Common Organic Compounds and Functional Groups

Class of Compound	General Structure	Functional Group	Example
alkanes	R—H	none	$CH_3CH_2CH_2CH_3$ butane
alkyl halides	R—X	X = F, Cl, Br, or I	$CH_3CH_2CH_2Cl$ 1-chloropropane
alkenes	R—CH=CH—R′	carbon–carbon double bond	$CH_3CH_2{-}CH{=}CH_2$ but-1-ene
alkynes	R—C≡C—R′	carbon–carbon triple bond	$CH_3{-}C{\equiv}C{-}CH_3$ but-2-yne
aromatic compounds		benzene ring, also drawn	C_6H_6 benzene
alcohols	R—OH	hydroxy group	$CH_3CH_2{-}OH$ ethanol
phenols	Ar—OH	hydroxy group on an aromatic ring	$C_6H_5{-}OH$ phenol
thiols	R—SH	sulfhydryl group	$CH_3{-}SH$ methanethiol
ethers	R—O—R′	oxygen between two alkyl groups	$CH_3CH_2{-}O{-}CH_2CH_3$ diethyl ether
thioethers	R—S—R′	sulfur between two alkyl groups	$CH_3{-}S{-}CH_3$ dimethyl sulfide
epoxides	C—C bridged by O	ether in a 3-membered ring	1,2-epoxycyclohexane
ketones	$R{-}C({=}O){-}R'$	carbonyl group	$CH_3{-}C({=}O){-}CH_3$ acetone
aldehydes	$R{-}C({=}O){-}H$	carbonyl group	$CH_3CH_2{-}C({=}O){-}H$ propanal
carboxylic acids	$R{-}C({=}O){-}OH$	carboxyl group	$CH_3{-}C({=}O){-}OH$ acetic acid
esters	$R{-}C({=}O){-}O{-}R'$	carboalkoxy group	$CH_3{-}C({=}O){-}O{-}CH_2CH_3$ ethyl acetate
amides	$R{-}C({=}O){-}NH_2$	carboxamide group	$H{-}C({=}O){-}N(CH_3)_2$ *N,N*-dimethylformamide
amines	$R{-}NH_2$	amino group	$CH_3CH_2{-}NH_2$ ethylamine
nitriles	R—C≡N	cyano group	$CH_3CH_2{-}C{\equiv}N$ propionitrile
nitroalkanes	$R{-}NO_2$	nitro group	$CH_3CH_2{-}NO_2$ nitroethane

Common Groups in Organic Chemistry

Organic Groups Abbreviation	Meaning	Structure
Ac	acetyl	$CH_3-C(=O)-R$
	allyl	$H_2C{=}CH-CH_2-R$
Ar	aryl group	unspecified aromatic
Boc	*tert*-butyloxycarbonyl	$(CH_3)_3C-O-C(=O)-R$
Bn	benzyl	$Ph-CH_2-R$
Bu	butyl (*n*-butyl)	$CH_3-CH_2-CH_2-CH_2-R$
i-Bu	isobutyl	$(CH_3)_2CH-CH_2-R$
s-Bu	*sec*-butyl	$CH_3-CH_2-CH(CH_3)-R$
t-Bu	*tert*-butyl	$(CH_3)_3C-R$
Bz	benzoyl	$Ph-C(=O)-R$
Cbz (or Z)	benzyloxycarbonyl	$Ph-CH_2-O-C(=O)-R$
Et	ethyl	CH_3-CH_2-R
c-Hx	cyclohexyl	cyclohexane ring–R
Fmoc	9-fluorenylmethoxycarbonyl	fluorene ring with CH_2–OR at C-9
Me	methyl	CH_3-R
Ph	phenyl	benzene ring–R
Pr	propyl	$CH_3-CH_2-CH_2-R$
i-Pr	isopropyl	$(CH_3)_2CH-R$
R	alkyl group	unspecified
Sia	*secondary* isoamyl	$(CH_3)_2CH-CH(CH_3)-R$
THP	tetrahydropyranyl	tetrahydropyran ring (O) with R at C-2
TIPS	triisopropylsilyl	$(i\text{-}Pr)_3Si-R$
Ts	*para*-toluenesulfonyl, "tosyl"	$CH_3-C_6H_4-S(=O)_2-R$
	vinyl	$H_2C{=}CH-R$

Not all of these abbreviations are used in this text, but they are provided for reference.

Common Reagents and Solvents

Abbreviation		Structure
Ac_2O	acetic anhydride	$CH_3-C(=O)-O-C(=O)-CH_3$
DCC	dicyclohexylcarbodiimide	$C_6H_{11}-N{=}C{=}N-C_6H_{11}$
DIBAL-H	diisobutylaluminum hydride	$[(CH_3)_2CHCH_2]_2AlH$
DME, "glyme"	1,2-dimethoxyethane	$CH_3-O-CH_2CH_2-O-CH_3$
diglyme	bis(2-methoxyethyl) ether	$(CH_3-O-CH_2CH_2)_2O$
DMP	Dess–Martin periodinane	AcO, OAc, OAc on I; benzo-fused ring with O and C=O
DMF	dimethylformamide	$H-C(=O)-N(CH_3)_2$
DMSO	dimethyl sulfoxide	$CH_3-S(=O)-CH_3$
EtOH	ethanol	CH_3CH_2OH
EtO^-	ethoxide ion	$CH_3CH_2-O^-$
Et_2O	diethyl ether	$CH_3CH_2-O-CH_2CH_3$
LAH	lithium aluminum hydride	$LiAlH_4$
LDA	lithium diisopropylamide	$[(CH_3)_2CH]_2N^-\ Li^+$
mCPBA	*meta*-chloroperoxybenzoic acid	Cl–C_6H_4–$C(=O)-O-O-H$
MeOH	methanol	CH_3OH
MeO^-	methoxide ion	CH_3-O^-
MVK	methyl vinyl ketone	$CH_3-C(=O)-CH{=}CH_2$
NBS	*N*-bromosuccinimide	succinimide ring, N–Br
PCC	pyridinium chlorochromate	$pyr \cdot CrO_3 \cdot HCl$
Py or Pyr	pyridine	pyridine ring, N:
t-BuOH	*tertiary* butyl alcohol	$(CH_3)_3C-OH$
t-BuOK	potassium *tertiary* butoxide	$(CH_3)_3C-O^-\ K^+$
TEMPO	2,2,6,6-tetramethylpiperidinyl-1-oxy	tetramethylpiperidine ring, $N-O^{\cdot}$
THF	tetrahydrofuran	five-membered ring with O
TMS	tetramethylsilane	$(CH_3)_4Si$

ORGANIC CHEMISTRY

NINTH EDITION

LEROY G. WADE, JR.

WHITMAN COLLEGE

CONTRIBUTING AUTHOR:

JAN WILLIAM SIMEK

CALIFORNIA POLYTECHNIC STATE UNIVERSITY

PEARSON

Editor-in-Chief: Jeanne Zalesky
Executive Editor: Terry Haugen
Product Marketing Manager: Elizabeth Ellsworth
Executive Field Marketing Manager: Chris Barker
Director of Development: Jennifer Hart
Development Editor: David Chelton
Project Manager: Elisa Mandelbaum
Program Manager: Lisa Pierce
Editorial Assistant: Fran Falk
Associate Content Producer: Lauren Layn
Marketing Assistant: Megan Riley
Team Lead, Project Management: David Zielonka
Team Lead, Program Management: Kristen Flathman
Compositor: GEX Publishing Services
Project Manager: GEX Publishing Services
Illustrator: Lachina
Photo and Text Research Manager: Maya Gomez
Photo and Text Researcher: Candice Velez
Design Manager: Mark Ong
Interior and Cover Design: Gary Hespenheide
Operations Specialist: Maura Zaldivar-Garcia
Cover Photo Credit: Klaus Rein / image/BROKER / Alamy Stock Photo
Spectra: ©Sigma-Aldrich Co.

Library of Congress Cataloging-in-Publication Data
Wade, L. G., 1947- | Simek, Jan William.
Organic chemistry / L.G. Wade, Jr., Whitman College; contributing author, Jan William Simek, California Polytechnic State University.
Ninth edition. | Glenview, IL : Pearson, [2017] | Includes index.
ISBN-13: 978-0-321-97137-1 | ISBN-10: 0-321-97137-X
LCSH: Chemistry, Organic--Textbooks.
LCC QD251.3 .W33 2017 | DDC 547--dc23
LC record available at
http://lccn.loc.gov/2015038705

LCCN 2015038705

ISBN 10: 0-321-97137-X; ISBN 13: 978-0-321-97137-1

www.pearsonhighered.com

3 2019

To my students and colleagues at Whitman College

About the Authors

L. G. "Skip" Wade decided to become a chemistry major during his sophomore year at Rice University, while taking organic chemistry from Professor Ronald M. Magid. After receiving his B.A. from Rice in 1969, Wade went on to Harvard University, where he did research with Professor James D. White. While at Harvard, he served as the Head Teaching Fellow for the organic laboratories and was strongly influenced by the teaching methods of two master educators, Professors Leonard K. Nash and Frank H. Westheimer.

After completing his Ph.D. at Harvard in 1974, Dr. Wade joined the chemistry faculty at Colorado State University. Over the course of fifteen years at Colorado State, Dr. Wade taught organic chemistry to thousands of students working toward careers in all areas of biology, chemistry, human medicine, veterinary medicine, and environmental studies. He also authored research papers in organic synthesis and in chemical education, as well as eleven books reviewing current research in organic synthesis. In 1989, Dr. Wade joined the chemistry faculty at Whitman College, where he continued to teach organic chemistry and pursue research interests in organic synthesis and forensic chemistry. Dr. Wade received the A. E. Lange Award for Distinguished Science Teaching at Whitman in 1993.

Dr. Wade's interest in forensic science has led him to testify as an expert witness in court cases involving drugs and firearms, and he has worked as a police firearms instructor, drug consultant, and boating safety officer. He also enjoys repairing and restoring old violins and bows, which he has done professionally for many years.

Jan Simek was born to humble, coal-mining parents who taught him to appreciate the importance of carbon at a very early age. At age 14, he was inspired to pursue a career teaching chemistry by his high school chemistry teacher, Joe Plaskas. Under the guidance of Professor Kurt Kaufman at Kalamazoo College, Dr. Simek began lab work in synthesis of natural products that turned into research in hop extracts for the Kalamazoo Spice Extraction Company. After receiving a master's degree from Stanford University, Dr. Simek worked in the pharmaceutical industry, synthesizing compounds designed to control diabetes and atherosclerosis, and assisted in the isolation of anti-cancer antibiotics from natural sources. Returning to Stanford University, Dr. Simek completed his Ph.D. with the legendary Professor Carl Djerassi, who developed the first synthesis of steroidal oral contraceptives.

Dr. Simek's 35-year teaching career was spent primarily at California Polytechnic State University, San Luis Obispo, where he received the university's Distinguished Teaching Award. Other teaching experiences include Albion College, the University of Colorado at Boulder, Kalamazoo College, and the University of California at Berkeley. In addition to his pharmaceutical research, he has industrial experience investigating dyes, surfactants, and liquid crystals, and he continues to consult for the biotechnology industry.

Although his outside interests include free climbing in Yosemite, performing in a reggae band, and parasailing over the Pacific, as close as he gets to any of those is tending his backyard garden with his wife Judy.

Brief Contents

Contents

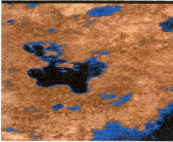

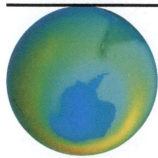

5 STEREOCHEMISTRY 201

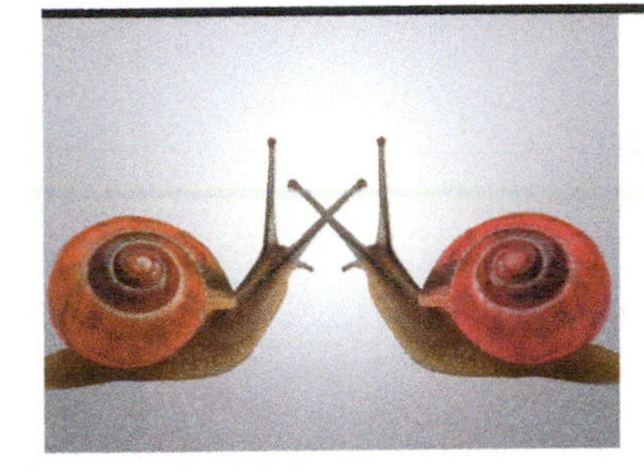

6 ALKYL HALIDES; NUCLEOPHILIC SUBSTITUTION 247

8 REACTIONS OF ALKENES 359

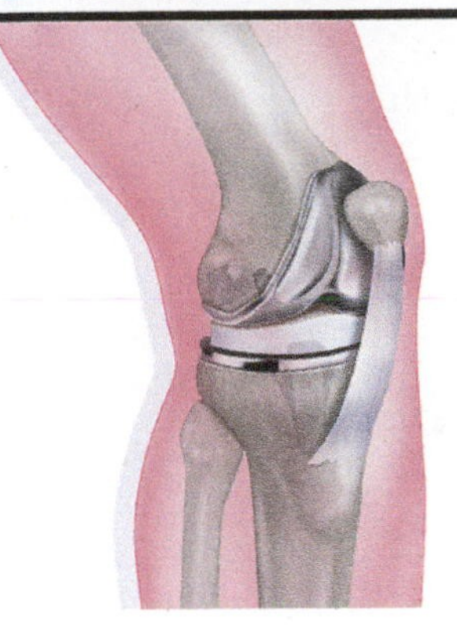

9 ALKYNES 428

10 STRUCTURE AND SYNTHESIS OF ALCOHOLS 460

11 REACTIONS OF ALCOHOLS 505

12 INFRARED SPECTROSCOPY AND MASS SPECTROMETRY 556

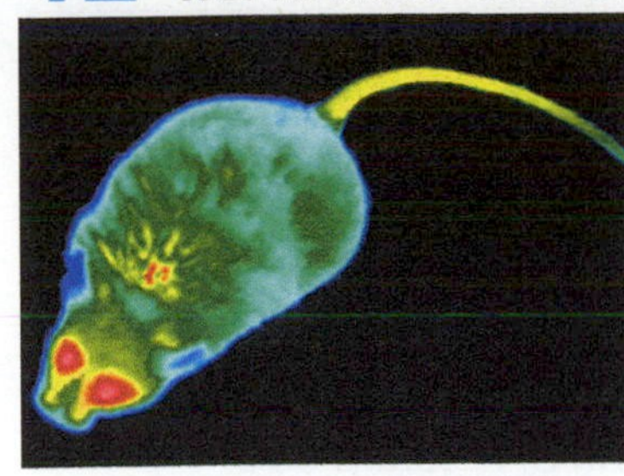

13 NUCLEAR MAGNETIC RESONANCE SPECTROSCOPY 607

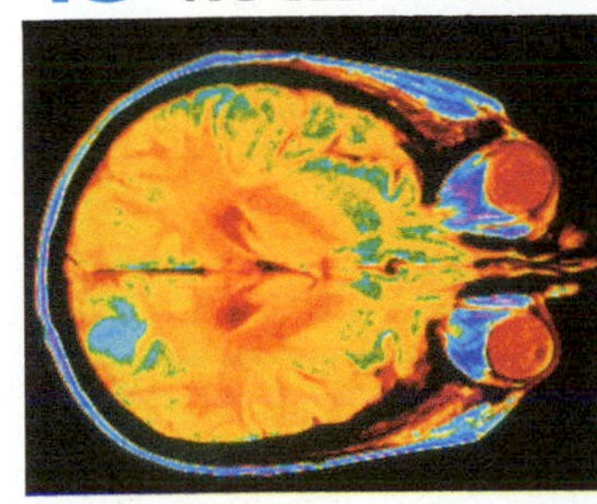

14 ETHERS, EPOXIDES, AND THIOETHERS 672

15 CONJUGATED SYSTEMS, ORBITAL SYMMETRY, AND ULTRAVIOLET SPECTROSCOPY 716

16 AROMATIC COMPOUNDS 764

17 REACTIONS OF AROMATIC COMPOUNDS 809

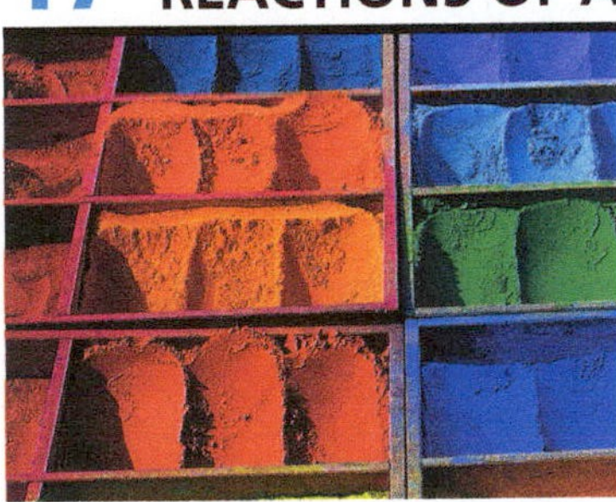

18 KETONES AND ALDEHYDES 876

19 AMINES 941

23 CARBOHYDRATES AND NUCLEIC ACIDS 1172

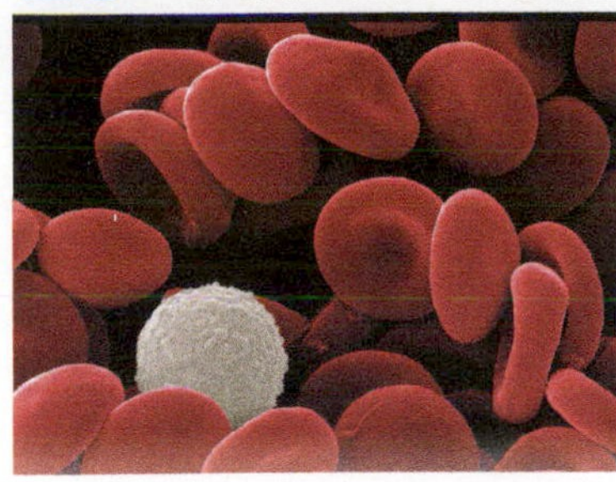

24 AMINO ACIDS, PEPTIDES, AND PROTEINS 1222

25 LIPIDS 1265

26 SYNTHETIC POLYMERS 1286

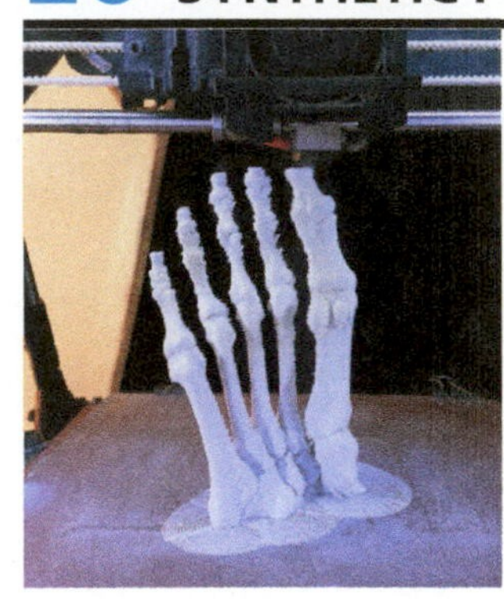

APPENDICES 1308

MECHANISMS

MECHANISMS (continued)

MECHANISMS (continued)

New to This Edition

1 **NEW!** Expanded coverage of **Acid/Base Chemistry** in chapter 2 and separation of the chapter on **Substitution and Elimination** into two distinct chapters allow students to build upon their existing knowledge and move through their first mechanisms with greater clarity and with more opportunities to test and apply their understanding without getting overwhelmed by organic chemistry. New problem-solving strategy spreads have been added to both corresponding chapters for additional support.

2 **NEW! Reaction Starbursts/Reaction Maps** appear before the end of every 'reaction-based' chapter to help students better understand and mentally organize reactive similarities and distinctions.

3 **NEW! Visual Guides to Organic Reactions** place the reactions covered in each chapter within the overall context of the reactions covered in the course.

4 **NEW! Problem Solving Strategies** have been added and explicitly highlighted in several chapters, including new strategies for resonance, acid-base equilibria, and multistep synthesis.

5 **NEW! Over 100 New Problems** include more synthesis problems and problems based on recent literature.

6 **NEW! Green Chemistry** is emphasized with presentation of less toxic, environmentally friendly reagents in many situations, such as oxidation of alcohols with bleach rather than with chromium reagents.

7 **NEW! Chapter Openers** focus on organic applications, with introductions and images for a more enticing, contemporary presentation.

8 **20 Key Mechanism Boxes** highlight the fundamental mechanistic principles that recur throughout the course and are the basis for some of the longer, more complex mechanisms. Each describes the steps of the reaction in detail with a specific example to reinforce the mechanism and a concluding problem to help students absorb these essential reactions.

9 **NEW! Explanations and Annotations to Mechanisms** help students better understand how each mechanism works.

Brief Chapter-by-Chapter Changes

Global Changes

Every chapter begins with a new chapter-opening photograph showing an interesting, real-world application of the material in that chapter. New Problem-Solving Hints and new Applications have been added to each chapter, and all of the chapters have gone through a careful revision process. All of the structures have been updated to the new IUPAC recommendations for showing stereochemistry. Green curved arrows are used to show the imaginary flow of electrons in resonance forms, in contrast to the red curved arrows used to show the actual flow in reactions.

Chapter 1 Structure and Bonding

- The material on structure, bonding, and molecular geometry has been consolidated into one chapter. A revised discussion of resonance includes a Problem-Solving Strategy, a Problem-Solving Hint on the types of arrows used in organic chemistry, and several new problems.

Chapter 2 Acids and Bases; Functional Groups

- The presentation of acids and bases has been moved from the previous Chapter 1 and greatly enhanced to become the main subject in the new Chapter 2. The new material includes sections on inductive, hybridization, resonance, and solvent effects on acidity and basicity; a section and Problem-Solving Strategy on predicting acid-base equilibrium positions; new Problem-Solving Hints; new figures; new applications; and 18 new problems.

Chapter 4 The Study of Chemical Reactions

- The values of bond dissociation enthalpies have been updated to the most recent experimental results throughout the chapter. A revised discussion of Hammond's postulate includes a figure that has been revised for clarity.

Chapter 5 Stereochemistry

- This chapter includes a revised summary of types of isomers, with revised figures for clarity. There are new Problem-Solving Hints on stereocenters, Fischer projections, and relative versus absolute configurations.

Chapter 6 Alkyl Halides; Nucleophilic Substitution

- The sections on E1 and E2 eliminations have been moved to Chapter 7. A new graphic showing the strengths of common nucleophiles has been added, and the summary of nucleophilic substitution conditions has been expanded. Several Problem-Solving Hints have been added on nucleophiles and bases, acid-base strength in the S_N1 reaction, and carbocation rearrangements.

Chapter 7 Structure and Synthesis of Alkenes; Elimination

- This chapter now contains expanded sections on E1 and E2 eliminations. Several Problem-Solving Hints have been added, as well as graphics on the competition between substitutions and eliminations. Several new problems have been added, including two solved problems.

Chapter 8 Reactions of Alkenes

- Several diagrams, applications, problems, and starburst summaries of reactions have been added. The new visual *Guide to Organic Reactions* is introduced in Chapter 8, and further updated in Chapters 11, 17, 18, 21, and 22.

Chapter 9 Alkynes

- New examples and a new starburst summary have been added. A new Problem-Solving Hint summarizes oxidative cleavages of alkynes.

Chapter 10 Structure and Synthesis of Alcohols

- The material on lithium dialkylcuprates has been expanded into a new section. New Problem-Solving Hints on Grignard reactions and organometallic reactions have also been added. A new starburst reaction summary has been added.

Chapter 11 Reactions of Alcohols

- A newly revised discussion of oxidizing agents emphasizes "green" reactions with sodium hypochlorite and acetic acid, or TEMPO, rather than toxic chromium reagents. A new interim summary compares alcohol oxidations with and without chromium reagents, and a new Problem-Solving Hint discusses ring-size changes and rearrangements. Two new starburst reaction summaries have been added.

Chapter 14 Ethers, Epoxides, and Thioethers

- New material and a new graphic have been added to clarify the regiochemistry of the opening of substituted epoxides. Several new problems have been added.

Chapter 15 Conjugated Systems, Orbital Symmetry, and Ultraviolet Spectroscopy

- Several figures have been revised for clarity, and new applications have been added.

Chapter 16 Aromatic Compounds

- New to this chapter are a Problem-Solving Hint on drawing energy diagrams for the MOs of cyclic systems, plus new applications and problems. A new starburst reaction summary has also been added.

Chapter 17 Reactions of Aromatic Compounds

- A new Problem-Solving Strategy has been added to explain multistep synthesis using electrophilic aromatic substitutions. The discussion of the Suzuki reaction has been expanded, including its mechanism. New applications, two new starburst reaction summaries, and several problems have also been added.

Chapter 18 Ketones and Aldehydes

- The discussion of syntheses of ketones and aldehydes has been revised to emphasize oxidations that use less toxic reagents such as bleach and TEMPO. Several new applications have been added, as well as a starburst reaction summary and several new problems.

Chapter 19 Amines

- A Problem-Solving Hint on pK_a of amines has been added, plus new applications and several new problems.

Chapter 20 Carboxylic Acids

- New problems and applications have been added as well as a starburst reaction summary.

Chapter 21 Carboxylic Acid Derivatives

- Several new problems and applications have been added, as well as a starburst reaction summary.

Chapter 22 Condensations and Alpha Substitutions of Carbonyl Compounds

- A new Problem-Solving Hint on ketone and ester carbonyl groups has been added, plus a new starburst reaction summary. Several applications and problems have been added as well.

Chapter 23 Carbohydrates and Nucleic Acids

- This chapter has been updated with a new application on glycoproteins. Some of the obsolete older reactions have been dropped.

Chapter 24 Amino Acids, Peptides, and Proteins

- The material on solid-phase peptide synthesis has been updated to use current techniques, and some of the obsolete, older methods have been deleted.

Chapter 26 Synthetic Polymers

- The organization of the chapter has been revised to emphasize chain-growth versus step-growth polymers, rather than addition versus condensation polymers. A new section has been added on the recycling of plastics, plus applications on 3D printing and PEX pipes.

Preface

To the Student

As you begin your study of organic chemistry, you might feel overwhelmed by the number of compounds, names, reactions, and mechanisms that confront you. You might even wonder whether you can learn all this material in a single year. The most important function of a textbook is to organize the material to show that most of organic chemistry consists of a few basic principles and many extensions and applications of these principles. Relatively little memorization is required if you grasp the major concepts and develop flexibility in applying those concepts. Frankly, I have a poor memory, and I hate memorizing lists of information. I don't remember the specifics of most of the reactions and mechanisms in this book, but I can work them out by remembering a few basic principles, such as "alcohol dehydrations usually go by E1 mechanisms."

Still, you'll have to learn some facts and fundamental principles to serve as the working "vocabulary" of each chapter. As a student, I learned this the hard way when I made a D on my second organic chemistry exam. I thought organic would be like general chemistry, where I could memorize a couple of equations and fake my way through the exams. For example, in the ideal gas chapter, I would memorize $PV = nRT$, and I was good to go. When I tried the same approach in organic, I got a D. We learn by making mistakes, and I learned a lot in organic chemistry.

In writing this book, I've tried to point out a small number of important facts and principles that should be learned to prepare for solving problems. For example, of the hundreds of reaction mechanisms shown in this book, about 20 are the fundamental mechanistic steps that combine into the longer, more complicated mechanisms. I've highlighted these fundamental mechanisms in *Key Mechanism* boxes to alert you to their importance. Similarly, the *Guide to Organic Reactions* appears in six chapters that contain large numbers of new reactions. This guide outlines the kinds of reactions we cover and shows how the reactions just covered fit into the overall organization. Spectroscopy is another area in which a student might feel pressured to memorize hundreds of facts, such as NMR chemical shifts and infrared vibration frequencies. I couldn't do that, so I've always gotten by with knowing about a dozen NMR chemical shifts and about a dozen IR vibration frequencies, and knowing how they are affected by other influences. I've listed those important infrared frequencies in Table 12-2 and the important NMR chemical shifts in Table 13-3.

Don't try to memorize your way through this course. It doesn't work; you have to know what's going on so you can apply the material. Also, don't think (like I did) that you can get by without memorizing *anything*. Read the chapter, listen carefully to the lectures, and *work the problems*. The problems will tell you whether or not you know the material. If you can do the problems, you should do well on the exams. If you can't do the problems, you probably won't be able to do the exams, either. If you keep having to look up an item to do the problems, that item is a good one to learn.

Here are some hints I give my students at the beginning of the course:

1. Read the material in the book before the lecture (expect 13–15 pages per lecture). Knowing what to expect and what is in the book, you can take fewer notes and spend more time listening and understanding the lecture.
2. After the lecture, review your notes and the book, and do the in-chapter problems. Also, read the material for the next lecture.
3. If you are confused about something, visit your instructor during office hours immediately, before you fall behind. Bring your attempted solutions to problems with you to show the instructor where you are having trouble.

4. To study for an exam, begin by reviewing each chapter and your notes, and reviewing any reaction summaries to make sure you can recognize and use those reactions. The "starburst" summaries are most useful for developing syntheses, since you can quickly glance at them and see the most useful conversions for that functional group. Then concentrate on the end-of-chapter problems. In each chapter, the Essential Problem-Solving Skills (EPSS) outline reviews the important concepts in the chapter and shows which problems can be used to review each concept. Also use old exams, if available, for practice. Many students find that working in a study group and posing problems for each other is particularly helpful.

Remember the two "golden rules" of organic chemistry.

1. ***Don't Get Behind!*** The course moves too fast, and it's hard to catch up.
2. ***Work Lots of Problems.*** Everyone needs the practice, and the problems show where you need more work.

I am always interested to hear from students using this book. If you have any suggestions about how the book might be improved, or if you've found an error, please let me know (L. G. Wade, Whitman College, Walla Walla, WA 99362: E-mail wadelg@whitman.edu). I take students' suggestions seriously, and hundreds of them now appear in this book. For example, Whitman student Brian Lian suggested Figure 21-9, and University of Minnesota student (and race-car driver) Jim Coleman gave me the facts on the fuels used at Indianapolis.

Good luck with your study of organic chemistry. I'm certain you will enjoy this course, especially if you let yourself relax and develop an interest in how organic compounds influence our lives. My goal in writing this book has been to make the process a little easier: to build the concepts logically on top of each other, so they flow naturally from one to the next. The hints and suggestions for problem solving have helped my students in the past, and I hope some of them will help you to learn and use the material. Even if your memory is worse than mine (highly unlikely), you should be able to do well in organic chemistry. I hope this will be a good learning experience for all of us.

To the Instructor

In writing the first edition of this text, my goal was to produce a modern, readable text that uses the most effective techniques of presentation and review. I wanted a book that presents organic chemistry at the level needed for chemistry and biochemistry majors, but one that presents and explains the material in ways that facilitate success for all the many different kinds of students who take the course. Subsequent editions have extended and refined these goals, with substantial rewriting and reorganizing and with many new features. This ninth edition adds several new features to help students organize types of reactions and mechanisms for easier learning and better understanding, as well as for reference.

New to This Edition

To help students organize functional group reactions, new **Starburst Summaries** have been added that provide visual links between synthetically related functional groups. This new feature is particularly useful when students are developing multistep syntheses, when the visual links help them to see the possible reactions moving forward from a reactant or synthetic intermediate. The new **Guides to Organic Reactions** will help students to organize mentally the many new reactions they are learning, and where those reactions fit within the overall scheme of the types of reactions we use in organic chemistry. **Chapter-opening photographs**, with captions that explain how the photograph relates to the chemistry presented in that chapter, have been added to all of the chapters. We have tried to select photos that are remarkable in some way or another and that grab the viewer's attention.

All of the features of the earlier editions have been retained in this ninth edition. In many cases, those that were introduced in earlier editions have been expanded and refined. Many **updated applications** have been added, including those relating to medicine, green chemistry, biochemistry, and other contemporary areas of interest. **Green chemistry** is emphasized in many areas, most notably in the use of methods that avoid chromium reagents, which are known to be toxic and carcinogenic. The older, more toxic reagents are mentioned, but they are no longer given as the first choice for a reagent. Mechanisms have been provided for the newest reactions, such as the Suzuki coupling, when they are relevant to the material and studied well enough to be confident they are correct.

Key Features

Expanded Coverage of Acids and Bases: After reviewing the basics of bonding, hybridization, and molecular structure in Chapter 1, Chapter 2 is centered around acids and bases and how these concepts apply to organic compounds. The Arrhenius, Brønsted-Lowry, and Lewis definitions are introduced and explained. The uses of pK_a and pK_b are described, followed by a discussion and a Problem-Solving Strategy feature on predicting the position of an acid-base equilibrium reaction. Factors that affect acidity and basicity are explained, including solvent effects, size, electronegativity, inductive effects, hybridization effects, and resonance effects. Lewis acid-base reactions are discussed, with a careful discussion of the correct use of the curved-arrow formalism.

Separation of Substitution and Elimination Reactions: The crucial chapters on substitution and elimination have been revised, with substitutions covered in Chapter 6 and eliminations in Chapter 7. This organization allows students to become more comfortable with the differences between S_N1 and S_N2 substitutions before the possible reaction pathways are expanded to include eliminations. Chapter 7 presents complete coverage of the competition between substitutions and eliminations, and how one can predict what mechanisms and products are most likely.

Organic Synthesis: Many new synthetic problems have been added, some of them coming from the recent literature. The material on organic synthesis and retrosynthetic analysis has been supplemented, with particular attention to multistep aromatic syntheses.

Nomenclature: We have tried to stay as current as possible with the constantly changing IUPAC nomenclature, and this edition reflects some of the most recent changes. Beginning with the eighth edition, we have used the 1993 IUPAC positioning of the locants in names (e.g., but-1-ene), while also showing the names using the older positions of the locants (e.g., 1-butene). We have also carefully defined stereochemical terms (such as *stereocenter* and *chiral center*) correctly and precisely, and we have endeavored to use the most precise term in each case.

In this edition, we have adopted three of the newest changes in the IUPAC rules:

1. In showing stereochemistry, IUPAC now recommends the "reverse perspective" (closer end is smaller) version of wedged dashed bonds. Wedged solid bonds are still drawn with normal perspective, with their closer end larger.
2. IUPAC now defines *hydroxyl* as referring only to the radical, not the functional group. The functional group is the *hydroxy group*. We have changed these terms where needed to conform to this rule.
3. At one time, the IUPAC banished the term *ketal*. It has now been reinstated as a subclass of acetals, and we have resumed using it.

This ninth edition also includes a new Nomenclature Appendix, which serves as a compact reference to the rules of naming organic compounds. This feature should make it easier for students to name compounds without always having to find the discussion pertaining to that particular functional group.

The Keys to Organic Chemistry

Wade & Simek's ninth edition of *Organic Chemistry* presents key principles of organic chemistry in the context of fundamental reasoning and problem solving. Written to reflect how today's students use textbooks, this text serves as a primary guide to organic chemistry, as well as a comprehensive study resource when working problems and preparing for exams.

The resources in this book include Problem-Solving Strategies throughout, plus Partially Solved Problems, Reaction Summaries, new Starburst Summaries, and new Reaction Guides. Through a careful, refined presentation and step-by-step guidance, this ninth edition gives students a contemporary overview of organic chemistry with tools for organizing and understanding reaction mechanisms and synthetic organic chemistry.

PROBLEM-SOLVING STRATEGY Predicting Substitutions and Eliminations

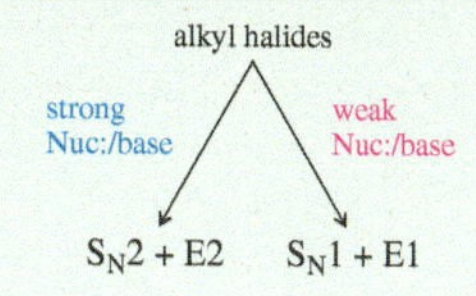

1° alkyl halide: R—CH₂—X (R—C(H)(H)—X)

strong Nuc:/base → S_N2

weak Nuc:/base → no reaction

Given a set of reagents and solvents, how can you predict what products will result and which mechanisms will be involved? Should you memorize all this theory about substitutions and eliminations? Students sometimes feel overwhelmed at this point.

Memorizing is not the best way to approach this material because the answers are not absolute and too many factors are involved. Besides, the real world with its real reagents and solvents is not as clean as our equations on paper. Most nucleophiles are also basic, and many solvents can solvate ions or react as nucleophiles or bases.

The first principle you must understand is that *you cannot always predict one unique product or one unique mechanism*. Often, the best you can do is to eliminate some of the possibilities and make some accurate predictions. Remembering this limitation, here are some general guidelines:

1. **The strength of the base or nucleophile determines the order of the reaction.** If a strong nucleophile (or base) is present, it will force second-order kinetics, either S_N2 or E2. If no strong base or nucleophile is present, you should consider first-order reactions, both S_N1 and E1. Addition of silver salts to the reaction can force some difficult ionizations.
2. **Primary halides usually undergo the S_N2 reaction, occasionally the E2 reaction.** Primary halides rarely undergo first-order reactions, unless the carbocation is resonance-stabilized. With good nucleophiles, S_N2 substitution is usually observed. With a strong base, E2 elimination may occasionally be observed.

SOLVED PROBLEM 7-7

(a) Propose a mechanism for the sulfuric acid-catalyzed dehydration of *tert*-butyl alcohol.

SOLUTION

The first step is protonation of the hydroxy group, which converts it to a good leaving group.

$$(CH_3)_3C-\ddot{O}-H + H_2SO_4 \rightleftarrows (CH_3)_3C-\ddot{O}^+H_2 + HSO_4^-$$

The second step is ionization of the protonated alcohol to give a carbocation.

$$(CH_3)_3C-\ddot{O}^+H_2 \rightleftarrows (CH_3)_3C^+ + H_2\ddot{O}:$$

Abstraction of a proton completes the mechanism.

$$H_2\ddot{O}: \;\; H-CH_2-C^+(CH_3)_2 \rightleftarrows H_2C=C(CH_3)_2 + H_3\ddot{O}^+$$

(b) Predict the products and propose a mechanism for the acid-catalyzed dehydration of 1-cyclohexylethanol.

PARTIAL SOLUTION:

Protonation of the hydroxy group, followed by loss of water, forms a carbocation.

1-cyclohexylethanol: $C_6H_{11}-CH(\ddot{O}H)-CH_3$ + H—OSO₃H ⇌ $C_6H_{11}-CH(\ddot{O}^+H_2)-CH_3$ + HSO_4^- ⇌ $C_6H_{11}-C^+H-CH_3$ (carbocation) + $H_2\ddot{O}:$

The carbocation can lose a proton, or it can rearrange to a more stable carbocation.

unrearranged 2° carbocation $\xrightarrow{\text{~H:}^-}$ rearranged 3° carbocation

PROBLEM-SOLVING HINT

Alcohol dehydrations usually go through E1 elimination of the protonated alcohol.

Reactivity is: 3° > 2° >> 1°

Rearrangements are common.

Protonated primary alcohols dehydrate at elevated temperatures with rearrangement (E1), or the adjacent carbon may lose a proton to a weak base at the same time water leaves (E2).

Principles, Preparation, and Problem Solving

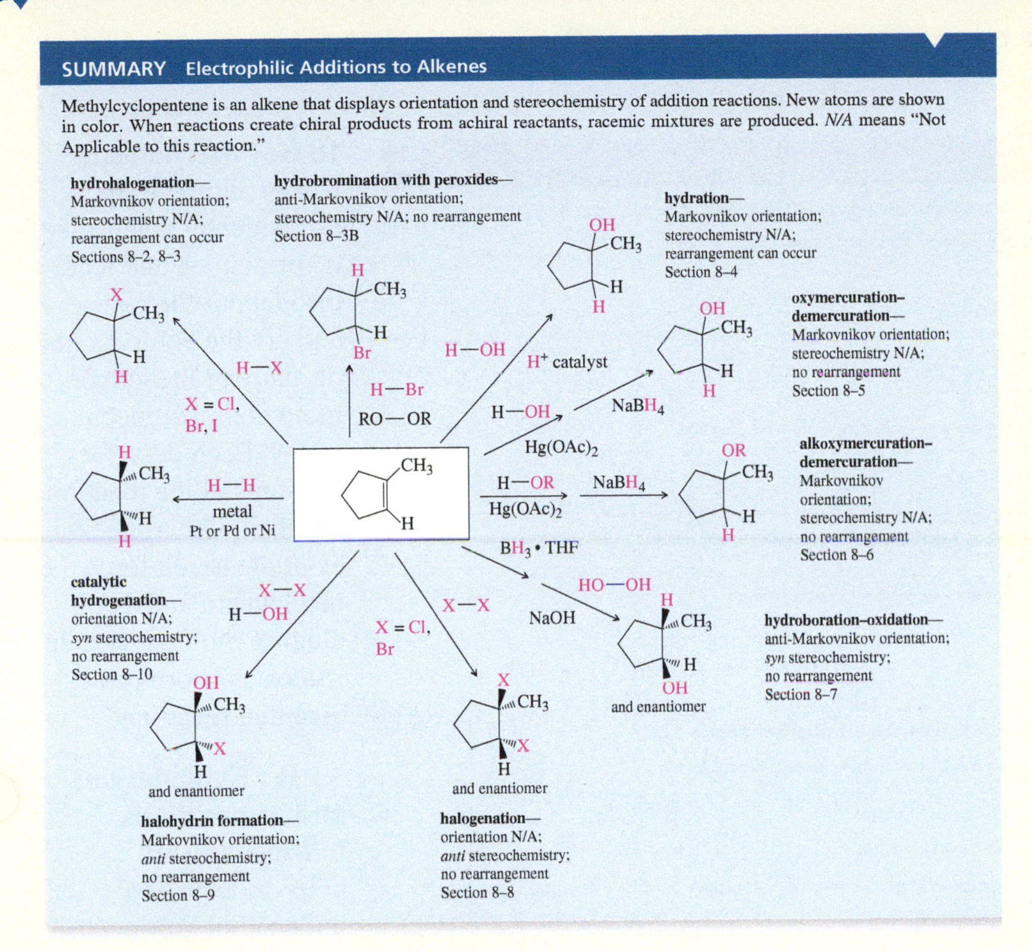

NEW! Starburst Reaction Summaries appear before the end-of-chapter material of "reaction-based" chapters to help students mentally organize the reactions and recognize their similarities and differences.

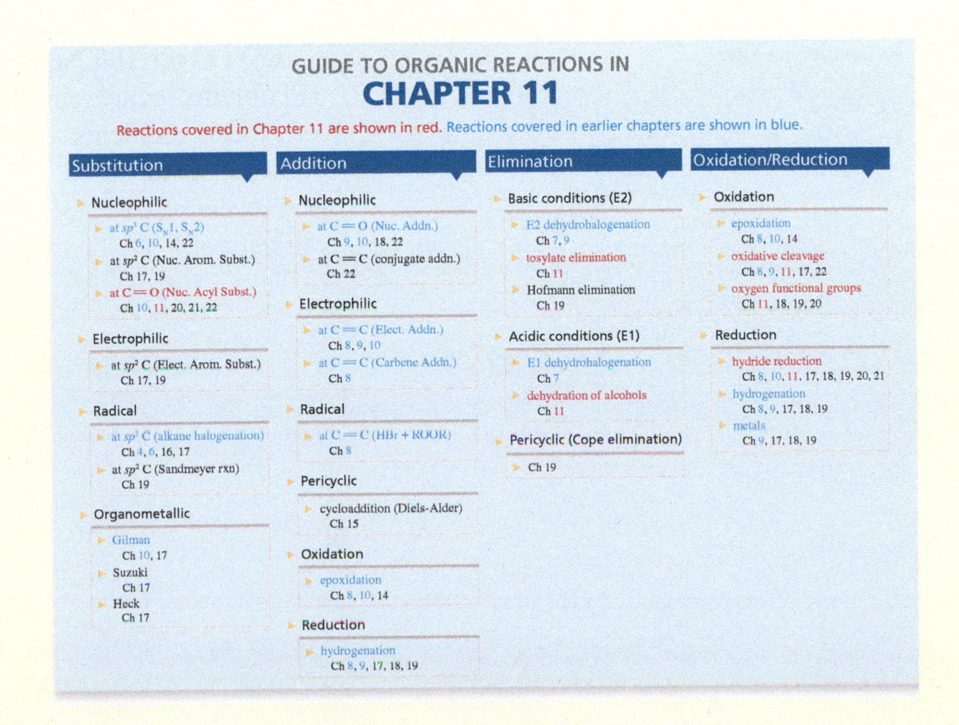

GUIDE TO ORGANIC REACTIONS IN
CHAPTER 11
Reactions covered in Chapter 11 are shown in red. Reactions covered in earlier chapters are shown in blue.
Substitution
Addition
Elimination
Oxidation/Reduction

NEW! Visual **Guides to Organic Reactions** place the reactions covered in each chapter within the overall context of the reactions covered in the course.

The Keys to Organic Chemistry

Over 80 Mechanism Boxes help students understand how specific reactions occur by zooming in on each individual step in detail.

18 Key Mechanism Boxes highlight the fundamental mechanistic principles that recur throughout the course and are the components of many of the longer, more complex mechanisms. Each describes the steps of the reaction in detail with a specific example to reinforce the mechanism and a concluding problem to help students absorb these essential reactions.

NEW! Explanations and Annotations to Mechanisms help students better understand how each mechanism works.

NEW! Over 100 New Problems include more synthesis problems and problems based on recent research literature.

KEY MECHANISM 7-5 Acid-Catalyzed Dehydration of an Alcohol

PROBLEM-SOLVING HINT

In acid-catalyzed mechanisms, the first step is often addition of H^+, and the last step is often loss of H^+.

Alcohol dehydrations usually involve E1 elimination of the protonated alcohol.

Step 1: Protonation of the hydroxy group (fast equilibrium).

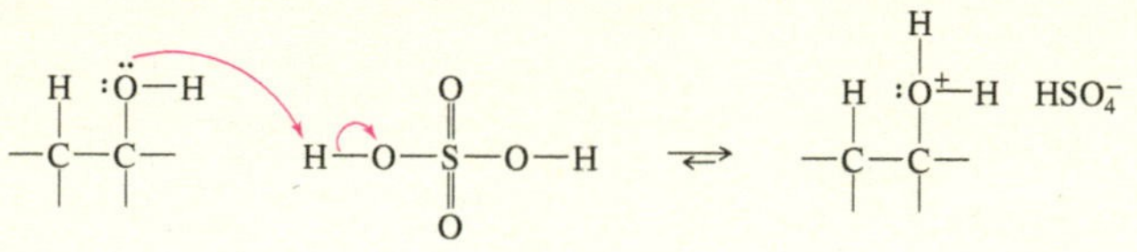

Step 2: Ionization to a carbocation (slow; rate limiting).

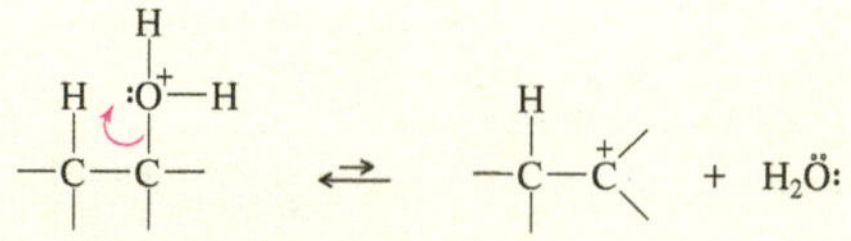

Step 3: Deprotonation to give the alkene (fast).

EXAMPLE: **Acid-catalyzed dehydration of butan-2-ol**

Step 1: Protonation of the hydroxy group (fast equilibrium).

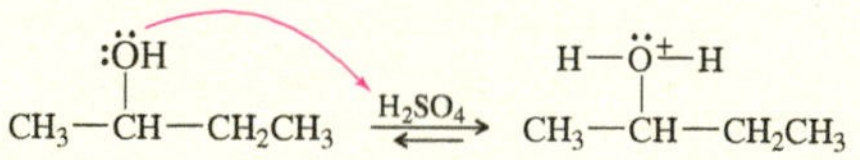

Step 2: Ionization to a carbocation (slow; rate limiting).

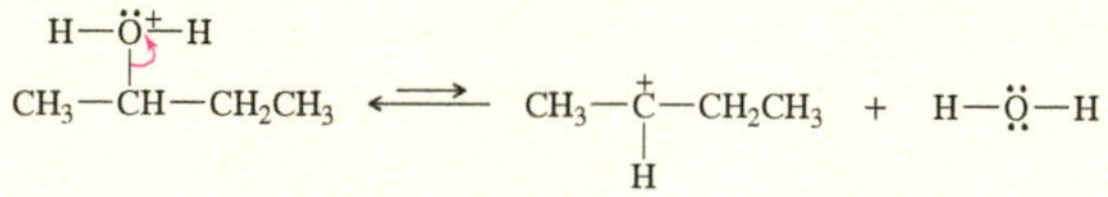

Step 3: Deprotonation to give the alkene (fast).

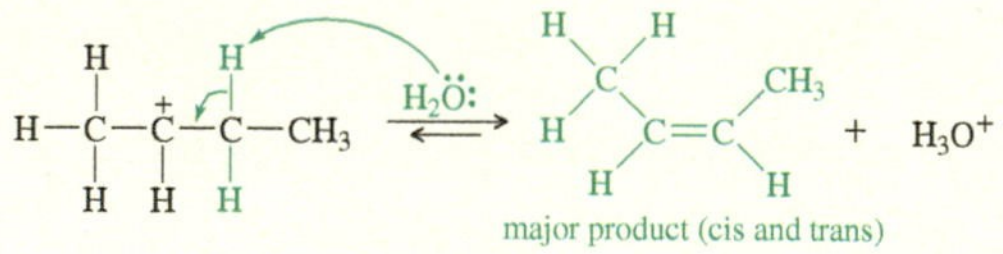

major product (cis and trans)

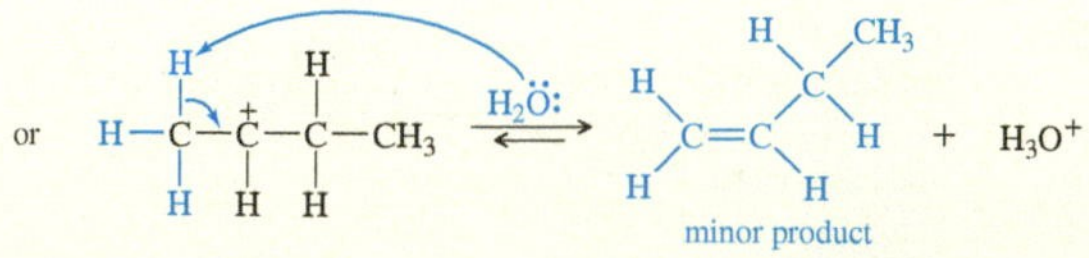

minor product

MasteringChemistry®

www.masteringchemistry.com

MasteringChemistry motivates students to practice organic chemistry outside of class and arrive prepared for lecture. The textbook works with MasteringChemistry to guide students toward what they need to know before testing them on the content. This edition continually engages students through pre-lecture, during-, and post-lecture activities that all include real-life applications.

Dynamic Study Modules

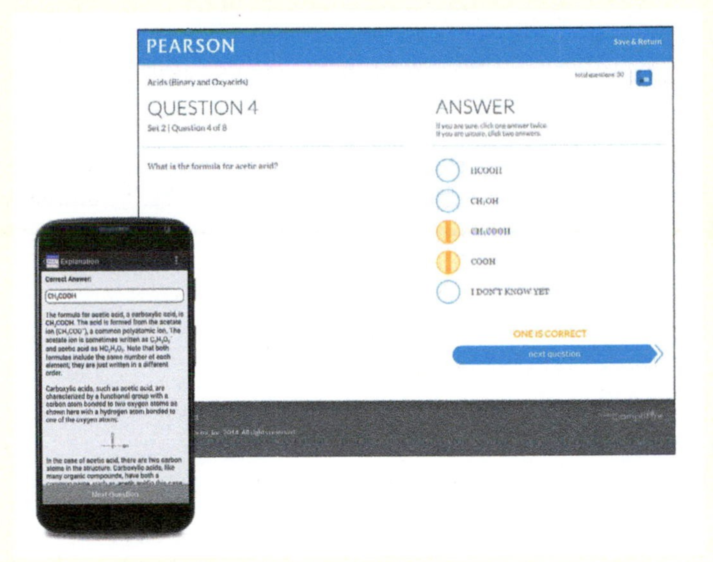

Help students understand the concepts more quickly! Now assignable, Dynamic Study Modules enable your students to study on their own and be better prepared with the fundamental concepts needed from general chemistry, as well as the problem solving skills and practice with nomenclature, functional groups, and key mechanisms skills needed to succeed in the organic chemistry course. The mobile app is available for iOS and Android devices for study on the go, and results can be tracked in the MasteringChemistry gradebook.

Spectroscopy Simulations

NEW! Six NMR/IR Spectroscopy simulations (a partnership with ACD labs) allow professors and students access to limitless spectral analysis with guided activities that can be used in lab, in the classroom, or after class to study how spectral information correlates with molecular structures. Activities authored by Mike Huggins, University of West Florida, prompt students to use the spectral simulator to solve analytical problems by drawing the right conclusions from the spectra and combining the NMR and IR data to propose a molecular structure.

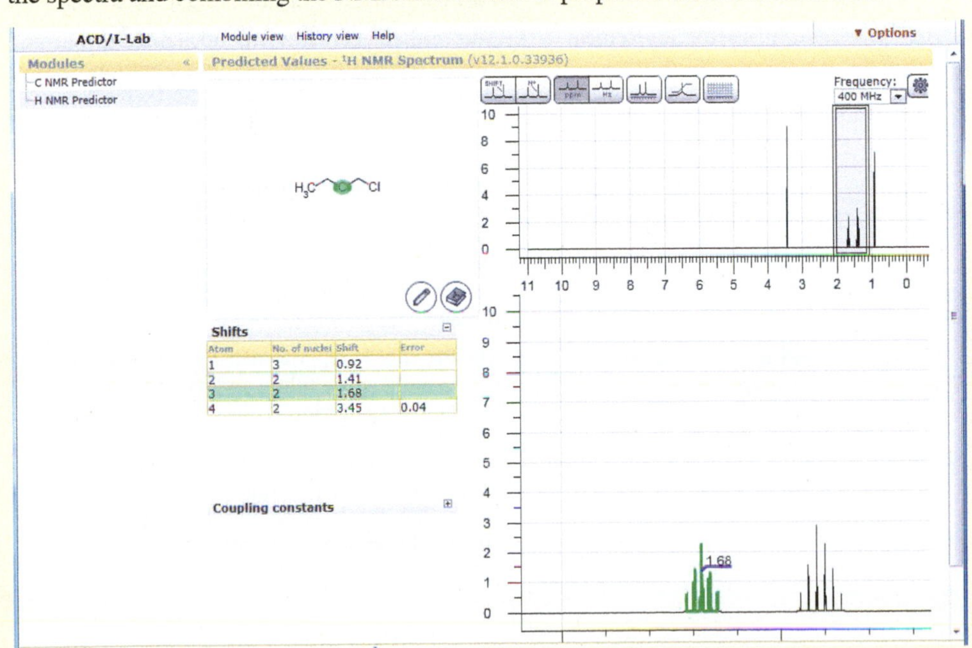

Resources in Print and Online

Supplement	Available in Print?	Available Online?	Instructor or Student Supplement	Description
MasteringChemistry®		✓	Instructor and Student Supplement	MasteringChemistry New! Organic Chemistry Dynamic Study modules help students efficiently prepare for lecture and exams by reinforcing understanding of general chemistry prerequisites, acid-base chemistry, functional groups, nomenclature, and key mechanisms. At the end of each personalized question set, Dynamic Study Modules provide feedback on whether the answer submitted was correct and give students an explanation of the correct and incorrect answers. The process repeats until the students answer all of the questions correctly and confidently. (Available at www.masteringchemistry.com)
Solutions Manual by Jan William Simek	✓		Instructor and Student Supplement	This Solutions Manual provides detailed solutions to all in-chapter as well as the end-of-chapter exercises in the text. It is now printed in two-color format to highlight details in the solutions and to improve consistency with the text.
Instructor Resources		✓	Instructor Supplement	This website provides an integrated collection of online resources to help instructors make efficient and effective use of their time. It includes all artwork from the text, including figures and tables in PDF format for high-resolution printing, as well as four pre-built PowerPoint™ presentations. The first presentation contains the images embedded within PowerPoint slides. The second includes a complete lecture outline that is modifiable by the user. Also available are powerpoints of the parent text "in chapter" sample exercises. It also includes the Test Bank. Access resources through http://www.pearsonhighered.com/.
Testbank		✓	Instructor Supplement	This testbank contains over 3000 multiple-choice, true/false, and matching questions. It is available in print format, in the TestGen program, in Word format, and is included in the item library of MasteringChemistry.
Organic Molecular Kit (Darling)	✓		Instructor and Student Supplement	Darling Models™ contain various pieces used to build atoms, bonds, and molecules. This model kit allows you to build molecules and see the three-dimensional aspects of organic chemistry that can only be imagined in a two-dimensional drawing.
Prentice Hall Molecular Model Kit for Organic Chemistry	✓		Instructor and Student Supplement	The Prentice Hall molecular model set allows you to build space-filling and ball-and-stick models of organic molecules. The components are precision-tooled from quality plastics, are virtually indestructible, and come in a sturdy plastic case for easy storage. Provides a useful Instruction Book—with photos, diagrams, and concise discussions of chemical principles.

Acknowledgments

I am pleased to thank the many talented people who helped with this revision. Certainly the largest contribution has come from Jan William Simek, long-time author of the Solutions Manual and now contributing author for this textbook. Jan has provided excellent advice and sound judgment through several editions of the book. In this edition, Jan has authored numerous new and revised sections, developed over 100 new problems, constructed the starburst reaction summaries, and authored the new Nomenclature Appendix.

Jan and I would also like to thank the many reviewers for their valuable insight and commentary. Although we did not adopt all their suggestions, we adopted most of them; they were helpful and contributed to the quality of the final product.

Ninth Edition Accuracy Reviewers

David A. Boyajian	Palomar College
Charles Kingsbury	University of Nebraska at Lincoln
Chris Scwhartz	Metropolitan Community College
Chad Synder	Grace College and Seminary

Ninth Edition Prescriptive Reviewers:

Brad Andersh	Bradley University
Anonymous	University of Nebraska-Lincoln
Anonymous	West Chester University
Ardeshir Azadnia	Michigan State University
David Boyajian	Palomar College
Xuefei Huang	Michigan State University
Rizalia Klausmeyer	Baylor University
Susan Lever	University of Missouri-Columbia
Richard Mullins	Xavier University
Helena Malinakova	University of Kansas
Alline Somlai	Delta State University
Maria Vogt	Bloomfield College

Additional Reviewers for 9e:

Angela Allen	Lenoir Community College
Anonymous	Emmanuel College
Anonymous	Illinois Wesleyan University
Stefan Bossman	Kansas State University
Russell Betts	Broward College Central
Adam Braunschweig	University of Miami
Jeffrey Charonnat	California State University Northridge
Joan M. Comar	Georgia State University
Joseph Cradlebaugh	Jacksonville University
Sean Curtis	University of Arkansas-Ft. Smith
Markus Etzkorn	University of North Carolina Charlotte
Ana Fraiman	Northeastern Illinois University
Andy Frazer	University of Central Florida
John Brent Friesen	Dominican University
John Hoberg	University of Wyoming, Laramie
Andrew Holland	Idaho State University, Pocatello
Jess Jones	University of South Florida
Raja Mani	University of Bridgeport
Joan Mutanyatta	Comar, Georgia State University
Chris Nicholson	University of West Florida
Oyindasola Oyelaran	Northeastern University
Ivana Peralta	Vincennes University
Suzanne Ruder	Virginia Commonwealth University
Joshua Ruppel	University of South Carolina-Upstate
Rekha Srinivasan	Case Western Reserve University
Eric Trump	Emporia State University

Reviewers of Previous Editions

Jung-Mo Ahn	University of Texas at Dallas
David Alonso	Andrews University
Merritt B. Andrus	Brigham Young University
Jon Antilla	University of South Florida
Arthur J. Ashe	University of Michigan
Bill Baker	University of South Florida

Dan Becker	Loyola University
John Berger	Montclair State University
Bob Bly	University of South Carolina
Mary Boyd	Loyola University, Chicago
Hindy Bronstein	Fordham College at Lincoln Center
David Brown	St. John's University
Eric Brown	Loyola University, Lake Shore
Philip Brown	North Carolina State University
Christine Brzezowski	University of Alberta
Patrick Buick	Florida Atlantic University
David Cantillo	Hillsborough Community College
Dee Ann Casteel	Bucknell University
Amber Charlebois	William Paterson University
Cai Chengzhi	University of Houston
Timothy B. Clark	Western Washington University
Barry Coddens	Northwestern University
Jamie Lee Cohen	Pace University
Barbara Colonna	University of Miami
Richard Conley	Middlesex County College
Robert Crow	St. Louis College of Pharmacy
Maria de GracaVicente	Louisiana State University
James Fletcher	Creighton University
Chris Gorman	North Carolina State University
Geneive Henry	Susquehanna University
William Jenks	Iowa State University
Przemyslaw Maslak	Pennsylvania State University
Kristen Meisenheimer	Cal Polytechnic at San Luis Obispo
Guillermo Moyna	University of the Sciences in Philadelphia
Rabi Musah	University at Albany
Stephen A. Miller	University of Florida
Hasan Palandoken	California Polytechnic State University
Keith Osbourne Pascoe	Georgia State University
Anthony J. Pearson	Case Western Reserve
Allan Pinhas	University of Cincinnati
Owen Priest	Northwestern University
Stanley Raucher	University ofWashington
Suzanne Ruder	Virginia Commonwealth University
K.C. Russell	Northern Kentucky University
Alline Somlai	Delta State University
Joseph B. Wachter	Michigan State University
David Son	Southern Methodist University
Solomon Weldegirma	University of South Florida

Finally, we want to thank the people at Pearson, whose dedication and commitment contributed to the completion of this project. Particular thanks are due to Developmental Editor David Chelton, who made thousands of useful suggestions throughout the writing and revision process, and who helped to shape this new edition. Special thanks are also due to Editor-in-Chief Jeanne Zalesky, who guided the project from start to finish and made many useful comments and suggestions that guided the direction of the revision. Director of Development Jennifer Hart and Program Manager Lisa Pierce kept the project moving and ensured the needed resources were made available. Project Managers Elisa Mandelbaum and Heidi Aguiar, and Text and Image Research Lead Maya Gomez kept the production process organized, on track, and on schedule. It has been a pleasure working with all these thoroughly professional and competent people.

We have enjoyed working on this new edition, and we hope that it is a scientific and pedagogical improvement over the eighth edition. We've tried to make this book as error-free as possible, but some errors may have slipped by. If you find errors, or have suggestions about how the book might be improved, please send those errors and suggestions to me at my e-mail address: wadelg@whitman.edu. Errors can be fixed quickly in the next printing. Please send any errors you find in the Solutions Manual, or suggestions for improvements, to Jan Simek at his e-mail address: jsimek@calpoly.edu.

We've already started a file of possible changes and improvements for the next edition, and we hope that many of the current users will contribute suggestions to this file. We hope this book makes the instructor's job easier and helps more students to succeed. That's the most important reason that we continue to work at improving it.

L. G. Wade, Jr.
Walla Walla, Washington
Jan William Simek
San Luis Obispo, California

1 Structure and Bonding

Goals for Chapter 1

1 Review concepts from general chemistry that are essential for success in organic chemistry, such as the electronic structure of the atom, Lewis structures and the octet rule, types of bonding, electronegativity, and formal charges.

2 Predict patterns of covalent and ionic bonding involving C, H, O, N, and the halogens.

3 Identify resonance-stabilized structures and compare the relative importance of their resonance forms.

4 Draw and interpret the types of structural formulas commonly used in organic chemistry, including condensed structural formulas and line–angle formulas.

5 Predict the hybridization and geometry of organic molecules based on their bonding.

6 Identify isomers and explain the differences between them.

◀ **Luciferin** is the light-emitting compound found in many firefly (Lampyridae) species. Luciferin reacts with atmospheric oxygen, under the control of an enzyme, to emit the yellow light that fireflies use to attract mates or prey.

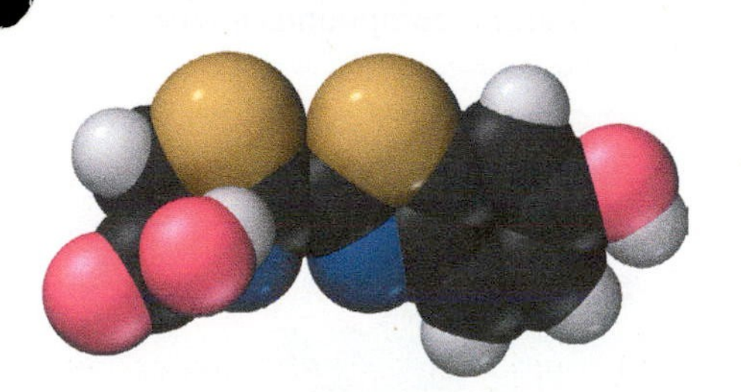

luciferin

S S OH N N O C OH

luciferin

1-1 The Origins of Organic Chemistry

The modern definition of organic chemistry is *the chemistry of carbon compounds.* What is so special about carbon that a whole branch of chemistry is devoted to its compounds? Unlike most other elements, carbon forms strong bonds to other carbon atoms and to a wide variety of other elements. Chains and rings of carbon atoms can be built up to form an endless variety of molecules. This diversity of carbon compounds provides the basis for life on Earth. Living creatures are composed largely of complex organic compounds that serve structural, chemical, or genetic functions.

The term **organic** literally means "derived from living organisms." Originally, the science of organic chemistry was the study of compounds extracted from living organisms and their natural products. Compounds such as sugar, urea, starch, waxes, and plant oils were considered "organic," and people accepted **vitalism,** the belief that natural products needed a "vital force" to create them. Organic chemistry, then, was the study of compounds having the vital force. Inorganic chemistry was the study of gases, rocks, and minerals, and the compounds that could be made from them.

Application: Drug Research

One of the reasons chemists synthesize derivatives of complex organic compounds such as morphine (shown below) is to discover new drugs that retain the good properties (potent pain-relieving) but not the bad properties (highly addictive).

In the 19th century, experiments showed that organic compounds could be synthesized from inorganic compounds. In 1828, the German chemist Friedrich Wöhler converted ammonium cyanate, made from ammonia and cyanic acid, to urea simply by heating it in the absence of oxygen.

$$\underset{\substack{\text{ammonium cyanate} \\ \text{(inorganic)}}}{NH_4^+ \ {}^-OCN} \xrightarrow{\text{heat}} \underset{\substack{\text{urea} \\ \text{(organic)}}}{H_2N{-}\overset{\overset{\displaystyle O}{\|}}{C}{-}NH_2}$$

Urea had always come from living organisms and was presumed to contain the vital force, yet ammonium cyanate is inorganic and thus lacks the vital force. Some chemists claimed that a trace of vital force from Wöhler's hands must have contaminated the reaction, but most recognized the possibility of synthesizing organic compounds from inorganics. Many other syntheses were carried out, and the vital force theory was eventually discarded.

Because vitalism was disproved in the early 19th century, you'd think it would be extinct by now. And you'd be wrong! Vitalism lives on today in the minds of those who believe that "natural" (plant-derived) vitamins, flavor compounds, etc., are somehow different and more healthful than the identical "artificial" (synthesized) compounds.

As chemists, we know that plant-derived compounds and the synthesized compounds are identical. Assuming they are pure, the only way to tell them apart is through ^{14}C dating: Compounds synthesized from petrochemicals have a lower content of radioactive ^{14}C and appear old because their ^{14}C has decayed over time. Plant-derived compounds are recently synthesized from CO_2 in the air. They have a higher content of radioactive ^{14}C. Some large chemical suppliers provide isotope-ratio analyses to show that their "naturals" have high ^{14}C content and are plant-derived. Such a sophisticated analysis lends a high-tech flavor to this 21st-century form of vitalism.

Even though organic compounds do not need a vital force, they are still distinguished from inorganic compounds. The distinctive feature of organic compounds is that they *all* contain one or more carbon atoms. Still, not all carbon compounds are organic;

FIGURE 1-1
(a) Venom of the blue-ringed octopus contains tetrodotoxin, which causes paralysis resulting in death. (b) Rose hips contain vitamin C, a radical inhibitor (Chapter 4) that prevents scurvy. (c) The prickly pear cactus is host to cochineal insects, used to prepare the red dye carmine (Chapter 15). (d) Opium poppies contain morphine, an addictive, pain-relieving alkaloid (Chapter 19).

substances such as diamond, graphite, carbon dioxide, ammonium cyanate, and sodium carbonate are derived from minerals and have typical inorganic properties. Most of the millions of carbon compounds are classified as organic, however.

We humans are composed largely of organic molecules, and we are nourished by the organic compounds in our food. The proteins in our skin, the lipids in our cell membranes, the glycogen in our livers, and the DNA in the nuclei of our cells are all organic compounds. Our bodies are also regulated and defended by complex organic compounds.

Chemists have learned to synthesize or simulate many of these complex molecules. The synthetic products serve as drugs, medicines, plastics, pesticides, paints, and fibers. Many of the most important advances in medicine are actually advances in organic chemistry. New synthetic drugs are developed to combat disease, and new polymers are molded to replace failing organs. Organic chemistry has gone full circle. It began as the study of compounds derived from "organs," and now it gives us the drugs and materials we need to save or replace those organs.

The AbioCor® self-contained artificial heart, which is used to sustain patients who are waiting for a heart transplant. The outer shell is polycarbonate, and the valves and inner bladder are polyurethane. Both of these durable substances are synthetic organic compounds.

1-2 Principles of Atomic Structure

Before we begin our study of organic chemistry, we must review some basic principles. These concepts of atomic and molecular structure are crucial to your understanding of the structure and bonding of organic compounds.

1-2A Structure of the Atom

Atoms are made up of protons, neutrons, and electrons. Protons are positively charged and are found together with (uncharged) neutrons in the nucleus. Electrons, which have a negative charge that is equal in magnitude to the positive charge on the proton, occupy the space surrounding the nucleus (Figure 1-2). Protons and neutrons have similar masses, about 1800 times the mass of an electron. Almost all the atom's mass is in the nucleus, but it is the electrons that take part in chemical bonding and reactions.

Each element is distinguished by the number of protons in the nucleus (the atomic number). The number of neutrons is usually similar to the number of protons, although the number of neutrons may vary. Atoms with the same number of protons but different numbers of neutrons are called **isotopes.** For example, the most common kind of carbon atom has six protons and six neutrons in its nucleus. Its mass number (the sum of the protons and neutrons) is 12, and we write its symbol as ^{12}C. About 1% of carbon atoms have seven neutrons; the mass number is 13, written ^{13}C. A very small fraction of carbon atoms have eight neutrons and a mass number of 14. The ^{14}C isotope is radioactive, with a half-life (the time it takes for half of the nuclei to decay) of 5730 years. The predictable decay of ^{14}C is used to determine the age of organic materials up to about 50,000 years old.

cloud of electrons

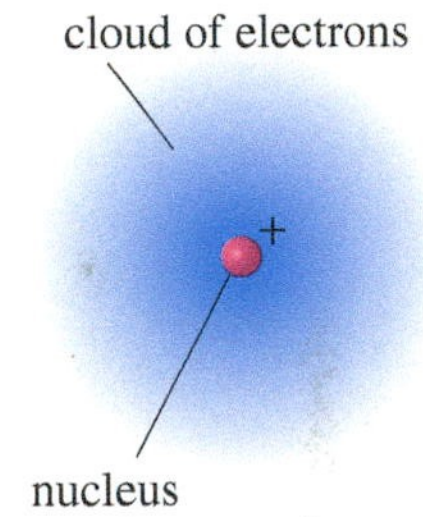

nucleus
(protons and neutrons)

FIGURE 1-2
Basic atomic structure. An atom has a dense, positively charged nucleus surrounded by a cloud of electrons.

1-2B Electron Shells and Orbitals

An element's chemical properties are determined by the number of protons in the nucleus and the corresponding number of electrons around the nucleus. The electrons form bonds and determine the structure of the resulting molecules. Because they are small and light, electrons show properties of both particles and waves; in many ways, the electrons in atoms and molecules behave more like waves than like particles.

Electrons that are bound to nuclei are found in **orbitals.** Orbitals are mathematical descriptions that chemists use to explain and predict the properties of atoms and molecules. The *Heisenberg uncertainty principle* states that we can never determine exactly where the electron is; nevertheless, we can determine the **electron density,** the probability of finding the electron in a particular part of the orbital. An orbital, then, is an allowed energy state for an electron, with an associated probability function that defines the distribution of electron density in space.

Relative orbital energies

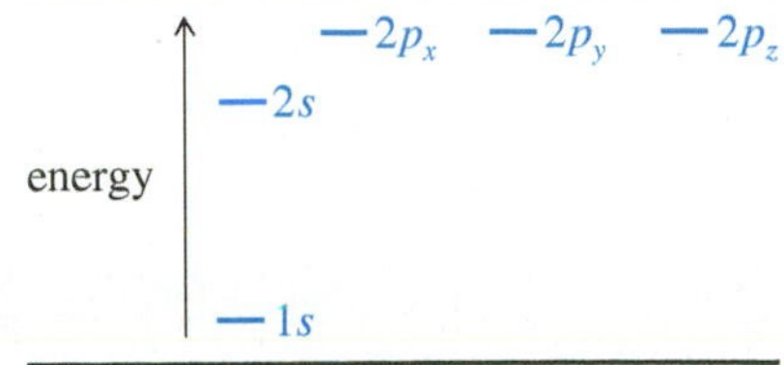

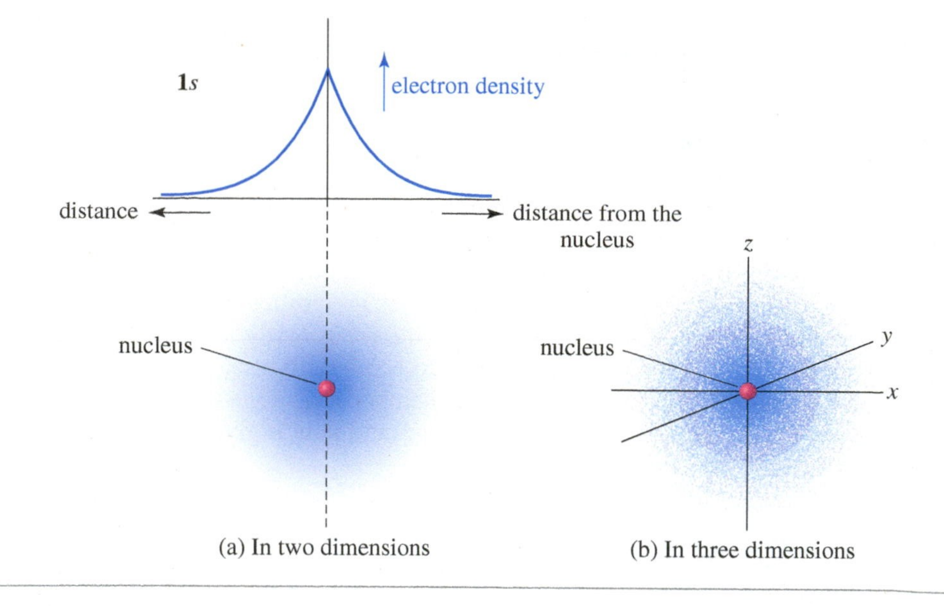

FIGURE 1-3
Graph and diagram of the 1*s* atomic orbital. The electron density is highest at the nucleus and drops off exponentially with increasing distance from the nucleus in any direction.

Atomic orbitals are grouped into different "shells" at different distances from the nucleus. Each shell is identified by a principal quantum number n, with $n = 1$ for the lowest-energy shell closest to the nucleus. As n increases, the shells are farther from the nucleus, are higher in energy, and can hold more electrons. Most of the common elements in organic compounds are found in the first two rows of the periodic table, indicating that their electrons are found in the first two electron shells. The first shell ($n = 1$) can hold two electrons, and the second shell ($n = 2$) can hold eight.

The first electron shell contains just the 1*s* orbital. All *s* orbitals are spherically symmetrical, meaning that they are nondirectional. The electron density is only a function of the distance from the nucleus. The electron density of the 1*s* orbital is graphed in Figure 1-3. Notice how the electron density is highest *at* the nucleus and falls off exponentially with increasing distance from the nucleus. The 1*s* orbital might be imagined as a cotton boll, with the cottonseed at the middle representing the nucleus. The density of the cotton is highest nearest the seed, and it becomes less dense at greater distances from this "nucleus."

The second electron shell consists of the 2*s* and 2*p* orbitals. The 2*s* orbital is spherically symmetrical like the 1*s* orbital, but its electron density is not a simple exponential function. The 2*s* orbital has a smaller amount of electron density close to the nucleus. Most of the electron density is farther away, beyond a region of zero electron density called a **node.** Because most of the 2*s* electron density is farther from the nucleus than that of the 1*s*, the 2*s* orbital is higher in energy. Figure 1-4 shows a graph of the 2*s* orbital.

In addition to the 2*s* orbital, the second shell also contains three 2*p* atomic orbitals, one oriented in each of the three spatial directions. These orbitals are called the $2p_x$, the $2p_y$, and the $2p_z$, according to their direction along the *x*, *y*, or *z* axis. The 2*p* orbitals are slightly higher in energy than the 2*s*, because the average location of the electron in a 2*p* orbital is farther from the nucleus. Each *p* orbital consists of two lobes, one on either side of the nucleus, with a **nodal plane** at the nucleus. The nodal plane is a flat (planar) region of space, including the nucleus, with zero electron density. The three 2*p* orbitals differ only in their spatial orientation, so they have identical energies. Orbitals with identical energies are called **degenerate orbitals.** Figure 1-5 shows the shapes of the three degenerate 2*p* atomic orbitals.

The *Pauli exclusion principle* tells us that each orbital can hold a maximum of two electrons, provided that their spins are paired. The first shell (one 1*s* orbital) can accommodate two electrons. The second shell (one 2*s* orbital and three 2*p* orbitals) can accommodate eight electrons, and the third shell (one 3*s* orbital, three 3*p* orbitals, and five 3*d* orbitals) can accommodate 18 electrons.

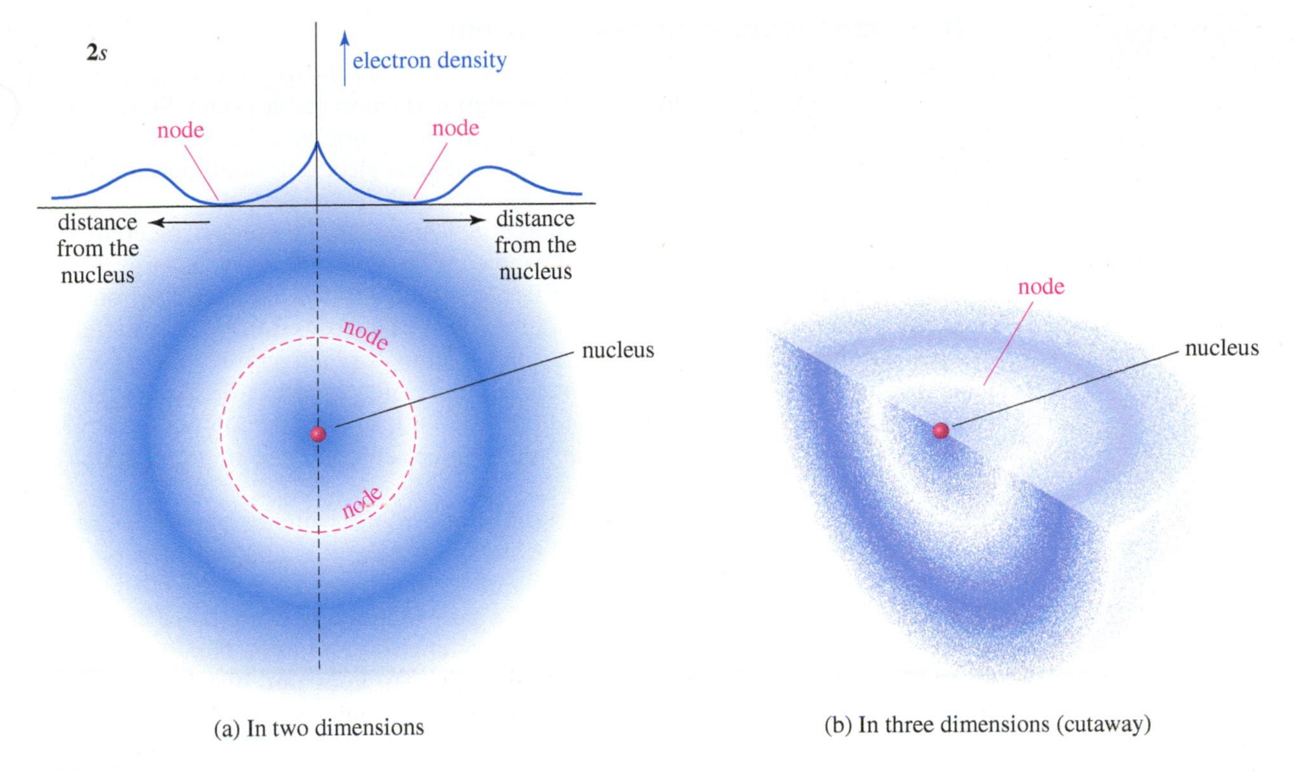

(a) In two dimensions

(b) In three dimensions (cutaway)

FIGURE 1-4
Graph and diagram of the 2*s* atomic orbital. The 2*s* orbital has a small region of high electron density close to the nucleus, but most of the electron density is farther from the nucleus, beyond a node, or region of zero electron density.

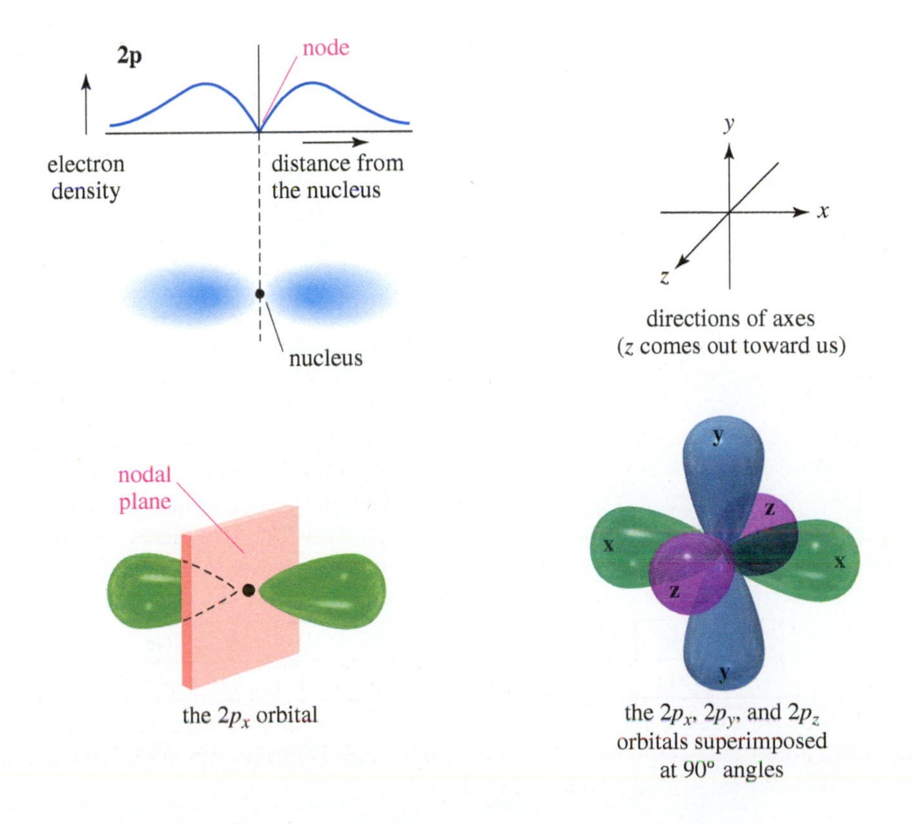

FIGURE 1-5
The 2*p* orbitals. Three 2*p* orbitals are oriented at right angles to each other. Each is labeled according to its orientation along the *x*, *y*, or *z* axis.

Application: Drugs

Lithium carbonate, a salt of lithium, is a mood-stabilizing agent used to treat the psychiatric disorder known as mania. Mania is characterized by behaviors such as elated mood, feelings of greatness, racing thoughts, and an inability to sleep. We don't know how lithium carbonate helps to stabilize these patients' moods.

1-2C Electronic Configurations of Atoms

Aufbau means "building up" in German, and the *aufbau principle* tells us how to build up the electronic configuration of an atom's ground (most stable) state. Starting with the lowest-energy orbital, we fill the orbitals in order until we have added the proper number of electrons. Table 1-1 shows the ground-state electronic configurations of the elements in the first two rows of the periodic table.

Two additional concepts are illustrated in Table 1-1. The **valence electrons** are those electrons in the outermost shell. Carbon has four valence electrons, nitrogen has five, and oxygen has six. Helium has a filled first shell with two valence electrons, and neon has has a filled second shell with eight valence electrons (ten electrons total). In general (for the representative elements), the column or group number of the periodic table corresponds to the number of valence electrons (Figure 1-6). Hydrogen and lithium have one valence electron, and they are both in the first column (group 1A) of the periodic table. Carbon has four valence electrons, and it is in group 4A of the periodic table.

Notice in Table 1-1 that carbon's third and fourth valence electrons are not paired; they occupy separate orbitals. Although the Pauli exclusion principle says that two electrons can occupy the same orbital, the electrons repel each other, and pairing requires additional energy. **Hund's rule** states that when there are two or more orbitals of the same energy, electrons go into *different* orbitals rather than pair up in the same orbital. The first $2p$ electron (boron) goes into one $2p$ orbital, the second $2p$ electron (carbon) goes into a different orbital, and the third $2p$ electron (nitrogen) occupies the last $2p$ orbital. The fourth, fifth, and sixth $2p$ electrons must pair up with the first three electrons.

TABLE 1-1
Electronic Configurations of the Elements of the First and Second Rows

Element	Configuration	Valence Electrons
H	$1s^1$	1
He	$1s^2$	2
Li	$1s^2 2s^1$	1
Be	$1s^2 2s^2$	2
B	$1s^2 2s^2 2p_x^1$	3
C	$1s^2 2s^2 2p_x^1 2p_y^1$	4
N	$1s^2 2s^2 2p_x^1 2p_y^1 2p_z^1$	5
O	$1s^2 2s^2 2p_x^2 2p_y^1 2p_z^1$	6
F	$1s^2 2s^2 2p_x^2 2p_y^2 2p_z^1$	7
Ne	$1s^2 2s^2 2p_x^2 2p_y^2 2p_z^2$	8

Partial periodic table

1A							noble gases 8A
H	2A	3A	4A	5A	6A	7A	He
Li	Be	B	C	N	O	F	Ne
Na	Mg	Al	Si	P	S	Cl	Ar

FIGURE 1-6
First three rows of the periodic table. The organization of the periodic table results from the filling of atomic orbitals in order of increasing energy. For these representative elements, the number of the column corresponds to the number of valence electrons.

PROBLEM 1-1

(a) Nitrogen has relatively stable isotopes (half-life greater than 1 second) of mass numbers 13, 14, 15, 16, and 17. (All except ^{14}N and ^{15}N are radioactive.) Calculate how many protons and neutrons are in each of these isotopes of nitrogen.

(b) Write the electronic configurations of the third-row elements shown in the partial periodic table in Figure 1-6.

1-3 Bond Formation: The Octet Rule

In 1915, G. N. Lewis proposed several new theories describing how atoms bond together to form molecules. One of these theories states that a filled shell of electrons is especially stable, and *atoms transfer or share electrons in such a way as to attain a filled shell of electrons.* A filled shell of electrons is simply the electron configuration of a noble gas, such as He, Ne, or Ar. This principle has come to be called the **octet rule** because a filled shell implies eight valence electrons for the elements in the second row of the periodic table. Elements in the third and higher rows (such as Al, Si, P, S, Cl, and above) can have an "expanded octet" of more than eight electrons because they have low-lying *d* orbitals available.

PROBLEM-SOLVING HINT

When we speak of a molecule having "all octets satisfied," we mean that all the second-row elements have octets. Hydrogen atoms have just two electrons (the He configuration) in their filled valence shell.

1-3A Ionic Bonding

There are two ways that atoms can interact to attain noble-gas configurations. Sometimes atoms attain noble-gas configurations by transferring electrons from one atom to another. For example, lithium has one electron more than the helium configuration, and fluorine has one electron less than the neon configuration. Lithium easily loses its valence electron, and fluorine easily gains one:

$$\underset{\text{electron transfer}}{Li\cdot \;\; \cdot\ddot{\underset{\cdot\cdot}{F}}:} \longrightarrow \underset{\text{He configuration}}{Li^+} + \underset{\text{Ne configuration}}{:\ddot{\underset{\cdot\cdot}{F}}:^-} \longrightarrow \underset{\text{ionic bond}}{Li^+ :\ddot{\underset{\cdot\cdot}{F}}:^-}$$

A transfer of one electron gives each of these two elements a noble-gas configuration. The resulting ions have opposite charges, and they attract each other to form an **ionic bond.** Ionic bonding usually results in the formation of a large crystal lattice rather than individual molecules. Ionic bonding is common in inorganic compounds but relatively uncommon in organic compounds.

1-3B Covalent Bonding

Covalent bonding, in which electrons are shared rather than transferred, is the most common type of bonding in organic compounds. Hydrogen, for example, needs a second electron to achieve the noble-gas configuration of helium. If two hydrogen atoms come together and form a bond, they "share" their two electrons, and each atom has two electrons in its valence shell.

$$H\cdot \; + \; H\cdot \longrightarrow \underset{\text{covalent bond}}{H:H} \qquad \text{each H shares two electrons (He configuration)}$$

We will study covalent bonding in more detail later in this chapter.

1-4 Lewis Structures

One way to symbolize the bonding in a covalent molecule is to use **Lewis structures.** In a Lewis structure, each valence electron is symbolized by a dot. A bonding pair of electrons is symbolized by a pair of dots or by a dash (—). We try to arrange all the atoms so that they have their appropriate noble-gas configurations: two electrons for hydrogen, and octets for the second-row elements.

Consider the Lewis structure of methane (CH_4).

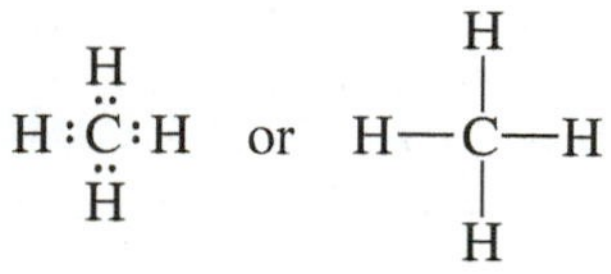

```
     H                H
   H:C:H    or     H—C—H
     H                H
```

methane

Carbon contributes four valence electrons, and each hydrogen contributes one, to give a total of eight electrons. All eight electrons surround carbon to give it an octet, and each hydrogen atom shares two of the electrons with the carbon atom.

The Lewis structure for ethane (C_2H_6) is more complex.

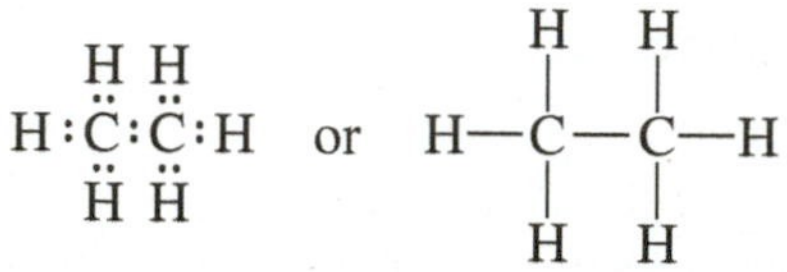

```
     H H                H  H
   H:C:C:H    or     H—C—C—H
     H H                H  H
```

ethane

Once again, we have computed the total number of valence electrons (14) and distributed them so that each carbon atom is surrounded by 8 and each hydrogen by 2. The only possible structure for ethane is the one shown, with the two carbon atoms sharing a pair of electrons and each hydrogen atom sharing a pair with one of the carbons. The ethane structure shows the most important characteristic of carbon—its ability to form strong carbon–carbon bonds.

Nonbonding electrons are valence-shell electrons that are *not* shared between two atoms. A pair of nonbonding electrons is often called a **lone pair.** Oxygen atoms, nitrogen atoms, and the halogens (F, Cl, Br, I) usually have nonbonding electrons in their stable compounds. These lone pairs of nonbonding electrons often serve as reactive sites in their parent compounds. The following Lewis structures show one lone pair of electrons on the nitrogen atom of methylamine and two lone pairs on the oxygen atom of ethanol. Halogen atoms usually have three lone pairs, as shown in the structure of chloromethane.

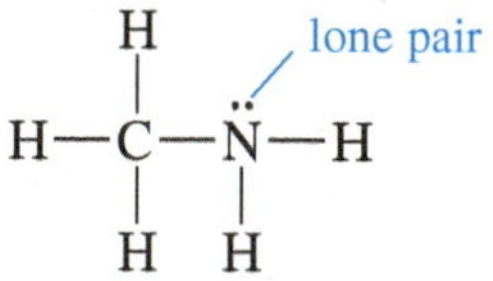

```
   H     lone pair         H  H    lone pairs          H
   |     ..                |  |      ..                |   ..
H—C—N—H              H—C—C—O:                   H—C—Cl:  lone pairs
   |  |                    |  |  |                     |   ..
   H  H                    H  H  H                     H
```

methylamine 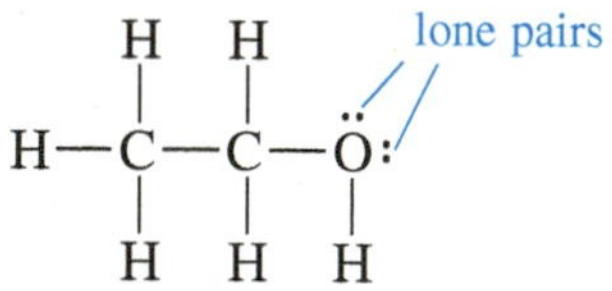 | ethanol 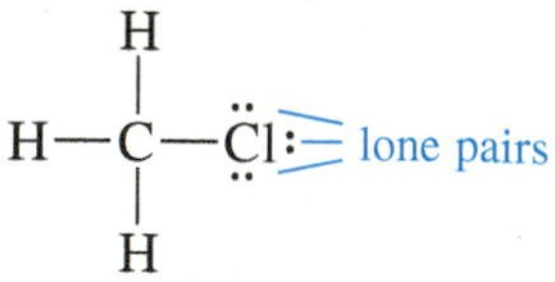 | chloromethane

A correct Lewis structure should show any lone pairs. Organic chemists often draw structures that omit most or all of the lone pairs. These are not true Lewis structures because you must imagine the correct number of nonbonding electrons.

PROBLEM-SOLVING HINT

Lewis structures are the way we write organic chemistry. Learning how to draw them quickly and correctly will help you throughout this course.

PROBLEM 1-2

Draw Lewis structures for the following compounds.

(a) ammonia, NH_3
(b) water, H_2O
(c) hydronium ion, H_3O^+
(d) propane, C_3H_8
(e) dimethylamine, CH_3NHCH_3
(f) diethyl ether, $CH_3CH_2OCH_2CH_3$
(g) 1-chloropropane, $CH_3CH_2CH_2Cl$
(h) propan-2-ol, $CH_3CH(OH)CH_3$
(i) borane, BH_3
(j) boron trifluoride, BF_3

Explain what is unusual about the bonding in the compounds in parts (i) and (j).

1-5 Multiple Bonding

In drawing Lewis structures in Section 1-4, we placed just one pair of electrons between any two atoms. The sharing of one pair between two atoms is called a **single bond.** Many molecules have adjacent atoms sharing two or even three electron pairs. The sharing of two pairs is called a **double bond,** and the sharing of three pairs is called a **triple bond.**

Ethylene (C_2H_4) is an organic compound with a double bond. When we draw a Lewis structure for ethylene, the only way to show both carbon atoms with octets is to draw them sharing two pairs of electrons. The following examples show organic compounds with double bonds. In each case, two atoms share four electrons (two pairs) to give them octets. A double dash (=) symbolizes a double bond.

$H_2C::CH_2$	$H_2C::\ddot{O}:$	$H_2C::\ddot{N}H$
or	or	or
$H_2C{=}CH_2$	$H_2C{=}\ddot{O}:$	$H_2C{=}\ddot{N}{-}H$
ethylene	formaldehyde	formaldimine

Acetylene (C_2H_2) has a triple bond. Its Lewis structure shows three pairs of electrons between the carbon atoms to give them octets. The following examples show organic compounds with triple bonds. A triple dash (≡) symbolizes a triple bond.

H:C:::C:H	H_3C:C:::C:CH_3	H_3C:C:::N:
or	or	or
H—C≡C—H	H_3C—C≡C—CH_3	H_3C—C≡N:
acetylene	dimethylacetylene	acetonitrile

All these Lewis structures show that carbon normally forms four bonds in neutral organic compounds. Nitrogen generally forms three bonds, and oxygen usually forms two. Hydrogen and the halogens usually form only one bond. The number of bonds an atom usually forms is called its **valence.** Carbon is tetravalent, nitrogen is trivalent, oxygen is divalent, and hydrogen and the halogens are monovalent. By remembering the usual number of bonds for these common elements, we can write organic structures more easily. If we draw a structure with each atom having its usual number of bonds, a correct Lewis structure usually results.

Application: FYI

Acetylene is a high-energy gaseous hydrocarbon that is explosive at high pressures. Combined with oxygen, acetylene burns with such a hot flame that it melts steel. Acetylene is commonly used in welding and cutting torches that work anywhere, even underwater. In gas cylinders, acetylene is dissolved in acetone to keep it from getting too concentrated and exploding.

SUMMARY Common Bonding Patterns (Uncharged)

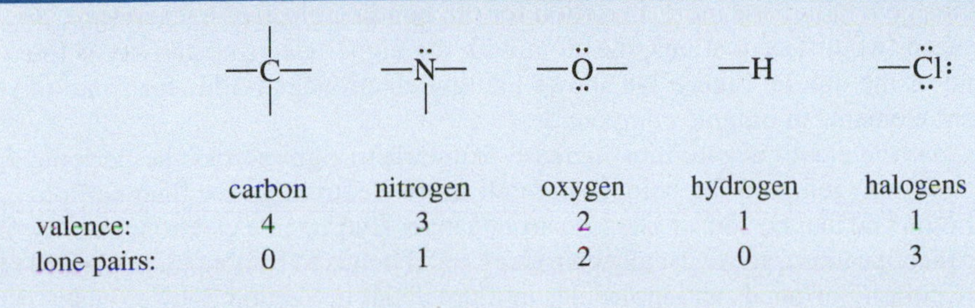

	carbon	nitrogen	oxygen	hydrogen	halogens
valence:	4	3	2	1	1
lone pairs:	0	1	2	0	3

PROBLEM-SOLVING HINT

These "usual numbers of bonds" might be single bonds, or they might be combined into double and triple bonds. For example, three bonds to nitrogen might be three single bonds, one single bond and one double bond, or one triple bond (:N≡N:). In working problems, consider all possibilities.

PROBLEM 1-3

Write Lewis structures for the following molecular formulas.

(a) N_2 **(b)** HCN **(c)** HONO
(d) CO_2 **(e)** CH_3CHNH **(f)** HCO_2H
(g) C_2H_3Cl **(h)** HNNH **(i)** C_3H_6 (one double bond)
(j) C_3H_4 (two double bonds) **(k)** C_3H_4 (one triple bond)

PROBLEM 1-4

Circle any lone pairs (pairs of nonbonding electrons) in the structures you drew for Problem 1-3.

1-6 Electronegativity and Bond Polarity

A bond with the electrons shared equally between the two atoms is called a **nonpolar covalent bond.** The bond in H_2 and the C—C bond in ethane are nonpolar covalent bonds. In most bonds between two different elements, the bonding electrons are attracted more strongly to one of the two nuclei. An unequally shared pair of bonding electrons is called a **polar covalent bond.**

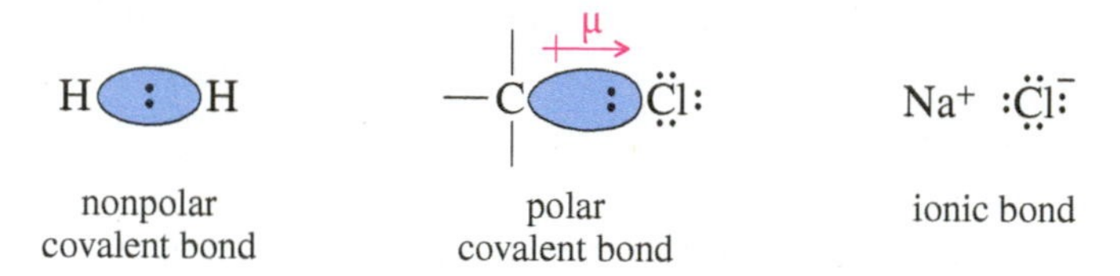

When carbon is bonded to chlorine, for example, the bonding electrons are attracted more strongly to the chlorine atom. The carbon atom bears a small partial positive charge, and the chlorine atom bears a partial negative charge. Figure 1-7 shows the polar carbon–chlorine bond in chloromethane. We symbolize the bond polarity using an arrow with its head at the negative end of the polar bond and a plus sign at the positive end. The bond polarity is measured by its **dipole moment** (μ), defined to be the amount of partial charge (δ^+ and δ^-) multiplied by the bond length (d). The symbol δ^+ means "a small amount of positive charge"; δ^- means "a small amount of negative charge." To symbolize a dipole moment, we use a crossed arrow pointing from the + charge toward the – charge.

$\mu = \delta \times d$
dipole moment

Figure 1-7 also shows an **electrostatic potential map (EPM)** for chloromethane, using color to represent the calculated charge distribution in the molecule. Red shows electron-rich regions. Blue and purple show electron-poor regions. Orange, yellow, and green show intermediate levels of electrostatic potential. In chloromethane, the red region shows the partial negative charge on chlorine, and the blue region shows the partial positive charges on carbon and the hydrogen atoms.

We often use **electronegativities** as a guide in predicting whether a given bond will be polar and the direction of its dipole moment. The Pauling electronegativity scale, most commonly used by organic chemists, is based on bonding properties, and it is useful for predicting the polarity of covalent bonds. Elements with higher electronegativities generally have more attraction for the bonding electrons. Therefore, in a bond between two different atoms, the atom with the higher electronegativity is the negative end of the dipole. Figure 1-8 shows Pauling electronegativities for some of the important elements in organic compounds.

Notice that the electronegativities increase from left to right across the periodic table. Nitrogen, oxygen, and the halogens are all more electronegative than carbon; sodium, lithium, and magnesium are less electronegative. Hydrogen's electronegativity is similar to that of carbon, so we usually consider C—H bonds to be nonpolar. We will consider the polarity of bonds and molecules in more detail in Section 2-1.

$\overset{\mu}{\longrightarrow}$

H$\diagdown^{\delta^+}$ δ^-
C—Cl
H/ |
H

chloromethane

chloromethane

FIGURE 1-7
Bond polarity. Chloromethane contains a polar carbon–chlorine bond with a partial negative charge on chlorine and a partial positive charge on carbon. The electrostatic potential map shows a red region (electron-rich) around the partial negative charge and a blue region (electron-poor) around the partial positive charge. Other colors show intermediate values of electrostatic potential.

H 2.2						
Li 1.0	Be 1.6	B 2.0	C 2.5	N 3.0	O 3.4	F 4.0
Na 0.9	Mg 1.3	Al 1.6	Si 1.9	P 2.2	S 2.6	Cl 3.2
K 0.8						Br 3.0
						I 2.7

FIGURE 1-8
The Pauling electronegativities of some of the elements found in organic compounds.

PROBLEM 1-5

Use electronegativities to predict the direction of the dipole moments of the following bonds.

(a) C—Cl **(b)** C—O **(c)** C—N **(d)** C—S **(e)** C—B
(f) N—Cl **(g)** N—O **(h)** N—S **(i)** N—B **(j)** B—Cl

1-7 Formal Charges

In polar bonds, the partial charges (δ^+ and δ^-) on the bonded atoms are *real*. **Formal charges** provide a method for keeping track of electrons, but they may or may not correspond to real charges. In most cases, if the Lewis structure shows that an atom has a formal charge, it actually bears at least part of that charge. The concept of formal charge helps us determine which atoms bear most of the charge in a charged molecule; it also helps us to see charged atoms in molecules that are neutral overall.

To calculate formal charges, count how many electrons contribute to the charge of each atom and compare that number with the number of valence electrons in the free, neutral atom (given by the group number in the periodic table on the inside back cover). The electrons that contribute to an atom's charge are

1. *all* its unshared (nonbonding) electrons; plus
2. *half* the (bonding) electrons it shares with other atoms, or one electron of each bonding pair.

The formal charge of a given atom can be calculated by the formula

$$\text{formal charge (FC)} = [\text{group number}] - [\text{nonbonding electrons}] - \tfrac{1}{2}[\text{shared electrons}]$$

SOLVED PROBLEM 1-1

Compute the formal charge (FC) on each atom in the following structures.

(a) Methane (CH_4)

H
H:C:H
H

SOLUTION

Each of the hydrogen atoms in methane has one bonding pair of electrons (two shared electrons). Half of two shared electrons is one electron, and one valence electron is what hydrogen needs to be neutral. Hydrogen atoms with one bond are formally neutral: FC = 1 − 0 − 1 = 0.

(continued)

The carbon atom has four bonding pairs of electrons (eight electrons). Half of eight shared electrons is four electrons, and four electrons are what carbon (group 4A) needs to be neutral. Carbon is formally neutral whenever it has four bonds: $FC = 4 - 0 - \frac{1}{2}(8) = 0$.

(b) The hydronium ion, H_3O^+

two nonbonding electrons

$H:\overset{+\,\cdot\cdot}{\underset{\cdot\cdot}{O}}:H$
H

three bonds, six bonding electrons

SOLUTION

In drawing the Lewis structure for this ion, we use eight electrons: six from oxygen plus three from the hydrogens, minus one because the ion has a positive charge. Each hydrogen has one bond and is formally neutral. Oxygen is surrounded by an octet, with six bonding electrons and two nonbonding electrons. Half the bonding electrons plus all the nonbonding electrons contribute to its charge: $\frac{6}{2} + 2 = 5$; but oxygen (group 6A) needs six valence electrons to be neutral. Consequently, the oxygen atom has a formal charge of +1: $FC = 6 - 2 - \frac{1}{2}(6) = +1$.

(c) $H_3N{-}BH_3$

Boron has four bonds, eight bonding electrons

H H
$H:\overset{+}{N}:\overset{-}{B}:H$
H H

Nitrogen has four bonds, eight bonding electrons

SOLUTION

This is a neutral compound where the individual atoms are formally charged. The Lewis structure shows that both nitrogen and boron have four shared bonding pairs of electrons. Both boron and nitrogen have $\frac{8}{2} = 4$ electrons contributing to their charges. Nitrogen (group 5A) needs five valence electrons to be neutral, so it bears a formal charge of +1. Boron (group 3A) needs only three valence electrons to be neutral, so it bears a formal charge of −1.

Nitrogen:	$FC = 5 - 0 - \frac{1}{2}(8) = +1$
Boron:	$FC = 3 - 0 - \frac{1}{2}(8) = -1$

(d) $[H_2CNH_2]^+$

$H_2C{=}\overset{+}{N}H_2$

SOLUTION

In this structure, both carbon and nitrogen have four shared pairs of bonding electrons. With four bonds, carbon is formally neutral; however, nitrogen is in group 5A, and it bears a formal positive charge: $FC = 5 - 0 - 4 = +1$.

This compound might also be drawn with the following Lewis structure:

$H_2\overset{+}{C}{-}\ddot{N}H_2$

In this structure, the carbon atom has three bonds with six bonding electrons. We calculate that $\frac{6}{2} = 3$ electrons, so carbon is one short of the four needed to be formally neutral: $FC = 4 - 0 - \frac{1}{2}(6) = +1$.

Nitrogen has six bonding electrons and two nonbonding electrons. We calculate that $\frac{6}{2} + 2 = 5$, so the nitrogen is uncharged in this second structure:

$$FC = 5 - 2 - \tfrac{1}{2}(6) = 0$$

The significance of these two Lewis structures is discussed in Section 1-9.

SUMMARY TABLE 1-1 Common Bonding Patterns in Organic Compounds and Ions

Atom	Valence Electrons	Positively Charged	Neutral	Negatively Charged
B	3		—B— (no octet)	—B⁻—
C	4	—C⁺— (no octet)	—C—	—C̈⁻—
N	5	—N⁺—	—N̈—	—N̈⁻—
O	6	—Ö⁺—	—Ö—	—Ö:⁻
halogens	7	—C̈l⁺—	—C̈l:	:C̈l:⁻

PROBLEM-SOLVING HINT

This is a very important table. Work enough problems to become familiar with these bonding patterns so that you can recognize other patterns as being either unusual or wrong.

Most organic compounds contain only a few common elements, usually with complete octets of electrons. Summary Table 1-1 shows the most commonly occurring bonding structures, using dashes to represent bonding pairs of electrons. Use the rules for calculating formal charges to verify the charges shown on these structures. A good understanding of the structures shown here will help you to draw organic compounds and their ions quickly and correctly.

1-8 Ionic Structures

Some organic compounds contain ionic bonds. For example, the structure of methylammonium chloride (CH_3NH_3Cl) cannot be drawn using just covalent bonds. That would require nitrogen to have five bonds, implying ten electrons in its valence shell. The correct structure shows the chloride ion ionically bonded to the rest of the structure.

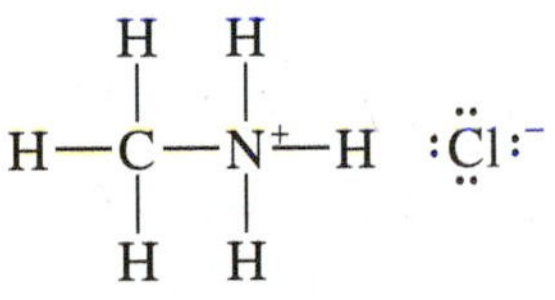

too many electrons around nitrogen

methylammonium chloride 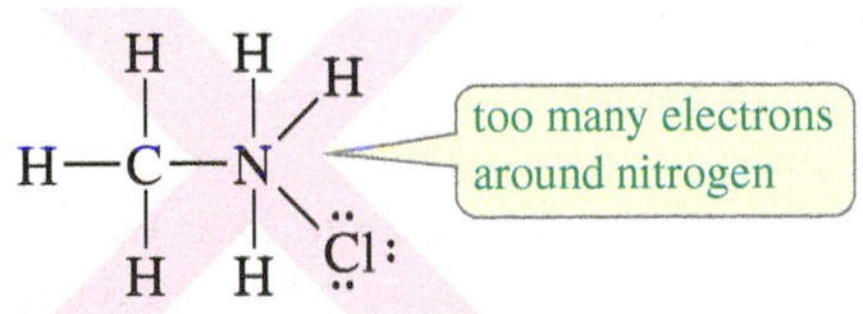   cannot be drawn covalently

Some molecules can be drawn either covalently or ionically. For example, sodium acetate ($NaOCOCH_3$) may be drawn with either a covalent bond or an ionic bond between sodium and oxygen. Because sodium generally forms ionic bonds with oxygen (as in NaOH), the ionically bonded structure is usually preferred. In general, bonds between atoms with very large electronegativity differences (about 2 or more) are usually drawn as ionic.

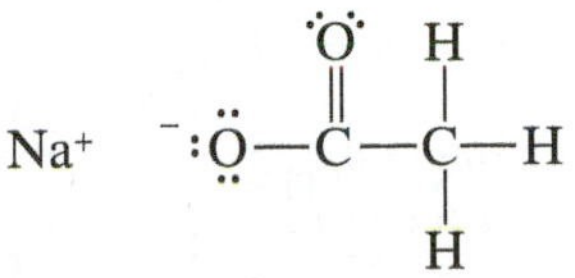

drawn as ionic (more common) 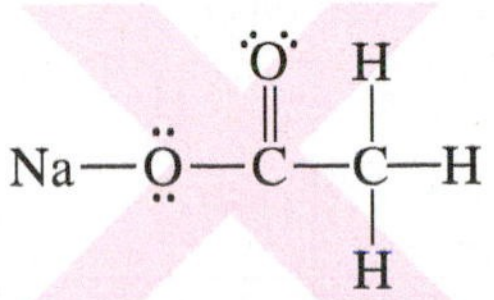   drawn as covalent (less common)

PROBLEM 1-6

Draw Lewis structures for the following compounds and ions, showing appropriate formal charges.

(a) $[CH_3OH_2]^+$	**(b)** NH_4Cl	**(c)** $(CH_3)_4NCl$
(d) $NaOCH_3$	**(e)** $^+CH_3$	**(f)** $^-CH_3$
(g) $NaBH_4$	**(h)** $NaBH_3CN$	**(i)** $(CH_3)_2O{-}BF_3$
(j) $[HONH_3]^+$	**(k)** $KOC(CH_3)_3$	**(l)** $[H_2C{=}OH]^+$

1-9 Resonance

Some compounds' structures are not adequately represented by a single Lewis structure. When two or more valence-bond structures are possible, differing only in the placement of electrons, the molecule will usually show characteristics of both structures. The different structures are called **resonance structures** or **resonance forms** because they are not different compounds, just different ways of drawing the same compound. The actual molecule is said to be a **resonance hybrid** of its resonance forms.

1-9A Resonance Hybrids

In Solved Problem 1-1(d), we saw that the ion $[H_2CNH_2]^+$ might be represented by either of the following resonance forms:

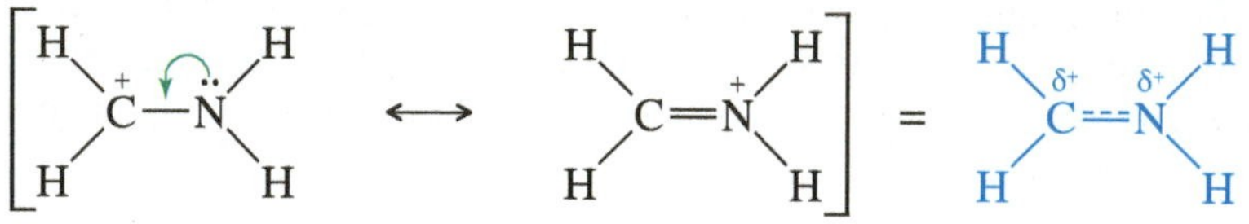

$$\left[H_2\overset{+}{C}{-}\ddot{N}H_2 \longleftrightarrow H_2C{=}\overset{+}{N}H_2 \right] = H_2\overset{\delta+}{C}{=\!=}\overset{\delta+}{N}H_2$$

resonance forms of a resonance hybrid

combined representation

The actual structure of this ion is a resonance hybrid of the two structures. In the actual molecule, the positive charge is **delocalized** (spread out) over both the carbon atom and the nitrogen atom. In the left resonance form, the positive charge is on carbon, but carbon does not have an octet. We can imagine moving nitrogen's nonbonding electrons into the bond (as indicated by the green arrow) to give the second structure, with a positive charge on nitrogen and an octet on carbon. The combined representation attempts to combine the two resonance forms into a single picture, with the charge shared by carbon and nitrogen.

Spreading the positive charge over two atoms makes the ion more stable than it would be if the entire charge were localized only on the carbon or only on the nitrogen. We call this a **resonance-stabilized** cation. Resonance is most important when it allows a charge to be delocalized over two or more atoms, as in this example.

Resonance stabilization plays a crucial role in organic chemistry, especially in the chemistry of compounds having double bonds. For example, the acidity of acetic acid (shown below) is enhanced by resonance effects. When acetic acid loses a proton, the resulting acetate ion has a negative charge delocalized over both of the oxygen atoms. Each oxygen atom bears half of the negative charge, and this delocalization stabilizes the ion. Each of the carbon–oxygen bonds is halfway between a single bond and a double bond, and they are said to have a *bond order* of $1\frac{1}{2}$.

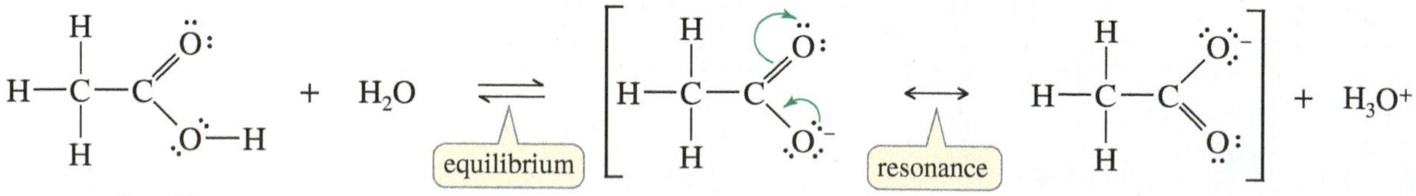

$$CH_3C(=\ddot{O}:)\ddot{O}{-}H + H_2O \rightleftharpoons \left[CH_3C(=\ddot{O}:)\ddot{O}:^- \longleftrightarrow CH_3C(\ddot{O}:^-)=\ddot{O}: \right] + H_3O^+$$

acetic acid — equilibrium — resonance — acetate ion

We use a single double-headed arrow between resonance forms (and often enclose them in brackets) to indicate that the actual structure is a hybrid of the Lewis structures we have drawn. By contrast, an equilibrium is represented by two arrows in opposite directions. Occasionally we use curved arrows (shown in green above) to help us see how we mentally move the electrons between one resonance form and another. The electrons do not actually move like these curved arrows show, and they do not "resonate" back and forth. They are delocalized over all the resonance forms at the same time.

Some uncharged molecules actually have resonance-stabilized structures with equal positive and negative formal charges. For example, we can draw two Lewis structures for nitromethane (CH_3NO_2), but both of them have a formal positive charge on nitrogen and a negative charge on one of the oxygens. Thus, nitromethane has a positive charge on the nitrogen atom and a negative charge spread equally over the two oxygen atoms. The N—O bonds are midway between single and double bonds, as indicated in the combined representation:

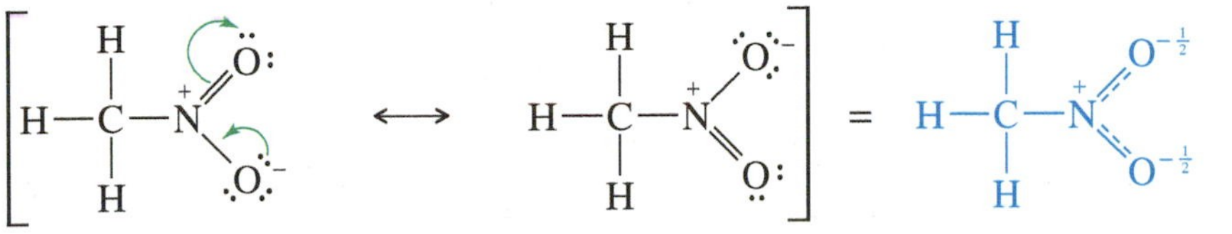

resonance forms

combined representation

Remember that individual resonance forms do not exist. The molecule does not "resonate" between these structures. It is a hybrid with some characteristics of both. An analogy is a mule, which is a hybrid of a horse and a donkey. The mule does not "resonate" between looking like a horse and looking like a donkey; it looks like a mule all the time, with the broad back of the horse and the long ears of the donkey.

1-9B Major and Minor Resonance Contributors

Two or more correct Lewis structures for the same compound may or may not represent electron distributions of equal energy. Although separate resonance forms do not exist, we can estimate their relative energies as if they did exist. More stable resonance forms are closer representations of the real molecule than less stable ones. The two resonance forms shown earlier for the acetate ion have similar bonding, and they are of identical energy. The same is true for the two resonance forms of nitromethane. The following resonance forms are bonded differently, however.

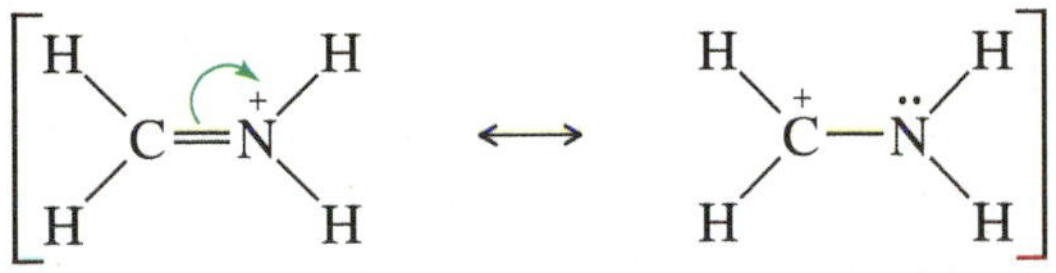

all octets (major contributor)

no octet on C (minor contributor)

These structures are not equal in estimated energy. The first structure has the positive charge on nitrogen. The second has the positive charge on carbon, and the carbon atom does not have an octet. The first structure is more stable because it has an additional bond and all the atoms have octets. Many stable ions have a positive charge on a nitrogen atom with four bonds (see Summary Table 1-1, page 6). We call the more stable resonance form the **major contributor,** and the less stable form is the **minor contributor.** Lower-energy (more stable) resonance forms are closer representations of the actual molecule or ion than are the higher-energy (less stable) ones.

Many organic molecules have major and minor resonance contributors. Formaldehyde ($H_2C{=}O$) can be written with a negative charge on oxygen, balanced by a positive charge on carbon. This polar resonance form is higher in estimated energy than the double-bonded structure because it has charge separation, fewer bonds, and

PROBLEM-SOLVING HINT

Second-row elements (B, C, N, O, F) cannot have more than eight electrons in their valence shells. The following is NOT a valid Lewis structure:

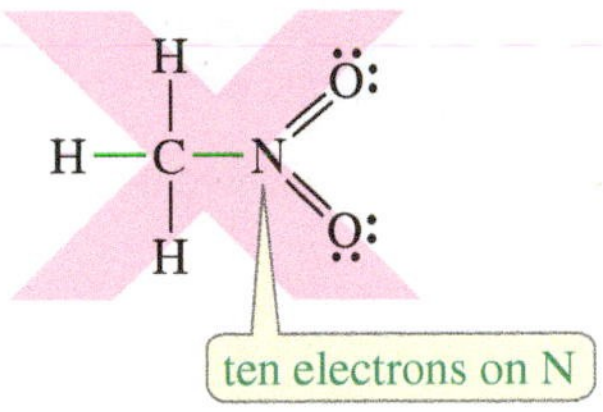

ten electrons on N

PROBLEM-SOLVING HINT

At this point, we are using four types of arrows in drawing structures and reactions:

- ⟶ Reaction arrows show the reactants converting to products.
- ⇌ Equilibrium arrows show equal rates of the forward and reverse reactions.
- ⟷ Resonance arrows connect forms that are actually the same structure but are drawn with the *delocalized* electrons arranged differently.
- Green curved arrows are a mental tool used to imagine moving electrons from one resonance form to another.

a positively charged carbon atom without an octet. The charge-separated structure is only a minor contributor, but it helps to explain why the formaldehyde C=O bond is very polar, with a partial positive charge on carbon and a partial negative charge on oxygen. The electrostatic potential map (EPM) of formaldehyde also shows an electron-rich region (red) around oxygen and an electron-poor region (blue) around carbon.

EPM of formaldehyde

all octets
more bonds
no charge separation
(major contributor)

no octet on C
fewer bonds
charge separation
(minor contributor)

dipole moment

In drawing resonance forms, we try to draw structures that are as low in energy as possible. The best candidates are those that have the maximum number of octets and the maximum number of bonds. Also, we look for structures with the minimum amount of charge separation.

Only electrons can be delocalized. Unlike electrons, nuclei cannot be delocalized. They must remain in the same places, with the same bond distances and angles, in all the resonance contributors. The following general rules will help us to draw realistic resonance structures:

1. All the resonance forms must be valid Lewis structures for the compound. Second-row elements (B, C, N, O, F) can never have more than eight electrons in their valence shells.
2. Only the placement of the electrons may be shifted from one structure to another. Electrons in double bonds and nonbonding electrons (lone pairs) are most commonly shifted.
3. Nuclei cannot be moved, and all bond angles must remain the same.

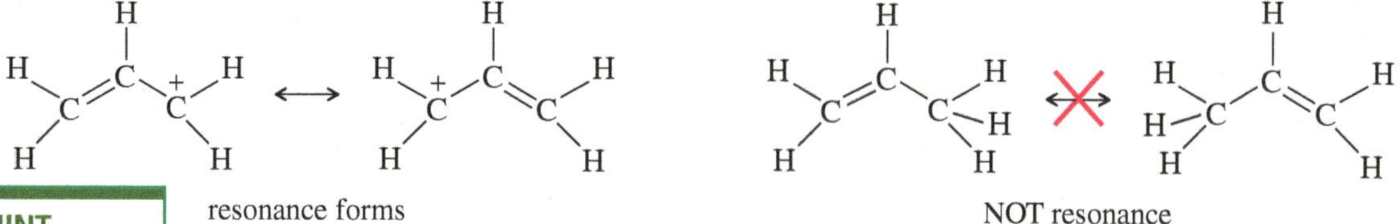

resonance forms

NOT resonance
a H nucleus has moved

4. Sigma bonds are very stable, and they are rarely involved in resonance.
5. The major resonance contributor is the one with the lowest energy. The best contributors generally have the most octets satisfied, as many bonds as possible, and as little charge separation as possible.
6. Electronegative atoms such as N, O, and halogens often help to delocalize positive charges, but they can bear a positive charge only if they have octets.
7. Resonance stabilization is most important when it serves to delocalize a charge or a radical over two or more atoms.

PROBLEM-SOLVING HINT

Resonance forms can be compared using the following criteria, beginning with the most important:

1. As many octets as possible
2. As many bonds as possible
3. As little charge separation as possible
4. Any negative charges on electronegative atoms

SOLVED PROBLEM 1-2

For each of the following compounds, draw the important resonance forms. Indicate which structures are major and minor contributors or whether they would have the same energy.

(a) $[CH_3OCH_2]^+$ **(b)** 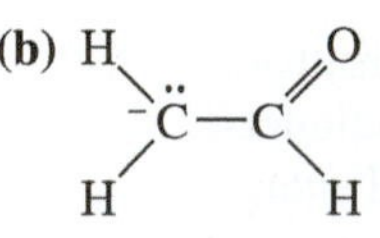 **(c)** H_2SO_4

SOLUTION (a)

$$\left[\text{H}-\overset{\text{H}}{\underset{\text{H}}{\text{C}}}-\ddot{\text{O}}-\text{C}^{+}\text{H}_2 \longleftrightarrow \text{H}-\overset{\text{H}}{\underset{\text{H}}{\text{C}}}-\overset{+}{\text{O}}=\text{CH}_2 \right]$$

minor contributor — major contributor

The first (minor) structure has a carbon atom with only six electrons around it. The second (major) structure has octets on all atoms and an additional bond.

SOLUTION (b)

$$\left[\text{H}_2\ddot{\text{C}}^{-}-\text{CH}=\ddot{\text{O}}: \longleftrightarrow \text{H}_2\text{C}=\text{CH}-\ddot{\text{O}}:^{-} \right]$$

minor contributor — major contributor

Both of these structures have octets on oxygen and both carbon atoms, and they have the same number of bonds. The first structure has the negative charge on carbon; the second has it on oxygen. Oxygen is the more electronegative element, so the second structure is the major contributor.

SOLUTION (c)

$$\left[\text{H}-\text{O}-\text{S}(=\text{O})_2-\text{O}-\text{H} \longleftrightarrow \text{H}-\text{O}-\overset{+}{\text{S}}(=\text{O})(\text{O}^{-})-\text{O}-\text{H} \longleftrightarrow \text{H}-\text{O}-\overset{+}{\text{S}}(=\text{O})(\text{O}^{-})-\text{O}-\text{H} \longleftrightarrow \text{H}-\text{O}-\overset{+2}{\text{S}}(\text{O}^{-})_2-\text{O}-\text{H} \right]$$

The first structure, with more bonds and less charge separation, is possible because sulfur is a third-row element with accessible *d* orbitals, giving it an expandable valence. For example, SF_6 is a stable compound with 12 electrons around sulfur. Theoretical calculations suggest that the last structure, with octets on all atoms, may be the major resonance contributor, however. Organic chemists mostly draw the first uncharged structure, whereas inorganic chemists mostly prefer the fourth structure with octets on all atoms.

PROBLEM 1-7

Draw the important resonance forms for the following molecules and ions.

(a) CO_3^{2-} **(b)** $H_2C{=}CH{-}\overset{+}{C}H_2$ **(c)** $H_2C{=}CH{-}\overset{-}{C}H_2$ **(d)** NO_3^-

(e) NO_2^- **(f)** $H{-}\overset{O}{\overset{\|}{C}}{-}CH{=}CH{-}\overset{-}{C}H_2$ **(g)** $[CH_3C(OCH_3)_2]^+$ **(h)** $B(OH)_3$

PROBLEM-SOLVING HINT

In drawing resonance forms for ions, see how you can delocalize the charge over several atoms. Try to spread a negative charge over electronegative elements such as oxygen and nitrogen. Try to spread a positive charge over as many carbons as possible, but especially over any atoms that can bear the positive charge and still have an octet, such as oxygen (with three bonds) or nitrogen (with four bonds).

PROBLEM 1-8 (PARTIALLY SOLVED)

For each of the following compounds, draw the important resonance forms. Indicate which structures are major and minor contributors or whether they have the same energy.

(a) H_2CNN

$$\left[\text{H}_2\text{C}=\overset{+}{\text{N}}=\ddot{\text{N}}:^{-} \longleftrightarrow \text{H}_2\overset{-}{\ddot{\text{C}}}-\overset{+}{\text{N}}\equiv\text{N}: \right] \qquad \text{H}_2\overset{-}{\ddot{\text{C}}}-\ddot{\text{N}}=\overset{+}{\text{N}}:$$

major — minor — not significant (no octet)

(continued)

SOLUTION

The first two resonance forms have all their octets satisfied, and they have the same number of bonds. They show that the central nitrogen atom bears a positive charge, and the outer carbon and nitrogen share a negative charge. A nitrogen atom is more electronegative than a carbon atom, so we expect the structure showing the minus charge on nitrogen to be the major contributor.

The third structure is not one we would normally draw, because it is much less significant than the other two. It has a nitrogen atom without an octet, and it has fewer bonds than the other two structures. Nitrogen often bears a positive charge when it has four bonds and an octet, but it rarely bears a positive charge without an octet.

(b) $[H_2CCN]^-$ **(c)** $[H_2COCH_3]^+$ **(d)** $[H_2CNO_2]^-$ **(e)** $[CH_3C(OH)_2]^+$

(f) $[CH_2CHNH]^-$ **(g)** H—C(=O)—$\bar{C}$H—C(=O)—H **(h)** CHO—$\bar{C}$H—CN

PROBLEM-SOLVING STRATEGY Drawing and Evaluating Resonance Forms

We commonly draw resonance forms (resonance structures) for one of two reasons:

1. To show how delocalization stabilizes a reactive species, such as a cation or anion, by spreading the charge over two or more atoms in a structure.
2. To explain the characteristics of a compound and why it reacts the way it does.

In drawing resonance forms to show how a cation or an anion is delocalized, you should look for double bonds and nonbonding electron pairs (lone pairs) next to the charged atom. A double bond next to a charged atom can delocalize the charge. In drawing resonance forms, you can use curved arrows to keep track of the electrons as you mentally move them from one place to another. Remember that this movement is imaginary. The actual molecule is a mixture of all the resonance forms all the time. The electrons and charges do not move back and forth, but are delocalized throughout the structure.

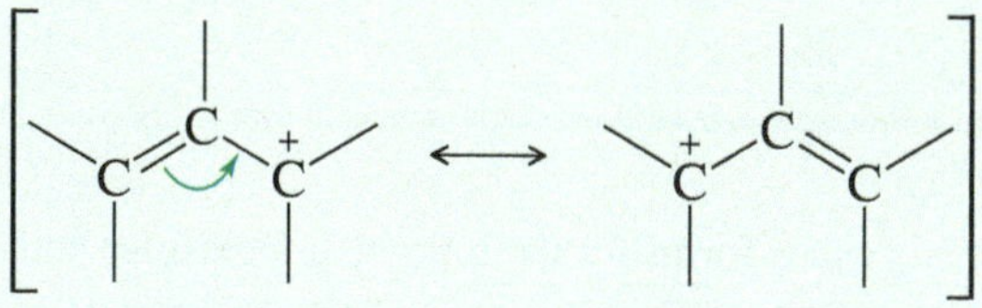

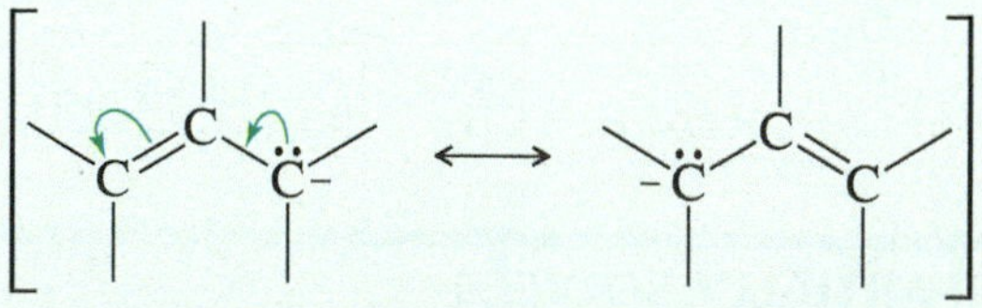

If the double bond allows an electronegative element to share a negative charge, that resonance form will be particularly stable. The most common examples are double bonds to oxygen and nitrogen.

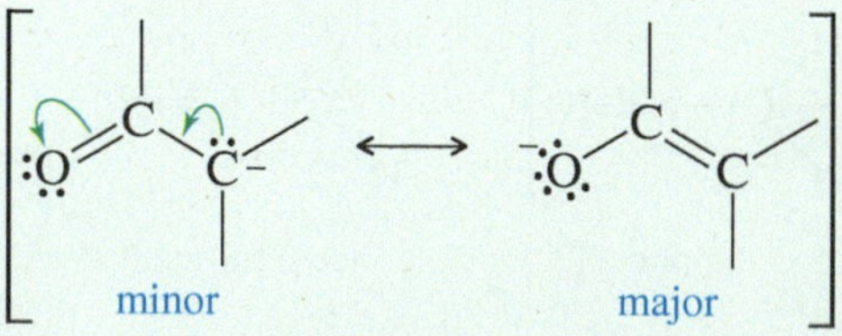

minor

major

The major resonance form has the negative charge on oxygen, an electronegative element.

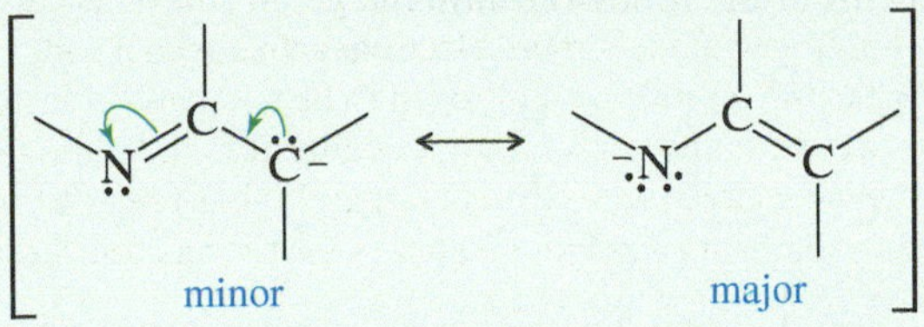

minor ↔ major

The major resonance form has the negative charge on nitrogen, an electronegative element.

These resonance forms show how the negative charge is spread over a carbon atom and the oxygen or nitrogen atom. All the atoms have octets in both structures, so the more important resonance form is the one with the negative charge on the more electronegative element, either oxygen or nitrogen. Still, these resonance forms show why this anion can react at either the carbon or the oxygen or nitrogen atom.

The resonance form shown below is also a valid Lewis structure, as well as a valid resonance form, but it is not an important resonance form (except in quantum-mechanical calculations) because it does not help to delocalize the negative charge, and it includes unnecessary charge separation and a carbon atom without an octet. You would not draw this resonance form in most cases.

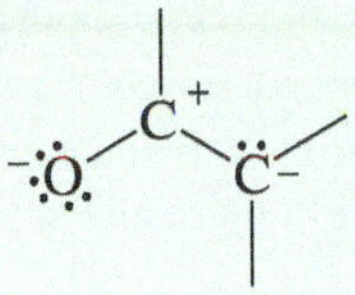

Not an important resonance form. Too much charge separation, too few bonds, and a carbon atom without an octet.

A carbon atom with a positive charge lacks an octet. If it is bonded to an atom with nonbonding electrons (often oxygen or nitrogen), the carbon atom can share those nonbonding electrons. This sharing allows it to gain an octet and delocalize the positive charge onto its neighbor. The following examples show how this delocalization affords more stability to the cation.

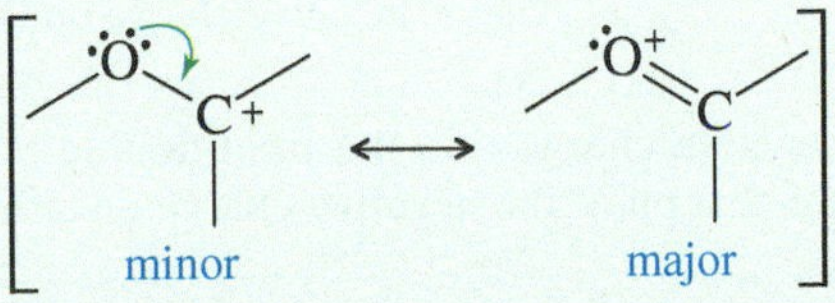

minor ↔ major

The major resonance form has all octets satisfied, and it has more bonds than the other form.

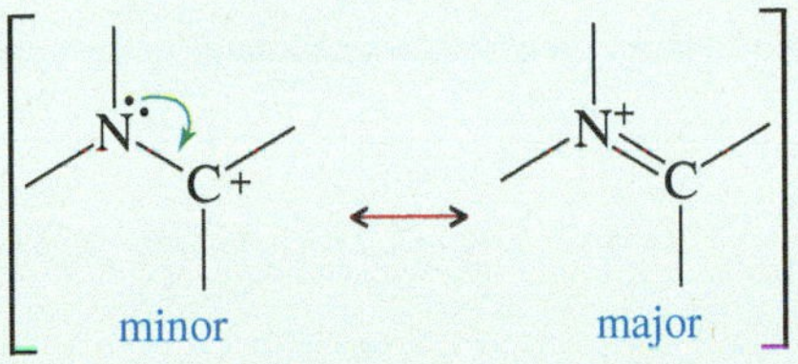

minor ↔ major

The major resonance form has all octets satisfied, and it has more bonds than the other form.

At first, it seems surprising that the major resonance form has the positive charge on oxygen or nitrogen, which are both more electronegative than carbon. But here, the comparison is between carbon with a positive charge and lacking an octet, versus oxygen or nitrogen with a positive charge and a complete octet. This situation is similar to the stable ions H_3O^+ and NH_4^+, which also bear a positive charge on oxygen or nitrogen, but with a complete octet.

Finally, let's see how you would approach problems that involve delocalizing positive and negative charges.

SOLVED PROBLEM 1-3

Draw the important resonance forms of the following anion:

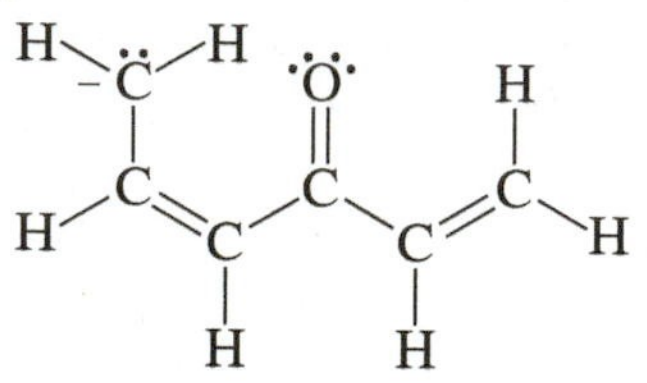

Looking ahead, you should try to put the negative charge on the electronegative oxygen atom of the C=O group. But first, the carbon with the negative charge has a C=C double bond next to it. The lone pair can delocalize into the double bond, sharing the negative charge with another carbon.

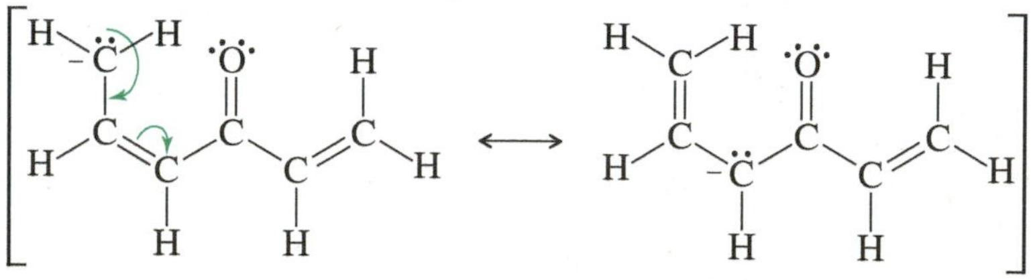

This second structure now has a C=O double bond next to the carbon bearing the negative charge. Moving the pair of electrons in that direction delocalizes the negative charge onto the oxygen atom, a favorable result.

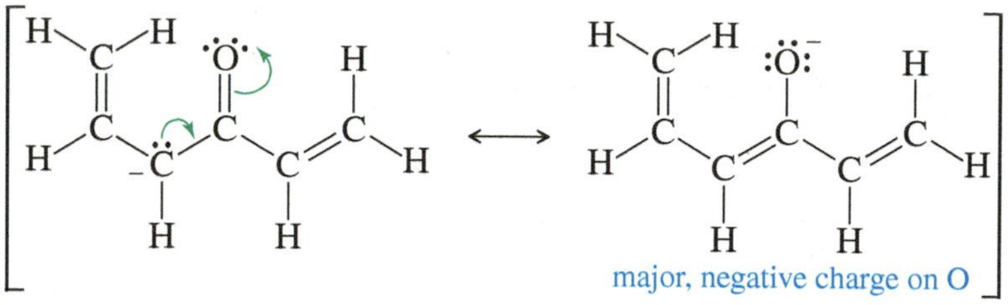

major, negative charge on O

Note that you cannot delocalize the negative charge into the final double bond on the right. There are no valid Lewis structures that place the negative charge on either of these two carbon atoms.

SOLVED PROBLEM 1-4

Draw the important resonance forms of the following cation:

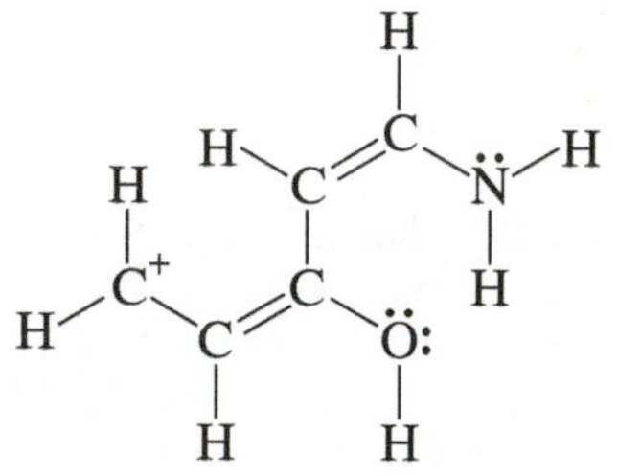

SOLUTION

Looking ahead, you should try to put the positive charge on oxygen (with three bonds) or nitrogen (with four bonds) to give resonance forms with octets. But first, notice the double bond next to the carbon with the positive charge. Moving the electrons in that double bond shares the charge with another carbon. This second carbon has a neighbor (oxygen) with lone pairs that can provide carbon with an octet, giving a relatively stable structure.

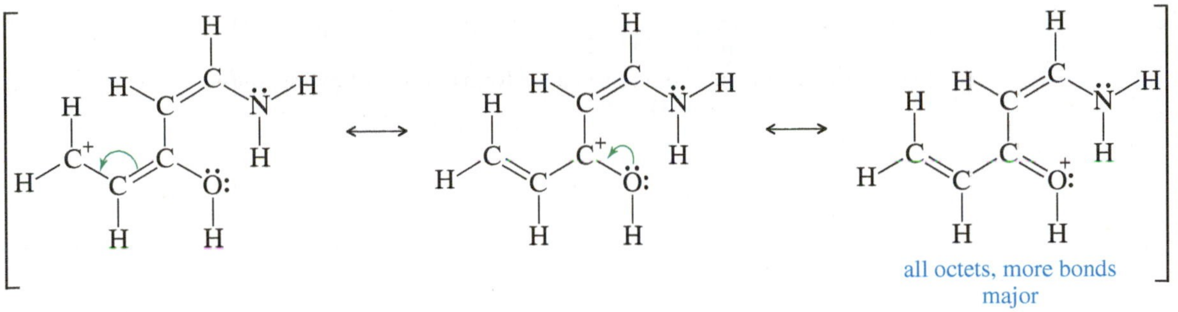

all octets, more bonds
major

This second carbon also has another double bond next to it. Moving the electrons in this double bond moves the charge to a third carbon. The third carbon with the positive charge has a neighbor (nitrogen) with lone pairs that can provide carbon with an octet, giving another relatively stable structure.

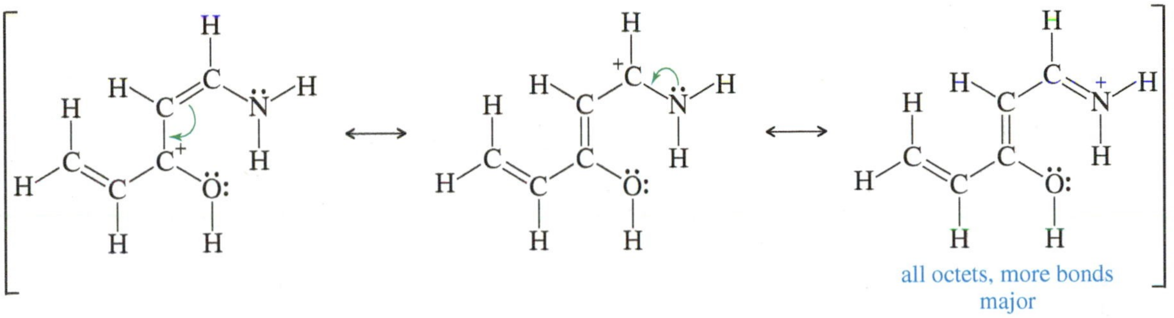

all octets, more bonds
major

These are all of the important resonance forms that you would normally draw. You could draw other valid Lewis structures, like the one below, but they do not help to delocalize the charge, and they have unnecessary charge separation. These are not important resonance forms. This one also shows a negative charge on a carbon atom, which is unrealistic in a cation where we seek to delocalize a positive charge. You would not normally draw this resonance form.

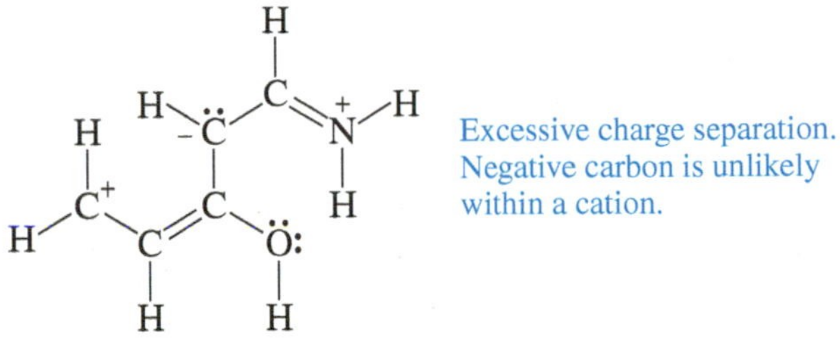

Excessive charge separation.
Negative carbon is unlikely
within a cation.

Finally, resonance forms can help you to predict where you should expect high or low electron density in a molecule. To do that, look for a highly polar area and see how resonance delocalizes those partial charges over the molecule. Atoms that bear a positive charge in one or more resonance forms are likely to be electron-poor, while those that bear a negative charge are likely to be electron-rich.

SOLVED PROBLEM 1-5

Use resonance structures to identify the areas of high and low electron density in the following compound:

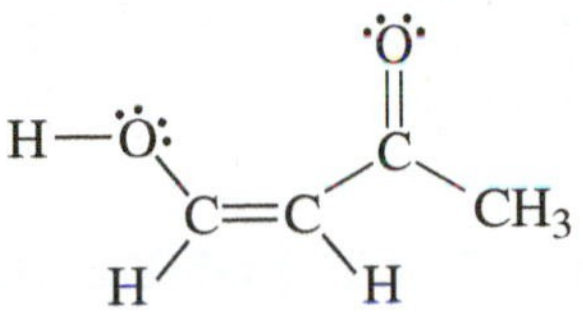

The C=O bond is highly polar, so you could begin by showing charges to represent its polarity. This resonance form is not nearly as stable as the original uncharged structure, but it helps to visualize the partial charges.

(continued)

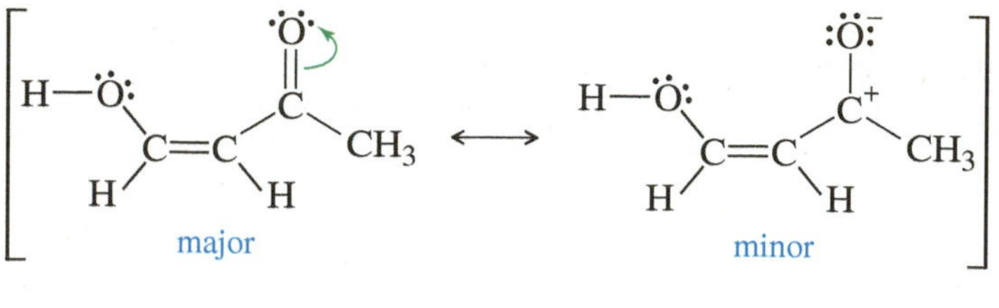

major ⟷ minor

Moving the bonds and nonbonding electrons next to these charges gives other important resonance forms. Note how the oxygen of the OH group is an electron donor, while the oxygen of the C═O group is an electron acceptor.

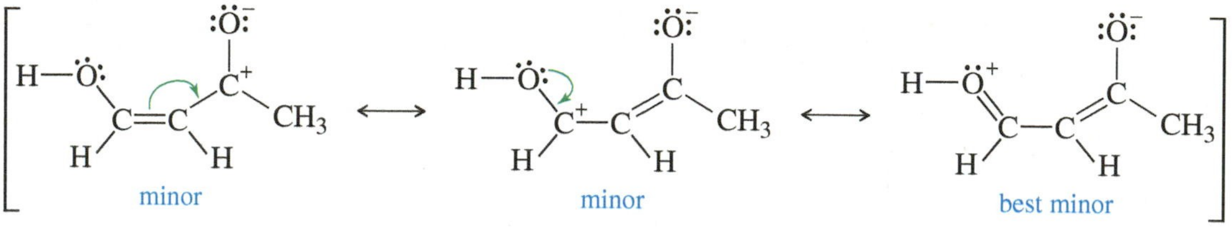

minor ⟷ minor ⟷ best minor

The final resonance form ("best minor") is the most stable of the structures with charge separation. The original uncharged structure labeled "major" is the best of all, with no charge separation, all octets satisfied, and the maximum number of bonds. It is more representative of the actual structure than any of the charge-separated resonance forms. Still, the charge-separated forms suggest where the molecule is likely to have electron-rich (negative) and electron-poor (positive) regions. We predict that the OH oxygen at left and the first and third carbons are electron-poor, while the oxygen at right (from the C═O double bond) is electron-rich.

PROBLEM 1-9

Draw the important resonance forms of the following cations and anions:

(a) $CH_3O—\overset{+}{C}H—CH═CH—NH(OH)$

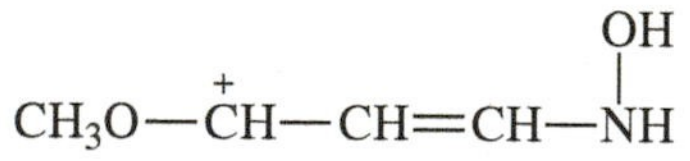

(b) $H_2N—CH═CH—C(═\overset{+}{O}H)—OCH_3$

(c) $H_2C═CH—C(NH_2)═\overset{+}{O}H$

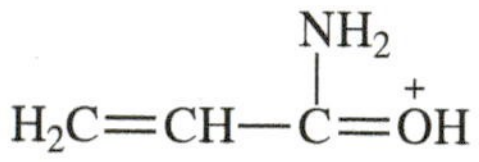

(d) $H_3C—\overset{-}{C}H—C(═O)—CH═CH—CN$

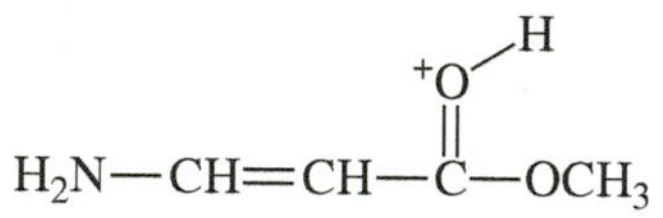

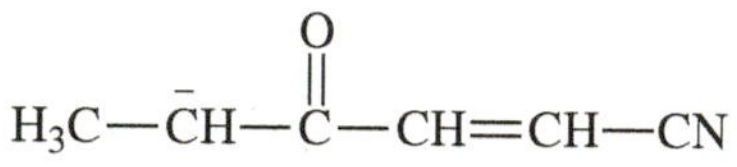

(e) $H_3C—C(═O)—\overset{-}{C}H—CH═CH—CN$

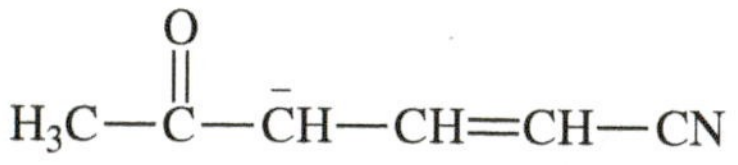

(f) $H_2C═CH—C(═NH)—\overset{-}{C}H_2$

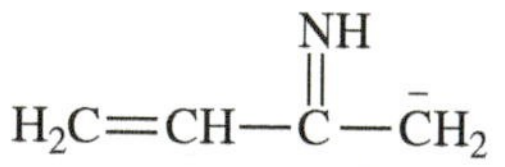

PROBLEM 1-10

Use resonance structures to identify the areas of high and low electron density in the following compounds:

(a) $H_2C═CH—NO_2$

(b) $H_2C═CH—C≡N$

(c) $H—C(═O)—NH_2$

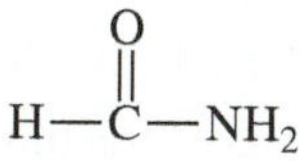

(d) $H—C(═O)—OCH_3$

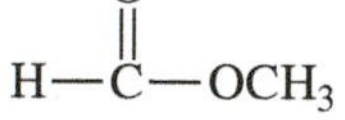

(e) $H—C(═O)—CH═CH—NH_2$

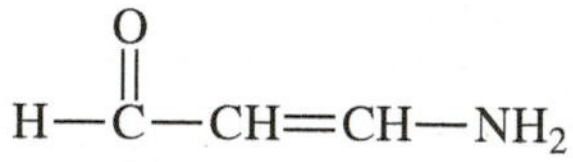

(f) $CH_3O—CH═CH—CN$

1-10 Structural Formulas

Organic chemists use several kinds of formulas to represent organic compounds. Some of these formulas involve a shorthand notation that requires some explanation. **Structural formulas** actually show which atoms are bonded to which. Structural formulas are of two major types: complete Lewis structures and condensed structural

TABLE 1-2
Examples of Condensed Structural Formulas

Compound	Lewis Structure	Condensed Structural Formula
ethane	H H H—C—C—H H H	CH_3CH_3
isobutane	H H H H—C——C——C—H H H—C—H H H	$(CH_3)_3CH$
n-hexane	H H H H H H H—C—C—C—C—C—C—H H H H H H H	$CH_3(CH_2)_4CH_3$
diethyl ether	H H H H H—C—C—Ö—C—C—H H H H H	$CH_3CH_2OCH_2CH_3$ or $CH_3CH_2—O—CH_2CH_3$ or $(CH_3CH_2)_2O$
ethanol	H H H—C—C—Ö—H H H	CH_3CH_2OH
isopropyl alcohol	H :Ö—H H H—C—C——C—H H H H	$(CH_3)_2CHOH$
dimethylamine	H H H—C—N̈—C—H H H H	$(CH_3)_2NH$

formulas. In addition, there are several ways of drawing condensed structural formulas. As we have seen, a Lewis structure symbolizes a bonding pair of electrons as a pair of dots or as a dash (—). Lone pairs of electrons are shown as pairs of dots.

1-10A Condensed Structural Formulas

Condensed structural formulas (Table 1-2) are written without showing all the individual bonds. In a condensed structural formula, each central atom is shown together with the atoms that are bonded to it. The atoms bonded to a central atom are often listed after the central atom (as in CH_3CH_3 rather than $H_3C—CH_3$) even if that is not their actual bonding order. In many cases, if there are two or more identical groups, parentheses and a subscript may be used to represent all the identical groups. Nonbonding electrons are rarely shown in condensed structural formulas.

When we write a condensed structural formula for a compound containing double or triple bonds, the multiple bonds are often drawn as they would be in a Lewis structure. Table 1-3 shows examples of condensed structural formulas containing multiple bonds. Notice that the —CHO group of an aldehyde and the —COOH group of a carboxylic acid are actually bonded differently from what the condensed notation suggests. Condensed structures are assumed to follow the octet rule even if the condensed notation does not show the bonding.

TABLE 1-3
Condensed Structural Formulas for Double and Triple Bonds

Compound	Lewis Structure	Condensed Structural Formula
but-2-ene	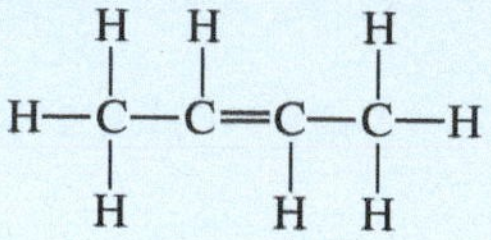 H—C(H)(H)—C(H)=C(H)—C(H)(H)—H	$CH_3CHCHCH_3$ or $CH_3CH{=}CHCH_3$
acetonitrile	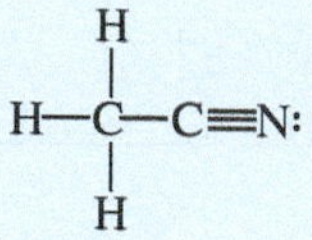 H—C(H)(H)—C≡N:	CH_3CN or $CH_3C{\equiv}N$
acetaldehyde	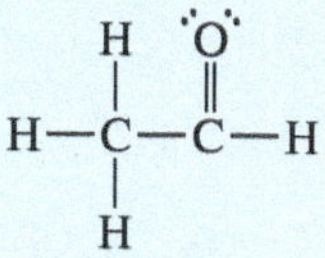 H—C(H)(H)—C(=Ö:)—H	CH_3CHO or $CH_3\overset{O}{\overset{\|}{C}}H$
acetone	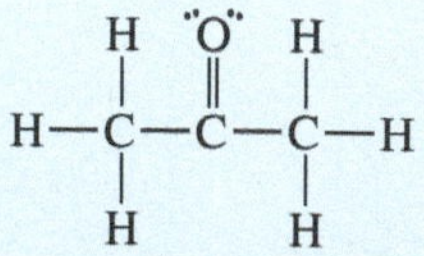 H—C(H)(H)—C(=Ö:)—C(H)(H)—H	CH_3COCH_3 or $CH_3\overset{O}{\overset{\|}{C}}CH_3$
acetic acid	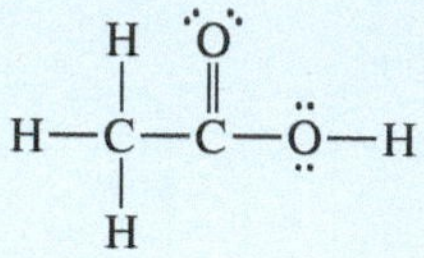 H—C(H)(H)—C(=Ö:)—Ö—H	CH_3COOH or $CH_3\overset{O}{\overset{\|}{C}}OH$ or CH_3CO_2H

As you can see from Tables 1-2 and 1-3, the distinction between a complete Lewis structural formula and a condensed structural formula can be blurry. Chemists often draw formulas with some parts condensed and other parts completely drawn out. You should work with these different types of formulas so that you understand what all of them mean.

PROBLEM 1-11

Draw complete Lewis structures for the following condensed structural formulas.

(a) $CH_3(CH_2)_3CH(CH_3)_2$
(b) $(CH_3)_2CHCH_2Cl$
(c) CH_3CH_2COCN
(d) CH_2CHCHO
(e) $(CH_3)_3CCOCHCH_2$
(f) $CH_3COCOOH$
(g) $(CH_3CH_2)_2CO$
(h) $(CH_3)_3COH$

PROBLEM-SOLVING HINT

In a line–angle formula, a carbon atom is implied at the end of every line and at every apex, unless another atom is specified.

1-10B Line–Angle Formulas

Another kind of shorthand used for organic structures is the **line–angle formula**, sometimes called a **skeletal structure** or a **stick figure.** Line–angle formulas are often used for cyclic compounds and occasionally for noncyclic ones. In a stick figure, bonds are represented by lines, and carbon atoms are assumed to be present wherever two lines meet or a line begins or ends. Nitrogen, oxygen, and halogen atoms are shown, but hydrogen atoms are not usually drawn unless they are bonded to an atom that is drawn. Each carbon atom is assumed to have enough hydrogen atoms to give it a total of four bonds. Nonbonding electrons are rarely shown. Table 1-4 shows some examples of line–angle drawings.

TABLE 1-4
Examples of Line–Angle Drawings

Compound	Condensed Structure	Line–Angle Formula
hexane	$CH_3(CH_2)_4CH_3$	
hex-2-ene	$CH_3CH{=}CHCH_2CH_2CH_3$	
hexan-3-ol	$CH_3CH_2CH(OH)CH_2CH_2CH_3$	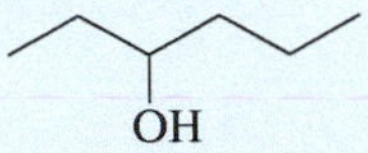 OH
cyclohex-2-en-1-one	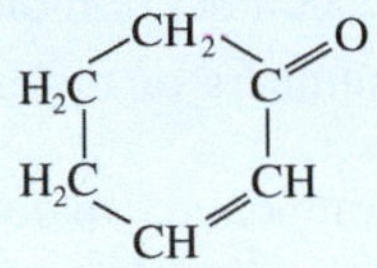 CH_2, C=O, H_2C, H_2C, CH, CH (ring)	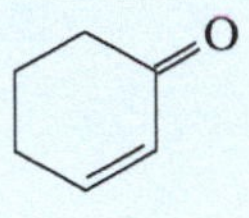 O
2-methylcyclohexan-1-ol (2-methylcyclohexanol)	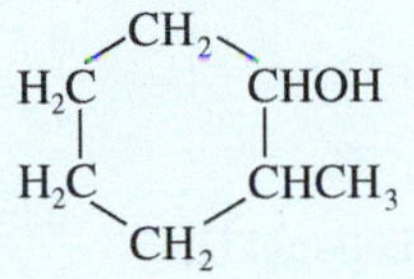 CH_2, H_2C, CHOH, H_2C, $CHCH_3$, CH_2 (ring)	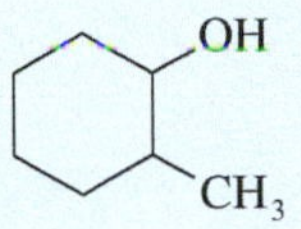 OH, CH_3 or OH 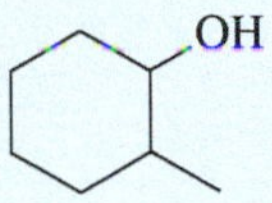
nicotinic acid (a vitamin, also called niacin)	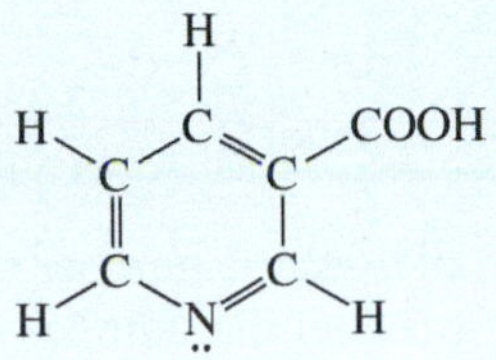 H, C, COOH, H, C, C, C, C, H, N, H (ring)	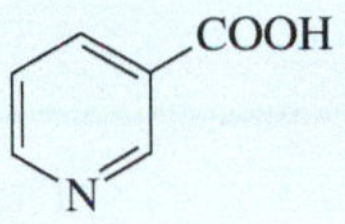 COOH, N or O, OH, N 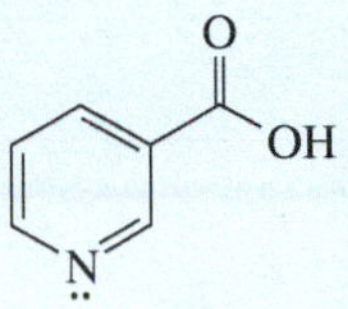

PROBLEM 1-12

Give Lewis structures corresponding to the following line–angle structures. Give the molecular formula for each structure.

(a) 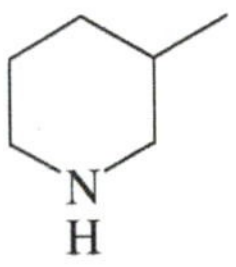 N, H

(b) 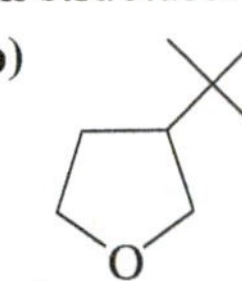

(c) 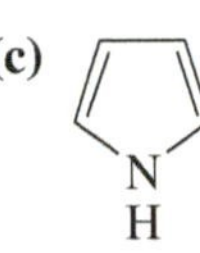

(d) 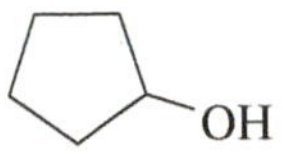 OH

(e) 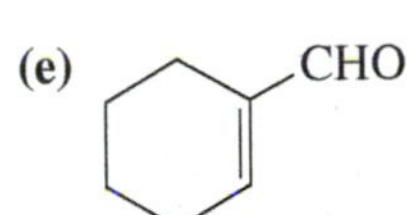

(f) 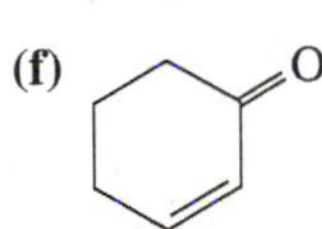

(g) 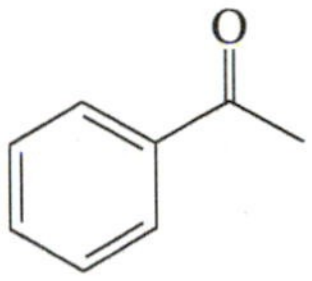 O

(h) 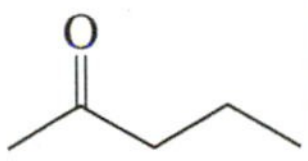 O

PROBLEM 1-13

Repeat Problem 1-11, this time drawing line–angle structures for compounds (a) through (h).

1-11 Molecular Formulas and Empirical Formulas

Before we can write possible structural formulas for a compound, we need to know its molecular formula. The **molecular formula** simply gives the number of atoms of each element in one molecule of the compound. For example, the molecular formula for butan-1-ol is $C_4H_{10}O$.

$$CH_3CH_2CH_2CH_2OH$$

butan-1-ol, molecular formula $C_4H_{10}O$

Calculation of the Empirical Formula Molecular formulas can be determined by a two-step process. The first step is the determination of an **empirical formula,** simply the relative ratios of the elements present. Suppose, for example, that an unknown compound was found by quantitative elemental analysis to contain 40.0% carbon and 6.67% hydrogen. The remainder of the weight (53.3%) is assumed to be oxygen. To convert these numbers to an empirical formula, we can follow a simple procedure.

1. Assume that the sample contains 100 g, so the percent value gives the number of grams of each element. Divide the number of grams of each element by the atomic weight to get the number of moles of that atom in the 100-g sample.
2. Divide each of these numbers of moles by the smallest one. This step should give recognizable ratios.

For the unknown compound, we do the following computations:

$$\frac{40.0 \text{ g C}}{12.0 \text{ g/mol}} = 3.33 \text{ mol C}; \qquad \frac{3.33 \text{ mol}}{3.33 \text{ mol}} = 1$$

$$\frac{6.67 \text{ g H}}{1.01 \text{ g/mol}} = 6.60 \text{ mol H}; \qquad \frac{6.60 \text{ mol}}{3.33 \text{ mol}} = 1.98 \cong 2$$

$$\frac{53.3 \text{ g O}}{16.0 \text{ g/mol}} = 3.33 \text{ mol O}; \qquad \frac{3.33 \text{ mol}}{3.33 \text{ mol}} = 1$$

The first computation divides the number of grams of carbon by 12.0, the number of grams of hydrogen by 1.0, and the number of grams of oxygen by 16.0. We compare the number of moles of C, H, and O by dividing them by the smallest number, 3.33. The final result is a ratio of one carbon to two hydrogens to one oxygen. This result gives the empirical formula $C_1H_2O_1$ or CH_2O, which simply shows the ratios of the elements. The molecular formula can be any multiple of this empirical formula, because any multiple also has the same ratio of elements. Possible molecular formulas are CH_2O, $C_2H_4O_2$, $C_3H_6O_3$, $C_4H_8O_4$, etc.

Calculation of the Molecular Formula How do we know the correct molecular formula? We can choose the right multiple of the empirical formula if we know the molecular weight. Molecular weights can be determined by methods that relate the freezing-point depression or boiling-point elevation of a solvent to the molal concentration of the unknown. If the compound is volatile, we can convert it to a gas and use its volume to determine the number of moles according to the gas law. Newer methods for determining molecular weights include *mass spectrometry*, which we will cover in Chapter 12.

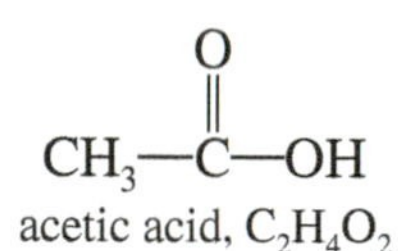

acetic acid, $C_2H_4O_2$

For our example (empirical formula CH_2O), let's assume that the molecular weight is determined to be about 60. The weight of one CH_2O unit is 30, so our unknown compound must contain twice this many atoms. The molecular formula must be $C_2H_4O_2$. The compound might be acetic acid or any of several other compounds with this molecular formula.

In Chapters 12, 13, and 15 we will use spectroscopic techniques to determine the complete structure for a compound once we know its molecular formula.

PROBLEM-SOLVING HINT

If an elemental analysis does not add up to 100%, the missing percentage is assumed to be oxygen.

PROBLEM 1-14

Compute the empirical and molecular formulas for each of the following elemental analyses. In each case, propose at least one structure that fits the molecular formula.

	C	H	N	Cl	MW
(a)	40.0%	6.67%	0	0	90
(b)	32.0%	6.67%	18.7%	0	75
(c)	25.6%	4.32%	15.0%	37.9%	93
(d)	38.4%	4.80%	0	56.8%	125

1-12 Wave Properties of Electrons in Orbitals

The chemistry we have covered so far does not explain the actual shapes and properties of organic molecules. To understand these aspects of molecular structure, we need to consider how the atomic orbitals on an atom mix to form *hybrid atomic orbitals* and how orbitals on different atoms combine to form *molecular orbitals*. Now we consider how combinations of orbitals account for the shapes and properties we observe in organic molecules.

1-12A Waveforms and Nodes

We like to picture the atom as a miniature solar system, with the electrons orbiting around the nucleus. This solar system picture satisfies our intuition, but it does not accurately reflect today's understanding of the atom. About 1923, Louis de Broglie suggested that the properties of electrons in atoms are better explained by treating the electrons as waves rather than as particles.

There are two general kinds of waves, *traveling waves* and *standing waves*. Examples of traveling waves are the sound waves that carry a thunderclap and the water waves that form the wake of a boat. Standing waves vibrate in a fixed location. Standing waves are found inside an organ pipe, where the rush of air creates a vibrating air column, and in the wave pattern of a guitar string when it is plucked. An electron in an atomic orbital is like a stationary, bound vibration: a standing wave.

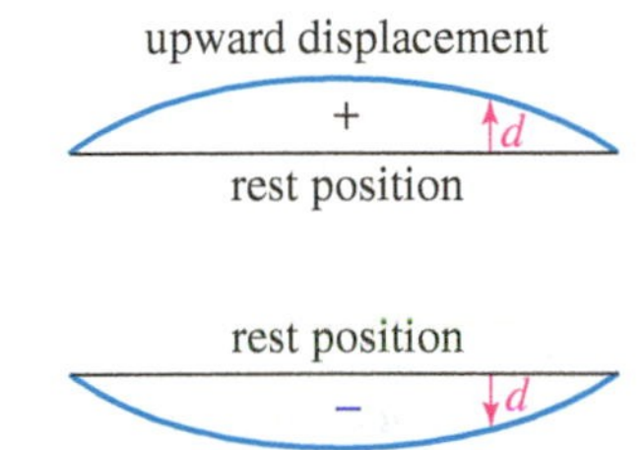

FIGURE 1-9
A standing wave. The fundamental frequency of a guitar string is a standing wave with the string alternately displaced upward and downward.

If we consider an electron in an orbital as a three-dimensional standing wave, what are the resulting properties? Let's first consider a one-dimensional analog of this standing wave; namely, the vibration of a guitar string (Figure 1-9). If you pluck a guitar string at its middle, a standing wave results. In this mode of vibration, all of the string is displaced upward for a fraction of a second, then downward for an equal time. An instantaneous picture of the waveform shows the string displaced in a smooth curve either upward or downward, depending on the exact instant of the picture.

The waveform of a 1*s* orbital is like this guitar string, except that it is three-dimensional. The orbital can be described by its **wave function,** ψ, which is the mathematical description of the shape of the wave as it vibrates. All of the wave is positive in sign for a brief instant; then it is negative in sign. The electron density at any point is given by ψ^2, the square of the wave function at that point. *The plus sign and the minus sign of these wave functions are not charges. The plus or minus sign is the instantaneous phase of the constantly changing wave function.* The 1*s* orbital is spherically symmetrical, and it is often represented by a circle (representing a sphere) with a nucleus in the center and with a plus or minus sign to indicate the instantaneous sign of the wave function (Figure 1-10).

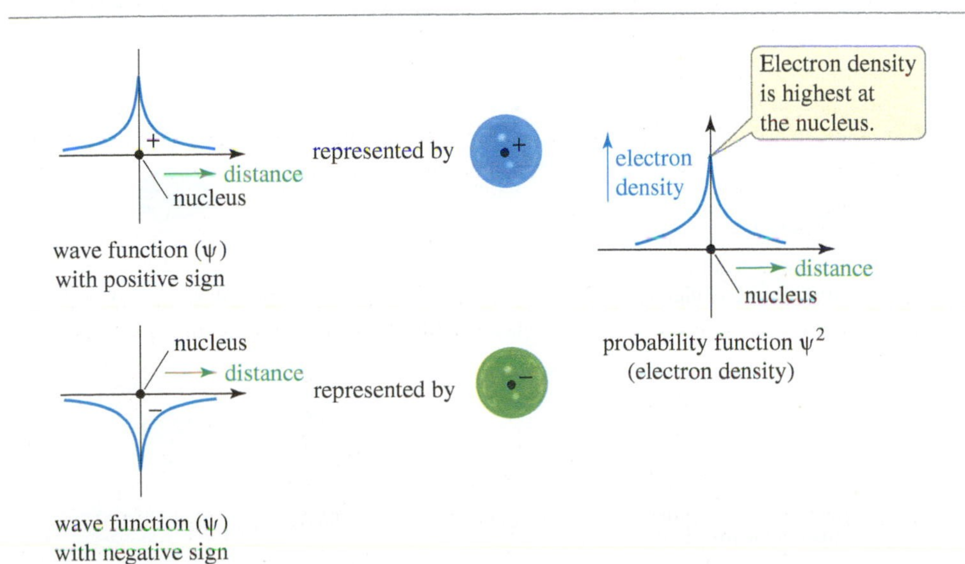

FIGURE 1-10
The 1*s* orbital. The 1*s* orbital is similar to the fundamental vibration of a guitar string. The wave function is instantaneously all positive or all negative. The square of the wave function gives the electron density. A circle with a nucleus is used to represent the spherically symmetrical *s* orbital.

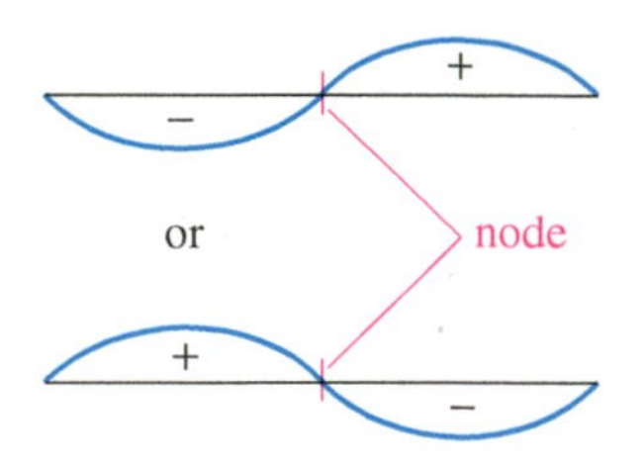

FIGURE 1-11
First harmonic of a guitar string. The two halves of the string are separated by a node, a point with zero displacement. The two halves vibrate out of phase with each other.

If you gently place a finger at the center of a guitar string while plucking the string, your finger keeps the midpoint of the string from moving. The displacement (movement + or −) at the midpoint is always zero; this point is a **node.** The string now vibrates in two parts, with the two halves vibrating in opposite directions. We say that the two halves of the string are *out of phase:* When one is displaced upward, the other is displaced downward. Figure 1-11 shows this first *harmonic* of the guitar string.

The first harmonic of the guitar string resembles the 2*p* orbital (Figure 1-12). We have drawn the 2*p* orbital as two "lobes," separated by a node (a nodal plane). The two lobes of the *p* orbital are out of phase with each other. Whenever the wave function has a plus sign in one lobe, it has a minus sign in the other lobe. When phase relationships are important, organic chemists often represent the phases with colors. Figures 1-10 and 1-12 use blue for regions with a positive phase and green for a negative phase.

1-12B Linear Combination of Atomic Orbitals

Atomic orbitals can combine and overlap to give more complex standing waves. We can add and subtract their wave functions to give the wave functions of new orbitals. This process is called **linear combination of atomic orbitals** (LCAO). The number of new orbitals generated always equals the number of starting orbitals.

PROBLEM-SOLVING HINT

When orbitals combine to form hybrid atomic orbitals or molecular orbitals, the number of orbitals formed always equals the number of orbitals that combine to form them.

1. When orbitals on *different* atoms interact, they produce **molecular orbitals** (MOs) that lead to bonding (or antibonding) interactions.
2. When orbitals on the *same* atom interact, they give **hybrid atomic orbitals** that define the geometry of the bonds.

We begin by looking at how atomic orbitals on different atoms interact to give molecular orbitals. Then we consider how atomic orbitals on the same atom can interact to give hybrid atomic orbitals.

1-13 Molecular Orbitals

The stability of a covalent bond results from a large amount of electron density in the bonding region, the space between the two nuclei (Figure 1-13). In the bonding region, the electrons are close to both nuclei, lowering the overall energy. The bonding electrons also mask the positive charges of the nuclei, so the nuclei do not repel each other as much as they would otherwise.

There is always an optimum distance for the two bonded nuclei. If they are too far apart, their attraction for the bonding electrons is diminished. If they are too close

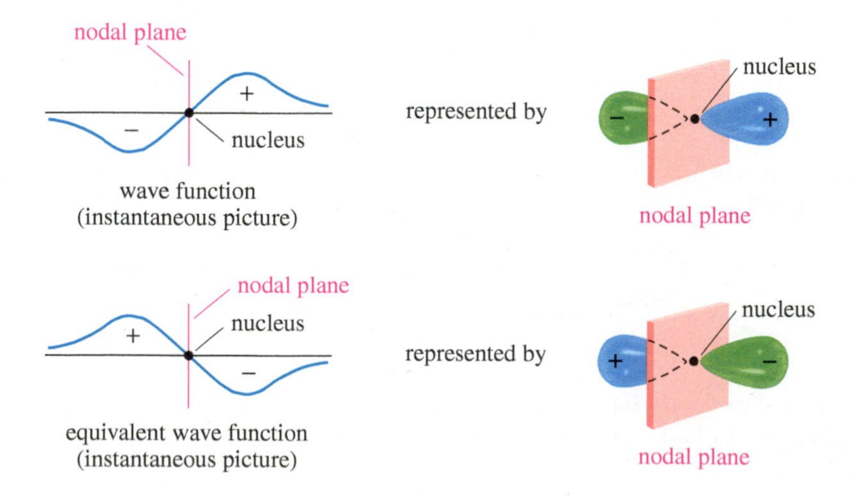

FIGURE 1-12
The 2*p* orbital. The 2*p* orbital has two lobes separated by a nodal plane. The two lobes are out of phase with each other. When one has a plus sign, the other has a minus sign.

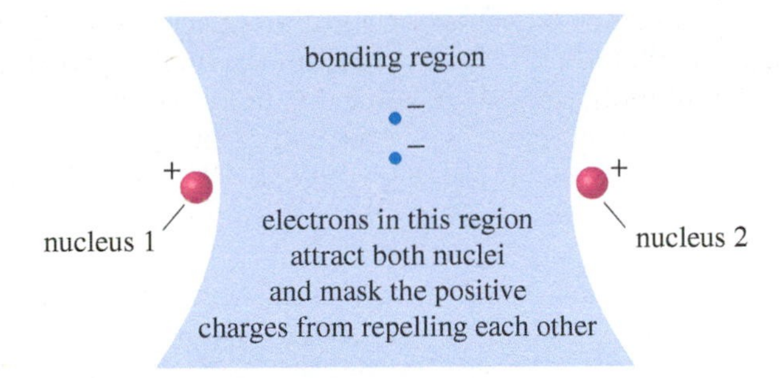

FIGURE 1-13
The bonding region. Electrons in the space between the two nuclei attract both nuclei and mask their positive charges. A bonding molecular orbital places a large amount of electron density in the bonding region.

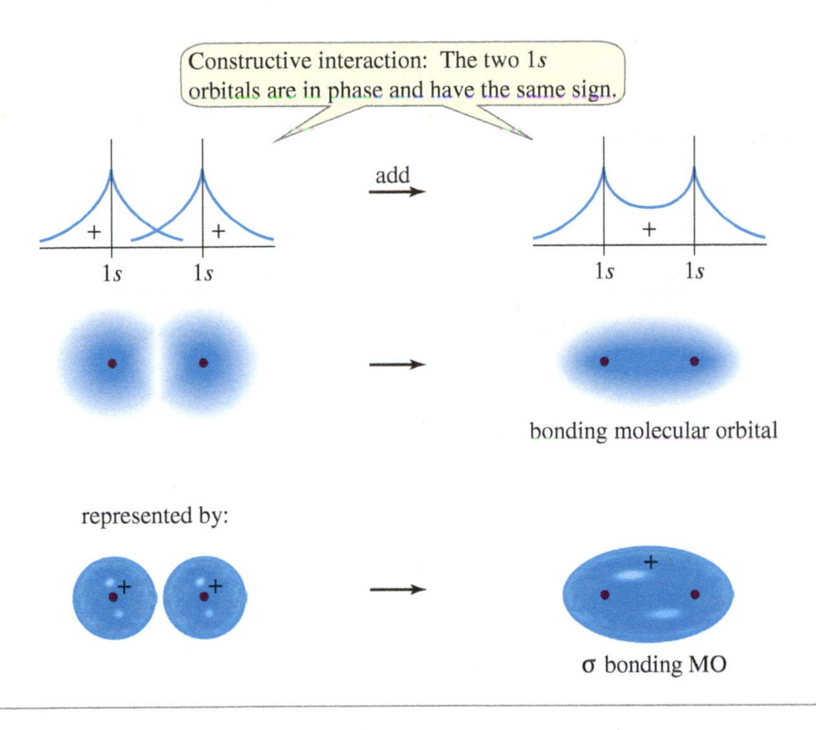

FIGURE 1-14
Formation of a σ bonding MO. When the 1*s* orbitals of two hydrogen atoms overlap in phase, they interact constructively to form a bonding MO. The electron density in the bonding region (between the nuclei) is increased. The result is a cylindrically symmetrical bond, or sigma (σ) bond.

together, their electrostatic repulsion pushes them apart. The internuclear distance where attraction and repulsion are balanced, which also gives the minimum energy (the strongest bond), is the *bond length.*

1-13A The Hydrogen Molecule; Sigma Bonding

The hydrogen molecule is the simplest example of covalent bonding. As two hydrogen atoms approach each other, their 1*s* wave functions can add *constructively* so that they reinforce each other or *destructively* so that they cancel out where they overlap. Figure 1-14 shows how the wave functions interact constructively when they are in phase and have the same sign in the region between the nuclei. The wave functions reinforce each other and increase the electron density in this bonding region. The result is a **bonding molecular orbital** (bonding MO).

The bonding MO depicted in Figure 1-14 has most of its electron density centered *along the line connecting the nuclei.* This type of bond is called a *cylindrically symmetrical bond* or a **sigma bond (σ bond).** Sigma bonds are the most common bonds in organic compounds. All single bonds in organic compounds are sigma bonds, and every double or triple bond contains one sigma bond. The electrostatic potential map (EPM) of H_2 shows its cylindrically symmetrical sigma bond, with the highest electron density (red) in the bonding region between the two protons.

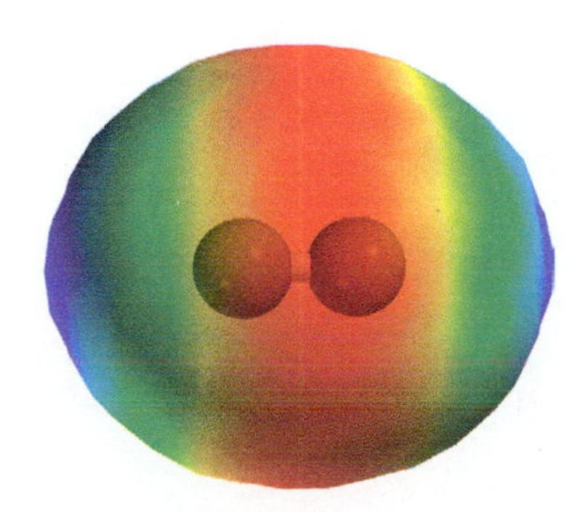

EPM of H_2

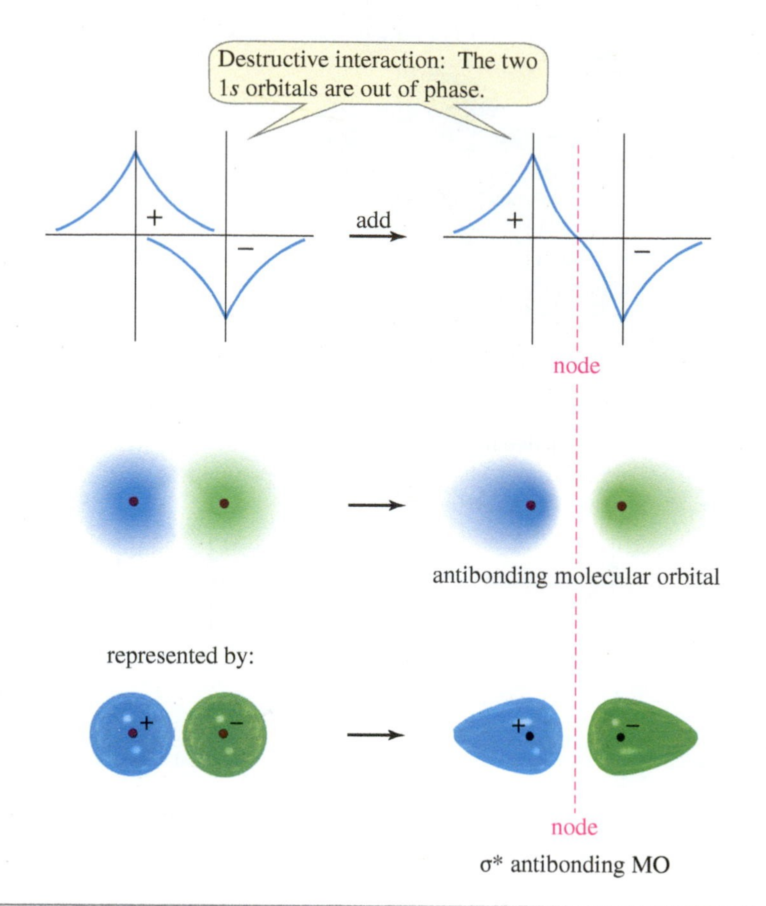

FIGURE 1-15
Formation of a σ^* antibonding MO. When two 1*s* orbitals overlap out of phase, they interact destructively to form an antibonding MO. The positive and negative values of the wave functions tend to cancel out in the region between the nuclei, and a node separates the nuclei. We use an asterisk (*) to designate antibonding orbitals such as this sigma antibonding orbital, σ^*.

When two hydrogen 1*s* orbitals overlap *out of phase* with each other, an **antibonding molecular orbital** results (Figure 1-15). The two 1*s* wave functions have opposite signs, so they tend to cancel out where they overlap. The result is a node (actually a nodal plane) separating the two atoms. The presence of a node separating the two nuclei usually indicates that the orbital is antibonding. The antibonding MO is designated σ^* to indicate an antibonding (*), cylindrically symmetrical (σ) molecular orbital.

Figure 1-16 shows the relative energies of the atomic orbitals and the molecular orbitals of the H_2 system. When the 1*s* orbitals are in phase, the resulting molecular orbital is a σ bonding MO, with lower energy than that of a 1*s* atomic orbital. When two 1*s* orbitals overlap out of phase, they form an antibonding (σ^*) orbital with higher energy than that of a 1*s* atomic orbital. The two electrons in the H_2 system are found with paired spins in the sigma bonding MO, giving a stable H_2 molecule. Both bonding and antibonding orbitals exist in all molecules, but the antibonding orbitals (such as σ^*) are usually vacant in stable molecules. Antibonding molecular orbitals often participate in reactions, however.

PROBLEM-SOLVING HINT

In stable compounds, all or most of the bonding orbitals will be filled, and all or most of the antibonding orbitals will be empty.

1-13B Sigma Overlap Involving *p* Orbitals

When two *p* orbitals overlap along the line between the nuclei, a bonding orbital and an antibonding orbital result. Once again, most of the electron density is centered along the line between the nuclei. This linear overlap is another type of sigma bonding MO. The constructive overlap of two *p* orbitals along the line joining the nuclei forms a σ bond represented as follows:

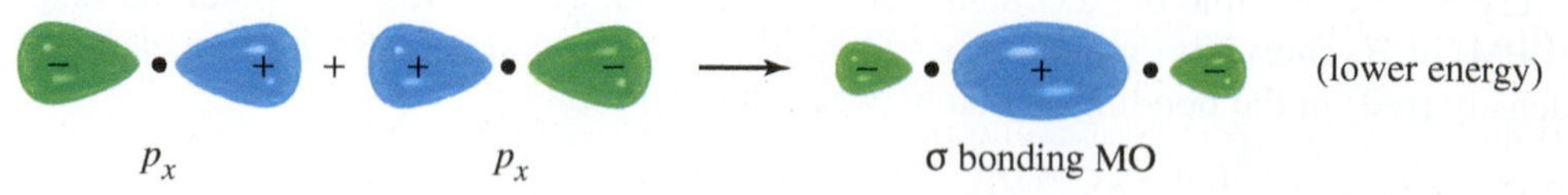

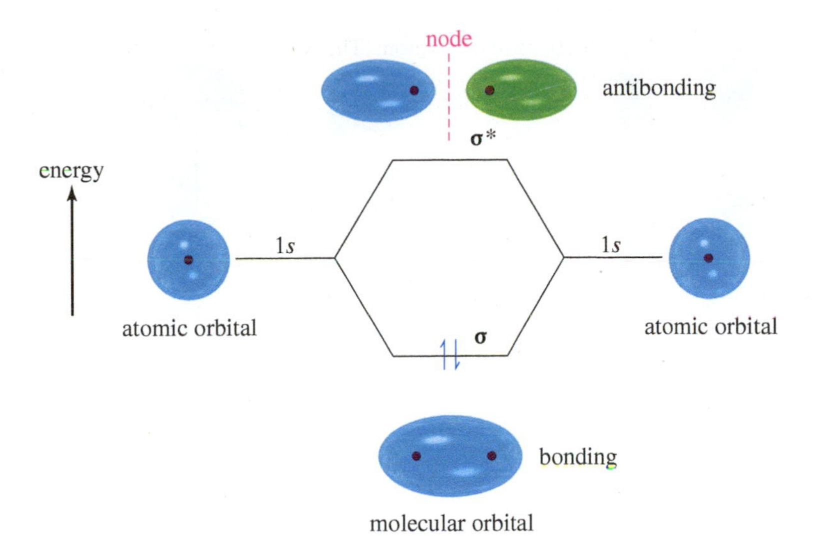

FIGURE 1-16
Relative energies of atomic and molecular orbitals. When the two hydrogen 1*s* orbitals overlap, a sigma bonding MO and a sigma antibonding MO result. The bonding MO is lower in energy than the atomic 1*s* orbital, and the antibonding orbital is higher in energy. Two electrons (represented by arrows) go into the bonding MO with opposite spins, forming a stable H_2 molecule. The antibonding orbital is vacant.

SOLVED PROBLEM 1-6

Draw the σ^* antibonding orbital that results from the destructive overlap of the two p_x orbitals just shown.

SOLUTION

This orbital results from the destructive overlap of lobes of the two *p* orbitals with opposite phases. If the signs are reversed on one of the orbitals, adding the two orbitals gives an antibonding orbital with a node separating the two nuclei:

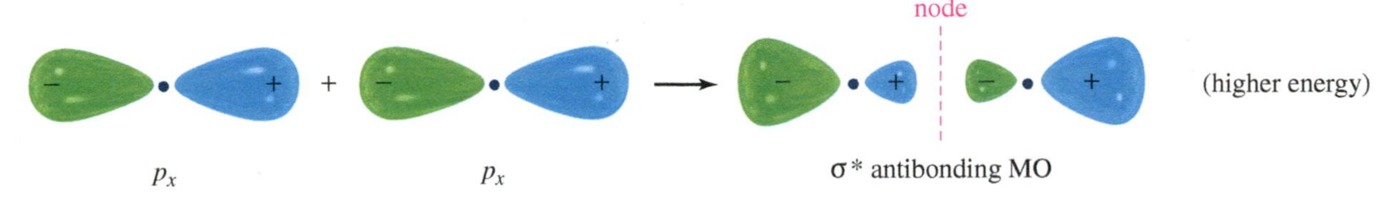

Overlap of an *s* orbital with a *p* orbital also gives a bonding MO and an antibonding MO, as shown in the following illustration. Constructive overlap of the *s* orbital with the p_x orbital gives a sigma bonding MO with its electron density centered along the line between the nuclei. Destructive overlap gives a σ^* antibonding orbital with a node separating the nuclei.

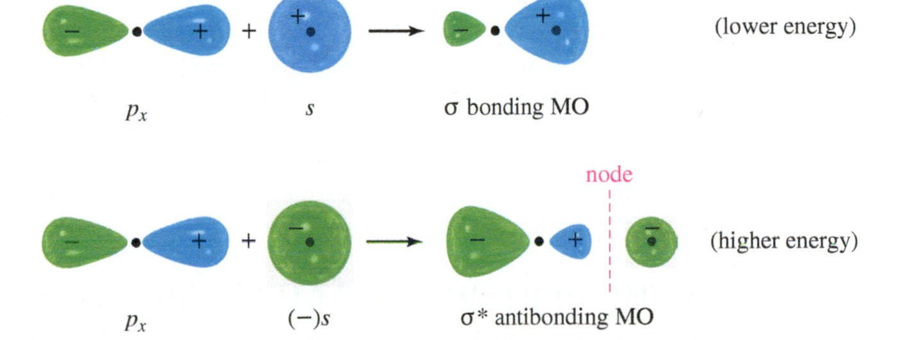

1-14 Pi Bonding

A **pi bond (π bond)** results from overlap between two *p* orbitals oriented perpendicular to the line connecting the nuclei (Figure 1-17). These parallel orbitals overlap sideways, with most of the electron density centered *above and below* the line connecting the nuclei. This overlap is parallel, not linear (a sigma bond is linear), so a pi molecular orbital is *not* cylindrically symmetrical. Figure 1-17 shows a π bonding MO and the corresponding π^* antibonding MO.

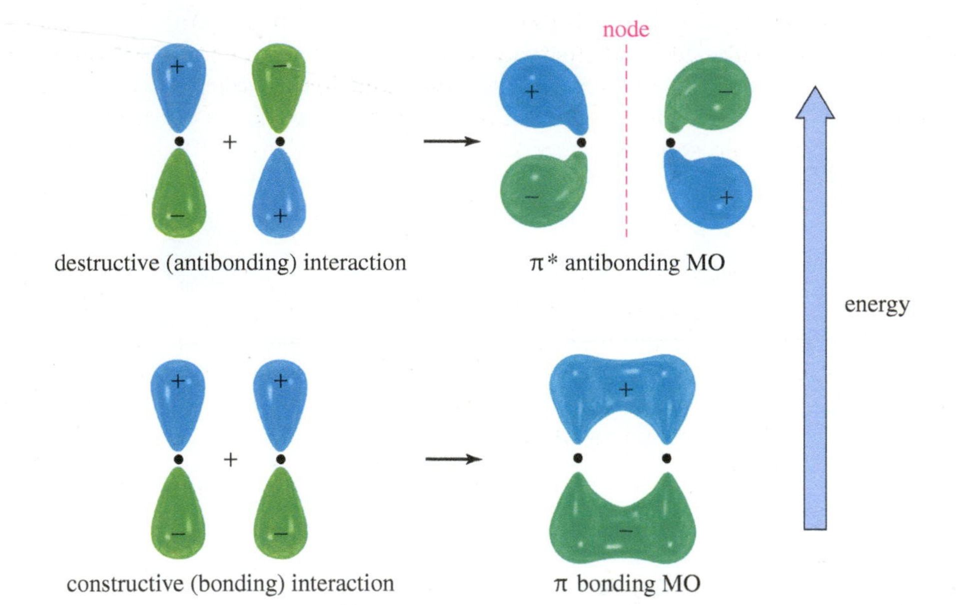

FIGURE 1-17
Pi bonding and antibonding molecular orbitals. The sideways overlap of two *p* orbitals leads to a π bonding MO and a π^* antibonding MO. A pi bond is not as strong as most sigma bonds.

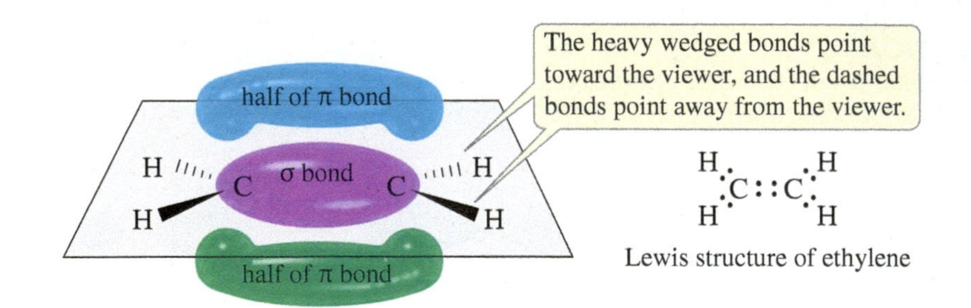

FIGURE 1-18
Structure of the double bond in ethylene. The first pair of electrons forms a σ bond. The second pair forms a π bond. The π bond has its electron density centered in two lobes above and below the σ bond. Together, the two lobes of the π bonding molecular orbital constitute one bond.

1-14A Single and Double Bonds

A **double bond** requires the presence of four electrons in the bonding region between the nuclei. The first pair of electrons goes into the sigma bonding MO, forming a strong sigma bond. The second pair of electrons cannot go into the same orbital or the same space. It goes into a pi bonding MO, with its electron density centered above and below the sigma bond.

This combination of one sigma bond and one pi bond is the normal structure of a double bond. Figure 1-18 shows the structure of ethylene, an organic molecule containing a carbon–carbon double bond.

1-15 Hybridization and Molecular Shapes

Thus far, we have discussed bonds involving overlap of simple *s* and *p* atomic orbitals. Although these simple bonds are occasionally seen in organic compounds, they are not as common as bonds formed using **hybrid atomic orbitals.** Hybrid atomic orbitals result from the mixing of orbitals on the *same* atom. The geometry of these hybrid orbitals helps us to account for the actual structures and bond angles observed in organic compounds.

If we predict the bond angles of organic molecules using just the simple *s* and *p* orbitals, we expect bond angles of about 90°. The *s* orbitals are nondirectional, and the *p* orbitals are oriented at 90° to one another (see Figure 1-5). Experimental evidence shows, however, that bond angles in organic compounds are usually close to 109°, 120°, or 180° (Figure 1-19). A common way of accounting for these bond angles is the **valence-shell electron-pair repulsion theory (VSEPR theory):** Electron pairs repel each other, and the bonds and lone pairs around a central atom generally are separated by the largest possible angles. An angle of 109.5° is the largest possible separation for four pairs of electrons; 120° is the largest separation for three pairs; and 180° is the largest separation for two pairs. All the structures in Figure 1-19 have bond angles that separate their bonds about as far apart as possible.

The shapes of these molecules cannot result from bonding between simple *s* and *p* atomic orbitals. Although *s* and *p* orbitals have the lowest energies for isolated atoms in space, they are not the best for forming bonds. To explain the shapes of common organic molecules, we conclude that the *s* and *p* orbitals combine to form hybrid atomic orbitals that separate the electron pairs more widely in space and place more electron density in the bonding region between the nuclei.

1-15A *sp* Hybrid Orbitals

Orbitals can interact to form new orbitals. We have used this principle to form molecular orbitals by adding and subtracting atomic orbitals on *different* atoms. We can also add and subtract orbitals on the *same* atom. Consider the result, shown in Figure 1-20, when we combine a *p* orbital and an *s* orbital on the same atom.

The resulting orbital is called an ***sp* hybrid orbital.** Its electron density is concentrated toward one side of the atom. We started with two orbitals (*s* and *p*), so we must finish with two *sp* hybrid orbitals. The second *sp* hybrid orbital results if we add the *p* orbital with the opposite phase (Figure 1-20).

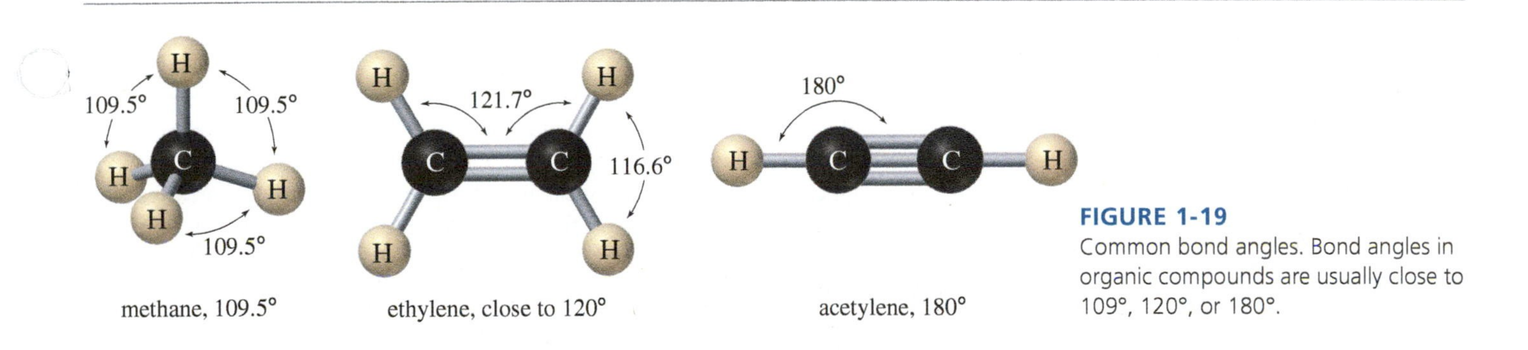

FIGURE 1-19 Common bond angles. Bond angles in organic compounds are usually close to 109°, 120°, or 180°.

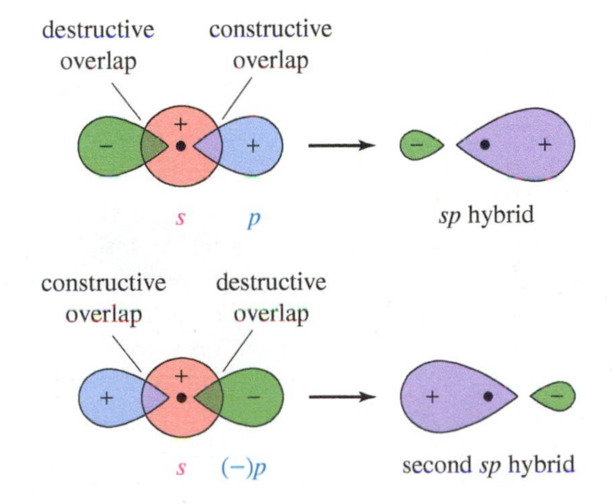

FIGURE 1-20 Formation of a pair of *sp* hybrid atomic orbitals. Addition of an *s* orbital to a *p* orbital gives an *sp* hybrid atomic orbital, with most of its electron density on one side of the nucleus. Adding the *p* orbital with opposite phase gives the other *sp* hybrid orbital, with most of its electron density on the opposite side of the nucleus from the first hybrid.

The result of this hybridization is a pair of directional *sp* hybrid orbitals pointed in opposite directions. These hybridized orbitals provide enhanced electron density in the bonding region for a sigma bond toward the left of the atom and for another sigma bond toward the right. They give a bond angle of 180°, separating the bonding electrons as much as possible. In general, *sp* hybridization results in this **linear** bonding arrangement.

SOLVED PROBLEM 1-7

Draw the Lewis structure for beryllium hydride, BeH_2. Draw the orbitals that overlap in the bonding of BeH_2, and label the hybridization of each orbital. Predict the H—Be—H bond angle.

SOLUTION

First, draw a Lewis structure for BeH_2.

H:Be:H

There are only four valence electrons in BeH_2 (two from Be and one from each H), so the Be atom cannot have an octet. The bonding must involve orbitals on Be that give the strongest bonds (the most electron density in the bonding region) and also allow the two pairs of electrons to be separated as far as possible.

Hybrid orbitals concentrate the electron density in the bonding region, and *sp* hybrids give 180° separation for two pairs of electrons. Hydrogen cannot use hybridized orbitals, because the closest available *p* orbitals are the 2*p*'s, and they are much higher in energy than the 1*s*. The bonding in BeH_2 results from overlap of *sp* hybrid orbitals on Be with the 1*s* orbitals on hydrogen. Figure 1-21 shows how this occurs.

1-15B *sp*² Hybrid Orbitals

For three bonds to be oriented as far apart as possible, bond angles of 120° are required. When an *s* orbital combines with two *p* orbitals, the resulting three hybrid orbitals are oriented at 120° angles to each other (Figure 1-22). These orbitals are called ***sp*² hybrid orbitals** because they are composed of one *s* and two *p* orbitals. The 120° arrangement is called **trigonal** geometry, in contrast to the linear geometry associated with *sp* hybrid orbitals. There remains an unhybridized *p* orbital (p_z) perpendicular to the plane of the three *sp*² hybrid orbitals.

First bond

Be (*sp*) H (1*s*)

Second bond

H (1*s*) Be (*sp*)

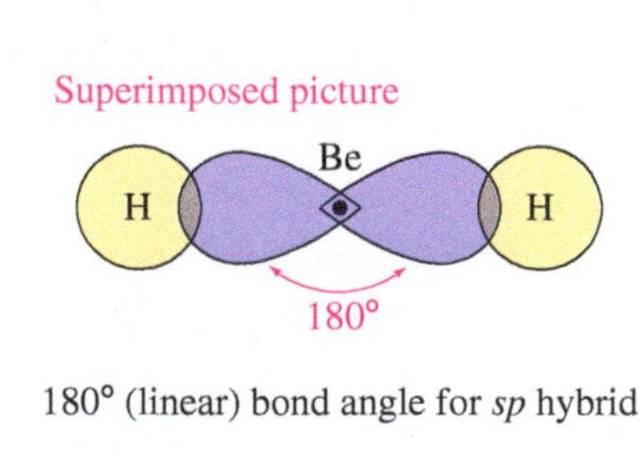

180° (linear) bond angle for *sp* hybrid

FIGURE 1-21 Linear geometry in the bonding of BeH_2. To form two sigma bonds, the two *sp* hybrid atomic orbitals on Be overlap with the 1*s* orbitals of hydrogen. The bond angle is 180° (linear).

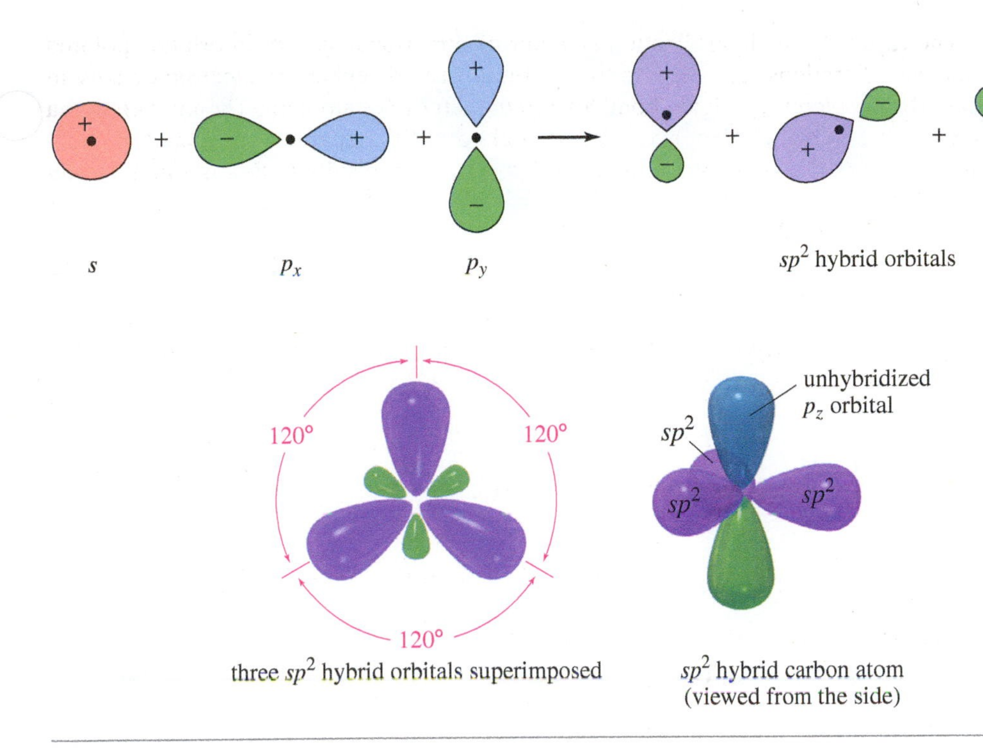

FIGURE 1-22
Trigonal geometry with sp^2 hybrid orbitals. Hybridization of an *s* orbital with two *p* orbitals gives a set of three sp^2 hybrid orbitals. This trigonal structure has bond angles of about 120°. The remaining *p* orbital is perpendicular to the plane of the three hybrid orbitals.

SOLVED PROBLEM 1-8

Borane (BH_3) is unstable under normal conditions, but it has been detected at low pressure.

(a) Draw the Lewis structure for borane.
(b) Draw a diagram of the bonding in BH_3, and label the hybridization of each orbital.
(c) Predict the H—B—H bond angle.

SOLUTION

There are only six valence electrons in borane, so the boron atom cannot have an octet. Boron has a single bond to each of the three hydrogen atoms.

```
   H
   ..
  .B.
 H   H
```

The best bonding orbitals are those that provide the greatest electron density in the bonding region while keeping the three pairs of bonding electrons as far apart as possible. Hybridization of an *s* orbital with two *p* orbitals gives three sp^2 hybrid orbitals directed 120° apart. Overlap of these orbitals with the hydrogen 1*s* orbitals gives a planar, trigonal molecule. (Note that the small back lobes of the hybrid orbitals have been omitted.)

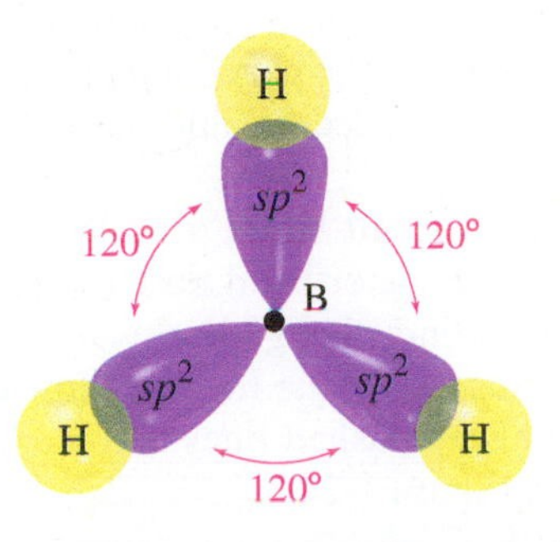

PROBLEM-SOLVING HINT

The number of hybrid orbitals formed is always the same as the total number of *s* and *p* orbitals hybridized.

Number of Orbitals	Hybrid	Angle
2	*sp*	180°
3	sp^2	120°
4	sp^3	109.5°

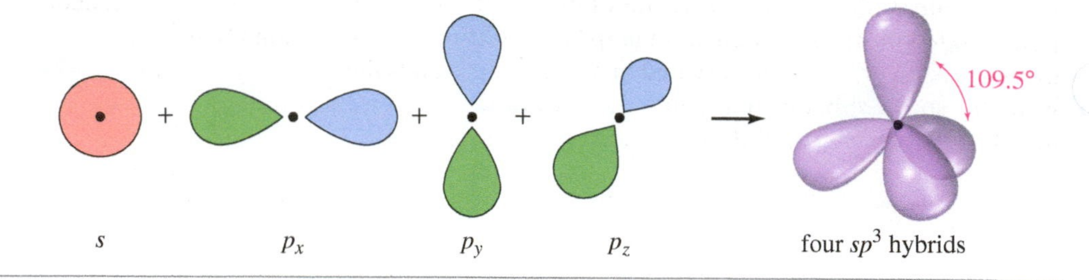

FIGURE 1-23
Tetrahedral geometry with sp^3 hybrid orbitals. Hybridization of an *s* orbital with all three *p* orbitals gives four sp^3 hybrid orbitals with tetrahedral geometry corresponding to 109.5° bond angles.

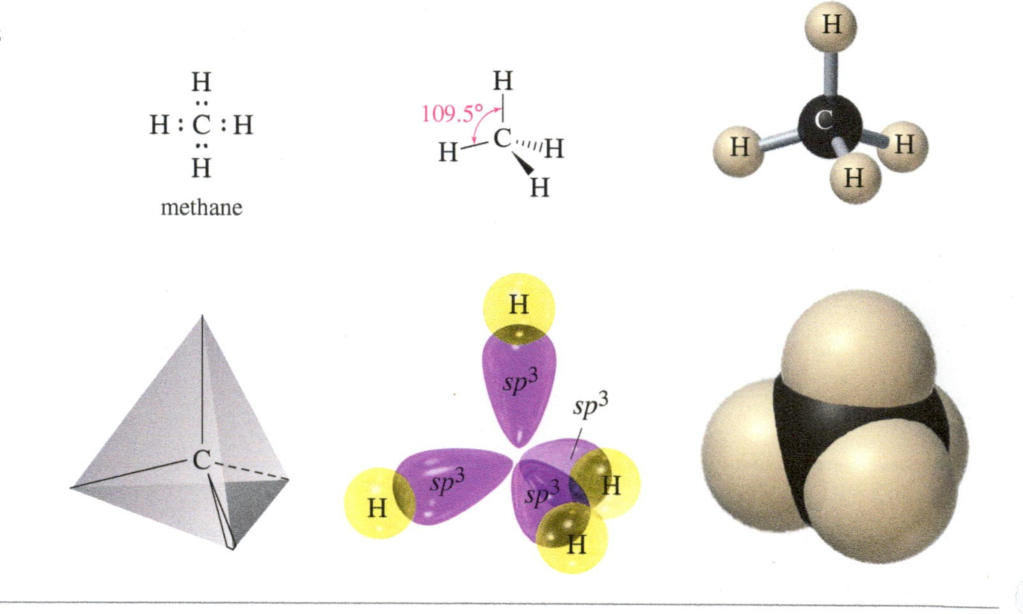

FIGURE 1-24
Several views of methane. Methane has tetrahedral geometry, using four sp^3 hybrid orbitals to form sigma bonds to the four hydrogen atoms.

1-15C sp^3 Hybrid Orbitals

Many organic compounds contain carbon atoms that are bonded to four other atoms. When four bonds are oriented as far apart as possible, they form a regular tetrahedron (109.5° bond angles), as pictured in Figure 1-23. This **tetrahedral** arrangement can be explained by combining the *s* orbital with all three *p* orbitals. The resulting four orbitals are called sp^3 **hybrid orbitals** because they are composed of one *s* and three *p* orbitals.

Methane (CH_4) is the simplest example of sp^3 hybridization (Figure 1-24). The Lewis structure for methane has eight valence electrons (four from carbon and one from each hydrogen), corresponding to four C—H single bonds. Tetrahedral geometry separates these bonds by the largest possible angle, 109.5°.

Application: Biology

Methanotrophs are bacteria or archaea that use methane as their source of carbon and energy. Those that live in the air use oxygen to oxidize methane to formaldehyde ($H_2C{=}O$) and CO_2. Those that live in anoxic marine sediments use sulfate (SO_4^{2-}) to oxidize methane to formaldehyde and CO_2, also reducing sulfate to H_2S.

1-16 Drawing Three-Dimensional Molecules

Figures 1-23 and 1-24 are more difficult to draw than the earlier figures because they depict three-dimensional objects on a two-dimensional surface. The p_z orbital should look like it points in and out of the page, and the tetrahedron should look three-dimensional. These drawings use perspective and the viewer's imagination to add the third dimension.

The use of perspective is difficult when a molecule is large and complicated. Organic chemists have developed a shorthand notation to simplify three-dimensional drawings. Dashed lines indicate bonds that go backward, away from the reader. Heavy wedge-shaped lines depict bonds that come forward, toward the reader. Straight lines are bonds in the plane of the page. Dashed lines and wedges show perspective in the second drawing of methane in Figure 1-24.

The three-dimensional structure of ethane, C_2H_6, has the shape of two tetrahedra joined together. Each carbon atom is sp^3 hybridized, with four sigma bonds formed by the four sp^3 hybrid orbitals. Dashed lines represent bonds that go away from the viewer, wedges represent bonds that come out toward the viewer, and other bond lines are in the plane of the page. All the bond angles are close to 109.5°.

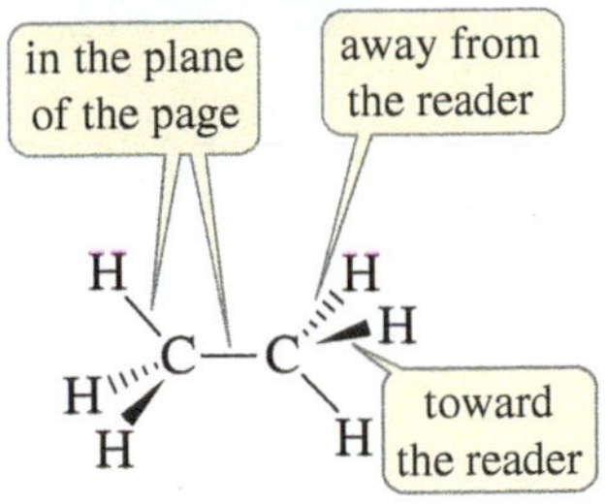

in the plane of the page
away from the reader
toward the reader

ethane 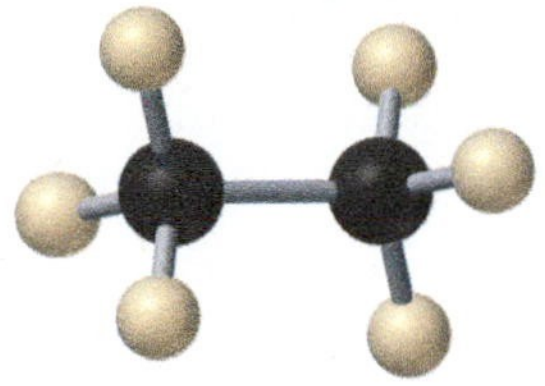 ethane  ethane

PROBLEM-SOLVING HINT

When showing perspective, do not draw another bond between the two bonds in the plane of the paper. Such a drawing shows an incorrect shape.

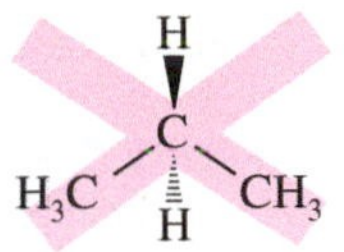

incorrect

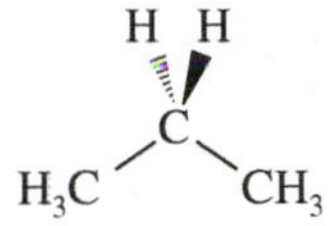

correct

correct

PROBLEM 1-15

(a) Use your molecular models to make ethane, and compare the model with the preceding structures.

(b) Make a model of propane (C_3H_8), and draw this model using dashed lines and wedges to represent bonds going back and coming forward.

1-17 General Rules of Hybridization and Geometry

At this point, we can consider some general rules for determining the hybridization of orbitals and the bond angles of atoms in organic molecules. After stating these rules, we solve some problems to show how to use these rules.

> Rule 1: Both sigma bonding electrons and lone pairs can occupy hybrid orbitals. The number of hybrid orbitals on an atom is computed by adding the number of sigma bonds and the number of lone pairs of electrons on that atom.

Because the first bond to another atom is always a sigma bond, the number of hybrid orbitals may be computed by adding the number of lone pairs to the number of atoms bonded to the central atom.

> Rule 2: Use the hybridization and geometry that give the widest possible separation of the calculated number of bonds and lone pairs.

SUMMARY Hybridization and Geometry

Hybrid Orbitals	Hybridization	Geometry	Approximate Bond Angles
2	$s + p = sp$	linear	180°
3	$s + p + p = sp^2$	trigonal	120°
4	$s + p + p + p = sp^3$	tetrahedral	109.5°

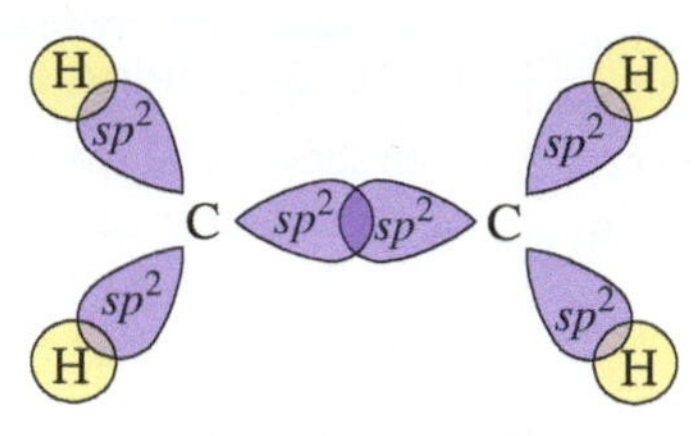

σ bond framework
(viewed from above the plane)

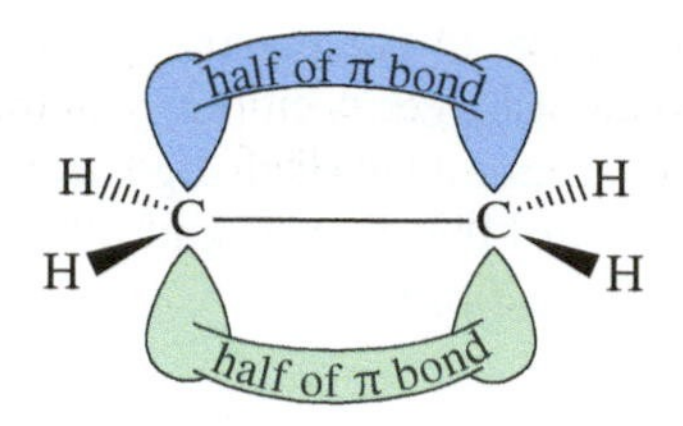

π bond
(viewed from alongside the plane)

ethylene

FIGURE 1-25
Planar geometry of ethylene. The carbon atoms in ethylene are sp^2 hybridized, with trigonal bond angles of about 120°. All the carbon and hydrogen atoms lie in the same plane.

FIGURE 1-26
Linear geometry of acetylene. The carbon atoms in acetylene are *sp* hybridized, with linear (180°) bond angles. The triple bond contains one sigma bond and two perpendicular pi bonds.

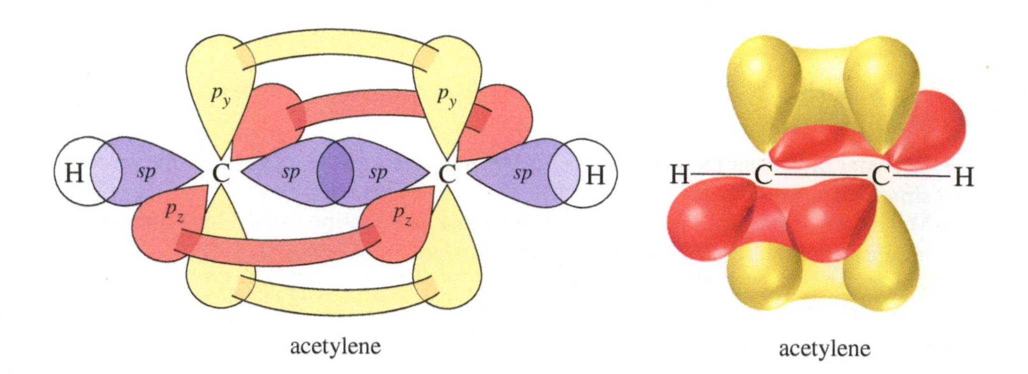

The number of hybrid orbitals obtained equals the number of atomic orbitals combined. Lone pairs of electrons take up more space than bonding pairs of electrons; thus, they compress the bond angles.

> Rule 3: If two or three pairs of electrons form a multiple bond between two atoms, the first bond is a sigma bond formed by a hybrid orbital. The second bond is a pi bond, consisting of two lobes above and below the sigma bond, formed by two unhybridized *p* orbitals (see the structure of ethylene in Figure 1-25). The third bond of a triple bond is another pi bond, perpendicular to the first pi bond (shown in Figure 1-26).

Solved Problems 1-9 through 1-13 show how to use these rules to predict the hybridization and bond angles in organic compounds.

SOLVED PROBLEM 1-9

Predict the hybridization of the nitrogen atom in ammonia, NH_3. Draw a picture of the three-dimensional structure of ammonia, and predict the bond angles.

SOLUTION

The hybridization depends on the number of sigma bonds plus lone pairs. A Lewis structure provides this information.

H:N̈: with H above and below (H:N: / H) or H—N: with H above and below — lone pair

In this structure, there are three sigma bonds and one pair of nonbonding electrons. Four hybrid orbitals are required, implying sp^3 hybridization and tetrahedral geometry around the nitrogen atom, with bond angles of about 109.5°. The resulting structure is much like that of methane, except that one of the sp^3 hybrid orbitals is occupied by a lone pair of electrons.

107.3°

The bond angles in ammonia (107.3°) are slightly smaller than the ideal tetrahedral angle, 109.5°. The nonbonding electrons are spread out more than a bonding pair of electrons, so they take up more space. The lone pair repels the electrons in the N—H bonds, compressing the bond angle.

PROBLEM 1-16

(a) Predict the hybridization of the oxygen atom in water, H_2O. Draw a picture of its three-dimensional structure, and explain why its bond angle is 104.5°.

(b) The electrostatic potential maps for ammonia and water are shown here. The structure of ammonia is shown within its EPM. Note how the lone pair creates a region of high electron potential (red), and the hydrogens are in regions of low electron potential (blue). Show how your three-dimensional structure of water corresponds with its EPM.

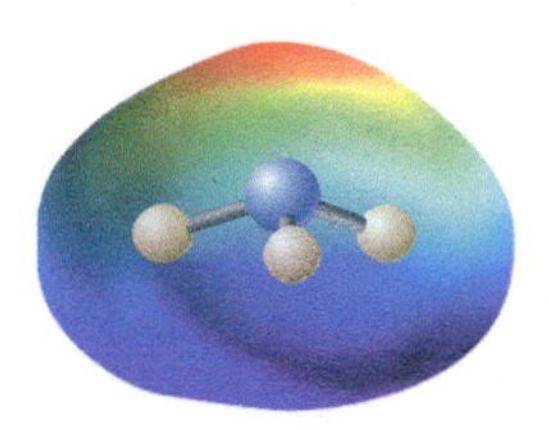

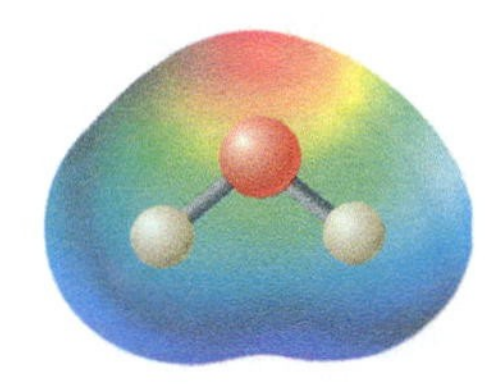

NH_3 H_2O

SOLVED PROBLEM 1-10

Predict the hybridization, geometry, and bond angles for ethylene (C_2H_4).

SOLUTION

The Lewis structure of ethylene is

H:C::C:H (with H above and below each C) or $H_2C{=}CH_2$

Each carbon atom has an octet, and there is a double bond between the carbon atoms. Each carbon is bonded to three other atoms (three sigma bonds), and there are no lone pairs. The carbon atoms are sp^2 hybridized, and the bond angles are trigonal: about 120°. The double bond is composed of a sigma bond formed by overlap of two sp^2 hybridized orbitals, plus a pi bond formed by overlap of the unhybridized p orbitals remaining on the carbon atoms. Because the pi bond requires parallel alignment of its two p orbitals, the ethylene molecule must be planar (Figure 1-25).

PROBLEM 1-17

Predict the hybridization, geometry, and bond angles for the central atoms in **(a)** but-2-ene, $CH_3CH{=}CHCH_3$. **(b)** $CH_3CH{=}NH$.

PROBLEM-SOLVING HINT

Begin with a valid Lewis structure, and use hybrid orbitals for the sigma bonds and lone pairs. Use pi bonds between unhybridized *p* orbitals for the second and third bonds of double and triple bonds.

SOLVED PROBLEM 1-11

Predict the hybridization, geometry, and bond angles for the carbon atoms in acetylene, C_2H_2.

SOLUTION

The Lewis structure of acetylene is

$$\text{H:C:::C:H} \quad \text{or} \quad \text{H—C≡C—H}$$

Both carbon atoms have octets, but each carbon is bonded to just two other atoms, requiring two sigma bonds. There are no lone pairs. Each carbon atom is *sp* hybridized and linear (180° bond angles). The *sp* hybrid orbitals are generated from the *s* orbital and the p_x orbital (the *p* orbital directed along the line joining the nuclei). The p_y orbitals and the p_z orbitals are unhybridized.

The **triple bond** is composed of one sigma bond, formed by overlap of *sp* hybrid orbitals, plus two pi bonds. One pi bond results from sideways overlap of the two p_y orbitals and another from sideways overlap of the two p_z orbitals (Figure 1-26).

PROBLEM 1-18

Predict the hybridization, geometry, and bond angles for the carbon and nitrogen atoms in acetonitrile (CH_3—C≡N:).

SOLVED PROBLEM 1-12

Predict the hybridization, geometry, and bond angles for the carbon and oxygen atoms in acetaldehyde (CH_3CHO).

SOLUTION

The Lewis structure for acetaldehyde is

H:C̈:C(H):Ö: (dot structure) or H—C(H)(H)—C(H)=Ö:

The oxygen atom and both carbon atoms have octets. The CH_3 carbon atom is sigma bonded to four atoms, so it is sp^3 hybridized (and tetrahedral). The C═O carbon is bonded to three atoms (no lone pairs), so it is sp^2 hybridized, and its bond angles are about 120°.

We predict that the oxygen atom is sp^2 hybridized because it is bonded to one atom (carbon) and has two lone pairs, requiring a total of three hybrid orbitals. We cannot experimentally measure the angles of the lone pairs on oxygen, however, so it is impossible to confirm whether the oxygen atom is really sp^2 hybridized.

The double bond between carbon and oxygen looks just like the double bond in ethylene. There is a sigma bond formed by overlap of sp^2 hybrid orbitals and a pi bond formed by overlap of the unhybridized *p* orbitals on carbon and oxygen (Figure 1-27).

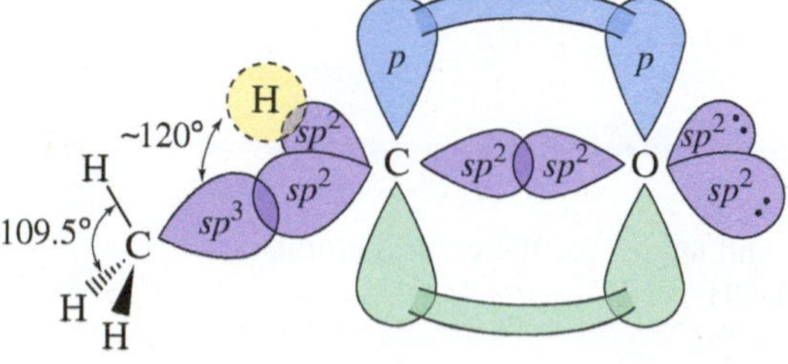

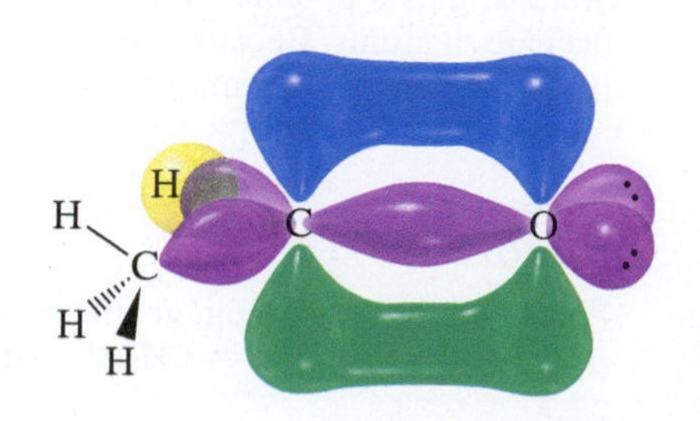

FIGURE 1-27
Structure of acetaldehyde. The CH_3 carbon in acetaldehyde is sp^3 hybridized, with tetrahedral bond angles of about 109.5°. The carbonyl (C═O) carbon is sp^2 hybridized, with bond angles of about 120°. The oxygen atom is probably sp^2 hybridized, but we cannot measure any bond angles to verify this prediction.

PROBLEM 1-19

1. Draw a Lewis structure for each compound.
2. Label the hybridization, geometry, and bond angles around each atom other than hydrogen.
3. Draw a three-dimensional representation (using wedges and dashed lines) of the structure.

(a) CO_2 **(b)** CH_3OCH_3 **(c)** $(CH_3)_3O^+$ **(d)** CH_3COOH
(e) CH_3CCH **(f)** CH_3CHNCH_3 **(g)** H_2CCO

PROBLEM 1-20

Allene, $CH_2{=}C{=}CH_2$, has the structure shown below. Explain how the bonding in allene requires the two $=CH_2$ groups at its ends to be at right angles to each other.

$$H_2C{=}C{=}CH_2$$

allene

SOLVED PROBLEM 1-13

In Sections 1-7 and 1-9, we considered the electronic structure of $[CH_2NH_2]^+$. Predict its hybridization, geometry, and bond angles.

SOLUTION

This is a tricky question. This ion has two important resonance forms:

$$\left[H_2\overset{+}{C}{-}\ddot{N}H_2 \longleftrightarrow H_2C{=}\overset{+}{N}H_2\right] = H_2\overset{\delta^+}{C}{\,\text{---}\,}\overset{\delta^+}{N}H_2$$

resonance forms combined representation

When resonance is involved, different resonance forms may suggest different hybridization and bond angles, but a real molecule can have only one set of bond angles, which must be compatible with all the important resonance forms. The bond angles imply the hybridization of the atoms, which also must be the same in all the resonance forms.

Looking at either resonance form for $[CH_2NH_2]^+$, we would predict sp^2 hybridization (120° bond angles) for the carbon atom; however, the first resonance form suggests sp^3 hybridization for nitrogen (109° bond angles), and the second suggests sp^2 hybridization (120° bond angles). Which is correct?

Experiments show that the bond angles on both carbon and nitrogen are about 120°, implying sp^2 hybridization. This nitrogen cannot be sp^3 hybridized because there must be an unhybridized *p* orbital available to form the pi bond in the second resonance form. In the first resonance form, we picture the lone pair residing in this unhybridized *p* orbital.

In general, resonance-stabilized structures have bond angles appropriate for the largest number of pi bonds needed at each atom—that is, with unhybridized *p* orbitals available for all the pi bonds shown in any important resonance form.

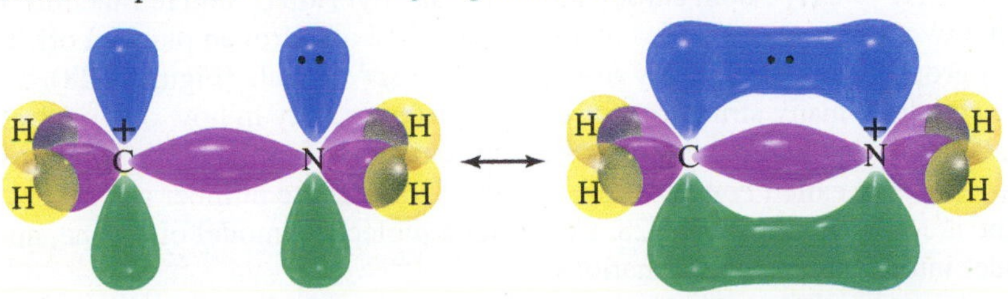

PROBLEM-SOLVING HINT

To predict the hybridization and geometry of an atom in a resonance hybrid, consider the resonance form with the most pi bonds to that atom. An atom involved in resonance generally will not be sp^3 hybridized because it needs at least one unhybridized *p* orbital for pi-bonding overlap.

PROBLEM 1-21 (PARTIALLY SOLVED)

1. Draw the important resonance forms for each compound.
2. Label the hybridization and bond angles around each atom other than hydrogen.
3. Use a three-dimensional drawing to show where the electrons are pictured to be in each resonance form.

(a) $HCONH_2$

SOLUTION

This compound has a carbonyl (C═O) group that is not obvious in the condensed formula. An important resonance form delocalizes the nonbonding electrons on nitrogen into a pi bond with carbon. This pi bonding requires the overlap of unhybridized *p* orbitals, requiring sp^2 hybridization on both carbon and nitrogen, and 120° bond angles. Then O is probably sp^2 hybridized, but we cannot confirm that assumption because there are no bond angles on O.

sp^2, 120°

major ⟷ minor

resonance forms

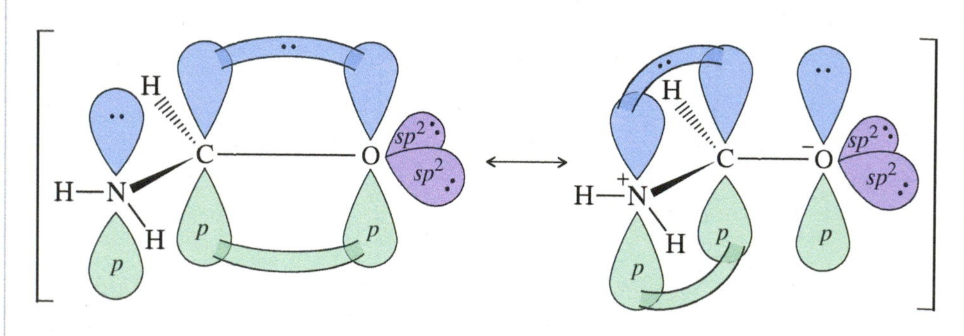

(b) $[CH_2OH]^+$ **(c)** $[CH_2CHO]^-$ **(d)** $[CH_3CHNO_2]^-$
(e) $[CH_2CN]^-$ **(f)** $B(OH)_3$ **(g)** ozone (O_3, bonded OOO)

PROBLEM-SOLVING HINT

Drawings that differ only by rotations of single bonds usually represent the same compound. For example, the following drawings all represent *n*-butane:

$CH_3CH_2CH_2CH_3$ H_3C—C(H)(H)—C(CH_3)(H)—H

H—C(CH_3)(H)—C(CH_3)(H)—H H—C(H)(CH_3)—C(CH_3)(H)—H

1-18 Bond Rotation

Some bonds rotate easily, but others do not. When we look at a structure, we must recognize which bonds rotate and which do not. If a bond rotates easily, each molecule can rotate through the different angular arrangements of atoms. If a bond cannot rotate, however, different angular arrangements may be distinct compounds with different properties.

1-18A Rotation of Single Bonds

In ethane (CH_3—CH_3), both carbon atoms are sp^3 hybridized and tetrahedral. Ethane looks like two methane molecules that have each had a hydrogen plucked off (to form a methyl group) and are joined by overlap of their sp^3 orbitals (Figure 1-28).

We can draw many structures for ethane, differing only in how one methyl group is twisted in relation to the other one. Such structures, differing only in rotations about a single bond, are called *conformations*. Two of the infinite number of conformations of ethane are shown in Figure 1-28. Construct a molecular model of ethane, and twist the model into these two conformations.

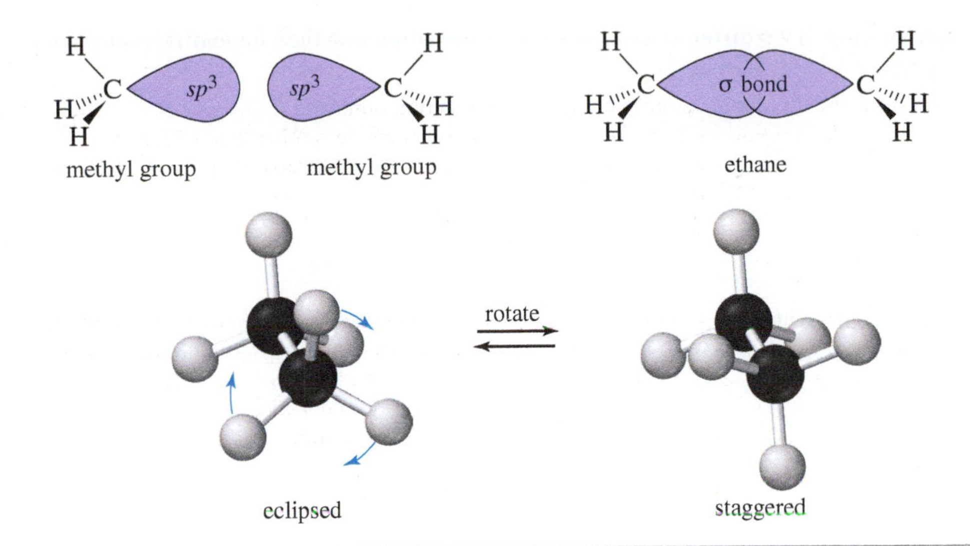

FIGURE 1-28
Rotation of single bonds. Ethane is composed of two methyl groups bonded by overlap of their sp^3 hybrid orbitals. These methyl groups may rotate with respect to each other.

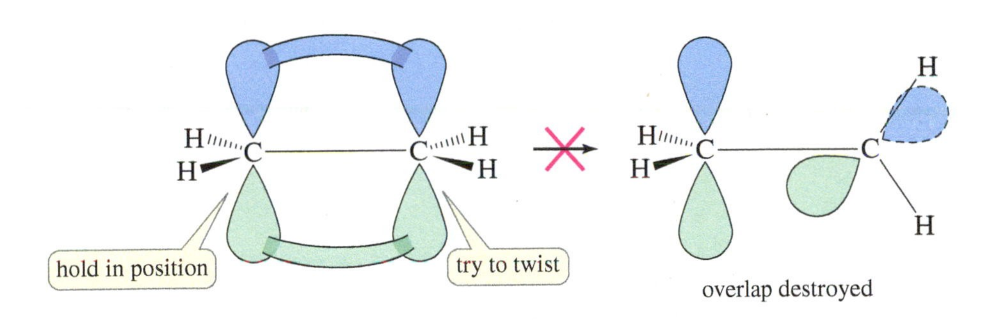

FIGURE 1-29
Rigidity of double bonds. Ethylene is not easily twisted because twisting destroys the overlap of its pi bond.

Which of these structures in Figure 1-28 for ethane is the "right" one? Are the two methyl groups lined up so that their C—H bonds are parallel (*eclipsed*), or are they *staggered*, as in the drawing on the right? The answer is that both structures, and all the possible structures in between, are correct structures for ethane, and a real ethane molecule rotates through all these conformations. The two carbon atoms are bonded by overlap of their sp^3 orbitals to form a sigma bond along the line between the carbons. The magnitude of this sp^3–sp^3 overlap remains nearly the same during rotation because the sigma bond is cylindrically symmetrical about the line joining the carbon nuclei. No matter how you turn one of the methyl groups, its sp^3 orbital still overlaps with the sp^3 orbital of the other carbon atom.

1-18B Rigidity of Double Bonds

Not all bonds allow free rotation; ethylene, for example, is quite rigid (Figure 1-29). In ethylene, the double bond between the two CH_2 groups consists of a sigma bond and a pi bond. When we twist one of the two CH_2 groups, the sigma bond is unaffected but the pi bond loses its overlap. The two *p* orbitals cannot overlap when the two ends of the molecule are at right angles, and the pi bond is effectively broken in this geometry.

We can make the following generalization:

> Rotation about single bonds is allowed, but double bonds are rigid and cannot be twisted under normal conditions.

Because double bonds are rigid, we can separate and isolate compounds that differ only in how their substituents are arranged on a double bond. For example, the double bond in but-2-ene (CH_3—CH=CH—CH_3) prevents the two ends of the molecule from rotating. Two different compounds are possible, and they have different physical properties:

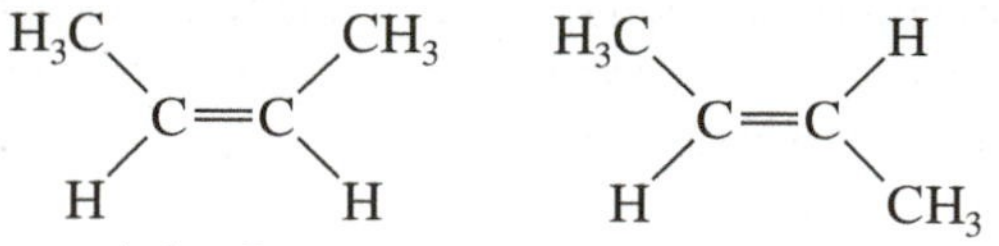

cis-but-2-ene: H_3C(H)C=C(H)CH_3 — *cis*-but-2-ene, bp = 3.7 °C

trans-but-2-ene: H_3C(H)C=C(CH_3)H — *trans*-but-2-ene, bp = 0.9 °C

The molecule with the methyl groups on the same side of the double bond is called *cis*-but-2-ene, and the one with the methyl groups on opposite sides is called *trans*-but-2-ene. These kinds of molecules are discussed further in Section 1-19B.

PROBLEM 1-22

For each pair of structures, determine whether they represent different compounds or a single compound.

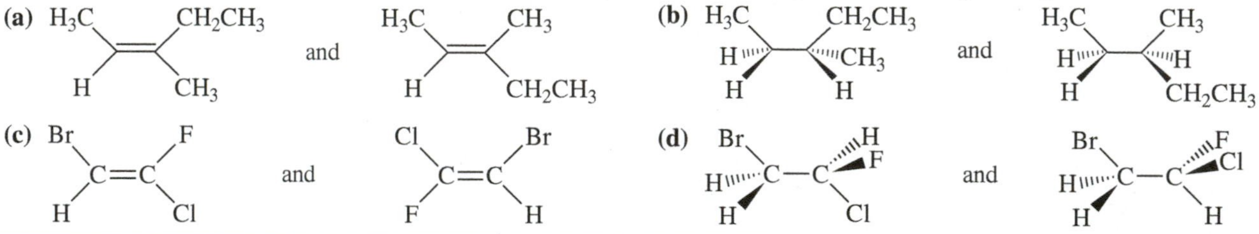

(a) H_3C, CH_2CH_3 / H, CH_3 and H_3C, CH_3 / H, CH_2CH_3

(b) H_3C, CH_2CH_3 / H, CH_3 / H, H and H_3C, CH_3 / H, H / H, CH_2CH_3

(c) Br, F / H, Cl (C=C) and Cl, Br / F, H (C=C)

(d) Br, H, H–C–C H, F, Cl and Br, H, H–C–C F, Cl, H

PROBLEM 1-23

Two compounds with the formula $CH_3—CH=N—CH_3$ are known.

(a) Draw a Lewis structure for this molecule, and label the hybridization of each carbon and nitrogen atom.

(b) What two compounds have this formula?

(c) Explain why only one compound with the formula $(CH_3)_2CNCH_3$ is known.

1-19 Isomerism

Isomers are different compounds with the same molecular formula. There are several types of isomerism in organic compounds, and we will cover them in detail in Chapter 5 (Stereochemistry). For now, we need to recognize the two large classes of isomers: constitutional isomers and stereoisomers.

1-19A Constitutional Isomerism

Constitutional isomers (or **structural isomers**) are isomers that differ in their bonding sequence; that is, their atoms are connected differently. Let's use butane as an example. If you were asked to draw a structural formula for C_4H_{10}, either of the following structures would be correct:

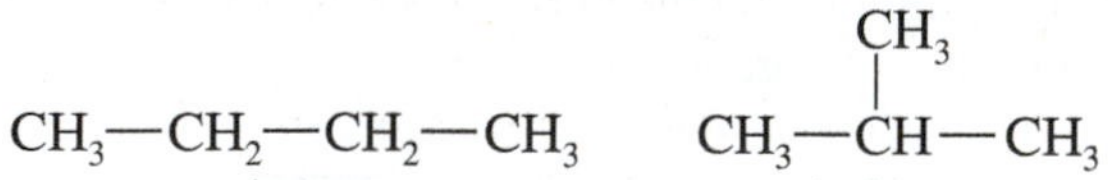

$$CH_3—CH_2—CH_2—CH_3$$
n-butane

$$CH_3—CH(CH_3)—CH_3$$
isobutane

These two compounds are isomers because they are different compounds with different properties, yet they have the same molecular formula. They are constitutional isomers because their atoms are connected differently. The first compound (*n*-butane for "normal" butane) has its carbon atoms in a straight chain four carbons long. The second compound ("isobutane" for "an isomer of butane") has a branched structure with a longest chain of three carbon atoms and a methyl side chain.

There are three constitutional isomers of pentane (C_5H_{12}), whose common names are *n*-pentane, isopentane, and neopentane. The number of isomers increases rapidly as the number of carbon atoms increases.

$CH_3—CH_2—CH_2—CH_2—CH_3$ (*n*-pentane)

$CH_3—CH(CH_3)—CH_2—CH_3$ (isopentane)

$CH_3—C(CH_3)_2—CH_3$ (neopentane)

Constitutional isomers may differ in ways other than the branching of their carbon chain. They may differ in the position of a double bond or other group or by having a ring or some other feature. Notice how the following constitutional isomers all differ by the ways in which atoms are bonded to other atoms. (Check the number of hydrogens bonded to each carbon.) These compounds are not isomers of the pentanes just shown, however, because these have a different molecular formula (C_5H_{10}).

PROBLEM-SOLVING HINT

Constitutional isomers (structural isomers) differ in the order in which their atoms are bonded.

$H_2C{=}CH—CH_2CH_2CH_3$ (pent-1-ene)

$CH_3—CH{=}CH—CH_2CH_3$ (pent-2-ene)

cyclopentane

CH_3 methylcyclobutane

1-19B Stereoisomers

Stereoisomers are isomers that differ only in how their atoms are oriented in *space*. Their atoms are bonded in the same order, however. For example, *cis*- and *trans*-but-2-ene have the same connections of bonds, so they are not constitutional isomers. They are stereoisomers because they differ only in the spatial orientation of the groups attached to the double bond. The cis isomer has the two methyl groups on the same side of the double bond, and the trans isomer has them on opposite sides. In contrast, but-1-ene is a constitutional isomer of *cis*- and *trans*-but-2-ene.

PROBLEM-SOLVING HINT

Stereoisomers are different compounds that differ only in how their atoms are oriented in space.

PROBLEM-SOLVING HINT

Similar groups on the same side of the double bond: *cis*. Similar groups on opposite sides of the double bond: *trans*.

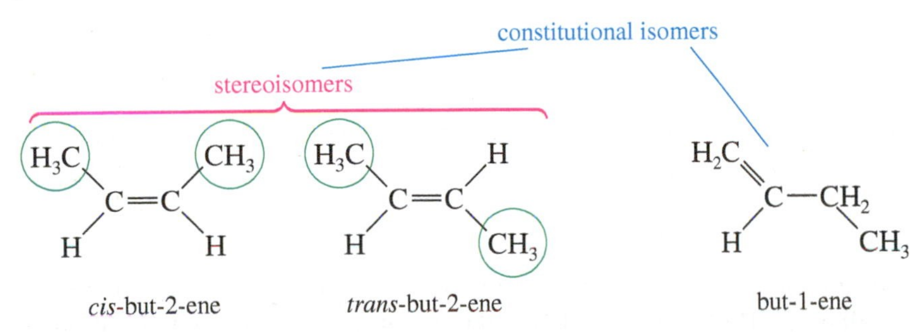

Cis and trans isomers are only one type of stereoisomerism. The study of the structure and chemistry of stereoisomers is called **stereochemistry.** We will encounter stereochemistry throughout our study of organic chemistry, and Chapter 5 is devoted entirely to this field.

Cis-trans isomers are also called **geometric isomers** because they differ in the geometry of the groups on a double bond. The cis isomer is always the one with similar groups on the same side of the double bond, and the trans isomer has similar groups on opposite sides of the double bond.

To have cis-trans isomerism, there must be two *different* groups on each end of the double bond. For example, but-1-ene has two identical hydrogens on one end of the double bond. Reversing their positions does not give a different compound. Similarly, 2-methylbut-2-ene has two identical methyl groups on one end of the double bond.

Application: Drugs

Stereoisomers can have different therapeutic effects. Quinine, a natural product isolated from the bark of the cinchona tree, was the first compound effective against malaria. Quinidine, a stereoisomer of quinine, is used to treat irregular heartbeat.

Reversing the methyl groups does not give a different compound. These compounds cannot show cis-trans isomerism.

identical H, H on C=C with H, CH_2CH_3 — but-1-ene, no cis or trans

identical CH_3, CH_3 on C=C with H, CH_3 — 2-methylbut-2-ene, no cis or trans

PROBLEM-SOLVING HINT

Two identical groups on one of the double-bonded carbons implies no cis-trans isomerism.

PROBLEM 1-24

Which of the following compounds show cis-trans isomerism? Draw the cis and trans isomers of those that do.

(a) CHF=CHF (b) $F_2C=CH_2$ (c) $CH_2=CH-CH_2-CH_3$

(d) cyclopentane=$CHCH_3$ (e) cyclopentane$-CHCHCH_3$ (f) 2,2-dimethylcyclopentane=$CHCH_3$

PROBLEM 1-25

Give the relationship between the following pairs of structures. The possible relationships are:

same compound *constitutional isomers (structural isomers)*
cis-trans isomers *not isomers (different molecular formula)*

(a) $CH_3CH_2CH(CH_2CH_3)CH_2CH_3$ and $CH_3CH_2CH(CH_3)CH_2CH_2CH_3$

(b) Br(H)C=C(H)(Br) and Br(H)C=C(Br)(H)

(c) Br(H)C=C(H)(Br) and $Br_2C=CH_2$

(d) Br(H)C=C(H)(Br) and H(Br)C=C(Br)(H)

(e) H—C(H)(Cl)—C(Cl)(H)—H and H—C(Cl)(H)—C(Cl)(H)—H

(f) H—C(H)(CH_3)—C(CH_3)(H)—H and H—C(CH_3)(H)—C(H)(H)—CH_3

(g) $CH_3-CH_2-CH_2-CH_3$ and $CH_3-CH=CH-CH_3$

(h) $CH_2=CH-CH_2CH_2CH_3$ and $CH_3-CH=CH-CH_2CH_3$

(i) $CH_2=CHCH_2CH_2CH_3$ and $CH_3CH_2CH_2CH=CH_2$

(j) 1,2-dimethylcyclohexane (CH_3, CH_3) and 1,3-dimethylcyclohexane (CH_3, CH_3)

(k) 1-methylcyclohexene (CH_3) and 3-methylcyclohexene (CH_3)

Essential Terms

Each chapter ends with a glossary that summarizes the most important new terms in the chapter. These glossaries are more than just a dictionary to look up unfamiliar terms as you encounter them (the index serves that purpose). The glossary is one of the tools for reviewing the chapter. You can read carefully through the glossary to see if you understand and remember all the terms and associated chemistry mentioned there. Anything that seems unfamiliar should be reviewed by turning to the page number given in the glossary listing.

cis-trans isomers — **(geometric isomers)** Stereoisomers that differ in their cis-trans arrangement on a double bond or on a ring. The cis isomer has similar groups on the same side, and the trans isomer has similar groups on opposite sides. (p. 45)

constitutional isomers — **(structural isomers)** Isomers whose atoms are connected differently; they differ in their bonding sequence. (p. 44)

covalent bonding — Bonding that occurs by the sharing of electrons in the region between two nuclei. (p. 7)

- **single bond:** A covalent bond that involves the sharing of one pair of electrons. (p. 9)
- **double bond:** A covalent bond that involves the sharing of two pairs of electrons. (p. 9)
- **triple bond:** A covalent bond that involves the sharing of three pairs of electrons. (p. 9)

degenerate orbitals — Orbitals with identical energies. (p. 4)

delocalized charge — A charge that is spread out over two or more atoms. We usually draw resonance forms to show how the charge can appear on each of the atoms sharing the charge. (p. 14)

dipole moment (μ) — A measure of the polarity of a bond (or a molecule), proportional to the product of the charge separation times the bond length. (p. 10)

double bond — A bond containing four electrons between two nuclei. One pair of electrons forms a sigma bond, and the other pair forms a pi bond. (p. 32)

electron density — The relative probability of finding an electron in a certain region of space. (p. 3)

electronegativity — A measure of an element's ability to attract electrons. Elements with higher electronegativities attract electrons more strongly. (p. 10)

electrostatic potential map (EPM) — A computer-calculated molecular representation that uses colors to show the charge distribution in a molecule. In most cases, the EPM uses red to show electron-rich regions (most negative electrostatic potential) and blue or purple to show electron-poor regions (most positive electrostatic potential). The intermediate colors orange, yellow, and green show regions with intermediate electrostatic potentials. (p. 10)

empirical formula — The ratios of atoms in a compound. (p. 26) See also **molecular formula**.

formal charges — A method for keeping track of charges, showing what charge would be on an atom in a particular Lewis structure. (p. 11)

geometric isomers — (IUPAC term: **cis-trans isomers**) Stereoisomers that differ in their cis-trans arrangement on a double bond or on a ring. (p. 45)

Hund's rule — When there are two or more unfilled orbitals of the same energy (degenerate orbitals), the lowest-energy configuration places the electrons in different orbitals (with parallel spins) rather than pairing them in the same orbital. (p. 6)

hybrid atomic orbital — A directional orbital formed from a combination of *s* and *p* orbitals on the same atom. (pp. 28–31)

- ***sp* hybrid orbitals** give two orbitals with a bond angle of 180° (**linear** geometry).
- ***sp^2* hybrid orbitals** give three orbitals with bond angles of 120° (**trigonal** geometry).
- ***sp^3* hybrid orbitals** give four orbitals with bond angles of 109.5° (**tetrahedral** geometry).

ionic bonding — Bonding that occurs by the attraction of oppositely charged ions. Ionic bonding usually results in the formation of a large, three-dimensional crystal lattice. (p. 7)

isomers — Different compounds with the same molecular formula. (p. 44)

- **Constitutional isomers (structural isomers)** are connected differently; they differ in their bonding sequence.
- **Cis-trans isomers (geometric isomers)** are stereoisomers that differ in their cis-trans arrangement on a double bond or on a ring.
- **Stereoisomers** differ only in how their atoms are oriented in space.
- **Stereochemistry** is the study of the structure and chemistry of stereoisomers.

isotopes — Atoms with the same number of protons but different numbers of neutrons; atoms of the same element but with different atomic masses. (p. 3)

LCAO — **(linear combination of atomic orbitals)** Wave functions can add to each other to produce the wave functions of new orbitals. The number of new orbitals generated equals the original number of orbitals. (p. 28)

Lewis structure — A structural formula that shows all valence electrons, with the bonds symbolized by dashes (—) or by pairs of dots, and nonbonding electrons symbolized by dots. (p. 8)

line–angle formula — **(skeletal structure, stick figure)** A shorthand structural formula with bonds represented by lines. Carbon atoms are implied wherever two lines meet or a line begins or bends. Atoms other than C and H are drawn in, but hydrogen atoms are not shown unless they are on an atom that is drawn. Each carbon atom is assumed to have enough hydrogens to give it four bonds. (p. 24)

Lewis structure of cyclohex-2-en-1-ol

cyclohex-2-en-1-ol equivalent line–angle formula

lone pair — A pair of nonbonding electrons. (p. 8)

molecular formula — The number of atoms of each element in one molecule of a compound. The **empirical formula** simply gives the ratios of atoms of the different elements. For example, the molecular formula of glucose is $C_6H_{12}O_6$. Its empirical formula is CH_2O. Neither the molecular formula nor the empirical formula gives structural information. (p. 25)

molecular orbital (MO) — An orbital formed by the overlap of atomic orbitals on different atoms. MOs can be either bonding or antibonding, but only the bonding MOs are filled in most stable molecules. (pp. 28–31)

A **bonding molecular orbital** places a large amount of electron density in the bonding region between the nuclei. The energy of an electron in a bonding MO is lower than it is in an atomic orbital.

An **antibonding molecular orbital** places most of the electron density outside the bonding region. The energy of an electron in an antibonding MO is higher than it is in an atomic orbital.

node — In an orbital, a region of space with zero electron density. (p. 4)

nodal plane — In an orbital, a flat (planar) region of space with zero electron density. (p. 4)

nonbonding electrons — Valence electrons that are not used for bonding. A pair of nonbonding electrons is often called a **lone pair.** (p. 8)

octet rule — Atoms generally form bonding arrangements that give them filled shells of electrons (noble-gas configurations). For the second-row elements, this configuration has eight valence electrons. (p. 7)

orbital — An allowed energy state for an electron bound to a nucleus; the probability function that defines the distribution of electron density in space. The *Pauli exclusion principle* states that up to two electrons can occupy each orbital if their spins are paired. (p. 3)

organic chemistry — New definition: The chemistry of carbon compounds. Old definition: The study of compounds derived from living organisms and their natural products. (p. 1)

pi bond (π bond) — A bond formed by sideways overlap of two *p* orbitals. A pi bond has its electron density in two lobes, one above and one below the line joining the nuclei. (p. 31)

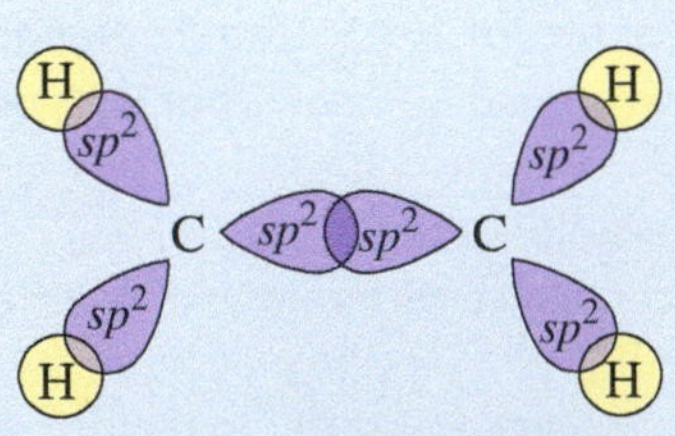

σ bond framework (viewed from above the plane)

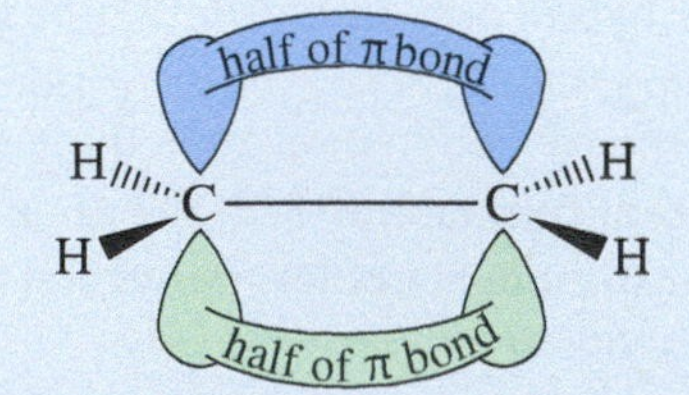

π bond (viewed from alongside the plane)

ethylene

polar covalent bond — A covalent bond in which electrons are shared unequally. A bond with equal sharing of electrons is called a **nonpolar covalent bond.** (p. 10)

resonance hybrid — A molecule or ion for which two or more valid Lewis structures can be drawn, differing only in the placement of the valence electrons. These Lewis structures are called **resonance forms** or

resonance structures. Individual resonance forms do not exist, but we can estimate their relative energies. The more important (lower-energy) structures are called **major contributors**, and the less important (higher-energy) structures are called **minor contributors**. When a charge is spread over two or more atoms by resonance, it is said to be **delocalized** and the molecule is said to be **resonance stabilized**. (pp. 14–15)

sigma bond (σ bond) A bond with most of its electron density centered along the line joining the nuclei; a cylindrically symmetrical bond. Single bonds are normally sigma bonds. (p. 29)

stereochemistry The study of the structure and chemistry of stereoisomers. (p. 45)

stereoisomers Isomers that differ only in how their atoms are oriented in space. (p. 45)

structural formulas A **complete structural formula** (such as a Lewis structure) shows all the atoms and bonds in the molecule. A **condensed structural formula** shows each central atom along with the atoms bonded to it. A **line–angle formula** (sometimes called a **skeletal structure** or **stick figure**) assumes that there is a carbon atom wherever two lines meet or a line begins or ends. See Section 1-10 for examples. (pp. 22–25)

structural isomers (IUPAC term: **constitutional isomers**) Isomers whose atoms are connected differently; they differ in their bonding sequence. (p. 44)

triple bond A bond containing six electrons between two nuclei. One pair of electrons forms a sigma bond, and the other two pairs form two pi bonds at right angles to each other. (p. 40)

valence The number of bonds an atom usually forms. (p. 9)

valence electrons Those electrons that are in the outermost shell. (p. 6)

vitalism The belief that syntheses of organic compounds require the presence of a "vital force." (p. 1)

VSEPR theory **(valence-shell electron-pair repulsion theory)** Bonds and lone pairs around a central atom tend to be separated by the largest possible angles: about 180° for two, 120° for three, and 109.5° for four. (p. 33)

wave function (ψ) The mathematical description of an orbital. The square of the wave function (ψ^2) is proportional to the electron density. (p. 27)

Essential Problem-Solving Skills in Chapter 1

Each skill is followed by problem numbers exemplifying that particular skill.

1 Write the electronic configurations for the elements hydrogen through neon. Explain how electronic configurations determine the electronegativities and bonding properties of these elements, and how the third row elements (e.g., Si, P, and S) differ from them.	Problems 1-26, 27, 28, and 30
2 Draw all possible structures corresponding to a given molecular formula.	Problems 1-34, 35, and 36
3 Draw and interpret Lewis, condensed, and line–angle structural formulas. Calculate formal charges.	Problems 1-30, 31, 32, 33, 34, 37, 38, 41, 42, and 45
4 Predict patterns of covalent and ionic bonding involving C, H, O, N, and the halogens. Identify resonance-stabilized structures and compare the relative importance of their resonance forms.	Problems 1-26, 29, 36, 40, 41, 42, 43, 44, 45, and 46
Problem-Solving Strategy: Drawing and Evaluating Resonance Forms	Problems 1-26, 41, 42, 43, 44, 45, 46, 53, and 54
5 Calculate empirical and molecular formulas from elemental composition.	Problems 1-47 and 48
6 Draw the structure of a single bond, a double bond, and a triple bond.	Problems 1-52, 55, 57, and 60
7 Predict the hybridization and geometry of the atoms in a molecule, and draw a three-dimensional representation of the molecule.	Problems 1-49, 51, 52, 53, 56, 57, and 60
8 Draw the resonance forms of a resonance hybrid, and predict its hybridization and geometry. Predict which resonance form is the major contributor.	Problems 1-43, 44, 52, 53, and 54
9 Identify constitutional isomers and stereoisomers, and predict which compounds can exist as constitutional isomers and as cis-trans (geometric) isomers.	Problems 1-49, 56, 58, and 59

Study Problems

It's easy to fool yourself into thinking you understand organic chemistry when you actually may not. As you read through this book, all the facts and ideas may make sense, yet you have not learned to combine and use those facts and ideas. An examination is a painful time to learn that you do not really understand the material.

The best way to learn organic chemistry is to use it. You will certainly need to read and reread all the material in the chapter, but this level of understanding is just the beginning. Problems are provided so that you can work with the ideas, applying them to new compounds and new reactions that you have never seen before. By working problems, you force yourself to use the material and fill in the gaps in your understanding. You also increase your level of self-confidence and your ability to do well on exams.

Several kinds of problems are included in each chapter. There are problems within the chapters, providing examples and drill for the material as it is covered. Work these problems as you read through the chapter to ensure your understanding as you go along. Answers to many of these in-chapter problems are found at the back of this book. Study Problems at the end of each chapter give you additional experience using the material, and they force you to think in depth about the ideas. Problems with red stars (*) are more difficult problems that require extra thought and perhaps some extension of the material presented in the chapter. Some of the Study Problems have short answers in the back of this book, and all of them have detailed answers in the accompanying Solutions Manual.

Taking organic chemistry without working the problems is like skydiving without a parachute. Initially there is a breezy sense of freedom and daring. But then, there is the inevitable jolt that comes at the end for those who went unprepared.

1-26 (a) Draw the resonance forms for SO_2 (bonded O—S—O).
(b) Draw the resonance forms for ozone (bonded O—O—O).
(c) Sulfur dioxide has one more resonance form than ozone. Explain why this structure is not possible for ozone.

1-27 Name the element that corresponds to each electronic configuration.
(a) $1s^2 2s^2 2p^2$ (b) $1s^2 2s^2 2p^4$ (c) $1s^2 2s^2 2p^6 3s^2 3p^3$ (d) $1s^2 2s^2 2p^6 3s^2 3p^5$

1-28 There is a small portion of the periodic table that you must know to do organic chemistry. Construct this part from memory, using the following steps.
(a) From memory, make a list of the elements in the first two rows of the periodic table, together with their numbers of valence electrons.
(b) Use this list to construct the first two rows of the periodic table.
(c) Organic compounds often contain sulfur, phosphorus, chlorine, bromine, and iodine. Add these elements to your periodic table.

1-29 For each compound, state whether its bonding is covalent, ionic, or a mixture of covalent and ionic.
(a) $NaCl$ (b) $NaOH$ (c) CH_3Li (d) CH_2Cl_2
(e) $NaOCH_3$ (f) HCO_2Na (g) CF_4

1-30 (a) Both PCl_3 and PCl_5 are stable compounds. Draw Lewis structures for these two compounds.
(b) NCl_3 is a known compound, but all attempts to synthesize NCl_5 have failed. Draw Lewis structures for NCl_3 and a hypothetical NCl_5, and explain why NCl_5 is an unlikely structure.

1-31 Draw a Lewis structure for each species.
(a) N_2H_4 (b) N_2H_2 (c) $(CH_3)_2NH_2Cl$ (d) CH_3CN
(e) CH_3CHO (f) $CH_3S(O)CH_3$ (g) H_2SO_4 (h) CH_3NCO
(i) $CH_3OSO_2OCH_3$ (j) $CH_3C(NH)CH_3$ (k) $(CH_3)_3CNO$

1-32 Draw a Lewis structure for each compound. Include all nonbonding pairs of electrons.
(a) $CH_3COCH_2CHCHCOOH$ (b) $NCCH_2COCH_2CHO$
(c) $CH_2CHCH(OH)CH_2CO_2H$ (d) $CH_2CHC(CH_3)CHCOOCH_3$

1-33 Draw a line–angle formula for each compound in Problem 1-32.

1-34 Draw Lewis structures for
(a) two compounds of formula C_4H_{10} (b) two compounds of formula C_2H_6O
(c) two compounds of formula C_2H_7N (d) three compounds of formula C_2H_7NO
(e) three compounds of formula $C_3H_8O_2$ (f) three compounds of formula C_2H_4O

1-35 Draw a complete structural formula and a condensed structural formula for
(a) three compounds of formula C_3H_8O (b) five compounds of formula C_3H_6O

1-36 Some of the following molecular formulas correspond to stable compounds. When possible, draw a stable structure for each formula.

CH_2 CH_3 CH_4 CH_5

C_2H_2 C_2H_3 C_2H_4 C_2H_5 C_2H_6 C_2H_7

C_3H_3 C_3H_4 C_3H_5 C_3H_6 C_3H_7 C_3H_8 C_3H_9

Propose a general rule for the numbers of hydrogen atoms in stable hydrocarbons.

1-37 Draw complete Lewis structures, including lone pairs, for the following compounds.

(a) 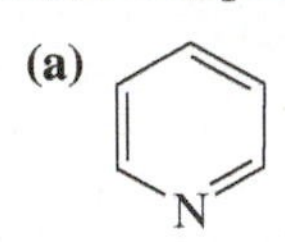
pyridine

(b) 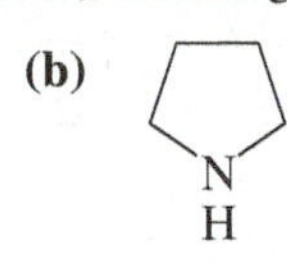
pyrrolidine

(c)  furan

(d)  H_2N–(CH₂)₃–COOH
γ-aminobutyric acid
(a neurotransmitter)

(e) 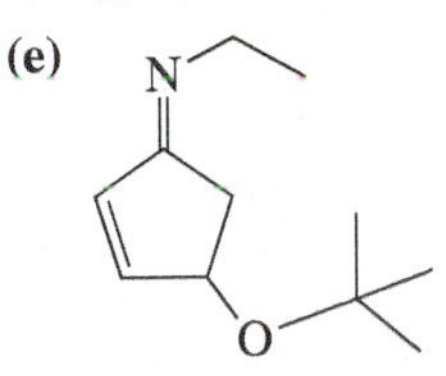

(f) 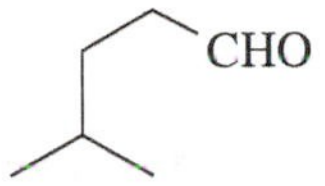 CHO

(g) 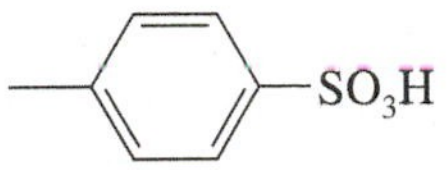 SO_3H

(h) 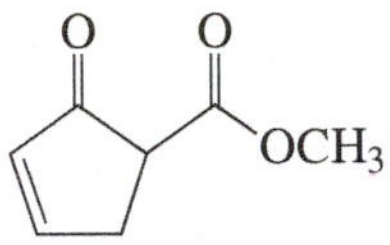 OCH_3

1-38 Give the molecular formula of each compound shown in Problem 1-37.

1-39 1. From what you remember of electronegativities, show the direction of the dipole moments of the following bonds.
2. In each case, predict whether the dipole moment is relatively large (electronegativity difference >0.5) or small.

(a) C—Cl (b) C—H (c) C—Li (d) C—N (e) C—O
(f) C—B (g) C—Mg (h) N—H (i) O—H (j) C—Br

1-40 For each of the following structures,
1. Draw a Lewis structure; fill in any nonbonding electrons.
2. Calculate the formal charge on each atom other than hydrogen.

(a) CH_3NO (nitrosomethane)
(b) $(CH_3)_3NO$ (trimethylamine oxide)
(c) $[N_3]^-$ (azide ion)
(d) $[(CH_3)_3O]^+$
(e) CH_3NC
(f) $(CH_3)_4NBr$

1-41 Determine whether the following pairs of structures are actually different compounds or simply resonance forms of the same compounds.

(a) 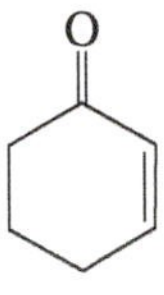 and 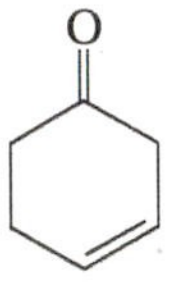
(b) 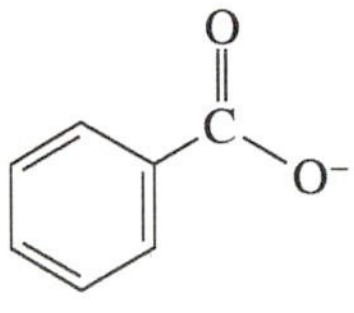 and 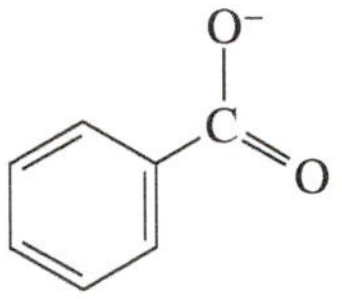
(c) 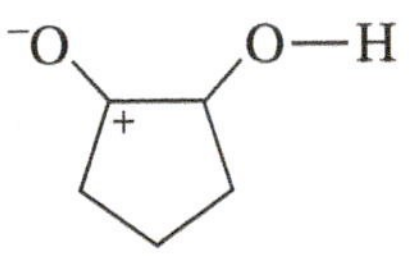 and 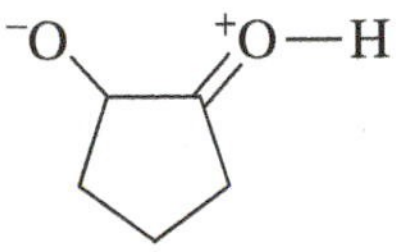
(d) 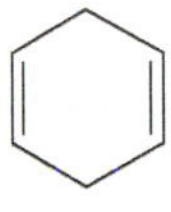 and

(e) 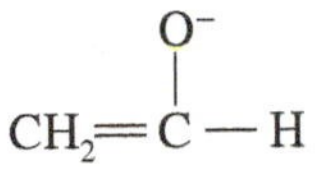 $CH_2{=}C(O^-){-}H$ and $^-CH_2{-}C({=}O){-}H$
(f) 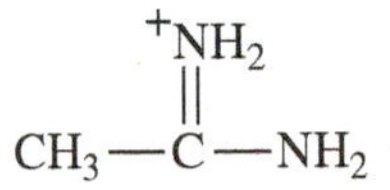 $CH_3{-}C({=}^+NH_2){-}NH_2$ and $CH_3{-}C(NH_2){=}^+NH_2$
(g) $CH_2{=}CH{-}\overset{+}{C}H_2$ and $\overset{+}{C}H_2{-}CH{=}CH_2$
(h) $CH_2{=}C{=}O$ and $H{-}C{\equiv}C{-}OH$
(i) 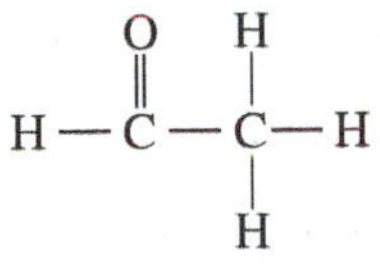 $H{-}C({=}O){-}CH_2{-}H$ and $H{-}C(OH){=}C(H){-}H$
(j) 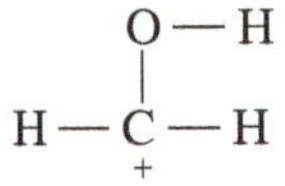 $H{-}\overset{+}{C}(OH){-}H$ and $H{-}C({=}^+O{-}H){-}H$
(k) 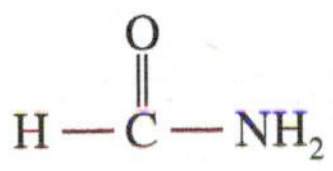 $H{-}C({=}O){-}NH_2$ and $H{-}C(O^-){=}\overset{+}{N}H_2$
(l) 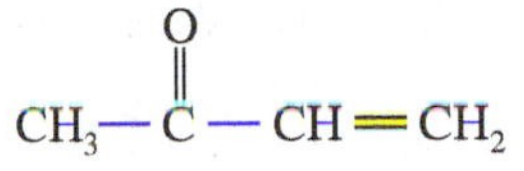 $CH_3{-}C({=}O){-}CH{=}CH_2$ and $CH_3{-}C(O^-){=}CH{-}\overset{+}{C}H_2$

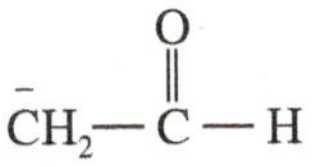

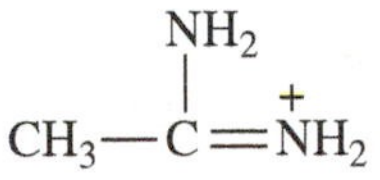

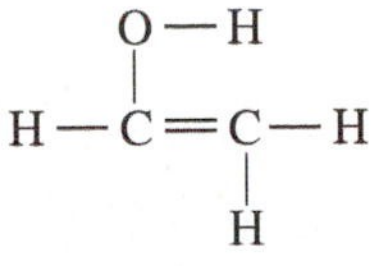

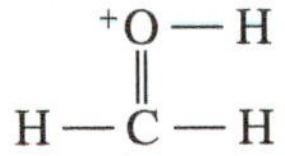

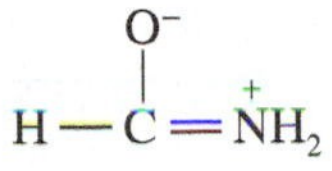

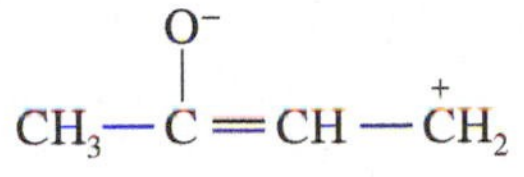

1-42 Draw the important resonance forms to show the delocalization of charges in the following ions. In each case, indicate the major resonance form(s).

(a) $CH_3-\overset{O}{\overset{\|}{C}}-\overset{-}{C}H_2$

(b) $H-\overset{O}{\overset{\|}{C}}-CH{=}CH-\overset{-}{C}H_2$

(c) 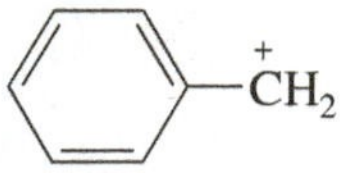 $\overset{+}{C}H_2$

(d) 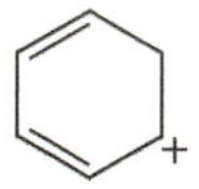 +

(e) 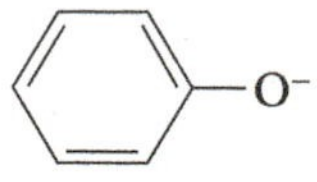 O^-

(f) 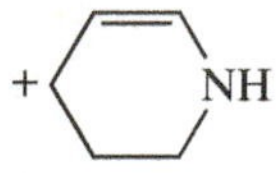 + NH

(g) 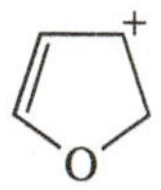 + O

(h) 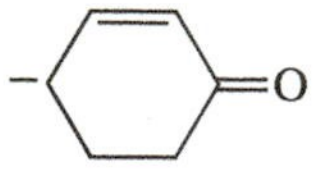 − =O

(i) $CH_2{=}CH-CH{=}CH-\overset{+}{C}H-CH_3$

(j) $CH_3-CH{=}CH-CH{=}CH-CH_2-\overset{+}{C}H_2$

(k) $\overset{+}{C}H_2-CH{=}CH-O-CH{=}CH_2$

1-43 In the following sets of resonance forms, label the major and minor contributors and state which structures would be of equal energy. Add any missing important resonance forms.

(a) $[CH_3-\overset{-}{\underset{\cdot\cdot}{C}}H-C{\equiv}N: \longleftrightarrow CH_3-CH{=}C{=}\ddot{N}:^-]$

(b) $[CH_3-\overset{O^-}{\overset{|}{C}}{=}CH-\overset{+}{C}H-CH_3 \longleftrightarrow CH_3-\overset{O^-}{\overset{|}{\underset{+}{C}}}-CH{=}CH-CH_3]$

(c) $[CH_3-\overset{O}{\overset{\|}{C}}-\overset{-}{C}H-\overset{O}{\overset{\|}{C}}-CH_3 \longleftrightarrow CH_3-\overset{O^-}{\overset{|}{C}}{=}CH-\overset{O}{\overset{\|}{C}}-CH_3]$

(d) $[CH_3-\overset{-}{C}H-CH{=}CH-NO_2 \longleftrightarrow CH_3-CH{=}CH-\overset{-}{C}H-NO_2]$

(e) $[CH_3-CH_2-\overset{NH_2}{\overset{|}{\underset{+}{C}}}-NH_2 \longleftrightarrow CH_3-CH_2-\overset{NH_2}{\overset{|}{C}}{=}\overset{+}{N}H_2]$

1-44 For each of these ions, draw the important resonance forms and predict which resonance form is likely to be the major contributor.

(a) 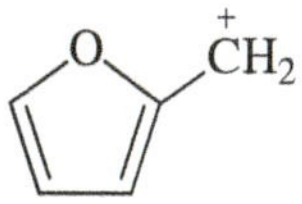 O $\overset{+}{C}H_2$

(b) O $\overset{-}{C}H_2$

(c) HN $\overset{+}{C}H_2$

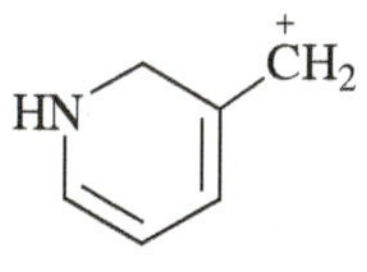

(d) 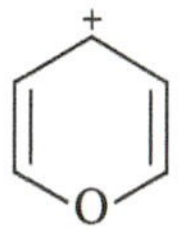 + O

(e)  −

(f) 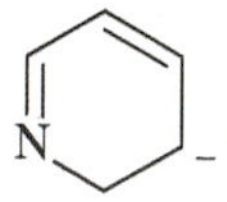 N −

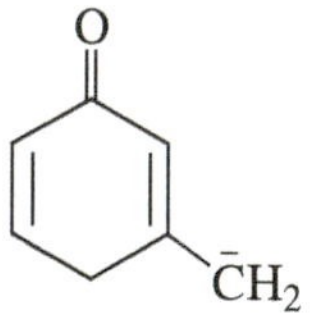

1-45 For each pair of ions, determine which ion is more stable. Use resonance forms to explain your answers.

(a) $CH_3-\overset{+}{C}H-CH_3$ or $CH_3-\overset{+}{C}H-OCH_3$

(b) CH_3-N-CH_3 / $CH_3-\underset{+}{C}-CH_3$ or $CH_3-CH-CH_3$ / $CH_3-\underset{+}{C}-CH_3$

(c) $CH_2{=}CH-\overset{+}{C}H-CH_3$ or $CH_2{=}CH-CH_2-\overset{+}{C}H_2$

(d) $\overset{-}{C}H_2-CH_3$ or $\overset{-}{C}H_2-C{\equiv}N:$

(e) 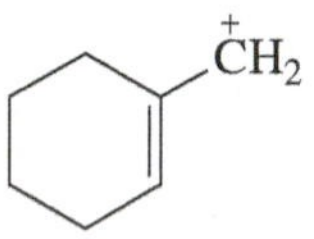 $\overset{+}{C}H_2$ or $\overset{+}{C}H_2$

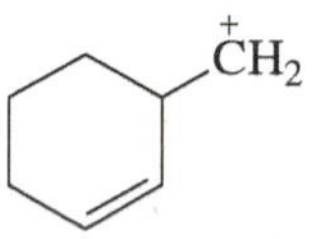

(f) 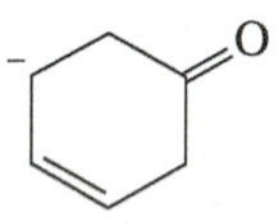 − O or O −

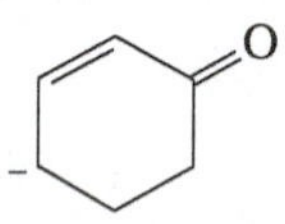

1-46 Use resonance structures to identify the areas of high and low electron density in the following compounds:

(a) 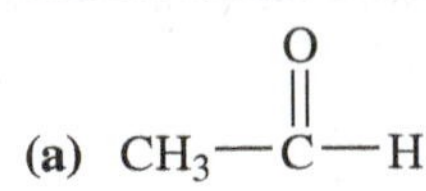

(b) $CH_3-\overset{O}{\overset{\|}{C}}-NH_2$

(c) $CH_3-\overset{NH}{\overset{\|}{C}}-H$

(d) $CH_3-\overset{O}{\overset{\|}{C}}-OCH_3$

(e) 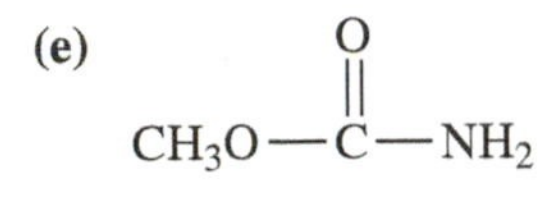

(f) NH, O (ring structure)

(g) CN, NH2 (ring structure)

(h) 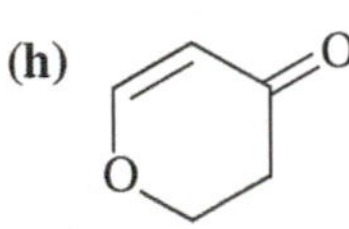

(i) H–N, O (ring structure)

(j) CH_3O–C$_6$H$_4$–C(=O)–H

1-47 Compound X, isolated from lanolin (sheep's wool fat), has the pungent aroma of dirty sweatsocks. A careful analysis showed that compound X contains 62.0% carbon and 10.4% hydrogen. No nitrogen or halogen was found.

(a) Compute an empirical formula for compound X.
(b) A molecular weight determination showed that compound X has a molecular weight of approximately 117. Find the molecular formula of compound X.
(c) Many possible structures have this molecular formula. Draw complete structural formulas for four of them.

***1-48** In 1934, Edward A. Doisy of Washington University extracted 3000 lb of hog ovaries to isolate a few milligrams of pure estradiol, a potent female hormone. Doisy burned 5.00 mg of this precious sample in oxygen and found that 14.54 mg of CO_2 and 3.97 mg of H_2O were generated.

(a) Determine the empirical formula of estradiol.
(b) The molecular weight of estradiol was later determined to be 272. Determine the molecular formula of estradiol.

1-49 If the carbon atom in CH_2Cl_2 were flat, there would be two stereoisomers. The carbon atom in CH_2Cl_2 is actually tetrahedral. Make a model of this compound, and determine whether there are any stereoisomers of CH_2Cl_2.

(H above; H—C—Cl; Cl below)   (H above; Cl—C—Cl; H below)

1-50 Cyclopropane (C_3H_6, a three-membered ring) is more reactive than most other cycloalkanes.

(a) Draw a Lewis structure for cyclopropane.
(b) Compare the bond angles of the carbon atoms in cyclopropane with those in an acyclic (noncyclic) alkane.
(c) Suggest why cyclopropane is so reactive.

1-51 For each of the following compounds,

1. Give the hybridization and approximate bond angles around each atom except hydrogen.
2. Draw a three-dimensional diagram, including any lone pairs of electrons.

(a) H_3O^+ (b) ^-OH (c) CH_2CHCN (d) $(CH_3)_3N$ (e) $[CH_3NH_3]^+$
(f) CH_3COOH (g) CH_3CHNH (h) CH_3OH (i) CH_2O

1-52 For each of the following compounds and ions,

1. Draw a Lewis structure.
2. Show the kinds of orbitals that overlap to form each bond.
3. Give approximate bond angles around each atom except hydrogen.

(a) $[NH_2]^-$ (b) $[CH_2OH]^+$ (c) $CH_2=N-CH_3$
(d) $CH_3-CH=CH_2$ (e) $HC\equiv C-CHO$ (f) H_2N-CH_2-CN
(g) $CH_3-\overset{O}{\overset{\|}{C}}-OH$ (h) (ring with O and =O) (i) + (ring with O)

1-53 In most amines, the nitrogen atom is sp^3 hybridized, with a pyramidal structure and bond angles close to 109°. In urea, both nitrogen atoms are found to be planar, with bond angles close to 120°. Explain this surprising finding. (*Hint*: Consider resonance forms and the overlap needed in them.)

$$H_2\ddot{N}-\overset{O}{\overset{\|}{C}}-\ddot{N}H_2$$

urea

1-54 Predict the hybridization and geometry of the carbon and nitrogen atoms in the following molecules and ions. (*Hint:* Resonance.)

(a) 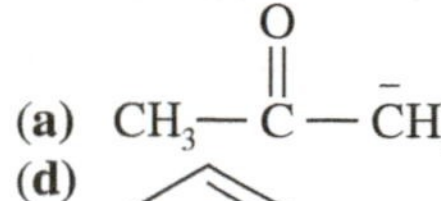
(b) $H_2N—CH{=}CH—\overset{+}{C}H_2$
(c) $\overset{-}{C}H_2—C{\equiv}N$

(d) 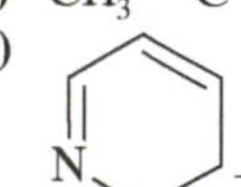
(e) 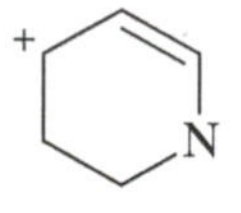
(f)

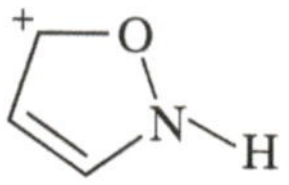

(g) 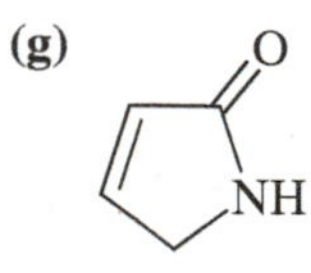
(h) 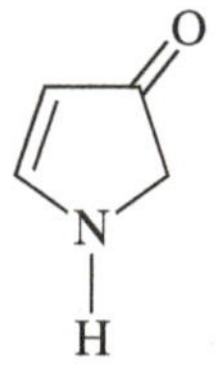
(i)

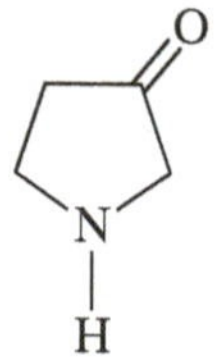

1-55 Draw orbital pictures of the pi bonding in the following compounds:

(a) CH_3COCH_3
(b) HCN
(c) $CH_2{=}CH—CHCHCN$
(d) $CH_3C{\equiv}CCHO$
(e) $CH_3CH{=}C{=}CHCH_3$
(f) $CH_3—CH{=}N—CH{=}C{=}O$

1-56 (a) Draw the structure of *cis*- $CH_3—CH{=}CH—CH_2CH_3$ showing the pi bond with its proper geometry.
(b) Circle the six coplanar atoms in this compound.
(c) Draw the trans isomer, and circle the coplanar atoms. Are there still six?
(d) Circle the coplanar atoms in the following structure:

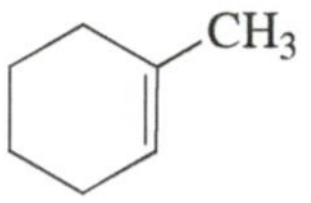

1-57 In pent-2-yne ($CH_3CCCH_2CH_3$), there are four atoms in a straight line. Use dashed lines and wedges to draw a three-dimensional representation of this molecule, and circle the four atoms that are in a straight line.

1-58 Which of the following compounds show cis-trans isomerism? Draw the cis and trans isomers of the ones that do.

(a) $CH_3CH{=}CHCH_3$
(b) $CH_3—C{\equiv}C—CH_3$
(c) $CH_2{=}C(CH_3)_2$
(d) cyclopentene, 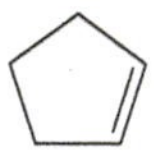
(e) $CH_3—CH{=}C(CH_2CH_2CH_3)—CH_2—CH_3$
(f) $CH_3—CH{=}N—CH_3$

1-59 Give the relationships between the following pairs of structures. The possible relationships are as follows: same compound, cis-trans isomers, constitutional (structural) isomers, and not isomers (different molecular formula).

(a) $CH_3CH_2CH_2CH_3$ and $(CH_3)_3CH$
(b) $CH_2{=}CH—CH_2Cl$ and $CHCl{=}CH—CH_3$
(c) 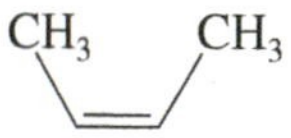 and 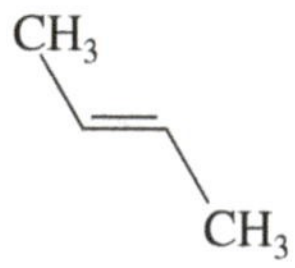
(d) 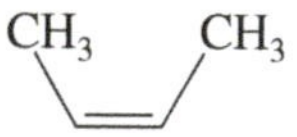 and 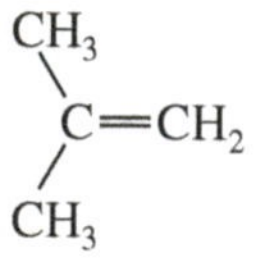
(e) 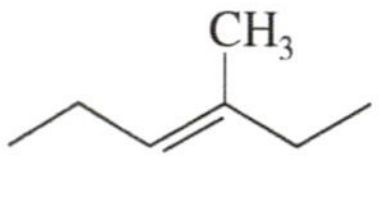 and 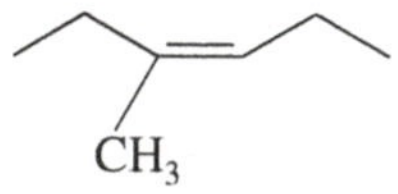
(f) 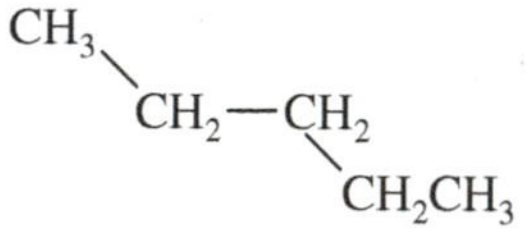 and 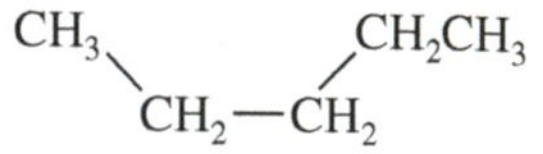
(g) 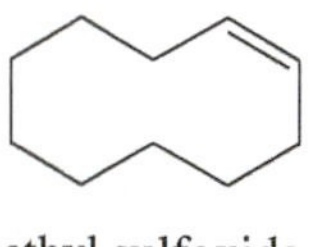 and 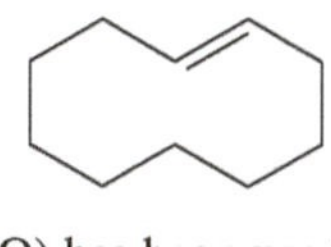
(h) 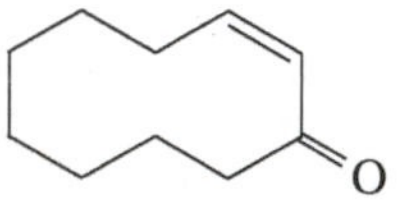 and

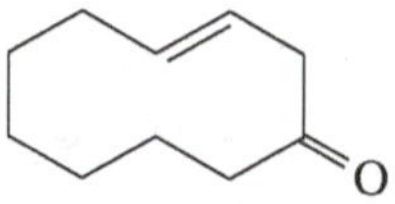

***1-60** Dimethyl sulfoxide (DMSO) has been used as an anti-inflammatory rub for race horses. DMSO and acetone appear to have similar structures, but the C=O carbon atom in acetone is planar, while the S=O sulfur atom in DMSO is pyramidal. Draw Lewis structures for DMSO and acetone, predict the hybridizations, and explain these observations.

DMSO

acetone

2 Acids and Bases; Functional Groups

Goals for Chapter 2

1 Identify the molecular features that cause compounds to be polar and to engage in hydrogen bonding.

2 Predict general trends in physical properties such as boiling points and solubilities.

3 Identify acids, bases, electrophiles, and nucleophiles. Compare their strengths and predict their reactions based on structure and bonding, as well as K_a and pK_a values.

4 Identify the nucleophiles and electrophiles in Lewis acid–base reactions and use curved arrows to show the flow of electrons.

5 Identify the general classes of organic compounds.

◀ Citric acid and ascorbic acid (vitamin C) are two water-soluble organic acids found in citrus fruits. Citric acid plays a central role in metabolism, and ascorbic acid is needed for growth and repair of collagen, a structural protein in connective tissue. Unlike most animals, humans cannot synthesize their own ascorbic acid. Ascorbic acid deficiency causes scurvy, with debilitating and often deadly symptoms resulting from damage to connective tissues.

Chemists categorize organic compounds into families based on characteristic groups of atoms called **functional groups.** These functional groups determine the properties of their parent compounds, and they serve as acidic or basic sites within the molecules. The acidic and basic properties of organic compounds are crucial to understanding their stability and reactivity. Most organic compounds can react as both acids in a basic environment and bases in an acidic environment. Nearly all the substances around you and inside you are acids and bases—from the food you eat to the products you use to clean up afterward.

In this chapter, we expand the ideas from Chapter 1 to consider more about the distribution of electrons in molecules and how these electron distributions affect the molecular properties. Ultimately, electron distributions cause the compounds to react as acids and bases. We cover the acid–base reactions of organic compounds and show that many reactions involve the transfer of protons. However, many acid–base reactions do not involve protons at all, but depend on changes in electron distribution. Then we go on to discuss the families of organic compounds, showing how the characteristic functional groups in each class determine the electron distribution in the molecules and their reactivity. We will refer to the relative acidity or basicity of molecules throughout this course and show how these concepts affect organic reactions.

2-1 Polarity of Bonds and Molecules

In Section 1-6, we reviewed the concept of polar covalent bonds between atoms with different electronegativities. Now we are ready to combine this concept with molecular geometry to study the polarity of entire molecules.

2-1A Bond Dipole Moments

Bond polarities can range from nonpolar covalent, through polar covalent, to totally ionic. In the following examples, ethane has a nonpolar covalent C—C bond. Methylamine, methanol, and chloromethane have increasingly polar (C—N, C—O, and C—Cl) covalent bonds. Methylammonium chloride ($CH_3NH_3^+ \, Cl^-$) has an ionic bond between the methylammonium ion and the chloride ion.

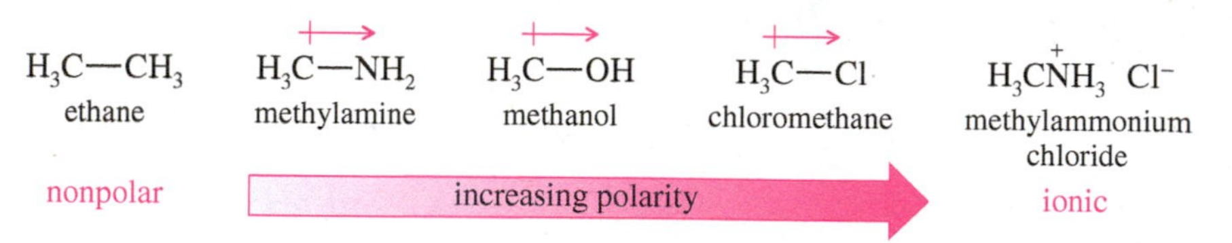

The polarity of an individual bond is measured as its **bond dipole moment,** μ, defined as

$$\mu = \delta \times d$$

where δ is the amount of charge at either end of the dipole and d is the distance between the charges.

Dipole moments are expressed in units of the debye (D), where 1 debye = 3.34×10^{-30} coulomb-meter. If a proton and an electron (charge 1.60×10^{-19} coulomb) were 1 Å apart (distance 10^{-10} meter), the dipole moment would be

$$\mu = (1.60 \times 10^{-19} \text{ coulomb}) \times (10^{-10} \text{ meter}) = 1.60 \times 10^{-29} \text{ coulomb-meter}$$

Expressed in debyes,

$$\mu = \frac{1.60 \times 10^{-29} \text{ C} \cdot \text{m}}{3.34 \times 10^{-30} \text{ C} \cdot \text{m/D}} = 4.8 \text{ D}$$

A simple rule of thumb, using common units, is that

$$\mu \text{ (in debyes)} = 4.8 \times \delta \text{ (electron charge)} \times d \text{ (in angstroms)}$$

Dipole moments are measured experimentally, and they can be used to calculate other information such as bond lengths and charge separations.

Bond dipole moments in organic compounds range from zero in symmetrical bonds to about 3.6 D for the strongly polar C≡N: triple bond. Table 2-1 shows typical dipole moments for some of the bonds common in organic molecules. Recall that the positive end of the crossed arrow corresponds to the less electronegative (partial positive charge) end of the dipole.

TABLE 2-1 BOND DIPOLE MOMENTS (DEBYE) FOR SOME COMMON COVALENT BONDS

Bond	Dipole Moment, μ	Bond	Dipole Moment, μ
$\overset{+\!\!\longrightarrow}{\text{C—N}}$	0.22 D	$\overset{+\!\!\longrightarrow}{\text{H—C}}$	0.3 D
$\overset{+\!\!\longrightarrow}{\text{C—O}}$	0.86 D	$\overset{+\!\!\longrightarrow}{\text{H—N}}$	1.31 D
$\overset{+\!\!\longrightarrow}{\text{C—F}}$	1.51 D	$\overset{+\!\!\longrightarrow}{\text{H—O}}$	1.53 D
$\overset{+\!\!\longrightarrow}{\text{C—Cl}}$	1.56 D	$\overset{+\!\!\longrightarrow}{\text{C=O}}$	2.4 D
$\overset{+\!\!\longrightarrow}{\text{C—Br}}$	1.48 D	$\overset{+\!\!\longrightarrow}{\text{C≡N}}$	3.6 D
$\overset{+\!\!\longrightarrow}{\text{C—I}}$	1.29 D		

SOLVED PROBLEM 2-1

Calculate the amount of charge separation for a typical C—O single bond, with a bond length of 1.43 Å and a dipole moment of 0.86 D.

SOLUTION

The bond is polarized with a partial negative charge on oxygen and an equal amount of positive charge on carbon.

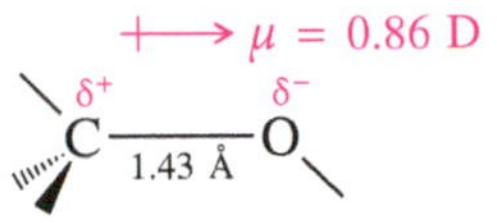

$\mu = 0.86$ D
δ^+ C—O δ^- (1.43 Å)

Using the formula for the dipole moment, we have

$$0.86 \text{ D} = 4.8 \times \delta \times 1.43\text{Å}$$

$$\delta = 0.125\ e$$

The amount δ of charge separation is about 0.125 electronic charge, so the carbon atom has about an eighth of a positive charge, and the oxygen atom has about an eighth of a negative charge.

PROBLEM 2-1

The C=O double bond has a dipole moment of about 2.4 D and a bond length of about 1.23 Å.

(a) Calculate the amount of charge separation in this bond.

(b) Use this information to evaluate the relative importance of the following two resonance contributors:

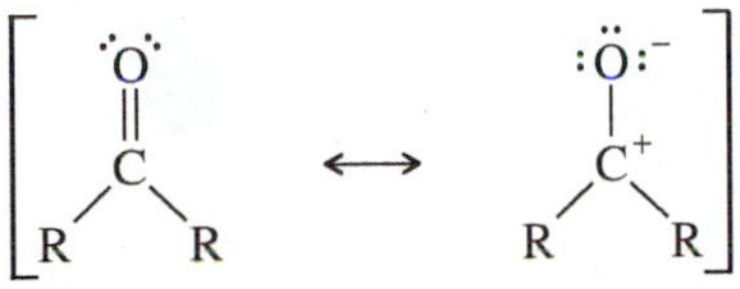

[R₂C=Ö: ↔ R₂C⁺—Ö:⁻]

(R is a general symbol for a carbon-containing group.)

2-1B Molecular Dipole Moments

A **molecular dipole moment** is the dipole moment of the molecule taken as a whole. It is a good indicator of a molecule's overall polarity. Molecular dipole moments can be measured directly, in contrast to bond dipole moments, which must be estimated by comparing various compounds. The value of the molecular dipole moment is equal to the *vector* sum of the individual bond dipole moments. This vector sum reflects both the magnitude and the direction of each individual bond dipole moment.

For example, formaldehyde has one strongly polar C═O bond, and carbon dioxide has two. We might expect CO_2 to have the larger dipole moment, but its dipole moment is actually zero. The symmetry of the carbon dioxide molecule explains this surprising result. The structures of formaldehyde and carbon dioxide are shown here, together with their electrostatic potential maps. These electrostatic potential maps show the directions of the bond dipole moments, with red at the negative ends and blue at the positive ends of the dipoles. In carbon dioxide, the bond dipole moments are oriented in opposite directions, so they cancel each other.

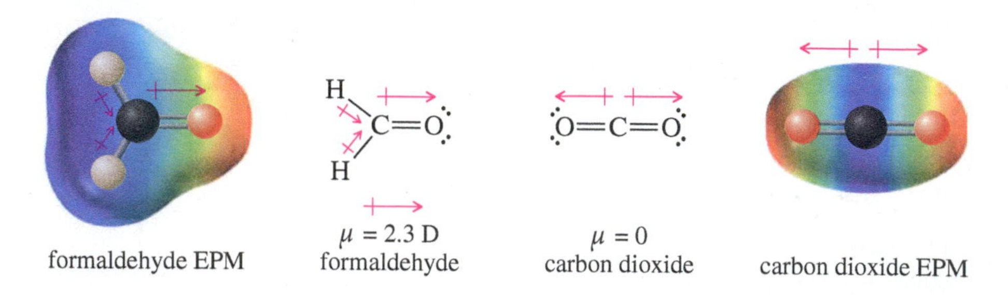

Figure 2-1 shows some examples of molecular dipole moments. Notice that the dipole moment of C—H bonds is small, so we often treat C—H bonds as nearly nonpolar. Also note that the tetrahedral symmetry of CCl_4 positions the four C—Cl dipole moments in directions so that they cancel. A partial canceling of the bond dipole moments explains why $CHCl_3$, with three C—Cl bonds, has a smaller molecular dipole moment than CH_3Cl, with only one.

Lone pairs of electrons contribute to the dipole moments of bonds and molecules. Each lone pair corresponds to a charge separation, with the nucleus having a partial positive charge balanced by the negative charge of the lone pair. Figure 2-2 shows four molecules with lone pairs and large dipole moments. Notice how the lone pairs contribute to the large dipole moments, especially in the C═O and C≡N bonds. Also notice the red areas in the electrostatic potential maps, indicating high negative potential in the electron-rich regions of the lone pairs.

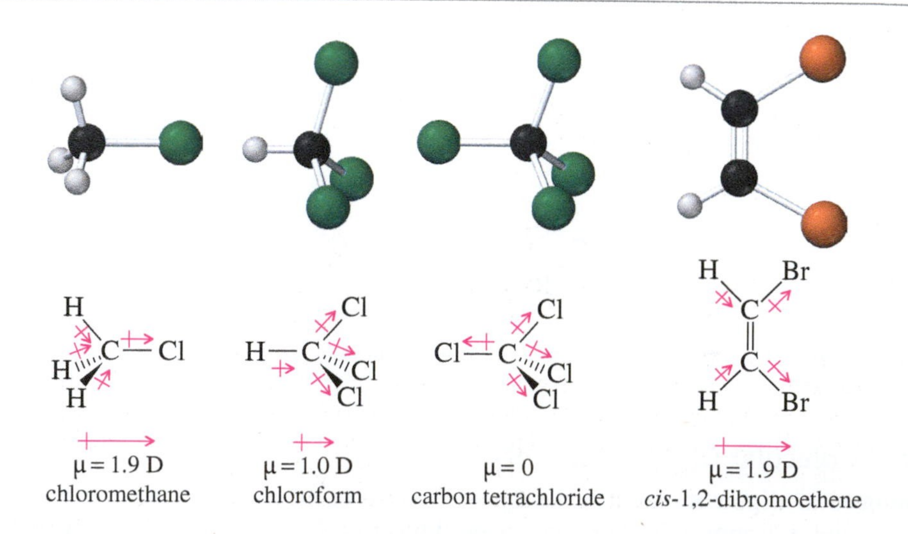

FIGURE 2-1
Molecular dipole moments. A molecular dipole moment is the vector sum of the individual bond dipole moments.

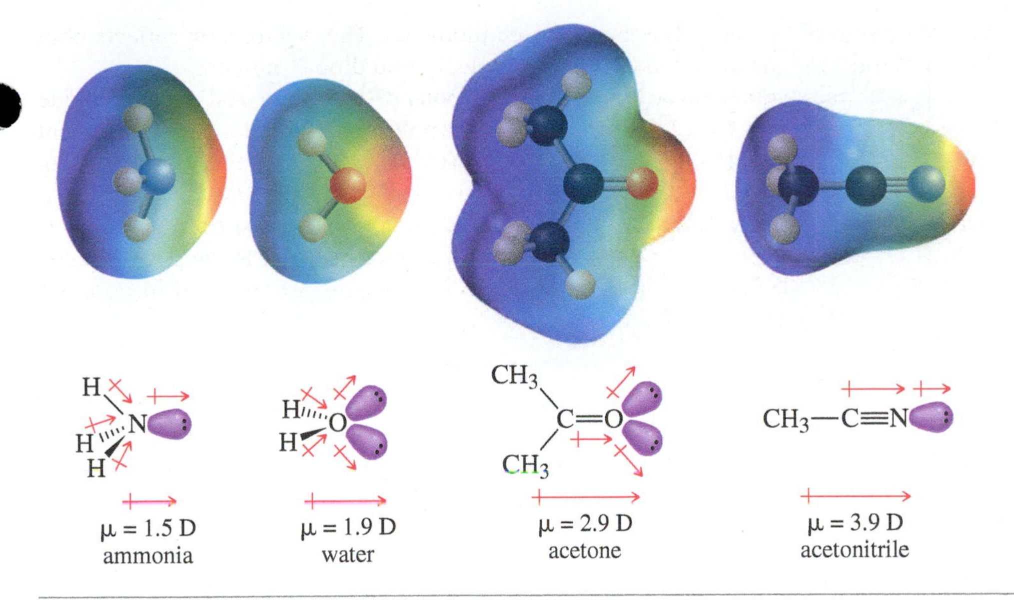

FIGURE 2-2
Effects of lone pairs on dipole moments. Lone pairs can make large contributions to molecular dipole moments.

PROBLEM 2-2

The N—F bond is more polar than the N—H bond, but NF_3 has a *smaller* dipole moment than NH_3. Explain this curious result.

NH_3	NF_3
$\mu = 1.5$ D	$\mu = 0.2$ D

PROBLEM 2-3 (PARTIALLY SOLVED)

For each of the following compounds:

1. Draw the Lewis structure.
2. Show how the bond dipole moments (and those of any nonbonding pairs of electrons) contribute to the molecular dipole moment.
3. Estimate whether the compound will have a large, small, or zero dipole moment.

(a) NH_4^+ **(b)** O_3

(a) NH_4^+ has four polar N—H bonds. These bonds are probably more polarized than a typical N—H bond, because the N in NH_4^+ bears a formal positive charge. Nevertheless, these four polar bonds have a symmetric tetrahedral arrangement so they cancel each other.

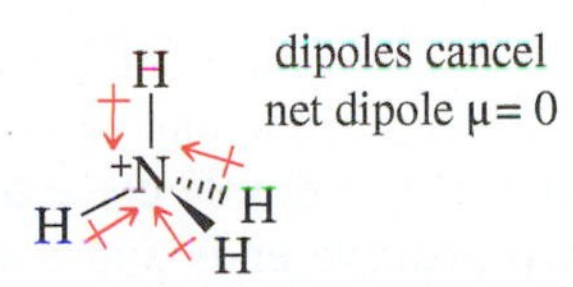

(b) Ozone (O_3) is an sp^2 hybrid structure, with a lone pair on the central oxygen atom. Therefore, O_3 must be bent. The resonance structures imply partial negative charges on the outer oxygens and a partial positive charge on the central oxygen. The lone pair on the central oxygen cancels part, but not all, of the vector sum of the two O—O dipoles. The resulting net dipole is relatively small.

(continued)

resonance forms of ozone

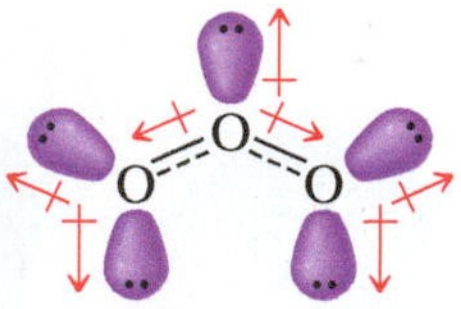

small net dipole (0.52 D)

(c) CH_2Cl_2
(d) CH_3F
(e) CF_4
(f) CH_3OH
(g) HCN
(h) CH_3CHO
(i) $H_2C{=}NH$
(j) $(CH_3)_3N$
(k) $CH_2{=}CHCl$
(l) BF_3
(m) $BeCl_2$

PROBLEM 2-4

Two isomers of 1,2-dichloroethene are known. One has a dipole moment of 2.4 D; the other has zero dipole moment. Draw the two isomers, and explain why one has zero dipole moment.

$CHCl{=}CHCl$
1,2-dichloroethene

2-2 Intermolecular Forces

When two molecules approach, they attract or repel each other. This interaction can be described fairly simply in the case of atoms (such as the noble gases) or simple molecules such as H_2 and Cl_2. In general, the forces are attractive until the molecules come so close that they infringe on each other's van der Waals radius. When this happens, the small attractive force quickly becomes a large repulsive force, and the molecules "bounce" off each other. With complicated organic molecules, these attractive and repulsive forces are more difficult to predict. We can still describe the nature of the forces, however, and we can show how they affect the physical properties of organic compounds.

Attractions between molecules are particularly important in solids and liquids. In these "condensed" phases, the molecules are continuously in contact with each other. The melting points, boiling points, and solubilities of organic compounds show the effects of these forces. Boiling points roughly indicate the strength of intermolecular forces, because those forces must be overcome to boil the compound. Three major kinds of attractive forces cause molecules to associate into solids and liquids:

1. the dipole–dipole forces of polar molecules;
2. the London dispersion forces that affect all molecules; and
3. the "hydrogen bonds" that link molecules having —OH or —NH groups.

2-2A Dipole–Dipole Forces

Most molecules have permanent dipole moments as a result of their polar bonds. Each molecular dipole moment has a positive end and a negative end. The most stable arrangement has the positive end of one dipole close to the negative end of another. When two negative ends or two positive ends approach each other, they repel, but they may turn and orient themselves in the more stable positive-to-negative arrangement. **Dipole–dipole forces,** therefore, are generally attractive intermolecular forces resulting from the attraction of the positive and negative ends of the dipole moments of polar molecules. Figure 2-3 shows the attractive and repulsive orientations of polar molecules, using chloromethane as the example.

Polar molecules are mostly oriented in the lower-energy positive-to-negative arrangement, and the net force is attractive. This attraction must be overcome when the liquid vaporizes, resulting in larger heats of vaporization and higher boiling points for strongly polar compounds.

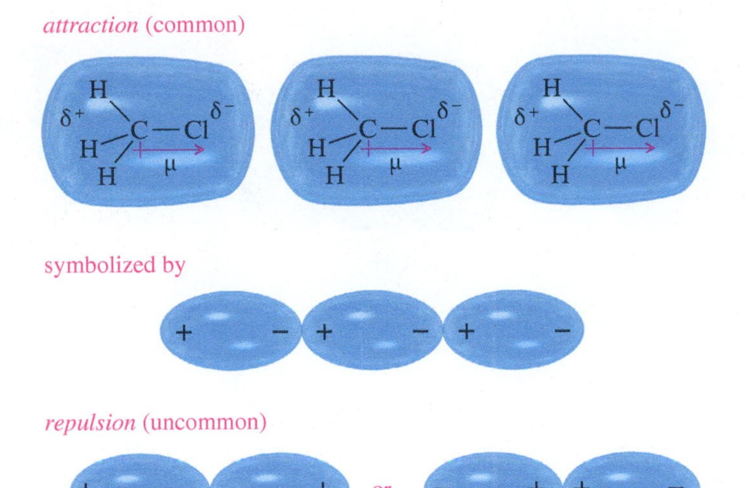

FIGURE 2-3
Dipole–dipole interactions. Dipole–dipole interactions result from the approach of two polar molecules. If their positive and negative ends approach, the interaction is attractive. If two negative ends or two positive ends approach, the interaction is repulsive. In a liquid or a solid, the molecules orient mostly with their positive and negative ends together, and the net force is attractive.

2-2B The London Dispersion Force

Carbon tetrachloride (CCl_4) has a zero dipole moment, yet its boiling point is higher than that of chloroform ($\mu = 1.0$ D). Clearly, there must be some kind of force other than dipole–dipole forces holding the molecules of carbon tetrachloride together.

$\mu = 0$
carbon tetrachloride, bp = 77 °C

$\mu = 1.0$ D
chloroform, bp = 62 °C

In nonpolar molecules such as carbon tetrachloride, the principal attractive force is the **London dispersion force,** one of the **van der Waals forces** (Figure 2-4). The London force arises from temporary dipole moments that are induced in a molecule by other nearby molecules. Even though carbon tetrachloride has no permanent dipole moment, the electrons are not always evenly distributed. A small temporary dipole moment is induced when one molecule approaches another molecule in which the electrons are slightly displaced from a symmetrical arrangement. The electrons in the approaching molecule are displaced slightly so that an attractive dipole–dipole interaction results.

These temporary dipoles last only a fraction of a second, and they constantly change; yet they are correlated so that their net force is attractive. This attractive force depends on close surface contact of two molecules, so it is roughly proportional to the molecular surface area. Carbon tetrachloride has a larger surface area than chloroform (a chlorine atom is much larger than a hydrogen atom), so the intermolecular London dispersion attractions between carbon tetrachloride molecules are stronger than they are between chloroform molecules.

We can see the effects of London forces in the boiling points of simple hydrocarbons. If we compare the boiling points of several isomers, the isomers with larger surface areas (and greater potential for London force attraction) have higher boiling points. The boiling points of three C_5H_{12} isomers are given here. The long-chain isomer (*n*-pentane) has the greatest surface area and the highest boiling point. As the amount of chain branching increases, the molecule becomes more spherical and its surface area decreases. The most highly branched isomer (neopentane) has the smallest surface area and the lowest boiling point.

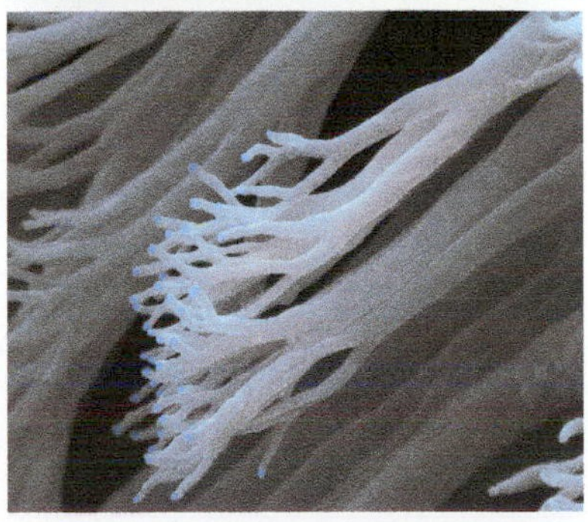

A gecko easily climbs a polished glass window, even though the glass has no crevices. A gecko's toe has thousands of tiny hairs, with each hair split into hundreds of tiny tips. Each tiny hair tip is attracted to the surface by van der Waals forces. Millions of hair tips provide a large surface area, so these weak intermolecular attractions easily support the gecko's weight.

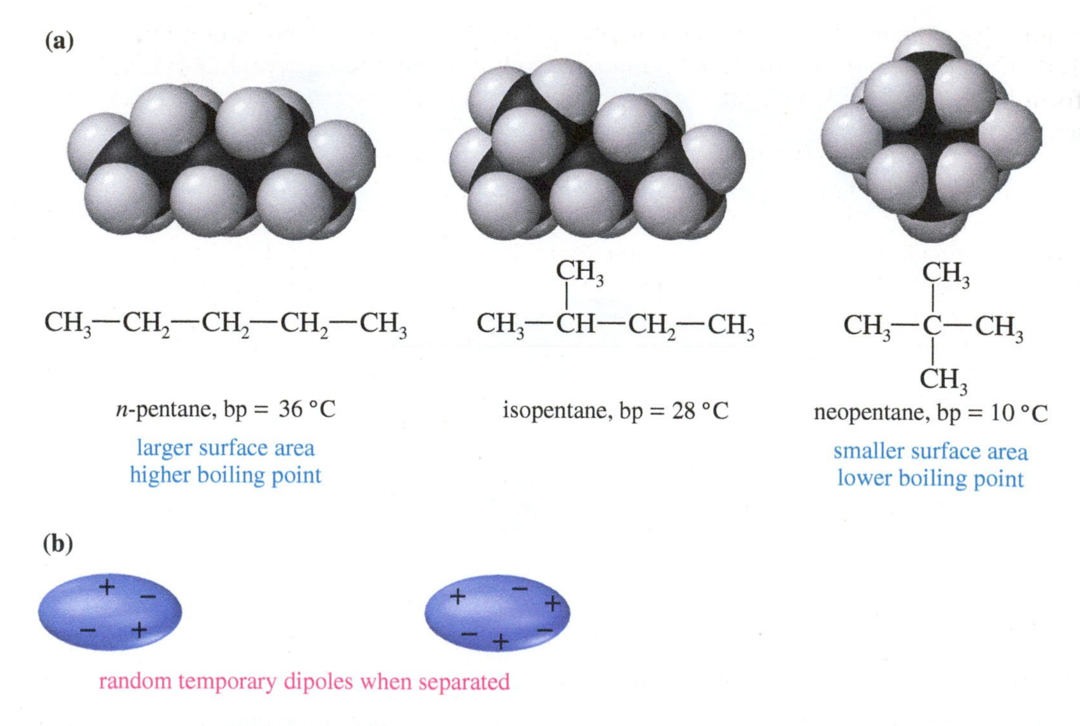

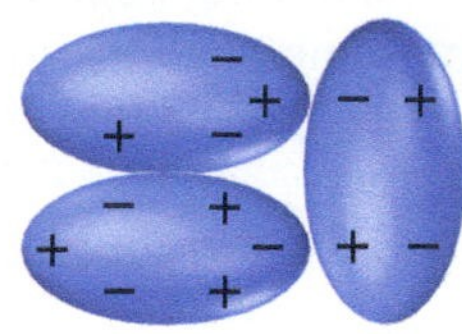

FIGURE 2-4
London dispersion forces between nonpolar molecules. London dispersion forces result from the attraction of correlated temporary dipoles.

2-2C Hydrogen Bonding

A **hydrogen bond** is not a true bond but a particularly strong dipole–dipole attraction. A hydrogen atom can participate in hydrogen bonding if it is bonded to oxygen, nitrogen, or fluorine. Organic compounds do not contain H—F bonds, so we consider only N—H and O—H hydrogens to be hydrogen bonded (Figure 2-5).

The O—H and N—H bonds are strongly polarized, leaving the hydrogen atom with a partial positive charge. This electrophilic hydrogen has a strong affinity for nonbonding electrons, and it forms intermolecular attachments with the nonbonding electrons on oxygen or nitrogen atoms.

Hydrogen bonding is a strong form of intermolecular attraction, yet it is much weaker than a normal C—H, N—H, or O—H covalent bond. Breaking a hydrogen bond requires about 20/kJ mol (5 kcal/mol), compared with about 400 kJ/mol (about 100 kcal/mol) required to break a C—H, N—H, or O—H bond.

Hydrogen bonding has a large effect on the physical properties of organic compounds, as shown by the boiling points of ethanol (ethyl alcohol) and dimethyl ether, two isomers of molecular formula C_2H_6O:

$CH_3—CH_2—OH$	$CH_3—O—CH_3$
ethanol, bp 78 °C	dimethyl ether, bp −25 °C

These two isomers have the same size and the same molecular weight. Alcohols such as ethanol have O—H hydrogens, however, so they are extensively hydrogen bonded. Dimethyl ether has no O—H hydrogen, so it cannot form hydrogen bonds. As a result of its hydrogen bonding, ethanol has a boiling point more than 100 °C higher than that of dimethyl ether. Both ethanol and dimethyl ether can form hydrogen bonds with the —OH groups of water, however, so both of them are soluble in water.

The effect of N—H hydrogen bonding on boiling points can be seen in the isomers of formula C_3H_9N shown below. Trimethylamine has no N—H hydrogens, so it is not hydrogen bonded. Ethylmethylamine has one N—H hydrogen atom, and the resulting hydrogen bonding raises its boiling point about 34 °C above that of trimethylamine. Propylamine, with two N—H hydrogens, is more extensively hydrogen bonded and has the highest boiling point of these three isomers.

Application: Energy Conversion

The joule is the SI unit for energy, corresponding to the energy of a mass of 1 kg moving at 1 meter per second. The calorie is the cgs unit for energy, corresponding to the energy required to raise the temperature of 1 gram of water from 14.5 °C to 15.5 °C. Both units are widely used. They are related by 1 cal = 4.184 J, or 1 kcal = 4.184 kJ.

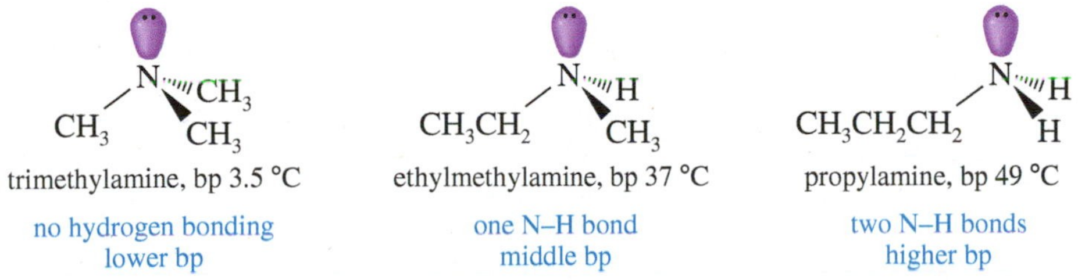

trimethylamine, bp 3.5 °C
no hydrogen bonding
lower bp

ethylmethylamine, bp 37 °C
one N–H bond
middle bp

propylamine, bp 49 °C
two N–H bonds
higher bp

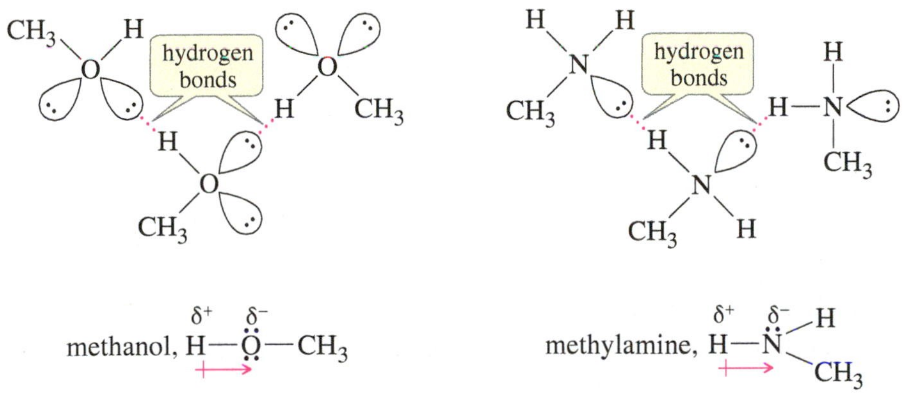

hydrogen bonds

methanol, H—Ö—CH_3 (δ^+ δ^-)

methylamine, H—N̈(CH_3)H (δ^+ δ^-)

FIGURE 2-5
Hydrogen bonding. Hydrogen bonding is a strong attraction between an electrophilic O—H or N—H hydrogen atom and a pair of nonbonding electrons.

Alcohols form stronger hydrogen bonds than amines because oxygen is more electronegative than nitrogen. Thus, the O—H bond is more strongly polarized than the N—H bond. This effect is seen in the boiling points of the preceding isomers, with more than 100 °C difference in the boiling points of ethanol and dimethyl ether, compared with a 34 °C difference for ethylmethylamine and trimethylamine.

PROBLEM 2-5

Draw the hydrogen bonding that takes place between
(a) two molecules of ethanol.
(b) two molecules of propylamine.
(c) a molecule of dimethyl ether and two molecules of water.
(d) two molecules of trimethylamine and a molecule of water.

SOLVED PROBLEM 2-2

Rank the following compounds in order of increasing boiling points. Explain the reasons for your chosen order.

$CH_3-C(CH_3)_2-CH_3$ neopentane   2-methylbutan-2-ol   2,3-dimethylbutane

pentan-1-ol   hexane

SOLUTION

To predict relative boiling points, we should look for differences in (1) hydrogen bonding, (2) molecular weight and surface area, and (3) dipole moments. Except for neopentane, these compounds have similar molecular weights. The two alcohols (*-ol* suffix) engage in hydrogen bonding, so they should be the highest-boiling compounds. Unbranched pentan-1-ol should boil at a higher temperature than branched 2-methylbutan-2-ol, which has less surface area for van der Waals forces.

Among the other compounds, neopentane has the smallest molecular weight, and it is a compact spherical structure that minimizes van der Waals attractions. Neopentane should be the lowest-boiling compound.

Comparing the last two compounds, 2,3-dimethylbutane is more highly branched (and has a smaller surface area) than hexane. Therefore, 2,3-dimethylbutane should have a lower boiling point than hexane. We predict the following order:

neopentane < 2,3-dimethylbutane < hexane < 2-methylbutan-2-ol < pentan-1-ol

10 °C   58 °C   69 °C   102 °C   138 °C

The actual boiling points are given here to show that our prediction is correct.

PROBLEM-SOLVING HINT

To predict relative boiling points, look for differences in

1. hydrogen bonding,
2. molecular weight and surface area, and
3. dipole moments.

PROBLEM 2-6

For each pair of compounds, circle the compound you expect to have the higher boiling point. Explain your reasoning.

(a) $(CH_3)_3C-C(CH_3)_3$ or $(CH_3)_2CH-CH_2CH_2-CH(CH_3)_2$
(b) $CH_3(CH_2)_6CH_3$ or $CH_3(CH_2)_5CH_2OH$
(c) $CH_3CH_2OCH_2CH_3$ or $CH_3CH_2CH_2CH_2OH$
(d) $HOCH_2-(CH_2)_4-CH_2OH$ or $(CH_3)_3CCH(OH)CH_3$
(e) $(CH_3CH_2CH_2)_2NH$ or $(CH_3CH_2)_3N$
(f) piperidine (ring NH) or cyclohexyl$-NH_2$

Application: Biochemistry

Most vitamins contain charged groups, making them water-soluble. As a result, they are rapidly eliminated and generally nontoxic. Vitamins A and D, however, are nonpolar and are stored in the fat tissue of the body, which is also nonpolar. Hence, these two vitamins are potentially toxic in large doses.

2-3 Polarity Effects on Solubilities

In addition to affecting boiling points and melting points, intermolecular forces determine the solubility properties of organic compounds. The general rule is that "*like dissolves like*." Polar substances dissolve in polar solvents, and nonpolar substances dissolve in nonpolar solvents. We discuss the reasons for this rule now, then apply the rule in later chapters when we discuss the solubility properties of organic compounds.

We consider four different cases: (1) a polar solute with a polar solvent, (2) a polar solute with a nonpolar solvent, (3) a nonpolar solute with a nonpolar solvent, and (4) a nonpolar solute with a polar solvent. We use sodium chloride and water as examples of polar solutes and solvents and paraffin "wax" and gasoline as examples of nonpolar solutes and solvents.

Polar Solute in a Polar Solvent (Dissolves) When you think about sodium chloride dissolving in water, it seems remarkable that the oppositely charged ions can be separated from each other. A great deal of energy is required to separate these ions. A polar solvent (such as water) can separate the ions because it *solvates* them (Figure 2-6). If water is the solvent, the solvation process is called *hydration.* As the salt dissolves, water molecules surround each ion, with the appropriate end of the water dipole moment next to the ion. The oxygen atoms of the water molecules approach the positively charged sodium ions. Water's hydrogen atoms approach the negatively charged chloride ions.

Because water molecules are strongly polar, a large amount of energy is released when the sodium and chloride ions are hydrated. This energy is nearly sufficient to overcome the lattice energy of the crystal. The salt dissolves, partly because of strong solvation by water molecules and partly because of the increase in entropy (randomness or freedom of movement) when it dissolves.

Polar Solute in a Nonpolar Solvent (Does Not Dissolve) If you stir sodium chloride with a nonpolar solvent such as turpentine or gasoline, you will find that the salt does not dissolve (Figure 2-7). The nonpolar molecules of these solvents do not solvate ions very strongly, and they cannot overcome the large lattice energy of the salt crystal. This is a case where the attractions of the ions in the solid for each other are much greater than their attractions for the solvent.

Nonpolar Solute in a Nonpolar Solvent (Dissolves) Paraffin "wax" dissolves in gasoline. Both paraffin and gasoline are mixtures of nonpolar hydrocarbons (Figure 2-8). The molecules of a nonpolar substance (paraffin) are weakly attracted to each other, and these van der Waals attractions are easily overcome by van der Waals attractions with the solvent. Although there is little change in energy when the nonpolar substance dissolves in a nonpolar solvent, there is a large increase in entropy.

Application: Biochemistry

Hydrogen bonding is essential for the structural integrity of many biological molecules. For example, the double helix structure of DNA is maintained, in part, by hydrogen bonds between the bases: Adenine (A) pairs with thymine (T), joined by two hydrogen bonds, and guanine (G) pairs with cytosine (C), joined by three hydrogen bonds. In the diagram below, the hydrogen bonds are represented by lines of red dots.

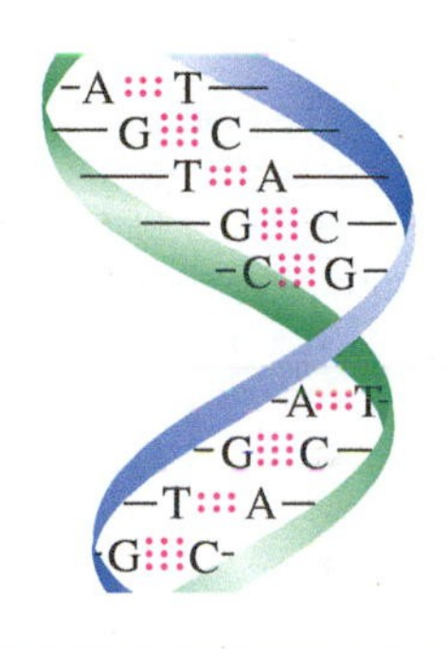

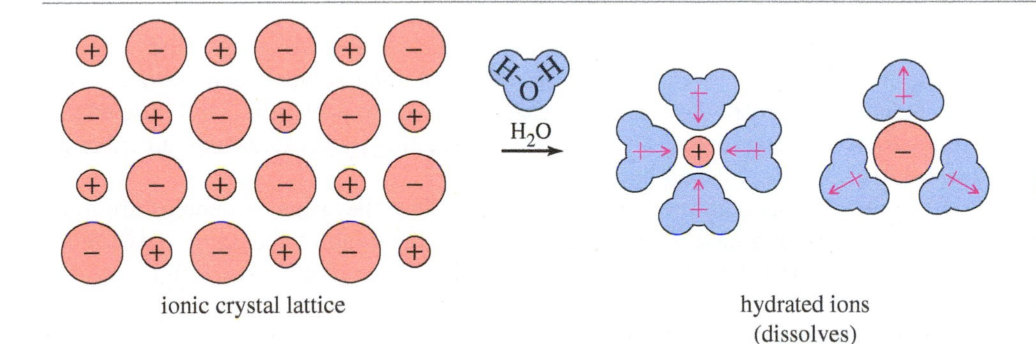

FIGURE 2-6
Polar solute in water (a polar solvent). The hydration of sodium and chloride ions by water molecules overcomes the lattice energy of sodium chloride. The salt dissolves.

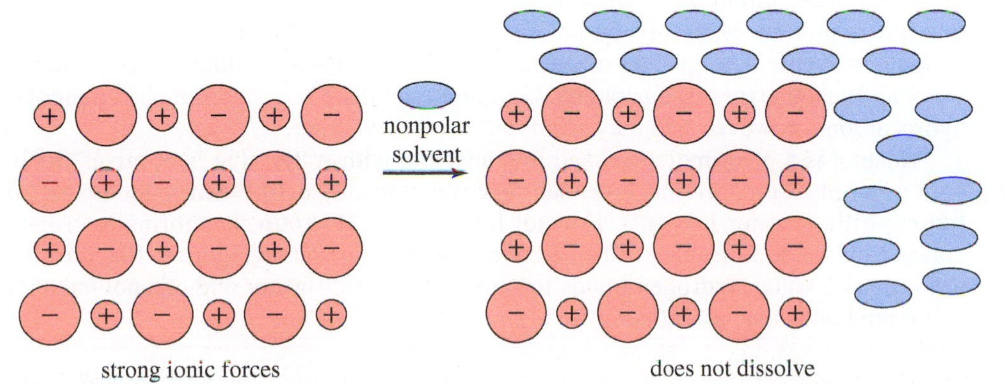

FIGURE 2-7
Polar solute in a nonpolar solvent. The intermolecular attractions of polar substances are stronger than their attractions for nonpolar solvent molecules. Thus, a polar substance does not dissolve in a nonpolar solvent.

FIGURE 2-8
Nonpolar solute in a nonpolar solvent. The weak intermolecular attractions of a nonpolar substance are overcome by the weak attractions for a nonpolar solvent. The nonpolar substance dissolves.

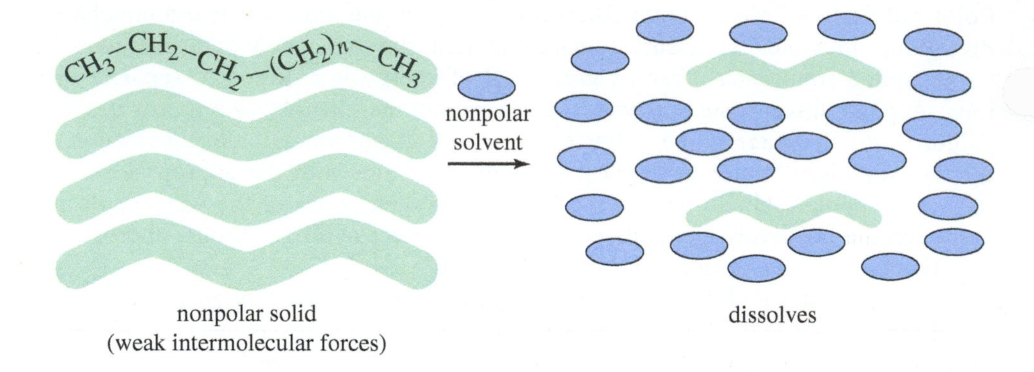

FIGURE 2-9
Nonpolar solute in a polar solvent (water). Nonpolar substances are hydrophobic. They do not dissolve in water because of the unfavorable entropy effects associated with forming a hydrogen-bonded shell of water molecules around a nonpolar molecule.

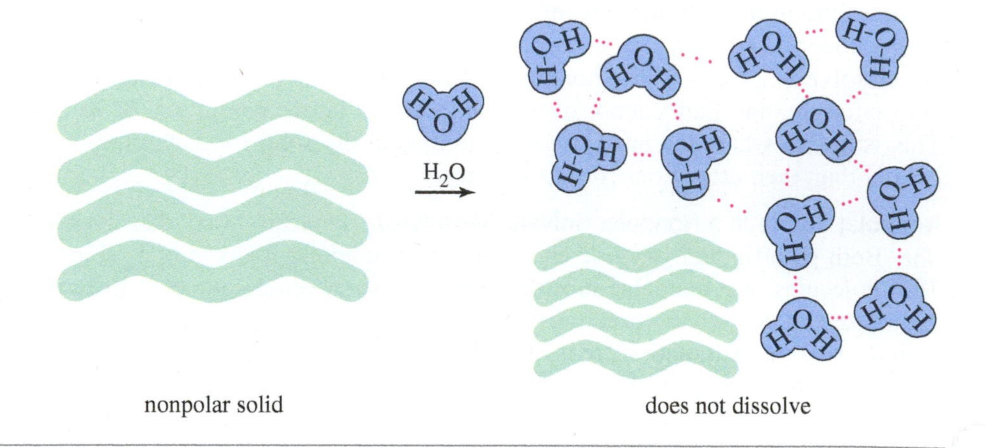

Motor oil and water do not mix because the nonpolar oil molecules cannot displace the strong intermolecular attractions between water molecules.

Nonpolar Solute in a Polar Solvent (Does Not Dissolve) Nonpolar solids such as paraffin "wax" and other waxes do not dissolve in a polar solvent such as water. Why not? The nonpolar molecules are only weakly attracted to each other, and little energy is required to separate them. The problem is that the water molecules are strongly attracted to each other by their hydrogen bonding.

If a nonpolar paraffin molecule were to dissolve, the water molecules around it would have to form a cavity. Water molecules at the edge of the cavity have fewer available neighbors for hydrogen bonding, resulting in a tighter, more rigid, ice-like structure around the cavity. This tighter structure results in an unfavorable decrease in entropy (Figure 2-9). Nonpolar substances that do not dissolve in water are called **hydrophobic** ("water-hating").

Figures 2-6 through 2-9 show why the saying "like dissolves like" is generally true. Polar substances dissolve in polar solvents, and nonpolar substances dissolve in nonpolar solvents. This general rule also applies to the mixing of liquids. For example, water and gasoline (or oil) do not mix. Gasoline and oil are both hydrophobic, nonpolar hydrocarbons, however, and they mix freely with each other.

Ethanol is a polar molecule, and it is miscible with water; that is, it mixes freely with water in all proportions. Polar substances that readily dissolve in water are called **hydrophilic** ("water-loving"). Ethanol has an O—H group that forms hydrogen bonds with water molecules. When ethanol dissolves in water (Figure 2-10), it forms new ethanol–water hydrogen bonds to replace the water–water and ethanol–ethanol hydrogen bonds that are broken:

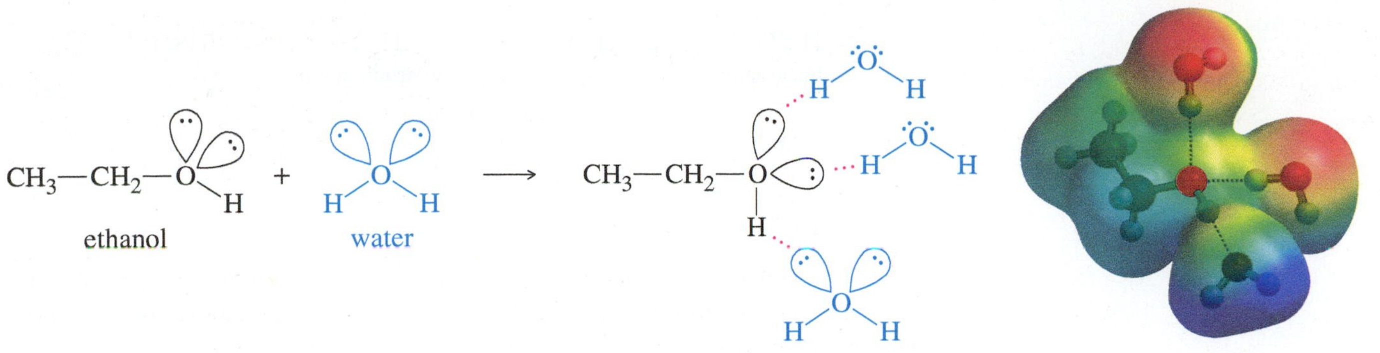

$CH_3—CH_2—OH$ + H_2O ⟶ $CH_3—CH_2—O—H$ hydrogen bonded to water

ethanol water

FIGURE 2-10
Ethanol is hydrophilic because its OH groups form hydrogen bonds with water.

In Sections 2-15 through 2-17, we will see many kinds of organic compounds with a wide variety of functional groups. As you encounter these new compounds, you should look to see whether the molecules are polar or nonpolar and whether they can engage in hydrogen bonding. Those that are nonpolar tend to be hydrophobic. Those that are strongly polar may be hydrophilic, especially if they can engage in hydrogen bonding.

PROBLEM 2-7

Circle the member of each pair that is more soluble in water.

(a) $CH_3CH_2OCH_2CH_3$ or $CH_3CH_2CH_2CH_2CH_3$
(b) $CH_3CH_2OCH_2CH_3$ or $CH_3CH_2CH_2OH$
(c) $CH_3CH_2NHCH_3$ or $CH_3CH_2CH_2CH_3$
(d) CH_3CH_2OH or $CH_3CH_2CH_2CH_2OH$
(e) 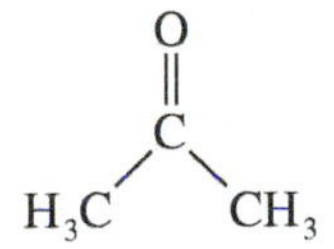 $H_3C—C(=O)—CH_3$ or cyclohexanone

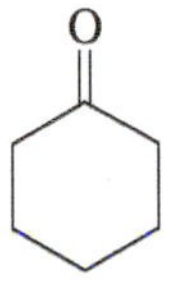

PROBLEM-SOLVING HINT

Hydrogen bonding is the most important factor in whether a compound is hydrophilic. In general, one hydrogen-bonding polar group can carry about four carbons into water. Most hydrogen-bonding compounds with three or fewer carbons are miscible with water. Multiple hydrogen-bonding groups in a molecule increase its solubility in water.

2-4 Arrhenius Acids and Bases

The properties and reactions of acids and bases are central to our study of organic chemistry. We need to consider exactly what is meant by the terms **acid** and **base.** Most people would agree that H_2SO_4 is an acid and NaOH is a base. Is BF_3 an acid or a base? Is ethylene ($H_2C{=}CH_2$) an acid or a base? To answer these questions, we need to understand the three different definitions of acids and bases: the Arrhenius definition, the Brønsted–Lowry definition, and the Lewis definition.

Acidic compounds were first classified on the basis of their sour taste. The Latin terms *acidus* (sour) and *acetum* (vinegar) gave rise to our modern terms *acid* and *acetic acid.* Alkaline compounds (bases) were substances that neutralize acids, such as limestone and plant ashes (*al kalai* in Arabic).

In 1887, Swedish chemist Svante Arrhenius introduced the **Arrhenius** theory, which defines acids as *substances that dissociate in water to give hydronium ions.* The stronger acids, such as sulfuric acid (H_2SO_4), were assumed to dissociate to a greater degree than weaker acids, such as acetic acid (CH_3COOH).

$$\underset{\text{sulfuric acid}}{H_2SO_4} + H_2O \rightleftharpoons \underset{\text{hydronium ion}}{H_3O^+} + HSO_4^-$$

$$\underset{\text{acetic acid}}{CH_3-\overset{O}{\overset{\|}{C}}-OH} + H_2O \rightleftharpoons \underset{\text{hydronium ion}}{H_3O^+} + CH_3-\overset{O}{\overset{\|}{C}}-O^-$$

According to the Arrhenius definition, bases are *substances that dissociate in water to give hydroxide ions.* Strong bases, such as NaOH, were assumed to dissociate more completely than weaker, sparingly soluble bases such as $Mg(OH)_2$.

$$NaOH \rightleftharpoons Na^+ + {}^-OH$$

$$Mg(OH)_2 \rightleftharpoons Mg^{2+} + 2\ {}^-OH$$

The acidity or basicity of an aqueous (water) solution is measured by the concentration of H_3O^+. This value also implies the concentration of ^-OH because these two concentrations are related by the water ion-product constant:

$$K_w = [H_3O^+][^-OH] = 1.00 \times 10^{-14}\ M^2 \text{ (at 25 °C)}$$

In a neutral solution, the concentrations of H_3O^+ and ^-OH are equal.

$$[H_3O^+] = [^-OH] = 1.00 \times 10^{-7}\ M \text{ in a neutral solution}$$

Acidic and basic solutions are defined by an excess of H_3O^+ or ^-OH.

$$\text{acidic: } [H_3O^+] > 10^{-7}\ M \text{ and } [^-OH] < 10^{-7}\ M$$

$$\text{basic: } [H_3O^+] < 10^{-7}\ M \text{ and } [^-OH] > 10^{-7}\ M$$

Because these concentrations can span a wide range of values, the acidity or basicity of a solution is usually measured on a logarithmic scale. The pH is defined as the negative logarithm (base 10) of the H_3O^+ concentration.

$$pH = -\log_{10}[H_3O^+]$$

A neutral solution has a pH of 7, an acidic solution has a pH less than 7, and a basic solution has a pH greater than 7.

PROBLEM 2-8

Calculate the pH of the following solutions.
(a) 5.00 g of HBr in 100 mL of aqueous solution
(b) 1.50 g of NaOH in 50 mL of aqueous solution

The Arrhenius definition was an important contribution to understanding many acids and bases, but it does not explain why a compound such as ammonia (NH_3) neutralizes acids, even though it has no hydroxide ion in its molecular formula. In the next section, we discuss a more versatile theory of acids and bases that includes ammonia and a wider variety of organic acids and bases.

2-5 Brønsted–Lowry Acids and Bases

In 1923, Danish chemist Johannes Brønsted and English chemist Thomas Lowry independently proposed defining acids and bases by following the transfer of protons. A **Brønsted–Lowry acid** is *any species that can donate a proton*, and a **Brønsted–Lowry base** is *any species that can accept a proton.* These definitions also include all the Arrhenius acids and bases because compounds that dissociate to give H_3O^+ are proton donors, and compounds that dissociate to give ^-OH are proton acceptors. (Hydroxide ion accepts a proton to form H_2O.)

In addition to Arrhenius acids and bases, the Brønsted–Lowry definition includes bases that have no hydroxide ions, yet can accept protons. Consider the following examples of acids donating protons to bases. NaOH is a base under either the Arrhenius or Brønsted–Lowry definition. The other three are Brønsted–Lowry bases but not Arrhenius bases, because they have no hydroxide ions.

$$\underset{\text{proton donor}}{HCl} + \underset{\text{proton acceptor}}{NaOH} \rightleftharpoons NaCl + H{-}OH$$

$$\underset{\text{proton donor}}{H_2SO_4} + \underset{\text{proton acceptor}}{:NH_3} \rightleftharpoons HSO_4^- + H{-}\overset{+}{N}H_3$$

$$\underset{\text{proton donor}}{HCl} + \underset{\text{proton acceptor}}{(H)_2C{=}C(H)(CH_3)} \rightleftharpoons Cl^- + H{-}CH_2{-}C^+(H)(CH_3)$$

$$\underset{\text{proton donor}}{HNO_3} + \underset{\text{proton acceptor}}{(H)_2C{=}\ddot{N}{-}H} \rightleftharpoons NO_3^- + (H)_2C{=}N^+(H)_2$$

When a base accepts a proton, it becomes an acid capable of returning that proton. When an acid donates its proton, it becomes a base capable of accepting that proton back. One of the most important principles of the Brønsted–Lowry definition is this concept of **conjugate acids and bases.** For example, NH_4^+ and NH_3 are a conjugate acid–base pair. NH_3 is the base; when it accepts a proton, it is transformed into its conjugate acid, NH_4^+. Many compounds (water, for instance) can react as either an acid or a base. Here are some additional examples of conjugate acid–base pairs.

$$\underset{\text{acid}}{H_2SO_4} + \underset{\text{base}}{H_2O} \rightleftharpoons \underset{\text{conjugate base}}{HSO_4^-} + \underset{\text{conjugate acid}}{H_3O^+}$$

$$\underset{\text{acid}}{H_2O} + \underset{\text{base}}{:NH_3} \rightleftharpoons \underset{\text{conjugate base}}{^-OH} + \underset{\text{conjugate acid}}{NH_4^+}$$

$$\underset{\text{acid}}{H{-}\overset{O}{\overset{\|}{C}}{-}OH} + \underset{\text{base}}{CH_3{-}\ddot{\underset{..}{O}}{:}^-} \rightleftharpoons \underset{\text{conjugate base}}{H{-}\overset{O}{\overset{\|}{C}}{-}\ddot{\underset{..}{O}}{:}^-} + \underset{\text{conjugate acid}}{CH_3{-}\ddot{\underset{..}{O}}{-}H}$$

2-6 Strengths of Acids and Bases

It seems obvious that the strength of an acid should reflect its tendency to protonate a base by donating a proton. Similarly, the strength of a base should reflect its tendency to deprotonate an acid by accepting a proton. But we need quantitative values to compare different acids and bases. For example, if we add both sulfuric acid and acetic acid to water, which one will protonate water? If we add water to ammonium sulfate, will water act as an acid or a base? The theory of acid strength and base strength allows us to answer these questions with confidence.

2-6A Acid Strength

The strength of a Brønsted–Lowry acid is expressed as it is in the Arrhenius definition, by the extent of its ionization in water. The general reaction of an acid (HA) with water is as follows:

$$\underset{\text{acid}}{HA} + H_2O \overset{K_a}{\rightleftharpoons} H_3O^+ + \underset{\text{base}}{A^-} \qquad K_a = \frac{[H_3O^+][A^-]}{[HA]}$$

conjugate acid–base pair

K_a is called the **acid-dissociation constant,** and its value indicates the relative strength of the acid. The stronger the acid, the more it dissociates, giving a larger value of K_a. Acid-dissociation constants vary over a wide range. Strong acids are almost completely ionized in water, and their dissociation constants are greater than 1. Most organic acids are weak acids, with K_a values less than 10^{-4}. Many organic compounds are extremely weak acids; for example, methane and ethane are essentially with K_a values less than 10^{-40}.

Because they span such a wide range, acid-dissociation constants are often expressed on a logarithmic scale. The pK_a of an acid is defined just like the pH of a solution: as the negative logarithm (base 10) of K_a.

$$pK_a = -\log_{10} K_a$$

PROBLEM-SOLVING HINT

Students sometimes forget that the K_a of water (1.8×10^{-16} *M*) is different from the ion-product constant $K_w = 1.00 \times 10^{-14}$ M^2. In calculating the K_a, you must divide by the concentration of the acid (water), which is 55.6 moles per liter in pure water.

SOLVED PROBLEM 2-3

Calculate K_a and pK_a for water.

SOLUTION

The equilibrium that defines K_a for water is

$$\underset{\text{acid (HA)}}{H_2O} + \underset{\text{solvent}}{H_2O} \overset{K_a}{\rightleftharpoons} \underset{\text{conjugate base } (A^-)}{H_3O^+ + {}^-OH}$$

Water serves as both the acid and the solvent in this dissociation. The equilibrium expression is

$$K_a = \frac{[H_3O^+][A^-]}{[HA]} = \frac{[H_3O^+][^-OH]}{H_2O}$$

We already know that $[H_3O^+][^-OH] = K_w = 1.00 \times 10^{-14}$ M^2, the ion-product constant for water.

The concentration of H_2O in water is simply the number of moles of water in 1 L (about 1 kg).

$$\frac{1000 \text{ g/L}}{18.0 \text{ g/mol}} = 55.6 \text{ mol/L}$$

Substitution gives

$$K_a = \frac{[H_3O^+][^-OH]}{[H_2O]} = \frac{1.00 \times 10^{-14}}{55.6} = 1.8 \times 10^{-16}\, M$$

The logarithm of 1.8×10^{-16} is -15.7, so the pK_a of water is 15.7.

Strong acids generally have pK_a values around 0 (or even negative), and weak acids, such as most organic acids, have pK_a values that are greater than 4. *Weaker acids have larger* pK_a *values.* Table 2-2 lists K_a and pK_a values for some common inorganic and organic compounds. Notice that the pK_a values increase as the K_a values decrease. A much larger table of pK_a values appears in Appendix 4.

TABLE 2-2 RELATIVE STRENGTH OF SOME COMMON ORGANIC AND INORGANIC ACIDS AND THEIR CONJUGATE BASES

(Acid strength: stronger at top → weaker at bottom; base strength: weaker bases at top → stronger bases at bottom)

	Acid			Conjugate Base	K_a	pK_a
strong acids	HCl hydrochloric acid	$+ H_2O$	$\rightleftarrows H_3O^+ +$	Cl^- chloride ion	1×10^7	−7
	H_3O^+ hydronium ion	$+ H_2O$	$\rightleftarrows H_3O^+ +$	H_2O water	55.6	−1.7
	HF hydrofluoric acid	$+ H_2O$	$\rightleftarrows H_3O^+ +$	F^- fluoride ion	6.8×10^{-4}	3.17
	$H-C(=O)-OH$ formic acid	$+ H_2O$	$\rightleftarrows H_3O^+ +$	$H-C(=O)-O^-$ formate ion	1.7×10^{-4}	3.76
	$CH_3-C(=O)-OH$ acetic acid	$+ H_2O$	$\rightleftarrows H_3O^+ +$	$CH_3-C(=O)-O^-$ acetate ion	1.8×10^{-5}	4.74
weak acids	$H-C\equiv N:$ hydrocyanic acid	$+ H_2O$	$\rightleftarrows H_3O^+ +$	$^-:C\equiv N:$ cyanide ion	6.0×10^{-10}	9.22
	$^+NH_4$ ammonium ion	$+ H_2O$	$\rightleftarrows H_3O^+ +$	$:NH_3$ ammonia	5.8×10^{-10}	9.24
	H_2O water	$+ H_2O$	$\rightleftarrows H_3O^+ +$	HO^- hydroxide ion	1.8×10^{-16}	15.7
	CH_3CH_2-OH ethyl alcohol	$+ H_2O$	$\rightleftarrows H_3O^+ +$	$CH_3CH_2O^-$ ethoxide ion	1.3×10^{-16}	15.9
very weak	NH_3 ammonia	$+ H_2O$	$\rightleftarrows H_3O^+ +$	$^-:\ddot{N}H_2$ amide ion	10^{-36}	36
not acidic	CH_4 methane	$+ H_2O$	$\rightleftarrows H_3O^+ +$	$^-:CH_3$ methyl anion	$<10^{-50}$	50

PROBLEM 2-9

Ammonia appears in Table 2-2 as both an acid and a conjugate base.

(a) Explain how ammonia can act as both an acid and a base. Which of these roles does it commonly fill in aqueous solutions?

(b) Show how water can serve as both an acid and a base.

(c) Calculate K_a and pK_a for the hydronium ion, H_3O^+.

(d) Show how methanol (CH_3OH) can serve as both an acid and a base. Write an equation for the reaction of methanol with sulfuric acid.

PROBLEM-SOLVING HINT

In most cases, the pK_a of an acid corresponds to the pH where the acid is about half dissociated. At a lower (more acidic) pH, the acid is mostly undissociated; at a higher (more basic) pH, the acid is mostly dissociated.

2-6B Base Strength

The strength of an acid is inversely related to the strength of its conjugate base. For an acid (HA) to be strong, its conjugate base (A^-) must be stable in its anionic form; otherwise, HA would not easily lose its proton. Therefore, the conjugate base of a strong acid must be a weak base. On the other hand, when an acid is weak, its conjugate base is a strong base.

$$\underset{\text{strong acid}}{HCl} + H_2O \rightleftarrows H_3O^+ + \underset{\text{weak base}}{Cl^-}$$

$$\underset{\text{weak acid}}{CH_3\text{—}\ddot{O}H} + H_2O \rightleftarrows H_3O^+ + \underset{\text{strong base}}{CH_3\ddot{O}:^-}$$

In the reaction of an acid with a base, the equilibrium generally favors the *weaker* acid and base. For example, in the preceding reactions, H_3O^+ is a weaker acid than HCl but a stronger acid than CH_3OH. It also follows that H_2O is a stronger base than Cl^- but a weaker base than CH_3O^-.

The strength of a base is measured much like the strength of an acid, by using the equilibrium constant of the hydrolysis reaction.

$$\underset{\text{conjugate base}}{A^-} + H_2O \overset{K_b}{\rightleftarrows} \underset{\text{conjugate acid}}{HA} + {}^-OH$$

The equilibrium constant (K_b) for this reaction is called the *base-dissociation constant* for the base A^-. Because this constant spans a wide range of values, it is often given in logarithmic form. The negative logarithm (base 10) of K_b is defined as pK_b.

$$K_b = \frac{[HA][^-OH]}{[A^-]} \qquad pK_b = -\log_{10} K_b$$

When we multiply K_a by K_b, we can see how the acidity of an acid is related to the basicity of its conjugate base.

$$(K_a)(K_b) = \frac{[H_3O^+][A^-]}{[HA]}\frac{[HA][^-OH]}{[A^-]} = \underset{\text{water ion-product constant}}{[H_3O^+][^-OH] = K_w = 1.0 \times 10^{-14}}$$

$$(K_a)(K_b) = 10^{-14}$$

Logarithmically,

$$pK_a + pK_b = -\log 10^{-14} = 14$$

Application: Drugs

The acid–base properties of many natural products are important for their isolation, distribution in the body, and therapeutic effects. For example, morphine (page 2) is isolated from the opium poppy and crosses into the brain as the free base, where the nitrogen is not charged. However, it exerts its pain-relieving effects as the protonated, charged species.

In an aqueous solution, the product of K_a and K_b must always equal the ion-product constant of water, $K_w = 10^{-14}$. If the value of K_a is large, the value of K_b must be small; that is, the stronger an acid, the weaker its conjugate base. Similarly, a small value of K_a (weak acid) implies a large value of K_b (strong base).

A stronger acid has a weaker conjugate base.
A weaker acid has a stronger conjugate base.
Acid–base reactions favor the weaker acid and the weaker base.

2-7 Equilibrium Positions of Acid–Base Reactions

A Brønsted–Lowry acid–base reaction transfers a positively charged proton from an acid to a base. The products are the conjugate base of the original acid and the conjugate acid of the original base. For the simple case where the acids are uncharged and the bases have negative charges, this is the general equation:

$$\underbrace{\underset{\text{acid}}{H{-}A} + \underset{\text{base}}{B{:}^-}}_{\text{reactants}} \rightleftharpoons \underbrace{\underset{\text{conjugate base}}{A{:}^-} + \underset{\text{conjugate acid}}{H{-}B}}_{\text{products}}$$

How do we know whether the reaction proceeds from left to right as written? Another way of asking that question is "Does the reaction favor the products at equilibrium?" How do we know which way the reaction goes and which species are favored by the equilibrium?

We can show the following three principles mathematically:

1. **The acid–base equilibrium favors formation of the weaker acid and the weaker base.**
2. **The weaker acid has the larger pK_a. The weaker base has the larger pK_b.**
3. **The weaker acid and the weaker base are always on the same side of the equation.** Both of them are reactants, or both of them are products.

We can predict which side of the equation is favored by comparing the strength of the two acids or by comparing the strength of the two bases. Either comparison gives the same result. The weaker acid is the one with the smaller K_a, which implies a larger pK_a. The weaker base is the one with the smaller K_b, which implies a larger pK_b. If we know the pK_a values or the pK_b values, then we can easily predict which side of the equation is favored.

For example, consider the reaction of acetic acid with sodium hydroxide in water. We need to identify which reactant is the acid, which is the base, and which proton is transferred. Once those are identified, we can predict the products.

proton transferred

$$\underset{\substack{\text{acid} \\ pK_a = 4.74 \\ \text{stronger acid}}}{CH_3COOH\,(aq)} + \underset{\text{base}}{NaOH\,(aq)} \rightleftharpoons \underset{\text{conjugate base}}{CH_3COO^-\,Na^+\,(aq)} + \underset{\substack{\text{conjugate acid} \\ pK_a = 15.7 \\ \text{weaker acid}}}{H{-}OH}$$

products are favored

From Table 2-2 we know that the pK_a for acetic acid is 4.74, and that the pK_a for water is 15.7. Water is the weaker acid by about 11 pK_a units. The equilibrium for this reaction lies far to the right and favors the products by 11 pK_a units. We can confidently say that this reaction goes to completion.

We could write this equation backward, interchanging the products and reactants. In that case, we would say that the reaction does not proceed, or that the equilibrium lies far to the left and favors the reactants.

$$\underset{\substack{pK_b = 9.26 \\ \text{weaker base}}}{CH_3COO^-\,Na^+\,(aq)} + \underset{\substack{pK_a = 15.7 \\ \text{weaker acid}}}{H{-}OH} \rightleftharpoons \underset{\substack{pK_a = 4.74 \\ \text{stronger acid}}}{CH_3COOH\,(aq)} + \underset{\substack{pK_b = -1.7 \\ \text{stronger base}}}{NaOH\,(aq)}$$

reactants are favored

Note that in this equation, the pK_b values of the bases have been included simply by subtracting the pK_a values from 14 (K_w, the ion-product constant for water): pK_b = 14 − pK_a of the conjugate acid. If we had compared the pK_b values instead of the pK_a values, we would have obtained the same result.

Consider another example: the possible acid–base reaction between phenol and aniline. Phenol is a weak acid with $pK_a = 10.0$, and aniline, with $pK_b = 9.4$, is a weak base. Does this reaction go toward the left or the right?

$$C_6H_5-\ddot{O}-H + C_6H_5-\ddot{N}H_2 \overset{?}{\rightleftharpoons} C_6H_5-\ddot{O}:^- \; Na^+ + C_6H_5-\overset{+}{N}H_3$$

phenol, acid, $pK_a = 10.0$; aniline, base, $pK_b = 9.40$; conjugate base; conjugate acid

To compare pK_a or pK_b values, we must first subtract from 14 to get the pK_a and pK_b of the products. We find that the conjugate base of phenol has $pK_b = 4.0$, and that the conjugate acid of aniline has $pK_a = 4.60$. The equilibrium lies far to the left, and this reaction will not go as written. The reverse reaction would go spontaneously.

$$C_6H_5-\ddot{O}-H + C_6H_5-\ddot{N}H_2 \rightleftharpoons C_6H_5-\ddot{O}:^- \; Na^+ + C_6H_5-\overset{+}{N}H_3$$

phenol, acid, $pK_a = 10.0$, weaker acid; aniline, base, $pK_b = 9.40$, weaker base; conjugate base, $pK_b = 4.0$; conjugate acid, $pK_a = 4.60$

reactants are favored

Predicting the position of an acid–base equilibrium becomes more challenging when no values are available for pK_a or pK_b. In most cases, we can estimate approximate values of pK_a or pK_b by comparing the compounds with similar compounds listed in the tables. For example, we can predict the following equilibrium using only the values in Table 2-2:

$$CH_3CH_2\overset{O}{\overset{\|}{C}}-OH + HN(CH_2CH_2)_2O \overset{?}{\rightleftharpoons}$$

propionic acid; morpholine

Propionic acid is a carboxylic acid, like acetic acid. The pK_a of acetic acid is 4.74, so we estimate the pK_a of propionic acid to be about 5. Morpholine should be a weak base like ammonia. We predict that propionic acid might lose a proton to morpholine. Table 2-2 shows that the pK_a of protonated ammonia (the ammonium ion, NH_4^+) is 9.26, so we estimate the pK_a of protonated morpholine to be about 9.

$$CH_3CH_2\overset{O}{\overset{\|}{C}}-OH + HN(CH_2CH_2)_2O \rightleftharpoons CH_3CH_2\overset{O}{\overset{\|}{C}}-O^- + H_2\overset{+}{N}(CH_2CH_2)_2O$$

propionic acid, stronger acid, $pK_a \cong 5$; morpholine, base; propionate ion; morpholinium ion, weaker acid, $pK_a \cong 9$

products are favored

The products contain the weaker acid, so the products are favored. We can be fairly confident of this result even though we used estimated pK_a values, because there is a large difference in pK_a between the reactants and the products. (The literature values of pK_a are 4.87 for propionic acid and 8.4 for the morpholinium ion.)

PROBLEM-SOLVING STRATEGY Predicting Acid–Base Equilibrium Positions

The preceding examples show how we can confidently predict the favored products or reactants in most acid–base reactions. Let's summarize the steps involved and then use these steps to work through a particularly difficult problem. The steps we used above (sometimes combined) are these:

Step 1: Complete the reaction (if necessary), and label the acids and bases on each side.

Step 2: Assign pK_a or pK_b values from tables or by analogy with similar compounds. Note that Appendix 4 gives an extensive list of pK_a values.

Step 3: Calculate any needed pK_a or pK_b values using the fact that pK_a + pK_b = 14 for a conjugate acid–base pair.

Step 4: Choose the side with the weaker acid or weaker base (larger value of pK_a or pK_b) as favored. The difference in pK_a or pK_b values shows how strongly that side is favored.

As an example, the following reaction is a strongly basic one that cannot take place in water because water would protonate the strong bases involved. In fact, this reaction is usually done in liquid ammonia solution.

$$\underset{\text{propyne}}{H_3C{-}C{\equiv}C{-}H} + \underset{\text{sodium amide}}{Na^+ \ {}^-\!:\!\ddot{N}H_2} \overset{?}{\rightleftharpoons} \underset{\text{sodium propynide}}{H_3C{-}C{\equiv}C\!:^- \ Na^+} + \underset{\text{ammonia}}{H{-}\ddot{N}H_2}$$

Step 1: In this reaction, a very weakly acidic proton is transferred from propyne to the amide ion (NH_2^-) to give the propynide anion and ammonia. Propyne serves as a weak acid, and ammonia is the conjugate acid of the amide ion, a powerful base.

Step 2: The pK_a of ammonia is shown in Table 2-2 and Appendix 4 as about 36. When a pK_a is this large, it depends on the solvent, and it is difficult to measure. The pK_a of propyne is not listed in Table 2-2 or in Appendix 4, but we can estimate it. Appendix 4 lists the pK_a of acetylene, H—C≡C—H, as 25. This should be a good approximation for the pK_a of propyne.

Step 3: The pK_b of the amide anion can be found by subtracting the pK_a of ammonia from 14: $14 - 36 = -22$. This large, negative pK_b indicates an unusually powerful base. The pK_b of the propynide ion is $14 - 25 = -11$. This is also a very strong base, but not as strong as the amide ion.

$$H_3C{-}C{\equiv}C{-}H + Na^+ \ {}^-\!:\!\ddot{N}H_2 \rightleftharpoons H_3C{-}C{\equiv}C\!:^- \ Na^+ + H{-}\ddot{N}H_2$$

propyne	sodium amide	sodium propynide	ammonia
acid, p$K_a \cong 25$	base, p$K_b = -22$	conjugate base	conjugate acid
stronger acid	stronger base	p$K_b = -11$	p$K_a = 36$
		weaker base	weaker acid

products are favored

Step 4: We can compare the pK_a values or the pK_b values. Either way, the products are the weaker acid and the weaker base, by about 11 pK_a units. The products are strongly favored in this reaction, and the reaction should proceed as written.

PROBLEM 2-10 (PARTIALLY SOLVED)

Write equations for the following acid–base reactions. Use the information in Table 2-2 or Appendix 4 to predict whether the equilibrium will favor the reactants or the products.

(a) $HCOOH + {}^{-}CN$

(b) $CH_3COO^- + CH_3OH$

(c) $(CH_3)_2CHOH + NaNH_2$

(d) $NaOCH_3 + HCN$

(e) $HCl + CH_3CH_2OH$

(f) $H_3O^+ + CH_3O^-$

(g) pyrrolidine (N—H) + phenol (—OH)

(h) pyrrolidine (N—H) + cyclopentanecarboxylic acid (—COOH)

(i) thiophenol (—SH) + $CH_3CH_2O^-$

(j) CH_3CH_2OH + phenyl—C≡C⁻ Na^+

SOLUTION

to (a): Cyanide is the conjugate base of HCN. It can accept a proton from formic acid:

H—C(=O)—O—H	+	$^-$:C≡N:	⇌	H—C(=O)—O:$^-$	+	H—C≡N:
formic acid, pK_a = 3.76 stronger acid		cyanide stronger base		formate weaker base		pK_a = 9.22 weaker acid

Reading from Table 2-2, formic acid (pK_a = 3.76) is a stronger acid than HCN (pK_a = 9.22), and cyanide is a stronger base than formate. The products (weaker acid and base) are favored.

PROBLEM-SOLVING HINT

An acid will donate a proton to the conjugate base of any weaker acid (smaller K_a or higher pK_a).

PROBLEM-SOLVING HINT

Use either acid strengths or base strengths to determine the favored side of an acid–base equilibrium. The side with weaker acid (larger pK_a) or the weaker base (larger pK_b) is favored.

PROBLEM 2-11

Ethanol, methylamine, and acetic acid are all amphoteric, reacting as either acids or bases depending on the conditions.

CH_3CH_2OH	CH_3NH_2	CH_3—C(=O)—OH
ethanol	methylamine	acetic acid

(a) Rank ethanol, methylamine, and acetic acid in decreasing order of acidity. In each case, show the equation for the reaction with a generic base (B:$^-$) to give the conjugate base.

(b) Rank ethanol, methylamine, and acetic acid in decreasing order of basicity. In each case, show the equation for the reaction with a generic acid (HA) to give the conjugate acid.

2-8 Solvent Effects on Acidity and Basicity

Water is amphoteric; it can react as both an acid and a base. Its conjugate acid is H_3O^+, and its conjugate base is ^-OH. When water is the solvent for a reaction, it is present in large excess. Any acid that is stronger than H_3O^+ (pK_a = −1.7) does not survive in water, but reacts with water to give H_3O^+. For example, if we add HCl gas (pK_a = −7) to water, it immediately reacts exothermically to give H_3O^+, a weaker acid.

HCl	+	H_2O	⟶	Cl^-	+	H_3O^+
stronger acid pK_a = −7		base		conjugate base		conjugate acid pK_a = −1.7 weaker acid

When we say that we add HCl (aq) to a reaction, we are actually adding H_3O^+ (aq) and Cl^- (aq) because HCl is too strong an acid to coexist with water. We should realize that we are not using an acid with $pK_a = -7$, but actually one with $pK_a = -1.7$.

The same principle applies to bases that are stronger than ^-OH, the conjugate base of water, with $pK_b = -1.7$. If we add some sodium amide ($NaNH_2$, $pK_b = -22$) to water, it immediately reacts exothermically to give ^-OH, a weaker base.

$$Na^+ \ {}^-{:}\ddot{N}H_2 + H_2O \longrightarrow {:}NH_3 + Na^+ \ {}^-{:}\ddot{O}H$$

$Na^+ \ {}^-{:}\ddot{N}H_2$	+	H_2O	$\longrightarrow$	$:NH_3$	+	$Na^+ \ {}^-{:}\ddot{O}H$
sodium amide base, $pK_b = -22$ stronger base		stronger acid $pK_a = 15.7$		conjugate acid $pK_a = 36$ weaker acid		conjugate base $pK_b = -1.7$ weaker base

Thus, water narrows ("levels") the possible ranges of acid strength and base strength to values of $pK_a > -1.7$ and values of $pK_b > -1.7$. Any acid with pK_a more negative than -1.7 will be leveled to H_3O^+, with $pK_a = -1.7$. Any base with pK_b more negative than -1.7 will be leveled to ^-OH, with $pK_b = -1.7$.

Alcohols (R—OH) react with both strong acids and strong bases just like water does. For example, strong acids protonate ethanol (CH_3CH_2OH) to give protonated ethanol ($CH_3CH_2OH_2^+$), an acid with $pK_a = -2.4$. Any acid with a pK_a that is more negative than -2.4 will donate a proton to ethanol. Strong bases deprotonate ethanol to give ethoxide ion ($CH_3CH_2O^-$), a base with $pK_b = -1.9$. Any base with a pK_b that is more negative than -1.9 will become protonated in ethanol solution.

Other solvents may or may not narrow the range of acidity and basicity as much as water and alcohols do. For example, simple ethers tend to level acids but not bases. Diethyl ether ($CH_3CH_2—O—CH_2CH_3$) is not an acid, so it can be used as a solvent for some exceedingly strong bases. Diethyl ether is a weak base, however, so it narrows the range of acids. It reacts with any acid that is stronger than protonated diethyl ether ($pK_a = -3.6$).

HCl	+	$CH_3CH_2—\ddot{O}—CH_2CH_3$	$\longrightarrow$	Cl^-	+	$CH_3CH_2—\overset{H}{\overset{\vert}{O^+}}—CH_2CH_3$
$pK_a = -7$ stronger acid		diethyl ether base		conjugate base		conjugate acid $pK_a = -3.6$ weaker acid

Most ethers are not acidic, so they are often used as solvents for reagents that are stronger bases than hydroxide ions. Some basic reactions are best run in liquid ammonia, which has a pK_a of 36, and accommodates bases with $pK_b > -22$. For example, reactions using sodium amide (the conjugate base of ammonia) are usually done in liquid ammonia, NH_3 (*l*).

$H_3C—C{\equiv}C—H$	+	$Na^+ \ {}^-{:}\ddot{N}H_2$	$\xrightarrow{NH_3\,(l)}$	$H_3C—C{\equiv}C{:}^- \ Na^+$	+	$H—\ddot{N}H_2$
propyne acid, $pK_a = 25$ stronger acid		sodium amide base, $pK_b = -22$ stronger base		sodium propynide conjugate base $pK_b = -11$ weaker base		ammonia conjugate acid $pK_a = 36$ weaker acid

Some organic solvents do not react appreciably with either acids or bases. Alkanes (saturated hydrocarbons) such as pentane and hexane are not acidic or basic, and we can use strong acids and bases in these solvents with little or no leveling of acidity or basicity.

$CH_3CH_2CH_2CH_2CH_3$	$CH_3CH_2CH_2CH_2CH_2CH_3$
pentane	hexane

SUMMARY Acidity and Basicity Limitations in Common Solvents

Solvent	Structure	Typical pK_a Limit	Limiting Species	Typical pK_b Limit	Limiting Species
water	H—O—H	−1.7	H_3O^+	−1.7	HO^-
alcohols	R—O—H	−2.4	ROH_2^+	−1.9	RO^-
ethers	R—O—R	−3.6	$R-\overset{+}{O}(H)-R$	none	none
ammonia	NH_3	9.2	NH_4^+	−22	NH_2^-
alkanes	R—H	none	none	none	none

Values of pK_a and pK_b are defined using equilibrium reactions of the acids and bases in water. Values of pK_a between −1.7 (H_3O^+) and 15.7 (H_2O) and values of pK_b between −1.7 (^-OH) and 15.7 (H_2O) are measured in water, and are most accurate. Values outside this range cannot be measured in water because of water's leveling effect. There is no standard method for measuring these difficult values of pK_a and pK_b, and they vary with the solvent. Therefore, different references give slightly different values. Appendix 4 gives pK_a values that are consistent with the relative strengths of acids and bases, even if the values differ from one source to another.

PROBLEM-SOLVING HINT

Water and alcohols (R—O—H) can react as both weak acids and weak bases, and they limit the strength of strong acids and bases dissolved in them. Ethers (R—O—R) can react as weak bases, and they limit the strength of strong acids (but not bases) dissolved in them. Alkanes do not react as either acids or bases, and they do not limit the strength of acids or bases dissolved in them.

PROBLEM 2-12

For each of the following reactions, suggest which solvent(s) would be compatible with the acids and bases involved. (We will ignore any other possible reactions for now.) Your choices of solvents are pentane, diethyl ether, ethanol, water, and ammonia. Refer to Appendix 4 for any needed values of pK_a, or estimate them.

(a) $CH_3Li + H-C\equiv C-H \longrightarrow CH_4 + H-C\equiv CLi$

(b) $CH_3Li + (CH_3)_3C-OH \longrightarrow CH_4 + (CH_3)_3C-OLi$

(c) 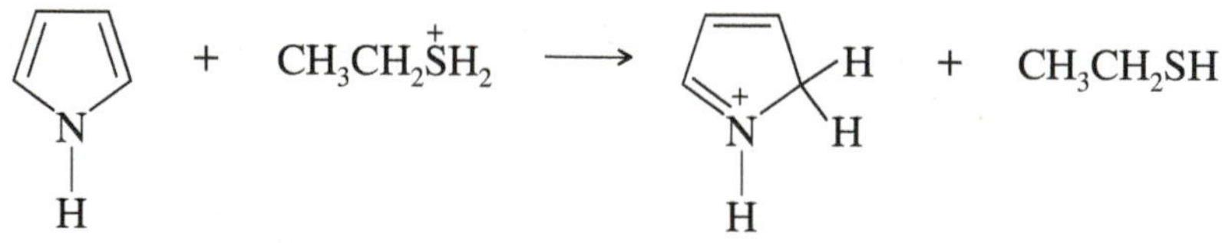 pyrrole (N—H) $+ CH_3CH_2\overset{+}{S}H_2 \longrightarrow$ protonated pyrrole ($\overset{+}{N}$—H, ring carbon bearing H, H) $+ CH_3CH_2SH$

(d) $NaNH_2 + (CH_3)_3C-OH \longrightarrow NH_3 + (CH_3)_3C-ONa$

(e) $HBr + CH_3-\overset{O}{\overset{\|}{C}}-CH_3 \longrightarrow CH_3-\overset{^+O-H}{\overset{\|}{C}}-CH_3 + Br^-$

(f) $CH_3-\overset{O}{\overset{\|}{C}}-NH_2 + HCl \longrightarrow CH_3-\overset{^+O-H}{\overset{\|}{C}}-NH_2 + Cl^-$

2-9 Effects of Size and Electronegativity on Acidity

How can we look at a structure and predict whether a compound will be a strong acid, a weak acid, or not an acid at all? To be a Brønsted–Lowry acid (HA), a compound must contain a hydrogen atom that can be lost as a proton. A strong acid must have a stable conjugate base (A:$^-$) after losing the proton.

The stability of the conjugate base is a good guide to acidity. More stable anions tend to be weaker bases, and their conjugate acids tend to be stronger acids. Some of the factors that affect the stability of conjugate bases are electronegativity, size, and resonance.

Electronegativity A more electronegative element bears a negative charge more easily, giving a more stable conjugate base and a stronger acid. Electronegativities increase from left to right in the periodic table:

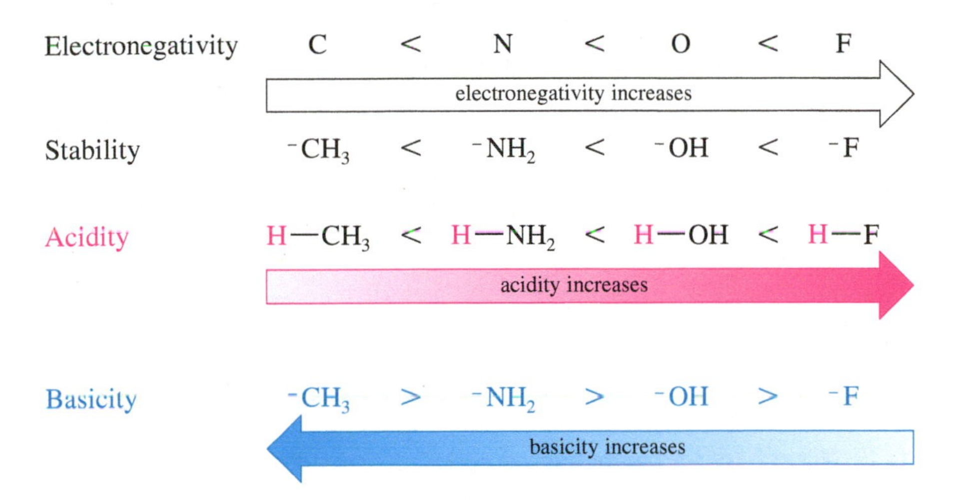

Size The negative charge of an anion is more stable if it is spread over a larger region of space. Within a column of the periodic table, acidity increases down the column, as the size of the element increases.

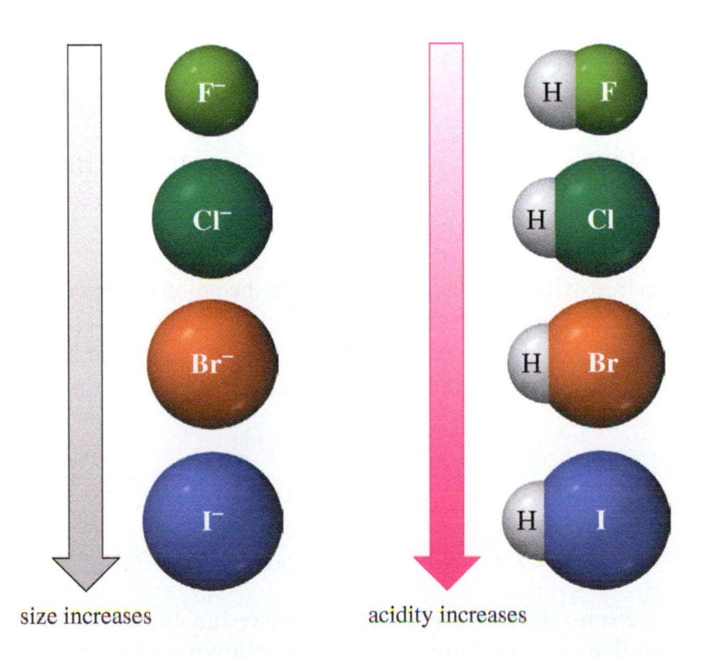

Figure 2-11 summarizes these two effects, using part of the periodic table to show the pK_a of the acid formed when each element is combined with enough hydrogen atoms to fill its valence. Acidity increases toward the right (increasing electronegativity) and downward (increasing size) on the periodic table. A smaller or more negative value of pK_a implies a stronger acid and a more stable conjugate base.

PROBLEM-SOLVING HINT

To evaluate or compare the acidity of acids, consider the stability of their conjugate bases.

To compare the basicity of bases, consider the stability of their conjugate acids.

PROBLEM-SOLVING HINT

In the comparison of similar bases, a negative charge is generally more stable on a larger atom or on a more electronegative atom. Enhanced stability of a base increases the strength of its conjugate acid.

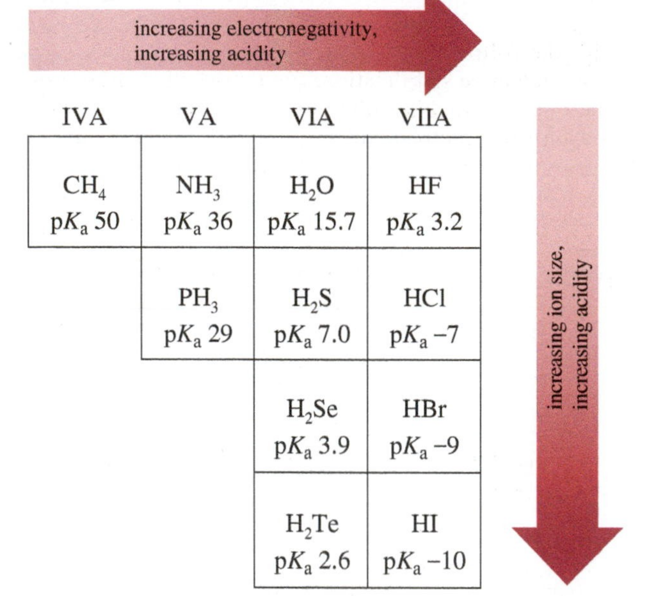

FIGURE 2-11
Acidity trends within the periodic table. Acidity depends on the stability of the conjugate base, which increases with the electronegativity and size of the anion. Electronegativity increases toward the right of the periodic table, and size increases down each column.

2-10 Inductive Effects on Acidity

> **PROBLEM-SOLVING HINT**
> When a problem asks you to compare acidities or explain an acidity, you should consider the structure and stability of the conjugate base. Factors that stabilize the conjugate base will increase the strength of the acid.

As we have seen, the strength of an acid is determined by the stability of its conjugate base. Electron-withdrawing atoms and groups can stabilize a conjugate base through the sigma bonds of the molecule. Stabilization of the conjugate base results in a stronger acid (lower value of pK_a). The magnitude of this **inductive effect** depends on the number of bonds between the electronegative element (or other electron-withdrawing group) and the site of the negative charge. Note how adding a chlorine atom to butanoic acid increases its acidity, and the effect is larger if the chlorine atom is closer to the acidic group.

$CH_3CH_2CH_2-C(=O)-OH$	$ClCH_2CH_2CH_2-C(=O)-OH$	$CH_3CHClCH_2-C(=O)-OH$	$CH_3CH_2CHCl-C(=O)-OH$
butanoic acid	4-chlorobutanoic acid	3-chlorobutanoic acid	2-chlorobutanoic acid
pK_a = 4.82	pK_a = 4.52	pK_a = 4.05	pK_a = 2.86

Stronger electron-withdrawing groups stabilize the anion of the conjugate base more than weaker groups do, leading to stronger acids. Fluorine is more electronegative and a stronger withdrawing group than chlorine, making fluoroacetic acid a stronger acid than chloroacetic acid. This trend continues through the other halogen-substituted acetic acids.

$ICH_2-C(=O)-OH$	$BrCH_2-C(=O)-OH$	$ClCH_2-C(=O)-OH$	$FCH_2-C(=O)-OH$
iodoacetic acid	bromoacetic acid	chloroacetic acid	fluoroacetic acid
pK_a = 3.2	pK_a = 2.9	pK_a = 2.8	pK_a = 2.7

Multiple electron-withdrawing groups stabilize the conjugate base and increase the acidity more than a single group does. The following examples show how each additional chlorine atom increases the acidity of the substituted acetic acid.

$CH_3-C(=O)-OH$	$ClCH_2-C(=O)-OH$	$Cl_2CH-C(=O)-OH$	$Cl_3CH-C(=O)-OH$
acetic acid	chloroacetic acid	dichloroacetic acid	trichloroacetic acid
pK_a = 4.76	pK_a = 2.8	pK_a = 1.3	pK_a = 0.6

PROBLEM 2-13

Write equations for the following acid–base reactions. Label the conjugate acids and bases, and show any inductive stabilization. Predict whether the equilibrium favors the reactants or products. Try to do this without using a table of pK_a values, but if you need a hint, you can consult Appendix 4.

(a) $CH_3CH_2OH + CH_3NH^-$
(b) $F_3CCOONa + Br_3C{-}COOH$
(c) $CH_3OH + H_2SO_4$
(d) $NaOH + H_2S$
(e) $CH_3NH_3^+ + CH_3O^-$
(f) $BrCH_2CH_2OH + F_3C{-}CH_2O^-$
(g) $NaOCH_2CH_3 + Cl_2CHCH_2OH$
(h) $H_2Se + NaNH_2$
(i) $CH_3CHFCOOH + FCH_2CH_2COO^-$
(j) $CF_3CH_2O^- + FCH_2CH_2OH$

PROBLEM-SOLVING HINT

Acidity is increased by
1. stronger electron-withdrawing substituents.
2. additional electron-withdrawing substituents.
3. closer proximity of the electron-withdrawing substituent to the acidic group.

PROBLEM 2-14

Rank the following acids in decreasing order of their acid strength. In each case, explain why the previous compound should be a stronger acid than the one that follows it.

O OH Cl   O OH Br   Br O OH   O OH Cl Cl

2-11 Hybridization Effects on Acidity

When an acid donates a proton, the electrons that were originally in the H—A bond remain with the conjugate base (A:$^-$) as a nonbonding electron pair (lone pair).

$$\underset{\text{acid}}{H{-}A} + \underset{\text{base}}{B:^-} \rightleftarrows \underset{\text{conjugate base}}{A:^-} + \underset{\text{conjugate acid}}{H{-}B}$$

The stability of the conjugate base, and thus the strength of the acid, is determined by the stability of this nonbonding electron pair. As we have seen in Sections 2-9 and 2-10, the stability of this lone pair depends on the electronegativity, size, charge, and inductive substituents of the atom bearing it. The stability of the lone pair also depends on the hybridization state of the orbital where it resides. For example, consider the compounds below. All of them can lose a proton from a positively charged nitrogen atom with four bonds to give a neutral nitrogen with three bonds and a lone pair. Yet their different hybridization states strongly influence their acidity. Note that the acidity of the N—H bond varies with hybridization in the order $sp^3 < sp^2 \ll sp$.

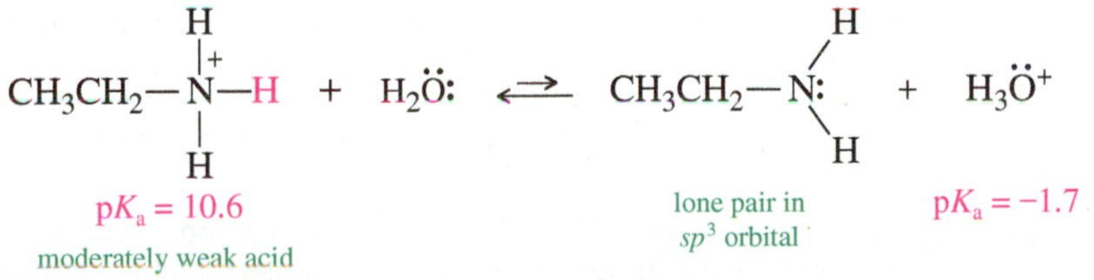

$$CH_3CH_2{-}\overset{+}{N}H_3 + H_2\ddot{O}: \rightleftarrows CH_3CH_2{-}\ddot{N}H_2 + H_3\ddot{O}^+$$

pK_a = 10.6, moderately weak acid; lone pair in sp^3 orbital; pK_a = −1.7

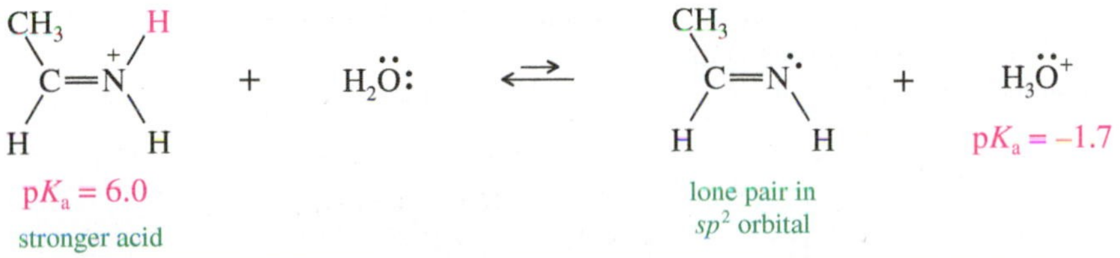

$$(CH_3)(H)C{=}\overset{+}{N}H_2 + H_2\ddot{O}: \rightleftarrows (CH_3)(H)C{=}\ddot{N}H + H_3\ddot{O}^+$$

pK_a = 6.0, stronger acid; lone pair in sp^2 orbital; pK_a = −1.7

$$CH_3{-}C{\equiv}\overset{+}{N}{-}H \;+\; H_2\ddot{O}: \;\rightleftharpoons\; CH_3{-}C{\equiv}N: \;+\; H_3\ddot{O}^+$$

$pK_a = -10.1$ very strong acid; lone pair in sp orbital; $pK_a = -1.7$

The analogous carbon compounds also show this hybridization dependence. The following carbon compounds are all very weak acids, yet their acidity varies in the same order as we saw with nitrogen: $sp^3 < sp^2 << sp$.

Compound	Conjugate Base	Hybridization	*s* Character	pK_a
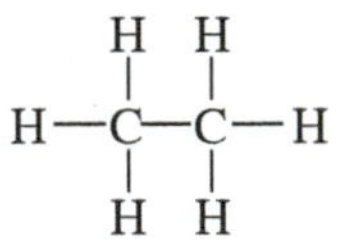 $H_3C{-}CH_3$	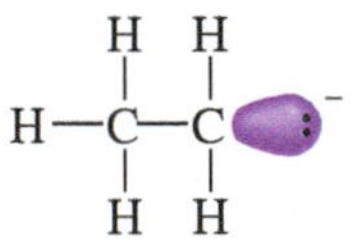 $H_3C{-}CH_2^-$	sp^3	25%	50
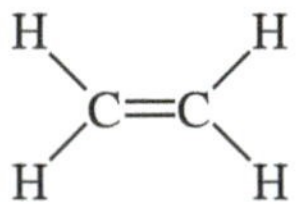 $H_2C{=}CH_2$	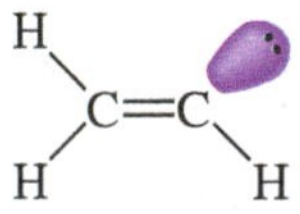 $H_2C{=}CH^-$	sp^2	33%	44
$H{-}C{\equiv}C{-}H$	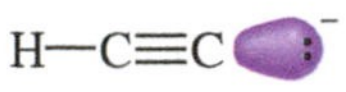 $H{-}C{\equiv}C^-$	sp	50%	25

We explain this hybridization dependence by considering the relative stability of the *s* and *p* components of these hybrid orbitals. In Section 1-2B, we saw that an electron in the 2*s* orbital is lower in energy than one in the 2*p* orbital. Therefore, a nonbonding pair of electrons remaining in a hybrid orbital (after an acid has donated a proton) is more stable in a hybrid using more of the *s* orbital and less of the *p* orbitals. The amount of the *s* orbital (*s* character) used in each hybrid orbital is 25% in the sp^3 hybrid, 33% in the sp^2 hybrid, and 50% in the *sp* hybrid. Thus, both the *s* character of the hybrid orbital, and the stability of a lone pair in the orbital, increase in the order $sp^3 < sp^2 < sp$.

The shapes of the orbitals also reflect how increasing the *s* character results in the electrons being held closer to the nucleus. The drawings below show lone pairs in sp^3, sp^2, and *sp* hybrid orbitals. Electrons in the sp^3 hybrid orbital are, on average, farthest from the nucleus. Electrons in the sp^2 hybrid orbital are held closer and more tightly. Electrons in the *sp* hybrid orbital are held the closest and tightest of all.

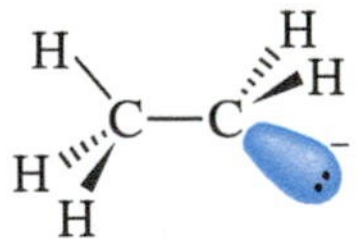

lone pair in sp^3 hybrid orbital 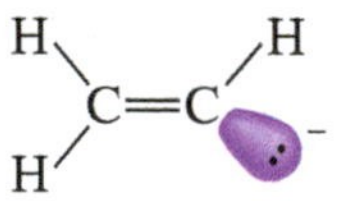 — lone pair in sp^2 hybrid orbital 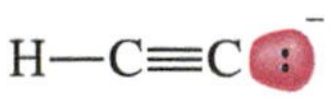 — lone pair in *sp* hybrid orbital

PROBLEM-SOLVING HINT

If we compare two similar acids that donate a proton from the same element with the same charge, differing only in the hybridization of the atom, acidity increases in the order $sp^3 < sp^2 << sp$.

PROBLEM 2-15

Like nitrogen and carbon, oxygen also shows this same hybridization effect on acidity. Both of the following compounds can lose a proton from a positively charged oxygen with three bonds to give a conjugate base containing a neutral oxygen with two bonds. One of these structures has $pK_a = -2.4$, while the other has $pK_a = -8.0$.

(a) Show the reaction of each compound with water.

(b) Match each structure with its pK_a, and explain your choice.

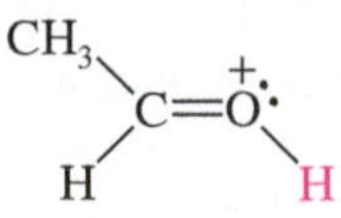

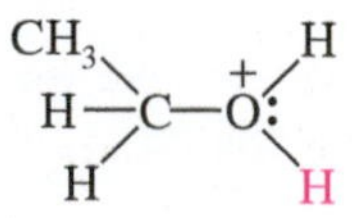

$CH_3(H)C{=}\overset{+}{O}{-}H$ and $CH_3CH_2{-}\overset{+}{O}H_2$

PROBLEM 2-16

Consider the type of orbitals involved, and rank the following nitrogen compounds in order of *decreasing basicity*. Give the structures of their conjugate acids, and estimate their pK_as from similar compounds in Appendix 4. Rank the conjugate acids in order of *increasing acidity*. (Hint: These two orders should be the same!)

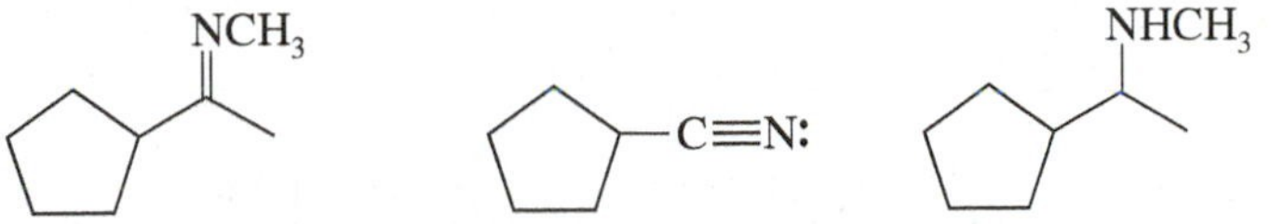

NCH_3 C≡N: $NHCH_3$

PROBLEM 2-17

Consider each pair of bases, and explain which one is more basic. Draw their conjugate acids, and show which one is a stronger acid.

(a) C≡C:⁻   **(b)** OH   O

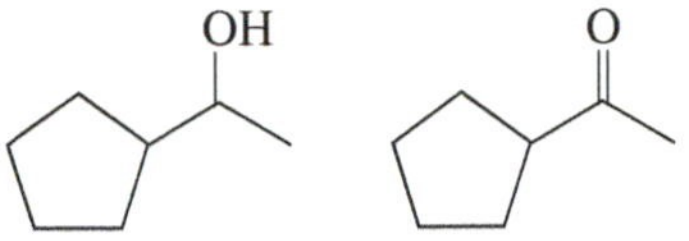

(c) N—H   N   **(d)** C≡N:

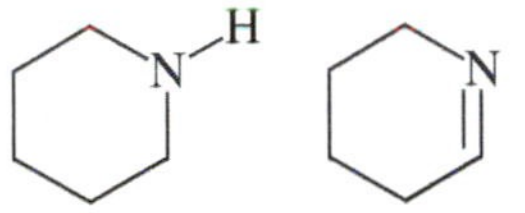

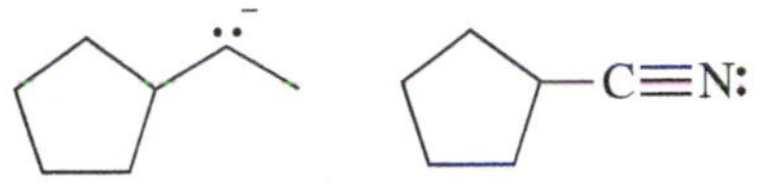

2-12 Resonance Effects on Acidity and Basicity

When an acid donates a proton, the conjugate base is left with the two electrons that once bonded it to the proton, and often a negative charge. This lone pair of electrons and the negative charge (if any) may be delocalized over two or more atoms by resonance. The amount of stabilization that occurs through resonance depends on how electronegative the atoms are and how many share the charge and the lone pair. In many cases, resonance delocalization is the dominant effect helping to stabilize the conjugate base. For example, acetic acid is over 11 pK_a units more acidic than ethanol (see below). This large difference is due to enhanced stability of the acetate ion, which is the conjugate base of acetic acid. The negative charge in the acetate ion is delocalized over two oxygen atoms, in contrast to the ethoxide ion, which has the negative charge localized on its single oxygen atom.

$$CH_3-CH_2-OH + H_2\ddot{O}: \rightleftarrows CH_3-CH_2-\ddot{O}:^- + H_3\ddot{O}^+$$

ethanol, pK_a = 15.9   ethoxide ion

$$CH_3-C(=\ddot{O}:)-\ddot{O}H + H_2\ddot{O}: \rightleftarrows \left[CH_3-C(=\ddot{O}:)-\ddot{O}:^- \longleftrightarrow CH_3-C(-\ddot{O}:^-)=\ddot{O}:\right] + H_3\ddot{O}^+$$

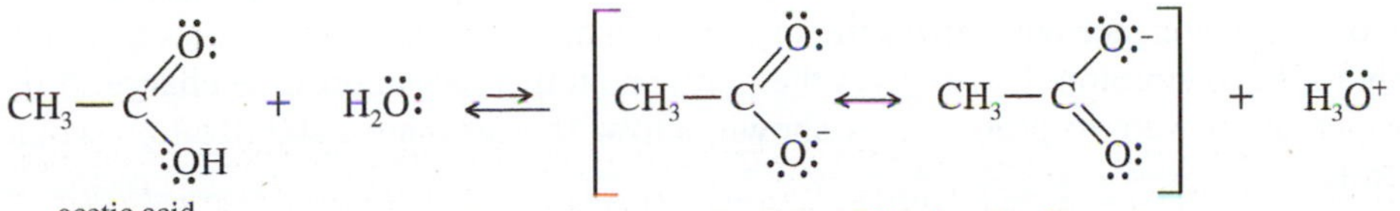

acetic acid, pK_a = 4.74   resonance-stabilized acetate ion

PROBLEM 2-18

Which is a stronger base: ethoxide ion or acetate ion? Give pK_b values (without looking them up) to support your choice.

Protons on carbon and nitrogen are also more acidic if they are located where the conjugate base is stabilized by resonance. The following examples show how resonance stabilization of the conjugate base can enhance their acidity. Each of these examples has a carbonyl (C=O) group next to the atom that donates the proton. These are both weak acids, but without the carbonyl group, they would be much weaker.

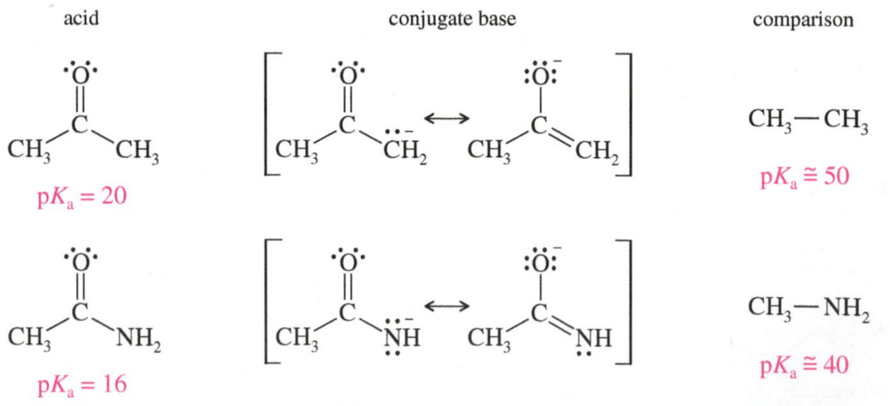

acid	conjugate base	comparison
$CH_3C(=O)CH_3$ $pK_a = 20$	[$CH_3C(=O)\ddot{C}H_2^-$ ⟷ $CH_3C(-O^-)=CH_2$]	CH_3-CH_3 $pK_a \cong 50$
$CH_3C(=O)NH_2$ $pK_a = 16$	[$CH_3C(=O)\ddot{N}H^-$ ⟷ $CH_3C(-O^-)=NH$]	CH_3-NH_2 $pK_a \cong 40$

Other electron-withdrawing groups can also stabilize a conjugate base through resonance. Two of the strongest stabilizing groups are the cyano (—C≡N) and nitro (—NO_2) groups.

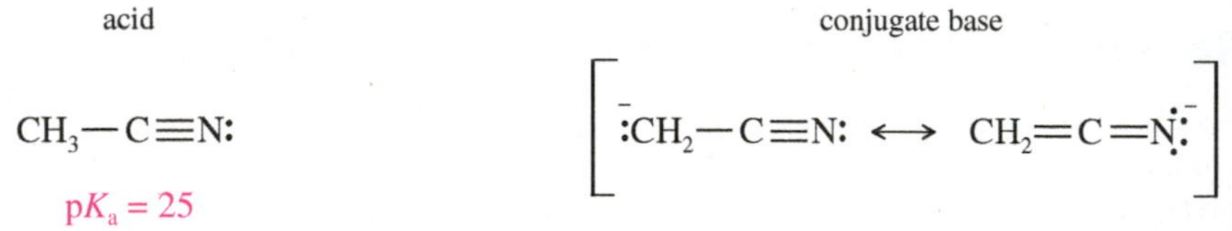

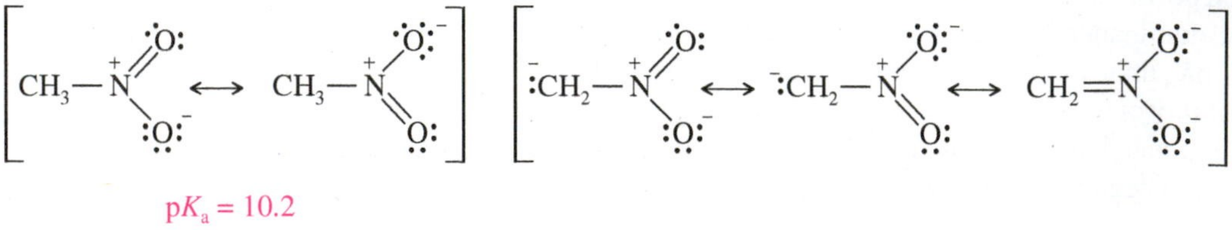

acid	conjugate base
$CH_3-C\equiv N:$ $pK_a = 25$	[$:\overset{-}{C}H_2-C\equiv N:$ ⟷ $CH_2=C=\ddot{N}:^-$]

[$CH_3-\overset{+}{N}(=O)-O^-$ ⟷ $CH_3-\overset{+}{N}(-O^-)=O$]
$pK_a = 10.2$

[$:\overset{-}{C}H_2-\overset{+}{N}(=O)-O^-$ ⟷ $:\overset{-}{C}H_2-\overset{+}{N}(-O^-)=O$ ⟷ $CH_2=\overset{+}{N}(-O^-)-O^-$]

Resonance stabilization also affects the basicity of common organic compounds. This effect is seen in amides, which are much less basic than amines. An amide is stabilized by resonance between the amino (—NH_2) group and the carbonyl (C=O) group. This resonance delocalizes the lone pair on nitrogen and makes it less available for bonding to a proton. It also gives the nitrogen atom a partial positive charge. If the nitrogen atom were to protonate as shown below, the resonance stabilization would be lost.

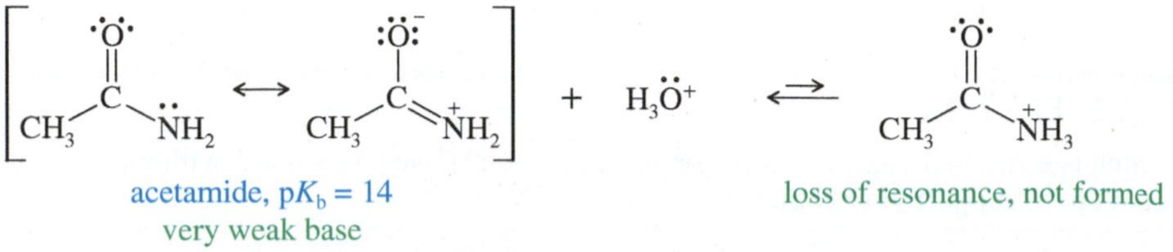

[$CH_3C(=O)\ddot{N}H_2$ ⟷ $CH_3C(-O^-)=\overset{+}{N}H_2$] + $H_3\ddot{O}^+$ ⇄ $CH_3C(=O)\overset{+}{N}H_3$

acetamide, $pK_b = 14$
very weak base

loss of resonance, not formed

Compare this amide's very weak basicity with the corresponding amine, which is over 10 pK_b units more basic.

$$CH_3{-}CH_2{-}\ddot{N}H_2 \quad + \quad H_3\ddot{O}^+ \quad \rightleftharpoons \quad CH_3{-}CH_2{-}\overset{+}{N}H_3$$

ethylamine, $pK_b = 3.3$ — ethylammonium ion, $pK_a = 10.7$; no loss of resonance

Resonance delocalization is the most common type of stabilization we see in the conjugate bases of common organic acids. Resonance is often the largest stabilizing factor as well, sometimes contributing 10 or more pK_a units to the acidity of the conjugate acid.

PROBLEM 2-19

The preceding equation for the protonation of acetamide shows a hypothetical product that is not actually formed. When acetamide is protonated by a strong enough acid, it does not protonate on nitrogen, but at a different basic site. Draw the structure of the actual conjugate acid of acetamide, and explain (resonance) why protonation occurs where it does rather than on nitrogen. Calculate the pK_a of this conjugate acid.

PROBLEM 2-20

Acetic acid can also react as a very weak base ($pK_b = 20$). Two different sites on acetic acid might become protonated to give the conjugate acid. Draw both of these possible conjugate acids, and explain (resonance) why the correct one is more stable. Calculate the pK_a of this conjugate acid.

PROBLEM-SOLVING HINT

Resonance delocalization of the charge(s) on a base helps to stabilize the base, making it a weaker base and strengthening its conjugate acid. The strongest common stabilizing groups are the carbonyl (C═O), cyano (C≡N), and nitro (NO_2) groups.

PROBLEM 2-21 (PARTIALLY SOLVED)

Choose the more acidic member of each pair of isomers, and show why the acid you chose is more acidic.

(a) 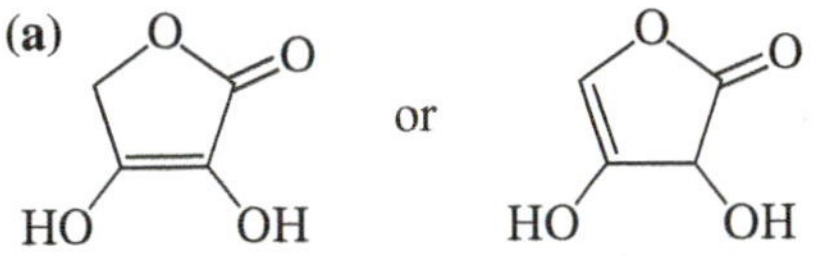 [structure: HO, OH, O, O] or [structure: HO, OH, O, O]

SOLUTION

to (a): The first (left) compound is more acidic because deprotonation of its left —OH group gives a resonance-stabilized anion that delocalizes the negative charge onto the carbonyl group. Compare this compound with vitamin C, shown at the beginning of the chapter. This resonance explains why vitamin C (ascorbic acid) is acidic even though it does not contain a carboxylic acid (—COOH) group.

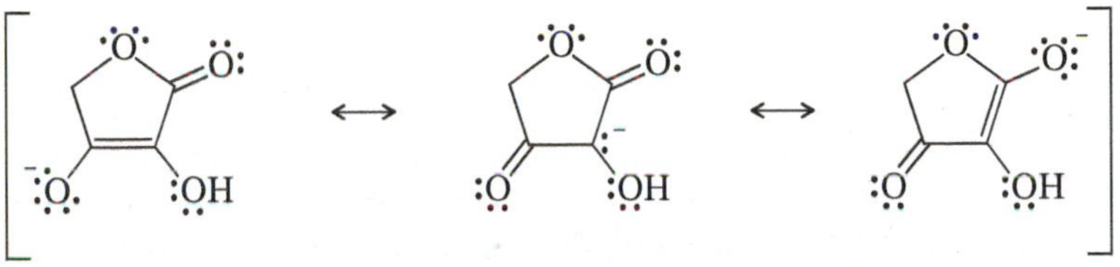

[resonance structures: :Ö:⁻ / :ÖH ↔ :O / :ÖH ↔ :O / :ÖH, with :Ö:⁻]

In the second compound, deprotonation of either —OH group fails to give an anion that delocalizes the negative charge onto the carbonyl group.

(b) 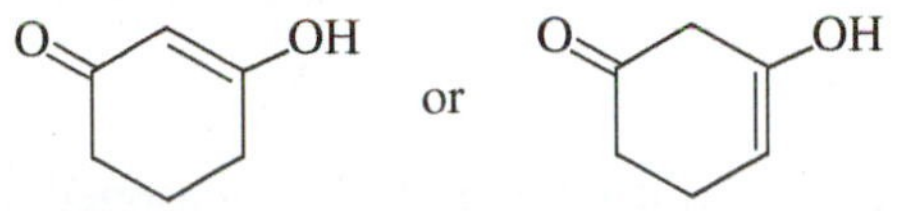 [O, OH] or [O, OH]

(c) 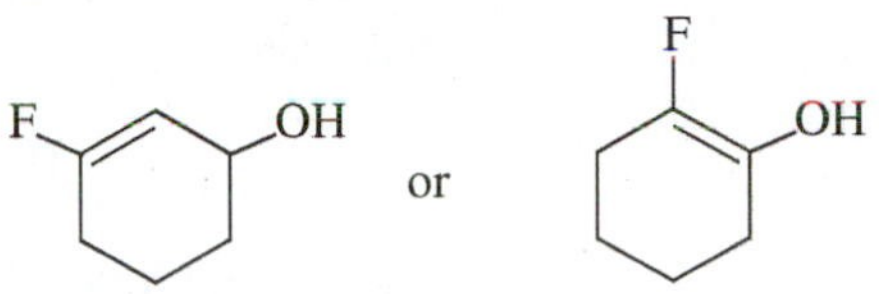 [F, OH] or [F, OH]

(d) 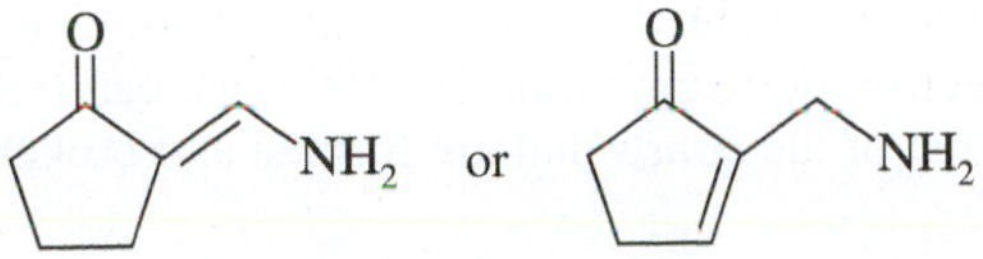 [O, NH_2] or [O, NH_2]

(e) 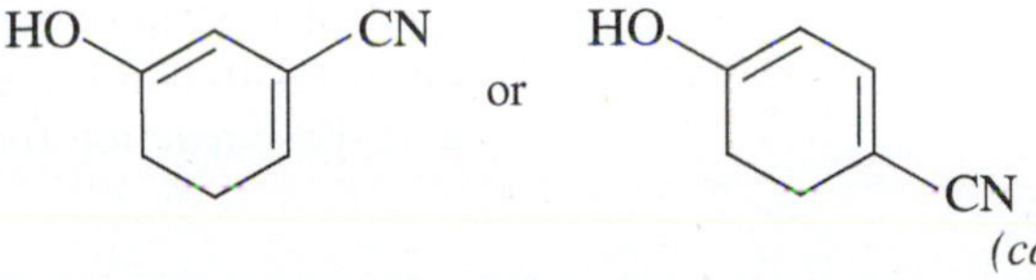 [HO, CN] or [HO, CN]

(continued)

(f) or

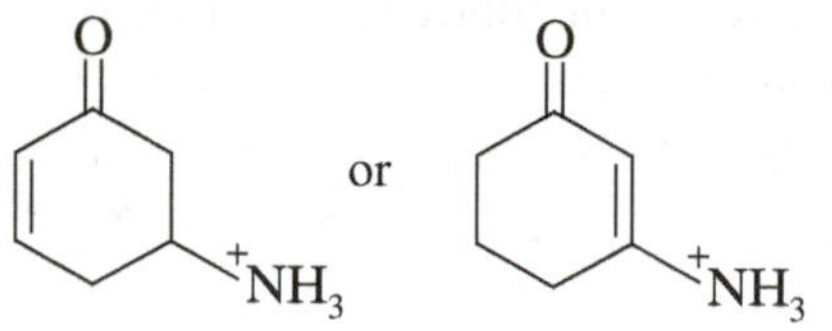

HO OH

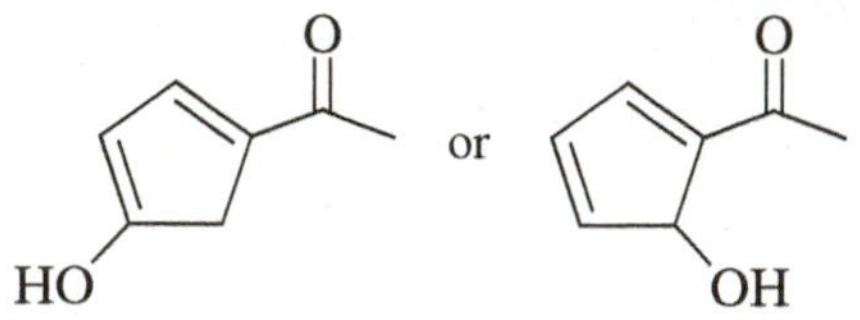

(g) or

$\overset{+}{N}H_3$ $\overset{+}{N}H_3$

(h) or

OH OH

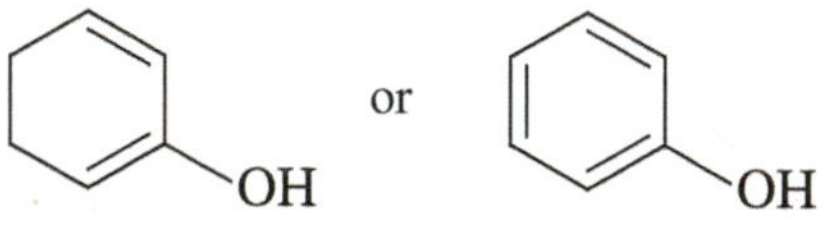

(i) or

H O H O

OH OH

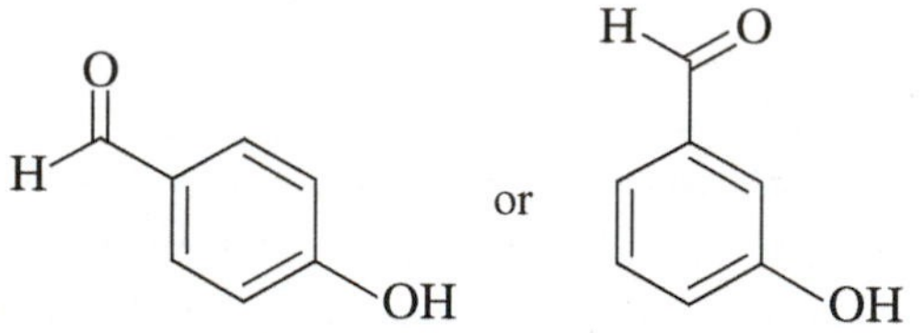

PROBLEM 2-22 (PARTIALLY SOLVED)

Choose the more basic member of each pair of isomers, and show why the base you chose is more basic.

(a) O O

NH_2 or NH_2

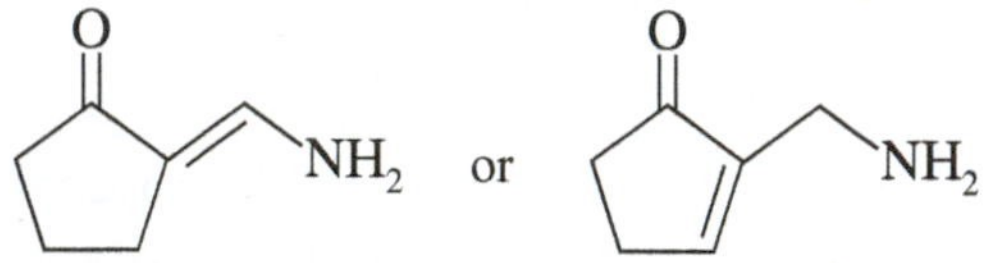

SOLUTION

to (a): The first (left) compound is less basic because it is resonance-stabilized (like an amide), and protonation of its $-NH_2$ group would destroy that resonance. Moreover, the lone pair on the $-NH_2$ group is shared with the carbonyl group, and the $-NH_2$ group carries a partial positive charge. The compound on the right is an amine, which we expect to be more basic. Protonation of its $-NH_2$ group would not cause loss of resonance stabilization.

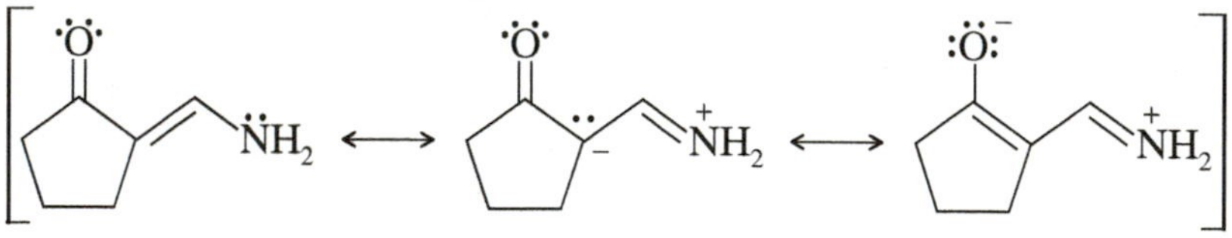

$\ddot{N}H_2 \longleftrightarrow \overset{+}{N}H_2 \longleftrightarrow \overset{+}{N}H_2$

(b) O O⁻ or O O⁻

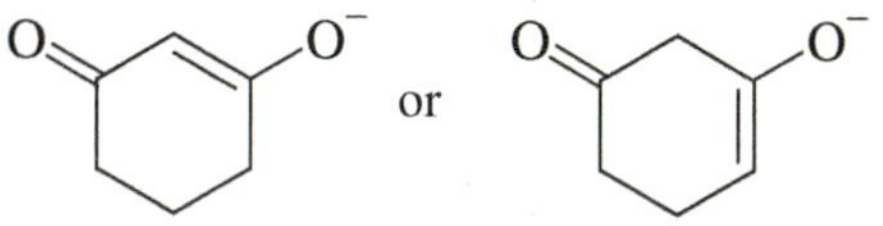

(c) O N O⁻ or O⁻ O NH

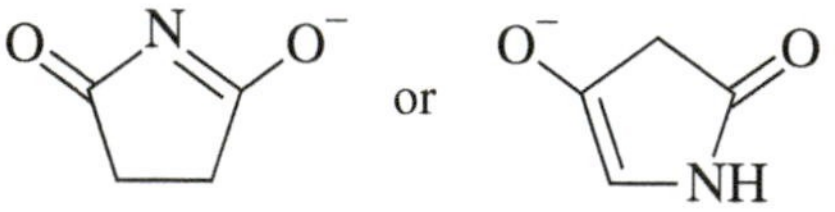

(d) NO_2 or NO_2 O⁻ O⁻

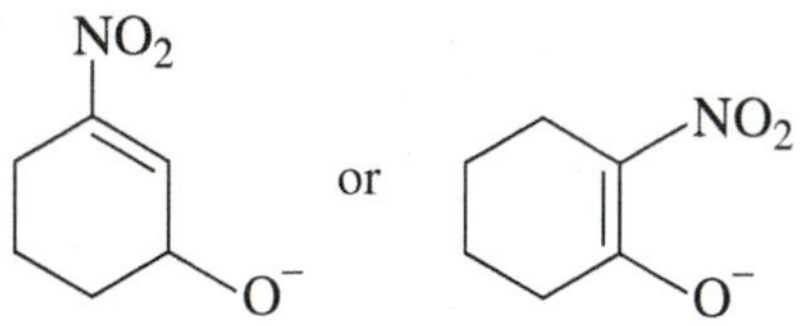

(e) or C≡N N≡C O⁻ O⁻

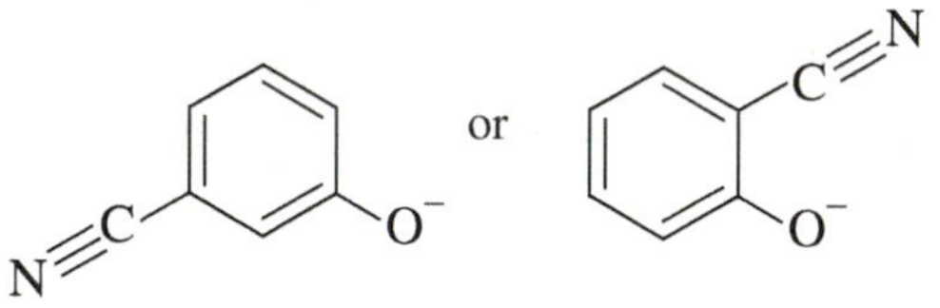

2-13 Lewis Acids and Bases

The Brønsted–Lowry definition of acids and bases depends on the transfer of a proton from the acid to the base. The base uses a pair of nonbonding electrons to form a bond to the proton.

$$B:^- + H:A \rightleftharpoons B:H + {}^-:A$$

In 1923, Gilbert N. Lewis (University of California, Berkeley) reasoned that this kind of reaction does not need a proton. Instead, a base could use its lone pair of electrons to bond to some other electron-deficient atom. In effect, we can look at an acid–base reaction from the viewpoint of the *bonds* that are formed and broken rather than a proton that is transferred. The following reaction shows the proton transfer, with emphasis on the bonds being broken and formed. Organic chemists routinely use curved arrows to show the movement of the participating electrons.

$$B:^- \longrightarrow H:A \rightleftarrows B:H + {}^-:A$$

PROBLEM-SOLVING HINT

A nucleophile donates electrons. An electrophile accepts electrons. Acidic protons may serve as electron acceptors.

Lewis bases are species with available electrons that can be donated to form new bonds. **Lewis acids** are species that can accept these electron pairs to form new bonds. Because a Lewis acid *accepts* a pair of electrons, it is called an **electrophile,** from the Greek words meaning "lover of electrons." A Lewis base is called a **nucleophile,** or "lover of nuclei," because it donates electrons to a nucleus with an empty (or easily vacated) orbital. In this book, we sometimes use colored type for emphasis: blue for nucleophiles, green for electrophiles, and occasionally red for acidic protons.

The Lewis acid–base definitions include reactions having nothing to do with protons. Following are some examples of Lewis acid–base reactions. Notice that the common Brønsted–Lowry acids and bases also fall under the Lewis definition, with a proton serving as the electrophile. Curved arrows (red) are used to show the movement of electrons, generally from the nucleophile to the electrophile.

PROBLEM-SOLVING HINT

Curved arrows (red) are used to show the movement of electrons from the nucleophile to the electrophile.

$B:^-$ + H^+ → B—H
nucleophile (Lewis base); electrophile (Lewis acid); bond formed

$H_3N:$ + BF_3 → $H_3\overset{+}{N}—\overset{-}{B}F_3$
nucleophile; electrophile; bond formed

$CH_3—\ddot{O}:^-$ + $H_3C—Cl$ → $CH_3—O—CH_3$ + $:\ddot{Cl}:^-$
nucleophile; electrophile; bond formed

$H_3N:$ + $H—O—C(=O)—CH_3$ → $H_3\overset{+}{N}—H$ + $^-:\ddot{O}—C(=O)—CH_3$
nucleophile base; electrophile acid; bond formed (conjugate acid); (conjugate base)

Some of the terms associated with acids and bases have evolved specific meanings in organic chemistry. When organic chemists use the term *base*, they usually mean a proton acceptor (a Brønsted–Lowry base). Similarly, the term *acid* usually means a proton donor (a Brønsted–Lowry acid). When the acid–base reaction involves formation of a bond to some other element (especially carbon), organic chemists refer to the electron donor as a *nucleophile* (Lewis base) and the electron acceptor as an *electrophile* (Lewis acid).

The following illustration shows electrostatic potential maps for the reaction of NH_3 (the nucleophile/electron donor) with BF_3 (the electrophile/electron acceptor). The electron-rich (red) region of NH_3 attacks the electron-poor (blue) region of BF_3. In contrast, the product shows high electron density on the boron atom and its three fluorine atoms and low electron density on nitrogen and its three hydrogen atoms.

PROBLEM-SOLVING HINT

So far, we have used several types of arrows in this course.

Reaction arrows: ⟶

Equilibrium arrows: ⇄

Resonance arrows: ⟷

Flow of electrons in Lewis acid–base reactions: ↷ (red)

Imaginary flow of electrons between resonance forms: ↷ (green)

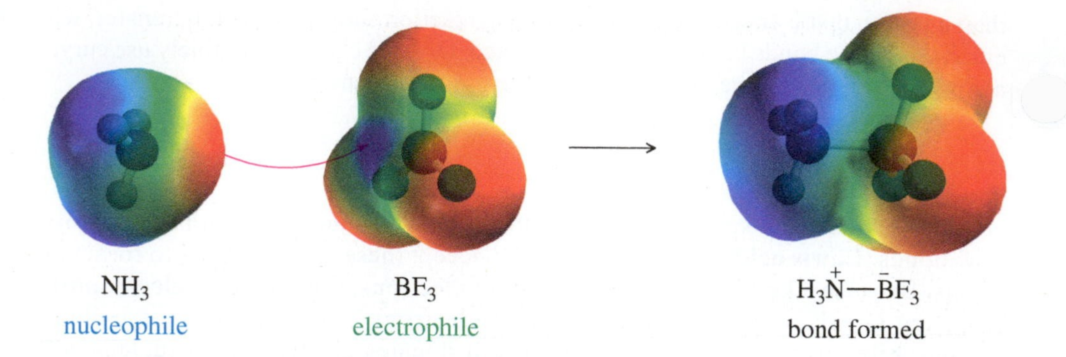

2-14 The Curved-Arrow Formalism

We often break down complicated organic reactions and biochemical reactions into series of steps involving simple Lewis acid–base reactions of nucleophiles with electrophiles. This section describes the mechanics of drawing these individual steps using curved arrows to show movements of the electrons.

The **curved-arrow formalism** is used to show the flow of an electron pair *from the electron donor to the electron acceptor*. The movement of each pair of electrons involved in making or breaking bonds is indicated by its own separate arrow, as shown in the preceding set of reactions. In this book, these curved arrows are always printed in red. In the preceding reaction of CH_3O^- with CH_3Cl, one curved arrow shows the lone pair on oxygen forming a bond to carbon. Another curved arrow shows that the C—Cl bonding pair detaches from carbon and becomes a lone pair on the Cl^- product.

PROBLEM-SOLVING HINT

Use one curved arrow for each pair of electrons participating in the reaction.

$$CH_3—\ddot{\underset{..}{O}}:^- \quad H—\underset{H}{\overset{H}{C}}—\ddot{\underset{..}{Cl}}: \longrightarrow CH_3—\ddot{\underset{..}{O}}—\underset{H}{\overset{H}{C}}—H + :\ddot{\underset{..}{Cl}}:^-$$

nucleophile (electron donor) electrophile (electron acceptor)

The curved-arrow formalism is universally used for keeping track of the flow of electrons in reactions. In this way, we show exactly what changes take place in a reaction, as well as the details of how those changes occur. We have also used this device (in Section 1-9, for example) to keep track of electrons in resonance structures as we imagined their "flow" in going from one resonance structure to another. We colored them green in that case, rather than red, because electrons do not actually "flow" in resonance structures; they are simply delocalized. Still, the curved-arrow formalism helps our *minds* flow from one resonance structure to another. We will find ourselves constantly using these curved arrows to keep track of electrons, both as reactants change to products and as we imagine additional resonance structures of a hybrid.

PROBLEM 2-23 (PARTIALLY SOLVED)

In the following acid–base reactions,

1. draw Lewis structures of the reactants and the products.
2. determine which species are acting as electrophiles (acids) and which are acting as nucleophiles (bases).
3. use the curved-arrow formalism to show the movement of electron pairs in these reactions, as well as the imaginary movement in the resonance hybrids of the products.

4. indicate which reactions are best termed Brønsted–Lowry acid–base reactions.

(a)

$$\underset{\text{acetaldehyde}}{CH_3-\overset{\overset{\displaystyle O}{\|}}{C}-H} + HCl \longrightarrow CH_3-\overset{\overset{\displaystyle \overset{+}{O}-H}{\|}}{C}-H + Cl^-$$

This reaction is a proton transfer from HCl to the C═O to the group of acetaldehyde. Therefore, it is a Brønsted–Lowry acid–base reaction, with HCl acting as the acid (proton donor) and acetaldehyde acting as the base (proton acceptor). Before drawing any curved arrows, remember that one arrow must show the movement of electrons *from* the electron-pair donor (the nucleophile/base) *to* the electron-pair acceptor (the electrophile/acid). Another curved arrow must show the electrons that once formed the H—Cl bond leaving with chlorine to give a chloride ion. Drawing these arrows is easier once we draw valid Lewis structures for all the reactants and products.

$$\underset{\text{base}}{CH_3-\overset{\overset{\displaystyle :O:}{\|}}{C}-H} \quad \underset{\text{acid}}{H-\ddot{\underset{\cdot\cdot}{Cl}}:} \longrightarrow \left[\underset{\text{major}}{CH_3-\overset{\overset{\displaystyle :\overset{+}{O}-H}{\|}}{C}-H} \longleftrightarrow \underset{\text{minor}}{CH_3-\overset{\overset{\displaystyle :\ddot{O}-H}{|}}{\overset{+}{C}}-H}\right] + :\ddot{\underset{\cdot\cdot}{Cl}}:^-$$

The resonance forms of the product show that a pair of electrons can be moved between the oxygen atom and the C═O pi bond. The positive charge is delocalized over the carbon and oxygen atoms, with most of the positive charge on oxygen because all octets are satisfied in that resonance structure.

(b)

$$\underset{\text{acetaldehyde}}{CH_3-\overset{\overset{\displaystyle O}{\|}}{C}-H} + CH_3-O^- \longrightarrow CH_3-\overset{\overset{\displaystyle O^-}{|}}{\underset{\underset{\displaystyle O-CH_3}{|}}{C}}-H$$

In this case, no proton has been transferred, so this is not a Brønsted–Lowry acid–base reaction. Instead, a bond has formed between the oxygen of the CH_3-O^- group and the C═O carbon atom. Drawing the Lewis structures helps to show that the CH_3-O^- group (the nucleophile in this reaction) donates the electrons to form the new bond to acetaldehyde (the electrophile). This result agrees with our intuition that a negatively charged ion is likely to be electron-rich and therefore an electron donor.

$$\underset{\text{electrophile}}{CH_3-\overset{\overset{\displaystyle :O:}{\|}}{C}-H} \quad \underset{\text{nucleophile}}{^-:\ddot{\underset{\cdot\cdot}{O}}-CH_3} \longrightarrow CH_3-\overset{\overset{\displaystyle :\ddot{O}:^-}{|}}{\underset{\underset{\displaystyle :\underset{\cdot\cdot}{O}-CH_3}{|}}{C}}-H$$

Notice that acetaldehyde acts as the nucleophile (Lewis base) in part (a) and as the electrophile (Lewis acid) in part (b). Like most organic compounds, acetaldehyde is both acidic and basic. It acts as a base if we add a strong enough acid to make it donate electrons or accept a proton. It acts as an acid if the base we add is strong enough to donate an electron pair or abstract a proton.

(c)

$$BH_3 + CH_3-O-CH_3 \longrightarrow CH_3-\overset{\overset{\displaystyle ^-BH_3}{|}}{\underset{+}{O}}-CH_3$$

(d)

$$CH_3-\overset{\overset{\displaystyle O}{\|}}{C}-H + {}^-OH \longrightarrow CH_3-\overset{\overset{\displaystyle O^-}{|}}{\underset{\underset{\displaystyle OH}{|}}{C}}-H$$

(continued)

PROBLEM-SOLVING HINT

The curved red arrows show movement of electrons. Each arrow must begin at a pair of electrons: either an unshared electron pair or a pair of electrons in a bond.

Each arrow should point toward where that pair of electrons will be in the products, either forming a bond to another atom or residing as an unshared pair on an atom.

(e) $CH_3-\overset{O}{\overset{\|}{C}}-H + {}^-OH \longrightarrow \left[H-\overset{H}{\overset{|}{\underset{-}{C}}}-\overset{O}{\overset{\|}{C}}-H \longleftrightarrow H-\overset{H}{\overset{|}{C}}=\overset{O^-}{\overset{|}{C}}-H \right] + H_2O$

(f) $CH_3-NH_2 + CH_3-Cl \rightarrow CH_3-\overset{+}{N}H_2-CH_3 + Cl^-$

(g) $CH_3-CH=CH_2 + BF_3 \rightarrow CH_3-\overset{+}{C}H-CH_2-\overset{-}{B}F_3$

(h) $CH_3-\overset{+}{C}H-CH_2-\overset{-}{B}F_3 + CH_3-CH=CH_2 \longrightarrow CH_3-\overset{+}{C}H-CH_2-\overset{CH_3}{\overset{|}{C}H}-CH_2-\overset{-}{B}F_3$

(i) [cyclic structure: O, Cl, OH] $\longrightarrow$ [ring structure: O^-, Cl, O^+, H]

(j) [cyclopentene with OCH_3] $+ O=\overset{+}{N}=O \longrightarrow$ [cyclopentyl cation with $\overset{+}{}OCH_3$ and NO_2]

2-15 Hydrocarbons

In later chapters, we will study many different types of organic compounds. The various kinds of compounds are briefly described here so that you can recognize them as you encounter them. For the purpose of this brief survey, we divide organic compounds into three classes: (1) hydrocarbons, (2) compounds containing oxygen, and (3) compounds containing nitrogen.

The **hydrocarbons** are compounds composed entirely of carbon and hydrogen. The major classes of hydrocarbons are alkanes, alkenes, alkynes, and aromatic hydrocarbons. Nearly all hydrocarbons are nonpolar or only slightly polar. Therefore, hydrocarbons (and hydrocarbon parts of larger molecules) tend to be hydrophobic.

2-15A Alkanes

Alkanes are hydrocarbons that contain only single bonds. Alkane names generally have the *-ane* suffix, and the first part of the name indicates the number of carbon atoms. Table 2-3 shows how the prefixes in the names correspond with the number of carbon atoms. Naming of compounds (**nomenclature**) is covered in more detail in the chapters devoted to the individual functional groups and summarized in Appendix 5 (*Summary of Organic Nomenclature*).

TABLE 2-3 CORRESPONDENCE OF PREFIXES AND NUMBERS OF CARBON ATOMS

Alkane Name	Number of Carbons	Alkane Name	Number of Carbons
*meth*ane	1	*hex*ane	6
*eth*ane	2	*hept*ane	7
*prop*ane	3	*oct*ane	8
*but*ane	4	*non*ane	9
*pent*ane	5	*dec*ane	10

CH_4 methane; CH_3-CH_3 ethane; $CH_3-CH_2-CH_3$ propane; $CH_3-CH_2-CH_2-CH_3$ butane; $CH_3-\overset{CH_3}{\overset{|}{C}H}-CH_3$ isobutane

The **cycloalkanes** are a special class of alkanes in the form of a ring. Figure 2-12 shows the Lewis structures and line–angle formulas of cyclopentane and cyclohexane, the cycloalkanes containing five and six carbons, respectively.

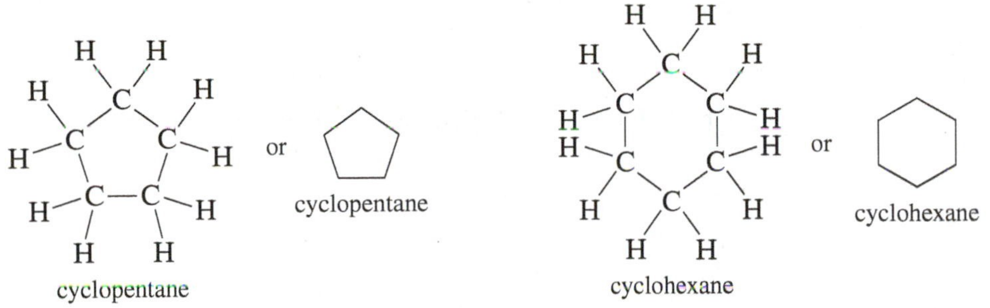

cyclopentane or cyclopentane

cyclohexane or cyclohexane

FIGURE 2-12
Cycloalkanes. Cycloalkanes are alkanes in the form of a ring.

Alkanes are the major components of heating gases (natural gas and liquefied petroleum gas), gasoline, jet fuel, diesel fuel, motor oil, fuel oil, and paraffin "wax." Other than combustion, alkanes undergo few reactions. In fact, when a molecule contains an alkane portion and a nonalkane portion, we often ignore the presence of the alkane portion because it is relatively unreactive. Alkanes undergo few reactions because they have no **functional group,** the part of the molecule where reactions usually occur. Functional groups are distinct chemical units, such as double bonds, hydroxy groups, or halogen atoms, that are reactive. Most organic compounds are characterized and classified by their functional groups.

An **alkyl group** is an alkane portion of a molecule, with one hydrogen atom removed to allow bonding to the rest of the molecule. Figure 2-13 shows an ethyl group (C_2H_5) attached to cyclohexane to give ethylcyclohexane. We might try to name this compound as "cyclohexylethane," but we should treat the larger fragment as the parent compound (cyclohexane) and the smaller group as the alkyl group (ethyl).

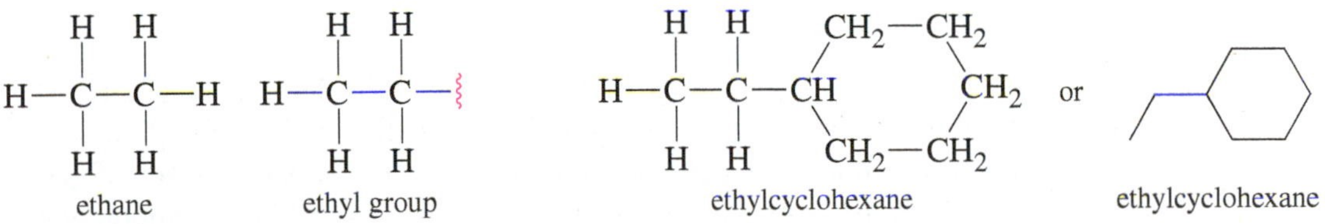

ethane, ethyl group, ethylcyclohexane or ethylcyclohexane

FIGURE 2-13
Naming alkyl groups. Alkyl groups are named like the alkanes they are derived from, with a *-yl* suffix.

We are often concerned primarily with the structure of the most important part of a molecule. In these cases, we can use the symbol R as a substituent to represent an alkyl group (or some other unreactive group). We presume that the exact nature of the R group is unimportant.

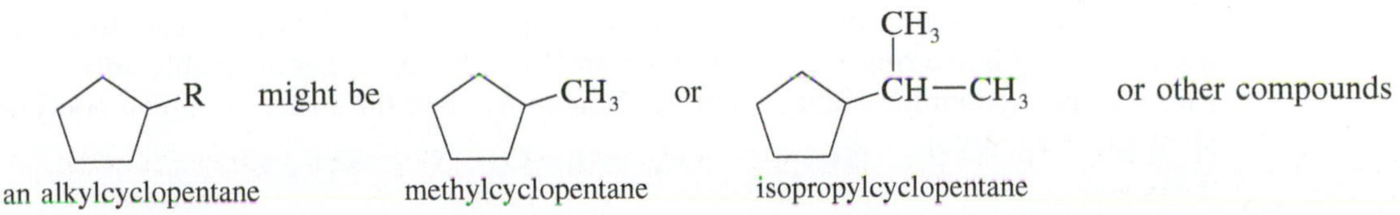

an alkylcyclopentane might be methylcyclopentane or isopropylcyclopentane or other compounds

2-15B Alkenes

Alkenes are hydrocarbons that contain carbon–carbon double bonds. A carbon–carbon double bond is the most reactive part of an alkene, so we say that the double bond is the *functional group* of the alkene. Alkene names end in the *-ene* suffix. If the double bond might be in more than one position, then the chain is numbered and the lower number of the two double-bonded carbons is added to the name to indicate the position of the double bond.

$CH_2{=}CH_2$
ethene (ethylene)

$CH_2{=}CH{-}CH_3$
propene (propylene)

$\overset{1}{CH_2}{=}\overset{2}{CH}{-}\overset{3}{CH_2}{-}\overset{4}{CH_3}$
but-1-ene

$\overset{1}{CH_3}{-}\overset{2}{CH}{=}\overset{3}{CH}{-}\overset{4}{CH_3}$
but-2-ene

Carbon–carbon double bonds cannot rotate, and many alkenes show geometric (cis-trans) isomerism (Sections 1-18B and 1-19B). Following are the cis-trans isomers of some simple alkenes:

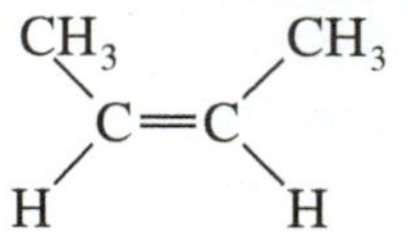

cis-but-2-ene

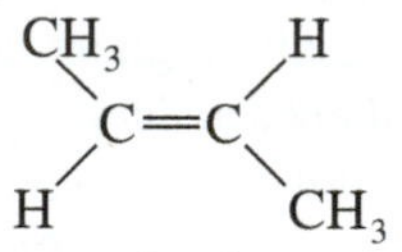

trans-but-2-ene

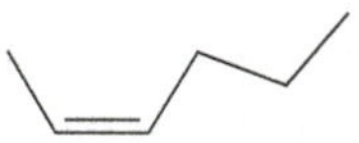

cis-hex-2-ene

trans-hex-2-ene

Cycloalkenes are also common. Unless the rings are very large, cycloalkenes are always the cis isomers, and the term *cis* is omitted from the names. In a large ring, a trans double bond may occur, giving a *trans*-cycloalkene.

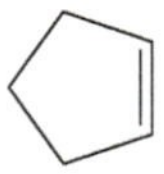

cyclopentene

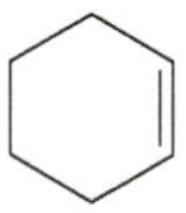

cyclohexene

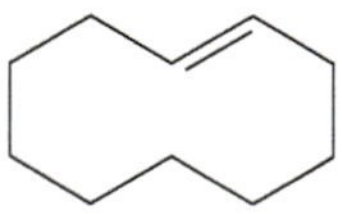

trans-cyclodecene

2-15C Alkynes

Alkynes are hydrocarbons with carbon–carbon triple bonds as their functional group. Alkyne names generally have the *-yne* suffix, although some of their common names (*acetylene*, for example) do not conform to this rule. The triple bond is linear, so there is no possibility of geometric (cis-trans) isomerism in alkynes.

$H{-}C{\equiv}C{-}H$
ethyne (acetylene)

$H{-}C{\equiv}C{-}CH_3$
propyne (methylacetylene)

$H{-}\overset{1}{C}{\equiv}\overset{2}{C}{-}\overset{3}{CH_2}{-}\overset{4}{CH_3}$
but-1-yne

$\overset{1}{CH_3}{-}\overset{2}{C}{\equiv}\overset{3}{C}{-}\overset{4}{CH_3}$
but-2-yne

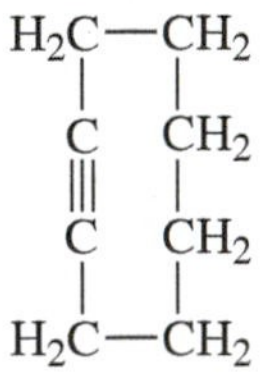

In an alkyne, four atoms must be in a straight line. These four collinear atoms are not easily bent into a ring, so cycloalkynes are rare. Cycloalkynes are stable only if the ring is large, containing eight or more carbon atoms. See the structure of cyclooctyne shown in the margin.

cyclooctyne

2-15D Aromatic Hydrocarbons

The following compounds may look like cycloalkenes, but their properties are different from those of simple alkenes. These **aromatic hydrocarbons** (also called **arenes**) are all derivatives of *benzene*, represented by a six-membered ring with three double bonds. This bonding arrangement is particularly stable, for reasons that we explain in Chapter 16.

or | or | or

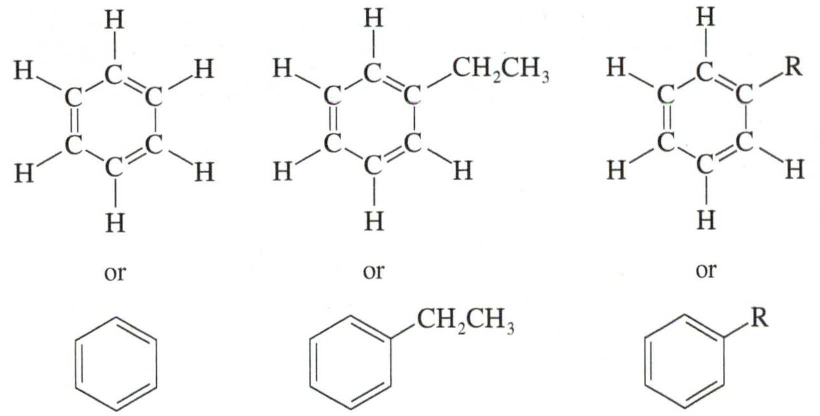

benzene | ethylbenzene | an alkylbenzene

Application: Toxicity

The presence of a methyl or ethyl group can make a big difference in biological systems. For example, benzene is quite toxic and causes leukemia, but methyl benzene (and ethyl benzene) are less toxic because enzymes can oxidize the methyl or ethyl group.

Just as a generic alkyl group substituent is represented by R, a generic aryl group is represented by Ar. When a benzene ring serves as a substituent, it is called a phenyl group, abbreviated Ph.

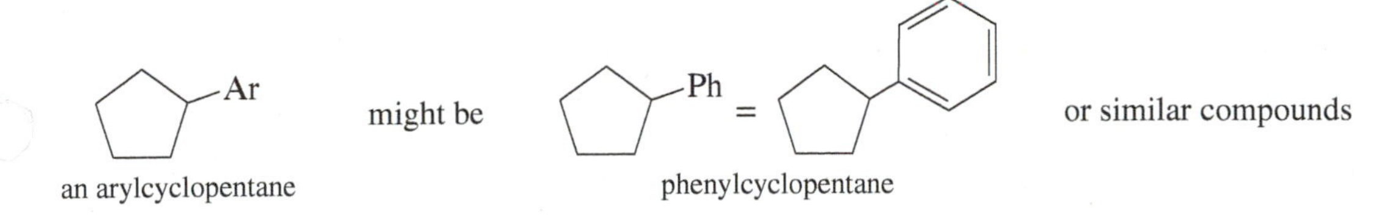

an arylcyclopentane | phenylcyclopentane

PROBLEM 2-24

Classify the following hydrocarbons, and draw a Lewis structure for each one. A compound may fit into more than one of the following classifications:

alkane	cycloalkane	aromatic hydrocarbon
alkene	cycloalkene	
alkyne	cycloalkyne	

(a) $(CH_3CH_2)_2CHCH(CH_3)_2$ **(b)** $CH_3CHCHCH_2CH_3$ **(c)** $CH_3CCCH_2CH_2CH_3$

(d) 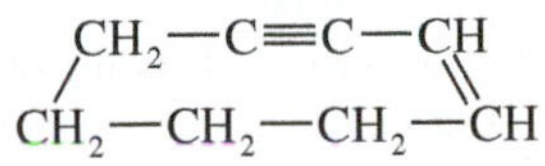 $CH_2-C\equiv C-CH$ / $CH_2-CH_2-CH_2-CH$ **(e)** 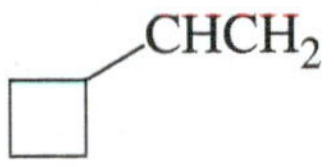 $CHCH_2$ **(f)** 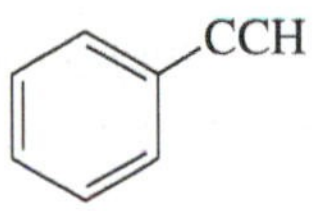 CCH

(g) 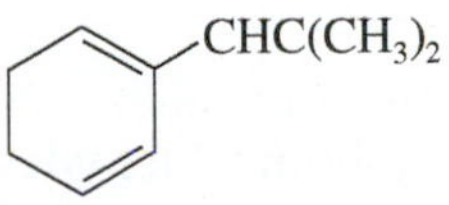 $CHC(CH_3)_2$ **(h)** 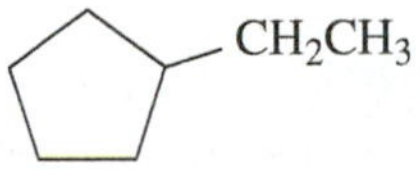 CH_2CH_3 **(i)**

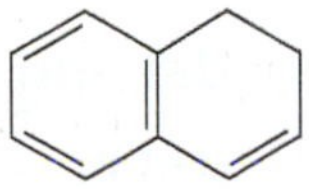

2-16 Functional Groups with Oxygen

Many organic compounds contain oxygen atoms bonded to alkyl groups. The major classes of oxygen-containing compounds are alcohols, ethers, ketones, aldehydes, carboxylic acids, and acid derivatives. We will cover their nomenclature in greater detail in upcoming chapters.

2-16A Alcohols

Alcohols are organic compounds that contain the **hydroxy group** (—OH) as their functional group. The general formula for an alcohol is R—OH. Alcohols are among the most polar organic compounds because the hydroxy group is strongly polar and can participate in hydrogen bonding. Thus, the hydroxy group is hydrophilic, and it confers solubility in water if the hydrophobic alkyl part of the molecule is not too large. Some of the simple alcohols such as ethanol and methanol are miscible (soluble in all proportions) with water, but alcohols with more than four carbons have limited solubility in water. Names of alcohols end in the *-ol* suffix from the word *alcohol*, as shown for the following common alcohols:

R—OH
an alcohol

CH_3—OH
methanol
(methyl alcohol)

CH_3CH_2—OH
ethanol
(ethyl alcohol)

CH_3—CH_2—CH_2—OH
propan-1-ol
(*n*-propyl alcohol)

CH_3—CH(OH)—CH_3
propan-2-ol
(isopropyl alcohol)

Application: Environment

Ethanol is widely used as an additive to gasoline. "Gasohol" is an "oxygenated" fuel that contains about 10% ethanol, which causes the fuel to burn at a slightly lower temperature, producing lower concentrations of pollutants such as nitrogen oxides. Ethanol has replaced MTBE (methyl *tert*-butyl ether) as an oxygenated fuel additive because of safety concerns about MTBE contaminating groundwater when it leaks from storage tanks.

Alcohols are some of the most common organic compounds. Methyl alcohol (methanol), also known as "wood alcohol," is used as an industrial solvent and as an automotive racing fuel. Ethyl alcohol (ethanol) is sometimes called "grain alcohol" because it is produced by the fermentation of grain or almost any other organic material. "Isopropyl alcohol" is the common name for propan-2-ol, used as "rubbing alcohol."

2-16B Ethers

Ethers are composed of two alkyl groups bonded to an oxygen atom. The general formula for an ether is R—O—R′. (The symbol R′ represents another alkyl group, either the same as or different from the first.) Like alcohols, ethers are much more polar than hydrocarbons, but ethers have no O—H hydrogens, so they cannot hydrogen bond with themselves. Ethers do form hydrogen bonds with hydrogen-bond donors such as alcohols, amines, and water, enhancing their solubility with these compounds. Ether names are often formed from the names of the alkyl groups and the word *ether*. Diethyl ether is the common "ether" used for starting engines in cold weather and once used for surgical anesthesia.

R—O—R′
ROR′, an ether

CH_3CH_2—O—CH_2CH_3
diethyl ether

furan

CH_3—O—C(CH_3)$_2$—CH_3
methyl *tert*-butyl ether

hydrogen bonding of dimethyl ether with water

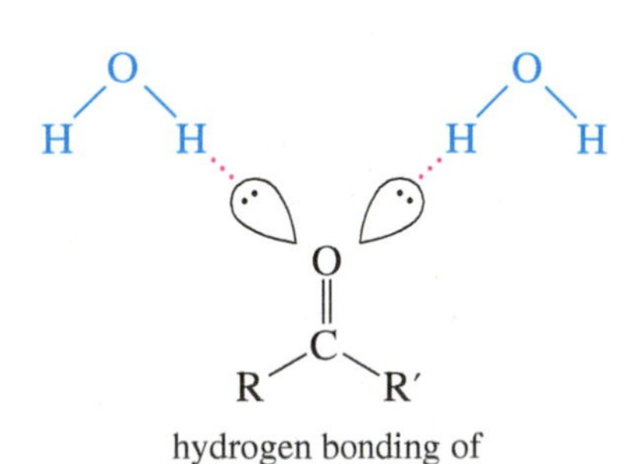

hydrogen bonding of a ketone with water

2-16C Aldehydes and Ketones

The **carbonyl group,** C=O, is the functional group for both aldehydes and ketones. A **ketone** has two alkyl groups bonded to the carbonyl group; an **aldehyde** has one alkyl group and a hydrogen atom bonded to the carbonyl group. Ketone names generally have the *-one* suffix; aldehyde names use either the *-al* suffix or the *-aldehyde* suffix.

The carbonyl group is strongly polar, and it can form hydrogen bonds with hydrogen-bond donors such as water, alcohols, and amines. The hydrophilic carbonyl group allows aldehydes and ketones containing up to four carbon atoms to be miscible with water. Those with more than four carbons have limited solubility in water.

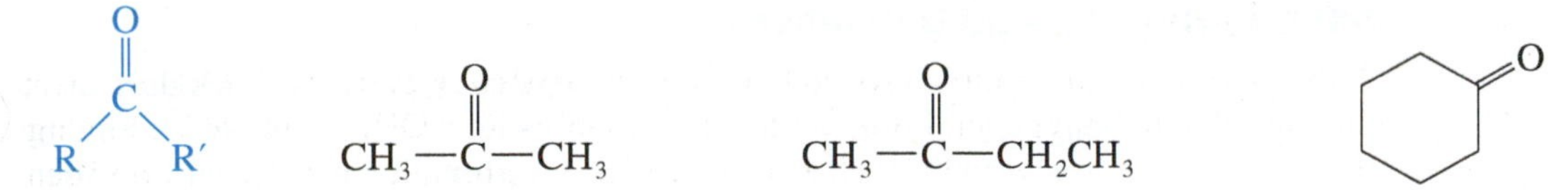

$$\mathrm{RCOR'}, \text{ a ketone} \qquad \mathrm{CH_3-\overset{O}{\overset{\|}{C}}-CH_3} \qquad \mathrm{CH_3-\overset{O}{\overset{\|}{C}}-CH_2CH_3}$$

propan-2-one (acetone) — butan-2-one (methyl ethyl ketone) — cyclohexanone

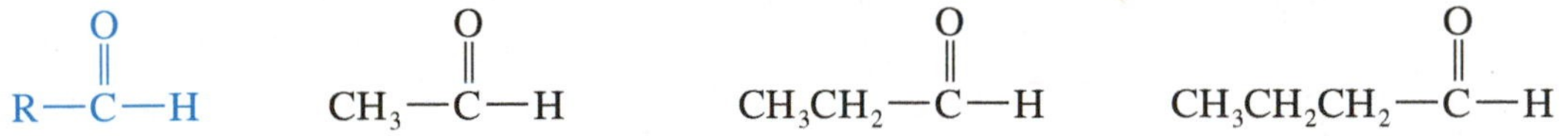

$$\mathrm{R-\overset{O}{\overset{\|}{C}}-H} \qquad \mathrm{CH_3-\overset{O}{\overset{\|}{C}}-H} \qquad \mathrm{CH_3CH_2-\overset{O}{\overset{\|}{C}}-H} \qquad \mathrm{CH_3CH_2CH_2-\overset{O}{\overset{\|}{C}}-H}$$

or RCHO an aldehyde — or CH_3CHO ethanal (acetaldehyde) — or CH_3CH_2CHO propanal (propionaldehyde) — or $CH_3CH_2CH_2CHO$ butanal (butyraldehyde)

2-16D Carboxylic Acids

Carboxylic acids contain the **carboxyl group,** —COOH, as their functional group. The general formula for a carboxylic acid is R—COOH (*or* RCO_2H) The carboxyl group is a combination of a carbonyl group and a hydroxy group, but this combination has different properties from those of ketones and alcohols. Carboxylic acids owe their acidity (pK_a of about 5) to the resonance-stabilized *carboxylate anions* formed by deprotonation. The following reaction shows the dissociation of a carboxylic acid:

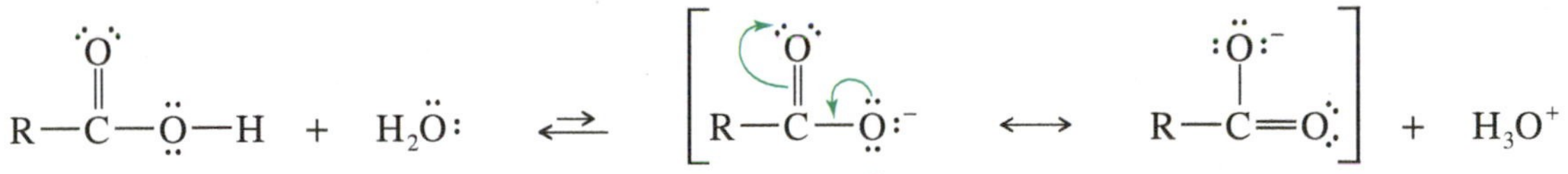

$$\mathrm{R-\overset{O}{\overset{\|}{C}}-\ddot{O}-H} + \mathrm{H_2\ddot{O}:} \rightleftharpoons \left[\mathrm{R-\overset{O}{\overset{\|}{C}}-\ddot{O}:^-} \longleftrightarrow \mathrm{R-\overset{:\ddot{O}:^-}{\overset{|}{C}}=\ddot{O}}\right] + \mathrm{H_3O^+}$$

carboxylic acid — carboxylate anion

Systematic names for carboxylic acids use the *-oic acid* suffix, but historical names are commonly used. Formic acid was first isolated from ants, genus *Formica.* Acetic acid, found in vinegar, gets its name from the Latin word for "sour" (*acetum*). Propionic acid gives the tangy flavor to sharp cheeses, and butyric acid provides the pungent aroma of rancid butter.

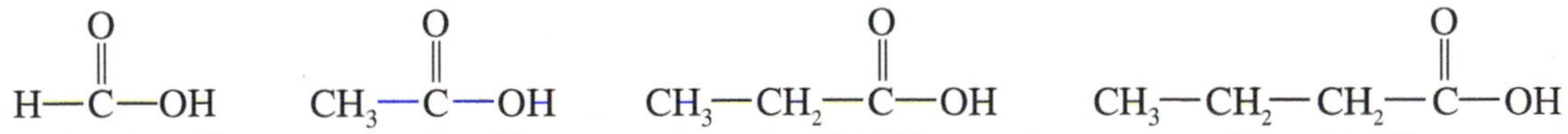

$$\mathrm{H-\overset{O}{\overset{\|}{C}}-OH} \qquad \mathrm{CH_3-\overset{O}{\overset{\|}{C}}-OH} \qquad \mathrm{CH_3-CH_2-\overset{O}{\overset{\|}{C}}-OH} \qquad \mathrm{CH_3-CH_2-CH_2-\overset{O}{\overset{\|}{C}}-OH}$$

methanoic acid (formic acid) — ethanoic acid (acetic acid) — propanoic acid (propionic acid) — butanoic acid (butyric acid)

Carboxylic acids are strongly polar, like ketones, aldehydes, and alcohols. The strongly hydrophilic carboxyl group enhances their solubility in water; in fact, all four of the carboxylic acids shown here are miscible (soluble in all proportions) with water.

PROBLEM 2-25

Draw a Lewis structure, and classify each of the following compounds. The possible classifications are as follows:

alcohol ketone carboxylic acid
ether aldehyde alkene

(a) CH_2CHCHO (b) $CH_3CH_2CH(OH)CH_3$ (c) $CH_3COCH_2CH_3$

(d) $CH_3CH_2OCHCH_2$ (e) cyclohexane–COOH (f) ring with O and double bond

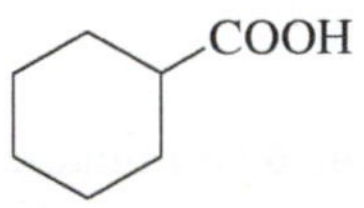

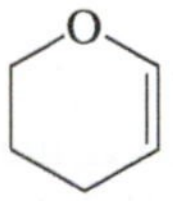

(g) cyclopentenone (=O) (h) cyclopentane–CHO (i) cyclopentane–CH_2OH

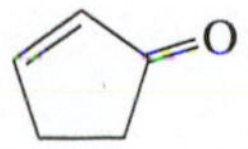

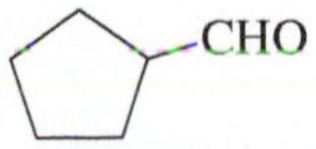

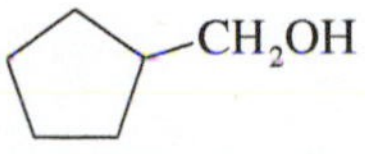

PROBLEM-SOLVING HINT

Condensed formulas are often confusing, especially when they involve carbonyl groups. Whenever you see a complicated condensed formula, convert it to a Lewis structure first for clarity.

2-16E Carboxylic Acid Derivatives

Carboxylic acids are easily converted to a variety of acid derivatives. Each derivative contains the carbonyl group bonded to an oxygen or another electron-withdrawing element. Among these functional groups are **acid chlorides, esters,** and **amides.** All of these groups can be converted back to carboxylic acids by acidic or basic hydrolysis.

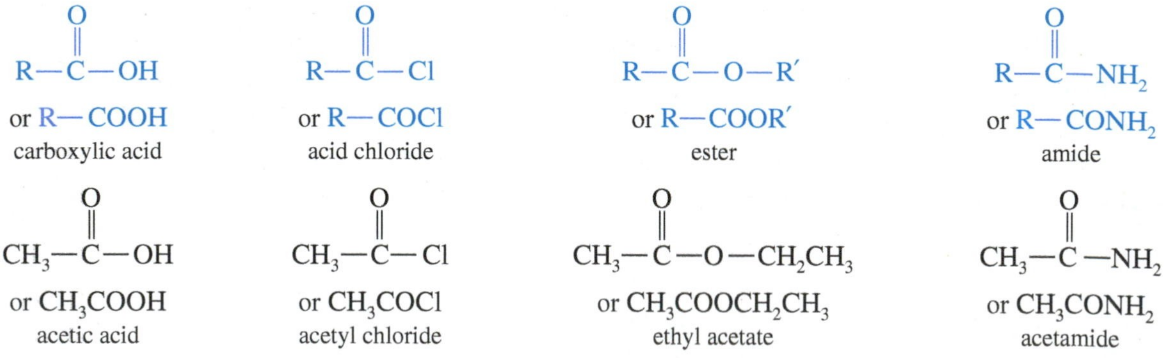

carboxylic acid	acid chloride	ester	amide
$R-C(=O)-OH$ or $R-COOH$	$R-C(=O)-Cl$ or $R-COCl$	$R-C(=O)-O-R'$ or $R-COOR'$	$R-C(=O)-NH_2$ or $R-CONH_2$
$CH_3-C(=O)-OH$ or CH_3COOH	$CH_3-C(=O)-Cl$ or CH_3COCl	$CH_3-C(=O)-O-CH_2CH_3$ or $CH_3COOCH_2CH_3$	$CH_3-C(=O)-NH_2$ or CH_3CONH_2
acetic acid	acetyl chloride	ethyl acetate	acetamide

2-17 Functional Groups with Nitrogen

Nitrogen is another element often found in the functional groups of organic compounds. The most common "nitrogenous" organic compounds are amines, amides, and nitriles.

2-17A Amines

Amines are alkylated derivatives of ammonia. Like ammonia, amines are basic.

$$R-\ddot{N}H_2 + H_2O \rightleftarrows R-\overset{+}{N}H_3 \ {}^{-}OH \qquad K_b \cong 10^{-4} \qquad pK_b \cong 4$$

Because of their basicity ("alkalinity"), naturally occurring amines are often called *alkaloids*. Simple amines are named by naming the alkyl groups bonded to nitrogen and adding the word "amine." The structures of some simple amines are shown below, together with the structure of nicotine, a toxic alkaloid found in tobacco leaves.

Amines: $R-\ddot{N}H_2$ or $R-\ddot{N}H-R'$ or $R-\ddot{N}(R')-R''$

$CH_3-\ddot{N}H_2$ methylamine

$CH_3-\ddot{N}H-CH_2CH_3$ ethylmethylamine

$(CH_3CH_2)_3N\!:$ triethylamine

piperidine

nicotine

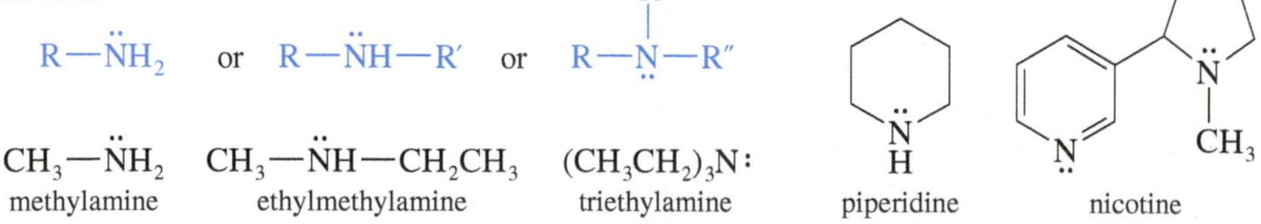

The nitrogen atom of an amine is hydrophilic because it can accept hydrogen bonds from water. If the amine has N—H bonds, these can donate hydrogen bonds to water. Therefore, many of the amines containing up to five carbon atoms are miscible with water.

2-17B Amides

Amides are acid derivatives that result from a combination of an acid with ammonia or an amine. Proteins have the structure of long-chain, complex amides.

$$R-\overset{\overset{\displaystyle O}{\|}}{C}-NH_2 \quad \text{or} \quad R-\overset{\overset{\displaystyle O}{\|}}{C}-NHR' \quad \text{or} \quad R-\overset{\overset{\displaystyle O}{\|}}{C}-NR'_2$$

amides

$$CH_3-\overset{\overset{\displaystyle O}{\|}}{C}-NH_2 \qquad CH_3-\overset{\overset{\displaystyle O}{\|}}{C}-NH-CH_3 \qquad C_6H_5-\overset{\overset{\displaystyle O}{\|}}{C}-N(CH_3)_2$$

acetamide — *N*-methylacetamide — *N*,*N*-dimethylbenzamide

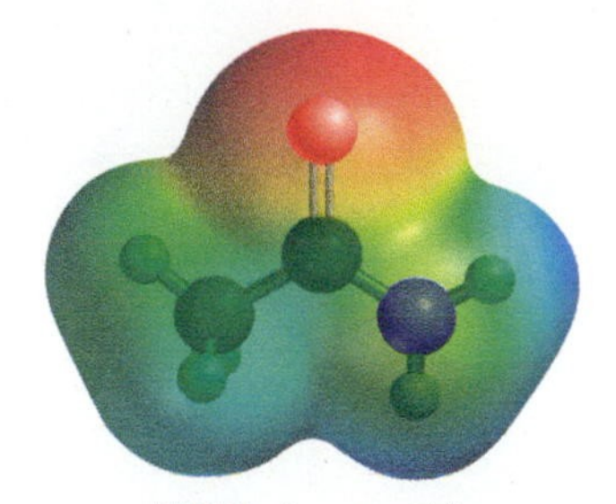
EPM of acetamide

Amides are among the most stable acid derivatives. The nitrogen atom of an amide is not as basic as the nitrogen of an amine because of the electron-withdrawing effect of the carbonyl group (see Section 2-12). The following resonance forms help to show why amides are very weak bases:

$$\left[R-\overset{\overset{\displaystyle :\ddot{O}\cdot}{\|}}{C}-\ddot{N}H_2 \longleftrightarrow R-\overset{\overset{\displaystyle :\ddot{O}:^-}{|}}{C}=\overset{+}{N}H_2\right] \quad \text{very weak base} \quad pK_b \cong 14$$

Amides form particularly strong hydrogen bonds, giving them high melting points and high boiling points. The strongly polarized amide N—H hydrogen forms unusually strong hydrogen bonds with the carbonyl oxygen that carries a partial negative charge in the polarized resonance form shown above. The following illustration shows this strong intermolecular hydrogen bonding.

$$\begin{array}{c} O^- \quad H \\ R-C=\overset{+}{N}-H \cdots O^- \\ R-C=\overset{+}{N}(H)-H \cdots O^--C(R)=\overset{+}{N}H_2 \end{array}$$

hydrogen bonding in amides

Application: Proteins

Proteins are specialized polymers of amides (Chapter 24). Their three-dimensional structures are defined and stabilized by the strong hydrogen bonding found in amides.

The alpha-helical structure shown below is held in position by hydrogen bonding between amide N—H groups and carbonyl groups in adjacent turns of the helix.

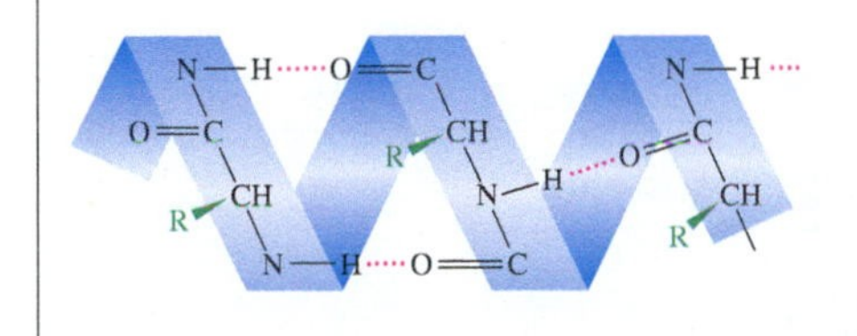

2-17C Nitriles

A **nitrile** is a compound containing the **cyano group,** —C≡N. The cyano group was introduced in Section 1-17 as an example of *sp* hybridized bonding. The cyano group is strongly polar by virtue of the C≡N triple bond, and most small nitriles are somewhat soluble in water. Acetonitrile is miscible with water.

$$R-C\equiv N: \qquad CH_3-C\equiv N: \qquad CH_3CH_2-C\equiv N: \qquad C_6H_5-C\equiv N:$$

a nitrile — acetonitrile — propionitrile — benzonitrile

All of these classes of compounds are summarized in the table of Common Organic Compounds and Functional Groups, given on the front inside cover for convenient reference.

PROBLEM 2-26

Draw a Lewis structure, and classify each of the following compounds:

(a) $CH_3CH_2CONHCH_3$ (b) $(CH_3CH_2)_2NH$ (c) $(CH_3)_2CHCOOCH_3$
(d) $CH_3CHCHCOCl$ (e) $(CH_3CH_2)_2O$ (f) $CH_3CH_2CH_2CN$
(g) $(CH_3)_3CCH_2CH_2COOH$ (h) 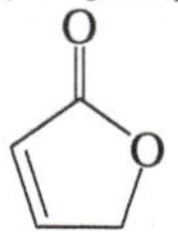 (i)

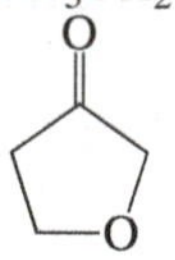

(j) 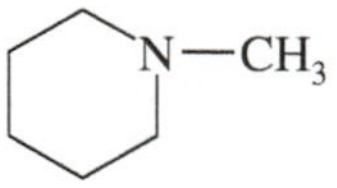 N—CH₃ (k) 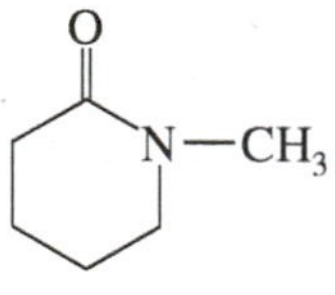 (l) 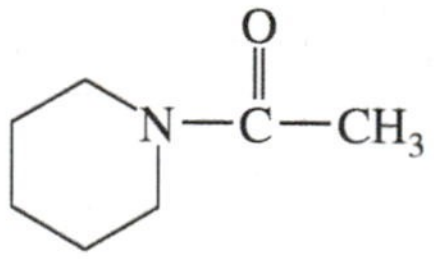 N—C—CH₃

(m) 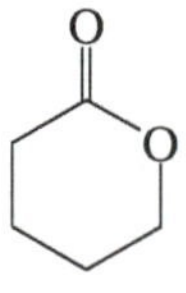 (n) 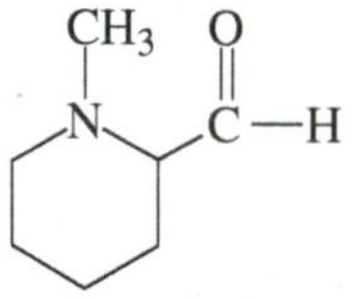 (o) 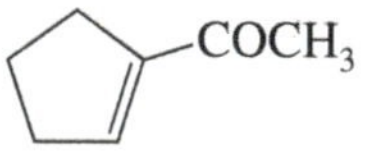 COCH₃

PROBLEM 2-27

Circle the functional groups in the following structures. State to which class (or classes) of compounds the structure belongs.

(a) $CH_2{=}CHCH_2COOCH_3$ (b) CH_3OCH_3 (c) CH_3CHO
(d) CH_3CONH_2 (e) CH_3NHCH_3 (f) RCOOH
(g) 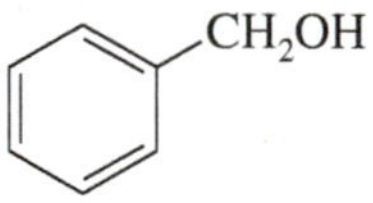 CH_2OH (h) 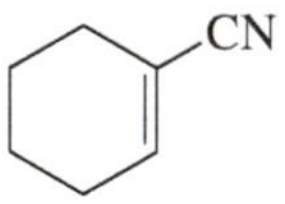 CN (i) 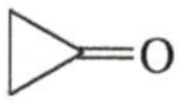 =O

(j) 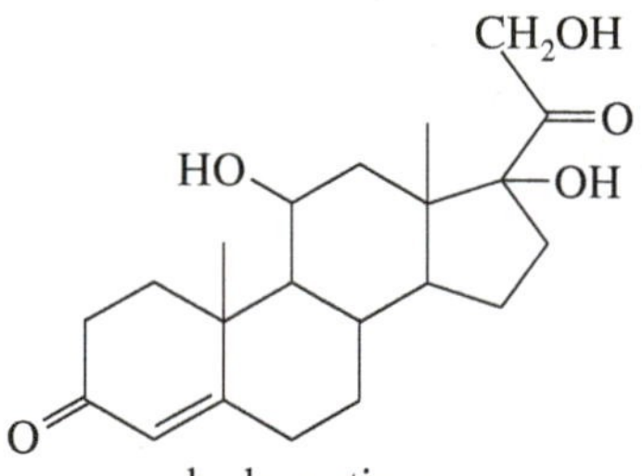 hydrocortisone (k) 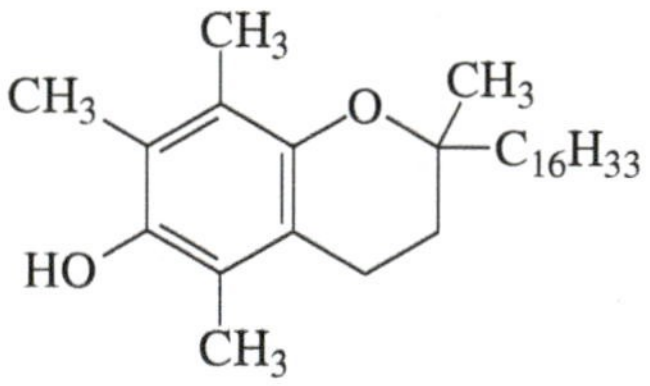 vitamin E

Essential Terms

acid chloride An acid derivative with a chlorine atom in place of the hydroxy group. (p. 96)

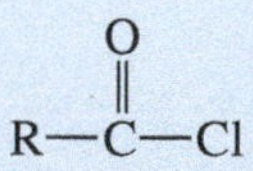

$$R{-}\overset{\overset{\displaystyle O}{\|}}{C}{-}Cl$$

acid-dissociation constant (K_a) The equilibrium constant for the reaction of the acid with water to generate H_3O^+. (p. 70)

$$\underset{\text{acid}}{HA} + H_2O \xrightleftharpoons{K_a} H_3O^+ + \underset{\text{base}}{A^-} \qquad K_a = \frac{[H_3O^+][A^-]}{[HA]}$$

conjugate acid–base pair

The negative logarithm of K_a is expressed as pK_a:

$$pK_a = -\log_{10} K_a$$

acids and bases (p. 67)

(Arrhenius definitions) **acid:** dissociates in water to give H_3O^+
base: dissociates in water to give ^-OH

(Brønsted–Lowry definitions) **acid:** proton donor
base: proton acceptor

(Lewis definitions) **acid:** electron-pair acceptor (electrophile)
base: electron-pair donor (nucleophile)

alcohol A compound that contains a hydroxy group bonded to a carbon atom; R—OH. (p. 94)

aldehyde A carbonyl group with one alkyl group and one hydrogen. (p. 94)

$$R-\overset{\overset{\displaystyle O}{\|}}{C}-H$$

alkanes Hydrocarbons containing only single bonds. (p. 90)

alkenes Hydrocarbons containing one or more C=C double bonds. (p. 92)

alkyl group A hydrocarbon group with only single bonds; an alkane with one hydrogen removed, to allow bonding to another group; symbolized by R. (p. 91)

alkynes Hydrocarbons containing one or more C≡C triple bonds. (p. 92)

amide An acid derivative that contains a nitrogen atom instead of the hydroxy group of the acid. (p. 96)

$$R-\overset{\overset{\displaystyle O}{\|}}{C}-NH_2 \qquad R-\overset{\overset{\displaystyle O}{\|}}{C}-NHR' \qquad R-\overset{\overset{\displaystyle O}{\|}}{C}-NR'_2$$

amine An alkylated analog of ammonia; $R-NH_2$, R_2NH, or R_3N. (p. 96)

aromatic hydrocarbons **(arenes)** Hydrocarbons containing a *benzene ring*, a six-membered ring with three double bonds. (p. 93)

bond dipole moment A measure of the polarity of an individual bond in a molecule, defined as $\mu = (4.8 \times d \times \delta)$, where μ is the dipole moment in **debyes** (10^{-10} esu-Å), d is the bond length in angstroms, and δ is the effective amount of charge separated, in units of the electronic charge. (p. 56)

carbonyl group The >C=O functional group, as in a ketone or an aldehyde. (p. 94)

carboxyl group The —COOH functional group, as in a carboxylic acid. (p. 95)

carboxylic acid A compound that contains the carboxyl group, —COOH. (p. 95)

$$R-\overset{\overset{\displaystyle O}{\|}}{C}-OH$$

conjugate acid The acid that results from protonation of a base. (p. 69)

conjugate base The base that results from loss of a proton from an acid. (p. 69)

curved-arrow formalism A method of drawing curved arrows to keep track of electron movement from nucleophile to electrophile (or within a molecule) during the course of a reaction. (p. 88)

cyano group The —C≡N functional group, as in a nitrile. (p. 97)

dipole–dipole forces Attractive intermolecular forces resulting from the attraction of the positive and negative ends of the permanent dipole moments of polar molecules. (p. 60)

dipole moment (μ) A measure of the polarity of a bond (or a molecule), proportional to the product of the charge separation times the bond length. (p. 56)

electrophile An electron-pair acceptor (Lewis acid). (p. 87)

ester An acid derivative with an alkyl group replacing the acid proton. (p. 96)

$$R-\overset{\overset{\displaystyle O}{\|}}{C}-OR'$$

ether A compound with an oxygen bonded between two alkyl (or aromatic) groups; R—O—R′. (p. 94)

functional group The reactive, nonalkane part of an organic molecule. (p. 91)

hydrocarbons Compounds composed exclusively of carbon and hydrogen.

- **alkanes:** Hydrocarbons containing only single bonds. (p. 90)
- **alkenes:** Hydrocarbons containing one or more C═C double bonds. (p. 92)
- **alkynes:** Hydrocarbons containing one or more C≡C triple bonds. (p. 92)
- **cycloalkanes, cycloalkenes, cycloalkynes:** Alkanes, alkenes, and alkynes in the form of a ring. (p. 91)
- **aromatic hydrocarbons:** Hydrocarbons containing a *benzene ring*, a six-membered ring with three double bonds. (p. 93)

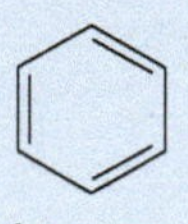

benzene

hydrogen bond A particularly strong attraction between a nonbonding pair of electrons and an electrophilic O—H or N—H hydrogen. Hydrogen bonds have bond energies of about 20 kJ/mol (5 kcal/mol), compared with about 400 kJ/mol (about 100 kcal/mol) for typical C—H bonds. (p. 62)

hydrophilic ("water-loving") Polar substances or groups that readily dissolve in water; water-attractive. (p. 66)

hydrophobic ("water-hating") Nonpolar substances or groups that do not readily dissolve in water; water-repellant. (p. 66)

hydroxy group **(hydroxyl group)** The —OH functional group, as in an alcohol. (p. 94)

isomers Different compounds with the same molecular formula. (p. 44)

ketone A carbonyl group bonded to two alkyl groups. (p. 94)

$$\mathrm{R-\overset{\overset{\displaystyle O}{\|}}{C}-R'}$$

Lewis acid An electron-pair acceptor (electrophile). (p. 87)

Lewis base An electron-pair donor (nucleophile). (p. 87)

London dispersion forces Intermolecular forces resulting from the attraction of correlated temporary dipole moments induced in adjacent molecules. (p. 61)

miscible Soluble in all proportions. (p. 66)

molecular dipole moment The vector sum of the bond dipole moments (and any nonbonding pairs of electrons) in a molecule; a measure of the polarity of a molecule. (p. 57)

nitrile A compound containing a cyano group, —C≡N. (p. 97)

nucleophile An electron-pair donor (Lewis base). (p. 87)

pH A measure of the acidity of a solution, defined as the negative logarithm (base 10) of the H_3O^+ concentration: $pH = -\log_{10}[H_3O^+]$. (p. 68)

van der Waals forces The attractive forces between neutral molecules, including dipole–dipole forces and London dispersion forces. (p. 61)

- **dipole–dipole forces:** The forces between polar molecules resulting from the attraction of their permanent dipole moments.
- **London forces:** Intermolecular forces resulting from the attraction of correlated temporary dipole moments induced in adjacent molecules.

Essential Problem-Solving Skills in Chapter 2

Each skill is followed by problem numbers exemplifying that particular skill.

1 Identify polar and nonpolar molecules. Identify those molecules that engage in hydrogen bonding, either between themselves or with other hydrogen-bond donors.	Problems 2-28, 29, 30, 31, and 32
2 Predict general trends in the boiling points and solubilities of compounds based on their size, polarity, and ability to form hydrogen bonds.	Problems 2-31, 32, 33, 34, and 35
3 Predict relative acidities and basicities based on the structure, bonding, and resonance of conjugate acid–base pairs.	Problems 2-36, 37, 38, 40, 45, 46, 47, 48, and 54
4 Calculate and interpret values of K_a, K_b, pK_a, and pK_b. Use these values, and values for similar compounds, to predict products of acid–base reactions.	Problems 2-39, 41, 42, 43, 44, and 48
Problem-Solving Strategy: Predicting Acid–Base Equilibrium Positions	Problems 2-39, 42, 43, and 49
5 Identify electrophiles (Lewis acids) and nucleophiles (Lewis bases), and write equations for Lewis acid–base reactions using curved arrows to show the flow of electrons.	Problems 2-50, 51, 52, and 53
6 Identify alkanes, alkenes, and aromatic hydrocarbons, and draw structural formulas for examples.	Problems 2-55, 56, and 57
7 Identify alcohols, ethers, aldehydes, ketones, carboxylic acids, and esters, and draw structural formulas for examples.	Problems 2-55, 56, and 57
8 Identify amines, amides, and nitriles, and draw structural formulas for examples.	Problems 2-55, 56, and 57

Study Problems

2-28 The C≡N triple bond in acetonitrile has a dipole moment of about 3.6 D and a bond length of about 1.16 Å. Calculate the amount of charge separation in this bond. How important is the charge-separated resonance form in the structure of acetonitrile?

$$\left[CH_3-C\equiv N: \longleftrightarrow CH_3-\overset{+}{C}=\overset{-}{\ddot{N}}: \right]$$

acetonitrile

2-29 For each of the following compounds,

1. draw the Lewis structure.
2. show how the bond dipole moments (and those of any nonbonding pairs of electrons) contribute to the molecular dipole moment.
3. estimate whether the compound will have a large, small, or zero dipole moment.

(a) $CH_3-CH=N-CH_3$ **(b)** CH_3-CH_2OH **(c)** CBr_4

(d) 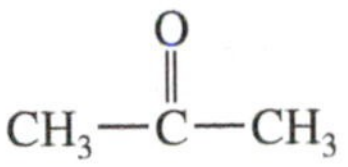 $CH_3-\overset{O}{\overset{\|}{C}}-CH_3$ **(e)** 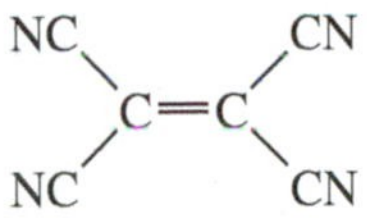 $(NC)_2C=C(CN)_2$ **(f)** 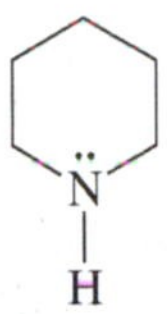 (piperidine ring with N–H)

(g) 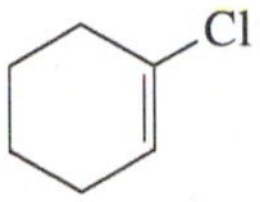 (1-chlorocyclohexene ring, Cl) **(h)** $CH_3-\overset{O}{\overset{\|}{C}}-NH_2$

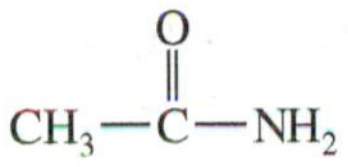

2-30 Sulfur dioxide has a dipole moment of 1.60 D. Carbon dioxide has a dipole moment of zero, even though C—O bonds are more polar than S—O bonds. Explain this apparent contradiction.

2-31 Which of the following pure compounds can form hydrogen bonds? Which can form hydrogen bonds with water? Which ones do you expect to be soluble in water?

(a) $(CH_3CH_2)_2NH$ **(b)** $(CH_3CH_2)_3N$ **(c)** $CH_3CH_2CH_2OH$

(d) $(CH_3CH_2CH_2)_2O$ **(e)** $CH_3(CH_2)_3CH_3$ **(f)** $CH_2{=}CH{-}CH_2CH_3$

(g) CH_3COCH_3 **(h)** CH_3CH_2COOH **(i)** CH_3CH_2CHO

(j) **(k)** **(l)** $CH_3-\overset{O}{\overset{\|}{C}}-NH_2$

2-32 Predict which member of each pair is more soluble in water. Explain your prediction.

(a) or **(b)** or **(c)** or CH_3

(d) or CH_2OH **(e)** or Br Br **(f)** or

2-33 Diethyl ether and butan-1-ol are isomers, and they have similar solubilities in water. Their boiling points are very different, however. Explain why these two compounds have similar solubility properties but dramatically different boiling points.

$CH_3CH_2-O-CH_2CH_3$	$CH_3CH_2CH_2CH_2-OH$
diethyl ether, bp 35 °C	butan-1-ol, bp 118 °C
8.4 mL dissolves in 100 mL H_2O	9.1 mL dissolves in 100 mL H_2O

2-34 *N*-Methylpyrrolidine has a boiling point of 81 °C, and piperidine has a boiling point of 106 °C.

(a) Explain the large difference (25 °C) in boiling point for these two isomers.

(b) Tetrahydropyran has a boiling point of 88 °C, and cyclopentanol has a boiling point of 141 °C. These two isomers have a boiling point difference of 53 °C. Explain why the two oxygen-containing isomers have a much larger boiling point difference than the two amine isomers.

(c) *N,N*-Dimethylformamide has a boiling point of 150 °C, and N-methylacetamide has a boiling point of 206 °C, for a difference of 56 °C. Explain why these two nitrogen-containing isomers have a much larger boiling point difference than the two amine isomers. Also explain why these two amides have higher boiling points than any of the other four compounds shown (two amines, an ether, and an alcohol).

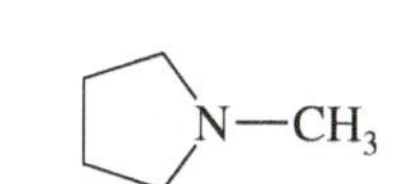

N-methylpyrrolidine, bp 81 °C

tetrahydropyran, bp 88 °C

$H-\overset{O}{\overset{\|}{C}}-N(CH_3)_2$

N,N-dimethylformamide, bp 150 °C

piperidine, bp 106 °C

cyclopentanol, bp 141 °C

$CH_3-\overset{O}{\overset{\|}{C}}-N(CH_3)H$

N-methylacetamide, bp 206 °C

2-35 Predict which compound in each pair has the higher boiling point. Explain your prediction.

(a) $CH_3CH_2OCH_3$ or $CH_3CH(OH)CH_3$ **(b)** $CH_3CH_2CH_2CH_3$ or $CH_3CH_2CH_2CH_2CH_3$

(c) $CH_3CH_2CH_2CH_2CH_3$ or $(CH_3)_2CHCH_2CH_3$ **(d)** $CH_3CH_2CH_2CH_2CH_3$ or $CH_3CH_2CH_2CH_2CH_2Cl$

(e)

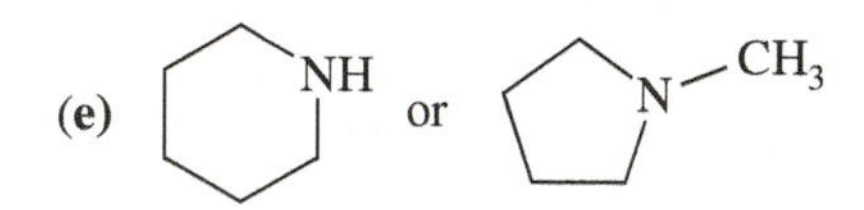

(f)

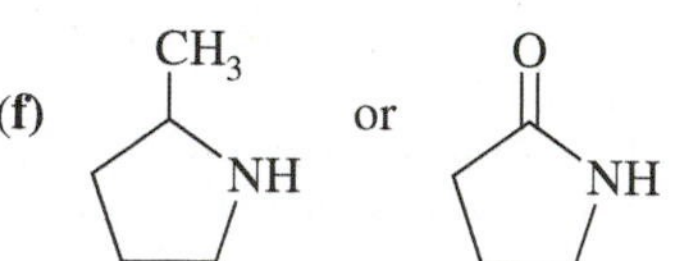

2-36 All of the following compounds can react as acids. Without using a table of acidities, rank them in order of increasing acidity. Explain your ranking.

(a) $CH_3CH_2SO_3H$ **(b)** CH_3CH_2OH **(c)** CH_3CH_2COOH

(d) $CH_3CHClCOOH$ **(e)** $ClCH_2CH_2COOH$

2-37 Rank the following species in order of increasing acidity. Explain your reasons for ordering them as you do.

$HF \quad NH_3 \quad H_2SO_4 \quad CH_3OH \quad CH_3COOH \quad H_3O^+ \quad H_2O$

2-38 Rank the following species in order of increasing basicity. Explain your reasons for ordering them as you do.

$NH_3 \quad CH_3O^- \quad H_2O \quad CH_3COO^- \quad NaOH \quad NH_2^- \quad HSO_4^-$

2-39 The K_a of phenylacetic acid is 5.2×10^{-5}, and the pK_a of propionic acid is 4.87.

$C_6H_5{-}CH_2{-}C(=O){-}OH$ phenylacetic acid, $K_a = 5.2 \times 10^{-5}$ $\qquad CH_3{-}CH_2{-}C(=O){-}OH$ propionic acid, $pK_a = 4.87$

(a) Calculate the pK_a of phenylacetic acid and the K_a of propionic acid.
(b) Which of these is the stronger acid? Calculate how much stronger an acid it is.
(c) Predict whether the following equilibrium will favor the reactants or the products.

$$C_6H_5{-}CH_2COO^- + CH_3CH_2COOH \rightleftharpoons C_6H_5{-}CH_2COOH + CH_3CH_2COO^-$$

***2-40** The following compound can become protonated on any of the three nitrogen atoms. One of these nitrogens is much more basic than the others, however.
(a) Draw the important resonance forms of the products of protonation on each of the three nitrogen atoms.
(b) Determine which nitrogen atom is the most basic.

$CH_3{-}NH{-}C({=}NH){-}NH_2$

2-41 The following compounds are listed in increasing order of acidity. In each case, the most acidic proton is shown in red.

$C_6H_5C(=O)CH_2{-}H$ **W**, $pKa = 25$ $\qquad C_6H_5C(=O)NH{-}H$ **X**, $pKa = 23$ $\qquad C_6H_5C(=O)N(OH){-}H$ **Y**, $pKa = 8.8$ $\qquad C_6H_5C(=O)O{-}H$ **Z**, $pKa = 4.2$

(a) Show the structure of the conjugate base of each acid, including any resonance forms.
(b) Explain why X is a stronger acid than W.
(c) Explain why Y is a stronger acid than X.
(d) Explain why Z is a stronger acid than Y.

2-42 Predict the products of the following acid–base reactions.
(a) $H_2SO_4 + CH_3COO^- \rightleftharpoons$
(b) $CH_3COOH + (CH_3)_3N: \rightleftharpoons$
(c) $C_6H_5{-}C(=O){-}OH + {}^-OH \rightleftharpoons$
(d) $HO{-}C(=O){-}OH + 2\ {}^-OH \rightleftharpoons$
(e) $H_2O + NH_3 \rightleftharpoons$
(f) $(CH_3)_3\overset{+}{N}H + {}^-OH \rightleftharpoons$
(g) $HCOOH + CH_3O^- \rightleftharpoons$
(h) $\overset{+}{N}H_3CH_2COOH + 2\ {}^-OH \rightleftharpoons$

2-43 Consider the following proposed Brønsted–Lowry acid–base reactions. In each case, draw the products of a transfer of the most acidic proton on the acid to the most basic site on the base. Use Appendix 4 to find or estimate the pK_a values for the acids and the pK_b values for the bases. Then determine which side of the reaction is favored, either reactants or products.

(a) pyridine + $O_2N{-}C_6H_4{-}OH$ (4-nitrophenol)
(b) pyridine + $O_2N{-}C_6H_4{-}COOH$ (4-nitrobenzoic acid)

(continued)

(c) 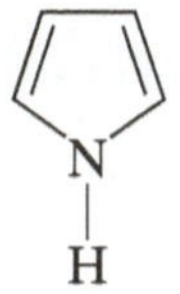 pyrrole (N—H) + $NaNH_2$

(d) $CH_3-C(=O)-NH_2$ + $NaOCH_2CH_3$

(e) $CH_3-C(=O)-NH_2$ + HCl (aq)

(f) $CH_3-C(=O)-CH_2-C(=O)-OCH_2CH_3$ + NaOH (aq)

2-44 Compare the relative acidity of 1-molar aqueous solutions of the following acids.

NH_4Cl	CH_3COOH	$BrCH_2COOH$	F_3CCOOH	H_2SO_4	HCl	HBr
$pK_a = 9.26$	$pK_a = 4.74$	$pK_a = 2.9$	$pK_a = 0.2$	$pK_a = -5$	$pK_a = -7$	$pK_a = -9$

2-45 The following compounds can all react as acids.

$CH_3-C(=O)-OH$ $\quad$ $CF_3-C(=O)-OH$ $\quad$ $CH_3-C(=O)-OOH$ $\quad$ $CF_3CH_2-C(=O)-OH$ $\quad$ CH_3CH_2OH

(a) For each compound, show its conjugate base. Show any resonance forms if applicable.
(b) Rank the conjugate bases in the order you would predict, from most stable to least stable.
(c) Rank the original compounds in order, from strongest acid to weakest acid.

2-46 The following compounds can all react as bases.

$CH_3CH_2-NH_2$ $\quad$ $CH_3-C(=O)-NH_2$ $\quad$ NaOH $\quad$ CH_3CH_2-OH $\quad$ $NaNH_2$

(a) For each compound, show its conjugate acid. Show any resonance forms if applicable.
(b) Rank the conjugate acids in the order you would predict, from most stable to least stable.
(c) Rank the original compounds in order, from strongest base to weakest base.

2-47 The following compounds can all react as acids.

$CH_3-C(=O)-OH$ $\quad$ $CH_3-C(=O)-NH_2$ $\quad$ $CH_3-S(=O)-OH$ $\quad$ $CH_3-S(=O)_2-OH$ $\quad$ $F-S(=O)_2-OH$

(a) For each compound, show its conjugate base. Show any resonance forms if applicable.
(b) Rank the conjugate bases in the order you would predict, from most stable to least stable.
(c) Rank the original compounds in order, from strongest acid to weakest acid.

2-48 Consider the following compounds that vary from nearly nonacidic to strongly acidic. Draw the conjugate bases of these compounds, and explain why the acidity increases so dramatically with substitution by nitro groups.

CH_4	CH_3NO_2	$CH_2(NO_2)_2$	$CH(NO_2)_3$
$pK_a \cong 50$	$pK_a = 10.2$	$pK_a = 3.6$	$pK_a = 0.17$

2-49 Methyllithium (CH_3Li) is often used as a base in organic reactions.
(a) Predict the products of the following acid–base reaction.

$$CH_3CH_2-OH + CH_3-Li \longrightarrow$$

(b) What is the conjugate acid of CH_3Li? Would you expect CH_3Li to be a strong base or a weak base?

2-50 Label the reactants in these acid–base reactions as Lewis acids (electrophiles) or Lewis bases (nucleophiles). Use curved arrows to show the movement of electron pairs in the reactions.

(a) $CH_3\ddot{\underset{..}{O}}:^- + CH_3-\ddot{\underset{..}{Cl}}: \longrightarrow CH_3-\ddot{\underset{..}{O}}-CH_3 + :\ddot{\underset{..}{Cl}}:^-$

(b) $CH_3-\ddot{O}^+(CH_3)-CH_3 + :\ddot{O}(H)-H \longrightarrow CH_3-\ddot{O}(CH_3): + CH_3-\ddot{O}^+(H)-H$

(c) 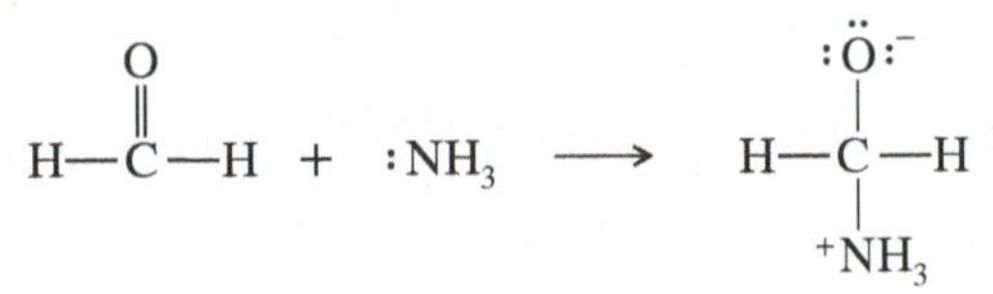 $H-\overset{O}{\overset{\|}{C}}-H + :NH_3 \longrightarrow H-\underset{^+NH_3}{\overset{:\ddot{O}:^-}{C}}-H$

(d) $CH_3-\ddot{N}H_2 + CH_3-CH_2-\ddot{Cl}: \longrightarrow CH_3-\overset{+}{N}H_2-CH_2CH_3 + :\ddot{Cl}:^-$

(e)  $CH_3-\overset{\ddot{O}}{\overset{\|}{C}}-CH_3 + H_2SO_4 \longrightarrow CH_3-\overset{\ddot{O}^+-H}{\overset{\|}{C}}-CH_3 + HSO_4^-$

(f) $(CH_3)_3CCl + AlCl_3 \longrightarrow (CH_3)_3C^+ + {}^-AlCl_4$

(g)  $CH_3-\overset{\ddot{O}}{\overset{\|}{C}}-CH_3 + {}^-:\ddot{O}-H \longrightarrow CH_3-\overset{:\ddot{O}:^-}{C}=CH_2 + H-\ddot{O}-H$

(h) $CH_2=CH_2 + BF_3 \longrightarrow \overset{-}{B}F_3-CH_2-\overset{+}{C}H_2$

(i) $\overset{-}{B}F_3-CH_2-\overset{+}{C}H_2 + CH_2=CH_2 \longrightarrow \overset{-}{B}F_3-CH_2-CH_2-CH_2-\overset{+}{C}H_2$

2-51 In each reaction, label the reactants as Lewis acids (electrophiles) or Lewis bases (nucleophiles). Use curved arrows to show the movement of electron pairs in the reactions. Draw any nonbonding electrons to show how they participate in the reactions.

(a) $(CH_3)_2NH + HCl \longrightarrow (CH_3)_2\overset{+}{N}H_2 + Cl^-$

(b) $(CH_3)_2NH + CH_3Cl \longrightarrow (CH_3)_3\overset{+}{N}H + Cl^-$

(c) 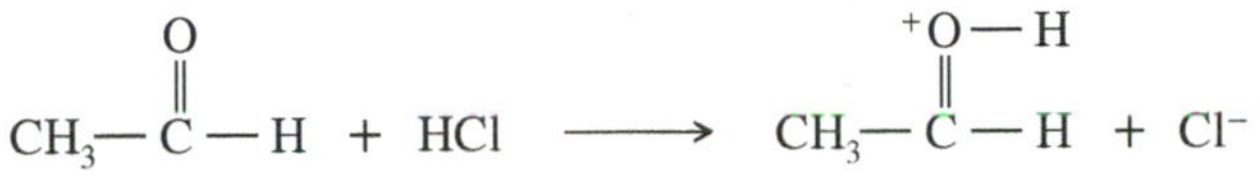 $CH_3-\overset{O}{\overset{\|}{C}}-H + HCl \longrightarrow CH_3-\overset{^+O-H}{\overset{\|}{C}}-H + Cl^-$

(d) 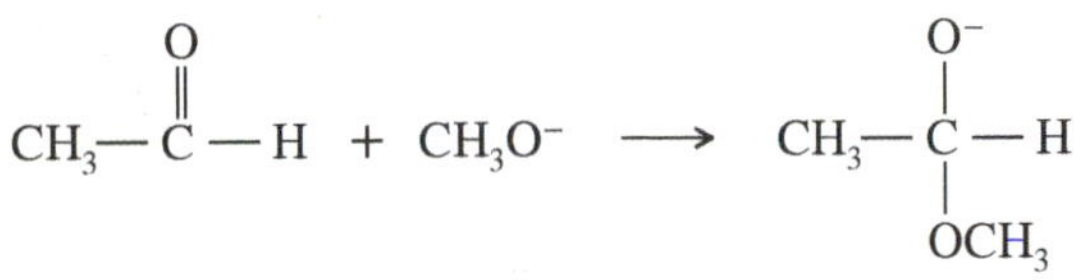 $CH_3-\overset{O}{\overset{\|}{C}}-H + CH_3O^- \longrightarrow CH_3-\underset{OCH_3}{\overset{O^-}{C}}-H$

(e) 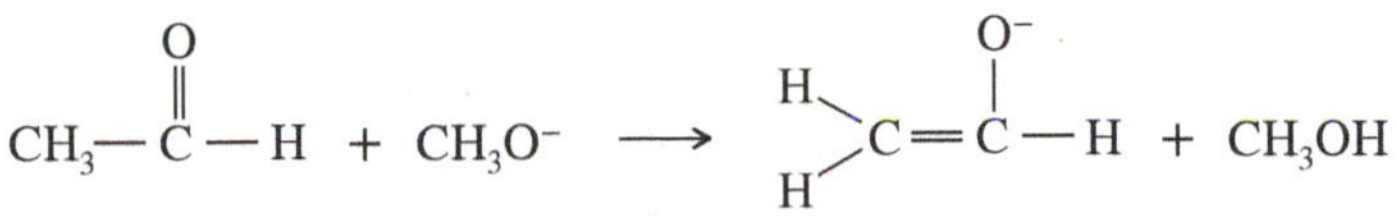 $CH_3-\overset{O}{\overset{\|}{C}}-H + CH_3O^- \longrightarrow (H)(H)C=\overset{O^-}{C}-H + CH_3OH$

2-52 Each of these compounds can react as an electrophile. In each case, use curved arrows to show how the electrophile would react with the strong nucleophile sodium ethoxide, $Na^+ {}^-OCH_2CH_3$.

(a) 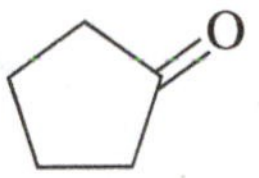 cyclopentanone (ring with =O)

(b) NH_4^+

(c) CH_3CH_2Br

(d) BH_3

(e) CH_3COOH

*(f) $H_2C=CH-CHO$ (two sites)

2-53 Each of these compounds can react as a nucleophile. In each case, use curved arrows to show how the nucleophile would react with the strong electrophile BF_3.

(a) 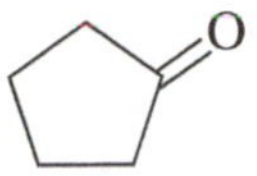 cyclopentanone (ring with =O)

(b)  cyclopentene

(c)  tetrahydrofuran (five-membered ring with O)

(d) $(CH_3)_3N$

(e) CH_3CH_2OH

(f) $(CH_3)_2S$

2-54 The pK_a of ascorbic acid (vitamin C, page 55) is 4.17, showing that it is slightly more acidic than acetic acid (CH_3COOH, pK_a 4.74).

(a) Show the four different conjugate bases that would be formed by deprotonation of the four different OH groups in ascorbic acid.

(b) Compare the stabilities of these four conjugate bases, and predict which OH group of ascorbic acid is the most acidic.

(c) Compare the most stable conjugate base of ascorbic acid with the conjugate base of acetic acid, and suggest why these two compounds have similar acidities, even though ascorbic acid lacks the carboxylic acid (COOH) group.

2-55 Give a definition and an example for each class of organic compounds.

(a) alkane **(b)** alkene **(c)** alkyne
(d) alcohol **(e)** ether **(f)** ketone
(g) aldehyde **(h)** aromatic hydrocarbon **(i)** carboxylic acid
(j) ester **(k)** amine **(l)** amide
(m) nitrile

2-56 Circle the functional groups in the following structures. State to which class (or classes) of compounds the structure belongs.

(a) 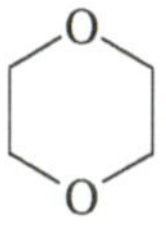 O, O (b) 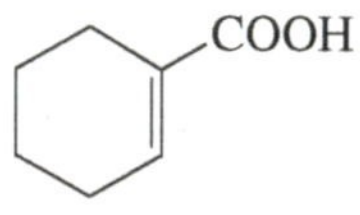 COOH (c) 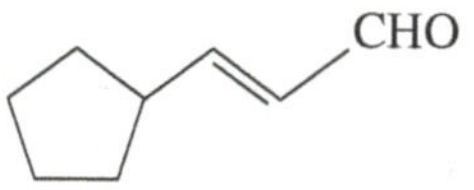 CHO

(d) 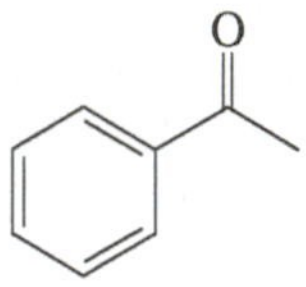 O (e) 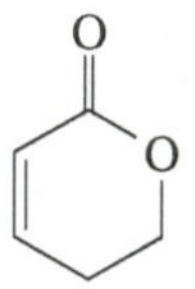 O, O (f) 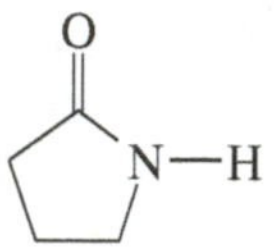 O, N—H

(g) 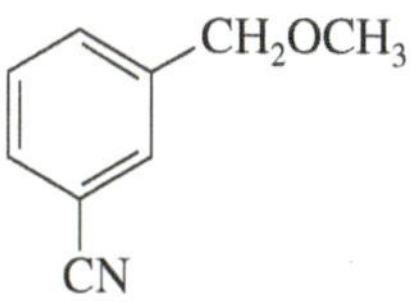 CH_2OCH_3, CN (h) 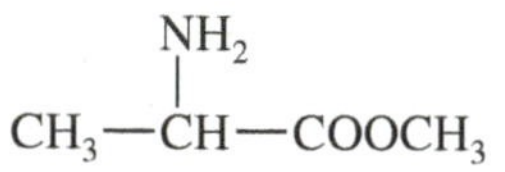 NH_2, CH_3—CH—$COOCH_3$ (i) 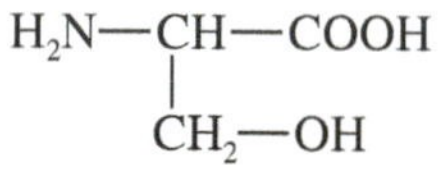 H_2N—CH—COOH, CH_2—OH

2-57 Many naturally occurring compounds contain more than one functional group. Identify the functional groups in the following compounds:

(a) Penicillin G is a naturally occurring antibiotic.

(b) Dopamine is the neurotransmitter that is deficient in Parkinson's disease.

(c) Capsaicin gives the fiery taste to chili peppers.

(d) Thyroxine is the principal thyroid hormone.

(e) Testosterone is a male sex hormone.

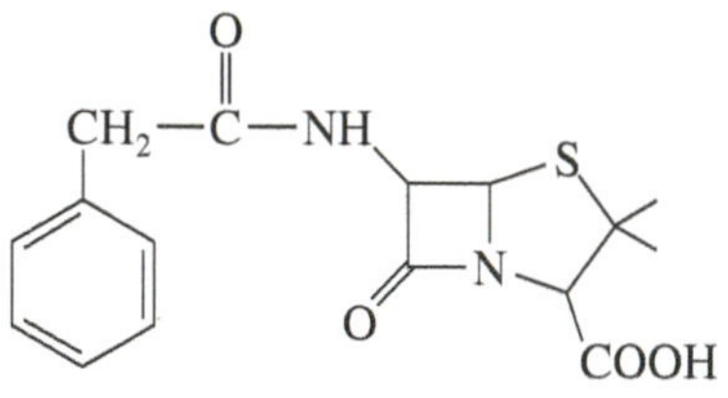

O, CH_2—C—NH, S, N, O, COOH

penicillin G

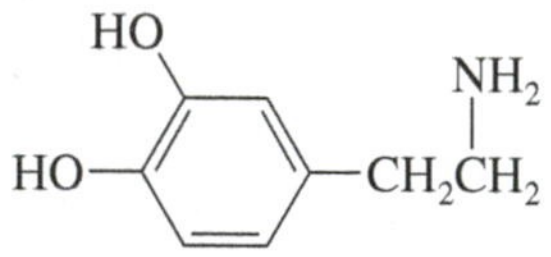

HO, NH_2, HO, CH_2CH_2

dopamine

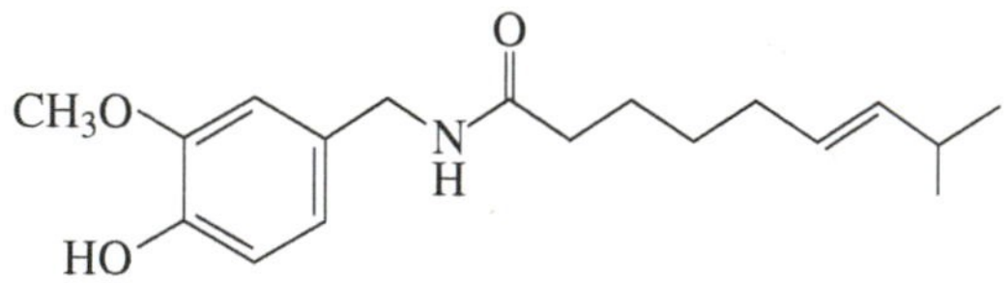

CH_3O, O, N, H, HO

capsaicin

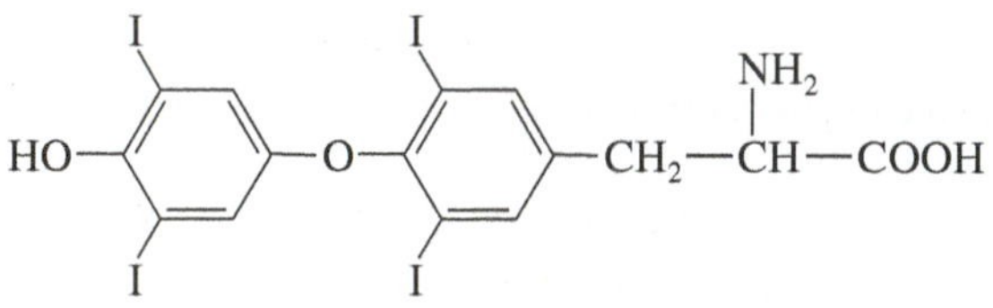

I, I, NH_2, HO, O, CH_2—CH—COOH, I, I

thyroxine

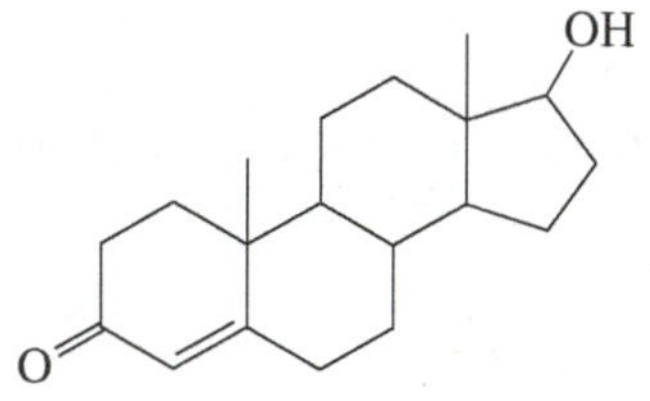

OH, O

testosterone

3 Structure and Stereochemistry of Alkanes

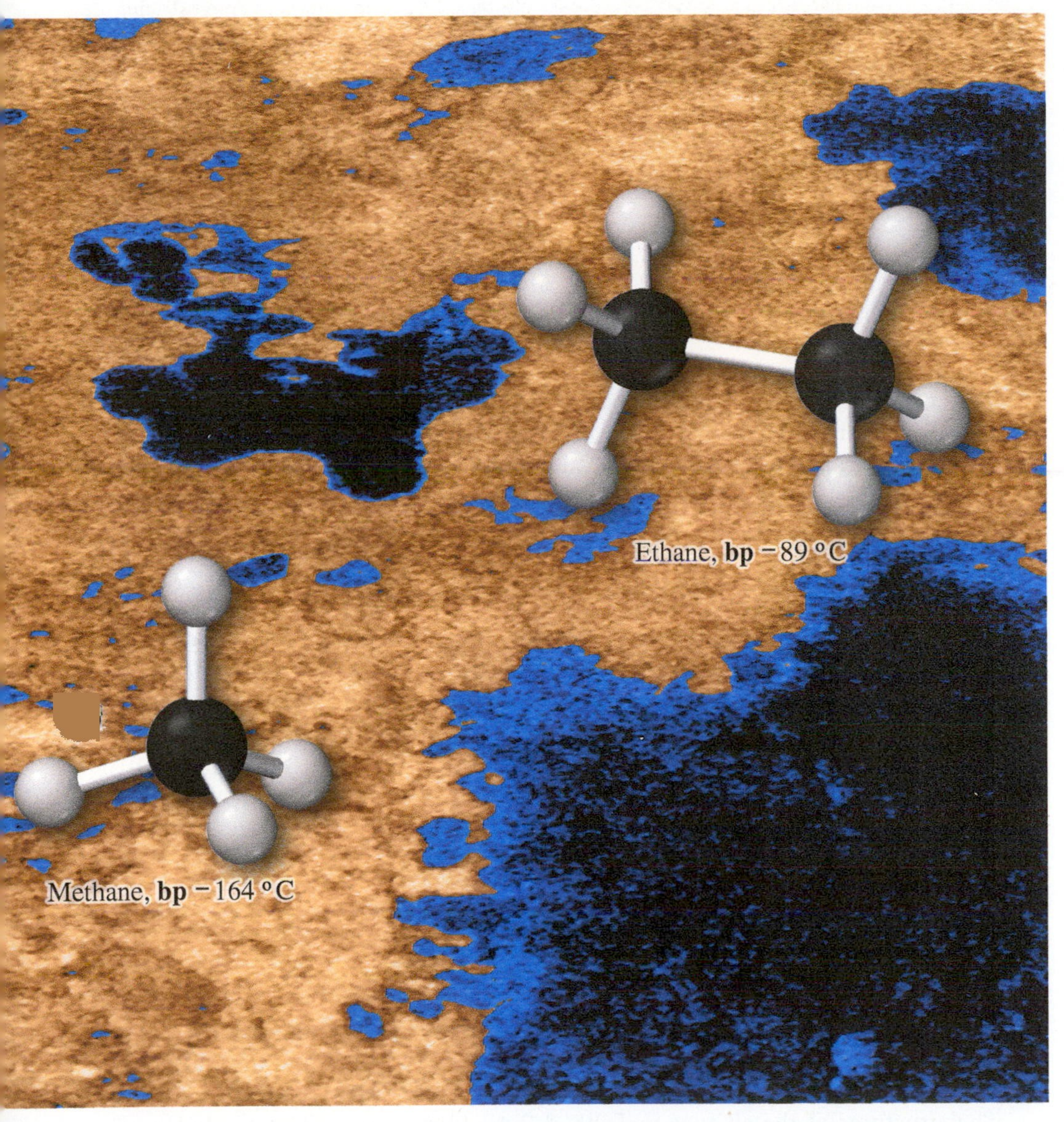

Goals for Chapter 3

1 Draw and name the isomers of alkanes, and explain the trends in their physical properties.

2 Draw alkane conformations, compare their energies, and predict the most stable conformations.

3 Draw and name the isomers of cycloalkanes, and explain ring strain.

4 Draw the conformations of cycloalkanes, compare their energies, and predict the most stable conformations.

◀ A colorized radar image of the surface of Titan, Saturn's largest moon, taken by the Cassini spacecraft. The light tan areas correspond to a solid surface with high radar reflectivity. The dark areas of low reflectivity are thought to be lakes of methane and ethane, which rain down on the moon and flow in rivers along its surface.

An **alkane** is a hydrocarbon that contains only single bonds. The alkanes are the simplest and least reactive class of organic compounds because they contain only hydrogen and sp^3 hybridized carbon, and they have no reactive functional groups. Alkanes contain no double or triple bonds and no heteroatoms (atoms other than carbon or hydrogen). They are poor acids and bases, and they are poor electrophiles and nucleophiles as well. Although alkanes undergo reactions such as cracking and combustion at high temperatures, they are much less reactive than other classes of compounds that have functional groups.

TABLE 3-1
Hydrocarbon Classifications

Compound Type	Functional Group	Example
alkanes	none (no double or triple bonds)	$CH_3-CH_2-CH_3$, propane
alkenes	$>C=C<$ double bond	$CH_2=CH-CH_3$, propene
alkynes	$-C\equiv C-$ triple bond	$H-C\equiv C-CH_3$, propyne
aromatics	benzene ring	CH_2CH_3 ethylbenzene

The study of organic compounds begins with alkanes because alkane structures form the hydrocarbon backbones of most organic compounds. Learning to name organic compounds begins with the names of alkanes because most organic compounds are named using the parent alkane as the root of the name.

3-1 Classification of Hydrocarbons (Review)

We classify hydrocarbons according to their bonding (Section 2-15), as shown in Table 3-1. Alkanes have only single bonds. A hydrocarbon with a carbon–carbon double bond (such as ethylene) is an *alkene*. If a hydrocarbon has a carbon–carbon triple bond (like acetylene), it is an *alkyne*. Hydrocarbons with aromatic rings (resembling benzene) are called *aromatic hydrocarbons*.

A hydrocarbon with no double or triple bonds is said to be **saturated** because it has the maximum number of bonded hydrogens. Another way to describe *alkanes*, then, is as the class of **saturated hydrocarbons.**

3-2 Molecular Formulas of Alkanes

Table 3-2 shows the structures and formulas of the first 20 unbranched alkanes, called the n-alkanes or "normal" alkanes. Any isomers of these compounds have the same molecular formulas even though their structures are different. Notice how the molecular formulas increase by two hydrogen atoms each time a carbon atom is added.

The structures of the alkanes in Table 3-2 are purposely written as chains of $-CH_2-$ groups (**methylene groups**), terminated at each end by a hydrogen atom. This is the general formula for the unbranched (straight-chain) alkanes. These alkanes differ only by the number of methylene groups in the chain. If the molecule contains n carbon atoms, it must contain $(2n + 2)$ hydrogen atoms. Figure 3-1 shows how this pattern appears in structures and how it leads to formulas of the form C_nH_{2n+2}.

A series of compounds such as the unbranched alkanes, that differ only by the number of $-CH_2-$ groups, is called a *homologous series*, and the individual members of the series are called **homologs.** For example, butane is a homolog of propane, and both of these are homologs of hexane and decane.

Although we have derived the C_nH_{2n+2} formula using the unbranched n-alkanes, it applies to branched alkanes as well. Any isomer of one of these n-alkanes has the same molecular formula. Just as butane and pentane follow the C_nH_{2n+2} rule, their branched isomers isobutane, isopentane, and neopentane also follow the rule.

TABLE 3-2
Formulas and Physical Properties of the Unbranched Alkanes, Called the *n*-Alkanes

Alkane	Number of Carbons	Structure	Formula	Boiling Point (°C)	Melting Point (°C)	Density[a]
methane	1	$H—CH_2—H$	CH_4	−164	−183	0.55
ethane	2	$H—(CH_2)_2—H$	C_2H_6	−89	−183	0.51
propane	3	$H—(CH_2)_3—H$	C_3H_8	−42	−189	0.50
butane	4	$H—(CH_2)_4—H$	C_4H_{10}	0	−138	0.58
pentane	5	$H—(CH_2)_5—H$	C_5H_{12}	36	−130	0.63
hexane	6	$H—(CH_2)_6—H$	C_6H_{14}	69	−95	0.66
heptane	7	$H—(CH_2)_7—H$	C_7H_{16}	98	−91	0.68
octane	8	$H—(CH_2)_8—H$	C_8H_{18}	126	−57	0.70
nonane	9	$H—(CH_2)_9—H$	C_9H_{20}	151	−51	0.72
decane	10	$H—(CH_2)_{10}—H$	$C_{10}H_{22}$	174	−30	0.73
undecane	11	$H—(CH_2)_{11}—H$	$C_{11}H_{24}$	196	−26	0.74
dodecane	12	$H—(CH_2)_{12}—H$	$C_{12}H_{26}$	216	−10	0.75
tridecane	13	$H—(CH_2)_{13}—H$	$C_{13}H_{28}$	235	−5	0.76
tetradecane	14	$H—(CH_2)_{14}—H$	$C_{14}H_{30}$	254	6	0.76
pentadecane	15	$H—(CH_2)_{15}—H$	$C_{15}H_{32}$	271	10	0.77
hexadecane	16	$H—(CH_2)_{16}—H$	$C_{16}H_{34}$	287	18	0.77
heptadecane	17	$H—(CH_2)_{17}—H$	$C_{17}H_{36}$	303	23	0.76
octadecane	18	$H—(CH_2)_{18}—H$	$C_{18}H_{38}$	317	28	0.76
nonadecane	19	$H—(CH_2)_{19}—H$	$C_{19}H_{40}$	330	32	0.78
eicosane	20	$H—(CH_2)_{20}—H$	$C_{20}H_{42}$	343	37	0.79
triacontane	30	$H—(CH_2)_{30}—H$	$C_{30}H_{62}$	>450	66	0.81

[a]Densities are given in g/mL at 20 °C, except for methane and ethane, whose densities are given at their boiling points.

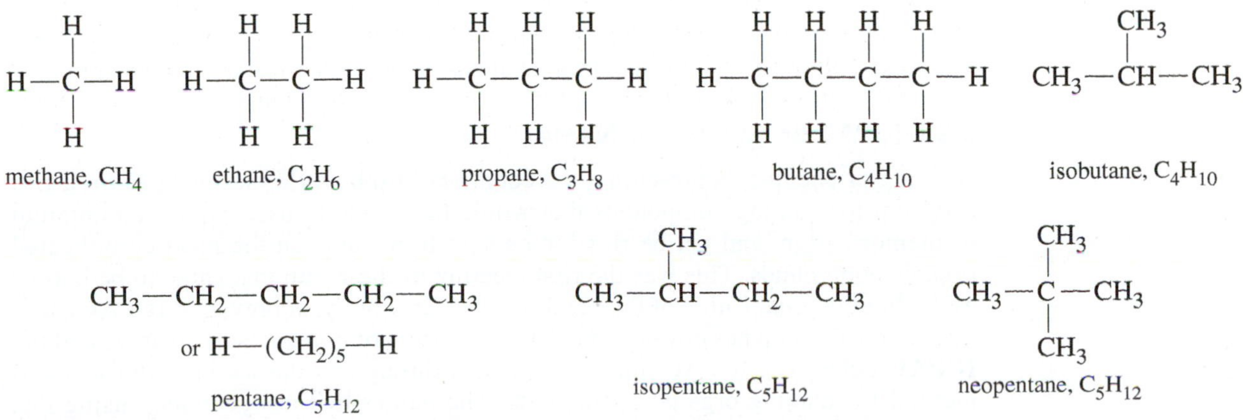

```
    H              H   H              H   H   H                H   H   H   H               CH3
    |              |   |              |   |   |                |   |   |   |               |
 H—C—H          H—C—C—H          H—C—C—C—H          H—C—C—C—C—H          CH3—CH—CH3
    |              |   |              |   |   |                |   |   |   |
    H              H   H              H   H   H                H   H   H   H
```

methane, CH_4 ethane, C_2H_6 propane, C_3H_8 butane, C_4H_{10} isobutane, C_4H_{10}

```
                                                  CH3                        CH3
                                                  |                          |
CH3—CH2—CH2—CH2—CH3                      CH3—CH—CH2—CH3              CH3—C—CH3
   or H—(CH2)5—H                                                            |
                                                                            CH3
```

pentane, C_5H_{12} isopentane, C_5H_{12} neopentane, C_5H_{12}

FIGURE 3-1
Examples of the general alkane molecular formula, C_nH_{2n+2}.

PROBLEM 3-1

Using the general molecular formula for alkanes:
(a) Predict the molecular formula of the C_{28} straight-chain alkane.
(b) Predict the molecular formula of the alkanes containing 44 carbon atoms with extensive branching.

3-3 Nomenclature of Alkanes

The names *methane, ethane, propane,* and *butane* have historical roots. From pentane on, alkanes are named using the Greek word for the number of carbon atoms, plus the suffix *-ane* to identify the molecule as an alkane. Table 3-2 gives the names and physical properties of the *n*-alkanes up to 20 carbon atoms.

3-3A Common Names

If all alkanes had unbranched (straight-chain) structures, their nomenclature would be simple. Most alkanes have structural isomers, however, and we need a way of naming all the different isomers. For example, there are two isomers of formula C_4H_{10}. The unbranched isomer is simply called *butane* (or *n-butane*, meaning "normal" butane), and the branched isomer is called *isobutane*, meaning an "isomer of butane."

$CH_3—CH_2—CH_2—CH_3$
butane (*n*-butane)

$CH_3—CH(CH_3)—CH_3$
isobutane

The three isomers of C_5H_{12} are called *pentane* (or *n-pentane*), *isopentane*, and *neopentane*.

$CH_3—CH_2—CH_2—CH_2—CH_3$
pentane (*n*-pentane)

$CH_3—CH(CH_3)—CH_2—CH_3$
isopentane

$CH_3—C(CH_3)_2—CH_3$
neopentane

Isobutane, isopentane, and *neopentane* are **common names** or **trivial names,** meaning historical names arising from common usage. Common names cannot easily describe the larger, more complicated molecules having many isomers, however. The number of isomers for any molecular formula grows rapidly as the number of carbon atoms increases. For example, there are 5 structural isomers of hexane, 18 isomers of octane, and 75 isomers of decane! We need a system of nomenclature that enables us to name complicated molecules without having to memorize hundreds of these historical common names.

3-3B IUPAC or Systematic Names

A group of chemists representing the countries of the world met in 1892 to devise a system for naming compounds that would be simple to use, require a minimum of memorization, and yet be flexible enough to name even the most complicated organic compounds. This was the first meeting of the group that came to be known as the International Union of Pure and Applied Chemistry, abbreviated **IUPAC.** This international group has developed a detailed system of nomenclature that we call the **IUPAC rules.** The IUPAC rules are accepted throughout the world as the standard method for naming organic compounds. The names that are generated using this system are called **IUPAC names** or **systematic names.** The IUPAC rules for naming organic compounds are summarized (for later reference) in Appendix 5, Summary of Organic Nomenclature.

The IUPAC system works consistently to name many different families of compounds. We will consider the naming of alkanes in detail, and later extend these rules to other kinds of compounds as we encounter them. The IUPAC system uses the longest chain of carbon atoms as the main chain, which is numbered to give the locations of side chains. Four rules govern this process.

Rule 1: The Main Chain The first rule of nomenclature gives the base name of the compound.

> Find the longest continuous chain of carbon atoms, and use the name of this chain as the base name of the compound.

For example, the longest chain of carbon atoms in the compound in the margin contains six carbons, so the compound is named as a *hexane* derivative. The longest chain is rarely drawn in a straight line; look carefully to find it.

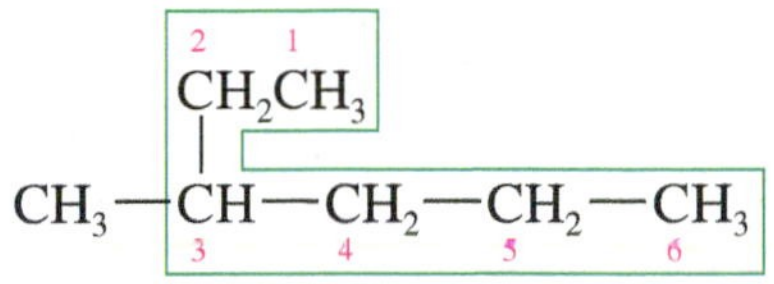

$CH_3-CH(CH_2CH_3)-CH_2-CH_2-CH_3$ (main chain numbered: CH_3 1, CH_2 2, CH 3, CH_2 4, CH_2 5, CH_3 6)

3-methyl*hexane*

The groups attached to the main chain are called **substituents** because they are substituted (in place of a hydrogen atom) on the main chain. *When there are two longest chains of equal length, use the chain with the greater number of substituents as the main chain.* The following compound contains two different seven-carbon chains and is named as a *heptane*. We choose the chain on the right as the main chain because it has more substituents (in red) attached to the chain.

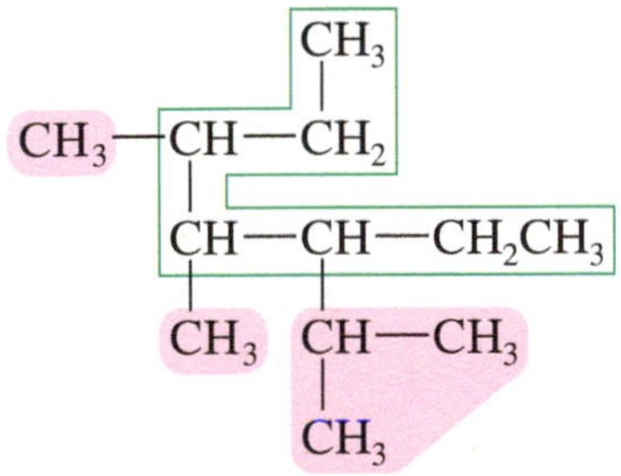

wrong
seven-carbon chain, but only three substituents

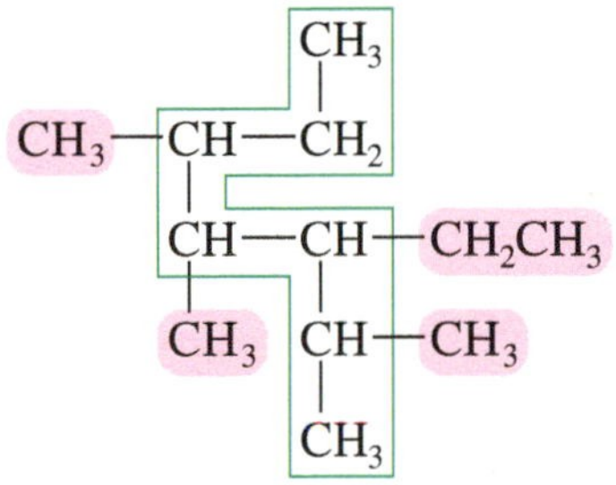

correct
seven-carbon chain, four substituents

PROBLEM-SOLVING HINT

When looking for the longest continuous chain (to give the base name), look for all the different chains of that length. Often, the longest chain with the most substituents is not obvious.

Rule 2: Numbering The Main Chain To give the locations of the substituents, assign a number to each carbon atom on the main chain.

> Number the longest chain, beginning with the end of the chain nearest a substituent.

We start the numbering from the end nearest a branch so that the numbers of the substituted carbons will be as low as possible. In the preceding heptane structure on the right, numbering from top to bottom gives the first branch at C3 (carbon atom 3), but numbering from bottom to top gives the first branch at C2. Numbering from bottom to top is correct. (If each end had a substituent the same distance in, we would start at the end nearer the second branch point.)

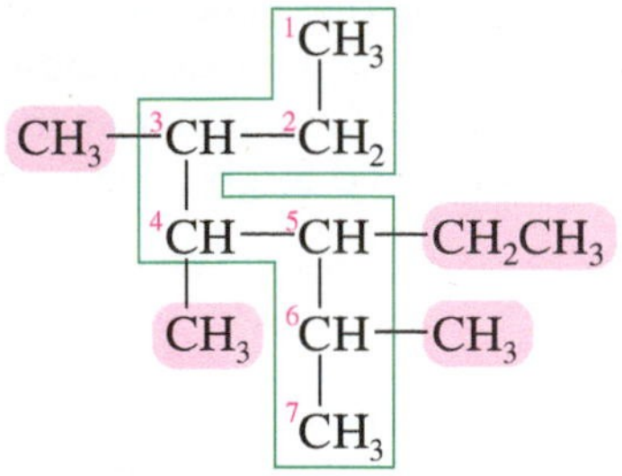

incorrect

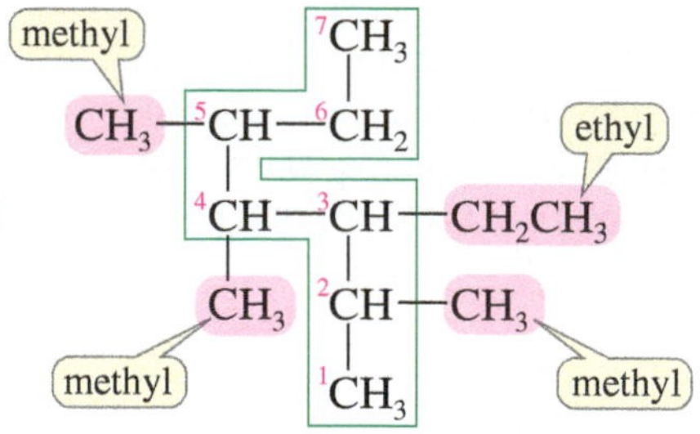

correct
3-ethyl-2,4,5-trimethylheptane

Rule 3: Naming Alkyl Groups Name the substituent groups.

> Name the substituent groups attached to the longest chain as **alkyl groups.** Give the location of each alkyl group by the number of the main-chain carbon atom to which it is attached.

Alkyl groups are named by replacing the *-ane* suffix of the alkane name with *-yl. Methane* becomes *methyl; ethane* becomes *ethyl.* You may encounter the word *amyl,* which is an archaic term for a pentyl (five-carbon) group.

Alkanes		Alkyl Groups	
CH_4	methane	CH_3-	methyl group
CH_3-CH_3	ethane	CH_3-CH_2-	ethyl group
$CH_3-CH_2-CH_3$	propane	$CH_3-CH_2-CH_2-$	propyl group
$CH_3-(CH_2)_2-CH_3$	butane	$CH_3-(CH_2)_2-CH_2-$	butyl group
$CH_3-(CH_2)_3-CH_3$	pentane	$CH_3-(CH_2)_3-CH_2-$	pentyl group (*n*-amyl group)

The following alkanes show the use of alkyl group nomenclature.

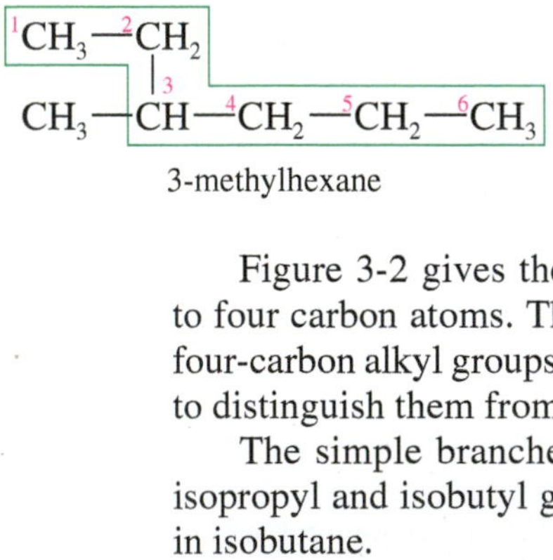

3-methylhexane

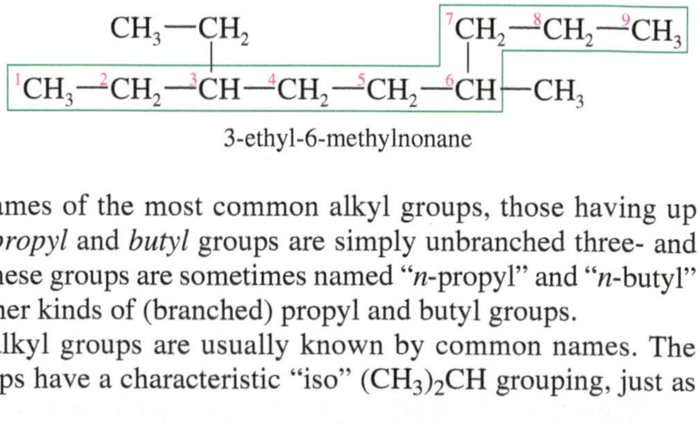

3-ethyl-6-methylnonane

Figure 3-2 gives the names of the most common alkyl groups, those having up to four carbon atoms. The *propyl* and *butyl* groups are simply unbranched three- and four-carbon alkyl groups. These groups are sometimes named "*n*-propyl" and "*n*-butyl" to distinguish them from other kinds of (branched) propyl and butyl groups.

The simple branched alkyl groups are usually known by common names. The isopropyl and isobutyl groups have a characteristic "iso" $(CH_3)_2CH$ grouping, just as in isobutane.

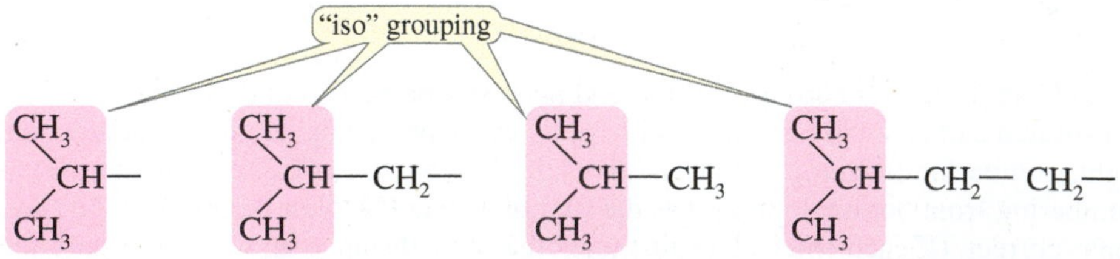

"iso" grouping

isopropyl group

isobutyl group

isobutane

isopentyl group
(isoamyl group)

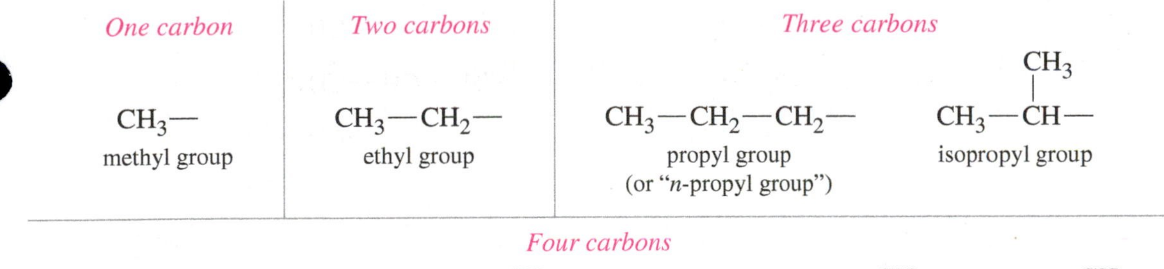

Four carbons

CH_3—CH_2—CH_2—CH_2—
butyl group
(or "*n*-butyl group")

CH_3—CH(CH_3)—CH_2—
isobutyl group

CH_3—CH_2—CH(CH_3)—
sec-butyl group

CH_3—C(CH_3)(CH_3)—
tert-butyl group
(or "*t*-butyl group")

FIGURE 3-2
Some common alkyl groups.

The names of the *secondary*-butyl (*sec*-butyl) and *tertiary*-butyl (*tert*-butyl or *t*-butyl) groups are based on the **degree of alkyl substitution** of the carbon atom attached to the main chain. In the *sec*-butyl group, the carbon atom bonded to the main chain is **secondary** (2°), or bonded to two other carbon atoms. In the *tert*-butyl group, it is **tertiary** (3°), or bonded to three other carbon atoms. In both the *n*-butyl group and the isobutyl group, the carbon atoms bonded to the main chain are **primary** (1°), bonded to only one other carbon atom.

R—C(H)(H)—
a *primary* (1°) carbon

R—C(R)(H)—
a *secondary* (2°) carbon

R—C(R)(R)—
a *tertiary* (3°) carbon

$CH_3CH_2CH_2$—C(H)(H)—
n-butyl group (1°)

CH_3CH_2—C(CH_3)(H)—
sec-butyl group (2°)

CH_3—C(CH_3)(CH_3)—
tert-butyl group (3°)

SOLVED PROBLEM 3-1

Give the structures of 4-isopropyloctane and 5-*tert*-butyldecane.

SOLUTION

The base name of 4-isopropyloctane is octane, so it has a main chain of eight carbons. There must be an isopropyl group on the fourth carbon from the end.

CH_3—CH—CH_3
|
CH_3—CH_2—CH_2—CH—CH_2—CH_2—CH_2—CH_3
4-isopropyloctane

The base name of 5-*tert*-butyldecane is decane, so it has a main chain of ten carbons. There must be a *tert*-butyl group on the fifth carbon from the end.

(continued)

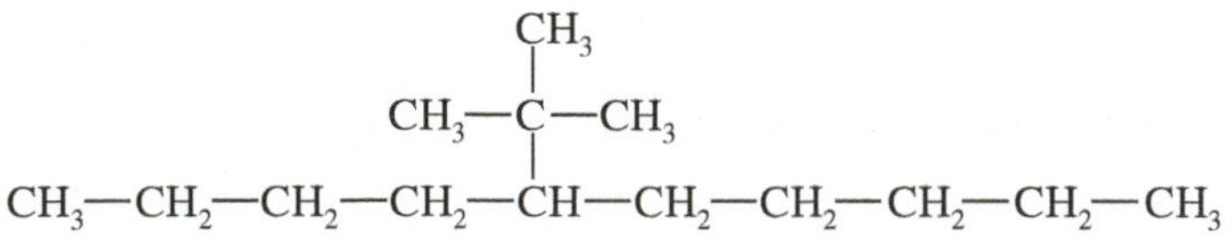

$$\begin{array}{c} CH_3 \\ | \\ CH_3-C-CH_3 \\ | \\ CH_3-CH_2-CH_2-CH_2-CH-CH_2-CH_2-CH_2-CH_2-CH_3 \end{array}$$

5-*tert*-butyldecane

Haloalkanes can be named just like alkanes, with the halogen atom treated as a substituent. Halogen substituents are named *fluoro-*, *chloro-*, *bromo-*, and *iodo-*.

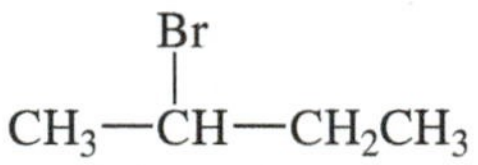

$CH_3-CH(Br)-CH_2CH_3$
2-bromobutane

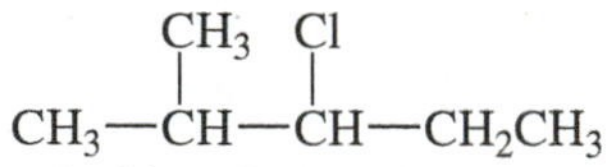

$CH_3-CH(CH_3)-CH(Cl)-CH_2CH_3$
3-chloro-2-methylpentane

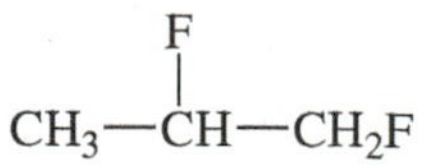

$CH_3-CH(F)-CH_2F$
1,2-difluoropropane

PROBLEM-SOLVING HINT

Please remember the alkyl groups in Figure 3-2. You will encounter them many times throughout this course.

PROBLEM 3-2

Name the following alkanes and haloalkanes. When two or more substituents are present, list them in alphabetical order.

(a) $CH_3-CH(CH_2-CH_3)-CH_2-CH_3$

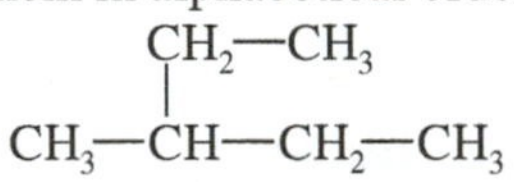

(b) $CH_3-CH(Br)-CH(CH_2CH_3)-CH_3$

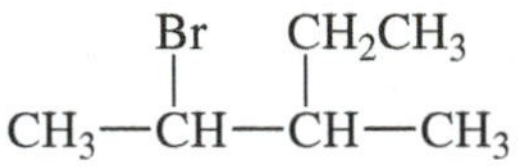

(c) $CH_3-CH_2-CH(CH_2-CH_3)-CH(CH_2CH(CH_3)_2)-CH_2-CH_2-CH_3$

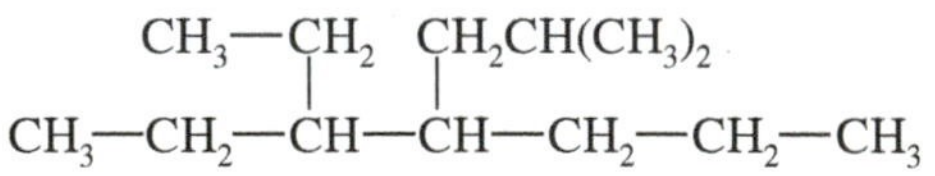

(d) $CH_3-CH_2-CH_2-CH_2-CH_2-CH_2-CH(CH_2-CH(CH_3)-CH_3)-CH(CH_3)-CH_3$

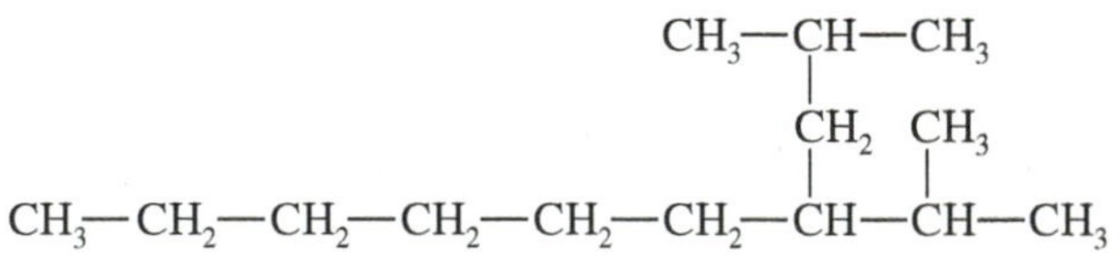

RULE 4: ORGANIZING MULTIPLE GROUPS The final rule deals with naming compounds with more than one substituent.

When two or more substituents are present, list them in alphabetical order. When two or more of the *same* alkyl substituent are present, use the prefixes *di-*, *tri-*, *tetra-*, etc., to avoid having to name the alkyl group twice. Include a position number for each substituent, even if it means repeating a number more than once.

di- means 2	*penta-* means 5	*octa-* means 8
tri- means 3	*hexa-* means 6	*nona-* means 9
tetra- means 4	*hepta-* means 7	*deca-* means 10

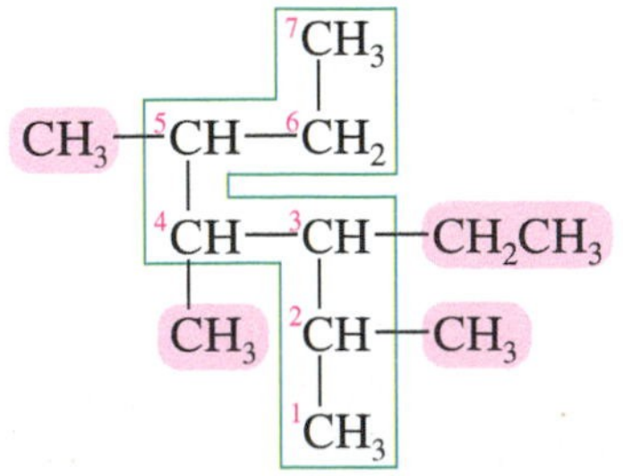

$$\begin{array}{l} \quad\quad\quad\;\; {}^7CH_3 \\ \quad\quad\quad\quad\;\; | \\ CH_3-{}^5CH-{}^6CH_2 \\ \quad\quad\;\; | \\ \quad\;\; {}^4CH-{}^3CH-CH_2CH_3 \\ \quad\quad\;\; | \quad\quad\; | \\ \quad\;\; CH_3\;\; {}^2CH-CH_3 \\ \quad\quad\quad\quad\;\; | \\ \quad\quad\quad\;\; {}^1CH_3 \end{array}$$

3-ethyl-2,4,5-trimethylheptane

Using this rule, we can construct names for some complicated structures. Let's finish naming the heptane shown again here in the margin. This compound has an ethyl group on C3 and three methyl groups on C2, C4, and C5. List the ethyl group alphabetically before the methyl groups, and give each of the four substituents a location number.

SOLVED PROBLEM 3-2

Give a systematic (IUPAC) name for the following compound.

```
                CH3
                |
                CH—CH3   CH2CH3
                |        |
CH3—CH—CH—CH2—CH—CH3
     |
CH3—C—CH3
     |
     CH3
```

SOLUTION

The longest carbon chain contains eight carbon atoms, so this compound is named as an octane. Numbering from left to right gives the first branch on C2; numbering from right to left gives the first branch on C3, so we number from left to right.

```
                CH3
                |
                CH—CH3   7CH2CH3 (8)
                |        |
CH3—3CH—4CH—5CH2—6CH—CH3
     |
1CH3—2C—CH3
     |
     CH3
```

There are four methyl groups: two on C2, one on C3, and one on C6. These four groups will be listed as "2,2,3,6-tetramethyl . . . ." There is an isopropyl group on C4. Listing the isopropyl group and the methyl groups alphabetically, we have

4-isopropyl-2,2,3,6-tetramethyloctane

PROBLEM-SOLVING HINT

When substituents are alphabetized, *iso-* is used as part of the alkyl group name, but the hyphenated prefixes are not. Thus, *isobutyl* is alphabetized with *i*, but *n*-butyl, *tert*-butyl, and *sec*-butyl are alphabetized with *b*. The number prefixes *di-*, *tri-*, *tetra-*, etc., are ignored in alphabetizing.

SUMMARY Rules for Naming Alkanes

To name an alkane, we follow four rules:

1. Find the longest continuous chain of carbon atoms, and use this chain as the base name.
2. Number the longest chain, beginning with the end nearest a branch.
3. Name the substituents on the longest chain (as alkyl groups). Give the location of each substituent by the number of the main-chain carbon atom to which it is attached.
4. When two or more substituents are present, list them in alphabetical order. When two or more of the *same* alkyl substituent are present, use the prefixes *di-*, *tri-*, *tetra-*, etc. (ignored in alphabetizing) to avoid having to name the alkyl group twice.

PROBLEM 3-3

Write structures for the following compounds.

(a) 3-ethyl-4-methylhexane
(b) 3-ethyl-5-isobutyl-3-methylnonane
(c) 4-*tert*-butyl-2-methylheptane
(d) 5-isopropyl-3,3,4-trimethyloctane

PROBLEM-SOLVING HINT

Always compare the total number of carbon atoms in the name with the number in the structure to make sure they match. For example, an *isopropyldimethyloctane* should have 3 + 2 + 8 carbon atoms.

PROBLEM 3-4

Provide IUPAC names for the following compounds.

(a) $(CH_3)_2CHCH_2CH_3$

(b) $CH_3-C(CH_3)_2-CH_3$

(c) $CH_3CH_2CH_2CH(CH_2CH_3)-CH(CH_3)_2$

(d) $CH_3-CH(CH_3)-CH_2-CH(CH_2CH_3)-CH_3$

(e) $CH_3CH_2CH(C(CH_3)_3)CH(CH(CH_3)_2)CH_3$

(f) $(CH_3)_3C-CH(CH(CH_3)CH_2CH_3)-CH_2CH_2CH_3$

PROBLEM-SOLVING HINT

To draw all alkanes of a given molecular formula, start with a straight chain, and then go to shorter chains with more branches. To draw all C_7H_{16} alkanes, for example, start with a C_7 chain, and then go to a C_6 chain with a methyl group in each possible position. Next, draw a C_5 chain with an ethyl group in each possible position, followed by two methyls in all possible combinations. Finally, draw a C_4 chain, which offers only one way to substitute three carbon atoms without extending the longest chain. To guard against duplicated structures, name each compound, and if a name appears twice, you have either duplicated a structure or named something wrong.

PROBLEM 3-5

All of the following names are incorrect or incomplete. In each case, draw the structure (or a possible structure) and name it correctly.

(a) 3-ethyl-4-methylpentane
(b) 2-ethyl-3-methylpentane
(c) 3-dimethylhexane
(d) 4-isobutylheptane
(e) 2-bromo-3-ethylbutane
(f) 2,3-diethyl-5-isopropylheptane

PROBLEM 3-6

Give structures and names for

(a) the five isomers of C_6H_{14}
(b) the nine isomers of C_7H_{16}

Complex Substituents Complex alkyl groups are named by a systematic method using the longest alkyl chain as the base alkyl group. The base alkyl group is numbered beginning with the carbon atom (the "head carbon") bonded to the main chain. The substituents on the base alkyl group are listed with appropriate numbers, and parentheses are used to set off the name of the complex alkyl group. The following examples illustrate the systematic method for naming complex alkyl groups.

$-\overset{1}{C}H(CH_2CH_3)-\overset{2}{C}H(CH_3)-\overset{3}{C}H_3$

a (1-ethyl-2-methylpropyl) group

$-\overset{1}{C}(CH_3)_2-\overset{2}{C}H_2-\overset{3}{C}H(CH_3)-\overset{4}{C}H_3$

a (1,1,3-trimethylbutyl) group

$\overset{1}{C}H_3\overset{2}{C}H_2-\overset{3}{C}H(CH_2CH_3)-\overset{4}{C}H_2-\overset{5}{C}H[\overset{1}{C}H(CH_2CH_3)-\overset{2}{C}H(CH_3)-\overset{3}{C}H_3]-\overset{6}{C}H_2\overset{7}{C}H_2\overset{8}{C}H_2\overset{9}{C}H_3$

3-ethyl-5-(1-ethyl-2-methylpropyl)nonane

1,1-dimethyl-3-(1,1,3-trimethylbutyl)cyclooctane

PROBLEM 3-7

Draw the structures of the following groups, and give their more common names.

(a) the (1-methylethyl) group
(b) the (2-methylpropyl) group
(c) the (1-methylpropyl) group
(d) the (1,1-dimethylethyl) group
(e) the (3-methylbutyl) group, sometimes called the "isoamyl" group

PROBLEM 3-8

Draw the structures of the following compounds.
(a) 4-(1,1-dimethylethyl)octane **(b)** 5-(1,2,2-trimethylpropyl)nonane
(c) 3,3-diethyl-4-(2,2-dimethylpropyl)octane

PROBLEM 3-9

Without looking at the structures, give molecular formulas for the compounds in Problem 3-8 (a) and (b). Use the names of the groups to determine the number of carbon atoms; then use the $(2n+2)$ rule.

3-4 Physical Properties of Alkanes

Alkanes are used primarily as fuels, solvents, and lubricants. Natural gas, gasoline, kerosene, heating oil, lubricating oil, and paraffin "wax" are all composed primarily of alkanes, with different physical properties resulting from different ranges of molecular weights.

3-4A Solubilities and Densities of Alkanes

Alkanes are nonpolar, so they dissolve in nonpolar or weakly polar organic solvents. Alkanes are said to be **hydrophobic** ("water-hating") because they do not dissolve in water. Alkanes are good lubricants and preservatives for metals because they keep water from reaching the metal surface and causing corrosion.

Densities of the *n*-alkanes are listed in Table 3-2 (page 109). Alkanes have densities around 0.7 g/mL, compared with a density of 1.0 g/mL for water. Because alkanes are less dense than water and insoluble in water, a mixture of an alkane (such as gasoline or oil) and water quickly separates into two phases, with the alkane on top.

3-4B Boiling Points of Alkanes

Table 3-2 also gives the boiling points and melting points of the unbranched alkanes. The boiling points increase smoothly with increasing numbers of carbon atoms and increasing molecular weights. Larger molecules have larger surface areas, resulting in increased intermolecular van der Waals attractions. These increased attractions must be overcome for vaporization and boiling to occur. Thus, a larger molecule, with greater surface area and greater van der Waals attractions, boils at a higher temperature.

A graph of *n*-alkane boiling points versus the number of carbon atoms (the blue line in Figure 3-3) shows that boiling points increase with increasing molecular weight. Each additional CH_2 group increases the boiling point by about 30 °C up to about ten carbons, and by about 20 °C in higher alkanes.

The green line in Figure 3-3 represents the boiling points of some branched alkanes. In general, a branched alkane boils at a lower temperature than the *n*-alkane with the same number of carbon atoms. This difference in boiling points arises because branched alkanes are more compact, with less surface area for London force interactions.

Oil floats on water. Note how the oil (from the 2010 blowout of the Macondo well in the Gulf of Mexico) floats on top of the water, where it can be burned off. Oil recovery booms, containing nonpolar fibers, are used to soak up and contain the floating oil. Note how most of the burning oil ends at the oil recovery boom.

3-4C Melting Points of Alkanes

The blue line in Figure 3-4 is a graph of the melting points of the *n*-alkanes. Like their boiling points, the melting points increase with increasing molecular weight. The melting point graph is not smooth, however. Alkanes with even numbers of carbon atoms pack better into a solid structure, so higher temperatures are needed to melt them. Alkanes with odd numbers of carbon atoms do not pack as well, and they melt at lower temperatures. The sawtooth-shaped graph of melting points is smoothed by drawing separate lines (green and red) for the alkanes with even and odd numbers of carbon atoms.

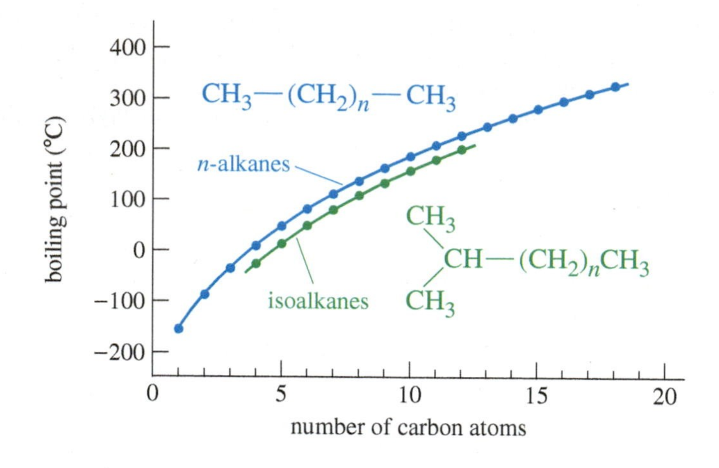

FIGURE 3-3
Alkane boiling points. The boiling points of the unbranched alkanes (blue) are compared with those of some branched alkanes (green). Because of their smaller surface areas, branched alkanes have lower boiling points than unbranched alkanes.

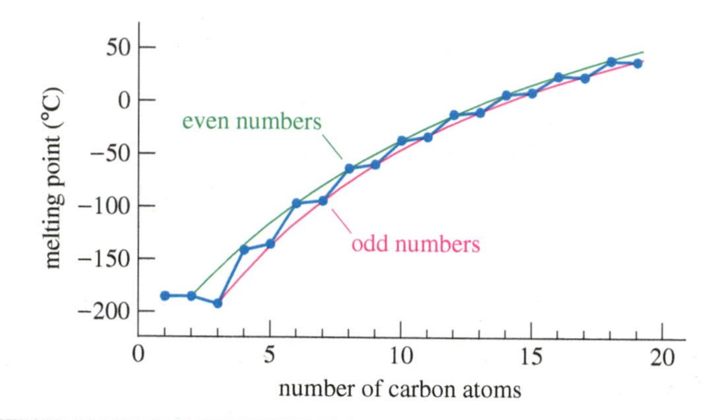

FIGURE 3-4
Alkane melting points. The melting-point curve for *n*-alkanes with even numbers of carbon atoms is slightly higher than the curve for alkanes with odd numbers of carbons.

PROBLEM 3-10

List each set of compounds in order of increasing boiling point.

(a) hexane, octane, and decane

(b) octane, $(CH_3)_3C—C(CH_3)_3$, and $CH_3CH_2C(CH_3)_2CH_2CH_2CH_3$

3-5 Uses and Sources of Alkanes

Distillation of petroleum separates alkanes into fractions with similar boiling points. These fractions are suited for different uses based on their physical properties, such as volatility and viscosity.

3-5A Major Uses of Alkanes

C_1–C_2 Methane and ethane are gases at room temperature and atmospheric pressure. Methane is the major constituent of *natural gas,* with a variable amount of ethane present as well. Natural gas is widely used as a clean-burning fuel for heating and power generation.

Both methane and ethane are difficult to liquefy, so they are usually handled as compressed gases. Upon cooling to cryogenic (very low) temperatures, however, methane and ethane become liquids. *Liquefied natural gas,* mostly methane, can be transported in special refrigerated tankers more easily than it can be transported as a compressed gas.

C_3–C_4 Propane and butane are also gases at room temperature and pressure, but they are easily liquefied at room temperature under modest pressure. These gases, often obtained along with liquid petroleum, are stored in low-pressure cylinders of *liquefied petroleum gas (LPG)*. Propane and butane are good fuels, both for heating and for internal combustion engines. They burn cleanly, and pollution-control equipment is rarely necessary. In many agricultural areas, propane and butane are more cost-effective tractor fuels than gasoline and diesel fuel. Propane and butane have largely replaced freons (see Section 6-3D) as propellants in aerosol cans. Unlike alkanes, the chlorofluorocarbon Freon propellants are implicated in damaging Earth's protective ozone layer.

Clean-burning vehicles powered by natural gas help to reduce air pollution in urban areas.

C_5–C_8 The next four alkanes are free-flowing, volatile liquids. Isomers of pentane, hexane, heptane, and octane are the primary constituents of gasoline. Their volatility is crucial for this use because the injection system simply squirts a stream of gasoline into the intake air as it rushes through. If gasoline did not evaporate easily, it would reach the cylinder in the form of droplets. Droplets cannot burn as efficiently as a vapor, so the engine would smoke and give low mileage.

In addition to being volatile, gasoline must resist the potentially damaging explosive combustion known as *knocking*. The antiknock properties of gasoline are rated by an **octane number** that is assigned by comparing the gasoline to a mixture of *n*-heptane (which knocks badly) and isooctane (2,2,4-trimethylpentane, which is not prone to knocking). The gasoline being tested is used in a test engine with a variable compression ratio. Higher compression ratios induce knocking, so the compression ratio is increased until knocking begins. Tables are available that show the percentage of isooctane in an isooctane/heptane blend that begins to knock at any given compression ratio. The octane number assigned to the gasoline is simply the percentage of isooctane in an isooctane/heptane mixture that begins to knock at that same compression ratio.

Application: Fuels

Tetraethyl lead (TEL), formula $(CH_3CH_2)_4Pb$, was once added to gasoline to increase the octane rating and lubricate the valves. In the 1970s, lead was banned from automotive gasoline because it inactivates catalytic converters and introduces lead into the environment. TEL is still used in 100LL, which is low-lead 100-octane aviation fuel for piston aircraft engines. No suitable replacement fuel has yet been certified for the old engines.

$CH_3CH_2CH_2CH_2CH_2CH_2CH_3$

n-heptane (0 octane)
prone to knocking

$CH_3-C(CH_3)_2-CH_2-CH(CH_3)-CH_3$

2,2,4-trimethylpentane (100 octane)
"isooctane," resists knocking

C_9–C_{16} The nonanes (C_9) through about the hexadecanes (C_{16}) are higher-boiling liquids that are somewhat viscous. These alkanes are used in kerosene, jet fuel, and diesel fuel. **Kerosene,** the lowest-boiling of these fuels, is less volatile than gasoline and less prone to forming explosive mixtures. Kerosene has been used in kerosene lamps and heaters, which use wicks to allow this heavier fuel to burn. Jet fuel is similar to kerosene, but more highly refined and less odorous.

Diesel fuel is not very volatile, so it does not evaporate in the intake air entering an engine. In a diesel engine, the fuel is sprayed directly into the cylinder right at the top of the compression stroke. The hot, highly compressed air in the cylinder causes the fuel to burn quickly, swirling and vaporizing as it burns. Some of the alkanes in diesel fuel have fairly high freezing points, and they may solidify in cold weather. This partial solidification causes the diesel fuel to turn into a waxy, semisolid mass. Owners of diesel engines in cold climates often mix a small amount of gasoline with their diesel fuel in the winter. The added gasoline dissolves the frozen alkanes, diluting the slush and allowing it to be pumped to the cylinders.

Application: Air Pollution

Incomplete combustion of gasoline and other motor fuels releases significant quantities of volatile organic compounds (VOCs) into the atmosphere. VOCs are composed of short-chained alkanes, alkenes, aromatic compounds, and a variety of other hydrocarbons. VOCs are components of air pollution and contribute to cardiac and respiratory diseases.

C_{16} and Up Alkanes with more than 16 carbon atoms are most often used as lubricating and heating oils. These are sometimes called "mineral" oils because they come from petroleum, which was once considered a mineral.

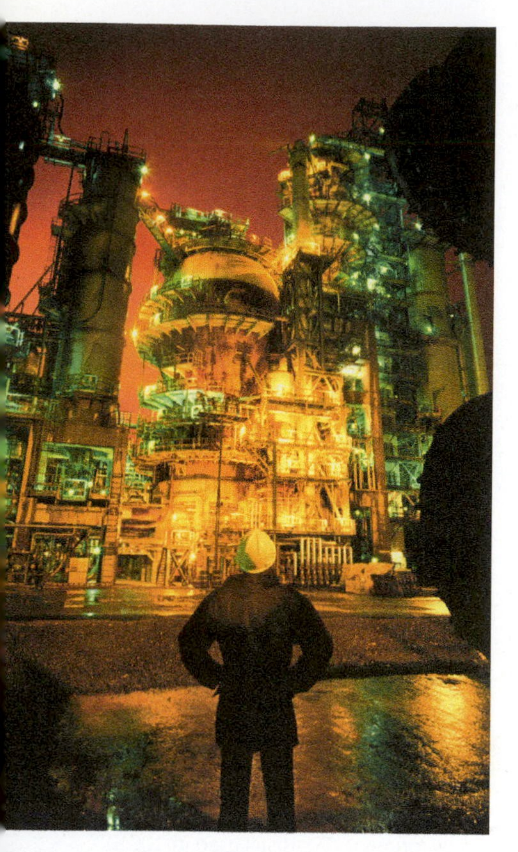

The large distillation tower at left is used to separate petroleum into fractions based on their boiling points. The "cat cracker" at right uses catalysts and high temperatures to crack large molecules into smaller ones.

Paraffin "wax" is not a true wax, but a purified mixture of high-molecular-weight alkanes with melting points well above room temperature. The true waxes are long-chain esters, discussed in Chapter 25.

3-5B Alkane Sources; Petroleum Refining

Alkanes are derived mostly from petroleum and petroleum by-products. *Petroleum*, often called *crude oil*, is pumped from wells that reach into pockets containing the remains of prehistoric plants. The principal constituents of crude oil are alkanes, some aromatics, and some undesirable compounds containing sulfur and nitrogen. The composition of petroleum and the amounts of contaminants vary from one source to another, and a refinery must be carefully adjusted to process a particular type of crude oil. Because of their different qualities, different prices are paid for light Arabian crude, West Texas crude, and other classes of crude petroleum.

The first step in refining petroleum is a careful fractional distillation. The products of that distillation are not pure alkanes, but mixtures of alkanes with useful ranges of boiling points. Table 3-3 shows the major fractions obtained from the distillation of crude petroleum.

After distillation, **catalytic cracking** converts some of the less valuable fractions to more valuable products. Catalytic cracking involves heating alkanes in the presence of materials that catalyze the cleavage of large molecules into smaller ones. Cracking is often used to convert higher-boiling fractions into mixtures that can be blended with gasoline. When cracking is done in the presence of hydrogen (**hydrocracking**), the product is a mixture of alkanes free of sulfur and nitrogen impurities. The following reaction shows the catalytic hydrocracking of a molecule of tetradecane into two molecules of heptane.

$$CH_3-(CH_2)_{12}-CH_3 + H_2 \xrightarrow[SiO_2 \text{ or } Al_2O_3 \text{ catalyst}]{\text{heat}} 2\,CH_3-(CH_2)_5-CH_3$$

3-5C Natural Gas; Methane

Natural gas was once treated as a waste product of petroleum production and destroyed by flaring it off. Now natural gas is an equally valuable natural resource, pumped and stored throughout the world. Natural gas is about 70% methane, 10% ethane, and 15% propane, depending on the source of the gas. Small amounts of other hydrocarbons and contaminants are also present. Natural gas is often found above pockets of petroleum or coal, although it is also found in places where there is little or no recoverable petroleum or coal. Natural gas is used primarily as a fuel to heat buildings and to generate electricity. It is also important as a starting material for the production of fertilizers.

TABLE 3-3
Major Fractions Obtained from Distillation of Crude Petroleum

Boiling Range (°C)	Number of Carbons	Fraction	Use
under 30°	2–4	petroleum gas	LP gas for heating
30°–180°	4–9	gasoline	motor fuel
160°–230°	8–16	kerosene	heating, jet fuel
200°–320°	10–18	diesel	motor fuel
300°–450°	16–30	heavy oil	heating, lubrication
>300° (vacuum)	>25	petroleum "jelly,"	
		paraffin "wax"	
residue	>35	asphalt	

Although the methane we burn as natural gas is millions of years old, another 300 million tons per year (estimated) of new methane is synthesized by microbes in diverse places such as the stomachs of plant-eating animals and the mud under the seafloor. Most of the undersea methane is eaten by other microbes, but some escapes at methane seeps. Under the sea, cold, high-pressure conditions may allow formation of **methane hydrate,** with individual methane molecules trapped inside cages of water molecules. When methane hydrate is brought to the surface, it quickly melts and the methane escapes. Methane is sometimes generated on a small scale from manure and landfill waste, but we have no practical methods for capturing and using naturally-occuring microbial methane or methane hydrate. Much of this methane escapes to the atmosphere, where it acts as a greenhouse gas and contributes to global warming.

Methane hydrate, consisting of methane molecules surrounded by water molecules, is formed under high pressure on the cold seafloor. When brought to the surface, it quickly melts and releases the methane. In this photo, the methane has been ignited to make it visible.

3-6 Reactions of Alkanes

Alkanes are the least reactive class of organic compounds. Their low reactivity is reflected in another term for alkanes: **paraffins.** The name *paraffin* comes from two Latin terms, *parum*, meaning "too little," and *affinis*, meaning "affinity." Chemists found that alkanes do not react with strong acids or bases or with most other reagents. They attributed this low reactivity to a lack of affinity for other reagents, so they coined the name "paraffins."

Most useful reactions of alkanes take place under energetic or high-temperature conditions. These conditions are inconvenient in a laboratory because they require specialized equipment, and the rate of the reaction is difficult to control. Alkane reactions often form mixtures of products that are difficult to separate. These mixtures may be commercially important for an industry, however, where the products may be separated and sold separately. Newer methods of selective functionalization may eventually change this picture. For now, however, the following alkane reactions are rarely seen in laboratory applications, but they are widely used in the chemical industry and even in your home and car.

3-6A Combustion

Combustion is a rapid oxidation that takes place at high temperatures, converting alkanes to carbon dioxide and water. Little control over the reaction is possible, except for moderating the temperature and controlling the fuel/air ratio to achieve efficient burning.

$$C_nH_{(2n+2)} \quad + \quad \text{excess } O_2 \quad \xrightarrow{\text{heat}} \quad n\,CO_2 \quad + \quad (n+1)\,H_2O$$

Example

$$CH_3CH_2CH_3 \quad + \quad 5\,O_2 \quad \xrightarrow{\text{heat}} \quad 3\,CO_2 \quad + \quad 4\,H_2O$$

Application: Global Warming

Methane is a stronger greenhouse gas than carbon dioxide. If all the methane trapped in methane hydrates were suddenly released into the atmosphere (by warming of the ocean, for example), the rate of global warming could increase dramatically.

Unfortunately, the burning of gasoline and fuel oil pollutes the air and depletes the petroleum resources needed for lubricants and chemical feedstocks. Solar and nuclear heat sources cause less pollution, and they do not deplete these important natural resources. Facilities that use these more environment-friendly heat sources are currently more expensive than those that rely on the combustion of alkanes.

3-6B Cracking and Hydrocracking

As discussed in Section 3-5B, catalytic **cracking** of large hydrocarbons at high temperatures produces smaller hydrocarbons. The cracking process usually operates under conditions that give the maximum yields of gasoline. In **hydrocracking,** hydrogen is added to give saturated hydrocarbons; cracking without hydrogen gives mixtures of alkanes and alkenes.

Combustion is the most common reaction of alkanes. Lightning initiated this fire in a tank containing 3 million gallons of gasoline at the Shell Oil storage facility in Woodbridge, New Jersey (June 11, 1996).

Catalytic hydrocracking

$C_{12}H_{26}$ long-chain alkane $\xrightarrow[\text{catalyst}]{H_2,\ \text{heat}}$ C_5H_{12} + C_7H_{16} shorter-chain alkanes

Catalytic cracking

$C_{12}H_{26}$ long-chain alkane $\xrightarrow[\text{catalyst}]{\text{heat}}$ C_5H_{10} + C_7H_{16} shorter-chain alkanes and alkenes

3-6C Halogenation

Alkanes can react with halogens (F_2, Cl_2, Br_2, I_2) to form alkyl halides. For example, methane reacts with chlorine (Cl_2) to form chloromethane (methyl chloride), dichloromethane (methylene chloride), trichloromethane (chloroform), and tetrachloromethane (carbon tetrachloride).

$$CH_4 + Cl_2 \xrightarrow{\text{heat or light}} CH_3Cl + CH_2Cl_2 + CHCl_3 + CCl_4 + HCl$$

Heat or light is usually needed to initiate this **halogenation.** Reactions of alkanes with chlorine and bromine proceed at moderate rates and are easily controlled. Reactions with fluorine are often too fast to control, however. Iodine reacts very slowly or not at all. We will discuss the halogenation of alkanes in Chapter 4.

3-7 Structure and Conformations of Alkanes

Although alkanes are not as reactive as other classes of organic compounds, they have many of the same structural characteristics. We will use simple alkanes as examples to study some of the properties of organic compounds, including the structure of sp^3 hybridized carbon atoms and properties of C—C and C—H single bonds.

3-7A Structure of Methane

The simplest alkane is *methane*, CH_4. Methane is perfectly tetrahedral, with the 109.5° bond angles predicted for an sp^3 hybridized carbon. Four hydrogen atoms are covalently bonded to the central carbon atom, with bond lengths of 1.09 Å.

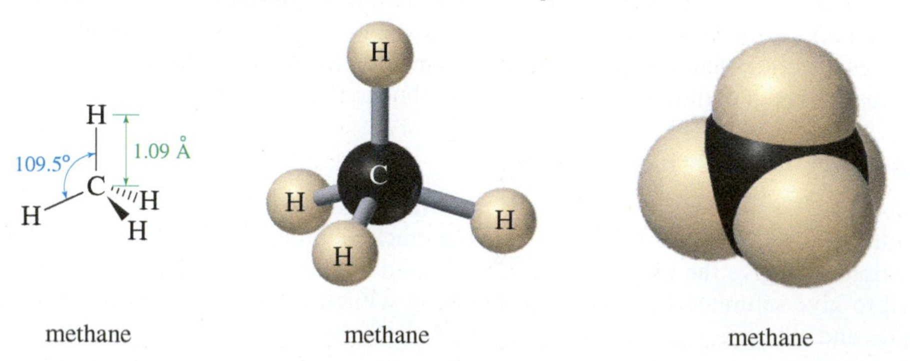

3-7B Conformations of Ethane

Ethane, the two-carbon alkane, is composed of two methyl groups with overlapping sp^3 hybrid orbitals forming a sigma bond between them.

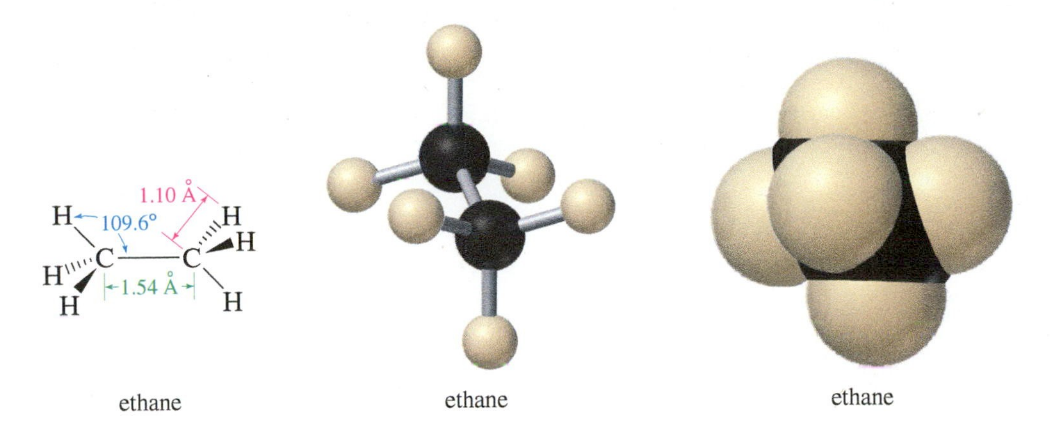

The two methyl groups are not fixed in a single position, but are relatively free to rotate about the sigma bond connecting the two carbon atoms. The bond maintains its linear bonding overlap as the carbon atoms turn. The different arrangements formed by rotations about a single bond are called **conformations,** and a specific conformation is called a **conformer** ("**conform**ational iso**mer**").* Pure conformers cannot be isolated in most cases, because the molecules are constantly rotating through all the possible conformations.

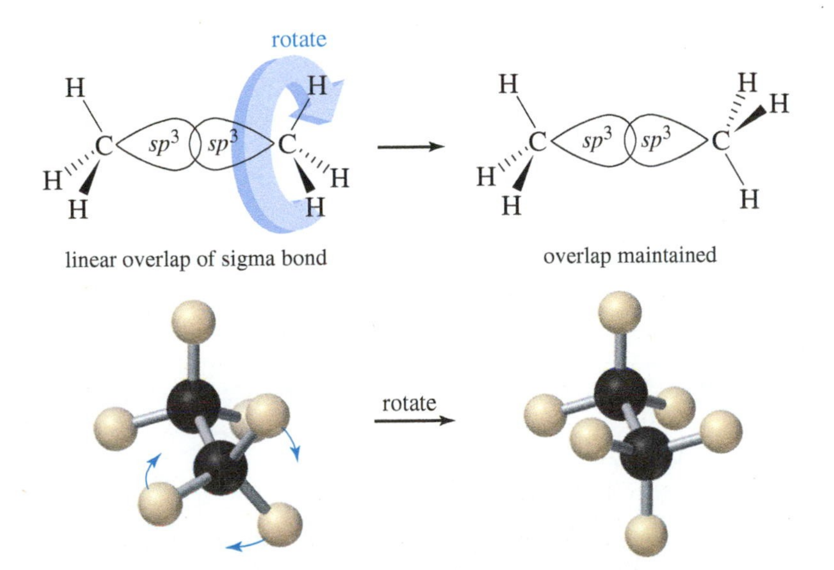

In drawing conformations, we often use **Newman projections,** a way of drawing a molecule looking straight down the bond connecting two carbon atoms (Figure 3-5). The front carbon atom is represented by three lines (three bonds) coming together in a Y shape. The back carbon is represented by a circle with three bonds pointing out from it. Until you become familiar with Newman projections, you should make models and compare your models with the drawings.

An infinite number of conformations are possible for ethane, because the angle between the hydrogen atoms on the front and back carbons can take on an infinite number of values. Figure 3-6 uses Newman projections and sawhorse structures to

*This is the common definition of conformers. The IUPAC definition also requires that a conformer correspond to a distinct potential energy minimum, such as the anti and gauche conformations of butane.

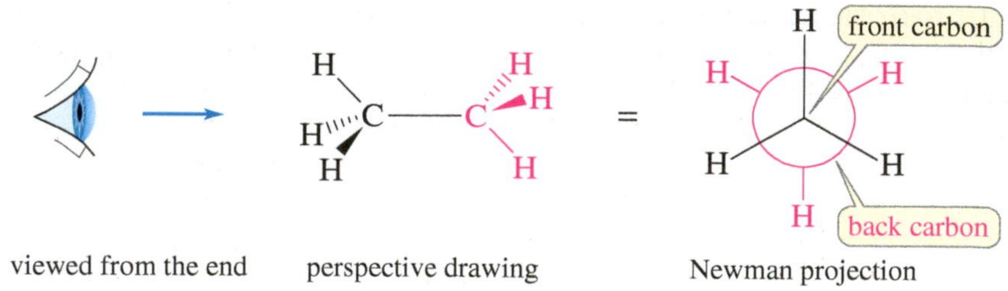

FIGURE 3-5
The Newman projection looks straight down the carbon–carbon bond.

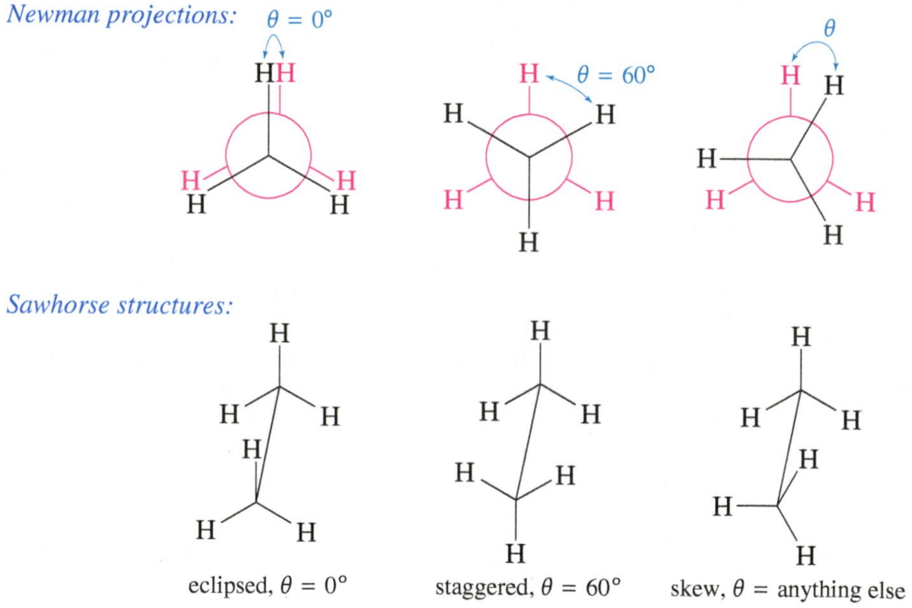

viewed from the end — perspective drawing — Newman projection (front carbon, back carbon)

FIGURE 3-6
Ethane conformations. The eclipsed conformation has a dihedral angle $\theta = 0°$, and the staggered conformation has $\theta = 60°$. Any other conformation is called a skew conformation.

Newman projections: $\theta = 0°$; $\theta = 60°$; θ

Sawhorse structures:

eclipsed, $\theta = 0°$ — staggered, $\theta = 60°$ — skew, θ = anything else

PROBLEM 3-11 (PARTIALLY SOLVED)

Draw Newman projections of the following molecules viewed from the direction of the blue arrows.

(a) (b) 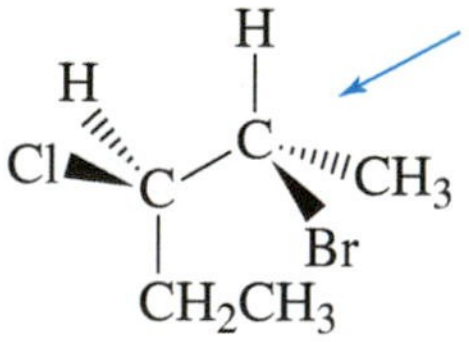 (c)

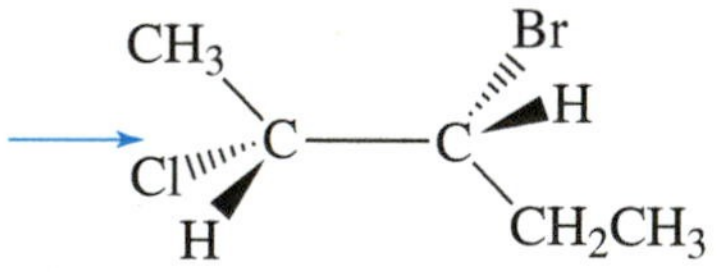

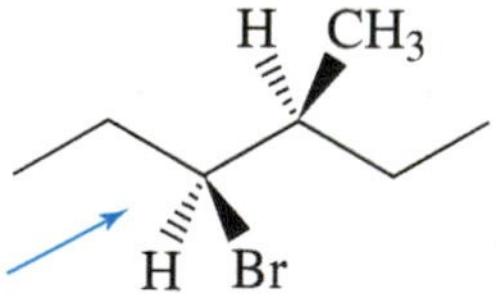

SOLUTION (a)

Looking from the direction of the arrow, the front carbon has a methyl group pointing straight up, a chlorine pointing down and to the left, and a hydrogen pointing down and to the right. The back carbon has an ethyl group pointing straight down, a bromine pointing up and to the left, and a hydrogen pointing up and to the right.

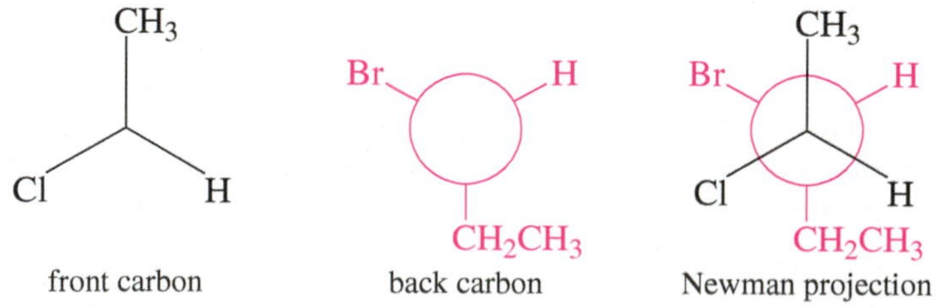

front carbon — back carbon — Newman projection

illustrate some of these ethane conformations. **Sawhorse structures** picture the molecule looking down at an angle toward the carbon–carbon bond. Sawhorse structures can be misleading, depending on how the eye sees them. We will generally use perspective or Newman projections to draw molecular conformations.

Any conformation can be specified by its **dihedral angle** (θ), the angle between the C—H bonds on the front carbon atom and the C—H bonds on the back carbon in the Newman projection. Two of the conformations have special names. The conformation with $\theta = 0°$ is called the **eclipsed conformation** because the Newman projection shows the hydrogen atoms on the back carbon to be hidden (eclipsed) by those on the front carbon. The **staggered conformation,** with $\theta = 60°$, has the hydrogen atoms on the back carbon staggered halfway between the hydrogens on the front carbon. Any other intermediate conformation is called a **skew conformation.**

In a sample of ethane gas at room temperature, the ethane molecules rotate millions of times per second, and their conformations are constantly changing. These conformations are not all equally favored, however. The lowest-energy conformation is the staggered conformation, with the electron clouds in the C—H bonds separated as much as possible. The interactions of the electrons in the bonds make the eclipsed conformation about 12.6 kJ/mol (3.0 kcal/mol) higher in energy than the staggered conformation. Three kilocalories is not a large amount of energy, and at room temperature, most molecules have enough kinetic energy to overcome this small rotational barrier.

Figure 3-7 shows how the potential energy of ethane changes as the carbon–carbon bond rotates. The y axis shows the potential energy relative to the most stable (staggered) conformation. The x axis shows the dihedral angle as it increases from 0° (eclipsed) through 60° (staggered) and on through additional eclipsed and staggered conformations as θ continues to increase. As ethane rotates toward an eclipsed conformation, its potential energy increases, and there is resistance to the rotation. This resistance to twisting (torsion) is called **torsional strain,** and the 12.6 kJ/mol (3.0 kcal/mol) of energy required is called **torsional energy.**

Conformational analysis is the study of the energetics of different conformations. Many reactions depend on a molecule's ability to twist into a particular conformation; conformational analysis can help to predict which conformations are favored and which reactions are more likely to take place. We will apply conformational analysis to propane and butane first, and later to some interesting cycloalkanes.

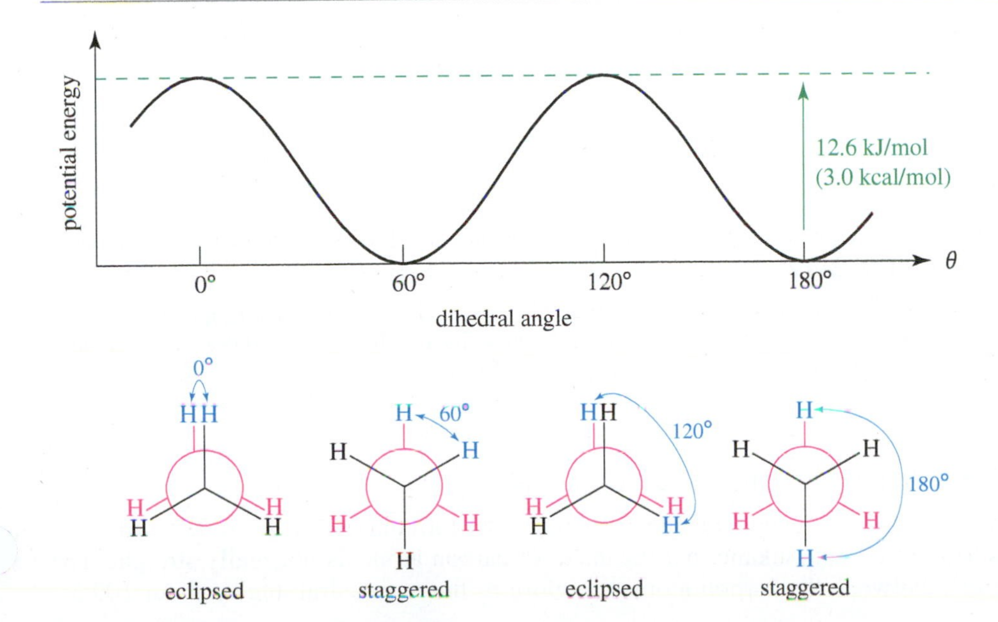

FIGURE 3-7
The torsional energy of ethane is lowest in the staggered conformation. The eclipsed conformation is about 12.6 kJ/mol (3.0 kcal/mol) higher in energy. At room temperature, this barrier is easily overcome and the molecules rotate constantly.

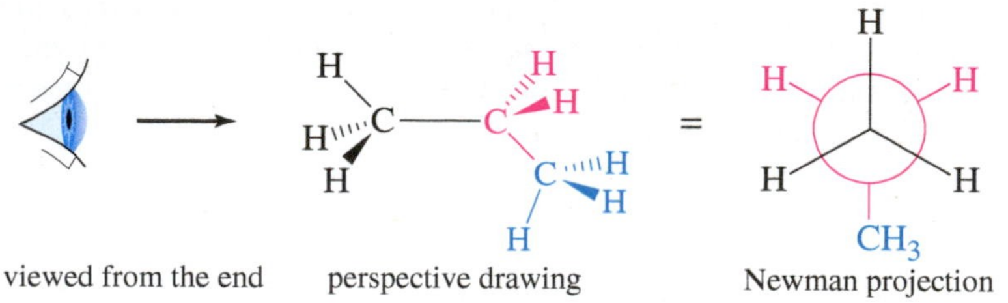

FIGURE 3-8
Propane is shown here as a perspective drawing and as a Newman projection looking down one of the carbon–carbon bonds.

viewed from the end — perspective drawing — Newman projection

FIGURE 3-9
Torsional energy of propane. When a C—C bond of propane rotates, the torsional energy varies much like it does in ethane, but with 13.8 kJ/mol (3.3 kcal/mol) torsional energy in the eclipsed conformation.

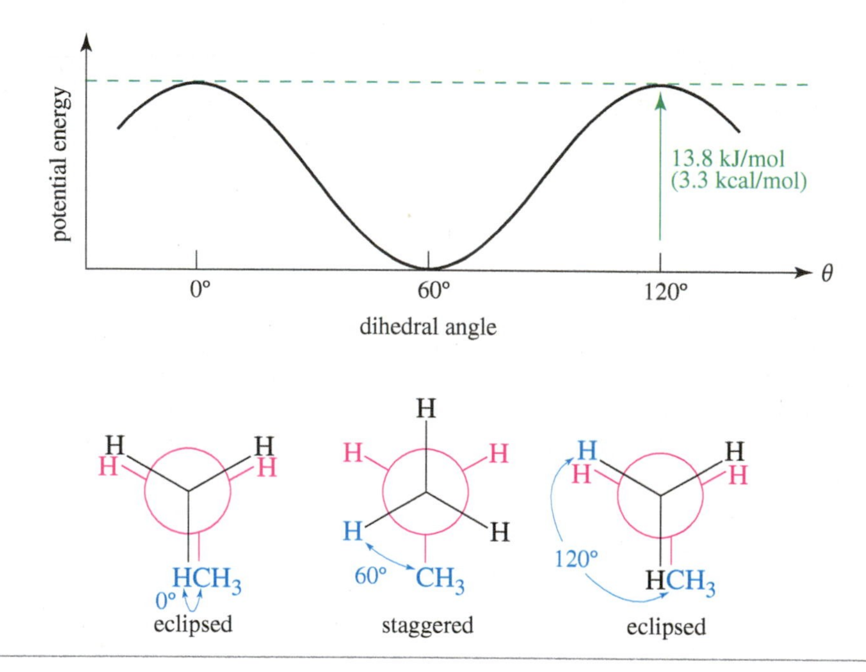

3-7C Conformations of Propane

PROBLEM-SOLVING HINT

A C—H bond eclipsed with another C—H bond contributes 4.2 kJ/mol (1.0 kcal/mol) torsional energy (one-third of eclipsed ethane). A C—H bond eclipsed with a C—CH_3 bond contributes 5.4 kJ/mol (1.3 kcal/mol).

Propane is the three-carbon alkane, with formula C_3H_8. Figure 3-8 shows a three-dimensional representation of propane and a Newman projection looking down one of the carbon–carbon bonds.

Figure 3-9 shows a graph of the torsional energy of propane as one of the carbon–carbon bonds rotates. The torsional energy of the eclipsed conformation is about 13.8 kJ/mol (3.3 kcal/mol), only 1.2 kJ (0.3 kcal) more than that required for ethane. Apparently, the torsional strain resulting from eclipsing a carbon–hydrogen bond with a carbon–methyl bond is only 1.2 kJ (0.3 kcal) more than the strain of eclipsing two carbon–hydrogen bonds.

PROBLEM 3-12

Draw a graph, similar to Figure 3-9, of the torsional strain of 2-methylpropane as it rotates about the bond between C1 and C2. Show the dihedral angle and draw a Newman projection for each staggered and eclipsed conformation.

3-8 Conformations of Butane

Butane is the four-carbon alkane, with molecular formula C_4H_{10}. We refer to *n*-butane as a straight-chain alkane, but the chain of carbon atoms is not really straight. The angles between the carbon atoms are close to the tetrahedral angle, about 109.5°.

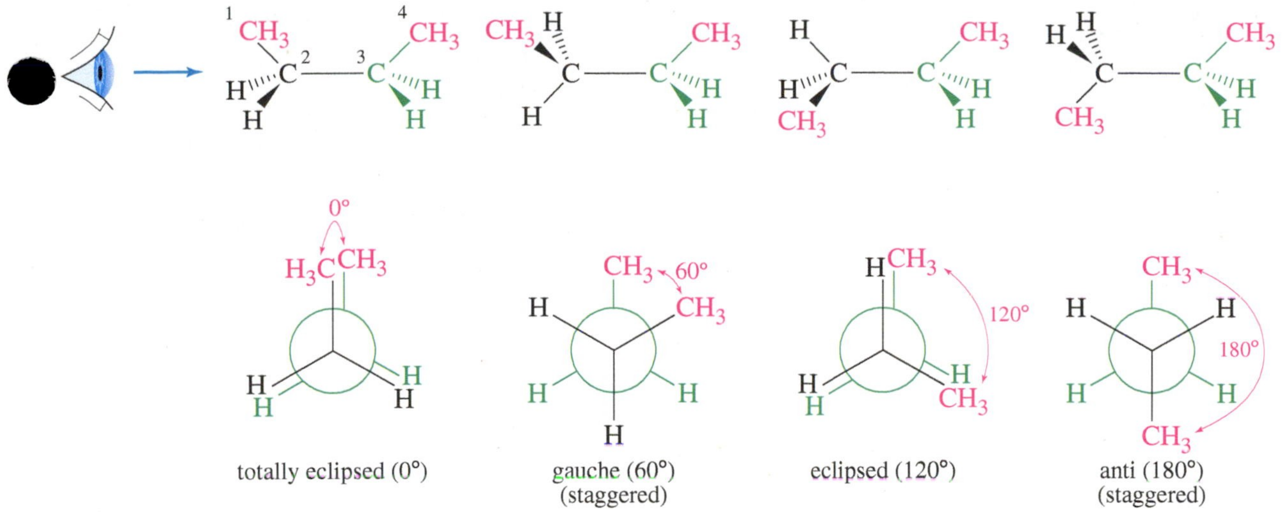

totally eclipsed (0°) gauche (60°) (staggered) eclipsed (120°) anti (180°) (staggered)

FIGURE 3-10
Butane conformations. Rotations about the center bond in butane give different molecular shapes. Three of these conformations have specific names.

Rotations about any of the carbon–carbon bonds are possible. Rotations about either of the end bonds (C1—C2 or C3—C4) just rotate a methyl group like they do in ethane and propane. Rotations about the central C2—C3 bond are more interesting, however. Figure 3-10 shows Newman projections, looking along the central C2—C3 bond, for four conformations of butane. Construct butane with your molecular models, and sight down the C2—C3 bond. Notice that we have defined the dihedral angle θ as the angle between the two end methyl groups.

Three of the conformations shown in Figure 3-10 are given special names. When the methyl groups are pointed in the same direction ($\theta = 0°$), they eclipse each other. This conformation is called **totally eclipsed,** to distinguish it from the other eclipsed conformations, such as the one at $\theta = 120°$. At $\theta = 60°$, the butane molecule is staggered and the methyl groups are toward the left and right of each other. This 60° conformation is called **gauche** (pronounced gōsh), a French word meaning "left" or "awkward."

Another staggered conformation occurs at $\theta = 180°$, with the methyl groups pointing in opposite directions. This conformation is called **anti** because the methyl groups are "opposed."

3-8A Torsional Energy of Butane

A graph of the relative torsional energies of the butane conformations is shown in Figure 3-11. All the staggered conformations (anti and gauche) are lower in energy than any of the eclipsed conformations. The anti conformation is lowest in energy because it places the bulky methyl groups as far apart as possible. The gauche conformations, with the methyl groups separated by just 60°, are 3.8 kJ (0.9 kcal) higher in energy than the anti conformation because the methyl groups are close enough that their electron clouds begin to repel each other. Use your molecular models to compare the crowding of the methyl groups in these conformations.

3-8B Steric Strain

The totally eclipsed conformation is about 6 kJ (1.4 kcal) higher in energy than the other eclipsed conformations because it forces the two end methyl groups so close together that their electron clouds experience a strong repulsion. This kind of

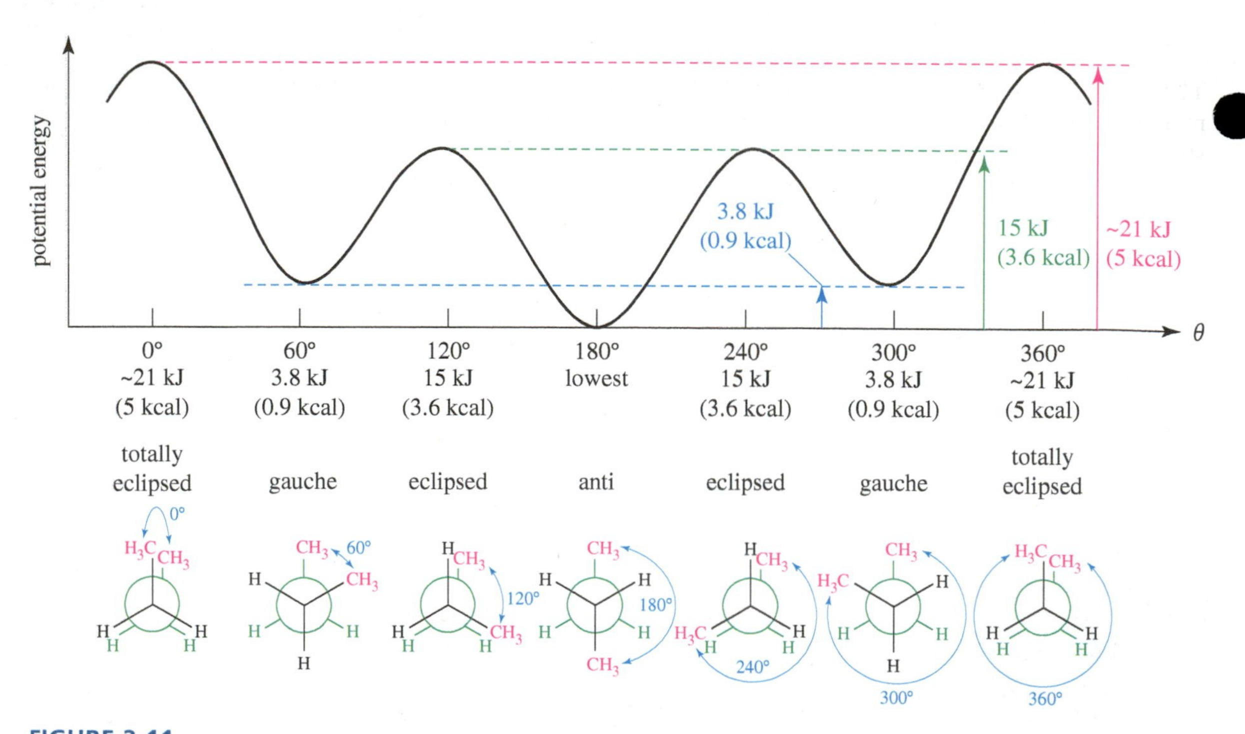

FIGURE 3-11
Torsional energy of butane. The anti conformation is lowest in energy, and the totally eclipsed conformation is highest in energy.

interference between two bulky groups is called **steric strain.*** The following structure shows the interference between the methyl groups in the totally eclipsed conformation.

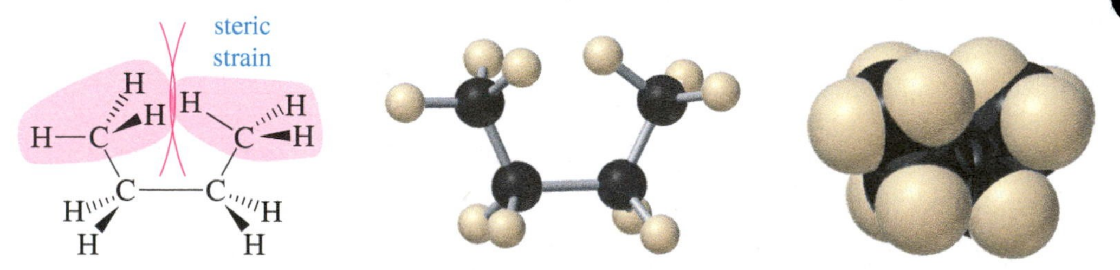

Totally eclipsed conformation of butane

Rotating the totally eclipsed conformation 60° to a gauche conformation releases most, but not all, of this steric strain. The gauche conformation is still 3.8 kJ (0.9 kcal) higher in energy than the most stable anti conformation.

What we have learned about the conformations of butane can be applied to other alkanes. We can predict that carbon–carbon single bonds will assume staggered conformations whenever possible to avoid eclipsing the groups attached to them. Among the staggered conformations, the anti conformation is preferred because it has the lowest torsional energy. We must remember, however, that enough thermal energy is present at room temperature for the molecules to rotate rapidly among all the different conformations. The relative stabilities are important because more molecules will be found in the more stable conformations than in the less stable ones.

PROBLEM-SOLVING HINT

A C—CH_3 bond eclipsed with another C—CH_3 bond contributes about 13 kJ/mol (3 kcal/mol) torsional energy.

Bond	Eclipsed with	Molar energy
C—H	C—H	4.2 kJ (1.0 kcal)
C—H	C—CH_3	5.4 kJ (1.3 kcal)
C—CH_3	C—CH_3	13 kJ (3 kcal)

PROBLEM 3-13

Draw a graph, similar to Figure 3-11, of the torsional energy of 2-methylbutane as it rotates about the C2—C3 bond.

*"Steric strain" is sometimes called "steric hindrance," a term that more appropriately refers to the slowing (hindrance) of a reaction because bulky groups interfere.

3-9 Conformations of Higher Alkanes

The higher alkanes resemble butane in their preference for anti and gauche conformations about the carbon–carbon bonds. The lowest-energy conformation for any straight-chain alkane is the one with all the internal carbon–carbon bonds in their anti conformations. These anti conformations give the chain a zigzag shape. At room temperature, the internal carbon–carbon bonds undergo rotation, and many molecules contain gauche conformations. Gauche conformations make kinks in the zigzag structure. Nevertheless, we frequently draw alkane chains in a zigzag structure to represent the most stable arrangement.

Conformations of long-chain alkyl groups are important in the chemistry of lipids (Chapter 25), with interesting implications for their melting points and for nutrition.

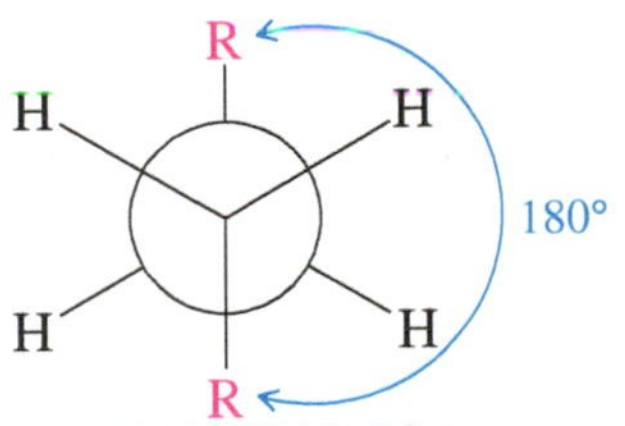

anti conformation

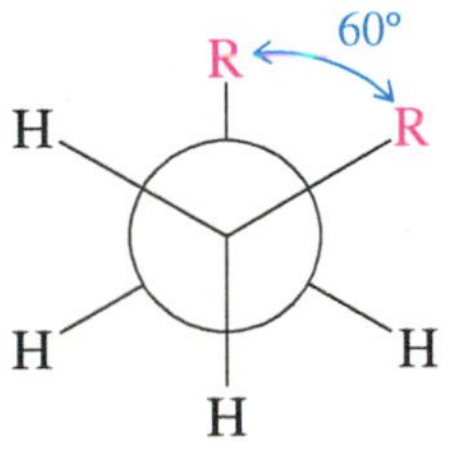

gauche conformation

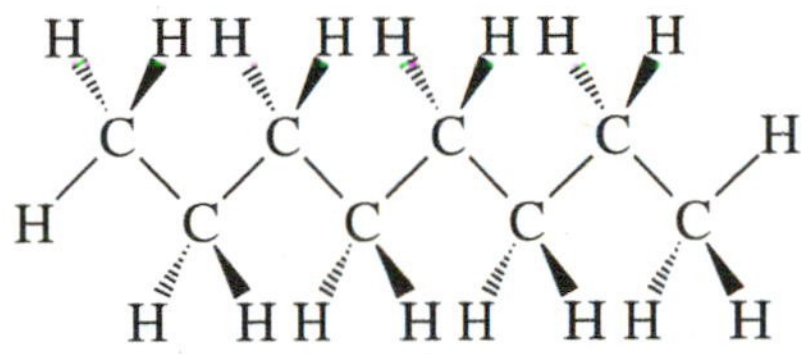

octane, all anti conformation

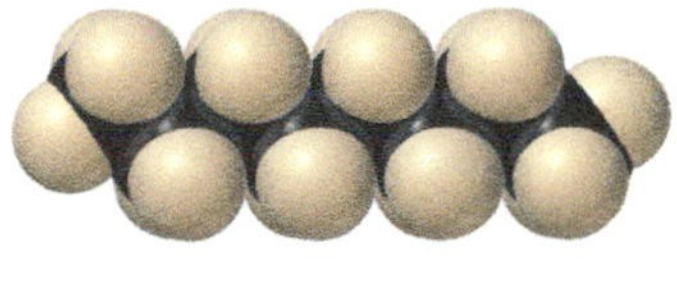

all anti, zigzag conformation

PROBLEM 3-14

Draw a perspective representation of the most stable conformation of 3-methylhexane.

3-10 Cycloalkanes

Many organic compounds are **cyclic:** they contain rings of atoms. The carbohydrates we eat are cyclic, the nucleotides that make up our DNA and RNA are cyclic, and the antibiotics we use to treat diseases are cyclic. In this chapter, we use the *cycloalkanes* to illustrate the properties and stability of cyclic compounds.

Cycloalkanes are alkanes that contain rings of carbon atoms. Simple cycloalkanes are named like acyclic (noncyclic) alkanes, with the prefix *cyclo-* indicating the presence of a ring. For example, the cycloalkane with four carbon atoms in a ring is called *cyclobutane.* The cycloalkane with seven carbon atoms in a ring is *cycloheptane.* Line–angle formulas are often used for drawing the rings of cycloalkanes (Figure 3-12).

3-10A General Molecular Formulas of Cycloalkanes

Simple cycloalkanes are rings of CH_2 groups (methylene groups). Each one has exactly twice as many hydrogen atoms as carbon atoms, giving the general molecular formula C_nH_{2n}. This general formula has two fewer hydrogen atoms than the $(2n + 2)$ formula for an acyclic alkane because a ring has no ends, and no hydrogens are needed to cap off the ends of the chain.

3-10B Physical Properties of Cycloalkanes

Most cycloalkanes resemble the **acyclic** (noncyclic), open-chain alkanes in their physical properties and in their chemistry. They are nonpolar, relatively inert compounds with boiling points and melting points that depend on their molecular weights. The cycloalkanes are held in a more compact cyclic shape, so their physical properties are similar to those of the compact, branched alkanes. The physical properties of some common cycloalkanes are listed in Table 3-4.

Application: Anesthesia

Cyclopropane was once used as general anesthetic because its vapors, like those of other simple alkanes and cycloalkanes, cause sleepiness and a loss of consciousness. After inhalation into the lungs, cyclopropane goes into the blood. Due to its nonpolar nature, it rapidly leaves the blood and passes through the nonpolar membranes surrounding the central nervous system, where it produces anesthesia. Cyclopropane is no longer used as an anesthetic because it is highly flammable (like ether) and can cause explosions when mixed with air.

TABLE 3-4
Physical Properties of Some Simple Cycloalkanes

Cycloalkane	Formula	Boiling Point (°C)	Melting Point (°C)	Density
cyclopropane	C_3H_6	−33	−128	0.72
cyclobutane	C_4H_8	−12	−50	0.75
cyclopentane	C_5H_{10}	49	−94	0.75
cyclohexane	C_6H_{12}	81	7	0.78
cycloheptane	C_7H_{14}	118	−12	0.81
cyclooctane	C_8H_{16}	148	14	0.83

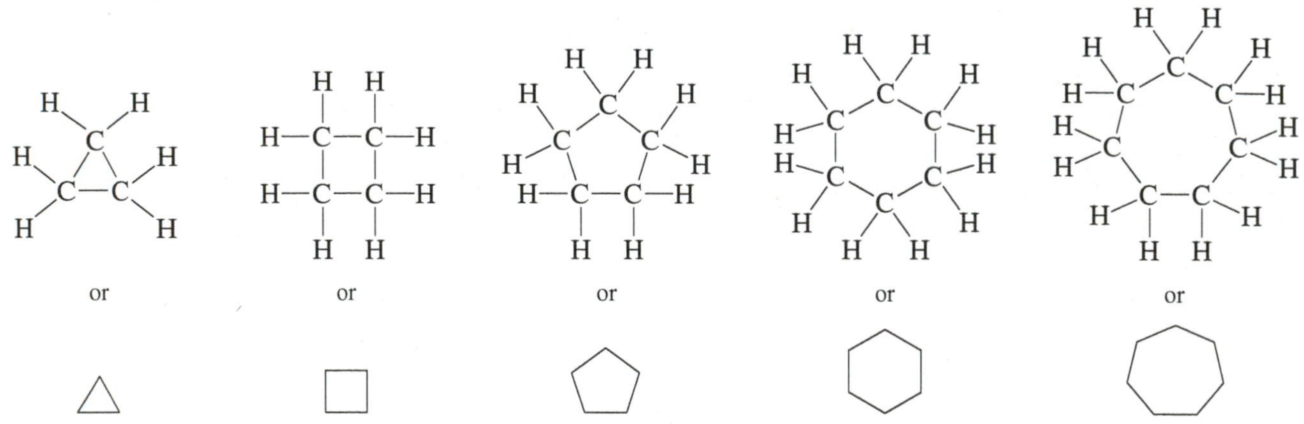

cyclopropane C_3H_6 cyclobutane C_4H_8 cyclopentane C_5H_{10} cyclohexane C_6H_{12} cycloheptane C_7H_{14}

FIGURE 3-12
Structures of some cycloalkanes.

3-10C Nomenclature of Cycloalkanes

Cycloalkanes are named much like acyclic alkanes. Substituted cycloalkanes use the cycloalkane for the base name, with the alkyl groups named as substituents. If there is just one substituent, no numbering is needed.

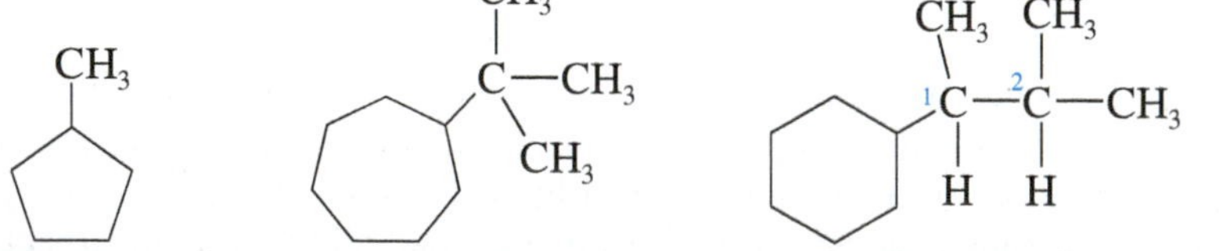

methylcyclopentane *tert*-butylcycloheptane (1,2-dimethylpropyl)cyclohexane

If there are two or more substituents on the ring, the ring carbons are numbered to give the lowest possible numbers for the substituted carbons. The numbering begins with one of the substituted ring carbons and continues in the direction that gives the lowest possible numbers to the other substituents. In the name, the substituents are listed in alphabetical order. When the numbering could begin with either of two substituted ring carbons (as in a disubstituted cycloalkane), begin with the one that has more substituents; otherwise, begin with the one that is first alphabetically.

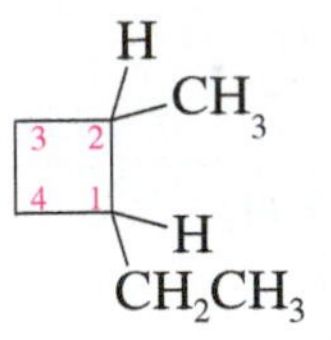

1-ethyl-2-methylcyclobutane 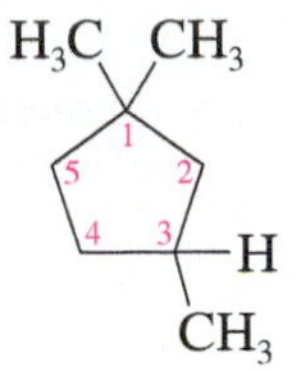   1,1,3-trimethylcyclopentane 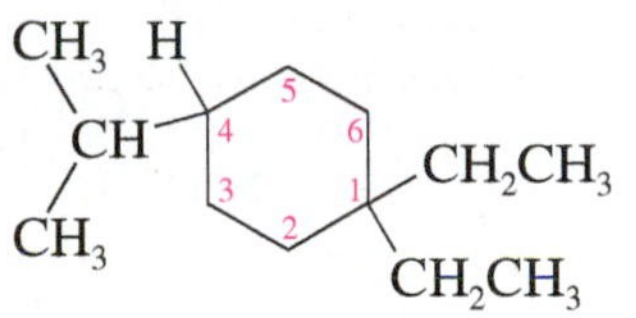   1,1-diethyl-4-isopropylcyclohexane

When the acyclic portion of the molecule contains more carbon atoms than the cyclic portion (or when it contains an important functional group), the cyclic portion is sometimes named as a cycloalkyl substituent.

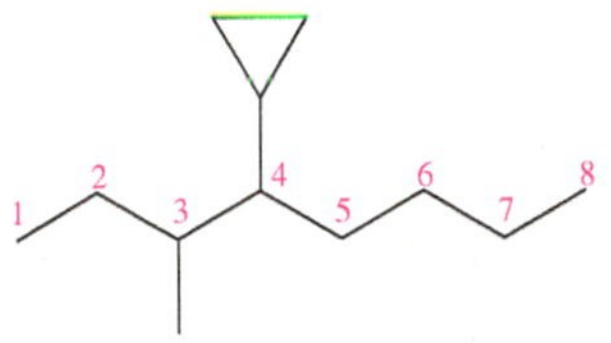

4-cyclopropyl-3-methyloctane 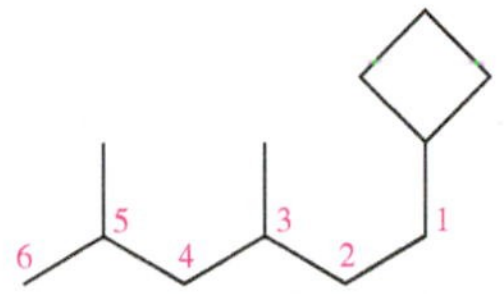   1-cyclobutyl-3,5-dimethylhexane 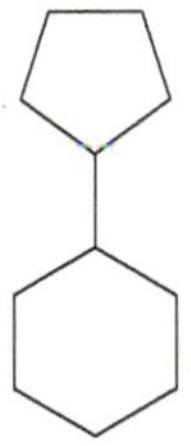   cyclopentylcyclohexane

PROBLEM 3-15

Give IUPAC names for the following compounds.

(a) $CH_3—CH—CH_2CH_3$ **(b)** 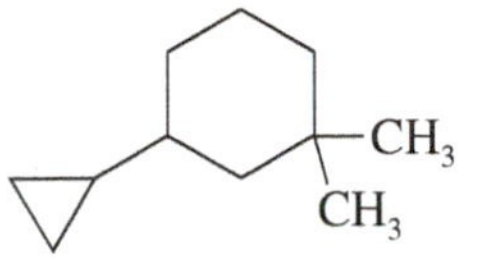 **(c)**

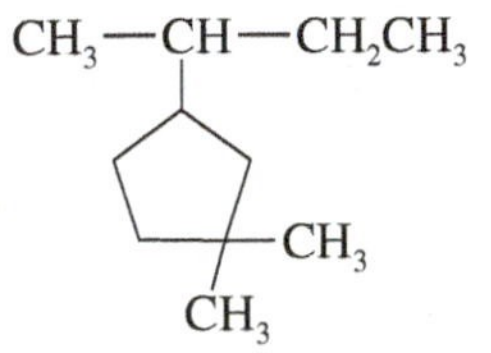

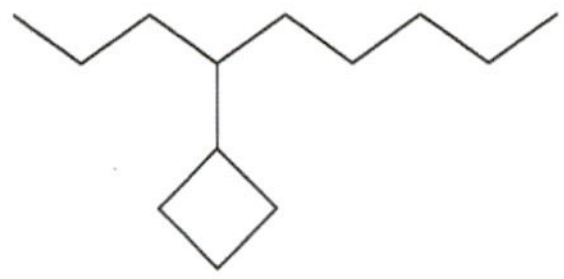

PROBLEM-SOLVING HINT

Students accidentally draw cyclic structures when acyclic structures are intended and vice versa. Always verify whether the name contains the prefix **cyclo-**.

PROBLEM 3-16

Draw the structure and give the molecular formula for each of the following compounds.

(a) 1-ethyl-3-methylcycloheptane
(b) isobutylcyclohexane
(c) cyclopropylcyclopentane
(d) 3-ethyl-1,1-dimethylcyclohexane
(e) 3-ethyl-2,4-dimethylhexane
(f) 1,1-diethyl-4-(3,3-dimethylbutyl)cyclohexane

3-11 Cis-trans Isomerism in Cycloalkanes

Open-chain (acyclic) alkanes undergo rotations about their carbon–carbon single bonds, so they are free to assume any of an infinite number of conformations. Alkenes have rigid double bonds that prevent rotation, giving rise to cis and trans isomers with different orientations of the groups on the double bond (Section 1-19B). Cycloalkanes are similar to alkenes in this respect. A cycloalkane has two distinct faces. If two substituents point toward the same face, they are **cis.** If they point toward opposite faces, they are **trans.** These **geometric isomers** cannot interconvert without breaking and re-forming bonds.

Figure 3-13 compares the cis-trans isomer of but-2-ene with those of 1,2-dimethylcyclopentane. Make models of these compounds to convince yourself that *cis*- and *trans*-1,2-dimethylcyclopentane cannot interconvert by simple rotations about the bonds. Note that in the perspective drawings, the thicker lines indicate the parts of the molecule that extend out toward the viewer.

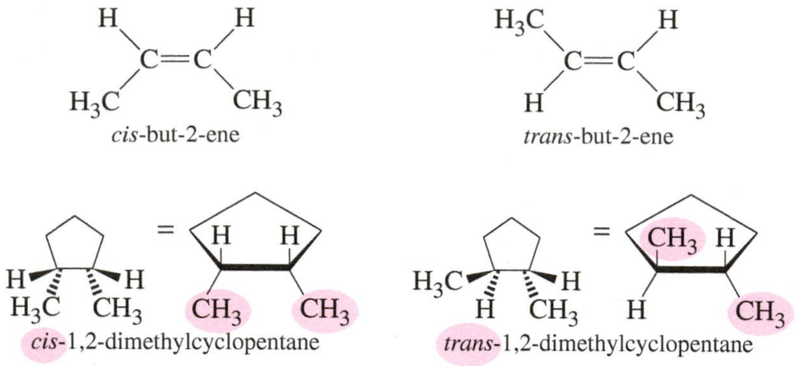

FIGURE 3-13
Cis-trans isomerism in cycloalkanes. Like alkenes, cycloalkane rings are restricted from free rotation. Two substituents on a cycloalkane must be on the same side (cis) or on opposite sides (trans) of the ring.

cis-but-2-ene

trans-but-2-ene

cis-1,2-dimethylcyclopentane

trans-1,2-dimethylcyclopentane

PROBLEM 3-17

Which of the following cycloalkanes are capable of geometric (cis-trans) isomerism? Draw the cis and trans isomers.

(a) 3-ethyl-1,1-dimethylcyclohexane
(b) 1-ethyl-3-methylcycloheptane
(c) 1-ethyl-3-methylcyclopentane
(d) 1-cyclopropyl-2-methylcyclohexane

PROBLEM 3-18

Give IUPAC names for the following cycloalkanes.

(a) **(b)** 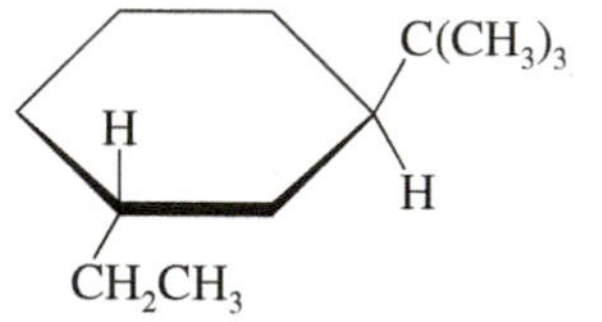 **(c)**

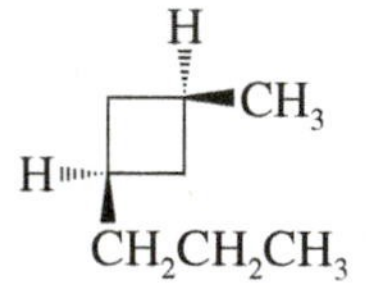

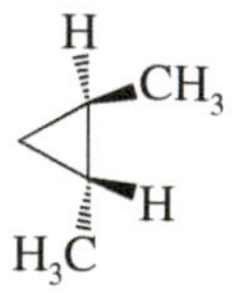

3-12 Stabilities of Cycloalkanes; Ring Strain

Although all the simple cycloalkanes (up to about C_{20}) have been synthesized, the most common rings contain five or six carbon atoms. We will study the stabilities and conformations of these rings in detail because they help to determine the properties of many important organic compounds.

Why are five-membered and six-membered rings more common than the other sizes? Adolf von Baeyer first attempted to explain the relative stabilities of cyclic molecules in the late 19th century, and he was awarded a Nobel Prize for this work in 1905. Baeyer reasoned that the carbon atoms in acyclic alkanes have bond angles of 109.5°. (We now explain this bond angle by the tetrahedral geometry of the sp^3 hybridized carbon atoms.)

If a cycloalkane requires bond angles other than 109.5°, the orbitals of its carbon–carbon bonds cannot achieve optimum overlap, and the cycloalkane must have some **angle strain** (sometimes called **Baeyer strain**) associated with it. Figure 3-14 shows that a planar cyclobutane, with 90° bond angles, is expected to have significant angle strain.

In addition to this angle strain, the Newman projection in Figure 3-14 shows that the bonds are eclipsed, resembling the *totally eclipsed* conformation of butane (Section 3-7). This eclipsing of bonds gives rise to torsional strain. Together, the angle strain and the torsional strain add to give what we call the **ring strain** of the cyclic compound. The amount of ring strain depends primarily on the size of the ring.

Before we discuss the ring strain of different cycloalkanes, we need to consider how ring strain is measured. In theory, we should measure the total amount of energy in the cyclic compound and subtract the amount of energy in a similar strain-free reference

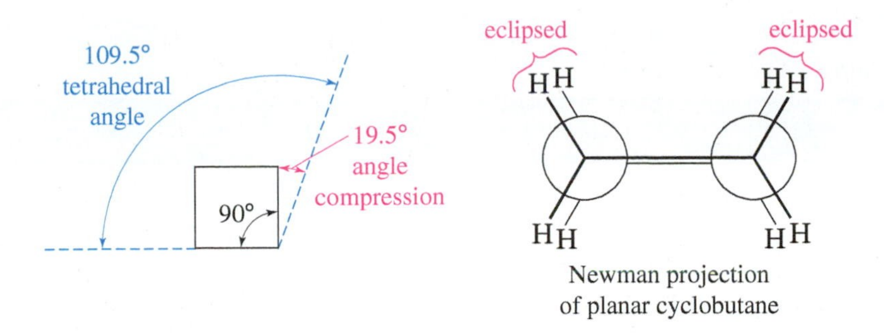

FIGURE 3-14
The ring strain of a planar cyclobutane results from two factors: angle strain from the compressing of the bond angles to 90° rather than the tetrahedral angle of 109.5°, and torsional strain from eclipsing of the bonds.

compound. The difference should be the amount of extra energy due to ring strain in the cyclic compound. These measurements are commonly made using *heats of combustion*.

3-12A Heats of Combustion

The **heat of combustion** is the amount of heat released when a compound is burned with an excess of oxygen in a sealed container called a *bomb calorimeter*. If the compound has extra energy as a result of ring strain, that extra energy is released in the combustion. The heat of combustion is usually measured by the temperature rise in the water bath surrounding the "bomb."

A cycloalkane can be represented by the molecular formula $(CH_2)_n$, so the general reaction in the bomb calorimeter is

$$\underset{\text{cycloalkane, }(CH_2)_n}{(CH_2)_n} + \tfrac{3}{2}n\,O_2 \longrightarrow n\,CO_2 + n\,H_2O + \underset{\text{heat of combustion}}{n(\text{energy per }CH_2)}$$

The molar heat of combustion of cyclohexane is nearly twice that of cyclopropane, simply because cyclohexane contains twice as many methylene (CH_2) groups per mole. To compare the relative stabilities of cycloalkanes, we divide the heat of combustion by the number of methylene (CH_2) groups. The result is the energy per CH_2 group. These normalized energies allow us to compare the relative amounts of ring strain (per methylene group) in the cycloalkanes.

Table 3-5 lists the heats of combustion for some simple cycloalkanes. The reference value of 658.6 kJ (157.4 kcal) per mole of CH_2 groups comes from an unstrained long-chain alkane. The values show large amounts of ring strain in cyclopropane and cyclobutane. Cyclopentane, cycloheptane, and cyclooctane have much smaller amounts of ring strain, and cyclohexane has no ring strain at all. We will discuss several of these rings in detail to explain this pattern of ring strain.

3-12B Cyclopropane

Table 3-5 shows that cyclopropane bears more ring strain per methylene group than any other cycloalkane. Two factors contribute to this large ring strain. First is the angle strain required to compress the bond angles from the tetrahedral angle of 109.5° to the 60° angles of cyclopropane. The bonding overlap of the carbon–carbon sp^3 orbitals is weakened when the bond angles differ so much from the tetrahedral angle. The sp^3 orbitals cannot point directly toward each other, and they overlap at an angle to form weaker "bent bonds" (Figure 3-15).

Torsional strain is the second factor in cyclopropane's large ring strain. The three-membered ring is planar, and all the bonds are eclipsed. A Newman projection of one of the carbon–carbon bonds (Figure 3-16) shows that the conformation resembles the

TABLE 3-5
Heats of Combustion (per Mole) for Some Simple Cycloalkanes

Ring Size	Cycloalkane	Molar Heat of Combustion	Heat of Combustion per CH_2 Group	Ring Strain per CH_2 Group	Total Ring Strain
3	cyclopropane	2091 kJ (499.8 kcal)	697.1 kJ (166.6 kcal)	38.5 kJ (9.2 kcal)	115 kJ (27.6 kcal)
4	cyclobutane	2744 kJ (655.9 kcal)	686.1 kJ (164.0 kcal)	27.5 kJ (6.6 kcal)	110 kJ (26.3 kcal)
5	cyclopentane	3320 kJ (793.5 kcal)	664.0 kJ (158.7 kcal)	5.4 kJ (1.3 kcal)	27 kJ (6.5 kcal)
6	cyclohexane	3951 kJ (944.4 kcal)	658.6 kJ (157.4 kcal)	0.0 kJ (0.0 kcal)	0.0 kJ (0.0 kcal)
7	cycloheptane	4637 kJ (1108.2 kcal)	662.4 kJ (158.3 kcal)	3.8 kJ (0.9 kcal)	27 kJ (6.4 kcal)
8	cyclooctane	5309 kJ (1268.9 kcal)	663.6 kJ (158.6 kcal)	5.1 kJ (1.2 kcal)	41 kJ (9.7 kcal)
reference: long-chain alkane			658.6 kJ (157.4 kcal)	0.0 kJ (0.0 kcal)	0.0 kJ (0.0 kcal)

All units are per mole.

totally eclipsed conformation of butane. The torsional strain in cyclopropane is not as great as its angle strain, but it helps to account for the large total ring strain.

Cyclopropane is generally more reactive than other alkanes. Reactions that open the cyclopropane ring release 115 kJ (27.6 kcal) per mole of ring strain, which provides an additional driving force for these reactions.

PROBLEM 3-19

The heat of combustion of *cis*-1,2-dimethylcyclopropane is larger than that of the trans isomer. Which isomer is more stable? Use drawings to explain this difference in stability.

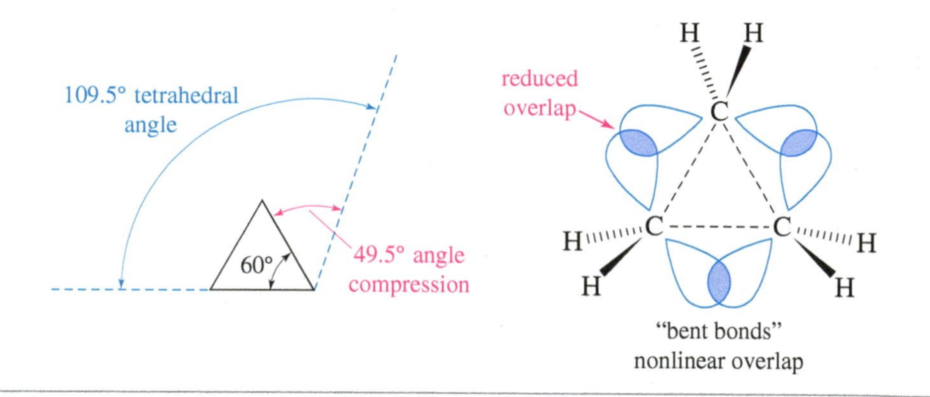

FIGURE 3-15
Angle strain in cyclopropane. The bond angles are compressed to 60° from the usual 109.5° bond angle of sp^3 hybridized carbon atoms. This severe angle strain leads to nonlinear overlap of the sp^3 orbitals and "bent bonds."

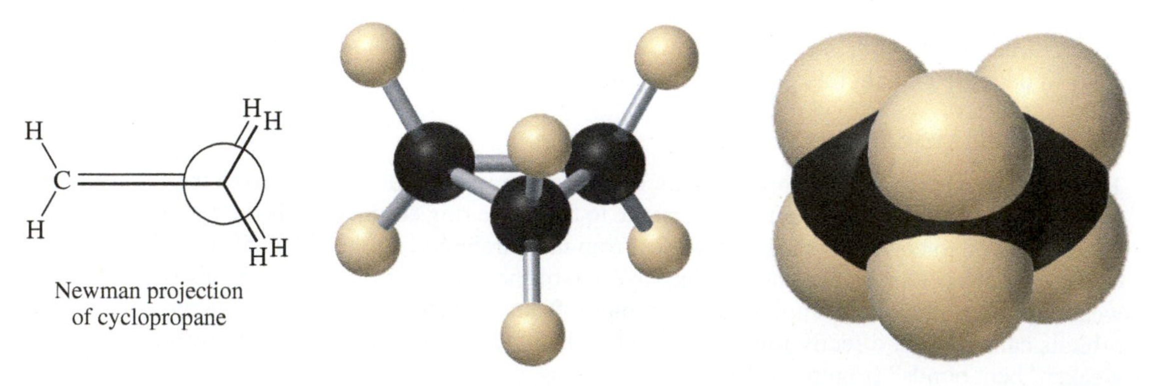

FIGURE 3-16
Torsional strain in cyclopropane. All the carbon–carbon bonds are eclipsed, generating torsional strain that contributes to the total ring strain.

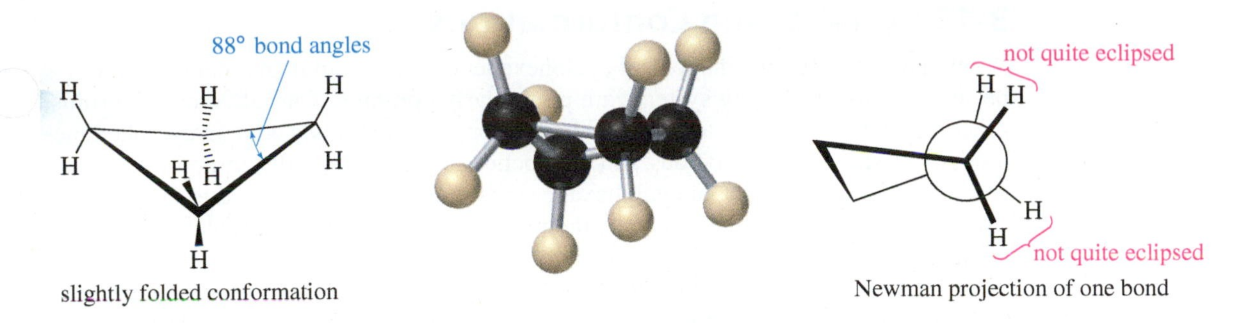

FIGURE 3-17
The conformation of cyclobutane is slightly folded. Folding gives partial relief from the eclipsing of bonds, as shown in the Newman projection. Compare this actual structure with the hypothetical planar structure in Figure 3-14.

3-12C Cyclobutane

The total ring strain in cyclobutane is almost as great as that in cyclopropane, but is distributed over four carbon atoms. If cyclobutane were perfectly planar and square, it would have 90° bond angles. A planar geometry requires eclipsing of all the bonds, however, as in cyclopropane. To reduce this torsional strain, cyclobutane actually assumes a slightly folded form, with bond angles of 88°. These smaller bond angles require slightly more angle strain than 90° angles, but the relief of some of the torsional strain appears to compensate for a small increase in angle strain (Figure 3-17).

PROBLEM 3-20

trans-1,2-Dimethylcyclobutane is more stable than *cis*-1,2-dimethylcyclobutane, but *cis*-1,3-dimethylcyclobutane is more stable than *trans*-1,3-dimethylcyclobutane. Use drawings to explain these observations.

3-12D Cyclopentane

If cyclopentane had the shape of a planar, regular pentagon, its bond angles would be 108°, close to the tetrahedral angle of 109.5°. A planar structure would require all the bonds to be eclipsed, however. Cyclopentane actually assumes a slightly puckered "envelope" conformation that reduces the eclipsing and lowers the torsional strain (Figure 3-18). This puckered shape is not fixed, but undulates by the thermal up-and-down motion of the five methylene groups. The "flap" of the envelope seems to move around the ring as the molecule undulates.

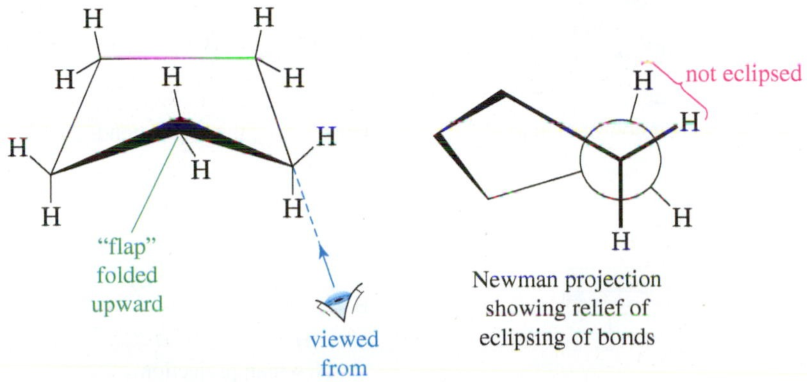

Application: Biochemistry

The conformation of cyclopentane is important because ribose and deoxyribose, the sugar components of RNA and DNA, respectively, assume cyclopentane-like ring conformations. These conformations are crucial to the properties and reactions of RNA and DNA.

FIGURE 3-18
The conformation of cyclopentane is slightly folded, like the shape of an envelope. This puckered conformation reduces the eclipsing of adjacent CH_2 groups, which reduces the torsional strain.

3-13 Cyclohexane Conformations

We will cover the conformations of cyclohexane in more detail than other cycloalkanes because cyclohexane ring systems are particularly common. Carbohydrates, steroids, plant products, pesticides, and many other important compounds contain cyclohexane-like rings whose conformations and stereochemistry are critically important to their reactivity. The abundance of cyclohexane rings in nature is probably due to both their stability and the selectivity offered by their predictable conformations. Nature probably forms more six-membered rings than all other ring sizes combined.

The combustion data (Table 3-5) show that cyclohexane has *no* ring strain. Cyclohexane must have bond angles that are near the tetrahedral angle (no angle strain) and also have no eclipsing of bonds (no torsional strain). A planar, regular hexagon would have bond angles of 120° rather than 109.5°, implying some angle strain. A planar ring would also have torsional strain because the bonds on adjacent CH_2 groups would be eclipsed. Therefore, the cyclohexane ring cannot be planar.

3-13A Chair and Boat Conformations

Cyclohexane achieves tetrahedral bond angles and staggered conformations by assuming a puckered conformation. The most stable conformation is the **chair conformation** shown in Figure 3-19. Build a molecular model of cyclohexane, and compare its shape with the drawings in Figure 3-19. In the chair conformation, the angles between the carbon–carbon bonds are all 109.5°. The Newman projection looking down the "seat" bonds shows all the bonds in staggered conformations.

The **boat conformation** of cyclohexane (Figure 3-20) also has bond angles of 109.5° and avoids angle strain. The boat conformation resembles the chair conformation except that the "footrest" methylene group is folded upward. The boat conformation suffers from torsional strain, however, because there is eclipsing of bonds.

This eclipsing forces two of the hydrogens on the ends of the "boat" to interfere with each other. These hydrogens are called **flagpole hydrogens** because they point upward from the ends of the boat like two flagpoles. The Newman projection in Figure 3-20 shows this eclipsing of the carbon–carbon bonds along the sides of the boat.

A cyclohexane molecule in the boat conformation actually exists as a slightly skewed **twist boat conformation,** also shown in Figure 3-20. If you assemble your molecular model in the boat conformation and twist it slightly, the flagpole hydrogens move away from each other and the eclipsing of the bonds is reduced. Even though the twist boat is lower in energy than the symmetrical boat, it is still about 23 kJ/mol (5.5 kcal/mol) higher in energy than the chair conformation. When someone refers to the "boat conformation," the twist boat (or simply twist) conformation is often intended.

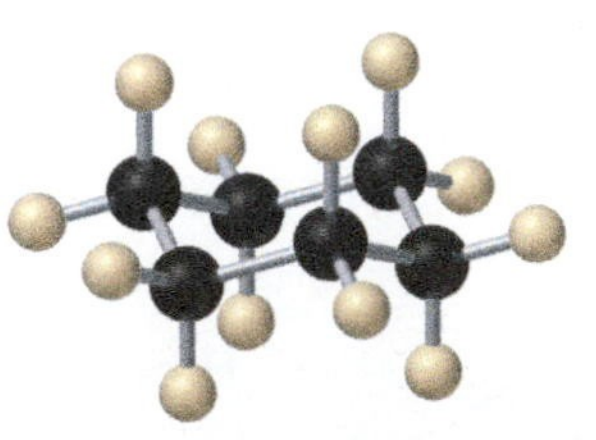

chair conformation

viewed along the "seat" bonds

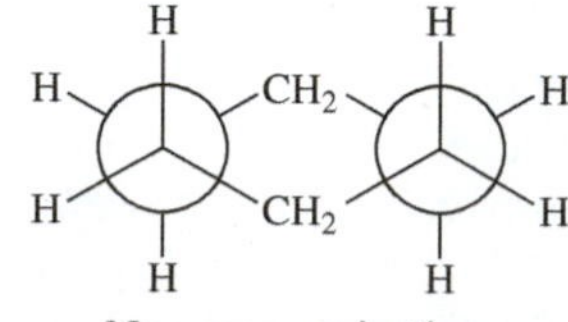

Newman projection

FIGURE 3-19
Viewed from the side, the chair conformation of cyclohexane appears to have one methylene group puckered upward and another puckered downward. Viewed from the Newman projection, the chair has no eclipsing of the carbon–carbon bonds. The bond angles are 109.5°.

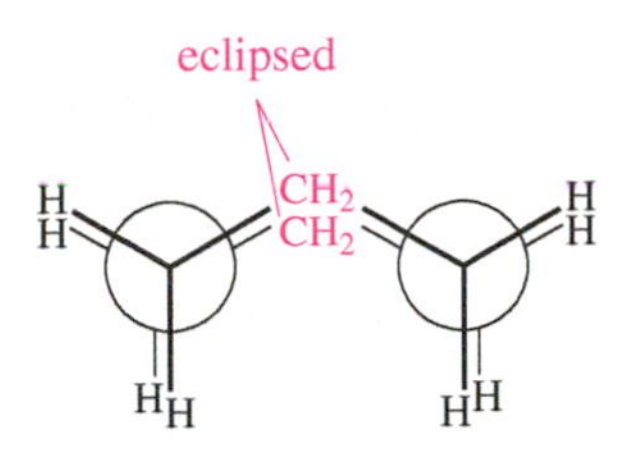

boat conformation

"flagpole" hydrogens

symmetrical boat

eclipsed

Newman projection

"twist" boat

FIGURE 3-20
In the symmetrical boat conformation of cyclohexane, eclipsing of bonds results in torsional strain. In the actual molecule, the boat is skewed to give the twist boat, a conformation with less eclipsing of bonds and less interference between the two flagpole hydrogens.

At any instant, most of the molecules in a cyclohexane sample are in chair conformations. The energy barrier between the boat and chair is sufficiently low, however, that the conformations interconvert many times each second. The interconversion from the chair to the boat takes place by the footrest of the chair flipping upward and forming the boat. The highest-energy point in this process is the conformation where the footrest is planar with the sides of the molecule. This unstable arrangement is called the **half-chair conformation.** Figure 3-21 shows how the energy of cyclohexane varies as it interconverts between the boat and chair forms.

Application: Biochemistry

The conformations of biological molecules are critical for their activities. For example, steroids fit into their receptors in only one conformation. The correct fit activates the receptor, resulting in a biological response.

3-13B Axial and Equatorial Positions

If we could freeze cyclohexane in a chair conformation, we would see that there are two different kinds of carbon–hydrogen bonds. Six of the bonds (one on each carbon atom) are directed up and down, parallel to the axis of the ring. These are called **axial bonds.** The other six bonds point out from the ring, along the "equator" of the ring. These are called **equatorial bonds.** The axial bonds and hydrogens are shown in red in Figure 3-22, and the equatorial bonds and hydrogens are shown in green.

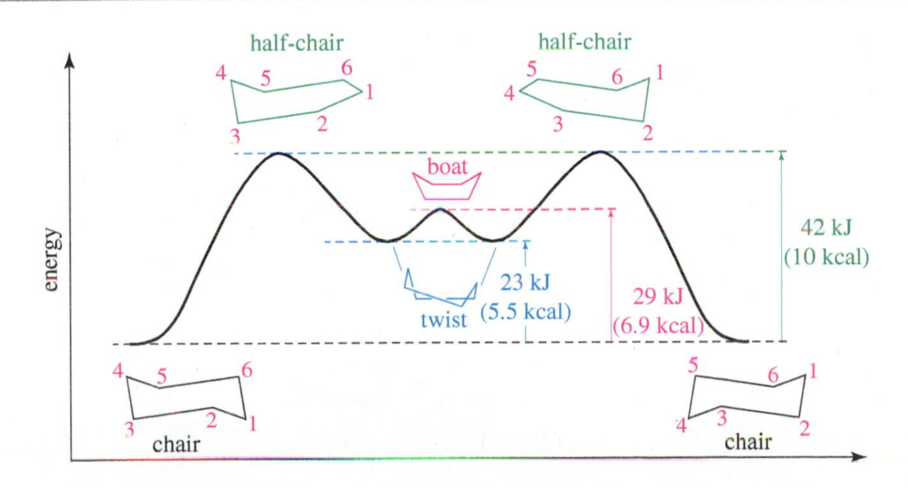

FIGURE 3-21
Conformational energy of cyclohexane. The chair conformation is most stable, followed by the twist boat. To convert between these two conformations, the molecule must pass through the unstable half-chair conformation.

FIGURE 3-22
Axial bonds are directed vertically, parallel to the axis of the ring. Equatorial bonds are directed outward, from the equator of the ring. As they are numbered here, the odd-numbered carbons have their *upward* bonds axial and their *downward* bonds equatorial. The even-numbered carbons have their *downward* bonds axial and their *upward* bonds equatorial.

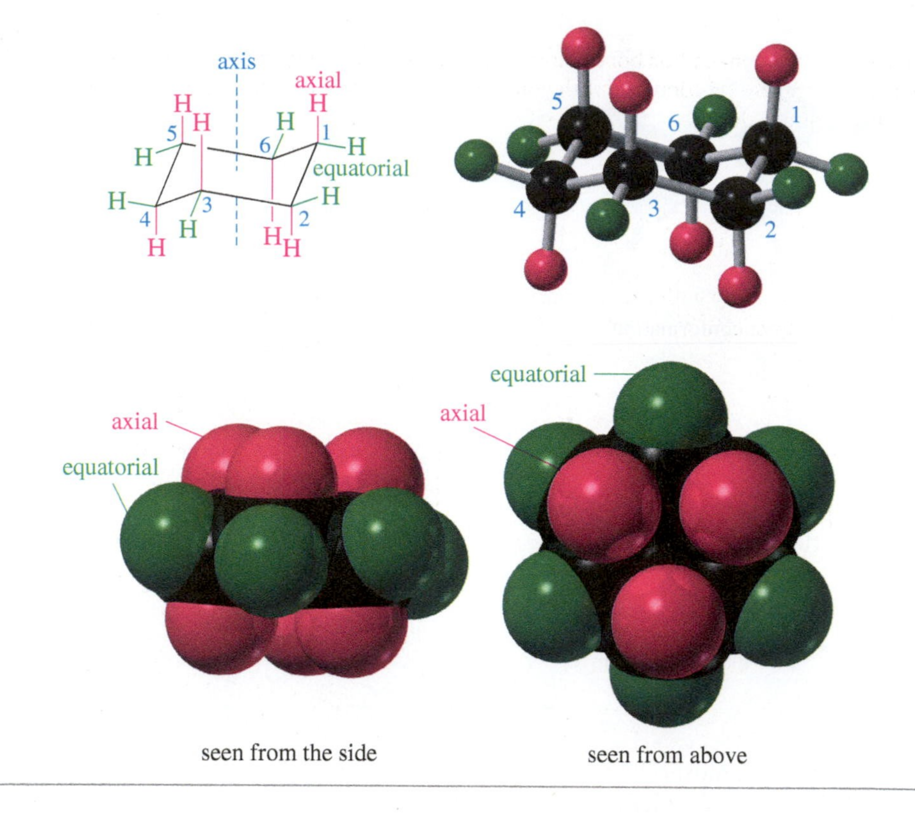

Each carbon atom in cyclohexane is bonded to two hydrogen atoms, one directed upward and one downward. As the carbon atoms are numbered in Figure 3-22, C1 has an axial bond upward and an equatorial bond downward. C2 has an equatorial bond upward and an axial bond downward. The pattern alternates. The odd-numbered carbon atoms have axial bonds up and equatorial bonds down, like C1. The even-numbered carbons have equatorial bonds up and axial bonds down, like C2. This pattern of alternating axial and equatorial bonds is helpful for predicting the conformations of substituted cyclohexanes, as we see in Sections 3-13 and 3-14.

PROBLEM-SOLVING STRATEGY Drawing Chair Conformations

Drawing realistic pictures of cyclohexane conformations is not difficult, but certain rules should be followed to show the actual positions and angles of the substituents on the ring. Make a cyclohexane ring with your models, put it in a chair conformation, and use it to follow along with this discussion. When you hold your model at the angle that corresponds to a drawing, the angles of the bonds in the model should correspond to the angles in the drawing.

To draw the carbon–carbon bond framework, first draw two parallel lines slightly slanted and slightly offset. The atoms at the ends of these bonds lie in a plane, and they define what will be the "armrests" of our chair.

Draw the headrest and footrest carbons, and draw the lines connecting them to the armrests. The two lines connecting the headrest carbon should be parallel to the two lines connecting the footrest.

headrest
in back
=
footrest
in front

Notice that the carbon–carbon bond framework uses lines with only three different slopes, labeled *a*, *b*, and *c*. Compare this drawing with your model, and notice the pairs of carbon–carbon bonds with three distinct slopes.

slope *a* slope *b* slope *c*

We can draw the chair with the headrest to the left and the footrest to the right or vice versa. Practice drawing it both ways.

Now fill in the axial and equatorial bonds. The axial bonds are drawn vertically, either up or down. When a vertex of the chair points upward, its axial bond also points upward. If the vertex points downward, its axial bond points downward. C1 is a downward-pointing vertex, and its axial bond also points downward. C2 points upward, and its axial bond points upward.

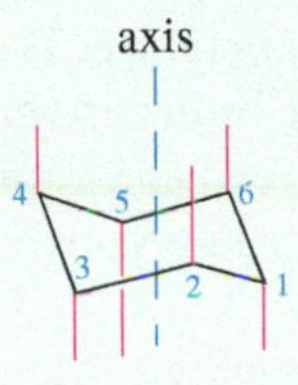

The equatorial bonds take more thought. Each carbon atom is represented by a vertex formed by two lines (bonds) having two of the possible slopes *a*, *b*, and *c*. Each equatorial bond should have the third slope: the slope that is *not* represented by the two lines forming the vertex.

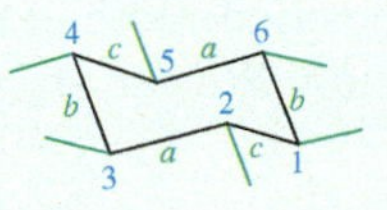

Look at your model as you add the equatorial bonds. The vertex C1 is formed by lines of slopes b and c, so its equatorial bond should have slope a. The equatorial bond at C2 should have slope b, and so on. Notice the W- and M-shaped patterns that result when these bonds are drawn correctly.

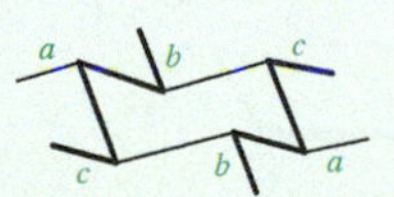

If your cyclohexane rings look awkward or slanted when using the analytical approach just shown, then try the artistic approach:* Draw a wide M, and draw a wide W below it, displaced about half a bond length to one side or the other. Connect the second atoms and the fourth atoms to give the cyclohexane ring with four equatorial bonds.

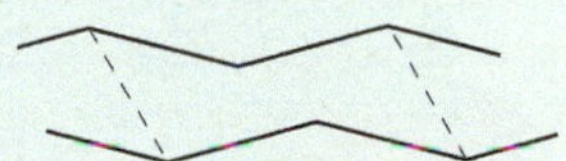

W displaced to the right W displaced to the left

The other two equatorial bonds are drawn parallel to the ring connections. The axial bonds are then drawn vertically.

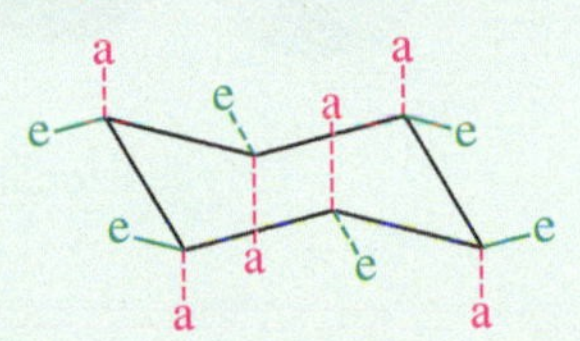

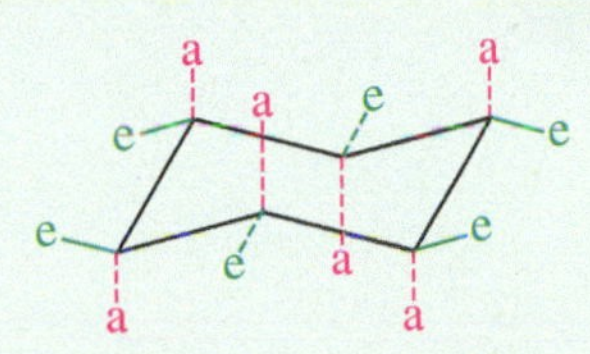

*See V. Dragojlovic, *J. Chem. Educ.* 2001, 78, 923.

PROBLEM 3-21

The cyclohexane chair shown in Figure 3-22 has the headrest to the right and the footrest to the left. Draw a cyclohexane chair with its axial and equatorial bonds, showing the headrest to the left and the footrest to the right.

PROBLEM 3-22

Draw 1,2,3,4,5,6-hexamethylcyclohexane with all the methyl groups
(a) in axial positions. **(b)** in equatorial positions.

3-14 Conformations of Monosubstituted Cyclohexanes

A substituent on a cyclohexane ring (in the chair conformation) can occupy either an axial or equatorial position. In many cases, the reactivity of the substituent depends on whether its position is axial or equatorial. The two possible chair conformations for methylcyclohexane are shown in Figure 3-23. These conformations are in equilibrium because they interconvert at room temperature. The boat (actually the twist boat) serves as an intermediate in this **chair–chair interconversion,** sometimes called a "ring-flip." Place different-colored atoms in the axial and equatorial positions of your cyclohexane model, and notice that the chair–chair interconversion changes axial to equatorial and equatorial to axial.

The two chair conformations of methylcyclohexane interconvert at room temperature, so the one that is lower in energy predominates. Careful measurements have shown that the chair with the methyl group in an equatorial position is the most stable conformation. It is about 7.6 kJ/mol (1.8 kcal/mol) lower in energy than the conformation with the methyl group in an axial position. Both of these chair conformations are lower in energy than any boat conformation. We can show how the 7.6 kJ energy difference

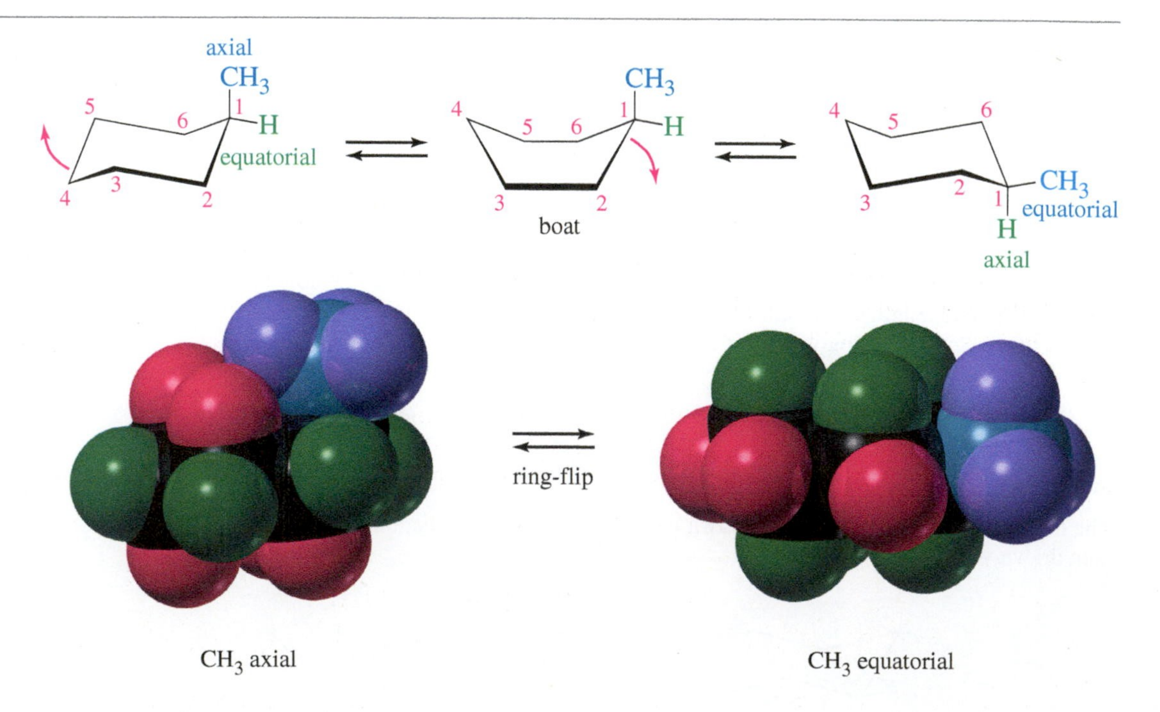

FIGURE 3-23
Chair–chair interconversion of methylcyclohexane. The methyl group is axial in one conformation, and equatorial in the other.

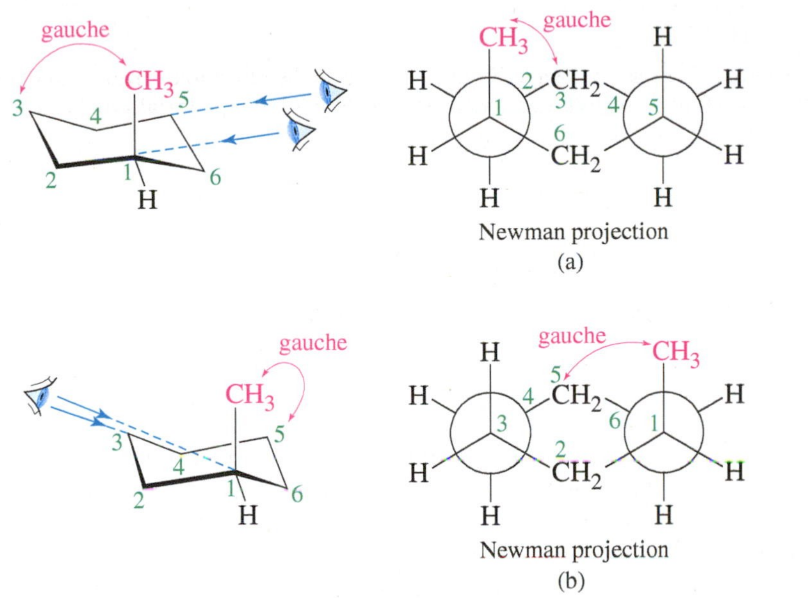

FIGURE 3-24
(a) When the methyl substituent is in an axial position on C1, it is gauche to C3. (b) The axial methyl group on C1 is also gauche to C5 of the ring.

between the axial and equatorial positions arises by examining molecular models and Newman projections of the two conformations. First, make a model of methylcyclohexane and use it to follow this discussion.

Consider a Newman projection looking along the armrest bonds of the conformation with the methyl group axial (Figure 3-24a): the methyl group is on C1, and we are looking from C1 toward C2. There is a 60° angle between the bond to the methyl group and the bond from C2 to C3, placing the axial methyl substituent and C3 in a gauche relationship. This axial methyl group is also gauche to C5, as you will see if you look along the C1—C6 bond in your model. Figure 3-24b shows this second gauche relationship.

The Newman projection for the conformation with the methyl group equatorial shows that the methyl group has an anti relationship to both C3 and C5. Figure 3-25 shows the Newman projection along the C1—C2 bond, with the anti relationship of the methyl group to C3.

PROBLEM 3-23

Draw a Newman projection, similar to Figure 3-25, down the C1—C6 bond in the equatorial conformation of methylcyclohexane. Show that the equatorial methyl group is also anti to C5. (Using your models will help.)

The axial methylcyclohexane conformation has two gauche interactions, each representing about 3.8 kJ (0.9 kcal) of additional energy. The equatorial methyl group has no gauche interactions. Therefore, we predict that the axial conformation is higher

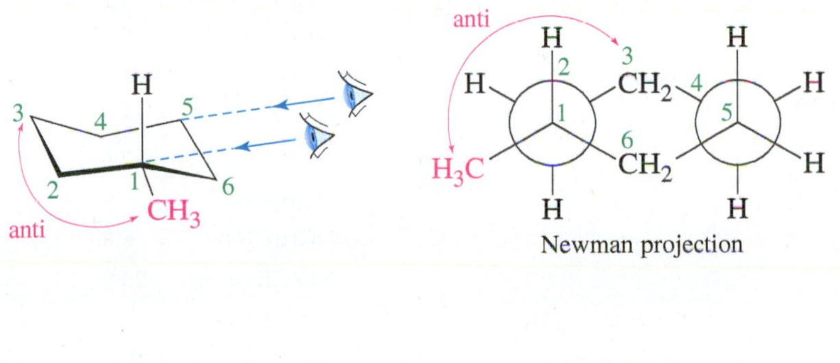

FIGURE 3-25
Looking down the C1—C2 bond of the equatorial conformation. Notice that the methyl group is anti to C3.

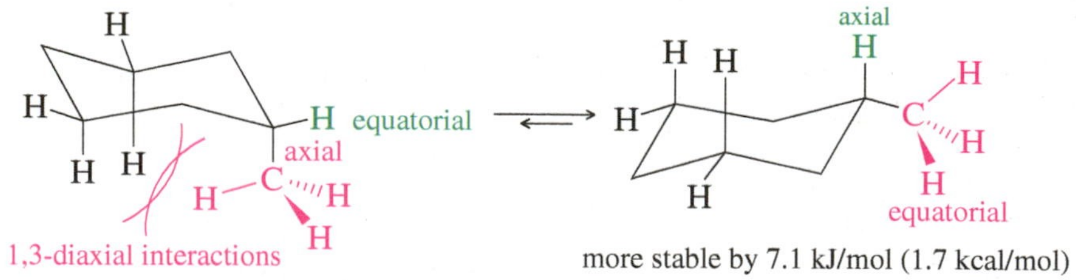

FIGURE 3-26
The axial substituent interferes with the axial hydrogens on C3 and C5. This interference is called a 1,3-diaxial interaction.

in energy by 7.6 kJ (1.8 kcal) per mole, in good agreement with the experimental value. Figure 3-26 shows that the gauche relationship of the axial methyl group with C3 and C5 places the methyl hydrogens close to the axial hydrogens on these carbons, causing their electron clouds to interfere. This form of steric strain is called a **1,3-diaxial interaction** because it involves substituents on carbon atoms of the ring that bear a 1,3 relationship. These 1,3-diaxial interactions are not present in the equatorial conformation.

A larger group usually has a larger energy difference between the axial and equatorial positions, because the 1,3-diaxial interaction shown in Figure 3-26 is stronger for larger groups. Table 3-6 lists the energy differences between the axial and equatorial positions for several alkyl groups and functional groups. The axial position is higher in energy in each case.

PROBLEM 3-24

Table 3-6 shows that the axial–equatorial energy difference for methyl, ethyl, and isopropyl groups increases gradually: 7.6, 7.9, and 8.8 kJ/mol (1.8, 1.9, and 2.1 kcal/mol). The *tert*-butyl group jumps to an energy difference of 23 kJ/mol (5.4 kcal/mol), over twice the value for the isopropyl group. Draw pictures of the axial conformations of isopropylcyclohexane and *tert*-butylcyclohexane, and explain why the *tert*-butyl substituent experiences such a large increase in axial energy over the isopropyl group.

PROBLEM 3-25

Draw the most stable conformation of
(a) ethylcyclohexane.
(b) 3-isopropyl-1,1-dimethylcyclohexane.
(c) *cis*-1-*tert*-butyl-4-isopropylcyclohexane.

TABLE 3-6
Energy Differences Between the Axial and Equatorial Conformations of Monosubstituted Cyclohexanes

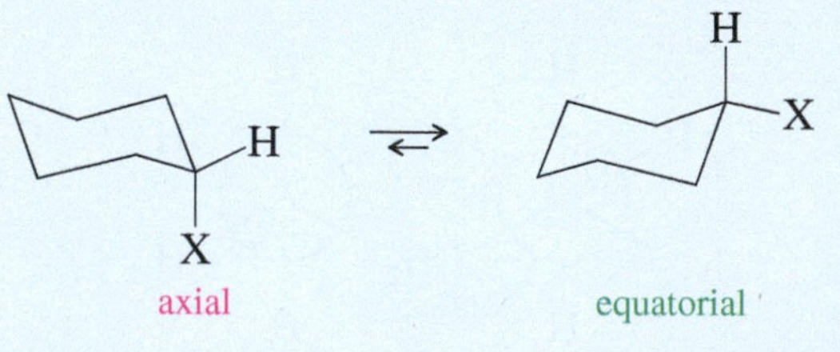

X	ΔG (axial–equatorial) (kJ/mol)	ΔG (axial–equatorial) (kcal/mol)
—F	0.8	0.2
—CN	0.8	0.2
—Cl	2.1	0.5
—Br	2.5	0.6
—OH	4.1	1.0
—COOH	5.9	1.4
$—CH_3$	7.6	1.8
$—CH_2CH_3$	7.9	1.9
$—CH(CH_3)_2$	8.8	2.1
$—C(CH_3)_3$	23	5.4

3-15 Conformations of Disubstituted Cyclohexanes

The steric interference between substituents in axial positions is particularly severe when there are large groups on two carbon atoms that bear a 1,3-diaxial relationship (*cis* on C1 and C3, or on C1 and C5), as in the two chair conformations of *cis*-1,3-dimethylcyclohexane shown here. The less stable conformation has both methyl groups in axial positions. The more stable conformation has both methyl groups in equatorial positions. Note the strongly unfavorable 1,3-diaxial interaction between the two methyl groups in the diaxial conformation. The molecule can relieve this 1,3-diaxial interference by flipping to the diequatorial conformation. Use your models to compare the diaxial and diequatorial forms of *cis*-1,3-dimethylcyclohexane.

1,3-diaxial interaction
axial
axial
equatorial
equatorial

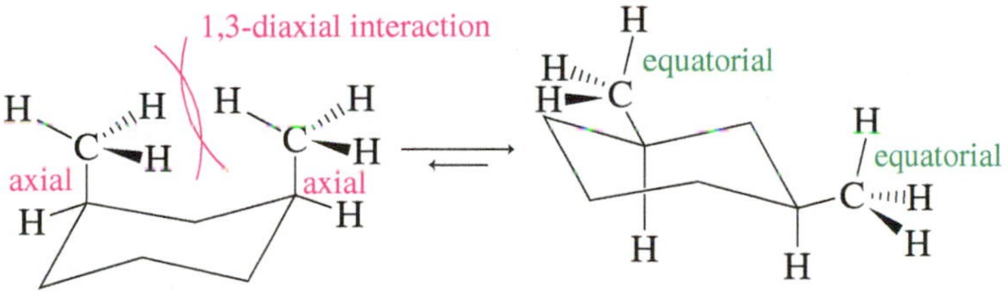

diaxial
very unfavorable, <0.1%

diequatorial
much more stable, >99.9%

trans-1,3-Dimethylcyclohexane does not have a conformation with a 1,3-diaxial interaction between two methyl groups. Either of its chair conformations places one methyl group in an axial position and one in an equatorial position. These conformations have equal energies, and they are present in equal amounts.

Chair conformations of trans-*1,3-dimethylcyclohexane*

axial CH_3 H
H CH_3 equatorial
same energy
equatorial CH_3
H H CH_3 axial

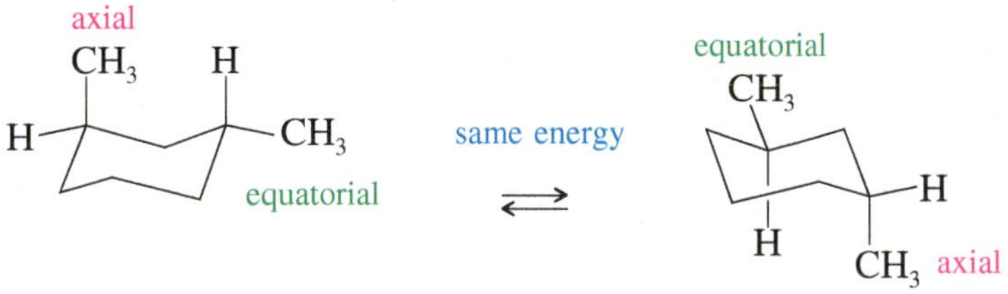

Now we can compare the relative stabilities of the cis and trans isomers of 1,3-dimethylcyclohexane. The most stable conformation of the cis isomer has both methyl groups in equatorial positions. Either conformation of the trans isomer places one methyl group in an axial position. The trans isomer is therefore higher in energy than the cis isomer by about 7.6 kJ/mol (1.8 kcal/mol), the energy difference between axial and equatorial methyl groups. Remember that the cis and trans isomers cannot interconvert, and there is no equilibrium between these isomers.

PROBLEM-SOLVING STRATEGY Recognizing Cis and Trans Isomers

Some students find it difficult to look at a chair conformation and tell whether a disubstituted cyclohexane is the cis isomer or the trans isomer. In the following drawing, the two methyl groups appear to be oriented in similar directions. They are actually trans but are often mistaken for cis.

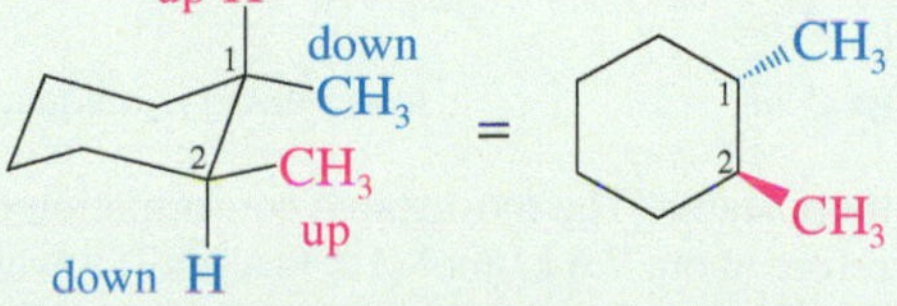

up H
down CH_3
CH_3 up
down H
=
CH_3
CH_3

trans-1,2-dimethylcyclohexane

(continued)

This ambiguity is resolved by recognizing that each of the ring carbons has two available bonds, one upward and one downward. In this drawing, the methyl group on C1 is on the downward bond, and the methyl on C2 is on the upward bond. Because one is down and one is up, their relationship is trans. A cis relationship would require both groups to be upward or both to be downward.

PROBLEM 3-26

Name the following compounds. Remember that two *up* bonds are cis; two *down* bonds are cis; one *up* bond and one *down* bond are trans.

(a) (b) 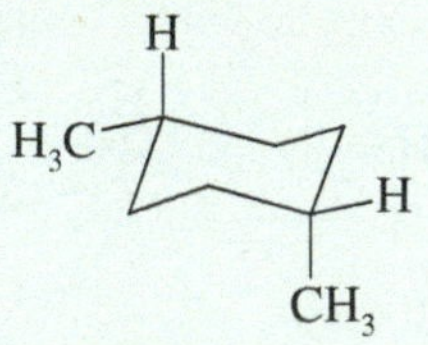 (c)

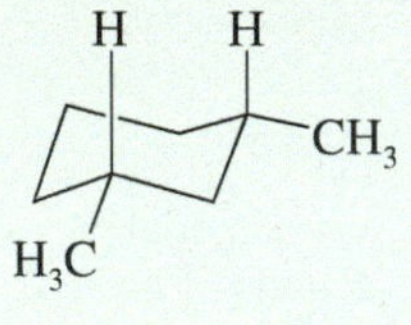

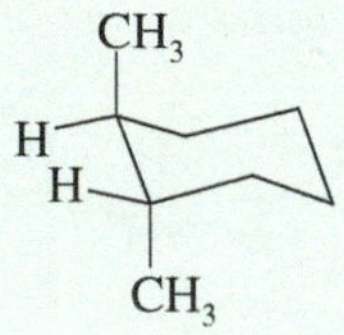

(d) 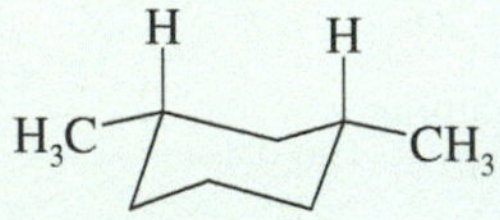 (e) 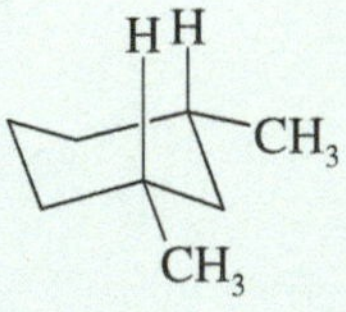 (f)

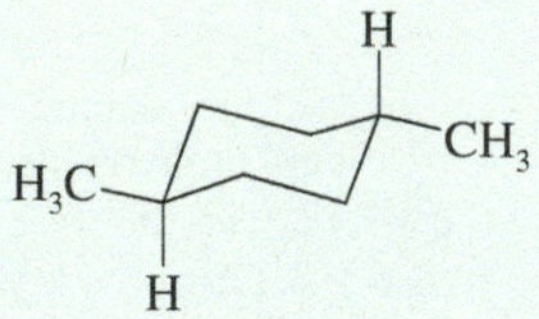

PROBLEM-SOLVING HINT

Ring-flips change the axial or equatorial positioning of groups, but they cannot change their cis-trans relationships. Converting cis into trans would require breaking and re-forming bonds.

PROBLEM-SOLVING HINT

If you number the carbons in a cyclohexane, the odd-numbered carbons are similar, as are the even-numbered carbons. If all the odd-numbered carbons have their up bond axial and their down bond equatorial, then all the even-numbered carbons will have their down bond axial and their up bond equatorial. For example, *cis*-1,3 (both up, both odd) will be both axial or both equatorial; *cis*-1,2 (both up, one odd, one even) will be one axial, one equatorial. This tip allows you to predict the answers before you draw them.

SOLVED PROBLEM 3-3

(a) Draw both chair conformations of *cis*-1,2-dimethylcyclohexane, and determine which conformer is more stable.

(b) Repeat for the trans isomer.

(c) Predict which isomer (cis or trans) is more stable.

SOLUTION

(a) The two possible chair conformations for the cis isomer interconvert at room temperature. Each of these conformations places one methyl group axial and one equatorial, giving them the same energy.

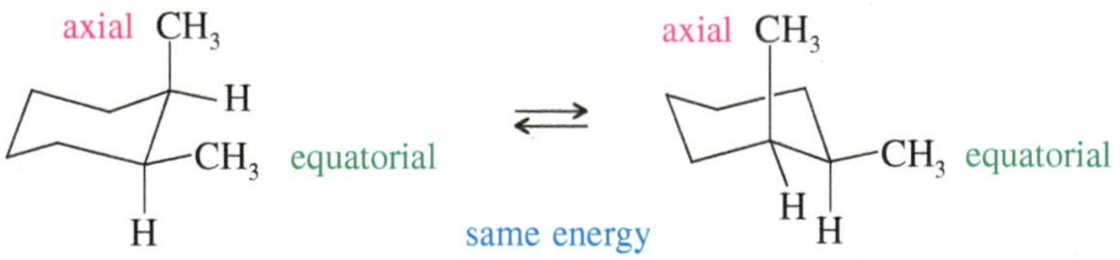

same energy

(b) The two chair conformations of the trans isomer interconvert at room temperature. Both methyl groups are axial in one, and both are equatorial in the other. The diequatorial conformation is more stable because neither methyl group occupies the more strained axial position.

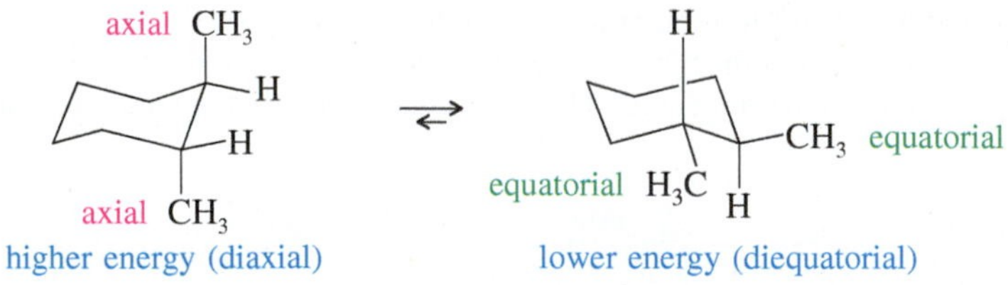

higher energy (diaxial) lower energy (diequatorial)

(c) The trans isomer is more stable. The most stable conformation of the trans isomer is diequatorial and therefore about 7.6 kJ/mol (1.8 kcal/mol) lower in energy than either conformation of the cis isomer, each having one methyl axial and one equatorial. Remember that cis and trans are distinct isomers and cannot interconvert.

PROBLEM 3-27

(a) Draw both chair conformations of *cis*-1,4-dimethylcyclohexane, and determine which conformer is more stable.
(b) Repeat for the trans isomer.
(c) Predict which isomer (cis or trans) is more stable.

PROBLEM 3-28

Use your results from Problem 3-27 to complete the following table. Each entry shows the positions of two groups arranged as shown. For example, two groups that are trans on adjacent carbons (*trans*-1,2) must be both equatorial (e,e) or both axial (a,a).

Positions	cis	trans
1,2	(e,a) or (a,e)	(e,e) or (a,a)
1,3		
1,4		

3-15A Substituents of Different Sizes

In many substituted cyclohexanes, the substituents are different sizes. As shown in Table 3-6 (page 142), the energy difference between the axial and equatorial positions for a larger group is greater than that for a smaller group. In general, if both groups cannot be equatorial, the most stable conformation has the larger group equatorial and the smaller group axial.

SOLVED PROBLEM 3-4

Draw the most stable conformation of *trans*-1-ethyl-3-methylcyclohexane.

SOLUTION

First, we draw the two conformations.

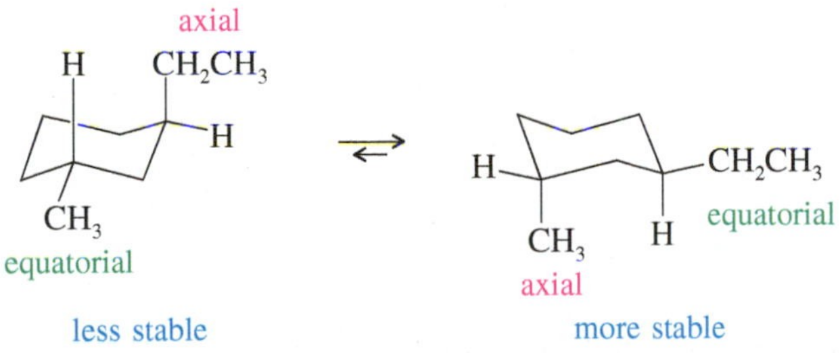

Both conformations require one group to be axial while the other is equatorial. The ethyl group is bulkier than the methyl group, so the conformation with the ethyl group equatorial is more stable. These chair conformations are in equilibrium at room temperature, and the one with the equatorial ethyl group predominates.

PROBLEM 3-29

Draw the two chair conformations of each of the following substituted cyclohexanes. In each case, label the more stable conformation.
(a) *cis*-1-ethyl-2-methylcyclohexane
(b) *trans*-1,2-diethylcyclohexane
(c) *cis*-1-ethyl-4-isopropylcyclohexane
(d) *trans*-1-ethyl-4-methylcyclohexane

3-15B Extremely Bulky Groups

Some groups, such as *tertiary*-butyl groups, are so bulky that they are extremely strained in axial positions. Regardless of the other groups present, cyclohexanes with *tert*-butyl substituents are most stable when the *tert*-butyl group is in an equatorial position. The following figure shows the severe steric interactions in a chair conformation with a *tert*-butyl group axial.

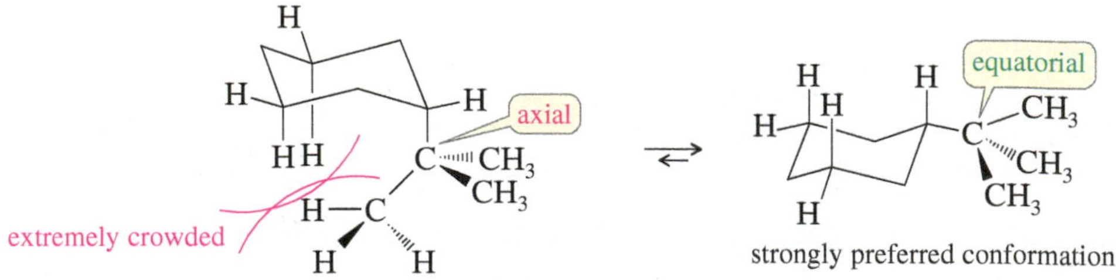

extremely crowded

axial

equatorial

strongly preferred conformation

If two *tert*-butyl groups are attached to the ring, both of them are much less strained in equatorial positions. When neither chair conformation allows both bulky groups to be equatorial, they may force the ring into a twist boat conformation. For example, both chair conformations of *cis*-1,4-di-*tert*-butylcyclohexane require one of the bulky *tert*-butyl groups to occupy an axial position. This compound is more stable in a twist boat conformation that allows both bulky groups to avoid axial positions.

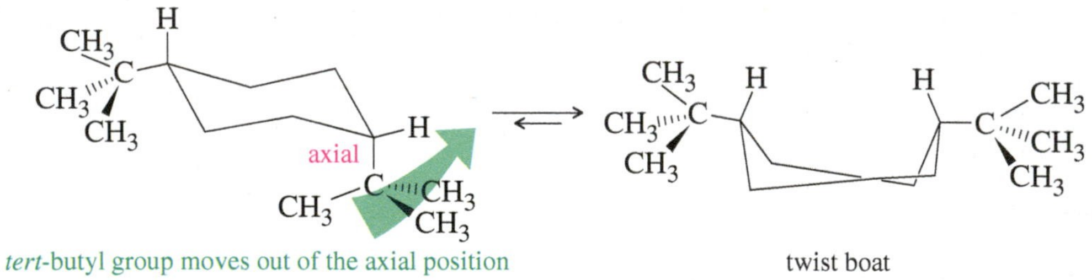

axial

tert-butyl group moves out of the axial position

twist boat

PROBLEM 3-30

Draw the most stable conformation of
(a) *cis*-1-*tert*-butyl-3-ethylcyclohexane.
(b) *trans*-1-*tert*-butyl-2-methylcyclohexane.
(c) *trans*-1-*tert*-butyl-3-(1,1-dimethylpropyl)cyclohexane.

3-16 Bicyclic Molecules

Two or more rings can be joined into *bicyclic* or *polycyclic* systems. There are three ways that two rings may be joined. **Fused rings** are most common, sharing two adjacent carbon atoms and the bond between them. **Bridged rings** are also common, sharing two nonadjacent carbon atoms (the **bridgehead carbons**) and one or more carbon atoms (the bridge) between them. **Spirocyclic compounds,** in which the two rings share only one carbon atom, are relatively rare. These ring systems may be drawn realistically puckered or flat, as shown below.

fused bicyclic

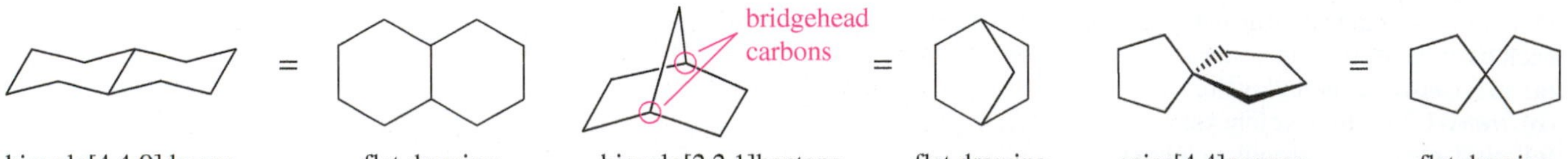

bicyclo[4.4.0]decane (decalin)

flat drawing of decalin

bridged bicyclic

bridgehead carbons

bicyclo[2.2.1]heptane (norbornane)

flat drawing of norbornane

spirocyclic

spiro[4.4]nonane

flat drawing of spiro[4.4]nonane

3-16A Nomenclature of Bicyclic Alkanes

The name of a bicyclic compound is based on the name of the alkane having the same number of carbons as there are in the ring system. This name follows the prefix *bicyclo* and a set of brackets enclosing three numbers. The following examples contain eight carbon atoms and are named bicyclo[4.2.0]octane and bicyclo[3.2.1]octane, respectively.

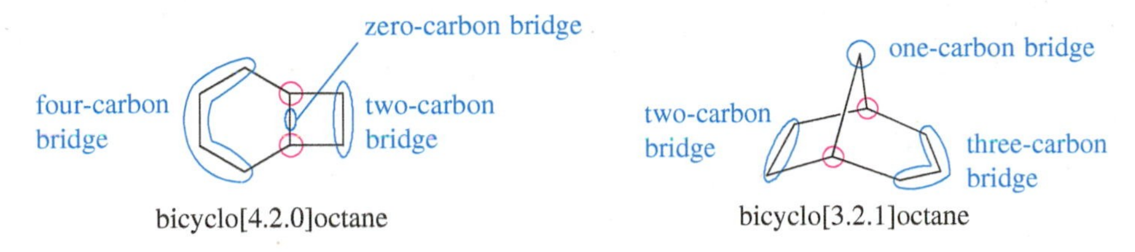

All fused and bridged bicyclic systems have three bridges connecting the two bridgehead atoms (red circles) where the rings connect. The numbers in the brackets give the number of carbon atoms in each of the three bridges connecting the bridgehead carbons, in order of decreasing size.

Application: Drugs

Bicyclic molecules are found in many natural product structures. Cocaine is a derivative of bicyclo[3.2.1]octane in which nitrogen replaces the carbon at the one-carbon bridge.

cocaine
in coca leaves

PROBLEM 3-31

Name the following compounds.

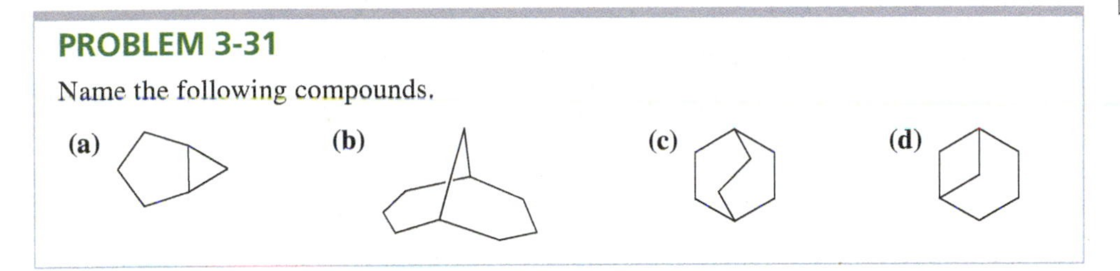

3-16B *cis*- and *trans*-Decalin

Decalin (bicyclo[4.4.0]decane) is the most common example of a fused ring system. Two geometric isomers of decalin exist, as shown in Figure 3-27. In one isomer, the rings are fused using two cis bonds, while the other is fused using two trans bonds. You should make a model of decalin to follow this discussion.

If we consider the left ring in the drawing of *cis*-decalin, the bonds to the right ring are both directed downward (and the attached hydrogens are directed upward). These bonds are therefore cis, and this is a cis ring fusion. One of the bonds to the right ring must be axial, and the other is equatorial. In *trans*-decalin, one of the bonds to the right ring is directed upward and the other downward. These bonds are trans, and this is a trans ring fusion. Both of the bonds to the right ring are equatorial. The six-membered rings in both isomers assume chair conformations, as shown in Figure 3-27.

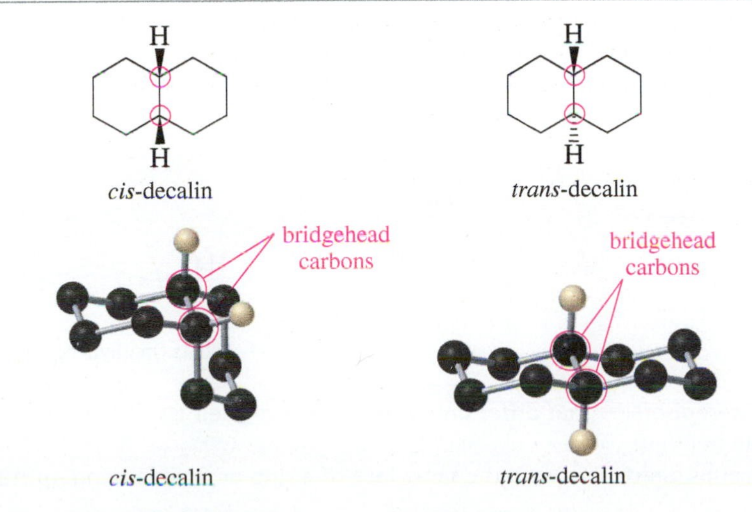

FIGURE 3-27
cis-Decalin has a ring fusion where the second ring is attached by two cis bonds. *trans*-Decalin is fused using two trans bonds. (The other hydrogens are omitted for clarity.)

The conformation of *cis*-decalin is somewhat flexible, but the trans isomer is quite rigid. If one of the rings in the trans isomer did a chair–chair interconversion, both bonds to the second ring would become axial and would be directed 180° apart. This is an impossible conformation, and it prevents any chair–chair interconversion in *trans*-decalin.

PROBLEM 3-32

Use your models to do a chair–chair interconversion on each ring of the conformation of *cis*-decalin shown in Figure 3-27. Draw the conformation that results.

Essential Terms

acyclic Not cyclic. (p. 129)

alkane A hydrocarbon having only single bonds; a **saturated hydrocarbon;** general formula: C_nH_{2n+2}. (p. 107)

alkyl group The group of atoms remaining after a hydrogen atom is removed from an alkane; an alkane-like substituent. Symbolized by R. (p. 112)

amyl An older common name for **pentyl.** (p. 112)

angle strain or Baeyer strain The strain associated with distorting bond angles to smaller (or larger) angles. (p. 132)

anti conformation A conformation with a 180° dihedral angle between the largest groups. Usually the lowest-energy conformation. (p. 127)

aromatic hydrocarbon A hydrocarbon having a benzene-like aromatic ring. (p. 108)

axial bond One of the six bonds (three up and three down) on the chair conformation of the cyclohexane ring that are parallel to the "axis" of the ring. The axial bonds are shown in red, and the equatorial bonds in green, in the drawing on page 137.

bridged bicyclic compound A compound containing two rings joined at nonadjacent carbon atoms. (p. 147)

bridged bicyclic systems (bridgeheads circled)

bridgehead carbons The carbon atoms shared by two or more rings. Three chains of carbon atoms (bridges) connect the bridgeheads. (p. 146)

chair–chair interconversion **(ring-flip)** The process of one chair conformation of a cyclohexane flipping into another one, with all the axial and equatorial positions reversed. The boat (or twist boat) conformation is an intermediate for the chair–chair interconversion. (p. 140)

chair (methyls axial) ⇄ boat ⇄ chair (methyls equatorial)

cis-trans isomers **(geometric isomers)** Stereoisomers that differ only with respect to their cis or trans arrangement on a ring or double bond. (p. 131)

cis: Having two similar groups directed toward the same face of a ring or double bond. (p. 131)

trans: Having two similar groups directed toward opposite faces of a ring or double bond. (p. 131)

cis-but-2-ene | *trans*-but-2-ene | *cis*-1,2-dimethylcyclopentane | *trans*-1,2-dimethylcyclopentane

combustion A rapid oxidation at high temperatures in the presence of air or oxygen. (p. 121)

common names The names that have developed historically, generally with a specific name for each compound; also called **trivial names.** (p. 110)

conformational analysis The study of the energetics of different conformations. (p. 125)

conformations and conformers Structures that are related by rotations about single bonds. Strictly speaking, a **conformer** is a conformation that corresponds to a relative minimum in energy, usually a staggered conformation. In most cases, conformations and conformers interconvert at room temperature, and they are not true isomers. (p. 123)

$\theta = 0°$ | $\theta = 60°$ | $\theta = 180°$

totally eclipsed conformation | gauche conformation | anti conformation

conformations of cyclohexanes (pp. 136–146)

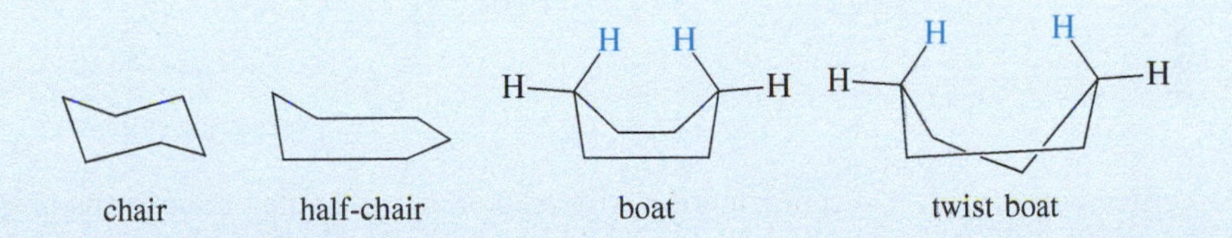

chair | half-chair | boat | twist boat

chair conformation: The most stable conformation of cyclohexane, with one part puckered upward and another part puckered downward. (p. 136)

boat conformation: The less stable puckered conformation of cyclohexane, with both parts puckered upward. The most stable boat is actually the **twist boat** (or simply **twist**) conformation. Twisting minimizes torsional strain and steric strain. (p. 136)

flagpole hydrogens: Two hydrogens (blue) in the boat conformation point upward like flagpoles. The twist boat reduces the steric repulsion of the flagpole hydrogens. (p. 136)

half-chair conformation: The unstable conformation halfway between the chair conformation and the boat conformation. Part of the ring is flat in the half-chair conformation. (p. 137)

constitutional isomers **(structural isomers)** Isomers whose atoms are connected differently; they differ in their bonding sequence. (p. 44)

cracking Heating large alkanes to cleave them into smaller molecules. (p. 121)

catalytic cracking: Cracking in the presence of a catalyst. (p. 120)

hydrocracking: Catalytic cracking in the presence of hydrogen to give mixtures of alkanes. (p. 120)

cyclic Containing a ring of atoms. (p. 129)

cycloalkane An alkane containing a ring of carbon atoms; general formula: C_nH_{2n}. (p. 129)

degree of alkyl substitution The number of alkyl groups bonded to a carbon atom in a compound or in an alkyl group. (p. 113)

$R{-}CH_2{-}H$	$R_2CH{-}H$	$R_3C{-}H$	$R_3C{-}R$
primary (1°) carbon atom	**secondary (2°)** carbon atom	**tertiary (3°)** carbon atom	**quaternary (4°)** carbon atom

1,3-diaxial interaction The strong steric strain between two axial groups on cyclohexane carbons with one carbon between them. (p. 142)

dihedral angle (θ) (see also **conformations**) The angle between two specified groups in a Newman projection. (p. 125)

eclipsed conformation Any conformation with bonds directly lined up with each other, one behind the other, in the Newman projection. The conformation with $\theta = 0°$ is an eclipsed conformation. See also **staggered conformation.** (p. 125)

equatorial bond One of the six bonds (three down and three up) on the cyclohexane ring that are directed out toward the "equator" of the ring. The equatorial bonds are shown in green in the drawing below. (p. 137)

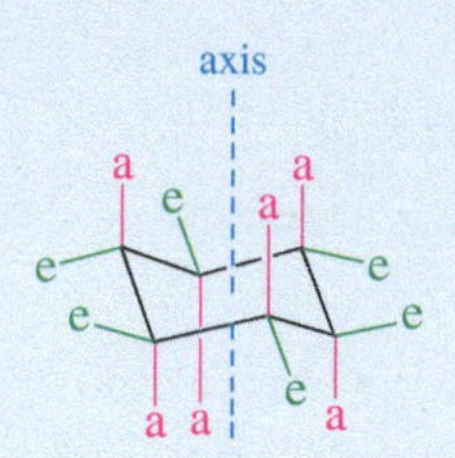

axial bonds in red;
equatorial bonds in green

fused ring system A molecule in which two or more rings share two adjacent carbon atoms. (p. 147)

fused ring systems

gauche conformation A conformation with a 60° dihedral angle between the largest groups. (p. 127)

geometric isomers See **cis-trans isomers**, the IUPAC term. (p. 131)

halogenation The reaction of alkanes with halogens, in the presence of heat or light, to give products with halogen atoms substituted for hydrogen atoms. (p. 122)

$$R{-}H + X_2 \xrightarrow{\text{heat or light}} R{-}X + XH \qquad X = F, Cl, Br$$

heat of combustion The heat given off when a mole of a compound is burned with excess oxygen to give CO_2 and H_2O in a *bomb calorimeter*. A measure of the energy content of a molecule. (p. 133)

homologs Two compounds that differ only by one or more $-CH_2-$ groups. (p. 108)

hydrophilic Attracted to water; soluble in water.

hydrophobic Repelled by water; insoluble in water. (p. 117)

IUPAC names The systematic names that follow the rules adopted by the International Union of Pure and Applied Chemistry. (p. 110)

kerosene A thin, volatile oil distilled from petroleum, with a boiling range higher than that of gasoline and lower than that of diesel fuel. Kerosene was once used in lanterns and heaters, but now most of this petroleum fraction is further refined for use as jet fuel. (p. 119)

methane hydrate An ice-like substance consisting of individual methane molecules trapped inside cages of water molecules. (p. 121)

methine group The —CH— group.

methylene group The $—CH_2—$ group. (p. 108)

methyl group The $—CH_3$ group. (p. 114)

***n*-alkane, normal alkane, or straight-chain alkane** An alkane with all its carbon atoms in a single chain, with no branching or alkyl substituents. (p. 108)

Newman projections A way of drawing the conformations of a molecule by looking straight down the bond connecting two carbon atoms. (p. 123)

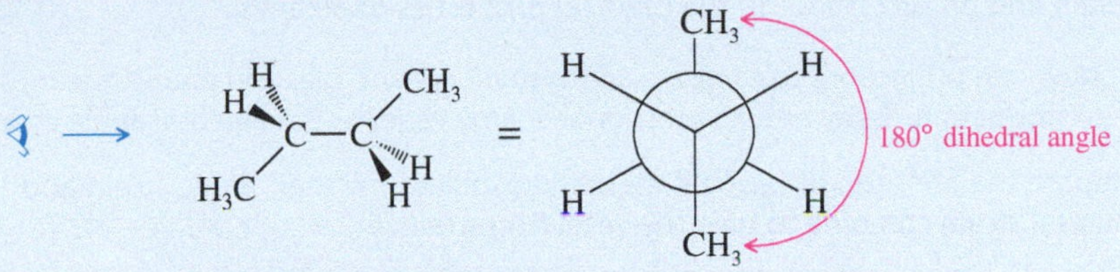

a Newman projection of butane in the anti conformation

octane number A rating of the antiknock properties of a gasoline blend. Its octane number is the percentage of isooctane (2,2,4-trimethylpentane) in an isooctane/heptane blend that begins to knock at the same compression ratio as the gasoline being tested. (p. 119)

paraffins Another term for alkanes. (p. 121)

ring strain The extra strain associated with the cyclic structure of a compound, as compared with a similar acyclic compound; composed of angle strain and torsional strain. (p. 132)

- **angle strain or Baeyer strain:** The strain associated with distorting bond angles to smaller (or larger) angles.
- **torsional strain:** The strain associated with eclipsing of bonds in the ring.

saturated Having no double or triple bonds. (p. 108)

sawhorse structures A way of picturing conformations by looking down at an angle toward the carbon–carbon bond. (p. 125)

skew conformation Any conformation that is not precisely staggered or eclipsed. (p. 125)

spirocyclic compounds Bicyclic compounds in which the two rings share only one carbon atom. (p. 146)

staggered conformation Any conformation with the bonds equally spaced in the Newman projection. The conformation with $\theta = 60°$ is a staggered conformation. (p. 125)

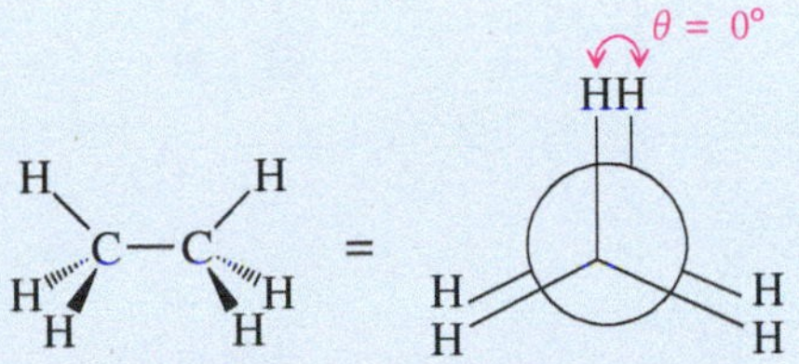

eclipsed conformation of ethane

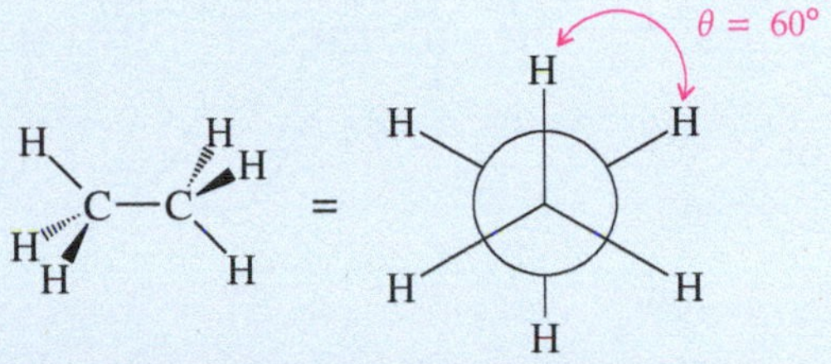

staggered conformation of ethane

steric strain The interference between two bulky groups that are so close together that their electron clouds experience a repulsion. (p. 128)

substituent A side chain or an appendage on the main chain. (p. 111)

systematic names Same as IUPAC names, the names that follow the rules adopted by the International Union of Pure and Applied Chemistry. (p. 110)

torsional energy or conformational energy The energy required to twist a bond into a specific conformation. (p. 125)

torsional strain The resistance to twisting about a bond. (p. 125)

totally eclipsed conformation A conformation with a 0° dihedral angle between the largest groups. Usually the highest-energy conformation. (p. 127)

Essential Problem-Solving Skills in Chapter 3

Each skill is followed by problem numbers exemplifying that particular skill.

1	Given the IUPAC name or common name of an alkane, draw the structure and give the molecular formula.	Problems 3-36, 37, 41, and 49
2	Draw isomers of alkanes and cycloalkanes, and use the IUPAC rules to name alkanes, cycloalkanes, and bicyclic alkanes.	Problems 3-33, 34, 35, 39, 43, and 52
3	Explain and predict trends in the physical properties of alkanes.	Problem 3-42
4	Use Newman projections to compare conformational energies and predict the most stable conformation. Show how the torsional energy varies as the dihedral angle changes.	Problems 3-40, 44, 48, and 49
5	Compare the energies of cycloalkanes, and explain how their angle strain and torsional strain combine to give the total ring strain.	Problem 3-45
6	Draw accurate cyclohexane chair conformations. Place axial and equatorial groups in their proper positions with correct bond angles.	Problems 3-45, 46, 47, 51, and 52
	Problem-Solving Strategy: Drawing Chair Conformations	Problems 3-45, 46, 47, 51, and 52
7	Predict the most stable conformations of mono and disubstituted cyclohexanes by placing the larger groups in equatorial rather than axial position.	Problems 3-45, 46, 47, and 51
	Problem-Solving Strategy: Recognizing Cis and Trans Isomers	Problems 3-45, 46, 47, 51, and 52

Study Problems

3-33 **(a)** There are 18 isomeric alkanes of molecular formula C_8H_{18}. Draw and name any eight of them.
(b) Draw and name the six isomeric cyclopentanes of molecular formula C_7H_{14}. These will include four constitutional isomers, of which two show geometric (cis-trans) stereoisomerism.

3-34 Which of the following structures represent the same compound? Which ones represent different compounds?

(a)

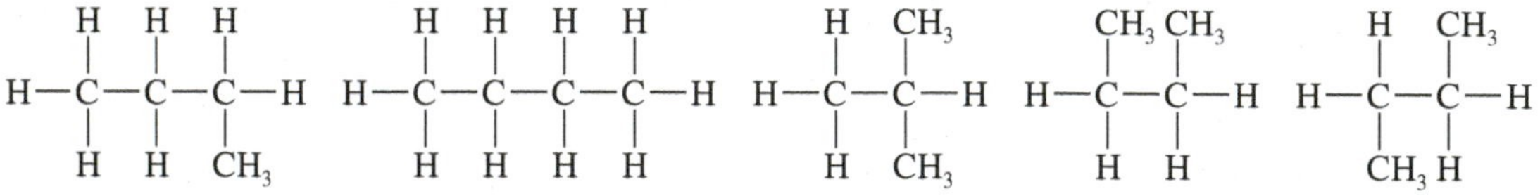

(b)

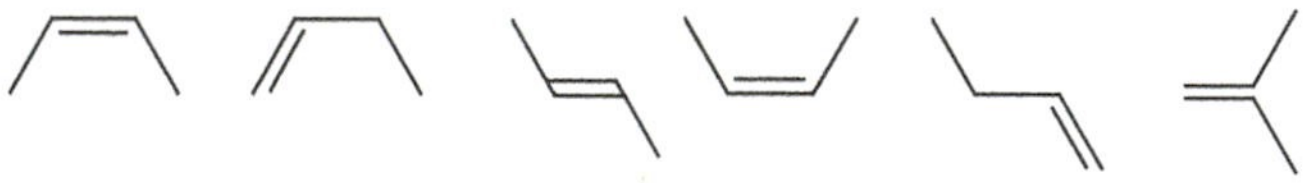

(c)

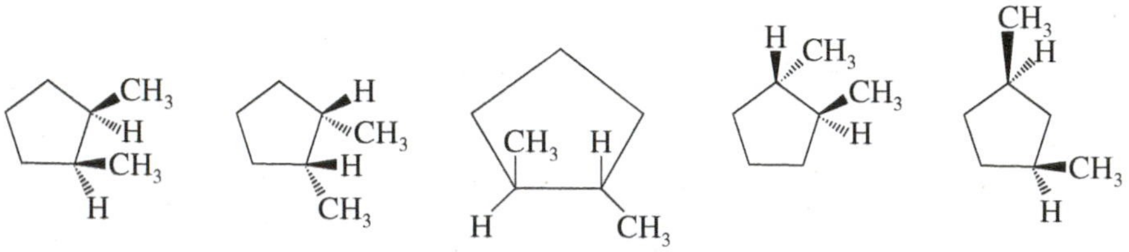

(d)

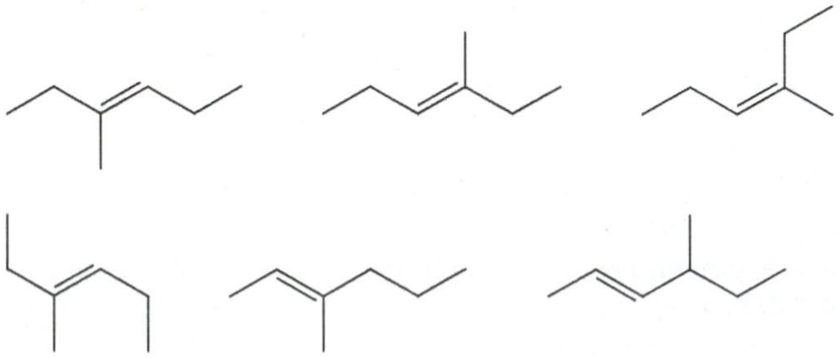

(e) CH_3, H, CH_3, H; CH_3, H, H, CH_3; H, CH_3, H, CH_3

CH_3, H, H, CH_3; H, CH_3, CH_3, H; H, CH_3, CH_3, H

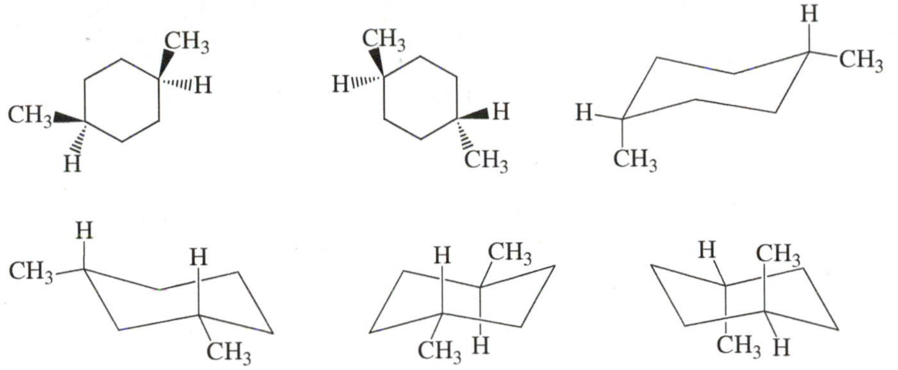

(f) Newman projections: CH_3, H, CH_3, H, CH_3, CH_3; H, H_3C, CH_3, H_3C, CH_3, H; CH_3, H, CH_3, H_3C, CH_3, H; H, H_3C, $CH(CH_3)_2$, H, H, H; H, H_3C, CH_2CH_3, H, H, CH_3

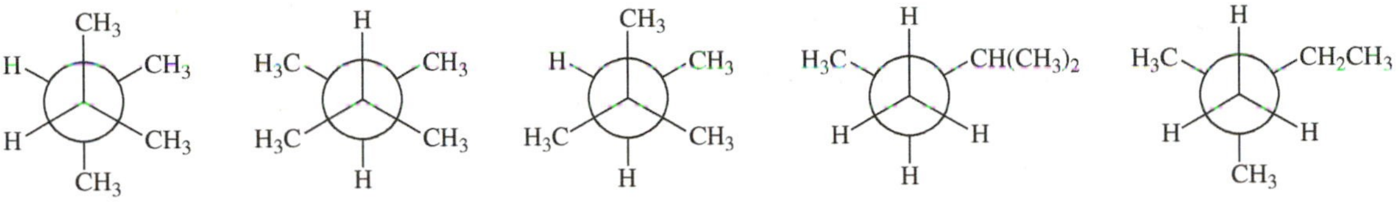

(g) Name the structures given in Problem 3-34, parts (a), (c), (e), and (f). Make sure your names are the same for structures that are the same, and different for structures that are different.

***3-35** (a) Draw and name the five cycloalkane structures of formula C_5H_{10}. Can any of these structures give rise to geometric (cis-trans) isomerism? If so, show the cis and trans stereoisomers.

(b) Draw and name the eight cycloalkane structures of formula C_6H_{12} that do not show geometric isomerism.

(c) Draw and name the four cycloalkanes of formula C_6H_{12} that do have cis-trans isomers.

3-36 Draw the structure that corresponds with each name.

(a) 3-ethyloctane
(b) 4-isopropyldecane
(c) *sec*-butylcycloheptane
(d) 2,3-dimethyl-4-propylnonane
(e) 2,2,4,4-tetramethylhexane
(f) *trans*-1,3-diethylcyclopentane
(g) *cis*-1-ethyl-4-methylcyclohexane
(h) isobutylcyclopentane
(i) *tert*-butylcyclohexane
(j) pentylcyclohexane
(k) cyclobutylcyclohexane
(l) *cis*-1-bromo-3-chlorocyclohexane

3-37 Each of the following descriptions applies to more than one alkane. In each case, draw and name two structures that match the description.

(a) an isopropylheptane
(b) a diethyldecane
(c) a *cis*-diethylcyclohexane
(d) a *trans*-dihalocyclopentane
(e) a (2,3-dimethylpentyl)cycloalkane
(f) a bicyclononane

3-38 Write structures for a homologous series of alcohols (R—OH) having from one to six carbons.

3-39 Give the IUPAC names of the following alkanes.

(a) $CH_3C(CH_3)_2CH(CH_2CH_3)CH_2CH_2CH(CH_3)_2$

(b) $CH_3CH_2—CH(CH_3CHCH_3)—CH_2CH_2—CH(CH_3CHCH_3)—CH_3$

(c)

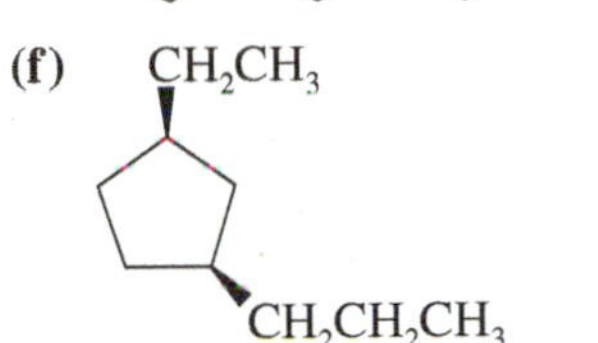

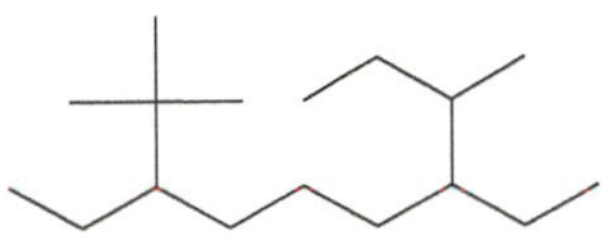

(d) cyclobutane with CH_2CH_3, CH_2CH_3, CH_3

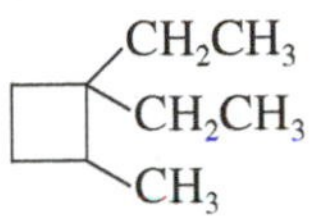

(e)

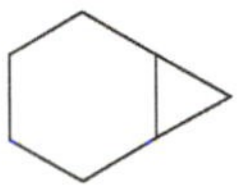

(f) CH_2CH_3, $CH_2CH_2CH_3$

(g) $C(CH_2CH_3)_3$

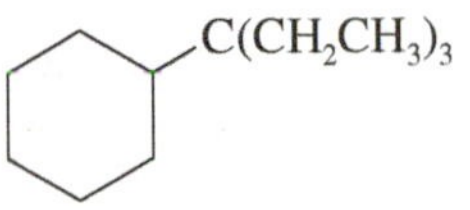

(h) CH_2CH_3, $CH(CH_3)_2$

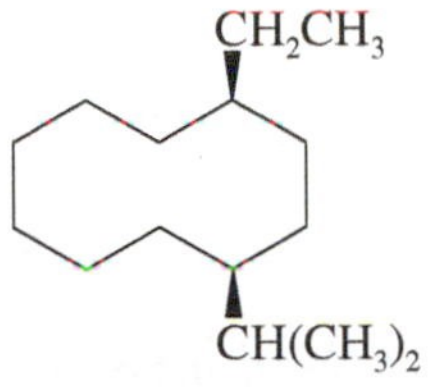

3-40 Construct a graph, similar to Figure 3-11, of the torsional energy of 3-methylpentane along the C2—C3 bond. Place C2 in front, represented by three bonds coming together in a Y shape, and C3 in back, represented by a circle with three bonds pointing out from it. Define the dihedral angle as the angle between the methyl group on the front carbon and the ethyl group on the back carbon. Begin your graph at the 0° dihedral angle and begin to turn the front carbon. Show the Newman projection and the approximate energy at each 60° of rotation. Indicate which conformations are the most stable (lowest energy) and the least stable (highest energy).

3-41 The following names are all incorrect or incomplete, but they represent real structures. Draw each structure and name it correctly.

(a) 2-ethylpentane
(b) 3-isopropylhexane
(c) 5-chloro-4-methylhexane
(d) 2-dimethylbutane
(e) 2-cyclohexylbutane
(f) 2,3-diethylcyclopentane

3-42 In each pair of compounds, which compound has the higher boiling point? Explain your reasoning.
(a) octane or 2,2,3-trimethylpentane **(b)** nonane or 2-methylheptane **(c)** 2,2,5-trimethylhexane or nonane

3-43 There are eight different five-carbon alkyl groups.
(a) Draw them. **(b)** Give them systematic names.
(c) In each case, label the degree of substitution (primary, secondary, or tertiary) of the head carbon atom bonded to the main chain.

3-44 Use a Newman projection about the indicated bond to draw the most stable conformer for each compound.
(a) 3-methylpentane about the C2—C3 bond **(b)** 3,3-dimethylhexane about the C3—C4 bond

3-45 **(a)** Draw the two chair conformations of *cis*-1,3-dimethylcyclohexane, and label all the positions as axial or equatorial.
(b) Label the higher-energy conformation and the lower-energy conformation.
(c) The energy difference in these two conformations has been measured to be about 23 kJ (5.4 kcal) per mole. How much of this energy difference is due to the torsional energy of gauche relationships?
(d) How much energy is due to the additional steric strain of the 1,3-diaxial interaction?

3-46 Draw the two chair conformations of each compound, and label the substituents as axial and equatorial. In each case, determine which conformation is more stable.
(a) *cis*-1-ethyl-2-isopropylcyclohexane **(b)** *trans*-1-ethyl-2-isopropylcyclohexane
(c) *cis*-1-ethyl-3-methylcyclohexane **(d)** *trans*-1-ethyl-3-methylcyclohexane
(e) *cis*-1-ethyl-4-methylcyclohexane **(f)** *trans*-1-ethyl-4-methylcyclohexane

3-47 Using what you know about the conformational energetics of substituted cyclohexanes, predict which of the two decalin isomers is more stable. Explain your reasoning.

3-48 Convert each Newman projection to the equivalent line–angle formula, and assign the IUPAC name.

(a) H; H, H; H, H; CH_2CH_3
(b) CH_2CH_3; H, CH_3; H, CH_3; CH_3
(c) Br; H, CH_3; H, H; CH_2CH_3
(d) CH_2CH_3; H, Cl; H, CH_3; $CH(CH_3)_2$
(e) $C(CH_3)_3$; Br, H; Cl, CH_3; CH_2CH_3
(f) CH_3; H, CH_3; H, H; $CH(CH_3)_2$
(g) F; H, CBr_3; F, F; H
(h) H; F, Cl; H, Br; CH_2CH_3
(i) $CH(CH_3)_2$; Br, H; Br, H; CH_2CH_3
(j) Br; H, CH_3; Cl, F; CH_2CH_3

***3-49** Draw Newman projections along the C3—C4 bond to show the most stable and least stable conformations of 3-ethyl-2,4,4-trimethylheptane.

***3-50** Conformational studies on ethane-1,2-diol ($HOCH_2$—CH_2OH) have shown the most stable conformation about the central C—C bond to be the gauche conformation, which is 9.6 kJ/mol (2.3 kcal/mol) more stable than the anti conformation. Draw Newman projections of these conformers, and explain this curious result.

3-51 The most stable form of the common sugar glucose contains a six-membered ring in the chair conformation with all the substituents equatorial. Draw this most stable conformation of glucose.

HO, O, CH_2OH, HO, OH, OH

glucose

3-52 This is a Newman projection of a substituted cyclohexane.
(a) Draw the equivalent chair form.
(b) Draw the equivalent structure using wedge and dash notation on a cyclohexane hexagon.
(c) Give the IUPAC name.

Br, H; H, H; H, CH_3; H, H

4 The Study of Chemical Reactions

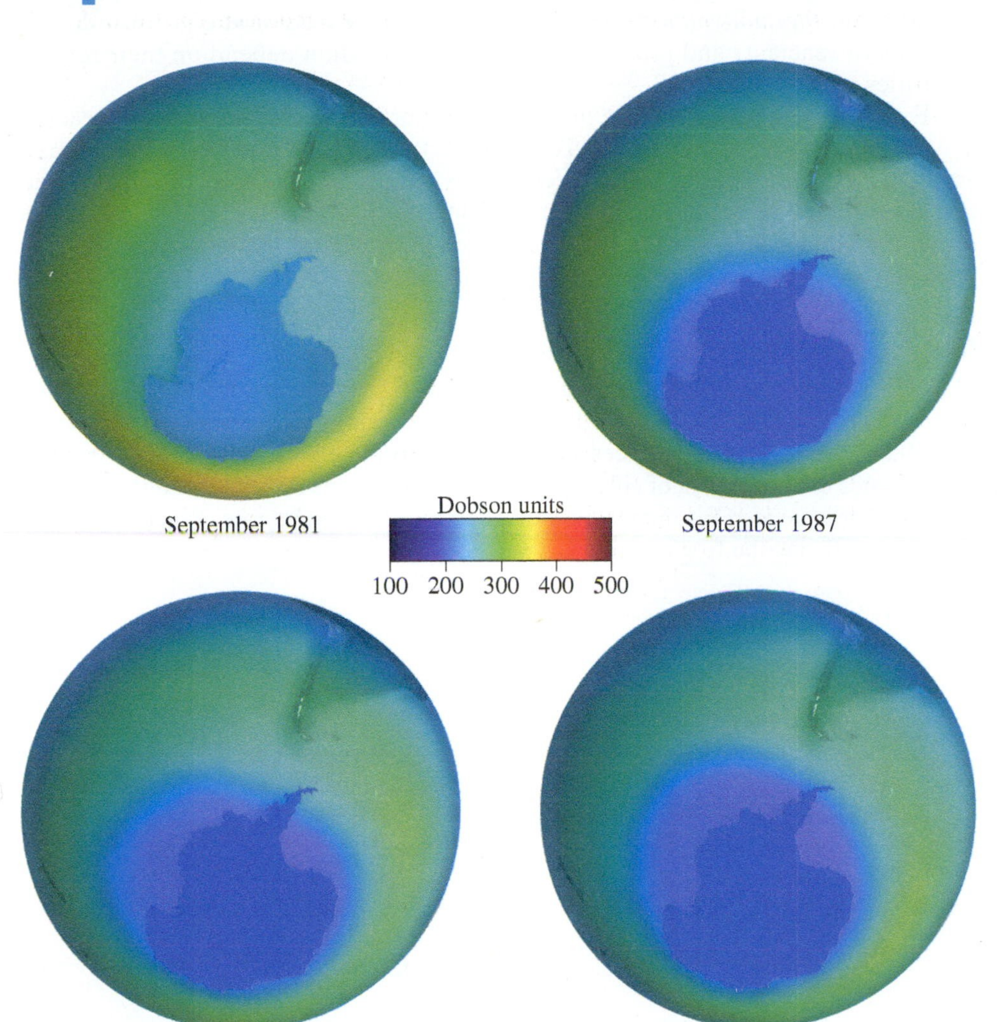

Goals for Chapter 4

1 Propose mechanisms and explain the steps for simple reactions such as free-radical halogenation.

2 Draw a reaction-energy diagram, and use it to identify the factors controlling the thermodynamics and kinetics of a reaction.

3 Use the mechanism, thermodynamics, and kinetics of a reaction to predict which of several possible products is the major product.

4 Identify reactive intermediates and explain their properties.

◀ This picture shows the "hole" in the protective ozone layer over Antarctica that resulted from the release of chlorofluorocarbons in refrigerants and aerosol spray cans. Chlorofluorocarbon refrigerants such as CF_2Cl_2 (Freon® 12) react with free radicals in the atmosphere to generate chlorine radicals, which catalyze decomposition of the ozone layer. The "hole" is now improving because the worst chlorofluorocarbons are no longer used.

4-1 Introduction

The most interesting part of organic chemistry begins now, with the study of making and breaking bonds. Organic chemists study reactions to learn how to use them to make interesting and useful products. We also want to understand how reactions occur so that we can predict new and useful reactions.

Nobody can remember thousands of specific organic reactions, but we can organize the reactions into logical groups based on how the reactions take place and what intermediates are involved. We begin our study by considering the **halogenation** of alkanes, a relatively simple substitution of a halogen for a hydrogen that can occur in the gas phase, without a solvent to complicate the reaction. In practice, alkanes are so unreactive that they are rarely used as starting materials for laboratory organic syntheses. We start with alkanes, however, because we have already studied their structure and properties, and their reactions are relatively uncomplicated. Once we have used alkanes to introduce the tools for studying reactions, we will apply those tools to a variety of more useful synthetic reactions.

Writing the overall equation, with the reactants on the left and the products on the right, is only the first step in our study of a reaction. If we truly want to understand a reaction, we must also know the *mechanism*, the step-by-step pathway from reactants to products. To predict how well the reaction goes to products, we study its *thermodynamics*, the energetics of the reaction at equilibrium. The amounts of reactants and products present at equilibrium depend on their relative stabilities.

Even though the equilibrium may favor the formation of a product, the reaction may not take place at a useful rate. To use a reaction in a realistic time period (and to keep the reaction from becoming violent), we study its *kinetics*, the variation of reaction rates with different conditions and concentrations of reagents. Understanding the reaction's kinetics helps us to propose reaction mechanisms that are consistent with the behavior we observe.

4-2 Chlorination of Methane

The chlorination of methane is an important industrial reaction, with a relatively simple mechanism that illustrates many of the important principles of a reaction. The reaction of methane with chlorine produces a mixture of chlorinated products, whose composition depends on the amount of chlorine added and also on the reaction conditions. Either light or heat is needed for the reaction to take place at a useful rate. When chlorine is added to methane, the first reaction is

$$\underset{\text{methane}}{CH_4} + \underset{\text{chlorine}}{Cl-Cl} \xrightarrow{\text{heat or light}} \underset{\text{chloromethane (methyl chloride)}}{CH_3Cl} + \underset{\text{hydrogen chloride}}{H-Cl}$$

This reaction may continue; heat (Δ) or light ($h\nu$) is needed for each step:

$$CH_3Cl \xrightarrow[(h\nu)]{Cl_2} CH_2Cl_2 + HCl \xrightarrow[(h\nu)]{Cl_2} CHCl_3 + HCl \xrightarrow[(h\nu)]{Cl_2} CCl_4 + HCl$$

This sequence raises several questions about the chlorination of methane. Why is heat or light needed for the reaction to go? Why do we get a mixture of products? Is there any way to modify the reaction to get just one pure product? Are the observed products formed because they are the most stable products possible? Or are they favored because they are formed faster than any other products?

The answers to these questions involve three aspects of the reaction: the mechanism, the thermodynamics, and the kinetics.

1. The **mechanism** is the complete, step-by-step description of exactly which bonds break and which bonds form, and in what order to give the observed products.
2. **Thermodynamics** is the study of the energy changes that accompany chemical and physical transformations. It allows us to compare the stability of reactants and products and predict which compounds are favored by the equilibrium.
3. **Kinetics** is the study of reaction rates, determining which products are formed fastest. Kinetics also helps to predict how the rate will change if we change the reaction conditions.

PROBLEM-SOLVING HINT

Free-radical reactions are some of the most common reactions in nature, but they are not the most common reactions used by organic chemists. The ionic reactions, covered in Chapters 6 and 7, are more common in the laboratory. We begin our study with free-radical reactions because they are not influenced by solvents or other variables that must be considered in the common ionic reactions.

We will use the chlorination of methane to show how to study a reaction. Before we can propose a detailed mechanism for the chlorination, we must learn everything we can about how the reaction works and what factors affect the reaction rate and the product distribution.

A careful study of the chlorination of methane has established three important characteristics:

1. *The chlorination does not occur at room temperature in the absence of light.* The reaction begins when light falls on the mixture or when it is heated. Thus, we know that this reaction requires some form of energy to *initiate* it.
2. *The most effective wavelength of light is a blue color that is strongly absorbed by chlorine gas.* This finding implies that absorbing light activates the chlorine molecule in some way so that it initiates the reaction with methane.
3. *The light-initiated reaction has a high quantum yield.* This means that many molecules of the product are formed for every photon of light absorbed. Our mechanism must explain how hundreds of individual reactions of methane with chlorine result from the absorption of a single photon by a single molecule of chlorine.

4-3 The Free-Radical Chain Reaction

A **chain reaction** mechanism has been proposed to explain the chlorination of methane. A chain reaction consists of three kinds of steps:

1. The **initiation step,** which generates a reactive intermediate.
2. **Propagation steps,** in which the reactive intermediate reacts with a stable molecule to form a product and another reactive intermediate, allowing the chain to continue until the supply of reactants is exhausted or the reactive intermediate is destroyed.
3. **Termination steps,** side reactions that destroy reactive intermediates and tend to slow or stop the reaction.

In studying the chlorination of methane, we will consider just the first reaction to form chloromethane (common name *methyl chloride*). This reaction is a **substitution:** Chlorine does not add to methane, but a chlorine atom *substitutes* for one of the hydrogen atoms, which becomes part of the HCl by-product.

$$\mathrm{CH_4} + \mathrm{Cl{-}Cl} \xrightarrow{\text{heat or light } (h\nu)} \mathrm{H_3C{-}Cl} + \mathrm{H{-}Cl}$$

methane | chlorine | chloromethane (methyl chloride) | substitution

4-3A The Initiation Step: Generation of Radicals

Initiation: Formation of reactive intermediates.

Blue light, absorbed by chlorine but not by methane, promotes this reaction. Therefore, initiation probably results from the absorption of light by a molecule of chlorine. Blue light has about the right energy to split a chlorine molecule (Cl_2) into two chlorine atoms, which requires 242 kJ/mol (58 kcal/mol).* The splitting of a chlorine molecule by absorption of a photon is shown as follows:

$$:\ddot{\mathrm{Cl}}:\ddot{\mathrm{Cl}}: + \text{photon } (h\nu) \longrightarrow :\ddot{\mathrm{Cl}}\cdot + \cdot\ddot{\mathrm{Cl}}:$$

radical electrons

*The energy of a photon of light is related to its frequency ν by the relationship $E = h\nu$, where h is Planck's constant. Blue light has an energy of about 250 kJ (60 kcal) per einstein (an einstein is a mole of photons).

Application: Aging

Free radicals may play a role in diseases and accelerate aging. In the course of everyday life, reactive oxygen species are encountered in the environment and produced in the body. These compounds break down into short-lived hydroxyl radicals, which can react with the body's proteins and DNA. The resulting damage accumulates and may result in heart disease, cancer, and premature aging.

Notice the fishhook-shaped half-arrows used to show the movement of single unpaired electrons. Just as we use curved arrows to represent the movement of electron *pairs*, we use these curved half-arrows to represent the movement of single electrons. These half-arrows show that the two electrons in the Cl—Cl bond separate, and one leaves with each chlorine atom.

The splitting of a Cl_2 molecule is an initiation step that produces two highly reactive chlorine atoms. A chlorine atom is an example of a **reactive intermediate,** a short-lived species that is never present in high concentration because it reacts as quickly as it is formed. Each Cl· atom has an odd number of valence electrons (seven), one of which is unpaired. The unpaired electron is called the *odd electron* or the *radical electron.* Species with unpaired electrons are called **radicals** or **free radicals.** Radicals are electron-deficient because they lack an octet. The odd electron readily combines with an electron in another atom to complete an octet and form a bond. Figure 4-1 shows the Lewis structures of some free radicals. Radicals are often represented by a structure with a single dot representing the unpaired odd electron.

PROBLEM 4-1

Draw Lewis structures for the following free radicals.

(a) The ethyl radical, $CH_3—\dot{C}H_2$ **(b)** The *tert*-butyl radical, $(CH_3)_3C\cdot$
(c) The isopropyl radical (2-propyl radical) **(d)** The iodine atom

4-3B Propagation Steps

Propagation: Formation of products with regeneration of reactive intermediates.

When a chlorine radical collides with a methane molecule, it abstracts (removes) a hydrogen atom from methane. One of the electrons in the C—H bond remains on carbon while the other combines with the odd electron on the chlorine atom to form the H—Cl bond.

First propagation step

$$H_3C{-}H \;+\; Cl\cdot \;\longrightarrow\; H_3C\cdot \;+\; H{-}Cl$$

methane + chlorine atom ⟶ methyl radical + hydrogen chloride

This step forms only one of the final products: the molecule of HCl. A later step must form chloromethane. Notice that the first propagation step begins with one free radical (the chlorine atom) and produces another free radical (the methyl radical). The regeneration of a free radical is characteristic of a propagation step of a chain reaction. The reaction can continue because another reactive intermediate is produced.

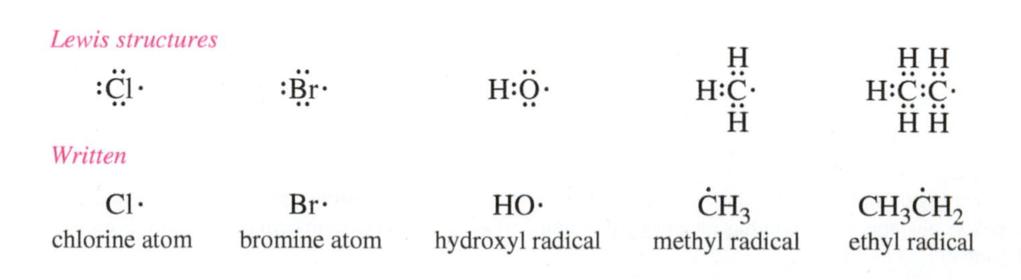

FIGURE 4-1
Free radicals are reactive species with odd numbers of electrons. The unpaired electron is represented by a single unpaired dot in the formula.

In the second propagation step, the methyl radical reacts with a molecule of chlorine to form chloromethane. The odd electron of the methyl radical combines with one of the two electrons in the Cl—Cl bond to give the Cl—CH_3 bond, and the chlorine atom is left with the odd electron.

Second propagation step

$$H_3C\cdot + Cl{-}Cl \longrightarrow H_3C{-}Cl + Cl\cdot$$

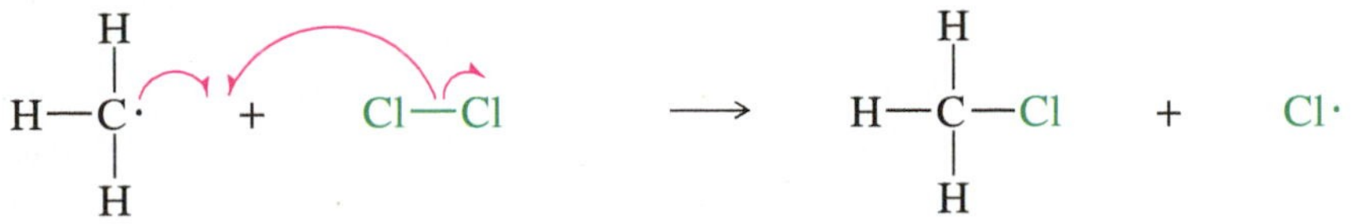

methyl radical · chlorine molecule · chloromethane · chlorine atom

In addition to forming chloromethane, the second propagation step produces another chlorine radical. The chlorine radical can react with another molecule of methane, giving HCl and a methyl radical, which reacts with Cl_2 to give chloromethane and regenerate yet another chlorine radical. In this way, the chain reaction continues until the supply of the reactants is exhausted or some other reaction consumes the radical intermediates. The chain reaction explains why hundreds of molecules of methyl chloride and HCl are formed by each photon of light that is absorbed. We can summarize the reaction mechanism as follows.

KEY MECHANISM 4-1 Free-Radical Halogenation

Like many other radical reactions, free-radical halogenation is a chain reaction. Chain reactions usually require one or more initiation steps to form radicals, followed by propagation steps that produce products and regenerate radicals.

Initiation: Radicals are formed.
Light supplies the energy to split a chlorine molecule.

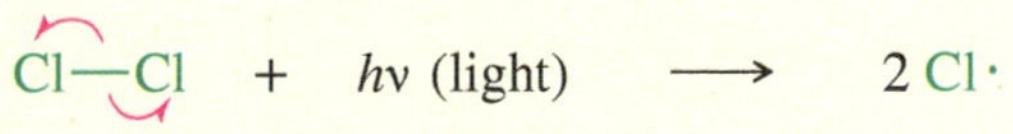

$$Cl{-}Cl + h\nu \text{ (light)} \longrightarrow 2\,Cl\cdot$$

Propagation: A radical reacts to generate another radical.
Step 1: A chlorine radical abstracts a hydrogen to generate an alkyl radical.

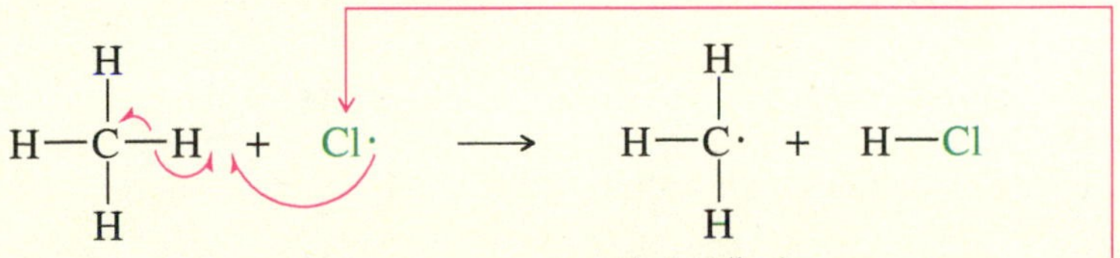

$$H_3C{-}H + Cl\cdot \longrightarrow H_3C\cdot + H{-}Cl$$

methyl radical

continues the chain

Step 2: The alkyl radical reacts with Cl_2 to generate the product and a chlorine radical.

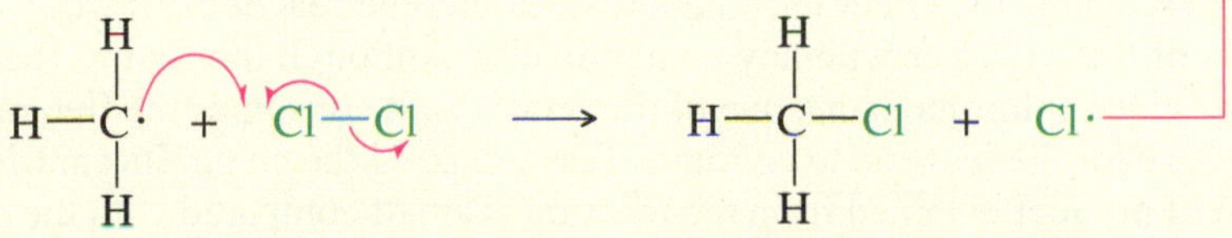

$$H_3C\cdot + Cl{-}Cl \longrightarrow H_3C{-}Cl + Cl\cdot$$

chloromethane

The chlorine radical generated in step 2 goes on to react in another step 1, continuing the chain.

(CONTINUED)

The overall reaction is simply the sum of the propagation steps:

$$\mathrm{H{-}\underset{H}{\overset{H}{\underset{|}{\overset{|}{C}}}}{-}H \;+\; Cl{-}Cl \;\longrightarrow\; H{-}\underset{H}{\overset{H}{\underset{|}{\overset{|}{C}}}}{-}Cl \;+\; H{-}Cl}$$

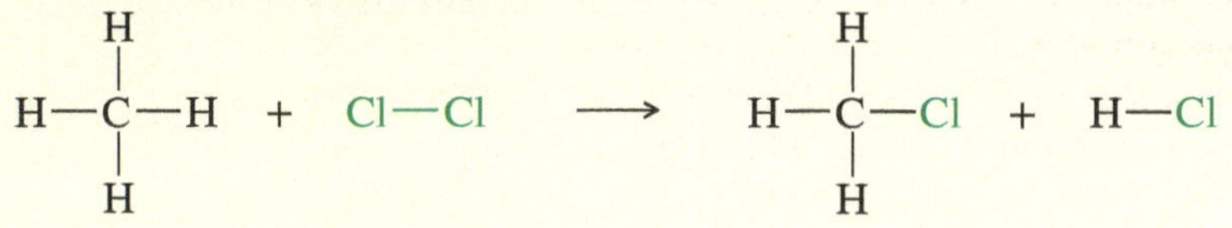

QUESTION: What factors characterize the propagation steps of a chain reaction?

PROBLEM 4-2

(a) Write the propagation steps leading to the formation of dichloromethane (CH_2Cl_2) from chloromethane.
(b) Explain why free-radical halogenation usually gives mixtures of products.
(c) How could an industrial plant control the proportions of methane and chlorine to favor production of CCl_4? To favor production of CH_3Cl?

4-3C Termination Reactions

Termination: Destruction of reactive intermediates.

If anything happens to consume some of the free-radical intermediates without generating new ones, the chain reaction will slow or stop. Such a side reaction is called a **termination reaction:** a step that produces fewer reactive intermediates (free radicals) than it consumes. Following are some of the possible termination reactions in the chlorination of methane:

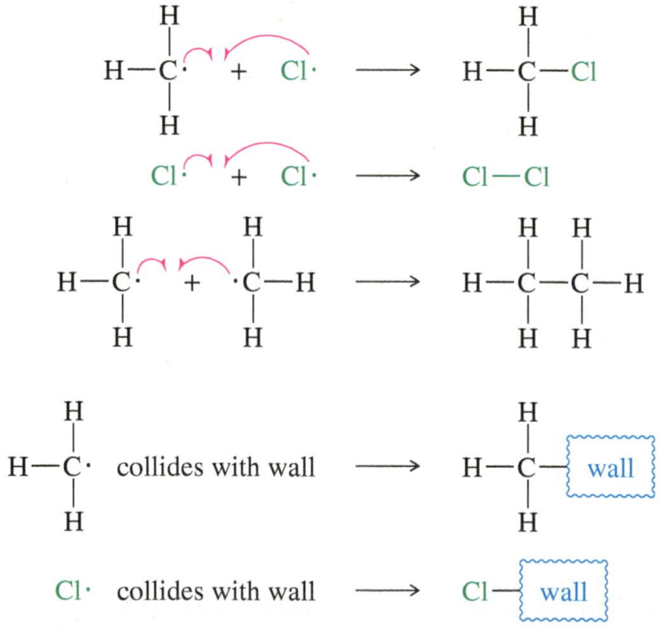

$$\mathrm{H{-}\underset{H}{\overset{H}{\underset{|}{\overset{|}{C}}}}\cdot \;+\; Cl\cdot \;\longrightarrow\; H{-}\underset{H}{\overset{H}{\underset{|}{\overset{|}{C}}}}{-}Cl}$$

$$\mathrm{Cl\cdot \;+\; Cl\cdot \;\longrightarrow\; Cl{-}Cl}$$

$$\mathrm{H{-}\underset{H}{\overset{H}{\underset{|}{\overset{|}{C}}}}\cdot \;+\; \cdot\underset{H}{\overset{H}{\underset{|}{\overset{|}{C}}}}{-}H \;\longrightarrow\; H{-}\underset{H}{\overset{H}{\underset{|}{\overset{|}{C}}}}{-}\underset{H}{\overset{H}{\underset{|}{\overset{|}{C}}}}{-}H}$$

$$\mathrm{H{-}\underset{H}{\overset{H}{\underset{|}{\overset{|}{C}}}}\cdot}\ \text{collides with wall} \;\longrightarrow\; \mathrm{H{-}\underset{H}{\overset{H}{\underset{|}{\overset{|}{C}}}}{-}}\text{wall}$$

$$\mathrm{Cl\cdot}\ \text{collides with wall} \;\longrightarrow\; \mathrm{Cl{-}}\text{wall}$$

The combination of any two free radicals is a termination step because it decreases the number of free radicals. Other termination steps involve reactions of free radicals with the walls of the vessel or with any contaminants. Although the first of these termination steps gives chloromethane, one of the products, it consumes the free radicals that are necessary for the reaction to continue, thus breaking the chain. Its contribution to the amount of product obtained from the reaction is small compared with the contribution of the propagation steps.

Application: Free-Radical Inhibitors

Some foods, additives, and supplements contain compounds that react with free radicals to terminate chain reactions. These radical inhibitors, called anti-oxidants, help to prevent the air-oxidation of foods by free-radical reactions. Current research is directed toward determining whether free-radical inhibitors in foods also can help cells to avoid damage caused by free radicals.

While a chain reaction is in progress, the concentration of radicals is very low. The probability that two radicals will combine in a termination step is lower than the probability that each will encounter a molecule of reactant and give a propagation step. The termination steps become important toward the end of the reaction, when there are relatively few molecules of reactants available. At this point, the free radicals are less likely to encounter a molecule of reactant than they are to encounter each other (or the wall of the container). The chain reaction quickly stops.

PROBLEM 4-3

Each of the following proposed mechanisms for the free-radical chlorination of methane is wrong. Explain how the experimental evidence disproves each mechanism.

(a) $Cl_2 + h\nu \rightarrow Cl_2^*$ (an "activated" form of Cl_2)
$Cl_2^* + CH_4 \rightarrow HCl + CH_3Cl$

(b) $CH_4 + h\nu \rightarrow \cdot CH_3 + H\cdot$
$\cdot CH_3 + Cl_2 \rightarrow CH_3Cl + Cl\cdot$
$Cl\cdot + H\cdot \rightarrow HCl$

PROBLEM-SOLVING HINT

In a free-radical chain reaction, initiation steps generally create new free radicals. Propagation steps usually combine a free radical and a reactant to give a product and another free radical. Termination steps generally decrease the number of free radicals.

PROBLEM 4-4

Free-radical chlorination of hexane gives very poor yields of 1-chlorohexane, while cyclohexane can be converted to chlorocyclohexane in good yield.

(a) How do you account for this difference?
(b) What ratio of reactants (cyclohexane and chlorine) would you use for the synthesis of chlorocyclohexane?

4-4 Equilibrium Constants and Free Energy

Now that we have determined a mechanism for the chlorination of methane, we can consider the energetics of the individual steps. Let's begin by reviewing some of the principles needed for this discussion.

Thermodynamics is the branch of chemistry that deals with the energy changes accompanying chemical and physical transformations. These energy changes are most useful for describing the properties of systems at **equilibrium.** Let's review how energy and entropy variables describe an equilibrium.

The equilibrium concentrations of reactants and products are governed by the **equilibrium constant** of the reaction. For example, if *a* moles of A and *b* moles of B react to give *c* moles of C and *d* moles of D, then the equilibrium constant K_{eq} is defined by the following equation:

$$aA + bB \rightleftharpoons cC + dD$$

$$K_{eq} = \frac{[\text{products}]}{[\text{reactants}]} = \frac{[C]^c[D]^d}{[A]^a[B]^b}$$

The value of K_{eq} tells us the position of the equilibrium: whether the products or the reactants are more stable, and therefore energetically favored. If K_{eq} is larger than 1, the reaction is favored as written from left to right. If K_{eq} is less than 1, the reverse reaction is favored (from right to left as written).

The chlorination of methane has a large equilibrium constant of about 1.1×10^{19}.

$$CH_4 + Cl_2 \rightleftharpoons CH_3Cl + HCl$$

$$K_{eq} = \frac{[CH_3Cl][HCl]}{[CH_4][Cl_2]} = 1.1 \times 10^{19}$$

The equilibrium constant for chlorination is so large that the remaining amounts of the reactants are close to zero at equilibrium. Such a reaction is said to *go to completion*, and the value of K_{eq} is a measure of the reaction's tendency to *go to completion*.

From the value of K_{eq}, we can calculate the change in **free energy** (sometimes called **Gibbs free energy**) that accompanies the reaction. Free energy is represented by G, and the change (Δ) in free energy associated with a reaction is represented by ΔG, the difference between the free energy of the products and the free energy of the reactants. ΔG is a measure of the amount of energy available to do work.

$$\Delta G = (\text{free energy of products}) - (\text{free energy of reactants})$$

Some chemists use the terms *exergonic* and *endergonic* to describe changes in free energy.
Exergonic = having a negative $\Delta G°$.
Endergonic = having a positive $\Delta G°$.

If the energy levels of the products are lower than the energy levels of the reactants (a "downhill" reaction), then the reaction is energetically favored; and this equation gives a negative value of ΔG, corresponding to a decrease in the energy of the system.

The **standard Gibbs free energy change,** $\Delta G°$, is most commonly used. The symbol ° designates a reaction involving reactants and products in their standard states (pure substances in their most stable states at 25 °C and 1 atm pressure). The relationship between $\Delta G°$ and K_{eq} is given by the expression

$$K_{eq} = e^{-\Delta G°/RT}$$

or, conversely, by

$$\Delta G° = -RT(\ln K_{eq}) = -2.303RT(\log_{10} K_{eq})$$

where

$R = 8.314$ J/kelvin-mol (1.987 cal/kelvin-mol), the gas constant
$T =$ absolute temperature, in kelvins*
$e = 2.718$, the base of natural logarithms

The value of RT at 25 °C is about 2.48 kJ/mol (0.592 kcal/mol).

The formula shows that a reaction is favored ($K_{eq} > 1$) if it has a *negative* value of $\Delta G°$ (energy is released). A reaction that has a positive value of $\Delta G°$ (energy must be added) is unfavorable. These predictions agree with our intuition that reactions should go from higher-energy states to lower-energy states, with a net decrease in free energy.

PROBLEM-SOLVING HINT

A reaction with a negative ΔG is favorable.
A reaction with a positive ΔG is unfavorable.
Compounds with lower free energy are more stable than those with higher free energy.

SOLVED PROBLEM 4-1

Calculate the value of $\Delta G°$ for the chlorination of methane.

SOLUTION

We know the value of K_{eq} for this reaction, so we can use the equation relating the free energy change with the equilibrium constant.

$$\Delta G° = -2.303RT(\log K_{eq})$$

K_{eq} for the chlorination is 1.1×10^{19}, and $\log K_{eq} = 19.04$

At 25 °C (about 298 °K), the value of RT is

$$RT = (8.314\ \text{J/kelvin-mol})(298\ \text{kelvin}) = 2478\ \text{J/mol, or } 2.48\ \text{kJ/mol}$$

Substituting, we have

$$\Delta G° = (-2.303)(2.478\ \text{kJ/mol})(19.04) = -108.7\ \text{kJ/mol}\ (-25.9\ \text{kcal/mol})$$

This is a large negative value for $\Delta G°$, showing that this chlorination has a large driving force that pushes it toward completion.

*Absolute temperatures (in kelvins) are correctly given without a degree sign, as in the equation 25 °C = 298 K. We will include the degree sign, however, to distinguish absolute temperatures (K) from equilibrium constants (K) as in 25 °C = 298 °K.

TABLE 4-1
Product Composition as a Function of ΔG° at 25 °C

G°			
kJ/mol	kcal/mol	K	Conversion to Products
+4.0	(+1.0)	0.20	17%
+2.0	(+0.5)	0.45	31%
0.0	(0.0)	1.0	50%
−2.0	(−0.5)	2.2	69%
−4.0	(−1.0)	5.0	83%
−8.0	(−1.9)	25	96%
−12.0	(−2.9)	127	99.2%
−16.0	(−3.8)	638	99.8%
−20.0	(−4.8)	3200	99.96%

In general, a reaction goes nearly to completion (>99%) for values of ΔG° that are more negative than about −12 kJ/mol or −3 kcal/mol. Table 4-1 shows what percentages of the starting materials are converted to products at equilibrium for reactions with various values of ΔG°.

PROBLEM 4-5

The following reaction has a value of $\Delta G^\circ = -2.1$ kJ/mol (−0.50 kcal/mol).

$$CH_3Br + H_2S \rightleftharpoons CH_3SH + HBr$$

(a) Calculate K_{eq} at room temperature (25 °C) for this reaction as written.
(b) Starting with a 1 M solution of CH_3Br and H_2S, calculate the final concentrations of all four species at equilibrium.

PROBLEM 4-6

Under base-catalyzed conditions, two molecules of acetone can condense to form diacetone alcohol. At room temperature (25 °C), about 5% of the acetone is converted to diacetone alcohol. Determine the value of ΔG° for this reaction.

$$2\ CH_3{-}C({=}O){-}CH_3 \xrightleftharpoons{^{-}OH} CH_3{-}C({=}O){-}CH_2{-}C(OH)(CH_3)_2$$

acetone ⇌ diacetone alcohol

4-5 Enthalpy and Entropy

Two factors contribute to the change in free energy: the change in **enthalpy** and the change in **entropy** multiplied by the temperature.

$$\Delta G^\circ = \Delta H^\circ - T\Delta S^\circ$$

$$\Delta G^\circ = \text{(free energy of products)} - \text{(free energy of reactants)}$$

$$\Delta H^\circ = \text{(enthalpy of products)} - \text{(enthalpy of reactants)}$$

$$\Delta S^\circ = \text{(entropy of products)} - \text{(entropy of reactants)}$$

At low temperatures, the enthalpy term (ΔH°) is usually much larger than the entropy term ($-T\Delta S^\circ$), and the entropy term is sometimes ignored.

4-5A Enthalpy

The **change in enthalpy** (ΔH°) is the heat of reaction—the amount of heat evolved or consumed in the course of a reaction, usually given in kilojoules (or kilocalories) per mole. The enthalpy change is a measure of the relative strength of bonding in the products and reactants. Reactions tend to favor products with the lowest enthalpy (those with the strongest bonds).

> **PROBLEM-SOLVING HINT**
>
> Breaking a bond is always endothermic and absorbs energy.
>
> Forming a bond is always exothermic and releases energy.

If weaker bonds are broken and stronger bonds are formed, heat is evolved and the reaction is **exothermic** (negative value of ΔH°). In an exothermic reaction, the enthalpy term makes a favorable negative contribution to ΔG°. If stronger bonds are broken and weaker bonds are formed, then energy is consumed in the reaction, and the reaction is **endothermic** (positive value of ΔH°). In an endothermic reaction, the enthalpy term makes an unfavorable positive contribution to ΔG°.

The value of ΔH° for the chlorination of methane is about -105.1 kJ/mol (-25.0 kcal/mol). This is a highly exothermic reaction, with the decrease in enthalpy serving as the primary driving force.

4-5B Entropy

Entropy is often described as a measure of randomness, disorder, or freedom of motion. Reactions tend to favor products with the greatest entropy. Notice the negative sign in the entropy term ($-T\Delta S^\circ$) of the free-energy expression. A positive value of the entropy change (ΔS°), indicating that the products have more freedom of motion than the reactants, makes a favorable (negative) contribution to ΔG°.

In many cases, the enthalpy change (ΔH°) is much larger than the entropy change (ΔS°), and the enthalpy term dominates the equation for ΔG°. Thus, a negative value of ΔS° does not necessarily mean that the reaction has an unfavorable value of ΔG°. The formation of strong bonds (the change in enthalpy) is usually the most important component in the driving force for a reaction.

In the chlorination of methane, the value of ΔS° is $+12.1$ J/kelvin-mole (2.89 cal/kelvin-mole). The $-T\Delta S^\circ$ term in the free energy is

$$-T\Delta S^\circ = -(298\ {}^\circ\text{K})(12.1\ \text{J/kelvin-mol}) = -3610\ \text{J/mol}$$
$$= -3.61\ \text{kJ/mol}\ (-0.86\ \text{kcal/mol})$$

The value of $\Delta G^\circ = -108.7$ kJ/mol is divided into enthalpy and entropy terms:

$$\Delta G^\circ = \Delta H^\circ - T\Delta S^\circ = -105.1\ \text{kJ/mol} - 3.61\ \text{kJ/mol}$$
$$= -108.7\ \text{kJ/mol}\ (-25.9\ \text{kcal/mol})$$

The enthalpy change is the largest factor in the driving force for chlorination. This is the case in most organic reactions: The entropy term is often small in relation to the enthalpy term. When we discuss chemical reactions involving the breaking and forming of bonds, we can often use the values of the enthalpy changes (ΔH°), under the assumption that $\Delta G^\circ \cong \Delta H^\circ$. We must be cautious in making this approximation, however, because some reactions have relatively small changes in enthalpy and larger changes in entropy.

SOLVED PROBLEM 4-2

Predict whether the value of ΔS° for the dissociation of Cl_2 is positive (favorable) or negative (unfavorable). What effect does the entropy term have on the sign of the value of ΔG° for this reaction?

$$Cl_2 \xrightarrow{h\nu} 2\ Cl\cdot$$

SOLUTION

Two isolated chlorine atoms have more freedom of motion than a single chlorine molecule. Therefore, the change in entropy is positive, and the entropy term ($-T\Delta S°$) is negative. This negative (favorable) value of ($-T\Delta S°$) is small, however, compared with the much larger, positive (unfavorable) value of $\Delta H°$ required to break the Cl—Cl bond. The chlorine molecule is much more stable than two chlorine atoms, showing that the positive enthalpy term predominates.

PROBLEM 4-7

When ethene is mixed with hydrogen in the presence of a platinum catalyst, hydrogen adds across the double bond to form ethane. At room temperature, the reaction goes to completion. Predict the signs of $\Delta H°$ and $\Delta S°$ for this reaction. Explain these signs in terms of bonding and freedom of motion.

$$H_2C{=}CH_2 + H_2 \xrightleftharpoons{\text{Pt catalyst}} H{-}CH_2{-}CH_2{-}H$$

ethene ethane

PROBLEM-SOLVING HINT

In general, two smaller molecules (or fragments, such as radicals) have more freedom of motion (greater entropy) than one larger molecule.

PROBLEM 4-8

For each reaction, estimate whether $\Delta S°$ for the reaction is positive, negative, or impossible to predict.

(a) $C_{10}H_{22}$ (*n*-decane) $\xrightarrow[\text{catalyst}]{\text{heat}}$ C_3H_6 (propene) + C_7H_{16} (heptane) (catalytic cracking)

(b) The formation of diacetone alcohol:

$$2\ CH_3{-}C({=}O){-}CH_3 \xrightleftharpoons{^-OH} CH_3{-}C({=}O){-}CH_2{-}C(OH)(CH_3)_2$$

(c) $$CH_3{-}C({=}O){-}OH + CH_3OH \xrightleftharpoons{H^+} CH_3{-}C({=}O){-}OCH_3 + H_2O$$

4-6 Bond-Dissociation Enthalpies

We can put known amounts of methane and chlorine into a bomb calorimeter and use a hot wire to initiate the reaction. The temperature rise in the calorimeter is used to calculate the precise value of the heat of reaction, $\Delta H°$. This measurement shows that 105 kJ (25 kcal) of heat is evolved (exothermic) for each mole of methane converted to chloromethane. Thus, $\Delta H°$ for the reaction is negative, and the heat of reaction is given as

$$\Delta H° = -105\ \text{kJ/mol}\ (-25\ \text{kcal/mol})$$

In many cases, we want to predict whether a particular reaction will be endothermic or exothermic, without actually measuring the heat of reaction. We can calculate an approximate heat of reaction by adding and subtracting the energies involved in the breaking and forming of bonds. To do this calculation, we need to know the energies of the affected bonds.

The **bond-dissociation enthalpy (BDE,** also called **bond-dissociation energy)** is the amount of enthalpy required to break a particular bond **homolytically**—that is, in such a way that each bonded atom retains one of the bond's two electrons. In contrast, when a bond is broken **heterolytically** (as in an acid–base reaction), one of the atoms retains both electrons.

Homolytic cleavage (free radicals result)

$$A{:}B \longrightarrow A\cdot + \cdot B \qquad \Delta H^\circ = \text{bond-dissociation enthalpy}$$

$$:\ddot{Cl}:\ddot{Cl}: \longrightarrow 2\,:\ddot{Cl}\cdot \qquad \Delta H^\circ = 242\text{ kJ/mol (58 kcal/mol)}$$

Heterolytic cleavage (ions result)

$$A{:}B \longrightarrow A^+ + {}^-{:}B$$

$$(CH_3)_3C{-}\ddot{Cl}: \longrightarrow (CH_3)_3C^+ + :\ddot{Cl}:^- \quad (\Delta H^\circ \text{ varies with solvent})$$

Homolytic cleavage (radical cleavage) forms free radicals, while **heterolytic cleavage (ionic cleavage)** forms ions. Enthalpies for heterolytic (ionic) cleavage depend strongly on the solvent's ability to solvate the ions that result. Homolytic cleavage is used to define bond-dissociation enthalpies because the values do not vary so much with different solvents or with no solvent. Note that a curved arrow is used to show the movement of the electron pair in an ionic cleavage and that curved half-arrows are used to show the separation of individual electrons in a homolytic cleavage.

Energy is released when bonds are formed, and energy is consumed to break bonds. Therefore, bond-dissociation enthalpies are always positive (endothermic). The overall enthalpy change for a reaction is the sum of the dissociation enthalpies of the bonds broken minus the sum of the dissociation enthalpies of the bonds formed.

$$\Delta H^\circ = \Sigma(\text{BDE of bonds broken}) - \Sigma(\text{BDE of bonds formed})$$

For the hypothetical reaction

$$A{-}B + C{-}D \rightleftharpoons A{-}C + B{-}D$$

$$\Delta H^\circ = (\text{BDE of } A{-}B) + (\text{BDE of } C{-}D) - (\text{BDE of } A{-}C) - (\text{BDE of } B{-}D)$$

By studying the heats of reaction for many different reactions, chemists have developed reliable tables of bond-dissociation enthalpies. Table 4-2 gives the bond-dissociation enthalpies for the homolysis of bonds in a variety of molecules.

4-7 Enthalpy Changes in Chlorination

We can use values from Table 4-2 to predict the heat of reaction for the chlorination of methane. This reaction involves the breaking (positive values) of a $CH_3{-}H$ bond and a $Cl{-}Cl$ bond, and the formation (negative values) of a $CH_3{-}Cl$ bond and a $H{-}Cl$ bond.

Overall reaction

$$CH_3{-}H + Cl{-}Cl \longrightarrow CH_3{-}Cl + H{-}Cl$$

Bonds broken	*ΔH° (per mole)*	*Bonds formed*	*ΔH° (per mole)*
$Cl{-}Cl$	+240 kJ (+57 kcal)	$H{-}Cl$	−432 kJ (−103 kcal)
$CH_3{-}H$	+439 kJ (+105 kcal)	$CH_3{-}Cl$	−350 kJ (−84 kcal)
Total	+679 kJ (+162 kcal)	Total	−782 kJ (−187 kcal)

$$\Delta H^\circ = +679\text{ kJ/mol} + (-782)\text{ kJ/mol} = -103\text{ kJ/mol } (-25\text{ kcal/mol})$$

TABLE 4-2
Bond-Dissociation Enthalpies for Homolytic Cleavages

$$A{:}B \longrightarrow A\cdot + \cdot B$$

Bond	Bond-Dissociation Enthalpy kJ/mol	Bond-Dissociation Enthalpy kcal/mol
H—X bonds and X—X bonds		
H—H	436	104
D—D	440	105
F—F	154	37
Cl—Cl	240	57
Br—Br	190	45
I—I	149	36
H—F	570	136
H—Cl	432	103
H—Br	366	88
H—I	298	71
HO—H	497	119
HS—H	382	91
HOO—H	367	88
HO—OH	213	51
H_2N—H	450	108
Methyl bonds		
CH_3—H	439	105
CH_3—F	481	115
CH_3—Cl	350	84
CH_3—Br	302	72
CH_3—I	241	58
CH_3—OH	385	92
CH_3—NH_2	356	85
Bonds to primary carbons		
CH_3CH_2—H	423	101
CH_3CH_2—F	464*	111*
CH_3CH_2—Cl	355	85
CH_3CH_2—Br	303	72
CH_3CH_2—I	238	57
CH_3CH_2—OH	393	94
$CH_3CH_2CH_2$—H	423	101
$CH_3CH_2CH_2$—F	464*	111*
$CH_3CH_2CH_2$—Cl	355	85
$CH_3CH_2CH_2$—Br	303	72
$CH_3CH_2CH_2$—I	238	57
Bonds to secondary carbons		
$(CH_3)_2CH$—H	413	99
$(CH_3)_2CH$—F	463	111
$(CH_3)_2CH$—Cl	356	85
$(CH_3)_2CH$—Br	309	74
$(CH_3)_2CH$—I	238	57
$(CH_3)_2CH$—OH	399	95
Bonds to tertiary carbons		
$(CH_3)_3C$—H	403	96
$(CH_3)_3C$—F	464*	111*
$(CH_3)_3C$—Cl	355	85
$(CH_3)_3C$—Br	304	73
$(CH_3)_3C$—I	233	56
$(CH_3)_3C$—OH	401	96
Other C—H bonds		
$PhCH_2$—H (benzylic)	376	90
$H_2C{=}CHCH_2$—H (allylic)	372	89
$H_2C{=}CH$—H (vinyl)	463	111
Ph—H (aromatic)	472	113
$HC{\equiv}C$—H (acetylenic)	558	133
C—C bonds		
CH_3—CH_3	377	90
CH_3CH_2—CH_3	372	89
CH_3CH_2—CH_2CH_3	368	88
$(CH_3)_2CH$—CH_3	371	89
$(CH_3)_3C$—CH_3	366	87
C=C and C=X bonds		
$H_2C{=}CH_2$	728	174
$H_2C{=}NH$	736*	176
$H_2C{=}O$	749	179

*approximate

The bond-dissociation enthalpies also provide the heat of reaction for each individual step:

First propagation step

$$Cl\cdot + CH_4 \longrightarrow \cdot CH_3 + HCl$$

Breaking a CH_3—H bond	+439 kJ/mol (+105 kcal/mol)
Forming a H—Cl bond	−432 kJ/mol (−103 kcal/mol)
Step total	+7 kJ/mol (+2 kcal/mol)

Second propagation step

$$\cdot CH_3 + Cl_2 \longrightarrow CH_3Cl + Cl\cdot$$

Breaking a Cl—Cl bond	+240 kJ/mol (+57 kcal/mol)
Forming a CH_3—Cl bond	−350 kJ/mol (−84 kcal/mol)
Step total	−110 kJ/mol (−27 kcal/mol)

Grand total = +7 kJ/mol + (−110 kJ/mol) = −103 kJ/mol (−25 kcal/mol)

PROBLEM-SOLVING HINT

Bond-dissociation enthalpies are for breaking bonds, which costs energy. In calculating values of $\Delta H°$, use positive BDE values for bonds that are broken and negative values for bonds that are formed.

The sum of the values of $\Delta H°$ for the individual propagation steps gives the overall enthalpy change for the reaction. The initiation step, $Cl_2 \rightarrow 2\ Cl\cdot$, is not added to give the overall enthalpy change because it is not necessary for each molecule of product formed. The first splitting of a chlorine molecule simply begins the chain reaction, which generates hundreds or thousands of molecules of chloromethane. The energy needed to break the Cl—Cl bond is already included in the second propagation step.

PROBLEM 4-9

(a) Propose a mechanism for the free-radical chlorination of ethane,

$$CH_3{-}CH_3 + Cl_2 \xrightarrow{h\nu} CH_3{-}CH_2Cl + HCl$$

(b) Calculate $\Delta H°$ for each step in this reaction.
(c) Calculate the overall value of $\Delta H°$ for this reaction.

Alternative Mechanism The mechanism we have used is not the only one that might be proposed to explain the reaction of methane with chlorine. We know that the initiating step must be the splitting of a molecule of Cl_2, but there are other propagation steps that would form the correct products:

(a) $Cl\cdot + CH_3{-}H \longrightarrow CH_3{-}Cl + H\cdot$ $\quad \Delta H° = +439\ kJ - 350\ kJ = +89\ kJ$ (+105 kcal − 84 kcal = +21 kcal)
(b) $H\cdot + Cl{-}Cl \longrightarrow H{-}Cl + Cl\cdot$ $\quad \Delta H° = +240\ kJ - 432\ kJ = -192\ kJ$ (+ 57 kcal − 103 kcal = −46 kcal)

Total −103 kJ (Total −25 kcal)

This alternative mechanism seems plausible, but step (a) is endothermic by 89 kJ/mol (21 kcal/mol). Reactions generally follow the lowest-energy pathway available, and the previous mechanism provides a lower-energy alternative. When a chlorine atom collides with a methane molecule, it will not react to give methyl chloride and a hydrogen atom ($\Delta H°$ = +89 kJ = +21 kcal); it will react to give HCl and a methyl radical ($\Delta H°$ = +7 kJ = +2 kcal), the first propagation step of the correct mechanism.

PROBLEM 4-10

(a) Using bond-dissociation enthalpies from Table 4-2 (page 167), calculate the heat of reaction for each step in the free-radical bromination of methane.

$$Br_2 + CH_4 \xrightarrow{\text{heat or light}} CH_3Br + HBr$$

(b) Calculate the overall heat of reaction.

4-8 Kinetics and the Rate Equation

Kinetics is the study of reaction rates. How fast a reaction goes is just as important as the position of its equilibrium. Just because thermodynamics favors a reaction (negative $\Delta G°$) does not necessarily mean the reaction will actually occur. For example, a mixture of gasoline and oxygen does not react without a spark or a catalyst. Similarly, a mixture of methane and chlorine does not react if it is kept cold and dark.

The **rate of a reaction** is a measure of how fast the products appear and the reactants disappear. We can determine the rate by measuring the increase in the concentrations of the products with time, or the decrease in the concentrations of the reactants with time.

Reaction rates depend on the concentrations of the reactants. The greater the concentrations, the more often the reactants collide and the greater the chance of reaction. A **rate equation** (sometimes called a **rate law**) is the relationship between the concentrations of the reactants and the observed reaction rate. Each reaction has its own rate equation, *determined experimentally* by changing the concentrations of the reactants and measuring the change in the rate. For example, consider the general reaction

$$A + B \longrightarrow C + D$$

The reaction rate is usually proportional to the concentrations of the reactants ([A] and [B]) raised to some powers, a and b. We can use a general rate expression to represent this relationship as

$$\text{rate} = k_r[A]^a[B]^b$$

where k_r is the **rate constant,** and the values of the powers (a and b) must be determined experimentally. We cannot guess or calculate the rate equation from just the stoichiometry of the reaction. The rate equation depends on the mechanism of the reaction and on the rates of the individual steps.

In the general rate equation, the power a is called the **order** of the reaction with respect to reactant A, and b is the order of the reaction with respect to B. The sum of these powers, $(a + b)$, is called the **overall order** of the reaction.

The following reaction has a simple rate equation:

$$CH_3{-}Br \quad + \quad ^-OH \xrightarrow{H_2O/\text{acetone}} CH_3{-}OH \quad + \quad Br^-$$

Experiments show that doubling the concentration of methyl bromide, $[CH_3Br]$, doubles the rate of reaction. Doubling the concentration of hydroxide ion, $[^-OH]$, also doubles the rate. Thus, the rate is proportional to both $[CH_3Br]$ and $[^-OH]$, so the rate equation has the following form:

$$\text{rate} = k_r[CH_3Br][^-OH]$$

This rate equation is *first order* in each of the two reagents because it is proportional to the first power of their concentrations. The rate equation is *second order overall* because the sum of the powers of the concentrations in the rate equation is 2; that is, (first order) + (first order) = second order overall.

Reactions of the same overall type do not necessarily have the same form of rate equation. For example, the following similar reaction has a different kinetic order:

$$(CH_3)_3C{-}Br \quad + \quad ^-OH \xrightarrow{H_2O/\text{acetone}} (CH_3)_3C{-}OH \quad + \quad Br^-$$

Doubling the concentration of *tert*-butyl bromide $[(CH_3)_3C{-}Br]$ causes the rate to double, but doubling the concentration of hydroxide ion $[^-OH]$ has no effect on the rate of this particular reaction. The rate equation is

$$\text{rate} = k_r[(CH_3)_3C{-}Br]$$

This reaction is first order in *tert*-butyl bromide, and zeroth order in hydroxide ion (proportional to [$^{-}$OH] to the zeroth power). It is first order overall. This reaction is zeroth order in hydroxide ion because the slow step involves only *tert*-butyl bromide and not hydroxide ion:

$$(CH_3)_3C{-}Br \rightleftharpoons (CH_3)_3C^+ + Br^-$$

The most important fact to remember is that *the rate equation must be determined experimentally*. We cannot predict the form of the rate equation from the stoichiometry of the reaction. We determine the rate equation experimentally, then use that information to propose consistent mechanisms.

SOLVED PROBLEM 4-3

Chloromethane reacts with dilute sodium cyanide ($Na^+\ {}^-C{\equiv}N$) according to the following equation:

$$\underset{\text{chloromethane}}{CH_3{-}Cl} + \underset{\text{cyanide}}{{}^-C{\equiv}N} \longrightarrow \underset{\text{acetonitrile}}{CH_3{-}C{\equiv}N} + \underset{\text{chloride}}{Cl^-}$$

When the concentration of chloromethane is doubled, the rate is observed to double. When the concentration of cyanide ion is tripled, the rate is observed to triple.

(a) What is the kinetic order with respect to chloromethane?
(b) What is the kinetic order with respect to cyanide ion?
(c) What is the kinetic order overall?
(d) Write the rate equation for this reaction.

SOLUTION

(a) When [CH_3Cl] is doubled, the rate doubles, which is 2 to the first power. The reaction is first order with respect to chloromethane.
(b) When [$^{-}$CN] is tripled, the reaction rate triples, which is 3 to the first power. The reaction is first order with respect to cyanide ion.
(c) First order plus first order equals second order overall.
(d) rate = $k_r[CH_3Cl][{}^-CN]$

PROBLEM 4-11

The reaction of *tert*-butyl chloride with methanol

$$\underset{\textit{tert}\text{-butyl chloride}}{(CH_3)_3C{-}Cl} + \underset{\text{methanol}}{CH_3{-}OH} \longrightarrow \underset{\text{methyl }\textit{tert}\text{-butyl ether}}{(CH_3)_3C{-}OCH_3} + HCl$$

is found to follow the rate equation

$$\text{rate} = k_r[(CH_3)_3C{-}Cl]$$

(a) What is the kinetic order with respect to *tert*-butyl chloride?
(b) What is the kinetic order with respect to methanol?
(c) What is the kinetic order overall?

PROBLEM 4-12

Under certain conditions, the bromination of cyclohexene follows an unusual rate law:

$$\text{cyclohexene} + Br_2 \longrightarrow \text{1,2-dibromocyclohexane} \qquad \text{rate} = k_r[\text{cyclohexene}][Br_2]^2$$

(a) What is the kinetic order with respect to cyclohexene?
(b) What is the kinetic order with respect to bromine?
(c) What is the overall kinetic order?

PROBLEM 4-13

When a small piece of platinum is added to a mixture of ethene and hydrogen, the following reaction occurs:

$$\mathrm{H_2C{=}CH_2} + \mathrm{H_2} \xrightarrow{\text{Pt catalyst}} \mathrm{H{-}CH_2{-}CH_2{-}H}$$

ethene + hydrogen → ethane

Doubling the concentration of hydrogen has no effect on the reaction rate. Doubling the concentration of ethene also has no effect.

(a) What is the kinetic order of this reaction with respect to ethene? With respect to hydrogen? What is the overall order?

(b) Write the unusual rate equation for this reaction.

(c) Explain this strange rate equation, and suggest what one might do to accelerate the reaction.

4-9 Activation Energy and the Temperature Dependence of Rates

Each reaction has its own characteristic rate constant, k_r. Its value depends on the conditions of the reaction, especially the temperature. This temperature dependence is expressed by the *Arrhenius equation*,

$$k_r = Ae^{-E_a/RT}$$

where

A = a constant (the "frequency factor")
E_a = activation energy
R = the gas constant, 8.314 J/kelvin-mole (1.987 cal/kelvin-mole)
T = the absolute temperature

The **activation energy,** E_a, is the minimum kinetic energy the molecules must have to overcome the repulsions between their electron clouds when they collide. The exponential term $e^{-E_a/RT}$ corresponds to the fraction of collisions in which the particles have the minimum energy E_a needed to react. We can calculate E_a for a reaction by measuring how k_r varies with temperature, and substituting into the Arrhenius equation.

The *frequency factor A* accounts for the frequency of collisions and the fraction of collisions with the proper orientation for the reaction to occur. In most cases, only a small fraction of collisions occur between molecules with enough speed and with just the right orientation for reaction to occur. Far more collisions occur without enough kinetic energy or without the proper orientation, and the molecules simply bounce off each other.

The Arrhenius equation implies that the rate of a reaction depends on the fraction of collisions with kinetic energy of at least E_a. Figure 4-2 shows how the distribution of kinetic energies in a sample of a gas depends on the temperature. The black curved line shows the molecular energy distribution at room temperature, and the dashed lines show the energy needed to overcome barriers of 4 kJ/mol (1 kcal/mol), 40 kJ/mol (10 kcal/mol), and 80 kJ (19 kcal/mol). The area under the curve to the right of each barrier corresponds to the fraction of molecules with enough energy to overcome that barrier.

The red curve shows how the energy distribution is shifted at 100 °C. At 100 °C, many more molecules have the energy needed to overcome the energy barriers, especially the 80 kJ/mol barrier. For smaller temperature changes, chemists often use an approximation: for reactions with typical activation energies of about 40 to 60 kJ/mol (10 to 15 kcal/mol), the reaction rate approximately doubles when the temperature is raised by 10 °C, as from 27 °C (near room temperature) to 37 °C (body temperature).

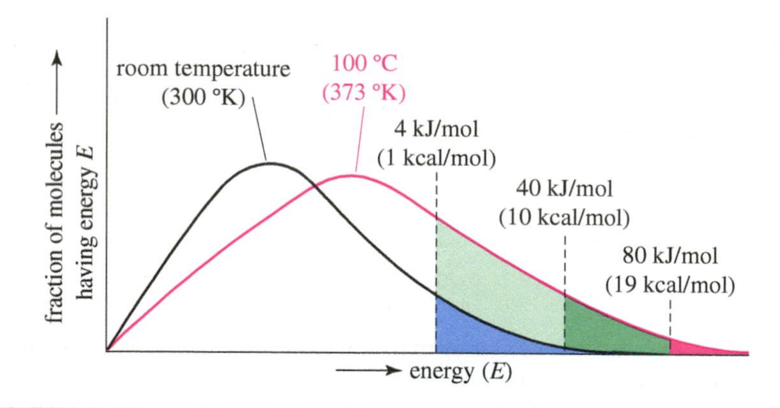

FIGURE 4-2
The dependence of kinetic energies on temperature. This graph shows how the fraction of molecules with a given activation energy decreases as the activation energy increases. At a higher temperature (red curve), more collisions have the needed energy.

Because the rate constant increases quickly when the temperature is raised, it might seem that raising the temperature would always be a good way to save time by making reactions go faster. The problem with raising the temperature is that *all* reactions are accelerated, including all the unwanted side reactions. We try to find a temperature that allows the desired reaction to go at a reasonable rate without producing unacceptable rates of side reactions.

4-10 Transition States

The activation energy E_a represents the energy difference between the reactants and the **transition state,** the highest-energy state in a molecular collision that leads to reaction. In effect, the activation energy is the barrier that must be overcome for the reaction to take place. The value of E_a is always positive, and its magnitude depends on the relative energy of the transition state. The term *transition state* implies that this configuration is the transition between the reactants and products, and the molecules can either go on to products or return to reactants.

Unlike the reactants or products, a transition state is unstable and cannot be isolated. It is not an intermediate, because an **intermediate** is a species that exists for some finite length of time, even if it is very short. An intermediate has at least some stability, but the transition state is a transient on the path from one intermediate to another. The transition state is often symbolized by a superscript double dagger (‡), and the changes in variables such as free energy, enthalpy, and entropy involved in achieving the transition state are symbolized $\Delta G^{\ddagger}$, $\Delta H^{\ddagger}$, and $\Delta S^{\ddagger}$. $\Delta G^{\ddagger}$ is similar to E_a, and the symbol $\Delta G^{\ddagger}$ is often used in speaking of the activation energy.

Transition states have high energies because bonds must begin to break before other bonds can form. The following equation shows the reaction of a chlorine radical with methane. The transition state shows the C—H bond partially broken and the H—Cl bond partially formed. Transition states are often enclosed by brackets to emphasize their transient nature. The double dagger next to the bracket means that this is a transition state, and the dot means that it contains a radical (unpaired) electron.

$$CH_4 + \cdot Cl \rightleftharpoons \left[H_3C\cdots H\cdots Cl \right]^{\ddagger} \longrightarrow H_3C\cdot + H—Cl$$

transition state

Reaction-Energy Diagrams The concepts of transition state and activation energy are easier to understand graphically. Figure 4-3 shows a **reaction-energy diagram** for a one-step exothermic reaction. The vertical axis of the energy diagram represents the total potential energy of all the species involved in the reaction. The horizontal axis is called the **reaction coordinate.** The reaction coordinate symbolizes the progress of the reaction, going from the reactants on the left to the products on the right.

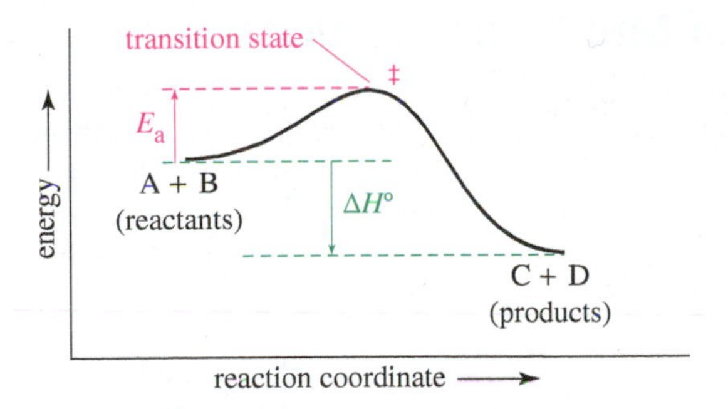

FIGURE 4-3
Reaction-energy diagram for a one-step exothermic reaction. The reactants are toward the left, and the products are toward the right. The vertical axis represents the potential energy. The transition state is the highest point on the graph, and the activation energy is the energy difference between the reactants and the transition state.

The transition state is the highest point on the graph, and the activation energy is the energy difference between the reactants and the transition state. The heat of reaction ($\Delta H°$) is the difference in energy between the reactants and the products.

If a **catalyst** were added to the reaction in Figure 4-3, it would create a transition state of lower energy, thereby lowering the activation energy and increasing the reaction rate. Addition of a catalyst would not change the energies of the reactants and products, however, so the heat of reaction and the equilibrium constant would be unaffected.

Application: Biochemistry

Enzymes serve as biological catalysts. They speed up reactions without changing the energies of the reactants (called substrates) and products. Without enzymes, most of the reactions in our cells would not go fast enough to keep us alive. The enzyme shown is carbonic anhydrase, which increases the rate of the reaction $CO_2 + H_2O \rightleftarrows H_2CO_3$ by a factor of 7.7×10^6. This rate enhancement corresponds to a reduction in E_a of 40 kJ/mol (9.5 kcal/mol).

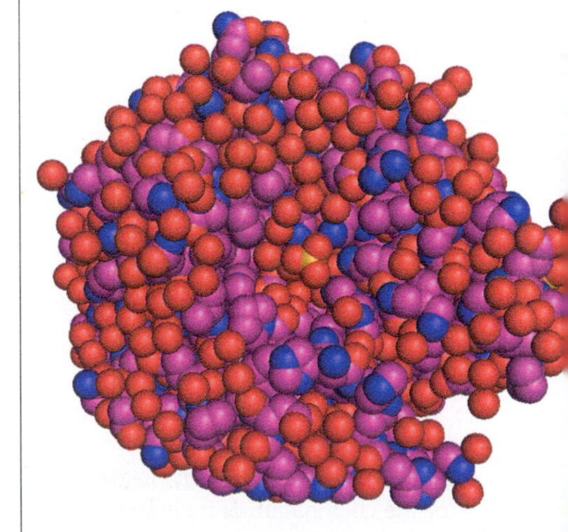

SOLVED PROBLEM 4-4

Consider the following reaction:

$$CH_4 \quad + \quad Cl\cdot \quad \longrightarrow \quad \cdot CH_3 \quad + \quad HCl$$

This reaction has an activation energy (E_a) of +17 kJ/mol (+4 kcal/mol) and a $\Delta H°$ of +7 kJ/mol (+2 kcal/mol). Draw a reaction-energy diagram for this reaction.

SOLUTION

We draw a diagram that shows the products to be 7 kJ *higher* in energy than the reactants. The barrier is made to be 17 kJ higher in energy than the reactants.

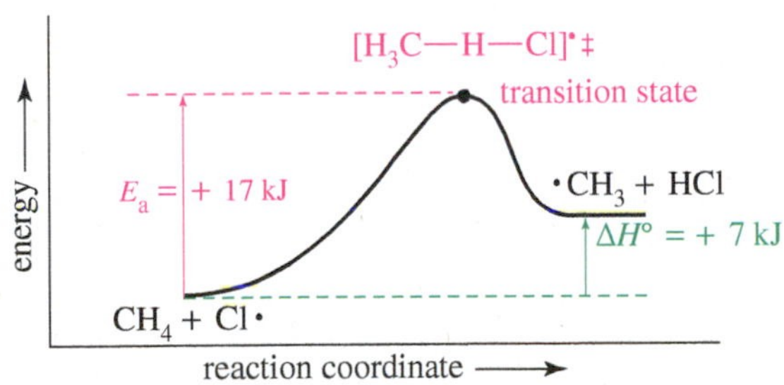

PROBLEM 4-14

(a) Draw the reaction-energy diagram for the following reverse reaction:

$$\cdot CH_3 \quad + \quad HCl \quad \longrightarrow \quad CH_4 \quad + \quad Cl\cdot$$

(b) What is the activation energy for this reverse reaction?
(c) What is the heat of reaction ($\Delta H°$) for this reverse reaction?

PROBLEM 4-15

(a) Draw a reaction-energy diagram for the following reaction:

$$\cdot CH_3 + Cl_2 \longrightarrow CH_3Cl + Cl\cdot$$

The activation energy is 4 kJ/mol (1 kcal/mol), and the overall $\Delta H°$ for the reaction is −110 kJ/mol (−27 kcal/mol).

(b) Give the equation for the reverse reaction.
(c) What is the activation energy for the reverse reaction?

4-11 Rates of Multistep Reactions

Many reactions proceed by mechanisms involving several steps and several intermediates. As we saw in Section 4-7, for example, the reaction of methane with chlorine goes through two propagation steps. The propagation steps are shown here, along with their heats of reaction and their activation energies. Just the propagation steps are shown because the rate of the initiation step is controlled by the amount of light or heat available to split chlorine molecules.

Step	$\Delta H°$ (per mole)	E_a (per mole)
$CH_4 + Cl\cdot \longrightarrow \cdot CH_3 + HCl$	+7 kJ (+2 kcal)	17 kJ (4 kcal)
$\cdot CH_3 + Cl_2 \longrightarrow CH_3Cl + Cl\cdot$	−110 kJ (−27 kcal)	4 kJ (1 kcal)

In this reaction, $Cl\cdot$ and $CH_3\cdot$ are *reactive intermediates*. Unlike transition states, reactive intermediates are stable as long as they do not collide with other atoms or molecules. As free radicals, however, $Cl\cdot$ and $\cdot CH_3$ are quite reactive toward other molecules. Figure 4-4 shows a single reaction-energy profile that includes both propagation steps of the chlorination. The energy maxima (high points) are the unstable transition states, and the energy minima (low points) are the intermediates. This complete energy profile provides most of the important information about the energetics of the reaction.

The Rate-Limiting Step In a multistep reaction, each step has its own characteristic rate. There can be only one overall reaction rate, however, and it is controlled by the **rate-limiting step** (also called the **rate-determining step**). In general, the *highest-energy* step of a multistep reaction is the "bottleneck," and it determines the overall rate. How can we tell which step is rate limiting? If we have the reaction-energy diagram, it is simple: The highest point in the energy diagram is the transition state with the highest energy—generally the transition state for the rate-limiting step.

The highest point in the energy diagram of the chlorination of methane (Figure 4-4) is the transition state for the reaction of methane with a chlorine radical. This step must be rate limiting. If we calculate a rate for this slow step, it will be the rate for the overall reaction. The second, faster step will consume the products of the slow step as fast as they are formed.

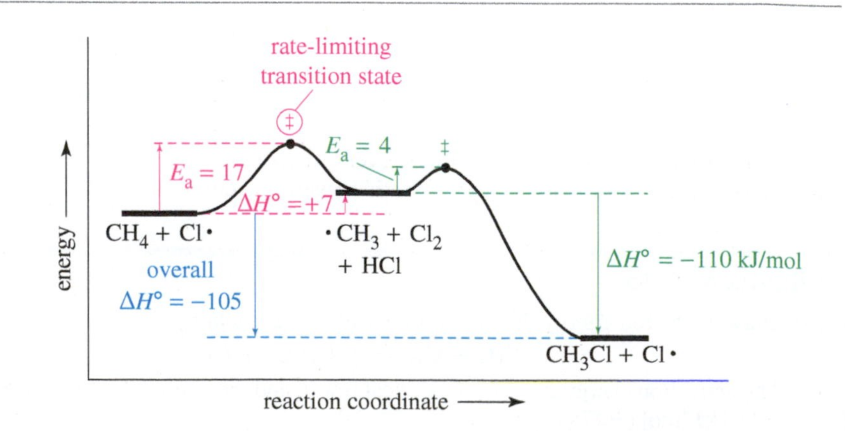

FIGURE 4-4
Combined reaction-energy diagram for the chlorination of methane. (Units are kJ/mol.) The energy maxima are transition states, and the energy minima are intermediates. $Cl\cdot$ and $\cdot CH_3$ are the reactive intermediates in this mechanism.

4-12 Temperature Dependence of Halogenation

We now apply what we know about rates to the reaction of methane with halogens. The rate-limiting step for chlorination is the endothermic reaction of the chlorine atom with methane to form a methyl radical and a molecule of HCl.

Rate-limiting step

$$CH_4 + Cl\cdot \longrightarrow \cdot CH_3 + HCl$$

The activation energy for this step is 17 kJ/mol (4 kcal/mol). At room temperature, the value of $e^{-E_a/RT}$ is 1300×10^{-6}. This value represents a rate that is fast but controllable.

In a free-radical chain reaction, every propagation step must occur quickly; otherwise, the free radicals will undergo unproductive collisions and participate in termination steps. We can predict how quickly the various halogen atoms react with methane given relative rates based on the measured activation energies of the slowest steps:

Reaction	E_a (per mole)	Relative Rate ($e^{-E_a/RT} \times 10^6$) 27 °C (300 °K)	227 °C (500 °K)
$F\cdot + CH_4 \longrightarrow HF + \cdot CH_3$	5 kJ (1.2 kcal)	140,000	300,000
$Cl\cdot + CH_4 \longrightarrow HCl + \cdot CH_3$	17 kJ (4 kcal)	1300	18,000
$Br\cdot + CH_4 \longrightarrow HBr + \cdot CH_3$	75 kJ (18 kcal)	9×10^{-8}	0.015
$I\cdot + CH_4 \longrightarrow HI + \cdot CH_3$	140 kJ (34 kcal)	2×10^{-19}	2×10^{-9}

These relative rates suggest how easily and quickly methane reacts with the different halogen radicals. The reaction with fluorine should be difficult to control because its rate is very high. Chlorine should react moderately at room temperature, but it may become difficult to control if the temperature rises much (the rate at 500 °K is rather high). The reaction with bromine is very slow, but heating might give an observable rate. Iodination is probably out of the question because its rate is exceedingly slow, even at 500 °K.

Laboratory halogenations show that our predictions are right. In fact, fluorine reacts explosively with methane, and chlorine reacts at a moderate rate. A mixture of bromine and methane must be heated to react, and iodine does not react at all.

PROBLEM 4-16

The bromination of methane proceeds through the following steps:

	$\Delta H°$ (per mole)	E_a (per mole)
$Br_2 \xrightarrow{h\nu} 2\,Br\cdot$	+190 kJ (45 kcal)	190 kJ (45 kcal)
$CH_4 + Br\cdot \rightarrow \cdot CH_3 + HBr$	+73 kJ (17 kcal)	79 kJ (19 kcal)
$\cdot CH_3 + Br_2 \rightarrow CH_3Br + Br\cdot$	−112 kJ (−27 kcal)	4 kJ (1 kcal)

(a) Draw a complete reaction-energy diagram for this reaction.
(b) Label the rate-limiting step.
(c) Draw the structure of each transition state.
(d) Compute the overall value of $\Delta H°$ for the bromination.

PROBLEM 4-17

(a) Using the BDEs in Table 4-2 (page 167), compute the value of $\Delta H°$ for each step in the iodination of methane.
(b) Compute the overall value of $\Delta H°$ for iodination.
(c) Suggest *two* reasons why iodine does not react well with methane.

4-13 Selectivity in Halogenation

Up to now, we have limited our discussions to the halogenation of methane. Beginning our study with such a simple compound allowed us to concentrate on the thermodynamics and kinetics of the reaction. Now we consider halogenation of the "higher" alkanes, meaning those of higher molecular weight.

4-13A Chlorination of Propane: Product Ratios

Halogenation is a substitution, where a halogen atom replaces a hydrogen.

$$\mathrm{R{-}H} + \mathrm{X_2} \longrightarrow \mathrm{R{-}X} + \mathrm{H{-}X}$$

In methane, all four hydrogen atoms are identical, and it does not matter which hydrogen is replaced. In the higher alkanes, replacement of different hydrogen atoms may lead to different products. In the chlorination of propane, for example, two monochlorinated (just one chlorine atom) products are possible. One has the chlorine atom on a primary carbon atom, and the other has the chlorine atom on the secondary carbon atom.

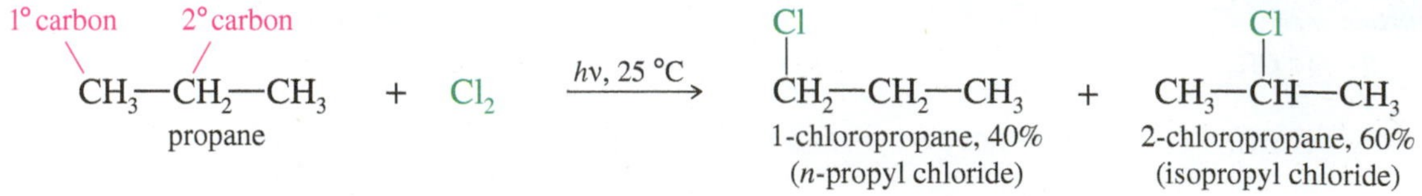

$$\underset{\text{propane}}{\mathrm{CH_3{-}CH_2{-}CH_3}} + \mathrm{Cl_2} \xrightarrow{h\nu,\ 25\ ^\circ\mathrm{C}} \underset{\substack{\text{1-chloropropane, 40\%} \\ (n\text{-propyl chloride})}}{\mathrm{CH_2(Cl){-}CH_2{-}CH_3}} + \underset{\substack{\text{2-chloropropane, 60\%} \\ \text{(isopropyl chloride)}}}{\mathrm{CH_3{-}CH(Cl){-}CH_3}}$$

(In propane, the 1° carbon and the 2° carbon are labeled.)

The product ratio shows that replacement of hydrogen atoms by chlorine is not random. Propane has six primary hydrogens (hydrogens bonded to primary carbons) and only two secondary hydrogens (bonded to the secondary carbon), yet the major product results from substitution of a secondary hydrogen. We can calculate how reactive each kind of hydrogen is by dividing the amount of product observed by the number of hydrogens that can be replaced to give that product.

Figure 4-5 shows the definition of primary, secondary, and tertiary hydrogens and the calculation of their relative reactivity. Replacing either of the two secondary hydrogens accounts for 60% of the product, and replacing any of the six primary hydrogens accounts for 40% of the product. We calculate that each secondary hydrogen is 4.5 times as reactive as each primary hydrogen.

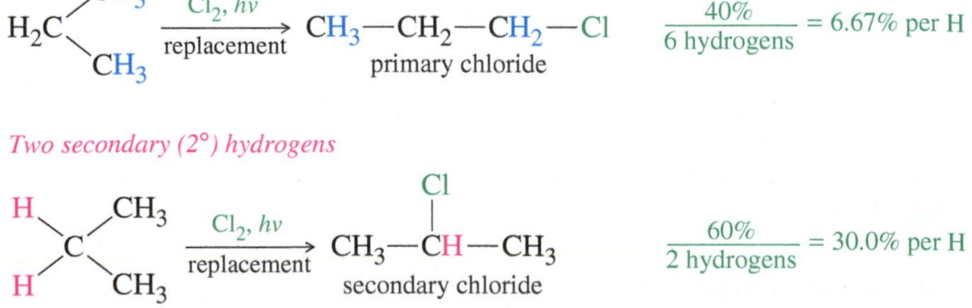

FIGURE 4-5
Definitions of primary, secondary, and tertiary hydrogens. There are six primary hydrogens in propane and only two secondary hydrogens, yet the major product results from replacement of a secondary hydrogen.

$\mathrm{R{-}CH_2{-}H}$ (C bonded to R, H, H, H): primary (1°) hydrogens

$\mathrm{R_2CH{-}H}$ (C bonded to R, R, H, H): secondary (2°) hydrogens

$\mathrm{R_3C{-}H}$ (C bonded to R, R, R, H): tertiary (3°) hydrogen

Six primary (1°) hydrogens

$$\mathrm{H_2C(CH_3)_2} \xrightarrow[\text{replacement}]{\mathrm{Cl_2},\ h\nu} \underset{\text{primary chloride}}{\mathrm{CH_3{-}CH_2{-}CH_2{-}Cl}} \qquad \text{relative reactivity: } \frac{40\%}{6\ \text{hydrogens}} = 6.67\%\ \text{per H}$$

Two secondary (2°) hydrogens

$$\mathrm{H_2C(CH_3)_2} \xrightarrow[\text{replacement}]{\mathrm{Cl_2},\ h\nu} \underset{\text{secondary chloride}}{\mathrm{CH_3{-}CH(Cl){-}CH_3}} \qquad \frac{60\%}{2\ \text{hydrogens}} = 30.0\%\ \text{per H}$$

The 2° hydrogens are $\dfrac{30.0}{6.67} = 4.5$ times as reactive as the 1° hydrogens.

Initiation: Splitting of the chlorine molecule

$$Cl_2 + h\nu \longrightarrow 2\ Cl\cdot$$

First propagation step: Abstraction (removal) of a primary or secondary hydrogen

$$CH_3{-}CH_2{-}CH_3 + Cl\cdot \longrightarrow \underset{\text{primary radical}}{\cdot CH_2{-}CH_2{-}CH_3} \text{ or } \underset{\text{secondary radical}}{CH_3{-}\dot{C}H{-}CH_3} + HCl$$

Second propagation step: Reaction with chlorine to form the alkyl chloride

$$\underset{\text{primary radical}}{\cdot CH_2{-}CH_2{-}CH_3} + Cl_2 \longrightarrow \underset{\substack{\text{primary chloride}\\ \text{(1-chloropropane)}}}{Cl{-}CH_2{-}CH_2{-}CH_3} + Cl\cdot$$

$$\text{or } \underset{\text{secondary radical}}{CH_3{-}\dot{C}H{-}CH_3} + Cl_2 \longrightarrow \underset{\substack{\text{secondary chloride}\\ \text{(2-chloropropane)}}}{CH_3{-}CH(Cl){-}CH_3} + Cl\cdot$$

FIGURE 4-6
The mechanism for free-radical chlorination of propane. The first propagation step forms either a primary radical or a secondary radical. This radical determines whether the final product will be the primary alkyl chloride or the secondary alkyl chloride.

To explain this preference for reaction at the secondary position, we must look carefully at the reaction mechanism, as shown in Figure 4-6. When a chlorine atom reacts with propane, abstraction of a hydrogen atom can give either a primary radical or a secondary radical. The structure of the radical formed in this step determines the structure of the observed product, either 1-chloropropane or 2-chloropropane. The product ratio shows that the secondary radical is formed preferentially. This preference for reaction at the secondary position results from the greater stability of the secondary free radical and the transition state leading to it. We explain this preference in more detail in the next section.

PROBLEM 4-18

What would be the product ratio in the chlorination of propane if all the hydrogens were abstracted at equal rates?

PROBLEM 4-19

Classify each hydrogen atom in the following compounds as primary (1°), secondary (2°), or tertiary (3°).

(a) butane **(b)** isobutane **(c)** 2-methylbutane
(d) cyclohexane **(e)** norbornane (bicyclo[2.2.1]heptane)

4-13B Free-Radical Stabilities

Figure 4-7 shows the energy required (the bond-dissociation enthalpy) to form a free radical by breaking a bond between a hydrogen atom and a carbon atom. This energy is greatest for a methyl carbon, and it decreases for a primary carbon, a secondary carbon, and a tertiary carbon. The more highly substituted the carbon atom, the less energy required to form the free radical.

From the information in Figure 4-7, we conclude that free radicals are more stable when they are more highly substituted. The following free radicals are listed in increasing order of stability.

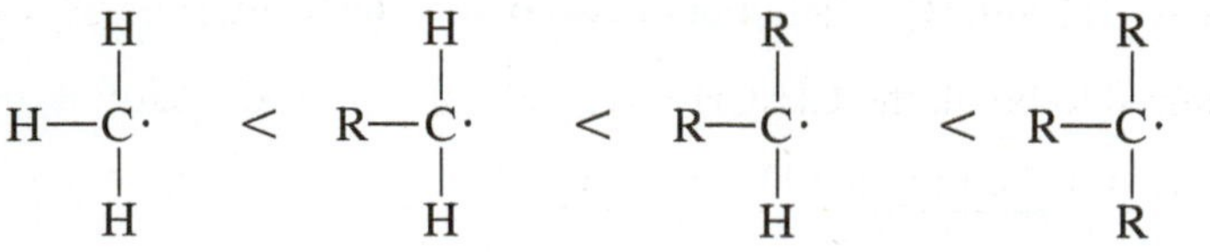

$$\begin{array}{ccccccc} \text{H} & & \text{H} & & \text{R} & & \text{R} \\ | & & | & & | & & | \\ \text{H—C·} & < & \text{R—C·} & < & \text{R—C·} & < & \text{R—C·} \\ | & & | & & | & & | \\ \text{H} & & \text{H} & & \text{H} & & \text{R} \end{array}$$

$$\text{Me·} < 1° < 2° < 3°$$

$$\text{methyl} < \text{primary} < \text{secondary} < \text{tertiary}$$

PROBLEM-SOLVING HINT

The first propagation step of chlorination is exothermic for all alkanes except methane. For methane, it is slightly endothermic, about +4 kJ/mol (+1 kcal/mol).

In the chlorination of propane, the secondary hydrogen atom is abstracted more often because the secondary radical and the transition state leading to it are lower in energy than the primary radical and its transition state. Using the bond-dissociation enthalpies in Table 4-2 (page 167), we can calculate $\Delta H°$ for each of the possible reaction steps. Abstraction of the secondary hydrogen is 13 kJ/mol (3 kcal/mol) more exothermic than abstraction of the primary hydrogen.

FIGURE 4-7
Enthalpy required to form a free radical. Bond-dissociation enthalpies show that more highly substituted free radicals are more stable than less highly substituted ones.

Bond-dissociation enthalpy

Formation of a methyl radical

$$CH_4 \longrightarrow H\cdot + \cdot CH_3 \qquad \Delta H° = 439\text{ kJ (105 kcal)}$$

Formation of a primary (1°) radical

$$CH_3—CH_2—CH_3 \longrightarrow H\cdot + CH_3—CH_2—\dot{C}H_2 \qquad \Delta H° = 423\text{ kJ (101 kcal)}$$

Formation of a secondary (2°) radical

$$CH_3—CH_2—CH_3 \longrightarrow H\cdot + CH_3—\dot{C}H—CH_3 \qquad \Delta H° = 413\text{ kJ (99 kcal)}$$

Formation of a tertiary (3°) radical

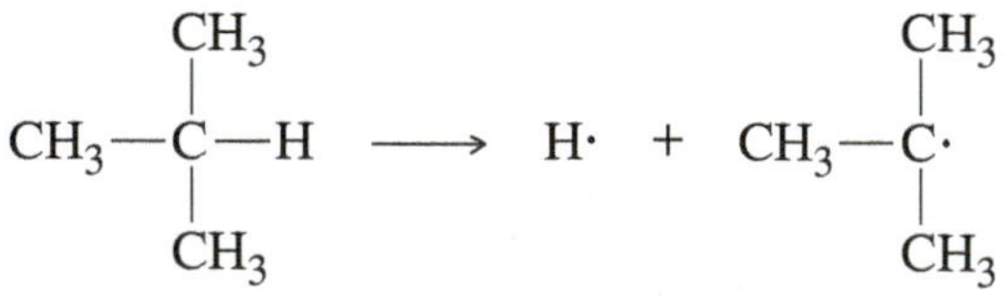

$$(CH_3)_3C—H \longrightarrow H\cdot + (CH_3)_3C\cdot \qquad \Delta H° = 403\text{ kJ (96 kcal)}$$

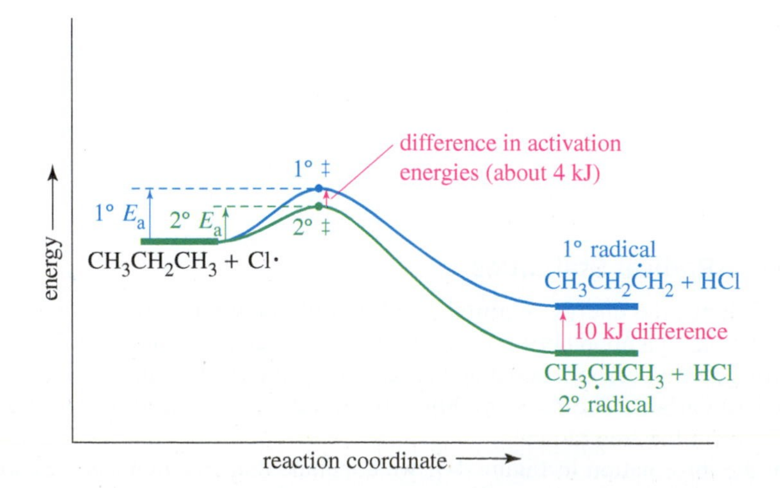

FIGURE 4-8
Reaction-energy diagram for the first propagation step in the chlorination of propane. Formation of the secondary radical has a lower activation energy than does formation of the primary radical.

$$1° \text{ H:} \quad CH_3{-}CH_2{-}CH_3 \ + \ Cl\cdot \ \longrightarrow \ CH_3{-}CH_2{-}CH_2\cdot \ + \ H{-}Cl$$

Energy required to break the $CH_3CH_2CH_2{-}H$ bond	+423 kJ/mol (+101 kcal/mol)
Energy released in forming the $H{-}Cl$ bond	−432 kJ/mol (−103 kcal/mol)
Total energy for reaction at the primary position:	−9 kJ/mol (−2 kcal/mol)

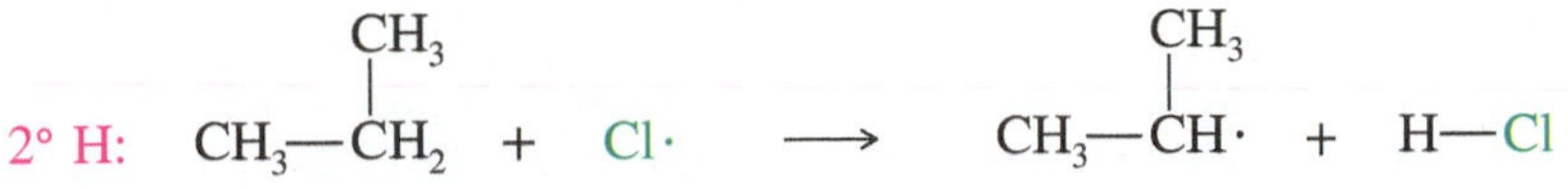

$$2° \text{ H:} \quad CH_3{-}CH_2(CH_3) \ + \ Cl\cdot \ \longrightarrow \ CH_3{-}\dot{C}H(CH_3) \ + \ H{-}Cl$$

Energy required to break the $CH_3{-}CH(CH_3){-}H$ bond	+413 kJ/mol (+99 kcal/mol)
Energy released in forming the $H{-}Cl$ bond	−432 kJ/mol (−103 kcal/mol)
Total energy for reaction at the secondary position:	−19 kJ/mol (−4 kcal/mol)

A reaction-energy diagram for this rate-limiting first propagation step appears in Figure 4-8. The activation energy to form the secondary radical is slightly lower, so the secondary radical is formed faster than the primary radical.

SOLVED PROBLEM 4-5

Tertiary hydrogen atoms react with Cl· about 5.5 times as fast as primary ones. Predict the product ratios for chlorination of isobutane.

SOLUTION

Isobutane has nine primary hydrogens and one tertiary hydrogen.

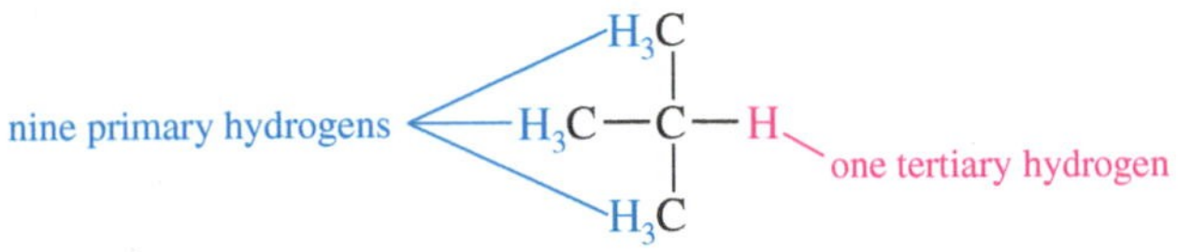

$$(H_3C)_3C{-}H$$

nine primary hydrogens (the three H_3C groups); one tertiary hydrogen (the C—H)

(9 primary hydrogens) × (reactivity 1.0) = 9.0 relative amount of reaction

(1 tertiary hydrogen) × (reactivity 5.5) = 5.5 relative amount of reaction

Even though the primary hydrogens are less reactive, there are so many of them that the primary product is the major product. The product ratio will be 9.0:5.5, or about 1.6:1.

$$\text{fraction of primary} = \frac{9.0}{9.0 + 5.5} = 62\%$$

$$\text{fraction of tertiary} = \frac{5.5}{9.0 + 5.5} = 38\%$$

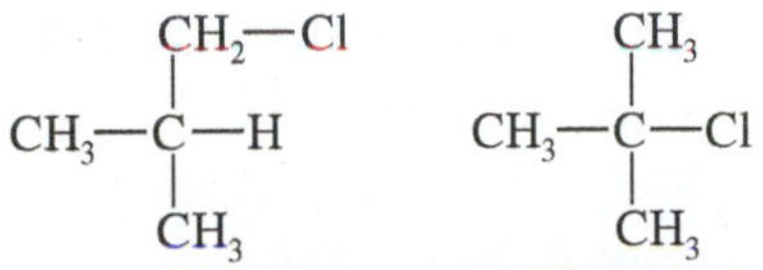

$CH_3{-}CH(CH_3){-}CH_2{-}Cl$ — major product 62%

$CH_3{-}C(CH_3)_2{-}Cl$ — minor product 38%

PROBLEM 4-20

Use the bond-dissociation enthalpies in Table 4-2 (page 167) to calculate the heats of reaction for the two possible first propagation steps in the chlorination of isobutane. Use this information to draw a reaction-energy diagram like Figure 4-8, comparing the activation energies for formation of the two radicals.

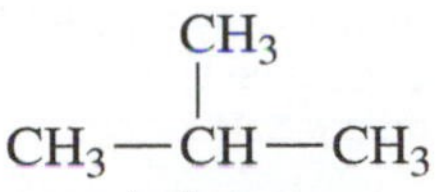

$CH_3{-}CH(CH_3){-}CH_3$

isobutane
2-methylpropane

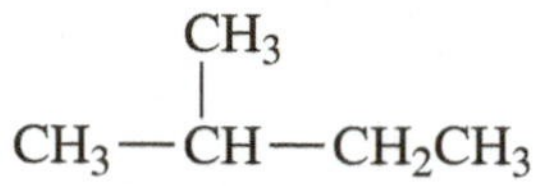

$$\begin{array}{c} CH_3 \\ | \\ CH_3-CH-CH_2CH_3 \end{array}$$

isopentane
2-methylbutane

PROBLEM 4-21

Predict the ratios of products that result from chlorination of isopentane (2-methylbutane).

$$\begin{array}{c} CH_3 \qquad\quad CH_3 \\ | \qquad\qquad | \\ CH_3-C-CH_2-CH-CH_3 \\ | \\ CH_3 \end{array}$$

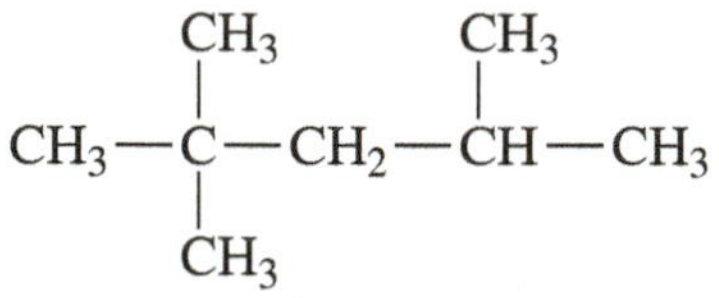

isooctane
2,2,4-trimethylpentane

PROBLEM 4-22

(a) When *n*-heptane burns in a gasoline engine, the combustion process takes place too quickly. The explosive detonation makes a noise called *knocking*. When 2,2,4-trimethylpentane (isooctane) is burned, combustion takes place in a slower, more controlled manner. Combustion is a free-radical chain reaction, and its rate depends on the reactivity of the free-radical intermediates. Explain why isooctane has less tendency to knock than does *n*-heptane.

(b) Alkoxy radicals ($R-O\cdot$) are generally more stable than alkyl ($R\cdot$) radicals. Write an equation showing an alkyl free radical (from burning gasoline) abstracting a hydrogen atom from *tert*-butyl alcohol, $(CH_3)_3COH$. Explain why *tert*-butyl alcohol works as an antiknock additive for gasoline.

(c) Use the information in Table 4-2 (page 167) to explain why toluene ($PhCH_3$) has a very high octane rating of 111. Write an equation to show how toluene reacts with an alkyl free radical to give a relatively stable radical.

4-13C Bromination of Propane

Figure 4-9 shows the free-radical reaction of propane with bromine. Notice that this reaction is both heated to 125 °C and irradiated with light to achieve a moderate rate. The secondary bromide (2-bromopropane) is favored by a 97:3 product ratio. From this product ratio, we calculate that each of the two secondary hydrogens is 97 times as reactive as one of the six primary hydrogens.

The 97:1 reactivity ratio for bromination is much larger than the 4.5:1 ratio for chlorination. We say that bromination is more *selective* than chlorination because the major reaction is favored by a larger amount. To explain this enhanced selectivity, we must consider the transition states and activation energies for the rate-limiting step.

As with chlorination, the rate-limiting step in bromination is the first propagation step: abstraction of a hydrogen atom by a bromine radical. The energetics of the two possible hydrogen abstractions are shown on the next page. Compare these numbers with the energetics of the first propagation step of chlorination shown on page 178. The bond dissociation enthalpies are taken from Table 4-2 (page 167).

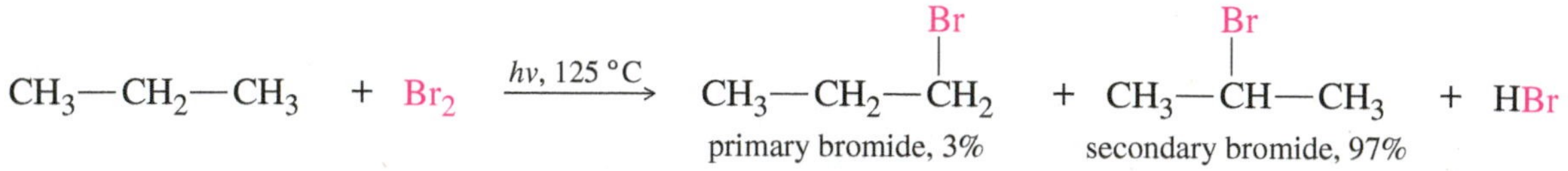

$$CH_3-CH_2-CH_3 + Br_2 \xrightarrow{h\nu,\ 125\ ^\circ C} \underset{\text{primary bromide, 3\%}}{CH_3-CH_2-\overset{\displaystyle Br}{\overset{|}{C}H_2}} + \underset{\text{secondary bromide, 97\%}}{CH_3-\overset{\displaystyle Br}{\overset{|}{C}H}-CH_3} + HBr$$

Relative reactivity

six primary hydrogens $\frac{3\%}{6} = 0.5\%$ per H

two secondary hydrogens $\frac{97\%}{2} = 48.5\%$ per H

The 2° hydrogens are $\frac{48.5}{0.5} = 97$ times as reactive as the 1° hydrogens.

FIGURE 4-9
The free-radical reaction of propane with bromine. This 97:3 ratio of products shows that bromine abstracts a secondary hydrogen 97 times as rapidly as a primary hydrogen. Bromination (reactivity ratio 97:1) is much more selective than chlorination (reactivity ratio 4.5:1).

1° H: $CH_3—CH_2—CH_3$ + Br· ⟶ $CH_3—CH_2—\dot{C}H_2$ + H—Br

Energy required to break the $CH_3CH_2CH_2$—H bond	+423 kJ/mol (+101 kcal/mol)
Energy released in forming the H—Br bond	−366 kJ/mol (−88 kcal/mol)
Total energy for reaction at the primary position:	+57 kJ/mol (+13 kcal/mol)

2° H: $CH_3—CH(CH_3)—H$ ($CH_3—CH_2$ with CH_3 substituent) + Br· ⟶ $CH_3—\dot{C}H(CH_3)$ + H—Br

Energy required to break the $CH_3—CH(CH_3)$—H bond	+413 kJ/mol (+99 kcal/mol)
Energy released in forming the H—Br bond	−366 kJ/mol (−88 kcal/mol)
Total energy for reaction at the secondary position:	+47 kJ/mol (+11 kcal/mol)

The energy differences between chlorination and bromination result from the difference in the bond-dissociation enthalpies of H—Cl (432 kJ) and H—Br (366 kJ). The HBr bond is weaker, and abstraction of a hydrogen atom by Br· is endothermic. This endothermic step explains why bromination is much slower than chlorination, but it still does not explain the enhanced selectivity observed with bromination.

Consider the reaction-energy diagram for the first propagation step in the bromination of propane (Figure 4-10). Although the difference in values of ΔH° between abstraction of a primary hydrogen and a secondary hydrogen is still 10 kJ/mol (2 kcal/mol), the energy diagram for bromination shows a much larger difference in activation energies for abstraction of the primary and secondary hydrogens than we saw for chlorination (Figure 4-8).

In bromination, the rate-limiting first propagation step is endothermic, and the energy maxima (corresponding to the activation energies) are closer to the products than to the reactants. A smooth graph (Figure 4-10) shows the activation energies nearly as far apart as the product energies. In chlorination, on the other hand, this first step is exothermic, and the energy maxima are closer to the reactants, which are the same and have the same energy for either route. The graph for chlorination (Figure 4-8) shows the activation energies separated by only a small fraction of the difference in product energies. This intuitive, graphic principle is formalized in **Hammond's postulate**.

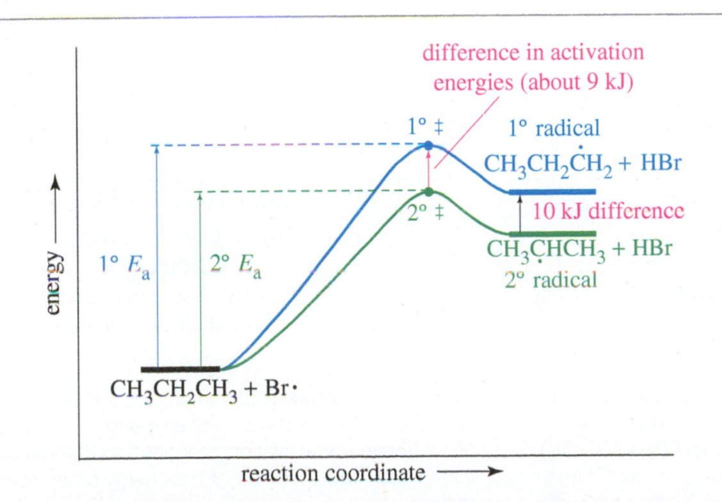

FIGURE 4-10
Reaction-energy diagram for the first propagation step in the bromination of propane. The energy difference in the transition states is nearly as large as the energy difference in the products.

4-14 Hammond's Postulate

Hammond's postulate is a general statement summarizing the differences we have seen in comparing the chlorination and bromination of propane. We call it a postulate, rather than a rule, because we cannot prove it, yet it appears to be self-evident.

> HAMMOND'S POSTULATE: Related species that are closer in energy are also closer in structure. The structure of a transition state resembles the structure of the closest stable species.

Figure 4-11 summarizes the energy diagrams for the first propagation steps in the bromination and chlorination of propane. Together, these energy diagrams explain the enhanced selectivity observed in bromination, and they show how Hammond's postulate applies to this reaction.

Two important differences are apparent in the reaction-energy diagrams for the first propagation steps of chlorination and bromination:

1. The first propagation step is endothermic for bromination but exothermic for chlorination.
2. The transition states forming the 1° and 2° radicals for the endothermic bromination have a larger energy difference than those for the exothermic chlorination, even though the energy difference of the products is the same (10 kJ, or 2 kcal) in both reactions.

Hammond's postulate tells us that the structures of the transition states in the endothermic bromination should resemble the products more than the reactants because they are closer in energy to the products. In the exothermic chlorination, the transition states are closer in energy to the reactants than the products; so we expect their structure to resemble the reactants more than the products. Figure 4-12 compares the transition states for bromination and chlorination.

In the product-like transition state for bromination, the C—H bond is nearly broken and the carbon atom has a great deal of radical character. The energy of this transition state reflects most of the energy difference of the radical products. In the reactant-like transition

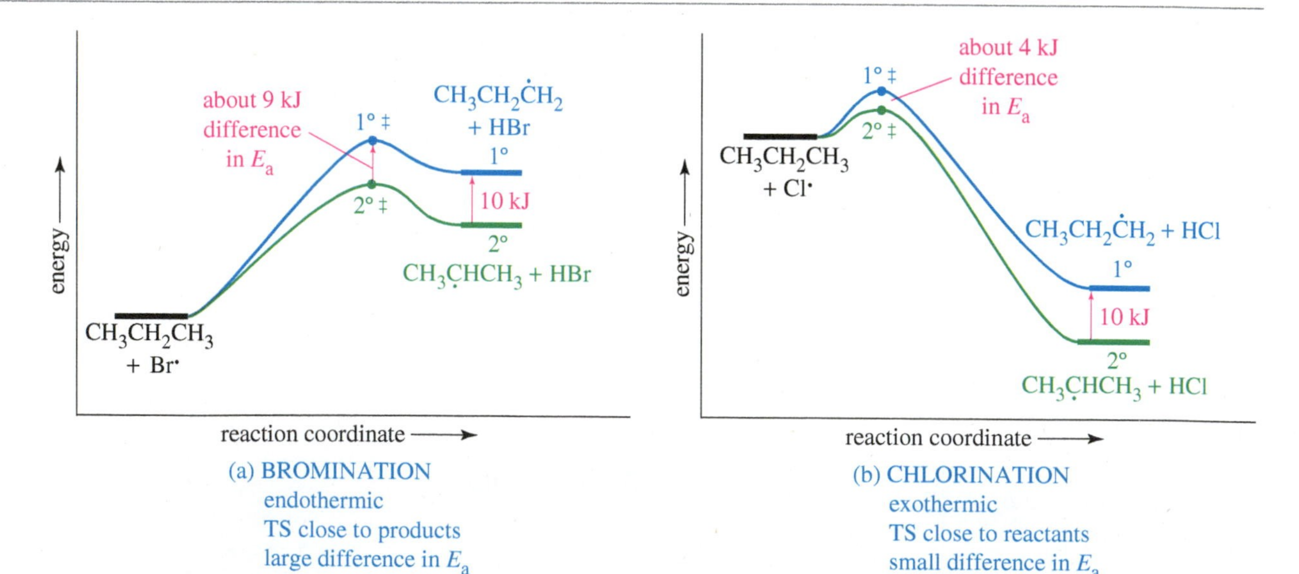

FIGURE 4-11
The energy diagrams for bromination and chlorination of propane. (**a**) In the endothermic bromination, the transition states are closer to the products (the radicals) in energy and in structure. The difference in the 1° and 2° activation energies is about 9 kJ (2.2 kcal), nearly the entire energy difference of the radicals. (**b**) In the exothermic chlorination, the transition states are closer to the reactants in energy and in structure. The difference in activation energies for chlorination is about 4 kJ (1 kcal), less than half of the energy difference of the radicals.

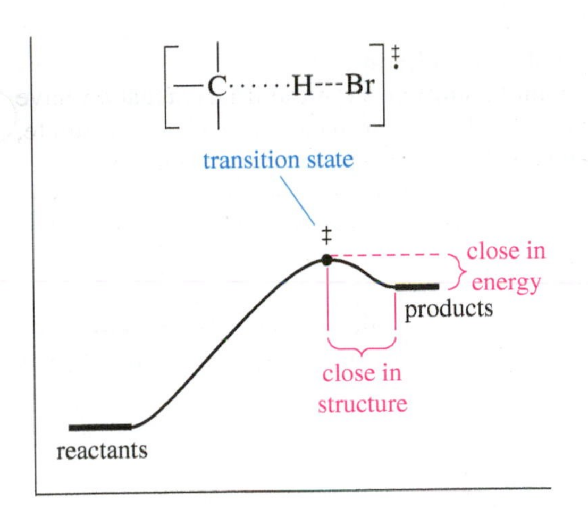

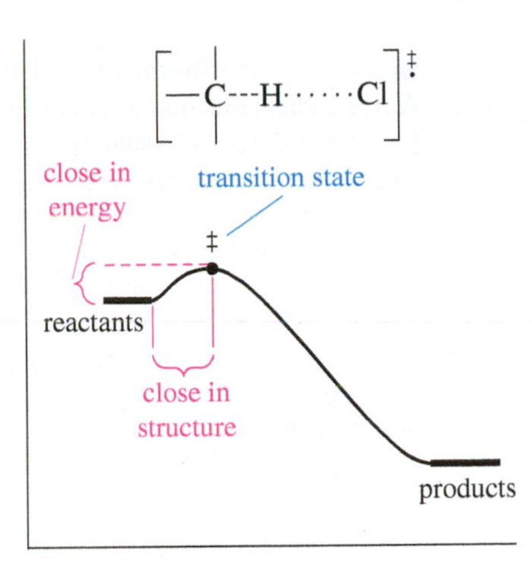

FIGURE 4-12
Comparison of transition states for bromination and chlorination. In the endothermic bromination, the transition state resembles the products (the free radical and HBr). In the exothermic chlorination, the free radical has just begun to form in the transition state, so the transition state resembles the reactants.

state for chlorination, the C—H bond is just beginning to break, and the carbon atom has little radical character. This transition state reflects only a small part (less than half) of the energy difference of the radical products. Therefore, chlorination is less selective. We can state this conclusion as a general rule:

> In an endothermic reaction, the transition state is closer to the products in energy and in structure. In an exothermic reaction, the transition state is closer to the reactants in energy and in structure.

The transition state is always the point of highest energy on the energy diagram. Its structure resembles either the reactants or the products, whichever are higher in energy. In an endothermic reaction, the products are higher in energy, and the transition state is product-like. In an exothermic reaction, the reactants are higher in energy, and the transition state is reactant-like. Thus, the Hammond postulate helps us understand why exothermic processes tend to be less selective than similar endothermic processes.

PROBLEM 4-23

(a) Compute the heats of reaction for abstraction of a primary hydrogen and a secondary hydrogen from propane by a fluorine radical.

$$CH_3-CH_2-CH_3 + F\cdot \longrightarrow CH_3-CH_2-\dot{C}H_2 + HF$$

$$CH_3-CH_2-CH_3 + F\cdot \longrightarrow CH_3-\dot{C}H-CH_3 + HF$$

(b) How selective do you expect free-radical fluorination to be?

(c) What product distribution would you expect to obtain from the free-radical fluorination of propane?

PROBLEM-SOLVING HINT

Free-radical bromination is highly selective, chlorination is moderately selective, and fluorination is nearly nonselective.

PROBLEM-SOLVING STRATEGY Proposing Reaction Mechanisms

Throughout this course, we will propose mechanisms to explain reactions. We will discuss methods for dealing with different types of mechanisms as we encounter them. These techniques for dealing with a variety of mechanisms are collected in Appendix 3A. At this point, however, we focus on free-radical mechanisms like those in this chapter.

Free-Radical Reactions

General principles: Free-radical reactions generally proceed by chain-reaction mechanisms, using an initiator with an easily broken bond (such as chlorine, bromine, or a peroxide) to start the chain reaction. In drawing the mechanism, expect free-radical intermediates (especially highly substituted or resonance-stabilized intermediates). Watch for the most stable free radicals, and avoid any high-energy radicals such as hydrogen atoms.

(continued)

1. **Draw a step that breaks the weak bond in the initiator.**
   A free-radical reaction usually begins with an initiation step in which the initiator undergoes homolytic (free-radical) cleavage to give two radicals.
2. **Draw a reaction of the initiator with one of the starting materials.**
   One of the initiator radicals reacts with one of the starting materials to give a free-radical version of the starting material. The initiator might abstract a hydrogen atom or add to a double bond, depending on what reaction leads to formation of the observed product. You might want to consider bond-dissociation enthalpies to see which reaction is energetically favored.
3. **Draw a reaction of the free-radical version of the starting material with another starting-material molecule to form a bond needed in the product and to generate a new radical intermediate.**
   Check your intermediates to be sure that you have used the most stable radical intermediates. For a realistic chain reaction, no new initiation steps should be required; a radical should be regenerated in each propagation step.
4. **Draw termination step(s).**
   The reaction ends with termination steps, which are side reactions rather than part of the product-forming mechanism. The reaction of any two free radicals to give a stable molecule is a termination step, as is a collision of a free radical with the reaction vessel.

Before we illustrate this procedure, let's consider a few common mistakes. Avoiding these mistakes will help you to draw correct mechanisms throughout this course.

Common Mistakes to Avoid

1. Do not use condensed or line–angle formulas for reaction sites. Draw all the bonds and all the substituents of each carbon atom affected throughout the mechanism. Three-bonded carbon atoms in intermediates are most likely to be radicals in the free-radical reactions we have studied. If you draw condensed formulas or line–angle formulas, you will likely misplace a hydrogen atom and show a reactive species on the wrong carbon.
2. Do not show more than one step occurring at once, unless they really do occur at once.

Sample Problem

Draw a mechanism for the reaction of methylcyclopentane with bromine under irradiation with light. Predict the major product.

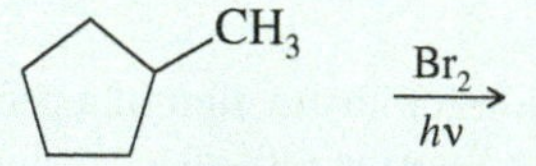

CH_3 $\xrightarrow[h\nu]{Br_2}$

In every mechanism problem, we first draw what we know, showing all the bonds and all the substituents of each carbon atom that may be affected throughout the mechanism.

H CH$_3$ H C H C C H H H—C—C—H H H $\xrightarrow[h\nu]{Br_2}$

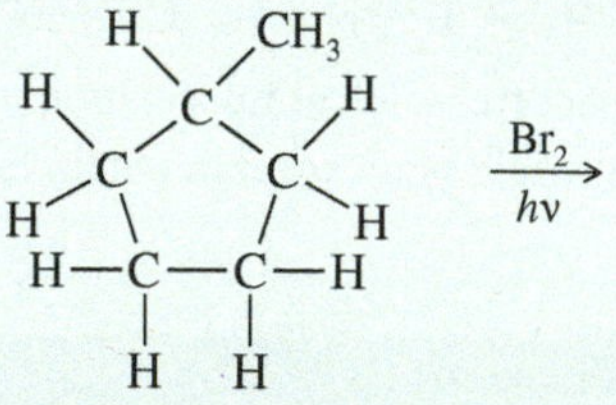

1. **Draw a step involving cleavage of the weak bond in the initiator.**
   The use of light with bromine suggests a free-radical reaction, with light providing the energy for dissociation of Br_2. This homolytic cleavage initiates the chain reaction by generating two Br· radicals.

   Initiation step $\quad Br—Br \xrightarrow{h\nu} Br\cdot + Br\cdot$

2. **Draw a reaction of the initiator with one of the starting materials.**
   One of these initiator radicals should react with methylcyclopentane to give a free-radical version of methylcyclopentane. As we have seen, a bromine or chlorine radical can abstract a hydrogen atom from an alkane to generate an alkyl radical. The bromine radical is highly selective, and the most stable alkyl radical should result. Abstraction of the tertiary hydrogen atom gives a tertiary radical.

First propagation step

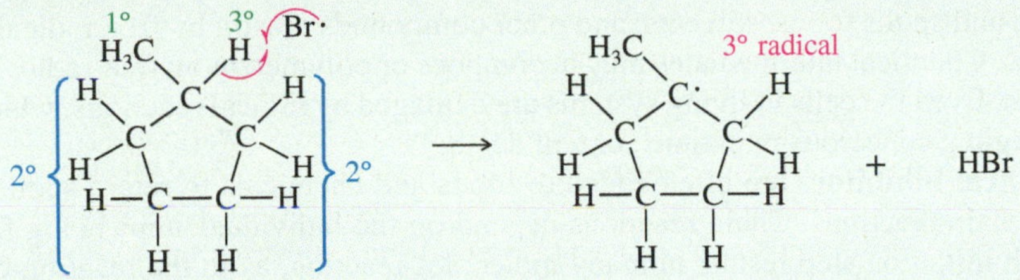

[Structure: methylcyclopentane with 1°, 2°, 3° H labels; Br· abstracts the 3° H] ⟶ [3° radical] + HBr

3. Draw a reaction of the free-radical version of the starting material with another starting-material molecule to form a bond needed in the product and to generate a new radical intermediate.

The alkyl radical should react with another starting-material molecule, in another propagation step, to generate a product and another radical. Reaction of the alkyl radical with Br_2 gives 1-bromo-1-methylcyclopentane (the major product) and another bromine radical to continue the chain.

Second propagation step

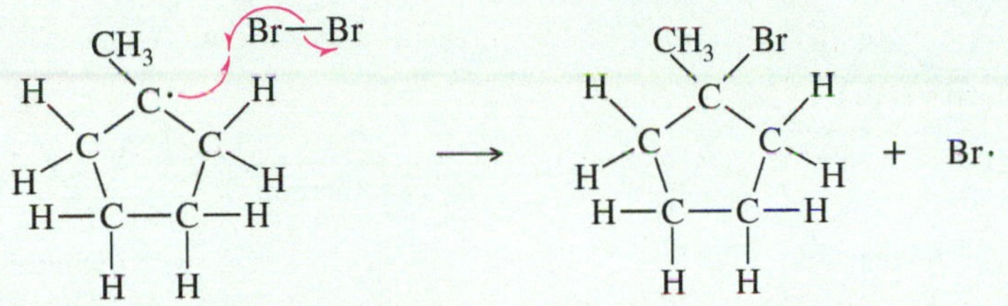

[Structure: 1-methylcyclopentyl radical + Br—Br] ⟶ [1-bromo-1-methylcyclopentane] + Br·

major product

4. Draw termination step(s).

It is left to you to add some possible termination steps and summarize the mechanism developed here.

As practice in using a systematic approach to proposing mechanisms for free-radical reactions, work Problem 4-24 by going through the four steps just outlined.

PROBLEM 4-24

2,3-Dimethylbutane reacts with bromine in the presence of light to give a monobrominated product. Further reaction gives a good yield of a dibrominated product. Predict the structures of these products, and propose a mechanism for the formation of the monobrominated product.

$CH_3-CH(CH_3)-CH(CH_3)-CH_3$

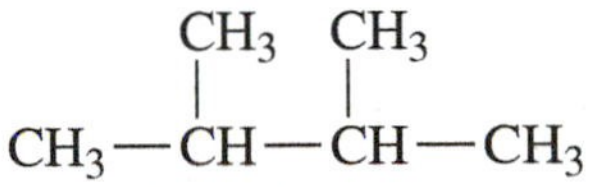

2,3-dimethylbutane

PROBLEM 4-25

In the presence of a small amount of bromine, cyclohexene undergoes the following light-promoted reaction:

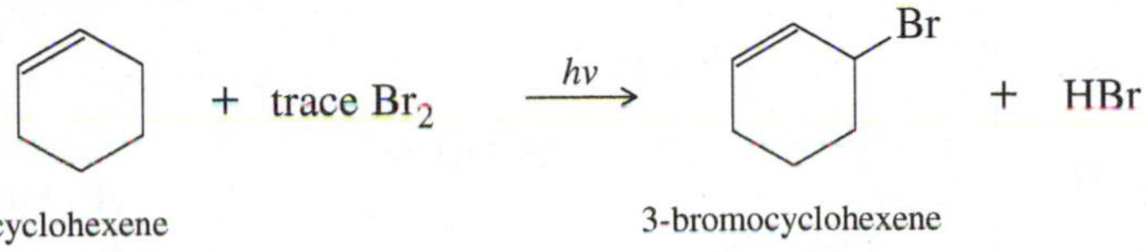

cyclohexene + trace Br_2 $\xrightarrow{h\nu}$ 3-bromocyclohexene + HBr

(a) Propose a mechanism for this reaction.
(b) Draw the structure of the rate-limiting transition state.
(c) Use Hammond's postulate to predict which intermediate most closely resembles this transition state.
(d) Explain why cyclohexene reacts with bromine much faster than cyclohexane, which must be heated to react.

4-15 Radical Inhibitors

We often want to prevent or retard free-radical reactions. For example, oxygen in the air oxidizes and spoils foods, solvents, and other compounds mostly by free-radical chain reactions. Chemical intermediates may decompose or polymerize by free-radical chain reactions. Even the cells in living systems are damaged by radical reactions, which can lead to aging, cancerous mutations, or cell death.

Radical inhibitors are often added to foods and chemicals to retard spoilage by radical chain reactions. Chain reactions depend on the individual steps being fast, so that each initiation step results in many molecules reacting, as in the reaction-energy diagram at the left in Figure 4-13. (Only the radicals are shown.)

The diagram at right in the figure shows how an inhibitor (I) can stop the chain by reacting with a radical intermediate in a fast, highly exothermic step to form an intermediate that is relatively stable. The next step in the chain becomes endothermic and very slow.

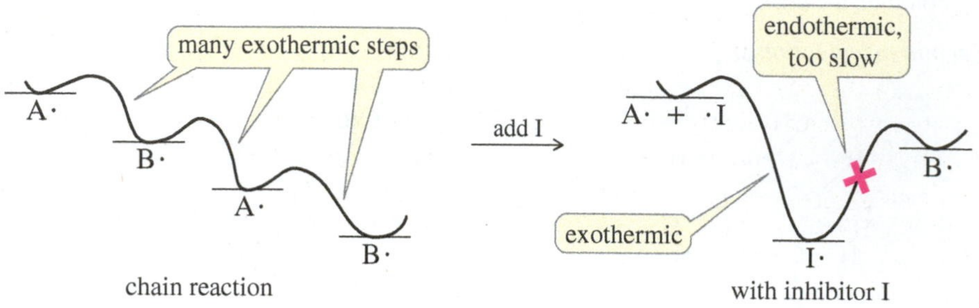

FIGURE 4-13
A radical chain reaction must be a series of exothermic (or mildly endothermic) steps because each step must go quickly, before a termination reaction occurs. An inhibitor reacts with a radical to give a relatively stable product that ends the chain, because it would require a strongly endothermic step for the chain reaction to continue.

"Butylated hydroxyanisole" (BHA) is often added to foods as an antioxidant. It stops oxidation by reacting with radical intermediates to form a relatively stable free-radical intermediate (BHA radical). The BHA radical can react with a second free radical to form an even more stable quinone with all its electrons paired.

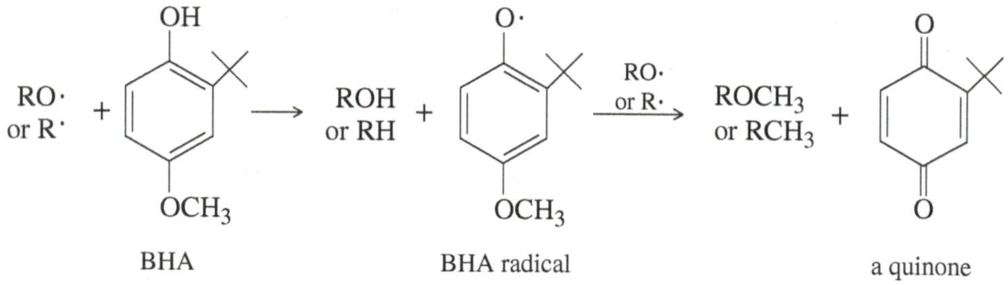

Application: Cancer Drugs

Some anti-cancer agents act by generating highly reactive hydroxyl radicals, which damage and degrade the DNA of the rapidly dividing tumor cells. As a result, the cells die and the tumor shrinks. One example of a radical generator is bleomycin, which is used for the treatment of testicular cancer.

Radical inhibitors also help to protect the cells of living systems. Like BHA, vitamin E is a *phenol* (an aromatic ring with an —OH group), and it is thought to react with radicals by losing the OH hydrogen atom as just shown for BHA. Ascorbic acid (vitamin C) is also thought to protect cells from free radicals, possibly by the following mechanism:

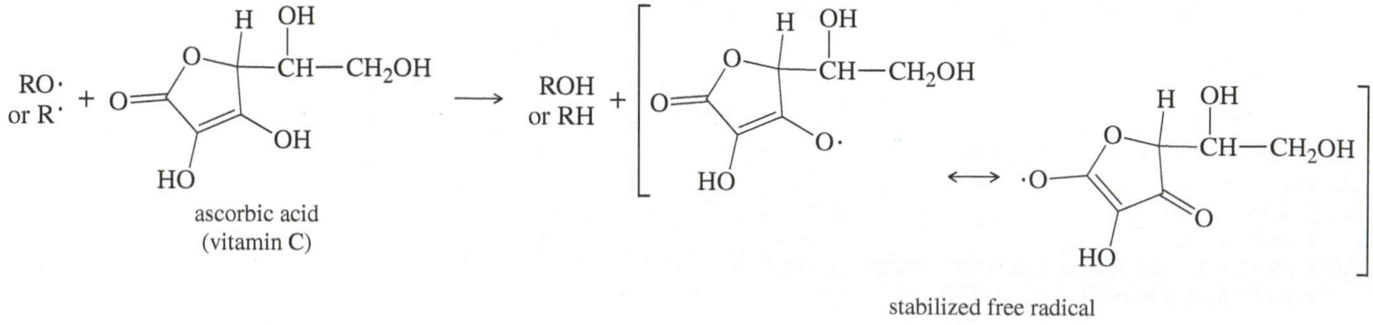

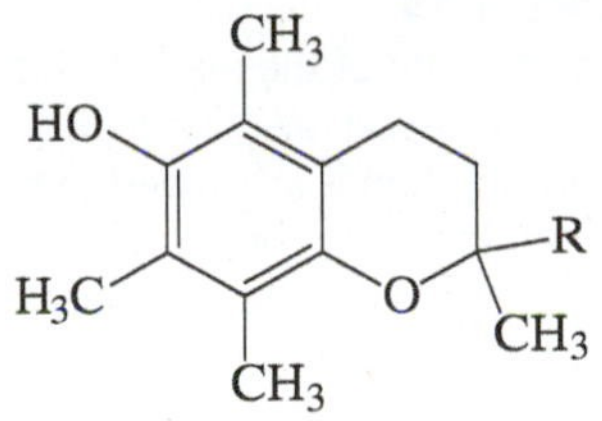

vitamin E
(R = alkyl chain)

PROBLEM 4-26

Draw resonance forms to show how the BHA radical is stabilized by delocalization of the radical electron over other atoms in the molecule.

PROBLEM 4-27

Write an equation for the reaction of vitamin E with an oxidizing radical (RO·) to give ROH and a less reactive free radical.

4-16 Reactive Intermediates

The free radicals we have studied are one class of reactive intermediates. **Reactive intermediates** are short-lived species that are never present in high concentrations because they react as quickly as they are formed. In most cases, reactive intermediates are fragments of molecules (like free radicals), often having atoms with unusual numbers of bonds. Some of the common reactive intermediates contain carbon atoms with only two or three bonds, compared with carbon's four bonds in its stable compounds. Such species react quickly with a variety of compounds to give more stable products with tetravalent carbon atoms.

Although reactive intermediates are not stable compounds, they are important to our study of organic chemistry. Most reaction mechanisms involve reactive intermediates. If you are to understand these mechanisms and propose mechanisms of your own, you need to know how reactive intermediates are formed and how they are likely to react. In this chapter, we consider their structure and stability. In later chapters, we see how they are formed and ways they react to give stable compounds.

Species with trivalent (three-bonded) carbon are classified according to their charge, which depends on the number of nonbonding electrons. The *carbocations* have no nonbonding electrons and are positively charged. The *radicals* have one nonbonding electron and are neutral. The *carbanions* have a pair of nonbonding electrons and are negatively charged.

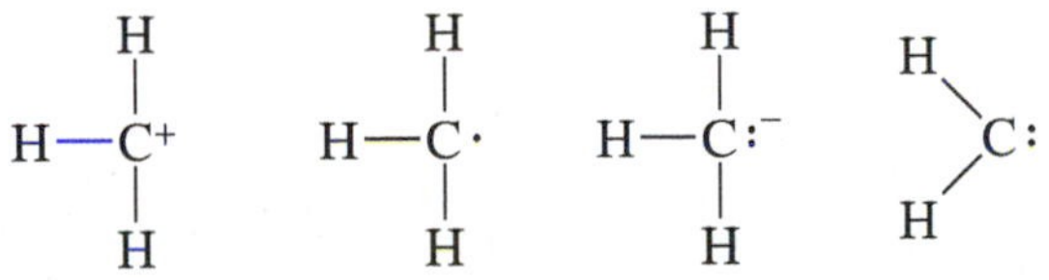

carbocation radical carbanion carbene

The most common intermediates with a divalent (two-bonded) carbon atom are the *carbenes*. A carbene has two nonbonding electrons on the divalent carbon atom, making it uncharged. Carbenes are not as common as carbocations, radicals, and carbanions.

4-16A Carbocations

A **carbocation** (also called a **carbonium ion** or a **carbenium ion**) is a species that contains a carbon atom bearing a positive charge. The positively charged carbon atom is bonded to three other atoms, and it has no nonbonding electrons, so it has only six electrons in its valence shell. It is sp^2 hybridized, with a planar structure and bond angles of about 120°. For example, the methyl cation ($^+CH_3$) is planar, with bond angles of exactly 120°. The unhybridized p orbital is vacant and lies perpendicular to the plane of the C—H bonds (Figure 4-14). The structure of $^+CH_3$ is similar to the structure of BH_3, discussed in Chapter 1.

With only six electrons in the positive carbon's valence shell, a carbocation is a powerful electrophile (Lewis acid), and it may react with any nucleophile it encounters. Like other strong acids, carbocations are unlikely to be found in basic solutions.

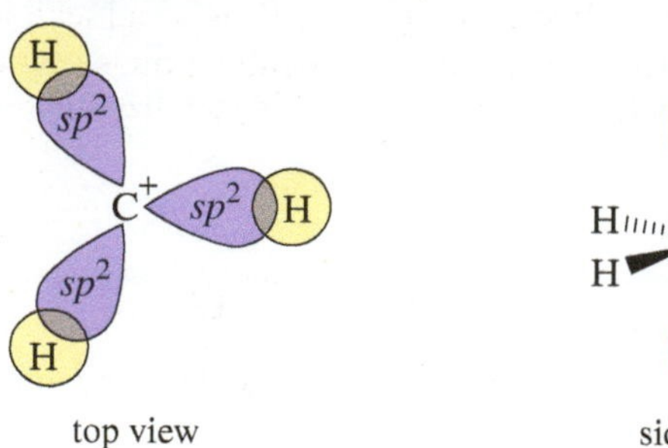

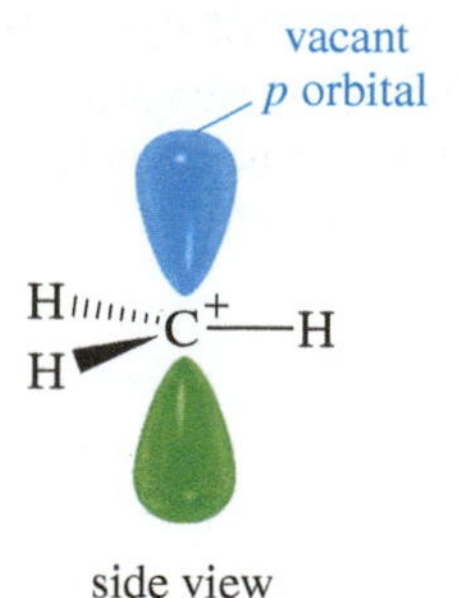

FIGURE 4-14
Orbital diagram of the methyl cation. The methyl cation is similar to BH_3. The carbon atom is σ bonded to three hydrogen atoms by overlap of its sp^2 hybrid orbitals with the *s* orbitals of hydrogen. A vacant *p* orbital lies perpendicular to the plane of the three C—H bonds.

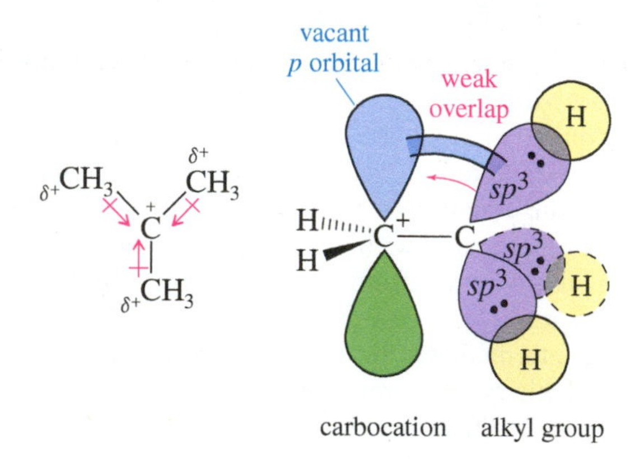

FIGURE 4-15
Effect of alkyl substituent on carbocation stability. A carbocation is stabilized by overlap of filled orbitals on an adjacent alkyl group with the vacant *p* orbital of the carbocation. Overlap between a σ bond and a *p* orbital is called *hyperconjugation*.

Carbocations are proposed as intermediates in many types of organic reactions, some of which we will encounter in Chapter 6.

Like free radicals, carbocations are *electron-deficient* species: They have fewer than eight electrons in the valence shell. Also like free radicals, carbocations are stabilized by alkyl substituents. An alkyl group stabilizes an electron-deficient carbocation in two ways: (1) through an inductive effect, and (2) through the partial overlap of filled orbitals with empty ones. The **inductive effect** is a donation of electron density through the sigma (σ) bonds of the molecule. The positively charged carbon atom withdraws some electron density from the polarizable alkyl groups bonded to it.

Alkyl substituents also have filled sp^3 orbitals that can overlap with the empty *p* orbital on the positively charged carbon atom, further stabilizing the carbocation (Figure 4-15). Even though the attached alkyl group rotates, one of its sigma bonds is always aligned with the empty *p* orbital on the carbocation. The pair of electrons in this σ bond spreads out into the empty *p* orbital, stabilizing the electron-deficient carbon atom. This type of overlap between a *p* orbital and a sigma bond is called *hyperconjugation*.

In general, more highly substituted carbocations are more stable.

Stability of carbocations

$$R_3C^+ > R_2CH^+ > RCH_2^+ > CH_3^+$$

most stable ... least stable

3° > 2° > 1° > methyl

Unsaturated carbocations are also stabilized by **resonance stabilization.** If a pi (π) bond is adjacent to a carbocation, the filled *p* orbitals of the π bond will overlap with the empty *p* orbital of the carbocation. The result is a delocalized ion, with the positive charge shared by two atoms. Resonance delocalization is particularly effective in stabilizing carbocations.

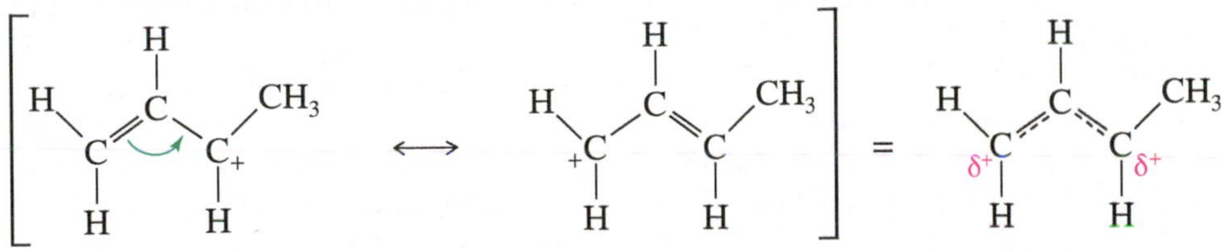

Carbocations are common intermediates in organic reactions. Highly substituted alkyl halides can ionize when they are heated in a polar solvent. The strongly electrophilic carbocation reacts with any available nucleophile, often the solvent.

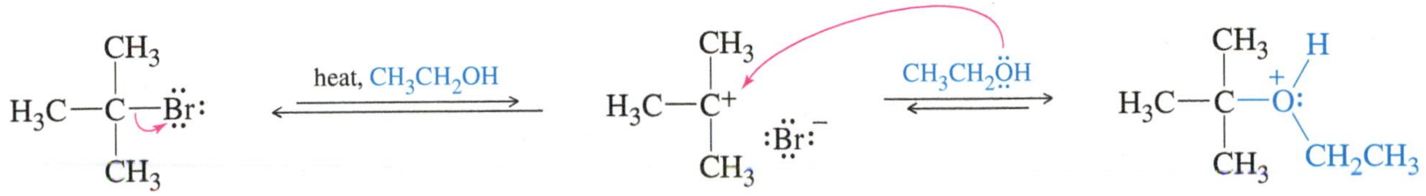

heat, CH_3CH_2OH

tert-butyl bromide

tert-butyl cation

Carbocations are also strong proton acids. The *tert*-butyl cation shown above can also lose a proton to a weak base, often the solvent.

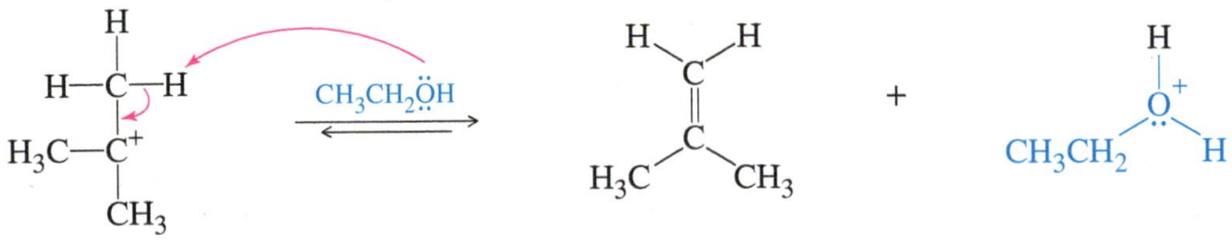

tert-butyl cation

PROBLEM 4-28

The triphenylmethyl cation is so stable that some of its salts can be stored for months. Explain why this cation is so stable.

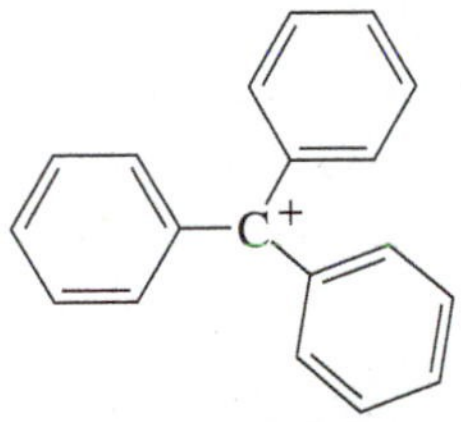

triphenylmethyl cation

PROBLEM 4-29

Rank the following carbocations in decreasing order of stability. Classify each as primary, secondary, or tertiary.

(a) The isopentyl cation, $(CH_3)_2CHCH_2—\overset{+}{C}H_2$

(b) The 3-methyl-2-butyl cation, $CH_3—\overset{+}{C}H—CH(CH_3)_2$

(c) The 2-methyl-2-butyl cation, $CH_3—\overset{+}{C}(CH_3)CH_2CH_3$

(d)

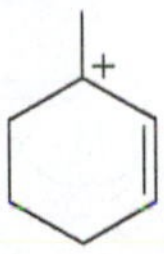

4-16B Free Radicals

Application: Biochemistry

Vitamin B_{12} is an essential dietary factor, and a deficiency results in anemia and neurological damage. The vitamin assists two different enzymes in the production and stabilization of methyl radicals. These methyl radicals are then used for the synthesis of important cellular components.

Like carbocations, **free radicals** are sp^2 hybridized and planar (or nearly planar). Unlike carbocations, however, the *p* orbital perpendicular to the plane of the C—H bonds of the radical is not empty; it contains the odd electron. Figure 4-16 shows the structure of the methyl radical.

Both radicals and carbocations are electron-deficient because they lack an octet around the carbon atom. Like carbocations, radicals are stabilized by the electron-donating effect of alkyl groups, making more highly substituted radicals more stable. This effect is confirmed by the bond-dissociation enthalpies shown in Figure 4-7: Less energy is required to break a C—H bond to form a more highly substituted radical.

Stability of radicals

$$R_3C\cdot \; > \; R_2HC\cdot \; > \; RH_2C\cdot \; > \; H_3C\cdot$$

most stable ... least stable

3° > 2° > 1° > methyl

PROBLEM-SOLVING HINT

We describe atoms as being electron-deficient when they lack a full octet, whether or not they bear a formal charge. Recall that the boron atoms in BH_3 and BF_3 are uncharged, yet they are electron-deficient and are powerful electrophiles (Lewis acids). Similarly, carbocations (6 electrons, positive charge) and free radicals (7 electrons, uncharged) are both electron-deficient and highly reactive.

Like carbocations, radicals can be stabilized by resonance. Overlap with the *p* orbitals of a π bond allows the odd electron to be delocalized over two carbon atoms. Resonance delocalization is particularly effective in stabilizing a radical.

$$\left[H\text{–}CH\!=\!CH\text{–}\dot{C}H\text{–}CH_3 \longleftrightarrow H\text{–}\dot{C}H\text{–}CH\!=\!CH\text{–}CH_3 \right] \quad \text{or} \quad H\text{–}\overset{\delta\cdot}{C}H\text{=-=}CH\text{=-=}\overset{\delta\cdot}{C}H\text{–}CH_3$$

PROBLEM 4-30

Rank the following radicals in decreasing order of stability. Classify each as primary, secondary, or tertiary.

(a) The isopentyl radical, $(CH_3)_2CHCH_2—\dot{C}H_2$

(b) The 3-methyl-2-butyl radical, $CH_3—\dot{C}H—CH(CH_3)_2$

(c) The 2-methyl-2-butyl radical, $CH_3—\dot{C}(CH_3)CH_2CH_3$

(d) 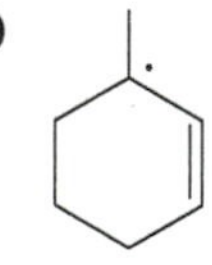

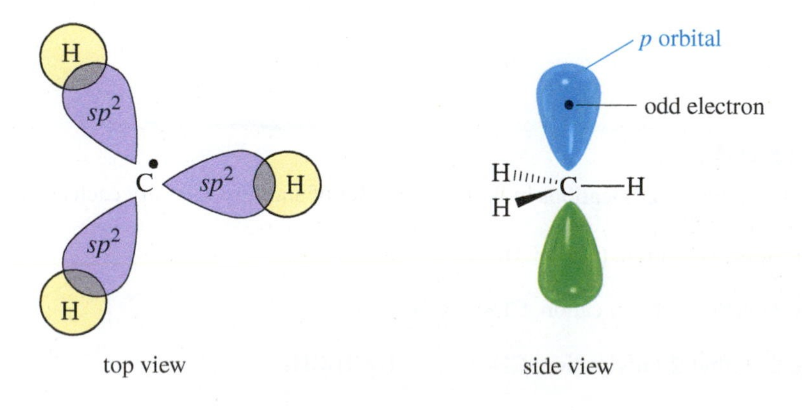

FIGURE 4-16
Orbital diagram of the methyl radical. The structure of the methyl radical is like that of the methyl cation (Figure 4-14), except there is an additional electron. The odd electron is in the *p* orbital perpendicular to the plane of the three C—H bonds.

4-16C Carbanions

A **carbanion** has a trivalent carbon atom that bears a negative charge. There are eight electrons around the carbon atom (three bonds and one lone pair), so it is not electron-deficient; rather, it is electron-rich and a strong nucleophile (Lewis base). A carbanion has the same electronic structure as an amine. Compare the structures of a methyl carbanion and ammonia:

$$\underset{\text{methyl anion}}{H_3C{:}^-} \qquad \underset{\text{ammonia}}{H_3N{:}}$$

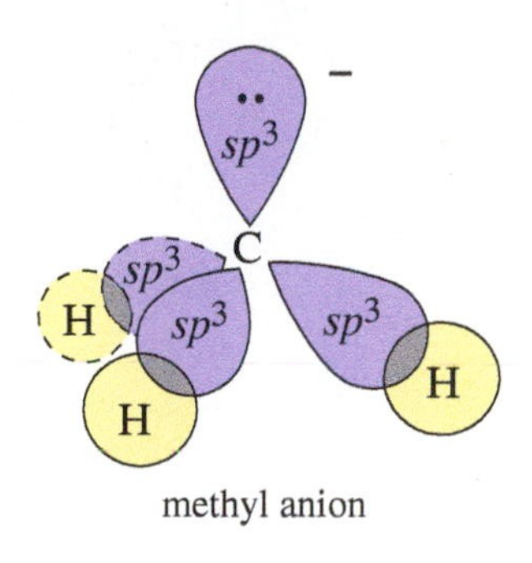

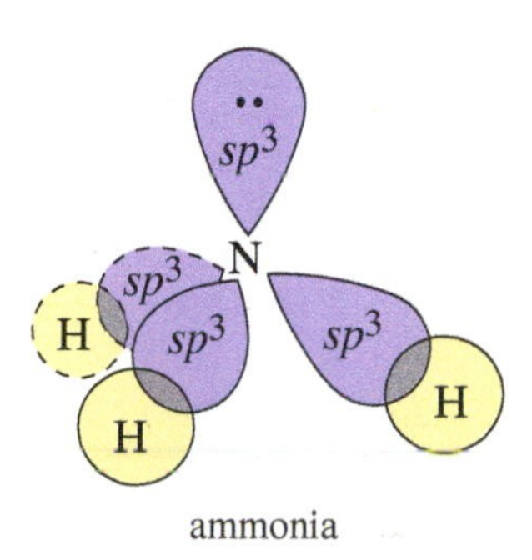

FIGURE 4-17
Comparison of orbital structures of the methyl anion and ammonia. Both the methyl anion and ammonia have an sp^3 hybridized central atom, with a nonbonding pair of electrons occupying one of the tetrahedral positions.

The hybridization and bond angles of a simple carbanion also resemble those of an amine. The carbon atom is sp^3 hybridized and tetrahedral. One of the tetrahedral positions is occupied by an unshared lone pair of electrons. Figure 4-17 compares the orbital structures and geometry of ammonia and the methyl anion.

Like amines, carbanions are nucleophilic and basic. A carbanion has a negative charge on its carbon atom, however, making it a more powerful base and a stronger nucleophile compared with an amine. For example, a carbanion is sufficiently basic to remove a proton from ammonia.

$$R_3C{:}^- \;+\; {:}NH_3 \;\longrightarrow\; R_3CH \;+\; ^-{:}\ddot{N}H_2$$

Like other strong bases, carbanions are unlikely to be found in acidic solutions. The stability order of carbanions reflects their high electron density. Alkyl groups and other electron-donating groups slightly destabilize a carbanion. The order of stability is usually the opposite of that for carbocations and free radicals.

Stability of carbanions

$$\underset{\text{least stable}}{R_3C{:}^-} < R_2CH{:}^- < RCH_2{:}^- < \underset{\text{most stable}}{H_3C{:}^-}$$

$$3° < 2° < 1° < \text{methyl}$$

Carbanions that occur as intermediates in organic reactions are almost always stabilized by neighboring groups. They can be stabilized either by inductive effects or by resonance. For example, halogen atoms are electron-withdrawing, so they stabilize carbanions through the inductive withdrawal of electron density. Resonance also plays an important role in stabilizing carbanions. A carbonyl group (C=O) stabilizes an adjacent carbanion by overlap of its π bond with the nonbonding electrons of the carbanion. The negative charge is delocalized onto the electronegative oxygen atom of the carbonyl group.

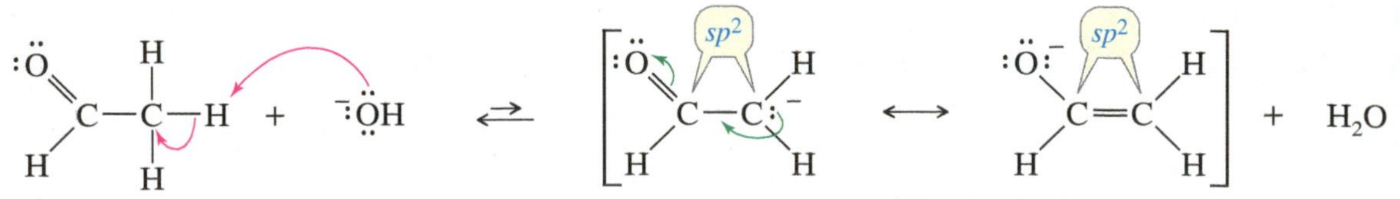

$$\text{H-C(=O)-CH}_3 + {}^{-}\text{:OH} \rightleftharpoons \left[\text{:O=C(H)-CH}_2^{:-} \longleftrightarrow {}^{-}\text{:O-C(H)=CH}_2 \right] + H_2O$$

resonance-stabilized carbanion

This resonance-stabilized carbanion must be sp^2 hybridized and planar for effective delocalization of the negative charge onto oxygen (Section 1-17). Resonance-stabilized carbanions are the most common type of carbanions we will encounter in organic reactions.

PROBLEM 4-31

Acetylacetone (pentane-2,4-dione) reacts with sodium hydroxide to give water and the sodium salt of a carbanion. Write a complete structural formula for the carbanion, and use resonance forms to show the stabilization of the carbanion.

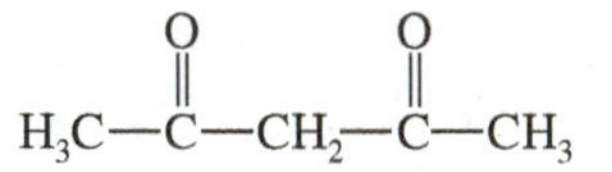

$$H_3C-C(=O)-CH_2-C(=O)-CH_3$$

acetylacetone (pentane-2,4-dione)

PROBLEM 4-32

Acetonitrile ($CH_3C{\equiv}N$) is deprotonated by very strong bases. Write resonance forms to show the stabilization of the carbanion that results.

4-16D Carbenes

Carbenes are uncharged reactive intermediates containing a divalent carbon atom. The simplest carbene has the formula $:CH_2$ and is called *methylene*, just as a $-CH_2-$ group in a molecule is called a *methylene group*. One way of generating carbenes is to form a carbanion that can expel a halide ion. For example, a strong base can abstract a proton from tribromomethane ($CHBr_3$) to give an inductively stabilized carbanion. This carbanion expels bromide ion to give dibromocarbene.

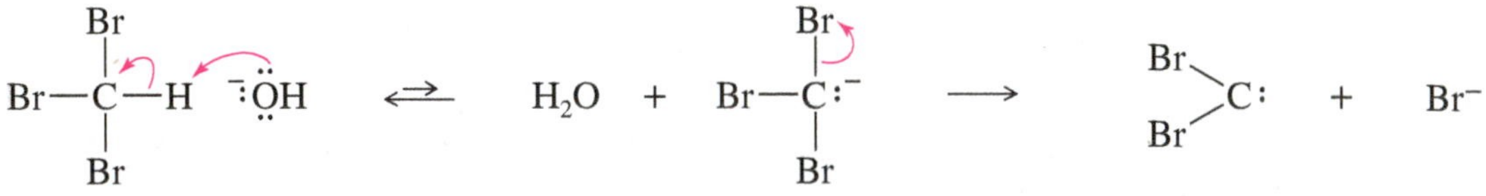

$$\underset{\text{tribromomethane}}{CHBr_3} + {}^{-}\text{:OH} \rightleftharpoons H_2O + \underset{\text{a carbanion}}{Br_3C:^{-}} \longrightarrow \underset{\text{dibromocarbene}}{Br_2C:} + Br^{-}$$

The carbon atom in dibromocarbene has only six electrons in its valence shell. It is sp^2 hybridized, with trigonal geometry. An unshared pair of electrons occupies one of the sp^2 hybrid orbitals, and there is an empty p orbital extending above and below the plane of the atoms. A carbene has both a lone pair of electrons and an empty p orbital, so it can react as a nucleophile or as an electrophile.

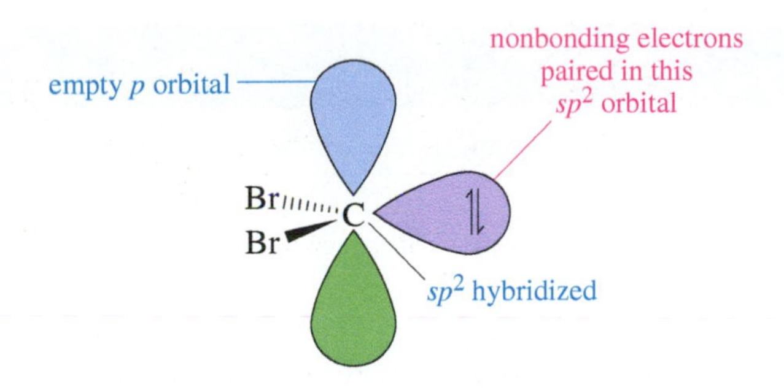

Methylene itself is formed when diazomethane (CH_2N_2) is heated or irradiated with light. The diazomethane molecule splits to form a stable nitrogen molecule and the very reactive carbene.

$$H_2\overset{-}{C}{-}\overset{+}{N}{\equiv}N: \xrightarrow{\text{heat or light}} H_2C: \; + \; :N{\equiv}N:$$

diazomethane — methylene — nitrogen

The most common synthetic reaction of carbenes is their addition to double bonds to form cyclopropane rings. For example, dibromocarbene adds to cyclohexene to give an interesting bicyclic compound.

$$\text{cyclohexene} + :CBr_2 \longrightarrow \text{7,7-dibromobicyclo product}$$

cyclohexene — dibromocarbene

No simple carbenes have ever been purified or even made in a high concentration, because when two carbenes collide, they immediately dimerize (two of them bond together) to give an alkene.

$$R_2C: \; + \; :CR_2 \xrightarrow{\text{very fast}} R_2C{=}CR_2$$

Carbenes and carbenoids (carbene-like reagents) are useful both for the synthesis of other compounds and for the investigation of reaction mechanisms. The carbene intermediate is generated in the presence of its target compound, so that it can react immediately, and the concentration of the carbene is always low. Reactions using carbenes are discussed in Chapter 8.

PROBLEM 4-33

When it is strongly heated, ethyl diazoacetate decomposes to give nitrogen gas and a carbene. Draw a Lewis structure of the carbene.

$$:N{\equiv}\overset{+}{N}{-}\overset{-}{C}H{-}\overset{\overset{\displaystyle O}{\|}}{C}{-}O{-}CH_2CH_3$$

ethyl diazoacetate

SUMMARY Reactive Intermediates

	Structure	*Stability*	*Properties*
carbocations	$-\overset{\vert}{\underset{\vert}{C}}^{+}$	3° > 2° > 1° > $^{+}CH_3$	electrophilic strong acids
radicals	$-\overset{\vert}{\underset{\vert}{C}}\cdot$	3° > 2° > 1° > $\cdot CH_3$	electron-deficient
carbanions	$-\overset{\vert}{\underset{\vert}{C}}:^{-}$	$^{-}:CH_3$ > 1° > 2° > 3°	nucleophilic strong bases
carbenes	$>C:$		both nucleophilic and electrophilic

Essential Terms

activation energy (E_a) The energy difference between the reactants and the transition state; the minimum energy the reactants must have for the reaction to occur. (p. 171)

bond-dissociation enthalpy (BDE) The amount of enthalpy required to break a particular bond homolytically, to give radicals. (p. 165)

$$A{:}B \longrightarrow A\cdot + \cdot B \qquad \Delta H^\circ = \text{BDE}$$

carbanion A strongly nucleophilic species with a negatively charged carbon atom having only three bonds. The carbon atom has a nonbonding pair of electrons. (p. 191)

carbene A highly reactive species with only two bonds to an uncharged carbon atom with a nonbonding pair of electrons. The simplest carbene is methylene, $:CH_2$. (p. 192)

carbocation **(carbonium ion, carbenium ion)** A strongly electrophilic species with a positively charged carbon atom having only three bonds. (p. 187)

catalyst A substance that increases the rate of a reaction (by lowering E_a) without being consumed in the reaction. (p. 173)

chain reaction A multistep reaction where a reactive intermediate formed in one step brings about a second step that generates the intermediate needed for the following step. (p. 157)

- **initiation step:** The preliminary step in a chain reaction, where the reactive intermediate is first formed.
- **propagation steps:** The steps in a chain reaction that are repeated over and over to form the product. The sum of the propagation steps should give the net reaction.
- **termination steps:** Any steps where a reactive intermediate is consumed without another one being generated.

enthalpy **(heat content; *H*)** A measure of the heat energy in a system. In a reaction, the heat absorbed or evolved is called the *heat of reaction*, ΔH°. A decrease in enthalpy (negative ΔH°) is favorable for a reaction. (p. 163)

- **endothermic:** Consuming heat (having a positive ΔH°).
- **exothermic:** Giving off heat (having a negative ΔH°).

entropy (*S*) A measure of disorder or freedom of motion. An increase in entropy (positive ΔS°) is favorable for a reaction. (p. 163)

equilibrium A state of a system such that no more net change is taking place; the rate of the forward reaction equals the rate of the reverse reaction. (p. 161)

equilibrium constant A quantity calculated from the relative amounts of the products and reactants present at equilibrium. (p. 161) For the reaction

$$aA + bB \rightleftharpoons cC + dD$$

the equilibrium constant is

$$K_{eq} = \frac{[C]^c[D]^d}{[A]^a[B]^b}$$

free energy (**Gibbs free energy; *G***) A measure of a reaction's tendency to go in the direction written. A decrease in free energy (negative ΔG) is favorable for a reaction. (p. 162)

Free-energy change is defined: $\Delta G = \Delta H - T\Delta S$

standard Gibbs free energy change: (**$\Delta G°$**) The free-energy change corresponding to reactants and products in their standard states (pure substances in their most stable states) at 25 °C and 1 atm pressure. $\Delta G°$ is related to K_{eq} by

$$K_{eq} = e^{-\Delta G°/RT}$$ (p. 162)

endergonic: Having a positive $\Delta G°$ (unfavorable).

exergonic: Having a negative $\Delta G°$ (favorable).

halogenation The reaction of a halogen (X_2) or halogen-containing reagent that incorporates one or more halogen atoms into a molecule. Free-radical halogenation of alkanes is an important industrial synthesis, but it is rarely used in a laboratory setting. We study the reaction primarily because it serves as an uncomplicated example for studying its thermodynamics and kinetics. (pp. 155, 159)

$$\text{R—H} \xrightarrow[\text{heat or light}]{X_2} \text{R—X} + \text{HX}$$

Hammond's postulate Related species (on a reaction-energy diagram) that are closer in energy are also closer in structure. In an exothermic reaction, the transition state is closer to the reactants in energy and in structure. In an endothermic reaction, the transition state is closer to the products in energy and in structure. (p. 181)

heterolytic cleavage (**ionic cleavage**) The breaking of a bond in such a way that one of the atoms retains both of the bond's electrons. A heterolytic cleavage forms two ions. (p. 166)

$$\text{A:B} \longrightarrow \text{A:}^- + \text{B}^+$$

homolytic cleavage (**radical cleavage**) The breaking of a bond in such a way that each atom retains one of the bond's two electrons. A homolytic cleavage produces two radicals. (p. 166)

$$\text{A:B} \longrightarrow \text{A}\cdot + \cdot\text{B}$$

inductive effect A donation (or withdrawal) of electron density through sigma bonds. (p. 188)

intermediate A molecule or a fragment of a molecule that is formed in a reaction and exists for a finite length of time before it reacts in the next step. An intermediate corresponds to a relative minimum (a low point) in the reaction-energy diagram. (pp. 157, 172)

reactive intermediate: A short-lived species that is never present in high concentration because it reacts as quickly as it is formed. (pp. 157, 158)

kinetics The study of reaction rates. (pp. 156, 169)

mechanism The step-by-step pathway from reactants to products, showing which bonds break and which bonds form in what order. The mechanism should include the structures of all intermediates and curved arrows to show the movement of electrons. (p. 156)

potential-energy diagram See **reaction-energy diagram.** (p. 172)

radical (**free radical**) A highly reactive species in which one of the atoms has an odd number of electrons. Most commonly, a radical contains a carbon atom with three bonds and an "odd" (unpaired) electron. (p. 158)

radical inhibitor A compound added to prevent the propagation of free-radical chain reactions. In most cases, the inhibitor reacts to form a radical that is too stable to propagate the chain. (p. 186)

rate equation (**rate law**) The relationship between the concentrations of the reagents and the observed reaction rate. (p. 169)

A general rate law for the reaction $A + B \longrightarrow C + D$ is

$$\text{rate} = k_r[A]^a[B]^b$$

order: (**kinetic order**) The power of a concentration term in the rate equation. The preceding rate equation is ath order in [A], bth order in [B], and $(a + b)$th order overall.

rate constant: The proportionality constant k_r in the rate equation.

rate-limiting step (**rate-determining step**) The slowest step in a multistep sequence of reactions. In general, the rate-limiting step is the step with the highest-energy transition state. (p. 174)

rate of a reaction The amount of product formed or reactant consumed per unit of time. (p. 169)

reaction-energy diagram (**potential-energy diagram**) A plot of potential-energy changes as the reactants are converted to products. The vertical axis is potential energy (usually free energy, but occasionally enthalpy). The horizontal axis is the **reaction coordinate,** a measure of the progress of the reaction. (p. 172)

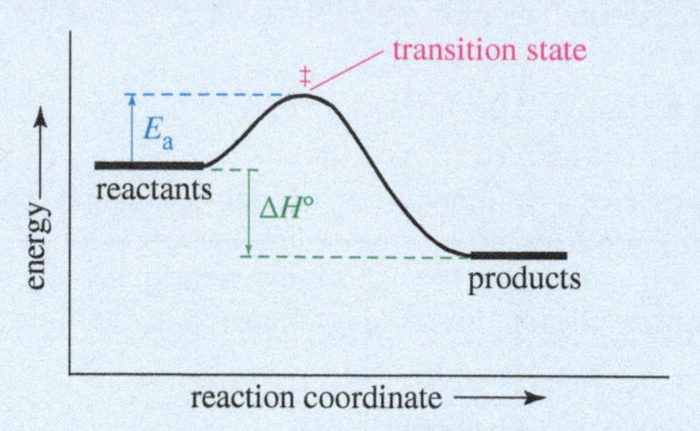

resonance stabilization Stabilization that takes place by delocalization of electrons in a π bonded system. Cations, radicals, and anions are often stabilized by resonance delocalization. (p. 188)

resonance-stabilized carbanion

substitution A reaction in which one atom replaces another, usually as a substituent on a carbon atom. (p. 157)

termination reaction A step that produces fewer reactive intermediates (usually free radicals) than it consumes. (p. 160)

thermodynamics The study of the energy changes accompanying chemical transformations. Thermodynamics is generally concerned with systems at equilibrium. (pp. 156, 161)

transition state (**activated complex**) The state of highest energy between reactants and products. A relative maximum (high point) on the reaction-energy diagram. (p. 172)

Essential Problem-Solving Skills in Chapter 4

Each skill is followed by problem numbers exemplifying that particular skill.

1	Propose a detailed mechanism for the free-radical halogenation of an alkane.	Problems 4-44, 46, 48, 51, 55, 56, and 58
2	Predict the major halogenation products based on the stability of the intermediates and the selectivity of the halogenation.	Problems 4-43, 47, 48, and 49
3	Draw a reaction-energy diagram for a mechanism, and point out the corresponding transition states, activation energies, intermediates, and rate-limiting steps.	Problems 4-35, 36, 37, and 38
4	Use bond-dissociation enthalpies to calculate the enthalpy change for each step of a reaction and the overall enthalpy change for the reaction.	Problems 4-41, 51, 54, 55, 56, and 58
5	Calculate free-energy changes from equilibrium constants, and calculate the position of an equilibrium from the free-energy changes.	Problems 4-53 and 58
6	Determine the kinetic order of a reaction based on its rate equation.	Problems 4-34 and 39
7	Use Hammond's postulate to predict whether a transition state will be reactant-like or product-like, and explain how this distinction affects the selectivity of a reaction.	Problems 4-50 and 57
	Problem-Solving Strategy: Proposing Reaction Mechanisms: Free-Radical Reactions	Problems 4-44, 46, 48, 51, 55, 56, and 58
8	Draw and describe the structures of carbocations, carbanions, free radicals, and carbenes and the structural features that stabilize them. Explain which are electrophilic and which are nucleophilic.	Problems 4-40, 42, 45, 46, 47, and 52

Study Problems

4-34 The following reaction is a common synthesis used in the organic chemistry laboratory course.

$$\underset{\text{1-bromobutane}}{CH_3CH_2CH_2CH_2Br} + \underset{\text{methoxide ion}}{CH_3O^-} \xrightarrow[\text{(methanol solvent)}]{CH_3OH} \underset{\text{methoxybutane}}{CH_3CH_2CH_2CH_2OCH_3} + \underset{\text{bromide ion}}{Br^-}$$

When we double the concentration of methoxide ion (CH_3O^-), we find that the reaction rate doubles. When we triple the concentration of 1-bromobutane, we find that the reaction rate triples.

(a) What is the order of this reaction with respect to 1-bromobutane? What is the order with respect to methoxide ion? Write the rate equation for this reaction. What is the overall order?

(b) One lab textbook recommends forming the sodium methoxide in methanol solvent, but before adding 1-bromobutane, it first distills off enough methanol to reduce the mixture to half of its original volume. What difference in rate will we see when we run the reaction (using the same amounts of reagents) in half the volume of solvent?

4-35 Consider the following reaction-energy diagram.

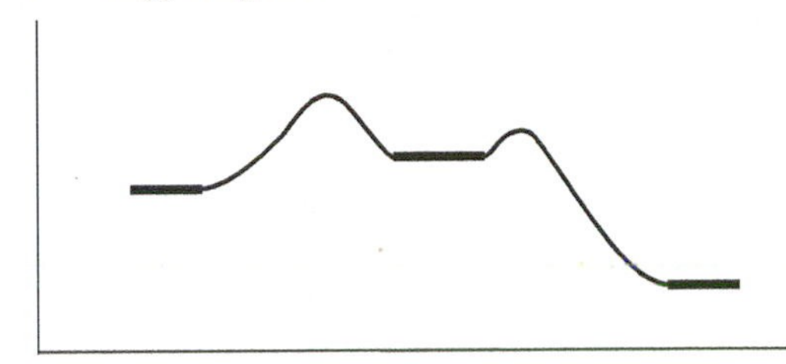

(a) Label the reactants and the products. Label the activation energy for the first step and the second step.

(b) Is the overall reaction endothermic or exothermic? What is the sign of $\Delta H°$?

(c) Which points in the curve correspond to intermediates? Which correspond to transition states?

(d) Label the transition state of the rate-limiting step. Does its structure resemble the reactants, the products, or an intermediate?

4-36 Draw a reaction-energy diagram for a one-step exothermic reaction. Label the parts that represent the reactants, products, transition state, activation energy, and heat of reaction.

4-37 Draw a reaction-energy diagram for a two-step endothermic reaction with a rate-limiting second step.

4-38 **(a)** Draw an approximate reaction-energy diagram for the acid–base reaction of phenol (see below) with 1-molar aqueous sodium hydroxide solution.

(b) On the same diagram, draw an approximate reaction-energy diagram for the acid–base reaction of *tert*-butyl alcohol (see below) with 1-molar aqueous sodium hydroxide solution.

C_6H_5—OH phenol, $pK_a = 10.0$

$(CH_3)_3C$—OH *tert*-butyl alcohol, $pK_a = 18.0$

4-39 Treatment of *tert*-butyl alcohol with concentrated HCl gives *tert*-butyl chloride.

$$\underset{\textit{tert}\text{-butyl alcohol}}{(CH_3)_3C{-}OH} + H^+ + Cl^- \longrightarrow \underset{\textit{tert}\text{-butyl chloride}}{(CH_3)_3C{-}Cl} + H_2O$$

When the concentration of H^+ is doubled, the reaction rate doubles. When the concentration of *tert*-butyl alcohol is tripled, the reaction rate triples. When the chloride ion concentration is quadrupled, however, the reaction rate is unchanged. Write the rate equation for this reaction.

4-40 Label each hydrogen atom in the following compounds as primary (1°), secondary (2°), or tertiary (3°).

(a) $CH_3CH_2CH(CH_3)_2$ (b) $(CH_3)_3CCH_2CH_3$ (c) $(CH_3)_2CHCH(CH_3)CH_2CH_3$

(d) 

(e) 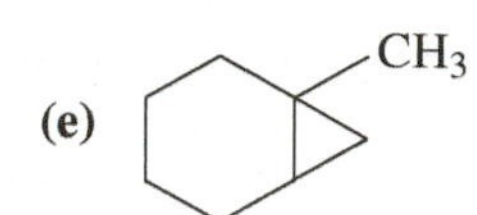

(f) 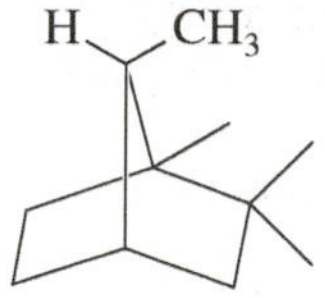 H CH3

4-41 Use bond-dissociation enthalpies (Table 4-2, p. 167) to calculate values of $\Delta H°$ for the following reactions.

(a) $CH_3{-}CH_3 + I_2 \longrightarrow CH_3CH_2I + HI$
(b) $CH_3CH_2Cl + HI \longrightarrow CH_3CH_2I + HCl$
(c) $(CH_3)_3C{-}OH + HCl \longrightarrow (CH_3)_3C{-}Cl + H_2O$
(d) $CH_3CH_2CH_3 + H_2 \longrightarrow CH_3CH_3 + CH_4$
(e) $CH_3CH_2OH + HBr \longrightarrow CH_3CH_2{-}Br + H_2O$

4-42 Use the information in Table 4-2 (p. 167) to rank the following radicals in decreasing order of stability.

$\cdot CH_3$ $CH_3\dot{C}H_2$ (benzene ring)$-\dot{C}H_2$ 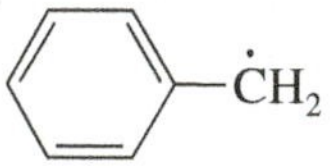 $(CH_3)_3C\cdot$ $(CH_3)_2\dot{C}H$ $CH_2{=}CH{-}\dot{C}H_2$

4-43 For each alkane,

1. draw all the possible monochlorinated derivatives.
2. determine whether free-radical chlorination would be a good way to make any of these monochlorinated derivatives. (Will the reaction give mostly one major product?)
3. which monobrominated derivatives could you form in good yield by free-radical bromination?

(a) cyclopentane (b) methylcyclopentane
(c) 2-methylpentane (d) 2,2,3,3-tetramethylbutane

4-44 Write a mechanism for the light-initiated reaction of cyclohexane with chlorine to give chlorocyclohexane. Label the initiation and propagation steps.

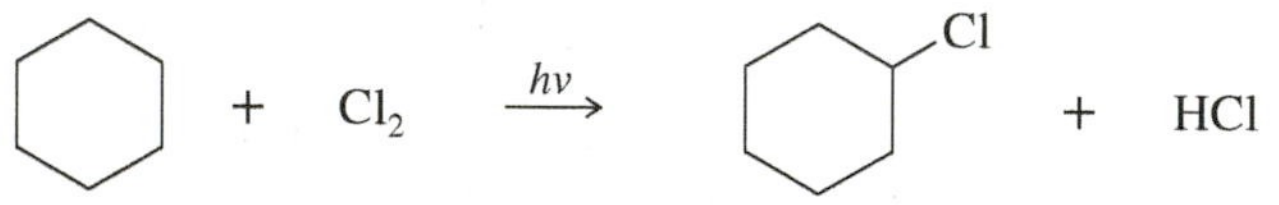

cyclohexane + Cl_2 $\xrightarrow{h\nu}$ chlorocyclohexane (Cl) + HCl

4-45 Draw the important resonance forms of the following free radicals.

(a) $CH_2{=}CH{-}\dot{C}H_2$ (b) 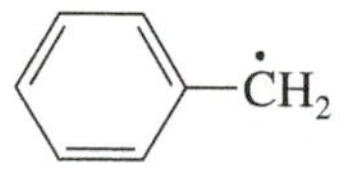 (benzene ring)$-\dot{C}H_2$ (c) $CH_3{-}\overset{O}{\overset{\|}{C}}{-}O\cdot$

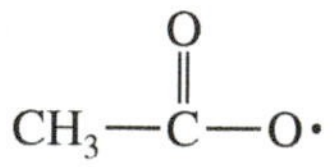

(d)  (e) 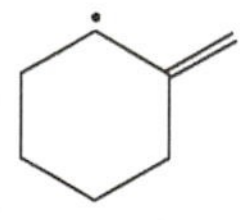 (f) 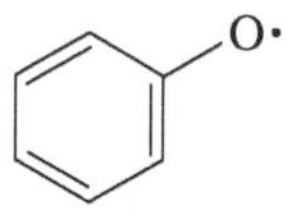 O·

***4-46** In the presence of a small amount of bromine, the following light-promoted reaction has been observed.

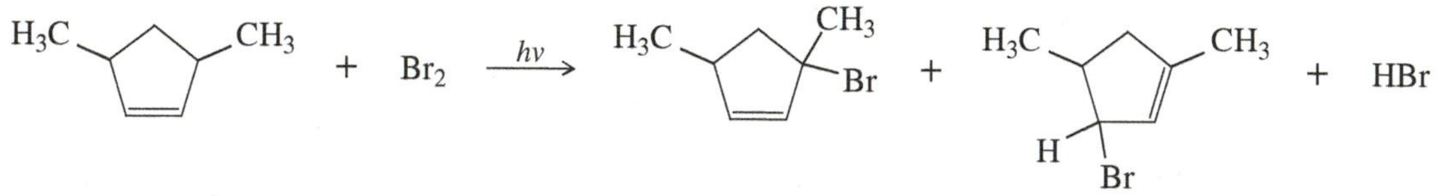

H3C, CH3 (cyclopentene) + Br_2 $\xrightarrow{h\nu}$ H3C, CH3, Br (product) + H3C, CH3, H, Br (product) + HBr

(a) Write a mechanism for this reaction. Your mechanism should explain how both products are formed. (*Hint*: Notice which H atom has been lost in both products.)
(b) Explain why only this one type of hydrogen atom has been replaced, in preference to any of the other hydrogen atoms in the starting material.

4-47 For each compound, predict the major product of free-radical bromination. Remember that bromination is highly selective, and only the most stable radical will be formed.

(a) cyclohexane (b) methylcyclopentane (c) decalin

(d) hexane (e) 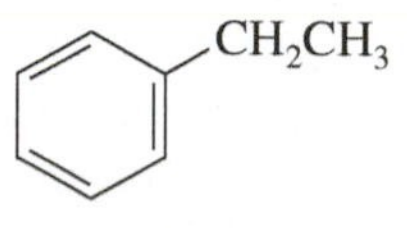 CH_2CH_3 ethylbenzene (f) 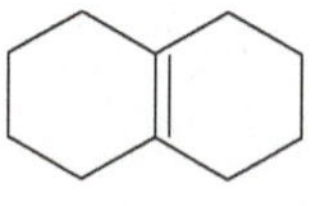 (2 products)

4-48 When exactly 1 mole of methane is mixed with exactly 1 mole of chlorine and light is shone on the mixture, a chlorination reaction occurs. The products are found to contain substantial amounts of di-, tri-, and tetrachloromethane, as well as unreacted methane.

(a) Explain how a mixture is formed from this stoichiometric mixture of reactants, and propose mechanisms for the formation of these compounds from chloromethane.

(b) How would you run this reaction to get a good conversion of methane to CH_3Cl? Of methane to CCl_4?

4-49 The chlorination of pentane gives a mixture of three monochlorinated products.

(a) Draw their structures.

(b) Predict the ratios in which these monochlorination products will be formed, remembering that a chlorine atom abstracts a secondary hydrogen about 4.5 times as fast as it abstracts a primary hydrogen.

4-50 **(a)** Draw the structure of the transition state for the second propagation step in the chlorination of methane.

$$\cdot CH_3 + Cl_2 \longrightarrow CH_3Cl + Cl\cdot$$

Show whether the transition state is product-like or reactant-like and which of the two partial bonds is stronger.

(b) Repeat for the second propagation step in the bromination of methane.

4-51 Peroxides are often added to free-radical reactions as initiators because the oxygen–oxygen bond cleaves homolytically rather easily. For example, the bond-dissociation enthalpy of the O—O bond in hydrogen peroxide (H—O—O—H) is only 213 kJ/mol (51 kcal/mol). Give a mechanism for the hydrogen peroxide-initiated reaction of cyclopentane with chlorine. The BDE for HO—Cl is 210 kJ/mol (50 kcal/mol).

***4-52** When dichloromethane is treated with strong NaOH, an intermediate is generated that reacts like a carbene. Draw the structure of this reactive intermediate, and propose a mechanism for its formation.

***4-53** When ethene is treated in a calorimeter with H_2 and a Pt catalyst, the heat of reaction is found to be −137 kJ/mol (−32.7 kcal/mol), and the reaction goes to completion. When the reaction takes place at 1400 °K, the equilibrium is found to be evenly balanced, with $K_{eq} = 1$. Compute the value of ΔS for this reaction.

$$CH_2{=}CH_2 + H_2 \underset{}{\overset{\text{Pt catalyst}}{\rightleftharpoons}} CH_3{-}CH_3 \qquad \Delta H = -137\ \text{kJ/mol}\ (-32.7\ \text{kcal/mol})$$

***4-54** When a small amount of iodine is added to a mixture of chlorine and methane, it prevents chlorination from occurring. Therefore, iodine is a *free-radical inhibitor* for this reaction. Calculate $\Delta H°$ values for the possible reactions of iodine with species present in the chlorination of methane, and use these values to explain why iodine inhibits the reaction. (The I—Cl bond-dissociation enthalpy is 211 kJ/mol or 50 kcal/mol.)

***4-55** Tributyltin hydride (Bu_3SnH) is used synthetically to reduce alkyl halides, replacing a halogen atom with hydrogen. Free-radical initiators promote this reaction, and free-radical inhibitors are known to slow or stop it. Your job is to develop a mechanism, using the following reaction as the example.

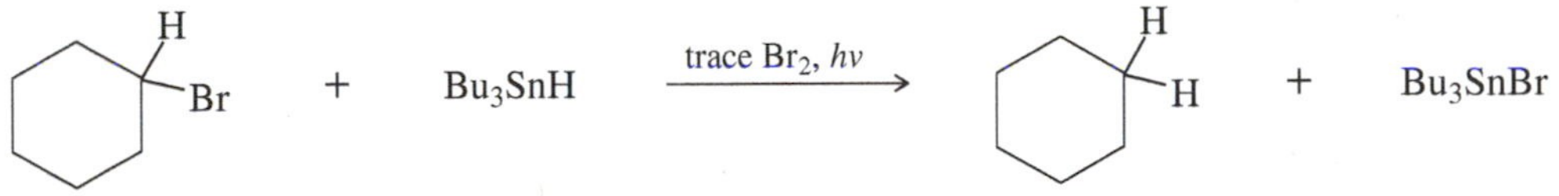

$$\text{(cyclohexyl bromide)} + Bu_3SnH \xrightarrow{\text{trace } Br_2,\ h\nu} \text{(cyclohexane)} + Bu_3SnBr$$

The following bond-dissociation enthalpies may be helpful:

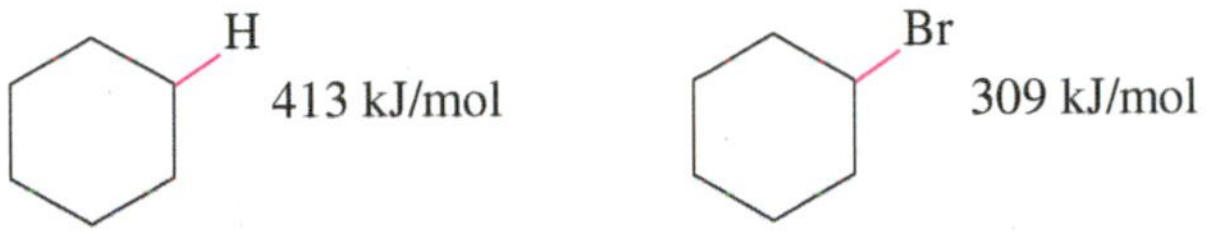

cyclohexyl—H	413 kJ/mol
cyclohexyl—Br	309 kJ/mol
Br—Br	190 kJ/mol
H—Br	366 kJ/mol
Bu_3Sn—H	310 kJ/mol
Bu_3Sn—Br	552 kJ/mol

(a) Propose initiation and propagation steps to account for this reaction.

(b) Calculate values of ΔH for your proposed steps to show that they are energetically feasible. (*Hint*: A trace of Br_2 and light suggests it's there only as an initiator, to create Br· radicals. Then decide which atom can be abstracted most favorably from the starting materials by the Br· radical. That should complete the initiation. Finally, decide what energetically favored propagation steps will accomplish the reaction.)

***4-56** When healthy, Earth's stratosphere contains a low concentration of ozone (O_3) that absorbs potentially harmful ultraviolet (UV) radiation by the cycle shown at right.

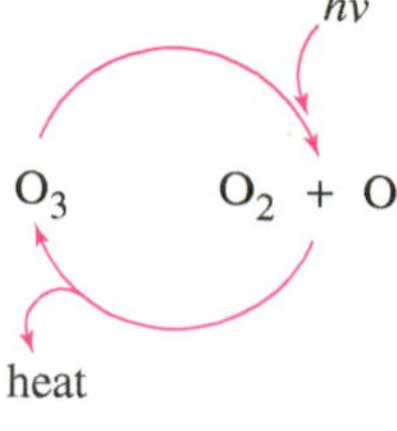

Chlorofluorocarbon refrigerants, such as Freon 12 (CF_2Cl_2), are stable in the lower atmosphere, but in the stratosphere they absorb high-energy UV radiation to generate chlorine radicals.

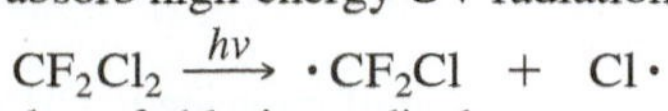

The presence of a small number of chlorine radicals appears to lower ozone concentrations dramatically. The following reactions are all known to be exothermic (except the one requiring light) and to have high rate constants. Propose two mechanisms to explain how a small number of chlorine radicals can destroy large numbers of ozone molecules. Which of the two mechanisms is more likely when the concentration of chlorine atoms is very small?

$$\mathrm{Cl{-}O{-}O{-}Cl} \xrightarrow{h\nu} \mathrm{O_2} + 2\,\mathrm{Cl}\cdot \qquad \mathrm{Cl{-}O}\cdot + \mathrm{O} \longrightarrow \mathrm{O_2} + \mathrm{Cl}\cdot$$

$$\mathrm{Cl}\cdot + \mathrm{O_3} \longrightarrow \mathrm{Cl{-}O}\cdot + \mathrm{O_2} \qquad 2\,\mathrm{Cl{-}O}\cdot \longrightarrow \mathrm{Cl{-}O{-}O{-}Cl}$$

***4-57** Deuterium (D) is the hydrogen isotope of mass number 2, with a proton and a neutron in its nucleus. The chemistry of deuterium is nearly identical to the chemistry of hydrogen, except that the C—D bond is slightly stronger than the C—H bond by 5.0 kJ/mol (1.2 kcal/mol). Reaction rates tend to be slower when a C—D bond (as opposed to a C—H bond) is broken in a rate-limiting step.

This effect, called a *kinetic isotope effect*, is clearly seen in the chlorination of methane. Methane undergoes free-radical chlorination 12 times as fast as tetradeuteriomethane (CD_4).

$$\textit{Faster:}\quad \mathrm{CH_4} + \mathrm{Cl}\cdot \longrightarrow \mathrm{CH_3Cl} + \mathrm{HCl} \quad \text{relative rate} = 12$$

$$\textit{Slower:}\quad \mathrm{CD_4} + \mathrm{Cl}\cdot \longrightarrow \mathrm{CD_3Cl} + \mathrm{DCl} \quad \text{relative rate} = 1$$

(a) Draw the transition state for the rate-limiting step of each of these reactions, showing how a bond to hydrogen or deuterium is being broken in this step.

(b) Monochlorination of deuterioethane (C_2H_5D) leads to a mixture containing 93% C_2H_4DCl and 7% C_2H_5Cl. Calculate the relative rates of abstraction per hydrogen and deuterium in the chlorination of deuterioethane.

(c) Consider the thermodynamics of the chlorination of methane and the chlorination of ethane, and use the Hammond postulate to explain why one of these reactions has a much larger isotope effect than the other.

***4-58** Iodination of alkanes using iodine (I_2) is usually an unfavorable reaction. (See Problem 4-17, for example.) Tetraiodomethane (CI_4) can be used as the iodine source for iodination in the presence of a free-radical initiator such as hydrogen peroxide. Propose a mechanism (involving mildly exothermic propagation steps) for the following proposed reaction. Calculate the value of ΔH for each of the steps in your proposed mechanism.

cyclopentane + CI_4 $\xrightarrow[\text{heat}]{H_2O_2}$ iodocyclopentane + HCI_3

The following bond-dissociation energies may be helpful:

Bond	Energy	Bond	Energy	Bond	Energy
$I_3C{-}I$	188 kJ/mol (45 kcal/mol)	$HO{-}I$	234 kJ/mol (56 kcal/mol)	$I_3C{-}H$	418 kJ/mol (100 kcal/mol)
$HO{-}OH$	213 kJ/mol (51 kcal/mol)	cyclopentyl—H	413 kJ/mol (99 kcal/mol)	cyclopentyl—I	238 kJ/mol (57 kcal/mol)

5 Stereochemistry

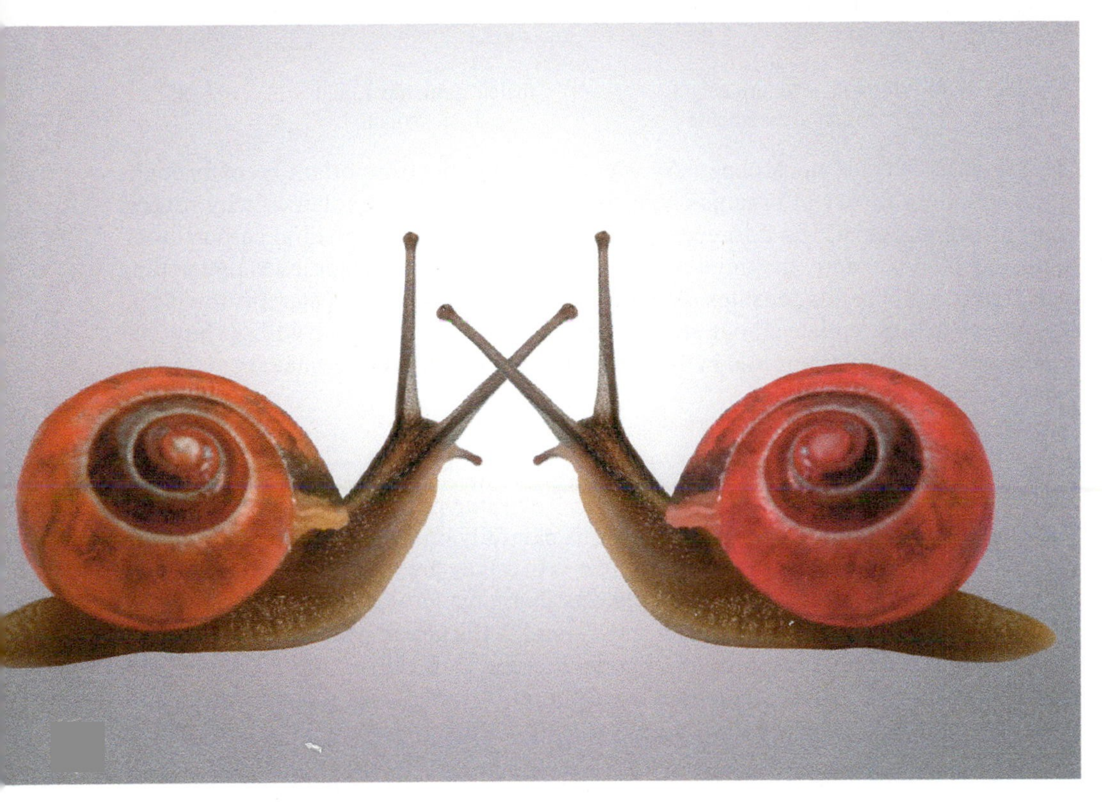

Goals for Chapter 5

1 Recognize structures that have stereoisomers, and identify the relationships between the stereoisomers.

2 Recognize chiral structures, draw their mirror images, and identify features that may suggest chirality.

3 Identify asymmetric carbon atoms and other stereocenters, and assign their configurations.

4 Explain the relationships between optical activity and chirality, optical purity, and enantiomeric excess.

5 Explain how the different types of stereoisomers differ in their physical and chemical properties.

◀ Chirality is not an obscure scientific concept. Almost all of nature is chiral, and all living things are chiral. Hands, feet, paws, plants, flowers, hooves, protein, DNA, and people are all chiral. This photo shows two otherwise identical snails with mirror-image shells (Section 5-2).

5-1 Introduction

Stereochemistry is the study of the three-dimensional structure of molecules. No one can understand organic chemistry, biochemistry, or biology without using stereochemistry. Biological systems are exquisitely selective, and they often discriminate between molecules with subtle stereochemical differences. We have seen (Section 1-19) that isomers are grouped into two broad classes: constitutional isomers and stereoisomers. **Constitutional isomers** (**structural isomers**) differ in their bonding sequence; their atoms are connected differently. **Stereoisomers** (also called **configurational isomers**) have the same bonding sequence, but they differ in the orientation of their atoms in space.

Differences in spatial orientation might seem unimportant, but stereoisomers often have remarkably different physical, chemical, and biological properties. For example, the cis and trans isomers of butenedioic acid (next page) are a special type of stereoisomers called *cis-trans isomers* (or *geometric isomers*). Both compounds have the formula HOOC—CH=CH—COOH, but they differ in how these atoms are arranged in space. The cis isomer is called *maleic acid*, and the trans isomer is called *fumaric acid*. Fumaric acid is an essential metabolic intermediate in both plants and animals, but maleic acid is toxic and irritating to tissues.

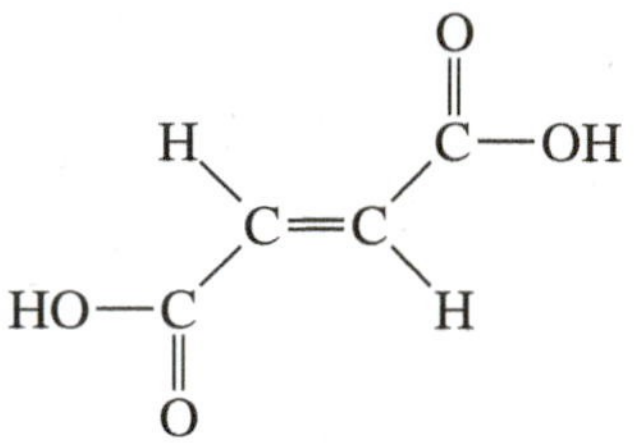

fumaric acid, mp 287 °C
essential metabolite

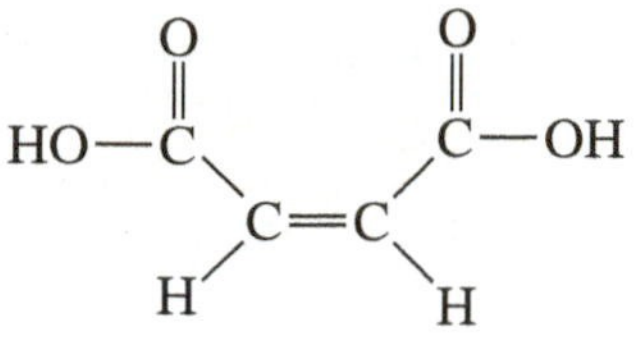

maleic acid, mp 138 °C
toxic irritant

The discovery of stereochemistry was one of the most important breakthroughs in the structural theory of organic chemistry. Stereochemistry explained why several types of isomers exist, and it forced scientists to propose the tetrahedral carbon atom. In this chapter, we study the three-dimensional structures of molecules to understand their stereochemical relationships. We compare the various types of stereoisomers and study ways to differentiate among stereoisomers. In later chapters, we will see how stereochemistry plays a major role in the properties and reactions of organic compounds.

Application: Mirror-Image People

Most people have their heart on their left side and their liver and appendix on the right. About 1 person in 10,000 is born with a congenital condition called *situs inversus,* in which their major organs are in the mirror image positions, with the heart on the right and the liver and appendix on the left. People with situs inversus are approximately (but not in minute detail) the mirror image of the majority of people.

5-2 Chirality

What is the difference between your left hand and your right hand? They look similar, yet a left-handed glove does not fit the right hand. The same principle applies to your feet. They look almost identical, yet the left shoe fits painfully on the right foot. The relationship between your two hands or your two feet is that they are nonsuperimposable (nonidentical) mirror images of each other. Objects that have left-handed and right-handed forms are called **chiral** (*kī′rel*, rhymes with "spiral"), the Greek word for "handed."

We can tell whether an object is chiral by looking at its mirror image (Figure 5-1). Every physical object (with the possible exception of a vampire) has a mirror image, but *a chiral object has a mirror image that is different from the original object.* For example, a chair and a spoon and a glass of water all look the same in a mirror. Such objects are called **achiral,** meaning "not chiral." A hand looks different in the mirror. If the original hand were the right hand, it would look like a left hand in the mirror.

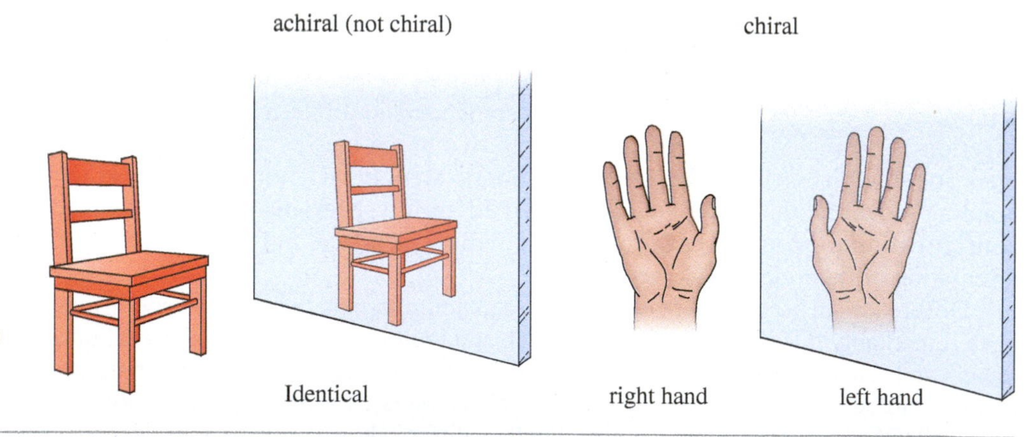

FIGURE 5-1
Use of a mirror to test for chirality. An object is chiral if its mirror image is different from the original object.

Besides shoes and gloves, we encounter many other chiral objects every day (Figure 5-2). What is the difference between an English car and an American car? The English car has the steering wheel on the right-hand side, while the American car has it on the left. To a first approximation, the English and American cars are nonsuperimposable mirror images. Most screws have right-hand threads and are turned clockwise to tighten. The mirror image of a right-handed screw is a left-handed screw, turned counterclockwise to tighten. Those of us who are left-handed realize that scissors are chiral. Most scissors are right-handed. If you use them in your left hand, they cut poorly, if at all. A left-handed person must go to a well-stocked store (or a specialized "Leftorium") to find a pair of left-handed scissors, the mirror image of the "standard" right-handed scissors.

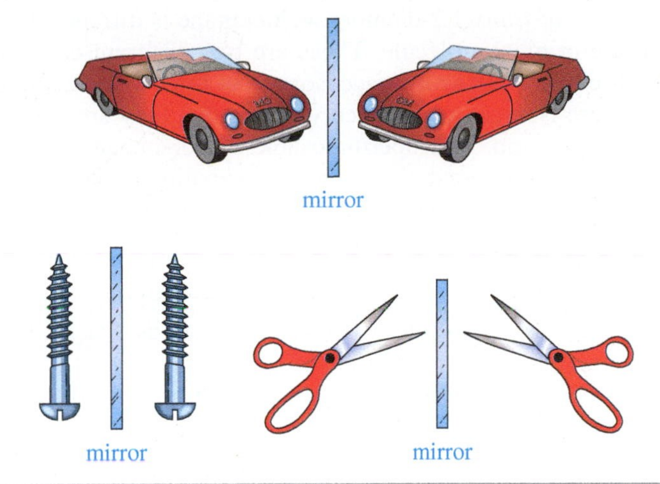

FIGURE 5-2
Common chiral objects. Many objects come in "left-handed" and "right-handed" versions.

PROBLEM 5-1

Determine whether the following objects are chiral or achiral.

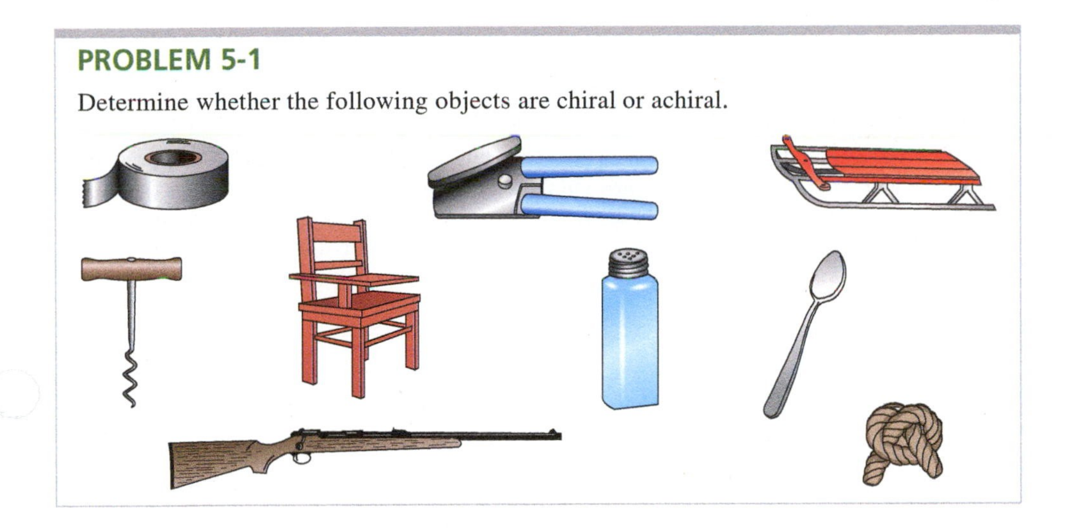

5-2A Chirality and Enantiomers in Organic Molecules

Like other objects, molecules are either chiral or achiral. For example, consider the two geometric isomers of 1,2-dichlorocyclopentane (Figure 5-3). The cis isomer is achiral because its mirror image is superimposable on the original molecule. Two molecules are said to be **superimposable** if they can be placed on top of each other and the three-dimensional position of each atom of one molecule coincides with the equivalent atom of the other molecule. To draw the mirror image of a molecule, simply draw the same structure with left and right reversed. The up-and-down and front-and-back directions are unchanged. The two mirror-image structures of the cis isomer are identical (superimposable), and *cis*-1,2-dichlorocyclopentane is achiral.

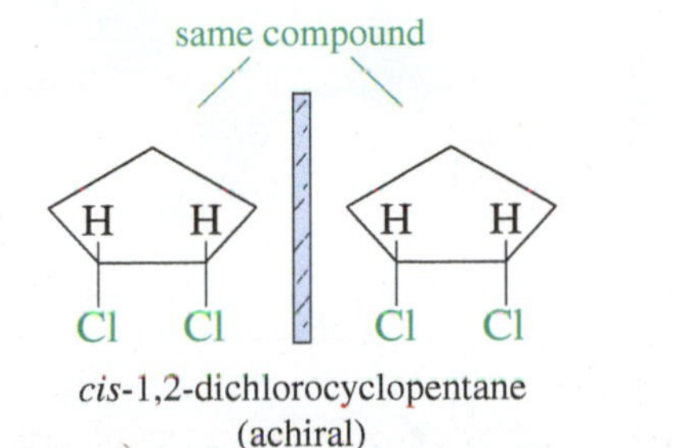

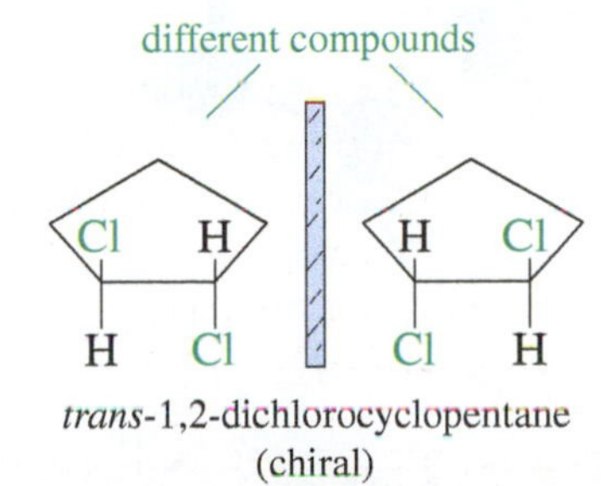

FIGURE 5-3
Stereoisomers of 1,2-dichlorocyclopentane. The cis isomer has no enantiomers; it is achiral. The trans isomer is chiral; it can exist in either of two nonsuperimposable enantiomeric forms.

PROBLEM-SOLVING HINT

Every object has a mirror image. Is its mirror image the same or different? Different: The object is chiral. Same: The object is achiral.

The mirror image of *trans*-1,2-dichlorocyclopentane is different from (nonsuperimposable with) the original molecule. These are two different compounds, and we should expect to discover two mirror-image isomers of *trans*-1,2-dichlorocyclopentane. Make models of these isomers to convince yourself that they are different no matter how you twist and turn them. Nonsuperimposable mirror-image molecules are called **enantiomers.** A chiral compound always has an enantiomer (a nonsuperimposable mirror image). An achiral compound always has a mirror image that is the same as the original molecule. Let's review the definitions of these words.

enantiomers:	mirror-image isomers; pairs of compounds that are nonsuperimposable mirror images
chiral:	("handed") different from its mirror image; having an enantiomer
achiral:	("not handed") identical with its mirror image; not chiral

Any compound that is chiral must have an enantiomer. Any compound that is achiral cannot have an enantiomer.

PROBLEM-SOLVING HINT

Stereochemistry is a difficult topic for many students. Use your models to help you see the relationships between structures. Once you have experience working with these three-dimensional relationships, you may be able to visualize them without constructing models.

PROBLEM 5-2

Make a model and draw a three-dimensional structure for each compound. Then draw the mirror image of your original structure and determine whether the mirror image is the same compound. Label each structure as being chiral or achiral, and label pairs of enantiomers.

(a) *cis*-1,2-dimethylcyclobutane
(b) *trans*-1,2-dimethylcyclobutane
(c) *cis*- and *trans*-1,3-dimethylcyclobutane
(d) 2-bromobutane
(e) O
(f)

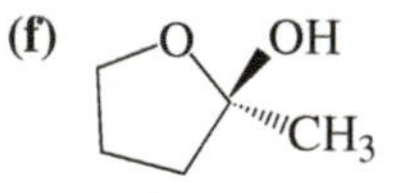

5-2B Asymmetric Carbon Atoms, Chiral Centers, and Stereocenters

The three-dimensional drawing of 2-bromobutane in Figure 5-4 shows that 2-bromobutane cannot be superimposed on its mirror image. This simple molecule is chiral, with two distinct enantiomers. What is it about a molecule that makes it chiral? The most common feature (but not the only one) that leads to chirality is a carbon atom that is bonded to four different groups. Such a carbon atom is called an **asymmetric carbon atom** or a **chiral carbon atom,** and is often designated by an asterisk (*). Carbon atom 2 of 2-bromobutane

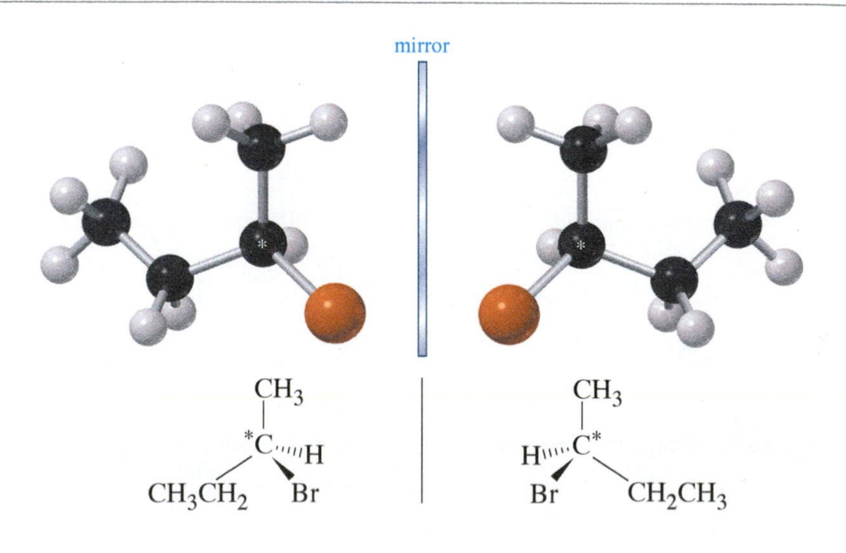

FIGURE 5-4
Enantiomers (nonsuperimposable mirror image isomers) of 2-bromobutane. 2-Bromobutane is chiral by virtue of an asymmetric carbon atom (chiral carbon atom), marked with an *.

is bonded to a hydrogen atom, a bromine atom, a methyl group, and an ethyl group. It is an asymmetric carbon atom, and it is responsible for the chirality of 2-bromobutane.

An asymmetric carbon atom is the most common example of a **chiral center** (or **chirality center**), the IUPAC term for any atom holding a set of ligands in a spatial arrangement that is not superimposable on its mirror image. Chiral centers belong to an even broader group called *stereocenters*. A **stereocenter** (or **stereogenic atom**) is any atom at which the interchange of two groups gives a stereoisomer.* Asymmetric carbons and the double-bonded carbon atoms in cis-trans isomers are the most common types of stereocenters. Figure 5-5 compares these successively broader definitions.

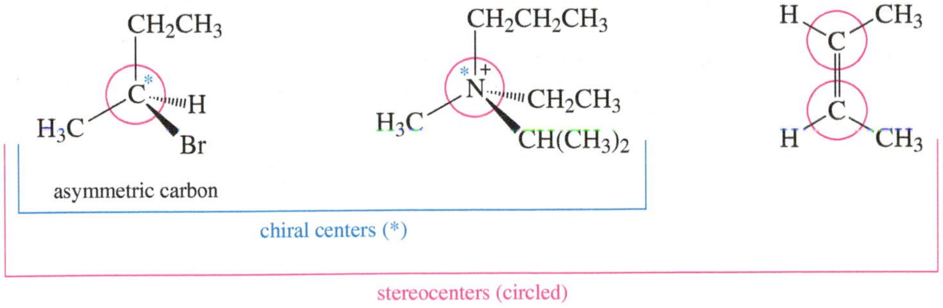

CH_2CH_3 — H_3C—C*—H, Br
asymmetric carbon

$CH_2CH_2CH_3$ — H_3C—N^{+}*—CH_2CH_3, $CH(CH_3)_2$

H, CH_3 C=C H, CH_3

chiral centers (*)

stereocenters (circled)

FIGURE 5-5
Asymmetric carbon atoms are examples of chiral centers, which are examples of stereocenters.

Make a model of an asymmetric carbon atom, bonded to four different-colored atoms. Also make its mirror image and try to superimpose the two (Figure 5-6). No matter how you twist and turn the models, they never look exactly the same.

If two of the four groups on a carbon atom are the same, however, the arrangement usually is not chiral. Figure 5-7 shows the mirror image of a tetrahedral structure with only three different groups; two of the four groups are the same. If the structure on the right is rotated 180°, it can be superimposed on the left structure.

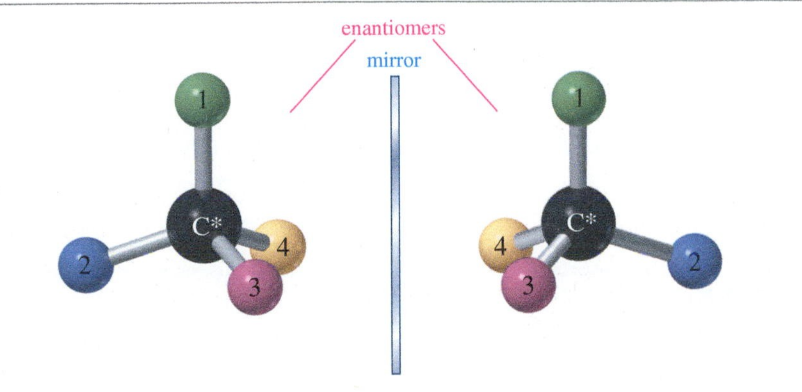

FIGURE 5-6
Enantiomers of an asymmetric carbon atom. These two mirror images are nonsuperimposable.

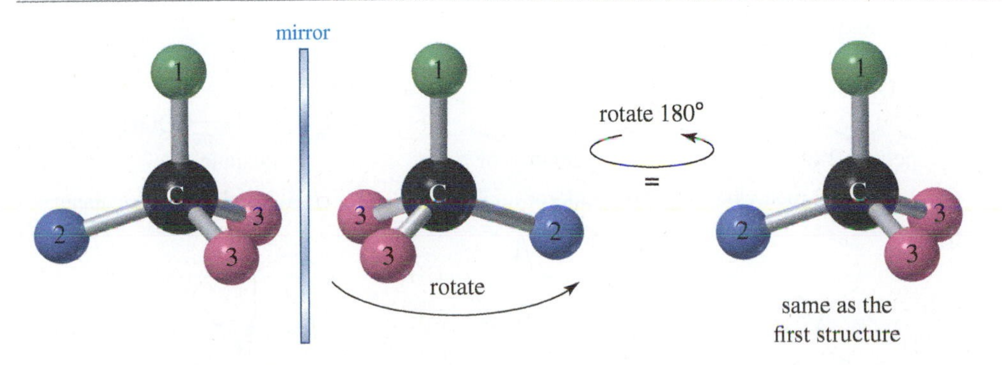

FIGURE 5-7
A carbon atom bonded to just three different types of groups is not chiral.

*The term *stereocenter* (*stereogenic atom*) is not consistently defined. The original (Mislow) definition is given here. Some sources simply define it as a synonym for an *asymmetric carbon* (*chiral carbon*) or for a *chiral center*.

COMMON TERMINOLOGY

Many organic chemists use the broad term *stereocenter* to mean "chiral center." You may need to determine the meaning from the context. Most of the rules that apply to asymmetric carbons and other chiral centers do not apply to the stereocenters in double bonds.

We can generalize at this point, but keep in mind that the ultimate test for chirality is always whether the molecule's mirror image is the same or different.

1. If a compound has no asymmetric carbon atom, it is usually achiral. (We will see exceptions in Section 5-9.)
2. If a compound has just one asymmetric carbon atom, it must be chiral.
3. If a compound has more than one asymmetric carbon, it may or may not be chiral. (We will see examples in Section 5-12.)

SOLVED PROBLEM 5-1

Identify each asymmetric carbon atom in the following structure:

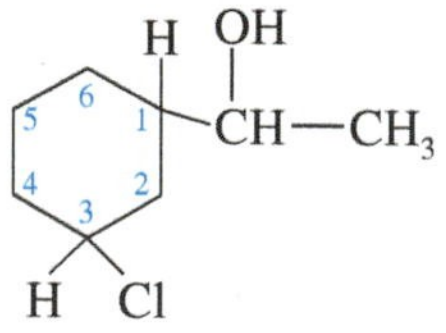

H OH
CH—CH_3
H Cl

SOLUTION

This structure contains three asymmetric carbons:

1. The (CHOH) carbon of the side chain is asymmetric. Its four substituents are the ring, a hydrogen atom, a hydroxy group, and a methyl group.
2. Carbon atom C1 of the ring is asymmetric. Its four substituents are the side chain, a hydrogen atom, the part of the ring closer to the chlorine atom (—CH_2—CHCl—), and the part of the ring farther from the chlorine atom (—CH_2—CH_2—CH_2—CHCl—).
3. The ring carbon C3 bearing the chlorine atom is asymmetric. Its four substituents are the chlorine atom, a hydrogen atom, the part of the ring closer to the side chain, and the part of the ring farther from the side chain.

Notice that different groups might be different in any manner. For example, the ring carbon bearing the chlorine atom is asymmetric even though two of its ring substituents initially appear to be —CH_2— groups. These two parts of the ring are different because one is closer to the side chain and one is farther away. The *entire* structure of the group must be considered.

PROBLEM-SOLVING HINT

To draw the mirror image of a structure, keep up-and-down and front-and-back aspects as they are in the original structure, but reverse left and right.

PROBLEM-SOLVING HINT

To determine whether a ring carbon is asymmetric, see if there is a difference in the path around the ring in each direction. If there is, then the two ring bonds are "different groups."

PROBLEM 5-3

Draw a three-dimensional structure for each compound, and star all asymmetric carbon atoms. Draw the mirror image for each structure, and state whether you have drawn a pair of enantiomers or just the same molecule twice. Build molecular models of any of these examples that seem difficult to you.

(a) OH — pentan-2-ol
(b) OH — pentan-3-ol
(c) NH_2 / CH_3—CH—COOH — alanine
(d) 1-bromo-2-methylbutane
(e) chlorocyclohexane
(f) *cis*-1,2-dichlorocyclobutane
(g) H, H, C=C, H, CH_3
(h) CH_3
(i) H, H

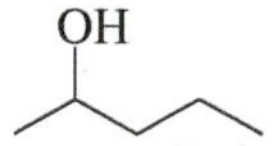

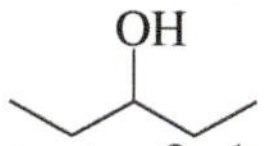

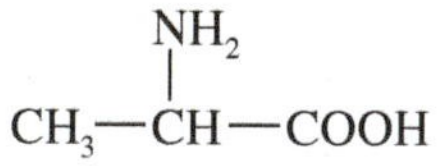

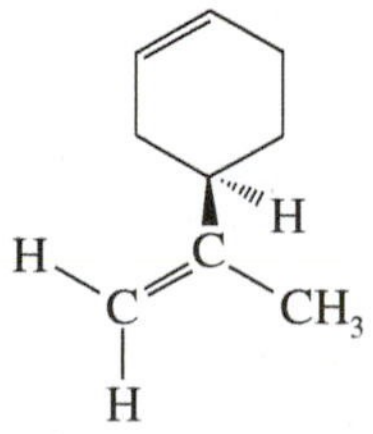

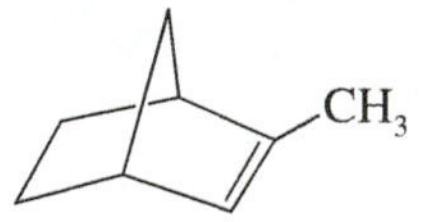

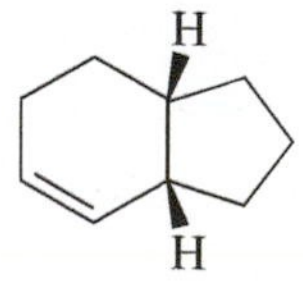

PROBLEM 5-4

For each of the stereocenters (circled) in Figure 5-5,
(a) draw the compound with two of the groups on the stereocenter interchanged.
(b) give the relationship of the new compound to the original compound.

5-2C Mirror Planes of Symmetry

In Figure 5-3, we saw that *cis*-1,2-dichlorocyclopentane is achiral. Its mirror image was found to be identical with the original molecule. Figure 5-8 illustrates a shortcut that often shows whether a molecule is chiral. If we draw a line down the middle of *cis*-1,2-dichlorocyclopentane, bisecting a carbon atom and its two hydrogen atoms, the part of the molecule that appears to the right of the line is the mirror image of the part on the left. This kind of symmetry is called an **internal mirror plane,** sometimes symbolized by the Greek lowercase letter sigma (σ). Because the right-hand side of the molecule is the reflection of the left-hand side, the molecule's mirror image is the same as the original molecule.

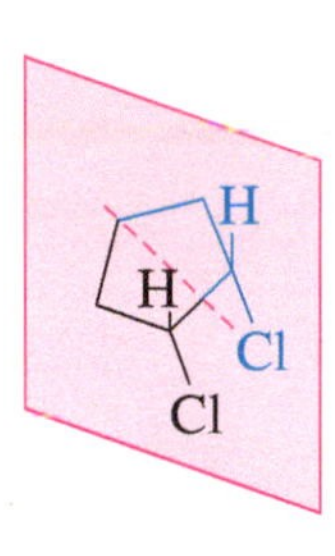

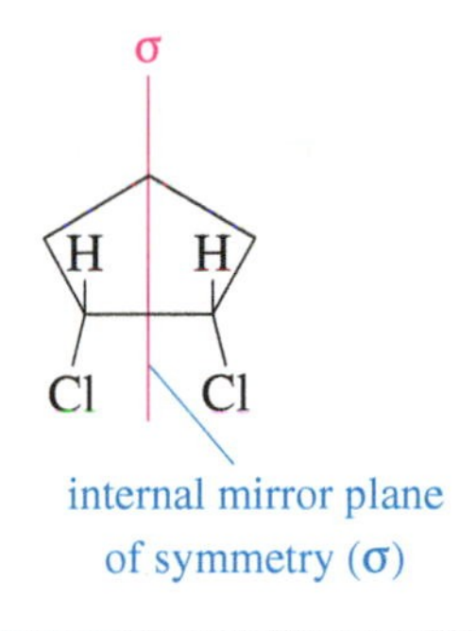

FIGURE 5-8
Internal mirror plane. *cis*-1,2-Dichlorocyclopentane has an internal mirror plane of symmetry. Any compound with an internal mirror plane of symmetry *cannot* be chiral.

Notice in the following figure that the chiral trans isomer of 1,2-dichlorocyclopentane does not have a mirror plane of symmetry. The chlorine atoms do not reflect into each other across our hypothetical mirror plane. One of them is directed up, and the other is directed down.

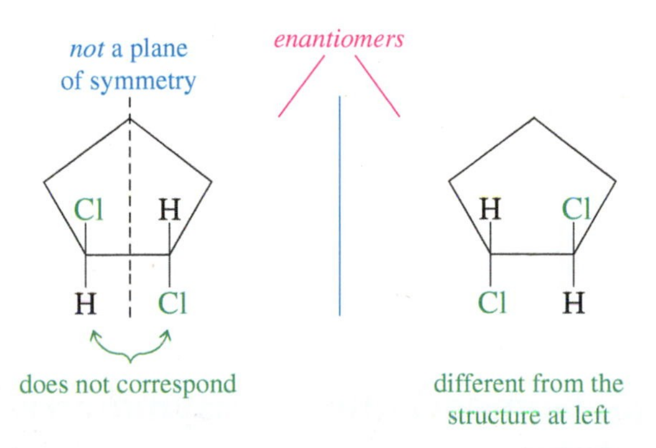

We can generalize from these and other examples to state the following principle:

Any molecule that has an internal mirror plane of symmetry cannot be chiral, even though it may contain asymmetric carbon atoms.

The converse is not true, however. When we cannot find a mirror plane of symmetry, that does not necessarily mean that the molecule must be chiral. The following example has no internal mirror plane of symmetry, yet the mirror image is superimposable on the original molecule. You may need to make models to show that these mirror images are just two drawings of the same compound.

(identical mirror images)

Using what we know about mirror planes of symmetry, we can see why a chiral (asymmetric) carbon atom is special. Figure 5-4 showed that an asymmetric carbon has a mirror image that is nonsuperimposable on the original structure; it has no internal mirror plane of symmetry. If a carbon atom has only three different kinds of substituents, however, it has an internal mirror plane of symmetry (Figure 5-9). Therefore, it cannot contribute to chirality in a molecule.

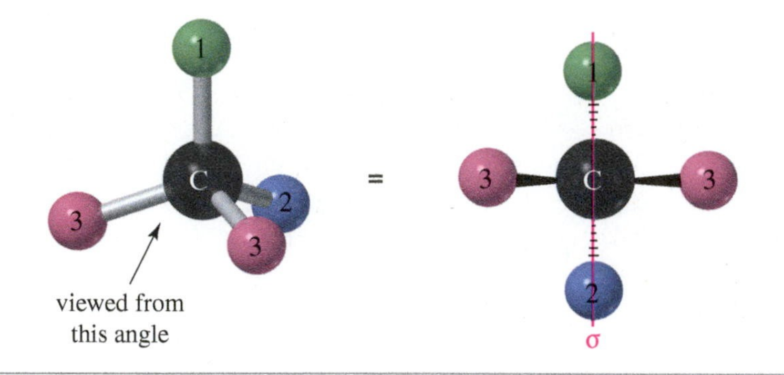

FIGURE 5-9 A carbon atom with two identical substituents (only three different substituents) usually has an internal mirror plane of symmetry. The structure is not chiral.

PROBLEM-SOLVING HINT

Various methods are used to show the stereochemistry of bonds. Bonds toward the viewer are shown as heavy or dark, often with perspective (the end toward the viewer appears larger). Bonds away from the viewer are shown as dashed or dotted, with or without perspective. The IUPAC has recently recommended the convention that we follow in this book. All four structures shown below depict exactly the same molecule.

perspective — no perspective — IUPAC convention — other options

PROBLEM 5-5

For each compound, determine whether the molecule has an internal mirror plane of symmetry. If it does, draw the mirror plane on a three-dimensional drawing of the molecule. If the molecule does not have an internal mirror plane, determine whether the structure is chiral.

(a) methane
(b) *cis*-1,2-dibromocyclobutane
(c) *trans*-1,2-dibromocyclobutane
(d) 1,2-dichloropropane
(e) $HOCH_2—CH(OH)—CHO$ glyceraldehyde
(f) $CH_3—CH(NH_2)—COOH$ alanine
(g) CH_3–cyclohexane–$CH(CH_3)_2$
(h)

5-3 (*R*) and (*S*) Nomenclature of Asymmetric Carbon Atoms

Alanine, from Problem 5-5(f), is one of the amino acids found in common proteins. Alanine has an asymmetric carbon atom, and it exists in two enantiomeric forms.

natural alanine — unnatural alanine

These mirror images are different, and this difference is reflected in their biochemistry. Only the enantiomer on the left can be metabolized by the usual enzyme; the one on the right is not recognized as a useful amino acid. Both are named alanine, however, or 2-aminopropanoic acid in the IUPAC system. We need a simple way to distinguish between enantiomers and to give each of them a unique name.

The difference between the two enantiomers of alanine lies in the three-dimensional arrangement of the four groups around the asymmetric carbon atom. Any asymmetric carbon has two possible (mirror-image) spatial arrangements, which we call **configurations.** The alanine enantiomers represent the two possible arrangements of its four groups around the asymmetric carbon atom. If we can name the two configurations of any asymmetric carbon atom, then we have a way of specifying and naming the enantiomers of alanine or any other chiral compound.

The **Cahn–Ingold–Prelog convention** is the most widely accepted system for naming the configurations of chiral centers. Each asymmetric carbon atom is assigned a letter (*R*) or (*S*) based on its three-dimensional configuration. To determine the name, we follow a two-step procedure that assigns "priorities" to the four substituents and then assigns the name based on the relative positions of these substituents. Here is the procedure:

1. *Assign a relative "priority" to each group bonded to the asymmetric carbon.* We speak of group 1 as having the highest priority, group 2 as having the second priority, group 3 as having the third priority, and group 4 as having the lowest priority.

(a) Look at the first atom of the group—the atom bonded to the asymmetric carbon. *Atoms with higher atomic numbers receive higher priorities.* For example, consider the structure shown at right. If the four groups bonded to an asymmetric carbon atom are H, CH_3, NH_2, and F, the fluorine atom (atomic number 9) receives the highest priority, followed by the nitrogen atom of the NH_2 group (atomic number 7), then by the carbon atom of the methyl group (atomic number 6). Note that we look only at the atomic number of the atom directly attached to the asymmetric carbon, not the entire group. Hydrogen would have the lowest priority.

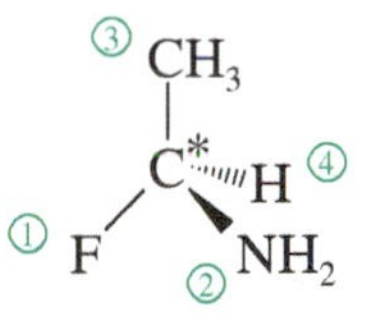

③ CH_3 — C* — ④ H, ① F, ② NH_2

With different isotopes of the same element, the heavier isotopes have higher priorities. For example, tritium (^3H) receives a higher priority than deuterium (^2H), followed by hydrogen (^1H).

Examples of priority for atoms bonded to an asymmetric carbon:

$$I > Br > Cl > S > F > O > N > {}^{13}C > {}^{12}C > Li > {}^{3}H > {}^{2}H > {}^{1}H$$

(b) *In case of ties, use the next atoms along the chain of each group as tiebreakers.* For example, in the structure at right, we assign a higher priority to isopropyl $-CH(CH_3)_2$ than to ethyl $-CH_2CH_3$ or bromoethyl $-CH_2CH_2Br$. The first carbon in the isopropyl group is bonded to two carbons, while the first carbon in the ethyl group (or the bromoethyl group) is bonded to only one carbon. An ethyl group and a $-CH_2CH_2Br$ have identical first atoms and second atoms, but the bromine atom in the third position gives $-CH_2CH_2Br$ a higher priority than $-CH_2CH_3$. One high-priority atom takes priority over any number of lower-priority atoms.

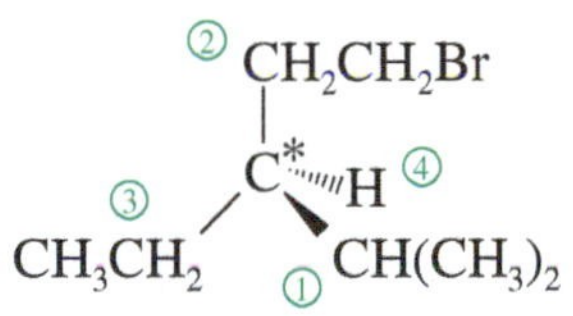

② CH_2CH_2Br — C* — ④ H, ③ CH_3CH_2, ① $CH(CH_3)_2$

Examples

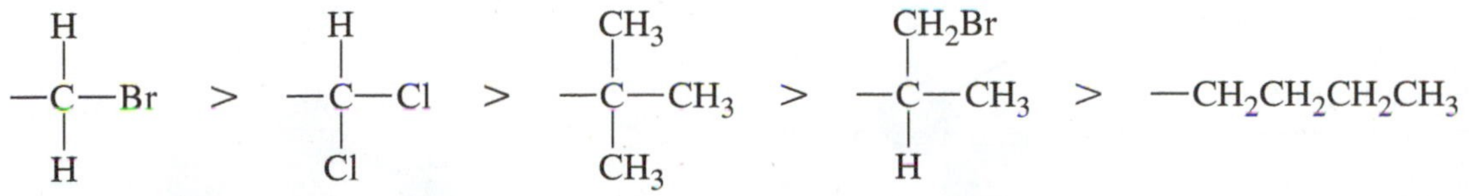

$$-CH_2Br > -CHCl_2 > -C(CH_3)_3 > -CH(CH_3)(CH_2Br)... $$

H–C(–Br)(H) > H–C(–Cl)(Cl) > CH_3–C(–CH_3)(CH_3) > CH_2Br–C(–CH_3)(H) > $-CH_2CH_2CH_2CH_3$

(c) *Treat double and triple bonds as if each were a bond to a separate atom.* For this method, imagine that each pi bond is broken and the atoms at both ends duplicated. Note that when you break a bond, you always add *two* imaginary atoms. (Imaginary atoms are circled below.) Examples of assigning priorities with double bonds are shown at left.

alanine

R—CH=CH₂ *break and duplicate* becomes R—CH(C)—CH₂(C)

R—CH=NH *break and duplicate* becomes R—CH(N)—NH(C)

R—C≡C—H *break and duplicate* becomes R—C(C)(C)—C(C)(C)—H

R—C(OH)=O *break and duplicate* becomes R—C(OH)(O)—O(C)

2. *Using a three-dimensional drawing or a model, put the fourth-priority group away from you and view the molecule with the first, second, and third priority groups radiating toward you like the spokes of a steering wheel* (Figure 5-10). *Draw an arrow from the first-priority group, through the second, to the third.* If the arrow points clockwise, the asymmetric carbon atom is called (*R*) (Latin, *rectus*, "upright"). If the arrow points counterclockwise, the chiral carbon atom is called (*S*) (Latin, *sinister*, "left").

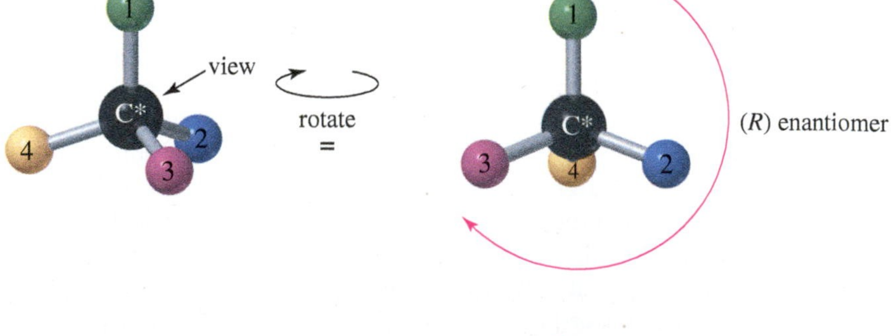

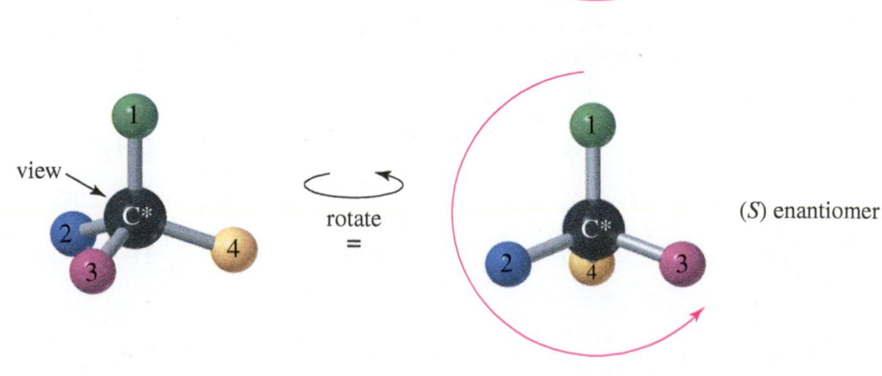

FIGURE 5-10
Assignment of (*R*) and (*S*) labels to asymmetric carbon atoms. View the molecule with the lowest- (fourth-) priority atom (often hydrogen) away from you. Then draw a circular arrow from group 1 to group 2 to group 3. A clockwise arrow (to the right) corresponds to the (*R*) configuration, and a counterclockwise arrow (to the left) corresponds to (*S*).

Alternatively, you can draw the arrow and imagine turning a car's steering wheel in that direction. If the car would go to the left, the asymmetric carbon atom is designated (*S*). If the car would go to the right, the asymmetric carbon atom is designated (*R*).

Let's use the enantiomers of alanine as an example. The naturally occurring enantiomer is the one on the left, determined to have the (*S*) configuration. Of the four atoms attached to the asymmetric carbon in alanine, nitrogen has the largest atomic number, giving it the highest priority. Next is the —COOH carbon atom, because it is bonded to oxygen atoms. Third is the methyl group, followed by the hydrogen atom. When we position the natural enantiomer with its hydrogen atom pointing away from us, the arrow from $-NH_2$ to —COOH to $-CH_3$ points counterclockwise. Thus, the naturally occurring enantiomer of alanine has the (*S*) configuration. Make models of these enantiomers to illustrate how they are named (*R*) and (*S*).

PROBLEM-SOLVING HINT

Until you become comfortable working with drawings, use models to help you assign (*R*) and (*S*) configurations.

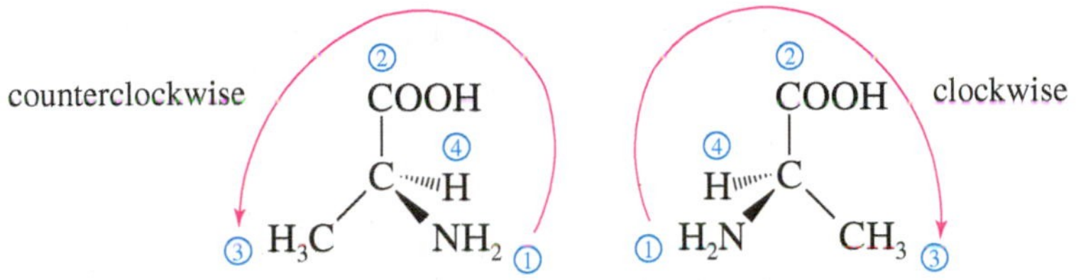

counterclockwise — ② COOH, ④ H, ① NH_2, ③ H_3C

natural (*S*)-alanine

clockwise — ② COOH, ④ H, ① H_2N, ③ CH_3

unnatural (*R*)-alanine

SOLVED PROBLEM 5-2

Draw the enantiomers of 1,3-dibromobutane and label them as (*R*) and (*S*). (Making a model is particularly helpful for this type of problem.)

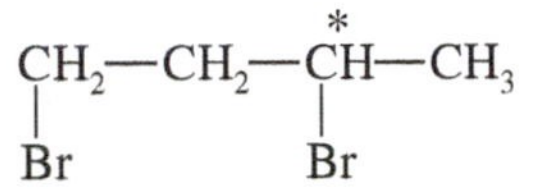

$$\underset{\text{Br}}{CH_2}-CH_2-\underset{\text{Br}}{\overset{*}{CH}}-CH_3$$

SOLUTION

The third carbon atom in 1,3-dibromobutane is asymmetric. The bromine atom receives first priority, the ($-CH_2CH_2Br$) group second priority, the methyl group third, and the hydrogen fourth. The following mirror images are drawn with the hydrogen atom back, ready to assign (*R*) or (*S*) as shown.

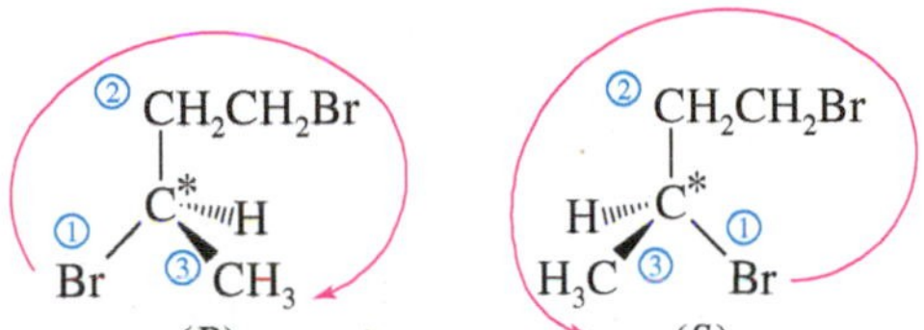

② CH_2CH_2Br, ① Br, ③ CH_3, H — (*R*)

② CH_2CH_2Br, ① Br, ③ H_3C, H — (*S*)

SOLVED PROBLEM 5-3

The structure of one of the enantiomers of carvone is shown here. Find the asymmetric carbon atom, and determine whether it has the (*R*) or the (*S*) configuration.

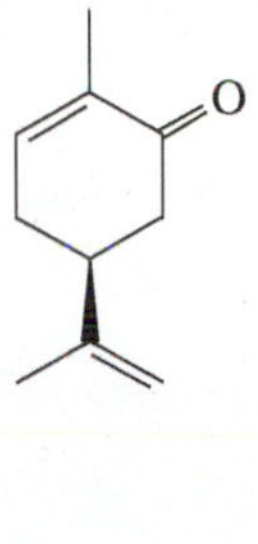

O

PROBLEM-SOLVING HINT

In assigning priorities for a ring carbon, go around the ring in each direction until you find a point of difference; then use the difference to determine which ring carbon has higher priority than the other.

Application: Biochemistry

Scientists frequently use the isotopes of hydrogen to assign the configuration of the products of biological reactions. Ethanol, made chiral by the presence of a deuterium (D or ^{2}H), is one of the early examples.

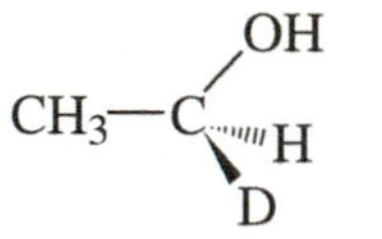

(S)-1-deuterioethanol

SOLUTION

The asymmetric carbon atom is one of the ring carbons, as indicated by the asterisk in the following structure. Although two $-CH_2-$ groups are bonded to the carbon, they are different $-CH_2-$ groups. One is a $-CH_2-CO-$ group, and the other is a $-CH_2-CH{=}C$ group. The groups are assigned priorities, and this is found to be the (S) enantiomer.

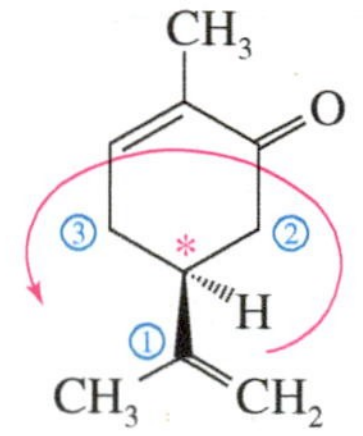

(S)-carvone

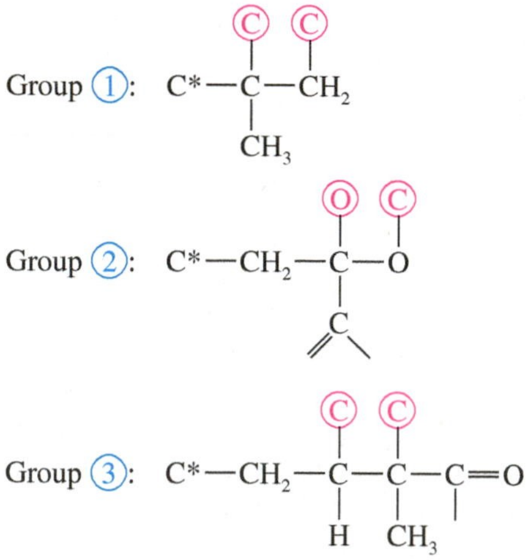

Group ①: $C^*-C(C)(C)-CH_2$, with CH_3

Group ②: $C^*-CH_2-C(O)(C)-O$

Group ③: $C^*-CH_2-C(C)(H)-C(C)(CH_3)-C{=}O$

PROBLEM-SOLVING HINT

If the lowest-priority atom (usually H) is oriented toward you, you don't need to turn the structure around. You can leave it as it is with the H toward you, find the (R) or (S) configuration, and reverse your answer.

PROBLEM 5-6

Star (*) each asymmetric carbon atom in the following examples, and determine whether it has the (R) or (S) configuration.

(a) (b) 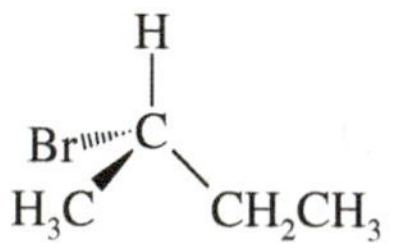 (c) 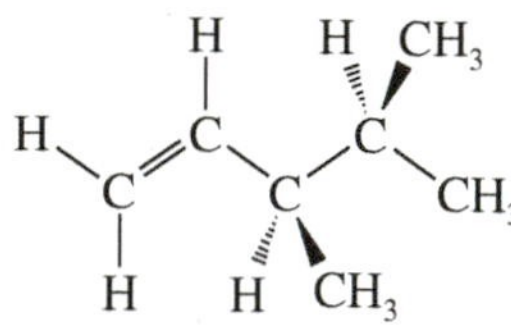 (d) 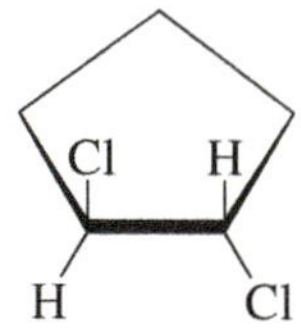 (e) 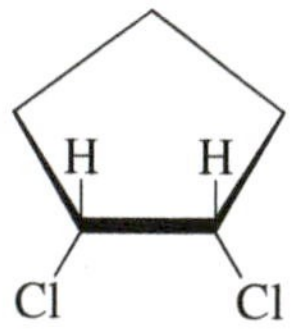 (f) 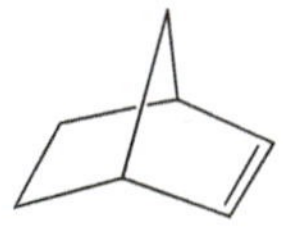 (g) 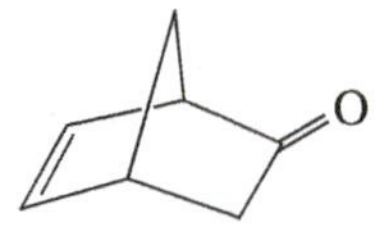 (h) 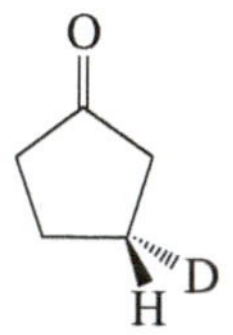 (*i)

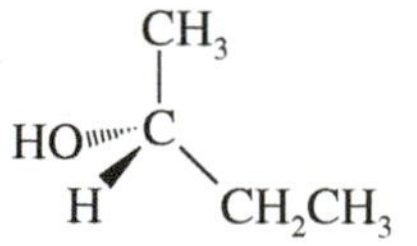

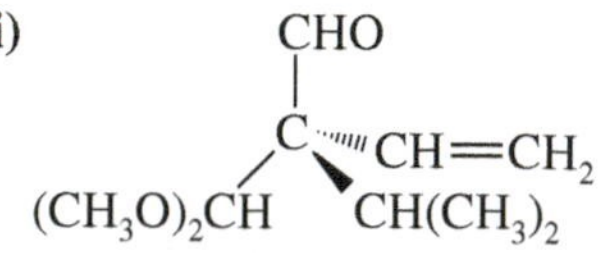

PROBLEM-SOLVING HINT

Interchanging any two substituents on an asymmetric carbon atom inverts its (R) or (S) configuration. If there is only one chiral center in a molecule, inverting its configuration gives the enantiomer.

PROBLEM 5-7

In Problem 5-3, you drew the enantiomers for a number of chiral compounds. Now go back and designate each asymmetric carbon atom as either (R) or (S).

5-4 Optical Activity

Mirror-image molecules have identical physical properties in an achiral environment. Compare the following properties of (*R*)-2-bromobutane and (*S*)-2-bromobutane.

	(*R*)-2-Bromobutane	(*S*)-2-Bromobutane
boiling point (°C)	91.2	91.2
melting point (°C)	−112	−112
refractive index	1.436	1.436
density	1.253	1.253

Differences in enantiomers become apparent in their interactions with other chiral molecules, such as enzymes. Still, we need a simple method to distinguish between enantiomers and measure their purity in the laboratory. **Polarimetry** is a common method used to distinguish between enantiomers, based on their ability to rotate the plane of polarized light in opposite directions. For example, the two enantiomers of thyroid hormone are shown below. The (*S*) enantiomer has a powerful effect on the metabolic rate of all the cells in the body. The (*R*) enantiomer is useless. In the laboratory, we distinguish between the enantiomers by observing that the active one rotates the plane of polarized light to the left.

thyroid hormone (*S*)
rotates polarized light to the left

wrong enantiomer (*R*)
rotates polarized light to the right

5-4A Plane-Polarized Light

Most of what we see is unpolarized light, consisting of waves vibrating randomly in all directions. **Plane-polarized light** is composed of waves that vibrate in only one plane. Although there are other types of "polarized light," the term usually refers to plane-polarized light.

When unpolarized light passes through a polarizing filter, the randomly vibrating light waves are filtered so that most of the light passing through is vibrating in one direction (Figure 5-11). The direction of vibration is called the *axis* of the filter. Polarizing filters may be made from carefully cut calcite crystals or from specially treated plastic sheets. Plastic polarizing filters are often used as lenses in sunglasses, because the axis of the filters can be positioned to filter out reflected glare.

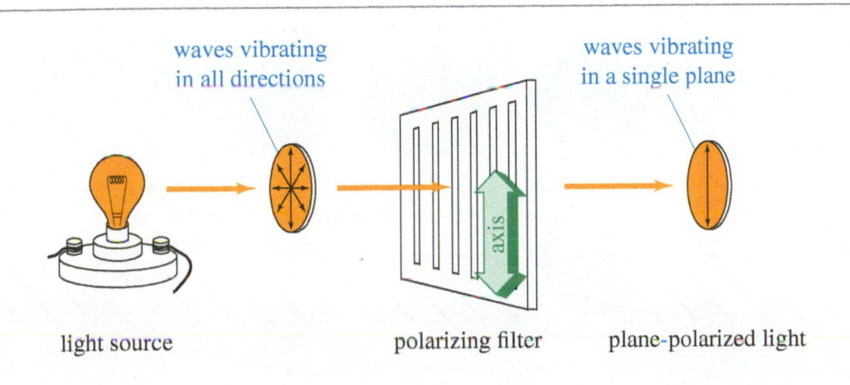

FIGURE 5-11
Function of a polarizing filter. The waves of plane-polarized light vibrate primarily in a single plane.

When light passes first through one polarizing filter and then through another, the amount of light emerging depends on the relative orientation of the axes of the two filters (Figure 5-12). If the axes of the two filters are lined up (parallel), then nearly all the light that passes through the first filter also passes through the second. If the axes of the two filters are perpendicular (*crossed poles*), however, all the polarized light that emerges from the first filter is stopped by the second. At intermediate angles of rotation, intermediate amounts of light pass through.

You can demonstrate this effect for yourself by wearing a pair of polarized sunglasses while looking at a light source through another pair (Figure 5-13). The second pair seems to be transparent, as long as its axis is lined up with the pair you are wearing. When the second pair is rotated to 90°, however, the lenses become opaque, as if they were covered with black ink.

5-4B Rotation of Plane-Polarized Light

When polarized light passes through a solution containing a chiral compound, the chiral compound causes the plane of vibration to rotate. Rotation of the plane of polarized light is called **optical activity,** and substances that rotate the plane of polarized light are said to be **optically active.**

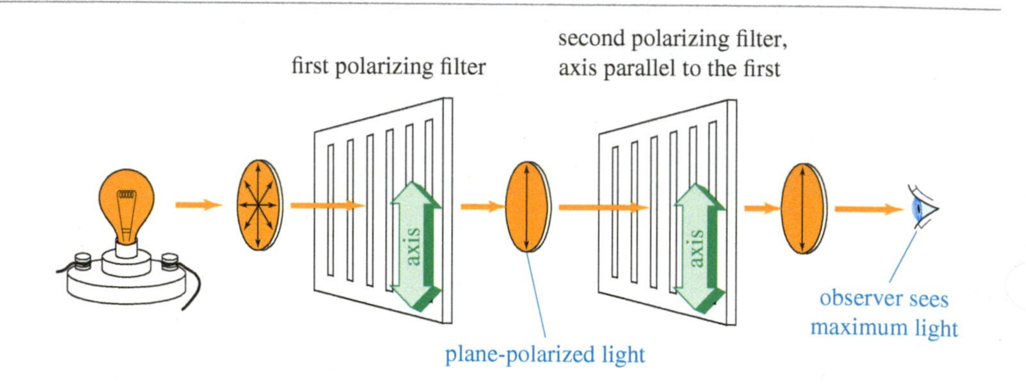

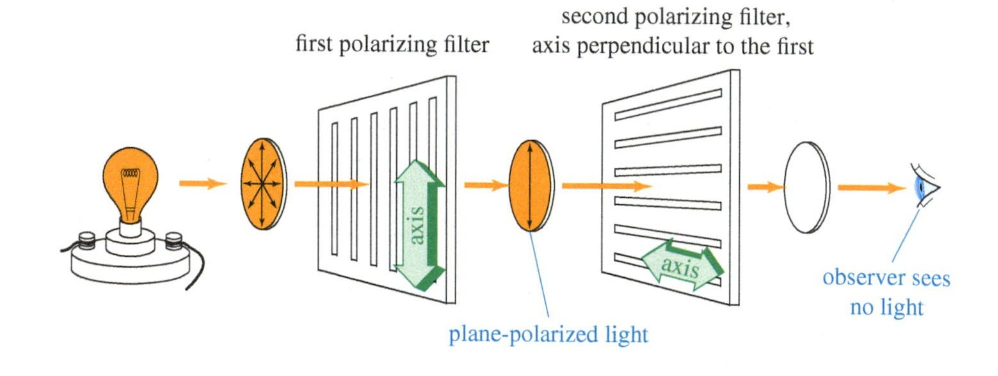

FIGURE 5-12
Crossed poles. When the axis of a second polarizing filter is parallel to the first, a maximum amount of light passes through. When the axes of the filters are perpendicular (crossed poles), no light passes through.

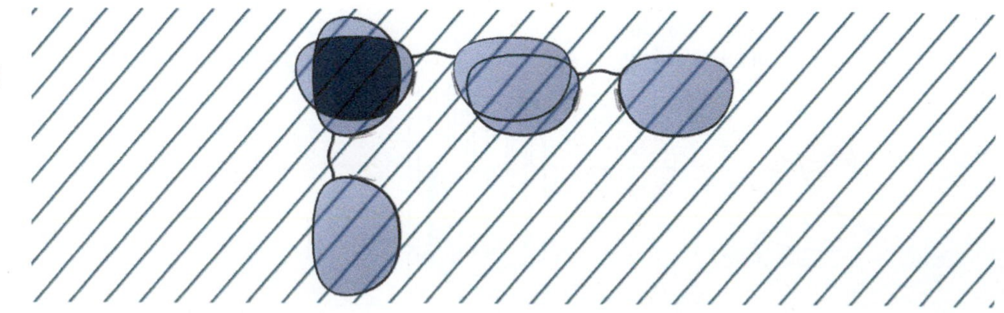

FIGURE 5-13
Using sunglasses to demonstrate parallel axes of polarization and crossed poles. When the pairs of sunglasses are parallel, a maximum amount of light passes through. When they are perpendicular, very little light passes through.

Before the relationship between chirality and optical activity was known, enantiomers were called **optical isomers** because they seemed identical except for their opposite optical activity. The term was loosely applied to more than one type of isomerism among optically active compounds, however, and this ambiguous term has been replaced by the well-defined term *enantiomers*.

Two enantiomers have identical physical properties, except for the direction they rotate the plane of polarized light.

> Enantiomeric compounds rotate the plane of polarized light by exactly the same amount but in opposite directions.

If the (*R*) enantiomer rotates the plane 30° clockwise, the (*S*) enantiomer will rotate it 30° counterclockwise. If the (*R*) enantiomer rotates the plane 5° counterclockwise, the (*S*) enantiomer will rotate it 5° clockwise.

> We cannot predict which direction a particular enantiomer [either (*R*) or (*S*)] will rotate the plane of polarized light.

(*R*) and (*S*) are simply names, but the direction and magnitude of rotation are physical properties that must be measured.

PROBLEM-SOLVING HINT

Don't confuse the process for *naming* a structure (*R*) or (*S*) with the process for *measuring* an optical rotation. Just because we use the terms *clockwise* and *counterclockwise* in naming (*R*) and (*S*) does not mean that light follows our naming rules.

5-4C Polarimetry

A **polarimeter** measures the rotation of polarized light. It has a tubular cell filled with a solution of the optically active material and a system for passing polarized light through the solution and measuring the rotation as the light emerges (Figure 5-14). The light from a sodium lamp is filtered so that it consists of just one wavelength (one color), because most compounds rotate different wavelengths of light by different amounts. The wavelength of light most commonly used for polarimetry is a yellow emission line in the spectrum of sodium, called the *sodium D line*.

Monochromatic (one-color) light from the source passes through a polarizing filter, then through the sample cell containing a solution of the optically active compound. Upon leaving the sample cell, the polarized light encounters another polarizing filter. This filter is movable, with a scale allowing the operator to read the angle between the axis of the second (analyzing) filter and the axis of the first (polarizing) filter. The operator rotates the analyzing filter until the maximum amount of light is transmitted, then reads the observed rotation from the protractor. The observed rotation is symbolized by α, the Greek letter alpha.

Compounds that rotate the plane of polarized light toward the right (clockwise as we look through the detector) are called **dextrorotatory,** from the Greek word

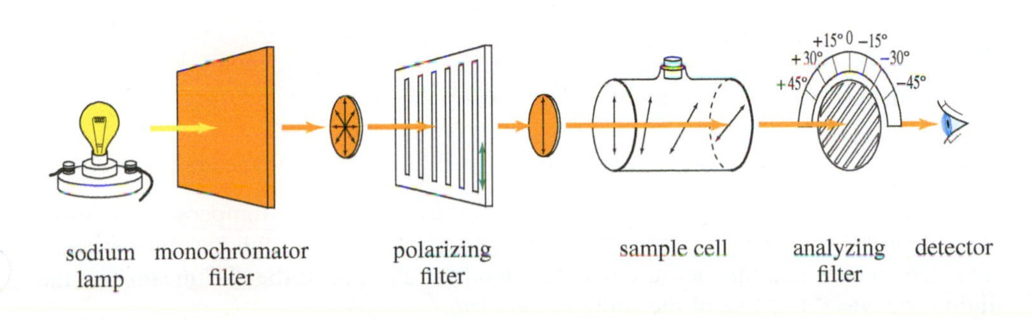

FIGURE 5-14
Schematic diagram of a polarimeter. The light originates at a source (usually a sodium lamp) and passes through a polarizing filter and the sample cell. An optically active solution rotates the plane of polarized light. The analyzing filter is another polarizing filter equipped with a protractor. It is turned until a maximum amount of light is observed, and the rotation is read from the protractor.

dexios, meaning "toward the right." Compounds that rotate the plane toward the left (counterclockwise) are called **levorotatory,** from the Latin word *laevus*, meaning "toward the left." These terms are sometimes abbreviated by a lowercase *d* or *l*. Using IUPAC notation, the direction of rotation is specified by the (+) or (−) sign of the rotation:

> Dextrorotatory (clockwise) rotations are (+) or (*d*).
> Levorotatory (counterclockwise) rotations are (−) or (*l*).

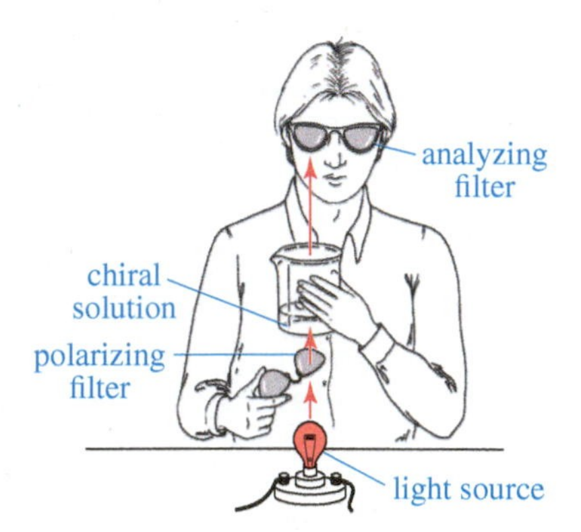

FIGURE 5-15
Apparatus for using a lightbulb and two pairs of polarized sunglasses as a simple polarimeter.

For example, the isomer of butan-2-ol that rotates the plane of polarized light clockwise is named (+)-butan-2-ol or *d*-butan-2-ol. Its enantiomer, (−)-butan-2-ol or *l*-butan-2-ol, rotates the plane counterclockwise by exactly the same amount.

You can see the principle of polarimetry by using two pairs of polarized sunglasses, a beaker, and some corn syrup or sugar solution. Wear one pair of sunglasses, look down at a light, and hold another pair of sunglasses above the light. Notice that the most light is transmitted through the two pairs of sunglasses when their axes are parallel. Very little light is transmitted when their axes are perpendicular.

Put syrup in the beaker, and hold the beaker above the bottom pair of sunglasses so that the light passes through one pair of sunglasses (the polarizing filter), then the beaker (the optically active sample), and then the other pair of sunglasses (the analyzing filter); see Figure 5-15. Again, check the angles giving maximum and minimum light transmission. Is the syrup solution dextrorotatory or levorotatory? Did you notice the color variation as you rotated the filter? You can see why just one color of light should be used for accurate work.

5-4D Specific Rotation

The angular rotation of polarized light by a chiral compound is a characteristic physical property of that compound, just like the boiling point or the density. The rotation (α) observed in a polarimeter depends on the concentration of the sample solution and the length of the cell, as well as the optical activity of the compound. For example, twice as concentrated a solution would give twice the original rotation. Similarly, a 20-cm cell gives twice the rotation observed using a similar concentration in a 10-cm cell.

To use the rotation of polarized light as a characteristic property of a compound, we must standardize the conditions for measurement. We define a compound's **specific rotation** $[\alpha]$ as the rotation found using a 10-cm (1-dm) sample cell and a concentration of 1 g/mL. Other cell lengths and concentrations may be used, as long as the observed rotation is divided by the path length of the cell (l) and the concentration (c).

$$[\alpha] = \frac{\alpha(\text{observed})}{c \cdot l}$$

where

α(observed) = rotation observed in the polarimeter
c = concentration in grams per mL
l = length of sample cell (path length) in decimeters (dm)

A rotation depends on the wavelength of light used and on the temperature, so these data are given together with the rotation. We often see specific rotations given as $[\alpha]_D^{25}$. The "25" means that the measurement was made at 25 °C, and the "D" means that the light used was the D line of the sodium spectrum.

SOLVED PROBLEM 5-4

When one of the enantiomers of butan-2-ol is placed in a polarimeter, the observed rotation is 4.05° counterclockwise. The solution was made by diluting 6.00 g of butan-2-ol to a total of 40.0 mL, and the solution was placed into a 200-mm polarimeter tube for the measurement. Determine the specific rotation for this enantiomer of butan-2-ol.

SOLUTION

Because it is levorotatory, this must be (−)-butan-2-ol. The concentration is 6.00 g per 40.0 mL = 0.150 g/mL, and the path length is 200 mm = 2.00 dm. The specific rotation is −13.5°.

$$[\alpha]_{\mathrm{D}}^{25} = \frac{-4.05^\circ}{(0.150)(2.00)} = -13.5^\circ$$

Without even measuring it, we can predict that the specific rotation of the other enantiomer of butan-2-ol will be

$$[\alpha]_{\mathrm{D}}^{25} = +13.5^\circ$$

where the (+) refers to the clockwise direction of the rotation. This enantiomer would be called (+)-butan-2-ol. We could refer to this pair of enantiomers as (+)-butan-2-ol and (−)-butan-2-ol or as (*R*)-butan-2-ol and (*S*)-butan-2-ol.

Does this mean that (*R*)-butan-2-ol is the dextrorotatory isomer because it is named (*R*) and that (*S*)-butan-2-ol is levorotatory because it is named (*S*)? Not at all! The rotation of a compound, (+) or (−), is something that we measure in the polarimeter, depending on how the molecule interacts with light. The (*R*) and (*S*) nomenclature is our own artificial way of describing how the atoms are arranged in space.

In the laboratory, we can measure a rotation and see whether a particular substance is (+) or (−). On paper, we can determine whether a particular drawing is named (*R*) or (*S*). But it is difficult to predict whether a structure we call (*R*) will rotate polarized light clockwise or counterclockwise. Similarly, it is difficult to predict whether a dextrorotatory substance in a flask has the (*R*) or (*S*) configuration.

PROBLEM 5-8

A solution of 2.0 g of (+)-glyceraldehyde, $HOCH_2CHOHCHO$, in 10.0 mL of water was placed in a 100-mm cell. Using the sodium D line, a rotation of +1.74° was found at 25 °C. Determine the specific rotation of (+)-glyceraldehyde.

PROBLEM 5-9

A solution of 0.50 g of (−)-epinephrine (see Figure 5-16) dissolved in 10.0 mL of dilute aqueous HCl was placed in a 20-cm polarimeter tube. Using the sodium D line, the rotation was found to be −5.1° at 25 °C. Determine the specific rotation of epinephrine.

PROBLEM 5-10

A chiral sample gives a rotation that is close to 180°. How can one tell whether this rotation is +180° or −180°?

5-5 Biological Discrimination of Enantiomers

If the direction of rotation of polarized light were the only difference between enantiomers, you might ask if the difference is important. However, biological systems commonly distinguish between enantiomers, and two enantiomers may have totally different biological properties. In fact, any **chiral probe** can distinguish between enantiomers, and a polarimeter is only one example of a chiral probe. Another example is your hand. If you needed to sort a box of gloves into right-handed gloves and left-handed gloves, you could distinguish between them by checking to see which ones fit your right hand.

Enzymes in living systems are chiral, and they are capable of distinguishing between enantiomers. Usually, only one enantiomer of a pair fits properly into the chiral active site of an enzyme. For example, the levorotatory form of epinephrine is one of the principal hormones secreted by the adrenal medulla. When synthetic epinephrine is given to a patient, the (−) form has the same stimulating effect as the natural hormone. The (+) form lacks this effect and is mildly toxic. Figure 5-16 shows a simplified picture of how only the (−) enantiomer fits into the enzyme's active site.

Biological systems are capable of distinguishing between the enantiomers of many different chiral compounds. In general, just one of the enantiomers produces the characteristic effect; the other either produces no effect or has a different effect. Even your nose is capable of distinguishing between some enantiomers. For example,

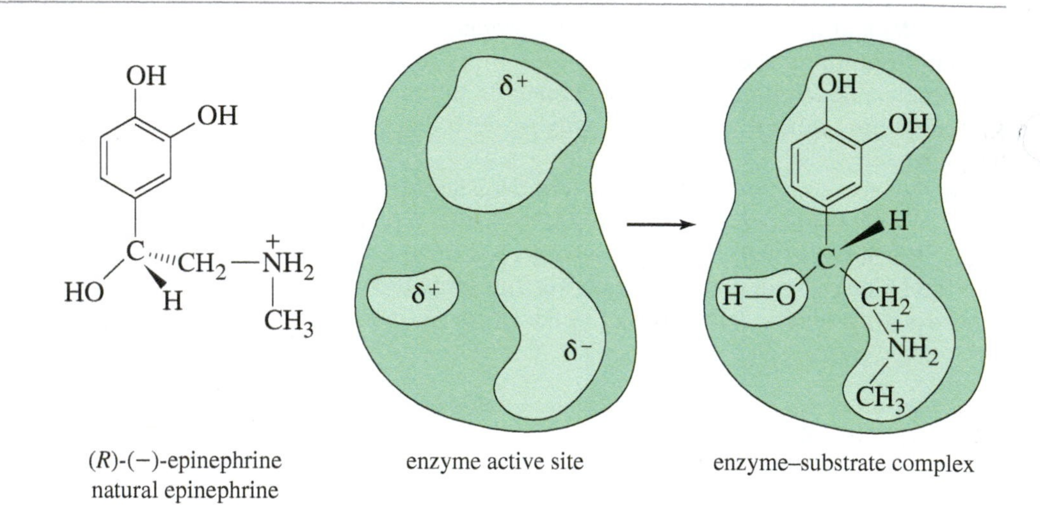

(S)-(+)-epinephrine
unnatural epinephrine

does not fit the enzyme's active site

FIGURE 5-16
Chiral recognition of epinephrine by an enzyme. Only the levorotatory enantiomer fits into the active site of the enzyme.

(−)-carvone is the fragrance associated with spearmint oil; (+)-carvone has the tangy odor of caraway seed. The receptor sites for the sense of smell must be chiral, therefore, just as the active sites in most enzymes are chiral. In general, enantiomers do not interact identically with other chiral molecules, whether or not they are of biological origin.

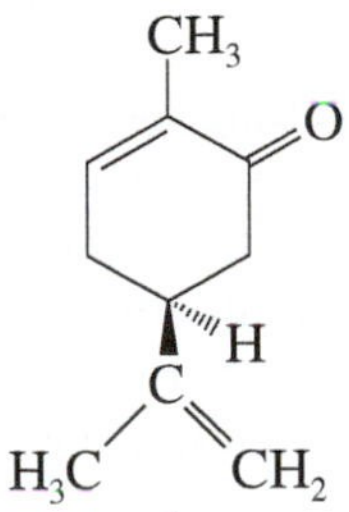

(+)-carvone (caraway seed) 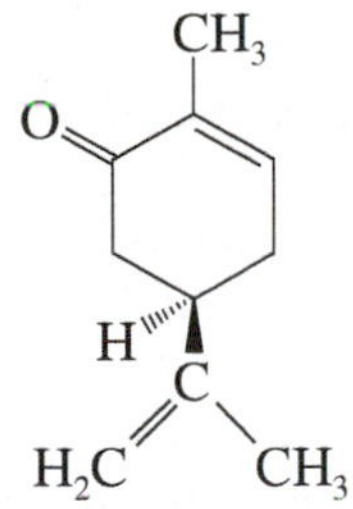 (−)-carvone (spearmint)

Application: Biochemistry

Enzymes can exist as two enantiomers, although only one enantiomer is found in nature. In 1992, Stephen Kent and coworkers reported the synthesis of both enantiomers of an enzyme that cuts peptide substrates, and showed for the first time that each enzyme acts only on the corresponding enantiomeric peptide substrate.

PROBLEM 5-11

If you had the two enantiomers of carvone in unmarked bottles, could you use just your nose and a polarimeter to determine

(a) whether it is the (+) or (−) enantiomer that smells like spearmint?
(b) whether it is the (*R*) or (*S*) enantiomer that smells like spearmint?
(c) With the information given in the drawings of carvone above, what can you add to your answers to (a) and (b)?

5-6 Racemic Mixtures

Suppose we had a mixture of equal amounts of (+)-butan-2-ol and (−)-butan-2-ol. The (+) isomer would rotate polarized light clockwise with a specific rotation of +13.5°, and the (−) isomer would rotate the polarized light counterclockwise by exactly the same amount. We would observe a rotation of zero, just as though butan-2-ol were achiral. A solution of equal amounts of two enantiomers, so that the mixture is optically inactive, is called a **racemic mixture.** Sometimes a racemic mixture is called a **racemate,** a **(±) pair,** or a **(*d,l*) pair.** A racemic mixture is symbolized by placing (±) or (*d,l*) in front of the name of the compound. For example, racemic butan-2-ol would be symbolized by "(±)-butan-2-ol" or "(*d,l*)-butan-2-ol."

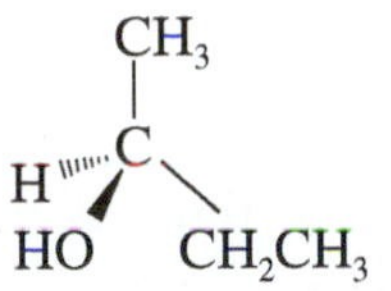

(*S*)-(+)-butan-2-ol, +13.5° rotation and (*R*)-(−)-butan-2-ol, −13.5° rotation

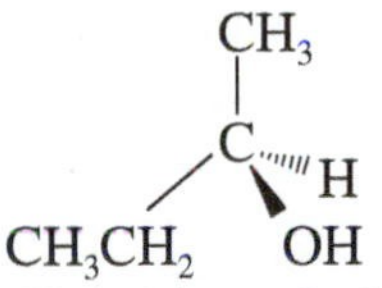

A racemic mixture contains equal amounts of the two enantiomers.

You might think that a racemic mixture would be unusual because it requires exactly equal amounts of the two enantiomers. This is not the case, however. Many reactions lead to racemic products, especially when an achiral molecule is converted to a chiral molecule.

A reaction that uses optically inactive reactants and catalysts cannot produce a product that is optically active. Any chiral product must be formed as a racemic mixture.

Application: Drugs

Many drugs currently on the market are racemic mixtures. Ketamine, for example, is a potent anesthetic agent, but its use is limited because it is hallucinogenic (making it a drug of abuse widely known as "K"). The (*S*) isomer is responsible for the anesthetic effects, and the (*R*) isomer causes the hallucinogenic effects.

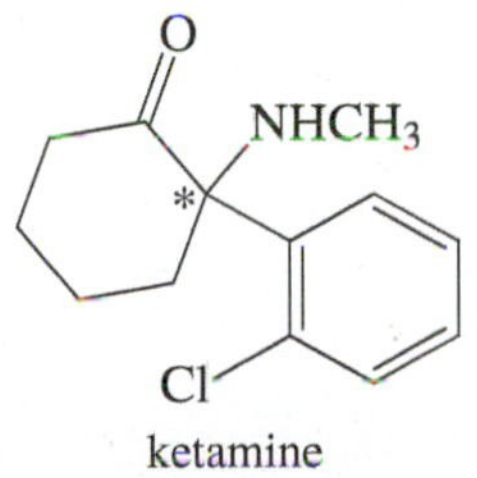

ketamine

For example, hydrogen adds across the C=O double bond of a ketone to produce an alcohol.

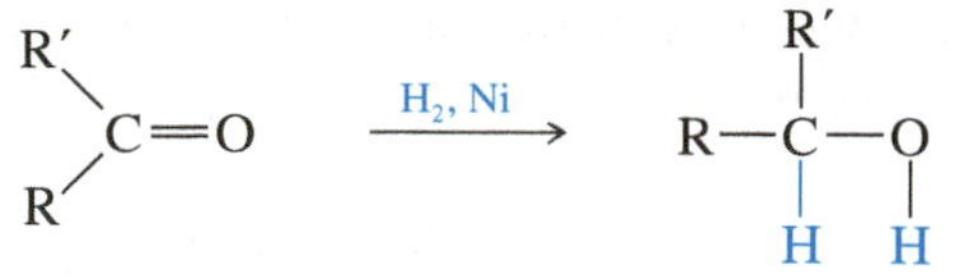

$$R'(R)C{=}O \xrightarrow{H_2,\ Ni} R{-}C(R')(H){-}O{-}H$$

Because the carbonyl group is flat, a simple ketone such as butan-2-one is achiral. Hydrogenation of butan-2-one gives butan-2-ol, a chiral molecule (Figure 5-17). This reaction involves adding hydrogen atoms to the C=O carbon atom and oxygen atom. If the hydrogen atoms are added to one face of the double bond, the (*S*) enantiomer results. Addition of hydrogen to the other face forms the (*R*) enantiomer. It is equally probable for hydrogen to add to either face of the double bond, and equal amounts of the (*R*) and (*S*) enantiomers are formed.

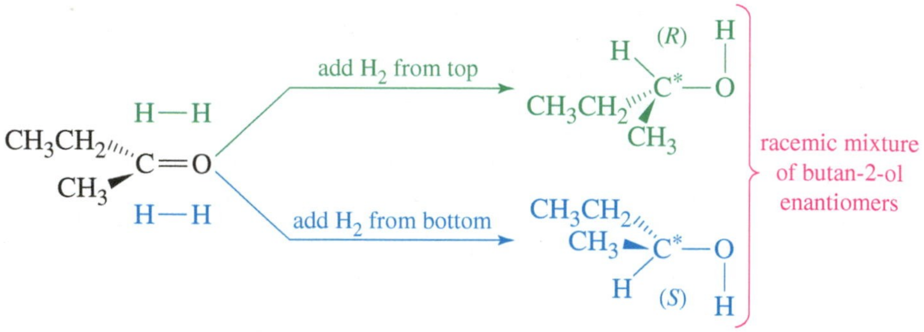

FIGURE 5-17
Hydrogenation of butan-2-one forms racemic butan-2-ol. Hydrogen adds to either face of the double bond. Addition of H_2 to one face gives the (*R*) product, while addition to the other face gives the (*S*) product.

Logically, it makes sense that optically inactive reagents and catalysts cannot form optically active products. If the starting materials and reagents are optically inactive, there is no reason for the dextrorotatory product to be favored over a levorotatory one or vice versa. The (+) product and the (−) product are favored equally, and they are formed in equal amounts: a racemic mixture.

5-7 Enantiomeric Excess and Optical Purity

Sometimes we deal with mixtures that are neither optically pure (all one enantiomer) nor racemic (equal amounts of two enantiomers). In these cases, we specify the **optical purity (o.p.)** of the mixture. The optical purity of a mixture is defined as the ratio of its rotation to the rotation of a pure enantiomer. For example, if we have some [mostly (+)] butan-2-ol with a specific rotation of +9.72°, we compare this rotation with the +13.5° rotation of the pure (+) enantiomer.

$$\text{o.p.} = \frac{\text{observed rotation}}{\text{rotation of pure enantiomer}} \times 100\% = \frac{9.72^\circ}{13.5^\circ} \times 100\% = 72.0\%$$

The **enantiomeric excess (e.e.)** is a more common method for expressing the relative amounts of enantiomers in a mixture. To compute the enantiomeric excess of a mixture, we calculate the *excess* of the predominant enantiomer as a percentage of the entire mixture. For a chemically pure compound, the calculation of enantiomeric excess generally gives the same result as the calculation of optical purity, and we often use the two terms interchangeably. Algebraically, we use the following formula:

$$\text{o.p.} = \text{e.e.} = \frac{|d - l|}{d + l} \times 100\% = \frac{\text{(excess of one over the other)}}{\text{(entire mixture)}} \times 100\%$$

The units cancel out in the calculation of either e.e. or o.p., so these formulas can be used whether the amounts of the enantiomers are expressed in concentrations, grams, or percentages. For the butan-2-ol mixture just described, the optical purity of 72% (+) implies that $d - l = 72\%$, and we know that $d + l = 100\%$. Adding the equations gives $2d = 172\%$. We conclude that the mixture contains 86% of the d or (+) enantiomer and 14% of the l or (−) enantiomer.

Application: Drugs

Several racemic drugs have become available as pure active enantiomers. For example, the drug Nexium® (for controlling acid reflux) contains just the active enantiomer of the racemic mixture in Prilosec®.

SOLVED PROBLEM 5-5

Calculate the e.e. and the specific rotation of a mixture containing 6.0 g of (+)-butan-2-ol and 4.0 g of (−)-butan-2-ol.

SOLUTION

In this mixture, there is a 2.0 g excess of the (+) isomer and a total of 10.0 g, for an e.e. of 20%. We can envision this mixture as 80% racemic [4.0 g(+) and 4.0 g(−)] and 20% pure (+).

$$\text{o.p.} = \text{e.e.} = \frac{|6.0 - 4.0|}{6.0 + 4.0} = \frac{2.0}{10.0} = 20\%$$

The specific rotation of enantiomerically pure (+)-butan-2-ol is +13.5°. The rotation of this mixture is

$$\begin{aligned}\text{observed rotation} &= (\text{rotation of pure enantiomer}) \times (\text{o.p.})\\ &= (+13.5^\circ) \times (20\%) = +2.7^\circ\end{aligned}$$

PROBLEM 5-12

When optically pure (*R*)-2-bromobutane is heated with water, butan-2-ol is the product. The reaction forms twice as much (*S*)-butan-2-ol as (*R*)-butan-2-ol. Calculate the e.e. and the specific rotation expected for the product.

PROBLEM 5-13

A chemist finds that the addition of (+)-epinephrine to the catalytic reduction of butan-2-one (Figure 5-17) gives a product that is slightly optically active, with a specific rotation of +0.45°. Calculate the percentages of (+)-butan-2-ol and (−)-butan-2-ol formed in this reaction.

5-8 Chirality of Conformationally Mobile Systems

Let's consider whether *cis*-1,2-dibromocyclohexane is chiral. If we did not know about chair conformations, we might draw a flat cyclohexane ring. With a flat ring, the molecule has an internal mirror plane of symmetry (σ), and it is achiral.

But we know the ring is puckered into a chair conformation with one bromine atom axial and one equatorial. A chair conformation of *cis*-1,2-dibromocyclohexane and its mirror image are shown below. These two mirror-image structures are nonsuperimposable. You may be able to see the difference more easily if you make models of these two chiral conformations.

H Br H Br ⇄ H Br H Br

fast equilibrium

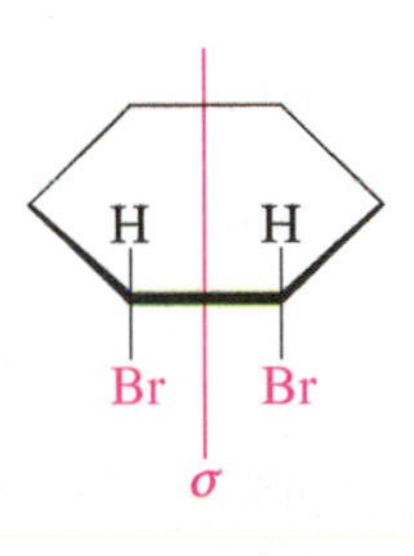

Does this mean that *cis*-1,2-dibromocyclohexane is chiral? No, it does not, because the chair–chair interconversion is rapid at room temperature. If we had a bottle of just the conformation on the left, the molecules would quickly undergo chair–chair interconversions. Because the two mirror-image conformations interconvert and have identical energies, any sample of *cis*-1,2-dibromocyclohexane must contain equal amounts of the two mirror images. Similarly, nearly all achiral compounds can exist in transient chiral conformations that are in equilibrium with their mirror-image conformations. For a compound to be considered chiral, it must be possible, at least in theory, to isolate a sample containing just one enantiomer.

> We consider a molecule to be achiral if its chiral conformations are in rapid equilibrium with their mirror-image conformations, such that we cannot purify a sample of just one enantiomer.

cis-1,2-Dibromocyclohexane appears to exist as a racemic mixture, but with a major difference: It is impossible to isolate an optically active sample of *cis*-1,2- dibromocyclohexane. We could have predicted the correct result by imagining that the cyclohexane ring is flat.

This finding leads to a general principle we can use with conformationally mobile systems:

> To determine whether a conformationally mobile molecule is chiral, consider its most symmetric conformation.

We can consider cyclohexane rings as though they were flat (the most symmetric conformation), and we should consider straight-chain compounds in their most symmetric conformations (often an eclipsed conformation).

Organic compounds commonly exist as rapidly interconverting chiral conformations. Even ethane is chiral in its skew conformations. When we speak of chirality, however, we intend to focus on observable, persistent properties rather than transient conformations. For example, butane exists in gauche conformations that are chiral, but they quickly interconvert. They are in equilibrium with the anti and totally eclipsed conformations, which are both symmetric, implying that butane must be achiral.

CH_3 / H, H, H, H / σ / CH_3 anti (achiral) ⇌ CH_3 / H_3C, H, H, H / H gauche (chiral) ⇌ CH_3, H_3C / H, H, H, H / σ totally eclipsed (achiral) ⇌ CH_3 / H, CH_3, H, H / H gauche (chiral)

In contrast, we can draw any of the conformations of (*R*)-2-bromobutane (Figure 5-4), and they are all chiral. None of them can interconvert with their mirror images because the (*R*) asymmetric carbon cannot interconvert with its (*S*) mirror image.

SOLVED PROBLEM 5-6

Draw each compound in its most stable conformation(s). Then draw it in its most symmetric conformation, and determine whether it is chiral.
(a) 2-methylbutane

SOLUTION

The most stable conformations of 2-methylbutane are two mirror-image conformations. These conformations are nonsuperimposable, but they can interconvert by rotation around the central bond. So they are not enantiomers.

chiral conformation   chiral conformation

2-Methylbutane has two symmetric conformations: Either of these conformations is sufficient to show that 2-methylbutane is achiral.

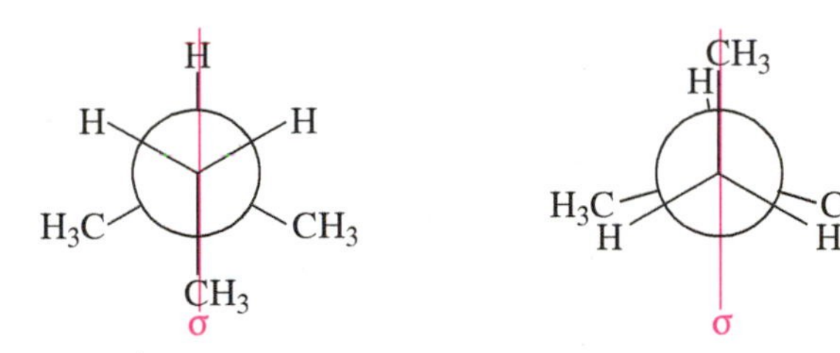

symmetric conformation   symmetric conformation

(b) *trans*-1,2-dibromocyclohexane

SOLUTION

We can draw two nonsuperimposable mirror images of the most stable chair conformation of *trans*-1,2-dibromocyclohexane with both bromines equatorial. These structures cannot interconvert by ring-flips or other rotations about bonds, however. They are mirror-image isomers: enantiomers.

This molecule's chirality is more apparent when drawn in its most symmetric conformation. Drawn flat, the two mirror-image structures of *trans*-1,2-dibromocyclohexane are still nonsuperimposable. This compound is inherently chiral, and no conformational changes can interconvert the two enantiomers.

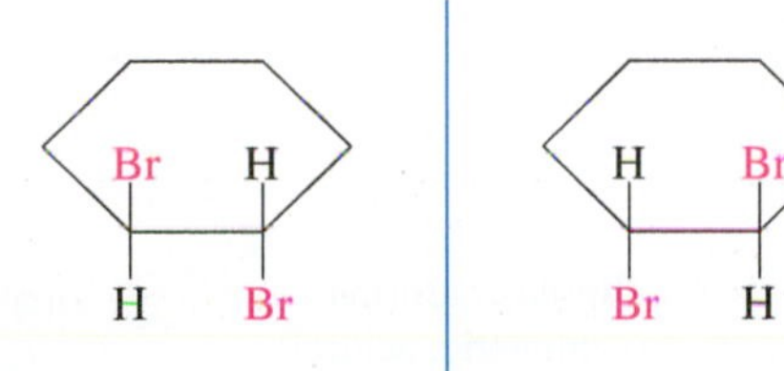

PROBLEM-SOLVING HINT

Nearly all organic compounds have chiral conformations.

A chiral compound has no symmetric (achiral) conformations, and none of its conformations can interconvert with their mirror images.

An achiral compound can (and usually does) rotate through chiral conformations, but the chiral conformations quickly interconvert with their mirror images.

PROBLEM-SOLVING HINT

Consider the most symmetric accessible conformation. You can also consider the most stable conformation and see if it can interconvert with its mirror image.

PROBLEM 5-14

1. Make a model of each compound, draw it in its most symmetric conformation, and determine whether it is capable of showing optical activity.
   (a) 1-bromo-1-chloroethane **(b)** 1-bromo-2-chloroethane
   (c) 1,2-dichloropropane **(d)** *cis*-1,3-dibromocyclohexane
   (e) *trans*-1,3-dibromocyclohexane **(f)** *trans*-1,4-dibromocyclohexane
2. Star (*) each asymmetric carbon atom in part (1), label each as (*R*) or (*S*), and compare your result from part (1) with the prediction you would make based on the asymmetric carbons.

5-9 Chiral Compounds Without Asymmetric Atoms

Most chiral organic compounds have at least one asymmetric carbon atom. Some compounds are chiral because they have another asymmetric atom, such as phosphorus, sulfur, or nitrogen, serving as a chiral center. Some compounds are chiral even though they have no asymmetric atoms at all. In these types of compounds, special characteristics of the molecules' shapes lead to chirality.

5-9A Conformational Enantiomerism

Some molecules are so bulky or so highly strained that they cannot easily convert from one chiral conformation to the mirror-image conformation. They cannot achieve the most symmetric conformation because it has too much steric strain or ring strain. Because these molecules are "locked" into a conformation, we must evaluate the individual locked-in conformation to determine whether the molecule is chiral.

Figure 5-18 shows three conformations of a sterically crowded derivative of biphenyl. The center drawing shows the molecule in its most symmetric conformation. This conformation is planar, and it has a mirror plane of symmetry. If the molecule could achieve this conformation, or even pass through it for an instant, it would not be optically active. This planar conformation is very high in energy, however, because the iodine and bromine atoms are too large to be forced so close together. The molecule is *conformationally locked*. It can exist only in one of the two staggered conformations shown on the left and right. These conformations are nonsuperimposable mirror images, and they do not interconvert. They are enantiomers, and they can be separated and isolated. Each of them is optically active, and they have equal and opposite specific rotations.

Even a simple strained molecule can show conformational enantiomerism (Figure 5-19). *trans*-Cyclooctene is the smallest stable *trans*-cycloalkene, and it is strained. If *trans*-cyclooctene existed as a planar ring, even for an instant, it could not be chiral. Make a molecular model of *trans*-cyclooctene, however, and you will see that it cannot exist as a planar ring. Its ring is folded into the three-dimensional structure pictured in Figure 5-19. The mirror image of this structure is different, and *trans*-cyclooctene is a chiral molecule. In fact, the enantiomers of *trans*-cyclooctene have been separated and characterized, and they are optically active.

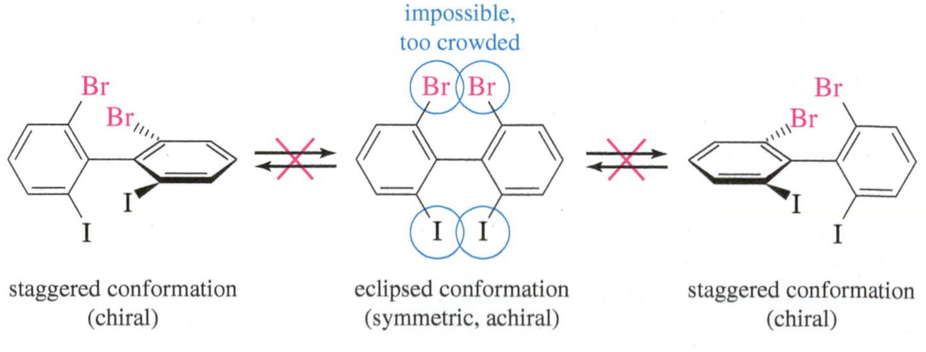

FIGURE 5-18 Three conformations of a biphenyl. This biphenyl cannot pass through its symmetric conformation because there is too much crowding of the iodine and bromine atoms. The molecule is "locked" into one of the two chiral, enantiomeric, staggered conformations.

5-9B Allenes

Allenes are compounds that contain the C=C=C unit, with two C=C double bonds meeting at a single carbon atom. The parent compound, propadiene, has the common name *allene.*

$$H_2C{=}C{=}CH_2$$

(central C: *sp* hybridized)

allene

In allene, the central carbon atom is *sp* hybridized and linear (Section 1-15), and the two outer carbon atoms are sp^2 hybridized and trigonal. We might imagine that the whole molecule lies in a plane, but this is not correct. The central *sp* hybrid carbon atom must use different *p* orbitals to form the pi bonds with the two outer carbon atoms. The two unhybridized *p* orbitals on the *sp* hybrid carbon atom are perpendicular, so the two pi bonds must also be perpendicular. Figure 5-20 shows the bonding and three-dimensional structure of allene. Allene itself is achiral. If you make a model of its mirror image, you will find it identical with the original molecule. If we add some substituents to allene, however, the molecule may be chiral.

Make a model of the following compound:

$$\overset{1}{CH_3}{-}\overset{2}{CH}{=}\overset{3}{C}{=}\overset{4}{CH}{-}\overset{5}{CH_3}$$

penta-2,3-diene

Carbon atom 3 is the *sp* hybrid allene-type carbon atom. Carbons 2 and 4 are both sp^2 and planar, but their planes are perpendicular to each other. None of the carbon atoms is attached to four different atoms, so there is no asymmetric carbon atom. Nevertheless, penta-2,3-diene is chiral, as you should see from your models and from the following drawings of the enantiomers.

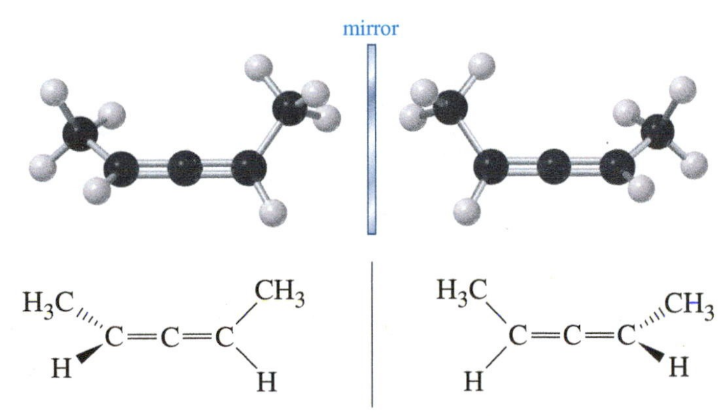

enantiomers of penta-2,3-diene

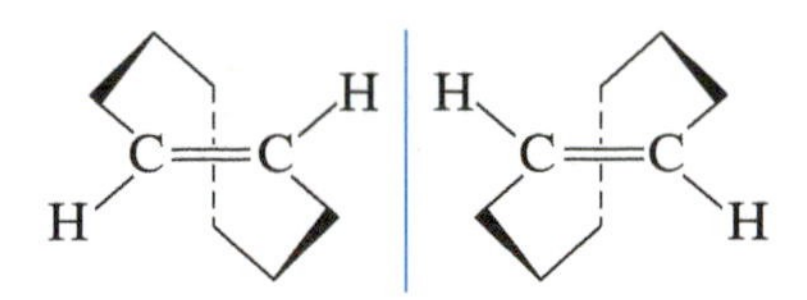

FIGURE 5-19
Conformational enantiomerism. *trans*-Cyclooctene is strained, unable to achieve a symmetric planar conformation. It is locked into one of these two enantiomeric conformations. Either pure enantiomer is optically active, with $[\alpha] = \pm 430°$.

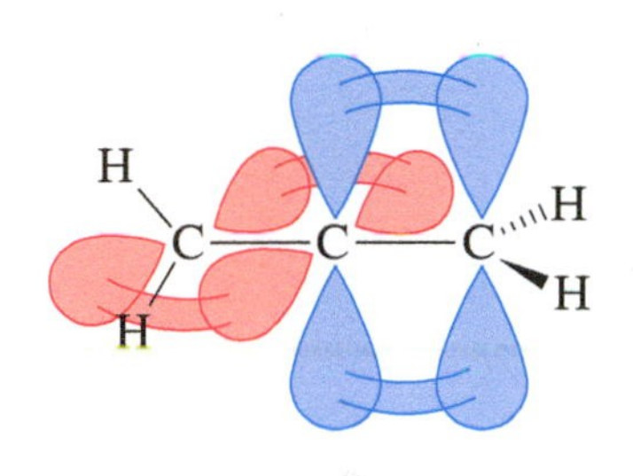

H, H–C=C=C–H, H

FIGURE 5-20
Structure of allene. The two ends of the allene molecule are perpendicular.

PROBLEM-SOLVING HINT

Dienes are compounds with two double bonds. In the name, each double bond is given the lower number of its two carbon atoms. *Allenes* are dienes with the two double bonds next to each other, joined at one carbon atom. An allene is chiral if each end has two distinct substituents.

PROBLEM 5-15

Draw three-dimensional representations of the following compounds. Which have asymmetric carbon atoms? Which have no asymmetric carbons but are chiral anyway? Use your models for parts (a) through (d) and any others that seem unclear.

(a) ClHC=C=CHCl
1,3-dichloropropadiene

(b) ClHC=C=CHCH$_3$
1-chlorobuta-1,2-diene

(c) ClHC=C=C(CH$_3$)$_2$
1-chloro-3-methylbuta-1,2-diene

(d) ClHC=CH—CH=CH$_2$
1-chlorobuta-1,3-diene

(e) Cl

(f)
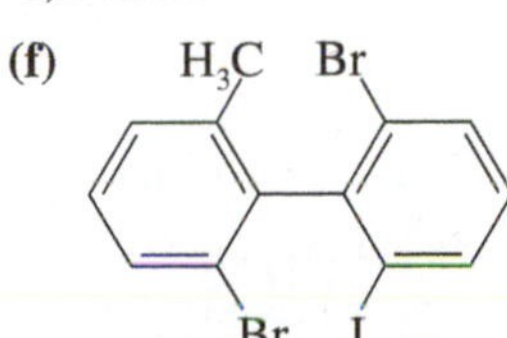

(g)
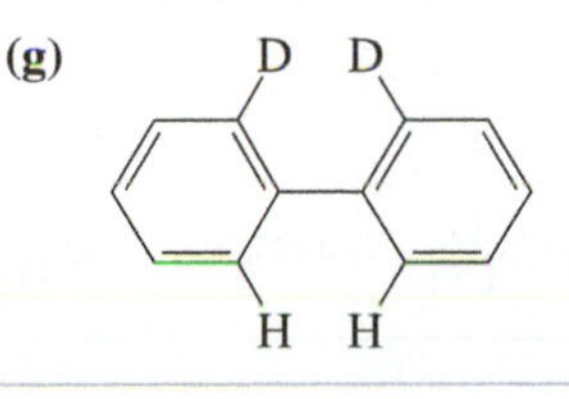

PROBLEM-SOLVING HINT

All asymmetric carbons are stereocenters, but not all stereocenters are asymmetric carbons.

5-10 Fischer Projections

We have been using dashed lines and wedges to indicate perspective in drawing the stereochemistry of asymmetric carbon atoms. When we draw molecules with several asymmetric carbons, perspective drawings become time-consuming and cumbersome. In addition, the complicated drawings make it difficult to see the similarities and differences in groups of stereoisomers.

At the turn of the 20th century, Emil Fischer was studying the stereochemistry of sugars (Chapter 23), which contain as many as seven asymmetric carbon atoms. To draw these structures in perspective would have been difficult, and to pick out minor stereochemical differences in the drawings would have been nearly impossible. Fischer developed a symbolic way of drawing asymmetric carbon atoms, allowing them to be drawn rapidly. The **Fischer projection** also facilitates comparison of stereoisomers, holding them in their most symmetric conformation and emphasizing any differences in stereochemistry. We learn to use Fischer projections because they clarify stereochemical differences and allow fast and accurate comparison of structures.

5-10A Drawing Fischer Projections

The Fischer projection looks like a cross, with the asymmetric carbon (usually not drawn in) at the point where the lines cross. The horizontal lines are taken to be wedges—that is, bonds that project out toward the viewer. The vertical lines are taken to project away from the viewer, as dashed lines. Figure 5-21 shows the perspective implied by the Fischer projection. The center drawing, with the wedged horizontal bonds looking like a bow tie, illustrates why this projection is sometimes called the "bow-tie convention." Problem 5-16 should help you to visualize how the Fischer projection is used.

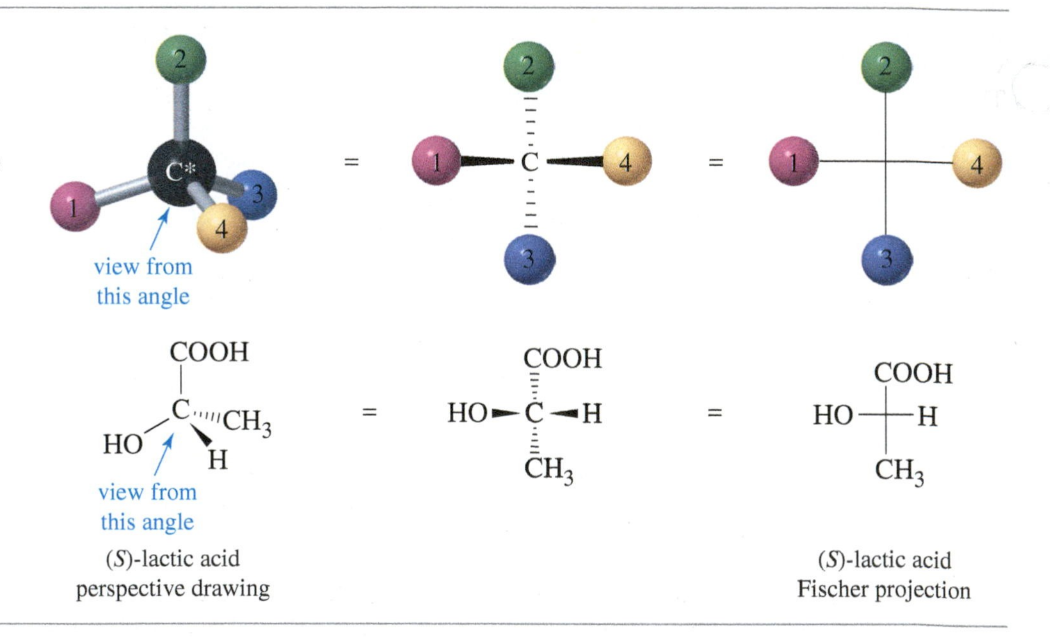

FIGURE 5-21
Perspective in a Fischer projection. The Fischer projection uses a cross to represent an asymmetric carbon atom. The horizontal lines project toward the viewer, and the vertical lines project away from the viewer.

PROBLEM 5-16

For each set of examples, make a model of the first structure, and indicate the relationship of each of the other structures to the first structure. Examples of relationships: same compound, enantiomer, structural isomer.

(a)

top	left	right	bottom
COOH	H	OH	CH_3
COOH	HO	H	CH_3
H	H_3C	COOH	OH
CH_3	HO	H	COOH

(b) CH_2CH_3 / H—Br / CH_3 | CH_3 / Br—H / CH_2CH_3 | CH_2CH_3 / Br—H / CH_3 | CH_3 / H—Br / CH_2CH_3

(c) (*R*)-butan-2-ol | CH_3 / H—OH / CH_2CH_3 | CH_3 / HO—H / CH_2CH_3 | CH_2CH_3 / H—OH / CH_3

In working Problem 5-16, you may have noticed that Fischer projections that differ by a 180° rotation are the same. When we rotate a Fischer projection by 180°, the vertical (dashed line) bonds still end up vertical, and the horizontal (wedged) lines still end up horizontal. The "horizontal lines forward, vertical lines back" convention is maintained.

Rotation by 180° is allowed.

COOH / H—OH / CH_3 = COOH / H►C◄OH / CH_3 rotate 180° = CH_3 / HO►C◄H / COOH = CH_3 / HO—H / COOH

On the other hand, if we were to rotate a Fischer projection by 90°, we would change the configuration and confuse the viewer. The original projection has the vertical groups back (dashed lines) and the horizontal groups forward. When we rotate the projection by 90°, the vertical bonds become horizontal and the horizontal bonds become vertical. The viewer assumes that the horizontal bonds come forward and that the vertical bonds go back. The viewer sees a different molecule (actually, the enantiomer of the original molecule).

A 90° rotation is NOT allowed.

COOH / H—OH / CH_3 = COOH / H►C◄OH / CH_3 90° = H / H_3C···C···COOH / OH (incorrect orientation) ✗ H / H_3C—COOH / OH (enantiomer)

In comparing Fischer projections, we cannot rotate them by 90°, and we cannot flip them over. Either of these operations gives an incorrect representation of the molecule. The Fischer projection must be kept in the plane of the paper, and it may be rotated only by 180°.

The final rule for drawing Fischer projections helps to ensure that we do not rotate the drawing by 90°. This rule is that the carbon chain is drawn along the vertical line of the Fischer projection, usually with the IUPAC numbering from top to bottom. In most cases, this numbering places the most highly oxidized carbon substituent at the top. For example, to represent (*R*)-propane-1,2-diol with a Fischer projection, we should arrange the three carbon atoms along the vertical. C1 is placed at the top, and C3 at the bottom.

$^1CH_2OH$ / ^2C ($^3CH_3$, H, OH) (viewed from this angle) = $^1CH_2OH$ / H►^{2}C◄OH / $^3CH_3$ = $^1CH_2OH$ / H—^{2}—OH / $^3CH_3$

(*R*)-propane-1,2-diol

PROBLEM-SOLVING HINT

Interchanging any two groups on a Fischer projection (or on a perspective drawing) inverts the configuration of that asymmetric carbon from (*R*) to (*S*) or from (*S*) to (*R*).

PROBLEM 5-17

Draw a Fischer projection for each compound. Remember that the cross represents an asymmetric carbon atom, and the carbon chain should be along the vertical, with the IUPAC numbering from top to bottom.

(a) (*S*)-propane-1,2-diol
(b) (*R*)-2-bromobutan-1-ol
(c) (*S*)-1,2-dibromobutane
(d) (*R*)-butan-2-ol
(e) (*R*)-glyceraldehyde, HO—$^{3}CH_2$—$^{2}CH(OH)$—^{1}CHO

COOH / H—+—OH / CH_3 (*R*)-lactic acid

COOH / HO—+—H / CH_3 (*S*)-lactic acid

5-10B Drawing Mirror Images of Fischer Projections

How should we draw the mirror image of a molecule drawn in Fischer projection? With our perspective drawings, the rule was to reverse left and right while keeping the other directions (up and down, front and back) in their same positions. This rule still applies to Fischer projections. Interchanging the groups on the horizontal part of the cross reverses left and right while leaving the other directions unchanged.

Testing for enantiomerism is particularly simple using Fischer projections. If the Fischer projections are properly drawn (carbon chain along the vertical), and if the mirror image cannot be made to look the same as the original structure with a 180° rotation in the plane of the paper, the two mirror images are enantiomers. In the following examples, any groups that fail to superimpose after a 180° rotation are circled in red.

	Original	*Mirror image*		*180° rotation*
propan-2-ol	CH_3 / σ---H—+—OH--- / CH_3	CH_3 / ---HO—+—H---σ / CH_3	180° =	CH_3 / H—+—OH / CH_3

These mirror images are the same. Propan-2-ol is achiral.

(*R*)-propane-1,2-diol	CH_2OH / H—+—OH / CH_3	CH_2OH / HO—+—H / CH_3	180° =	(CH_3) / H—+—OH / (CH_2OH)

These mirror images are different. Propane-1,2-diol is chiral.

(2*S*,3*S*)-2,3-dibromobutane	CH_3 / H—+—Br / Br—+—H / CH_3	CH_3 / Br—+—H / H—+—Br / CH_3	180° =	CH_3 / (Br)—+—(H) / (H)—+—(Br) / CH_3

These mirror images are different. This structure is chiral.

Mirror planes of symmetry are particularly easy to identify from the Fischer projection because this projection is normally the most symmetric conformation. In the first preceding example (propan-2-ol) and in the following example [(2*S*,3*R*)-2,3-dibromobutane], the symmetry planes are indicated in red; these molecules with symmetry planes cannot be chiral.

(2*S*,3*R*)-2,3-dibromobutane	CH_3 / H—+—Br / σ- - - - - - / H—+—Br / CH_3	CH_3 / Br—+—H / - - - - - -σ / Br—+—H / CH_3	180° =	CH_3 / H—+—Br / H—+—Br / CH_3

These mirror images are the same. This structure is achiral.

PROBLEM 5-18

For each Fischer projection,

1. make a model.
2. draw the mirror image.
3. determine whether the mirror image is the same as, or different from, the original structure.
4. draw any mirror planes of symmetry that are apparent from the Fischer projections.

(a) CHO; H—|—OH; CH_2OH

(b) CH_2OH; H—|—Br; CH_2OH

(c) CH_2Br; Br—|—Br; CH_3

(d) CHO; H—|—OH; H—|—OH; CH_2OH

(e) CH_2OH; H—|—OH; H—|—OH; CH_2OH

(f) CH_2OH; HO—|—H; H—|—OH; CH_2OH

PROBLEM-SOLVING HINT

Using Fischer projections, we can easily solve difficult stereochemical problems that would be intractable using perspective drawings. Most problems involving two or more asymmetric carbons are more easily solved using Fischer projections.

5-10C Assigning (*R*) and (*S*) Configurations from Fischer Projections

The Cahn–Ingold–Prelog convention (Section 5-3) can be applied to structures drawn using Fischer projections. Let's review the two rules for assigning (*R*) and (*S*): (1) Assign priorities to the groups bonded to the asymmetric carbon atom; (2) put the lowest-priority group (usually H) in back, and draw an arrow from group 1 to group 2 to group 3. Clockwise is (*R*), and counterclockwise is (*S*).

The (*R*) or (*S*) configuration can also be determined directly from the Fischer projection, without having to convert it to a perspective drawing. The lowest-priority atom is usually hydrogen. In the Fischer projection, the carbon chain is along the vertical line, so the hydrogen atom is usually on the horizontal line and projects out in front. Once we have assigned priorities, we can draw an arrow from group 1 to group 2 to group 3 and see which way it goes. If the molecule were turned around so that the hydrogen would be in back [as in the definition of (*R*) and (*S*)], the arrow would rotate in the other direction. By mentally turning the arrow around (or simply applying the rule backward), we can assign the configuration.

As an example, consider the Fischer projection formula of one of the enantiomers of glyceraldehyde. First priority goes to the —OH group, followed by the —CHO group and the —CH_2OH group. The hydrogen atom receives the lowest priority. The arrow from group 1 to group 2 to group 3 appears counterclockwise in the Fischer projection. If the molecule is turned over so that the hydrogen is in back, the arrow becomes clockwise, making this the (*R*) enantiomer of glyceraldehyde.

counterclockwise

② CHO; H—|—OH ①; CH_2OH ③

Fischer projection
(*R*)-(+)-glyceraldehyde

=

CHO; H—C—OH; CH_2OH

hydrogen in front

=

clockwise

② CHO; H—C; ① HO; ③ CH_2OH

perspective drawing
(*R*)-(+)-glyceraldehyde

PROBLEM-SOLVING HINT

When naming (*R*) and (*S*) from Fischer projections with the hydrogen on a horizontal bond (toward you instead of away from you), just apply the normal rules backward.

PROBLEM 5-19

For each Fischer projection, label each asymmetric carbon atom as (*R*) or (*S*).
(a)–(f) the structures in Problem 5-18

(g) Fischer projection: top CH_2CH_3; left H; right Br; bottom CH_3

(h) Fischer projection: top COOH; left H_2N; right H; bottom CH_3

(i) Fischer projection: top CH_2OH; left Br; right Cl; bottom CH_3
(careful—no hydrogen)

SUMMARY Fischer Projections and Their Use

1. They are most useful for compounds with two or more asymmetric carbon atoms.
2. Asymmetric carbons are at the centers of crosses.
3. The vertical lines project away from the viewer; the horizontal lines project toward the viewer (like a bow tie ).
4. The carbon chain is placed along the vertical, with the IUPAC numbering from top to bottom. In most cases, this places the more oxidized end (the carbon with the most bonds to O or halogen) at the top.
5. The entire projection can be rotated 180° (but not 90°) in the plane of the paper without changing its stereochemistry.
6. Interchanging any two groups on an asymmetric carbon (for example, those on the horizontal line) inverts its stereochemistry.

5-11 Diastereomers

We have defined *stereoisomers* as isomers whose atoms are bonded together in the same order but differ in how the atoms are directed in space. We have also considered enantiomers (mirror-image isomers) in detail. All other stereoisomers are classified as **diastereomers,** which are defined as *stereoisomers that are not mirror images.* Most diastereomers are either geometric isomers or compounds containing two or more chiral centers.

5-11A Cis-trans Isomerism on Double Bonds

We have already seen one class of diastereomers: the **cis-trans isomers,** or **geometric isomers.** For example, there are two isomers of but-2-ene:

H_3C and CH_3 on the same side, H and H below: $C{=}C$

cis-but-2-ene

H_3C and H above, H and CH_3 below: $C{=}C$

trans-but-2-ene

These stereoisomers are not mirror images of each other, so they are not enantiomers. They are diastereomers.

5-11B Cis-trans Isomerism on Rings

Cis-trans isomerism is also possible when a ring is present. *Cis-* and *trans*-1,2-dimethylcyclopentane are geometric isomers, and they are diastereomers. The trans diastereomer has an enantiomer, but because the cis diastereomer has an internal mirror plane of symmetry, it must be achiral.

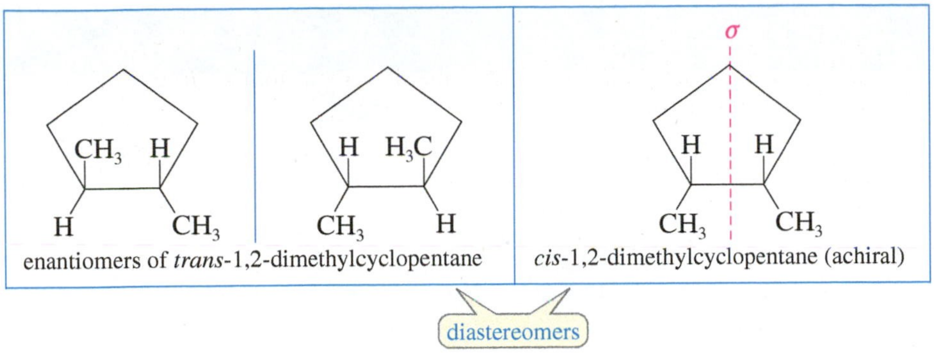

CH$_3$ H / H CH$_3$ | H H$_3$C / CH$_3$ H
enantiomers of *trans*-1,2-dimethylcyclopentane

σ
H H / CH$_3$ CH$_3$
cis-1,2-dimethylcyclopentane (achiral)

diastereomers

5-11C Diastereomers of Molecules with Two or More Chiral Centers

Apart from geometric isomers, most other compounds that show diastereomerism have two or more chiral centers, usually asymmetric carbon atoms. For example, 2-bromo-3-chlorobutane has two asymmetric carbon atoms, and it exists in two diastereomeric forms (shown next). Make molecular models of these two stereoisomers.

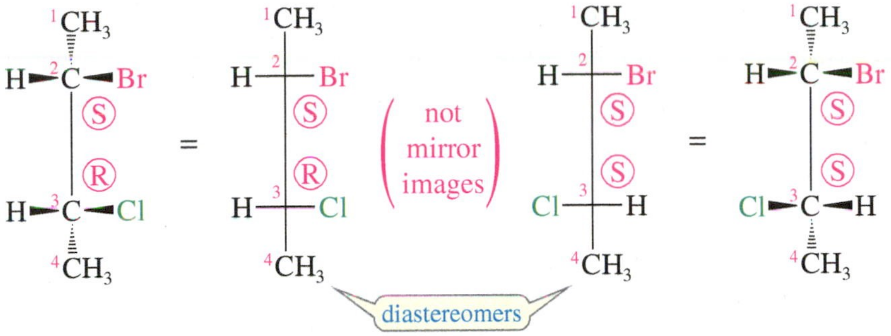

(S) (R) = (S) (R) (not mirror images) (S) (S) = (S) (S)

diastereomers

These two structures are not the same; they are stereoisomers because they differ in the orientation of their atoms in space. They are not enantiomers, however, because they are not mirror images of each other: C2 has the (*S*) configuration in both structures, while C3 is (*R*) in the structure on the left and (*S*) in the structure on the right. The C3 carbon atoms are mirror images of each other, but the C2 carbon atoms are not. If these two compounds were mirror images of each other, both asymmetric carbons would have to be mirror images of each other.

Because these compounds are stereoisomers but not enantiomers, they must be diastereomers. In fact, both of these diastereomers are chiral and each has an enantiomer. Thus, there is a total of four stereoisomeric 2-bromo-3-chlorobutanes: two pairs of enantiomers. Either member of one pair of enantiomers is a diastereomer of either member of the other pair.

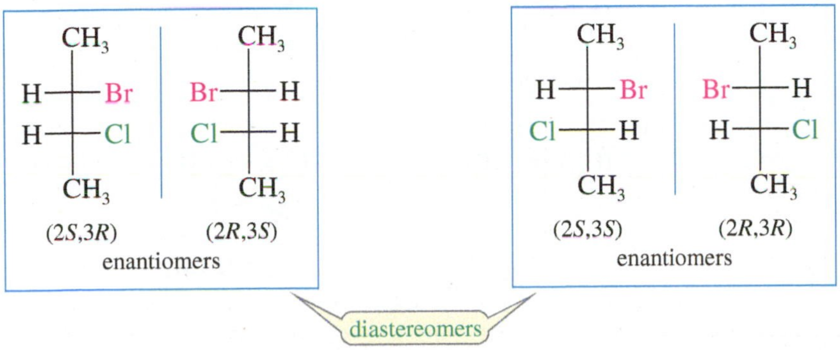

(2*S*,3*R*) (2*R*,3*S*)
enantiomers

(2*S*,3*S*) (2*R*,3*R*)
enantiomers

diastereomers

PROBLEM-SOLVING HINT

When a compound contains two or more chiral centers (usually asymmetric carbons), consider drawing it as a Fischer projection. The stereochemical differences we see in the structures at left are not as clear in perspective drawings, which are also much harder to draw.

We have now seen all the types of isomers we need to study, and we can diagram their relationships and summarize their definitions.

SUMMARY Types of Isomers

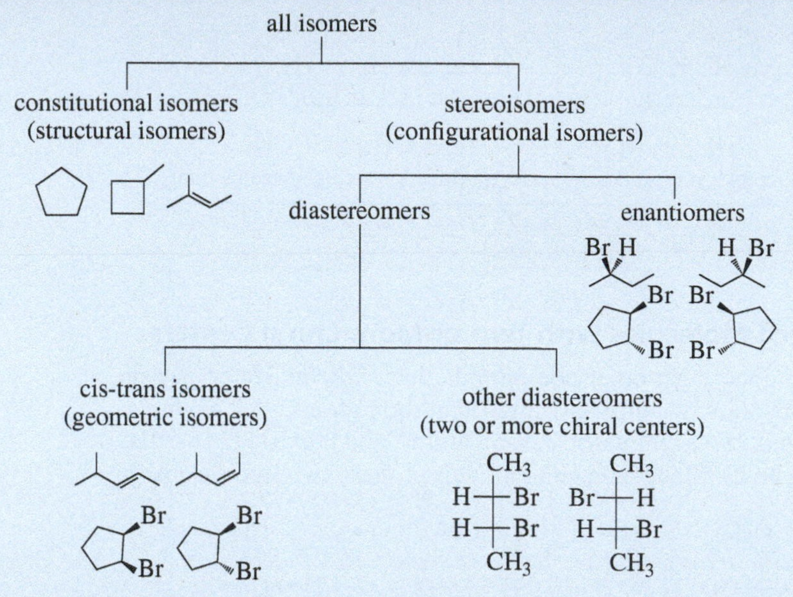

Isomers are different compounds with the same molecular formula.

Constitutional isomers are isomers that differ in the order in which atoms are bonded together. Constitutional isomers are sometimes called **structural isomers** because they have different connections among their atoms.

Stereoisomers are isomers that differ only in the orientation of the atoms in space.

Enantiomers are mirror-image isomers.

Diastereomers are stereoisomers that are not mirror images of each other.

Cis-trans isomers (geometric isomers) are diastereomers that differ in their cis-trans arrangement on a ring or double bond.

PROBLEM 5-20

For each pair, give the relationship between the two compounds. Making models will be helpful.

(a) (2*R*,3*S*)-2,3-dibromohexane and (2*S*,3*R*)-2,3-dibromohexane

(b) (2*R*,3*S*)-2,3-dibromohexane and (2*R*,3*R*)-2,3-dibromohexane

(c) H_3C, H, Br on C — C with Br, H, CH_3 and H_3C, Br, H on C — C with H, CH_3, Br

(d) 3-bromocyclohexene (Br wedge) and (Br hashed)

(e) Cl-substituted bicyclic alkenes (Cl on bridge) and (Cl on bridgehead)

(f) Fischer projections: CHO; H—OH; H—OH; CH_2OH and CHO; HO—H; H—OH; CH_2OH

(g) Fischer projections: CHO; HO—H; H—OH; H—OH; CH_2OH and CHO; H—OH; HO—H; HO—H; CH_2OH

(h) Newman projections: front CH_3, H, H; back H, Br, CH_2CH_3 and front CH_3, H, H; back Br, H, CH_2CH_3

(i) CH_2CH_3, H, H_3C on C — $\overset{+}{N}$ with CH_3, CH_2CH_3, $CH(CH_3)_2$ and CH_2CH_3, H, H_3C on C — $\overset{+}{N}$ with CH_2CH_3, CH_3, $CH(CH_3)_2$

5-12 Stereochemistry of Molecules with Two or More Asymmetric Carbons

In the preceding section, we saw that there are four stereoisomers (two pairs of enantiomers) of 2-bromo-3-chlorobutane. These four isomers are simply all the permutations of (*R*) and (*S*) configurations at the two asymmetric carbon atoms, C2 and C3:

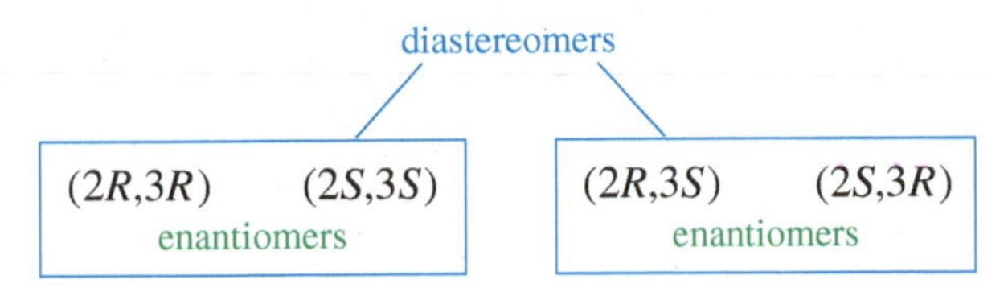

A compound with *n* asymmetric carbon atoms might have as many as has 2^n stereoisomers. This formula is called the **2^n rule,** where *n* is the number of chiral centers (usually asymmetric carbon atoms). The 2^n rule suggests that we should look for a *maximum* of 2^n stereoisomers. We may not always find 2^n isomers, especially when two of the asymmetric carbon atoms have identical substituents.

2,3-Dibromobutane has fewer than 2^n stereoisomers. It has two asymmetric carbons (C2 and C3), so the 2^n rule predicts a maximum of four stereoisomers. The four permutations of (*R*) and (*S*) configurations at C2 and C3 are shown next. Make molecular models of these structures to compare them.

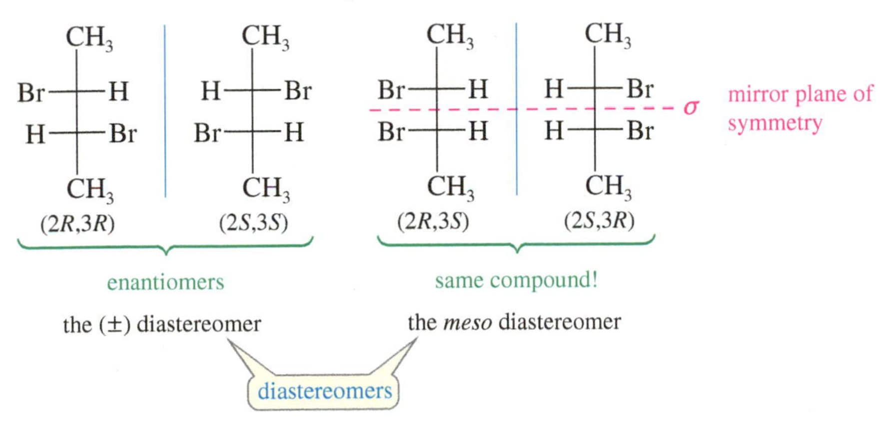

There are only three stereoisomers of 2,3-dibromobutane because two of the four structures are identical. The diastereomer on the right is achiral, having a mirror plane of symmetry. The asymmetric carbon atoms have identical substituents, and the one with (*R*) configuration reflects into the other having (*S*) configuration. It seems almost as though the molecule were a racemic mixture within itself.

5-13 Meso Compounds

Compounds that are achiral even though they have asymmetric carbon atoms are called **meso compounds.** The (2*R*,3*S*) isomer of 2,3-dibromobutane is a meso compound; most meso compounds have this kind of symmetric structure, with two similar halves of the molecule having opposite configurations. In speaking of the two diastereomers of 2,3-dibromobutane, the symmetric one is called the *meso diastereomer*, and the chiral one is called the (±) *diastereomer*, because one enantiomer is (+) and the other is (−).

> MESO COMPOUND: An achiral compound that has chiral centers (usually asymmetric carbons).

The term *meso* (Greek, "middle") was used to describe an achiral member of a set of diastereomers, some of which are chiral. The optically inactive isomer seemed to be in the

PROBLEM-SOLVING HINT

A meso compound with two chiral centers will be (*R*,*S*) or (*S*,*R*) because the chiral centers must be mirror images of each other, reflected across the internal mirror plane.

"middle" between the dextrorotatory and levorotatory isomers. The definition just given ("an achiral compound with chiral centers") is nearly as complete, and more easily applied, especially when you remember that chiral centers are usually asymmetric carbon atoms.

We have already seen other meso compounds, although we have not yet called them that. For example, the cis isomer of 1,2-dichlorocyclopentane has two asymmetric carbon atoms, yet it is achiral. Thus, it is a meso compound. *cis*-1,2-Dibromocyclohexane is not symmetric in its chair conformation, but it consists of equal amounts of two enantiomeric chair conformations in a rapid equilibrium. We are justified in looking at the molecule in its symmetric flat conformation to show that it is achiral and meso. For acyclic compounds, the Fischer projection helps to show the symmetry of meso compounds.

σ H H Cl Cl

cis-1,2-dichlorocyclopentane

σ H H Br Br

cis-1,2-dibromocyclohexane

CH_3 Br—H σ Br—H CH_3

meso-2,3-dibromobutane

COOH σ H—OH H—OH COOH

meso-tartaric acid

SOLVED PROBLEM 5-7

Determine which of the following compounds are chiral. Star (*) any asymmetric carbon atoms, and draw in any mirror planes. Label any meso compounds. (Use your molecular models to follow along.)

(a) CH_3 H—OH HO—H CH_3

(b) CH_2OH Br—Cl Br—Cl CH_2OH

(c) CH_3 H—OH HO—H HO—H H—OH CH_3

(d) H Br H Cl

SOLUTION

(a) CH_3 H—*—OH HO—*—H CH_3 | CH_3 HO—*—H H—*—OH CH_3

This compound does *not* have a plane of symmetry, and we suspect that it is chiral. Drawing the mirror image shows that it is nonsuperimposable on the original structure. These are the enantiomers of a chiral compound.

(b) CH_2OH Br—*—Cl σ Br—*—Cl CH_2OH meso

(c) CH_3 H—*—OH HO—*—H σ HO—*—H H—*—OH CH_3 meso

Both (b) and (c) have mirror planes of symmetry and are achiral. Because they have asymmetric carbon atoms yet are achiral, they are meso.

(d) H H * * Cl Br | H H * * Br Cl

Drawing this compound in its most symmetric conformation (flat) shows that it does not have a mirror plane of symmetry. When we draw the mirror image, we find it to be an enantiomer.

SOLVED PROBLEM 5-8

One source defines a meso compound as "an achiral compound with stereocenters." Why is this a poor definition?

SOLUTION

A stereocenter is an atom at which the interchange of two groups gives a stereoisomer. Stereocenters include both chiral centers and double-bonded carbons giving rise to cis-trans isomers. For example, the isomers of but-2-ene are achiral and they contain stereocenters (circled), so they meet this definition. They have no chiral diastereomers, however, so they are not correctly called meso.

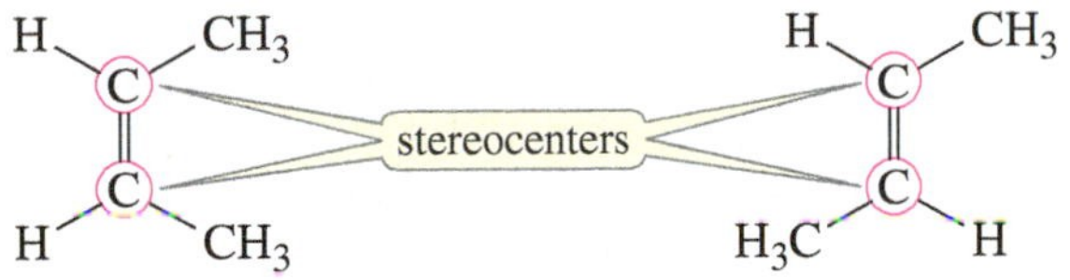

PROBLEM-SOLVING HINT

To draw all the stereoisomers of a given structure, first identify all the stereocenters. Then systematically go through all the permutations that result when the stereocenters are in their two configurations. For example, if there are three chiral centers, you write RRR, RRS, RSR, SRR, RSS, SRS, SSR, and SSS. In a Fischer projection, you can start with all the substituents on one side and go through the same process moving them to the other side. If the molecule has symmetry that creates meso structures, you must eliminate the duplicates.

PROBLEM 5-21

Which of the following compounds are chiral? Draw each compound in its most symmetric conformation, star (*) any asymmetric carbon atoms, and draw any mirror planes. Label any meso compounds. You may use Fischer projections if you prefer.

(a) *meso*-2,3-dibromo-2,3-dichlorobutane
(b) (±)-2,3-dibromo-2,3-dichlorobutane
(c) (2*R*,3*S*)-2-bromo-3-chlorobutane
(d) (2*R*,3*S*)-2,3-dibromobutane
(e) (*R*,*R*)-2,3-dibromobutane
(f) **(g)** 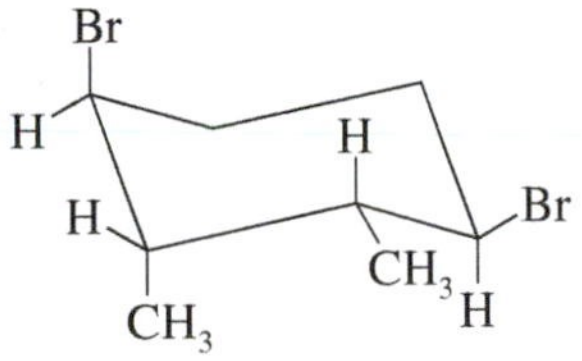 **(*h)**

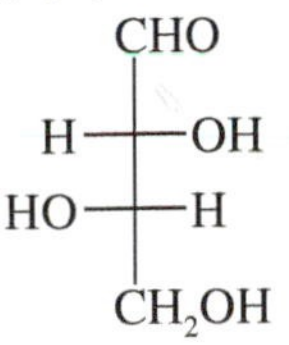

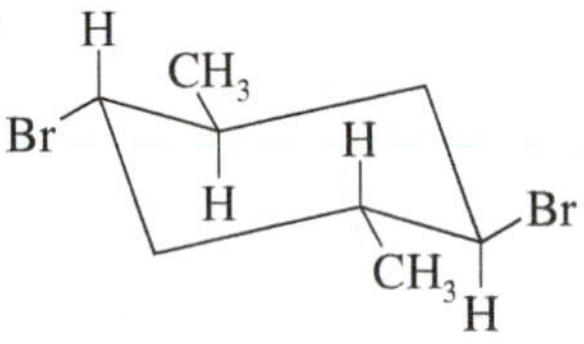

PROBLEM 5-22

Draw all the distinct stereoisomers for each structure. Show the relationships (enantiomers, diastereomers, etc.) between the isomers. Label any meso isomers, and draw any mirror planes of symmetry.

(a) CH_3—CHCl—CHOH—COOH
(b) tartaric acid, HOOC—CHOH—CHOH—COOH
(c) HOOC—CHBr—CHOH—CHOH—COOH
(d) **(e)**

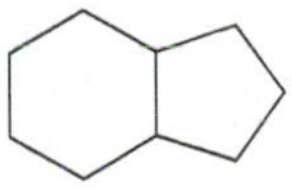

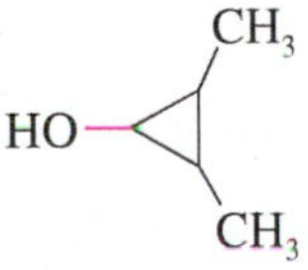

PROBLEM-SOLVING HINT

In part (e), the carbon bearing the OH group is not asymmetric, but it can be a stereocenter if the methyl groups are cis to each other.

5-14 Absolute and Relative Configuration

Throughout our study of stereochemistry, we have drawn three-dimensional representations, and we have spoken of asymmetric carbons having the (*R*) or (*S*) configuration. These ways of describing the configuration of a chiral center are *absolute*; that is, they give the actual orientation of the atoms in space. We say that these methods specify the **absolute configuration** of the molecule. For example, given the name "(*R*)-butan-2-ol," any chemist can construct an accurate molecular model or draw a three-dimensional representation.

ABSOLUTE CONFIGURATION: The detailed stereochemical picture of a molecule, including how the atoms are arranged in space. Alternatively, the (*R*) or (*S*) configuration at each chiral center.

Chemists have determined the absolute configurations of many chiral compounds since 1951, when X-ray crystallography was first used to find the orientation of atoms in space. Before 1951, there was no way to link the stereochemical drawings with the actual enantiomers and their observed rotations. No absolute configurations were known. It was possible, however, to correlate the configuration of one compound with another and to show that two compounds had the same or opposite configurations. When we convert one compound into another using a reaction that does not break bonds at the asymmetric carbon atom, we know that the product must have the same **relative configuration** as the reactant, even if we cannot determine the absolute configuration of either compound.

RELATIVE CONFIGURATION: The experimentally determined relationship between the configurations of two molecules, even though we may not know the absolute configuration of either.

For example, optically active 2-methylbutan-1-ol reacts with PBr_3 to give optically active 1-bromo-2-methylbutane. None of the bonds to the asymmetric carbon atom are broken in this reaction, so the product must have the same configuration at the asymmetric carbon as the starting material does.

$$CH_3CH_2\overset{*}{C}H(CH_3)CH_2OH + PBr_3 \longrightarrow CH_3CH_2\overset{*}{C}H(CH_3)CH_2Br$$

(+)-2-methylbutan-1-ol $[\alpha]_D^{25} = +5.8°$ — (−)-1-bromo-2-methylbutane $[\alpha]_D^{25} = -4.0°$

We say that (+)-2-methylbutan-1-ol and (−)-1-bromo-2-methylbutane have the same relative configuration, even though we don't have the foggiest idea whether either of these is (*R*) or (*S*) unless we relate them to a compound whose absolute configuration has been established by X-ray crystallography.

Before the advent of X-ray crystallography, several systems were used to compare the relative configurations of chiral compounds with those of standard compounds. Only one of these systems is still in common use today: the **D–L system,** also known as the *Fischer–Rosanoff convention*. The configurations of sugars and amino acids were related to the enantiomers of glyceraldehyde. Compounds with the same relative configuration as (+)-glyceraldehyde were assigned the D prefix, and those with the relative configuration of (−)-glyceraldehyde were given the L prefix.

PROBLEM-SOLVING HINT

With chiral cyclic compounds, *cis* and *trans* prefixes give relative configurations. In the name *cis*-1-bromo-2-methylcyclohexane, *cis* indicates the relative positions of the methyl group and the bromine atom. It does not specify the absolute configuration of either chiral center, however, and we might correctly draw either cis enantiomer. The name (1*S*,2*R*)-1-bromo-2-methylcyclohexane specifies the absolute configuration of one of the cis enantiomers with only one possible structure.

CHO	COOH	COOH	COOH
HO—+—H (*S*)	H_2N—+—H	H_2N—+—H	H_2N—+—H
CH_2OH	R	CH_2OH	CH_2CH_2COOH
L-(−)-glyceraldehyde	an L-amino acid	L-(−)-serine	L-(+)-glutamic acid

We now know the absolute configurations of the glyceraldehyde enantiomers: The (+) enantiomer has the (*R*) configuration, with the hydroxy (OH) group on the right in the Fischer projection. The (−) enantiomer has the (*S*) configuration, with the hydroxy group on the left. Most naturally occurring amino acids have the L configuration, with the amino (NH_2) group on the left in the Fischer projection.

Sugars have several asymmetric carbons, but they can all be degraded to glyceraldehyde by oxidizing them from the aldehyde end. (We discuss these reactions in Chapter 23.) Most naturally occurring sugars degrade to (+)-glyceraldehyde, so they are given the D prefix. This means that the bottom asymmetric carbon of the sugar has its hydroxy (OH) group on the right in the Fischer projection.

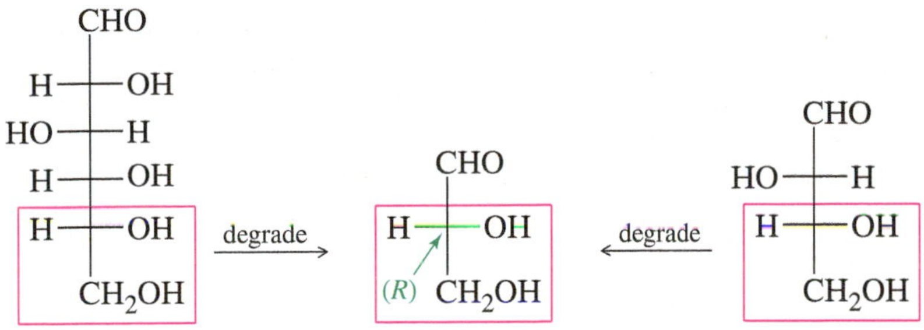

D-(+)-glucose —degrade→ D-(+)-glyceraldehyde ←degrade— D-(−)-threose

5-15 Physical Properties of Diastereomers

We have seen that enantiomers have identical physical properties except for the direction in which they rotate polarized light. Diastereomers, on the other hand, generally have different physical properties. For example, consider the diastereomers of but-2-ene (shown next). The symmetry of *trans*-but-2-ene causes the dipole moments of the bonds to cancel. The dipole moments in *cis*-but-2-ene do not cancel but add together to create a molecular dipole moment. The dipole–dipole attractions of *cis*-but-2-ene give it a higher boiling point than *trans*-but-2-ene.

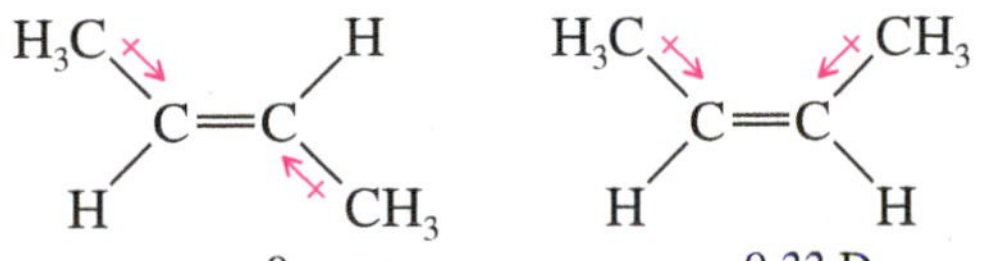

$\mu = 0$
trans-but-2-ene
bond dipoles cancel
bp = 0.9 °C

$\mu = 0.33$ D
cis-but-2-ene
vector sum dipole
bp = 3.7 °C

Diastereomers that are not geometric isomers also have different physical properties. The two diastereomers of 2,3-dibromosuccinic acid have melting points that differ by nearly 100 °C!

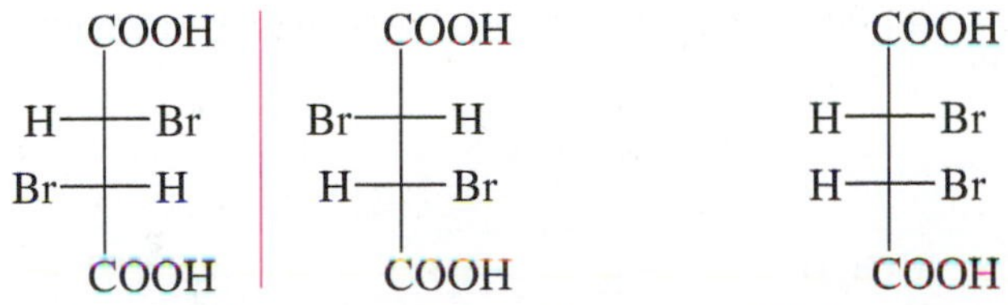

(+) and (−)-2,3-dibromosuccinic acid
mp for either is 158 °C

meso-2,3-dibromosuccinic acid
mp 256 °C

Most of the common sugars are diastereomers of glucose. All these diastereomers have different physical properties. For example, glucose and galactose are diastereomeric sugars that differ only in the stereochemistry of one asymmetric carbon atom, C4.

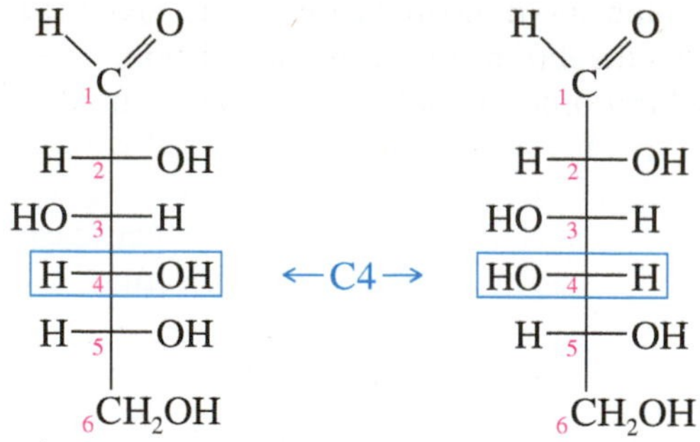

D-(+)-glucose, mp 148 °C ←C4→ D-(+)-galactose, mp 167 °C

Because diastereomers have different physical properties, we can separate them by ordinary means such as distillation, recrystallization, and chromatography. As we will see in the next section, the separation of enantiomers is a more difficult process.

PROBLEM 5-23

Which of the following pairs of compounds could be separated by recrystallization or distillation?

(a) *meso*-tartaric acid and (±)-tartaric acid (HOOC—CHOH—CHOH—COOH)

(b) [structure] and [structure]

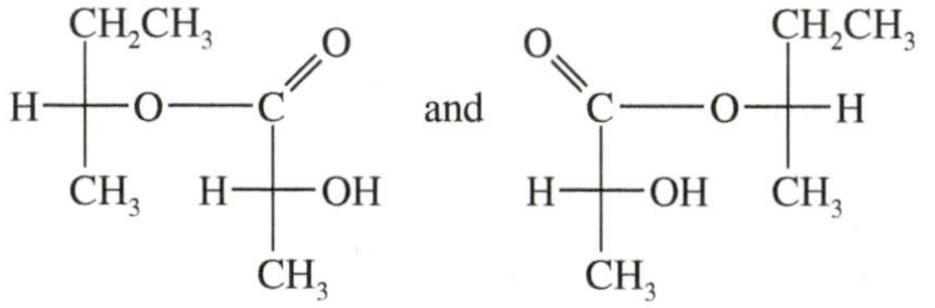

(c) [structure] and [structure]

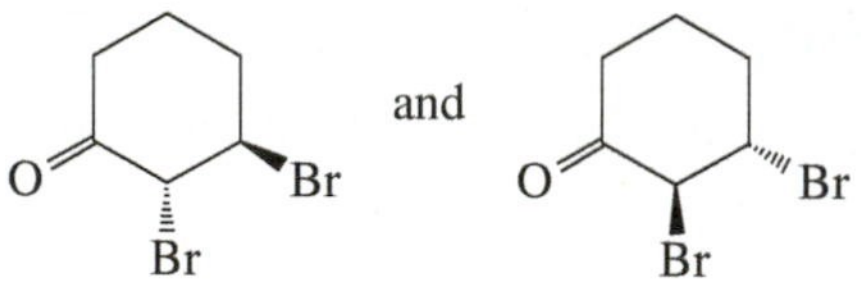

(d)* [structure] and [structure]

(an acid–base salt)

Ph = [phenyl ring]

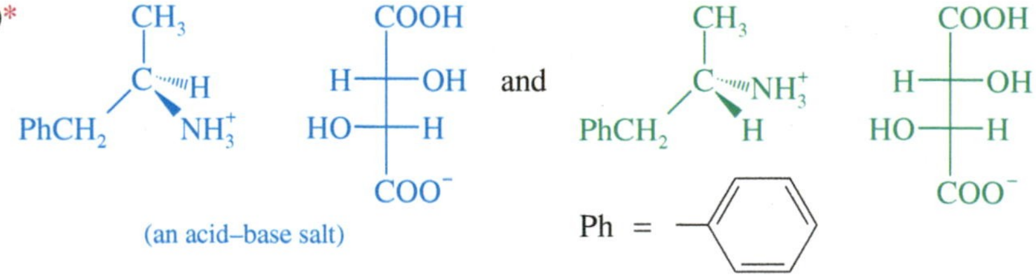

5-16 Resolution of Enantiomers

Pure enantiomers of optically active compounds are often obtained by isolation from biological sources. Most optically active molecules are found as only one enantiomer in living organisms. For example, pure (+)-tartaric acid can be isolated from the precipitate formed by yeast during the fermentation of wine. Pure (+)-glucose is obtained from many different sugar sources, such as grapes, sugar beets, sugarcane, and honey. Alanine is a common amino acid found in protein as the pure (+) enantiomer.

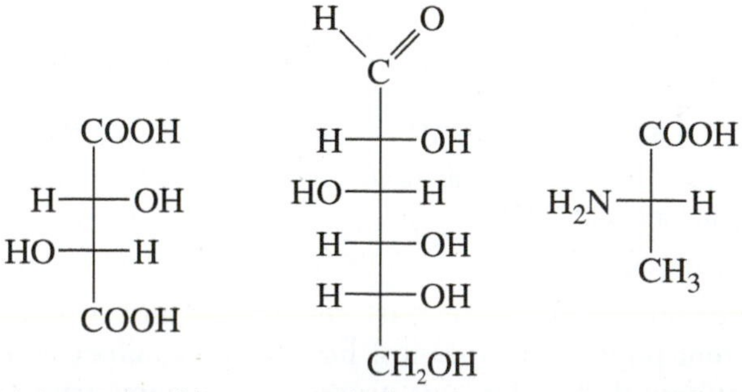

L-(+)-tartaric acid
(2*R*,3*R*)-tartaric acid

D-(+)-glucose

L-(+)-alanine

When a chiral compound is synthesized from achiral reagents, however, a racemic mixture of enantiomers results. For example, we saw that the reduction of butan-2-one (achiral) to butan-2-ol (chiral) gives a racemic mixture:

$$H_3C-\overset{\overset{O}{\|}}{C}-CH_2-CH_3 \xrightarrow{H_2,\ Pt} \underset{(R)\text{-butan-2-ol}}{(H)(OH)C(CH_3)(CH_2CH_3)} + \underset{(S)\text{-butan-2-ol}}{(HO)(H)C(CH_3)(CH_2CH_3)}$$

butan-2-one

If we need one pure enantiomer of butan-2-ol, we must find a way of separating it from the other enantiomer. The separation of enantiomers is called **resolution,** and it is a different process from the usual physical separations. A chiral probe is necessary for the resolution of enantiomers; such a chiral compound or apparatus is called a **resolving agent.**

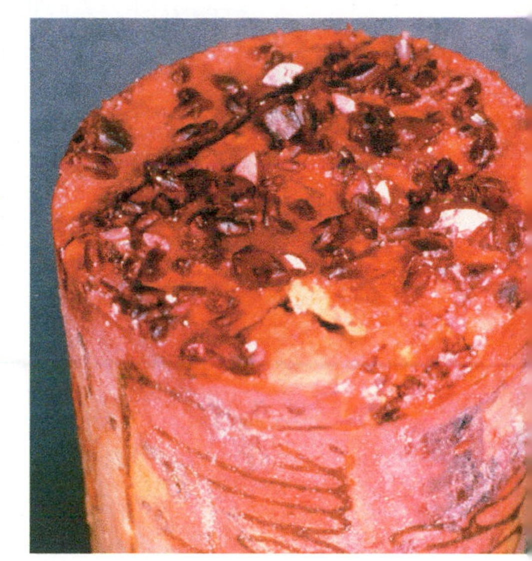

Wine sediments often contain salts of tartaric acid. This wine cork bears crystals of the potassium salt of L-(+)-tartaric acid.

In 1848, Louis Pasteur noticed that a salt of racemic (±)-tartaric acid crystallizes into mirror-image crystals. Using a microscope and a pair of tweezers, he physically separated the enantiomeric crystals. He found that solutions made from the "left-handed" crystals rotate polarized light in one direction and solutions made from the "right-handed" crystals rotate polarized light in the opposite direction. Pasteur had accomplished the first artificial resolution of enantiomers. Unfortunately, few racemic compounds crystallize as separate enantiomers, and other methods of separation are required.

5-16A Chemical Resolution of Enantiomers

The traditional method for resolving a racemic mixture into its enantiomers is to use an enantiomerically pure natural product that bonds with the compound to be resolved. When the enantiomers of the racemic compound bond to the pure resolving agent, a pair of diastereomers results. The diastereomers are separated, then the resolving agent is cleaved from the separated enantiomers.

Let's consider how we might resolve a racemic mixture of (*R*)- and (*S*)-butan-2-ol. We need a resolving agent that reacts with an alcohol and that is readily available in an enantiomerically pure state. A carboxylic acid combines with an alcohol to form an ester. Although we have not yet studied the chemistry of esters (Chapter 21), the following equation shows how an acid and an alcohol can combine with the loss of water to form an ester.

$$\underset{\text{acid}}{R-\overset{\overset{O}{\|}}{C}-OH} + \underset{\text{alcohol}}{R'-OH} \overset{(H^+\ \text{catalyst})}{\rightleftharpoons} \underset{\text{ester}}{R-\overset{\overset{O}{\|}}{C}-O-R'} + \underset{\text{water}}{H-O-H}$$

For our resolving agent, we need an optically active chiral acid to react with butan-2-ol. Any winery can provide large amounts of pure (+)-tartaric acid. Figure 5-22 shows that diastereomeric esters are formed when (*R*)- and (*S*)-butan-2-ol react with (+)-tartaric acid. We can represent the reaction schematically as follows:

An illustration of Louis Pasteur working in the laboratory. He is, no doubt, contemplating the implications of enantiomerism in tartaric acid crystals.

[(*R*)- and (*S*)-butan-2-ol] plus [(*R,R*)-tartaric acid] $\xrightarrow{H^+}$ [(*R*)-2-butyl (*R,R*)-tartrate] + [(*S*)-2-butyl (*R,R*)-tartrate]

diastereomers, *not* mirror images

The diastereomers of 2-butyl tartrate have different physical properties, and they can be separated by conventional distillation, recrystallization, or chromatography. Separation of the diastereomers leaves us with two flasks, each containing one of the diastereomeric esters. The resolving agent is then cleaved from the separated

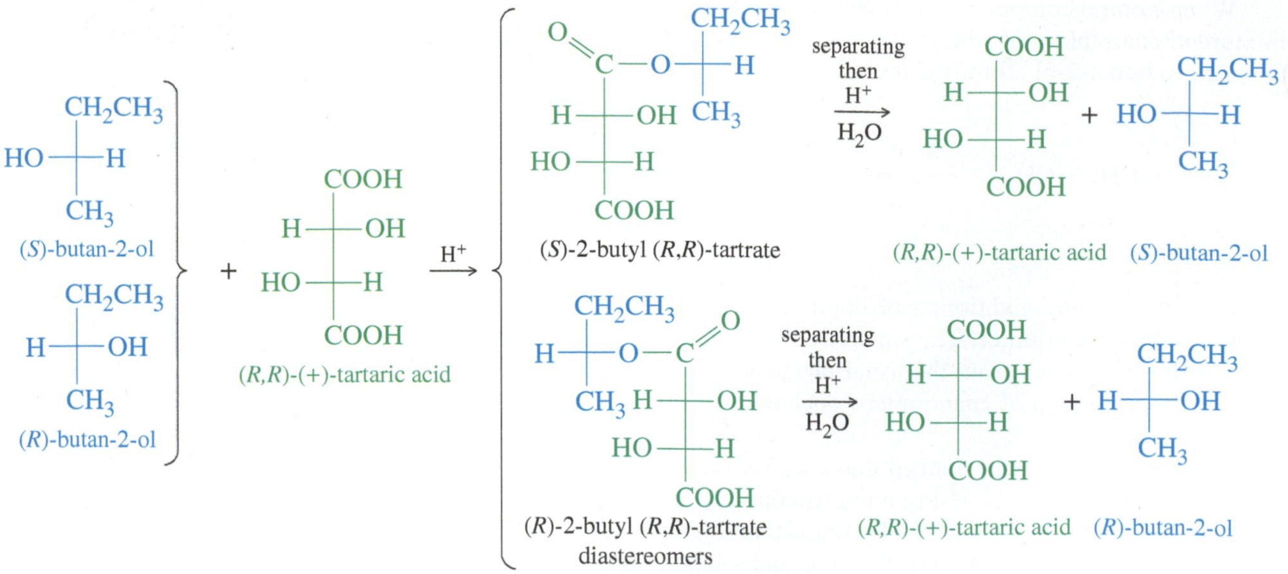

FIGURE 5-22
Formation of (R)- and (S)-2-butyl tartrate. The reaction of a pure enantiomer of one compound with a racemic mixture of another compound produces a mixture of diastereomers. Separation of the diastereomers, followed by hydrolysis, gives the resolved enantiomers.

enantiomers of butan-2-ol by the reverse of the reaction used to make the ester. Adding an acid catalyst and an excess of water to an ester drives the equilibrium toward the acid and the alcohol:

$$\underset{\text{ester}}{\text{R—C(=O)—O—R}'} + \underset{\text{water}}{\text{H—O—H}} \xrightleftharpoons{(\text{H}^+ \text{ catalyst})} \underset{\text{acid}}{\text{R—C(=O)—OH}} + \underset{\text{alcohol}}{\text{R}'\text{—OH}}$$

Hydrolysis of (R)-butan-2-ol tartrate gives (R)-butan-2-ol and (+)-tartaric acid, and hydrolysis of (S)-butan-2-ol tartrate gives (S)-butan-2-ol and (+)-tartaric acid. The recovered tartaric acid would probably be thrown away, because it is cheap and nontoxic. Many other chiral resolving agents are expensive, so they must be carefully recovered and recycled.

PROBLEM 5-24

To show that (R)-2-butyl (R,R)-tartrate and (S)-2-butyl (R,R)-tartrate are not enantiomers, draw and name the mirror images of these compounds.

5-16B Chromatographic Resolution of Enantiomers

Chromatography is a powerful method for separating compounds. One type of chromatography involves passing a solution through a column containing particles whose surface tends to adsorb organic compounds. Compounds that are adsorbed strongly spend more time on the stationary particles; they come off the column later than less strongly adsorbed compounds, which spend more time in the mobile solvent phase.

In some cases, enantiomers may be resolved by passing the racemic mixture through a column containing particles whose surface is coated with chiral molecules (Figure 5-23). As the solution passes through the column, the enantiomers form weak complexes, usually through hydrogen bonding, with the chiral column packing.

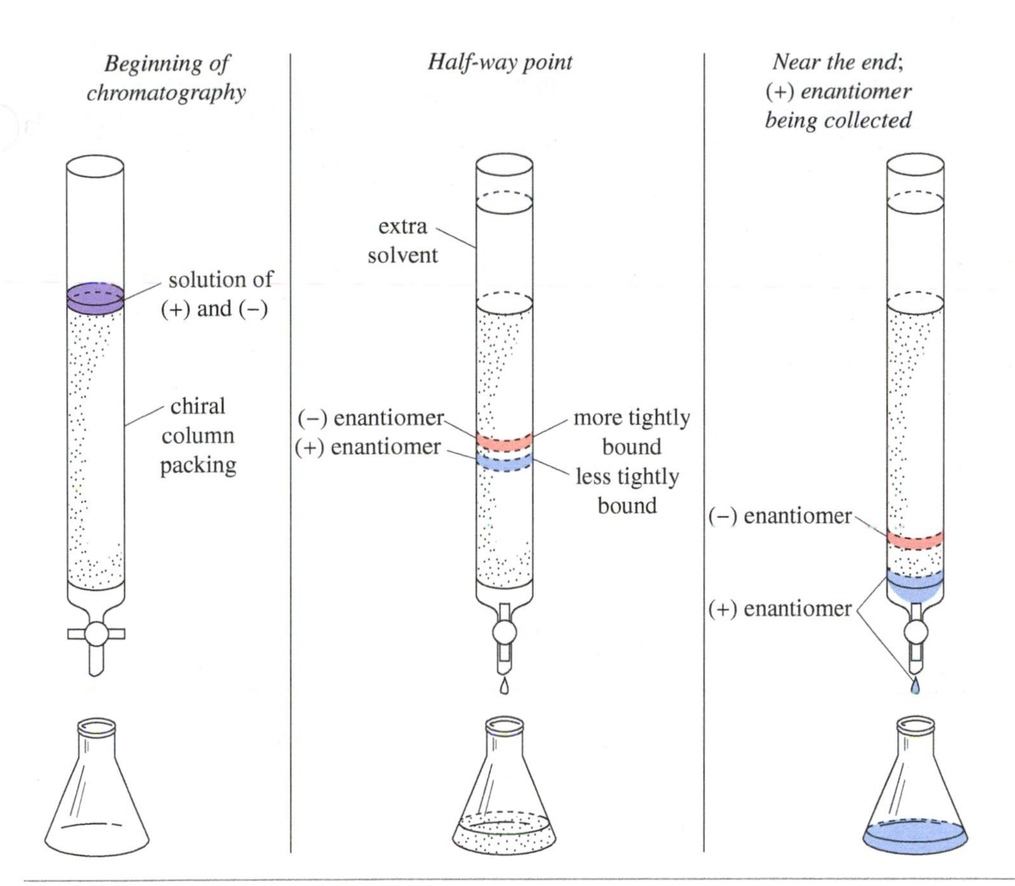

FIGURE 5-23
Chromatographic resolution of enantiomers. The enantiomers of the racemic compound form diastereomeric complexes with the chiral material on the column packing. One of the enantiomers binds more tightly than the other, so it moves more slowly through the column.

The solvent flows continually through the column, and the dissolved enantiomers gradually move along, retarded by the time they spend complexed with the column packing.

The special feature of this chromatography is the fact that the enantiomers form diastereomeric complexes with the chiral column packing. These diastereomeric complexes have different physical properties. They also have different binding energies and different equilibrium constants for complexation. One of the two enantiomers will spend more time complexed with the chiral column packing. The more strongly complexed enantiomer passes through the column more slowly and emerges from the column after the faster-moving (more weakly complexed) enantiomer.

Application: Biochemistry

Enzymes can also be used to eliminate an undesired stereoisomer. The enzyme will process only one isomer in a racemic mixture and leave the other stereoisomer untouched.

Essential Terms

absolute configuration The detailed stereochemical picture of a molecule, including how the atoms are arranged in space. Alternatively, the (*R*) or (*S*) configuration at each asymmetric carbon atom. (p. 235)

achiral Not chiral. (p. 202)

allenes Compounds having two C=C double bonds that meet at a single carbon atom, C=C=C. The two outer carbon atoms are trigonal planar, with their planes perpendicular to each other. Many substituted allenes are chiral. (p. 225)

asymmetric carbon atom **(chiral carbon atom)** A carbon atom that is bonded to four different groups. (p. 204)

Cahn–Ingold–Prelog convention The accepted method for designating the absolute configuration of a chiral center (usually an asymmetric carbon) as either (*R*) or (*S*). (p. 209)

chiral Different from its mirror image. (p. 202)

chiral carbon atom **(asymmetric carbon atom)** A carbon atom that is bonded to four different groups. (p. 204)

chiral center **(chirality center)** The IUPAC term for an atom holding a set of ligands in a spatial arrangement that is not superimposable on its mirror image. Asymmetric carbon atoms are the most common chiral centers. (p. 205)

chiral probe A molecule or an object that is chiral and can use its own chirality to differentiate between mirror images. (p. 218)

cis On the same side of a ring or double bond. (p. 230)

cis-trans isomers **(geometric isomers)** Isomers that differ in their geometric arrangement on a ring or double bond; cis-trans isomers are a subclass of diastereomers. (p. 232)

configurational isomers (see **stereoisomers**)

configurations The two possible spatial arrangements around a chiral center or other stereocenter. (p. 209)

conformations Structures that differ only by rotations about single bonds. In most cases, conformations interconvert at room temperature; thus, they are not different compounds and not true isomers. (p. 221)

constitutional isomers **(structural isomers)** Isomers that differ in the order in which their atoms are bonded together. (p. 201)

D–L configurations **(Fischer–Rosanoff convention)** D has the same relative configuration as (+)-glyceraldehyde. L has the same relative configuration as (−)-glyceraldehyde. (p. 236)

dextrorotatory, (+), or (*d*) Rotating the plane of polarized light clockwise. (p. 215)

diastereomers Stereoisomers that are not mirror images. (p. 230)

enantiomeric excess (e.e.) The excess of one enantiomer in a mixture of enantiomers expressed as a percentage of the mixture. Similar to optical purity. (p. 220) Algebraically,

$$\text{e.e.} = \frac{|R - S|}{R + S} \times 100\%$$

enantiomers A pair of nonsuperimposable mirror-image molecules: mirror-image isomers. (p. 204)

Fischer projection A method for drawing an asymmetric carbon atom as a cross. The carbon chain is kept along the vertical, with the IUPAC numbering from top to bottom. Vertical bonds project away from the viewer, and horizontal bonds project toward the viewer. (p. 226)

geometric isomers (see **cis-trans isomers**) (p. 230)

internal mirror plane (σ) A plane of symmetry through the middle of a molecule, dividing the molecule into two mirror-image halves. A molecule with an internal mirror plane of symmetry cannot be chiral. (p. 207)

isomers Different compounds with the same molecular formula. (p. 232)

Leftorium A store that sells the enantiomers of everyday chiral objects such as scissors, rifles, can openers, etc. (p. 202)

levorotatory, (−), or (*l*) Rotating the plane of polarized light counterclockwise. (p. 216)

meso compound An achiral compound that contains chiral centers (usually asymmetric carbon atoms). Originally, an achiral compound that has chiral diastereomers. (p. 233)

optical activity Rotation of the plane of polarized light. (p. 214)

optically active Capable of rotating the plane of polarized light. (p. 214)

optical isomers (archaic; see **enantiomers)** Compounds with identical properties except for the direction in which they rotate polarized light. (p. 215)

optical purity (o.p.) The specific rotation of a mixture of two enantiomers, expressed as a percentage of the specific rotation of one of the pure enantiomers. Similar to enantiomeric excess. (p. 220) Algebraically,

$$\text{o.p.} = \frac{\text{observed rotation}}{\text{rotation of pure enantiomer}} \times 100\%$$

plane-polarized light Light composed of waves that vibrate in only one plane. (p. 213)

polarimeter An instrument that measures the rotation of plane-polarized light by an optically active compound. (p. 215)

racemic mixture **[racemate, racemic modification, (±) pair, (*d,l*) pair]** A mixture of equal quantities of enantiomers, such that the mixture is optically inactive. (p. 219)

relative configuration The experimentally determined relationship between the configurations of two molecules, even though the absolute configuration of either may not be known. (p. 236)

resolution The process of separating a racemic mixture into the pure enantiomers. Resolution requires a chiral resolving agent. (p. 239)

resolving agent A chiral compound (or chiral material on a chromatographic column) used for separating enantiomers. (p. 239)

2^n rule A molecule with *n* chiral centers might have as many as 2^n stereoisomers. (p. 233)

specific rotation A measure of a compound's ability to rotate the plane of polarized light, given by

$$[\alpha]_D^{25} = \frac{\alpha(\text{observed})}{c \cdot l}$$

where *c* is concentration in g/mL and *l* is length of sample cell (path length) in decimeters. (p. 216)

stereocenter **(stereogenic atom)** An atom that gives rise to stereoisomers when its groups are interchanged. Asymmetric carbon atoms and double-bonded carbons in cis-trans alkenes are the most common stereocenters. (p. 205)

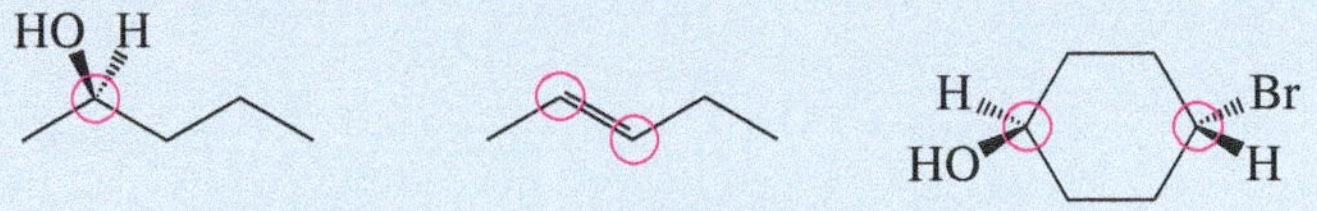

examples of stereocenters (circled)

stereochemistry The study of the three-dimensional structure of molecules. (p. 201)

stereogenic Giving rise to stereoisomers. Characteristic of an atom or a group of atoms such that interchanging any two groups creates a stereoisomer. (p. 205)

stereoisomers **(configurational isomers)** Isomers whose atoms are bonded together in the same order but differ in how the atoms are oriented in space. (p. 201)

structural isomers (see **constitutional isomers**) Isomers that differ in the order in which their atoms are bonded together. (p. 201)

superimposable **(congruent)** Identical in all respects. The three-dimensional positions of all atoms coincide when the molecules are placed on top of each other. (p. 203)

trans On opposite sides of a ring or double bond. (p. 230)

Essential Problem-Solving Skills in Chapter 5

Each skill is followed by problem numbers exemplifying that particular skill.

1 Draw all the stereoisomers of a given structure, and identify the relationships between the stereoisomers.	Problems 5-27, 30, 35, and 37
2 Classify molecules as chiral or achiral, and draw their mirror images. Identify and draw any mirror planes of symmetry.	Problems 5-26, 27, 31, 35, and 37
3 Identify asymmetric carbon atoms and other chiral centers and name them using the (*R*) and (*S*) nomenclature.	Problems 5-25, 26, 27, 36, 37, and 38
4 Calculate specific rotations from polarimetry data, and use specific rotations to determine the optical purity and the enantiomeric excess of mixtures.	Problems 5-32, 33, and 34
5 Use Fischer projections to represent the stereochemistry of compounds with one or more asymmetric carbon atoms.	Problems 5-28, 29, 30, and 37
6 Identify pairs of enantiomers, diastereomers, and meso compounds, and explain how they differ in their physical and chemical properties.	Problems 5-30, 31, and 36
7 Explain how different types of stereoisomers can be separated.	Problems 5-30 and 36

Study Problems

5-25 The following four structures are naturally occurring optically active compounds. Star (*) the asymmetric carbon atoms in these structures.

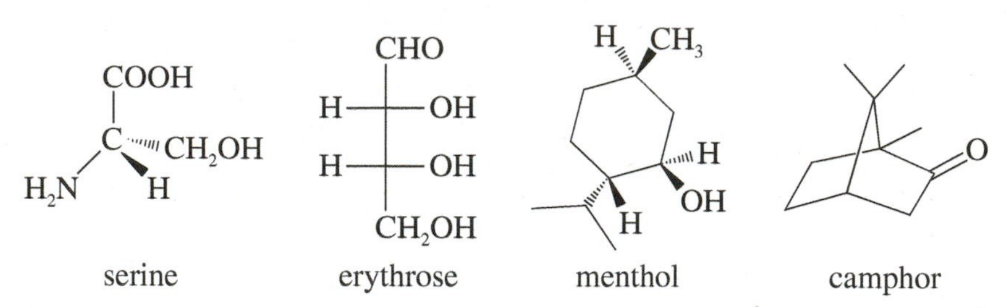

5-26 For each structure,

1. star (*) any asymmetric carbon atoms.
2. label each asymmetric carbon as (*R*) or (*S*).
3. draw any internal mirror planes of symmetry.
4. label the structure as chiral or achiral.
5. label any meso structures.

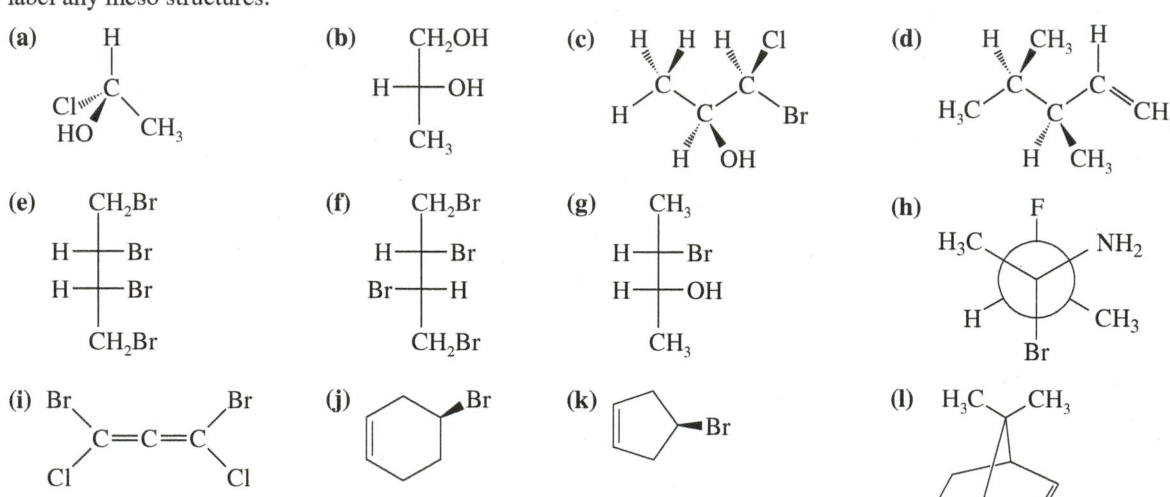

5-27 For each of the compounds described by the following names,

1. draw a three-dimensional representation.
2. star (*) each chiral center.
3. draw any planes of symmetry.
4. draw any enantiomer.
5. draw any diastereomers.
6. label each structure you have drawn as chiral or achiral.

 (a) (*S*)-2-chlorobutane
 (b) (*R*)-1,1,2-trimethylcyclohexane
 (c) (2*R*,3*S*)-2,3-dibromohexane
 (d) (1*R*,2*R*)-1,2-dibromocyclohexane
 (e) *meso*-hexane-3,4-diol, $CH_3CH_2CH(OH)CH(OH)CH_2CH_3$
 (f) (±)-hexane-3,4-diol

5-28 Convert the following perspective formulas to Fischer projections.

(a) H, OH, C, CH_3, CH_2OH **(b)** CH_3, C, Br, H, CHO **(c)** Br, H, C—C, H, OH, $HOCH_2$, CH_3 **(d)** H, OH, C, CH_3, $HOCH_2$, C, H, OH

5-29 Convert the following Fischer projections to perspective formulas.

(a) COOH; H_2N—H; CH_3 **(b)** CHO; H—OH; CH_2OH **(c)** CH_2OH; Br—Cl; CH_3 **(d)** CH_2OH; H—Br; H—Cl; CH_3

5-30 Give the stereochemical relationships between each pair of structures. Examples are same compound, structural isomers, enantiomers, diastereomers. Which pairs could you (theoretically) separate by distillation or recrystallization?

(a) **(b)** 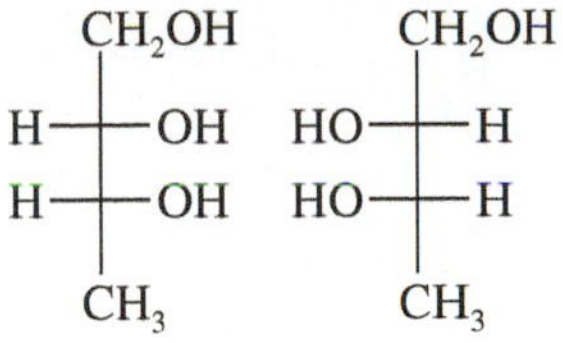 **(c)** 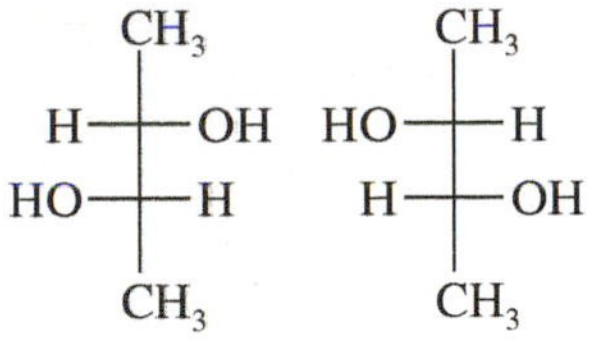 **(d)** 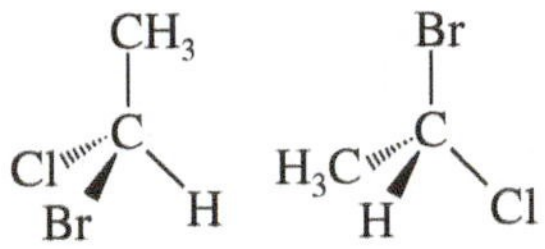 **(e)** 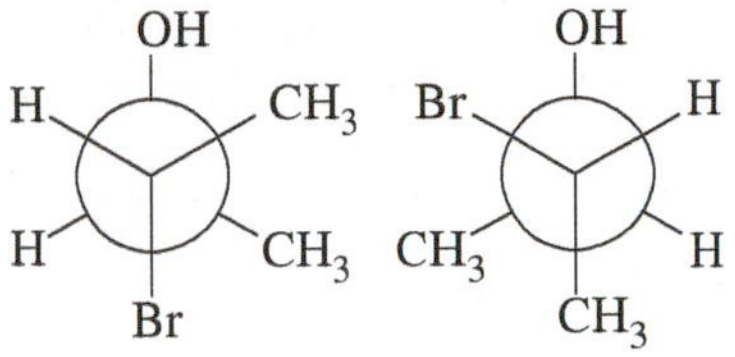 **(f)** 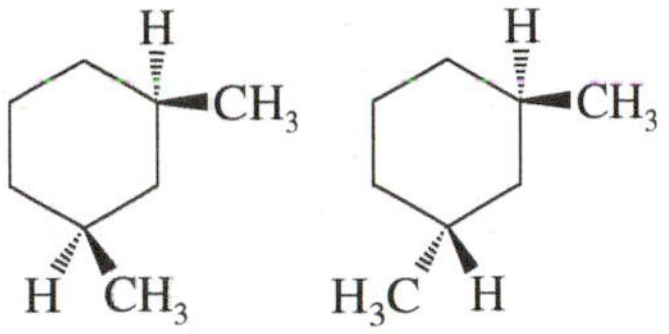 **(g)** 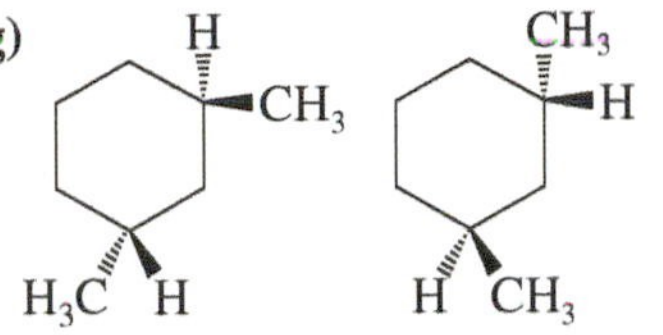 ***(h)** 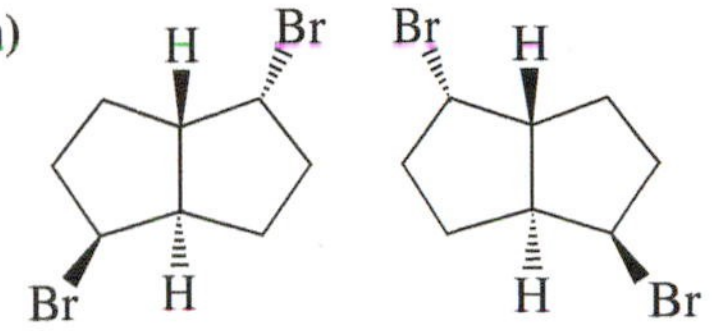 **(i)**

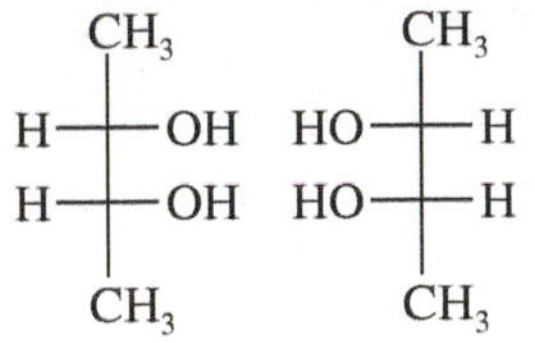

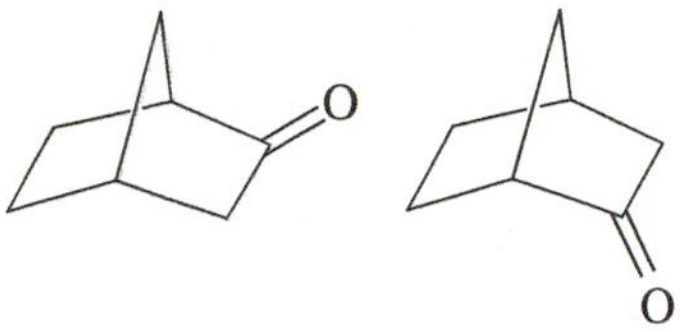

5-31 Draw the enantiomer, if any, for each structure.

(a) **(b)** 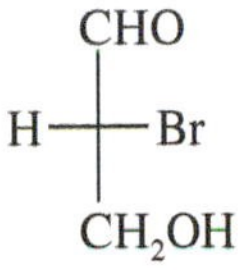 **(c)** 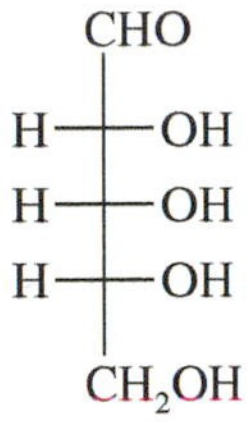 **(d)** 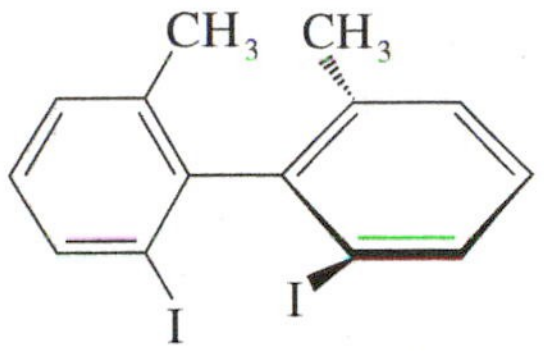 **(e)** 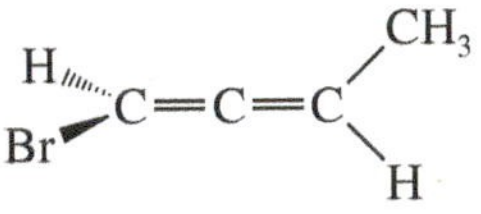 **(f)** 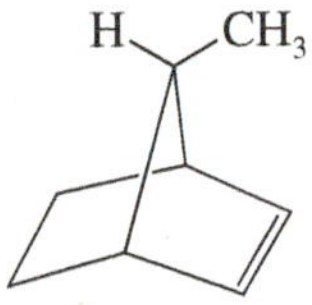 **(g)** 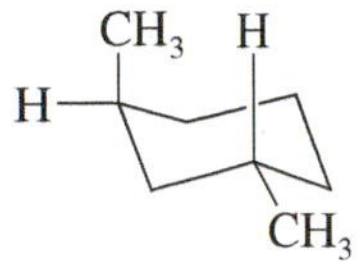 **(h)**

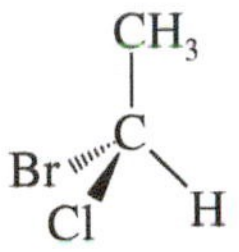

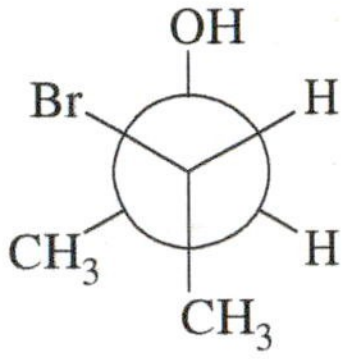

5-32 Calculate the specific rotations of the following samples taken at 25 °C using the sodium D line.

(a) 1.00 g of sample is dissolved in 20.0 mL of ethanol. Then 5.00 mL of this solution is placed in a 20.0-cm polarimeter tube. The observed rotation is 1.25° counterclockwise.

(b) 0.050 g of sample is dissolved in 2.0 mL of ethanol, and this solution is placed in a 2.0-cm polarimeter tube. The observed rotation is clockwise 0.043°.

5-33 (+)-Tartaric acid has a specific rotation of +12.0°. Calculate the specific rotation of a mixture of 68% (+)-tartaric acid and 32% (−)-tartaric acid.

5-34 The specific rotation of (*S*)-2-iodobutane is +15.90°.

(a) Draw the structure of (*S*)-2-iodobutane.

(b) Predict the specific rotation of (*R*)-2-iodobutane.

(c) Determine the percentage composition of a mixture of (*R*)- and (*S*)-2-iodobutane with a specific rotation of −7.95°.

5-35 For each structure,

1. draw all the stereoisomers.
2. label each structure as chiral or achiral.
3. give the relationships between the stereoisomers (enantiomers, diastereomers).

(a) ***(b)** 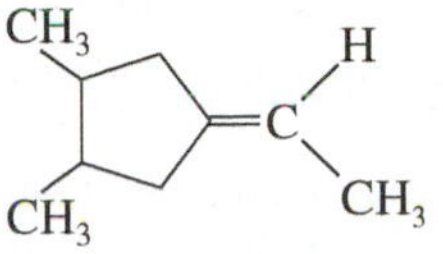 ****(c)**

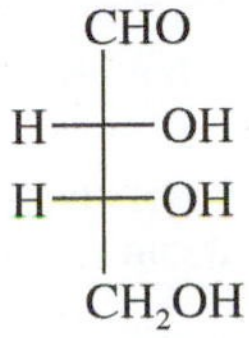

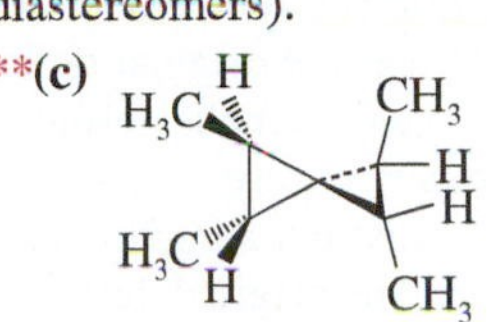

***5-36** Free-radical bromination of the following compound introduces bromine primarily at the benzylic position next to the aromatic ring. If the reaction stops at the monobromination stage, two stereoisomers result.

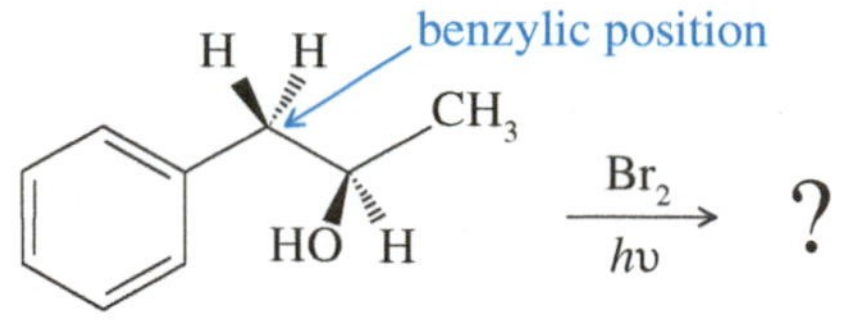

H H benzylic position CH_3 HO H $\xrightarrow[h\nu]{Br_2}$?

(a) Propose a mechanism to show why free-radical halogenation occurs almost exclusively at the benzylic position.
(b) Draw the two stereoisomers that result from monobromination at the benzylic position.
(c) Assign *R* and *S* configurations to the asymmetric carbon atoms in the products.
(d) What is the relationship between the two isomeric products?
(e) Will these two products be produced in identical amounts? That is, will the product mixture be exactly 50:50?
(f) Will these two stereoisomers have identical physical properties such as boiling point, melting point, solubility, etc.? Could they be separated (theoretically, at least) by distillation or recrystallization?

***5-37** If you think you know your definitions, try this difficult problem.
(a) Draw all the stereoisomers of 2,3,4-tribromopentane. (Using Fischer projections may be helpful.) You should find two meso structures and one pair of enantiomers.
(b) Star (*) the asymmetric carbon atoms, and label each as (*R*) or (*S*).
(c) In the meso structures, show how C3 is not asymmetric, nor is it a chiral center, yet it is a stereocenter.
(d) In the enantiomers, show how C3 is not a stereocenter.

***5-38** 3,4-Dimethylpent-1-ene has the formula $CH_2{=}CH{-}CH(CH_3){-}CH(CH_3)_2$. When pure (*R*)-3,4-dimethylpent-1-ene is treated with hydrogen over a platinum catalyst, the product is (*S*)-2,3-dimethylpentane.
(a) Draw the equation for this reaction. Show the stereochemistry of the reactant and the product.
(b) Has the chiral center retained its configuration during this hydrogenation, or has it been inverted?
(c) The reactant is named (*R*), but the product is named (*S*). Does this name change imply a change in the spatial arrangement of the groups around the chiral center? So why does the name switch from (*R*) to (*S*)?
(d) How useful is the (*R*) or (*S*) designation for predicting the sign of an optical rotation? Can you predict the sign of the rotation of the reactant? Of the product? (*Hint* from Juliet Capulet: "What's in a name? That which we call a rose/By any other name would smell as sweet.")

***5-39** A graduate student was studying enzymatic reductions of cyclohexanones when she encountered some interesting chemistry. When she used an enzyme and NADPH to reduce the following ketone, she was surprised to find that the product was optically active. She carefully repurified the product so that no enzyme, NADPH, or other contaminants were present. Still, the product was optically active.

O= COOH H $\xrightarrow[\text{enzyme}]{\text{NADPH}}$ HO H COOH H

optically active

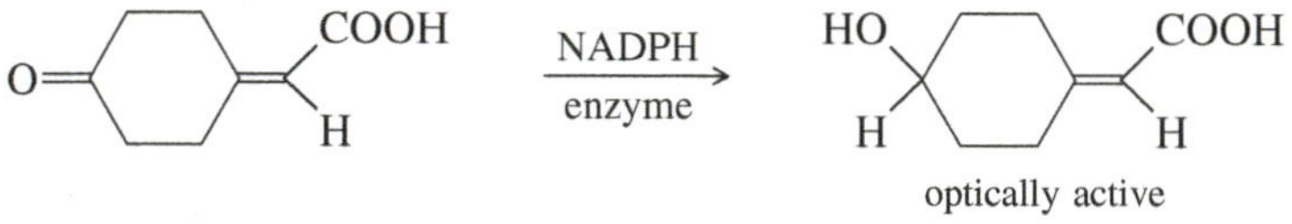

(a) Does the product have any asymmetric carbon atoms or other stereocenters?
(b) Is the product capable of showing optical activity? If it is, explain how.
(c) If this reaction could be accomplished using H_2 and a nickel catalyst, would the product be optically active? Explain.

***5-40** D-(−)-Erythrose has the formula $HOCH_2{-}CH(OH){-}CH(OH){-}CHO$, and the D in its name implies that it can be degraded to D-(+)-glyceraldehyde. The (−) in its name implies that D-(−)-erythrose is optically active (levorotatory). When D-(−)-erythrose is reduced (using H_2 and a nickel catalyst), it gives an optically inactive product of formula $HOCH_2{-}CH(OH){-}CH(OH){-}CH_2OH$. Knowing the absolute configuration of D-(+)-glyceraldehyde (Section 5-14), determine the absolute configuration of D-(−)-erythrose.

***5-41** The original definition of *meso* is "an achiral compound that has chiral diastereomers." Our working definition of *meso* is "an achiral compound that has chiral centers (usually asymmetric carbon atoms)." The working definition is much easier to apply, because we don't have to envision all possible chiral diastereomers of the compound. Still, the working definition is not quite as complete as the original definition.
(a) Show how *cis*-cyclooctene is defined as a *meso* compound under the original definition, but not under our working definition. (Review Figure 5-19.)
(b) See if you can construct a double allene that is achiral, although it has chiral diastereomers, and is therefore a *meso* compound under the original definition. The allene structure is not a chiral center, but it can be a *chiral axis*.

6 Alkyl Halides; Nucleophilic Substitution

cymbal bubble snail
Haminoea cymbalum

kumepaloxane

Goals for Chapter 6

1 Name alkyl halides, explain their physical properties, and describe their common uses.

2 Identify what a substitution is and show how it differs from an elimination.

3 Predict the products of substitution reactions.

4 Identify the differences between unimolecular and bimolecular substitutions and explain what factors determine the order of the reaction.

5 Given a set of reaction conditions, identify the possible substitution mechanism(s) and predict which product(s) are most likely.

◀ Halogen-containing organic compounds are not very common in nature. Exceptions occur in ocean plants and invertebrates, where organic halides are common in toxic plants and venoms. The Cymbal bubble snail *Haminoea cymbalum* discharges **kumepaloxane**, a toxic defense chemical containing both chlorine and bromine, when it is disturbed by predatory fishes.

6-1 Introduction

Our study of organic chemistry is organized into families of compounds classified by their functional groups. In this chapter, we consider the properties and reactions of alkyl halides. We use alkyl halides to introduce substitution and elimination, two of the most important types of reactions in organic chemistry. Stereochemistry (Chapter 5) will play a major role in our study of these reactions. Many other reactions show similarities to substitution and elimination, and the techniques introduced in this chapter will be used throughout our study of organic reactions.

There are three major classes of halogenated organic compounds: the alkyl halides, the vinyl halides, and the aryl halides. An **alkyl halide** simply has a halogen atom bonded to one of the sp^3 hybrid carbon atoms of an alkyl group. A **vinyl halide** has a halogen atom bonded to one of the sp^2 hybrid carbon atoms of an alkene. An **aryl halide** has a halogen atom bonded to one of the sp^2 hybrid carbon atoms of an aromatic ring. The chemistry of vinyl halides and aryl halides is different from that of alkyl halides because their bonding and hybridization are different. We consider the reactions of vinyl halides and aryl halides in later chapters. The structures of some representative alkyl halides, vinyl halides, and aryl halides are shown here, with their most common names and uses.

Alkyl halides

$CHCl_3$	$CHClF_2$	$CCl_3—CH_3$	$CF_3—CHClBr$
solvent	refrigerant	cleaning fluid	nonflammable anesthetic
chloroform	Freon-22®	1,1,1-trichloroethane	Halothane

Vinyl halides

$$\text{Cl(H)C=CH}_2$$

vinyl chloride
monomer for poly(vinyl chloride)

$$\text{F}_2\text{C=CF}_2$$

tetrafluoroethylene (TFE)
monomer for Teflon®

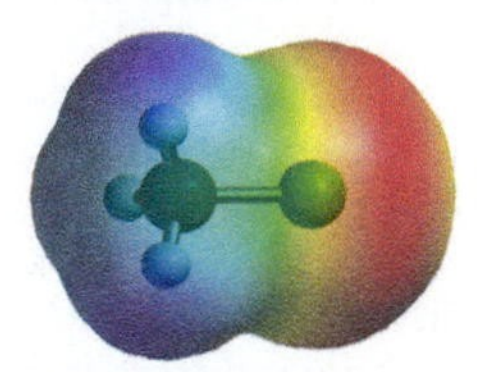

chloromethane

EPM of chloromethane

FIGURE 6-1
Chloromethane and its electrostatic potential map (EPM). The polar C—Cl bond is seen in the EPM as an electron-rich (red) region around chlorine and an electron-poor (blue) region around carbon and the hydrogen atoms.

Aryl halides

para-dichlorobenzene
mothballs

thyroxine
thyroid hormone

The carbon–halogen bond in an alkyl halide is polar because halogen atoms are more electronegative than carbon atoms. Most reactions of alkyl halides result from breaking this polarized bond. The electrostatic potential map of chloromethane (Figure 6-1) shows higher electron density (red) around the chlorine atom and relatively low electron density (blue) around the carbon and hydrogen atoms. The carbon atom has a partial positive charge, making it somewhat electrophilic. A nucleophile can attack this electrophilic carbon, and the halogen atom can leave as a halide ion, taking the bonding pair of electrons with it. By serving as a leaving group, the halogen can be eliminated from the alkyl halide, or it can be replaced (substituted for) by a wide variety of functional groups. This versatility allows alkyl halides to serve as intermediates in the synthesis of many other functional groups.

PROBLEM 6-1

Classify each compound as an alkyl halide, a vinyl halide, or an aryl halide.

(a) $CH_3CHCFCH_3$ **(b)** $(CH_3)_3CBr$ **(c)** CH_3CCl_3

(d) bromocyclohexane **(e)** 1-bromocyclohexene **(f)** a PCB (polychlorinated biphenyl)

6-2 Nomenclature of Alkyl Halides

There are two ways of naming alkyl halides. The systematic (IUPAC) nomenclature treats an alkyl halide as an alkane with a *halo-* substituent: Fluorine is *fluoro-*, chlorine is *chloro-*, bromine is *bromo-*, and iodine is *iodo-*. The result is a systematic **haloalkane** name, as in 1-chlorobutane and 2-bromopropane. Common or "trivial" names are constructed by naming the alkyl group and then the halide, as in "isopropyl bromide." This is the origin of the term *alkyl halide*. Common names are useful only for simple alkyl halides such as the following:

	$CH_3CH_2—F$	$CH_2(Cl)CH_2CH_2CH_3$	$CH_3—CH(Br)—CH_3$
IUPAC name:	fluoroethane	1-chlorobutane	2-bromopropane
common name:	ethyl fluoride	*n*-butyl chloride	isopropyl bromide

	(iodocyclohexane structure)	(cyclopentane structure)
IUPAC name:	iodocyclohexane	*trans*-1-chloro-3-methylcyclopentane
common name:	cyclohexyl iodide	(none)

	$CH_3CH_2—CH(CH_2—I)—CH_2CH_3$	$CH_3CH_2CH_2—CH(CH_2CH_2—F)—CH_2CH_2CH_3$
IUPAC name:	3-(iodomethyl)pentane	4-(2-fluoroethyl)heptane

PROBLEM-SOLVING HINT

A summary of organic nomenclature is given in Appendix 5.

Some of the halomethanes have acquired common names that are not clearly related to their structures. A compound of formula CH_2X_2 (a methylene group with two halogens) is called a *methylene halide*, a compound of formula CHX_3 is called a *haloform*, and a compound of formula CX_4 is called a *carbon tetrahalide*.

	CH_2Cl_2	$CHCl_3$	CCl_4
IUPAC name:	dichloromethane	trichloromethane	tetrachloromethane
common name:	methylene chloride	chloroform	carbon tetrachloride

PROBLEM 6-2

Give the structures of the following compounds.

(a) methylene iodide
(b) carbon tetrabromide
(c) 3-bromo-2-methylpentane
(d) iodoform
(e) 2-bromo-3-ethyl-2-methylhexane
(f) isobutyl bromide
(g) *cis*-1-fluoro-3-(fluoromethyl)cyclohexane
(h) *tert*-butyl chloride

Alkyl halides are classified according to the nature of the carbon atom bonded to the halogen. If the halogen-bearing carbon is bonded to one carbon atom, it is primary (1°) and the alkyl halide is a **primary halide.** If two carbon atoms are bonded to the halogen-bearing carbon, it is secondary (2°) and the compound is a **secondary halide.** A **tertiary halide** (3°) has three other carbon atoms bonded to the halogen-bearing carbon atom. If the halogen-bearing carbon atom is a methyl group (bonded to no other carbon atoms), the compound is a *methyl halide*.

$CH_3—X$	$R—CH_2—X$	$R—CH(R)—X$	$R—C(R)(R)—X$
methyl halide	primary (1°) halide	secondary (2°) halide	tertiary (3°) halide

Examples

	$CH_3—Br$	$CH_3CH_2CH_2—F$ (1°)	$CH_3—CH(I)—CH_2CH_3$ (2°)	$(CH_3)_3C—Cl$ (3°)
IUPAC name:	bromomethane	1-fluoropropane	2-iodobutane	2-chloro-2-methylpropane
common name:	methyl bromide	*n*-propyl fluoride	*sec*-butyl iodide	*tert*-butyl chloride

A **geminal dihalide** (Latin, *geminus*, "twin") has the two halogen atoms bonded to the same carbon atom. A **vicinal dihalide** (Latin, *vicinus*, "neighboring") has the two halogens bonded to adjacent carbon atoms.

$Br_2C<$	$—C(Cl)—C(Cl)—$
a geminal dibromide	a vicinal dichloride

PROBLEM 6-3

For each of the following compounds,

1. give the IUPAC name.
2. give the common name (if possible).
3. classify the compound as a methyl, primary, secondary, or tertiary halide.

(continued)

(a) $(CH_3)_2CHCH_2Cl$

(b) $(CH_3)_3CBr$

(c) 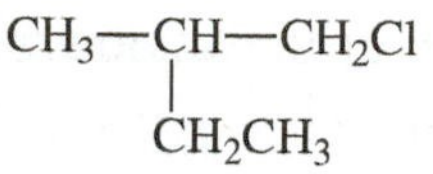 $CH_3-CH(CH_2CH_3)-CH_2Cl$

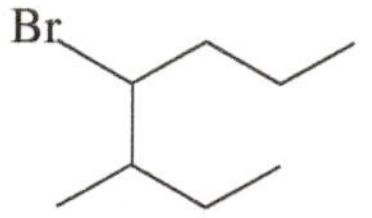

(d) 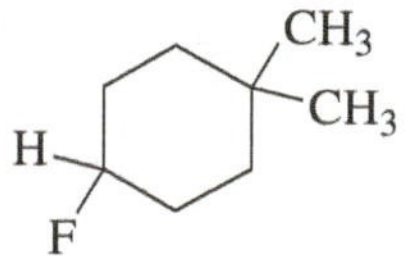 [cyclohexane ring bearing CH_3, CH_3 on one carbon and H, F on the opposite carbon]

(e) [skeletal structure with Br]

(f) [cyclobutane with Br, H, Cl, H wedge/dash substituents]

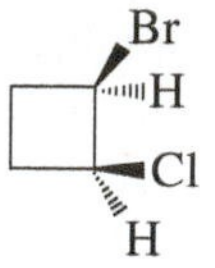

6-3 Common Uses of Alkyl Halides

6-3A Solvents

Alkyl halides are used primarily as industrial and household solvents. Carbon tetrachloride (CCl_4) was once used for dry cleaning, spot removing, and other domestic cleaning. Carbon tetrachloride is toxic and carcinogenic (causes cancer), however, so dry cleaners now use 1,1,1-trichloroethane and other solvents instead.

Methylene chloride (CH_2Cl_2) and chloroform ($CHCl_3$) are also good solvents for cleaning and degreasing work. Methylene chloride was once used to dissolve the caffeine from coffee beans to produce decaffeinated coffee. Concerns about the safety of coffee with residual traces of methylene chloride prompted coffee producers to use supercritical carbon dioxide instead. Chloroform is more toxic and carcinogenic than methylene chloride; it has been replaced by methylene chloride and other solvents in most industrial degreasers and paint removers.

Even the safest halogenated solvents, such as methylene chloride and 1,1,1-trichloroethane, should be used carefully. They are all potentially toxic and carcinogenic, and they dissolve the fatty oils that protect skin, causing a form of dermatitis.

Application: FYI

Chloroform can be converted to phosgene, a very reactive acid chloride, in the presence of oxygen at room temperature. Phosgene is extremely toxic because it reacts with and deactivates many biological molecules. A small amount of alcohol is sometimes added to bottles of chloroform to destroy any phosgene that might be formed.

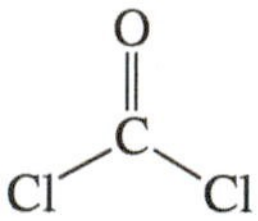

$Cl-C(=O)-Cl$

phosgene

6-3B Reagents

Many syntheses use alkyl halides as starting materials for making more complex molecules. The conversion of alkyl halides to organometallic reagents (compounds containing carbon–metal bonds) is a particularly important tool for organic synthesis. We discuss the formation of organometallic compounds in Section 10-8.

6-3C Anesthetics

In the 1840s, chloroform ($CHCl_3$) was found to produce general anesthesia, opening new possibilities for careful surgery with a patient who is unconscious and relaxed. Chloroform is toxic and carcinogenic, however, and it was soon abandoned in favor of safer anesthetics such as diethyl ether. A less toxic halogenated anesthetic is a mixed alkyl halide, $CF_3CHClBr$, which goes by the trade name Halothane. Ethyl chloride is often used as a topical anesthetic for minor procedures. When sprayed on the skin, its evaporation (bp 12 °C) cools the area and enhances the numbing effect.

Chemists have developed other alkyl halides and haloethers that are much less toxic than chloroform, and they are still among the most common anesthetics. Some of the recent ones are Isoflurane, Desflurane, and Sevoflurane. Their structures are compared with Halothane below.

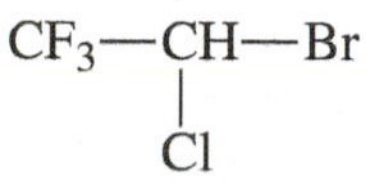

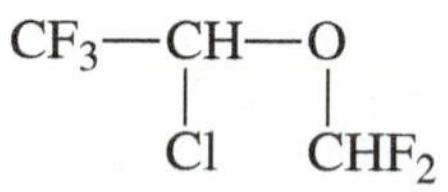

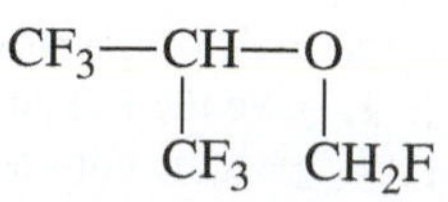

$CF_3-CH(Cl)-Br$ halothane

$CF_3-CH(Cl)-O-CHF_2$ isoflurane

$CF_3-CH(F)-O-CHF_2$ desflurane

$CF_3-CH(CF_3)-O-CH_2F$ sevoflurane

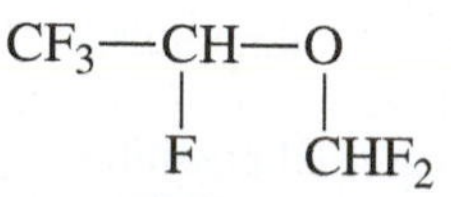

6-3D Freons: Refrigerants and Foaming Agents

The **freons** (also called *chlorofluorocarbons*, or CFCs) are fluorinated haloalkanes that were developed to replace ammonia as a refrigerant gas. Ammonia is toxic, and leaking refrigerators often killed people who were working or sleeping nearby. Freon-12®, CF_2Cl_2, was at one time the most widely used refrigerant. Low-boiling freons (such as Freon-11®, CCl_3F) were once used as *foaming agents* that were added to a plastic to vaporize and form a froth that hardens into a plastic foam.

Freons released into the atmosphere began to destroy Earth's protective ozone layer. CFCs gradually diffuse up into the stratosphere, where the chlorine atoms catalyze the decomposition of ozone (O_3) into oxygen (O_2). Most scientists blame the freon-catalyzed depletion of ozone for the "hole" in the ozone layer that has been detected over the South Pole. Paul Crutzen, Mario Molina, and F. Sherwood Rowland received the 1995 Nobel Prize in Chemistry for their work showing how UV light forms the ozone layer in the atmosphere and how chlorine radicals catalyze its destruction. The opening page of Chapter 4 shows the extent of the ozone hole over the South Pole.

International treaties have limited the future production and use of the ozone-destroying freons. Freon-12 has been replaced in aerosol cans by low-boiling hydrocarbons or carbon dioxide. In refrigerators and automotive air conditioners, Freon-12 has been replaced by Freon-22®, $CHClF_2$. Freons with C—H bonds (such as Freon-22), called HCFCs, are generally destroyed at lower altitudes before they reach the stratosphere. Propane, CO_2, and HCFC-123 ($CHCl_2CF_3$) are used as substitutes for Freon-11 in making plastic foams. Freon-134a (1,1,1,2-tetrafluoroethane), used in automotive air conditioners, contains no chlorine.

Application: Environment

Halogenated compounds persist in the environment because they resist breakdown by soil bacteria. Many are chemically unreactive and insoluble in water, hindering bacterial degradation. Nevertheless, there are strains of bacteria that can use halogenated compounds as their source of food.

6-3E Pesticides

Alkyl halides have contributed to human health through their use as insecticides. Since antiquity, people have died from famine and disease caused or carried by mosquitoes, fleas, lice, and other vermin. The "Black Death" of the Middle Ages wiped out nearly a third of the population of Europe through infection by the flea-borne bubonic plague. Whole regions of Africa and tropical America were uninhabited and unexplored because people could not survive insect-borne diseases such as malaria, yellow fever, and sleeping sickness.

Arsenic compounds, nicotine, and other crude insecticides were developed in the nineteenth century, but these compounds are just as toxic to birds, animals, and people as they are to insects. Their use is extremely hazardous, but a hazardous insecticide was still preferable to certain death by disease or starvation.

The war against insects changed dramatically in 1939 with the discovery of DDT (Figure 6-2). DDT is extremely toxic to insects, but its toxicity in mammals is quite low. About an ounce of DDT is required to kill a person, but that same amount of insecticide protects an acre of land against locusts or mosquitoes. In 1970, the U.S. National Academy of Sciences reported, "in little more than two decades DDT has prevented 500 million deaths due to malaria." Similar advances were made against the mosquitoes carrying yellow fever and the tsetse flies carrying sleeping sickness. Using DDT as a body dust protected people against louse-borne typhus, and dusting rodent burrows controlled the threat of plague.

As with many inventions, DDT showed undesired side effects. It is a long-lasting insecticide, and its residues accumulate in the environment. The widespread use of DDT as an agricultural insecticide led to the development of substantial DDT concentrations in wildlife, causing declines in several species. In 1972, DDT was banned by the U.S. Environmental Protection Agency for use as an agricultural insecticide. It is still used, however, in places where insect-borne diseases threaten human life. DDT-treated bed netting is still the most cost-effective protection against malaria, and careful spraying of DDT around dwellings and in rodent burrows has helped to control the spread of deadly diseases.

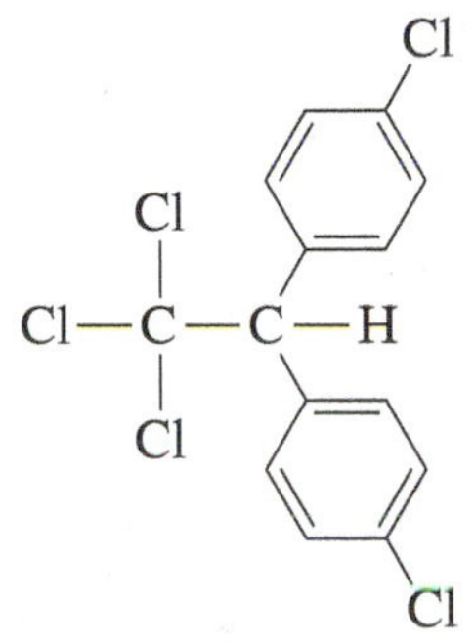

DDT

FIGURE 6-2
Structure of DDT. DDT is DichloroDiphenylTrichloroethane, or 1,1,1-trichloro-2,2-bis-(*p*-chlorophenyl) ethane. DDT was the first chlorinated insecticide. Its use rendered large parts of the world safe from insect-borne disease and starvation, but it accumulated in the environment and nearly caused the extinction of several species of birds.

lindane   kepone   aldrin   chlordane

FIGURE 6-3
Four chlorinated insecticides. These pesticides are rarely used because they accumulate in the environment. Lindane is still used in small quantities to control body lice.

Many other chlorinated insecticides have been developed. Most of them also accumulate in the environment, gradually producing toxic effects in wildlife. Others can be used with little adverse impact if they are applied properly. Because of their persistent toxic effects, chlorinated insecticides are rarely used in agriculture. They are generally used when a potent insecticide is needed to protect life or property. For example, lindane is used in shampoos to kill lice, and chlordane was once used to protect wooden buildings from termites. The structures of some chlorinated insecticides are shown in Figure 6-3.

Body lice transmit a variety of human diseases, including typhus, trench fever, and bubonic plague. Typhus killed millions of people in Europe and Russia during World War II. The epidemic was contained when DDT became readily available for body dusting. Lotions containing lindane (Kwell® and others) are still used as topical treatments for lice.

PROBLEM 6-4

Kepone, aldrin, and chlordane are synthesized from hexachlorocyclopentadiene and other five-membered-ring compounds. Show how these three pesticides are composed of two five-membered rings.

hexachlorocyclopentadiene

6-4 Structure of Alkyl Halides

In an alkyl halide, the halogen atom is bonded to an sp^3 hybrid carbon atom. The halogen is more electronegative than carbon, and the C—X bond is polarized with a partial positive charge on carbon and a partial negative charge on the halogen.

The dipole moment (μ) is given in debyes (D):
$\mu = 4.8 \times \delta \times d$
where δ is the amount of charge separation and d is the bond length.

The electronegativities of the halogens *increase* in the order

	I	<	Br	<	Cl	<	F
electronegativity:	2.7		3.0		3.2		4.0

The carbon–halogen bond lengths *increase* as the halogen atoms become bigger (larger atomic radii) in the order

	C—F	<	C—Cl	<	C—Br	<	C—I
bond length:	1.38 Å		1.78 Å		1.94 Å		2.14 Å

TABLE 6-1
Molecular Dipole Moments of Methyl Halides

X	CH_3X	CH_2X_2	CHX_3	CX_4
F	1.82 D	1.97 D	1.65 D	0
Cl	1.94 D	1.60 D	1.03 D	0
Br	1.79 D	1.45 D	1.02 D	0
I	1.64 D	1.11 D	1.00 D	0

carbon tetrachloride
The bond dipole moment vectors add to zero.

These two effects oppose each other, with the larger halogens having longer bonds but weaker electronegativities. The overall result is that the bond dipole moments increase in the order

	C—I	<	C—Br	<	C—F	<	C—Cl
dipole moment, μ:	1.29 D		1.48 D		1.51 D		1.56 D

A *molecular* dipole moment is the vector sum of the individual bond dipole moments. Molecular dipole moments are not easy to predict because they depend on the bond angles and other factors that vary with the specific molecule. Table 6-1 lists the experimentally measured dipole moments of the halogenated methanes. Notice how the four symmetrically oriented polar bonds of the carbon tetrahalides cancel to give a molecular dipole moment of zero.

PROBLEM 6-5

For each pair of compounds, predict which one has the higher molecular dipole moment, and explain your reasoning.

(a) ethyl chloride or ethyl iodide
(b) 1-bromopropane or cyclopropane
(c) *cis*-2,3-dibromobut-2-ene or *trans*-2,3-dibromobut-2-ene
(d) *cis*-1,2-dichlorocyclobutane or *trans*-1,3-dichlorocyclobutane

6-5 Physical Properties of Alkyl Halides

6-5A Boiling Points

Two types of intermolecular forces influence the boiling points of alkyl halides. The London force (Section 2-2B) is the strongest intermolecular attraction in alkyl halides. London forces are *surface* attractions, resulting from coordinated temporary dipoles. Molecules with larger surface areas have larger London attractions, resulting in higher boiling points. Dipole–dipole attractions (arising from the polar C—X bond) also affect the boiling points, but to a smaller extent.

Molecules with higher molecular weights generally have higher boiling points because they are heavier (and therefore slower moving), and they have greater surface area. The surface areas of the alkyl halides vary with the surface areas of halogens. We can get an idea of the relative surface areas of halogen atoms by considering their van der Waals radii. Figure 6-4 shows that an alkyl fluoride has nearly the same surface area as the corresponding alkane; thus, its London attractive forces are similar. The alkyl fluoride has a larger dipole moment, however, so the total attractive forces are slightly greater in the alkyl fluoride, giving it a higher boiling point. For example, the boiling point of *n*-butane is 0 °C, while that of *n*-butyl fluoride is 33 °C.

The other halogens are considerably larger than fluorine, giving them more surface area and raising the boiling points of their alkyl halides. With a boiling point of 78 °C, *n*-butyl chloride shows the influence of chlorine's much larger surface area. This trend continues with *n*-butyl bromide (bp 102 °C) and *n*-butyl iodide (bp 131 °C). Table 6-2

Halogen	van der Waals Radius (10^{-8} cm)
F	1.35
Cl	1.8
Br	1.95
I	2.15
H (for comparison)	1.2

FIGURE 6-4
Space-filling drawings of the ethyl halides. The heavier halogens are larger, with much greater surface areas. As a result, the boiling points of the ethyl halides increase in the order F < Cl < Br < I.

ethyl fluoride, bp −38 °C
smallest surface area, lowest bp

ethyl chloride, bp 12 °C

ethyl bromide, bp 38 °C

ethyl iodide, bp 72 °C
largest surface area, highest bp

lists the boiling points and densities of some simple alkyl halides. Notice that compounds with branched, more spherical shapes have lower boiling points as a result of their smaller surface areas. For example, *n*-butyl bromide has a boiling point of 102 °C, while the more spherical *tert*-butyl bromide has a boiling point of only 73 °C. This effect is similar to the one we saw with alkanes.

PROBLEM 6-6

For each pair of compounds, predict which compound has the higher boiling point. Check Table 6-2 to see if your prediction was right; then explain why that compound has the higher boiling point.
(a) isopropyl bromide and *n*-butyl bromide
(b) isopropyl chloride and *tert*-butyl bromide
(c) 1-bromobutane and 1-chlorobutane

6-5B Densities

Table 6-2 also lists the densities of common alkyl halides. Like their boiling points, their densities follow a predictable trend. Alkyl fluorides and alkyl chlorides (those with just one chlorine atom) are less dense than water (1.00 g/mL). Alkyl chlorides with two or more chlorine atoms are denser than water, and all alkyl bromides and alkyl iodides are denser than water.

PROBLEM 6-7

When water is shaken with hexane, the two liquids separate into two phases. Which compound is present in the top phase, and which is present in the bottom phase? When water is shaken with chloroform, a similar two-phase system results. Again, which compound is present in each phase? Explain the difference in the two experiments. What do you expect to happen when water is shaken with ethanol (CH_3CH_2OH)?

TABLE 6-2
Physical Properties of Alkyl Halides

Compound	Molecular Weight	Boiling Point (°C)	Density (g/mL)
$CH_3—F$	34	−78	
$CH_3—Cl$	50.5	−24	0.92
$CH_3—Br$	95	4	1.68
$CH_3—I$	142	42	2.28
CH_2Cl_2	85	40	1.34
$CHCl_3$	119	61	1.50
CCl_4	154	77	1.60
$CH_3CH_2—F$	48	−38	0.72
$CH_3CH_2—Cl$	64.5	12	0.90
$CH_3CH_2—Br$	109	38	1.46
$CH_3CH_2—I$	156	72	1.94
$CH_3CH_2CH_2—F$	62	3	0.80
$CH_3CH_2CH_2—Cl$	78.5	47	0.89
$CH_3CH_2CH_2—Br$	123	71	1.35
$CH_3CH_2CH_2—I$	170	102	1.75
$(CH_3)_2CH—Cl$	78.5	36	0.86
$(CH_3)_2CH—Br$	123	59	1.31
$(CH_3)_2CH—I$	170	89	1.70
$CH_3CH_2CH_2CH_2—F$	76	33	0.78
$CH_3CH_2CH_2CH_2—Cl$	92.5	78	0.89
$CH_3CH_2CH_2CH_2—Br$	137	102	1.28
$CH_3CH_2CH_2CH_2—I$	184	131	1.62
$(CH_3)_3C—Cl$	92.5	52	0.84
$(CH_3)_3C—Br$	137	73	1.23
$(CH_3)_3C—I$	184	100	1.54

6-6 Preparation of Alkyl Halides

Most syntheses of alkyl halides exploit the chemistry of functional groups we have not yet covered. For now, we review free-radical halogenation and only summarize other, often more useful, syntheses of alkyl halides. The other syntheses are discussed in subsequent chapters.

6-6A Free-Radical Halogenation

Although we discussed its mechanism at length in Section 4-3, free-radical halogenation is rarely an effective method for the synthesis of alkyl halides. It usually produces mixtures of products because there are different kinds of hydrogen atoms that can be abstracted. Also, more than one halogen atom may react, giving multiple substitutions. For example, the chlorination of propane can give a messy mixture of products.

$$CH_3—CH_2—CH_3 \quad + \quad Cl_2 \xrightarrow{h\nu} \begin{cases} CH_3—CH_2—CH_2Cl \quad + \quad CH_3—CHCl—CH_3 \\ + \quad CH_3—CHCl—CH_2Cl \quad + \quad CH_3—CCl_2—CH_3 \\ + \quad CH_3—CH_2—CHCl_2 \quad + \quad \text{others} \end{cases}$$

In industry, free-radical halogenation is sometimes useful because the reagents are cheap, the mixture of products can be separated by distillation, and each of the individual products is sold separately. In a laboratory, however, we need a good yield of one particular product. Free-radical halogenation rarely provides good selectivity and yield, so it is seldom used in the laboratory. Laboratory syntheses using free-radical halogenation are generally limited to specialized compounds that give a single major product, such as the following examples.

$$\text{cyclohexane} + Cl_2 \xrightarrow{h\nu} \text{chlorocyclohexane (50\%)} + H{-}Cl$$

$$\underset{\text{isobutane}}{(CH_3)_3C{-}H} + Br_2 \xrightarrow{h\nu} \underset{\textit{tert}\text{-butyl bromide (90\%)}}{(CH_3)_3C{-}Br} + H{-}Br$$

All the hydrogen atoms in cyclohexane are equivalent, and free-radical chlorination gives a usable yield of chlorocyclohexane. Formation of dichlorides and trichlorides is possible, but these side reactions are controlled by using only a small amount of chlorine and an excess of cyclohexane. Free-radical bromination is highly selective (Section 4-14), and it gives good yields of products that have one type of hydrogen atom that is more reactive than the others. Isobutane has only one tertiary hydrogen atom, and this atom is preferentially abstracted to give a tertiary free radical. In general, however, we are not inclined to use free-radical halogenation in the laboratory because it tends to be plagued by mixtures of products.

6-6B Allylic Bromination

Although free-radical halogenation is a poor synthetic method in most cases, free-radical bromination of alkenes can be carried out in a highly selective manner. An **allylic** position is a carbon atom next to a carbon–carbon double bond. Allylic intermediates (cations, radicals, and anions) are stabilized by resonance with the double bond, allowing the charge or radical to be delocalized. The following bond dissociation enthalpies show that less energy is required to form a resonance-stabilized primary allylic radical than a typical secondary radical.

$$H_3C{-}CH(CH_3){-}H \longrightarrow \underset{\text{(2° radical)}}{H_3C{-}\dot{C}(CH_3){-}H} + H\cdot \qquad \Delta H = +413 \text{ kJ/mol} \quad (99 \text{ kcal/mol})$$

$$H_2C{=}CH{-}CH_3 \longrightarrow \underset{\text{(1° allylic radical)}}{\left[H_2C{=}CH{-}\dot{C}H_2 \longleftrightarrow H_2\dot{C}{-}CH{=}CH_2\right]} + H\cdot \qquad \Delta H = +372 \text{ kJ/mol} \quad (89 \text{ kcal/mol})$$

Recall from Section 4-13C that bromination is highly selective, with only the most stable radical being formed. If there is an allylic hydrogen, the allylic radical is usually the most stable of the radicals that might be formed. For example, consider the free-radical bromination of cyclohexene. Under the right conditions, free-radical bromination of cyclohexene can give a good yield of 3-bromocyclohexene, where bromine has substituted for an allylic hydrogen on the carbon atom next to the double bond.

allylic positions

cyclohexene

allylic hydrogens

$$\text{cyclohexene} + Br_2 \xrightarrow{h\nu} \text{3-bromocyclohexene} + HBr$$

3-bromocyclohexene (80%)

The mechanism is similar to other free-radical halogenations. A bromine radical abstracts an allylic hydrogen atom to give a resonance-stabilized allylic radical. This radical reacts with Br_2, regenerating a bromine radical that continues the chain reaction.

abstraction of allylic H

$$\text{cyclohexene} \xrightarrow{Br\cdot} Br{-}H + [\text{allylic radical}] \xrightarrow{Br{-}Br} \text{allylic bromide} + Br\cdot$$

cyclohexene | allylic radical | allylic bromide

The general mechanism for allylic bromination shows that either end of the resonance-stabilized allylic radical can react with bromine to give products. In one of the products, the bromine atom appears in the same position where the hydrogen atom was abstracted. The other product results from reaction at the carbon atom that bears the radical in the second resonance form of the allylic radical. This second compound is said to be the product of an **allylic shift.**

For efficient allylic bromination, a large concentration of bromine must be avoided because bromine can also add to the double bond (Chapter 8). ***N*-Bromosuccinimide (NBS)** is often used as the bromine source in free-radical brominations because it combines with the HBr by-product to regenerate a constant low concentration of bromine. No additional bromine is needed because most samples of NBS contain traces of Br_2 to initiate the reaction.

$$\text{N-Br} + HBr \longrightarrow \text{N-H} + Br_2$$

N-bromosuccinimide (NBS) | regenerates a low concentration of Br_2

NBS also works well for brominating benzylic positions, next to an aromatic ring (see Problem 6-10). Allylic and benzylic halogenations are discussed in more detail in Chapter 15.

MECHANISM 6-1 Allylic Bromination

Initiation Step: Bromine absorbs light, causing formation of radicals.

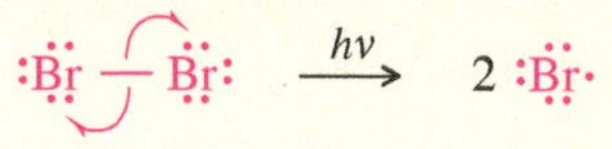

$$:\ddot{Br}-\ddot{Br}: \xrightarrow{h\nu} 2\ :\ddot{Br}\cdot$$

First Propagation Step: A bromine radical abstracts an allylic hydrogen.

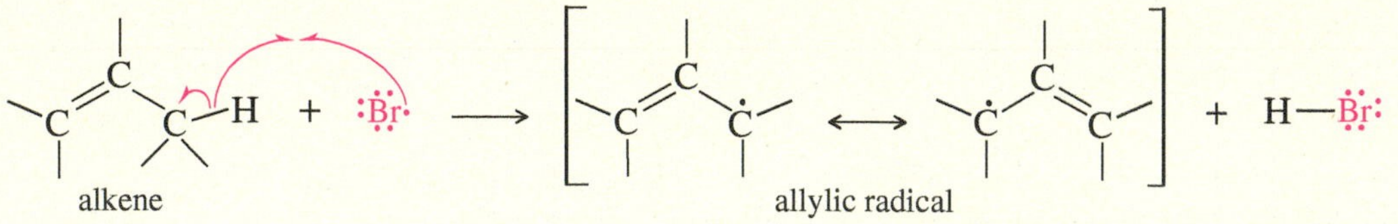

alkene + $:\ddot{Br}\cdot$ ⟶ [allylic radical ⟷ allylic radical] + $H-\ddot{Br}:$

alkene

allylic radical

Second Propagation Step: Either radical carbon can react with bromine.

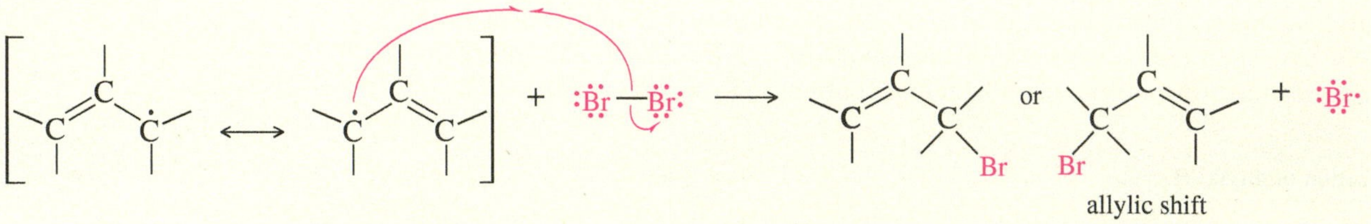

[allylic radical ⟷ allylic radical] + $:\ddot{Br}-\ddot{Br}:$ ⟶ allylic bromide or allylic bromide + $:\ddot{Br}\cdot$

Br Br

allylic shift

Overall reaction

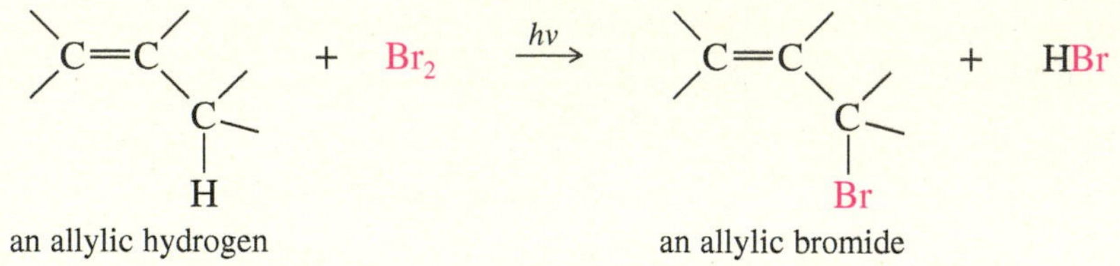

$$\text{C=C-C-H} + Br_2 \xrightarrow{h\nu} \text{C=C-C-Br} + HBr$$

H

an allylic hydrogen

Br

an allylic bromide

PROBLEM 6-8

(a) Propose a mechanism for the following reaction:

$$H_2C{=}CH{-}CH_3 + Br_2 \xrightarrow{h\nu} H_2C{=}CH{-}CH_2Br + HBr$$

(b) Use the bond-dissociation enthalpies given in Table 4-2 (page 167) to calculate the value of $\Delta H°$ for each step shown in your mechanism. (The BDE for $CH_2{=}CHCH_2{-}Br$ is about 280 kJ/mol, or 67 kcal/mol.) Calculate the overall value of $\Delta H°$ for the reaction. Are these values consistent with a rapid free-radical chain reaction?

PROBLEM 6-9

The light-initiated reaction of 2,3-dimethylbut-2-ene with *N*-bromosuccinimide (NBS) gives two products:

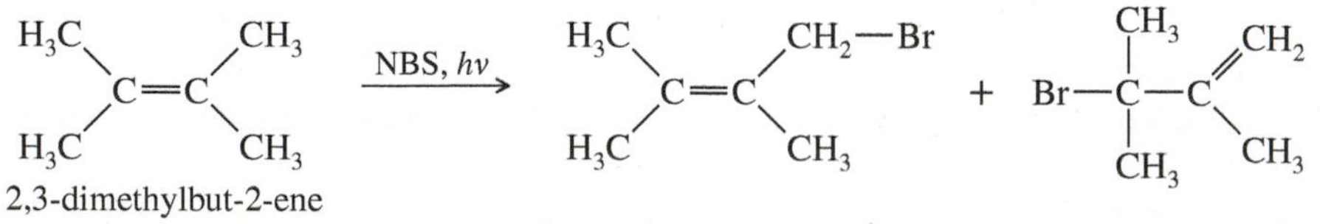

$$(CH_3)_2C{=}C(CH_3)_2 \xrightarrow{NBS,\ h\nu} (CH_3)_2C{=}C(CH_3){-}CH_2{-}Br + Br{-}C(CH_3)_2{-}C(CH_3){=}CH_2$$

2,3-dimethylbut-2-ene

(a) Give a mechanism for this reaction, showing how the two products arise as a consequence of the resonance-stabilized intermediate.

(b) The bromination of cyclohexene using NBS gives only one major product, as shown on the previous page. Explain why there is no second product from an allylic shift.

PROBLEM 6-10

Show how free-radical halogenation might be used to synthesize the following compounds. In each case, explain why we expect to get a single major product.

(a) 1-chloro-2,2-dimethylpropane (neopentyl chloride)

(b) 2-bromo-2-methylbutane

(c) 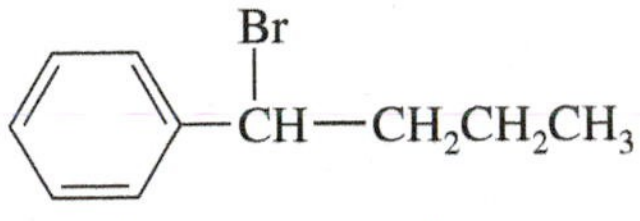 $C_6H_5-CH(Br)-CH_2CH_2CH_3$

1-bromo-1-phenylbutane

(d) 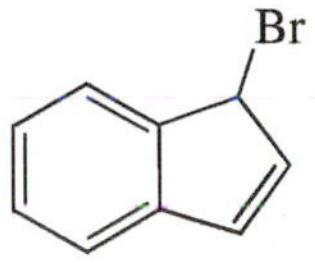 (1-bromoindene structure: Br on the sp^3 carbon of the five-membered ring fused to a benzene ring)

Following is a brief summary of the most important methods of making alkyl halides. Many of them are more general and more useful than free-radical halogenation. Several of these methods are not discussed until later in the text (note the appropriate section references). They are listed here so that you can use this summary for reference throughout the course.

SUMMARY Methods for Preparing Alkyl Halides

1. *From alkanes: free-radical halogenation (synthetically useful only in certain cases)* (Sections 4-13 and 6-6)

$$R-H \xrightarrow[\text{heat or light}]{X_2} R-X \;+\; H-X$$

2. *From alkenes and alkynes* (covered later as shown)

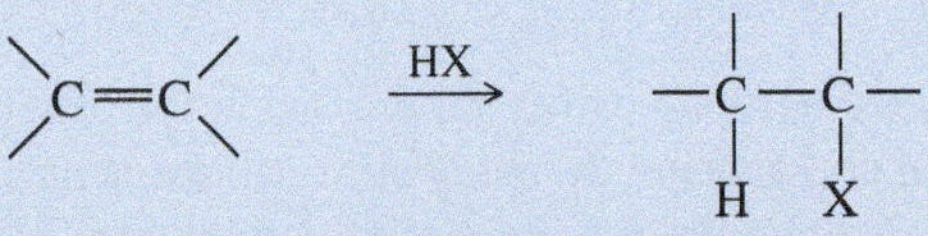

$$\gt C=C\lt \xrightarrow{HX} -\underset{H}{\overset{|}{C}}-\underset{X}{\overset{|}{C}}- \quad \text{(Section 8-8)}$$

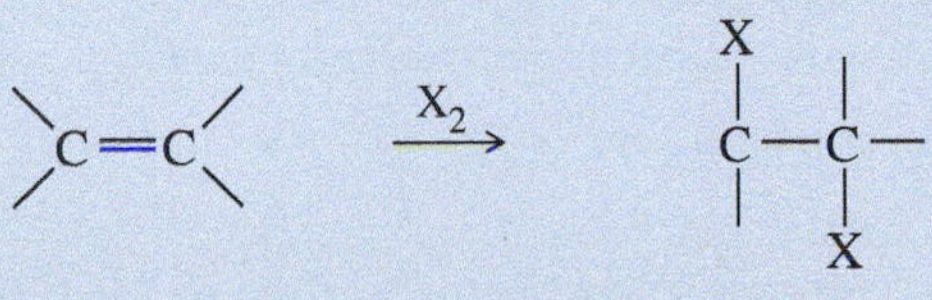

$$\gt C=C\lt \xrightarrow{X_2} -\underset{|}{\overset{X}{C}}-\underset{X}{\overset{|}{C}}- \quad \text{(Section 8-8)}$$

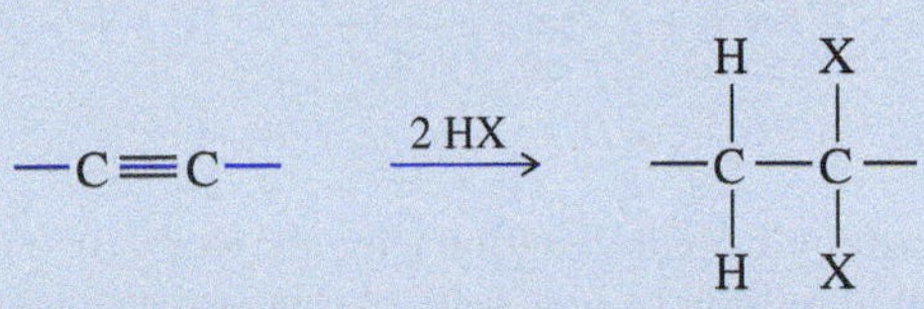

$$-C\equiv C- \xrightarrow{2\,HX} -\underset{H}{\overset{H}{C}}-\underset{X}{\overset{X}{C}}- \quad \text{(Section 9-9)}$$

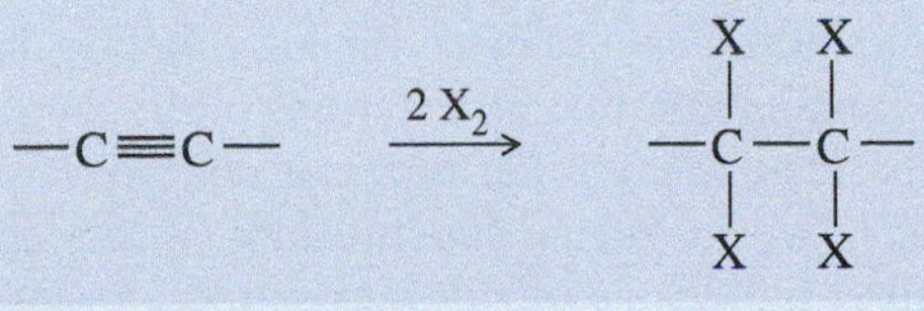

$$-C\equiv C- \xrightarrow{2\,X_2} -\underset{X}{\overset{X}{C}}-\underset{X}{\overset{X}{C}}- \quad \text{(Section 9-9)}$$

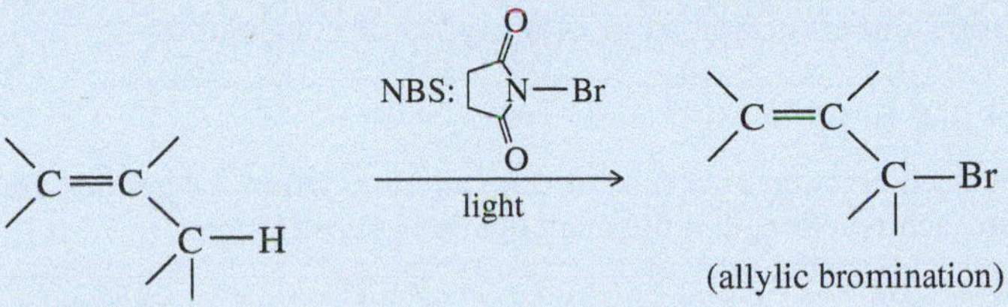

$$\gt C=C\lt_{\;C-H} \xrightarrow[\text{light}]{\text{NBS: (N-bromosuccinimide) } N-Br} \gt C=C\lt_{\;C-Br} \quad \text{(Sections 6-6 and 15-7)}$$

(allylic bromination)

(continued)

3. *From alcohols* (covered later in sections 11-7, 11-8 and 11-9)

$$R—OH \xrightarrow{HX,\ PX_3,\ \text{or others}} R—X$$

4. *From other halides* (Section 6-9)

$$R—X + I^- \xrightarrow[\text{acetone}]{} R—I + X^-$$

$$R—Cl + KF \xrightarrow[CH_3CN]{\text{18-crown-6}} R—F$$

6-7 Reactions of Alkyl Halides: Substitution and Elimination

Alkyl halides are easily converted to many other functional groups. The halogen atom can leave with its bonding pair of electrons to form a stable halide ion; we say that a halide is a good **leaving group.** When another atom replaces the halide ion, the reaction is a **substitution.**

In a **nucleophilic substitution,** a nucleophile (Nuc:$^-$) replaces a leaving group (:X:$^-$) from a carbon atom, using its lone pair of electrons to form a new bond to the carbon atom.

Nucleophilic substitution

$$—C(H)—C(X)— + Nuc:^- \longrightarrow —C(H)—C(Nuc)— + :X:^-$$

When the halide ion leaves along with another atom or ion (often H^+), a new pi bond results, and the reaction is an **elimination.** In many eliminations, a molecule of H—X is lost from the alkyl halide to give an alkene. These eliminations are called **dehydrohalogenations** because a hydrogen halide has been removed from the alkyl halide.

Elimination (covered in Chapter 7)

$$—C(H)—C(X)— + B:^- \longrightarrow B—H + \;>C=C<\; + :X:^-$$

In the elimination (a dehydrohalogenation), the reagent (B:$^-$) reacts as a base, abstracting a proton from the alkyl halide. Most nucleophiles are also basic and can engage in either substitution or elimination, depending on the alkyl halide and the reaction conditions. Therefore, substitution and elimination reactions often compete with each other.

Besides alkyl halides, many other types of compounds undergo substitution and elimination reactions. We introduce substitutions in this chapter and eliminations in Chapter 7, using alkyl halides as our first examples. In later chapters, we encounter substitutions and eliminations of other types of compounds.

PROBLEM 6-11

Classify each reaction as a substitution, an elimination, or neither. Identify the leaving group in each reaction, and the nucleophile in substitutions.

(a) cyclohexane(H)(Br) $\xrightarrow{Na^+\ ^-OCH_3}$ cyclohexane(H)(OCH_3) + NaBr

(b) 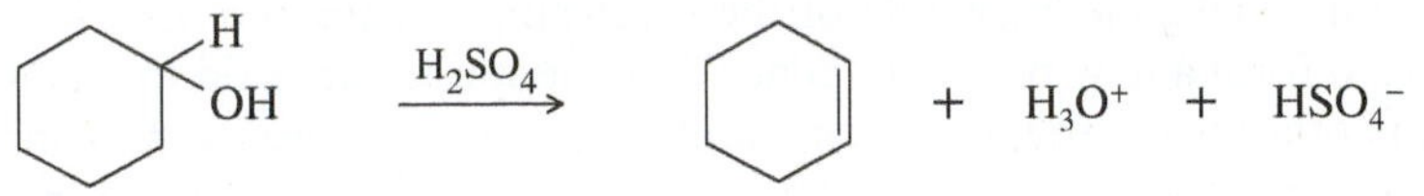 cyclohexanol $\xrightarrow{H_2SO_4}$ cyclohexene + H_3O^+ + HSO_4^-

(c) 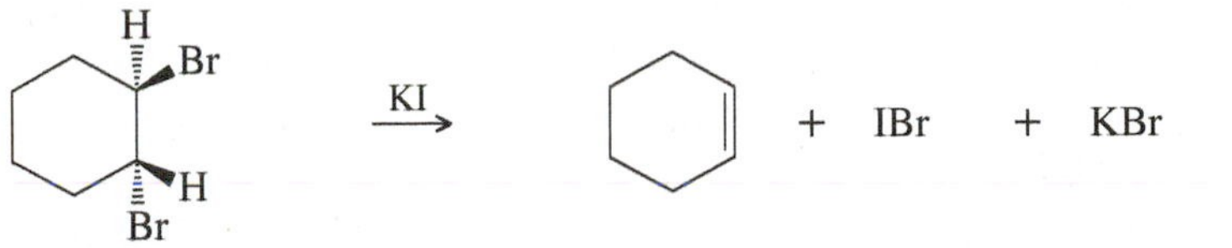 trans-1,2-dibromocyclohexane $\xrightarrow{KI}$ cyclohexene + IBr + KBr

PROBLEM-SOLVING HINT

Many common nucleophiles are also bases, so the same species might react both as Nuc:⁻ (giving substitution) and B:⁻ (giving elimination) in the same reaction. We study substitutions in Chapter 6, then add eliminations in Chapter 7.

PROBLEM 6-12

Give the structures of the substitution products expected when 1-bromohexane reacts with
(a) $NaOCH_2CH_3$ **(b)** KCN **(c)** NaOH

6-8 Bimolecular Nucleophilic Substitution: The S_N2 Reaction

A nucleophilic substitution has the general form

General nucleophilic substitution:

$$\text{Nuc:}^- + -\overset{|}{\underset{|}{C}}-\ddot{X}: \longrightarrow \text{Nuc}-\overset{|}{\underset{|}{C}}- + :\ddot{X}:^-$$

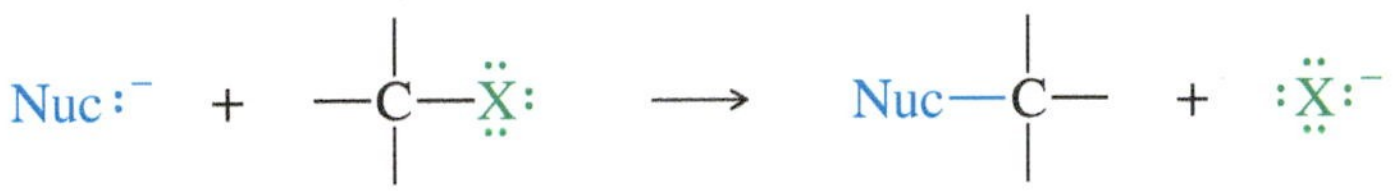

nucleophile substrate product leaving group

where Nuc:⁻ is the nucleophile and $:\ddot{X}:^-$ is the leaving halide ion. An example is the reaction of iodomethane (CH_3I) with hydroxide ion. The product is methanol.

Example:

$$H-\ddot{O}:^- + H-\overset{H}{\underset{H}{C}}-\ddot{I}: \longrightarrow H-\ddot{O}-\overset{H}{\underset{H}{C}}-H + :\ddot{I}:^-$$

hydroxide (nucleophile) iodomethane (substrate) methanol (product) iodide (leaving group)

Hydroxide ion is a strong **nucleophile** (donor of an electron pair) because the oxygen atom has unshared pairs of electrons and a negative charge. Iodomethane is called the **substrate,** meaning the compound that is attacked by the reagent. The carbon atom of iodomethane is electrophilic because it is bonded to an electronegative iodine atom. Electron density is drawn away from carbon by the halogen atom, giving the carbon atom a partial positive charge. The negative charge of hydroxide ion is attracted to this partial positive charge.

$$H\ddot{O}:^- + H_3C^{\delta+}-\ddot{I}:^{\delta-} \longrightarrow \left[H\ddot{O}\text{---}CH_3\text{---}\ddot{I}:\right]^{\ddagger -} \longrightarrow H\ddot{O}-CH_3 + :\ddot{I}:^-$$

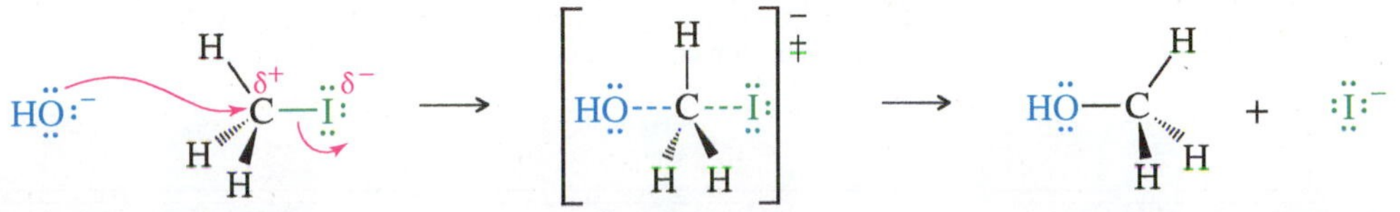

nucleophile electrophile (substrate) transition state product leaving group

Hydroxide ion attacks the back side of the electrophilic carbon atom, donating a pair of electrons to form a new bond. (In general, nucleophiles are said to attack electrophiles, not the other way around.) Notice that curved arrows are used to show the movement of electron pairs—from the electron-rich nucleophile to the electron-poor carbon atom of the electrophile. Carbon can accommodate only eight electrons in its valence shell, so the carbon–iodine bond must begin to break as the carbon–oxygen bond begins to form. Iodide ion is the leaving group; it leaves with the pair of electrons that once bonded it to the carbon atom.

This one-step mechanism is supported by kinetic information. We can vary the concentrations of the reactants and observe the effects on the reaction rate (how much methanol is formed per second). The rate is found to double when the concentration of *either* reactant is doubled. The reaction is therefore first order in each of the reactants and second order overall. The rate equation has the following form:

$$\text{second-order rate} = k_r[CH_3I][^{-}OH]$$

This rate equation is consistent with a mechanism that requires a collision between a molecule of methyl iodide and a hydroxide ion. Both of these species are present in the transition state, and the collision frequency is proportional to both concentrations. The rate constant k_r depends on several factors, including the energy of the transition state and the temperature (Section 4-9).

This one-step nucleophilic substitution is an example of the **S_N2 mechanism.** The abbreviation S_N2 stands for *Substitution, Nucleophilic, bimolecular*. The term *bimolecular* means that the transition state of the rate-limiting step (the only step in this reaction) involves the collision of *two* molecules. Bimolecular reactions usually have rate equations that are second order overall.

The S_N2 reaction of methyl iodide (iodomethane) with hydroxide ion is a **concerted reaction,** taking place in a single step with bonds breaking and forming at the same time. The middle structure is a **transition state,** a point of maximum energy, rather than an intermediate. In this transition state, the bond to the nucleophile (hydroxide) is partially formed, and the bond to the leaving group (iodide) is partially broken. Remember that a transition state is not a discrete molecule that can be isolated; it exists for only an instant.

PROBLEM-SOLVING HINT

A transition state is unstable and cannot be isolated. It exists for only an instant. In contrast, an intermediate exists for a finite length of time.

The reaction-energy diagram for this substitution (Figure 6-5) shows only one transition state and no intermediates between the reactants and the products. The

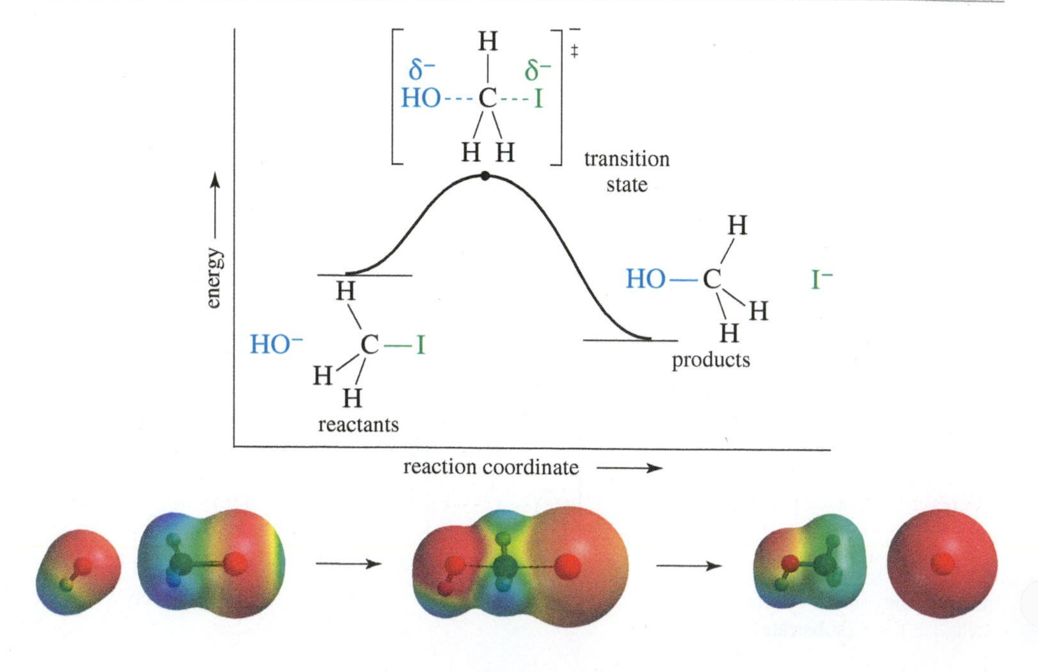

FIGURE 6-5
The reaction-energy diagram for the S_N2 reaction of methyl iodide with hydroxide shows only one energy maximum: the transition state. There are no intermediates.

The electrostatic potential maps of the reactants, transition state, and products show that the negatively charged nucleophile (red) attacks the electrophilic (blue) region of the substrate. In the transition state, the negative charge (red) is delocalized over the nucleophile and the leaving group. The negative charge leaves with the leaving group.

reactants are shown slightly higher in energy than the products because this reaction is known to be exothermic. The transition state is much higher in energy because it involves a five-coordinate carbon atom with two partial bonds.

Key Mechanism 6-2 shows a general S_N2 reaction. A nucleophile attacks the substrate to give a transition state in which a bond to the nucleophile is forming at the same time the bond to the leaving group is breaking.

KEY MECHANISM 6-2 The S_N2 Reaction

The S_N2 reaction takes place in a single (concerted) step. A strong nucleophile attacks the electrophilic carbon, forcing the leaving group to leave.

Nuc:⁻ + C—X: → [Nuc---C---X:]‡⁻ → Nuc—C + :X:⁻

nucleophile | substrate (electrophile) | transition state | product | leaving group

The order of reactivity for substrates is $CH_3X > 1° > 2°$. (3° alkyl halides cannot react by this mechanism.)

EXAMPLE: Reaction of 1-bromobutane with sodium methoxide gives 1-methoxybutane.

$$\underset{\text{sodium methoxide}}{NaOCH_3} + \underset{\text{1-bromobutane}}{CH_3CH_2CH_2CH_2Br} \longrightarrow \underset{\text{1-methoxybutane}}{CH_3CH_2CH_2CH_2OCH_3} + NaBr$$

$CH_3—\ddot{O}:^-$ + $CH_3CH_2CH_2$(H)(H)C—Br: → [$CH_3—\ddot{O}$---C($CH_2CH_2CH_3$)(H)(H)---Br:]‡ → $CH_3—\ddot{O}—C(CH_2CH_2CH_3)(H)(H)$ + Na^+ :Br:⁻

nucleophile | electrophile (substrate) | transition state | product | leaving group

PROBLEM 6-13

(a) Under certain conditions, the reaction of 0.5 M 1-bromobutane with 1.0 M sodium methoxide forms 1-methoxybutane at a rate of 0.05 mol/L per second. What would be the rate if 0.1 M 1-bromobutane and 2.0 M $NaOCH_3$ were used?

(b) Consider the reaction of 1-bromobutane with a large excess of ammonia (NH_3). Draw the reactants, the transition state, and the products. Note that the initial product is the salt of an amine ($RNH_3^+\ Br^-$), which is deprotonated by the excess ammonia to give the amine.

(c) Show another S_N2 reaction using a different combination of an alkoxide and an alkyl bromide that also produces 1-methoxybutane.

6-9 Generality of the S_N2 Reaction

Many useful reactions take place by the S_N2 mechanism. The reaction of an alkyl halide, such as methyl iodide, with hydroxide ion gives an alcohol. Other nucleophiles convert alkyl halides to a wide variety of functional groups. The following table summarizes some of the types of compounds that can be formed by nucleophilic displacement of alkyl halides.

SUMMARY S_N2 Reactions of Alkyl Halides

$$\text{Nuc:}^- + \text{R—X} \longrightarrow \text{Nuc—R} + \text{X}^-$$

Nucleophile		Product	Class of Product
R—X + $^-$:I:	⟶	R—I:	alkyl halide
R—X + $^-$:ÖH	⟶	R—ÖH	alcohol
R X + $^-$:ÖR′	⟶	R—ÖR′	ether
R—X + $^-$:SH	⟶	R—SH	thiol (mercaptan)
R—X + $^-$:SR′	⟶	R—SR′	thioether (sulfide)
R—X + :NH_3	⟶	R—NH_3^+ X^-	amine salt
R—X + $^-$:N=$\overset{+}{N}$=N:$^-$	⟶	R—N=$\overset{+}{N}$=N:$^-$	azide
R—X + $^-$:C≡C—R′	⟶	R—C≡C—R′	alkyne
R—X + $^-$:C≡N:	⟶	R—C≡N:	nitrile
R—X + $^-$:O—C(=O)—R′	⟶	R—O—C(=O)—R′	ester
R—X + :PPh_3	⟶	$[R—PPh_3]^+$ X^-	phosphonium salt

Halogen Exchange Reactions The S_N2 reaction provides a useful method for synthesizing alkyl iodides and fluorides, which are more difficult to make than alkyl chlorides and bromides. Halides can be converted to other halides by **halogen exchange reactions,** in which one halide displaces another.

Iodide is a good nucleophile, and many alkyl chlorides react with sodium iodide to give alkyl iodides. Alkyl fluorides are difficult to synthesize directly, and they are often made by treating alkyl chlorides or bromides with KF under conditions that use a crown ether (Section 14-2D) to dissolve the fluoride salt in an aprotic solvent, which enhances the normally weak nucleophilicity of the fluoride ion (see Section 6-10).

$$\text{R—X} + \text{I}^- \longrightarrow \text{R—I} + \text{X}^-$$

$$\text{R—X} + \text{KF} \xrightarrow[\text{CH}_3\text{CN}]{\text{18-crown-6}} \text{R—F} + \text{KX}$$

Examples

$$\underset{\text{allyl chloride}}{H_2C{=}CH{-}CH_2Cl} + NaI \longrightarrow \underset{\text{allyl iodide}}{H_2C{=}CH{-}CH_2I} + NaCl$$

$$\underset{\text{ethyl chloride}}{CH_3CH_2Cl} + KF \xrightarrow[\text{CH}_3\text{CN}]{\text{18-crown-6}} \underset{\text{ethyl fluoride}}{CH_3CH_2F} + KCl$$

PROBLEM 6-14

Predict the major products of the following substitutions.

(a) $\underset{\text{ethyl bromide}}{CH_3CH_2Br} + \underset{\text{potassium } tert\text{-butoxide}}{(CH_3)_3CO^- K^+} \longrightarrow$

(b) $\underset{\text{sodium acetylide}}{HC{\equiv}C{:}^- Na^+} + \underset{\text{1-chlorobutane}}{CH_3CH_2CH_2CH_2Cl} \longrightarrow$

(c) $(CH_3)_2CHCH_2Br$ + excess NH_3 ⟶
(d) $CH_3CH_2CH_2I$ + NaCN ⟶
(e) 1-chloropentane + NaI ⟶
(f) 1-chloropentane + KF $\xrightarrow[CH_3CN]{\text{18-crown-6}}$

PROBLEM 6-15

Show how you might use S_N2 reactions to convert 1-chlorobutane into the following compounds.

(a) butan-1-ol
(b) 1-fluorobutane
(c) 1-iodobutane
(d) $CH_3—(CH_2)_3—CN$
(e) $CH_3—(CH_2)_3—C{\equiv}CH$
(f) $CH_3CH_2—O—(CH_2)_3—CH_3$
(g) $CH_3—(CH_2)_3—NH_2$

6-10 Factors Affecting S_N2 Reactions: Strength of the Nucleophile

We will use the S_N2 reaction as an example of how we study the factors that affect the rates and products of organic reactions. Both the nucleophile and the substrate (the alkyl halide) are important, as well as the type of solvent used. We begin by considering what makes a good nucleophile.

A "stronger" nucleophile is an ion or a molecule that reacts faster in the S_N2 reaction than a "weaker" nucleophile under the same conditions. A strong nucleophile is much more effective than a weak one in attacking an electrophilic carbon atom. For example, both methanol (CH_3OH) and methoxide ion (CH_3O^-) have easily shared pairs of nonbonding electrons, but methoxide ion reacts with electrophiles in the S_N2 reaction about 1 million times faster than methanol. It is generally true that a species with a negative charge is a stronger nucleophile than a similar, neutral species.

Methoxide ion has nonbonding electrons that are readily available for bonding. In the transition state, the negative charge is shared by the oxygen of methoxide ion and by the halide leaving group. Methanol, however, has no negative charge; the transition state has a partial negative charge on the halide but a partial positive charge on the methanol oxygen atom. We can generalize the case of methanol and the methoxide ion to say that

> a base is always a stronger nucleophile than its conjugate acid.

$CH_3—\ddot{O}:^-$ + $CH_3—I$ ⟶ $[CH_3—O^{\delta-}\cdots CH_3 \cdots I^{\delta-}]^{\ddagger}$ ⟶ $CH_3—O—CH_3$ + $:I:^-$

base (stronger nucleophile)

lower E_a

$CH_3—\ddot{O}(H):$ + $CH_3—I$ ⟶ $[CH_3—O^{\delta+}(H)\cdots CH_3 \cdots I^{\delta-}]^{\ddagger}$ ⟶ $CH_3—O^+(H)—CH_3$ + $:I:^-$

conjugate acid (weaker nucleophile)

higher E_a

TABLE 6-3
Some Common Nucleophiles
Listed in increasing order of nucleophilicity in polar solvents such as water and alcohols

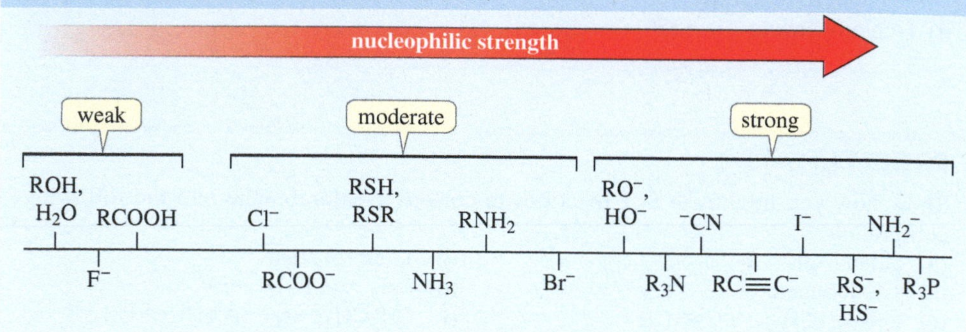

We might be tempted to say that methoxide is a much better nucleophile because it is much more basic. This would be a mistake because basicity and nucleophilicity are different properties. **Basicity** is defined by the *equilibrium constant* for abstracting a *proton.* **Nucleophilicity** is defined by the *rate* of attack on an electrophilic *carbon atom.* In both cases, the nucleophile (or base) forms a new bond. If it forms a new bond to a proton, it has reacted as a **base;** if it forms a new bond to carbon, it has reacted as a **nucleophile.** Predicting which way a species will react may be difficult; most (but not all) good nucleophiles are also strong bases, and vice versa.

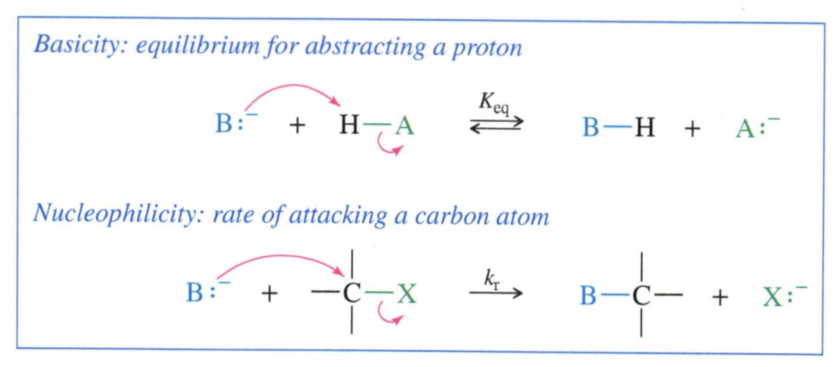

Table 6-3 lists some common nucleophiles in increasing order of their nucleophilicity in polar solvents such as water and alcohols. The strength of nucleophiles in these solvents shows three major trends, as described in the Summary below.

SUMMARY Trends in Nucleophilicity

1. A species with a negative charge is a stronger nucleophile than a similar neutral species. In particular, a base is a stronger nucleophile than its conjugate acid.

$$^{-}{:}\ddot{O}H > H_2\ddot{O}{:} \qquad ^{-}{:}\ddot{S}H > H_2\ddot{S}{:} \qquad ^{-}{:}\ddot{N}H_2 > {:}NH_3$$

2. Nucleophilicity decreases from left to right in the periodic table, following the increase in electronegativity from left to right. The more electronegative elements have more tightly held nonbonding electrons that are less reactive toward forming new bonds.

$$^{-}{:}\ddot{N}H_2 > {}^{-}{:}\ddot{O}H > {:}\ddot{F}{:}^{-} \qquad {:}NH_3 > H_2\ddot{O} \qquad (CH_3CH_2)_3P{:} > (CH_3CH_2)_2\ddot{S}{:}$$

3. Nucleophilicity increases down the periodic table, following the increase in size and polarizability and the decrease in electronegativity.

$$ {:}\ddot{I}{:}^{-} > {:}\ddot{Br}{:}^{-} > {:}\ddot{Cl}{:}^{-} > {:}\ddot{F}{:}^{-} \qquad ^{-}{:}\ddot{Se}H > {}^{-}{:}\ddot{S}H > {}^{-}{:}\ddot{O}H \qquad (CH_3CH_2)_3P{:} > (CH_3CH_2)_3N{:}$$

sp^3 orbital back lobe

F^-

"hard," small valence shell

little bonding

transition state

I^-

"soft," large valence shell

more bonding

transition state

FIGURE 6-6 Comparison of fluoride ion and iodide ion as nucleophiles in the S_N2 reaction. Fluoride has tightly bound electrons that cannot begin to form a C—F bond until the atoms are close together. Iodide has more loosely bound outer electrons that begin bonding earlier in the reaction.

The third trend (size and polarizability) reflects an atom's ability to engage in partial bonding as it begins to attack an electrophilic carbon atom. As we go down a column in the periodic table, the atoms become larger, with more electrons at a greater distance from the nucleus. The electrons are more loosely held, and the atom is more **polarizable:** Its electrons can move more freely toward a positive charge, resulting in stronger bonding in the transition state. The increased mobility of its electrons enhances the atom's ability to begin to form a bond at a relatively long distance.

Figure 6-6 illustrates this polarizability effect by comparing the attack of iodide ion and fluoride ion on a methyl halide. The outer shell of the fluoride ion is the second shell. These electrons are tightly held, close to the nucleus. Fluoride is a "hard" (low-polarizability) nucleophile, and its nucleus must approach the carbon nucleus quite closely before the electrons can begin to overlap and form a bond. In the transition state, there is little bonding between fluorine and carbon. In contrast, the outer shell of the iodide ion is the fifth shell. These electrons are loosely held, making the iodide ion a "soft" (high-polarizability) nucleophile. The outer electrons begin to shift and overlap with the carbon atom from farther away. There is a great deal of bonding between iodine and carbon in the transition state, which lowers the energy of the transition state.

6-10A Steric Effects on Nucleophilicity

To serve as a nucleophile, an ion or a molecule must get close to a carbon atom to attack it. Bulky groups on the nucleophile hinder this close approach, and they slow the reaction rate. For example, *tert*-butoxide ion is a stronger *base* (for abstracting protons) than ethoxide ion, but *tert*-butoxide ion has three methyl groups that hinder any close approach to a more crowded carbon atom. Therefore, ethoxide ion is a stronger nucleophile than *tert*-butoxide ion. When bulky groups interfere with a reaction by virtue of their size, we call the effect **steric hindrance.**

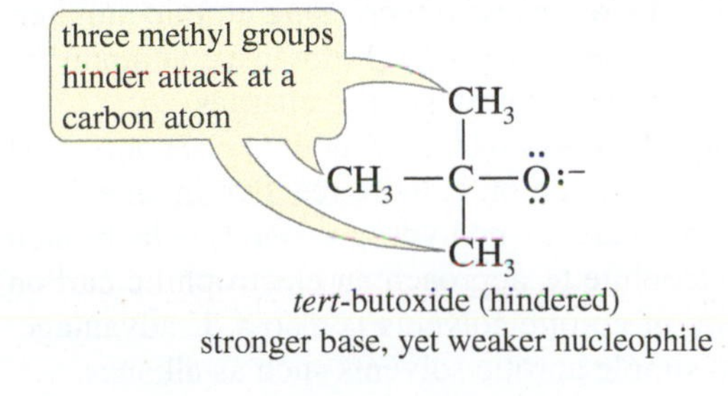

tert-butoxide (hindered)
stronger base, yet weaker nucleophile

$CH_3—CH_2—\ddot{O}:^-$
ethoxide (unhindered)
weaker base, yet stronger nucleophile

Application: Biochemistry

The "nitrogen mustard" anticancer drugs are believed to alkylate DNA using two S_N2 reactions. First, the nitrogen nucleophile displaces chloride on the primary alkyl chloride portion to generate a reactive intermediate. Second, the intermediate alkylates a nitrogen atom of DNA. The process is repeated, linking the two strands of the double-helix DNA, and thereby preventing replication of the DNA.

nitrogen mustard

intermediate

alkylates DNA

Steric hindrance has little effect on *basicity* because basicity involves attack on an unhindered proton. When a *nucleophile* attacks a carbon atom, however, a bulky nucleophile cannot approach the carbon atom so easily. Most bases are also nucleophiles, capable of attacking either a proton or an electrophilic carbon atom. If we want a species to act as a base, we use a bulky reagent such as *tert*-butoxide ion. If we want it to react as a nucleophile, we use a less hindered reagent such as ethoxide.

PROBLEM-SOLVING HINT

Steric hindrance (bulkiness) hinders nucleophilicity (S_N2) more than it hinders basicity.

PROBLEM 6-16

For each pair, predict the stronger nucleophile in the S_N2 reaction (using an alcohol as the solvent). Explain your prediction.

(a) $(CH_3CH_2)_3N$ or $(CH_3CH_2)_2NH$
(b) $(CH_3)_2O$ or $(CH_3)_2S$
(c) NH_3 or PH_3
(d) CH_3S^- or H_2S
(e) $(CH_3)_3N$ or $(CH_3)_2O$
(f) CH_3COO^- or CF_3COO^-
(g) $(CH_3)_2CHO^-$ or $CH_3CH_2CH_2O^-$
(h) I^- or Cl^-

6-10B Solvent Effects on Nucleophilicity

Another factor affecting the nucleophilicity of these ions is their solvation, particularly in protic solvents. A **protic solvent** has acidic protons, usually in the form of O—H or N—H groups. Those with O—H groups are often called **hydroxylic solvents.** These groups form hydrogen bonds to negatively charged nucleophiles. Protic solvents, especially alcohols, are convenient solvents for nucleophilic substitutions because the reagents (alkyl halides, nucleophiles, etc.) tend to be quite soluble.

Small anions are solvated more strongly than large anions in a protic solvent because the solvent molecules approach a small anion more closely and form stronger hydrogen bonds. When an anion reacts as a nucleophile, energy is required to "strip off" some of the solvent molecules, breaking some of the hydrogen bonds that stabilized the solvated anion. More energy is required to strip off solvent from a small, strongly solvated ion such as fluoride than from a large, diffuse, less strongly solvated ion such as iodide.

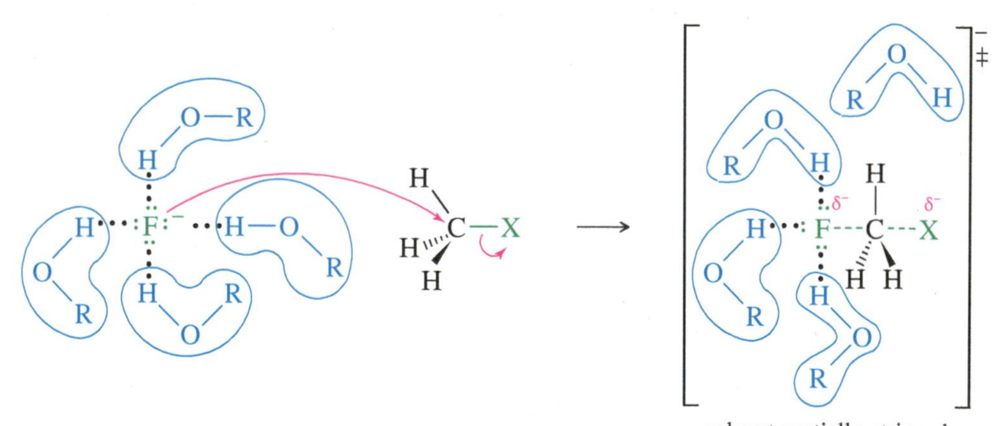

solvent partially stripped off in the transition state

The enhanced solvation of smaller anions in protic solvents, requiring more energy to strip off their solvent molecules, reduces their nucleophilicity. This trend reinforces the trend in polarizability: The polarizability increases with increasing atomic number, and the solvation energy (in protic solvents) decreases with increasing atomic number. Therefore, nucleophilicity (in protic solvents) generally increases down a column in the periodic table, as long as we compare similar species with similar charges.

In contrast with protic solvents, **aprotic solvents** (solvents without O—H or N—H groups) enhance the nucleophilicity of anions. An anion is more reactive in an aprotic solvent because it is not so strongly solvated. There are no hydrogen bonds to be broken when solvent must make way for the nucleophile to approach an electrophilic carbon atom. The relatively weak solvating ability of aprotic solvents is also a disadvantage: Most polar, ionic reagents are insoluble in simple aprotic solvents such as alkanes.

Polar aprotic solvents have strong dipole moments to enhance solubility, yet they have no O—H or N—H groups to form hydrogen bonds with anions. Examples of useful polar aprotic solvents (shown below) are acetonitrile, dimethylformamide (DMF), acetone, and dimethyl sulfoxide (DMSO). Changing from a hydroxylic solvent to a polar aprotic solvent often enhances the rate of an S_N2 reaction, sometimes by a factor of 1000 or more. In some cases, we can add specific solvating reagents to enhance solubility without affecting the reactivity of the nucleophile. For example, the "crown ether" 18-crown-6 solvates potassium ions. Using the potassium salt of a nucleophile and solvating the potassium ions causes the nucleophilic anion to be dragged along into solution.

Examples of polar aprotic solvents:

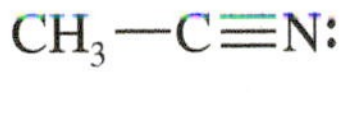

$CH_3—C{\equiv}N{:}$ acetonitrile

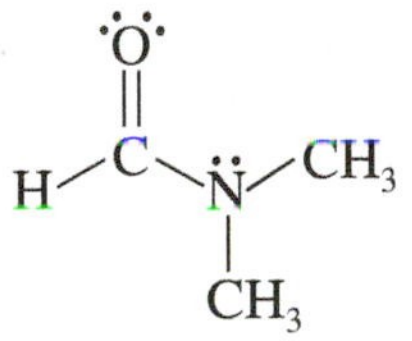

dimethylformamide (DMF)

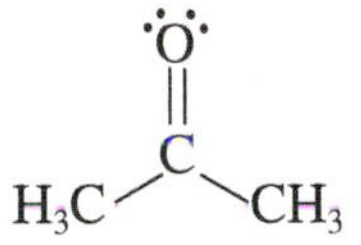

acetone

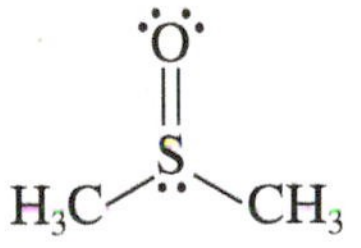

dimethyl sulfoxide (DMSO)

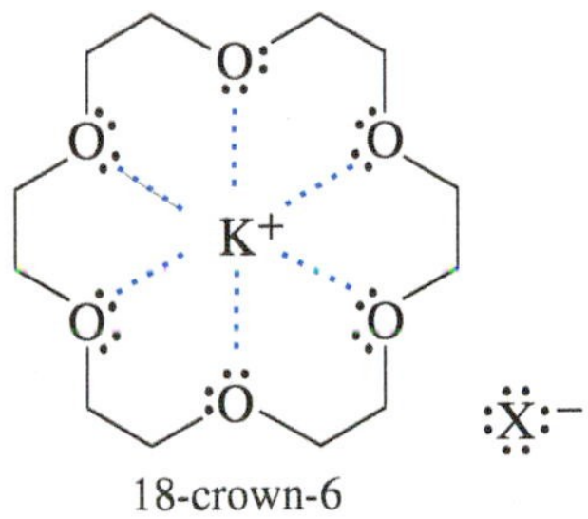

18-crown-6 solvates K^+ ions

The following example shows how fluoride ion, normally a poor nucleophile in hydroxylic (protic) solvents, can be a good nucleophile in an aprotic solvent. Although KF is not very soluble in acetonitrile, 18-crown-6 solvates the potassium ions, and the poorly solvated fluoride ion follows. This poorly solvated fluoride ion is nucleophilic, and it attacks carbon more readily than it would in a protic solvent.

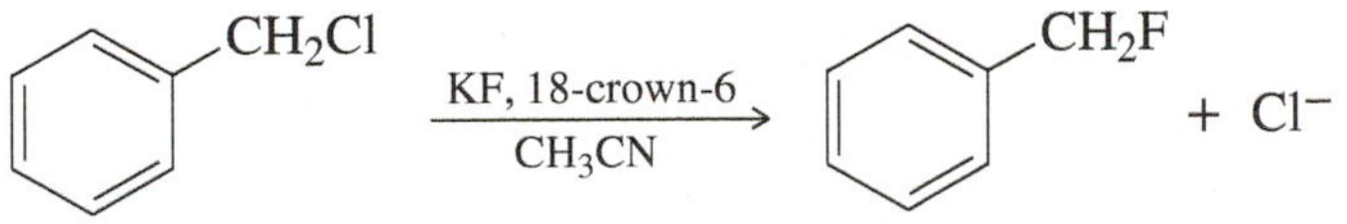

$$C_6H_5CH_2Cl \xrightarrow[CH_3CN]{KF,\ 18\text{-crown-6}} C_6H_5CH_2F + Cl^-$$

6-11 Reactivity of the Substrate in S_N2 Reactions

Just as the nucleophile is important in the S_N2 reaction, the structure of the alkyl halide is equally important. We will often refer to the alkyl halide as the **substrate:** literally, the compound that is being attacked by the reagent. Besides alkyl halides, a variety of other types of compounds serve as substrates in S_N2 reactions. To be a good substrate for S_N2 attack by a nucleophile, a molecule must have an electrophilic carbon atom with a good leaving group, and that carbon atom must not be too sterically hindered for a nucleophile to attack.

6-11A Leaving-Group Effects on the Substrate

A leaving group serves two purposes in the S_N2 reaction:

1. It polarizes the C—X bond, making the carbon atom electrophilic.
2. It leaves with the pair of electrons that once bonded it to the electrophilic carbon atom.

To fill these roles, a good leaving group should be

1. electron-withdrawing, to polarize the carbon atom,
2. stable (not a strong base) once it has left, and
3. polarizable, to stabilize the transition state.

1. The leaving group must be *electron-withdrawing* to create a partial positive charge on the carbon atom, making the carbon electrophilic. An electron-withdrawing leaving group also stabilizes the negatively charged transition state. Halogen atoms are strongly electronegative, so alkyl halides are common substrates for S_N2 reactions. Oxygen, nitrogen, and sulfur also form strongly polarized bonds with carbon; given the right substituents, they can form the basis for excellent leaving groups.

Strongly polarized

$\overset{+\!\!\!\rightarrow}{C-X}$ (X = halogen) $\overset{+\!\!\!\rightarrow}{C-O}$ $\overset{+\!\!\!\rightarrow}{C-N}$ $\overset{+\!\!\!\rightarrow}{C-S}$

2. The leaving group must be *stable* once it has left with the pair of electrons that bonded it to carbon. A stable leaving group is needed for favorable energetics. The leaving group is leaving in the transition state; a reactive leaving group would raise the energy of the transition state, slowing the reaction. Also, the energy of the leaving group is reflected in the energy of the products. A reactive leaving group would raise the energy of the products, driving the equilibrium toward the reactants.

$$\text{Nuc}:^- + \text{C—X} \longrightarrow \left[\overset{\delta^-}{\text{Nuc}}\text{---C---}\overset{\delta^-}{\text{X}}\right]^{\ddagger} \longrightarrow \text{Nuc—C} + :\text{X}^-$$

bond forming   bond breaking

transition state

Good leaving groups should be *weak bases*; therefore, they are the conjugate bases of strong acids. The hydrohalic acids HCl, HBr, and HI are strong, and their conjugate bases (Cl^-, Br^-, and I^-) are all weak bases. Other weak bases, such as sulfate ions, sulfonate ions, and phosphate ions, can also serve as good leaving groups. Table 6-4 lists examples of good leaving groups.

Hydroxide ions, alkoxide ions, and other strong bases are poor leaving groups for S_N2 reactions. For example, the —OH group of an alcohol is a poor leaving group because it would have to leave as hydroxide ion.

$$Na^+ \; :\ddot{Br}:^- \; CH_3-\ddot{O}H \;\not\longrightarrow\; Br-CH_3 + Na^+ \; ^-:\ddot{O}H \;\text{(strong base)}$$

Ions that are strong bases and poor leaving groups:

$^-:\ddot{O}H$	$^-:\ddot{O}R$	$^-:\ddot{N}H_2$
hydroxide	alkoxide	amide

Table 6-4 also lists some neutral molecules that can be good leaving groups. A neutral molecule often serves as the leaving group from a *positively charged* species. For example, if an alcohol is placed in an acidic solution, the hydroxy group is protonated. Water then serves as the leaving group. Note that the need to protonate the alcohol (requiring acid) limits the choice of nucleophiles to those few that are weak bases, such as bromide and iodide. A strongly basic nucleophile would become protonated in acid.

$$CH_3-\ddot{O}H + H^+ \underset{}{\overset{HBr}{\rightleftharpoons}} :\ddot{Br}:^- \; CH_3-\overset{H}{\overset{|}{O^+}}-H \longrightarrow :\ddot{Br}-CH_3 + :\overset{H}{\overset{|}{\ddot{O}}}-H$$

protonated alcohol   water

3. Finally, a good leaving group should be *polarizable* to maintain partial bonding with the carbon atom in the transition state. This bonding helps stabilize the transition state and reduce the activation energy. The departure of a leaving group is much like the attack of a nucleophile, except that the bond is breaking rather than forming. Polarizable nucleophiles and polarizable leaving groups both stabilize the transition state by engaging in more bonding at a longer distance. Iodide ion, one of the most polarizable ions, is both a good nucleophile and a good leaving group. In contrast, fluoride ion is a small, "hard" ion. Fluoride is both a poor nucleophile (in protic solvents) and a poor leaving group in S_N2 reactions.

TABLE 6-4
Weak Bases That Are Common Leaving Groups

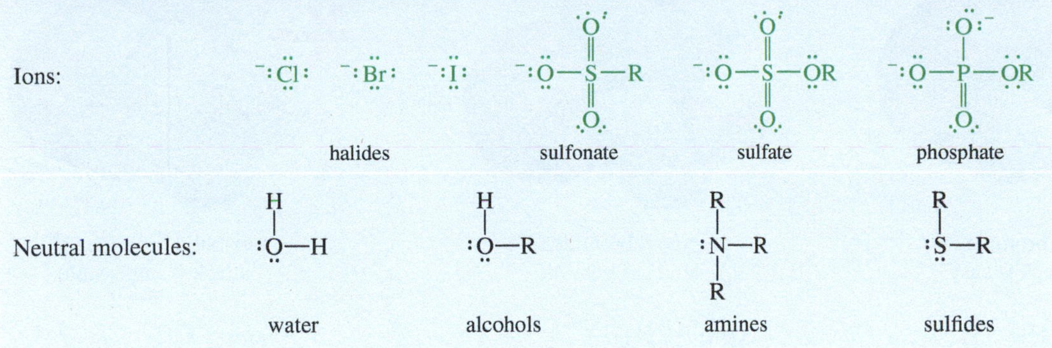

Ions:	$^-:\ddot{Cl}:$ $^-:\ddot{Br}:$ $^-:\ddot{I}:$	$^-:\ddot{O}-S(=O)_2-R$	$^-:\ddot{O}-S(=O)_2-\ddot{O}R$	$^-:\ddot{O}-P(=O)(O^-)-\ddot{O}R$
	halides	sulfonate	sulfate	phosphate
Neutral molecules:	$H-\ddot{O}-H$	$H-\ddot{O}-R$	$R_2\ddot{N}-R$	$R-\ddot{S}-R$
	water	alcohols	amines	sulfides

PROBLEM 6-17

When diethyl ether ($CH_3CH_2OCH_2CH_3$) is treated with concentrated HBr, the initial products are CH_3CH_2Br and CH_3CH_2OH. Propose a mechanism to account for this reaction.

6-11B Steric Effects on the Substrate

Different alkyl halides undergo S_N2 reactions at vastly different rates. The structure of the substrate is the most important factor in its reactivity toward S_N2 displacement. The reaction goes rapidly with methyl halides and with most primary substrates. It is more sluggish with secondary halides. Tertiary halides fail to react at all by the S_N2 mechanism. Table 6-5 shows the effect of alkyl substitution on the rate of S_N2 displacements.

For simple alkyl halides, the relative rates for S_N2 displacement are

$$\text{Relative rates for } S_N2: \quad CH_3X > 1° > 2° \gg 3°$$

The physical explanation for this order of reactivity is suggested by the information in Table 6-5. All the slow-reacting compounds have one property in common: The back side of the electrophilic carbon atom is crowded by the presence of bulky groups. Tertiary halides are more hindered than secondary halides, which are more hindered

TABLE 6-5
Effect of Substituents on the Rates of SN2 Reactions

Class of Halide	Example	Structure	Relative Rate
methyl	CH_3-Br		>1000
primary (1°)	CH_3CH_2-Br	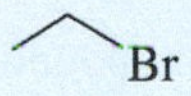	50
secondary (2°)	$(CH_3)_2CH-Br$	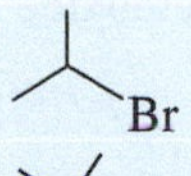	1
tertiary (3°)	$(CH_3)_3C-Br$	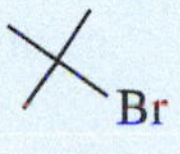	<0.001
n-butyl (1°)	$CH_3CH_2CH_2CH_2-Br$	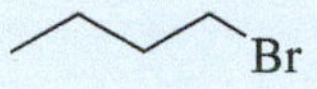	20
isobutyl (1°)	$(CH_3)_2CHCH_2-Br$	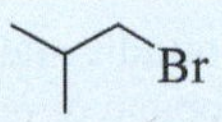	2
neopentyl (1°)	$(CH_3)_3CCH_2-Br$	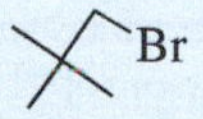	0.0005

Note: Two or three alkyl groups, or even a single bulky alkyl group, slow the reaction rate. The rates listed are compared to the secondary case (isopropyl bromide), assigned a relative rate of 1.

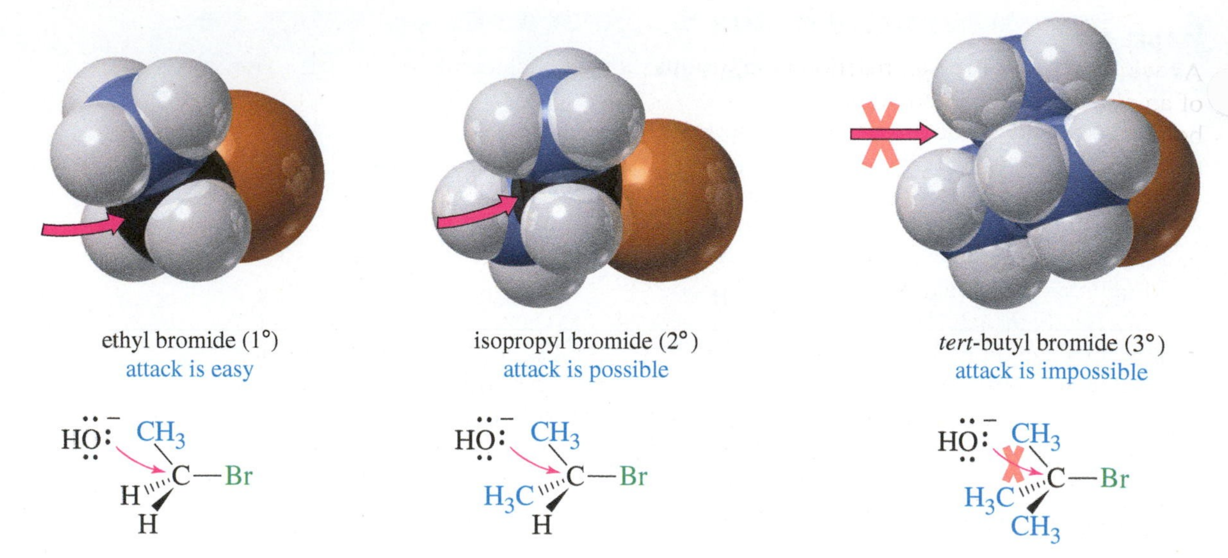

FIGURE 6-7
S_N2 attack on a simple primary alkyl halide is unhindered. Attack on a secondary halide is hindered, and attack on a tertiary halide is impossible.

than primary halides. Even a bulky primary halide (like neopentyl bromide) undergoes S_N2 reaction at a rate similar to that of a tertiary halide. The relative rates show that it is the bulk of the alkyl groups, rather than an electronic effect, that hinders the reactivity of bulky alkyl halides in the S_N2 displacement.

This effect on the rate is another example of **steric hindrance.** When the nucleophile approaches the back side of the electrophilic carbon atom, it must come within bonding distance of the back lobe of the C—X sp^3 orbital. If two alkyl groups are bonded to the carbon atom, this process is difficult. Three alkyl groups make it impossible. Just one alkyl group can produce a large amount of steric hindrance if it is unusually bulky, such as the *tert*-butyl group of neopentyl bromide.

Figure 6-7 shows the S_N2 reaction of hydroxide ion with ethyl bromide (1°), isopropyl bromide (2°), and *tert*-butyl bromide (3°). The nucleophile can easily approach the electrophilic carbon atom of ethyl bromide. In isopropyl bromide, the approach is hindered, but still possible. In contrast, S_N2 approach to the tertiary carbon of *tert*-butyl bromide is impossible because of the steric hindrance of the three methyl groups. Make models of ethyl bromide, isopropyl bromide, and *tert*-butyl bromide, and compare the ease of bringing in an atom for a back-side attack.

PROBLEM-SOLVING HINT

Do not write S_N2 reactions occurring on tertiary alkyl halides.

PROBLEM 6-18

Rank the following compounds in decreasing order of their reactivity toward the S_N2 reaction with sodium ethoxide ($Na^+ \, ^-OCH_2CH_3$) in ethanol.

methyl chloride	*tert*-butyl iodide	neopentyl bromide
isopropyl bromide	methyl iodide	ethyl chloride

PROBLEM 6-19

For each pair of compounds, state which compound is the better S_N2 substrate.

(a) 2-methyl-1-iodopropane or *tert*-butyl iodide
(b) cyclohexyl bromide or 1-bromo-1-methylcyclohexane
(c) 2-bromobutane or isopropyl bromide
(d) 1-chloro-2,2-dimethylbutane or 2-chlorobutane
(e) 1-iodobutane or 2-iodopropane

6-12 Stereochemistry of the S_N2 Reaction

As we have seen, the S_N2 reaction requires attack by a nucleophile on the back side of an electrophilic carbon atom (Figure 6-8). A carbon atom can have only four filled bonding orbitals (an octet), so the leaving group must leave as the nucleophile bonds to the carbon atom. The nucleophile's electrons insert into the back lobe of carbon's sp^3 hybrid orbital in its antibonding combination with the orbital of the leaving group (because the bonding MO is already filled). These electrons in the antibonding MO help to weaken the C—Br bond as bromide leaves. The transition state shows partial bonding to both the nucleophile and the leaving group.

Back-side attack literally turns the tetrahedron of the carbon atom inside out, like an umbrella caught by the wind (Figure 6-8). In the product, the nucleophile assumes a stereochemical position opposite the position the leaving group originally occupied. We call this result an **inversion of configuration** at the carbon atom.

In the case of an asymmetric carbon atom, back-side attack gives the opposite configuration of the carbon atom (Mechanism 6-3). The S_N2 displacement is the most common example of a **Walden inversion,** a step (in a reaction sequence) where an

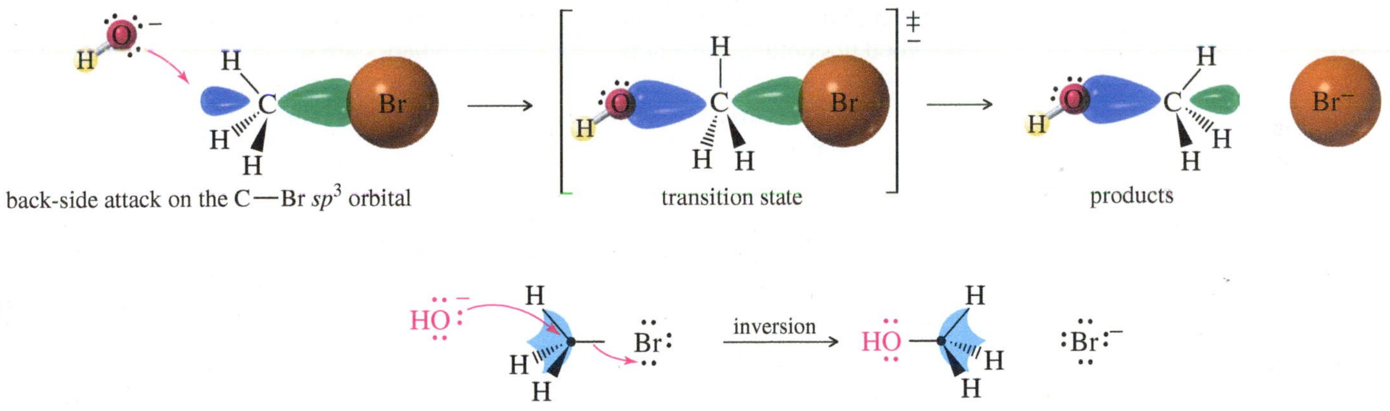

FIGURE 6-8
Back-side attack in the S_N2 reaction. The S_N2 reaction takes place through nucleophilic attack on the back lobe of carbon's sp^3 hybrid orbital. This back-side attack inverts the carbon atom's tetrahedron, like a strong wind inverts an umbrella.

MECHANISM 6-3 Inversion of Configuration in the S_N2 Reaction

Back-side attack inverts the configuration of the carbon atom.

$$\text{Nuc:}^- + \text{C—X} \longrightarrow [\text{Nuc}\cdots\text{C}\cdots\text{X}]^{\ddagger} \longrightarrow \text{Nuc—C} \quad \text{X:}^-$$

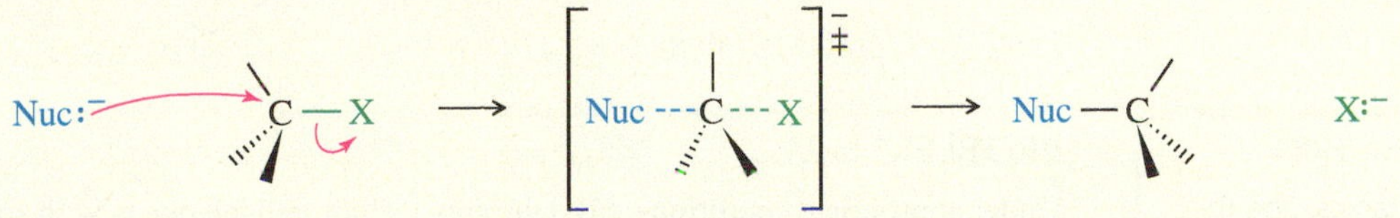

EXAMPLE:

$$\text{HO:}^- + \text{(S)-2-bromobutane} \longrightarrow [\text{HO}\cdots\text{C}\cdots\text{Br}]^{\ddagger} \longrightarrow \text{(R)-butan-2-ol} + \text{:Br:}^-$$

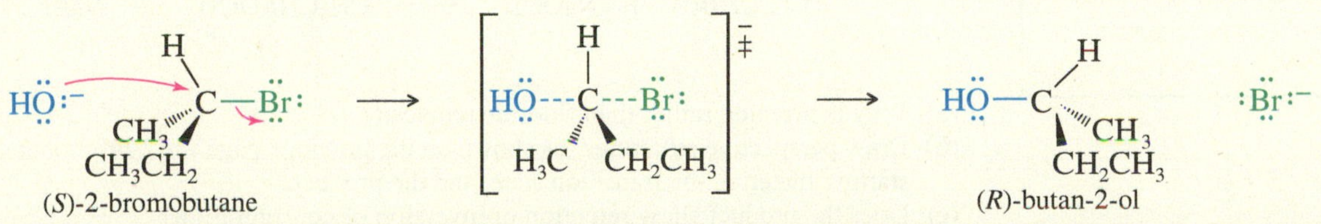

asymmetric carbon atom undergoes inversion of configuration. In the 1890s, Paul Walden, of the University of Tübingen (Germany), was one of the first to study reactions giving inversion of configuration.

In some cases, inversion of configuration is readily apparent. For example, when *cis*-1-bromo-3-methylcyclopentane undergoes S_N2 displacement by hydroxide ion, inversion of configuration gives *trans*-3-methylcyclopentanol.

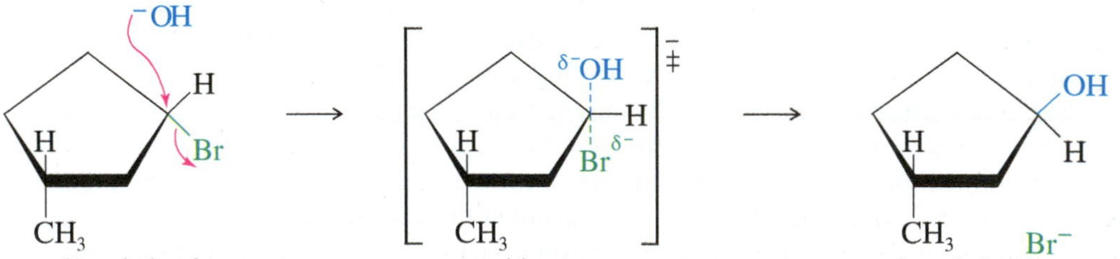

$^-$OH, H, H, Br, CH_3 → [δ^-OH, H, H, δ^- Br, CH_3]‡ → OH, H, H, CH_3, Br^-

cis-1-bromo-3-methylcyclopentane — transition state — *trans*-3-methylcyclopentanol

The S_N2 displacement is a good example of a **stereospecific reaction:** one in which different stereoisomers react to give different stereoisomers of the product. To study the mechanism of a nucleophilic substitution, we often look at the product to see if the reaction is stereospecific, with inversion of configuration. If it is, the S_N2 mechanism is a good possibility, especially if the reaction kinetics are second order. In many cases (no asymmetric carbon or ring, for example), it is impossible to determine whether inversion has occurred. In these cases, we use kinetics and other evidence to help determine the reaction mechanism.

PROBLEM 6-20

Draw a perspective structure or a Fischer projection for the products of the following S_N2 reactions.

(a) *trans*-1-bromo-3-methylcyclopentane + KOH

(b) (*R*)-2-bromopentane + KCN

(c) Fischer projection: CH_3 (top); Br—C—H; H—C—CH_3; CH_2CH_3 (bottom) + NaI $\xrightarrow{\text{acetone}}$

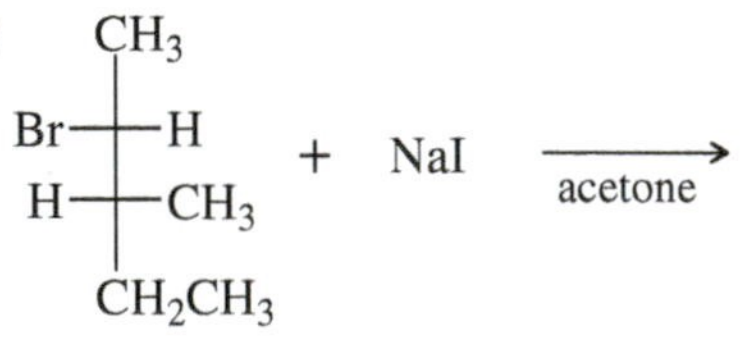

(d) Br, C, H, F, CH_3 + NaSH

(e) Fischer projection: CH_3 (top); H—C—Br; $CH_2CH_2CH_3$ (bottom) + $NaOCH_3$ $\xrightarrow{CH_3OH}$

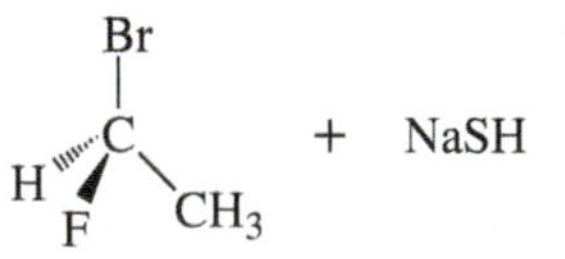

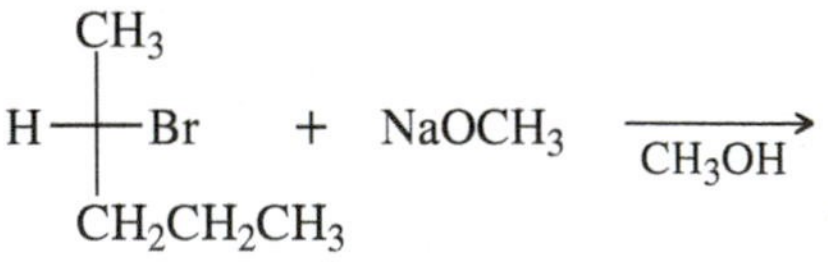

(f) H, C, D, Cl (attached to benzene ring) + NH_3 excess

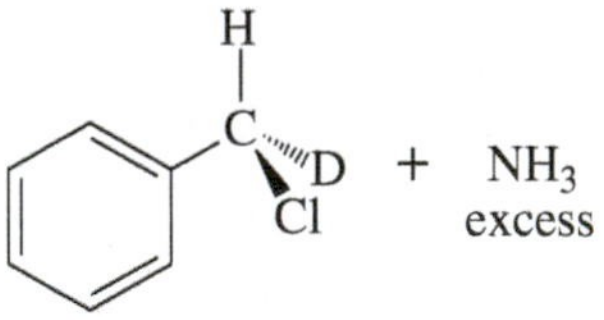

PROBLEM-SOLVING HINT

(*R*) and (*S*) are just names, and they vary with changing priorities of the substituents. Don't rely on names to determine the stereochemistry of a reaction.

PROBLEM 6-21

Under appropriate conditions, (*S*)-1-bromo-1-fluoroethane reacts with sodium methoxide to give pure (*S*)-1-fluoro-1-methoxyethane.

$$\underset{(S)}{CH_3CHBrF} + NaOCH_3 \longrightarrow \underset{(S)}{CH_3CHFOCH_3} + NaBr$$

(a) Why is bromide rather than fluoride replaced?

(b) Draw perspective structures (as shown on the previous page for 2-bromobutane) for the starting material, the transition state, and the product.

(c) Does the product show retention or inversion of configuration?

(d) Is this result consistent with reaction by the S_N2 mechanism?

6-13 Unimolecular Nucleophilic Substitution: The S_N1 Reaction

When *tert*-butyl bromide is placed in boiling methanol, methyl *tert*-butyl ether can be isolated from the reaction mixture. Because this reaction takes place with the solvent acting as the nucleophile, it is called a **solvolysis** (*solvo* for "solvent," plus *lysis*, meaning "cleavage").

$$\underset{\textit{tert}\text{-butyl bromide}}{(CH_3)_3C{-}Br} + \underset{\text{methanol}}{CH_3{-}OH} \xrightarrow{\text{boil}} \underset{\text{methyl } \textit{tert}\text{-butyl ether}}{(CH_3)_3C{-}O{-}CH_3} + HBr$$

This solvolysis is a substitution because methoxide has replaced bromide on the *tert*-butyl group. It does not go through the S_N2 mechanism, however. The S_N2 requires a strong nucleophile and a substrate that is not too hindered. Methanol is a weak nucleophile, and *tert*-butyl bromide is a hindered tertiary halide—a poor S_N2 substrate.

If this substitution cannot go by the S_N2 mechanism, what kind of mechanism might be involved? An important clue is kinetic: Its rate does not depend on the concentration of methanol, the nucleophile. The rate depends only on the concentration of the substrate, *tert*-butyl bromide.

$$\text{first-order rate} = k_r[(CH_3)_3C{-}Br]$$

This rate equation is first order overall: first order in the concentration of the alkyl halide and zeroth order in the concentration of the nucleophile. Because the rate does not depend on the concentration of the nucleophile, we infer that the nucleophile is not present in the transition state of the rate-limiting step. The nucleophile must react *after* the slow step.

This type of substitution is called an **S_N1 reaction,** for *Substitution, Nucleophilic, unimolecular*. The term *unimolecular* means there is only one molecule involved in the transition state of the rate-limiting step. The mechanism of the S_N1 reaction of *tert*-butyl bromide with methanol is shown here. Ionization of the alkyl halide (first step) is the rate-limiting step.

Step 1: Formation of carbocation (rate limiting)

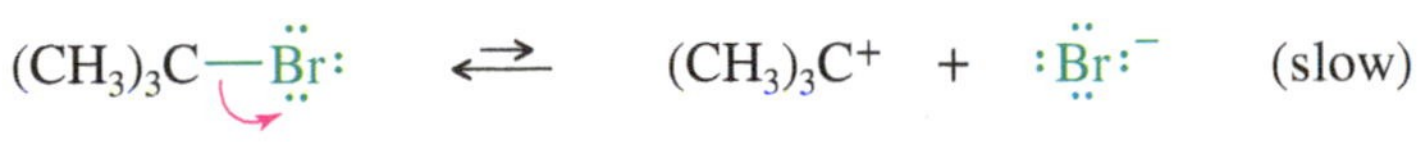

$$(CH_3)_3C{-}\ddot{\underset{..}{Br}}{:} \rightleftharpoons (CH_3)_3C^+ + {:}\ddot{\underset{..}{Br}}{:}^- \quad \text{(slow)}$$

Step 2: Nucleophilic attack on the carbocation

$$(CH_3)_3C^+ \;\; {:}\ddot{O}(H){-}CH_3 \rightleftharpoons (CH_3)_3C{-}\overset{+}{\ddot{O}}(H){-}CH_3 \quad \text{(fast)}$$

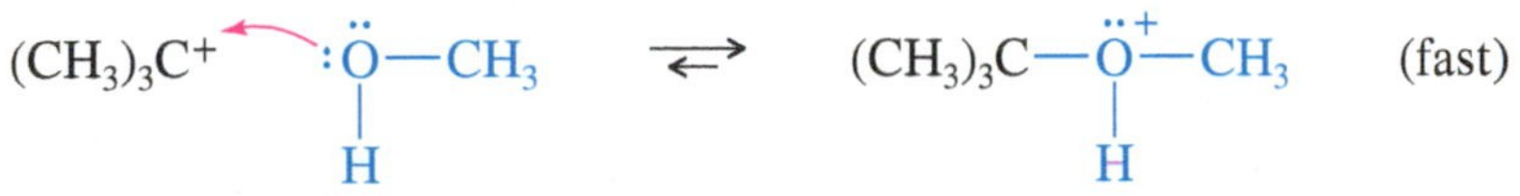

Final Step: Loss of proton to solvent

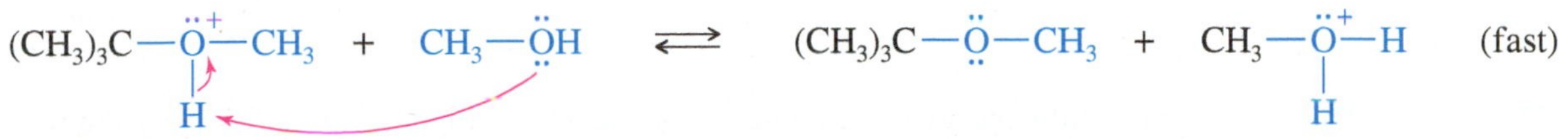

$$(CH_3)_3C{-}\overset{+}{\ddot{O}}(H){-}CH_3 + CH_3{-}\ddot{\underset{..}{O}}H \rightleftharpoons (CH_3)_3C{-}\ddot{\underset{..}{O}}{-}CH_3 + CH_3{-}\overset{+}{\ddot{O}}(H){-}H \quad \text{(fast)}$$

The S_N1 mechanism is a multistep process. The first step is a slow ionization to form a carbocation. The second step is a fast attack on the carbocation by a nucleophile. The carbocation is a strong electrophile; it reacts very quickly with nucleophiles, including weak nucleophiles. The nucleophile in S_N1 reactions is usually weak, because a strong nucleophile would be more likely to attack the substrate and force some kind of second-order reaction. If the nucleophile is an uncharged molecule such as water or an alcohol, the positively charged product must lose a proton to give the final uncharged product. The general mechanism for the S_N1 reaction is summarized in Key Mechanism 6-4.

PROBLEM-SOLVING HINT

Strong acids such as HCl, HBr, and HI are completely dissociated in solutions of alcohols or water. Consider the pK_a values:

HI	−10
HBr	−9
HCl	−7
$CH_3OH_2^+$	−2.5
$CH_3CH_2OH_2^+$	−2.4
H_3O^+	−1.7

These pK_a values show that the equilibrium

$$ROH + HX \rightleftharpoons ROH_2^+ + X^-$$

lies far to the right. When we write "HBr (aq)" or "HBr in ethanol," we actually mean the dissociated ions. The values also show that alcohols and water are much more basic than halide ions, so they are more likely than halide ion to serve as the base in the final deprotonation in the S_N1 reaction and similar reactions under acidic conditions.

KEY MECHANISM 6-4 The S_N1 Reaction

The S_N1 reaction involves a two-step mechanism. A slow ionization gives a carbocation that reacts quickly with a (usually weak) nucleophile. Reactivity: 3° > 2° > 1°.

***Step 1*:** Formation of the carbocation (rate-limiting).

$$R{-}\ddot{X}: \rightleftharpoons R^+ + :\ddot{X}:^-$$

***Step 2*:** Nucleophilic attack on the carbocation (fast).

$$R^+ + Nuc:^- \longrightarrow R{-}Nuc$$

If the nucleophile is water or an alcohol, a third step is needed to deprotonate the product.

EXAMPLE: Solvolysis of 1-iodo-1-methylcyclohexane in methanol.

***Step 1*:** Formation of a carbocation (rate-limiting).

***Step 2*:** Nucleophilic attack by the solvent (methanol).

***Step 3*:** Deprotonation to form the product.

product (protonated methanol)

PROBLEM 6-22

Propose an S_N1 mechanism for the solvolysis of 3-bromo-2,3-dimethylpentane in ethanol.

PROBLEM-SOLVING HINT

Never show a proton falling off into thin air. Show a possible base (often the solvent) abstracting the proton.

The reaction-energy diagram of the S_N1 reaction (Figure 6-9) shows why the rate does not depend on the strength or concentration of the nucleophile. The ionization (first step) is highly endothermic, and its large activation energy determines the overall reaction rate. The nucleophilic attack (second step) is strongly exothermic, with a lower-energy transition state. In effect, a nucleophile reacts with the carbocation almost as soon as it forms.

The reaction-energy diagrams of the S_N1 mechanism and the S_N2 mechanism are compared in Figure 6-9. The S_N1 has a true intermediate, the carbocation. The intermediate appears as a relative minimum (a low point) in the reaction-energy diagram. Reagents and conditions that favor formation of the carbocation (the slow step) accelerate the S_N1 reaction; reagents and conditions that hinder its formation retard the reaction.

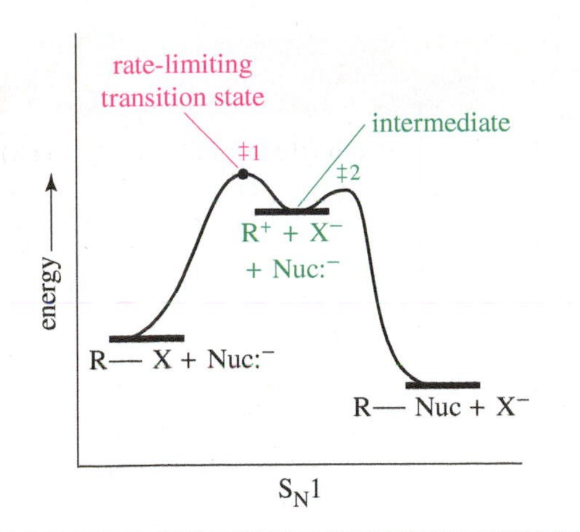

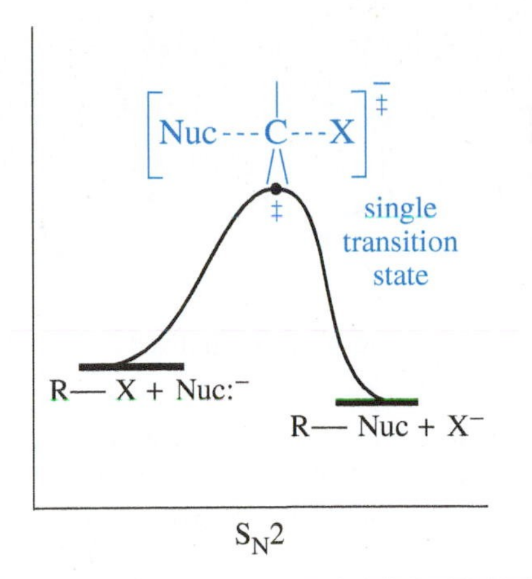

FIGURE 6-9
Reaction-energy diagrams of the S_N1 and S_N2 reactions. The S_N1 is a two-step mechanism with two transition states (‡1 and ‡2) and a carbocation intermediate. The S_N2 has only one transition state and no intermediate.

6-13A Substituent Effects

The rate-limiting step of the S_N1 reaction is ionization to form a carbocation, a strongly endothermic process. The first transition state resembles the carbocation (Hammond's postulate, Section 4-14); consequently, rates of S_N1 reactions depend strongly on carbocation stability. In Section 4-16A, we saw that alkyl groups stabilize carbocations by donating electrons through sigma bonds (the *inductive effect*) and through overlap of filled orbitals with the empty *p* orbital of the carbocation *(hyperconjugation)*. Highly substituted carbocations are therefore more stable.

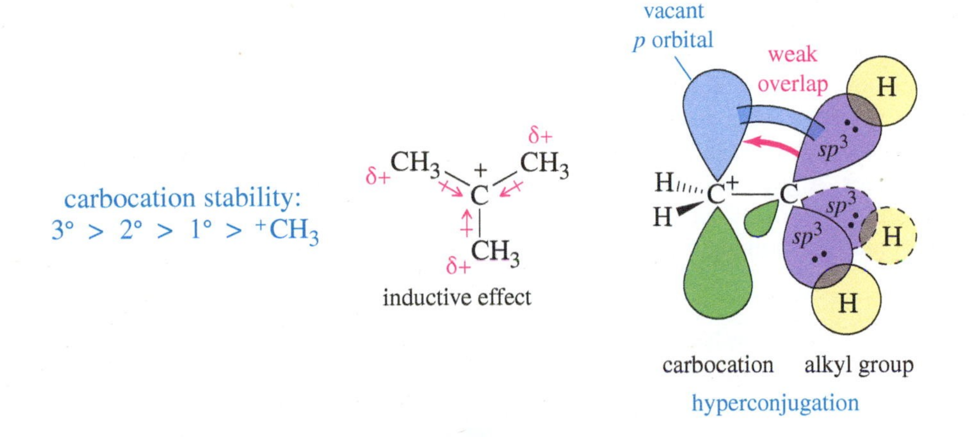

Reactivity toward S_N1 substitution mechanisms follows the stability of carbocations:

S_N1 reactivity: 3° > 2° > 1° > CH_3X

This order is *opposite* that of the S_N2 reaction. Alkyl groups hinder the S_N2 by blocking attack of the strong nucleophile, but alkyl groups enhance the S_N1 by stabilizing the carbocation intermediate. Simple primary and methyl carbocations are not normally seen in solution.

Resonance stabilization of the carbocation can also promote the S_N1 reaction. For example, allyl bromide is a primary halide, but it undergoes the S_N1 reaction about as fast as a secondary halide. The carbocation formed by ionization is resonance-stabilized, with the positive charge spread equally over two carbon atoms.

allyl bromide ⇄ [resonance-stabilized carbocation] $\xrightarrow{Nuc:^-}$

Vinyl and aryl halides generally do not undergo S_N1 or S_N2 reactions. An S_N1 reaction would require ionization to form a vinyl or aryl cation, either of which is less stable than most alkyl carbocations. An S_N2 reaction would require back-side attack by the nucleophile, which is made impossible by the repulsion of the electrons in the double bond or aromatic ring.

a vinyl halide or an aryl halide → no S_N1, no S_N2

6-13B Leaving-Group Effects

The leaving group is breaking its bond to carbon in the rate-limiting ionization step of the S_N1 mechanism. A highly polarizable leaving group helps stabilize the rate-limiting transition state through partial bonding as it leaves. The leaving group should be a weak base—very stable after it leaves with the pair of electrons that bonded it to carbon.

Figure 6-10 shows the transition state of the ionization step of the S_N1 reaction. Notice how the leaving group is taking on a negative charge while it stabilizes the new carbocation through partial bonding. The leaving group should be stable as it takes on this negative charge, and it should be polarizable to engage in effective partial bonding as it leaves. A good leaving group is just as necessary in the S_N1 reaction as it is in the S_N2, and similar leaving groups are effective for either reaction. Table 6-4 (page 271) lists some common leaving groups for either reaction.

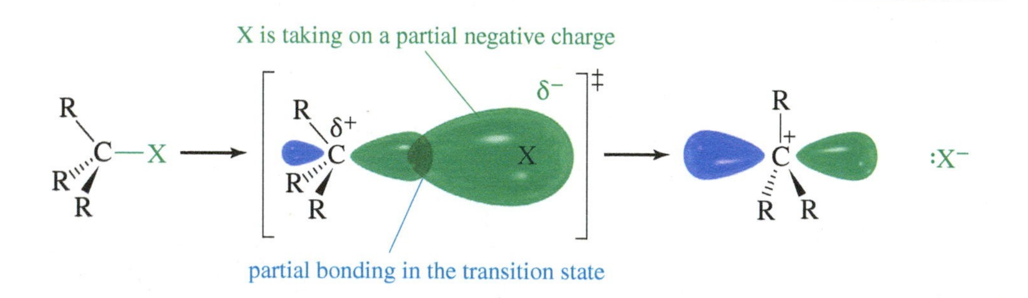

FIGURE 6-10
In the transition state of the S_N1 ionization, the leaving group is taking on a negative charge. The C—X bond is breaking, and a polarizable leaving group can still maintain substantial overlap.

PROBLEM-SOLVING HINT

Primary cations are rarely formed in solution unless they are resonance-stabilized.

PROBLEM 6-23

Choose the member of each pair that will react faster by the S_N1 mechanism.

(a) 1-bromopropane or 2-bromopropane
(b) 2-bromo-2-methylbutane or 2-bromo-3-methylbutane
(c) *n*-propyl bromide or allyl bromide
(d) 1-bromo-2,2-dimethylpropane or 2-bromopropane
(e) 2-iodo-2-methylbutane or *tert*-butyl chloride
(f) 2-bromo-2-methylbutane or ethyl iodide

PROBLEM 6-24

3-Bromocyclohexene is a secondary halide, and benzyl bromide is a primary halide. Both halides undergo S_N1 substitution about as fast as most tertiary halides. Use resonance structures to explain this enhanced reactivity.

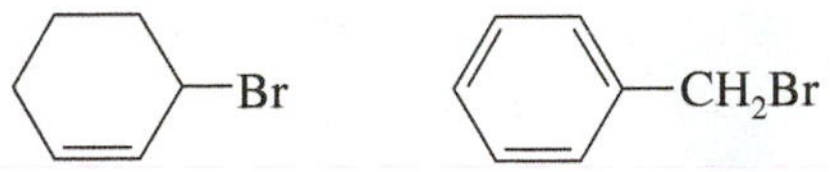

3-bromocyclohexene benzyl bromide

6-13C Solvent Effects on S_N1 Reactions

The S_N1 reaction goes much more readily in polar solvents that stabilize ions. The rate-limiting step forms two ions, and ionization is taking place in the transition state. Polar solvents solvate these ions by an interaction of the solvent's dipole moment with the charge of the ion. Protic solvents such as alcohols and water are even more effective solvents because anions form hydrogen bonds with the —OH hydrogen atom, and cations complex with the nonbonding electrons of the —OH oxygen atom.

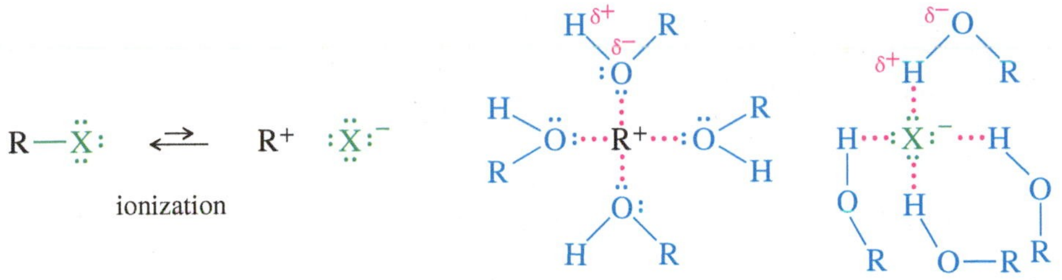

$$R{-}\ddot{X}: \rightleftharpoons R^+ \; :\ddot{X}:^-$$

ionization

solvated ions

Ionization of an alkyl halide requires formation and separation of positive and negative charges, similar to what happens when sodium chloride dissolves in water. Therefore, S_N1 reactions require highly polar solvents that strongly solvate ions. One measure of a solvent's ability to solvate ions is its *dielectric constant* (ε), a measure of the solvent's polarity. Table 6-6 lists the dielectric constants of some common solvents and the relative ionization rates for *tert*-butyl chloride in these solvents. Note that ionization occurs much faster in highly polar solvents such as water and alcohols. Although most alkyl halides are not soluble in water, they often dissolve in highly polar mixtures of acetone and alcohols with water.

TABLE 6-6 Dielectric Constants (ε) and Ionization Rates of *tert*-Butyl Chloride in Common Solvents

Solvent	ε	Relative Rate
water	78	8000
methanol	33	1000
ethanol	24	200
acetone	21	1
diethyl ether	4.3	0.001
hexane	2.0	<0.0001

6-14 Stereochemistry of the S_N1 Reaction

Recall from Section 6-12 that the S_N2 reaction is stereospecific: the nucleophile attacks from the back side of the electrophilic carbon atom, giving inversion of configuration. In contrast, the S_N1 reaction is not stereospecific. In the S_N1 mechanism, the carbocation intermediate is sp^2 hybridized and planar. A nucleophile can attack the carbocation from either face. Figure 6-11 shows the S_N1 solvolysis of a chiral compound, (*S*)-3-bromo-2,3-dimethylpentane, in ethanol. The carbocation is planar and achiral; attack from both faces gives both enantiomers of the product. Such a process, giving both enantiomers of the product (whether or not the two enantiomers are produced in equal amounts), is called **racemization.** The product is either racemic or at least less optically pure than the starting material.

If a nucleophile attacks the carbocation in Figure 6-11 from the front side (the side the leaving group left), the product molecule shows **retention of configuration.** Attack from the back side gives a product molecule showing **inversion of configuration.** Racemization is simply a combination of retention and inversion. When racemization occurs, the product is rarely completely racemic, however; there is often more inversion than retention of configuration. As the leaving group leaves, it partially blocks the front side of the carbocation. The back side is unhindered, so attack is more likely there.

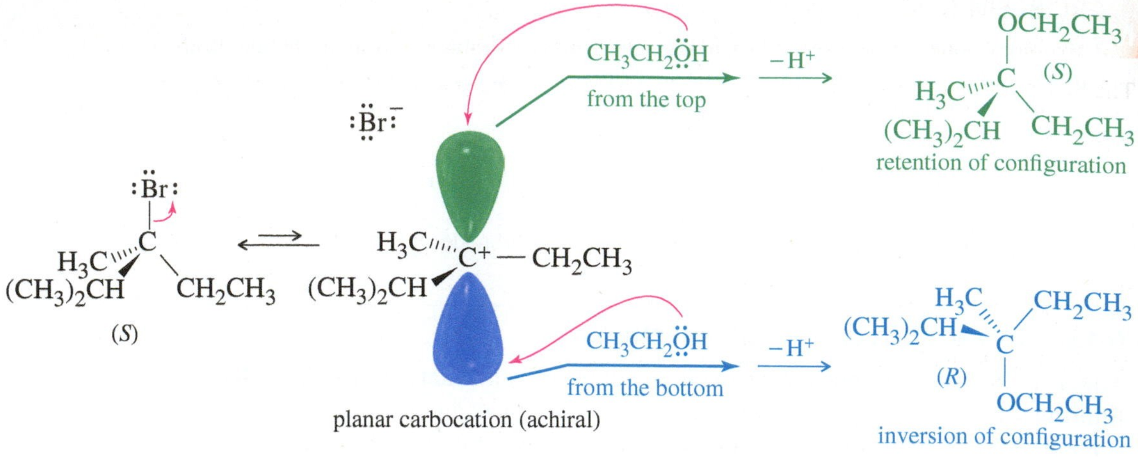

FIGURE 6-11
Racemization. An asymmetric carbon atom undergoes racemization when it ionizes to a planar, achiral carbocation. A nucleophile can attack the carbocation from either face, giving either enantiomer of the product.

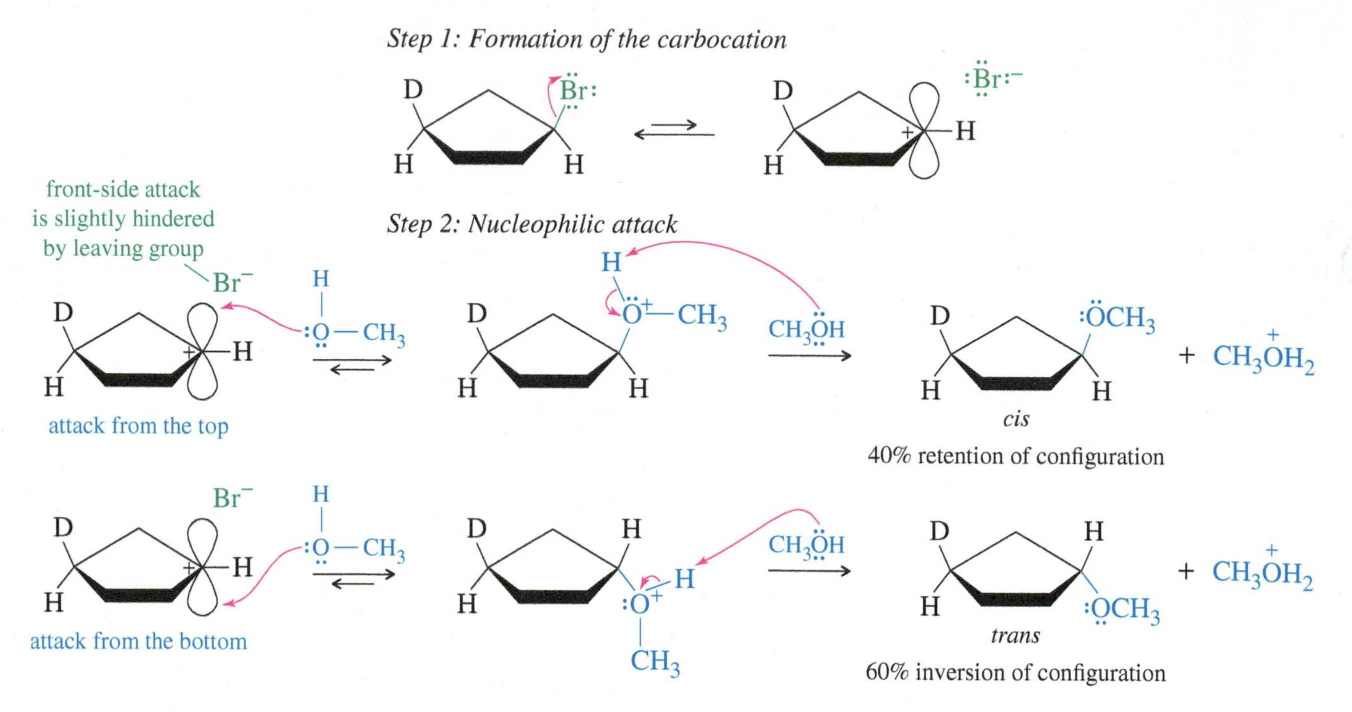

FIGURE 6-12
In the S_N1 reaction of *cis*-1-bromo-3-deuteriocyclopentane with methanol, the carbocation can be attacked from either face. Because the leaving group (bromide) partially blocks the front side as it leaves, back-side attack (inversion of configuration) is slightly favored.

Figure 6-12 shows a cyclic case where one of the faces of a cyclopentane ring has been "labeled" by a deuterium atom. Deuterium has the same size and shape as hydrogen, and it undergoes the same reactions. It distinguishes between the two faces of the ring: the bromine atom is cis to the deuterium in the reactant, so the nucleophile is cis to the deuterium in the retention product. The nucleophile is trans to the deuterium in the inversion product. The product mixture contains both cis and trans isomers, with the trans isomer slightly favored because the leaving group hinders approach of the nucleophilic solvent from the front side.

MECHANISM 6-5 Racemization in the S_N1 Reaction

The S_N1 reaction involves ionization to a flat carbocation, which can be attacked from either side.

Step 1: Ionization of a tetrahedral carbon gives a flat carbocation.

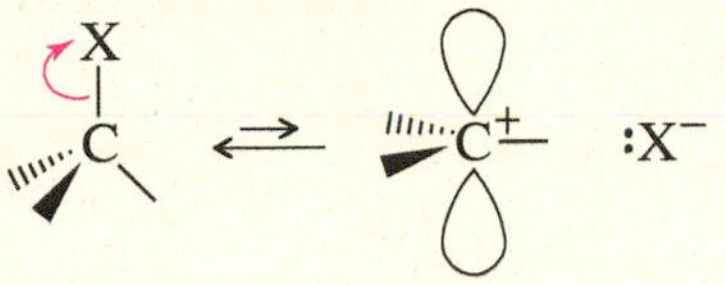

:X^-

Step 2: A nucleophile may attack either side of the carbocation.

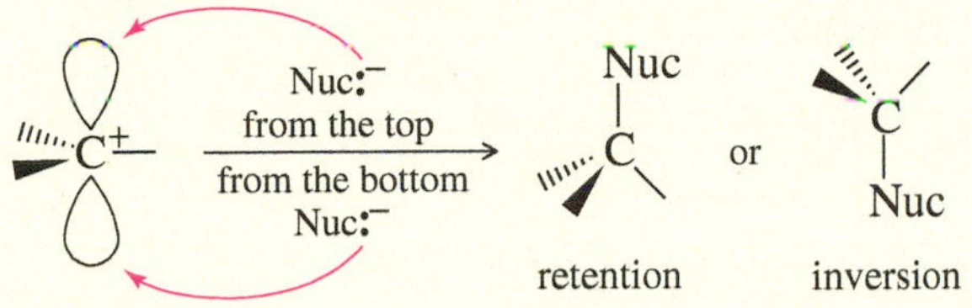

Nuc:$^-$ from the top → from the bottom Nuc:$^-$

retention or inversion

Racemization: Formation of products showing both retention and inversion of configuration.

These two products may be different if the carbon atom is stereogenic.

6-15 Rearrangements in S_N1 Reactions

Carbocations frequently undergo structural changes, called **rearrangements,** to form more stable ions. A rearrangement may occur after a carbocation has formed, or it may occur as the leaving group is leaving. Rearrangements are not seen in S_N2 reactions, where no carbocation is formed and the one-step mechanism allows no opportunity for rearrangement.

An example of a reaction with rearrangement is the S_N1 reaction of 2-bromo-3-methylbutane in boiling ethanol. The product is a mixture of 2-ethoxy-3-methylbutane (not rearranged) and 2-ethoxy-2-methylbutane (rearranged).

$$CH_3-CH(Br)-CH(CH_3)-CH_3 \xrightarrow[\text{heat}]{CH_3CH_2OH} CH_3-CH(OCH_2CH_3)-CH(CH_3)-CH_3 + CH_3-CH_2-C(OCH_2CH_3)(CH_3)-CH_3 + HBr$$

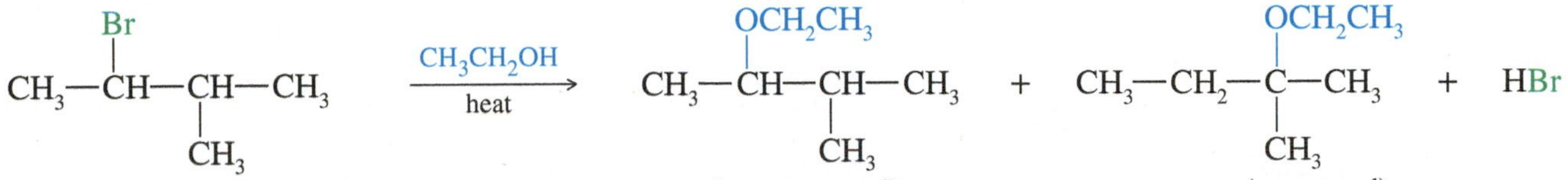

2-bromo-3-methylbutane; (not rearranged) 2-ethoxy-3-methylbutane; (rearranged) 2-ethoxy-2-methylbutane

PROBLEM 6-25

Give the S_N1 mechanism for the formation of 2-ethoxy-3-methylbutane, the unrearranged product in this reaction.

PROBLEM-SOLVING HINT

A carbocation usually rearranges just once. Most rearrangements convert the carbocation to a more stable carbocation, typically
2° ⟶ 3°
or 2° ⟶ resonance-stabliized
or 3° ⟶ resonance-stabilized.

The rearranged product, 2-ethoxy-2-methylbutane, results from a **hydride shift,** the movement of a hydrogen atom with its bonding pair of electrons. A hydride shift is represented by the symbol ~H. In this case, the hydride shift converts the initially formed secondary carbocation to a more stable tertiary carbocation. Attack by the solvent gives the rearranged product.

MECHANISM 6-6 Hydride Shift in an S_N1 Reaction

Carbocations often rearrange to form more stable carbocations. This may occur when a hydrogen atom moves with its bonding pair of electrons. Formally, this is the movement of a hydride ion (H:$^-$), although no actual free hydride ion is involved.

Step 1: Unimolecular ionization gives a carbocation.

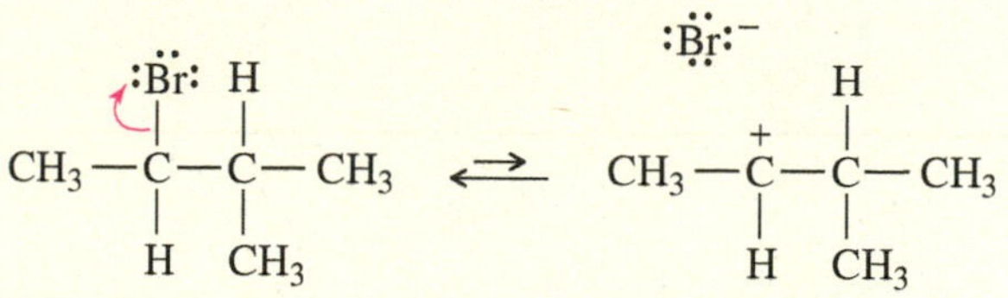

$$CH_3-CH(Br)-CH(CH_3)-CH_3 \rightleftarrows CH_3-\overset{+}{C}H-CH(CH_3)-CH_3 + :\ddot{Br}:^-$$

2° carbocation

Step 2: A hydride shift forms a more stable carbocation.

hydrogen moves with pair of electrons

$$CH_3-\overset{+}{C}H-CH(CH_3)-CH_3 \xrightarrow{\sim H} CH_3-CH_2-\overset{+}{C}(CH_3)-CH_3$$

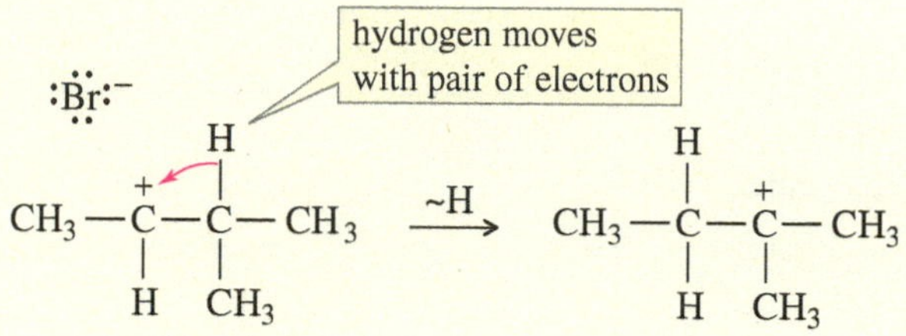

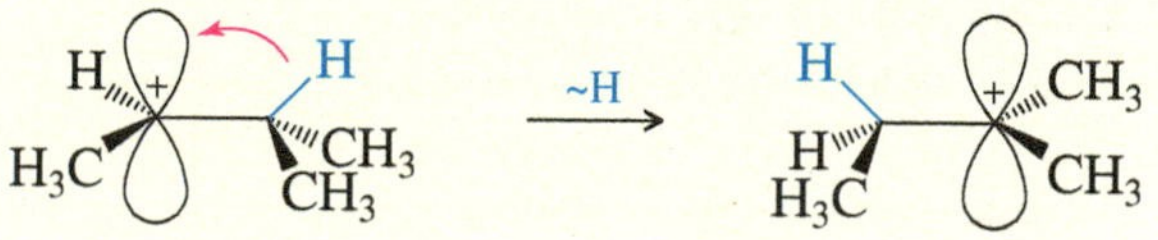

2° carbocation → 3° carbocation

This rearrangement involves movement of a hydrogen atom with its bonding pair of electrons over to the empty *p* orbital of the carbocation. In three dimensions, the rearrangement looks like this:

2° carbocation $\xrightarrow{\sim H}$ 3° carbocation

Step 3: Solvent (a weak nucleophile) attacks the rearranged carbocation.

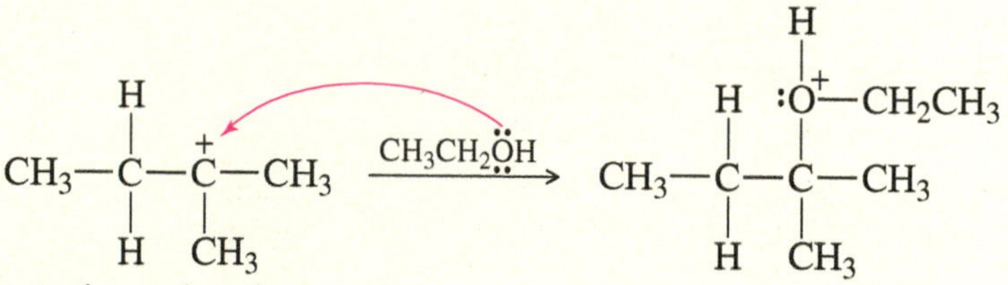

$$CH_3-CH_2-\overset{+}{C}(CH_3)-CH_3 \xrightarrow{CH_3CH_2\ddot{O}H} CH_3-CH_2-C(CH_3)(\overset{+}{O}H-CH_2CH_3)-CH_3$$

tertiary carbocation

Step 4: Deprotonation gives the rearranged product.

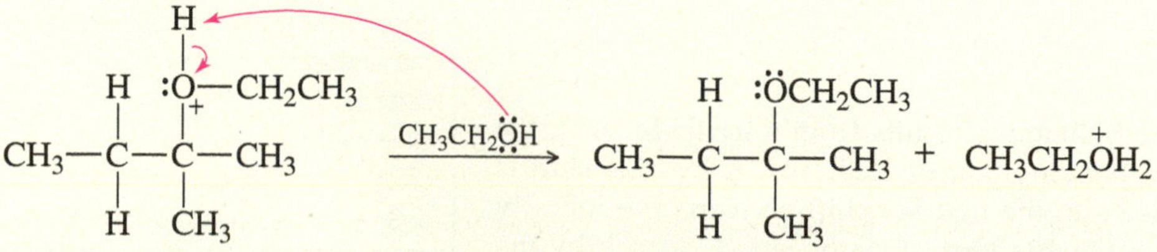

$$CH_3-CH_2-C(CH_3)(\overset{+}{O}H-CH_2CH_3)-CH_3 \xrightarrow{CH_3CH_2\ddot{O}H} CH_3-CH_2-C(CH_3)(\ddot{O}CH_2CH_3)-CH_3 + CH_3CH_2\overset{+}{O}H_2$$

rearranged product

When neopentyl bromide is boiled in ethanol, it gives *only* a rearranged substitution product. This product results from a **methyl shift** (represented by the symbol $\sim CH_3$), the migration of a methyl group together with its pair of electrons. Without rearrangement, ionization of neopentyl bromide would give a very unstable primary carbocation.

PROBLEM-SOLVING HINT

Primary halides and methyl halides rarely ionize to carbocations in solution. If a primary halide ionizes, it will likely ionize with rearrangement.

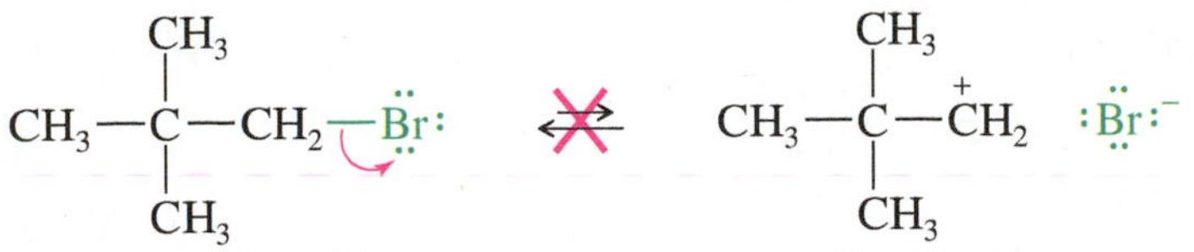

$$CH_3-C(CH_3)_2-CH_2-\ddot{Br}: \;\not\rightleftarrows\; CH_3-C(CH_3)_2-\overset{+}{C}H_2 \quad :\ddot{Br}:^-$$

neopentyl bromide — 1° carbocation

The methyl shift occurs *while* bromide ion is leaving, so that only the more stable tertiary carbocation is formed.

MECHANISM 6-7 Methyl Shift in an S_N1 Reaction

An alkyl group can rearrange to make a carbocation more stable.

Step 1: Ionization occurs with a methyl shift.

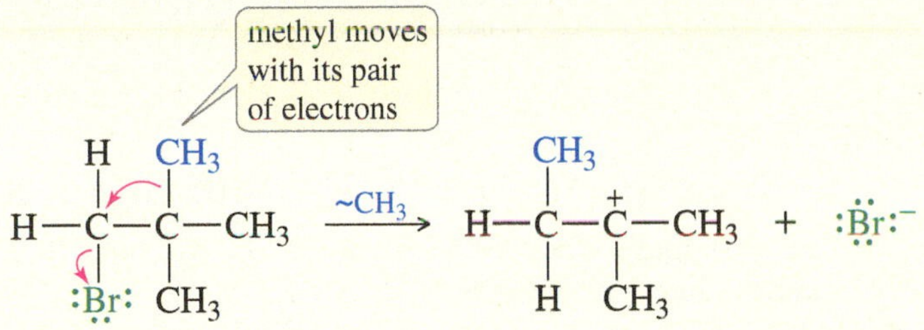

methyl moves with its pair of electrons

$$H-C(H)(Br)-C(CH_3)_2-CH_3 \xrightarrow{\sim CH_3} H-C(H)(CH_3)-\overset{+}{C}(CH_3)-CH_3 + :\ddot{Br}:^-$$

In three dimensions,

$$\xrightarrow{\sim CH_3} \text{3° carbocation} + Br^-$$

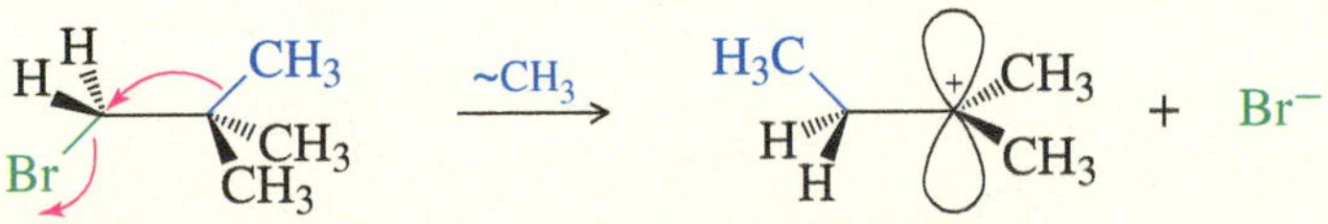

3° carbocation

Step 2: Attack by ethanol gives a protonated version of the rearranged product.

$$H-C(H)(CH_3)-\overset{+}{C}(CH_3)-CH_3 \xrightarrow{CH_3CH_2\ddot{O}H} H-C(H)(H_3C)-C(CH_3)(\overset{+}{O}(H)-CH_2CH_3)-CH_3$$

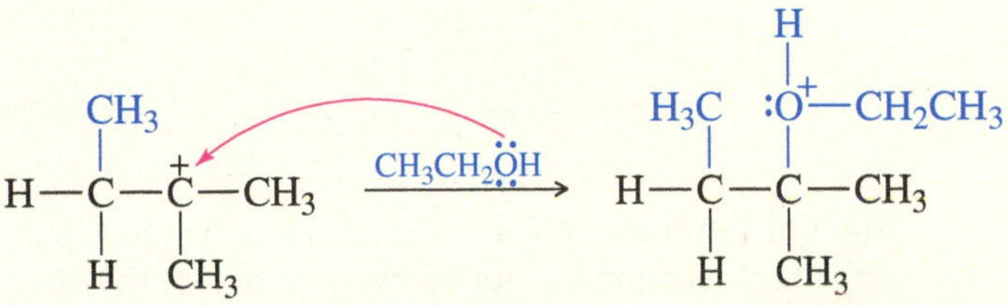

protonated product

Step 3: Deprotonation gives the rearranged product.

$$H-C(H)(H_3C)-C(CH_3)(\overset{+}{O}(H)-CH_2CH_3)-CH_3 \xrightarrow{CH_3CH_2\ddot{O}H} H-C(H)(H_3C)-C(CH_3)(\ddot{O}-CH_2CH_3)-CH_3 + CH_2CH_3\overset{+}{O}H_2$$

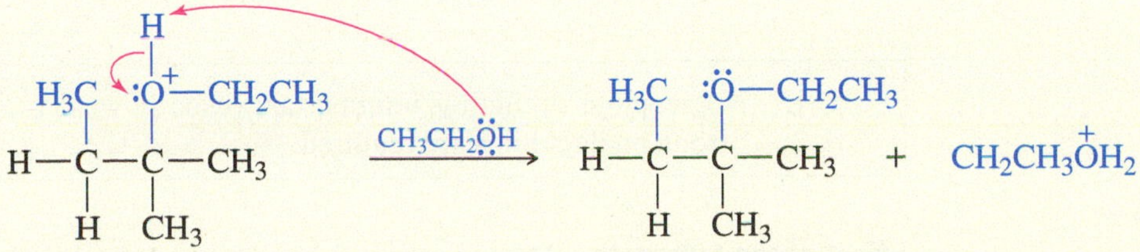

rearranged product

Because rearrangement is required for ionization of this primary halide, only rearranged products are observed.

In general, we should expect rearrangements in reactions involving carbocations whenever a hydride shift or an **alkyl shift** can form a more stable carbocation. Most rearrangements convert 2° (or incipient 1°) carbocations to 3° or resonance-stabilized carbocations.

PROBLEM-SOLVING HINT

Most rearrangements convert 2° (or incipient 1°) carbocations to 3° or resonance-stabilized carbocations.

PROBLEM 6-26

Propose a mechanism involving a hydride shift or an alkyl shift for each solvolysis reaction. Explain how each rearrangement forms a more stable intermediate.

(a) 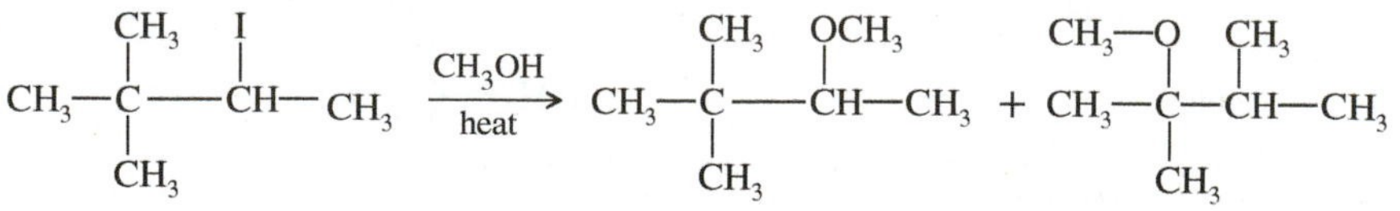 $(CH_3)_3C{-}CHI{-}CH_3 \xrightarrow[\text{heat}]{CH_3OH} (CH_3)_3C{-}CH(OCH_3){-}CH_3 + CH_3{-}C(CH_3)(OCH_3){-}CH(CH_3){-}CH_3$

(b) 1-chloro-2-methylcyclohexane $\xrightarrow[\text{heat}]{CH_3CH_2OH}$ OCH_2CH_3 / CH_3 (2-methylcyclohexyl ethyl ether) + OCH_2CH_3 / CH_3 (1-methylcyclohexyl ethyl ether)

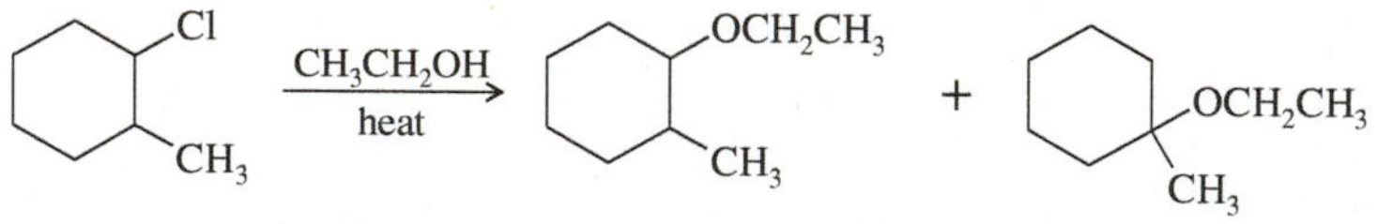

(c) 3-iodocyclohexene + $CH_3{-}\overset{O}{\overset{\|}{C}}{-}OH \xrightarrow{\text{heat}}$ cyclohex-2-enyl $O{-}\overset{O}{\overset{\|}{C}}{-}CH_3$ + cyclohex-2-enyl $O{-}\overset{O}{\overset{\|}{C}}{-}CH_3$

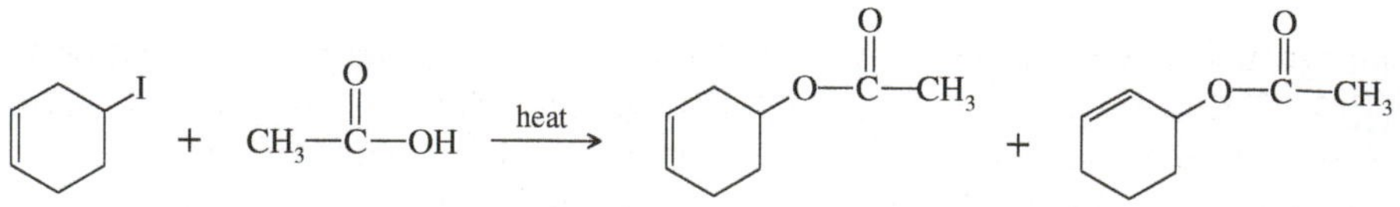

(d) 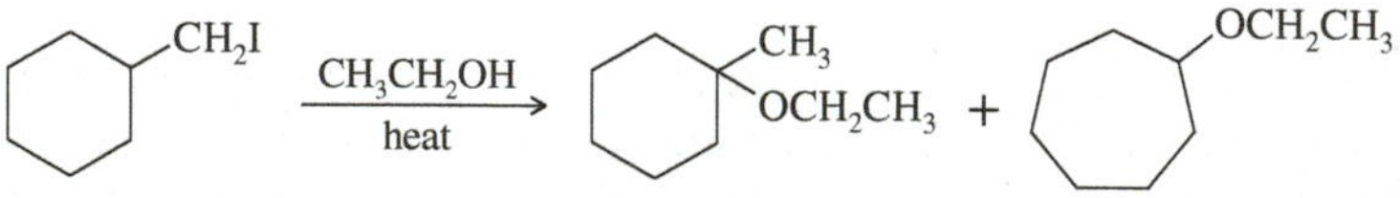 cyclohexyl-$CH_2I \xrightarrow[\text{heat}]{CH_3CH_2OH}$ 1-methylcyclohexyl (CH_3, OCH_2CH_3) + cycloheptyl-OCH_2CH_3

6-16 Comparison of S_N1 and S_N2 Reactions

Let's compare what we know about the S_N1 and S_N2 reactions, then organize this material into a brief table.

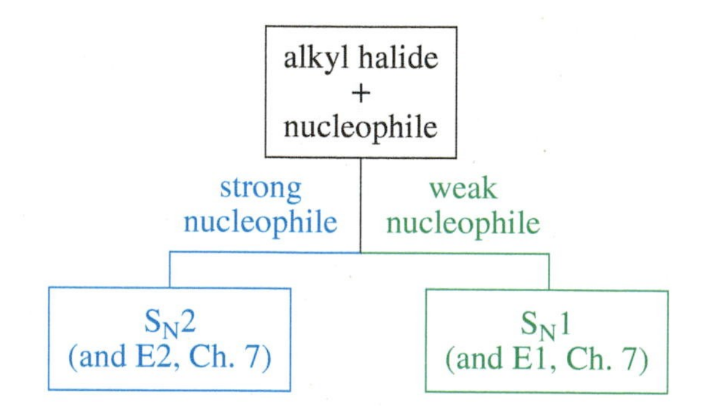

Effect of the Nucleophile The nucleophile takes part in the slow step (the only step) of the S_N2 reaction but not in the slow step of the S_N1. Therefore, a strong nucleophile promotes the S_N2 but not the S_N1. Weak nucleophiles fail to promote the S_N2 reaction; therefore, reactions with weak nucleophiles often go by the S_N1 mechanism if the substrate is secondary or tertiary.

S_N1: Nucleophile strength is unimportant (usually weak).
S_N2: Strong nucleophiles are required.

Effect of the Substrate The structure of the substrate (the alkyl halide) is an important factor in determining which of these substitution mechanisms might operate. Most methyl halides and primary halides are poor substrates for S_N1 substitutions because they cannot easily ionize to high-energy methyl and primary carbocations. They are relatively unhindered, however, so they make good S_N2 substrates.

Tertiary halides are too hindered to undergo S_N2 displacement, but they can ionize to form tertiary carbocations. Tertiary halides undergo substitution exclusively through the S_N1 mechanism. Secondary halides can undergo substitution by either mechanism, depending on the conditions.

S_N1 substrates:	$3° > 2°$	($1°$ and CH_3X are unlikely)
S_N2 substrates:	$CH_3X > 1° > 2°$	($3°$ is unsuitable)

If silver nitrate ($AgNO_3$) is added to an alkyl halide in a good ionizing solvent, the silver ion removes the halide ion to give a carbocation. This technique can force some unlikely ionizations, often giving interesting rearrangements (see Problem 6-29).

Effect of the Solvent The slow step of the S_N1 reaction involves formation of two ions. Solvation of these ions is crucial to stabilizing them and lowering the activation energy for their formation. Very polar ionizing solvents such as water and alcohols are needed for the S_N1. The solvent may be heated to reflux (boiling) to provide the energy needed for ionization.

Less charge separation is generated in the transition state of the S_N2 reaction. Strong solvation may weaken the strength of the nucleophile because of the energy needed to strip off the solvent molecules. Thus, the S_N2 reaction often goes faster in less polar solvents if the nucleophile will dissolve. Polar aprotic solvents may enhance the strength of weak nucleophiles.

S_N1:	Good ionizing solvent required.
S_N2:	May go faster in a less polar solvent.

Kinetics The rate of the S_N1 reaction is proportional to the concentration of the alkyl halide but not the concentration of the nucleophile. It follows a first-order rate equation.

The rate of the S_N2 reaction is proportional to the concentrations of both the alkyl halide [R—X] and the nucleophile [Nuc:$^-$]. It follows a second-order rate equation.

$$S_N1 \text{ rate} = k_r[\text{R—X}]$$
$$S_N2 \text{ rate} = k_r[\text{R—X}][\text{Nuc:}^-]$$

Stereochemistry The S_N1 reaction involves a flat carbocation intermediate that can be attacked from either face. Therefore, the S_N1 usually gives a mixture of inversion and retention of configuration.

The S_N2 reaction takes place through a back-side attack, which inverts the stereochemistry of the carbon atom. Complete inversion of configuration is the result.

S_N1 stereochemistry:	Mixture of retention and inversion; racemization.
S_N2 stereochemistry:	Complete inversion.

Rearrangements The S_N1 reaction involves a carbocation intermediate. This intermediate can rearrange, usually by a hydride shift or an alkyl shift, to give a more stable carbocation.

The S_N2 reaction takes place in one step with no intermediates. No rearrangement is possible in the S_N2 reaction.

> S_N1: Rearrangements are common.
> S_N2: Rearrangements are impossible.

SUMMARY Nucleophilic Substitutions

	S_N1	S_N2
Promoting factors		
nucleophile	weak nucleophiles are OK	strong nucleophile needed
substrate (RX)	3° > 2°	CH_3X > 1° > 2°
solvent	good ionizing solvent needed	wide variety of solvents
leaving group	good one required	good one required
other	$AgNO_3$ forces ionization	
Characteristics		
kinetics	first order, k_r[RX]	second order, k_r[RX][Nuc:$^-$]
stereochemistry	mixture of inversion and retention	complete inversion
rearrangements	common	impossible

	S_N1 conditions (weak nucleophile)	S_N2 conditions (strong nucleophile)
methyl halides CH_3X	No reaction. A methyl cation is too unstable.	S_N2 is unhindered and favored.
primary halides RCH_2X	Rarely any reaction unless it can form a resonance-stabilized cation.	S_N2 is favored unless the R group is exceptionally bulky.
secondary halides R_2CHX	S_N1, often with rearrangement, can occur in a good solvent.	S_N2 can occur unless the alkyl groups or the nucleophile are bulky.
tertiary halides R_3CX	S_N1 occurs readily in a good solvent.	S_N2 cannot occur because of steric hindrance.

PROBLEM-SOLVING HINT

The strength of the nucleophile (or base) usually determines the order of the reaction. Strong nucleophiles encourage bimolecular, second-order reactions, and weak nucleophiles more commonly react by unimolecular, first-order mechanisms. Also, S_N2 is unlikely with 3° halides, and S_N1 is unlikely with 1° halides unless they are resonance-stabilized.

PROBLEM 6-27

For each reaction, give the expected substitution product, and predict whether the mechanism will be predominantly first order (S_N1) or second order (S_N2).

(a) 2-chloro-2-methylbutane + CH_3COOH
(b) isobutyl bromide + sodium methoxide
(c) 1-iodo-1-methylcyclohexane + ethanol
(d) cyclohexyl bromide + methanol
(e) cyclohexyl bromide + sodium ethoxide

PROBLEM 6-28

Under certain conditions, when (*R*)-2-bromobutane is heated with water, the S_N1 substitution proceeds twice as fast as the S_N2. Calculate the e.e. and the specific rotation expected for the product. The specific rotation of (*R*)-butan-2-ol is −13.5°. Assume that the S_N1 gives equal amounts of the two enantiomers.

PROBLEM 6-29

A reluctant first-order substrate can be forced to ionize by adding some silver nitrate (one of the few soluble silver salts) to the reaction. Silver ion reacts with the halogen to form a silver halide (a highly exothermic reaction), generating the cation of the alkyl group.

$$R{-}X + Ag^+ \longrightarrow R^+ + AgX\downarrow$$

Give mechanisms for the following silver-promoted rearrangements.

(a) $CH_3{-}C(CH_3)_2{-}CH_2{-}I \xrightarrow{AgNO_3,\ H_2O} CH_3{-}C(CH_3)(OH){-}CH_2{-}CH_3$

(b) (bicyclic compound with CH_2I) $\xrightarrow{AgNO_3,\ H_2O/CH_3CH_2OH}$ (ring-expanded bicyclic compound with OH)

The Rest of the Story Elimination reactions often accompany nucleophilic substitutions of alkyl halides and other compounds containing good leaving groups. Under conditions that favor them, elimination reactions often predominate over substitutions. In Chapter 6, we have concentrated on the factors that favor S_N1 and S_N2 substitutions. In Chapter 7, we will cover the factors that favor eliminations. At that point, we can begin to predict the products and mechanisms of many reactions that are subject to competition between substitutions and eliminations.

SUMMARY Reactions of Alkyl Halides

Some of these reactions have not yet been covered, but they are included here for completeness and for later reference. The reactions that we have not yet covered are shown in the summary with a gray background. Notice the section numbers, indicating where each reaction is covered.

1. *Nucleophilic substitutions* (Section 6-9)

a. *Alcohol formation*

$$R{-}X + {}^{-}\!:\ddot{O}H \longrightarrow R{-}OH + :X^-$$

b. *Halide exchange*

$$R{-}X + :\ddot{I}:^- \longrightarrow R{-}I + :X^-$$

$$R{-}Cl + KF \xrightarrow[CH_3CN]{18\text{-crown-6}} R{-}F + KCl$$

c. *Williamson ether synthesis*

$$R{-}X + R'\ddot{O}:^- \longrightarrow R{-}\ddot{O}{-}R' + :X^- \quad \text{ether synthesis}$$

$$R{-}X + R'\ddot{S}:^- \longrightarrow R{-}\ddot{S}{-}R' + :X^- \quad \text{thioether synthesis}$$

d. *Amine synthesis*

$$R{-}X + \underset{\text{excess}}{:NH_3} \longrightarrow R{-}NH_3^+\ :X^- \xrightarrow{:NH_3} \underset{\text{amine}}{R{-}\ddot{N}H_2} + NH_4^+\ :X^-$$

(continued)

e. *Nitrile synthesis*

$$R{-}X \quad + \quad {}^{-}\!:\!C{\equiv}N\!: \quad \longrightarrow \quad R{-}C{\equiv}N\!: \quad + \quad :\!X^{-}$$

cyanide → nitrile

f. *Alkyne synthesis* (covered later in Section 9-7)

$$R{-}C{\equiv}C\!:^{-} \quad + \quad R'{-}X \quad \longrightarrow \quad R{-}C{\equiv}C{-}R' \quad + \quad :\!X^{-}$$

acetylide ion → alkyne

2. *Dehydrohalogenation* (covered later in Chapter 7)

$$\text{H}{-}\overset{|}{\underset{|}{C}}{-}\overset{|}{\underset{|}{C}}{-}\text{X} \xrightarrow{\text{KOH}} \;>\!C{=}C\!<\; + \; :\!X^{-}$$

3. *Formation of organometallic reagents* (covered later in Section 10-8)

a. *Grignard reagents*

$$R{-}X \quad + \quad Mg \xrightarrow{CH_3CH_2{-}O{-}CH_2CH_3} R{-}Mg{-}X$$

(X = Cl, Br, or I) → organomagnesium halide (Grignard reagent)

b. *Organolithium reagents*

$$R{-}X \quad + \quad 2\,Li \quad \longrightarrow \quad R{-}Li \quad + \quad Li^{+}\;X^{-}$$

(X = Cl, Br, or I) → organolithium

4. *Coupling of organocopper reagents* (covered later in Section 10-9)

$$2\,R{-}Li \quad + \quad CuI \quad \longrightarrow \quad R_2CuLi \quad + \quad LiI$$

$$R_2CuLi \quad + \quad R'{-}X \quad \longrightarrow \quad R{-}R' \quad + \quad R{-}Cu \quad + \quad LiX$$

5. *Reduction* (covered later in Section 10-10)

$$R{-}X \xrightarrow[\text{(2) } H_2O]{\text{(1) Mg or Li}} R{-}H$$

Essential Terms

acid A species that can donate a proton. (p. 67)

acidity: **(acid strength)** The thermodynamic reactivity of an acid, expressed quantitatively by the acid-dissociation constant K_a. (p. 70)

Lewis acid: **(electrophile)** A species that can accept an electron pair from a nucleophile, forming a bond. (pp. 87, 271)

alkyl halide **(haloalkane)** A derivative of an alkane in which one (or more) of the hydrogen atoms has been replaced by a halogen. (p. 247)

alkyl shift (symbolized **~R**) Movement of an alkyl group with a pair of electrons from one atom (usually carbon) to another. Alkyl shifts are examples of rearrangements that convert carbocations into more stable carbocations. (p. 283)

allylic The saturated position adjacent to a carbon–carbon double bond. (p. 256)

allylic halogenation Substitution of a halogen for a hydrogen at the allylic position. (p. 256)

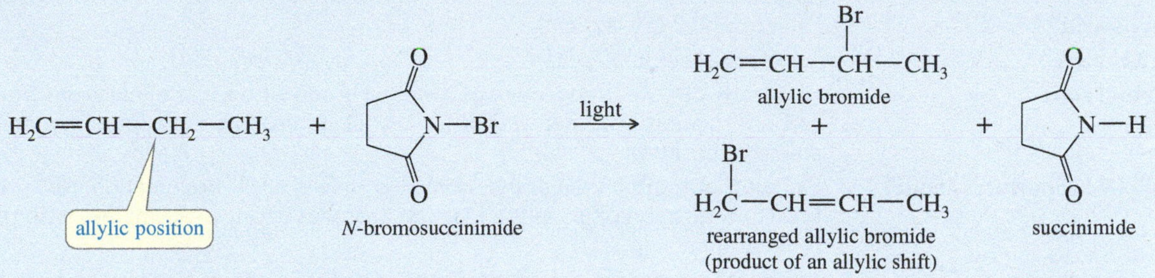

$$H_2C{=}CH{-}CH_2{-}CH_3 + \text{N-bromosuccinimide} \xrightarrow{\text{light}} H_2C{=}CH{-}CH(Br){-}CH_3 \text{ (allylic bromide)} + BrH_2C{-}CH{=}CH{-}CH_3 \text{ (rearranged allylic bromide)} + \text{succinimide}$$

allylic position

N-bromosuccinimide

allylic bromide

rearranged allylic bromide (product of an allylic shift)

succinimide

allylic shift A rearrangement that results from reaction at either end of a resonance-stabilized allylic intermediate. (p. 257)

aprotic solvent A solvent that has no acidic protons; a solvent with no O—H or N—H groups. (p. 268)

aryl halide An aromatic compound (benzene derivative) in which a halogen is bonded to one of the carbon atoms of the aromatic ring. (p. 247)

base An electron-rich species that can abstract a proton. (pp. 68, 266)

basicity: **(base strength)** The thermodynamic reactivity of a base, expressed quantitatively by the base-dissociation constant K_b. (pp. 72, 266)

Lewis base: **(nucleophile)** An electron-rich species that can donate a pair of electrons to form a bond. (p. 87)

concerted reaction A reaction in which the breaking of bonds and the formation of new bonds occur at the same time (in one step). (p. 262)

dehydrohalogenation An elimination in which the two atoms lost are a hydrogen atom and a halogen atom. (p. 260)

electrophile **(Lewis acid)** A species that can accept an electron pair from a nucleophile, forming a bond. (p. 87)

electrophilicity: **(electrophile strength)** The kinetic reactivity of an electrophile. (p. 271)

elimination A reaction that involves the loss of two atoms or groups from the substrate, usually resulting in the formation of a pi bond. (p. 260)

freons A generic name for a group of chlorofluorocarbons used as refrigerants, propellants, and solvents. Freon-12® is CF_2Cl_2, and Freon-22® is $CHClF_2$. (p. 251)

geminal dihalide A dihalide with both halogens on the same carbon atom. (p. 249)

$$CH_3{-}CH_2{-}CBr_2{-}CH_3$$

a geminal dibromide

haloalkane **(alkyl halide)** A derivative of an alkane in which one (or more) of the hydrogen atoms has been replaced by a halogen. (p. 248)

halogen exchange reaction A substitution where one halogen atom replaces another; commonly used to form fluorides and iodides. (p. 264)

hydride shift (symbolized **~H**) Movement of a hydrogen atom with a pair of electrons from one atom (usually carbon) to another. Hydride shifts are examples of rearrangements that convert carbocations into more stable carbocations. (p. 281)

hydroxylic solvent A solvent containing OH groups (the most common type of protic solvents). (p. 268)

inversion of configuration (see also **Walden inversion**) A process in which the groups around an asymmetric carbon atom are changed to the opposite spatial configuration, usually as a result of **back-side attack.** (p. 273)

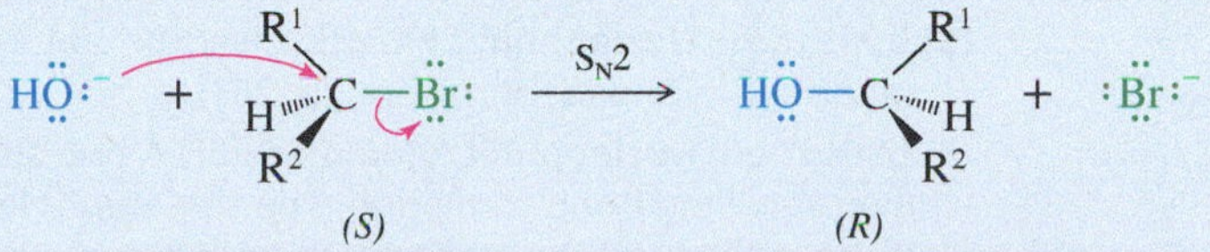

$$HO^- + (S)\text{-}C(R^1)(R^2)(H){-}Br \xrightarrow{S_N2} (R)\text{-}HO{-}C(R^1)(R^2)(H) + Br^-$$

The S_N2 reaction goes with *inversion of configuration.*

leaving group The atom or group of atoms that departs during a substitution or elimination. The leaving group can be charged or uncharged, but it leaves with the pair of electrons that originally bonded the group to the remainder of the molecule. (p. 260)

Lewis acid See **electrophile.** (pp. 87, 271)

Lewis base See **nucleophile.** (pp. 87, 261)

methyl shift (symbolized ~CH_3) Rearrangement of a methyl group with a pair of electrons from one atom (usually carbon) to another. A methyl shift (or any alkyl shift) in a carbocation generally results in a more stable carbocation. (p. 283)

NBS (*N*-bromosuccinimide) A reagent that is often used as the bromine source in allylic bromination. NBS provides a consistently low concentration of Br_2 as the reaction proceeds because it combines with the HBr byproduct to regenerate Br_2. (p. 257)

nucleophile **(Lewis base)** An electron-rich species that can donate a pair of electrons to form a bond. (pp. 87, 261)

nucleophilicity: **(nucleophile strength)** The kinetic reactivity of a nucleophile; a measure of the rate of substitution in a reaction with a standard substrate. (p. 266)

nucleophilic substitution A reaction where a nucleophile replaces another group or atom (the leaving group) in a molecule. (p. 260)

organic synthesis The preparation of desired organic compounds from readily available starting materials. (p. 250)

polarizable Having electrons that are easily displaced toward a positive charge. Polarizable atoms can begin to form a bond at a relatively long distance. (p. 267)

primary halide, secondary halide, tertiary halide These terms specify the substitution of the halogen-bearing carbon atom (sometimes called the ***head carbon***). If the head carbon is bonded to one other carbon, it is **primary;** if it is bonded to two carbons, it is **secondary;** and if bonded to three carbons, it is **tertiary.** (p. 249)

$CH_3—CH_2—Br$ — a primary halide (1°)

$CH_3—CH(CH_3)—Br$ — a secondary halide (2°)

$CH_3—C(CH_3)_2—Br$ — a tertiary halide (3°)

protic solvent A solvent containing acidic protons, usually O—H or N—H groups. (p. 268)

racemization The loss of optical activity that occurs when a reaction shows neither clean retention of configuration nor clean inversion of configuration. (p. 279)

reagent The compound that serves as the attacking species in a reaction. (p. 260)

rearrangement A reaction involving a change in the bonding sequence within a molecule. Rearrangements are common in reactions such as the S_N1 and E1 involving carbocation intermediates. (p. 281)

retention of configuration Formation of a product with the same configuration as the reactant. In a nucleophilic substitution, retention of configuration occurs when the nucleophile assumes the same stereochemical position in the product as the leaving group occupied in the reactant. (p. 279)

solvolysis A nucleophilic substitution or elimination where the solvent serves as the attacking reagent. Solvolysis literally means "cleavage by the solvent." (p. 275)

$$(CH_3)_3C—Br \xrightarrow{CH_3OH,\ heat} (CH_3)_3C—OCH_3 + (CH_3)_2C{=}CH_2 + HBr$$

stereospecific reaction A reaction in which different stereoisomers of the starting material react to give different stereoisomers of the product. (p. 274)

steric hindrance Interference by bulky groups that slow a reaction or prevent it from occurring. (p. 267)

substitution (displacement) A reaction in which an attacking species (nucleophile, electrophile, or free radical) replaces another group. (p. 260)

S_N2 reaction: **(Substitution, Nucleophilic, bimolecular)** The concerted displacement of one nucleophile by another on an sp^3 hybrid carbon atom. (p. 261)

S_N1 reaction: **(Substitution, Nucleophilic, unimolecular)** A two-step interchange of nucleophiles, with bond breaking preceding bond formation. The first step is ionization to form a carbocation. The second step is the reaction of the carbocation with a nucleophile. (p. 275)

substrate The compound that is attacked by the reagent. (p. 261)

transition state In each individual step of a reaction, the state of highest energy between reactants and products. The transition state is a relative maximum (high point) on the reaction-energy diagram. (p. 262)

vicinal dihalide A dihalide with the halogens on adjacent carbon atoms. (p. 249)

$CH_3—CHBr—CHBr—CH_3$
a vicinal dibromide

$CH_3—CH{=}C(Br)—CH_3$
a vinyl bromide

vinyl halide A derivative of an alkene in which one (or more) of the hydrogen atoms on the double-bonded carbon atoms has been replaced by a halogen. (p. 247)

Walden inversion (see also **inversion of configuration**) A step in a reaction sequence in which an asymmetric carbon atom undergoes inversion of configuration. (p. 273)

Essential Problem-Solving Skills in Chapter 6

Each skill is followed by problem numbers exemplifying that particular skill.

1 Correctly name alkyl halides; summarize their physical properties; and identify them as 1°, 2°, or 3°.	Problems 6-31, 32, and 47
2 Show how free-radical halogenation might be used for the synthesis of some alkyl halides, especially for making allylic and benzylic alkyl halides.	Problems 6-30, 46, 51, 57, 58, and 59
3 Predict the products of S_N1 and S_N2 reactions, including stereochemistry.	Problems 6-35, 40, 41, 44, 45, 46, and 52
4 Draw the mechanisms and reaction-energy diagrams of S_N1 and S_N2 reactions.	Problems 6-40, 41, 48, 49, 52, 53, 54, 55, and 56
5 Predict and explain the relative stabilities and rearrangements of cations involved in first-order reactions.	Problems 6-40, 41, 42, 43, and 54
6 Predict which substitutions will be faster, based on differences in substrate, nucleophile, leaving group, and solvent.	Problems 6-33, 34, 36, 37, 38, and 39
7 Given a set of reaction conditions, predict whether the reaction will be unimolecular (first-order) or bimolecular (second-order), and predict which product(s) are most likely.	Problems 6-37, 38, 54, and 56
8 Show how substitutions of alkyl halides might be used to synthesize other types of compounds.	Problems 6-30, 35, 36, 39, 40, 45, 46, 53, and 54

Study Problems

6-30 Show how you would convert (in one or two steps) 1-phenylpropane to the three products shown below. In each case, explain what unwanted reactions might produce undesirable impurities in the product.

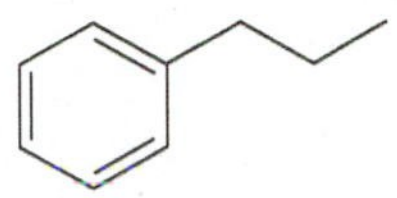

1-phenylpropane

(a) Br — 1-bromo-1-phenylpropane

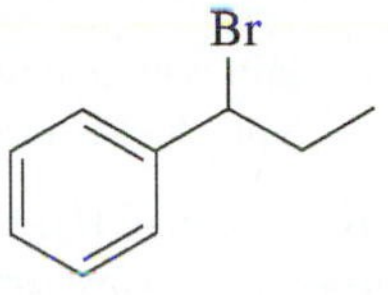

(b) OCH_3 — 1-methoxy-1-phenylpropane

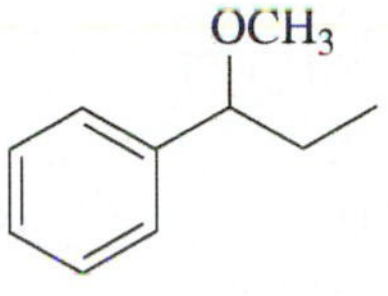

(c) C≡N — 2-phenylbutanenitrile

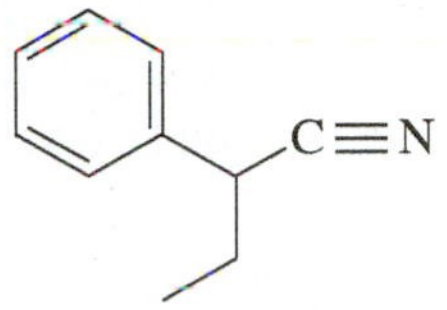

6-31 Draw the structures of the following compounds.

(a) *sec*-butyl chloride **(b)** isobutyl bromide **(c)** 1,2-dibromo-3-methylpentane
(d) 2,2,2-trichloroethanol **(e)** *trans*-1-chloro-2-methylcyclohexane **(f)** methylene chloride
(g) chloroform **(h)** 1-chloro-1-isopropylcyclopentane **(h)** *tert*-pentyl iodide

6-32 Give systematic (IUPAC) names for the following compounds.

(a) 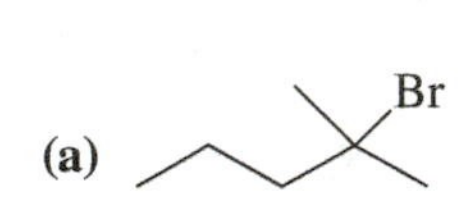

(b) 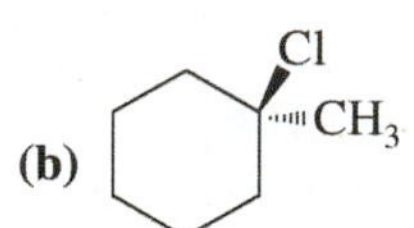

(c) 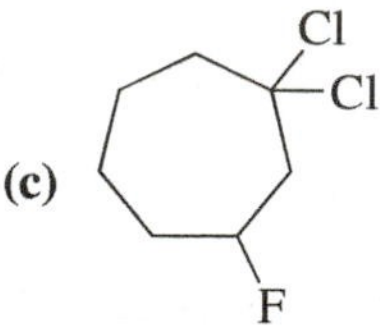

(d) 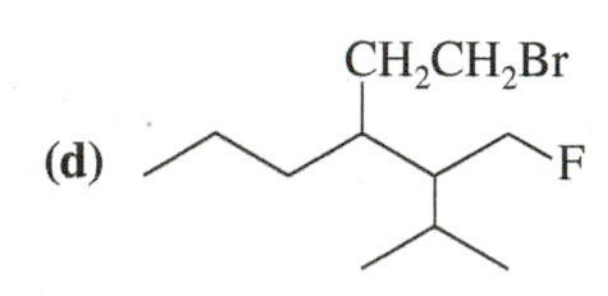

(e) 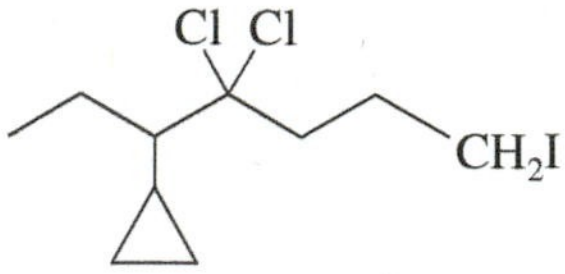 Cl Cl CH_2I

(f) 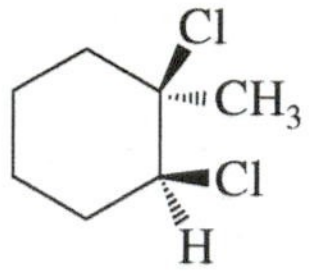 Cl CH_3 Cl H

6-33 Predict the compound in each pair that will undergo the S_N2 reaction faster.

(a) 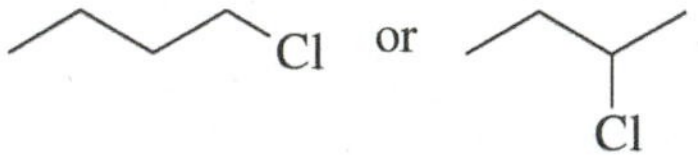 Cl or Cl

(b) 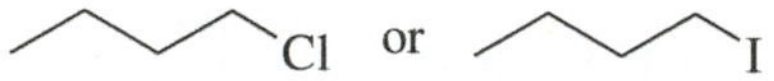 Cl or I

(c) 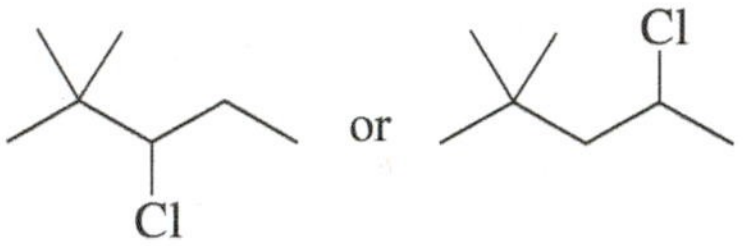 Cl or Cl

(d) 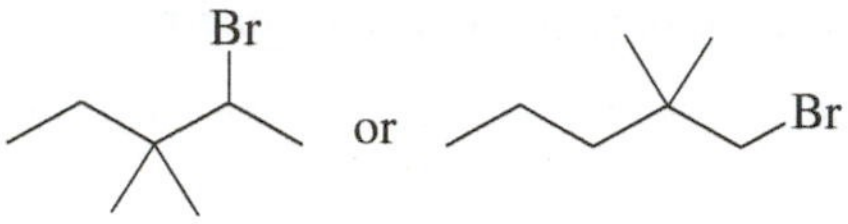 Br or Br

(e) 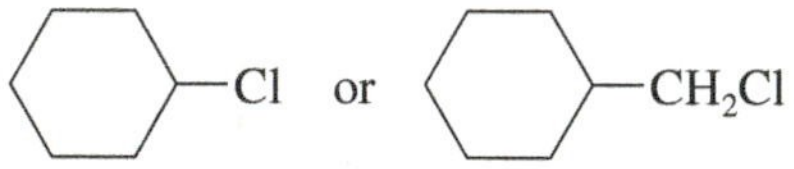 Cl or CH_2Cl

(f) 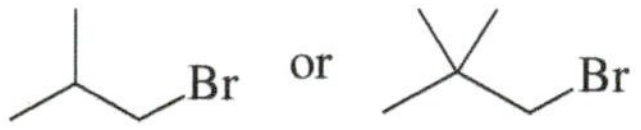 Br or Br

6-34 Predict the compound in each pair that will undergo solvolysis (in aqueous ethanol) more rapidly.

(a) $(CH_3CH_2)_2CH—Cl$ or $(CH_3)_3C—Cl$

(b) 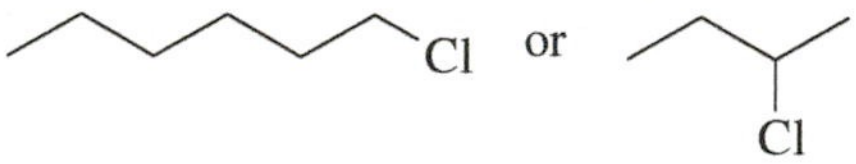 Cl or Cl

(c) 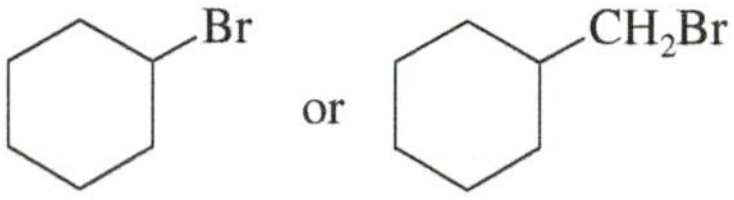 Br or CH_2Br

(d) 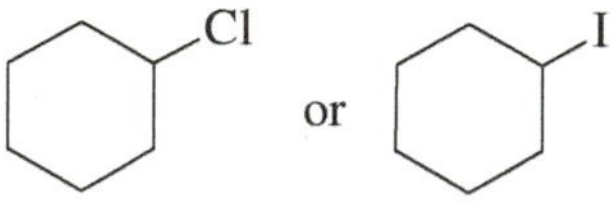 Cl or I

(e) 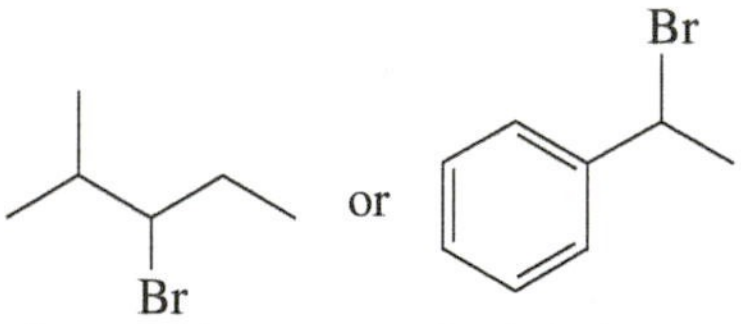 Br or Br

(f) 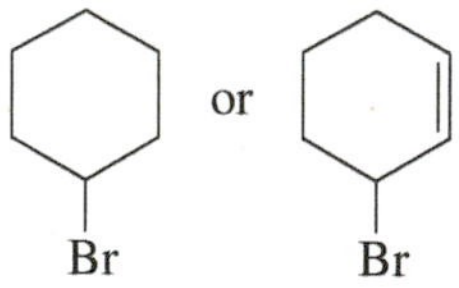 Br or Br

6-35 Show how each compound might be synthesized by the S_N2 displacement of an alkyl halide.

(a) 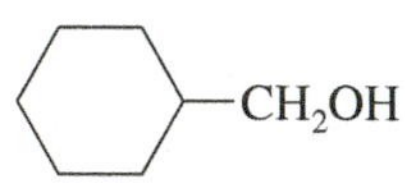 CH_2OH

(b) 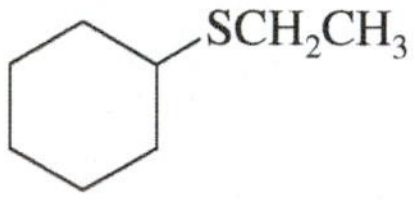 SCH_2CH_3

(c)  O

(d) 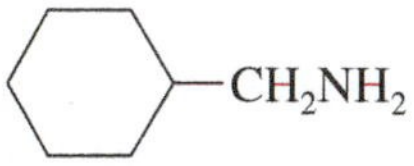 CH_2NH_2

(e) $H_2C=CH—CH_2CN$

(f) $H—C≡C—CH_2CH_2CH_3$

6-36 Give two syntheses for $(CH_3)_2CH—O—CH_2CH_3$, and explain which synthesis is better.

6-37 When ethyl bromide is added to potassium *tert*-butoxide, the product is ethyl *tert*-butyl ether.

$$\underset{\text{ethyl bromide}}{CH_3CH_2—Br} + \underset{\text{potassium } tert\text{-butoxide}}{(CH_3)_3C—O^- \ K^+} \longrightarrow \underset{\text{ethyl } tert\text{-butyl ether}}{(CH_3)_3C—O—CH_2CH_3}$$

(a) What happens to the reaction rate if the concentration of ethyl bromide is doubled?
(b) What happens to the rate if the concentration of potassium *tert*-butoxide is tripled and the concentration of ethyl bromide is doubled?
(c) What happens to the rate if the temperature is raised?

6-38 When *tert*-butyl bromide is heated with an equal amount of ethanol in an inert solvent, one of the products is ethyl *tert*-butyl ether.
(a) What happens to the reaction rate if the concentration of ethanol is doubled?
(b) What happens to the rate if the concentration of *tert*-butyl bromide is tripled and the concentration of ethanol is doubled?
(c) What happens to the rate if the temperature is raised?

6-39 Chlorocyclohexane reacts with sodium cyanide (NaCN) in ethanol to give cyanocyclohexane. The rate of formation of cyanocyclohexane increases when a small amount of sodium iodide is added to the solution. Explain this acceleration in the rate.

6-40 Give the substitution products expected from solvolysis of each compound by heating in ethanol.

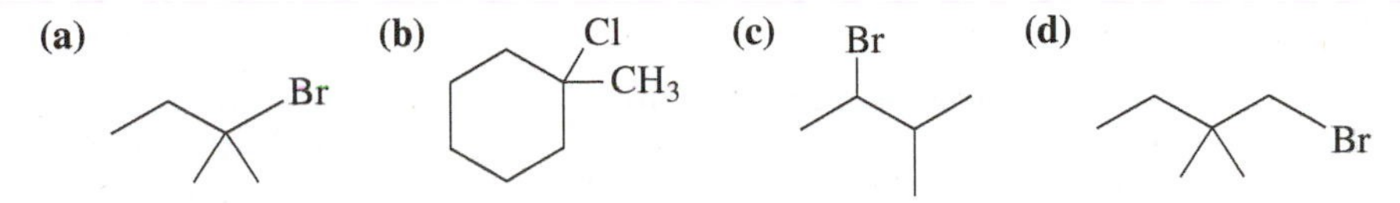

6-41 Allylic halides have the structure

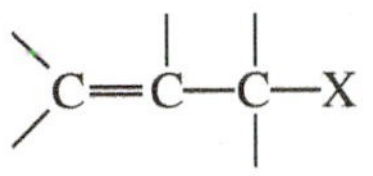

C=C—C—X

(a) Show how the first-order ionization of an allylic halide leads to a resonance-stabilized cation.
(b) Draw the resonance structures of the allylic cations formed by ionization of the following halides.
(c) Show the products expected from S_N1 solvolysis of these halides in ethanol.

(i) Br (ii) CH_2Br (iii) Br (iv) Br

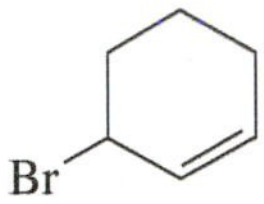

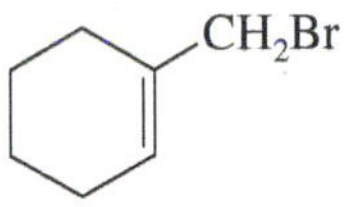

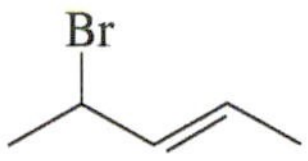

6-42 List the following carbocations in decreasing order of their stability.

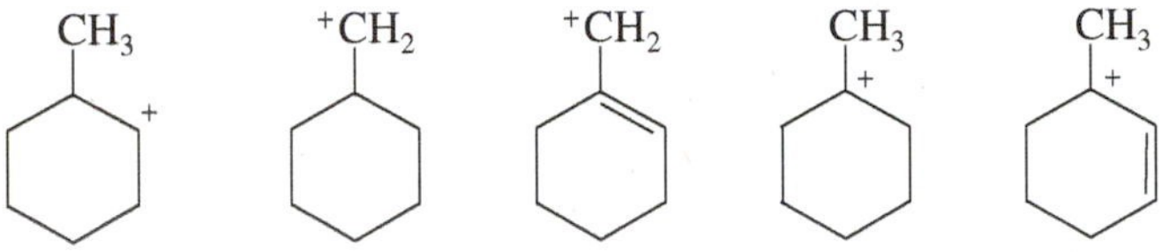

CH_3 (+) $^+CH_2$ $^+CH_2$ CH_3 (+) CH_3 (+)

6-43 Two of the carbocations in Problem 6-42 are prone to rearrangement. Show how they might rearrange to more stable carbocations.

6-44 Draw perspective structures or Fischer projections for the substitution products of the following reactions.

(a) CH_3; H—Br; CH_2CH_3 + NaCN $\xrightarrow{\text{acetone}}$

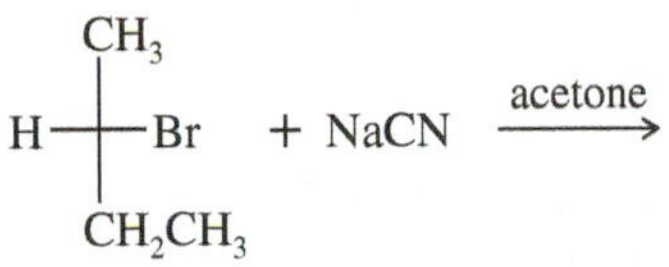

(b) CH_3; Br—H; H—CH_3; CH_2CH_3 + NaOH $\xrightarrow{\text{water/acetone}}$

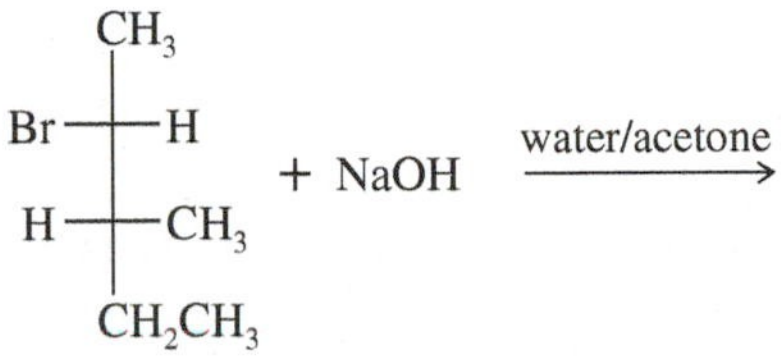

(c) CH_2CH_3; Br—CH_3; $CH(CH_3)_2$ $\xrightarrow{\text{EtOH, heat}}$

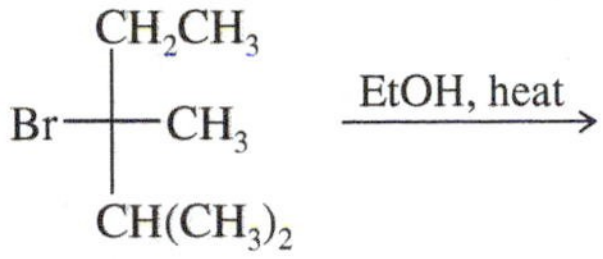

6-45 Predict the products of the following S_N2 reactions.

(a) $CH_3CH_2ONa + CH_3CH_2Cl \longrightarrow$

(b) (benzene)—CH_2CH_2Br + NaCN $\longrightarrow$

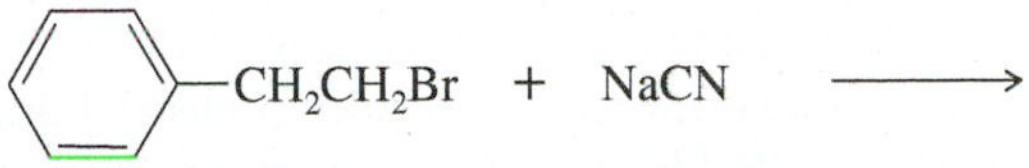

(c) (benzene)—S^- Na^+ + $CH_3CH_2Br \longrightarrow$

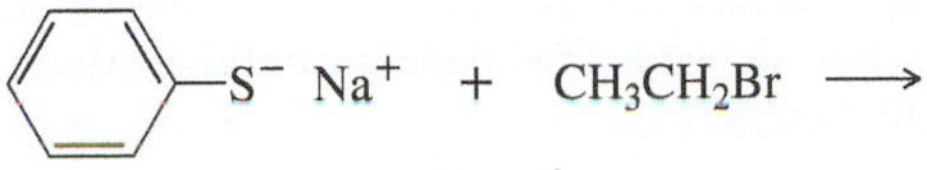

(d) $CH_3(CH_2)_8CH_2Cl + Na^+ \ ^-{:}C{\equiv}CH \longrightarrow$

(e) (pyridine) N: + $CH_3I \longrightarrow$

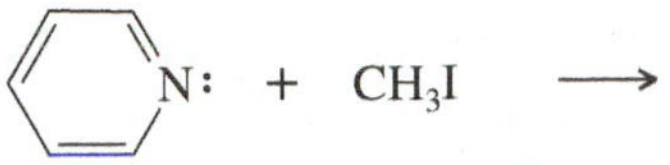

(f) $(CH_3)_3C—CH_2CH_2Br$ + excess $NH_3 \longrightarrow$

(g) Cl, OH + NaOH $\longrightarrow$

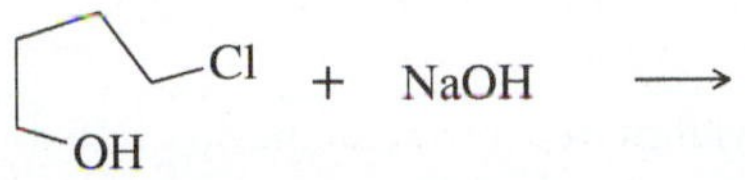

(h) Br—(cyclohexane)—CH_3 $\xrightarrow[CH_3OH]{NaOH}$

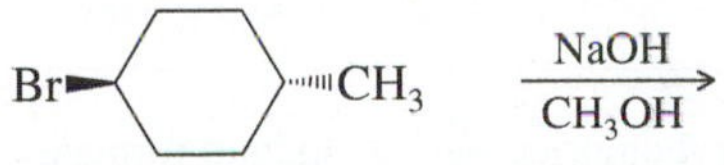

6-46 Using cyclohexane as one of your starting materials, show how you would synthesize the following compounds.

(a) 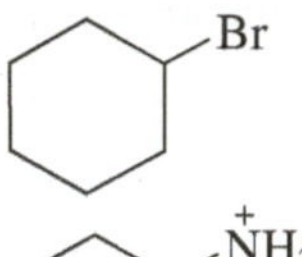 cyclohexyl–Br (b) 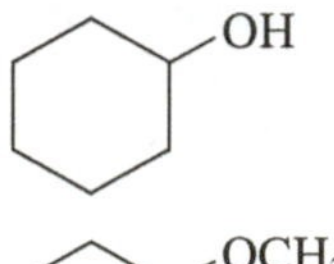 cyclohexyl–OH (c) 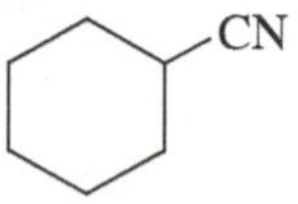 cyclohexyl–CN

(d) 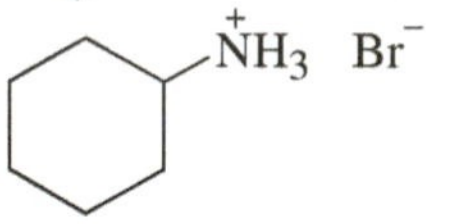 cyclohexyl–$\overset{+}{N}H_3$ Br^- (e) 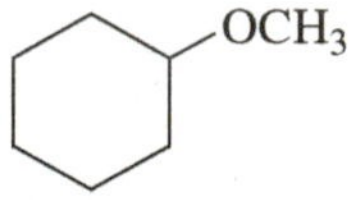 cyclohexyl–OCH_3

6-47 Strawberry growers have used large quantities of methyl bromide (b.p. 4 °C) to sterilize the soil before planting their crops. Like some of the freons, methyl bromide can diffuse up into the stratosphere, where it damages the protective ozone layer. Agricultural chemists have suggested using methyl iodide (b.p. 43 °C) as a replacement for methyl bromide. Why is methyl iodide likely to be more toxic to agricultural pests (and people) than methyl bromide? Why is methyl iodide less likely to reach the stratosphere than methyl bromide?

6-48 A solution of pure (*S*)-2-iodobutane ($[\alpha] = +15.90°$) in acetone is allowed to react with radioactive iodide, $^{131}I^-$, until 1.0% of the iodobutane contains radioactive iodine. The specific rotation of this recovered iodobutane is found to be +15.58°.

(a) Determine the percentages of (*R*)- and (*S*)-2-iodobutane in the product mixture.

(b) What does this result suggest about the mechanism of the reaction of 2-iodobutane with iodide ion?

6-49 **(a)** Optically active 2-bromobutane undergoes racemization on treatment with a solution of KBr. Give a mechanism for this racemization.

(b) In contrast, optically active butan-2-ol does not racemize on treatment with a solution of KOH. Explain why a reaction like that in part (a) does not occur.

(c) Optically active butan-2-ol racemizes in dilute acid. Propose a mechanism for this racemization.

6-50 Give a mechanism to explain the two products formed in the following reaction.

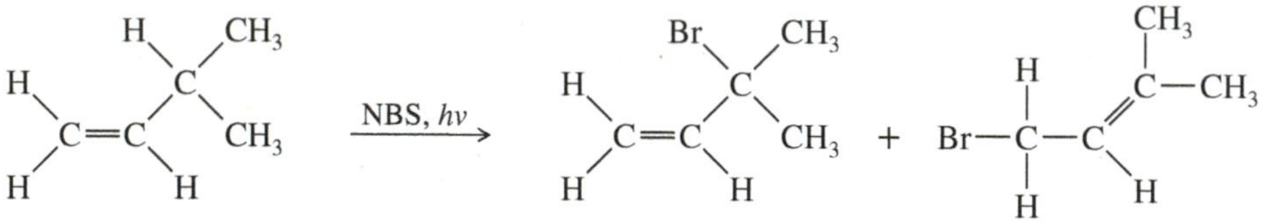

$H_2C{=}CH{-}CH(CH_3)_2$ (3-methylbut-1-ene) $\xrightarrow{\text{NBS, } h\nu}$ $H_2C{=}CH{-}CBr(CH_3)_2$ (not rearranged) + $BrCH_2{-}CH{=}C(CH_3)_2$ (rearranged)

6-51 Predict the major product of the following reaction, and give a mechanism to support your prediction.

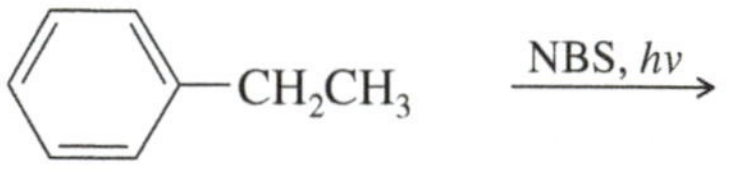

$C_6H_5{-}CH_2CH_3$ (ethylbenzene) $\xrightarrow{\text{NBS, } h\nu}$

6-52 Because the S_N1 reaction goes through a flat carbocation, we might expect an optically active starting material to give a completely racemized product. In most cases, however, S_N1 reactions actually give more of the *inversion* product. In general, as the stability of the carbocation increases, the excess inversion product decreases. Extremely stable carbocations give completely racemic products. Explain these observations.

6-53 Triethyloxonium tetrafluoroborate, $(CH_3CH_2)_3O^+\ BF_4^-$, is a solid with melting point 91–92 °C. Show how this reagent can transfer an ethyl group to a nucleophile (Nuc:$^-$) in an S_N2 reaction. What is the leaving group? Why might this reagent be preferred to using an ethyl halide? (Consult Table 6-2.)

6-54 Furfuryl chloride can undergo substitution by both S_N2 and S_N1 mechanisms. Because it is a 1° alkyl halide, we expect S_N2 but not S_N1 reactions. Draw a mechanism for the S_N1 reaction shown below, paying careful attention to the structure of the intermediate. How can this primary halide undergo S_N1 reactions?

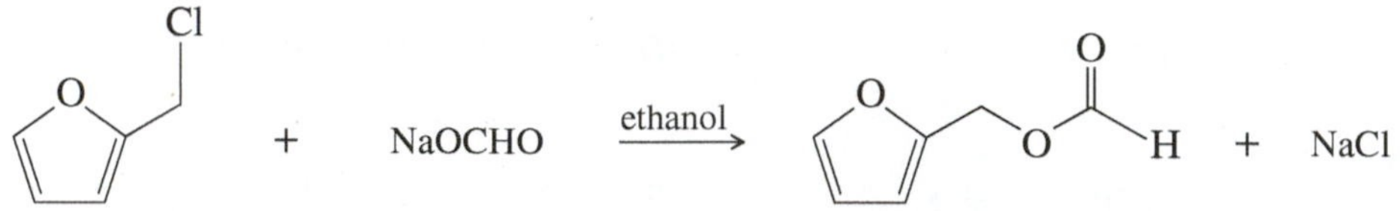

furfuryl chloride + NaOCHO (sodium formate) $\xrightarrow{\text{ethanol}}$ furfuryl formate + NaCl

6-55 The reaction of an amine with an alkyl halide gives an ammonium salt.

$$\underset{\text{amine}}{R_3N} + \underset{\text{alkyl halide}}{R'{-}X} \longrightarrow \underset{\text{ammonium salt}}{R_3\overset{+}{N}{-}R'\ X^-}$$

The rate of this S_N2 reaction is sensitive to the polarity of the solvent. Draw an energy diagram for this reaction in a nonpolar solvent and another in a polar solvent. Consider the nature of the transition state, and explain why this reaction should be sensitive to the polarity of the solvent. Predict whether it will be faster or slower in a more polar solvent.

6-56 The following reaction takes place under second-order conditions (strong nucleophile), yet the structure of the product shows rearrangement. Also, the rate of this reaction is several thousand times faster than the rate of substitution of hydroxide ion on 2-chlorobutane under similar conditions. Propose a mechanism to explain the enhanced rate and rearrangement observed in this unusual reaction. ("Et" is the abbreviation for ethyl.)

$$\underset{\text{Et}_2\text{N:}\qquad\quad\ \ \text{Cl}\qquad\qquad}{H_2C{-}CH{-}CH_2CH_3} \xrightarrow{^-OH} \underset{\qquad\quad\text{:NEt}_2\ \ (\text{on CH}),\ \ \text{OH}\ (\text{on H}_2\text{C})}{H_2C{-}CH{-}CH_2CH_3} + Cl^-$$

6-57 Propose mechanisms to account for the observed products in the following reactions.

(a) $\xrightarrow[h\nu]{\text{NBS}}$ Br + Br

(b) $\xrightarrow{\text{NBS, } h\nu}$ Br + CH_2Br

6-58 Show the products you expect when each compound reacts with NBS with light shining on the reaction.

(a)

(b)

(c) O, OCH_3

(d)

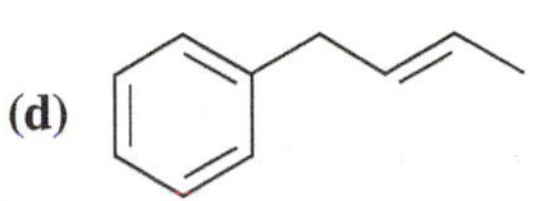

6-59 A student adds NBS to a solution of 1-methylcyclohexene and irradiates the mixture with a sunlamp until all the NBS has reacted. After a careful distillation, the product mixture contains two major products of formula $C_7H_{11}Br$.

(a) Draw the resonance forms of the three possible allylic free radical intermediates.

(b) Rank these three intermediates from most stable to least stable.

(c) Draw the products obtained from each free-radical intermediate.

7 Structure and Synthesis of Alkenes; Elimination

Goals for Chapter 7

1 Draw and name alkenes and cycloalkenes.

2 Given a molecular formula, calculate the number of pi bonds and rings.

3 Explain how the stability of alkenes depends on their structures.

4 Write equations to show how alkenes can be synthesized by eliminations from alkyl halides and alcohols.

5 Predict the products of substitution and elimination reactions, and explain what factors favor each type of reaction.

6 Identify the differences between unimolecular and bimolecular eliminations and explain what factors determine the order of the reaction.

7 Given a set of reaction conditions, identify the possible mechanisms, and predict which mechanism(s) and product(s) are most likely.

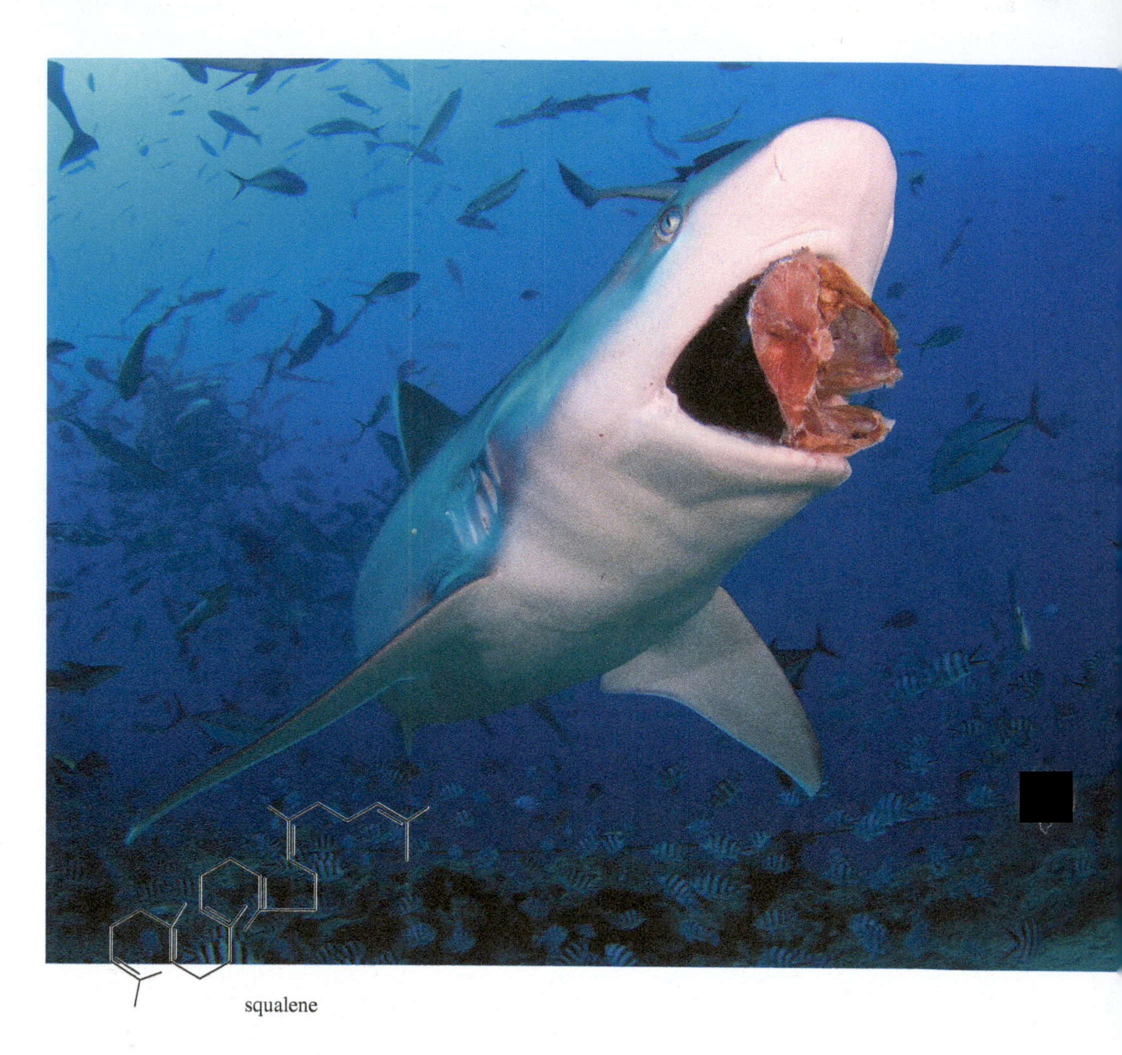

▲ **Squalene** is a 30-carbon alkene with six carbon–carbon double bonds. Squalene occurs in all plants and animals to make the steroid compounds used as hormones, vitamins, emulsifiers, and cell membrane components (Chapter 25). Squalene is used in cosmetics and as an adjuvant (makes the immune system respond more strongly) in vaccines. Squalene was originally obtained from shark livers, where sharks use it as an incompressible low-density liquid to make them neutrally buoyant in salt water. Plant sources, such as olive oil, are now used as cheaper and more sustainable sources of squalene.

7-1 Introduction

Alkenes are hydrocarbons with carbon–carbon double bonds. Alkenes are central to organic chemistry because they are manufactured in large quantities from crude petroleum, and they are readily converted to many of the other functional groups. In this chapter, we consider the characteristics of alkenes and some of the ways they are synthesized. In studying the syntheses of alkenes, we begin with the elimination reactions of alkyl halides, which compete with the substitution reactions covered in Chapter 6. In Chapter 8, we continue the study of alkenes, emphasizing the wide variety of reactions they can undergo.

Alkenes are sometimes called **olefins,** a term derived from *olefiant gas*, meaning "oil-forming gas." This term originated with early experimentalists who noticed the oily appearance of alkene derivatives. Alkenes are among the most important industrial compounds (see Section 7-6), and many alkenes are also synthesized by plants and animals. *Ethylene* is the largest-volume industrial organic compound, used to make polyethylene and a variety of other industrial and consumer chemicals. Ethylene is also a fruit-ripening hormone found in the air released by plants. *Pinene* is a major component of *turpentine*, the paint solvent distilled from extracts of evergreen trees. *Muscalure* (*cis*-tricos-9-ene) is the sex attractant of the common housefly.

$H_2C{=}CH_2$ — ethylene (ethene)

α-pinene

$CH_3(CH_2)_7CH{=}CH(CH_2)_{12}CH_3$ — *cis*-tricos-9-ene, "muscalure"

A carbon–carbon double bond consists of a very stable sigma bond and a less stable pi bond. The total bond energy of a carbon–carbon double bond is about 611 kJ/mol (146 kcal/mol), compared with the single-bond energy of about 347 kJ/mol (83 kcal/mol). From these energies, we can calculate the approximate energy of a pi bond:

double-bond dissociation energy	611 kJ/mol	(146 kcal/mol)
subtract sigma bond dissociation energy	(−)347 kJ/mol	(−)(83 kcal/mol)
pi bond dissociation energy	264 kJ/mol	(63 kcal/mol)

This value of 264 kJ/mol is much less than the sigma bond energy of 347 kJ/mol, showing that pi bonds should be more reactive than sigma bonds.

Because a carbon–carbon double bond is relatively reactive, it is considered to be a *functional group*, and alkenes are characterized by the reactions of their double bonds. In Chapter 6, we considered substitution reactions of alkyl halides in depth and briefly showed that elimination reactions can lead to alkenes. Here in Chapter 7, we study alkenes in more detail, with particular emphasis on the mechanisms of elimination reactions.

7-2 The Orbital Description of the Alkene Double Bond

In a Lewis structure, the double bond of an alkene is represented by two pairs of electrons between the carbon atoms. The Pauli exclusion principle tells us that two pairs of electrons can go into the region of space between the carbon nuclei only if each pair has its own molecular orbital. Using ethylene as an example, let's consider how the electrons are distributed in the double bond.

7-2A The Sigma Bond Framework

In Section 1-15, we saw how we can visualize the sigma bonds of organic molecules using hybrid atomic orbitals. In ethylene, each carbon atom is bonded to three other atoms (one carbon and two hydrogens), and there are no nonbonding electrons. Three hybrid orbitals are needed, implying sp^2 hybridization. Recall from Section 1-15 that sp^2 hybridization corresponds to bond angles of about 120°, giving optimum separation of the three atoms bonded to the carbon atom.

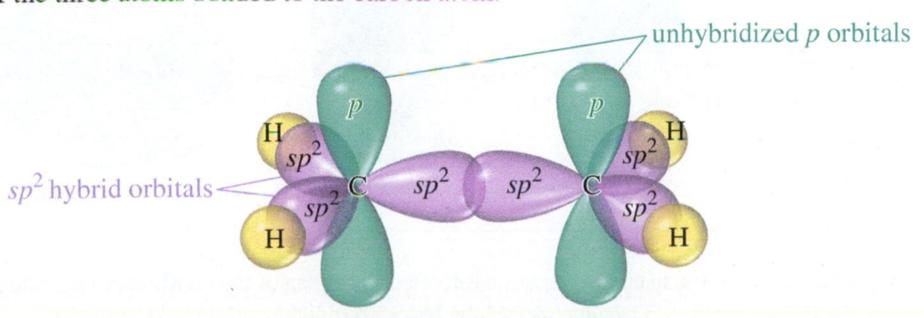

sigma bonding orbitals of ethylene

Each of the carbon–hydrogen sigma bonds is formed by overlap of an sp^2 hybrid orbital on carbon with the 1*s* orbital of a hydrogen atom. The C—H bond length in ethylene (1.08 Å) is slightly shorter than the C—H bond in ethane (1.09 Å) because the sp^2 orbital in ethylene has more *s* character (one-third *s*) than an sp^3 orbital (one-fourth *s*). The *s* orbital is closer to the nucleus than the *p* orbital, contributing to shorter bonds.

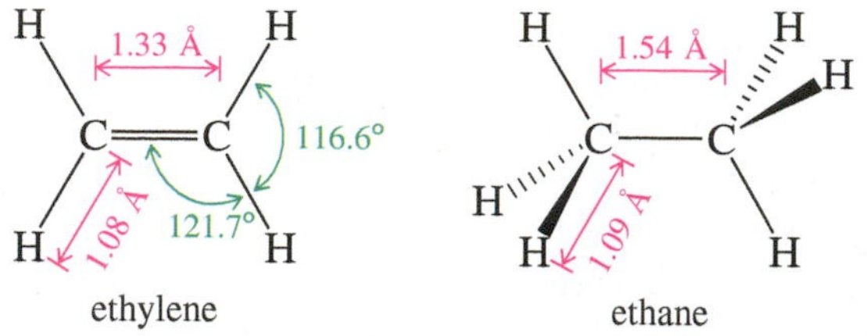

The remaining sp^2 orbitals overlap in the region between the carbon nuclei, providing a bonding orbital. The pair of electrons in this bonding orbital forms one of the bonds between the carbon atoms. This bond is a sigma bond because its electron density is centered along the line joining the nuclei. The C═C bond in ethylene (1.33 Å) is much shorter than the C—C bond (1.54 Å) in ethane, partly because the sigma bond of ethylene is formed from sp^2 orbitals (with more *s* character) and partly because two bonds draw the atoms closer together.

7-2B The Pi Bond

Two more electrons must go into the carbon–carbon bonding region to form the double bond in ethylene. Each carbon atom still has an unhybridized *p* orbital, and these overlap to form a pi-bonding molecular orbital. The two electrons in this orbital form the second bond between the double-bonded carbon atoms. For pi overlap to occur, these *p* orbitals must be parallel, which requires that the two carbon atoms be oriented with all their C—H bonds in a single plane (Figure 7-1). Half of the pi-bonding orbital is above the C—C sigma bond, and the other half is below the sigma bond. The pi-bonding electrons give rise to regions of high electron density (red) in the electrostatic potential map of ethylene shown in Figure 7-1.

Figure 7-2 shows that the two ends of the ethylene molecule cannot be twisted with respect to each other without disrupting the pi bond. Unlike single bonds, a carbon–carbon double bond does not permit rotation. Six atoms, including the double-bonded carbon atoms and the four atoms bonded to them, must remain in the same plane. This is the origin of cis-trans isomerism. If two groups are on the same side of a double bond (cis), they cannot rotate to opposite sides (trans) without breaking the pi bond. Figure 7-2 shows the two distinct isomers of but-2-ene: *cis*-but-2-ene and *trans*-but-2-ene.

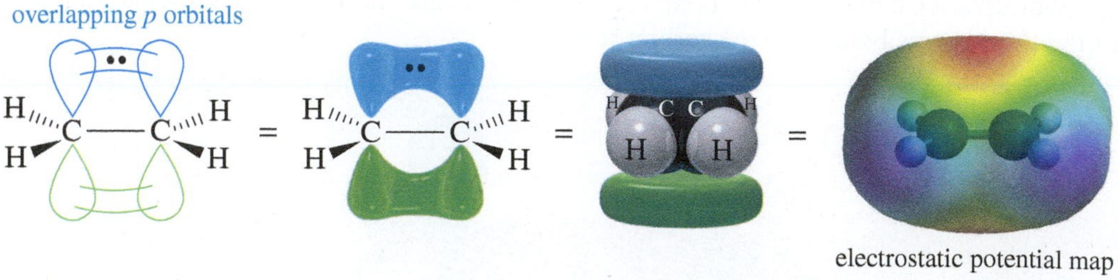

FIGURE 7-1
Parallel *p* orbitals in ethylene. The pi bond in ethylene is formed by overlap of the unhybridized *p* orbitals on the sp^2 hybrid carbon atoms. This overlap requires the two ends of the molecule to be coplanar.

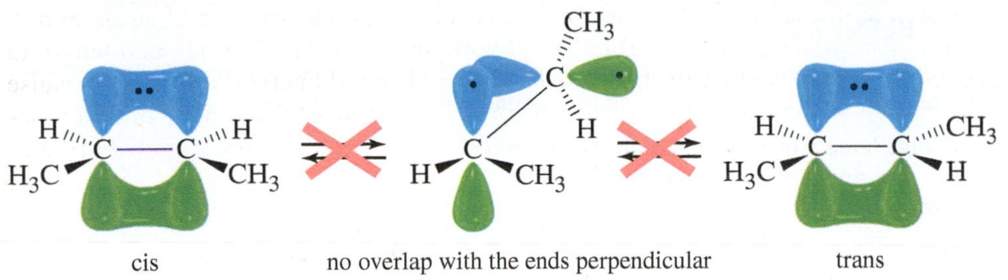

cis — no overlap with the ends perpendicular — trans

FIGURE 7-2
Distinct isomers resulting from C—C double bonds. The two isomers of but-2-ene cannot interconvert by rotation about the carbon–carbon double bond without breaking the pi bond.

7-3 Elements of Unsaturation

7-3A Elements of Unsaturation in Hydrocarbons

Alkenes are said to be **unsaturated** because they are capable of adding hydrogen in the presence of a catalyst. The product, an alkane, is called **saturated** because it cannot react with any more hydrogen. The presence of a pi bond of an alkene (or an alkyne) or the ring of a cyclic compound decreases the number of hydrogen atoms in a molecular formula. These structural features are called **elements of unsaturation.** Each element of unsaturation corresponds to two fewer hydrogen atoms than in the "saturated" formula.

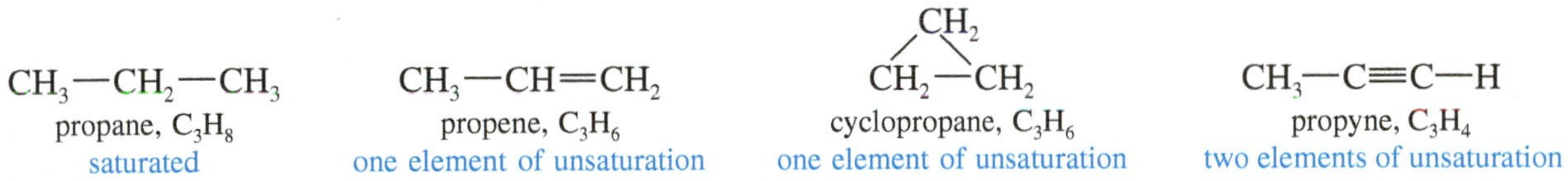

$CH_3—CH_2—CH_3$	$CH_3—CH=CH_2$	cyclo-(CH_2–CH_2–CH_2)	$CH_3—C\equiv C—H$
propane, C_3H_8	propene, C_3H_6	cyclopropane, C_3H_6	propyne, C_3H_4
saturated	one element of unsaturation	one element of unsaturation	two elements of unsaturation

Consider, for example, the formula C_4H_8. A saturated alkane would have a $C_nH_{(2n+2)}$ formula, or C_4H_{10}. The formula C_4H_8 is missing two hydrogen atoms, so it has one element of unsaturation, either a pi bond or a ring. There are five constitutional isomers of formula C_4H_8:

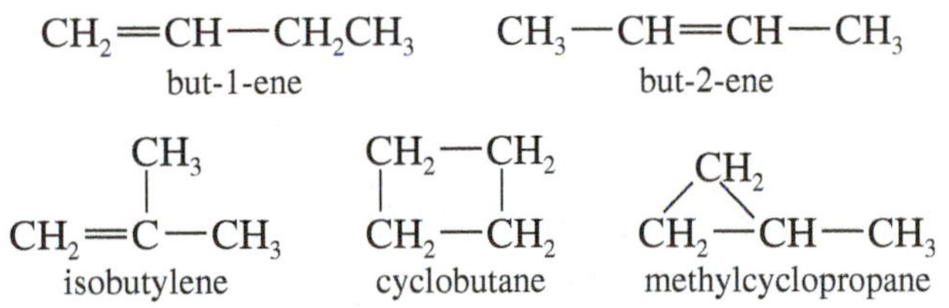

$CH_2=CH—CH_2CH_3$ — but-1-ene

$CH_3—CH=CH—CH_3$ — but-2-ene

$CH_2=C(CH_3)—CH_3$ — isobutylene

cyclo-(CH_2–CH_2–CH_2–CH_2) — cyclobutane

cyclo-(CH_2–CH_2–CH)$—CH_3$ — methylcyclopropane

PROBLEM-SOLVING HINT

Elements of Unsaturation, Degree of Unsaturation, Unsaturation Number, and *Index of Hydrogen Deficiency* are all equivalent terms.

When you need a structure for a particular molecular formula, it helps to find the number of elements of unsaturation. Calculate the maximum number of hydrogen atoms from the saturated formula, $C_nH_{(2n+2)}$, and see how many are missing. The number of elements of unsaturation is simply half the number of missing hydrogens. This simple calculation allows you to consider possible structures quickly, without having to check for the correct molecular formula.

PROBLEM 7-1

(a) If a hydrocarbon has nine carbon atoms, three double bonds, and one ring, how many hydrogen atoms must it have?
(b) Calculate the number of elements of unsaturation implied by the molecular formula C_6H_{12}.
(c) Give five examples of structures with this formula (C_6H_{12}). At least one should contain a ring, and at least one should contain a double bond.

PROBLEM-SOLVING HINT

If you prefer to use a formula, elements of unsaturation

$$= \frac{1}{2}(2C + 2 - H)$$

C = number of carbons
H = number of hydrogens

PROBLEM 7-2

Determine the number of elements of unsaturation in the molecular formula C_4H_6. Give all nine possible structures having this formula. Remember that

a double bond = one element of unsaturation
a ring = one element of unsaturation
a triple bond = two elements of unsaturation

7-3B Elements of Unsaturation with Heteroatoms

Heteroatoms (*hetero*, "different") are any atoms other than carbon and hydrogen. The rule for calculating elements of unsaturation in hydrocarbons can be extended to include heteroatoms. Let's consider how the addition of a heteroatom affects the number of hydrogen atoms in the formula.

Halogens Halogens simply substitute for hydrogen atoms in the molecular formula. The formula C_2H_6 is saturated, so the formula $C_2H_4F_2$ is also saturated. C_4H_8 has one element of unsaturation, and $C_4H_5Br_3$ also has one element of unsaturation. In calculating the number of elements of unsaturation, simply *count halogen atoms as hydrogen atoms.*

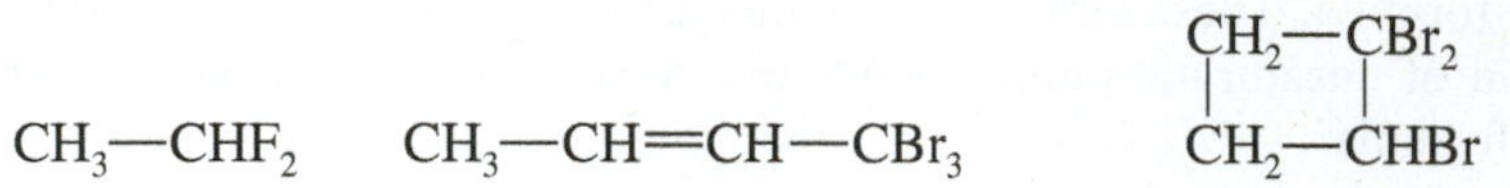

$CH_3—CHF_2$	$CH_3—CH=CH—CBr_3$	$CH_2—CBr_2$ / $CH_2—CHBr$ (ring)
$C_2H_4F_2$	$C_4H_5Br_3$	$C_4H_5Br_3$
saturated	one element of unsaturation	one element of unsaturation

Oxygen An oxygen atom can be added to the chain (or added to a C—H bond to make a C—OH group) without changing the number of hydrogen atoms or carbon atoms. In calculating the number of elements of unsaturation, *ignore the number of oxygen atoms.*

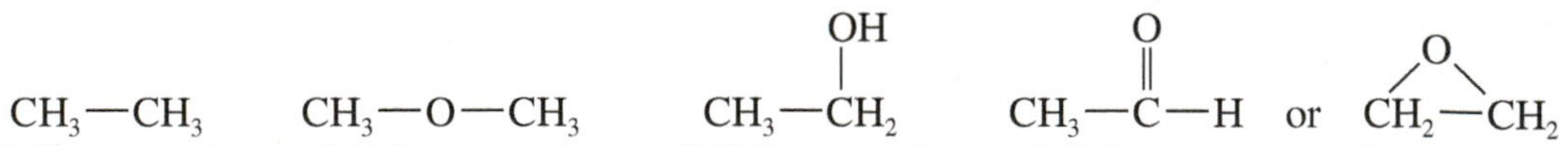

$CH_3—CH_3$	$CH_3—O—CH_3$	$CH_3—CH_2—OH$	$CH_3—C(=O)—H$ or $CH_2—CH_2$ (epoxide, O bridging)
C_2H_6, saturated	C_2H_6O, saturated	C_2H_6O, saturated	C_2H_4O, one element of unsaturation

PROBLEM-SOLVING HINT

If you prefer to use a formula, the number of elements of unsaturation

$$= \frac{1}{2}[2C + 2 + N - (H + Hal)]$$

C = number of carbons
H = number of hydrogens
N = number of nitrogens
Hal = number of halogens

Nitrogen A nitrogen atom can take the place of a carbon atom in the chain, but nitrogen is trivalent. A nitrogen atom has only one additional hydrogen atom, compared with two hydrogens for each additional carbon atom. In computing the elements of unsaturation, *count nitrogen as half a carbon atom.*

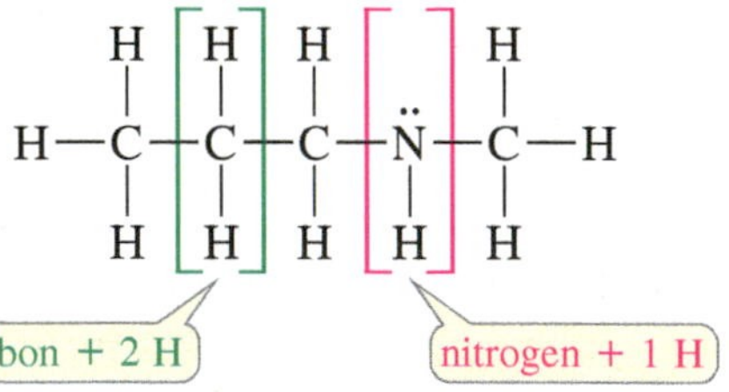

$$H—CH_2—[CH_2]—CH_2—[\ddot{N}H]—CH_2—H$$

carbon + 2 H   nitrogen + 1 H

The formula C_4H_9N is like a formula with $4\frac{1}{2}$ carbon atoms, with saturated formula $C_{4.5}H_{9+2}$. The formula C_4H_9N has one element of unsaturation, because it is two hydrogen atoms short of the saturated formula.

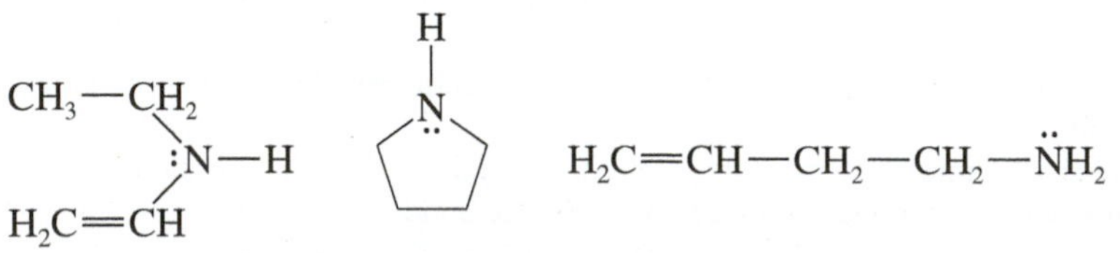

$CH_3—CH_2—\ddot{N}H—CH=CH_2$ pyrrolidine (N—H ring) $H_2C=CH—CH_2—CH_2—\ddot{N}H_2$

examples of formula C_4H_9N, one element of unsaturation

SOLVED PROBLEM 7-1

Draw at least four compounds of formula C_4H_6NOCl.

SOLUTION

Counting the nitrogen as $\frac{1}{2}$ carbon, ignoring the oxygen, and counting chlorine as a hydrogen shows that the formula is equivalent to $C_{4.5}H_7$. The saturated formula for 4.5 carbon atoms is $C_{4.5}H_{11}$, so C_4H_6NOCl has two elements of unsaturation. These could be two double bonds, two rings, one triple bond, or a ring and a double bond. There are many possibilities, four of which are listed here.

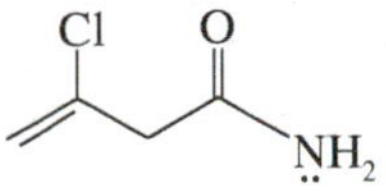

two double bonds 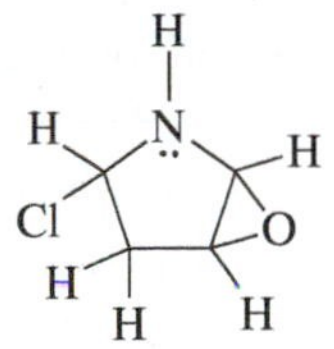 — two rings — $H-C{\equiv}C-CH(OCH_2Cl)(NH_2)$ one triple bond 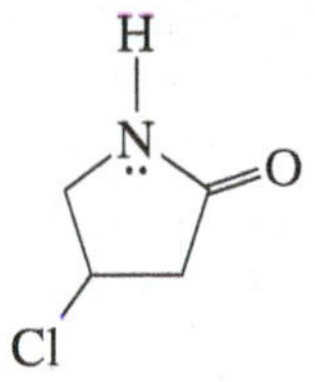 — one ring, one double bond

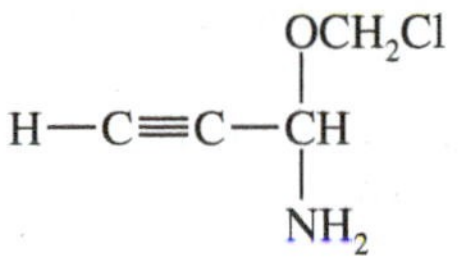

PROBLEM-SOLVING HINT

In figuring elements of unsaturation:
- Count halogens as hydrogens.
- Ignore oxygen.
- Count nitrogen as half a carbon.

PROBLEM 7-3

Draw five more compounds of formula C_4H_6NOCl.

PROBLEM 7-4

For each of the following molecular formulas, determine the number of elements of unsaturation, and draw three examples.

(a) $C_4H_4Cl_2$ **(b)** C_4H_8O **(c)** $C_6H_8O_2$ **(d)** $C_5H_5NO_2$ **(e)** C_6H_3NClBr

7-4 Nomenclature of Alkenes

Simple alkenes are named much like alkanes, using the root name of the longest chain containing the double bond. The ending is changed from *-ane* to *-ene*. For example, "ethane" becomes "ethene," "propane" becomes "propene," and "cyclohexane" becomes "cyclohexene."

	$CH_2{=}CH_2$	$CH_2{=}CH-CH_3$	 (cyclohexene ring)
IUPAC names:	ethene	propene	cyclohexene
Common names:	ethylene	propylene	

PROBLEM-SOLVING HINT

The nomenclature of alkenes is reviewed in Appendix 5, the Summary of Organic Nomenclature.

When the chain contains more than three carbon atoms, a number is used to give the location of the double bond. The chain is numbered starting from the end closest to the double bond, and the double bond is given the *lower* number of its two double-bonded carbon atoms. Cycloalkenes are assumed to have the double bond in the number 1 position.

	$\overset{1}{C}H_2{=}\overset{2}{C}H-\overset{3}{C}H_2-\overset{4}{C}H_3$	$\overset{1}{C}H_2{=}\overset{2}{C}H-\overset{3}{C}H_2-\overset{4}{C}H_2-\overset{5}{C}H_3$
old IUPAC names:	1-butene	1-pentene
new IUPAC names:	but-1-ene	pent-1-ene

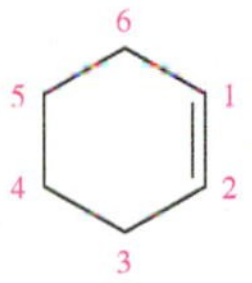

cyclohexene (ring numbered 1–6, double bond between C1 and C2)

	$\overset{1}{C}H_3-\overset{2}{C}H{=}\overset{3}{C}H-\overset{4}{C}H_3$	$\overset{1}{C}H_3-\overset{2}{C}H{=}\overset{3}{C}H-\overset{4}{C}H_2-\overset{5}{C}H_3$
old IUPAC names:	2-butene	2-pentene
new IUPAC names:	but-2-ene	pent-2-ene

Old and New IUPAC Names

In 1993, the IUPAC recommended a logical change in the positions of the numbers used in names. Instead of placing the numbers before the root name (1-butene), it recommended placing them immediately before the part of the name they locate (but-1-ene). The new placement is helpful for clarifying the names of compounds containing multiple functional groups. You should be prepared to recognize names using either placement of the numbers, because both are widely used. In this section, names using the old number placement are printed in blue, and those using the new number placement are printed in green. Throughout this book, we will generally use the new number placement.

A compound with two double bonds is a **diene.** A **triene** has three double bonds, and a **tetraene** has four. Numbers are used to specify the locations of the double bonds.

$CH_2{=}CH{-}CH{=}CH_2$
old IUPAC names: 1,3-butadiene
new IUPAC names: buta-1,3-diene

$CH_3{-}CH{=}CH{-}CH{=}CH{-}CH{=}CH_2$
1,3,5-heptatriene
hepta-1,3,5-triene

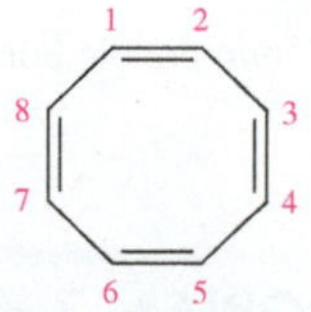

1,3,5,7-cyclooctatetraene
cycloocta-1,3,5,7-tetraene

Each alkyl group attached to the main chain is listed with a number to give its location. Note that the double bond is still given preference in numbering, however.

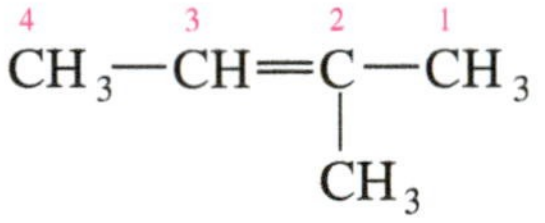

$CH_3{-}CH{=}C(CH_3){-}CH_3$
2-methyl-2-butene
2-methylbut-2-ene

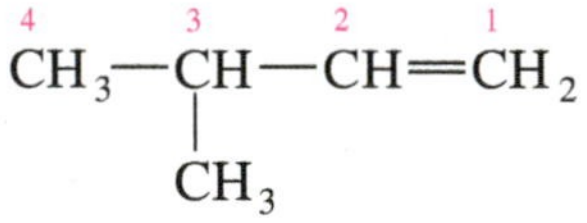

$CH_3{-}CH(CH_3){-}CH{=}CH_2$
3-methyl-1-butene
3-methylbut-1-ene

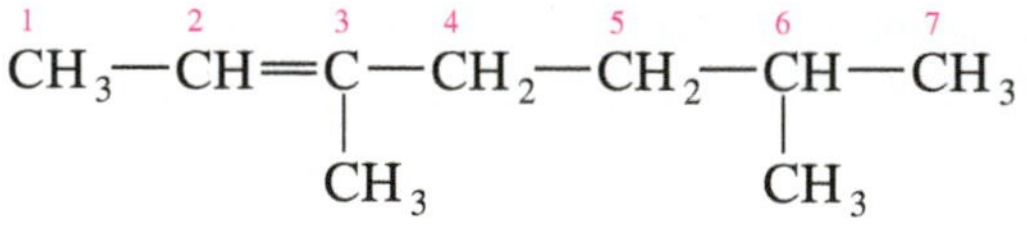

$CH_3{-}CH{=}C(CH_3){-}CH_2{-}CH_2{-}CH(CH_3){-}CH_3$
3,6-dimethyl-2-heptene
3,6-dimethylhept-2-ene

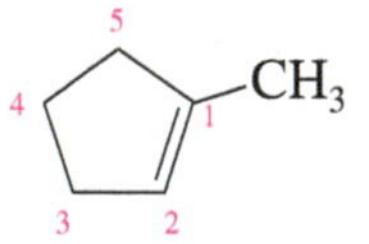

1-methylcyclopentene
1-methylcyclopentene

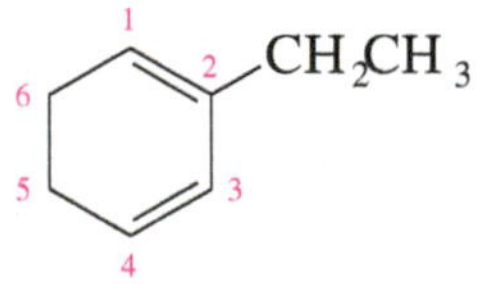

2-ethyl-1,3-cyclohexadiene
2-ethylcyclohexa-1,3-diene

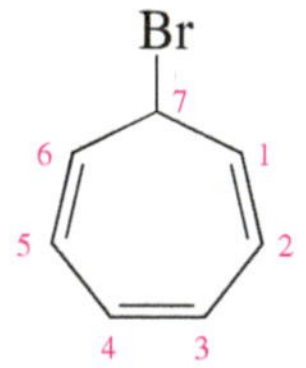

7-bromo-1,3,5-cycloheptatriene
7-bromocyclohepta-1,3,5-triene

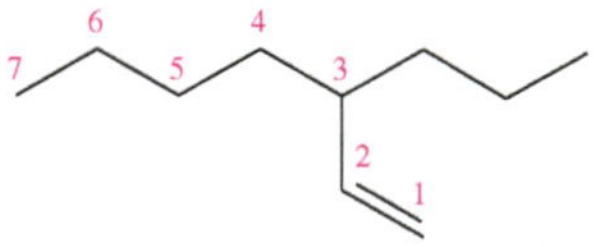

3-propyl-1-heptene
3-propylhept-1-ene

Alkenes as Substituents When a double bond is not part of the main chain, we name the group containing the double bond as a substituent called an *alkenyl group*. Alkenyl groups can be named systematically (ethenyl, propenyl, etc.) or by common names. Common alkenyl substituents are the vinyl, allyl, methylene, and phenyl groups. The phenyl group (Ph) is different from the others because it is aromatic (see Chapter 16) and does not undergo the typical reactions of alkenes.

$\{{=}CH_2$
methylene group
(methylidene group)

$\{{-}CH{=}CH_2$
vinyl group
(ethenyl group)

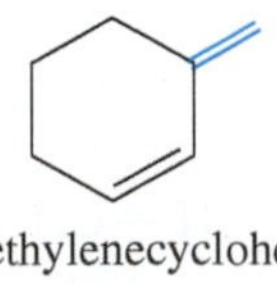

3-methylenecyclohexene

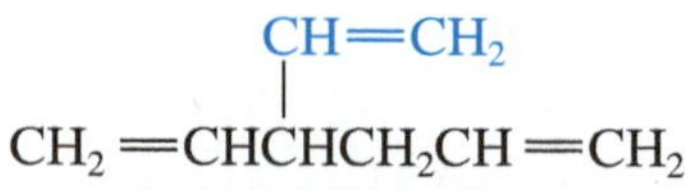

$CH_2{=}CHCH(CH{=}CH_2)CH_2CH{=}CH_2$
3-vinyl-1,5-hexadiene
3-vinylhexa-1,5-diene

$\sim CH_2-CH{=}CH_2$
allyl group
2-propenyl group

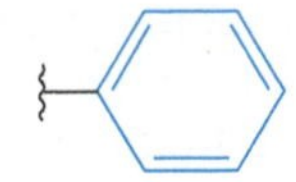

phenyl group
(Ph)

$CH_2{=}CH-CH_2-Cl$
allyl chloride
3-chloropropene

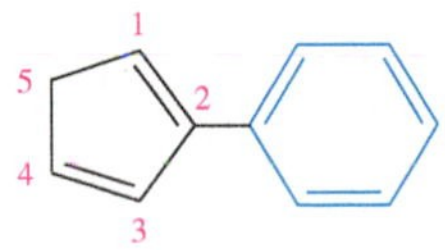

2-phenyl-1,3-cyclopentadiene
2-phenylcyclopenta-1,3-diene

Common Names Most alkenes are conveniently named by the IUPAC system, but common names are sometimes used for the simplest compounds.

	$CH_2{=}CH_2$	$CH_2{=}CH-CH_3$
IUPAC names:	ethene	propene
common name:	ethylene	propylene

$CH_2{=}C(CH_3)-CH_3$
2-methylpropene
isobutylene

$CH_2{=}C(CH_3)-CH{=}CH_2$
2-methylbuta-1,3-diene
isoprene

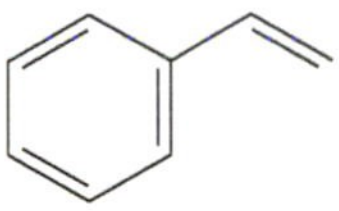

ethenylbenzene
styrene
(vinylbenzene)

PROBLEM 7-5

Give the systematic (IUPAC) names of the following alkenes.

(a) $CH_2{=}CH-CH_2-CH(CH_3)_2$

(b) $CH_3(CH_2)_3-C(=CH_2)-CH_2CH_3$

(c) $CH_2{=}CH-CH_2-CH{=}CH_2$

(d) $CH_2{=}C{=}CH-CH{=}CH_2$

(e) [structure with CH_3 and CH_3]

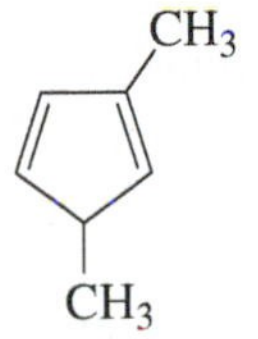

(f) [structure]

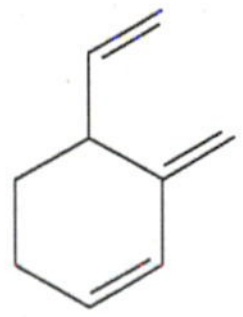

(g) [structure]

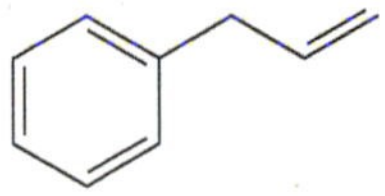

> **Application: Antifungal Drugs**
>
> The polyene antifungals are a group of drugs with a nonpolar region consisting of four to seven sets of alternating single and double bonds. They insert themselves in the cell membranes of fungi, causing disruption and leakiness that results in fungal cell death.
>
> The best-known polyene antifungal drug is Amphotericin B, whose structure is shown below.
>
> Amphotericin B 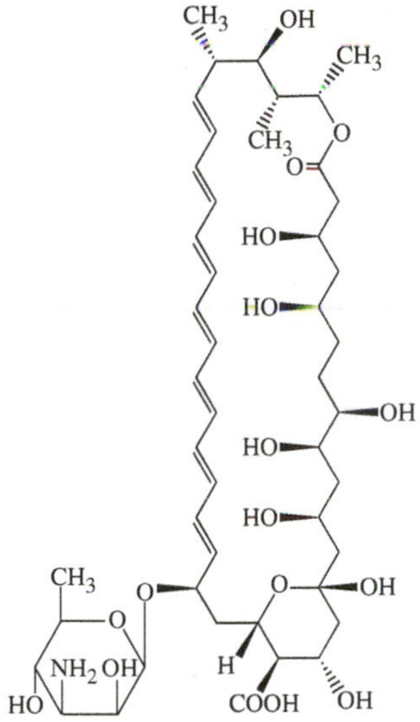

7-5 Nomenclature of Cis-Trans Isomers

Depending on their substituents, the double-bonded carbon atoms of an alkene may be stereogenic, giving rise to cis-trans stereoisomers. In that case, a complete name should also specify which of the possible stereoisomers is intended.

7-5A Cis-Trans Nomenclature

In Chapters 2 and 5, we saw how the rigidity and lack of rotation of carbon–carbon double bonds give rise to **cis-trans isomerism,** also called **geometric isomerism.** A double bond that gives rise to cis-trans stereoisomerism is called **stereogenic,** and it has two possible configurations. If two of the same groups (often hydrogen) are bonded to the same side of the carbons of the double bond, the alkene is the **cis** isomer.

If the same groups are on opposite sides of the double bond, the alkene is **trans.** Not all alkenes are capable of showing cis-trans isomerism. If either carbon of the double bond holds two identical groups, the molecule cannot have cis and trans forms.

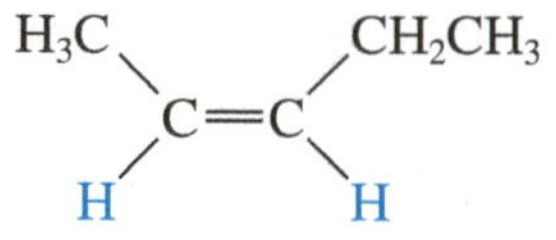

H_3C, CH_2CH_3, H, H: *cis*-2-pentene, *cis*-pent-2-ene

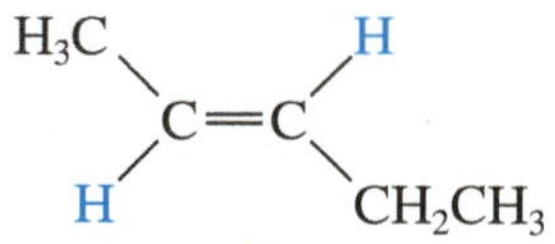

H_3C, H, H, CH_2CH_3: *trans*-2-pentene, *trans*-pent-2-ene

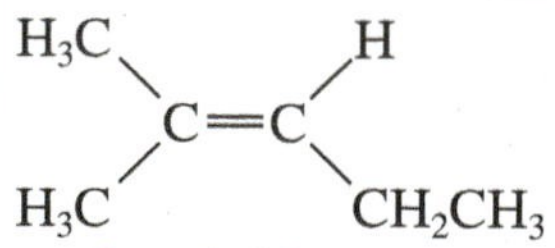

H_3C, H, H_3C, CH_2CH_3: 2-methyl-2-pentene, 2-methylpent-2-ene

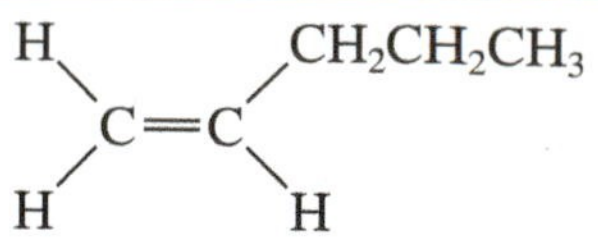

H, $CH_2CH_2CH_3$, H, H: 1-pentene, pent-1-ene

(neither cis nor trans)

Application: Biochemistry

A double bond in rhodopsin, a visual pigment found in your eyes that enables you to see at night, is converted from the cis isomer to the trans isomer when light strikes the eye. As a result, a nerve impulse travels to the brain and you see the source of the light.

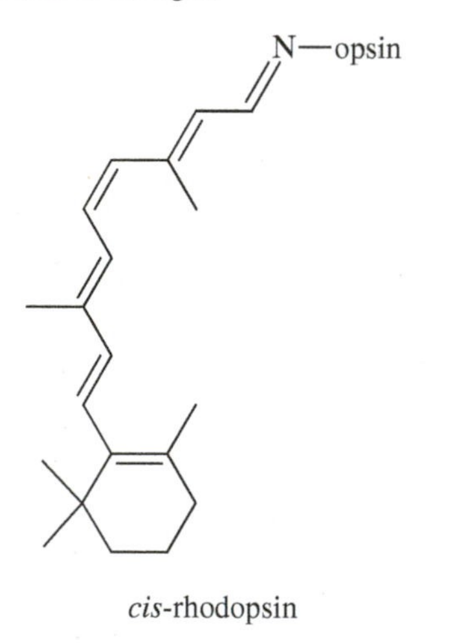

cis-rhodopsin

Trans cycloalkenes are unstable unless the ring is large enough (at least eight carbon atoms) to accommodate the trans double bond (Section 7-8D). Therefore, all cycloalkenes are assumed to be cis unless they are specifically named trans. The cis name is rarely used with cycloalkenes, except to distinguish a large cycloalkene from its trans isomer.

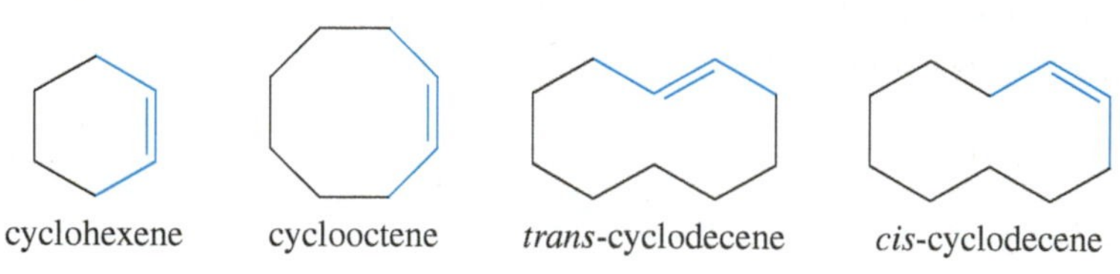
cyclohexene   cyclooctene   *trans*-cyclodecene   *cis*-cyclodecene

7-5B *E-Z* Nomenclature

The cis-trans nomenclature for geometric isomers sometimes gives an ambiguous name. For example, the isomers of 1-bromo-1-chloropropene are not clearly cis or trans because it is not obvious which substituents are referred to as being cis or trans.

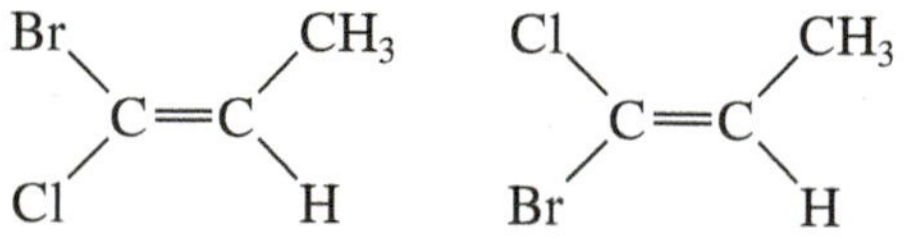

Br, CH_3, Cl, H   Cl, CH_3, Br, H

geometric isomers of 1-bromo-1-chloropropene

To deal with this problem, we use the ***E-Z* system** of nomenclature (pun intended) for stereogenic double bonds. The *E-Z* system is patterned after the Cahn–Ingold–Prelog convention for asymmetric carbon atoms (Section 5-3). It assigns a unique configuration of either *E* or *Z* to any stereogenic double bond.

To name an alkene by the *E-Z* system, mentally separate the double bond into its two ends. Recall how you used the Cahn–Ingold–Prelog rules (page 209) to assign relative priorities to groups on an asymmetric carbon atom so that you could name it (*R*) or (*S*). Consider each end of the double bond separately, and use those same rules to assign first and second priorities to the two substituent groups on that end. Do the same for the other end of the double bond. If the two first-priority atoms are *together (cis)* on the same side of the double bond, you have the ***Z*** isomer, from the German word *zusammen*, "together." If the two first-priority atoms are on *opposite (trans)* sides of the double bond, you have the ***E*** isomer, from the German word *entgegen*, "opposite."

PROBLEM-SOLVING HINT

Those who are unfamiliar with German terms can remember that *Z* = "zame zide."

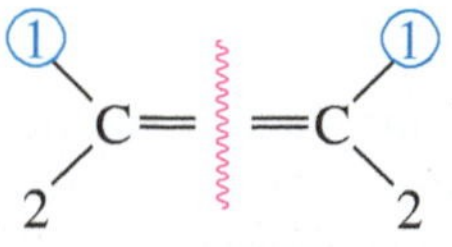

Zusammen (1s on same side)   Entgegen (1s on opposite sides)

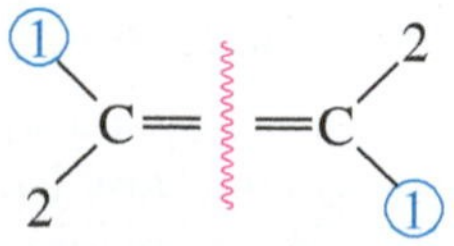

For example,

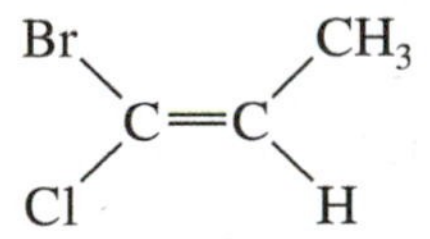

(Z)-1-bromo-1-chloropropene

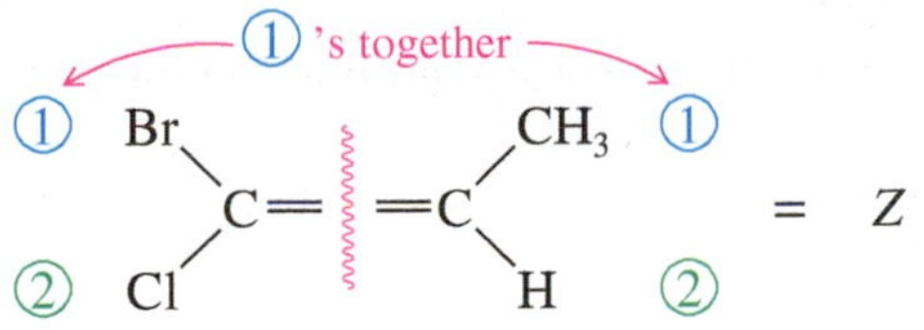

The other isomer is named similarly:

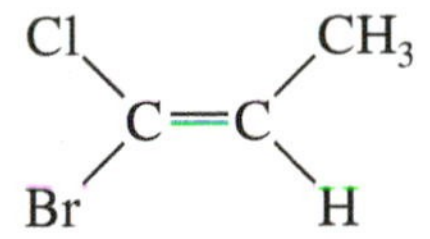

(E)-1-bromo-1-chloropropene

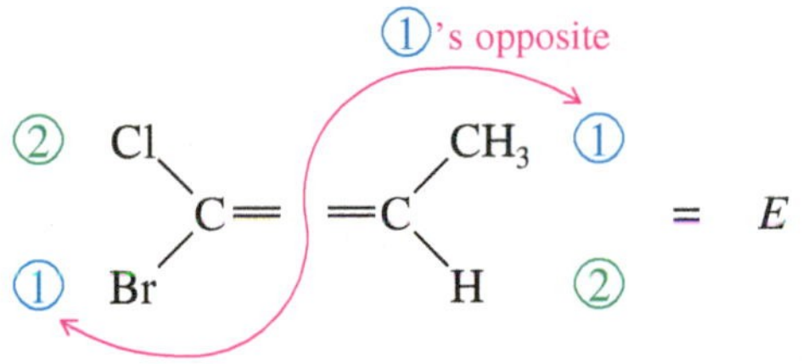

The following example shows the use of the *E-Z* nomenclature with cyclic stereoisomers that are not clearly cis or trans.

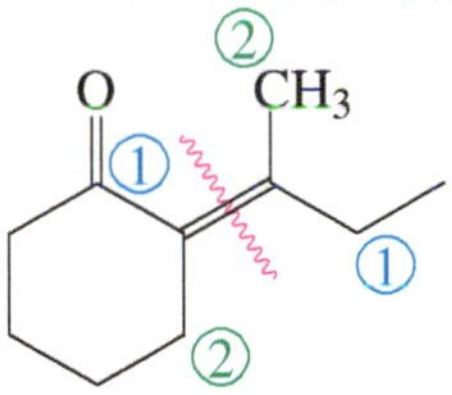

(*E*) isomer 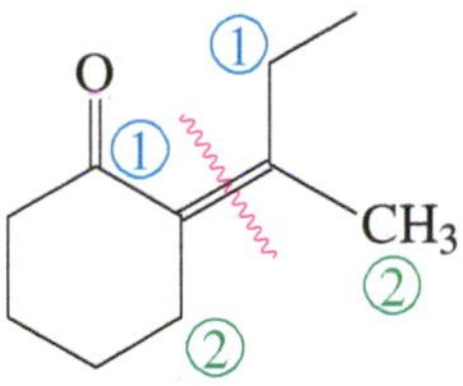 (*Z*) isomer

If the alkene has more than one double bond, the stereochemistry about each double bond should be specified. The following compound is properly named (3*Z*,5*E*)-3-bromoocta-3,5-diene:

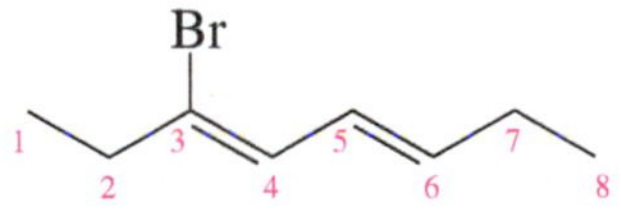

(3*Z*,5*E*)-3-bromoocta-3,5-diene

The use of *E-Z* names (rather than cis and trans) is always an option, but it is required whenever a double bond is not clearly cis or trans. Most trisubstituted and tetrasubstituted double bonds are more clearly named *E* or *Z* rather than cis or trans.

PROBLEM-SOLVING HINT

For a compound with multiple double bonds, the maximum possible number of cis-trans stereoisomers is given by 2^n, where n = the number of stereogenic double bonds. Fewer than 2^n stereoisomers will exist if the molecule has symmetry that leads to duplications among the possible isomers. (See Problem 7-9.)

SUMMARY Rules for Naming Alkenes

The following rules summarize the IUPAC system for naming alkenes:

1. Select the longest chain or largest ring that contains the *largest possible number of double bonds*, and name it with the *-ene* suffix. If there are two double bonds, the suffix is *-diene*; for three, *-triene*; for four, *-tetraene*; etc.
2. Number the chain from the end closest to the double bond(s). Number a ring so that the double bond is between carbons 1 and 2. Place the numbers giving the locations of the double bonds in front of the root name (old system) or in front of the suffix *-ene, -diene*, etc. (new system).
3. Name substituent groups as in alkanes, indicating their locations by the number of the main-chain carbon to which they are attached. The ethenyl group and the propenyl group are usually called the *vinyl* group and the *allyl* group, respectively.
4. For compounds that show geometric isomerism, add the appropriate prefix: *cis-* or *trans-*, or *E-* or *Z-*. Cycloalkenes are assumed to be cis unless named otherwise.

PROBLEM-SOLVING HINT

To see whether a compound can have cis and trans isomers, draw the structure, and then draw it again with the groups on one end of the double bond reversed. See if you can describe a difference between the two.

PROBLEM 7-6

1. Determine which of the following compounds show cis-trans isomerism.
2. Draw and name the cis and trans (or Z and E) isomers of those that do.
   (a) hex-3-ene **(b)** buta-1,3-diene **(c)** hexa-2,4-diene
   (d) 3-methylpent-2-ene **(e)** 2,3-dimethylpent-2-ene **(f)** 3,4-dibromocyclopentene

PROBLEM 7-7

The following names are all incorrect. Draw the structure represented by the incorrect name (or a consistent structure if the name is ambiguous), and give your drawing the correct name.

(a) *cis*-dimethylpent-2-ene **(b)** 3-vinylhex-4-ene
(c) 2-methylcyclopentene **(d)** 6-chlorocyclohexadiene
(e) 2,5-dimethylcyclohexene **(f)** *cis*-2,5-dibromo-3-ethylpent-2-ene

PROBLEM 7-8

Some of the following examples can show geometric isomerism, and some cannot. For the ones that can, draw all the geometric isomers, and assign complete names using the *E-Z* system.

(a) 3-bromo-2-chloropent-2-ene **(b)** 3-ethylhexa-2,4-diene
(c) 3-bromo-2-methylhex-3-ene **(d)** penta-1,3-diene
(e) 3-ethyl-5-methyloct-3-ene **(f)** 3,7-dichloroocta-2,5-diene

(g)

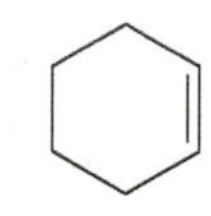

cyclohexene

(h)

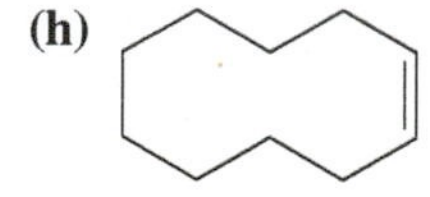

cyclodecene

(i)

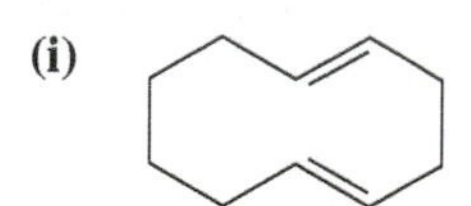

cyclodeca-1,5-diene

Application: Drug Side Effects

A derivative of *trans*-stilbene known as diethylstilbestrol, or DES, was once taken by women during pregnancy to prevent miscarriages. The use of DES was discontinued because studies showed that DES increases the risk of cervical cancer in the children of women who take it.

HO, CH_2CH_3, C=C, CH_3CH_2, OH

diethylstilbestrol (DES)

PROBLEM 7-9*

(a) How many stereogenic double bonds are in octa-1,3,5-triene? How many stereocenters are there? Draw and name the four stereoisomers of octa-1,3,5-triene.

(b) Draw the six stereoisomers of octa-2,4,6-triene. Explain why there are only six stereoisomers, rather than the eight we might expect for a compound with three stereogenic double bonds.

7-6 Commercial Importance of Alkenes

Because the carbon–carbon double bond is readily converted to other functional groups, alkenes are important intermediates in the synthesis of plastics, drugs, pesticides, and other valuable chemicals.

Ethylene is the organic compound produced in the largest volume, at over 300 billion pounds per year worldwide. Most of this ethylene is polymerized to form about 160 billion pounds of polyethylene per year. The remainder is used to synthesize a wide variety of organic chemicals including ethanol, acetic acid, ethylene glycol, and vinyl chloride (Figure 7-3). Ethylene also serves as a plant hormone, accelerating the ripening of fruit. For example, tomatoes are harvested and shipped while green, then treated with ethylene to make them ripen and turn red just before they are placed on display.

Propylene is produced at the rate of about 160 billion pounds per year worldwide, with most of that going to make about 100 billion pounds of polypropylene. The rest is used to make propylene glycol, acetone, isopropyl alcohol, and a variety of other useful organic chemicals (Figure 7-4).

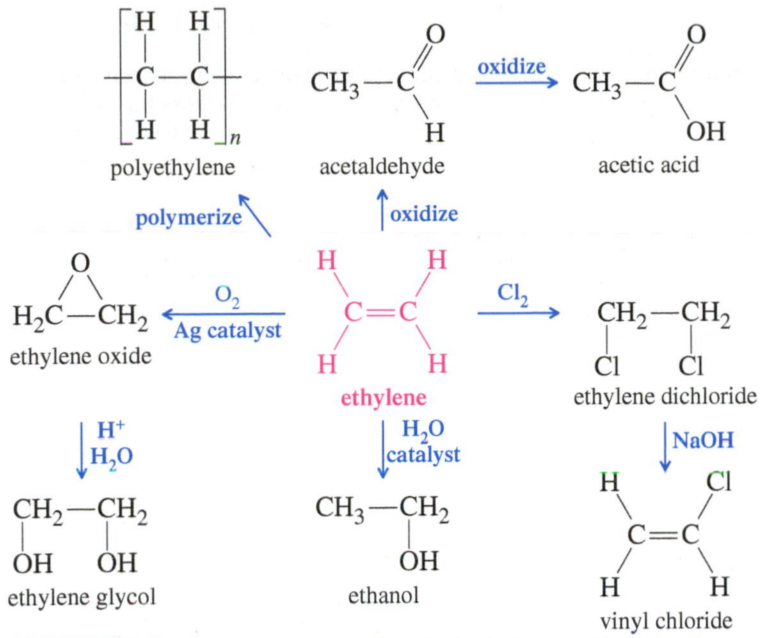

FIGURE 7-3
Uses of ethylene. Ethylene is the largest-volume industrial organic chemical. Most of the ethylene produced each year is polymerized to make polyethylene. Much of the rest is used to produce a variety of useful two-carbon compounds.

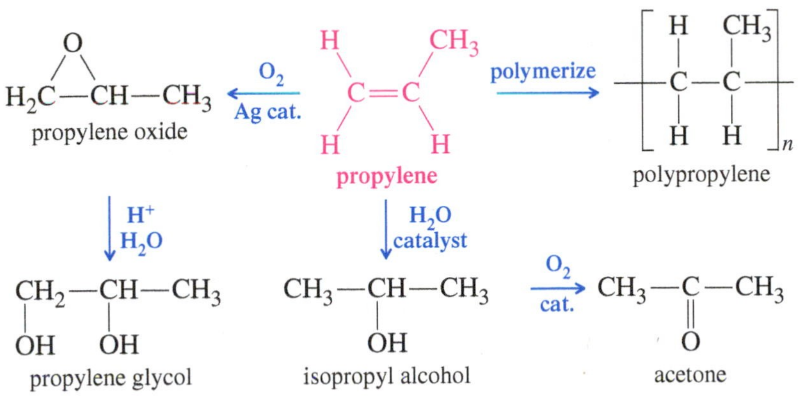

FIGURE 7-4
Uses of propylene. Most propylene is polymerized to make polypropylene. It is also used to make several important three-carbon compounds.

Many common polymers are made by polymerizing alkenes. These polymers are used in consumer products from shoes to plastic bags to car bumpers. A **polymer** (Greek, *poly*, "many," and *meros*, "parts") is a large molecule made up of many **monomer** (Greek, *mono*, "one") molecules. An alkene monomer can **polymerize** by a chain reaction where additional alkene molecules add to the end of the growing polymer chain. Because these polymers result from addition of many individual alkene units, they are called **addition polymers. Polyolefins** are polymers made from monofunctional (single functional group) alkenes such as ethylene and propylene. Figure 7-5 shows some addition polymers made from simple alkenes and haloalkenes. We discuss polymerization reactions in more detail in Chapters 8 and 26.

PROBLEM 7-10

Teflon-coated frying pans routinely endure temperatures that would cause polyethylene or polypropylene to oxidize and decompose. Decomposition of polyethylene is initiated by free-radical abstraction of a hydrogen atom by O_2. Bond-dissociation energies of C—H bonds are about 400 kJ/mol, and C—F bonds are about 460 kJ/mol. The BDE of the H—OO bond is about 192 kJ/mol, and the F—OO bond is about 63 kJ/mol. Show why Teflon (Figure 7-5) is much more resistant to oxidation than polyethylene is.

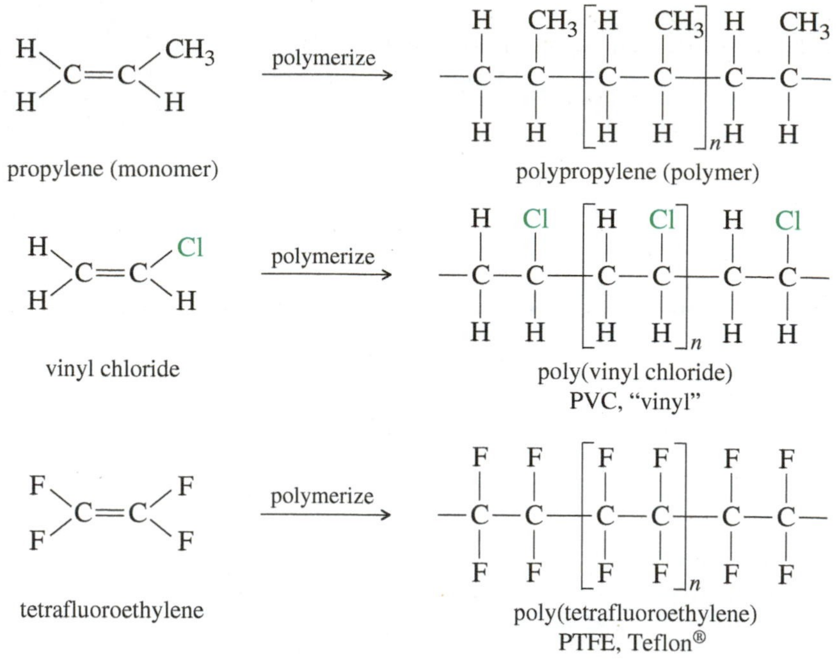

FIGURE 7-5
Addition polymers. Alkenes polymerize in this way to form many common plastics. Uses of these materials are listed in Table 26-1.

7-7 Physical Properties of Alkenes

7-7A Boiling Points and Densities

Most physical properties of alkenes are similar to those of the corresponding alkanes. For example, the boiling points of but-1-ene, *cis*-but-2-ene, *trans*-but-2-ene, and *n*-butane are all close to 0 °C. Also like the alkanes, alkenes have densities around 0.6 or 0.7 g/cm^3. The boiling points and densities of some representative alkenes are listed in Table 7-1. The table shows that boiling points of alkenes increase smoothly with molecular weight. As with alkanes, increased branching leads to greater volatility and lower boiling points. For example, 2-methylpropene (isobutylene) has a boiling point of −7 °C, which is lower than the boiling point of any of the unbranched butenes.

7-7B Polarity

Like alkanes, alkenes are relatively nonpolar. They are insoluble in water but soluble in nonpolar solvents such as hexane, gasoline, halogenated solvents, and ethers. Alkenes tend to be slightly more polar than alkanes, however, for two reasons: The more weakly held electrons in the pi bond are more polarizable (contributing to instantaneous dipole moments), and the vinylic bonds tend to be slightly polar (contributing to a permanent dipole moment).

Alkyl groups are slightly electron-donating toward a double bond, helping to stabilize it. This donation slightly polarizes the vinylic bond, with a small partial positive charge on the alkyl group and a small negative charge on the double-bond carbon atom. For example, propene has a small dipole moment of 0.35 D.

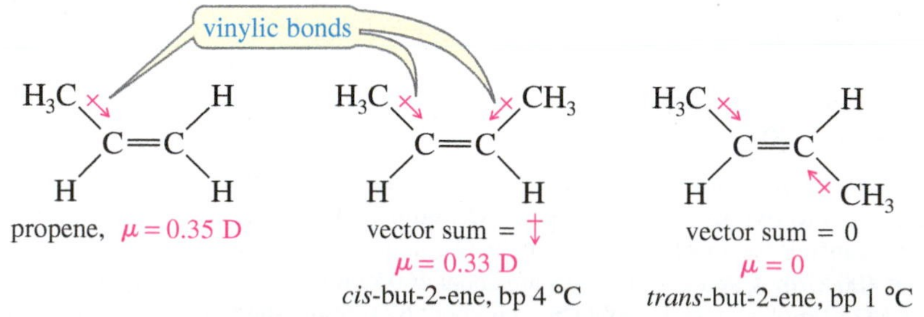

In a cis-disubstituted alkene, the vector sum of the two dipole moments is directed perpendicular to the double bond. In a trans-disubstituted alkene, the two dipole

TABLE 7-1
Physical Properties of Some Representative Alkenes

Name	Structure	Carbons	Boiling Point (°C)	Density (g/cm3)
ethene (ethylene)	$CH_2{=}CH_2$	2	−104	
propene (propylene)	$CH_3CH{=}CH_2$	3	−47	0.52
2-methylpropene (isobutylene)	$(CH_3)_2C{=}CH_2$	4	−7	0.59
but-1-ene	$CH_3CH_2CH{=}CH_2$	4	−6	0.59
trans-but-2-ene	$H_3C(H)C{=}C(H)CH_3$ (trans)	4	1	0.60
cis-but-2-ene	$H_3C(H)C{=}C(H)CH_3$ (cis)	4	4	0.62
3-methylbut-1-ene	$(CH_3)_2CH{-}CH{=}CH_2$	5	25	0.65
pent-1-ene	$CH_3CH_2CH_2{-}CH{=}CH_2$	5	30	0.64
trans-pent-2-ene	$H_3C(H)C{=}C(H)CH_2CH_3$ (trans)	5	36	0.65
cis-pent-2-ene	$H_3C(H)C{=}C(H)CH_2CH_3$ (cis)	5	37	0.66
2-methylbut-2-ene	$(CH_3)_2C{=}CH{-}CH_3$	5	39	0.66
hex-1-ene	$CH_3(CH_2)_3{-}CH{=}CH_2$	6	64	0.68
2,3-dimethylbut-2-ene	$(CH_3)_2C{=}C(CH_3)_2$	6	73	0.71
hept-1-ene	$CH_3(CH_2)_4{-}CH{=}CH_2$	7	93	0.70
oct-1-ene	$CH_3(CH_2)_5{-}CH{=}CH_2$	8	122	0.72
non-1-ene	$CH_3(CH_2)_6{-}CH{=}CH_2$	9	146	0.73
dec-1-ene	$CH_3(CH_2)_7{-}CH{=}CH_2$	10	171	0.74

moments tend to cancel out. If an alkene is symmetrically trans disubstituted, the dipole moment is zero. For example, *cis*-but-2-ene has a nonzero dipole moment, but *trans*-but-2-ene has no measurable dipole moment.

Compounds with permanent dipole moments engage in dipole–dipole attractions, while those without permanent dipole moments engage only in van der Waals attractions. *cis*-But-2-ene and *trans*-but-2-ene have similar van der Waals attractions, but only the cis isomer has dipole–dipole attractions. Because of its increased intermolecular attractions, *cis*-but-2-ene must be heated to a slightly higher temperature (4 °C versus 1 °C) before it begins to boil.

PROBLEM-SOLVING HINT

Boiling points are rough indicators of the strength of intermolecular forces.

The effect of bond polarity is even more apparent in the 1,2-dichloroethenes, with their strongly polar carbon–chlorine bonds. The cis isomer has a large dipole moment (2.4 D), giving it a boiling point 12 degrees higher than the trans isomer, with zero dipole moment.

cis: $Cl(H)C{=}C(H)Cl$
vector sum = ↑
μ = 2.4 D
bp = 60 °C

trans: $Cl(H)C{=}C(H)Cl$
vector sum = 0
μ = 0
bp = 48 °C

PROBLEM 7-11

For each pair of compounds, predict the one with a higher boiling point. Which compounds have zero dipole moments?

(a) *cis*-1,2-dichloroethene or *cis*-1,2-dibromoethene
(b) *cis*- or *trans*-2,3-dichlorobut-2-ene
(c) cyclohexene or 1,2-dichlorocyclohexene

7-8 Stability of Alkenes

In making alkenes, we often find that the major product is the most stable alkene. Many reactions also provide opportunities for alkenes to rearrange to more stable isomers by movement of the double bonds. Therefore, we need to know how the stability of an alkene depends on its structure.

We can compare stabilities by converting different compounds to a common product and comparing the amounts of heat given off. One possibility would be to measure heats of combustion from converting alkenes to CO_2 and H_2O. However, heats of combustion are large numbers (thousands of kJ per mole), and measuring small differences in these large numbers is difficult. Instead, alkene energies are often compared by measuring the **heat of hydrogenation:** the heat given off ($\Delta H°$) during catalytic hydrogenation. Heats of hydrogenation can be measured about as easily as heats of combustion, yet they are smaller numbers and provide more accurate energy differences.

7-8A Heats of Hydrogenation

When an alkene is treated with hydrogen in the presence of a platinum catalyst, hydrogen adds to the double bond, reducing the alkene to an alkane. Hydrogenation is mildly exothermic, evolving about 110 to 136 kJ (26 to 33 kcal) of heat per mole of hydrogen consumed. Consider the hydrogenation of but-1-ene and *trans*-but-2-ene:

$$H_2C{=}CH{-}CH_2{-}CH_3 + H_2 \xrightarrow{Pt} CH_2(H){-}CH(H){-}CH_2{-}CH_3 \qquad \Delta H° = -125.9\ kJ/mol\ (-30.1\ kcal/mol)$$

but-1-ene (monosubstituted) → butane

$$(H_3C)HC{=}CH(CH_3) + H_2 \xrightarrow{Pt} CH_3{-}CH(H){-}CH(H){-}CH_3 \qquad \Delta H° = -114.6\ kJ/mol\ (-27.4\ kcal/mol)$$

trans-but-2-ene (disubstituted) → butane

Figure 7-6 shows these heats of hydrogenation on a reaction-energy diagram. The difference in the stabilities of but-1-ene and *trans*-but-2-ene is the difference in their heats of hydrogenation. *trans*-But-2-ene is more stable by

$$125.9\ kJ/mol - 114.6\ kJ/mol = 11.3\ kJ/mol\ (2.7\ kcal/mol)$$

7-8B Substitution Effects

An 11.3 kJ/mol (2.7 kcal/mol) stability difference is typical between a monosubstituted alkene (but-1-ene) and a trans-disubstituted alkene (*trans*-but-2-ene). In the following equations, we compare the monosubstituted double bond of 3-methylbut-1-ene with the trisubstituted double bond of 2-methylbut-2-ene. The trisubstituted alkene is more stable by 14.7 kJ/mol (3.5 kcal/mol).

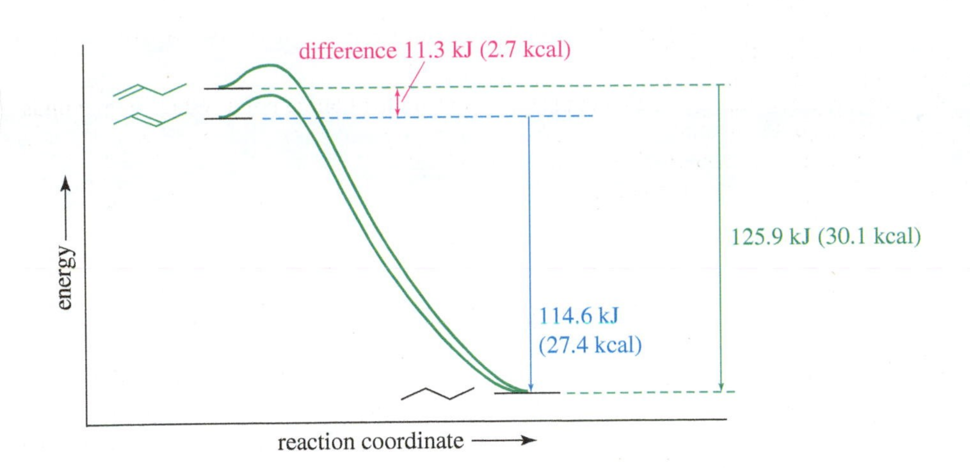

FIGURE 7-6
Relative heats of hydrogenation per mole of alkene. *trans*-But-2-ene is more stable than but-1-ene by 11.3 kJ/mol (2.7 kcal/mol).

$$CH_2{=}CH{-}CH(CH_3){-}CH_3 \xrightarrow{H_2,\ Pt} CH_3{-}CH_2{-}CH(CH_3){-}CH_3 \qquad \Delta H° = -126.3\ \text{kJ}\ (-30.2\ \text{kcal})$$

3-methylbut-1-ene (monosubstituted) → 2-methylbutane

$$CH_3{-}CH{=}C(CH_3){-}CH_3 \xrightarrow{H_2,\ Pt} CH_3{-}CH_2{-}CH(CH_3){-}CH_3 \qquad \Delta H° = -111.6\ \text{kJ}\ (-26.7\ \text{kcal})$$

2-methylbut-2-ene (trisubstituted) → 2-methylbutane

To be completely correct, we should compare heats of hydrogenation only for compounds that give the same alkane, as 3-methylbut-1-ene and 2-methylbut-2-ene do. However, most alkenes with similar substitution patterns give similar heats of hydrogenation. For example, 3,3-dimethylbut-1-ene (below) hydrogenates to give a different alkane than does 3-methylbut-1-ene or but-1-ene (above); yet these three monosubstituted alkenes have similar heats of hydrogenation because the alkanes formed have similar energies. In effect, the heat of hydrogenation is a measure of the energy content of the pi bond.

$$CH_2{=}CH{-}C(CH_3)_2{-}CH_3 \xrightarrow{H_2,\ Pt} CH_3{-}CH_2{-}C(CH_3)_2{-}CH_3 \qquad \Delta H° = -125.8\ \text{kJ}\ (-30.1\ \text{kcal})$$

3,3-dimethylbut-1-ene (monosubstituted) → 2,2-dimethylbutane

In practice, we can use heats of hydrogenation to compare the stabilities of different alkenes as long as they hydrogenate to give alkanes of similar energies. Most acyclic alkanes and unstrained cycloalkanes have similar energies, and we can use this approximation. Table 7-2 shows the heats of hydrogenation of a variety of alkenes with different substitution. The compounds are ranked in decreasing order of their heats of hydrogenation, that is, from the least stable double bonds to the most stable. Note that the values are similar for alkenes with similar substitution patterns.

The most stable double bonds are those with the most alkyl groups attached. For example, hydrogenation of ethylene (no alkyl groups attached) evolves 136.0 kJ/mol, while but-1-ene and pent-1-ene (one alkyl group for each) give off about 126 kJ/mol. Double bonds with two alkyl groups hydrogenate to produce about 114–119 kJ/mol. Three or four alkyl substituents further stabilize the double bond, as with 2-methylbut-2-ene (trisubstituted, 111.6 kJ/mol) and 2,3-dimethylbut-2-ene (tetrasubstituted, 110.4 kJ/mol).

TABLE 7-2
Molar Heats of Hydrogenation of Alkenes

Name	Structure	Molar Heat of Hydrogenation ($-\Delta H°$) kJ	kcal	General Structure
ethene (ethylene)	$H_2C{=}CH_2$	136.0	32.5	unsubstituted
propene (propylene)	$CH_3{-}CH{=}CH_2$	123.4	29.5	monosubstituted $R{-}CH{=}CH_2$
but-1-ene	$CH_3{-}CH_2{-}CH{=}CH_2$	125.9	30.1	
pent-1-ene	$CH_3{-}CH_2{-}CH_2{-}CH{=}CH_2$	126.0	30.1	
hex-1-ene	$CH_3{-}(CH_2)_3{-}CH{=}CH_2$	125.0	29.9	
3-methylbut-1-ene	$(CH_3)_2CH{-}CH{=}CH_2$	126.3	30.2	
3,3-dimethylbut-1-ene	$(CH_3)_3C{-}CH{=}CH_2$	125.8	30.1	
cis-but-2-ene	$(CH_3)(H)C{=}C(CH_3)(H)$ (CH₃ groups cis)	118.5	28.3	disubstituted (cis) $(R)(H)C{=}C(R)(H)$ (R groups cis)
cis-pent-2-ene	$(CH_3{-}CH_2)(H)C{=}C(CH_3)(H)$ (alkyl groups cis)	117.7	28.1	
2-methylpropene (isobutylene)	$(CH_3)_2C{=}CH_2$	117.8	28.2	disubstituted (geminal) $R_2C{=}CH_2$
2-methylbut-1-ene	$CH_3{-}CH_2{-}C(CH_3){=}CH_2$	118.2	28.3	
2,3-dimethylbut-1-ene	$(CH_3)_2CH{-}C(CH_3){=}CH_2$	116.3	27.8	
trans-but-2-ene	$(CH_3)(H)C{=}C(H)(CH_3)$ (CH₃ groups trans)	114.6	27.4	disubstituted (trans) $(R)(H)C{=}C(H)(R)$ (R groups trans)
trans-pent-2-ene	$(CH_3{-}CH_2)(H)C{=}C(H)(CH_3)$ (alkyl groups trans)	113.8	27.2	
2-methylbut-2-ene	$CH_3{-}C(CH_3){=}CH{-}CH_3$	111.6	26.7	trisubstituted $R_2C{=}CHR$
2,3-dimethylbut-2-ene	$(CH_3)_2C{=}C(CH_3)_2$	110.4	26.4	tetrasubstituted $R_2C{=}CR_2$

Note: A lower heat of hydrogenation corresponds to lower energy and greater stability of the alkene.

The values in Table 7-2 confirm **Zaitsev's rule:**

> *Alkenes with more highly substituted double bonds are usually more stable.*

In other words, the alkyl groups attached to the double-bonded carbons stabilize the alkene.

Two factors are probably responsible for the stabilizing effect of alkyl groups on a double bond. Alkyl groups are mildly electron-donating, and they contribute electron density to the pi bond. In addition, bulky substituents such as alkyl groups are best situated as far apart as possible. In an alkane, they are separated by the tetrahedral bond angle, about 109.5°. A double bond increases this separation to about 120°. In general, alkyl groups are separated best by the most highly substituted double bond. This steric effect is illustrated in Figure 7-7 for two **double-bond isomers** (isomers that differ

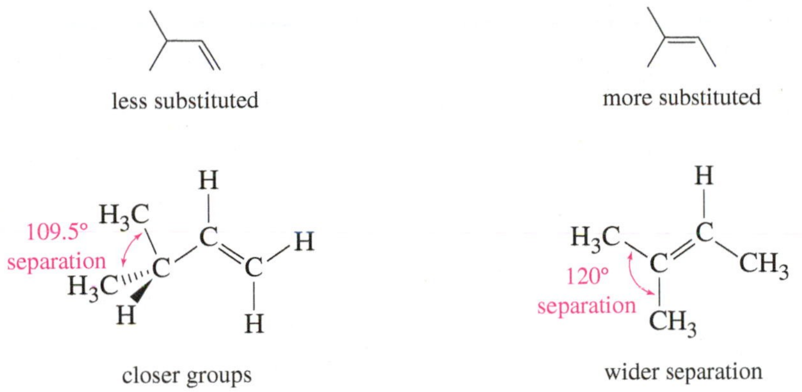

less substituted

more substituted

$$\underset{\text{closer groups}}{\text{(H}_3\text{C)}_2\text{CH—CH=CH}_2}\qquad 109.5° \text{ separation}$$

$$\underset{\text{wider separation}}{\text{(H}_3\text{C)}_2\text{C=CH—CH}_3}\qquad 120° \text{ separation}$$

FIGURE 7-7
Bond angles in double-bond isomers. The isomer with the more substituted double bond has a larger angular separation between the bulky alkyl groups.

only in the position of the double bond). The isomer with the monosubstituted double bond separates the alkyl groups by only 109.5°, while the trisubstituted double bond separates them by about 120°.

PROBLEM 7-12

Use the data in Table 7-2 to predict the energy difference between 2,3-dimethylbut-1-ene and 2,3-dimethylbut-2-ene. Which of these double-bond isomers is more stable?

PROBLEM-SOLVING HINT

Heats of hydrogenation are usually exothermic. A larger amount of heat given off implies a less stable alkene, because the less stable alkene starts from a higher potential energy.

7-8C Energy Differences in cis-trans Isomers

The heats of hydrogenation in Table 7-2 show that trans isomers are generally more stable than the corresponding cis isomers. This trend seems reasonable because the alkyl substituents are separated farther in trans isomers than they are in cis isomers. The greater stability of the trans isomer is evident in the pent-2-enes, which show nearly a 4 kJ/mol (1 kcal/mol) difference between the cis and trans isomers.

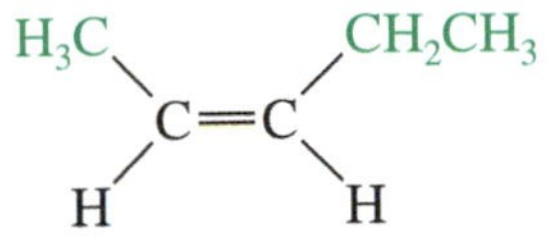

$$\underset{\text{cis, less stable}}{\text{(H}_3\text{C)HC=CH(CH}_2\text{CH}_3)} + H_2 \xrightarrow{Pt} CH_3—CH_2—CH_2—CH_2CH_3 \qquad \Delta H° = -117.7 \text{ kJ/mol } (-28.1 \text{ kcal/mol})$$

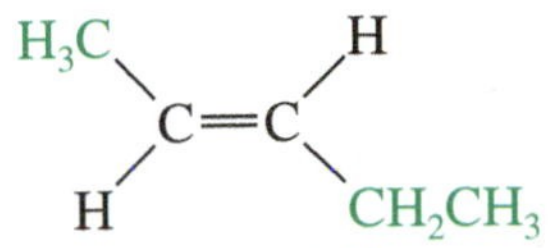

$$\underset{\text{trans, more stable}}{\text{(H}_3\text{C)HC=CH(CH}_2\text{CH}_3)} + H_2 \xrightarrow{Pt} CH_3—CH_2—CH_2—CH_2CH_3 \qquad \Delta H° = -113.8 \text{ kJ/mol } (-27.2 \text{ kcal/mol})$$

A 4 kJ/mol difference between cis and trans isomers is typical for disubstituted alkenes. Figure 7-8 summarizes the relative stabilities of alkenes, comparing them with ethylene, the least stable of the simple alkenes. Geminal isomers, $CR_2{=}CH_2$, tend to fall between the cis and trans isomers in energy, but the differences are not as predictable as the cis/trans differences.

PROBLEM 7-13

Using Table 7-2 as a guide, predict which member of each pair is more stable, as well as by about how many kJ/mol or kcal/mol.

(a) *cis,cis*-hexa-2,4-diene or *trans,trans*-hexa-2,4-diene
(b) 2-methylbut-1-ene or 3-methylbut-1-ene
(c) 2-methylbut-1-ene or 2-methylbut-2-ene
(d) *cis*-4-methylpent-2-ene or 2-methylpent-2-ene

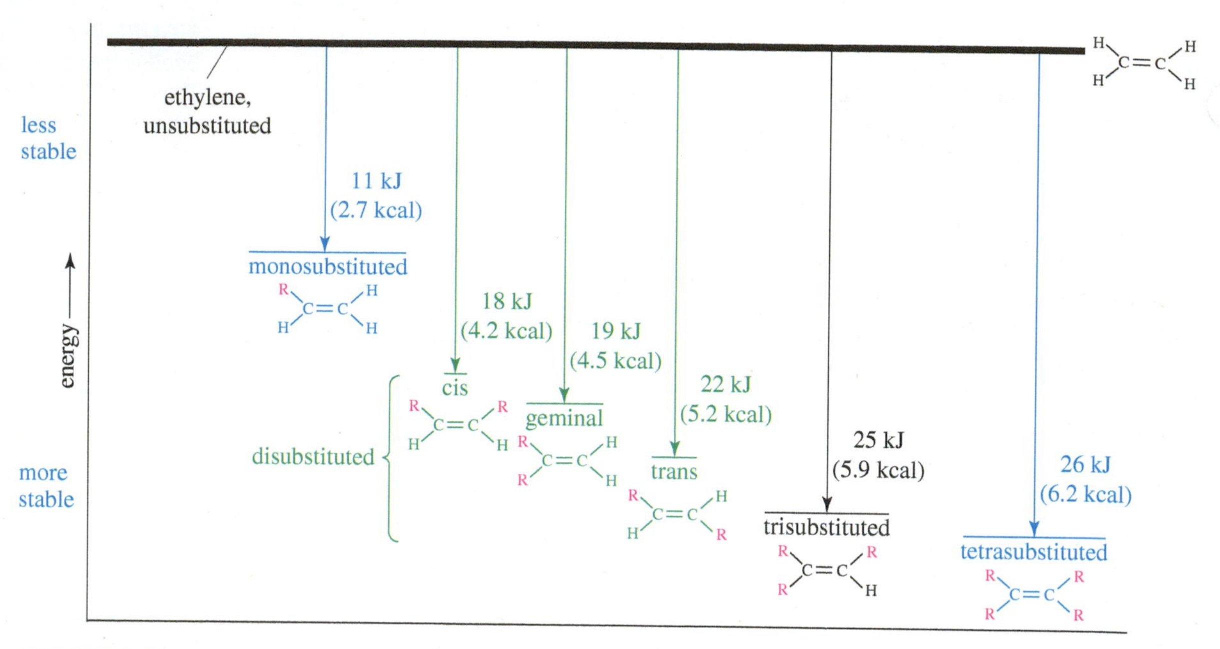

FIGURE 7-8
Relative energies of typical π bonds compared with ethylene. (The numbers are approximate.)

7-8D Stability of Cycloalkenes

Most cycloalkenes react like acyclic (noncyclic) alkenes. The presence of a ring makes a major difference only if there is ring strain, either because of a small ring or because of a trans double bond. Rings that are five-membered or larger can easily accommodate double bonds, and these cycloalkenes react much like straight-chain alkenes. Three- and four-membered rings show evidence of ring strain, however.

Cyclobutene Cyclobutene has a heat of hydrogenation of −129.7 kJ/mol (−31.0 kcal/mol), compared with −112.7 kJ/mol (−26.9 kcal/mol) for cyclopentene.

cyclobutene + H_2 $\xrightarrow{\text{Pt}}$ cyclobutane $\quad \Delta H° = -129.7$ kJ/mol (−31.0 kcal/mol)

cyclopentene + H_2 $\xrightarrow{\text{Pt}}$ cyclopentane $\quad \Delta H° = -112.7$ kJ/mol (−26.9 kcal/mol)

The double bond in cyclobutene has about 17 kJ/mol of *extra* ring strain (in addition to the ring strain in cyclobutane) by virtue of the small ring. The 90° bond angles in cyclobutene compress the angles of the sp^2 hybrid carbons (normally 120°) more than they compress the sp^3 hybrid angles (normally 109.5°) in cyclobutane. The extra ring strain in cyclobutene makes its double bond more reactive than a typical double bond.

Cyclopropene Cyclopropene has bond angles of about 60°, compressing the bond angles of the carbon–carbon double bond to half their usual value of 120°. The double bond in cyclopropene is highly strained.

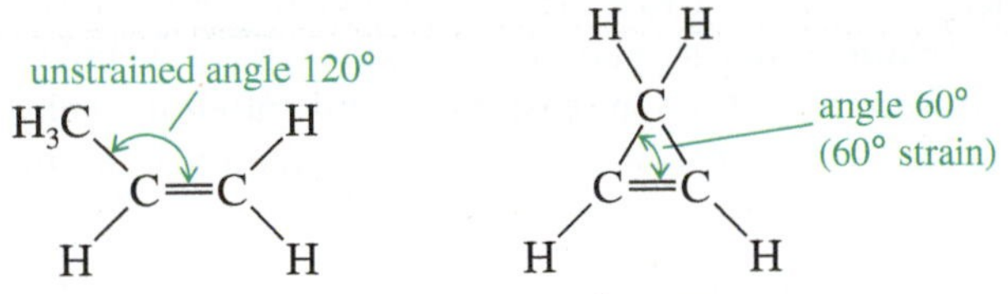
unstrained angle 120°
angle 60°
(60° strain)

propene cyclopropene

Many chemists once believed that a cyclopropene could never be made because it would snap open (or polymerize) immediately from the large ring strain. Cyclopropene was eventually synthesized, however, and it can be stored in the cold. Cyclopropenes were still considered to be strange, highly unusual compounds. Natural-product chemists were surprised when they found that the kernel oil of *Sterculia foelida*, a tropical tree, contains *sterculic acid*, a carboxylic acid with a cyclopropene ring.

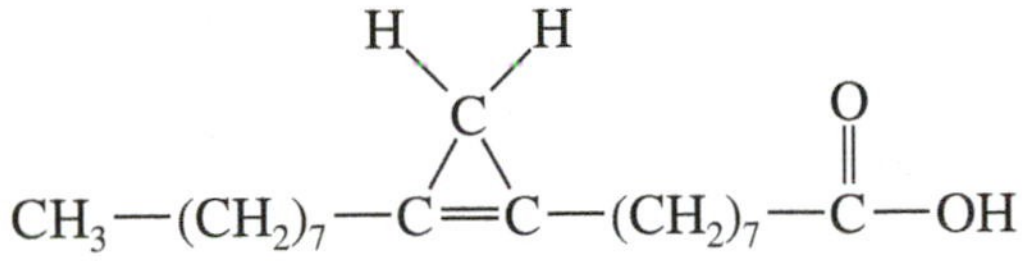

sterculic acid

Application: Toxicology

Sterculic acid is a potent inhibitor of several enzymes that are responsible for the formation of double bonds in long-chain fatty acids. These unsaturated fatty acids are needed in the cells for energy sources and membrane components and to make other critical biological molecules.

Consequently, vegetable oils containing sterculic acid must be hydrogenated or processed at high temperatures to reduce or destroy the cyclopropene ring.

Trans Cycloalkenes Another difference between cyclic and acyclic alkenes is the relationship between cis and trans isomers. In acyclic alkenes, the trans isomers are usually more stable; but the trans isomers of small cycloalkenes are rare, and those with fewer than eight carbon atoms are unstable at room temperature. The problem with making a trans cycloalkene lies in the geometry of the trans double bond. The two alkyl groups on a trans double bond are so far apart that several carbon atoms are needed to complete the ring.

Try to make a model of *trans*-cyclohexene, being careful that the large amount of ring strain does not break your models. *trans*-Cyclohexene is too strained to be isolated, but *trans*-cycloheptene can be isolated at low temperatures. *trans*-Cyclooctene is stable at room temperature, although its cis isomer is still more stable.

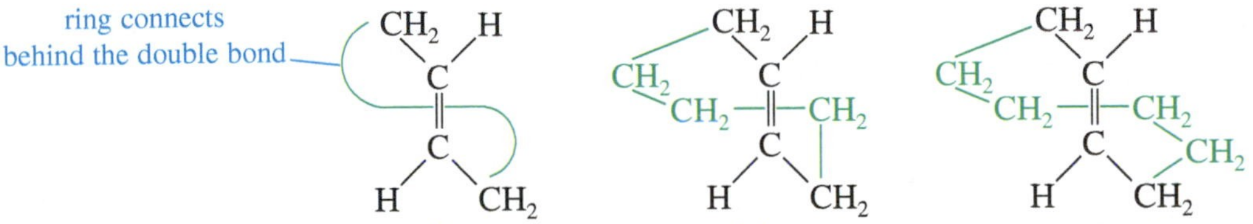
ring connects
behind the double bond

trans cyclic system — *trans*-cycloheptene, marginally stable — *trans*-cyclooctene, stable — *cis*-cyclooctene, more stable

Once a cycloalkene contains at least ten or more carbon atoms, it can easily accommodate a trans double bond. For cyclodecene and larger cycloalkenes, the trans isomer is nearly as stable as the cis isomer.

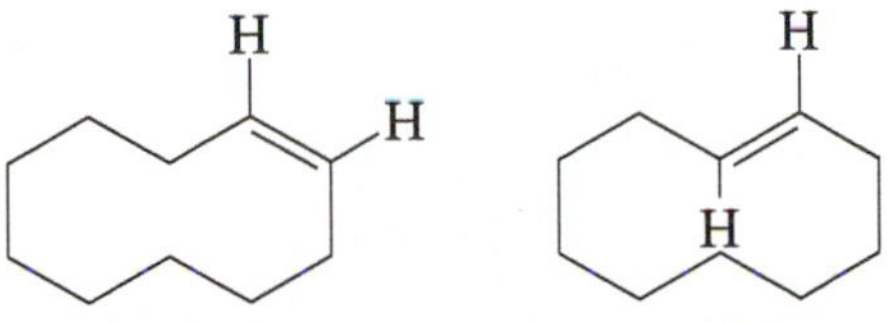

cis-cyclodecene *trans*-cyclodecene

7-8E Bredt's Rule

We have seen that a trans cycloalkene is not stable unless there are at least eight carbon atoms in the ring. An interesting extension of this principle is called **Bredt's rule.**

> BREDT'S RULE: A bridged bicyclic compound cannot have a double bond at a bridgehead position unless one of the rings contains at least eight carbon atoms.

Let's review exactly what Bredt's rule means. A **bicyclic** compound contains two rings. The **bridgehead carbon atoms** are part of both rings, with three links connecting them. A **bridged bicyclic** compound has at least one carbon atom in each of the three links between the bridgehead carbons. In the following examples, the bridgehead carbon atoms are circled in red.

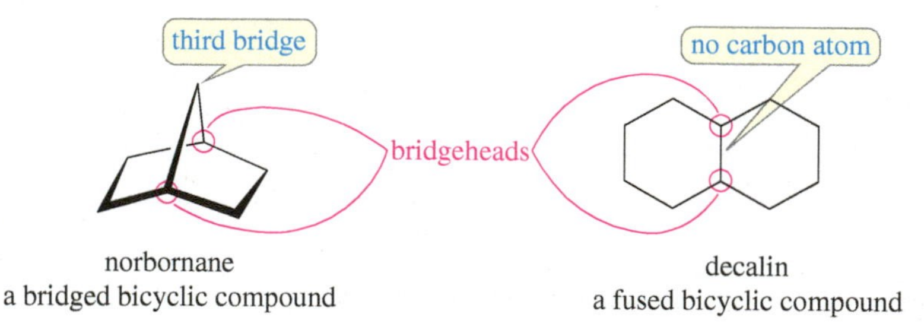

norbornane
a bridged bicyclic compound

decalin
a fused bicyclic compound

If there is a double bond at the bridgehead carbon of a bridged bicyclic system, one of the two rings contains a cis double bond and the other must contain a trans double bond. For example, the following structures show that the norbornane ring system contains a five-membered ring and a six-membered ring. If there is a double bond at the bridgehead carbon atom, the five-membered ring contains a cis double bond and the six-membered ring contains a trans double bond. This unstable arrangement is called a "Bredt's rule violation." If the larger ring contains at least eight carbon atoms, then it can contain a trans double bond and the bridgehead double bond is stable.

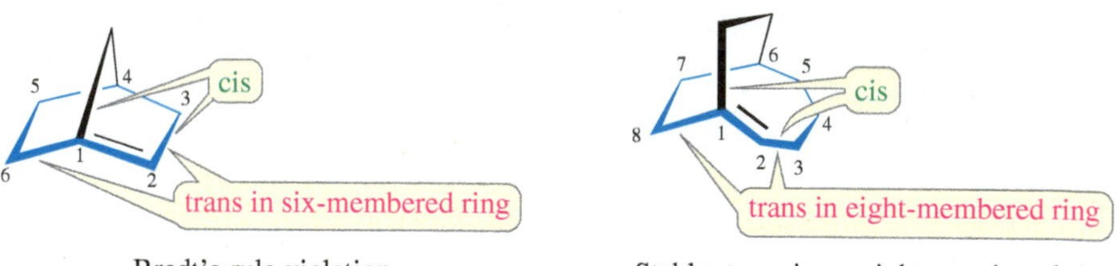

Bredt's rule violation

Stable: trans in an eight-membered ring

In general, compounds that violate Bredt's rule are not stable at room temperature. In a few cases, such compounds (usually with seven carbon atoms in the largest ring) have been synthesized at low temperatures.

SOLVED PROBLEM 7-2

Which of the following alkenes are stable?

(a) cis, trans

(b) trans

(c) cis

(d) trans

SOLUTION

Compound (a) is stable. Although the double bond is at a bridgehead, it is not a bridged bicyclic system. The trans double bond is in a ten-membered ring. Compound (b) is a Bredt's rule violation and is not stable. The largest ring contains six carbon atoms, and the trans double bond cannot be stable in this bridgehead position.

Compound (c) (norbornene) is stable. The (cis) double bond is not at a bridgehead carbon.

Compound (d) is stable. Although the double bond is at the bridgehead of a bridged bicyclic system, there is an eight-membered ring to accommodate the trans double bond.

PROBLEM 7-14

Explain why each of the following alkenes is stable or unstable.

(a) 1,2-dimethylcyclopentene **(b)** *trans*-1,2-dimethylcyclopentene

(c) *trans*-3,4-dimethylcyclopentene **(d)** *trans*-1,2-dimethylcyclodecene

(e)  **(f)** 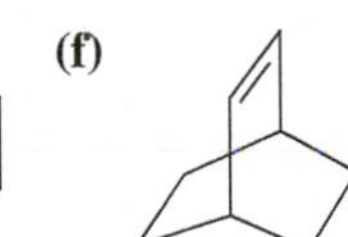 **(g)** 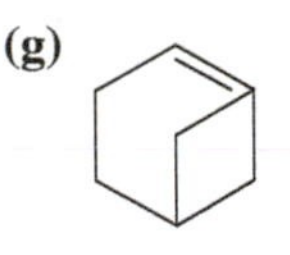 **(h)** 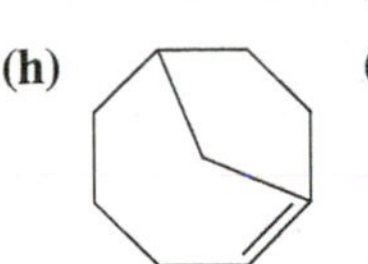 **(i)** 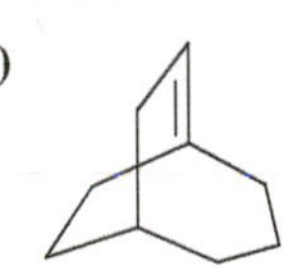

7-8F Enhanced Stability of Conjugated Double Bonds

Double bonds can interact with each other if they are separated by just one single bond. Such interacting double bonds are said to be **conjugated.** Double bonds with two or more single bonds separating them have little interaction and are called **isolated double bonds.** For example, *trans*-hexa-1,3-diene has conjugated double bonds, while *trans*-hexa-1,4-diene has isolated double bonds.

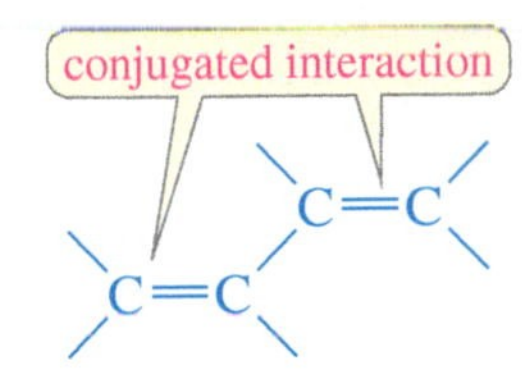

conjugated double bonds
(more stable than isolated double bonds)

no interaction

C=C CH$_2$ C=C

isolated double bonds

trans-hexa-1,3-diene
conjugated, more stable

trans-hexa-1,4-diene
isolated

The pi-bonding electrons of conjugated double bonds are delocalized over both bonds, enhancing the stability of the conjugated system over a similar compound with isolated double bonds. In the examples above, *trans*-hexa-1,3-diene is conjugated, and it is more stable than *trans*-hexa-1,4-diene by about 17 kJ/mol (4.0 kcal/mol). This extra stability of the conjugated molecule is called its **resonance energy.** We will discuss the reasons for the stability of conjugated systems and the origin of the resonance energy in Chapter 15.

Conjugation effects are particularly important in aromatic systems, with three double bonds conjugated in a six-membered ring (in the simplest case). Benzene (below) has a huge resonance energy, 151 kJ/mol (36 kcal/mol) of enhanced stability over three isolated bonds. We will discuss the reasons for the remarkable stability of aromatic systems in Chapter 16. For now, you should recognize that aromatic systems are particularly stable, and they are not likely to react as easily as typical alkenes.

PROBLEM-SOLVING HINT

Benzene can be represented by either of two resonance forms, which are equivalent.

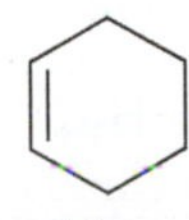

cyclohexene
isolated

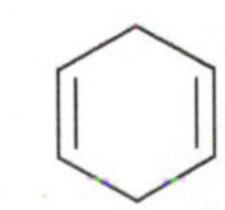

cyclohexa-1,4-diene
isolated

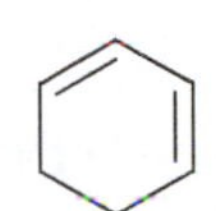

cyclohexa-1,3-diene
conjugated

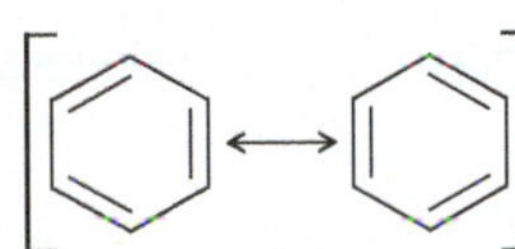

benzene
aromatic

resonance energy:	cyclohexene	cyclohexa-1,4-diene	cyclohexa-1,3-diene	benzene
	0	0	8 kJ (1.9 kcal)	151 kJ (36 kcal)

PROBLEM 7-15

For each set of isomers, choose the isomer that you expect to be most stable and the isomer you expect to be least stable.

(a)

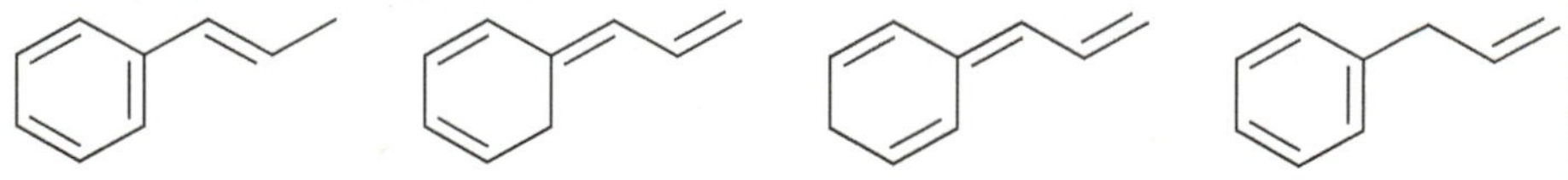

(b)

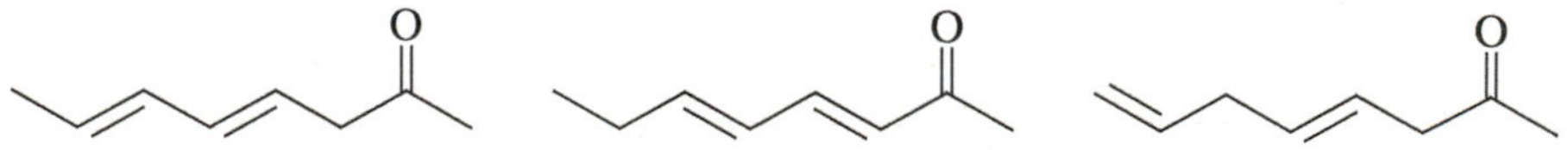

(c)

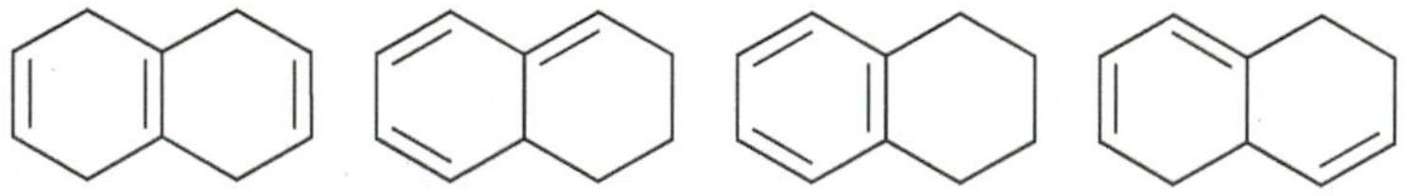

7-9 Formation of Alkenes by Dehydrohalogenation of Alkyl Halides

Elimination reactions are some of the most common methods of synthesizing alkenes. Alkyl halides undergo eliminations to alkenes under conditions that are similar to those covered in Chapter 6 for their nucleophilic substitutions. In fact, nucleophilic substitutions and eliminations often take place under the same conditions in direct competition with each other.

An **elimination** involves the loss of two atoms or groups from the substrate, usually with formation of a new π bond. Elimination of a proton and a halide ion is called **dehydrohalogenation,** and the product is an alkene.

Elimination

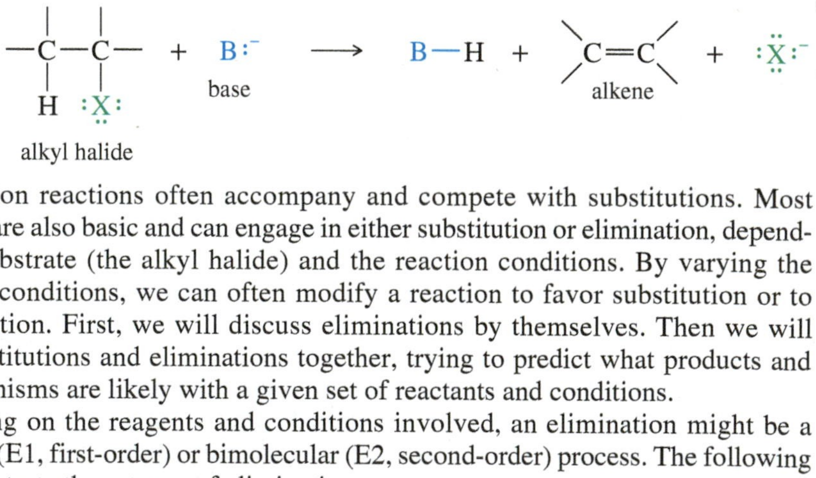

$$\underset{\text{alkyl halide}}{-\underset{\text{H}}{\overset{|}{\text{C}}}-\underset{:\ddot{\text{X}}:}{\overset{|}{\text{C}}}-} + \underset{\text{base}}{\text{B}:^-} \longrightarrow \text{B}-\text{H} + \underset{\text{alkene}}{\text{C}=\text{C}} + :\ddot{\text{X}}:^-$$

Elimination reactions often accompany and compete with substitutions. Most nucleophiles are also basic and can engage in either substitution or elimination, depending on the substrate (the alkyl halide) and the reaction conditions. By varying the reagents and conditions, we can often modify a reaction to favor substitution or to favor elimination. First, we will discuss eliminations by themselves. Then we will consider substitutions and eliminations together, trying to predict what products and which mechanisms are likely with a given set of reactants and conditions.

Depending on the reagents and conditions involved, an elimination might be a unimolecular (E1, first-order) or bimolecular (E2, second-order) process. The following examples illustrate these types of eliminations.

E1:

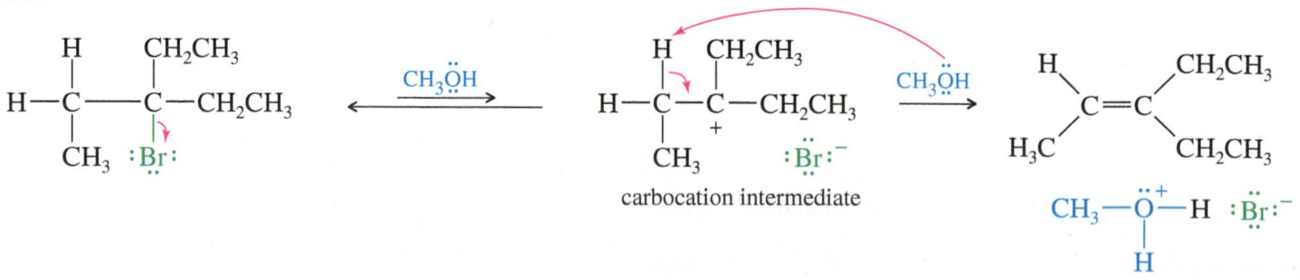

$$\text{H}-\underset{\text{CH}_3}{\overset{\text{H}}{\text{C}}}-\underset{:\text{Br}:}{\overset{\text{CH}_2\text{CH}_3}{\text{C}}}-\text{CH}_2\text{CH}_3 \xrightarrow{\text{CH}_3\ddot{\text{O}}\text{H}} \underset{\text{carbocation intermediate}}{\text{H}-\underset{\text{CH}_3}{\overset{\text{H}}{\text{C}}}-\overset{\text{CH}_2\text{CH}_3}{\underset{+}{\text{C}}}-\text{CH}_2\text{CH}_3 \quad :\ddot{\text{Br}}:^-} \xrightarrow{\text{CH}_3\ddot{\text{O}}\text{H}} \underset{\text{H}_3\text{C}}{\overset{\text{H}}{}}\text{C}=\text{C}\underset{\text{CH}_2\text{CH}_3}{\overset{\text{CH}_2\text{CH}_3}{}} \quad \text{CH}_3-\underset{\text{H}}{\ddot{\text{O}}^+}-\text{H} \quad :\ddot{\text{Br}}:^-$$

E2:

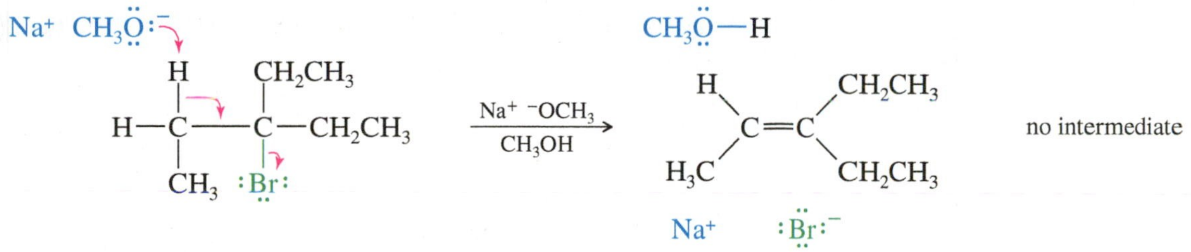

$Na^+\ CH_3O^-$ + $H-C(H)(CH_3)-C(CH_2CH_3)(Br)-CH_2CH_3$ $\xrightarrow[CH_3OH]{Na^+\ {}^-OCH_3}$ CH_3O-H + $(H)(H_3C)C=C(CH_2CH_3)_2$ + Na^+ Br^-

no intermediate

7-10 Unimolecular Elimination: The E1 Reaction

The abbreviation **E1** stands for *Elimination, unimolecular*. The mechanism is called *unimolecular* because the rate-limiting transition state involves a single molecule rather than a collision between two molecules. The slow step of an E1 reaction is the same as in the S_N1 reaction: unimolecular ionization to form a carbocation. In a fast second step, a base abstracts a proton from the carbon atom adjacent to the C^+. The electrons that once formed the carbon–hydrogen bond now form a pi bond between two carbon atoms. The general mechanism for the E1 reaction is shown in Key Mechanism 7-1.

KEY MECHANISM 7-1 The E1 Reaction

The E1 reaction requires ionization to a carbocation intermediate like the S_N1. The E1 and S_N1 reactions follow the same order of reactivity based on carbocation stability: 3° >> 2° >> 1°. Tertiary alkyl halides react readily by the E1 mechanism, secondary halides react much slower, and simple primary halides do not undergo E1 or S_N1 reactions.

Step 1: Unimolecular ionization to give a carbocation (rate-limiting).

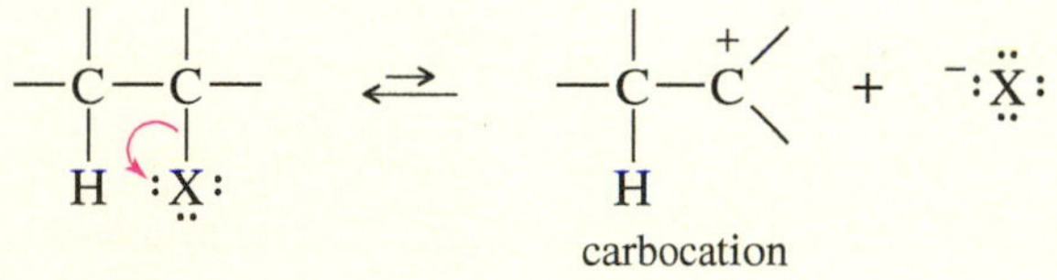

$$-\overset{|}{\underset{H}{C}}-\overset{|}{\underset{X}{C}}- \rightleftharpoons -\overset{|}{\underset{H}{C}}-\overset{+}{C}\langle \quad + \quad {}^-:\ddot{X}:$$

carbocation

Step 2: Deprotonation by a weak base (often the solvent) gives the alkene (fast).

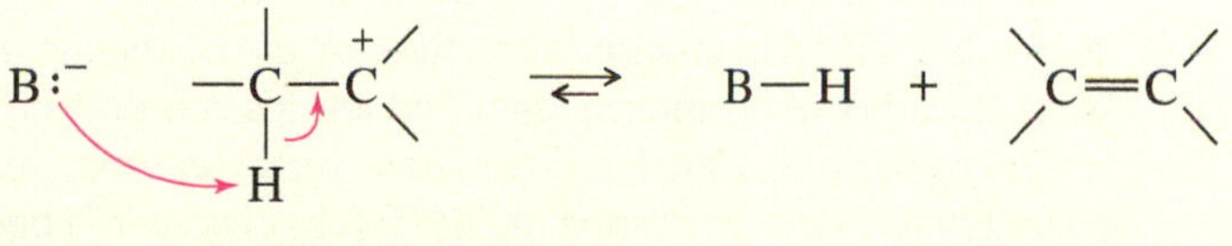

$$B:^- \quad -\overset{|}{\underset{H}{C}}-\overset{+}{C}\langle \rightleftharpoons B-H \quad + \quad \rangle C=C\langle$$

EXAMPLE: E1 elimination of bromocyclohexane in methanol.

Step 1: Ionization gives a carbocation and bromide ion in a slow step.

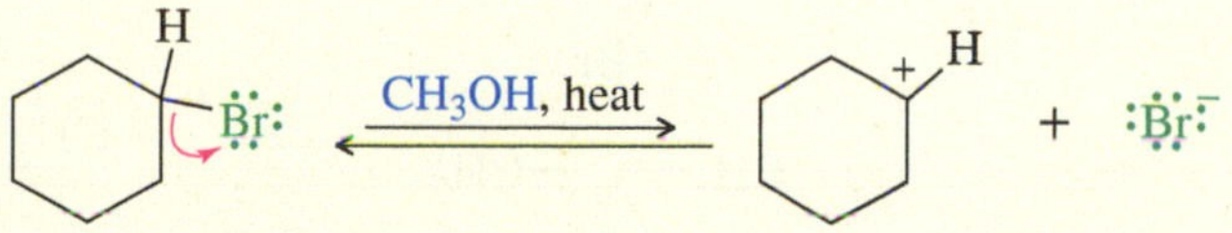

$$C_6H_{11}Br \underset{}{\xrightleftharpoons{CH_3OH,\ heat}} C_6H_{11}^+ \quad + \quad :\ddot{Br}:^-$$

(CONTINUED)

Step 2: Methanol abstracts a proton to give cyclohexene in a fast step.

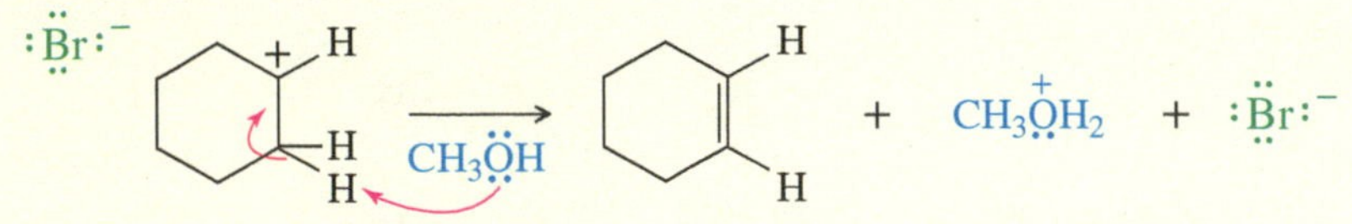

$$:\ddot{Br}:^- \quad \text{(cyclohexyl cation with H's)} \xrightarrow{CH_3\ddot{O}H} \text{cyclohexene} + CH_3\overset{+}{O}H_2 + :\ddot{Br}:^-$$

PROBLEM 7-16

Show what happens in step 2 of the example if the solvent acts as a *nucleophile* (forming a bond to carbon) rather than as a *base* (removing a proton).

The rate-limiting step involves ionization of the alkyl halide. Because the step is unimolecular (involves only one molecule), the rate equation is first-order. The rate depends only on the concentration of the alkyl halide, and not on the strength or concentration of the base.

$$\text{E1 rate} = k_r[\text{RX}]$$

The weak base (often the solvent) takes part in the fast second step of the reaction.

Unimolecular (E1) dehydrohalogenations usually take place in a good ionizing solvent (such as an alcohol or water), without a strong base present to force a bimolecular (E2) reaction. The substrate for E1 dehydrohalogenation is usually a tertiary alkyl halide. If secondary halides undergo E1 reactions, they require stronger conditions and react much more slowly.

7-10A Competition between the E1 and S_N1 Reactions

The E1 elimination almost always competes with the S_N1 substitution. Whenever a carbocation is formed, it can undergo either substitution or elimination, and mixtures of products often result. The following reaction shows the formation of both elimination and substitution products in the reaction of *tert*-butyl bromide with boiling ethanol.

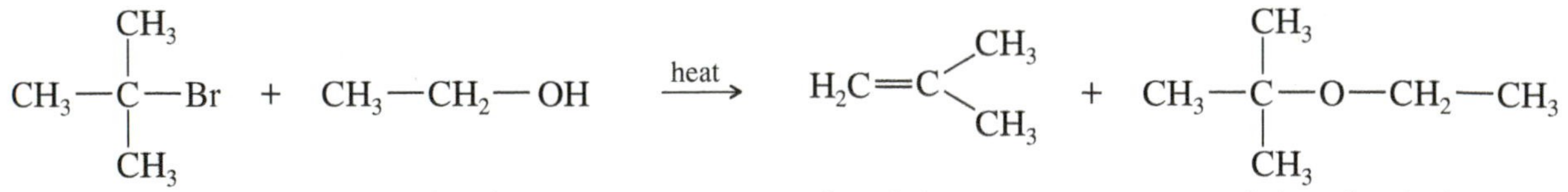

$$(CH_3)_3C{-}Br + CH_3{-}CH_2{-}OH \xrightarrow{\text{heat}} H_2C{=}C(CH_3)_2 + (CH_3)_3C{-}O{-}CH_2{-}CH_3$$

tert-butyl bromide; ethanol; 2-methylpropene (E1 product); ethyl *tert*-butyl ether (S_N1 product)

The 2-methylpropene product results from dehydrohalogenation, an elimination of hydrogen and a halogen atom. Under these first-order conditions (the absence of a strong base), dehydrohalogenation takes place by the E1 mechanism: Ionization of the alkyl halide gives a carbocation intermediate, which loses a proton to give the alkene. Substitution results from an S_N1 nucleophilic attack on the carbocation. This kind of reaction is called a **solvolysis** (*solvo* for "solvent," *lysis* for "cleavage") because the solvent reacts with the substrate. In this example, ethanol (the solvent) serves as a base (removing a proton) in the elimination and as a nucleophile (forming a bond to carbon) in the substitution.

Step 1: Ionization to form a carbocation.

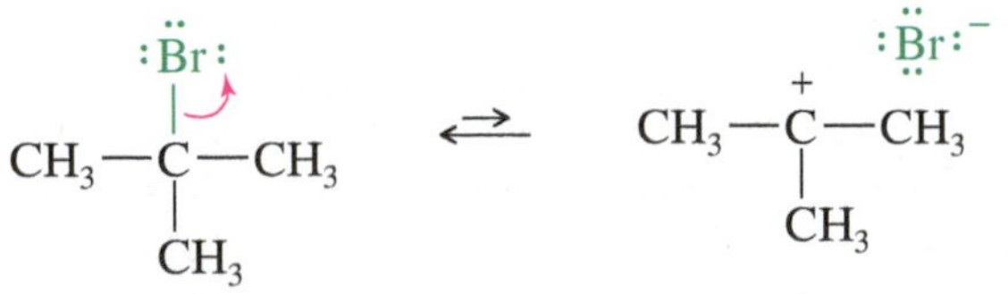

$$(CH_3)_3C{-}\ddot{Br}: \rightleftarrows (CH_3)_3C^+ + :\ddot{Br}:^-$$

tert-butyl bromide; carbocation intermediate

Step 2 (by the E1 mechanism): Basic attack by the solvent abstracts a proton to give an alkene.

Elimination: $CH_3CH_2-\ddot{O}H$ + $H-CH_2-C^+(CH_3)-CH_3$ (carbocation) $\rightleftarrows$ $(H)_2C=C(CH_3)_2$ (2-methylpropane) + $CH_3CH_2-\ddot{O}^+H_2$ $:\ddot{Br}:^-$

or step 2 (by the S_N1 mechanism): Nucleophilic attack by the solvent on the carbocation.

Substitution: $:\ddot{Br}:^-$ $CH_3-C^+(CH_3)-CH_3$ (carbocation) $\xrightarrow{CH_3CH_2-\ddot{O}-H}$ $:\ddot{Br}:^-$ $CH_3CH_2-\ddot{O}^+(H)-C(CH_3)_2-CH_3$ $\xrightarrow{CH_3CH_2-\ddot{O}-H}$ $CH_3-C(:\ddot{O}CH_2CH_3)(CH_3)-CH_3$ (2-ethoxy-2-methylpropane) $:\ddot{Br}:^-$ $CH_3CH_2\overset{+}{O}H_2$

Under ideal conditions, one of these first-order reactions provides a good yield of one product or the other. Often, however, carbocation intermediates react in two or more ways to give mixtures of products. For this reason, S_N1 and E1 reactions of alkyl halides are not frequently used for organic synthesis. They have been studied in great detail to learn about the properties of carbocations, however.

PROBLEM 7-17

S_N1 substitution and E1 elimination frequently compete in the same reaction.

(a) Propose a mechanism and predict the products for the solvolysis of 2-bromo-2,3,3-trimethylbutane in methanol.

(b) Compare the function of the solvent (methanol) in the E1 and S_N1 reactions.

7-10B Orbitals and Energetics of the E1 Reaction

In the second step of the E1 mechanism, the carbon atom next to the C^+ must rehybridize to sp^2 as the base attacks the proton and electrons flow into the new pi bond.

B:⁻ H overlap sp^3 C $^+$C CH_3 CH_3 H H empty p ⟶ B—H sp^2 H H C C CH_3 CH_3 π bond

The potential-energy diagram for the E1 reaction (Figure 7-9) is similar to that for the S_N1 reaction. The ionization step is strongly endothermic, with a rate-limiting transition state. The second step is a fast exothermic deprotonation by a base. The base is not involved in the reaction until *after* the rate-limiting step, so the rate depends only on the concentration of the alkyl halide. Weak bases are common in E1 reactions.

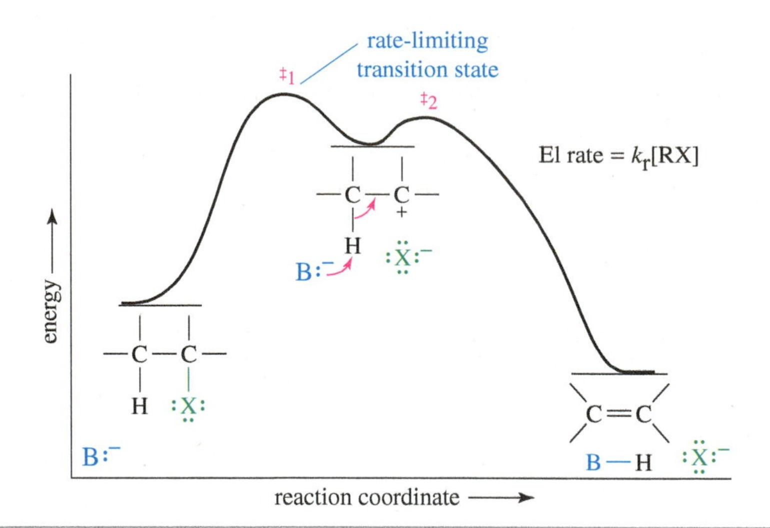

FIGURE 7-9
Reaction-energy diagram of the E1 reaction. The first step is a rate-limiting ionization. Compare this energy profile with that of the S_N1 reaction in Figure 6-8.

7-10C Rearrangements in E1 Reactions

Like other carbocation reactions, the E1 may be accompanied by rearrangement. Compare the following E1 reaction (with rearrangement) with the S_N1 reaction of the same substrate, shown in Mechanism 6-6. Note that the solvent acts as a *base* in the E1 reaction and a *nucleophile* in the S_N1 reaction.

MECHANISM 7-2 Rearrangement in an E1 Reaction

Like other reactions involving carbocations, the E1 may be accompanied by rearrangements such as hydride shifts and alkyl shifts.

Step 1: Ionization to form a carbocation. (slow)

CH_3—CH(Br)—CH(CH_3)—CH_3 ⇌ CH_3—$\overset{+}{C}$H—CH(CH_3)—CH_3 + :Br:⁻

2-bromo-3-methylbutane → 2° carbocation

Step 2: A hydride shift forms a more stable carbocation. (fast)

hydrogen moves with pair of electrons

CH_3—$\overset{+}{C}$H—CH(CH_3)—CH_3 $\xrightarrow{\sim H}$ CH_3—CH_2—$\overset{+}{C}$(CH_3)—CH_3

2° carbocation → 3° carbocation

Step 3: The weakly basic solvent removes either adjacent proton. (fast)

$CH_3CH_2\ddot{O}H$:Br:⁻

CH_3—CH(H)—$\overset{+}{C}$(CH_3)—CH_2—H ⟶ $(CH_3)C(H)$=C$(CH_3)_2$ + CH_3—CH_2—C(CH_3)=CH_2 + $CH_3CH_2\overset{+}{O}H_2$:Br:⁻

2-methylbut-2-ene + 2-methylbut-1-ene

SOLVED PROBLEM 7-3

When the following compound is heated in methanol, several different products are formed. Propose mechanisms to account for the four products shown.

(cyclobutyl)CH_2Br $\xrightarrow[CH_3OH]{heat}$ methylenecyclobutane + 1-methoxy-1-methylcyclobutane (CH_3, OCH_3) + cyclopentene + methoxycyclopentane (OCH_3)

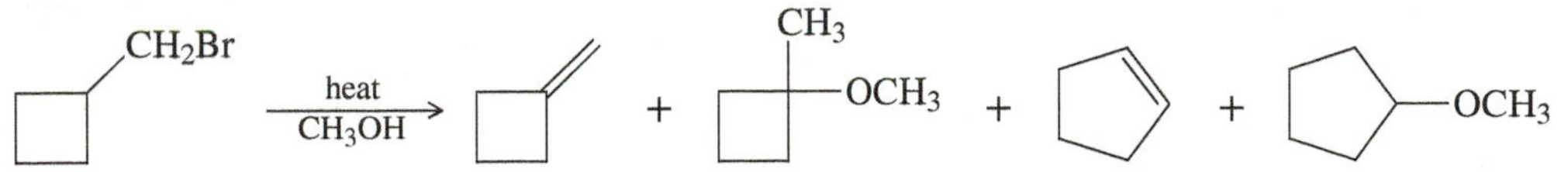

PARTIAL SOLUTION

With no strong base and a good ionizing solvent, we would expect a first-order reaction. But this is a primary alkyl halide, so ionization is difficult unless it rearranges. It might rearrange as it forms, but we'll imagine the cation forming then rearranging.

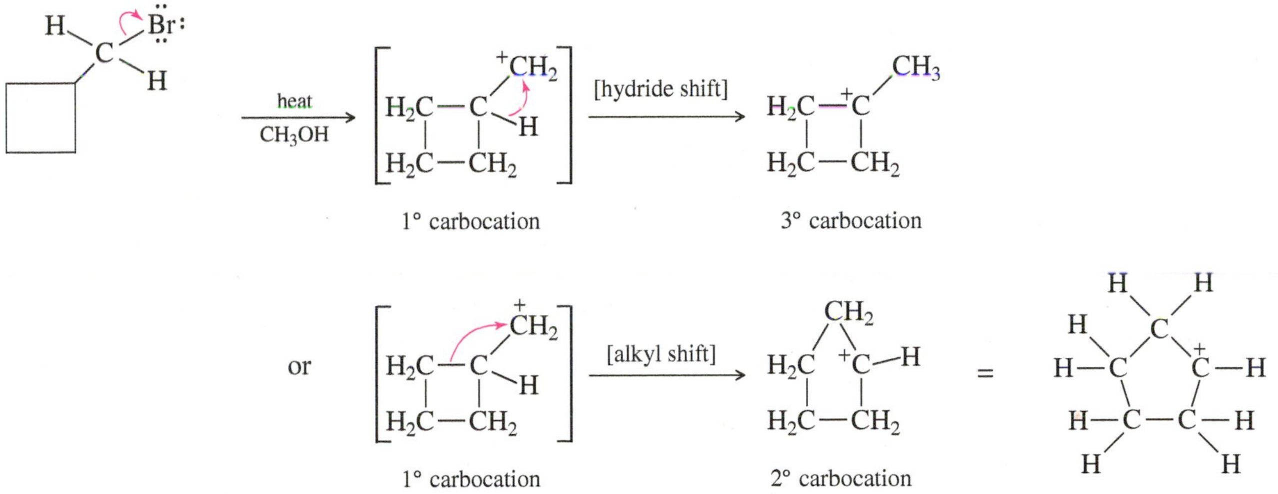

H–C(–H)(–Br:)–cyclobutane $\xrightarrow[CH_3OH]{heat}$ [$^+CH_2$–C(H) in ring H_2C—C / H_2C—CH_2] 1° carbocation $\xrightarrow{\text{[hydride shift]}}$ CH_3–^+C in ring H_2C—^+C / H_2C—CH_2 3° carbocation

or [$^+CH_2$–C(H) in ring H_2C—C / H_2C—CH_2] 1° carbocation $\xrightarrow{\text{[alkyl shift]}}$ CH_2, H_2C ^+C—H, H_2C—CH_2 2° carbocation = cyclopentyl cation drawn with all H atoms (H, H on C; H—C; $\overset{+}{C}$—H; H—C—C—H; H, H)

From these rearranged intermediates, either loss of a proton (E1) or attack by the solvent (S_N1) gives the observed products. Note that the actual reaction may give more than just these products, but the other products are not required for the problem.

PROBLEM 7-18

Finish Solved Problem 7-3 by showing how the rearranged carbocations give the four products shown in the problem. Be careful when using curved arrows to show deprotonation and/or nucleophilic attack by the solvent. The curved arrows always show movement of electrons, not movement of protons or other species.

PROBLEM 7-19

The solvolysis of 2-bromo-3-methylbutane potentially can give several products, including both E1 and S_N1 products from both the unrearranged carbocation and the rearranged carbocation. Mechanisms 6-6 (page 282) and 7-2 (previous page) show the products from the rearranged carbocation. Summarize all the possible products, showing which carbocation they come from and whether they are the products of E1 or S_N1 reactions.

We can now summarize four ways that a carbocation can react to become more stable.

SUMMARY Carbocation Reactions

A carbocation can

1. React with its own leaving group to return to the reactant: $R^+ + :X^- \longrightarrow R—X$
2. React with a nucleophile to form a substitution product (S_N1): $R^+ + Nuc:^- \longrightarrow R—Nuc$

(continued)

3. Lose a proton to form an elimination product (an alkene) (E1):

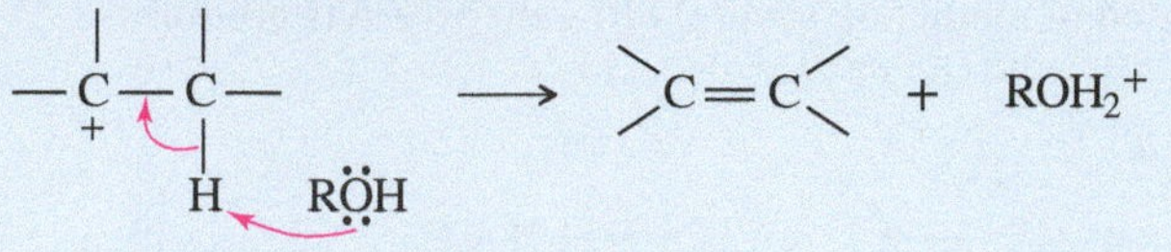

$$-\overset{+}{C}-\underset{H}{C}- \quad R\ddot{O}H \longrightarrow \;>C=C< \; + \; ROH_2^+$$

4. Rearrange to a more stable carbocation and then react further.
The order of stability of carbocations is: resonance-stabilized, 3° > 2° > 1°.

PROBLEM 7-20

Give the substitution and elimination products you would expect from the following reactions.
(a) 3-bromo-3-ethylpentane heated in methanol
(b) 1-iodo-1-phenylcyclopentane heated in ethanol
(c) 1-bromo-2-methylcyclohexane + silver nitrate in water ($AgNO_3$ forces ionization)

7-11 Positional Orientation of Elimination: Zaitsev's Rule

We have seen that many compounds can eliminate in more than one way, to give mixtures of alkenes. In many cases, we can predict which elimination product will predominate. In the example shown in Mechanism 7-2, the carbocation can lose a proton on either of two adjacent carbon atoms.

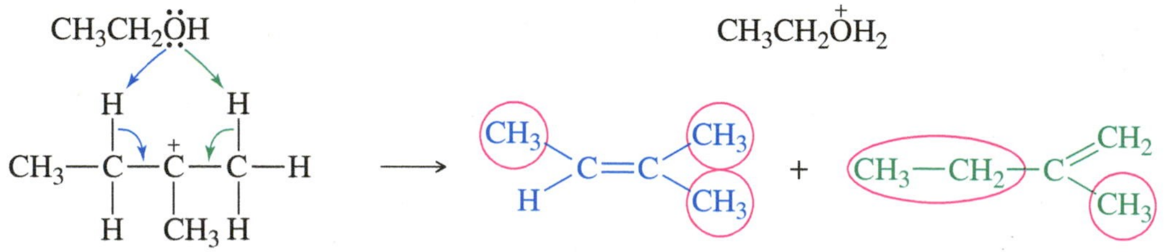

$$CH_3CH_2\ddot{O}H + CH_3-CH_2-\overset{+}{C}(CH_3)-CH_3 \longrightarrow CH_3CH_2\overset{+}{O}H_2 + (CH_3)_2C=CH(CH_3) + CH_3-CH_2-C(CH_3)=CH_2$$

2-methylbut-2-ene
trisubstituted, major (90%)

2-methylbut-1-ene
disubstituted, minor (10%)

The first product has a *trisubstituted* double bond, with three substituents (circled) on the doubly bonded carbons. It has the general formula $R_2C{=}CHR$. The second product has a *disubstituted* double bond, with general formula $R_2C{=}CH_2$ (or $R-CH{=}CH-R$). In most E1 and E2 eliminations where there are two or more possible elimination products, *the product with the most substituted double bond will predominate*. This general principle is the applied form of **Zaitsev's rule,** which we first encountered in Section 7-8B. Reactions that give the most substituted alkene are said to follow **Zaitsev orientation.**

PROBLEM-SOLVING HINT

Zaitsev is transliterated from the Russian name, and may also be spelled ***Saytzeff.***

ZAITSEV'S RULE: In elimination reactions, the most substituted alkene usually predominates.

$$\underset{\text{tetrasubstituted}}{R_2C{=}CR_2} > \underset{\text{trisubstituted}}{R_2C{=}CHR} > \underset{\text{disubstituted}}{RHC{=}CHR \text{ and } R_2C{=}CH_2} > \underset{\text{monosubstituted}}{RHC{=}CH_2}$$

This order of preference is the same as the order of stability of alkenes.

SOLVED PROBLEM 7-4

When 3-iodo-2,2-dimethylbutane is treated with silver nitrate in ethanol, three elimination products are formed. Give their structures, and predict which ones are formed in larger amounts.

SOLUTION

Silver nitrate forces ionization of the alkyl iodide to give solid silver iodide and a cation.

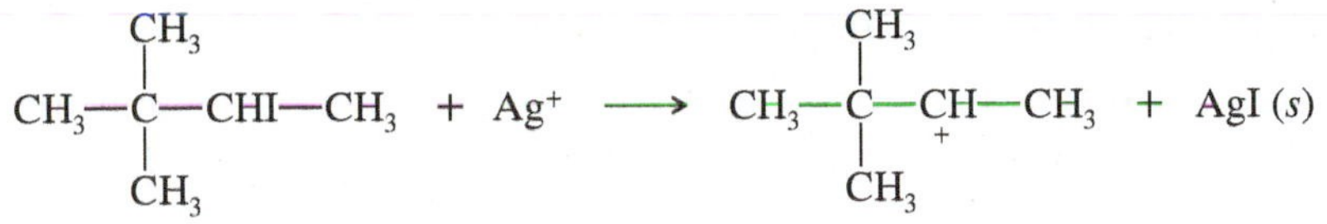

$$CH_3-C(CH_3)_2-CHI-CH_3 + Ag^+ \longrightarrow CH_3-C(CH_3)_2-\overset{+}{C}H-CH_3 + AgI\,(s)$$

This secondary carbocation can lose a proton to give an unrearranged alkene (**A**), or it can rearrange to a more stable tertiary cation.

Loss of a proton

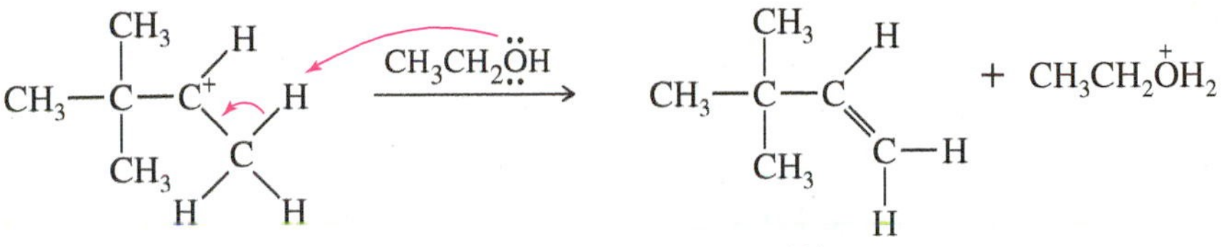

$$CH_3-C(CH_3)_2-\overset{+}{C}H-CH_3 \xrightarrow{CH_3CH_2\ddot{O}H} CH_3-C(CH_3)_2-CH=CH_2 + CH_3CH_2\overset{+}{O}H_2$$

(A)

Rearrangement

$$CH_3-C(CH_3)_2-\overset{+}{C}H-CH_3 \xrightarrow[\text{(methyl shift)}]{\sim CH_3} CH_3-\overset{+}{C}(CH_3)-CH(CH_3)-CH_3$$

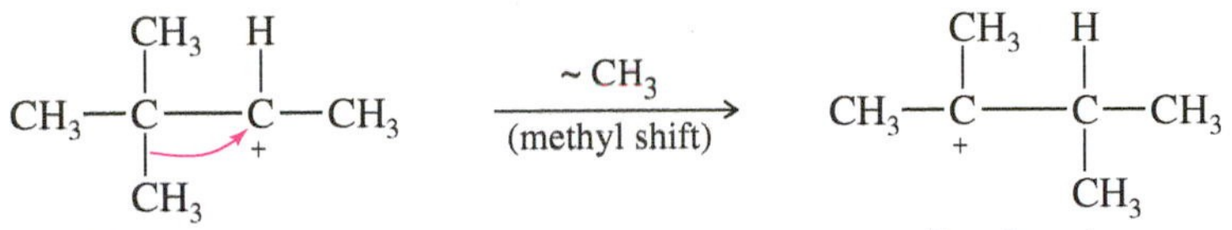

2° carbocation → 3° carbocation

The tertiary cation can lose a proton in either of two positions. One of the products (**B**) is a tetrasubstituted alkene, and the other (**C**) is disubstituted.

Formation of a tetrasubstituted alkene

$$CH_3-\overset{+}{C}(CH_3)-CH(CH_3)-CH_3 \xrightarrow{H\ddot{O}CH_2CH_3} (H_3C)_2C=C(CH_3)_2 + CH_3CH_2\overset{+}{O}H_2$$

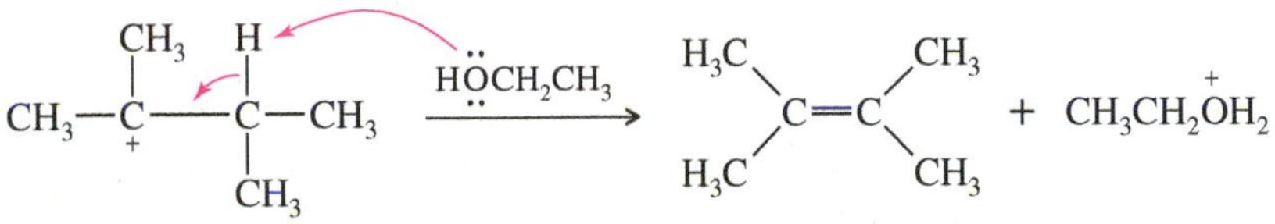

(B)
(tetrasubstituted)

Formation of a disubstituted alkene

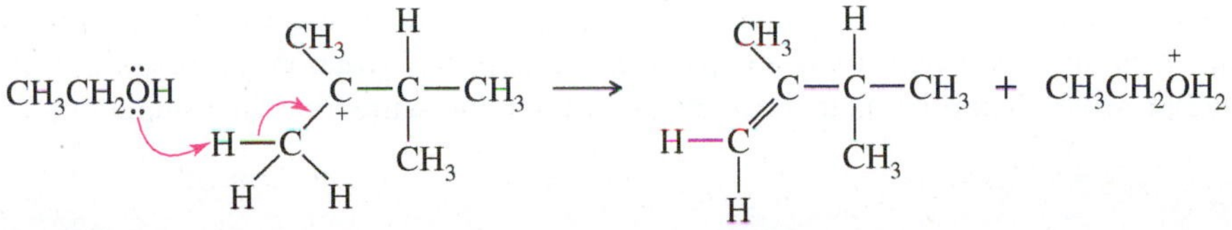

$$CH_3CH_2\ddot{O}H + H-CH_2-\overset{+}{C}(CH_3)-CH(CH_3)-CH_3 \longrightarrow H_2C=C(CH_3)-CH(CH_3)-CH_3 + CH_3CH_2\overset{+}{O}H_2$$

(C)
(disubstituted)

Product **B** predominates over product **C** because the double bond in **B** is more substituted. Whether product **A** is a major product will depend on the specific reaction conditions and whether proton loss or rearrangement occurs faster.

PROBLEM-SOLVING HINT

Whenever a carbocation has the positive charge on a carbon atom next to a more highly substituted carbon, consider whether a rearrangement might occur.

PROBLEM 7-21

Each of the two carbocations in Solved Problem 7-4 can also react with ethanol to give a substitution product. Give the structures of the two substitution products formed in this reaction.

PROBLEM 7-22

When (1-bromoethyl)cyclohexane is heated in methanol for an extended period of time, five products result: two ethers and three alkenes. Predict the products of this reaction, and propose mechanisms for their formation. Predict which of the three alkenes is the major elimination product.

7-12 Bimolecular Elimination: The E2 Reaction

Eliminations can also take place under second-order conditions with a strong base present. As an example, consider the reaction of *tert*-butyl bromide with methoxide ion in methanol. Methoxide ion is a strong base as well as a strong nucleophile. It attacks the alkyl halide faster than the halide can ionize to give a first-order reaction. No substitution product (methyl *tert*-butyl ether) is observed, however. The S_N2 mechanism is blocked because the tertiary alkyl halide is too hindered. The observed product is 2-methylpropene, resulting from elimination of HBr and formation of a double bond.

S_N2 is too hindered

$CH_3\ddot{O}:^-$ + $(CH_3)_3C-Br$ $\xrightarrow{E2}$ $CH_3-\ddot{O}-H$ + $H_2C=C(CH_3)_2$ + $:\ddot{Br}:^-$

Rate = $k_r[(CH_3)_3C-Br][^-OCH_3]$

The rate of this elimination is proportional to the concentrations of both the alkyl halide and the base, giving a second-order rate equation. This is a *bimolecular* process, with both the base and the alkyl halide participating in the transition state, so this mechanism is abbreviated **E2** for *Elimination, bimolecular.*

$$\text{E2 rate} = k_r[\text{RX}]\,[\text{B:}^-]$$

In the E2 reaction just shown, methoxide reacts as a *base* (removing a proton) rather than as a *nucleophile* (forming a bond to carbon). Most strong nucleophiles are also strong bases, and elimination commonly results when a strong base/nucleophile is used with a poor S_N2 substrate such as a 3° or hindered 2° alkyl halide. Instead of attacking the back side of the hindered electrophilic carbon, methoxide abstracts a proton from one of the methyl groups. This reaction takes place in one step, with bromide leaving as the base abstracts a proton.

In the general mechanism of the E2 reaction, a strong base abstracts a proton on a carbon atom adjacent to the one with the leaving group. As the base abstracts a proton, a double bond forms and the leaving group leaves. Like the S_N2 reaction, the E2 is a *concerted* reaction in which bonds break and new bonds form at the same time, in a single step.

KEY MECHANISM 7-3 The E2 Reaction

The concerted E2 reaction takes place in a single step. A strong base abstracts a proton on a carbon next to the leaving group, and the leaving group leaves. The product is an alkene.

$B:^-$ + H–C–C–X $\longrightarrow$ $[\overset{\delta-}{B}\text{---}H\text{---}C{=}C\text{---}\overset{\delta-}{X}]^\ddagger$ $\longrightarrow$ B—H + C=C + $X:^-$

transition state

EXAMPLE: E2 elimination of 3-bromopentane with sodium ethoxide.

$$Na^+ \ {}^-:\ddot{O}-CH_2CH_3 + \text{3-bromopentane} \xrightarrow{E2} H-\ddot{O}-CH_2CH_3 + \text{pent-2-ene (mostly trans)} + Na^+ \ :\ddot{Br}:^-$$

3-bromopentane

pent-2-ene (mostly trans)

The order of reactivity for alkyl halides in E2 reactions is 3° > 2° > 1°.

PROBLEM 7-23

Under second-order conditions (strong base/nucleophile), S_N2 and E2 reactions may occur simultaneously and compete with each other. Show what products might be expected from the reaction of 2-bromo-3-methylbutane (a moderately hindered 2° alkyl halide) with sodium ethoxide.

Reactivity of the Substrate in the E2 The order of reactivity of alkyl halides toward E2 dehydrohalogenation is found to be

$$3° > 2° > 1°$$

This reactivity order reflects the greater stability of highly substituted double bonds. Elimination of a secondary halide gives a more substituted alkene than elimination of a primary halide. Elimination of a tertiary halide gives an even more substituted alkene than a secondary halide. The stabilities of the alkene products are reflected in the transition states, giving lower activation energies and higher rates for elimination of alkyl halides that lead to highly substituted alkenes.

Mixtures of Products in the E2 The E2 reaction requires abstraction of a proton on a carbon atom next to the carbon bearing the halogen. If there are two or more possibilities, mixtures of products may result. In most cases, Zaitsev's rule predicts which of the possible products will be the major product: the most substituted alkene. For example, the E2 reaction of 2-bromobutane with potassium hydroxide gives a mixture of two products, but-1-ene (a monosubstituted alkene) and but-2-ene (a disubstituted alkene). As predicted by Zaitsev's rule, the disubstituted isomer but-2-ene is the major product.

$$\text{2-bromobutane} \xrightarrow[H_2O]{KOH} \text{but-1-ene} + \text{but-2-ene} \qquad (CH_3OH + KBr)$$

2-bromobutane

19% but-1-ene
monosubstituted

81% but-2-ene
disubstituted

Similarly, the reaction of 1-bromo-1-methylcyclohexane with sodium ethoxide gives a mixture of a disubstituted alkene and a trisubstituted alkene. The trisubstituted alkene is the major product.

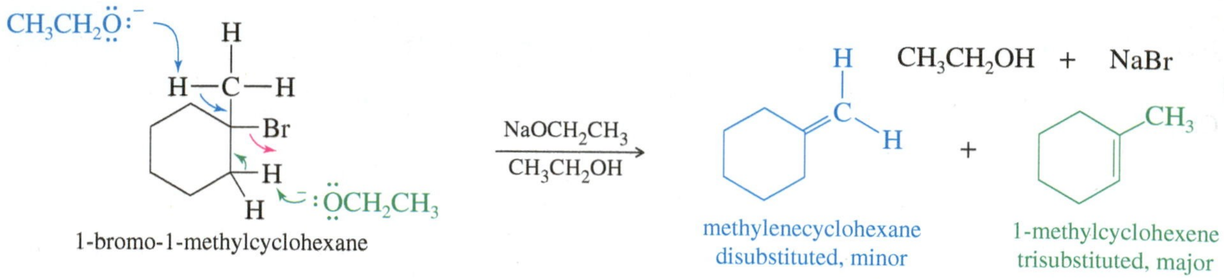

$CH_3CH_2\ddot{O}:^-$ ... 1-bromo-1-methylcyclohexane $\xrightarrow[CH_3CH_2OH]{NaOCH_2CH_3}$ methylenecyclohexane (disubstituted, minor) + 1-methylcyclohexene (trisubstituted, major) + CH_3CH_2OH + NaBr

PROBLEM-SOLVING HINT

Zaitsev's rule usually applies in E2 reactions unless the base and/or the leaving group are unusually bulky.

PROBLEM 7-24

1. Predict the elimination products of the following reactions. When two alkenes are possible, predict which one will be the major product. Explain your answers, showing the degree of substitution of each double bond in the products.
2. Which of these reactions are likely to produce both elimination and substitution products?
   (a) 2-bromopentane + $NaOCH_3$
   (b) 3-bromo-3-methylpentane + NaOMe (Me = methyl, CH_3)
   (c) 2-bromo-3-ethylpentane + NaOH

7-13 Bulky Bases in E2 Eliminations; Hofmann Orientation

Use of a Bulky Base If the substrate in an E2 elimination is prone to substitution, we can minimize the amount of substitution by using a bulky base. Large alkyl groups on a bulky base hinder its approach to attack a carbon atom (substitution), yet it can easily abstract a proton (elimination). Some of the bulky strong bases commonly used for elimination are *tert*-butoxide ion, diisopropylamine, triethylamine, and 2,6-dimethylpyridine.

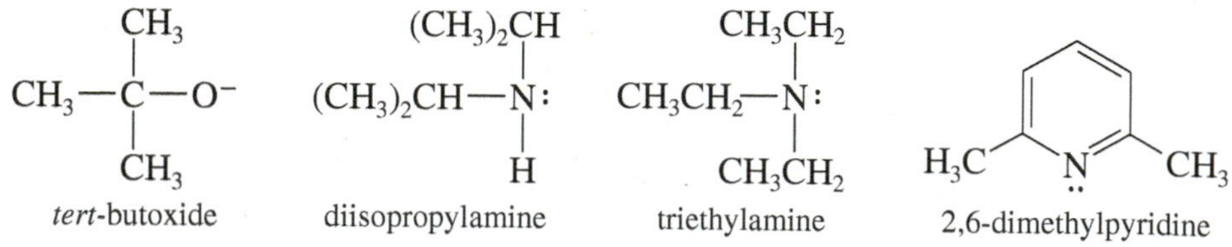

$(CH_3)_3C{-}O^-$ *tert*-butoxide; $[(CH_3)_2CH]_2N{-}H$ diisopropylamine; $(CH_3CH_2)_3N:$ triethylamine; 2,6-dimethylpyridine

The dehydrohalogenation of bromocyclohexane illustrates the use of a bulky base for elimination. Bromocyclohexane, a secondary alkyl halide, can undergo both substitution and elimination. Elimination (E2) is favored over substitution (S_N2) by using a bulky base such as diisopropylamine. Diisopropylamine is too bulky to be a good nucleophile, but it acts as a strong base to abstract a proton.

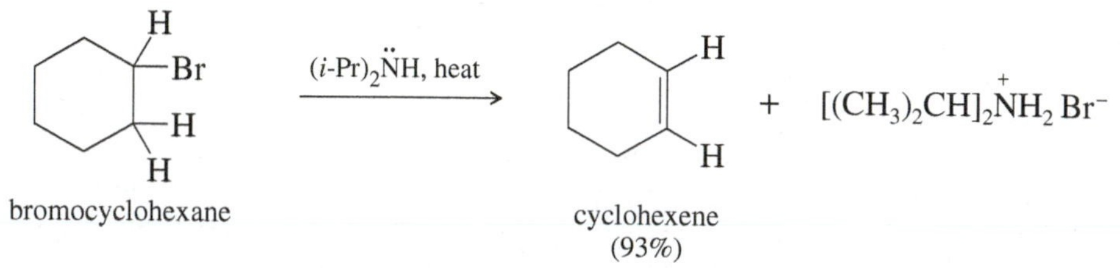

bromocyclohexane $\xrightarrow{(i\text{-Pr})_2\ddot{N}H,\ \text{heat}}$ cyclohexene (93%) + $[(CH_3)_2CH]_2\overset{+}{N}H_2\ Br^-$

Formation of the Hofmann Product Bulky bases can also accomplish dehydrohalogenations that do not follow the Zaitsev rule. Steric hindrance often prevents a bulky base from abstracting the proton that leads to the most highly substituted alkene. In these cases, it abstracts a less hindered proton, often the one that leads to formation of the least highly substituted product, called the **Hofmann product.** The following reaction gives mostly the **Zaitsev product** with the relatively unhindered ethoxide ion, but mostly the Hofmann product with the bulky *tert*-butoxide ion.

Zaitsev product | *Hofmann product*

$CH_3-CH(H)-C(CH_3)(Br)-CH_2H$ $\xrightarrow[CH_3CH_2OH]{^{-}OCH_2CH_3}$ $(H_3C)(H)C{=}C(CH_3)(CH_3)$ 71% + $(CH_3-CH_2)(H_3C)C{=}C(H)(H)$ 29%

$CH_3-CH(H)-C(CH_3)(Br)-CH_2H$ $\xrightarrow[(CH_3)_3COH \text{ (bulky base)}]{^{-}OC(CH_3)_3}$ $(H_3C)(H)C{=}C(CH_3)(CH_3)$ 28% + $(CH_3-CH_2)(H_3C)C{=}C(H)(H)$ 72%

less hindered

PROBLEM 7-25

For each reaction, decide whether substitution or elimination (or both) is possible, and predict the products you expect. Label the major products.

(a) 1-bromo-1-methylcyclohexane + NaOH in acetone
(b) 1-bromo-1-methylcyclohexane + triethylamine (Et_3N:)
(c) chlorocyclohexane + $NaOCH_3$ in CH_3OH
(d) chlorocyclohexane + $NaOC(CH_3)_3$ in $(CH_3)_3COH$

7-14 Stereochemistry of the E2 Reaction

Like the S_N2 reaction, the E2 follows a **concerted mechanism:** Bond breaking and bond formation take place at the same time, and the partial formation of new bonds lowers the energy of the transition state. Concerted mechanisms require specific geometric arrangements so that the orbitals of the bonds being broken can overlap with those being formed and the electrons can flow smoothly from one bond to another. The geometric arrangement required by the S_N2 reaction is a back-side attack; with the E2 reaction, a coplanar arrangement of the orbitals is needed.

E2 elimination requires partial formation of a new pi bond, with its parallel *p* orbitals, in the transition state. The electrons that once formed a C—H bond must begin to overlap with the orbital that the leaving group is vacating. Formation of this new pi bond implies that these two sp^3 orbitals must be parallel so that pi overlap is possible as the hydrogen and halogen leave and the orbitals rehybridize to the *p* orbitals of the new pi bond.

7-14A Anti-Coplanar and Syn-Coplanar E2 Eliminations

Figure 7-10 shows two conformations that provide the necessary coplanar alignment of the leaving group, the departing hydrogen, and the two carbon atoms. When the hydrogen and the halogen are **anti** to each other ($\theta = 180°$), their orbitals are aligned. This is called the **anti-coplanar** conformation. When the hydrogen and the halogen eclipse each other ($\theta = 0°$), their orbitals are once again aligned. This is called the **syn-coplanar** conformation. Make a model corresponding to Figure 7-10, and use it to follow along with this discussion.

Of these possible conformations, the anti-coplanar arrangement is most commonly seen in E2 reactions. The transition state for the anti-coplanar arrangement is a staggered conformation, with the base far away from the leaving group. In most cases, this transition state is lower in energy than that for the syn-coplanar elimination.

The transition state for syn-coplanar elimination is an eclipsed conformation. In addition to the higher energy resulting from eclipsing interactions, the transition state suffers from interference between the attacking base and the leaving group. To abstract the proton, the base must approach quite close to the leaving group. In most cases, the leaving group is bulky and negatively charged, and the repulsion between the base and the leaving group raises the energy of the syn-coplanar transition state.

Application: Biochemistry

Enzyme-catalyzed eliminations generally proceed by E2 mechanisms and produce only one stereoisomer. Two catalytic groups are involved: One abstracts the hydrogen, and the other assists in the departure of the leaving group. The groups are positioned appropriately to allow an anti-coplanar elimination.

For example, one of the steps in cellular respiration involves elimination of water from citric acid to give only the *Z* isomer of the product, commonly called *cis*-aconitic acid.

This proton is removed. — H; COOH; CH_2COOH; HOOC; OH; H — This proton is not removed.

citric acid

$\xrightarrow[\text{(an enzyme)}]{\text{aconitase}}$ $(HOOC)(H)C{=}C(COOH)(CH_2COOH)$ + H_2O

cis-aconitic acid (100%)

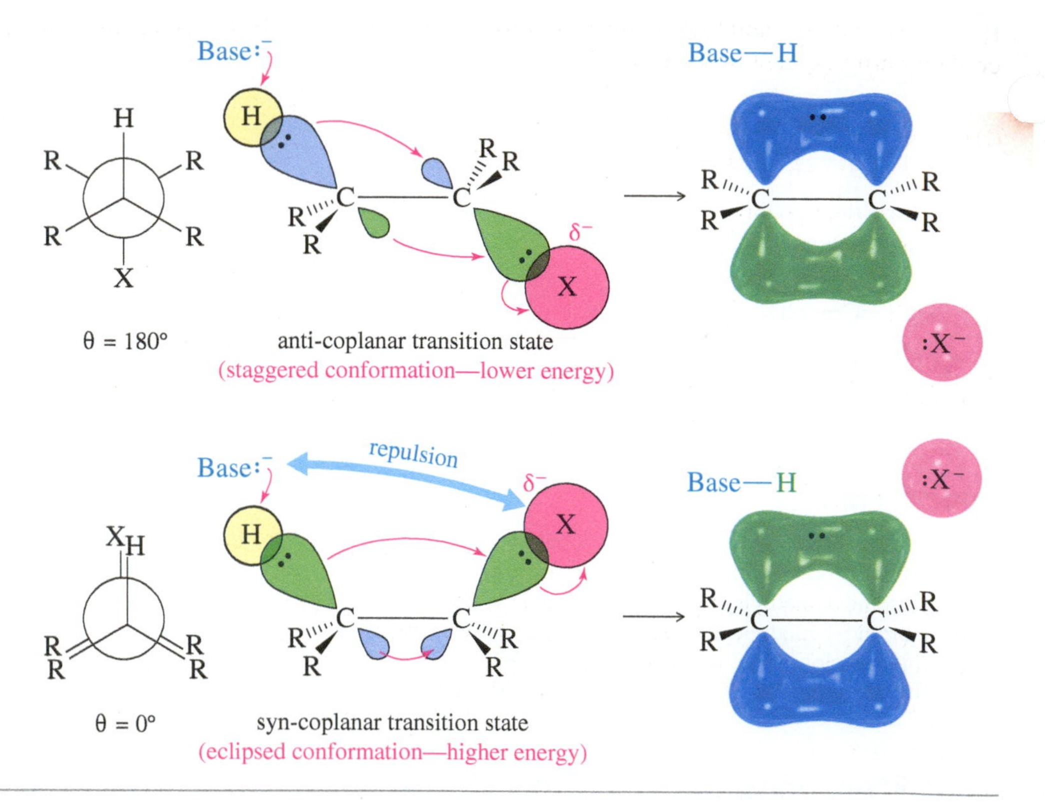

FIGURE 7-10
Concerted transition states of the E2 reaction. The orbitals of the hydrogen atom and the halide must be aligned so that they can begin to form a pi bond in the transition state.

Some molecules are rigidly held in eclipsed (or nearly eclipsed) conformations, with a hydrogen atom and a leaving group in a syn-coplanar arrangement. Such compounds are likely to undergo E2 elimination by a concerted syn-coplanar mechanism. Deuterium labeling (using D, the hydrogen isotope with mass number 2) is used in the following reaction to show which atom is abstracted by the base. Only the hydrogen atom is abstracted, because it is held in a syn-coplanar position with the bromine atom. Remember that syn-coplanar eliminations are unusual, however, and anti-coplanar eliminations are more common.

[Reaction scheme: bicyclic deuterated bromide (Br and H syn-coplanar) + $Na^+ \; ^-OCH_3$ → bicyclic alkene (D retained) + $HOCH_3$ + NaBr]

7-14B Stereospecific E2 Reactions

Like the S_N2 reaction (Section 6-12), the E2 is **stereospecific:** Different stereoisomers of the reactant give different stereoisomers of the product. The E2 is stereospecific because it normally goes through an anti-coplanar transition state. The products are alkenes, and different diastereomers of starting materials commonly give different diastereomers of alkenes. The following example shows why the E2 elimination of one diastereomer of 1-bromo-1,2-diphenylpropane gives only the trans isomer of the alkene product.

[Reaction scheme: Fischer projection (1S,2R)-1-bromo-1,2-diphenylpropane = (viewed from the side) → (rotate) → H and Br anti-coplanar, Base: → E2 → phenyl groups trans, Base—H + :Br:⁻ ; (Ph = phenyl group)]

If we make a model and look at this reaction from the left end of the molecule, the anti-coplanar arrangement of the H and Br is apparent.

MECHANISM 7-4 Stereochemistry of the E2 Reaction

Most E2 reactions go through an anti-coplanar transition state. This geometry is most apparent if we view the reaction sighting down the carbon–carbon bond between the hydrogen and leaving group. Viewed from the left:

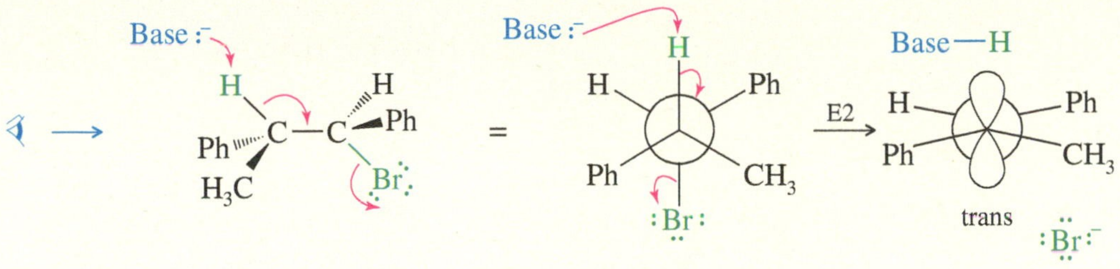

trans

The following reaction shows how the anti-coplanar elimination of the other diastereomer (*R,R*) gives only the cis isomer of the product. In effect, the two different diastereomers of the reactant give two different diastereomers of the product: a stereospecific result.

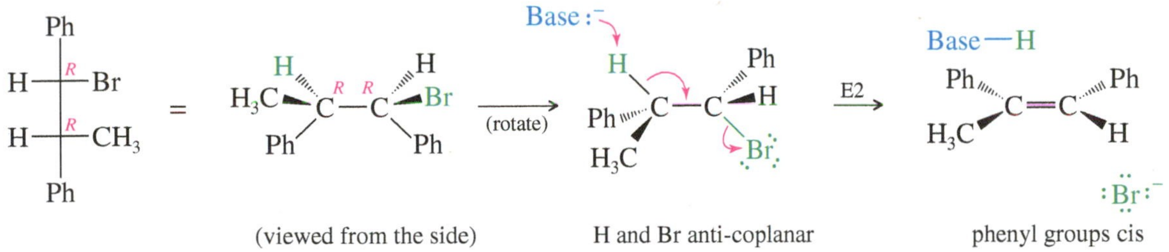

(viewed from the side) H and Br anti-coplanar phenyl groups cis

Viewed from the left end of the molecule:

cis

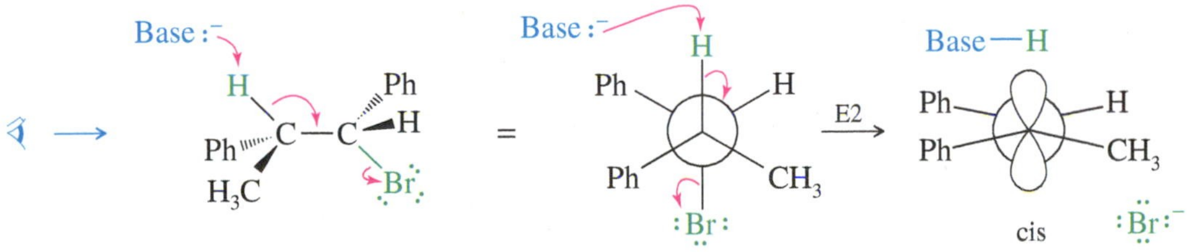

PROBLEM 7-26

Show that the (*S,S*) enantiomer of this (*R,R*) diastereomer of 1-bromo-1,2-diphenylpropane also undergoes E2 elimination to give the cis diastereomer of the product. (We do not expect these achiral reagents to distinguish between enantiomers.)

PROBLEM-SOLVING HINT

Don't try to memorize your way through these reactions. Look at each one, and consider what it might do. Use your models for the ones that involve stereochemistry.

PROBLEM 7-27

Make models of the following compounds, and predict the products formed when they react with the strong bases shown.

(a) + KOH ⟶ (substitution and elimination)

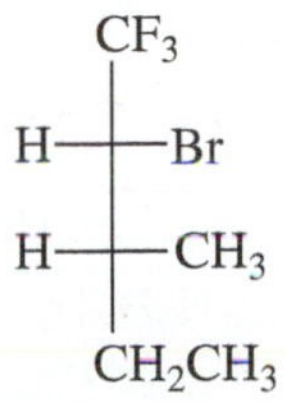

(continued)

PROBLEM-SOLVING HINT

Anti-coplanar E2 eliminations are common.

Syn-coplanar E2 eliminations are rare, usually occurring when free rotation is not possible.

(b) *meso*-1,2-dibromo-1,2-diphenylethane + $(CH_3CH_2)_3N$:
(c) (*d,l*)-1,2-dibromo-1,2-diphenylethane + $(CH_3CH_2)_3N$:
(d) 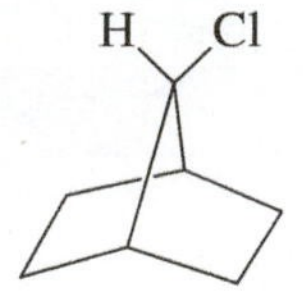 + NaOH in acetone
(e) 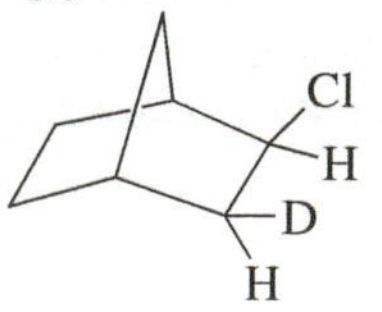 + $(CH_3)_3CO^-$
(f) 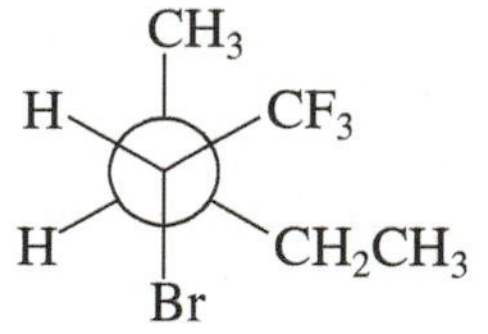 + $NaOCH(CH_3)_2$

7-15 E2 Reactions in Cyclohexane Systems

Nearly all cyclohexanes are most stable in chair conformations. In the chair, all the carbon–carbon bonds are staggered, and any two adjacent carbon atoms have axial bonds in an anti-coplanar conformation, ideally oriented for the E2 reaction. (As drawn in the following figure, the axial bonds are vertical.) On any two adjacent carbon atoms, one has its axial bond pointing up and the other has its axial bond pointing down. These two bonds are trans to each other, and we refer to their geometry as **trans-diaxial.**

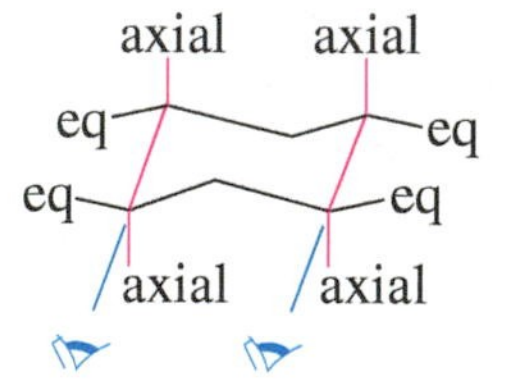

perspective view

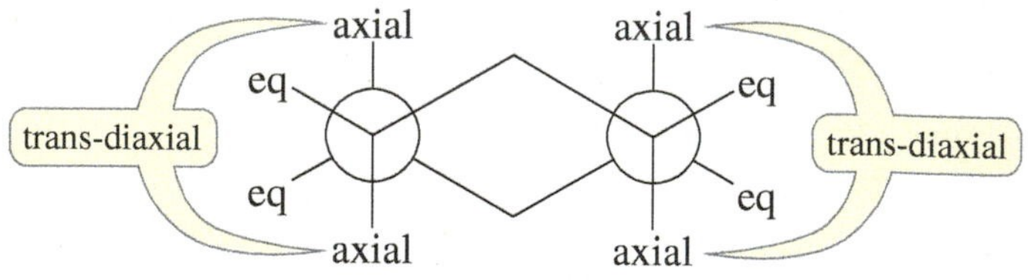

Newman projection

An E2 elimination can take place on this chair conformation only if the proton and the leaving group can get into a trans-diaxial arrangement. Figure 7-11 shows the E2 dehydrohalogenation of bromocyclohexane. The molecule must flip into the chair conformation with the bromine atom axial before elimination can occur.

(You should make models of the structures in the following examples and problems so that you can follow along more easily.)

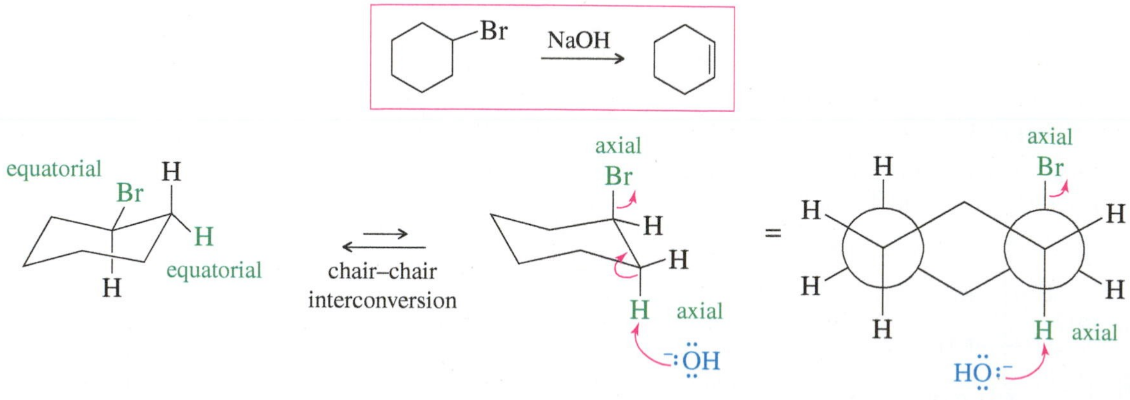

FIGURE 7-11
E2 eliminations on cyclohexane rings. E2 elimination of bromocyclohexane requires that both the proton and the leaving group be trans and both be axial.

SOLVED PROBLEM 7-5

Explain why the following deuterated 1-bromo-2-methylcyclohexane undergoes dehydrohalogenation by the E2 mechanism to give only the indicated product. Two other alkenes are not observed.

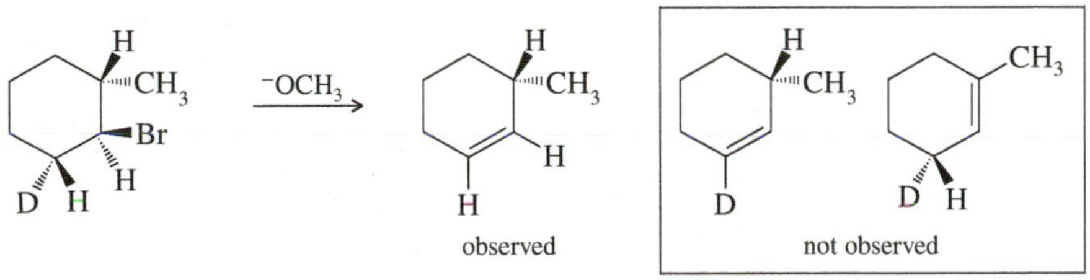

$^-OCH_3$ → observed | not observed

SOLUTION

In an E2 elimination, the hydrogen atom and the leaving group must have a trans-diaxial relationship. In this compound, only one hydrogen atom—the deuterium—is trans to the bromine atom. When the bromine atom is axial, the adjacent deuterium is also axial, providing a trans-diaxial arrangement.

E2 → ($CH_3\ddot{O}:^-$)

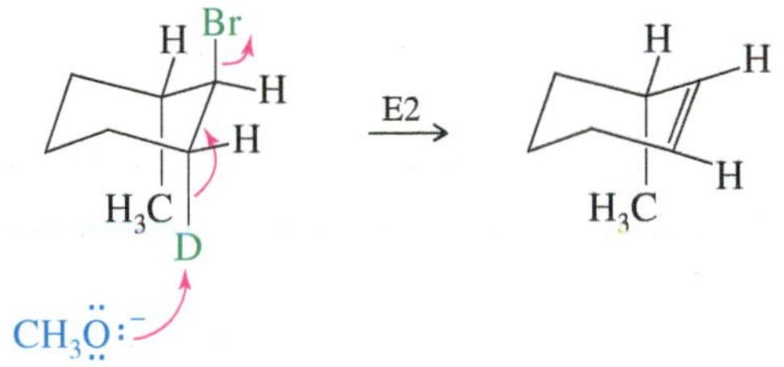

PROBLEM 7-28

Predict the elimination products of the following reactions, and label the major products.

(a) *cis*-1-bromo-2-methylcyclohexane + $NaOCH_3$ in CH_3OH

(b) *trans*-1-bromo-2-methylcyclohexane + $NaOCH_3$ in CH_3OH

PROBLEM-SOLVING HINT

In a chair conformation of a cyclohexane ring, a trans-diaxial arrangement places the two groups anti and coplanar.

PROBLEM 7-29

When the following stereoisomer of 2-bromo-1,3-dimethylcyclohexane is treated with sodium methoxide, no E2 reaction is observed. Explain why this compound cannot undergo the E2 reaction in the chair conformation.

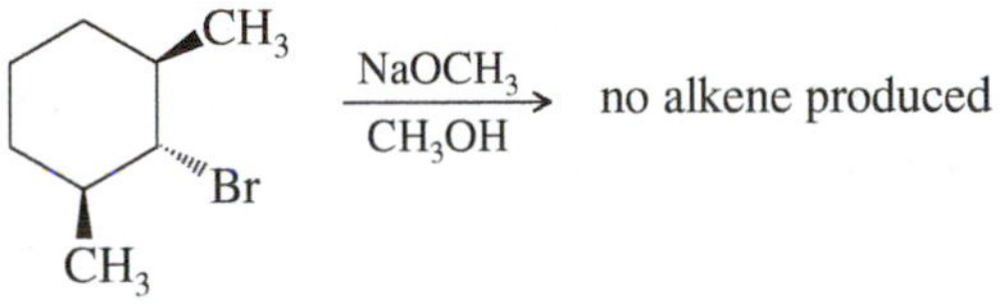

$\xrightarrow[CH_3OH]{NaOCH_3}$ no alkene produced

PROBLEM-SOLVING HINT

Look for a hydrogen trans to the leaving group; then see if the hydrogen and the leaving group can become diaxial.

PROBLEM 7-30

(a) Two stereoisomers of a bromodecalin are shown. Although the difference between these stereoisomers may seem trivial, one isomer undergoes elimination with KOH much faster than the other. Predict the products of these eliminations, and explain the large difference in the ease of elimination.

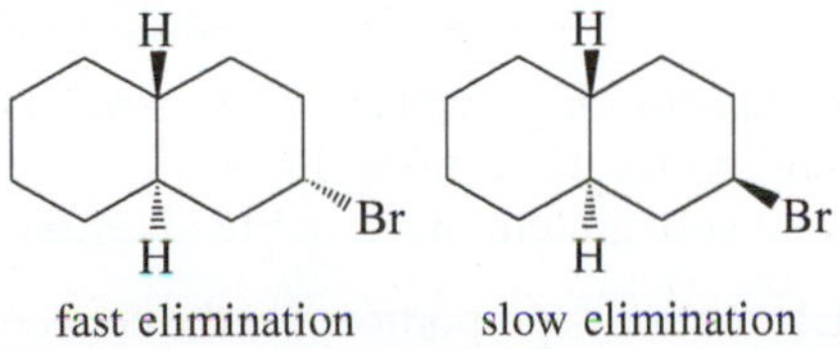

fast elimination   slow elimination

(continued)

(b) Predict which of the following compounds will undergo elimination with KOH faster, and explain why. Predict the major product that will be formed.

$CH(CH_3)_2$ $CH(CH_3)_2$

Br   Br

PROBLEM 7-31

Give the expected product(s) of E2 elimination for each reaction. (*Hint*: Use models!)

(a) CH_3, H, Br, H, H_3C, H $\xrightarrow{NaOCH_3}$ one product

(b)

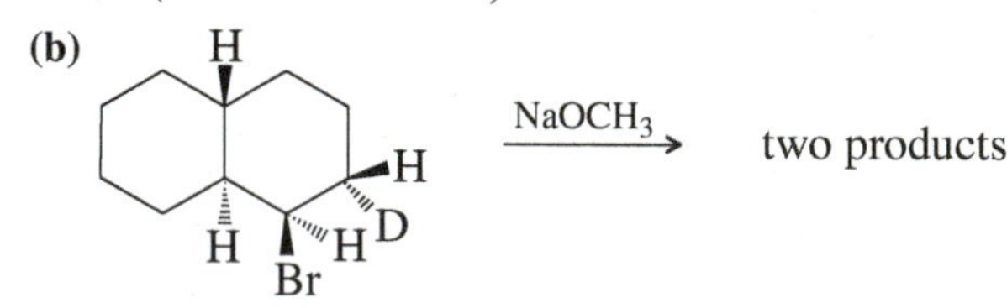

7-16 Comparison of E1 and E2 Elimination Mechanisms

Let's summarize the major points to remember about the E1 and E2 reactions, focusing on the factors that help us predict which of these mechanisms will operate under a given set of experimental conditions.

Effect of the Base The nature of the base is the single most important factor in determining whether an elimination will go by the E1 or E2 mechanism. If a strong base is present, the rate of the bimolecular reaction will be greater than the rate of ionization, and the E2 reaction will predominate (perhaps accompanied by the S_N2).

If no strong base is present, then a good solvent makes a unimolecular ionization likely. Subsequent loss of a proton to a weak base (such as the solvent) leads to elimination. Under these conditions, the E1 reaction usually predominates, usually accompanied by the S_N1.

E1:	Base strength is unimportant (usually weak).
E2:	Requires strong bases.

Effect of the Solvent The slow step of the E1 reaction is the formation of two ions. Like the S_N1, the E1 reaction critically depends on polar ionizing solvents such as water and the alcohols.

In the E2 reaction, the transition state spreads out the negative charge of the base over the entire molecule. There is no more need for solvation in the E2 transition state than in the reactants. The E2 is therefore less sensitive to the solvent; in fact, some reagents are stronger bases in less polar solvents.

E1:	Requires a good ionizing solvent.
E2:	Solvent polarity is not so important.

Effect of the Substrate For both the E1 and the E2 reactions, the order of reactivity is

E1, E2:	$3° > 2° > 1°$ (1° usually will not go E1)

In the E1 reaction, the rate-limiting step is formation of a carbocation, and the reactivity order reflects the stability of carbocations. In the E2 reaction, the more substituted halides generally form more substituted, more stable alkenes.

Kinetics The rate of the E1 reaction is proportional to the concentration of the alkyl halide [RX] but not to the concentration of the weak base. It follows a first-order rate equation.

The rate of the E2 reaction is proportional to the concentrations of both the alkyl halide [RX] and the strong base [B:$^-$]. It follows a second-order rate equation.

E1 rate $= k_r$[RX]
E2 rate $= k_r$[RX][B:$^-$]

Orientation of Elimination In most E1 and E2 eliminations with two or more possible products, the product with the most substituted double bond (the most stable product) predominates. This principle is called **Zaitsev's rule,** and the most highly substituted product is called the **Zaitsev product.**

Br — E1 or E2 → major (cis + trans) + minor

E1, E2: Usually Zaitsev orientation.

Stereochemistry The E1 reaction begins with an ionization to give a flat carbocation. No particular geometry is required for ionization.

The E2 reaction takes place through a concerted mechanism that requires a coplanar arrangement of the bonds to the atoms being eliminated. The transition state is usually anti-coplanar, although it may be syn-coplanar in rigid systems.

E1: No particular geometry required for the slow step.
E2: Coplanar arrangement (usually anti) required for the transition state.

Rearrangements The E1 reaction involves a carbocation intermediate. This intermediate can rearrange, usually by the shift of a hydride or an alkyl group, to give a more stable carbocation.

The E2 reaction takes place in one step with no intermediates. No rearrangement is possible in the E2 reaction.

E1: Rearrangements are common.
E2: No rearrangements.

SUMMARY Elimination Reactions

	E1	E2
Promoting factors		
base	weak bases work	strong base required
solvent	good ionizing solvent	wide variety of solvents
substrate	3° > 2°	3° > 2° > 1°
leaving group	good one required	good one required
Characteristics		
kinetics	first order, k_r[RX]	second order, k_r[RX][B:$^-$]
orientation	most substituted alkene	most substituted alkene
stereochemistry	no special geometry	coplanar transition state required
rearrangements	common	impossible

PROBLEM 7-32

Predict the major and minor elimination products of the following proposed reactions (ignoring any possible substitutions for now). In each case, explain whether you expect the mechanism of the elimination to be E1 or E2.

(a) 1-bromo-1-cyclopentylethane (Br, CH_3) $\xrightarrow[\text{CH}_3\text{CH}_2\text{OH, heat}]{\text{NaOCH}_2\text{CH}_3}$

(b) 1-bromo-1-cyclopentylethane (Br, CH_3) $\xrightarrow[\text{CH}_3\text{OH, heat}]{\text{AgNO}_3}$

7-17 Competition Between Substitutions and Eliminations

Substitutions require the substrate to react with a nucleophile, and eliminations require it to react with a base. Most bases are also nucleophilic, however, and most nucleophiles are also basic; so both substitution and elimination may be possible in the very same reaction. To predict the products of a reaction, we need to predict both the mechanism and whether substitution or elimination will predominate.

7-17A Bimolecular Substitution Versus Elimination (Strong Base or Nucleophile)

In most cases, the order of a reaction is determined by the strength of the reagent that serves as the base or nucleophile. Strong reagents are likely to attack the substrate faster than the substrate ionizes, so strong reagents favor bimolecular, second-order reactions: S_N2 and E2. For example, sodium methoxide ($NaOCH_3$) is both a strong base and a strong nucleophile. Sodium methoxide narrows the likely reaction mechanisms to S_N2 and E2, but it does not imply which of these will predominate. The following examples show substrates that give almost entirely substitution (1°), entirely elimination (3°), and a mixture of the two reactions (2°).

1-bromobutane (Br) (1° alkyl halide) $\xrightarrow[\text{CH}_3\text{OH}]{\text{Na}^+ \ ^-\text{OCH}_3}$ 1-methoxybutane (OCH_3) >95% — almost entirely S_N2

2-bromobutane (Br) (2° alkyl halide) $\xrightarrow[\text{CH}_3\text{OH}]{\text{Na}^+ \ ^-\text{OCH}_3}$ 2-butene (E2) + 2-methoxybutane (OCH_3) (S_N2)

(proportions depend on conditions)

$Br—C(CH_3)_2—CH_3$ (3° alkyl halide) $\xrightarrow[\text{CH}_3\text{OH}]{\text{Na}^+ \ ^-\text{OCH}_3}$ $H_2C{=}C(CH_3)_2$ 100% — entirely E2

These examples show the principles that apply to bimolecular substitutions and eliminations. Primary halides are most reactive toward S_N2 reactions, and they are least reactive toward E2 reactions. Primary halides usually react with strong nucleophiles to give substitution by the S_N2 mechanism. If the reagent is also a strong base, a small amount of elimination might also occur. Tertiary halides cannot undergo S_N2 reactions, so we can safely predict that a strong base will react with a tertiary halide to give only elimination by the E2 mechanism.

Secondary halides are hardest to predict. They react with strong bases by the E2 mechanism and with strong nucleophiles by the S_N2 mechanism. But most strong nucleophiles are also strong bases, so the proportions of substitution and elimination may be controlled by other factors such as temperature, solvent, and structural effects such as steric hindrance. A few strong nucleophiles are not strong bases, however, and they can promote S_N2 substitution over E2 elimination. Halide ions are the most common examples of strong nucleophiles that are weak bases.

Br $\xrightarrow{Na^+ \ I^-}$ I (major product) S_N2

Bulky bases are often used to promote E2 elimination over S_N2 substitution. Bulky bases are hindered from approaching the back side of a carbon atom to do the S_N2 displacement, but they can easily approach a hydrogen atom on a neighboring carbon and abstract a proton.

Br $\xrightarrow{Na^+ \ ^-OC(CH_3)_3}$ (major product) E2

7-17B Unimolecular Substitution Versus Elimination (Weak Base or Nucleophile)

In the absence of a strong base or strong nucleophile and in the presence of a good ionizing solvent (very polar, often hydroxylic), the reaction rate is first-order, limited by the rate of unimolecular ionization to form a carbocation. Under these conditions, we have little control over the fast second step, when the reagent either attacks the carbocation to give substitution (S_N1) or removes an adjacent proton to give elimination (E1). The carbocation intermediate can also rearrange, leading to even more products. This lack of control explains why unimolecular substitutions and eliminations of alkyl halides are rarely used for synthesis.

CH_3 Br $\xrightarrow[CH_3CH_2OH]{heat}$ CH_3 OCH_2CH_3 + CH_3 + CH_2

S_N1 E1

Figure 7-12 compares the substitution and elimination mechanisms under second-order (bimolecular) conditions with those that occur under first-order (unimolecular) conditions. A strong reagent is likely to displace the leaving group from carbon (S_N2) or remove a proton from a neighboring carbon (E2). A weak reagent is unlikely to react until ionization occurs. Then it may form a bond to the carbocation (S_N1) or remove a proton from a neighboring carbon (E1).

Temperature is one variable we can control to influence the competition between substitution and elimination. Unlike substitutions, eliminations produce more products than reactants.

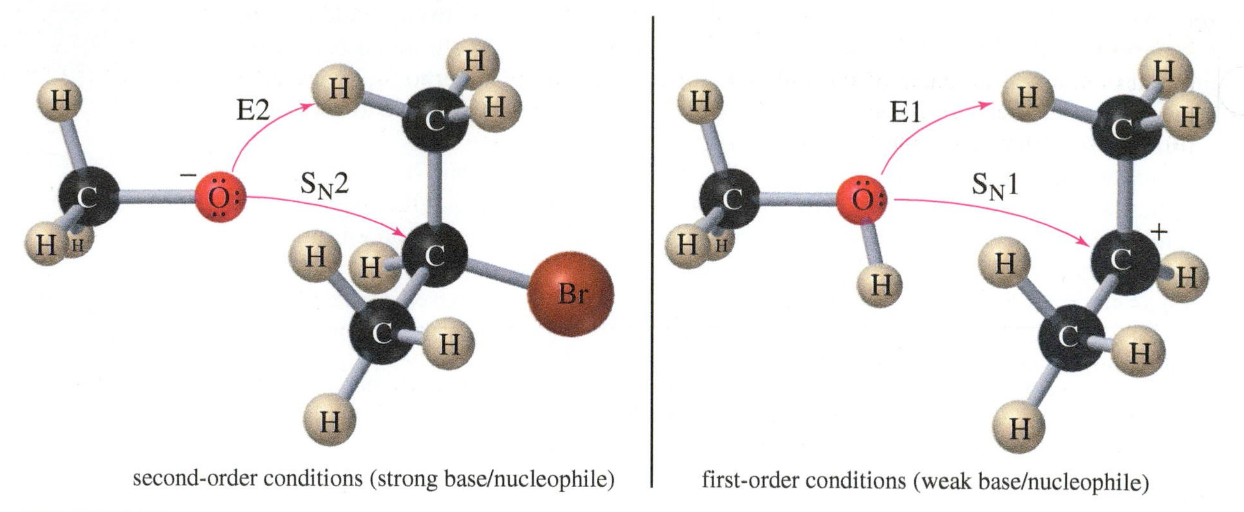

FIGURE 7-12
Under second-order conditions (strong base/nucleophile), a secondary alkyl halide might undergo either substitution (S_N2) or elimination (E2). Under first-order conditions (weak base/nucleophile), S_N1 and E1 are possible.

$$-\overset{|}{\underset{\underset{H}{|}}{C}}-\overset{|}{\underset{\underset{:\ddot{X}:}{|}}{C}}- \; + \; B:^- \longrightarrow B-H \; + \; \rangle C=C\langle \; + \; :\ddot{X}:^-$$

Reactions that produce more products than reactants generally increase the entropy of the system, meaning $\Delta S > 0$. By raising the temperature, we can increase the $(T\Delta S)$ term in the free-energy equation, $\Delta G = \Delta H - T\Delta S$, contributing to a more favorable (negative) value of ΔG. Therefore, an increase in temperature usually favors elimination, while a decrease in temperature favors substitution. This general principle applies to both unimolecular and bimolecular reactions when other factors are evenly balanced.

SUMMARY TABLE: Substitution and Elimination Reactions of Alkyl Halides

The reaction products and mechanisms depend on the type of alkyl halide and on the strength of the nucleophile/base.

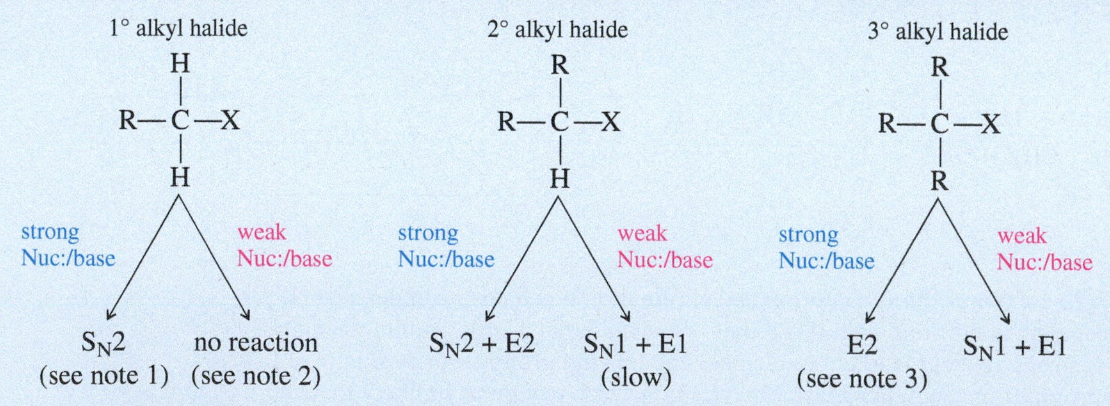

Note 1: In the presence of a strong nucleophile, unhindered 1° alkyl halides usually give good yields of substitution products with little elimination.
Note 2: 1° alkyl halides can be forced to ionize by $AgNO_3$/EtOH and heat, usually with rearrangement, to give S_N1 and E1 products.
Note 3: Any strongly hindered alkyl halide (even if 1° or 2°) will undergo S_N2 reactions very slowly or not at all.

SOLVED PROBLEM 7-6

Predict the major products of the following reactions, and indicate the mechanisms (S_N1, S_N2, E1, or E2) that are likely responsible for their formation.

(a) 1-bromo-1-methylcyclohexane reacts with sodium methoxide in methanol.

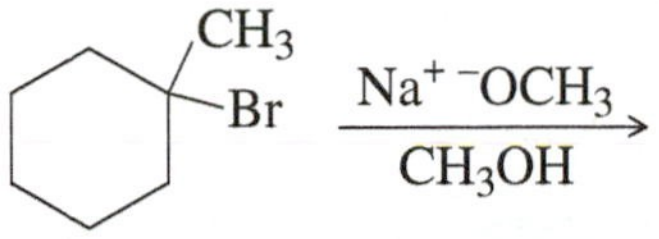

CH_3, Br; $\xrightarrow[CH_3OH]{Na^{+}\ ^{-}OCH_3}$

SOLUTION

(a) $NaOCH_3$ is a strong base and a strong nucleophile, suggesting that S_N2 and E2 are the likely reactions. This 3° alkyl halide cannot undergo S_N2 reactions, so the only remaining possibility is elimination by the E2 mechanism. The elimination can take place by abstraction of a methyl proton (circled below) or a ring proton on one of the CH_2 groups (circled) next to the carbon bonded to Br. $NaOCH_3$ is not a bulky base, so Zaitsev's rule probably applies.

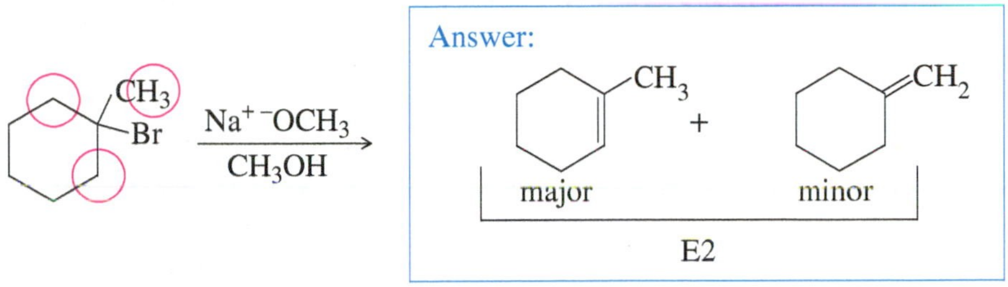

CH_3, Br; $\xrightarrow[CH_3OH]{Na^{+}\ ^{-}OCH_3}$

Answer: CH_3 (major) + CH_2 (minor); E2

Follow-up question: Draw a chair transition state for this elimination, showing how a trans-diaxial arrangement of the H and Br allows E2 elimination to give the major product.

(b) 1-iodo-2-methylcyclohexane is heated in methanol for an extended period.

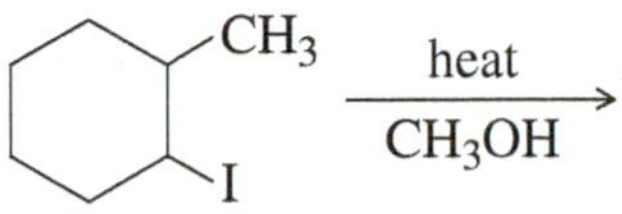

CH_3, I; $\xrightarrow[CH_3OH]{heat}$

SOLUTION

(b) A slow ionization of this secondary alkyl halide leads to formation of a 2° carbocation, which can lose a proton in either of two directions to give products of E1 elimination. Alternatively, methanol can attack the 2° carbocation, followed by loss of a proton, to give a product of S_N1 substitution. The 2° carbocation can also rearrange by a hydride shift to give a more stable 3° carbocation. That 3° carbocation can lose a proton in either of two directions to give products of E1 elimination, or it can react with methanol, followed by deprotonation, to give a product of S_N1 substitution. We don't know what the major product will be or even if there will be one major product.

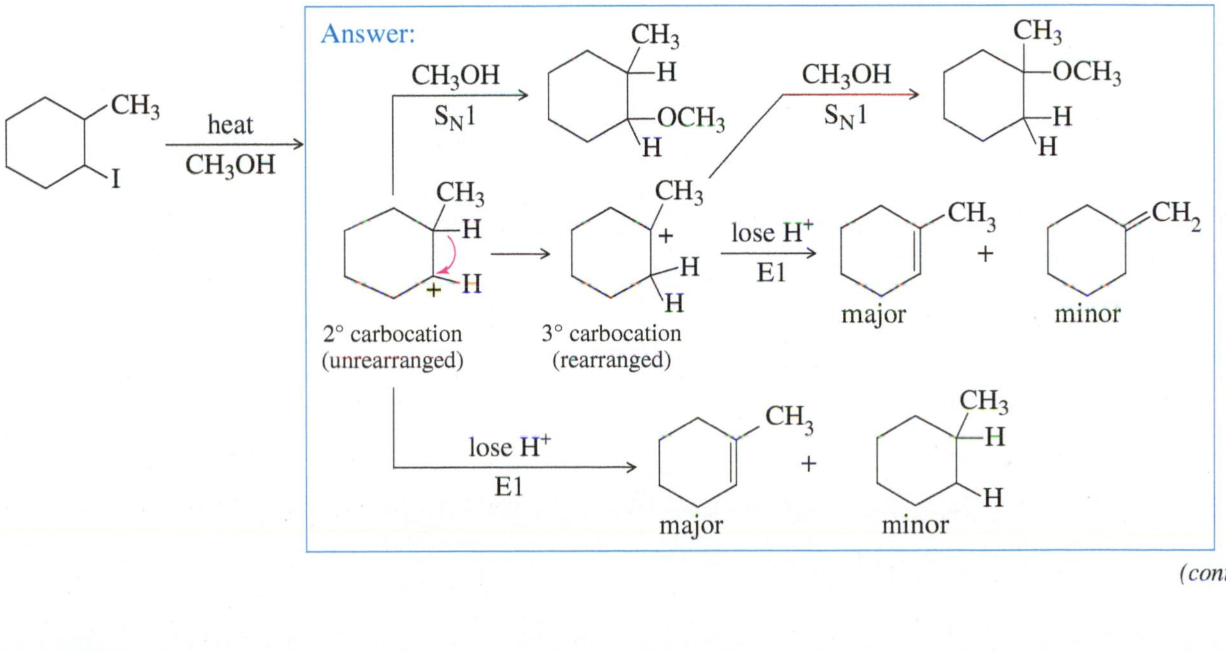

Answer:

CH_3, I; $\xrightarrow[CH_3OH]{heat}$

2° carbocation (unrearranged) (CH_3, H, +, H) → 3° carbocation (rearranged) (CH_3, +, H, H)

2° carbocation $\xrightarrow[S_N1]{CH_3OH}$ (CH_3, H, OCH_3, H)

3° carbocation $\xrightarrow[S_N1]{CH_3OH}$ (CH_3, OCH_3, H, H)

3° carbocation $\xrightarrow[E1]{lose\ H^{+}}$ CH_3 (major) + CH_2 (minor)

2° carbocation $\xrightarrow[E1]{lose\ H^{+}}$ CH_3 (major) + (CH_3, H, H) (minor)

(continued)

Among the elimination products, Zaitsev's rule tells us that the product with the most substituted double bond will predominate. That is the same product for both the initial carbocation and the rearranged carbocation. However, a substitution product might be the major product. Which one depends on whether the rearrangement is faster than attack by the methanol solvent. If the rearrangement is faster, then we see more rearranged product. This complicated reaction shows why E1 and S_N1 reactions of alkyl halides are rarely used for synthesis.

Follow-up question: Show the mechanistic steps that are omitted in the solution to (b) shown above.

PROBLEM-SOLVING STRATEGY Predicting Substitutions and Eliminations

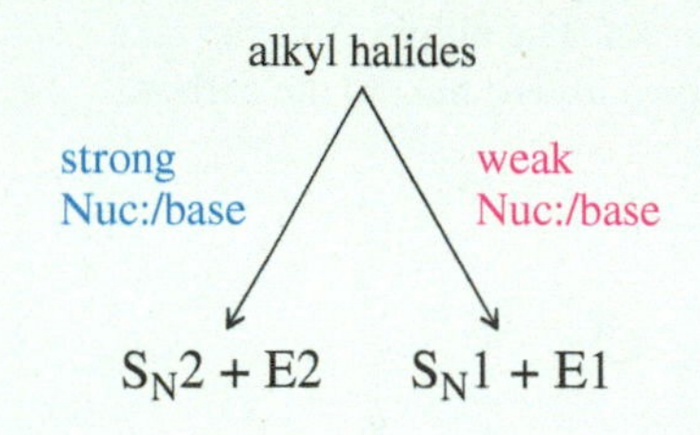

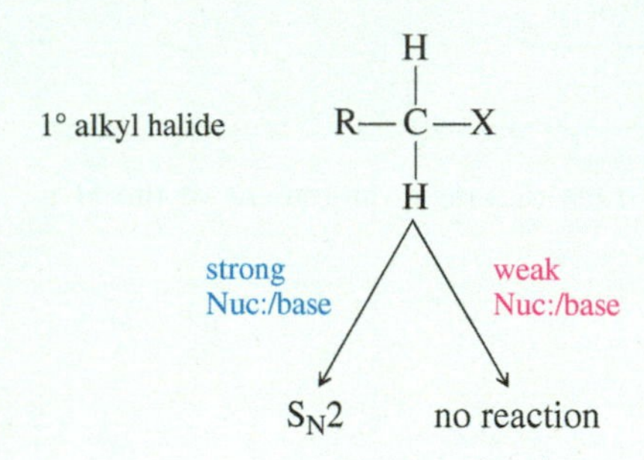

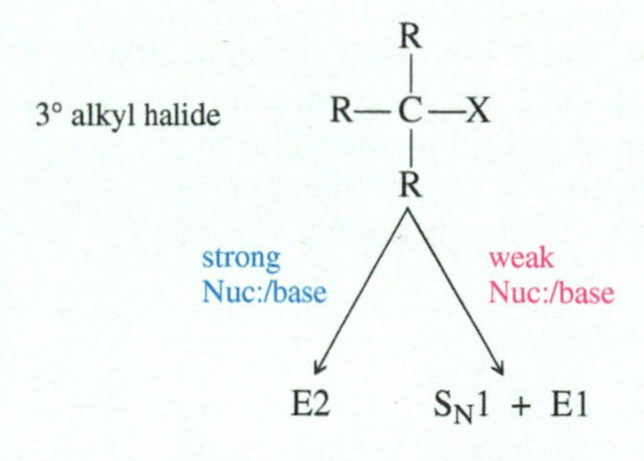

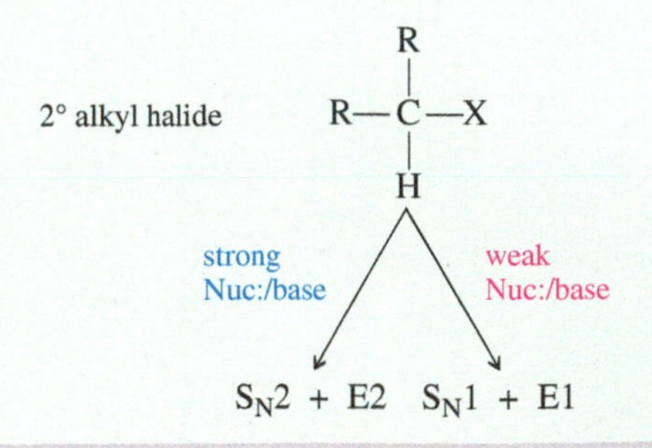

Given a set of reagents and solvents, how can you predict what products will result and which mechanisms will be involved? Should you memorize all this theory about substitutions and eliminations? Students sometimes feel overwhelmed at this point.

Memorizing is not the best way to approach this material because the answers are not absolute and too many factors are involved. Besides, the real world with its real reagents and solvents is not as clean as our equations on paper. Most nucleophiles are also basic, and many solvents can solvate ions or react as nucleophiles or bases.

The first principle you must understand is that *you cannot always predict one unique product or one unique mechanism.* Often, the best you can do is to eliminate some of the possibilities and make some accurate predictions. Remembering this limitation, here are some general guidelines:

1. **The strength of the base or nucleophile determines the order of the reaction.**
   If a strong nucleophile (or base) is present, it will force second-order kinetics, either S_N2 or E2. If no strong base or nucleophile is present, you should consider first-order reactions, both S_N1 and E1. Addition of silver salts to the reaction can force some difficult ionizations.
2. **Primary halides usually undergo the S_N2 reaction, occasionally the E2 reaction.**
   Primary halides rarely undergo first-order reactions, unless the carbocation is resonance-stabilized. With good nucleophiles, S_N2 substitution is usually observed. With a strong base, E2 elimination may occasionally be observed.
3. **Tertiary halides usually undergo the E2 reaction (strong base) or a mixture of S_N1 and E1 (weak base).**
   Tertiary halides cannot undergo the S_N2 reaction. A strong base forces second-order kinetics, resulting in elimination by the E2 mechanism. In the absence of a strong base, tertiary halides react by first-order processes, usually a mixture of S_N1 and E1. The specific reaction conditions determine the ratio of substitution to elimination.
4. **The reactions of secondary halides are the most difficult to predict.**
   With a strong base, either the S_N2 or the E2 reaction is possible. With a weak base and a good ionizing solvent, both the S_N1 and E1 reactions are possible, but both are slow. Mixtures of products are common.
5. **Some nucleophiles and bases favor substitution or elimination.**
   To promote elimination, the base should readily abstract a proton but not readily attack a carbon atom. A bulky strong base, such as *tert*-butoxide [$^{-}OC(CH_3)_3$], enhances elimination. Higher temperatures also favor elimination in most cases. To promote substitution, you need a good nucleophile with limited basicity: a highly polarizable species that is the conjugate base of a strong acid. Bromide (Br^-) and iodide (I^-) are examples of good nucleophiles that are weak bases and favor substitution.

PROBLEM 7-33

Predict the products and mechanisms of the following reactions. When more than one product or mechanism is possible, explain which are most likely.

(a) 1-bromohexane + sodium ethoxide in ethanol
(b) 2-chlorohexane + $NaOCH_3$ in methanol
(c) 2-chloro-2-methylbutane + $NaOCH_2CH_3$ in ethanol
(d) 2-chloro-2-methylbutane heated in ethanol
(e) isobutyl iodide + KOH in ethanol/water
(f) isobutyl chloride + $AgNO_3$ in ethanol/water
(g) 1-bromo-1-methylcyclopentane + NaOEt in ethanol
(h) 1-bromo-1-methylcyclopentane heated in methanol

7-18 Alkene Synthesis by Dehydration of Alcohols

Thus far, we have studied the eliminations that convert alkyl halides to alkenes. Alkenes are also made from a variety of other types of compounds. A common laboratory synthesis of alkenes involves the elimination of the elements of water from alcohols under acidic conditions. This reaction is called **dehydration,** which literally means "removal of water."

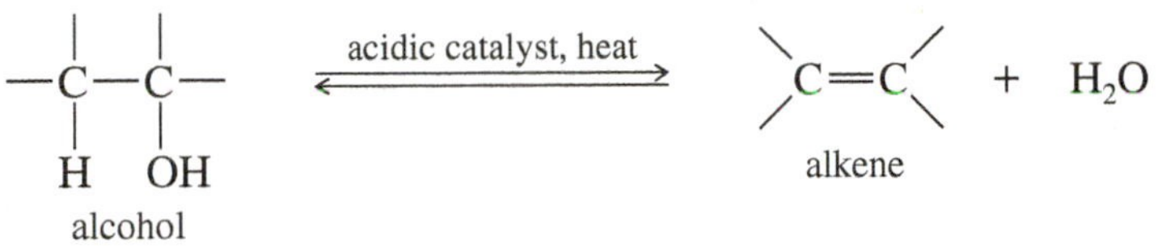

$$\text{—C(H)—C(OH)— (alcohol)} \xrightleftharpoons{\text{acidic catalyst, heat}} \text{C=C (alkene)} + H_2O$$

Dehydration is reversible, and in most cases, the equilibrium constant is not large. In fact, the reverse reaction (hydration) is a method for converting alkenes to alcohols (see Section 8-4). Dehydration can be forced to completion by removing the products from the reaction mixture as they form. The alkene boils at a lower temperature than the alcohol because the alcohol is hydrogen-bonded. A carefully controlled distillation removes the alkene while leaving the alcohol in the reaction mixture.

Concentrated sulfuric acid and/or concentrated phosphoric acid are often used as reagents for dehydration because these acids act both as acidic catalysts and as dehydrating agents. Hydration of these acids is strongly exothermic. The overall reaction (using sulfuric acid) is

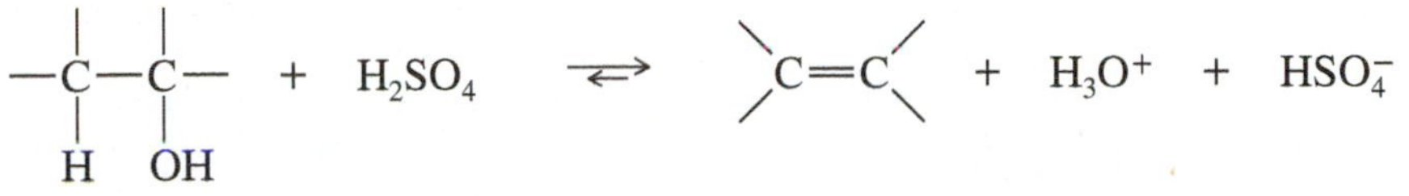

$$\text{—C(H)—C(OH)—} + H_2SO_4 \rightleftharpoons \text{C=C} + H_3O^+ + HSO_4^-$$

The mechanism of dehydration resembles the E1 mechanism (Key Mechanism 7-1, page 319). The hydroxy group of the alcohol is a poor leaving group (^-OH), but protonation by the acidic catalyst in the first step converts it to a good leaving group (H_2O). In the second step, loss of water from the protonated alcohol gives a carbocation. The carbocation is a very strong acid: Any weak base such as H_2O or HSO_4^- can abstract the proton in the final step to give the alkene.

Application: Biochemistry

Hydration and dehydration reactions are common in biological pathways. In one of the steps in cellular respiration, the enzyme fumarase catalyzes the reversible addition of water to the double bond of fumarate to form malate. In contrast to the harsh acidic conditions used in the laboratory reaction, the enzymatic reaction takes place at neutral pH and at 37 °C. The enzymatic reaction forms only one enantiomer of the product.

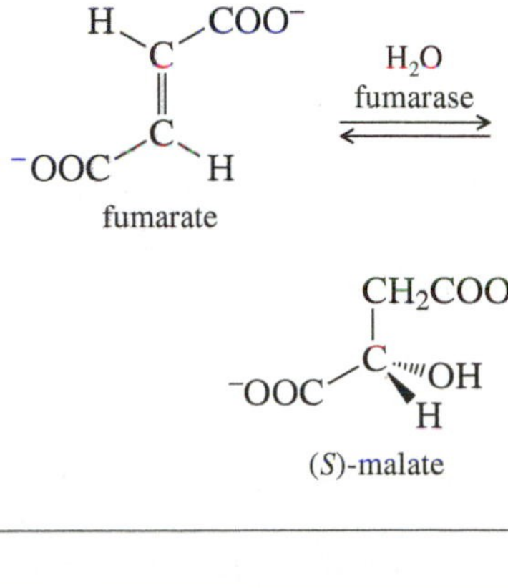

$$\underset{\text{fumarate}}{(H)(^-OOC)C{=}C(COO^-)(H)} \xrightleftharpoons[\text{fumarase}]{H_2O} \underset{(S)\text{-malate}}{^-OOC\text{—}C(H)(OH)\text{—}CH_2COO^-}$$

KEY MECHANISM 7-5 Acid-Catalyzed Dehydration of an Alcohol

PROBLEM-SOLVING HINT

In acid-catalyzed mechanisms, the first step is often addition of H^+, and the last step is often loss of H^+.

Alcohol dehydrations usually involve E1 elimination of the protonated alcohol.

Step 1: Protonation of the hydroxy group (fast equilibrium).

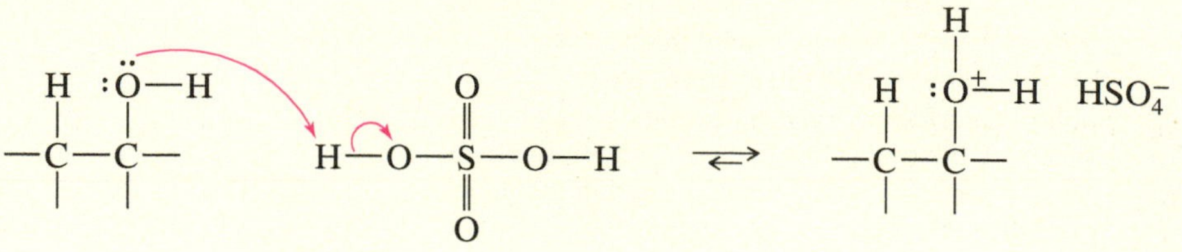

$$-\overset{|}{\underset{|}{C}}(H)-\overset{|}{\underset{|}{C}}(\ddot{O}H)- + H-O-SO_2-O-H \rightleftharpoons -\overset{|}{\underset{|}{C}}(H)-\overset{|}{\underset{|}{C}}(\ddot{O}^+H_2)- \; HSO_4^-$$

Step 2: Ionization to a carbocation (slow; rate limiting).

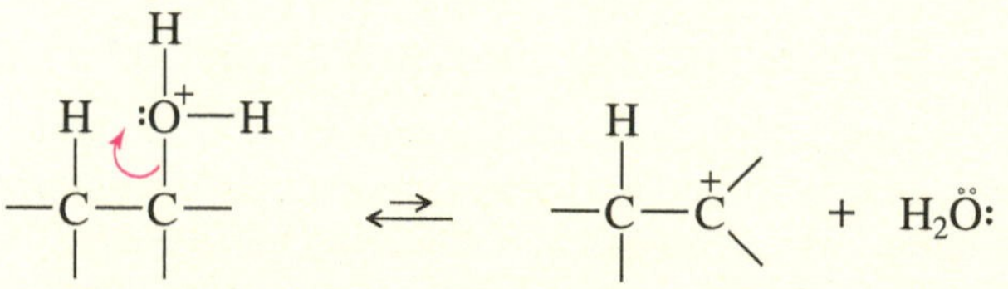

$$-\overset{|}{\underset{|}{C}}(H)-\overset{|}{\underset{|}{C}}(\ddot{O}^+H_2)- \rightleftharpoons -\overset{|}{\underset{|}{C}}(H)-\overset{+}{C}< + H_2\ddot{O}:$$

Step 3: Deprotonation to give the alkene (fast).

$$-\overset{|}{\underset{|}{C}}(H)-C^+< + H_2\ddot{O}: \longrightarrow >C=C< + H_3\ddot{O}^+$$

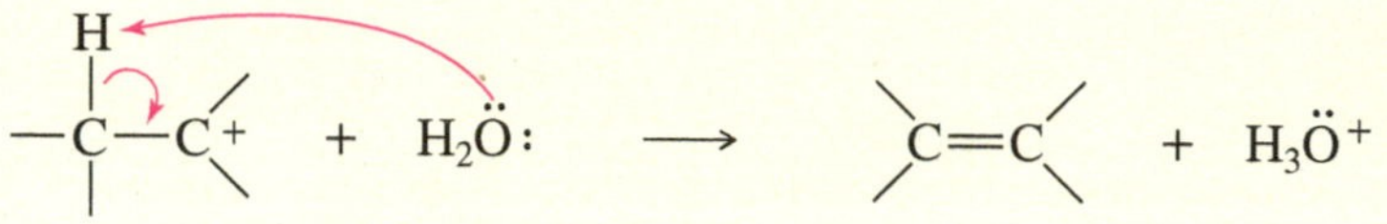

EXAMPLE: Acid-catalyzed dehydration of butan-2-ol

Step 1: Protonation of the hydroxy group (fast equilibrium).

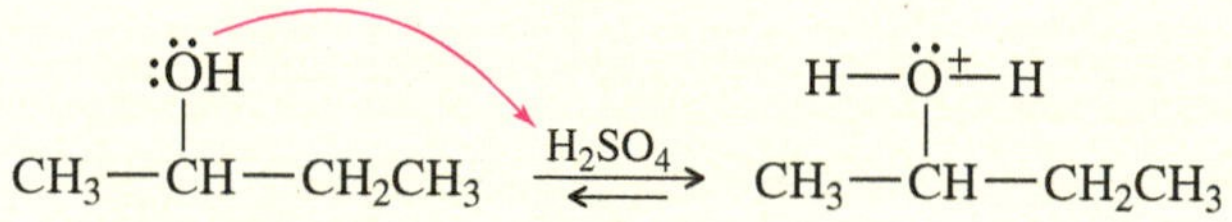

$$CH_3-CH(\ddot{O}H)-CH_2CH_3 \underset{}{\overset{H_2SO_4}{\rightleftharpoons}} CH_3-CH(H-\ddot{O}^+-H)-CH_2CH_3$$

Step 2: Ionization to a carbocation (slow; rate limiting).

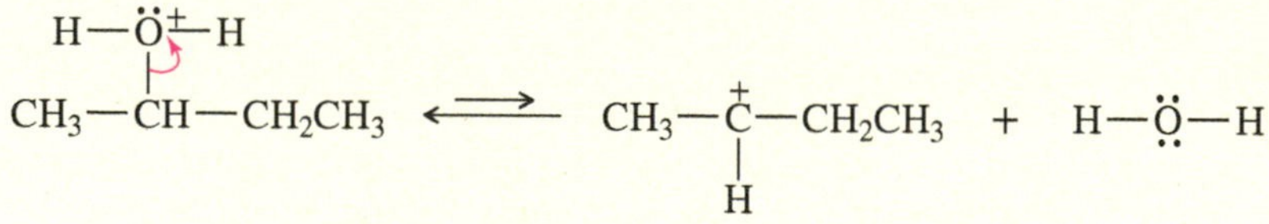

$$CH_3-CH(H-\ddot{O}^+-H)-CH_2CH_3 \rightleftharpoons CH_3-\overset{+}{C}H-CH_2CH_3 + H-\ddot{O}-H$$

Step 3: Deprotonation to give the alkene (fast).

$$H_3C-\overset{+}{C}H-CH_2-CH_3 \overset{H_2\ddot{O}:}{\rightleftharpoons} CH_3-CH=CH-CH_3 + H_3O^+$$

major product (cis and trans)

or

$$H_3C-\overset{+}{C}H-CH_2-CH_3 \overset{H_2\ddot{O}:}{\rightleftharpoons} H_2C=CH-CH_2-CH_3 + H_3O^+$$

minor product

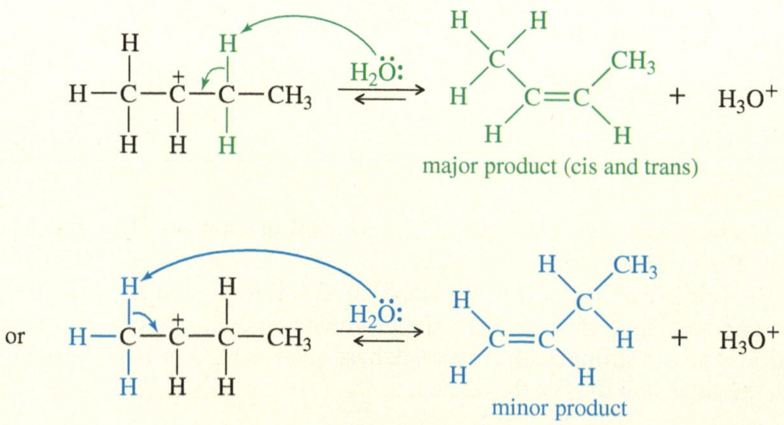

Like other E1 reactions, alcohol dehydration follows an order of reactivity that reflects carbocation stability: 3° alcohols react faster than 2° alcohols, and 1° alcohols are the least reactive. Rearrangements of the carbocation intermediates are common in alcohol dehydrations. In most cases, Zaitsev's rule applies: The major product is usually the one with the most substituted double bond.

SOLVED PROBLEM 7-7

(a) Propose a mechanism for the sulfuric acid-catalyzed dehydration of *tert*-butyl alcohol.

SOLUTION

The first step is protonation of the hydroxy group, which converts it to a good leaving group.

$$(CH_3)_3C{-}\ddot{O}{-}H + H_2SO_4 \rightleftarrows (CH_3)_3C{-}\ddot{O}^{+}H_2 + HSO_4^-$$

The second step is ionization of the protonated alcohol to give a carbocation.

$$(CH_3)_3C{-}\ddot{O}^{+}H_2 \rightleftarrows (CH_3)_2C^{+}{-}CH_3 + H_2\ddot{O}:$$

Abstraction of a proton completes the mechanism.

$$H_2\ddot{O}: \quad H{-}CH_2{-}C^{+}(CH_3)_2 \rightleftarrows H_2C{=}C(CH_3)_2 + H_3\ddot{O}^{+}$$

(b) Predict the products and propose a mechanism for the acid-catalyzed dehydration of 1-cyclohexylethanol.

PARTIAL SOLUTION:

Protonation of the hydroxy group, followed by loss of water, forms a carbocation.

$$\text{1-cyclohexylethanol} \xrightarrow{H{-}OSO_3H} \text{protonated alcohol} + HSO_4^- \rightleftarrows \text{carbocation} + H_2\ddot{O}:$$

1-cyclohexylethanol carbocation

The carbocation can lose a proton, or it can rearrange to a more stable carbocation.

$$\text{unrearranged 2° carbocation} \xrightarrow{\sim H:^-} \text{rearranged 3° carbocation}$$

unrearranged 2° carbocation rearranged 3° carbocation

Complete this problem by showing how the unrearranged 2° carbocation can lose either of two protons to give two of the following products. Also show how the rearranged 3° carbocation can lose either of two protons to give two of the following products.

PROBLEM-SOLVING HINT

Alcohol dehydrations usually go through E1 elimination of the protonated alcohol.

Reactivity is: 3° > 2° >> 1°

Rearrangements are common.

Protonated primary alcohols dehydrate at elevated temperatures with rearrangement (E1), or the adjacent carbon may lose a proton to a weak base at the same time water leaves (E2).

PROBLEM 7-34

Propose mechanisms for the following reactions.

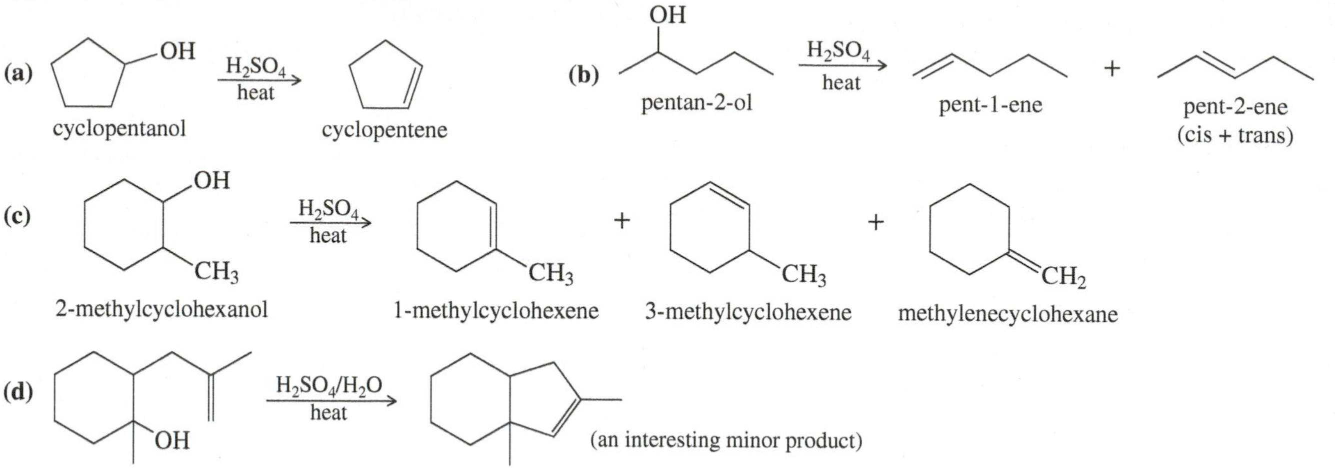

(a) cyclopentanol $\xrightarrow[\text{heat}]{H_2SO_4}$ cyclopentene

(b) pentan-2-ol $\xrightarrow[\text{heat}]{H_2SO_4}$ pent-1-ene + pent-2-ene (cis + trans)

(c) 2-methylcyclohexanol $\xrightarrow[\text{heat}]{H_2SO_4}$ 1-methylcyclohexene + 3-methylcyclohexene + methylenecyclohexane

(d) $\xrightarrow[\text{heat}]{H_2SO_4/H_2O}$ (an interesting minor product)

PROBLEM 7-35

Show the product(s) you expect from dehydration of the following alcohols when they are heated in sulfuric or phosphoric acid. In each case, use a mechanism to show how the products are formed.

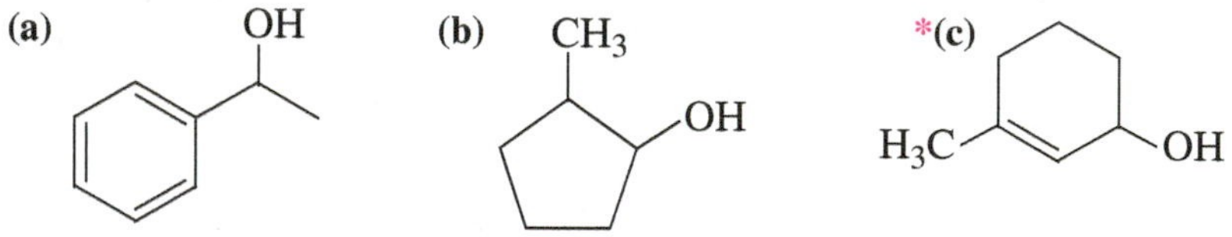

(a) (b) *(c)

7-19 Alkene Synthesis by High-Temperature Industrial Methods

7-19A Catalytic Cracking of Alkanes

The least expensive way to make alkenes on a large scale is by the **catalytic cracking** of petroleum: heating a mixture of alkanes in the presence of a catalyst (usually aluminosilicates). Alkenes are formed by bond cleavage to give an alkene and a shortened alkane.

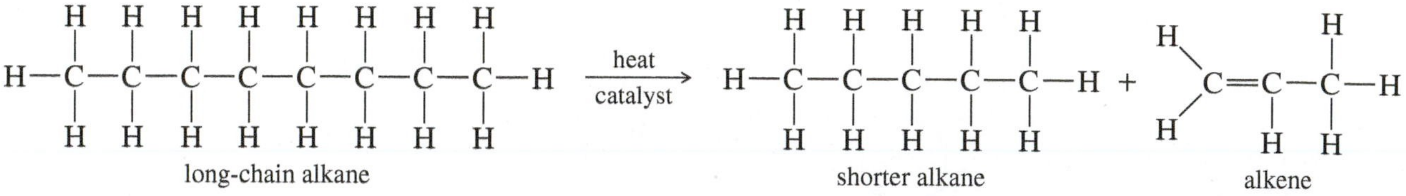

$$CH_3(CH_2)_6CH_3 \xrightarrow[\text{catalyst}]{\text{heat}} CH_3(CH_2)_3CH_3 + CH_2{=}CH{-}CH_3$$

long-chain alkane → shorter alkane + alkene

Cracking is used primarily to make small alkenes, up to about six carbon atoms. Its value depends on having a market for all the different alkenes and alkanes produced. The average molecular weight and the relative amounts of alkanes and alkenes can be controlled by varying the temperature, catalyst, and concentration of hydrogen in the cracking process. A careful distillation on a huge column separates the mixture into its pure components, ready to be packaged and sold.

Because the products are always mixtures, catalytic cracking is unsuitable for laboratory synthesis of alkenes. Better methods are available for synthesizing relatively pure alkenes from a variety of other functional groups. Dehydrohalogenation of alkyl halides and dehydration of alcohols are better methods shown in this chapter, and additional methods are included in the Summary on page 349.

This plant in Tokuyama, Japan, passes ethane rapidly over a hot catalyst. The products are ethylene and hydrogen.

7-19B Dehydrogenation of Alkanes

Dehydrogenation is the removal of H_2 from a molecule, just the reverse of hydrogenation. Dehydrogenation of an alkane gives an alkene. This reaction has an unfavorable enthalpy change but a favorable entropy change.

$$\underset{\text{alkane}}{-\overset{H}{\underset{|}{\overset{|}{C}}}-\overset{H}{\underset{|}{\overset{|}{C}}}-} \xrightarrow[\text{catalyst}]{\text{heat}} \underset{\text{alkene}}{>C=C<} + H_2$$

$\Delta H° = +80$ to $+120$ kJ/mol ($+20$ to $+30$ kcal/mol) $\quad \Delta S° = +125$ J/kelvin-mol

$$CH_3CH_2CH_2CH_3 \xrightarrow{\text{Pt, 500 °C}} \left\{ \begin{array}{l} \textit{cis}\text{-}H_3C(H)C{=}C(H)CH_3 + \textit{trans}\text{-}H_3C(H)C{=}C(H)CH_3 \\ + H_2C{=}CH{-}CH_2CH_3 + H_2C{=}CH{-}CH{=}CH_2 + H_2 \end{array} \right.$$

The hydrogenation of alkenes (Section 7-8) is exothermic, with values of $\Delta H°$ around -80 to -120 kJ/mol (-20 to -30 kcal/mol). Therefore, dehydrogenation is endothermic and has an unfavorable (positive) value of $\Delta H°$. The entropy change for dehydrogenation is strongly favorable ($\Delta S° = +120$ J/kelvin-mol), however, because one alkane molecule is converted into two molecules (the alkene and hydrogen), and two molecules are more disordered than one.

The equilibrium constant for the hydrogenation–dehydrogenation equilibrium depends on the change in free energy, $\Delta G = \Delta H - T\Delta S$. At room temperature, the enthalpy term predominates and hydrogenation is favored. When the temperature is raised, however, the entropy term ($-T\Delta S$) becomes larger and eventually dominates the expression. At a sufficiently high temperature, dehydrogenation is favored.

PROBLEM 7-36

The dehydrogenation of butane to *trans*-but-2-ene has $\Delta H° = +116$ kJ/mol (+27.6 kcal/mol) and $\Delta S° = +117$ J/kelvin-mol (+28.0 cal/kelvin-mol).

(a) Compute the value of $\Delta G°$ for dehydrogenation at room temperature (25 °C or 298 °K). Is dehydrogenation favored or disfavored?

(b) Compute the value of ΔG for dehydrogenation at 1000 °C, assuming that ΔS and ΔH are constant. Is dehydrogenation favored or disfavored?

PROBLEM-SOLVING HINT

When you are doing synthesis problems, avoid using these high-temperature industrial methods. They require specialized equipment, and they produce variable mixtures of products.

In many ways, dehydrogenation is similar to catalytic cracking. In both cases, a catalyst lowers the activation energy, and both reactions use high temperatures to increase a favorable entropy term ($-T\Delta S$) and overcome an unfavorable enthalpy term (ΔH). Unfortunately, dehydrogenation and catalytic cracking also share a tendency to produce mixtures of products, and neither reaction is well suited for the laboratory synthesis of alkenes.

PROBLEM-SOLVING STRATEGY Proposing Reaction Mechanisms

At this point, we have seen examples of three major classes of reaction mechanisms:

- Those involving strong bases and strong nucleophiles
- Those involving strong acids and strong electrophiles
- Those involving free radicals

Many students have difficulty proposing mechanisms. You can use some general principles to approach this process, however, by breaking it down into a series of logical steps. Using a systematic approach, you can usually come up with a mechanism that is at least possible and that explains the products, without requiring any unusual steps. Appendix 3A contains more complete methods for approaching mechanism problems.

First, Classify the Reaction

Before you begin to propose a mechanism, you must determine what kind of reaction you are dealing with. Examine what you know about the reactants and the reaction conditions:

A free-radical initiator such as chlorine, bromine, or a peroxide (with heat or light) suggests that a free-radical chain reaction is most likely. Free-radical reactions were discussed in Chapter 4.

Strong bases or strong nucleophiles suggest mechanisms such as the S_N2 or E2, involving attack by the strong base or nucleophile on a substrate.

Strong acids or strong electrophiles (or a reactant that can dissociate to give a strong electrophile) suggest mechanisms such as the S_N1, E1, and alcohol dehydration that involve carbocations and other strongly acidic intermediates.

General Principles for Drawing Mechanisms

Once you have decided which type of mechanism is most likely (acidic, basic, or free-radical), some general principles can guide you in proposing the mechanism. Some principles for free-radical reactions were discussed in Chapter 4. Now we consider reactions that involve either strong nucleophiles or strong electrophiles as intermediates. In later chapters, we will apply these principles to more complex mechanisms.

Whenever you start to work out a mechanism, **draw all the bonds** and all the substituents of each carbon atom affected throughout the mechanism. Three-bonded carbon atoms are likely to be the reactive intermediates. If you attempt to draw condensed formulas or line–angle formulas, you will likely misplace a hydrogen atom and show the wrong carbon atom as a radical, a cation, or an anion.

Show only one step at a time; never combine steps, unless two or more bonds really do change position in one step (as in the E2 reaction, for example). Protonation of an alcohol and loss of water to give a carbocation, for example, must be shown as two steps. You must not simply circle the hydroxy and the proton to show water falling off.

Use curved arrows to show the movement of electrons in each step of the reaction. This movement is always *from* the nucleophile (electron donor) to the electrophile (electron acceptor). For example, protonation of an alcohol must show the arrow going from the electrons of the hydroxy oxygen to the proton—never from the proton to the hydroxy group. *Don't* use curved arrows to try to "point out" where the proton (or other reagent) goes.

Reactions Involving Strong Nucleophiles

When a strong base or nucleophile is present, we expect to see intermediates that are also strong bases and strong nucleophiles; anionic intermediates are common. Acids and electrophiles in such a reaction are generally weak. Avoid drawing carbocations, H_3O^+, and other strong acids. They are unlikely to coexist with strong bases and strong nucleophiles.

Functional groups are often converted to alkoxides, carbanions, or other strong nucleophiles by deprotonation or reaction with a strong nucleophile. Then the carbanion or other strong nucleophile reacts with a weak electrophile such as a carbonyl group or an alkyl halide.

Consider, for example, the mechanism for the dehydrohalogenation of 3-bromopentane.

$$CH_3-CH_2-\underset{\displaystyle Br}{\underset{|}{CH}}-CH_2-CH_3 \xrightarrow{CH_3CH_2O^-} CH_3-CH=CH-CH_2-CH_3$$

Someone who has not read Chapter 7 or these guidelines might propose an ionization, followed by loss of a proton:

Incorrect mechanism

$$CH_3-\overset{H}{\underset{H}{C}}-\overset{H}{\underset{Br}{C}}-CH_2-CH_3 \xrightleftharpoons{CH_3CH_2O^-} CH_3-\overset{H}{\underset{H}{C}}-\overset{H}{\overset{+}{C}}-CH_2-CH_3 \quad Br^- \longrightarrow CH_3-\overset{H}{C}=\overset{H}{C}-CH_2-CH_3 \quad Br^-$$

unlikely (carbocation) — unlikely (H^+)

This mechanism would violate several general principles of proposing mechanisms. First, in the presence of ethoxide ion (a strong base), both the carbocation and the H^+ ion are unlikely. Second, the mechanism fails to explain why the strong base is required; the rate of ionization would be unaffected by the presence of ethoxide ion. Also, H^+ doesn't just fall off (even in an acidic reaction); it must be removed by a base.

The presence of ethoxide ion (a strong base and a strong nucleophile) in the reaction suggests that the mechanism involves only strong bases and nucleophiles and not any strongly acidic intermediates. As shown in Section 7-12, the reaction occurs by the E2 mechanism, an example of a reaction involving a strong base. In this concerted reaction, ethoxide ion removes a proton as the electron pair left behind forms a pi bond and expels bromide ion.

Correct mechanism

$$CH_3CH_2\ddot{O}:^- + CH_3-CH_2-CH(Br)-CH_2-CH_3 \longrightarrow CH_3CH_2\ddot{O}-H + CH_3-CH=CH-CH_2-CH_3 + :\ddot{Br}:^-$$

Reactions Involving Strong Electrophiles

When a strong acid or electrophile is present, expect to see intermediates that are also strong acids and strong electrophiles. Cationic intermediates are common, but avoid drawing any species with more than one + charge. Bases and nucleophiles in such a reaction are generally weak. Avoid drawing carbanions, alkoxide ions, and other strong bases. They are unlikely to coexist with strong acids and strong electrophiles.

Functional groups are often converted to carbocations or other strong electrophiles by protonation or by reaction with a strong electrophile; then the carbocation or other strong electrophile reacts with a weak nucleophile such as an alkene or the solvent.

For example, consider the dehydration of 2,2-dimethylpropan-1-ol:

$$CH_3-C(CH_3)_2-CH_2-OH \xrightarrow{H_2SO_4,\ 150\ °C} (CH_3)_2C=CH-CH_3$$

The presence of sulfuric acid indicates that the reaction is acidic and should involve strong electrophiles. The carbon skeleton of the product is different from the reactant. Under these acidic conditions, formation and rearrangement of a carbocation would be likely. The hydroxy group is a poor leaving group; it certainly cannot ionize to give a carbocation and ^-OH (and we do not expect to see a strong base such as ^-OH in this acidic reaction). The hydroxy group is weakly basic, however, and it can become protonated in the presence of a strong acid. The protonated OH group becomes a good leaving group.

Step 1: Protonation of the hydroxy group

$$CH_3-C(CH_3)_2-CH_2-\ddot{O}-H + H_2SO_4 \longrightarrow CH_3-C(CH_3)_2-CH_2-\dot{O}^+H_2 + HSO_4^-$$

starting alcohol — protonated alcohol

The protonated hydroxy group $-\overset{+}{O}H_2$ is a good leaving group. A simple ionization to a carbocation would form a primary carbocation. Primary carbocations, however, are very unstable. Thus, a methyl shift occurs as water leaves, so a primary carbocation is never formed. A tertiary carbocation results. (You can visualize this as two steps if you prefer.)

Step 2: Ionization with rearrangement

$$CH_3-C(CH_3)_2-CH_2-\dot{O}^+H_2 \xrightarrow[\text{with } CH_3 \text{ shift } (\sim CH_3)]{H_2O \text{ leaves}} CH_3-\overset{+}{C}(CH_3)-CH_2-CH_3 + H_2\ddot{O}:$$

protonated alcohol — tertiary carbocation

(continued)

The final step is loss of a proton to a weak base, such as H_2O or HSO_4^- (but *not* ^-OH, which is incompatible with the acidic solution). Either of two types of protons, labeled 1 and 2 in the following figure, could be lost to give alkenes. Loss of proton 2 gives the required product.

Step 3: Abstraction of a proton to form the required product

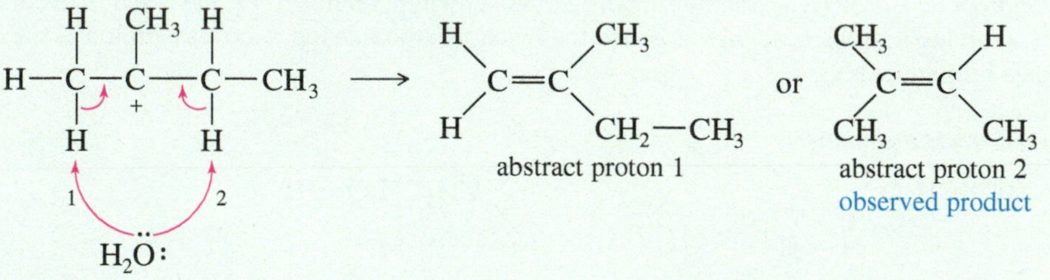

abstract proton 1

or

abstract proton 2
observed product

Because abstraction of proton 2 gives the more highly substituted (therefore more stable) product, Zaitsev's rule predicts it will be the major product. Note that in other problems, however, you may be asked to propose mechanisms to explain unusual compounds that are only minor products.

PROBLEM 7-37

For practice in recognizing mechanisms, classify each reaction according to the type of mechanism you expect:

1. Free-radical chain reaction
2. Reaction involving strong bases and strong nucleophiles
3. Reaction involving strong acids and strong electrophiles

(a) 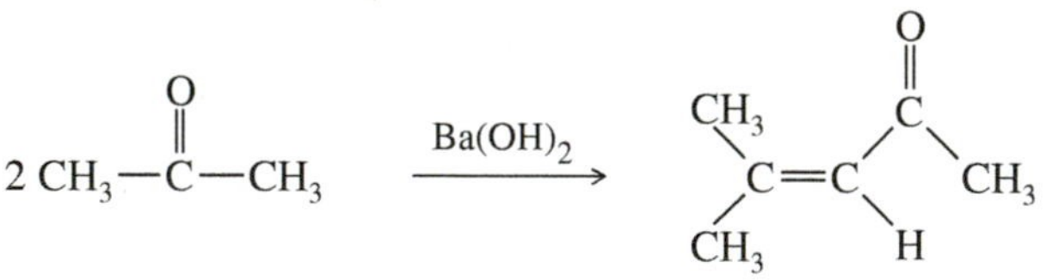 $2\ CH_3-CO-CH_3 \xrightarrow{Ba(OH)_2} (CH_3)_2C=CH-CO-CH_3$

(b) 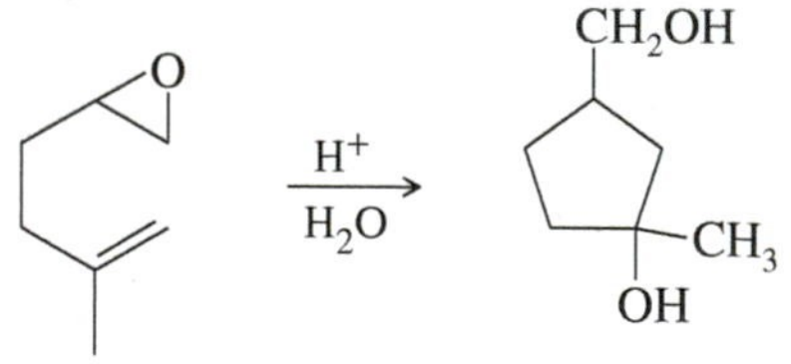 $\xrightarrow[H_2O]{H^+}$

(c) 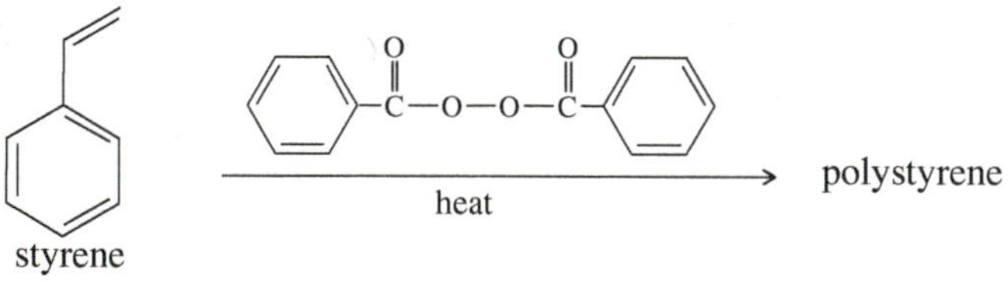 styrene $\xrightarrow[\text{heat}]{C_6H_5-CO-O-O-CO-C_6H_5}$ polystyrene

(d) ethylene $\xrightarrow{BF_3}$ polyethylene

PROBLEM 7-38

Propose mechanisms for the following reactions. Additional products may be formed, but your mechanism only needs to explain the products shown.

(a) $CH_3-CH_2-CH_2-CH_2-OH \xrightarrow{H_2SO_4,\ 140\ °C} CH_3-CH=CH-CH_3 + CH_2=CH-CH_2CH_3$

(*Hint*: Hydride shift)

(b) 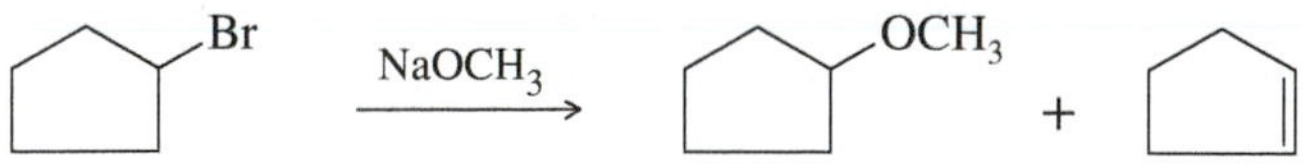 $\xrightarrow{NaOCH_3}$ +

(c) 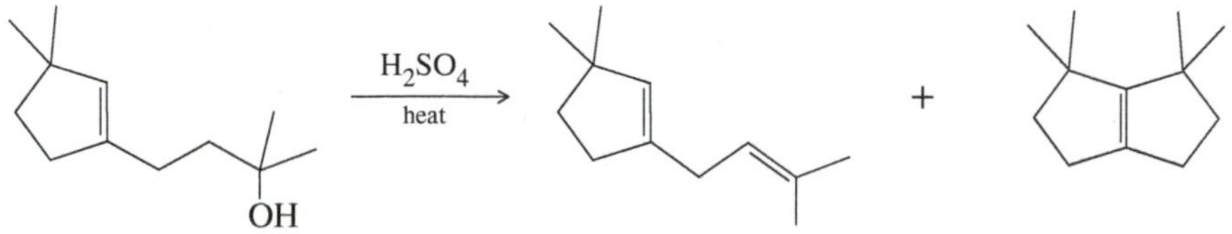 $\xrightarrow[\text{heat}]{H_2SO_4}$ +

PROBLEM 7-39

Propose mechanisms for the following reactions.

(a) 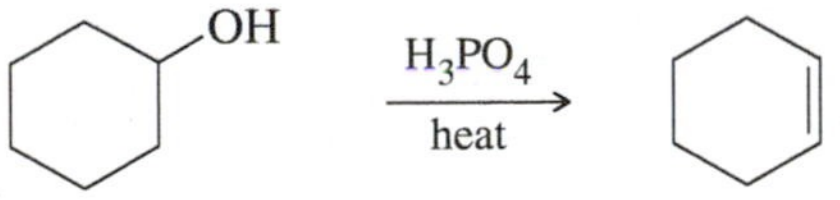 cyclohexanol $\xrightarrow[\text{heat}]{H_3PO_4}$ cyclohexene

(b) 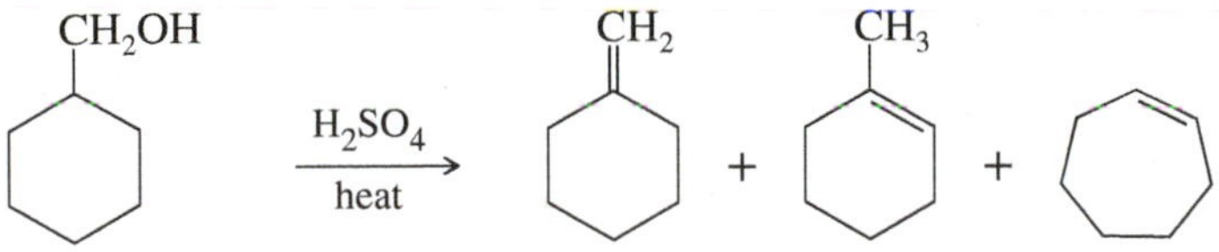 cyclohexylmethanol (CH_2OH) $\xrightarrow[\text{heat}]{H_2SO_4}$ methylenecyclohexane (CH_2) + 1-methylcyclohexene (CH_3) + cycloheptene

(c) 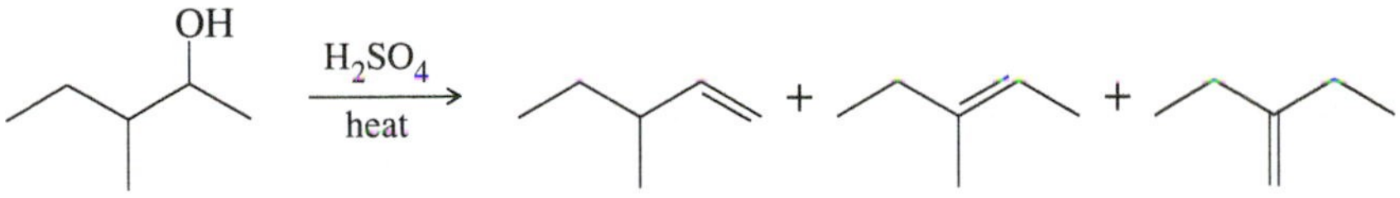 3-methyl-2-butanol (OH) $\xrightarrow[\text{heat}]{H_2SO_4}$ 3-methyl-1-butene + 2-methyl-2-butene + 2-methyl-1-butene

(d) 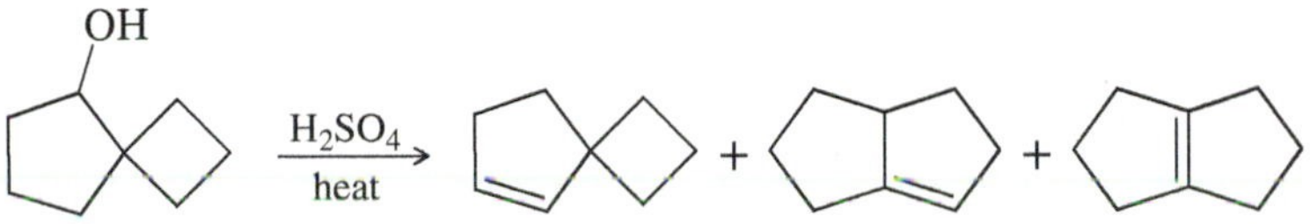 spiro alcohol (OH) $\xrightarrow[\text{heat}]{H_2SO_4}$ + +

PROBLEM-SOLVING HINT

Alcohol dehydrations usually go through E1 elimination of the protonated alcohol, with a carbocation intermediate. Rearrangements are common.

SUMMARY Methods for Synthesis of Alkenes

1. *Dehydrohalogenation of alkyl halides* (Section 7-9)

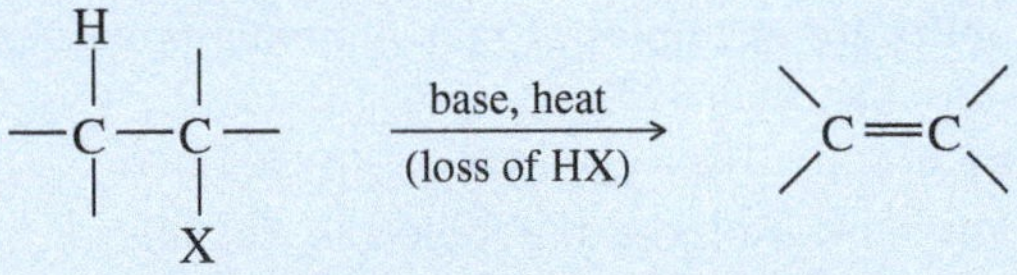

$$\text{H}-\overset{|}{\underset{|}{C}}-\overset{|}{\underset{\text{X}}{C}}- \xrightarrow[\text{(loss of HX)}]{\text{base, heat}} \;\; >C{=}C<$$

Requires an anti-coplanar conformation for the E2 reaction.

2. *Dehydration of alcohols* (Section 7-18)

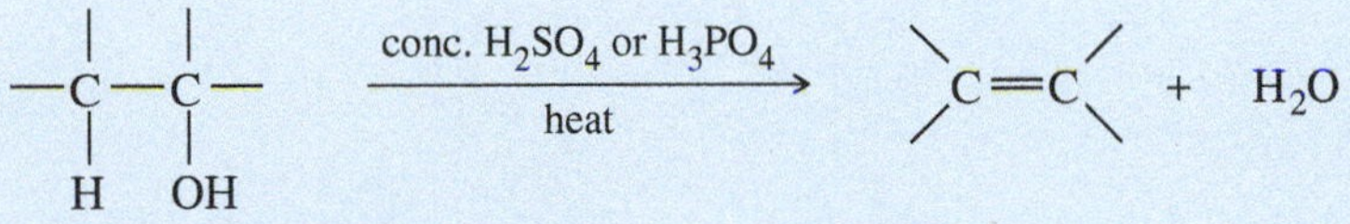

$$-\underset{\text{H}}{\overset{|}{C}}-\underset{\text{OH}}{\overset{|}{C}}- \xrightarrow[\text{heat}]{\text{conc. } H_2SO_4 \text{ or } H_3PO_4} \;\; >C{=}C< \; + \; H_2O$$

Rearrangements are common.

3. *Dehydrogenation of alkanes* (Section 7-19B)

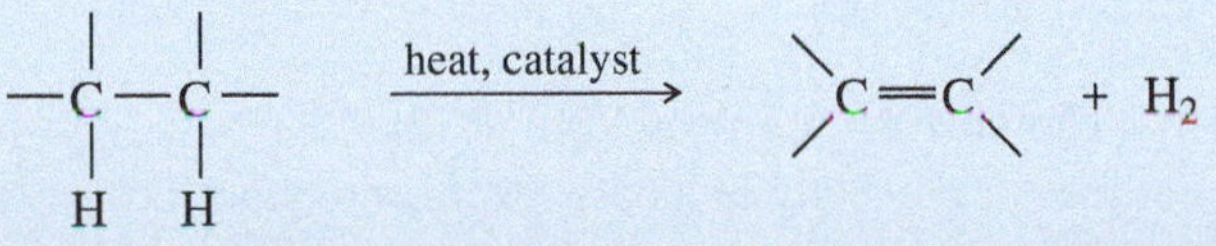

$$-\underset{\text{H}}{\overset{|}{C}}-\underset{\text{H}}{\overset{|}{C}}- \xrightarrow{\text{heat, catalyst}} \;\; >C{=}C< \; + \; H_2$$

Industrial prep, not used in the lab.
Useful only for small alkenes.
Commonly gives mixtures.

4. *Hofmann and Cope eliminations* (Covered later in Sections 19-14 and 19-15)

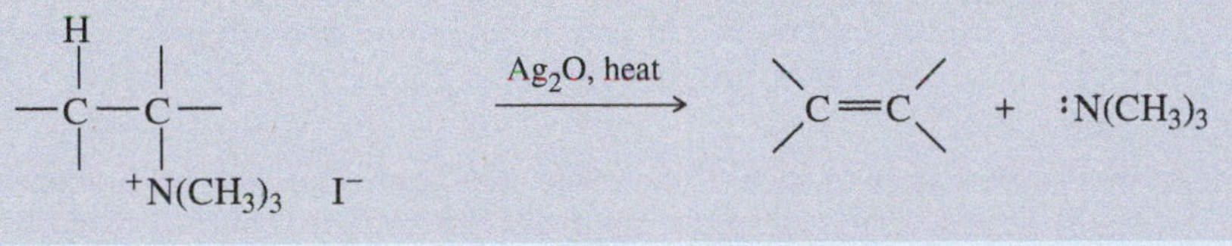

$$\text{H}-\overset{|}{\underset{|}{C}}-\overset{|}{\underset{{}^{+}N(CH_3)_3}{C}}- \;\; I^- \xrightarrow{Ag_2O, \text{ heat}} \;\; >C{=}C< \; + \; :N(CH_3)_3$$

Usually gives the least substituted alkene.

(continued)

5. *Reduction of alkynes* (Covered later in Section 9-9)

$$R-C\equiv C-R' \xrightarrow[\text{quinoline}]{H_2,\ Pd/BaSO_4} (R)(H)C=C(R')(H) \quad \text{Forms the cis alkene}$$

$$R-C\equiv C-R' \xrightarrow{Na,\ NH_3} (R)(H)C=C(H)(R') \quad \text{Forms the trans alkene}$$

6. *Wittig reaction* (Covered later in Section 18-18)

$$(R')(R)C=O \ + \ Ph_3P=CHR'' \longrightarrow (R')(R)C=CHR'' \ + \ Ph_3P=O$$

Essential Terms

alkene (olefin) A hydrocarbon with one or more carbon–carbon double bonds. (p. 296)

- **diene:** A compound with two carbon–carbon double bonds. (p. 302)
- **triene:** A compound with three carbon–carbon double bonds. (p. 302)
- **tetraene:** A compound with four carbon–carbon double bonds. (p. 302)

alkyl shift (symbolized ~**R**) Movement of an alkyl group with a pair of electrons from one atom (usually carbon) to another. Alkyl shifts are examples of rearrangements that convert carbocations into more stable carbocations. (pp. 283, 323)

allyl group A vinyl group plus a methylene group: $CH_2=CH-CH_2-$ (p. 303)

anti Adding to (or eliminating from) opposite faces of a molecule. (p. 329)

- **anti-coplanar** Having a dihedral angle of 180°.
- **syn-coplanar** Having a dihedral angle of 0°.

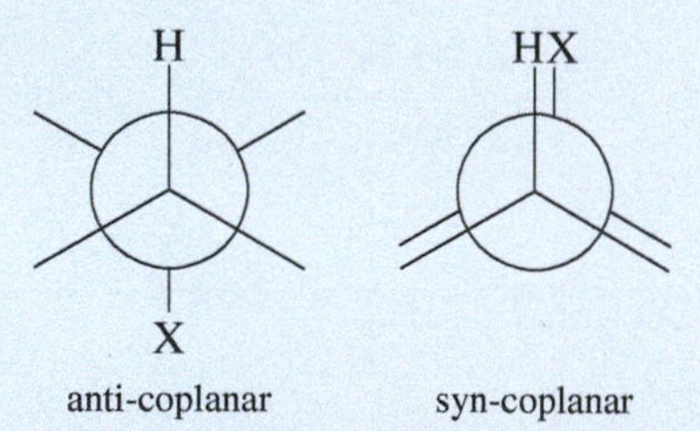

base An electron-rich species that can abstract a proton. (pp. 68, 266)

- **basicity:** **(base strength)** The thermodynamic reactivity of a base, expressed quantitatively by the base-dissociation constant K_b. (p. 72)
- **Lewis base:** **(nucleophile)** An electron-rich species that can donate a pair of electrons to form a bond. (pp. 87, 261)

Bredt's rule A stable bridged bicyclic compound cannot have a double bond at a bridgehead position unless one of the rings contains at least eight carbon atoms. (p. 315)

- **bicyclic:** Containing two rings.
- **bridged bicyclic:** Having at least one carbon atom in each of the three links connecting the bridgehead carbons. (p. 316)

5. *Reduction of alkynes* (Covered later in Section 9-9)

$$R-C\equiv C-R' \xrightarrow[\text{quinoline}]{H_2,\ Pd/BaSO_4} \text{(R)(H)C=C(R')(H)} \quad \text{Forms the cis alkene}$$

$$R-C\equiv C-R' \xrightarrow{Na,\ NH_3} \text{(R)(H)C=C(H)(R')} \quad \text{Forms the trans alkene}$$

6. *Wittig reaction* (Covered later in Section 18-18)

$$R'RC=O + Ph_3P=CHR'' \longrightarrow R'RC=CHR'' + Ph_3P=O$$

Essential Terms

alkene — **(olefin)** A hydrocarbon with one or more carbon–carbon double bonds. (p. 296)

- **diene:** A compound with two carbon–carbon double bonds. (p. 302)
- **triene:** A compound with three carbon–carbon double bonds. (p. 302)
- **tetraene:** A compound with four carbon–carbon double bonds. (p. 302)

alkyl shift — (symbolized ~R) Movement of an alkyl group with a pair of electrons from one atom (usually carbon) to another. Alkyl shifts are examples of rearrangements that convert carbocations into more stable carbocations. (pp. 283, 323)

allyl group — A vinyl group plus a methylene group: $CH_2=CH-CH_2-$ (p. 303)

anti — Adding to (or eliminating from) opposite faces of a molecule. (p. 329)

- **anti-coplanar** Having a dihedral angle of 180°.
- **syn-coplanar** Having a dihedral angle of 0°.

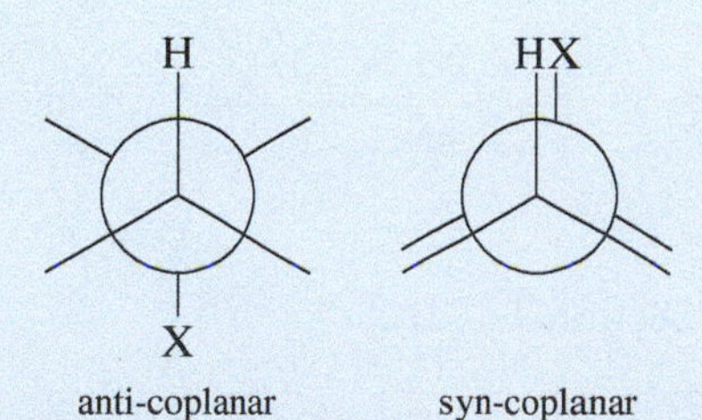

base — An electron-rich species that can abstract a proton. (pp. 68, 266)

- **basicity:** **(base strength)** The thermodynamic reactivity of a base, expressed quantitatively by the base-dissociation constant K_b. (p. 72)
- **Lewis base:** **(nucleophile)** An electron-rich species that can donate a pair of electrons to form a bond. (pp. 87, 261)

Bredt's rule — A stable bridged bicyclic compound cannot have a double bond at a bridgehead position unless one of the rings contains at least eight carbon atoms. (p. 315)

- **bicyclic:** Containing two rings.
- **bridged bicyclic:** Having at least one carbon atom in each of the three links connecting the bridgehead carbons. (p. 316)

PROBLEM 7-39

Propose mechanisms for the following reactions.

(a) cyclohexanol $\xrightarrow[\text{heat}]{H_3PO_4}$ cyclohexene

(b) cyclohexylmethanol (CH_2OH) $\xrightarrow[\text{heat}]{H_2SO_4}$ methylenecyclohexane (CH_2) + 1-methylcyclohexene (CH_3) + cycloheptene

(c) OH $\xrightarrow[\text{heat}]{H_2SO_4}$ + +

(d) OH $\xrightarrow[\text{heat}]{H_2SO_4}$ + +

PROBLEM-SOLVING HINT

Alcohol dehydrations usually go through E1 elimination of the protonated alcohol, with a carbocation intermediate. Rearrangements are common.

SUMMARY Methods for Synthesis of Alkenes

1. *Dehydrohalogenation of alkyl halides* (Section 7-9)

$$\underset{}{\overset{H}{-\overset{|}{\underset{|}{C}}}}-\underset{X}{\overset{|}{\underset{|}{C}}}- \xrightarrow[\text{(loss of HX)}]{\text{base, heat}} \gt C=C \lt$$

Requires an anti-coplanar conformation for the E2 reaction.

2. *Dehydration of alcohols* (Section 7-18)

$$-\underset{H}{\overset{|}{\underset{|}{C}}}-\underset{OH}{\overset{|}{\underset{|}{C}}}- \xrightarrow[\text{heat}]{\text{conc. } H_2SO_4 \text{ or } H_3PO_4} \gt C=C \lt + H_2O$$

Rearrangements are common.

3. *Dehydrogenation of alkanes* (Section 7-19B)

$$-\underset{H}{\overset{|}{\underset{|}{C}}}-\underset{H}{\overset{|}{\underset{|}{C}}}- \xrightarrow{\text{heat, catalyst}} \gt C=C \lt + H_2$$

Industrial prep, not used in the lab. Useful only for small alkenes. Commonly gives mixtures.

4. *Hofmann and Cope eliminations* (Covered later in Sections 19-14 and 19-15)

$$\overset{H}{-\overset{|}{\underset{|}{C}}}-\underset{^{+}N(CH_3)_3 \;\; I^-}{\overset{|}{\underset{|}{C}}}- \xrightarrow{Ag_2O, \text{ heat}} \gt C=C \lt + :N(CH_3)_3$$

Usually gives the least substituted alkene.

(continued)

bridgehead carbons:	Those carbon atoms that are part of both rings, with three bridges of bonds connecting them.

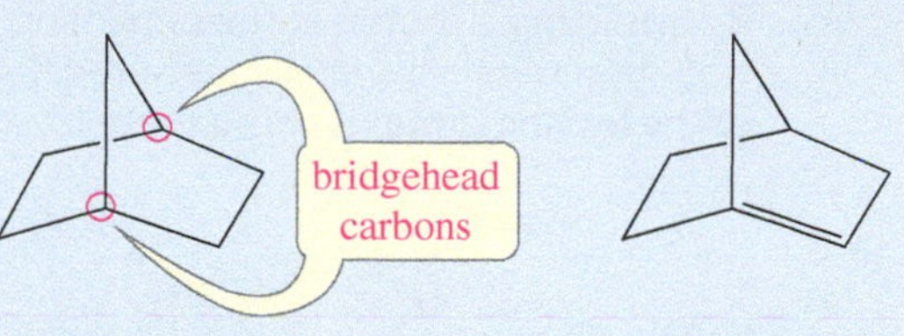

a bridged bicyclic compound a Bredt's rule violation

catalytic cracking	The heating of petroleum products in the presence of a catalyst (usually an aluminosilicate mineral), causing bond cleavage to form alkenes and alkanes of lower molecular weight. (p. 344)
cis-trans isomers	**(geometric isomers)** Isomers that differ in their cis-trans arrangement on a ring or double bond. Cis-trans isomers are a subclass of diastereomers. (p. 303)
cis:	Having similar groups on the same side of a double bond or a ring.
trans:	Having similar groups on opposite sides of a double bond or a ring.
***Z*:**	Having the higher-priority groups on the same side of a double bond.
***E*:**	Having the higher-priority groups on opposite sides of a double bond.
concerted reaction	A reaction in which the breaking of bonds and the formation of new bonds occur at the same time (in one step). (p. 326)
conjugated double bonds	Double bonds that are separated by only one single bond. The interaction between the pi bonds contributes to extra stability in the conjugated system. (p. 317)
dehydration	The elimination of water from a compound; usually acid-catalyzed. (p. 341)

$$-\overset{\displaystyle H}{\underset{|}{\overset{|}{C}}}-\overset{\displaystyle OH}{\underset{|}{\overset{|}{C}}}- \quad \underset{}{\overset{H^+}{\rightleftharpoons}} \quad >C=C< \; + \; H_2O$$

dehydrogenation	The elimination of hydrogen (H_2) from a compound; usually done in the presence of a catalyst. (p. 345)

$$-\overset{\displaystyle H}{\underset{|}{\overset{|}{C}}}-\overset{\displaystyle H}{\underset{|}{\overset{|}{C}}}- \quad \overset{\text{Pt, high temperature}}{\rightleftharpoons} \quad >C=C< \; + \; H_2$$

dehydrohalogenation	The elimination of a hydrogen halide (HX) from a compound; usually base-promoted. (p. 318)

$$-\overset{\displaystyle H}{\underset{|}{\overset{|}{C}}}-\overset{\displaystyle X}{\underset{|}{\overset{|}{C}}}- \quad \xrightarrow{\text{KOH}} \quad >C=C< \; + \; H_2O \; + \; K^+X^-$$

double-bond isomers	Constitutional isomers that differ only in the position of a double bond. Double-bond isomers hydrogenate to give the same alkane. (p. 312)
element of unsaturation	**(degree of unsaturation, unsaturation number, index of hydrogen deficiency)** A structural feature that results in two fewer hydrogen atoms in the molecular formula. A double bond or a ring is one element of unsaturation; a triple bond is two elements of unsaturation. (p. 299)
elimination	A reaction that involves the loss of two atoms or groups from the substrate, usually resulting in the formation of a pi bond. (p. 318)
E1 reaction:	**(elimination, unimolecular)** A multistep elimination where the leaving group is lost in a slow ionization step, then a proton is lost in a second step. Zaitsev orientation is generally preferred. (p. 319)

$$H_3C-C(CH_3)_2-:\ddot{C}l: \;\rightleftharpoons\; \overset{B:^-}{H}-CH_2-\overset{+}{C}(CH_3)_2 \;\; :\ddot{C}l:^- \;\overset{E1}{\rightleftharpoons}\; B-H \;\; H_2C=C(CH_3)_2 \;\; :\ddot{C}l:^-$$

E2 reaction: **(elimination, bimolecular)** A concerted elimination involving a transition state where the base is abstracting a proton at the same time the leaving group is leaving. The anti-coplanar transition state is generally preferred. Zaitsev orientation is usually preferred, unless the base or the leaving group is unusually bulky. (p. 326)

$$\mathrm{B:^-} + \mathrm{H_3C{-}C(CH_3)_2Cl} \xrightarrow{E2} \mathrm{B{-}H} + \mathrm{H_2C{=}C(CH_3)_2} + \mathrm{:\ddot{C}l:^-}$$

geminal dihalide A compound with two halogen atoms on the same carbon atom.

geometric isomers See **cis-trans isomers.** (p. 303)

heteroatom Any atom other than carbon or hydrogen. (p. 300)

Hofmann product The least highly substituted alkene product. (p. 328)

hydride shift (symbolized ~H) Movement of a hydrogen atom with a pair of electrons from one atom (usually carbon) to another. Hydride shifts are examples of rearrangements that convert carbocations into more stable carbocations. (pp. 281, 323)

hydrogenation Addition of hydrogen to a molecule. The most common hydrogenation is the addition of H_2 across a double bond in the presence of a catalyst (*catalytic hydrogenation*). The value of $(-\Delta H°)$ for this reaction is called the **heat of hydrogenation.** (p. 310)

$$\mathrm{>C{=}C<} + \mathrm{H_2} \xrightarrow{Pt} \mathrm{-\overset{H}{C}-\overset{H}{C}-} \qquad -\Delta H° = \text{heat of hydrogenation}$$

isolated double bonds Double bonds separated by two or more single bonds. Isolated double bonds react independently, as they do in a simple alkene. (p. 317)

methyl shift (symbolized ~CH_3) Rearrangement of a methyl group with a pair of electrons from one atom (usually carbon) to another. A methyl shift (or any alkyl shift) in a carbocation generally results in a more stable carbocations. (p. 325)

nucleophile **(Lewis base)** An electron-rich species that can donate a pair of electrons to form a bond. (pp. 87, 261)

nucleophilicity: **(nucleophile strength)** The kinetic reactivity of a nucleophile; a measure of the rate of substitution in a reaction with a standard substrate. (p. 266)

olefin An alkene. (p. 296)

polymer A substance of high molecular weight made by linking many small molecules, called **monomers.** (p. 307)

addition polymer: A polymer formed by simple addition of monomer units.

polyolefin: A type of addition polymer with an olefin (alkene) serving as the monomer.

rearrangement A reaction involving a change in the bonding sequence within a molecule. Rearrangements are common in reactions such as the S_N1 and E1 involving carbocation intermediates. (p. 322)

resonance energy The extra stability of a conjugated system compared with the energy of a compound with an equivalent number of isolated double bonds. (p. 317)

saturated Having only single bonds; incapable of undergoing addition reactions. (p. 299)

Saytzeff Alternate spelling of Zaitsev.

solvolysis A nucleophilic substitution or elimination where the solvent serves as the attacking reagent. Solvolysis literally means "cleavage by the solvent." (p. 320)

$$(CH_3)_3C{-}Br \xrightarrow{CH_3OH.\ heat} (CH_3)_3C{-}OCH_3 + (CH_3)_2C{=}CH_2 + HBr$$

stereogenic Giving rise to stereoisomers. Characteristic of an atom or a group of atoms such that interchanging any two groups creates a stereoisomer. (p. 303)

stereospecific reaction A reaction in which different stereoisomers of the starting material react to give different stereoisomers of the product. (p. 330)

syn Adding to (or eliminating from) the same face of a molecule. (p. 329)

syn-coplanar: Having a dihedral angle of 0°. See **anti-coplanar** for a diagram.

trans-diaxial An anti and coplanar arrangement allowing E2 elimination of two adjacent substituents on a cyclohexane ring. The substituents must be trans to each other, and both must be in axial positions on the ring. (p. 332)

unsaturated Having multiple bonds that can undergo addition reactions. (p. 299)

vicinal dihalide A compound with two halogens on adjacent carbon atoms. (p. 249)

vinyl group An ethenyl group, CH=CH—. (p. 302)

Zaitsev's rule **(Saytzeff's rule)** Alkenes with more highly substituted double bonds are usually more stable. An elimination usually gives the most stable alkene product, commonly the most substituted alkene. Zaitsev's rule does not always apply, especially with a bulky base or a bulky leaving group. (pp. 312, 324, 335)

Zaitsev elimination: An elimination that gives the Zaitsev product.

Zaitsev product: The most substituted alkene product.

Essential Problem-Solving Skills in Chapter 7

Each skill is followed by problem numbers exemplifying that particular skill.

1	Draw and name alkenes, and calculate their elements of unsaturation. Name geometric isomers using both the *E-Z* and *cis-trans* systems.	Problems 7-40, 41, 42, 43, 44, 45, 46, and 47
2	Predict the relative stabilities of alkenes and cycloalkenes based on their structure and stereochemistry.	Problems 7-47, 48, and 70
3	Propose logical mechanisms for dehydrohalogenation of alkyl halides and dehydration of alcohols.	Problems 7-59, 61, 62, 63, 64, 67, 69, 72, 73, 74, 75, and 76
4	Predict the products of dehydrohalogenation and dehydration reactions, and use Zaitsev's rule to predict the major and minor products.	Problems 7-49, 50, 51, 52, 53, 54, 55, 67, 68, and 72
5	Given a set of reaction conditions, predict whether a reaction will be unimolecular or bimolecular, and identify the possible mechanisms and the likely products.	Problems 7-50, 54, 73, 75, and 76
	Problem-Solving Strategy: Predicting Substitutions and Eliminations	Problems 7-61, 62, 63, 67, 68, and 75
6	Draw and compare the mechanisms and reaction-energy diagrams of S_N1, S_N2, E1, and E2 reactions. Predict which substitutions or eliminations will be faster based on differences in substrate, leaving group, solvent, and base or nucleophile.	Problems 7-60, 71, 75, and 76
7	Explain the stereochemistry of E2 eliminations to form alkenes, and predict the products of E1 reactions on stereoisomers and on cyclohexane systems.	Problems 7-55, 65, 66, 67, 68, 69, and 76
	Problem-Solving Strategy: Proposing Reaction Mechanisms	Problems 7-59, 61, 62, 63, 64, 67, 69, 72, 73, 74, 75, and 76
8	Propose effective syntheses of alkenes from alkyl halides and alcohols.	Problems 7-56, 57, 58, 65, and 66

Study Problems

7-40 Draw a structure for each compound (includes old and new names).

(a) 3-methylpent-1-ene **(b)** *cis*-3-methyl-3-hexene **(c)** 3,4-dibromobut-1-ene
(d) 1,3-cyclohexadiene **(e)** cycloocta-1,4-diene **(f)** (*Z*)-3-methyl-2-octene
(g) vinylcyclopropane **(h)** (*Z*)-2-bromo-2-pentene **(i)** (*3Z,6E*)-1,3,6-octatriene

7-41 Label each structure as *Z*, *E*, or neither.

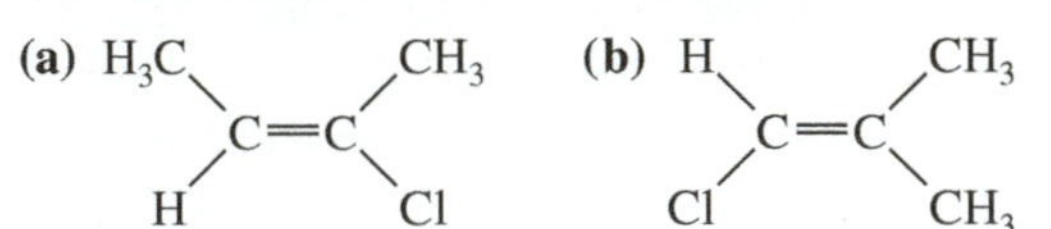

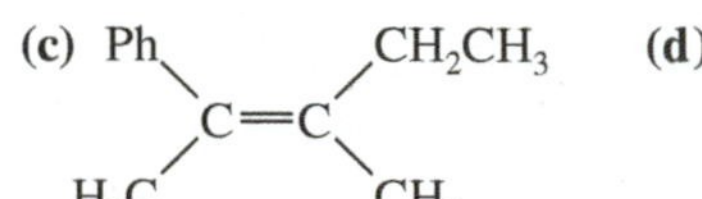

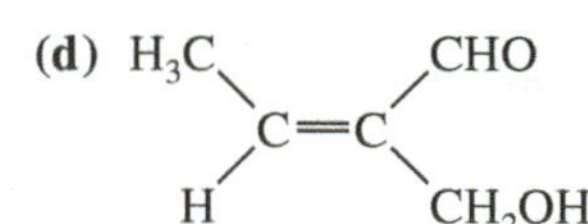

7-42 Determine which compounds show cis-trans isomerism. Draw and label the isomers, using both the cis-trans and *E*-*Z* nomenclatures where applicable.

(a) pent-1-ene
(b) pent-2-ene
(c) hex-3-ene
(d) 1,1-dibromopropene
(e) 1,2-dibromopropene
(f) 1-bromo-1-chlorohexa-1,3-diene

7-43 Give a correct name for each compound.

(a) $CH_3—CH_2—C(=CH_2)—CH_2—CH_2—CH_3$

(b) 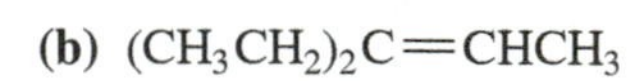

(c) 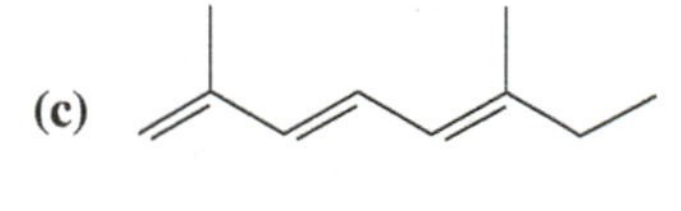

(d)

(e)

(f) 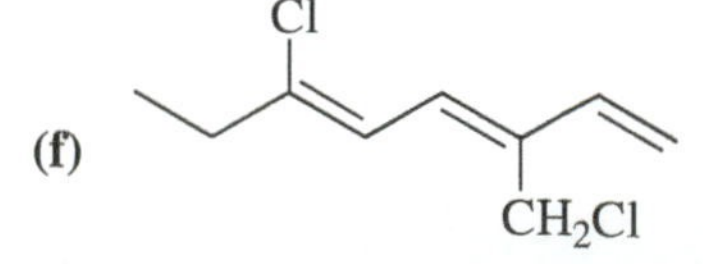

7-44 **(a)** Draw and name all five isomers of formula C_3H_5F.
(b) Draw all 12 acyclic (no rings) isomers of formula C_4H_7Br. Include stereoisomers.
(c) Cholesterol, $C_{27}H_{46}O$, has only one pi bond. With no additional information, what else can you say about its structure?

7-45 Draw and name all stereoisomers of 3-chlorohepta-2,4-diene
(a) using the cis-trans nomenclature.
(b) using the *E*-*Z* nomenclature.

7-46 For each alkene, indicate the direction of the dipole moment. For each pair, determine which compound has the larger dipole moment.
(a) *cis*-1,2-difluoroethene or *trans*-1,2-difluoroethene
(b) *cis*-1,2-dibromoethene or *trans*-2,3-dibromobut-2-ene
(c) *cis*-1,2-dibromo-1,2-dichloroethene or *cis*-1,2-dichloroethene

7-47 The energy difference between *cis*- and *trans*-but-2-ene is about 4 kJ/mol; however, the trans isomer of 4,4-dimethylpent-2-ene is nearly 16 kJ/mol more stable than the cis isomer. Explain this large difference.

7-48 A double bond in a six-membered ring is usually more stable in an endocyclic position than in an exocyclic position. Hydrogenation data on two pairs of compounds follow. One pair suggests that the energy difference between endocyclic and exocyclic double bonds is about 9 kJ/mol. The other pair suggests an energy difference of about 5 kJ/mol. Which number do you trust as being more representative of the actual energy difference? Explain your answer.

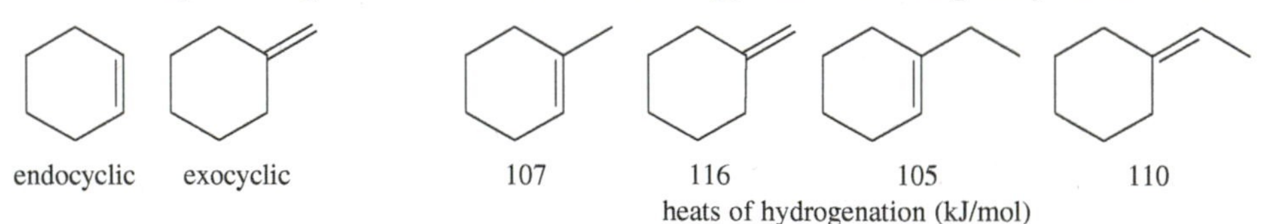

7-49 Predict the products of E1 elimination of the following compounds. Label the major products.

(a) 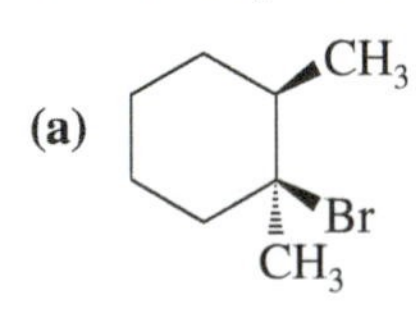

(b) 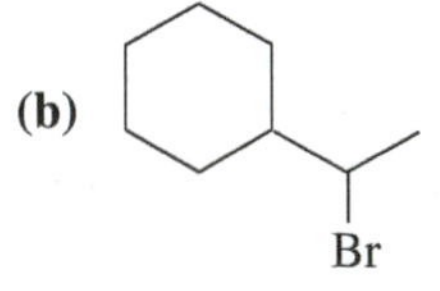

(c) 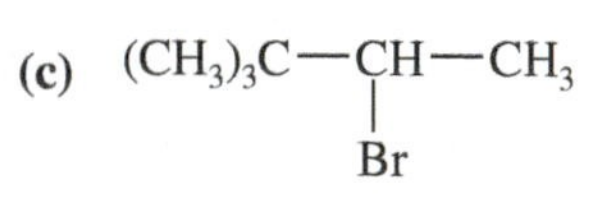

7-50 Predict the products formed by sodium hydroxide-promoted dehydrohalogenation of the following compounds. In each case, predict which will be the major product.
(a) 1-bromobutane
(b) 2-chlorobutane
(c) 3-bromopentane
(d) *cis*-1-bromo-2-methylcyclohexane
(e) *trans*-1-bromo-2-methylcyclohexane

7-51 What halides would undergo E2 dehydrohalogenation to give the following pure alkenes?
(a) hex-1-ene
(b) isobutylene
(c) pent-2-ene
(d) methylenecyclohexane
(e) 4-methylcyclohexene

7-52 Predict the major products of acid-catalyzed dehydration of the following alcohols.

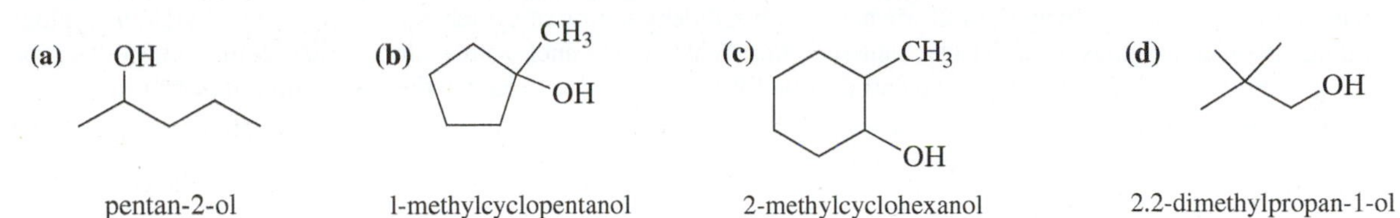

7-53 Predict the products of the following reactions. When more than one product is expected, predict which will be the major product.

(a) cyclopentanol (OH) $\xrightarrow[\text{heat}]{H_2SO_4}$ **(b)** (OH) $\xrightarrow[\text{heat}]{H_3PO_4}$ **(c)** (CH_3, OH) $\xrightarrow[\text{heat}]{H_2SO_4}$

7-54 Write a balanced equation for each reaction, showing the major product you expect.

(a) $CH_3{-}CH_2{-}CH(OH){-}CH_3 \xrightarrow{H_2SO_4,\ \text{heat}}$

(b) (H, Br) $\xrightarrow{NaOC(CH_3)_3}$

(c) (OH, OH) $\xrightarrow[\text{heat}]{H_2SO_4}$

(d) $CH_3{-}CH(CH_3){-}C(CH_3)(Br){-}CH_3 \xrightarrow{NaOH,\ \text{heat}}$

7-55 Predict the dehydrohalogenation product(s) that result when the following alkyl halides are heated in alcoholic KOH. When more than one product is formed, predict the major and minor products.

(a) $(CH_3)_2CH{-}C(CH_3)_2$ (Br) **(b)** $(CH_3)_2CH{-}CH(Br){-}CH_3$ **(c)** $(CH_3)_2C(Br){-}CH_2{-}CH_3$

(d) (CH_3, Cl, H, D) **(e)** $(CH_3)_3C$ (Cl, Cl) **(f)** (H, Br, Br, H)

7-56 Using cyclohexane as your starting material, show how you would synthesize each of the following compounds. (Once you have shown how to synthesize a compound, you may use it as the starting material in any later parts of this problem.)
(a) bromocyclohexane **(b)** cyclohexene **(c)** ethoxycyclohexane
(d) 3-bromocyclohex-1-ene **(e)** cyclohexa-1,3-diene **(f)** cyclohexanol

7-57 Show how you would prepare cyclopentene from each compound.
(a) cyclopentanol
(b) cyclopentyl bromide
(c) cyclopentane (not by dehydrogenation)

7-58 Show how you would convert (in one or two steps) 1-phenylpropane to the three products shown below. In each case, explain what unwanted reactions might produce undesirable impurities in the product.

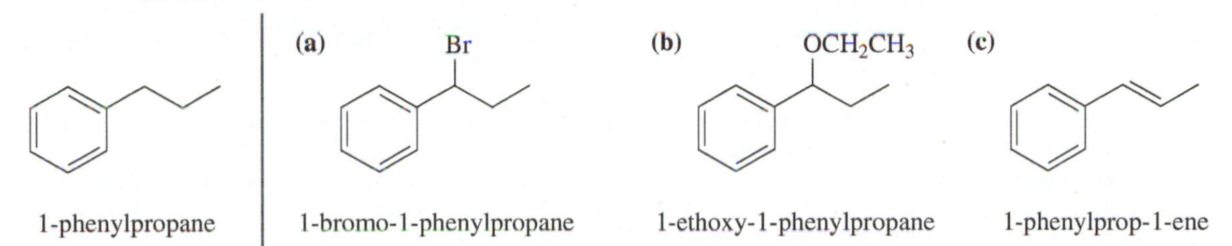

7-59 E1 eliminations of alkyl halides are rarely useful for synthetic purposes because they give mixtures of substitution and elimination products. Explain why the sulfuric acid-catalyzed dehydration of cyclohexanol gives a good yield of cyclohexene even though the reaction goes by an E1 mechanism. (*Hint*: What are the nucleophiles in the reaction mixture? What products are formed if these nucleophiles attack the carbocation? What further reactions can these substitution products undergo?)

7-60 Propose mechanisms and draw reaction-energy diagrams for the following reactions. Pay particular attention to the structures of any transition states and intermediates. Compare the reaction-energy diagrams for the two reactions, and explain the differences.

(a) 2-Bromo-2-methylbutane reacts with sodium methoxide in methanol to give 2-methylbut-2-ene (among other products).
(b) 2-Bromo-2-methylbutane reacts in boiling methanol to give 2-methylbut-2-ene (among other products).

7-61 Propose mechanisms for the following reactions. Additional products may be formed, but your mechanism only needs to explain the products shown.

(a) Br; CH_3OH, heat; OCH_3 + +

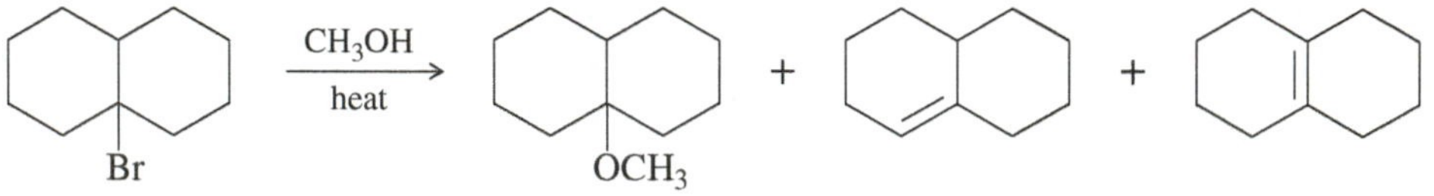

(b) CH_2OH; H_2SO_4, heat; + +

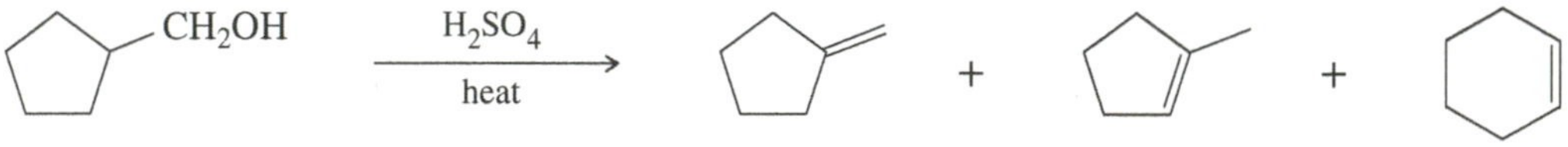

(c) Br; EtOH, heat; OEt + OEt +

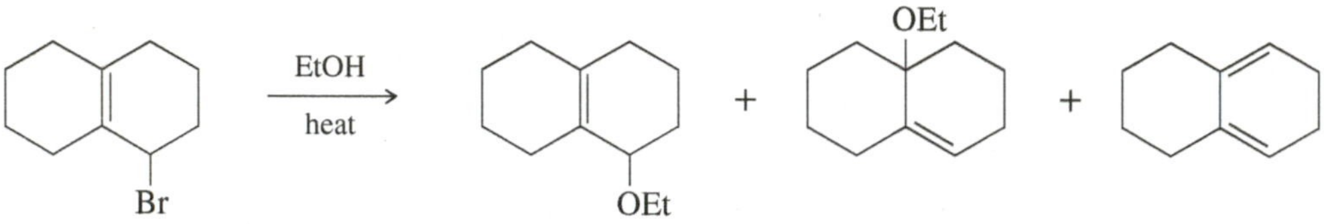

7-62 Propose mechanisms to account for the observed products in the following reactions. In some cases, more products are formed, but you only need to account for the ones shown here.

(a) Br, CH_2; CH_3OH, heat; CH_2OCH_3 + + OCH_3

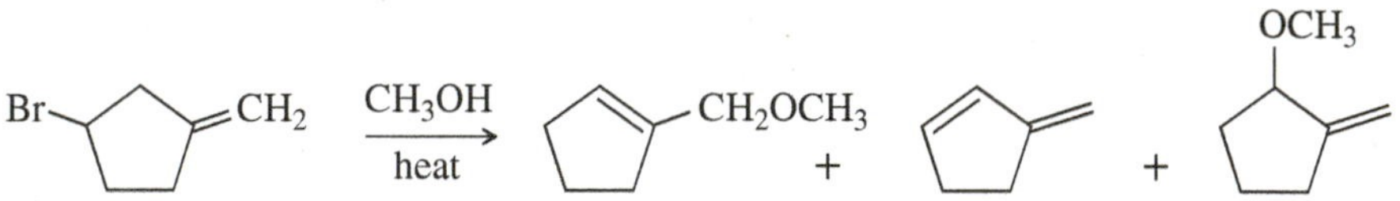

(b) Br; CH_3OH, heat; OCH_3 + OCH_3 +

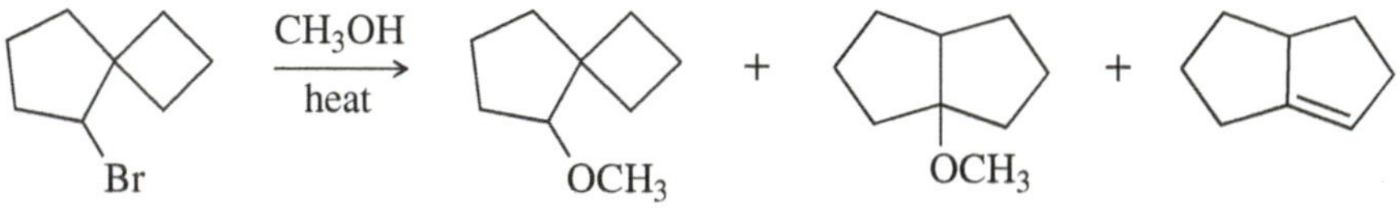

7-63 Silver-assisted solvolysis of bromomethylcyclopentane in methanol gives a complex product mixture of the following five compounds. Propose mechanisms to account for these products.

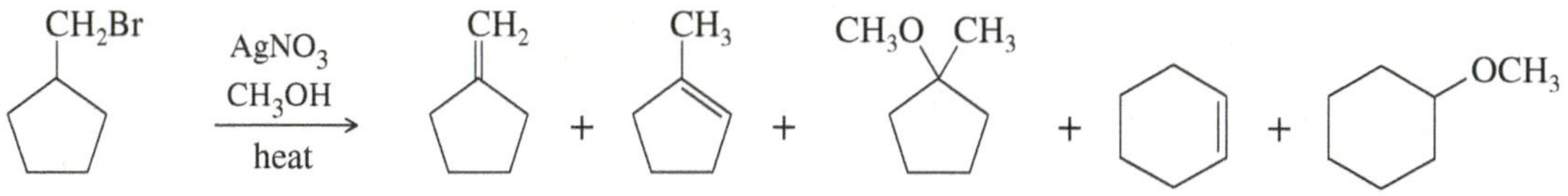

CH_2Br; $AgNO_3$, CH_3OH, heat; CH_2 + CH_3 + CH_3O CH_3 + + OCH_3

7-64 Protonation converts the hydroxy group of an alcohol to a good leaving group. Suggest a mechanism for each reaction.

(a) OH; HBr, heat (S_N2 or S_N1); Br + H_2O

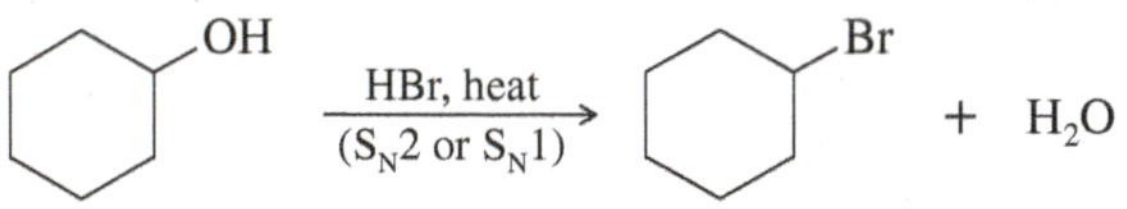

(b) OH; H_2SO_4, heat

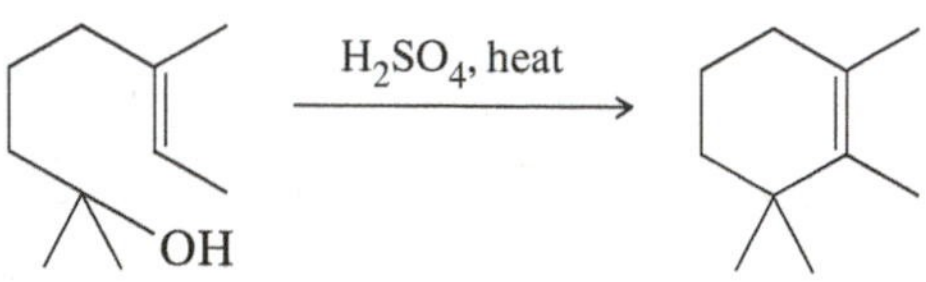

7-65 **(a)** Design an alkyl halide that will give *only* 2,4-diphenylpent-2-ene upon treatment with potassium *tert*-butoxide (a bulky base that promotes E2 elimination).
(b) What stereochemistry is required in your alkyl halide so that *only* the following stereoisomer of the product is formed?

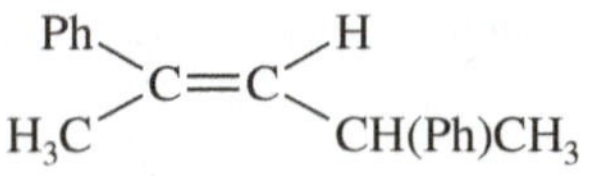

Ph, H, H_3C, C=C, CH(Ph)CH_3

7-66 A chemist allows some pure (2*S*,3*R*)-3-bromo-2,3-diphenylpentane to react with a solution of sodium ethoxide ($NaOCH_2CH_3$) in ethanol. The products are two alkenes: **A** (cis-trans mixture) and **B,** a single pure isomer. Under the same conditions, the reaction of (2*S*,3*S*)-3-bromo-2,3-diphenylpentane gives two alkenes, **A** (cis-trans mixture) and **C.** Upon catalytic hydrogenation, all three of these alkenes (**A, B,** and **C**) give 2,3-diphenylpentane. Determine the structures of **A, B,** and **C;** give equations for their formation; and explain the stereospecificity of these reactions.

***7-67** Pure (*S*)-2-bromo-2-fluorobutane reacts with methoxide ion in methanol to give a mixture of (*S*)-2-fluoro-2-methoxybutane and three fluoroalkenes.

(a) Use mechanisms to show which three fluoroalkenes are formed.

(b) Propose a mechanism to show how (*S*)-2-bromo-2-fluorobutane reacts to give (*S*)-2-fluoro-2-methoxybutane. Has this reaction gone with retention or inversion of configuration?

***7-68** When (±)-2,3-dibromobutane reacts with potassium hydroxide, some of the products are (2*S*,3*R*)-3-bromobutan-2-ol and its enantiomer and *trans*-2-bromobut-2-ene. Give mechanisms to account for these products. Why is no *cis*-2-bromobut-2-ene formed?

CH_3; H—OH; H—Br; CH_3 | CH_3; HO—H; Br—H; CH_3

(2*S*,3*R*) (2*R*,3*S*)

3-bromobutan-2-ol

H_3C, Br, C=C, H, CH_3

trans-2-bromobut-2-ene

***7-69** When 2-bromo-3-phenylbutane is treated with sodium methoxide, two alkenes result (by E2 elimination). The Zaitsev product predominates.

(a) Draw the reaction, showing the major and minor products.

(b) When one pure stereoisomer of 2-bromo-3-phenylbutane reacts, one pure stereoisomer of the major product results. For example, when (2*R*,3*R*)-2-bromo-3-phenylbutane reacts, the product is the stereoisomer with the methyl groups cis. Use your models to draw a Newman projection of the transition state to show why this stereospecificity is observed.

(c) Use a Newman projection of the transition state to predict the major product of elimination of (2*S*,3*R*)-2-bromo-3-phenylbutane.

(d) Predict the major product from elimination of (2*S*,3*S*)-2-bromo-3-phenylbutane. This prediction can be made without drawing any structures, by considering the results in part (b).

***7-70** Explain the dramatic difference in rotational energy barriers of the following three alkenes. (*Hint:* Consider what the transition states must look like.)

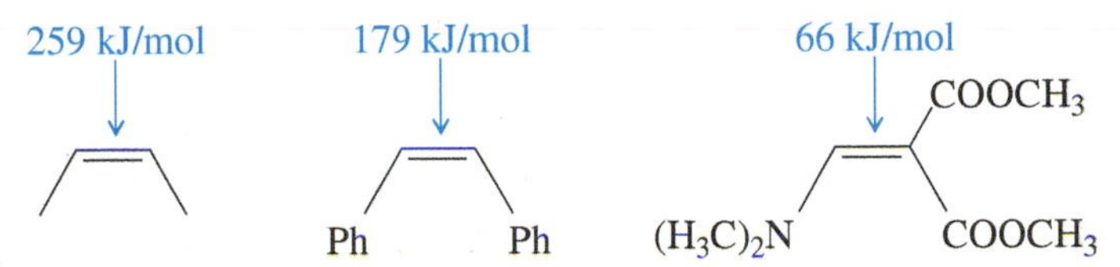

***7-71** One of the following dichloronorbornanes undergoes elimination much faster than the other. Determine which one reacts faster, and explain the large difference in rates.

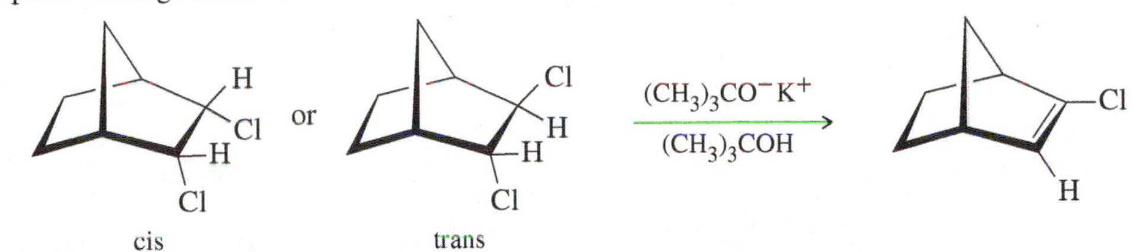

***7-72** A graduate student wanted to make methylenecyclobutane, and he tried the following reaction. Propose structures for the other products, and give mechanisms to account for their formation.

cyclobutyl—CH_2OH $\xrightarrow[\text{heat}]{H_2SO_4}$ methylenecyclobutane + other products

methylenecyclobutane (minor)

***7-73** Write a mechanism that explains the formation of the following product. In your mechanism, explain the cause of the rearrangement, and explain the failure to form the Zaitsev product.

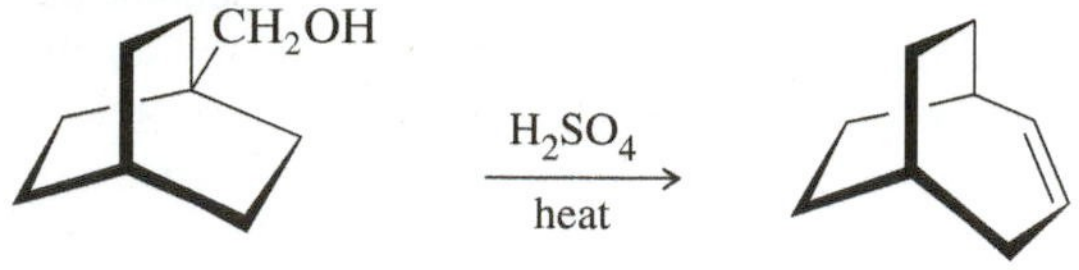

CH_2OH — $\xrightarrow[\text{heat}]{H_2SO_4}$

***7-74** The following reaction is called the *pinacol rearrangement*. The reaction begins with an acid-promoted ionization to give a carbocation. This carbocation undergoes a methyl shift to give a more stable, resonance-stabilized cation. Loss of a proton gives the observed product. Propose a mechanism for the pinacol rearrangement.

$$CH_3-C(CH_3)(OH)-C(CH_3)(OH)-CH_3 \xrightarrow{H_2SO_4,\ \text{heat}} CH_3-C(=O)-C(CH_3)_2-CH_3$$

pinacol → pinacolone

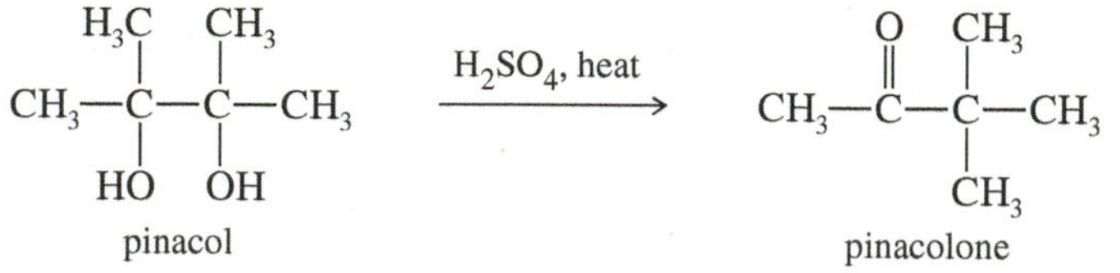

***7-75** Deuterium (D) is the isotope of hydrogen of mass number 2, with a proton and a neutron in its nucleus. The chemistry of deuterium is nearly identical to the chemistry of hydrogen, except that the C—D bond is slightly (5.0 kJ/mol, or 1.2 kcal/mol) stronger than the C—H bond. Reaction rates tend to be slower if a C—D bond (as opposed to a C—H bond) is broken in a rate-limiting step. This effect on the rate is called a *kinetic isotope effect*. (Review Problem 4-57.)

(a) Propose a mechanism to explain each product in the following reaction.

$$CH_3-CH(Br)-CH_3 \xrightarrow{\text{KOH, alcohol}} CH_2{=}CH-CH_3 + CH_3-CH(OH)-CH_3$$

elimination product; substitution product

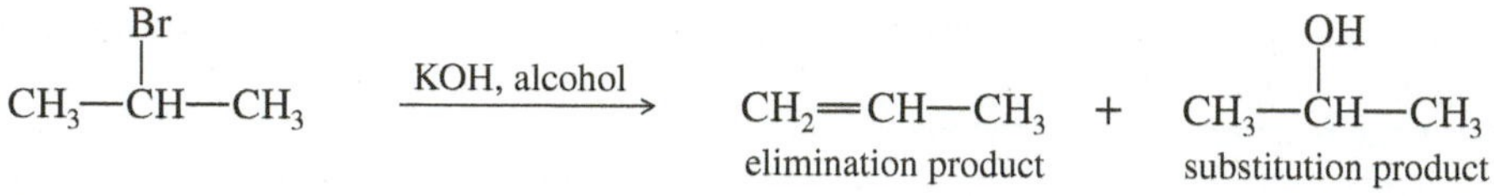

(b) When the following deuterated compound reacts under the same conditions, the rate of formation of the substitution product is unchanged, while the rate of formation of the elimination product is slowed by a factor of 7.

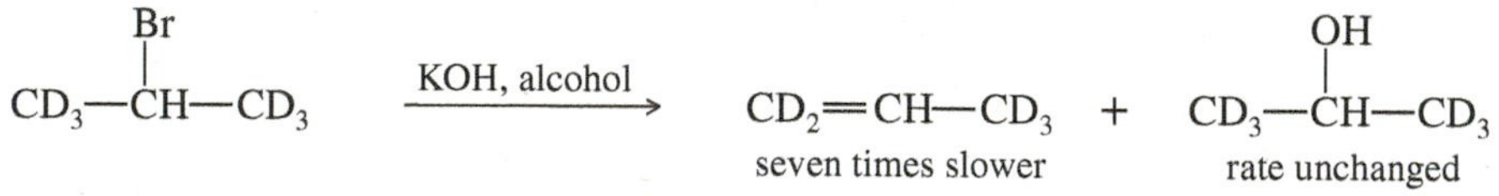

$$CD_3-CH(Br)-CD_3 \xrightarrow{\text{KOH, alcohol}} CD_2{=}CH-CD_3 + CD_3-CH(OH)-CD_3$$

seven times slower; rate unchanged

Explain why the elimination rate is slower, but the substitution rate is unchanged.

(c) A similar reaction takes place on heating the alkyl halide in an acetone/water mixture.

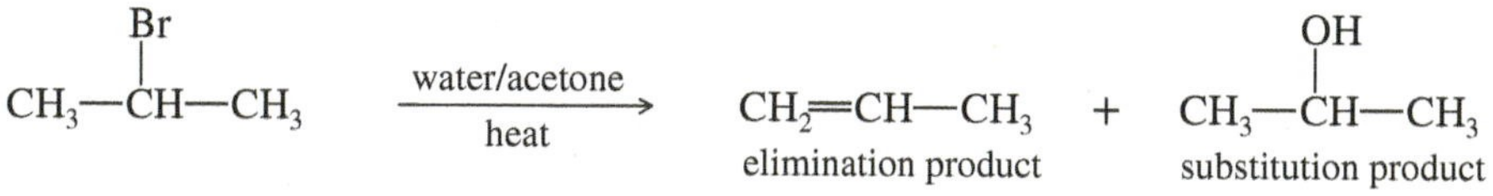

$$CH_3-CH(Br)-CH_3 \xrightarrow[\text{heat}]{\text{water/acetone}} CH_2{=}CH-CH_3 + CH_3-CH(OH)-CH_3$$

elimination product; substitution product

Give a mechanism for the formation of each product under these conditions, and predict how the rate of formation of each product will change when the deuterated halide reacts. Explain your prediction.

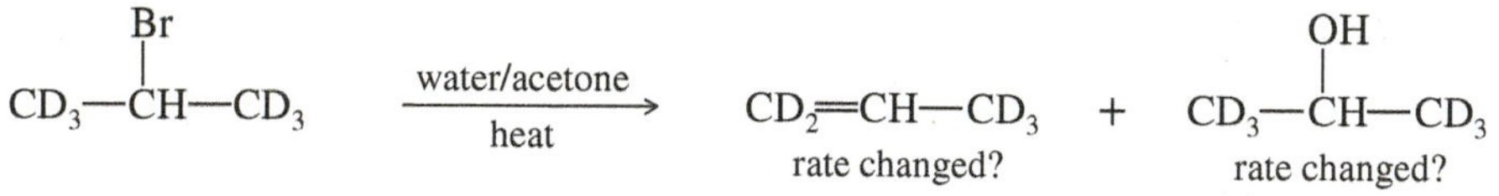

$$CD_3-CH(Br)-CD_3 \xrightarrow[\text{heat}]{\text{water/acetone}} CD_2{=}CH-CD_3 + CD_3-CH(OH)-CD_3$$

rate changed?; rate changed?

***7-76** When the following compound is treated with sodium methoxide in methanol, two elimination products are possible. Explain why the deuterated product predominates by about a 7:1 ratio (refer to Problem 7-75).

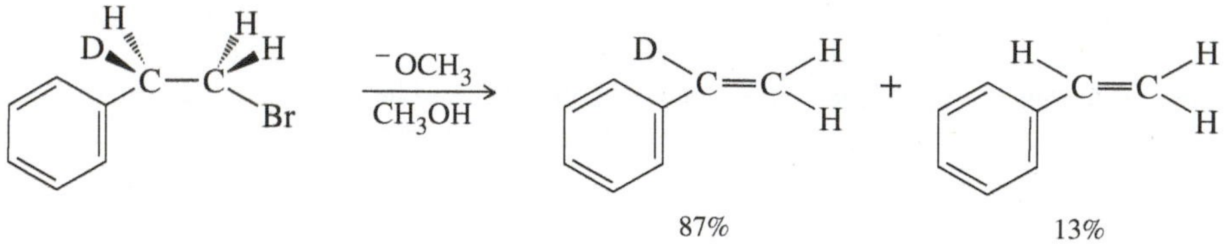

$$PhCHD-CH_2Br \xrightarrow[CH_3OH]{^-OCH_3} PhC(D){=}CH_2 + PhCH{=}CH_2$$

87% ; 13%

8 Reactions of Alkenes

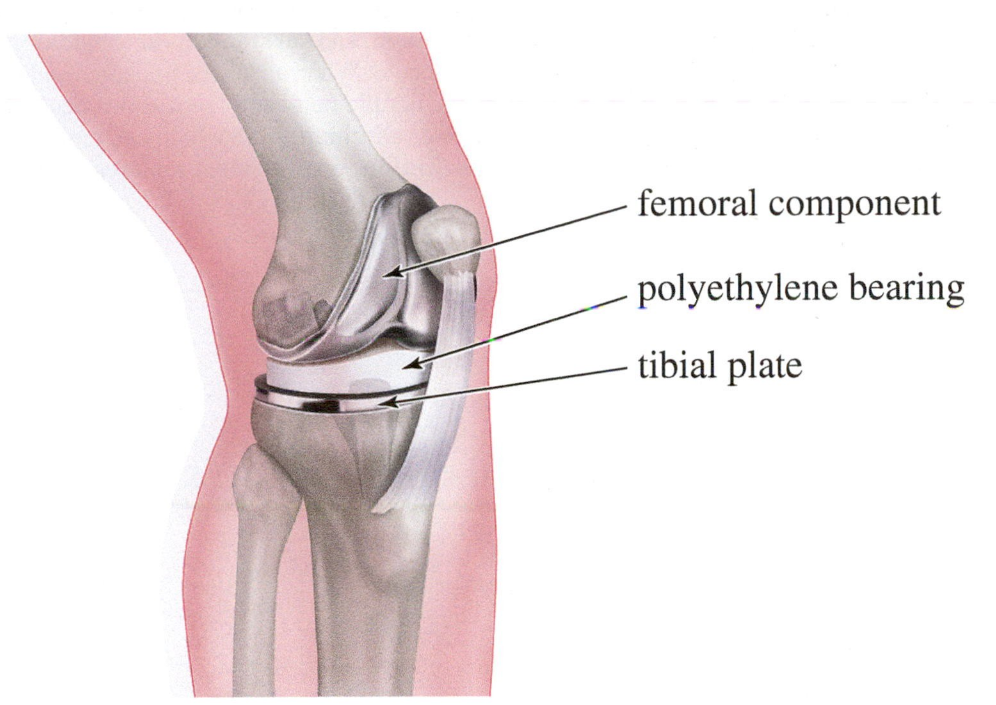

▲ Polyethylene provides a self-lubricating surface for movement of the metal parts of an artificial knee replacement. The highly cross-linked polyethylene used in these implants is fabricated to be exceptionally tough: It wears only about 0.1 mm per year. Polyethylene is compatible with the human body, and in most cases, it does not cause a foreign body reaction even after years of constant movement in the joint.

Goals for Chapter 8

1 Explain why electrophilic additions are among the most common reactions of alkenes.

2 Predict the products of the reactions of alkenes, including the orientation of the reaction (regiochemistry) and the stereochemistry.

3 Propose mechanisms to explain the observed products of alkene reactions.

4 Use retrosynthetic analysis to solve multistep synthesis problems with alkenes as reagents, intermediates, or products.

The alkene double bond is a gateway functional group. Alkene reactions lead to many other functional groups that lay the foundation for the rest of your study of organic chemistry. You can convert alkenes to alkyl halides, epoxides, alcohols, aldehydes, ketones, carboxylic acids, and other functional groups. The reactions of alkenes arise from the reactivity of their carbon–carbon double bonds. Organic chemists enjoy the challenge of taking a simple carbon–carbon double bond and manipulating it in all possible ways to produce other compounds, often mimicking biological reactions that occur in cells. This chapter covers the most common alkene reactions, including their mechanisms, reactivity, orientation, and stereochemistry.

Most reactions of alkenes involve addition of atoms or groups across the double bond, with one atom or group adding to each end. By studying the typical mechanisms involved when reagents add to double bonds, you will be able to predict the outcomes of alkene addition reactions that you have not seen before. You will also find it easier to understand the mechanisms of addition reactions we see in other families of compounds.

8-1 Reactivity of the Carbon–Carbon Double Bond

Because single bonds (sigma bonds) are more stable than pi bonds, the most common reactions of double bonds transform the pi bond into a sigma bond. For example, catalytic hydrogenation converts the C═C pi bond and the H—H sigma bond into two C—H sigma bonds (Section 7-8). The reaction is exothermic (ΔH° = about -80 to -120 kJ/mol or about -20 to -30 kcal/mol), showing that the product is more stable than the reactants.

$$\mathrm{>C{=}C<} \; + \; \mathrm{H{-}H} \xrightarrow{\text{catalyst}} \mathrm{-\underset{H}{\overset{|}{C}}-\underset{H}{\overset{|}{C}}-} \; + \; \text{energy}$$

Hydrogenation of an alkene is an example of an **addition,** one of the three major reaction types we have studied: addition, elimination, and substitution. In an addition, two molecules combine to form one product molecule. When an alkene undergoes addition, two groups add to the carbon atoms of the double bond and the carbons become saturated. In many ways, addition is the reverse of **elimination,** in which one molecule splits into two fragment molecules. In a **substitution,** one fragment replaces another fragment in a molecule.

Addition

$$\mathrm{>C{=}C<} \; + \; \mathrm{X{-}Y} \longrightarrow \mathrm{-\underset{X}{\overset{|}{C}}-\underset{Y}{\overset{|}{C}}-}$$

Elimination

$$\mathrm{-\underset{X}{\overset{|}{C}}-\underset{Y}{\overset{|}{C}}-} \longrightarrow \mathrm{>C{=}C<} \; + \; \mathrm{X{-}Y}$$

Substitution

$$\mathrm{-\underset{|}{\overset{|}{C}}-X} \; + \; \mathrm{Y^-} \longrightarrow \mathrm{-\underset{|}{\overset{|}{C}}-Y} \; + \; \mathrm{X^-}$$

Addition is the most common reaction of alkenes, and in this chapter, we consider additions to alkenes in detail. A wide variety of functional groups can be formed by adding suitable reagents to the double bonds of alkenes.

8-2 Electrophilic Addition to Alkenes

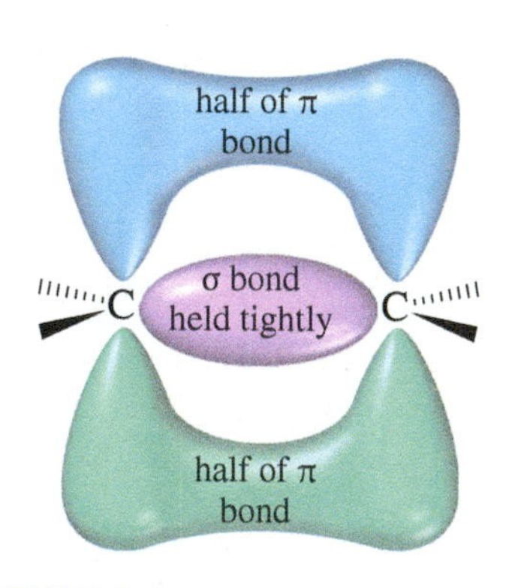

FIGURE 8-1
The electrons in the pi bond are spread farther from the carbon nuclei and are more loosely held than the sigma electrons.

In principle, many different reagents could add to a double bond to form more stable products; that is, the reactions are energetically favorable. Not all of these reactions have convenient rates, however. For example, the reaction of ethylene with hydrogen (to give ethane) is strongly exothermic, but the rate is very slow. A mixture of ethylene and hydrogen can remain for years without appreciable reaction. Adding a catalyst such as platinum, palladium, or nickel allows the reaction to take place at a rapid rate.

Some reagents react with carbon–carbon double bonds without the aid of a catalyst. To understand what types of reagents react with double bonds, consider the structure of the pi bond. Although the electrons in the sigma bond framework are tightly held, the pi bond is delocalized above and below the sigma bond (Figure 8-1). The pi-bonding electrons are spread farther from the carbon nuclei, and they are more loosely held. A strong electrophile has an affinity for these loosely held electrons. It can pull them away to form a new bond (Figure 8-2), leaving one of the carbon atoms with only three bonds and a positive charge: a carbocation. In effect, the double bond has reacted as a nucleophile, donating a pair of electrons to the electrophile.

Most addition reactions involve a second step in which a nucleophile attacks the carbocation (as in the second step of the S_N1 reaction), forming a stable addition product. In the product, both the electrophile and the nucleophile are bonded to the carbon atoms that were connected by the double bond. This reaction is outlined in Key Mechanism 8-1, identifying the electrophile as E^+ and the nucleophile as $Nuc:^-$. This type of reaction requires a strong electrophile to attract the electrons of the pi bond and generate a carbocation in the rate-limiting step. Most alkene reactions fall into this large class of **electrophilic additions** to alkenes.

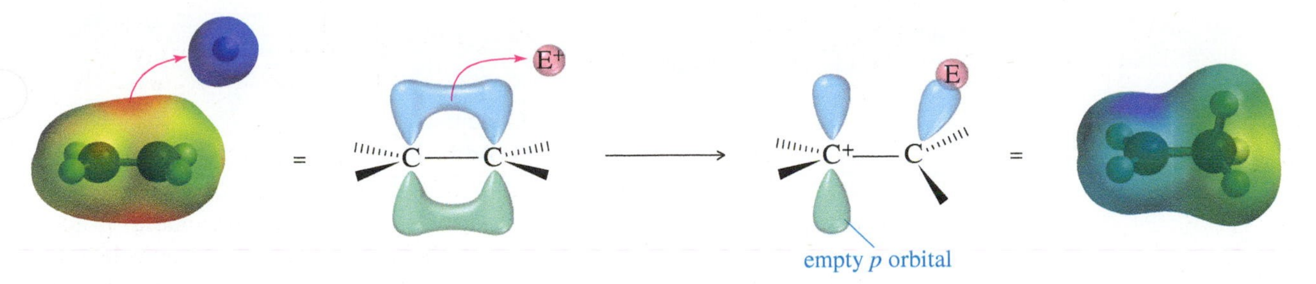

FIGURE 8-2
The pi bond as a nucleophile. A strong electrophile attracts the electrons out of the pi bond to form a new sigma bond, generating a carbocation. The (red) curved arrow shows the movement of electrons, from the electron-rich pi bond to the electron-poor electrophile.

KEY MECHANISM 8-1 Electrophilic Addition to Alkenes

A wide variety of electrophilic additions involve similar mechanisms. First, a strong electrophile attracts the loosely held electrons from the pi bond of an alkene. The electrophile forms a sigma bond to one of the carbons of the (former) double bond, while the other carbon becomes a carbocation. The carbocation (a strong electrophile) reacts with a nucleophile (often a weak nucleophile) to form another sigma bond. We symbolize the electrophile as E^+ and the nucleophile as Nuc:$^-$ because electrophilic additions often involve ionic reagents, with the electrophile having a positive charge and the nucleophile having a negative charge.

Step 1: Attack of the pi bond on the electrophile forms a carbocation.

$$>C=C< \; + \; E^+ \; \longrightarrow \; -\underset{E}{\overset{|}{C}}-C^+<$$

+ on the more substituted carbon

Step 2: Attack by a nucleophile gives the addition product.

$$-\underset{E}{\overset{|}{C}}-C^+< \; + \; \text{Nuc}:^- \; \longrightarrow \; -\underset{E}{\overset{|}{C}}-\underset{\text{Nuc}}{\overset{|}{C}}-$$

EXAMPLE: Ionic addition of HBr to but-2-ene

This example shows what happens when gaseous HBr adds to but-2-ene. The proton in HBr is electrophilic; it reacts with the alkene to form a carbocation. Bromide ion reacts rapidly with the carbocation to give a stable product in which the elements of HBr have added to the ends of the double bond.

Step 1: Protonation of the double bond forms a carbocation.

$$CH_3-\overset{H}{\overset{|}{C}}=\overset{H}{\overset{|}{C}}-CH_3 \; + \; H-\ddot{Br}: \; \rightleftharpoons \; CH_3-\underset{H}{\overset{H}{\overset{|}{\underset{|}{C}}}}-\overset{H}{\overset{|}{C^+}}-CH_3 \; + \; :\ddot{Br}:^-$$

Step 2: Bromide ion attacks the carbocation.

$$CH_3-\underset{H}{\overset{H}{\overset{|}{\underset{|}{C}}}}-\overset{H}{\overset{|}{C^+}}-CH_3 \; + \; :\ddot{Br}:^- \; \rightleftharpoons \; CH_3-\underset{H}{\overset{H}{\overset{|}{\underset{|}{C}}}}-\underset{:Br:}{\overset{H}{\overset{|}{\underset{|}{C}}}}-CH_3$$

PROBLEM: Explain why the + charge of the carbocation always appears at the carbon of the (former) double bond that has NOT bonded to the electrophile.

TABLE 8-1
Types of Additions to Alkenes

$\rangle C{=}C\langle$ → Type of Addition [Elements Added][a] → Product			
hydration $[H_2O]$	$-\overset{H}{C}-\overset{OH}{C}-$	halogenation $[X_2]$, an oxidation	$-\overset{X}{C}-\overset{X}{C}-$
hydrogenation $[H_2]$, a reduction	$-\overset{H}{C}-\overset{H}{C}-$	halohydrin formation $[HOX]$, an oxidation	$-\overset{X}{C}-\overset{OH}{C}-$
dihydroxylation [HOOH], an oxidation	$-\overset{OH}{C}-\overset{OH}{C}-$	HX addition [HX] (hydrohalogenation)	$-\overset{H}{C}-\overset{X}{C}-$
oxidative cleavage $[O_2]$, an oxidation	$\rangle C{=}O \quad O{=}C\langle$		
epoxidation [O], an oxidation	$-C-C-$ bridged by O (epoxide)	cyclopropanation $[CH_2]$	$-C-C-$ bridged by CH_2 (cyclopropane)

[a]These are not the reagents used, but simply the groups that appear in the product.

We will consider several types of additions to alkenes, using a wide variety of reagents: water, borane, hydrogen, carbenes, halogens, oxidizing agents, and even other alkenes. Most, but not all, of these will be electrophilic additions. Table 8-1 summarizes the classes of additions we will cover. Note that the table shows what elements have added across the double bond in the final product, but it says nothing about reagents or mechanisms. As we study these reactions, you should note the **regiochemistry** of each reaction, also called the *orientation of addition*, meaning which part of the reagent adds to which end of the double bond. Also note the *stereochemistry* when the reaction is stereospecific.

8-3 Addition of Hydrogen Halides to Alkenes

8-3A Orientation of Addition: Markovnikov's Rule

The simple mechanism shown for addition of HBr to but-2-ene applies to a large number of electrophilic additions. We can use this mechanism to predict the outcome of some fairly complicated reactions. For example, the addition of HBr to 2-methylbut-2-ene could lead to either of two products, yet only one is observed.

$$CH_3-\overset{CH_3}{\overset{|}{C}}=CH-CH_3 + H-Br \longrightarrow \underset{\text{observed}}{CH_3-\overset{CH_3}{\underset{Br}{C}}-\underset{H}{CH}-CH_3} \text{ or } \underset{\text{not observed}}{CH_3-\overset{CH_3}{\underset{H}{C}}-\underset{Br}{CH}-CH_3}$$

The first step is protonation of the double bond. If the proton adds to the secondary carbon, the product will be different from the one formed if the proton adds to the tertiary carbon.

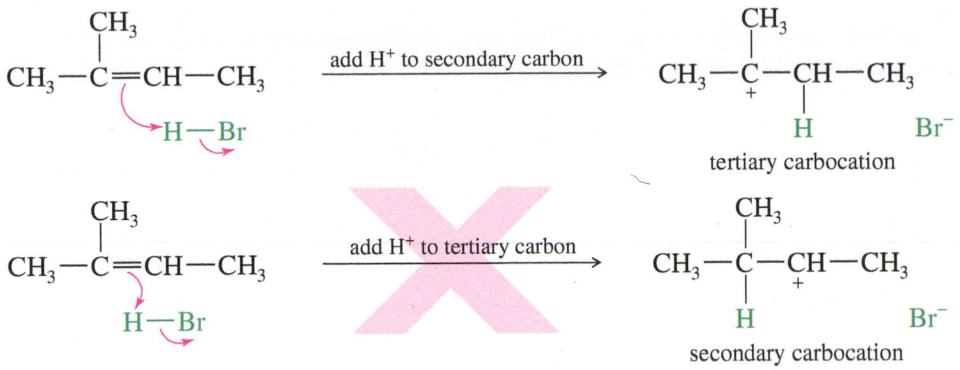

$$CH_3-\underset{|}{\overset{CH_3}{C}}=CH-CH_3 + H-Br \xrightarrow{\text{add } H^+ \text{ to secondary carbon}} CH_3-\overset{CH_3}{\underset{+}{C}}-\underset{H}{CH}-CH_3 \quad Br^-$$

tertiary carbocation

$$CH_3-\overset{CH_3}{C}=CH-CH_3 + H-Br \xrightarrow{\text{add } H^+ \text{ to tertiary carbon}} CH_3-\underset{H}{\overset{CH_3}{C}}-\underset{+}{CH}-CH_3 \quad Br^-$$

secondary carbocation

PROBLEM-SOLVING HINT

Stability of carbocations:

$$3° > 2° > 1° > {}^+CH_3$$

An electrophile adds to a double bond to give the most stable carbocation in the intermediate.

When the proton adds to the secondary carbon, a tertiary carbocation results. When the proton adds to the tertiary carbon atom, a secondary carbocation results. The tertiary carbocation is more stable (see Section 4-16A), so the first reaction is favored.

The second half of the mechanism produces the final product of the addition of HBr to 2-methylbut-2-ene.

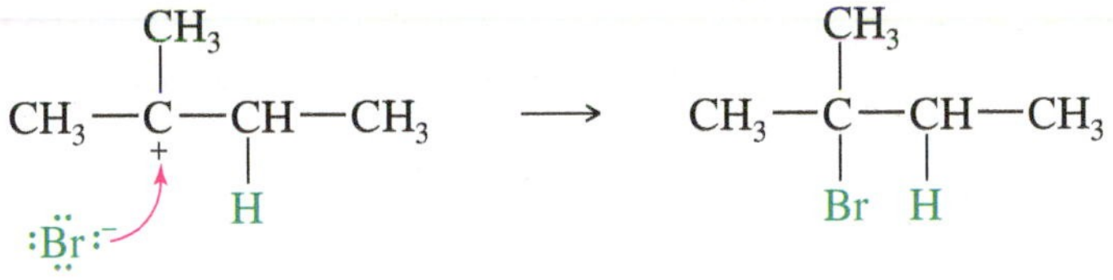

$$CH_3-\overset{CH_3}{\underset{+}{C}}-\underset{H}{CH}-CH_3 + :\ddot{Br}:^- \longrightarrow CH_3-\underset{Br}{\overset{CH_3}{C}}-\underset{H}{CH}-CH_3$$

Note that protonation of one carbon atom of a double bond gives a carbocation on the carbon atom that was *not* protonated. Therefore, the proton adds to the end of the double bond that is *less* substituted to give the *more substituted carbocation* (the more stable carbocation).

MECHANISM 8-2 Ionic Addition of HX to an Alkene

Step 1: Protonation of the pi bond forms a carbocation.

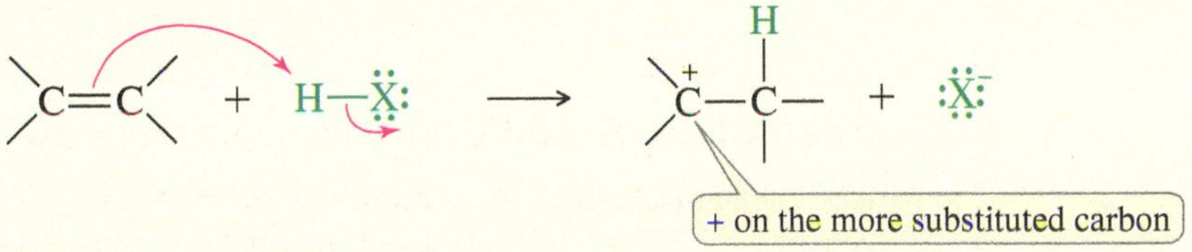

$$>C=C< + H-\ddot{X}: \longrightarrow >\overset{+}{C}-\underset{|}{\overset{H}{C}}- + :\ddot{X}:^-$$

+ on the more substituted carbon

Step 2: Attack by the halide ion gives the addition product.

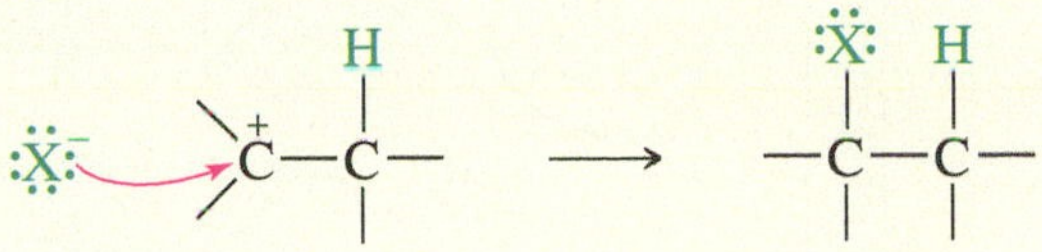

$$:\ddot{X}:^- + >\overset{+}{C}-\underset{|}{\overset{H}{C}}- \longrightarrow -\underset{|}{\overset{:\ddot{X}:}{C}}-\underset{|}{\overset{H}{C}}-$$

(CONTINUED)

EXAMPLE: The ionic addition of HBr to propene shows protonation of the less substituted carbon to give the more substituted carbocation. Reaction with bromide ion completes the addition.

less substituted

product

Positive charge on less substituted carbon. Less stable; *not formed.*

Figure 8-3 shows how addition of the proton to the less substituted end of the double bond gives the more substituted carbocation, which is more stable. Hammond's postulate predicts that the transition states resemble the carbocations, so the transition state leading to the more stable carbocation will be lower in energy than the one leading to the less stable carbocation. Therefore, the more highly substituted carbocation will be formed faster.

There are many examples of reactions where the proton adds to the less substituted carbon atom of the double bond to produce the more substituted carbocation. The addition of HBr (and other hydrogen halides) is said to be **regioselective** because in each case, one of the two possible orientations of addition results preferentially over the other.

Markovnikov's Rule A Russian chemist, Vladimir Markovnikov, first showed the orientation of addition of HBr to alkenes in 1869.

> MARKOVNIKOV'S RULE: The addition of a proton acid to the double bond of an alkene results in a product with the acid proton bonded to the carbon atom that already holds the greater number of hydrogen atoms.

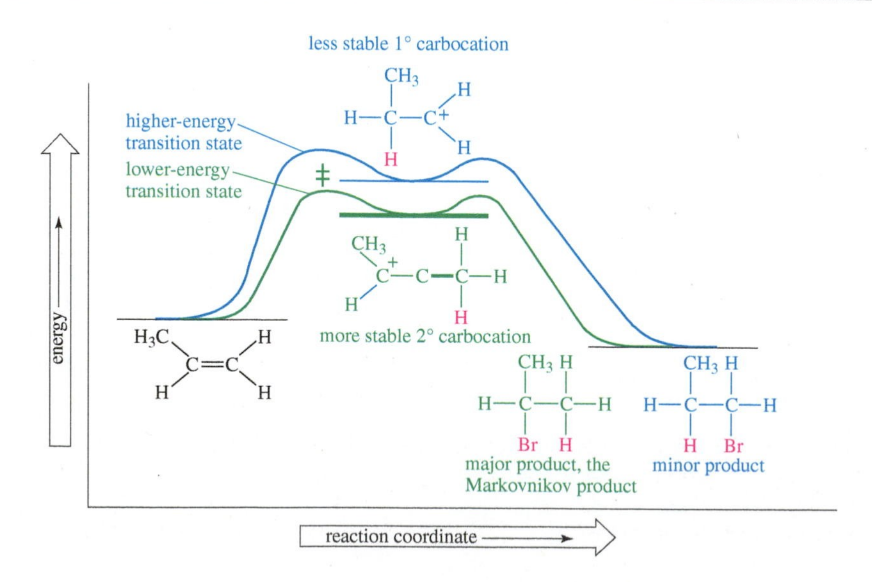

FIGURE 8-3
The reaction-energy diagram shows that the first step is rate-determining in the electrophilic addition to an alkene. The transition states resemble the intermediates (Hammond postulate), and the more highly substituted intermediate is formed faster because it has a lower-energy transition state leading to it. The more stable, more highly substituted intermediate generally leads to the Markovnikov product.

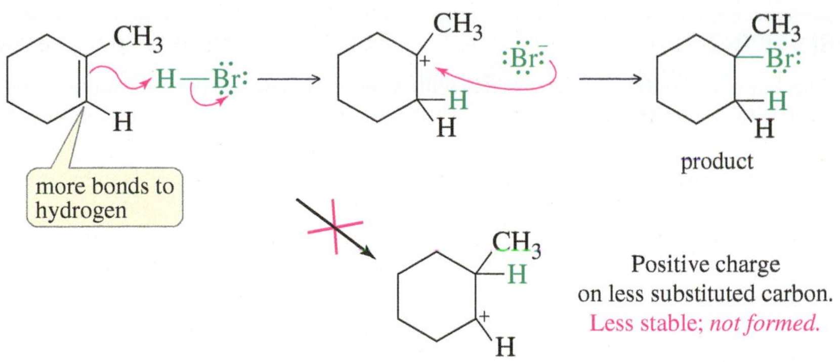

more bonds to hydrogen

product

Positive charge on less substituted carbon. Less stable; *not formed.*

FIGURE 8-4
An electrophile adds to the less substituted end of the double bond to give the more substituted (and therefore more stable) carbocation.

This is the original statement of **Markovnikov's rule.** Reactions that follow this rule are said to follow **Markovnikov orientation** and give the **Markovnikov product.** We are often interested in adding electrophiles other than proton acids to the double bonds of alkenes. Markovnikov's rule can be extended to include a wide variety of other additions, based on the addition of the electrophile in such a way as to produce the most stable carbocation.

MARKOVNIKOV'S RULE (extended): In an electrophilic addition to an alkene, the electrophile adds in such a way as to generate the most stable intermediate.

Figure 8-4 shows how HBr adds to 1-methylcyclohexene to give the product with an additional hydrogen bonded to the carbon that already had the most bonds to hydrogen (one) in the alkene. Note that this orientation results from addition of the proton in the way that generates the more stable carbocation.

Like HBr, both HCl and HI add to the double bonds of alkenes, and they also follow Markovnikov's rule; for example,

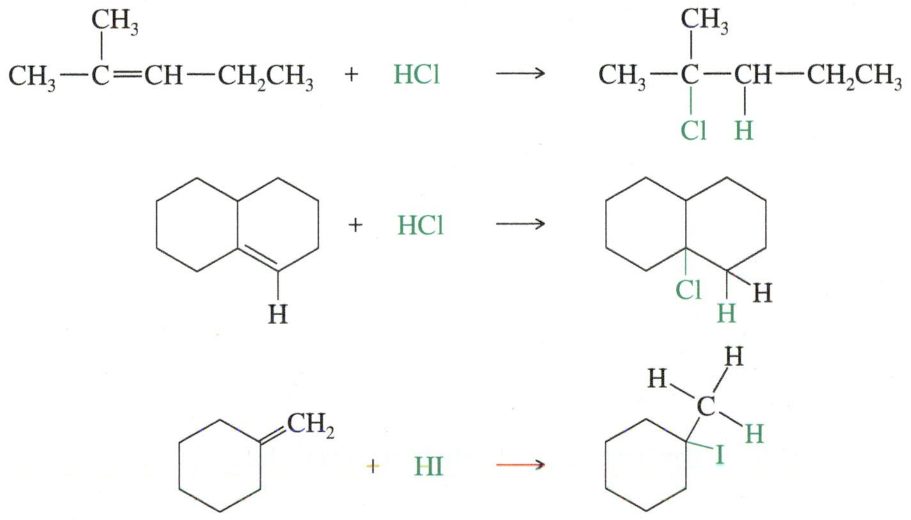

$$CH_3-C(CH_3)=CH-CH_2CH_3 + HCl \longrightarrow CH_3-C(CH_3)(Cl)-CH(H)-CH_2CH_3$$

$$\text{(bicyclic alkene)} + HCl \longrightarrow \text{(Cl, H product)}$$

$$\text{methylenecyclohexane} + HI \longrightarrow \text{1-iodo-1-methylcyclohexane}$$

PROBLEM-SOLVING HINT

The *orientation* or *regiochemistry* of an electrophilic addition is important when the reaction combines an unsymmetrical reagent with an unsymmetrical alkene. The first chemical species that reacts with the double bond is the electrophile (H^+ in the case of adding HBr). Where the electrophile reacts determines where the second part of the reagent (the nucleophile, Br^- in this case) will react. Understanding the factors that determine where the electrophile reacts allows us to predict the products. This is the basis of Markovnikov's rule.

PROBLEM 8-1

Predict the major products of the following reactions, and propose mechanisms to support your predictions.

(a) pent-1-ene + HCl
(b) 2-methylpropene + HCl
(c) 1-methylcyclohexene + HI
(d) 4-methylcyclohexene + HBr

PROBLEM 8-2

(a) When 1 mole of buta-1,3-diene reacts with 1 mole of HBr, both 3-bromobut-1-ene and 1-bromobut-2-ene are formed. Propose a mechanism to account for this mixture of products.

(b) When 1-chlorocyclohexene reacts with HBr, the major product is 1-bromo-1-chlorocyclohexane. Propose a mechanism for this reaction, and explain why your proposed intermediate is more stable than the other possible intermediate.

8-3B Free-Radical Addition of HBr: Anti-Markovnikov Addition

In 1933, M. S. Kharasch and F. W. Mayo found that some additions of HBr (but not HCl or HI) to alkenes gave products that were opposite those expected from Markovnikov's rule. These **anti-Markovnikov** reactions were most likely when the reagents or solvents came from old supplies that had accumulated peroxides from exposure to the air. Peroxides give rise to free radicals that initiate the addition, causing it to occur by a radical mechanism. The oxygen–oxygen bond in peroxides is rather weak, so it can break to give two alkoxyl radicals.

$$\text{R—}\ddot{\underset{..}{\text{O}}}\text{—}\ddot{\underset{..}{\text{O}}}\text{—R} \xrightarrow{\text{heat}} \text{R—}\ddot{\underset{..}{\text{O}}}\cdot + \cdot\ddot{\underset{..}{\text{O}}}\text{—R} \qquad \Delta H^\circ = +150 \text{ kJ } (+36 \text{ kcal})$$

Alkoxyl radicals (R—O·) initiate the anti-Markovnikov addition of HBr. Radicals are electrophilic because they do not have octets. In most cases, a radical needs one more electron for its octet. When a radical reacts with a double bond, one electron from the pi bond fills the octet, leaving the other electron as a radical at the other end of the bond. The mechanism of this free-radical chain reaction is shown in Mechanism 8-3.

MECHANISM 8-3 Free-Radical Addition of HBr to Alkenes

Initiation: Formation of radicals.

$$\text{R—}\ddot{\underset{..}{\text{O}}}\text{—}\ddot{\underset{..}{\text{O}}}\text{—R} \xrightarrow{\text{heat}} \text{R—}\ddot{\underset{..}{\text{O}}}\cdot + \cdot\ddot{\underset{..}{\text{O}}}\text{—R}$$

$$\text{R—}\ddot{\underset{..}{\text{O}}}\cdot + \text{H—}\ddot{\underset{..}{\text{Br}}}\text{:} \longrightarrow \text{R—}\ddot{\underset{..}{\text{O}}}\text{—H} + \text{:}\ddot{\underset{..}{\text{Br}}}\cdot$$

Propagation: A radical reacts to generate another radical.

Step 1: A bromine radical adds to the double bond to generate an alkyl radical on the more substituted carbon atom.

$$\text{:}\ddot{\underset{..}{\text{Br}}}\cdot + \text{C=C} \longrightarrow \text{—C(Br)—C}\cdot$$

radical on the more substituted carbon

Step 2: The alkyl radical abstracts a hydrogen atom from HBr to generate the product and a bromine radical.

$$\text{—C(Br)—C}\cdot + \text{H—}\ddot{\underset{..}{\text{Br}}}\text{:} \longrightarrow \text{—C(Br)—C(H)—} + \text{:}\ddot{\underset{..}{\text{Br}}}\cdot$$

The bromine radical generated in Step 2 goes on to react with another molecule of alkene in Step 1, continuing the chain.

EXAMPLE: Free-radical addition of HBr to propene.

Initiation: Radicals are formed.

$$\text{R—O—O—R} \xrightarrow{\text{heat}} \text{R—O}\cdot + \cdot\text{O—R}$$

$$\text{R—O}\cdot + \text{H—Br} \longrightarrow \text{R—O—H} + \text{Br}\cdot$$

Propagation: A radical reacts to generate another radical.

Step 1: A bromine radical adds to the double bond to generate an alkyl radical on the secondary carbon atom.

$$(\text{H})(\text{H}_3\text{C})\text{C}{=}\text{CH}_2 + \text{Br}\cdot \longrightarrow (\text{H})(\text{H}_3\text{C})\dot{\text{C}}\text{—CH}_2\text{Br}$$

· on the 2° carbon

Step 2: The alkyl radical abstracts a hydrogen atom from HBr to generate the product and a bromine radical.

$$(\text{H})(\text{H}_3\text{C})\dot{\text{C}}\text{—CH}_2\text{Br} + \text{H—Br} \longrightarrow \text{H—CH(CH}_3\text{)—CH}_2\text{Br} + \text{Br}\cdot$$

The bromine radical generated in Step 2 goes on to react with another molecule of the alkene in another Step 1, continuing the chain.

Let's consider the individual steps. In the initiation step, free radicals generated from the peroxide react with HBr to form bromine radicals.

$$\text{R—}\ddot{\text{O}}\cdot + \text{H—}\ddot{\text{Br}}: \longrightarrow \text{R—}\ddot{\text{O}}\text{—H} + :\ddot{\text{Br}}\cdot \qquad \Delta H° = -63 \text{ kJ } (-15 \text{ kcal})$$

The bromine radical lacks an octet of electrons in its valence shell, making it electron-deficient and electrophilic. It adds to a double bond, forming a new free radical with the odd electron on a carbon atom.

$$:\ddot{\text{Br}}\cdot + \text{C}{=}\text{C} \longrightarrow \text{—C(:Br:)—}\dot{\text{C}} \qquad \Delta H° = -12 \text{ kJ } (-3 \text{ kcal})$$

This free radical reacts with an HBr molecule to form a C—H bond and generate another bromine radical.

$$\text{—C(Br)—}\dot{\text{C}} + \text{H—}\ddot{\text{Br}}: \longrightarrow \text{—C(Br)—C(H)—} + :\ddot{\text{Br}}\cdot \qquad \Delta H° = -25 \text{ kJ } (-6 \text{ kcal})$$

The regenerated bromine radical reacts with another molecule of the alkene, continuing the chain reaction. Both of the propagation steps are moderately exothermic, allowing them to proceed faster than the termination steps. Note that each propagation step starts with one free radical and ends with another free radical. The number of free radicals is constant, until the reactants are consumed, and free radicals come together and terminate the chain reaction.

Radical Addition of HBr to Unsymmetrical Alkenes Now we must explain the anti-Markovnikov orientation found in the products of the peroxide-catalyzed reaction. With an unsymmetrical alkene such as 2-methylbut-2-ene, adding the bromine radical to the secondary end of the double bond forms a tertiary radical.

$$CH_3-\overset{CH_3}{\overset{|}{C}}=CH-CH_3 + Br\cdot \longrightarrow CH_3-\overset{CH_3}{\overset{|}{\dot{C}}}-\underset{Br}{\underset{|}{CH}}-CH_3 \quad \text{but not} \quad CH_3-\overset{CH_3}{\overset{|}{\underset{Br}{\underset{|}{C}}}}-\dot{C}H-CH_3$$

tertiary radical (more stable) — secondary radical (less stable)

As we saw in the protonation of an alkene, the electrophile (in this case, Br ·) adds to the less substituted end of the double bond, and the unpaired electron appears on the more substituted carbon to give the more stable free radical. This intermediate reacts with HBr to give the anti-Markovnikov product, in which H has added to the more substituted end of the double bond: the end that started with *fewer* hydrogens.

$$CH_3-\overset{CH_3}{\overset{|}{\dot{C}}}-\underset{Br}{\underset{|}{CH}}-CH_3 + H-Br \longrightarrow CH_3-\overset{CH_3}{\overset{|}{\underset{H}{\underset{|}{C}}}}-\underset{Br}{\underset{|}{CH}}-CH_3 + Br\cdot$$

anti-Markovnikov product

PROBLEM-SOLVING HINT

Stability of radicals:

3° > 2° > 1° > $\cdot CH_3$

A radical adds to a double bond to give the most stable radical in the intermediate.

Note that *both* mechanisms for the addition of HBr to an alkene (with and without peroxides) follow our extended statement of Markovnikov's rule: In both cases, the electrophile adds to the less substituted end of the double bond to give the more stable intermediate, either a carbocation or a free radical. In the ionic reaction, the electrophile is H^+. In the peroxide-catalyzed free-radical reaction, Br· is the electrophile.

Many students wonder why the reaction with Markovnikov orientation does not take place in the presence of peroxides, together with the free-radical chain reaction. It actually does take place, but the peroxide-catalyzed reaction is faster. If just a tiny bit of peroxide is present, a mixture of Markovnikov and anti-Markovnikov products results. If an appreciable amount of peroxide is present, the radical chain reaction is so much faster than the uncatalyzed ionic reaction that only the anti-Markovnikov product is observed.

The reversal of orientation in the presence of peroxides is called the **peroxide effect.** It occurs only with the addition of HBr to alkenes. The peroxide effect is not seen with HCl because the second step, the reaction of an alkyl radical with HCl, is strongly endothermic.

Proposed free-radical addition of HCl fails:

$$Cl-\overset{|}{\underset{|}{C}}-\dot{C}\!\!< + H-Cl \longrightarrow Cl-\overset{|}{\underset{|}{C}}-\overset{|}{\underset{|}{C}}-H + Cl\cdot \qquad \Delta H^\circ = +42\text{ kJ }(+10\text{ kcal})$$

Similarly, the peroxide effect is not observed with HI because the reaction of an iodine atom with an alkene is strongly endothermic. Only HBr has just the right reactivity for each step of the free-radical chain reaction to take place.

Proposed free-radical addition of HI fails:

$$I\cdot + >C=C< \longrightarrow I-\overset{|}{\underset{|}{C}}-\dot{C}\!\!< \qquad \Delta H^\circ = +54\text{ kJ }(+13\text{ kcal})$$

PROBLEM 8-3

Predict the major products of the following reactions, and propose mechanisms to support your predictions.

(a) 1-methylcyclopentene + HBr + $CH_3-\overset{\overset{\large O}{\|}}{C}-O-O-\overset{\overset{\large O}{\|}}{C}-CH_3$

(b) 1-phenylpropene + HBr + di-*tert*-butyl peroxide (phenyl = Ph = 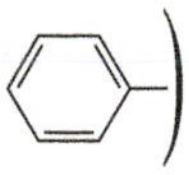)

PROBLEM-SOLVING HINT

Remember to write out complete structures, including all bonds and charges, when writing a mechanism or determining the course of a reaction.

SOLVED PROBLEM 8-1

Show how you would accomplish the following synthetic conversions.

(a) Convert 1-methylcyclohexene to 1-bromo-1-methylcyclohexane.

SOLUTION

This synthesis requires the addition of HBr to an alkene with Markovnikov orientation. Ionic addition of HBr gives the correct product.

1-methylcyclohexene + HBr ⟶ 1-bromo-1-methylcyclohexane

(b) Convert 1-methylcyclohexanol to 1-bromo-2-methylcyclohexane.

SOLUTION

This synthesis requires the conversion of an alcohol to an alkyl bromide with the bromine atom at the neighboring carbon atom. This is the anti-Markovnikov product, which could be formed by the radical-catalyzed addition of HBr to 1-methylcyclohexene.

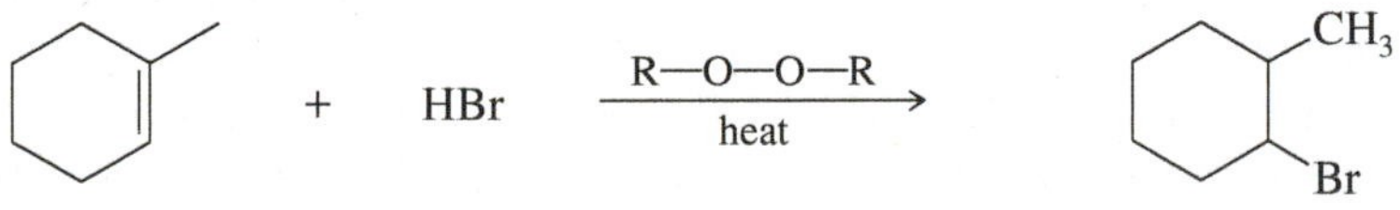

1-methylcyclohexene + HBr $\xrightarrow[\text{heat}]{R-O-O-R}$ 1-bromo-2-methylcyclohexane

1-Methylcyclohexene is easily synthesized by the dehydration of 1-methylcyclohexanol. The most substituted alkene is the desired product.

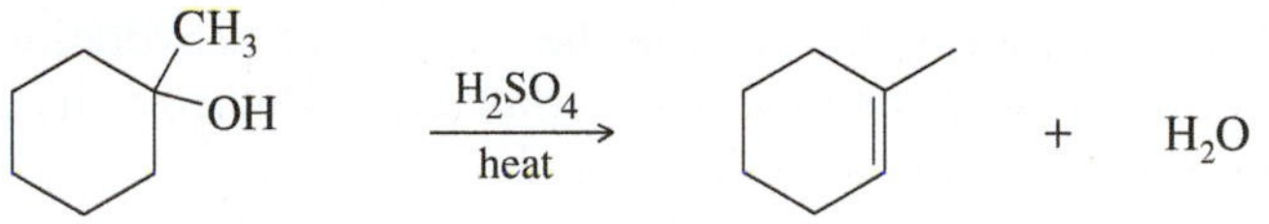

1-methylcyclohexanol $\xrightarrow[\text{heat}]{H_2SO_4}$ 1-methylcyclohexene + H_2O

The two-step synthesis is summarized as follows:

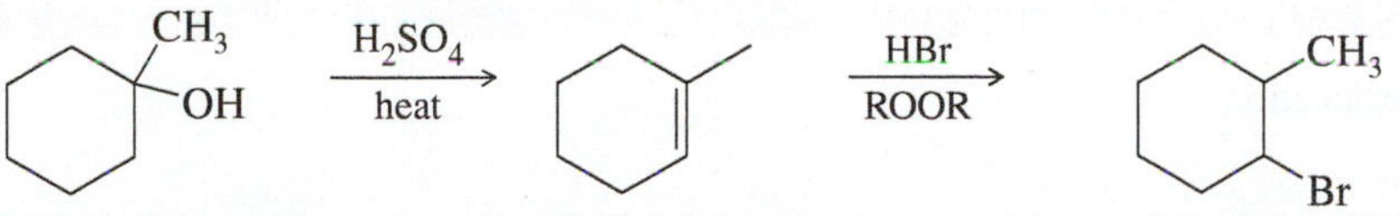

1-methylcyclohexanol $\xrightarrow[\text{heat}]{H_2SO_4}$ 1-methylcyclohexene $\xrightarrow[\text{ROOR}]{\text{HBr}}$ 1-bromo-2-methylcyclohexane

PROBLEM 8-4

Show how you would accomplish the following synthetic conversions.

(a) but-1-ene ⟶ 1-bromobutane

(b) but-1-ene ⟶ 2-bromobutane

(c) 2-methylcyclohexanol ⟶ 1-bromo-1-methylcyclohexane

(d) 2-methylbutan-2-ol ⟶ 2-bromo-3-methylbutane

8-4 Addition of Water: Hydration of Alkenes

An alkene may react with water in the presence of a strongly acidic catalyst to form an alcohol. Formally, this reaction is a **hydration** (the addition of water), with a hydrogen atom adding to one carbon and a hydroxy group adding to the other. Hydration of an alkene is the reverse of the dehydration of alcohols we studied in Section 7-18.

Hydration of an alkene

$$>C{=}C< \; + \; H_2O \;\underset{}{\overset{H^+}{\rightleftharpoons}}\; -\overset{H}{\underset{|}{\overset{|}{C}}}-\overset{OH}{\underset{|}{\overset{|}{C}}}-$$

alkene — alcohol (Markovnikov orientation)

Dehydration of an alcohol

$$-\overset{H}{\underset{|}{\overset{|}{C}}}-\overset{OH}{\underset{|}{\overset{|}{C}}}- \;\overset{H^+}{\rightleftharpoons}\; >C{=}C< \; + \; H_2O$$

alcohol — alkene

For dehydrating alcohols, a concentrated dehydrating acid (such as H_2SO_4 or H_3PO_4) is used to drive the equilibrium to favor the alkene. Hydration of an alkene, on the other hand, is accomplished by adding excess water to drive the equilibrium toward the alcohol.

8-4A Mechanism of Hydration

The *Principle of Microscopic Reversibility* states that a forward reaction and a reverse reaction taking place under the same conditions (as in an equilibrium) must follow the same reaction pathway in microscopic detail. The hydration and dehydration reactions are the two complementary reactions in an equilibrium; therefore, they must follow the same reaction pathway. It makes sense that the lowest-energy transition states and intermediates for the reverse reaction are the same as those for the forward reaction, except in reverse order.

According to the Principle of Microscopic Reversibility, we can write the hydration mechanism by reversing the order of the steps of the dehydration (Section 7-18). Protonation of the double bond forms a carbocation. Nucleophilic attack by water, followed by loss of a proton, gives the alcohol.

MECHANISM 8-4 Acid-Catalyzed Hydration of an Alkene

Step 1: Protonation of the double bond forms a carbocation.

$$>C{=}C< \; + \; H-\overset{H}{\overset{|}{\ddot{O}^+}}-H \;\rightleftharpoons\; >\overset{+}{C}-\overset{H}{\underset{|}{\overset{|}{C}}}- \; + \; H_2\ddot{O}:$$

+ on the more substituted carbon

Step 2: Nucleophilic attack by water gives a protonated alcohol.

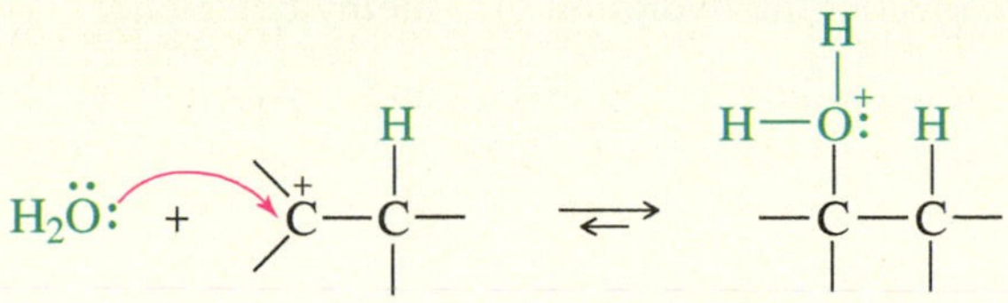

Step 3: Deprotonation gives the alcohol.

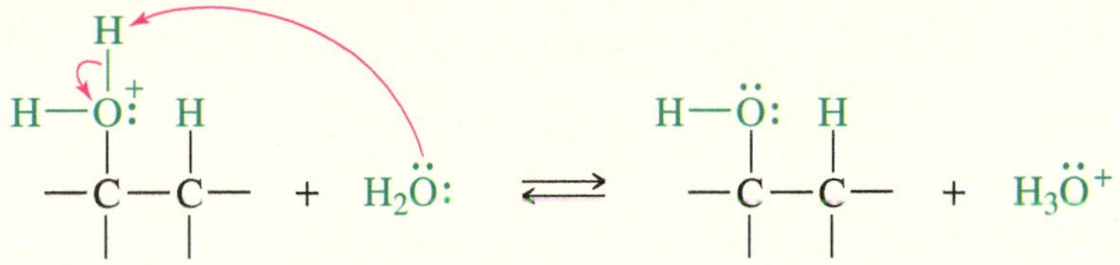

EXAMPLE: Acid-catalyzed hydration of propene.

Step 1: Protonation of the double bond forms a secondary carbocation.

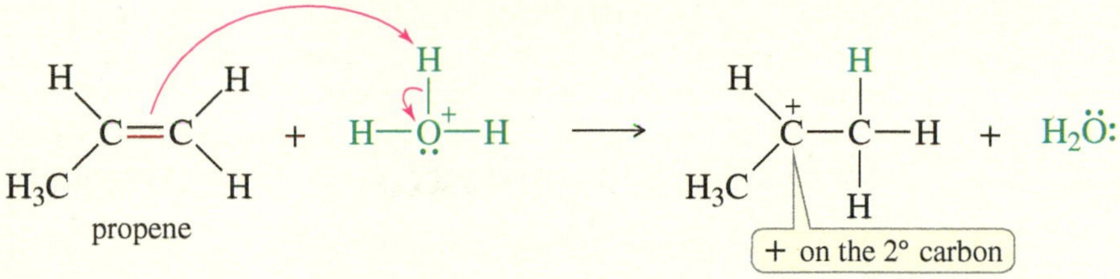

propene

+ on the 2° carbon

Step 2: Nucleophilic attack by water gives a protonated alcohol.

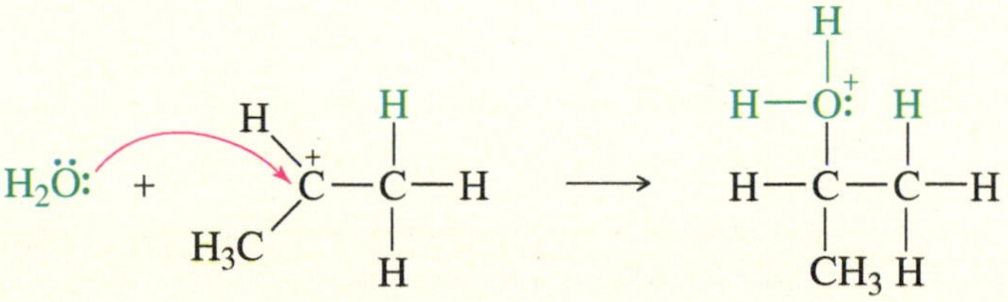

Step 3: Deprotonation gives the alcohol.

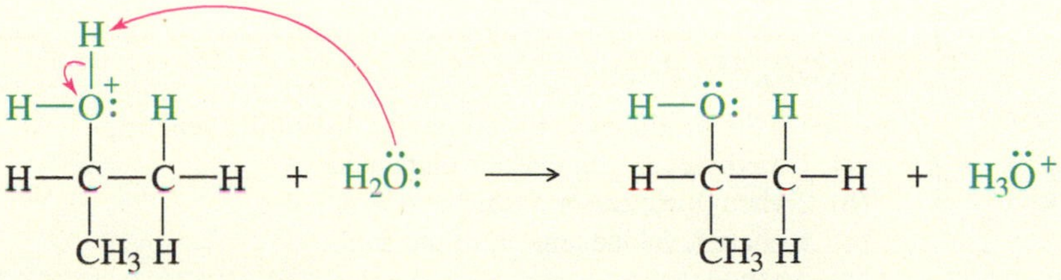

propan-2-ol

8-4B Orientation of Hydration

Step 1 of the hydration mechanism is similar to the first step in the addition of HBr. The proton adds to the *less* substituted end of the double bond to form the *more* substituted carbocation. Water attacks the carbocation to give (after loss of a proton) the alcohol with the —OH group on the more substituted carbon. Like the addition of hydrogen halides, hydration is *regioselective*: It follows Markovnikov's rule, giving a product in which the new hydrogen has added to the less substituted end of the double bond. Consider the hydration of 2-methylbut-2-ene:

$$CH_3-\overset{\overset{CH_3}{|}}{C}=CH-CH_3 + H-\overset{\overset{H}{|}}{\underset{..}{O^+}}-H \rightleftharpoons CH_3-\overset{\overset{CH_3}{|}}{\underset{+}{C}}-\underset{\underset{H}{|}}{CH}-CH_3 \text{ but not } CH_3-\overset{\overset{CH_3}{|}}{\underset{\underset{H}{|}}{C}}-\underset{+}{CH}-CH_3$$

3°, more stable 2°, less stable

The proton adds to the less substituted end of the double bond, so the positive charge appears at the more substituted end. Water attacks the carbocation to give the protonated alcohol.

$$CH_3-\overset{\overset{CH_3}{|}}{\underset{+}{C}}-\underset{\underset{H}{|}}{CH}-CH_3 \;\; H_2\ddot{O}: \rightleftharpoons CH_3-\overset{\overset{CH_3}{|}}{\underset{\underset{^{+}O:(H)(H)}{|}}{C}}-\underset{\underset{H}{|}}{CH}-CH_3 \;\; H_2\ddot{O}: \rightleftharpoons CH_3-\overset{\overset{CH_3}{|}}{\underset{\underset{OH}{|}}{C}}-\underset{\underset{H}{|}}{CH}-CH_3 \;\; H_3\ddot{O}^+$$

The reaction follows Markovnikov's rule. The proton has added to the end of the double bond that already had more hydrogens (that is, the less substituted end), and the —OH group has added to the more substituted end.

Like other reactions that involve carbocation intermediates, hydration may take place with rearrangement. For example, when 3,3-dimethylbut-1-ene undergoes acid-catalyzed hydration, the major product results from rearrangement of the carbocation intermediate.

$$CH_3-\overset{\overset{CH_3}{|}}{\underset{\underset{CH_3}{|}}{C}}-CH=CH_2 \xrightarrow{50\%\ H_2SO_4} CH_3-\overset{\overset{CH_3}{|}}{\underset{\underset{OH}{|}}{C}}-\underset{\underset{CH_3}{|}}{CH}-CH_3$$

3,3-dimethylbut-1-ene 2,3-dimethylbutan-2-ol (major product)

PROBLEM 8-5

Propose a mechanism to show how 3,3-dimethylbut-1-ene reacts with dilute aqueous H_2SO_4 to give 2,3-dimethylbutan-2-ol and a small amount of 2,3-dimethylbut-2-ene.

PROBLEM-SOLVING HINT

When predicting products for electrophilic additions, first draw the structure of the carbocation (or other intermediate) that results from electrophilic attack.

PROBLEM 8-6

Predict the products of the following hydration reactions.

(a) 1-methylcyclopentene + dilute acid
(b) 2-phenylpropene + dilute acid
(c) 1-phenylcyclohexene + dilute acid

8-5 Hydration by Oxymercuration–Demercuration

Many alkenes do not easily undergo hydration in aqueous acid. Some alkenes are nearly insoluble in aqueous acid, and others undergo side reactions such as rearrangement, polymerization, or charring under these strongly acidic conditions. In some cases, the overall equilibrium favors the alkene rather than the alcohol. No amount of catalysis can cause a reaction to occur if the energetics are unfavorable.

Oxymercuration–demercuration is another method for converting alkenes to alcohols with Markovnikov orientation. Oxymercuration–demercuration works with many alkenes that do not easily undergo direct hydration, and it takes place under milder conditions. No free carbocation is formed, so there is no opportunity for rearrangements or polymerization.

Oxymercuration–Demercuration

$$>C{=}C< \; + \; \ddot{H}g(OAc)_2 \xrightarrow{H_2O} -\underset{HO}{C}-\underset{:HgOAc}{C}- \xrightarrow{NaBH_4} -\underset{HO}{C}-\underset{H}{C}-$$

(Markovnikov orientation)

The reagent for mercuration is mercuric acetate, $Hg(OCOCH_3)_2$, abbreviated $Hg(OAc)_2$. There are several theories as to how this reagent acts as an electrophile; the simplest one is that mercuric acetate dissociates slightly to form a positively charged mercury species, $^+Hg(OAc)$.

$$\underset{Hg(OAc)_2}{CH_3-\overset{O}{\overset{\|}{C}}-O-\ddot{H}g-O-\overset{O}{\overset{\|}{C}}-CH_3} \rightleftharpoons \underset{^+Hg(OAc)}{CH_3-\overset{O}{\overset{\|}{C}}-O-\ddot{H}g^+} + \underset{^-OAc}{CH_3-\overset{O}{\overset{\|}{C}}-O^-}$$

Oxymercuration involves an electrophilic attack on the double bond by the positively charged mercury species. The product is a *mercurinium ion*, an organometallic cation containing a three-membered ring. In the second step, water from the solvent attacks the mercurinium ion to give (after deprotonation) an organomercurial alcohol. A subsequent reaction is **demercuration,** to remove the mercury. Sodium borohydride ($NaBH_4$, a reducing agent) replaces the mercuric acetate fragment with a hydrogen atom.

Application: Toxicology

Mercury and its compounds were used for centuries as ingredients in antibacterial drugs, skin creams, and antiseptics. Mercury compounds are quite toxic, however. In the body, mercury combines with the thiol groups of critical enzymes, inactivating them. Mercury poisoning causes brain and kidney damage, often leading to death.

We cover these reactions using mercury-based reagents because they are particularly interesting and the concepts are useful. For example, oxymercuration–demercuration adds water to alkenes without the rearrangements that are common in the acid-catalyzed hydration. When possible, we prefer to use other, less toxic reagents.

MECHANISM 8-5 Oxymercuration of an Alkene

Step 1: Electrophilic attack forms a mercurinium ion.

$$>C{=}C< \; + \; ^+\ddot{H}g(OAc) \longrightarrow \text{mercurinium ion: } -C-C- \text{ bridged by } Hg^+\text{–OAc}$$

mercurinium ion

Step 2: Water opens the ring to give an organomercurial alcohol.

$$\text{mercurinium ion} + H_2\ddot{O}: \longrightarrow -\underset{H-\overset{+}{O}:-H}{C}-\overset{\ddot{H}g(OAc)}{C}- \xrightarrow{H_2\ddot{O}:} -\underset{:\ddot{O}H}{C}-\overset{\ddot{H}g(OAc)}{C}- + H_3O^+$$

organomercurial alcohol

Demercuration replaces the mercuric fragment with hydrogen to give the alcohol.

(CONTINUED)

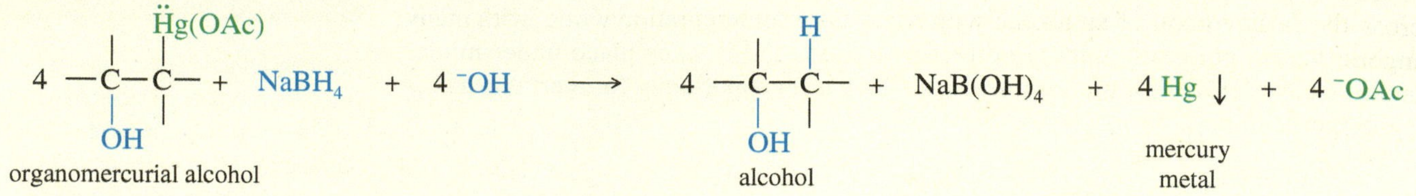

$$4\ \text{organomercurial alcohol (R–C(OH)–C–Hg(OAc))} + NaBH_4 + 4\ {}^{-}OH \longrightarrow 4\ \text{alcohol (R–C(OH)–C–H)} + NaB(OH)_4 + 4\ Hg\downarrow\ (\text{mercury metal}) + 4\ {}^{-}OAc$$

organomercurial alcohol — alcohol — mercury metal

EXAMPLE: Oxymercuration–demercuration of propene.

Step 1: Electrophilic attack forms a mercurinium ion.

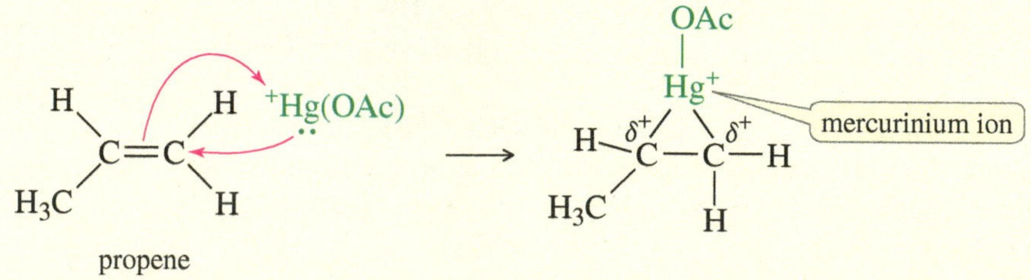

$$H_3C(H)C{=}CH_2 \ (\text{propene}) + {}^{+}Hg(OAc) \longrightarrow \text{mercurinium ion}$$

propene

mercurinium ion

Step 2: Water opens the ring to give an organomercurial alcohol.

$$\text{mercurinium ion} + H_2\ddot{O}: \longrightarrow H_3C{-}CH(\overset{+}{O}H_2){-}CH_2{-}Hg(OAc) \longrightarrow H_3C{-}CH(OH){-}CH_2{-}Hg(OAc) + H_3O^+$$

Water attacks the more substituted carbon.

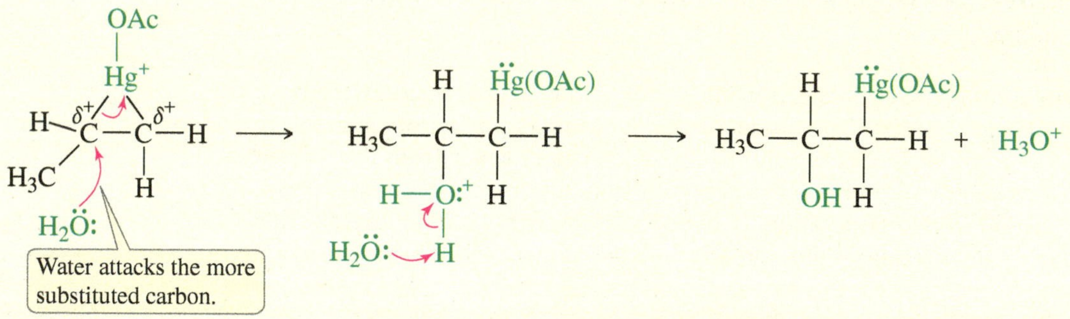

Demercuration replaces the mercuric fragment with hydrogen to give the alcohol.

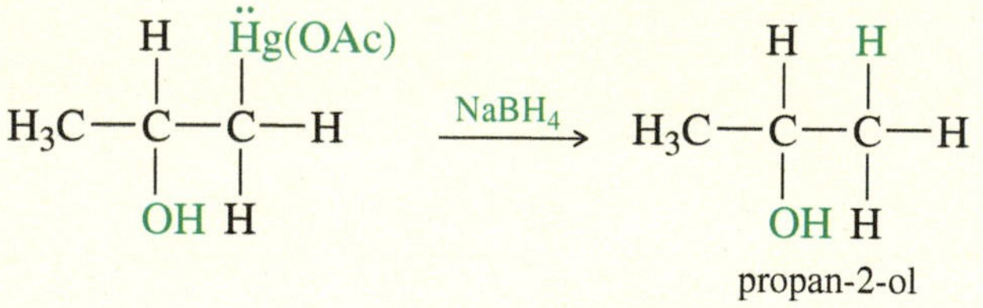

$$H_3C{-}CH(OH){-}CH_2{-}Hg(OAc) \xrightarrow{NaBH_4} H_3C{-}CH(OH){-}CH_2{-}H$$

propan-2-ol

Oxymercuration–demercuration of an unsymmetrical alkene generally gives Markovnikov orientation of addition, as shown by the oxymercuration of propene in the preceding example. The mercurinium ion has a considerable amount of positive charge on both of its carbon atoms, but there is more of a positive charge on the more substituted carbon atom, where it is more stable. Attack by water occurs on this more electrophilic carbon, giving Markovnikov orientation. The electrophile, ${}^{+}Hg(OAc)$, remains bonded to the less substituted end of the double bond. Reduction of the organomercurial alcohol gives the Markovnikov alcohol: propan-2-ol.

Similarly, oxymercuration–demercuration of 3,3-dimethylbut-1-ene gives the Markovnikov product, 3,3-dimethylbutan-2-ol, in excellent yield. Contrast this unrearranged product with the rearranged product formed in the acid-catalyzed hydration of the same alkene in Section 8-4B. Oxymercuration–demercuration reliably adds water across the double bond of an alkene with Markovnikov orientation and without rearrangement.

$$\underset{\text{3,3-dimethylbut-1-ene}}{(CH_3)_3C(H)C{=}CH_2} \xrightarrow[H_2O]{Hg(OAc)_2} \underset{\text{mercurinium ion}}{(CH_3)_3C{-}\overset{\delta+}{C}H{-}CH_2{-}\overset{\delta+}{Hg}(OAc)} \xrightarrow{-H^+} \underset{\text{Markovnikov product}}{(CH_3)_3C{-}CH(OH){-}CH_2{-}Hg(OAc)}$$

$$\underset{\text{Markovnikov product}}{(CH_3)_3C{-}CH(OH){-}CH_2{-}Hg(OAc)} \xrightarrow{NaBH_4} \underset{\substack{\text{3,3-dimethylbutan-2-ol}\\ \text{(94\% overall)}}}{(CH_3)_3C{-}CH(OH){-}CH_2H}$$

Of the methods we have seen for Markovnikov hydration of alkenes, oxymercuration–demercuration is most commonly used in the laboratory. It gives better yields than direct acid-catalyzed hydration, it avoids the possibility of rearrangements, and it does not involve harsh conditions. There are also disadvantages, however. Organomercurial compounds are highly toxic. They must be used with great care and then disposed of properly.

8-6 Alkoxymercuration–Demercuration

When mercuration takes place in an alcohol solvent, the alcohol serves as a nucleophile to attack the mercurinium ion. The resulting product contains an alkoxy (—O—R) group. In effect, **alkoxymercuration**–demercuration converts alkenes to ethers by adding an alcohol across the double bond of the alkene.

$$>C{=}C< \; + \; Hg(OAc)_2 \xrightarrow{ROH} -\underset{RO}{C}-\underset{HgOAc}{C}-$$

$$-\underset{RO}{C}-\underset{HgOAc}{C}- \xrightarrow{NaBH_4} \underset{\text{(Markovnikov orientation)}}{-\underset{RO}{C}-\underset{H}{C}-}$$

As we have seen, an alkene reacts to form a mercurinium ion that is attacked by the nucleophilic solvent. Attack by an alcohol solvent gives an organomercurial ether that can be reduced to the ether.

$$\text{mercurinium ion} + R{-}\ddot{O}H \longrightarrow -\underset{R-\ddot{O}^+-H}{C}-\underset{}{C}(Hg(OAc))- \;\; (+\; H{-}\ddot{O}{-}R) \longrightarrow \underset{\text{organomercurial ether}}{-\underset{R-\ddot{O}:}{C}-C(Hg(OAc))-} \xrightarrow{NaBH_4} \underset{\text{an ether}}{-\underset{R-O}{C}-\overset{H}{C}-}$$

The solvent attacks the mercurinium ion at the *more* substituted end of the double bond (where there is more δ^+ charge), giving Markovnikov orientation of addition. The Hg(OAc) group appears at the *less* substituted end of the double bond. Reduction gives the Markovnikov product, with hydrogen at the less substituted end of the double bond.

SOLVED PROBLEM 8-2

Show the intermediates and products that result from alkoxymercuration–demercuration of 1-methylcyclopentene, using methanol as the solvent.

SOLUTION

Mercuric acetate adds to 1-methylcyclopentene to give the cyclic mercurinium ion. This ion has a considerable amount of positive charge on the more substituted tertiary carbon atom. Methanol attacks this carbon from the opposite side, leading to ***anti*** **addition:** The reagents (HgOAc and OCH_3) have added to opposite faces of the double bond.

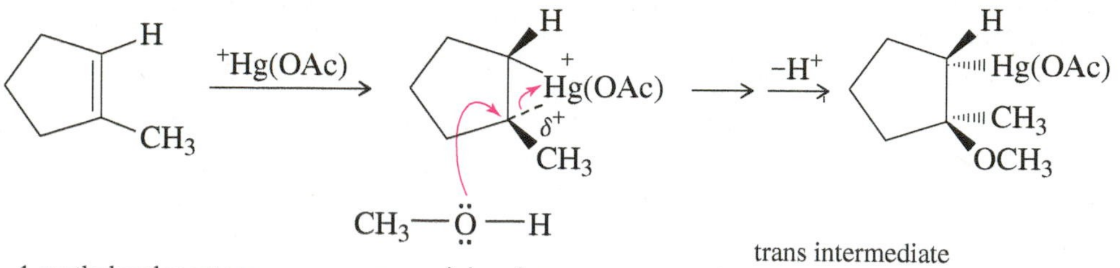

H, CH_3 — $\xrightarrow{^{+}Hg(OAc)}$ — H, $^{+}$Hg(OAc), δ^{+}, CH_3; $CH_3—\ddot{O}—H$ — $\longrightarrow \xrightarrow{-H^{+}}$ — H, Hg(OAc), CH_3, OCH_3

1-methylcyclopentene — mercurinium ion — trans intermediate (product of anti addition)

Reduction of the intermediate gives the Markovnikov product, 1-methoxy-1-methylcyclopentane.

H, Hg(OAc), CH_3, OCH_3 — $\xrightarrow{NaBH_4}$ — H, H, CH_3, OCH_3

intermediate — 1-methoxy-1-methylcyclopentane

PROBLEM 8-7

(a) Propose a mechanism for the following reaction.

$$CH_3—\underset{|}{\overset{CH_3}{\overset{|}{C}}}=CH—CH_3 \xrightarrow{Hg(OAc)_2,\ CH_3CH_2OH} CH_3—\underset{CH_3CH_2O}{\underset{|}{\overset{CH_3}{\overset{|}{C}}}}—\underset{Hg(OAc)}{\underset{|}{CH}}—CH_3$$

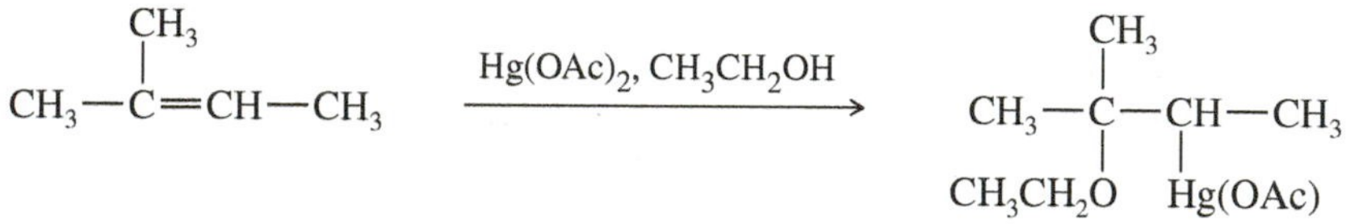

(90%)

(b) Give the structure of the product that results when this intermediate is reduced by sodium borohydride.

PROBLEM 8-8

Predict the major products of the following reactions.

(a) 1-methylcyclohexene + aqueous $Hg(OAc)_2$
(b) the product from part (a), treated with $NaBH_4$
(c) 4-chlorocycloheptene + $Hg(OAc)_2$ in CH_3OH
(d) the product from part (c), treated with $NaBH_4$

PROBLEM 8-9

Show how you would accomplish the following synthetic conversions.

(a) but-1-ene → 2-methoxybutane
(b) 1-iodo-2-methylcyclopentane → 1-methylcyclopentanol
(c) 3-methylpent-1-ene → 3-methylpentan-2-ol

Explain why acid-catalyzed hydration would be a poor choice for the reaction in (c).

8-7 Hydroboration of Alkenes

We have seen two methods for hydrating an alkene with Markovnikov orientation. What if we need to convert an alkene to the anti-Markovnikov alcohol? For example, the following transformation cannot be accomplished using the hydration procedures covered thus far.

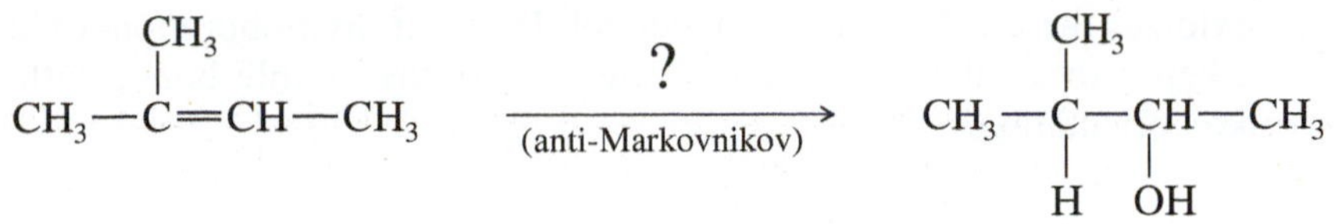

$$CH_3-\underset{\displaystyle CH_3}{\overset{|}{C}}=CH-CH_3 \xrightarrow[\text{(anti-Markovnikov)}]{?} CH_3-\underset{H}{C}(CH_3)-\underset{OH}{CH}-CH_3$$

2-methylbut-2-ene → 3-methylbutan-2-ol

Such an anti-Markovnikov hydration was impossible until H. C. Brown, of Purdue University, discovered that diborane (B_2H_6) adds to alkenes with anti-Markovnikov orientation to form alkylboranes, which can be oxidized to give anti-Markovnikov alcohols. This discovery led to the development of a large field of borane chemistry, and Brown received the Nobel Prize in Chemistry in 1979.

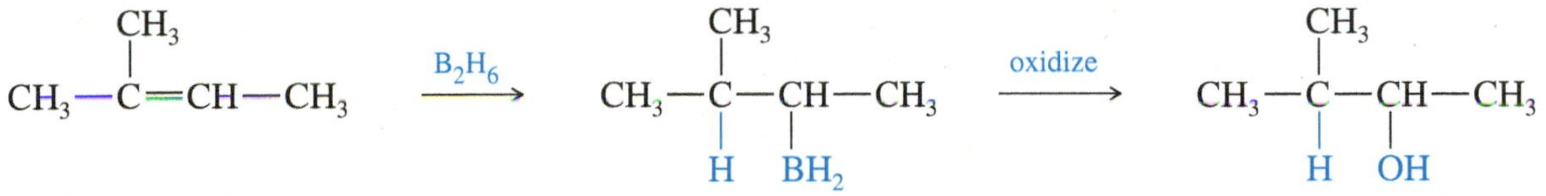

$$CH_3-C(CH_3)=CH-CH_3 \xrightarrow{B_2H_6} CH_3-\underset{H}{C}(CH_3)-\underset{BH_2}{CH}-CH_3 \xrightarrow{\text{oxidize}} CH_3-\underset{H}{C}(CH_3)-\underset{OH}{CH}-CH_3$$

2-methylbut-2-ene; an alkylborane; 3-methylbutan-2-ol (>90%)

Diborane (B_2H_6) is a dimer composed of two molecules of borane (BH_3). The bonding in diborane is unconventional because the boron atom in borane has only six electrons around it, and it is a powerful Lewis acid. Its tendency to acquire an additional electron pair leads it to share two hydrogens with the other borane molecule in diborane. We can depict this structure using two resonance forms showing two bridging hydrogens that are shared by the boron atoms.

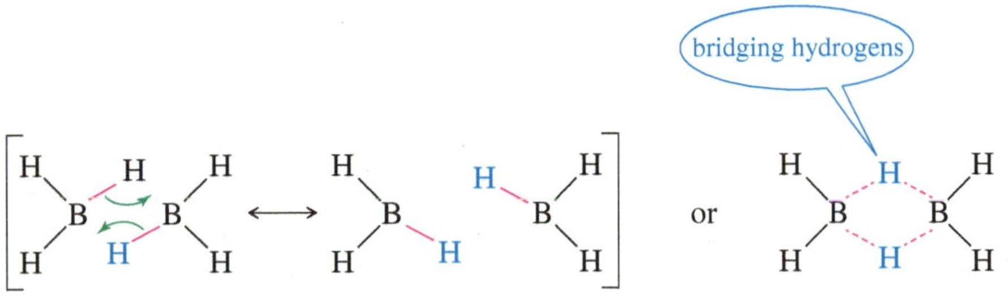

[Resonance forms of diborane: $H_2B(\mu\text{-}H)BH_2 \leftrightarrow$ …] or [bridged structure]

bridging hydrogens

diborane diborane

Diborane is in equilibrium with a very small amount of borane (BH_3), a highly reactive Lewis acid with only six electrons.

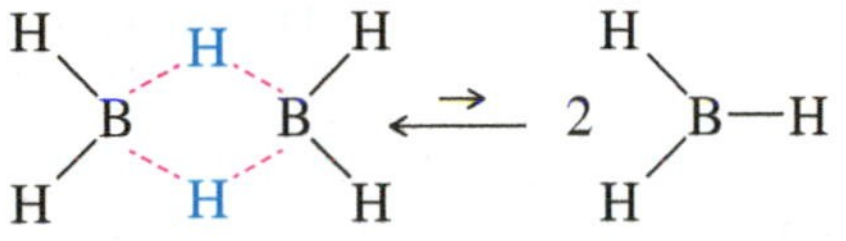

$$B_2H_6 \rightleftarrows 2\ BH_3$$

diborane borane

Diborane is an inconvenient reagent. It is a toxic, flammable, and explosive gas. It is more easily used as a complex with tetrahydrofuran (THF), a cyclic ether. This complex reacts like diborane, yet the solution is easily measured and transferred.

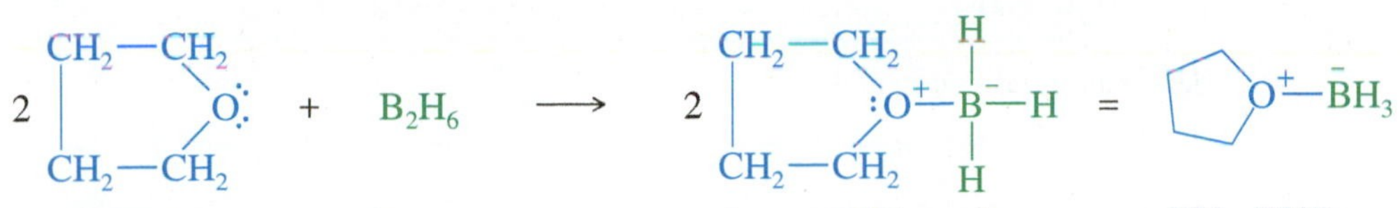

$$2\ (CH_2)_4O: \; + \; B_2H_6 \longrightarrow 2\ (CH_2)_4\overset{+}{O}-\overset{-}{B}H_3 = BH_3\cdot THF$$

tetrahydrofuran (THF); diborane; borane–THF complex = $BH_3 \cdot THF$

The $BH_3 \cdot THF$ reagent is the form of borane commonly used in organic reactions. BH_3 adds to the double bond of an alkene to give an alkylborane. Basic hydrogen peroxide oxidizes the alkylborane to an alcohol. In effect, hydroboration–oxidation converts alkenes to alcohols by adding water across the double bond, with anti-Markovnikov orientation.

Hydroboration–oxidation:

$$\text{>C=C<} + BH_3 \cdot THF \longrightarrow \text{—C(H)—C(BH}_2\text{)—} \xrightarrow{H_2O_2,\ ^-OH} \text{—C(H)—C(OH)—}$$

anti-Markovnikov orientation
(syn stereochemistry)

8-7A Mechanism of Hydroboration

Borane is an electron-deficient compound. It has only six valence electrons, so the boron atom in BH_3 cannot have an octet. Acquiring an octet is the driving force for the unusual bonding structures (bridging hydrogens, for example) found in boron compounds. As an electron-deficient compound, BH_3 is a strong electrophile, capable of adding to a double bond. This **hydroboration** of the double bond is thought to occur in one step, with the boron atom adding to the less substituted end of the double bond, as shown in Mechanism 8-6.

In the transition state, the electrophilic boron atom withdraws electrons from the pi bond, and the carbon at the other end of the double bond acquires a partial positive charge. This partial charge is more stable on the more substituted carbon atom. The product shows boron bonded to the less substituted end of the double bond and hydrogen bonded to the more substituted end. Also, steric hindrance favors boron adding to the less hindered, less substituted end of the double bond.

MECHANISM 8-6 Hydroboration of an Alkene

Borane adds to the double bond in a single step. Boron adds to the less hindered, less substituted carbon, and hydrogen adds to the more substituted carbon.

$$(CH_3)_2C{=}CH(CH_3) + H{-}BH_2 \nearrow CH_3{-}\overset{\delta+}{C}(CH_3)\text{=-=}CH{-}CH_3 \ [\text{H---}\underset{\delta-}{BH_2}] \longrightarrow CH_3{-}C(CH_3)(H){-}CH(BH_2){-}CH_3$$

more stable transition state

anti-Markovnikov product

$$\not\searrow \ CH_3{-}C(CH_3)\text{=-=}\overset{\delta+}{C}H{-}CH_3 \ [\underset{\delta-}{H_2B}\text{---H}]$$

less stable transition state

The boron atom is removed by oxidation, using aqueous sodium hydroxide and hydrogen peroxide (HOOH or H_2O_2) to replace the boron atom with a hydroxy (—OH) group. The oxidation does not affect the orientation of the product, because the anti-Markovnikov orientation was established in the first step, the addition of BH_3.

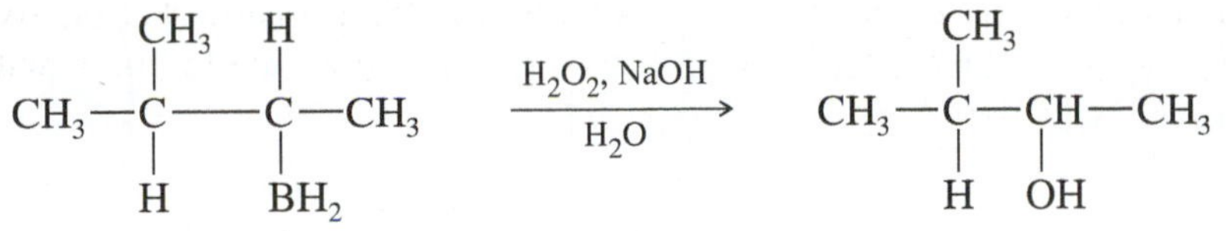

$$CH_3-\underset{H}{\overset{CH_3}{C}}-\underset{BH_2}{\overset{H}{C}}-CH_3 \xrightarrow[H_2O]{H_2O_2,\ NaOH} CH_3-\underset{H}{\overset{CH_3}{C}}-\underset{OH}{CH}-CH_3$$

This hydration of an alkene by hydroboration–oxidation is another example of a reaction that does not follow the original statement of Markovnikov's rule (the product is anti-Markovnikov), but still follows our understanding of the reasoning behind Markovnikov's rule. The electrophilic boron atom adds to the *less* substituted end of the double bond, placing the positive charge (and the hydrogen atom) at the more substituted end.

SOLVED PROBLEM 8-3

Show how you would convert 1-methylcyclopentanol to 2-methylcyclopentanol.

SOLUTION

Working backward, use hydroboration–oxidation to form 2-methylcyclopentanol from 1-methylcyclopentene. The use of (1) and (2) above and below the reaction arrow indicates individual steps in a two-step sequence.

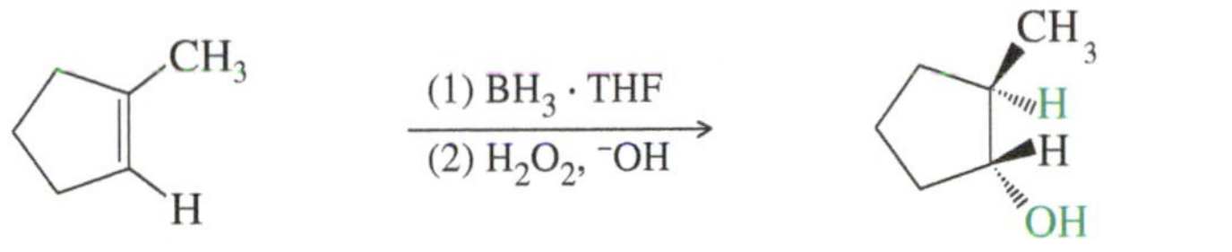

1-methylcyclopentene $\xrightarrow[(2)\ H_2O_2,\ ^-OH]{(1)\ BH_3 \cdot THF}$ *trans*-2-methylcyclopentanol

The 2-methylcyclopentanol that results from this synthesis is the pure trans isomer. This stereochemical result is discussed in Section 8-7C.

1-Methylcyclopentene is the most substituted alkene that results from dehydration of 1-methylcyclopentanol. Dehydration of the alcohol would give the correct alkene.

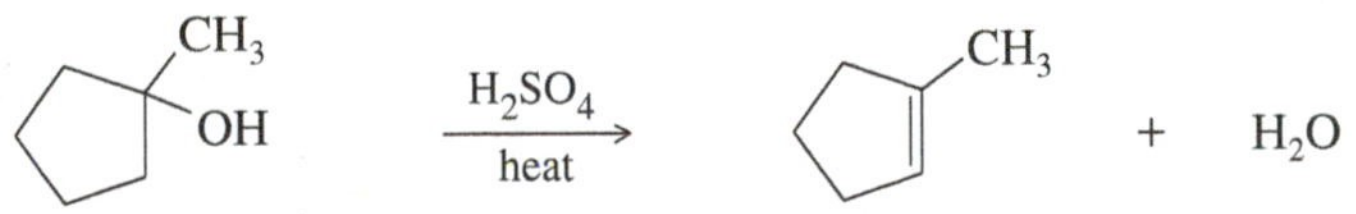

1-methylcyclopentanol $\xrightarrow[heat]{H_2SO_4}$ 1-methylcyclopentene + H_2O

PROBLEM-SOLVING HINT

Work backward on multistep syntheses.

PROBLEM 8-10

Predict the major products of the following reactions.

(a) propene + $BH_3 \cdot THF$
(b) the product from part (a) + H_2O_2/OH^-
(c) 2-methylpent-2-ene + $BH_3 \cdot THF$
(d) the product from part (c) + H_2O_2/OH^-
(e) 1-methylcyclohexene + $BH_3 \cdot THF$
(f) the product from part (e) + H_2O_2/OH^-

PROBLEM 8-11

Show how you would accomplish the following synthetic conversions.

(a) but-1-ene → butan-1-ol
(b) but-1-ene → butan-2-ol
(c) 2-bromo-2,4-dimethylpentane → 2,4-dimethylpentan-3-ol

8-7B Stoichiometry of Hydroboration

For simplicity, we have neglected the fact that 3 moles of an alkene can react with each mole of BH_3. Each B—H bond in BH_3 can add across the double bond of an alkene. The first addition forms an alkylborane, the second a dialkylborane, and the third a trialkylborane.

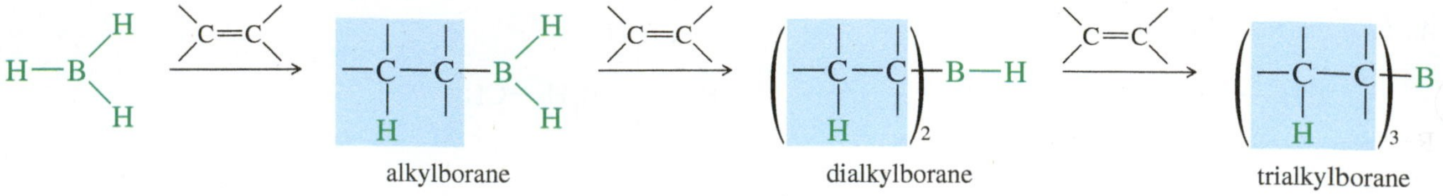

alkylborane dialkylborane trialkylborane

Summary

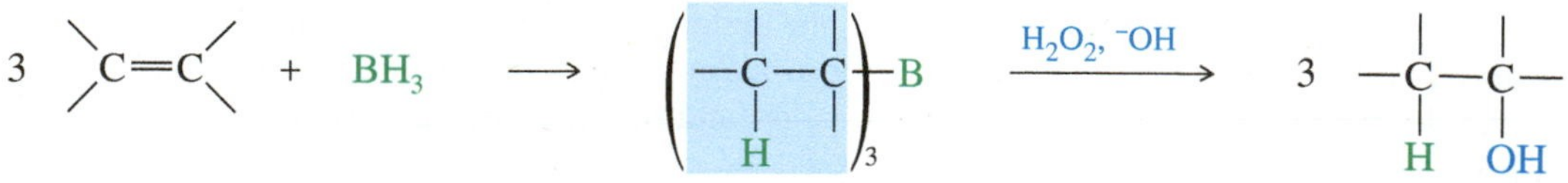

$$3\ C{=}C + BH_3 \longrightarrow (-C-C-)_3B \xrightarrow{H_2O_2,\ ^-OH} 3\ -C(H)-C(OH)-$$

Trialkylboranes react exactly as we have discussed, and they oxidize to give anti-Markovnikov alcohols. Trialkylboranes are quite bulky, further reinforcing the preference for boron to add to the less hindered carbon atom of the double bond. Boranes are often drawn as the 1:1 monoalkylboranes to simplify their structure and emphasize the organic part of the molecule.

8-7C Stereochemistry of Hydroboration

The simultaneous addition of boron and hydrogen to the double bond (as shown in Mechanism 8-6) leads to a **syn addition:** Boron and hydrogen add across the double bond on the *same side* of the molecule. (If they added to opposite sides of the molecule, the process would be an **anti addition.**)

The stereochemistry of the hydroboration–oxidation of 1-methylcyclopentene is that boron and hydrogen add to the same face of the double bond (syn) to form a trialkylborane. Oxidation of the trialkylborane replaces boron with a hydroxy group in the same stereochemical position. The product is *trans*-2-methylcyclopentanol. A racemic mixture is expected because a chiral product is formed from achiral reagents.

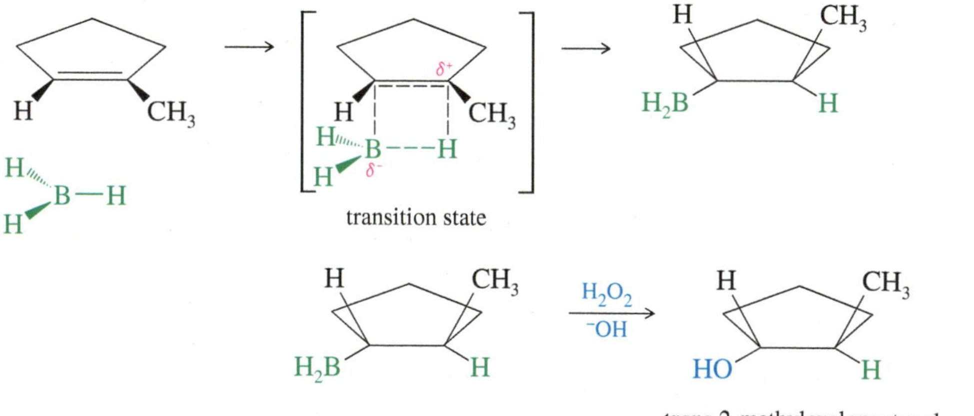

transition state

H_2O_2, ^-OH

trans-2-methylcyclopentanol
(85% overall)
(racemic mixture of enantiomers)

The second step (oxidation of the borane to the alcohol) takes place with retention of configuration. Hydroperoxide ion adds to the borane, causing the alkyl group to migrate from boron to oxygen. The alkyl group migrates with retention of configuration because it moves with its electron pair and does not alter the tetrahedral structure of the migrating carbon atom. Hydrolysis of the borate ester gives the alcohol.

Formation of hydroperoxide ion

$$H-\ddot{O}-\ddot{O}-H + :\ddot{O}H^- \rightleftharpoons H-\ddot{O}-\ddot{O}:^- + H_2\ddot{O}:$$

Addition of hydroperoxide and migration of the alkyl group

$$R_3B + {}^{-}\ddot{O}{-}\ddot{O}{-}H \longrightarrow R_3\bar{B}{-}\ddot{O}{-}\ddot{O}{-}H \longrightarrow R_2B{-}\ddot{O}{-}R + {}^{-}\ddot{O}H$$

hydroperoxide R migrates borate ester

Twice more to oxidize the other two alkyl groups

$$R_2B{-}O{-}R \xrightarrow{{}^{-}OOH} \xrightarrow{{}^{-}OOH} B(OR)_3$$

trialkyl borate ester

Hydrolysis of the borate ester

$$B(OR)_3 + {}^{-}OH \rightleftarrows (RO)_3\bar{B}{-}OH \rightleftarrows (RO)_2B{-}OH + {}^{-}OR \xrightarrow{H_2O} (RO)_2B{-}OH + R{-}OH + {}^{-}OH$$

(The other two OR groups hydrolyze similarly.)

Hydroboration of alkenes is another example of a **stereospecific reaction,** in which different stereoisomers of the starting compound react to give different stereoisomers of the product. Problem 8-14 considers the different products formed by the hydroboration–oxidation of two acyclic diastereomers.

SOLVED PROBLEM 8-4

A norbornene molecule labeled with deuterium is subjected to hydroboration–oxidation. Give the structures of the intermediates and products.

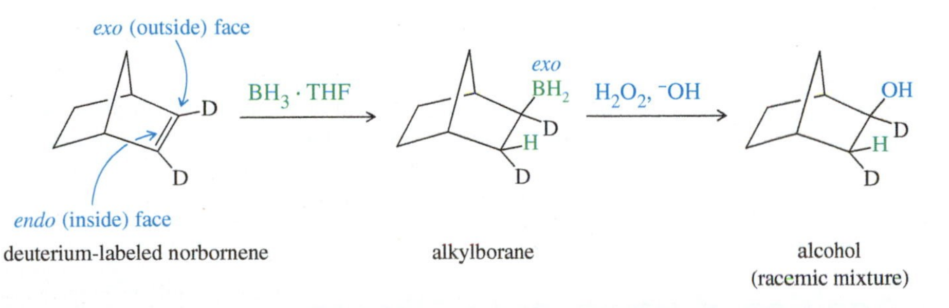

deuterium-labeled norbornene alkylborane alcohol (racemic mixture)

SOLUTION

The syn addition of BH_3 across the double bond of norbornene takes place mostly from the more accessible outside (exo) face of the double bond. Oxidation gives a product with both the hydrogen atom and the hydroxy group in exo positions. (The less accessible inner face of the double bond is called the endo face.)

PROBLEM 8-12

In the hydroboration of 1-methylcyclopentene shown in Solved Problem 8-3, the reagents are achiral, and the products are chiral. The product is a racemic mixture of *trans*-2-methylcyclopentanol, but only one enantiomer is shown. Show how the other enantiomer is formed.

PROBLEM 8-13

Predict the major products of the following reactions. Include stereochemistry where applicable.

(a) 1-methylcycloheptene + $BH_3 \cdot THF$, then H_2O_2, OH^-

(b) *trans*-4,4-dimethylpent-2-ene + $BH_3 \cdot THF$, then H_2O_2, OH^-

(c)

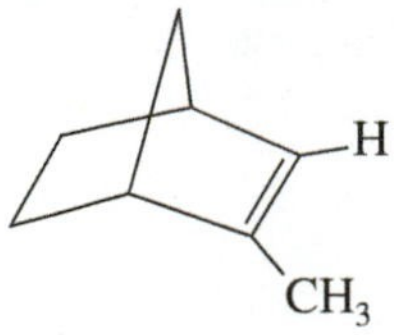

+ $BH_3 \cdot THF$, then H_2O_2, OH^-

PROBLEM 8-14

(a) When (*Z*)-3-methylhex-3-ene undergoes hydroboration–oxidation, two isomeric products are formed. Give their structures, and label each asymmetric carbon atom as (*R*) or (*S*). What is the relationship between these isomers?

(b) Repeat part (a) for (*E*)-3-methylhex-3-ene. What is the relationship between the products formed from (*Z*)-3-methylhex-3-ene and those formed from (*E*)-3-methylhex-3-ene?

PROBLEM 8-15

Show how you would accomplish the following transformations.

(a)

(b)

(c) 1-methylcycloheptanol ⟶ 2-methylcycloheptanol

PROBLEM 8-16

(a) When HBr adds across the double bond of 1,2-dimethylcyclopentene, the product is a mixture of the cis and trans isomers. Show why this addition is not stereospecific.

(b) When 1,2-dimethylcyclopentene undergoes hydroboration–oxidation, one diastereomer of the product predominates. Show why this addition is stereospecific, and predict the stereochemistry of the major product.

8-8 Addition of Halogens to Alkenes

Halogens add to alkenes to form vicinal dihalides.

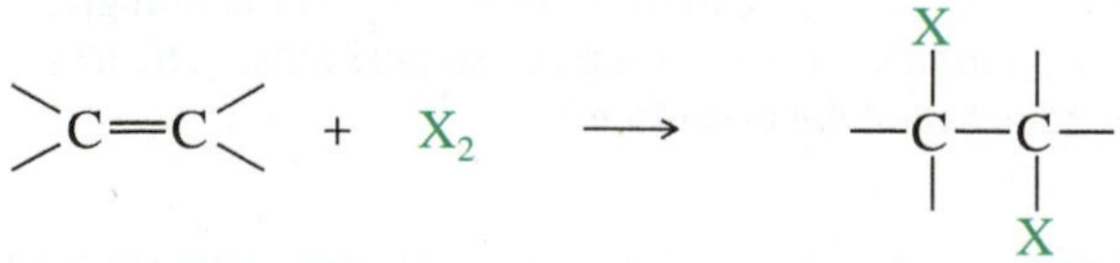

(X_2 = Cl_2, Br_2, sometimes I_2)

usually anti addition

8-8A Mechanism of Halogen Addition

A halogen molecule (Br_2, Cl_2, or I_2) is electrophilic; a nucleophile can react with a halogen, displacing a halide ion:

$$\text{Nuc:}^- + :\ddot{\text{Br}}-\ddot{\text{Br}}: \longrightarrow \text{Nuc}-\ddot{\text{Br}}: + :\ddot{\text{Br}}:^-$$

In this example, the nucleophile attacks the electrophilic nucleus of one bromine atom, and the other bromine serves as the leaving group, departing as bromide ion. Many reactions fit this general pattern; for example:

$$\text{H}\ddot{\text{O}}:^- + :\ddot{\text{Br}}-\ddot{\text{Br}}: \longrightarrow \text{HO}-\ddot{\text{Br}}: + :\ddot{\text{Br}}:^-$$

$$\text{H}_3\text{N}: + :\ddot{\text{Cl}}-\ddot{\text{Cl}}: \longrightarrow \text{H}_3\overset{+}{\text{N}}-\ddot{\text{Cl}}: + :\ddot{\text{Cl}}:^-$$

$$>\text{C}=\text{C}< + :\ddot{\text{Br}}-\ddot{\text{Br}}: \longrightarrow \text{bromonium ion} + :\ddot{\text{Br}}:^-$$

bromonium ion

In the last reaction, the pi electrons of an alkene attack the bromine molecule, expelling bromide ion. A **bromonium ion** results, containing a three-membered ring with a positive charge on the bromine atom. This bromonium ion is similar in structure to the mercurinium ion discussed in Section 8-5. Similar reactions with other halogens form other **halonium ions,** including *chloronium ions* and *iodonium ions.*

Examples

chloronium ion   bromonium ion   iodonium ion

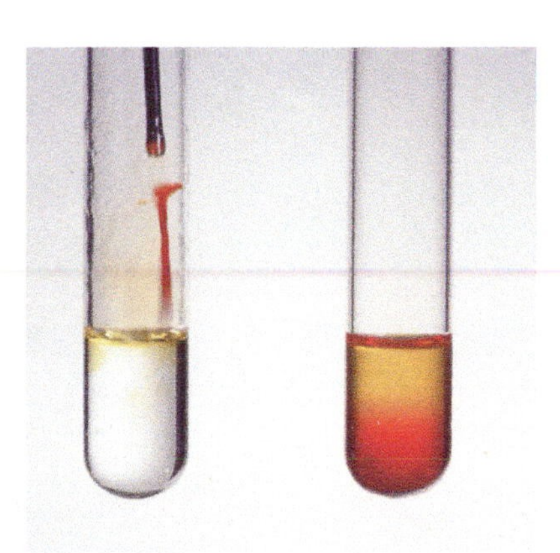

When a solution of bromine (red-brown) is added to cyclohexene, the bromine color quickly disappears because bromine adds across the double bond. When bromine is added to cyclohexane (at right), the color persists because no reaction occurs.

Unlike a normal carbocation, all the atoms in a halonium ion have filled octets. The three-membered ring has considerable ring strain, however, which, combined with a positive charge on an electronegative halogen atom, makes the halonium ion strongly electrophilic. Attack by a nucleophile, such as a halide ion, opens the halonium ion to give a stable product.

MECHANISM 8-7 Addition of Halogens to Alkenes

Step 1: Electrophilic attack forms a halonium ion.

$$>\text{C}=\text{C}< + :\ddot{\text{X}}-\ddot{\text{X}}: \longrightarrow \text{halonium ion} + :\ddot{\text{X}}:^-$$

halonium ion

Step 2: The halide ion opens the halonium ion.

X^- attacks from the back side

(CONTINUED)

EXAMPLE: Addition of Br_2 to propene.

Step 1: Electrophilic attack forms a bromonium ion.

propene → bromonium ion + $:\ddot{Br}:^-$

Step 2: Bromide ion opens the bromonium ion.

→ 1,2-dibromopropane

Chlorine and bromine commonly add to alkenes by the halonium ion mechanism. Iodination is used less frequently because diiodide products decompose easily. Any solvents used must be inert to the halogens; methylene chloride (CH_2Cl_2), chloroform ($CHCl_3$), and carbon tetrachloride (CCl_4) are the most frequent choices.

The addition of bromine has been used as a simple chemical test for the presence of olefinic double bonds. A solution of bromine in carbon tetrachloride is a clear, deep red color. When this red solution is added to an alkene, the red bromine color disappears (we say it is "decolorized"), and the solution becomes clear and colorless. (Although there are other functional groups that decolorize bromine, few do it as quickly as alkenes.)

8-8B Stereochemistry of Halogen Addition

The addition of bromine to cyclopentene is a stereospecific **anti addition.**

cyclopentene $\xrightarrow{Br_2}$ *trans*-1,2-dibromocyclopentane (92%) (both enantiomers) but not *cis*-1,2-dibromocyclopentane (not formed)

Anti stereochemistry results from the bromonium ion mechanism. When a nucleophile attacks a halonium ion, it must do so from the back side, in a manner similar to the S_N2 displacement. This back-side attack assures anti stereochemistry of addition.

→ → trans + enantiomer

Halogen addition is another example of a stereospecific reaction, in which different stereoisomers of the starting material give different stereoisomers of the product. Figure 8-5 shows additional examples of the anti addition of halogens to alkenes.

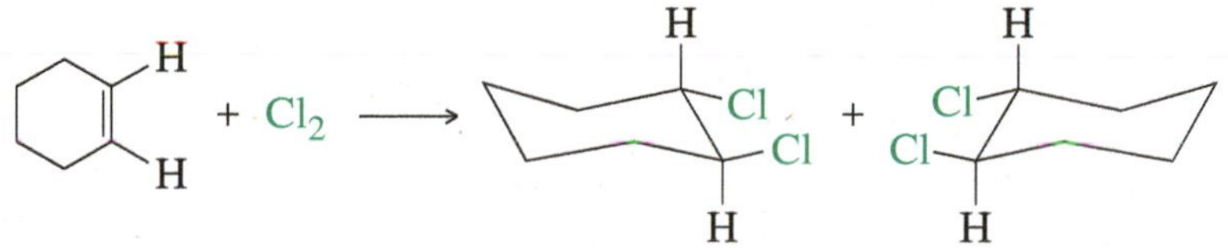

cyclohexene + Cl_2 ⟶ racemic *trans*-1,2-dichlorocyclohexane

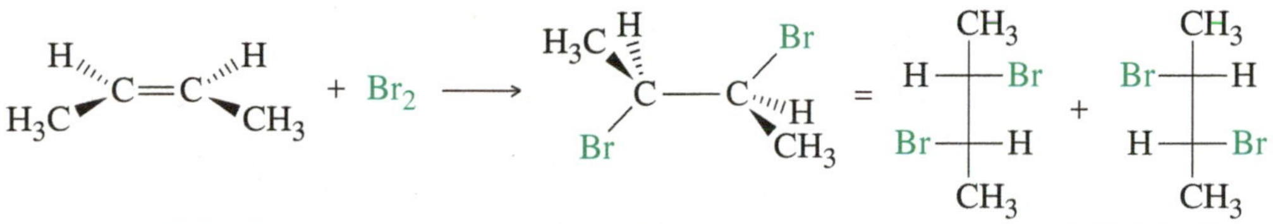

cis-but-2-ene + Br_2 ⟶ (both enantiomers) = (±)-2,3-dibromobutane

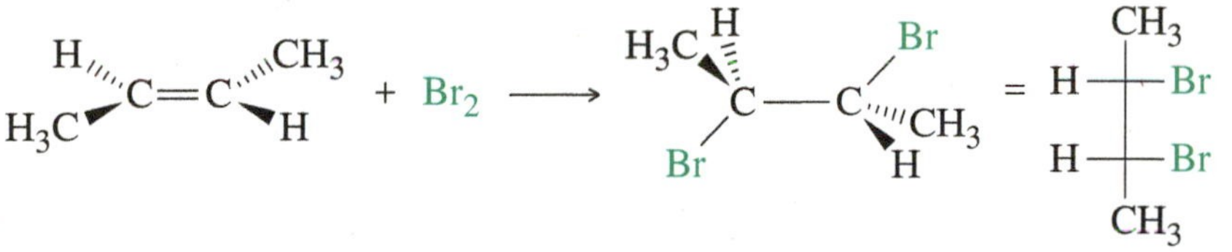

trans-but-2-ene + Br_2 ⟶ = *meso*-2,3-dibromobutane

FIGURE 8-5
Examples of the anti addition of halogens to alkenes. The stereospecific anti addition gives predictable stereoisomers of the products.

PROBLEM 8-17

Give mechanisms to account for the stereochemistry of the products observed from the addition of bromine to *cis*- and *trans*-but-2-ene (Figure 8-5). Why are two products formed from the cis isomer but only one from the trans? (Making models will be helpful.)

PROBLEM 8-18

Propose mechanisms and predict the major products of the following reactions. Include stereochemistry where appropriate.

(a) cycloheptene + Br_2 in CH_2Cl_2

(b) 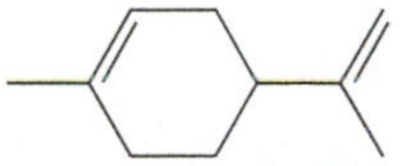 + 2 Cl_2 in CCl_4

(c) (*E*)-dec-3-ene + Br_2 in CCl_4

(d) (*Z*)-dec-3-ene + Br_2 in CCl_4

PROBLEM-SOLVING HINT

Models may be helpful whenever stereochemistry is involved. Write complete structures, including all bonds and charges, when writing mechanisms.

8-9 Formation of Halohydrins

A **halohydrin** is an alcohol with a halogen on the adjacent carbon atom. In the presence of water, halogens add to alkenes to form halohydrins. The electrophilic halogen adds to the alkene to give a halonium ion, which is also electrophilic. Water acts as a nucleophile to open the halonium ion and form the halohydrin.

MECHANISM 8-8 Formation of Halohydrins

Step 1: Electrophilic attack forms a halonium ion.

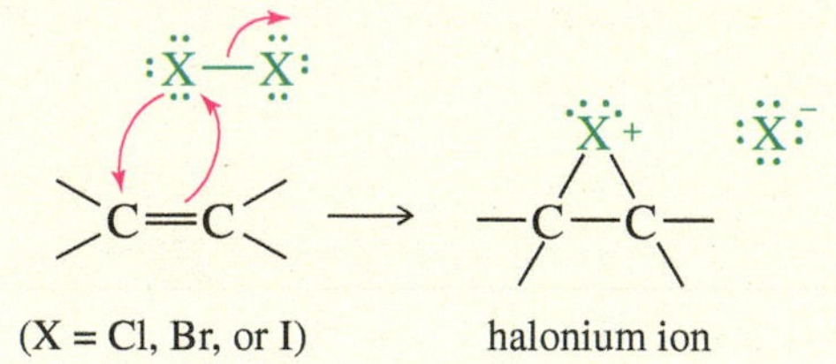

(X = Cl, Br, or I)

halonium ion

Step 2: Water opens the halonium ion; deprotonation gives the halohydrin.

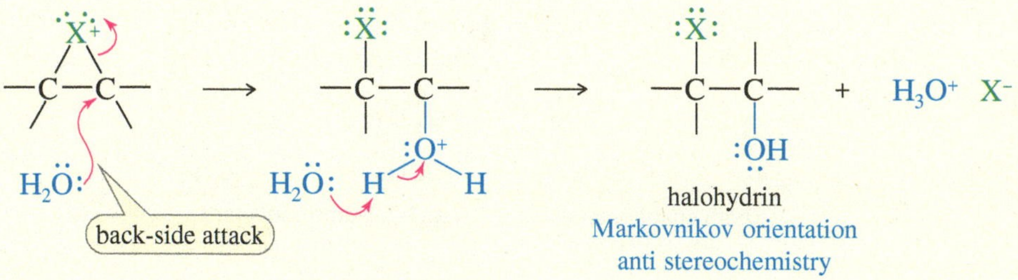

$H_2\ddot{O}$: back-side attack

$+ \; H_3O^+ \; X^-$

halohydrin
Markovnikov orientation
anti stereochemistry

EXAMPLE: Addition of Cl_2 to propene in water.

Step 1: Electrophilic attack forms a chloronium ion.

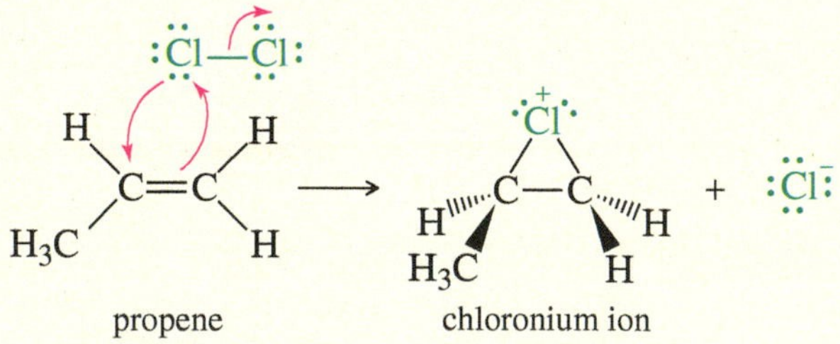

propene

chloronium ion

Step 2: Back-side attack by water opens the chloronium ion.

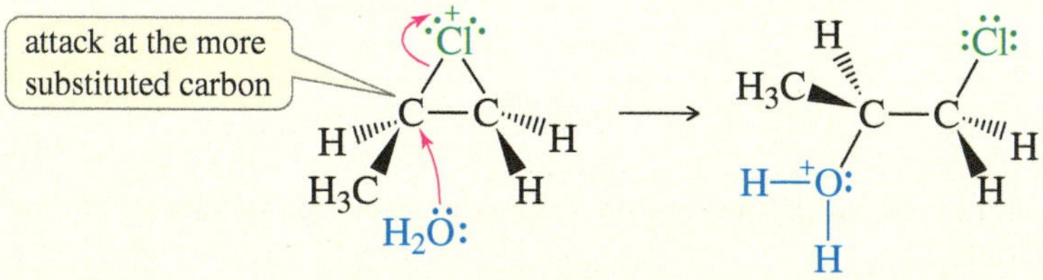

attack at the more substituted carbon

Step 3: Water removes a proton to give the chlorohydrin.

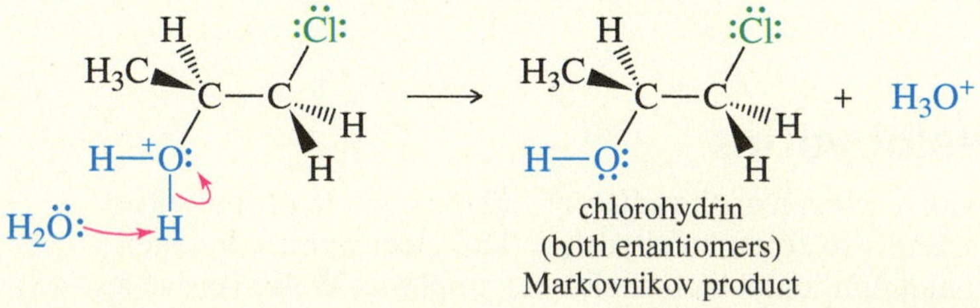

$+ \; H_3O^+$

chlorohydrin
(both enantiomers)
Markovnikov product

When halogenation takes place with no solvent or with an inert solvent such as carbon tetrachloride (CCl_4) or chloroform ($CHCl_3$), only the halide ion is available as a nucleophile to attack the halonium ion. A dihalide results. But when an alkene reacts with a halogen in the presence of a nucleophilic solvent such as water, a solvent molecule is the most likely nucleophile to attack the halonium ion. When a water molecule attacks the halonium ion, the final product is a halohydrin, with a halogen on one carbon atom and a hydroxy group on the adjacent carbon. The product may be a *chlorohydrin*, a *bromohydrin*, or an *iodohydrin*, depending on the halogen.

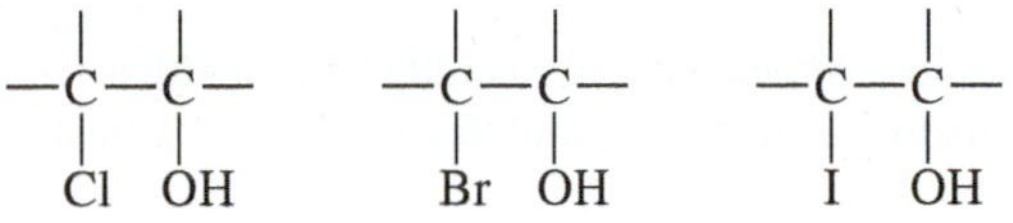

chlorohydrin (—C(Cl)—C(OH)—)   bromohydrin (—C(Br)—C(OH)—)   iodohydrin (—C(I)—C(OH)—)

Stereochemistry of Halohydrin Formation Because the mechanism involves a halonium ion, the stereochemistry of addition is anti, as in halogenation. For example, the addition of bromine water to cyclopentene gives *trans*-2-bromocyclopentanol, the product of anti addition across the double bond.

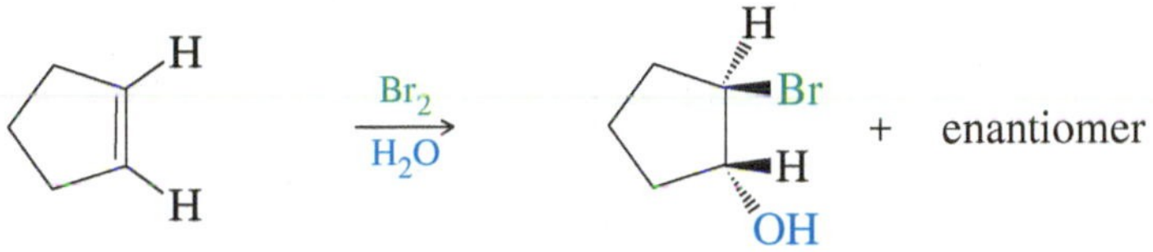

cyclopentene $\xrightarrow[H_2O]{Br_2}$ *trans*-2-bromocyclopentanol (cyclopentene bromohydrin) + enantiomer

PROBLEM 8-19

Propose a mechanism for the addition of bromine water to cyclopentene, being careful to show why the trans product results and how both enantiomers are formed.

Orientation of Halohydrin Formation Even though a halonium ion is involved, rather than a carbocation, the extended version of Markovnikov's rule applies to halohydrin formation. When propene reacts with chlorine water, the major product has the electrophile (the chlorine atom) bonded to the less substituted carbon of the double bond. The nucleophile (the hydroxy group) is bonded to the more substituted carbon.

$$H_2C{=}CH{-}CH_3 + Cl_2 + H_2O \longrightarrow H_2C(Cl){-}CH(OH){-}CH_3 + HCl$$

The Markovnikov orientation observed in halohydrin formation is explained by the structure of the halonium ion intermediate. The two carbon atoms bonded to the halogen have partial positive charges, with a larger charge (and a weaker bond to the halogen) on the more substituted carbon atom (Figure 8-6). The nucleophile (water) attacks this more substituted, more electrophilic carbon atom. The result is both anti stereochemistry and Markovnikov orientation.

larger δ^+ on the more substituted carbon

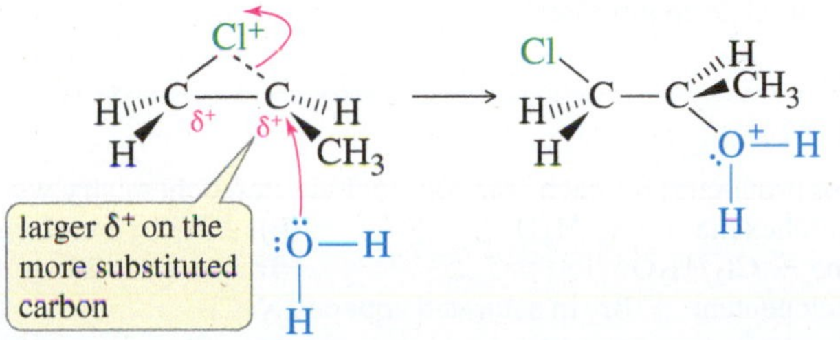

FIGURE 8-6
Orientation of halohydrin formation. The more substituted carbon of the chloronium ion bears more positive charge than the less substituted carbon. Attack by water occurs on the more substituted carbon to give the Markovnikov product.

This halonium ion mechanism can be used to explain and predict a wide variety of reactions in both nucleophilic and non-nucleophilic solvents. The halonium ion mechanism is similar to the mercurinium ion mechanism for oxymercuration of an alkene, and both give Markovnikov orientation (Section 8-5).

SOLVED PROBLEM 8-5

Propose a mechanism for the reaction of 1-methylcyclopentene with bromine water.

SOLUTION

1-Methylcyclopentene reacts with bromine to give a bromonium ion. Attack by water could occur at either the secondary carbon or the tertiary carbon of the bromonium ion. Attack actually occurs at the more substituted carbon, which bears more of the positive charge. The product is formed as a racemic mixture.

CH_3 :Br—Br: 3° CH_3 Br^+ 2° H :Br:⁻ $H_2\ddot{O}$: H ^+O—H CH_3 Br H $H_2\ddot{O}$: :ÖH CH_3 Br H + H_3O^+

(both enantiomers)

SOLVED PROBLEM 8-6

When cyclohexene is treated with bromine in saturated aqueous sodium chloride, a mixture of *trans*-2-bromocyclohexanol and *trans*-1-bromo-2-chlorocyclohexane results. Propose a mechanism to account for these two products.

SOLUTION

Cyclohexene reacts with bromine to give a bromonium ion, which will react with any available nucleophile. The most abundant nucleophiles in saturated aqueous sodium chloride solution are water and chloride ions. Attack by water gives the bromohydrin, and attack by chloride gives the dihalide. Either of these attacks gives anti stereochemistry.

cyclohexene $\xrightarrow{Br_2}$ bromonium ion (H, Br^+, H) — :Cl:⁻ → Cl, H, Br, H + enantiomer *trans*-1-bromo-2-chlorocyclohexane; — $H_2\ddot{O}$: → OH, H, Br, H + enantiomer *trans*-2-bromocyclohexanol

PROBLEM 8-20

The solutions to Solved Problem 8-5 and Solved Problem 8-6 showed only how one enantiomer of the product is formed. For each product, show how an equally probable reaction forms the other enantiomer.

PROBLEM 8-21

Predict the major product(s) for each reaction. Include stereochemistry where appropriate.

(a) 1-methylcyclohexene + Cl_2/H_2O
(b) 2-methylbut-2-ene + Br_2/H_2O
(c) *cis*-but-2-ene + Cl_2/H_2O
(d) *trans*-but-2-ene + Cl_2/H_2O
(e) 1-methylcyclopentene + Br_2 in saturated aqueous NaCl

PROBLEM 8-22

Show how you would accomplish the following synthetic conversions.
(a) 3-methylpent-2-ene → 2-chloro-3-methylpentan-3-ol
(b) chlorocyclohexane → *trans*-2-chlorocyclohexanol
(c) 1-methylcyclopentanol → 2-chloro-1-methylcyclopentanol

PROBLEM-SOLVING HINT

The opening of a halonium ion is driven by its electrophilic nature. The weak nucleophile attacks the carbon bearing more positive charge.

8-10 Catalytic Hydrogenation of Alkenes

Although we mentioned **catalytic hydrogenation** before (Sections 7-8A and 8-1), we now consider the mechanism and stereochemistry in more detail. Hydrogenation of an alkene is formally a reduction, with H_2 adding across the double bond to give an alkane. The process usually requires a catalyst containing Pt, Pd, or Ni.

$$\text{>C=C<} \;+\; H_2 \xrightarrow[\text{(Pt, Pd, Ni)}]{\text{catalyst}} \text{—C(H)—C(H)—}$$

alkene → alkane

Example

$$CH_3{-}CH{=}CH{-}CH_3 \;+\; H_2 \xrightarrow{Pt} CH_3{-}CH_2{-}CH_2{-}CH_3$$

For most alkenes, hydrogenation takes place at room temperature, using hydrogen gas at atmospheric pressure. The alkene is usually dissolved in an alcohol, an alkane, or acetic acid. A small amount of platinum, palladium, or nickel catalyst is added, and the container is shaken or stirred while the reaction proceeds. Hydrogenation actually takes place at the surface of the metal, where the liquid solution of the alkene comes into contact with hydrogen and the catalyst.

Hydrogen gas is adsorbed onto the surface of these metal catalysts, and the catalyst weakens the H—H bond. In fact, if H_2 and D_2 are mixed in the presence of a platinum catalyst, the two isotopes quickly scramble to produce a random mixture of HD, H_2, and D_2. (No scrambling occurs in the absence of the catalyst.) Hydrogenation is an example of **heterogeneous catalysis,** because the (solid) catalyst is in a different phase from the reactant solution. In contrast, **homogeneous catalysis** involves reactants and catalyst in the same phase, as in the acid-catalyzed dehydration of an alcohol.

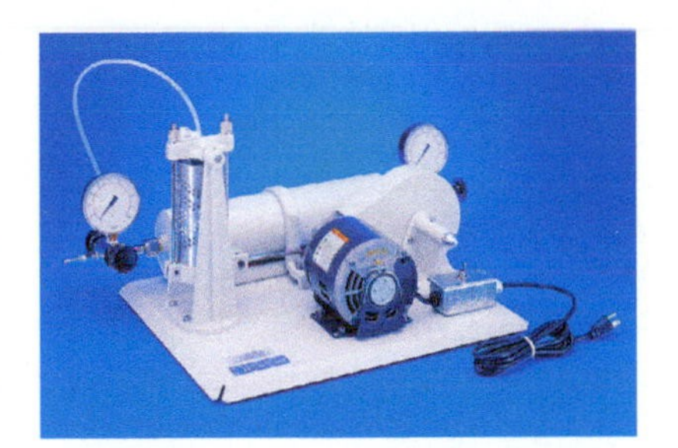

The Parr hydrogenation apparatus shakes the reaction vessel (containing the alkene and the solid catalyst) while a pressurized cylinder supplies hydrogen.

Because the two hydrogen atoms add from a solid surface, they add with **syn** stereochemistry. For example, when 1,2-dideuteriocyclohexene is treated with hydrogen gas over a catalyst, the product is the cis isomer resulting from syn addition (Figure 8-7).

$$\text{1,2-dideuteriocyclohexene} \xrightarrow[\text{Pt}]{H_2} \text{cis-1,2-dideuteriocyclohexane}$$

cis isomer

One face of the alkene pi bond binds to the catalyst, which has hydrogen adsorbed on its surface. Hydrogen inserts into the pi bond, and the product is freed from the catalyst. Both hydrogen atoms add to the face of the double bond that is complexed with the catalyst.

Soluble homogeneous catalysts, such as *Wilkinson's catalyst*, also catalyze the hydrogenation of carbon–carbon double bonds.

$$\text{>C=C<} \;+\; H_2 \xrightarrow[\text{(Wilkinson's catalyst)}]{(Ph_3P)_2Rh(Cl)(PPh_3)} \text{—C(H)—C(H)—}$$

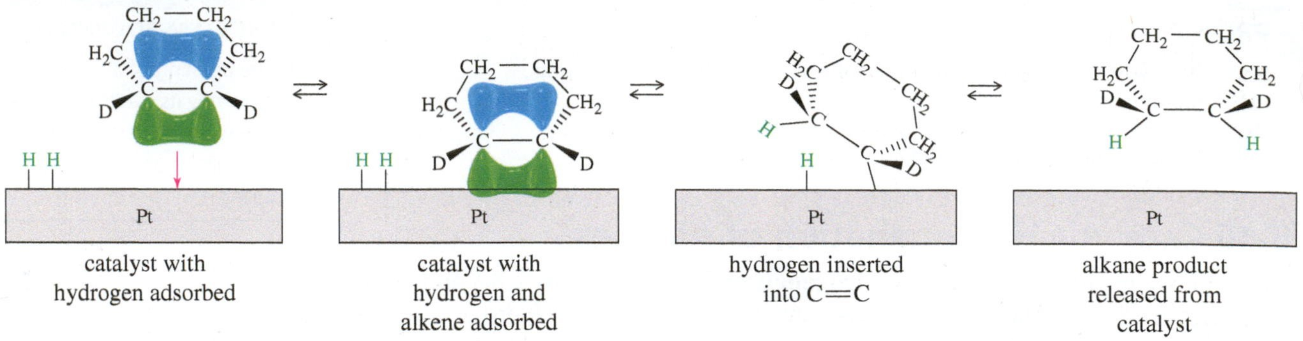

FIGURE 8-7
Syn stereochemistry in catalytic hydrogenation. A solid heterogeneous catalyst adds two hydrogen atoms to the same face of the pi bond (syn stereochemistry).

Application: Trans Fats

The double bonds in naturally occurring vegetable oils are almost exclusively cis. Catalytic hydrogenation reduces some of the double bonds, raising the melting point and giving a white, creamy shortening that resembles lard. Catalytic hydrogenation is a reversible process, however, and some of the molecules undergo hydrogenation followed by dehydrogenation, leaving double bonds in random positions and often with trans stereochemistry. Our metabolism is not well equipped to deal with trans double bonds, and trans fats are associated with increased serum cholesterol levels and increased incidence of heart disease. The FDA has recently recommended the removal from food of all partially hydrogenated vegetable oils containing trans fats. Also see Section 25-3A.

Wilkinson's catalyst is not chiral, but its triphenylphosphine (PPh_3) ligands can be replaced by chiral ligands to give chiral catalysts that are capable of converting optically inactive starting materials to optically active products. Such a process is called **asymmetric induction** or **enantioselective synthesis.** For example, Figure 8-8 shows a chiral ruthenium complex catalyzing an enantioselective hydrogenation of a carbon–carbon double bond to give a large excess of one enantiomer. Because the catalyst is chiral, the transition states leading to the two enantiomers of product are diastereomeric. They have different energies, and the transition state leading to the (*R*) enantiomer is favored. Ryoji Noyori and William Knowles shared the 2001 Nobel Prize in Chemistry for their work on chirally catalyzed hydrogenation reactions.

Enantioselective synthesis is particularly important in the pharmaceutical industry, because only one enantiomer of a chiral drug is likely to have the desired effect. For example, levodopa [(−)-dopa or *l*-dopa] is used in patients with Parkinson's disease to counteract a deficiency of dopamine, one of the neurotransmitters in the brain. Dopamine itself is useless as a drug because it cannot cross the "blood–brain barrier"; that is, it cannot get into the cerebrospinal fluid from the bloodstream. (−)-Dopa, on the other hand, is an amino acid related to tyrosine. It crosses the blood–brain barrier into the cerebrospinal fluid, where it undergoes enzymatic conversion to dopamine. Only the (−) enantiomer of dopa can be transformed into dopamine; the other enantiomer, (+)-dopa, is toxic to the patient.

The correct enantiomer can be synthesized from an achiral starting material by catalytic hydrogenation using a complex of rhodium with a chiral ligand called DIOP. Such an enantioselective synthesis is more efficient than making a racemic mixture, resolving it into enantiomers, and discarding the unwanted enantiomer.

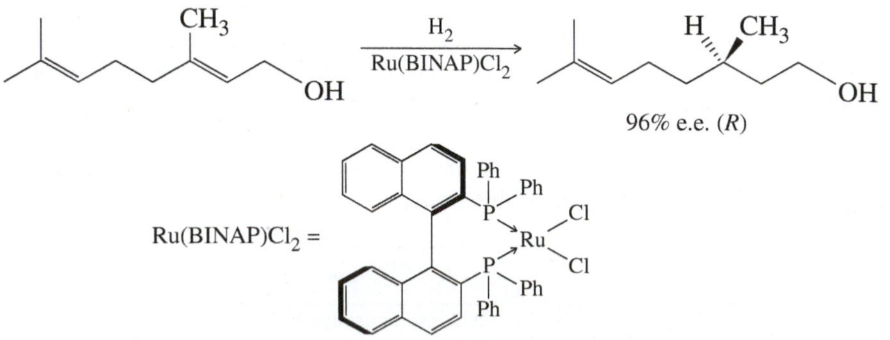

FIGURE 8-8
Chiral hydrogenation catalysts. Rhodium and ruthenium phosphines are effective homogeneous catalysts for hydrogenation. Chiral ligands can be attached to accomplish asymmetric induction, the creation of a new asymmetric carbon as mostly one enantiomer.

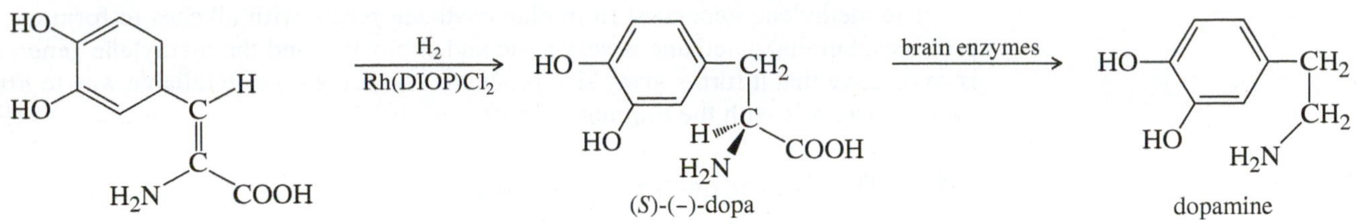

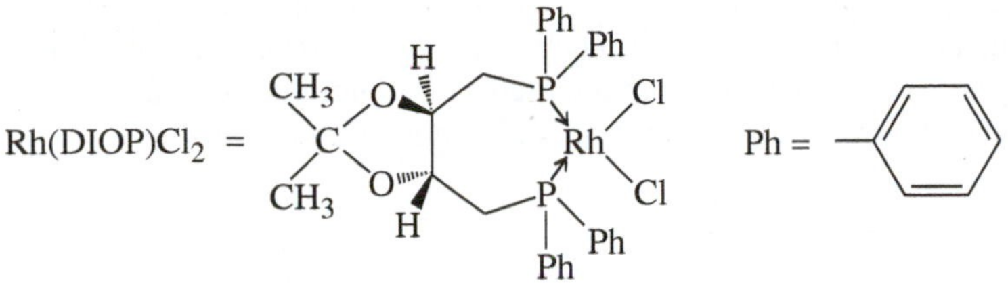

$$\text{HO, HO-aryl-CH}=\text{C(NH}_2\text{)COOH} \xrightarrow[\text{Rh(DIOP)Cl}_2]{H_2} \text{(S)-(−)-dopa} \xrightarrow{\text{brain enzymes}} \text{dopamine}$$

(*S*)-(−)-dopa

dopamine

Rh(DIOP)Cl_2 = Ph =

PROBLEM 8-23

Give the expected major product for each reaction, including stereochemistry where applicable.

(a) but-1-ene + H_2/Pt **(b)** *cis*-but-2-ene + H_2/Ni

(c) 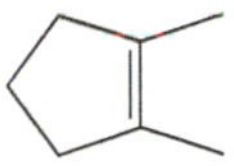 + H_2/Pt **(d)** 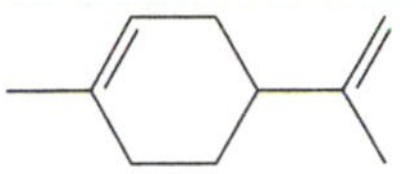 + excess H_2/Pt

PROBLEM 8-24

One of the principal components of lemongrass oil is *limonene*, $C_{10}H_{16}$. When limonene is treated with excess hydrogen and a platinum catalyst, the product is an alkane of formula $C_{10}H_{20}$. What can you conclude about the structure of limonene?

PROBLEM 8-25

The chiral BINAP ligand shown in Figure 8-8 contains no asymmetric carbon atoms. Explain how this ligand is chiral.

Application: Antibiotics

The enzymatic reduction of a double bond is a key step in the formation of a fatty acid that is ultimately incorporated into the cell wall of the bacterium that causes tuberculosis. The antituberculosis drug isoniazid blocks this enzyme, preventing reduction of the double bond. Without an intact cell wall, the bacteria die.

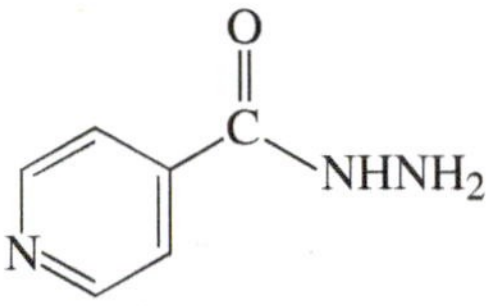

isoniazid

8-11 Addition of Carbenes to Alkenes

Methylene (:CH_2) is the simplest of the **carbenes:** uncharged, reactive intermediates that have a carbon atom with two bonds and two nonbonding electrons. Like borane (BH_3), methylene is a potent electrophile because it has an unfilled octet. It adds to the electron-rich pi bond of an alkene to form a cyclopropane.

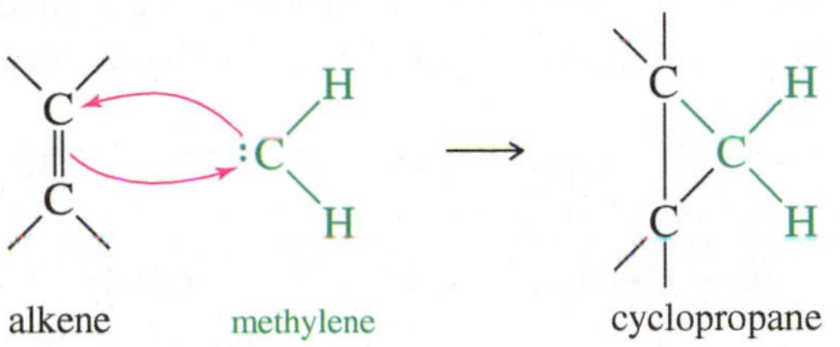

alkene methylene cyclopropane

Heating or photolysis of diazomethane gives nitrogen gas and methylene:

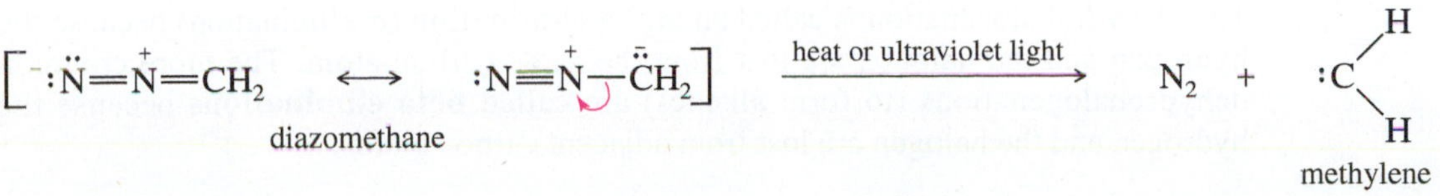

$$\left[\,^{-}\!:\ddot{N}=\overset{+}{N}=CH_2 \longleftrightarrow :N\equiv\overset{+}{N}-\ddot{C}H_2^{-}\right] \xrightarrow{\text{heat or ultraviolet light}} N_2 + :CH_2$$

diazomethane methylene

The methylene generated from diazomethane reacts with alkenes to form cyclopropanes, but diazomethane is very toxic and explosive, and the methylene generated is so reactive that it forms many side products. A safer and more reliable way to make cyclopropanes is with the Simmons–Smith reagent.

8-11A The Simmons–Smith Reaction

The **Simmons–Smith reagent,** named for the two DuPont chemists who discovered it, is made by adding methylene iodide to the "zinc–copper couple" (zinc dust that has been activated with an impurity of copper). The reagent probably resembles iodomethyl zinc iodide, ICH_2ZnI. This kind of reagent is called a *carbenoid* because it reacts much like a carbene, but it does not actually contain a divalent carbon atom.

$$CH_2I_2 + Zn(Cu) \longrightarrow ICH_2ZnI$$

Simmons–Smith reagent
(a carbenoid)

The Simmons-Smith reagent adds to double bonds much like methylene ($:CH_2$) would, except that the reagent is easier to use, is much more stable, and gives better yields without as many side reactions.

$$\text{>C=C<} + ICH_2ZnI \longrightarrow \text{cyclopropane } (-C—C-, \text{bridged by } CH_2) + ZnI_2$$

Example

$$\text{cyclohexene} \xrightarrow[\text{Zn, CuCl}]{CH_2I_2} \text{bicyclo[4.1.0]heptane}$$

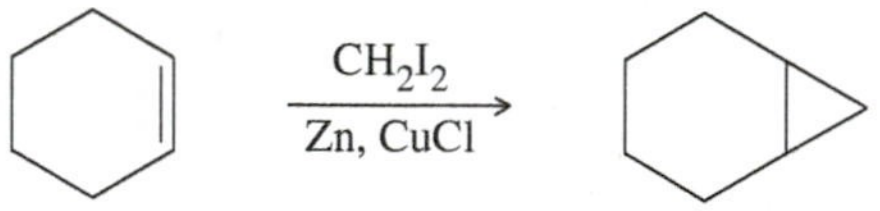

(59%)

PROBLEM 8-26

Predict the carbenoid addition products of the following reactions.

(a) *trans*-hex-3-ene + CH_2I_2, Zn(Cu) **(b)** *cis*-hept-2-ene + CH_2I_2, Zn(Cu)

8-11B Formation of Carbenes by Alpha Elimination

Carbenes are also formed by reactions of halogenated compounds with bases. If a carbon atom has bonds to at least one hydrogen and to enough halogen atoms to make the hydrogen slightly acidic, it may be possible to form a carbene. For example, bromoform ($CHBr_3$) reacts with a 50% aqueous solution of potassium hydroxide to form dibromocarbene.

$$\underset{\text{bromoform}}{CHBr_3} + K^+ \ {}^-OH \rightleftharpoons {}^-:CBr_3 \ K^+ + H_2O$$

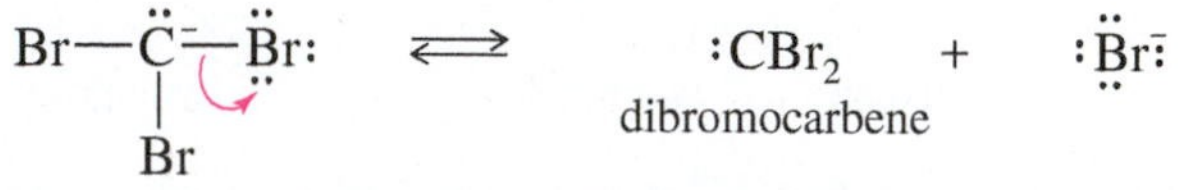

$$Br_2\ddot{C}^-—\ddot{Br}: \rightleftharpoons \underset{\text{dibromocarbene}}{:CBr_2} + :\ddot{Br}:^-$$

This dehydrohalogenation is called an **alpha elimination** (α elimination) because the hydrogen and the halogen are lost from the same carbon atom. The more common dehydrohalogenations (to form alkenes) are called **beta eliminations** because the hydrogen and the halogen are lost from adjacent carbon atoms.

Dibromocarbene formed from $CHBr_3$ can add to a double bond to form a dibromocyclopropane.

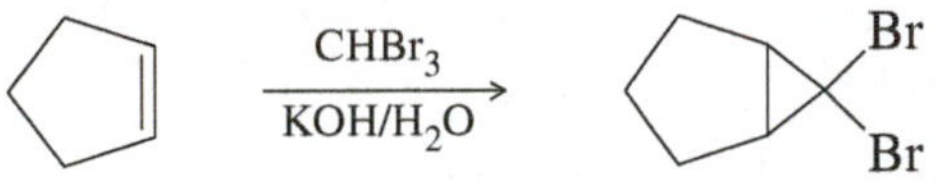

$$\text{cyclopentene} \xrightarrow[\text{KOH/H}_2\text{O}]{\text{CHBr}_3} \text{6,6-dibromobicyclo[3.1.0]hexane}$$

The products of these cyclopropanations retain any cis or trans stereochemistry of the reactants.

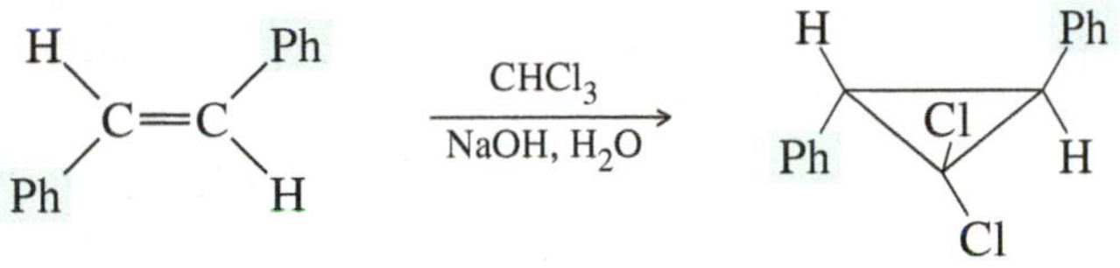

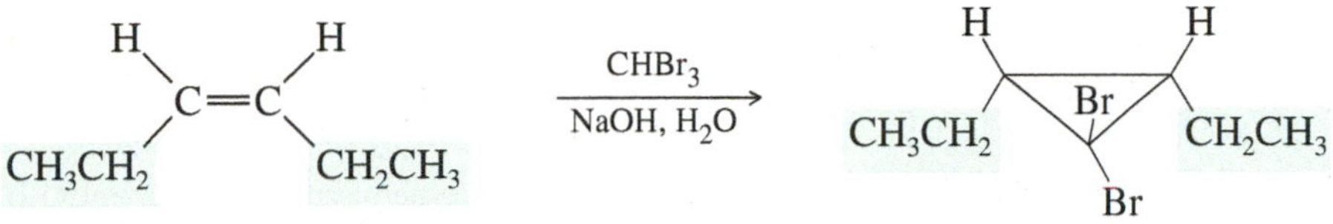

$$\text{trans-PhCH=CHPh} \xrightarrow[\text{NaOH, H}_2\text{O}]{\text{CHCl}_3} \text{trans-1,1-dichloro-2,3-diphenylcyclopropane}$$

$$\text{cis-CH}_3\text{CH}_2\text{CH=CHCH}_2\text{CH}_3 \xrightarrow[\text{NaOH, H}_2\text{O}]{\text{CHBr}_3} \text{cis-1,1-dibromo-2,3-diethylcyclopropane}$$

PROBLEM 8-27

Predict the carbene addition products of the following reactions.

(a) cyclohexene + $CHCl_3$, 50% $NaOH/H_2O$

(b) 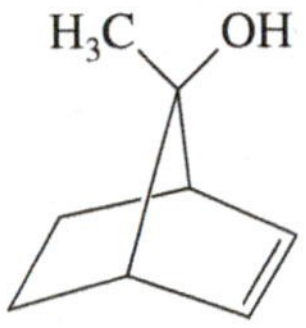 (7-methylbicyclo[2.2.1]hept-2-en-7-ol; H_3C, OH) + CH_2I_2, Zn(Cu)

(c) 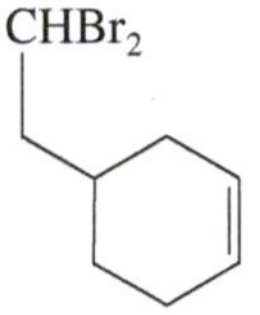 ($CHBr_2CH_2$-substituted cyclohexene) + 50% $NaOH/H_2O$

PROBLEM 8-28

Show how you would accomplish each of the following synthetic conversions.

(a) *trans*-but-2-ene ⟶ *trans*-1,2-dimethylcyclopropane

(b) cyclopentene ⟶ (6-bromobicyclo[3.1.0]hexane; H, Br)

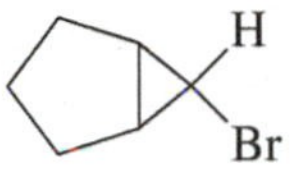

(c) cyclohexanol ⟶ (7,7-dichlorobicyclo[4.1.0]heptane; Cl, Cl)

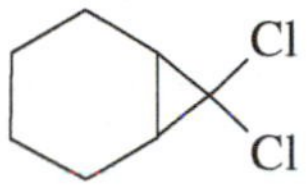

8-12 Epoxidation of Alkenes

Some of the most important reactions of alkenes involve oxidation. When we speak of oxidation, we usually mean reactions that form carbon–oxygen bonds. (Halogens are oxidizing agents, and the addition of a halogen molecule across a double bond is formally an oxidation as well.) Oxidations are particularly important because many common functional groups contain oxygen, and alkene oxidations are some of the best methods for introducing oxygen into organic molecules. We will consider methods for epoxidation, dihydroxylation, and oxidative cleavage of alkene double bonds.

An **epoxide** is a three-membered cyclic ether, also called an **oxirane.** Epoxides are valuable synthetic intermediates used for converting alkenes to a variety of other functional groups. An alkene is converted to an epoxide by a **peroxyacid,** a carboxylic acid that has an extra oxygen atom in a —O—O— (peroxy) linkage.

alkene + peroxyacid → epoxide (oxirane) + acid

The **epoxidation** of an alkene is clearly an oxidation, because an oxygen atom is added. Peroxyacids are highly selective oxidizing agents. Some simple peroxyacids (sometimes called *peracids*) and their corresponding carboxylic acids are shown next.

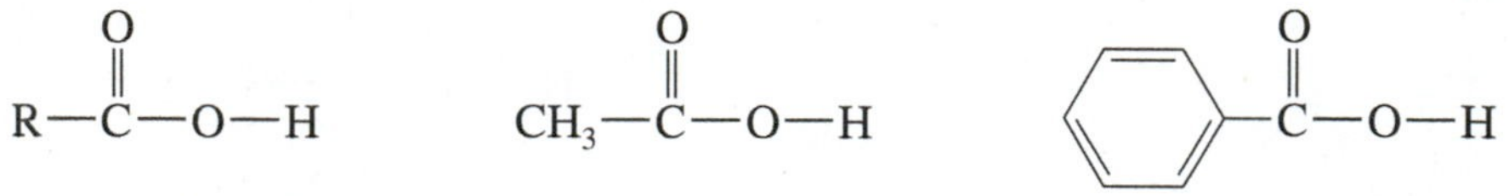

a carboxylic acid — acetic acid — benzoic acid, $PhCO_2H$

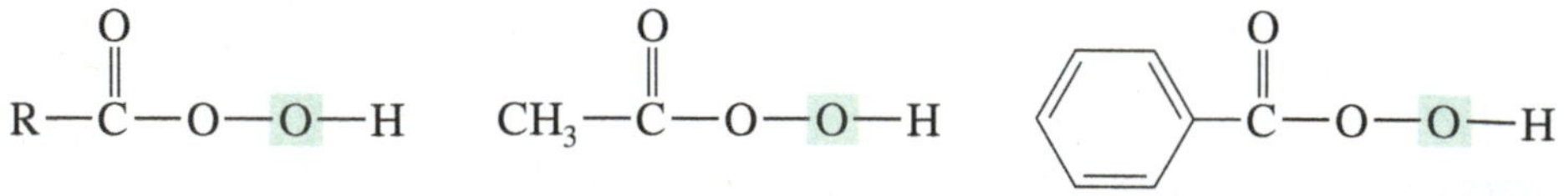

a peroxyacid — peroxyacetic acid — peroxybenzoic acid, $PhCO_3H$

Application: Insecticides

Epoxides are often found as components of natural products that plants use as defense mechanisms against insects. Typically, the epoxide reacts with critical cellular enzymes or DNA, preventing them from carrying out their normal functions. As a result, the insect is killed.

A peroxyacid epoxidizes an alkene by a concerted electrophilic reaction where several bonds are broken and several are formed at the same time. Starting with the alkene and the peroxyacid, a one-step reaction gives the epoxide and the acid directly, without any intermediates.

MECHANISM 8-9 Epoxidation of Alkenes

Peroxyacids epoxidize alkenes in a one-step (concerted) process.

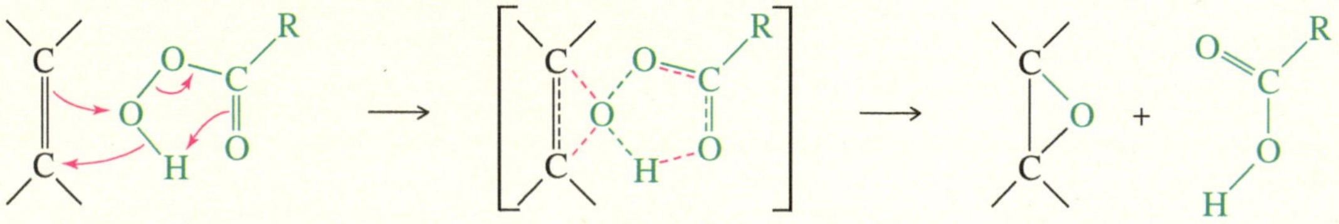

alkene peroxyacid → transition state → epoxide + acid

EXAMPLE: Epoxidation of propene by peroxyacetic acid.

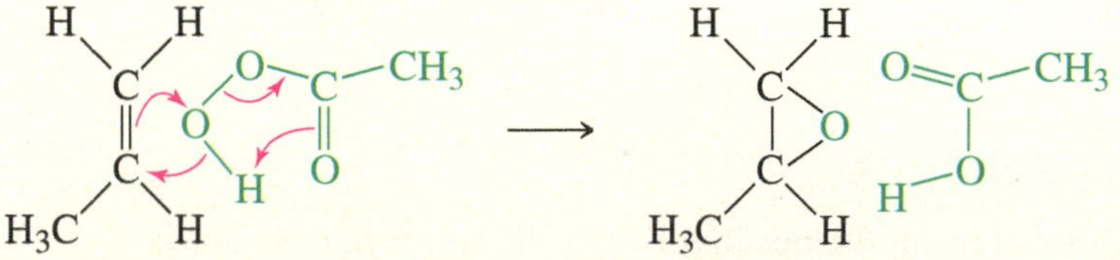

propene (propylene) + peroxyacetic acid → epoxypropane (propylene oxide) + acetic acid

Because the epoxidation takes place in one step, there is no opportunity for the alkene molecule to rotate and change its cis or trans geometry. The epoxide retains whatever stereochemistry is present in the alkene.

The following examples use *m*-chloroperoxybenzoic acid (mCPBA), a common epoxidizing reagent, to convert alkenes to epoxides having the same cis or trans stereochemistry. mCPBA is used for its desirable solubility properties: The peroxyacid dissolves, then the spent acid precipitates out of solution.

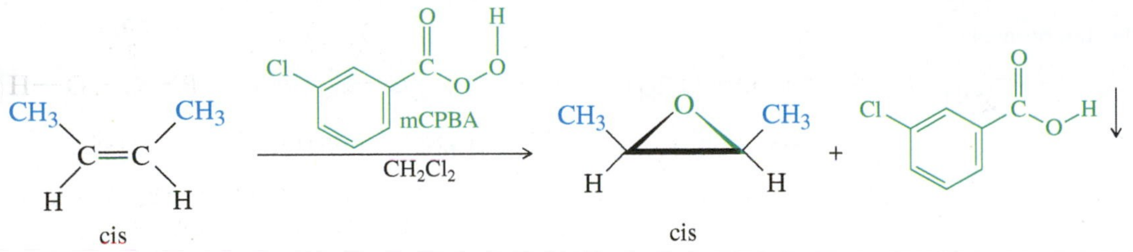

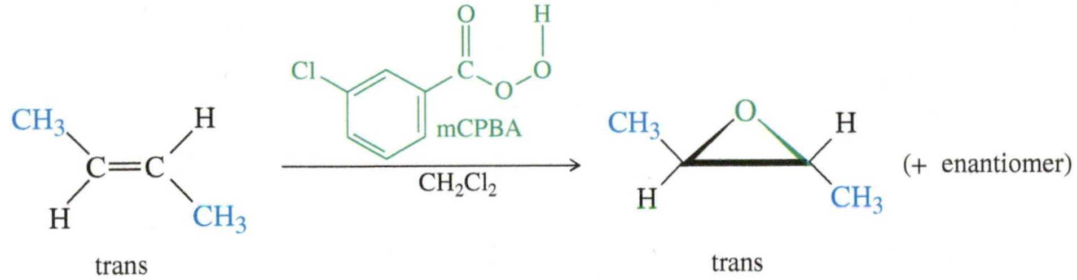

cis → (mCPBA, CH_2Cl_2) → cis + m-chlorobenzoic acid

trans → (mCPBA, CH_2Cl_2) → trans (+ enantiomer)

PROBLEM 8-29

Predict the products, including stereochemistry where appropriate, for the *m*-chloroperoxybenzoic acid epoxidations of the following alkenes.

(a) *cis*-hex-2-ene **(b)** *trans*-hex-2-ene
(c) *cis*-cyclodecene **(d)** *trans*-cyclodecene

8-13 Acid-Catalyzed Opening of Epoxides

Most epoxides are easily isolated as stable products if the solution is not too acidic. Any moderately strong acid protonates the epoxide, however. Water attacks the protonated epoxide, opening the ring and forming a 1,2-diol, commonly called a **glycol.**

MECHANISM 8-10 Acid-Catalyzed Opening of Epoxides

The crucial step is a back-side attack by the solvent on the protonated epoxide.

Step 1: Protonation of the epoxide activates it toward nucleophilic attack.

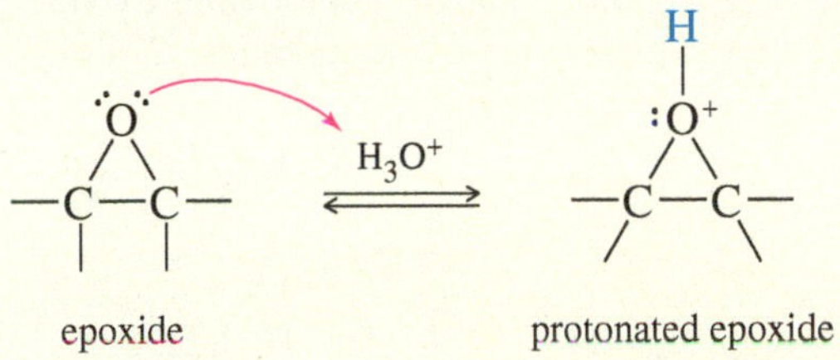

epoxide ⇌ (H_3O^+) protonated epoxide

Step 2: Back-side attack by the solvent (water) opens the ring.

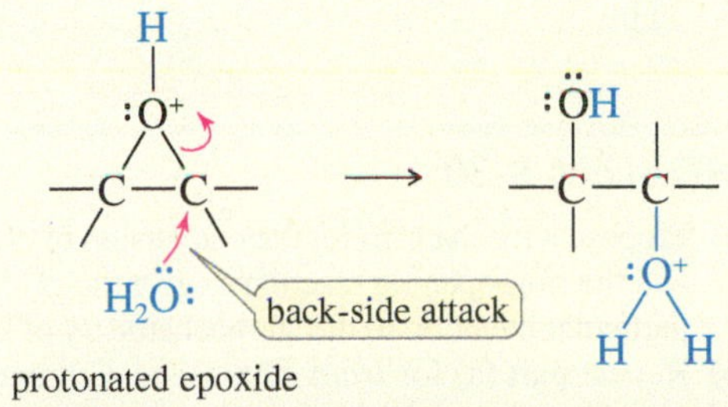

protonated epoxide + $H_2\ddot{O}$: (back-side attack) →

(CONTINUED)

Step 3: Deprotonation gives the diol product.

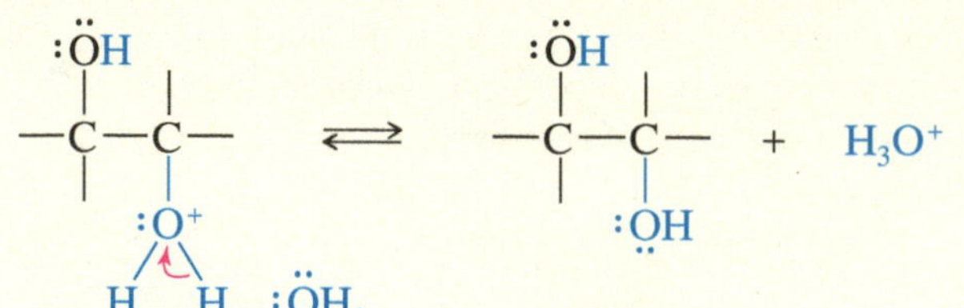

a glycol
(anti orientation)

EXAMPLE: Acid-catalyzed hydrolysis of propylene oxide (epoxypropane).

Step 1: Protonation of the epoxide.

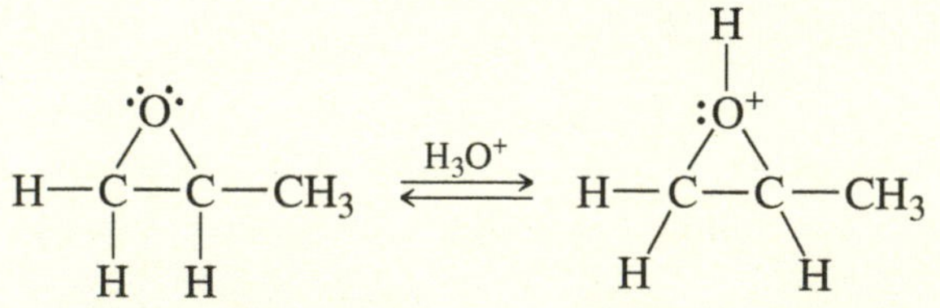

epoxypropane
(propylene oxide)

protonated epoxide

Steps 2 and 3: Back-side attack by water, then deprotonation of the product.

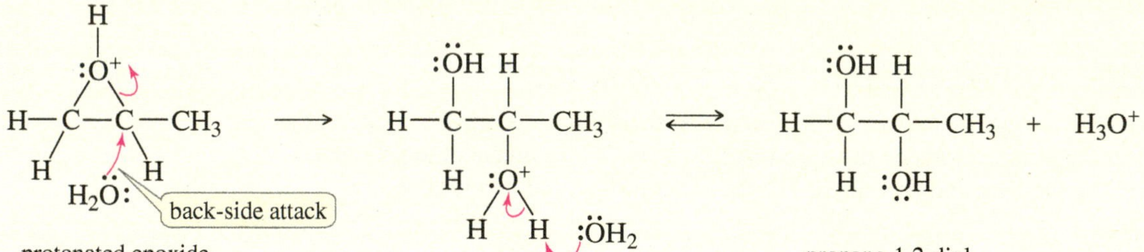

back-side attack

protonated epoxide

propane-1,2-diol
(propylene glycol)

Because glycol formation involves a back-side attack on a protonated epoxide, the result is anti orientation of the hydroxy groups on the double bond. For example, when 1,2-epoxycyclopentane ("cyclopentene oxide") is treated with dilute mineral acid, the product is pure *trans*-cyclopentane-1,2-diol.

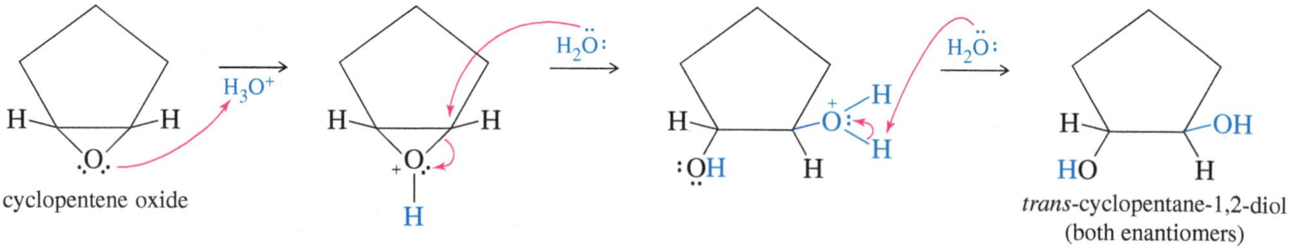

cyclopentene oxide

trans-cyclopentane-1,2-diol
(both enantiomers)

PROBLEM 8-30

(a) Propose a mechanism for the conversion of *cis*-hex-3-ene to the epoxide (3,4-epoxyhexane) and the ring-opening reaction to give the glycol, hexane-3,4-diol. In your mechanism, pay particular attention to the stereochemistry of the intermediates and products.

(b) Repeat part (a) for *trans*-hex-3-ene. Compare the products obtained from *cis*- and *trans*-hex-3-ene. Is this reaction sequence stereospecific?

Epoxidation reagents can be chosen to favor either the epoxide or the glycol. Peroxyacetic acid is used in strongly acidic water solutions. The acidic solution protonates the epoxide and converts it to the glycol. Peroxybenzoic acids are weak acids that can be used in non-nucleophilic solvents such as carbon tetrachloride. *m*-Chloroperoxybenzoic acid in CCl_4 generally gives good yields of epoxides. Figure 8-9 compares the uses of these reagents.

Application: Biochemistry

The body oxidizes the alkene components of drugs and other substances to epoxides, which are then hydrolyzed to diols by an epoxide hydrolase enzyme. The more reactive epoxides are rapidly converted to water-soluble diols and eliminated in the urine. Epoxide hydrolase enzymes are sometimes used in organic synthesis to produce chiral diols.

PROBLEM 8-31

Magnesium monoperoxyphthalate (MMPP) epoxidizes alkenes much like mCPBA. MMPP is more stable, however, and it may be safer to use for large-scale and industrial reactions. Propose a mechanism for the reaction of *trans*-2-methylhept-3-ene with MMPP, and predict the structure of the product(s).

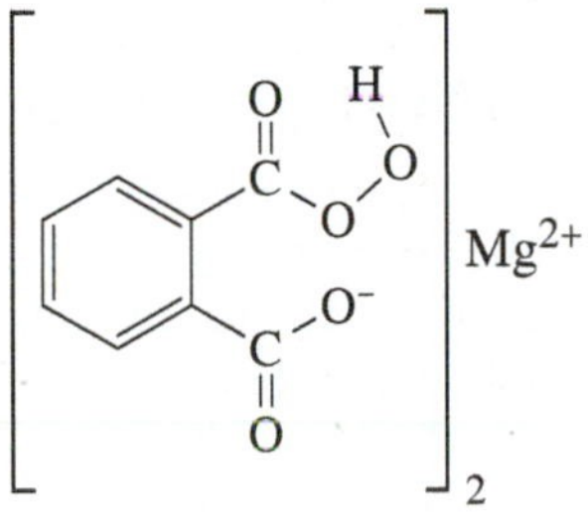

Mg^{2+}

Magnesium monoperoxyphthalate, MMPP

Ph—C(=O)—OOH peroxybenzoic acid, CCl_4 (100%)

CH_3—C(=O)—OOH peroxyacetic acid, H_3O^+ — not isolated — CH_3—C(=O)—OH, H_3O^+ — stronger acid, nucleophilic solvent — + enantiomer (75%)

mCPBA, CCl_4 — + enantiomer (90%)

CH_3—C(=O)—OOH peroxyacetic acid, H_3O^+ — (meso) (70%)

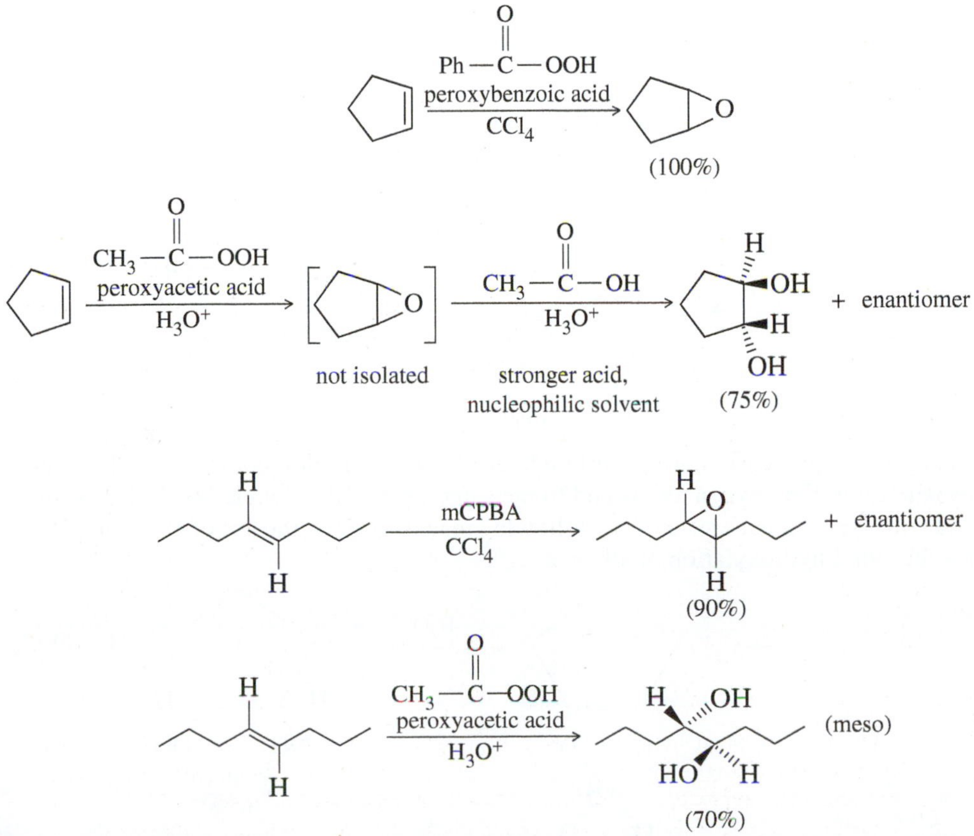

FIGURE 8-9
Reagents for epoxidation. Peroxyacetic acid is used in strongly acidic aqueous solutions. Alkenes are epoxidized, then opened to glycols in one step. Weakly acidic peroxyacids, such as peroxybenzoic acid or mCPBA, can be used in nonaqueous solutions to give good yields of epoxides.

PROBLEM 8-32

Predict the major products of the following reactions.
(a) *cis*-hex-2-ene + mCPBA in chloroform
(b) *trans*-hex-3-ene + peroxyacetic acid (CH_3CO_3H) in water
(c) 1-methylcyclohexene + MMPP in ethanol
(d) *trans*-cyclodecene + peroxyacetic acid in acidic water
(e) *cis*-cyclodecene + mCPBA in CH_2Cl_2, then dilute aqueous acid

PROBLEM 8-33

When 1,2-epoxycyclohexane (cyclohexene oxide) is treated with anhydrous HCl in methanol, the principal product is *trans*-2-methoxycyclohexanol. Propose a mechanism to account for the formation of this product.

Application: Catalysis

Older dihydroxylation procedures once used a full equivalent of OsO_4 followed by a reducing agent such as $NaHSO_3$ to reduce the osmate ester. These old procedures have been supplanted by methods using only catalytic amounts of OsO_4, an exceptionally toxic and expensive reagent.

8-14 Syn Dihydroxylation of Alkenes

Converting an alkene to a glycol requires adding a hydroxy group to each end of the double bond. This addition is called **dihydroxylation** (or **hydroxylation**) of the double bond. We have seen that epoxidation of an alkene, followed by acidic hydrolysis, gives *anti* dihydroxylation of the double bond. Reagents are also available for the dihydroxylation of alkenes with *syn* stereochemistry. The two most common reagents for this purpose are osmium tetroxide and potassium permanganate.

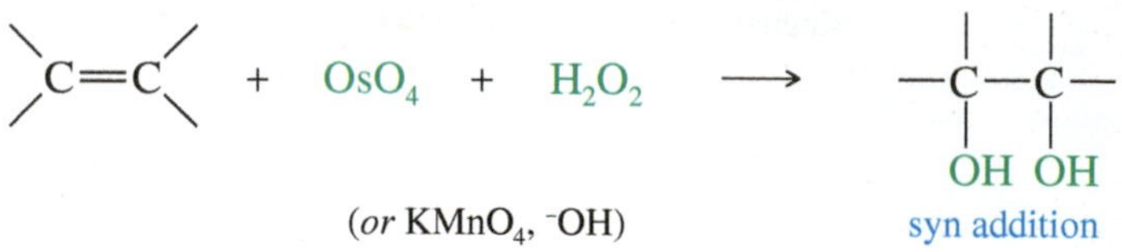

$$>C=C< \; + \; OsO_4 \; + \; H_2O_2 \longrightarrow -\underset{OH}{\overset{|}{C}}-\underset{OH}{\overset{|}{C}}-$$

(*or* $KMnO_4$, ^-OH) syn addition

8-14A Osmium Tetroxide Dihydroxylation

Osmium tetroxide (OsO_4, sometimes called *osmic acid*) reacts with alkenes in a concerted step to form a cyclic osmate ester. Oxidizing agents such as hydrogen peroxide (H_2O_2) or tertiary amine oxides ($R_3N^+{-}O^-$) are used to hydrolyze the osmate ester and reoxidize osmium to osmium tetroxide. The regenerated osmium tetroxide catalyst continues to hydroxylate more molecules of the alkene.

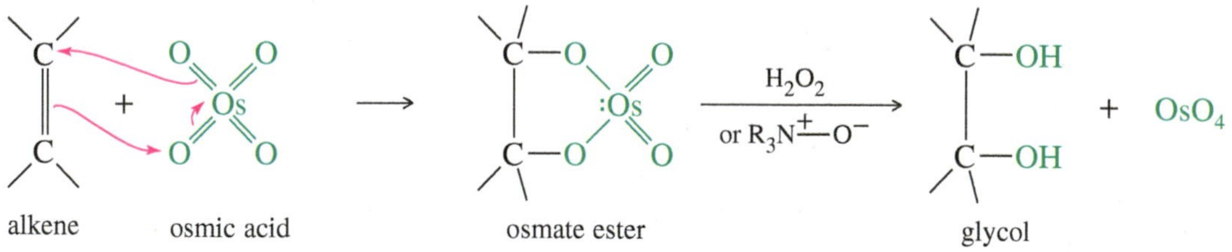

alkene + osmic acid ⟶ osmate ester $\xrightarrow[\text{or } R_3N^+{-}O^-]{H_2O_2}$ glycol + OsO_4

Because the two carbon–oxygen bonds are formed simultaneously with the cyclic osmate ester, the oxygen atoms add to the same face of the double bond; that is, they add with syn stereochemistry. The following reactions show the use of OsO_4 and H_2O_2 for the syn dihydroxylation of alkenes.

cyclopentene + OsO_4 ⟶ osmate ester $\xrightarrow{H_2O_2}$ *cis*-glycol (65%) + OsO_4

concerted formation of osmate ester

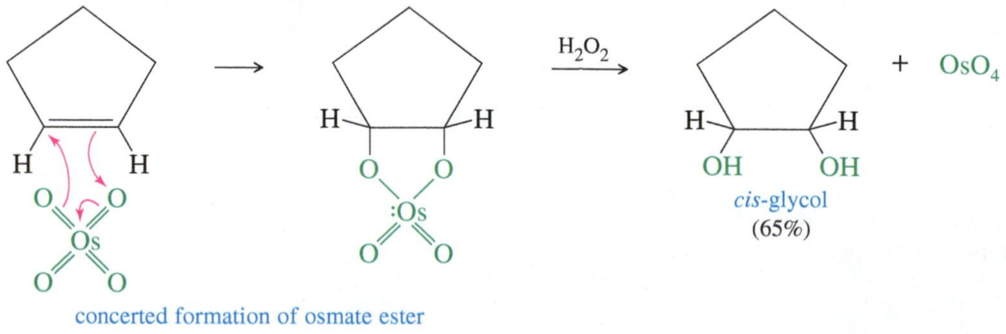

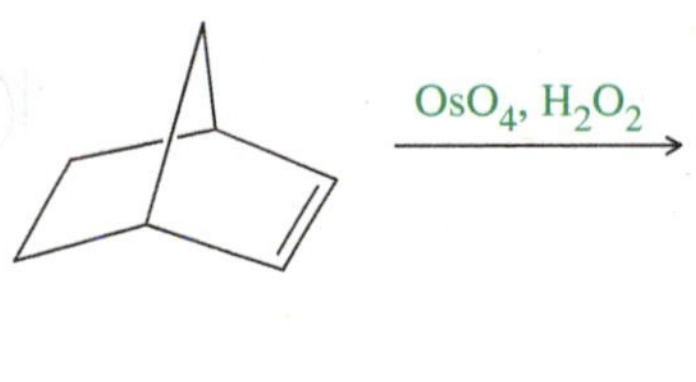

OsO_4, H_2O_2 → OH, H, OH, H

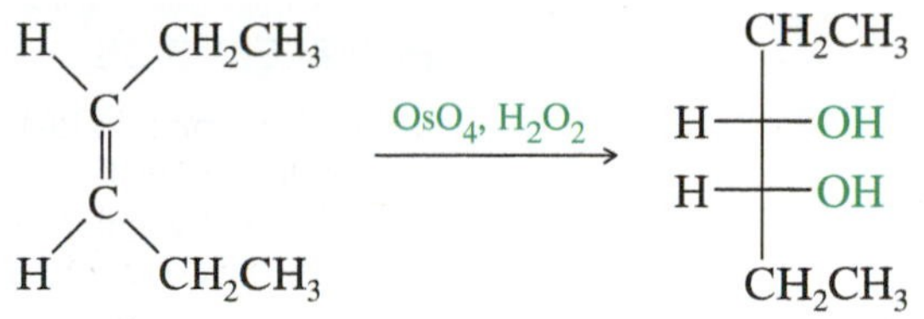

H, CH_2CH_3, C, C, H, CH_2CH_3

cis-hex-3-ene

OsO_4, H_2O_2 →

CH_2CH_3, H—OH, H—OH, CH_2CH_3

meso-hexane-3,4-diol

8-14B Permanganate Dihydroxylation

Osmium tetroxide is expensive, highly toxic, and volatile. A cold, dilute solution of potassium permanganate ($KMnO_4$) also hydroxylates alkenes with syn stereochemistry, with slightly reduced yields in most cases. Like osmium tetroxide, permanganate adds to the alkene double bond to form a cyclic ester: a manganate ester in this case. The basic solution hydrolyzes the manganate ester, liberating the glycol and producing a brown precipitate of manganese dioxide, MnO_2.

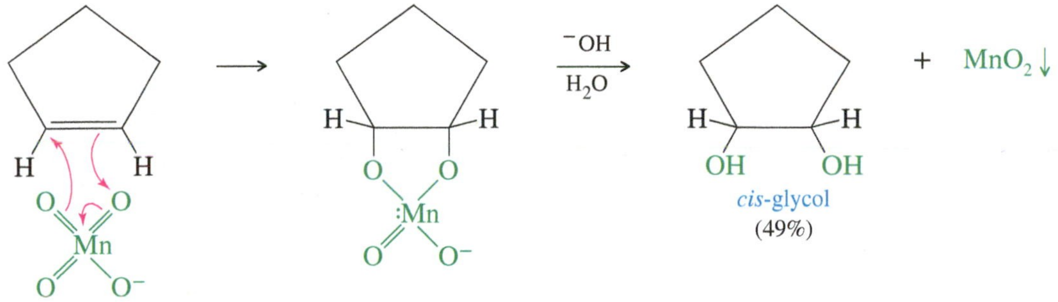

H, H, O, O, Mn, O, O^- → H, H, O, O, :Mn, O, O^- $\xrightarrow[H_2O]{^-OH}$ H, H, OH, OH + $MnO_2\downarrow$

concerted formation of manganate ester

cis-glycol
(49%)

In addition to its synthetic value, the permanganate oxidation of alkenes provides a simple chemical test for the presence of an alkene. When an alkene is added to a clear, deep purple aqueous solution of potassium permanganate, the solution loses its purple color and becomes the murky, opaque brown color of MnO_2. (Although there are other functional groups that decolorize permanganate, few do it as quickly as alkenes.)

8-14C Choosing a Reagent

To dihydroxylate an alkene with syn stereochemistry, which is the better reagent: osmium tetroxide or potassium permanganate? Osmium tetroxide gives better yields, but permanganate is cheaper and safer to use. The answer depends on the circumstances.

If the starting material is only 2 mg of a compound 15 steps along in a difficult synthesis, we use osmium tetroxide. The better yield is crucial because the starting material is precious and expensive, and little osmic acid is needed. If the dihydroxylation is the first step in a synthesis and involves 5 kg of the starting material, we use potassium permanganate. The cost of buying enough osmium tetroxide would be prohibitive, and dealing with such a large amount of a volatile, toxic reagent would be inconvenient. On such a large scale, we can accept the lower yield of the permanganate oxidation.

PROBLEM 8-34

Predict the major products of the following reactions, including stereochemistry.

(a) cyclohexene + $KMnO_4/H_2O$ (cold, dilute)
(b) cyclohexene + peroxyacetic acid in water
(c) *cis*-pent-2-ene + OsO_4/H_2O_2
(d) *cis*-pent-2-ene + peroxyacetic acid in water
(e) *trans*-pent-2-ene + OsO_4/H_2O_2
(f) *trans*-pent-2-ene + peroxyacetic acid in water

PROBLEM 8-35

Show how you would accomplish the following conversions.

(a) *cis*-hex-3-ene to *meso*-hexane-3,4-diol
(b) *cis*-hex-3-ene to (*d,l*)-hexane-3,4-diol
(c) *trans*-hex-3-ene to *meso*-hexane-3,4-diol
(d) *trans*-hex-3-ene to (*d,l*)-hexane-3,4-diol

8-15 Oxidative Cleavage of Alkenes

We just described two methods for oxidizing alkenes to glycols. Stronger conditions can further oxidize glycols to cleave the bond that was originally the double bond. Potassium permanganate and ozone are two of the most common reagents for such **oxidative cleavages.**

8-15A Cleavage by Permanganate

In a potassium permanganate dihydroxylation, if the solution is warm or acidic or too concentrated, oxidative cleavage of the glycol may occur. In effect, the double bond is cleaved to two carbonyl groups. The products are initially ketones and aldehydes, but aldehydes are oxidized to carboxylic acids under these strong oxidizing conditions. If the molecule contains a terminal $=CH_2$ group, that group is oxidized all the way to CO_2 and water.

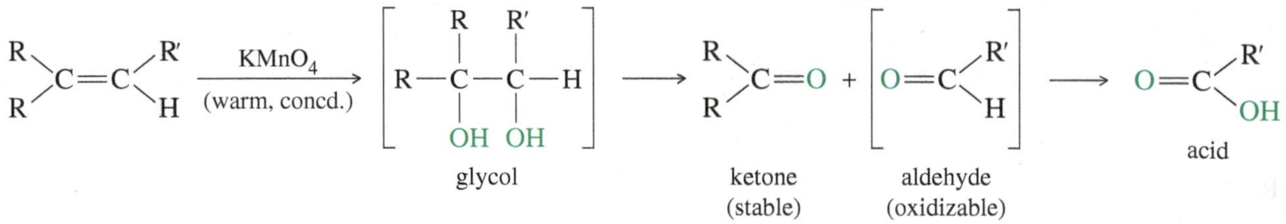

$R_2C=CHR'$ $\xrightarrow[\text{(warm, concd.)}]{KMnO_4}$ [R–C(R)(OH)–C(R')(OH)–H] glycol ⟶ $R_2C=O$ ketone (stable) + [O=C(R')H] aldehyde (oxidizable) ⟶ O=C(R')OH acid

Examples

(warm, concd.) $KMnO_4$ → +

$KMnO_4$ (warm, concd.) → COOH COOH O + CO_2

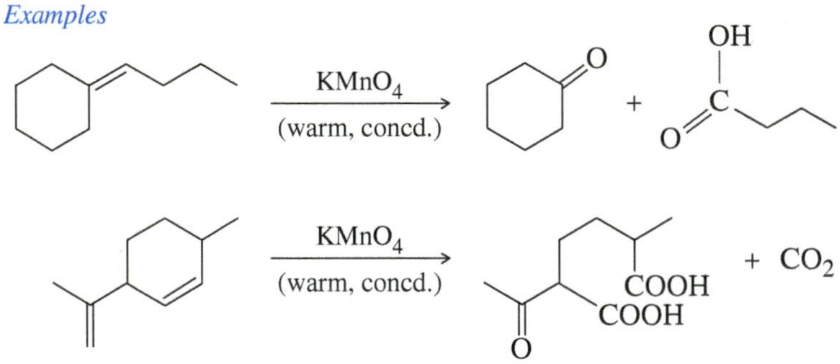

8-15B Ozonolysis

Like permanganate, ozone cleaves double bonds to give ketones and aldehydes. However, ozonolysis is milder, and both ketones and aldehydes can be recovered without further oxidation.

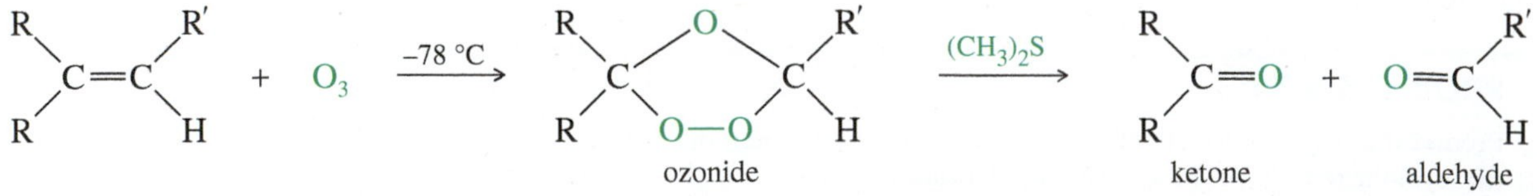

$R_2C=CHR'$ + O_3 $\xrightarrow{-78\ °C}$ ozonide $\xrightarrow{(CH_3)_2S}$ $R_2C=O$ ketone + O=CHR' aldehyde

Ozone (O_3) is a high-energy form of oxygen produced when ultraviolet light or an electrical discharge passes through oxygen gas. Ultraviolet light from the Sun converts oxygen to ozone in the upper atmosphere. This "ozone layer" shields Earth from some of the high-energy ultraviolet radiation it would otherwise receive.

$$\tfrac{3}{2}O_2 + 142\ \text{kJ (34 kcal)} \longrightarrow O_3$$

Ozone has 142 kJ/mol of excess energy over oxygen, and it is much more reactive. A Lewis structure of ozone shows that the central oxygen atom bears a positive charge, and that each of the outer oxygen atoms bears half a negative charge.

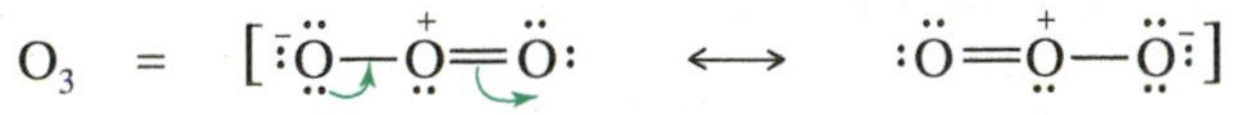

$$O_3 = [\,:\ddot{O}-\overset{+}{O}=\ddot{O}: \longleftrightarrow :\ddot{O}=\overset{+}{O}-\ddot{O}:\,]$$

Ozone reacts with an alkene to form a cyclic compound called a *primary ozonide* or *molozonide* (because 1 mole of ozone has been added). This reaction is usually cooled with dry ice (−78 °C) to minimize overoxidation and other side reactions. The molozonide has two peroxy (—O—O—) linkages, so it is quite unstable. It rearranges rapidly, even at low temperatures, to form an ozonide.

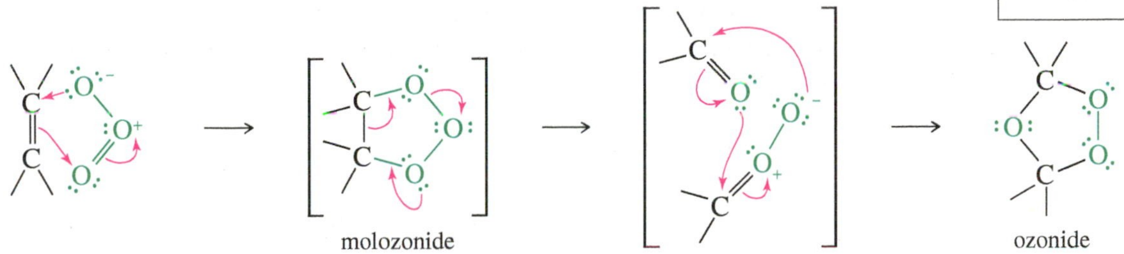

molozonide (primary ozonide)

ozonide

Ozonides are not very stable, and they are rarely isolated. In most cases, they are immediately reduced by a mild reducing agent such as zinc or (more recently) dimethyl sulfide. The products of this reduction are ketones and aldehydes.

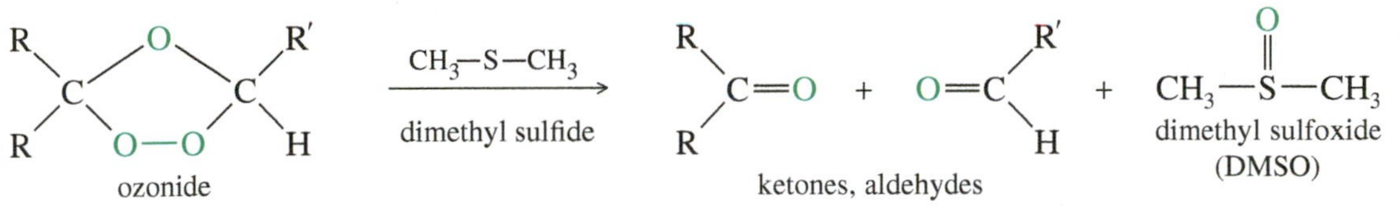

ozonide $\xrightarrow[\text{dimethyl sulfide}]{CH_3-S-CH_3}$ $R_2C=O + O=CHR'$ (ketones, aldehydes) + $CH_3-\overset{O}{\overset{\|}{S}}-CH_3$ dimethyl sulfoxide (DMSO)

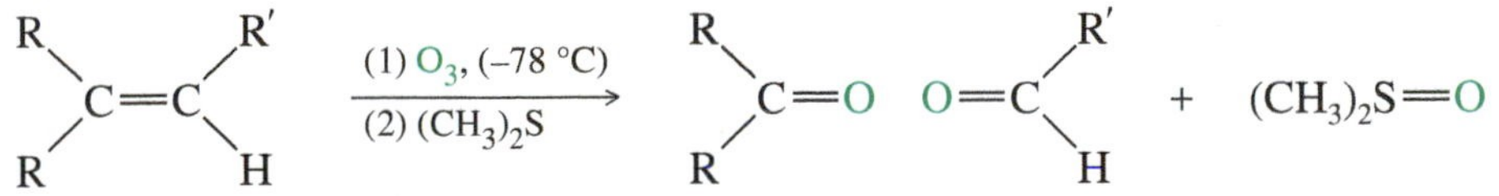

$$R_2C=CHR' \xrightarrow[(2)\ (CH_3)_2S]{(1)\ O_3,\ (-78\ °C)} R_2C=O \quad O=CHR' + (CH_3)_2S=O$$

The following reactions show the products obtained from ozonolysis of some representative alkenes. Note how (1) and (2) are used with a single reaction arrow to denote the steps in a two-step sequence.

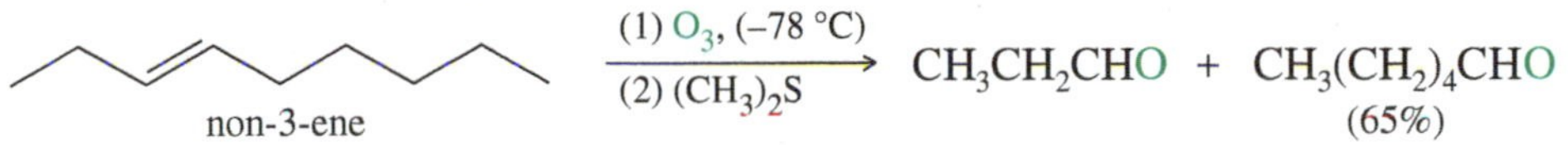

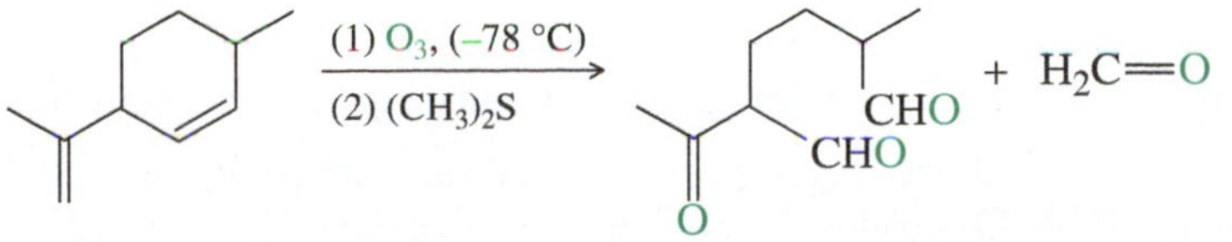

non-3-ene $\xrightarrow[(2)\ (CH_3)_2S]{(1)\ O_3,\ (-78\ °C)}$ $CH_3CH_2CHO + CH_3(CH_2)_4CHO$ (65%)

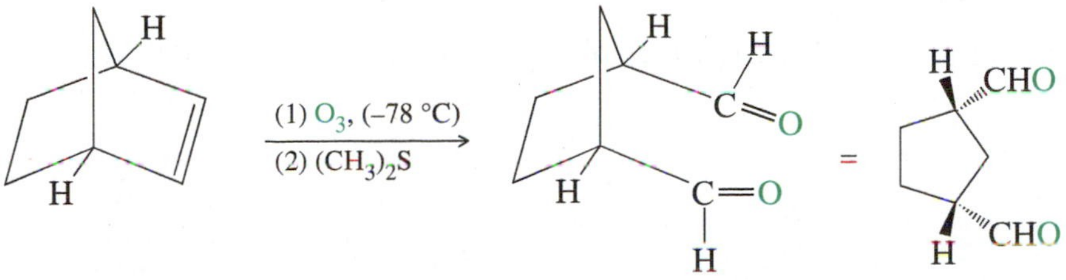

$\xrightarrow[(2)\ (CH_3)_2S]{(1)\ O_3,\ (-78\ °C)}$ = cyclopentane-1,3-dicarbaldehyde (H, CHO; H, CHO)

$\xrightarrow[(2)\ (CH_3)_2S]{(1)\ O_3,\ (-78\ °C)}$ CHO, CHO + $H_2C=O$

One of the most common uses of ozonolysis has been for determining the positions of double bonds in alkenes. For example, if we were uncertain of the position of the

Application: Air Pollution

Ozone is a powerful lung irritant, causing a cough, a sore throat, and tiredness. It can also increase a person's sensitivity to allergens. The mechanism may involve oxidation of the double bonds of the fatty acids that make up the surfactants and the membranes of the cells lining the bronchial airways and lungs.

methyl group in a methylcyclopentene, the products of ozonolysis–reduction would confirm the structure of the original alkene.

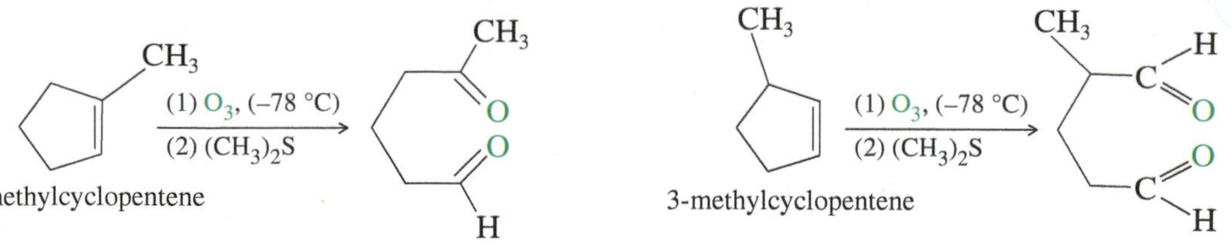

1-methylcyclopentene $\xrightarrow[(2)\ (CH_3)_2S]{(1)\ O_3,\ (-78\ °C)}$

3-methylcyclopentene $\xrightarrow[(2)\ (CH_3)_2S]{(1)\ O_3,\ (-78\ °C)}$

PROBLEM-SOLVING HINT

To predict the products from ozonolysis of an alkene, erase the double bond and add two oxygen atoms as carbonyl (C=O) groups where the double bond used to be.

SOLVED PROBLEM 8-7

Ozonolysis–reduction of an unknown alkene gives an equimolar mixture of cyclohexanecarbaldehyde and butan-2-one. Determine the structure of the original alkene.

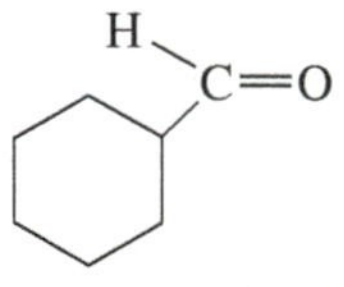

cyclohexanecarbaldehyde

$CH_3-\overset{O}{\overset{\|}{C}}-CH_2-CH_3$

butan-2-one

SOLUTION

We can reconstruct the alkene by removing the two oxygen atoms of the carbonyl groups (C=O) and connecting the remaining carbon atoms with a double bond. One uncertainty remains, however: The original alkene might be either of two possible geometric isomers.

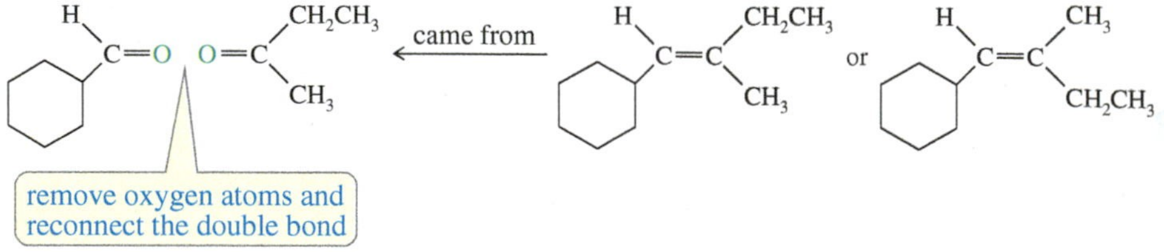

came from

or

remove oxygen atoms and reconnect the double bond

Application: Disinfectant

Ozone is a strong oxidizing agent that can be used instead of chlorine to disinfect the water in swimming pools. Ozone oxidizes organic matter, and it kills bacteria and algae. Ozone is used instead of chlorine because it can be generated on-site (rather than storing and using toxic chemicals such as chlorine gas or sodium hypochlorite) and because it doesn't produce as many harmful by-products.

PROBLEM 8-36

Give structures of the alkenes that would give the following products upon ozonolysis–reduction.

(a) $CH_3-\overset{O}{\overset{\|}{C}}-CH_2-CH_2-CH_2-\overset{O}{\overset{\|}{C}}-CH_2-CH_3$

(b) cyclohexanone and $CH_3-CH_2-CH_2-\overset{O}{\overset{\|}{C}}-H$

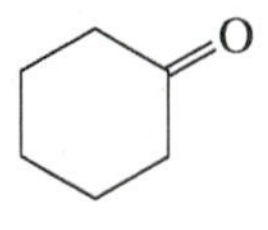

(c) $CH_3-CH_2-\overset{O}{\overset{\|}{C}}-CH_2-CH_2-CH_2-CH_3$ and $CH_3-CH_2-\overset{O}{\overset{\|}{C}}-H$

8-15C Comparison of Permanganate Cleavage and Ozonolysis

Both permanganate and ozonolysis break the carbon–carbon double bond and replace it with carbonyl (C=O) groups. In the permanganate cleavage, any aldehyde products are further oxidized to carboxylic acids. In the ozonolysis–reduction procedure, the aldehyde products are generated in the dimethyl sulfide reduction step (and not in the presence of ozone), and they are not oxidized.

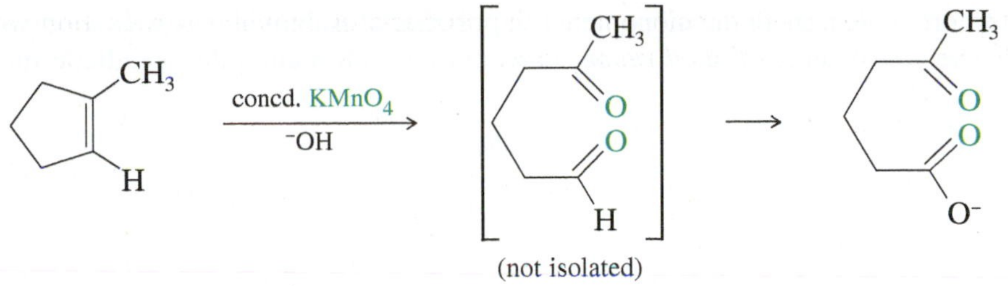

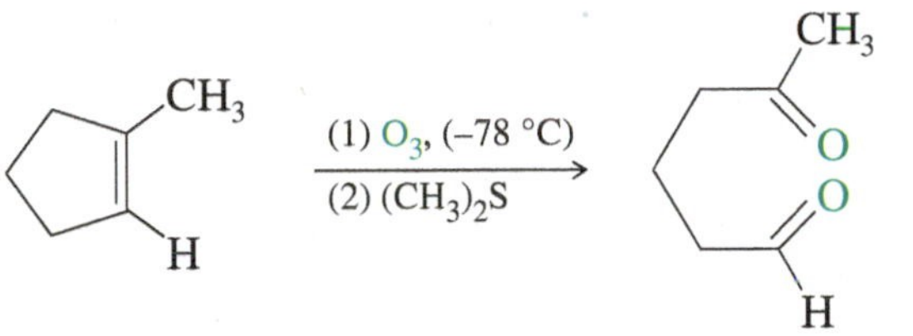

concd. $KMnO_4$, ^-OH → (not isolated) →

(1) O_3, (−78 °C)
(2) $(CH_3)_2S$

PROBLEM 8-37

Predict the major products of the following reactions.
(a) (*E*)-3-methyloct-3-ene + ozone, then $(CH_3)_2S$
(b) (*Z*)-3-methyloct-3-ene + warm, concentrated $KMnO_4$
(c) 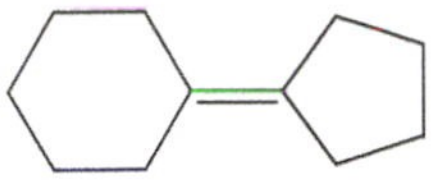 + O_3, then $(CH_3)_2S$
(d) 1-ethylcycloheptene + ozone, then $(CH_3)_2S$
(e) 1-ethylcycloheptene + warm, concentrated $KMnO_4$
(f) 1-ethylcycloheptene + cold, dilute $KMnO_4$

PROBLEM-SOLVING HINT

Three reagents oxidize the pi bond of an alkene but leave the sigma bond intact: (a) cold, dilute $KMnO_4$; (b) OsO_4-H_2O_2; and (c) RCO_3H epoxidation.
Two reagents break the double bond entirely, giving carbonyl groups: (a) warm, concentrated $KMnO_4$ and (b) O_3 followed by $(CH_3)_2S$.

8-16 Polymerization of Alkenes

A **polymer** is a large molecule composed of many smaller repeating units (the **monomers**) bonded together. Alkenes serve as monomers for some of the most common polymers, such as polyethylene, polypropylene, polystyrene, poly(vinyl chloride), and many others. Alkenes polymerize to give **addition polymers** resulting from repeated addition reactions across their double bonds.

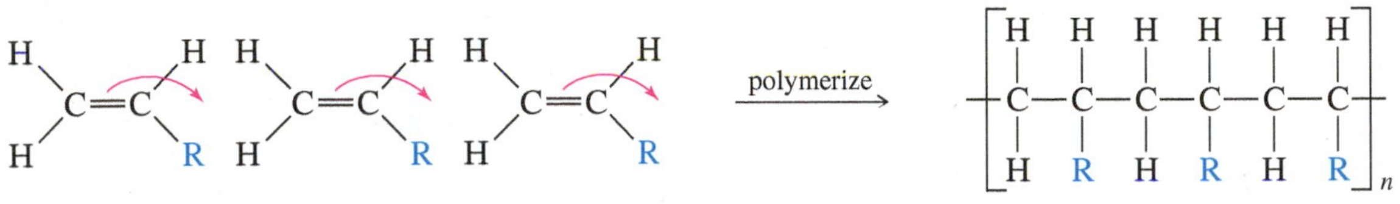

monomers — polymerize → addition polymer

R = H, CH_3, Ph, Cl, etc.

Addition polymers generally form by **chain-growth polymerization,** the rapid addition of one molecule at a time to a growing polymer chain. There is generally a reactive intermediate (cation, anion, or radical) at the growing end of the chain. The chain-growth mechanism involves addition of the reactive end of the chain across the double bond of the alkene monomer. The next sections show how this occurs with cations, radicals, and anions at the growing ends of the chains.

8-16A Cationic Polymerization

Alkenes that easily form carbocations are good candidates for **cationic polymerization,** which is just another example of electrophilic addition to an alkene. Consider what happens when pure isobutylene is treated with a trace of concentrated sulfuric acid. Protonation of the alkene forms a carbocation. If a large concentration of isobutylene is available, another molecule of the alkene may act as the nucleophile and attack the carbocation to form the *dimer* (two monomers joined together) and give another carbocation. If the conditions are right, the growing cationic end of the chain will keep adding across more molecules of the monomer. The polymer of isobutylene is *polyisobutylene*, one of the constituents of *butyl rubber* used in inner tubes and other synthetic rubber products.

Protonation *Attack by the second molecule of isobutylene*

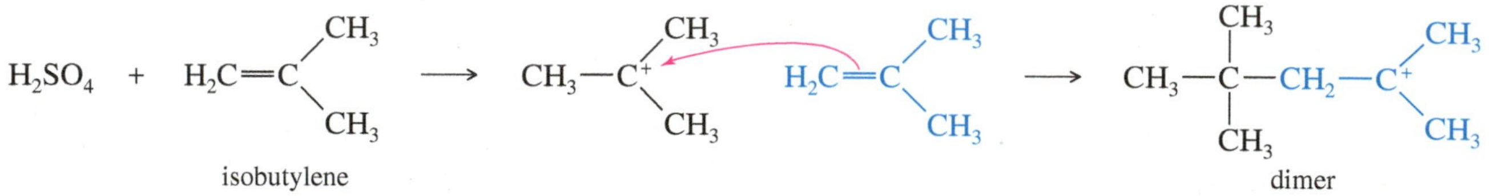

$$H_2SO_4 + H_2C{=}C(CH_3)_2 \longrightarrow CH_3{-}\overset{+}{C}(CH_3)_2 \quad H_2C{=}C(CH_3)_2 \longrightarrow CH_3{-}C(CH_3)_2{-}CH_2{-}\overset{+}{C}(CH_3)_2$$

isobutylene; dimer

Attack by a third molecule to give a trimer

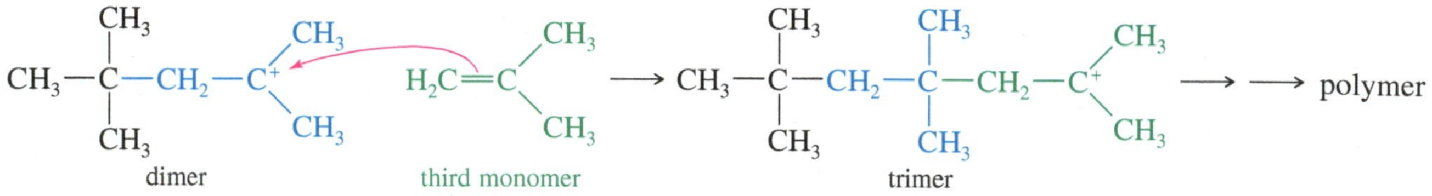

$$CH_3{-}C(CH_3)_2{-}CH_2{-}\overset{+}{C}(CH_3)_2 \quad H_2C{=}C(CH_3)_2 \longrightarrow CH_3{-}C(CH_3)_2{-}CH_2{-}C(CH_3)_2{-}CH_2{-}\overset{+}{C}(CH_3)_2 \longrightarrow \longrightarrow \text{polymer}$$

dimer; third monomer; trimer

Loss of a proton is the most common side reaction that terminates chain growth:

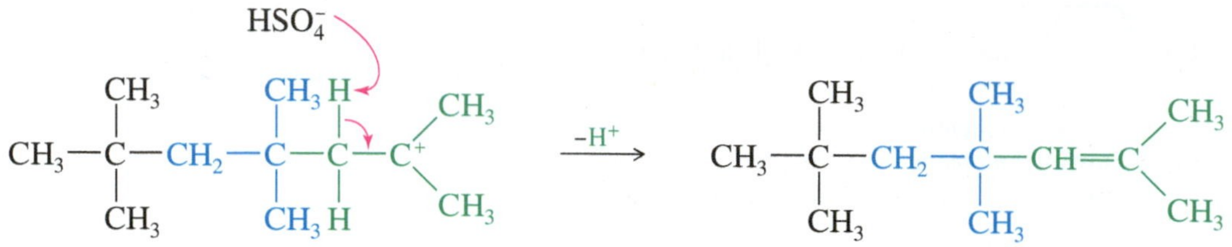

$$HSO_4^- \quad CH_3{-}C(CH_3)_2{-}CH_2{-}C(CH_3)_2{-}CH_2{-}\overset{+}{C}(CH_3)_2 \xrightarrow{-H^+} CH_3{-}C(CH_3)_2{-}CH_2{-}C(CH_3)_2{-}CH{=}C(CH_3)_2$$

Boron trifluoride (BF_3) is an excellent catalyst for cationic polymerization because it leaves no good nucleophile that might attack a carbocation intermediate and end the polymerization. Boron trifluoride is electron-deficient and a strong Lewis acid. It usually contains a trace of water that acts as a co-catalyst by adding to BF_3 and then protonating the monomer. Protonation occurs at the less substituted end of an alkene double bond to give the more stable carbocation. Each additional monomer molecule adds with the same orientation, always giving the more stable carbocation. The following reaction shows the polymerization of styrene (vinylbenzene) using BF_3 as the catalyst.

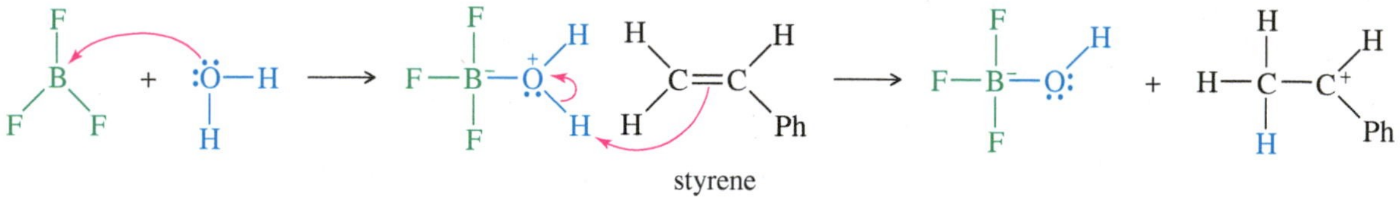

$$BF_3 + H_2\ddot{O} \longrightarrow F_3\overset{-}{B}{-}\overset{+}{O}H_2 \quad H_2C{=}CH(Ph) \longrightarrow F_3\overset{-}{B}{-}\ddot{O}H + H{-}CH_2{-}\overset{+}{C}H(Ph)$$

styrene

First chain-lengthening step

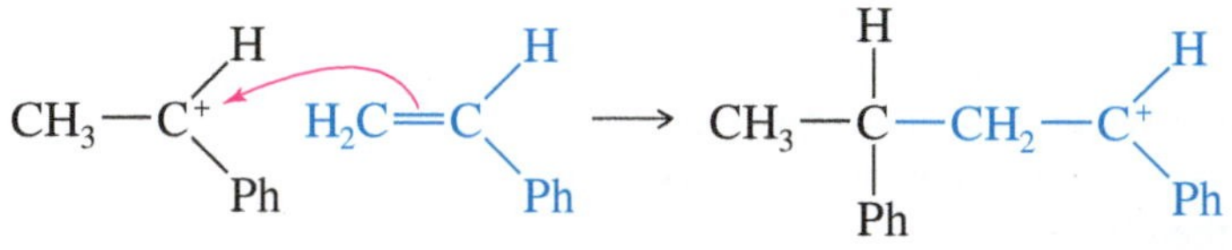

$$CH_3{-}\overset{+}{C}H(Ph) \quad H_2C{=}CH(Ph) \longrightarrow CH_3{-}CH(Ph){-}CH_2{-}\overset{+}{C}H(Ph)$$

After many steps, the polymerization continues

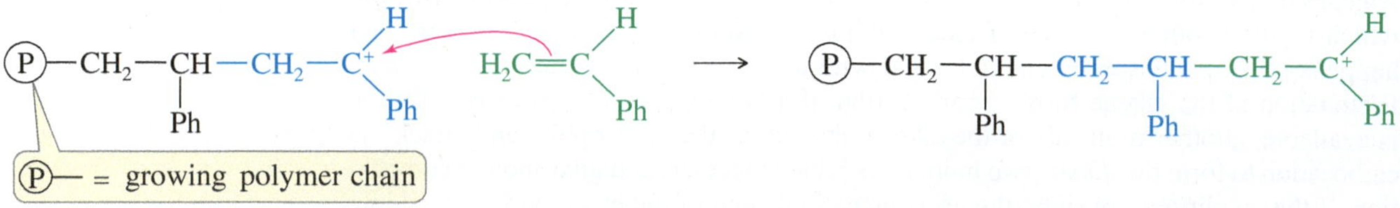

$$Ⓟ{-}CH_2{-}CH(Ph){-}CH_2{-}\overset{+}{C}H(Ph) \quad H_2C{=}CH(Ph) \longrightarrow Ⓟ{-}CH_2{-}CH(Ph){-}CH_2{-}CH(Ph){-}CH_2{-}\overset{+}{C}H(Ph)$$

Ⓟ— = growing polymer chain

The most likely ending of this BF_3-catalyzed polymerization is the loss of a proton from the carbocation at the end of the chain. This side reaction terminates one chain, but it also protonates another molecule of styrene, initiating a new chain.

Termination of a polymer chain

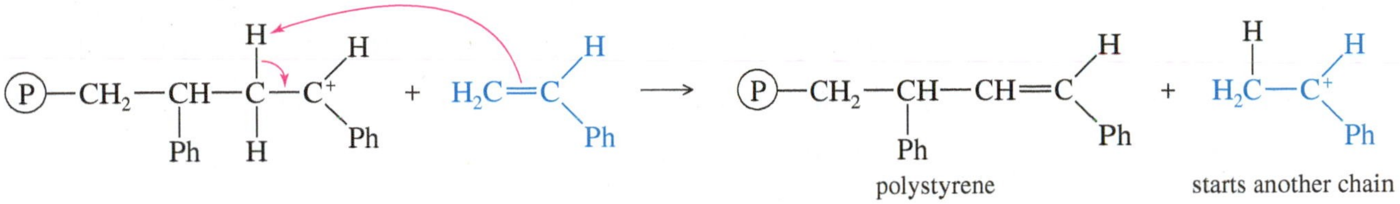

$$\text{Ⓟ}-CH_2-CH(Ph)-CH_2-C^+(H)(Ph) + H_2C{=}C(H)(Ph) \longrightarrow \text{Ⓟ}-CH_2-CH(Ph)-CH{=}C(H)(Ph) + H_3C-C^+(H)(Ph)$$

polystyrene; starts another chain

The product of this polymerization is polystyrene: a clear, brittle plastic that is often used for inexpensive lenses, transparent containers, and Styrofoam® insulation. Polystyrene is also the major component of the resin beads that are used to make synthetic proteins. (See Section 24-11.)

PROBLEM 8-38

(a) Propose a mechanism for the following reaction.

$$2\ (CH_3)_2C{=}CH{-}CH_3 + \text{cat. } H^+ \longrightarrow \text{2,3,4,4-tetramethylhex-2-ene}$$

(b) Show the first three steps (as far as the tetramer) in the BF_3-catalyzed polymerization of propylene to form polypropylene.

PROBLEM 8-39

When cyclohexanol is dehydrated to cyclohexene, a gummy green substance forms on the bottom of the flask. Suggest what this residue might be, and propose a mechanism for its formation (as far as the dimer).

8-16B Free-Radical Polymerization

Many alkenes undergo **free-radical polymerization** when they are heated with radical initiators. For example, styrene polymerizes to polystyrene when it is heated to 100 °C with a peroxide initiator. A radical adds to styrene to give a resonance-stabilized radical, which then attacks another molecule of styrene to give an elongated radical.

Initiation step $\quad ROOR \xrightarrow{\text{heat}} 2\ RO\cdot$

Propagation step

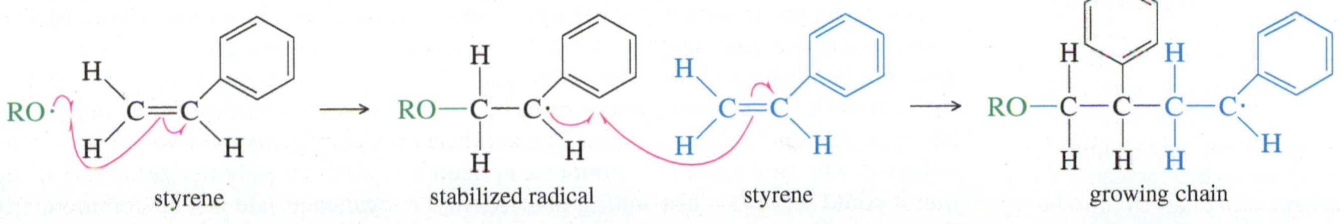

$$RO\cdot + H_2C{=}CH(C_6H_5) \longrightarrow RO-CH_2-\dot{C}H(C_6H_5) \xrightarrow{H_2C{=}CH(C_6H_5)} RO-CH_2-CH(C_6H_5)-CH_2-\dot{C}H(C_6H_5)$$

styrene; stabilized radical; styrene; growing chain

Each propagation step adds another molecule of styrene to the radical end of the growing chain. This addition always takes place with the orientation that gives another resonance-stabilized benzylic (next to a benzene ring) radical.

Propagation step

growing chain + styrene → elongated chain —add many more styrene molecules→ polystyrene

polystyrene
n = about 100 to 10,000

Chain growth may continue with addition of several hundred or several thousand styrene units. Eventually, the chain reaction stops, either by the coupling of two chains or by reaction with an impurity (such as oxygen) or simply by running out of monomer.

PROBLEM 8-40

Show the intermediate that would result if the growing chain added to the other end of the styrene double bond. Explain why the final polymer has phenyl groups substituted on alternate carbon atoms rather than randomly distributed.

Ethylene is also polymerized by free-radical chain-growth polymerization. With ethylene, the free-radical intermediates are less stable, so stronger reaction conditions are required. Ethylene is commonly polymerized by free-radical initiators at pressures around 3000 atm and temperatures of about 200 °C. The product, called *low-density polyethylene*, is the material commonly used in polyethylene bags.

This 1949 DuPont promotional photo shows how a Teflon rod (on our right, in the model's left hand) resists a hot acid solution compared with a rod made from another plastic. In World War II, Teflon was used for insulation in aircraft wiring and for seals in the equipment used to enrich corrosive uranium hexafluoride.

PROBLEM 8-41

The structures of three monomers are shown. In each case, show the structure of the polymer that would result from polymerization of the monomer. Vinyl chloride is polymerized to "vinyl" plastics and PVC pipe. Tetrafluoroethylene polymerizes to Teflon®, used as non-stick coatings and PTFE valves and gaskets. Acrylonitrile is polymerized to Orlon®, used in sweaters and carpets.

$H_2C{=}CHCl$ vinyl chloride; $F_2C{=}CF_2$ tetrafluoroethylene; $H_2C{=}CH{-}C{\equiv}N$ acrylonitrile

8-16C Anionic Polymerization

Like cationic polymerization, **anionic polymerization** depends on the presence of a stabilizing group. To stabilize anions, the double bond should have a strong electron-withdrawing group such as a carbonyl group, a cyano group, or a nitro group. Methyl α-cyanoacrylate contains two powerful electron-withdrawing groups, and it undergoes nucleophilic additions very easily. If this liquid monomer is spread in a thin film between two surfaces, traces of basic impurities (metal oxides, for example) can catalyze its rapid polymerization. The solidified polymer joins the two surfaces. The chemists who first made this monomer noticed how easily it polymerizes and realized that it could serve as a fast-setting glue. Methyl α-cyanoacrylate is sold commercially as "super" glue.

Initiation step

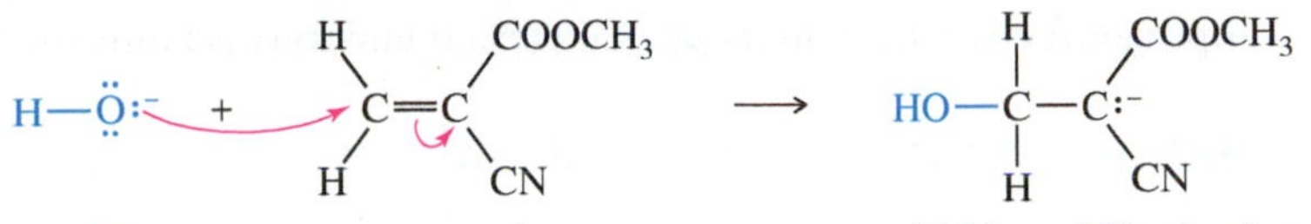

$$H{-}\ddot{\underset{..}{O}}{:}^{-} + \underset{H}{\overset{H}{>}}C{=}C\underset{CN}{\overset{COOCH_3}{<}} \longrightarrow HO{-}\underset{H}{\overset{H}{C}}{-}C{:}^{-}\underset{CN}{\overset{COOCH_3}{<}}$$

trace of base — super glue — highly stabilized anion

Chain-lengthening step

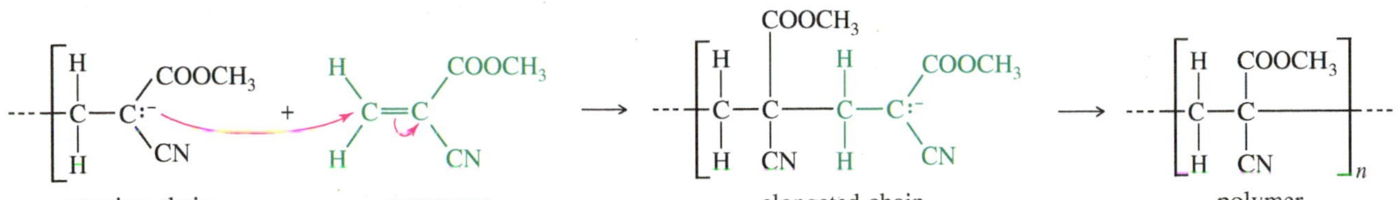

$$\cdots{-}\underset{H}{\overset{H}{C}}{-}C{:}^{-}\underset{CN}{\overset{COOCH_3}{<}} + \underset{H}{\overset{H}{>}}C{=}C\underset{CN}{\overset{COOCH_3}{<}} \longrightarrow \cdots{-}\underset{H}{\overset{H}{C}}{-}\underset{CN}{\overset{COOCH_3}{C}}{-}\underset{H}{\overset{H}{C}}{-}C{:}^{-}\underset{CN}{\overset{COOCH_3}{<}} \longrightarrow \cdots\left[\underset{H}{\overset{H}{C}}{-}\underset{CN}{\overset{COOCH_3}{C}}\right]_n\cdots$$

growing chain — monomer — elongated chain — polymer

PROBLEM 8-42

Draw a mechanism for a base-catalyzed polymerization of methyl α-methacrylate to give the Plexiglas® polymer.

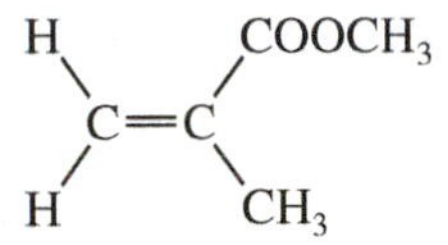

$$\underset{H}{\overset{H}{>}}C{=}C\underset{CH_3}{\overset{COOCH_3}{<}}$$

methyl α-methacrylate

8-17 Olefin Metathesis

The double bond is the strongest bond in an alkene, yet it is also the most reactive bond. Imagine how useful it would be if we could break molecules at their double bonds and reassemble them as we please. That is the goal of olefin metathesis. We can think of an alkene as two **alkylidene groups** (=CHR) held together by the double bond, and mentally divide it up just like we divide the molecule when we go to name it as *E* or *Z* (Section 7-5B). **Olefin metathesis** is any reaction that trades and interchanges these alkylidene groups. The word *metathesis* comes from the Greek words *meta* (change) and *thesis* (position), meaning that the alkylidene groups change their positions in the products. Figure 8-10 shows the trading of alkylidene groups that takes place during olefin metathesis.

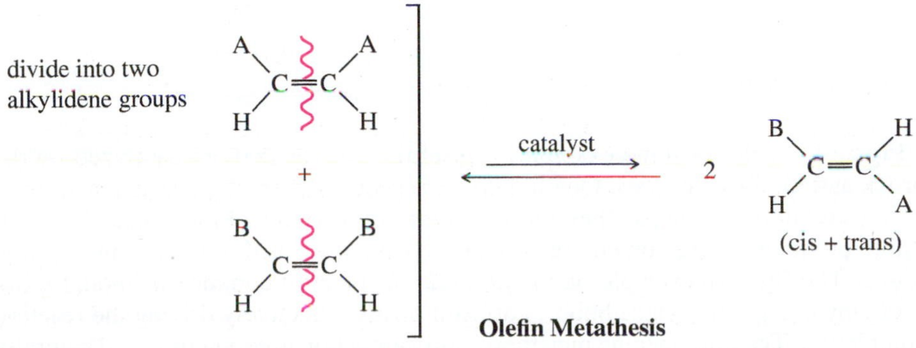

divide into two alkylidene groups

$$\left[\underset{H}{\overset{A}{>}}C{=}C\underset{H}{\overset{A}{<}} + \underset{H}{\overset{B}{>}}C{=}C\underset{H}{\overset{B}{<}}\right] \underset{}{\overset{\text{catalyst}}{\rightleftharpoons}} 2\ \underset{H}{\overset{B}{>}}C{=}C\underset{A}{\overset{H}{<}}$$

(cis + trans)

Olefin Metathesis

FIGURE 8-10 Olefin metathesis. During metathesis, the alkylidene groups of the reactant olefins trade partners and rearrange to give new combinations of alkenes in the products.

The 2005 Nobel Prize in Chemistry was awarded to Yves Chauvin (French Petroleum Institute), Robert Grubbs (Caltech), and Richard Schrock (MIT) for developing effective ways to induce alkenes to undergo metathesis.

8-17A Catalysts for Olefin Metathesis

Olefin metathesis was first observed in the 1950s, and was used in industry to convert propylene to a mixture of but-2-ene and ethylene. This *Phillips Triolefin Process* used an aluminum/molybdenum catalyst whose exact structure was unknown.

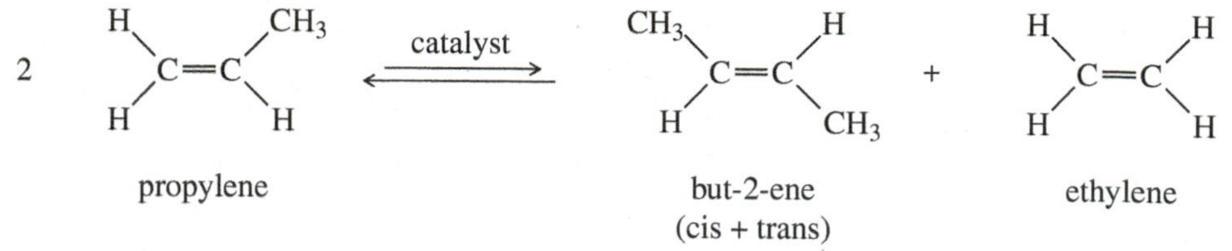

$$2\ H_2C{=}CHCH_3 \underset{}{\overset{\text{catalyst}}{\rightleftharpoons}} CH_3CH{=}CHCH_3 + H_2C{=}CH_2$$

propylene — but-2-ene (cis + trans) — ethylene

Around 1990, Richard Schrock developed versatile molybdenum and tungsten catalysts for olefin metathesis that tolerate a wide range of functional groups in the alkylidene fragments of the olefins. The Schrock catalyst shown in Figure 8-11a is now commercially available. The Schrock catalysts tend to be air- and moisture-sensitive, which limits their use in commercial processes.

In 1992, Robert Grubbs developed a ruthenium phosphine catalyst (Figure 8-11b) that is less sensitive to oxygen and moisture than the Schrock catalysts, and tolerates even more functional groups in the alkylidene fragments of the olefins. Both the Schrock and Grubbs catalysts have a metal atom that is double-bonded to an alkylidene (=CHR) group. They can be symbolized [M]=CHR, where the [M] in brackets signifies that the metal atom has other ligands that fine-tune its reactivity.

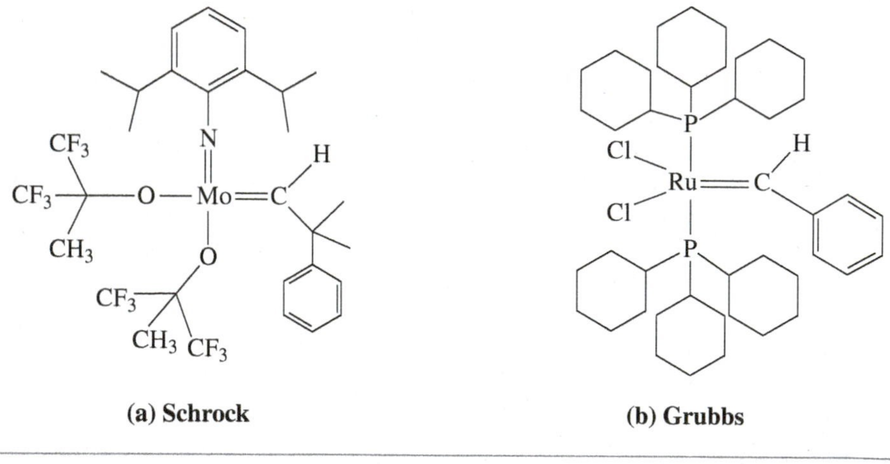

FIGURE 8-11
(a) One of the Schrock molybdenum metathesis catalysts. (b) One of the Grubbs ruthenium metathesis catalysts.

Figure 8-12 shows some examples of useful reactions that are catalyzed by the Schrock and Grubbs catalysts. One important aspect of these metathesis reactions is that they are all reversible, so they form equilibrium mixtures of the reactants and all possible products unless something is done to drive the reaction toward the desired products. The first two examples in Figure 8-12 use the most common method, formation of ethylene gas. Ethylene bubbles off as it forms, effectively driving the reaction to completion. The ring-opening metathesis polymerization is exothermic and naturally goes to products because the ring strain in the bicyclic norbornene is released when the ring opens to form the polymer.

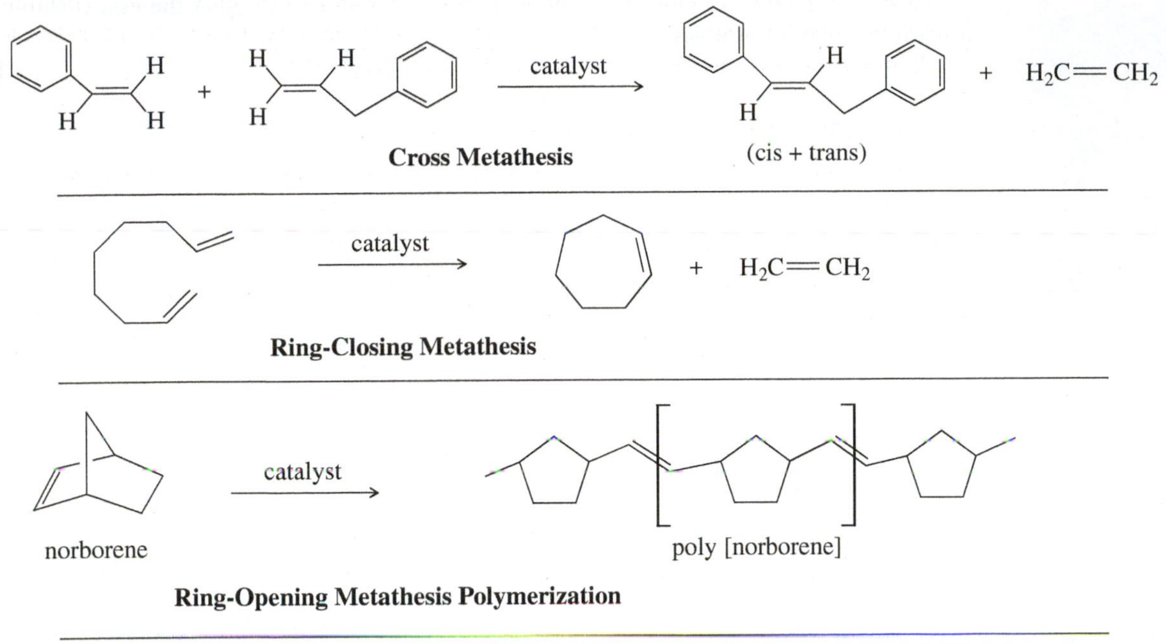

Cross Metathesis

catalyst

(cis + trans)

+ $H_2C{=}CH_2$

catalyst

+ $H_2C{=}CH_2$

Ring-Closing Metathesis

catalyst

norborene

poly [norborene]

Ring-Opening Metathesis Polymerization

FIGURE 8-12
Useful examples of metathesis reactions.

8-17B Mechanism of Olefin Metathesis

Several mechanisms were proposed to explain the catalytic metathesis reactions, but the mechanism published by Yves Chauvin in 1971 has come to be accepted as correct. We can think of an alkene as two alkylidene groups bonded together. Similarly, the Schrock and Grubbs catalysts are like a metal atom bonded to one alkylidene group.

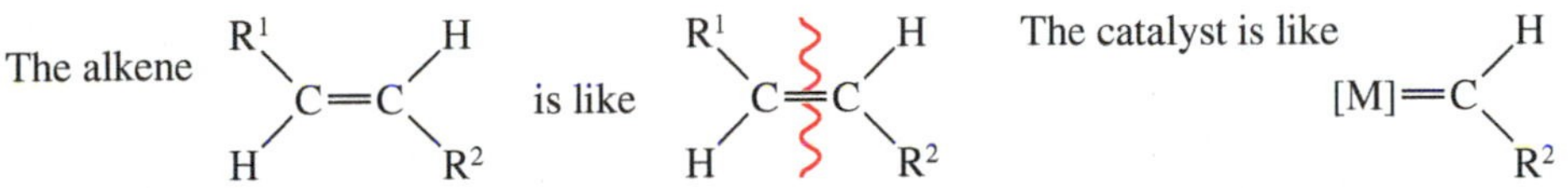

The alkene $R^1(H)C{=}C(H)R^2$ is like $R^1(H)C \wr C(H)R^2$ The catalyst is like $[M]{=}C(H)R^2$

Chauvin proposed that the metal-alkylidene catalyst forms an intermediate four-membered ring with an alkene, as shown in Mechanism 8-11. Then the ring breaks apart, either to give the starting alkene and catalyst or to give a new alkene that has traded one alkylidene group with the catalyst.

This mechanism allows the alkylidene groups to change partners back and forth with the catalytic metal until a thermodynamic equilibrium is reached. As we saw earlier, good yields of products result if there is an effective driving force (such as

MECHANISM 8-11 Olefin Metathesis

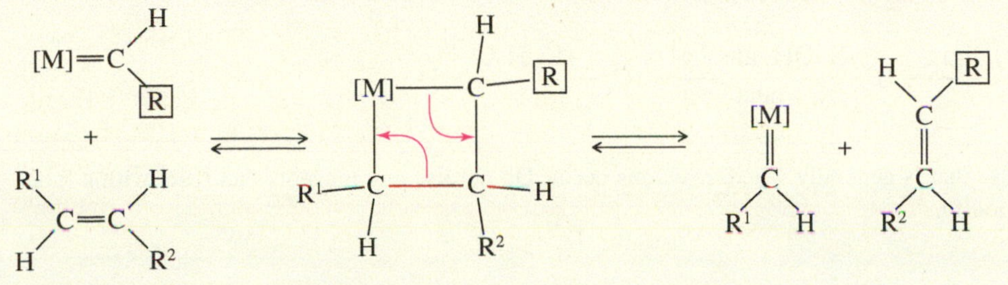

$[M]{=}C(H)R$ + $R^1(H)C{=}C(H)R^2$ ⇌ [four-membered ring: [M]—C(H)(R)—C(H)(R^2)—C(H)(R^1)] ⇌ $[M]{=}C(R^1)(H)$ + $R(H)C{=}C(R^2)(H)$

The omnivorous leafroller (OLR) feeds on a wide variety of fruits, vegetables, and ornamentals. Vineyards use pheromone traps to monitor the OLR populations and determine when control methods are needed.

formation of a gaseous by-product or release of ring strain) to push the equilibrium toward the desired products.

PROBLEM 8-43

Propose a mechanism for the triolefin process using a metal alkylidene as the catalyst.

$$2\ \text{H}_2\text{C{=}CHCH}_3 \underset{}{\overset{[\text{M}]{=}\text{CHCH}_3}{\rightleftharpoons}} \text{CH}_3\text{CH{=}CHCH}_3 + \text{H}_2\text{C{=}CH}_2$$

propylene — but-2-ene (cis + trans) — ethylene

PROBLEM 8-44

Show what reagents would be needed to synthesize the pheromone of the omnivorous leafroller (OLR) using olefin metathesis to assemble the molecule at the double bond.

OLR pheromone (cis + trans)

PROBLEM-SOLVING STRATEGY Organic Synthesis

Alkyl halides and alkenes are readily made from other compounds, and they are easily converted to other functional groups. This flexibility makes them useful as reagents and intermediates for organic synthesis. Alkenes are particularly important for industrial syntheses because they are inexpensive and available in large quantities from cracking and dehydrogenation of petroleum fractions.

Organic synthesis is the preparation of desired compounds from readily available materials. Synthesis is one of the major areas of organic chemistry, and nearly every chapter of this book involves organic synthesis in some way. A synthesis may be a simple one-step reaction, or it may involve many steps and incorporate a subtle strategy for assembling the correct carbon skeleton with all the functional groups in the right positions.

Many of the problems in this book are synthesis problems. In some synthesis problems, you are asked to show how to convert a given starting material to the desired product. There are obvious one-step answers to some of these problems, but others may require several steps, and there may be many correct answers. In solving multistep synthetic problems, it is often helpful to analyze the problem backward: Begin with the desired product (called the *target compound*) and see how it might be mentally changed or broken down to give the starting materials. This backward approach to synthesis is called a **retrosynthetic analysis.**

Some problems allow you to begin with any compounds that meet a certain restriction. For example, you might be allowed to use any alcohols containing no more than four carbon atoms. A retrosynthetic analysis can be used to break down the target compound into fragments no larger than four carbon atoms; then those fragments could be formed from the appropriate alcohols by functional group chemistry.

The following suggestions should help you solve synthesis problems:

1. Do not guess a starting material and try every possible reaction to convert it to the target compound. Rather, begin with the target compound and use a retrosynthetic analysis to simplify it.
2. Use simple equations, with reagents written above and below the arrows, to show the reactions. The equations do not need to be balanced, but they should include all the reagents and conditions that are important to the success of the reaction.

$$\text{A} \xrightarrow{\text{Br}_2,\ \text{light}} \text{B} \xrightarrow[\text{heat}]{\text{NaOH, alcohol}} \text{C} \xrightarrow{\text{H}^+,\ \text{H}_2\text{O}} \text{D}$$

3. Focus on the functional groups, because that is generally where reactions occur. Do not use any reagents that react with a functional group that you don't intend to modify.

In solving multistep synthesis problems, you will rarely be able to "see" the solution immediately. These problems are best approached systematically, working backward and considering alternative routes. To illustrate a systematic approach that can guide you in solving synthesis problems, we will work through the synthesis of a complex ether starting from alkenes. The problem-solving method described here will be extended in future chapters to multistep syntheses based on the reactions of additional functional groups.

A systematic retrosynthetic analysis begins with an examination of the structure of the product. We will consider the synthesis of the following compound from alkenes containing up to five carbon atoms.

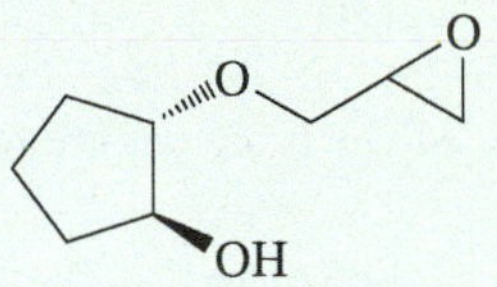

1. **Review the functional groups and carbon skeleton of the target compound.**
   The target compound is an ether. One alkyl group is a five-carbon cyclopentane ring with two oxygen atoms situated trans. The other group has three carbons containing a reactive epoxide ring.
2. **Review the functional groups and carbon skeletons of the starting materials (if specified), and see how their skeletons might fit together in the target compound.**
   The synthesis is to begin with alkenes containing up to five carbon atoms, so all the functional groups in the product must be derived from alkenes. Most likely, we will start with cyclopentene to give the five-carbon ring and propene to give the three-carbon chain.
3. **Compare methods for synthesizing the functional groups in the target compound, and select the reactions that are most likely to give the correct product.**
   This step may require writing several possible reactions and evaluating them.
   Ethers can be synthesized by nucleophilic reactions between alkyl halides and alkoxides (Section 6-9). The target compound might be formed by S_N2 attack of an alkoxide ion on an alkyl halide in either of two ways shown below:

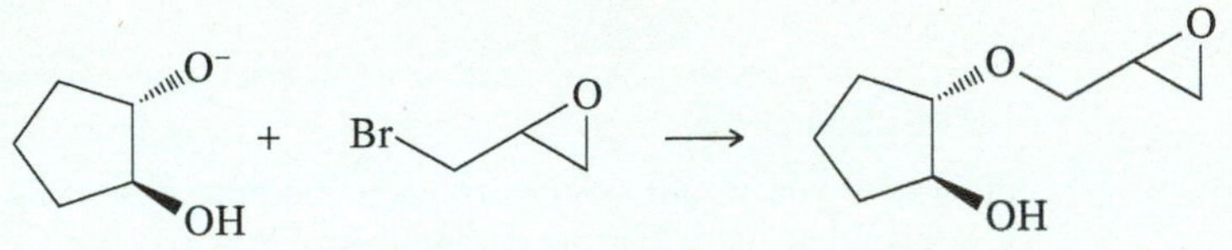

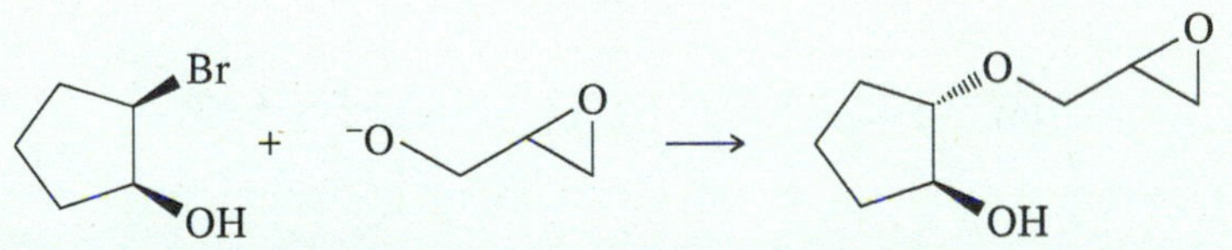

   The first reaction is better because the S_N2 attack is on a primary alkyl halide, while the second is on a secondary halide. Also, in the second reaction, the alkoxide might simply deprotonate the alcohol on the left and cause the reaction to fail.
4. **In general, reactive functional groups are best put into place toward the end of a synthesis.**
   The target compound contains a reactive epoxide ring. Epoxides react with acids and bases, and the epoxide might not survive the crucial ether-forming reaction just shown. Perhaps the epoxide is best added after formation of the ether. That gives us the following final two steps in the synthesis:

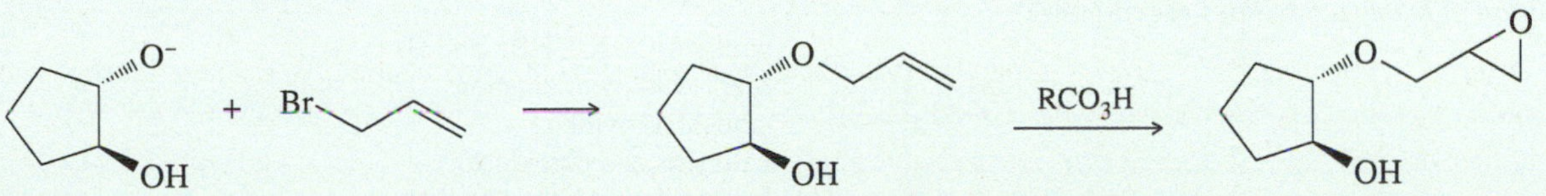

   RCO_3H

5. **Working backward through as many steps as necessary, compare methods for synthesizing the reactants needed for the final step.**
   This process may require writing several possible reaction sequences and evaluating them, keeping in mind the specified starting materials.
   Two reactants are needed to form the ether: an allylic halide and an alkoxide ion. Alkoxide ions are commonly formed by the reaction of an alcohol with sodium metal:

$$R{-}O{-}H + Na \longrightarrow Na^+ \ ^-O{-}R + \tfrac{1}{2}H_2\uparrow$$

(continued)

The alkoxide needed to make the ether is formed by adding sodium to a trans diol as shown below. Trans diols are formed by epoxidation and hydrolysis of alkenes (Section 8-13).

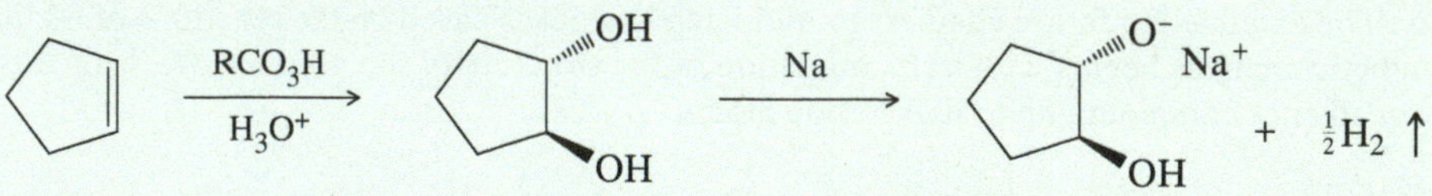

$$\text{cyclopentene} \xrightarrow[H_3O^+]{RCO_3H} \text{trans-cyclopentane-1,2-diol (OH, OH)} \xrightarrow{Na} \text{alkoxide } (O^-\ Na^+,\ OH) + \tfrac{1}{2}H_2\uparrow$$

The other piece we need is an allylic bromide. Allylic bromides are formed by allylic bromination of alkenes (Section 6-6B).

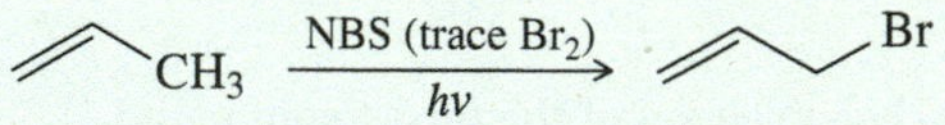

$$CH_2{=}CH{-}CH_3 \xrightarrow[hv]{\text{NBS (trace } Br_2)} CH_2{=}CH{-}CH_2Br$$

6. Summarize the complete synthesis in the forward direction, including all steps and all reagents, and check it for errors and omissions.
This summary is left to you as a review of both the chemistry involved in the synthesis and the method used to develop multistep syntheses.

PROBLEM: Summarize the synthesis outlined in the problem-solving strategy. This summary should be in the synthetic (forward) direction, showing each step and all reagents.

Problem 8-45 requires devising several multistep syntheses. As practice in working such problems, we suggest that you proceed in order through the five steps just outlined.

PROBLEM 8-45

Show how you would synthesize each compound, starting with alkenes or cycloalkenes that contain no more than six carbon atoms. You may use any additional reagents you need.

(a) 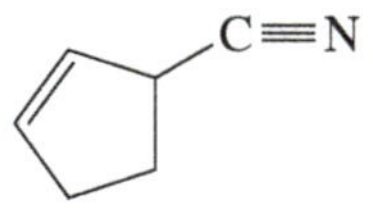 [structure: cyclopentenyl–CH₂–C≡N] (b) 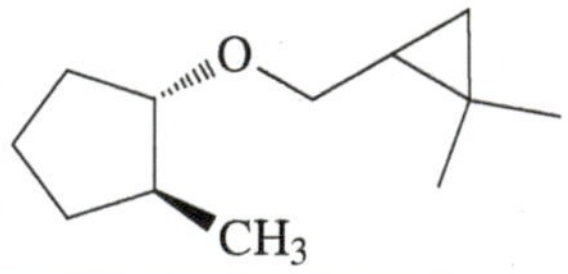 [structure: trans-2-methylcyclopentyl ether with O, CH₃] (c) [structure: cyclopentenyl ether with O, OH]

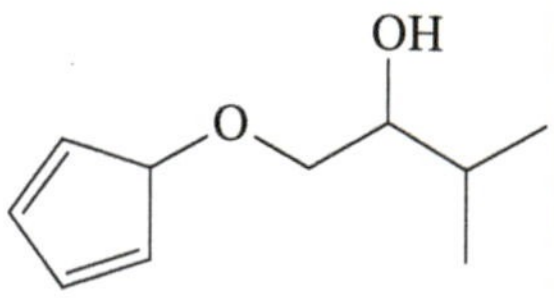

SUMMARY Reactions of Alkenes

1. *Electrophilic Additions*

a. *Addition of hydrogen halides* (Section 8-3)

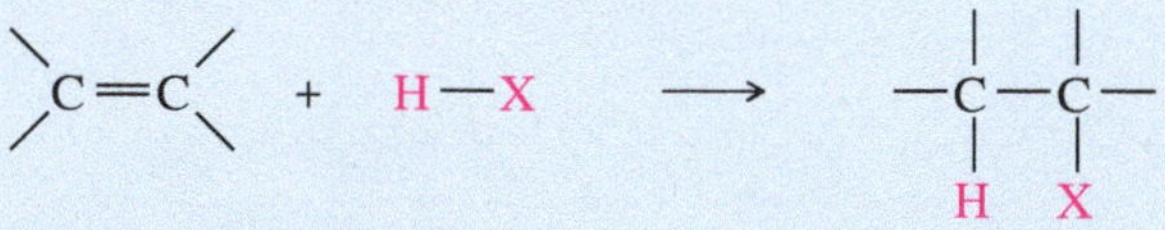

$$>C{=}C< \; + \; H{-}X \longrightarrow -\underset{H}{C}-\underset{X}{C}-$$

(HX = HCl, HBr, or HI)

works with HCl, HBr, and HI
Markovnikov orientation
(anti-Markovnikov with HBr and peroxides)
Rearrangements are possible

b. *Acid-catalyzed hydration* (Section 8-4)

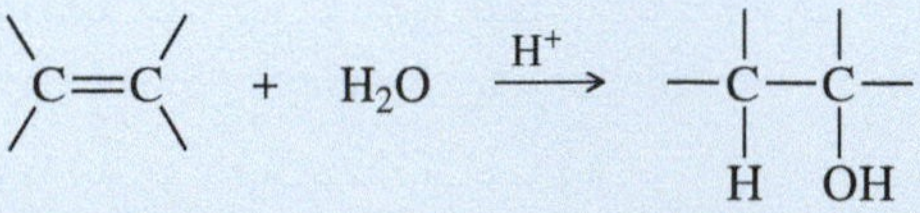

$$>C{=}C< \; + \; H_2O \xrightarrow{H^+} -\underset{H}{C}-\underset{OH}{C}-$$

Markovnikov orientation
Rearrangements are possible

c. *Oxymercuration–demercuration* (Section 8-5)

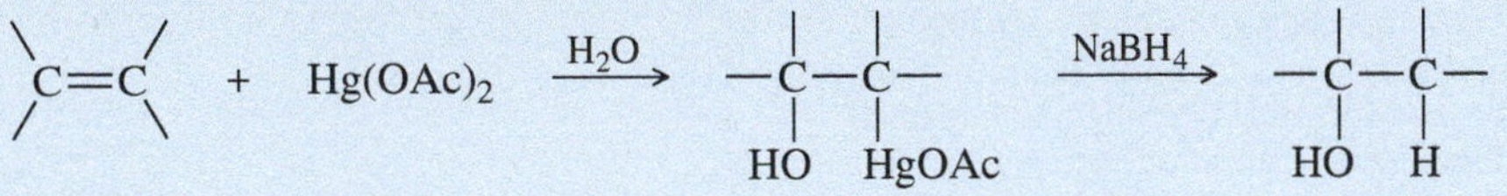

$$>C=C< + Hg(OAc)_2 \xrightarrow{H_2O} -\underset{HO}{C}-\underset{HgOAc}{C}- \xrightarrow{NaBH_4} -\underset{HO}{C}-\underset{H}{C}-$$

Markovnikov orientation
no rearrangements

d. *Alkoxymercuration–demercuration* (Section 8-6)

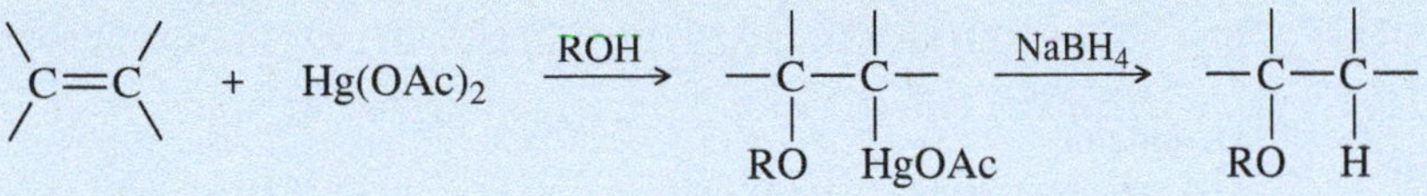

$$>C=C< + Hg(OAc)_2 \xrightarrow{ROH} -\underset{RO}{C}-\underset{HgOAc}{C}- \xrightarrow{NaBH_4} -\underset{RO}{C}-\underset{H}{C}-$$

Markovnikov orientation

e. *Hydroboration–oxidation* (Section 8-7)

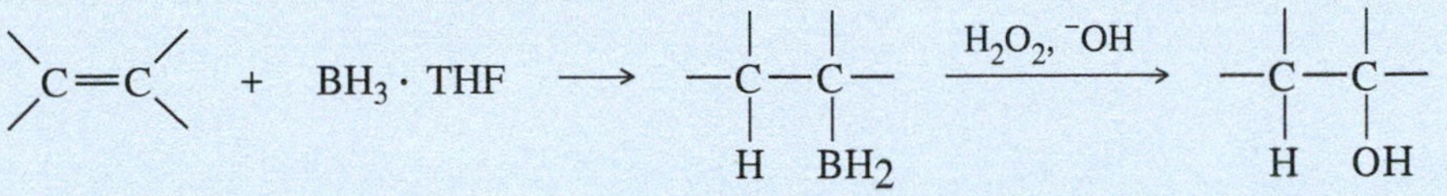

$$>C=C< + BH_3 \cdot THF \longrightarrow -\underset{H}{C}-\underset{BH_2}{C}- \xrightarrow{H_2O_2,\ ^-OH} -\underset{H}{C}-\underset{OH}{C}-$$

anti-Markovnikov orientation
(syn addition)
no rearrangements

f. *Polymerization* (Section 8-16)

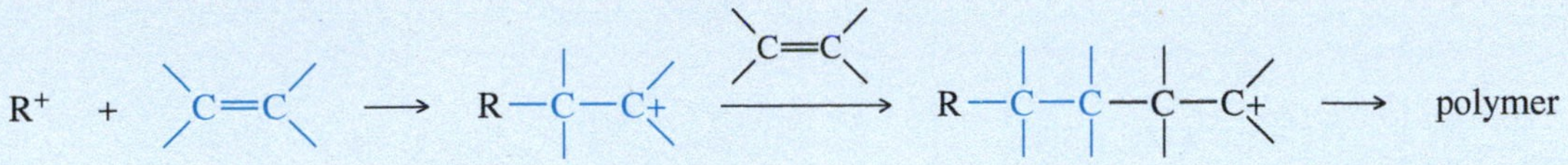

$$R^+ + >C=C< \longrightarrow R-C-C^+ \xrightarrow{>C=C<} R-C-C-C-C^+ \longrightarrow \text{polymer}$$

cationic, radical, or anionic polymerization

2. *Reduction: Catalytic Hydrogenation* (Section 8-10)

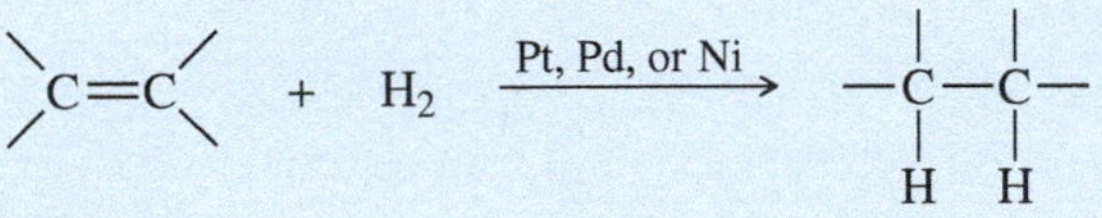

$$>C=C< + H_2 \xrightarrow{\text{Pt, Pd, or Ni}} -\underset{H}{C}-\underset{H}{C}-$$

syn addition

3. *Addition of Carbenes: Cyclopropanation* (Section 8-11)

a. *Simmons-Smith Reaction*

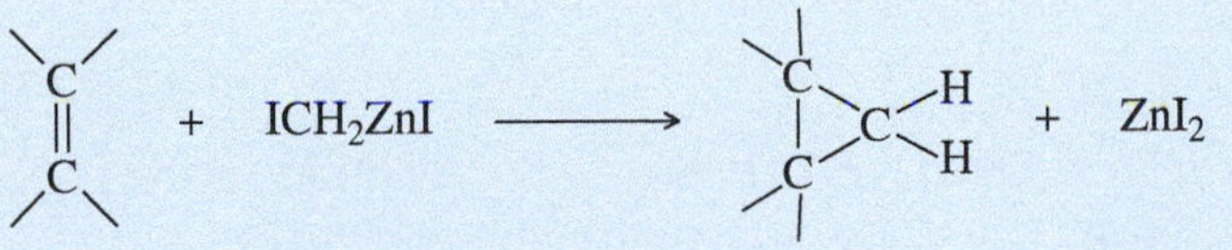

$$>C=C< + ICH_2ZnI \longrightarrow \text{cyclopropane } (CH_2) + ZnI_2$$

syn addition

b. *Alpha elimination*

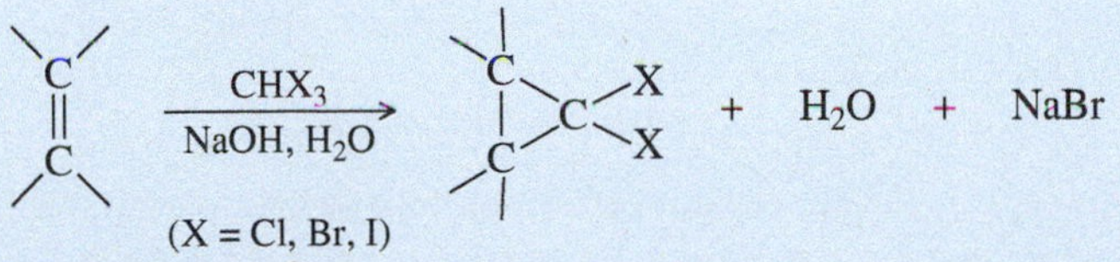

$$>C=C< \xrightarrow[NaOH,\ H_2O]{CHX_3} \text{cyclopropane } (CX_2) + H_2O + NaBr$$

(X = Cl, Br, I)

syn addition

4. *Oxidative Additions*

a. *Addition of halogens* (Section 8-8)

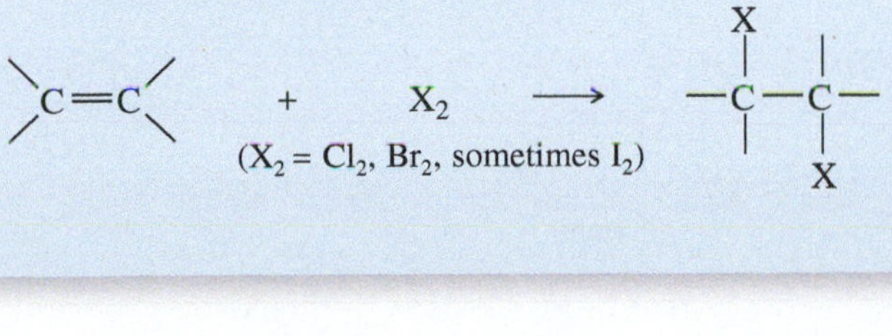

$$>C=C< + X_2 \longrightarrow -\overset{X}{C}-\underset{X}{C}-$$

($X_2 = Cl_2, Br_2$, sometimes I_2)

anti addition

(continued)

(or substitution)

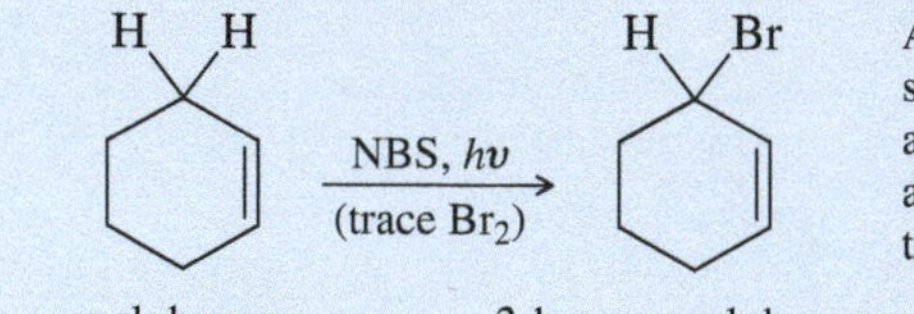

cyclohexene $\xrightarrow[\text{(trace } Br_2)]{\text{NBS, } h\nu}$ 3-bromocyclohexene

At low concentrations of Br_2, an allylic substitution may be observed. NBS provides a trace of Br_2 that (with light as initiator) allows radical substitution to proceed faster than the ionic addition. (Section 6-6B)

b. *Halohydrin formation* (Section 8-9)

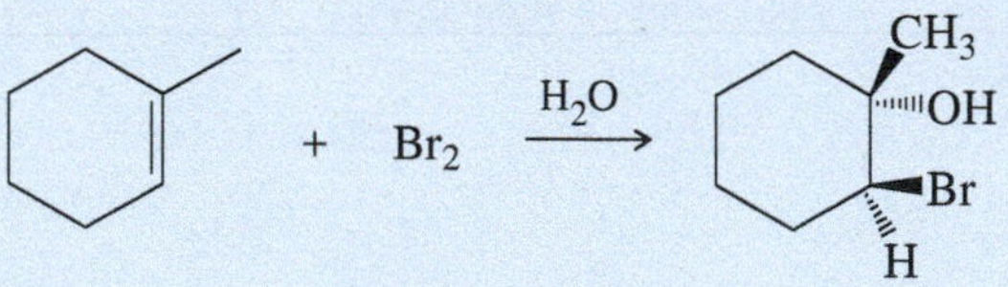

1-methylcyclohexene + Br_2 $\xrightarrow{H_2O}$ trans-2-bromo-1-methylcyclohexanol (CH3, OH, Br, H)

anti addition
Markovnikov orientation
Br^+ adds like H^+ would

c. *Epoxidation* (Section 8-12)

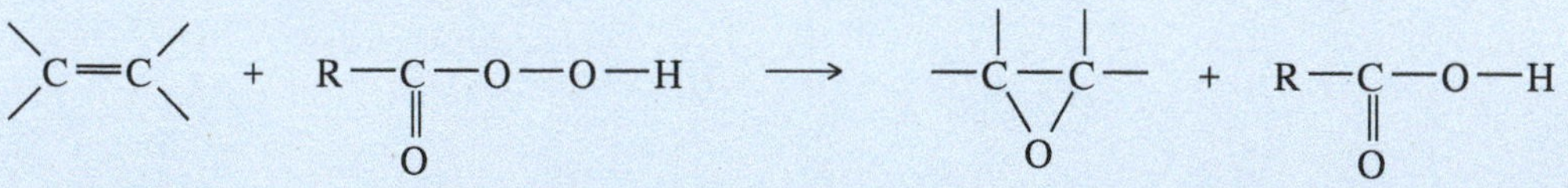

$$\text{>C=C<} + R-\underset{\|}{\overset{}{C}}(=O)-O-O-H \longrightarrow \text{epoxide} + R-C(=O)-O-H$$

alkene, peroxyacid, epoxide — syn addition

d. *Anti dihydroxylation* (Section 8-13)

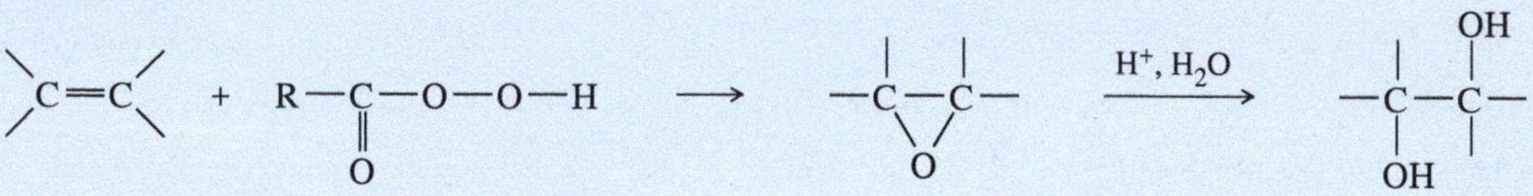

$$\text{>C=C<} + R-C(=O)-O-O-H \longrightarrow \text{epoxide} \xrightarrow{H^+,\ H_2O} -\underset{OH}{C}-\overset{OH}{C}-$$

anti addition

e. *Syn dihydroxylation* (Section 8-14)

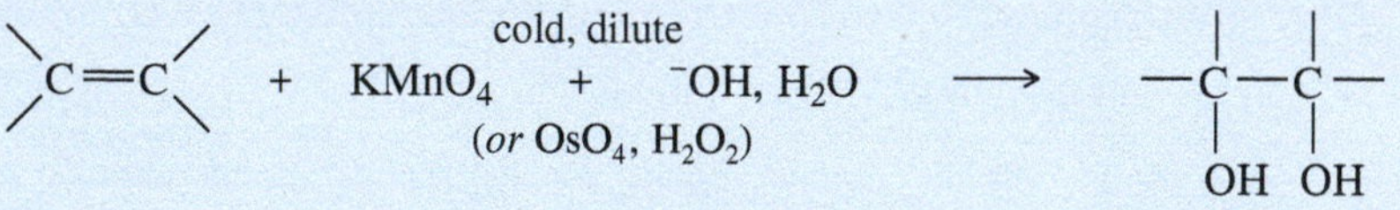

$$\text{>C=C<} + KMnO_4 + {}^-OH,\ H_2O \longrightarrow -\underset{OH}{C}-\underset{OH}{C}-$$

cold, dilute (*or* OsO_4, H_2O_2)

syn addition

5. *Oxidative Cleavage of Alkenes* (Section 8-15)

a. *Ozonolysis*

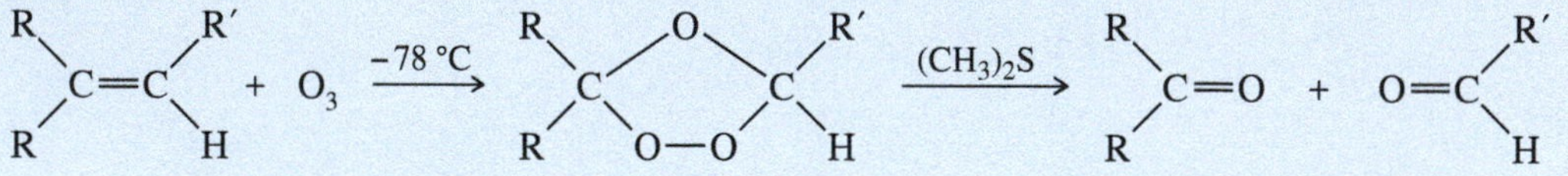

$$R_2C=CHR' + O_3 \xrightarrow{-78\,°C} \text{ozonide} \xrightarrow{(CH_3)_2S} R_2C=O + O=CHR'$$

ozonide; ketones and aldehydes — Aldehydes survive

b. *Potassium permanganate*

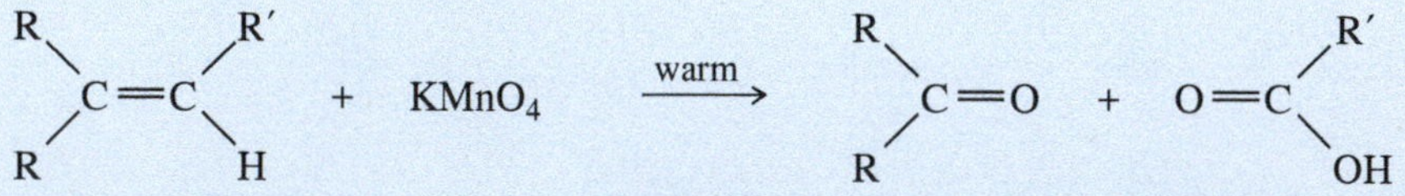

$$R_2C=CHR' + KMnO_4 \xrightarrow{\text{warm}} R_2C=O + O=C(OH)R'$$

ketones and acids — Aldehydes are oxidized to acids

6. *Olefin (Alkene) Metathesis* (Section 8-17)

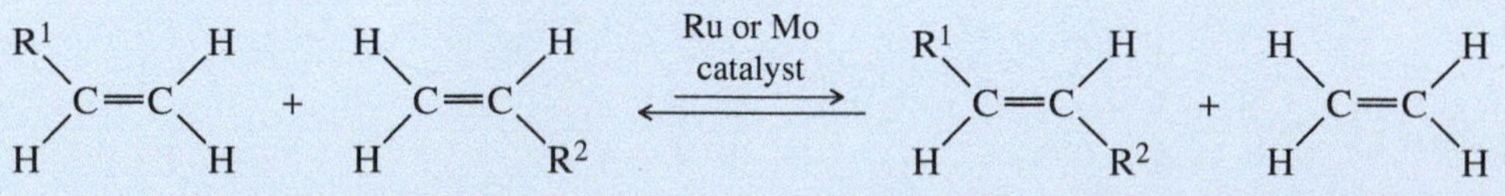

$$R^1CH=CH_2 + CH_2=CHR^2 \xrightleftharpoons{\text{Ru or Mo catalyst}} R^1CH=CHR^2 + CH_2=CH_2$$

(cis + trans); ethylene

SUMMARY Electrophilic Additions to Alkenes

Methylcyclopentene is an alkene that displays orientation and stereochemistry of addition reactions. New atoms are shown in color. When reactions create chiral products from achiral reactants, racemic mixtures are produced. *N/A* means "Not Applicable to this reaction."

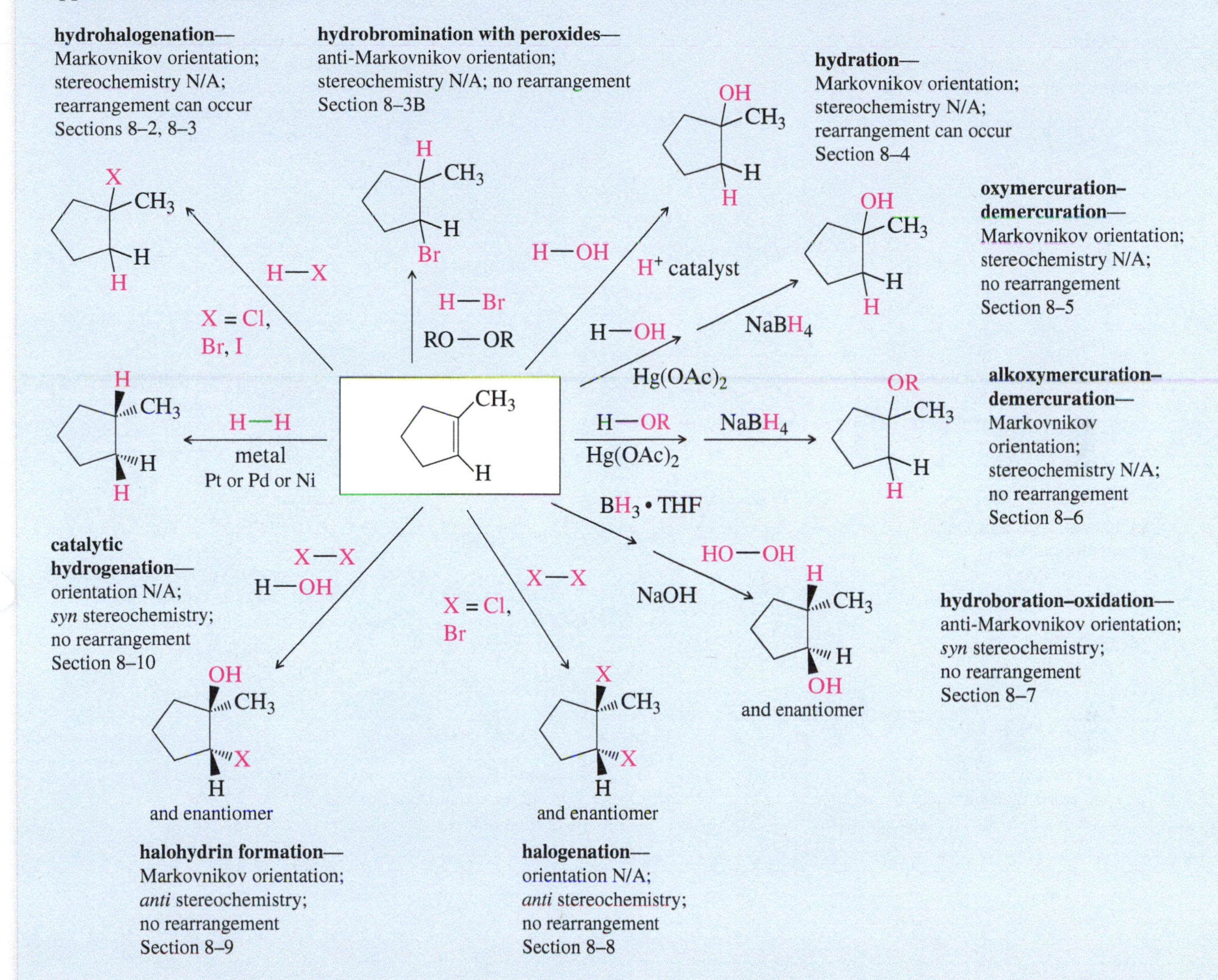

SUMMARY Oxidation and Cyclopropanation Reactions of Alkenes

Methylcyclopentene displays important features of oxidations such as orientation, stereochemistry, and functional groups. New atoms are shown in color. When reactions create chiral products from achiral reactants, racemic mixtures are produced. *N/A* means "Not Applicable to this reaction."

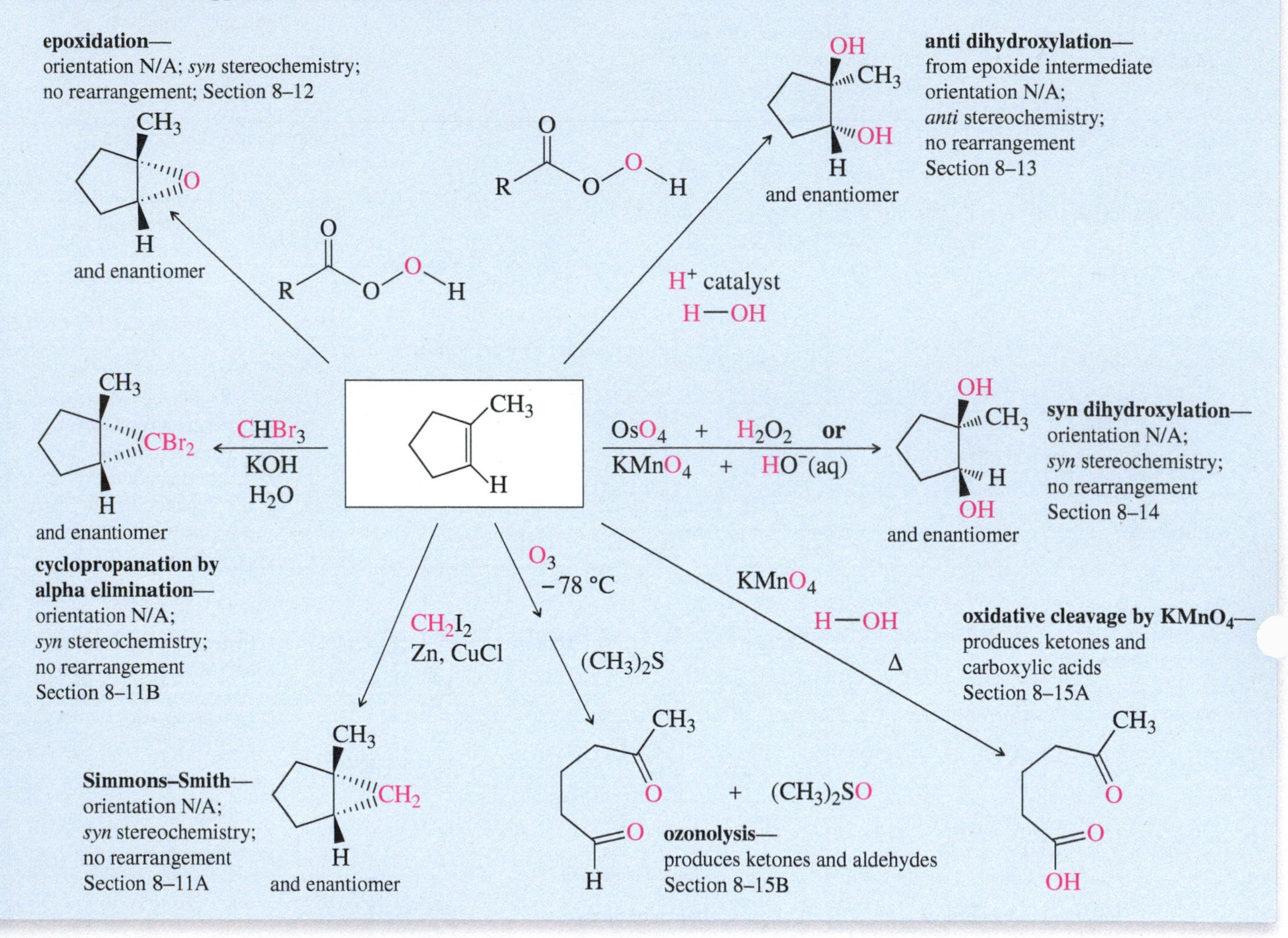

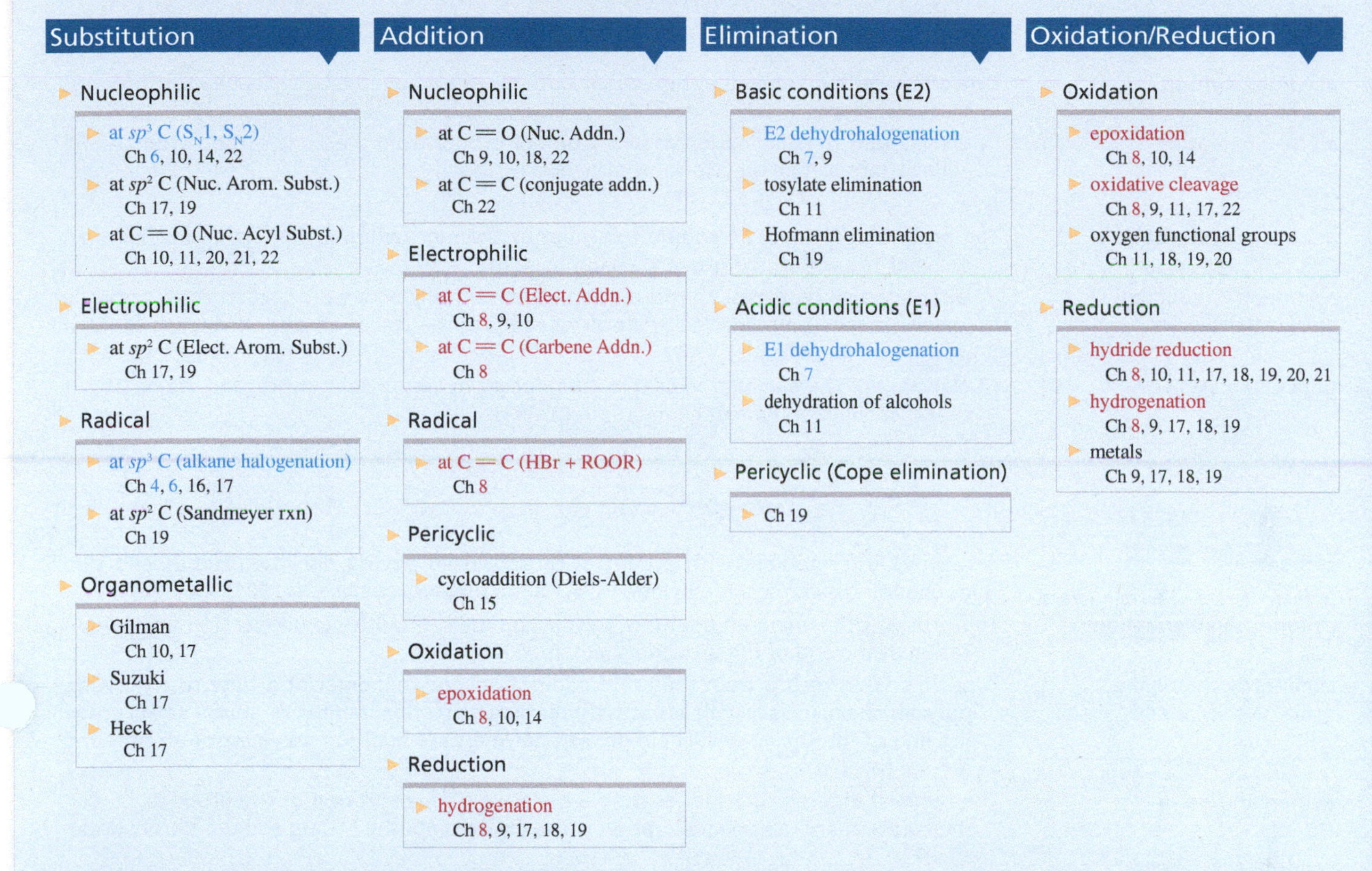

GUIDE TO ORGANIC REACTIONS IN
CHAPTER 8
Reactions covered in Chapter 8 are shown in red. Reactions covered in earlier chapters are shown in blue.
Substitution
Nucleophilic
at sp³ C (SN1, SN2) Ch 6, 10, 14, 22
at sp² C (Nuc. Arom. Subst.) Ch 17, 19
at C═O (Nuc. Acyl Subst.) Ch 10, 11, 20, 21, 22
Electrophilic
at sp² C (Elect. Arom. Subst.) Ch 17, 19
Radical
at sp³ C (alkane halogenation) Ch 4, 6, 16, 17
at sp² C (Sandmeyer rxn) Ch 19
Organometallic
Gilman Ch 10, 17
Suzuki Ch 17
Heck Ch 17
Addition
Nucleophilic
at C═O (Nuc. Addn.) Ch 9, 10, 18, 22
at C═C (conjugate addn.) Ch 22
Electrophilic
at C═C (Elect. Addn.) Ch 8, 9, 10
at C═C (Carbene Addn.) Ch 8
Radical
at C═C (HBr + ROOR) Ch 8
Pericyclic
cycloaddition (Diels-Alder) Ch 15
Oxidation
epoxidation Ch 8, 10, 14
Reduction
hydrogenation Ch 8, 9, 17, 18, 19
Elimination
Basic conditions (E2)
E2 dehydrohalogenation Ch 7, 9
tosylate elimination Ch 11
Hofmann elimination Ch 19
Acidic conditions (E1)
E1 dehydrohalogenation Ch 7
dehydration of alcohols Ch 11
Pericyclic (Cope elimination)
Ch 19
Oxidation/Reduction
Oxidation
epoxidation Ch 8, 10, 14
oxidative cleavage Ch 8, 9, 11, 17, 22
oxygen functional groups Ch 11, 18, 19, 20
Reduction
hydride reduction Ch 8, 10, 11, 17, 18, 19, 20, 21
hydrogenation Ch 8, 9, 17, 18, 19
metals Ch 9, 17, 18, 19

This *Guide to Organic Reactions* will appear at the end of six of the chapters in which large numbers of new reactions are introduced. The purposes are to show the overview of major reaction categories, to indicate where that chapter's reactions fit in the larger scheme, and to serve as a study tool by indicating which reaction types are new and which have been previously covered. The new reactions in the current chapter will appear in red type. Those reactions that have been covered in previous chapters will appear in blue type. As you progress through *Organic Chemistry*, you will gradually see more reactions appear in blue type because you have previously covered them.

Essential Terms

addition — A reaction involving an increase in the number of groups attached to the alkene and a decrease in the number of elements of unsaturation. (p. 360)

- **anti addition:** An addition in which two groups add to opposite faces of the double bond (as in addition of Br_2). (p. 380)
- **electrophilic addition:** An addition in which the electrophile (electron-pair acceptor) bonds to one of the double-bonded carbons first, followed by the nucleophile. (p. 360)
- **syn addition:** An addition in which two groups add to the same face of the double bond (as in osmium tetroxide **dihydroxylation**). (p. 380)

addition polymer — A polymer that results from the addition reactions of alkenes, dienes, or other compounds with double and triple bonds. Most addition polymers form by a chain-growth process. (p. 403)

alkoxy group — (**alkoxyl group**) (—O—R) An alkyl group bonded through an oxygen atom, as in an ether. (p. 375)

alkoxymercuration The addition of mercuric acetate to an alkene in an alcohol solution, forming an alkoxymercurial intermediate. Demercuration gives an ether. (p. 375)

$$>C=C< + Hg(OAc)_2 \xrightarrow{R-OH} -\underset{|}{\overset{R-O}{\overset{|}{C}}}-\underset{HgOAc}{\overset{|}{\underset{|}{C}}}- \xrightarrow{NaBH_4} -\underset{|}{\overset{R-O}{\overset{|}{C}}}-\underset{H}{\overset{|}{\underset{|}{C}}}-$$

alkylidene group One of the carbon atoms at either end of a double bond, together with its two substituents. Most commonly $=CH_2$ or $=CHR$ or $=CR_2$. (p. 407)

alpha elimination (**α elimination**) The elimination of two atoms or groups from the same carbon atom. Alpha eliminations can be used to form carbenes. (p. 392)

$$CHBr_3 + KOH \longrightarrow :CBr_2 + H_2O + KBr$$

anionic polymerization The process of forming an addition polymer by chain-growth polymerization involving an anion at the end of the growing chain. (p. 406)

asymmetric induction (**enantioselective synthesis**) The formation of an optically active product from an optically inactive starting material. Such a process requires the use of an optically active reagent or catalyst. (p. 390)

beta elimination (**β elimination**) The elimination of two atoms or groups from adjacent carbon atoms. This is the most common type of elimination. (p. 392)

$$-\underset{|}{\overset{H}{\overset{|}{C}}}-\underset{|}{\overset{Br}{\overset{|}{C}}}- + KOH \longrightarrow >C=C< + H_2O + KBr$$

carbene A reactive intermediate with a neutral carbon atom having only two bonds and two nonbonding electrons. Methylene ($:CH_2$) is the simplest carbene. (p. 391)

cationic polymerization The process of forming an addition polymer by chain-growth polymerization involving a cation at the end of the growing chain. (p. 403)

chain-growth polymer A polymer that results from the rapid addition of one monomer at a time to a growing polymer chain, usually with a reactive intermediate (cation, radical, or anion) at the growing end of the chain. Most chain-growth polymers are addition polymers of alkenes and dienes. (p. 403)

demercuration The removal of a mercury species from a molecule. Demercuration of the products of oxymercuration and alkoxymercuration is usually accomplished using sodium borohydride. (p. 373)

dihydroxylation (**hydroxylation**) The addition of two hydroxy groups, one at each carbon of the double bond; formally, an oxidation. (p. 398)

$$>C=C< + H_2O \xrightarrow{H^+} -\underset{|}{\overset{H}{\overset{|}{C}}}-\underset{|}{\overset{OH}{\overset{|}{C}}}-$$

epoxide (**oxirane**) A three-membered cyclic ether. (p. 393)

epoxidation: Formation of an epoxide, usually from an alkene. A peroxyacid is generally used for alkene epoxidations. (p. 394)

free-radical polymerization The process of forming an addition polymer by chain-growth polymerization involving a free radical at the end of the growing chain. (p. 405)

glycol A 1,2-diol. (p. 395)

halogenation The addition of a halogen (X_2) to a molecule, or the free-radical substitution of a halogen for a hydrogen. (p. 382)

halohydrin A beta-haloalcohol, with a halogen and a hydroxy group on adjacent carbon atoms. (p. 385)

$$>C=C< \xrightarrow[H_2O]{Cl_2} -\underset{Cl}{\overset{|}{\underset{|}{C}}}-\underset{OH}{\overset{|}{\underset{|}{C}}}- + HCl$$

a chlorohydrin

halonium ion A reactive, cationic intermediate with a three-membered ring containing a halogen atom; usually, a **chloronium ion,** a **bromonium ion,** or an **iodonium ion.** (p. 383)

heterogeneous catalysis Use of a catalyst that is in a separate phase from the reactants. For example, a platinum hydrogenation catalyst is a solid, a separate phase from the liquid alkene. (p. 389)

homogeneous catalysis Use of a catalyst that is in the same phase as the reactants. For example, the acid catalyst in hydration is in the liquid phase with the alkene. (p. 389)

hydration The addition of water to a molecule. Hydration of an alkene forms an alcohol. (p. 370)

$$\mathrm{>C{=}C<} \; + \; H_2O_2 \xrightarrow{OsO_4} \mathrm{-\underset{|}{\overset{HO}{\overset{|}{C}}}-\underset{|}{\overset{OH}{\overset{|}{C}}}-}$$

hydroboration The addition of borane (BH_3) or one of its derivatives ($BH_3 \cdot THF$, for example) to a molecule. (p. 378)

hydrogenation The addition of hydrogen to a molecule. The most common hydrogenation is the addition of H_2 across a double bond in the presence of a catalyst (**catalytic hydrogenation** or **catalytic reduction**). (p. 389)

hydroxylation See **dihydroxylation.**

Markovnikov's rule (*original statement*) When a proton acid adds to the double bond of an alkene, the proton bonds to the carbon atom that already has more hydrogen atoms. (p. 365)

Markovnikov's rule (*extended statement*) In an electrophilic addition to an alkene, the electrophile adds in such a way as to generate the most stable intermediate. (p. 365)

$$\text{1-methylcyclohexene } (CH_3, H) \; + \; HCl \longrightarrow \text{1-chloro-1-methylcyclohexane } (CH_3, Cl, H, H)$$

Markovnikov product

Markovnikov orientation: An orientation of addition that obeys the original statement of Markovnikov's rule; one that gives the **Markovnikov product.** (p. 365)

anti-Markovnikov orientation: An orientation of addition that is the opposite of that predicted by the original statement of Markovnikov's rule; one that gives the **anti-Markovnikov product.** (p. 366)

mCPBA (*meta*-chloroperoxybenzoic acid) A common reagent for epoxidizing alkenes. mCPBA dissolves in common solvents such as dichloromethane. As the epoxidation takes place, the *m*-chlorobenzoic acid by-product precipitates out of solution. (p. 394)

metathesis (**olefin metathesis**) Any reaction that trades and interchanges the alkylidene groups of an alkene. (p. 407)

$$R^1HC{=}CH_2 \; + \; H_2C{=}CHR^2 \xrightarrow[\text{olefin metathesis}]{\text{Ru or Mo catalyst}} R^1HC{=}CHR^2 \; (\text{cis + trans}) \; + \; H_2C{=}CH_2 \; (\text{ethylene})$$

monomer One of the small molecules that bond together to form a polymer. (p. 403)

organic synthesis The preparation of desired organic compounds from readily available materials. (p. 410)

oxidative cleavage The cleavage of a carbon–carbon bond through oxidation. Carbon–carbon double bonds are commonly cleaved by ozonolysis/reduction or by warm, concentrated permanganate. (p. 400)

oxymercuration The addition of aqueous mercuric acetate to an alkene. (p. 373)

$$\mathrm{>C{=}C<} \; + \; Hg(OAc)_2 \xrightarrow{H_2O} \mathrm{-\underset{|}{\overset{HO}{\overset{|}{C}}}-\underset{HgOAc}{\underset{|}{\overset{|}{C}}}-} \; + \; HOAc$$

ozonolysis The use of ozone, usually followed by reduction, to cleave a double bond. (p. 400)

peroxide effect The reversal of orientation of HBr addition to alkenes in the presence of peroxides. A free-radical mechanism is responsible for the peroxide effect. (p. 368)

peroxyacid (peracid) A carboxylic acid with an extra oxygen atom and a peroxy (—O—O—) linkage. The general formula is RCO_3H. (p. 393)

polymer A high-molecular-weight compound composed of many molecules of a smaller, simpler compound called the **monomer.** (p. 403)

polymerization: The reaction of monomer molecules to form a polymer. (p. 403)

regiochemistry The orientation of a chemical reaction on an unsymmetrical substrate. In additions to alkenes, the regiochemistry of the addition involves which part of the reagent adds to which end of an unsymmetrical alkene. (p. 362)

regioselective reaction A reaction in which one direction of bond making or bond breaking occurs preferentially over all other directions. For example, the addition of HCl is regioselective, predicted by Markovnikov's rule. Hydroboration–oxidation is regioselective because it consistently gives anti-Markovnikov orientation. (p. 364)

retrosynthetic analysis A method of working backward to solve multistep synthetic problems. (p. 410)

Simmons–Smith reaction A cyclopropanation of an alkene using the carbenoid reagent generated from diiodomethane and the zinc–copper couple. (p. 392)

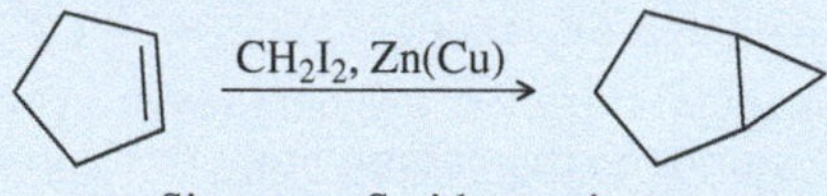

Simmons–Smith reaction

stereospecific reaction A reaction that converts different stereoisomers of the starting material into different stereoisomers of the product. (p. 381)

Essential Problem-Solving Skills in Chapter 8

Each skill is followed by problem numbers exemplifying that particular skill.

1	Use the extended version of Markovnikov's rule to predict the regiochemistry (orientation) of electrophilic additions to alkenes.	Problems 8-46, 47, and 50
2	Show how to control the stereochemistry and regiochemistry (orientation) of additions to alkenes to obtain the products we want.	Problems 8-47, 49, 50, 55, 61, 65, and 76
3	Show how to control the hydration of alkenes to give alcohols with either Markovnikov or anti-Markovnikov orientation, depending on the reagents.	Problems 8-46, 49, 50, 51, and 79
4	Predict the products of halogenations, oxidations, reductions, and cleavages of alkenes, including the orientation (regiochemistry) and the stereochemistry of the reaction.	Problems 8-46, 47, 48, 54, 55, 61, 65, 66, and 75
5	Predict the stereochemistry observed in the hydroboration, halogenation, and dihydroxylation reactions of alkenes.	Problems 8-46, 50, 51, 55, and 66
6	Propose logical mechanisms to explain the observed products of alkene reactions, including regiochemistry and stereochemistry.	Problems 8-56, 57, 58, 59, 64, 67, 70, 71, 76, 77, 78, and 79
7	Use retrosynthetic analysis to solve multistep synthesis problems with alkenes as reagents, intermediates, or products.	Problems 8-49, 50, 51, 55, and 61
	Problem-Solving Strategy: Organic Synthesis	Problems 8-49, 50, 51, 55, and 61
8	Use clues provided by the products of reactions, such as ozonolysis, to determine the structure of an unknown alkene.	Problems 8-48, 54, 55, 60, 61, 62, 63, and 68
9	Show how metathesis interchanges the alkylidene (=CHR) groups in alkenes, forming new and different alkenes.	Problems 8-46, 52, and 53
10	Given a particular monomer unit, draw the structure of the resulting polymer.	Problems 8-72, 73, and 74

In studying these reaction-intensive chapters, students ask whether they should "memorize" all the reactions. Doing organic chemistry is like speaking a foreign language, and the reactions are your vocabulary. Without knowing the words, how can you construct sentences? Making flash cards often helps students learn the reactions.

In organic chemistry, the mechanisms, regiochemistry, and stereochemistry are your grammar. You must develop *facility* with the reactions, as you develop facility with the words and grammar you use in speaking. Problems and multistep syntheses are the sentences of organic chemistry. You must practice combining all aspects of your vocabulary in solving these problems.

Students who fail organic chemistry exams often do so because they have memorized the vocabulary, but they have not practiced enough problems. Others fail because they think they can do problems, but they lack the vocabulary. If you understand the reactions and can do the end-of-chapter problems without referring to the chapter, then you should do well on your exams.

Study Problems

8-46 Predict the major products of the following reactions, and give the structures of any intermediates. Include stereochemistry where appropriate.

(a) 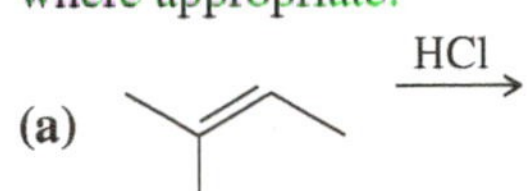 $\xrightarrow{HCl}$

(b) $\xrightarrow[CCl_4]{Br_2}$

(c) $\xrightarrow[(2)\ H_2O_2,\ ^-OH]{(1)\ BH_3 \cdot THF}$

(d) $\xrightarrow[(2)\ (CH_3)_2S]{(1)\ O_3\ (-78\ °C)}$

(e) $\xrightarrow[ROOR]{HBr}$

(f) $\xrightarrow[ROOR]{HCl}$

(g) $\xrightarrow{PhCO_3H}$

(h) $\xrightarrow[H_2O_2]{OsO_4}$

(i) $\xrightarrow[(cold,\ dil.)]{KMnO_4,\ ^-OH}$

(j) $\xrightarrow[H^+,\ H_2O]{CH_3CO_3H}$

(k) $\xrightarrow[(warm,\ concd.)]{KMnO_4,\ ^-OH}$

(l) $\xrightarrow[(2)\ (CH_3)_2S]{(1)\ O_3\ (-78\ °C)}$

(m) $\xrightarrow[Pt]{H_2}$

(n) $\xrightarrow{H^+,\ H_2O}$

(o) $\xrightarrow{[M]{=}CHR}$

(p) $\xrightarrow[(2)\ NaBH_4]{(1)\ Hg(OAc)_2,\ H_2O}$

(q) $\xrightarrow[H_2O]{Cl_2}$

8-47 Limonene is one of the compounds that give lemons their tangy odor. Show the structures of the products expected when limonene reacts with an excess of each of these reagents.

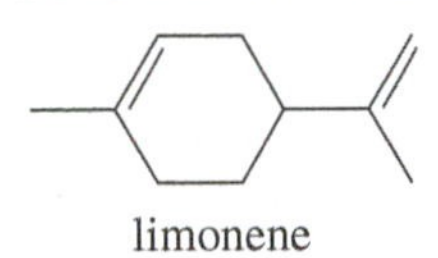
limonene

(a) borane in tetrahydrofuran, followed by basic hydrogen peroxide
(b) *m*-chloroperoxybenzoic acid
(c) ozone, then dimethyl sulfide
(d) a mixture of osmic acid and hydrogen peroxide
(e) hot, concentrated potassium permanganate
(f) peroxyacetic acid in acidic water
(g) hydrogen and a platinum catalyst
(h) hydrogen bromide gas
(i) hydrogen bromide gas in a solution containing dimethyl peroxide
(j) bromine water
(k) chlorine gas
(l) mercuric acetate in methanol, followed by sodium borohydride
(m) $CHBr_3$ and 50% aq. NaOH

8-48 Give the products expected when the following compounds are ozonized and reduced.

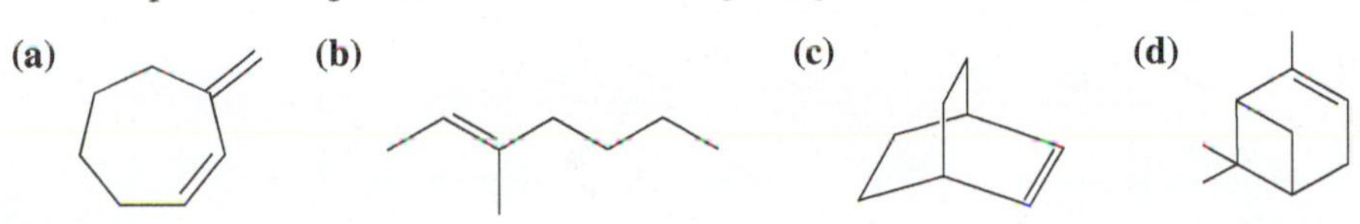

8-49 Show how you would make the following compounds from a suitable cyclic alkene.

(a)
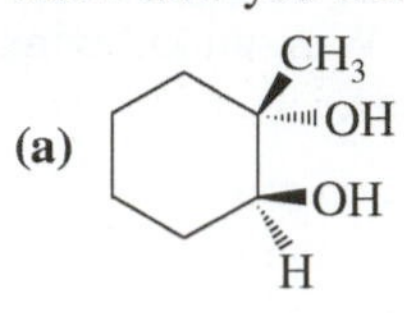

(b)
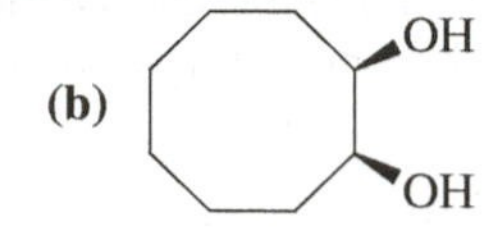

(c) Br Br

(d) Cl OH

(e) OH

(f) OCH_3

8-50 Using 1,2-dimethylcyclohexene as your starting material, show how you would synthesize the following compounds. (Once you have shown how to synthesize a compound, you may use it as the starting material in any later parts of this problem.) If a chiral product is shown, assume that it is part of a racemic mixture.

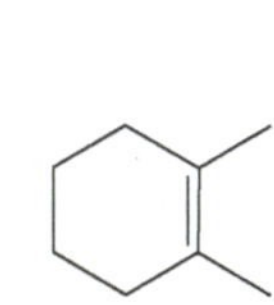
1,2-dimethylcyclohexene

(a)
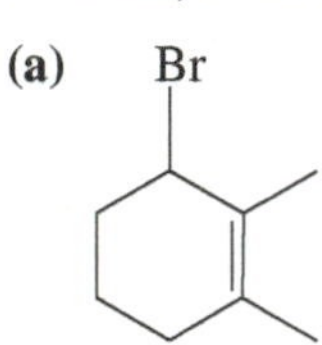

(b)
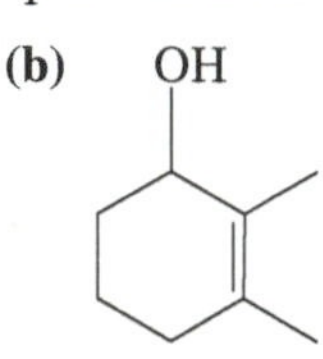

(c)
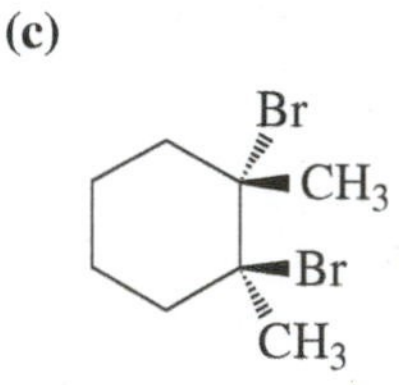

(d)
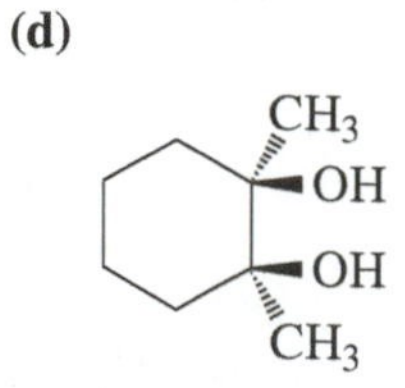

(e)
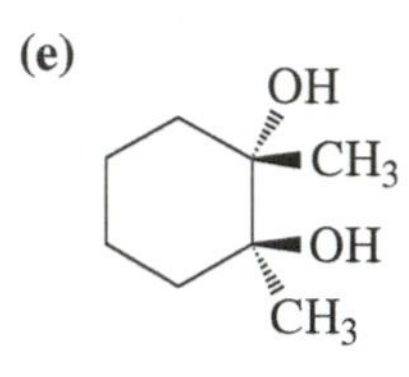

(f)
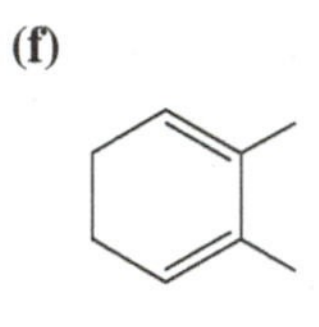

(g)
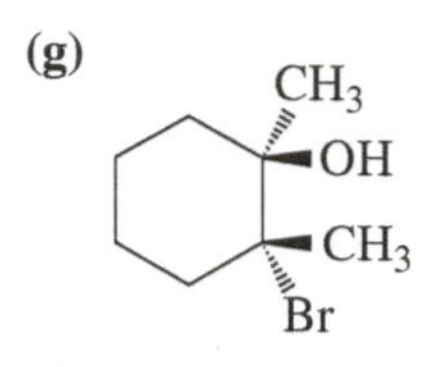

(h)
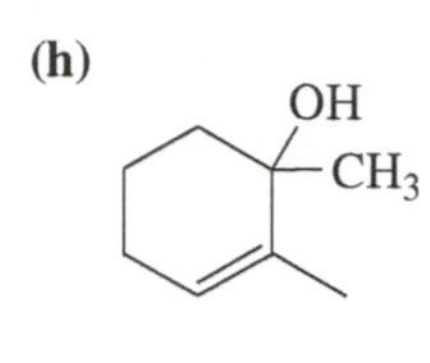

(i)
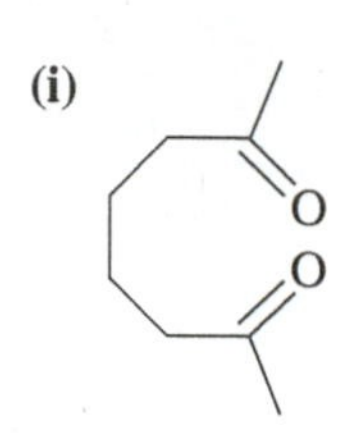

8-51 Show how you would synthesize each compound using methylenecyclohexane as your starting material.

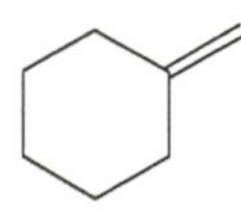
methylenecyclohexane

(a)
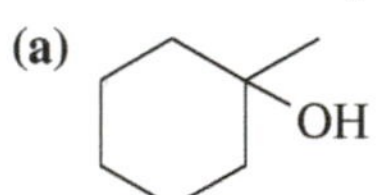

(b)
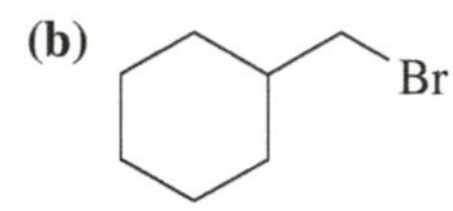

(c)
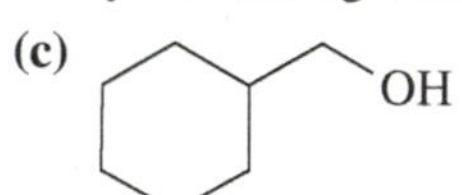

(d)
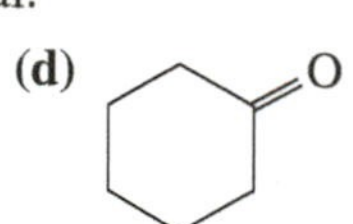

(e)
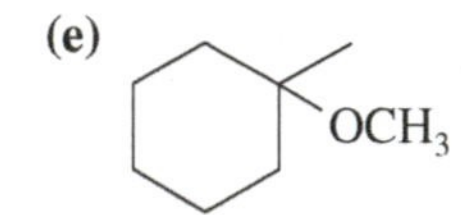

(f)
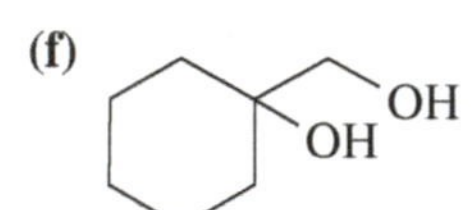

(g)
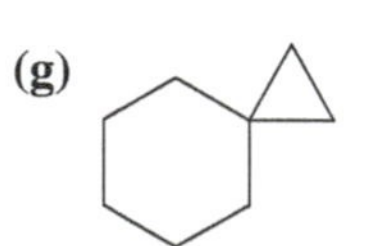

(h)
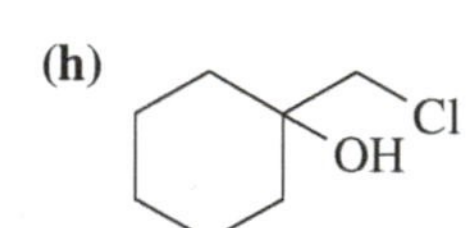

(i)
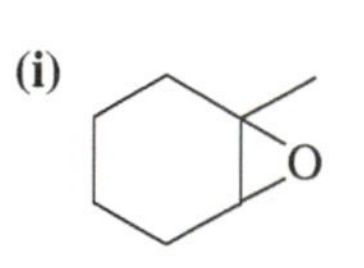

8-52 Show what products you would expect from the following metathesis reactions, using the Schrock or Grubbs catalysts.

(a) O $\xrightarrow{[M]=CHR}$

(b)
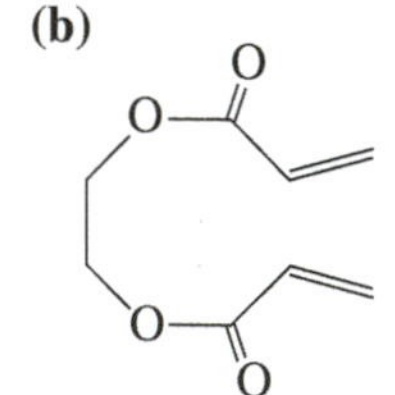

$\xrightarrow{[M]=CHR}$

(c) OCH_3 HO eugenol +
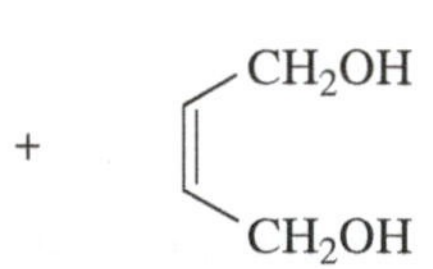

$\xrightarrow{[M]=CHR}$

8-53 Show how you might use olefin metathesis to assemble the following alkenes from smaller units:

(a)
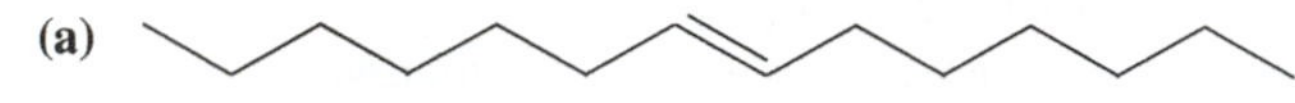

(b)

8-54 Professor Patrick Dussault (University of Nebraska at Lincoln) has developed an alternative to the standard two-step ozonolysis procedure requiring reduction of the ozonide in a second step. He uses 2 to 3 equivalents of pyridine, a mildly basic organic solvent, in a one-step process (*Organic Letters*, **2012**, *14*, 2242). Show the products you expect from the following examples.

(a) $\xrightarrow[-78\ °C]{O_3,\ \text{pyridine}}$? 93% yield

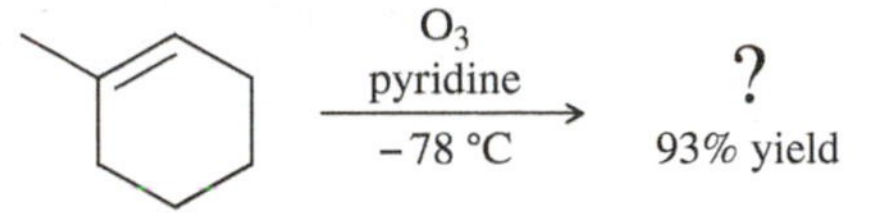

(b) t-Bu $\xrightarrow[-78\ °C]{O_3,\ \text{pyridine}}$? 85% yield

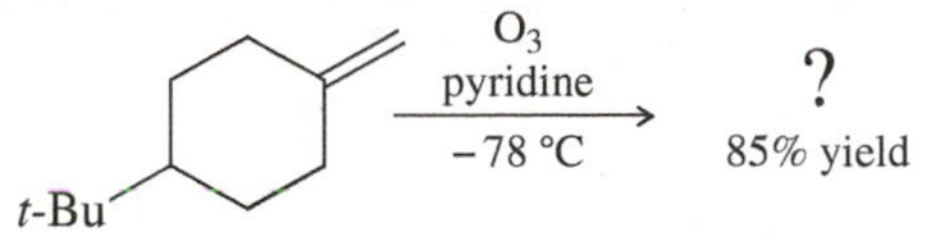

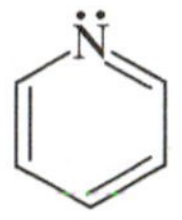

pyridine

(c) $(CH_2)_7COCH_3$ (O) / $(CH_2)_7CH_3$ $\xrightarrow[-78\ °C]{O_3,\ \text{pyridine}}$? 78% yield

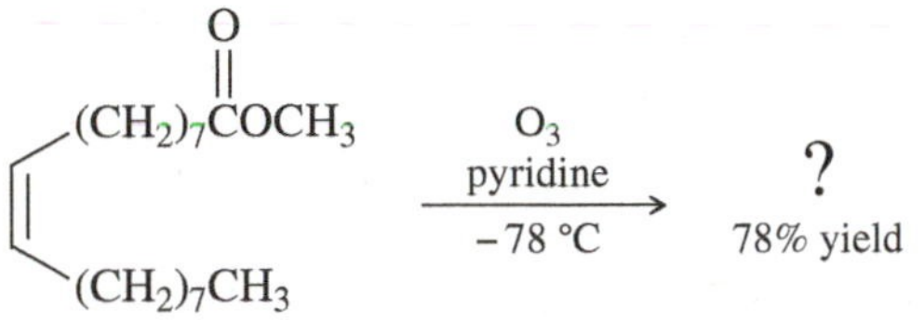

(d) CH_2, O $\xrightarrow[-78\ °C]{O_3,\ \text{pyridine}}$? 81% yield

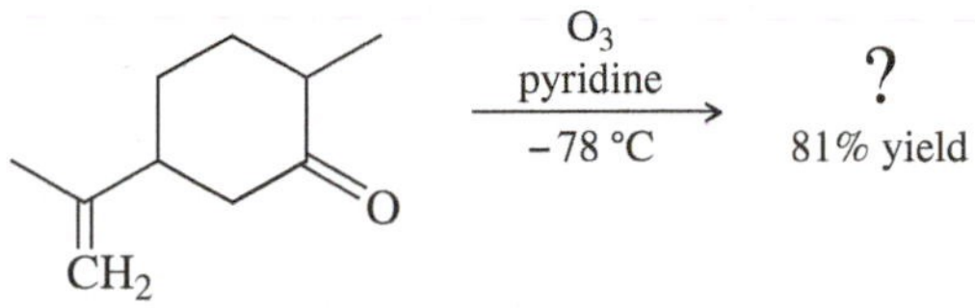

8-55 Complete each synthesis by providing the structure of the major product at each step, including any important stereochemistry.

(a) $\xrightarrow[H_2O]{Br_2}$ A $\xrightarrow{\text{excess } NH_3}$ B

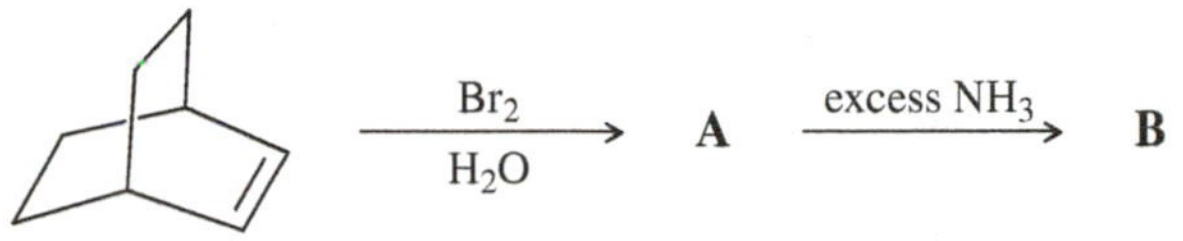

(b) $\xrightarrow[(2)\ NaBH_4]{(1)\ Hg(OAc)_2,\ H_2O}$ C $\xrightarrow{H_2SO_4}$ D $\xrightarrow[H_2O_2]{OsO_4}$ E

(c) $\xrightarrow[ROOR]{\text{excess HBr}}$ F $\xrightarrow[\Delta]{\text{excess KO-}t\text{-Bu}}$ G $\xrightarrow[2)\ (CH_3)_2S]{1)\ \text{excess } O_3,\ -78\ °C}$ H

(d) $\xrightarrow{Br_2}$ J $\xrightarrow[\Delta]{KOH}$ K ($C_{13}H_{16}$) $\xrightarrow[2)\ H_2O_2,\ NaOH]{1)\ \text{excess } BH_3 \cdot THF}$ L (ignore stereoisomers)

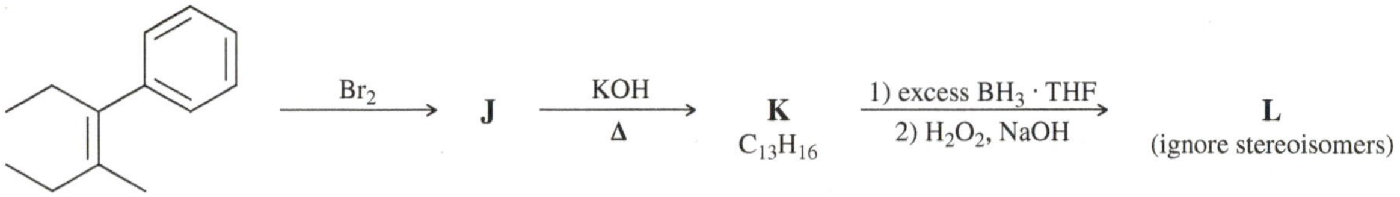

(Hint: Recall that benzene does not undergo most addition reactions.)

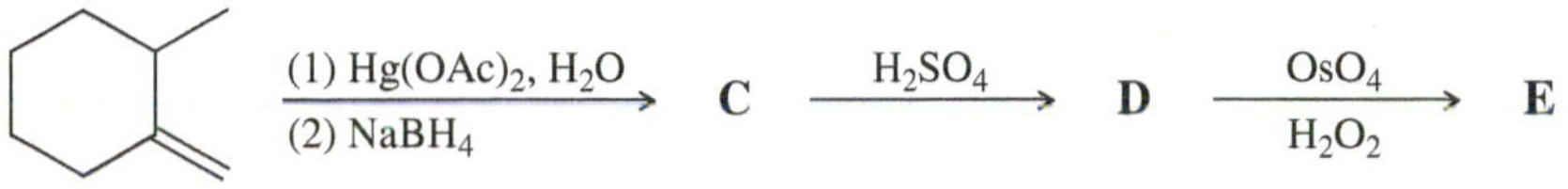

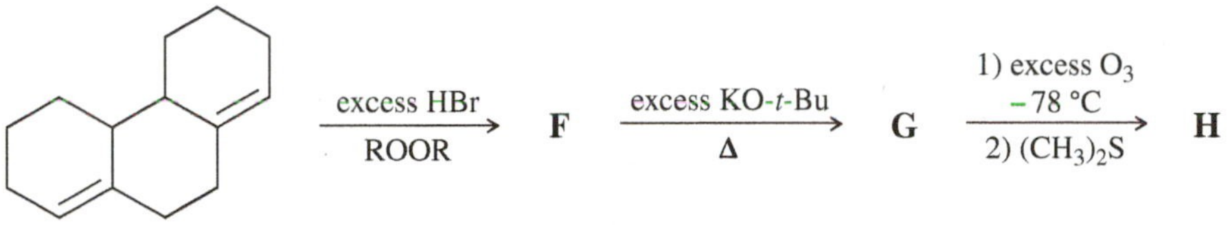

8-56 Propose mechanisms consistent with the following reactions.

(a) $\xrightarrow[ROOR]{HBr}$ Br

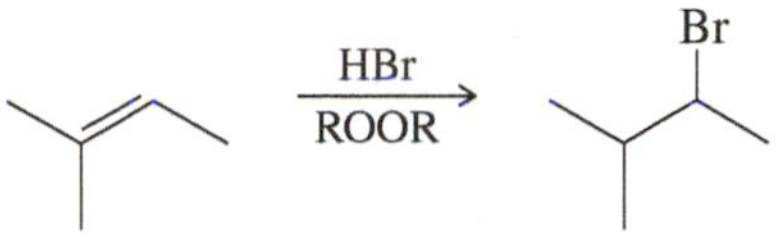

(b) $\xrightarrow[H_2O]{H_2SO_4}$ OH

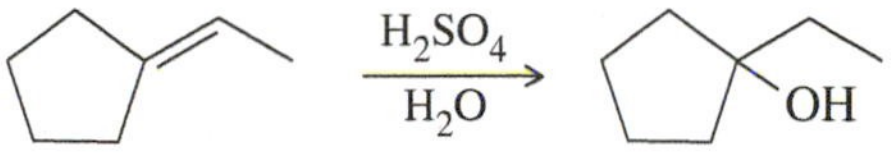

(c) $\xrightarrow{HBr}$ Br + Br

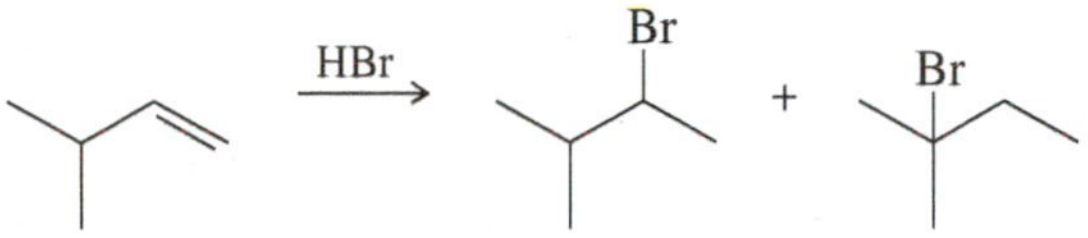

(d) $\xrightarrow[NaOH]{CHBr_3}$ Br, Br

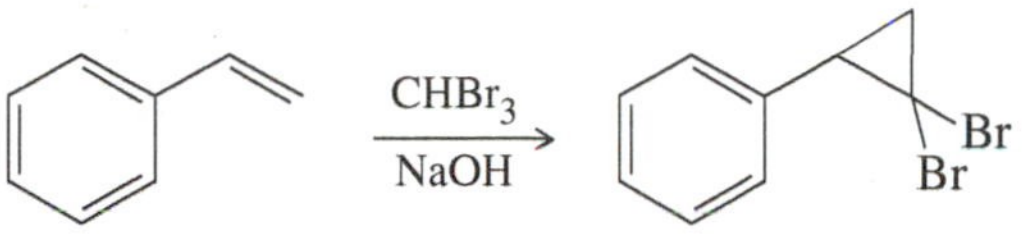

(e) Br $\xrightarrow[CH_3CH_2OH]{HCl}$ Cl, Br + OCH_2CH_3, Br

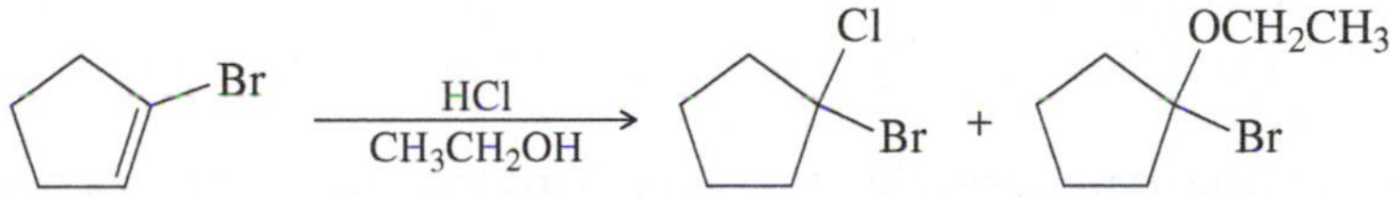

(f) CH_3CH_2, O, CH_2CH_3 $\xrightarrow{H^+,\ H_2O}$ CH_2CH_3; H—OH; H—OH; CH_2CH_3

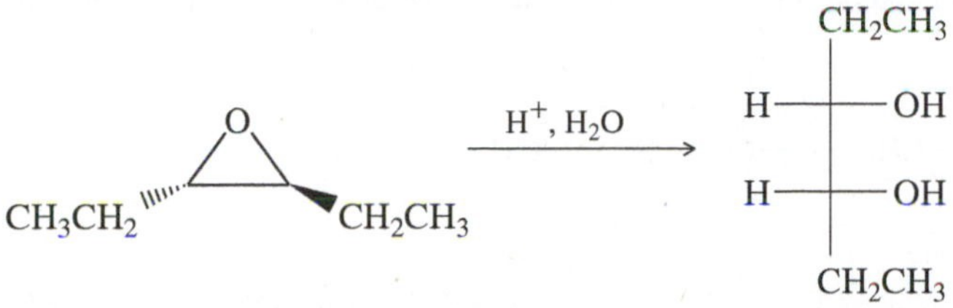

(continued)

(g) 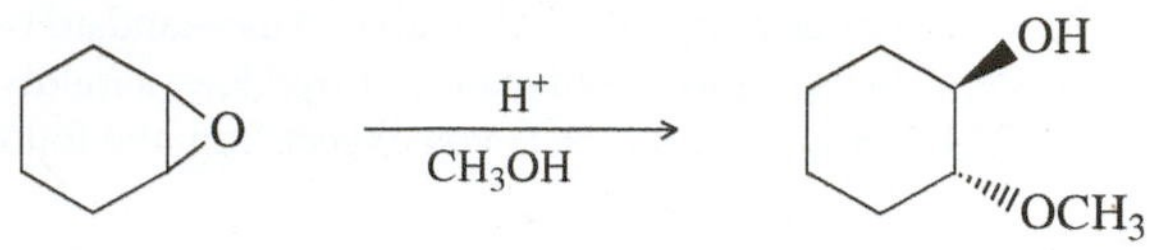 cyclohexene oxide $\xrightarrow[CH_3OH]{H^+}$ OH / OCH$_3$

(h) $H_2C{=}CH{-}C_6H_5 \xrightarrow{H^+} CH_3{-}CH(C_6H_5){-}CH{=}CH{-}C_6H_5$

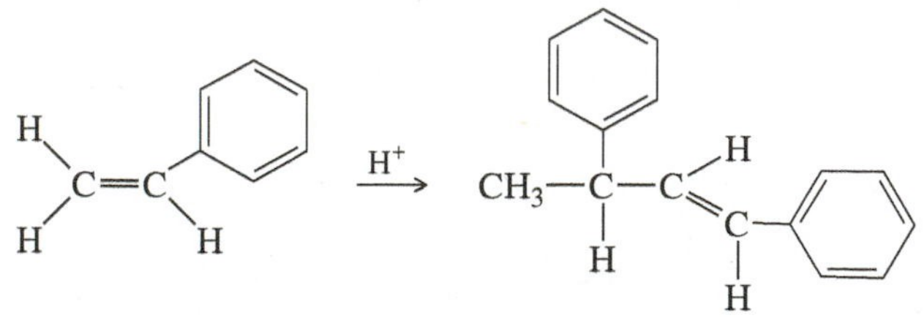

8-57 Draw an approximate reaction-energy diagram showing the curves for the two possible pathways for ionic addition of HBr to 1-methylcyclohexene. (a) Formation of the major product, 1-bromo-1-methylcyclohexane, and (b) formation of the minor product, 1-bromo-2-methylcyclohexane. Point out how these curves show that 1-bromo-1-methylcyclohexane should be formed faster.

8-58 Cyclohexene is dissolved in a solution of lithium chloride in chloroform. To this solution is added one equivalent of bromine. The material isolated from this reaction contains primarily a mixture of *trans*-1,2-dibromocyclohexane and *trans*-1-bromo-2-chlorocyclohexane. Propose a mechanism to show how these compounds are formed.

8-59 Draw a reaction-energy diagram for the propagation steps of the free-radical addition of HBr to isobutylene. Draw curves representing the reactions leading to both the Markovnikov and the anti-Markovnikov products. Compare the values of $\Delta G°$ and E_a for the rate-limiting steps, and explain why only one of these products is observed.

8-60 Unknown **X,** C_5H_9Br, does not react with bromine or with dilute $KMnO_4$. Upon treatment with potassium *tert*-butoxide, **X** gives only one product, **Y,** C_5H_8. Unlike **X, Y** decolorizes bromine and changes $KMnO_4$ from purple to brown. Catalytic hydrogenation of **Y** gives methylcyclobutane. Ozonolysis–reduction of **Y** gives dialdehyde **Z,** $C_5H_8O_2$. Propose consistent structures for **X, Y,** and **Z.** Is there any aspect of the structure of **X** that is still unknown?

8-61 One of the constituents of turpentine is α-pinene, formula $C_{10}H_{16}$. The following scheme (called a "road map") gives some reactions of α-pinene. Determine the structure of α-pinene and of the reaction products **A** through **E.**

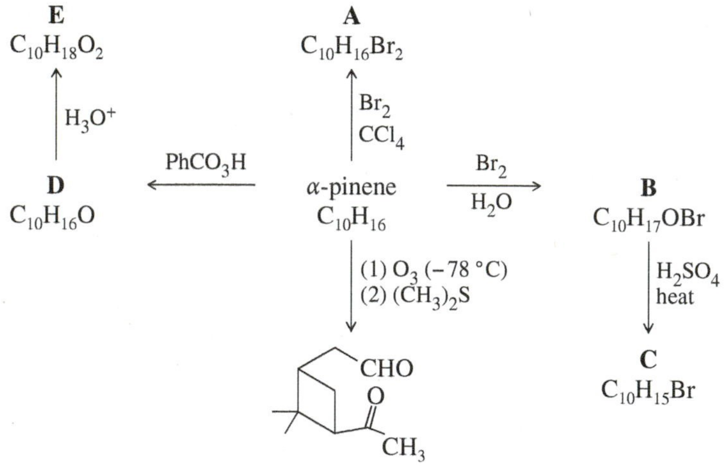

E $C_{10}H_{18}O_2$ ← (H_3O^+) ← **D** $C_{10}H_{16}O$ ← ($PhCO_3H$) ← α-pinene $C_{10}H_{16}$

α-pinene → (Br_2, CCl_4) → **A** $C_{10}H_{16}Br_2$

α-pinene → (Br_2, H_2O) → **B** $C_{10}H_{17}OBr$ → (H_2SO_4, heat) → **C** $C_{10}H_{15}Br$

α-pinene → (1) O_3 (−78 °C) (2) $(CH_3)_2S$ → keto-aldehyde (CHO, $C(=O)CH_3$)

8-62 The sex attractant of the housefly has the formula $C_{23}H_{46}$. When treated with warm potassium permanganate, this pheromone gives two products: $CH_3(CH_2)_{12}COOH$ and $CH_3(CH_2)_7COOH$. Suggest a structure for this sex attractant. Explain which part of the structure is uncertain.

8-63 In contact with a platinum catalyst, an unknown alkene reacts with three equivalents of hydrogen gas to give 1-isopropyl-4-methylcyclohexane. When the unknown alkene is ozonized and reduced, the products are the following:

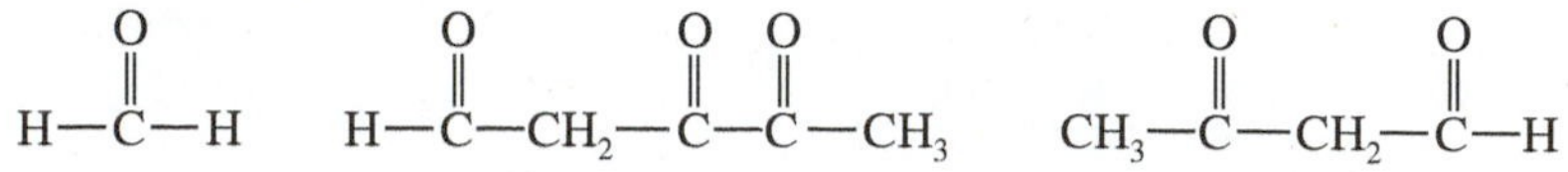

$$H{-}\overset{O}{\overset{\|}{C}}{-}H \qquad H{-}\overset{O}{\overset{\|}{C}}{-}CH_2{-}\overset{O}{\overset{\|}{C}}{-}\overset{O}{\overset{\|}{C}}{-}CH_3 \qquad CH_3{-}\overset{O}{\overset{\|}{C}}{-}CH_2{-}\overset{O}{\overset{\|}{C}}{-}H$$

Deduce the structure of the unknown alkene.

***8-64** Propose a mechanism for the following reaction.

H_2SO_4, H_2O

HO

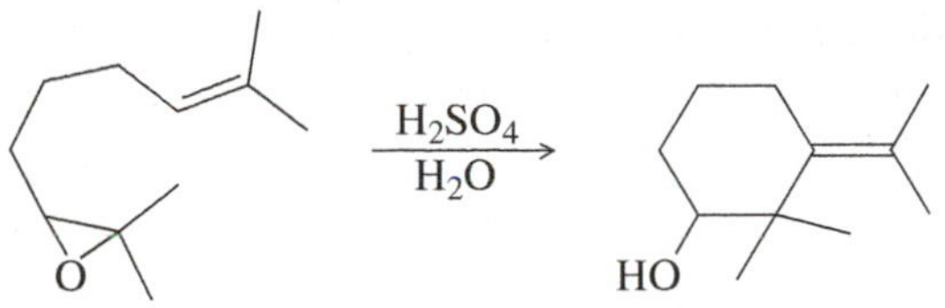

8-65 The two butenedioic acids are called *fumaric acid* (trans) and *maleic acid* (cis). 2,3-Dihydroxybutanedioic acid is called *tartaric acid.*

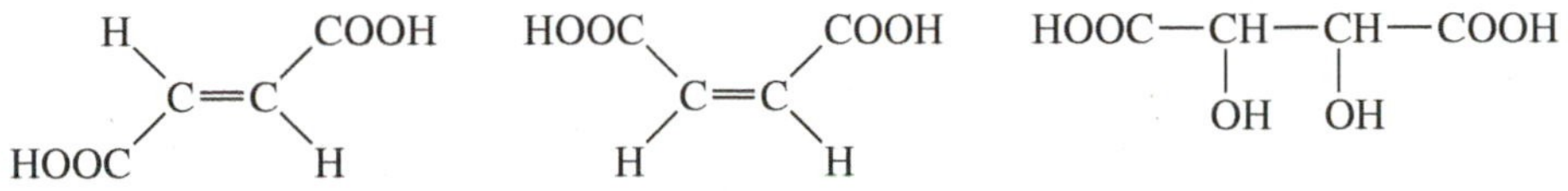

fumaric acid: H(COOH)C=C(H)HOOC

maleic acid: HOOC(H)C=C(COOH)H

tartaric acid: HOOC—CH(OH)—CH(OH)—COOH

Show how you would convert
(a) fumaric acid to (±)-tartaric acid.
(b) fumaric acid to *meso*-tartaric acid.
(c) maleic acid to (±)-tartaric acid.
(d) maleic acid to *meso*-tartaric acid.

8-66 The compound BD_3 is a deuterated form of borane. Predict the product formed when 1-methylcyclohexene reacts with $BD_3 \cdot THF$, followed by basic hydrogen peroxide.

8-67 A routine addition of HBr across the double bond of a vinylcyclopentane gave a small amount of an unexpected rearranged product. Propose a mechanism for the formation of this product, and explain why the rearrangement occurs.

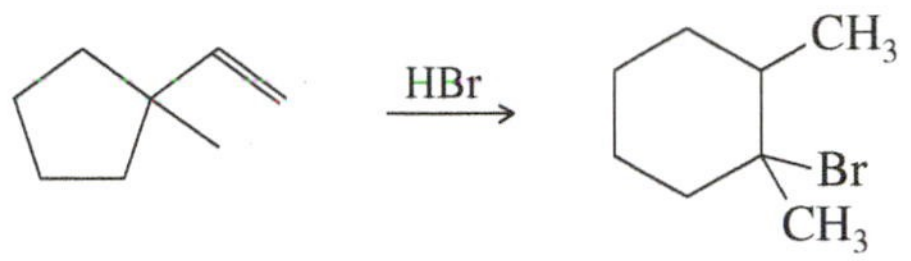

HBr

CH_3 Br CH_3

8-68 An unknown compound decolorizes bromine in carbon tetrachloride, and it undergoes catalytic reduction to give decalin. When treated with warm, concentrated potassium permanganate, this compound gives *cis*-cyclohexane-1,2-dicarboxylic acid and oxalic acid. Propose a structure for the unknown compound.

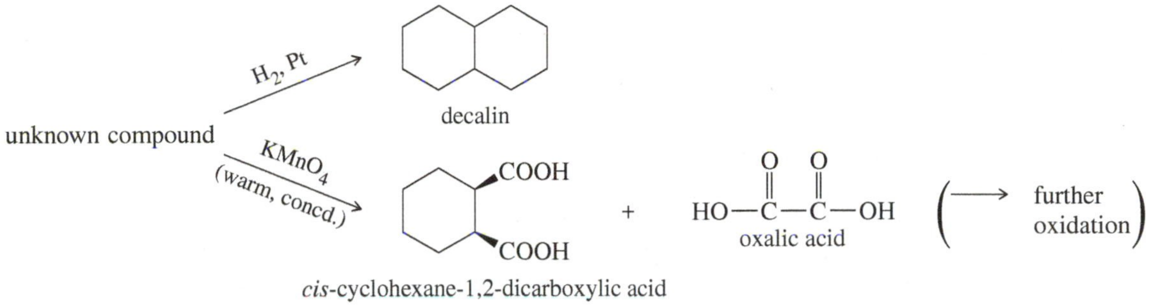

unknown compound —(H_2, Pt)→ decalin

unknown compound —($KMnO_4$ (warm, concd.))→ *cis*-cyclohexane-1,2-dicarboxylic acid (COOH, COOH) + HO—C(=O)—C(=O)—OH oxalic acid (→ further oxidation)

***8-69** Many enzymes catalyze reactions that are similar to reactions we might use for organic synthesis. Enzymes tend to be stereospecific in their reactions, and asymmetric induction is common. The following reaction, part of the tricarboxylic acid cycle of cell respiration, resembles a reaction we might use in the laboratory; however, the enzyme-catalyzed reaction gives only the (*S*) enantiomer of the product, malic acid.

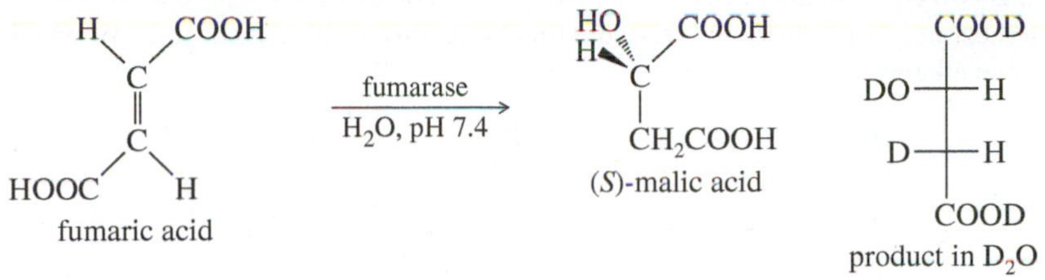

fumaric acid (H, COOH, HOOC, H) —(fumarase, H_2O, pH 7.4)→ (*S*)-malic acid (HO, H, COOH, CH_2COOH)

product in D_2O: COOD; DO—H; D—H; COOD

(a) What type of reaction does fumarase catalyze?
(b) Is fumaric acid chiral? Is malic acid chiral? In the enzyme-catalyzed reaction, is the product (malic acid) optically active?
(c) If we could run the preceding reaction in the laboratory using sulfuric acid as the catalyst, would the product (malic acid) be optically active?
(d) Do you expect the fumarase enzyme to be a chiral molecule?
(e) When the enzyme-catalyzed reaction takes place in D_2O, the *only* product is the stereoisomer just pictured. No enantiomer or diastereomer of this compound is formed. Is the enzyme-catalyzed reaction a syn or anti addition?
(f) Assume that we found conditions to convert fumaric acid to deuterated malic acid using hydroboration with $BD_3 \cdot THF$, followed by oxidation with D_2O_2 and NaOD. Use Fischer projections to show the stereoisomer(s) of deuterated malic acid you would expect to be formed.

***8-70** **(a)** The following cyclization has been observed in the oxymercuration–demercuration of this unsaturated alcohol. Propose a mechanism for this reaction.

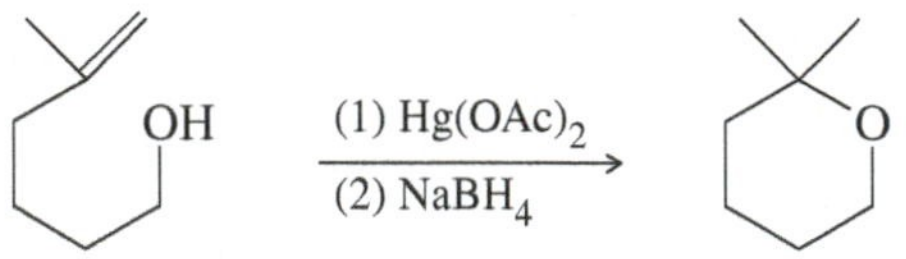

OH — (1) $Hg(OAc)_2$ (2) $NaBH_4$ → O

(b) Predict the product of formula $C_7H_{13}BrO$ from the reaction of this same unsaturated alcohol with bromine. Propose a mechanism to support your prediction.

8-71 A graduate student attempted to form the iodohydrin of the alkene shown below. Her analysis of the products showed a good yield of an unexpected product. Propose a mechanism to explain the formation of this product.

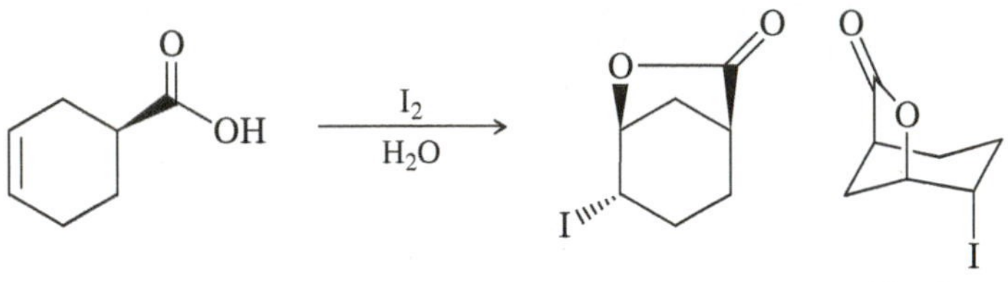

O, OH — I_2 / H_2O → O, O, I; O, O, I

side view

8-72 Propose a mechanism for reaction of the first three propylene units in the polymerization of propylene in the presence of a peroxide.

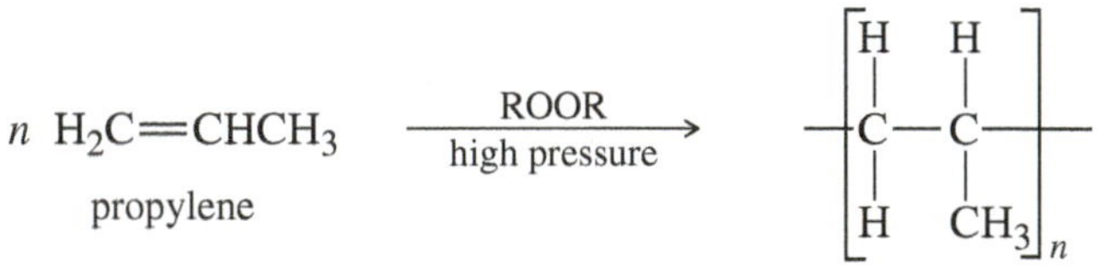

$n\ H_2C{=}CHCH_3$ (propylene) $\xrightarrow[\text{high pressure}]{\text{ROOR}}$ $\left[\begin{array}{cc} H & H \\ | & | \\ -C- & -C- \\ | & | \\ H & CH_3 \end{array} \right]_n$ (polypropylene)

***8-73** When styrene (vinylbenzene) is commercially polymerized, about 1–3% of 1,4-divinylbenzene is often added to the styrene. The incorporation of some divinylbenzene gives a polymer with more strength and better resistance to organic solvents. Explain how a very small amount of divinylbenzene has a marked effect on the properties of the polymer.

8-74 The cationic polymerization of isobutylene (2-methylpropene) is shown in Section 8-16A. Isobutylene is often polymerized under free-radical conditions. Propose a mechanism for the free-radical polymerization of isobutylene.

8-75 Ozonolysis can be applied selectively to different types of carbon–carbon double bonds. The compound shown below contains two vinyl ether double bonds, which are electron-rich because of the electron-donating alkoxy groups. Ozone reacts more quickly with electron-rich double bonds and more slowly with hindered double bonds. At –78 °C, this compound quickly adds two equivalents of ozone. Immediate reduction of the ozonide gives a good yield of a single product. Show the expected ozonolyis product, and label the functional groups produced, some of which are not typical from ozonolysis of simple alkenes.

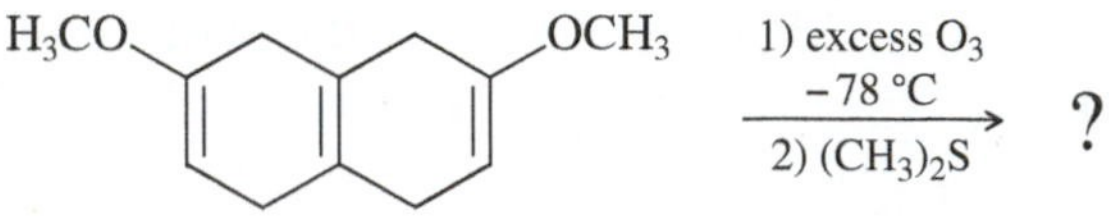

H_3CO, OCH_3 — 1) excess O_3, –78 °C; 2) $(CH_3)_2S$ → ?

(continued)

8-76 Propose mechanisms to explain the opposite regiochemistry observed in the following two reactions.

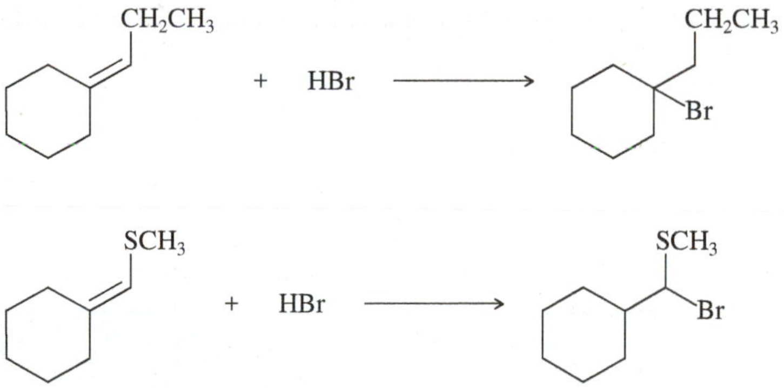

CH_2CH_3 + HBr → CH_2CH_3 Br

SCH_3 + HBr → SCH_3 Br

***8-77** An inexperienced graduate student treated dec-5-ene with borane in THF, placed the flask in a refrigerator, and left for a party. When he returned from the party, he discovered that the refrigerator was broken and that it had gotten quite warm inside. Although all the THF had evaporated from the flask, he treated the residue with basic hydrogen peroxide. To his surprise, he recovered a fair yield of decan-1-ol. Use a mechanism to show how this reaction might have occurred. (*Hint*: The addition of BH_3 is reversible.)

***8-78** We have seen many examples where halogens add to alkenes with anti stereochemistry via the halonium ion mechanism. However, when 1-phenylcyclohexene reacts with chlorine in carbon tetrachloride, a mixture of the cis and trans isomers of the product is recovered. Propose a mechanism, and explain this lack of stereospecificity.

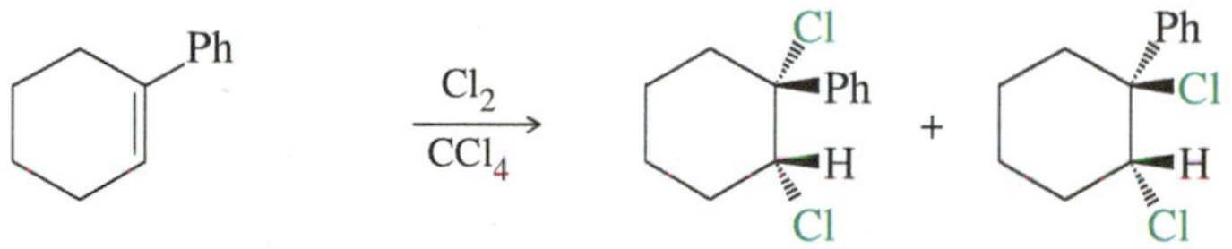

Ph $\xrightarrow[CCl_4]{Cl_2}$ Cl Ph H Cl + Ph Cl H Cl

1-phenylcyclohexene *cis*- and *trans*- 1,2-dichloro-1-phenylcyclohexane

8-79 The bulky borane 9-BBN was developed to enhance the selectivity of hydroboration. In this example, 9-BBN adds to the less hindered carbon with 99.3% regioselectivity, compared with only 57% for diborane.

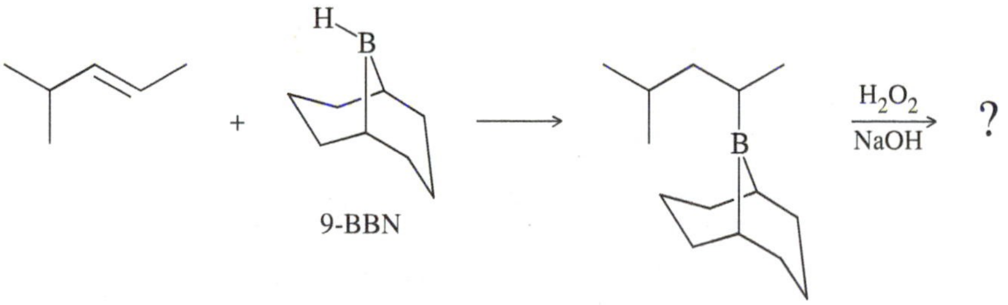

+ H B (9-BBN) → B (99.3% this isomer) $\xrightarrow[NaOH]{H_2O_2}$?

(a) Show the two organic products generated when the trialkylborane is oxidized with H_2O_2/NaOH.

(b) 9-BBN is synthesized by adding BH_3 across a symmetric, cyclic diene. What is the structure of the diene?

9 Alkynes

Goals for Chapter 9

1 Draw and name alkynes and their derivatives.

2 Show how to synthesize alkynes by eliminations from alkyl halides and by the additions and substitutions of acetylide ions.

3 Predict the products of the reactions of alkynes, including the orientation of the reaction (regiochemistry) and the stereochemistry.

4 Propose mechanisms to explain the observed products of alkyne reactions.

5 Use retrosynthetic analysis to solve multistep synthesis problems with alkynes as reagents, intermediates, or products.

▶ The harlequin poison frog (*Dendrobates histrionicus*) protects itself from attack by secreting a fluid containing the alkaloid histrionicotoxin from glands in its back. Histrionicotoxin has a unique structure with a spirocyclic center and two *cis*-enyne side chains that are not found in any other natural products. It is not known whether the frog synthesizes histrionicotoxin itself or accumulates it from its diet.

9-1 Introduction

Alkynes are hydrocarbons that contain carbon–carbon triple bonds. Alkynes are also called **acetylenes** because they are derivatives of acetylene, the simplest alkyne.

$H-C\equiv C-H$	$CH_3CH_2-C\equiv C-H$	$CH_3-C\equiv C-CH_3$
acetylene	ethylacetylene	dimethylacetylene
ethyne	but-1-yne	but-2-yne

The chemistry of the carbon–carbon triple bond is similar to that of the double bond. In this chapter, we see that alkynes undergo most of the same reactions as alkenes, especially the additions and the oxidations. We also consider reactions that are specific to alkynes: some that depend on the unique characteristics of the $C\equiv C$ triple bond, and others that depend on the unusual acidity of the acetylenic $\equiv C-H$ bond.

A triple bond gives an alkyne four fewer hydrogens than the corresponding alkane. Its molecular formula is like that of a molecule with two double bonds: C_nH_{2n-2}. Therefore, the triple bond contributes two elements of unsaturation (eu) (Section 7-3).

H_3C-CH_3 (H–C(H)(H)–C(H)(H)–H)	$H_2C=CH_2$	$H-C\equiv C-H$
ethane, C_2H_6	ethene, C_2H_4	ethyne, C_2H_2
0 eu, C_nH_{2n+2}	1 eu, C_nH_{2n}	2 eu, C_nH_{2n-2}

Alkynes are not as common in nature as alkenes, but some plants and insects use alkynes to protect themselves against disease or predators. Some interesting examples are shown:

- Cicutoxin is a toxic compound found in poisonous water hemlock.
- A substituted panaxytriol, a diacetylenic compound found in ginseng, may alleviate the side effects of cancer chemotherapy.
- Soldier beetles secrete DHMA to ward off predators and pathogens.
- Capillin is used by a plant found near the Red Sea to protect itself against fungal diseases.
- Parsalmide is used as an analgesic.
- Ethynyl estradiol, a synthetic female hormone, is a common ingredient in birth control pills.
- Dynemicin A is an antibacterial compound that is also being tested as an antitumor agent.

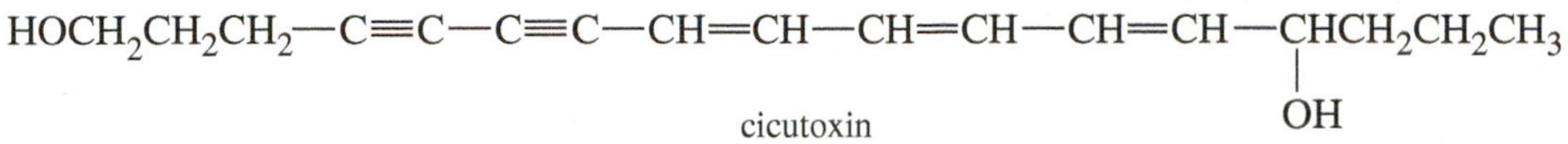

$$HOCH_2CH_2CH_2{-}C{\equiv}C{-}C{\equiv}C{-}CH{=}CH{-}CH{=}CH{-}CH{=}CH{-}\underset{\displaystyle OH}{CH}CH_2CH_2CH_3$$

cicutoxin

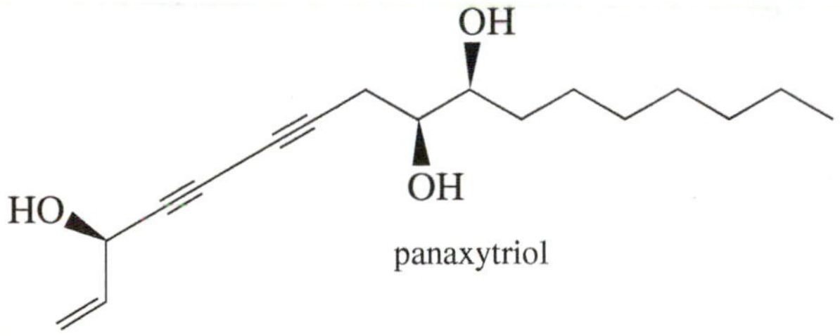

panaxytriol

8Z-dihydromatricaria acid (DHMA)

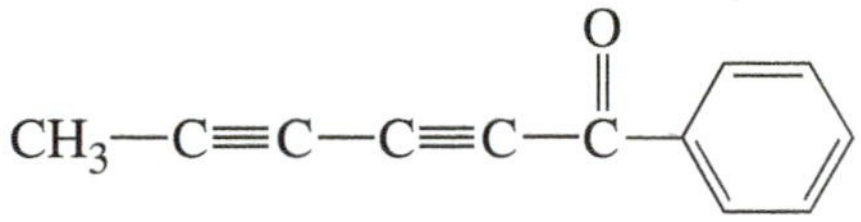

$$CH_3{-}C{\equiv}C{-}C{\equiv}C{-}\overset{\displaystyle O}{\overset{\|}{C}}{-}C_6H_5$$

capillin

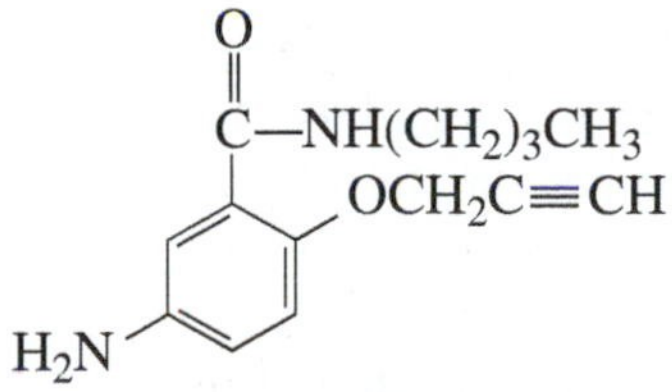

parsalmide

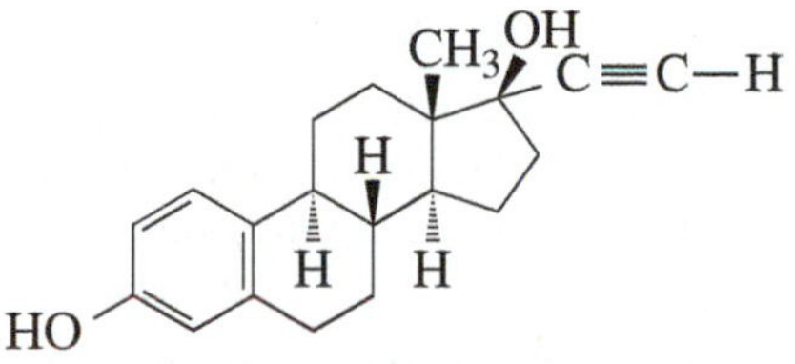

ethynyl estradiol

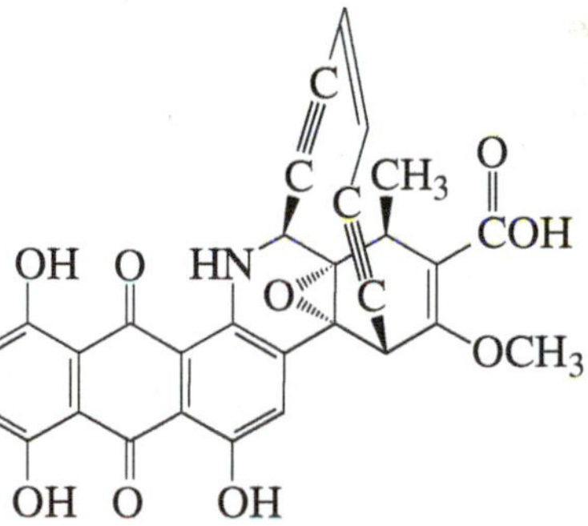

dynemicin A

PROBLEM 9-1

(a) Count the elements of unsaturation in parsalmide, ethynyl estradiol, and dynemicin A.

(b) Draw structural formulas of at least two alkynes of each molecular formula.

(1) C_6H_{10} **(2)** C_8H_{12} **(3)** C_7H_8

9-2 Nomenclature of Alkynes

IUPAC Names The IUPAC nomenclature for alkynes is similar to that for alkenes. We find the longest continuous chain of carbon atoms that includes the triple bond and change the *-ane* ending of the parent alkane to *–yne*. The chain is numbered from the end closest to the triple bond, and the position of the triple bond is designated by its lower-numbered carbon atom. Substituents are given numbers to indicate their locations.

	$H-C\equiv C-H$	$CH_3-C\equiv C-H$	$CH_3-C\equiv C-CH_3$	$CH_3-CH(CH_3)-C\equiv C-CH_2-CHBr-CH_3$
old IUPAC name:	ethyne	propyne	2-butyne	6-bromo-2-methyl-3-heptyne
new IUPAC name:	ethyne (acetylene)	propyne	but-2-yne	6-bromo-2-methylhept-3-yne

When additional functional groups are present, the suffixes are combined to produce the compound names of the *alkenynes* (a double bond and a triple bond), *alkynols* (a triple bond and an alcohol), etc. The new IUPAC system (placing the number right before the group) helps to clarify these names. The IUPAC rules give alcohols higher priority than alkenes or alkynes (which are given equal priority), so the numbering begins at the end closer to an alcohol. The priorities of functional groups in naming organic compounds are listed in Table 9-1. If the double bond and the triple bond are equidistant from the ends of the chain, number the chain so that the double bond receives a lower number than the triple bond (because "ene" comes before "yne" in the alphabet).

	$H_2C=C(CH_3)-C\equiv C-CH_3$	$CH_3-CH(OH)-C\equiv C-H$	$CH_2=CH-CH_2-C\equiv CH$
old IUPAC name:	2-methyl-1-penten-3-yne	3-butyn-2-ol	1-penten-4-yne
new IUPAC name:	2-methylpent-1-en-3-yne	but-3-yn-2-ol	pent-1-en-4-yne

TABLE 9-1 Priority of Functional Groups in Naming Organic Compounds

acids (highest)
esters
aldehydes
ketones
alcohols
amines
alkenes, alkynes
alkanes
ethers
alkyl halides (lowest)

Common Names The common names of alkynes describe them as derivatives of acetylene. Most alkynes can be named as a molecule of acetylene with one or two alkyl substituents. This nomenclature is like the common nomenclature for ethers, where we name the two alkyl groups bonded to oxygen.

$H-C\equiv C-H$ acetylene	$R-C\equiv C-H$ an alkylacetylene	$R-C\equiv C-R'$ a dialkylacetylene
$CH_3-C\equiv C-H$ methylacetylene	$Ph-C\equiv C-H$ phenylacetylene	$CH_3-C\equiv C-CH_2CH_3$ ethylmethylacetylene
$(CH_3)_2CH-C\equiv C-CH(CH_3)_2$ diisopropylacetylene	$Ph-C\equiv C-Ph$ diphenylacetylene	$H-C\equiv C-CH_2OH$ hydroxymethylacetylene (propargyl alcohol)

Many of an alkyne's chemical properties depend on whether there is an acetylenic hydrogen ($H-C\equiv C$), that is, whether the triple bond comes at the end of a carbon chain. Such an alkyne is called a **terminal alkyne** or a **terminal acetylene.** If the triple bond is located somewhere other than the end of the carbon chain, the alkyne is called an **internal alkyne** or an **internal acetylene.**

acetylenic hydrogen

$H-C\equiv C-CH_2CH_3$

but-1-yne, a *terminal* alkyne

(no acetylenic hydrogen)

$CH_3-C\equiv C-CH_3$

but-2-yne, an *internal* alkyne

PROBLEM-SOLVING HINT

The nomenclature of alkynes is reviewed in Appendix 5, the Summary of Organic Nomenclature.

PROBLEM 9-2

For each molecular formula, draw all the isomeric alkynes, and give their IUPAC names. Circle the acetylenic hydrogen of each terminal alkyne.

(a) C_5H_8 (three isomers) **(b)** C_6H_{10} (seven isomers)

9-3 Physical Properties of Alkynes

The physical properties of alkynes (Table 9-2) are similar to those of alkanes and alkenes of similar molecular weights. Alkynes are relatively nonpolar and nearly insoluble in water. They are quite soluble in most organic solvents, including acetone, ether, methylene chloride, chloroform, and alcohols. Many alkynes have characteristic, mildly offensive odors. Ethyne, propyne, and the butynes are gases at room temperature, just like the corresponding alkanes and alkenes. In fact, the boiling points of alkynes are nearly the same as those of alkanes and alkenes with similar carbon skeletons.

TABLE 9-2
Physical Properties of Selected Alkynes

Name	Structure	mp (°C)	bp (°C)	Density (g/cm^3)
ethyne (acetylene)	$H-C\equiv C-H$	−82	−84	0.62
propyne	$H-C\equiv C-CH_3$	−101	−23	0.67
but-1-yne	$H-C\equiv C-CH_2CH_3$	−126	8	0.67
but-2-yne	$CH_3-C\equiv C-CH_3$	−32	27	0.69
pent-1-yne	$H-C\equiv C-CH_2CH_2CH_3$	−90	40	0.70
pent-2-yne	$CH_3-C\equiv C-CH_2CH_3$	−101	55	0.71
3-methylbut-1-yne	$CH_3-CH(CH_3)-C\equiv C-H$		28	0.67
hex-1-yne	$H-C\equiv C-(CH_2)_3-CH_3$	−132	71	0.72
hex-2-yne	$CH_3-C\equiv C-CH_2CH_2CH_3$	−90	84	0.73
hex-3-yne	$CH_3CH_2-C\equiv C-CH_2CH_3$	−101	82	0.73
3,3-dimethylbut-1-yne	$(CH_3)_3C-C\equiv C-H$	−81	38	0.67
hept-1-yne	$H-C\equiv C-(CH_2)_4CH_3$	−81	100	0.73
oct-1-yne	$H-C\equiv C-(CH_2)_5CH_3$	−79	125	0.75
non-1-yne	$H-C\equiv C-(CH_2)_6CH_3$	−50	151	0.76
dec-1-yne	$H-C\equiv C-(CH_2)_7CH_3$	−36	174	0.77

9-4 Commercial Importance of Alkynes

Of the alkynes that are used commercially, acetylene is used in much greater amounts than all of the other alkynes combined: about 5 million tons per year. Acetylene is used both as the fuel for the oxyacetylene welding torch and as an important industrial feedstock for making other compounds.

9-4A Uses of Acetylene and Methylacetylene

Acetylene is a colorless, foul-smelling gas that burns in air with a yellow, sooty flame. When the flame is supplied with pure oxygen, however, the color turns to light blue, and the flame temperature increases dramatically. A comparison of the heat of combustion for acetylene with those of ethene and ethane shows why this gas makes an excellent fuel for a high-temperature flame.

$$CH_3CH_3 + \tfrac{7}{2}\,O_2 \longrightarrow 2\,CO_2 + 3\,H_2O \qquad \Delta H° = -1561\text{ kJ } (-373\text{ kcal})$$

$$-1561\text{ kJ divided by 5 moles of products} = -312\text{ kJ/mol of products } (-75\text{ kcal/mol of products})$$

$$H_2C{=}CH_2 + 3\,O_2 \longrightarrow 2\,CO_2 + 2\,H_2O \qquad \Delta H° = -1410\text{ kJ } (-337\text{ kcal})$$

$$-1410\text{ kJ divided by 4 moles of products} = -352\text{ kJ/mol of products } (-84\text{ kcal/mol of products})$$

(continued)

An oxygen–acetylene flame is hot enough to melt steel for welding. A cutting torch uses an extra jet of oxygen to burn away hot steel.

$$\mathrm{HC{\equiv}CH} + \tfrac{5}{2}\mathrm{O_2} \longrightarrow 2\,\mathrm{CO_2} + 1\,\mathrm{H_2O} \qquad \Delta H° = -1326\text{ kJ } (-317\text{ kcal})$$

$$-1326\text{ kJ divided by 3 moles of products} = -442\text{ kJ/mol of products } (-106\text{ kcal/mol of products})$$

If we were simply heating a house by burning one of these fuels, we might choose ethane as our fuel because it produces the most heat per mole of gas consumed. In the welding torch, we want the highest possible *temperature* of the gaseous products. The heat of reaction must raise the temperature of the products to the flame temperature. Roughly speaking, the increase in temperature of the products is proportional to the heat given off *per mole of products* formed. This rise in temperature is largest with acetylene, which gives off the most heat per mole of products. The oxyacetylene flame reaches temperatures as high as 3500 °C.

When acetylene was first used for welding, it was considered a dangerous, explosive gas. Acetylene is thermodynamically unstable. When the compressed gas is subjected to thermal or mechanical shock, it decomposes to its elements, releasing 234 kJ (56 kcal) of energy per mole. This initial decomposition often splits the container, allowing the products (hydrogen and finely divided carbon) to burn in the air.

$$\mathrm{H{-}C{\equiv}C{-}H} \longrightarrow 2\,\mathrm{C} + \mathrm{H_2} \qquad \Delta H° = -234\text{ kj/mol} \quad (-56\text{ kcal/mol})$$

$$2\,\mathrm{C} + \mathrm{H_2} \xrightarrow{\frac{5}{2}\mathrm{O_2}} 2\,\mathrm{CO_2} + \mathrm{H_2O} \qquad \Delta H° = -1090\text{ kj/mol } (-261\text{ kcal/mol})$$

Acetylene is safely stored and handled in cylinders that are filled with crushed firebrick wet with acetone. Acetylene dissolves freely in acetone, and the dissolved gas is not so prone to decomposition. Firebrick helps to control the decomposition by minimizing the free volume of the cylinder, cooling and controlling any decomposition before it gets out of control.

Methylacetylene also is used in welding torches. Methylacetylene does not decompose as easily as acetylene, and it burns better in air (as opposed to pure oxygen). Methylacetylene is well suited for household soldering and brazing that requires higher temperatures than propane torches can reach. The industrial synthesis of methylacetylene gives a mixture with its isomer, propadiene (allene). This mixture is sold commercially under the name MAPP® gas (**M**ethyl**A**cetylene-**P**ro**P**adiene).

$$\underset{\text{methylacetylene}}{\mathrm{CH_3{-}C{\equiv}C{-}H}} \qquad \underset{\text{propadiene (allene)}}{\mathrm{H_2C{=}C{=}CH_2}}$$

9-4B Manufacture of Acetylene

Acetylene, one of the cheapest organic chemicals, is made from coal or from natural gas. The synthesis from coal involves heating lime and coke (roasted coal) in an electric furnace to produce calcium carbide. Addition of water to calcium carbide produces acetylene and hydrated lime.

$$\underset{\text{coke}}{3\,\mathrm{C}} + \underset{\text{lime}}{\mathrm{CaO}} \xrightarrow{\text{electric furnace, 2500 °C}} \underset{\text{calcium carbide}}{\mathrm{CaC_2}} + \mathrm{CO}$$

$$\mathrm{CaC_2} + 2\,\mathrm{H_2O} \longrightarrow \underset{\text{acetylene}}{\mathrm{H{-}C{\equiv}C{-}H}} + \underset{\text{hydrated lime}}{\mathrm{Ca(OH)_2}}$$

This second reaction once served as a light source in coal mines until battery-powered lights became available. A miner's lamp works by allowing water to drip slowly onto some calcium carbide. Acetylene is generated, feeding a small flame where the gas burns in air with a yellow flickering light. Unfortunately, this flame ignites the methane gas commonly found in coal seams, causing explosions. Battery-powered miner's lamps have replaced carbide lamps in most mines, although some cave explorers still prefer carbide lamps because of their reliability and their usefulness as hand warmers.

The synthesis of acetylene from natural gas is a simple process. Natural gas consists mostly of methane, which forms acetylene when it is heated for a very short period of time.

$$2\ CH_4 \xrightarrow[0.01\ \text{sec}]{1500\ °\text{C}} H{-}C{\equiv}C{-}H \ + \ 3\ H_2$$

Although this reaction is endothermic, there are twice as many moles of products as reactants. The increase in the number of moles results in an increase in entropy, and the $(-T\Delta S)$ term in the free energy $(\Delta G = \Delta H - T\Delta S)$ predominates at this high temperature.

A miner's carbide lamp. Water in the upper chamber slowly drips onto calcium carbide in the lower chamber, generating acetylene.

PROBLEM 9-3

What reaction would acetylene likely undergo if it were kept at 1500 °C for too long?

9-5 Electronic Structure of Alkynes

In Section 1-15, we studied the electronic structure of a triple bond. Let's review this structure, using acetylene as the example. The Lewis structure of acetylene shows three pairs of electrons in the region between the carbon nuclei:

H:C:::C:H

Each carbon atom is bonded to two other atoms, and there are no nonbonding valence electrons. Each carbon atom needs two hybrid orbitals to form the sigma bond framework. Hybridization of the *s* orbital with one *p* orbital gives two hybrid orbitals, directed 180° apart, for each carbon atom. Overlap of these *sp* hybrid orbitals with each other and with the hydrogen *s* orbitals gives the sigma bond framework. Experimental results have confirmed this linear (180°) structure.

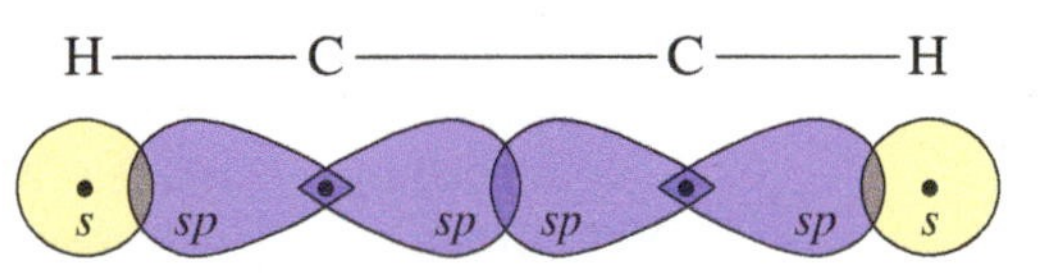

Two pi bonds result from overlap of the two remaining unhybridized *p* orbitals on each carbon atom. These orbitals overlap at right angles to each other, forming one pi bond with electron density above and below the C—C sigma bond, and the other with electron density in front and in back of the sigma bond. The shape of these pi bonds is such that they blend to form a cylinder of electron density encircling the sigma bond between the two carbon atoms (Figure 9-1).

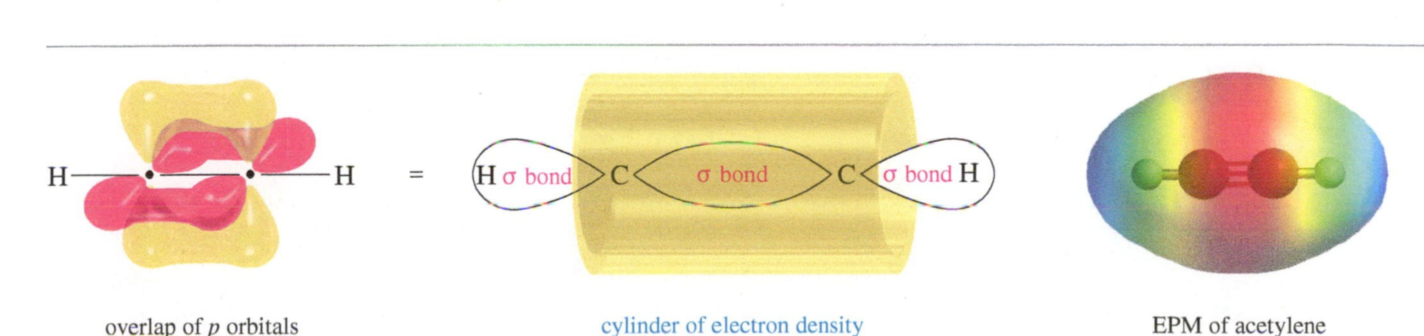

FIGURE 9-1
Bonding in acetylene. The C≡C triple bond consists of a central sigma bond (formed by overlap of two *sp* hybrid orbitals on the carbon atoms) plus two pi bonds formed by overlap of the remaining two *p* orbitals on each carbon. A calculation shows that the two pi bonds add to form a cylinder of electron density surrounding the sigma bond.

The carbon–carbon bond length in acetylene is 1.20 Å, and each carbon–hydrogen bond is 1.06 Å. Both bonds are shorter than the corresponding bonds in ethane and in ethene.

ethane	ethene	ethyne
C—C 1.54 Å; C—H 1.09 Å	C=C 1.33 Å; C—H 1.08 Å	C≡C 1.20 Å; C—H 1.06 Å

The triple bond is relatively short because of the attractive overlap of three bonding pairs of electrons and the high ***s* character** of the *sp* hybrid orbitals. The *sp* hybrid orbitals are about one-half *s* character (as opposed to one-third *s* character of sp^2 hybrids and one-fourth of sp^3 hybrids), using more of the closer, tightly held *s* orbital. The *sp* hybrid orbitals also account for the slightly shorter C—H bonds in acetylene compared with ethylene.

9-6 Acidity of Alkynes; Formation of Acetylide Ions

Terminal alkynes are much more acidic than other hydrocarbons. Removal of an acetylenic proton forms an acetylide ion, which plays a central role in alkyne chemistry.

The acidity of an acetylenic hydrogen stems from the nature of the *sp* hybrid ≡C—H bond. Table 9-3 shows how the acidity of a C—H bond varies with its hybridization, increasing with the increasing *s* character of the orbitals: $sp^3 < sp^2 < sp$. (Recall that a *smaller* value of pK_a corresponds to a stronger acid.) The acetylenic proton is about 10^{19} times more acidic than a vinyl proton.

Abstraction of an acetylenic proton gives a carbanion that has the lone pair of electrons in the *sp* hybrid orbital. Electrons in this orbital are close to the nucleus, and there is less charge separation than in carbanions with the lone pair in sp^2 or sp^3 hybrid orbitals. Ammonia and alcohols are included for comparison; note that acetylene can be deprotonated by the amide ($^-NH_2$) ion, but not by an alkoxide ion (^-OR).

Very strong bases (such as sodium amide, $NaNH_2$) deprotonate terminal acetylenes to form carbanions called **acetylide ions** (or **alkynide ions**). Hydroxide ion and alkoxide ions are not strong enough bases to deprotonate alkynes. Internal alkynes do not have acetylenic protons, so they do not react.

TABLE 9-3

Compound	Conjugate Base	Hybridization	*s* Character	pK_a	
$H_3C—CH_3$	$H_3C—CH_2^-$	sp^3	25%	50	weakest acid
$H_2C=CH_2$	$H_2C=CH^-$	sp^2	33%	44	↓
$:NH_3$	$^-:\ddot{N}H_2$	(ammonia)		36	↓
H—C≡C—H	H—C≡C:⁻	*sp*	50%	25	↓
R—OH	$R—\ddot{O}:^-$	(alcohols)		16–18	stronger acid

acidic proton

$$CH_3CH_2{-}C{\equiv}C{-}H + Na^+ \ {}^-{:}\ddot{N}H_2 \longrightarrow CH_3CH_2{-}C{\equiv}C{:}^- \ Na^+ + NH_3$$

but-1-yne, a terminal alkyne, $pK_a = 25$; sodium amide; sodium butynide; NH_3 $pK_a = 36$

$$C_6H_{11}{-}C{\equiv}C{-}H + NaNH_2 \longrightarrow C_6H_{11}{-}C{\equiv}C{:}^- \ Na^+ + NH_3$$

cyclohexylacetylene, $pK_a = 25$; sodium amide; sodium cyclohexylacetylide; NH_3 $pK_a = 36$

$$CH_3{-}C{\equiv}C{-}CH_3 \xrightarrow{NaNH_2} \text{no reaction}$$

(no acetylenic proton)
but-2-yne, an internal alkyne

Sodium amide ($Na^+ \ {}^-{:}\ddot{N}H_2$) is frequently used as the base in forming acetylide salts. The amide ion (${}^-NH_2$) is the conjugate base of ammonia, a compound that is itself a base. Ammonia is also a very weak acid, however, with $K_a = 10^{-36}$ ($pK_a = 36$). One of its hydrogens can be reduced by sodium metal to give the sodium salt of the amide ion, a very strong conjugate base.

$$NH_3 + Na \xrightarrow{Fe^{3+}\text{ catalyst}} Na^+ \ {}^-{:}\ddot{N}H_2 + \tfrac{1}{2}H_2\uparrow$$

ammonia; sodium amide ("sodamide")

$$R{-}C{\equiv}C{-}H + Na^+ \ {}^-{:}\ddot{N}H_2 \longrightarrow R{-}C{\equiv}C{:}^- \ Na^+ + {:}NH_3$$

a sodium acetylide

Acetylide ions are strong nucleophiles. In fact, one of the best methods for synthesizing substituted alkynes is a nucleophilic attack by an acetylide ion on an unhindered alkyl halide. We consider this displacement reaction in detail in Section 9-7A.

$$CH_3CH_2{-}C{\equiv}C{:}^- \ Na^+ + H_3C{-}I \longrightarrow CH_3CH_2{-}C{\equiv}C{-}CH_3 + NaI$$

PROBLEM 9-4

The boiling points of hex-1-ene (64 °C) and hex-1-yne (71 °C) are sufficiently close that it is difficult to achieve a clean separation by distillation. Show how you might use the acidity of hex-1-yne to remove the last trace of it from a sample of hex-1-ene.

PROBLEM 9-5

Predict the products of the following acid–base reactions, or indicate if no significant reaction would take place.

(a) $H{-}C{\equiv}C{-}H + NaNH_2$
(b) $H{-}C{\equiv}C{-}H + CH_3Li$
(c) $H{-}C{\equiv}C{-}H + NaOCH_3$
(d) $H{-}C{\equiv}C{-}H + NaOH$
(e) $H{-}C{\equiv}C{:}^- \ Na^+ + CH_3OH$
(f) $H{-}C{\equiv}C{:}^- \ Na^+ + H_2O$
(g) $H{-}C{\equiv}C{:}^- \ Na^+ + H_2C{=}CH_2$
(h) $H_2C{=}CH_2 + NaNH_2$
(i) $CH_3OH + NaNH_2$

9-7 Synthesis of Alkynes from Acetylides

Two different approaches are commonly used for the synthesis of alkynes. In the first, an appropriate electrophile undergoes nucleophilic attack by an acetylide ion. The electrophile may be an unhindered primary alkyl halide (undergoes S_N2), or it may be a carbonyl compound (undergoes addition to give an alcohol). Either reaction joins two fragments and gives a product with a lengthened carbon skeleton. This approach is used in many laboratory syntheses of alkynes.

The second approach forms the triple bond by a double dehydrohalogenation of a dihalide. This reaction does not enlarge the carbon skeleton. Isomerization of the triple bond may occur (see Section 9-8), so dehydrohalogenation is useful only when the desired product has the triple bond in a thermodynamically favored position.

9-7A Alkylation of Acetylide Ions

An acetylide ion is a strong base and a powerful nucleophile. It can displace a halide ion from a suitable substrate, giving a substituted acetylene.

PROBLEM-SOLVING HINT

Alkylation of acetylide ions is an excellent way of lengthening a carbon chain. The triple bond can later be reduced (to an alkane or an alkene) if needed.

$$R{-}C{\equiv}C{:}^- + R'{-}X \xrightarrow{S_N2} R{-}C{\equiv}C{-}R' + X^-$$

($R'{-}X$ must be a primary alkyl halide)

If this S_N2 reaction is to produce a good yield, the alkyl halide must be an excellent S_N2 substrate: It must be methyl or primary, with no bulky substituents or branches close to the reaction center. In the following examples, acetylide ions displace primary halides to form elongated alkynes.

$$H{-}C{\equiv}C{:}^-\ Na^+ + CH_3CH_2CH_2CH_2{-}Br \longrightarrow H{-}C{\equiv}C{-}CH_2CH_2CH_2CH_3 + NaBr$$

sodium acetylide; 1-bromobutane; hex-1-yne (butylacetylene) (75%)

$$C_6H_{11}{-}C{\equiv}C{-}H \xrightarrow[\text{(2) ethyl bromide}]{\text{(1) } NaNH_2} C_6H_{11}{-}C{\equiv}C{-}CH_2CH_3$$

ethynylcyclohexane (cyclohexylacetylene); 1-cyclohexylbut-1-yne (ethylcyclohexylacetylene) (70%)

If the back-side approach is hindered, the acetylide ion may abstract a proton, giving elimination by the E2 mechanism.

$$CH_3CH_2{-}C{\equiv}C{:}^- + H_3C{-}CH(Br){-}CH_3 \xrightarrow{E2} CH_3CH_2{-}C{\equiv}C{-}H + H_2C{=}CH{-}CH_3 + Br^-$$

butynide ion; isopropyl bromide; but-1-yne; propene

SOLVED PROBLEM 9-1

Show how to synthesize dec-3-yne from acetylene and any necessary alkyl halides.

SOLUTION

Another name for dec-3-yne is ethyl *n*-hexyl acetylene. It can be made by adding an ethyl group and a hexyl group to acetylene. This can be done in either order. We begin by adding the hexyl group.

$$H{-}C{\equiv}C{-}H \xrightarrow[\text{(2) } CH_3(CH_2)_5Br]{\text{(1) } NaNH_2} CH_3(CH_2)_5{-}C{\equiv}C{-}H$$

acetylene; oct-1-yne

$$CH_3(CH_2)_5{-}C{\equiv}C{-}H \xrightarrow[\text{(2) } CH_3CH_2Br]{\text{(1) } NaNH_2} CH_3(CH_2)_5{-}C{\equiv}C{-}CH_2CH_3$$

oct-1-yne; dec-3-yne

PROBLEM 9-6

Solved Problem 9-1 showed the synthesis of dec-3-yne by adding the hexyl group first, then the ethyl group. Show the reagents and intermediates involved in the other order of synthesis of dec-3-yne, by adding the ethyl group first and the hexyl group last.

PROBLEM 9-7

Show how you might synthesize the following compounds, using acetylene and any suitable alkyl halides as your starting materials. If the compound given cannot be synthesized by this method, explain why.

(a) hex-1-yne
(b) hex-2-yne
(c) hex-3-yne
(d) 4-methylhex-2-yne
(e) 5-methylhex-2-yne
(f) cyclodecyne

9-7B Addition of Acetylide Ions to Carbonyl Groups

Like other carbanions, acetylide ions are strong nucleophiles and strong bases. In addition to displacing halide ions in S_N2 reactions, they can add to carbonyl (C═O) groups. Figure 9-2 shows the structure of the carbonyl group. Because oxygen is more electronegative than carbon, the C═O double bond is polarized. The oxygen atom has a partial negative charge balanced by an equal amount of positive charge on the carbon atom.

The positively charged carbon is electrophilic; attack by a nucleophile places a negative charge on the electronegative oxygen atom.

$$\text{Nuc:}^- + \overset{\delta+}{\text{C}}=\overset{\delta-}{\ddot{\text{O}}}: \longrightarrow \text{Nuc}-\text{C}-\ddot{\text{O}}:^-$$

alkoxide ion

The product of this nucleophilic attack is an alkoxide ion, a strong base. (An **alkoxide ion** is the conjugate base of an alcohol, a weak acid.) Addition of water or a dilute acid protonates the alkoxide to give the alcohol.

$$\text{Nuc}-\text{C}-\ddot{\text{O}}:^- \quad \text{H}-\ddot{\text{O}}-\text{H} \longrightarrow \text{Nuc}-\text{C}-\ddot{\text{O}}\text{H} + {}^-:\ddot{\text{O}}\text{H}$$

alkoxide (or H_3O^+)

Application: Drugs

The addition of an acetylide ion to a carbonyl group is used in the synthesis of ethchlorvynol, a drug used to cause drowsiness and induce sleep. Ethchlorvynol is relatively nonpolar, enhancing its distribution into the fatty tissue of the central nervous system.

$$CH_3CH_2-C(OH)(CH{=}CHCl)-C\equiv CH$$

ethchlorvynol

PROBLEM-SOLVING HINT

Numbers (1), (2), (3), etc. are used to show a sequence of separate reactions over a single arrow. If the numbers were omitted, it would incorrectly imply mixing all these reagents together, rather than adding them in separate steps.

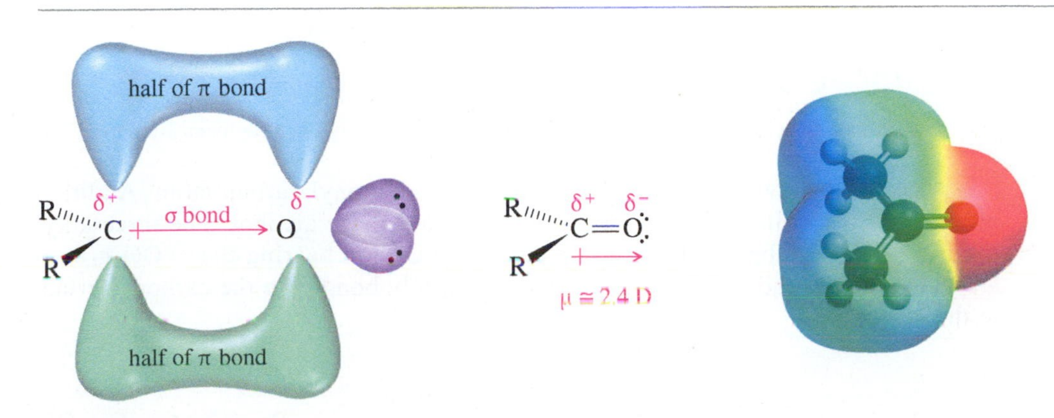

FIGURE 9-2
The C═O double bond of a carbonyl group resembles the C═C double bond of an alkene; however, the carbonyl double bond is strongly polarized. The oxygen atom bears a partial negative charge, and the carbon atom bears a partial positive charge.

An acetylide ion can serve as the nucleophile in this addition to a carbonyl group. The acetylide ion adds to the carbonyl group to form an alkoxide ion. Addition of dilute acid (in a separate step) protonates the alkoxide to give the alcohol.

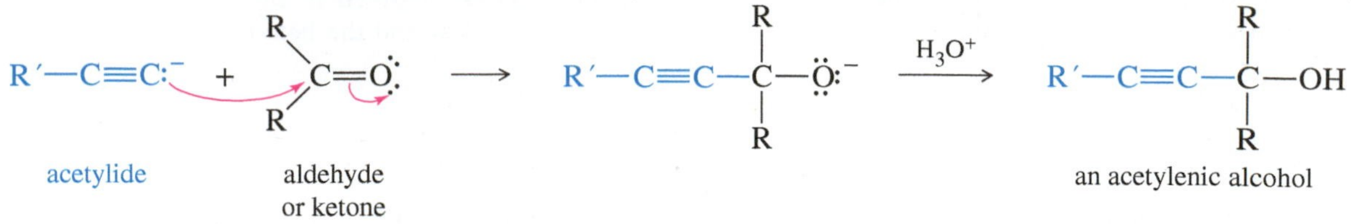

$$R'-C\equiv C:^- + R_2C=\ddot{O}: \longrightarrow R'-C\equiv C-CR_2-\ddot{O}:^- \xrightarrow{H_3O^+} R'-C\equiv C-CR_2-OH$$

acetylide; aldehyde or ketone; an acetylenic alcohol

An acetylide adds to formaldehyde ($H_2C{=}O$) to give (after the protonation step) a primary alcohol with one more carbon atom than there was in the acetylide.

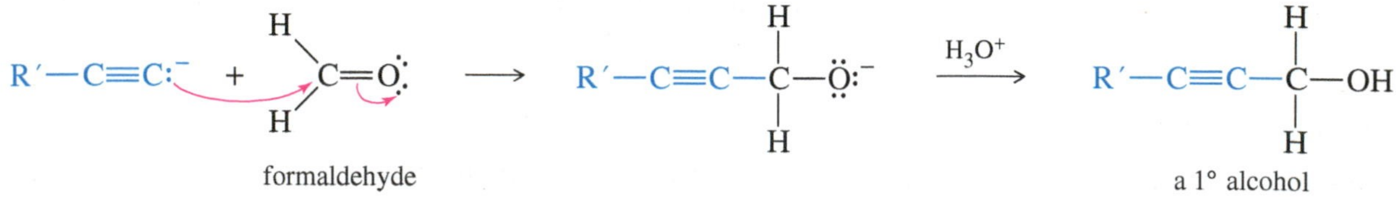

$$R'-C\equiv C:^- + H_2C=\ddot{O}: \longrightarrow R'-C\equiv C-CH_2-\ddot{O}:^- \xrightarrow{H_3O^+} R'-C\equiv C-CH_2-OH$$

formaldehyde; a 1° alcohol

Example

$$CH_3-C\equiv C-H \xrightarrow[(3)\ H_3O^+]{(1)\ NaNH_2\quad (2)\ H_2C=O} CH_3-C\equiv C-CH_2-OH$$

propyne; but-2-yn-1-ol (1°)

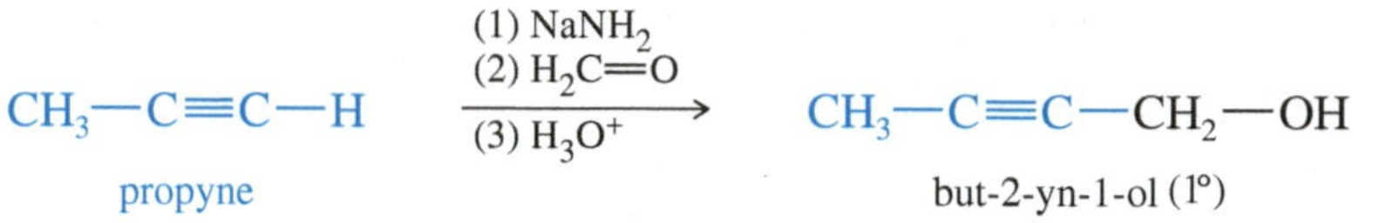

An acetylide adds to an aldehyde to give, after protonation, a secondary alcohol. The two groups of the secondary alcohol are the acetylide and the alkyl group that was bonded to the carbonyl group of the aldehyde.

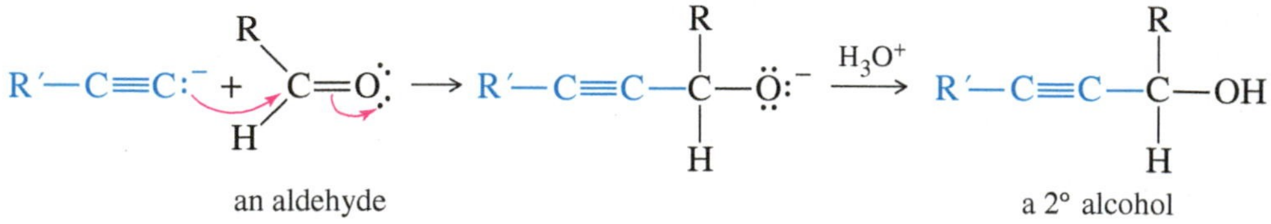

$$R'-C\equiv C:^- + RHC=\ddot{O}: \longrightarrow R'-C\equiv C-CHR-\ddot{O}:^- \xrightarrow{H_3O^+} R'-C\equiv C-CHR-OH$$

an aldehyde; a 2° alcohol

Example

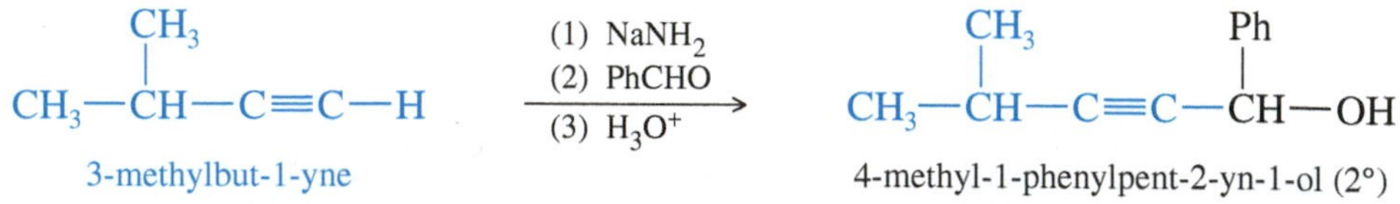

$$CH_3-CH(CH_3)-C\equiv C-H \xrightarrow[(3)\ H_3O^+]{(1)\ NaNH_2\quad (2)\ PhCHO} CH_3-CH(CH_3)-C\equiv C-CH(Ph)-OH$$

3-methylbut-1-yne; 4-methyl-1-phenylpent-2-yn-1-ol (2°)

A ketone has two alkyl groups bonded to its carbonyl carbon atom. Addition of an acetylide, followed by protonation, gives a tertiary alcohol. The three alkyl groups bonded to the carbinol carbon atom (the carbon bearing the —OH group) are the acetylide and the two alkyl groups originally bonded to the carbonyl group in the ketone.

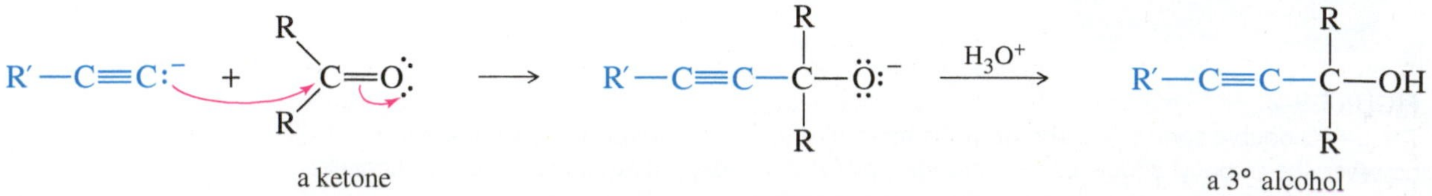

$$R'-C\equiv C:^- + R_2C=\ddot{O}: \longrightarrow R'-C\equiv C-CR_2-\ddot{O}:^- \xrightarrow{H_3O^+} R'-C\equiv C-CR_2-OH$$

a ketone; a 3° alcohol

Example

$$\text{cyclohexanone} \xrightarrow[(2)\ H_3O^+]{(1)\ Na^+\ ^-\!:C{\equiv}C{-}H} \text{1-ethynylcyclohexanol (3°)}$$

cyclohexanone

1-ethynylcyclohexanol (3°)

SOLVED PROBLEM 9-2

Show how you would synthesize the following compound, beginning with acetylene and any necessary additional reagents.

$$\text{cyclopentyl}{-}CH(OH){-}C{\equiv}C{-}CH_2CH_3$$

SOLUTION

We need to add two groups to acetylene: an ethyl group and a six-carbon aldehyde (to form the secondary alcohol). If we formed the alcohol group first, the weakly acidic —OH group would interfere with the alkylation by the ethyl group. Therefore, we should add the less reactive ethyl group first, and then add the alcohol group later in the synthesis.

$$H{-}C{\equiv}C{-}H \xrightarrow[(2)\ CH_3CH_2Br]{(1)\ NaNH_2} H{-}C{\equiv}C{-}CH_2CH_3$$

The ethyl group is not acidic, and it does not interfere with the addition of the second group:

$$H{-}C{\equiv}C{-}CH_2CH_3 \xrightarrow{NaNH_2} Na^+\ ^-\!:C{\equiv}C{-}CH_2CH_3 \xrightarrow[(2)\ H_3O^+]{(1)\ \text{cyclopentyl}{-}CHO} \text{cyclopentyl}{-}CH(OH){-}C{\equiv}C{-}CH_2CH_3$$

PROBLEM 9-8

Show how you would synthesize each compound, beginning with acetylene and any necessary additional reagents.

(a) prop-2-yn-1-ol (propargyl alcohol)

$H{-}C{\equiv}C{-}CH_2OH$

(b) hept-2-yn-4-ol

$CH_3{-}C{\equiv}C{-}CH(OH){-}CH_2CH_2CH_3$

(c) 2-phenylbut-3-yn-2-ol

$CH_3{-}C(OH)(Ph){-}C{\equiv}C{-}H$

(d) 3-methylhex-4-yn-3-ol

$CH_3CH_2{-}C(OH)(CH_3){-}C{\equiv}C{-}CH_3$

PROBLEM-SOLVING HINT

If a synthesis requires both alkylation of an acetylide and addition to a carbonyl, add the less reactive group first: alkylate, then add to a carbonyl. In general, you should add reactive functional groups late in a synthesis.

PROBLEM 9-9

Show how you would synthesize 2-phenylhex-3-yn-2-ol, starting with acetophenone ($PhCOCH_3$) and any other reagents you need. ("2-ol" means there is an OH group on C2.)

9-8 Synthesis of Alkynes by Elimination Reactions

In some cases, we can generate a carbon–carbon triple bond by eliminating two molecules of HX from a dihalide. Dehydrohalogenation of a *geminal* or *vicinal* dihalide gives a vinyl halide. Under strongly basic conditions, a second dehydrohalogenation may occur to form an alkyne.

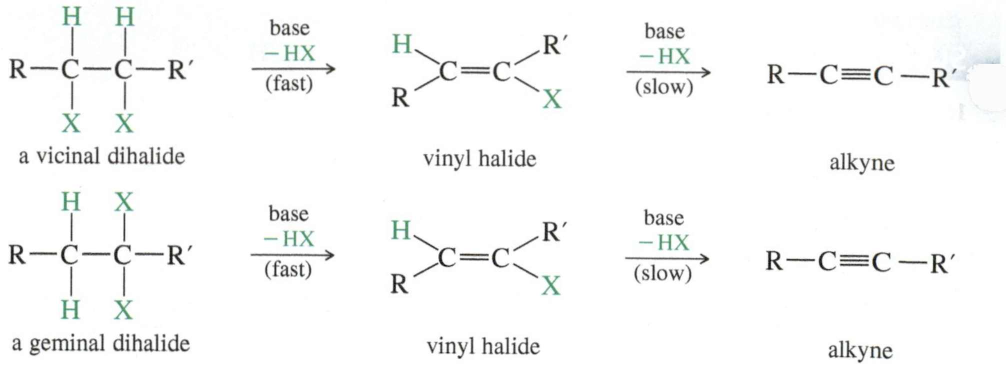

$$R{-}CH(X){-}CH(X){-}R' \xrightarrow[\text{(fast)}]{\text{base, }-HX} H(R)C{=}C(R')X \xrightarrow[\text{(slow)}]{\text{base, }-HX} R{-}C{\equiv}C{-}R'$$

a vicinal dihalide — vinyl halide — alkyne

$$R{-}CH_2{-}CX_2{-}R' \xrightarrow[\text{(fast)}]{\text{base, }-HX} H(R)C{=}C(R')X \xrightarrow[\text{(slow)}]{\text{base, }-HX} R{-}C{\equiv}C{-}R'$$

a geminal dihalide — vinyl halide — alkyne

We have already seen (Section 7-9) many examples of dehydrohalogenation of alkyl halides. The second step is new, however, because it involves dehydrohalogenation of a vinyl halide to give an alkyne. This second dehydrohalogenation occurs only under extremely basic conditions—for example, fused (molten) KOH or alcoholic KOH in a sealed tube, usually heated to temperatures close to 200 °C. Sodium amide is also used for the double dehydrohalogenation. Because the amide ion ($^{-}\!:\!\ddot{N}H_2$) is a much stronger base than hydroxide ion, the amide reaction takes place at a lower temperature.

Using either KOH or sodium amide at these elevated temperatures implies brutal reaction conditions, encouraging side reactions and rearrangements. Yields are often poor. The following reactions are carefully chosen to form products that are not prone to side reactions. The KOH elimination tends to give the most stable, most highly substituted internal alkyne.

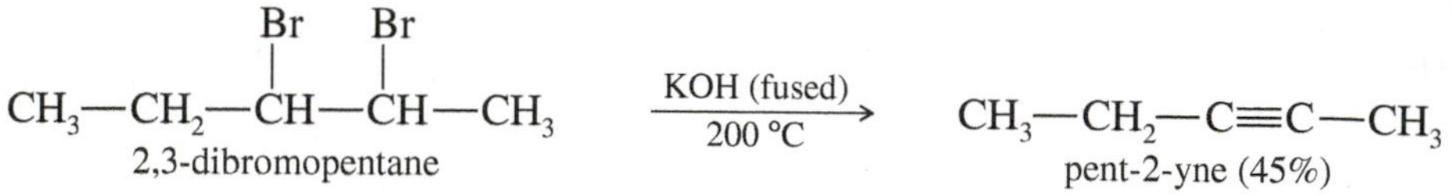

$$CH_3{-}CH_2{-}CH(Br){-}CH(Br){-}CH_3 \xrightarrow[\text{200 °C}]{\text{KOH (fused)}} CH_3{-}CH_2{-}C{\equiv}C{-}CH_3$$

2,3-dibromopentane — pent-2-yne (45%)

The sodium amide elimination tends to give a terminal alkyne (where possible) because the acetylenic hydrogen is deprotonated by the amide ion, giving an acetylide ion as the initial product. Protonation of the acetylide ion gives the terminal alkyne.

$$CH_3CH_2CH_2CH_2{-}CHCl_2 \xrightarrow[\text{150 °C}]{NaNH_2} CH_3CH_2CH_2C{\equiv}C{:}^{-}\,Na^{+} \xrightarrow{H_2O} CH_3CH_2CH_2C{\equiv}CH$$

1,1-dichloropentane — acetylide ion — pent-1-yne (55%)

PROBLEM-SOLVING HINT

$R(H)C{=}C(X)R'$

KOH, 200 °C → most stable internal alkyne

(1) $NaNH_2$, 150 °C (2) H_2O → terminal alkyne

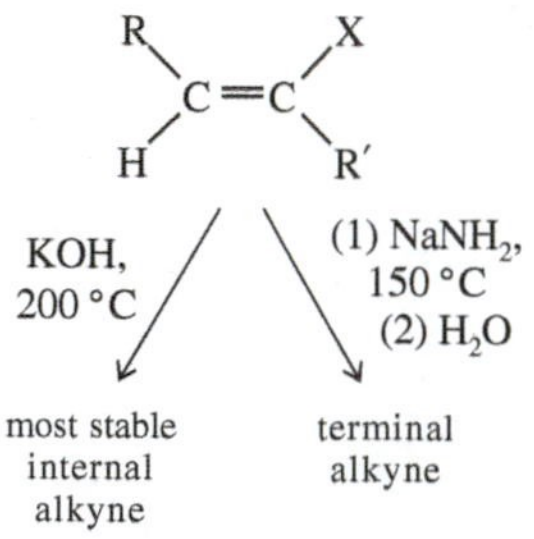

PROBLEM 9-10

When 2,2-dibromo-1-phenylpropane is heated overnight in fused KOH at 200 °C, the major product is a foul-smelling compound of formula C_9H_8. Propose a structure for this product, and give a mechanism to account for its formation.

PROBLEM 9-11

When 2,2-dibromo-1-phenylpropane is heated overnight with sodium amide at 150 °C, the major product (after addition of water) is a different foul-smelling compound of formula C_9H_8. Propose a structure for this product, and give a mechanism to account for its formation.

SUMMARY Syntheses of Alkynes

1. *Alkylation of acetylide ions* (Section 9-7A)

$$\text{R—C≡C:}^- + \text{R'—X} \xrightarrow{S_N2} \text{R—C≡C—R'} + X^-$$

R′—X must be an unhindered primary halide.

2. *Additions to carbonyl groups* (Section 9-7B)

$$\text{R—C≡C:}^- + \text{R'}_2\text{C=O} \longrightarrow \text{R—C≡C—CR'}_2\text{—O:}^- \xrightarrow[\text{(or } H_3O^+)]{H_2O} \text{R—C≡C—CR'}_2\text{—OH}$$

3. *Double dehydrohalogenation of alkyl dihalides* (Section 9-8)

$$\text{R—CX(H)—CX(H)—R'} \text{ or } \text{R—CH(H)—CX(X)—R'} \xrightarrow[\substack{or\ NaNH_2 \\ \text{(severe conditions)}}]{\text{fused KOH}} \text{R—C≡C—R'}$$

KOH forms internal alkynes; $NaNH_2$ forms terminal alkynes.

9-9 Addition Reactions of Alkynes

We have already discussed some of the most important reactions of alkynes. The nucleophilic attack of acetylide ions on electrophiles, for example, is one of the best methods for making more complicated alkynes (Section 9-7). Now we consider reactions that involve transformations of the carbon–carbon triple bond itself.

Many of the reactions of alkynes are similar to the corresponding reactions of alkenes because both involve pi bonds between two carbon atoms. Like the pi bond of an alkene, the pi bonds of an alkyne are electron-rich, and they readily undergo addition reactions. Table 9-4 shows how the energy differences between the kinds of carbon–carbon bonds can be used to estimate how much energy it takes to break a particular bond. The bond energy of the alkyne triple bond is only about 226 kJ (54 kcal) more than the bond energy of an alkene double bond. This is the energy needed to break one of the pi bonds of an alkyne.

TABLE 9-4
Approximate Bond Energies of Carbon–Carbon Bonds

Bond	Total Energy	Class of Bond	Approximate Energy
C—C	347 kJ (83 kcal)	alkane sigma bond	347 kJ (83 kcal)
C=C	611 kJ (146 kcal)	alkene pi bond	264 kJ (63 kcal)
C≡C	837 kJ (200 kcal)	second alkyne pi bond	226 kJ (54 kcal)

Reagents add across the triple bonds of alkynes just as they add across the double bonds of alkenes. In effect, this reaction converts a pi bond of the alkyne and a sigma bond of the reagent into two new sigma bonds. Because sigma bonds are generally stronger than pi bonds, the reaction is usually exothermic. Alkynes have two pi bonds, so up to two molecules can add across the triple bond, depending on the reagents and the conditions.

$$\underset{\pi\text{ bond}}{\text{R—C≡C—R'}} + \underset{\sigma\text{ bond}}{\text{A—B}} \longrightarrow \underset{2\ \sigma\text{ bonds}}{\text{A(R)C=C(B)R'}} \xrightarrow{\text{A—B}} \text{R—CA}_2\text{—CB}_2\text{—R'}$$

We must consider the possibility of a double addition whenever a reagent adds across the triple bond of an alkyne. Some conditions may allow the reaction to stop after a single addition, while other conditions give double addition.

9-9A Catalytic Hydrogenation to Alkanes

In the presence of a suitable catalyst, hydrogen adds to an alkyne, reducing it to an alkane. For example, when either of the butyne isomers reacts with hydrogen and a platinum catalyst, the product is *n*-butane. Platinum, palladium, and nickel catalysts are commonly used in this reduction.

$$\mathrm{R{-}C{\equiv}C{-}R'} + 2\,\mathrm{H_2} \xrightarrow{\text{Pt, Pd, or Ni}} \mathrm{R{-}CH_2{-}CH_2{-}R'}$$

Examples

$$\underset{\text{but-1-yne}}{\mathrm{H{-}C{\equiv}C{-}CH_2CH_3}} + 2\,\mathrm{H_2} \xrightarrow{\text{Pt}} \underset{\text{butane (100\%)}}{\mathrm{H{-}CH_2{-}CH_2{-}CH_2CH_3}}$$

$$\underset{\text{but-2-yne}}{\mathrm{CH_3{-}C{\equiv}C{-}CH_3}} + 2\,\mathrm{H_2} \xrightarrow{\text{Pt}} \underset{\text{butane (100\%)}}{\mathrm{CH_3{-}CH_2{-}CH_2{-}CH_3}}$$

Catalytic hydrogenation takes place in two stages, with an alkene intermediate. With efficient catalysts such as platinum, palladium, or nickel, it is usually impossible to stop the reaction at the alkene stage.

$$\mathrm{R{-}C{\equiv}C{-}R'} \xrightarrow{\mathrm{H_2,\ Pt}} \left[\mathrm{(R)(H)C{=}C(R')(H)}\right] \xrightarrow{\mathrm{H_2,\ Pt}} \mathrm{R{-}CH_2{-}CH_2{-}R'}$$

9-9B Partial Catalytic Hydrogenation to cis Alkenes

Hydrogenation of an alkyne can be stopped at the alkene stage by using a "poisoned" (partially deactivated) catalyst made by treating a good catalyst with a compound that makes the catalyst less effective. **Lindlar's catalyst*** is a poisoned palladium catalyst, composed of powdered barium sulfate coated with palladium, and poisoned with quinoline. Nickel boride (Ni_2B) is a newer alternative to Lindlar's catalyst that is more easily made and often gives better yields.

$$\underset{\text{alkyne}}{\mathrm{R{-}C{\equiv}C{-}R'}} \xrightarrow[\text{quinoline, } \mathrm{CH_3OH}\ \text{(Lindlar's catalyst)}]{\mathrm{H_2,\ Pd/BaSO_4}} \underset{\text{cis alkene}}{\mathrm{(R)(H)C{=}C(R')(H)}}$$

The partial catalytic hydrogenation of alkynes is similar to the hydrogenation of alkenes, and both proceed with syn stereochemistry. In catalytic hydrogenation, the face of a pi bond contacts the solid catalyst, and the catalyst weakens the pi bond, allowing two hydrogen atoms to add (Figure 9-3). This simultaneous (or nearly simultaneous) addition of two hydrogen atoms on the same face of the alkyne ensures syn stereochemistry.

*Lindlar's catalyst was originally Pd on $CaCO_3$, deactivated by $Pb(OAc)_2$. Cram and Allinger modified the procedure to use Pd on $BaSO_4$, deactivated by quinoline.

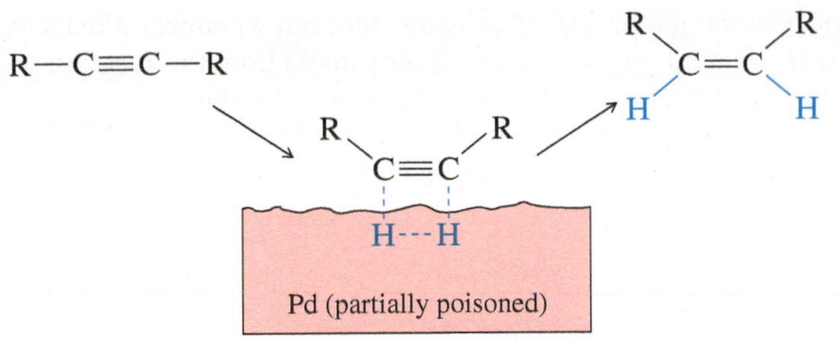

$$R{-}C{\equiv}C{-}R \longrightarrow \underset{\text{Pd (partially poisoned)}}{\text{R–C≡C–R on H---H}} \longrightarrow \text{RHC=CHR (cis)}$$

FIGURE 9-3
Catalytic hydrogenation of alkynes using Lindlar's catalyst.

In an internal alkyne, syn addition gives a cis product. For example, when hex-2-yne is hydrogenated using Lindlar's catalyst, the product is *cis*-hex-2-ene.

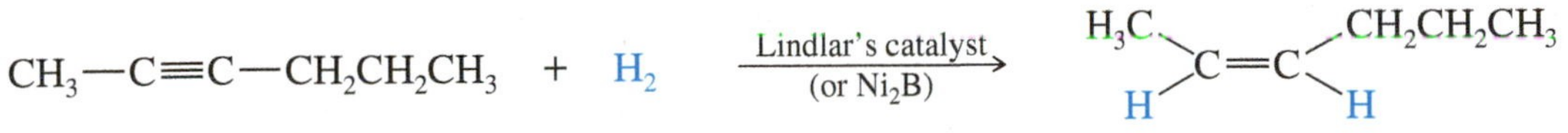

$$\underset{\text{hex-2-yne}}{CH_3{-}C{\equiv}C{-}CH_2CH_2CH_3} + H_2 \xrightarrow[\text{(or } Ni_2B)]{\text{Lindlar's catalyst}} \underset{\textit{cis}\text{-hex-2-ene}}{(H_3C)(H)C{=}C(CH_2CH_2CH_3)(H)}$$

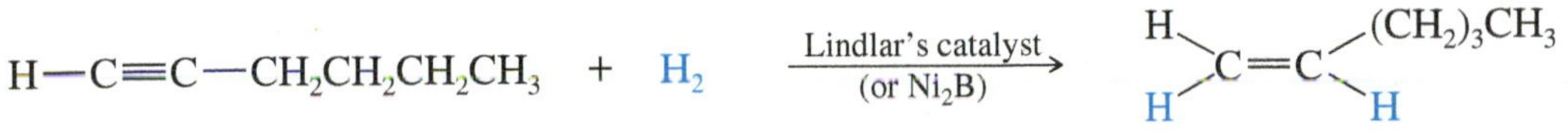

$$\underset{\text{hex-1-yne}}{H{-}C{\equiv}C{-}CH_2CH_2CH_2CH_3} + H_2 \xrightarrow[\text{(or } Ni_2B)]{\text{Lindlar's catalyst}} \underset{\text{hex-1-ene}}{(H)(H)C{=}C((CH_2)_3CH_3)(H)}$$

9-9C Metal–Ammonia Reduction to trans Alkenes

To form a trans alkene, two hydrogens must be added to the alkyne with anti stereochemistry. Sodium metal in liquid ammonia reduces alkynes with anti stereochemistry, so this reduction is used to convert alkynes to trans alkenes.

$$\underset{\text{alkyne}}{R{-}C{\equiv}C{-}R'} + Na/NH_3 \longrightarrow \underset{\text{trans alkene}}{(R)(H)C{=}C(H)(R')}$$

Example

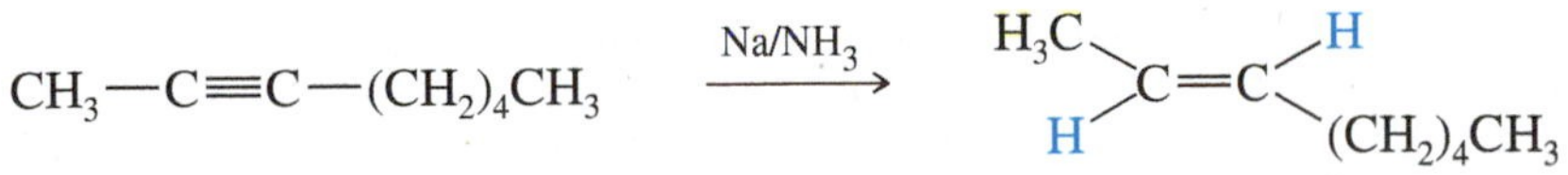

$$\underset{\text{oct-2-yne}}{CH_3{-}C{\equiv}C{-}(CH_2)_4CH_3} \xrightarrow{Na/NH_3} \underset{\textit{trans}\text{-oct-2-ene (80\%)}}{(H_3C)(H)C{=}C(H)((CH_2)_4CH_3)}$$

Ammonia (bp −33 °C) is a gas at room temperature, but it is kept liquid by using dry ice to cool the reaction vessel. As sodium dissolves in liquid ammonia, it gives up electrons, which produce a deep blue color. It is these solvated electrons that actually reduce the alkyne.

$$NH_3 + Na \longrightarrow \underset{\text{solvated electron}}{NH_3 \cdot e^-} \text{ (deep blue solution)} + Na^+$$

The metal–ammonia reduction proceeds by addition of an electron to the alkyne to form a radical anion, followed by protonation to give a neutral radical. Protons are provided by the ammonia solvent or by an alcohol added as a cosolvent. Addition of another electron, followed by another proton, gives the product.

MECHANISM 9-1 Metal–Ammonia Reduction of an Alkyne

This mechanism involves addition of an electron, followed by a proton, then addition of a second electron, followed by a second proton.

Step 1: An electron adds to the alkyne, forming a radical anion.

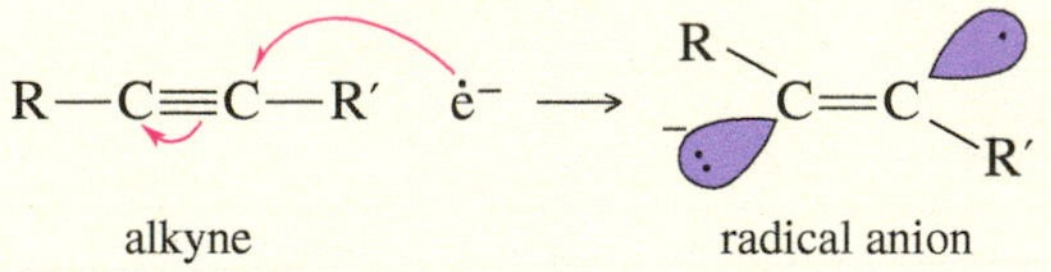

R—C≡C—R′ $\dot{e}^-$ ⟶ radical anion

alkyne radical anion

Step 2: The radical anion is protonated to give a radical.

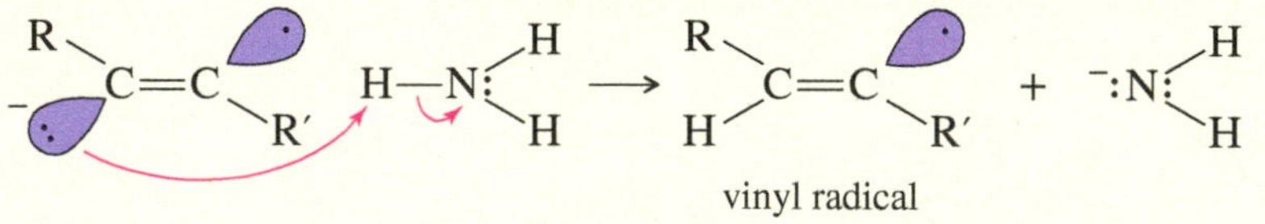

+ H—NH_2 ⟶ vinyl radical + $^-:NH_2$

vinyl radical

Step 3: An electron adds to the radical, forming an anion.

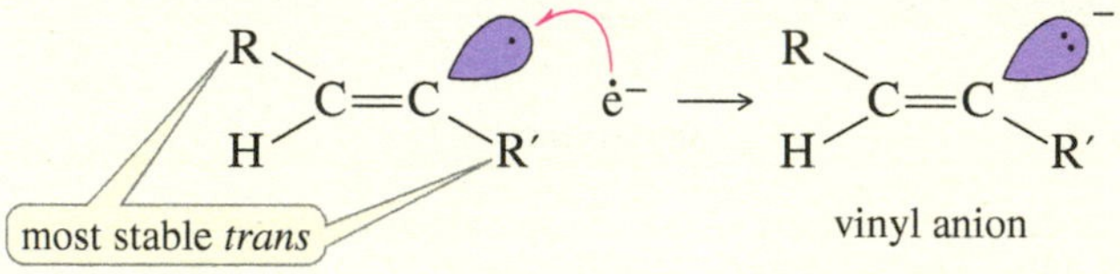

most stable *trans*

vinyl anion

Step 4: Protonation of the anion gives an alkene.

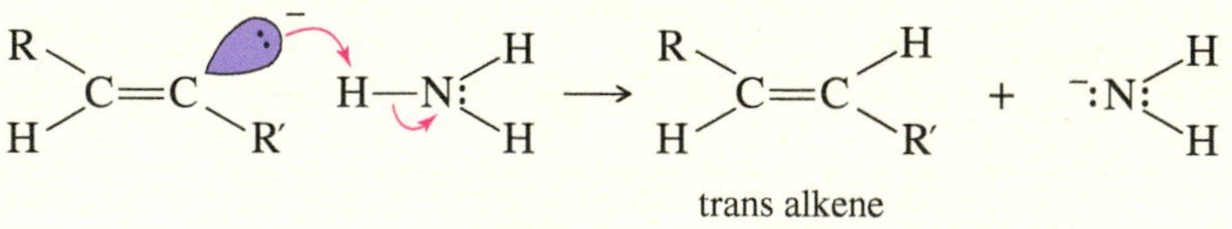

trans alkene

The anti stereochemistry of the sodium–ammonia reduction appears to result from the greater stability of the vinyl radical in the trans configuration, where the alkyl groups are farther apart. An electron is added to the trans radical to give a trans vinyl anion, which is quickly protonated to the trans alkene.

PROBLEM 9-12

Show how you would convert
(a) oct-3-yne to *cis*-oct-3-ene.
(b) pent-2-yne to *trans*-pent-2-ene.
(c) *cis*-cyclodecene to *trans*-cyclodecene.
(d) but-1-yne to *cis*-hex-3-ene.

PROBLEM 9-13

The fragrance of (*Z*)-1-phenylhex-2-en-1-ol resembles that of roses, with a delicate citrus edge. Show how you would synthesize this compound from benzaldehyde (PhCHO) and any other reagents you need.

9-9D Addition of Halogens

Bromine and chlorine add to alkynes just as they add to alkenes. If 1 mole of halogen adds to 1 mole of an alkyne, the product is a dihaloalkene. The stereochemistry of addition may be either syn or anti, and the products are often mixtures of cis and trans isomers.

$$R{-}C{\equiv}C{-}R' + X_2 \longrightarrow \underset{X}{\overset{R}{>}}C{=}C\underset{R'}{\overset{X}{<}} + \underset{X}{\overset{R}{>}}C{=}C\underset{X}{\overset{R'}{<}}$$
$(X_2 = Cl_2 \text{ or } Br_2)$

Example

$$CH_3(CH_2)_3{-}C{\equiv}C{-}H + Br_2 \longrightarrow \underset{Br}{\overset{CH_3(CH_2)_3}{>}}C{=}C\underset{H}{\overset{Br}{<}} + \underset{Br}{\overset{CH_3(CH_2)_3}{>}}C{=}C\underset{Br}{\overset{H}{<}}$$
(72%) (28%)

If 2 moles of halogen add to 1 mole of an alkyne, a tetrahalide results. Sometimes it is difficult to keep the reaction from proceeding all the way to the tetrahalide even when we want it to stop at the dihalide.

$$R{-}C{\equiv}C{-}R' + 2\,X_2 \longrightarrow R{-}\underset{X}{\overset{X}{C}}{-}\underset{X}{\overset{X}{C}}{-}R'$$
$(X_2 = Cl_2 \text{ or } Br_2)$

Example

$$CH_3(CH_2)_3{-}C{\equiv}C{-}H + 2\,Cl_2 \longrightarrow CH_3(CH_2)_3{-}\underset{Cl}{\overset{Cl}{C}}{-}\underset{Cl}{\overset{Cl}{C}}{-}H$$
(100%)

PROBLEM 9-14

In the addition of just 1 mole of bromine to 1 mole of hex-1-yne, should the hex-1-yne be added to a bromine solution or should the bromine be added to the hex-1-yne? Explain your answer.

9-9E Addition of Hydrogen Halides

Hydrogen halides add across the triple bond of an alkyne in much the same way they add across the alkene double bond. The initial product is a vinyl halide. When a hydrogen halide adds to a terminal alkyne, the product has the orientation predicted by Markovnikov's rule. A second molecule of HX can add, usually with the same orientation as the first.

$$R{-}C{\equiv}C{-}H + \boxed{H}{-}X \longrightarrow \underset{X}{\overset{R}{>}}C{=}C\underset{\boxed{H}}{\overset{H}{<}} \xrightarrow{\boxed{H}-X} R{-}\underset{X}{\overset{X}{C}}{-}\underset{\boxed{H}}{\overset{\boxed{H}}{C}}{-}H$$
$(HX = HCl, HBr, \text{ or } HI)$

For example, the reaction of pent-1-yne with HBr gives the Markovnikov product. In an internal alkyne such as pent-2-yne, however, the acetylenic carbon atoms are equally substituted, and a mixture of products results.

$$H{-}C{\equiv}C{-}CH_2CH_2CH_3 + HBr \longrightarrow \underset{H}{\overset{H}{>}}C{=}C\underset{Br}{\overset{CH_2CH_2CH_3}{<}}$$
pent-1-yne → 2-bromopent-1-ene (Markovnikov product)

$$CH_3{-}C{\equiv}C{-}CH_2CH_3 + HBr \longrightarrow CH_3{-}\overset{Br}{C}{=}\overset{H}{C}{-}CH_2CH_3 + CH_3{-}\overset{H}{C}{=}\overset{Br}{C}{-}CH_2CH_3$$
pent-2-yne → 2-bromopent-2-ene (*E* and *Z* isomers) + 3-bromopent-2-ene (*E* and *Z* isomers)

The mechanism is similar to the mechanism of hydrogen halide addition to alkenes. The short-lived **vinyl cation** formed in the first step is more stable with the positive charge on the more highly substituted carbon atom. Attack by halide ion completes the reaction.

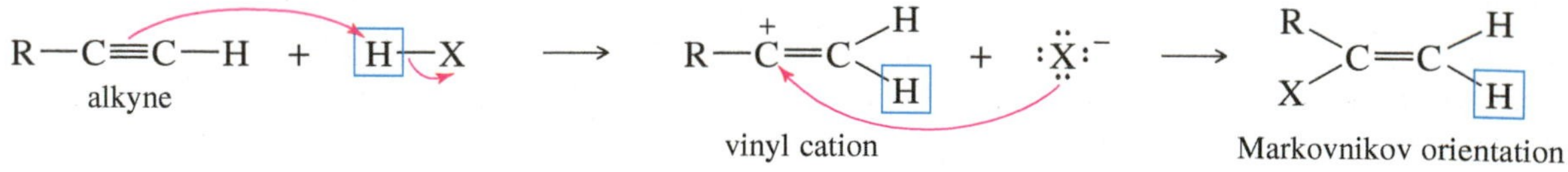

$$\text{R—C≡C—H} + \text{H—X} \longrightarrow \text{R—}\overset{+}{\text{C}}\text{=CH}_2 + :\ddot{\text{X}}:^- \longrightarrow \text{R(X)C=CH}_2$$

alkyne; vinyl cation; Markovnikov orientation

When 2 moles of a hydrogen halide add to an alkyne, the second mole usually adds with the same orientation as the first. This consistent orientation leads to a geminal dihalide. For example, a double Markovnikov addition of HBr to pent-1-yne gives 2,2-dibromopentane.

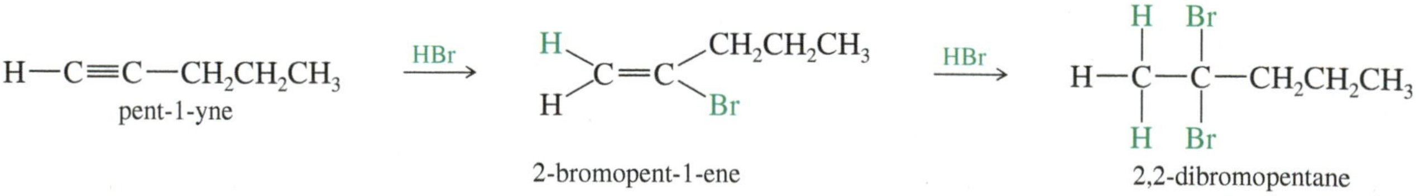

$$\text{H—C≡C—CH}_2\text{CH}_2\text{CH}_3 \xrightarrow{\text{HBr}} \text{H}_2\text{C=C(Br)CH}_2\text{CH}_2\text{CH}_3 \xrightarrow{\text{HBr}} \text{H—CH}_2\text{—CBr}_2\text{—CH}_2\text{CH}_2\text{CH}_3$$

pent-1-yne; 2-bromopent-1-ene; 2,2-dibromopentane

PROBLEM 9-15

Propose a mechanism for the entire reaction of pent-1-yne with 2 moles of HBr. Show why Markovnikov's rule should be observed in both the first and second additions of HBr.

PROBLEM 9-16

Predict the major product(s) of the following reactions:

(a) phenylacetylene + 2 HBr
(b) hex-1-yne + 2 HCl
(c) cyclooctyne + 2 HBr
*__(d)__ hex-2-yne + 2 HCl

In Section 8-3B, we saw the effect of peroxides on the addition of HBr to alkenes. Peroxides catalyze a free-radical chain reaction that adds HBr across the double bond of an alkene in the anti-Markovnikov sense. A similar reaction occurs with alkynes, with HBr adding with anti-Markovnikov orientation.

$$\text{H—C≡C—CH}_2\text{CH}_2\text{CH}_3 + \text{H—Br} \xrightarrow{\text{ROOR}} \text{H(Br)C=C(H)CH}_2\text{CH}_2\text{CH}_3$$

pent-1-yne; 1-bromopent-1-ene (mixture of *E* and *Z* isomers)

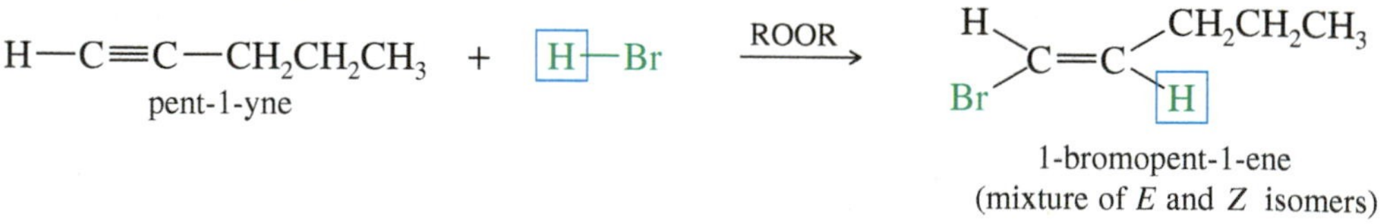

PROBLEM 9-17

Propose a mechanism for the reaction of pent-1-yne with HBr in the presence of peroxides. Show why anti-Markovnikov orientation results.

PROBLEM 9-18

Show how hex-1-yne might be converted to

(a) 1,2-dichlorohex-1-ene.
(b) 1-bromohex-1-ene.
(c) 2-bromohex-1-ene.
(d) 1,1,2,2-tetrabromohexane.
(e) 2-bromohexane.
(f) 2,2-dibromohexane.

9-9F Hydration of Alkynes to Ketones and Aldehydes

Mercuric Ion-Catalyzed Hydration Alkynes undergo acid-catalyzed addition of water across the triple bond in the presence of mercuric ion as a catalyst. A mixture of mercuric sulfate in aqueous sulfuric acid is commonly used as the reagent. The hydration of alkynes is similar to the hydration of alkenes, and it also goes with Markovnikov orientation. The products are not the alcohols we might expect, however. Hydration of an alkyne initially gives a vinyl alcohol that cannot be isolated, because it quickly rearranges to a ketone as described below.

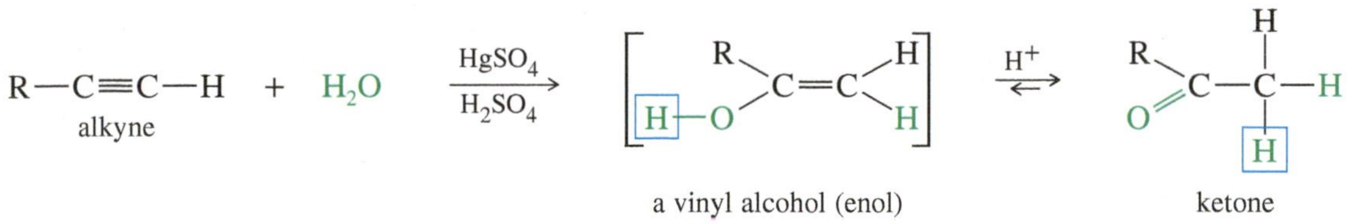

R—C≡C—H + H_2O $\xrightarrow[H_2SO_4]{HgSO_4}$ [H—O(R)C=CH₂] $\overset{H^+}{\rightleftarrows}$ R(O=)C—CH₂—H

alkyne; a vinyl alcohol (enol); ketone

Electrophilic addition of mercuric ion gives a vinyl cation, which reacts with water and loses a proton to give an organomercurial alcohol.

R—C≡C—H $\xrightarrow{Hg^{2+}}$ R—C⁺=CH(Hg⁺) $\xrightarrow{H_2\ddot{O}:}$ H—O⁺(H)—C(R)=CH(Hg⁺) + $H_2\ddot{O}:$ ⟶ HO(R)C=CH(Hg⁺) + H_3O^+

alkyne; vinyl cation; organomercurial alcohol

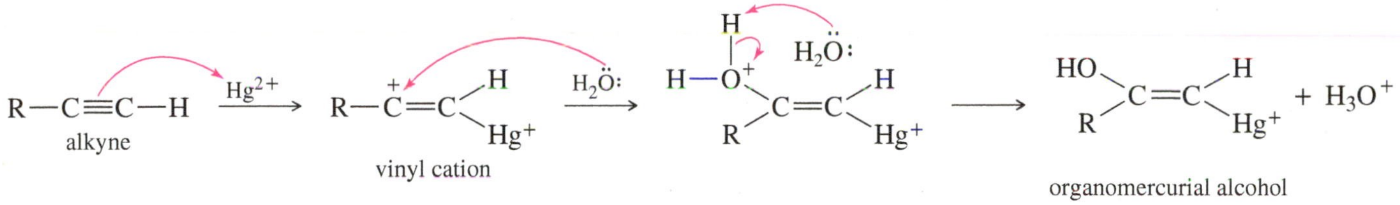

Under the acidic reaction conditions, mercury is replaced by hydrogen to give a vinyl alcohol, called an **enol.**

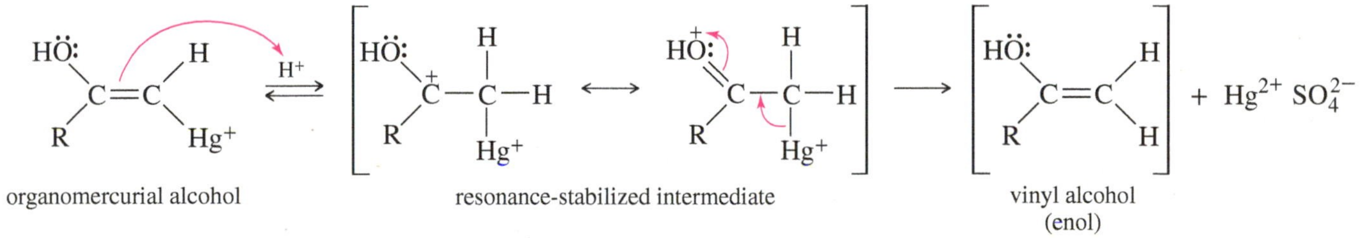

HÖ:(R)C=CH(Hg⁺) $\overset{H^+}{\rightleftarrows}$ [HÖ:(R)C⁺—CH(H)(Hg⁺) ⟷ HO⁺(R)=C—CH(H)(Hg⁺)] ⟶ [HÖ:(R)C=CH₂] + Hg^{2+} SO_4^{2-}

organomercurial alcohol; resonance-stabilized intermediate; vinyl alcohol (enol)

Enols tend to be unstable and isomerize to the ketone form. As shown next, this isomerization involves the shift of a proton and a double bond. The (boxed) hydroxy proton is lost, and a proton is regained at the methyl position, while the pi bond shifts from the C=C position to the C=O position. This type of rapid equilibrium is called a **tautomerism.** The one shown is the **keto–enol tautomerism,** which is covered in more detail in Chapter 22. In general, a C=O double bond is more stable than a C=C double bond, so the keto form usually predominates.

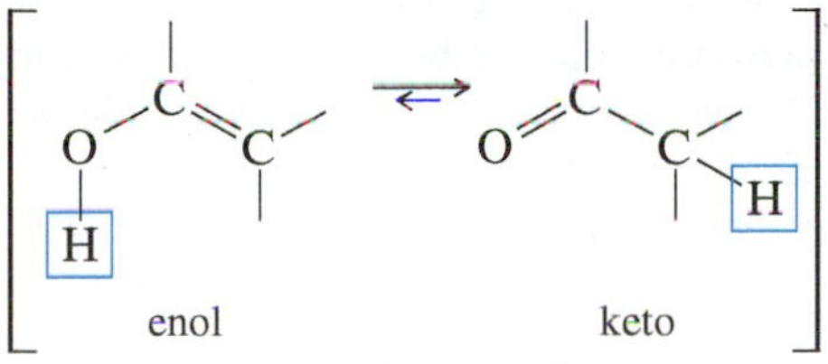

[H—O—C=C (enol) ⇄ O=C—C—H (keto)]

keto–enol tautomerism

Application: Biochemistry

A number of biological reactions involve the formation of an enol. Researchers are focusing on ways to use these reactions for therapeutic purposes. Several investigators have synthesized stable enols by placing bulky substituents around the double bond.

PROBLEM-SOLVING HINT

To move a proton (as in a tautomerism) under acidic conditions, try adding a proton in the new position, then removing it from the old position.

In acidic solution, the keto–enol tautomerism takes place by addition of a proton to the adjacent carbon atom, followed by loss of the hydroxy proton from oxygen.

MECHANISM 9-2 Acid-Catalyzed Keto–Enol Tautomerism

Under acidic conditions, the proton first adds at its new position on the adjacent carbon atom, and then is removed from its old position in the hydroxy group.

Step 1: Addition of a proton at the methylene group.

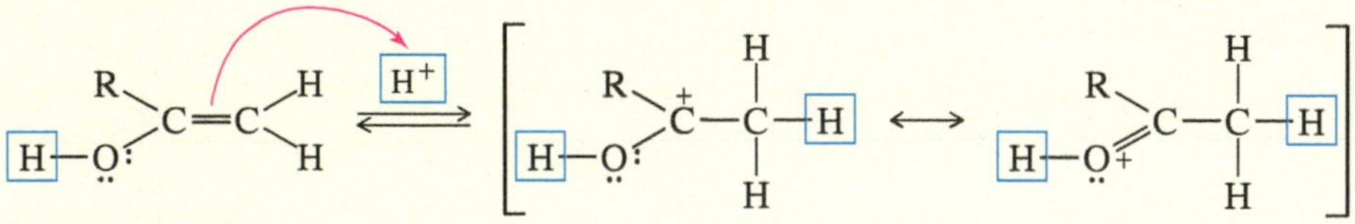

enol form ⇌ H^+ [resonance-stabilized intermediate]

Step 2: Loss of the hydroxy proton.

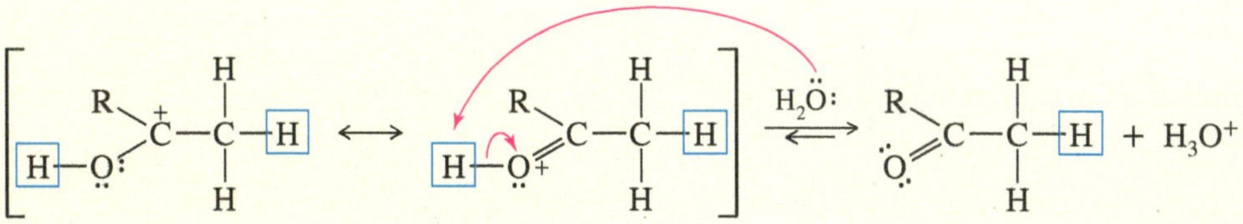

[resonance-stabilized intermediate] ⇌ $H_2\ddot{O}$: keto form + H_3O^+

For example, the mercuric-catalyzed hydration of but-1-yne gives but-1-en-2-ol as an intermediate. In the acidic solution, the intermediate quickly equilibrates to its more stable keto tautomer, butan-2-one.

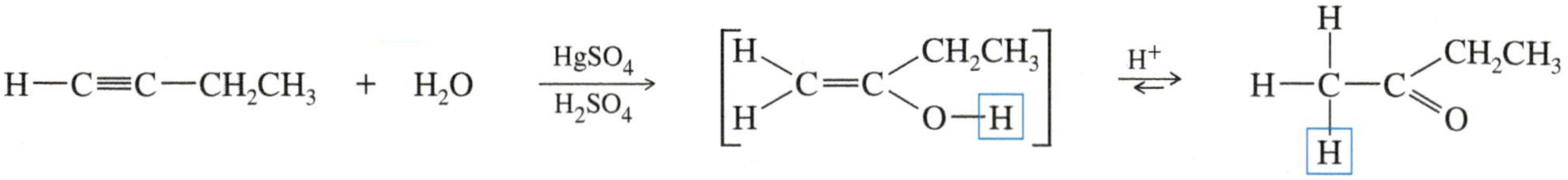

$$H-C\equiv C-CH_2CH_3 + H_2O \xrightarrow[H_2SO_4]{HgSO_4} [H_2C=C(OH)CH_2CH_3] \overset{H^+}{\rightleftharpoons} CH_3-C(=O)-CH_2CH_3$$

but-1-yne — but-1-en-2-ol — butan-2-one

PROBLEM 9-19

When pent-2-yne reacts with mercuric sulfate in dilute sulfuric acid, the product is a mixture of two ketones. Give the structures of these products, and use mechanisms to show how they are formed.

Hydroboration–Oxidation In Section 8-7, we saw that hydroboration–oxidation adds water across the double bonds of alkenes with anti-Markovnikov orientation. A similar reaction takes place with alkynes, except that a hindered dialkylborane must be used to prevent addition of two molecules of borane across the triple bond. Di(secondary isoamyl)borane, called "disiamylborane," adds to the triple bond only once to give a vinylborane. (**Amyl** is an older common name for pentyl.) In a terminal alkyne, the boron atom bonds to the terminal carbon atom.

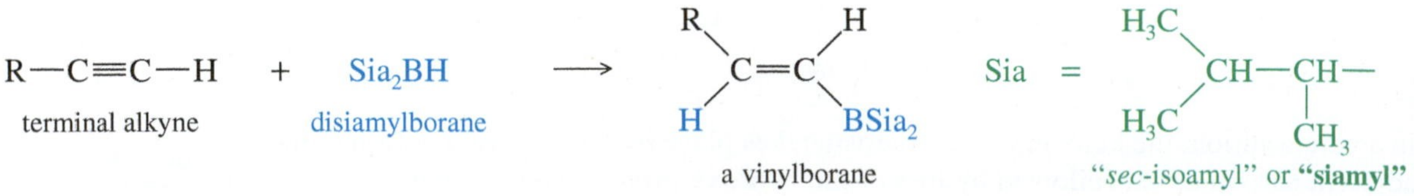

$$R-C\equiv C-H + Sia_2BH \longrightarrow R(H)C=C(H)BSia_2$$

terminal alkyne — disiamylborane — a vinylborane

$Sia = (H_3C)_2CH-CH(CH_3)-$

"*sec*-isoamyl" or "**siamyl**"

Oxidation of the vinylborane (using basic hydrogen peroxide) gives a vinyl alcohol (enol), resulting from anti-Markovnikov addition of water across the triple bond. This enol quickly tautomerizes to its more stable carbonyl (keto) form. In the case of a terminal alkyne, the keto product is an aldehyde. This sequence is an excellent method for converting terminal alkynes to aldehydes.

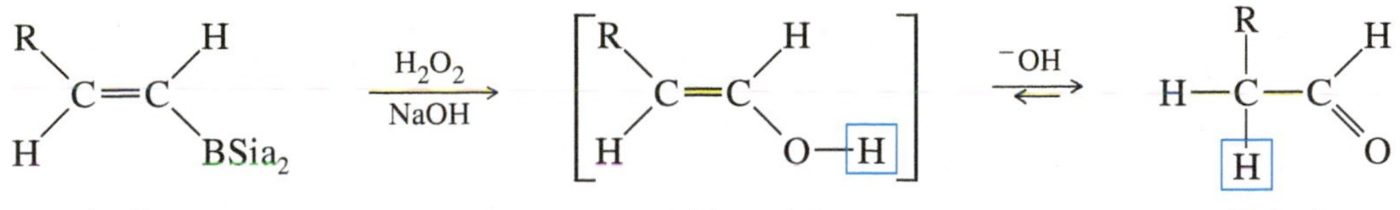

$$\underset{\text{vinylborane}}{(R)(H)C{=}C(H)(BSia_2)} \xrightarrow[NaOH]{H_2O_2} \underset{\text{unstable enol form}}{\left[(R)(H)C{=}C(H)(O{-}\boxed{H})\right]} \overset{^-OH}{\rightleftharpoons} \underset{\text{aldehyde}}{H{-}C(R)(\boxed{H}){-}C(H){=}O}$$

Under basic conditions, the keto–enol tautomerism operates by a different mechanism than it does in acid. In base, the proton is first removed from its old position in the OH group, and then replaced on carbon. In acid, the proton was first added on carbon, and then removed from the hydroxy group.

MECHANISM 9-3 Base-Catalyzed Keto–Enol Tautomerism

Under basic conditions, the proton is first removed from its old position in the enol, and then replaced in its new position on the adjacent carbon atom of the ketone or aldehyde.

Step 1: Loss of the hydroxy proton.

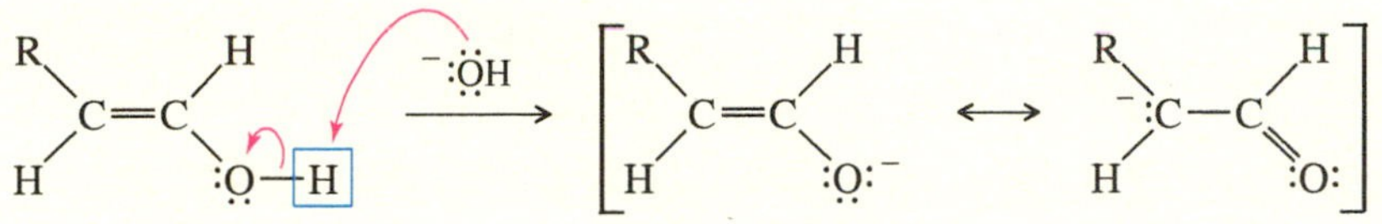

$$\underset{\text{enol form}}{(R)(H)C{=}C(H)(\ddot{O}{-}\boxed{H})} \xrightarrow{^-:\ddot{O}H} \underset{\text{stabilized "enolate" ion}}{\left[(R)(H)C{=}C(H)(:\ddot{O}:^-) \longleftrightarrow (R)(H)\,^-\!:C{-}C(H){=}\ddot{O}:\right]}$$

Step 2: Reprotonation on the adjacent carbon atom.

$$\underset{\text{stabilized "enolate" ion}}{\left[(R)(H)C{=}C(H)(:\ddot{O}:^-) \longleftrightarrow (R)(H)\,^-\!:C{-}C(H){=}\ddot{O}:\right]} \xrightarrow{H{-}\ddot{O}{-}H} \underset{\text{keto form}}{H{-}C(R)(\boxed{H}){-}C(H){=}\ddot{O}:} + {}^-OH$$

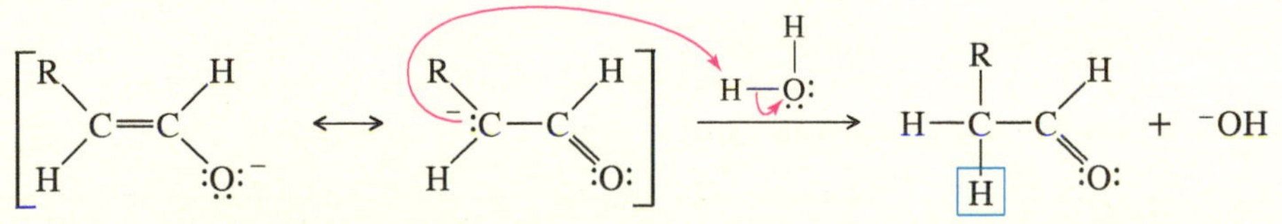

PROBLEM-SOLVING HINT

To move a proton (as in a tautomerism) under basic conditions, try removing the proton from its old position, then adding it to the new position.

Hydroboration of hex-1-yne, for example, gives the vinylborane with boron on the less highly substituted carbon. Oxidation of this intermediate gives an enol that quickly tautomerizes to hexanal.

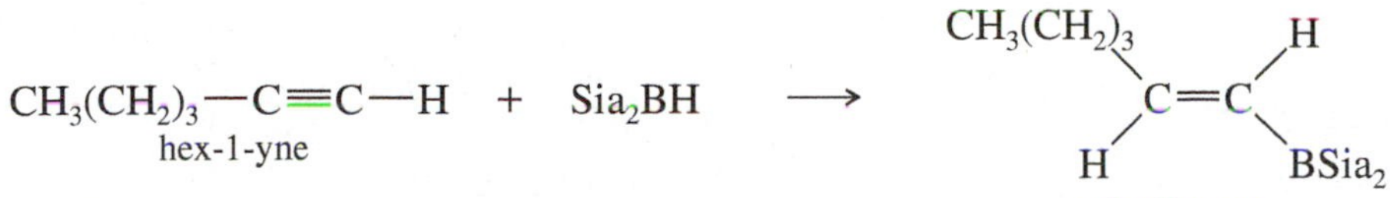

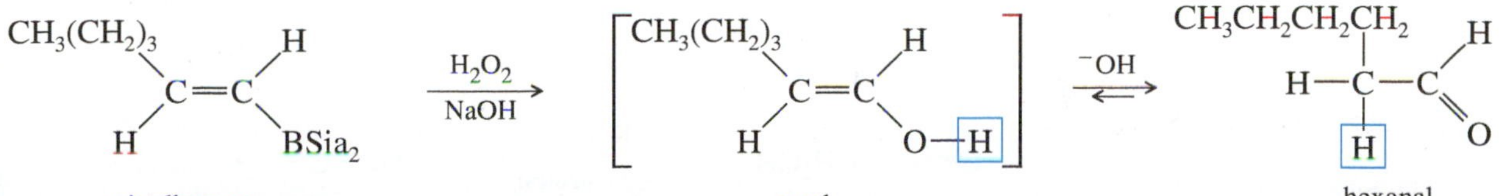

$$\underset{\text{hex-1-yne}}{CH_3(CH_2)_3{-}C{\equiv}C{-}H} + Sia_2BH \longrightarrow \underset{\text{a vinylborane}}{(CH_3(CH_2)_3)(H)C{=}C(H)(BSia_2)}$$

$$\underset{\text{vinylborane}}{(CH_3(CH_2)_3)(H)C{=}C(H)(BSia_2)} \xrightarrow[NaOH]{H_2O_2} \underset{\text{enol}}{\left[(CH_3(CH_2)_3)(H)C{=}C(H)(O{-}\boxed{H})\right]} \overset{^-OH}{\rightleftharpoons} \underset{\text{hexanal (65\%)}}{CH_3CH_2CH_2CH_2{-}C(H)(\boxed{H}){-}C(H){=}O}$$

PROBLEM 9-20

The hydroboration–oxidation of internal alkynes produces ketones.

(a) When hydroboration–oxidation is applied to but-2-yne, a single pure product is obtained. Determine the structure of this product, and show the intermediates in its formation.

(b) When hydroboration–oxidation is applied to pent-2-yne, two products are obtained. Show why a mixture of products should be expected with any unsymmetrical internal alkyne.

PROBLEM 9-21

For each compound, give the product(s) expected from (1) $HgSO_4/H_2SO_4$-catalyzed hydration and (2) hydroboration–oxidation.

(a) hex-1-yne
(b) hex-2-yne
(c) hex-3-yne
(d) cyclodecyne

PROBLEM 9-22

Disiamylborane adds only once to alkynes by virtue of its two bulky secondary isoamyl groups. Disiamylborane is prepared by the reaction of $BH_3 \cdot THF$ with an alkene.

(a) Draw the structural formulas of the reagents and the products in the preparation of disiamylborane.

(b) Explain why the reaction in part (a) goes only as far as the dialkylborane. Why is Sia_3B not formed?

9-10 Oxidation of Alkynes

Oxidation of alkynes occurs with the same reagents used to oxidize alkenes: permanganate and ozone. In the case of alkynes, however, two pi bonds are subject to oxidation, leading to different types of products.

9-10A Permanganate Oxidations

Under mild conditions, potassium permanganate oxidizes alkenes to glycols, compounds with two —OH groups on adjacent carbon atoms (Section 8-14B). Recall that this oxidation involves adding a hydroxy group to each end of the double bond (hydroxylation). A similar reaction occurs with alkynes. If an alkyne is treated with cold aqueous potassium permanganate under nearly neutral conditions, an α-diketone results. This is conceptually the same as hydroxylating each of the two pi bonds of the alkyne, then losing two molecules of water to give the diketone.

$$\mathrm{R{-}C{\equiv}C{-}R'} \xrightarrow[\mathrm{H_2O,\ neutral}]{\mathrm{KMnO_4}} \left[\mathrm{R{-}C(OH)_2{-}C(OH)_2{-}R'}\right] \xrightarrow{(-2\,\mathrm{H_2O})} \mathrm{R{-}\overset{O}{\overset{\|}{C}}{-}\overset{O}{\overset{\|}{C}}{-}R'}$$

diketone

For example, when pent-2-yne is treated with a cold, dilute solution of neutral permanganate, the product is pentane-2,3-dione.

$$\underset{\text{pent-2-yne}}{\mathrm{CH_3{-}C{\equiv}C{-}CH_2CH_3}} \xrightarrow[\mathrm{H_2O,\ neutral}]{\mathrm{KMnO_4}} \underset{\text{pentane-2,3-dione (90\%)}}{\mathrm{CH_3{-}\overset{O}{\overset{\|}{C}}{-}\overset{O}{\overset{\|}{C}}{-}CH_2CH_3}}$$

Terminal alkynes probably give a keto-aldehyde at first, but the aldehyde quickly oxidizes to an acid under these conditions.

$$\underset{\text{terminal alkyne}}{R-C\equiv C-H} \xrightarrow[H_2O,\ \text{neutral}]{KMnO_4} \underset{\text{keto-aldehyde}}{\left[R-\overset{O}{\overset{\|}{C}}-\overset{O}{\overset{\|}{C}}-H\right]} \xrightarrow{KMnO_4} \underset{\text{keto-acid}}{R-\overset{O}{\overset{\|}{C}}-\overset{O}{\overset{\|}{C}}-OH}$$

If the reaction mixture becomes warm or too basic, the diketone undergoes oxidative cleavage. The products are the carboxylate salts of carboxylic acids, which can be converted to the free acids by adding dilute acid.

$$R-C\equiv C-R' \xrightarrow[H_2O,\ \text{heat}]{KMnO_4,\ KOH} \underset{\text{carboxylate salts}}{R-\overset{O}{\overset{\|}{C}}-O^- + {}^-O-\overset{O}{\overset{\|}{C}}-R'} \xrightarrow[H_2O]{HCl} R-\overset{O}{\overset{\|}{C}}-OH + HO-\overset{O}{\overset{\|}{C}}-R'$$

For example, warm basic permanganate cleaves the triple bond of pent-2-yne to give acetate and propionate ions. Acidification reprotonates these anions to acetic acid and propionic acid.

$$\underset{\text{pent-2-yne}}{CH_3-C\equiv C-CH_2CH_3} \xrightarrow[H_2O,\ \text{heat}]{KMnO_4,\ KOH} \underset{\text{acetate}}{CH_3-\overset{O}{\overset{\|}{C}}-O^-} + \underset{\text{propionate}}{{}^-O-\overset{O}{\overset{\|}{C}}-CH_2CH_3} \xrightarrow{H^+} \underset{\text{acetic acid}}{CH_3-\overset{O}{\overset{\|}{C}}-OH} + \underset{\text{propionic acid}}{HO-\overset{O}{\overset{\|}{C}}-CH_2CH_3}$$

Terminal alkynes are cleaved similarly to give a carboxylate ion and formate ion. Under these oxidizing conditions, formate oxidizes further to carbonate, which becomes CO_2 after protonation.

$$\underset{\text{hex-1-yne}}{CH_3(CH_2)_3-C\equiv C-H} \xrightarrow[H_2O]{KMnO_4,\ KOH} \underset{\text{pentanoate}}{CH_3(CH_2)_3-\overset{O}{\overset{\|}{C}}-O^-} + \underset{\text{formate}}{\left[{}^-O-\overset{O}{\overset{\|}{C}}-H\right]}$$

$$\underset{\text{formate}}{\left[{}^-O-\overset{O}{\overset{\|}{C}}-H\right]} \xrightarrow[H_2O]{KMnO_4,\ KOH} \underset{\text{carbonate}}{{}^-O-\overset{O}{\overset{\|}{C}}-O^-} \xrightarrow{H^+} \underset{\text{carbonic acid}}{\left[HO-\overset{O}{\overset{\|}{C}}-OH\right]} \rightleftarrows CO_2 + H_2O$$

The overall reaction is as follows:

$$\underset{\text{hex-1-yne}}{CH_3(CH_2)_3-C\equiv C-H} \xrightarrow[(2)\ H^+,\ H_2O]{(1)\ KMnO_4,\ KOH,\ H_2O} \underset{\text{pentanoic acid}}{CH_3(CH_2)_3-\overset{O}{\overset{\|}{C}}-OH} + CO_2\uparrow$$

PROBLEM 9-23

Predict the product(s) you would expect from treatment of each compound with (1) dilute, neutral $KMnO_4$ and (2) warm basic $KMnO_4$, then dilute acid.

(a) hex-1-yne **(b)** hex-2-yne **(c)** hex-3-yne
(d) 2-methylhex-3-yne **(e)** cyclodecyne

PROBLEM-SOLVING HINT

Oxidative cleavage of an internal alkyne gives two carboxylic acids. A terminal alkyne gives one carboxylic acid and CO_2.

9-10B Ozonolysis

Ozonolysis of an alkyne, followed by hydrolysis, cleaves the triple bond and gives two carboxylic acids. Either permanganate cleavage or ozonolysis can be used to determine the position of the triple bond in an unknown alkyne (see Problem 9-24).

$$R{-}C{\equiv}C{-}R' \xrightarrow[(2)\ H_2O]{(1)\ O_3\ (-78\ ^\circ C)} R{-}COOH + R'{-}COOH$$

Examples

$$\underset{\text{pent-2-yne}}{CH_3{-}C{\equiv}C{-}CH_2CH_3} \xrightarrow[(2)\ H_2O]{(1)\ O_3\ (-78\ ^\circ C)} \underset{\text{acetic acid}}{CH_3{-}COOH} + \underset{\text{propionic acid}}{CH_2CH_3{-}COOH}$$

$$\underset{\text{hex-1-yne}}{CH_3(CH_2)_3{-}C{\equiv}C{-}H} \xrightarrow[(2)\ H_2O]{(1)\ O_3\ (-78\ ^\circ C)} \underset{\text{pentanoic acid}}{CH_3(CH_2)_3{-}\overset{\overset{O}{\|}}{C}{-}OH} + \underset{\text{formic acid}}{HO{-}\overset{\overset{O}{\|}}{C}{-}H}$$

PROBLEM 9-24

Oxidative cleavages can help to determine the positions of the triple bonds in alkynes.

(a) An unknown alkyne undergoes oxidative cleavage to give adipic acid and two equivalents of acetic acid. Propose a structure for the alkyne.

$$\text{unknown alkyne} \xrightarrow[(2)\ H_2O]{(1)\ O_3\ (-78\ ^\circ C)} \underset{\text{adipic acid}}{HOOC{-}(CH_2)_4{-}COOH} + 2\ CH_3COOH$$

(b) An unknown alkyne undergoes oxidative cleavage to give the following triacid plus one equivalent of propionic acid. Propose a structure for the alkyne.

$$\text{unknown alkyne} \xrightarrow[(2)\ H_2O]{(1)\ O_3\ (-78\ ^\circ C)} \underset{\text{a triacid}}{HOOC{-}(CH_2)_7{-}\overset{\overset{COOH}{|}}{CH}{-}COOH} + \underset{\text{propionic acid}}{CH_3CH_2COOH}$$

PROBLEM-SOLVING STRATEGY Multistep Synthesis

Multistep synthesis problems are useful for exercising your knowledge of organic reactions. Chapter 8 illustrated a systematic approach to synthesis that we now apply to a fairly difficult problem emphasizing alkyne chemistry. The compound to be synthesized is *cis*-2-methylhex-4-en-3-ol. (The "3-ol" means there is an alcohol —OH group on C3.)

$$\underset{\text{cis-2-methylhex-4-en-3-ol}}{\overset{6}{H_3C}{-}\overset{5}{CH}{=}\overset{4}{CH}{-}\overset{3}{CH}(OH){-}\overset{2}{CH}(CH_3){-}\overset{1}{CH_3}\ \ (\text{H on C5 and C4 cis})}$$

The starting materials are acetylene and compounds containing no more than four carbon atoms. In this problem, you should consider not only how to assemble the carbon skeleton and how to introduce the functional groups, but also when it is best to put in the functional groups. Begin with an examination of the target compound, and then examine possible intermediates and synthetic routes.

1. **Review the functional groups and carbon skeleton of the target compound.**
   The target compound contains seven carbon atoms and two functional groups: a cis carbon–carbon double bond and an alcohol. The best method for generating a cis double bond is the catalytic hydrogenation of a triple bond (Section 9-9B).

$$H_3C{-}C{\equiv}C{-}\overset{\overset{OH}{|}}{CH}{-}CH(CH_3)_2 \xrightarrow[\text{Lindlar's catalyst}]{H_2} H_3C{-}CH{=}CH{-}\underset{\underset{OH}{|}}{CH}{-}\underset{\underset{CH_3}{|}}{CH}{-}CH_3\ \ (\text{cis})$$

   Using this hydrogenation as the final step simplifies the problem to a synthesis of this acetylenic alcohol. You know how to form carbon–carbon bonds next to triple bonds, and you have seen the formation of acetylenic alcohols (Section 9-7B).

2. Review the functional groups and carbon skeletons of the starting materials, and see how their skeletons might fit together in the target compound.
Acetylene is listed as one of the starting materials, and you have good methods (Section 9-7) for making carbon–carbon bonds next to triple bonds by using acetylide ions as nucleophiles. You can break the target structure into three pieces, each containing no more than four carbon atoms.

$$\underset{\text{1 carbon}}{H_3C-} \qquad \underset{\text{acetylene}}{-C\equiv C-} \qquad \underset{\text{4 carbons (functionalized)}}{-\overset{OH}{\overset{|}{C}H}-\overset{CH_3}{\overset{|}{C}H}-CH_3}$$

3. Compare methods for assembling the carbon skeleton of the target compound. Which ones provide a key intermediate with the correct carbon skeleton and functional groups correctly positioned for conversion to the functionality in the target molecule?
Acetylenic alcohols result when acetylides add to ketones and aldehydes (Section 9-7B). Reaction of the acetylide ion with 2-methylpropanal gives one of the groups needed on the triple bond.

$$H-C\equiv C:^- \;+\; \underset{\text{2-methylpropanal}}{H-\overset{O}{\overset{\|}{C}}-CH(CH_3)_2} \longrightarrow \xrightarrow{H_3O^+} H-C\equiv C-\underset{H}{\overset{OH}{C}}-CH(CH_3)_2$$

A methyl group is needed on the other end of the double bond of the target compound. Methylation requires formation of an acetylide, however (Section 9-7A):

$$CH_3I \;+\; ^-:C\equiv C-R \longrightarrow H_3C-C\equiv C-R \;+\; I^-$$

Because the hydroxy group in the acetylenic alcohol is much more acidic than the acetylenic proton, any attempt to form the acetylide would fail.

$$H-C\equiv C-\overset{OH}{\overset{|}{C}H}-CH(CH_3)_2 \;+\; NaNH_2 \longrightarrow H-C\equiv C-\overset{O^-}{\overset{|}{C}H}-CH(CH_3)_2 \;+\; NH_3$$

This problem can be overcome by adding the methyl group first and then the alcohol portion. *In general, try to add less reactive groups earlier in a synthesis, and more reactive groups later.* In this case, you make the alcohol group after adding the alkyl group because the alkyl group is less likely to be affected by subsequent reactions.

$$H-C\equiv C-H \xrightarrow[\text{(2) } CH_3I]{\text{(1) } NaNH_2} H_3C-C\equiv C-H \xrightarrow{NaNH_2} H_3C-C\equiv C:^-\ Na^+$$

$$H_3C-C\equiv C:^- \;+\; H-\overset{O}{\overset{\|}{C}}-CH(CH_3)_2 \longrightarrow \xrightarrow{H_3O^+} H_3C-C\equiv C-\overset{OH}{\overset{|}{C}H}-CH(CH_3)_2$$

4. Working backward through as many steps as necessary, compare methods for synthesizing the reactants needed for assembly of the key intermediate with the correct carbon skeleton and functionality.
These compounds are all allowed as starting materials. Later, when we have covered more synthetic reactions, you will encounter problems that require you to evaluate how to make the compounds needed to assemble the key intermediates.

5. Summarize the complete synthesis in the forward direction, including all steps and all reagents, and check it for errors and omissions.
This final step is left to you as an exercise. Try to do it without looking at this solution, reviewing each thought process as you summarize the synthesis.

Now practice using a systematic approach with the syntheses in Problem 9-25.

PROBLEM 9-25

Develop syntheses for the following compounds, using acetylene and compounds containing no more than four carbon atoms as your organic starting materials.

(a) 3-methylnon-4-yn-3-ol ("3-ol" means there is an OH group on C3.)
(b) *cis*-1-ethyl-2-methylcyclopropane
(c) epoxide: CH_3CH_2 and H on one carbon, H and $CH_2CH_2CH_3$ on the other (ring O)
(d) meso-hexane-3,4-diol

PROBLEM-SOLVING HINT

Add less reactive groups earlier in a synthesis and more reactive groups later.

SUMMARY Reactions of Alkynes

I. *ACETYLIDE CHEMISTRY*

1. *Formation of acetylide anions (alkynides)* (Section 9-6)

$$R{-}C{\equiv}C{-}H + NaNH_2 \longrightarrow R{-}C{\equiv}C{:}^- \; Na^+ + NH_3$$
$$R{-}C{\equiv}C{-}H + R'Li \longrightarrow R{-}C{\equiv}CLi + R'H$$
$$R{-}C{\equiv}C{-}H + R'MgX \longrightarrow R{-}C{\equiv}CMgX + R'{-}H$$

2. *Alkylation of acetylide ions* (Section 9-7A)

$$R{-}C{\equiv}C{:}^- + R'{-}X \longrightarrow R{-}C{\equiv}C{-}R'$$

($R'{-}X$ must be an unhindered primary halide or tosylate.)

3. *Reactions with carbonyl groups* (Section 9-7B)

$$R{-}C{\equiv}C{:}^- + R'_2C{=}\ddot{O}{:} \longrightarrow R{-}C{\equiv}C{-}C(R')_2{-}\ddot{O}{:}^- \xrightarrow[\text{(or } H_3O^+)]{H_2O} R{-}C{\equiv}C{-}C(R')_2{-}OH$$

II. *ADDITIONS TO THE TRIPLE BOND*

1. *Reduction to alkanes* (Section 9-9A)

$$R{-}C{\equiv}C{-}R' + 2\,H_2 \xrightarrow{\text{Pt, Pd, or Ni}} R{-}CH_2{-}CH_2{-}R'$$

2. *Reduction to alkenes* (Sections 9-9B and 9-9C)

$$R{-}C{\equiv}C{-}R' + H_2 \xrightarrow[\text{(Lindlar's catalyst)}]{Pd/BaSO_4,\ \text{quinoline}} \underset{cis}{(R)(H)C{=}C(R')(H)}$$

$$R{-}C{\equiv}C{-}R' \xrightarrow{Na,\ NH_3} \underset{trans}{(R)(H)C{=}C(H)(R')}$$

3. *Addition of halogens* ($X_2 = Cl_2, Br_2$) (Section 9-9D)

$$R{-}C{\equiv}C{-}R' \xrightarrow{X_2} R{-}CX{=}CX{-}R' \xrightarrow{X_2} R{-}CX_2{-}CX_2{-}R'$$

5. *Addition of water* (Section 9-9F)

a. *Catalyzed by $HgSO_4/H_2SO_4$*

$$R{-}C{\equiv}C{-}H + H_2O \xrightarrow{HgSO_4,\ H_2SO_4} \left[(R)(HO)C{=}CH_2\right] \rightleftarrows R{-}C({=}O){-}CH_3$$

(Markovnikov orientation) — vinyl alcohol (unstable) — ketone (stable)

b. *Hydroboration–oxidation*

$$R{-}C{\equiv}C{-}H \xrightarrow[(2)\ H_2O_2,\ NaOH]{(1)\ Sia_2BH \cdot THF} \left[(R)(H)C{=}C(H)(OH)\right] \rightleftarrows R{-}CH_2{-}C({=}O){-}H$$

(anti-Markovnikov orientation) — vinyl alcohol (unstable) — aldehyde (stable)

III. *OXIDATION OF ALKYNES (SECTION 9-10)*

1. *Oxidation to α-diketones* (Section 9-10A)

$$\text{R—C}\equiv\text{C—R}' \xrightarrow[\text{H}_2\text{O, neutral}]{\text{KMnO}_4} \text{R—C(=O)—C(=O)—R}'$$

2. *Oxidative cleavage* (Section 9-10B)

$$\text{R—C}\equiv\text{C—R}' \xrightarrow[\text{(or O}_3\text{, then H}_2\text{O)}]{\text{(1) KMnO}_4,\ ^-\text{OH} \quad \text{(2) H}^+} \text{R—C(=O)—OH} + \text{HO—C(=O)—R}'$$

SUMMARY Reactions of Terminal Alkynes

A terminal alkyne displays orientation and stereochemistry of addition reactions. New atoms are shown in color.

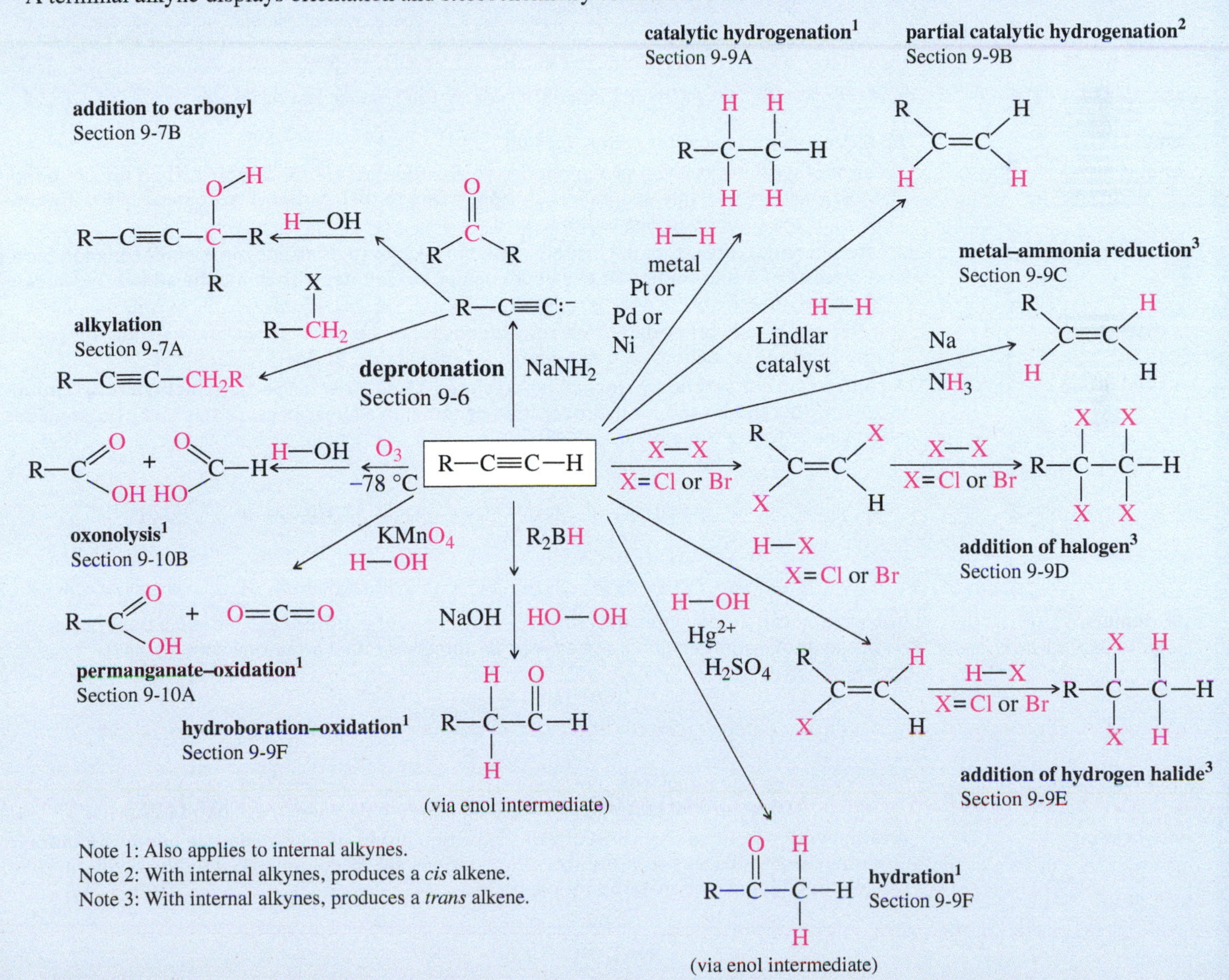

Note 1: Also applies to internal alkynes.
Note 2: With internal alkynes, produces a *cis* alkene.
Note 3: With internal alkynes, produces a *trans* alkene.

Essential Terms

acetylene The simplest alkyne, H—C≡C—H. Also used as a synonym for *alkyne,* a generic term for a compound containing a C≡C triple bond. (p. 428)

acetylide ion **(alkynide ion)** The anionic salt of a terminal alkyne. Metal acetylides are organometallic compounds with a metal atom in place of the weakly acidic acetylenic hydrogen of a terminal alkyne. (p. 434)

$$\text{R—C}\equiv\text{C—H} \;+\; \text{Na}^+ \;^-\text{:}\ddot{\text{N}}\text{H}_2 \longrightarrow \text{R—C}\equiv\text{C:}^-\,\text{Na}^+ \;+\; \text{:NH}_3$$

a sodium acetylide

alkoxide ion R—O⁻, the conjugate base of an alcohol. (p. 437)

$$\underset{\text{alkoxide}}{\text{R—}\ddot{\underset{\cdot\cdot}{\text{O}}}\text{:}^-} \;+\; \text{H}_2\text{O} \rightleftharpoons \underset{\text{alcohol}}{\text{R—}\ddot{\underset{\cdot\cdot}{\text{O}}}\text{—H}} \;+\; ^-\text{OH}$$

alkyne Any compound containing a carbon–carbon triple bond. (p. 428)

A **terminal alkyne** has a triple bond at the end of a chain, with an **acetylenic hydrogen.**

An **internal alkyne** has the triple bond somewhere other than at the end of the chain.

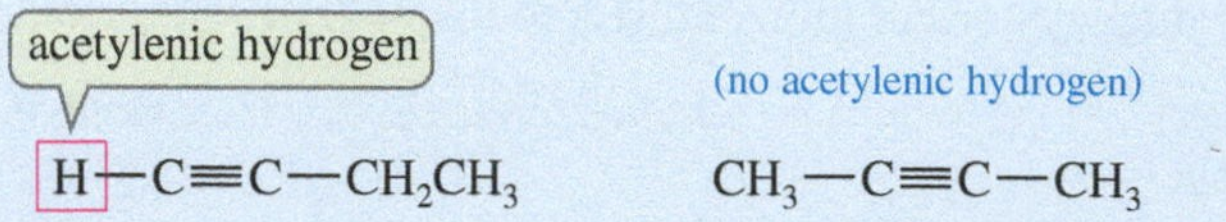

acetylenic hydrogen

$\text{H—C}\equiv\text{C—CH}_2\text{CH}_3$

but-1-yne, a terminal alkyne

(no acetylenic hydrogen)

$\text{CH}_3\text{—C}\equiv\text{C—CH}_3$

but-2-yne, an internal alkyne

amyl An older common name for pentyl. (p. 448)

enol An alcohol with the hydroxy group bonded to a carbon atom of a carbon–carbon double bond. Most enols are unstable, spontaneously isomerizing to their carbonyl tautomers, called the **keto** form of the compound. See **tautomers.** (p. 447)

Lindlar's catalyst A heterogeneous catalyst for the hydrogenation of alkynes to cis alkenes. In its most common form, it consists of a thin coating of palladium on barium sulfate, with quinoline added to decrease the catalytic activity. (p. 442)

***s* character** The fraction of a hybrid orbital that corresponds to an *s* orbital; about one-half for *sp* hybrids, one-third for sp^2 hybrids, and one-fourth for sp^3 hybrids. (p. 434)

siamyl group A contraction for secondary isoamyl, abbreviated "Sia." This is the 1,2-dimethylpropyl group. Disiamylborane is used for hydroboration of terminal alkynes because this bulky borane adds only once to the triple bond. (p. 448)

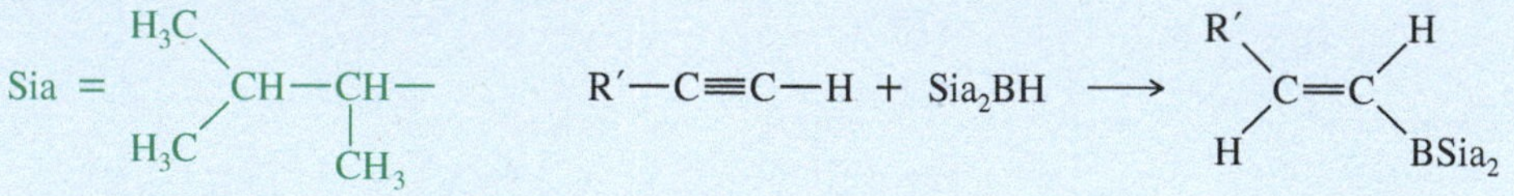

Sia = $(\text{H}_3\text{C})_2\text{CH—CH(CH}_3\text{)—}$

"*sec*-isoamyl" or "siamyl"

$$\underset{\text{alkyne}}{\text{R}'\text{—C}\equiv\text{C—H}} + \underset{\text{disiamylborane}}{\text{Sia}_2\text{BH}} \longrightarrow \underset{\text{a vinylborane}}{(\text{R}')(\text{H})\text{C=C}(\text{H})(\text{BSia}_2)}$$

tautomers Isomers that can quickly interconvert by the movement of a proton (and a double bond) from one site to another. An equilibrium between tautomers is called a **tautomerism.** (p. 447)

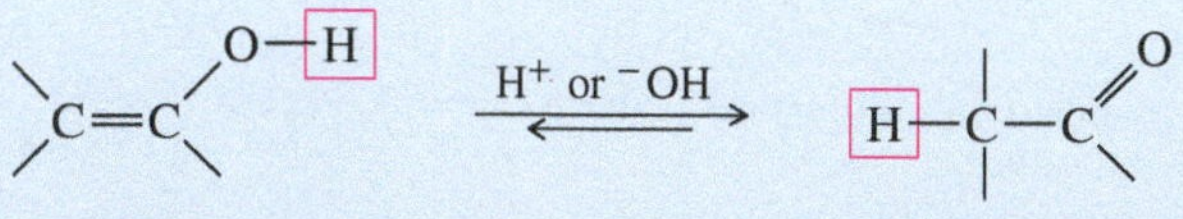

$$\underset{\text{enol form}}{\text{>C=C<}^{\text{O—H}}} \;\underset{}{\overset{\text{H}^+ \text{ or } ^-\text{OH}}{\rightleftharpoons}}\; \underset{\text{keto form}}{\text{H—}\overset{|}{\underset{|}{\text{C}}}\text{—C}(\text{=O})\text{—}}$$

The **keto–enol tautomerism** is the equilibrium between these two tautomers.

vinyl cation A cation with a positive charge on one of the carbon atoms of a C=C double bond. The cationic carbon atom is usually *sp* hybridized. Vinyl cations are often generated by the addition of an electrophile to a carbon–carbon triple bond. (p. 446)

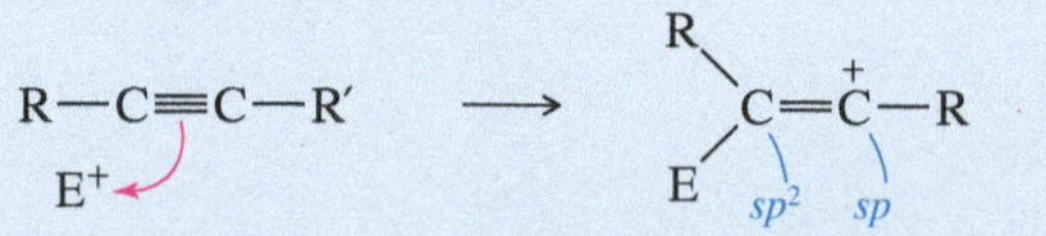

$$\text{R—C}\equiv\text{C—R}' \;\xrightarrow{\text{E}^+}\; (\text{R})(\text{E})\underset{sp^2}{\text{C}}\text{=}\overset{+}{\underset{sp}{\text{C}}}\text{—R}$$

a vinyl cation

Essential Problem-Solving Skills in Chapter 9

Each skill is followed by problem numbers exemplifying that particular skill.

1	Name alkynes and their derivatives, and draw correct structures when given their names.	Problems 9-26, 27, and 28
2	Explain why alkynes are more acidic than alkanes and alkenes. Show how to generate nucleophilic acetylide ions and use them in syntheses.	Problems 9-31, 34, 35, 36, 37, 40, and 42
3	Propose mechanisms to explain the observed products of alkyne reactions.	Problem 9-41
4	Propose effective single-step and multistep syntheses of alkynes by eliminations from alkyl halides and by the additions and substitutions of acetylide ions.	Problems 9-29, 31, 34, 35, 37, 40, and 42
5	Predict the products of additions, oxidations, reductions, and cleavages of alkynes, including the orientation of the reaction (regiochemistry) and the stereochemistry.	Problems 9-30, 32, 33, 34, 35, and 36
6	Use alkynes as starting materials and intermediates in one-step and multistep syntheses.	Problems 9-30, 32, 34, 35, 36, 37, and 42
	Problem-Solving Strategy: Multistep Synthesis	Problems 9-29, 31, 32, 34, 35, 37, 40, and 42
7	Determine the structure of an unknown alkyne from the products it forms in reactions such as ozonolysis.	Problems 9-38, 39, and 43

Study Problems

9-26 Write structural formulas for the following compounds (includes both old- and new-style names).

(a) 2-octyne **(b)** ethylisopentylacetylene **(c)** ethynylbenzene
(d) cyclohexylacetylene **(e)** 5-methyl-3-octyne **(f)** *trans*-3,5-dibromocyclodecyne
(g) 5,5-dibromo-4-phenylcyclooct-1-yne **(h)** (*E*)-6-ethyloct-2-en-4-yne **(i)** 1,4-heptadiyne
(j) vinylacetylene **(k)** (*S*)-3-methyl-1-penten-4-yne

9-27 Give common names for the following compounds.

(a) $CH_3-C\equiv C-CH_2CH_3$ **(b)** $Ph-C\equiv C-H$
(c) 3-methyloct-4-yne **(d)** $(CH_3)_3C-C\equiv C-CH(CH_3)CH_2CH_3$

9-28 Give IUPAC names for the following compounds.

(a) $CH_3-C\equiv C-CH(Ph)-CH_3$

(b) $(H_3C)(H)C=C(CH_3)(C\equiv C-CH_2CH_3)$ (H₃C and CH₃ on the same side; H and $C\equiv C-CH_2CH_3$ on the same side)

(c) $(CH_3)_3C-C\equiv C-CH(CH_3)CH_2CH_3$

(d) $CH_3-CBr_2-C\equiv C-CH_3$

(e) $CH_3-C\equiv C-C(CH_3)(CH_2CH_3)-OH$

(f) cyclopentyl$-CH_2-C\equiv C-CH_3$

9-29 Using hex-1-ene as your starting material, show how you would synthesize the following compounds. (Once you have shown how to synthesize a compound, you may use it as the starting material in any later parts of this problem.)

(a) 1,2-dibromohexane **(b)** hex-1-yne **(c)** 2,2-dibromohexane
(d) hex-2-yne **(e)** hexan-2-one **(f)** hexanal
(g) pentanoic acid **(h)** pentanal **(i)** undec-6-yn-5-ol

9-30 Using cyclooctyne as your starting material, show how you would synthesize the following compounds. (Once you have shown how to synthesize a compound, you may use it as the starting material in any later parts of this problem.)

(a) *cis*-cyclooctene **(b)** cyclooctane **(c)** *trans*-1,2-dibromocyclooctane
(d) cyclooctanone **(e)** 1,1-dibromocyclooctane **(f)** 3-bromocyclooctene
(g) cyclooctane-1,2-dione **(h)** $H-C(=O)-(CH_2)_6-C(=O)-H$ **(i)** $HO-C(=O)-(CH_2)_6-C(=O)-OH$

9-31 The application box in the margin of page 437 states, "The addition of an acetylide ion to a carbonyl group is used in the synthesis of ethchlorvynol, a drug used to cause drowsiness and induce sleep." Show how you would accomplish this synthesis from acetylene and a carbonyl compound.

OH

$CH_3CH_2—C—C{\equiv}CH$

C

H CHCl

ethchlorvynol

9-32 *Muscalure*, the sex attractant of the common housefly, is *cis*-tricos-9-ene. Most syntheses of alkenes give the more stable trans isomer as the major product. Devise a synthesis of muscalure from acetylene and other compounds of your choice. Your synthesis must give specifically the cis isomer of muscalure.

$CH_3(CH_2)_7$ $(CH_2)_{12}CH_3$

C=C

H H

cis-tricos-9-ene, "muscalure"

9-33 Predict the products of reaction of pent-1-yne with the following reagents.

(a) 1 equivalent of HCl
(b) 2 equivalents of HCl
(c) excess H_2, Ni
(d) H_2, $Pd/BaSO_4$, quinoline
(e) 1 equivalent of Br_2
(f) 2 equivalents of Br_2
(g) cold, dilute $KMnO_4$
(h) warm, concd. $KMnO_4$, NaOH
(i) Na, liquid ammonia
(j) $NaNH_2$
(k) $H_2SO_4/HgSO_4$, H_2O
(l) Sia_2BH, then H_2O_2, ^-OH

9-34 Show how you would accomplish the following synthetic transformations. Show all intermediates.

(a) 2,2-dibromobutane ⟶ but-1-yne
(b) 2,2-dibromobutane ⟶ but-2-yne
(c) but-1-yne ⟶ oct-3-yne
(d) *trans*-hex-2-ene ⟶ hex-2-yne
(e) 2,2-dibromohexane ⟶ hex-1-yne
(f) cyclodecyne ⟶ *cis*-cyclodecene
(g) cyclodecyne ⟶ *trans*-cyclodecene
(h) hex-1-yne ⟶ hexan-2-one, $CH_3COCH_2CH_2CH_2CH_3$
(i) hex-1-yne ⟶ hexanal, $CH_3(CH_2)_4CHO$
(j) *trans*-hex-2-ene ⟶ *cis*-hex-2-ene

9-35 Show how you would synthesize the following compounds from acetylene and any other needed reagents:

(a) 6-phenylhex-1-en-4-yne
(b) *cis*-1-phenylpent-2-ene
(c) *trans*-1-phenylpent-2-ene
(d)

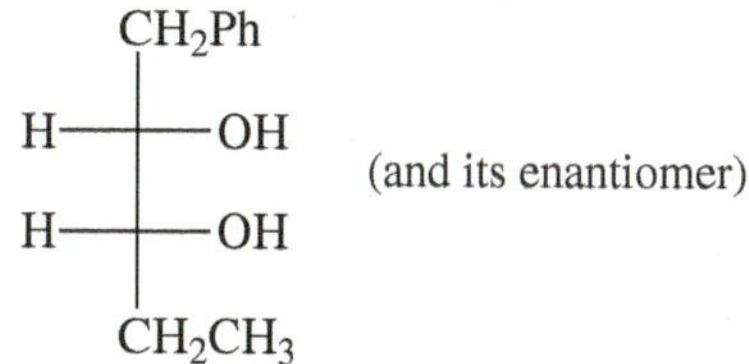

(and its enantiomer)

9-36 Predict the products formed when $CH_3CH_2—C{\equiv}C{:}^-\ Na^+$ reacts with the following compounds.

(a) ethyl bromide
(b) *tert*-butyl bromide
(c) formaldehyde
(d) cyclohexanone
(e) $CH_3CH_2CH_2CHO$
(f) cyclohexanol
(g) butan-2-one, $CH_3CH_2COCH_3$

9-37 Show how you would synthesize the following compounds, starting with acetylene and any compounds containing no more than four carbon atoms.

(a) hex-1-yne
(b) hex-2-yne
(c) *cis*-hex-2-ene
(d) *trans*-hex-2-ene
(e) 1,1-dibromohexane
(f) 2,2-dibromohexane
(g) pentanal, $CH_3CH_2CH_2CH_2CHO$
(h) pentan-2-one, $CH_3—CO—CH_2CH_2CH_3$
(i) (±)-3,4-dibromohexane
(j) *meso*-butan-2,3-diol
(k) 2-methylhex-3-yn-2-ol

9-38 When treated with hydrogen and a platinum catalyst, an unknown compound **X** absorbs 5 equivalents of hydrogen to give *n*-butylcyclohexane. Treatment of **X** with an excess of ozone, followed by dimethyl sulfide and water, gives the following products:

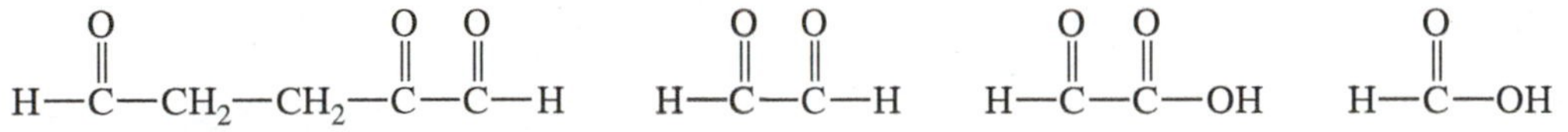

$$H-\overset{O}{\overset{\|}{C}}-CH_2-CH_2-\overset{O}{\overset{\|}{C}}-\overset{O}{\overset{\|}{C}}-H \qquad H-\overset{O}{\overset{\|}{C}}-\overset{O}{\overset{\|}{C}}-H \qquad H-\overset{O}{\overset{\|}{C}}-\overset{O}{\overset{\|}{C}}-OH \qquad H-\overset{O}{\overset{\|}{C}}-OH$$

Propose a structure for the unknown compound **X.** Is there any uncertainty in your structure?

9-39 When compound **Z** is treated with ozone, followed by dimethyl sulfide and washing with water, the products are formic acid, 3-oxobutanoic acid, and hexanal.

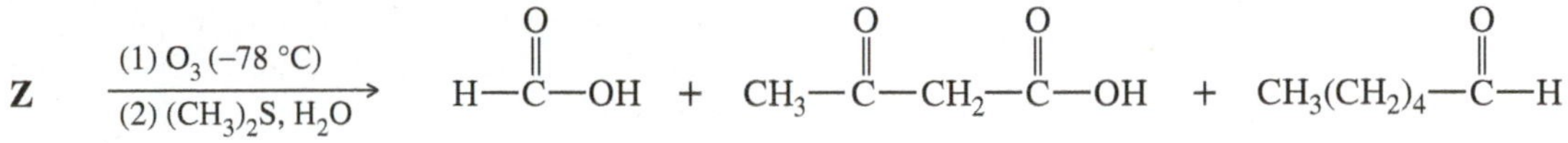

$$\mathbf{Z} \xrightarrow[(2)\ (CH_3)_2S,\ H_2O]{(1)\ O_3\ (-78\ ^\circ C)} H-\overset{O}{\overset{\|}{C}}-OH + CH_3-\overset{O}{\overset{\|}{C}}-CH_2-\overset{O}{\overset{\|}{C}}-OH + CH_3(CH_2)_4-\overset{O}{\overset{\|}{C}}-H$$

Propose a structure for compound **Z.** What uncertainty is there in the structure you have proposed?

***9-40** Show how you would convert the following starting materials into the target compound. You may use any additional reagents you need.

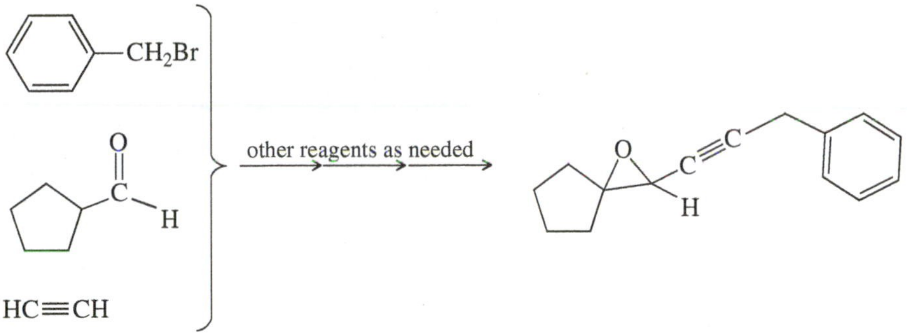

Starting materials: benzyl bromide ($C_6H_5-CH_2Br$), cyclopentanecarbaldehyde (cyclopentyl–CHO), $HC{\equiv}CH$ $\xrightarrow{\text{other reagents as needed}}$ target: spiro epoxide of cyclopentane bearing H and $-C{\equiv}C-CH_2-C_6H_5$ on the epoxide carbon.

***9-41** The following functional-group interchange is a useful synthesis of aldehydes.

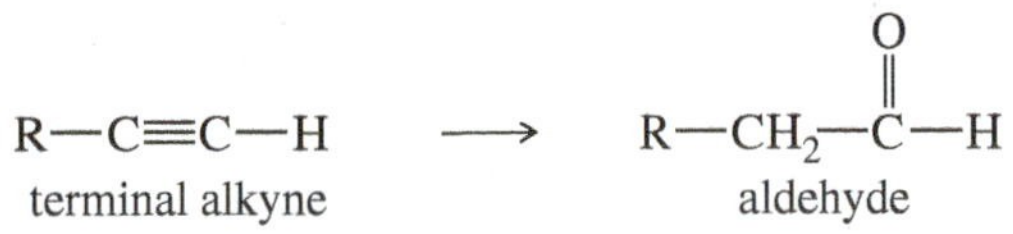

$$\underset{\text{terminal alkyne}}{R-C{\equiv}C-H} \longrightarrow \underset{\text{aldehyde}}{R-CH_2-\overset{O}{\overset{\|}{C}}-H}$$

(a) What reagents were used in this chapter for this transformation? Give an example to illustrate this method.

(b) This functional-group interchange can also be accomplished using the following sequence.

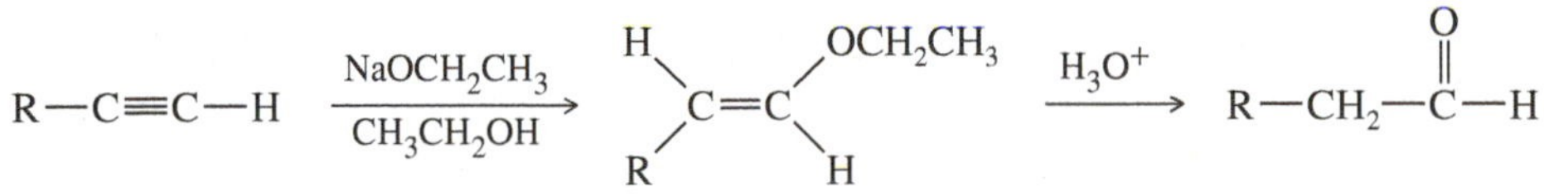

$$R-C{\equiv}C-H \xrightarrow[CH_3CH_2OH]{NaOCH_2CH_3} \text{(H)(R)}C{=}C\text{(OCH}_2\text{CH}_3\text{)(H)} \xrightarrow{H_3O^+} R-CH_2-\overset{O}{\overset{\|}{C}}-H$$

Propose mechanisms for these steps.

(c) Explain why a nucleophilic reagent such as ethoxide adds to an alkyne more easily than it adds to an alkene.

***9-42** Using any necessary inorganic reagents, show how you would convert acetylene and isobutyl bromide to

(a) *meso*-2,7-dimethyloctane-4,5-diol, $(CH_3)_2CHCH_2CH(OH)CH(OH)CH_2CH(CH_3)_2$.

(b) $(\pm)$-2,7-dimethyloctane-4,5-diol.

9-43 Deduce the structure of each compound from the information given. All unknowns in this problem have molecular formula C_8H_{12}.

(a) Upon catalytic hydrogenation, unknown **W** gives cyclooctane. Ozonolysis of **W,** followed by reduction with dimethyl sulfide, gives octanedioic acid, $HOOC-(CH_2)_6-COOH$. Draw the structure of **W.**

(b) Upon catalytic hydrogenation, unknown **X** gives cyclooctane. Ozonolysis of **X,** followed by reduction with dimethyl sulfide, gives two equivalents of butanedial, $O{=}CH-CH_2CH_2-CH{=}O$. Draw the structure of **X.**

(c) Upon catalytic hydrogenation, unknown **Y** gives cyclooctane. Ozonolysis of **Y,** followed by reduction with dimethyl sulfide, gives a three-carbon dialdehyde and a five-carbon dialdehyde. Draw the structure of **Y.**

***(d)** Upon catalytic hydrogenation, unknown **Z** gives *cis*-bicyclo[4.2.0]octane. Ozonolysis of **Z,** followed by reduction with dimethyl sulfide, gives a cyclobutane with a three-carbon aldehyde ($-CH_2-CH_2-CHO$) group on C1 and a one-carbon aldehyde ($-CHO$) group on C2. Draw the structure of **Z.**

10 Structure and Synthesis of Alcohols

Goals for Chapter 10

1 Draw structures and assign names for alcohols, phenols, diols, and thiols.

2 Predict relative boiling points, acidities, and solubilities of alcohols and thiols.

3 Show how to convert alkenes, alkyl halides, and carbonyl compounds to alcohols.

4 Use organometallic reagents for the synthesis of primary, secondary, and tertiary alcohols with the needed carbon skeletons.

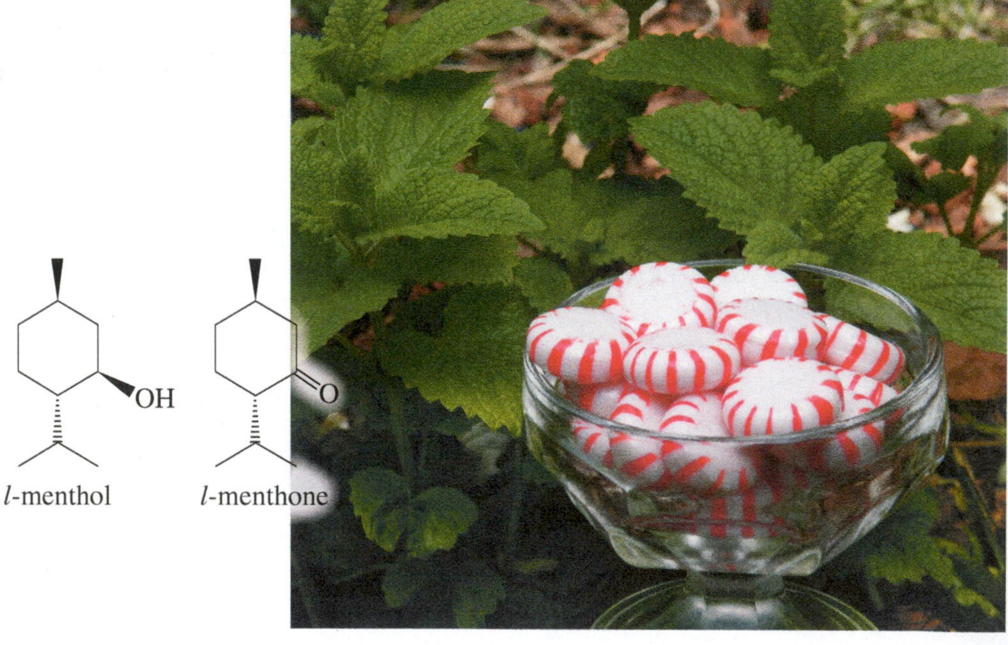

▶ Extraction or steam distillation of the peppermint plant (*Mentha piperita*) gives a water-insoluble liquid called peppermint oil. The two main components of peppermint oil are the alcohol *l*-menthol and the ketone *l*-menthone, both produced by the plant as single enantiomers. People have used peppermint for millennia, both as a spice and as a medicine. Peppermint oil is still used to treat nausea, vomiting, and other digestive illnesses. Peppermint candies are among the most popular in the world.

10-1 Introduction

Alcohols are organic compounds containing hydroxy (—OH) groups. They are some of the most common and useful compounds in nature, in industry, and around the house. The word *alcohol* is one of the oldest chemical terms, derived from the early Arabic *al-kuhl*. Originally it meant "the powder," and later "the essence." Ethyl alcohol, distilled from wine, was considered to be "the essence" of wine. Ethyl alcohol (grain alcohol) is found in alcoholic beverages, cosmetics, and drug preparations. Methyl alcohol (wood alcohol) is used as a fuel and solvent. Isopropyl alcohol (rubbing alcohol) is used as a skin cleanser for injections and minor cuts.

$CH_3—CH_2—OH$	$CH_3—OH$	$CH_3—CH(OH)—CH_3$
ethyl alcohol	methyl alcohol	isopropyl alcohol
ethanol	methanol	propan-2-ol

Alcohols are synthesized by a wide variety of methods, and the hydroxy group may be converted to most other functional groups. For these reasons, alcohols are versatile synthetic intermediates. In this chapter, we discuss the physical properties of alcohols and summarize the methods used to synthesize them. In the next chapter (Reactions of Alcohols), we continue our study of the central role that alcohols play in organic chemistry as reagents, solvents, and synthetic intermediates.

10-2 Structure and Classification of Alcohols

The structure of an alcohol resembles the structure of water, with an alkyl group replacing one of the hydrogen atoms of water. Figure 10-1 compares the structures of water and methanol. Both have sp^3-hybridized oxygen atoms, but the C—O—H bond

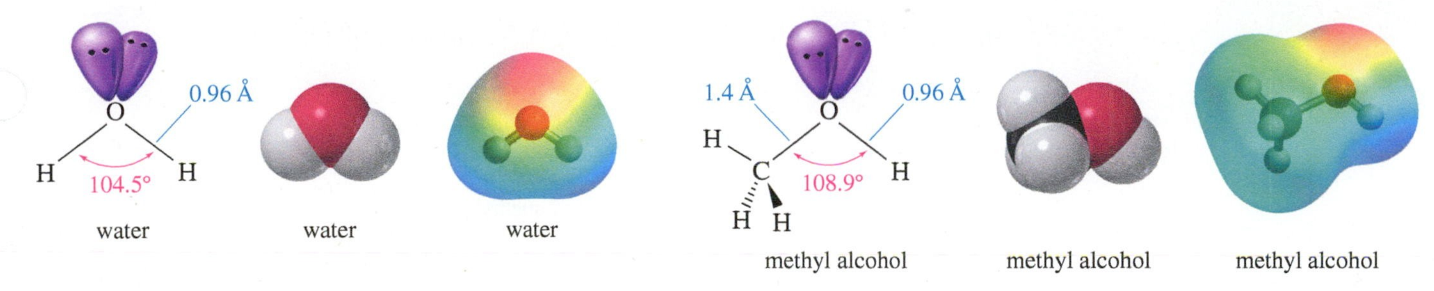

FIGURE 10-1
Comparison of the structures of water and methyl alcohol.

angle in methanol (108.9°) is larger than the H—O—H bond angle in water (104.5°) because the methyl group is much larger than a hydrogen atom. The bulky methyl group counteracts the bond angle compression caused by oxygen's nonbonding pairs of electrons. The O—H bond lengths are about the same in water and methanol (0.96 Å), but the C—O bond is considerably longer (1.4 Å), reflecting the larger covalent radius of carbon compared to hydrogen.

One way of organizing the alcohol family is to classify each alcohol according to the type of **carbinol carbon atom:** the one bonded to the —OH group. If this carbon atom is primary (bonded to one other carbon atom), the compound is a **primary alcohol.** A **secondary alcohol's** —OH group is attached to a secondary carbon atom, and a **tertiary alcohol's** —OH group is bonded to a tertiary carbon. When we studied alkyl halides (Chapter 6), we saw that primary, secondary, and tertiary halides react differently. The same is true for alcohols. We need to learn how these classes of alcohols are similar and under what conditions they react differently. Figure 10-2 shows examples of primary, secondary, and tertiary alcohols.

Compounds with a hydroxy group bonded directly to an aromatic (benzene) ring are called **phenols.** Phenols have many properties similar to those of alcohols, whereas other properties derive from their aromatic character. In this chapter, we consider the properties of phenols that are similar to those of alcohols and note some of the differences. In Chapter 16, we consider the aromatic nature of phenols and the reactions that result from their aromaticity.

10-3 Nomenclature of Alcohols and Phenols

10-3A IUPAC Names ("Alkanol" Names)

The IUPAC system provides unique names for alcohols, based on rules that are similar to those for other classes of compounds. In general, the name carries the *-ol* suffix, together with a number to give the location of the hydroxy group. The formal rules are summarized in the following three steps:

1. Name the longest carbon chain that contains the carbon atom bearing the —OH group. Drop the final *-e* from the alkane name and add the suffix *-ol* to give the root name.
2. Number the longest carbon chain starting at the end nearest the hydroxy group, and use the appropriate number to indicate the position of the —OH group. (The hydroxy group takes precedence over double and triple bonds.)
3. Name all the substituents and give their numbers, as you would for an alkane or an alkene.

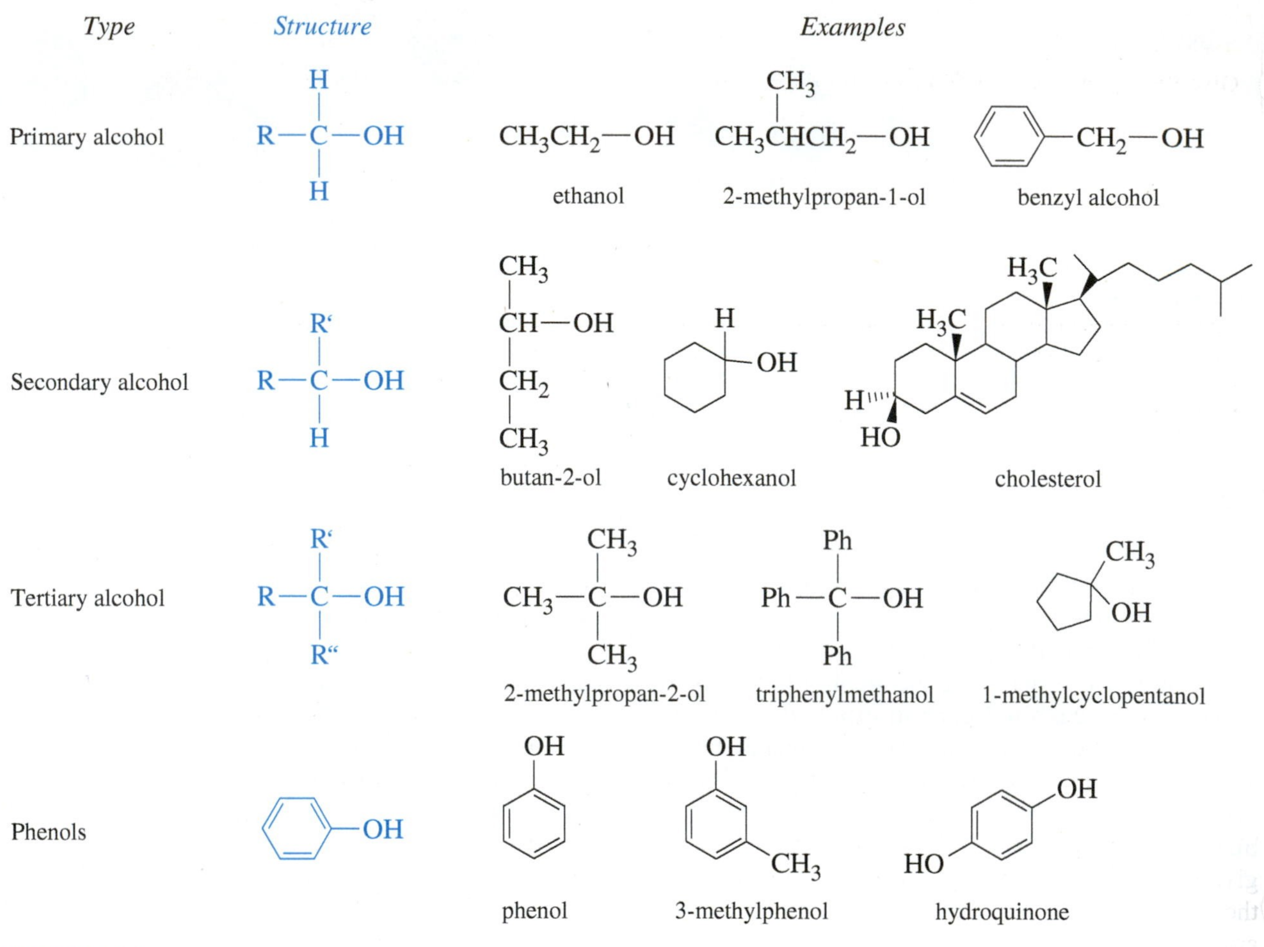

FIGURE 10-2
Classification of alcohols. Alcohols are classified according to the type of carbon atom (primary, secondary, or tertiary) bonded to the hydroxy group. Phenols have a hydroxy group bonded to a carbon atom in a benzene ring.

In the following example, the longest carbon chain has four carbons, so the root name is *butanol*. The —OH group is on the second carbon atom, so this is a butan-2-ol. The complete IUPAC name is 1-bromo-3,3-dimethylbutan-2-ol.

$$\overset{4}{CH_3}-\overset{3}{C}(CH_3)_2-\overset{2}{CH}(OH)-\overset{1}{CH_2}-Br$$

Cyclic alcohols are named using the prefix *cyclo-*; the hydroxy group is assumed to be on C1.

IUPAC name: *trans*-2-bromocyclohexanol 1-ethylcyclopropanol

SOLVED PROBLEM 10-1

Give the systematic (IUPAC) name for the following alcohol.

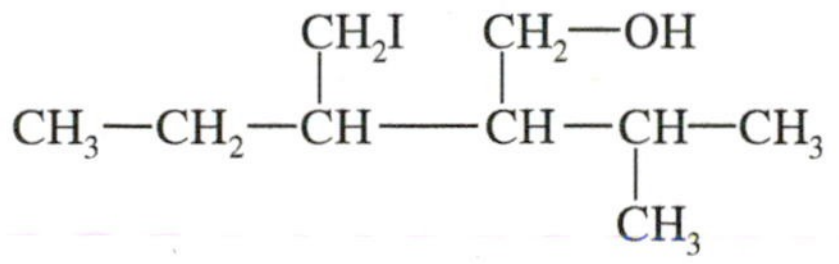

$$CH_3{-}CH_2{-}CH(CH_2I){-}CH(CH_2{-}OH){-}CH(CH_3){-}CH_3$$

SOLUTION

The longest chain contains six carbon atoms, but it does not contain the carbon bonded to the hydroxy group. The longest chain containing the carbon bonded to the —OH group is the one outlined by the green box, containing five carbon atoms. This chain is numbered from right to left to give the hydroxy-bearing carbon atom the lowest possible number.

$$^5CH_3{-}^4CH_2{-}^3CH(CH_2I){-}^2CH(^1CH_2{-}OH){-}CH(CH_3){-}CH_3$$

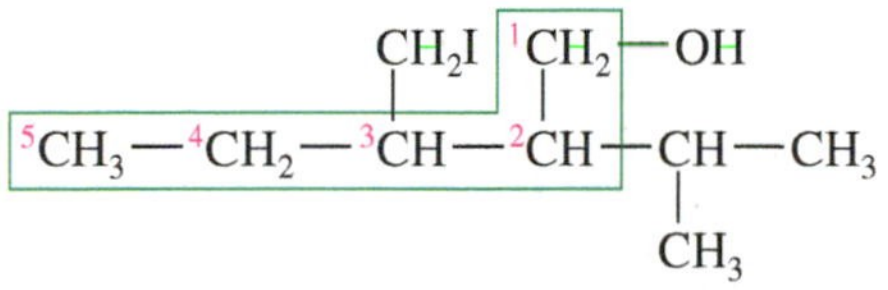

The correct name for this compound is 3-(iodomethyl)-2-isopropylpentan-1-ol.

In naming alcohols containing double and triple bonds, use the *-ol* suffix after the alkene or alkyne name. The alcohol functional group takes precedence over double and triple bonds; so the chain is numbered to give the lowest possible number to the carbon atom bonded to the hydroxy group. The position of the —OH group is given by putting its number before the *-ol* suffix. Numbers for the multiple bonds were once given early in the name, but the 1993 revision of the IUPAC rules puts them next to the *-en* or *-yn* suffix they describe. Both the new and old placements of the numbers are shown in the following figure.

PROBLEM-SOLVING HINT

The nomenclature of alcohols is reviewed in Appendix 5, the Summary of Organic Nomenclature.

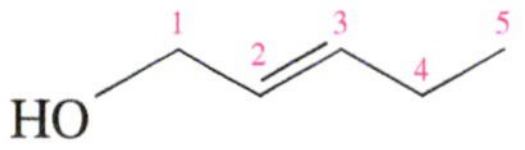

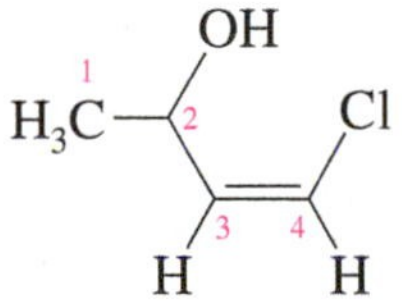

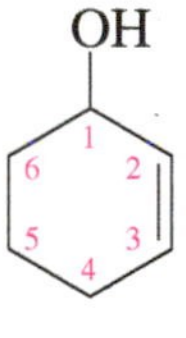

	HO–CH₂–CH=CH–CH₂–CH₃	H₃C–CH(OH)–CH=CH–Cl	cyclohexenol
old IUPAC name:	*trans*-2-penten-1-ol	(*Z*)-4-chloro-3-buten-2-ol	2-cyclohexen-1-ol
new IUPAC name:	*trans*-pent-2-en-1-ol	(*Z*)-4-chlorobut-3-en-2-ol	cyclohex-2-en-1-ol

Table 10-1 is a partial table showing the order of precedence of functional groups for assigning IUPAC names. A more complete table, titled "Summary of Functional Group Nomenclature," appears inside the back cover. In general, the highest-priority functional group is considered the "main" group, and the others are treated as substituents.

The —OH functional group is named as a **hydroxy** substituent* when it appears on a structure with a higher-priority functional group or when the structure is too difficult to name as a simple alcohol.

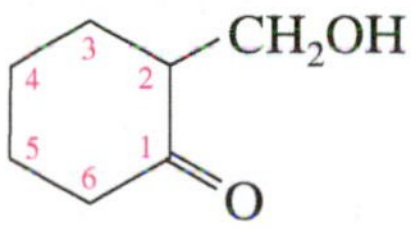

2-(hydroxymethyl)cyclohexanone

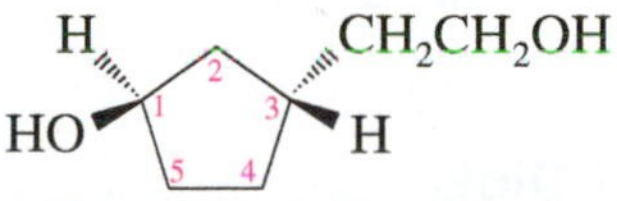

trans-3-(2-hydroxyethyl)cyclopentanol
(1*R*,3*R*)-3-(2-hydroxyethyl)cyclopentanol

$$^4CH_3{-}^3CH(OH){-}^2CH_2{-}^1C(=O){-}OH$$

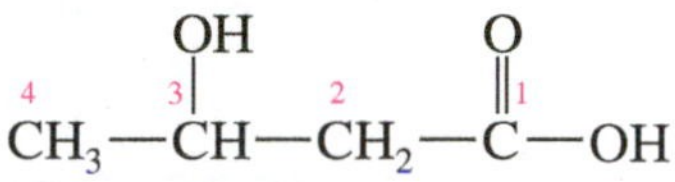

3-hydroxybutanoic acid

*In the past, the terms *hydroxy* and *hydroxyl* were both used to describe any OH group. The IUPAC has recently decided that the term "hydroxyl" should refer only to the hydroxyl radical, ·OH. This new rule is often disregarded in common usage.

TABLE 10-1
Priority of Functional Groups in Naming Organic Compounds

acids (highest)
esters
aldehydes
ketones
alcohols
amines
alkenes, alkynes
alkanes
ethers
halides (lowest)

PROBLEM 10-1

Give the IUPAC names of the following alcohols.

(a) 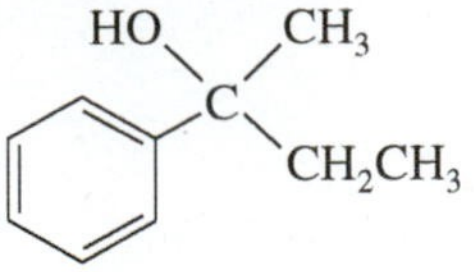 (b) 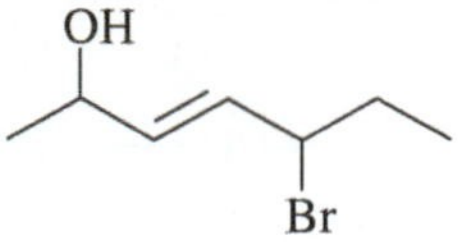 (c)

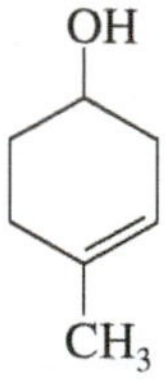

(d) 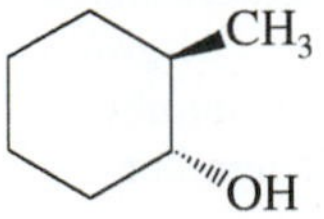 (e) 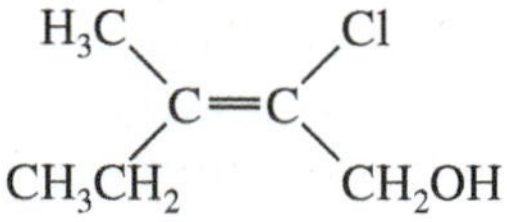 (f)

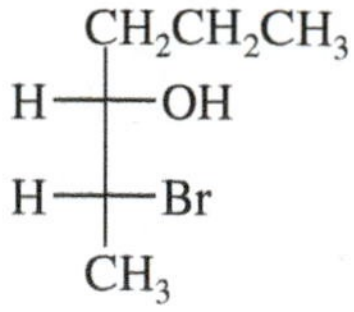

10-3B Common Names of Alcohols

The common name of an alcohol is derived from the common name of the alkyl group and the word *alcohol*. This system pictures an alcohol as a molecule of water with an alkyl group replacing one of the hydrogen atoms. If the structure is complex, the common nomenclature becomes awkward, and the IUPAC nomenclature should be used.

	CH_3—OH	$CH_3CH_2CH_2$—OH	CH_3—CH(OH)—CH_3	H_2C=CH—CH_2—OH
common name:	methyl alcohol	*n*-propyl alcohol	isopropyl alcohol	allyl alcohol
IUPAC name:	methanol	propan-1-ol	propan-2-ol	prop-2-en-1-ol

	$CH_3CH_2CH_2CH_2$—OH	CH_3—CH(OH)—CH_2CH_3	CH_3—C(CH_3)$_2$—OH	CH_3—CH(CH_3)—CH_2—OH
common name:	*n*-butyl alcohol	*sec*-butyl alcohol	*tert*-butyl alcohol	isobutyl alcohol
IUPAC name:	butan-1-ol	butan-2-ol	2-methylpropan-2-ol	2-methylpropan-1-ol

PROBLEM 10-2

Give both the IUPAC name and the common name for each alcohol.

(a) $CH_3CH_2CH(OH)CH_3$ (b)  (c)

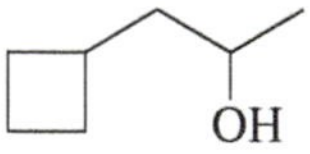

(d) $(CH_3)_2CHCH_2CH_2OH$

PROBLEM 10-3

For each molecular formula, draw all the possible constitutional isomers of alcohols with that formula. Give the IUPAC name for each alcohol.

(a) C_3H_8O (b) $C_4H_{10}O$ (c) C_3H_6O (d) C_3H_4O

10-3C Names of Diols

Alcohols with two —OH groups are called **diols** or **glycols.** They are named like other alcohols except that the suffix *diol* is used and two numbers are needed to tell where the two hydroxy groups are located. This is the preferred, systematic (IUPAC) method for naming diols.

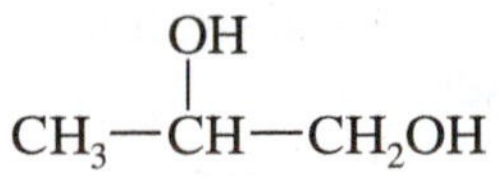
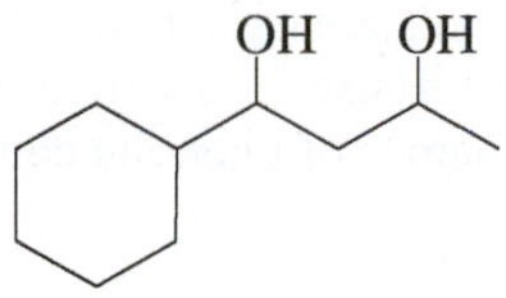
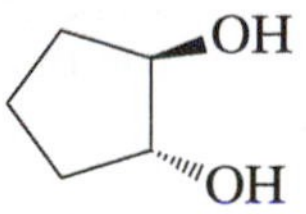

IUPAC name: propane-1,2-diol 1-cyclohexylbutane-1,3-diol *trans*-cyclopentane-1,2-diol

The term *glycol* generally means a 1,2-diol, or **vicinal diol,** with its two hydroxy groups on adjacent carbon atoms. Glycols are usually synthesized by the hydroxylation of alkenes, using peroxyacids, osmium tetroxide, or potassium permanganate (Section 8-14).

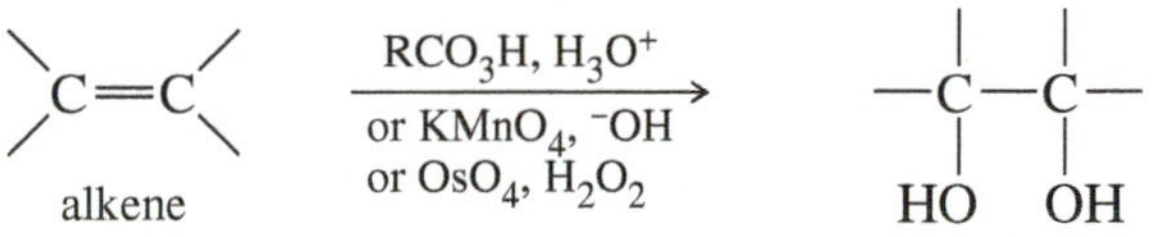

alkene → vicinal diol (glycol)

This synthesis of glycols is reflected in their common names. The glycol is named for the alkene from which it is synthesized:

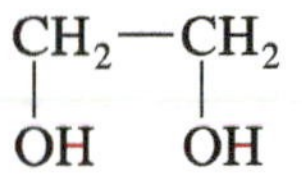
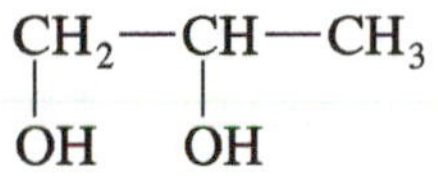
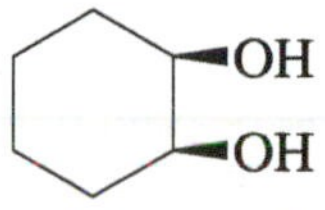

IUPAC name:	ethane-1,2-diol	propane-1,2-diol	*cis*-cyclohexane-1,2-diol
common name:	ethylene glycol	propylene glycol	*cis*-cyclohexene glycol

The common names of glycols can be awkward and confusing because the *-ene* portion of the name implies the presence of an alkene double bond, but the glycol does not contain a double bond. We will generally use the IUPAC "diol" nomenclature for diols, but be aware that the names "ethylene glycol" (automotive antifreeze) and "propylene glycol" (used in medicines and foods) are universally accepted for these common diols.

PROBLEM 10-4

Give a systematic (IUPAC) name for each diol.

(a) $CH_3CH(OH)(CH_2)_4CH(OH)C(CH_3)_3$

(b) $HO{-}(CH_2)_8{-}OH$

(c)

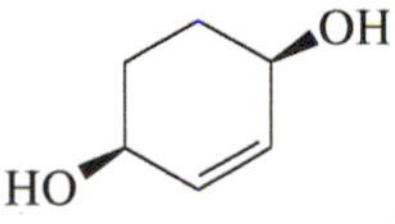

(d)

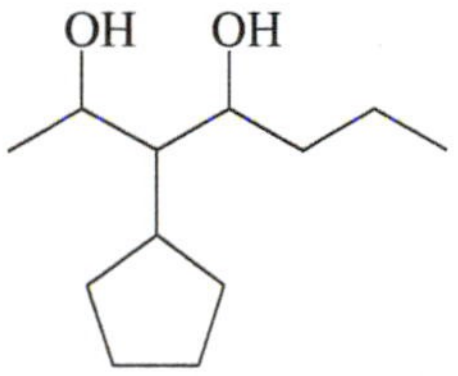

(e)

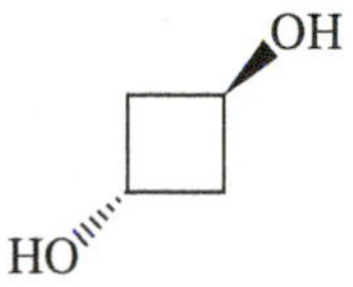

10-3D Names of Phenols

Because the phenol structure involves a benzene ring, the terms *ortho* (1,2-disubstituted), *meta* (1,3-disubstituted), and *para* (1,4-disubstituted) are often used in the common names. The following examples illustrate the systematic names and the common names of some simple phenols.

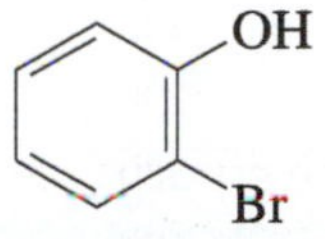
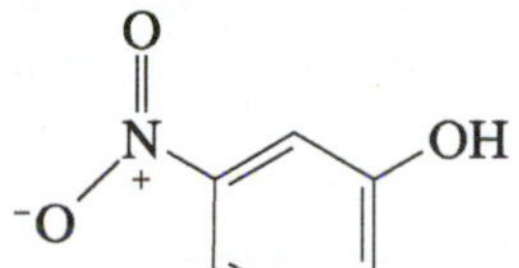
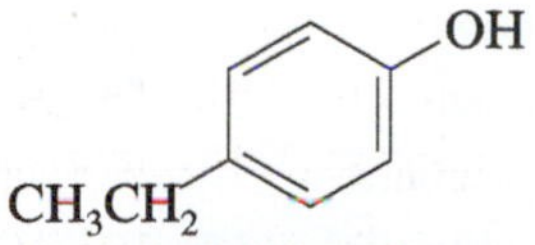

IUPAC name:	2-bromophenol	3-nitrophenol	4-ethylphenol
common name:	*ortho*-bromophenol	*meta*-nitrophenol	*para*-ethylphenol

Application: Antiseptic

Phenol was given the common name "carbolic acid" because it was originally isolated from the acidic fraction of coal tar. Concentrated phenol is a potent neurotoxin, and skin exposure causes deep, painless burns. Weak aqueous solutions of phenol are effective antiseptics.

The methylphenols are called *cresols*, whereas the names of the benzenediols are based on their historical uses and sources rather than their structures. We will generally use the systematic names of phenolic compounds.

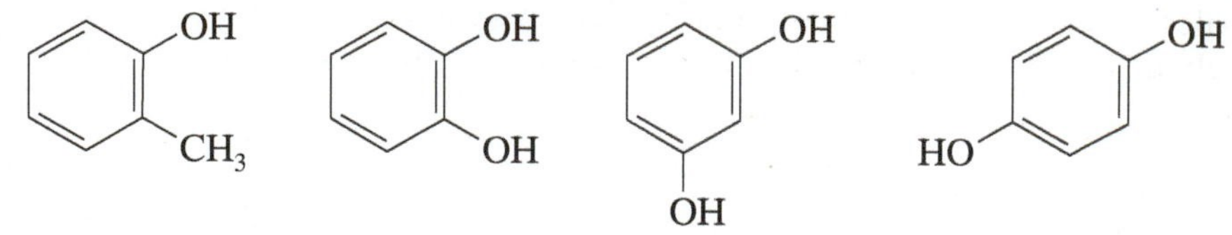

IUPAC name:	2-methylphenol	benzene-1,2-diol	benzene-1,3-diol	benzene-1,4-diol
common name:	*ortho*-cresol	catechol	resorcinol	hydroquinone

Application: Environment

Soil bacteria convert many aromatic compounds to catechol. Then they cleave the ring and degrade the products further to water and carbon dioxide. The soil around highways is often enriched in these bacteria feeding on the runoff of hydrocarbons that vehicles deposit on the road.

10-4 Physical Properties of Alcohols

Most of the common alcohols, up to about 11 or 12 carbon atoms, are liquids at room temperature. Methanol and ethanol are free-flowing volatile liquids with characteristic fruity odors. The higher alcohols (the butanols through the decanols) are somewhat viscous, and some of the highly branched isomers are solids at room temperature. These higher alcohols have heavier but still fruity odors. Propan-1-ol and propan-2-ol (rubbing alcohol) fall in the middle, with a barely noticeable viscosity and a characteristic odor often associated with a physician's office. Table 10-2 lists the physical properties of some common alcohols.

TABLE 10-2
Physical Properties of Selected Alcohols

IUPAC Name	Common Name	Formula	mp (°C)	bp (°C)	Density (g/mL)
methanol	methyl alcohol	CH_3OH	−97	65	0.79
ethanol	ethyl alcohol	CH_3CH_2OH	−114	78	0.79
propan-1-ol	*n*-propyl alcohol	$CH_3CH_2CH_2OH$	−126	97	0.80
propan-2-ol	isopropyl alcohol	$(CH_3)_2CHOH$	−89	82	0.79
butan-1-ol	*n*-butyl alcohol	$CH_3(CH_2)_3OH$	−90	118	0.81
butan-2-ol	*sec*-butyl alcohol	$CH_3CH(OH)CH_2CH_3$	−114	100	0.81
2-methylpropan-1-ol	isobutyl alcohol	$(CH_3)_2CHCH_2OH$	−108	108	0.80
2-methylpropan-2-ol	*tert*-butyl alcohol	$(CH_3)_3COH$	25	83	0.79
pentan-1-ol	*n*-pentyl alcohol	$CH_3(CH_2)_4OH$	−79	138	0.82
3-methylbutan-1-ol	isopentyl alcohol	$(CH_3)_2CHCH_2CH_2OH$	−117	132	0.81
2,2-dimethylpropan-1-ol	neopentyl alcohol	$(CH_3)_3CCH_2OH$	52	113	0.81
cyclopentanol	cyclopentyl alcohol	*cyclo*-C_5H_9OH	−19	141	0.95
hexan-1-ol	*n*-hexyl alcohol	$CH_3(CH_2)_5OH$	−52	156	0.82
cyclohexanol	cyclohexyl alcohol	*cyclo*-$C_6H_{11}OH$	25	162	0.96
heptan-1-ol	*n*-heptyl alcohol	$CH_3(CH_2)_6OH$	−34	176	0.82
octan-1-ol	*n*-octyl alcohol	$CH_3(CH_2)_7OH$	−16	194	0.83
nonan-1-ol	*n*-nonyl alcohol	$CH_3(CH_2)_8OH$	−6	214	0.83
decan-1-ol	*n*-decyl alcohol	$CH_3(CH_2)_9OH$	6	233	0.83
prop-2-en-1-ol	allyl alcohol	$H_2C{=}CH{-}CH_2OH$	−129	97	0.86
phenylmethanol	benzyl alcohol	$Ph{-}CH_2OH$	−15	205	1.05
diphenylmethanol	diphenylcarbinol	Ph_2CHOH	69	298	
triphenylmethanol	triphenylcarbinol	Ph_3COH	162	380	1.20
ethane-1,2-diol	ethylene glycol	$HOCH_2CH_2OH$	−13	198	1.12
propane-1,2-diol	propylene glycol	$CH_3CH(OH)CH_2OH$	−59	188	1.04
propane-1,2,3-triol	glycerol	$HOCH_2CH(OH)CH_2OH$	18	290	1.26

10-4A Boiling Points of Alcohols

Because we often deal with liquid alcohols, we forget how surprising it *should* be that the lower-molecular-weight alcohols are liquids instead of gases. For example, ethyl alcohol and propane have similar molecular weights, yet their boiling points differ by about 120 °C. Dimethyl ether has an intermediate boiling point.

$\mu = 1.69$ D	$\mu = 1.30$ D	$\mu = 0.08$ D
$H-O-CH_2CH_3$	$H_3C-O-CH_3$	$H_3C-CH_2-CH_3$
ethanol, MW 46	dimethyl ether, MW 46	propane, MW 44
bp 78 °C	bp −25 °C	bp −42 °C

Such a large difference in boiling points suggests that ethanol molecules are attracted to each other much more strongly than propane molecules are. Two important intermolecular forces are responsible: hydrogen bonding and dipole–dipole attractions (Section 2-2).

Hydrogen bonding is the major intermolecular attraction responsible for ethanol's high boiling point. The hydroxy hydrogen of ethanol is strongly polarized by its bond to oxygen, and it forms a hydrogen bond with a pair of nonbonding electrons from the oxygen atom of another alcohol molecule (Section 2-2C). Ethers have two alkyl groups bonded to their oxygen atoms, so they have no O—H hydrogen atoms to form hydrogen bonds. Hydrogen bonds have a strength of about 21 kJ (5 kcal) per mole, which is weaker than typical covalent bonds of 300 to 500 kJ, but much stronger than dipole–dipole attractions.

Dipole–dipole attractions also contribute to the relatively high boiling points of alcohols and ethers. The polarized C—O and H—O bonds and the nonbonding electrons add to produce a dipole moment of 1.69 D in ethanol, compared with a dipole moment of only 0.08 D in propane. In liquid ethanol, the positive and negative ends of these dipoles align to produce attractive interactions.

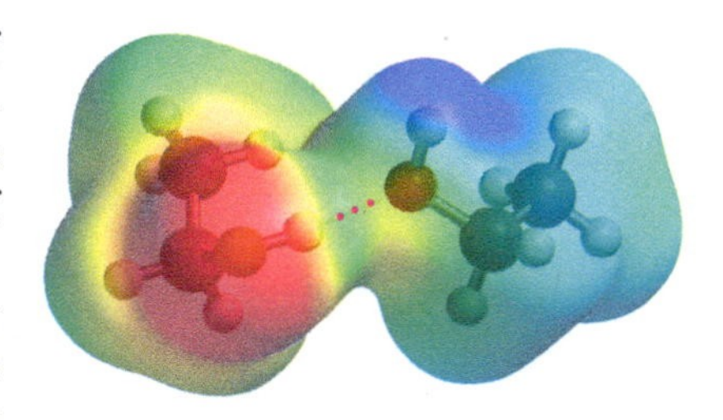
hydrogen bonding in ethanol

We can compare the effects of hydrogen bonding and dipole–dipole attractions by comparing ethanol with dimethyl ether. Like ethanol, dimethyl ether has a large dipole moment (1.30 D), but dimethyl ether cannot engage in hydrogen bonding because it has no —O—H hydrogens.

Alcohol	*Ether*
R—O—H···O(H)—R	R—O—R O(R)—R
hydrogen bond	no hydrogen bond

The boiling point of dimethyl ether is −25 °C, which is about 17° higher than that of propane, but still 103° lower than that of ethanol. Hydrogen bonds are clearly much stronger intermolecular attractions than dipole–dipole attractions.

10-4B Solubility Properties of Alcohols

Water and alcohols have similar properties because they all contain hydroxy groups that can form hydrogen bonds. Alcohols form hydrogen bonds with water, and several of the lower-molecular-weight alcohols are **miscible** (soluble in any proportions) with water. Similarly, alcohols are much better solvents than hydrocarbons for polar substances. Significant amounts of ionic compounds such as sodium chloride can dissolve in some of the lower alcohols. We call the hydroxy group **hydrophilic,** meaning "water loving," because of its affinity for water and other polar substances.

The alcohol's alkyl group is called **hydrophobic** ("water hating") because it acts like an alkane: It disrupts the network of hydrogen bonds and the dipole–dipole attractions of a polar solvent such as water. The alkyl group makes the alcohol less

TABLE 10-3
Solubility of Alcohols in Water (at 25 °C)

Alcohol	Solubility in Water
methyl	miscible
ethyl	miscible
n-propyl	miscible
tert-butyl	miscible
isobutyl	10.0%
n-butyl	9.1%
n-pentyl	2.7%
cyclohexyl	3.6%
n-hexyl	0.6%
phenol	9.3%
hexane-1,6-diol	miscible

hydrophilic, yet it lends solubility in nonpolar organic solvents. As a result, most alcohols are miscible with a wide range of nonpolar organic solvents.

Table 10-3 lists the solubility of some simple alcohols in water. The water solubility decreases as the alkyl group becomes larger. Alcohols with one-, two-, or three-carbon alkyl groups are miscible with water. A four-carbon alkyl group is large enough that some isomers are not miscible, yet *tert*-butyl alcohol, with a compact spherical shape, is miscible. In general, each hydroxy group or other hydrogen-bonding group can carry about four carbon atoms into water. Hexan-1-ol, with six carbon atoms, is only slightly soluble in water, but hexane-1,6-diol, with two hydrogen-bonding groups, is miscible with water. Phenol is unusually soluble for a six-carbon alcohol because of its compact shape and the particularly strong hydrogen bonds formed between phenolic —OH groups and water molecules.

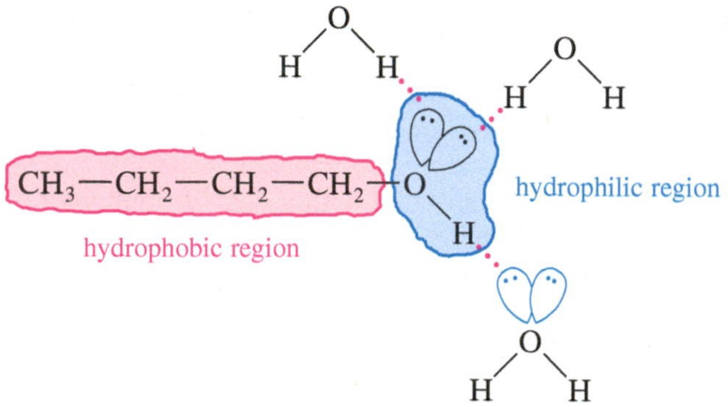

$CH_3—CH_2—CH_2—CH_2—O—H$

hydrophobic region

hydrophilic region

PROBLEM 10-5

Predict which member of each pair will be more soluble in water. Explain the reasons for your answers.

(a) hexan-1-ol or cyclohexanol
(b) heptan-1-ol or 4-methylphenol
(c) 3-ethylhexan-3-ol or octan-2-ol
(d) hexan-2-ol or cyclooctane-1,4-diol
(e) [decalin-diol structure] or [decalin-diol structure]

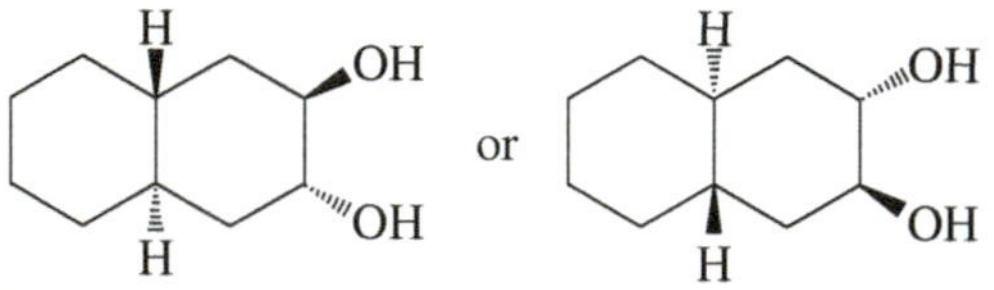

PROBLEM 10-6

Dimethylamine, $(CH_3)_2NH$, has a molecular weight of 45 and a boiling point of 7.4 °C. Trimethylamine, $(CH_3)_3N$, has a higher molecular weight (59) but a *lower* boiling point (3.5 °C). Explain this apparent discrepancy.

10-5 Commercially Important Alcohols

Alcohols are among the most important chemical commodities worldwide. Millions of tons of methanol, ethanol, isopropyl alcohol (propan-2-ol), and ethylene glycol (ethane-1,2-diol) are produced and used each year. These common alcohols are also among the most versatile chemicals, used as fuels, solvents, antiseptics, beverages, and starting materials for chemical syntheses.

10-5A Methanol

Methanol (methyl alcohol) was originally produced by the destructive distillation of wood chips in the absence of air. This source led to the name **wood alcohol.** During Prohibition (1919–1933), when the manufacture of alcoholic beverages was prohibited in the United States, anything called "alcohol" was often used for mixing drinks. Since methanol is more toxic than ethanol, this practice resulted in many cases of blindness and death.

Today, most methanol is synthesized by a catalytic reaction of carbon monoxide with hydrogen. This reaction uses high temperatures and pressures and requires large, complicated industrial reactors.

$$\underset{\text{synthesis gas}}{CO + 2\,H_2} \xrightarrow[\text{CuO–ZnO/Al}_2\text{O}_3]{300\text{–}400\ ^\circ\text{C, }200\text{–}300\text{ atm H}_2} CH_3OH$$

Synthesis gas, containing the hydrogen and carbon monoxide needed to make methanol, can be generated by the partial burning of coal in the presence of water. Careful regulation of the amount of water added allows production of synthesis gas with the correct ratio of carbon monoxide to hydrogen.

$$3\,C + 4\,H_2O \xrightarrow{\text{high temperature}} CO_2 + \underset{\text{synthesis gas}}{2\,CO + 4\,H2}$$

Methanol is one of the most common industrial solvents. It is cheap, it is relatively less toxic (compared with halogenated solvents), and it dissolves a wide variety of polar and nonpolar substances. Methanol is also a starting material for a wide variety of methyl ethers, methyl esters, and other compounds used in plastics, medicines, fuels, and solvents.

Methanol is a good fuel for internal combustion engines. From 1965–2006, all the cars at the Indianapolis 500 used methanol-fueled engines. The switch from gasoline to methanol was driven by a bad fire after a crash in 1964. Methanol is less flammable than gasoline, and water is effective against methanol fires (water mixes with and dilutes methanol). As with any alternative fuel, there are advantages and disadvantages to the use of methanol. Its high octane rating, low pollutant emissions, and lower flammability must be weighed against its lower energy content (smaller ΔH of combustion per gram), requiring 1.7 g of methanol to produce the same energy as 1 g of gasoline. Because of its excellent solvent properties, methanol is hard on rings, seals, and plastic fuel-system parts. Its tendency to burn with little or no visible flame can allow dangerous methanol fires to go undetected.

10-5B Ethanol

The prehistoric discovery of ethanol probably occurred when rotten fruit was consumed and found to have an intoxicating effect. This discovery presumably led to the intentional fermentation of fruit juices. The primitive wine that resulted could be stored (in a sealed container) without danger of decomposition, and it also served as a safe, unpolluted source of water to drink. Many societies, including the ancient Romans, have also used beer as a safe source of drinking water. Wine containers have been found from about 7000 BC, and beer containers from about 5000 BC.

Ethanol can be produced by the fermentation of sugars and starches from many different sources. Grains such as corn, wheat, rye, and barley are common sources, resulting in the name **grain alcohol** for ethanol. Cooking the grain and then adding sprouted barley, called *malt*, converts some of the starches to simpler sugars. Brewer's yeast is then added, and the solution is incubated while the yeast cells convert simple sugars, such as glucose, to ethanol and carbon dioxide.

$$\underset{\text{glucose}}{C_6H_{12}O_6} \xrightarrow{\text{yeast enzymes}} \underset{\text{ethanol}}{2C_2H_5OH} + 2\,CO_2$$

The alcoholic solution that results from fermentation contains only 12–15% alcohol, because yeast cells cannot survive higher concentrations. Distillation increases the alcohol concentration to about 40–50% (80 to 100 "proof") for "hard" liquors.

Experience at Indianapolis has proved that methanol (derived from coal) is an excellent fuel for automotive engines. The racers switched to ethanol as their primary fuel in 2006, when the ethanol industry agreed to supply ethanol free of charge. Since 2013, the racers have used E85, containing 85% ethanol and 15% gasoline.

Application: Biofuels

Government subsidies have encouraged the fermentation of food grains (primarily corn) to produce ethanol for fuel. The major effect has been to increase the price of food grains, while having little or no impact on fuel supplies.

Fermentation is not the most efficient way to produce ethanol, and the growing of corn and converting it to ethanol consumes about as much fuel as it produces. In general, foods are more valuable commodities than fuels, and these "food to fuel" schemes are not viable unless they are subsidized.

The economic values of chemical commodities fall into a pattern, with medicines most valuable and waste products least valuable:

	↑ value (per kg)
medicines	
plastics	
foods	
fuels	
waste products	

A viable process should convert less valuable starting materials into more valuable products. Waste materials, available at little or no expense, can be economically converted to fuels in some cases. Brazil has become independent of foreign oil by converting its sugarcane waste to ethanol for use as a motor fuel.

Distillation of ethanol–water solutions cannot increase the ethanol concentration above 95% because a solution of 95% ethanol and 5% water boils at a lower temperature (78.15 °C) than either pure water (100 °C) or pure ethanol (78.3 °C). Such a mixture of liquids that boils at a lower temperature than either of its components is called a minimum-boiling **azeotrope.**

The 95% alcohol produced by distillation is well suited for use as a solvent and a reagent when traces of water do not affect the reaction. When **absolute alcohol** (100% ethanol) is required, the 95% azeotrope is passed through a dehydrating agent such as anhydrous calcium oxide (CaO), which removes the final 5% of water.

Since World War II, most industrial ethanol has been synthesized directly by the catalyzed high-temperature, high-pressure, gas-phase reaction of water with ethylene. This process uses catalysts such as P_2O_5, tungsten oxide, or various specially treated clays.

$$H_2C{=}CH_2 \quad + \quad H_2O \quad \xrightarrow[\text{catalyst}]{\text{100–300 atm, 300 °C}} \quad CH_3{-}CH_2{-}OH$$

Like methanol, ethanol is an excellent solvent of low toxicity that is cheap to produce. Unfortunately, the liquor tax makes ethanol relatively expensive. Use of untaxed ethanol is possible, but it requires extensive record keeping and purchase of a special license. **Denatured alcohol** is ethanol that contains impurities that make it undrinkable. Denatured ethanol is untaxed, but the impurities (methanol, methyl isobutyl ketone, aviation gasoline, etc.) also make it unsuitable for many laboratory uses.

Like methanol, ethanol is a good motor fuel, with similar advantages and disadvantages. A car's carburetor must be adjusted (for a richer mixture) and fitted with alcohol-resistant seals if it is to run on pure ethanol. Solutions of about 10% ethanol in gasoline ("gasohol") work well without any adjustments, however.

Application: Toxicity

Everything is toxic in large enough amounts—even water. Toxicity depends on amounts and concentrations. The important questions are "How much?" and "Is that much harmful?"

Many people imagine ethanol to be nontoxic, and methanol to be horribly toxic. Actually, methanol is about twice as toxic as ethanol: Typical fatal doses for adults are about 100 mL of methanol or about 200 mL of ethanol, although smaller doses of methanol may damage the optic nerve. Many people die each year from underestimating ethanol's toxicity. In the lab, we would never ingest even a tiny fraction of these amounts. Therefore, we consider these solvents to be relatively nontoxic compared with truly hazardous solvents such as benzene and chloroform.

10-5C Propan-2-ol

Application: Antiseptics

Both ethanol and isopropyl alcohol are effective topical antiseptics. Ethanol is also an ingredient in many mouthwashes. These alcohols kill the microorganisms on the wound surface and in the mouth, but their low toxicity does not kill the cells of the skin or mouth tissues.

Propan-2-ol (2-propanol, isopropyl alcohol) is made by the catalytic hydration of propylene. Isopropyl alcohol is commonly used as **rubbing alcohol** (rather than ethanol) because it has less of a drying effect on the skin, and it is not regulated and taxed by the government. Rubbing alcohol usually contains about 70% isopropyl alcohol and 30% water. People have used rubbing alcohol as a safe and effective antiseptic since the early 1920s. It is about as toxic as methanol when taken orally, but it is safer for use on the skin because it does not pass through skin as easily as methanol.

$$\underset{\text{propylene}}{CH_3{-}CH{=}CH_2} \quad + \quad H_2O \quad \xrightarrow[\text{catalyst}]{\text{100–300 atm, 300 °C}} \quad \underset{\text{propan-2-ol}}{CH_3{-}\underset{\substack{|\\ OH}}{CH}{-}CH_3}$$

10-6 Acidity of Alcohols and Phenols

Like the hydroxy proton of water, the hydroxy proton of an alcohol is weakly acidic. A strong base can remove the hydroxy proton to give an **alkoxide ion.**

$$\underset{\text{alcohol}}{R-\ddot{\underset{\cdot\cdot}{O}}-H} + B:^- \rightleftharpoons \underset{\text{alkoxide ion}}{R-\ddot{\underset{\cdot\cdot}{O}}:^-} + B-H$$

Example

$$\underset{\text{ethanol}}{CH_3CH_2-\ddot{\underset{\cdot\cdot}{O}}-H} + B:^- \rightleftharpoons \underset{\text{ethoxide ion}}{CH_3CH_2-\ddot{\underset{\cdot\cdot}{O}}:^-} + B-H$$

The acidities of alcohols vary widely, from alcohols that are about as acidic as water to some that are much less acidic. The acid-dissociation constant, K_a, of an alcohol is defined by the equilibrium

$$R-O-H + H_2O \overset{K_a}{\rightleftharpoons} R-O^- + H_3O^+$$

$$K_a = \frac{[H_3O^+][RO^-]}{[ROH]} \qquad pK_a = -\log(K_a)$$

Table 10-4 compares the acid-dissociation constants of some alcohols with those of water and other acids.

10-6A Effects on Acidity

The acid-dissociation constants for alcohols vary according to their structure, from about 10^{-16} for methanol down to about 10^{-18} for most tertiary alcohols. The acidity decreases as the substitution on the alkyl group increases, because a more highly substituted alkyl group inhibits solvation of the alkoxide ion, decreasing the stability of the alkoxide ion and driving the dissociation equilibrium toward the left.

Table 10-4 shows that substitution by electron-withdrawing halogen atoms enhances the acidity of alcohols. For example, 2-chloroethanol is more acidic than ethanol because the electron-withdrawing chlorine atom helps to stabilize the 2-chloroethoxide ion.

$$\underset{\text{ethanol}}{CH_3-CH_2-OH} + H_2O \rightleftharpoons \underset{\substack{\text{ethoxide ion} \\ \text{(less stable)}}}{CH_3-CH_2-O^-} + H_3O^+ \qquad K_a = 1.3 \times 10^{-16}$$

$$\underset{\text{2-chloroethanol}}{Cl-CH_2-CH_2-OH} + H_2O \rightleftharpoons \underset{\substack{\text{2-chloroethoxide ion} \\ \text{(stabilized by Cl)}}}{Cl-CH_2-CH_2-O^-} + H_3O^+ \qquad K_a = 5.0 \times 10^{-15}$$

TABLE 10-4
Acid-Dissociation Constants of Representative Alcohols

Alcohol	Structure	K_a	pK_a
methanol	CH_3-OH	3.2×10^{-16}	15.5
ethanol	CH_3CH_2-OH	1.3×10^{-16}	15.9
2-chloroethanol	$Cl-CH_2CH_2-OH$	5.0×10^{-15}	14.3
2,2,2-trichloroethanol	Cl_3C-CH_2-OH	6.3×10^{-13}	12.2
isopropyl alcohol	$(CH_3)_2CH-OH$	3.2×10^{-17}	16.5
tert-butyl alcohol	$(CH_3)_3C-OH$	1.0×10^{-18}	18.0
cyclohexanol	$C_6H_{11}-OH$	1.0×10^{-18}	18.0
phenol	C_6H_5-OH	1.0×10^{-10}	10.0
	Comparison with Other Acids		
water	H_2O	1.8×10^{-16}	15.7
acetic acid	CH_3COOH	1.6×10^{-5}	4.74
hydrochloric acid	HCl	$1 \times 10^{+7}$	−7

PROBLEM 10-7

Predict which member of each pair will be more acidic. Explain your answers.
(a) methanol or *tert*-butyl alcohol
(b) 2-chloropropan-1-ol or 3-chloropropan-1-ol
(c) 2-chloroethanol or 2,2-dichloroethanol
(d) 2,2-dichloropropan-1-ol or 2,2-difluoropropan-1-ol

PROBLEM 10-8

Without looking them up, rank the following compounds in decreasing order of acidity. These examples represent large classes of compounds that differ widely in acidity.

water, ethanol, 2-chloroethanol, *tert*-butyl alcohol, ammonia, sulfuric acid, hexane, hex-1-yne, acetic acid

10-6B Formation of Sodium and Potassium Alkoxides

Alkoxide ions are strong nucleophiles and strong bases, and we have already seen many of their useful reactions. When an alkoxide ion is needed in a synthesis, it is often formed by the reaction of sodium or potassium metal with the alcohol. This is an oxidation–reduction, with the metal being oxidized and the hydrogen ion reduced to form hydrogen gas. Hydrogen bubbles out of the solution, leaving the sodium or potassium salt of the alkoxide ion.

$$\text{R—O—H} + \text{Na} \longrightarrow \text{R—O}^-\,\text{Na}^+ + \tfrac{1}{2}\,\text{H}_2\uparrow$$

Example

$$\underset{\text{ethanol}}{\text{CH}_3\text{CH}_2\text{OH}} + \underset{\text{sodium metal}}{\text{Na}} \longrightarrow \underset{\text{sodium ethoxide}}{\text{CH}_3\text{CH}_2\text{O}^-\,\text{Na}^+} + \underset{\text{hydrogen gas}}{\tfrac{1}{2}\,\text{H}_2\uparrow}$$

The more acidic alcohols, such as methanol and ethanol, react rapidly with sodium to form sodium methoxide and sodium ethoxide. Secondary alcohols, such as propan-2-ol, react more slowly. Tertiary alcohols, such as *tert*-butyl alcohol, react very slowly with sodium. Potassium is often used with secondary and tertiary alcohols because it is more reactive than sodium, and the reaction can be completed in a convenient amount of time.

$$\underset{\textit{tert}\text{-butyl alcohol}}{(\text{CH}_3)_3\text{C—OH}} + \underset{\text{potassium metal}}{\text{K}} \longrightarrow \underset{\text{potassium }\textit{tert}\text{-butoxide}}{(\text{CH}_3)_3\text{C—O}^-\text{K}^+} + \tfrac{1}{2}\,\text{H}_2\uparrow$$

Some alcohols react slowly with both sodium and potassium. In these cases, a useful alternative is sodium hydride, often in tetrahydrofuran (THF) solution. Sodium hydride reacts quickly to form the alkoxide, even with difficult compounds.

$$\underset{\text{alcohol}}{\text{R—}\ddot{\underset{..}{\text{O}}}\text{—H}} + \underset{\text{sodium hydride}}{\text{NaH}} \xrightarrow[\text{THF}]{} \underset{\text{sodium alkoxide}}{\text{R—}\ddot{\underset{..}{\text{O}}}\text{:}^-\ \ \text{Na}^+} + \underset{\text{hydrogen}}{\text{H}_2\uparrow}$$

10-6C Acidity of Phenols

We might expect that phenol would have about the same acidity as cyclohexanol, since their structures are similar. This prediction is wrong: Phenol is nearly 100 million (10^8) times more acidic than cyclohexanol.

$$H_2O + \text{cyclohexanol} \underset{}{\overset{K_a = 10^{-18}}{\rightleftharpoons}} \text{alkoxide ion} + H_3O^+$$

$$H_2O + \text{phenol} \overset{K_a = 10^{-10}}{\rightleftharpoons} \text{phenoxide ion (phenolate ion)} + H_3O^+$$

Application: Antiseptic

British surgeon Joseph Lister (1827–1912) used dilute solutions of phenol to kill germs in wounds. He found that antiseptic surgery using phenol reduced the mortality rate from amputations by nearly 50%. Within the next decade, Lister's phenol-based antiseptic techniques revolutionized surgery.

Cyclohexanol is a typical secondary alcohol, with a typical acid-dissociation constant for an alcohol. There must be something special about phenol that makes it unusually acidic. The phenoxide ion is more stable than a typical alkoxide ion because the negative charge is not confined to the oxygen atom but is delocalized over the oxygen and three carbon atoms of the ring.

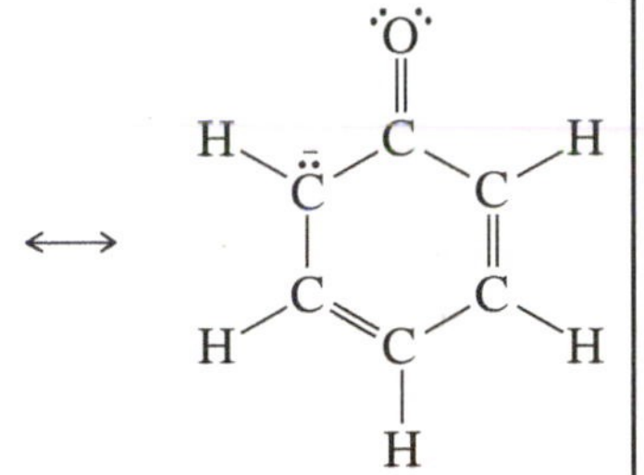

A large part of the negative charge in the resonance hybrid still resides on the oxygen atom, since it is the most electronegative of the four atoms sharing the charge. But the ability to spread the negative charge over four atoms rather than concentrating it on just one atom produces a more stable ion. The reaction of phenol with sodium hydroxide is exothermic, and the following equilibrium lies to the right.

$$\text{phenol, p}K_a = 10.0 \quad + \quad Na^+ \; ^-{:}\ddot{O}H \rightleftharpoons \text{sodium phenoxide} \quad + \quad H_2O \;\; \text{p}K_a = 15.7$$

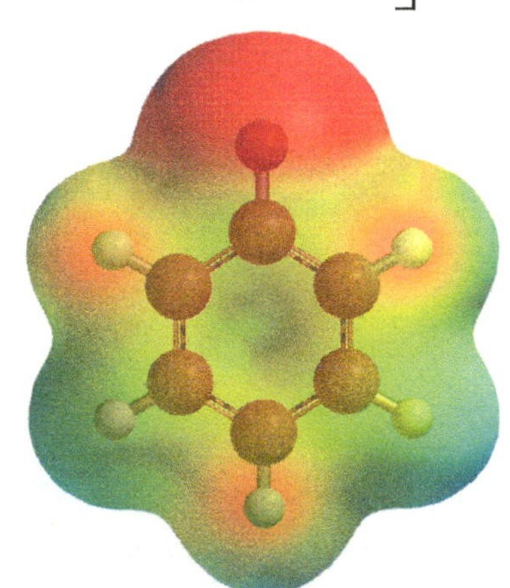

EPM of phenoxide ion

Phenoxide anions are prepared simply by adding the phenol to an aqueous solution of sodium hydroxide or potassium hydroxide. There is no need to use sodium or potassium metal. Phenol was once called *carbolic acid* because of its ability to neutralize common bases.

Application: Biochemistry

Phenol is commonly used to separate cellular DNA from proteins, one step in the process that leads to the cloning of a gene. An aqueous phenol solution dissolves DNA, but it complexes with the proteins and causes them to precipitate. Centrifugation removes the precipitated proteins. Finally, addition of ethanol causes the purified DNA to precipitate so that it can be collected.

PROBLEM 10-9

A nitro group ($-NO_2$) effectively stabilizes a negative charge on an adjacent carbon atom through resonance:

$$R_2\ddot{C}^-{-}\overset{+}{N}(=O){-}O^- \;\longleftrightarrow\; R_2\ddot{C}^-{-}\overset{+}{N}(-O^-)=O \;\longleftrightarrow\; R_2C=\overset{+}{N}(-O^-)_2$$

minor minor major

Two of the following nitrophenols are much more acidic than phenol itself. The third compound is only slightly more acidic than phenol. Use resonance structures of the appropriate phenoxide ions to show why two of these anions should be unusually stable.

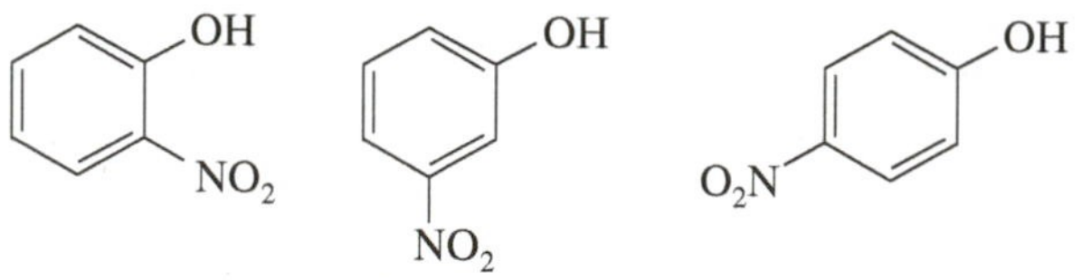

OH, NO_2 — 2-nitrophenol; OH, NO_2 — 3-nitrophenol; OH, O_2N — 4-nitrophenol

PROBLEM 10-10

The following compounds are only slightly soluble in water, but one of them is very soluble in a dilute aqueous solution of sodium hydroxide. The other is still only slightly soluble.

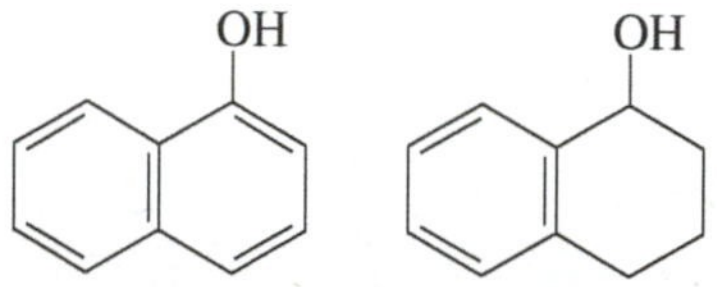

OH OH

(a) Explain the difference in solubility of these compounds in dilute sodium hydroxide.
(b) Show how this difference might be exploited to separate a mixture of these two compounds using a separatory funnel.

10-7 Synthesis of Alcohols: Introduction and Review

One of the reasons alcohols are important synthetic intermediates is that they can be synthesized directly from a wide variety of other functional groups. In Chapters 6 and 8, we examined the conversion of alkyl halides to alcohols by substitution and the conversion of alkenes to alcohols by hydration, hydroboration, and dihydroxylation. These reactions are summarized here, with references for review if needed.

Following this review, we will consider the largest and most versatile group of alcohol syntheses: nucleophilic additions to carbonyl compounds.

SUMMARY Previous Alcohol Syntheses

Nucleophilic Substitution on an Alkyl Halide **(Chapter 6)**
Usually via the S_N2 mechanism; competes with elimination.

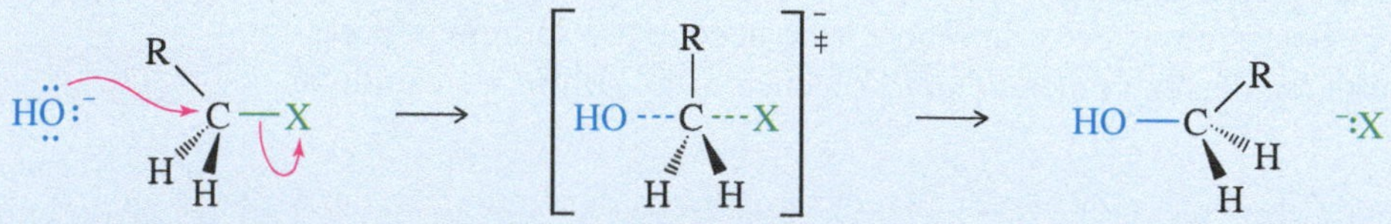

$HO:^- + RCH_2-X \longrightarrow [HO\cdots C\cdots X]^{\ddagger-} \longrightarrow HO-CH_2R + {}^-:X$

transition state

Synthesis of Alcohols from Alkenes **(Chapter 8)**

1. Acid-catalyzed hydration (Section 8-4)

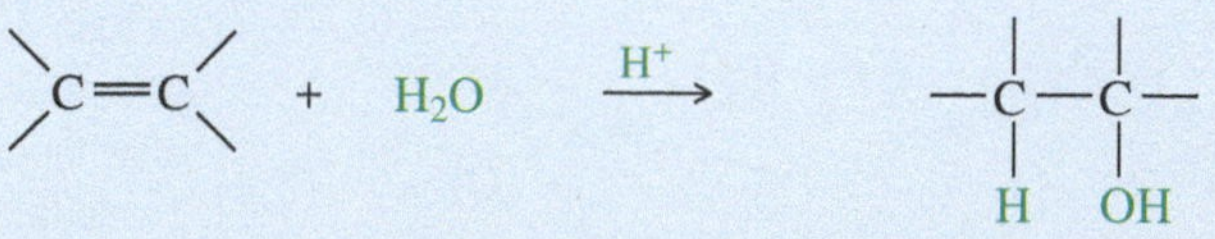

$$>C=C< \; + \; H_2O \xrightarrow{H^+} -\overset{|}{\underset{H}{C}}-\overset{|}{\underset{OH}{C}}-$$

Markovnikov orientation

2. *Oxymercuration–demercuration* (Section 8-5)

$$\mathrm{C{=}C} + \mathrm{Hg(OAc)_2} \xrightarrow{H_2O} \mathrm{-C(HgOAc)-C(OH)-} \xrightarrow{NaBH_4} \mathrm{-C(H)-C(OH)-}$$

Markovnikov orientation

3. *Hydroboration–oxidation* (Section 8-7)

$$\mathrm{C{=}C} \xrightarrow[(2)\ H_2O_2,\ NaOH]{(1)\ BH_3\cdot THF} \mathrm{-C(H)-C(OH)-}$$

syn addition, anti-Markovnikov orientation

4. *Dihydroxylation: synthesis of 1,2-diols from alkenes* (Sections 8-13 and 8-14)

Syn Dihydroxylation

$$\mathrm{C{=}C} \xrightarrow[\text{or } KMnO_4,\ ^-OH \text{ (cold, dilute)}]{OsO_4,\ H_2O_2} \mathrm{-C(HO)-C(OH)-}$$

syn dihydroxylation

Anti Dihydroxylation

$$\mathrm{C{=}C} \xrightarrow{R-C(=O)-OOH,\ H_3O^+} \mathrm{-C(OH)-C(OH)-}$$

anti dihydroxylation

5. *Addition of acetylides to carbonyl compounds* (Section 9-7)

$$\underset{\text{acetylide}}{R-C\equiv C:^-} + \underset{\text{ketone or aldehyde}}{R'_2C{=}\ddot{O}:} \longrightarrow \underset{\text{alkoxide}}{R-C\equiv C-CR'_2-\ddot{O}:^-} \xrightarrow{H_3O^+} \underset{\text{acetylenic alcohol}}{R-C\equiv C-CR'_2-OH}$$

10-8 Organometallic Reagents for Alcohol Synthesis

Organometallic compounds contain covalent bonds between carbon atoms and metal atoms. Organometallic reagents are useful because they have nucleophilic carbon atoms, in contrast to the electrophilic carbon atoms of alkyl halides. Most metals (M) are more electropositive than carbon, and the C—M bond is polarized with a partial positive charge on the metal and a partial negative charge on carbon. The following partial periodic table shows the electronegativities of some metals used in making organometallic compounds.

Electronegativities

Li	1.0					C	2.5
Na	0.9	Mg	1.3	Al	1.6		
K	0.8						

C—M bond

$\overset{\delta^-}{C}-\overset{\delta^+}{Li}$

$\overset{\delta^-}{C}-\overset{\delta^+}{Mg}$

We have already encountered one type of organometallic compound with a negative charge on carbon: sodium acetylides, covered in Section 9-7. Terminal alkynes are weakly acidic, and they are converted to sodium acetylides by treatment with an unusually strong base, sodium amide. These sodium acetylides are useful nucleophiles, reacting with alkyl halides and carbonyl compounds to form new carbon–carbon bonds.

$$\underset{\text{terminal alkyne}}{R{-}C{\equiv}C{-}H} + \underset{\text{sodium amide}}{NaNH_2} \longrightarrow \underset{\text{a sodium acetylide}}{R{-}C{\equiv}C{:}^-\ Na^+} + \underset{\text{ammonia}}{NH_3}$$

$$\underset{\text{acetylide}}{R{-}C{\equiv}C{:}^-} + \underset{\text{alkyl halide}}{R'{-}CH_2{-}X} \longrightarrow \underset{\text{substituted alkyne}}{R{-}C{\equiv}C{-}CH_2{-}R'} + X^-$$

$$\underset{\text{acetylide}}{R{-}C{\equiv}C{:}^-} + \underset{\text{ketone or aldehyde}}{R'_2C{=}\ddot{O}{:}} \longrightarrow \underset{\text{alkoxide}}{R{-}C{\equiv}C{-}CR'_2{-}\ddot{O}{:}^-} \xrightarrow{H_3O^+} \underset{\text{acetylenic alcohol}}{R{-}C{\equiv}C{-}CR'_2{-}OH}$$

Most alkyl and alkenyl groups are not acidic enough to be deprotonated by sodium amide, but they can be made into Grignard reagents and organolithium reagents. These reagents are extremely versatile, providing some of our best ways of forming carbon–carbon bonds.

10-8A Grignard Reagents

PROBLEM-SOLVING HINT

Grignard reactions are some of the most valuable tools ever discovered for use in synthetic organic chemistry. We will use Grignard reactions often to form carbon–carbon bonds in many different types of syntheses.

Organometallic compounds of lithium and magnesium are frequently used for the synthesis of alcohols. The organomagnesium halides, of empirical formula R—Mg—X, are called **Grignard reagents** in honor of the French chemist Victor Grignard, who discovered their utility around 1905 and received the Nobel Prize in Chemistry in 1912. Grignard reagents result from the reaction of an alkyl halide with magnesium metal. This reaction is always carried out in a dry (anhydrous) ether solvent, which is needed to solvate and stabilize the Grignard reagent as it forms. Although we write the Grignard reagent as R—Mg—X, the actual species in solution usually contains two, three, or four of these units associated together with several molecules of the ether solvent. Diethyl ether, $CH_3CH_2{-}O{-}CH_2CH_3$, is the most common solvent for these reactions, although other ethers are also used.

$$\underset{(X = Cl,\ Br,\ or\ I)}{R{-}X} + Mg \xrightarrow{CH_3CH_2OCH_2CH_3} \underset{\substack{\text{organomagnesium halide} \\ \text{(Grignard reagent)}}}{\overset{\delta-}{R}{-}\overset{\delta+}{Mg}{-}X} \qquad \text{reacts like } R{:}^-\ \overset{+}{Mg}X$$

Grignard reagents may be made from primary, secondary, and tertiary alkyl halides, as well as from vinyl and aryl halides. Alkyl iodides are the most reactive halides, followed by bromides and chlorides. Alkyl fluorides generally do not react.

reactivity: $R{-}I > R{-}Br > R{-}Cl \gg R{-}F$

The following reactions show the formation of some typical Grignard reagents.

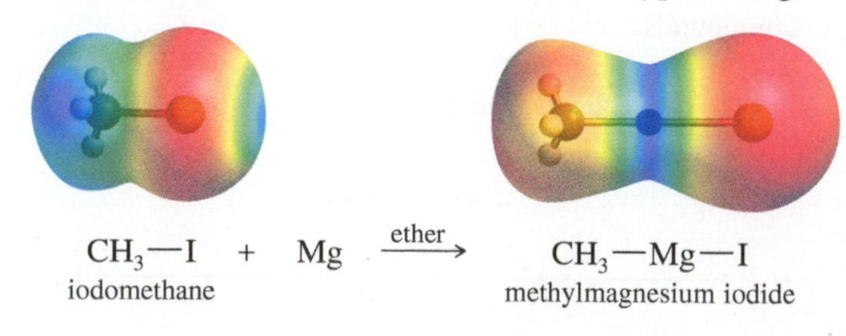

$$\underset{\text{iodomethane}}{CH_3{-}I} + Mg \xrightarrow{\text{ether}} \underset{\text{methylmagnesium iodide}}{CH_3{-}Mg{-}I}$$

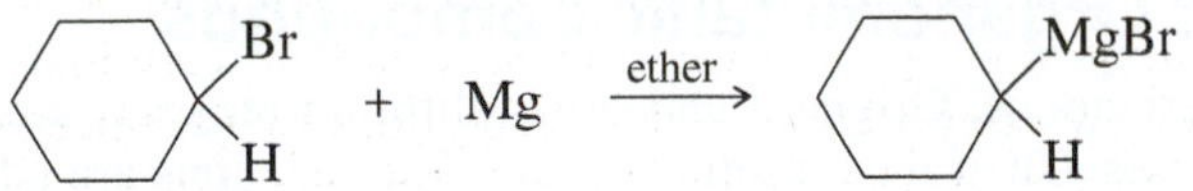

$$\text{bromocyclohexane} + Mg \xrightarrow{\text{ether}} \text{cyclohexylmagnesium bromide}$$

bromocyclohexane — cyclohexylmagnesium bromide

$$H_2C{=}CH{-}CH_2{-}Br + Mg \xrightarrow{\text{ether}} H_2C{=}CH{-}CH_2{-}MgBr$$

allyl bromide — allylmagnesium bromide

10-8B Organolithium Reagents

Like magnesium, lithium reacts with alkyl halides, vinyl halides, and aryl halides to form organometallic compounds. Ether is not necessary for this reaction. **Organolithium reagents** are made and used in a wide variety of solvents, including alkanes.

$$R{-}X + 2\,Li \longrightarrow Li^{+}\,{}^{-}X + R{-}Li \quad \text{reacts like } R{:}^{-}\,Li^{+}$$

(X = Cl, Br, or I) — organolithium

Examples

$$CH_3CH_2CH_2CH_2{-}Br + 2\,Li \xrightarrow{\text{hexane}} CH_3CH_2CH_2CH_2{-}Li + LiBr$$

n-butyl bromide — *n*-butyllithium

$$H_2C{=}CH{-}Cl + 2\,Li \xrightarrow{\text{pentane}} H_2C{=}CH{-}Li + LiCl$$

vinyl chloride — vinyllithium

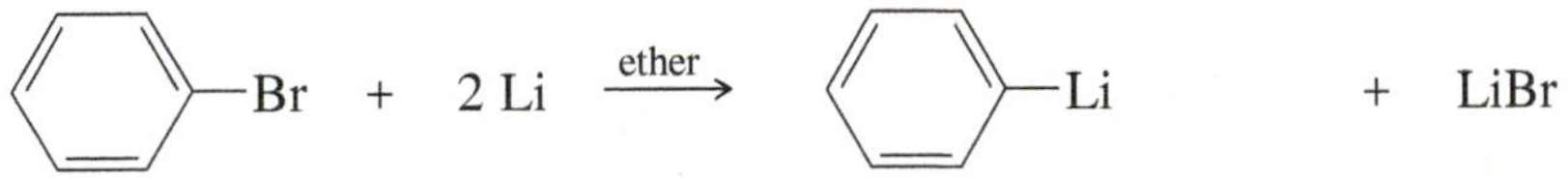

$$C_6H_5{-}Br + 2\,Li \xrightarrow{\text{ether}} C_6H_5{-}Li + LiBr$$

bromobenzene — phenyllithium

The electrostatic potential map (EPM) of methyllithium is shown at right. The blue (electron-poor) color of the metal results from its partial positive charge, and the red (electron-rich) color of the methyl group shows its partial negative charge.

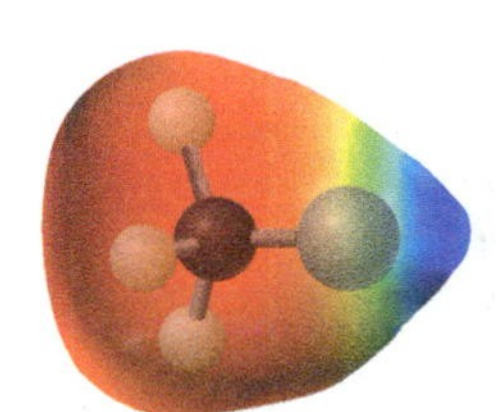

EPM of CH_3Li

PROBLEM 10-11

Which of the following compounds are suitable solvents for Grignard reactions?

(a) *n*-hexane **(b)** $CH_3{-}O{-}CH_3$ **(c)** $CHCl_3$
(d) cyclohexane **(e)** benzene **(f)** $CH_3OCH_2CH_2OCH_3$

(g) 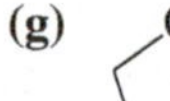

THF (tetrahydrofuran)

(h) 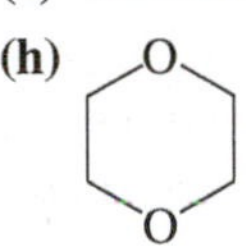
1,4-dioxane

PROBLEM 10-12

Predict the products of the following reactions.

(a) $CH_3CH_2Br + Mg \xrightarrow{\text{ether}}$

(b) isobutyl iodide + 2Li $\xrightarrow{\text{hexane}}$

(c) 1-bromo-4-fluorocyclohexane + Mg $\xrightarrow{\text{THF}}$

(d) $CH_2{=}CCl{-}CH_2CH_3 + 2\,Li \xrightarrow{\text{ether}}$

10-9 Reactions of Organometallic Compounds

Because they resemble carbanions, Grignard and organolithium reagents are strong nucleophiles and strong bases. Their most useful nucleophilic reactions are additions to carbonyl (C═O) groups, much like we saw with acetylide ions (Section 9-7B). The carbonyl group is polarized, with a partial positive charge on carbon and a partial negative charge on oxygen. The positively charged carbon is electrophilic; attack by a nucleophile places a negative charge on the electronegative oxygen atom.

$$R:^- + \overset{\delta+}{C}{=}\overset{\delta-}{\ddot{O}}: \longrightarrow R{-}\underset{|}{\overset{|}{C}}{-}\ddot{\underset{..}{O}}:^-$$

The product of this nucleophilic attack is an alkoxide ion, a strong base. Addition of water or a dilute acid in a second step protonates the alkoxide to give the alcohol.

$$\underset{\text{alkoxide}}{R{-}\underset{|}{\overset{|}{C}}{-}\ddot{\underset{..}{O}}:^-} \quad \underset{(\text{or } H_3O^+)}{H{-}\ddot{\underset{..}{O}}{-}H} \longrightarrow R{-}\underset{|}{\overset{|}{C}}{-}\ddot{O}H \; + \; ^-:\ddot{\underset{..}{O}}H$$

Either a Grignard reagent or an organolithium reagent can serve as the nucleophile in this addition to a carbonyl group. The following discussions refer to Grignard reagents, but they also apply to organolithium reagents. Key Mechanism 10-1 shows that the Grignard reagent first adds to the carbonyl group to form an alkoxide ion. Addition of dilute acid (in a separate step) protonates the alkoxide to give the alcohol.

We are interested primarily in the reactions of Grignard reagents with ketones and aldehydes. **Ketones** are compounds with two alkyl groups bonded to a carbonyl group. **Aldehydes** have one alkyl group and one hydrogen atom bonded to the carbonyl group. **Formaldehyde** has two hydrogen atoms bonded to the carbonyl group.

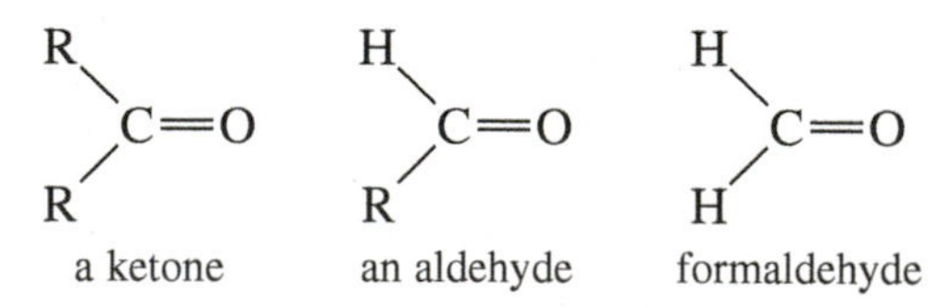

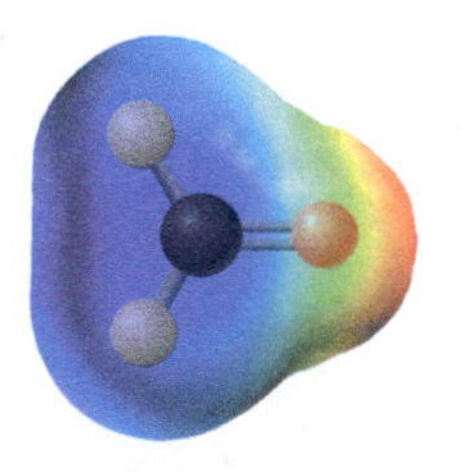

EPM of formaldehyde

The electrostatic potential map (EPM) of formaldehyde shows the polarization of the carbonyl group, with an electron-rich (red) region around oxygen and an electron-poor (blue) region near carbon.

KEY MECHANISM 10-1 Grignard Reactions

Grignard and organolithium reagents provide some of the best methods for assembling a carbon skeleton. These strong nucleophiles add to ketones and aldehydes to give alkoxide ions, which are protonated to give alcohols.

Formation of the Grignard reagent: Magnesium reacts with an alkyl halide in an anhydrous ether solution.

$$R'{-}X \; + \; Mg \xrightarrow[\text{ether}]{} R'{-}MgX$$

Reaction 1: The Grignard reagent attacks a carbonyl compound to form an alkoxide salt.

$$\overset{\delta-}{R'}{-}\overset{\delta+}{MgX} \; + \; R_2C{=}\ddot{O}: \xrightarrow[\text{ether}]{} R'{-}\underset{R}{\overset{R}{C}}{-}\ddot{\underset{..}{O}}:^- \; ^+MgX$$

magnesium alkoxide salt

Reaction 2: After the first reaction is complete, water or dilute acid is added to protonate the alkoxide and give the alcohol.

$$R'\!-\!CR_2\!-\!\ddot{O}:^{-}\ {}^{+}MgX \xrightarrow{H-\overset{+}{O}H-H} R'\!-\!CR_2\!-\!\ddot{O}\!-\!H + XMgOH$$

magnesium alkoxide salt — alcohol

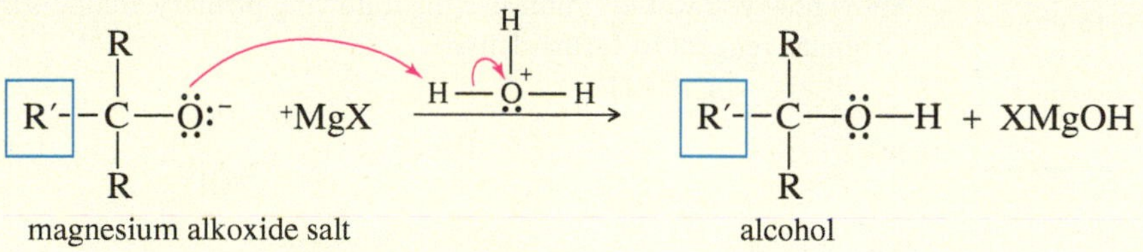

EXAMPLE: Addition of phenylmagnesium bromide to acetone

Formation of the Grignard reagent: Magnesium reacts with bromobenzene in an ether solution to give phenylmagnesium bromide.

$$C_6H_5\!-\!Br + Mg \xrightarrow{\text{ether}} C_6H_5\!-\!MgBr$$

phenylmagnesium bromide

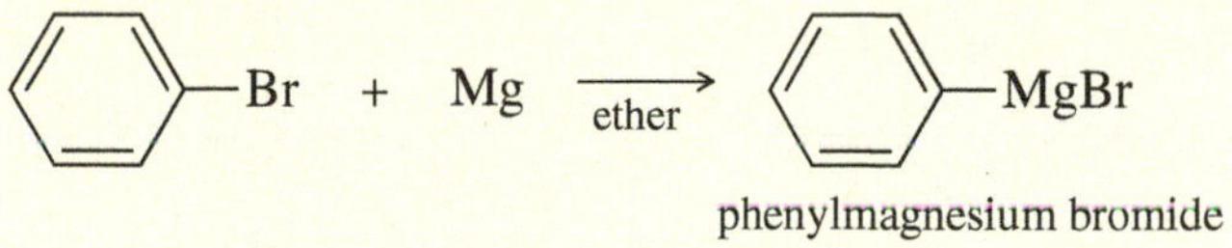

Reaction 1: The Grignard reagent attacks a carbonyl compound to form an alkoxide salt.

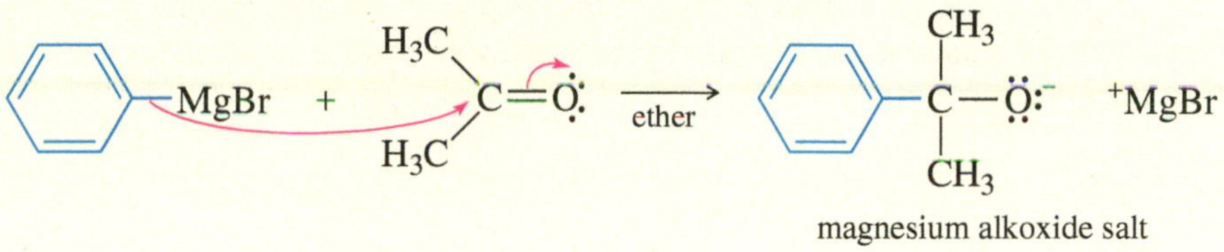

$$C_6H_5\!-\!MgBr + (H_3C)_2C\!=\!\ddot{O}: \xrightarrow{\text{ether}} C_6H_5\!-\!C(CH_3)_2\!-\!\ddot{O}:^{-}\ {}^{+}MgBr$$

magnesium alkoxide salt

Reaction 2: After the first reaction is complete, water or dilute acid is added to protonate the alkoxide and give the alcohol.

$$C_6H_5\!-\!C(CH_3)_2\!-\!\ddot{O}:^{-}\ {}^{+}MgBr \xrightarrow{H-\overset{+}{O}H-H} C_6H_5\!-\!C(CH_3)_2\!-\!OH + BrMgOH$$

magnesium alkoxide salt — 2-phenylpropan-2-ol

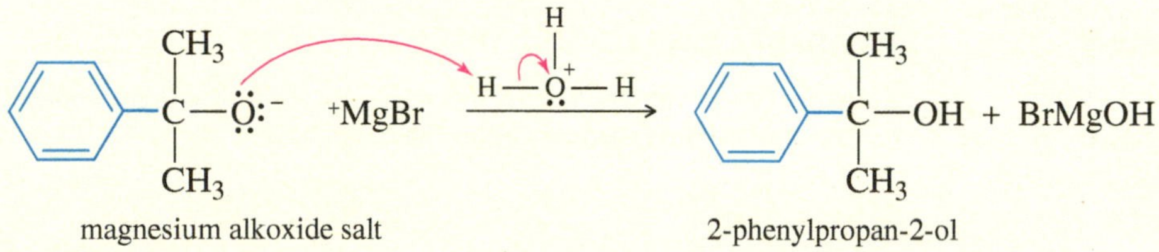

QUESTION: What would be the result if water were accidentally added in the first reaction with the Grignard reagent and the carbonyl compound?

10-9A Addition to Formaldehyde: Formation of Primary Alcohols

Addition of a Grignard reagent to formaldehyde, followed by protonation, gives a primary alcohol with one more carbon atom than in the Grignard reagent.

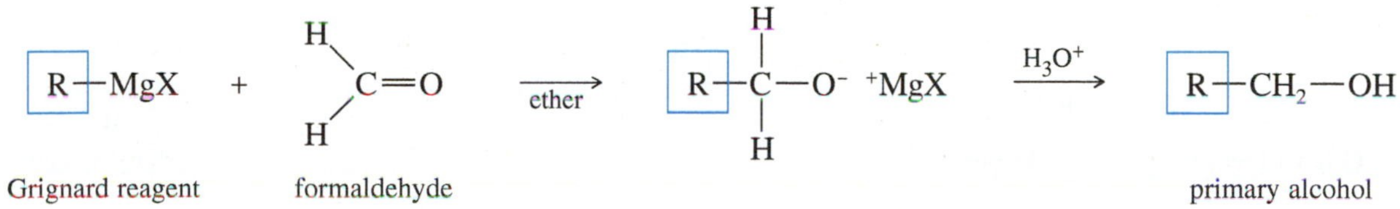

$$R\!-\!MgX + H_2C\!=\!O \xrightarrow{\text{ether}} R\!-\!CH_2\!-\!O^{-}\ {}^{+}MgX \xrightarrow{H_3O^{+}} R\!-\!CH_2\!-\!OH$$

Grignard reagent — formaldehyde — primary alcohol

For example,

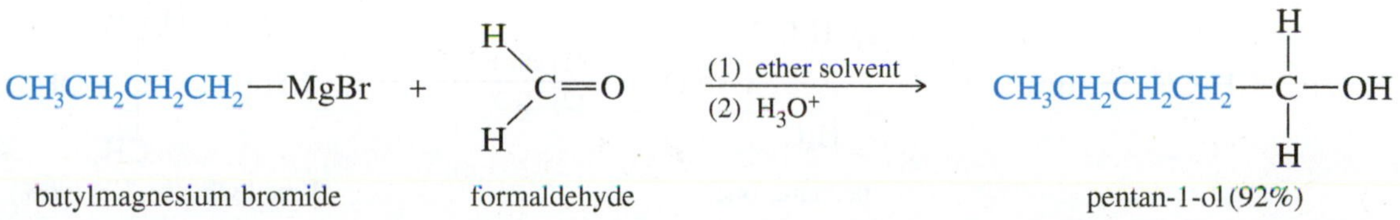

$$CH_3CH_2CH_2CH_2\!-\!MgBr + H_2C\!=\!O \xrightarrow[(2)\ H_3O^{+}]{(1)\ \text{ether solvent}} CH_3CH_2CH_2CH_2\!-\!CH_2\!-\!OH$$

butylmagnesium bromide — formaldehyde — pentan-1-ol (92%)

PROBLEM-SOLVING HINT

Note the use of $\xrightarrow[(2)]{(1)}$ to show separate reactions with one reaction arrow.

PROBLEM 10-13

Show how you would synthesize the following primary alcohols by adding an appropriate Grignard reagent to formaldehyde.

(a) cyclohexyl–CH_2OH (b) 3-methylbutan-1-ol (skeletal, OH) (c) cyclopentenyl–CH_2OH

10-9B Addition to Aldehydes: Formation of Secondary Alcohols

Grignard reagents add to aldehydes to give, after protonation, secondary alcohols.

$$\boxed{R}\!-\!MgX + R'(H)C{=}O \xrightarrow{\text{ether}} \boxed{R}\!-\!C(R')(H)\!-\!O^- \; ^+MgX \xrightarrow{H_3O^+} \boxed{R}\!-\!C(R')(H)\!-\!OH$$

Grignard reagent aldehyde secondary alcohol

The two alkyl groups of the secondary alcohol are the alkyl group from the Grignard reagent and the alkyl group that was bonded to the carbonyl group of the aldehyde.

$$CH_3CH_2{-}MgBr + H_3C(H)C{=}O \xrightarrow{\text{ether}} CH_3{-}CH_2{-}C(CH_3)(H){-}O^- \; ^+MgBr \xrightarrow{H_3O^+} CH_3CH_2{-}C(CH_3)(H){-}OH$$

acetaldehyde butan-2-ol (85%)

PROBLEM-SOLVING HINT

A secondary alcohol has two groups on the carbinol carbon atom. Consider two possible reactions, with either group added as the Grignard reagent.

PROBLEM 10-14

Show two ways you could synthesize each of the following secondary alcohols by adding an appropriate Grignard reagent to an aldehyde.

(a) (skeletal structure, OH) (b) (phenyl, OH) (c) (cyclohexyl, H, C, OH, vinyl)

10-9C Addition to Ketones: Formation of Tertiary Alcohols

A ketone has two alkyl groups bonded to its carbonyl carbon atom. Addition of a Grignard reagent, followed by protonation, gives a tertiary alcohol, with three alkyl groups bonded to the carbinol carbon atom.

$$\boxed{R}\!-\!MgX + R'(R'')C{=}O \xrightarrow{\text{ether}} \boxed{R}\!-\!C(R')(R'')\!-\!O^- \; ^+MgX \xrightarrow{H_3O^+} \boxed{R}\!-\!C(R')(R'')\!-\!OH$$

Grignard reagent ketone tertiary alcohol

Two of the alkyl groups are the two originally bonded to the ketone carbonyl group. The third alkyl group comes from the Grignard reagent.

$$CH_3CH_2{-}MgBr + CH_3CH_2CH_2(H_3C)C{=}O \xrightarrow[(2)\ H_3O^+]{(1)\ \text{ether solvent}} CH_3CH_2{-}C(CH_2CH_2CH_3)(CH_3){-}OH$$

pentan-2-one 3-methylhexan-3-ol (90%)

SOLVED PROBLEM 10-2

Show how you would synthesize the following alcohol from compounds containing no more than five carbon atoms.

$$\text{cyclopentyl}-\underset{\substack{|\\ CH_2CH_3}}{\overset{\substack{CH_3\\ |}}{C}}-OH$$

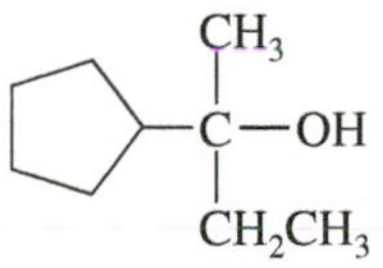

SOLUTION

This is a tertiary alcohol; any one of the three alkyl groups might be added in the form of a Grignard reagent. We can propose three combinations of Grignard reagents with ketones:

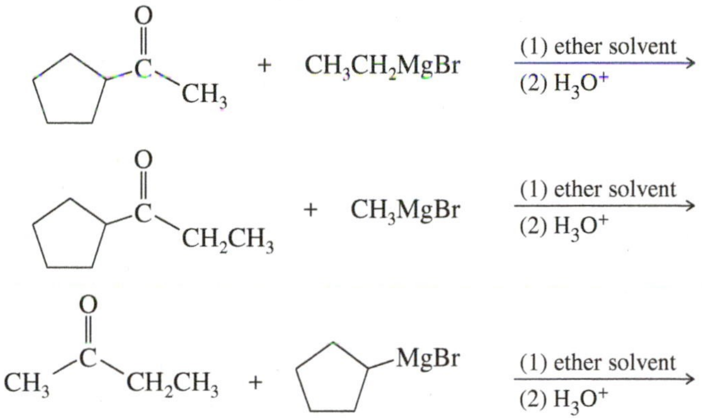

$$\text{cyclopentyl}-\overset{O}{\overset{\|}{C}}-CH_3 \; + \; CH_3CH_2MgBr \xrightarrow[(2)\; H_3O^+]{(1)\text{ ether solvent}}$$

$$\text{cyclopentyl}-\overset{O}{\overset{\|}{C}}-CH_2CH_3 \; + \; CH_3MgBr \xrightarrow[(2)\; H_3O^+]{(1)\text{ ether solvent}}$$

$$CH_3-\overset{O}{\overset{\|}{C}}-CH_2CH_3 \; + \; \text{cyclopentyl}-MgBr \xrightarrow[(2)\; H_3O^+]{(1)\text{ ether solvent}}$$

Any of these three syntheses would probably work, but only the third begins with fragments containing no more than five carbon atoms. The other two syntheses would require further steps to generate the ketones from compounds containing no more than five carbon atoms.

PROBLEM 10-15

Show how you would synthesize each tertiary alcohol by adding an appropriate Grignard reagent to a ketone.

(a) 3-phenylhexan-3-ol (3 ways) **(b)** Ph_3COH

(c) 1-ethylcyclopentanol **(d)** 2-cyclopentylpentan-2-ol

PROBLEM-SOLVING HINT

A tertiary alcohol has three groups on the carbinol carbon atom. Consider three possible reactions (as in Solved Problem 10-2), with each of these groups added as the Grignard reagent.

10-9D Addition to Acid Chlorides and Esters

Acid chlorides and **esters** are derivatives of carboxylic acids. In such **acid derivatives,** the —OH group of a carboxylic acid is replaced by other electron-withdrawing groups. In acid chlorides, the hydroxy group of the acid is replaced by a chlorine atom. In esters, the hydroxy group is replaced by an alkoxy (—O—R) group.

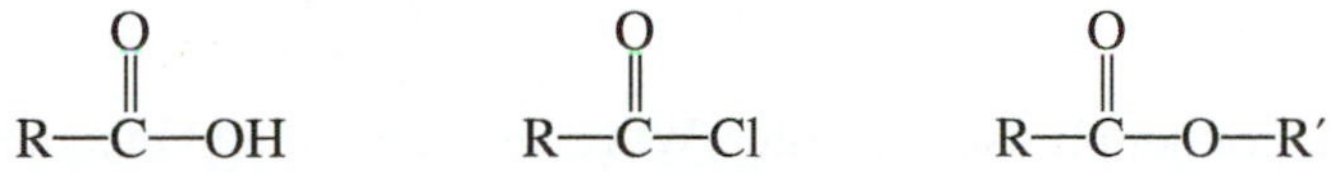

$$\underset{\text{carboxylic acid}}{R-\overset{O}{\overset{\|}{C}}-OH} \qquad \underset{\text{acid chloride}}{R-\overset{O}{\overset{\|}{C}}-Cl} \qquad \underset{\text{ester}}{R-\overset{O}{\overset{\|}{C}}-O-R'}$$

Acid chlorides and esters react with two equivalents of Grignard reagents to give (after protonation) tertiary alcohols.

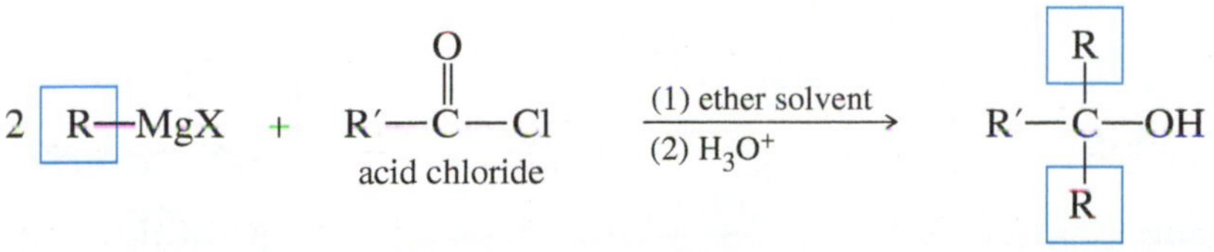

$$2\; \boxed{R}-MgX \; + \; \underset{\text{acid chloride}}{R'-\overset{O}{\overset{\|}{C}}-Cl} \xrightarrow[(2)\; H_3O^+]{(1)\text{ ether solvent}} \underset{\text{tertiary alcohol}}{R'-\underset{\substack{|\\ \boxed{R}}}{\overset{\substack{\boxed{R}\\ |}}{C}}-OH}$$

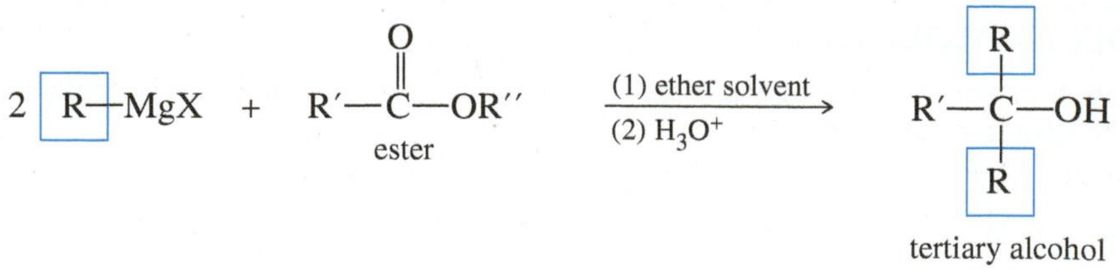

$$2\,\boxed{R}\!-\!MgX + R'-\overset{O}{\overset{\|}{C}}-OR'' \xrightarrow[(2)\ H_3O^+]{(1)\ \text{ether solvent}} R'-\underset{\boxed{R}}{\overset{\boxed{R}}{C}}-OH$$

ester — tertiary alcohol

Addition of the first equivalent of the Grignard reagent produces an unstable intermediate that expels a chloride ion (in the acid chloride) or an alkoxide ion (in the ester), to give a ketone. The alkoxide ion is a suitable leaving group in this reaction because its leaving stabilizes a negatively charged intermediate in a fast, strongly exothermic step.

Attack on an acid chloride

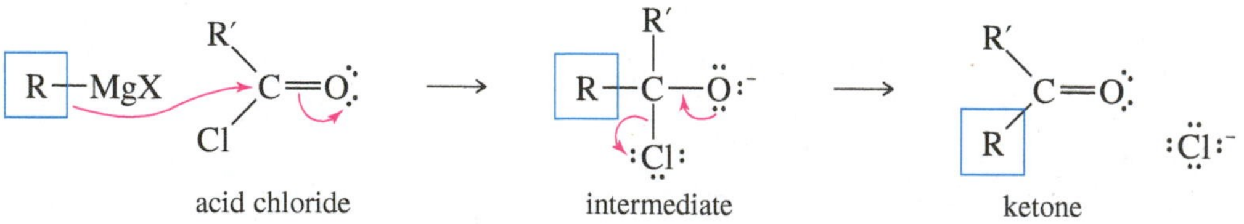

R—MgX + R′(Cl)C=O: ⟶ R—C(R′)(Cl:)—O:⁻ ⟶ R′(R)C=O: + :Cl:⁻

acid chloride — intermediate — ketone

Attack on an ester

R—MgX + R′(R″—O:)C=O: ⟶ R—C(R′)(:O—R″)—O:⁻ ⟶ R′(R)C=O + ⁻:O—R″

ester — intermediate — ketone

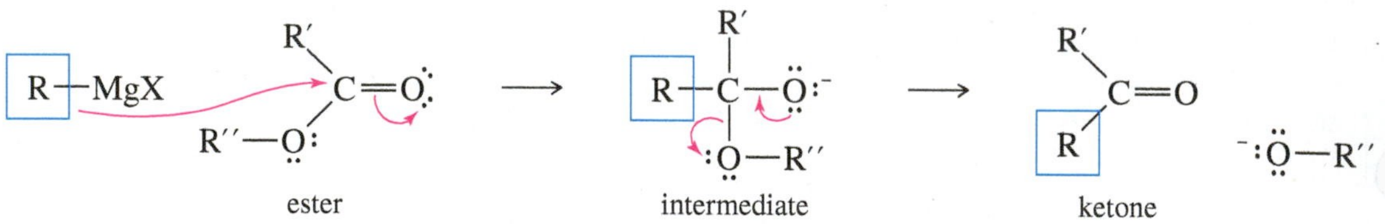

The ketone reacts with a second equivalent of the Grignard reagent, forming the magnesium salt of a tertiary alkoxide. Protonation gives a tertiary alcohol with one of its alkyl groups derived from the acid chloride or ester, and the other two derived from the Grignard reagent.

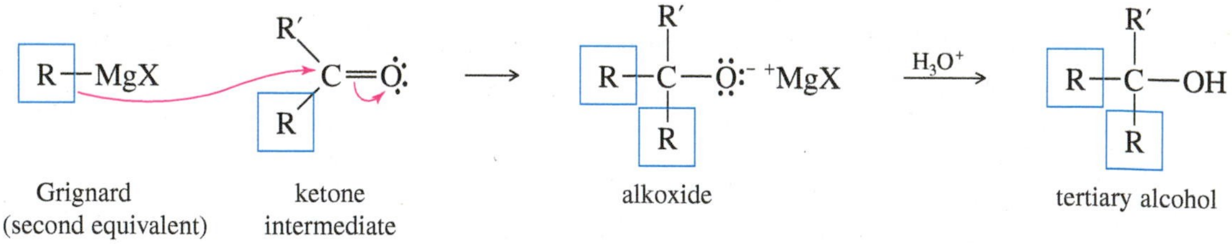

R—MgX + R′(R)C=O: ⟶ R—C(R′)(R)—O:⁻ ⁺MgX $\xrightarrow{H_3O^+}$ R—C(R′)(R)—OH

Grignard (second equivalent) — ketone intermediate — alkoxide — tertiary alcohol

PROBLEM-SOLVING HINT

When making a tertiary alcohol with two identical alkyl groups, consider using an acid chloride or ester.

Consider an example using an ester. When an excess of ethylmagnesium bromide is added to methyl benzoate, the first equivalent adds and methoxide is expelled, giving propiophenone. Addition of a second equivalent, followed by protonation, gives a tertiary alcohol: 3-phenylpentan-3-ol.

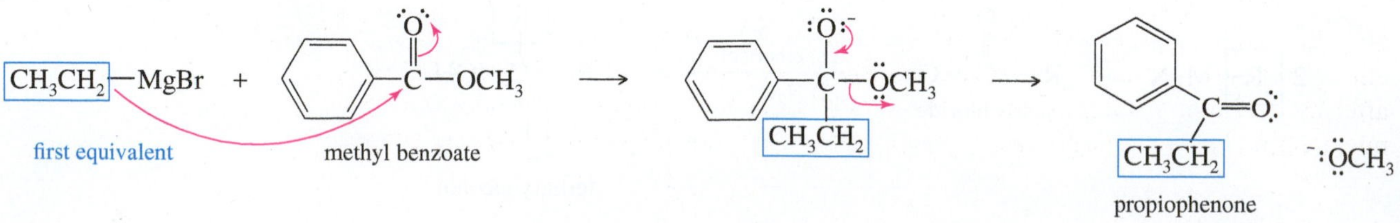

CH_3CH_2—MgBr + Ph—C(=O:)—OCH_3 ⟶ Ph—C(CH_3CH_2)(:O:⁻)—$\ddot{O}CH_3$ ⟶ Ph(C=O:)CH_3CH_2 + ⁻:$\ddot{O}CH_3$

first equivalent — methyl benzoate — propiophenone

$$CH_3CH_2\text{—}MgBr + \underset{\text{propiophenone}}{Ph(CH_3CH_2)C{=}\ddot{O}:} \longrightarrow CH_3CH_2\text{—}C(Ph)(CH_2CH_3)\text{—}\ddot{O}:^-\ ^+MgBr \xrightarrow{H_3O^+} \underset{\text{3-phenylpentan-3-ol (82\%)}}{CH_3CH_2\text{—}C(Ph)(CH_2CH_3)\text{—}OH}$$

second equivalent

PROBLEM 10-16

Propose a mechanism for the reaction of acetyl chloride with phenylmagnesium bromide to give 1,1-diphenylethanol.

$$\underset{\text{acetyl chloride}}{CH_3\text{—}\overset{O}{\overset{\|}{C}}\text{—}Cl} + 2\ \underset{\text{phenylmagnesium bromide}}{C_6H_5\text{—}MgBr} \xrightarrow[\text{(2) } H_3O^+]{\text{(1) ether solvent}} \underset{\text{1,1-diphenylethanol}}{CH_3\text{—}\overset{OH}{\overset{|}{C}}(C_6H_5)_2}$$

PROBLEM 10-17

Show how you would add Grignard reagents to acid chlorides or esters to synthesize the following alcohols.

(a) $Ph_3C\text{—}OH$ **(b)** 3-ethyl-2-methylpentan-3-ol
(c) dicyclohexylphenylmethanol

PROBLEM-SOLVING HINT

When making a secondary alcohol with identical alkyl groups, consider using a formate ester.

PROBLEM 10-18

A formate ester, such as ethyl formate, reacts with an excess of a Grignard reagent to give (after protonation) secondary alcohols with two identical alkyl groups.

$$2\,R\text{—}MgX + \underset{\text{ethyl formate}}{H\text{—}\overset{O}{\overset{\|}{C}}\text{—}O\text{—}CH_2CH_3} \xrightarrow[\text{(2) } H_3O^+]{\text{(1) ether solvent}} \underset{\text{secondary alcohol}}{R\text{—}\overset{OH}{\overset{|}{C}H}\text{—}R}$$

(a) Propose a mechanism to show how the reaction of ethyl formate with an excess of allylmagnesium bromide gives, after protonation, hepta-1,6-dien-4-ol.

$$\underset{\text{allylmagnesium bromide}}{2\,H_2C{=}CH\text{—}CH_2MgBr} + \underset{\text{ethyl formate}}{H\text{—}\overset{O}{\overset{\|}{C}}\text{—}OCH_2CH_3} \xrightarrow[\text{(2) } H_3O^+]{\text{(1) ether solvent}} \underset{\text{hepta-1,6-dien-4-ol (80\%)}}{(H_2C{=}CH\text{—}CH_2)_2CH\text{—}OH}$$

(b) Show how you would use reactions of Grignard reagents with ethyl formate to synthesize the following secondary alcohols.
(i) pentan-3-ol **(ii)** diphenylmethanol **(iii)** *trans,trans*-nona-2,7-dien-5-ol

10-9E Ring-Opening with Ethylene Oxide

Grignard reagents usually do not react with ethers, but **epoxides** are unusually reactive ethers because of their ring strain. Ethylene oxide reacts with Grignard reagents to give, after protonation, primary alcohols with *two* additional carbon atoms. Notice that the nucleophilic attack by the Grignard reagent opens the ring and relieves the ring strain.

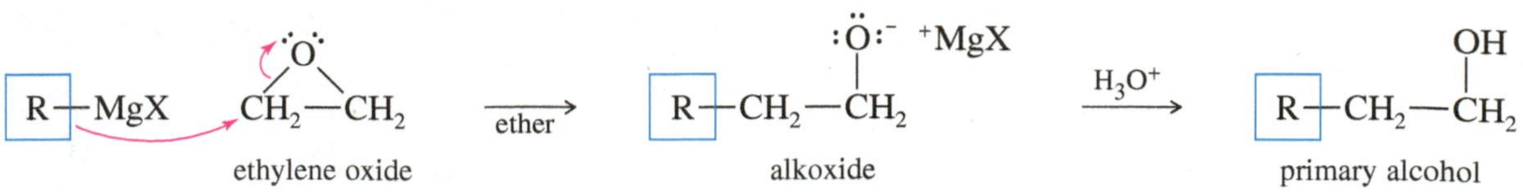

$$R{-}MgX + \underset{\text{ethylene oxide}}{CH_2{-}CH_2\ (\text{O bridge})} \xrightarrow{\text{ether}} \underset{\text{alkoxide}}{R{-}CH_2{-}CH_2{-}O^{-}\ {}^{+}MgX} \xrightarrow{H_3O^+} \underset{\text{primary alcohol}}{R{-}CH_2{-}CH_2{-}OH}$$

Example

$$\underset{\text{butylmagnesium bromide}}{CH_3(CH_2)_3{-}MgBr} + \underset{\text{ethylene oxide}}{CH_2{-}CH_2\ (\text{O bridge})} \longrightarrow C_4H_9{-}CH_2{-}CH_2{-}O^{-}\ {}^{+}MgBr \xrightarrow{H_3O^+} \underset{\text{hexan-1-ol (61\%)}}{C_4H_9{-}CH_2{-}CH_2{-}OH}$$

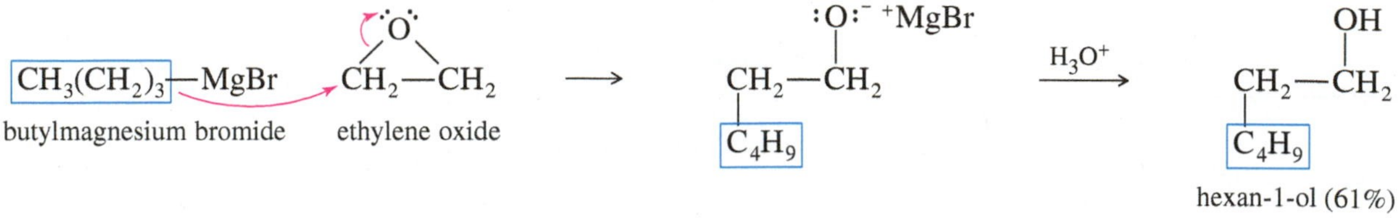

PROBLEM-SOLVING HINT

The reaction of a Grignard reagent with an epoxide is the only Grignard reaction we have seen where the new OH group is NOT on the same carbon atom where the Grignard formed a new bond. In this case, the new OH group appears on the second carbon from the new bond.

PROBLEM 10-19

Show how you would synthesize the following alcohols by adding Grignard reagents to ethylene oxide.

(a) 2-phenylethanol **(b)** 4-methylpentan-1-ol **(c)** cyclohexane bearing CH_2CH_2OH and CH_3 on adjacent carbons

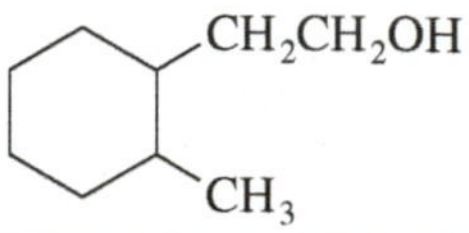

PROBLEM 10-20

In Section 9-7B, we saw how acetylide ions add to carbonyl groups in much the same way as Grignard and organolithium reagents. Acetylide ions also add to ethylene oxide much like Grignard and organolithium reagents. Predict the products obtained by adding the following acetylide ions to ethylene oxide, followed by a dilute acid workup.

(a) $HC{\equiv}C{:}^{-}$ **(b)** $CH_3CH_2{-}C{\equiv}C{:}^{-}$

SUMMARY Grignard Reactions

1. Nucleophilic Additions to Carbonyl Compounds

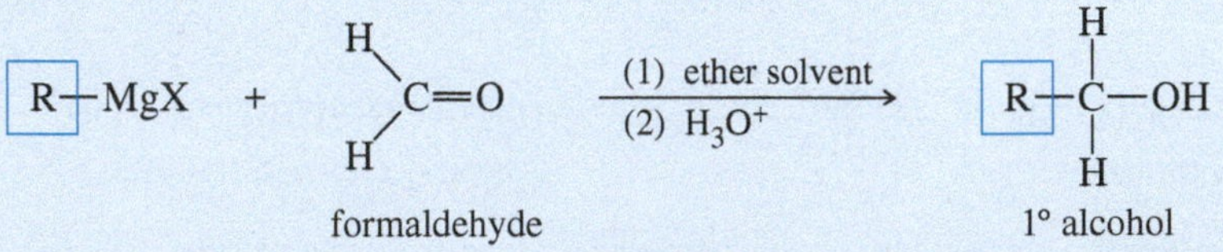

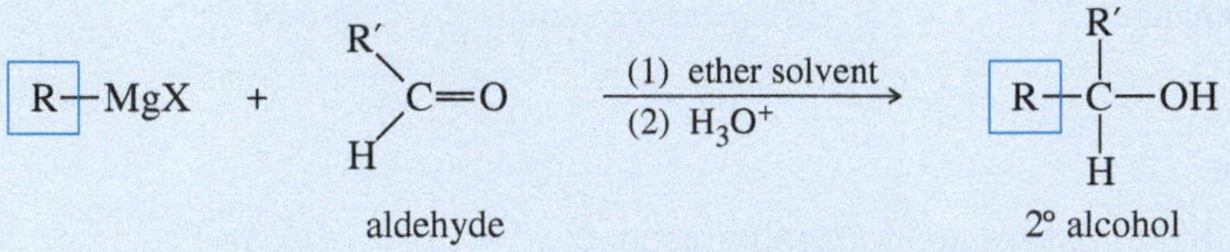

$$R{-}MgX + \underset{\text{formaldehyde}}{H_2C{=}O} \xrightarrow[(2)\ H_3O^+]{(1)\ \text{ether solvent}} \underset{1^\circ\ \text{alcohol}}{R{-}CH_2{-}OH}$$

$$R{-}MgX + \underset{\text{aldehyde}}{R'(H)C{=}O} \xrightarrow[(2)\ H_3O^+]{(1)\ \text{ether solvent}} \underset{2^\circ\ \text{alcohol}}{R{-}CR'H{-}OH}$$

$$R{-}MgX + R'R''C{=}O \xrightarrow[(2)\ H_3O^+]{(1)\ \text{ether solvent}} R{-}C(R')(R''){-}OH$$

ketone → 3° alcohol

$$2\ R{-}MgX + R'{-}C(=O){-}OR'' \xrightarrow[(2)\ H_3O^+]{(1)\ \text{ether solvent}} R'{-}C(R)(R){-}OH$$

ester or acid chloride → 3° alcohol, two groups added

2. Nucleophilic Displacement of Epoxides

$$R{-}MgX + CH_2{-}CH_2\ (\text{epoxide O}) \xrightarrow[(2)\ H_3O^+]{(1)\ \text{ether solvent}} R{-}CH_2CH_2{-}OH$$

ethylene oxide → 1° alcohol, two groups added

10-9F Lithium Dialkylcuprates

Recall from Section 9-7 how acetylide ions are alkylated by displacing unhindered alkyl halides.

$$H{-}C{\equiv}C{:}^- \quad R{-}CH_2{-}Br \longrightarrow H{-}C{\equiv}C{-}CH_2{-}R + Br^-$$

Like acetylide ions, Grignard and organolithium reagents are strong bases and strong nucleophiles. Fortunately, they do not displace halides as easily as acetylide ions do. If they did displace alkyl halides, it would be impossible to form Grignard reagents from alkyl halides because whenever a molecule of reagent formed, it would react with a molecule of the halide starting material. All that would be formed is a coupling product. In fact, coupling is a side reaction that hurts the yield of many Grignard reactions.

undesirable coupling

$$R{-}Br + Mg \xrightarrow{\text{ether}} R{-}Mg{-}Br \xrightarrow{R{-}Br} R{-}R + MgBr_2$$

If we *want* to couple two groups together efficiently, we can do it by using an organocopper reagent, a **lithium dialkylcuprate,** to couple with an alkyl halide.

$$R_2CuLi + R'{-}X \longrightarrow R{-}R' + R{-}Cu + LiX$$

a lithium dialkylcuprate

The lithium dialkylcuprate (also called a *Gilman reagent*) is formed by the reaction of two equivalents of the corresponding organolithium reagent (Section 10-8B) with cuprous iodide:

$$2\,R{-}Li + CuI \longrightarrow R_2CuLi + LiI$$

a Gilman reagent

The coupling takes place as if a carbanion ($R{:}^-$) were present and the carbanion attacked the alkyl halide to displace the halide ion. This is not the actual mechanism, however, because dialkylcuprates also couple with vinyl halides and aryl halides, which are incapable of undergoing S_N2 substitutions. The actual mechanism depends on the structure of the cuprate and the alkyl halide, and likely involves additional steps.

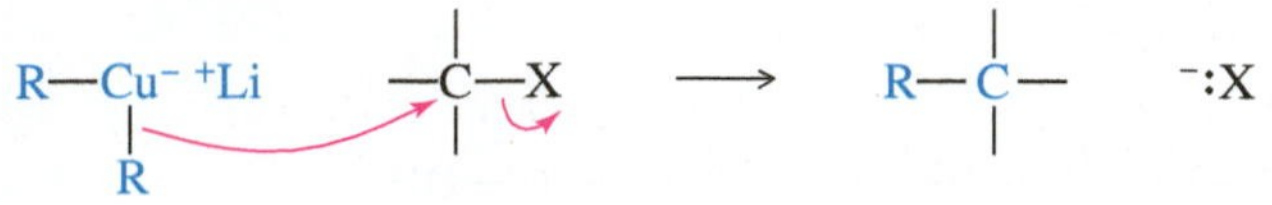

$$R{-}Cu^{-}\ {}^{+}Li\ (R) + {-}C{-}X \longrightarrow R{-}C{-} \quad {}^{-}{:}X$$

(hypothetical simplified mechanism)

Example

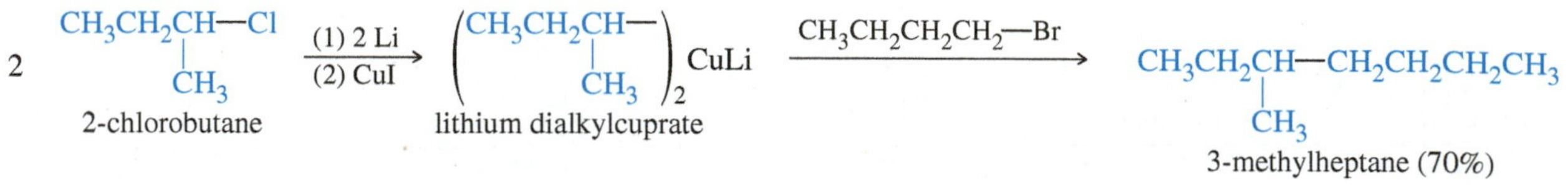

$$2\ CH_3CH_2CH(CH_3){-}Cl \xrightarrow[(2)\ CuI]{(1)\ 2\ Li} (CH_3CH_2CH(CH_3){-})_2CuLi \xrightarrow{CH_3CH_2CH_2CH_2{-}Br} CH_3CH_2CH(CH_3){-}CH_2CH_2CH_2CH_3$$

2-chlorobutane; lithium dialkylcuprate; 3-methylheptane (70%)

We know that these coupling reactions cannot actually involve an S_N2 mechanism, because they also take place on vinyl halides, which cannot undergo S_N2 reactions. Reactions of lithium dialkylcuprates with vinyl halides usually preserve the stereochemistry of the vinyl halide, as shown in the following example.

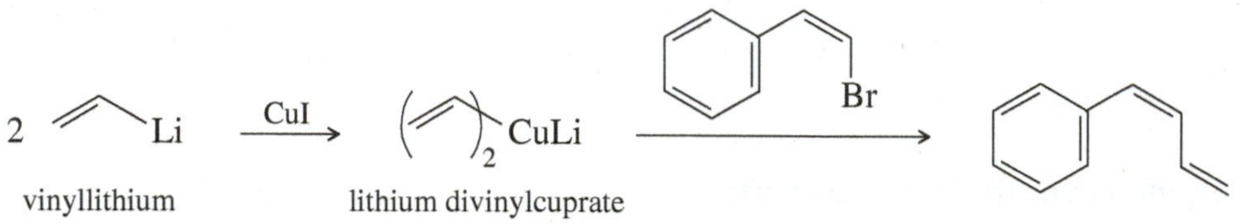

2 vinyllithium (Li) $\xrightarrow{CuI}$ lithium divinylcuprate (CuLi, subscript 2) $\xrightarrow{\text{PhCH=CHBr}}$ PhCH=CH–CH=CH$_2$

vinyllithium; lithium divinylcuprate

Lithium dialkylcuprates are not as reactive toward carbonyl groups as Grignard reagents are. The Gilman reagents react much slower with ketones than with acid chlorides, so the reaction can be stopped at the ketone stage. We use this selectivity to synthesize ketones in Chapter 18.

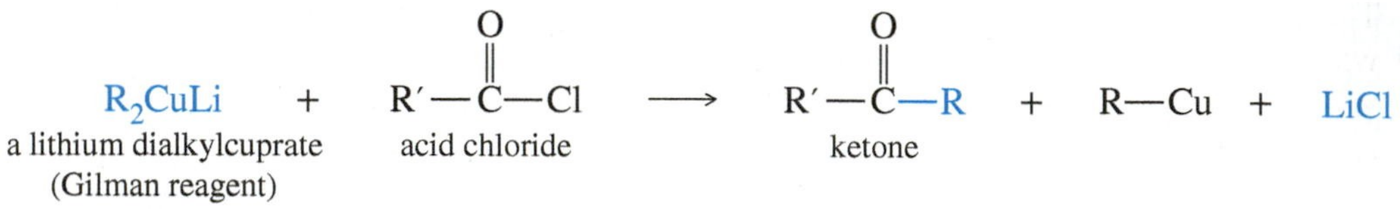

$$R_2CuLi + R'{-}C(=O){-}Cl \longrightarrow R'{-}C(=O){-}R + R{-}Cu + LiCl$$

a lithium dialkylcuprate (Gilman reagent); acid chloride; ketone

Example

lithium divinylcuprate (CuLi, subscript 2) + benzoyl chloride (Cl) ⟶ phenyl vinyl ketone

lithium divinylcuprate; benzoyl chloride

PROBLEM 10-21

Show how you would synthesize the following compounds from alkyl halides, vinyl halides, and aryl halides containing no more than six carbon atoms.

(a) octane **(b)** vinylcyclohexane
(c) *trans*-oct-3-ene **(d)** cyclopentyl propyl ketone

10-10 Side Reactions of Organometallic Reagents: Reduction of Alkyl Halides

Grignard and organolithium reagents are strong nucleophiles and strong bases. Besides their additions to carbonyl compounds, they react with other acidic or electrophilic compounds. In some cases, these are useful reactions, but they are often seen as annoying side reactions where a small impurity of water or an alcohol destroys the reagent.

10-10A Reactions with Acidic Compounds

Grignard and organolithium reagents react vigorously and irreversibly with water. Therefore, all reagents and solvents used in these reactions must be dry.

$$\overset{\delta^-}{R}-\overset{\delta^+}{MgX} + H-O-H \longrightarrow R-H + XMgOH$$
$$pK_a = 15.7 \qquad pK_a > 45$$

For example, consider the reaction of ethyllithium with water:

$$\underset{\text{ethyllithium}}{CH_3-\overset{\delta^-}{CH_2}-\overset{\delta^+}{Li}} + H-O-H \longrightarrow \underset{\text{ethane}}{CH_3-CH_2-H} + Li^+ \ ^-OH$$
$$pK_a = 15.7 \qquad pK_a \approx 50$$

PROBLEM-SOLVING HINT

Acid–base reactions in which the pK_a values of the acids differ by about 10 or more are essentially irreversible.

The products are strongly favored in this reaction. Ethane is a *very* weak acid (K_a of about 10^{-50}), so the reverse reaction (abstraction of a proton from ethane by lithium hydroxide) is unlikely. When ethyllithium is added to water, ethane instantly bubbles to the surface.

Why would we ever want to add an organometallic reagent to water? This is a method for reducing an alkyl halide to an alkane:

$$R-X + Mg \xrightarrow{\text{ether}} R-MgX \xrightarrow{H_2O} R-H + XMgOH$$

$$R-X + 2Li \longrightarrow R-Li + LiX \xrightarrow{H_2O} R-H + LiOH$$

The overall reaction is a *reduction* because it replaces the electronegative halogen atom with a hydrogen atom. In particular, this reaction provides a way to "label" a compound with deuterium (D or ^{2}H, a heavy isotope of hydrogen) at any position where a halogen is present.

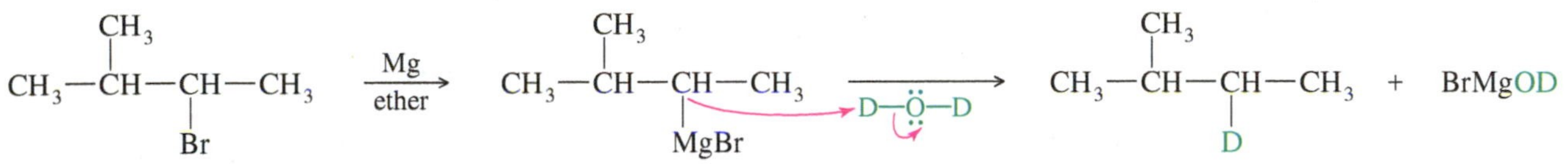

$$CH_3-\underset{}{\overset{CH_3}{CH}}-\underset{Br}{CH}-CH_3 \xrightarrow[\text{ether}]{Mg} CH_3-\overset{CH_3}{CH}-\underset{MgBr}{CH}-CH_3 \xrightarrow{D-\ddot{O}-D} CH_3-\overset{CH_3}{CH}-\underset{D}{CH}-CH_3 + BrMgOD$$

In addition to O—H groups, the protons of N—H and S—H groups and the hydrogen atom of a terminal alkyne, —C≡C—H, are sufficiently acidic to protonate Grignard and organolithium reagents. Unless we want to protonate the reagent, compounds with these groups are considered incompatible with Grignard and organolithium reagents.

$$CH_3CH_2CH_2CH_2Li + (CH_3CH_2)_2NH \longrightarrow CH_3CH_2CH_2CH_3 + (CH_3CH_2CH)_2N^- \ Li^+$$
$$pK_a = 40 \qquad pK_a > 45$$

$$CH_3CH_2CH_2CH_2Li + CH_3(CH_2)_4-C{\equiv}C-H \longrightarrow CH_3CH_2CH_2CH_3 + CH_3(CH_2)_4-C{\equiv}C-Li$$
$$pK_a = 25 \qquad pK_a > 45$$

PROBLEM 10-22

Predict the products of the following reactions.

(a) *sec*-butylmagnesium iodide + D_2O ⟶

(b) *n*-butyllithium + CH_3CH_2OH ⟶

(c) isobutylmagnesium bromide + but-1-yne ⟶

(d) cyclohexyllithium + $CH_3-\overset{O}{\overset{\|}{C}}-OH$ (acetic acid) $\longrightarrow$

(e) phenylmagnesium bromide + D_2O $\longrightarrow$

PROBLEM-SOLVING HINT

Grignard reagents are incompatible with water or acid. Dilute acid is used *in a separate step* to hydrolyze the magnesium alkoxide.

correct: $\xrightarrow[(2)\ H_3O^+]{(1)\ RMgX}$

correct: $\xrightarrow[ether]{RMgX} \xrightarrow{H_3O^+}$

incorrect: $\xrightarrow[H_3O^+]{RMgX}$

(The incorrect example means using a Grignard in aqueous acid.)

10-10B Reactions with Electrophilic Multiple Bonds

Grignard reagents are useful because they add to the electrophilic double bonds of carbonyl groups. However, we must make sure that the *only* electrophilic double bond in the solution is the one we want the reagent to attack. There must be no electrophilic double (or triple) bonds in the solvent or in the Grignard reagent itself; otherwise, they will be attacked as well. Any multiple bond involving a strongly electronegative element is likely to be attacked, including C=O, S=O, C=N, N=O, and C≡N bonds.

In later chapters, we will encounter methods for *protecting* susceptible groups to prevent the reagent from attacking them. For now, simply remember that the following groups react with Grignard and organolithium reagents; avoid compounds containing these groups except for the one carbonyl group that gives the desired reaction.

Protonate the Grignard or organolithium: O—H, N—H, S—H, C≡C—H

Attacked by the Grignard or organolithium: C=O, C=N, C≡N, S=O, N=O

PROBLEM 10-23

Point out the flaws in the following incorrect Grignard syntheses.

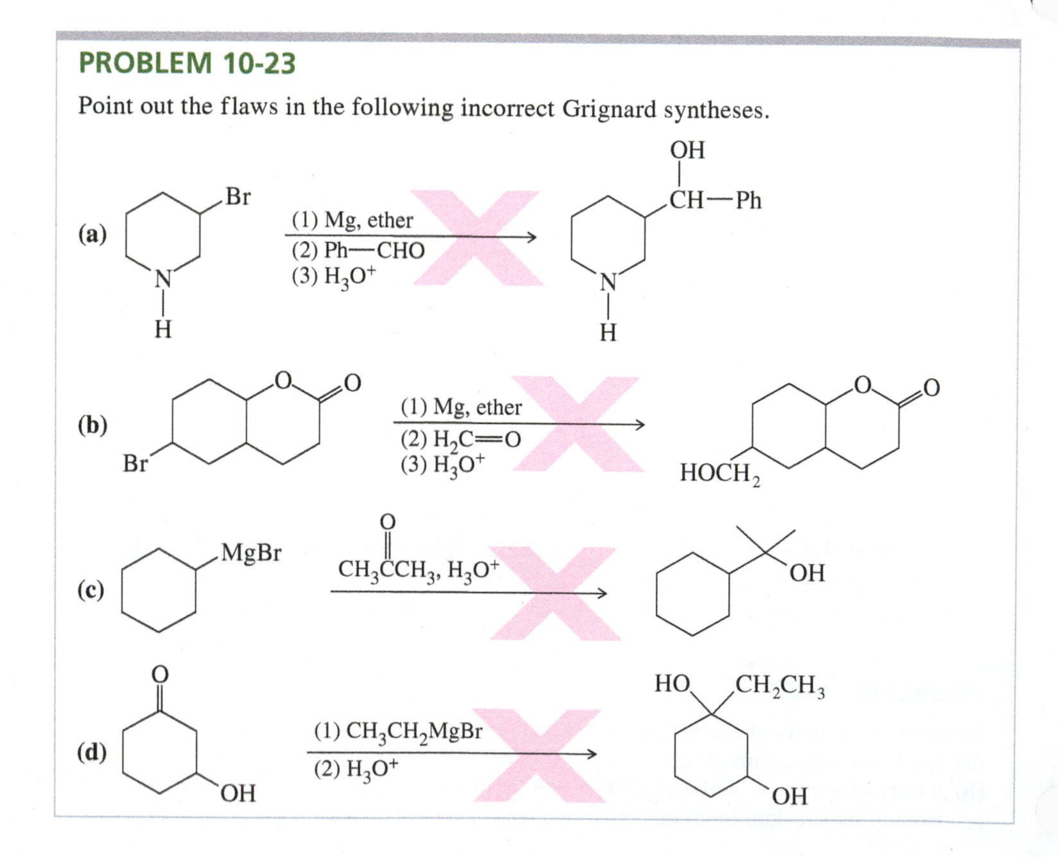

10-11 Reduction of the Carbonyl Group: Synthesis of 1° and 2° Alcohols

Grignard reagents convert carbonyl compounds to alcohols by adding alkyl groups. **Hydride reagents** add a hydride ion ($H:^-$), reducing the carbonyl group to an alkoxide ion with no additional carbon atoms. Subsequent protonation gives the alcohol. Converting a ketone or an aldehyde to an alcohol involves adding two hydrogen atoms across the C=O bond: a reduction. Mechanism 10-2 shows the mechanism for this reduction.

The two most useful hydride reagents, sodium borohydride ($NaBH_4$) and lithium aluminum hydride ($LiAlH_4$), reduce carbonyl groups in excellent yields. These reagents are called *complex hydrides* because they do not have a simple hydride structure such as Na^+H^- or Li^+H^-. Instead, their hydrogen atoms, bearing partial negative charges, are covalently bonded to boron and aluminum atoms. This arrangement makes the hydride a better nucleophile while reducing its basicity.

Na^+ BH_4^- — sodium borohydride

Li^+ AlH_4^- — lithium aluminum hydride

MECHANISM 10-2 Hydride Reduction of a Carbonyl Group

Sodium borohydride and lithium aluminum hydride (LAH) reduce ketones and aldehydes to alcohols. Either reagent can transfer a hydride ion ($H:^-$) to the electrophilic end of the carbonyl group.

Lithium aluminum hydride (LAH) is a strong reducing agent that cannot be used in hydroxylic solvents.

Reaction 1: LAH transfers a hydride ion ($H:^-$) to carbon, forming an alkoxide ion.

$$Li^+\ H{-}AlH_3^- + R_2C{=}O \xrightarrow{\text{ether}} [H{-}CR_2{-}O:^-]\ Li^+ + AlH_3$$

lithium aluminum hydride — alkoxide ion — (forms complexes)

Reaction 2: After the first reaction is complete, water or dilute acid is carefully added to protonate the alkoxide.

$$H{-}CR_2{-}O:^- \xrightarrow{H{-}O{-}H} H{-}CR_2{-}OH$$

alkoxide ion — alcohol

EXAMPLE: Borohydride reduction of cyclopentanone to cyclopentanol.
Sodium borohydride can conveniently be used in alcohol solutions.

Reaction 1: Borohydride transfers a hydride ion ($H:^-$) to carbon, forming an alkoxide ion.

$$Na^+\ H{-}BH_3^- + \text{cyclopentanone} \xrightarrow{R{-}OH} \text{alkoxide ion} + Na^+ + BH_3$$

sodium borohydride — cyclopentanone — alkoxide ion (forms complexes)

Reaction 2: The alcohol solvent protonates the alkoxide.

alkoxide ion $\xrightarrow{H-\ddot{O}-R}$ cyclopentanol + ^{-}OR

Comparison of electronegativities

B—H: 2.0 2.2 Al—H: 1.6 2.2

Aluminum is less electronegative than boron, so more of the negative charge in the AlH_4^- ion is borne by the hydrogen atoms. Therefore, lithium aluminum hydride (LAH) is a much stronger reducing agent, and it is much more difficult to work with than sodium borohydride. LAH reacts explosively with water and alcohols, liberating hydrogen gas and sometimes starting fires. Sodium borohydride reacts slowly with alcohols and with water as long as the pH is high (basic). Sodium borohydride is a convenient and highly selective reducing agent.

10-11A Uses of Sodium Borohydride

Sodium borohydride ($NaBH_4$) reduces aldehydes to primary alcohols, and ketones to secondary alcohols. The reactions take place in a wide variety of solvents, including alcohols, ethers, and water. The yields are generally excellent.

R—C(=O)—H aldehyde; R—C(=O)—R′ ketone; R—C(=O)—OR′ ester; R—C(=O)—O⁻ acid (anion)

$NaBH_4$ reduces (aldehyde, ketone); $LiAlH_4$ reduces (aldehyde, ketone, ester, acid anion)

ease of reduction

cyclohexanecarbaldehyde $\xrightarrow{NaBH_4,\ CH_3CH_2OH}$ cyclohexylmethanol (95%)

$$CH_3-C(=O)-CH_2CH_3 \xrightarrow{NaBH_4,\ CH_3OH} CH_3-CH(OH)-CH_2CH_3$$

butan-2-one → (±)-butan-2-ol (100%)

Sodium borohydride is selective; it usually does not react with carbonyl groups that are less reactive than ketones and aldehydes. For example, carboxylic acids and esters are unreactive toward borohydride reduction. Thus, sodium borohydride can reduce a ketone or an aldehyde in the presence of an acid or an ester.

$$O{=}C_6H_{10}-CH_2-C(=O)-OCH_3 \xrightarrow{NaBH_4} HO(H)C_6H_{10}-CH_2-C(=O)-OCH_3$$

PROBLEM-SOLVING HINT

LAH and water are incompatible. Water is added in a *separate* hydrolysis step. An explosion and fire would result from the process indicated by $\xrightarrow[H_3O^+]{LiAlH_4}$.

10-11B Uses of Lithium Aluminum Hydride

Lithium aluminum hydride ($LiAlH_4$, abbreviated LAH) is a much stronger reagent than sodium borohydride. It easily reduces ketones and aldehydes and also the less reactive carbonyl groups: those in acids, esters, and other acid derivatives (see Chapter 21). LAH reduces ketones to secondary alcohols, and it reduces aldehydes, acids, and esters to primary alcohols. The lithium salt of the alkoxide ion is initially formed, after which the (cautious!) addition of dilute acid protonates the alkoxide. For example, LAH reduces both functional groups of the keto ester in the previous example.

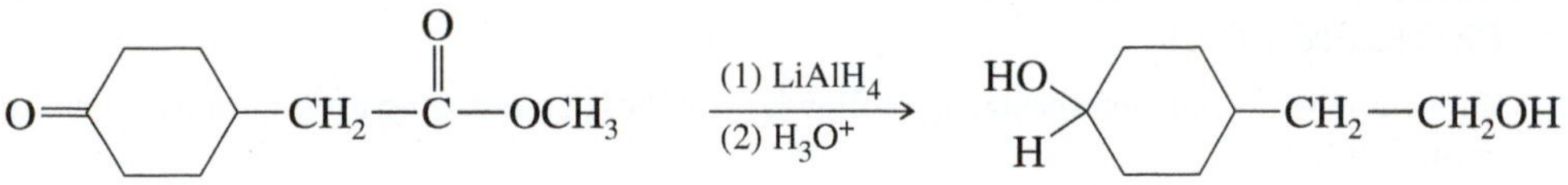

$$\text{O}{=}\text{C}_6\text{H}_{10}{-}\text{CH}_2{-}\overset{\text{O}}{\overset{\|}{\text{C}}}{-}\text{OCH}_3 \xrightarrow[(2)\ \text{H}_3\text{O}^+]{(1)\ \text{LiAlH}_4} \text{HO(H)C}_6\text{H}_{10}{-}\text{CH}_2{-}\text{CH}_2\text{OH}$$

In summary, sodium borohydride is usually the best reagent for reduction of a simple ketone or aldehyde. Using $NaBH_4$, we can reduce a ketone or an aldehyde in the presence of an acid or an ester, but we do not have a method (so far) for reducing an acid or an ester in the presence of a ketone or an aldehyde. The sluggish acid or ester requires the use of $LiAlH_4$, and this reagent also reduces the ketone or aldehyde.

SUMMARY Reactions of $LiAlH_4$ and $NaBH_4$

		$NaBH_4$	*$LiAlH_4$*
aldehyde	$\text{R}{-}\overset{\text{O}}{\overset{\|}{\text{C}}}{-}\text{H}$	$\text{R}{-}\text{CH}_2{-}\text{OH}$	$\text{R}{-}\text{CH}_2{-}\text{OH}$
ketone	$\text{R}{-}\overset{\text{O}}{\overset{\|}{\text{C}}}{-}\text{R}'$	$\text{R}{-}\underset{}{\overset{\text{OH}}{\overset{\vert}{\text{C}}}}\text{H}{-}\text{R}'$	$\text{R}{-}\overset{\text{OH}}{\overset{\vert}{\text{C}}}\text{H}{-}\text{R}'$
alkene	>C=C<	no reaction	no reaction
acid anion	$\text{R}{-}\overset{\text{O}}{\overset{\|}{\text{C}}}{-}\text{O}^-$ anion in base	no reaction	$\text{R}{-}\text{CH}_2{-}\text{OH}$
ester	$\text{R}{-}\overset{\text{O}}{\overset{\|}{\text{C}}}{-}\text{OR}'$	no reaction	$\text{R}{-}\text{CH}_2{-}\text{OH}$

Note: The products shown are the final products, after hydrolysis of the alkoxide.

PROBLEM 10-24

Predict the products you would expect from the reaction of $NaBH_4$ with the following compounds. You may assume that these reactions take place in methanol as the solvent.

(a) $CH_3-(CH_2)_8-CHO$ **(b)** $\text{CH}_3\text{CH}_2{-}\overset{\text{O}}{\overset{\|}{\text{C}}}{-}\text{OCH}_3$ **(c)** Ph—COOH

(d) 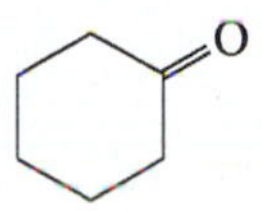 cyclohexanone (ring with =O) **(e)** 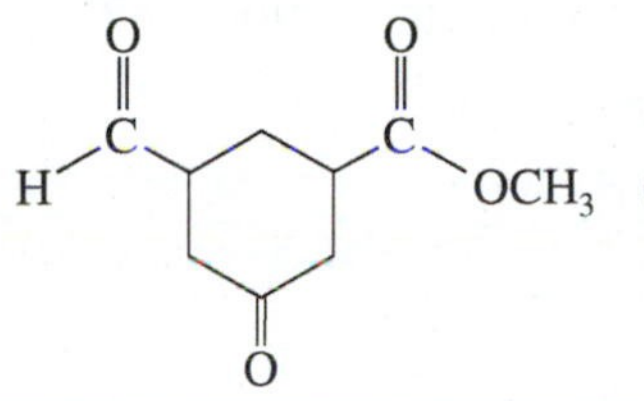 H—C(=O)— cyclohexanone ring —C(=O)—OCH₃ (ring bearing =O) **(f)** 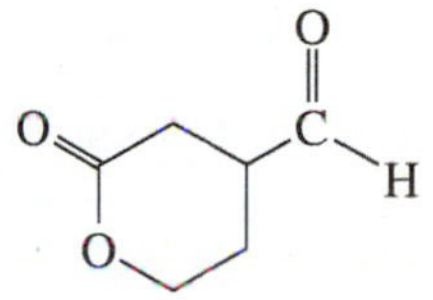 O= lactone ring (O in ring) bearing —C(=O)—H

PROBLEM-SOLVING HINT

When making a primary or secondary alcohol, you can consider adding an alkyl group last (as a Grignard reagent) or adding a hydrogen last (by reducing a ketone or aldehyde).

PROBLEM 10-25

Repeat Problem 10-24 using $LiAlH_4$ (followed by hydrolysis) as the reagent.

PROBLEM 10-26

Show how you would synthesize the following alcohols by reducing appropriate carbonyl compounds.

(a) heptan-1-ol **(b)** heptan-2-ol **(c)** 2-methylhexan-3-ol **(d)** [bicyclic lactone structure with OH]

10-11C Catalytic Hydrogenation of Ketones and Aldehydes

Reducing a ketone or an aldehyde to an alcohol involves adding two hydrogen atoms across the C═O bond. This addition can be accomplished by catalytic hydrogenation, commonly using Raney nickel as the catalyst.

$$-\overset{O}{\overset{\|}{C}}- \;+\; H_2 \xrightarrow{\text{Raney Ni}} -\overset{OH}{\overset{|}{C}H}-$$

Raney nickel is a finely divided hydrogen-bearing form of nickel made by treating a nickel–aluminum alloy with a strong sodium hydroxide solution. The aluminum in the alloy reacts to form hydrogen, leaving behind a finely divided nickel powder saturated with hydrogen. Raney nickel is an effective catalyst for the hydrogenation of ketones and aldehydes to alcohols. Carbon–carbon double bonds are also reduced under these conditions, however, so any alkene double bonds in the starting material will also be reduced. In most cases, sodium borohydride is more convenient for reducing simple ketones and aldehydes.

$$H_2C{=}CH{-}CH_2{-}C(CH_3)_2{-}CHO \;+\; 2\,H_2 \xrightarrow{\text{Raney Ni}} CH_3{-}CH_2{-}CH_2{-}C(CH_3)_2{-}CH_2OH$$

2,2-dimethylpent-4-enal → 2,2-dimethylpentan-1-ol (94%)

$$\text{2,2-dimethylpent-4-enal} \xrightarrow[\text{(for comparison)}]{NaBH_4} H_2C{=}CH{-}CH_2{-}C(CH_3)_2{-}CH_2OH$$

2,2-dimethylpent-4-en-1-ol

SUMMARY Alcohol Syntheses by Nucleophilic Additions to Carbonyl Groups

(Alcohol syntheses from alkenes and alkyl halides were summarized in Section 10-7.)

1. *Addition of a Grignard or organolithium reagent* (Section 10-9)

$$-\overset{O}{\overset{\|}{C}}- \;+\; R{-}MgX \xrightarrow{\text{ether}} -\overset{O^-\;{}^+MgX}{\overset{|}{C}}(R)- \xrightarrow{H_3O^+} -\overset{OH}{\overset{|}{C}}(R)-$$

a. *Addition to formaldehyde gives a primary alcohol*

$$\underset{\text{ethylmagnesium bromide}}{CH_3CH_2MgBr} \;+\; H_2C{=}O \xrightarrow[\text{(2) } H_3O^+]{\text{(1) ether solvent}} \underset{\text{propan-1-ol}}{CH_3CH_2{-}CH_2{-}OH}$$

b. *Addition to an aldehyde gives a secondary alcohol*

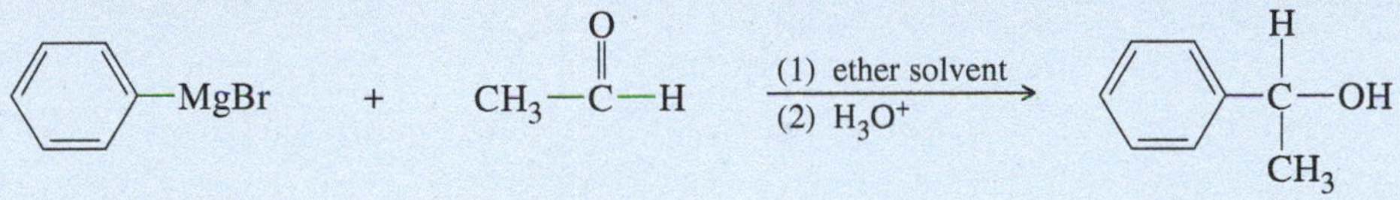

$$C_6H_5\text{-MgBr} + CH_3\text{-}\overset{O}{\overset{\|}{C}}\text{-H} \xrightarrow[(2)\ H_3O^+]{(1)\ \text{ether solvent}} C_6H_5\text{-}\underset{CH_3}{\overset{H}{C}}\text{-OH}$$

phenylmagnesium bromide acetaldehyde 1-phenylethanol

c. *Addition to a ketone gives a tertiary alcohol*

$$CH_3CH_2MgCl + \text{cyclohexanone} \xrightarrow[(2)\ H_3O^+]{(1)\ \text{ether}} \text{1-ethylcyclohexanol (OH, } CH_2CH_3\text{)}$$

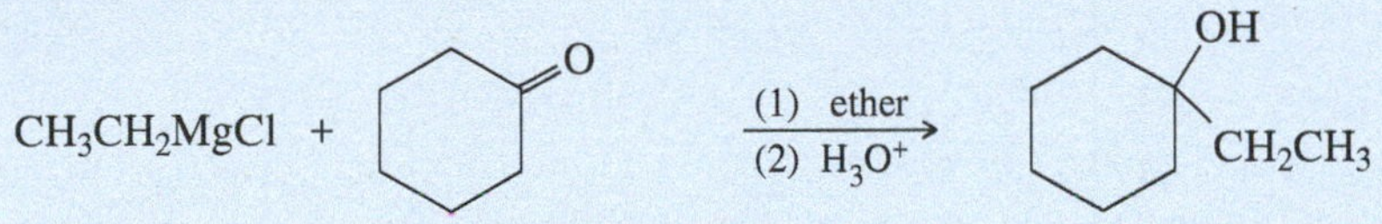

cyclohexanone 1-ethylcyclohexanol

d. *Addition to an acid halide or an ester gives a tertiary alcohol*

$$\left.\begin{array}{c} CH_3\text{-}\overset{O}{\overset{\|}{C}}\text{-Cl} \\ \text{acetyl chloride} \\ \text{or} \\ CH_3\text{-}\overset{O}{\overset{\|}{C}}\text{-}OCH_3 \\ \text{methyl acetate} \end{array}\right\} \xrightarrow[(2)\ H_3O^+]{(1)\ 2\ C_6H_{11}\text{-MgBr (cyclohexylmagnesium bromide)}} CH_3\text{-}\overset{OH}{C}(C_6H_{11})_2$$

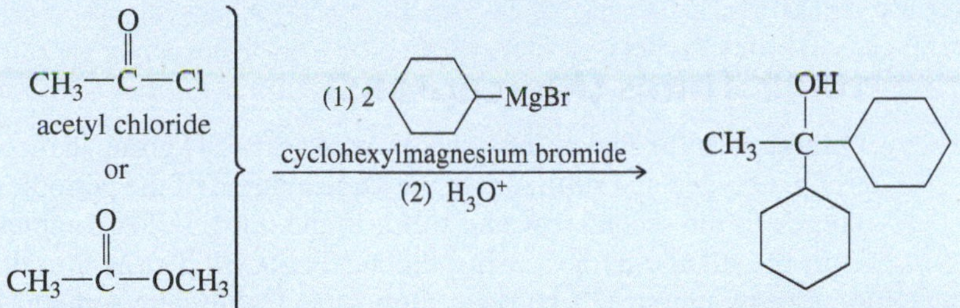

1,1-dicyclohexylethanol

e. *Addition to ethylene oxide gives a primary alcohol (with two additional carbon atoms added)*

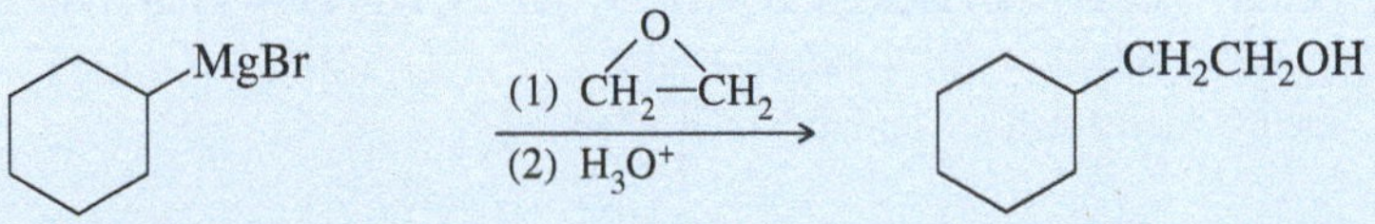

$$C_6H_{11}\text{-MgBr} \xrightarrow[(2)\ H_3O^+]{(1)\ CH_2\text{-}CH_2\ (\text{O bridged})} C_6H_{11}\text{-}CH_2CH_2OH$$

cyclohexylmagnesium bromide 2-cyclohexylethanol

2. *Reduction of carbonyl compounds* (Section 10-11)

a. *Catalytic hydrogenation of aldehydes and ketones*

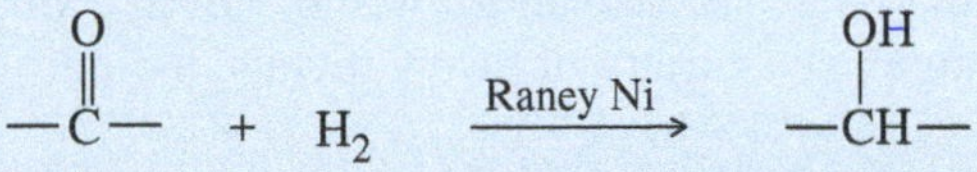

$$-\overset{O}{\overset{\|}{C}}- + H_2 \xrightarrow{\text{Raney Ni}} -\overset{OH}{\overset{|}{C}H}-$$

This method is usually not as selective or as effective as the use of hydride reagents.

b. *Use of hydride reagents*

(1) Reduction of an aldehyde gives a primary alcohol

$$C_6H_5\text{-}\overset{O}{\overset{\|}{C}}\text{-H} \xrightarrow{NaBH_4} C_6H_5\text{-}CH_2OH$$

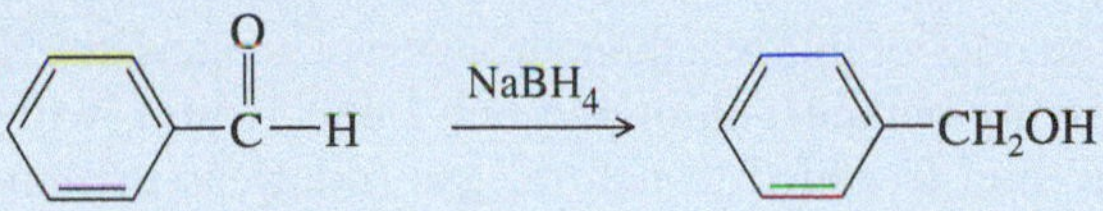

benzaldehyde benzyl alcohol

(2) Reduction of a ketone gives a secondary alcohol

$$\text{cyclohexanone} \xrightarrow{NaBH_4} \text{cyclohexanol}$$

cyclohexanone → cyclohexanol

(3) Reduction of an acid or ester gives a primary alcohol

$$\left.\begin{array}{c} CH_3-(CH_2)_8-\overset{\overset{O}{\|}}{C}-OH \\ \text{decanoic acid} \\ CH_3-(CH_2)_8-\overset{\overset{O}{\|}}{C}-OCH_3 \\ \text{methyl decanoate} \end{array}\right\} \xrightarrow[(2)\ H_3O^+]{(1)\ LiAlH_4} \begin{array}{c} CH_3-(CH_2)_8-CH_2-OH \\ \text{decan-1-ol} \end{array}$$

10-12 Thiols (Mercaptans)

Thiols are sulfur analogs of alcohols, with an —SH group in place of the alcohol —OH group. Oxygen and sulfur are in the same column of the periodic table (group 6A), with oxygen in the second row and sulfur in the third. IUPAC names for thiols are derived from the alkane names, using the suffix *-thiol*. Thiols are also called **mercaptans** ("captures mercury") because they form stable heavy-metal derivatives. Common names are formed like those of alcohols, using the name of the alkyl group with the word *mercaptan*. The —SH group itself is called a *mercapto* group.

	CH_3-SH	$CH_3CH_2CH_2CH_2-SH$	$CH_3CH{=}CHCH_2-SH$	$HS-CH_2CH_2-OH$
IUPAC name:	methanethiol	butane-1-thiol	but-2-ene-1-thiol	2-mercaptoethanol
common name:	methyl mercaptan	*n*-butyl mercaptan		

Offensive use of thiols. Skunks spray thiols to protect themselves from people, dogs, and other animals.

Thiols' ability to complex heavy metals has proved useful for making antidotes to heavy-metal poisoning. For example, in World War II the Allies were concerned that the Germans would use lewisite, a volatile arsenic compound, as a chemical warfare agent. Thiols complex strongly with arsenic, and British scientists developed dimercaprol (2,3-dimercaptopropan-1-ol) as an effective antidote. The Allies came to refer to this compound as "British anti-lewisite" (BAL), a name that is still used. Dimercaprol is useful against a variety of heavy metals, including arsenic, mercury, and gold.

$$\text{lewisite: } (Cl)(H)C{=}C(H)(AsCl_2) \qquad \text{dimercaprol: } \underset{SH}{CH_2}-\underset{SH}{CH}-\underset{OH}{CH_2}$$

lewisite

dimercaprol
British anti-lewisite (BAL)

The odor of thiols is their strongest characteristic. **Skunk** scent is composed mainly of 3-methylbutane-1-thiol and but-2-ene-1-thiol, with small amounts of other thiols. Ethanethiol is added to natural gas (odorless methane) to give it the characteristic "gassy" odor for detecting leaks.

Although oxygen is more electronegative than sulfur, thiols are more acidic than alcohols. Their enhanced acidity results from two effects: First, S—H bonds are generally weaker than O—H bonds, making S—H bonds easier to break. Second, the **thiolate ion** ($R-S^-$) has its negative charge on sulfur, which allows the charge to be delocalized over a larger region than the negative charge of an alkoxide ion, borne on a smaller oxygen atom. Thiolate ions are easily formed by treating the thiol with aqueous sodium hydroxide.

$$CH_3{-}CH_2{-}SH + {}^{-}OH \rightleftharpoons CH_3{-}CH_2{-}S^{-} + H_2O$$

ethanethiol, $pK_a = 10.5$; ethanethiolate; H_2O, $pK_a = 15.7$

$$C_6H_5{-}SH + {}^{-}OH \rightleftharpoons C_6H_5{-}S^{-} + H_2O$$

benzenethiol (thiophenol), $pK_a = 7.8$; benzenethiolate; H_2O, $pK_a = 15.7$

For comparison

$$CH_3{-}CH_2{-}OH + {}^{-}OH \rightleftharpoons CH_3{-}CH_2{-}O^{-} + H_2O$$

ethanol, $pK_a = 15.9$; ethoxide; H_2O, $pK_a = 15.7$

PROBLEM 10-27

Arrange the following compounds in order of decreasing acidity.
CH_3COOH CH_3OH CH_3CH_3 CH_3SO_3H CH_3NH_2 CH_3SH $CH_3C{\equiv}CH$

Thiols can be prepared by S_N2 reactions of sodium hydrosulfide with unhindered alkyl halides. The thiol product is still nucleophilic, so a large excess of hydrosulfide is used to prevent the product from undergoing a second alkylation to give a sulfide (R—S—R).

$$Na^{+}\ H{-}\ddot{S}{:}^{-} + R{-}X \longrightarrow R{-}SH + Na^{+}\ {}^{-}X$$

sodium hydrosulfide; alkyl halide; thiol

Unlike alcohols, thiols are easily oxidized to give a dimer called a **disulfide.** The reverse reaction, reduction of the disulfide to the thiol, takes place under reducing conditions. Formation and cleavage of disulfide linkages is an important aspect of protein chemistry (Chapter 24), where disulfide "bridges" between cysteine amino acid residues hold the protein chain in its active conformation. More examples of disulfide bridges appear in Section 24-8C.

$$R{-}SH + HS{-}R \underset{Zn,\ HCl}{\overset{Br_2}{\rightleftharpoons}} R{-}S{-}S{-}R + 2\,HBr$$

two molecules of thiol; disulfide

Example

$$2\ H_2N{-}CH(CH_2SH){-}C(=O){-}OH \underset{[H]\ (reduce)}{\overset{[O]\ (oxidize)}{\rightleftharpoons}} H_2N{-}CH({-}C(=O)OH){-}CH_2{-}S{-}S{-}CH_2{-}CH(NH_2){-}C(=O){-}OH + H_2O$$

two cysteine residues; cystine disulfide bridge

Application: Herbs

Garlic has served throughout history as a remedy for numerous diseases. Movies have depicted its mystic power to repel werewolves and vampires. The characteristic garlic odor derives from the many sulfur compounds it contains. One of the major constituents is allicin, a compound with antibacterial properties.

$$CH_2{=}CH{-}CH_2{-}S(=O){-}S{-}CH_2{-}CH{=}CH_2$$

allicin

Application: Enzymes

A protein can be made to withstand higher temperatures by the introduction of additional disulfide bonds. For example, laundry detergents frequently contain enzymes to remove protein and blood stains. By replacing strategically located amino acid residues with cysteines, a modified enzyme results that stays in its active conformation in hot water.

Just as mild oxidation converts thiols to disulfides, vigorous oxidation converts them to **sulfonic acids.** $KMnO_4$ or nitric acid (HNO_3), or even bleach (NaOCl), can be used as the oxidant for this reaction. Any Lewis structure of a sulfonic acid requires either separation of formal charges or more than 8 electrons around sulfur. Sulfur can have an expanded octet, as it does in SF_4 (10 electrons) and SF_6 (12 electrons). The three resonance forms shown here are most commonly used. Organic chemists tend to use the form with an expanded octet, and inorganic chemists tend to use the forms with charge separation.

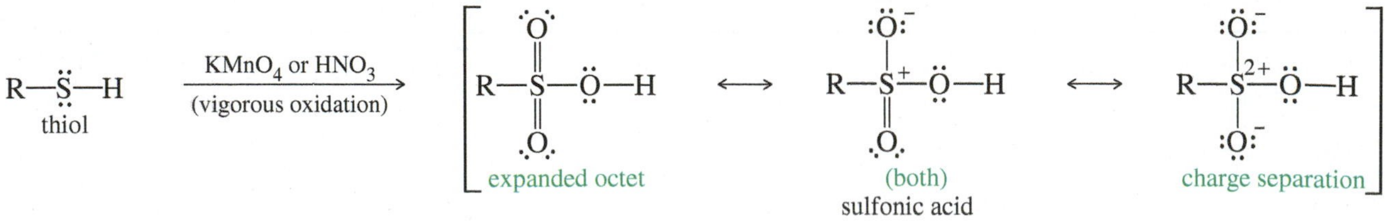

R—S—H (thiol) $\xrightarrow[\text{(vigorous oxidation)}]{KMnO_4 \text{ or } HNO_3}$ [R—S(=O)(=O)—O—H (expanded octet) ⟷ R—S^{+}(—O^{-})(=O)—O—H (both) ⟷ R—S^{2+}(—O^{-})(—O^{-})—O—H (charge separation)] sulfonic acid

Example

benzenethiol (C₆H₅—SH) $\xrightarrow[\text{(boil)}]{HNO_3}$ benzenesulfonic acid (C₆H₅—S(=O)(=O)—OH)

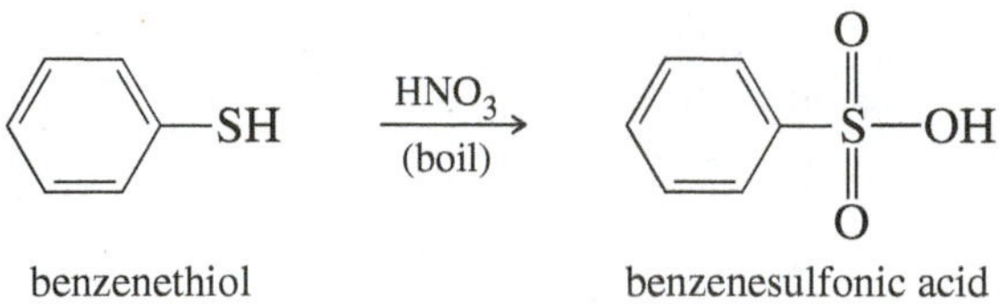

Application: Biochemistry

Glutathione, a tripeptide containing a thiol group, serves as a mild reducing agent to detoxify peroxides and maintain the cysteine residues of hemoglobin in the reduced state. Glutathione can also detoxify alkylating agents. For example, the thiol of glutathione reacts with methyl iodide by an S_N2 reaction, rendering the methyl iodide harmless and preventing its reaction with other molecules in the body.

PROBLEM 10-28

Give IUPAC names for the following compounds.

(a) 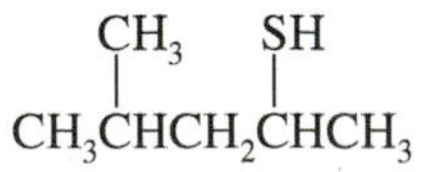 $CH_3CH(CH_3)CH_2CH(SH)CH_3$

(b) 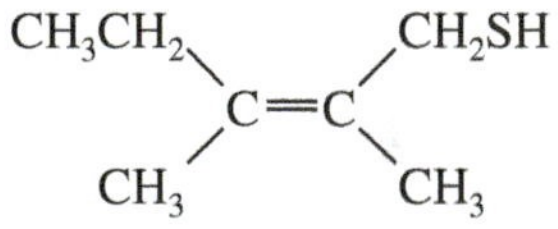 $(CH_3CH_2)(CH_3)C{=}C(CH_2SH)(CH_3)$

(c) cyclohex-2-ene ring bearing SH

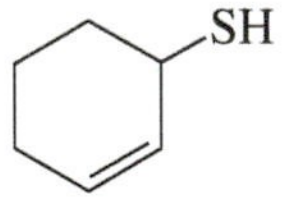

PROBLEM 10-29

Authentic skunk spray has become valuable for use in scent-masking products. Show how you would synthesize the two major components of skunk spray (3-methylbutane-1-thiol and but-2-ene-1-thiol) from any of the readily available butenes or from buta-1,3-diene.

SUMMARY Synthesis of Alcohols from Carbonyl Compounds

From aldehydes and ketones

aldehyde when R′ = H;
ketone when R and R′ are alkyl or aryl

$R{-}C({=}O){-}R' \xrightarrow{NaBH_4,\ ROH} R{-}C(OH)(H){-}R'$ — 1° or 2° alcohol depending on R′; Section 10-11A

$R{-}C({=}O){-}R' \xrightarrow{LiAlH_4} \xrightarrow{H{-}OH} R{-}C(OH)(H){-}R'$ — 1° or 2° alcohol depending on R′; Section 10-11B

$R{-}C({=}O){-}R' \xrightarrow{R''{-}C{\equiv}C:^-} \xrightarrow{H{-}OH} R{-}C(OH)(C{\equiv}C{-}R''){-}R'$ — 2° or 3° alcohol depending on R′; Section 9-7B

$R{-}C({=}O){-}R' \xrightarrow{R''{-}Li \text{ or } R''{-}MgX} \xrightarrow{H{-}OH} R{-}C(OH)(R''){-}R'$ — 2° or 3° alcohol depending on R′; Section 10-9ABC

From esters and acid chlorides

$R{-}C({=}O){-}Cl \xrightarrow{NaBH_4} \xrightarrow{H{-}OH} R{-}C(OH)(H){-}H$ — 1° alcohol. Section 10-11A

$NaBH_4$ reduces acid chlorides only, not esters.

$R{-}C({=}O){-}Cl$ or $R{-}C({=}O){-}OR' \xrightarrow{LiAlH_4} \xrightarrow{H{-}OH} R{-}C(OH)(H){-}H$ — 1° alcohol. Section 10-11B

$R{-}C({=}O){-}Cl$ or $R{-}C({=}O){-}OR' \xrightarrow{R''{-}Li \text{ or } R''{-}MgX} \xrightarrow{H{-}OH} R{-}C(OH)(R''){-}R''$ — 1° alcohol. Section 10-9D

Essential Terms

acid derivatives Compounds that are related to carboxylic acids but have other electron-withdrawing groups in place of the —OH group of the acid. Three examples are **acid chlorides, esters,** and **amides.** (p. 481)

$R{-}C({=}O){-}OH$ carboxylic acid; $R{-}C({=}O){-}Cl$ acid chloride; $R{-}C({=}O){-}O{-}R'$ ester; $R{-}C({=}O){-}NH_2$ amide

alcohol A compound in which a hydrogen atom of a hydrocarbon has been replaced by a hydroxy group, —OH. (p. 460) Alcohols are classified as **primary, secondary,** or **tertiary** depending on whether the hydroxy group is bonded to a primary, secondary, or tertiary carbon atom. (p. 461)

$R{-}C(OH)(H){-}H$ primary alcohol; $R{-}C(OH)(H){-}R$ secondary alcohol; $R{-}C(OH)(R){-}R$ tertiary alcohol

aldehyde A carbonyl compound with one alkyl group and one hydrogen on the carbonyl group. **Formaldehyde** has two hydrogens on the carbonyl group. (p. 478)

alkoxide ion The anion (R—Ö:⁻) formed by deprotonation of an alcohol. (p. 470)

azeotrope A mixture of two or more liquids that distills at a constant temperature and gives a distillate of definite composition. For example, a mixture of 95% ethanol and 5% water has a lower boiling point than pure ethanol or pure water. (p. 470)

carbinol carbon atom In an alcohol, the carbon atom bonded to the hydroxy group. (p. 461)

denatured alcohol A form of ethanol containing toxic impurities, making it unfit for drinking. (p. 470)

diol A compound with two alcohol —OH groups. (p. 464)

disulfide The oxidized dimer of a thiol, R—S—S—R. (p. 495)

epoxides **(oxiranes)** Compounds containing oxygen in a three-membered ring. (p. 483)

ester An acid derivative in which the hydroxy group of the carboxylic acid is replaced by an alkoxy (—OR′) group. (p. 481)

$$\underset{\text{acid}}{R-C(=O)-OH} + \underset{\text{alcohol}}{R'-OH} \rightleftarrows \underset{\text{ester}}{R-C(=O)-OR'} + \underset{\text{water}}{H_2O}$$

glycol Synonymous with **diol.** The term *glycol* is most commonly applied to the 1,2-diols, also called **vicinal diols.** (pp. 464–465)

grain alcohol Ethanol, ethyl alcohol. **Absolute alcohol** is 100% ethanol. (pp. 469–470)

Grignard reagent An organomagnesium halide, written in the form R—Mg—X. The actual reagent is more complicated in structure, usually a dimer or trimer complexed with several molecules of ether. (p. 476)

hydride reagent A compound of hydrogen with a less-electronegative element, so the hydrogen can be donated with its pair of electrons. Simple hydrides such as NaH and LiH tend to be more basic than nucleophilic, often reacting with H^+ to give H_2 gas. The complex hydrides $NaBH_4$ and $LiAlH_4$ tend to form new H—C bonds, reducing carbonyl groups as shown in the following general reaction. (p. 489)

$$\underset{\text{hydride reagent}}{M^+ \; H:^-} + \;>C{=}\ddot{O}: \longrightarrow \underset{\text{reduced}}{H-\overset{|}{\underset{|}{C}}-\ddot{O}:^- \; ^+M}$$

hydrophilic ("water loving") Attracted to water, water-soluble. (p. 467)

hydrophobic ("water hating") Repelled by water, water-insoluble. (p. 467)

hydroxy group **(hydroxyl group)** The —OH group, as in an alcohol. (Strictly speaking, the term *hydroxyl* should apply only to the hydroxyl radical, ·OH.) (pp. 460, 463)

ketone A carbonyl compound with two alkyl groups bonded to the carbonyl group. (p. 478)

lithium dialkylcuprate **(Gilman reagent)** An organometallic reagent used to couple with an alkyl halide. (p. 485)

$$\underset{\text{lithium dialklcuprate}}{R_2CuLi} + R'-X \longrightarrow R-R' + R-Cu + LiX$$

mercaptan **(thiol)** The sulfur analogue of an alcohol, R—SH. The —SH group is called a **mercapto** group. (p. 494)

miscible Mutually soluble in any proportions. (p. 467)

organolithium reagent An organometallic reagent of the form R—Li. (p. 477)

organometallic compounds (organometallic reagents) Compounds containing metal atoms directly bonded to carbon. (p. 475)

phenol A compound with a hydroxy group bonded directly to an aromatic ring. (p. 461)

Raney nickel A finely divided nickel/aluminum alloy that has been treated with NaOH to dissolve out most of the aluminum. Used as a catalyst for the hydrogenation of ketones and aldehydes to alcohols. (p. 492)

rubbing alcohol Propan-2-ol, isopropyl alcohol. (p. 470)

skunk (*noun*) A plantigrade omnivorous quadruped that effectively synthesizes thiols; (*verb*) to prevent from scoring in a game or contest. (p. 494)

sulfonic acid A strongly acidic compound of formula $R-SO_3H$, formed by vigorous oxidation of a thiol. (p. 496)

thiol **(mercaptan)** The sulfur analogue of an alcohol, R—SH. (p. 494)

thiolate ion **(mercaptide)** The R—S̈:⁻ anion, formed by deprotonation of a thiol. (pp. 494–495)

wood alcohol Methanol, methyl alcohol. (p. 468)

Essential Problem-Solving Skills in Chapter 10

Each skill is followed by problem numbers exemplifying that particular skill.

	Skill	Problems
1	Draw structures and assign names for alcohols, phenols, diols, and thiols.	Problems 10-30, 31, and 32
2	Predict relative boiling points, acidities, and solubilities of alcohols.	Problems 10-33, 34, 35, 42, and 53
3	Show how to convert alkenes, alkyl halides, and carbonyl compounds to alcohols.	Problems 10-37, 38, 40, 41, 43, 52, 56, 57, and 57
4	Predict the alcohol products of the hydration, hydroboration, and dihydroxylation of alkenes.	Problems 10-36, 47, 49, and 51
5	Use Grignard and organolithium reagents to synthesize primary, secondary, and tertiary alcohols with the required carbon skeletons.	Problems 10-36, 37, 38, 39, 41, 44, 47, 55, 56, 57, and 58
6	Propose syntheses and oxidation products for simple thiols.	Problems 10-37, 40, and 50

Study Problems

10-30 Give a systematic (IUPAC) name for each alcohol. Classify each as primary, secondary, or tertiary.

(a) 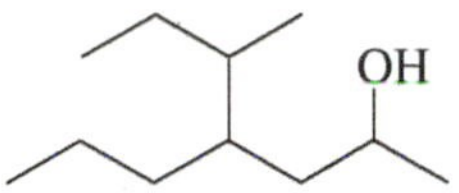 (b) 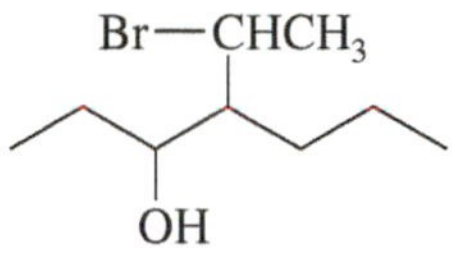 (c) 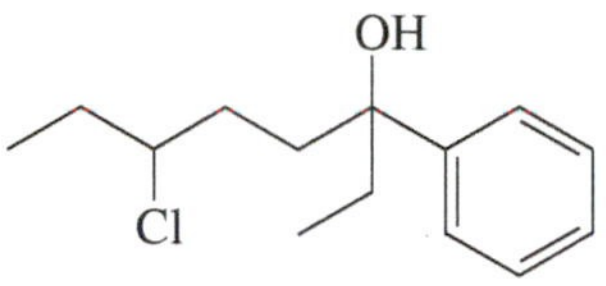 (d) 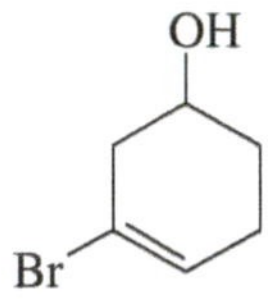 (e)  f) 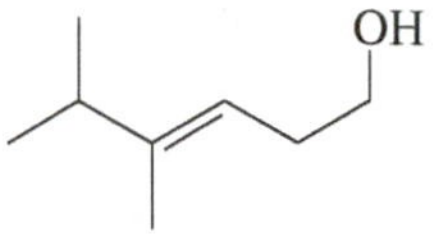 (g)

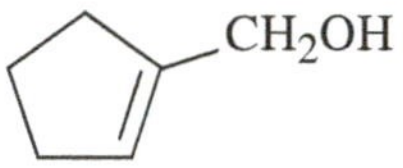

10-31 Give systematic (IUPAC) names for the following diols and phenols.

(a) 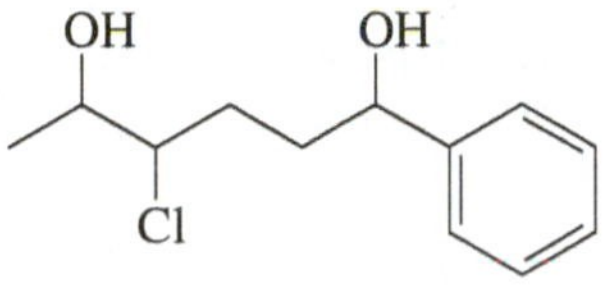 (b) 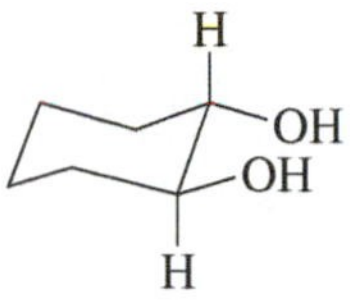 (c) 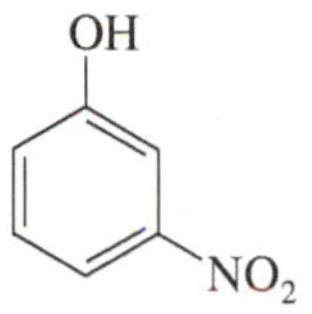 (d)

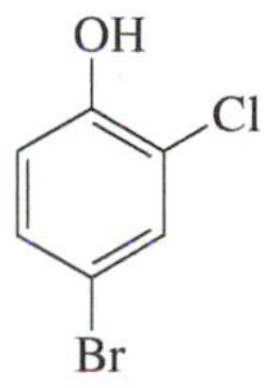

10-32 Draw the structures of the following compounds. (Includes both new and old names.)
(a) triphenylmethanol
(b) 4-(chloromethyl)heptan-3-ol
(c) 2-cyclohexen-1-ol
(d) 3-cyclopentylhexan-3-ol
(e) *meso*-2,4-pentanediol
(f) cyclopentene glycol
(g) 3-(iodomethyl)phenol
(h) (2*R*,3*R*)-2,3-hexanediol
(i) cyclopent-3-ene-1-thiol
(j) dimethyl disulfide
(k) 3-methylhex-4-yn-2-ol

10-33 Predict which member of each pair has the higher boiling point, and explain the reasons for your predictions.
(a) hexan-1-ol or 3,3-dimethylbutan-1-ol
(b) hexan-2-one or hexan-2-ol
(c) hexan-2-ol or hexane-1,5-diol
(d) pentan-2-ol or hexan-2-ol

10-34 Predict which member of each pair is more acidic, and explain the reasons for your predictions.
(a) cyclopentanol or 3-chlorophenol
(b) cyclohexanol or cyclohexanethiol
(c) cyclohexanol or cyclohexanecarboxylic acid
(d) butan-1-ol or 2,2-dichlorobutan-1-ol

10-35 Predict which member of each group is most soluble in water, and explain the reasons for your predictions.
(a) butan-1-ol, pentan-1-ol, or propan-2-ol
(b) chlorocyclohexane, cyclohexanol, or cyclohexane-1,2-diol
(c) phenol, cyclohexanol, or 4-methylcyclohexanol

10-36 Draw the organic products you would expect to isolate from the following reactions (after hydrolysis).

(a) 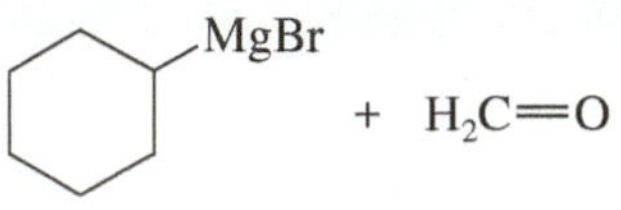 MgBr + $H_2C{=}O$ **(b)** MgCl + O **(c)** $CH_3{-}CH(MgI){-}CH_3$ + Ph—CHO

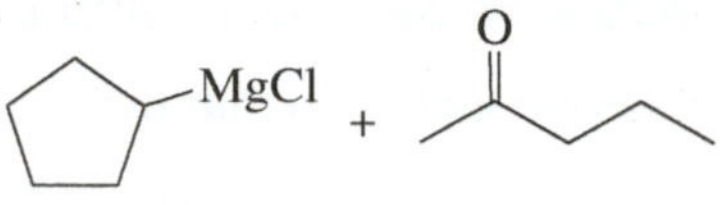

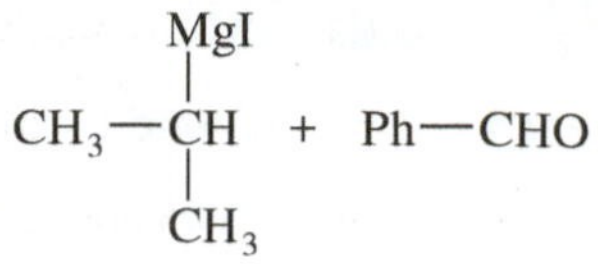

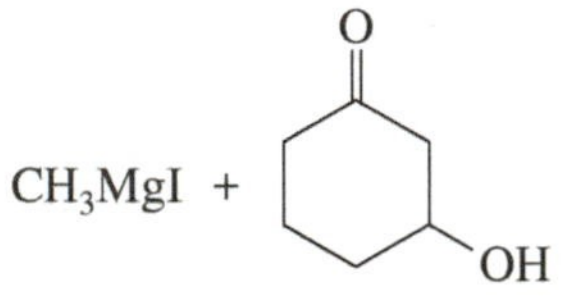

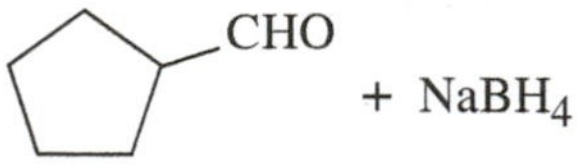

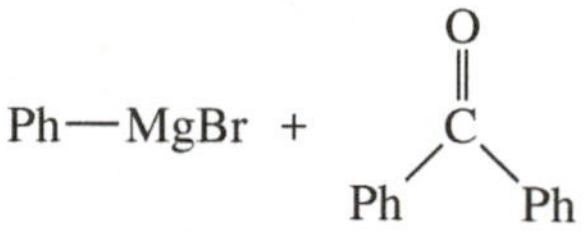

(d) CH_3MgI + O OH **(e)** CHO + $NaBH_4$ **(f)** Ph—MgBr + Ph—C(=O)—Ph

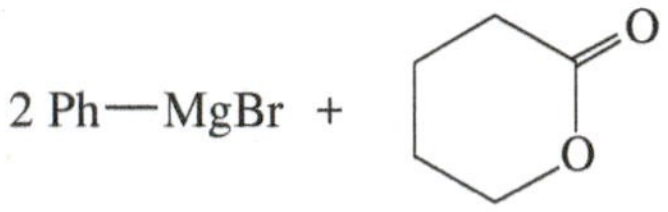

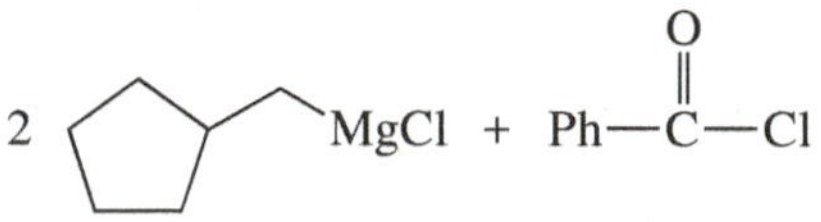

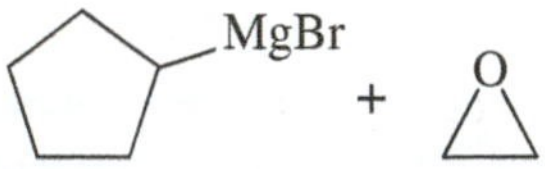

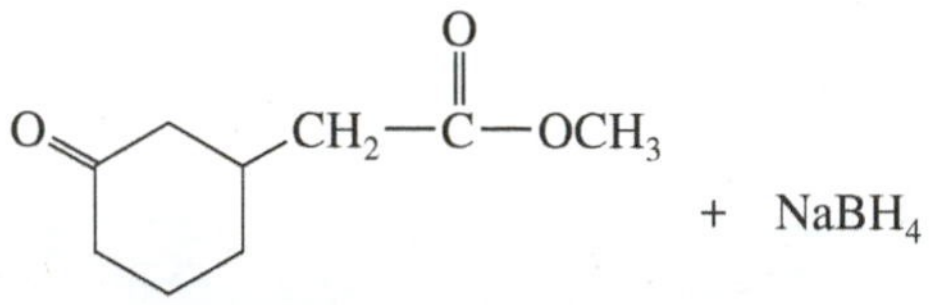

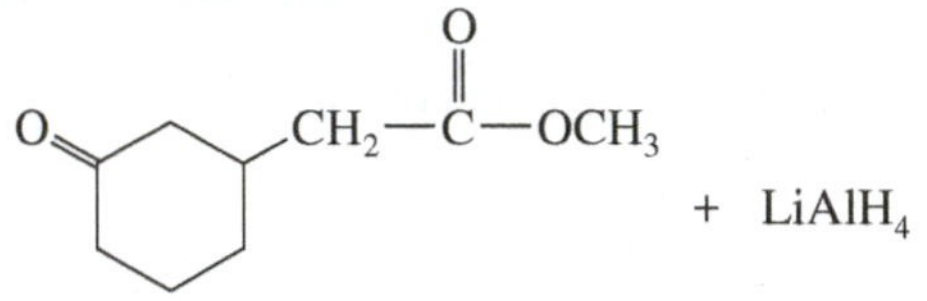

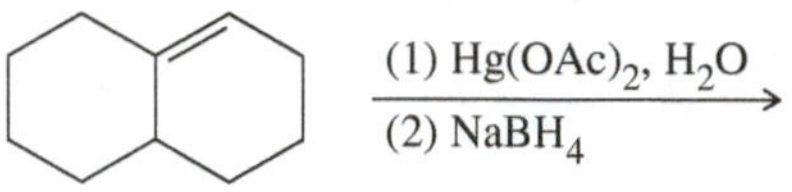

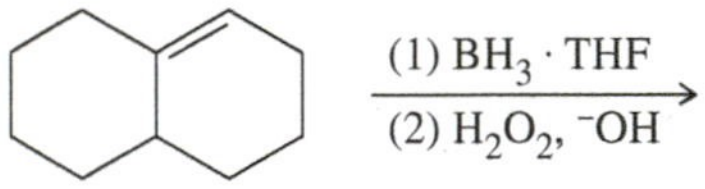

(g) 2 Ph—MgBr + O O **(h)** 2 MgCl + Ph—C(=O)—Cl

(i) MgBr + O **(j)** O $CH_2{-}C(=O){-}OCH_3$ + $NaBH_4$

(k) O $CH_2{-}C(=O){-}OCH_3$ + $LiAlH_4$ **(l)** (1) $Hg(OAc)_2$, H_2O (2) $NaBH_4$

(m) (1) $BH_3 \cdot THF$ (2) H_2O_2, ^-OH **(n)** H_3C, H, C=C, H, $CH_2CH_2CH_3$ cold, dilute $KMnO_4$ ^-OH

(o) H_3C, H, C=C, H, $CH_2CH_2CH_3$ HCO_3H H_3O^+ **(p)** $(CH_2{=}CH)_2CuLi$ + $CH_3CH_2CH{=}CHCH_2Br$

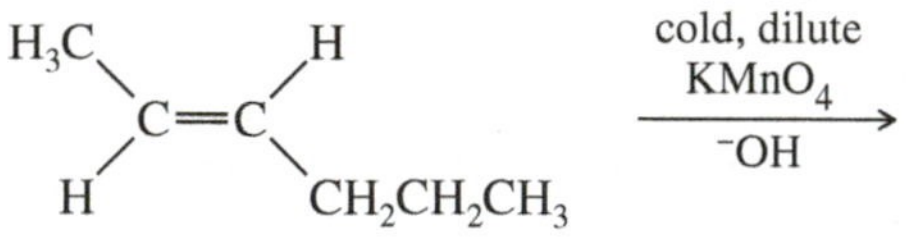

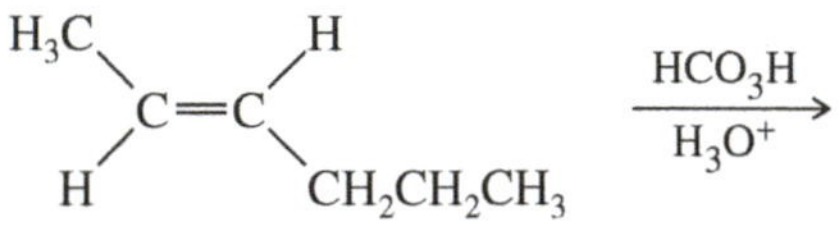

10-37 Starting from bromobenzene and any other reagents and solvents you need, show how you would synthesize the following compounds. Any of these products may be used as starting materials in subsequent parts of this problem.

(a) 1-phenylpropan-1-ol **(b)** 1-phenylpropene **(c)** 1-phenylpropan-2-ol
(d) 3-phenylprop-2-en-1-ol **(e)** 2-phenylbutan-2-ol **(f)** 2-phenylbut-2-ene

10-38 Show how you would synthesize the following alcohols from appropriate alkenes.

(a) 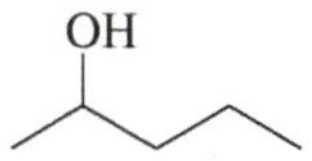 OH **(b)** 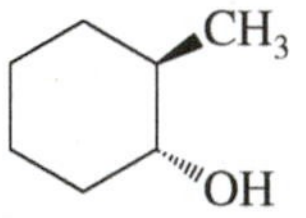 CH_3 OH

(c) 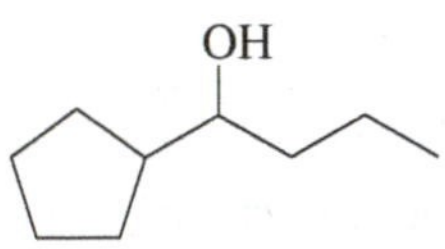 OH **(d)** 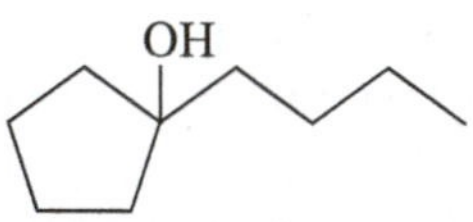 OH

10-39 Show how you would use Grignard syntheses to prepare the following alcohols from the indicated starting materials and any other necessary reagents.

(a) octan-3-ol from hexanal, $CH_3(CH_2)_4CHO$
(b) octan-1-ol from 1-bromoheptane
(c) 1-cyclohexylethanol from acetaldehyde, CH_3CHO
(d) 2-cyclohexylethanol from bromocyclohexane
(e) benzyl alcohol (Ph—CH_2 —OH) from bromobenzene (Ph—Br)

(f) 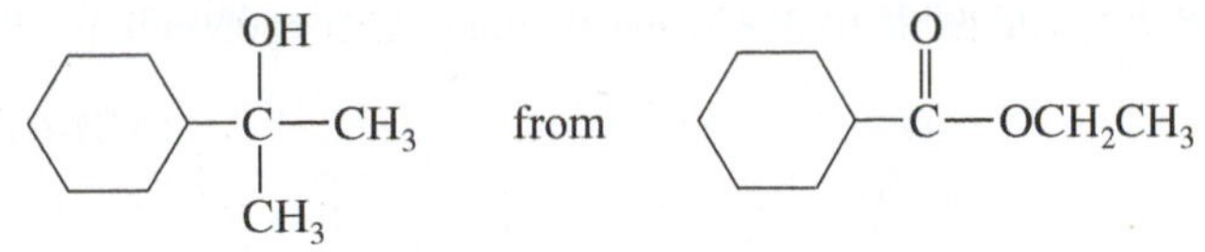 [cyclohexyl–C(OH)(CH$_3$)–CH$_3$] from [cyclohexyl–C(=O)–OCH$_2$CH$_3$]

(g) cyclopentylphenylmethanol from benzaldehyde (Ph—CHO)
(h) octan-1-ol from 1-bromohexane

10-40 Show how you would accomplish the following transformations. You may use any additional reagents you need.

(a) 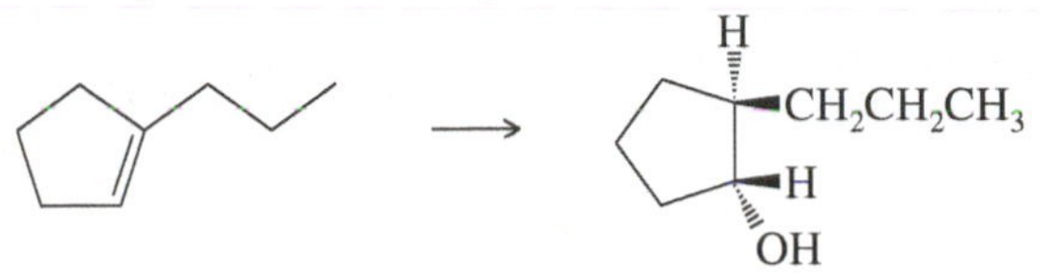 [1-propylcyclopentene] ⟶ [cyclopentane with H, $CH_2CH_2CH_3$, H, OH] (racemic)

(b) 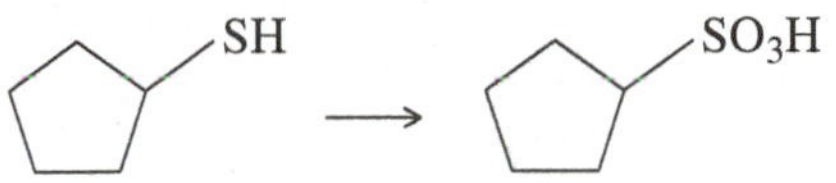 [cyclopentyl–SH] ⟶ [cyclopentyl–SO_3H]

(c)  [cyclohex-2-en-1-one] ⟶ [cyclohex-2-en-1-ol]

(d)  [cyclohex-2-en-1-one] ⟶ [cyclohexanone]

(e)  $CH_3—C(=O)—CH_2CH_2—C(=O)—OCH_2CH_3$ ⟶ $CH_3—CH(OH)—CH_2CH_2—C(=O)—OCH_2CH_3$

(f) 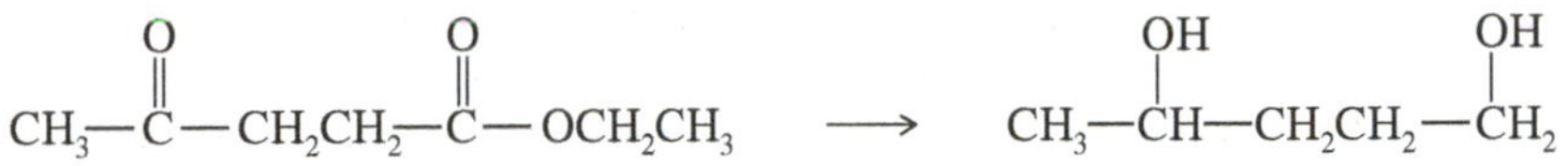 $CH_3—C(=O)—CH_2CH_2—C(=O)—OCH_2CH_3$ ⟶ $CH_3—CH(OH)—CH_2CH_2—CH_2(OH)$

10-41 Show how you would synthesize the following:
(a) 2-phenylethanol by the addition of formaldehyde to a suitable Grignard reagent
(b) 2-phenylethanol from a suitable alkene
(c) cyclohexylmethanol from an alkyl halide using an S_N2 reaction
(d) 3-cyclohexylpropan-1-ol by the addition of ethylene oxide to a suitable Grignard reagent
(e) *cis*-pent-2-en-1-thiol from a suitable alkenyl halide
(f) 2,5-dimethylhexane from a four-carbon alkyl halide

10-42 Complete the following acid–base reactions. In each case, indicate whether the equilibrium favors the reactants or the products, and explain your reasoning.

(a) 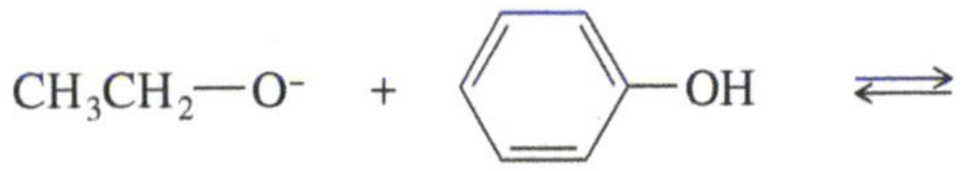 $CH_3CH_2—O^-$ + [phenol, Ph–OH] $\rightleftharpoons$

(b) 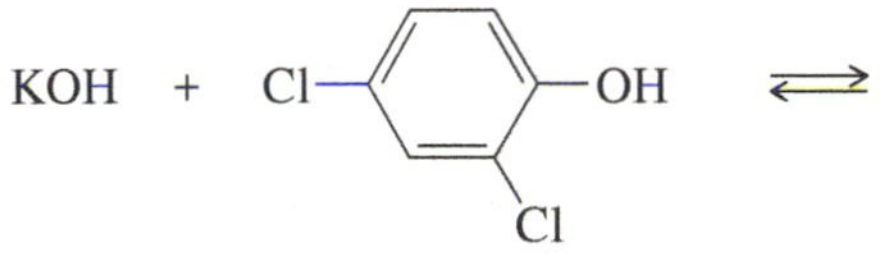 KOH + [2,4-dichlorophenol: Cl–C$_6$H$_3$(Cl)–OH] $\rightleftharpoons$

(c) [1-naphthol, OH] + CH_3O^- $\rightleftharpoons$

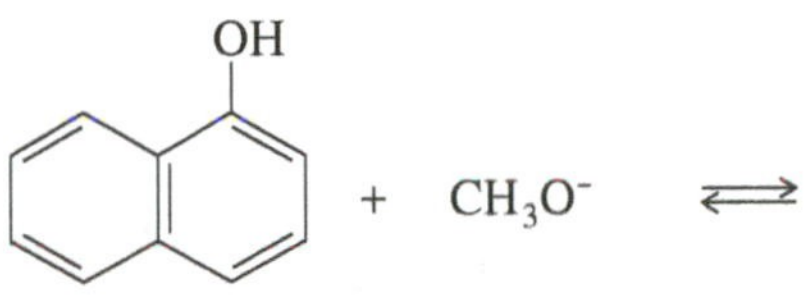

(d) 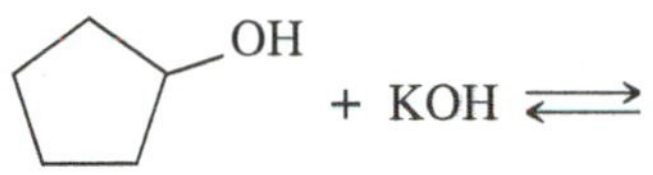 [cyclopentanol, OH] + KOH $\rightleftharpoons$

(e) $(CH_3)_3C—O^-$ + CH_3CH_2OH $\rightleftharpoons$

(f) $(CH_3)_3C—O^-$ + H_2O $\rightleftharpoons$

(g) KOH + CH_3CH_2OH $\rightleftharpoons$

10-43 Suggest carbonyl compounds and reducing agents that might be used to form the following alcohols.
(a) octan-1-ol
(b) 1-cyclohexylpropan-1-ol
(c) 1-phenylbutan-1-ol
(d) 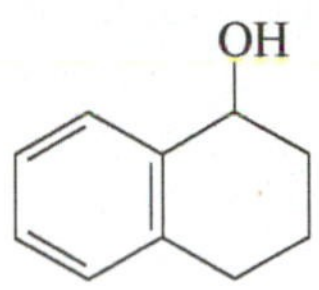 [1,2,3,4-tetrahydronaphthalen-1-ol, OH]
(e) 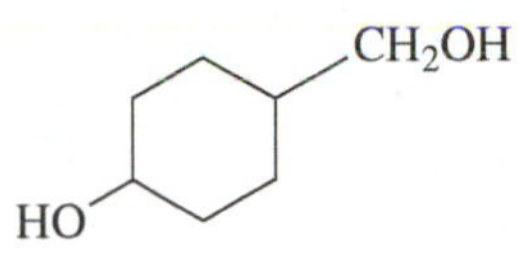 [4-(hydroxymethyl)cyclohexanol: HO–C$_6$H$_{10}$–CH_2OH]
(f) 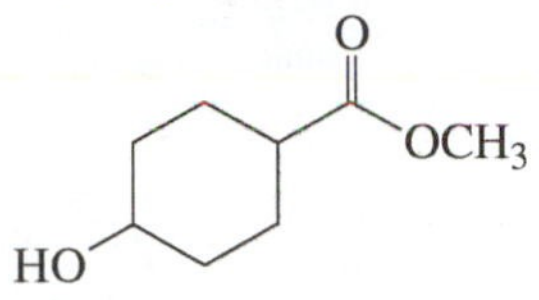 [methyl 4-hydroxycyclohexanecarboxylate: HO–C$_6$H$_{10}$–C(=O)OCH_3]

10-44 Show how you would synthesize the following compounds from any starting materials containing no more than six carbon atoms.

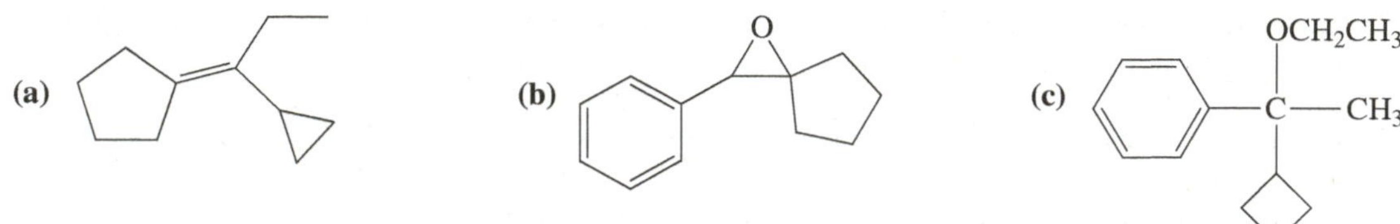

***10-45** Geminal diols, or 1,1-diols, are usually unstable, spontaneously losing water to give carbonyl compounds. Therefore, geminal diols are regarded as hydrated forms of ketones and aldehydes. Propose a mechanism for the acid-catalyzed loss of water from propane-2,2-diol to give acetone.

$$CH_3-C(OH)_2-CH_3 \underset{}{\overset{H^+}{\rightleftharpoons}} CH_3-\overset{O}{\overset{\|}{C}}-CH_3 + H_2O$$

propane-2,2-diol — acetone

10-46 Vinyl alcohols are generally unstable, quickly isomerizing to carbonyl compounds. Propose mechanisms for the following isomerizations.

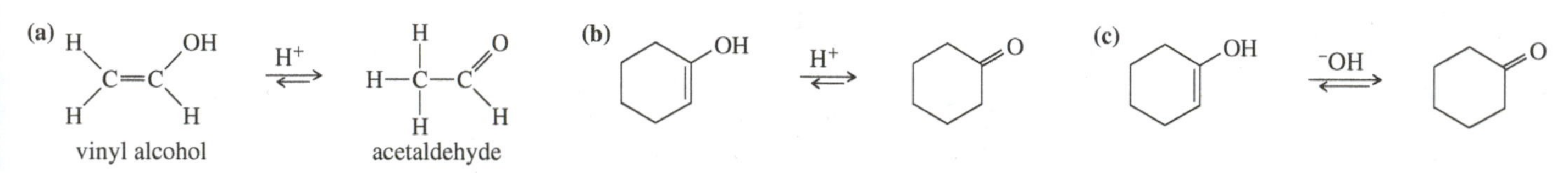

10-47 Compound A ($C_7H_{11}Br$) is treated with magnesium in ether to give **B** ($C_7H_{11}MgBr$), which reacts violently with D_2O to give 1-methylcyclohexene with a deuterium atom on the methyl group (**C**). Reaction of **B** with acetone (CH_3COCH_3) followed by hydrolysis gives **D** ($C_{10}H_{18}O$). Heating **D** with concentrated H_2SO_4 gives **E** ($C_{10}H_{16}$), which decolorizes two equivalents of Br_2 to give **F** ($C_{10}H_{16}Br_4$). **E** undergoes hydrogenation with excess H_2 and a Pt catalyst to give isobutylcyclohexane. Determine the structures of compounds **A** through **F**, and show your reasoning throughout.

***10-48** Grignard reagents react slowly with oxetane to produce primary alcohols. Propose a mechanism for this reaction, and suggest why oxetane reacts with Grignard reagents even though most ethers do not.

$$R-Mg-X \quad + \quad \text{oxetane} \quad \longrightarrow \quad R-CH_2CH_2CH_2-O^- \; ^+MgX$$

Grignard reagent — oxetane — salt of 1° alcohol

***10-49** Determine the structures of compounds **A** through **G**, including stereochemistry where appropriate.

Cyclopentanone —(1) CH_3MgI, (2) H_3O^+→ $C_6H_{12}O$ **A** —H_2SO_4, heat→ C_6H_{10} **B** —H_2, Pt→ methylcyclopentane

$C_6H_{12}O$ **D** —H_2SO_4, heat→ **B**

Cyclopentene —$PhCO_3H$→ C_5H_8O **C** —(1) CH_3MgI, (2) H_3O^+→ **D**

Cyclopentene —HBr→ C_5H_9Br **E** —(1) Mg, ether; (2) cyclopentanone; (3) H_3O^+→ $C_{10}H_{18}O$ **F** —H_2SO_4, heat→ $C_{10}H_{16}$ **G** —$KMnO_4$, warm, concd.→ cyclopentanone

***10-50** Many hunting dogs enjoy standing nose-to-nose with a skunk while barking furiously, oblivious to the skunk spray directed toward them. One moderately effective way of lessening the amount of odor is to wash the dog in a bath containing dilute hydrogen peroxide, sodium bicarbonate, and some mild dish detergent. Use chemical reactions to describe how this mixture helps to remove the skunk spray from the dog. The two major components of skunk oil are 3-methylbutane-1-thiol and but-2-ene-1-thiol.

10-51 Propose structures for intermediates and products (**A**) through (**K**).

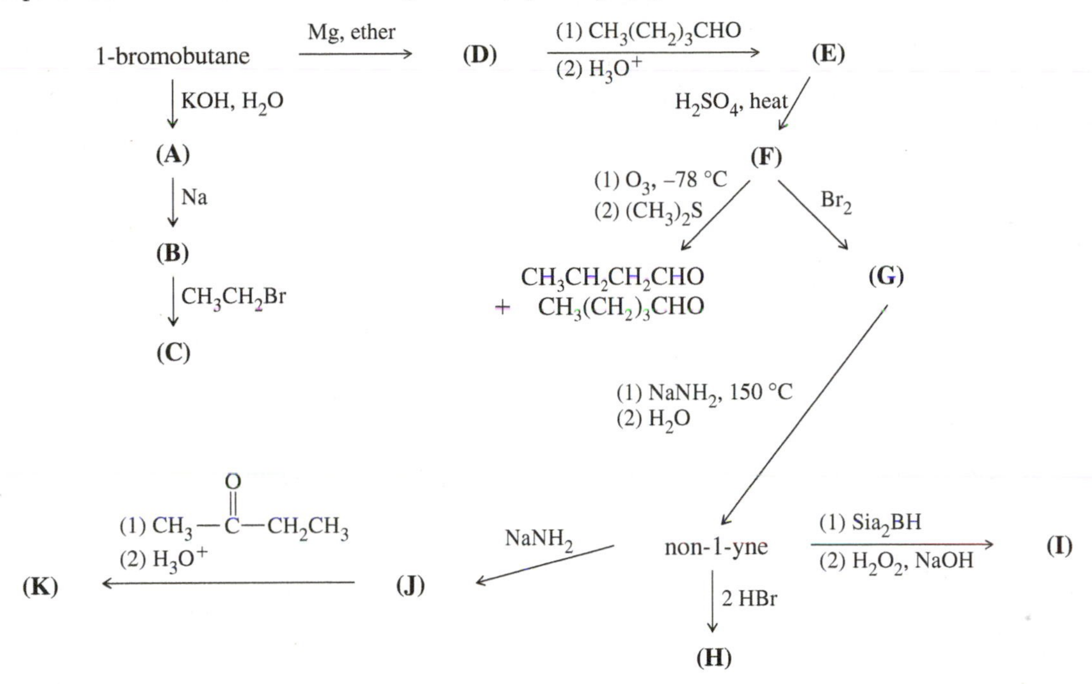

10-52 Devise a synthesis for each compound, starting with methylenecyclohexane and any other reagents you need.
(a) 1-methylcyclohexanol
(b) cyclohexylmethanol
(c) 1-(hydroxymethyl)cyclohexanol
(d) *trans*-2-methylcyclohexanol
(e) 2-chloro-1-methylcyclohexanol
(f) 1-(phenylmethyl)cyclohexanol

10-53 Compare the properties of propan-2-ol (**I**) and the hexafluoro analog (**II**).

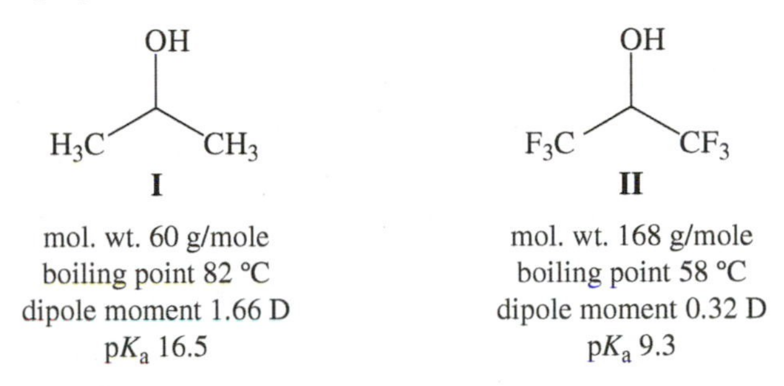

I	**II**
mol. wt. 60 g/mole	mol. wt. 168 g/mole
boiling point 82 °C	boiling point 58 °C
dipole moment 1.66 D	dipole moment 0.32 D
pK_a 16.5	pK_a 9.3

(a) Compound **II** has almost triple the molecular weight of **I**, but **II** has a *lower* boiling point. Explain.
(b) Explain why the dipole moment of Compound **II** is much lower than the dipole moment of **I**, despite the presence of the six electronegative fluorine atoms.
(c) Why is **II** a stronger acid than **I**?

***10-54** Compounds containing deuterium (D = ^2H) are useful for kinetic studies and metabolic studies with new pharmaceuticals. One way to introduce deuterium is by using the reagent $LiAlD_4$, equivalent in reactivity to $LiAlH_4$. Show how to make these deuterium-labeled compounds, using $LiAlD_4$ and D_2O as your sources of deuterium, and any non-deuterated starting materials you wish.
(a) CH_3CHDOH
(b) CH_3CD_2OH
(c) CH_3CD_2OD

***10-55** Show how to make these deuterium-labeled compounds, using CD_3MgBr and D_2O as your sources of deuterium, and any non-deuterated starting materials you wish.
(a) $CH_3CH(OD)CD_3$
(b) $CH_3C(OH)(CD_3)_2$
(c) $CD_3CH_2CH_2OH$
(d) $Ph(CD_3)_2COD$

10-56 Often, compounds can be synthesized by more than one method. Show how this 3° alcohol can be made from the following:

(a) two different ketones
(b) two different alkenes
(c) an ester
(d) a 3° alkyl bromide

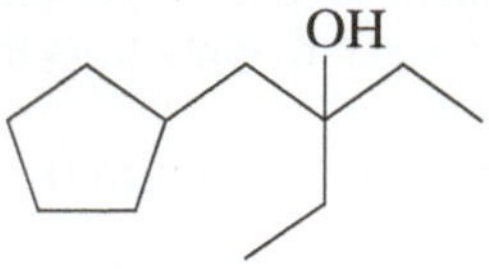

OH

10-57 Show how this 1° alcohol can be made from the following:

(a) a 1° alkyl bromide
(b) formaldehyde
(c) a 7-carbon aldehyde
(d) a carboxylic acid
(e) an alkene
(f) ethylene oxide

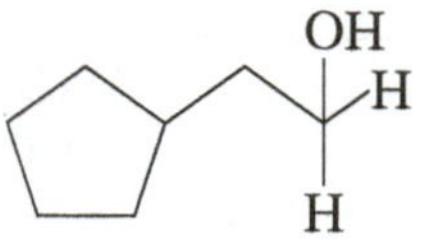

OH H H

10-58 For each synthesis, start with bromocyclohexane and predict the products. Assume that an excess of each reactant is added so that all possible reactions that can happen will happen.

(a) Br-cyclohexane $\xrightarrow[\text{EtOH, }\Delta]{\text{KOH}}$ **A** $\xrightarrow[\text{(2) Me}_2\text{S}]{\text{(1) O}_3\text{, –78 °C}}$ **B** $\xrightarrow[\text{ether}]{\text{CH}_2{=}\text{CHCH}_2\text{MgBr}}$ $\xrightarrow{\text{dilute H}_3\text{O}^+}$ **C**

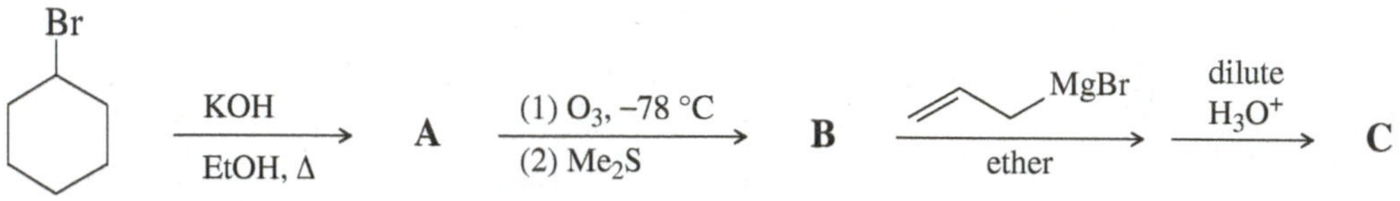

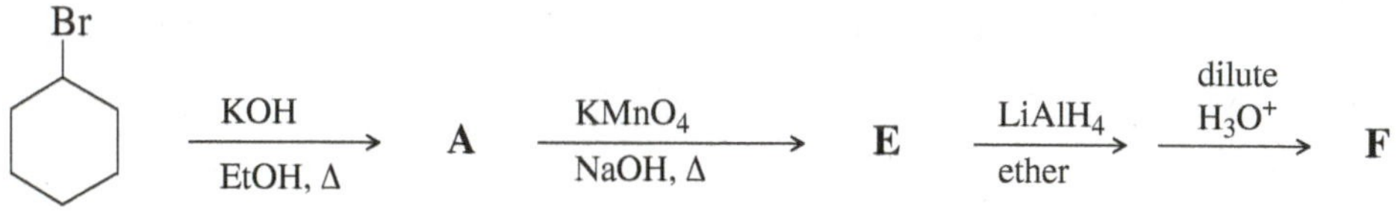

(b) Br-cyclohexane $\xrightarrow[\text{EtOH, }\Delta]{\text{KOH}}$ **A** $\xrightarrow[\text{NaOH, }\Delta]{\text{KMnO}_4}$ **E** $\xrightarrow[\text{ether}]{\text{LiAlH}_4}$ $\xrightarrow{\text{dilute H}_3\text{O}^+}$ **F**

(c) Br-cyclohexane $\xrightarrow[\text{EtOH, }\Delta]{\text{KOH}}$ **A** $\xrightarrow{\text{mCPBA}}$ **H** $\xrightarrow[\text{ether}]{\text{PhMgBr}}$ $\xrightarrow{\text{dilute H}_3\text{O}^+}$ **I** $\xrightarrow[\Delta]{\text{H}_2\text{SO}_4}$ **J**

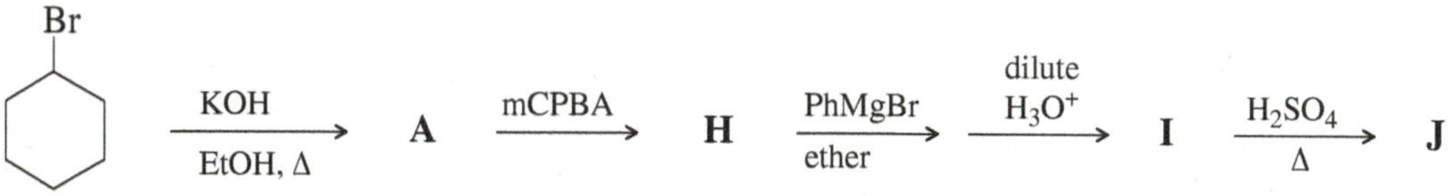

10-59 Ever since the Grignard reaction was discovered to react with ethylene oxide, chemists have experimented with other epoxides to test the limits of this reaction (which chemists call the *scope* of the reaction). The conclusion is that the success of the Grignard reaction is limited by steric hindrance in the epoxide. One group published results of several Grignard reagents on one epoxide; one of their products is shown here. What Grignard reagent and what epoxide were used to make this compound? Comment on the steric requirements of the epoxide and the stereochemistry of the product. (Reference: Schomaker, J. M., et al., *Org. Lett.*, **2003**, *5*(17), 3089.)

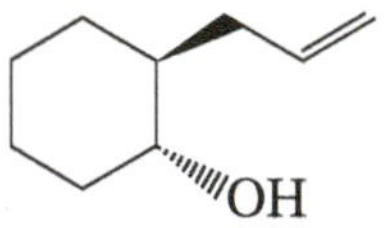

OH

10-60 Problem 8-54 describes a new method to perform ozonolysis reactions that used pyridine (py) to generate the final aldehydes and ketones in a non-aqueous reaction medium. In a subsequent publication (*J. Org. Chem.*, **2013**, *78*, 42), Professor Dussault (U. of Nebraska at Lincoln) described a "tandem" process in which two reactions are performed sequentially without having to isolate the intermediate aldehyde or ketone. Show the final product from each sequence. (Hint: The isolated products were from the larger part of the structure. Ignore stereochemistry.)

(a) $\xrightarrow[\text{(2) PhMgBr; (3) H}_3\text{O}^+]{\text{(1) O}_3\text{, py, –78 °C}}$

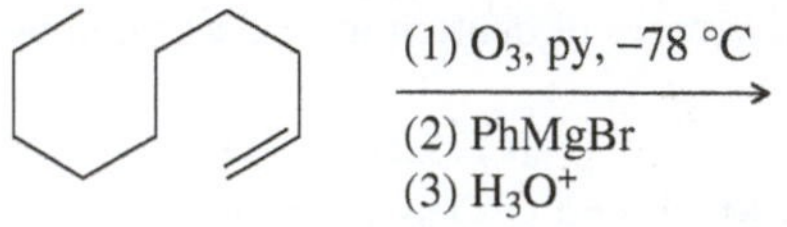

(b) $\xrightarrow[\text{(2) PhMgBr; (3) H}_3\text{O}^+]{\text{(1) O}_3\text{, py, –78 °C}}$

(c) OCH$_3$ $\xrightarrow[\text{(2) CH}_3\text{MgBr; (3) H}_3\text{O}^+]{\text{(1) O}_3\text{, py, –78 °C}}$

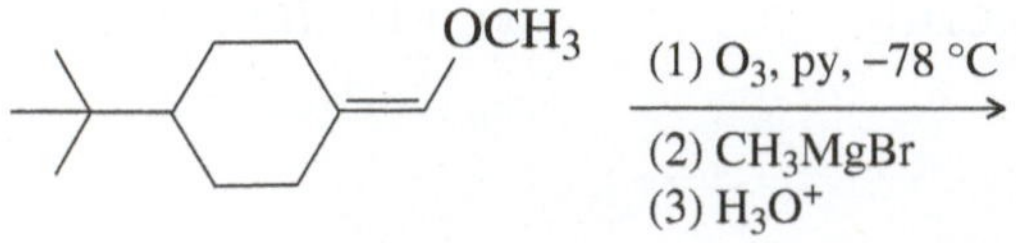

(d) $\xrightarrow[\text{(2) PhMgBr; (3) H}_3\text{O}^+]{\text{(1) O}_3\text{, py, –78 °C}}$

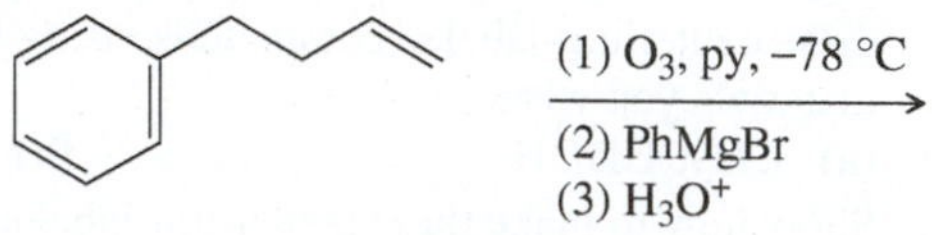

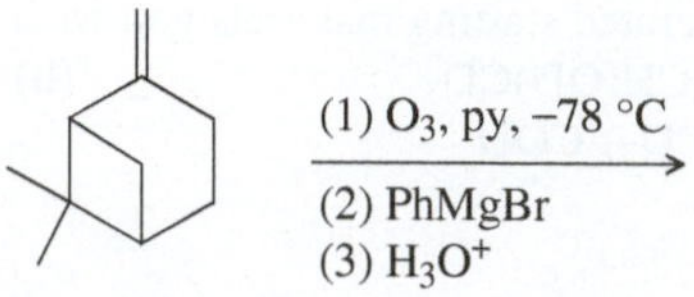

11 Reactions of Alcohols

isopentyl acetate

$CH_3(CH_2)_7CH{=}CH(CH_2)_9CH_2OH$

(*Z*)-eicos-11-en-1-ol

▲ Honey bees (*Apis millifera*) within a colony communicate largely through the use of pheromones, chemicals they secrete that other bees recognize and respond to. The alarm pheromone, which causes other bees to attack and sting, consists of at least 40 compounds, including isopentyl acetate and the alcohols butan-1-ol, hexan-1-ol, octan-1-ol, nonan-2-ol, and (*Z*)-eicos-11-en-1-ol.

Goals for Chapter 11

1 In an oxidation-reduction reaction of organic compounds, identify the species that have undergone oxidation and those that have undergone reduction.

2 Predict the products of the reactions of alcohols with oxidants, hydrohalic acids, dehydrating agents, and alkali metals.

3 Predict the products of the reactions of alkoxide ions.

4 Propose mechanisms to explain alcohol reactions such as dehydration and oxidation.

5 Plan multistep syntheses using alcohols as starting materials and intermediates.

Alcohols are important organic compounds because the hydroxy group is easily converted to almost any other functional group. In Chapter 10, we studied reactions that form alcohols. In this chapter, we seek to understand how alcohols react and which reagents are best for converting them to other kinds of compounds. Table 11-1 summarizes the types of reactions alcohols undergo and the products that result.

TABLE 11-1
Types of Reactions of Alcoholss

R—OH	type of reaction →	Product	R—OH	type of reaction →	Product
R—OH	dehydration →	alkenes	R—OH	esterification →	R—O—C(=O)—R′ esters
R—OH	oxidation →	ketones, aldehydes, acids	R—OH	tosylation →	R—OTs tosylate esters (good leaving group)
R—OH	substitution →	R—X halides			
R—OH	reduction →	R—H alkanes	R—OH	(1) form alkoxide (2) R′X →	R—O—R′ ethers

11-1 Oxidation States of Alcohols and Related Functional Groups

Oxidation of alcohols leads to ketones, aldehydes, and carboxylic acids. These functional groups, in turn, undergo a wide variety of additional reactions. For these reasons, alcohol oxidations are some of the most common organic reactions.

In inorganic chemistry, we think of oxidation as a loss of electrons and reduction as a gain of electrons. This picture works well for inorganic ions, as when Cr^{6+} is reduced to Cr^{3+}. Most organic compounds are uncharged, however, and gain or loss of electrons is not obvious. Organic chemists tend to think of **oxidation** as the result of adding an oxidizing agent (O_2, Br_2, etc.), and **reduction** as the result of adding a reducing agent (H_2, $NaBH_4$, etc.). Most organic chemists habitually use the following simple rules, based on the change in the formula of the substance:

> **OXIDATION:** addition of O or O_2; addition of X_2 (halogens); loss of H_2
> **REDUCTION:** addition of H_2 (or H^-); loss of O or O_2; loss of X_2
> **NEITHER:** addition or loss of H^+, ^-OH, H_2O, HX, etc. is neither an oxidation nor a reduction.

We can tell that an oxidation or a reduction of an alcohol has taken place by counting the number of C—O bonds to the carbon atom. Oxidation usually converts C—H bonds to C—O bonds. The first row of structures in Figure 11-1 shows that a primary alcohol is more oxidized than an alkane because the carbinol (C—OH) carbon atom

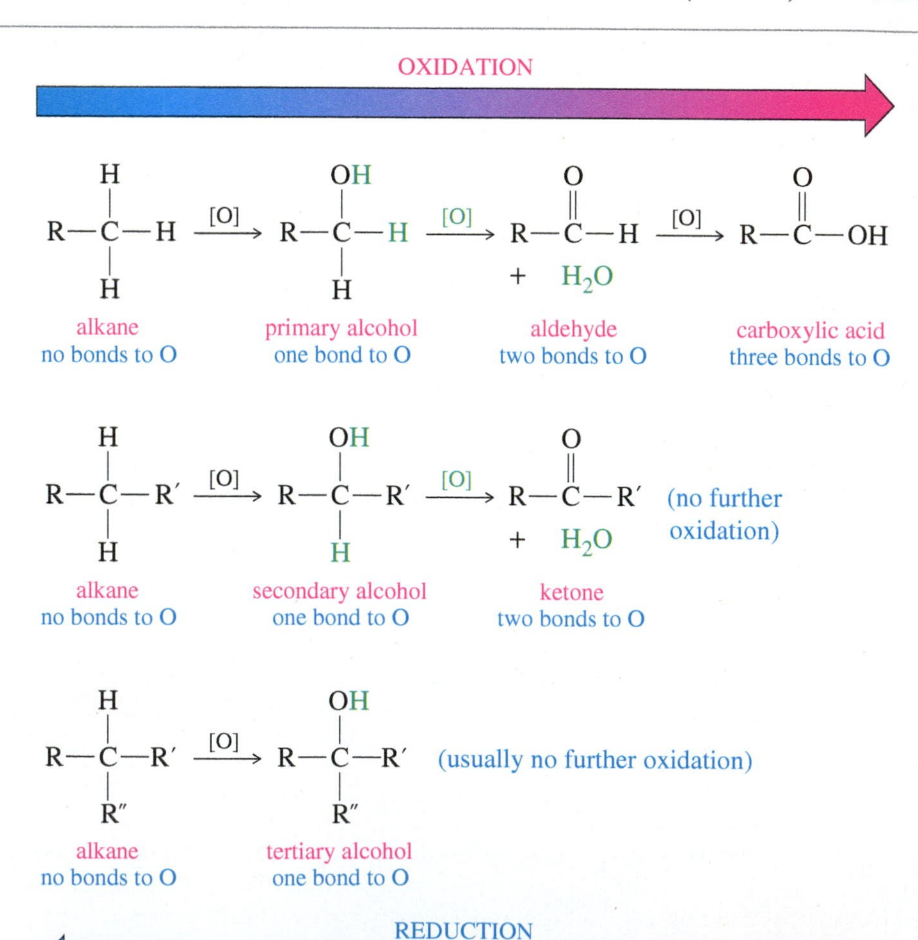

FIGURE 11-1
Oxidation states of alcohols. An alcohol is more oxidized than an alkane, yet less oxidized than carbonyl compounds such as ketones, aldehydes, and acids. Oxidation of a primary alcohol leads to an aldehyde, and further oxidation leads to an acid. Secondary alcohols are oxidized to ketones. Tertiary alcohols cannot be oxidized without breaking carbon–carbon bonds.

has one bond to oxygen, whereas the alkane has no bonds to oxygen. Oxidation of a primary alcohol gives an aldehyde with a carbonyl carbon having two bonds to oxygen. Oxidation of the aldehyde to an acid adds another bond to oxygen, for a total of three. Further oxidation would require breaking a carbon–carbon bond to give four bonds to oxygen, the oxidation state of carbon dioxide.

Figure 11-1 compares the oxidation states of primary, secondary, and tertiary alcohols with those obtained by oxidation or reduction. The symbol [O] indicates an unspecified oxidizing agent. Notice that oxidation of a primary or secondary alcohol forms a carbonyl (C=O) group by the removal of two hydrogen atoms: one from the carbinol carbon and one from the hydroxy group. A tertiary alcohol cannot easily oxidize because there is no hydrogen atom available on the carbinol carbon.

PROBLEM 11-1

Classify each reaction as an oxidation, a reduction, or neither.

(a) $CH_3-CH_2OH \xrightarrow{CrO_3\cdot pyridine} CH_3-C(=O)-H \xrightarrow{H_2CrO_4} CH_3-C(=O)-OH$

(b) $CH_4 \longrightarrow CH_3OH \longrightarrow H-C(=O)-OH \longrightarrow H-C(=O)-H \longrightarrow HO-C(=O)-OH$

(c) $CH_3-C(CH_3)(OH)-C(CH_3)(OH)-CH_3 \xrightarrow{H^+} CH_3-C(=O)-C(CH_3)_2-CH_3 + H_2O$

(d) $CH_3-CH_2-OH \xrightarrow{LiAlH_4/TiCl_4} CH_3-CH_3$

(e) cyclohexene $\xrightarrow{Br_2}$ trans-1,2-dibromocyclohexane (+ enantiomer)

(f) cyclopentanecarbaldehyde $\xrightarrow{H^+,\ CH_3OH}$ cyclopentyl-CH(OCH_3)$_2$ + H_2O

(g) cyclohexene $\xrightarrow{HBr}$ bromocyclohexane

(h) cyclohexanol $\xrightarrow[(-H_2O)]{H_2SO_4}$ cyclohexene $\xrightarrow[Pt]{H_2}$ cyclohexane

(i) cyclohexene $\xrightarrow[H_2O_2]{OsO_4}$ cis-cyclohexane-1,2-diol

(j) cyclohexene $\xrightarrow{RCO_3H}$ cyclohexene oxide $\xrightarrow{H_3O^+}$ trans-cyclohexane-1,2-diol (+ enantiomer)

(k) cyclohexene $\xrightarrow[H_2O]{Cl_2}$ trans-2-chlorocyclohexanol (+ enantiomer)

(l) cyclohexene $\xrightarrow{(1)\ BH_3\cdot THF\quad (2)\ H_2O_2,\ ^-OH}$ cyclohexanol

11-2 Oxidation of Alcohols

Alcohol oxidations commonly form aldehydes, ketones, and carboxylic acids. A wide variety of oxidants are used, including chromium oxides, permanganate, nitric acid, and sodium hypochlorite (NaOCl, household bleach). Many of the traditional oxidants for laboratory use are based on chromium(VI) compounds. These chromium reagents are highly toxic, and they are known to cause cancer. They are also difficult to dispose of legally. Chemists prefer to use less toxic oxidants. In this section, we will cover the

PROBLEM-SOLVING HINT

Phenols form esters and alkoxides like other alcohols do, but their other reactions involve the aromatic ring. Those reactions are covered in Chapter 17.

traditional uses of the chromium reagents and also discuss less toxic alternatives. In many cases, the cheapest, least toxic, and most easily available oxidant is household bleach, which is a 6–8% aqueous solution of NaOCl.

All of the common oxidants have an element (Cr, Cl, I, S, Mn, or others) in a high oxidation state bonded to oxygen. Moreover, they follow similar mechanisms. The following mechanism shows the oxidation of a secondary alcohol by bleach (NaOCl), with acetic acid used as a mild acid catalyst:

Protonation of hypochlorite to form a reactive intermediate

$Na^+ \; ^-$:Ö—C̈l: ⇌ (CH_3COOH, acetic acid) H—Ö—C̈l: ⇌ (CH_3COOH, acetic acid) :C̈l—O^+(H)(H)

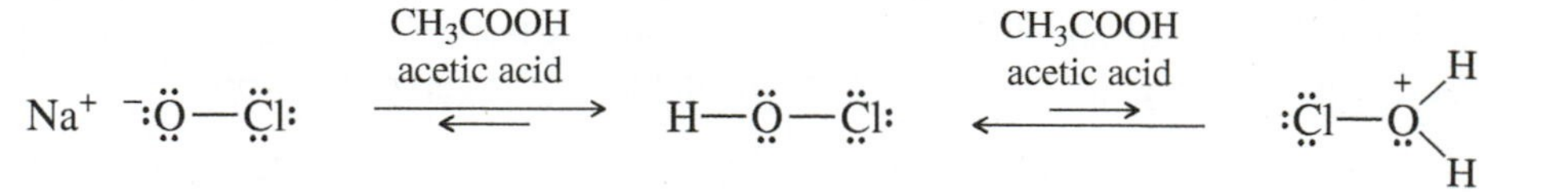

sodium hypochlorite — hypochlorous acid — (reactive intermediate)

Formation of an alkyl hypochlorite derivative

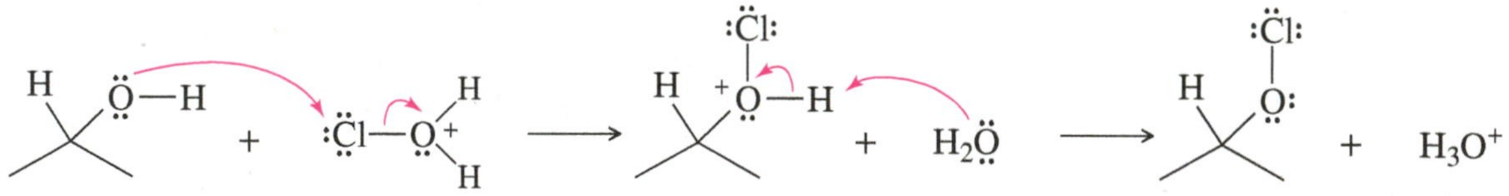

(alcohol) + :C̈l—O^+H₂ ⟶ (protonated alkyl hypochlorite) + H_2Ö ⟶ (alkyl hypochlorite) + H_3O^+

an alkyl hypochlorite

Up to this point, no oxidation states have changed.

Elimination of HCl oxidizes C and reduces Cl

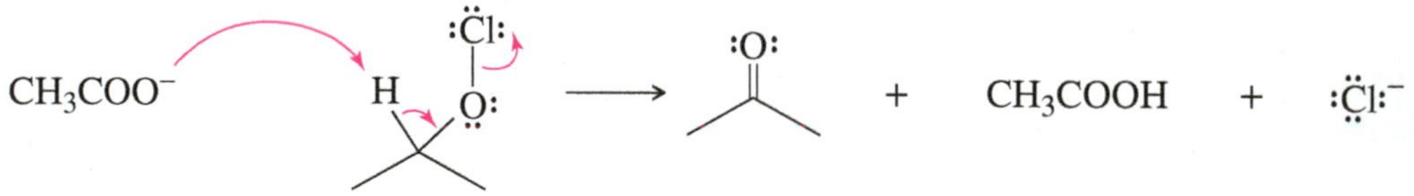

CH_3COO^- + (alkyl hypochlorite) ⟶ (ketone) + CH_3COOH + :C̈l:$^-$

In the final step, a base such as acetate ion or water removes a proton from the carbinol carbon atom, giving it a double bond to oxygen, which leaves it oxidized. The oxidant (chlorine in this case) leaves with an additional pair of electrons and fewer bonds to oxygen, giving it a lower (reduced) oxidation state.

11-2A Oxidation of Secondary Alcohols

Secondary alcohols are easily oxidized to give excellent yields of ketones. Sodium hypochlorite, often used with acetic acid, is an inexpensive and relatively safe oxidant.

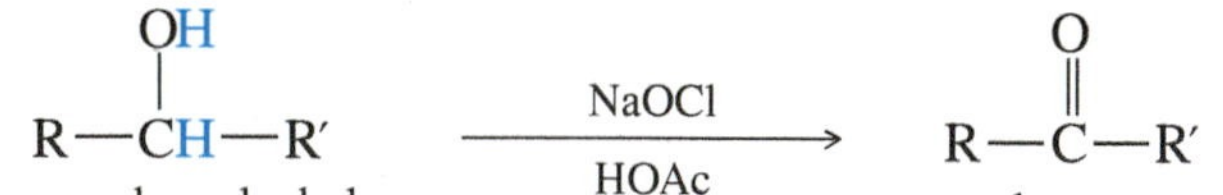

$$R-CH(OH)-R' \xrightarrow[\text{HOAc}]{\text{NaOCl}} R-C(=O)-R'$$

secondary alcohol — ketone

Example

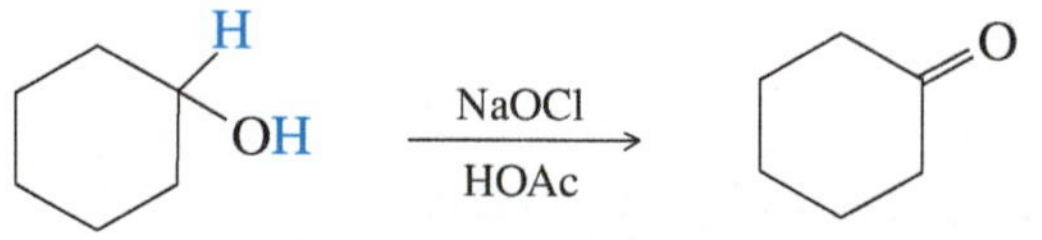

cyclohexanol $\xrightarrow[\text{HOAc}]{\text{NaOCl}}$ cyclohexanone (92%)

The traditional reagent for these laboratory oxidations is the **chromic acid reagent,** which is occasionally used when less hazardous reagents fail to give good yields. The chromic acid reagent is prepared by dissolving either sodium dichromate ($Na_2Cr_2O_7$) or chromium trioxide (CrO_3) in a mixture of sulfuric acid and water. Great care must be exercised in the formation and use of the chromic acid reagent, which is extremely toxic and corrosive, and tends to spatter.

$$\underset{\substack{\text{sodium dichromate}\\ \text{(or } CrO_3)}}{Na_2Cr_2O_7} + H_2O + 2\,H_2SO_4 \longrightarrow \underset{\text{chromic acid } (H_2CrO_4)}{2\,HO{-}Cr(=O)_2{-}OH} + 2\,Na^+ + 2\,HSO_4^-$$

The mechanism of chromic acid oxidation probably involves the formation of a chromate ester. The final step, elimination of the chromate ester, is similar to the final step of the hypochlorite oxidation. In this step, oxidation of the carbon atom and reduction of the chromium atom take place. A weak base removes a proton from carbon, giving it a double bond to oxygen. Chromium leaves with an additional pair of electrons, going from Cr(VI) to Cr(IV).

Formation of the chromate ester

$$\underset{\text{alcohol}}{R{-}CHR'{-}O{-}H} + \underset{\text{chromic acid}}{H{-}O{-}Cr(=O)_2{-}OH} \longrightarrow \underset{\text{chromate ester}}{R{-}CHR'{-}O{-}Cr(=O)_2{-}OH} + H_2O$$

Elimination of the chromate ester and oxidation of the carbinol carbon

$$H_2\ddot{O}: + R{-}CHR'{-}\ddot{O}{-}\underset{\text{Cr(VI)}}{Cr(=O)_2}{-}OH \longrightarrow R{-}C(R')=\ddot{O}: + H_3O^+ + \underset{\text{Cr(IV)}}{(O=)(^-O)Cr{-}OH}$$

The chromium(IV) species formed reacts further to give the stable reduced form, chromium(III). Both sodium dichromate and chromic acid are orange, but chromic ion (Cr^{3+}) is a deep blue. One can follow the progress of a chromic acid oxidation by observing the color change from orange through various shades of green to a greenish blue. In fact, the color change observed with chromic acid can be used as a test for the presence of an oxidizable alcohol.

11-2B Oxidation of Primary Alcohols

Oxidation of a primary alcohol initially forms an aldehyde. Unlike a ketone, however, an aldehyde is easily oxidized further to give a carboxylic acid.

$$\underset{\text{primary alcohol}}{R{-}CH(OH){-}H} \xrightarrow{[O]} \underset{\text{aldehyde}}{R{-}C(=O){-}H} \xrightarrow{[O]} \underset{\text{carboxylic acid}}{R{-}C(=O){-}OH}$$

Obtaining the aldehyde is often difficult, because most oxidizing agents strong enough to oxidize primary alcohols also oxidize aldehydes. Chromic acid, a powerful oxidant, usually oxidizes a primary alcohol all the way to the carboxylic acid.

Application: Oxidation Catalyst

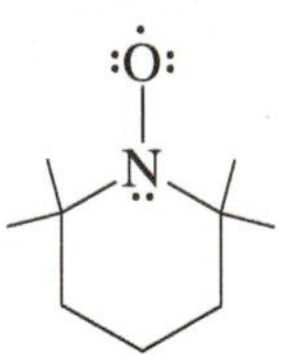

TEMPO

TEMPO (2,2,6,6-tetramethylpiperidinyl-1-oxy) is a stable free radical that catalyzes many types of oxidations. For example, a catalytic amount of TEMPO added to a hypochlorite oxidation of an alcohol increases the rate by enabling lower-energy reaction mechanisms involving reversible oxidation of the N—O bond.

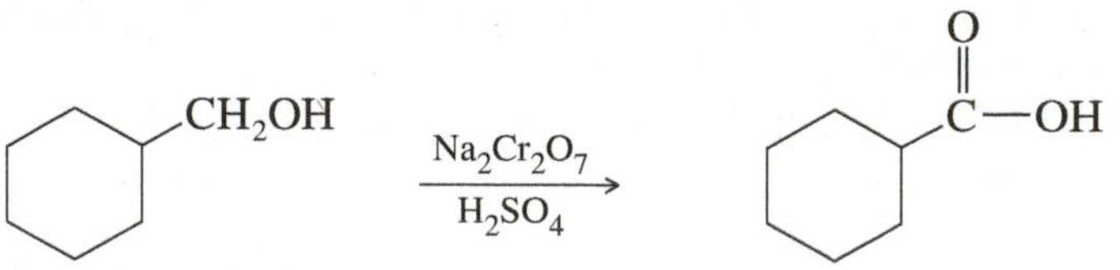

cyclohexylmethanol $\xrightarrow[H_2SO_4]{Na_2Cr_2O_7}$ cyclohexanecarboxylic acid (92%)

Sodium hypochlorite (NaOCl) can serve as a more selective oxidant, especially when it is used with TEMPO, a stable free radical that catalyzes a variety of oxidations (see structure at left). The combination of sodium hypochlorite with TEMPO can selectively oxidize primary alcohols to aldehydes under carefully controlled conditions.

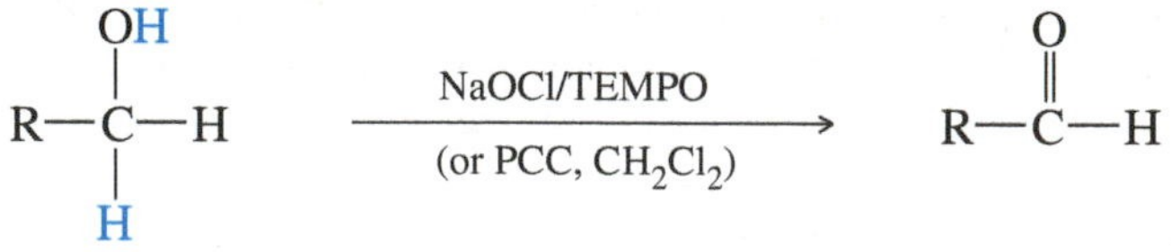

primary alcohol R—CH(OH)—H $\xrightarrow[\text{(or PCC, } CH_2Cl_2)]{\text{NaOCl/TEMPO}}$ R—C(=O)—H aldehyde

Examples

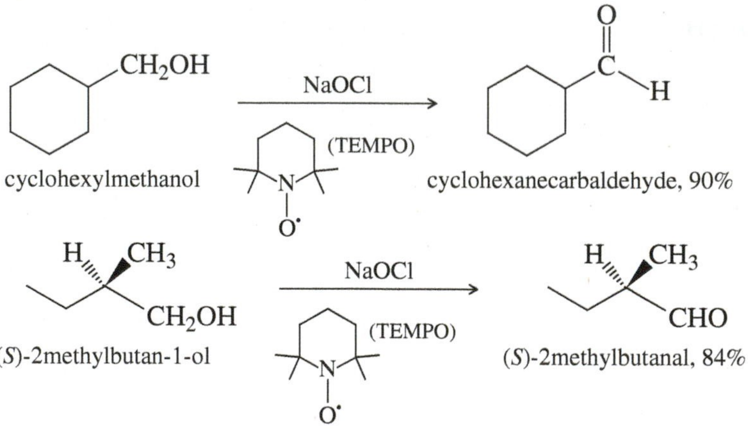

cyclohexylmethanol $\xrightarrow[\text{(TEMPO)}]{\text{NaOCl}}$ cyclohexanecarbaldehyde, 90%

(*S*)-2methylbutan-1-ol $\xrightarrow[\text{(TEMPO)}]{\text{NaOCl}}$ (*S*)-2methylbutanal, 84%

The combination of sodium hypochlorite with TEMPO can also be used to oxidize primary alcohols all the way to carboxylic acids, simply by using an excess of hypochlorite. In many cases, this hypochlorite/TEMPO combination gives better yields of carboxylic acids than the brute-force technique using hot chromic acid.

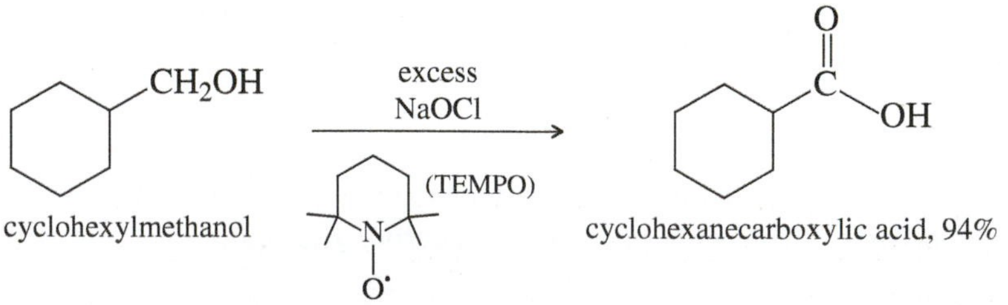

cyclohexylmethanol $\xrightarrow[\text{(TEMPO)}]{\text{excess NaOCl}}$ cyclohexanecarboxylic acid, 94%

Pyridinium chlorochromate (PCC):

$C_5H_5N^+$—H CrO_3Cl^-

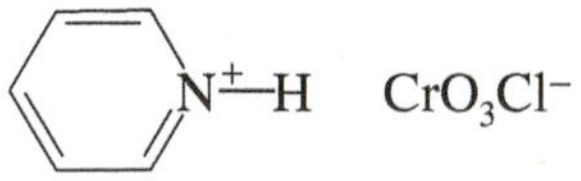

$CrO_3 \cdot$ pyridine $\cdot$ HCl
or $pyH^+ \ CrO_3Cl^-$

The traditional reagent for the limited oxidation of primary alcohols to aldehydes is **pyridinium chlorochromate (PCC),** a complex of chromium trioxide with pyridine and HCl. PCC oxidizes most primary alcohols to aldehydes in excellent yields. Unlike most other oxidants, PCC is soluble in nonpolar solvents such as dichloromethane (CH_2Cl_2), which is an excellent solvent for most organic compounds. PCC can also serve as a mild reagent for oxidizing secondary alcohols to ketones. Using PCC is not entirely simple, however. The reagent is relatively difficult to make, and it involves chromium reagents that are hazardous to use and to dispose of.

$$\underset{\text{heptan-1-ol}}{CH_3(CH_2)_5{-}CH_2OH} \xrightarrow[CH_2Cl_2]{PCC} \underset{\text{heptanal (78\%)}}{CH_3(CH_2)_5{-}\overset{\overset{\displaystyle O}{\|}}{C}{-}H}$$

11-2C Resistance of Tertiary Alcohols to Oxidation

Oxidation of tertiary alcohols is not an important reaction in organic chemistry. Tertiary alcohols have no hydrogen atoms on the carbinol carbon atom, so oxidation must take place by breaking carbon–carbon bonds. These oxidations require severe conditions and result in mixtures of products.

The **chromic acid test** for primary and secondary alcohols exploits the resistance of tertiary alcohols to oxidation. When a primary or secondary alcohol is added to the chromic acid reagent, the orange color changes to green or blue. When a nonoxidizable substance (such as a tertiary alcohol, a ketone, or an alkane) is added to the reagent, no immediate color change occurs.

Application: Portable Breath Tester

A small fuel cell in this portable breath tester catalyzes the oxidation of ethanol by oxygen in the air. The oxidation generates an electric current that is proportional to the concentration of ethanol in the sample.

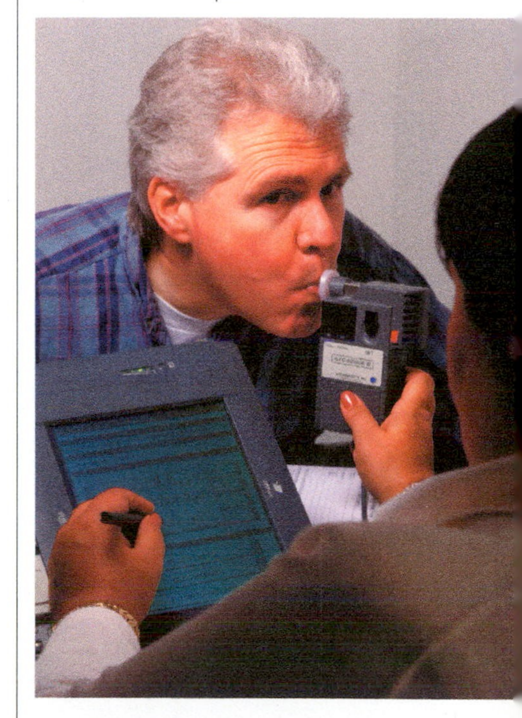

PROBLEM 11-2

Predict the products of the reactions of the following compounds with:
(1) chromic acid or excess sodium hypochlorite with acetic acid.
(2) PCC or NaOCl (1 equivalent) with TEMPO.

(a) cyclohexanol
(b) 1-methylcyclohexanol
(c) cyclopentylmethanol
(d) cyclohexanone
(e) cyclohexane
(f) 1-phenylpropan-1-ol
(g) hexan-1-ol
(h) acetaldehyde, CH_3CHO

11-3 Additional Methods for Oxidizing Alcohols

Many other reagents and procedures have been developed for oxidizing alcohols. Some are simply modifications of the procedures we have seen. For example, the *Collins reagent* is a complex of chromium trioxide and pyridine, the original version of PCC. The *Jones reagent* is a milder form of chromic acid, a solution of diluted chromic acid in acetone.

Two other strong oxidants are potassium permanganate and nitric acid. Both of these reagents are less expensive than the chromium reagents, and both of them give by-products that are less environmentally hazardous than spent chromium reagents. Both permanganate and nitric acid oxidize secondary alcohols to ketones and primary alcohols to carboxylic acids. Used primarily in industry, these strong oxidants can form explosive mixtures and cleave carbon–carbon bonds if the temperature and concentrations are not precisely controlled.

The **Swern oxidation** uses dimethyl sulfoxide (DMSO) as the oxidizing agent to convert alcohols to ketones and aldehydes. DMSO and oxalyl chloride are added to the alcohol at low temperature, followed by a hindered base such as triethylamine. The reactive species $(CH_3)_2\overset{+}{S}Cl$, formed in the solution, is thought to act as the oxidant in the Swern oxidation. Secondary alcohols are oxidized to ketones, and primary alcohols are oxidized only as far as the aldehyde. The by-products of this reaction are all volatile and are easily separated from the organic products.

$$\underset{\text{alcohol}}{{-}\overset{\overset{\displaystyle OH}{|}}{\underset{|}{C}}{-}H} + \underset{\substack{\text{DMSO}\\\text{dimethyl sulfoxide}}}{H_3C{-}\overset{\overset{\displaystyle O}{\|}}{S}{-}CH_3} + \underset{\substack{(COCl)_2\\\text{oxalyl chloride}}}{Cl{-}\overset{\overset{\displaystyle O}{\|}}{C}{-}\overset{\overset{\displaystyle O}{\|}}{C}{-}Cl} \xrightarrow[CH_2Cl_2]{(CH_3CH_2)_3N:} \underset{\substack{\text{ketone}\\\text{or aldehyde}}}{{-}C(=O){-}} + \underset{\text{dimethyl sulfide}}{H_3C{-}S{-}CH_3} \begin{array}{l} + CO_2\\ + CO\\ + 2\,HCl\end{array}$$

Examples

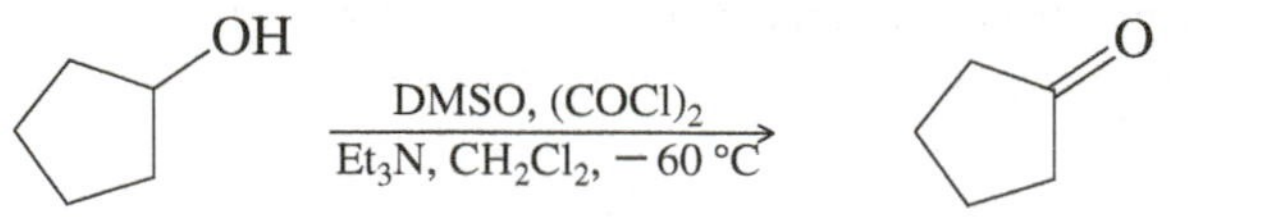

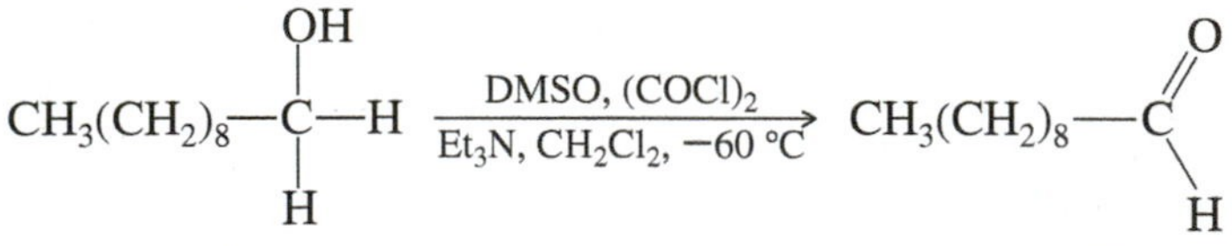

cyclopentanol $\xrightarrow[\text{Et}_3\text{N, CH}_2\text{Cl}_2,\ -60\ ^\circ\text{C}]{\text{DMSO, (COCl)}_2}$ cyclopentanone (90%)

decan-1-ol $CH_3(CH_2)_8-CH_2-OH$ $\xrightarrow[\text{Et}_3\text{N, CH}_2\text{Cl}_2,\ -60\ ^\circ\text{C}]{\text{DMSO, (COCl)}_2}$ $CH_3(CH_2)_8-CHO$ decanal (85%)

Like the Swern oxidation, the **Dess–Martin periodinane (DMP)** reagent oxidizes primary alcohols to aldehydes and secondary alcohols to ketones without using chromium or other heavy-metal compounds. The reaction with DMP takes place under mild conditions (room temperature, neutral pH) and gives excellent yields. The DMP reagent, which owes its oxidizing ability to a high-valence iodine atom, is a commercially available solid that is easily stored.

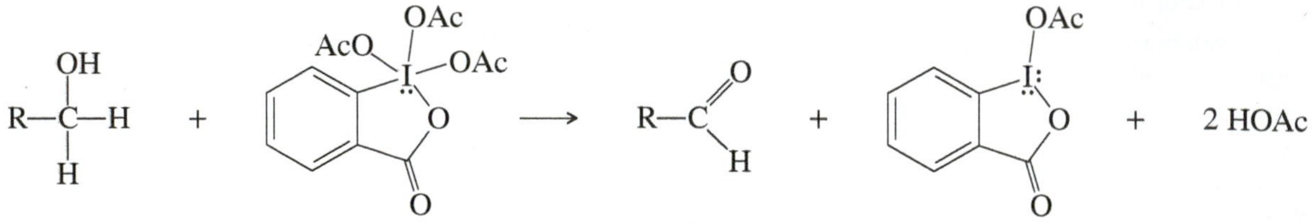

primary alcohol (or secondary) R–CH(OH)–H + DMP reagent ⟶ aldehyde (or ketone) R–CHO + reduced form + 2 HOAc acetic acid

SUMMARY OF ALCOHOL OXIDATIONS

To Oxidize	Product	Chromium-free Reagent	Chromium Reagent
2° alcohol	ketone	NaOCl-HOAc, Swern, DMP	chromic acid (or PCC)
1° alcohol	aldehyde	NaOCl-TEMPO, Swern, DMP	PCC
1° alcohol	carboxylic acid	NaOCl (excess)-TEMPO	chromic acid

PROBLEM 11-3

We have covered several oxidants that use a multi-valent atom (Cr, Cl, S, or I) as their active species, going from a higher oxidation state before the oxidation to a lower oxidation state after oxidizing the alcohol. Draw the Lewis structures of the following atoms, before and after the oxidation of an alcohol to a ketone or aldehyde. How many bonds to oxygen does each atom have before and after the oxidation?

(a) the Cr in chromic acid
(b) the Cl in sodium hypochlorite
(c) the S in the Swern oxidation
(d) the I in the DMP reagent
(e) the carbinol C in the alcohol that is oxidized

PROBLEM 11-4

Give the structure of the principal product(s) when each of the following alcohols reacts with (1) $Na_2Cr_2O_7/H_2SO_4$, (2) PCC, (3) DMP, and (4) 1 equiv NaOCl-TEMPO.

(a) octan-1-ol
(b) octan-3-ol
(c) 4-hydroxydecanal
(d) 1-methylcyclohexan-1,4-diol

PROBLEM 11-5

Predict the products you expect when the following starting material undergoes oxidation with an excess of each of the reagents shown below.

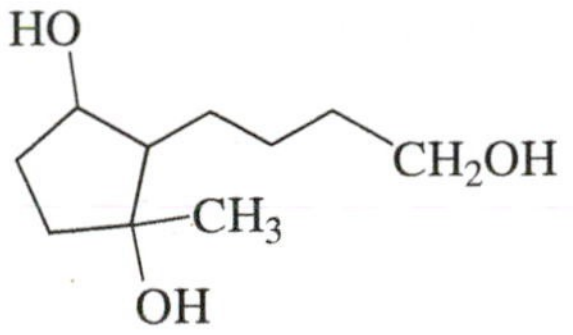

HO / CH_2OH / CH_3 / OH

(a) chromic acid
(b) PCC (pyridinium chlorochromate)
(c) sodium hypochlorite/acetic acid
(d) DMSO and oxalyl chloride
(e) DMP (periodinane) reagent

SOLVED PROBLEM 11-1

Suggest the most appropriate method for each of the following laboratory syntheses.

(a) cyclopentanol ⟶ cyclopentanone

SOLUTION

Many reagents are available to oxidize a simple secondary alcohol to a ketone. Most labs would have chromium trioxide or sodium dichromate available, and the chromic acid oxidation would work. Bleach (sodium hypochlorite) and acetic acid (HOAc) would be a cheaper and less polluting alternative to the chromium reagents. DMP and the Swern oxidation would also work.

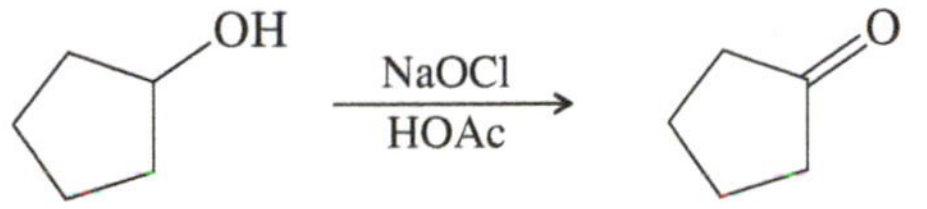

OH — NaOCl / HOAc ⟶ O

cyclopentanol cyclopentanone

(b) oct-2-en-1-ol ⟶ oct-2-enal

SOLUTION

This synthesis requires more finesse. The aldehyde is easily over-oxidized to a carboxylic acid, and the double bond reacts with oxidants such as $KMnO_4$. Our choices are limited to bleach with TEMPO, PCC, DMP, or the Swern oxidation.

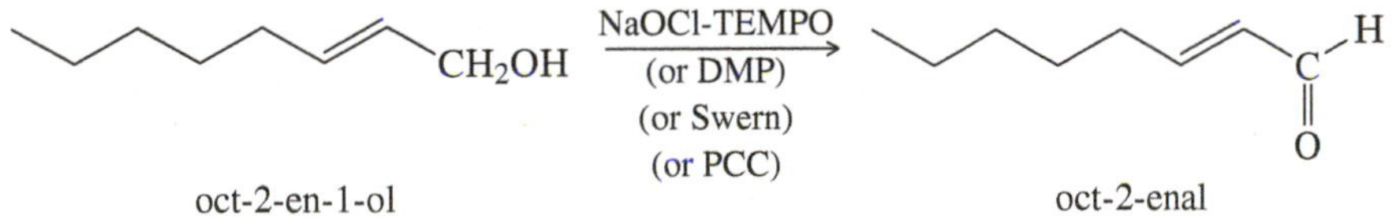

CH_2OH — NaOCl-TEMPO (or DMP) (or Swern) (or PCC) ⟶ C(=O)H

oct-2-en-1-ol oct-2-enal

PROBLEM 11-6

Suggest the most appropriate method for each of the following *laboratory* syntheses. In each case, suggest both a chromium reagent and a chromium-free reagent.

(a) butan-1-ol ⟶ butanal, $CH_3CH_2CH_2CHO$
(b) but-2-en-1-ol ⟶ but-2-enoic acid, $CH_3CH{=}CH{-}COOH$
(c) butan-2-ol ⟶ butan-2-one, $CH_3COCH_2CH_3$
(d) cyclopentanol ⟶ 1-ethylcyclopentanol (two steps)
(e) cyclopentylmethanol ⟶ 1-cyclopentylpropan-1-ol (two steps)
(f) 1-methylcyclohexanol ⟶ 2-methylcyclohexanone (several steps)

11-4 Biological Oxidation of Alcohols

Although it is the least toxic alcohol, ethanol is still a poisonous substance. When someone is suffering from a mild case of ethanol poisoning, we say that he or she is in*toxic*ated. Animals often consume food that has fermented and contains alcohol.

Their bodies must detoxify any alcohol in the food to keep it from building up in the blood and poisoning the brain. To detoxify ethanol, the liver produces an enzyme called **alcohol dehydrogenase (ADH).**

Alcohol dehydrogenase catalyzes an oxidation: the removal of two hydrogen atoms from the alcohol molecule. The oxidizing agent is **nicotinamide adenine dinucleotide (NAD).** NAD exists in two forms: the oxidized form, called NAD^+, and the reduced form, called NADH. The following equation shows that ethanol is oxidized to acetaldehyde, and NAD^+ is reduced to NADH.

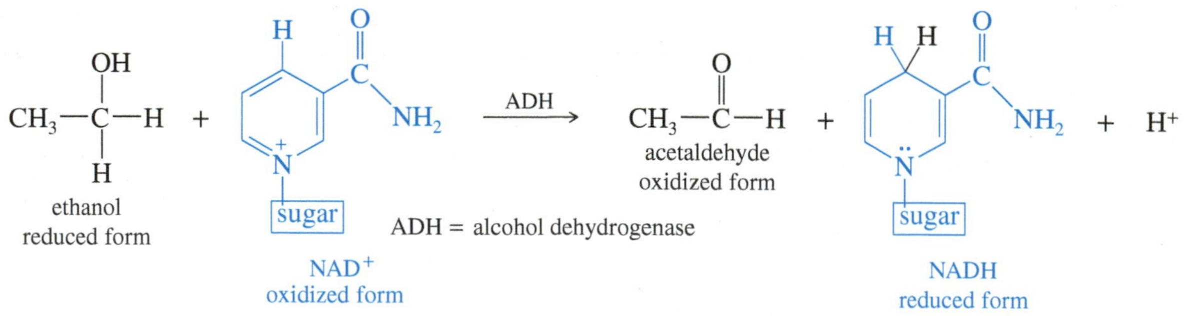

$$CH_3-CH(OH)-H + NAD^+ \xrightarrow{\text{ADH}} CH_3-C(=O)-H + NADH + H^+$$

ethanol (reduced form); NAD^+ (oxidized form); acetaldehyde (oxidized form); NADH (reduced form)

ADH = alcohol dehydrogenase

A subsequent oxidation, catalyzed by **aldehyde dehydrogenase (ALDH),** converts acetaldehyde to acetic acid, a normal metabolite.

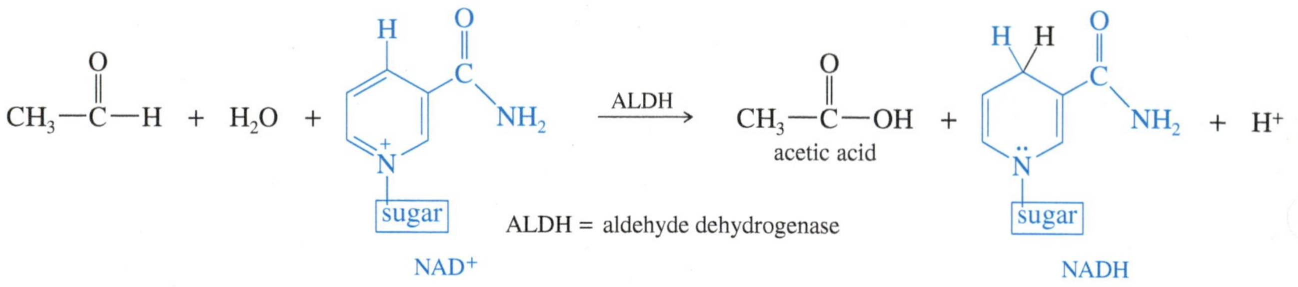

$$CH_3-C(=O)-H + H_2O + NAD^+ \xrightarrow{\text{ALDH}} CH_3-C(=O)-OH + NADH + H^+$$

acetic acid

ALDH = aldehyde dehydrogenase

These oxidations take place with most small primary alcohols. Unfortunately, the oxidation products of some other alcohols are more toxic than acetic acid. Methanol is oxidized first to formaldehyde and then to formic acid. Both of these compounds are more toxic than methanol itself.

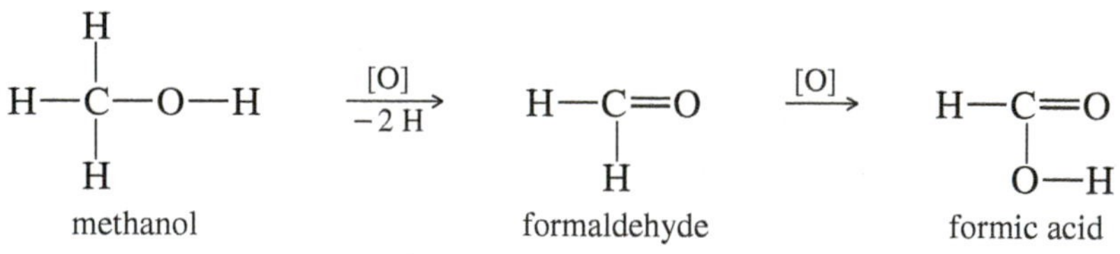

$$H-CH_2-O-H \xrightarrow[-2\,H]{[O]} H-C(H){=}O \xrightarrow{[O]} H-C(=O)-O-H$$

methanol → formaldehyde → formic acid

Ethylene glycol is a toxic diol. Its oxidation product is oxalic acid, the toxic compound found in the leaves of rhubarb and many other plants.

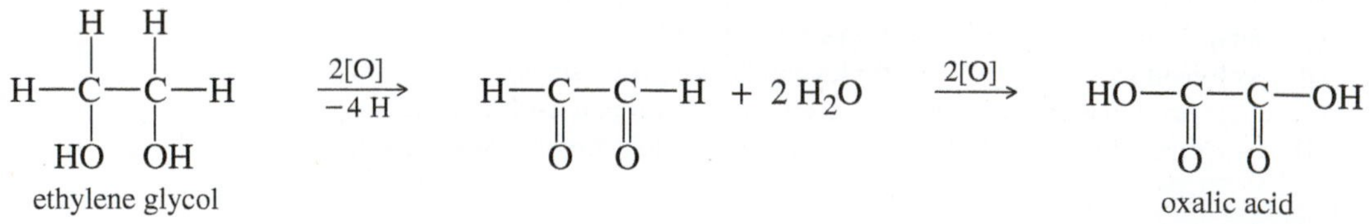

$$H-CH(OH)-CH(OH)-H \xrightarrow[-4\,H]{2[O]} H-C(=O)-C(=O)-H + 2\,H_2O \xrightarrow{2[O]} HO-C(=O)-C(=O)-OH$$

ethylene glycol → oxalic acid

Many poisonings by methanol and ethylene glycol occur each year. Alcoholics occasionally drink ethanol that has been denatured by the addition of methanol. Methanol is oxidized to formic acid, which may cause blindness and death. Dogs are often poisoned by sweet-tasting ethylene glycol when antifreeze is left in an open container. Once the glycol is metabolized to oxalic acid, the dog's kidneys fail, causing death.

The treatment for methanol or ethylene glycol poisoning is the same. The patient is given intravenous infusions of diluted ethanol. The ADH enzyme is swamped by all the ethanol, allowing time for the kidneys to excrete most of the methanol (or ethylene glycol) before it can be oxidized to formic acid (or oxalic acid). This is an example of *competitive inhibition* of an enzyme. The enzyme catalyzes oxidation of both ethanol and methanol, but a large quantity of ethanol ties up the enzyme, allowing time for excretion of most of the methanol before it is oxidized.

Application: Vitamins

NAD^+ is derived from vitamin B_3, known as niacin. There is a misconception that taking large supplements of niacin before a night of drinking ethanol will lessen the severity of the "hangover" the next morning. Metabolism is more complex than this simple idea suggests.

One common effect of taking large amounts of niacin is flushing caused by dilation of the capillaries in the skin. This flushing looks like a severe sunburn and can be quite uncomfortable, but it usually subsides in a few hours.

COOH

N

nicotinic acid
(niacin)

PROBLEM 11-7

A chronic alcoholic requires a much larger dose of ethanol as an antidote to methanol poisoning than does a nonalcoholic patient. Suggest a reason why a larger dose of the competitive inhibitor is required for an alcoholic.

PROBLEM 11-8

Unlike ethylene glycol, propylene glycol (propane-1,2-diol) is nontoxic because it oxidizes to a common metabolic intermediate. Give the structures of the biological oxidation products of propylene glycol.

11-5 Alcohols as Nucleophiles and Electrophiles; Formation of Tosylates

One reason alcohols are such versatile chemical intermediates is that they react as both nucleophiles and electrophiles. The following scheme shows an alcohol reacting as a weak nucleophile, bonding to a strong electrophile (in this case, a carbocation).

$$R{-}\ddot{O}{:}(H) + {>}C^{+}{-} \longrightarrow R{-}\ddot{O}^{+}(H){-}C{-} \longrightarrow R{-}\ddot{O}{-}C{-}$$

weak nucleophile; strong electrophile; $R{-}\ddot{O}{-}H$; $R{-}\overset{+}{O}H_2$

An alcohol is easily converted to a strong nucleophile by forming its alkoxide ion. The alkoxide ion can attack a weaker electrophile, such as an alkyl halide.

$$R{-}\ddot{O}{-}H \xrightarrow{Na} R{-}\ddot{O}{:}^{-}\ Na^+ \quad {-}C{-}X \longrightarrow R{-}\ddot{O}{-}C{-} \quad X^-\ Na^+$$

weak nucleophile; strong nucleophile; weak electrophile

The O—H bond is broken when alcohols react as nucleophiles, both when an alcohol reacts as a weak nucleophile, or when an alcohol is converted to its alkoxide that then reacts as a strong nucleophile. In contrast, when an alcohol reacts as an electrophile, the C—O bond is broken.

This bond is broken when alcohols react as nucleophiles.

$${-}C{-}O{\wr}H$$

This bond is broken when alcohols react as electrophiles.

$${-}C{\wr}O{-}H$$

An alcohol is a weak electrophile because the hydroxy group is a poor leaving group. The hydroxy group becomes a good leaving group (H_2O) when it is protonated. For example, HBr reacts with a primary alcohol by an S_N2 attack of bromide on the protonated alcohol. Note that the C—O bond is broken in this reaction.

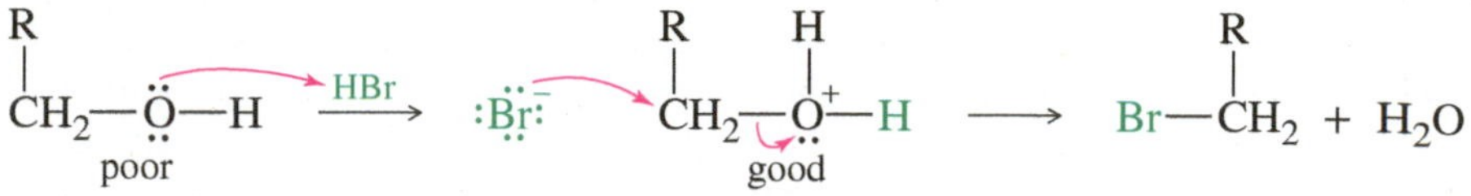

$$R{-}CH_2{-}\ddot{O}{-}H \xrightarrow{HBr} :\ddot{Br}:^- \; R{-}CH_2{-}\overset{+}{O}H{-}H \longrightarrow Br{-}CH_2{-}R + H_2O$$

poor electrophile — good electrophile

The disadvantage of using a protonated alcohol is that a strongly acidic solution is required to protonate the alcohol. Although halide ions are stable in acid, few other good nucleophiles are stable in strongly acidic solutions. Most strong nucleophiles are also basic and will abstract a proton in acid. Once protonated, the reagent is no longer nucleophilic. For example, an acetylide ion would instantly become protonated if it were added to a protonated alcohol.

$$R{-}\overset{+}{O}(H){-}H + {}^-:C{\equiv}C{-}H \longrightarrow R{-}\ddot{O}{-}H + H{-}C{\equiv}C{-}H$$

no S_N2

PROBLEM-SOLVING HINT

Ts = $-SO_2-C_6H_4-CH_3$ (tosyl group)

ROTs = $R-O-SO_2-C_6H_4-CH_3$ (tosylate ester)

TsOH = $H-O-SO_2-C_6H_4-CH_3$ (tosic acid)

TsCl = $Cl-SO_2-C_6H_4-CH_3$ (tosyl chloride)

$^-$OTs = $^-O-SO_2-C_6H_4-CH_3$ (tosylate ion)

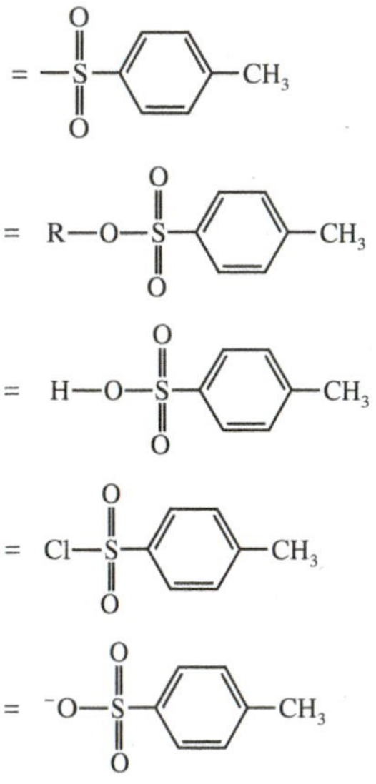

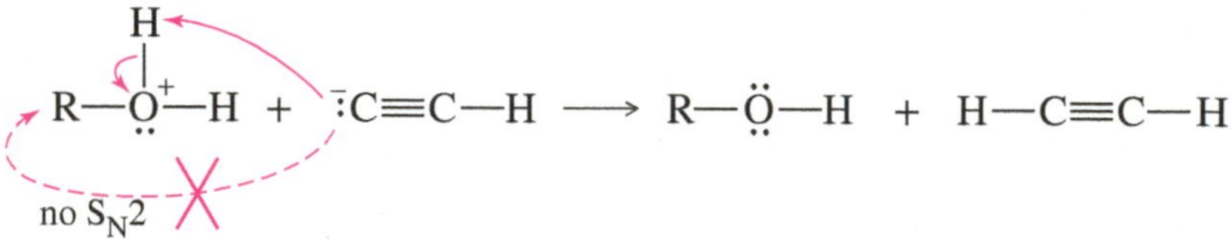

How can we convert an alcohol to an electrophile that is compatible with basic nucleophiles? We can convert it to an alkyl halide, or we can simply make its tosylate ester. A **tosylate ester** (symbolized ROTs) is the product of condensation of an alcohol with *p*-toluenesulfonic acid (symbolized TsOH).

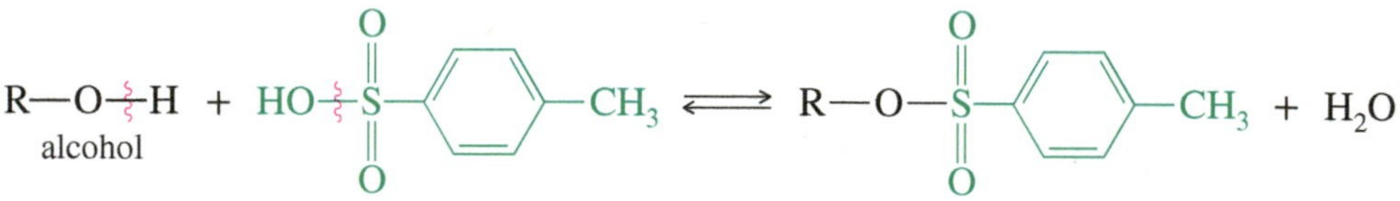

$$R{-}O{-}H + HO{-}SO_2{-}C_6H_4{-}CH_3 \rightleftharpoons R{-}O{-}SO_2{-}C_6H_4{-}CH_3 + H_2O$$

alcohol; TsOH, *p*-toluenesulfonic acid; alkyl tosylate, ROTs, a *p*-toluenesulfonate ester

The tosylate group is an excellent leaving group, and alkyl tosylates undergo substitution and elimination much like alkyl halides. In many cases, a tosylate is more reactive than the equivalent alkyl halide.

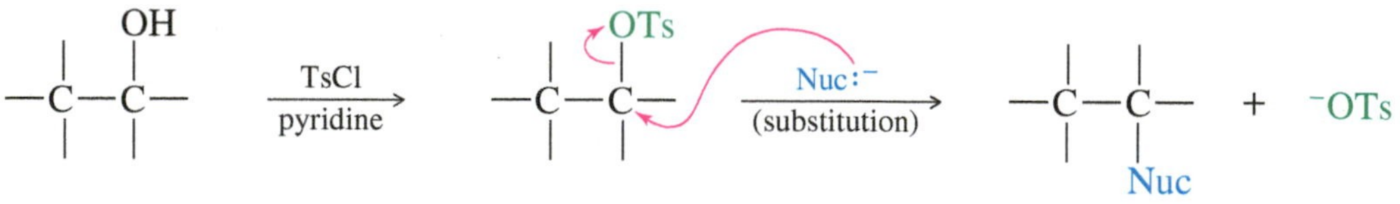

$$-C-C(OH)- \xrightarrow[\text{pyridine}]{TsCl} -C-C(OTs)- \xrightarrow[\text{(substitution)}]{Nuc:^-} -C-C(Nuc)- + {}^-OTs$$

or elimination:

$$B:^- + H{-}C{-}C{-}OTs \xrightarrow{\text{(elimination)}} C{=}C + B{-}H + {}^-OTs$$

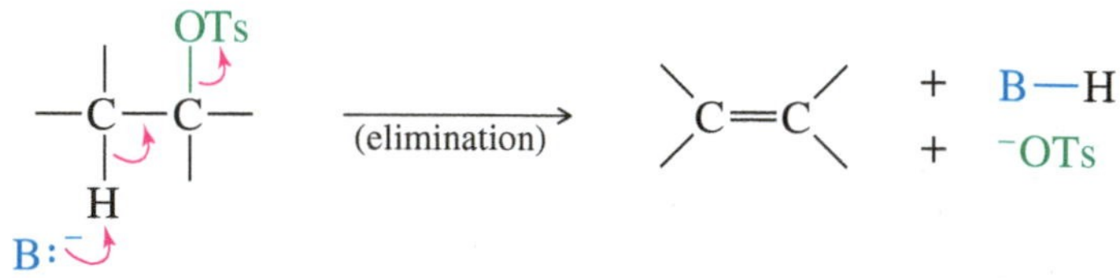

Tosylates are made from alcohols using tosyl chloride (TsCl) in pyridine, as shown next. This reaction gives much higher yields than the reaction with TsOH itself. The mechanism of tosylate formation shows that the C—O bond of the alcohol remains intact throughout the reaction, and the alcohol retains its stereochemical configuration. Pyridine is a stable aromatic amine that serves as an organic base to remove the HCl formed in the reaction, preventing it from protonating the alcohol and causing side reactions.

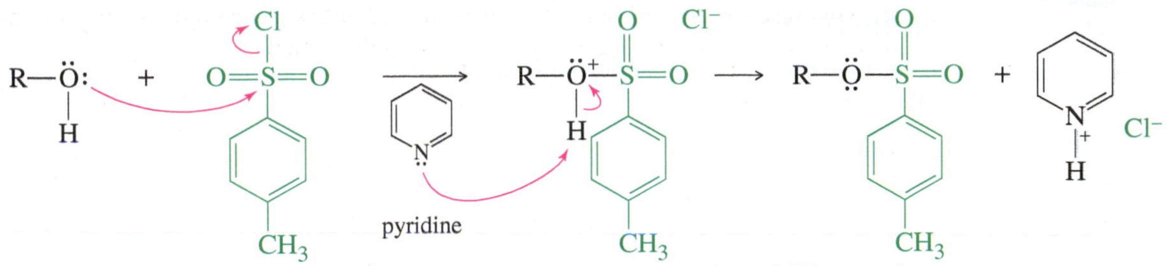

p-toluenesulfonyl chloride
TsCl, "tosyl chloride"

pyridine

ROTs, a tosylate ester

The following reaction shows the S_N2 displacement of tosylate ion (^-OTs) from (S)-2-butyl tosylate with inversion of configuration. The tosylate ion is a particularly stable anion, with its negative charge delocalized over three oxygen atoms.

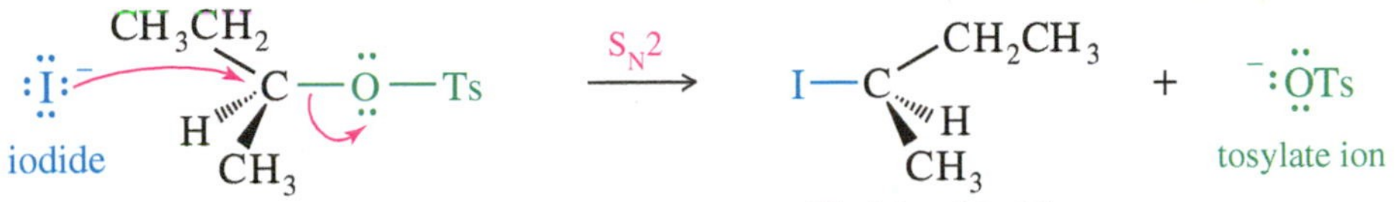

iodide

(S)-2-butyl tosylate

(R)-2-butyl iodide

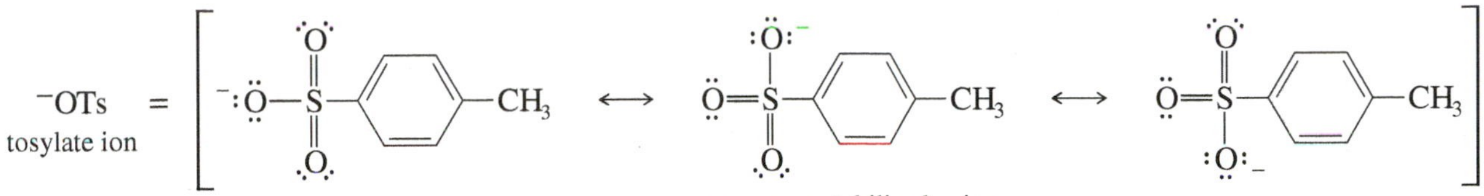

tosylate ion

^-OTs = tosylate ion

resonance-stabilized anion

Like halides, the tosylate leaving group is displaced by a wide variety of nucleophiles. The S_N2 mechanism (strong nucleophile) is more commonly used in synthetic preparations than the S_N1. The following reactions show the generality of S_N2 displacements of tosylates. In each case, R must be an unhindered primary or secondary alkyl group if substitution is to predominate over elimination.

SUMMARY S_N2 Reactions of Tosylate Esters

$R—OTs + {}^-OH$ (hydroxide) $\longrightarrow$ $R—OH$ (alcohol) $+ {}^-OTs$

$R—OTs + {}^-C{\equiv}N$ (cyanide) $\longrightarrow$ $R—C{\equiv}N$ (nitrile) $+ {}^-OTs$

$R—OTs + Br^-$ (halide) $\longrightarrow$ $R—Br$ (alkyl halide) $+ {}^-OTs$

$R—OTs + R'—O^-$ (alkoxide) $\longrightarrow$ $R—O—R'$ (ether) $+ {}^-OTs$

$R—OTs + :NH_3$ (ammonia) $\longrightarrow$ $R—NH_3^+\ {}^-OTs$ (amine salt)

$R—OTs + LiAlH_4$ (LAH) $\longrightarrow$ $R—H$ (alkane) $+ {}^-OTs$

PROBLEM-SOLVING HINT

Tosylate esters are particularly useful: They are great leaving groups, often better than halides. Grignard reactions build alcohols, which are easily converted to tosylates for substitution or elimination.

PROBLEM 11-9

Predict the major products of the following reactions.

(a) ethyl tosylate + potassium *tert*-butoxide
(b) isobutyl tosylate + NaI
(c) (*R*)-2-hexyl tosylate + NaCN
(d) the tosylate of cyclohexylmethanol + excess NH_3
(e) *n*-butyl tosylate + sodium acetylide, $H-C\equiv C:^{-}\ {}^{+}Na$

PROBLEM 11-10

Show how you would convert propan-1-ol to the following compounds using tosylate intermediates. You may use whatever additional reagents are needed.

(a) 1-bromopropane
(b) propan-1-amine, $CH_3CH_2CH_2NH_2$
(c) $CH_3CH_2CH_2OCH_2CH_3$ ethyl propyl ether
(d) $CH_3CH_2CH_2CN$ butyronitrile

11-6 Reduction of Alcohols

The reduction of alcohols to alkanes is not a common reaction because it removes a functional group, leaving fewer options for further reactions. Still, sometimes we use Grignard or similar reactions to generate the carbon skeleton, but we no longer want the product's alcohol group. In those cases, we reduce the —OH group to a hydrogen atom.

$$R-OH \xrightarrow{\text{reduction}} R-H \text{ (rare)}$$

We can reduce an alcohol in two steps, by dehydrating it to an alkene and then hydrogenating the alkene.

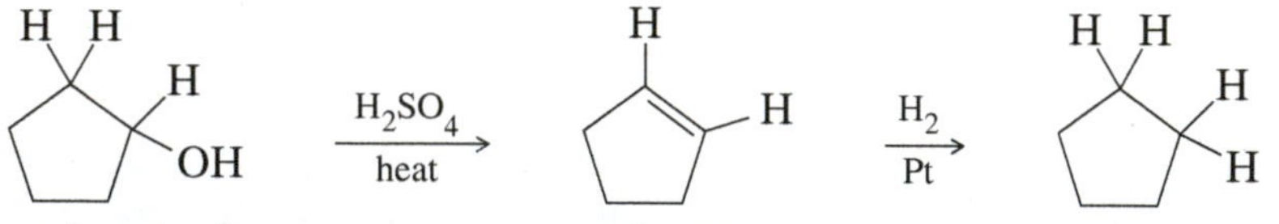

$$\text{cyclopentanol} \xrightarrow[\text{heat}]{H_2SO_4} \text{cyclopentene} \xrightarrow[\text{Pt}]{H_2} \text{cyclopentane}$$

Another method for reducing an alcohol involves converting the alcohol to the tosylate ester and then using a hydride reducing agent to displace the tosylate leaving group. This reaction works with most primary and secondary alcohols.

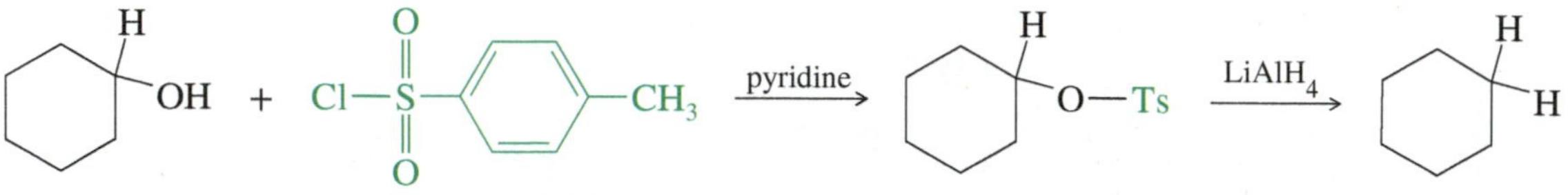

$$\text{cyclohexanol} + Cl-SO_2-C_6H_4-CH_3 \text{ (tosyl chloride, TsCl)} \xrightarrow{\text{pyridine}} \text{cyclohexyl tosylate (O—Ts)} \xrightarrow{LiAlH_4} \text{cyclohexane (75\%)}$$

PROBLEM 11-11

Predict the products of the following reactions.

(a) cyclohexylmethanol + TsCl/pyridine
(b) product of (a) + $LiAlH_4$
(c) 1-methylcyclohexanol + H_2SO_4, heat
(d) product of (c) + H_2, Pt

11-7 Reactions of Alcohols with Hydrohalic Acids

Tosylation of an alcohol, followed by displacement of the tosylate by a halide ion, converts an alcohol to an alkyl halide. This is not the most common method for converting alcohols to alkyl halides, however, because simple, one-step reactions are available. A common method is to treat the alcohol with a hydrohalic acid, either HBr, HCl, or HI.

In acidic solution, an alcohol is in equilibrium with its protonated form. Protonation converts the hydroxy group from a poor leaving group (^-OH) to a good leaving group (H_2O). Once the alcohol is protonated, all the usual substitution and elimination reactions are feasible, depending on the structure (1°, 2°, 3°) of the alcohol.

$$R-\ddot{O}-H \quad + \quad H^+ \quad \rightleftharpoons \quad R-\overset{H}{\overset{|}{O^+}}-H \quad \xrightarrow[S_N1 \text{ or } S_N2]{X^-} \quad R-X$$

poor leaving group — good leaving group

Most good nucleophiles are basic, becoming protonated and losing their nucleophilicity in acidic solutions. Halide ions are exceptions, however. Halides are anions of strong acids, so they are weak bases. Solutions of HBr, HCl, or HI contain nucleophilic Br^-, Cl^-, or I^- ions. These acids are commonly used to convert alcohols to the corresponding alkyl halides.

11-7A Reactions with Hydrobromic Acid

$$R-OH \quad + \quad HBr/H_2O \quad \longrightarrow \quad R-Br$$

Concentrated hydrobromic acid rapidly converts *tert*-butyl alcohol to *tert*-butyl bromide. The strong acid protonates the hydroxy group, converting it to a good leaving group. The hindered tertiary carbon atom cannot undergo S_N2 displacement, but it can ionize to a tertiary carbocation. Attack by bromide gives the alkyl bromide. The mechanism is similar to other S_N1 mechanisms we have studied, except that water serves as the leaving group from the protonated alcohol.

MECHANISM 11-1 Reaction of a Tertiary Alcohol with HBr (S_N1)

A tertiary alcohol reacts with HBr by the S_N1 mechanism.

EXAMPLE: Conversion of *tert*-butyl alcohol to *tert*-butyl bromide.

Step 1: Protonation converts the hydroxy group to a good leaving group.

$$(CH_3)_3C-\ddot{O}-H \quad \xrightleftharpoons{H-\ddot{B}r:} \quad (CH_3)_3C-\overset{+}{O}H_2 \quad + \quad :\ddot{B}r:^-$$

tert-butyl alcohol

Step 2: Water leaves, forming a carbocation.

$$(CH_3)_3C-\overset{+}{O}H_2 \quad \rightleftharpoons \quad (CH_3)_3C^+ \quad + \quad H_2O$$

carbocation

(CONTINUED)

Step 3: Bromide ion attacks the carbocation.

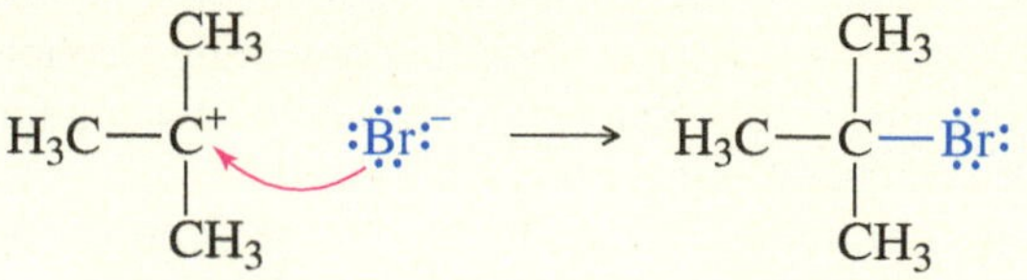

$$(CH_3)_3C^+ \;\; :\ddot{Br}:^- \longrightarrow (CH_3)_3C-\ddot{Br}:$$

carbocation → *tert*-butyl bromide

Many other alcohols react with HBr, with the reaction mechanism depending on the structure of the alcohol. For example, butan-1-ol reacts with sodium bromide in concentrated sulfuric acid to give 1-bromobutane by an S_N2 displacement. The sodium bromide/sulfuric acid reagent generates HBr in the solution.

$$CH_3(CH_2)_2-CH_2OH \xrightarrow{NaBr,\ H_2SO_4} CH_3(CH_2)_2-CH_2Br$$

butan-1-ol → 1-bromobutane (90%)

Protonation converts the hydroxy group to a good leaving group, but ionization to a primary carbocation is unfavorable. The protonated primary alcohol is well suited for the S_N2 displacement, however. Back-side attack by bromide ion gives 1-bromobutane.

MECHANISM 11-2 Reaction of a Primary Alcohol with HBr (S_N2)

A primary alcohol reacts with HBr by the S_N2 mechanism.

EXAMPLE: Conversion of butan-1-ol to 1-bromobutane.

Step 1: Protonation converts the hydroxy group to a good leaving group.

$$CH_3CH_2CH_2-CH_2-\ddot{O}-H + H-\ddot{Br}: \rightleftharpoons CH_3CH_2CH_2-CH_2-\overset{+}{O}H_2 + :\ddot{Br}:^-$$

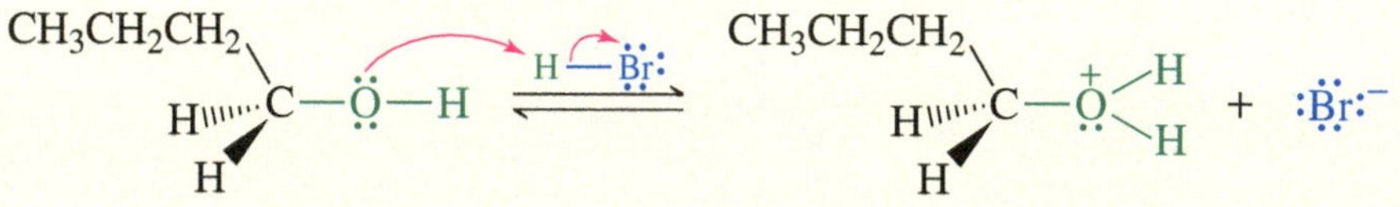

butan-1-ol

Step 2: Bromide displaces water to give the alkylbromide.

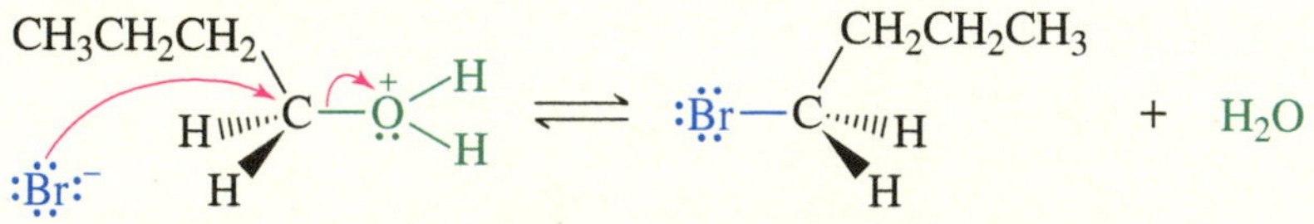

$$:\ddot{Br}:^- + CH_3CH_2CH_2-CH_2-\overset{+}{O}H_2 \rightleftharpoons :\ddot{Br}-CH_2CH_2CH_2CH_3 + H_2O$$

1-bromobutane

Secondary alcohols also react with HBr to form alkyl bromides, usually by the S_N1 mechanism. For example, cyclohexanol is converted to bromocyclohexane using HBr as the reagent.

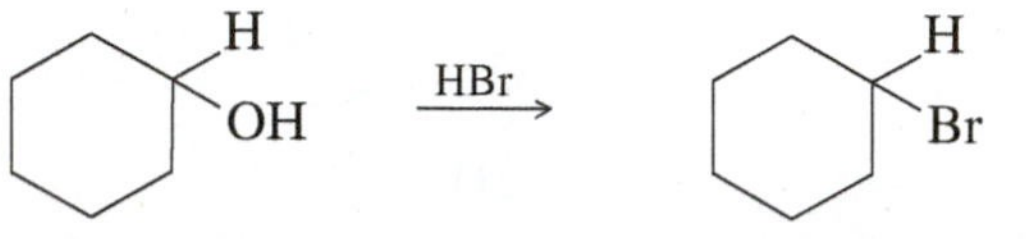

$$C_6H_{11}OH \xrightarrow{HBr} C_6H_{11}Br$$

cyclohexanol → bromocyclohexane (80%)

PROBLEM 11-12

Propose a mechanism for the reaction of
(a) 1-methylcyclohexanol with HBr to form 1-bromo-1-methylcyclohexane.
(b) 2-cyclohexylethanol with HBr to form 1-bromo-2-cyclohexylethane.

PROBLEM-SOLVING HINT

Memorizing all these mechanisms is not the best way to study this material. Depending on the substrate, these reactions can go by more than one mechanism. Gain experience working problems, and then consider each individual case to propose a likely mechanism.

11-7B Reactions with Hydrochloric Acid

$$\text{R—OH} + \text{HCl/H}_2\text{O} \xrightarrow{\text{ZnCl}_2} \text{R—Cl}$$

Hydrochloric acid (HCl) reacts with alcohols in much the same way that hydrobromic acid does. For example, concentrated aqueous HCl reacts with *tert*-butyl alcohol to give *tert*-butyl chloride.

$$\underset{\textit{tert}\text{-butyl alcohol}}{(CH_3)_3C\text{—}OH} + HCl/H_2O \longrightarrow \underset{\textit{tert}\text{-butyl chloride (98\%)}}{(CH_3)_3C\text{—}Cl} + H_2O$$

PROBLEM 11-13

The reaction of *tert*-butyl alcohol with concentrated HCl goes by the S_N1 mechanism. Write a mechanism for this reaction.

Chloride ion is a weaker nucleophile than bromide ion because it is smaller and less polarizable. An additional Lewis acid, such as zinc chloride ($ZnCl_2$), is sometimes necessary to promote the reaction of HCl with primary and secondary alcohols. Zinc chloride coordinates with the oxygen of the alcohol in the same way a proton does—except that zinc chloride coordinates more strongly.

The reagent composed of HCl and $ZnCl_2$ is called the **Lucas reagent.** Secondary and tertiary alcohols react with the Lucas reagent by the S_N1 mechanism.

S_N1 reaction with the Lucas reagent (fast)

$$(CH_3)_2CH\text{—}\ddot{O}\text{—}H \overset{ZnCl_2}{\rightleftharpoons} (CH_3)_2CH\text{—}\overset{+}{O}(H)\text{—}\bar{Z}nCl_2 \rightleftharpoons (CH_3)_2\overset{+}{C}H$$

alcohol–zinc chloride complex; carbocation

$$(CH_3)_2\overset{+}{C}H \xrightarrow{Cl^-} (CH_3)_2CH\text{—}Cl + H\ddot{O}\text{—}\bar{Z}nCl_2$$

When a primary alcohol reacts with the Lucas reagent, ionization is not possible—the primary carbocation is too unstable. Primary substrates react by an S_N2 mechanism, which is slower than the S_N1 reaction of secondary and tertiary substrates. For example, when butan-1-ol reacts with the Lucas reagent, the chloride ion attacks the complex from the back, displacing the leaving group.

S_N2 reaction with the Lucas reagent (slow)

$$:\ddot{C}l:^- + CH_3CH_2CH_2\text{—}CH_2\text{—}\overset{+}{O}(H)\text{—}\bar{Z}nCl_2 \longrightarrow \left[\overset{\delta-}{Cl}\text{---}C(H)(H)(CH_2CH_2CH_3)\text{---}\overset{\delta+}{O}(H)\text{—}\bar{Z}nCl_2\right]^{\ddagger} \longrightarrow Cl\text{—}CH_2CH_2CH_2CH_3 + :\ddot{O}(H)\text{—}\bar{Z}nCl_2$$

transition state

The Lucas Test The Lucas reagent reacts with primary, secondary, and tertiary alcohols at predictable rates, and these rates can distinguish among the three types of alcohols. When the reagent is first added to the alcohol, the mixture forms a single homogeneous phase: The concentrated HCl solution is very polar, and the polar alcohol–zinc chloride complex dissolves. Once the alcohol has reacted to form the alkyl halide, the relatively nonpolar halide separates into a second phase. (R—OH dissolves, but R—Cl does not.)

The **Lucas test** involves adding the Lucas reagent to an unknown water-soluble alcohol and watching for the second phase to separate (see Table 11-2). Tertiary alcohols react and show a second phase almost instantly because they form relatively stable tertiary carbocations. Secondary alcohols react in about 1 to 5 minutes because their secondary carbocations are less stable than tertiary ones. Primary alcohols react very slowly. Because the activated primary alcohol cannot form a carbocation, it simply remains in solution until it is attacked by the chloride ion. With a primary alcohol, the reaction may take from 10 minutes to several days.

TABLE 11-2
Reactions of Alcohols With the Lucas Reagent

Alcohol Type	Time to React (min)
primary	>6
secondary	1–5
tertiary	<1

PROBLEM 11-14

Show how you would use a simple chemical test to distinguish between the following pairs of compounds. Tell what you would observe with each compound.

(a) isopropyl alcohol and *tert*-butyl alcohol
(b) isopropyl alcohol and butan-2-one, $CH_3COCH_2CH_3$
(c) hexan-1-ol and cyclohexanol
(d) allyl alcohol and propan-1-ol
(e) butan-2-one and *tert*-butyl alcohol

11-7C Limitations on the Use of Hydrohalic Acids with Alcohols

The reactions of alcohols with hydrohalic acids do not always give good yields of the expected alkyl halides. Four principal limitations restrict the generality of this technique.

1. *Poor yields of alkyl chlorides from primary and secondary alcohols.* Primary and secondary alcohols react with HCl much more slowly than tertiary alcohols, even with zinc chloride added. Under these conditions, side reactions may prevent good yields of the alkyl halides.
2. *Eliminations.* Heating an alcohol in a concentrated acid such as HCl or HBr often leads to elimination. Once the hydroxy group of the alcohol has been protonated and converted to a good leaving group, it becomes a candidate for both substitution and elimination.
3. *Rearrangements.* Carbocation intermediates are always prone to rearrangements. We have seen (Section 6-15) that hydrogen atoms and alkyl groups can migrate from one carbon atom to another to form a more stable carbocation. This rearrangement may occur as the leaving group leaves, or it may occur once the cation has formed.
4. *Limited ability to make alkyl iodides.* Many alcohols do not react with HI to give acceptable yields of alkyl iodides. Alkyl iodides are valuable intermediates, however, because iodides are the most reactive of the alkyl halides. We will discuss another technique for making alkyl iodides in the next section.

SOLVED PROBLEM 11-2

When 3-methylbutan-2-ol is treated with concentrated HBr, the major product is 2-bromo-2-methylbutane. Propose a mechanism for the formation of this product.

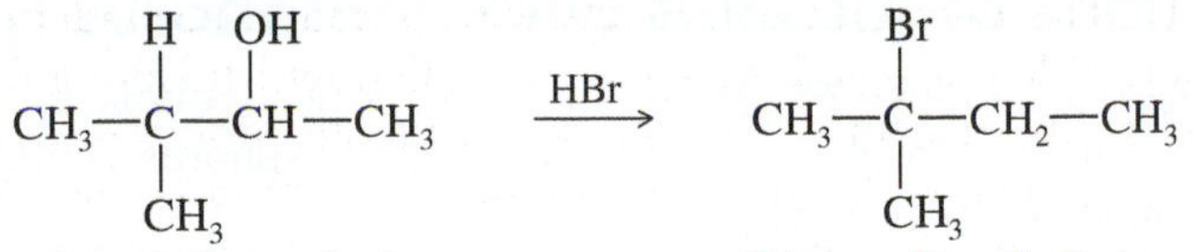

$$CH_3-C(H)(CH_3)-CH(OH)-CH_3 \xrightarrow{HBr} CH_3-C(Br)(CH_3)-CH_2-CH_3$$

3-methylbutan-2-ol → 2-bromo-2-methylbutane

SOLUTION

The alcohol is protonated by the strong acid. This protonated secondary alcohol loses water to form a secondary carbocation.

$$CH_3-C(H)(CH_3)-CH(\ddot{O}H)-CH_3 \overset{H^+}{\rightleftharpoons} CH_3-C(H)(CH_3)-CH(\overset{+}{O}H_2)-CH_3 \rightleftharpoons CH_3-C(H)(CH_3)-\overset{+}{C}H-CH_3 + H_2\ddot{O}:$$

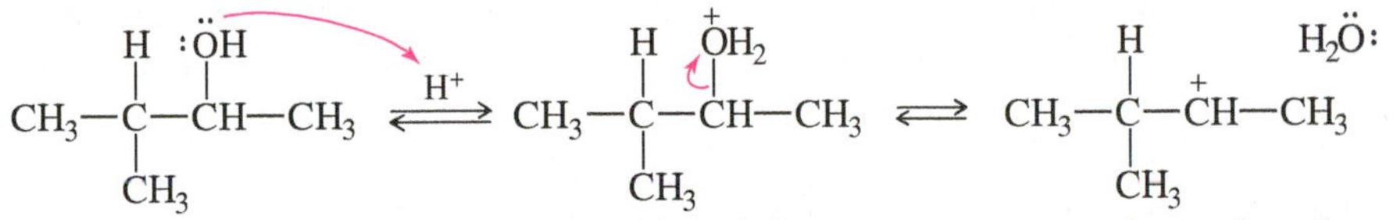

protonated alcohol; secondary carbocation

A hydride shift transforms the secondary carbocation into a more stable tertiary cation. Attack by bromide leads to the observed product.

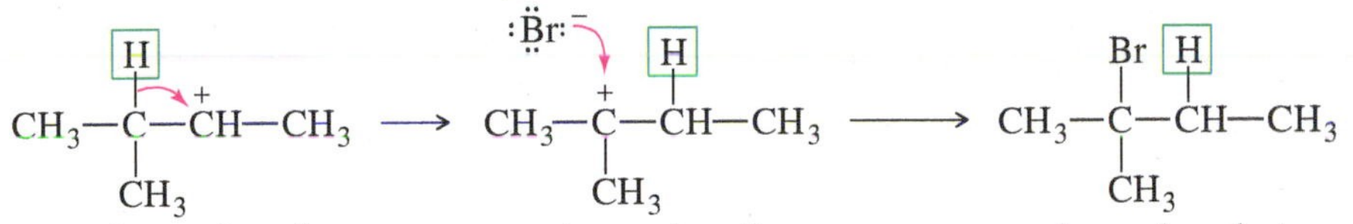

$$CH_3-C(H)(CH_3)-\overset{+}{C}H-CH_3 \longrightarrow CH_3-\overset{+}{C}(CH_3)-CH(H)-CH_3 \xrightarrow{:\ddot{Br}:^-} CH_3-C(Br)(CH_3)-CH(H)-CH_3$$

secondary carbocation; tertiary carbocation; observed product

Although rearrangements are usually seen as annoying side reactions, a clever chemist can use a rearrangement to accomplish a synthetic goal. Problem 11-15 shows how an alcohol substitution with rearrangement might be used in a synthesis.

PROBLEM 11-15

Neopentyl alcohol, $(CH_3)_3CCH_2OH$, reacts with concentrated HBr to give 2-bromo-2-methylbutane, a rearranged product. Propose a mechanism for the formation of this product.

PROBLEM 11-16

Explain the products observed in the following reaction of an alcohol with the Lucas reagent.

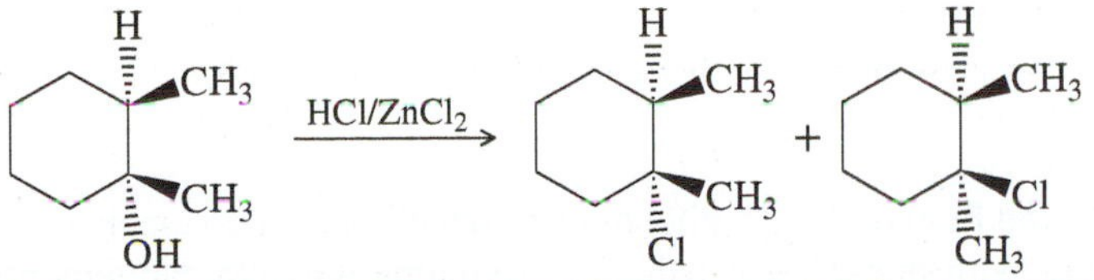

1,2-dimethylcyclohexanol (H, CH_3; CH_3, OH) $\xrightarrow{HCl/ZnCl_2}$ (H, CH_3; CH_3, Cl) + (H, CH_3; Cl, CH_3)

PROBLEM 11-17

When *cis*-2-methylcyclohexanol reacts with the Lucas reagent, the major product is 1-chloro-1-methylcyclohexane. Propose a mechanism to explain the formation of this product.

11-8 Reactions of Alcohols with Phosphorus Halides

Several phosphorus halides are useful for converting alcohols to alkyl halides. Phosphorus tribromide, phosphorus trichloride, and phosphorus pentachloride work well and are commercially available.

$$3\,R{-}OH + PCl_3 \longrightarrow 3\,R{-}Cl + P(OH)_3$$
$$3\,R{-}OH + PBr_3 \longrightarrow 3\,R{-}Br + P(OH)_3$$
$$R{-}OH + PCl_5 \longrightarrow R{-}Cl + POCl_3 + HCl$$

Phosphorus triiodide is not sufficiently stable to be stored, but it can be generated in situ (in the reaction mixture) by the reaction of phosphorus with iodine.

$$2\,P + 3\,I_2 \rightleftharpoons 2\,PI_3$$
$$6\,R{-}OH + 2\,P + 3\,I_2 \longrightarrow 6\,R{-}I + 2\,P(OH)_3$$

Phosphorus halides produce good yields of most primary and secondary alkyl halides, but none works well with tertiary alcohols. The two phosphorus halides used most often are PBr_3 and the phosphorus/iodine combination. Phosphorus tribromide is often the best reagent for converting a primary or secondary alcohol to the alkyl bromide, especially if the alcohol might rearrange in strong acid. A phosphorus and iodine combination is one of the best reagents for converting a primary or secondary alcohol to the alkyl iodide. For the synthesis of alkyl chlorides, thionyl chloride (discussed in the next section) generally gives better yields than PCl_3 or PCl_5, especially with tertiary alcohols.

The following examples show the conversion of primary and secondary alcohols to bromides and iodides by treatment with PBr_3 and P/I_2.

$$(CH_3)_3C{-}CH_2OH + PBr_3 \longrightarrow (CH_3)_3C{-}CH_2Br$$

neopentyl alcohol → neopentyl bromide (60%)

$$CH_3(CH_2)_{14}{-}CH_2OH + P/I_2 \longrightarrow CH_3(CH_2)_{14}{-}CH_2I$$

(85%)

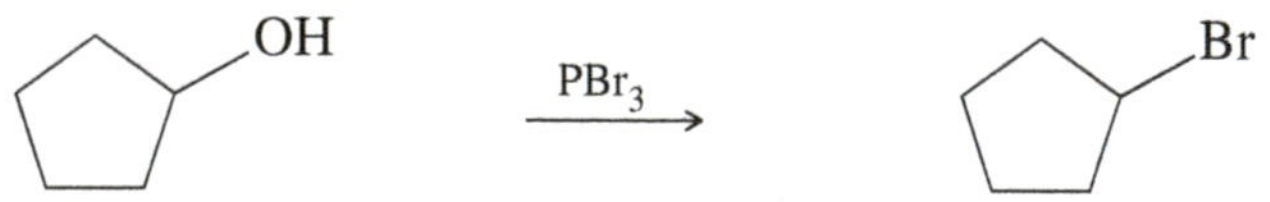

cyclopentanol (OH) $\xrightarrow{PBr_3}$ cyclopentyl bromide (Br) (90%)

PROBLEM 11-18

Write balanced equations for the three preceding reactions.

Mechanism of the Reaction with Phosphorus Trihalides The mechanism of the reaction of alcohols with phosphorus trihalides explains why rearrangements are uncommon and why phosphorus halides work poorly with tertiary alcohols. The mechanism is shown here using PBr_3 as the reagent; PCl_3 and PI_3 (generated from phosphorus and iodine) react in a similar manner.

Rearrangements are uncommon because no carbocation is involved, so there is no opportunity for rearrangement. This mechanism also explains the poor yields with tertiary alcohols. The final step is an S_N2 displacement where bromide attacks the back side of the alkyl group. This attack is hindered if the alkyl group is tertiary. In the case of a tertiary alcohol, an ionization to a carbocation is needed. This ionization is slow, and it invites side reactions.

MECHANISM 11-3 Reaction of Alcohols with PBr_3

Step 1: PBr_3 is a strong electrophile. An alcohol displaces bromide ion from PBr_3 to give an excellent leaving group.

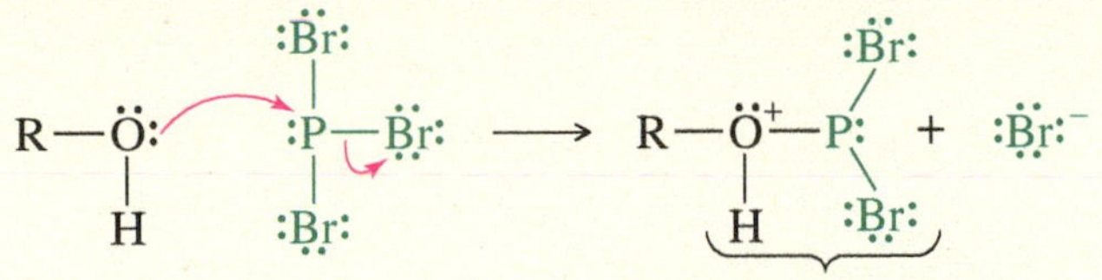

$$R{-}\ddot{O}(H){:} + PBr_3 \longrightarrow R{-}\ddot{O}^{+}(H){-}PBr_2 + {:}\ddot{Br}{:}^{-}$$

excellent leaving group

Step 2: Bromide displaces the leaving group to give the alkyl bromide.

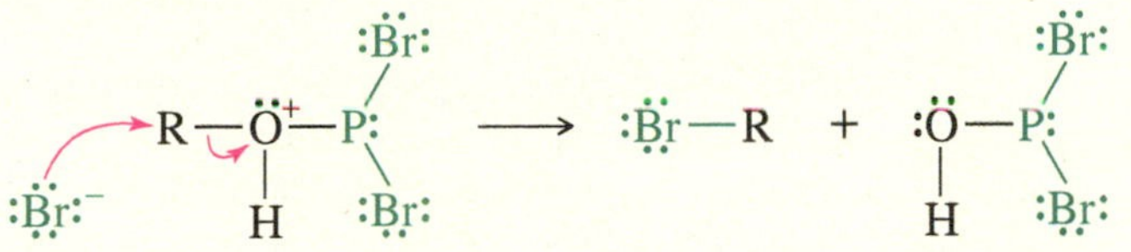

$${:}\ddot{Br}{:}^{-} + R{-}\ddot{O}^{+}(H){-}PBr_2 \longrightarrow {:}\ddot{Br}{-}R + {:}\ddot{O}(H){-}PBr_2$$

leaving group

EXAMPLE: Reaction of (*R*)-pentan-2-ol with PBr_3.

Step 1: Displacement of bromide and formation of a leaving group

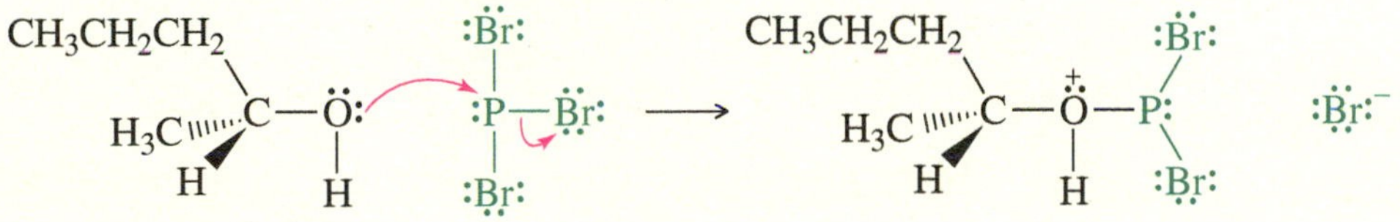

$$CH_3CH_2CH_2(H_3C)(H)C{-}\ddot{O}{-}H + PBr_3 \longrightarrow CH_3CH_2CH_2(H_3C)(H)C{-}\ddot{O}^{+}(H){-}PBr_2 + {:}\ddot{Br}{:}^{-}$$

(*R*)-pentan-2-ol

Step 2: Bromide displaces the leaving group to give (*S*)-2-bromopentane.

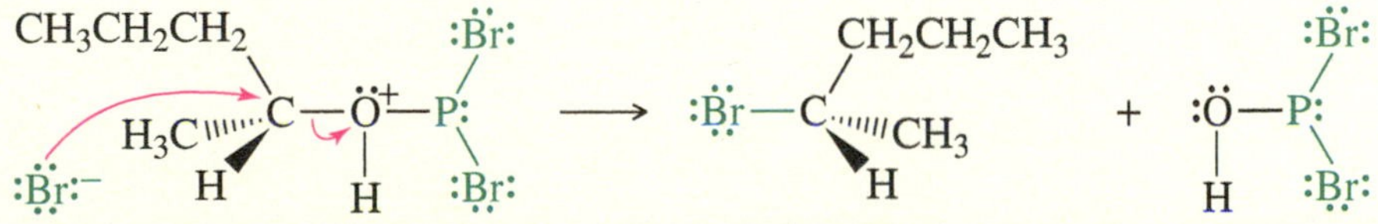

$${:}\ddot{Br}{:}^{-} + CH_3CH_2CH_2(H_3C)(H)C{-}\ddot{O}^{+}(H){-}PBr_2 \longrightarrow {:}\ddot{Br}{-}C(H)(CH_3)CH_2CH_2CH_3 + {:}\ddot{O}(H){-}PBr_2$$

(*S*)-2-bromopentane

11-9 Reactions of Alcohols with Thionyl Chloride

Thionyl chloride ($SOCl_2$) is often the best reagent for converting an alcohol to an alkyl chloride. The by-products (gaseous SO_2 and HCl) leave the reaction mixture and ensure there can be no reverse reaction.

$$R{-}OH + Cl{-}\overset{\overset{O}{\|}}{S}{-}Cl \xrightarrow{\text{heat}} R{-}Cl + SO_2 + HCl$$

Under the proper conditions, thionyl chloride reacts by the interesting mechanism summarized next. In the first step, the nonbonding electrons of the hydroxy oxygen atom attack the electrophilic sulfur atom of thionyl chloride. A chloride ion is expelled, and a proton is lost to give a chlorosulfite ester. In the next step, the chlorosulfite ester ionizes (when R = 2° or 3°), and the sulfur atom quickly delivers chloride to the carbocation. When R is primary, chloride probably bonds to carbon at the same time that the C—O bond is breaking.

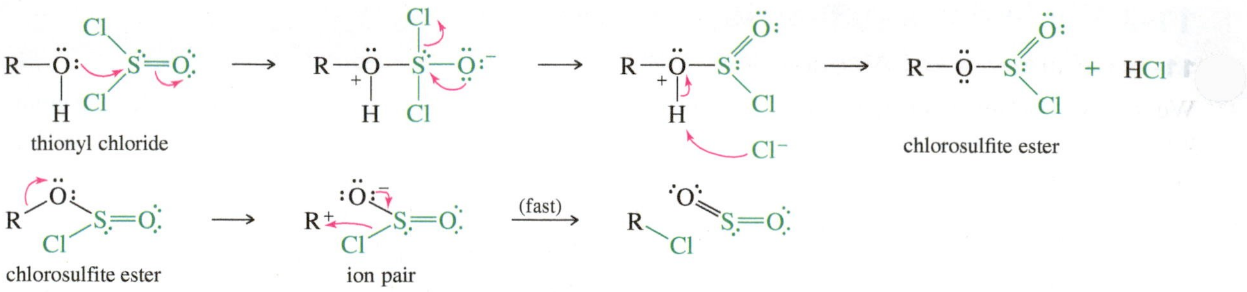

thionyl chloride

chlorosulfite ester

chlorosulfite ester

ion pair

(fast)

This mechanism resembles the S_N1, except that the nucleophile is delivered to the carbocation by the leaving group, usually giving retention of configuration as shown in the following example. (Under different conditions, retention of configuration might not be observed.)

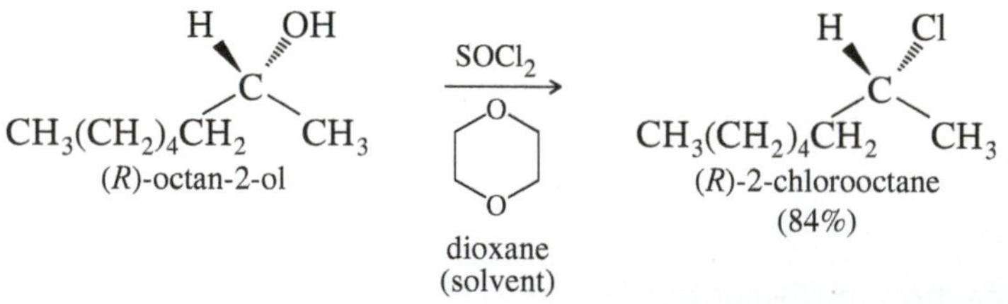

(*R*)-octan-2-ol $\xrightarrow{SOCl_2}$ (*R*)-2-chlorooctane (84%)

dioxane (solvent)

SUMMARY OF THE BEST REAGENTS FOR CONVERTING ALCOHOLS TO ALKYL HALIDES

Class of Alcohol	Chloride	Bromide	Iodide
primary	$SOCl_2$	PBr_3 or HBr^*	P/I_2
secondary	$SOCl_2$	PBr_3	P/I_2^*
tertiary	HCl	HBr	HI^*

*Works only in selected cases.

PROBLEM 11-19

Suggest how you would convert *trans*-4-methylcyclohexanol to
(a) *trans*-1-chloro-4-methylcyclohexane.
(b) *cis*-1-chloro-4-methylcyclohexane.

PROBLEM-SOLVING HINT

Thionyl chloride reacts with alcohols by various mechanisms that depend on the substrate, the solvent, and the temperature. Be cautious in predicting the structure and stereochemistry of a product unless you know the actual mechanism.

PROBLEM 11-20

Two products are observed in the following reaction.

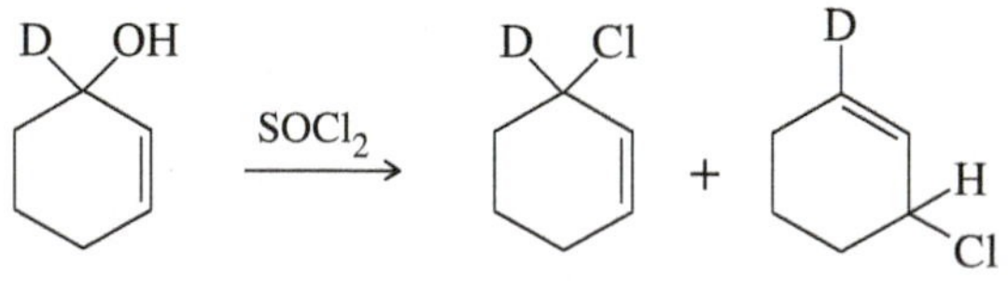

$\xrightarrow{SOCl_2}$ +

(a) Suggest a mechanism to explain how these two products are formed.
(b) Your mechanism for part (a) should be different from the usual mechanism of the reaction of $SOCl_2$ with alcohols. Explain why the reaction follows a different mechanism in this case.

PROBLEM 11-21

Give the structures of the products you would expect when each alcohol reacts with (1) HCl, $ZnCl_2$; (2) HBr; (3) PBr_3; (4) P/I_2; and (5) $SOCl_2$.
(a) butan-1-ol **(b)** 2-methylbutan-2-ol
(c) 2,2-dimethylbutan-1-ol **(d)** *cis*-3-methylcyclopentanol

11-10 Dehydration Reactions of Alcohols

11-10A Formation of Alkenes

We studied the mechanism for dehydration of alcohols to alkenes in Section 7-18, together with other syntheses of alkenes. Dehydration requires an acidic catalyst to protonate the hydroxy group of the alcohol and convert it to a good leaving group. Loss of water, followed by loss of a proton, gives the alkene. An equilibrium is established between reactants and products.

MECHANISM 11-4 (Review): Acid-Catalyzed Dehydration of an Alcohol

Dehydration results from E1 elimination of the protonated alcohol.

Step 1: Protonation converts the hydroxy group to a good leaving group.

$$\text{H}-\text{C}-\text{C}-\ddot{\text{O}}-\text{H} \;\overset{\text{H}^+}{\rightleftharpoons}\; \text{H}-\text{C}-\text{C}-\overset{+}{\text{O}}\text{H}_2$$

Step 2: Water leaves, forming a carbocation.

$$\text{H}-\text{C}-\text{C}-\overset{+}{\text{O}}\text{H}_2 \;\rightleftharpoons\; \text{H}-\text{C}-\overset{+}{\text{C}}- \;+\; \text{H}-\ddot{\text{O}}-\text{H}$$

Step 3: Loss of a proton gives the alkene.

$$\text{H}_2\ddot{\text{O}}: \;+\; \text{H}-\text{C}-\overset{+}{\text{C}}- \;\rightleftharpoons\; \text{C}=\text{C} \;+\; \text{H}_3\text{O}^+$$

To drive this equilibrium to the right, we remove one or both of the products as they form, either by distilling the products out of the reaction mixture or by adding a dehydrating agent to remove water. In practice, we often use a combination of distillation and a dehydrating agent. The alcohol is mixed with a dehydrating acid, and the mixture is heated to boiling. The alkene boils at a lower temperature than the alcohol (because the alcohol is hydrogen-bonded), and the alkene distills out of the mixture. For example,

$$\text{cyclohexanol} \;\underset{}{\overset{H_2SO_4}{\rightleftharpoons}}\; \text{cyclohexene} + H_2O$$

cyclohexanol, bp 161 °C

cyclohexene, bp 83 °C (80%)
(distilled from the mixture)

Alcohol dehydrations generally take place through the E1 mechanism. Protonation of the hydroxy group converts it to a good leaving group. Water leaves, forming a carbocation. Loss of a proton gives the alkene.

Figure 11-2 shows the reaction-energy diagram for the E1 dehydration of an alcohol. The first step is a mildly exothermic protonation, followed by an endothermic, rate-limiting ionization. A fast, strongly exothermic deprotonation gives the alkene. Because the rate-limiting step is formation of a carbocation, the ease of dehydration follows from the ease of formation of carbocations: 3° > 2° > 1°. As in other carbocation reactions, rearrangements are common.

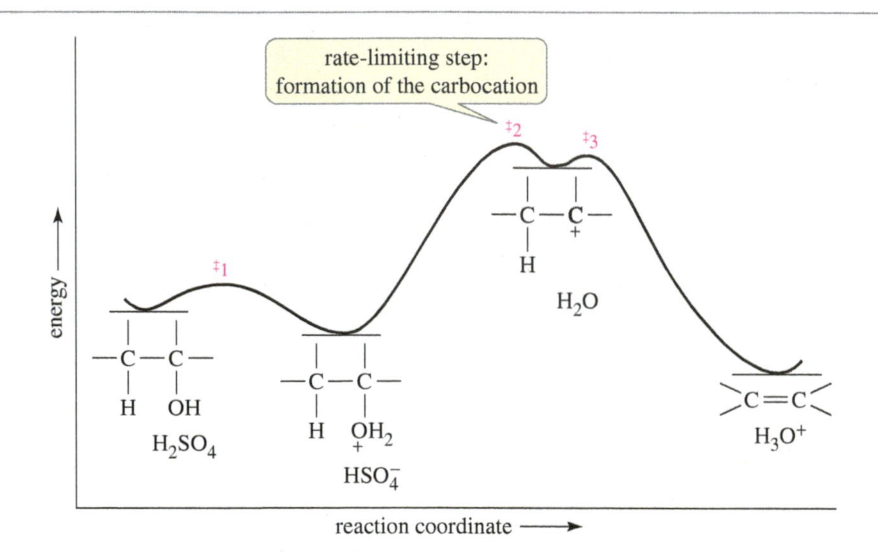

FIGURE 11-2
Reaction-energy diagram for dehydration of an alcohol.

With primary alcohols, the reaction conditions must be severe. Rearrangement and isomerization of the products are so common that acid-catalyzed dehydration is rarely a good method for converting them to alkenes. The following mechanism shows how butan-1-ol undergoes dehydration with rearrangement to give a mixture of but-1-ene and but-2-ene. The more highly substituted product, but-2-ene, is the major product, in accordance with Zaitsev's rule (Section 7-11).

Ionization of the protonated alcohol, with rearrangement

H_2O leaves
$H:^-$ migrates

secondary carbocation

Loss of either proton to give two products

loss of H_a^+
but-2-ene (major, 70%)
a disubstituted alkene

loss of H_b^+
but-1-ene (minor, 30%)
a monosubstituted alkene

Let's review the utility of dehydration and give guidelines for predicting the products:

1. Dehydration usually goes by the E1 mechanism. Rearrangements may occur to form more stable carbocations.

2. Dehydration works best with tertiary alcohols and almost as well with secondary alcohols. Rearrangements and poor yields are common with primary alcohols.
3. (Zaitsev's rule) If two or more alkenes might be formed by deprotonation of the carbocation, the most substituted alkene usually predominates.

Solved Problem 11-3 shows how these rules are used to predict the products of dehydrations. The carbocations are drawn to show how rearrangements occur and how more than one product may result.

SOLVED PROBLEM 11-3

Predict the products of sulfuric acid-catalyzed dehydration of the following alcohols.
(a) 1-methylcyclohexanol
(b) neopentyl alcohol

SOLUTION

(a) 1-Methylcyclohexanol reacts to form a tertiary carbocation. A proton may be abstracted from any one of three carbon atoms. The two secondary atoms are equivalent, and abstraction of a proton from one of them leads to the trisubstituted double bond of the major product. Abstraction of a methyl proton leads to the disubstituted double bond of the minor product.

1-methylcyclohexanol —H^+→ protonated —$-H_2O$→ cation

cation —$H_2\ddot{O}$:→ loss of H_a major product (trisubstituted) + loss of H_b minor product (disubstituted) + H_3O^+

SOLUTION

(b) Neopentyl alcohol cannot simply ionize to form a primary cation. Rearrangement occurs as the leaving group leaves, giving a tertiary carbocation. Loss of a proton from the adjacent secondary carbon gives the trisubstituted double bond of the major product. Loss of a proton from the methyl group gives the disubstituted double bond of the minor product.

neopentyl alcohol (2,2-dimethylpropan-1-ol) —H^+→ ionization with rearrangement → 3° cation + $:\ddot{O}H_2$

—$-H^+$→ loss of H_a^+ major product (trisubstituted) + loss of H_b^+ minor product (disubstituted)

PROBLEM-SOLVING HINT

Most alcohol dehydrations go by E1 mechanisms involving protonation of the OH group, followed by loss of water.

PROBLEM-SOLVING HINT

Draw the carbocation, look for possible rearrangements, and then consider all the ways that the original carbocation and any rearranged carbocation might lose protons to give alkenes. Zaitsev's rule usually predicts the major product.

PROBLEM 11-22

Predict the products of the sulfuric acid-catalyzed dehydration of the following alcohols. When more than one product is expected, label the major and minor products.
(a) 2-methylbutan-2-ol **(b)** pentan-1-ol **(c)** pentan-2-ol
(d) 1-isopropylcyclohexanol **(e)** 2-methylcyclohexanol

PROBLEM 11-23

Some alcohols undergo rearrangement or other unwanted side reactions when they dehydrate in acid. Alcohols may be dehydrated under mildly *basic* conditions using phosphorus oxychloride ($POCl_3$) in pyridine. The alcohol reacts with phosphorus oxychloride much like it reacts with tosyl chloride (Section 11-5), displacing a chloride ion from phosphorus to give an alkyl dichlorophosphate ester. The dichlorophosphate group is an outstanding leaving group. Pyridine reacts as a base with the dichlorophosphate ester to give an E2 elimination. Propose a mechanism for the dehydration of cyclohexanol by $POCl_3$ in pyridine.

O, P, Cl, Cl, Cl, N

phosphorus oxychloride pyridine

11-10B Bimolecular Condensation to Form Ethers (Industrial)

In some cases, a protonated primary alcohol may be attacked by another molecule of the alcohol and undergo an S_N2 displacement. The net reaction is a bimolecular dehydration to form an ether. For example, the attack by ethanol on a protonated molecule of ethanol gives diethyl ether.

$CH_3CH_2-\ddot{O}:$ (H) + CH_3, H, H–C–O^+(H)(H) $\xrightarrow{S_N2}$ $CH_3CH_2-\ddot{O}^+-C$ (CH_3, H, H) (H) + $:\ddot{O}:$ (Et)(H) + H_2O $\longrightarrow$ $CH_3CH_2-\ddot{O}-C$ (CH_3, H, H) + $Et\overset{+}{O}H_2$ + H_2O

nucleophilic electrophilic protonated ether diethyl ether

This bimolecular dehydration of alcohols is a type of **condensation,** a reaction that joins two (or more) molecules, often with the loss of a small molecule such as water. This method is used for the industrial synthesis of diethyl ether ($CH_3CH_2-O-CH_2CH_3$) and dimethyl ether (CH_3-O-CH_3). Under the acidic dehydration conditions, two reactions compete: Elimination (dehydration to give an alkene) competes with substitution (condensation to give an ether).

Elimination to give the alkene, a unimolecular dehydration

$$\underset{\text{ethanol}}{CH_3CH_2OH} \xrightarrow{H_2SO_4,\ 180\ °C} \underset{\text{ethylene}}{CH_2{=}CH_2} + H_2O$$

Substitution to give the ether, a bimolecular condensation

$$\underset{\text{ethanol}}{2\ CH_3CH_2OH} \xrightarrow{H_2SO_4,\ 140\ °C} \underset{\text{diethyl ether}}{CH_3CH_2-O-CH_2CH_3} + H_2O$$

PROBLEM 11-24

Contrast the mechanisms of the two preceding reactions, the dehydration and condensation of ethanol.

How can we control these two competing reactions? The ether synthesis (substitution) shows two molecules of alcohol giving two product molecules, one of diethyl ether and one of water. The elimination shows one molecule of alcohol giving two molecules, one of ethylene and one of water. The elimination results in an increase in the number of molecules and therefore an *increase* in the randomness (entropy) of the system. The elimination has a more positive change in entropy (ΔS) than the substitution, and the $-T\Delta S$ term in the Gibbs free energy becomes more favorable for the elimination as the temperature increases. Substitution (condensation to give the ether) is favored around 140 °C and below, and elimination is favored around 180 °C and above. Diethyl ether is produced industrially by heating ethanol with an acidic catalyst at around 140 °C.

PROBLEM 11-25

Explain why the acid-catalyzed condensation is a poor method for the synthesis of an unsymmetrical ether such as ethyl methyl ether, $CH_3CH_2{-}O{-}CH_3$.

PROBLEM-SOLVING STRATEGY Proposing Reaction Mechanisms

In view of the large number of reactions we've covered, proposing mechanisms for reactions you have never seen may seem nearly impossible. As you gain experience in working mechanism problems, you will start to see similarities to known reactions. Let's consider how an organic chemist systematically approaches a mechanism problem. (A more complete version of this method appears in Appendix 3A.) Although this stepwise approach cannot solve all mechanism problems, it should provide a starting point to begin building your experience and confidence.

Determining the Type of Mechanism

First, determine what kinds of conditions and catalysts are involved. In general, reactions may be classified as involving (a) strong electrophiles (including acid-catalyzed reactions), (b) strong nucleophiles (including base-catalyzed reactions), or (c) free radicals. These three types of mechanisms are quite distinct, and you should first try to determine which type is involved.

(a) In the presence of a strong acid or a reactant that can dissociate to give a strong electrophile, the mechanism probably involves strong electrophiles as intermediates. Acid-catalyzed reactions and reactions involving carbocations (such as the S_N1, the E1, and most alcohol dehydrations) fall into this category.

(b) In the presence of a strong base or a strong nucleophile, the mechanism probably involves strong nucleophiles as intermediates. Base-catalyzed reactions and those depending on base strength (such as the S_N2 and the E2) generally fall into this category.

(c) Free-radical reactions usually require a free-radical initiator such as chlorine, bromine, NBS, or a peroxide. In most free-radical reactions, there is no need for a strong acid or base.

Once you have determined which type of mechanism you will write, use a systematic approach to the problem. At this point, we consider mostly the electrophilic reactions covered in recent chapters. Suggestions for drawing the mechanisms of reactions involving strong nucleophiles and free-radical reactions are collected in Appendix 3A.

Reactions Involving Strong Electrophiles

When a strong acid or electrophile is present, expect to see intermediates that are strong acids and strong electrophiles. Cationic intermediates are common. Bases and nucleophiles in such a reaction are generally weak, however. Avoid drawing carbanions, hydroxide ions, alkoxide ions, and other strong bases. They are unlikely to co-exist with strong acids and strong electrophiles.

Functional groups are often converted to carbocations or other strong electrophiles by protonation or reaction with a strong electrophile. Then the carbocation or other strong electrophile reacts with a weak nucleophile such as an alkene or the solvent.

1. **Consider the carbon skeletons of the reactants and products, and decide which carbon atoms in the products are most likely derived from which carbon atoms in the reactants.**
2. **Consider whether any of the reactants is a strong enough electrophile to react without being activated. If not, consider how one of the reactants might be converted to a strong electrophile by protonation of a basic site, or complexation with a Lewis acid or ionization.**

(continued)

Protonation of an alcohol, for example, converts it to a strong electrophile, which can undergo attack or lose water to give a carbocation, an even stronger electrophile. Protonation of an alkene converts it to a carbocation.

3. **Consider how a nucleophilic site on another reactant (or, in a cyclization in another part of the same molecule) can attack the strong electrophile to form a bond needed in the product. Draw the product of this bond formation.**
If the intermediate is a carbocation, consider whether it is likely to rearrange to form a bond in the product. If there isn't any possible nucleophilic attack that leads in the direction of the product, consider other ways of converting one of the reactants to a strong electrophile.
4. **Consider how the product of a nucleophilic attack might be converted to the final product (if it has the right carbon skeleton) or reactivated to form another bond needed in the product.**
To move a proton from one atom to another under acidic conditions (as in an isomerization), try adding a proton to the new position and then removing it from the old position.
5. **Draw out all the steps of the mechanism using curved arrows to show the movement of electrons.**
Be careful to show only one step at a time.

Common Mistakes to Avoid in Drawing Mechanisms

1. Do not use condensed or line–angle formulas for reaction sites. Draw all the bonds and all the substituents of each carbon atom affected throughout the mechanism. In reactions involving strong electrophiles and acidic conditions, three-bonded carbon atoms are likely to be carbocations. If you draw condensed formulas or line–angle formulas, you will likely misplace a hydrogen atom and show a reactive species on the wrong carbon.
2. Do not show more than one step occurring at once. Do not show two or three bonds changing position in one step unless the changes really are concerted (take place simultaneously). For example, protonation of an alcohol and loss of water to give a carbocation are two steps. You must not show the hydroxy group "jumping" off the alcohol to join up with an anxiously waiting proton.
3. Remember that curved arrows show *movement of electrons*, always from the nucleophile (electron donor) to the electrophile (electron acceptor). For example, protonation of a double bond must show the arrow going from the electrons of the double bond to the proton—never from the proton to the double bond. Resist the urge to use an arrow to "point out" where the proton (or other reagent) goes.

Sample Problem

To illustrate the stepwise method for reactions involving strong electrophiles, we will develop a mechanism to account for the following cyclization:

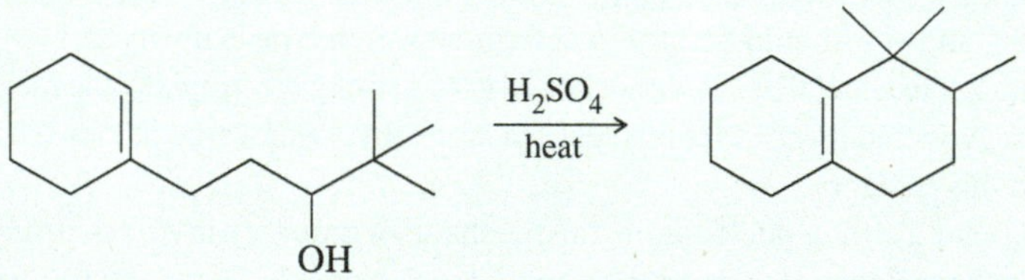

OH $\xrightarrow[\text{heat}]{H_2SO_4}$

The cyclized product is a minor product in this reaction. Note that a mechanism problem is different from a synthesis problem: In a mechanism problem, we are limited to the reagents given and are asked to explain how these reactants form these products under the conditions shown. Also, a mechanism problem may deal with how an unusual or unexpected minor product is formed.

In the presence of sulfuric acid, this is clearly an acid-catalyzed mechanism. We expect strong electrophiles, cationic intermediates (possibly carbocations), and strong acids. Carbanions, hydroxide ions, alkoxide ions, and other strong bases and strong nucleophiles are unlikely.

1. **Consider the carbon skeletons of the reactants and products, and decide which carbon atoms in the products are most likely derived from which carbon atoms in the reactants.**
Drawing the starting material and the product with all the substituents of the affected carbon atoms, we see the major changes shown here. A vinyl hydrogen must be lost, a =C—C bond must be formed, a methyl group must move over one carbon atom, and the hydroxy group must be lost.

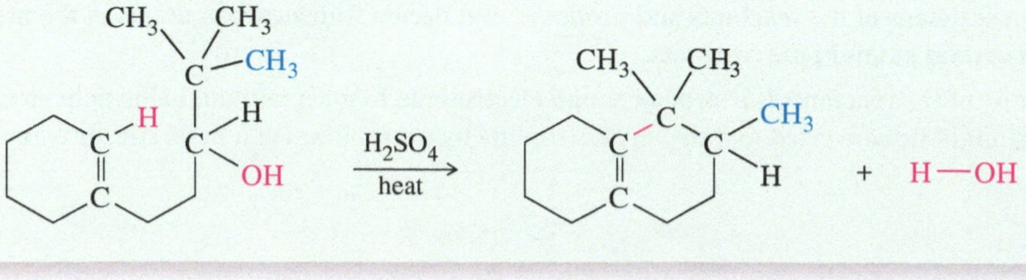

CH_3 CH_3 CH_3 C H C H C OH C $\xrightarrow[\text{heat}]{H_2SO_4}$ CH_3 CH_3 C CH_3 C C H C + H—OH

2. Consider whether any of the reactants is a strong enough electrophile to react without being activated. If not, consider how one of the reactants might be converted to a strong electrophile by protonation of a basic site, or complexation with a Lewis acid, or ionization.

The starting material is not a strong electrophile, so it must be activated. Sulfuric acid could generate a strong electrophile either by protonating the double bond or by protonating the hydroxy group. Protonating the double bond would form the tertiary carbocation, activating the wrong end of the double bond. Also, there is no good nucleophilic site on the side chain to attack this carbocation to form the correct ring. Protonating the double bond is a dead end.

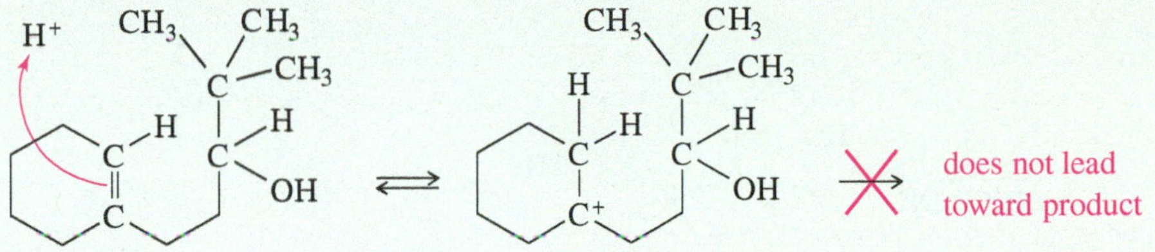

does not lead toward product

The other basic site is the hydroxy group. An alcohol can protonate on the hydroxy group and lose water to form a carbocation.

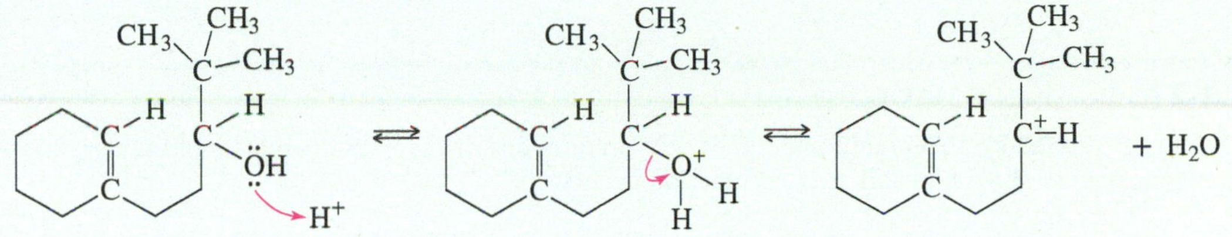

3. Consider how a nucleophilic site on another reactant (or, in a cyclization, in another part of the same molecule) can attack the strong electrophile to form a bond needed in the product. Draw the product of this bond formation.

The carbocation can be attacked by the electrons in the double bond to form a ring, but the positive charge is on the wrong carbon atom to give a six-membered ring. A favorable rearrangement of the secondary carbocation to a tertiary one shifts the positive charge to the correct carbon atom and accomplishes the methyl shift we identified in step 1. Attack by the (weakly) nucleophilic electrons in the double bond gives the correct six-membered ring.

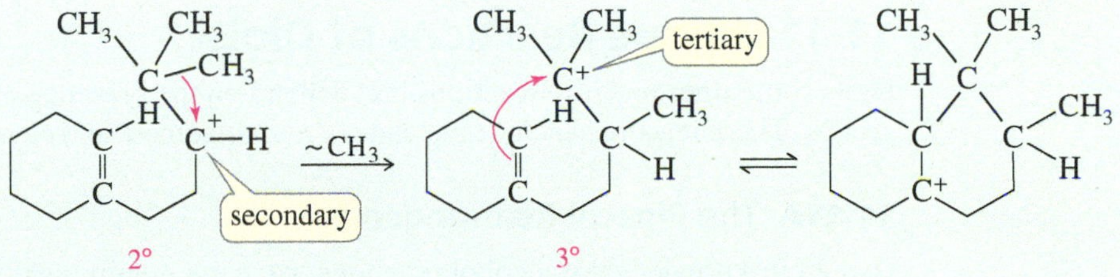

secondary 2° ~CH_3 tertiary 3°

4. Consider how the product of nucleophilic attack might be converted to the final product (if it has the right carbon skeleton) or reactivated to form another bond needed in the product.

Loss of a proton (to HSO_4^- or H_2O, but *not* to ^-OH, which is not compatible with acid) gives the observed product.

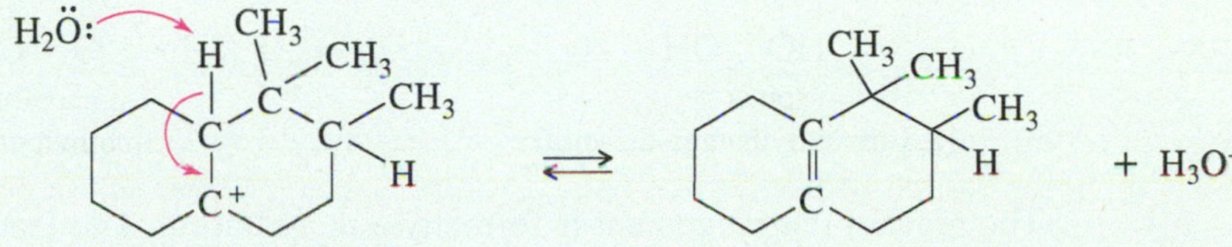

$$H_2\ddot{O}: \rightleftarrows \; + H_3O^+$$

5. Draw out all the steps of the mechanism using curved arrows to show the movement of electrons.

Combining the equations written immediately above gives the complete mechanism for this reaction.

The following problems require proposing mechanisms for reactions involving strong electrophiles. Work each one by completing the five steps just described.

PROBLEM 11-26

Propose a mechanism for each reaction.

(a) 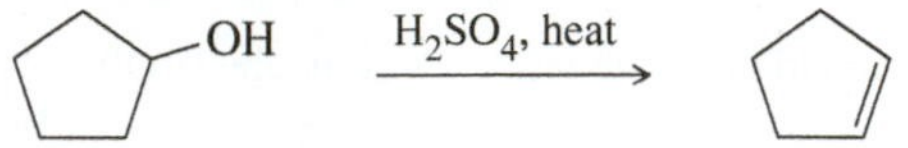 cyclopentanol $\xrightarrow{H_2SO_4,\ heat}$ cyclopentene

(b) 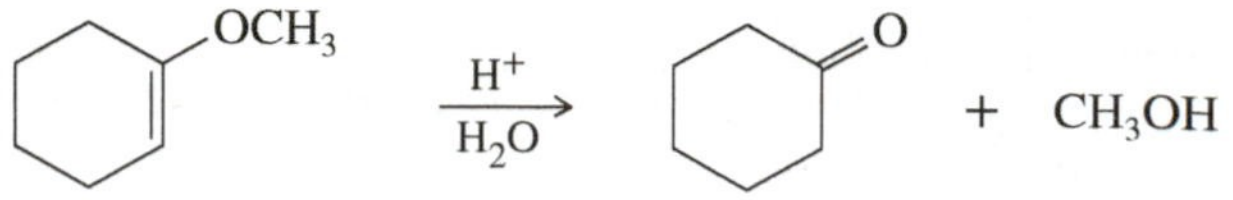 1-methoxycyclohexene (OCH_3) $\xrightarrow[H_2O]{H^+}$ cyclohexanone (O) + CH_3OH

(c) 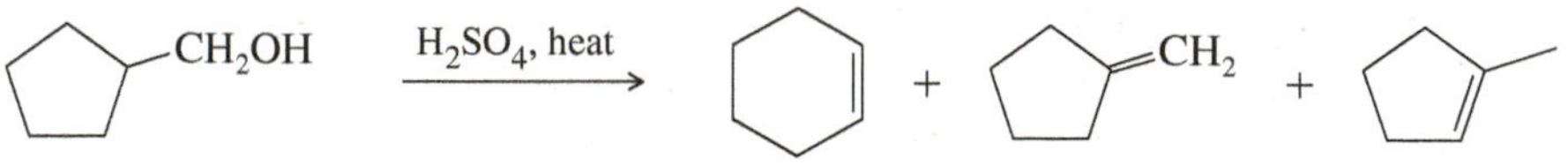 cyclopentylmethanol (CH_2OH) $\xrightarrow{H_2SO_4,\ heat}$ cyclohexene + methylenecyclopentane (CH_2) + 1-methylcyclopentene

(d) 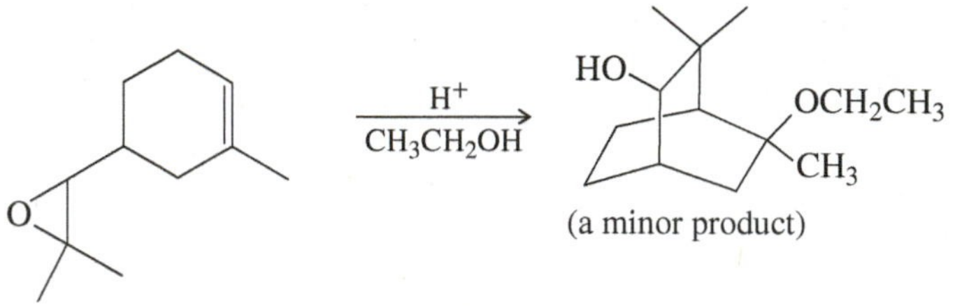 epoxide (O) $\xrightarrow[CH_3CH_2OH]{H^+}$ HO, OCH_2CH_3, CH_3 bicyclic product
(a minor product)

PROBLEM 11-27

When the following substituted cycloheptanol undergoes dehydration, one of the minor products has undergone a ring contraction. Propose a mechanism to show how this ring contraction occurs.

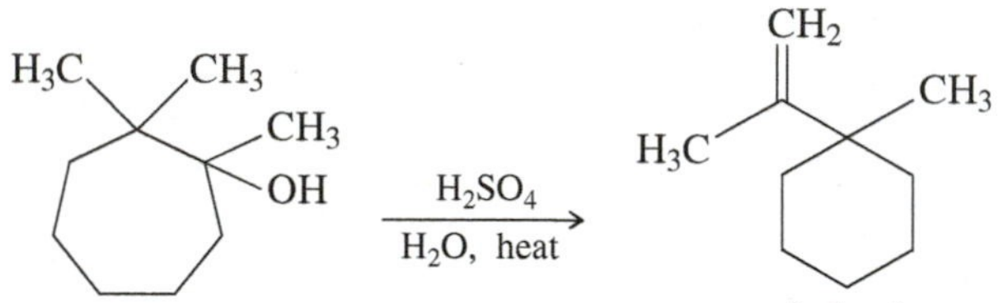

H_3C, CH_3, CH_3, OH (substituted cycloheptanol) $\xrightarrow[H_2O,\ heat]{H_2SO_4}$ CH_2, CH_3, H_3C (substituted cyclohexane)
(minor)

11-11 Unique Reactions of Diols

Diols sometimes undergo reactions that depend on the presence of two adjacent —OH groups. These reactions include the pinacol rearrangement and cleavage by periodic acid.

11-11A The Pinacol Rearrangement

Using our knowledge of alcohol reactions, we can explain results that seem strange at first glance. The following dehydration is an example of the **pinacol rearrangement:**

$$(CH_3)_2C(OH)-C(OH)(CH_3)_2 \xrightarrow[100\ °C]{H_2SO_4} CH_3-C(=O)-C(CH_3)_3 + H_2O$$

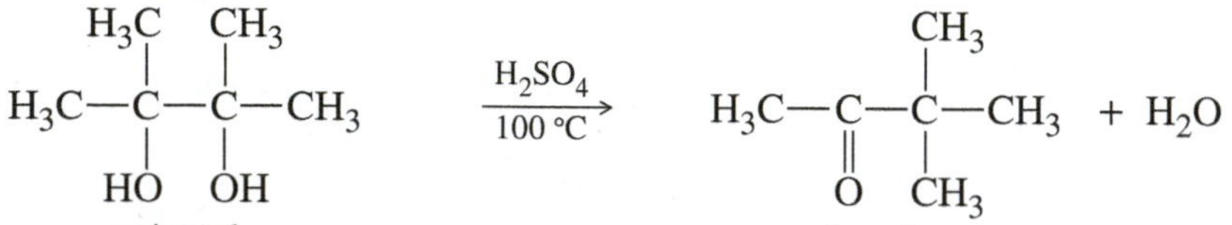

pinacol
(2,3-dimethylbutane-2,3-diol)

pinacolone
(3,3-dimethylbutan-2-one)

The pinacol rearrangement is formally a dehydration. The reaction is acid-catalyzed, and the first step is protonation of one of the hydroxy oxygens. Loss of water gives a tertiary carbocation, as expected for any tertiary alcohol. Migration of a methyl group places the positive charge on the carbon atom bearing the second —OH group, where oxygen's nonbonding electrons help to stabilize the positive charge through resonance. This extra stability is the driving force for the rearrangement, which converts a relatively stable 3° carbocation into an even better resonance-stabilized carbocation. Deprotonation of the resonance-stabilized cation gives the product, pinacolone.

MECHANISM 11-5 The Pinacol Rearrangement

Step 1: Protonation of a hydroxy group.

Step 2: Loss of water gives a carbocation.

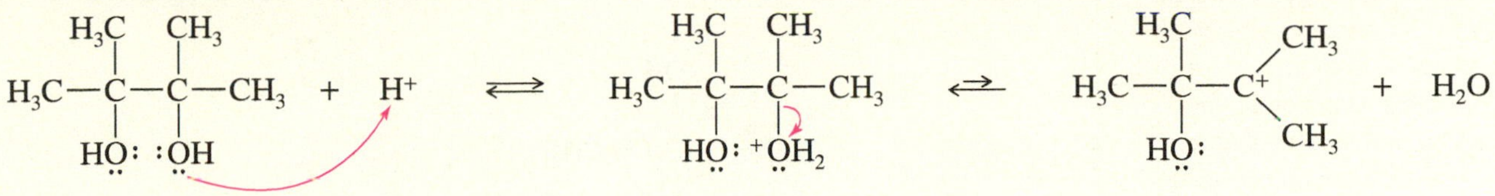

$$(CH_3)_2C(OH)-C(OH)(CH_3)_2 + H^+ \rightleftharpoons (CH_3)_2C(OH)-C(CH_3)_2(\overset{+}{O}H_2) \rightleftharpoons (CH_3)_2C(OH)-\overset{+}{C}(CH_3)_2 + H_2O$$

Step 3: Methyl migration forms a resonance-stabilized carbocation.

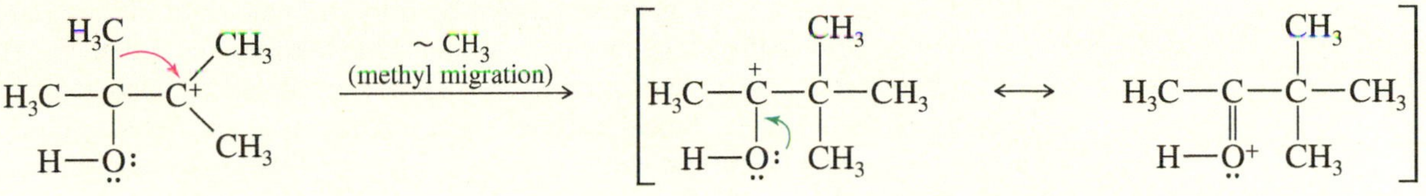

$$(CH_3)_2C(OH)-\overset{+}{C}(CH_3)_2 \xrightarrow[\text{(methyl migration)}]{\sim CH_3} \left[H_3C-\overset{+}{C}(OH)-C(CH_3)_3 \longleftrightarrow H_3C-C(=\overset{+}{O}H)-C(CH_3)_3 \right]$$

resonance-stabilized carbocation

Step 4: Deprotonation gives the product.

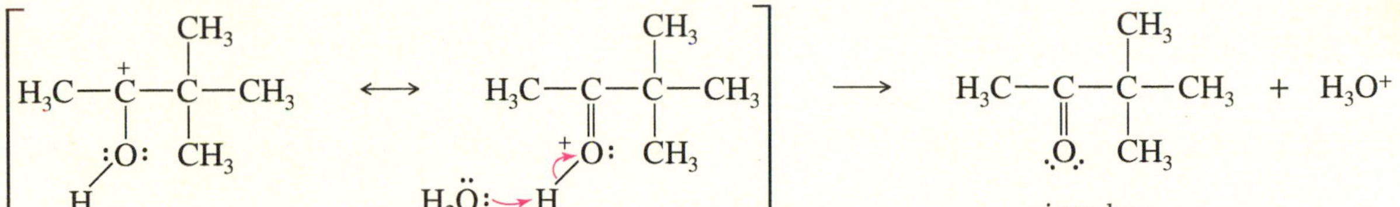

$$\left[H_3C-\overset{+}{C}(OH)-C(CH_3)_3 \longleftrightarrow H_3C-C(=\overset{+}{O}H)-C(CH_3)_3 \right] + H_2O \longrightarrow H_3C-C(=O)-C(CH_3)_3 + H_3O^+$$

resonance-stabilized carbocation

pinacolone

Pinacol-like rearrangements are common in acid-catalyzed reactions of diols. One of the hydroxy groups protonates and leaves as water, forming a carbocation. Rearrangement gives a resonance-stabilized cation with the remaining hydroxy group helping to stabilize the positive charge. Problem 11-28 shows some additional examples of pinacol rearrangements.

PROBLEM 11-28

Propose a mechanism for each reaction.

(a) 1,2-dimethylcyclohexane-1,2-diol (H_3C, CH_3, HO OH) $\xrightarrow{H^+}$ 1-acetyl-1-methylcyclopentane (O) + 2,2-dimethylcyclohexanone (O)

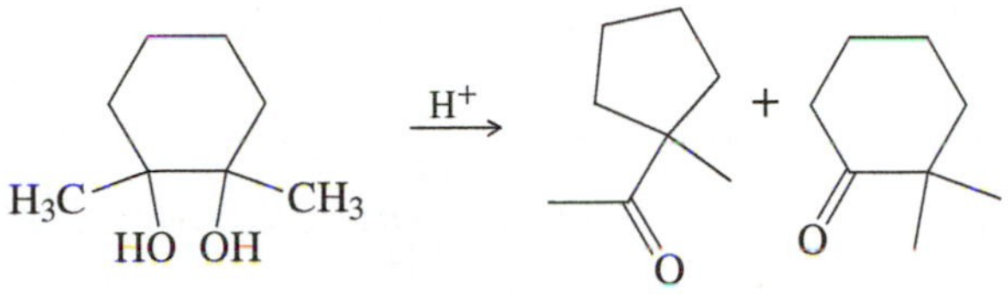

(b) 1-(hydroxydiphenylmethyl)cyclopentanol (OH; C—OH; Ph Ph) $\xrightarrow{H_2SO_4}$ 2,2-diphenylcyclohexanone (O; Ph; Ph)

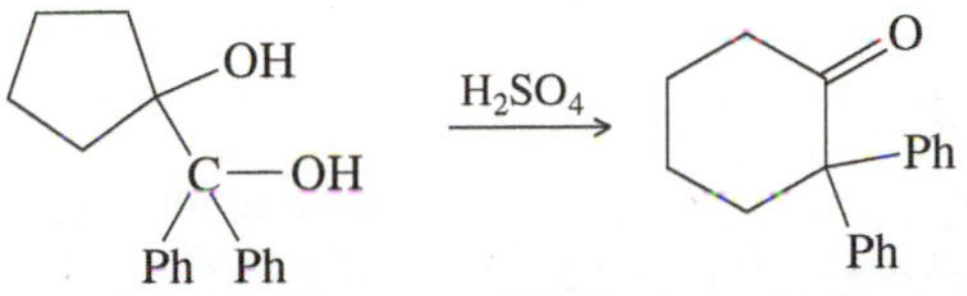

PROBLEM-SOLVING HINT

If a ring changes size during a reaction, a rearrangement has occurred; specifically, a ring atom has migrated. In most cases, migration occurs to form a more stable intermediate, or it may release ring strain in three- and four-membered rings.

PROBLEM-SOLVING HINT

By analogy with the pinacol rearrangement, watch for carbocation rearrangements that move the + charge to a carbinol carbon atom. Rearrangements often move a + charge to any carbon atom that is bonded to an oxygen or nitrogen atom having lone pairs that help to stabilize the positive charge.

PROBLEM 11-29

The following reaction involves a starting material with a double bond and a hydroxy group, yet its mechanism resembles a pinacol rearrangement. Propose a mechanism, and point out the part of your mechanism that resembles a pinacol rearrangement.

OH, CH=CH$_2$ $\xrightarrow{H_2SO_4}$ O, CH$_3$

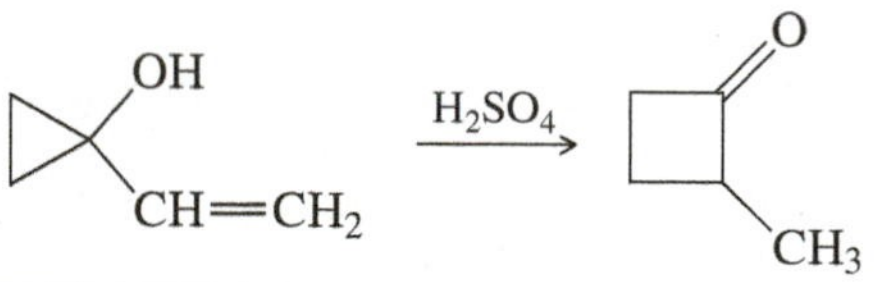

11-11B Periodic Acid Cleavage of Glycols

1,2-Diols (glycols), such as those formed by dihydroxylation of alkenes, are cleaved by periodic acid (HIO_4). The products are the same ketones and aldehydes that would be formed by ozonolysis–reduction of the alkene. Periodic acid cleavage of 1,2-diols is commonly used to determine the structures of unknown sugars, because it only cleaves a bond between two carbon atoms if there is a free —OH group on each carbon.

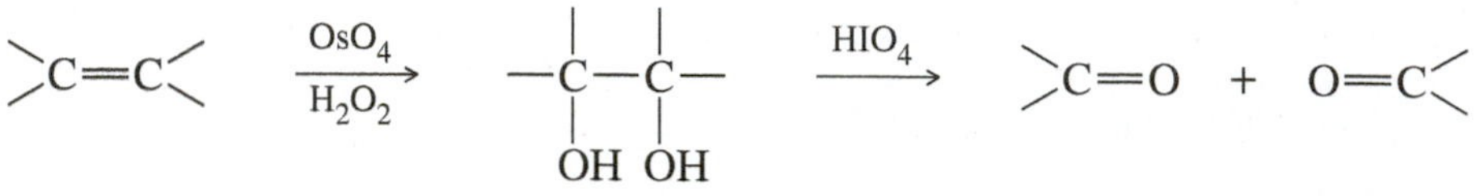

$$>C=C< \xrightarrow[H_2O_2]{OsO_4} -\underset{OH}{\overset{|}{C}}-\underset{OH}{\overset{|}{C}}- \xrightarrow{HIO_4} >C=O \; + \; O=C<$$

alkene — glycol — ketones and aldehydes

Periodic acid cleavage of a glycol probably involves a cyclic periodate intermediate like that shown here.

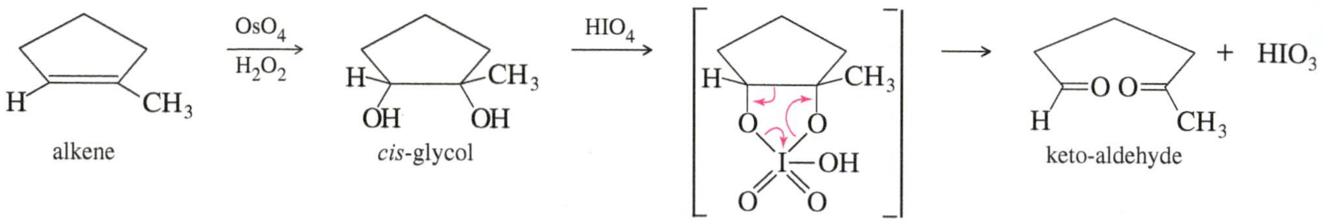

H, CH$_3$ — alkene $\xrightarrow[H_2O_2]{OsO_4}$ H, CH$_3$, OH, OH — *cis*-glycol $\xrightarrow{HIO_4}$ [H, CH$_3$, O, O, I—OH, O, O] — cyclic periodate intermediate $\longrightarrow$ H, =O O=, CH$_3$ — keto-aldehyde + HIO_3

PROBLEM-SOLVING HINT

Periodic acid cleaves a 1,2-diol, breaking the carbon–carbon bond and oxidizing the two —OH groups to carbonyl groups.

PROBLEM 11-30

Predict the products formed by periodic acid cleavage of the following diols.

(a) $CH_3CH(OH)CH(OH)CH_3$

(b) CH$_2$OH, OH

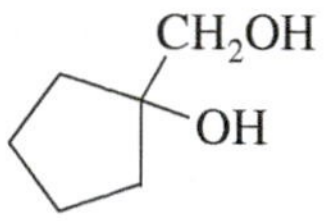

(c) Ph—C(OH)(CH$_3$)—CH(OH)CH$_2$CH$_3$

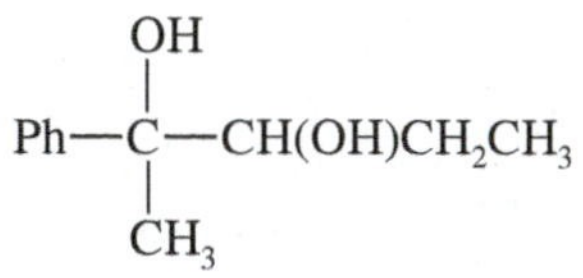

(d) OH, H, OH, H

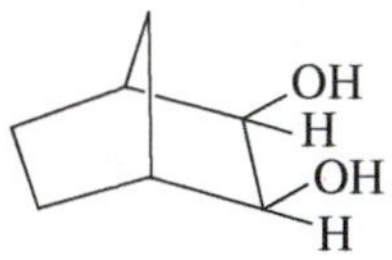

11-12 Esterification of Alcohols

To an organic chemist, the term **ester** normally means an ester of a carboxylic acid, unless some other kind of ester is specified. Replacing the —OH group of a carboxylic acid with the —OR group of an alcohol gives a carboxylic ester. The following condensation, called the **Fischer esterification,** shows the relationship between the alcohol and the acid on the left and the ester and water on the right.

$$\underset{\text{alcohol}}{R{-}O{-}H} + \underset{\text{acid}}{H{-}O{-}\overset{O}{\overset{\|}{C}}{-}R'} \xrightleftharpoons{H^+} \underset{\text{ester}}{R{-}O{-}\overset{O}{\overset{\|}{C}}{-}R'} + H{-}O{-}H$$

For example, if we mix isopropyl alcohol with acetic acid and add a drop of sulfuric acid as a catalyst, the following equilibrium results.

$$\underset{\text{isopropyl alcohol}}{(CH_3)_2CH{-}O{-}H} + \underset{\text{acetic acid}}{H{-}O{-}\overset{O}{\overset{\|}{C}}{-}CH_3} \xrightleftharpoons{H_2SO_4} \underset{\text{isopropyl acetate}}{(CH_3)_2CH{-}O{-}\overset{O}{\overset{\|}{C}}{-}CH_3} + \underset{\text{water}}{H{-}O{-}H}$$

Because the Fischer esterification is an equilibrium (often with an unfavorable equilibrium constant), clever techniques are often required to achieve good yields of esters. For example, we can use a large excess of the alcohol or the acid. Adding a dehydrating agent removes water (one of the products), driving the reaction to the right. There is a more powerful way to form an ester, however, without having to deal with an unfavorable equilibrium. An alcohol reacts with an acid chloride in an exothermic reaction to give an ester. Pyridine is often added to these reactions to neutralize the HCl by-product, and it may function as a catalyst as well.

Application: Drugs

The alcohol groups of unpleasant-tasting drugs are often converted to esters to mask the taste. In most cases, the ester has a less unpleasant taste than the free alcohol.

$$\underset{\text{alcohol}}{R{-}O{-}H} + \underset{\text{acid chloride}}{Cl{-}\overset{O}{\overset{\|}{C}}{-}R'} \xrightarrow{\text{pyridine}} \underset{\text{ester}}{R{-}O{-}\overset{O}{\overset{\|}{C}}{-}R'} + C_5H_5N^+{-}H\ Cl^-$$

The mechanisms of these reactions that form acid derivatives are covered with similar mechanisms in Chapter 21.

PROBLEM 11-31

Show the alcohol and the acid chloride that combine to make the following esters.

(a) $CH_3CH_2CH_2\overset{O}{\overset{\|}{C}}{-}OCH_2CH_2CH_3$ *n*-propyl butyrate

(b) $CH_3(CH_2)_3{-}O{-}\overset{O}{\overset{\|}{C}}{-}CH_2CH_3$ *n*-butyl propionate

(c) $H_3C{-}C_6H_4{-}O{-}\overset{O}{\overset{\|}{C}}{-}CH(CH_3)_2$ *p*-tolyl isobutyrate

(d) $C_3H_5{-}O{-}\overset{O}{\overset{\|}{C}}{-}C_6H_5$ cyclopropyl benzoate

PROBLEM-SOLVING HINT

An ester is a composite of an acid plus an alcohol, with loss of water.

11-13 Esters of Inorganic Acids

In addition to forming esters with carboxylic acids, alcohols form **inorganic esters** with inorganic acids such as nitric acid, sulfuric acid, and phosphoric acid. In each type of ester, the alkoxy (—OR) group of the alcohol replaces a hydroxy group of the acid, with loss of water. We have already studied tosylate esters, composed of *para*-toluenesulfonic acid and alcohols (but made using tosyl chloride, Section 11-5). Tosylate esters are analogous to sulfate esters (Section 11-13A), which are composed of sulfuric acid and alcohols.

$$\underset{\text{alcohol}}{R{-}O{-}H} + \underset{\substack{para\text{-toluenesulfonic acid}\\ (TsOH)}}{HO{-}SO_2{-}C_6H_4{-}CH_3} \rightleftarrows \underset{\substack{para\text{-toluenesulfonate ester}\\ (ROTs)}}{R{-}O{-}SO_2{-}C_6H_4{-}CH_3} + H_2O$$

Made using tosyl chloride

$$\underset{\text{alcohol}}{R{-}O{-}H} + \underset{\substack{para\text{-toluenesulfonyl chloride}\\ (TsCl)}}{Cl{-}SO_2{-}C_6H_4{-}CH_3} \xrightarrow{\text{pyridine}} \underset{\substack{\text{tosylate ester}\\ (ROTs)}}{R{-}O{-}SO_2{-}C_6H_4{-}CH_3} + HCl$$

11-13A Sulfate Esters

A **sulfate ester** is like a sulfonate ester, except there is no alkyl group directly bonded to the sulfur atom. In an alkyl sulfate ester, alkoxy groups are bonded to sulfur through oxygen atoms. Using methanol as the alcohol,

$$\underset{\text{sulfuric acid}}{HO{-}SO_2{-}OH} \underset{}{\overset{CH_3O{-}H}{\rightleftarrows}} \underset{\text{methyl sulfate}}{CH_3{-}O{-}SO_2{-}OH} + H_2O \overset{CH_3OH}{\rightleftarrows} \underset{\text{dimethyl sulfate}}{CH_3{-}O{-}SO_2{-}O{-}CH_3} + H_2O$$

Sulfate ions are excellent leaving groups. Like sulfonate esters, sulfate esters are good electrophiles. Nucleophiles react with sulfate esters to give alkylated products. For example, the reaction of dimethyl sulfate with ammonia gives a sulfate salt of methylamine, $CH_3NH_3^+\ CH_3OSO_3^-$.

$$\underset{\text{ammonia}}{H_3N\!:} + \underset{\text{dimethyl sulfate}}{CH_3{-}O{-}SO_2{-}O{-}CH_3} \longrightarrow \underset{\text{methylammonium ion}}{H_3\overset{+}{N}{-}CH_3} \quad \underset{\text{methylsulfate ion}}{{}^{-}O{-}SO_2{-}O{-}CH_3}$$

Application: Biochemistry

The body converts the hydroxy groups of some drugs to their sulfate derivatives to produce water-soluble compounds that are readily excreted. The reaction is not as common as it might be because of limited availability of inorganic sulfate in the body.

PROBLEM 11-32

Use resonance forms of the conjugate bases to explain why methanesulfonic acid (CH_3SO_3H, $pK_a = -2.6$) is a much stronger acid than acetic acid (CH_3COOH, $pK_a = 4.8$).

11-13B Nitrate Esters

Nitrate esters are formed from alcohols and nitric acid.

$$\underset{\text{alcohol}}{R-O-H} + \underset{\text{nitric acid}}{H-O-N^{+}(=O)O^{-}} \longrightarrow \underset{\text{alkyl nitrate ester}}{R-O-N^{+}(=O)O^{-}} + H-O-H$$

The best-known nitrate ester is "nitroglycerin," whose systematic name is *glyceryl trinitrate*. Glyceryl trinitrate results from the reaction of glycerol (1,2,3-propanetriol) with three molecules of nitric acid.

$$\underset{\text{glycerol (glycerin)}}{\begin{array}{l} CH_2-O-H \\ CH-O-H \\ CH_2-O-H \end{array}} + 3\ \underset{\text{nitric acid}}{HO-NO_2} \longrightarrow \underset{\text{glyceryl trinitrate (nitroglycerin)}}{\begin{array}{l} CH_2-O-NO_2 \\ CH-O-NO_2 \\ CH_2-O-NO_2 \end{array}} + 3\ H_2O$$

Illustration of Alfred Nobel around 1860, operating the apparatus he used to make nitroglycerin. The temperature must be monitored and controlled carefully during this process; therefore, the operator's stool has only one leg to ensure that he stays awake.

First made in 1847, nitroglycerin was found to be a much more powerful explosive than black powder, which is a physical mixture of potassium nitrate, sulfur, and charcoal. In black powder, potassium nitrate is the oxidizer, and sulfur and charcoal provide the fuel to be oxidized. The rate of a black powder explosion is limited by how fast oxygen from the grains of heated potassium nitrate can diffuse to the grains of sulfur and charcoal. A black powder explosion does its work by the rapid increase in pressure resulting from the reaction. The explosion must be confined, as in a cannon or a firecracker, to be effective.

In nitroglycerin, the nitro groups are the oxidizer and the CH and CH_2 groups are the fuel to be oxidized. This intimate association of fuel and oxidizer allows the explosion to proceed at a much faster rate, forming a shock wave that propagates through the explosive and initiates the reaction. The explosive shock wave can shatter rock or other substances without the need for confinement. Because of its unprecedented explosive power, nitroglycerin was called a *high explosive*. Many other high explosives have been developed, including picric acid, TNT (trinitrotoluene), PETN (pentaerythritol tetranitrate), and RDX (research department explosive). Nitroglycerin and PETN are nitrate esters. Picric acid and TNT are nitrobenzene derivatives, not esters.

Pure nitroglycerin is hazardous to make, use, and transport. Alfred Nobel's family were experts at making and using nitroglycerin, yet his brother and several workers were killed by an explosion. In 1866, Nobel found that nitroglycerin soaks into diatomaceous earth to give a pasty mixture that can be molded into sticks that do not detonate so easily. He called the sticks *dynamite* and founded the firm Dynamit Nobel, which is still one of the world's leading ammunition and explosives manufacturers. The Nobel prizes are funded from an endowment that originated with Nobel's profits from the dynamite business.

Application: Drugs

Nitroglycerin is used to relieve angina, a condition in which the heart is not receiving enough oxygen. Angina is characterized by severe pain in the chest, often triggered by stress or exercise. Workers with angina at the Nobel dynamite plant discovered this effect. They noticed that their symptoms improved when they went to work. In the mitochondria, nitroglycerin is metabolized to NO (nitric oxide), which regulates and controls many metabolic processes.

[Structures: picric acid (benzene ring with OH, O_2N, NO_2, NO_2); TNT (benzene ring with CH_3, O_2N, NO_2, NO_2); PETN, $C(CH_2ONO_2)_4$: $O_2NOCH_2-C-CH_2ONO_2$ with CH_2ONO_2 above and below; RDX (six-membered ring of alternating N and CH_2 with three N–NO_2 groups)]

picric acid | TNT | PETN | RDX

Application: Biochemistry

By controlling the formation of phosphate esters on key proteins, the body is able to regulate many cellular processes. Any disruption of these phosphorylation processes can result in numerous health problems, including cancer, diabetes, and obesity.

11-13C Phosphate Esters

Alkyl phosphates are composed of 1 mole of phosphoric acid combined with 1, 2, or 3 moles of an alcohol. For example, methanol forms three **phosphate esters.**

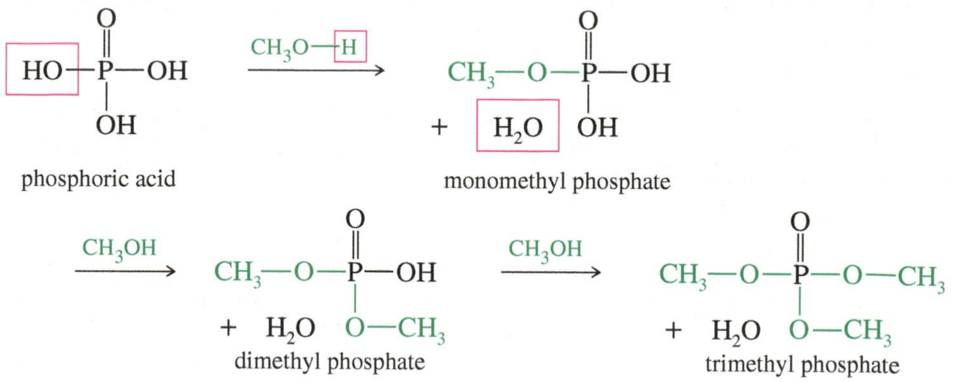

$$\underset{\text{phosphoric acid}}{HO-P(=O)(OH)-OH} \xrightarrow{CH_3O-H} \underset{\text{monomethyl phosphate}}{CH_3-O-P(=O)(OH)-OH} + H_2O$$

$$\xrightarrow{CH_3OH} \underset{\text{dimethyl phosphate}}{CH_3-O-P(=O)(O-CH_3)-OH} + H_2O \xrightarrow{CH_3OH} \underset{\text{trimethyl phosphate}}{CH_3-O-P(=O)(O-CH_3)-O-CH_3} + H_2O$$

Phosphate esters play a central role in biochemistry. Figure 11-3 shows how phosphate ester linkages compose the backbone of the nucleic acids RNA (ribonucleic acid) and DNA (deoxyribonucleic acid). These nucleic acids, which carry the genetic information in the cell, are discussed in Chapter 23.

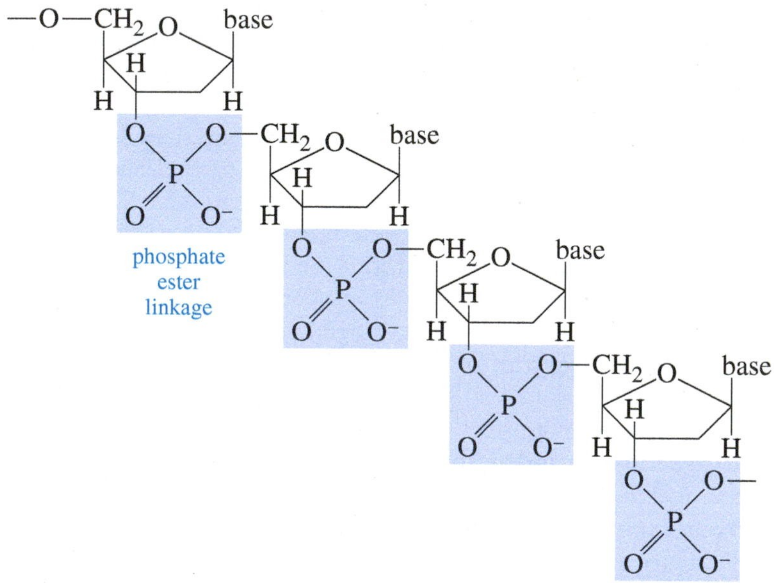

FIGURE 11-3 Phosphate ester groups bond the individual nucleotides together in DNA. The "base" on each of the nucleotides corresponds to one of the four heterocyclic bases of DNA (see Section 23-15).

11-14 Reactions of Alkoxides

In Section 10-6B, we learned to remove the hydroxy proton from an alcohol by reduction with an "active" metal such as sodium or potassium. This reaction generates a sodium or potassium salt of an **alkoxide ion** and hydrogen gas.

$$R-\ddot{\underset{\cdot\cdot}{O}}-H \quad + \quad Na \quad \longrightarrow \quad R-\ddot{\underset{\cdot\cdot}{O}}:^- \; Na^+ \quad + \quad \tfrac{1}{2}H_2\uparrow$$

$$R-\ddot{\underset{\cdot\cdot}{O}}-H \quad + \quad K \quad \longrightarrow \quad R-\ddot{\underset{\cdot\cdot}{O}}:^- \; K^+ \quad + \quad \tfrac{1}{2}H_2\uparrow$$

The reactivity of alcohols toward sodium and potassium decreases in the order: methyl > 1° > 2° > 3°. Sodium reacts quickly with primary alcohols and some secondary alcohols. Potassium is more reactive than sodium and is commonly used with tertiary alcohols and some secondary alcohols.

Some alcohols react sluggishly with both sodium and potassium. In these cases, a useful alternative is sodium hydride, usually in tetrahydrofuran solution. Sodium hydride reacts quickly to form the alkoxide, even with difficult compounds.

$$\underset{\substack{\text{alcohol} \\ pK_a \approx 16}}{R{-}\ddot{\underset{\cdot\cdot}{O}}{-}H} + \underset{\text{sodium hydride}}{NaH} \xrightarrow{\text{THF}} \underset{\text{sodium alkoxide}}{R{-}\ddot{\underset{\cdot\cdot}{O}}{:}^-\ Na^+} + \underset{\substack{\text{hydrogen} \\ pK_a = 36}}{H_2\uparrow}$$

Sodium metal reacts vigorously with simple primary alcohols such as ethanol.

The alkoxide ion is a strong nucleophile as well as a powerful base. Unlike the alcohol itself, the alkoxide ion reacts with primary alkyl halides and tosylates to form **ethers.** This general reaction, called the **Williamson ether synthesis,** is an S_N2 displacement. The alkyl halide (or tosylate) must be primary so that a back-side attack is not hindered. When the alkyl halide is not primary, elimination usually results.

KEY MECHANISM 11-6 The Williamson Ether Synthesis

This is the most important method for making ethers.

Step 1: Form the alkoxide of the alcohol having the more hindered group.

$$R{-}\ddot{\underset{\cdot\cdot}{O}}{-}H + Na\ (\text{or NaH or K}) \longrightarrow \underset{\text{alkoxide ion}}{R{-}\ddot{\underset{\cdot\cdot}{O}}{:}^-\ Na^+} + \tfrac{1}{2} H_2\uparrow$$

Step 2: The alkoxide displaces the leaving group of a good S_N2 substrate.

$$\underset{\text{alkoxide ion}}{R{-}\ddot{\underset{\cdot\cdot}{O}}{:}^-\ {}^+Na} \quad \underset{\text{primary halide or tosylate}}{R'{-}CH_2{-}X} \longrightarrow \underset{\text{ether}}{R{-}\ddot{\underset{\cdot\cdot}{O}}{-}CH_2{-}R'} + NaX$$

EXAMPLE: Synthesis of cyclopentyl ethyl ether.

Step 1: Form the alkoxide of the alcohol with the more hindered group.

$$\text{cyclopentyl–OH} + NaH \longrightarrow \text{cyclopentyl–}O^-\ Na^+ + H_2\uparrow$$

Step 2: The alkoxide displaces the leaving group of a good S_N2 substrate.

$$\text{cyclopentyl–}O^-\ Na^+ \quad H_3C{-}CH_2{-}Br \longrightarrow \text{cyclopentyl–}O{-}CH_2{-}CH_3 + Na^+\ Br^-$$

QUESTION: Why is the cyclopentyl group chosen for the alkoxide and the ethyl group chosen for the halide? Why not use cyclopentyl bromide and sodium ethoxide to make cyclopentyl ethyl ether?

In the Williamson ether synthesis, the alkyl halide (or tosylate) must be a good S_N2 substrate (usually primary). In proposing a Williamson synthesis, we usually choose the less hindered alkyl group to be the halide (or tosylate) and the more hindered group to be the alkoxide ion.

PROBLEM 11-33

A good Williamson synthesis of ethyl methyl ether would be

$$\underset{\text{sodium ethoxide}}{CH_3CH_2{-}O^-\ Na^+} + \underset{\text{methyl iodide}}{CH_3I} \longrightarrow \underset{\text{ethyl methyl ether}}{CH_3CH_2{-}O{-}CH_3} + NaI$$

What is wrong with the following proposed synthesis of ethyl methyl ether? First, ethanol is treated with acid to protonate the hydroxy group (making it a good leaving group), and then sodium methoxide is added to displace water.

$$CH_3CH_2{-}OH + H^+ \longrightarrow CH_3CH_2{-}\overset{+}{O}H_2 \xrightarrow[\times]{Na^+\ {}^-OCH_3} CH_3CH_2{-}O{-}CH_3$$

(incorrect synthesis of ethyl methyl ether)

PROBLEM-SOLVING HINT

When using the Williamson ether synthesis to make R—O—R′, choose the less hindered alkyl group to serve as the alkyl halide (R′—X), because it will make a better S_N2 substrate. Choose the more hindered alkyl group to form the alkoxide (R—O⁻), because it is less sensitive to steric hindrance in the reaction.

$$R{-}O^- + R'{-}X \longrightarrow R{-}O{-}R'$$

PROBLEM 11-34

(a) Show how ethanol and cyclohexanol may be used to synthesize cyclohexyl ethyl ether (tosylation followed by the Williamson ether synthesis).

(b) Why can't we synthesize this product simply by mixing the two alcohols, adding some sulfuric acid, and heating?

PROBLEM 11-35

A student wanted to use the Williamson ether synthesis to make (*R*)-2-ethoxybutane. He remembered that the Williamson synthesis involves an S_N2 displacement, which takes place with inversion of configuration. He ordered a bottle of (*S*)-butan-2-ol for his chiral starting material. He also remembered that the S_N2 goes best on primary halides and tosylates, so he made ethyl tosylate and sodium (*S*)-but-2-oxide. After warming these reagents together, he obtained an excellent yield of 2-ethoxybutane.

(a) What enantiomer of 2-ethoxybutane did he obtain? Explain how this enantiomer results from the S_N2 reaction of ethyl tosylate with sodium (*S*)-but-2-oxide.

(b) What would have been the best synthesis of (*R*)-2-ethoxybutane?

(c) How can this student convert the rest of his bottle of (*S*)-butan-2-ol to (*R*)-2-ethoxybutane?

PROBLEM 11-36

Phenols ($pK_a \approx 10$) are more acidic than other alcohols, so they are easily deprotonated by sodium hydroxide or potassium hydroxide. The anions of phenols (phenoxide ions) can be used in the Williamson ether synthesis, especially with very reactive alkylating reagents such as dimethyl sulfate. Using phenol, dimethyl sulfate, and other necessary reagents, show how you would synthesize methyl phenyl ether.

PROBLEM-SOLVING STRATEGY Multistep Synthesis

Chemists use organic syntheses both to make larger amounts of useful natural compounds and to invent totally new compounds in search of improved properties and biological effects. Solving synthesis problems also serves as one of the best methods for developing a firm command of organic chemistry. Planning a practical multistep synthesis requires a working knowledge of the applications and the limitations of a variety of organic reactions. We will often use synthesis problems for reviewing and reinforcing the reactions we have covered.

We use a systematic approach to solving multistep synthesis problems, working backward, in the "retrosynthetic" direction. We begin by studying the target molecule and considering what final reactions might be used to create it from simpler intermediate compounds. Most syntheses require comparison of two or more pathways and the intermediates involved. Eventually, this retrosynthetic analysis should lead back to starting materials that are readily available or meet the requirements defined in the problem.

We can now extend our systematic analysis to problems involving alcohols and Grignard reactions. As examples, we consider the syntheses of an acyclic diol and a disubstituted cyclohexane, concentrating on the crucial steps that assemble the carbon skeletons and generate the final functional groups.

Sample Problem

Our first problem is to synthesize 3-ethylpentane-2,3-diol from compounds containing no more than three carbon atoms.

$$CH_3-CH(OH)-C(OH)(CH_2CH_3)-CH_2-CH_3$$

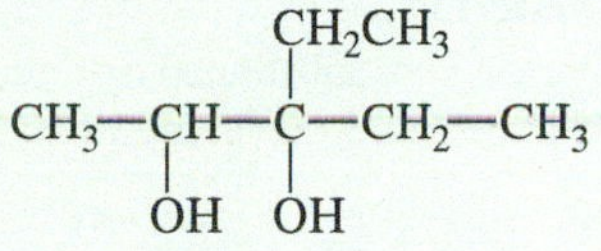

3-ethylpentane-2,3-diol

1. **Review the functional groups and carbon skeleton of the target compound.**
The compound is a vicinal diol (glycol) containing seven carbon atoms. Glycols are commonly made by dihydroxylation of alkenes, and this glycol would be made by dihydroxylation of 3-ethylpent-2-ene, which effectively becomes the target compound.

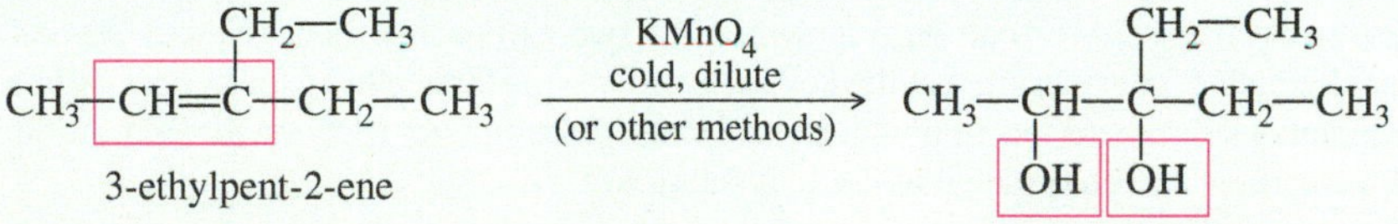

$$CH_3-CH=C(CH_2-CH_3)-CH_2-CH_3 \xrightarrow[\text{(or other methods)}]{KMnO_4\text{, cold, dilute}} CH_3-CH(OH)-C(OH)(CH_2-CH_3)-CH_2-CH_3$$

3-ethylpent-2-ene → 3-ethylpentane-2,3-diol

2. **Review the functional groups and carbon skeletons of the starting materials (if specified), and see how their skeletons might fit together into the target compound.**
The limitation is that the starting materials must contain no more than three carbon atoms. To form a 7-carbon product requires at least three fragments, probably a 3-carbon fragment and two 2-carbon fragments. A functional group that can be converted to an alkene will be needed on either C2 or C3 of the chain, because 3-ethylpent-2-ene has a double bond between C2 and C3. We can circle some likely 2-carbon and 3-carbon fragments as a possible way to construct the molecule.

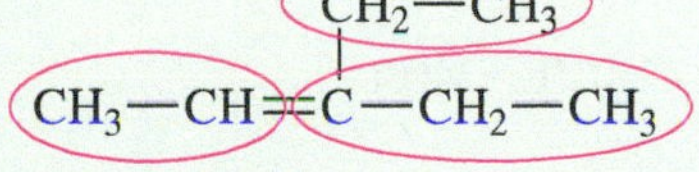

$$CH_3-CH=C(CH_2-CH_3)-CH_2-CH_3$$

3. **Compare methods for assembling the carbon skeleton of the target compound. Which ones provide a key intermediate with the correct carbon skeleton and functional groups correctly positioned for conversion to the functionality in the target molecule?**
At this point, the Grignard reaction is our most powerful method for assembling a carbon skeleton, and Grignards can be used to make primary, secondary, and tertiary alcohols (Section 10-9). The secondary alcohol 3-ethylpentan-2-ol has its functional group on C2, and the tertiary alcohol 3-ethylpentan-3-ol has it on C3. Either of these alcohols can be synthesized by an appropriate Grignard reaction, but 3-ethylpentan-2-ol may dehydrate to give a mixture of products. Because of its symmetry, 3-ethylpentan-3-ol dehydrates to give only the desired alkene, 3-ethylpent-2-ene. It also dehydrates more easily because it is a tertiary alcohol.

(continued)

$$\underset{\text{3-ethylpentan-2-ol}}{CH_3-\underset{\underset{OH}{|}}{CH}-\overset{\overset{CH_2-CH_3}{|}}{CH}-CH_2-CH_3} \xrightarrow{H_2SO_4} \left\{ \underset{\substack{\text{(major)}\\ \text{3-ethylpent-2-ene}}}{CH_3-CH=\overset{\overset{CH_2-CH_3}{|}}{C}-CH_2-CH_3} + \underset{\substack{\text{(minor)}\\ \text{3-ethylpent-1-ene}}}{CH_2=CH-\overset{\overset{CH_2-CH_3}{|}}{CH}-CH_2-CH_3} \right\}$$

Preferred synthesis:

$$\underset{\text{3-ethylpentan-3-ol}}{CH_3-CH_2-\overset{\overset{CH_2-CH_3}{|}}{\underset{\underset{OH}{|}}{C}}-CH_2-CH_3} \xrightarrow{H_2SO_4} \underset{\substack{\text{(only product)}\\ \text{3-ethylpent-2-ene}}}{CH_3-CH=\overset{\overset{CH_2-CH_3}{|}}{C}-CH_2-CH_3}$$

4. Working backward through as many steps as necessary, compare methods for synthesizing the reactants needed for assembly of the key intermediate. (This process may require writing several possible reaction sequences and evaluating them, keeping in mind the specified starting materials.)

The key intermediate, 3-ethylpentan-3-ol, is simply methanol substituted by three ethyl groups. The last step in its synthesis must add an ethyl group. Addition of ethyl magnesium bromide to pentan-3-one gives 3-ethylpentan-3-ol.

$$\underset{\text{pentan-3-one}}{CH_3-CH_2-\underset{\underset{O}{\|}}{C}-CH_2-CH_3} \xrightarrow[\text{(2) } H_3O^+]{\text{(1) } CH_3CH_2-MgBr} \underset{\text{3-ethylpentan-3-ol}}{CH_3-CH_2-\overset{\overset{CH_2-CH_3}{|}}{\underset{\underset{OH}{|}}{C}}-CH_2-CH_3}$$

The synthesis of pentan-3-one from a three-carbon fragment and a two-carbon fragment requires several steps (see Problem 11-37). Perhaps there is a better alternative, considering that the key intermediate has three ethyl groups on a carbinol carbon atom. Two similar alkyl groups can be added in one Grignard reaction with an acid chloride or an ester (Section 10-9D). Addition of 2 moles of ethyl magnesium bromide to a three-carbon acid chloride gives 3-ethylpentan-3-ol.

$$\underset{\text{propionyl chloride}}{CH_3-CH_2-\overset{\overset{O}{\|}}{C}-Cl} \xrightarrow[\text{(2) } H_3O^+]{\text{(1) 2 } CH_3CH_2-MgBr} \underset{\text{3-ethylpentan-3-ol}}{CH_3-CH_2-\overset{\overset{CH_2-CH_3}{|}}{\underset{\underset{OH}{|}}{C}}-CH_2-CH_3}$$

5. Summarize the complete synthesis in the forward direction, including all steps and all reagents, and check it for errors and omissions.

$$\underset{\text{propionyl chloride}}{CH_3-CH_2-\overset{\overset{O}{\|}}{C}-Cl} \xrightarrow[\text{(2) } H_3O^+]{\text{(1) 2 } CH_3CH_2-MgBr} \underset{\text{3-ethylpentan-3-ol}}{CH_3-CH_2-\overset{\overset{CH_2-CH_3}{|}}{\underset{\underset{OH}{|}}{C}}-CH_2-CH_3} \xrightarrow{H_2SO_4}$$

$$\underset{\text{3-ethylpent-2-ene}}{CH_3-CH=\overset{\overset{CH_2-CH_3}{|}}{C}-CH_2-CH_3} \xrightarrow[\text{(cold, dilute)}]{KMnO_4} \underset{\text{3-ethylpentane-2,3-diol}}{CH_3-\underset{\underset{OH}{|}}{CH}-\overset{\overset{CH_2-CH_3}{|}}{\underset{\underset{OH}{|}}{C}}-CH_2-CH_3}$$

PROBLEM 11-37

To practice working through the early parts of a multistep synthesis, devise syntheses of
(a) pentan-3-one from alcohols containing no more than three carbon atoms.
(b) 3-ethylpentan-2-one from compounds containing no more than three carbon atoms.

Sample Problem

As another example of the systematic approach to multistep synthesis, let's consider the synthesis of 1-bromo-2-methylcyclohexane from cyclohexanol.

1. **Review the functional groups and carbon skeleton of the target compound.**
The skeleton has seven carbon atoms: a cyclohexyl ring with a methyl group. It is an alkyl bromide, with the bromine atom on a ring carbon that is one atom away from the one bearing the methyl group.
2. **Review the functional groups and carbon skeletons of the starting materials (if specified), and see how their skeletons might fit together into the target compound.**
The starting compound has only six carbon atoms. So the methyl group must be added, presumably at the functional group. There are no restrictions on the methylating reagent, but it must provide a product with a functional group that can be converted to an adjacent halide.
3. **Compare methods for assembling the carbon skeleton of the target compound to determine which methods provide a key intermediate with the correct carbon skeleton and functional groups at the correct positions for being converted to the functionality in the target molecule.**
Once again, the best choice is a Grignard reaction, but there are two possible reactions that give the methylcyclohexane skeleton. A cyclohexyl Grignard reagent can add to formaldehyde, or a methyl Grignard reagent can add to cyclohexanone. (There are other possibilities, but none that are more direct.)

(1) ether, (2) H_3O^+ → alcohol A (to be functionalized)

CH_3MgBr + cyclohexanone, (1) ether, (2) H_3O^+ → alcohol B (to be functionalized)

Neither product has its alcohol functional group on the carbon atom that is functionalized in the target compound. Alcohol **A** needs its functional group moved two carbon atoms, but alcohol **B** needs it moved only one carbon atom. Converting alcohol **B** to an alkene functionalizes the correct carbon atom. Anti-Markovnikov addition of HBr converts the alkene to an alkyl halide with the bromine atom on the correct carbon atom.

alcohol B → (H_2SO_4, heat) → alkene → (HBr, ROOR) → target compound

(continued)

4. Working backward through as many steps as necessary, compare methods for synthesizing the reactants needed for assembly of the key intermediate.

All that remains is to make cyclohexanone by oxidation of cyclohexanol.

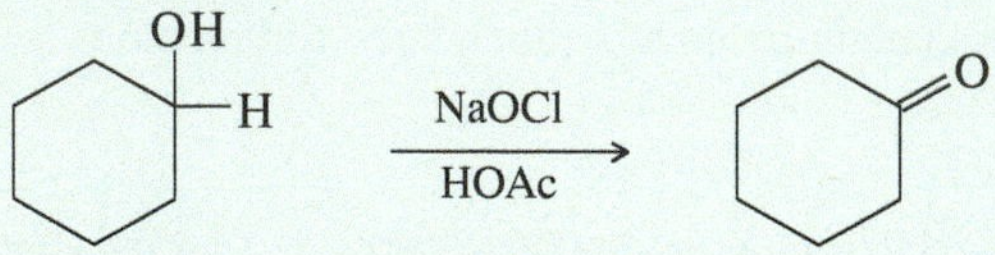

$$\text{cyclohexanol} \xrightarrow[\text{HOAc}]{\text{NaOCl}} \text{cyclohexanone}$$

5. Summarize the complete synthesis in the forward direction, including all steps and all reagents, and check it for errors and omissions.

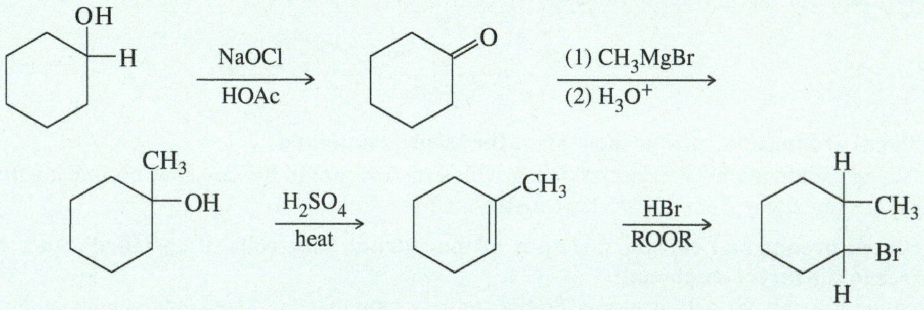

$$\text{cyclohexanol} \xrightarrow[\text{HOAc}]{\text{NaOCl}} \text{cyclohexanone} \xrightarrow[(2)\ H_3O^+]{(1)\ CH_3MgBr} \text{1-methylcyclohexanol}$$

$$\text{1-methylcyclohexanol} \xrightarrow[\text{heat}]{H_2SO_4} \text{1-methylcyclohexene} \xrightarrow[\text{ROOR}]{\text{HBr}} \text{trans-1-bromo-2-methylcyclohexane}$$

Problem 11-38 provides practice in multistep syntheses and using alcohols as intermediates.

PROBLEM 11-38

Develop syntheses for the following compounds. As starting materials, you may use cyclopentanol, alcohols containing no more than four carbon atoms, and any common reagents and solvents.

(a) *trans*-cyclopentane-1,2-diol **(b)** 1-chloro-1-ethylcyclopentane

(c) 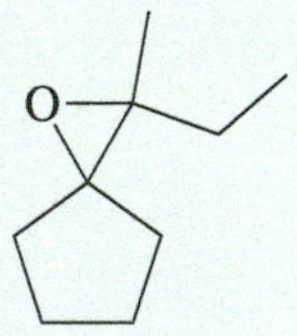 (epoxide: cyclopentane spiro-fused to oxirane bearing methyl and ethyl) **(d)** 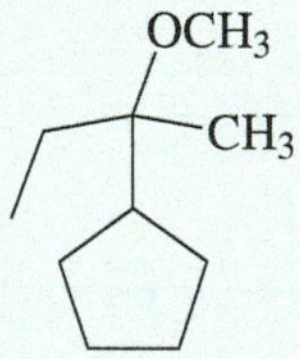 (cyclopentyl–C(OCH_3)(CH_3)–ethyl) **(e)** 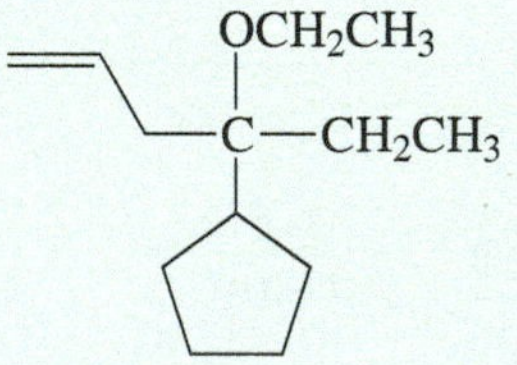 (allyl–C(OCH_2CH_3)(CH_2CH_3)–cyclopentyl) **(f)** 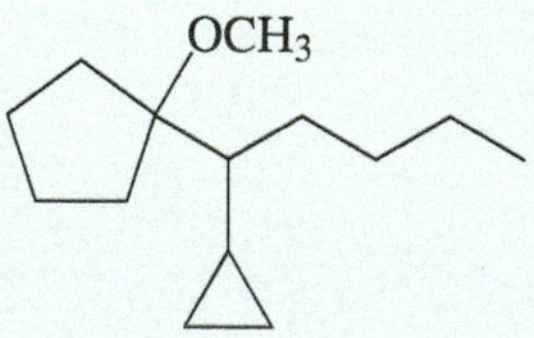 (1-methoxycyclopentyl–CH(cyclopropyl)–butyl)

SUMMARY Reactions of Alcohols

1. Oxidation–reduction reactions

a. *Oxidation of secondary alcohols to ketones* (Sections 11-2A and 11-3)

$$R{-}CH(OH){-}R' \xrightarrow{\text{NaOCl/HOAc}} R{-}C(=O){-}R'$$

H_2CrO_4, the DMP reagent, and the Swern oxidation are alternatives.

b. *Oxidation of primary alcohols to carboxylic acids* (Sections 11-2B and 11-3)

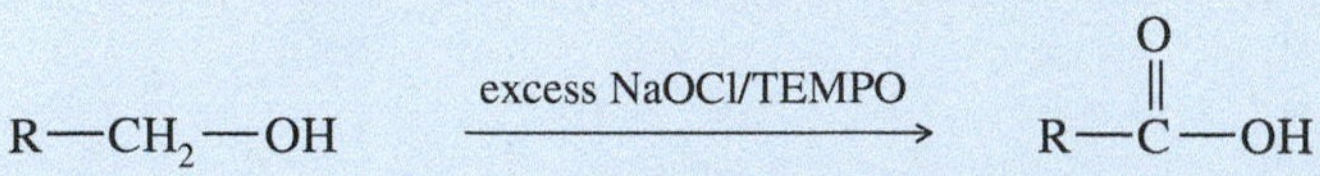

$$R{-}CH_2{-}OH \xrightarrow{\text{excess NaOCl/TEMPO}} R{-}C(=O){-}OH$$

H_2CrO_4 is the chromium alternative.

c. *Oxidation of primary alcohols to aldehydes* (Sections 11-2B and 11-3)

$$R{-}CH_2{-}OH \xrightarrow{\text{NaOCl/TEMPO}} R{-}\overset{\displaystyle O}{\overset{\|}{C}}{-}H$$

The DMP reagent, PCC, and the Swern oxidation are alternatives.

d. *Reduction of alcohols to alkanes* (Section 11-6)

$$R{-}OH \xrightarrow[\text{(2) LiAlH}_4]{\text{(1) TsCl/pyridine}} R{-}H$$

2. Cleavage of the alcohol hydroxy group C‡O—H

a. *Conversion of alcohols to alkyl halides* (Section 11-7 through 11-9)

$$R{-}OH \xrightarrow{\text{HCl or SOCl}_2\text{/pyridine}} R{-}Cl$$

$$R{-}OH \xrightarrow{\text{HBr or PBr}_3} R{-}Br$$

$$R{-}OH \xrightarrow{\text{HI or P/I}_2} R{-}I$$

HCl, HBr, and HI work well for most 3°.

$SOCl_2$, PBr_3, and P/I_2 work well for most 1° and 2°.

b. *Dehydration of alcohols to form alkenes* (Section 11-10A)

$$\underset{|}{\overset{\displaystyle H}{\overset{|}{-C}}}{-}\underset{|}{\overset{\displaystyle OH}{\overset{|}{C-}}} \underset{}{\overset{H_2SO_4 \text{ or } H_3PO_4}{\rightleftharpoons}} \;\rangle C{=}C\langle \; + \; H_2O$$

c. *Industrial condensation of alcohols to form ethers* (Section 11-10B)

$$2\,R{-}OH \overset{H^+}{\rightleftharpoons} R{-}O{-}R \;+\; H_2O$$

3. *Cleavage of the hydroxy proton* $-\overset{|}{\underset{|}{C}}-O\mathord{\wr}H$

a. *Tosylation* (Section 11-5)

$$R{-}OH \;+\; Cl{-}\overset{\displaystyle O}{\overset{\|}{\underset{\underset{\displaystyle O}{\|}}{S}}}{-}C_6H_4{-}CH_3 \xrightarrow{\text{pyridine}} R{-}O{-}\overset{\displaystyle O}{\overset{\|}{\underset{\underset{\displaystyle O}{\|}}{S}}}{-}C_6H_4{-}CH_3 \;+\; HCl$$

alcohol — tosyl chloride (TsCl) — alkyl tosylate

b. *Acylation to form esters* (Section 11-12)

$$R{-}OH \xrightarrow[\text{(acyl chloride)}]{R'{-}\overset{\displaystyle O}{\overset{\|}{C}}{-}Cl} R{-}O{-}\overset{\displaystyle O}{\overset{\|}{C}}{-}R' \;+\; HCl$$

ester

c. *Deprotonation to form an alkoxide* (Section 11-14)

$$R{-}OH \;+\; \text{Na (or K)} \longrightarrow R{-}O^-\ Na^+ \;+\; \tfrac{1}{2}\ H_2\uparrow$$

$$R{-}OH \;+\; NaH \longrightarrow R{-}O^-\ Na^+ \;+\; H_2\uparrow$$

d. *Williamson ether synthesis* (Sections 11-14 and 14-5)

$$R{-}O^- \;+\; R'X \longrightarrow R{-}O{-}R' \;+\; X^-$$

(R′ must be unhindered, usually primary)

SUMMARY Reactions of Alcohols: O—H Cleavage

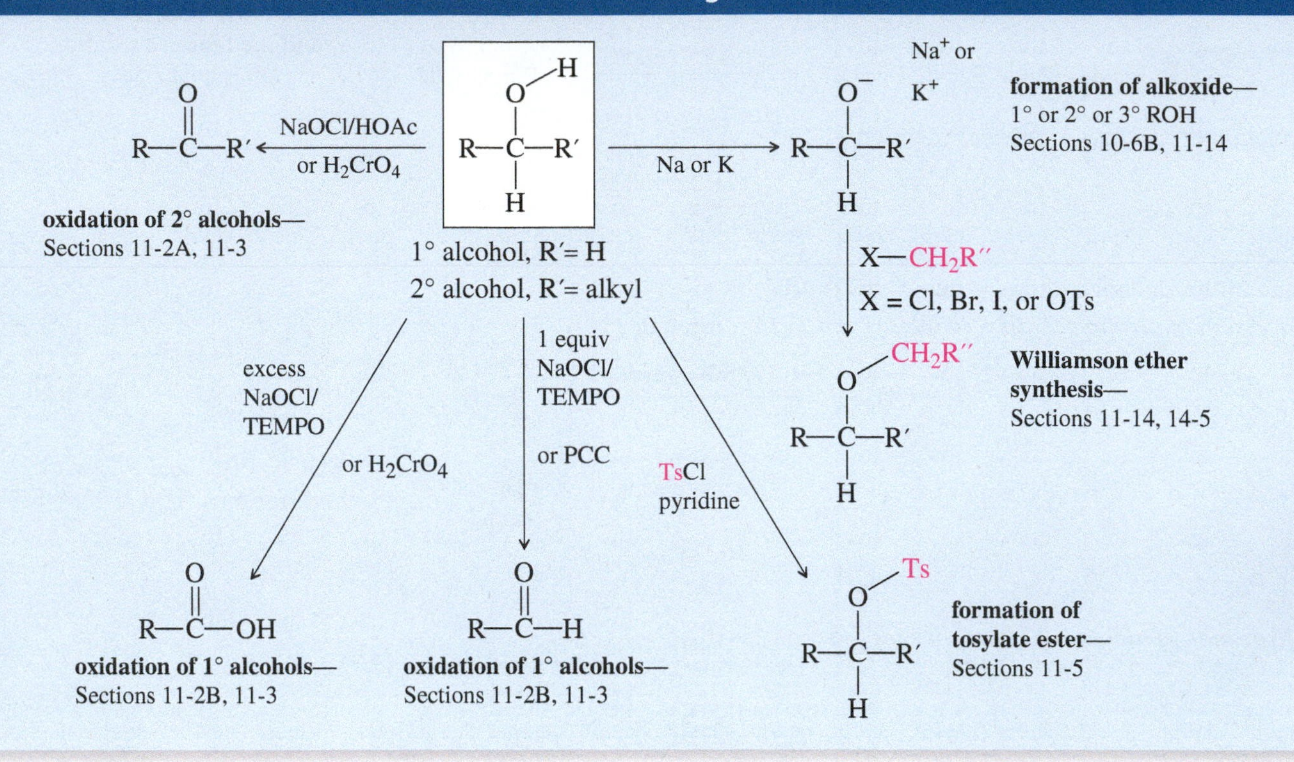

SUMMARY Reactions of Alcohols: C—O Cleavage

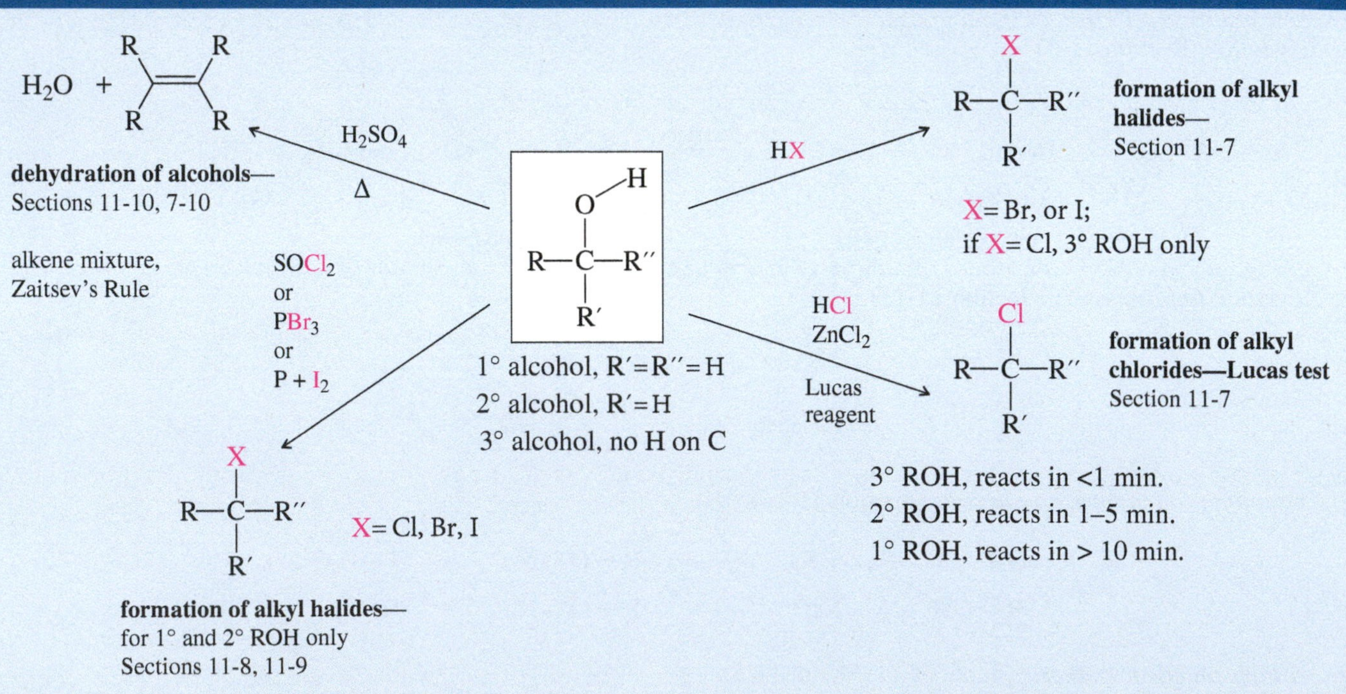

GUIDE TO ORGANIC REACTIONS IN CHAPTER 11

Reactions covered in Chapter 11 are shown in red. Reactions covered in earlier chapters are shown in blue.

Substitution

- Nucleophilic
  - at sp^3 C (S_N1, S_N2) Ch 6, 10, 14, 22
  - at sp^2 C (Nuc. Arom. Subst.) Ch 17, 19
  - at C=O (Nuc. Acyl Subst.) Ch 10, 11, 20, 21, 22
- Electrophilic
  - at sp^2 C (Elect. Arom. Subst.) Ch 17, 19
- Radical
  - at sp^3 C (alkane halogenation) Ch 4, 6, 16, 17
  - at sp^2 C (Sandmeyer rxn) Ch 19
- Organometallic
  - Gilman Ch 10, 17
  - Suzuki Ch 17
  - Heck Ch 17

Addition

- Nucleophilic
  - at C=O (Nuc. Addn.) Ch 9, 10, 18, 22
  - at C=C (conjugate addn.) Ch 22
- Electrophilic
  - at C=C (Elect. Addn.) Ch 8, 9, 10
  - at C=C (Carbene Addn.) Ch 8
- Radical
  - at C=C (HBr + ROOR) Ch 8
- Pericyclic
  - cycloaddition (Diels-Alder) Ch 15
- Oxidation
  - epoxidation Ch 8, 10, 14
- Reduction
  - hydrogenation Ch 8, 9, 17, 18, 19

Elimination

- Basic conditions (E2)
  - E2 dehydrohalogenation Ch 7, 9
  - tosylate elimination Ch 11
  - Hofmann elimination Ch 19
- Acidic conditions (E1)
  - E1 dehydrohalogenation Ch 7
  - dehydration of alcohols Ch 11
- Pericyclic (Cope elimination)
  - Ch 19

Oxidation/Reduction

- Oxidation
  - epoxidation Ch 8, 10, 14
  - oxidative cleavage Ch 8, 9, 11, 17, 22
  - oxygen functional groups Ch 11, 18, 19, 20
- Reduction
  - hydride reduction Ch 8, 10, 11, 17, 18, 19, 20, 21
  - hydrogenation Ch 8, 9, 17, 18, 19
  - metals Ch 9, 17, 18, 19

Essential Terms

alcohol dehydrogenase (ADH) An enzyme used by living cells to catalyze the oxidation of ethyl alcohol to acetaldehyde. (p. 514)

aldehyde dehydrogenase (ALDH) An enzyme used by living cells to catalyze the oxidation of acetaldehyde to acetic acid. (p. 514)

alkoxide ion The anion with structure R—O⁻, bearing the negative charge on oxygen. Commonly formed by deprotonating an alcohol. (p. 540)

$$\text{R—}\ddot{\text{O}}\text{—H} + \text{Na} \longrightarrow \text{R—}\ddot{\text{O}}\text{:}^- \ \text{Na}^+ + \tfrac{1}{2}\text{H}_2\uparrow$$

chromic acid reagent (H_2CrO_4) The solution formed by adding sodium or potassium dichromate (and a small amount of water) to concentrated sulfuric acid. (p. 508)

chromic acid test: When a primary or secondary alcohol is warmed with the chromic acid reagent, the orange color changes to green or blue. A nonoxidizable compound (such as a tertiary alcohol, a ketone, or an alkane) produces no color change. (p. 511)

condensation A reaction that joins two (or more) molecules, often with the loss of a small molecule such as water or an alcohol. (p. 530)

DMP reagent The Dess–Martin periodinane reagent, used to oxidize primary alcohols to aldehydes and secondary alcohols to ketones. The DMP reagent uses a high-valence iodine atom as the oxidizing species. (p. 512)

(continued)

ester — An acid derivative formed by the reaction of an acid with an alcohol with loss of water. The most common esters are carboxylic esters (or carboxylate esters), composed of carboxylic acids and alcohols. (p. 536)

Fischer esterification: The acid-catalyzed reaction of an alcohol with a carboxylic acid to form an ester. (p. 537)

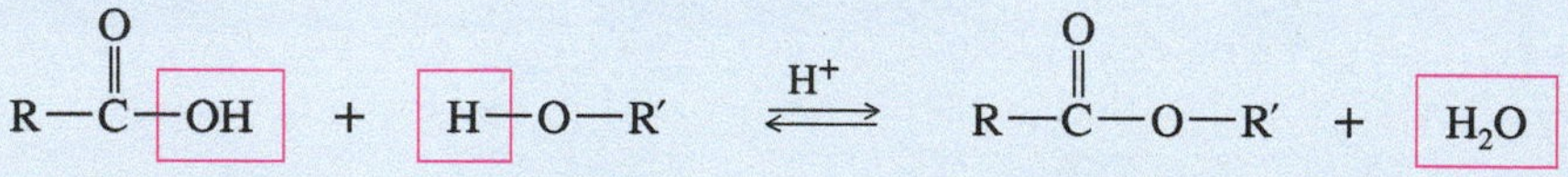

$$\underset{\text{carboxylic acid}}{R{-}C({=}O){-}OH} + \underset{\text{alcohol}}{H{-}O{-}R'} \overset{H^+}{\rightleftharpoons} \underset{\text{carboxylic ester}}{R{-}C({=}O){-}O{-}R'} + H_2O$$

inorganic esters: Compounds derived from alcohols and inorganic acids with loss of water. (p. 537) Examples are

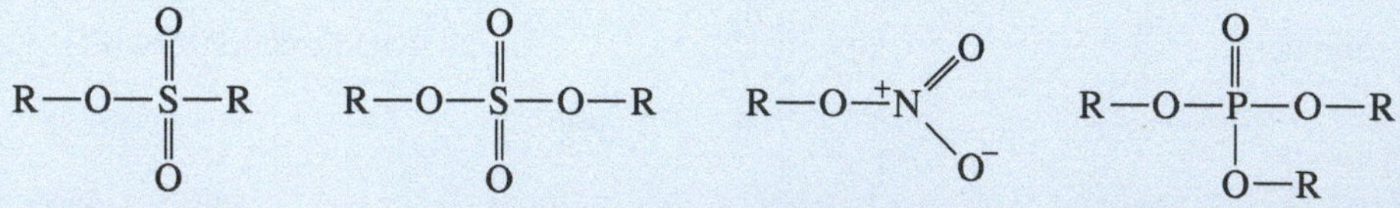

$$\underset{\text{sulfonate esters}}{R{-}O{-}S(O_2){-}R} \qquad \underset{\text{sulfate esters}}{R{-}O{-}S(O_2){-}O{-}R} \qquad \underset{\text{nitrate esters}}{R{-}O{-}\overset{+}{N}({=}O){-}O^-} \qquad \underset{\text{phosphate esters}}{R{-}O{-}P({=}O)(O{-}R){-}O{-}R}$$

ether — A compound containing an oxygen atom bonded to two alkyl or aryl groups. (p. 541)

glycol — Synonymous with **diol.** The term "glycol" is most commonly applied to the 1,2-diols, also called **vicinal diols.** (p. 464)

Lucas test — A test used to determine whether an alcohol is primary, secondary, or tertiary. The test measures the rate of reaction with the **Lucas reagent,** $ZnCl_2$ in concentrated HCl. Tertiary alcohols react fast (seconds), secondary alcohols react more slowly (minutes), and primary alcohols react very slowly (hours). (p. 522)

nicotinamide adenine dinucleotide — **(NAD)** A biological oxidizing/reducing reagent that operates in conjunction with enzymes such as alcohol dehydrogenase. (p. 514)

oxidation — Loss of H_2; addition of O or O_2; addition of X_2 (halogens). Alternatively, an increase in the number of bonds to oxygen or halogens or a decrease in the number of bonds to hydrogen. (p. 506)

pinacol rearrangement — Dehydration of a glycol in which one of the groups migrates to give a ketone. (p. 534)

pyridinium chlorochromate — **(PCC)** A complex of chromium trioxide with pyridine and HCl. PCC oxidizes primary alcohols to aldehydes without over-oxidizing them to carboxylic acids. (p. 510)

reduction — Addition of H_2 (or H^-); loss of O or O_2; loss of X_2 (halogens). Alternatively, a reduction in the number of bonds to oxygen or halogens or an increase in the number of bonds to hydrogen. (p. 506)

Swern oxidation — A mild oxidation, using DMSO and oxalyl chloride, that can oxidize primary alcohols to aldehydes and secondary alcohols to ketones. (p. 511)

tosylate ester — **(R—OTs)** An ester of an alcohol with *para*-toluenesulfonic acid. Like halide ions, the tosylate anion is an excellent leaving group. (p. 516)

Williamson ether synthesis — The S_N2 reaction between an alkoxide ion and a primary alkyl halide or tosylate. The product is an ether. (p. 541)

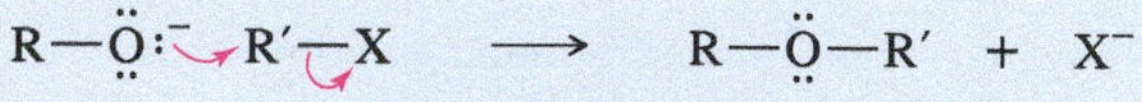

$$R{-}\ddot{\underset{\cdot\cdot}{O}}{:}^- \;\; R'{-}X \longrightarrow R{-}\ddot{\underset{\cdot\cdot}{O}}{-}R' + X^-$$

Essential Problem-Solving Skills in Chapter 11

Each skill is followed by problem numbers exemplifying that particular skill.

Skill	Problems
1 Identify whether oxidation or reduction is needed to interconvert alkanes, alcohols, aldehydes, ketones, and acids, and identify reagents that will accomplish the conversion.	Problems 11-40, 48, and 56
2 Predict the products of the reactions of alcohols with (a) oxidizing and reducing agents. (b) carboxylic acids and acid chlorides. (c) dehydrating reagents, especially H_2SO_4 and H_3PO_4. (d) hydrohalic acids (HCl, HBr, and HI) and the phosphorus halides. (e) sodium metal, potassium metal, and sodium hydride.	Problems 11-39, 43, 44, 49, 50, and 53
3 Predict the products of the reactions of alkoxide ions.	Problems 11-39, 48, 51, 56, and 60
4 Use your knowledge of alcohol and diol reactions to propose mechanisms and products for similar reactions that are new.	Problems 11-45, 47, 54, 55, 60, 61, 62, 63, and 65
Problem-Solving Strategy: Proposing Reaction Mechanisms	Problems 11-45, 47, 54, 55, 60, 61, 62, 63, and 65
5 Show how to convert an alcohol to a related compound with a different functional group.	Problems 11-40, 41, 42, 46, 47, and 48
6 Use retrosynthetic analysis to propose single-step and multistep syntheses of compounds using alcohols as starting materials and intermediates. Show how Grignard and organolithium reagents can be used to assemble the carbon skeletons.	Problems 11-42, 46, 48, 56, and 57
Problem-Solving Strategy: Multistep Synthesis	Problems 11-40, 46, 48, 56, and 57

Study Problems

11-39 Predict the major products of the following reactions, including stereochemistry where appropriate.

(a) (*R*)-butan-2-ol + TsCl in pyridine
(b) (*S*)-2-butyl tosylate + NaBr
(c) cyclooctanol + NaOCl/HOAc
(d) cyclopentylmethanol + $CrO_3 \cdot$ pyridine $\cdot$ HCl
(e) cyclopentylmethanol + $Na_2Cr_2O_7/H_2SO_4$
(f) cyclopentanol + $HCl/ZnCl_2$
(g) *n*-butanol + HBr
(h) cyclooctylmethanol + CH_3CH_2MgBr
(i) potassium *tert*-butoxide + methyl iodide
(j) sodium methoxide + *tert*-butyl iodide
(k) cyclopentanol + H_2SO_4/heat
(l) product from (k) + OsO_4/H_2O_2, then HIO_4
(m) sodium ethoxide + 1-bromobutane
(n) sodium ethoxide + 2-methyl-2-bromobutane
(o) octan-1-ol + DMSO + oxalyl chloride
(p) 4-cyclopentylhexan-1-ol + DMP reagent

11-40 Show how you would convert 2-methylcyclopentanol to the following products. Any of these products may be used as the reactant in any subsequent part of this problem.

(a) 1-methylcyclopentene
(b) 2-methylcyclopentyl tosylate
(c) 2-methylcyclopentanone
(d) 1-methylcyclopentanol
(e) 1,2-dimethylcyclopentanol
(f) 1-bromo-2-methylcyclopentane
(g) 2-methylcyclopentyl acetate
(h) 1-bromo-1-methylcyclopentane

11-41 In each case, show how you would synthesize the chloride, bromide, and iodide from the corresponding alcohol.

(a) 1-halobutane (halo = chloro, bromo, iodo)
(b) halocyclopentane
(c) 1-halo-1-methylcyclohexane
(d) 1-halo-2-methylcyclohexane

11-42 Show how you would accomplish the following synthetic conversions.

(a) OH, H → OCH_2CH_3, H

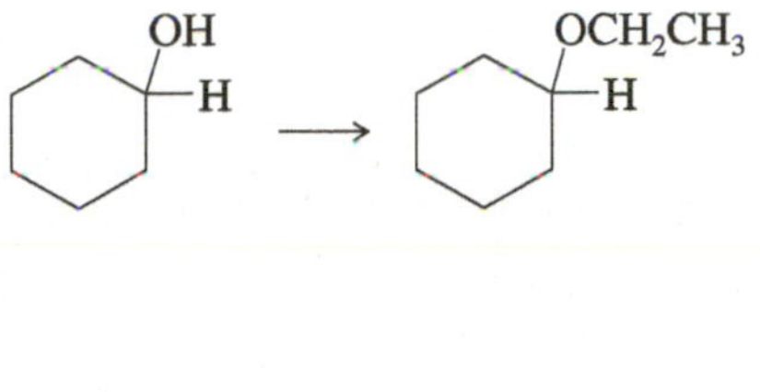

(b) 2 CH_2Br →

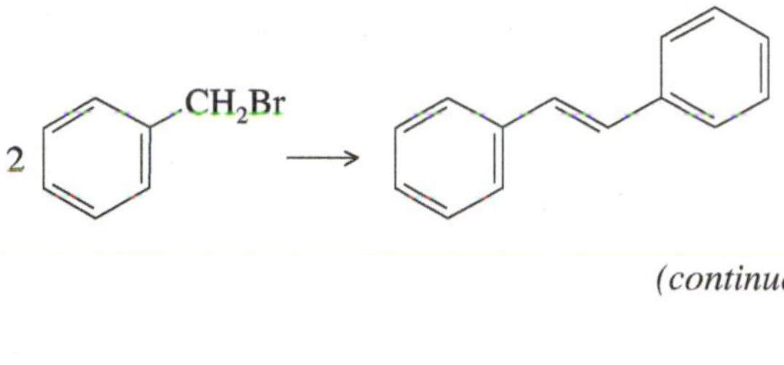

(continued)

(c) (bromocyclohexane) $\longrightarrow$ C_6H_{11}–$C(=O)$–$CH_2CH_2CH_3$

(d) C_6H_{11}–CH_2OH $\longrightarrow$ C_6H_{11}–$CH(OH)CH_2CH_3$

11-43 Predict the major products of dehydration catalyzed by sulfuric acid.
(a) hexan-1-ol
(b) hexan-2-ol
(c) pentan-3-ol
(d) 1-methylcyclopentanol
(e) cyclopentylmethanol
(f) 2-methylcyclopentanol

11-44 Predict the esterification products of the following acid/alcohol pairs.
(a) $CH_3CH_2CH_2COOH + CH_3OH$
(b) $CH_3OH + HNO_3$
(c) $2\ CH_3CH_2OH + H_3PO_4$
(d) C_6H_5COOH + CH_3CH_2OH
(e) C_6H_5OH + CH_3–$C(=O)$–OH

11-45 Both *cis*- and *trans*-2-methylcyclohexanol undergo dehydration in warm sulfuric acid to give 1-methylcyclohexene as the major alkene product. These alcohols can also be converted to alkenes by tosylation using TsCl and pyridine, followed by elimination using $KOC(CH_3)_3$ as a strong base. Under these basic conditions, the tosylate of *cis*-2-methylcyclohexanol eliminates to give mostly 1-methylcyclohexene, but the tosylate of *trans*-2-methylcyclohexanol eliminates to give only 3-methylcyclohexene. Explain how this stereochemical difference in reactants controls a regiochemical difference in the products of the basic elimination, but not in the acid-catalyzed elimination.

11-46 Show how you would convert (*S*)-hexan-2-ol to
(a) (*S*)-2-chlorohexane.
(b) (*R*)-2-bromohexane.
(c) (*R*)-hexan-2-ol.

11-47 When 1-cyclohexylethanol is treated with concentrated aqueous HBr, the major product is 1-bromo-1-ethylcyclohexane.

C_6H_{11}–$CH(OH)CH_3$ $\xrightarrow[H_2O]{HBr}$ 1-bromo-1-ethylcyclohexane (Br)

(a) Propose a mechanism for this reaction.
(b) How would you convert 1-cyclohexylethanol to (1-bromoethyl)cyclohexane in good yield?

C_6H_{11}–$CH(OH)CH_3$ $\xrightarrow{?}$ C_6H_{11}–$CH(Br)CH_3$

11-48 Show how you would make each compound, beginning with an alcohol of your choice.

(a)
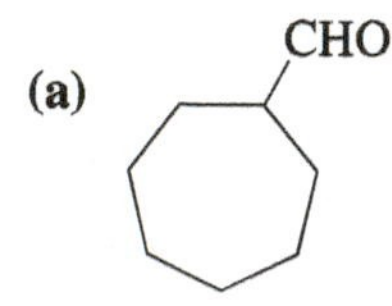

(b) C_6H_{11}–CH_2Br

(c) C_6H_{11}–OCH_3

(d)
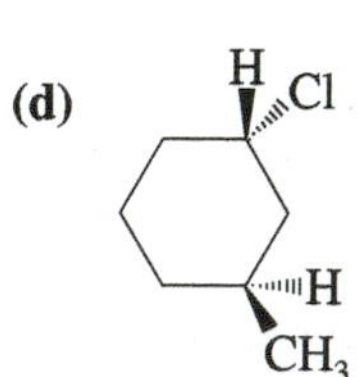

(e)
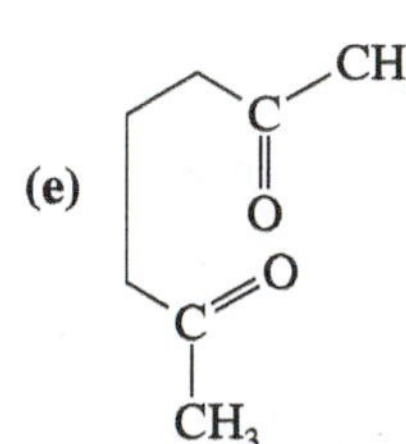

(f) C_5H_9–$C(=O)$–OH

(g) 4-*tert*-butylcyclohexanone (O)

(h) cyclopentane with H, CH_3 and H, OTs

11-49 Predict the major products (including stereochemistry) when *cis*-3-methylcyclohexanol reacts with the following reagents.
(a) PBr_3
(b) $SOCl_2$
(c) Lucas reagent
(d) concentrated HBr
(e) TsCl/pyridine, then NaBr

11-50 Show how you would use simple chemical tests to distinguish between the following pairs of compounds. In each case, describe what you would do and what you would observe.

(a) butan-1-ol and butan-2-ol **(b)** butan-2-ol and 2-methylbutan-2-ol
(c) cyclohexanol and cyclohexene **(d)** cyclohexanol and cyclohexanone
(e) cyclohexanone and 1-methylcyclohexanol

11-51 The compound shown below has three different types of OH groups, all with different acidities. Show the structure produced after this compound is treated with different amounts of NaH followed by a methylating reagent. Add a brief explanation.

(a) 1 equivalent of NaH, followed by 1 equivalent of CH_3I and heat
(b) 2 equivalents of NaH, followed by 2 equivalents of CH_3I and heat
(c) 3 equivalents of NaH, followed by 3 equivalents of CH_3I and heat

HO COOH OH

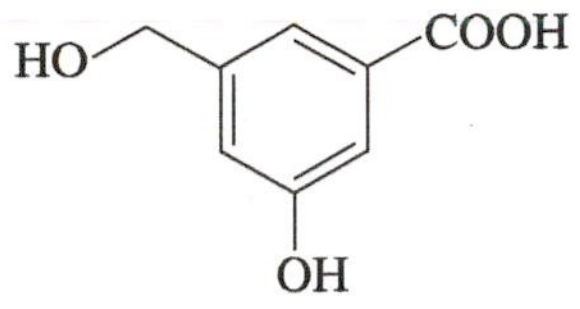

11-52 Compound **A** is an optically active alcohol. Treatment with chromic acid converts **A** into a ketone, **B.** In a separate reaction, **A** is treated with PBr_3, converting **A** into compound **C.** Compound **C** is purified, and then it is allowed to react with magnesium in ether to give a Grignard reagent, **D.** Compound **B** is added to the resulting solution of the Grignard reagent. After hydrolysis of the initial product (**E**), this solution is found to contain 3,4-dimethylhexan-3-ol. Propose structures for compounds **A, B, C, D,** and **E.**

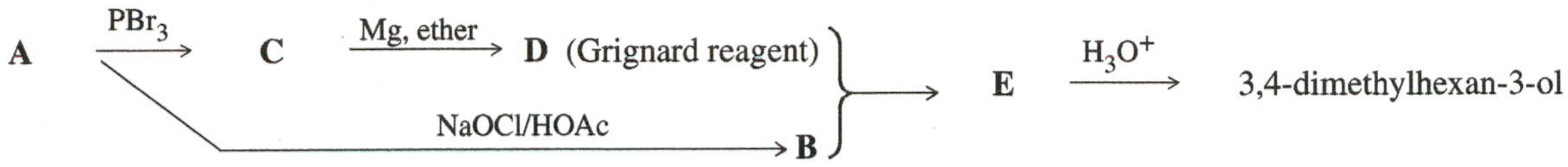

A $\xrightarrow{PBr_3}$ C $\xrightarrow{\text{Mg, ether}}$ D (Grignard reagent); A $\xrightarrow{\text{NaOCl/HOAc}}$ B; D + B → E $\xrightarrow{H_3O^+}$ 3,4-dimethylhexan-3-ol

11-53 Give the structures of the intermediates and products **V** through **Z.**

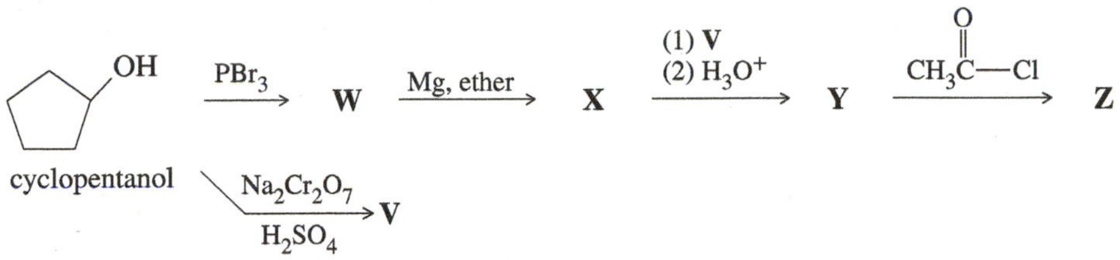

cyclopentanol $\xrightarrow{PBr_3}$ **W** $\xrightarrow{\text{Mg, ether}}$ **X** $\xrightarrow{(1)\ V\ (2)\ H_3O^+}$ **Y** $\xrightarrow{CH_3C(=O)-Cl}$ **Z**

cyclopentanol $\xrightarrow[H_2SO_4]{Na_2Cr_2O_7}$ **V**

11-54 Under acid catalysis, tetrahydrofurfuryl alcohol reacts to give surprisingly good yields of dihydropyran. Propose a mechanism to explain this useful synthesis.

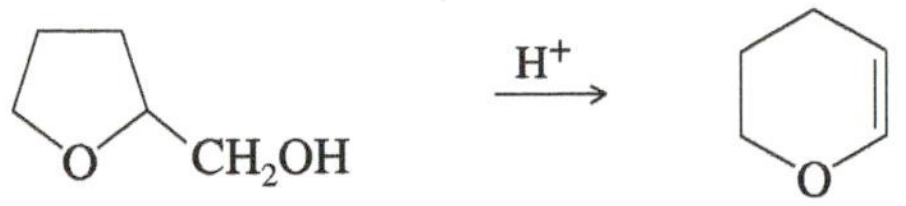

tetrahydrofurfuryl alcohol (CH_2OH) $\xrightarrow{H^+}$ dihydropyran

11-55 Propose mechanisms for the following reactions. In most cases, more products are formed than are shown here. You only need to explain the formation of the products shown, however.

(a) cyclohexyl–CH_2-OH $\xrightarrow{HCl,\ ZnCl_2}$ cycloheptyl–Cl (a minor product)

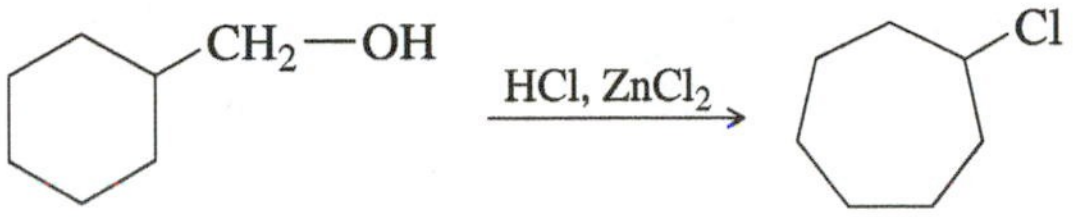

(b) H_3C, OH $\xrightarrow[\text{heat}]{H_2SO_4}$ CH_3 + CH_3

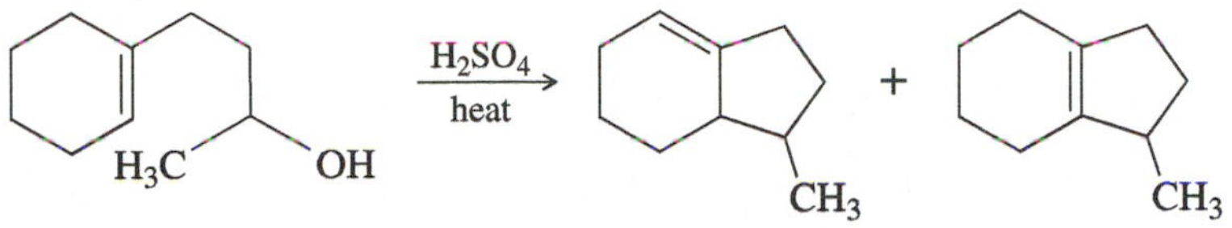

(c) OH, OH $\xrightarrow[\text{heat}]{H_2SO_4}$ + + O

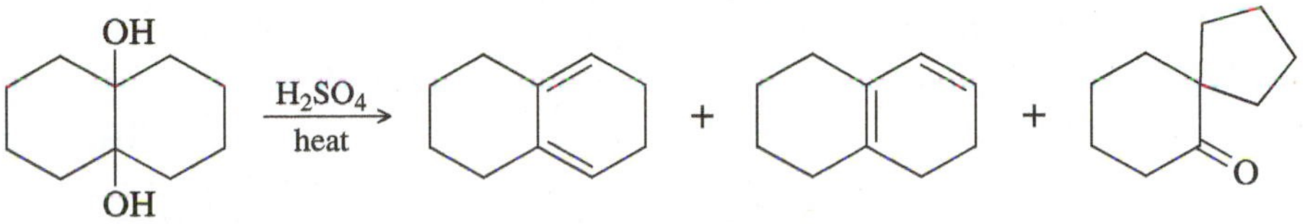

11-56 Show how you would synthesize the following compounds. As starting materials, you may use any alcohols containing four or fewer carbon atoms, cyclohexanol, and any necessary solvents and inorganic reagents.

(a) (b) (c) (d)

(e) (f) (g) (h)

***11-57** Show how you would synthesize the following compound. As starting materials, you may use any alcohols containing five or fewer carbon atoms and any necessary solvents and inorganic reagents.

11-58 The following pseudo-syntheses (guaranteed not to work) exemplify a common conceptual error.

$$(CH_3)_3C{-}Br \xrightarrow[CH_3OH]{heat} (CH_3)_3C^+ \; Br^- \xrightarrow{Na^+ \; ^-OCH_3} (CH_3)_3C{-}OCH_3$$

(a) What is the conceptual error implicit in these syntheses?
(b) Propose syntheses that are more likely to succeed.

11-59 Two unknowns, **X** and **Y**, both having the molecular formula C_4H_8O, give the following results with four chemical tests. Propose structures for **X** and **Y** consistent with this information.

	Bromine	Na Metal	Chromic Acid	Lucas Reagent
Compound **X**	decolorizes	bubbles	orange to green	no reaction
Compound **Y**	no reaction	no reaction	no reaction	no reaction

11-60 The Williamson ether synthesis involves the displacement of an alkyl halide or tosylate by an alkoxide ion. Would the synthesis shown be possible by making a tosylate and displacing it? If so, show the sequence of reactions. If not, explain why not and show an alternative synthesis that would be more likely to work.

make the tosylate and displace?

***11-61** Chromic acid oxidation of an alcohol (Section 11-2A) occurs in two steps: formation of the chromate ester, followed by an elimination of H^+ and chromium. Which step do you expect to be rate-limiting? Careful kinetic studies have shown that Compound **A** undergoes chromic acid oxidation over 10 times as fast as Compound **B**. Explain this large difference in rates.

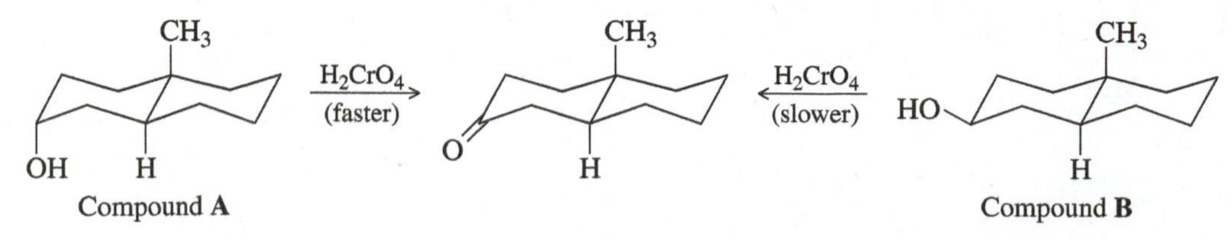

***11-62** **(a)** The reaction of butan-2-ol with concentrated aqueous HBr goes with partial racemization, giving more inversion than retention of configuration. Propose a mechanism that accounts for racemization with excess inversion.

(b) Under the same conditions, an optically active sample of *trans*-2-bromocyclopentanol reacts with concentrated aqueous HBr to give an optically inactive product, (racemic) *trans*-1,2-dibromocyclopentane. Propose a mechanism to show how this reaction goes with apparently complete retention of configuration, yet with racemization. (*Hint*: Draw out the mechanism of the reaction of cyclopentene with Br_2 in water to give the starting material, *trans*-2- bromocyclopentanol. Consider how parts of this mechanism might be involved in the reaction with HBr.)

***11-63** Alcohols combine with ketones and aldehydes to form interesting derivatives, which we will discuss in Chapter 18. The following reactions show the hydrolysis of two such derivatives. Propose mechanisms for these reactions.

(a) $\xrightarrow[H_2O]{H^+}$ + CH_3OH

(b) $\xrightarrow[H_2O]{H^+}$

***11-64** Unknown **Q** is determined to have a molecular formula of $C_6H_{12}O$. **Q** is not optically active, and passing it through a chiral column does not separate it into enantiomers. **Q** does not react with Br_2, nor with cold, dilute $KMnO_4$, nor does it take up H_2 under catalytic hydrogenation. Heating of **Q** with H_2SO_4 gives product **R**, of formula C_6H_{10}, which can be separated into enantiomers. Ozonolysis of a single enantiomer of **R** produces **S**, an acyclic, optically active ketoaldehyde of formula $C_6H_{10}O_2$. Propose structures for compounds **Q, R,** and **S,** and show how your structures would react appropriately to give these results.

11-65 Trichloroisocyanuric acid, TCICA, also known as "swimming pool chlorine," is a stable solid that oxidizes alcohols, following a mechanism similar to oxidation by HOCl. No reaction occurs between TCICA and the alcohol (in a solvent such as acetonitrile) until one drop of HCl(aq) is added, whereupon the reaction is over in a few minutes. Write the mechanism for this oxidation that shows the key role of the acid catalyst. Show the oxidation of just one alcohol, not three. (*Hint*: The carbonyls are the most basic sites on TCICA.)

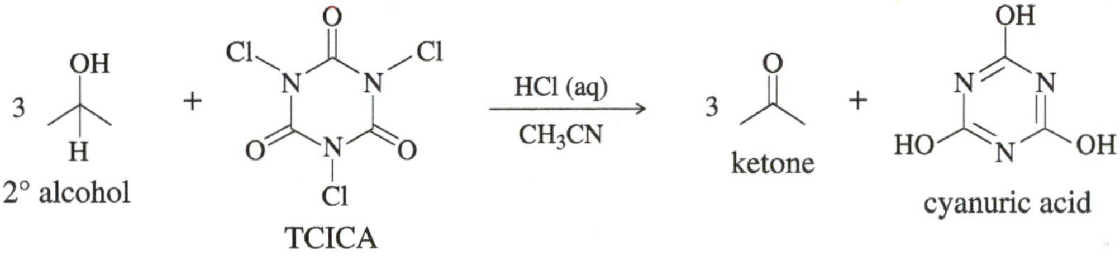

3 (2° alcohol) + TCICA $\xrightarrow[CH_3CN]{HCl\ (aq)}$ 3 (ketone) + cyanuric acid

11-66 Under normal circumstances, tertiary alcohols are not oxidized. However, when the tertiary alcohol is allylic, it can undergo a migration of the double bond (called an allylic shift) and subsequent oxidation of the alcohol. A particularly effective reagent for this reaction is Bobbitt's reagent, similar to TEMPO used in many oxidations. (M. Shibuya *et al.*, *J. Org. Chem.*, **2008**, *73*, 4750.)

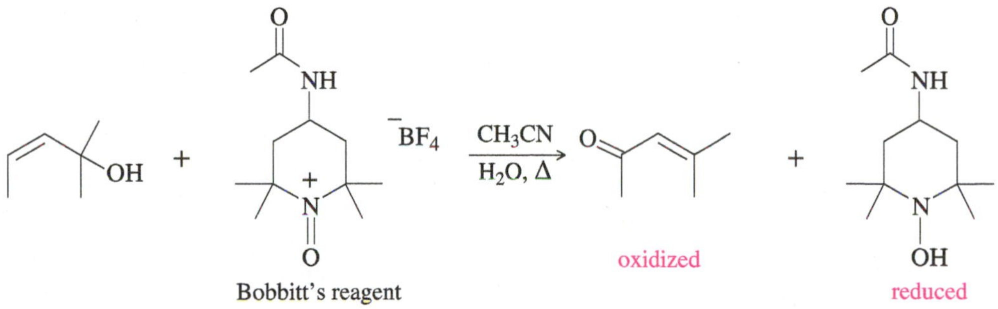

(3° allylic alcohol) + Bobbitt's reagent ($^-BF_4$) $\xrightarrow[H_2O,\ \Delta]{CH_3CN}$ (oxidized) + (reduced)

Show the expected product when each of these 3° allylic alcohols is oxidized by Bobbitt's reagent.

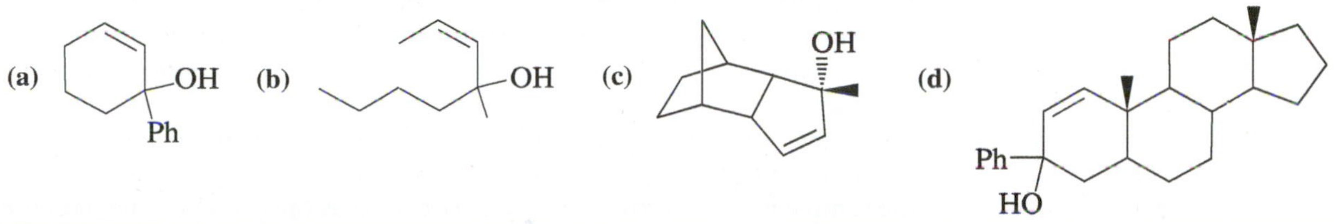

(a) (b) (c) (d)

12 Infrared Spectroscopy and Mass Spectrometry

Goals for Chapter 12

1 Identify the reliable characteristic absorptions in an infrared spectrum, and propose which functional groups are likely to be present in the molecule.

2 Explain which functional groups cannot be present in a molecule because their characteristic peaks are absent from the IR spectrum.

3 Use a mass spectrum to determine a compound's molecular weight, and propose which elements are likely to be present.

4 Given a structure, predict the major ions that will be observed in the mass spectrum from fragmentation of the molecular ion. Use these predictions to determine whether a proposed structure is consistent with the spectrum.

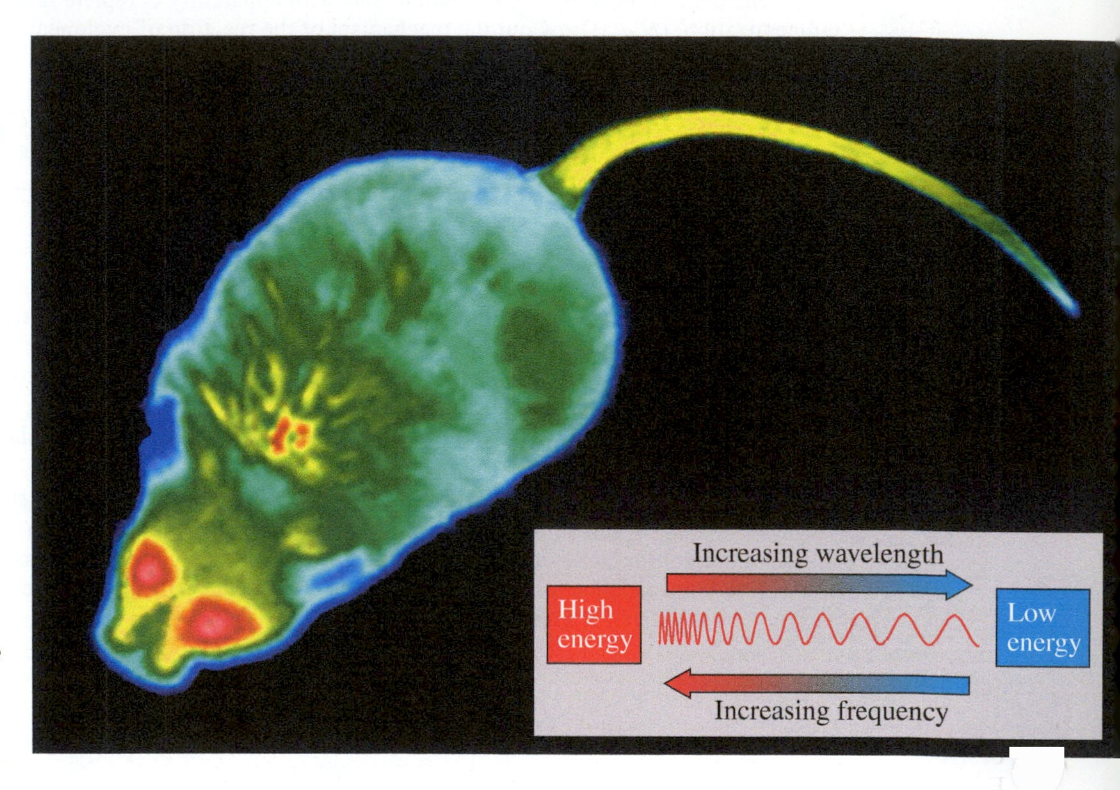

▶ This infrared image of a mouse depicts the wavelengths of infrared light as visible colors: The shorter wavelengths (higher energy) of infrared are shown as warmer, brighter colors. The image shows where the mouse loses the most heat, mostly through the eyes and the tail. This mouse is losing excessive heat through its tail because it has defective thyroid hormone receptors and is unable to control the constriction of the blood vessels in its tail.

12-1 Introduction

One of the most important tasks of organic chemistry is the determination of organic structures. When an interesting compound is isolated from a natural source, its structure must be completely determined before a synthesis can begin. Whenever we run a reaction, we must determine whether the product has the desired structure. The structure of an unwanted product must be known so that the reaction conditions can be altered to favor the desired product.

In many cases, a compound can be identified by chemical means. We find the molecular formula by analyzing the elemental composition and determining the molecular weight. If the compound has been characterized before, we can compare its physical properties (melting point, boiling point, etc.) with the published values. Chemical tests can suggest the functional groups and narrow the range of possible structures before the physical properties are used to make an identification.

These procedures are insufficient, however, for complex compounds that have never been synthesized and characterized. They are also impractical with compounds that are difficult to obtain, because a relatively large sample is required to complete the elemental analysis and all the functional group tests. We need analytical techniques that work with tiny samples and that do not destroy the sample.

Spectroscopic techniques often meet these requirements. **Absorption spectroscopy** is the measurement of the amount of light absorbed by a compound as a function of the wavelength of light. In general, a spectrometer irradiates the sample with light, measures the amount of light transmitted as a function of wavelength, and plots the results on a graph. Unlike chemical tests, most spectroscopic techniques are *nondestructive*; that is, the sample is not destroyed. Many different kinds of spectra can be measured with little or no loss of sample.

In this book, we cover four spectroscopic or related techniques that serve as powerful tools for structure determination in organic chemistry:

Infrared (IR) spectroscopy, covered in this chapter, observes the vibrations of bonds and provides evidence of the functional groups present.

Mass spectrometry (MS), also covered in this chapter, is not a *spectroscopic* technique, because it does not measure absorption or emission of light. A mass spectrometer bombards molecules with electrons and breaks the molecules into fragments. Analysis of the masses of the fragments gives the molecular weight, possibly the molecular formula, and clues to the structure and functional groups. Less than a milligram of sample is destroyed in this analysis.

Nuclear magnetic resonance (NMR) spectroscopy, covered in Chapter 13, observes the chemical environments of the hydrogen atoms or the carbon atoms and provides evidence for the structure of the alkyl groups and clues to the functional groups.

Ultraviolet (UV) spectroscopy, covered in Chapter 15, observes electronic transitions and provides information on the electronic bonding in the sample.

These spectroscopic techniques are complementary, and they are most powerful when used together. In many cases, an unknown compound cannot be completely identified from one spectrum without additional information, yet the structure can be determined with confidence using two or more different types of spectra. In Chapter 13, we consider how clues from different types of spectroscopy are combined to provide a reliable structure.

12-2 The Electromagnetic Spectrum

Visible light, infrared light, ultraviolet light, microwaves, and radio waves are examples of electromagnetic radiation. They all travel at the speed of light, about 3×10^{10} cm/second, but they differ in frequency and wavelength. The **frequency** of a wave is the number of complete wave cycles that pass a fixed point in a second. Frequency, represented by the Greek letter ν (nu), is usually given in hertz (Hz), meaning "cycles per second." The **wavelength,** represented by the Greek letter λ (lambda), is the distance between any two peaks (or any two troughs) of the wave.

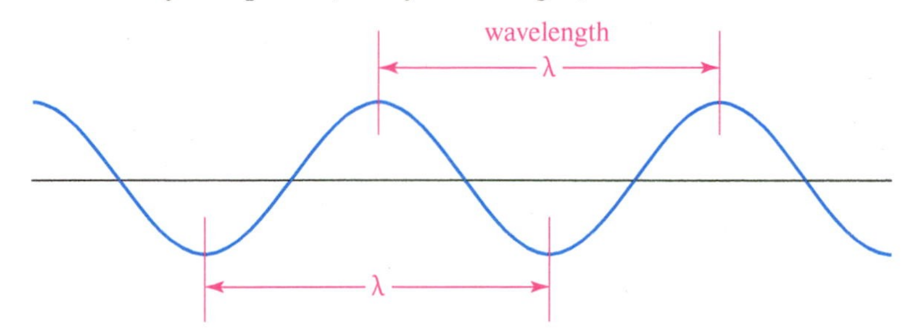

The wavelength and frequency, which are inversely proportional, are related by the equation

$$\nu\lambda = c \quad \text{or} \quad \lambda = \frac{c}{\nu}$$

where

c = speed of light (3×10^{10} cm/sec)
ν = frequency in hertz
λ = wavelength in centimeters

Electromagnetic waves travel as **photons,** which are massless packets of energy. The energy of a photon is proportional to its frequency and inversely proportional to its wavelength. A photon of frequency ν (or wavelength λ) has an energy given by

$$E = h\nu = \frac{hc}{\lambda}$$

where h is Planck's constant, 6.62×10^{-37} kJ • sec or 1.58×10^{-37} kcal • sec. Under certain conditions, a molecule struck by a photon may absorb the photon's energy. In this case, the molecule's energy is increased by an amount equal to the photon's energy, $h\nu$. For this reason, we often represent the irradiation of a reaction mixture by the symbol $h\nu$.

The **electromagnetic spectrum** is the range of all possible frequencies, from zero to infinity. In practice, the spectrum ranges from the very low radio frequencies used to communicate with submarines to the very high frequencies of gamma rays. Figure 12-1 shows the wavelength and energy relationships of the various parts of the electromagnetic spectrum.

The electromagnetic spectrum is continuous, and the exact positions of the dividing lines between the different regions are somewhat arbitrary. Toward the top of the spectrum in Figure 12-1 are the higher frequencies, shorter wavelengths, and higher energies. Toward the bottom are the lower frequencies, longer wavelengths, and lower energies. X-rays (very high energy) are so energetic that they excite electrons past all the energy levels, causing ionization. Energies in the ultraviolet–visible range excite electrons to higher energy levels within molecules. Infrared energies excite molecular vibrations, and microwave energies excite rotations. Radio-wave frequencies (very low energy) excite the nuclear spin transitions observed in NMR spectroscopy.

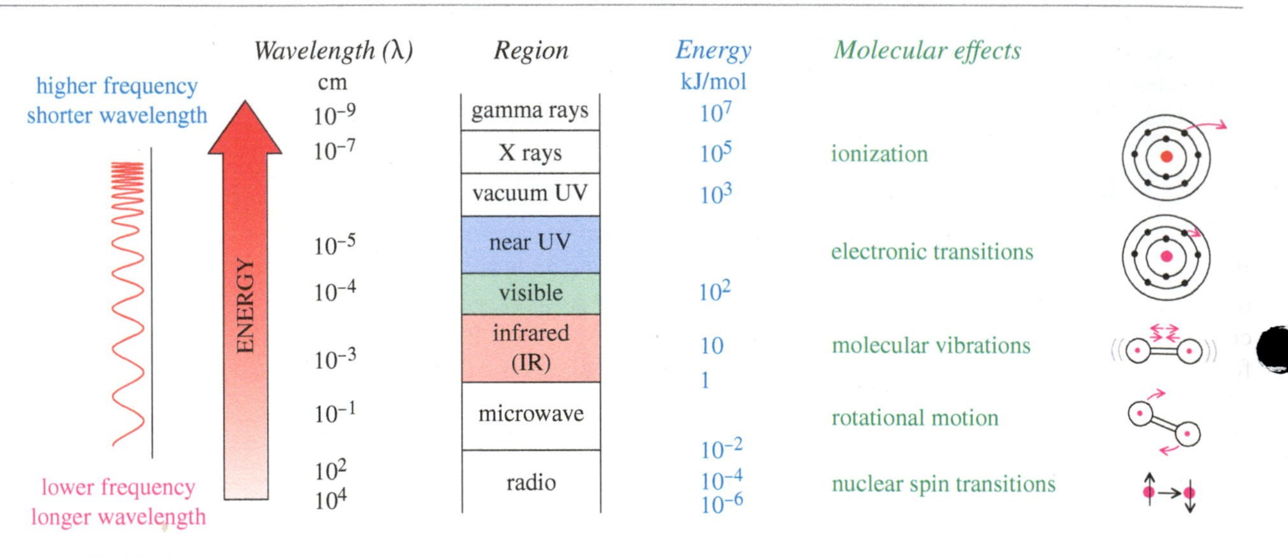

FIGURE 12-1
The electromagnetic spectrum and the resultant molecular effects.

12-3 The Infrared Region

The infrared (from the Latin, *infra*, meaning "below" red) region of the spectrum corresponds to frequencies from just below the visible frequencies to just above the highest microwave and radar frequencies: wavelengths of about 8×10^{-5} cm to 1×10^{-2} cm. Common infrared spectrometers operate in the middle of this region, at wavelengths between 2.5×10^{-4} cm and 25×10^{-4} cm, corresponding to energies of about 4.6 to 46 kJ/mol (1.1 to 11 kcal/mol). Infrared photons do not have enough energy to cause electronic transitions, but they can cause groups of atoms to vibrate with respect to the bonds that connect them. Like electronic transitions, these vibrational transitions correspond to distinct energies, and molecules absorb infrared radiation only at certain wavelengths and frequencies.

The position of an infrared band can be specified by its wavelength (λ), which is measured in *microns* (μm). A micron (or *micrometer*) corresponds to one millionth (10^{-6}) of a meter, or 10^{-4} cm. A more common unit, however, is the **wavenumber** ($\bar{\nu}$), which corresponds to the number of cycles (wavelengths) in a centimeter. The wavenumber is the reciprocal of the wavelength (in centimeters). Because 1 cm = 10,000 μm, the wavenumber can be calculated by dividing 10,000 by the wavelength in microns. The units of the wavenumber are cm^{-1} (*reciprocal centimeters*).

$$\bar{\nu}\,(\text{cm}^{-1}) = \frac{1}{\lambda\,(\text{cm})} = \frac{10{,}000\ \mu\text{m/cm}}{\lambda\,(\mu\text{m})} \quad \text{or} \quad \lambda\,(\mu\text{m}) = \frac{10{,}000\ \mu\text{m/cm}}{\bar{\nu}\,(\text{cm}^{-1})}$$

For example, an absorption at a wavelength of 4 μm corresponds to a wavenumber of 2500 cm^{-1}.

$$\bar{\nu} = \frac{10{,}000\ \mu\text{m/cm}}{4\ \mu\text{m}} = 2500\ \text{cm}^{-1} \quad \text{or} \quad \lambda = \frac{10{,}000\ \mu\text{m/cm}}{2500\ \text{cm}^{-1}} = 4\ \mu\text{m}$$

Wavenumbers (in cm^{-1}) have become the most common method for specifying IR absorptions, and we will use wavenumbers throughout this book. The wavenumber is proportional to the frequency (ν) of the wave, so it is also proportional to the energy of a photon of this frequency ($E = h\nu$). Some references still use microns, however, so you should know how to convert these units.

PROBLEM 12-1

Complete the following conversion table.

$\bar{\nu}$(cm^{-1})	4000				1700	1640	1600	400
λ (μm)	2.50	3.03	3.33	4.55				25.0

12-4 Molecular Vibrations

Before discussing characteristic infrared absorptions, it's helpful to understand some theory about the vibrational energies of molecules. The following drawing shows how a covalent bond between two atoms acts like a spring. If the bond is stretched, a restoring force pulls the two atoms together toward their equilibrium bond length. If the bond is compressed, the restoring force pushes the two atoms apart. If the bond is stretched or compressed and then released, the atoms vibrate.

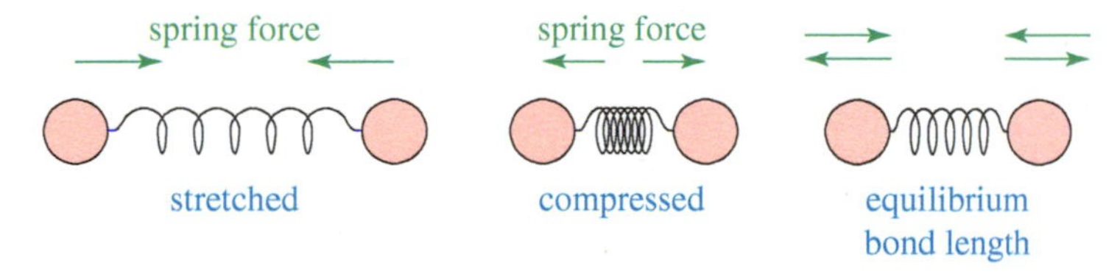

The frequency of the stretching vibration depends on the masses of the atoms and the stiffness of the bond. Heavier atoms vibrate more slowly than lighter ones; for example, the characteristic frequency of a C—D bond is lower than that of a C—H bond. In a group of bonds with similar bond energies, the *frequency decreases with increasing atomic weight.*

Stronger bonds are generally stiffer, requiring more force to stretch or compress them. Thus, stronger bonds usually vibrate faster than weaker bonds (assuming the atoms have similar masses). For example, O—H bonds are stronger than C—H bonds, so O—H bonds vibrate at higher frequencies. Triple bonds are stronger than double bonds, so triple bonds vibrate at higher frequencies than double bonds. Similarly, double bonds vibrate at higher frequencies than single bonds. In a group of bonds having atoms of similar masses, the *frequency increases with bond energy*.

Table 12-1 lists some common types of bonds, together with their stretching frequencies, to show how frequency varies with the masses of the atoms and the strengths of the bonds.

An **infrared spectrum** is a graph of the energy absorbed by a molecule as a function of the frequency or wavelength of light. The IR spectrum of methanol is shown in Figure 12-2. In the infrared region, absorptions generally result from exciting the vibrational modes of the bonds in the molecule. Even with simple compounds, infrared spectra contain many different absorptions, not just one absorption for each bond.

Use *spectrum* (singular) and *spectra* (plural) correctly: "This spectrum is..." "These spectra are...."

TABLE 12-1
Bond Stretching Frequencies

In a group of bonds with similar bond energies, *the frequency decreases with increasing atomic weight.* In a group of bonds between similar atoms, the *frequency increases with bond energy.* The bond energies and frequencies listed here are approximate.

Bond		Bond Energy [kJ (kcal)]		Stretching Frequency (cm^{-1})	
Frequency decreases with increasing atomic mass					
C—H		420 (100)		3000	
C—D	heavier atoms ↓	420 (100)		2100	$\bar{\nu}$ decreases ↓
C—C		350 (83)		1200	
Frequency increases with bond energy					
C—C		350 (83)		1200	
C=C		611 (146)	stronger bond ↓	1660	$\bar{\nu}$ increases ↓
C≡C		840 (200)		2200	
C—N		305 (73)		1200	
C=N		615 (147)	stronger bond ↓	1650	$\bar{\nu}$ increases ↓
C≡N		891 (213)		2200	
C—O		360 (86)	stronger bond ↓	1100	$\bar{\nu}$ increases ↓
C=O		745 (178)		1700	

The methanol spectrum (Figure 12-2) is a good example. We can see the broad O—H stretch around 3300 cm^{-1}, the C—H stretch just below 3000 cm^{-1}, and the C—O stretch just above 1000 cm^{-1}. We also see absorptions resulting from bending vibrations, including scissoring and twisting vibrations. In a bending vibration, the bond lengths stay constant, but the bond angles vibrate about their equilibrium values.

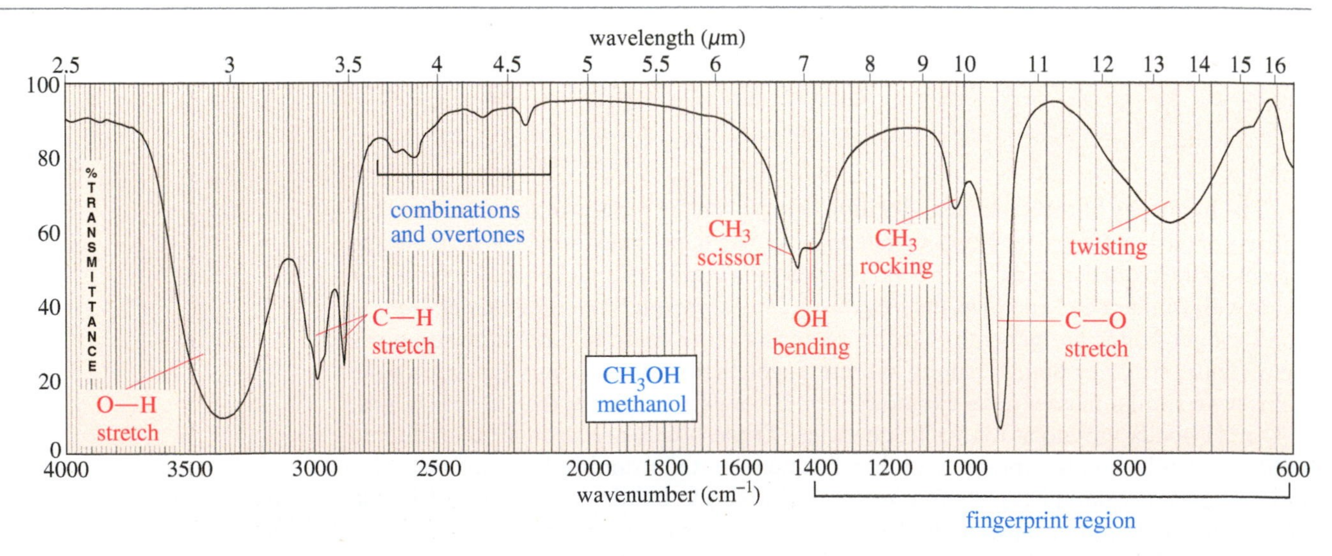

FIGURE 12-2
The infrared spectrum of methanol shows O—H, C—H, and C—O stretching absorptions, together with absorptions from several bending modes.

Consider the fundamental vibrational modes of a water molecule in the diagram on the next page. The two O—H bonds can stretch in phase with each other (*symmetric stretching*), or they can stretch out of phase (*antisymmetric stretching*). The H—O—H bond angle can also change in a bending vibration, making a *scissoring* motion.

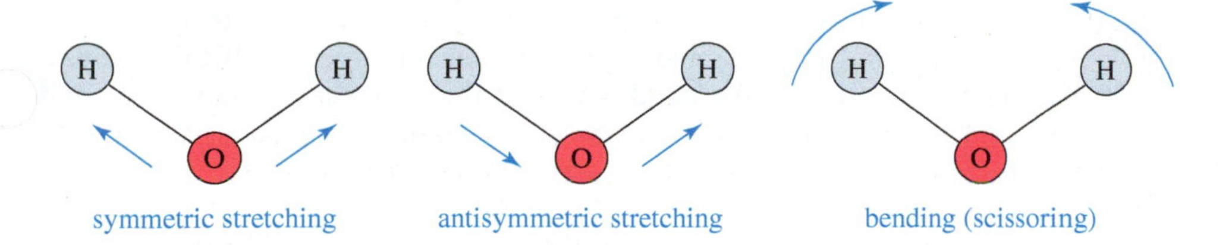

A nonlinear molecule with n atoms generally has $3n - 6$ fundamental vibrational modes. Water (3 atoms) has $3(3) - 6 = 3$ fundamental modes, as shown in the preceding figure. Methanol has $3(6) - 6 = 12$ fundamental modes, and ethanol has $3(9) - 6 = 21$ fundamental modes. We also observe combinations and multiples (overtones) of these simple fundamental vibrational modes. As you can see, the number of absorptions in an infrared spectrum can be quite large, even for simple molecules.

It is highly unlikely that the IR spectra of two different compounds (except enantiomers) will show the same frequencies for all their various complex vibrations. For this reason, the infrared spectrum provides a "fingerprint" of a molecule. In fact, the region of the IR spectrum containing most of these complex vibrations (600 to 1400 cm^{-1}) is commonly called the **fingerprint region** of the spectrum.

The simple stretching vibrations in the 1600 to 3500 cm^{-1} region are the most characteristic and predictable. Our study of infrared spectroscopy will concentrate on them. Although our introductory study of IR spectra will largely ignore bending vibrations, you should remember that these absorptions generally appear in the 600 to 1400 cm^{-1} region of the spectrum. Experienced spectroscopists can tell a great deal about the structure of a molecule from the various kinds of bending vibrations known as "wagging," "scissoring," "rocking," and "twisting" that appear in the fingerprint region (see Figure 12-2). The reference table of IR frequencies (Appendix 2) lists both stretching and bending characteristic frequencies.

12-5 IR-Active and IR-Inactive Vibrations

Not all molecular vibrations absorb infrared radiation. To understand which ones do and which do not, we need to consider how an electromagnetic field interacts with a molecular bond. The key to this interaction lies with the polarity of the bond, measured as its dipole moment. A bond with a dipole moment can be visualized as a positive charge and a negative charge separated by a spring. If this bond is placed in an electric field (Figure 12-3), it is either stretched or compressed, depending on the direction of the field.

One of the components of an electromagnetic wave is a rapidly reversing electric field ($\vec{E}$). This field alternately stretches and compresses a polar bond, as shown in Figure 12-3. When the electric field is in the same direction as the dipole moment, the

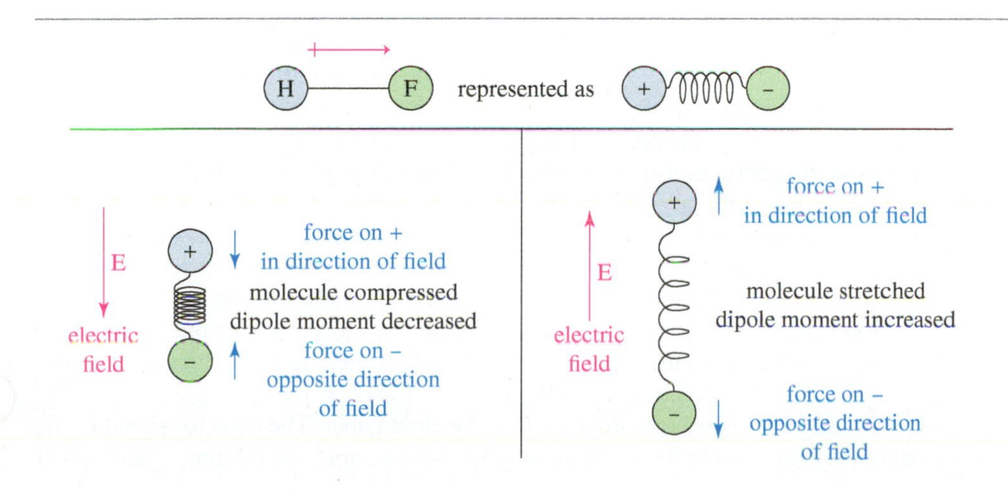

FIGURE 12-3
Effect of an electric field on a polar bond. A bond with a dipole moment (as in HF, for example) is either stretched or compressed by an electric field, depending on the direction of the field. Notice that the force on the positive charge is in the direction of the electric field ($\vec{\mathbf{E}}$), and the force on the negative charge is in the opposite direction.

bond is compressed and its dipole moment decreases. When the field is opposite the dipole moment, the bond stretches and its dipole moment increases. If this alternate stretching and compressing of the bond occurs at the frequency of the molecule's natural rate of vibration, energy may be absorbed. Vibrations of bonds with dipole moments generally result in IR absorptions and are said to be **IR-active.**

If a bond is symmetrical and has zero dipole moment, the electric field does not interact with the bond. For example, the triple bond of acetylene (H—C≡C—H) has zero dipole moment, and the dipole moment remains zero if the bond is stretched or compressed. Because the vibration produces no change in the dipole moment, there is no absorption of energy. This vibration is said to be **IR-inactive,** and it produces no absorption in the IR spectrum. The key to an IR-active vibration is that *the vibration must change the dipole moment of the molecule.*

In general, if a bond has a dipole moment, its stretching frequency causes an absorption in the IR spectrum. If a bond is symmetrically substituted and has zero dipole moment, its stretching vibration is weak or absent in the spectrum. Bonds with zero dipole moments sometimes produce absorptions (usually weak) because molecular collisions, rotations, and vibrations make them unsymmetrical part of the time. Strongly polar bonds (C=O groups, for example) may absorb so strongly that they also produce **overtone** peaks, which are relatively small peaks at a multiple (usually double) of the fundamental vibration frequency.

PROBLEM 12-2

Which of the bonds shown in red are expected to have IR-active stretching frequencies?

(a) H—C≡C—H **(b)** H—C≡C—H **(c)** H—C≡C—CH_3

(d) H_3C—C≡C—CH_3 **(e)** H_3C—C≡C—CH_3 **(f)** H_3C—CH_3

(g) H_3C—CH_2—H (central C bearing two H, shown vertically)

(h) $H_3C(H)C{=}C(H)CH_3$ (trans)

(i) $H_3C(H)C{=}C(H)CH_3$ (cis)

(j) $H_2C{=}C(H)CH_3$

12-6 Measurement of the IR Spectrum

Infrared spectra can be measured using liquid, solid, or gaseous samples that are placed in the beam of infrared light. A drop of a liquid can be placed as a thin film between two salt plates made of NaCl or KBr, which are transparent to infrared light at most important frequencies. A solid can be ground with KBr and pressed into a disk that is placed in the light beam. Alternatively, a solid sample can be ground into a pasty *mull* with paraffin oil. As with a liquid, the mull is placed between two salt plates. Solids can also be dissolved in common solvents such as CH_2Cl_2, CCl_4, or CS_2 that do not have absorptions in the areas of interest. Gases are placed in a longer cell with polished salt windows. These *gas cells* often contain mirrors that reflect the beam through the cell several times for stronger absorption.

An **infrared spectrometer** measures the frequencies of infrared light absorbed by a compound. In a simple infrared spectrometer (Figure 12-4), two beams of light are used. The *sample beam* passes through the sample cell, while the *reference beam* passes through a reference cell that contains only the solvent. A rotating mirror alternately allows light from each of the two beams to enter the monochromator.

The monochromator uses prisms or diffraction gratings to allow only one frequency of light to enter the detector at a time. It scans the range of infrared frequencies as a pen moves along the corresponding frequencies on the *x* axis of the chart paper. Higher frequencies (shorter wavelengths) appear toward the left of the chart paper. The detector signal is proportional to the *difference* in the intensity of light in the sample and reference beams, with

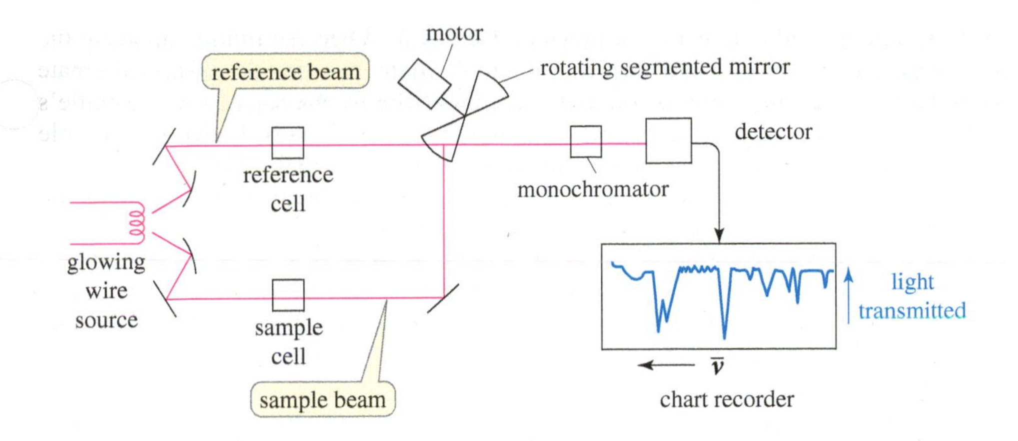

FIGURE 12-4
Block diagram of a dispersive infrared spectrometer. The *sample beam* passes through the sample cell while the *reference beam* passes through a reference cell that contains only the solvent. A rotating mirror alternately allows light from each of the two beams to enter the monochromator where they are compared. The chart recorder graphs the difference in light transmittance between the two beams.

the reference beam compensating for any absorption by air or by the solvent. The detector signal controls movement of the pen along the *y* axis, with 100% transmittance (no absorption) at the top of the paper, and 0% transmittance (absorption of all the light) at the bottom.

The spectrometer shown in Figure 12-4 is called a *dispersive* instrument because it disperses light into all the different frequencies and measures them individually. Dispersive instruments require expensive prisms and diffraction gratings, and they must be manually aligned and calibrated on a regular basis. Since only one frequency is observed at a time, dispersive instruments require strong IR sources, and they require 2 to 10 minutes to scan through a complete spectrum. Dispersive infrared spectrometers are being replaced by Fourier transform infrared (FT–IR) spectrometers for most uses.

A **Fourier transform infrared spectrometer (FT–IR)** uses an **interferometer,** like that shown in Figure 12-5, to measure an IR spectrum. The infrared light goes from the glowing source to a beamsplitter, usually made of polished KBr, placed at a 45° angle. Part of the beam passes through the beamsplitter, and part is reflected at a right angle. The reflected beam strikes a stationary mirror, while the transmitted beam strikes a mirror that moves at a constant speed. The beams return from the mirrors to recombine at the beamsplitter. The beam from the moving mirror has traveled a different distance than the beam from the fixed mirror, and the two beams combine to create an interference pattern called an **interferogram.** This interferogram, which simultaneously contains all frequencies, passes through the sample compartment to reach the detector.

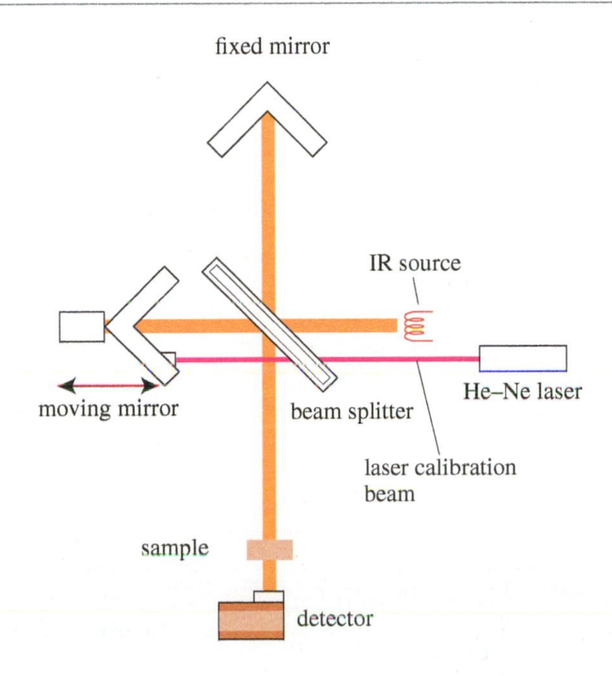

FIGURE 12-5
Block diagram of an interferometer in an FT–IR spectrometer. The light beams reflected from the fixed and moving mirrors are combined to form an interferogram, which passes through the sample to enter the detector.

The interferogram shown in the upper half of Figure 12-6 contains all the information contained in the spectrum shown in the lower half. The interferogram is said to be in the *time domain*, corresponding to the energy seen by the detector as the mirror moves through the signal. A standard computer algorithm called a Fourier transform converts the time domain to the *frequency domain* spectrum that allows us to see the strength of absorption as a function of the frequency (or wavelength). Figure 12-6 shows both the interferogram and the IR spectrum of *n*-octane.

The FT–IR spectrometer has several major advantages over the dispersive instrument: Its sensitivity is better because it measures all frequencies simultaneously rather than scanning through the individual frequencies. Less energy is needed from the source, and less time (typically 1 to 2 seconds) is needed for a scan. Several scans can be completed in a few seconds and averaged to improve the signal. Resolution and accuracy are also improved because a laser beam is used alongside the IR beam to control the speed of the moving mirror and to time the collection of data points. The laser beam is a precise frequency reference that keeps the spectrometer accurately calibrated.

In the infrared spectrum of *n*-octane [Figure 12-6(b)] are four major absorption bands. The broad band between 2800 and 3000 cm^{-1} results from C—H stretching vibrations, and the band at 1467 cm^{-1} results from a scissoring vibration of the CH_2 groups. The absorptions at 1378 and 722 cm^{-1} result from the bending vibrations (rocking) of CH_3 and CH_2 groups, respectively. Because most organic compounds contain at least some saturated C—H bonds and some CH_2 and CH_3 groups, all these bands are common. In fact, without an authentic spectrum for comparison, *we could not look at this spectrum and conclude that the compound is octane*. We could be fairly certain that it is an alkane, however, because we see no absorption bands corresponding to other functional groups.

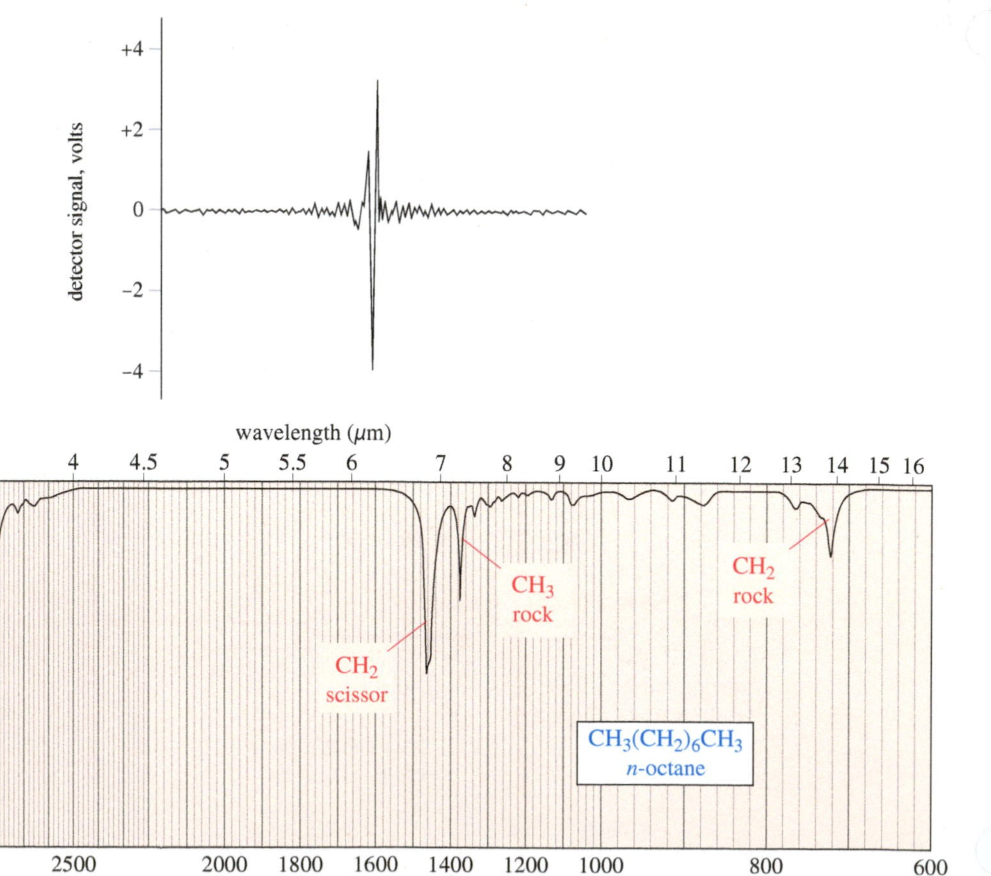

FIGURE 12-6
(a) Interferogram generated by *n*-octane. (b) Infrared spectrum of *n*-octane. Notice that the frequencies shown in a routine IR spectrum range from about 600 cm^{-1} to about 4000 cm^{-1}.

Another characteristic in the octane spectrum is the absence of any identifiable C—C stretching absorptions. (Table 12-1 shows that C—C stretching absorptions occur around 1200 cm^{-1}.) Although there are seven C—C bonds in octane, their dipole moments are small, and their absorptions are weak and indistinguishable. This result is common for alkanes with no functional groups to polarize the C—C bonds.

12-7 Infrared Spectroscopy of Hydrocarbons

Hydrocarbons contain only carbon–carbon bonds and carbon–hydrogen bonds. An infrared spectrum does not provide enough information to identify a structure conclusively (unless an authentic spectrum is available to compare "fingerprints"), but the absorptions of the carbon–carbon and carbon–hydrogen bonds can indicate the presence of double and triple bonds.

12-7A Carbon–Carbon Bond Stretching

Stronger bonds generally absorb at higher frequencies because of their greater stiffness. Carbon–carbon single bonds absorb around 1200 cm^{-1}, C=C double bonds absorb around 1660 cm^{-1}, and C≡C triple bonds absorb around 2200 cm^{-1}.

Carbon–carbon bond stretching frequencies	
C—C	1200 cm^{-1}
C=C	1660 cm^{-1}
C≡C	<2200 cm^{-1}

As discussed for the octane spectrum, C—C single bond absorptions (and most other absorptions in the fingerprint region) are not very reliable. We use the fingerprint region primarily to confirm the identity of an unknown compound by comparison with an authentic spectrum.

The absorptions of C=C double bonds, however, are useful for structure determination. Most unsymmetrically substituted double bonds produce observable stretching absorptions in the region of 1600 to 1680 cm^{-1}. The specific frequency of the double-bond stretching vibration depends on whether there is another double bond nearby. When two double bonds are one bond apart (as in cyclohexa-1,3-diene on the following figure) they are said to be **conjugated.** As we will see in Chapter 15, conjugated double bonds are slightly more stable than isolated double bonds because there is a small amount of pi bonding between them. This overlap between the pi bonds leaves a little less electron density in the double bonds themselves. As a result, they are a little less stiff and vibrate a little more slowly than an isolated double bond. Isolated double bonds absorb around 1640 to 1680 cm^{-1}, while conjugated double bonds absorb around 1620 to 1640 cm^{-1}.

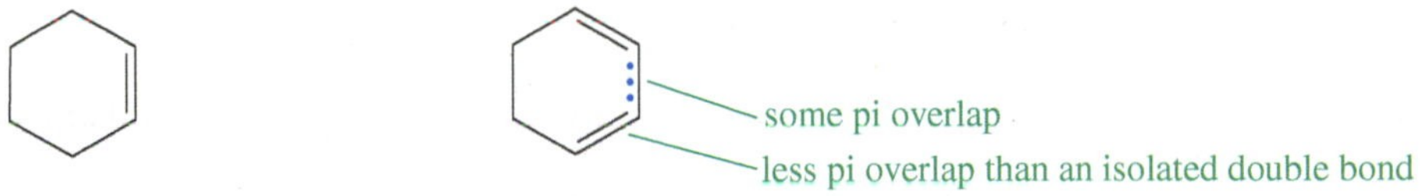

some pi overlap
less pi overlap than an isolated double bond

1645 cm^{-1}
cyclohexene (isolated)

1620 cm^{-1}
cyclohexa-1,3-diene (conjugated)

The effect of conjugation is even more pronounced in aromatic compounds, which have three conjugated double bonds in a six-membered ring. Aromatic C=C bonds are more like $1\frac{1}{2}$ bonds than true double bonds, and their reduced pi bonding results in less stiff bonds with lower stretching frequencies, around 1600 cm^{-1}.

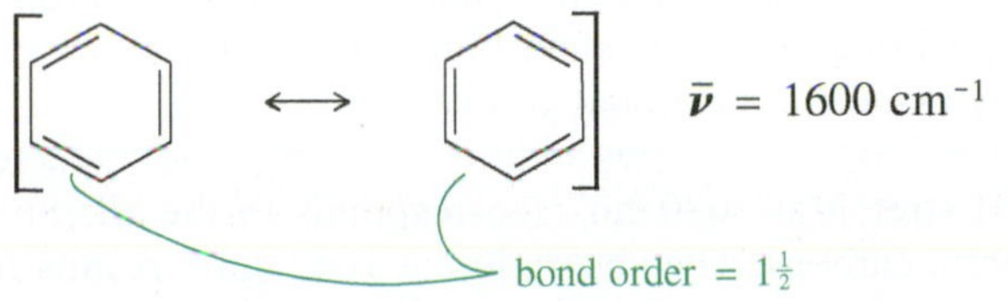

$\bar{\nu} = 1600\ cm^{-1}$

bond order = $1\frac{1}{2}$

PROBLEM-SOLVING HINT

The resonance delocalization of electrons in aromatic rings can be represented by a circle. This aromatic circle is a common shorthand among organic chemists.

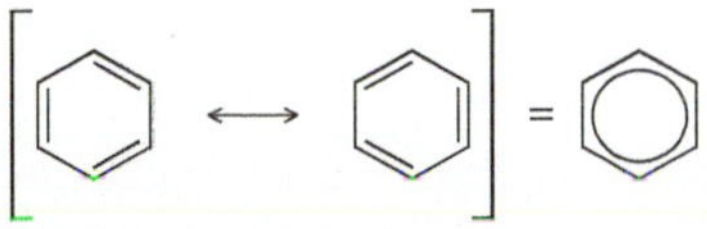

Characteristic C=C stretching frequencies	
isolated C=C	1640–1680 cm^{-1}
conjugated C=C	1620–1640 cm^{-1}
aromatic C=C	approx. 1600 cm^{-1}

Carbon–carbon triple bonds in alkynes are stronger and stiffer than carbon–carbon single or double bonds, and they absorb infrared light at higher frequencies. Most alkyne C≡C triple bonds have stretching frequencies between 2100 and 2200 cm^{-1}. Terminal alkynes usually give sharp C≡C stretching signals of moderate intensity. The C≡C stretching absorption of an internal alkyne may be weak or absent, however, because of the symmetry of the disubstituted triple bond with a very small or zero dipole moment.

μ

R—C≡C—H — C≡C stretch observed around 2100 to 2200 cm^{-1}
terminal alkyne

μ is small or zero

R—C≡C—R′ — C≡C stretch may be weak or absent
internal alkyne

12-7B Carbon–Hydrogen Bond Stretching

Alkanes, alkenes, and alkynes also have characteristic C—H stretching frequencies. Carbon–hydrogen bonds involving sp^3 hybrid carbon atoms generally absorb at frequencies just *below* (to the right of) 3000 cm^{-1}. Those involving sp^2 hybrid carbons absorb just *above* (to the left of) 3000 cm^{-1}. We explain this difference by the amount of *s* character in the carbon orbital used to form the bond. The *s* orbital is closer to the nucleus than the *p* orbitals, and stronger, stiffer bonds result from orbitals with more *s* character. Even if an alkene's C=C absorption is weak or absent, the unsaturated C—H stretch above 3000 cm^{-1} reveals the presence of the double bond.

An sp^3 orbital is one-fourth *s* character, and an sp^2 orbital is one-third *s* character. We expect the bond using the sp^2 orbital to be slightly stronger, with a higher vibration frequency. The C—H bond of a terminal alkyne is formed using an *sp* hybrid orbital, with about one-half *s* character. This bond is stiffer than a C—H bond using an sp^3 or sp^2 hybrid carbon, and it absorbs at a higher frequency: about 3300 cm^{-1}.

C—H bond stretching frequencies: sp > sp^2 > sp^3

—C—C—H (alkane)	sp^3 hybridized, one-fourth *s* character	2800–3000 cm^{-1}
C=C—H (alkene)	sp^2 hybridized, one-third *s* character	3000–3100 cm^{-1}
—C≡C—H	*sp* hybridized, one-half *s* character	3300 cm^{-1} (sharp)

12-7C Interpreting the IR Spectra of Hydrocarbons

Figure 12-7 compares the IR spectra of hexane, hex-1-ene, and *cis*-oct-2-ene. The hexane spectrum is similar to that of *n*-octane (Figure 12-6). The C—H stretching frequencies form a band between 2800 and 3000 cm^{-1}, and the bands in the fingerprint region are due to the bending vibrations discussed for Figure 12-6. This spectrum simply indicates the *absence* of any IR-active functional groups.

The spectrum of hex-1-ene shows additional absorptions characteristic of a double bond. The C—H stretch at 3080 cm^{-1} corresponds to the alkene =C—H bonds involving sp^2 hybrid carbons. The absorption at 1642 cm^{-1} results from stretching of

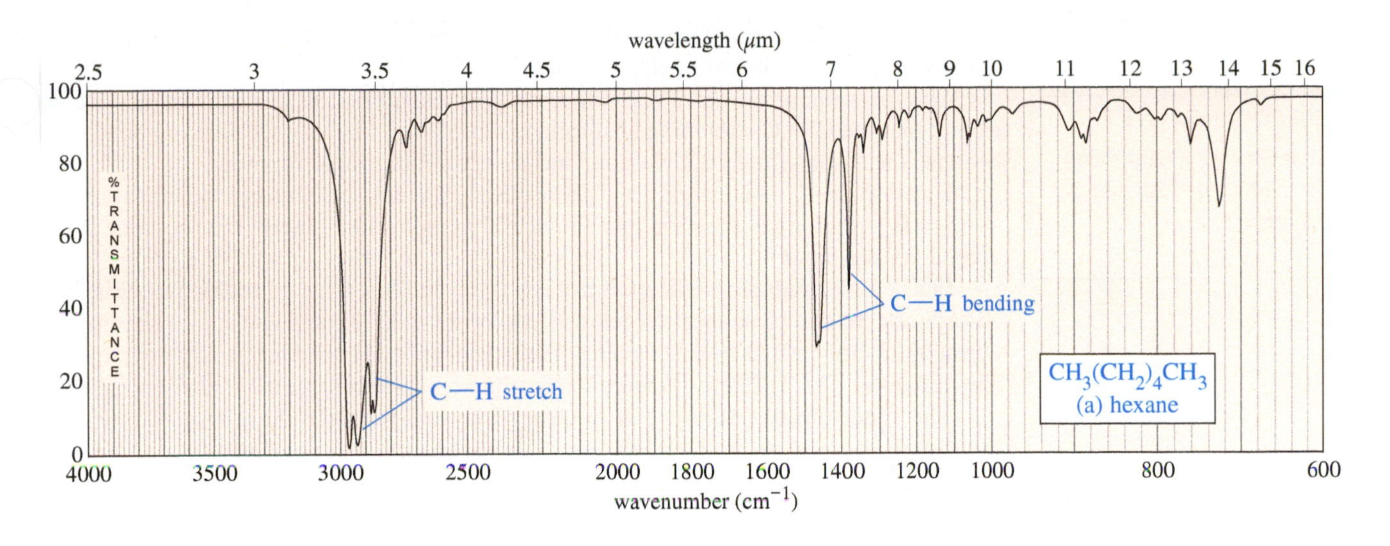

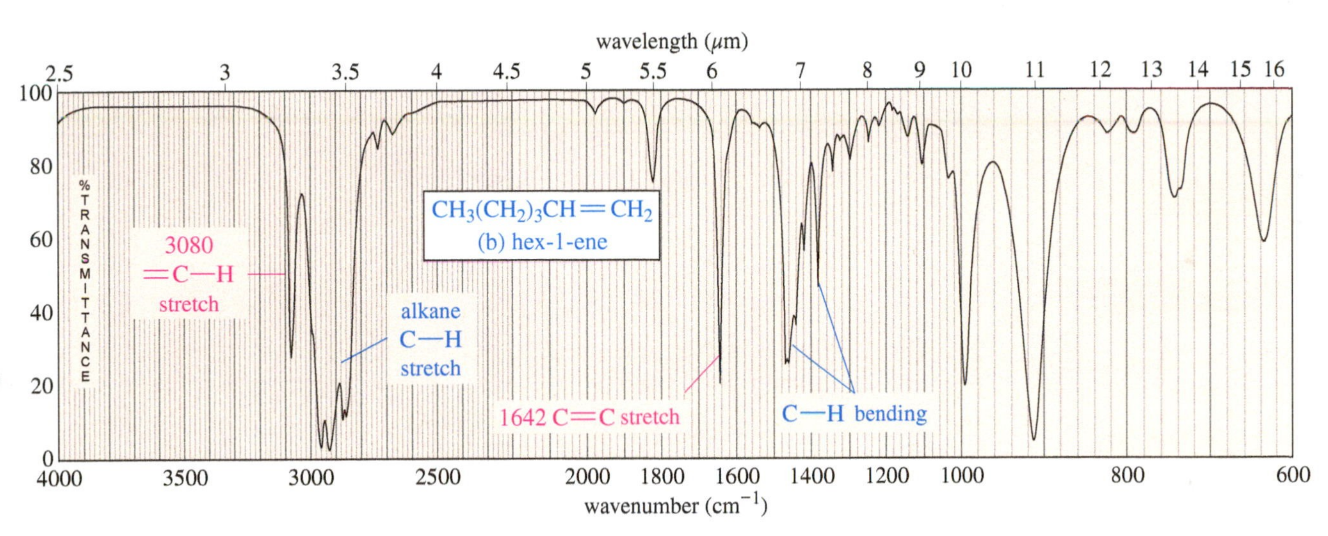

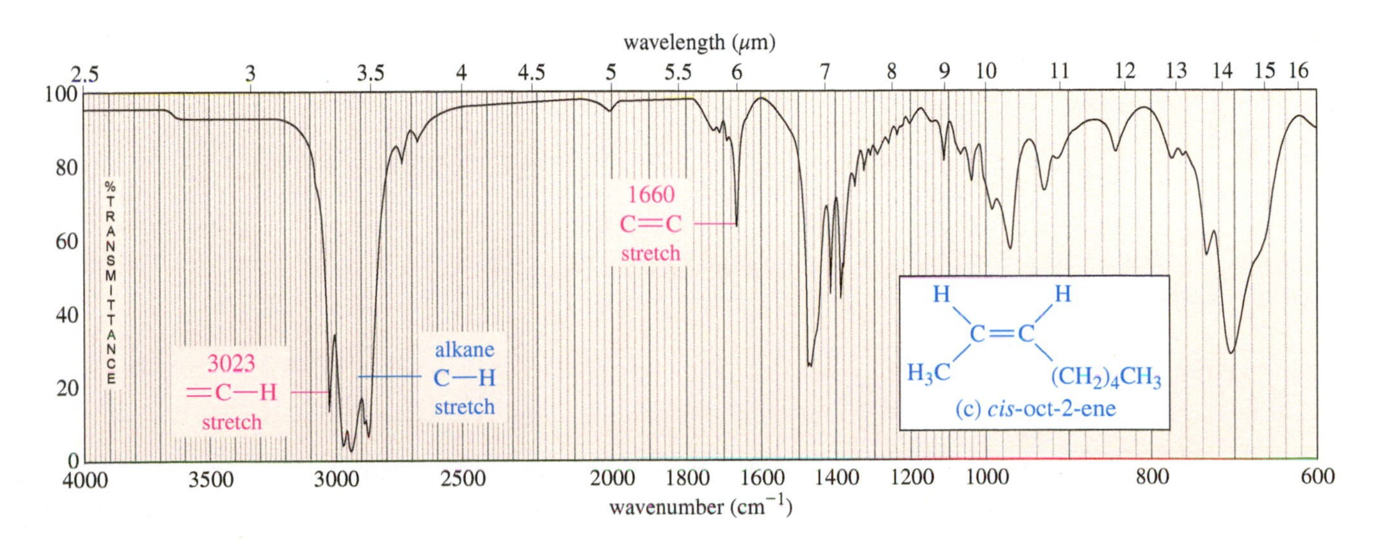

FIGURE 12-7
Comparison of the IR spectra of (a) hexane, (b) hex-1-ene, and (c) *cis*-oct-2-ene. The most characteristic absorptions in the hex-1-ene spectrum are the C=C stretch at 1642 cm^{-1} and the unsaturated =C—H stretch at 3080 cm^{-1}. The nearly symmetrically substituted double bond in *cis*-oct-2-ene gives a weak C=C absorption at 1660 cm^{-1}. The unsaturated =C—H stretch at 3023 cm^{-1} is still apparent, however.

the C═C double bond. (The small peak at 1820 cm^{-1} is likely an overtone at double the frequency of the intense peak at 910 cm^{-1}.)

The spectrum of *cis*-oct-2-ene (Figure 12-7c) resembles the spectrum of hex-1-ene, except that the C═C stretching absorption at 1660 cm^{-1} is very weak in *cis*-oct-2-ene because the disubstituted double bond has a very small dipole moment.

Even if the C═C stretching absorption is weak or absent, the unsaturated ═C—H stretching absorption just above 3000 cm^{-1} still suggests the presence of an alkene double bond.

PROBLEM-SOLVING HINT

The unsaturated ═C—H stretch, to the left of 3000 cm^{-1}, should alert you to look for a weak C═C stretch.

Figure 12-8 compares the IR spectra of oct-1-yne and oct-4-yne. In addition to the alkane absorptions, the oct-1-yne spectrum shows sharp peaks at 3313 and 2119 cm^{-1}. The absorption at 3313 cm^{-1} results from stretching of the stiff ≡C—H bond formed by the *sp* hybrid alkyne carbon. The 2119 cm^{-1} absorption results from stretching of the C≡C triple bond.

The spectrum of oct-4-yne is not very helpful. Because there is no acetylenic hydrogen, there is no ≡C—H stretching absorption around 3300 cm^{-1}. There is no visible C≡C stretching absorption around 2100 to 2200 cm^{-1} either, because the disubstituted triple bond has a very small dipole moment. This spectrum fails to alert us to the presence of a triple bond.

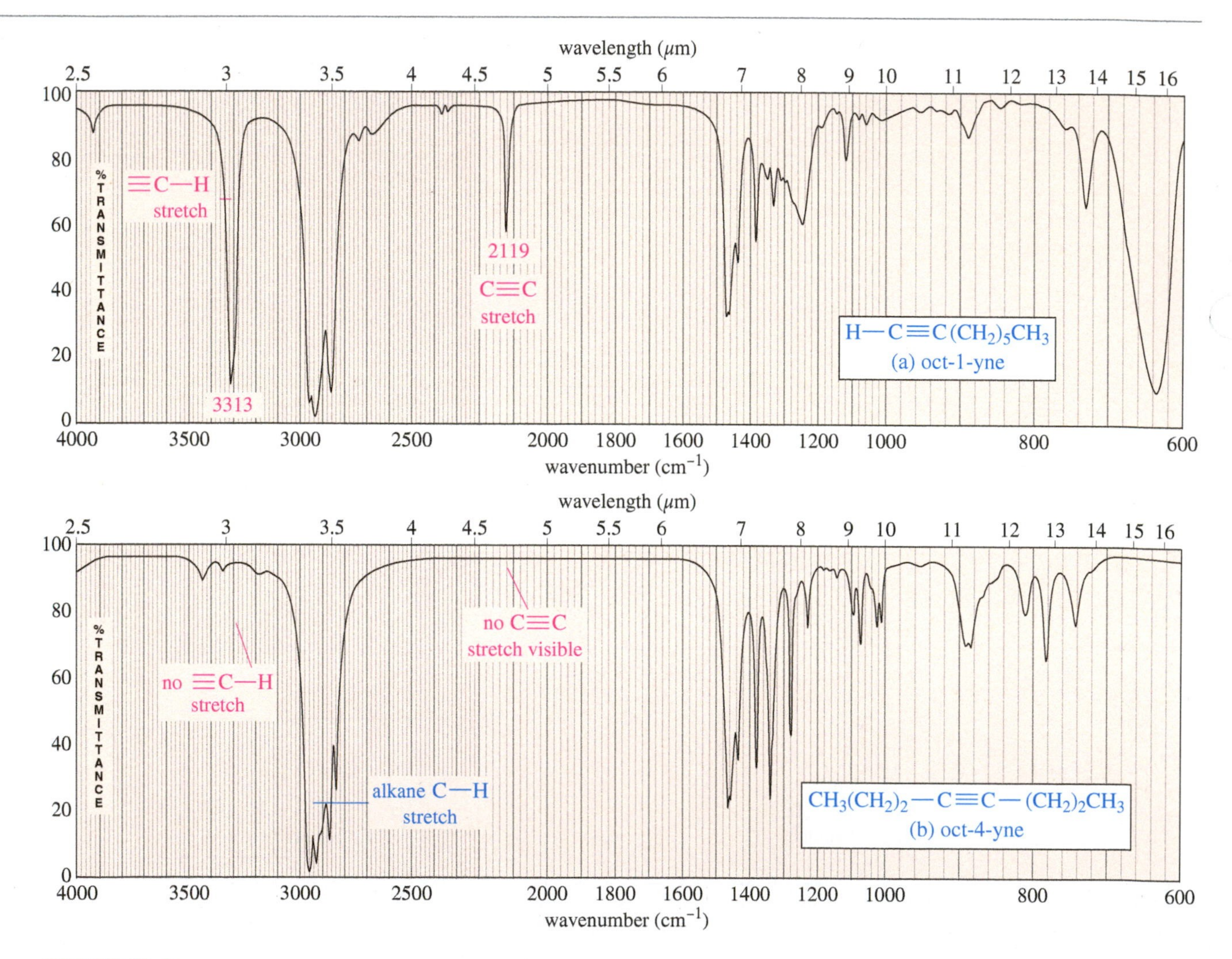

FIGURE 12-8
Comparison of the IR spectra of oct-1-yne and oct-4-yne.
(a) The IR spectrum of oct-1-yne shows characteristic absorptions at 3313 cm^{-1} (alkynyl ≡C—H stretch) and at 2119 cm^{-1} (C≡C stretch).
(b) We cannot tell that oct-4-yne is an alkyne from its IR spectrum because it displays neither of the characteristic absorptions seen in (a). There is no alkynyl ≡C—H bond, and its symmetrically substituted triple bond has too small a dipole moment to produce the C≡C stretching absorption seen in the spectrum of oct-1-yne.

PROBLEM 12-3

For each hydrocarbon spectrum, determine whether the compound is an alkane, an alkene, an alkyne, or an aromatic hydrocarbon, and assign the major peaks above (to the left of) 1600 cm^{-1}. More than one unsaturated group may be present.

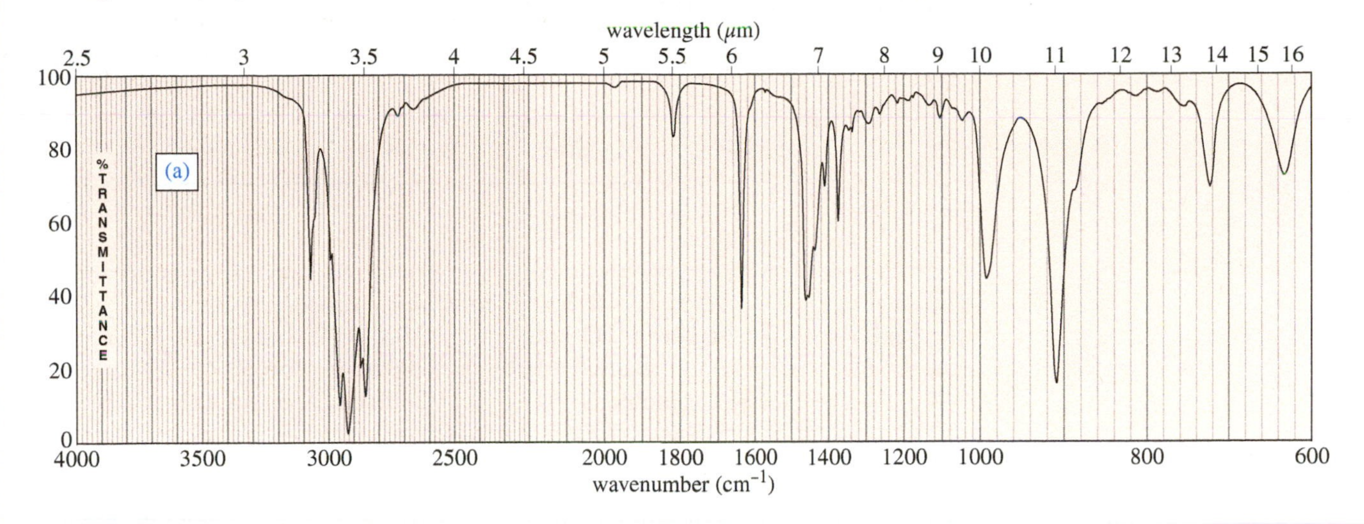

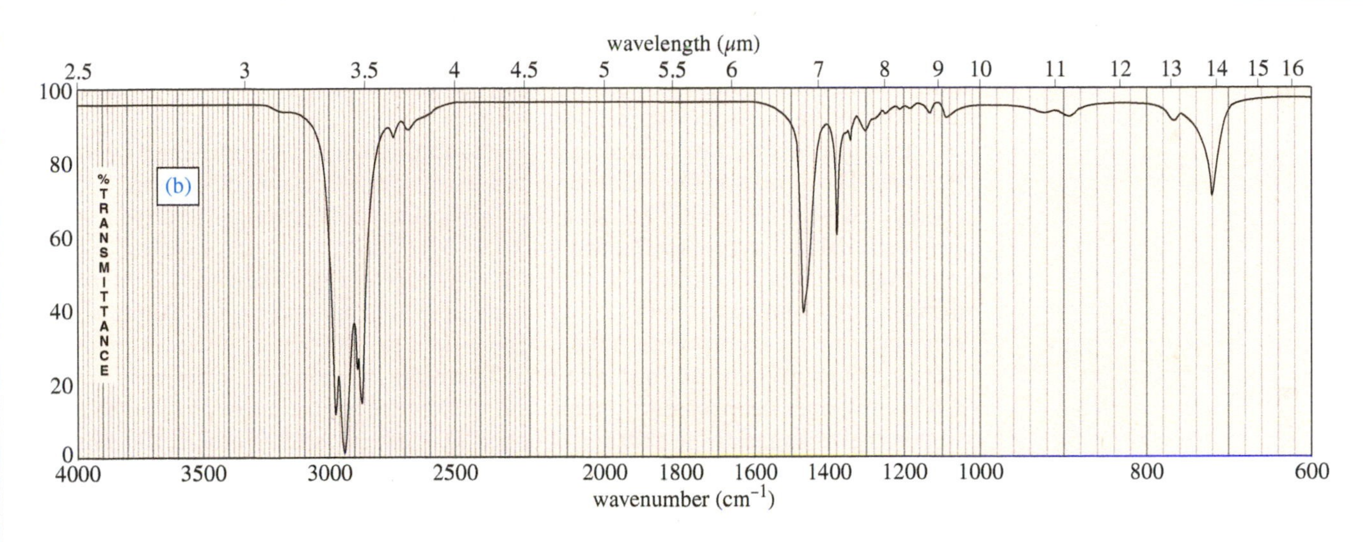

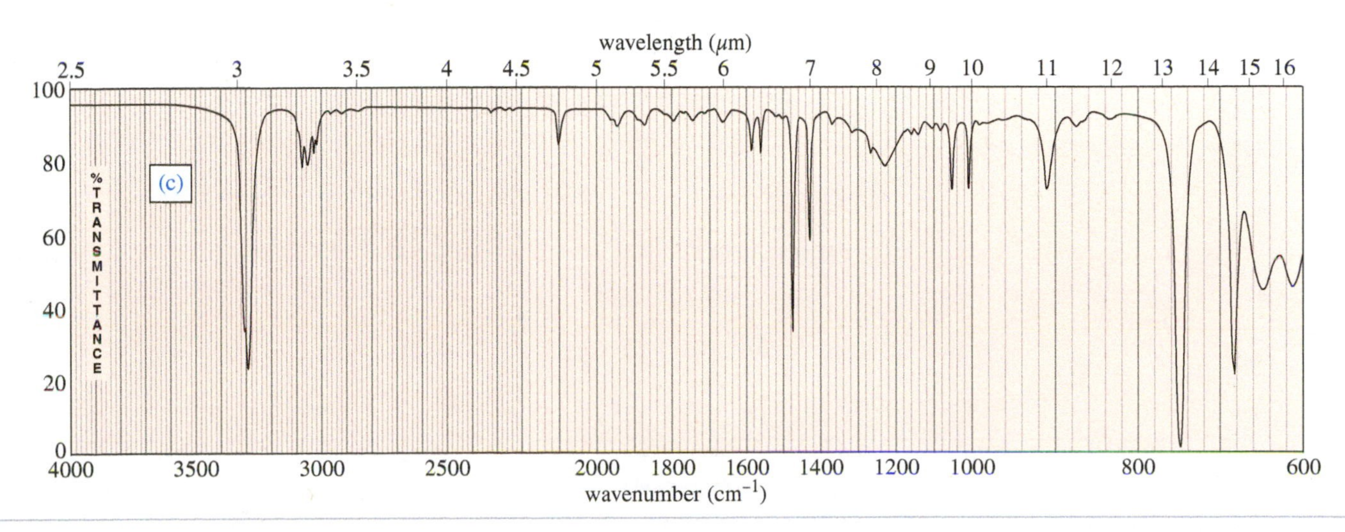

12-8 Characteristic Absorptions of Alcohols and Amines

The O—H bonds of alcohols and the N—H bonds of amines are strong and stiff. The vibration frequencies of O—H and N—H bonds therefore occur at higher frequencies than those of most C—H bonds (except for alkynyl ≡C—H bonds).

R—O—H	R—NH—H	R—NH—R′	R—NR″—R′
alcohol	primary	secondary	tertiary
	amines		

O—H and N—H stretching frequencies	
alcohol O—H	3300 cm^{-1}, broad
acid O—H	3000 cm^{-1}, broad
amine N—H	3300 cm^{-1}, broad with spikes

Alcohol O—H bonds absorb over a wide range of frequencies, centered around 3300 cm^{-1}. Alcohol molecules are involved in hydrogen bonding, with different molecules having different instantaneous arrangements. The O—H stretching frequencies reflect this diversity of hydrogen-bonding arrangements, resulting in very broad absorptions. Notice the broad O—H absorption centered around 3300 cm^{-1} in the infrared spectrum of butan-1-ol (Figure 12-9).

Like alcohols, carboxylic acids give O—H absorptions that are broadened by hydrogen bonding. The broad acid O—H absorption is usually centered around 3000 cm^{-1}, however (compared with 3300 cm^{-1} for an alcohol), because of the stronger hydrogen bonding between acid molecules (see Section 12-9A).

Figure 12-9 also shows a strong C—O stretching absorption centered near 1060 cm^{-1}. Compounds with C—O bonds (alcohols and ethers, for example) generally show strong absorptions in the range of 1000 to 1200 cm^{-1}; however, other functional groups also absorb in this region. Therefore, a strong peak between 1000 and 1200 cm^{-1} does not necessarily imply a C—O bond, but the *absence* of an absorption in this region suggests the absence of a C—O bond. For simple ethers, this unreliable C—O absorption is usually the only clue that the compound might be an ether.

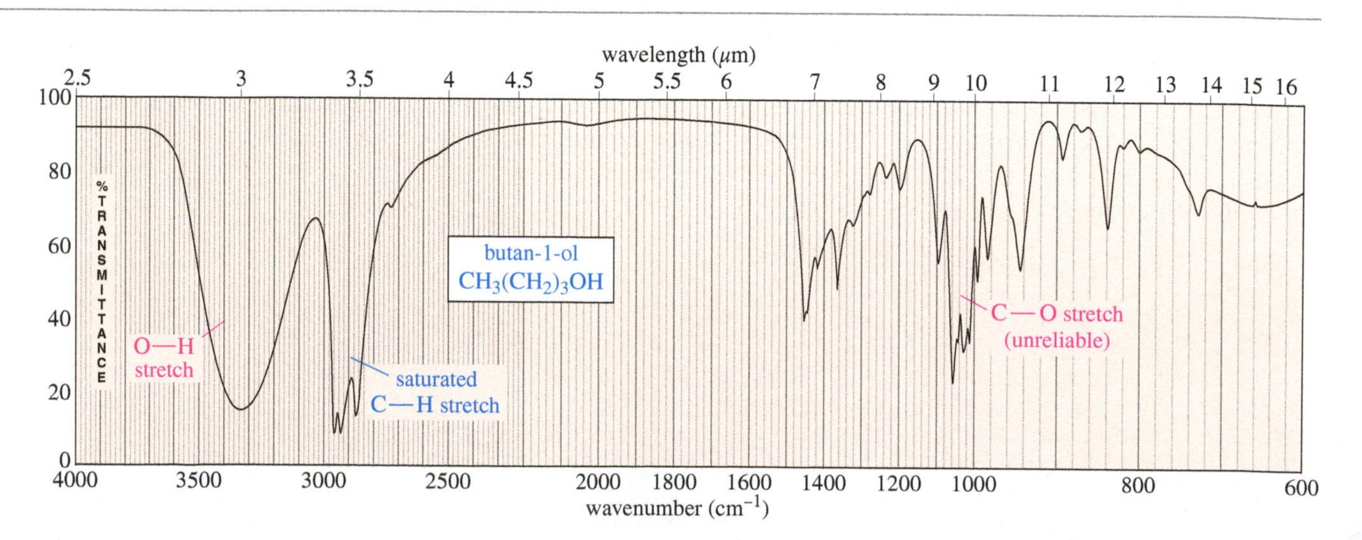

FIGURE 12-9
The IR spectrum of butan-1-ol shows a broad, intense O—H stretching absorption centered around 3300 cm^{-1}. The broad shape is due to the diverse nature of the hydrogen-bonding interactions of alcohol molecules.

Amine N—H bonds also have stretching frequencies that are in the 3300 cm^{-1} region, or even slightly higher. Like alcohols, amines participate in hydrogen bonding that can broaden the N—H absorptions. With amines, however, the absorption is somewhat weaker, and there may be one or more sharp spikes superimposed on the broad N—H stretching absorption—often one N—H spike for the single N—H bond of a secondary amine (R_2NH) and two N—H spikes for the symmetric and antisymmetric stretch of the two N—H bonds in a primary amine (RNH_2). These sharp spikes, combined with the presence of nitrogen in the molecular formula, help to distinguish amines from alcohols. Tertiary amines (R_3N) have no N—H bonds, and they do not give rise to N—H stretching absorptions in the IR spectrum. Figure 12-10 shows the spectrum of dipropylamine, a secondary amine.

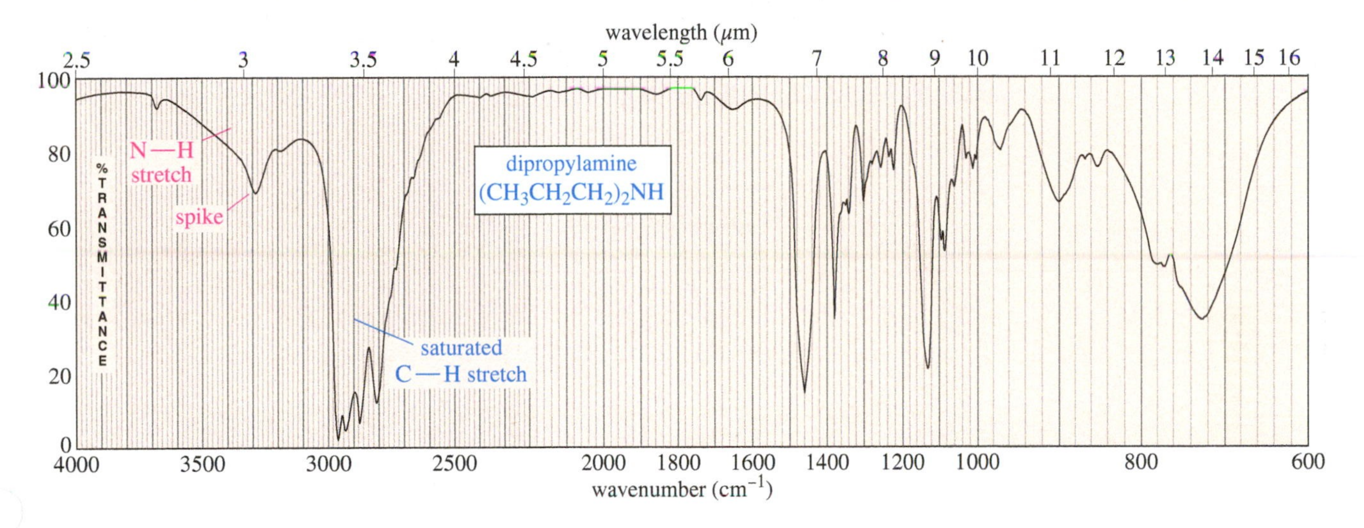

FIGURE 12-10
The IR spectrum of dipropylamine shows a broad N—H stretching absorption centered around 3300 cm^{-1}. Notice the spike in this broad absorption.

12-9 Characteristic Absorptions of Carbonyl Compounds

Because it has a large dipole moment, the C=O double bond produces intense infrared stretching absorptions. Carbonyl groups absorb at frequencies around 1700 cm^{-1}, but the exact frequency depends on the specific functional group and the rest of the molecule. For these reasons, infrared spectroscopy is often the best method for detecting and identifying the type of carbonyl group in an unknown compound. To simplify our discussion of carbonyl absorptions, we first consider the "normal" stretching frequencies for simple ketones, aldehydes, and carboxylic acids, and then we examine the types of carbonyl groups that deviate from this frequency.

12-9A Simple Ketones, Aldehydes, and Acids

The C=O stretching vibrations of simple ketones and carboxylic acids occur at frequencies around 1710 cm^{-1}. Aldehydes are a little higher, about 1725 cm^{-1}. These frequencies are higher than those for C=C double bonds because the C=O double bond is stronger and stiffer. Carbonyl absorptions may be so intense that they produce small overtone peaks around 3400 cm^{-1}, double their fundamental frequency.

R—C(=O)—R′ (C=O: 1710 cm^{-1})
ketone

R—C(=O)—H (C=O: 1725 cm^{-1}; C—H: 2700, 2800 cm^{-1})
aldehyde

R—C(=O)—O—H (C=O: 1710 cm^{-1}; O—H: broad, 2500–3500 cm^{-1})
acid

PROBLEM-SOLVING HINT

Real spectra are rarely perfect. Samples often contain traces of water, giving weak absorptions in the O—H region.

Many compounds oxidize in air. For example, alcohols often give weak C=O absorptions from oxidized impurities.

In addition to the strong C=O stretching absorption, an aldehyde shows a characteristic set of two low-frequency C—H stretching frequencies around 2700 and 2800 cm^{-1}. Neither a ketone nor an acid produces absorptions at these positions. Figure 12-11 compares the IR spectra of a ketone and an aldehyde. Notice the characteristic carbonyl stretching absorptions in both spectra, as well as the aldehyde C—H absorptions at 2720 and 2820 cm^{-1} in the butyraldehyde spectrum. Both spectra in Figure 12-11 also show small overtone peaks around 3400 cm^{-1}, double their carbonyl frequencies.

A carboxylic acid produces a characteristic broad O—H absorption in addition to the intense carbonyl stretching absorption (Figure 12-12). Because of the unusually strong hydrogen bonding in carboxylic acids, the broad O—H stretching frequency is shifted to about 3000 cm^{-1}, centered on top of the usual C—H absorption. This broad O—H absorption (which may have a shoulder or small spikes around 2500–2700 cm^{-1}) gives a characteristic overinflated shape to the peaks in the C—H stretching region. Participation of the acid carbonyl group in hydrogen bonding frequently results in broadening of the strong carbonyl absorption as well.

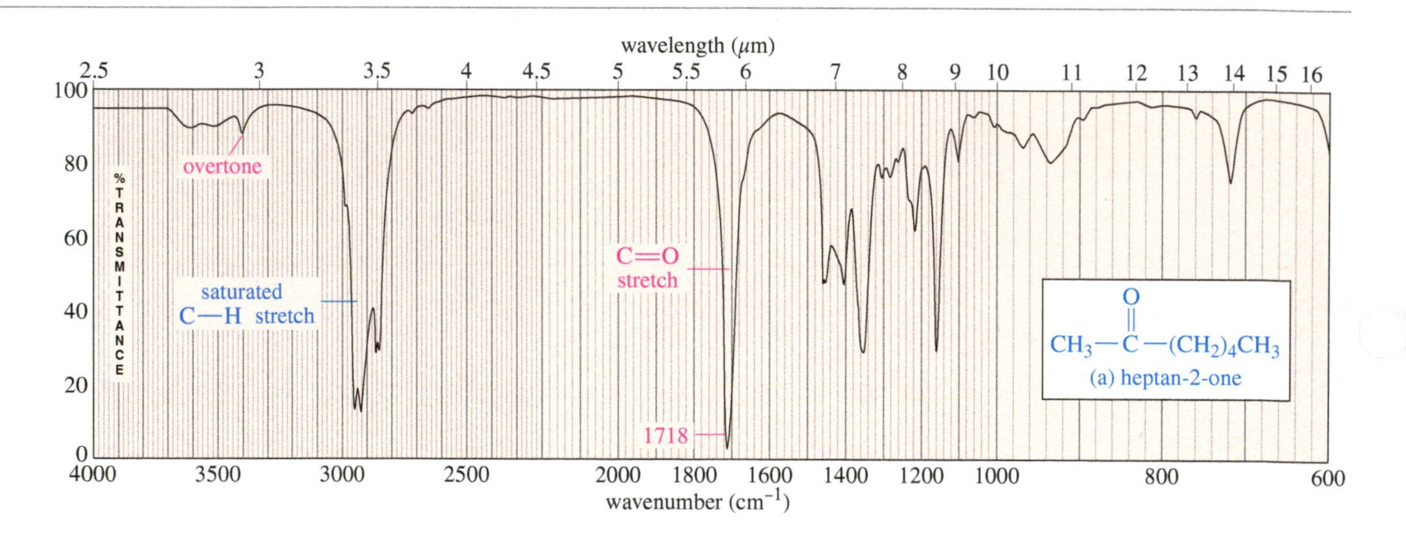

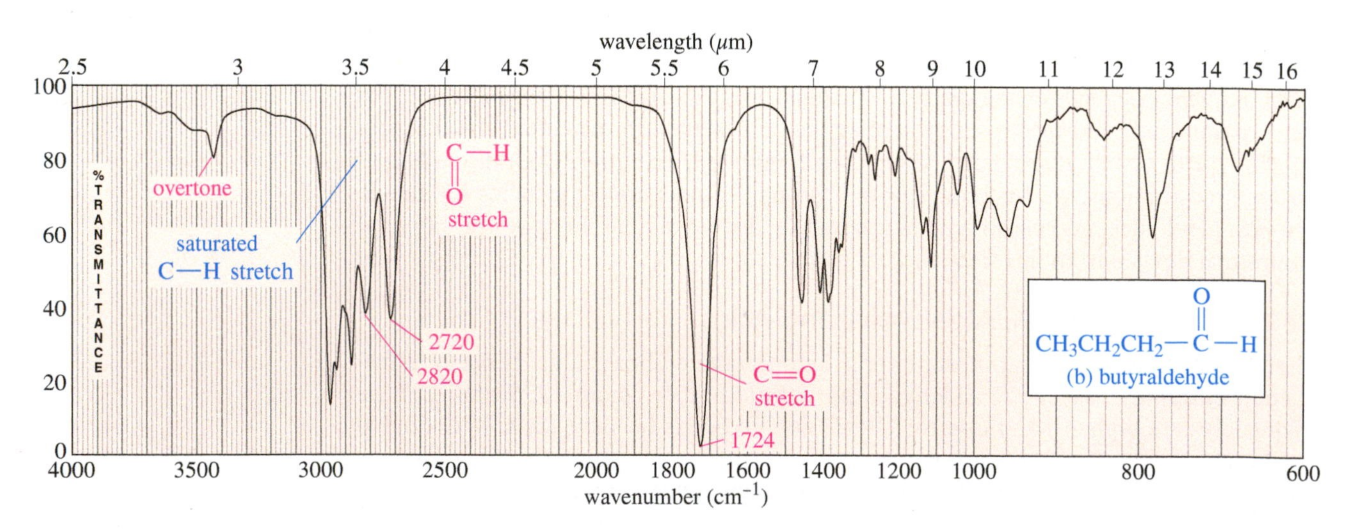

FIGURE 12-11
Infrared spectra of (a) heptan-2-one and (b) butyraldehyde. Both the ketone and the aldehyde show intense carbonyl absorptions near 1720 cm^{-1}. In the aldehyde spectrum, there are two peaks (2720 and 2820 cm^{-1}) characteristic of the aldehyde C—H stretch.

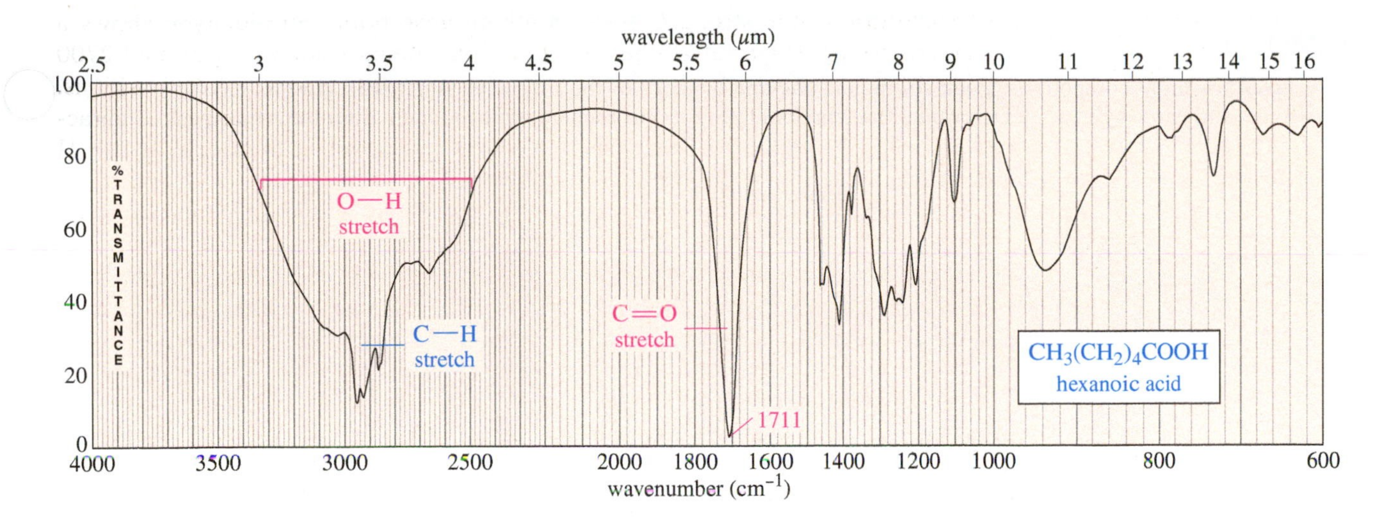

1710 cm^{-1}

broad, about 3000 cm^{-1}

FIGURE 12-12
Infrared spectrum of hexanoic acid. Carboxylic acids show a broad O—H absorption from about 2500 to 3500 cm^{-1}. This broad absorption gives the entire C—H stretching region a broad appearance, punctuated by sharper C—H stretching absorptions.

SOLVED PROBLEM 12-1

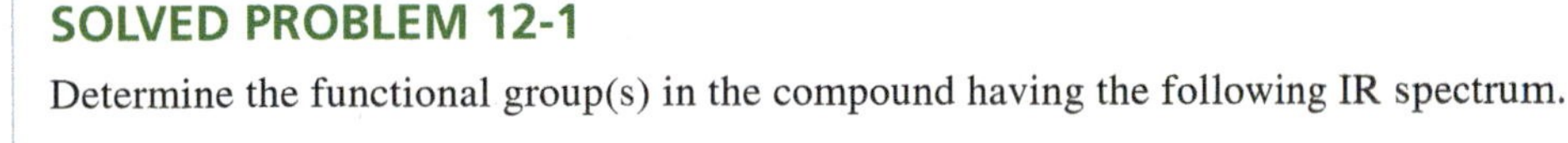

Determine the functional group(s) in the compound having the following IR spectrum.

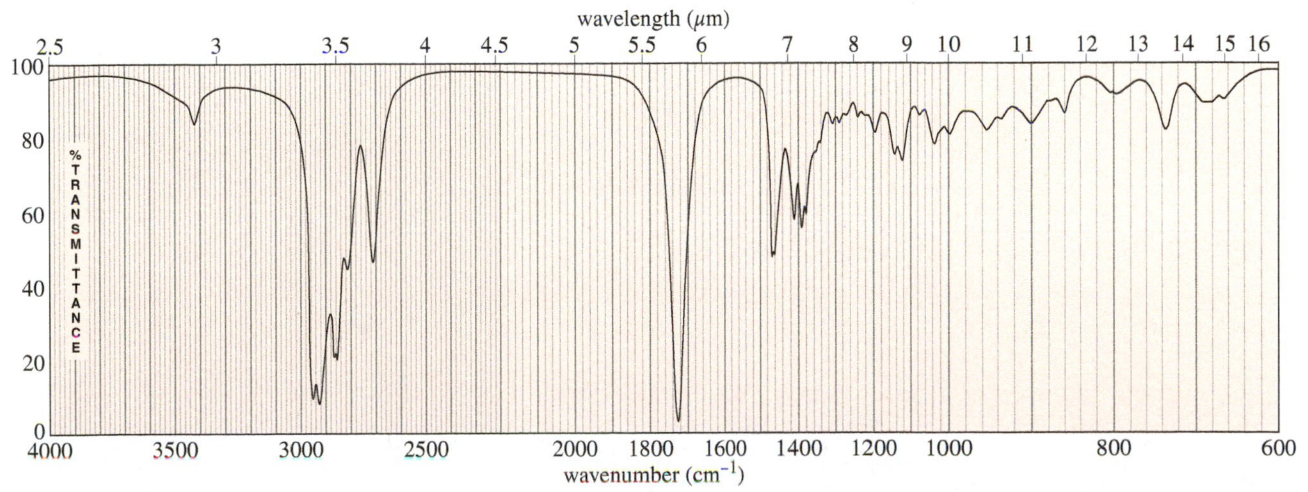

SOLUTION
First, look at the spectrum and see what peaks (outside the fingerprint region) don't look like alkane peaks: a weak peak around 3420 cm^{-1}, a strong peak about 1725 cm^{-1}, and an unusual C—H stretching region. The C—H region has two additional peaks around 2720 and 2820 cm^{-1}. The strong peak at 1725 cm^{-1} must be a C=O, and the peaks at 2720 and 2820 cm^{-1} suggest an aldehyde. The weak peak around 3420 cm^{-1} might be mistaken for an alcohol O—H. From experience, we know alcohols give much stronger, smoother O—H absorptions. This small peak is probably an overtone of the intense C=O absorption. Many IR spectra show small absorptions in the O—H region from overtones, water, or other impurities.

PROBLEM 12-4

Spectra are given for three compounds. Each compound has one or more of the following functional groups: alcohol, amine, ketone, aldehyde, and carboxylic acid. Determine the functional group(s) in each compound, and assign the major peaks above 1600 cm^{-1}.

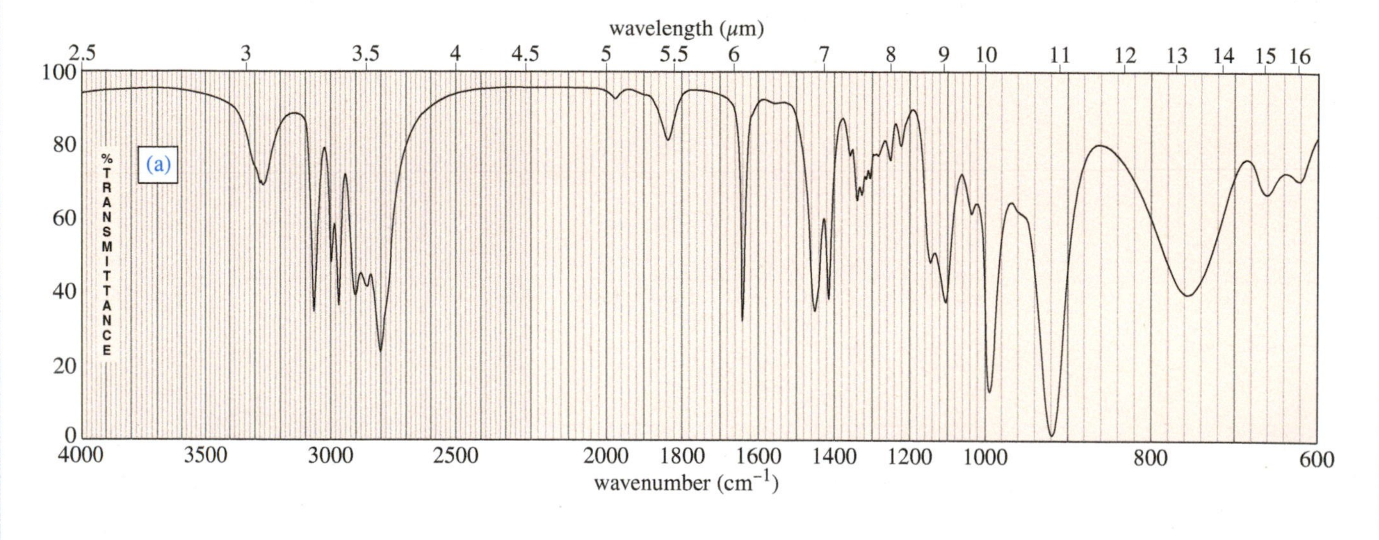

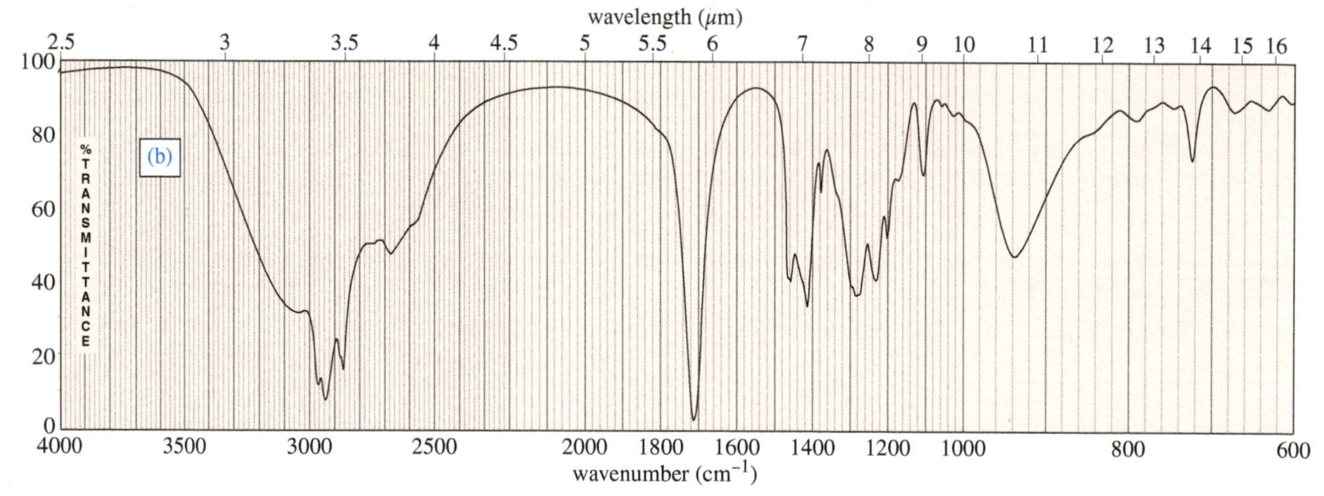

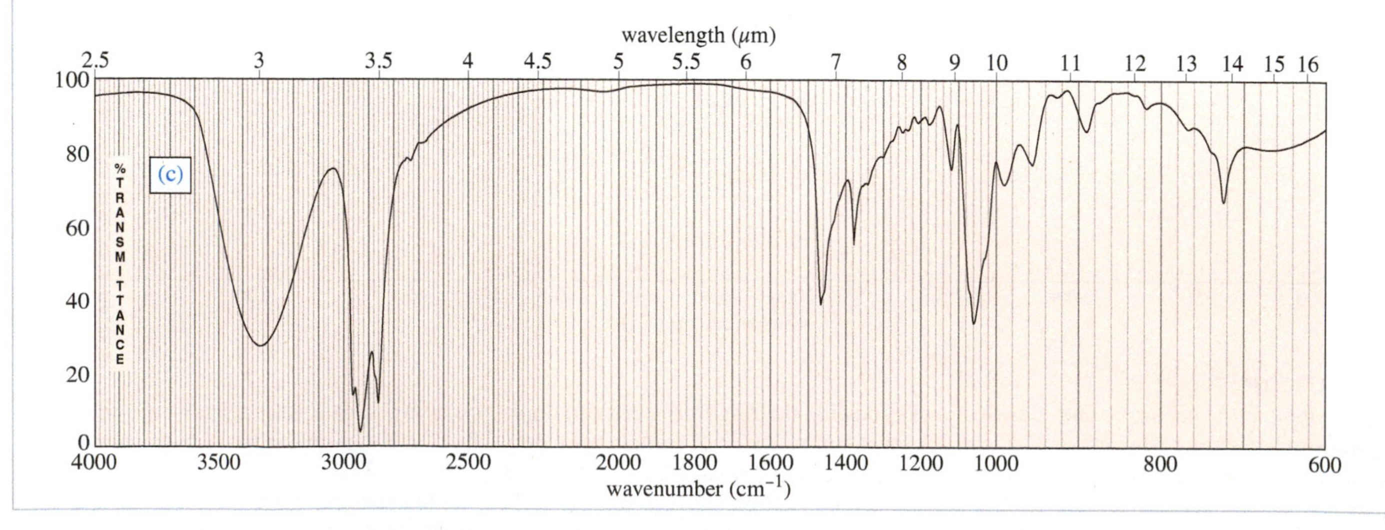

12-9B Resonance Lowering of Carbonyl Frequencies

In Section 12-7A we saw that conjugation of a C=C double bond lowers its stretching frequency. This is also true of conjugated carbonyl groups. Delocalization of the pi electrons reduces the electron density of the carbonyl double bond, weakening it and lowering the stretching frequency from about 1710 cm^{-1} to about 1685 cm^{-1} for conjugated ketones, aldehydes, and acids.

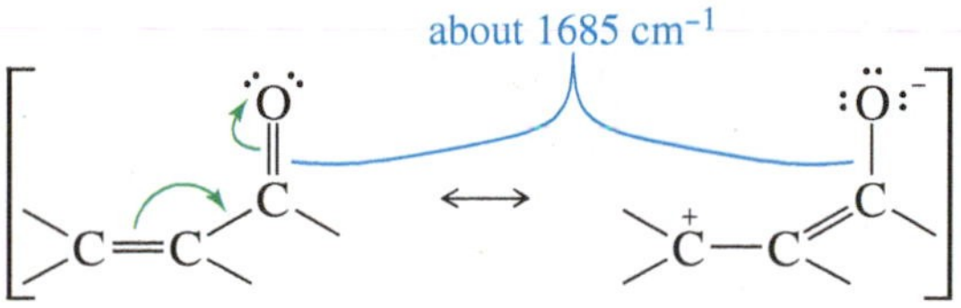

about 1685 cm^{-1}

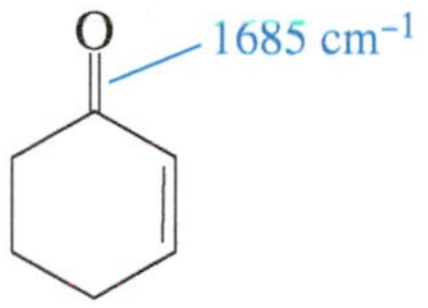

1685 cm^{-1} — cyclohex-2-enone

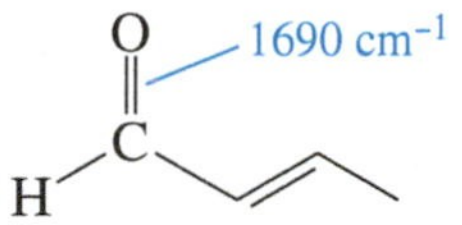

1690 cm^{-1} — (*E*)-but-2-enal

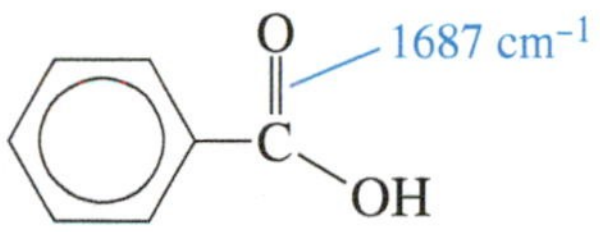

1687 cm^{-1} — benzoic acid

The C=C absorption of a conjugated carbonyl compound may not be apparent in the IR spectrum because it is so much weaker than the C=O absorption. The presence of the C=C double bond can still be inferred from its effect on the C=O frequency and the presence of unsaturated =C—H absorptions above 3000 cm^{-1}.

The carbonyl groups of amides absorb at particularly low IR frequencies: about 1640 to 1680 cm^{-1} (Figure 12-13). The dipolar resonance structure (shown next) places part of the pi bond between carbon and nitrogen, leaving less than a full C=O double bond.

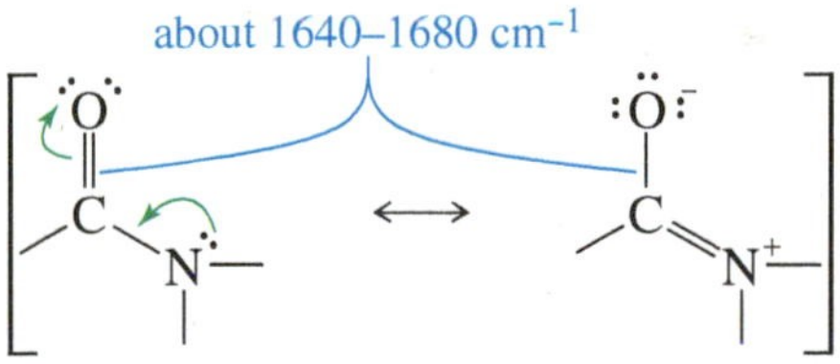

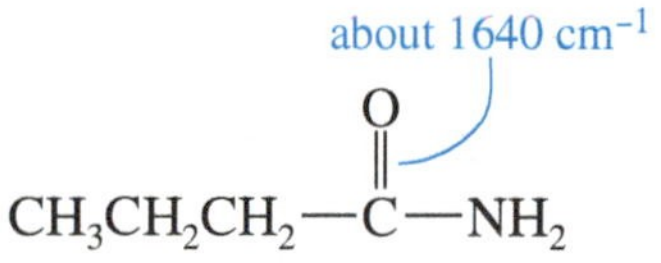

Application: Biochemistry

The IR absorption frequency of the amide N—H group is sensitive to the strength of hydrogen bonding. Therefore, IR spectroscopy provides structural information about peptide and protein conformations, which are stabilized by the hydrogen bonding of amide groups.

about 1640–1680 cm^{-1}

about 1640 cm^{-1}

$CH_3CH_2CH_2—C(=O)—NH_2$

The very low frequency of the amide carbonyl might be mistaken for an alkene C=C stretch. For example, consider the spectra of butyramide (C=O about 1640 cm^{-1}) and 1-methylcyclopentene (C=C at 1658 cm^{-1}) in Figure 12-13. Three striking differences are evident in these spectra: (1) The amide carbonyl absorption is much stronger and broader (from hydrogen bonding) than the absorption of the alkene double bond; (2) there are prominent N—H stretching absorptions in the amide spectrum; and (3) there is an unsaturated C—H stretching (just to the left of 3000 cm^{-1}) in the alkene spectrum. These examples show that we can distinguish between C=O and C=C absorptions, even when they appear in the same part of the spectrum.

Like primary amines, most primary amides show two spikes in the N—H stretching region (about 3300 cm^{-1}), as in the butyramide spectrum (Figure 12-13). Secondary amides (like secondary amines) generally show one N—H spike.

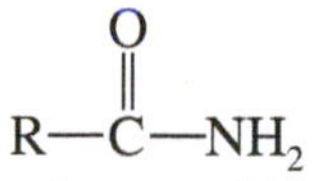

$R—C(=O)—NH_2$
primary amide

$R—NH_2$
primary amine

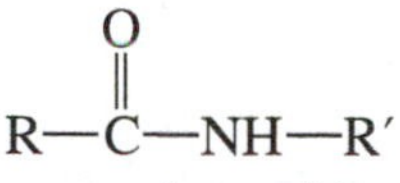

$R—C(=O)—NH—R'$
secondary amide

$R—NH—R'$
secondary amine

12-9C Carbonyl Absorptions Above 1725 cm^{-1}

Some carbonyl groups absorb at frequencies *higher* than 1725 cm^{-1}. For example, simple carboxylic esters absorb around 1735 cm^{-1}. These higher-frequency absorptions are also seen in strained cyclic ketones (in a five-membered ring or smaller). In a small ring, the angle strain on the carbonyl group forces more electron density into the C=O double bond, resulting in a stronger, stiffer bond.

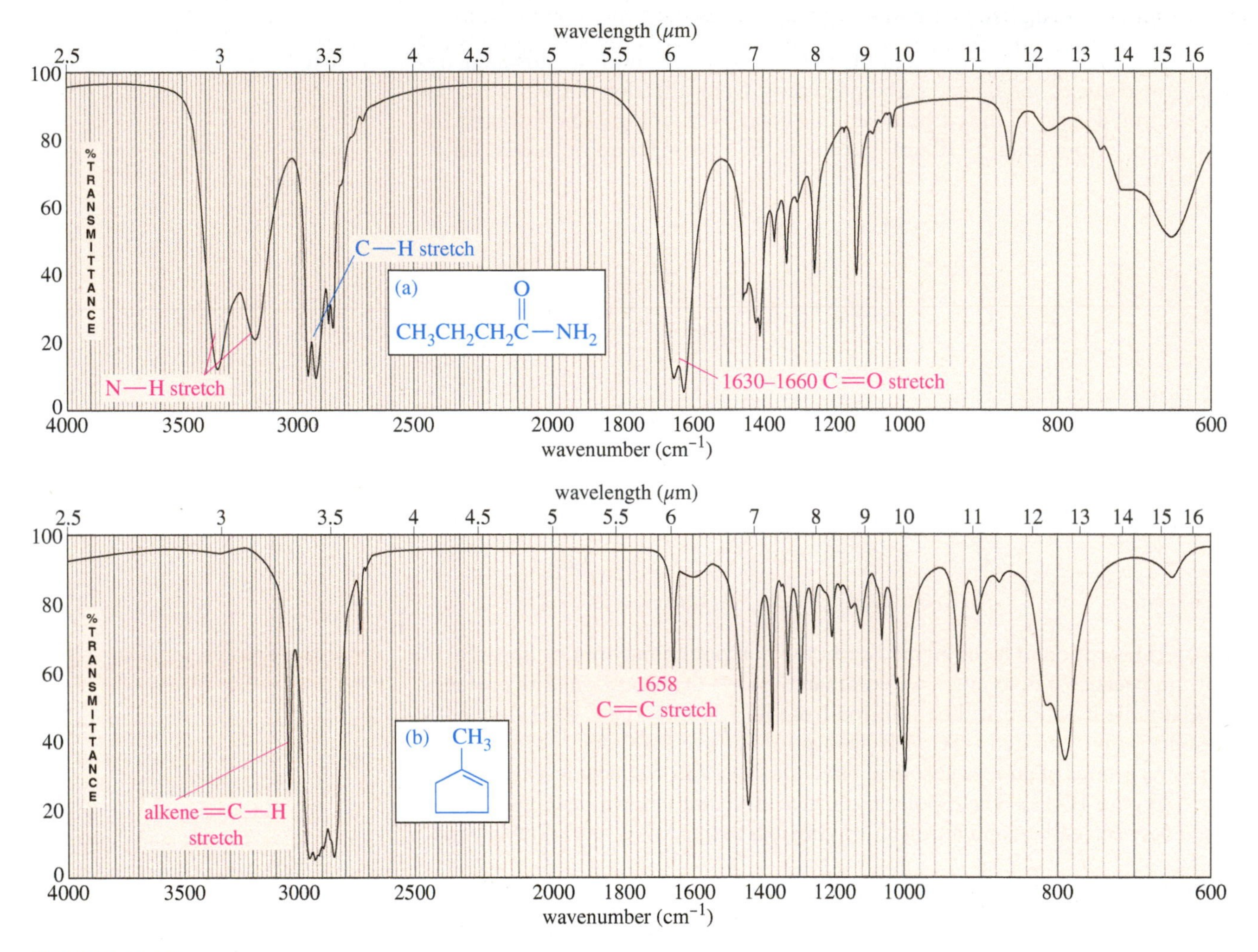

FIGURE 12-13
Characteristic IR spectra of amides. The carbonyl group of butyramide (a) and the C=C double bond of 1-methylcyclopentene (b) absorb in the same region, but three clues distinguish the alkene from the amide: (1) The amide C=O absorption is much stronger and broader than the C=C; (2) there are N—H absorptions (near 3300 cm^{-1}) in the amide; and (3) there is an unsaturated =C—H absorption in the alkene.

Application: Biochemistry

IR spectroscopy can also be used to monitor the progress of biological reactions. For example, the hydrolysis of complex lipids (esters of glycerol) causes a characteristic decrease in intensity of the ester carbonyl absorption at 1735 cm^{-1}, with a corresponding appearance of a carboxylic acid absorption near 1710 cm^{-1}.

R—C(=O)—O—R′ (about 1735 cm^{-1})
a carboxylic ester

$CH_3(CH_2)_6C(=O)—OCH_2CH_3$ (1738 cm^{-1})
ethyl octanoate

cyclobutanone (1785 cm^{-1})

12-10 Characteristic Absorptions of C—N Bonds

Infrared absorptions of carbon–nitrogen bonds are similar to those of carbon–carbon bonds, except that carbon–nitrogen bonds are more polar and give stronger absorptions. Carbon–nitrogen single bonds absorb around 1200 cm^{-1}, in a region close to many C—C and C—O absorptions. Therefore, the C—N single bond stretch is rarely useful for structure determination.

Carbon–nitrogen double bonds absorb in the same region as C=C double bonds, around 1660 cm^{-1}; however, the C=N bond gives rise to stronger absorptions because of its greater dipole moment. The C=N stretch often resembles a carbonyl absorption in intensity.

The most readily recognized carbon–nitrogen bond is the triple bond of a nitrile (Figure 12-14). The stretching frequency of the nitrile C≡N bond is close to that of an acetylenic C≡C triple bond, about 2200 cm^{-1}; however, nitriles generally absorb *above* 2200 cm^{-1} (2200 to 2300 cm^{-1}), while alkynes absorb *below* 2200 cm^{-1}. Also, nitrile triple bonds are more polar than C≡C triple bonds, so nitriles usually produce stronger absorptions than alkynes.

C—N *bond stretching frequencies*

	C—N	1200 cm^{-1}	usually strong
	C=N	1660 cm^{-1}	
	C≡N	>2200 cm^{-1}	
for comparison:	C≡C	<2200 cm^{-1}	(usually moderate or weak)

Figure 12-15 shows the regions of the IR spectrum that correspond with bonds to hydrogen, followed by triple bonds, double bonds, and single bonds between heavier elements.

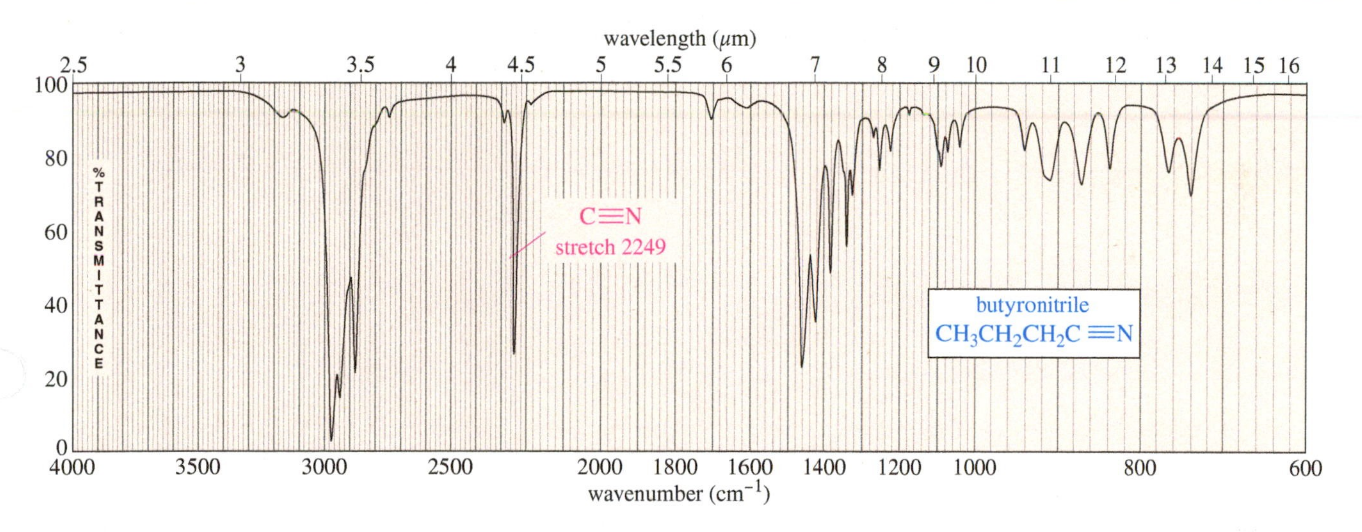

FIGURE 12-14
Nitrile triple bond stretching absorptions are at slightly higher frequencies (and usually more intense) than those of alkyne triple bonds. Compare this spectrum of butyronitrile with that of oct-1-yne in Figure 12-8.

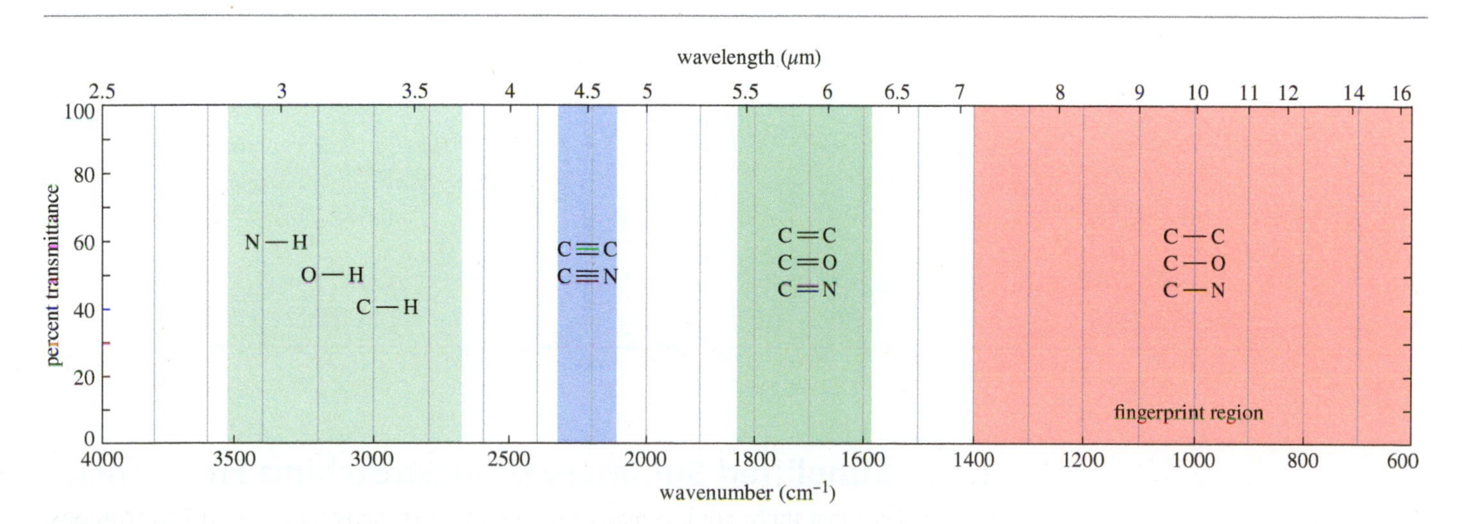

FIGURE 12-15
The highest-frequency (shortest wavelength) region of the IR spectrum corresponds to bonds to hydrogen. Next highest are triple bonds, followed by double bonds, followed by single bonds.

PROBLEM 12-5

The infrared spectra for three compounds are provided. Each compound has one or more of the following functional groups: conjugated ketone, ester, amide, nitrile, and alkyne. Determine the functional group(s) in each compound, and assign the major peaks above 1600 cm^{-1}.

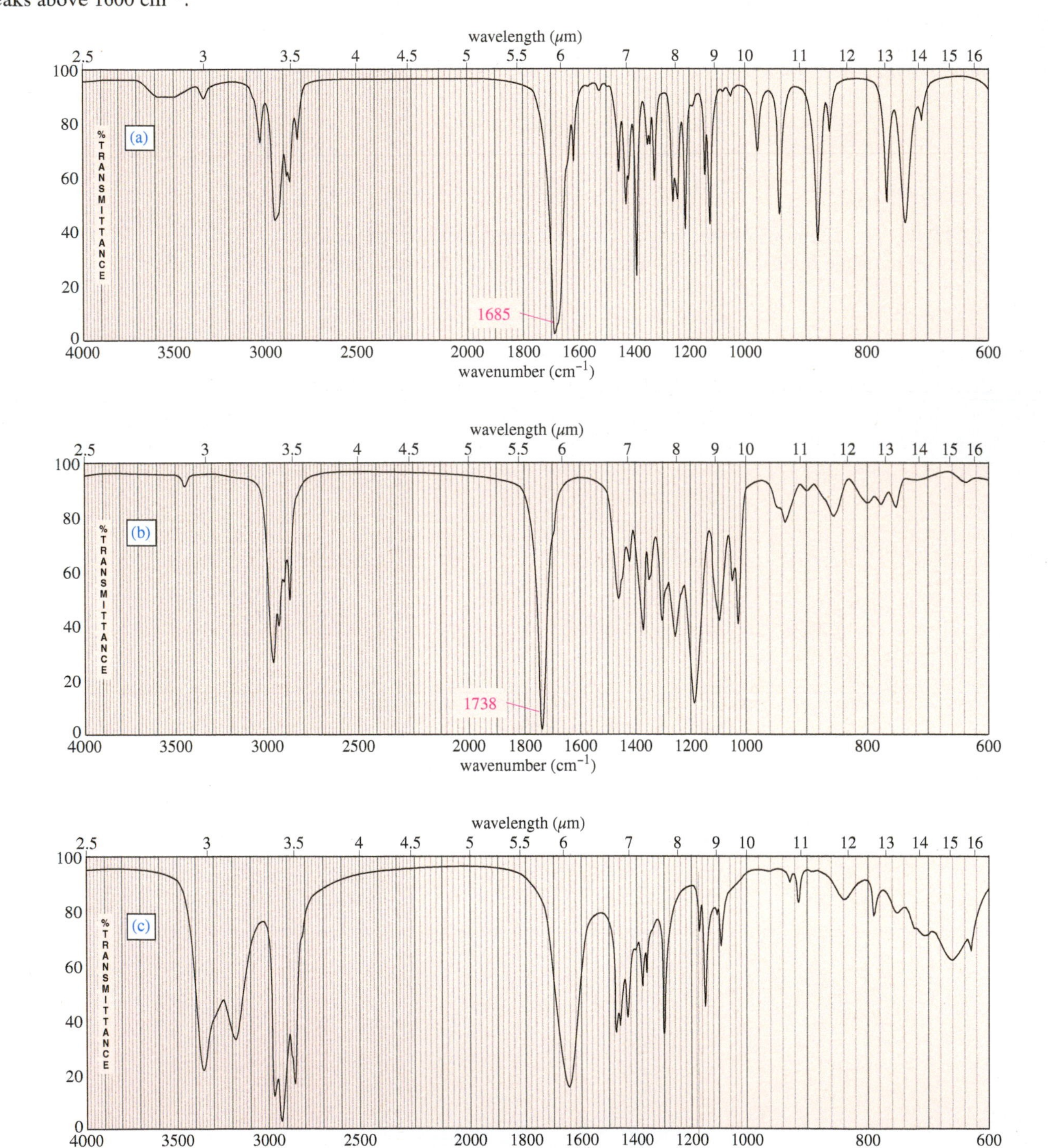

12-11 Simplified Summary of IR Stretching Frequencies

It may seem that there are too many numbers to memorize in infrared spectroscopy. Hundreds of characteristic absorptions for different kinds of compounds are listed in Appendix 2. Please glance at Appendix 2, and note that Appendix 2A is organized visually, while Appendix 2B is organized by functional groups. For everyday use, we can get by with only a few stretching frequencies, shown in Table 12-2. When using this table, remember that the numbers are approximate and they do not give ranges to cover all the unusual cases. Also, remember how frequencies change as a result of conjugation, ring strain, and other factors.

Strengths and Limitations of Infrared Spectroscopy The most useful aspect of infrared spectroscopy is its ability to identify functional groups. IR does not provide much information about the carbon skeleton or the alkyl groups in the compound, however. These aspects of the structure are more easily determined by NMR, as we will see in Chapter 13. Even an expert spectroscopist can rarely determine a structure based only on the IR spectrum.

Ambiguities often arise in the interpretation of IR spectra. For example, a strong absorption at 1680 cm^{-1} might arise from an amide, an isolated double bond, a conjugated ketone, a conjugated aldehyde, or a conjugated carboxylic acid. Familiarity with other regions of the spectrum usually enables us to determine which of these functional groups is present. In some cases, we cannot be entirely certain of the functional group without additional information, usually provided by other types of spectroscopy.

Infrared spectroscopy *can* provide conclusive proof that two compounds are either the same or different. The peaks in the fingerprint region depend on complex vibrations involving the entire molecule, and it is highly improbable for any two compounds (except enantiomers) to have precisely the same infrared spectrum.

To summarize, an infrared spectrum is valuable in three ways:

1. It indicates the functional groups in the compound.
2. It shows the *absence* of other functional groups that would give strong absorptions if they were present.
3. It can confirm the identity of a compound by comparison with a known sample.

TABLE 12-2
Summary of IR Stretching Frequencies

Frequency (cm^{-1})	Functional Group		Comments
3300	alcohol	O—H	always broad
	amine, amide	N—H	may be broad, sharp, or broad with spikes
	alkyne	≡C—H	always sharp, usually strong
3000	alkane	—C—H	just below 3000 cm^{-1}
	alkene	=C—H	just above 3000 cm^{-1}
	acid	O—H	very broad
2200	alkyne	—C≡C—	just below 2200 cm^{-1}
	nitrile	—C≡N	just above 2200 cm^{-1}
1710 (very strong)	carbonyl	>C=O	ketones, acids about 1710 cm^{-1} aldehydes about 1725 cm^{-1} esters higher, about 1735 cm^{-1} amides lower, about 1650 cm^{-1} conjugation lowers frequency
1660	alkene	>C=C<	conjugation lowers frequency aromatic C=C about 1600 cm^{-1}
	imine	>C=N—	stronger than C=C
	amide	>C=O	stronger than C=C (see above)

Ethers, esters, and alcohols also show C—O stretching between 1000 and 1200 cm^{-1}.

PROBLEM-SOLVING HINT

Table 12-2 provides the *numbers* but not the understanding and practice needed to work most IR problems. Learn to use the material in this table, and then practice doing problems until you feel confident.

SOLVED PROBLEM 12-2

You have an unknown with an absorption at 1680 cm^{-1}; it might be an amide, an isolated double bond, a conjugated ketone, a conjugated aldehyde, or a conjugated carboxylic acid. Describe what spectral characteristics you would look for to help you determine which of these possible functional groups might be causing the 1680 peak.

SOLUTION

Amide: (1680 peak is strong) Look for N—H absorptions (with spikes) around 3300 cm^{-1}.
Isolated double bond: (1680 peak is weak or moderate) Look for =C—H absorptions just above 3000 cm^{-1}.
Conjugated ketone: (1680 peak is strong) There must be a double bond nearby, conjugated with the C=O, to lower the C=O frequency to 1680 cm^{-1}. Look for the C=C of the nearby double bond (moderate, 1620 to 1640 cm^{-1}) and its =C—H above 3000 cm^{-1}.
Conjugated aldehyde: (1680 peak is strong) Look for the aldehyde C—H stretch about 2700 and 2800 cm^{-1}. Also look for the C=C of the nearby double bond (1620 to 1640 cm^{-1}) and its =C—H (just above 3000 cm^{-1}).
Conjugated carboxylic acid: (1680 peak is strong) Look for the characteristic acid O—H stretch centered *on top of* the C—H stretch around 3000 cm^{-1}. Also look for the C=C of the nearby double bond (1620 to 1640 cm^{-1}) and its =C—H (just above 3000 cm^{-1}).

12-12 Reading and Interpreting IR Spectra (Solved Problems)

Many students are unsure how much information they should be able to obtain from an infrared spectrum. In Chapter 13, we will use IR together with NMR and other information to determine the entire structure. For the present, concentrate on getting as much information as you can from the IR spectrum by itself. Several solved problems are included in this section to show what information can be inferred. An experienced spectroscopist could obtain more information from these spectra, but we will concentrate on the major, most reliable, features.

Study this section by looking at each spectrum and writing down the important frequencies and your proposed functional groups. Then look at the solution and compare it with your solution. The actual structures of these compounds are shown at the end of this section. They are not given with the solutions because *you cannot determine these structures using only the infrared spectra*, so a complete structure is not a part of a realistic solution.

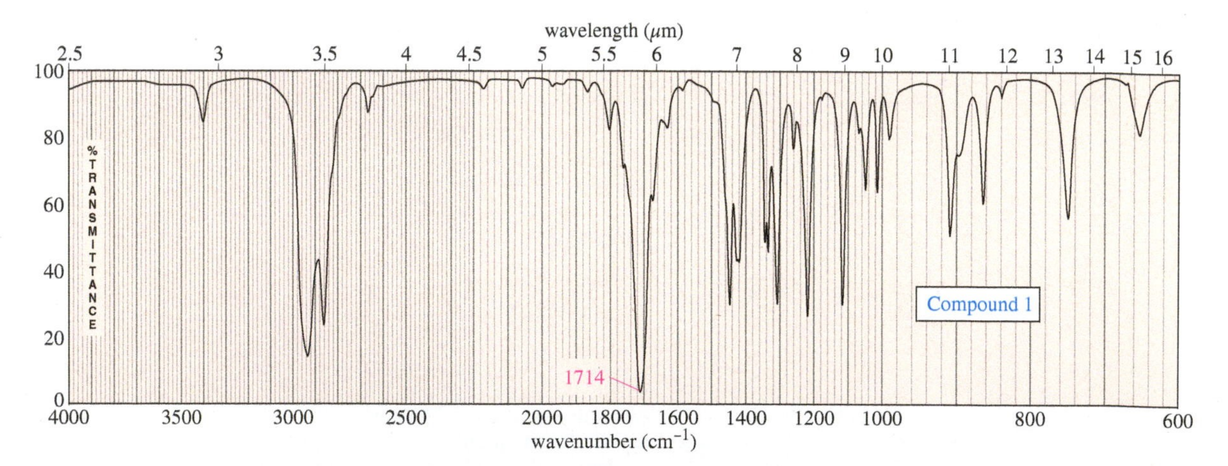

Compound 1 This spectrum is most useful for what it does *not* show. There is a carbonyl absorption at 1714 cm^{-1} and little else. There is no aldehyde C—H, no hydroxy O—H, and no N—H. The weak absorption at 3400 cm^{-1} is probably an overtone of the strong C=O absorption. The carbonyl absorption could indicate an aldehyde, ketone, or acid, except that the lack of aldehyde C—H stretch eliminates an aldehyde, and the lack of O—H stretch eliminates an acid. There is no visible C=C stretch and no unsaturated C—H absorption above 3000 cm^{-1}, so the compound appears to be otherwise saturated. The compound is probably a simple ketone.

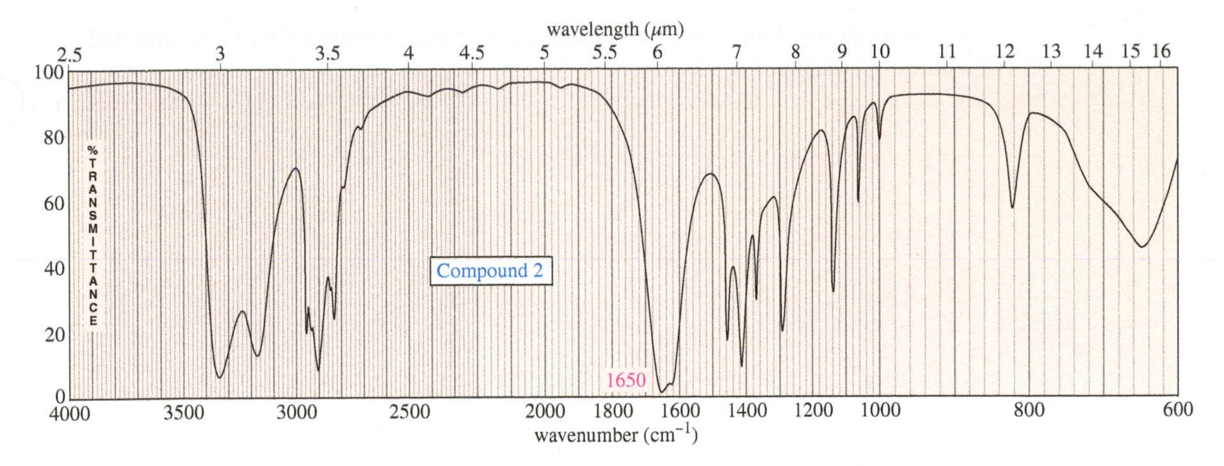

Compound 2 The absorption at 1650 cm^{-1} is so intense that it probably indicates a carbonyl group. A carbonyl group at this low frequency suggests an amide. The doublet (a pair of peaks) of N—H absorption around 3300 cm^{-1} also suggests a primary amide, R—$CONH_2$. Since there is no C—H absorption above 3000 cm^{-1}, this is probably a saturated amide.

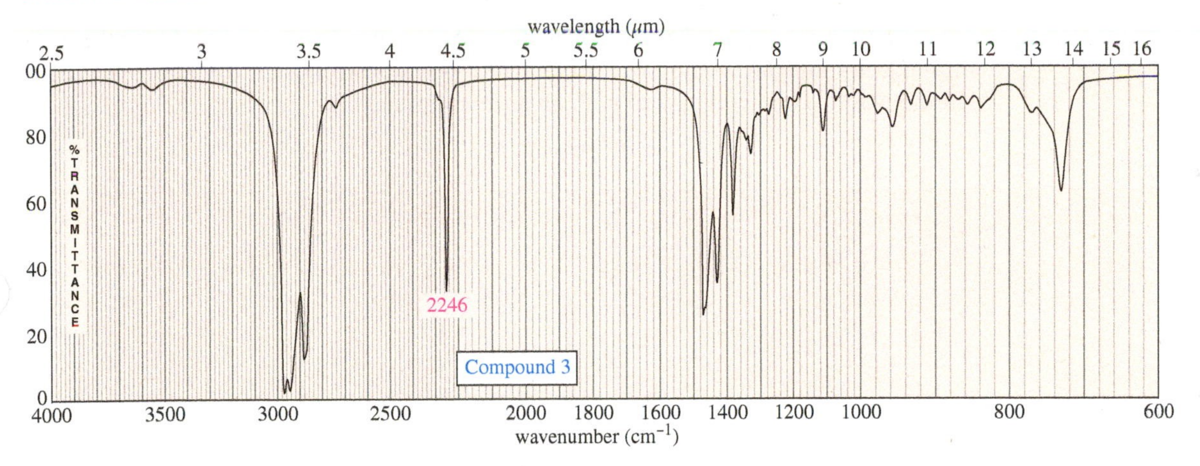

Compound 3 The sharp peak at 2246 cm^{-1} results from a nitrile C≡N stretch. (An alkyne C≡C absorption would be weaker and below 2200 cm^{-1}.) The absence of C=C stretch or C—H stretch above 3000 cm^{-1} suggests that this nitrile is otherwise saturated.

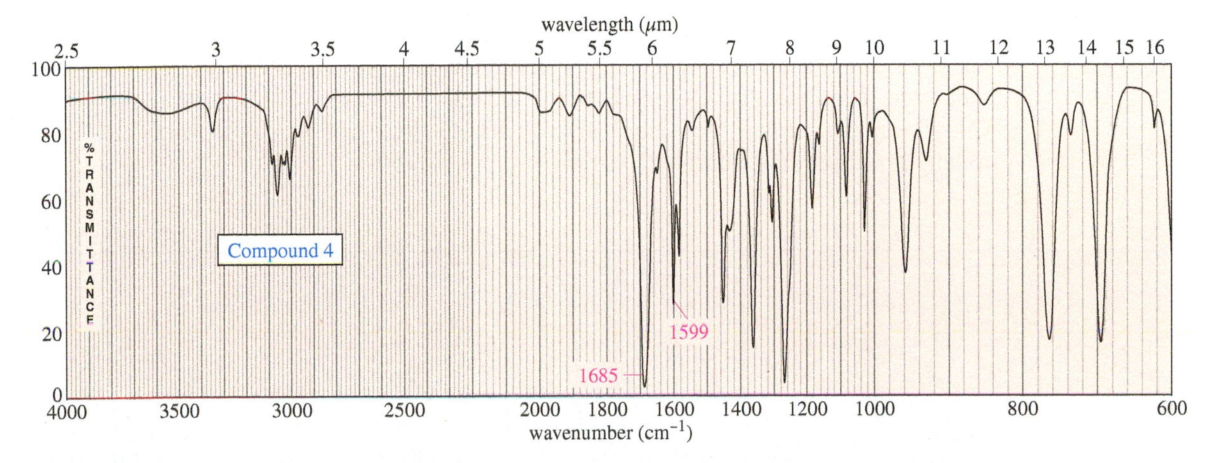

Compound 4 The carbonyl absorption at 1685 cm^{-1} is about right for a conjugated ketone, aldehyde, or acid. (An amide would be lower in frequency, and a C=C double bond would not be so strong.) The absence of any N—H stretch, O—H stretch, or aldehyde C—H stretch leaves a conjugated ketone as the best possibility. The C=C stretch at 1599 cm^{-1} indicates an aromatic ring, confirmed by the unsaturated C—H

absorption above 3000 cm^{-1}. We presume that the aromatic ring is conjugated with the carbonyl group of the ketone.

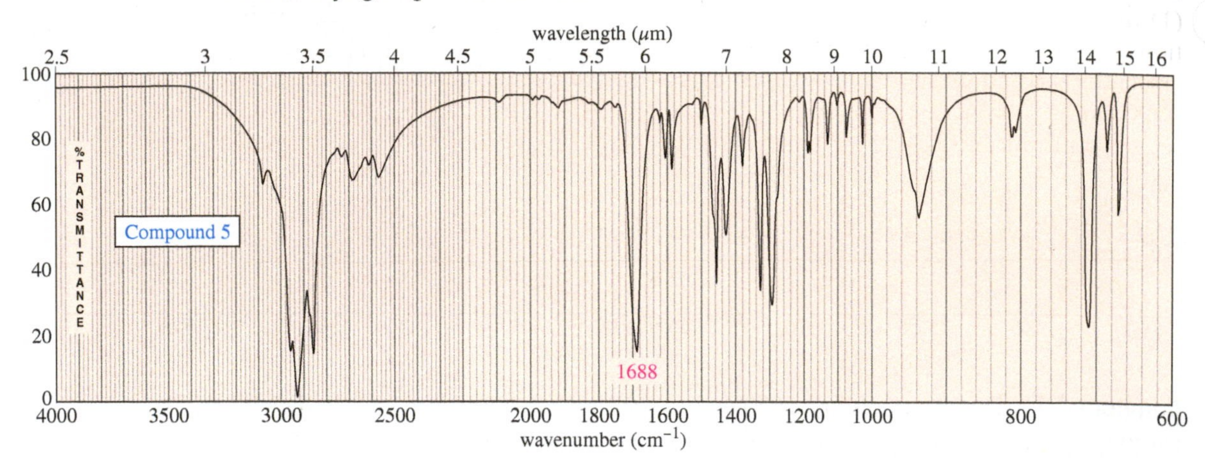

Compound 5 The broad O—H stretch that spans most of the C—H stretching region suggests a carboxylic acid. (This acid is a solid, and its O—H absorption is weaker than that of the liquid shown in Figure 12-12.) This acid O—H also has a shoulder with spikes about 2500–2700 cm^{-1}. The C=O stretch is low for an acid (1688 cm^{-1}), implying a conjugated acid. The aromatic C=C absorption at 1600 cm^{-1} suggests that the acid may be conjugated with an aromatic ring.

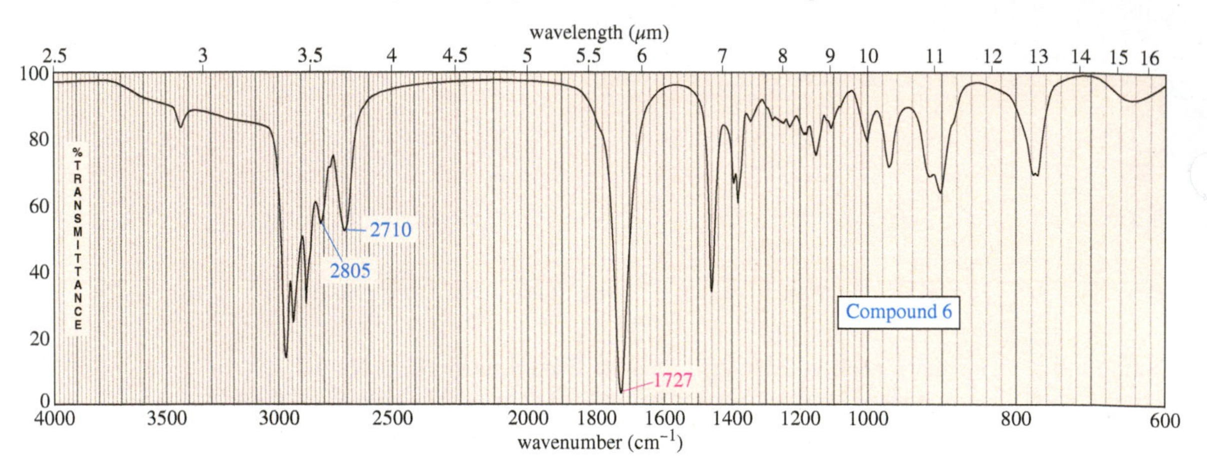

Compound 6 The carbonyl absorption at 1727 cm^{-1} suggests an aldehyde, or possibly a ketone or an acid. The C—H stretching at 2710 and 2805 cm^{-1} confirms an aldehyde. Because all the C—H stretch is below 3000 cm^{-1} and there is no visible C=C stretch around 1660 cm^{-1} or aromatic C=C stretch around 1600 cm^{-1}, the aldehyde is probably saturated.

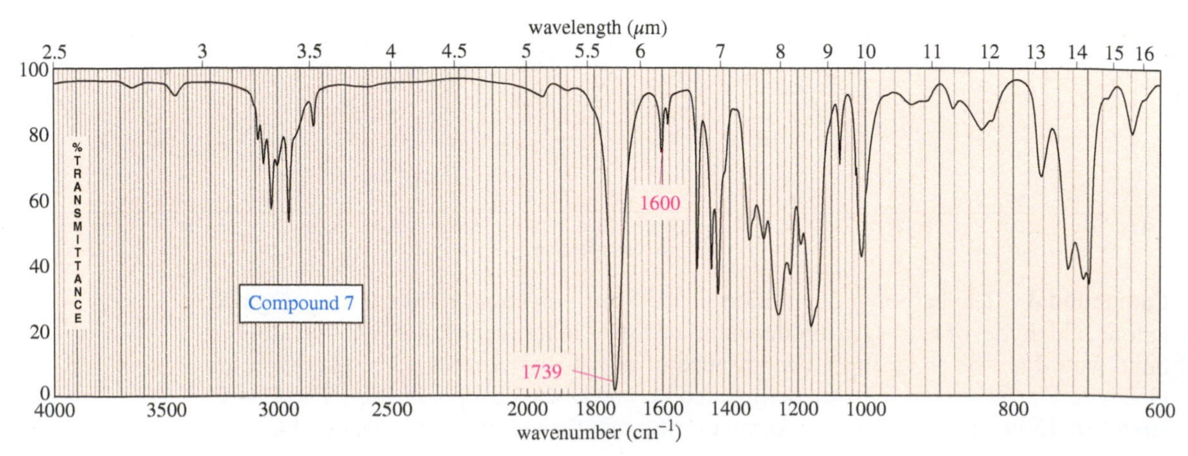

Compound 7 The carbonyl absorption at 1739 cm^{-1} suggests an ester. The weak peak at 1600 cm^{-1} indicates an aromatic ring, but it is not conjugated with the ester because (1) the ester absorption is close to its usual (unconjugated) position, and (2) conjugation with a polar carbonyl group would polarize the aromatic ring and give a stronger aromatic absorption than what we see here. The presence of both saturated (below 3000 cm^{-1}) and unsaturated (above 3000 cm^{-1}) C—H stretching in the 3000 cm^{-1} region confirms the presence of both alkyl and unsaturated portions of the molecule.

> **PROBLEM-SOLVING HINT**
>
> Appendices 2A and 2B are valuable tables for precise interpretation and prediction of infrared spectra. These tables will continue to be useful in your further studies in chemistry.

Structures of the compounds

(These structures cannot be determined from their IR spectra alone.)

compound 1 (cyclohexanone, ring with =O)

$CH_3CH_2—C(=O)—NH_2$
compound 2

$CH_3(CH_2)_4—C{\equiv}N$
compound 3

(benzene ring)$—C(=O)—CH_3$
compound 4

(benzene ring)$—C(=O)—OH$
compound 5

$CH_3CH_2—CH(CH_3)—C(=O)—H$
compound 6

(benzene ring)$—CH_2—C(=O)—OCH_3$
compound 7

PROBLEM 12-6

For each spectrum, interpret all the significant stretching frequencies above 1580 cm^{-1}.

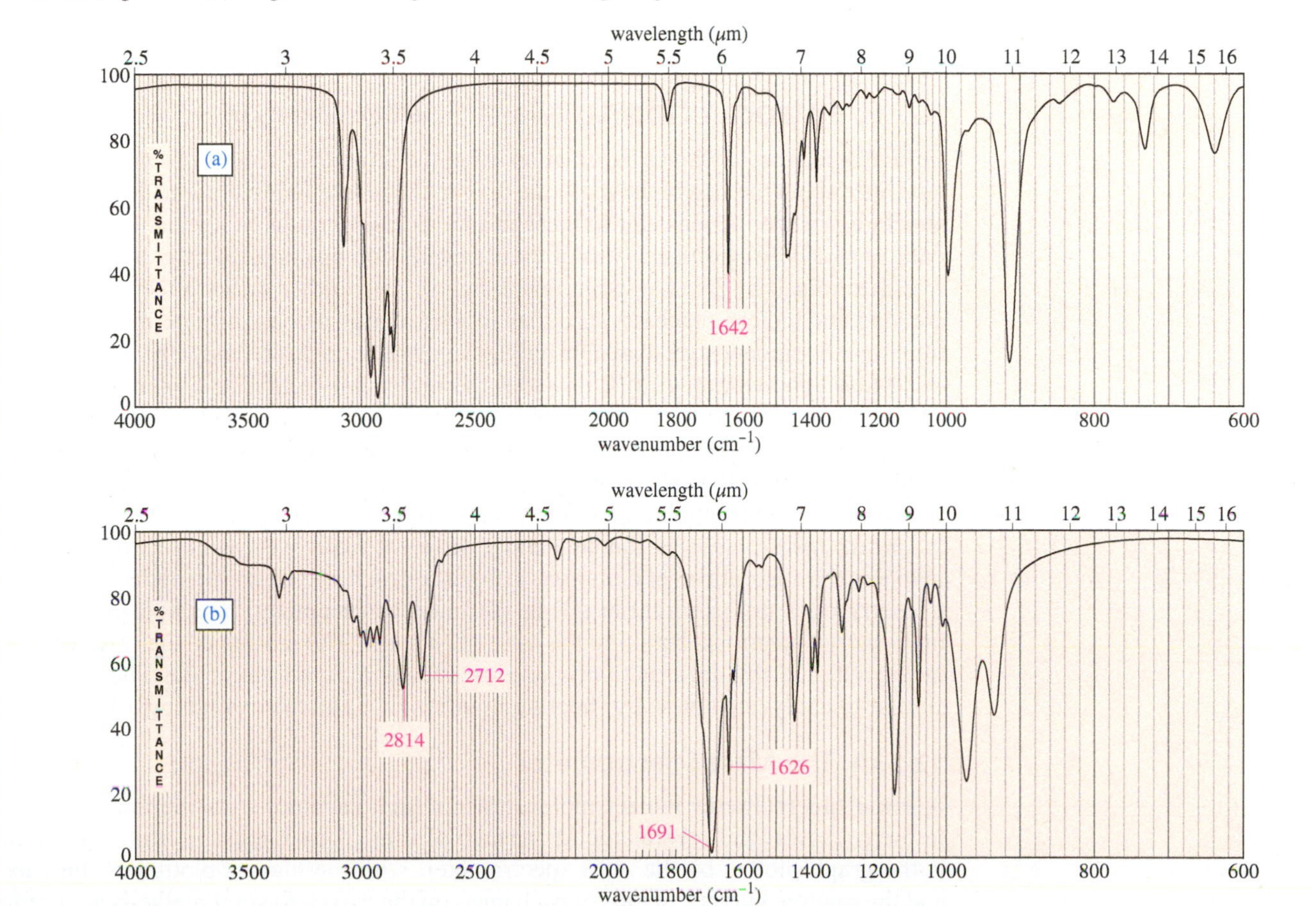

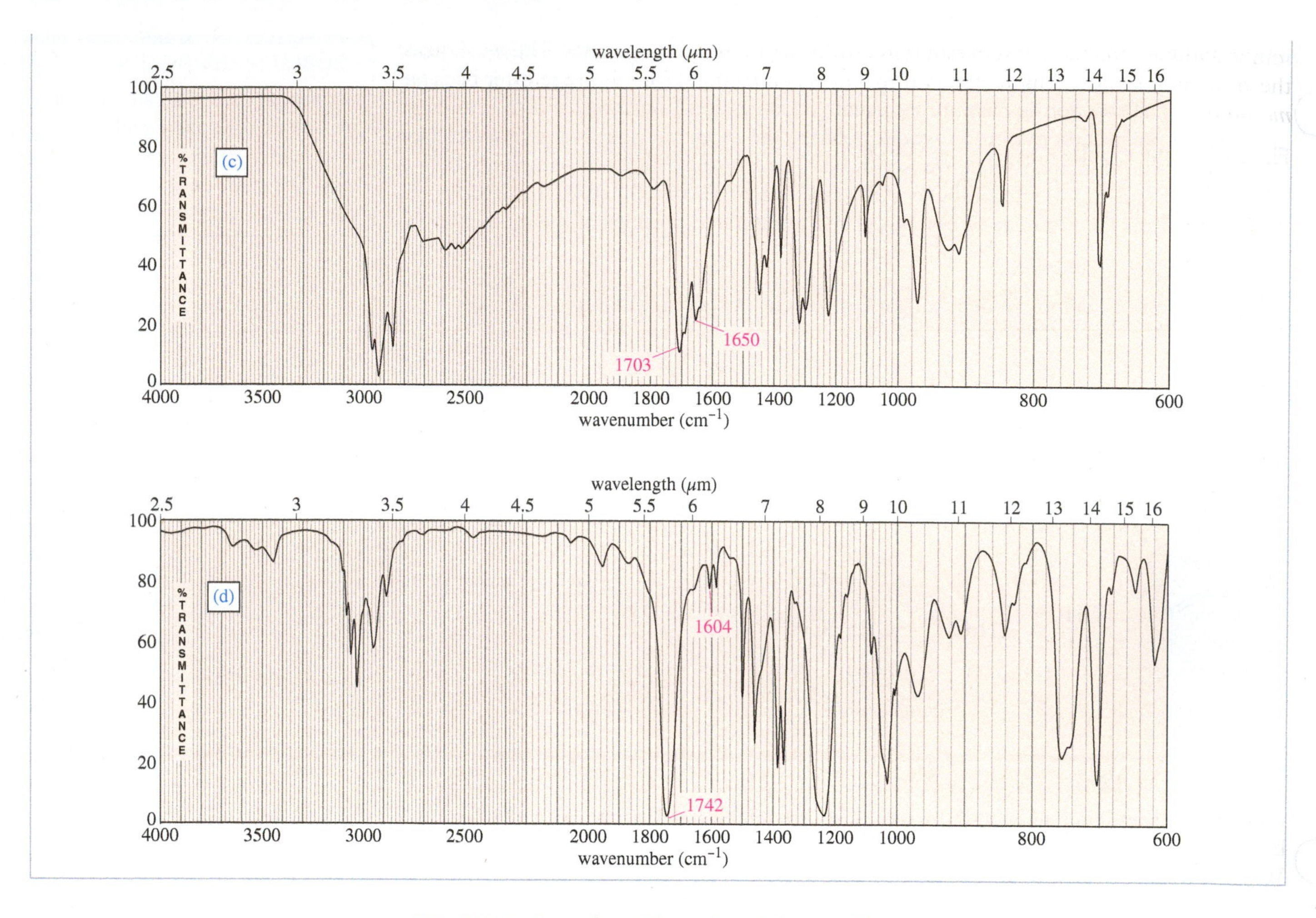

12-13 Introduction to Mass Spectrometry

Infrared spectroscopy gives information about the functional groups in a molecule, but it tells little about the size of the molecule or what heteroatoms are present. To determine a structure, we need a molecular weight and a molecular formula. Molecular formulas were once obtained by careful analysis of the elemental composition, and a molecular weight was determined by freezing-point depression or some other difficult technique. These are long and tedious processes, and they require a large amount of pure material. Many important compounds are available only in small quantities, and they may be impure.

Mass spectrometry (MS) provides the molecular weight and valuable information about the molecular formula, using a very small sample. High-resolution mass spectrometry (HRMS) can provide an accurate molecular formula, even for an impure sample. The mass spectrum also provides structural information that can confirm a structure derived from NMR and IR spectroscopy.

Mass *spectrometry* is fundamentally different from *spectroscopy*. Spectroscopy involves the absorption (or emission) of light over a range of wavelengths. Mass spectrometry does not use light at all. In the mass spectrometer, a sample is struck by high-energy electrons, breaking the molecules apart. The masses of the fragments are measured, and this information is used to reconstruct the molecule. The process is similar to analyzing a vase by shooting it with a rifle and then weighing all the pieces.

12-13A The Mass Spectrometer

A **mass spectrometer** ionizes molecules in a high vacuum, sorts the ions according to their masses, and records the abundance of ions of each mass. A **mass spectrum** is the graph plotted by the mass spectrometer, with the masses plotted as the *x* axis and the relative number of ions of each mass on the *y* axis. Several methods are used to ionize samples and then to separate ions according to their masses. We will emphasize the most common techniques: *electron impact ionization* for forming the ions, and *magnetic deflection* for separating the ions.

Electron Impact Ionization In the **ion source,** the sample is bombarded by a beam of electrons. When an electron strikes a neutral molecule, it may ionize that molecule by knocking out an additional electron.

$$e^- + M \longrightarrow [M]^{+\cdot} + 2\,e^-$$

When a molecule loses one electron, it then has a positive charge and one unpaired electron. The ion is therefore a **radical cation.** The electron impact ionization of methane is

$$\underset{\text{electron}}{e^-} + \underset{\text{methane}}{CH_4} \longrightarrow 2\,e^- + \underset{M^{+\cdot},\ \text{radical cation}}{[CH_4]^{+\cdot}}$$

(unpaired electron)

Most carbocations have a three-bonded carbon atom with six paired electrons in its valence shell. The radical cation just shown is not a normal carbocation. The carbon atom has seven electrons around it, and they bond it to four other atoms. This unusual cation is represented by the formula $[CH_4]^{+\cdot}$, with the + indicating the positive charge and the · indicating the unpaired electron.

In addition to ionizing a molecule, the impact of an energetic electron may break it apart. This **fragmentation** process gives a characteristic mixture of ions. The radical cation corresponding to the mass of the original molecule is called the **molecular ion,** abbreviated $M^{+\cdot}$. The ions of smaller molecular weights are called *fragments.* Bombardment of ethane molecules by energetic electrons, for example, produces the molecular ion and several fragments. Both charged and uncharged fragments are formed, but *only the positively charged fragments are detected by the mass spectrometer.* We will often use green type for the "invisible" uncharged fragments.

Application: Protein MS

Mass spectrometry can be used to determine the precise mass of a protein. Because of their large size and low volatility, proteins require specialized mass spectral techniques, such as electrospray ionization: spraying a charged, heated stream of droplets into the evacuated source chamber. The solvent evaporates to leave ions of the compound to be analyzed.

$$e^- + CH_3{-}CH_3 \longrightarrow \text{can give } \underset{\substack{\text{molecular ion, } M^{+\cdot} \\ m/z = 30}}{[CH_3{-}CH_3]^{+\cdot}} \text{ or } \underset{m/z = 29}{CH_3{-}CH_2^+} + H\cdot$$

$$\text{or } \underset{m/z = 15}{CH_3^+} + \cdot CH_3 \quad \text{or various other combinations of radicals and ions}$$

We discuss the common modes of fragmentation in Section 12-15.

Separation of Ions of Different Masses Once ionization and fragmentation have formed a mixture of ions, these ions are separated and detected. The most common type of mass spectrometer, shown in Figure 12-16, separates ions by *magnetic deflection.*

After ionization, the positively charged ions are attracted to a negatively charged accelerator plate, which has a narrow slit to allow some of the ions to pass through. The ion beam enters an evacuated flight tube, with a curved portion positioned between the poles of a large magnet. When a charged particle passes through a magnetic field, a transverse force bends its path. The path of a heavier ion bends less than the path of a lighter ion.

The exact radius of curvature of an ion's path depends on its mass-to-charge ratio, symbolized by $\boldsymbol{m/z}$ (or by $\boldsymbol{m/e}$ in the older literature). In this expression, m is the mass of the ion (in amu) and z is its charge in units of the electronic charge. The vast majority of ions have a charge of +1, so we consider their path to be curved by an amount that depends only on their mass.

A double-focusing mass spectrometer. This one is combined with a gas chromatograph to use as a GC–MS. The gas chromatograph separates a mixture into its components and injects the purified components into the ion source of the mass spectrometer.

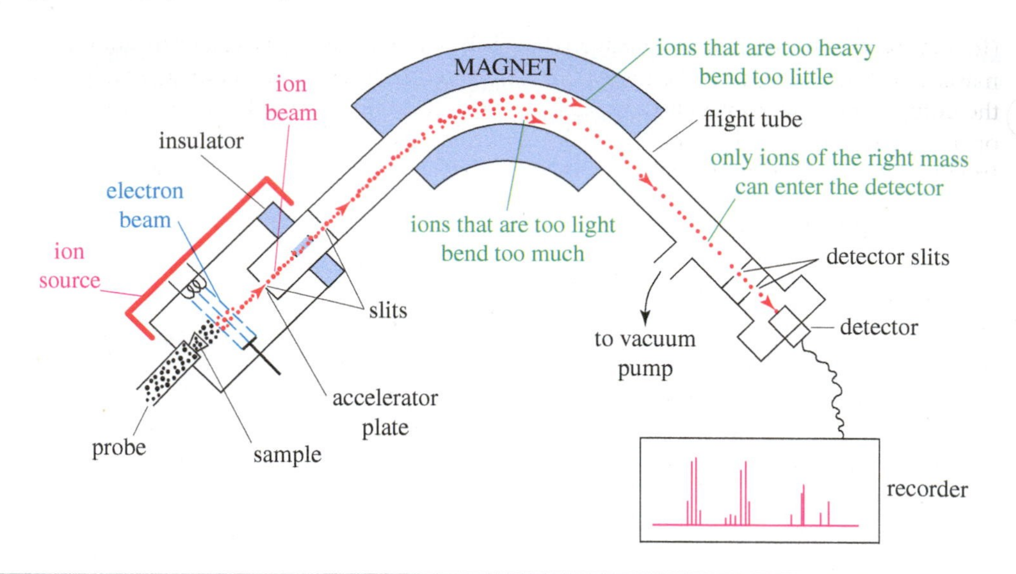

FIGURE 12-16
Diagram of a mass spectrometer. A beam of electrons causes molecules to ionize and fragment. The mixture of ions is accelerated and passes through a magnetic field, where the paths of lighter ions are bent more than those of heavier ions. By varying the magnetic field, the spectrometer plots the abundance of ions of each mass.

Application: Biochemistry

New techniques for vaporizing and ionizing large (but fragile) biomolecules, such as proteins and DNA fragments, have made mass spectrometry an important analytical tool for biochemistry and molecular biology. In the MALDI (Matrix-Assisted Laser Desorption/Ionization) technique, for example, the biomolecule is embedded in the crystal of a different, UV-absorbing compound. A UV laser vaporizes this matrix, gently ionizing the biomolecule and carrying it into the gas phase, where its mass spectrum can be measured.

At the end of the flight tube is another slit, followed by an ion detector connected to an amplifier. At any given magnetic field, only ions of one particular mass are bent exactly the right amount to pass through the slit and enter the detector. The detector signal is proportional to the number of ions striking it. By varying the magnetic field, the spectrometer scans through all the possible ion masses and produces a graph of the number of ions of each mass.

12-13B The Mass Spectrum

The mass spectrometer usually plots the spectrum as a graph on a computer screen. This information is tabulated, and the spectrum is printed as a bar graph or as a table of relative abundances (Figure 12-17). In the printed mass spectrum, all the masses are rounded to the nearest whole-number mass unit. The peaks are assigned abundances as percentages of the strongest peak, called the **base peak.** Notice that *the base peak does not necessarily correspond to the mass of the molecular ion.* It is simply the strongest peak, making it easy for other peaks to be expressed as percentages.

A molecular ion peak (also called the *parent peak*) is observed in most mass spectra, meaning that a detectable number of molecular ions (M^+) reach the detector without fragmenting. These molecular ions are usually the particles of highest mass in the spectrum and

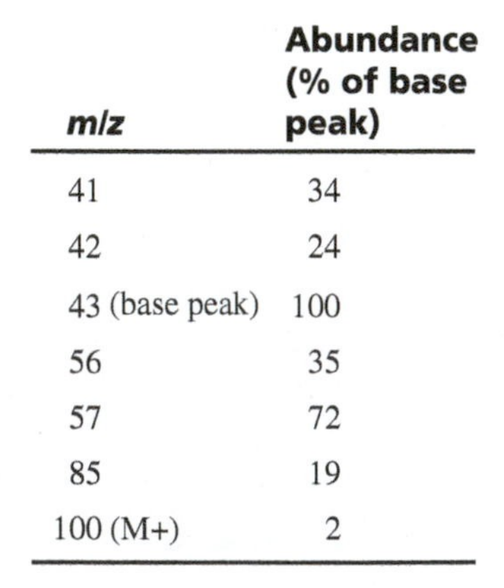

m/z	**Abundance (% of base peak)**
41	34
42	24
43 (base peak)	100
56	35
57	72
85	19
100 (M+)	2

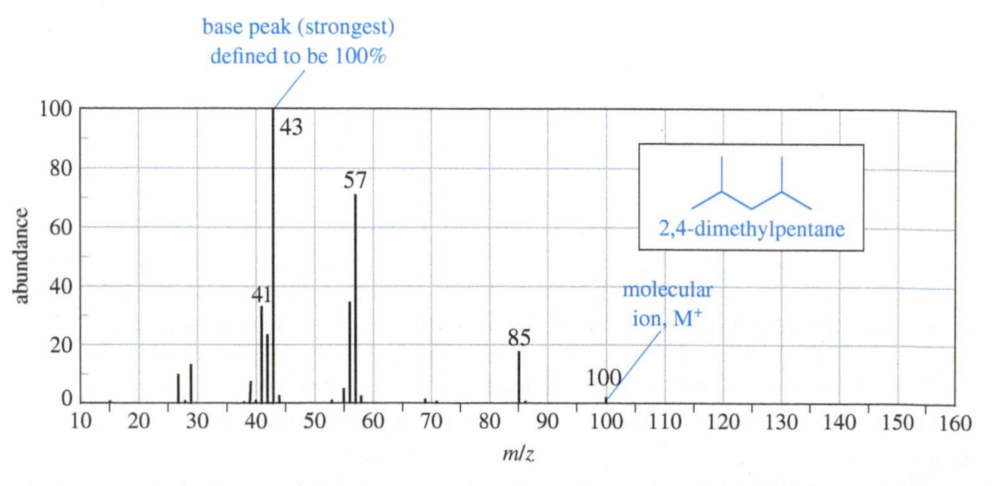

FIGURE 12-17
Mass spectrum of 2,4-dimethylpentane, given both as a bar graph and in tabular form. Abundances are given as percentages of the strongest peak (the base peak). In this example, the base peak is at m/z 43 and the molecular ion peak (parent peak) is at m/z 100. Notice that the molecular ion has an even mass number, while most of the fragments are odd.

(for compounds not containing nitrogen) the molecular ion usually has an even-numbered mass. The value of m/z for the molecular ion immediately gives the molecular weight of the compound. If no molecular ion peak is observed in the standard mass spectrum, the operator can use a gentler ionization. The energy of the electron beam can be decreased from the typical 70 electron volts (eV) to 20–25 eV, where much less fragmentation occurs.

12-13C Mass Spectrometry of Mixtures: The GC–MS

Mass spectrometry is combined with gas chromatography for routine analysis of mixtures of compounds, such as reaction mixtures or environmental samples. Figure 12-18 shows a simplified diagram of a common type of GC–MS. The **gas chromatograph** uses a heated capillary column coated on the inside with silicone rubber (or other *stationary phase*) to separate the components of the mixture. A small amount of sample (about 10^{-6} gram is enough) is injected into a heated injector, where a gentle flow of helium sweeps it into the column. As the sample passes through the column, the more volatile components (that interact less with the stationary phase) move through the column faster than the less volatile components. The separated components leave the column at different times, passing through a transfer line into the ion source of the mass spectrometer, where the molecules are ionized and allowed to fragment.

Most gas chromatograph–mass spectrometer systems use a *quadrupole mass filter* to separate the ions. In a high vacuum, the ions pass down the length of four rods, which have varying voltages applied to them. (Figure 12-18 shows two of the four rods.) The varying electric fields cause the ions to follow complex orbits, and only one mass reaches the detector at any instant. By scanning the voltages, a wide range of masses can be measured in less than 1 second. In this way, many mass spectra are taken and stored on a computer disk as the components of the sample pass from the chromatograph column into the mass spectrometer. This powerful GC–MS combination allows many components of a mixture to be separated by the gas chromatograph and later identified by their mass spectra.

Application: Drug Testing

We can establish the identity of a drug by infrared spectroscopy (IR) or gas chromatography–mass spectrometry (GC–MS). IR provides clues to the functional groups, and it can confirm the structure by comparing the spectrum with that of an authentic sample.

GC–MS provides a characteristic retention time that suggests possible compounds, and it confirms the structure by comparison of the retention time and the mass spectrum with those of an authentic sample. GC–MS includes a GC separation, so it works well with impure compounds or mixtures.

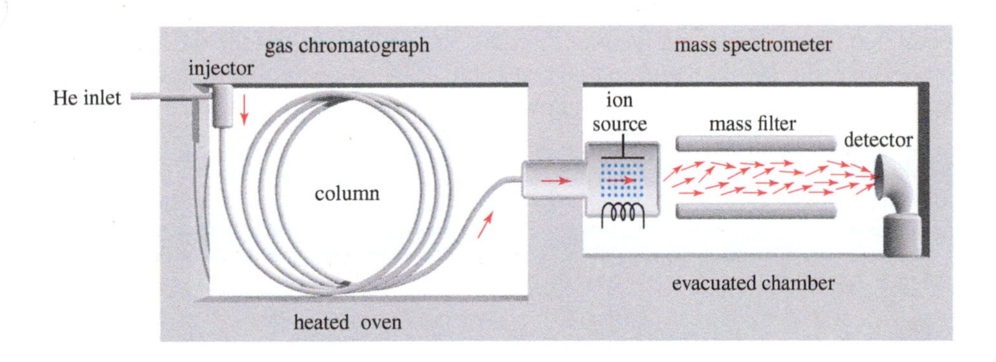

FIGURE 12-18
Block diagram of a gas chromatograph–mass spectrometer (GC–MS). The gas chromatograph column separates the mixture into its components. The quadrupole mass spectrometer scans mass spectra of the components as they leave the column.

12-14 Determination of the Molecular Formula by Mass Spectrometry

Mass spectrometry usually provides a reliable molecular weight for an unknown compound. In some cases, it provides information about the molecular formula as well. A very precise mass measurement can give the molecular formula in many cases. Some elements are apparent in the mass spectrum because of their characteristic isotopic mixtures.

12-14A High-Resolution Mass Spectrometry

Although mass spectra usually show the particle masses rounded to the nearest whole number, the masses are not really integral. The ^{12}C nucleus is *defined* to have a mass of exactly 12 atomic mass units (amu), and all other nuclei have masses based on this standard. For example, a proton has a mass of about 1, but not exactly: Its mass is 1.007825 amu. Table 12-3 shows the atomic masses for the most common isotopes found in organic compounds.

TABLE 12-3
"Exact" Masses of Common Isotopes

Isotope	Atomic Mass (amu)
^{12}C	12.000000
^{1}H	1.007825
^{16}O	15.994914
^{14}N	14.003050

Determination of a molecular formula is possible using a **high-resolution mass spectrometer** (HRMS), one that uses extra stages of electrostatic or magnetic focusing to form a very precise beam and to detect particle masses to an accuracy of about 1 part in 20,000. A mass determined to several significant figures using an HRMS is called an *exact mass*. Although it is not really exact, it is much more accurate than the usual integral mass numbers. Comparing the exact mass with masses calculated by molecular formula makes it possible to identify the correct formula.

Consider a molecular ion with a mass of 44. This approximate molecular weight might correspond to C_3H_8 (propane), C_2H_4O (acetaldehyde), CO_2, or CN_2H_4. Each of these molecular formulas corresponds to a different exact mass:

C_3H_8		C_2H_4O		CO_2		CN_2H_4	
3 C	36.00000	2 C	24.00000	1 C	12.00000	1 C	12.00000
8 H	8.06260	4 H	4.03130			4 H	4.03130
		1 O	15.99491	2 O	31.98983	2 N	28.00610
	44.06260		44.02621		43.98983		44.03740

If the HRMS measured the exact mass of this ion as 44.029 mass units, we would conclude that the compound has a molecular formula of C_2H_4O, because the mass corresponding to this formula is closest to the observed value. Published tables of exact masses are available for comparison with values obtained from the HRMS. Depending on the completeness of the tables, they may include sulfur, halogens, or other elements.

Application: Drug Testing

Isotope ratios can help to identify banned substances in Olympic athletes. For example, mass spectrometry can distinguish between synthetic testosterone and the naturally occurring hormone by detecting differences in the isotope ratios of ^{13}C and ^{12}C.

12-14B Use of Heavier Isotope Peaks

Whether or not a high-resolution mass spectrometer is available, molecular ion peaks often provide information about the molecular formula. Most elements do not consist of a single isotope, but contain heavier isotopes in varying amounts. These heavier isotopes give rise to small peaks at higher mass numbers than the major M^+ molecular ion peak. A peak that is one mass unit heavier than the M^+ peak is called the **M+1 peak;** two units heavier, the **M+2 peak;** and so on. Table 12-4 gives the isotopic compositions of some common elements, showing how they contribute to M+1 and M+2 peaks.

TABLE 12-4
Isotopic Composition of Some Common Elements

Element	M^+		M + 1		M + 2	
hydrogen	^1H	100.0%				
carbon	^{12}C	98.9%	^{13}C	1.1%		
nitrogen	^{14}N	99.6%	^{15}N	0.4%		
oxygen	^{16}O	99.8%			^{18}O	0.2%
sulfur	^{32}S	95.0%	^{33}S	0.8%	^{34}S	4.2%
chlorine	^{35}Cl	75.5%			^{37}Cl	24.5%
bromine	^{79}Br	50.5%			^{81}Br	49.5%
iodine	^{127}I	100.0%				

Ideally, we could use the isotopic compositions in Table 12-4 to determine the entire molecular formula of a compound, by carefully measuring the abundances of the M^+, M+1, and M+2 peaks. In practice, however, there are background peaks at every mass number. These background peaks are often similar in intensity to the M+1 peak, preventing an accurate measurement of the M+1 peak. High-resolution mass spectrometry is much more reliable.

Some elements (particularly S, Cl, Br, I, and N) are recognizable from molecular ion peaks, however, as the spectra shown next illustrate. A typical compound with no sulfur, chlorine, or bromine has a small M+1 peak and an even smaller (or no visible)

M+2 peak. If a compound contains sulfur, the M+2 peak is larger than the M+1 peak: about 4% of the M^+ peak. If chlorine is present, the M+2 peak (containing ^{37}Cl) is about a third as large as the M^+ peak (containing ^{35}Cl). If bromine is present, the M^+ and M+2 ions have about equal abundances; the molecular ion appears as a doublet separated by two mass units, with one mass corresponding to ^{79}Br and one to ^{81}Br.

Iodine is recognized by the presence of the iodonium ion, I^+, at m/z 127. This clue is combined with a characteristic 127-unit gap in the spectrum corresponding to loss of the iodine radical. Nitrogen (or an odd number of nitrogen atoms) gives an odd molecular weight, and usually gives some major even-numbered fragments. Stable compounds containing only carbon, hydrogen, and oxygen have even molecular weights, and most of their major fragments are odd-numbered.

Recognizable elements in the mass spectrum

Br	M+2 as large as M^+
Cl	M+2 a third as large as M^+
I	I^+ at 127; large gap
N	odd M^+, some even fragments
S	M+2 larger than usual (4% of M^+)

The following spectra show compounds containing sulfur, chlorine, and bromine.

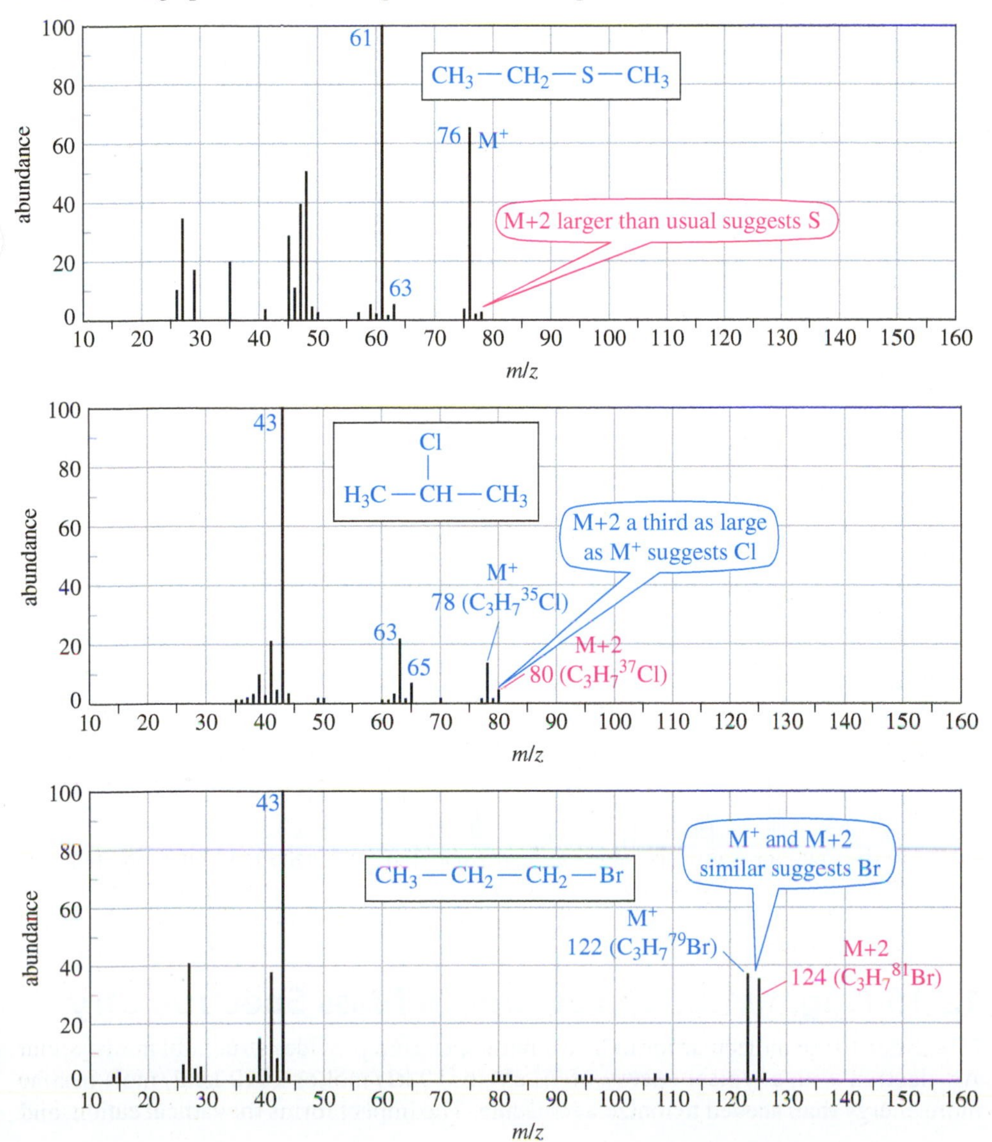

PROBLEM 12-7

Identify which of these four mass spectra indicate the presence of sulfur, chlorine, bromine, iodine, or nitrogen. Suggest a molecular formula for each.

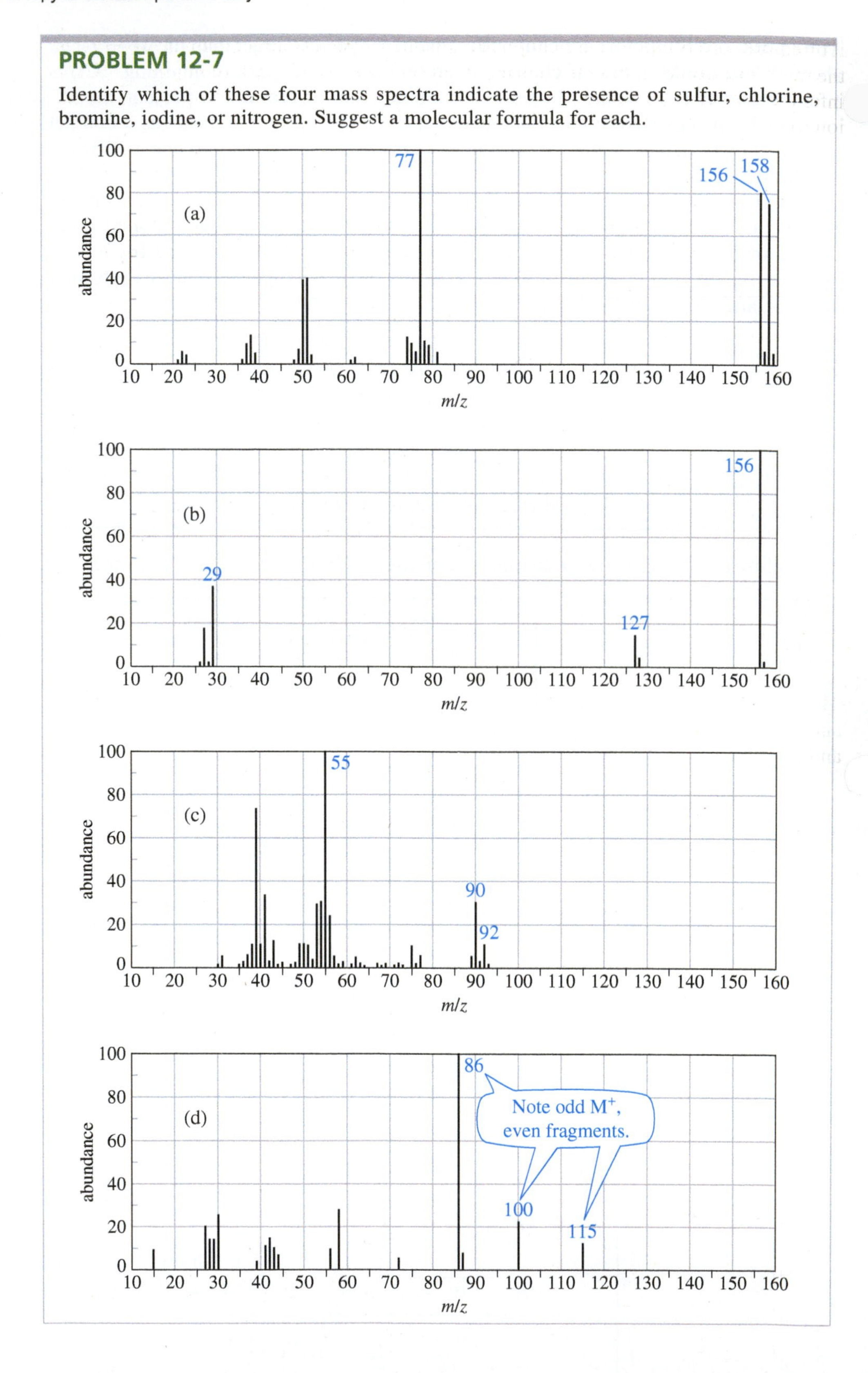

12-15 Fragmentation Patterns in Mass Spectrometry

In addition to the molecular formula, the mass spectrum provides structural information. An electron with a typical energy of 70 eV (6740 kJ/mol or 1610 kcal/mol) has far more energy than needed to ionize a molecule. The impact forms the radical cation, and it often breaks a bond to give a cation and a radical. The resulting cation is observed by the mass spectrometer, but the uncharged radical is not accelerated or detected. We can infer the mass of the uncharged radical from the amount of mass lost from the molecular ion to give the observed cation fragment.

Ionization

$$\text{R:R}' \;+\; e^- \longrightarrow \underset{\substack{\text{radical cation}\\\text{(molecular ion)}}}{[\text{R}\cdot\text{R}']^{+\cdot}} \;+\; 2\,e^-$$

Fragmentation

$$[\text{R}\cdot\text{R}']^{+\cdot} \longrightarrow \underset{\substack{\text{cation fragment}\\\text{(observed)}}}{\text{R}^+} \;+\; \underset{\substack{\text{radical fragment}\\\text{(not observed)}}}{\cdot\text{R}'}$$

This bond breaking does not occur randomly; it tends to form the most stable fragments. By knowing what stable fragments result from different kinds of compounds, we can recognize structural features and use the mass spectrum to confirm a proposed structure.

12-15A Mass Spectra of Alkanes

The mass spectrum of hexane (Figure 12-19) shows several characteristics typical of straight-chain alkanes. Like other compounds not containing nitrogen, the molecular ion (M^+) has an even-numbered mass, and most of the fragments are odd-numbered. The base peak (m/z 57) corresponds to loss of an ethyl group, giving an ethyl radical and a butyl cation. The neutral ethyl radical is not detected, because it is not charged and is not accelerated or deflected.

m/z of the charged fragment on this side of the broken bond: 57

$$\underset{\substack{\text{hexane radical cation}\\ M^+ \ 86}}{[CH_3CH_2CH_2CH_2{-}CH_2CH_3]^{+\cdot}} \longrightarrow \underset{\substack{\text{1-butyl cation}\\ \text{detected at } m/z \ 57}}{CH_3CH_2CH_2CH_2^+} \;+\; \underset{\substack{\text{ethyl radical (29)}\\ \text{not detected}}}{\cdot CH_2CH_3}$$

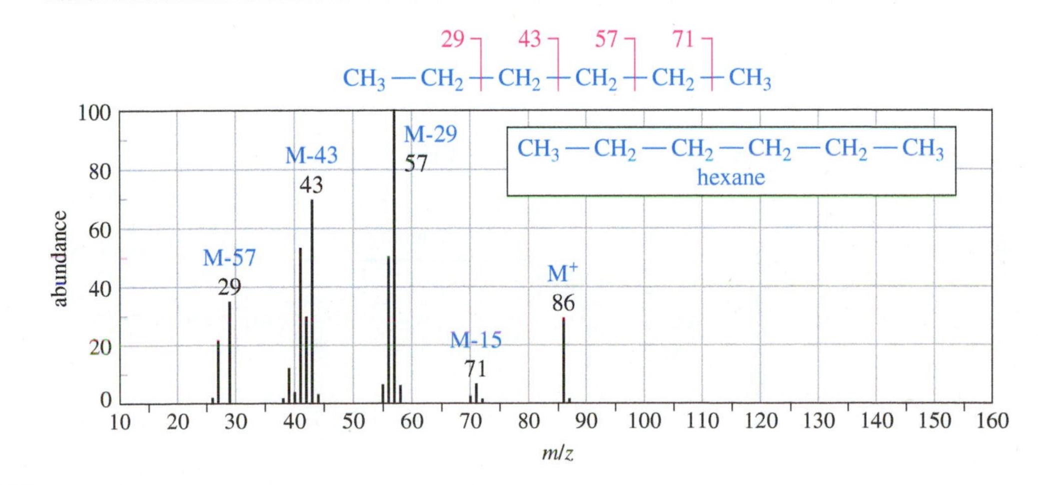

FIGURE 12-19 Mass spectrum of hexane. Groups of ions correspond to loss of one-, two-, three-, and four-carbon fragments.

A similar fragmentation gives an ethyl cation and a butyl radical. In this case, the ethyl fragment (m/z 29) is detected.

$$[CH_3CH_2CH_2CH_2 \vdash CH_2CH_3]^{+\cdot} \longrightarrow CH_3CH_2CH_2CH_2\cdot + {}^+CH_2CH_3$$

(cleavage at 29) hexane radical cation M^+ 86 → 1-butyl radical (57), not detected + ethyl cation, detected at m/z 29

Symmetric cleavage of hexane gives a propyl cation and a propyl radical.

$$[CH_3CH_2CH_2 \vdash CH_2CH_2CH_3]^{+\cdot} \longrightarrow CH_3CH_2CH_2^+ + \cdot CH_2CH_2CH_3$$

(cleavage at 43) hexane radical cation M^+ 86 → propyl cation, detected at m/z 43 + propyl radical (43), not detected

PROBLEM-SOLVING HINT

Most molecular ions have even mass numbers. Most fragments have odd mass numbers. (With a nitrogen atom, the molecular ion is odd and most fragments containing N are even.)

Cleavage to give a pentyl cation (m/z 71) and a methyl radical is weak because the methyl radical is less stable than a substituted radical. Cleavage to give a methyl cation (m/z 15) and a pentyl radical is not visible because the methyl cation is less stable than a substituted cation. The stability of the cation is apparently more important than the stability of the radical, since a weak peak appears corresponding to loss of a methyl radical, but we see no cleavage to give a methyl cation.

$$[CH_3CH_2CH_2CH_2CH_2 \vdash CH_3]^{+\cdot} \longrightarrow CH_3CH_2CH_2CH_2CH_2^+ + \cdot CH_3$$

(cleavage at 71) hexane radical cation M^+ 86 → pentyl cation, weak at m/z 71 + methyl radical (15), not detected

$$[CH_3CH_2CH_2CH_2CH_2 \vdash CH_3]^{+\cdot} \not\longrightarrow CH_3CH_2CH_2CH_2CH_2\cdot + {}^+CH_3$$

(15 (not formed)) hexane radical cation M^+ 86 → pentyl radical (71), not detected + methyl cation (too unstable)

Cation and radical stabilities help to explain the mass spectra of branched alkanes as well. Figure 12-20 shows the mass spectrum of 2-methylpentane. Fragmentation of a branched alkane commonly occurs at a branch carbon atom to give the most highly

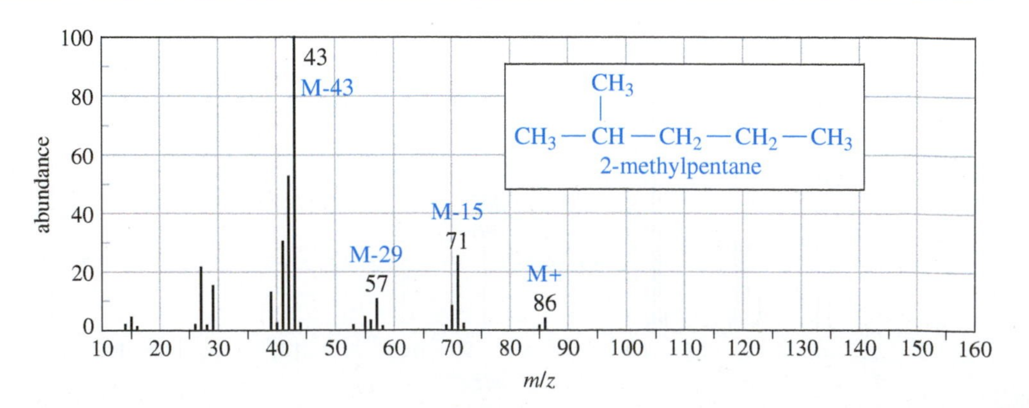

FIGURE 12-20
Mass spectrum of 2-methylpentane. The base peak corresponds to loss of a propyl radical to give an isopropyl cation.

substituted cation and radical. Fragmentation of 2-methylpentane at the branched carbon atom can give a secondary carbocation in either of two ways:

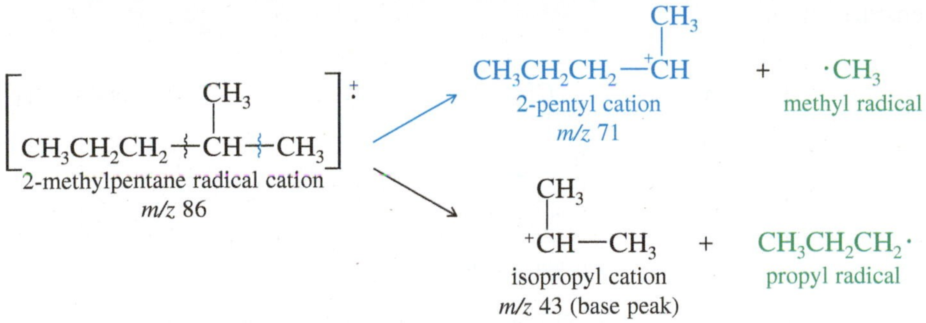

$$\left[\mathrm{CH_3CH_2CH_2{-}CH(CH_3){-}CH_3}\right]^{+\cdot} \longrightarrow \mathrm{CH_3CH_2CH_2{-}\overset{+}{C}H(CH_3)} + \cdot\mathrm{CH_3}$$

2-methylpentane radical cation, m/z 86 → 2-pentyl cation, m/z 71 + methyl radical

$$\left[\mathrm{CH_3CH_2CH_2{-}CH(CH_3){-}CH_3}\right]^{+\cdot} \longrightarrow \mathrm{{}^{+}CH(CH_3){-}CH_3} + \mathrm{CH_3CH_2CH_2\cdot}$$

isopropyl cation, m/z 43 (base peak) + propyl radical

Both fragmentations give secondary cations, but the second gives a primary radical instead of a methyl radical. Therefore, the second fragmentation accounts for the base (largest) peak, while the first accounts for another large peak at m/z 71. Other fragmentations (to give primary cations) account for the weaker peaks.

PROBLEM-SOLVING HINT

The guidelines we used to predict carbocation stability in E1 and S_N1 reactions are also useful for interpreting mass spectra. Relatively stable carbocations are generally more abundant in the mass spectrum.

PROBLEM 12-8

Show the fragmentation that accounts for the cation at m/z 57 in the mass spectrum of 2-methylpentane. Explain why this ion is less abundant than those at m/z 71 and 43.

PROBLEM 12-9

Show the fragmentations that give rise to the peaks at m/z 43, 57, and 85 in the mass spectrum of 2,4-dimethylpentane (Figure 12-17).

12-15B Fragmentation Giving Resonance-Stabilized Cations

Fragmentation in the mass spectrometer gives resonance-stabilized cations whenever possible. The most common fragmentation of alkenes is cleavage of an allylic bond to give a resonance-stabilized allylic cation.

$$[\mathrm{R{-}CH{=}CH{-}CH_2{-}R'}]^{+\cdot} \longrightarrow [\mathrm{R{-}CH{=}CH{-}\overset{+}{C}H_2} \longleftrightarrow \mathrm{R{-}\overset{+}{C}H{-}CH{=}CH_2}] + \cdot\mathrm{R'}$$

allylic cation

Figure 12-21 shows how the radical cation of *trans*-hex-2-ene undergoes allylic cleavage to give the resonance-stabilized cation responsible for the base peak at m/z 55.

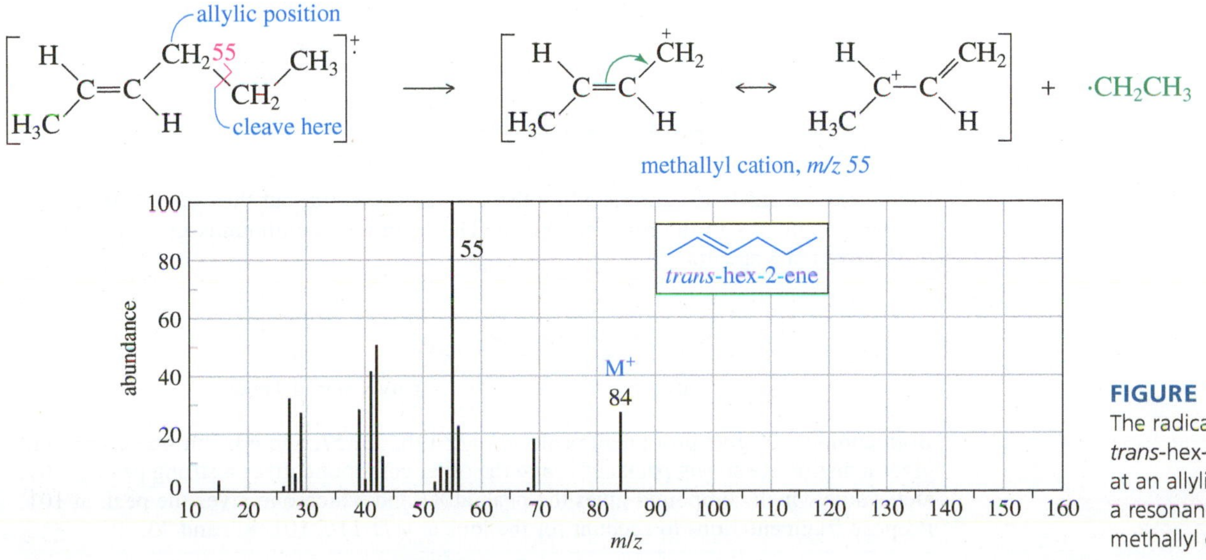

$$[\mathrm{H_3C{-}CH{=}CH{-}CH_2{-}CH_2CH_3}]^{+\cdot} \longrightarrow [\mathrm{H_3C{-}CH{=}CH{-}\overset{+}{C}H_2} \longleftrightarrow \mathrm{H_3C{-}\overset{+}{C}H{-}CH{=}CH_2}] + \cdot\mathrm{CH_2CH_3}$$

allylic position; cleave here

methallyl cation, m/z 55

abundance vs. m/z (mass spectrum of *trans*-hex-2-ene): base peak 55; M^+ 84

FIGURE 12-21 The radical cation of *trans*-hex-2-ene cleaves at an allylic bond to give a resonance-stabilized methallyl cation, m/z 55.

Compounds containing aromatic rings tend to fragment at the carbon (called a *benzylic* carbon) next to the aromatic ring. Such a cleavage forms a resonance-stabilized benzylic cation.

$$[C_6H_5-CH_2-R]^{+\cdot} \longrightarrow [C_6H_5-\overset{+}{C}H_2 \longleftrightarrow \ldots \longleftrightarrow \ldots \longleftrightarrow \ldots] + R^{\cdot}$$

benzylic cation

Ethers, amines, and carbonyl compounds can also fragment to give resonance-stabilized cations. The oxygen and nitrogen atoms in these compounds have nonbonding electrons that can stabilize the positive charge of a cation through resonance forms with octets on all the atoms. Common fragmentations often cleave the bond next to the carbon atom bearing the oxygen or nitrogen. We will see examples of these favorable fragmentations in later chapters covering the chemistry of these functional groups.

PROBLEM-SOLVING HINT

Fragmentations often split off simple alkyl groups, so it helps to recognize their mass numbers:

methyl	CH_3	15
ethyl	C_2H_5	29
propyl	C_3H_7	43
butyl	C_4H_9	57

Note that the numbers increase by 14, the mass of a CH_2 group.

Ketones and aldehydes: loss of alkyl groups to give acylium ions

$$[R-\overset{O}{\overset{\|}{C}}-R']^{+\cdot} \longrightarrow R-C\equiv O^+ + \cdot R'$$

m/z is even — acylium ion (odd m/z)

Ethers: α cleavage

$$[R-CH_2-O-R']^{+\cdot} \xrightarrow{\alpha \text{ cleavage}} H_2C=\overset{+}{O}-R' + \cdot R$$

m/z is even — stabilized cation (odd m/z)

or loss of an alkyl group

$$[R-CH_2-O-R']^{+\cdot} \longrightarrow R-CH=\overset{+}{O}H + \cdot R'$$

m/z is even — stabilized cation (odd m/z)

Amines: α cleavage to give stabilized cations

$$[R_2N-CH_2-R']^{+\cdot} \longrightarrow R_2\overset{+}{N}=CH_2 + \cdot R'$$

m/z is odd — iminium ion (even m/z)

PROBLEM 12-10

Ethers are not easily differentiated by their infrared spectra, but they tend to form predictable fragments in the mass spectrum. The following compounds give similar but distinctive mass spectra.

butyl propyl ether — butyl isopropyl ether

Both compounds give prominent peaks at m/z 116, 73, 57, and 43. But one compound gives a distinctive strong peak at 87, and the other compound gives a strong peak at 101. Determine which compound gives the peak at 87 and which one gives the peak at 101. Propose fragmentations to account for the ions at m/z 116, 101, 87, and 73.

12-15C Fragmentation Splitting Out a Small Molecule; Mass Spectra of Alcohols

Mass spectral peaks are often seen corresponding to loss of small, stable molecules. Loss of a small molecule is usually indicated by a fragment peak with an even mass number, corresponding to loss of an even mass number. A radical cation may lose water (mass 18), CO (28), CO_2 (44), and even ethene (28) or other alkenes. The most common example is the loss of water from alcohols, which occurs so readily that the molecular ion is often weak or absent. The peak corresponding to loss of water (the M–18 peak) is usually strong, however.

Alcohols often lose water.

$$\left[-\overset{\text{H}}{\underset{|}{\overset{|}{\text{C}}}}-\overset{\text{OH}}{\underset{|}{\overset{|}{\text{C}}}}- \right]^{+\cdot} \longrightarrow \left[>\text{C}=\text{C}< \right]^{+\cdot} + H_2O$$

even m/z even m/z loss of 18

The mass spectrum of 3-methylbutan-1-ol (Figure 12-22) shows a favorable loss of water. The even-numbered peak at m/z 70 that *appears* to be the molecular ion is actually the intense M–18 peak. The molecular ion (m/z 88) is not observed because it loses water very readily. The base peak at m/z 55 corresponds to loss of water and a methyl group.

In addition to losing water, alcohols commonly fragment next to the carbinol carbon atom to give a resonance-stabilized carbocation. This fragmentation is called an *alpha cleavage* because it breaks the bond next to the carbon bearing the hydroxy group.

α cleavage of an alcohol

$$\left[-\overset{\text{OH}}{\underset{|}{\overset{|}{\text{C}}}}-\underset{|}{\overset{|}{\text{C}}}- \right]^{+\cdot} \longrightarrow \left[-\overset{:\ddot{\text{O}}\text{H}}{\underset{|}{\overset{|}{\text{C}^+}}} \longleftrightarrow -\overset{^+\ddot{\text{O}}\text{H}}{\underset{|}{\overset{\|}{\text{C}}}} \right] + \cdot\underset{|}{\overset{|}{\text{C}}}-$$

resonance-stabilized

An alpha cleavage is prominent in the spectrum of 2,6-dimethylheptan-4-ol shown in Problem 12-11.

$$\left[CH_3-\underset{}{\overset{CH_3}{\overset{|}{CH}}}-CH_2-CH_2-OH \right]^{+\cdot} \nearrow H_2O + \left[CH_3-\overset{CH_3}{\overset{|}{CH}}-CH=CH_2 \right]^{+\cdot} \quad m/z\ 70$$

$$\searrow H_2O + \cdot CH_3 + \left[CH_3-\overset{H}{\overset{|}{C^+}}-CH=CH_2 \longleftrightarrow CH_3-\overset{H}{\overset{|}{C}}=CH-\overset{+}{C}H_2 \right]$$

allylic cation m/z 55 (base peak)

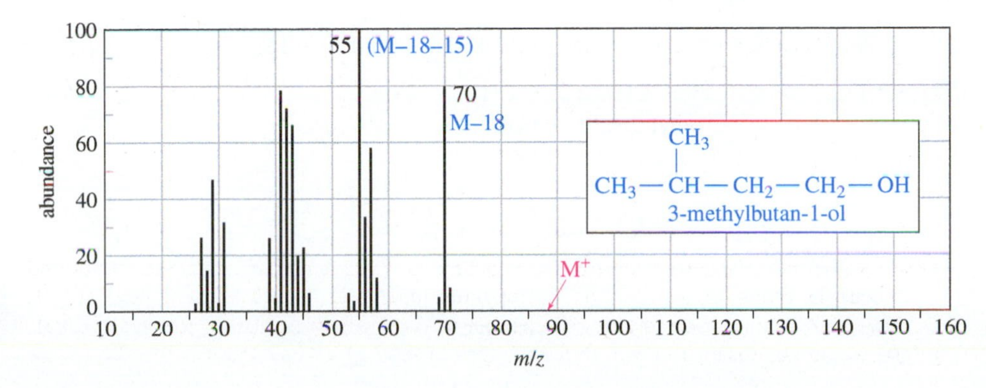

FIGURE 12-22 The mass spectrum of 3-methylbutan-1-ol. The strong peak at m/z 70 is actually the M–18 peak, corresponding to loss of water. The molecular ion is not visible because it loses water easily.

PROBLEM-SOLVING HINT

In general, you should be able to propose favorable fragmentations for two or three of the largest peaks in a spectrum. Also, the spectrum should contain large peaks corresponding to the most favorable fragmentations of your proposed structure. You shouldn't expect to account for all the peaks, however.

PROBLEM 12-11

Account for the peaks at m/z 87, 111, and 126 in the mass spectrum of 2,6-dimethylheptan-4-ol.

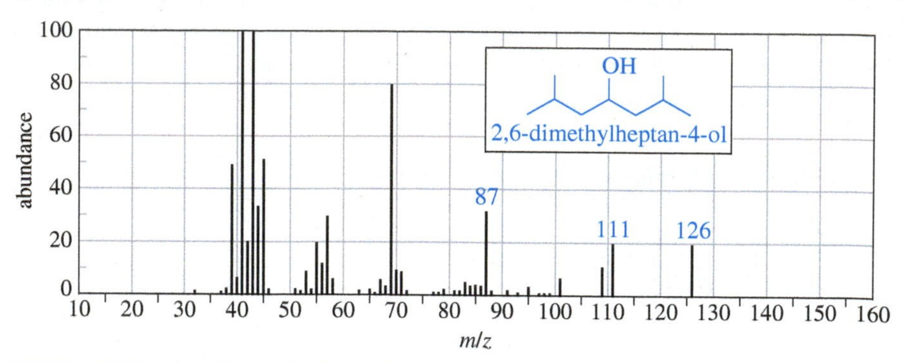

SUMMARY Common Fragmentation Patterns

This summary is provided for rapid reference to the common fragmentation patterns of simple functional groups. Some of these functional groups are discussed in more detail in later chapters.

1. *Alkanes:* cleavage to give the most stable carbocations (Section 12-15A)

$$\left[R-\underset{H}{\overset{R'}{C}}-R'' \right]^{+\cdot} \longrightarrow R-\underset{H}{\overset{R'}{C^+}} \; + \; \cdot R''$$

m/z is even → m/z is odd

2. *Alcohols:* loss of water (Section 12-15C)

$$\left[-\overset{H}{C}-\overset{OH}{C}- \right]^{+\cdot} \longrightarrow \left[>C=C< \right]^{+\cdot} \; + \; H_2O$$

m/z is even → m/z is even

or α cleavage (Section 12-15C)

$$\left[-\overset{OH}{C}-C- \right]^{+\cdot} \longrightarrow \left[-\overset{OH}{C^+} \longleftrightarrow -\overset{^+OH}{\overset{\|}{C}} \right] \; + \; \cdot C-$$

m/z is even → m/z is odd

3. *Alkenes and aromatics:* cleavage to give allylic and benzylic carbocations (Section 12-15B and Section 16-15)

$$[R-CH=CH-CH_2-R']^{+\cdot} \longrightarrow R-CH=CH-\overset{+}{C}H_2 \; + \; \cdot R'$$

allylic cation (odd m/z)

$$\left[C_6H_5-CH_2-R \right]^{+\cdot} \longrightarrow C_6H_5-\overset{+}{C}H_2 \; + \; \cdot R \longrightarrow C_7H_7^+ \; + \; \cdot R$$

benzylic cation m/z 91 — tropylium ion m/z 91

4. *Amines:* α cleavage next to the carbon bearing the nitrogen to give stabilized cations (Section 12-15B and 19-8D)

$$[R_2N—CH_2 | R']^{+\cdot} \longrightarrow R_2\overset{+}{N}{=}CH_2 + \cdot R'$$

m/z is odd — iminium ion (even *m/z*)

5. *Ethers:* loss of an alkyl group (Sections 12-15B and 14-4)

$$[R—CH_2—O | R']^{+\cdot} \longrightarrow R—CH{=}\overset{+}{O}H + \cdot R'$$

m/z is even — stabilized cation (odd *m/z*)

or

$$[R—CH_2—O | R']^{+\cdot} \longrightarrow R—CH_2—\ddot{O}\cdot + {}^{+}R$$

m/z is even — alkyl cation (odd *m/z*)

or α cleavage next to the carbon bearing the oxygen

$$[R | CH_2—O—R']^{+\cdot} \xrightarrow{\alpha\ \text{cleavage}} H_2C{=}\overset{+}{O}—R' + \cdot R$$

m/z is even — stabilized cation (odd *m/z*)

6. *Ketones and aldehydes:* loss of alkyl groups next to the carbon bearing the oxygen to give acylium ions (Section 12-15B and 18-5)

$$[R—C({=}O) | R']^{+\cdot} \longrightarrow R—C{\equiv}O^{+} + \cdot R'$$

m/z is even — acylium ion (odd *m/z*)

The McLafferty rearrangement splits out alkenes (covered in Section 18-5).

γ-hydrogen

$$[R—C({=}O)—C_{\alpha}—C_{\beta}—C_{\gamma}(H)—R']^{+\cdot} \longrightarrow [R—C(OH){=}C_{\alpha}]^{+\cdot} + C_{\beta}{=}C_{\gamma}R'$$

m/z is even — *m/z* is even

Essential Terms

absorption spectroscopy The measurement of the amount of light absorbed by a compound as a function of the wavelength. (p. 556)

base peak The strongest peak in a mass spectrum. (p. 586)

conjugated double bonds Double bonds that alternate with single bonds, allowing their pi bonding orbitals to overlap with each other. (p. 565)

electromagnetic spectrum The range of all possible electromagnetic frequencies from zero to infinity. In practice, it ranges from radio waves up to gamma rays. (p. 558)

fingerprint region The portion of the infrared spectrum between 600 and 1400 cm^{-1}, where many complex vibrations occur. So named because no two different compounds (except enantiomers) have exactly the same absorptions in this region. (p. 561)

FT–IR **(Fourier-transform infrared spectrometer)** Infrared light passes through both the sample and a scanning interferometer to give an interference pattern (interferogram). The interferogram is digitized, and the Fourier-transformed spectrum is calculated. (p. 563)

fragmentation The breaking apart of a molecular ion upon ionization in a mass spectrometer. (p. 585)

frequency (ν) The number of complete wave cycles that pass a fixed point in a second; or the number of reversals of the electromagnetic field per second. (p. 557)

gas chromatograph (GC) An instrument that vaporizes a mixture, passes the vapor through a column to separate the components, and detects the components as they emerge from the column. Mass spectrometry is one of the methods used to detect the components. (p. 587)

HRMS **(high-resolution mass spectrometer)** A mass spectrometer that measures masses very accurately, usually to 1 part in 20,000. This high precision allows calculation of molecular formulas using the known atomic masses of the elements. (p. 588)

infrared spectrometer A device that measures a compound's absorption of infrared light as a function of frequency or wavelength. (p. 562)

infrared spectrum A graph of the infrared energy absorbed by a sample as a function of the frequency ($\bar{\nu}$, expressed as a wavenumber, cm^{-1}) or the wavelength (λ, expressed in μm). (p. 559)

interferometer The light-measuring portion of an FT–IR spectrometer. The light is split into two beams. One beam is reflected from a stationary mirror, and the other from a moving mirror. The beams are recombined to form an interference pattern called an **interferogram.** Fourier transformation of the interferogram gives the spectrum. (p. 563)

IR-active A vibration that changes the dipole moment of the molecule and thus can absorb infrared light. (p. 562)

IR-inactive A vibration that does not change the dipole moment of the molecule and thus cannot absorb infrared light. (p. 562)

mass spectrometer An instrument that ionizes molecules, sorts the ions according to their masses, and records the abundance of ions of each mass. (p. 584)

mass spectrum The graph produced by a mass spectrometer, showing the masses along the *x* axis and their abundance along the *y* axis. (p. 584)

- ***m*/*z* (formerly *m*/*e*):** The mass-to-charge ratio of an ion. Most ions have a charge of +1, and *m*/*z* simply represents their masses. (p. 585)

molecular ion, M^+ **(parent ion)** In mass spectrometry, the ion with the same mass as the molecular weight of the original compound; no fragmentation has occurred. (p. 585)

- **M + 1 peak:** An isotopic peak that is one mass unit heavier than the major molecular ion peak. (p. 588)
- **M + 2 peak:** An isotopic peak that is two mass units heavier than the major molecular ion peak. (p. 588)

overtone A relatively weak absorption at a multiple of (usually double) the fundamental vibration frequency. Occurs with very strong absorptions, such as those of carbonyl (C=O) groups. (p. 562)

photon A massless packet of electromagnetic energy. (p. 557)

radical cation A positively charged ion with an unpaired electron; commonly formed by electron impact ionization, when the impinging electron knocks out an additional electron. (p. 585)

$$\mathrm{R:R} + e^- \longrightarrow \underset{\text{radical cation}}{[\mathrm{R\cdot R}]^+} + 2e^-$$

source (ion source) The part of a mass spectrometer where the sample is ionized and undergoes fragmentation. (p. 585)

wavelength (λ) The distance between any two peaks (or any two troughs) of a wave. (p. 557)

wavenumber ($\bar{\nu}$) The number of wavelengths that fit into one centimeter (cm^{-1} or reciprocal centimeters); proportional to the frequency. The product of the wavenumber (in cm^{-1}) and the wavelength (in μm) is 10,000. (p. 558)

Essential Problem-Solving Skills in Chapter 12

Each skill is followed by problem numbers exemplifying that particular skill.

1 Identify the reliable characteristic peaks in an infrared spectrum, and propose which functional groups are likely to be present in the molecule.	Problems 12-14, 15, 16, 23, 24, 25, and 30
2 Explain which functional groups cannot be present in a molecule because their characteristic peaks are absent in the IR spectrum.	Problems 12-15, 16, 19, 23, 24, and 25
3 Explain why some characteristic infrared absorptions are usually strong, and why others may be weak or absent.	Problems 12-14, 16, 19, 25, and 28
4 Identify conjugated and strained C=O bonds and conjugated and aromatic C=C bonds from their absorptions in the IR spectrum.	Problems 12-15, 16, 24, 25, 28, and 30
5 Determine a compound's molecular weight from its mass spectrum.	Problems 12-20, 23, 24, and 29
6 Recognize the presence of Br, Cl, I, N, and S atoms, based on the mass spectrum.	Problems 12-20, 26, 27, 29, and 31
7 Given a structure, predict the major ions that will be observed in the mass spectrum from fragmentation of the molecular ion.	Problems 12-17, 18, 22, 29, and 31
8 Use the fragmentation pattern to determine whether a proposed structure is consistent with the mass spectrum.	Problems 12-20, 23, 26, 29, and 31

Study Problems

12-12 Predict the characteristic infrared absorptions of the functional groups in the following molecules.

(a) cyclohexene **(b)** pentan-2-ol **(c)** pentan-2-one
(d) pent-1-yne **(e)** diethylamine **(f)** pentanoic acid
(g) pentanenitrile **(h)** ethyl acetate **(i)** pentanamide

12-13 Convert the following infrared wavelengths to cm^{-1}.

(a) 6.24 μm, typical for an aromatic C=C
(b) 3.38 μm, typical for a saturated C—H bond
(c) 5.85 μm, typical for a ketone carbonyl
(d) 5.75 μm, typical for an ester carbonyl
(e) 4.52 μm, typical for a nitrile
(f) 3.03 μm, typical for an alcohol O—H

12-14 All of the following compounds absorb infrared radiation between 1600 and 1800 cm^{-1}. In each case,

1. show which bonds absorb in this region.
2. predict the approximate absorption frequencies.
3. predict which compound of each pair absorbs more strongly in this region.

(a) $H_2C{=}CH(CH_2CH_3)$ or $O{=}CH(CH_2CH_3)$

(b) $H_2C{=}CH(CH_2CH_3)$ or $H_2C{=}CH(OCH_2CH_3)$

(c) $H\ddot{N}{=}CH(CH_2CH_3)$ or $H_2C{=}CH(CH_2CH_3)$

(d) $(H)(H_3C)C{=}C(CH_3)(H)$ or $H_2C{=}CH(CH_2CH_3)$

12-15 Describe the characteristic infrared absorption frequencies that would allow you to distinguish between the following pairs of compounds.

(a) 2,3-dimethylbut-2-ene and 2,3-dimethylbut-1-ene

(b) cyclohexa-1,3-diene and cyclohexa-1,4-diene

(c) $CH_3(CH_2)_3—C(=O)—H$ (pentanal) and $CH_3(CH_2)_2—C(=O)—CH_3$ (pentan-2-one)

(d) $CH_3CH_2CH_2—C(=O)—NH_2$ (butanamide) and $CH_3CH_2—C(=O)—CH_2CH_3$ (pentan-3-one)

(e) $CH_3(CH_2)_5—C{\equiv}C—H$ (oct-1-yne) and $CH_3(CH_2)_6—C{\equiv}N$ (octanenitrile)

(f) $CH_3CH_2CH_2—C(=O)—OH$ (butanoic acid) and $CH_3—CH(OH)—CH_2—C(=O)—H$ (3-hydroxybutanal)

(g) cyclohexanol and cyclohexanone

(h) cyclohex-2-enone and cyclohex-3-enone

12-16 Four infrared spectra are shown, corresponding to four of the following compounds. For each spectrum, determine the structure and explain how the peaks in the spectrum correspond to the structure you have chosen.

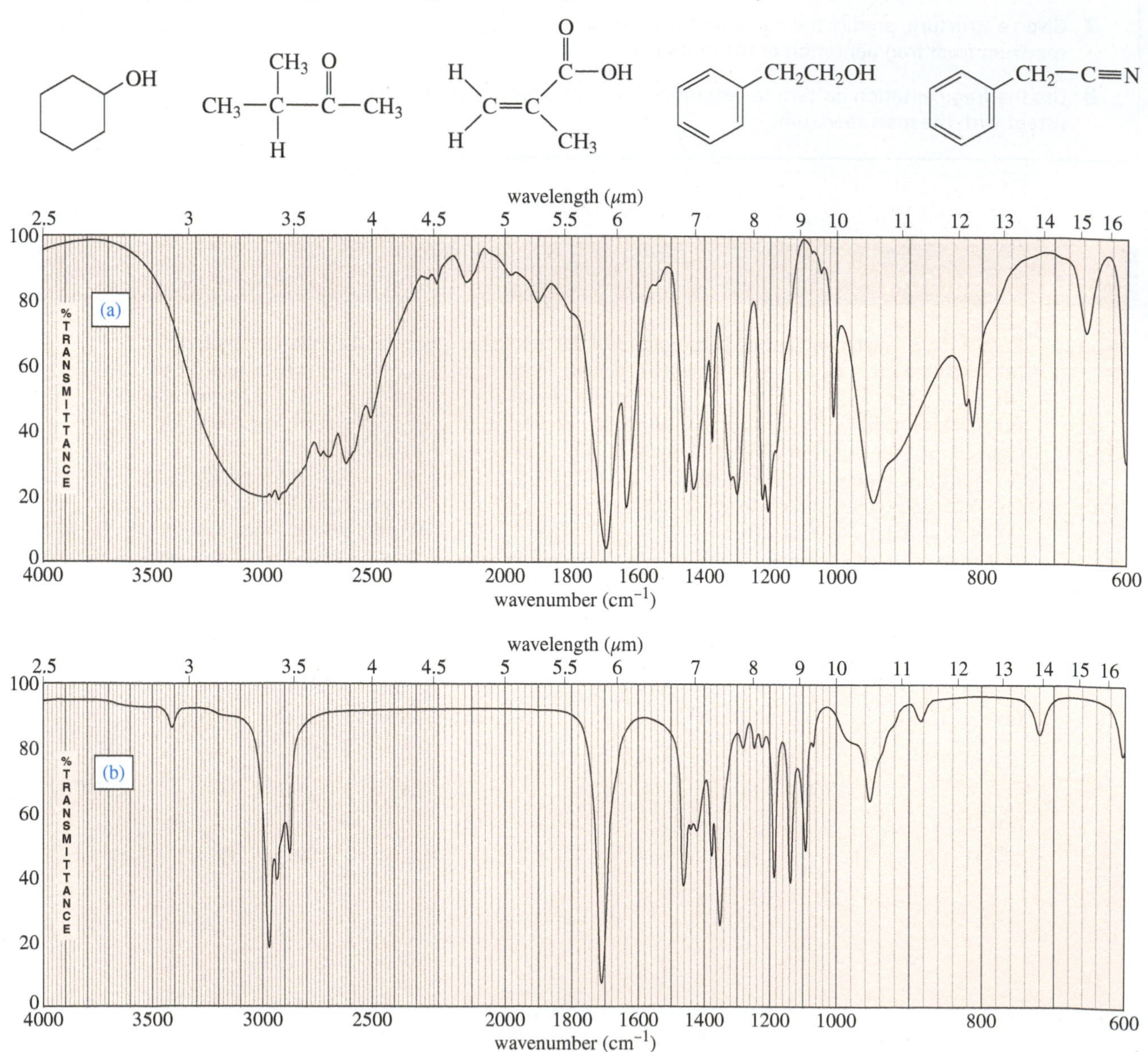

(continued)

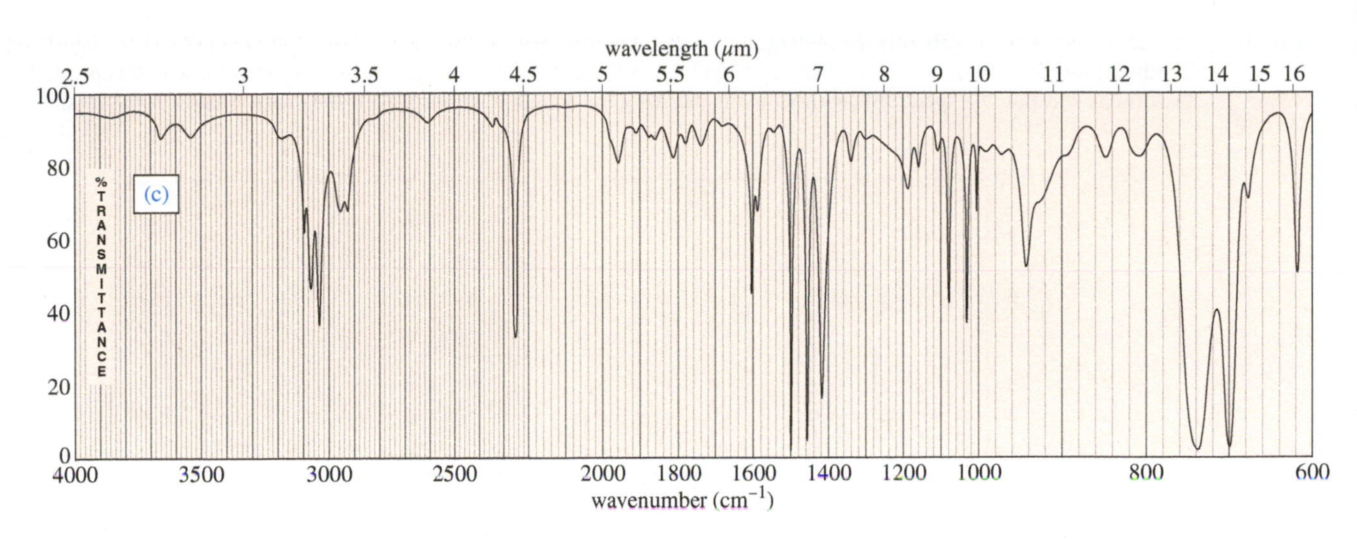

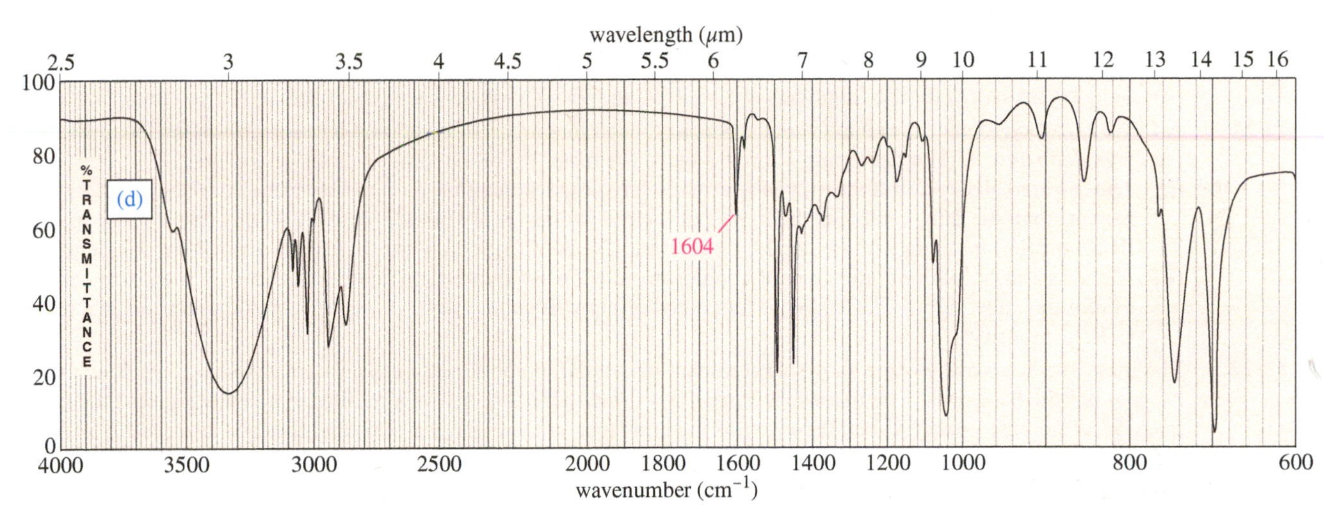

12-17 Predict the masses and the structures of the most abundant fragments observed in the mass spectra of the following compounds.

(a) 2-methylpentane
(b) 3-methylhex-2-ene
(c) 4-methylpentan-2-ol
(d) 2-methyl-1-phenylpropane
(e) cyclohexyl isopropyl ether [cyclohexyl—O—$CH(CH_3)_2$]
(f) $CH_3CH_2CH_2NHC(CH_3)_3$ *tert*-butyl propyl amine
(g) C_6H_5—C(=O)—CH_3 acetophenone
***(h)** 3-bromo-2-methylpentane

12-18 Give logical fragmentation reactions to account for the following ions observed in these mass spectra.

(a) *n*-octane: 114, 85, 71, 57
(b) methylcyclohexane: 98, 83
(c) 2-methylpent-2-ene: 84, 69
(d) pentan-1-ol: 70, 55, 41, 31
(e) *N*-ethylaniline ($PhNHCH_2CH_3$): 121, 106, 77
***(f)** 1-bromo-2-methylbutane: 152, 150, 123, 121, 71 (base)

12-19 A common lab experiment is the dehydration of cyclohexanol to cyclohexene.

(a) Explain how you could tell from the IR spectrum whether your product was pure cyclohexene, pure cyclohexanol, or a mixture of cyclohexene and cyclohexanol. Give approximate frequencies for distinctive peaks.
(b) Explain why mass spectrometry might not be a good way to distinguish cyclohexene from cyclohexanol.

12-20 (A true story.) While organizing the undergraduate stockroom, a new chemistry professor found a half-gallon jug containing a cloudy liquid (bp 100–105 °C), marked only "STUDENT PREP." She ran a quick mass spectrum, which is printed below. As soon as she saw the spectrum (without even checking the actual mass numbers), she said, "I know what it is."

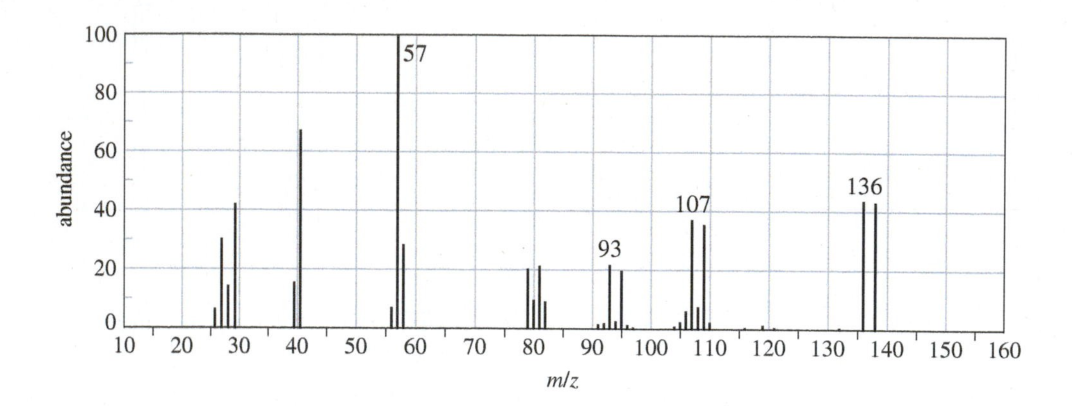

(a) What compound is the "student prep"? Any uncertainty in the structure?

(b) Suggest structures for the fragments at 136, 107, and 93. Why is the base peak (at m/z 57) so strong?

***12-21** A C—D (carbon–deuterium) bond is electronically much like a C—H bond, and it has a similar stiffness, measured by the *spring constant, k*. The deuterium atom has twice the mass (m) of a hydrogen atom, however.

(a) The infrared absorption frequency is approximately proportional to $\sqrt{k/m}$, when one of the bonded atoms is much heavier than the other, and m is the lighter of the two atoms (H or D in this case). Use this relationship to calculate the IR absorption frequency of a typical C—D bond. Use 3000 cm^{-1} as a typical C—H absorption frequency.

(b) A chemist dissolves a sample in deuterochloroform ($CDCl_3$) and then decides to take the IR spectrum and simply evaporates most of the $CDCl_3$. What functional group will *appear* to be present in this IR spectrum as a result of the $CDCl_3$ impurity?

***12-22** The mass spectrum of n-octane shows a prominent molecular ion peak (m/z 114). There is also a large peak at m/z 57, but it is not the base peak. The mass spectrum of 3,4-dimethylhexane shows a smaller molecular ion, and the peak at mass 57 is the base peak. Explain these trends in abundance of the molecular ions and the ions at mass 57, and predict the intensities of the peaks at masses 57 and 114 in the spectrum of 2,2,3,3-tetramethylbutane.

12-23 An unknown, foul-smelling hydrocarbon gives the mass spectrum and infrared spectrum shown.

(a) Use the mass spectrum to propose a molecular formula. How many elements of unsaturation are there?

(b) Use the IR spectrum to determine the functional group(s), if any.

(c) Propose one or more structures for this compound. What parts of the structure are uncertain? If you knew that hydrogenation of the compound gives n-octane, would the structure still be uncertain?

(d) Propose structures for the major fragments at 39, 67, 81, and 95 in the mass spectrum.

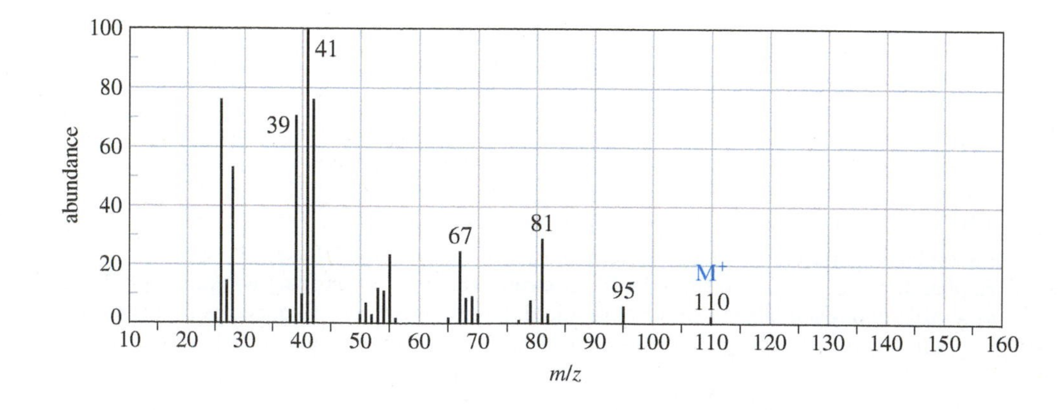

(continued)

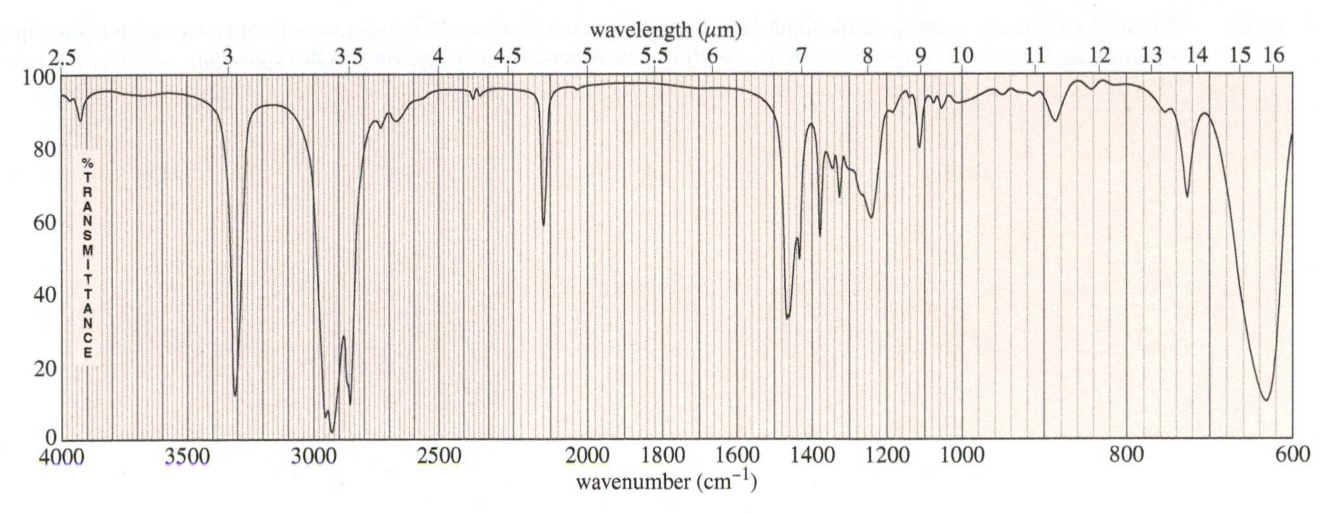

***12-24** Chapter 9 covered a synthesis of alkynes by a double dehydrohalogenation of dihalides. A student tried to convert *trans*-2,5-dimethylhex-3-ene to 2,5-dimethylhex-3-yne by adding bromine across the double bond and then doing a double elimination. The infrared and mass spectra of the major product are shown here.

$\xrightarrow{Br_2}$ (Br, Br dibromide) $\xrightarrow[\text{heat}]{\text{KOH}}$? C≡C

(a) Do the spectra confirm the right product? If not, what is it?
(b) Explain the important peaks in the IR spectrum.

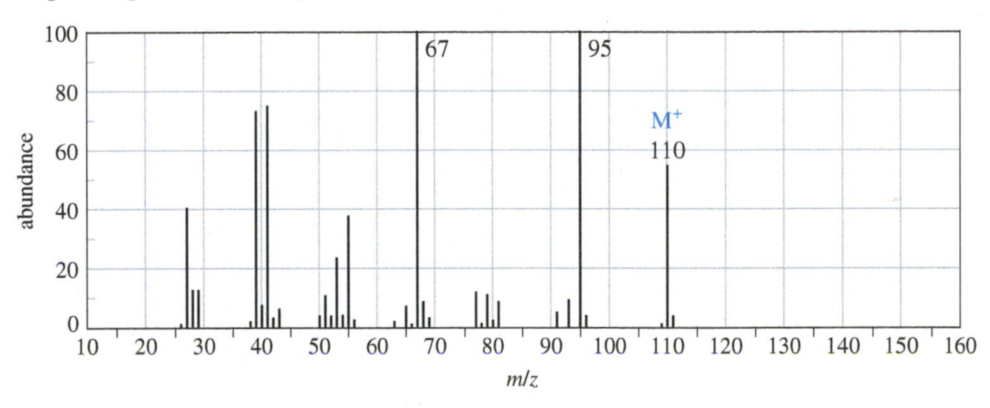

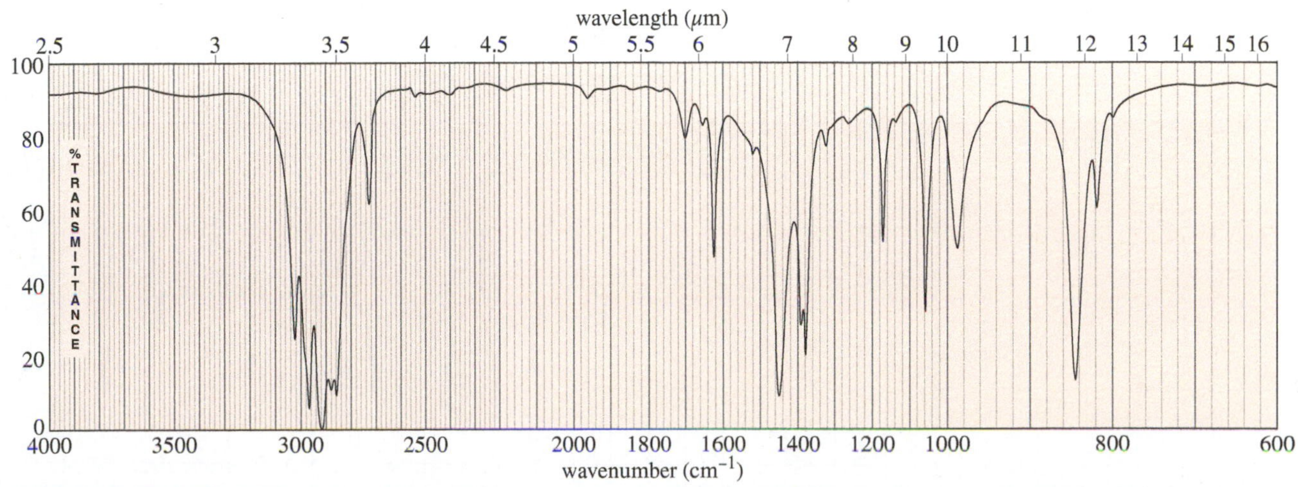

12-25 Three IR spectra are shown, corresponding to three of the following compounds. For each spectrum, determine the structure and explain how the peaks in the spectrum correspond to the structure you have chosen.

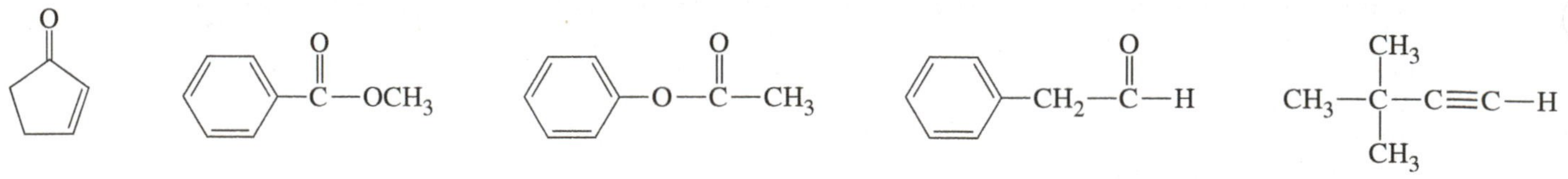

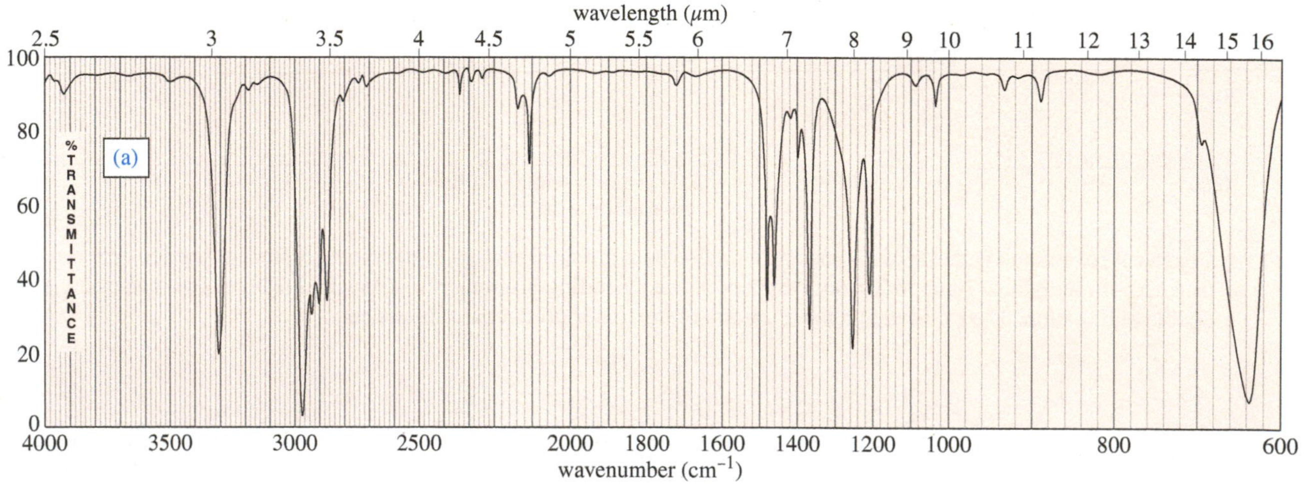

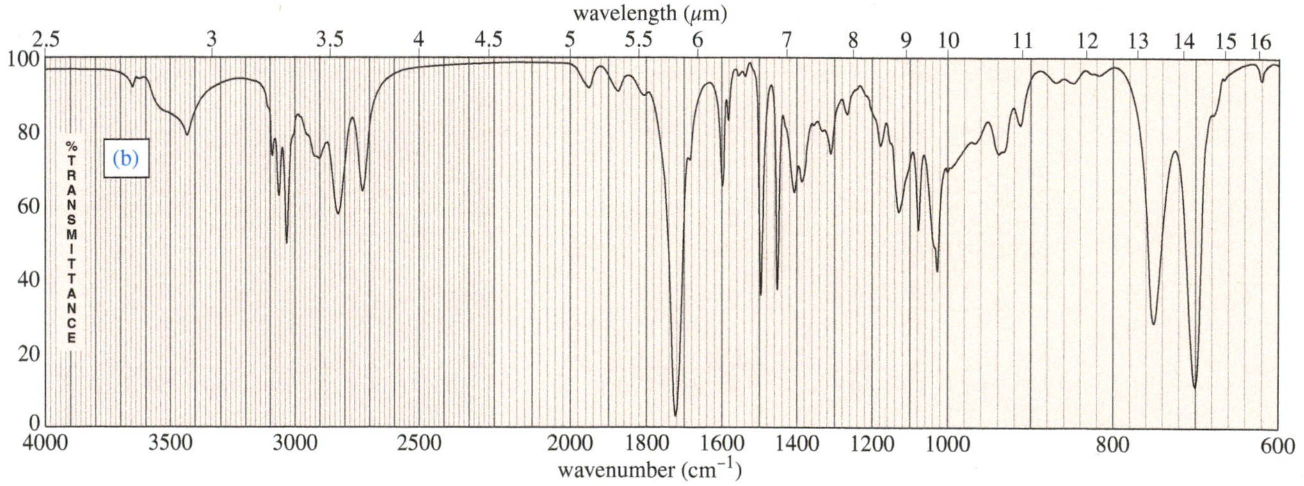

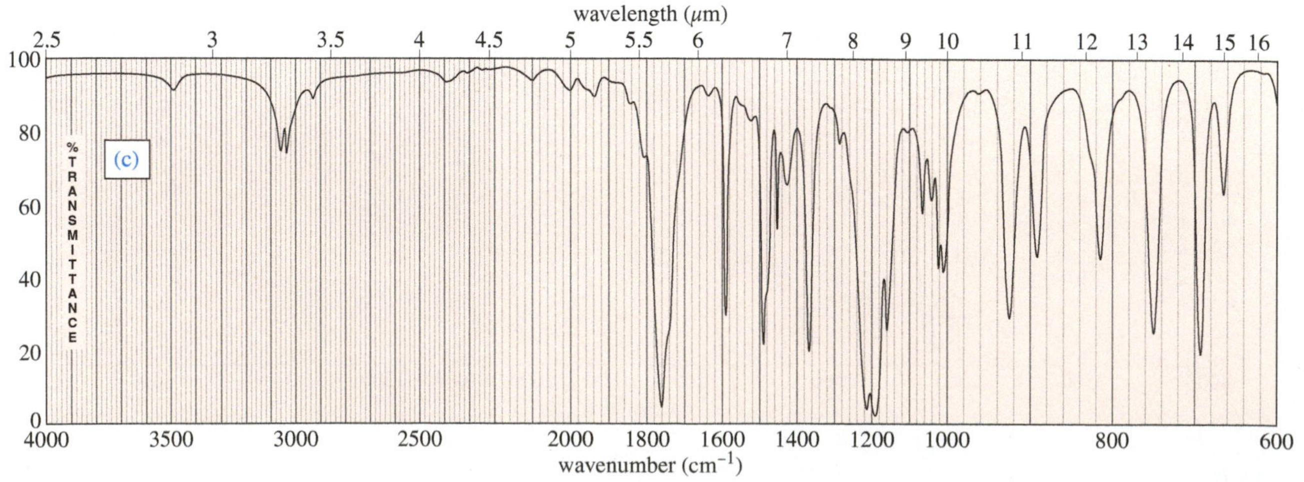

O (cyclopentenone) Ph—C(=O)—OCH$_3$ Ph—O—C(=O)—CH$_3$ Ph—CH$_2$—C(=O)—H CH$_3$—C(CH$_3$)$_2$—C≡C—H

wavelength (μm)
2.5 3 3.5 4 4.5 5 5.5 6 7 8 9 10 11 12 13 14 15 16
100 80 60 40 20 0
% TRANSMITTANCE
(a)
4000 3500 3000 2500 2000 1800 1600 1400 1200 1000 800 600
wavenumber (cm^{-1})

wavelength (μm)
2.5 3 3.5 4 4.5 5 5.5 6 7 8 9 10 11 12 13 14 15 16
100 80 60 40 20 0
% TRANSMITTANCE
(b)
4000 3500 3000 2500 2000 1800 1600 1400 1200 1000 800 600
wavenumber (cm^{-1})

wavelength (μm)
2.5 3 3.5 4 4.5 5 5.5 6 7 8 9 10 11 12 13 14 15 16
100 80 60 40 20 0
% TRANSMITTANCE
(c)
4000 3500 3000 2500 2000 1800 1600 1400 1200 1000 800 600
wavenumber (cm^{-1})

12-26 A laboratory student added 1-bromobutane to a flask containing dry ether and magnesium turnings. An exothermic reaction resulted, and the ether boiled vigorously for several minutes. Then she added acetone to the reaction mixture and the ether boiled even more vigorously. She added dilute acid to the mixture and separated the layers. She evaporated the ether layer, and distilled a liquid that boiled at 143 °C. GC–MS analysis of the distillate showed one major product with a few minor impurities. The mass spectrum of the major product is shown here.

(a) Draw out the reactions that took place and show the product that was formed.

(b) Explain why the molecular ion is or is not visible in the mass spectrum, and show what ions are likely to be responsible for the strong peaks at m/z 59 and 101.

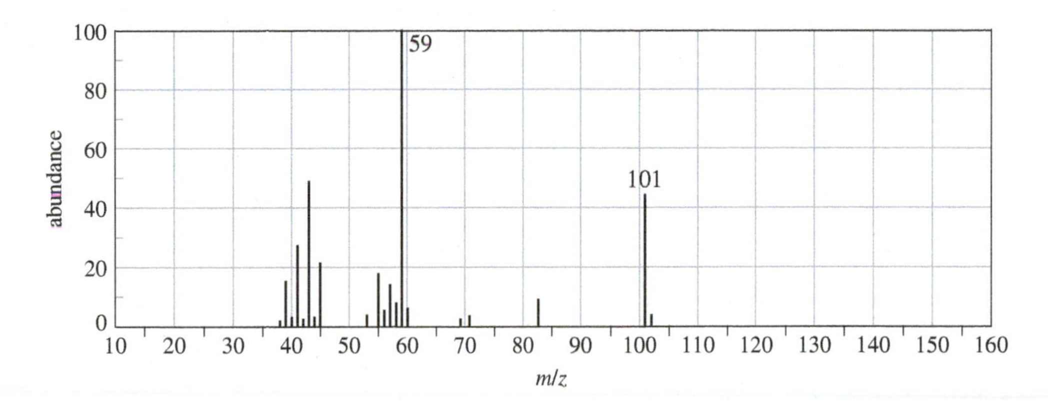

12-27 (Another true story.) A student who was checking into her lab desk found an unlabeled sample from a previous student. She was asked to identify the sample. She did an IR spectrum and declared, "It looks like an alkane." But it seemed too reactive to be an alkane, so she did a GC–MS. The mass spectrum is shown next. Identify the compound as far as you can, and state what part of your identification is uncertain. Propose fragments corresponding to the numbered peaks.

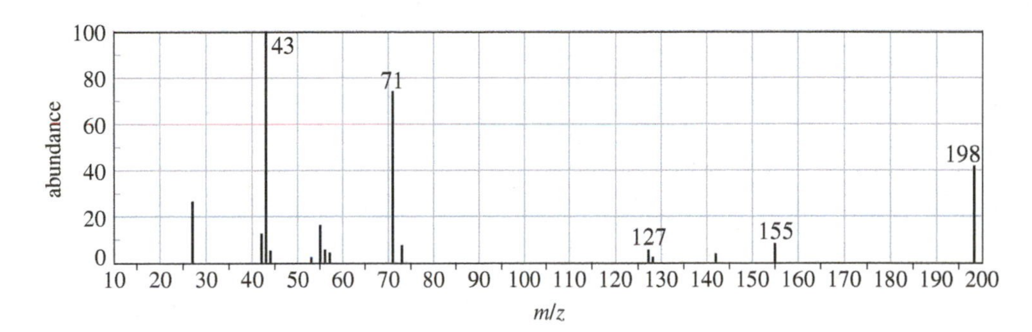

12-28 Three common lab experiments are shown. In each case, describe how the IR spectrum of the product would differ from that of the reactant. Give approximate frequencies for distinctive peaks in the IR spectrum of the reactant and also that of the product.

$H_3C-C(CH_3)(OH)-C(CH_3)(OH)-CH_3 \xrightarrow[\text{heat}]{H_2SO_4} H_3C-C(=O)-C(CH_3)_2-CH_3$

pinacol → pinacolone

salicylic acid (2-hydroxybenzoic acid, $C_6H_4(OH)COOH$) $\xrightarrow{CH_3OH,\ H^+}$ methyl salicylate (wintergreen) ($C_6H_4(OH)COOCH_3$)

cinnamaldehyde ($C_6H_5CH=CH-CHO$) $\xrightarrow[CH_3OH]{NaBH_4}$ cinnamyl alcohol ($C_6H_5CH=CH-CH_2OH$)

***12-29** The ultimate test of fluency in MS and IR is whether you can determine a moderately complex structure from just the MS and the IR, with no additional information. The IR and MS of a compound are shown below. Use everything you know about IR and MS, plus reasoning and intuition, to determine a likely structure. Then show how your proposed structure is consistent with these spectra.

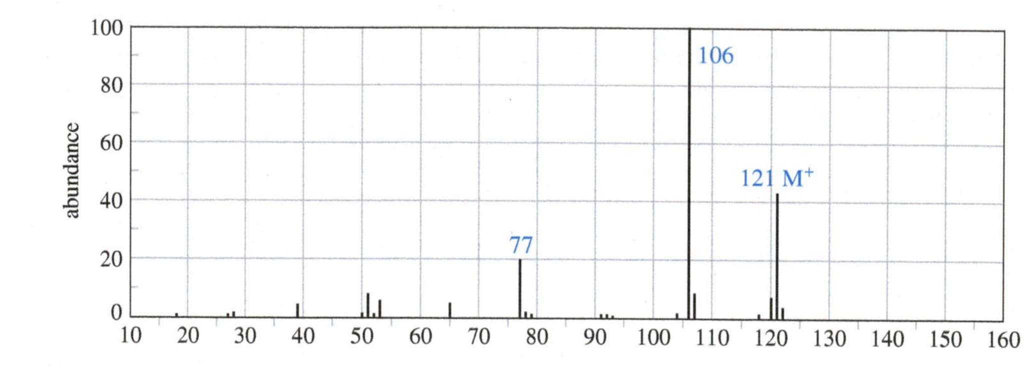

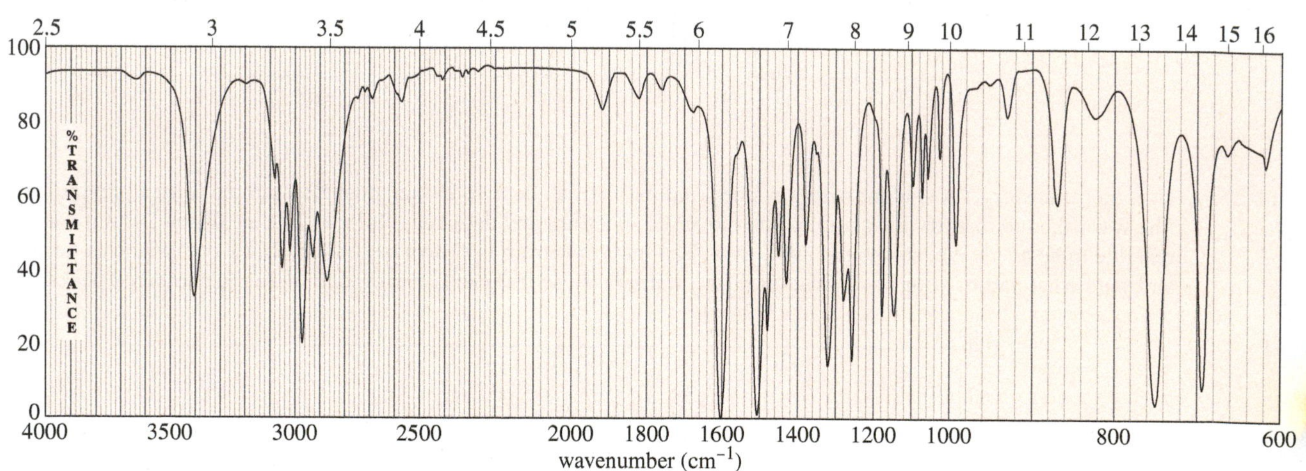

12-30 These five structures all have distinguishing absorptions in the IR. Match each structure with its characteristic absorption.

OH — Structure 1 | O — Structure 2 | NH_2 — Structure 3 | CN — Structure 4 | O, N — Structure 5

(a) sharp, 2254 cm^{-1}
(b) very broad, centered about 3330 cm^{-1}
(c) strong, slightly broadened, 1645 cm^{-1}
(d) broad with spikes at 3367 and 3292 cm^{-1}
(e) strong, sharp 1717 cm^{-1}

12-31 Consider the following four structures, followed by mass spectral data. Match each structure with its characteristic molecular ion or fragment. In each case, give a likely structure of the ion responsible for the base peak.

Br — Structure 1 | O — Structure 2 | O — Structure 3 | NH_2 — Structure 4

(a) base peak at 105
(b) base peak at 72
(c) M^+ doublet at 198 and 200, base peak at 91
(d) base peak at 91, large peak at 43

13 Nuclear Magnetic Resonance Spectroscopy

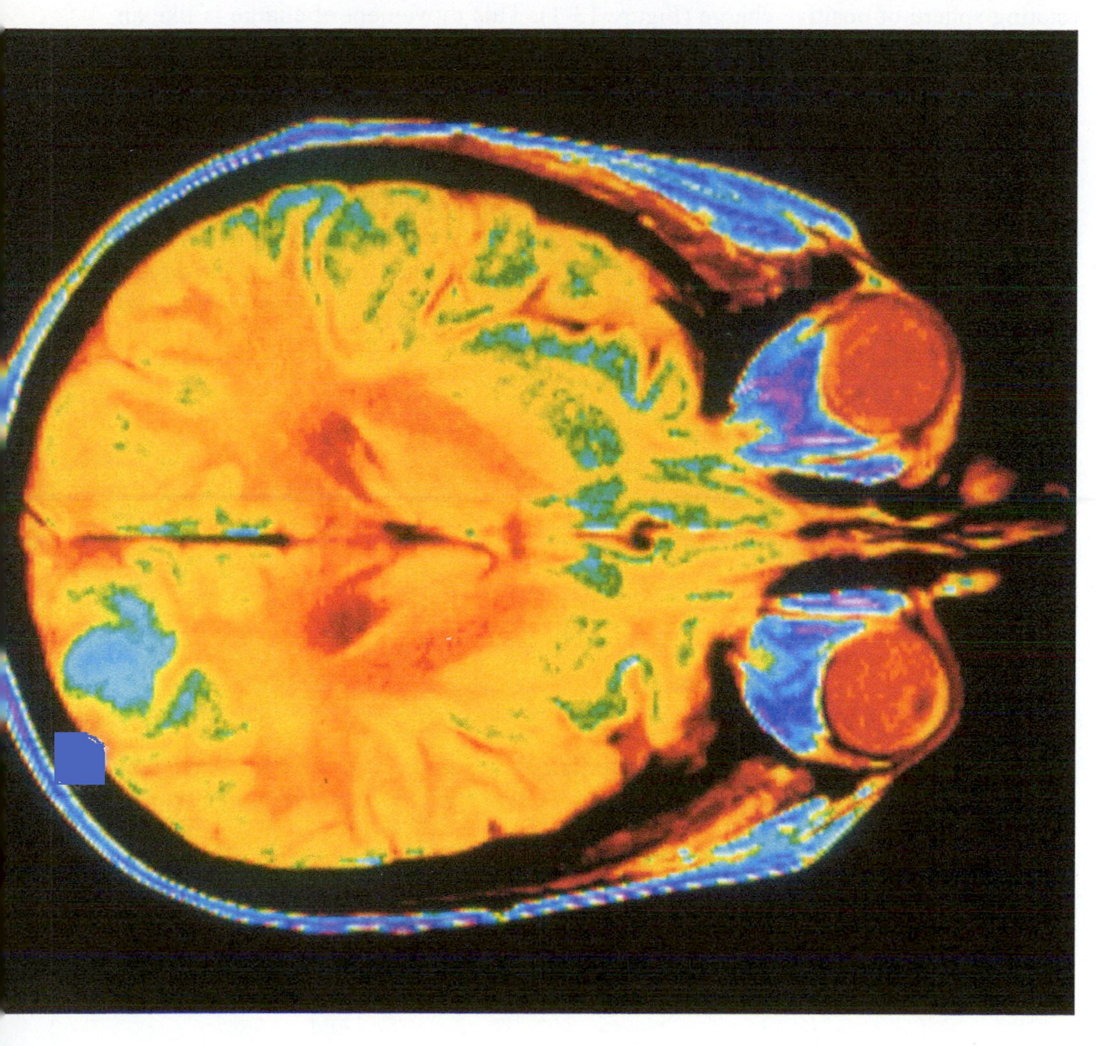

◀ This MRI scan of a human brain (viewed from below) shows a metastatic tumor, appearing as bright blue, in the left occipital lobe. The MRI (nuclear magnetic resonance imaging) instrument is a large-bore, broadband NMR instrument that produces a rapidly varying field gradient to obtain spatial information that is used to form an image.

Goals for Chapter 13

1 Use the chemical shifts, splitting patterns, and integrations shown in a proton NMR spectrum to propose structures for possible compounds.

2 Use the number of peaks and their chemical shifts in a ^{13}C NMR spectrum to determine the number of types of carbon atoms in the compound and what functional groups they might represent.

3 Given a chemical structure, predict the major features of its proton and ^{13}C NMR spectra.

4 Combine the information from NMR spectra, IR spectra, and mass spectra to determine the structures of unknown organic compounds.

13-1 Introduction

Nuclear magnetic resonance spectroscopy (NMR) is the most powerful tool available for organic structure determination. Like infrared spectroscopy, NMR can be used with a very small sample, and it does not harm the sample. The NMR spectrum provides a great deal of information about the structure of the compound, and many structures can be determined using only the NMR spectrum. More commonly, however, NMR spectroscopy is used in conjunction with other forms of spectroscopy and chemical analysis to determine the structures of complicated organic molecules.

NMR is used to study a wide variety of nuclei, including ^{1}H, ^{13}C, ^{15}N, ^{19}F, and ^{31}P. Organic chemists find **proton (^{1}H)** and **carbon-13 (^{13}C) NMR** to be most useful because hydrogen and carbon are major components of organic compounds. Historically, NMR was first used to study protons (the nuclei of hydrogen atoms), and proton magnetic resonance (^{1}H NMR) spectrometers have been the most common. "Nuclear magnetic resonance" is assumed to mean "proton magnetic resonance" unless a different nucleus is specified. We begin our study of NMR with ^{1}H NMR and conclude with a discussion of ^{13}C NMR.

13-2 Theory of Nuclear Magnetic Resonance

A nucleus with an odd atomic number or an odd mass number has a *nuclear spin* that can be observed by the NMR spectrometer. A proton is the simplest nucleus, and its odd atomic number of 1 implies it has a spin. We can visualize a spinning proton as a rotating sphere of positive charge (Figure 13-1). This movement of charge is like an electric current in a loop of wire. It generates a magnetic field (symbolized by B), called the **magnetic moment,** that looks like the field of a small bar magnet.

When a small bar magnet is placed in the field of a larger magnet (Figure 13-2), it twists to align itself with the field of the larger magnet—a lower energy arrangement than an orientation against the field. The same effect is seen when a proton is placed in an external magnetic field (B_0), as shown here. Quantum mechanics requires the proton's magnetic moment to be aligned either *with* the external field or *against* the field. The lower-energy state with the proton aligned with the field is called the *alpha-spin* (α-spin) *state*. The higher-energy state with the proton aligned against the external magnetic field is called the *beta-spin* (β-spin) *state*.

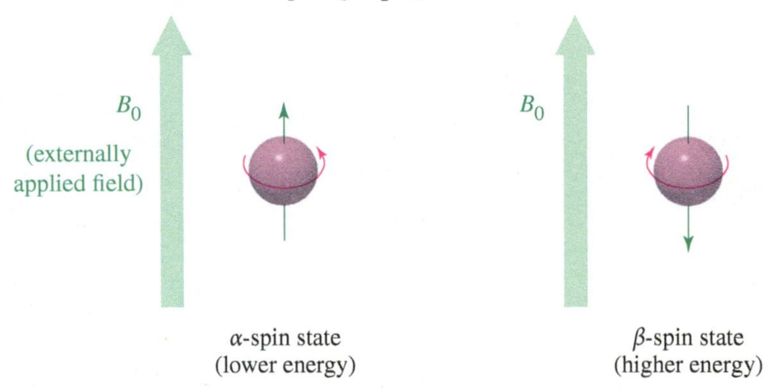

In the absence of an external magnetic field, proton magnetic moments have random orientations. When an external magnetic field is applied, each proton in a sample

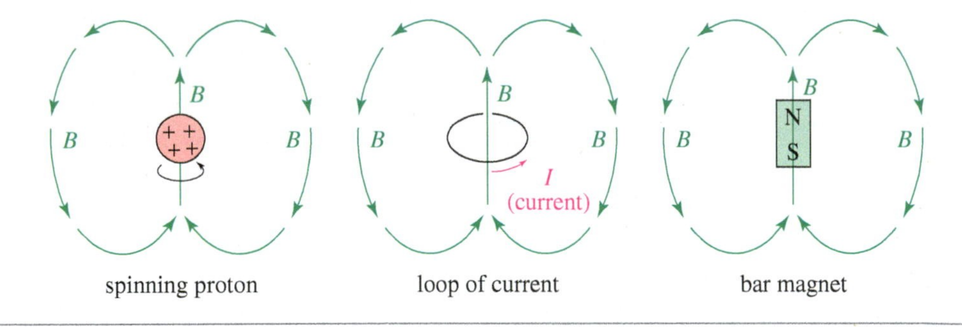

FIGURE 13-1
The magnetic moment. A spinning proton generates a magnetic field, called its magnetic moment. This magnetic field (B) resembles that of a small loop of current or bar magnet.

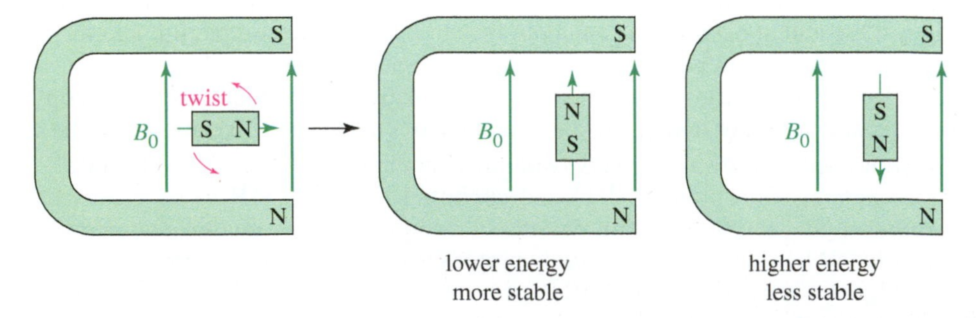

FIGURE 13-2
The effect of an external magnetic field. An external magnetic field (B_0) applies a force to a small bar magnet, twisting the bar magnet to align it with the external field. The arrangement of the bar magnet aligned *with* the field is lower in energy than the arrangement aligned *against* the field.

assumes the α state or the β state. Because the α-spin state is lower in energy, there are more α spins than β spins.

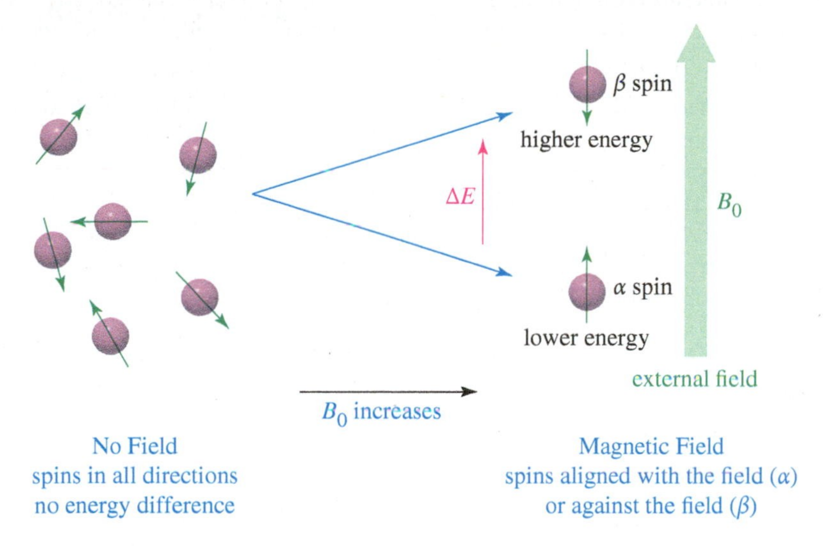

In a strong magnetic field, the energy difference between the two spin states is larger than it is in a weaker field. In fact, the energy difference is proportional to the strength of the magnetic field, as expressed in the equation

$$\Delta E = \gamma \frac{h}{2\pi} B_0$$

where

ΔE = energy difference between α and β states
h = Planck's constant
B_0 = strength of the external magnetic field
γ = gyromagnetic ratio, 26,753 sec^{-1} gauss^{-1} for a proton

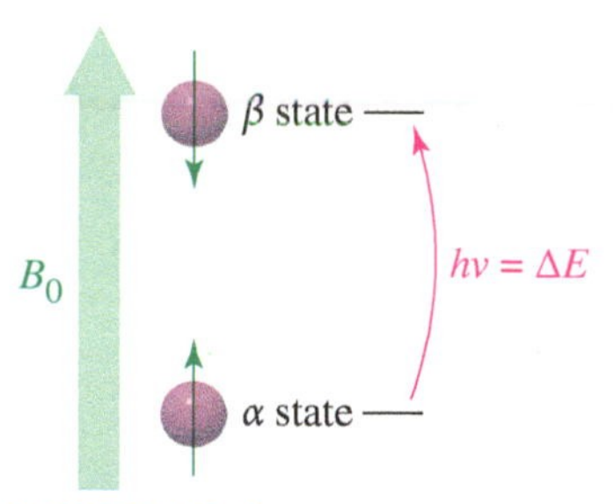

FIGURE 13-3
NMR absorption. A nucleus is "in resonance" when it is irradiated with radio-frequency photons having energy equal to the energy difference between the spin states. Under these conditions, a proton in the α-spin state can absorb a photon and flip to the β-spin state.

The **gyromagnetic ratio** (γ) is a constant that depends on the magnetic moment of the nucleus under study. Magnetic fields are measured in *gauss*; for example, the strength of the Earth's magnetic field is about 0.57 gauss. The SI unit for magnetic field is the *tesla* (T), which is simply 10,000 gauss.

The energy difference between a proton's two spin states is small. For a strong external magnetic field of 25,000 gauss (2.5 T), it is only about 10^{-5} kcal/mol (4×10^{-5} kJ/mol). Even this small energy difference can be detected by NMR. When a proton interacts with a photon with just the right amount of electromagnetic energy, the proton's spin can flip from α to β or from β to α. A nucleus aligned with the field can absorb the energy needed to flip and become aligned against the field.

When a nucleus is subjected to the right combination of magnetic field and electromagnetic radiation to flip its spin, it is said to be "in resonance" (Figure 13-3), and its absorption of energy is detected by the NMR spectrometer. This is the origin of the term *nuclear magnetic resonance.*

As we saw in Chapter 12, a photon's energy is given by $E = h\nu$, meaning that the energy, E, is proportional to ν, the frequency of the electromagnetic wave. This equation can be combined with the equation for the energy difference between the spin states:

$$\Delta E = h\nu = \gamma \frac{h}{2\pi} B_0$$

Rearranging to solve for ν shows that the resonance frequency ν is proportional to the applied magnetic field (B_0) and the gyromagnetic ratio (γ):

$$\nu = \frac{1}{2\pi} \gamma B_0$$

For a proton, $\gamma = 26{,}753\ \text{sec}^{-1}\ \text{gauss}^{-1}$, and

$$\nu = \frac{(26{,}753\ \text{sec}^{-1}\ \text{gauss}^{-1})}{2\pi} \times B_0 = (4257.8\ \text{sec}^{-1}\ \text{gauss}^{-1}) \times B_0$$

For the fields of currently available magnets, proton resonance frequencies occur in the radio-frequency (RF) region of the spectrum. NMR spectrometers are usually designed for the most powerful magnet that is practical for the price range of the spectrometer, and the radio frequency needed for resonance is calculated based on the field. A more powerful magnet makes ΔE larger and more easily detected, and it increases the frequency difference between signals, giving spectra that are more clearly resolved and easier to interpret. In the past, the most common operating frequency for student spectrometers has been 60 MHz (megahertz; 1 million cycles per second), corresponding to a magnetic field of 14,092 gauss. Higher-resolution instruments commonly operate at frequencies of 200 to 600 MHz (and higher), corresponding to fields of 46,972 to 140,918 gauss.

SOLVED PROBLEM 13-1

Calculate the magnetic fields that correspond to proton resonance frequencies of 60.00 MHz and 300.00 MHz.

SOLUTION

We substitute into the equation $\nu = (1/2\pi)\gamma B_0$.

$$60.00\ \text{MHz} = 60.00 \times 10^6\ \text{sec}^{-1} = (4257.8\ \text{sec}^{-1}\ \text{gauss}^{-1}) \times B_0$$
$$B_0 = 14{,}092\ \text{gauss}\ (1.4092\ \text{tesla})$$

$$300.00\ \text{MHz} = 300.00 \times 10^6\ \text{sec}^{-1} = (4257.8\ \text{sec}^{-1}\ \text{gauss}^{-1}) \times B_0$$
$$B_0 = 70{,}459\ \text{gauss}\ (7.0459\ \text{tesla})$$

13-3 Magnetic Shielding by Electrons

Up to now, we have considered the resonance of a naked proton in a magnetic field, but real protons in organic compounds are not naked. They are surrounded by electrons that partially shield them from the external magnetic field. The electrons circulate and generate a small **induced magnetic field** that opposes the externally applied field.

A similar effect occurs when a loop of wire is moved into a magnetic field. The electrons in the wire are induced to flow around the loop in the direction shown in Figure 13-4; this is the principle of the electric generator. The induced electric current creates a magnetic field that opposes the external field.

In a molecule, the electron cloud around each nucleus acts like a loop of wire, rotating in response to the external field. This induced rotation is a circular current whose magnetic field opposes the external field. The result is that the magnetic field at the nucleus is weaker than the external field, and we say the nucleus is **shielded.** The effective magnetic field *at the shielded proton* is always weaker than the external field, so the applied field must be increased for resonance to occur at a given frequency (Figure 13-5).

$$B_{\text{effective}} = B_{\text{external}} - B_{\text{shielding}}$$

At 300 MHz, an unshielded naked proton absorbs at 70,459 gauss, but a shielded proton requires a stronger field. For example, if a proton is shielded by 1 gauss when the external field is 70,459 gauss, the effective magnetic field *at the proton* is 70,458 gauss. If the external field is increased to 70,460 gauss, the effective magnetic field at the proton is increased to 70,459 gauss, which brings this proton into resonance.

If all protons were shielded by the same amount, they would all be in resonance at the same combination of frequency and magnetic field. Fortunately, protons in different chemical environments are shielded by different amounts. In methanol, for example, the electronegative oxygen atom withdraws some electron density from around the

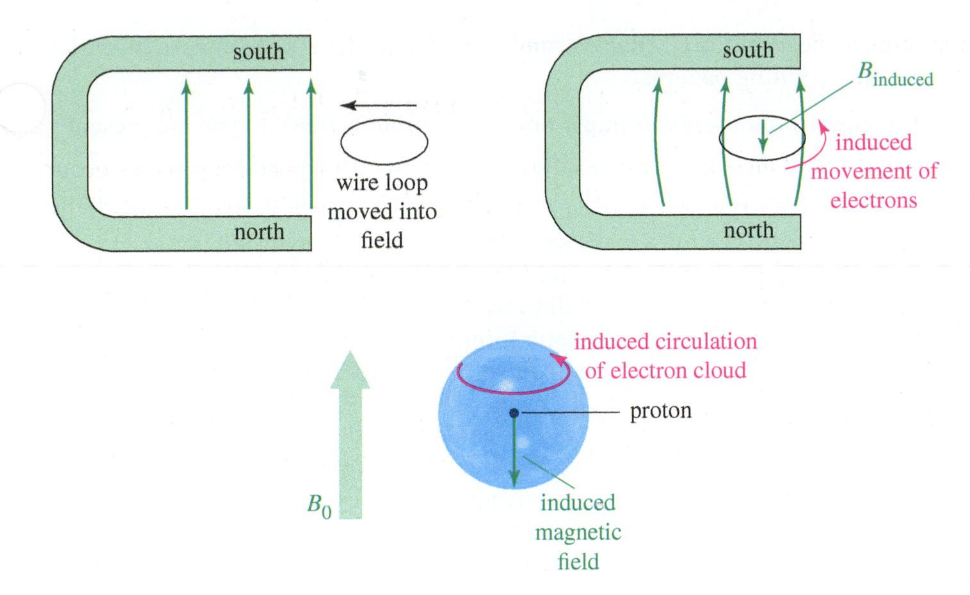

FIGURE 13-4
Induced magnetic field. Moving a loop of wire into a magnetic field induces a current in the wire. This current produces its own smaller magnetic field in the direction opposite the applied field. In a molecule, electrons can circulate around a nucleus. The resulting "current" sets up a magnetic field that opposes the external field, so the nucleus feels a slightly weaker field.

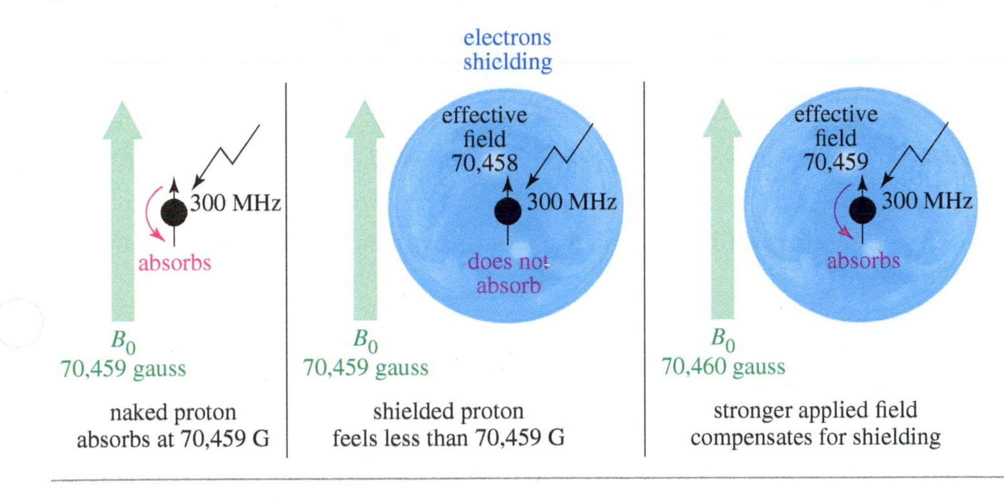

FIGURE 13-5
A proton shielded by electrons.
The magnetic field must be increased slightly above 70,459 gauss (at 300 MHz) for resonance of a shielded proton.

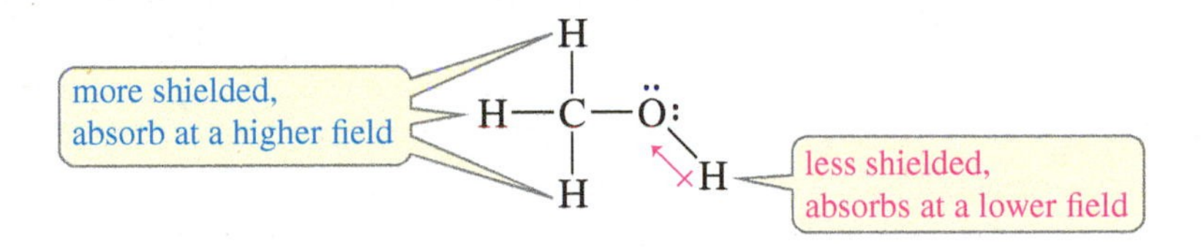

hydroxy proton. The hydroxy proton is not shielded as much as the methyl protons, so the hydroxy proton absorbs at a lower field than the methyl protons (but still at a higher field than a naked proton). We say that the hydroxy proton is **deshielded** somewhat by the presence of the electronegative oxygen atom.

Because of the diverse and complex structures of organic molecules, the shielding effects of electrons at various positions are generally different. A careful measurement of the field strengths required for resonance of the various protons in a molecule provides us with two important types of information:

1. The *number of different absorptions* (also called *signals* or *peaks*) implies how many different types of protons are present.
2. The *amount of shielding* shown by these absorptions implies the electronic structure of the molecule close to each type of proton.

Two other aspects of the NMR spectrum we will consider are the intensities of the signals and their splitting patterns:

3. The *intensities of the signals* imply how many protons of each type are present.
4. The *splitting of the signals* gives information about other nearby protons.

Before discussing the design of spectrometers, let's review what happens in an NMR spectrometer. Protons (in the sample compound) are placed in a magnetic field, where they align either with the field or against it. While still in the magnetic field, the protons are subjected to radiation of a frequency they can absorb by changing the orientation of their magnetic moment relative to the field. If protons were isolated, they would all absorb at the same frequency, proportional to the magnetic field.

But protons in a molecule are partially shielded from the magnetic field, and this shielding depends on each proton's environment. Thus, protons in different environments within a molecule exposed to a constant frequency absorb the radiation at different magnetic field strengths. The NMR spectrometer was originally developed to vary the magnetic field and plot a graph of energy absorption as a function of the magnetic field strength. Such a graph is called a **nuclear magnetic resonance spectrum.**

13-4 The NMR Spectrometer

The original, simplest type of NMR spectrometer (Figure 13-6) consisted of four parts:

1. A stable magnet, with a sensitive controller to produce a precise magnetic field
2. A radio-frequency (RF) transmitter, emitting a precise frequency (continuous wave or CW)
3. A detector to measure the sample's absorption of RF energy
4. A recorder to plot the output from the detector versus the applied magnetic field

The printer records a graph of absorption (on the y axis) as a function of the applied magnetic field (on the x axis). Higher values of the magnetic field are toward the right (**upfield**), and lower values are toward the left (**downfield**). The absorptions of more **shielded** protons appear upfield, toward the right of the spectrum, and more **deshielded** protons appear downfield, toward the left. The NMR spectrum of methanol is shown in Figure 13-7.

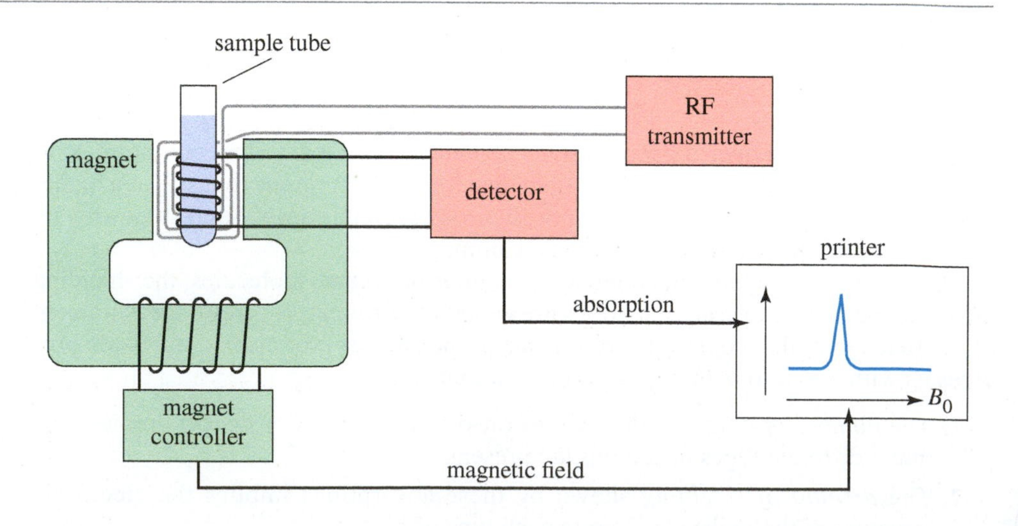

FIGURE 13-6
Simplified block diagram of a nuclear magnetic resonance spectrometer.

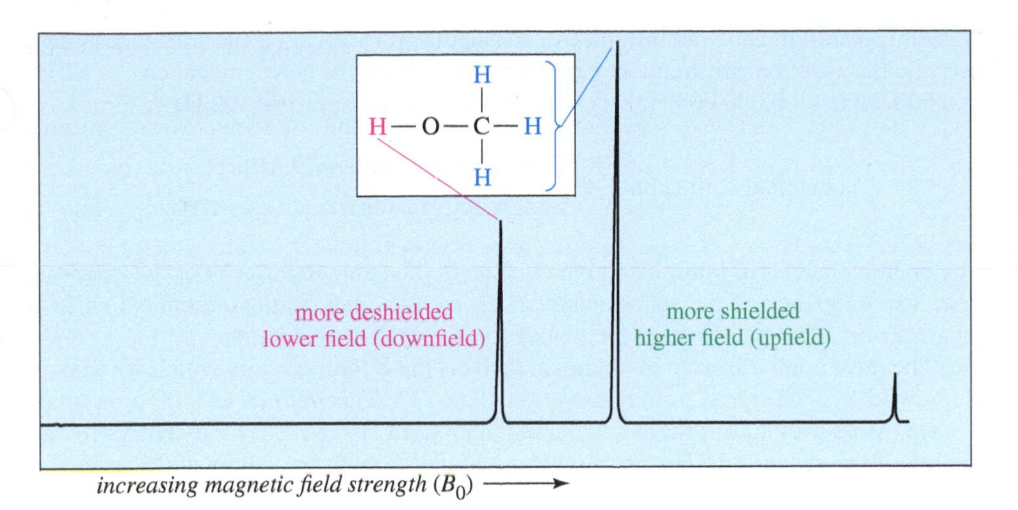

FIGURE 13-7
Proton NMR spectrum of methanol. The more shielded methyl protons appear toward the right of the spectrum (higher field); the less shielded hydroxy proton appears toward the left (lower field).

13-5 The Chemical Shift

An NMR spectrum provides information about the electronic environment of each type of proton in a molecule. Each proton environment is characterized by the relative strength of the magnetic field that causes that proton to absorb energy from the RF transmitter. The variations in the positions of NMR absorptions, arising from electronic shielding and deshielding, are called **chemical shifts.**

Chemical shift The difference (in parts per million) between the resonance field or frequency of the proton being observed and that of tetramethylsilane (TMS).

13-5A Measurement of Chemical Shifts

In practice, it is difficult to measure the absolute field where a proton absorbs with enough accuracy to distinguish individual protons, because the signals often differ by only a few thousandths of a gauss at an applied field of 70,459 gauss. A more accurate method for expressing chemical shifts is to determine the value relative to a reference compound added to the sample. The *difference* in the magnetic field strength between the resonances of the sample protons and the reference protons can be measured very accurately.

CH_3
CH_3—Si—CH_3
CH_3
tetramethylsilane (TMS)

The most common NMR reference compound is *tetramethylsilane* $(CH_3)_4Si$, abbreviated **TMS.** Because silicon is less electronegative than carbon, the methyl groups of TMS are relatively electron-rich, and their protons are well shielded. They absorb at a higher field strength than most hydrogens bonded to carbon or other elements, so most NMR signals appear *downfield* (to the left, *deshielded*) of the TMS signal. All 12 protons in TMS absorb at exactly the same applied magnetic field, giving one strong absorption.

A small amount of TMS is added to the sample, and the instrument measures the difference in magnetic field between where the protons in the sample absorb and where those in TMS absorb. For each type of proton, the distance downfield of TMS is the chemical shift of those protons. Newer spectrometers operate at a constant magnetic field, and they measure the chemical shift as a frequency difference between the resonances of the protons in the sample and those in TMS. Remember that frequency units (ν) and magnetic field units (B_0) are always proportional in NMR, with $\nu = \gamma B_0/2\pi$.

Chemical shifts are measured in *parts per million* (ppm), a dimensionless fraction of either the total applied field or the total radio frequency. By custom, the difference (the chemical shift) between the NMR signal of a proton and that of TMS is shown on the horizontal axis of the NMR spectrum calibrated in frequency units (hertz or Hz).

A chemical shift in parts per million can be calculated by dividing the shift measured in hertz by the spectrometer frequency measured in millions of hertz (megahertz or MHz). In a 300-MHz (300,000,000 Hz) spectrum, for example, 1 ppm = 300 Hz.

$$\text{chemical shift (ppm)} = \frac{\text{shift downfield from TMS (Hz)}}{\text{total spectrometer frequency (MHz)}}$$

The chemical shift (in ppm) of a given proton is the same regardless of the operating field and frequency of the spectrometer. The use of dimensionless chemical shifts to locate absorptions standardizes the values for all NMR spectrometers.

The most common scale of chemical shifts is the δ (delta) scale, which we will use (Figure 13-8). The signal from tetramethylsilane (TMS) is *defined* as 0.00 ppm on the δ scale. Most protons are more deshielded than TMS, so the δ scale increases toward the left of the spectrum. The spectrum is calibrated in both frequency and ppm δ.

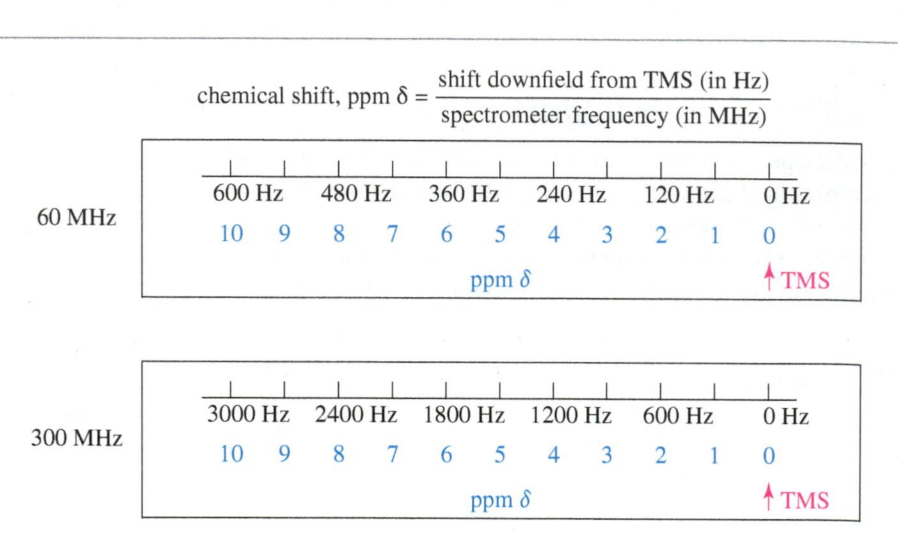

FIGURE 13-8
Use of the δ scale with 60- and 300-MHz spectrometers. The absorption of TMS is defined as 0, with the scale increasing from right to left (downfield, toward more deshielded protons). Each δ unit is 1 ppm difference from TMS: 60 Hz at 60 MHz and 300 Hz at 300 MHz.

PROBLEM-SOLVING HINT

The chemical shift (in ppm) of a given proton is the same in any NMR spectrometer, regardless of the operating field and frequency of the spectrometer. The frequency shift (in Hz) is proportional to the operating frequency of the spectrometer.

SOLVED PROBLEM 13-2

A 300-MHz spectrometer records a proton that absorbs at a frequency 2130 Hz downfield (deshielded) from TMS.
(a) Determine its chemical shift.
(b) Predict this proton's chemical shift at 60 MHz. In a 60-MHz spectrometer, how far downfield (in hertz) from TMS would this proton absorb?

SOLUTION

(a) The chemical shift is the fraction

$$\frac{\text{shift downfield (Hz)}}{\text{spectrometer frequency (MHz)}} = \frac{2130\ \text{Hz}}{300\ \text{MHz}} = 7.10\ \text{ppm}$$

(b) The chemical shift is unchanged at 60 MHz: δ 7.10. The frequency shift is

$$60.00\ \text{MHz} \times (7.10 \times 10^{-6}) = 426\ \text{Hz}$$

PROBLEM 13-1

In a 300-MHz spectrometer, the protons in iodomethane absorb at a position 650 Hz downfield from TMS.
(a) What is the chemical shift of these protons?
(b) What is the chemical shift of the iodomethane protons in a 60-MHz spectrometer?
(c) How many hertz downfield from TMS would they absorb at 60 MHz?

The 300-MHz NMR spectrum of methanol (Figure 13-9) shows the two signals from methanol together with the TMS reference peak at δ 0.0. The methyl protons absorb 1025 Hz downfield from TMS. Their chemical shift is 3.4 ppm, so we say that the methyl protons absorb at δ 3.4. The hydroxy proton absorbs farther downfield, at a position around 1450 Hz from TMS. Its chemical shift is δ 4.8.

Both the hydroxy proton and the methyl protons of methanol show the deshielding effects of the electronegative oxygen atom. The chemical shift of a methyl group in an alkane is about δ 0.9. Therefore, the methanol oxygen deshields the methyl protons by an additional 2.5 ppm. Other electronegative atoms produce similar deshielding effects. Table 13-1 compares the chemical shifts of methanol with those of the methyl halides. Notice that the chemical shift of the methyl protons depends on the electronegativity of the substituent, with more electronegative substituents deshielding more and giving larger chemical shifts.

The effect of an electronegative group on the chemical shift also depends on its distance from the protons. In methanol, the hydroxy proton is separated from oxygen by one bond, and its chemical shift is δ 4.8. The methyl protons are separated from oxygen by two bonds, and their chemical shift is δ 3.4. In general, the effect of an electron-withdrawing substituent decreases with increasing distance, and the effects are usually negligible on protons that are separated from the electronegative group by four or more bonds.

This decreasing effect can be seen by comparing the chemical shifts of the various protons in 1-bromobutane with those in butane. The deshielding effect of an electronegative substituent drops off rapidly with distance. In 1-bromobutane, protons on the α carbon are deshielded by about 2.5 ppm, and the β protons are deshielded by about 0.4 ppm. Protons that are more distant than the β protons are deshielded by a negligible amount.

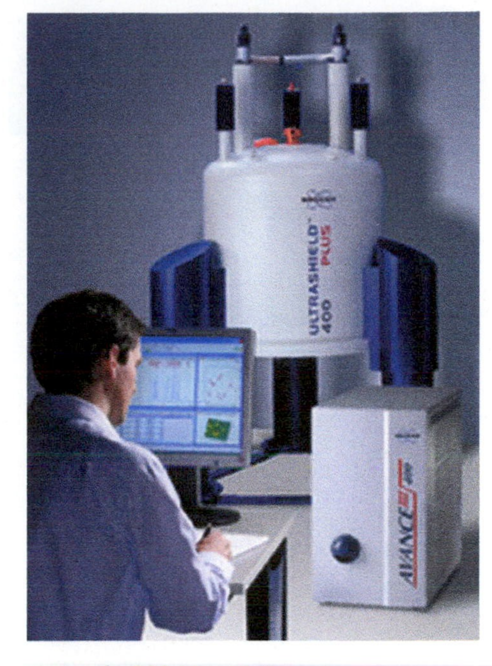

Photo of a 300-MHz NMR spectrometer. The metal container at the back contains the superconducting magnet, cooled by a liquid helium bath. The electronics used to control the spectrometer and to calculate spectra are at right.

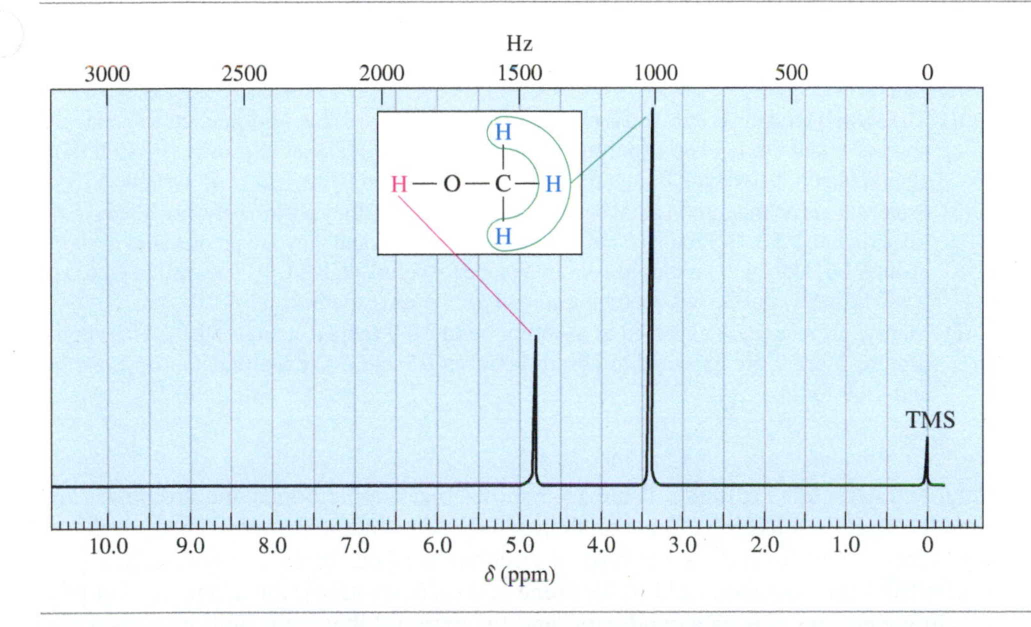

FIGURE 13-9
A 300-MHz proton NMR spectrum of methanol. The methyl protons absorb at δ 3.4, and the hydroxy proton absorbs at δ 4.8.

TABLE 13-1
Variation of Chemical Shift with Electronegativity

	X in CH_3—X				
	F	OH	Cl	Br	I
electronegativity of X	4.0	3.4	3.2	3.0	2.7
chemical shift of CH_3—X	δ 4.3	δ 3.4	δ 3.0	δ 2.7	δ 2.2

TABLE 13-2
Chemical Shifts of the Chloromethanes

Compound	Chemical Shift	Difference
CH_4 (H—C—H)	δ 0.2	
		2.8 ppm
CH_3Cl (H—C—Cl)	δ 3.0	
		2.3 ppm
CH_2Cl_2 (H—C—Cl)	δ 5.3	
		1.9 ppm
$CHCl_3$ (H—C—Cl)	δ 7.2	

Note: Each chlorine atom added changes the chemical shift of the remaining methyl protons by 2 to 3 ppm. These changes are nearly additive.

	butane				1-bromobutane			
structure	$H-CH_2-CH_2-CH_2-CH_2-H$				$H-CH_2^{\delta}-CH_2^{\gamma}-CH_2^{\beta}-CH_2^{\alpha}-Br$			
chemical shift:	0.9	1.3	1.3	0.9	0.9	1.3	1.7	3.4
deshielding resulting from Br, ppm:					0.0	0.0	0.4	2.5

If more than one electron-withdrawing group is present, the deshielding effects are nearly (but not quite) additive. In the chloromethanes (Table 13-2), the addition of the first chlorine atom causes a shift to δ 3.0, the second chlorine shifts the absorption further to δ 5.3, and the third chlorine moves the chemical shift to δ 7.2 for chloroform. The chemical shift *difference* is about 2 to 3 ppm each time another chlorine atom is added, but each additional chlorine moves the peak slightly less than the previous one did.

13-5B Characteristic Values of Chemical Shifts

Because the chemical shift of a proton is determined by its environment, we can construct a table of approximate chemical shifts for many types of compounds. Let's begin with a short table of representative chemical shifts (Table 13-3) and consider the reasons for some of the more interesting and unusual values. A comprehensive table of chemical shifts appears in Appendix 1.

SOLVED PROBLEM 13-3

Using Table 13-3, predict the chemical shifts of the protons in the following compounds.

(a) $CH_3-C(=O)-OH$ **(b)** $Cl-CH_2^{a}-CH_2^{b}-CH_3^{c}$ **(c)** $(CH_3^{a})_3CCH^{b}=CH_2^{c}$

SOLUTION

(a) The methyl group in acetic acid is next to a carbonyl group; Table 13-3 predicts a chemical shift of about δ 2.1. (The experimental value is δ 2.10.) The acid proton (—COOH) should absorb between δ 10 and δ 12. (The experimental value is δ 11.4, variable.)

(b) Protons *a* are on the carbon atom bearing the chlorine, and they absorb between δ 3 and δ 4 (experimental: δ 3.7). Protons *b* are one carbon removed, and they are predicted to absorb about δ 1.7, like the β protons in 1-bromobutane (experimental: δ 1.8). The methyl protons *c* will be nearly unaffected, absorbing around δ 0.9 ppm (experimental: δ 1.0).

(c) Methyl protons *a* are expected to absorb around δ 0.9 (experimental: δ 1.0). The vinyl protons *b* and *c* are expected to absorb between δ 5 and δ 6 (experimental: δ 5.8 for *b* and δ 4.9 for *c*).

Vinyl and Aromatic Protons Table 13-3 shows that double bonds and aromatic rings produce large deshielding effects on their vinyl and aromatic protons. These deshielding effects result from the same type of circulation of electrons that normally shields nuclei from the magnetic field. In benzene and its derivatives, the aromatic ring of pi bonding electrons acts as a conductor, and the external magnetic field induces a *ring current* (Figure 13-10). At the center of the ring, the induced field acts to oppose the external field. These induced field lines curve around, however, and on the edge of the ring the induced field *adds to* the external field. As a result, the aromatic protons are strongly *deshielded*, resulting in a large chemical shift. Benzene absorbs at δ 7.2, and most aromatic protons absorb in the range of δ 7 to δ 8.

The benzene molecule is not always lined up in the position shown in Figure 13-10. Because benzene is constantly tumbling in the solution, the chemical shift observed for its protons is an average of all the possible orientations. If we could hold a benzene

TABLE 13-3
Typical Values of Chemical Shifts

Type of Proton	Approximate δ	Type of Proton	Approximate δ
alkane (—CH_3) methyl	0.9	>C=C< CH_3 allylic	1.7
alkane (—CH_2—) methylene	1.3	Ph—H aromatic	7.2
alkane (—CH— with branch) methine	1.4	Ph—CH_3 benzylic	2.3
		R—C(=O)—H aldehyde	9–10
—C(=O)—CH_3 methyl ketone	2.1		
—C≡C—H acetylenic	2.5	R—C(=O)—OH acid	10–12
		R—OH alcohol	variable, about 2–5
R—CH_2—X (X = halogen, O)	3–4	Ar—OH phenol	variable, about 4–7
>C=C< H vinyl	5–6	R—NH_2 amine	variable, about 1.5–4

Note: These values are approximate, as all chemical shifts are affected by neighboring substituents. The numbers given here assume that alkyl groups are the only other substituents present. A more complete table of chemical shifts appears in Appendix 1.

PROBLEM-SOLVING HINT

Table 13-3 provides the numbers but not the understanding and practice needed to work most NMR problems. Learn the material in this table, and then practice doing problems until you feel confident.

molecule in the position shown in Figure 13-10, its protons would absorb at a field even lower than δ 7.2. Other orientations, such as the one with the benzene ring edge-on to the magnetic field, would be less deshielded and would absorb at a higher field. It is the *average* of all these orientations that is observed by the resonance at δ 7.2.

Figure 13-11 shows the NMR spectrum of toluene (methylbenzene). The aromatic protons absorb around δ 7.2. The methyl protons are deshielded by a smaller amount, absorbing at δ 2.3.

The pi electrons of an alkene deshield the vinyl protons in the same way that an aromatic ring of electrons deshields the aromatic protons. The effect is not as large in the alkene, however, because there is not such a large, effective ring of electrons as there is in benzene. Once again, the motion of the pi electrons generates an induced magnetic field that opposes the applied field at the middle of the double bond. The vinyl protons are on the periphery of this field, however, where the induced field bends around and reinforces the external field (Figure 13-12). As a result of this deshielding effect, most vinyl protons absorb in the range of δ 5 to δ 6.

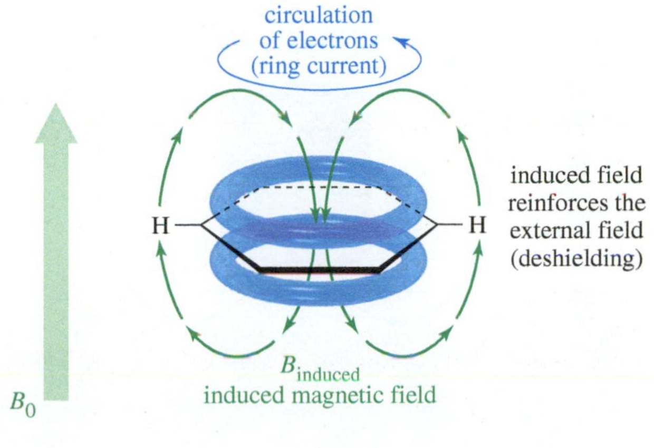

FIGURE 13-10
Aromatic ring current. The induced magnetic field of the circulating aromatic electrons opposes the applied magnetic field along the axis of the ring. The aromatic hydrogens are on the equator of the ring, where the induced field lines curve around and reinforce the applied field.

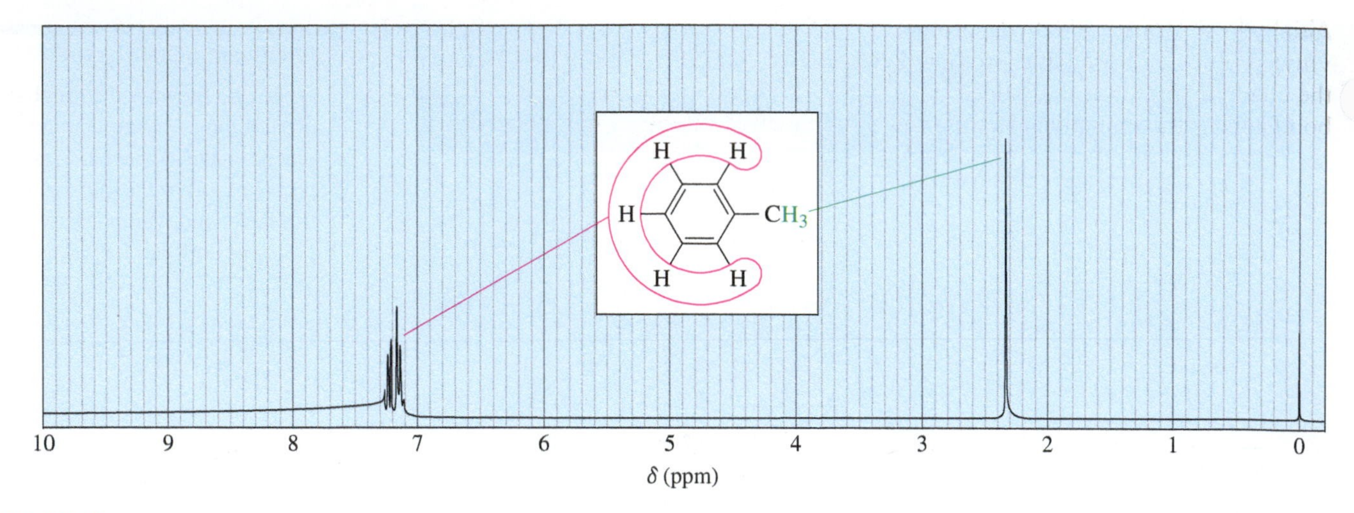

FIGURE 13-11
300-MHz proton NMR spectrum of toluene. The aromatic protons absorb at chemical shifts near δ 7.2, and the methyl protons absorb at δ 2.3.

FIGURE 13-12
Deshielding by a pi bond. Vinyl protons are positioned on the periphery of the induced magnetic field of the pi electrons. In this position, they are deshielded by the induced magnetic field.

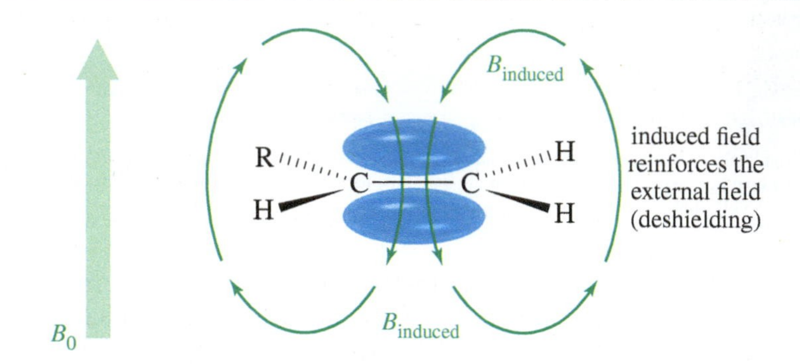

Acetylenic Hydrogens Because the pi bond of an alkene deshields the vinyl protons, we might expect an acetylenic hydrogen ($-C\equiv C-H$) to be even more deshielded by the two pi bonds of the triple bond. The opposite is true: Acetylenic hydrogens absorb around δ2.5, compared with δ5 to δ6 for vinyl protons. Figure 13-13 shows that the triple bond has a cylinder of electron density surrounding the sigma bond. As the molecules tumble in solution, in some orientations this cylinder of electrons can circulate to produce an induced magnetic field. The acetylenic proton lies along the *axis* of this induced field, which is a shielded region. When this shielded orientation is averaged with all other possible (mostly deshielded) orientations, the result is a resonance around δ2.5.

FIGURE 13-13
Partial shielding by a triple bond. When the acetylenic triple bond is aligned with the magnetic field, the cylinder of electrons circulates to create an induced magnetic field. The acetylenic proton lies along the axis of this field, which opposes the external field.

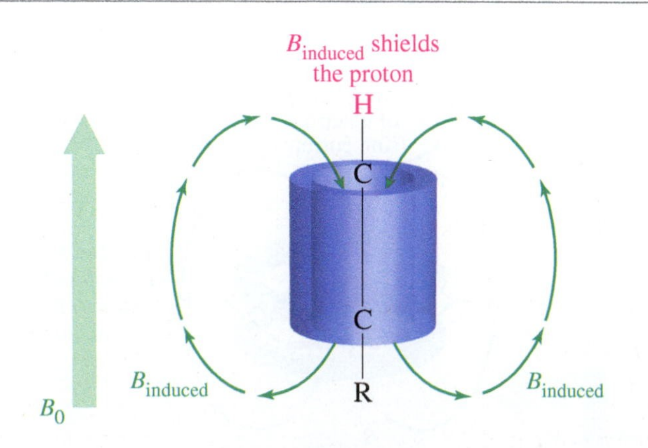

Aldehyde Protons Aldehyde protons (—CHO) absorb at even lower fields than vinyl protons and aromatic protons, between $\delta 9$ and $\delta 10$. Figure 13-14 shows that the aldehyde proton is deshielded both by the circulation of the electrons in the double bond and by the inductive electron-withdrawing effect of the carbonyl oxygen atom.

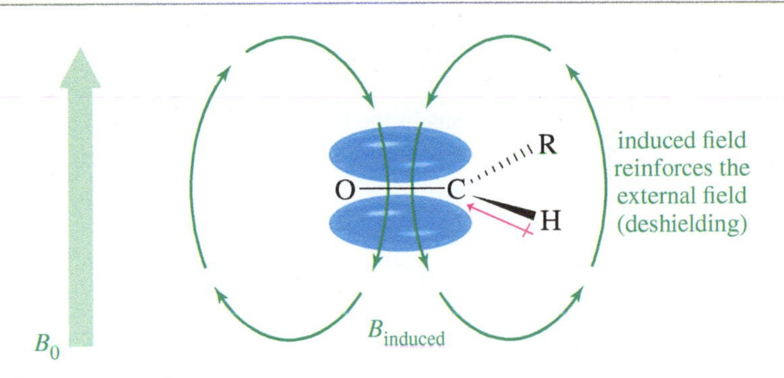

FIGURE 13-14
Deshielding of the aldehyde proton. Like a vinyl proton, the aldehyde proton is deshielded by the circulation of electrons in the pi bond. It is also deshielded by the electron-withdrawing effect of the carbonyl (C═O) group, giving a resonance between $\delta 9$ and $\delta 10$.

Hydrogen-Bonded Protons The chemical shifts of O—H protons in alcohols and N—H protons in amines depend on the concentration. In concentrated solutions, these protons are deshielded by hydrogen bonding, and they absorb at a relatively low field, about $\delta 3.5$ for an amine N—H and about $\delta 4.5$ for an alcohol O—H. When the alcohol or amine is diluted with a non-hydrogen-bonding solvent such as CCl_4, hydrogen bonding becomes less important. In dilute solutions, these signals are observed around $\delta 2$.

Hydrogen bonding and the proton exchange that accompanies it may contribute to a broadening of the peak from an O—H or N—H proton. A broad peak appears because protons exchange from one molecule to another during the NMR resonance (see Section 13-12). The protons pass through a variety of environments during this exchange, absorbing over a wider range of frequencies and field strengths.

Carboxylic Acid Protons Because carboxylic acid protons are bonded to an oxygen next to a carbonyl group, they have considerable positive character. They are strongly deshielded and absorb at chemical shifts greater than $\delta 10$. Carboxylic acids frequently exist as hydrogen-bonded dimers (shown at right), with moderate rates of proton exchange that broaden the absorption of the acid proton.

carboxylic acid dimer

The proton NMR spectrum of acetic acid is shown in Figure 13-15. As we expect, the methyl group next to the carbonyl absorbs at a chemical shift of $\delta 2.1$. The acid

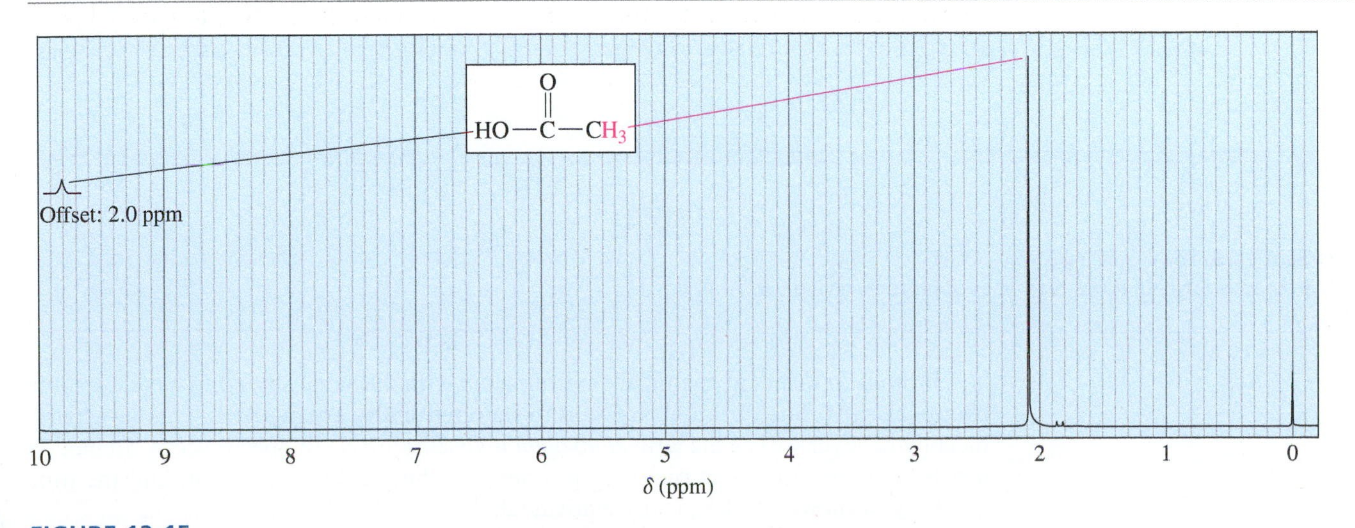

FIGURE 13-15
Proton NMR spectrum of acetic acid. The methyl protons of acetic acid are deshielded to about $\delta 2.1$ by the adjacent carbonyl group. The acid proton appears at $\delta 11.8$, shown on an offset trace.

proton signal appears at a chemical shift that is not scanned in the usual range of the NMR spectrum. It is seen in a second trace with a 2.0 ppm offset, meaning that this trace corresponds to frequencies with chemical shifts 2.0 ppm *larger* than shown on the trace. The acid proton appears around δ 11.8, the sum of δ 9.8 read from the trace, plus the δ 2.0 offset.

PROBLEM 13-2

Predict the chemical shifts of the protons in the following compounds.

(a) $(CH_3)_3C(H)C{=}C(H)C(CH_3)_3$ (b) benzene ring with H, CH_3, CH_3, H, CH_3, CH_3 (c) CH_3O–benzene ring (H, H, H, H)–OCH_3

(d) $CH_3{-}C(CH_3)(OH){-}C{\equiv}C{-}H$ (e) $C_6H_5{-}CH_2CH_2{-}C(=O){-}OH$ (f) $CH_3{-}C(CH_3)(Br){-}CH_2Br$

13-6 The Number of Signals

In general, the number of NMR signals corresponds to the number of different kinds of protons present in the molecule. For example, methyl *tert*-butyl ether has two types of protons (Figure 13-16). The three methoxy protons are chemically identical, and they give rise to a single absorption at δ 3.2. The *tert*-butyl protons are chemically different from the methoxy protons, absorbing at δ 1.2.

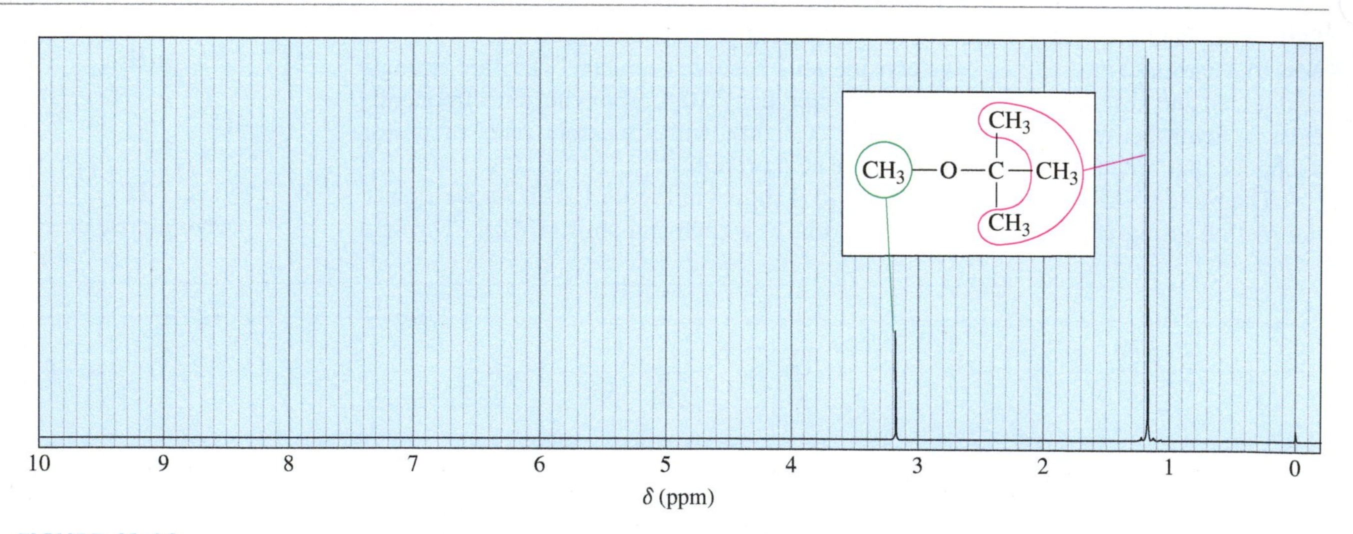

FIGURE 13-16
Methyl *tert*-butyl ether has two types of protons, giving two NMR signals.

Protons in identical chemical environments with the same shielding have the same chemical shift. Such protons are said to be **chemically equivalent.** This is what is meant whenever we use the term *equivalent* in discussing NMR spectroscopy. In methyl *tert*-butyl ether, the three methoxy protons are chemically equivalent, and the nine *tert*-butyl protons are chemically equivalent.

The spectrum of *tert*-butyl acetoacetate (Figure 13-17) shows three types of protons: the *tert*-butyl protons (*a*), with a chemical shift of δ 1.5; the methyl protons (*b*),

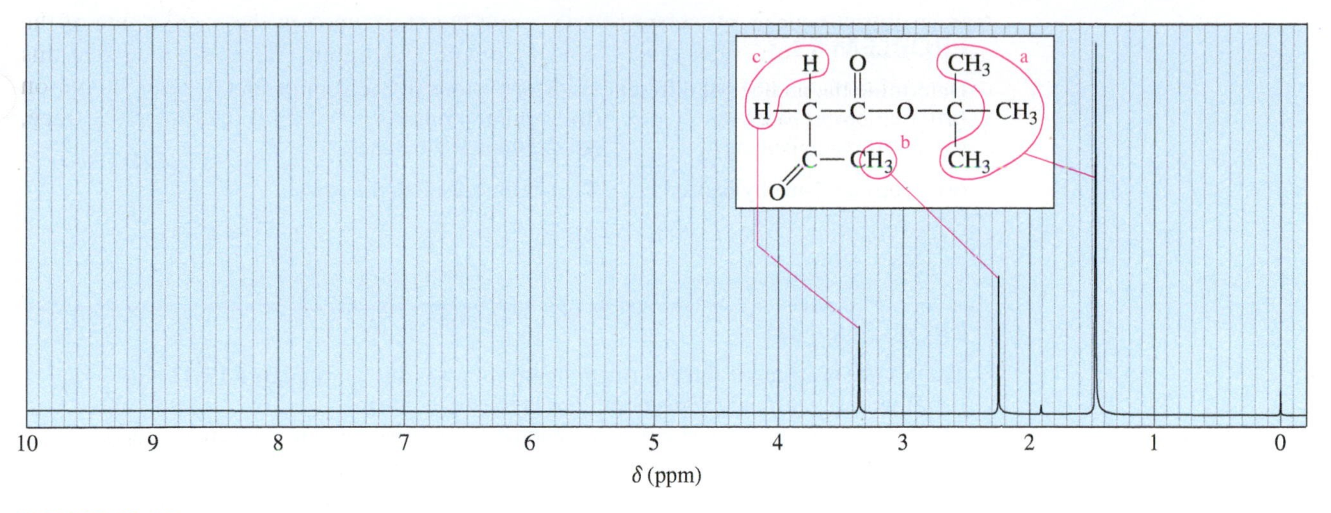

FIGURE 13-17
tert-Butyl acetoacetate has three types of protons, giving three signals in the NMR spectrum.

deshielded by an adjacent carbonyl group, with a chemical shift of $\delta 2.25$; and the methylene protons (c), deshielded by two adjacent carbonyl groups, at $\delta 3.35$.

In some cases, there may be fewer signals in the NMR spectrum than there are different types of protons in the molecule. For example, Figure 13-18 shows the structure and spectrum of *o*-xylene (1,2-dimethylbenzene). There are three different types of protons, labeled a for the two equivalent methyl groups, b for the protons adjacent to the methyl groups, and c for the protons two carbons removed. The spectrum shows only two distinct signals, however.

The upfield signal at $\delta 2.3$ corresponds to the six methyl protons, H^a. The absorption at $\delta 7.2$ corresponds to all four of the aromatic protons, H^b and H^c. Although the two types of aromatic protons are different, the methyl groups do not strongly influence the electron density of the ring or the amount of shielding felt by any of the substituents on the ring. The aromatic protons produce two signals, but these signals happen to occur at nearly the same chemical shift. Protons that are not chemically equivalent but happen to absorb at the same chemical shift are said to be **accidentally equivalent.**

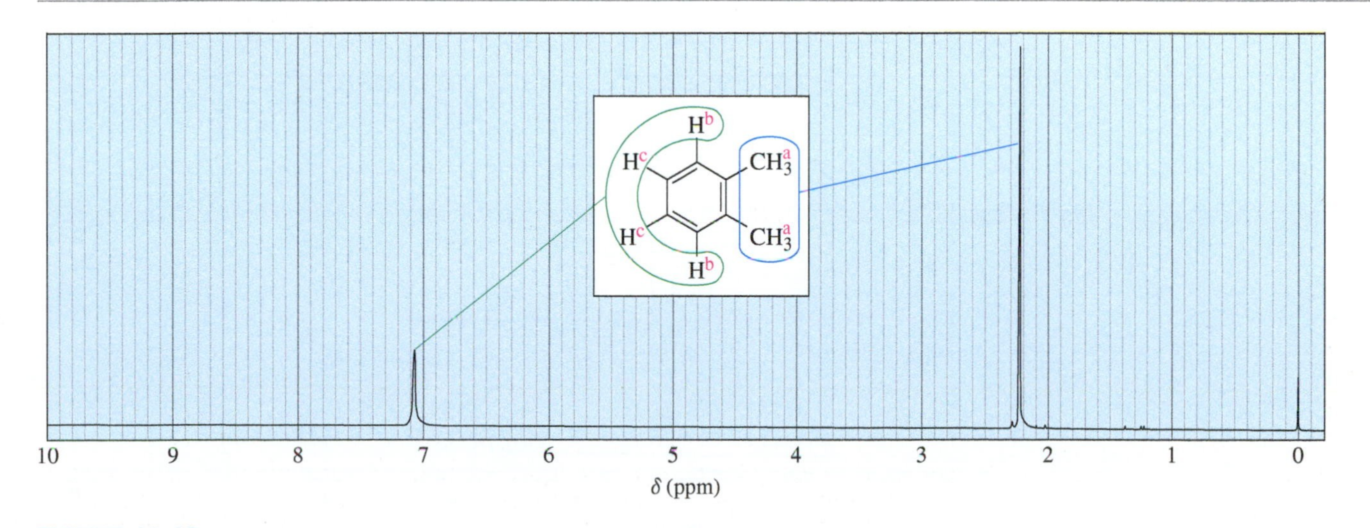

FIGURE 13-18
Proton NMR spectrum of *o*-xylene. There are three types of protons in *o*-xylene, but only two absorptions are seen in the spectrum. The aromatic protons H^b and H^c are accidentally equivalent, producing a broadened peak at δ 7.1.

PROBLEM 13-3

Determine the number of different kinds of protons in each compound.

(a) 1-chloropropane **(b)** 2-chloropropane
(c) 2,2-dimethylbutane **(d)** 2,3-dimethylbutane
(e) 1-bromo-4-methylbenzene **(f)** 1-bromo-2-methylbenzene,

PROBLEM 13-4

The NMR spectrum of toluene (methylbenzene) was shown in Figure 13-11.

(a) How many different kinds of protons are there in toluene?
(b) Explain why the aromatic region around δ7.2 is broad, with more than one sharp absorption.

PROBLEM-SOLVING HINT

1. To measure the heights of the rises in the integral trace, use a ruler to measure the integrals in millimeters.
2. You don't know the total number of hydrogens, so try setting the smallest integral equal to one hydrogen and the others proportionally. If some of the other integrals are not whole numbers of hydrogens, then set the smallest equal to 2 or 3 as required. For example, 1 : 1.3 : 2 would become 3 : 4 : 6 and you would look for a compound with this ratio or 6 : 8 : 12 or 9 : 12 : 18, etc.

13-7 Areas of the Peaks

The area under a peak is proportional to the number of hydrogens contributing to that peak. For example, in the methyl *tert*-butyl ether spectrum (Figure 13-19), the absorption of the *tert*-butyl protons is larger and stronger than that of the methoxy protons because there are three times as many *tert*-butyl protons as methoxy protons. We cannot simply compare peak heights, however; the *area* under the peak is proportional to the number of protons.

NMR spectrometers have **integrators** that compute the relative areas of peaks. The integrator draws a second trace (the integral trace) that rises when it goes over a peak. The amount the integral trace rises is proportional to the area of that peak. You can measure these integrals using a millimeter ruler. Newer digital instruments also print a number representing the area of each peak. These numbers correspond to the heights of the rises in the integral trace.

Neither an integral trace (shown in blue in Figure 13-19) nor a digital integral can specifically indicate that methyl *tert*-butyl ether has three methyl hydrogens and nine *tert*-butyl hydrogens. Each simply shows that about three times as many hydrogens are represented by the peak at δ1.2 as are represented by the peak at δ3.2. We must interpret what the 3 : 1 ratio means in terms of the structure.

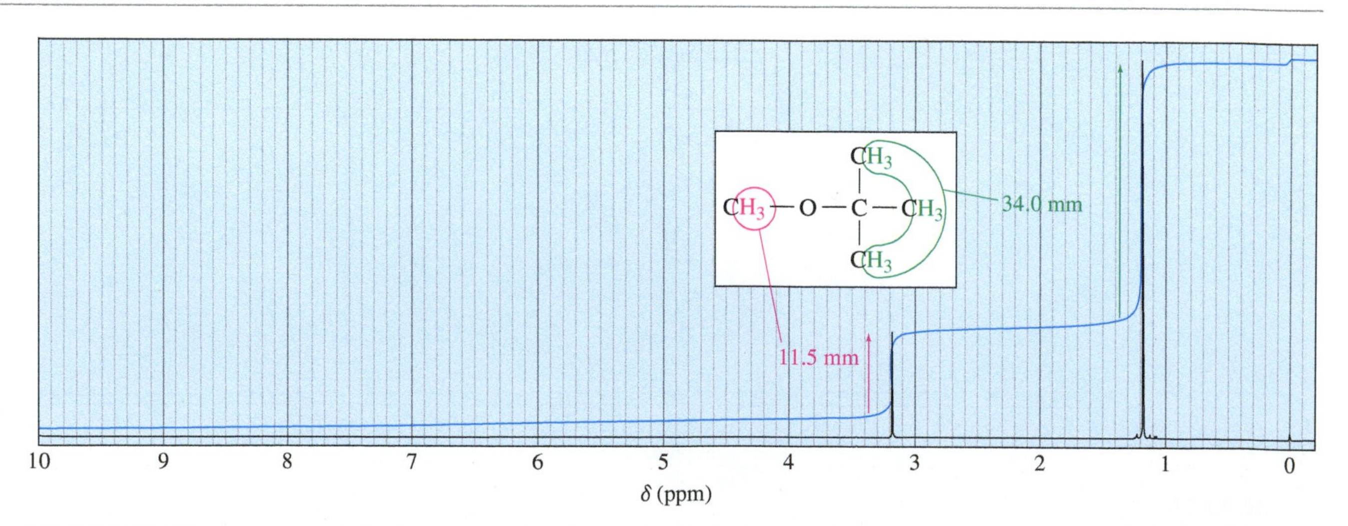

FIGURE 13-19
Integrated proton NMR spectrum of methyl *tert*-butyl ether. In going over a peak, the integrator trace (blue) rises by an amount that is proportional to the area under the peak.

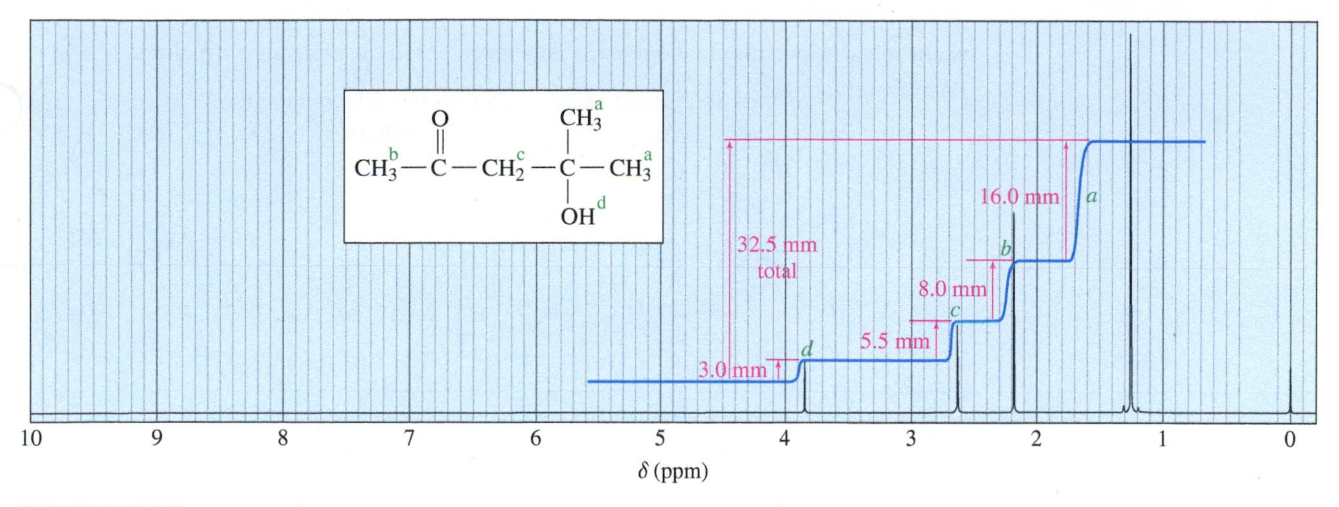

FIGURE 13-20
Proton NMR spectrum for a compound of molecular formula $C_6H_{12}O_2$.

Figure 13-20 shows the integrated spectrum of a compound with molecular formula $C_6H_{12}O_2$. Because we know the molecular formula, we can use the integral trace to determine exactly how many protons are responsible for each peak. The integrator has moved a total of 32.5 mm vertically in integrating the 12 protons in the molecule. Each proton is represented by

$$\frac{32.5\text{ mm}}{12\text{ hydrogens}} = \text{about 2.7 mm per hydrogen}$$

The signal at $\delta\,3.8$ has an integral of 3.0 mm, so it must represent one proton. At $\delta\,2.6$, the integrator moves 5.5 mm, corresponding to two protons. The signal at $\delta\,2.2$ has an integral of 8.0 mm, for three protons, and the signal at $\delta\,1.2$ (16.0 mm) corresponds to six protons. Considering the expected chemical shifts together with the information provided by the integrator leaves no doubt which protons are responsible for which signals in the spectrum.

O
CH$_3^a$ — area = 6, δ 1.2
H$_3^b$C—C—CH$_2^c$—C—CH$_3^a$
OH^d
area = 3, δ 2.2
area = 2, δ 2.6
area = 1, δ 3.8

PROBLEM-SOLVING HINT

Oxygen atoms are σ-withdrawers and π-donors of electron density. They deshield protons on the adjacent carbon atom to δ 3-4.

H δ 3–4
—C—H
OR
σ-withdrawing

When attached to aromatic rings, however, O—H and O—R groups donate electron density into the π system of the ring. Protons that are ortho or para to oxygen absorb *upfield* of the usual δ 7.2 for benzene (often around δ 6.8).

C
C H
:OR
π-donating

para, δ 6.8 (shielded)
meta, δ 7.2 (unaffected)
ortho, δ 6.8 (shielded)
H H H H H
OCH$_3$
shielding of ortho and para protons in anisole

PROBLEM 13-5

Draw the integral trace expected for the NMR spectrum of *tert*-butyl acetoacetate, shown in Figure 13-17.

PROBLEM 13-6

Determine the ratios of the peak areas in the following spectra. Then use this information, together with the chemical shifts, to pair up the compounds with their spectra. Assign the

(continued)

peaks in each spectrum to the protons they represent in the molecular structure.

Possible structures:

CH_3–C_6H_4–CH_3 (para) $\quad$ CH_3O–C_6H_4–OCH_3 (para) $\quad$ C_6H_5–C≡C–H

CH_3–C(OH)(CH_3)–C≡C–H $\quad$ CH_3–CH(Br)–CH_2–CH_3 $\quad$ Br–CH(H)–C(CH_3)(Br)–CH_3

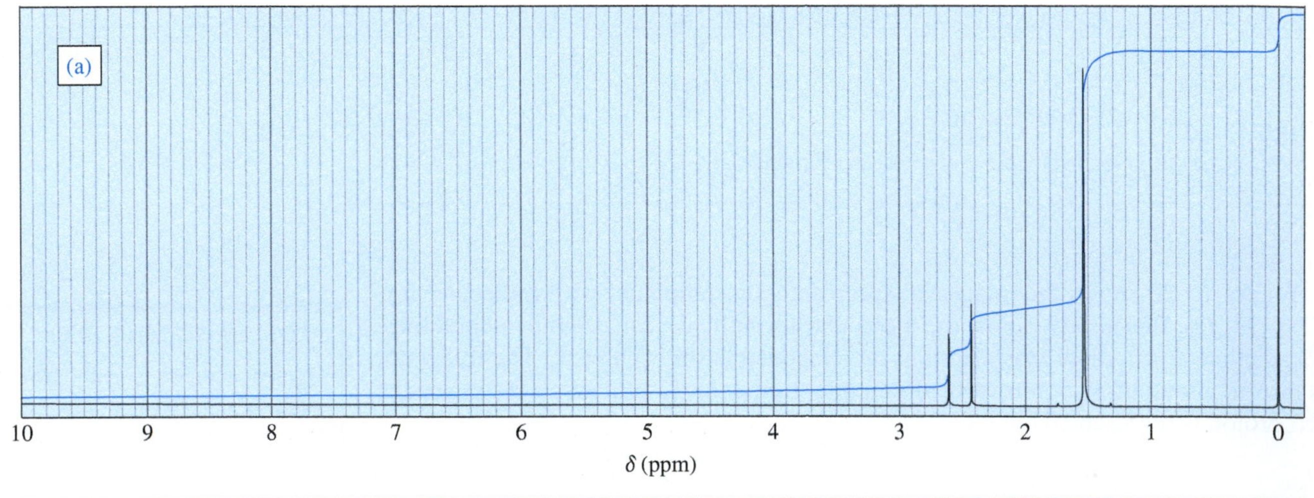

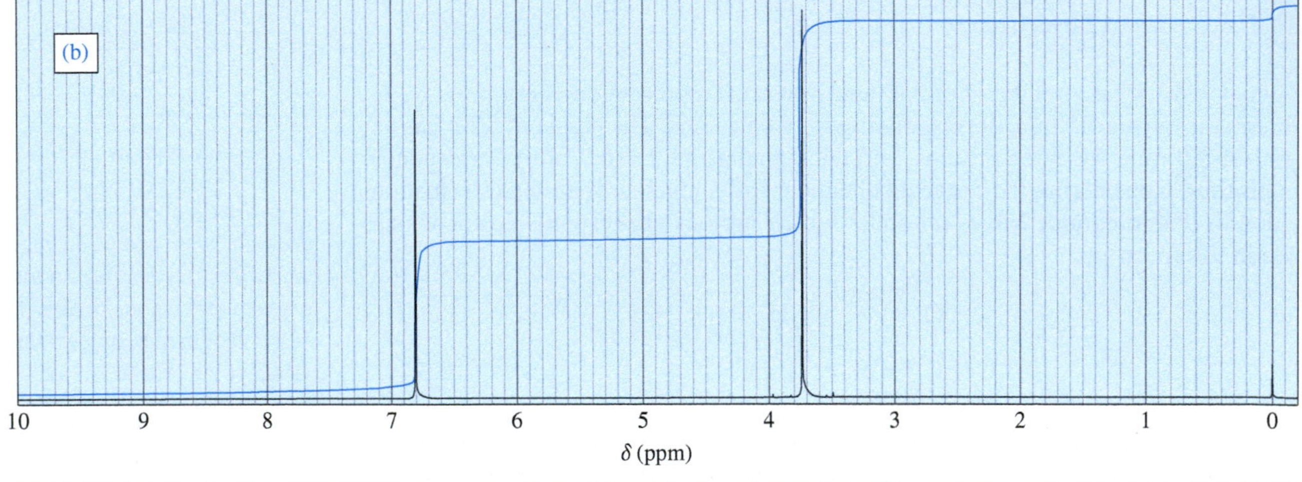

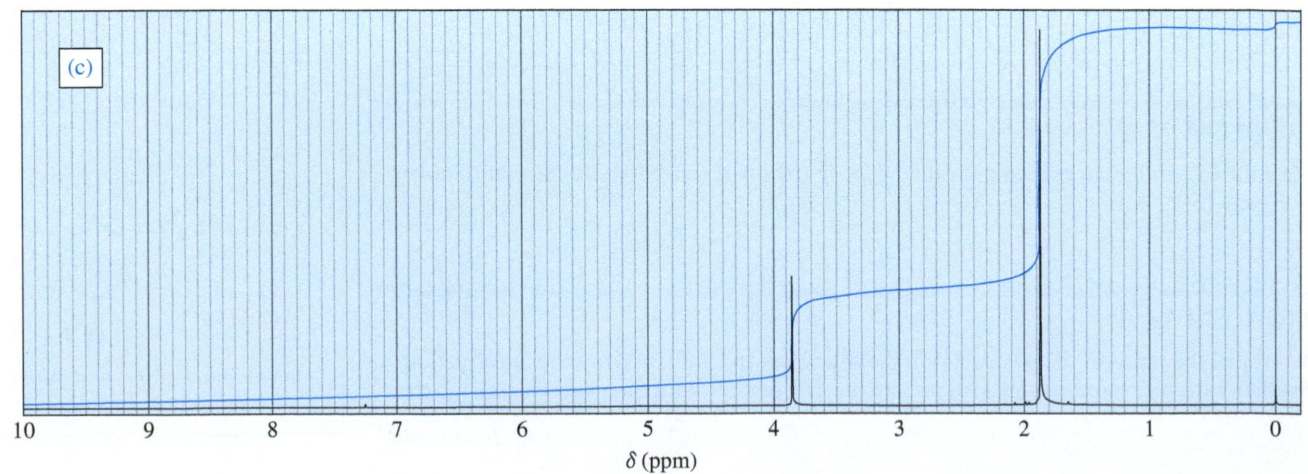

13-8 Spin-Spin Splitting

A proton in the NMR spectrometer is subjected to both the external magnetic field and the induced field of the shielding electrons. If there are other protons nearby, their small magnetic fields also affect the absorption frequencies of the protons we are observing. These small magnetic fields affect nearby protons in predictable ways that provide additional structural information.

13-8A Theory of Spin-Spin Splitting

Consider the spectrum of 1,1,2-tribromoethane (Figure 13-21). As expected, there are two signals with areas in the ratio of 1 : 2. The smaller signal (H^a) appears at $\delta 5.7$, deshielded by the two adjacent bromine atoms. The larger signal (H^b) appears at $\delta 4.1$. These signals do not appear as single peaks but as a triplet (three peaks) and a doublet (two peaks), respectively. This splitting of signals into multiplets, called **spin-spin splitting,** results when two different types of protons are close enough that their magnetic fields influence each other. Such protons are said to be **magnetically coupled.**

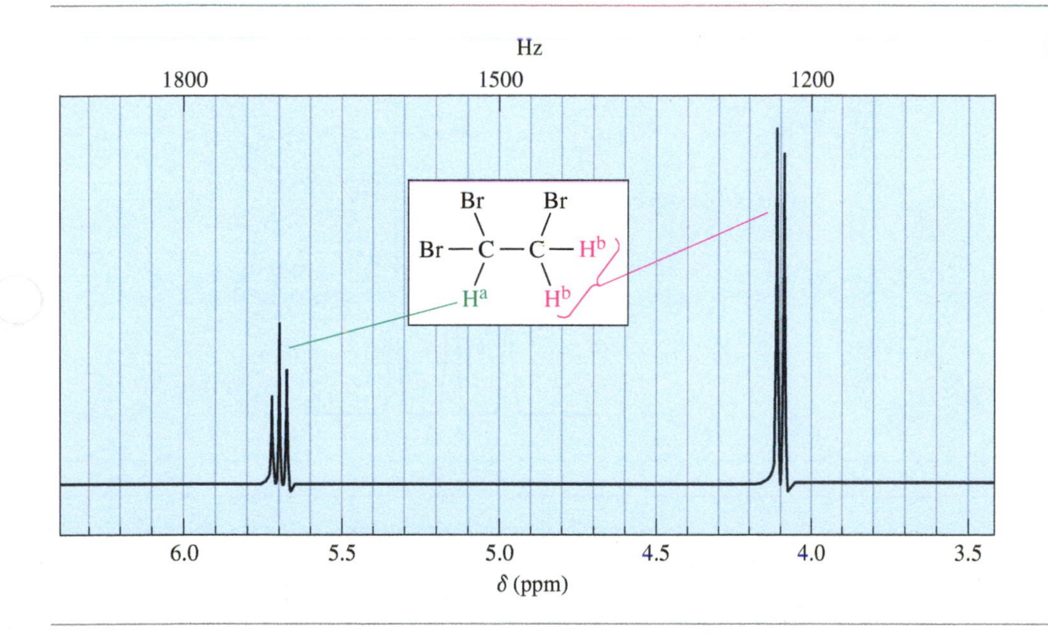

FIGURE 13-21
The proton NMR spectrum of 1,1,2-tribromoethane shows a triplet of area 1 at $\delta 5.7$ ($-CHBr_2$) and a doublet of area 2 at $\delta 4.1$ ($-CH_2Br$).

Spin-spin splitting can be explained by considering the individual spins of the magnetically coupled protons. Assume that our spectrometer is scanning the signal for the H^b protons of 1,1,2-tribromoethane at $\delta 4.1$ (Figure 13-22). These protons are under the influence of the small magnetic field of the adjacent proton, H^a. The orientation of H^a is not the same for every molecule in the sample. In some molecules, H^a is aligned with the external magnetic field, and in others, it is aligned against the field.

When H^a is aligned with the field, the H^b protons feel a slightly stronger total field: They are effectively deshielded, and they absorb at a lower field. When the magnetic moment of the H^a proton is aligned against the external field, the H^b protons are shielded, and they absorb at a higher field. These are the two absorptions of the doublet seen for the H^b protons. About half of the molecules have H^a aligned with the field and about half against the field, so the two absorptions of the doublet are nearly equal in area.

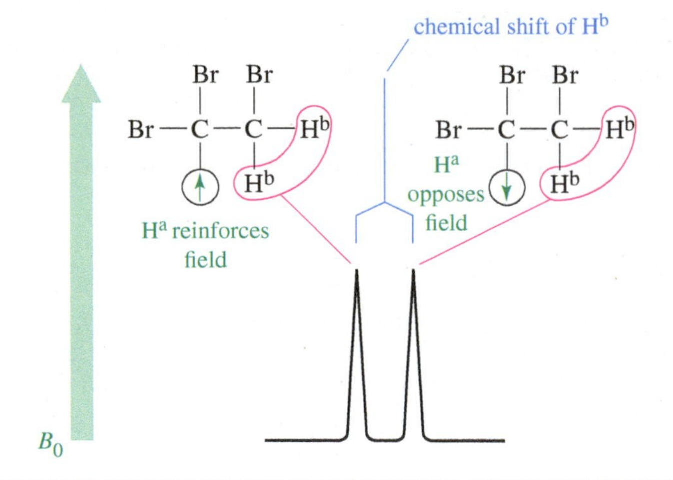

FIGURE 13-22 Splitting of the $—CH_2Br$ group in 1,1,2-tribromoethane. When the nearby H^a proton is aligned with the external magnetic field, it deshields H^b; when H^a is aligned against the field, it shields H^b.

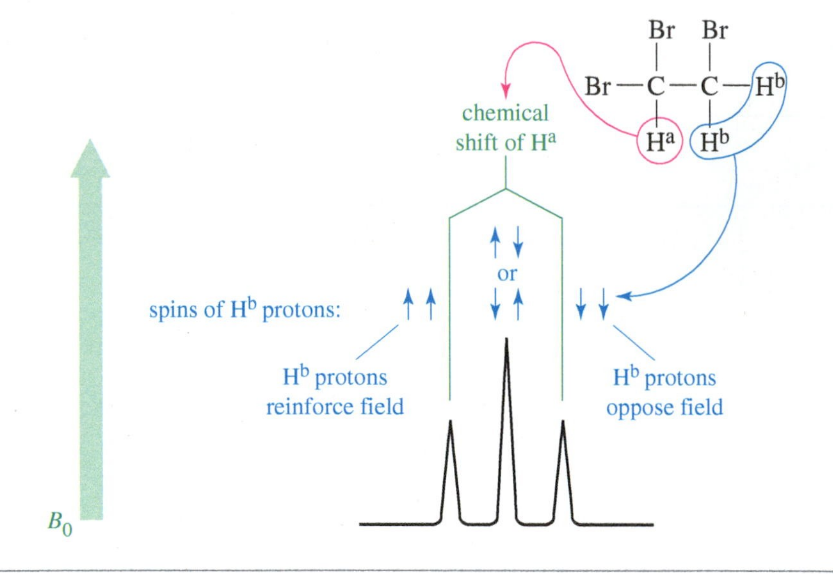

FIGURE 13-23 Splitting of the $—CHBr_2$ group in 1,1,2-tribromoethane. The H^a absorption is affected by the three combinations of H^b spins. When the H^b spins reinforce the external field, the H^a absorption occurs at a lower field. When the H^b spins oppose the external field, the H^a absorption occurs at a higher field. Two permutations, where the H^b proton spins cancel each other, allow H^a to absorb at its "normal" position. The peak area ratios are 1 : 2 : 1.

Spin-spin splitting is a reciprocal property. If one proton splits another, the second proton must split the first.

Proton a in Figure 13-21 appears as a triplet (at $\delta 5.7$) because there are four permutations of the two H^b proton spins, with two of them giving the same magnetic field (Figure 13-23). When both H^b spins are aligned with the applied field, proton a is deshielded; when both H^b spins are aligned against the field, proton a is shielded; and when the two H^b spins are opposite each other (two possible permutations), they cancel each other out. Three signals result, with the middle signal twice as large as the others because it corresponds to two possible spin permutations.

The two H^b protons do not split each other because they are chemically equivalent and absorb at the same chemical shift. Protons that absorb at the same chemical shift cannot split each other because they are in resonance at the same combination of frequency and field strength.

13-8B The $N + 1$ Rule

The preceding analysis for the splitting of 1,1,2-tribromoethane can be extended to more complicated systems. In general, the multiplicity (number of peaks) of an NMR signal is given by the **$N + 1$ rule:**

> *N+1* rule: If a signal is split by *N* neighboring equivalent protons, it will be split into *N+1* peaks.

The relative areas of the *N*+1 **multiplet** that results are approximately given by the appropriate line of *Pascal's triangle*:

RELATIVE PEAK INTENSITIES OF SYMMETRIC MULTIPLETS

Number of Equivalent Protons Causing Splitting (*N*)	Number of Peaks (multiplicity, *N* + 1)	Area Ratios (Pascal's triangle)
0	1 (singlet)	1
1	2 (doublet)	1 1
2	3 (triplet)	1 2 1
3	4 (quartet)	1 3 3 1
4	5 (quintet)	1 4 6 4 1
5	6 (sextet)	1 5 10 10 5 1
6	7 (septet)	1 6 15 20 15 6 1

Consider the splitting of the signals for the ethyl group in ethylbenzene (Figure 13-24). The methyl protons are split by two adjacent protons, and they appear upfield as a triplet of areas 1 : 2 : 1. The methylene (CH_2) protons are split by three protons, appearing farther downfield as a quartet of areas 1 : 3 : 3 : 1. This splitting pattern is typical for an ethyl group. Because ethyl groups are common, you should learn to recognize this familiar pattern. All five aromatic protons absorb close to 7.2 ppm because the alkyl substituent has only a small effect on the chemical shifts of the aromatic protons. The aromatic protons split each other in a complicated manner in this high-resolution spectrum (Section 13-9). At a lower field with less resolution, these aromatic protons would not be resolved, and they would appear as a slightly broadened single peak.

PROBLEM-SOLVING HINT

When you see splitting, look for nonequivalent protons on neighboring carbon atoms.

13-8C The Range of Magnetic Coupling

In ethylbenzene, there is no spin-spin splitting between the aromatic protons and the protons of the ethyl group. These protons are not on adjacent carbon atoms, so they are too far away to be magnetically coupled.

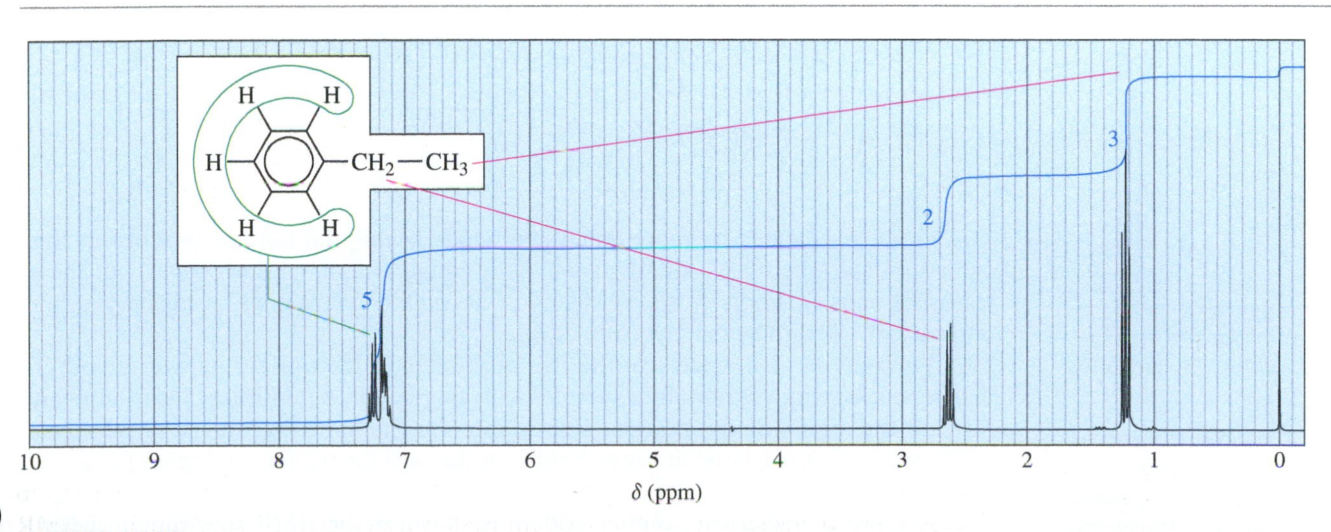

FIGURE 13-24
Proton NMR spectrum of ethylbenzene. The ethyl group appears as a triplet at δ1.2 (—CH_3) and as a quartet at δ2.6 (—CH_2—). The aromatic protons appear as a multiplet near δ7.2.

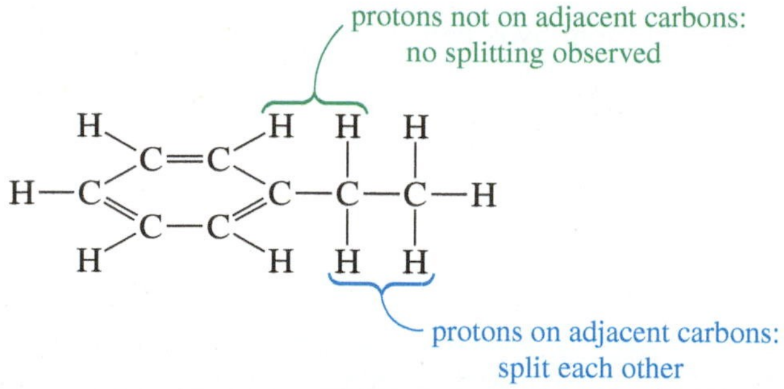

protons not on adjacent carbons: no splitting observed

protons on adjacent carbons: split each other

The magnetic coupling that causes spin-spin splitting takes place primarily through the bonds of the molecule. Most examples of spin-spin splitting involve coupling between protons that are separated by three bonds, so they are bonded to adjacent carbon atoms (vicinal protons).

PROBLEM-SOLVING HINT

In most cases, protons on the same carbon atom are equivalent and do not split each other. If they are nonequivalent, however (Section 13-10), they may split each other.

> Most spin-spin splitting is between protons on adjacent carbon atoms.

Protons bonded to the same carbon atom (geminal protons) can split each other *only if they are nonequivalent.* In most cases, protons on the same carbon atom are equivalent, and equivalent protons cannot split each other.

Bonded to the same carbon: two bonds between protons

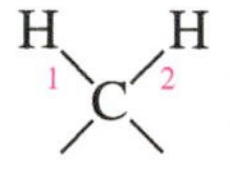

Spin-spin splitting is normally observed (if nonequivalent).

Bonded to adjacent carbons: three bonds between protons

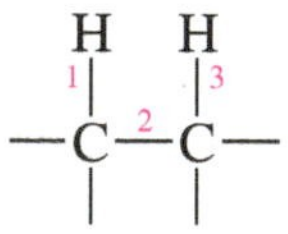

Spin-spin splitting is normally observed. (This is the most common case).

Bonded to nonadjacent carbons: four or more bonds between protons

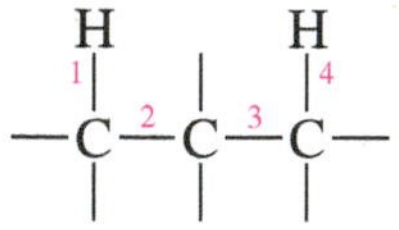

Spin-spin splitting is *not* normally observed.

Protons separated by more than three bonds usually do not produce observable spin-spin splitting. Occasionally, such "long-range coupling" does occur, but these cases are unusual. For now, we consider only nonequivalent protons on adjacent carbon atoms (or closer) to be magnetically coupled.

You may have noticed that the two multiplets in the upfield part of the ethylbenzene spectrum are not quite symmetrical. In general, a multiplet "leans" upward toward the signal of the protons responsible for the splitting. In the ethyl signal (Figure 13-25) the quartet at lower field leans toward the triplet at a higher field, and vice versa.

Another characteristic splitting pattern is shown in the NMR spectrum of methyl isopropyl ketone (3-methylbutan-2-one) in Figure 13-26.

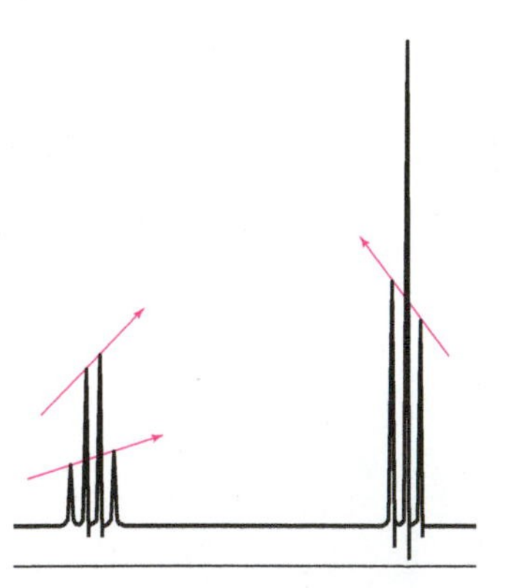

FIGURE 13-25
Leaning of a multiplet. A multiplet often "leans" upward toward the protons that are causing the splitting. The ethyl multiplets in ethylbenzene lean toward each other.

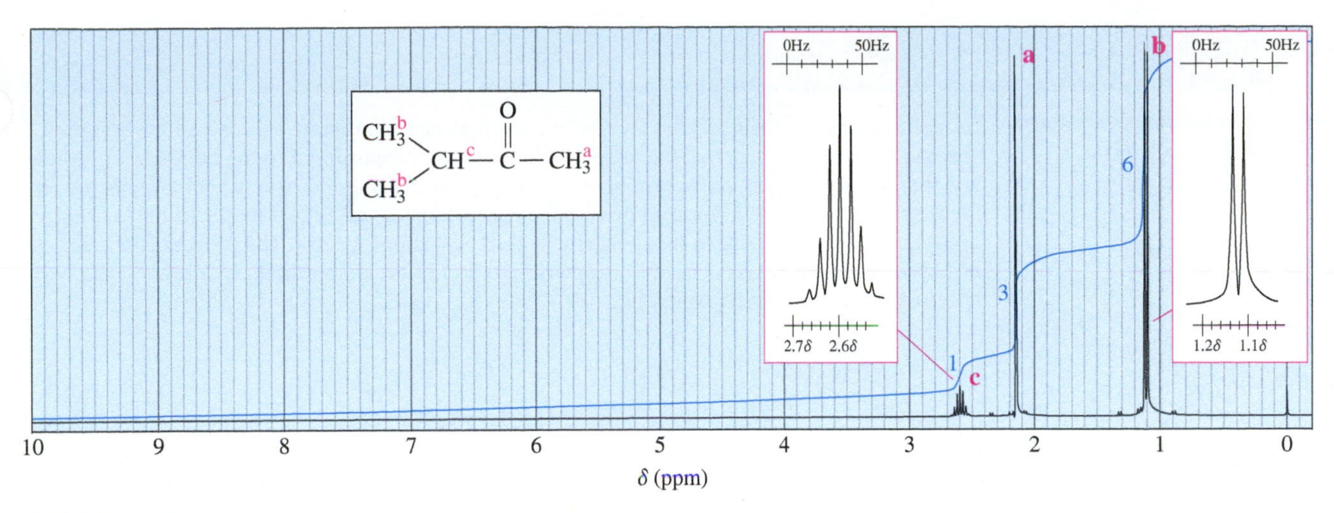

FIGURE 13-26
Proton NMR spectrum of methyl isopropyl ketone. The isopropyl group appears as a characteristic pattern of a strong doublet at a higher field and a weak multiplet (a septet) at a lower field. The methyl group appears as a singlet at δ2.1.

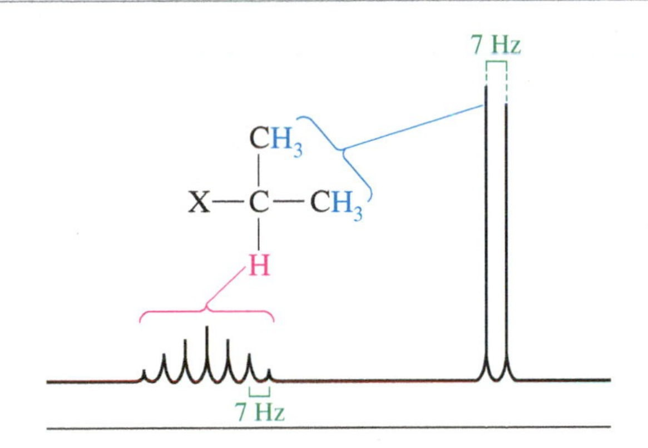

FIGURE 13-27
Characteristic isopropyl group pattern.

The three equivalent protons (a) of the methyl group bonded to the carbonyl appear as a singlet of relative area 3, near δ2.1. Methyl ketones and acetate esters characteristically give such singlets around δ2.1, because there are no protons on the adjacent carbon atom.

singlet, δ 2.1 → CH_3—C(=O)—R
a methyl ketone

singlet, δ 2.1 → CH_3—C(=O)—O—R
an acetate ester

The six methyl protons (b) of the isopropyl group are equivalent. They appear as a doublet of relative area 6 at about δ1.1, slightly deshielded by the carbonyl group two bonds away. This doublet leans downfield because these protons are magnetically coupled to the methine proton (c). This doublet is also shown in an insert box with the horizontal scale expanded for clarity. The vertical scale is adjusted so that the peaks fit in the box.

The methine proton H^c appears as a multiplet of relative area 1, at δ2.6. This absorption is a septet (seven peaks) because it is coupled to the six adjacent methyl protons (b). Some small peaks of this septet may not be visible unless the spectrum is amplified, as it is in the expanded insert box. The pattern seen in this spectrum, summarized in Figure 13-27, is typical for an isopropyl group: The methyl protons give a strong doublet at a higher field, and the methine proton gives a weak multiplet (usually difficult to count the peaks) at a lower field. Isopropyl groups are easily recognized from this characteristic pattern.

PROBLEM-SOLVING HINT

Ethyl and isopropyl groups are common. Learn to recognize them from their splitting patterns.

PROBLEM-SOLVING STRATEGY Drawing an NMR Spectrum

PROBLEM-SOLVING HINT

To estimate the chemical shift of protons that are deshielded by two groups, add the chemical shifts you would expect with each deshielding group individually, and subtract 1.3 (the δ for an alkane CH_2 group) from the result.

$C_6H_5-(CH_2)-C(=O)-CH_3$

phenylacetone

For the CH_2 group of phenylacetone, we calculate the following:

CH_2 next to phenyl	δ 2.5
CH_2 next to C=O	δ 2.3
Total	4.8
subtract	1.3
predict	δ 3.5

(The experimental value is δ 3.7.) This estimation will generally predict a chemical shift within ±0.5 to 1 ppm of the correct value.

In learning about NMR spectra, we have seen that chemical shift values can be assigned to specific types of protons, that the areas under peaks are proportional to the numbers of protons, and that nearby protons cause spin-spin splitting. By analyzing the structure of a molecule with these principles in mind, you can predict the characteristics of an NMR spectrum. Learning to draw spectra will help you to recognize the features of actual spectra. The process is not difficult if a systematic approach is used. A stepwise method is illustrated here, by drawing the NMR spectrum of the following compound.

$$(CH_3^a)_2CH^b-O-C(=O)-CH_2^c-CH_3^d$$

1. **Determine how many types of protons are present, together with their proportions.** In the example, there are four types of protons, labeled *a*, *b*, *c*, and *d*. The area ratios should be 6 : 1 : 2 : 3.
2. **Estimate the chemical shifts of the protons. (Table 13-3 and Appendix 1 serve as guides.)** Proton *b* is on a carbon atom bonded to oxygen; it should absorb around δ3 to δ4. Protons *a* are less deshielded by the oxygen, probably around δ1 to δ2. Protons *c* are on a carbon bonded to a carbonyl group; they should absorb around δ2.1 to δ2.5. Protons *d*, one carbon removed from a carbonyl, will be deshielded less than protons *c* and also less than protons *a*, which are next to a more strongly deshielded carbon atom. Protons *d* should absorb around δ1.0.
3. **Determine the splitting patterns.** Protons *a* and *b* split each other into a doublet and a septet, respectively (a typical isopropyl group pattern). Protons *c* and *d* split each other into a quartet and a triplet, respectively (a typical ethyl group pattern).
4. **Summarize each absorption in order, from the lowest field to the highest.**

	Proton *b*	Protons *c*	Protons *a*	Protons *d*
area	1	2	6	3
chemical shift	3–4	2.1–2.5	1–2	1
splitting	septet	quartet	doublet	triplet

5. **Draw the spectrum, using the information from your summary.** Work through the following problem to become comfortable with predicting NMR spectra.

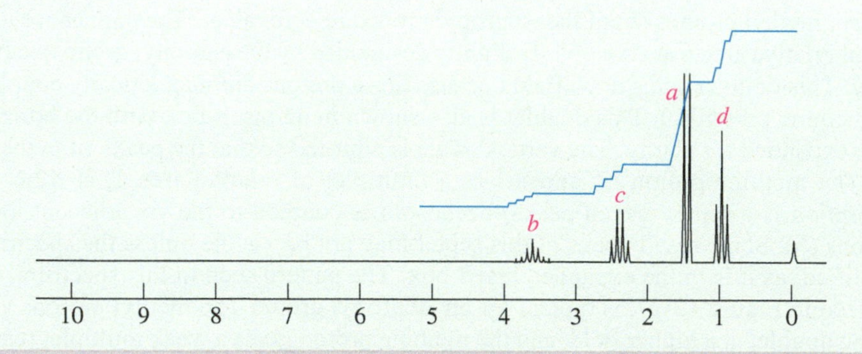

PROBLEM 13-7

Draw the NMR spectra you would expect for the following compounds.

(a) $(CH_3)_2CH-O-CH(CH_3)_2$

(b) $Cl-CH_2-CH_2-C(=O)-O-CH_3$

(c) $Ph-CH(CH_3)_2$

(d)  $CH_3CH_2O-C_6H_4-OCH_2CH_3$

(e) $CH_2-COOCH_2CH_3$ | $CH_2-COOCH_2CH_3$

13-8D Coupling Constants

The distances between the peaks of multiplets can provide additional structural information. These distances are all about 7 Hz in the methyl isopropyl ketone spectrum (Figures 13-26 and 13-27). These splittings are equal because *any two magnetically coupled protons must have equal effects on each other.* The distance between adjacent peaks of the H^c multiplet (split by H^b) must equal the distance between the peaks of the H^b doublet (split by H^c).

The distance between the peaks of a multiplet (measured in hertz) is called the **coupling constant.** Coupling constants are represented by J, and the coupling constant between H^a and H^b is represented by J_{ab}. In complicated spectra with many types of protons, groups of neighboring protons can sometimes be identified by measuring their coupling constants. Multiplets that have the same coupling constant may arise from adjacent groups of protons that split each other.

The magnetic effect that one proton has on another depends on the bonds connecting the protons, but it does not depend on the strength of the external magnetic field. For this reason, the coupling constant (measured in hertz) does not vary with the field strength of the spectrometer. A spectrometer operating at 300 MHz records the same coupling constants as a 60-MHz instrument.

Figure 13-28 shows typical values of coupling constants. The most commonly observed coupling constant is the 7-Hz splitting of protons on adjacent carbon atoms in freely rotating alkyl groups.

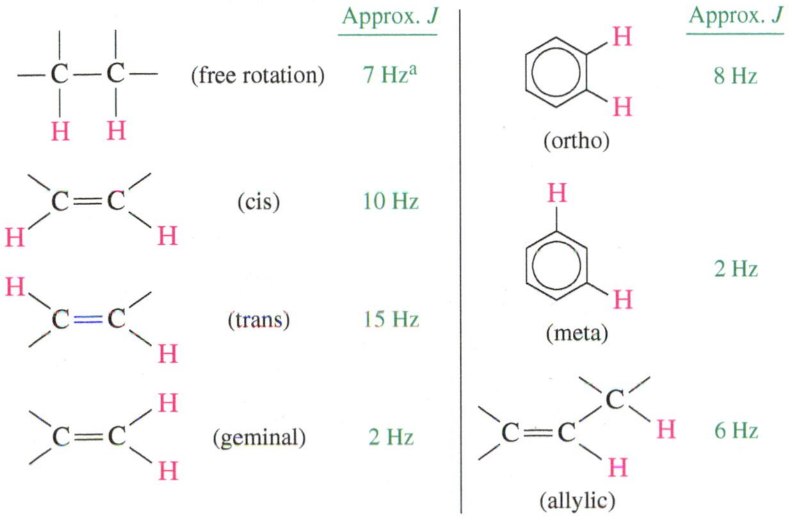

Structure		Approx. J	Structure	Approx. J
H–C–C–H	(free rotation)	7 Hz[a]	(ortho)	8 Hz
H–C=C–H	(cis)	10 Hz	(meta)	2 Hz
H–C=C–H	(trans)	15 Hz	(allylic)	6 Hz
C=CH₂	(geminal)	2 Hz		

[a]The value of 7 Hz in an alkyl group is averaged for rapid rotation about the carbon–carbon bond. If rotation is hindered by a ring or bulky groups, other splitting constants may be observed.

FIGURE 13-28
Typical values of proton coupling constants.

Splitting patterns and their coupling constants help to distinguish among the possible isomers of a compound, as in the spectrum of *p*-nitrotoluene (Figure 13-29). The methyl protons (*c*) absorb as a singlet near $\delta 2.5$, and the aromatic protons appear as a pair of doublets. The doublet centered around $\delta 7.3$ corresponds to the two aromatic protons closest to the methyl group (*a*). The doublet centered around $\delta 8.1$ corresponds to the two protons closest to the electron-withdrawing nitro group (*b*).

Each proton *a* is magnetically coupled to one *b* proton, splitting the H^a absorption into a doublet. Similarly, each proton *b* is magnetically coupled to one proton *a*, splitting the H^b absorption into a doublet. The coupling constant is 8 Hz, just a little wider than the 7-Hz grid in the insert box. This 8-Hz coupling suggests that the magnetically coupled protons H^a and H^b are ortho to each other.

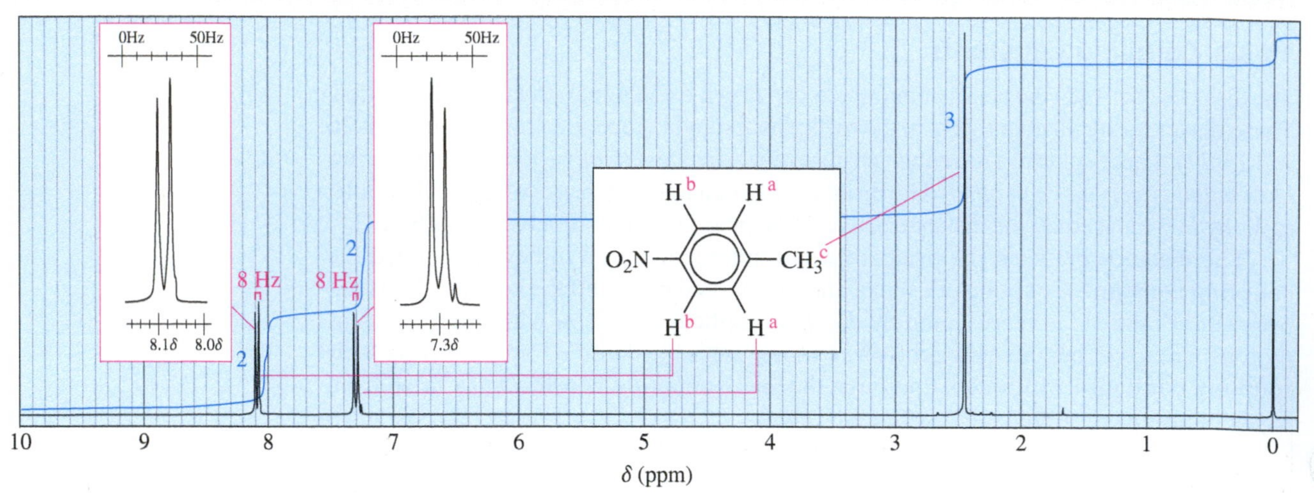

FIGURE 13-29
Proton NMR spectrum of *p*-nitrotoluene.

Both the ortho and meta isomers of nitrotoluene have four distinct types of aromatic protons, and the spectra for these isomers are more complex. Figure 13-29 must correspond to the para isomer of nitrotoluene.

Coupling constants also help to distinguish stereoisomers. In Figure 13-30(a), the 9-Hz coupling constant between the two vinyl protons shows that they are cis to one another. In Figure 13-30(b), the 15-Hz coupling constant shows that the two

PROBLEM-SOLVING HINT

Protons on the β carbon of an α,β-unsaturated carbonyl compound absorb at very low fields (about $\delta 7$) because of the electron-withdrawing resonance effect of the carbonyl group.

H(β)C=C(α)–C=O ⟷ H(β)$^{+}$C–C(α)=C–O^{-}

PROBLEM 13-8

Draw the NMR spectra you expect for the following compounds.

(a) Ph(H)C=C(H)C(CH$_3$)$_3$ — Ph and H on one carbon; H and C(CH$_3$)$_3$ on the other

(b) CH$_3$O(Cl)C=C(CH$_3$)(H)

(c) (CH$_3$)$_3$C(H)C=C(H)–C(=O)–OCH$_2$CH$_2$ — (CH$_3$)$_3$C, H, H, C(=O)OCH$_2$CH$_3$

(d) benzene ring with CH$_3$, four H, and C(=O)OH (para)

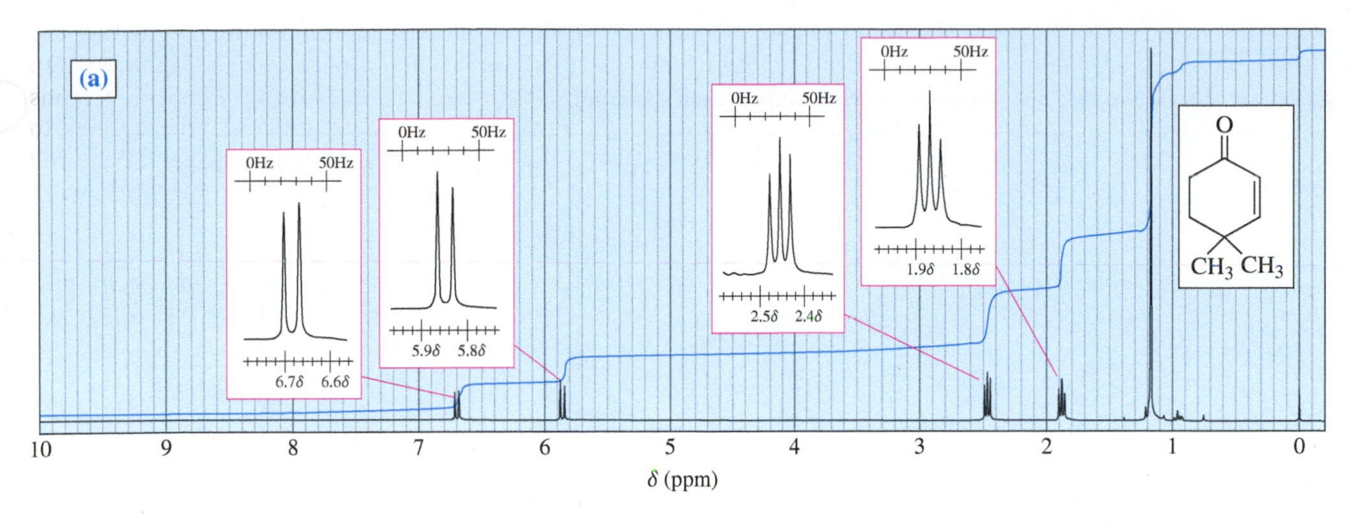

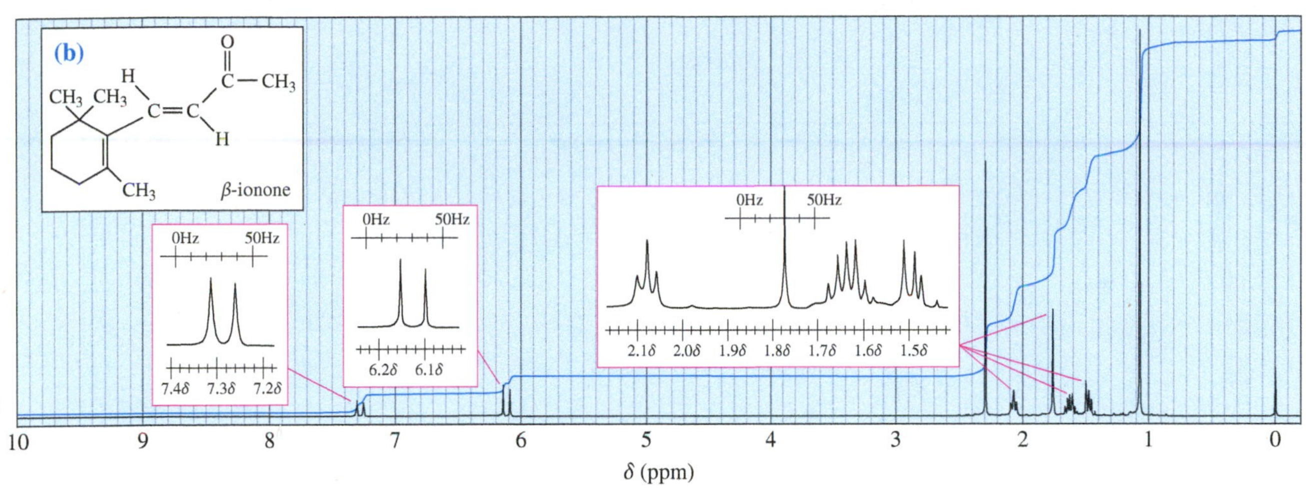

FIGURE 13-30
Proton NMR spectra of (a) 4,4-dimethylcyclohex-2-en-1-one and (b) β-ionone.

vinyl protons are trans. Notice that the 9-Hz coupling looks too wide for common alkyl group splitting, represented by the 7-Hz grid in the insert box. The 15-Hz coupling looks about double the grid spacing corresponding to the common 7-Hz alkyl group splitting.

PROBLEM-SOLVING HINT

Watch for unusually large coupling constants, especially in the vinyl region, where they may indicate the stereochemistry about a double bond.

PROBLEM 13-9

(a) Assign protons to the peaks in the NMR spectrum of 4,4-dimethylcyclohex-2-en-1-one in Figure 13-30(a). Explain the splitting that gives the triplets at δ1.8 and δ2.4.

(b) Assign protons to the peaks in the NMR spectrum of β-ionone in Figure 13-30(b). Explain the splitting seen in the three multiplets at δ1.5, δ1.65, and δ2.1. Explain how you know which of these multiplets corresponds to which methylene groups in the molecule.

PROBLEM 13-10

An unknown compound (C_3H_2NCl) shows moderately strong IR absorptions around 1650 cm^{-1} and 2200 cm^{-1}. Its NMR spectrum consists of two doublets ($J = 14$ Hz) at δ5.9 and δ7.1. Propose a structure consistent with these data.

PROBLEM 13-11

Two spectra are shown. Propose a structure that corresponds to each spectrum.

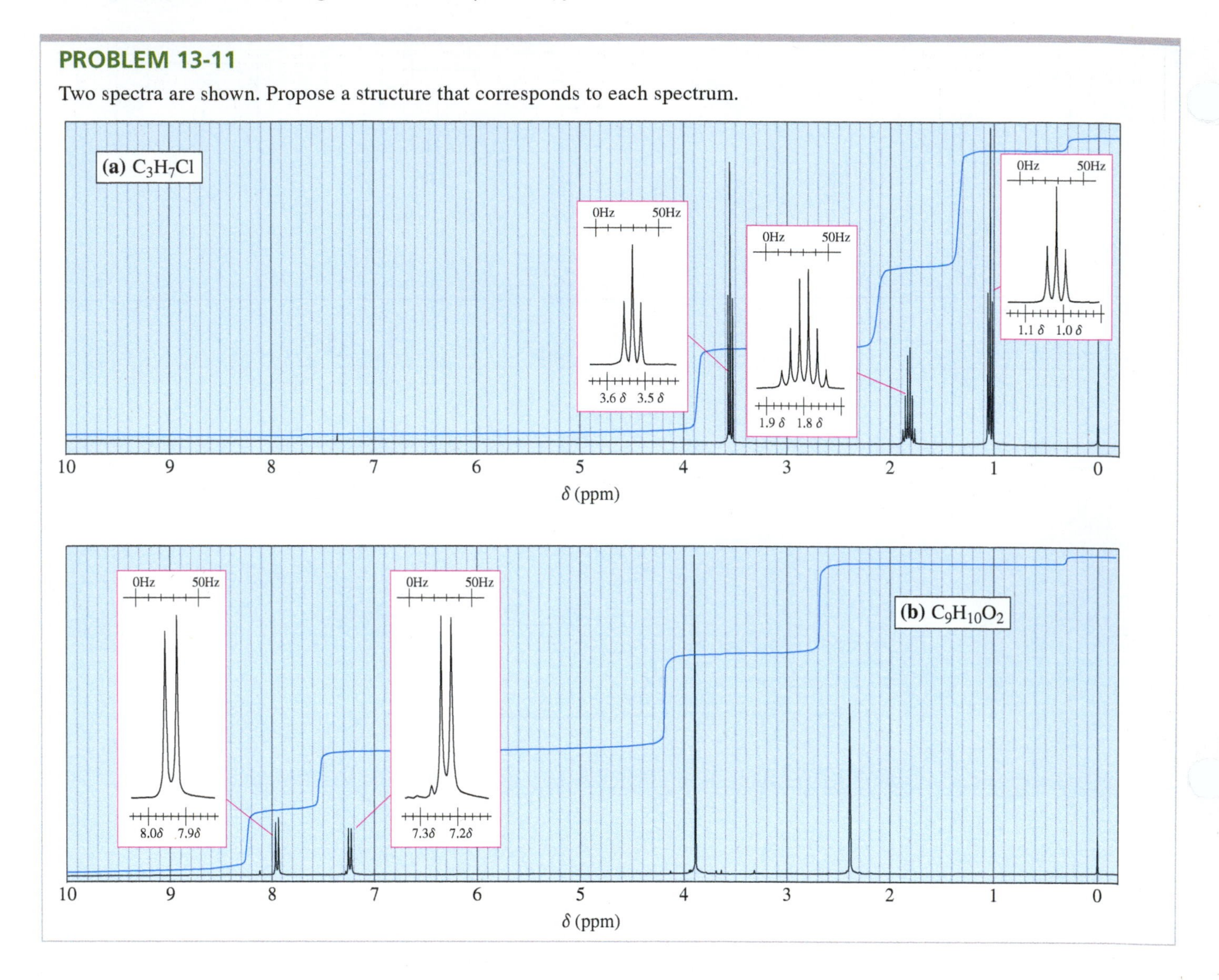

13-9 Complex Splitting

> **PROBLEM-SOLVING HINT**
>
> We could estimate the chemical shift of H^a by using the formula suggested in the Hint on page 630.
>
> | | |
> |---|---|
> | vinyl protons | δ5 to 6 |
> | Ph—CH protons | δ2.3 to 2.5 |
> | Total | 7.3 to 8.5 |
> | subtract | 1.3 |
> | Estimated chem shift | δ6 to 7 |

There are many cases of **complex splitting,** where signals are split by adjacent protons of more than one type, with different coupling constants. The proton NMR spectrum of styrene (Figure 13-31) includes several absorptions that show the results of complex splitting.

Consider the vinyl proton H^a, adjacent to the phenyl ring of styrene.

H^a H^c C=C H^b

styrene

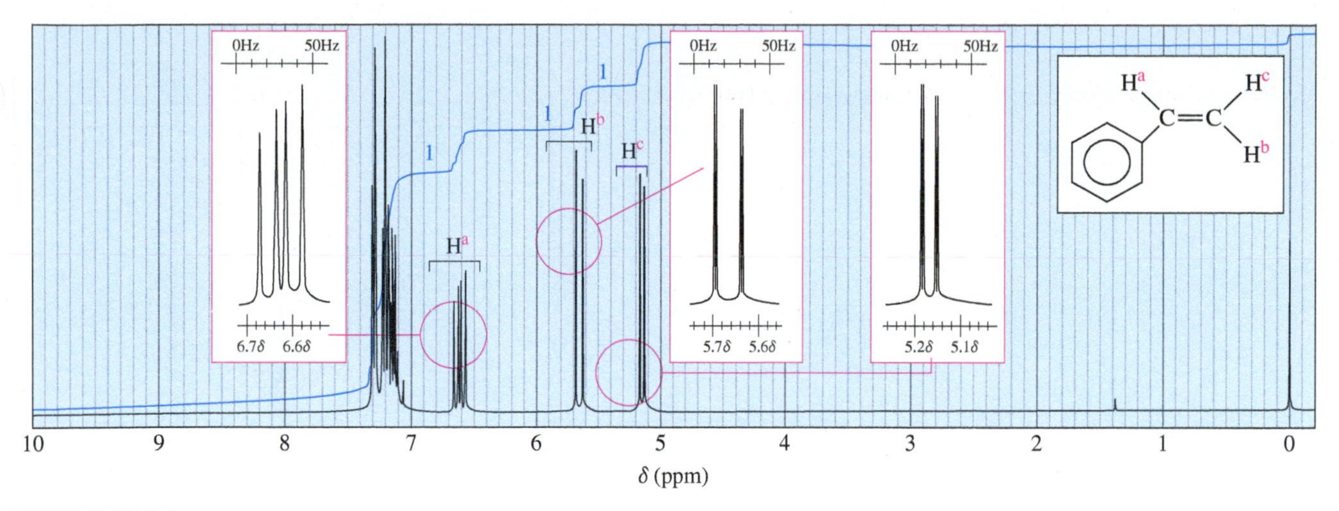

FIGURE 13-31
Proton NMR spectrum of styrene.

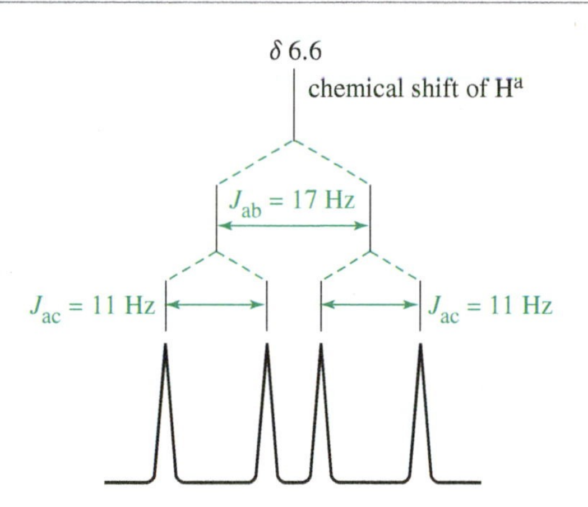

FIGURE 13-32
A splitting tree for the H^a proton in styrene. The H^a signal is split ($J_{ab} = 17$ Hz) by coupling with H^b and further split ($J_{ac} = 11$ Hz) by coupling with H^c.

The chemical shift of H^a is $\delta 6.6$ because it is deshielded by both the vinyl group and the aromatic ring. H^a is coupled to H^b with a typical trans coupling constant, $J_{ab} = 17$ Hz. It is also coupled to proton H^c with $J_{ac} = 11$ Hz. The H^a signal is therefore split into a doublet of spacing 17 Hz, and each of those peaks is further split into a doublet of spacing 11 Hz, for a total of four peaks. This complex splitting, called a *doublet of doublets*, can be analyzed by a diagram called a *splitting tree*, shown in Figure 13-32.

Proton H^b is farther from the deshielding influence of the phenyl group, giving rise to the multiplet centered at $\delta 5.65$ in the styrene NMR spectrum. H^b is also split by two nonequivalent protons: It is split by H^a with a trans coupling constant $J_{ac} = 17$ Hz and further split by H^c with a geminal coupling constant $J_{bc} = 1.4$ Hz. The splitting tree for H^b, showing a doublet of narrow doublets, is shown in Figure 13-33.

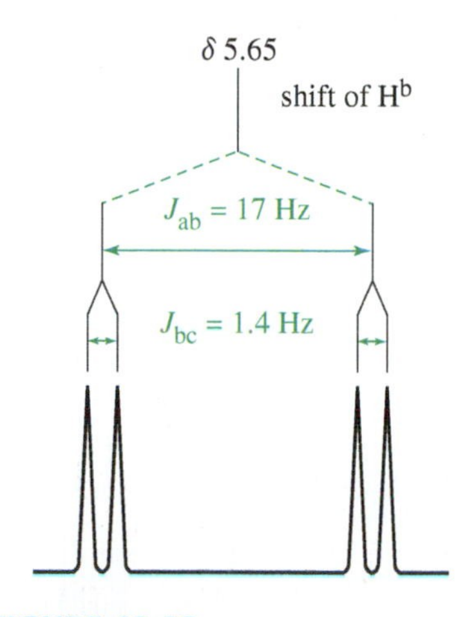

FIGURE 13-33
A splitting tree for the H^b proton in styrene. The H^b signal is split by coupling with H^a ($J_{ab} = 17$ Hz) and further split by coupling with H^c ($J_{bc} = 1.4$ Hz).

PROBLEM 13-12

Draw a splitting tree, similar to Figures 13-32 and 13-33, for proton H^c in styrene. What is the chemical shift of proton H^c?

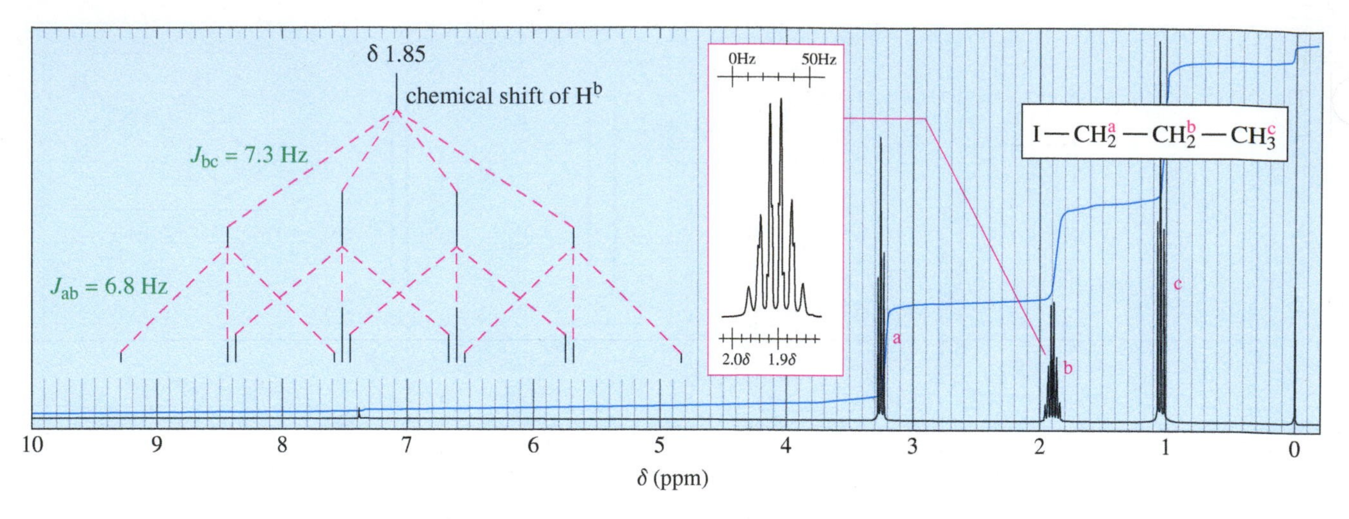

FIGURE 13-34
The NMR spectrum of 1-iodopropane. The H^b signal appears to be split into a sextet by the five hydrogens on the adjacent carbon atoms. On closer inspection, the multiplet is seen to be an imperfect sextet, the result of complex splitting by two sets of protons (a and c) with similar splitting constants.

Sometimes a signal is split by two or more different kinds of protons with similar coupling constants. Consider 1-iodopropane (Figure 13-34), where the *b* protons on the middle carbon atom are split by two types of protons: the methyl protons (H^c) and the CH_2I protons (H^a).

The coupling constants for these two interactions are similar: $J_{ab} = 6.8$ Hz, and $J_{bc} = 7.3$ Hz. The spectrum shows the H^b signal as a sextet, almost as though there were five equivalent protons coupled with H^b. The trace in the insert box, enlarged and offset, shows that the pattern is not a perfect sextet. The analysis of the splitting pattern serves as a reminder that the $N+1$ rule gives a perfect multiplet only when the signal is split by N protons with exactly the same coupling constant.

PROBLEM 13-13

The spectrum of *trans*-hex-2-enoic acid follows.

(a) Assign peaks to show which protons give rise to which peaks in the spectrum.

(b) Draw a tree to show the complex splitting of the vinyl proton centered around 7 ppm. Estimate the values of the coupling constants.

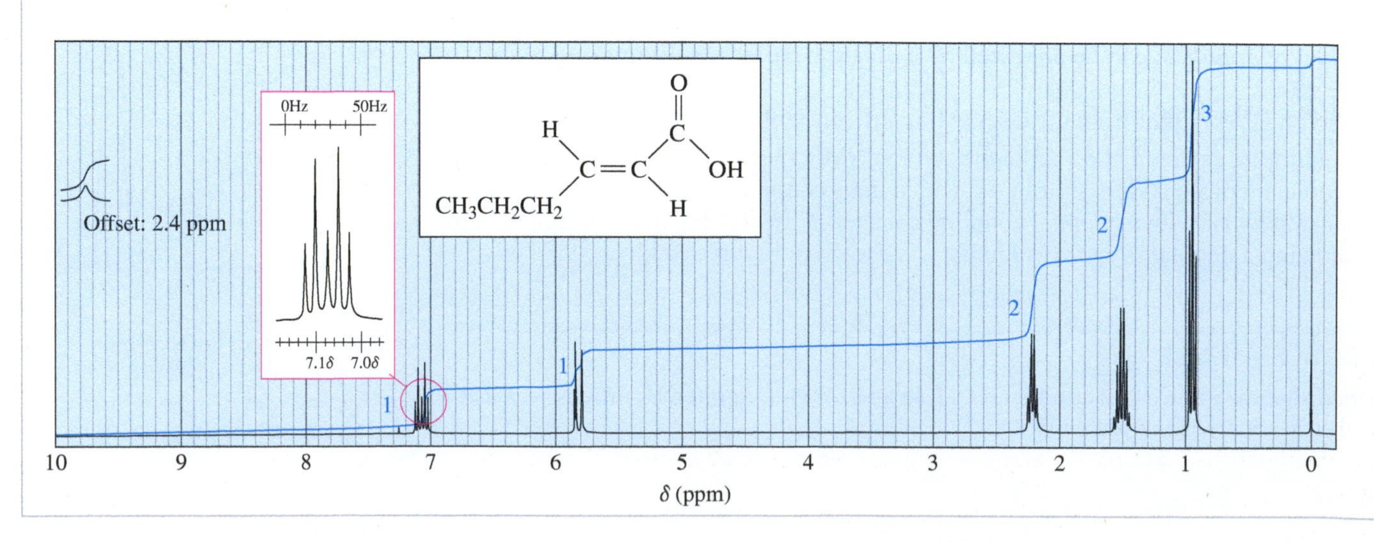

PROBLEM 13-14

The NMR spectrum of cinnamaldehyde follows.

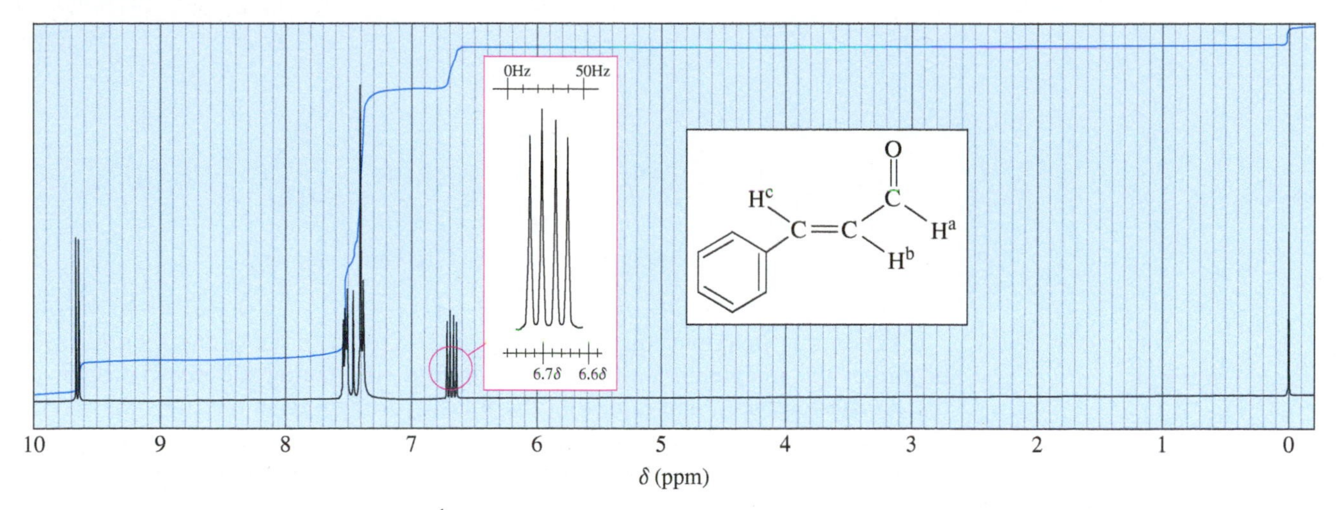

(a) Determine the chemical shifts of H^a, H^b, and H^c. The absorption of one of these protons is difficult to see; look carefully at the integrals.
(b) Estimate the coupling constants J_{ab} and J_{bc}.
(c) Draw a tree to analyze the complex splitting of the proton centered at δ 6.7.

PROBLEM 13-15

Consider the proton NMR spectrum of the following ketone.
(a) Predict the approximate chemical shift of each type of proton.
(b) Predict the number of NMR peaks for each type of proton.
(c) Draw a tree to show the splitting predicted for the absorption of the circled proton.

13-10 Stereochemical Nonequivalence of Protons

Stereochemical differences often result in different chemical shifts for protons on the same carbon atom. For example, the two protons on C_1 of allyl bromide (3-bromopropene) are not equivalent. H^a is cis to the $-CH_2Br$ group, and H^b is trans. H^a absorbs at δ5.3; H^b absorbs at δ5.1. There are four different (by NMR) types of protons in allyl bromide, as shown in the structure at right in the margin.

To determine whether similar-appearing protons are equivalent, mentally substitute another atom for each of the protons in question. *If the same product is formed by imaginary replacement of either of two protons, those protons are chemically equivalent* (also called *homotopic*). For example, replacing any of the three methyl protons in ethanol by an imaginary *Z* atom gives the same compound. These three hydrogens are chemically equivalent, and they will all appear at the same chemical shift. Freely rotating CH_3 groups always have chemically equivalent protons.

allyl bromide

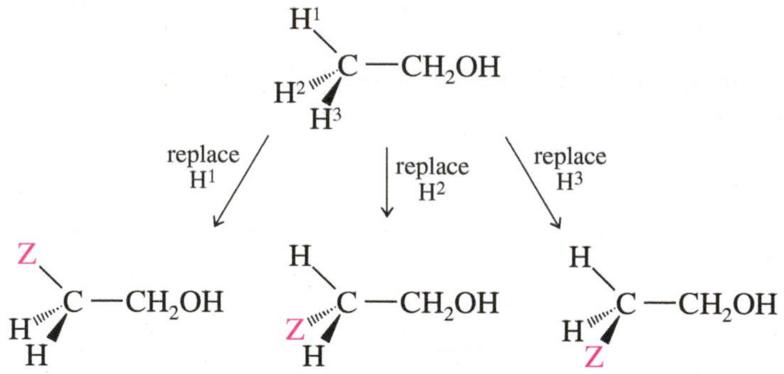

different conformations of the same compound

When this imaginary replacement test is applied to the protons on C_1 of allyl bromide, the imaginary products are different. Replacing the cis hydrogen gives the cis diastereomer, and replacing the trans hydrogen gives the trans diastereomer. Because the two imaginary products are diastereomers, these protons on C_1 are called **diastereotopic** protons. Diastereotopic protons appear in the NMR at different chemical shifts, and they can split each other.

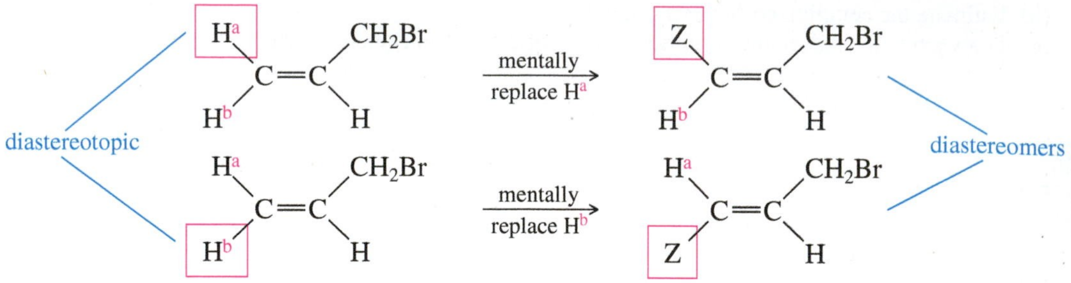

Cyclobutanol shows these stereochemical relationships in a cyclic system. The hydroxy proton H^a is clearly unique; it absorbs between $\delta 3$ and $\delta 5$, depending on the solvent and concentration. H^b is also unique, absorbing between $\delta 3$ and $\delta 4$. Protons H^e and H^f are diastereotopic (and absorb at different fields) because H^e is cis to the hydroxy group, whereas H^f is trans.

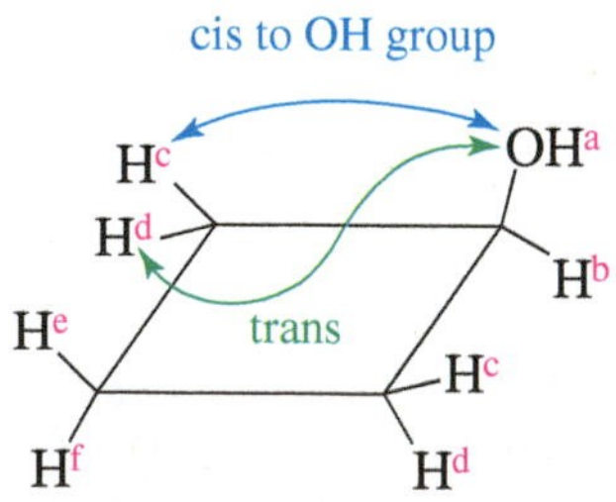

cyclobutanol

To distinguish among the other four protons, notice that cyclobutanol has an internal mirror plane of symmetry. Protons H^c are cis to the hydroxy group, whereas protons H^d are trans. Therefore, protons H^c are diastereotopic to protons H^d and the two sets of protons absorb at different magnetic fields and are capable of splitting each other.

PROBLEM 13-16

Use the imaginary replacement technique to show that protons H^c and H^d in cyclobutanol are diastereotopic.

*PROBLEM 13-17

If the imaginary replacement of either of two protons forms enantiomers, then those protons are said to be *enantiotopic*. The NMR is not a chiral probe, and it cannot distinguish between enantiotopic protons. They are seen to be "equivalent by NMR."

(a) Use the imaginary replacement technique to show that the two allylic protons (those on C_3) of allyl bromide are enantiotopic.

(b) Similarly, show that the two H^c protons in cyclobutanol are enantiotopic.

(c) What other protons in cyclobutanol are enantiotopic?

PROBLEM-SOLVING HINT

To compare two protons, use the imaginary replacement test. Draw the compound, replacing one of the protons with an imaginary atom (Z, for example). Then draw the compound again, replacing the other proton. Compare the two imaginary products. If they are the same, then the protons are chemically equivalent (homotopic). If they are diastereomers, then the protons are diastereotopic. If they are enantiomers, then the protons are enantiotopic. Under most circumstances, enantiotopic protons cannot be distinguished by NMR.

Diastereomerism also occurs in saturated, acyclic compounds; for example, 1,2-dichloropropane is a simple compound that contains diastereotopic protons. The two protons on the $—CH_2Cl$ group are diastereotopic; their imaginary replacement gives diastereomers.

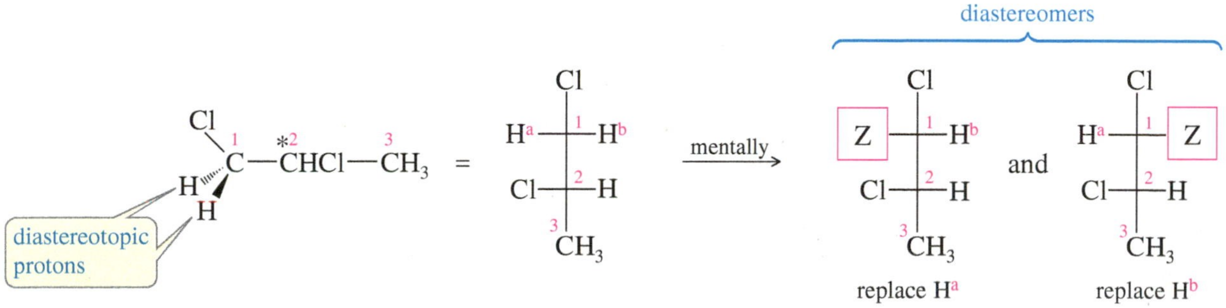

The most stable conformation of 1,2-dichloropropane (in the margin) shows that the diastereotopic protons on C1 exist in different chemical environments. They experience different magnetic fields and are nonequivalent by NMR. The NMR spectrum of 1,2-dichloropropane is shown in Figure 13-35. The methyl protons appear as a doublet at δ1.6, and the single proton on C2 appears as a complex multiplet at δ4.15. The two protons on C1 appear as distinct absorptions at δ3.60 and δ3.77. They are split by the proton on C2, and they also split each other.

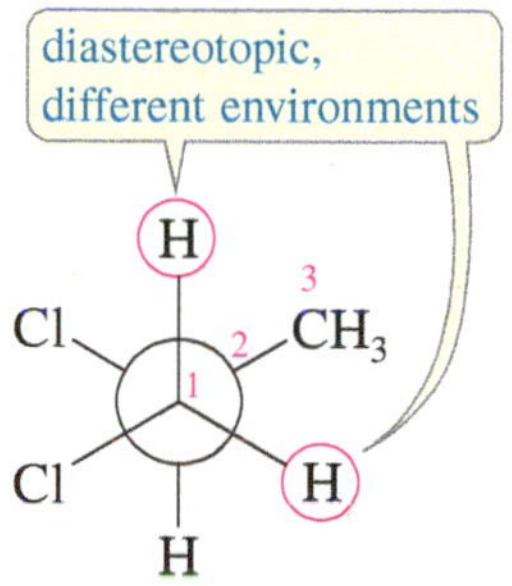

The presence of an asymmetric carbon atom adjacent to the CH_2Cl group gives rise to the different chemical environments of these diastereotopic protons. When a molecule contains an asymmetric carbon atom, the protons on any methylene group are usually diastereotopic. They may or may not be resolved in the NMR, however, depending on the differences in their environments.

*PROBLEM 13-18

Predict the theoretical number of different NMR signals produced by each compound, and give approximate chemical shifts. Point out any diastereotopic relationships.

(a) 2-bromobutane
(b) cyclopentanol
(c) $Ph—CHBr—CH_2Br$
(d) vinyl chloride

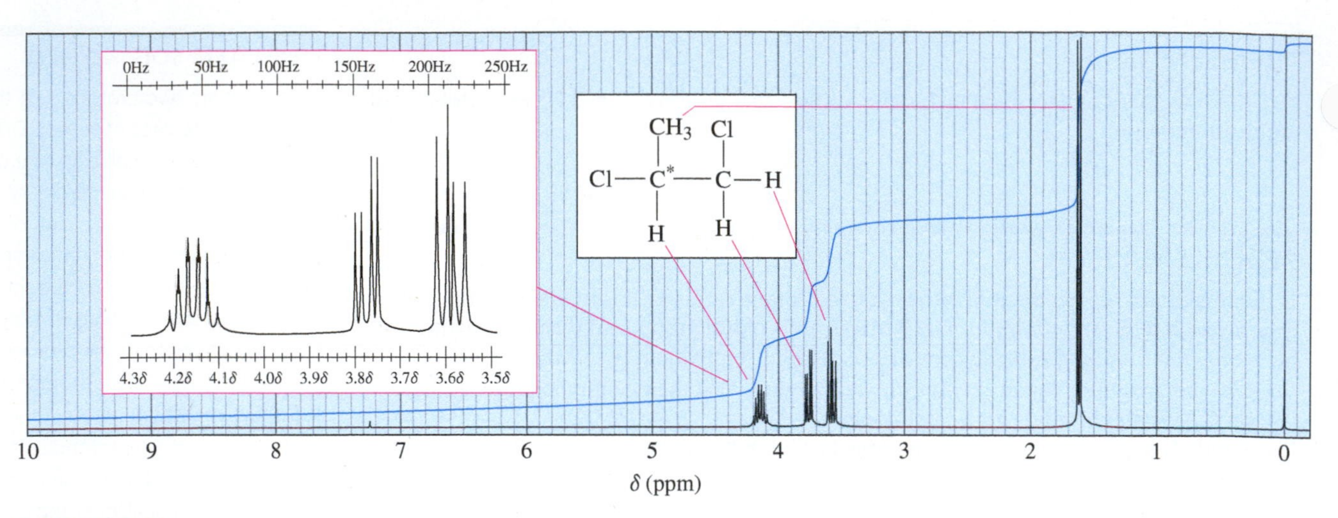

FIGURE 13-35
Diastereotopic protons in 1,2-dichloropropane. The proton NMR spectrum of 1,2-dichloropropane shows distinct absorptions for the methylene protons on C1. These hydrogen atoms are diastereotopic and are chemically nonequivalent.

13-11 Time Dependence of NMR Spectroscopy

We have already seen evidence that NMR does not provide an instantaneous picture of a molecule. For example, a terminal alkyne does not give a spectrum where the molecules oriented along the field absorb at a high field and those oriented perpendicular to the field absorb at a lower field. What we see is one signal whose position is averaged over the chemical shifts of all the orientations of a rapidly tumbling molecule. In general, any type of movement or change that takes place faster than about a hundredth of a second will produce an averaged NMR spectrum.

13-11A Conformational Changes

This principle is illustrated by the cyclohexane spectrum. In the chair conformation, there are two kinds of protons: the axial hydrogens and the equatorial hydrogens. The axial hydrogens become equatorial and the equatorial hydrogens become axial by chair–chair interconversions. These interconversions are fast on an NMR time scale at room temperature. The NMR spectrum of cyclohexane shows only one sharp, averaged peak (at δ 1.4) at room temperature.

H_1 (ax), H_2 (eq) ⇌ H_1 (eq), H_2 (ax)

Low temperatures retard the chair–chair interconversion of cyclohexane. The NMR spectrum at −89 °C shows two nonequivalent types of protons that split each other, giving two broad bands corresponding to the absorptions of the axial and equatorial protons. The broadening of the bands results from spin-spin splitting between axial and equatorial protons on the same carbon atom and on adjacent carbons. This technique of using low temperatures to stop conformational interconversions is called *freezing out* the conformations.

Application: Biochemistry

NMR spectroscopy is useful for determining the conformations of biological molecules such as neurotransmitters. For example, information about the conformations of acetylcholine has been used to design rigid analogs that are used as drugs to treat Alzheimer's disease.

13-11B Fast Proton Transfers

Hydroxy Protons Like conformational interconversions, chemical processes often occur faster than the NMR technique can observe them. Figure 13-36 shows two NMR spectra for ethanol.

Figure 13-36(a) shows coupling between the hydroxy (—OH) proton and the adjacent methylene (—CH_2—) protons, with a coupling constant of about 5 Hz. This is an ultrapure sample of ethanol with no contamination of acid, base, or water. Part (b) shows

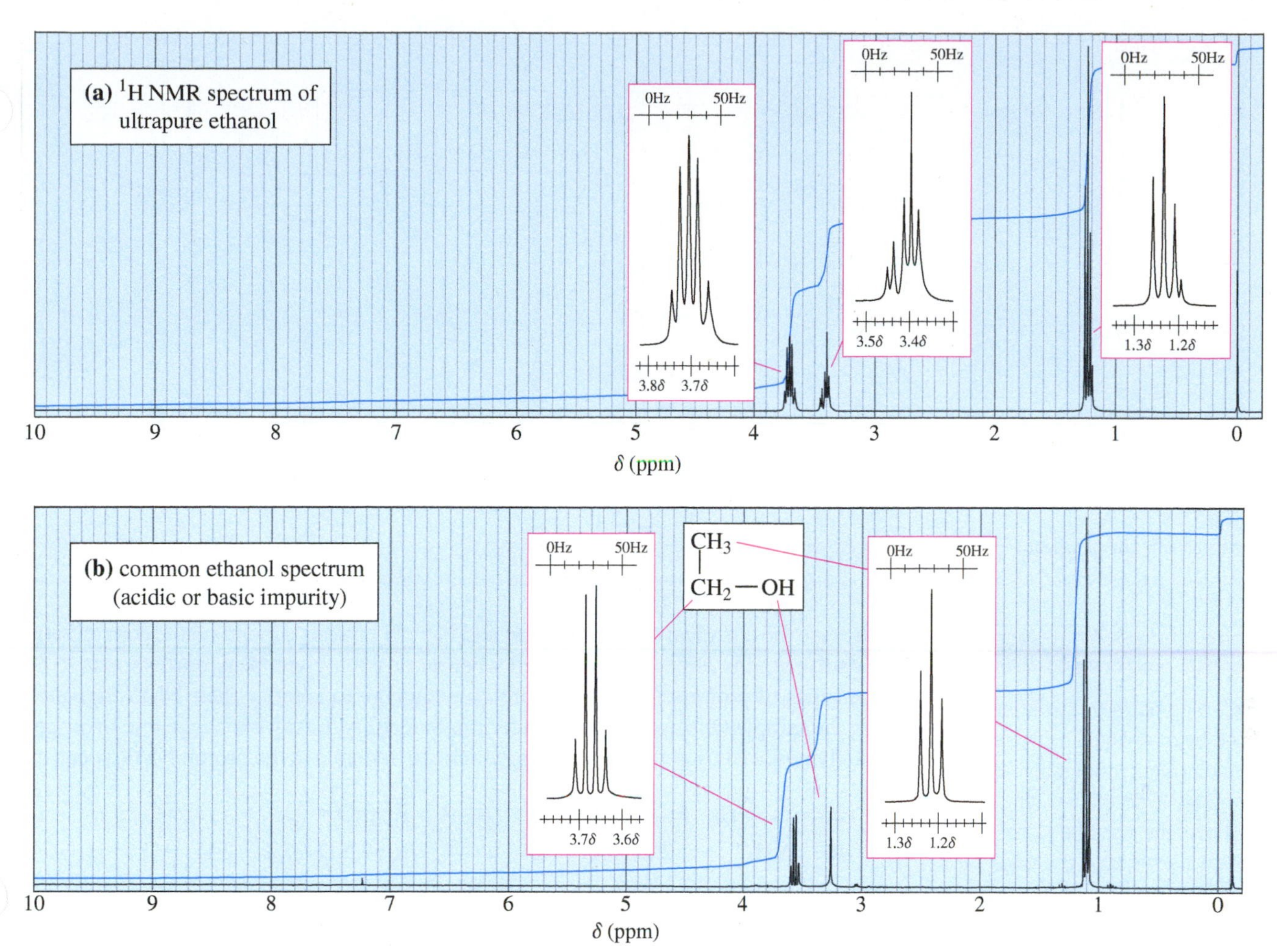

FIGURE 13-36
Comparison of the NMR spectrum of unusually pure ethanol and the spectrum of ethanol with a trace of an acidic (or basic) impurity. The impurity catalyzes a fast exchange of the —OH proton from one ethanol molecule to another. This rapidly exchanging proton produces a single, unsplit absorption at an averaged field.

a typical sample of ethanol, with some acid or base present to catalyze the interchange of the hydroxy protons. No splitting is seen between the hydroxy proton and the methylene protons. During the NMR measurement, each hydroxy proton becomes attached to a large number of different ethanol molecules and experiences all possible spin arrangements of the methylene group. What we see is a single, unsplit hydroxy absorption corresponding to the averaged field the proton experiences from bonding to many different ethanol molecules.

Proton exchange occurs in most alcohols and carboxylic acids, and in many amines and amides. If the exchange is fast (as it usually is for —OH protons), we see one sharp averaged signal. If the exchange is very slow, we see splitting. If the exchange is moderately slow, we may see a broadened peak that is neither cleanly split nor cleanly averaged.

PROBLEM 13-19

Propose mechanisms to show the interchange of protons between ethanol molecules under
(a) acid catalysis. **(b)** base catalysis.

N—H Protons Protons on nitrogen often show broadened signals in the NMR, both because of moderate rates of exchange and because of the magnetic properties of the nitrogen nucleus. Depending on the rate of exchange and other factors, N—H protons may give absorptions that are sharp and cleanly split, sharp and unsplit (averaged), or broad and shapeless. Figure 13-37 illustrates an NMR spectrum where the $—NH_2$ protons produce a very broad absorption, the shapeless peak centered at δ 5.2.

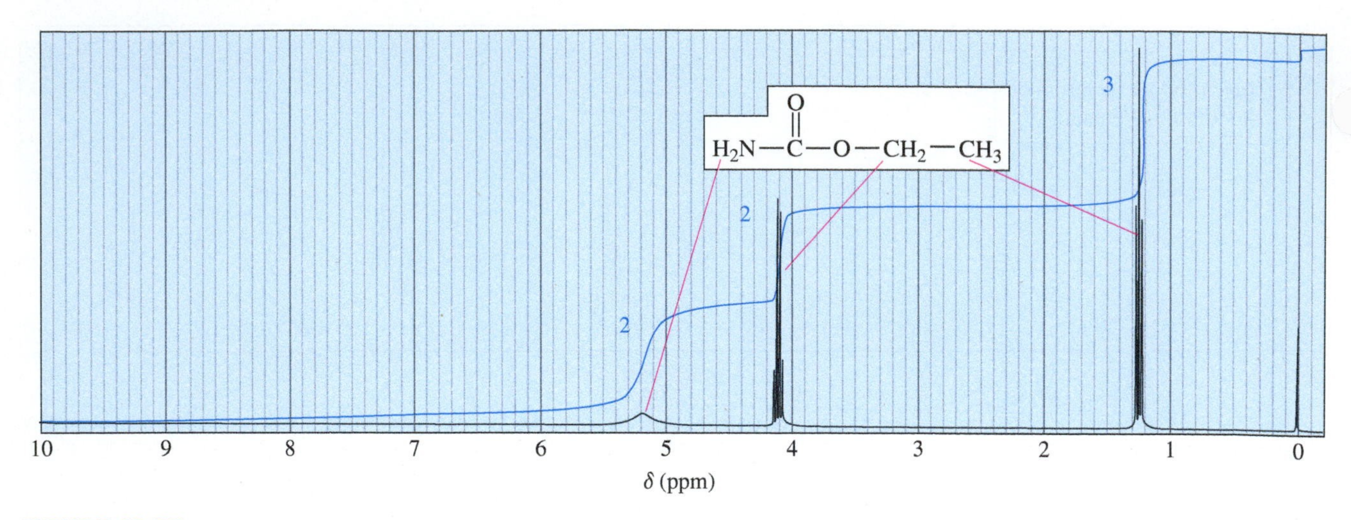

FIGURE 13-37
Proton NMR spectrum of ethyl carbamate, showing a broad N—H absorption.

Application: Deuteration

Complex splitting patterns can often be simplified by replacing a hydrogen with deuterium. Deuterium is invisible in the proton NMR, so the resulting spectrum shows the loss of a signal and simplified signals from the adjacent hydrogens.

Because the chemical shifts of O—H and N—H protons depend on the concentration and the solvent, it is often difficult to tell whether or not a given peak corresponds to one of these types of protons. We can use proton exchange to identify their NMR signals by shaking the sample with an excess of deuterium oxide, D_2O. Any exchangeable hydrogens are quickly replaced by deuterium atoms, which are invisible in the proton NMR spectrum.

$$R-O-H + D-O-D \rightleftharpoons R-O-D + D-O-H$$
$$R-NH_2 + 2\,D-O-D \rightleftharpoons R-ND_2 + 2\,D-O-H$$

When a second NMR spectrum is recorded (after shaking with D_2O), the signals from any exchangeable protons are either absent or much less intense.

PROBLEM 13-20

Draw the NMR spectrum expected from ethanol that has been shaken with a drop of D_2O.

PROBLEM 13-21

Propose chemical structures consistent with the following NMR spectra and molecular formulas. In spectrum (a), explain why the peaks around δ 1.65 and δ 3.75 are not clean multiplets, but show complex splitting. In spectrum (b), explain why some of the protons are likely to be missed.

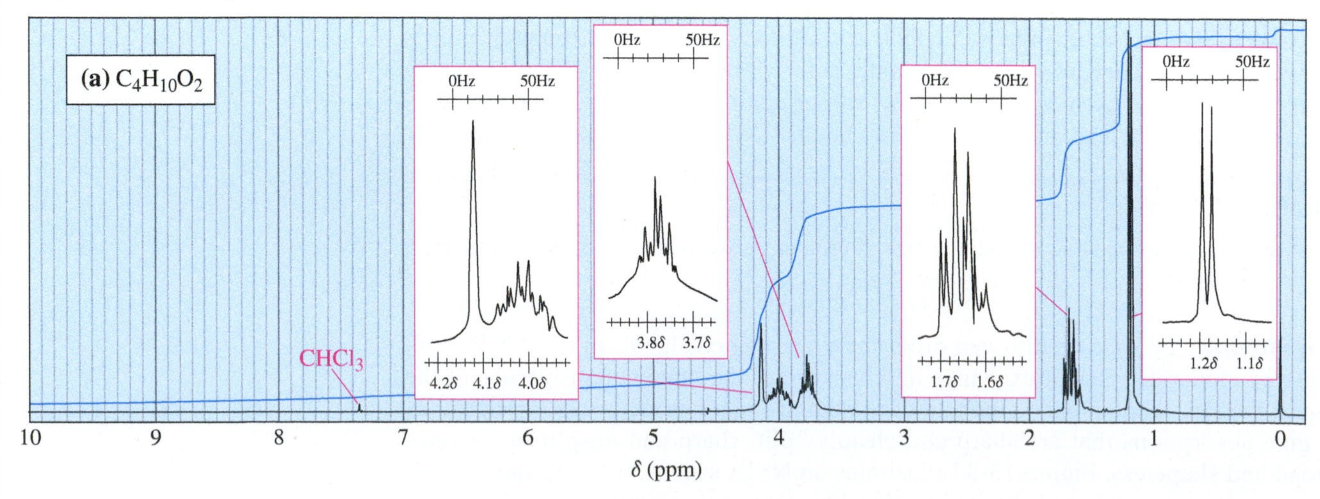

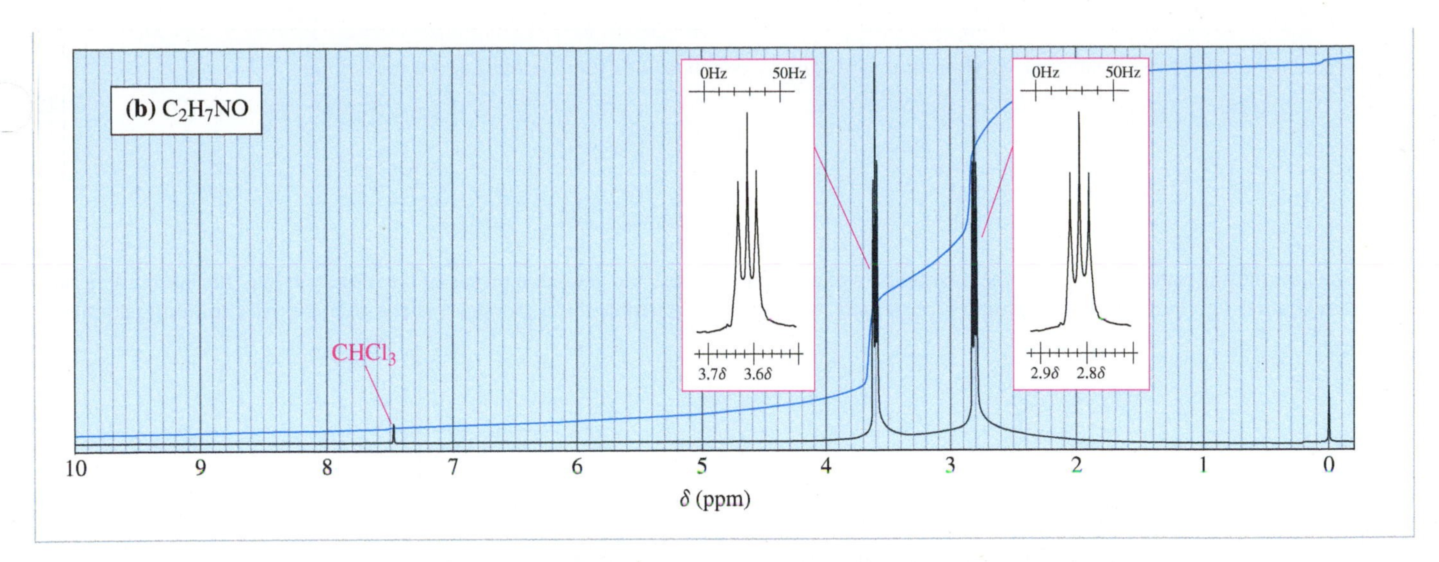

PROBLEM-SOLVING STRATEGY Interpreting Proton NMR Spectra

Learning to interpret NMR spectra requires practice with a large number of examples and problems. The problems at the end of this chapter should help you gain confidence in your ability to assemble a structure from the NMR spectrum combined with other information. This section provides some hints that can help make spectral analysis a little easier.

When you first look at a spectrum, consider the major features before getting bogged down in the minor details. Here are a few major characteristics you might watch for:

> **PROBLEM-SOLVING HINT**
>
> Remember to look for structural information based on
> 1. number of absorptions,
> 2. chemical shifts,
> 3. areas of peaks, and
> 4. spin-spin splitting.

1. If the molecular formula is known, use it to determine the number of elements of unsaturation (see Section 7-3). The elements of unsaturation suggest rings, double bonds, or triple bonds. Matching the integrated peak areas with the number of protons in the formula gives the numbers of protons represented by the individual peaks.
2. Any broadened singlets in the spectrum might be due to —OH or —NH protons. If the broad singlet is deshielded past 10 ppm, an acid —OH group is likely.

   —O—H —N—H (frequently broad singlets) —C(=O)—OH (broad (or sharp) singlet, $\delta > 10$)

3. A signal around δ 3 to δ 4 suggests protons on a carbon bearing an electronegative element such as oxygen or a halogen. Protons that are more distant from the electronegative atom will be less strongly deshielded.

   —O—C—H Br—C—H Cl—C H I—C—H

   around δ 3–δ 4 for hydrogens on carbons bearing oxygen or halogen

4. Signals around δ 7 to δ 8 suggest the presence of an aromatic ring. If some of the aromatic absorptions are farther downfield than δ 7.2, an electron-withdrawing substituent may be attached.

   (benzene ring bearing X, Y, and H) around δ 7–δ 8 H

(continued)

5. Signals around δ 5 to δ 6 suggest vinyl protons. Splitting constants can differentiate cis and trans isomers.

around δ 5–δ 6; J = 10 Hz   around δ 5–δ 6; J = 15 Hz

6. Use the splitting patterns to determine the numbers of adjacent protons, and assemble pieces of the molecule in a trial structure. Learn to recognize ethyl groups and isopropyl groups (and structures that resemble these groups) by their characteristic splitting patterns.

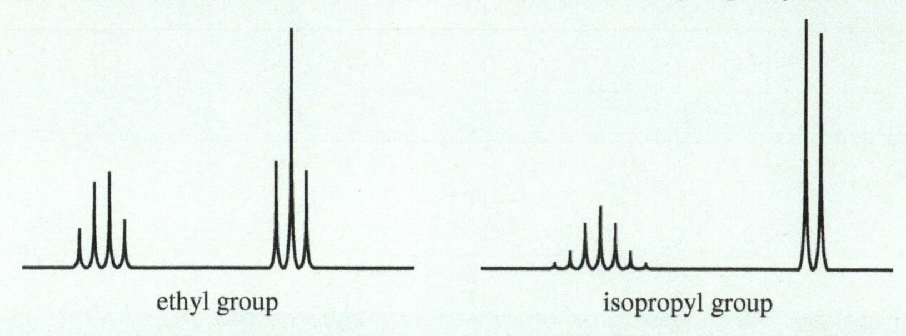

7. Signals around δ 2.1 to δ 2.5 may suggest protons adjacent to a carbonyl group or next to an aromatic ring. A singlet at δ 2.1 often results from a methyl group bonded to a carbonyl group.

around δ 2.1–δ 2.5   singlet, δ 2.1   singlet, δ 2.3

8. Signals in the range δ 9 to δ 10 suggest an aldehyde.

aldehyde, δ 9–δ 10

9. A sharp singlet around δ 2.5 suggests a terminal alkyne.

—C≡C—H
around δ 2.5

These hints are neither exact nor complete. They are simple methods for making educated guesses about the major features of a compound from its NMR spectrum. The hints can be used to draw partial structures to examine all the possible ways they might be combined to give a molecule that corresponds with the spectrum. Figure 13-38 gives a graphic presentation of some of the most common chemical shifts. A more complete table of chemical shifts appears in Appendix 1.

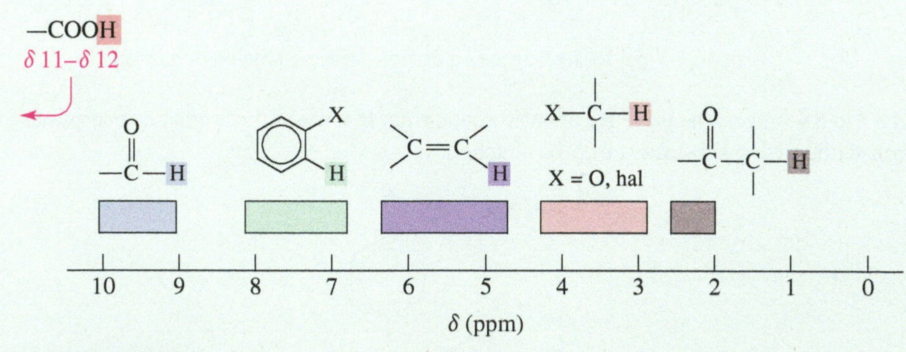

FIGURE 13-38
Common chemical shifts in the ^{1}H NMR spectrum.

Sample Problem

Consider how you might approach the NMR spectrum shown in Figure 13-39. The molecular formula is known to be $C_4H_8O_2$, implying one element of unsaturation (the saturated formula would be $C_4H_{10}O_2$). Three types of protons appear in this spectrum. The signals at δ4.1 and δ1.3 resemble an ethyl group—confirmed by the 2:3 ratio of the integrals.

partial structure: $—CH_2—CH_3$

The ethyl group is probably bonded to an electronegative element, because its methylene ($—CH_2—$) protons absorb close to δ4. The molecular formula contains oxygen, so an ethoxy group is suggested.

partial structure: $—O—CH_2—CH_3$

The singlet at δ2.1 (area = 3) might be a methyl group bonded to a carbonyl group. A carbonyl group would also account for the element of unsaturation.

partial structure: $—C(=O)—CH_3$

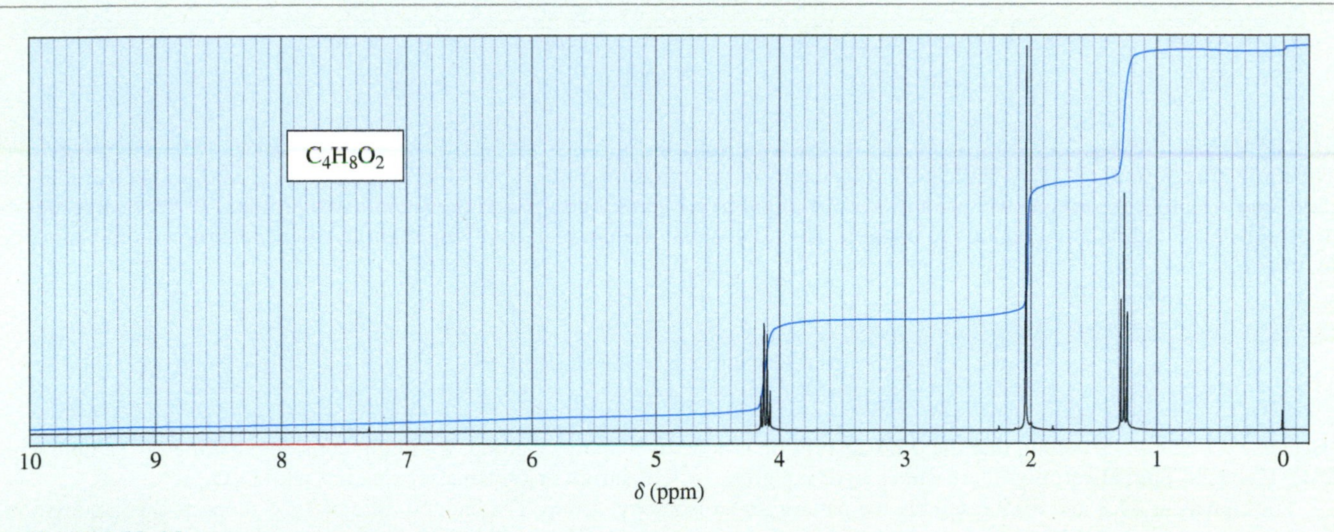

FIGURE 13-39
Proton NMR spectrum for a compound of formula $C_4H_{10}O_2$.

We have accounted for all eight hydrogen atoms in the spectrum. Putting together all the clues, we arrive at a proposed structure.

$CH_3^a—CH_2^b—O—C(=O)—CH_3^c$

ethyl acetate

At this point, the structure should be rechecked to make sure it is consistent with the molecular formula, the proton ratios given by the integrals, the chemical shifts of the signals, and the spin-spin splitting. In ethyl acetate, the H^a protons give a triplet (split by the adjacent CH_2 group, J = 7 Hz) of area 3 at δ1.3; the H^b protons give a quartet (split by the adjacent CH_3 group, J = 7 Hz) of area 2 at δ4.1; and the H^c protons give a singlet of area 3 at δ2.1.

PROBLEM 13-22

Draw the expected NMR spectrum of methyl propionate, and point out how it differs from the spectrum of ethyl acetate.

$CH_3—O—C(=O)—CH_2—CH_3$

methyl propionate

SOLVED PROBLEM 13-4

Propose a structure for the compound of molecular formula $C_4H_{10}O$ whose proton NMR spectrum follows.

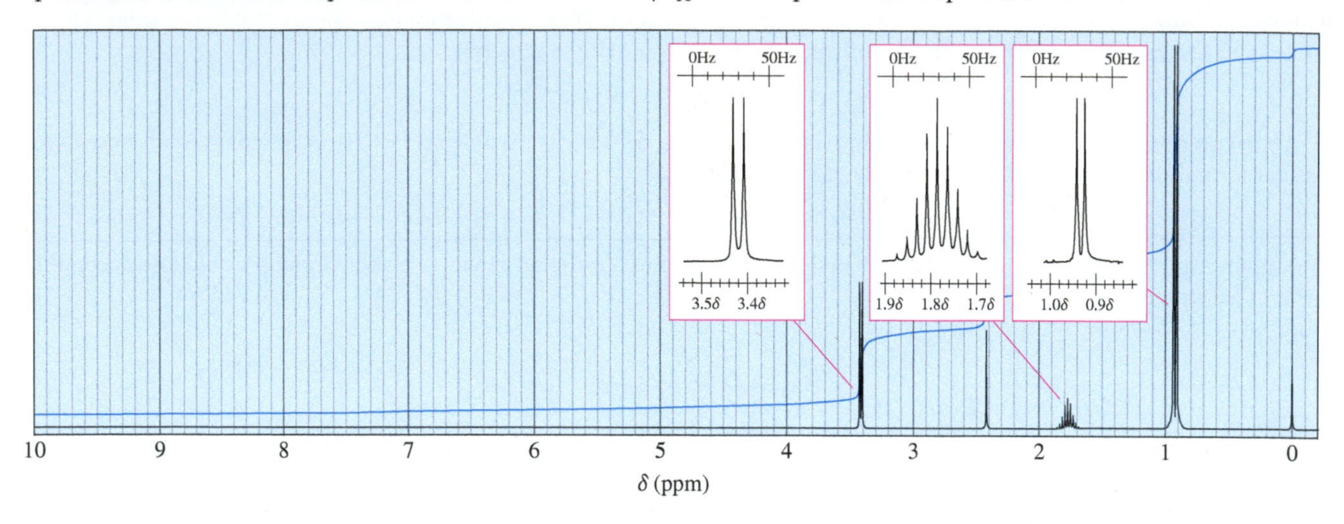

SOLUTION

The molecular formula $C_4H_{10}O$ indicates no elements of unsaturation. Four types of hydrogens appear in this spectrum, in the ratio 2 : 1 : 1 : 6. The singlet (one proton) at δ2.4 might be a hydroxy group, and the signal (two protons) at δ3.4 corresponds to protons on a carbon atom bonded to oxygen. The δ3.4 signal is a doublet, implying that the adjacent carbon atom bears one hydrogen.

partial structure: H—O—CH_2—C(H)(—)—

(Because we cannot be certain that the δ2.4 absorption is actually a hydroxy group, we might consider shaking the sample with D_2O. If the 2.4 ppm absorption represents a hydroxy group, it will shrink or vanish after shaking with D_2O.)

The signals at δ1.8 and δ0.9 resemble the pattern for an isopropyl group. The integral ratio of 1 : 6 supports this assumption. Because the methine (—CH—) proton of the isopropyl group absorbs at a fairly high field, the isopropyl group must be bonded to a carbon atom rather than an oxygen.

partial structure: —C(—)(—)—CH(CH_3)(CH_3)

Our two partial structures add to a total of six carbon atoms (compared with the four in the molecular formula) because two of the carbon atoms appear in both partial structures. Drawing the composite of the partial structures, we have isobutyl alcohol:

$$H^a—O—CH_2^b—CH^c(CH_3^d)(CH_3^d)$$

This structure must be rechecked to make sure that it has the correct molecular formula and that it accounts for all the structural evidence provided by the spectrum (Problem 13-23).

PROBLEM 13-23

Give the spectral assignments for the protons in isobutyl alcohol (Solved Problem 13-4). For example,

H^a is a singlet, area = 1, at δ2.4

PROBLEM 13-24

Five proton NMR spectra are given here, together with molecular formulas. In each case, propose a structure that is consistent with the spectrum.

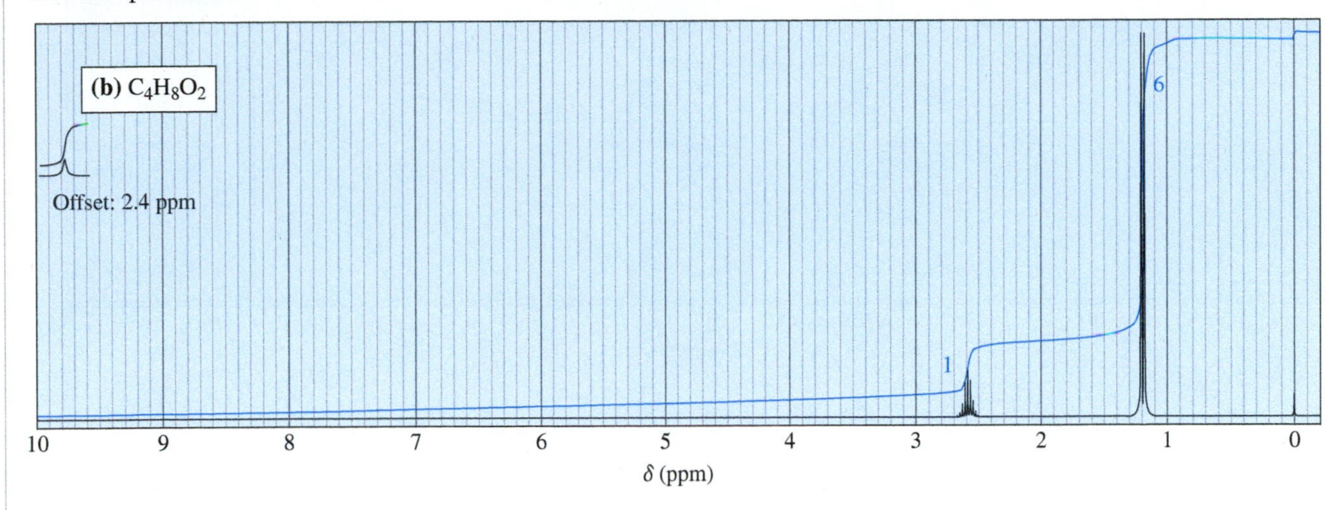

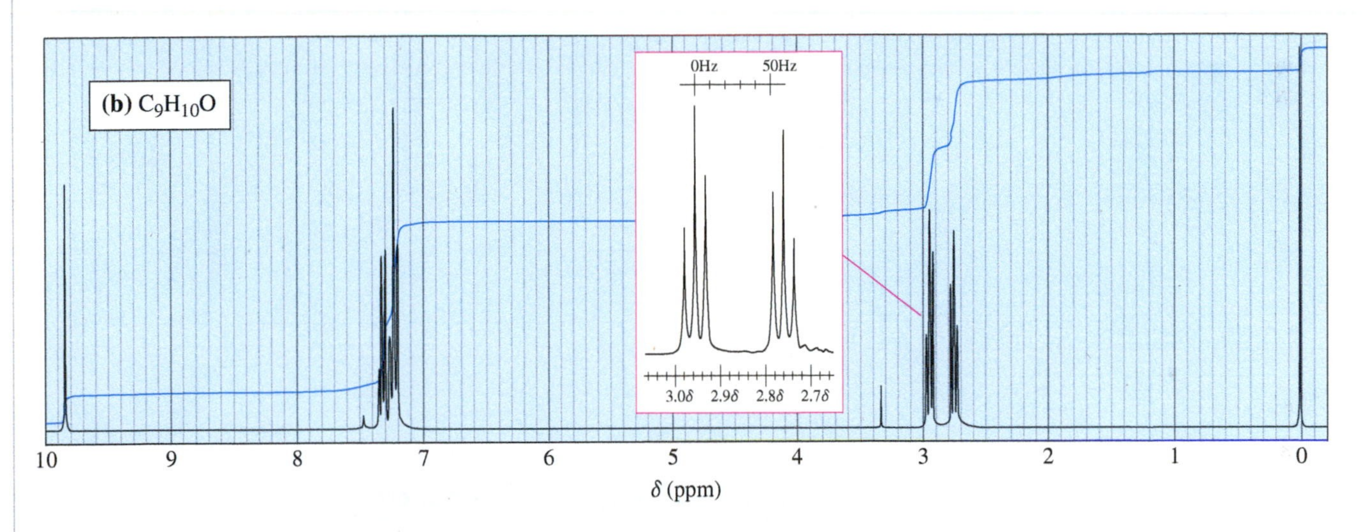

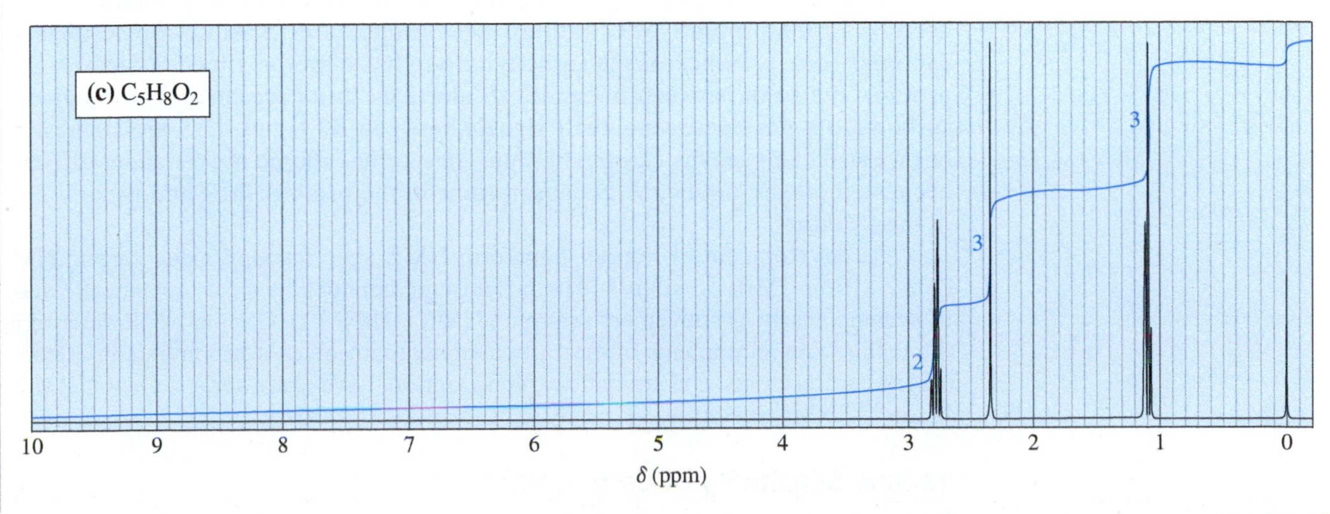

(continued)

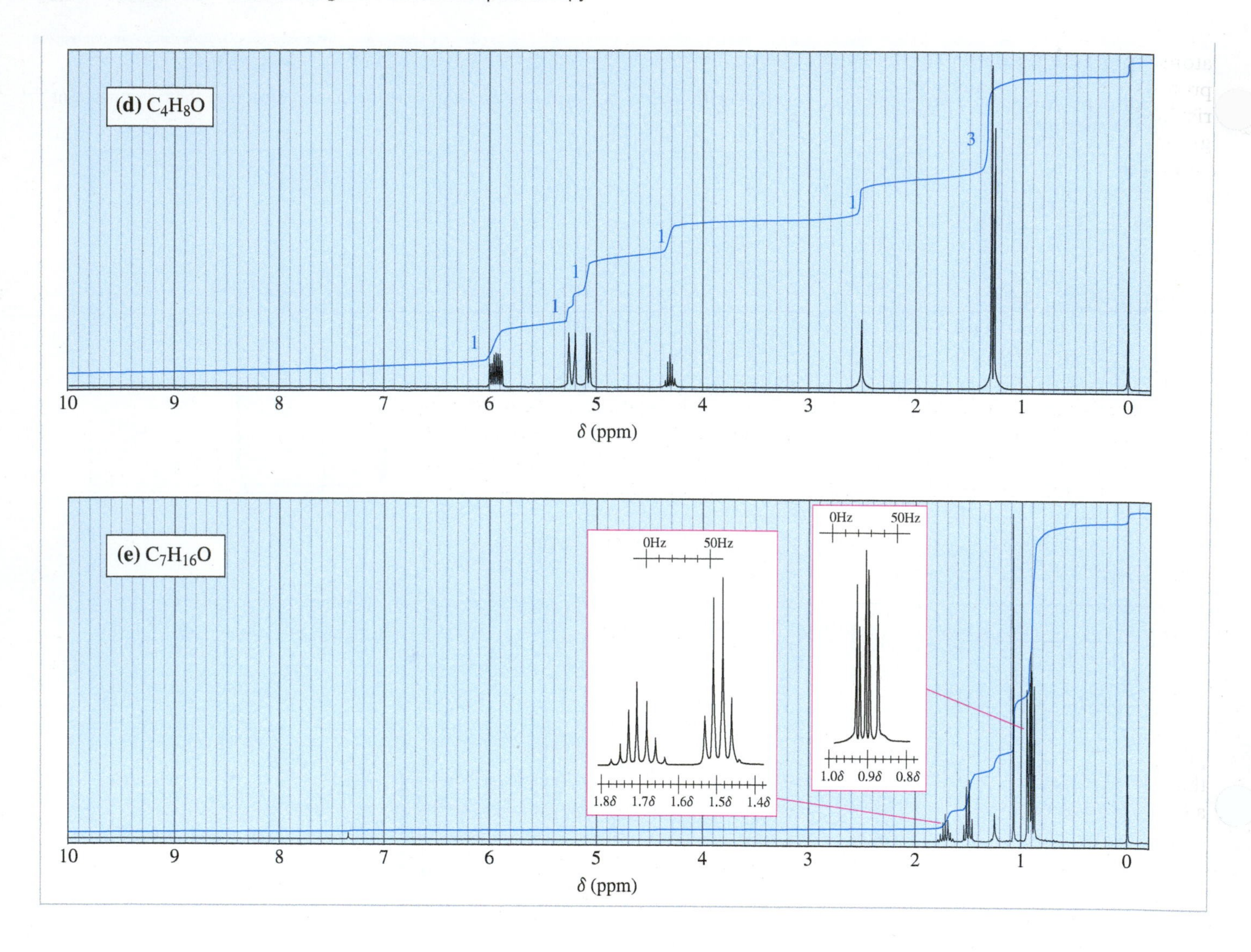

13-12 Carbon-13 NMR Spectroscopy

Where does a carbonyl group absorb in the NMR? Where does an internal alkyne absorb? In the proton NMR, both of these groups are invisible. Sometimes we can *infer* their presence: If the carbonyl group has a proton attached (an aldehyde proton), the peak between δ 9 and δ 10 alerts us to its presence. If the adjacent carbon atom has hydrogens, their signals between δ2.1 and δ 2.5 are suggestive, but we still can't see the carbonyl group. An internal alkyne is even more difficult, because there are no distinctive absorptions in the proton NMR and usually none in the IR either.

The development of Fourier transform NMR spectroscopy made carbon NMR (^{13}C NMR or CMR) possible, and high-field superconducting spectrometers allowed it to become nearly as convenient as proton NMR (^{1}H NMR). Carbon NMR determines the magnetic environments of the carbon atoms themselves. Carbonyl carbon atoms, alkyne carbon atoms, and aromatic carbon atoms all have characteristic chemical shifts in the ^{13}C NMR spectrum.

13-12A Sensitivity of Carbon NMR

Carbon NMR took longer than proton NMR to become a routine technique because carbon NMR signals are much weaker than proton signals, and the electronics in the early instruments could not detect the weak carbon signal. About 99% of the carbon atoms in a natural sample are the isotope ^{12}C. This isotope has an even number of protons and an even number of neutrons, so it has no magnetic spin and cannot give rise to NMR signals. The less abundant isotope ^{13}C has an odd number of neutrons, giving it a magnetic spin of one-half, just like a proton. Because only 1% of the carbon atoms in a sample are the magnetic ^{13}C isotope, the sensitivity of ^{13}C NMR is decreased by a factor of 100. In addition, the gyromagnetic ratio of ^{13}C is only one-fourth that of the proton, so the ^{13}C resonance frequency (at a given magnetic field) is only one-fourth of that for ^{1}H NMR. The smaller gyromagnetic ratio leads to a further decrease in sensitivity.

Because ^{13}C NMR is less sensitive than ^{1}H NMR, special techniques are needed to obtain a spectrum. The original type of NMR spectrometer shown in Figure 13-6 (called a CW or *continuous wave* spectrometer) produces ^{13}C signals that are very weak and become lost in the noise. When many spectra are averaged, however, the random noise tends to cancel while the desired signals are reinforced. If several spectra are taken and stored in a computer, they can be averaged and the accumulated spectrum plotted by the computer. Because the ^{13}C NMR technique is much less sensitive than the ^{1}H NMR technique, hundreds of spectra are commonly averaged to produce a usable result. Several minutes are required to scan each CW spectrum, and this averaging procedure is long and tedious. Fortunately, there is a better way.

13-12B Fourier Transform NMR Spectroscopy

When magnetic nuclei are placed in a uniform magnetic field and irradiated with a pulse of radio frequency close to their resonant frequency, the nuclei absorb some of the energy and precess like little tops at their resonant frequencies (Figure 13-40). This precession of many nuclei at slightly different frequencies produces a complex signal that decays as the nuclei lose the energy they gained from the pulse. This signal is called a **free induction decay** (or **transient**) and it contains all the information needed to calculate a spectrum. The free induction decay (FID) can be recorded by a radio receiver and a computer in 1 to 2 seconds, and many FIDs can be averaged in a few minutes. A computer converts the averaged transients into a spectrum.

A *Fourier transform* is the mathematical technique used to compute the spectrum from the free induction decay, and this technique of using pulses and collecting transients is called **Fourier transform spectroscopy.** A Fourier transform spectrometer requires sophisticated electronics capable of generating precise pulses and accurately receiving the complicated transients. A good ^{13}C NMR instrument usually has the capability to do ^{1}H NMR spectra as well. When used with proton spectroscopy, the Fourier transform technique produces good spectra with very small amounts (less than a milligram) of sample.

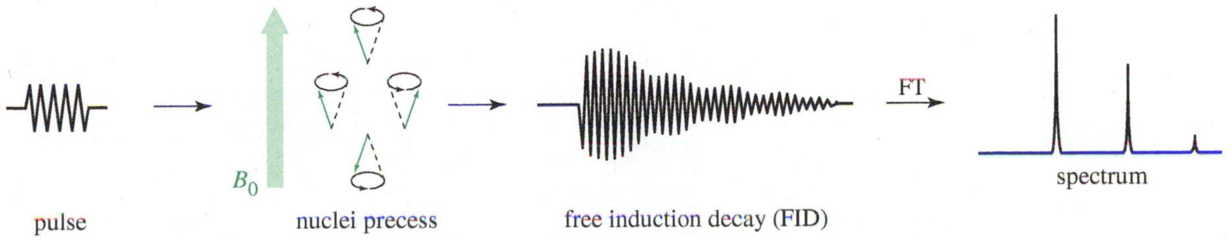

pulse — nuclei precess (B_0) — free induction decay (FID) —FT→ spectrum

FIGURE 13-40
Fourier transform NMR spectroscopy. The FT NMR spectrometer delivers a radio-frequency pulse close to the resonance frequency of the nuclei. Each nucleus precesses at its own resonance frequency, generating a free induction decay (FID). Many of these transient FIDs are accumulated and averaged in a short period of time. A computer does a Fourier transform (FT) on the averaged FID, producing the spectrum recorded on the printer.

Application: Biochemistry

NMR spectroscopy is an important method for determining the three-dimensional structures of proteins in solution. Advanced techniques using both ^{13}C and ^{1}H spectroscopy give a family of lowest-energy structures (a), from which a computer calculates a final molecular structure (b). Diagram (a) shows carbons in light blue, nitrogen in dark blue, oxygens in red, and amide protons in white.

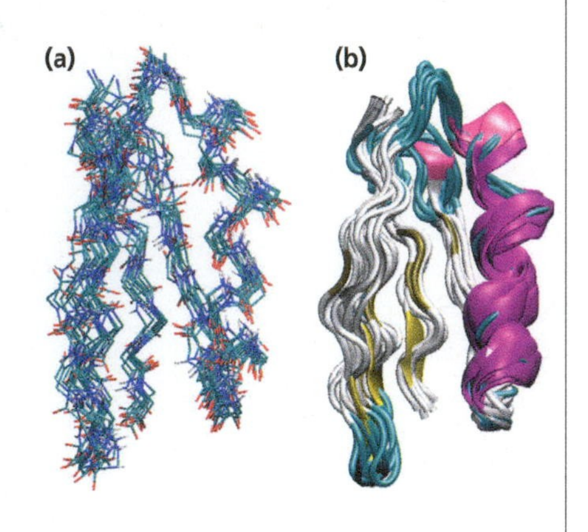

13-12C Carbon Chemical Shifts

Figure 13-41 gives typical ranges of chemical shifts for carbon atoms in organic molecules. A more detailed table of carbon chemical shifts is provided as Appendix 1C. As in proton NMR, many ^{13}C signals are deshielded by electron-withdrawing substituents. Carbon chemical shifts are usually about 15 to 20 times larger than comparable proton chemical shifts. This difference makes sense because the carbon atom is one atom closer to a shielding or deshielding group than its attached hydrogen. For example, an aldehyde proton (bonded to the oxygen through the carbon) absorbs around $\delta 9.4$ in the ^{1}H NMR spectrum, and the carbonyl carbon atom (directly bonded to the oxygen) absorbs around 180 ppm downfield from TMS in the ^{13}C spectrum. Figure 13-42 compares the proton and carbon spectra of a complex aldehyde to show this relationship between proton and carbon chemical shifts.

The proton (lower) and carbon (upper) spectra in Figure 13-42 are calibrated so that the full width of the proton spectrum is 10 ppm, whereas the width of the ^{13}C spectrum is 200 ppm (20 times as large). Notice how the corresponding peaks in the two spectra almost line up vertically. This proportionality between ^{13}C NMR and ^{1}H NMR chemical shifts is an approximation that allows us to make a first estimate of a carbon atom's chemical shift. For example, the peak for the aldehyde *proton* is at $\delta 9.5$ in the proton spectrum, so we expect the peak for the aldehyde *carbon* to appear at a chemical shift between 15 and 20 times as large (between $\delta 144$ and $\delta 192$) in the carbon spectrum. The actual position is at $\delta 180$.

Notice also the triplet at $\delta 77$ in the ^{13}C NMR spectrum in Figure 13-42. This is the carbon signal for deuterated chloroform ($CDCl_3$), split into three equal-sized peaks by coupling with the deuterium atom. Chloroform-*d* ($CDCl_3$) is a common solvent for ^{13}C NMR because the spectrometer can "lock" onto the signal from deuterium at a different frequency from carbon. The $CDCl_3$ solvent signal is a common feature of carbon NMR spectra, and it can be used as an internal reference instead of TMS if desired.

Because chemical shift effects are larger in ^{13}C NMR, an electron-withdrawing group has a substantial effect on the chemical shift of a carbon atom beta (one carbon removed) to the group. For example, Figure 13-43 shows the ^{1}H NMR and ^{13}C NMR spectra of 1,2,2-trichloropropane. The methyl (CH_3) carbon absorbs at 33 ppm downfield from TMS because the two chlorine atoms on the adjacent $-CCl_2-$ carbon have a substantial effect on the methyl carbon. The chemical shift of this methyl carbon is about 15 times that of its attached protons ($\delta 2.1$), in accordance with our prediction. Similarly, the chemical shift of the $-CH_2Cl$ carbon (56 ppm) is about 15 times that of its protons ($\delta 4.0$). Although the CCl_2 carbon has no protons, the proton in a $-CHCl_2$ group generally absorbs around $\delta 5.8$. The carbon absorption at 87 ppm is about 15 times this proton shift.

13-12D Important Differences Between Proton and Carbon Techniques

Most of the characteristics of ^{13}C NMR spectroscopy are similar to those of the ^{1}H NMR technique. There are some important differences, however.

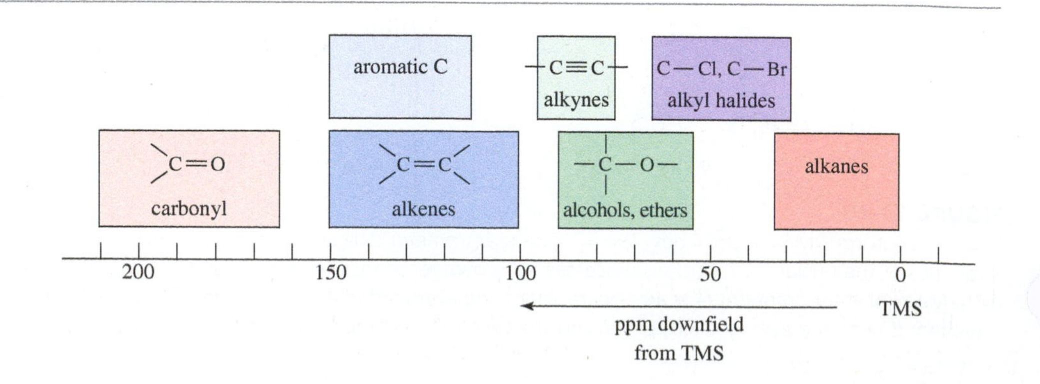

FIGURE 13-41
Table of approximate chemical shift values for ^{13}C NMR. Most of these values for a carbon atom are about 15 to 20 times the chemical shift of a proton if it were bonded to the carbon atom.

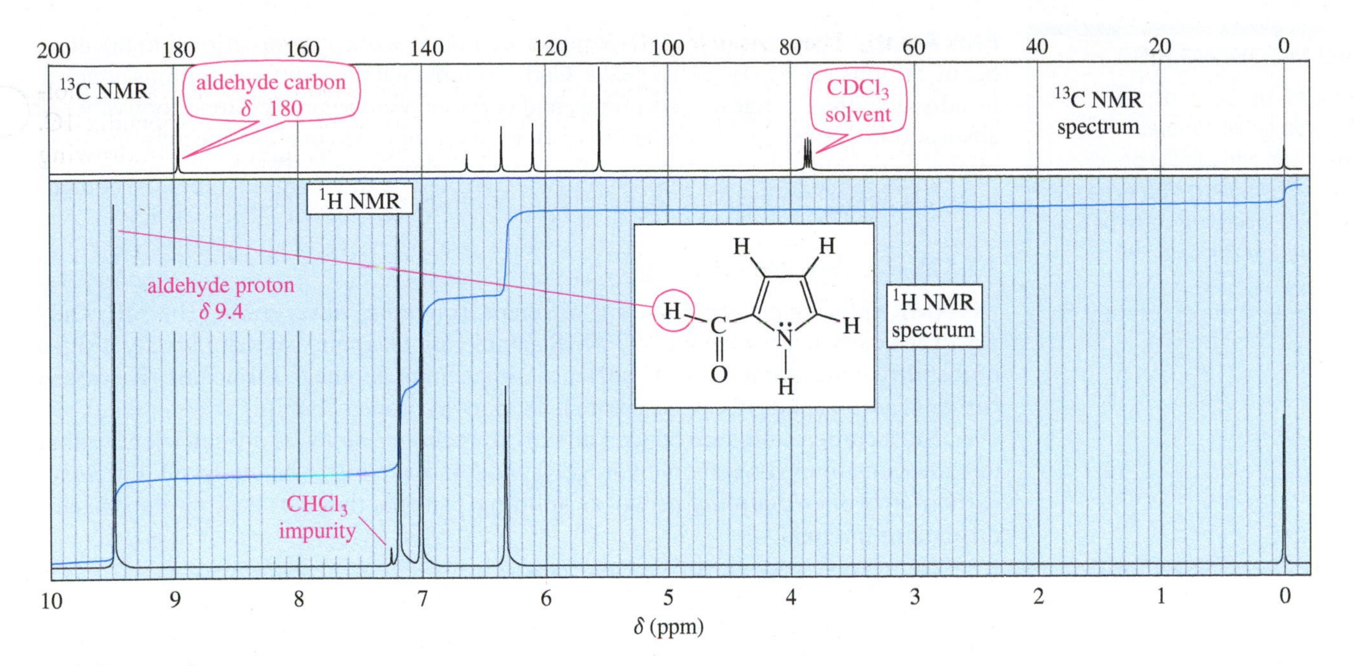

FIGURE 13-42
^{1}H NMR (lower) and ^{13}C NMR (upper) spectra of a heterocyclic aldehyde. Notice the correlation between the chemical shifts in the two spectra. The proton spectrum has a sweep width of 10 ppm, and the carbon spectrum has a width of 200 ppm.

Operating Frequency The gyromagnetic ratio for ^{13}C is about one-fourth that of the proton, so the resonance frequency is also about one-fourth. A spectrometer with a 70,459-gauss magnet needs a 300-MHz transmitter for protons and a 75.6-MHz transmitter for ^{13}C. A spectrometer with a 14,092-gauss magnet needs a 60-MHz transmitter for protons and a 15.1-MHz transmitter for ^{13}C.

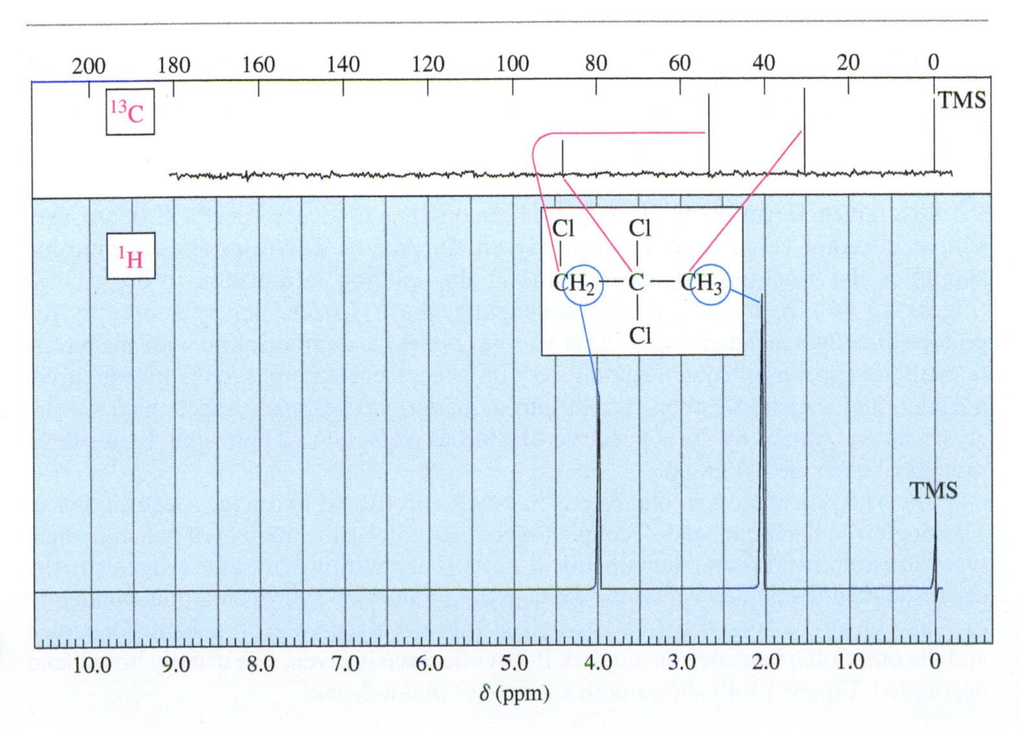

FIGURE 13-43
Proton and ^{13}C NMR spectra of 1,2,2-trichloropropane.

PROBLEM-SOLVING HINT

Appendices 1A, 1B, and 1C are valuable tables for precise interpretation and prediction of proton and carbon NMR spectra. These tables will continue to be useful in your further studies in chemistry.

Peak Areas The areas of ^{13}C NMR peaks are not necessarily proportional to the number of carbons giving rise to the peaks. Carbon atoms with two or three protons attached usually give the strongest absorptions, and carbons with no protons tend to give weak absorptions. Newer spectrometers have an integrating mode that uses decoupling techniques to equalize the absorptions of different carbon atoms. This mode makes peak integrals nearly proportional to the relative numbers of carbon atoms.

13-12E Spin-Spin Splitting

^{13}C NMR splitting patterns are quite different from those observed in ^{1}H NMR. Only 1% of the carbon atoms in the ^{13}C NMR sample are magnetic, so there is only a small probability that an observed ^{13}C nucleus is adjacent to another ^{13}C nucleus. Therefore, carbon–carbon splitting can be ignored. Carbon–hydrogen coupling is common, however. Most carbon atoms are bonded directly to hydrogen atoms or are sufficiently close to hydrogen atoms for carbon–hydrogen spin-spin coupling to be observed. Extensive carbon–hydrogen coupling produces splitting patterns that can be complicated and difficult to interpret.

Proton Spin Decoupling To simplify ^{13}C NMR spectra, they are commonly recorded using **proton spin decoupling,** where the protons are continuously irradiated with a broadband ("noise") proton transmitter. As a result, all the protons are continuously in resonance, and they rapidly flip their spins. The carbon nuclei see an *average* of the possible combinations of proton spin states. Each carbon signal appears as a single, unsplit peak because any carbon–hydrogen splitting has been eliminated. The spectra in Figures 13-42 and 13-43 were generated in this manner.

PROBLEM 13-25

Draw the expected broadband-decoupled ^{13}C NMR spectra of the following compounds. Use Figure 13-41 (page 650) to estimate the chemical shifts.

(a) O

(b)

(c) H_3C–C(=O)–C(=CH_2)–CH_3

(d) H(H)C=C(H)–C(=O)–H

Off-Resonance Decoupling Proton spin decoupling produces spectra that are very simple, but some valuable information is lost in the process. **Off-resonance decoupling** simplifies the spectrum but allows some of the splitting information to be retained (Figure 13-44). With off-resonance decoupling, the ^{13}C nuclei are split only by the protons directly bonded to them. The $N+1$ rule applies, so a carbon atom with one proton (a methine) appears as a doublet, a carbon with two attached protons (a methylene) gives a triplet, and a methyl carbon is split into a quartet. Off-resonance-decoupled spectra are easily recognized by the appearance of TMS as a quartet at 0 ppm, split by the three protons of each methyl group.

The best procedure for obtaining a ^{13}C NMR spectrum is to run the spectrum twice: The singlets in the broadband-decoupled spectrum indicate the number of nonequivalent types of carbon atoms and their chemical shifts. The multiplicities of the signals in the off-resonance-decoupled spectrum indicate the number of hydrogen atoms bonded to each carbon atom. ^{13}C spectra are often given with two traces, one broadband decoupled and the other off-resonance decoupled. If just one trace is given, it is usually broadband decoupled. Figure 13-45 shows both spectra for butan-2-one.

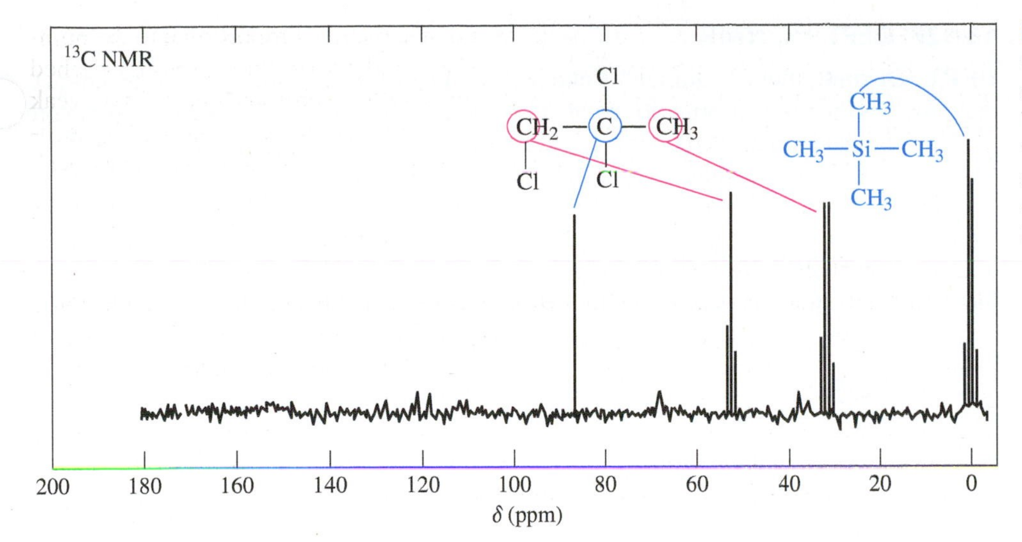

FIGURE 13-44
Off-resonance-decoupled ^{13}C NMR spectrum of 1,2,2-trichloropropane. The CCl_2 group appears as a singlet, the CH_2Cl group appears as a triplet, and the CH_3 group appears as a quartet. Compare this spectrum with Figure 13-43.

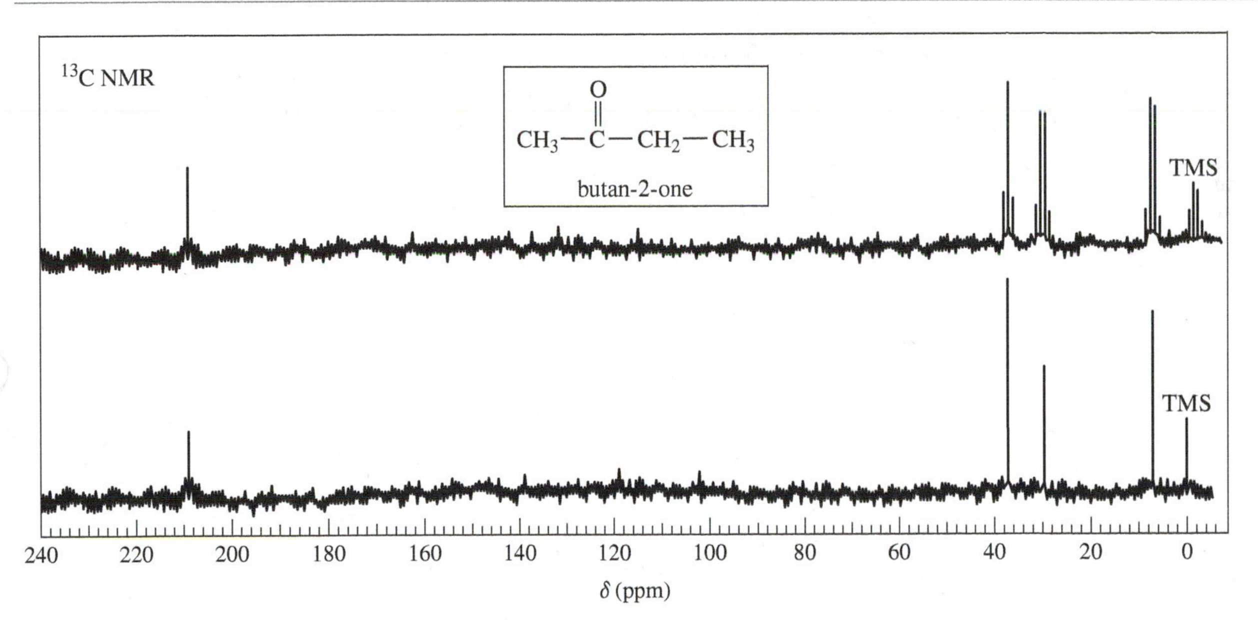

FIGURE 13-45
Off-resonance-decoupled (upper) and broadband-decoupled (lower) ^{13}C NMR spectra of butan-2-one.

PROBLEM 13-26

(a) Show which carbon atoms correspond with which peaks in the ^{13}C NMR spectrum of butan-2-one (Figure 13-45).

(b) Draw the proton NMR spectrum you would expect for butan-2-one. How well do the proton chemical shifts predict the carbon chemical shifts using the "15 to 20 times as large" rule of thumb?

PROBLEM 13-27

Repeat Problem 13-25, sketching the off-resonance-decoupled ^{13}C spectra of the compounds.

13-12F DEPT ^{13}C NMR

DEPT (Distortionless Enhanced Polarization Transfer) is a more recent technique that provides the same information as off-resonance decoupling. DEPT is easier to run on modern, computer-controlled Fourier transform spectrometers. DEPT gives better sensitivity, and it avoids overlapping multiplets because all the peaks remain decoupled singlets.

Each ^{13}C nucleus is magnetically coupled to the protons bonded to it. Under the right circumstances, this magnetic coupling allows the transfer of polarization from the protons to the carbon nucleus. The number of protons bonded to the ^{13}C nucleus determines how this polarization transfer occurs. A DEPT experiment usually includes three spectral scans:

1. the normal decoupled scan, in which each type of ^{13}C nucleus appears as a singlet.
2. the DEPT-90 scan, in which *only* the CH (methine) carbons bonded to exactly one proton appear.
3. the DEPT-135 scan, in which the CH_3 (methyl) groups and CH (methine) groups appear normally, and the CH_2 groups give negative peaks. Carbons that are bonded to no protons do not appear.

As shown graphically in Table 13-4, this information allows us to distinguish among carbons bonded to 0, 1, 2, or 3 hydrogen atoms:

- Carbons with no H's appear only in the normal spectrum, but not in either DEPT spectrum.
- Methine carbons (CH) give normal positive peaks in all three spectra.
- Methylene (CH_2) carbons give normal peaks in the normal spectrum, no peaks in the DEPT-90 spectrum, and negative peaks in the DEPT-135 spectrum.
- Methyl (CH_3) carbons give normal peaks in the normal spectrum, no peaks in the DEPT-90 spectrum, and normal peaks in the DEPT-135 spectrum.

Figure 13-46 shows the normal decoupled ^{13}C NMR spectrum of but-3-en-2-one (1), plus the DEPT-90 spectrum (2), and the DEPT-135 spectrum (3). Note that the carbonyl carbon (C_b, no protons) appears only in the regular spectrum. C_c, with 1 proton, appears normally in all the spectra. C_d, with two protons, appears as a negative peak

TABLE 13-4
Summary of DEPT Spectra

	Type of ^{13}C	*Protons*	*Normal ^{13}C NMR*	*DEPT-90*	*DEPT-135*
quaternary	$-\overset{\vert}{\underset{\vert}{C}}-$	0	positive peak	no peak	no peak
methine	$-\overset{\vert}{\underset{\vert}{C}}-H$	1	positive peak	positive peak	positive peak
methylene	$-\overset{\vert}{C}H_2$	2	positive peak	no peak	negative peak
methyl	$-CH_3$	3	positive peak	no peak	positive peak

in the DEPT-135 spectrum. C_a, the methyl carbon with three protons, vanishes in the DEPT-90 spectrum but appears as a normal peak in the DEPT-135 spectrum.

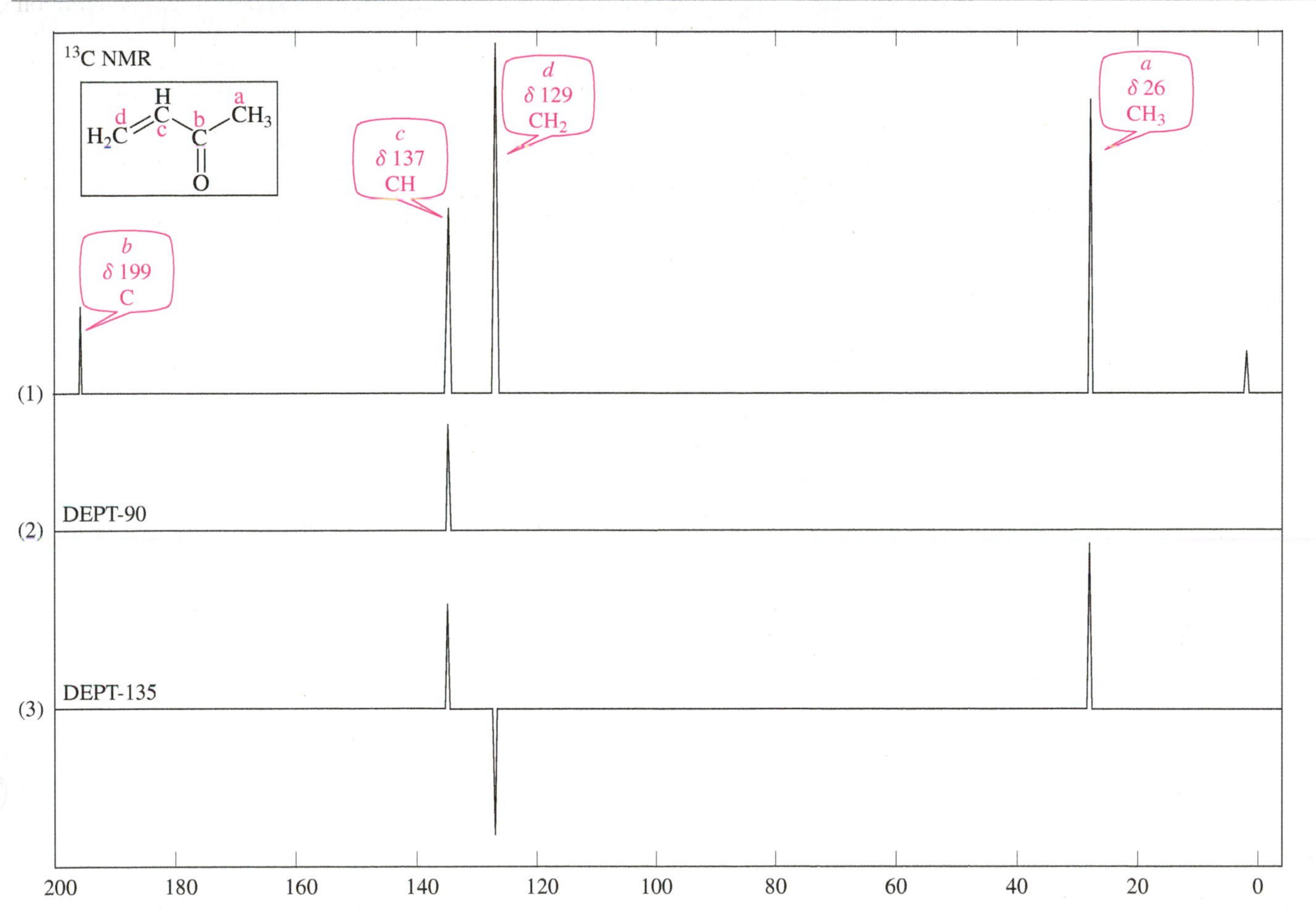

FIGURE 13-46
^{13}C NMR spectrum and DEPT spectra of but-3-en-2-one.

PROBLEM 13-28

The standard ^{13}C NMR spectrum of phenyl propanoate is shown here. Predict the appearance of the DEPT-90 and DEPT-135 spectra.

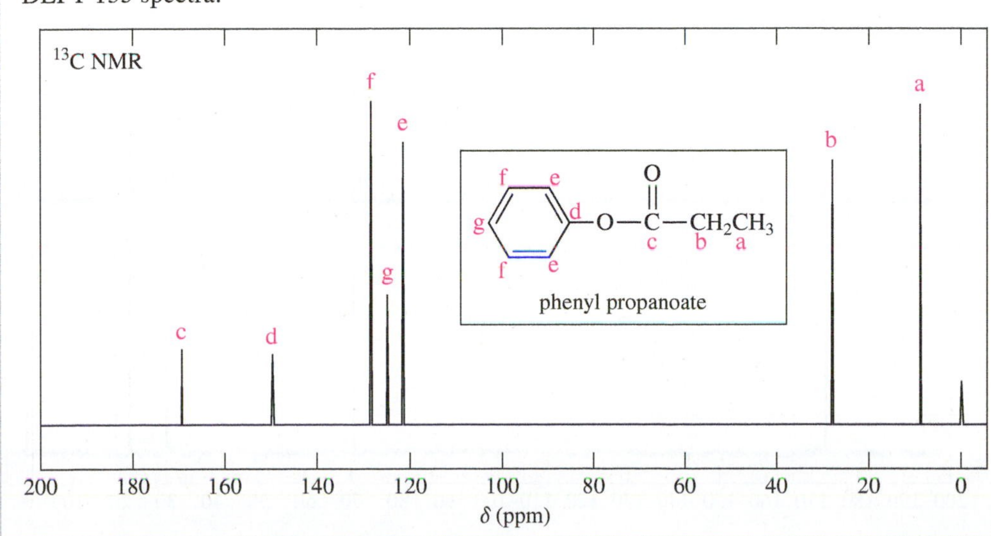

13-13 Interpreting Carbon NMR Spectra

Interpreting ^{13}C NMR spectra uses the same principles as interpreting ^{1}H NMR spectra. In fact, carbon spectra are often easier to interpret. The ^{13}C NMR spectrum provides the following information:

1. The *number of different signals* implies how many different types of carbons are present.
2. The *chemical shifts* of those signals suggest what types of functional groups contain those carbon atoms.
3. The *splitting of signals* in the off-resonance-decoupled spectrum or the DEPT-90 and DEPT-135 spectra indicate how many protons are bonded to each carbon atom.

For example, in the ^{13}C NMR spectrum of δ-valerolactone (Figure 13-47), the CH_2 groups in the upper (off-resonance-decoupled) spectrum are split into triplets, but they appear as singlets in the lower (broadband-decoupled) spectrum.

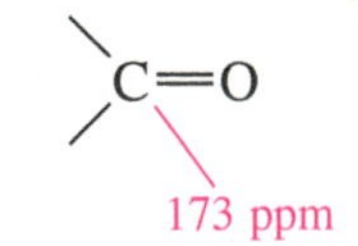

Let's consider how we might solve this structure, given only the ^{13}C NMR spectrum and the molecular formula. As we have seen in Figures 13-41 and 13-42, the signal at 173 ppm is appropriate for a carbonyl carbon. The off-resonance-decoupled spectrum shows a singlet at 173 ppm, implying that no hydrogens are bonded to the carbonyl carbon.

The chemical shift of the next absorption is about 70 ppm. This is about 20 times the chemical shift of a proton on a carbon bonded to an electronegative element. The molecular formula implies that the electronegative element must be oxygen. Because the signal at 70 ppm is a triplet in the off-resonance-decoupled spectrum, this carbon must be a methylene (—CH_2—) group.

H δ 3–δ 4
—C—O—
70 ppm

partial structures: —CH_2—O— —C(=O)—

δ 2.1 H O
30 ppm —C—C—

The signal at 30 ppm corresponds to a carbon atom bonded to a carbonyl group. Remember that a proton on a carbon adjacent to a carbonyl group absorbs around 2.1 ppm, and we expect the carbon to have a chemical shift about 15 to 20 times as large. This carbon atom is a methylene group, as shown by the triplet in the off-resonance-decoupled spectrum.

partial structures: —CH_2—O— —CH_2—C(=O)—

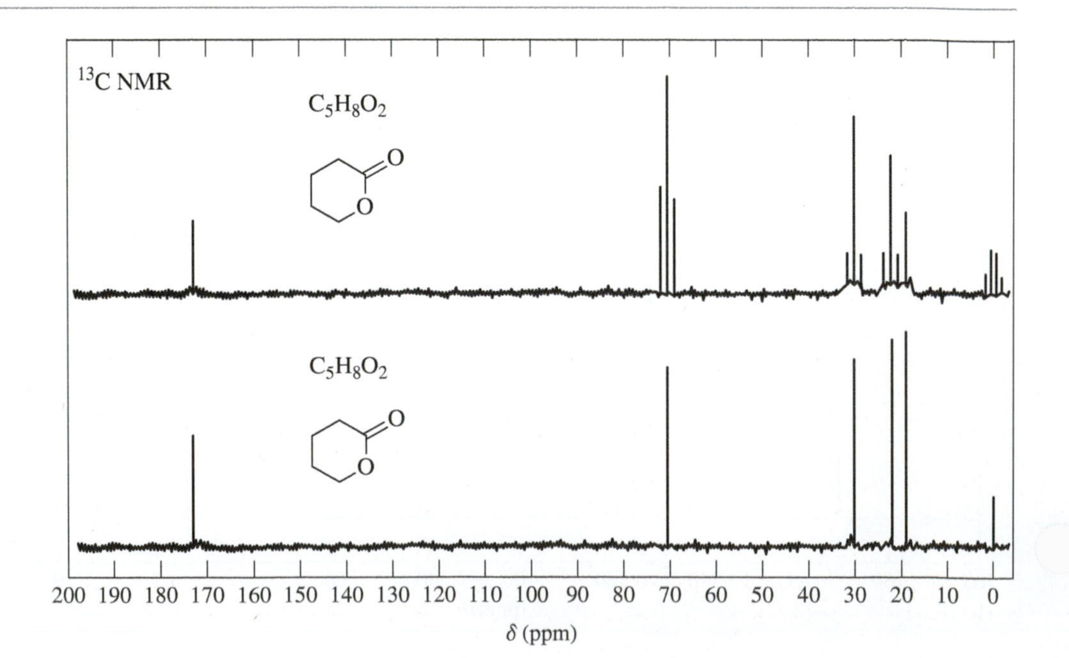

FIGURE 13-47
Off-resonance-decoupled and broadband-decoupled spectra of δ-valerolactone, molecular formula $C_5H_8O_2$.

The two signals at 19 and 22 ppm are from carbon atoms that are not directly bonded to any deshielding group, although the carbon at 22 ppm is probably closer to one of the oxygen atoms. These are also triplets in the off-resonance-decoupled spectrum and they correspond to methylene groups. We can propose the following:

partial structures: $-\overset{|}{\underset{|}{C}}-CH_2-CH_2-CH_2-O-$ $\qquad -CH_2-\overset{O}{\overset{\|}{C}}-$

The molecular formula $C_5H_8O_2$ implies the presence of two elements of unsaturation. The carbonyl (C=O) group accounts for one, but there are no more carbonyl groups and no double-bonded alkene carbon atoms. The other element of unsaturation must be a ring. Combining the partial structures into a ring gives the complete structure.

In the following problems, only the broadband-decoupled spectra are provided. In cases where off-resonance or DEPT spectra are available, the number of protons is given for each carbon atom: either zero (C), one (CH), two (CH_2), or three (CH_3).

PROBLEM 13-29

A bottle of allyl bromide was found to contain a large amount of an impurity. A careful distillation separated the impurity, which has the molecular formula C_3H_6O. The following ^{13}C NMR spectrum of the impurity was obtained:

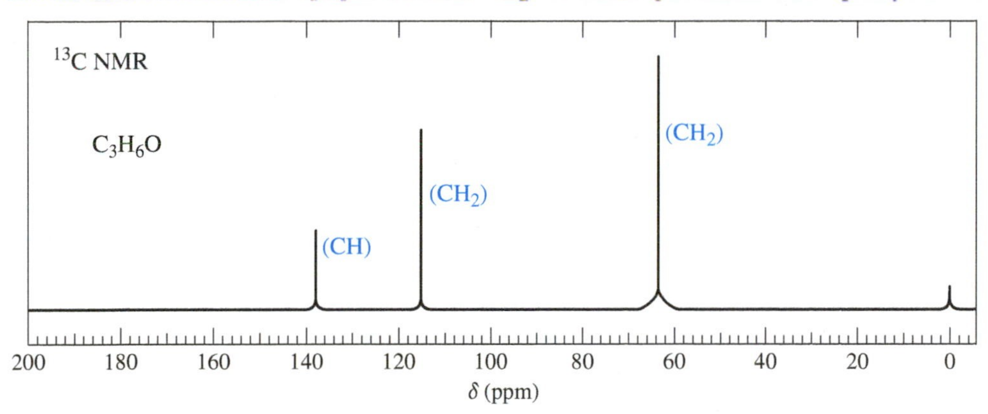

(a) Propose a structure for this impurity.
(b) Assign the peaks in the ^{13}C NMR spectrum to the carbon atoms in the structure.
(c) Suggest how this impurity arose in the allyl bromide sample.

PROBLEM 13-30

An inexperienced graduate student was making some 4-hydroxybutanoic acid. He obtained an excellent yield of a different compound, whose ^{13}C NMR spectrum is shown here.

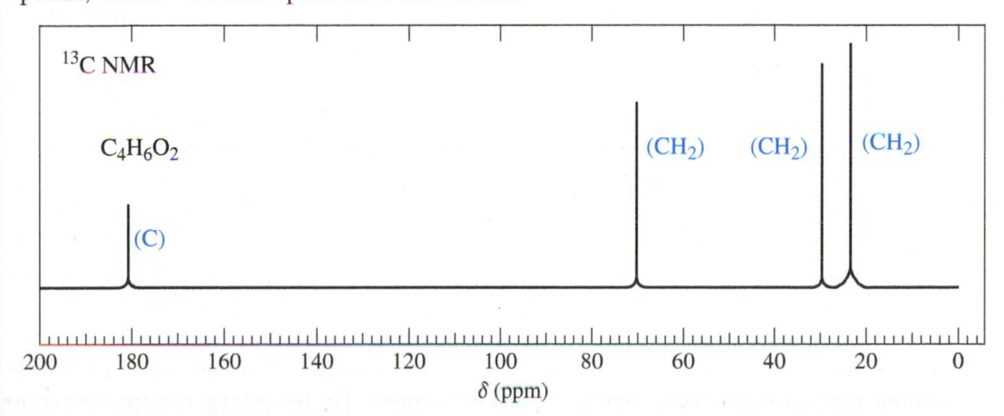

(a) Propose a structure for this product.
(b) Assign the peaks in the ^{13}C NMR spectrum to the carbon atoms in the structure.

PROBLEM 13-31

A laboratory student was converting cyclohexanol to cyclohexyl bromide by using one equivalent of sodium bromide in a large excess of concentrated sulfuric acid. The major product she recovered was not cyclohexyl bromide, but a compound of formula C_6H_{10} that gave the following ^{13}C NMR spectrum:

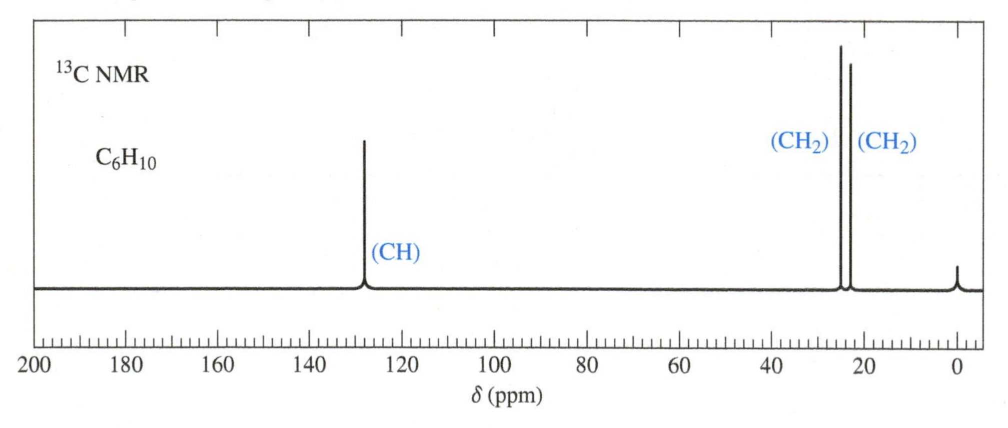

(a) Propose a structure for this product.
(b) Assign the peaks in the ^{13}C NMR spectrum to the carbon atoms in the structure.
(c) Suggest modifications in the reaction to obtain a better yield of cyclohexyl bromide.

13-14 Nuclear Magnetic Resonance Imaging

When chemists use NMR spectroscopy, they take great pains to get the most uniform magnetic field possible (often homogeneous to within one part per billion). They place small tubes of homogeneous solutions in the magnetic field and spin the tubes to average out any remaining variations in the magnetic field. Their goal is to have the sample behave as if it were all at a single point in the magnetic field, with every molecule subjected to exactly the same external magnetic field.

Nuclear magnetic resonance imaging uses the same physical effect, but its goals are almost the opposite of chemical NMR. In NMR imaging, a heterogeneous sample (commonly a living human body) is placed in the magnetic field of a large-bore superconducting magnet. The magnetic field is purposely nonuniform, with a gradient that allows just the protons in one plane of the sample to be in resonance at any one time. By using a combination of field gradients and sophisticated Fourier transform techniques, the instrument can look selectively at one point within the sample, a line within the sample, or a plane within the sample. The computer generates an image of a two-dimensional slice through the sample. A succession of slices can be accumulated in the computer to give a three-dimensional plot of the proton resonances within the bulk of the sample.

Medical NMR imaging is commonly called **magnetic resonance imaging (MRI)** to avoid the common fear of the word *nuclear* and the misconception that *nuclear* means *radioactive*. There is nothing radioactive about an NMR spectrometer. In fact, MRI is the least invasive, least hazardous method available for imaging the interior of the body. The only common side effect is claustrophobia from being confined within the ring of the wide-bore magnet.

Application: ^{31}P NMR

The ^{31}P signal of phosphates in cells and tissues can be followed by ^{31}P NMR spectroscopy. This technique has been used to study the effects of exercise and oxygen starvation on the metabolism of phosphate esters such as ATP.

The MRI image can easily distinguish watery tissues, fatty tissues, bone, air spaces, blood, etc. by their differences in composition and movement. By using proton **relaxation times,** the technique becomes even more useful. In a strong magnetic field, slightly more proton spins are aligned with the field (the lower-energy state) than against it. A radio-frequency pulse of just the right duration inverts some spins, increasing the number of spins oriented against the magnetic field. The spins gradually relax to their normal state over a period of a few seconds. By following the free-induction decay, the spectrometer measures how quickly spin relaxation occurs in each pixel of the sample.

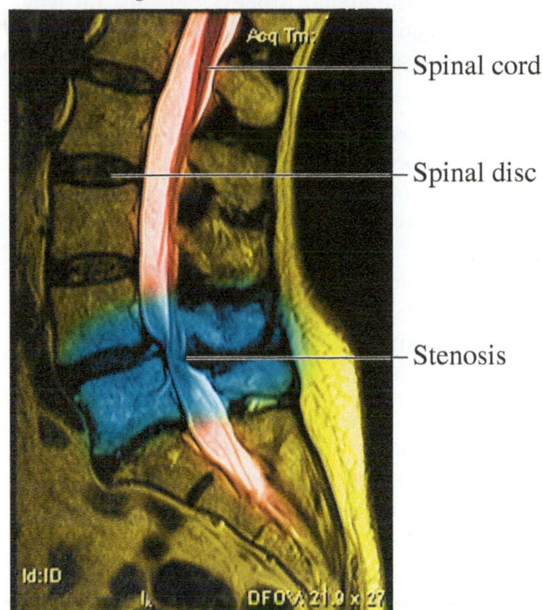

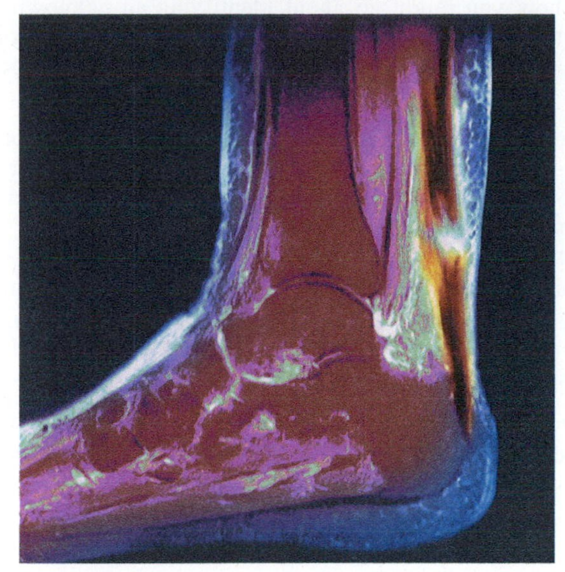

FIGURE 13-48
(a) This MRI image shows how spinal stenosis at lumbar vertebrae L4 and L5 has pinched the spinal cord in this patient. It also shows that the intervertebral disc between L4 and L5 has degenerated and no longer supports the vertebrae in their proper positions.
(b) This MRI image shows a foot with a torn Achilles tendon. Rupture of the Achilles tendon is the most common tendon break injury, often resulting from overstress in athletics.

Differing relaxation times are coded by color or intensity in the image, giving valuable information about the tissues involved. For example, cancerous tissues tend to have longer relaxation times than the corresponding normal tissues, so tumors are readily apparent in the NMR image. Figure 13-48 shows two actual MRI images: The first image is a slice through a patient's head and torso showing spinal stenosis and damage to a disc, explaining why the patient feels severe lower back pain. The second image is a slice through a patient's knee showing a torn ACL, which will require surgical repair.

PROBLEM-SOLVING STRATEGY Spectroscopy Problems

We have now learned to use both IR and NMR spectroscopy, as well as mass spectrometry, to determine the structures of unknown organic compounds. These techniques usually provide a unique structure with little chance of error. A major part of successful spectral interpretation is using an effective strategy rather than simply looking at the spectra, hoping that something obvious will jump out. A systematic approach should take into account the strengths and weaknesses of each technique. The following table summarizes the information provided by each spectroscopic technique.

SUMMARY OF THE INFORMATION PROVIDED BY EACH TYPE OF SPECTROSCOPY

	MS	IR	NMR
molecular weight	√		
molecular formula	√ (HRMS)		
heteroatoms	√	H	S
functional groups	S	√	H
alkyl substituents	S		√

Notes: √, usually provides this information
H, usually provides helpful information
S, sometimes provides helpful information

(continued)

We can summarize how you might go about identifying an unknown compound, but the actual process depends on what you already know about the chemistry of the compound and what you learn from each spectrum. Always go through the process with scratch paper and a pencil so that you can keep track of mass numbers, formulas, possible functional groups, and carbon skeletons.

1. *Mass spectrum.* Look for a molecular ion, and determine a tentative molecular weight. Most compounds not containing nitrogen will show an even-numbered molecular ion and mostly odd-numbered fragments. Remember that some compounds (alcohols, for example) may fail to give a visible molecular ion. If the molecular weight is odd, with some major even-numbered fragments, consider a nitrogen atom. If an HRMS is available, compare the "exact" mass with the tables to find a molecular formula with a mass close to the experimental value.

   Look for anything unusual or characteristic about the mass spectrum: Does the M+2 peak of the parent ion look larger than the M+1 peak? It might contain S, Cl, or Br. Is there a large gap and a peak at 127 characteristic of iodine?

   Although you might look at the MS fragmentation patterns to help determine the structure, this is more time-consuming than going on to other spectra. You can verify the fragmentation patterns more easily once you have a proposed structure.

2. *Infrared spectrum.* Look for O—H, N—H, or ≡C—H peaks in the 3300 cm^{-1} region. Are there saturated C—H peaks to the right of 3000 cm^{-1}? Unsaturated =C—H peaks to the left of 3000 cm^{-1}? Also look for C≡C or C≡N stretch around 2200 cm^{-1}, and for C=O, C=C, or C=N stretch between 1600 and 1800 cm^{-1}. The exact position of the peak, plus other characteristics (intensity, broadening), should help to determine the functional groups. For example, a broad O—H band centered over the C—H stretch at 3000 cm^{-1} might imply a carboxylic acid group, —COOH.

   The combination of IR and an odd molecular ion in the mass spectrum should confirm amines, amides, and nitriles. A strong alcohol —OH absorption in the IR might suggest that the apparent molecular ion in the mass spectrum could be low by 18 units from loss of water.

3. *Nuclear magnetic resonance spectrum.* First, consider the number of signals and their chemical shifts. Look for strongly deshielded protons, such as carboxylic acids (δ10 to δ12), aldehydes (δ9 to δ10), and aromatic protons (δ7 to δ8). Moderately deshielded peaks might be vinyl protons (δ5 to δ6) or protons on a carbon bonded to an electronegative atom such as oxygen or halogen (δ3 to δ4). A peak around δ2.1 to δ2.5 might be an acetylenic proton or a proton on a carbon next to a carbonyl group, a benzene ring, or a vinyl group.

   These possibilities should be checked to see which are consistent with the IR spectrum. The peak integrations should reveal the relative numbers of protons responsible for these signals. Finally, the spin-spin splitting patterns should be analyzed to suggest the structures of the alkyl groups present.

   If the ^{13}C NMR spectrum is available, use the number of signals and their chemical shifts to provide information on how many types of carbon atoms are present, and their possible chemical environments, consistent with the functional groups suggested by the IR spectrum.

   Once you have considered all the spectra, there should be one or two tentative structures. Each structure should be checked to see whether it accounts for the major characteristics of all the spectra.

- Are the molecular weight and formula of the tentative structure consistent with the appearance (or the absence) of the molecular ion in the mass spectrum? Are there peaks in the mass spectrum corresponding to the expected fragmentation products?
- Does the tentative structure explain each of the characteristic stretching frequencies in the infrared spectrum? Does it account for any shifting of frequencies from their usual positions?
- Does the tentative structure account for each proton (or carbon) in the NMR spectrum? Does it also account for the observed chemical shifts and spin-spin splitting patterns?

If the tentative structure successfully accounts for all these features of the spectra, you can be confident that it is correct.

PROBLEM 13-32 (PARTIALLY SOLVED)

Sets of spectra are given for two compounds. For each set,

(1) look at each spectrum individually, and list the structural characteristics you can determine from that spectrum.
(2) look at the set of spectra as a group, and propose a tentative structure.
(3) verify that your proposed structure accounts for the major features of each spectrum. The solution for compound 1 is given after the problem, but go as far as you can before looking at the solution.

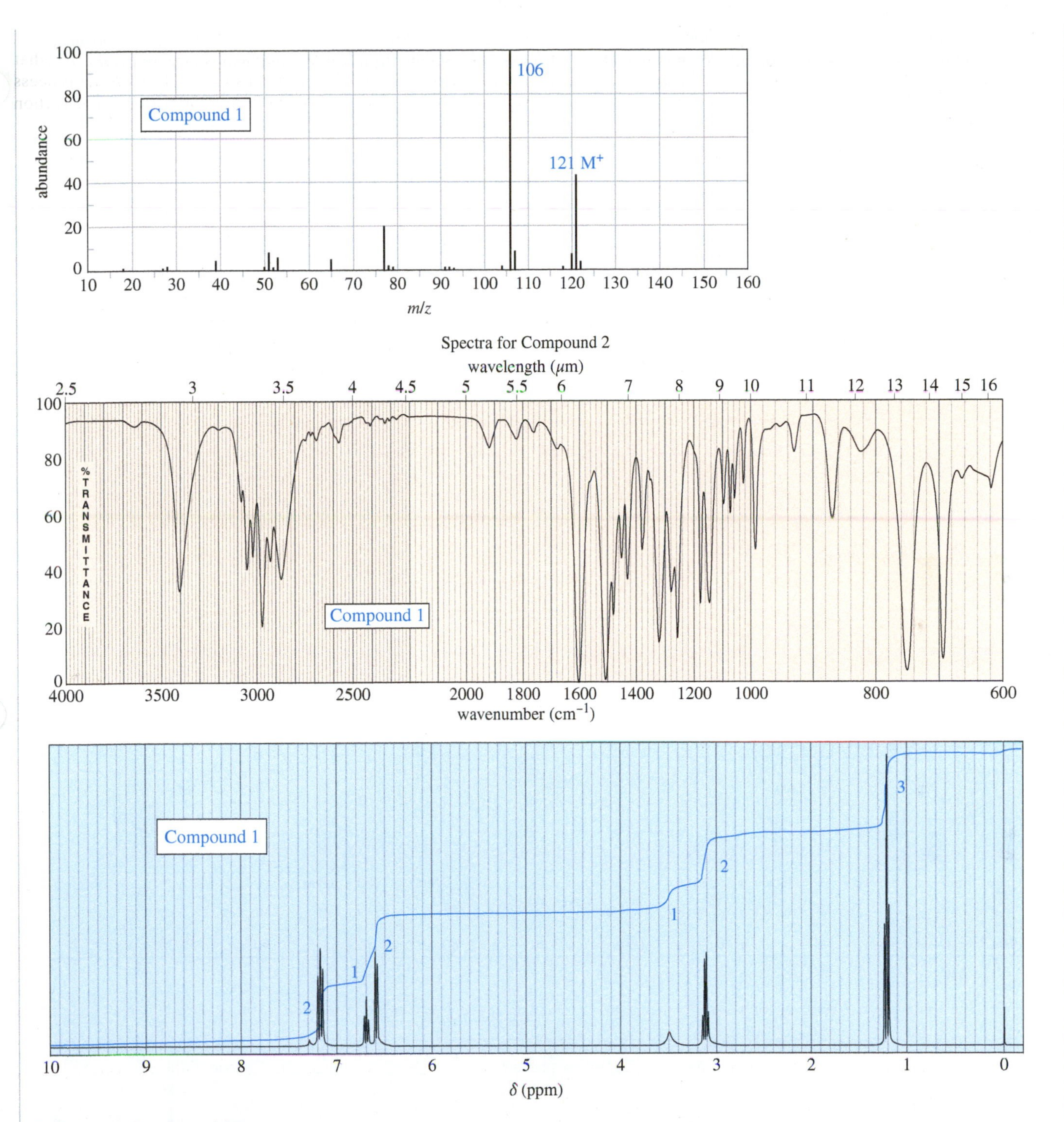

Solution to Compound 1:

Mass spectrum: The MS shows an odd molecular weight at 121 and a large even-numbered fragment at 106. These features may indicate the presence of a nitrogen atom.

Infrared spectrum: The IR shows a sharp peak around 3400 cm^{-1}, possibly the N—H of an amine or the ≡C—H of a terminal alkyne. Because the MS suggests a nitrogen atom, and there is no other evidence for an alkyne (no C≡C stretch around 2200 cm^{-1}), the 3400 cm^{-1} absorption is probably an N—H bond. The unsaturated =C—H absorptions above 3000 cm^{-1}, combined with an aromatic C=C stretch around 1600 cm^{-1}, indicate an aromatic ring.

NMR spectrum: The NMR shows complex splitting in the aromatic region, probably from a benzene ring: The total integral of 5 suggests the ring is monosubstituted. Part of the aromatic absorption is shifted upfield of δ 7.2, suggesting that the substituent on the benzene ring is a pi electron-donating group such as an amine or an ether. An ethyl group (total area 5) is seen at δ 1.2 and δ 3.1, appropriate for protons on a carbon atom bonded to nitrogen. A broad singlet of area 1 appears at δ 3.5, probably resulting from the N—H seen

(continued)

in the IR spectrum. Combining this information, we propose a nitrogen atom bonded to a hydrogen atom, a benzene ring, and an ethyl group. The total molecular weight for this structure would be 121, in agreement with the molecular ion in the mass spectrum.

Proposed structure for compound 1

$$C_6H_5-\ddot{N}(H)-CH_2-CH_3$$

The proposed structure shows an aromatic ring with 5 protons, which explains the aromatic signals in the NMR, the C=C at 1600 cm^{-1}, and the =C—H above 3000 cm^{-1} in the IR. The aromatic ring is bonded to an electron-donating —NHR group, which explains the odd molecular weight, the N—H absorption in the IR, and the aromatic signals shifted above δ7.2 in the NMR. The ethyl group bonded to nitrogen explains the ethyl signals in the NMR, deshielded to δ3.1 by the nitrogen atom. The base peak in the MS (M − 15 = 106) is explained by the loss of a methyl group to give a resonance-stabilized cation:

$$\left[Ph-\ddot{N}(H)-\underbrace{CH_2}_{106}\!+\!CH_3\right]^{+\cdot} \longrightarrow \left[Ph-\ddot{N}(H)-\overset{+}{C}H_2 \longleftrightarrow Ph-\overset{+}{N}(H)=CH_2\right] + \cdot CH_3$$

m/z 106 — loss of 15

Spectra for Compound 2

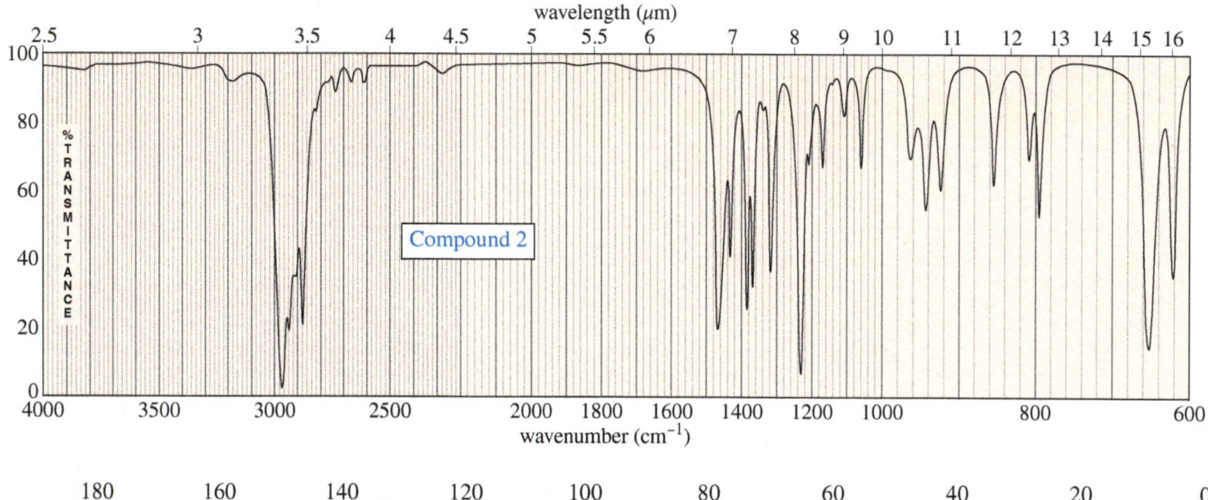

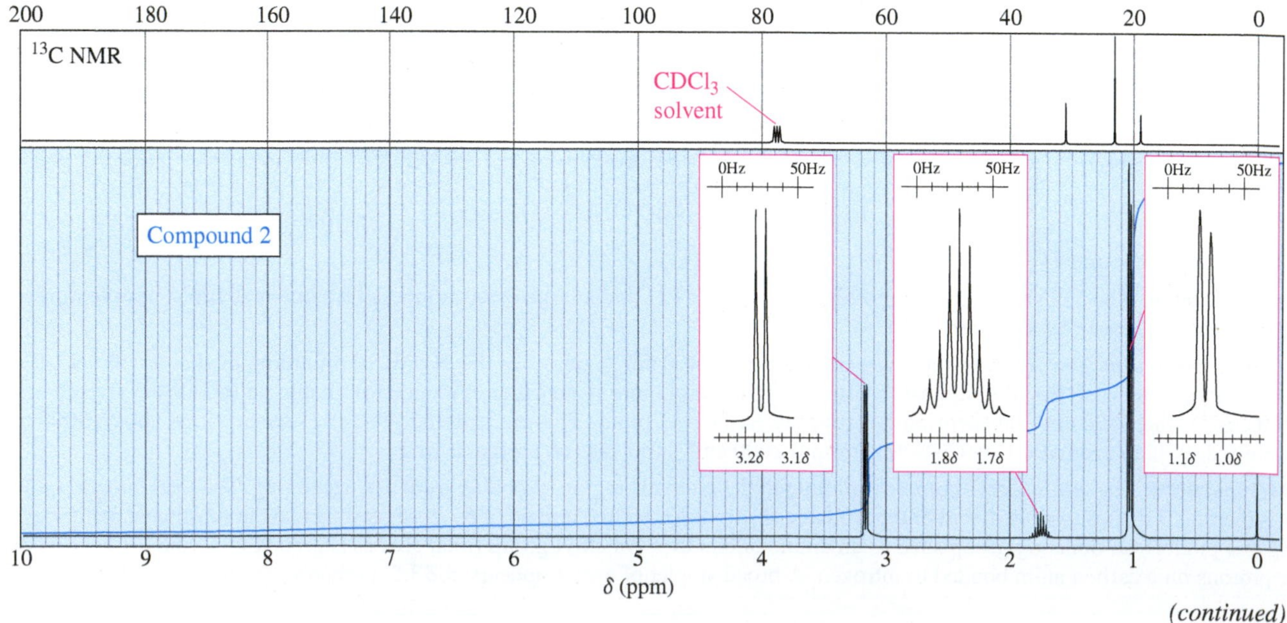

(continued)

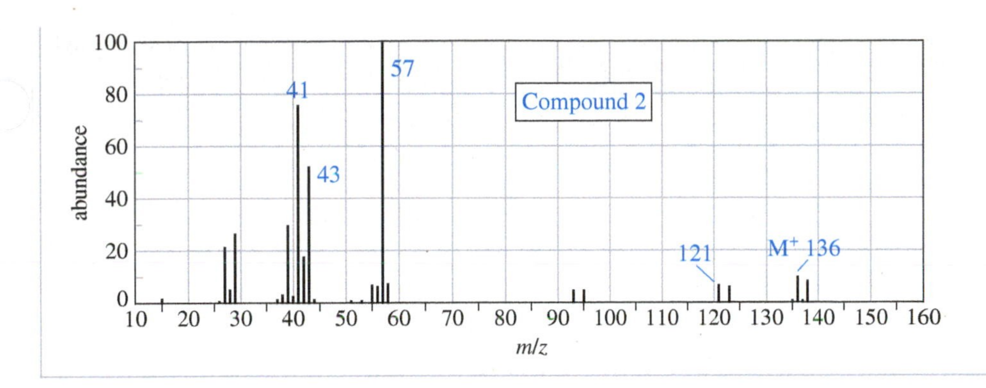

Essential Terms

accidentally equivalent nuclei Nuclei that are not chemically equivalent, yet absorb at nearly the same chemical shift and are not resolved. Nuclei that absorb at the same chemical shift cannot split each other, whether they are chemically equivalent or accidentally equivalent. (p. 621)

chemically equivalent atoms Atoms that cannot be distinguished chemically. The replacement test for chemically equivalent atoms gives identical compounds. (p. 620)

chemical shift The difference (in ppm) between the resonance field or frequency of the proton (or carbon nucleus) being observed and that of tetramethylsilane (TMS). Chemical shifts are usually given on the δ (delta) scale, in parts per million downfield from TMS. (p. 613)

complex splitting Splitting by two or more different kinds of protons with different coupling constants. (p. 634)

coupling constant (*J*) The distance (in hertz) between two adjacent peaks of a multiplet. (p. 631)

DEPT **(Distortionless Enhanced Polarization Transfer)** A method of running several ^{13}C experiments with different pulse sequences so that the carbon atoms appear differently depending on whether they are bonded to zero, one, two, or three protons. (p. 654)

deshielded Bonded to a group that withdraws part of the electron density from around the nucleus. The absorptions of deshielded nuclei are moved downfield, resulting in larger chemical shifts. (p. 611)

diastereotopic atoms Nuclei that occupy diastereomeric positions. The replacement test for diastereotopic atoms gives diastereomers. Diastereotopic nuclei can be distinguished by NMR, and they can split each other unless they are **accidentally equivalent.** (p. 638)

downfield At a lower value of the applied magnetic field, toward the left (higher values of δ) on the NMR spectrum. The more deshielded a nucleus is, the farther downfield it absorbs. (p. 612)

Fourier transform spectroscopy Spectroscopy that involves collecting transients (containing all the different resonance frequencies) and converting the averaged transients into a spectrum using the mathematical Fourier transform. (p. 649)

transient **(free induction decay, or FID)** The signal that results when many nuclei are irradiated by a pulse of energy and precess at their resonance frequencies. (p. 649)

gyromagnetic ratio (γ) A measure of the magnetic properties of a nucleus. The resonance frequency (ν) is given by the equation $\nu = \gamma B_{eff}/2\pi$, where B_{eff} is the effective magnetic field at the nucleus. The gyromagnetic ratio of a proton is 26,753 sec^{-1} $gauss^{-1}$. The gyromagnetic ratio of a ^{13}C nucleus is 6728 sec^{-1} $gauss^{-1}$. (p. 609)

induced magnetic field The magnetic field set up by the motion of electrons in a molecule (or in a wire) in response to the application of an external magnetic field. (p. 610)

integration The measurement of the area under a peak, proportional to the number of protons giving rise to that peak. (p. 622)

magnetically coupled Nuclei that are close enough that their magnetic fields influence each other, resulting in spin-spin splitting. (p. 625)

magnetic moment The magnitude of a nuclear magnetic field, characterized by the gyromagnetic ratio γ. (p. 608)

magnetic resonance imaging (MRI) The medical term for NMR imaging, avoiding the word *nuclear*. Use of field gradients in a large-bore magnet to scan two-dimensional slices of a patient's body. (p. 658)

multiplet A group of peaks resulting from the spin-spin splitting of the signal from a single type of nucleus. A **doublet** has two peaks, a **triplet** has three peaks, a **quartet** has four peaks, etc. (p. 627)

$N+1$ rule A signal that is being split by N neighboring equivalent protons is split into a multiplet with $N+1$ individual peaks. (p. 626)

NMR **(nuclear magnetic resonance spectroscopy)** A form of spectroscopy that measures the absorption of radio-frequency energy by nuclei in a magnetic field. The energy absorbed causes nuclear spin transitions. (p. 607)

- **carbon magnetic resonance** (**^{13}C NMR, CMR**) NMR of the ^{13}C isotope of carbon. (p. 607)
- **proton magnetic resonance** (**^{1}H NMR, PMR**) NMR of protons. (p. 607)

off-resonance decoupling A technique used with ^{13}C NMR, in which only the protons directly bonded to a carbon atom cause spin-spin splitting. (p. 652)

relaxation time A measure of how slowly the nuclear spins return to their normal state after an RF pulse near their resonance frequency. Alternatively, the evening after a chemistry exam. (p. 658)

shielded Surrounded by electrons whose induced magnetic field opposes the externally applied magnetic field. The effective magnetic field at the shielded nucleus is less than the applied magnetic field. (p. 610)

spin decoupling Elimination of spin-spin splitting by constantly irradiating one type of nuclei at its resonance frequency. (p. 652)

spin-spin splitting (magnetic coupling) The interaction of the magnetic fields of two or more nuclei, usually through the bonds connecting them. Spin-spin splitting converts a single signal to a **multiplet,** a set of smaller peaks. (p. 625)

TMS Tetramethylsilane, an NMR standard whose absorption is defined as δ 0.00. (p. 613)

upfield At a higher value of the applied magnetic field, toward the right (lower values of δ) on the NMR spectrum. The more shielded a nucleus is, the farther upfield it absorbs. (p. 612)

Essential Problem-Solving Skills in Chapter 13

Each skill is followed by problem numbers exemplifying that particular skill.

1 Given a structure, explain which protons are equivalent and which are nonequivalent. Predict the number of signals in the proton NMR and their approximate chemical shifts.	Problems 13-34, 35, 50, 56, and 57
2 Given the chemical shifts of absorptions, suggest likely types of protons. Use the integrations to determine the relative numbers of the different types of protons.	Problems 13-33, 36, 38, 43, 44, and 54
Problem-Solving Strategy: Drawing an NMR Spectrum	Problem 13-39
3 Explain which protons in a spectrum are magnetically coupled, and use the spin-spin splitting patterns to determine the structures of alkyl and other groups.	Problems 13-36, 38, 43, 44, and 45
Problem-Solving Strategy: Interpreting Proton NMR Spectra	Problems 13-33, 36, 38, 43, 44, and 54
4 Draw the general features of the proton and ^{13}C NMR spectra of a given compound.	Problems 13-34, 35, 39, 40, and 41
5 Predict the number of signals and approximate chemical shifts of carbon atoms in a given compound.	Problems 13-41, 49, 50, 51, 53, 55, and 57
6 Given the chemical shifts of ^{13}C absorptions, suggest likely types of carbons. Use either the off-resonance-decoupled spectrum or the DEPT ^{13}C spectra to determine the number of hydrogen atoms bonded to a given carbon atom.	Problems 13-42, 43, 45, and 48
7 Combine the chemical shifts, integrals, and spin-spin splitting patterns in NMR spectra with information from infrared and mass spectra to determine the structures of organic compounds.	Problems 13-47 and 48
Problem-Solving Strategy: Spectroscopy Problems	Problems 13-43, 45, 47, 48, 52, and 57

Study Problems

13-33 An unknown compound has the molecular formula $C_9H_{11}Br$. Its proton NMR spectrum shows the following absorptions:
singlet, δ 7.1, integral 44 mm
singlet, δ 2.3, integral 130 mm
singlet, δ 2.2, integral 67 mm
Propose a structure for this compound.

13-34 Predict the multiplicity (the number of peaks as a result of splitting) and the chemical shift for each shaded proton in the following compounds.

(a) $CH_3-CH_2-CCl_2-CH_3$ **(b)** $CH_3-CH(CH_3)-OH$ **(c)** $CH_3-CH(CH_3)-CH_3$

(d) $H_3C-C_6H_4$ (toluene, ring H's shown) **(e)** $CH_3-CH_2-C(=O)-O-CH_2-CH_3$ ***(f)** 2-methyltetrahydropyran (H and CH_3 on C-2)

13-35 Predict the approximate chemical shifts of the protons in the following compounds.

(a) benzene **(b)** cyclohexane **(c)** $CH_3-O-CH_2CH_2CH_2Cl$

(d) $CH_3CH_2-C\equiv C-H$ **(e)** $CH_3CH_2-C(=O)-CH_3$ **(f)** $(CH_3)_2CH-O-CH_2CH_2OH$

(g) $C_6H_5-C(=O)-H$ **(h)** $CH_3-CH=CH-CHO$ **(i)** $HO-C(=O)-CH_2CH_2-C(=O)-OCH(CH_3)_2$

(j) methylenecyclohexane **(k)** indane **(l)** indene

13-36 The following proton NMR spectrum is of a compound of molecular formula C_3H_8O.

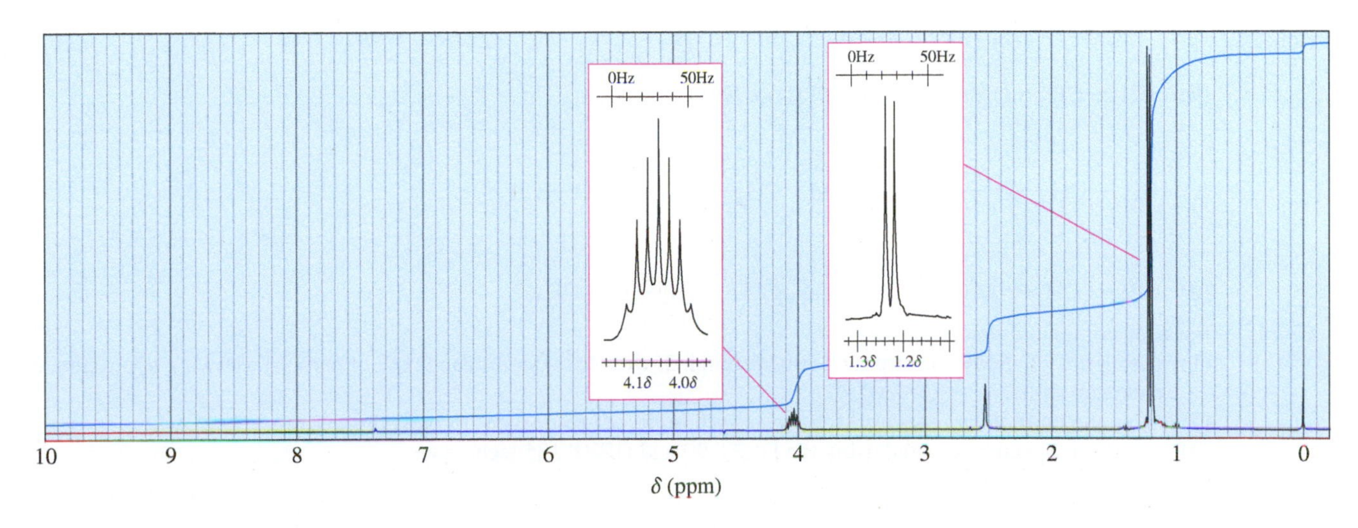

(a) Propose a structure for this compound.
(b) Assign peaks to show which protons give rise to which signals in the spectrum.

13-37 Using a 60-MHz spectrometer, a chemist observes the following absorption:

doublet, $J = 7$ Hz, at δ 4.00

(a) What would the chemical shift (δ) be in the 300-MHz spectrum?
(b) What would the splitting value J be in the 300-MHz spectrum?
(c) How many hertz from the TMS peak is this absorption in the 60-MHz spectrum? In the 300-MHz spectrum?

13-38 A compound ($C_{10}H_{12}O_2$) whose spectrum is shown here was isolated from a reaction mixture containing 2-phenylethanol and acetic acid.

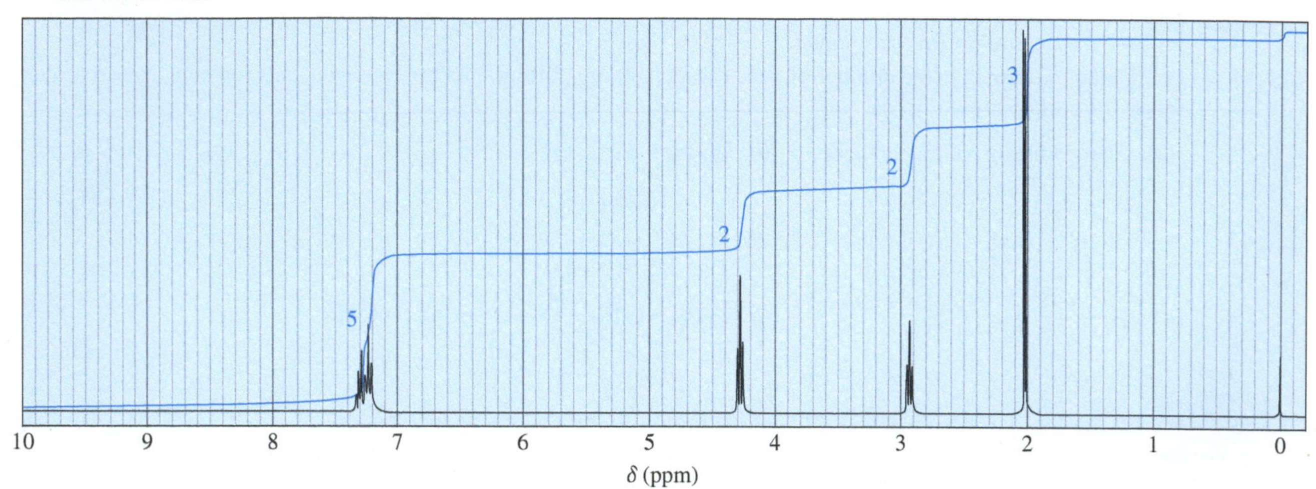

(a) Propose a structure for this compound.
(b) Assign peaks to show which protons give rise to which signals in the spectrum.

13-39 Sketch your predictions of the proton NMR spectra of the following compounds.

(a) $CH_3—O—CH_2CH_3$ **(b)** $(CH_3)_2CH—C(=O)—CH_3$ **(c)** $Cl—CH_2—CH_2—CH_2—Cl$

(d) 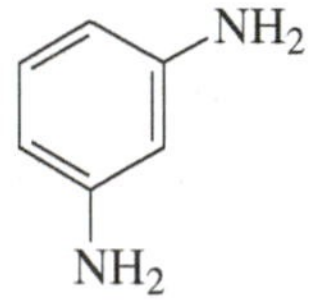 1,3-diaminobenzene (benzene ring with NH_2 groups) **(e)** 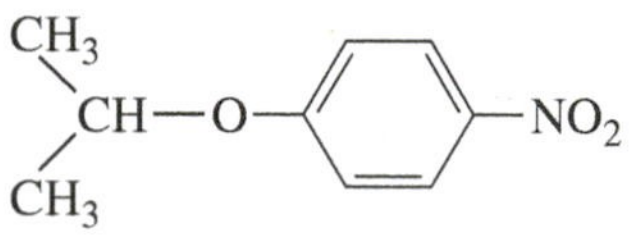 $(CH_3)_2CH—O—C_6H_4—NO_2$ **(f)** 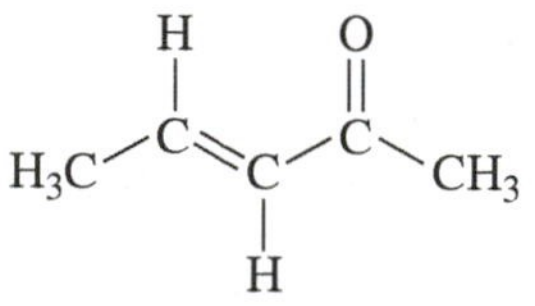 $H_3C—CH=CH—C(=O)—CH_3$

13-40 Tell precisely how you would use the proton NMR spectra to distinguish between the following pairs of compounds.

(a) 1-bromopropane and 2-bromopropane

(b) $CH_3—CH_2—C(=O)—CH_3$ and $(CH_3)_2CH—C(=O)—CH_3$

(c) $CH_3—CH_2—O—C(=O)—CH_3$ and $CH_3—CH_2—C(=O)—O—CH_3$

(d) $CH_3—CH_2—C\equiv C—CH_3$ and $CH_3—CH_2—C(=O)—CH_3$

13-41 For each compound shown below,

(1) sketch the ^{13}C NMR spectrum (totally decoupled, with a singlet for each type of carbon), showing approximate chemical shifts.
(2) show the multiplicity expected for each signal in the off-resonance-decoupled spectrum.
(3) sketch the spectra expected using the DEPT-90 and DEPT-135 techniques.

(a) $CH_3—C(=O)—O—CH_2—CH_3$ ethyl acetate **(b)** $H_2C=CH—CH_2Cl$ 3-chloropropene

(c) 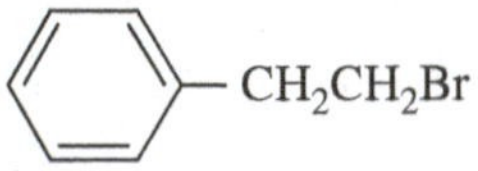 $C_6H_5—CH_2CH_2Br$ **(d)** $CH_3—C\equiv C—C(=O)—CH_2CH_3$

13-42 The following off-resonance-decoupled carbon NMR was obtained from a compound of formula $C_3H_5Cl_3$. Propose a structure for this compound, and show which carbon atoms give rise to which peaks in the spectrum.

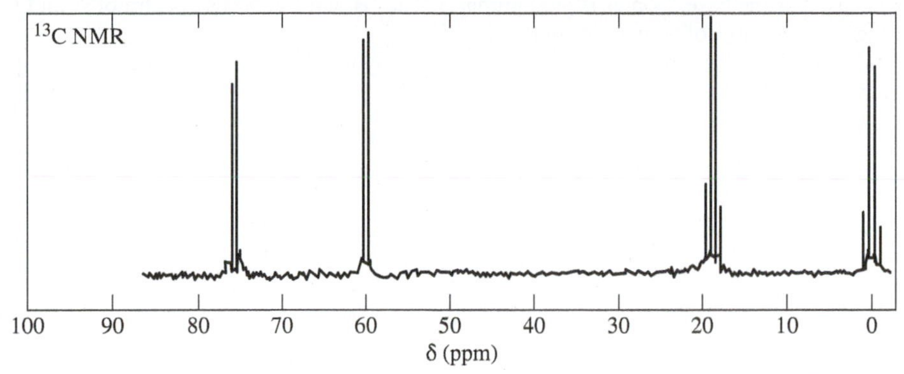

13-43 A small pilot plant was adding bromine across the double bond of but-2-ene to make 2,3-dibromobutane. A controller malfunction allowed the reaction temperature to rise beyond safe limits. A careful distillation of the product showed that several impurities had formed, including the one having the NMR spectra that appear below. Determine its structure, and assign the peaks to the protons in your structure.

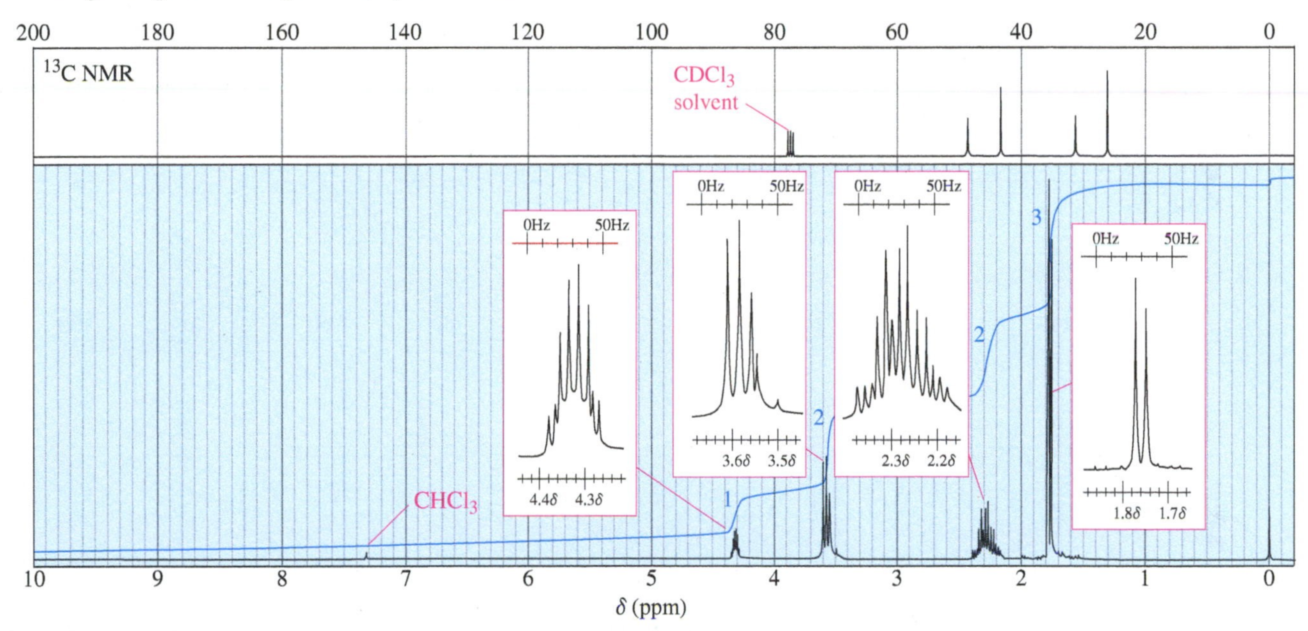

13-44 A new chemist moved into an industrial lab where work was being done on oxygenated gasoline additives. Among the additives that had been tested, she found an old bottle containing a clear, pleasant-smelling liquid that was missing its label. She took the quick NMR spectrum shown and was able to determine the identity of the compound without any additional information. Propose a structure, and assign the peaks. (*Hint:* This is a very pure sample.)

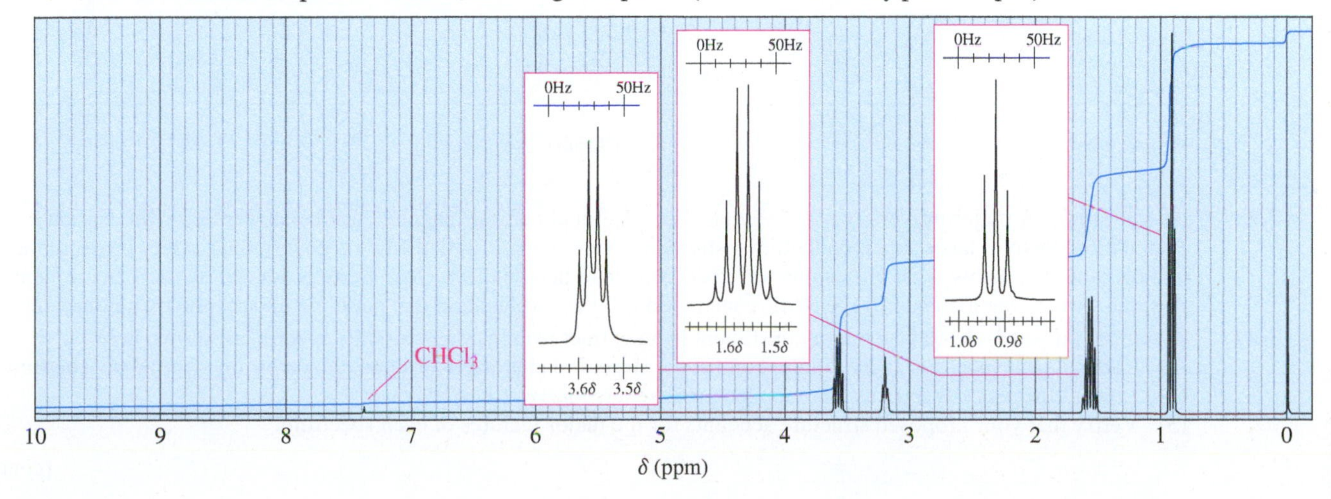

***13-45** When 2-chloro-2-methylbutane is treated with a variety of strong bases, the products always seem to contain two isomers (A and B) of formula C_5H_{10}. When sodium hydroxide is used as the base, isomer A predominates. When potassium *tert*-butoxide is used as the base, isomer B predominates. The ^1H and ^{13}C NMR spectra of A and B are given below.

(a) Determine the structures of isomers A and B.

(b) Explain why A is the major product when using sodium hydroxide as the base and why B is the major product when using potassium *tert*-butoxide as the base.

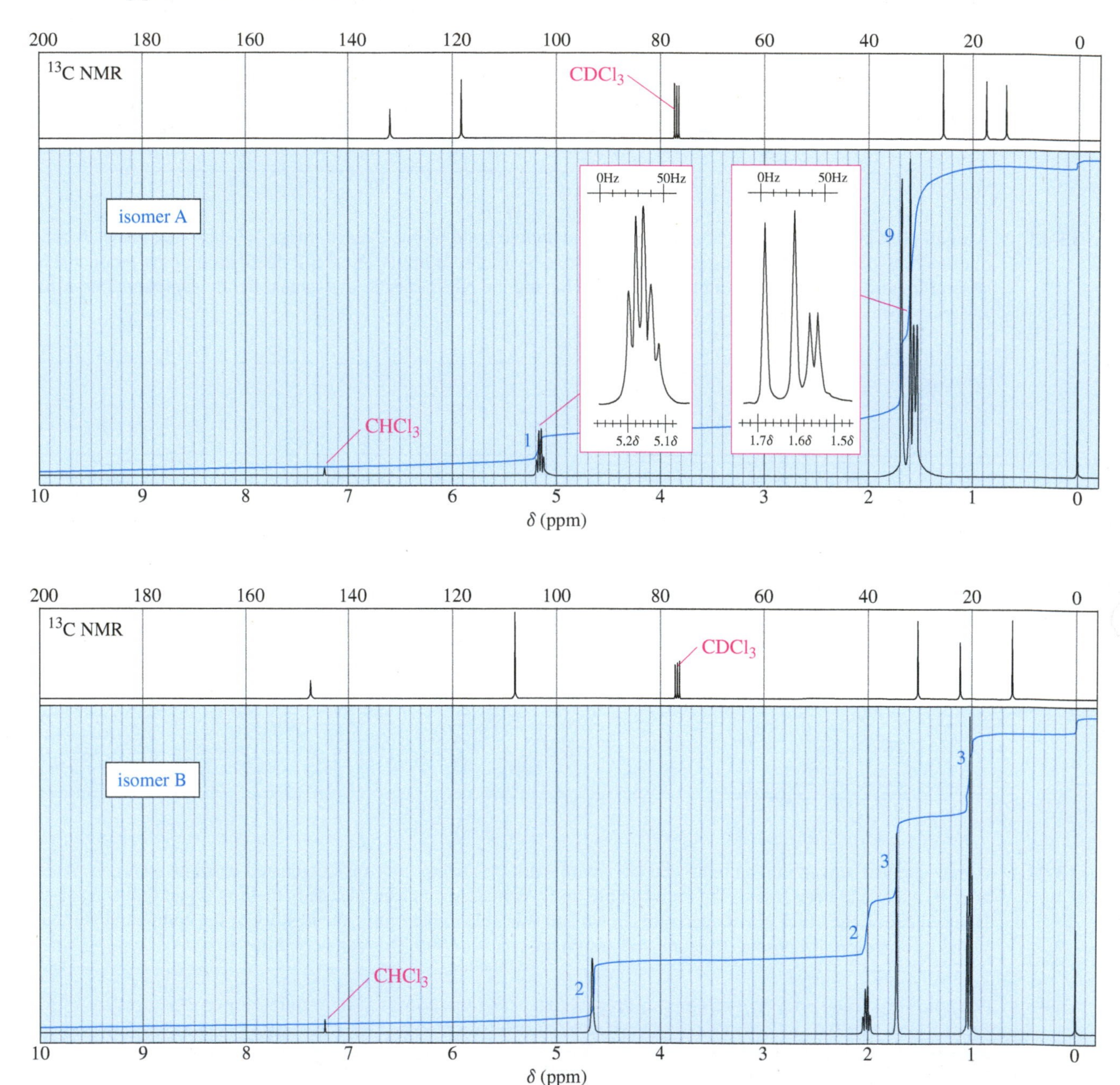

13-46 (A true story.) A major university was designated as a national nuclear magnetic resonance center by the National Science Foundation. Several large superconducting instruments were being installed when a government safety inspector appeared and demanded to know what provisions were being made to handle the nuclear waste produced by these instruments. Assume you are the manager of the NMR center, and offer an explanation that could be understood by a nonscientist.

13-47 A compound was isolated as a minor constituent in an extract from garden cress. Its spectra are shown here.

(1) Look at each spectrum individually, and list the structural characteristics you can determine from that spectrum.

(2) Look at the set of spectra as a group, and propose a tentative structure.

(3) Verify that your proposed structure accounts for the major features of each spectrum.

(continued)

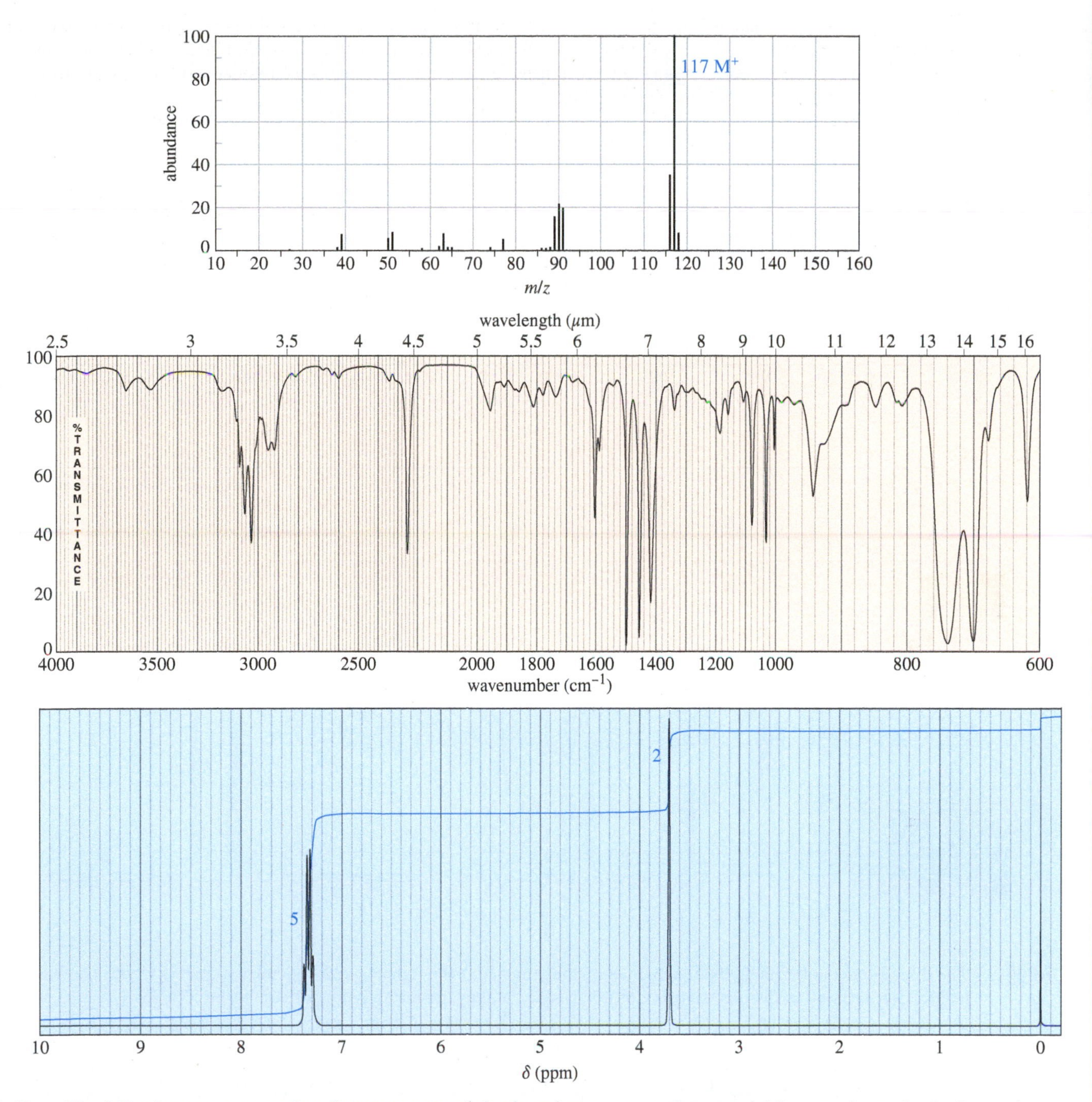

13-48 The following spectra are taken from a compound that is an important starting material for organic synthesis. Determine the structure, first by considering each spectrum individually, and then by considering all the spectra together. Assign peaks to show that your proposed structure accounts for all the major features of each spectrum. DEPT information is given in blue on the carbon NMR.

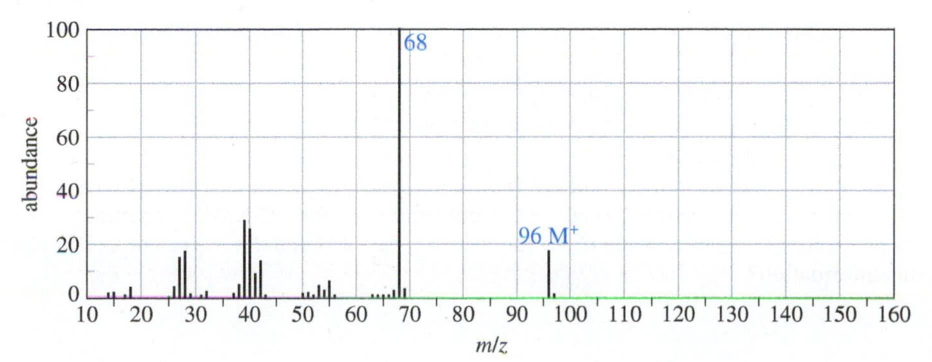

(continued)

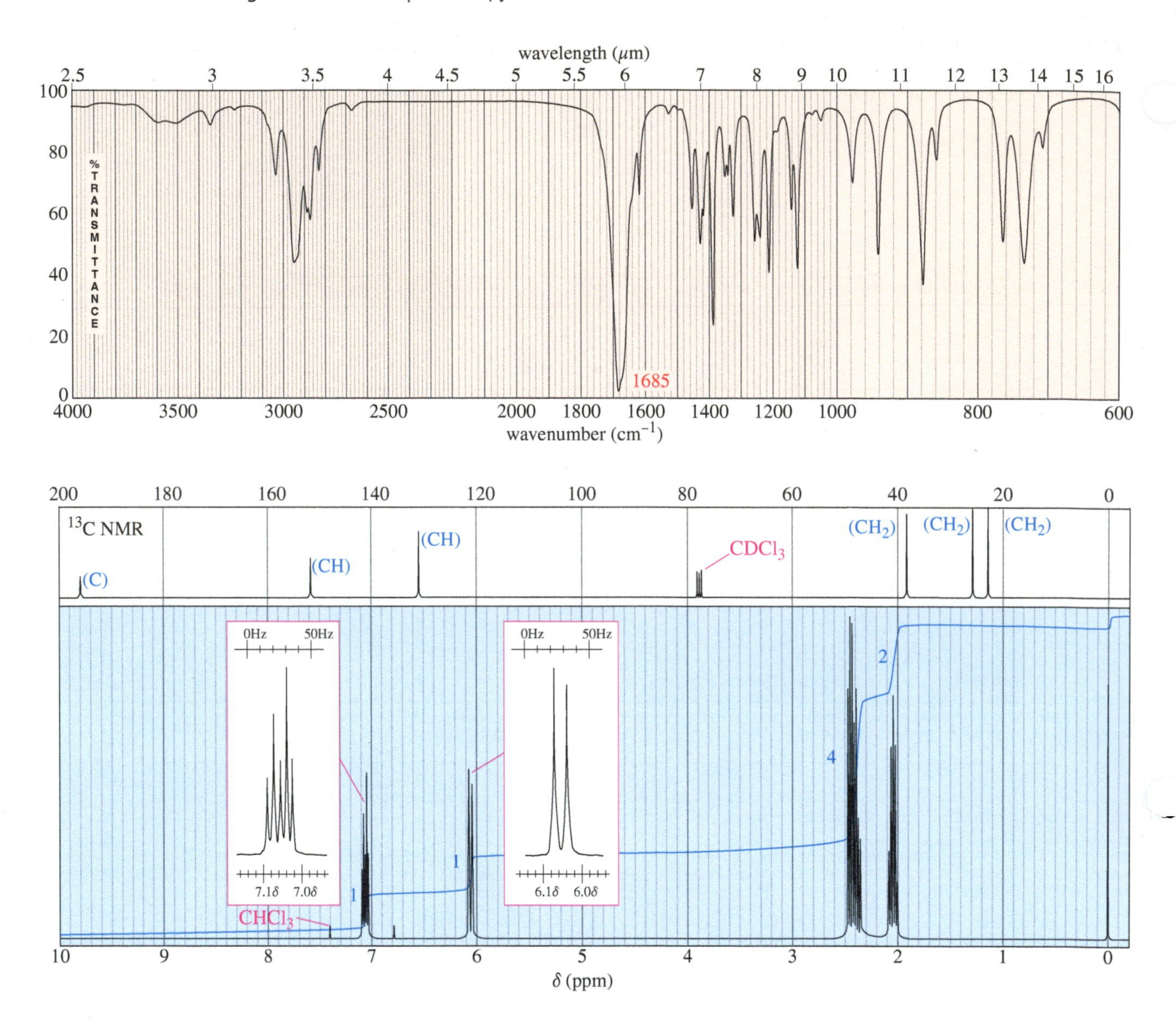

13-49 The three isomers of dimethylbenzene are commonly named *ortho*-xylene, *meta*-xylene, and *para*-xylene. These three isomers are difficult to distinguish using proton NMR, but they are instantly identifiable using ^{13}C NMR.

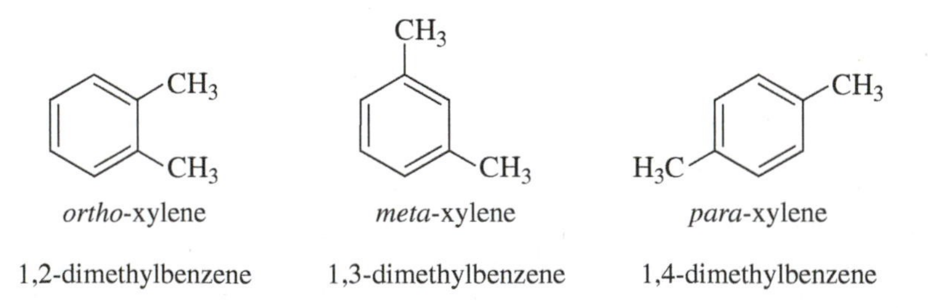

(a) Describe how carbon NMR distinguishes these three isomers.
(b) Explain why they are difficult to distinguish using proton NMR.

13-50 **(a)** Draw all six isomers of formula C_4H_8 (including stereoisomers).
(b) For each structure, show how many types of H would appear in the proton NMR spectrum.
(c) For each structure, show how many types of C would appear in the ^{13}C NMR spectrum.
(d) If an unknown compound of formula C_4H_8 shows two types of H and three types of C, can you determine its structure from this information?

13-51 Different types of protons and carbons in alkanes tend to absorb at similar chemical shifts, making structure determination difficult. Explain how the ^{13}C NMR spectrum, including the DEPT technique, would allow you to distinguish among the following four isomers.

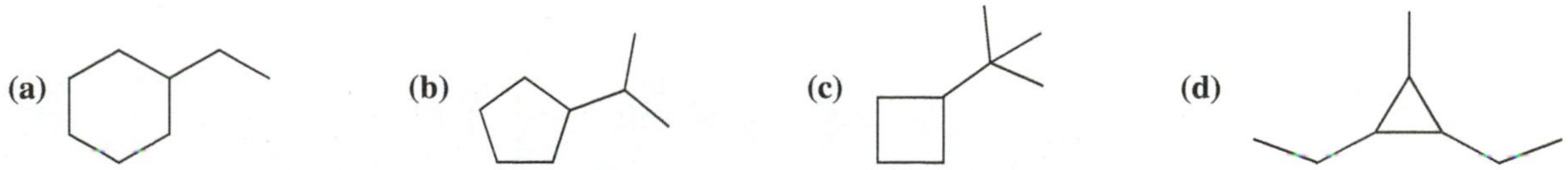

13-52 Hexamethylbenzene undergoes free-radical chlorination to give one monochlorinated product ($C_{12}H_{17}Cl$) and four dichlorinated products ($C_{12}H_{16}Cl_2$). These products are easily separated by GC-MS, but the dichlorinated products are difficult to distinguish by their mass spectra. Draw the monochlorinated product and the four dichlorinated products, and explain how ^{13}C NMR would easily distinguish among these compounds.

13-53 Each of these four structures has molecular formula $C_4H_8O_2$. Match the structure with its characteristic proton NMR signals. (Not all of the signals are listed in each case.)

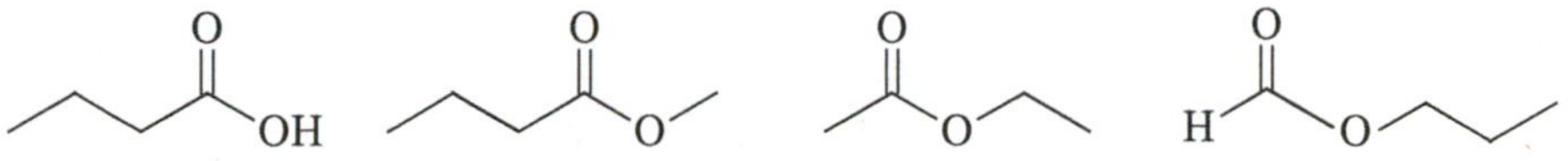

structure 1 structure 2 structure 3 structure 4

(a) sharp 1H singlet at δ 8.0 and 2H triplet at δ 4.0
(b) sharp 3H singlet at δ 2.0 and 2H quartet at δ 4.1
(c) sharp 3H singlet at δ 3.7 and 2H quartet at δ 2.3
(d) broad 1H singlet at δ 11.5 and 2H triplet at δ 2.3

13-54 How many signals would you expect to see in the ^{13}C NMR of the following compounds? In each case, show which carbon atoms are equivalent in the ^{13}C NMR.

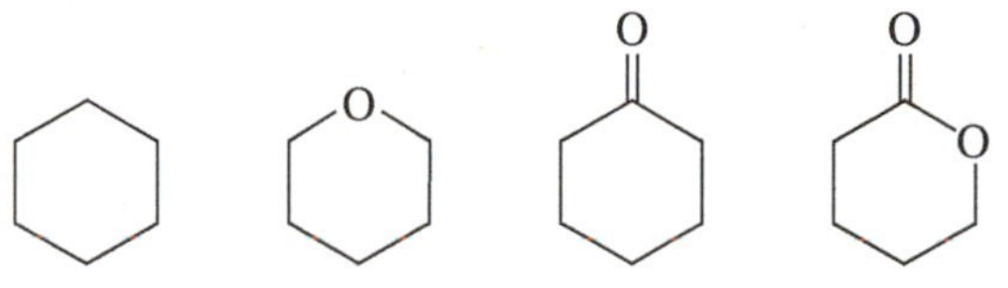

structure 1 structure 2 structure 3 structure 4

***13-55** Phenyl Grignard reagent adds to 2-methylpropanal to give the secondary alcohol shown. The proton NMR of 2-methylpropanal shows the two methyl groups as equivalent (one doublet at δ 1.1), yet the product alcohol, a racemic mixture, shows two different 3H doublets, one at δ 0.75 and one around δ 1.0.

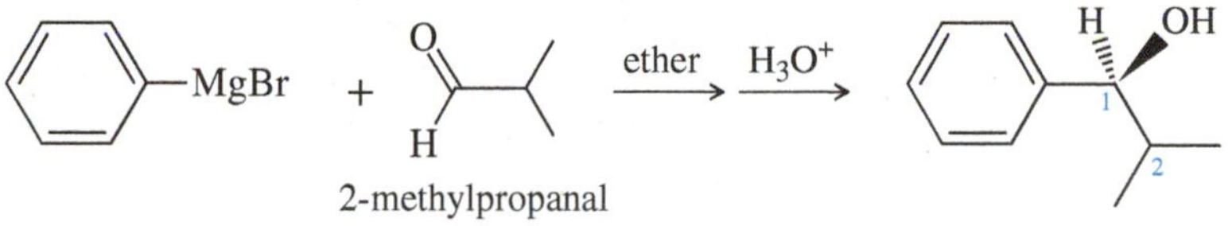

2-methylpropanal

(a) Draw a Newman projection of the product along the C1–C2 axis.
(b) Explain why the two methyl groups have different NMR chemical shifts. What is the term applied to protons such as these?

***13-56** Show how you would distinguish among the following three compounds
(a) using infrared spectroscopy and no other information.
(b) using proton NMR spectroscopy and no other information.
(c) using ^{13}C NMR, including DEPT, and no other information.

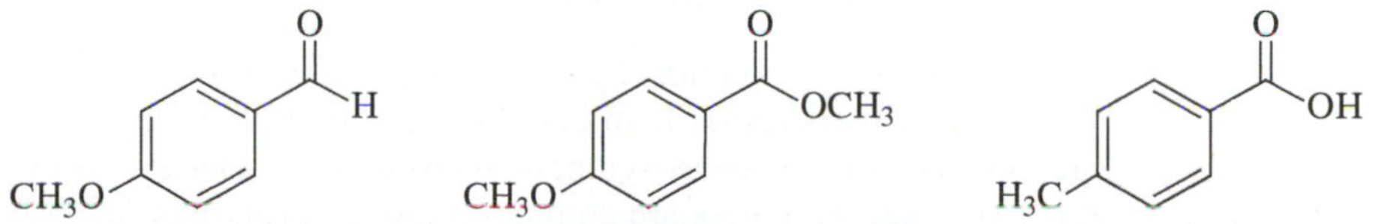

compound 1 compound 2 compound 3

14 Ethers, Epoxides, and Thioethers

Goals for Chapter 14

1 Draw and name ethers and heterocyclic ethers, including epoxides. Explain the trends in their boiling points, solubilities, and solvent properties.

2 Determine the structures of ethers from their spectra, and explain their characteristic absorptions and fragmentations.

3 Devise efficient laboratory syntheses of ethers and epoxides.

4 Predict the products of reactions of ethers and epoxides.

5 Propose mechanisms showing the formation and reactions of ethers and epoxides.

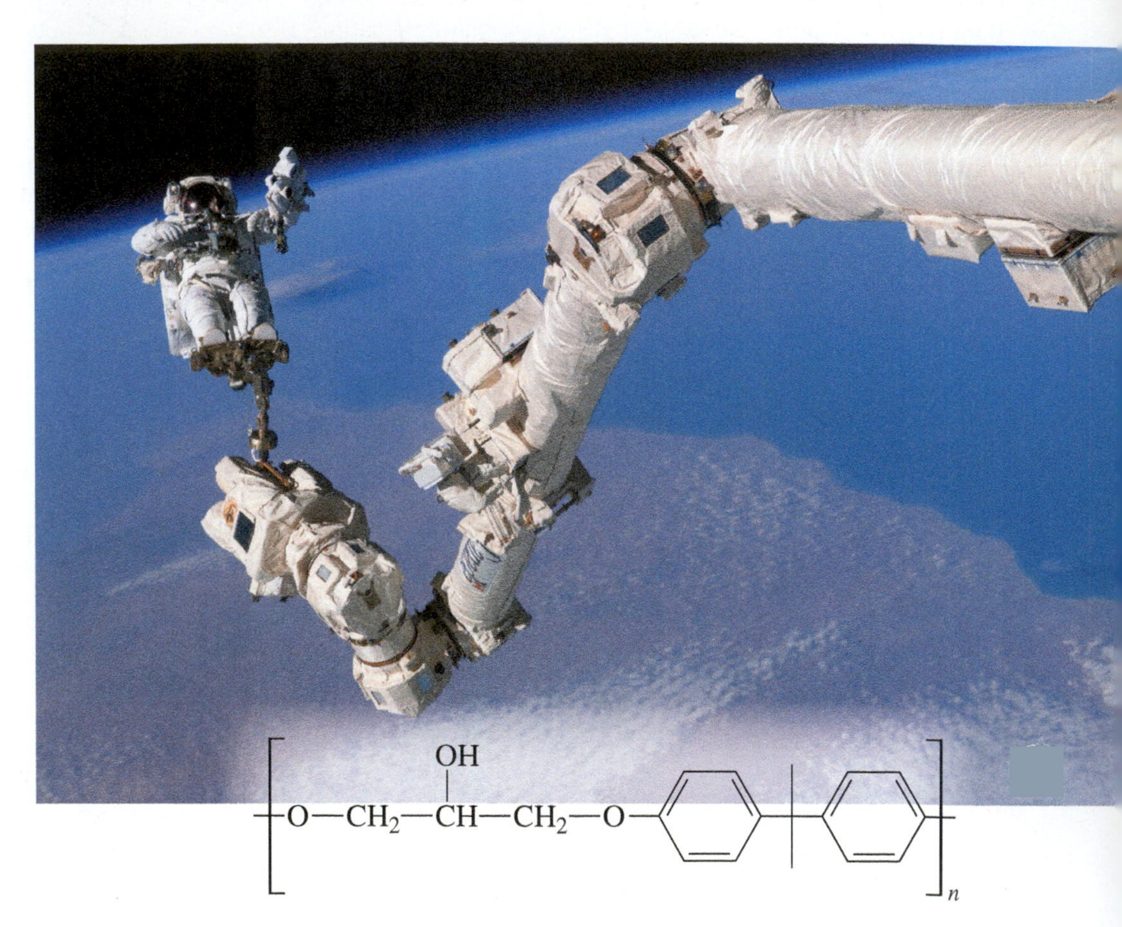

▶ Canadarm2 is the primary component of the Mobile Servicing System (MSS) on the International Space Station (ISS). The boom sections of Canadarm2 use composites of epoxy resins cured over a carbon fiber matrix (Section 14-16) for their high strength and stiffness combined with low weight.

14-1 Introduction

Ethers are compounds of formula R—O—R′, where R and R′ may be alkyl groups or aryl (benzene ring) groups. Like alcohols, ethers are related to water, with alkyl groups replacing the hydrogen atoms. In an alcohol, one hydrogen atom of water is replaced by an alkyl group. In an ether, both hydrogens are replaced by alkyl groups. The two alkyl groups are the same in a **symmetrical ether** and different in an **unsymmetrical ether.**

H—O—H	R—O—H	R—O—R′
water	alcohol	ether

Examples of ethers

$CH_3CH_2—O—CH_2CH_3$	$C_6H_5—O—CH_3$	(five-membered ring containing O)
diethyl ether (a symmetrical ether)	methyl phenyl ether (an unsymmetrical ether)	tetrahydrofuran (a symmetrical, cyclic ether)

As with other functional groups, we will discuss how ethers are formed and how they react. Ethers (other than epoxides) are relatively unreactive, however, and they are not frequently used as synthetic intermediates. Because they are stable with many types of reagents, ethers are commonly used as solvents for organic reactions. In this chapter, we consider the properties of ethers and how these properties make ethers such valuable solvents.

The most important commercial ether is diethyl ether, often called "ethyl ether," or simply "ether." Ether is a good solvent for reactions and extractions, and it is used as a volatile starting fluid for diesel and gasoline engines. Ether was used as a surgical anesthetic for over a hundred years (starting in 1842), but it is highly flammable, and patients often vomited as they regained consciousness. Several compounds that are less flammable and more easily tolerated are now in use, including nitrous oxide (N_2O) and halothane ($CF_3—CHClBr$).

Application: Anesthetics

Diethyl ether was found to be a much safer anesthetic than chloroform. Like chloroform, ether is more soluble in fatty tissue than in water, so it passes into the central nervous system and takes effect quickly. Ether is also volatile, making it easy to administer. But ether is much less toxic than chloroform because ether degrades to ethanol, which the body can oxidize.

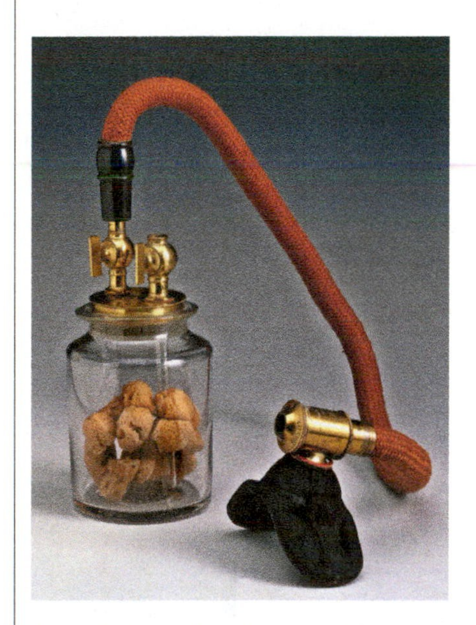

An early inhaler for ether anesthesia, about 1847. The glass bottle, containing ether-soaked sponges, is connected to a face mask with valves so that air is drawn through the bottle and over the sponges as the patient inhales.

14-2 Physical Properties of Ethers

14-2A Structure and Polarity of Ethers

Like water, ethers have a bent structure, with an sp^3 hybrid oxygen atom giving a nearly tetrahedral bond angle. In water, the nonbonding electrons compress the H—O—H bond angle to 104.5°, but in a typical ether, the bulk of the alkyl groups enlarges the bond angle. Figure 14-1 shows the structure of dimethyl ether, with a tetrahedral bond angle of 110°.

Although ethers lack the polar hydroxy group of alcohols, they are still strongly polar compounds. The dipole moment of an ether is the vector sum of two polar C—O bonds, with a substantial contribution from the two lone pairs of electrons. Table 14-1 compares the dipole moments of dimethyl ether, diethyl ether, and tetrahydrofuran (THF) with those of alkanes and alcohols of similar molecular weights. An ether such as THF provides a strongly polar solvent without the reactivity of a hydroxy group.

14-2B Boiling Points of Ethers; Hydrogen Bonding

Table 14-1 compares the boiling points of several ethers, alcohols, and alkanes. Notice that the boiling points of dimethyl ether and diethyl ether are nearly 100 °C lower than those of alcohols having similar molecular weights. This large difference results mostly from hydrogen bonding in the alcohols. Pure ethers cannot engage in hydrogen bonding because they have no O—H groups. Ethers do have large dipole moments, resulting in dipole–dipole attractions, but these attractions appear to have relatively little effect on their boiling points.

Although pure ethers have no hydroxy groups to engage in hydrogen bonding, they can hydrogen bond with other compounds that do have O—H or N—H groups. Figure 14-2 shows that a hydrogen bond requires both a hydrogen bond *donor* and a hydrogen bond *acceptor*. The donor is the molecule with an O—H or N—H group. The acceptor is the molecule whose lone pair of electrons forms a weak partial bond to the hydrogen atom provided by the donor. An ether molecule has the lone pair to form a hydrogen bond with an alcohol (or other hydrogen bond donor), but it cannot form a hydrogen bond with another ether molecule. As a result, ethers are much more volatile than alcohols having similar molecular weights. Table 14-2 lists the physical properties of a representative group of common ethers.

14-2C Ethers as Polar Solvents

Ethers are ideally suited as solvents for many organic reactions. They dissolve a wide range of polar and nonpolar substances, and their relatively low boiling points simplify

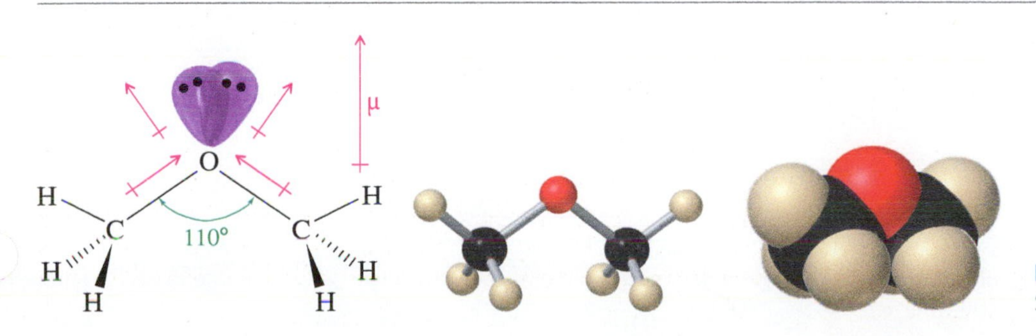

FIGURE 14-1
Structure of dimethyl ether.

TABLE 14-1
Comparison of the Boiling Points of Ethers, Alkanes, and Alcohols of Similar Molecular Weights

Compound	Formula	MW	bp (°C)	Dipole Moment (D)
water	H_2O	18	100	1.9
ethanol	$CH_3CH_2—OH$	46	78	1.7
dimethyl ether	$CH_3—O—CH_3$	46	−25	1.3
propane	$CH_3CH_2CH_3$	44	−42	0.1
n-butanol	$CH_3CH_2CH_2CH_2—OH$	74	118	1.7
tetrahydrofuran	(five-membered ring with O)	72	66	1.6
diethyl ether	$CH_3CH_2—O—CH_2CH_3$	74	35	1.2
pentane	$CH_3CH_2CH_2CH_2CH_3$	72	36	0.1

Note: The alcohols are hydrogen bonded, giving them much higher boiling points. The ethers have boiling points that are closer to those of alkanes with similar molecular weights.

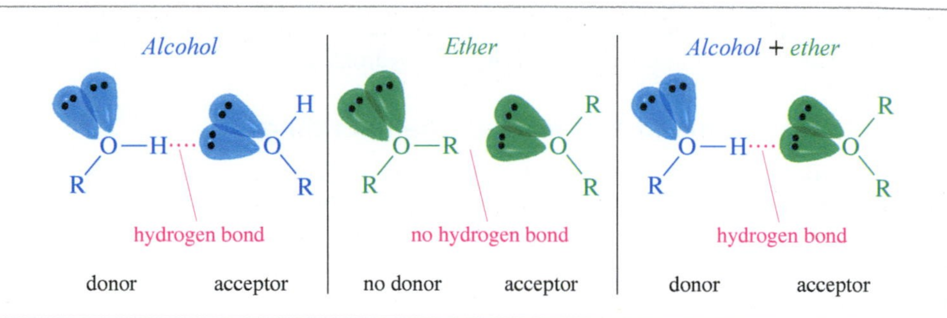

FIGURE 14-2
Hydrogen bonding in alcohols and ethers. A molecule of water or an alcohol can serve as both a hydrogen bond donor and acceptor. Ether molecules have no hydroxy groups, so they are not hydrogen bond donors. If a hydrogen bond donor is present, ethers can serve as hydrogen bond acceptors.

TABLE 14-2
Physical Properties of Ethers

Name	Structure	mp (°C)	bp (°C)	Density (g/mL)
dimethyl ether	$CH_3—O—CH_3$	−140	−25	0.66
ethyl methyl ether	$CH_3CH_2—O—CH_3$		8	0.72
diethyl ether	$CH_3CH_2—O—CH_2CH_3$	−116	35	0.71
di-*n*-propyl ether	$CH_3CH_2CH_2—O—CH_2CH_2CH_3$	−122	91	0.74
diisopropyl ether	$(CH_3)_2CH—O—CH(CH_3)_2$	−86	68	0.74
1,2-dimethoxyethane (DME)	$CH_3—O—CH_2CH_2—O—CH_3$	−58	83	0.86
methyl phenyl ether (anisole)	$CH_3—O—C_6H_5$	−37	154	0.99
diphenyl ether	$C_6H_5—O—C_6H_5$	27	259	1.07
furan	(five-membered aromatic ring with O)	− 86	32	0.94
tetrahydrofuran (THF)	(five-membered ring with O)	−108	65	0.89
1,4-dioxane	(six-membered ring with two O)	11	101	1.03

their evaporation from the reaction products. Nonpolar substances tend to be more soluble in ethers than in alcohols because ethers have no hydrogen-bonding network to be broken up by the nonpolar solute.

Polar substances tend to be nearly as soluble in ethers as in alcohols because ethers have large dipole moments as well as the ability to serve as hydrogen bond acceptors. The nonbonding electron pairs of an ether effectively solvate cations, as shown in Figure 14-3. Ethers do not solvate anions as well as alcohols do, however. Ionic substances with small, "hard" anions requiring strong solvation to overcome their ionic bonding are often insoluble in ether solvents. Substances with large, diffuse anions, such as iodides, acetates, and other organic anions, tend to be more soluble in ether solvents than substances with smaller, harder anions such as fluorides.

Alcohols cannot be used as solvents for reagents that are more strongly basic than the alkoxide ion. The hydroxy group quickly protonates the base, destroying the basic reagent.

$$\underset{\text{strong base}}{B:^-} + \underset{\text{alcohol}}{R{-}\ddot{\underset{..}{O}}H} \rightleftarrows \underset{\text{protonated base}}{B{-}H} + \underset{\text{alkoxide ion}}{R{-}\ddot{\underset{..}{O}}:^-}$$

Ethers are nonhydroxylic (no hydroxy group), and they are normally unreactive toward strong bases. For this reason, ethers are frequently used as solvents for very strong polar bases (like the Grignard reagent) that require polar solvents. The four ethers shown here are common solvents for organic reactions. DME, THF, and dioxane are miscible with water, and diethyl ether is sparingly soluble in water.

$CH_3CH_2{-}O{-}CH_2CH_3$ — diethyl ether, "ether", bp 35 °C

$CH_3{-}O{-}CH_2CH_2{-}O{-}CH_3$ — 1,2-dimethoxyethane, DME, "glyme", bp 82 °C

tetrahydrofuran, THF, oxolane, bp 65 °C

1,4-dioxane, dioxane, bp 101 °C

Application: Green Chemistry

2-Methyltetrahydrofuran is another polar aprotic solvent like THF that is considered a "green" solvent because it comes from corncobs and other waste plant sources rather than petroleum.

2-methyltetrahydrofuran

PROBLEM 14-1

Rank the given solvents in decreasing order of their ability to dissolve each compound.

Solutes

(a) NaOAc
(b) naphthalene
(c) 2-naphthol

Solvents
ethyl ether
water
ethanol
dichloromethane

14-2D Stable Complexes of Ethers with Reagents

The special properties of ethers (polarity, lone pairs, but relatively unreactive) enhance the formation and use of many reagents. For example, Grignard reagents cannot form unless an ether is present, possibly to share its lone pairs of electrons

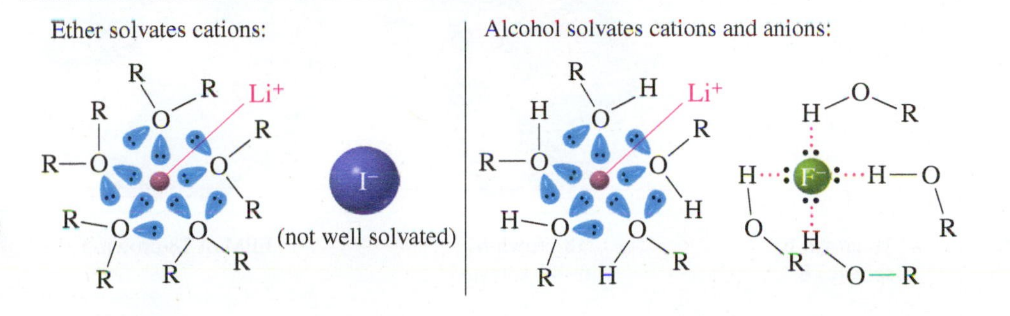

FIGURE 14-3
Ethers solvate cations. An ionic substance such as lithium iodide (LiI) is moderately soluble in ethers because the small lithium cation is strongly solvated by the ether's lone pairs of electrons. Unlike alcohols, ethers cannot serve as hydrogen bond donors, so they do not solvate anions well.

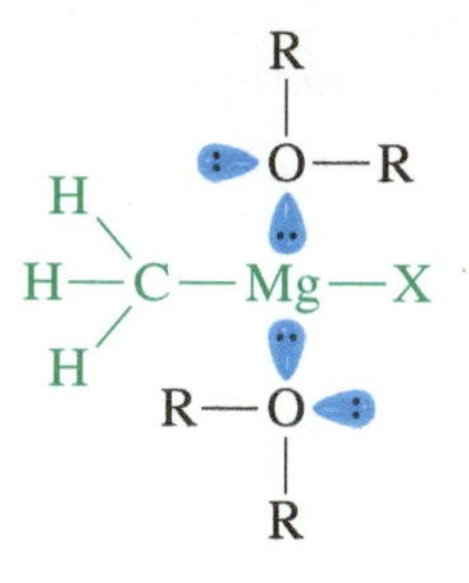

FIGURE 14-4
Complexation of an ether with a Grignard reagent stabilizes the reagent and helps keep it in solution.

with the magnesium atom. This sharing of electrons stabilizes the reagent and helps keep it in solution (Figure 14-4).

Complexes with Electrophiles An ether's nonbonding electrons also stabilize borane, BH_3. Pure borane exists as a dimer called diborane, B_2H_6. Diborane is a toxic, flammable, and explosive gas, whose use is both dangerous and inconvenient. Borane forms a stable complex with tetrahydrofuran. The $BH_3 \cdot$ THF complex is commercially available as a 1 M solution, easily measured and transferred like any other air-sensitive liquid reagent. The availability of $BH_3 \cdot$ THF has contributed greatly to the convenience of hydroboration (Section 8-7).

$$B_2H_6 \;(\text{diborane}) + 2\;\text{tetrahydrofuran} \rightleftharpoons 2\;BH_3 \cdot \text{THF}$$

Boron trifluoride is used as a Lewis acid catalyst in a wide variety of reactions. Like diborane, BF_3 is a toxic gas, but BF_3 forms a stable complex with ethers, allowing it to be conveniently stored, measured, and even distilled without breaking the complex apart. The complex of BF_3 with diethyl ether is called "boron trifluoride etherate."

$$BF_3 \;(\text{boron trifluoride}) + :O(CH_2CH_3)_2 \;(\text{diethyl ether}) \rightleftharpoons F_3B^{-}\text{—}O^{+}(CH_2CH_3)_2 \;\; BF_3 \cdot OEt_2 \;\text{"boron trifluoride etherate"}$$

PROBLEM 14-2

Aluminum trichloride ($AlCl_3$) dissolves in ether with the evolution of a large amount of heat. (In fact, this reaction can become rather violent if it gets too warm.) Show the structure of the resulting aluminum chloride etherate complex.

Application: Radioactive Cleanup

Crown ethers are also used to remove radioactive elements from radioactive waste. For example, radioactive cesium and strontium can be extracted using specialized derivatives of 18-crown-6.

Crown Ether Complexes In Chapter 6, we encountered the use of **crown ethers,** large cyclic polyethers that specifically solvate metal cations by complexing the metal in the center of the ring. Different crown ethers solvate different cations, depending on the relative sizes of the crown ether and the cation and the number of binding sites around the cation. The EPM of 18-crown-6 shows that the cavity in the center of the molecule is surrounded by electron-rich oxygen atoms that complex with the guest potassium cation.

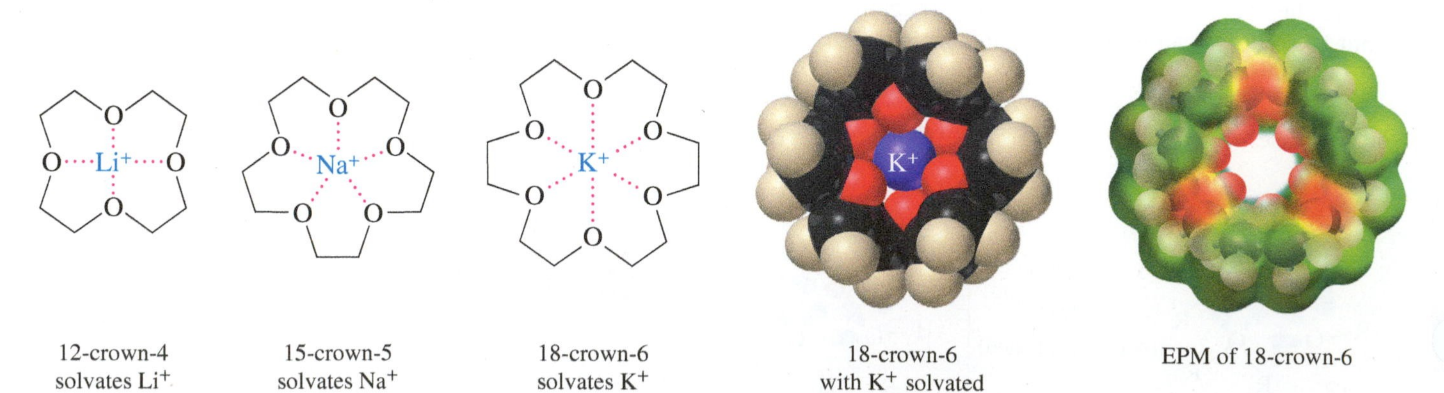

12-crown-4 solvates Li^+ | 15-crown-5 solvates Na^+ | 18-crown-6 solvates K^+ | 18-crown-6 with K^+ solvated | EPM of 18-crown-6

Complexation by crown ethers often helps polar inorganic salts to dissolve in nonpolar organic solvents. This enhanced solubility allows polar salts to be used under aprotic conditions, where the uncomplexed anions may show greatly enhanced reactivity. For example, in Section 6-10B, we discussed using 18-crown-6 to dissolve potassium fluoride in acetonitrile (CH_3CN), where the poorly solvated fluoride ion is a moderately strong nucleophile. Many other salts, including carboxylate salts ($RCOO^- K^+$), cyanides (KCN), and permanganates ($KMnO_4$), can be dissolved in aprotic (and often nonpolar) organic solvents using crown ethers. In each case, the crown ether complexes only the cation, leaving the anion bare and highly reactive.

PROBLEM 14-3

In the presence of 18-crown-6, potassium permanganate dissolves in benzene to give "purple benzene," a useful reagent for oxidizing alkenes in an aprotic environment. Use a drawing of the complex to show why $KMnO_4$ dissolves in benzene and why the reactivity of the permanganate ion is enhanced.

14-3 Nomenclature of Ethers

We have been using the common nomenclature of ethers, which is sometimes called the *alkyl alkyl ether* system. The IUPAC system, generally used with more complicated ethers, is sometimes called the *alkoxy alkane* system. Common names are almost always used for simple ethers.

14-3A Common Names (Alkyl Alkyl Ether Names)

Common names of ethers are formed by naming the two alkyl groups on oxygen and adding the word *ether*. Under the current system, the alkyl groups should be named in alphabetical order, but many people still use the old system, which named the groups in order of increasing complexity. For example, if one of the alkyl groups is methyl and the other is *tert*-butyl, the current common name should be "*tert*-butyl methyl ether," but most chemists use the older common name, "methyl *tert*-butyl ether" (or MTBE). If both groups are methyl, the name is "dimethyl ether." If just one alkyl group is described in the name, it implies that the ether is symmetrical, as in "ethyl ether."

Application: Oxygenated Gasoline

The Clean Air Act of 1990 requires the use of "oxygenated gasoline" in areas with severe air pollution. The preferred "oxygenate" was often MTBE because it blends well with gasoline, lowers the amounts of pollutants in the exhaust, burns well without engine modifications, and has a low toxicity. In 1999, California began a phaseout of MTBE from the gasoline in that state because of concerns that it was polluting groundwater. Since that time, many other states have banned the use of MTBE in gasoline.

14-3B IUPAC Names (Alkoxy Alkane Names)

IUPAC names use the more complex alkyl group as the root name, and the rest of the ether as an **alkoxy group.** For example, cyclohexyl methyl ether is named "methoxycyclohexane." This systematic nomenclature is often the only clear way to name complex ethers.

	$CH_3—O—CH_2CH_3$	(benzene ring)–OCH_3	$Cl—CH_2—O—CH_3$
IUPAC name:	methoxyethane	methoxybenzene	chloromethoxymethane
common name:	ethyl methyl ether	methyl phenyl ether, or anisole	chloromethyl methyl ether

	(cyclohexane: H_3C, CH_3, H, OCH_2CH_3)	(cyclobutane: Cl, H, OCH_3, H)	$CH_2—OH$ / $CH_2—O—CH_2CH_3$
IUPAC name:	3-ethoxy-1,1-dimethylcyclohexane	*trans*-1-chloro-2-methoxycyclobutane	2-ethoxyethanol

PROBLEM 14-4

Give a common name (when possible) and a systematic name for each compound.

(a) $CH_3OCH{=}CH_2$ **(b)** $CH_3CH_2OCH(CH_3)_2$ **(c)** $ClCH_2CH_2OCH_3$

(d) 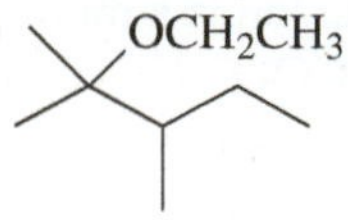 OCH_2CH_3 **(e)** 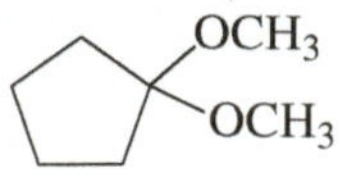 OCH_3, OCH_3 **(f)** 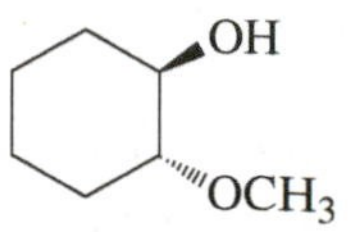 OH, OCH_3

(g) 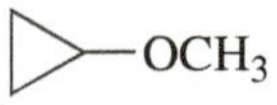 OCH_3 **(h)** $CH_3C{\equiv}CCH_2OCH_3$ **(i)** 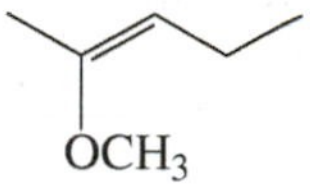 OCH_3

14-3C Nomenclature of Cyclic Ethers

Cyclic ethers are our first examples of **heterocyclic compounds,** containing a ring in which a ring atom is an element other than carbon. This atom, called the **heteroatom,** is numbered 1 when numbering the ring atoms. Heterocyclic ethers are especially important and useful ethers.

Application: Fumigant

Ethylene oxide has been used as a fumigant for foods, textiles, and soil, and for sterilizing biomedical instruments. It readily diffuses through materials without damaging them. Its antibacterial effect is probably due to its ability to alkylate critical cellular enzymes.

Epoxides (Oxiranes) We have already encountered some of the chemistry of epoxides in Section 8-12. **Epoxides** are three-membered cyclic ethers, usually formed by peroxyacid oxidation of the corresponding alkenes. The common name of an epoxide is formed by adding "oxide" to the name of the alkene that is oxidized. The following reactions show the synthesis and common names of two simple epoxides.

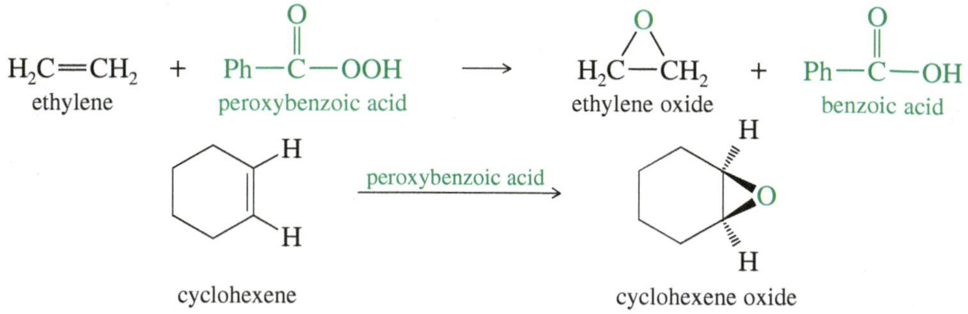

$$H_2C{=}CH_2 + Ph{-}\overset{O}{\overset{\|}{C}}{-}OOH \longrightarrow H_2C{-}CH_2\,(O) + Ph{-}\overset{O}{\overset{\|}{C}}{-}OH$$

ethylene + peroxybenzoic acid → ethylene oxide + benzoic acid

cyclohexene —peroxybenzoic acid→ cyclohexene oxide

One systematic method for naming epoxides is to name the rest of the molecule and use the term *epoxy* as a substituent, giving the numbers of the two carbon atoms bonded to the epoxide oxygen.

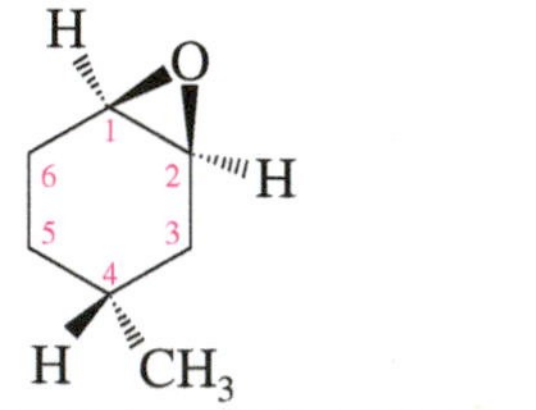

trans-1,2-epoxy-4-methylcyclohexane

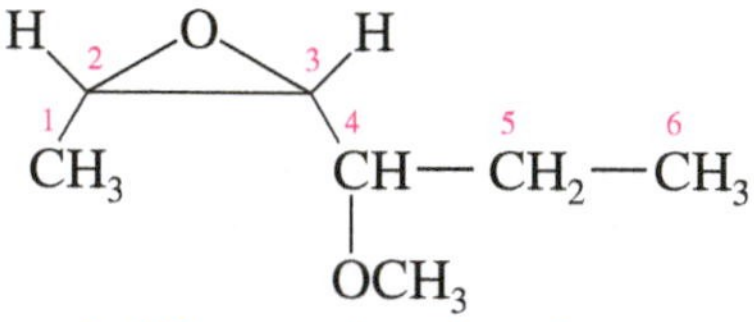

cis-2,3-epoxy-4-methoxyhexane

Another systematic method names epoxides as derivatives of the parent compound, ethylene oxide, using "oxirane" as the systematic name for ethylene oxide. In this system, the ring atoms of a heterocyclic compound are numbered starting with the heteroatom and going in the direction to give the lowest substituent numbers. The "epoxy" system names are also listed (in blue) for comparison. Note that the numbering is different for the "epoxy" system names, which number the longest chain rather than the ring.

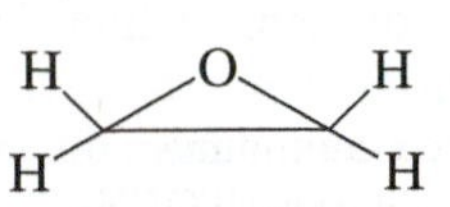

oxirane
1,2-epoxyethane

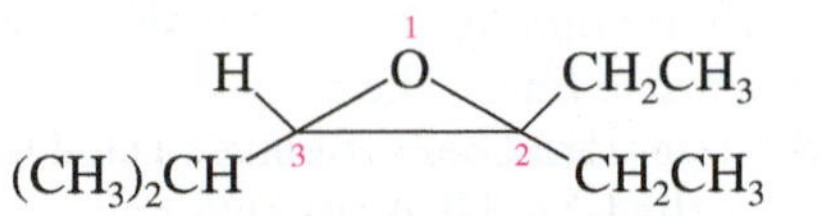

2,2-diethyl-3-isopropyloxirane
3,4-epoxy-4-ethyl-2-methylhexane

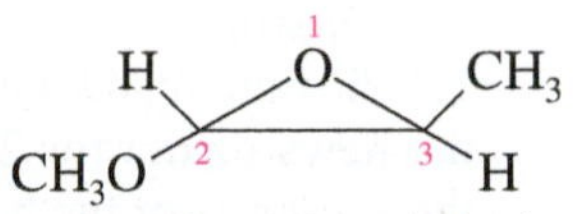

trans-2-methoxy-3-methyloxirane
1,2-epoxy-1-methoxypropane

Oxetanes The least common cyclic ethers are the four-membered **oxetanes.** Because these four-membered rings are strained, they are more reactive than larger cyclic ethers and open-chain ethers. They are not as reactive as the highly strained oxiranes (epoxides), however.

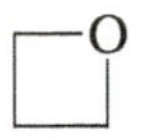

oxetane

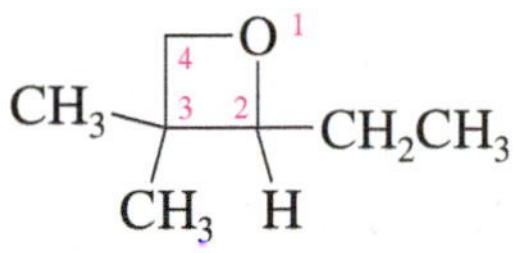

2-ethyl-3,3-dimethyloxetane

Furans (Oxolanes) The five-membered cyclic ethers are commonly named after an aromatic member of this group, **furan.** We consider the aromaticity of furan and other heterocycles in Chapter 16. The systematic term **oxolane** is also used for a five-membered ring containing an oxygen atom.

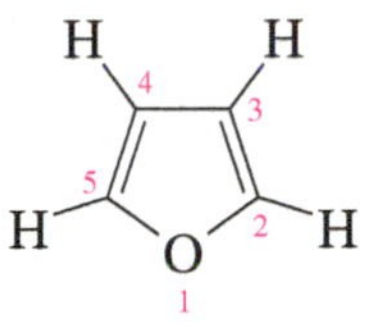

furan

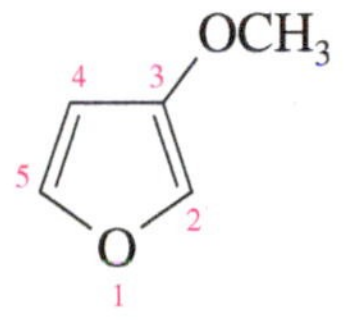

3-methoxyfuran

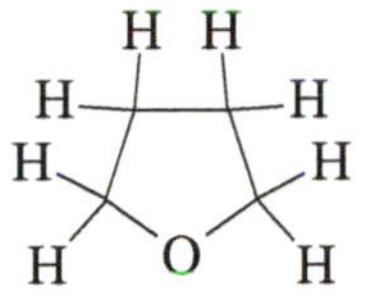

tetrahydrofuran (THF)
(oxolane)

The saturated five-membered cyclic ether resembles furan but has four additional hydrogen atoms. Therefore, it is called *tetrahydrofuran* (THF). One of the most polar ethers, tetrahydrofuran is an excellent nonhydroxylic organic solvent for polar reagents. Grignard reactions sometimes succeed in THF even when they fail in diethyl ether.

Pyrans (Oxanes) The six-membered cyclic ethers are commonly named as derivatives of **pyran,** an unsaturated ether. The saturated compound has four more hydrogen atoms, so it is called *tetrahydropyran* (THP). The systematic term **oxane** is also used for a six-membered ring containing an oxygen atom.

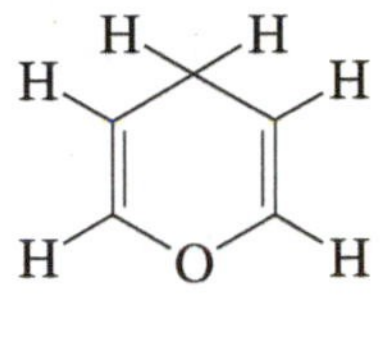

pyran

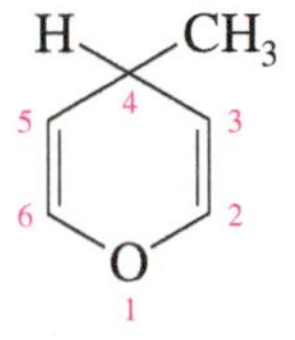

4-methylpyran

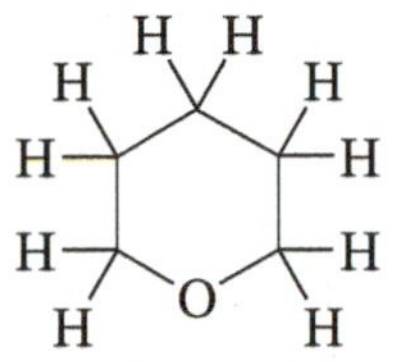

tetrahydropyran (THP)
(oxane)

Dioxanes Heterocyclic ethers with two oxygen atoms in a six-membered ring are called **dioxanes.** The most common form of dioxane is the one with the two oxygen atoms in a 1,4-relationship. 1,4-Dioxane is miscible with water, and it is widely used as a polar solvent for organic reactions.

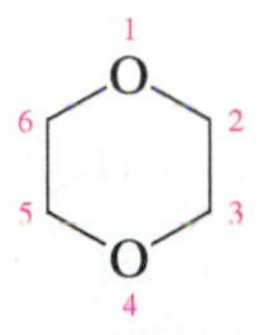

1,4-dioxane

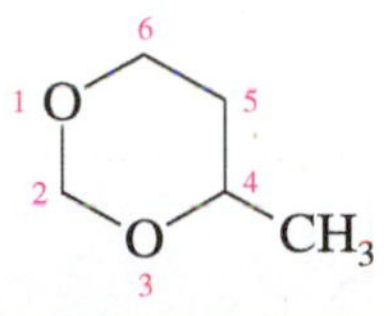

4-methyl-1,3-dioxane

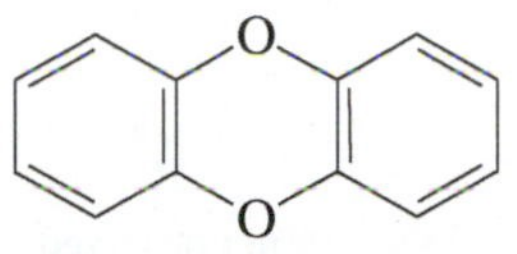

dibenzo-1,4-dioxane (dioxin)

Dioxin (previous page) is a common name for dibenzo-1,4-dioxane, which is 1,4-dioxane fused with two benzene rings. The name *dioxin* is often used incorrectly in the news media for 2,3,7,8-tetrachlorodibenzodioxin (TCDD), a toxic contaminant in the synthesis of the herbicide called 2,4,5-T or Agent Orange. Surprisingly, TCDD has been in the environment for many millions of years because it is also formed in forest fires. Most dioxins are toxic and carcinogenic (cause cancer) because they associate with DNA and cause a misreading of the genetic code that can result in a genetic mutation.

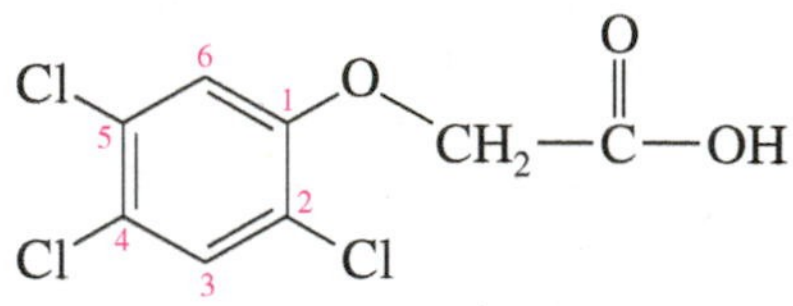

2,4,5-trichlorophenoxyacetic acid (2,4,5-T or Agent Orange)

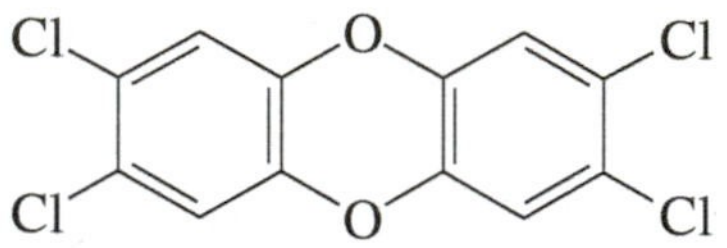

2,3,7,8-tetrachlorodibenzodioxin (TCDD, incorrectly "dioxin")

PROBLEM 14-5

1,4-Dioxane is made commercially by the acid-catalyzed condensation of an alcohol.
(a) Show what alcohol will undergo condensation, with loss of water, to give 1,4-dioxane.
(b) Propose a mechanism for this reaction.

PROBLEM 14-6

Name the following heterocyclic ethers.

(a) (b) 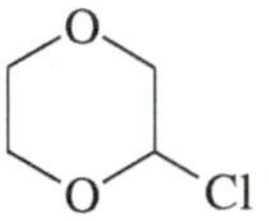 (c) $CH(CH_3)_2$

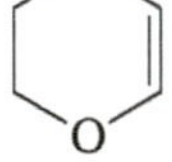

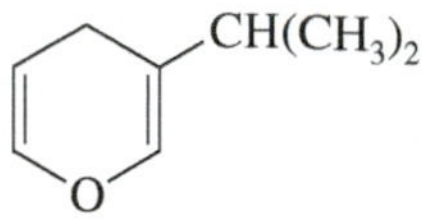

(d) H, CH_2CH_3, CH_3CH_2, H (e) Br, OCH_2CH_3 (f) CH_3, Br, CH_3

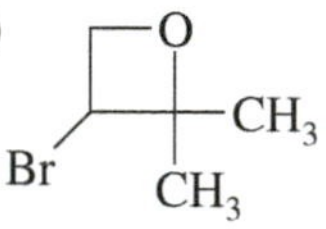

14-4 Spectroscopy of Ethers

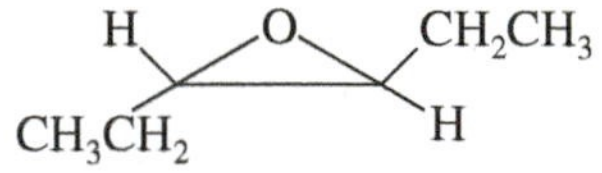

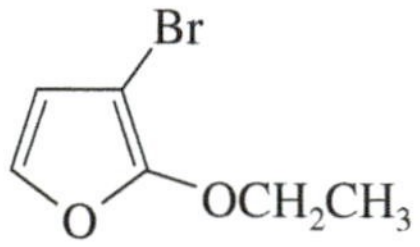

Infrared Spectroscopy of Ethers Infrared spectra do not show obvious or reliable absorptions for ethers. Most ethers give a moderate to strong C—O stretch around 1000 to 1200 cm^{-1} (in the fingerprint region), but many compounds other than ethers give similar absorptions. Nevertheless, the IR spectrum can be useful because it shows the *absence* of carbonyl (C═O) groups and hydroxy (O—H) groups. If the molecular formula contains an oxygen atom, the lack of carbonyl or hydroxy absorptions in the IR suggests an ether.

Mass Spectrometry of Ethers The most common fragmentation of ethers is cleavage next to one of the carbon atoms bonded to oxygen. Because this carbon is *alpha* to the oxygen atom, this fragmentation is called **α cleavage.** The resulting *oxonium ion* (oxygen with three bonds and a positive charge) is resonance-stabilized by the nonbonding electrons on oxygen.

α Cleavage

$$[R-CH_2-O-R']^{+\cdot} \longrightarrow R\cdot + \left[H_2C^{+}-\ddot{O}-R' \longleftrightarrow H_2C=\ddot{O}^{+}-R' \right]$$

R· not observed

oxonium ion

Another common cleavage is the loss of either of the two alkyl groups to give another oxonium ion or an alkyl cation.

Loss of an alkyl group

$$[\mathrm{R{-}CH_2{-}O}\,|\,\mathrm{R'}]^{+\cdot} \xrightarrow{\sim\mathrm{H}} \left[\mathrm{R{-}\overset{+}{C}H{-}\ddot{O}{-}H} \longleftrightarrow \mathrm{R{-}CH{=}\overset{+}{\ddot{O}}{-}H}\right] + \cdot\mathrm{R'}$$

oxonium ion; $\cdot R'$ not observed

or

$$[\mathrm{R{-}CH_2{-}O}\,|\,\mathrm{R'}]^{+\cdot} \longrightarrow \mathrm{R{-}CH_2{-}O\cdot} + {}^{+}\mathrm{R'}$$

$R{-}CH_2{-}O\cdot$ not observed; $^{+}R'$ alkyl cation

The mass spectrum of diethyl ether appears in Figure 14-5. The four most abundant ions correspond to the molecular ion, loss of an ethyl group, α cleavage, and loss of an ethylene molecule combined with α cleavage. All these modes of cleavage form resonance-stabilized oxonium ions.

NMR Spectroscopy of Ethers In the ^{13}C NMR spectrum, a carbon atom bonded to oxygen generally absorbs between δ 65 and δ 90. Protons on carbon atoms bonded to oxygen usually absorb at chemical shifts between δ 3.5 and δ 4 in the ^{1}H NMR spectrum. Both alcohols and ethers have resonances in this range. See, for example, the NMR spectra of methyl *tert*-butyl ether (page 620) and ethanol (page 641). If a compound containing C, H, and O has resonances in the correct range, and if there is no O—H stretch or C=O stretch in the IR spectrum, an ether is the most likely functional group.

$H{-}\underset{|}{C}(H){-}O{-}$: ^{13}C $\delta 65{-}\delta 90$; ^{1}H $\delta 3.5{-}\delta 4$

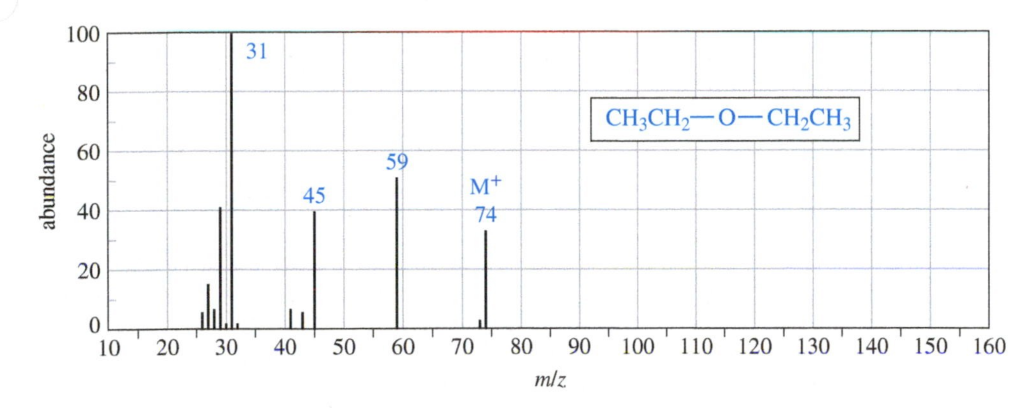

Loss of an ethyl group

$$[\mathrm{CH_3{-}CH_2}\,|^{45}\,\mathrm{O{-}CH_2{-}CH_3}]^{+\cdot} \longrightarrow \mathrm{H{-}\overset{+}{O}{=}CH{-}CH_3} + \cdot\mathrm{CH_2CH_3}$$

m/z 74 → *m/z* 45 + loss of 29

α Cleavage

$$[\mathrm{CH_3{-}CH_2{-}O{-}CH_2}\,|^{59}\,\mathrm{CH_3}]^{+\cdot} \longrightarrow \mathrm{CH_3{-}CH_2{-}\overset{+}{O}{=}CH_2} + \cdot\mathrm{CH_3}$$

m/z 74 → *m/z* 59 + loss of 15

α Cleavage combined with loss of an ethylene molecule

$$\mathrm{CH_3{-}CH_2{-}\overset{+}{O}{=}CH_2} \longrightarrow \mathrm{H{-}\overset{+}{O}{=}CH_2} + \mathrm{CH_2{=}CH_2}$$

m/z 59 → *m/z* 31 + loss of 28

FIGURE 14-5
The mass spectrum of diethyl ether shows major peaks for the molecular ion, loss of an ethyl group, α cleavage, and α cleavage combined with loss of a molecule of ethylene.

PROBLEM 14-7

Propose a fragmentation to account for each numbered peak in the mass spectrum of *n*-butyl isopropyl ether.

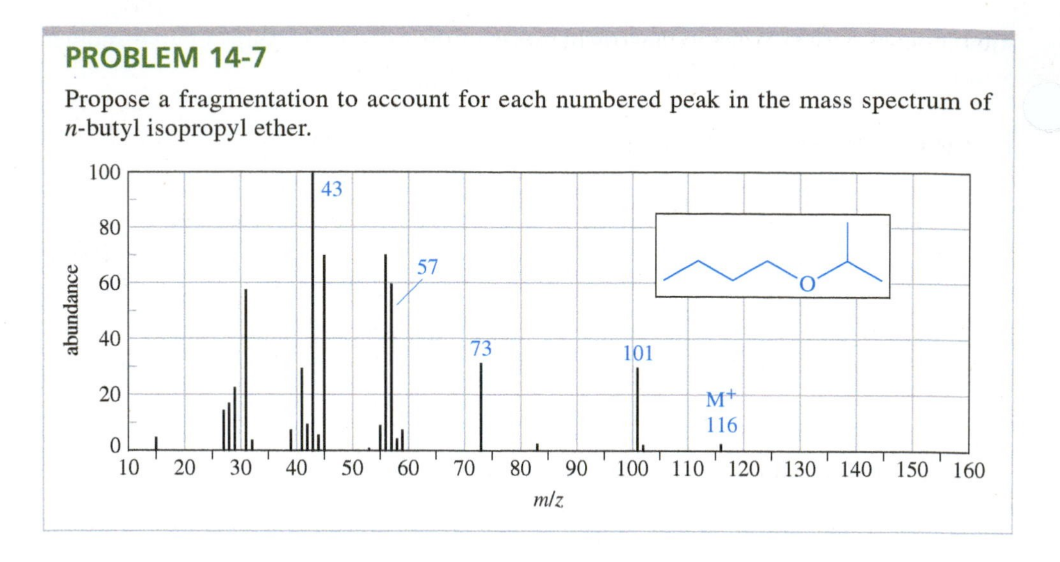

14-5 The Williamson Ether Synthesis

We have already seen most of the common methods for synthesizing ethers. We review them at this time, looking more closely at the mechanisms to see which methods are most suitable for preparing various kinds of ethers. The **Williamson ether synthesis** (Section 11-14) is the most reliable and versatile ether synthesis. This method involves the S_N2 attack of an alkoxide ion on an unhindered primary alkyl halide or tosylate. Secondary alkyl halides and tosylates are occasionally used in the Williamson synthesis, but elimination competes, and the yields are often poor.

$$\mathrm{R{-}\ddot{\underset{..}{O}}{:}^-}\quad \mathrm{R'{-}\ddot{\underset{..}{X}}{:}} \longrightarrow \mathrm{R{-}\ddot{\underset{..}{O}}{-}R'} \quad \mathrm{{:}\ddot{\underset{..}{X}}{:}^-}$$

alkoxide   alkyl halide   ether   halide

The alkoxide is commonly made by adding Na, K, or NaH to the alcohol (Section 11-14).

Examples

cyclohexanol $\xrightarrow[(2)\ CH_3CH_2OTs]{(1)\ Na}$ ethoxycyclohexane (92%)

3,3-dimethylpentan-2-ol $\xrightarrow[(2)\ CH_3I]{(1)\ NaH}$ 2-methoxy-3,3-dimethylpentane (90%)

SOLVED PROBLEM 14-1

(a) Why is the following reaction a poor method for the synthesis of *tert*-butyl propyl ether?

(b) What would be the major product from this reaction?

(c) Propose a better synthesis of *tert*-butyl propyl ether.

$$CH_3CH_2CH_2{-}\ddot{\underset{..}{O}}{:}^-\ Na^+ + CH_3{-}C(CH_3)_2{-}Br \xrightarrow[\text{give}]{\text{does } not} \times\ CH_3{-}C(CH_3)_2{-}O{-}CH_2CH_2CH_3$$

sodium propoxide — *tert*-butyl bromide — *tert*-butyl propyl ether

SOLUTION

(a) The desired S_N2 reaction cannot occur on the tertiary alkyl halide.

(b) The alkoxide ion is a strong base as well as a nucleophile, and elimination prevails.

$$CH_3CH_2CH_2{-}\ddot{\underset{..}{O}}{:}^-\ Na^+ + H{-}CH_2{-}C(CH_3)(Br){-}CH_3 \xrightarrow{E2} H_2C{=}C(CH_3)_2 + CH_3CH_2CH_2OH + NaBr$$

sodium propoxide — *tert*-butyl bromide — isobutylene

(c) A better synthesis would use the less hindered alkyl group as the S_N2 substrate and the alkoxide of the more hindered alkyl group.

$$CH_3{-}C(CH_3)_2{-}\ddot{\underset{..}{O}}{:}^-\ Na^+ + CH_3CH_2{-}CH_2{-}Br \xrightarrow{S_N2} CH_3{-}C(CH_3)_2{-}O{-}CH_2CH_2CH_3$$

sodium *tert*-butoxide — 1-bromopropane — *tert*-butyl propyl ether

PROBLEM 14-8

Propose a Williamson synthesis of 3-butoxy-1,1-dimethylcyclohexane from 3,3-dimethylcyclohexanol and butan-1-ol.

Synthesis of Phenyl Ethers A phenol (aromatic alcohol) can be used as the alkoxide fragment, but not the halide fragment, for the Williamson ether synthesis. Phenols are more acidic than aliphatic alcohols (Section 10-6), and sodium hydroxide is sufficiently basic to form the phenoxide ion. As with other alkoxides, the electrophile should have an unhindered primary alkyl group and a good leaving group.

$$\text{2-nitrophenol} \xrightarrow[(2)\ CH_3CH_2CH_2CH_2{-}I]{(1)\ NaOH} \text{2-butoxynitrobenzene } (80\%)$$

2-nitrophenol (OH, NO_2) — 2-butoxynitrobenzene ($O{-}CH_2CH_2CH_2CH_3$, NO_2) (80%)

PROBLEM 14-9

Show how you would use the Williamson ether synthesis to prepare the following ethers. You may use any alcohols or phenols as your organic starting materials.

(a) cyclohexyl propyl ether
(b) isopropyl methyl ether
(c) 1-methoxy-4-nitrobenzene
(d) ethyl *n*-propyl ether (two ways)
(e) benzyl *tert*-butyl ether (benzyl = $Ph{-}CH_2{-}$)

PROBLEM-SOLVING HINT

To convert two alcohols to an ether, convert the more hindered alcohol to its alkoxide. Convert the less hindered alcohol to its tosylate (or an alkyl halide). Make sure the tosylate (or halide) is a good S_N2 substrate.

14-6 Synthesis of Ethers by Alkoxymercuration–Demercuration

The **alkoxymercuration–demercuration** process adds a molecule of an alcohol across the double bond of an alkene (Section 8-6). The product is an ether, as shown here.

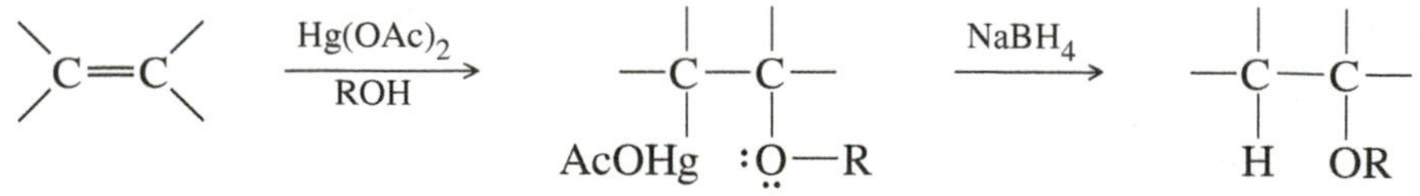

$$\mathrm{>C{=}C<} \xrightarrow[\mathrm{ROH}]{\mathrm{Hg(OAc)_2}} \underset{\text{mercurial ether}}{\mathrm{-\overset{|}{\underset{AcOHg}{\underset{|}{C}}}-\overset{|}{\underset{:\ddot{O}-R}{\underset{|}{C}}}-}} \xrightarrow{\mathrm{NaBH_4}} \mathrm{-\overset{|}{\underset{H}{\underset{|}{C}}}-\overset{|}{\underset{OR}{\underset{|}{C}}}-}$$

Example

$$\underset{\text{hex-1-ene}}{CH_3(CH_2)_3-CH{=}CH_2} \xrightarrow[\text{(2) } NaBH_4]{\text{(1) } Hg(OAc)_2,\ CH_3OH} \underset{\substack{\text{2-methoxyhexane, 80\%} \\ \text{(Markovnikov product)}}}{CH_3(CH_2)_3-\overset{\overset{OCH_3}{|}}{C}H-CH_3}$$

PROBLEM-SOLVING HINT

Alkoxymercuration adds the —OR group of the alcohol to the more substituted carbon atom of the C=C double bond.

PROBLEM 14-10

Show how the following ethers might be synthesized using (1) alkoxymercuration–demercuration and (2) the Williamson synthesis. (When one of these methods cannot be used for the given ether, point out why it will not work.)

(a) 2-methoxybutane
(b) ethyl cyclohexyl ether
(c) 1-methoxy-2-methylcyclopentane
(d) 1-methoxy-1-methylcyclopentane
(e) 1-isopropoxy-1-methylcyclopentane
(f) *tert*-butyl phenyl ether

14-7 Industrial Synthesis: Bimolecular Condensation of Alcohols

The least expensive method for synthesizing simple symmetrical ethers is the acid-catalyzed bimolecular **condensation** (joining of two molecules, often with loss of a small molecule such as water), discussed in Section 11-10B. Unimolecular dehydration (to give an alkene) competes with bimolecular condensation. To form an ether, the alcohol must have an unhindered primary alkyl group, and the temperature must not be allowed to rise too high. If the alcohol is hindered or the temperature is too high, the delicate balance between substitution and elimination shifts in favor of elimination, and very little ether is formed. Bimolecular condensation is used in industry to make symmetrical ethers from primary alcohols. Because the condensation is so limited in its scope, it finds little use in the laboratory synthesis of ethers.

Bimolecular condensation

$$2\ R-OH \overset{H^+}{\rightleftarrows} R-O-R + H_2O$$

Examples

$$\underset{\text{methyl alcohol}}{2\ CH_3OH} \xrightarrow{H_2SO_4,\ 140\ °C} \underset{\substack{\text{dimethyl ether} \\ (100\%)}}{CH_3-O-CH_3} + H_2O$$

$$\underset{\text{ethyl alcohol}}{2\ CH_3CH_2OH} \xrightarrow{H_2SO_4,\ 140\ °C} \underset{\substack{\text{diethyl ether} \\ (88\%)}}{CH_3CH_2-O-CH_2CH_3} + H_2O$$

$$2\ CH_3CH_2CH_2OH \xrightarrow{H_2SO_4,\ 140\ ^\circ C} CH_3CH_2CH_2{-}O{-}CH_2CH_2CH_3 + H_2O$$

n-propyl alcohol → *n*-propyl ether (75%)

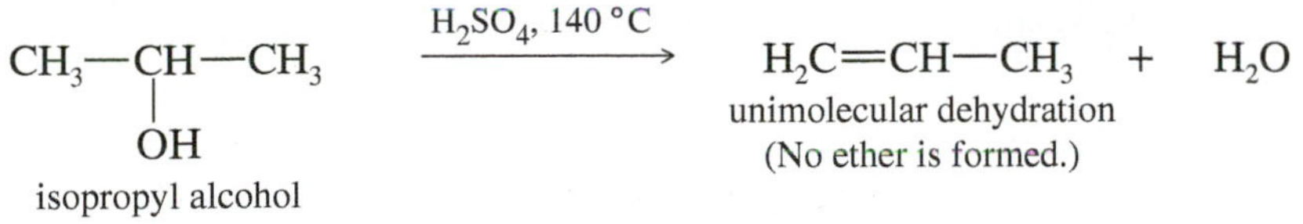

$$CH_3{-}\underset{\underset{OH}{|}}{CH}{-}CH_3 \xrightarrow{H_2SO_4,\ 140\ ^\circ C} H_2C{=}CH{-}CH_3 + H_2O$$

isopropyl alcohol → unimolecular dehydration (No ether is formed.)

If the conditions are carefully controlled, bimolecular condensation is a cheap synthesis of diethyl ether. In fact, this is the industrial method used to produce millions of gallons of diethyl ether each year.

PROBLEM 14-11

Explain why bimolecular condensation is a poor method for making unsymmetrical ethers such as ethyl methyl ether.

PROBLEM-SOLVING HINT

Bimolecular condensation of alcohols is generally a poor synthetic method.

PROBLEM 14-12

Propose a mechanism for the acid-catalyzed condensation of *n*-propyl alcohol to *n*-propyl ether, as shown above. When the temperature is allowed to rise too high, propene is formed. Propose a mechanism for the formation of propene, and explain why it is favored at higher temperatures.

PROBLEM 14-13

Which of the following ethers can be formed in good yield by condensation of the corresponding alcohols? For those that cannot be formed by condensation, suggest an alternative method that will work.

(a) dibutyl ether **(b)** ethyl *n*-propyl ether **(c)** di-*sec*-butyl ether

SUMMARY Syntheses of Ethers (Review)

1. *The Williamson ether synthesis* (Sections 11-14 and 14-5)

$$R{-}\ddot{\underset{..}{O}}{:}^- + R'{-}X \longrightarrow R{-}\ddot{\underset{..}{O}}{-}R' + X^-$$

$X = Cl, Br, I, OTs$, etc. — R' must be primary.

2. *Addition of an alcohol across a double bond: alkoxymercuration–demercuration* (Sections 8-6 and 14-6)

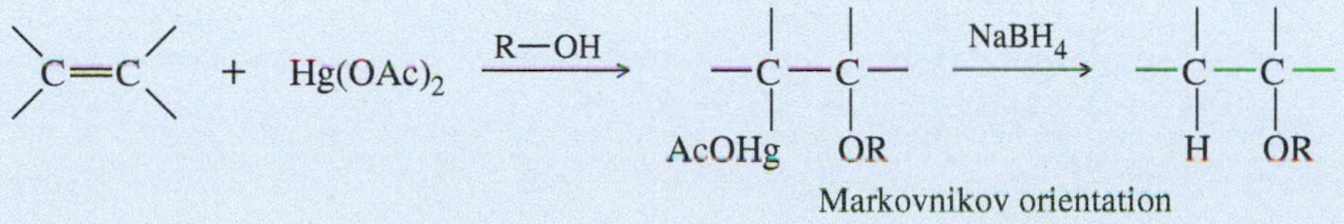

$$\gt C{=}C\lt + Hg(OAc)_2 \xrightarrow{R-OH} -\underset{\underset{AcOHg}{|}}{\overset{|}{C}}-\underset{\underset{OR}{|}}{\overset{|}{C}}- \xrightarrow{NaBH_4} -\underset{\underset{H}{|}}{\overset{|}{C}}-\underset{\underset{OR}{|}}{\overset{|}{C}}-$$

Markovnikov orientation

3. *Bimolecular condensation of alcohols: industrial synthesis* (Sections 11-10B and 14-7)

$$2\ R{-}OH \overset{H^+}{\rightleftharpoons} R{-}O{-}R + H_2O$$

R must be primary.

14-8 Cleavage of Ethers by HBr and HI

Unlike alcohols, ethers are not commonly used as synthetic intermediates because they do not undergo many reactions. This unreactivity makes ethers attractive as solvents. Even so, ethers do undergo a limited number of characteristic reactions.

Ethers are cleaved by heating with HBr or HI to give alkyl bromides or alkyl iodides.

$$\underset{\text{ether}}{R{-}O{-}R'} \xrightarrow[(X = \text{Br or I})]{\text{excess HX}} \underset{\text{alkyl halide}}{R{-}X} + \underset{\text{alkyl halide}}{R'{-}X}$$

Ethers are unreactive toward most bases, but they can react under acidic conditions. A protonated ether can undergo substitution or elimination with an alcohol serving as a neutral leaving group. Ethers react with concentrated HBr and HI because these reagents are sufficiently acidic to protonate the ether, whereas bromide and iodide are good nucleophiles for the substitution. Under these conditions, the alcohol leaving group usually reacts further with HX to give another alkyl halide.

$$\underset{\text{ether}}{R{-}\ddot{O}{-}R'} + \underset{(X = \text{Br or I})}{H^+ \; X^-} \rightleftharpoons \underset{\text{protonated ether}}{R{-}\overset{H}{\overset{|}{O^+}}{-}R'} \longrightarrow \underset{\text{alkyl halide}}{X{-}R} + \underset{\text{alcohol}}{:\overset{H}{\overset{|}{\ddot{O}}}{-}R'} \xrightarrow{HX} X{-}R + X{-}R'$$

In effect, this reaction converts a dialkyl ether into two alkyl halides. The conditions are very strong, however, and the molecule must not contain any acid-sensitive functional groups.

Iodide and bromide ions are good nucleophiles but weak bases, so they are more likely to substitute by the S_N2 mechanism than to promote elimination by the E2 mechanism. Mechanism 14-1 shows how bromide ion cleaves the protonated ether by displacing an alcohol. In the following example, cyclopentyl ethyl ether reacts with HBr to produce cyclopentanol by this displacement. Cyclopentanol reacts further with HBr, though, so the final products are ethyl bromide and bromocyclopentane.

MECHANISM 14-1 Cleavage of an Ether by HBr or HI

Ethers are cleaved by a nucleophilic substitution of Br^- or I^- on the protonated ether.

Step 1: Protonation of the ether to form a good leaving group.

$$R{-}\ddot{O}{-}R' + H{-}\ddot{Br}: \rightleftharpoons R{-}\overset{+}{O}(H){-}R' + :\ddot{Br}:^-$$

Step 2: S_N2 cleavage of the protonated ether.

$$:\ddot{Br}:^- + R{-}\overset{+}{O}(H){-}R' \longrightarrow :\ddot{Br}{-}R + :\ddot{O}(H){-}R'$$

Step 3: Conversion of the alcohol fragment to the alkyl halide. (Does not occur with phenols.)

$$R'{-}O{-}H \xrightarrow{HBr} R'{-}Br + H_2O$$

This conversion can occur by either of the two mechanisms shown in Section 11-7, depending on the structure of the alcohol and the reaction conditions. The protonated alcohol undergoes either S_N1 or S_N2 substitution by bromide ion.

EXAMPLE: Cleavage of cyclopentyl ethyl ether by HBr.

Step 1: Protonation of the ether to form a good leaving group.

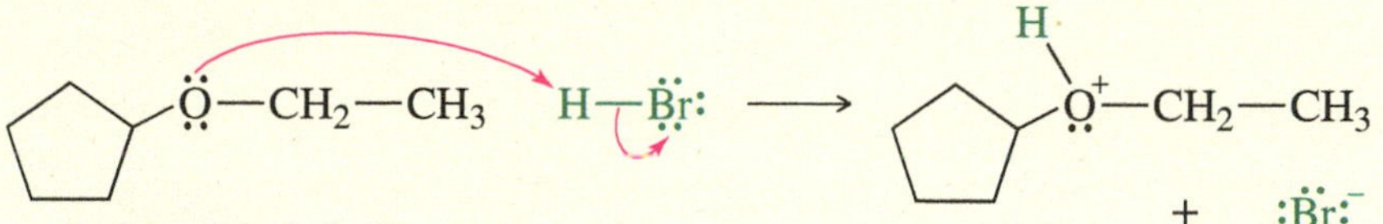

cyclopentyl ethyl ether

Step 2: Cleavage of the protonated ether.

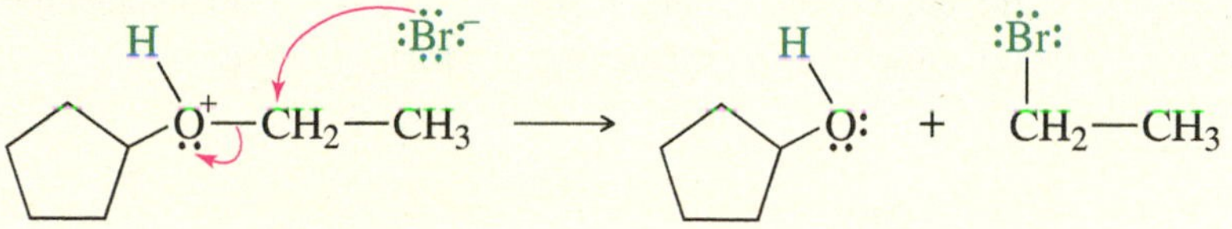

Step 3: Conversion of the alcohol fragment to the alkyl halide. First, the alcohol is protonated to form a good leaving group.

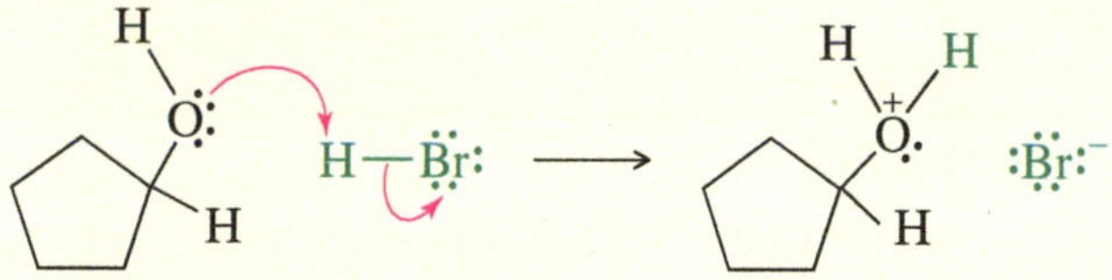

The protonated alcohol undergoes S_N1 or S_N2 substitution by bromide ion.

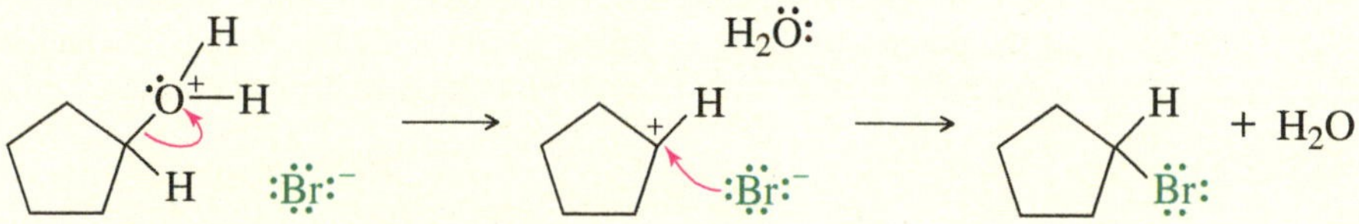

Hydroiodic acid (HI) reacts with ethers the same way HBr does. Aqueous iodide is a stronger nucleophile than aqueous bromide, and iodide reacts at a faster rate. We can rank the hydrohalic acids in order of their reactivity toward the cleavage of ethers:

$$\mathrm{HI} > \mathrm{HBr} \gg \mathrm{HCl}$$

PROBLEM 14-14

Propose a mechanism for the following reaction.

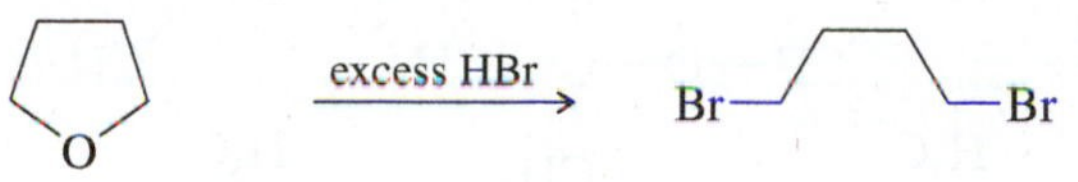

tetrahydrofuran —excess HBr→ 1,4-dibromobutane

Phenyl Ethers Phenyl ethers (one of the groups bonded to oxygen is a benzene ring) react with HBr or HI to give alkyl halides and phenols. Phenols do not react further to give halides because the sp^2-hybridized carbon atom of the phenol cannot undergo the S_N2 (or S_N1) reaction needed for conversion to the halide.

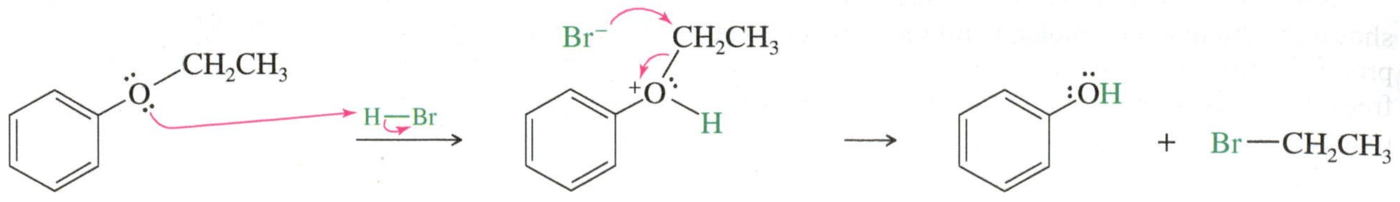

ethyl phenyl ether $\xrightarrow{H-Br}$ protonated ether $\longrightarrow$ phenol (no further reaction) + $Br-CH_2CH_3$ ethyl bromide

PROBLEM-SOLVING HINT

HBr and HI convert both alkyl groups (but not aromatic groups) of an ether to alkyl halides. Phenolic products are unreactive, however.

PROBLEM 14-15

Predict the products of the following reactions. An excess of acid is available in each case.

(a) ethoxycyclohexane + HBr
(b) tetrahydropyran + HI
(c) anisole (methoxybenzene) + HBr
(d) 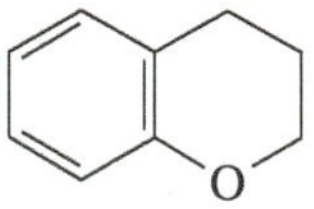 (chromane, benzo-fused six-membered ring with O) + HI
(e) $PhOCH_2CH_2-CH(CH_3)-CH_2OCH_2CH_3$ + HBr

PROBLEM 14-16

Boron tribromide (BBr_3) cleaves ethers to give alkyl halides and alcohols.

$$R-O-R' + BBr_3 \longrightarrow R-O-BBr_2 + R'Br$$
$$R-O-BBr_2 + 3\,H_2O \longrightarrow ROH + B(OH)_3 + 2\,HBr$$

The reaction is thought to involve attack by a bromide ion on the Lewis acid–base adduct of the ether with BBr_3 (a strong Lewis acid). Propose a mechanism for the reaction of butyl methyl ether with BBr_3 to give (after hydrolysis) butan-1-ol and bromomethane.

14-9 Autoxidation of Ethers

When ethers are stored in the presence of atmospheric oxygen, they slowly oxidize to produce **hydroperoxides** and **dialkyl peroxides,** both of which are explosive. Such a spontaneous oxidation by atmospheric oxygen is called **autoxidation.**

$$\underset{\text{ether}}{R-O-CH_2-R'} \xrightarrow[\text{(slow)}]{\text{excess } O_2} \underset{\text{hydroperoxide}}{R-O-CH(OOH)-R'} + \underset{\text{dialkyl peroxide}}{R-O-O-CH_2-R'}$$

Example

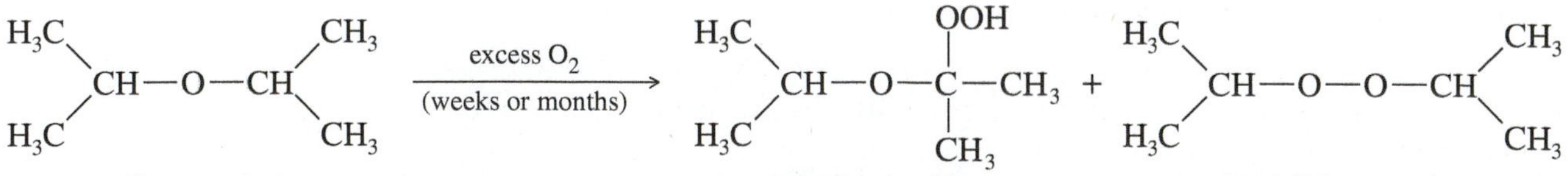

$$\underset{\text{diisopropyl ether}}{(H_3C)_2CH-O-CH(CH_3)_2} \xrightarrow[\text{(weeks or months)}]{\text{excess } O_2} \underset{\text{hydroperoxide}}{(H_3C)_2CH-O-C(OOH)(CH_3)-CH_3} + \underset{\text{diisopropyl peroxide}}{(H_3C)_2CH-O-O-CH(CH_3)_2}$$

Organic chemists often buy large containers of ethers and use small quantities over several months. Once a container has been opened, it contains atmospheric oxygen, and the autoxidation process begins. After several months, a large amount of peroxide may be present. Distillation or evaporation concentrates the peroxides, and an explosion may occur.

Such an explosion may be avoided by taking a few simple precautions. Ethers should be bought in small quantities, kept in tightly sealed containers, and used promptly. Any procedure requiring evaporation or distillation should use only peroxide-free ether. Any ether that might be contaminated with peroxides should be discarded or treated to destroy the peroxides.

SUMMARY Reactions of Ethers

1. *Cleavage by HBr and HI* (Section 14-8)

$$\text{R—O—R}' \xrightarrow[(\text{X} = \text{Br, I})]{\text{excess HX}} \text{R—X} + \text{R}'\text{—X}$$

$$\text{Ar—O—R} \xrightarrow[(\text{X} = \text{Br, I})]{\text{excess HX}} \text{Ar—OH} + \text{R—X}$$

Ar = aromatic ring

2. *Autoxidation* (Section 14-9)

$$\underset{\text{ether}}{\text{R—O—CH}_2\text{—R}'} \xrightarrow[(\text{slow})]{\text{excess O}_2} \underset{\text{hydroperoxide}}{\text{R—O—CH(OOH)—R}'} + \underset{\text{dialkyl peroxide}}{\text{R—O—O—CH}_2\text{—R}'}$$

14-10 Thioethers (Sulfides) and Silyl Ethers

Thioethers, also called **sulfides,** are ethers with a sulfur atom replacing the oxygen atom of an ether, just like the sulfur in a thiol replaces the oxygen atom of an alcohol. The chemistry of thioethers is much like the chemistry of ethers, except that thioethers can undergo oxidation and alkylation of the sulfur atom.

$$\underset{\text{ether}}{\text{R—O—R}'} \qquad \underset{\text{thioether (sulfide)}}{\text{R—S—R}'} \qquad \underset{\text{silyl ether}}{\text{R—O—SiR}'_3}$$

Silyl ethers are ethers with a substituted silicon atom replacing one of the alkyl groups of an ether. Silyl ethers share some of the properties of ethers (resistant to some acids, bases, and oxidizing agents), but they are more easily formed and more easily hydrolyzed. These properties make them useful as protecting groups, and silyl ethers are frequently used to protect alcohols.

14-10A Thioethers (Sulfides)

Like thiols, thioethers have strong characteristic odors: The odor of dimethyl sulfide is reminiscent of oysters that have been kept in the refrigerator for too long. Sulfides are named like ethers, with "sulfide" replacing "ether" in the common names. In the IUPAC (alkoxy alkane) names, "alkylthio" replaces "alkoxy."

$\text{CH}_3\text{—S—CH}_3$ dimethyl sulfide

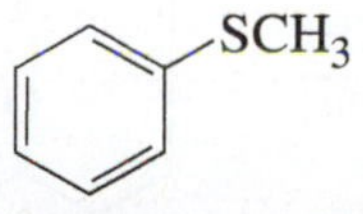

$\text{C}_6\text{H}_5\text{—SCH}_3$ methyl phenyl sulfide

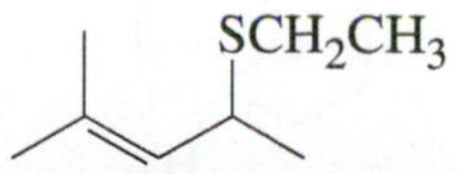

$\text{(CH}_3)_2\text{C=CH—CH(SCH}_2\text{CH}_3)\text{—CH}_3$ 4-ethylthio-2-methylpent-2-ene

Thioethers are easily synthesized by the Williamson ether synthesis, using a thiolate ion as the nucleophile.

$$CH_3CH_2{-}\ddot{\underset{..}{S}}{:}^- + CH_3CH_2CH_2{-}\ddot{\underset{..}{Br}}{:} \longrightarrow CH_3CH_2CH_2{-}\ddot{\underset{..}{S}}{-}CH_2CH_3 + {:}\ddot{\underset{..}{Br}}{:}^-$$

ethanethiolate; 1-bromopropane; ethyl propyl sulfide

Thiols are more acidic than water. Therefore, thiolate ions are easily generated by treating thiols with aqueous sodium hydroxide.

$$\underset{pK_a = 10.5}{CH_3CH_2{-}SH} + Na^{+}\,{}^{-}OH \rightleftharpoons \underset{\text{sodium ethanethiolate}}{CH_3CH_2{-}S^{-}\,Na^{+}} + \underset{pK_a = 15.7}{H_2O}$$

Because sulfur is larger and more polarizable than oxygen, thiolate ions are even better nucleophiles than alkoxide ions. Thiolates are such effective nucleophiles that secondary alkyl halides often react to give good yields of S_N2 products.

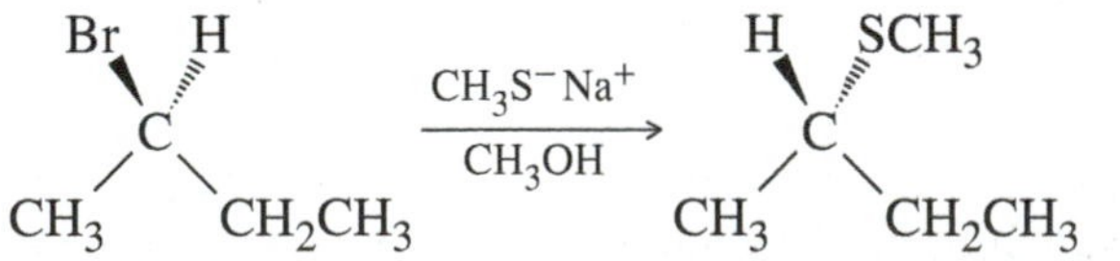

$$\text{(}R\text{)-2-bromobutane} \xrightarrow[CH_3OH]{CH_3S^{-}\,Na^{+}} \text{(}S\text{)-2-(methylthio)butane}$$

PROBLEM 14-17

Show how you would synthesize butyl isopropyl sulfide using butan-1-ol, propan-2-ol, and any solvents and reagents you need.

Sulfides are much more reactive than ethers. In a sulfide, sulfur valence is not necessarily filled: Sulfur can form additional bonds with other atoms. Sulfur forms particularly strong bonds with oxygen, and sulfides are easily oxidized to sulfoxides and sulfones. **Sulfoxides** and **sulfones** are drawn using either hypervalent double-bonded structures or formally charged single-bonded structures as shown here.

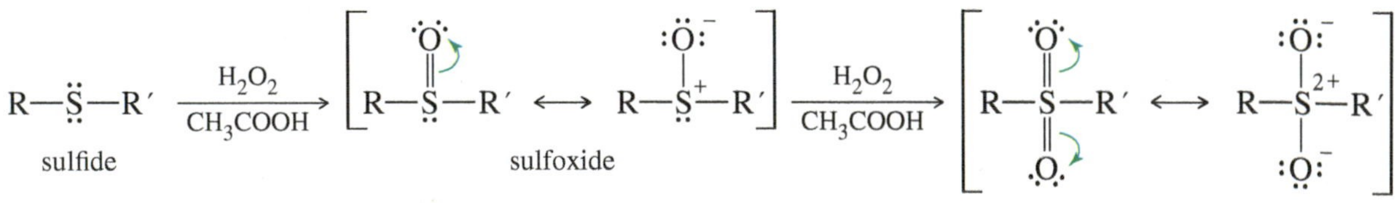

$$\underset{\text{sulfide}}{R{-}\ddot{S}{-}R'} \xrightarrow[CH_3COOH]{H_2O_2} \underset{\text{sulfoxide}}{\left[R{-}\underset{..}{S}({=}O){-}R' \longleftrightarrow R{-}\underset{..}{S^{+}}(O^{-}){-}R'\right]} \xrightarrow[CH_3COOH]{H_2O_2} \underset{\text{sulfone}}{\left[R{-}S({=}O)_2{-}R' \longleftrightarrow R{-}S^{2+}(O^{-})_2{-}R'\right]}$$

The hydrogen peroxide/acetic acid combination is a good oxidant for sulfides. One equivalent of peroxide gives the sulfoxide, and a second equivalent further oxidizes the sulfoxide to the sulfone. This reagent combination probably reacts via the peroxyacid, which is formed in equilibrium with hydrogen peroxide.

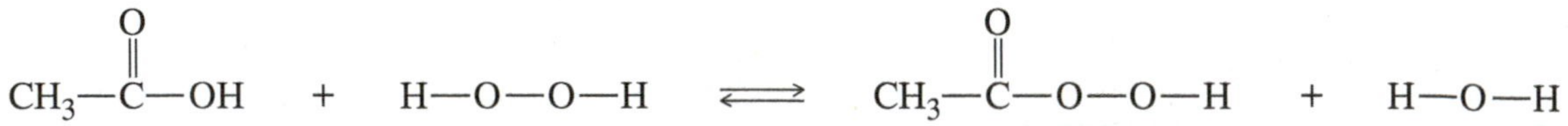

$$\underset{\text{acid}}{CH_3{-}C({=}O){-}OH} + H{-}O{-}O{-}H \rightleftharpoons \underset{\text{peroxyacid}}{CH_3{-}C({=}O){-}O{-}O{-}H} + H{-}O{-}H$$

Because they are easily oxidized, sulfides are often used as mild reducing agents. For example, we have used dimethyl sulfide to reduce the potentially explosive ozonides that result from ozonolysis of alkenes (Section 8-15).

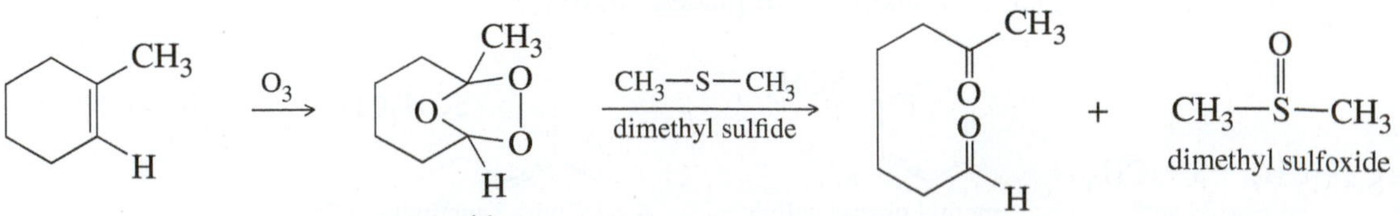

$$\text{1-methylcyclohexene} \xrightarrow{O_3} \text{ozonide} \xrightarrow[\text{dimethyl sulfide}]{CH_3{-}S{-}CH_3} \text{6-oxoheptanal} + \underset{\text{dimethyl sulfoxide}}{CH_3{-}S({=}O){-}CH_3}$$

Sulfur compounds are more nucleophilic than the corresponding oxygen compounds, because sulfur is larger and more polarizable and its electrons are less tightly held in orbitals that are farther from the nucleus. Although ethers are weak nucleophiles, sulfides are relatively strong nucleophiles. Sulfides attack unhindered alkyl halides to give **sulfonium salts.**

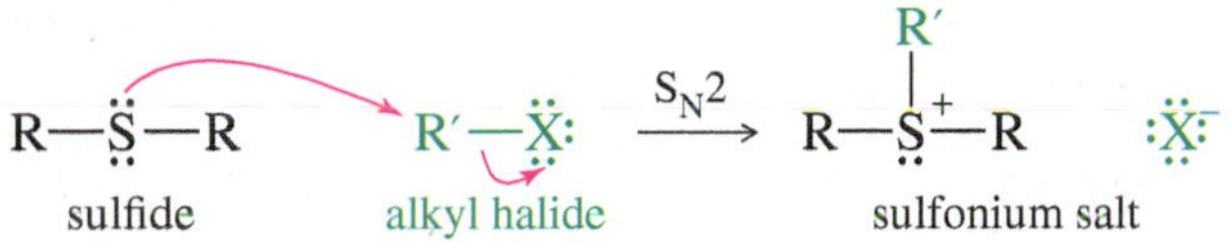

$$R-\ddot{S}-R \quad + \quad R'-X: \xrightarrow{S_N2} R-\overset{R'}{\overset{|}{S^+}}-R \quad :X:^-$$

sulfide — alkyl halide — sulfonium salt

Example

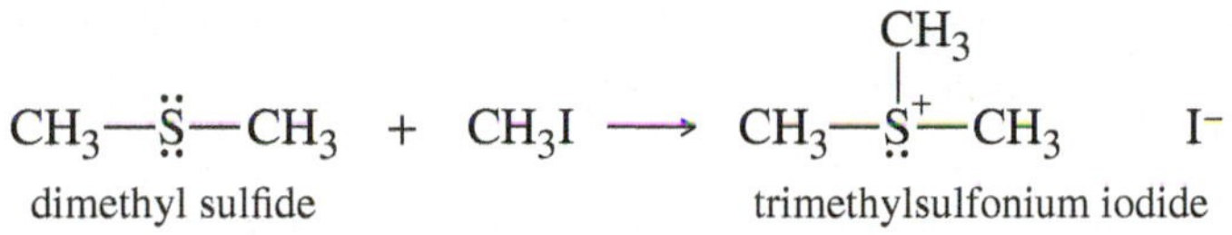

$$CH_3-\ddot{S}-CH_3 \; + \; CH_3I \longrightarrow CH_3-\overset{CH_3}{\overset{|}{S^+}}-CH_3 \quad I^-$$

dimethyl sulfide — trimethylsulfonium iodide

Sulfonium salts are strong alkylating agents. The sulfonium salt polarizes the carbon atom, making it electrophilic. Then, attack by a nucleophile expels an uncharged sulfide, which is an excellent leaving group. Sulfur's polarizability enhances partial bonding in the transition state, lowering its energy.

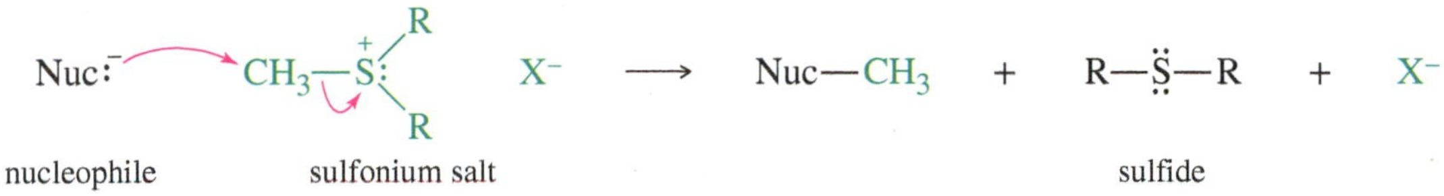

$$Nuc:^- \quad CH_3-\overset{+}{S}R_2 \quad X^- \longrightarrow Nuc-CH_3 \; + \; R-\ddot{S}-R \; + \; X^-$$

nucleophile — sulfonium salt — sulfide

Example

$$\text{pyridine } (N:) \; + \; CH_3-\overset{CH_3}{\overset{|}{S^+}}-CH_3 \quad I^- \longrightarrow \text{pyridinium } N^+-CH_3 \; I^- \; + \; CH_3-\ddot{S}-CH_3$$

pyridine — trimethylsulfonium iodide — *N*-methylpyridinium iodide — dimethyl sulfide

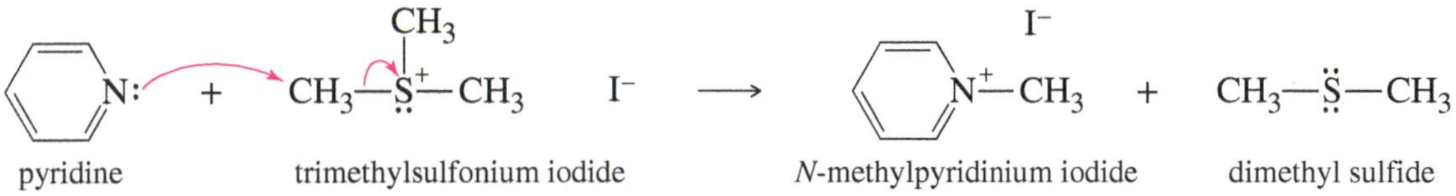

Sulfonium salts are common alkylating agents in biological systems. For example, ATP activation of methionine forms the sulfonium salt *S*-adenosylmethionine (SAM), a biological methylating agent.

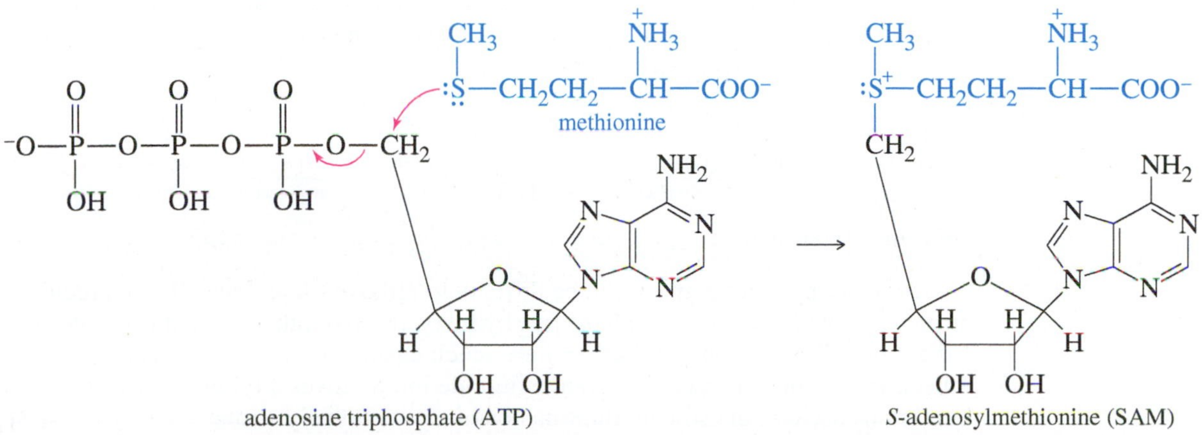

$$^-O-\overset{O}{\overset{\|}{P}}(OH)-O-\overset{O}{\overset{\|}{P}}(OH)-O-\overset{O}{\overset{\|}{P}}(OH)-O-CH_2\text{-adenosine} \; + \; :\overset{CH_3}{\overset{|}{S}}-CH_2CH_2-\overset{\overset{+}{N}H_3}{\overset{|}{CH}}-COO^- \longrightarrow :\overset{CH_3}{\overset{|}{S^+}}(CH_2\text{-adenosine})-CH_2CH_2-\overset{\overset{+}{N}H_3}{\overset{|}{CH}}-COO^-$$

methionine

adenosine triphosphate (ATP) — *S*-adenosylmethionine (SAM)

SAM converts norepinephrine to epinephrine (adrenaline) in the adrenal glands.

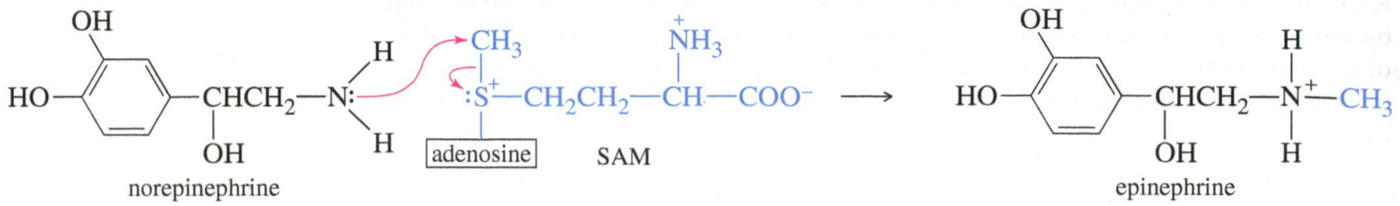

norepinephrine SAM adenosine epinephrine

Application: Cancer Chemotherapy

The sulfur mustards gave rise to the nitrogen mustards, which are less reactive alkylating agents that are used as anticancer drugs. Nitrogen mustards alkylate DNA, which prevents its reproduction and ultimately kills the cells.

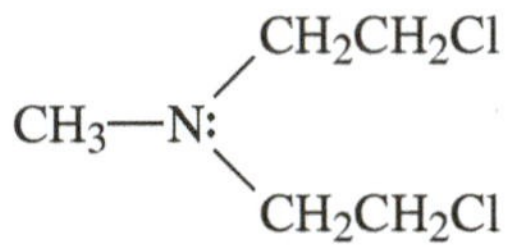

$CH_3-N(CH_2CH_2Cl)_2$

nitrogen mustard

PROBLEM 14-18

Mustard gas, $Cl-CH_2CH_2-S-CH_2CH_2-Cl$, was used as a poisonous chemical agent in World War I. Mustard gas is much more toxic than a typical primary alkyl chloride. Its toxicity stems from its ability to alkylate amino groups on important metabolic enzymes, rendering the enzymes inactive.

(a) Propose a mechanism to explain why mustard gas is an exceptionally potent alkylating agent.

(b) Bleach (sodium hypochlorite, NaOCl, a strong oxidizing agent) neutralizes and inactivates mustard gas. Bleach is also effective on organic stains because it oxidizes colored compounds to colorless compounds. Propose products that might be formed by the reaction of mustard gas with bleach.

14-10B Silyl Ethers as Alcohol-Protecting Groups

If we have a compound with two or more functional groups, and we would like to modify just one of those functional groups, we often must protect any other functional groups to prevent them from reacting as well. For example, if we wanted to add a Grignard reagent to the carbonyl group of a keto-alcohol, the alcohol group would protonate the Grignard reagent, and the reaction would fail.

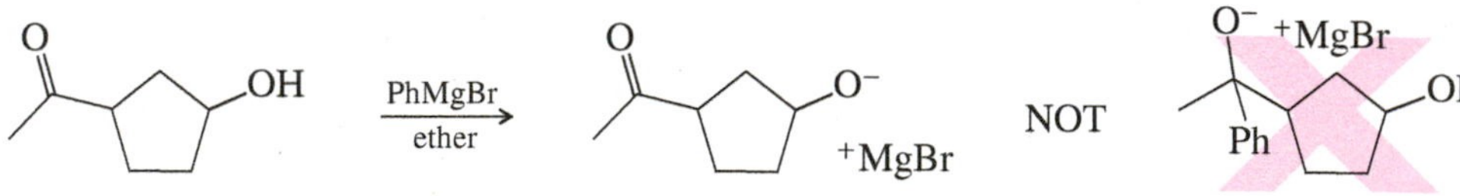

PhMgBr, ether → O^- ^+MgBr NOT O^- ^+MgBr Ph OH

Alcohol functional groups are common and useful, but they react with acids, bases, and oxidizing agents. Alcohols must be protected if they are to survive a reaction at another functional group on the molecule. A good **protecting group** must be easy to add to the group it protects, and then it must be resistant to the reagents used to modify other parts of the molecule. Finally, a good protecting group must be easy to remove to regenerate the original functional group. To accomplish the Grignard reaction shown above, we would need to convert the hydroxy group to something that is resistant to Grignard reagents. For example, we might consider using an ether to protect a hydroxy group in a Grignard reaction.

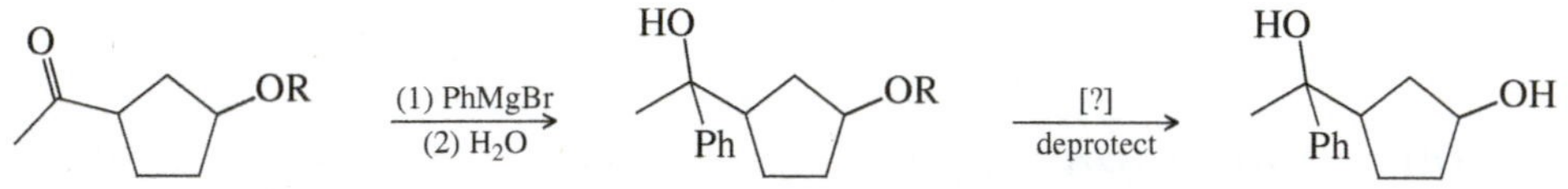

OR (1) PhMgBr (2) H_2O → HO Ph OR [?] deprotect → HO Ph OH

(OH converted to an ether)

An ether protecting group can be difficult to remove (deprotect). It often requires strong acid, which can react with the free hydroxy group or other parts of the molecule. Ethers based on silicon (silyl ethers) are much easier to remove than carbon-based ethers. In aqueous or organic solvents, fluoride ion removes silyl ethers under gentle conditions because the silicon–fluorine bond is exceptionally strong.

Synthetic organic chemists have developed many different silyl protecting groups that vary widely in their reactivity and are carefully chosen for a specific use. We will use the **triisopropylsilyl** (Tri-Iso-Propyl-Silyl or **TIPS**) protecting group, of structure

R-O-Si(i-Pr)$_3$ as our example. The three bulky isopropyl groups stabilize this silyl ether by hindering attack by nucleophiles. Silyl ethers are commonly formed by the reaction of alcohols with chlorosilanes in the presence of tertiary amines. We can form a TIPS ether by a reaction of chlorotriisopropylsilane (TIPSCl) with a tertiary amine such as triethylamine (Et_3N:).

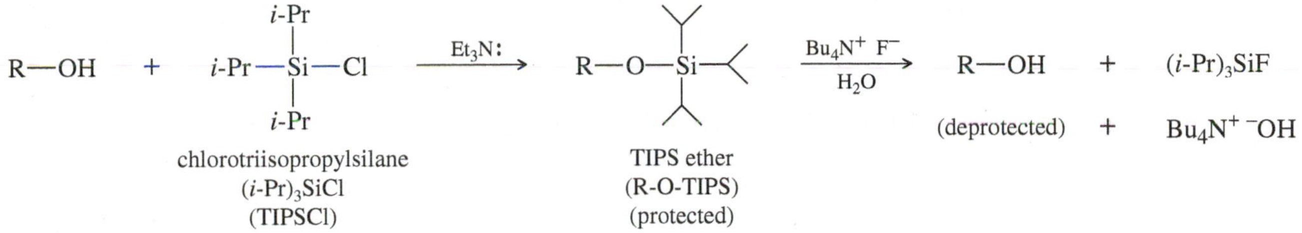

$$R{-}OH + (i\text{-Pr})_3Si{-}Cl \xrightarrow{Et_3N:} R{-}O{-}Si(i\text{-Pr})_3 \xrightarrow[H_2O]{Bu_4N^+\ F^-} R{-}OH + (i\text{-Pr})_3SiF + Bu_4N^+\ {}^-OH$$

chlorotriisopropylsilane (i-Pr)$_3$SiCl (TIPSCl); TIPS ether (R-O-TIPS) (protected); (deprotected)

TIPS ethers are stable to most acids and bases and oxidizing and reducing agents. Our keto-alcohol shown above would react with TIPS chloride (TIPSCl) and triethylamine (Et_3N:) to give a protected alcohol. In our example, we can add a Grignard reagent to the carbonyl group in the presence of the protected alcohol.

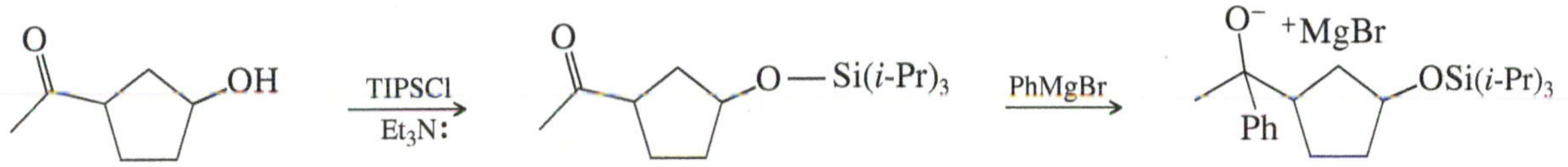

TIPSCl, Et_3N: → O—Si(i-Pr)$_3$; PhMgBr → O$^-$ $^+$MgBr, Ph, OSi(i-Pr)$_3$

After the Grignard reaction is completed, protonation of the magnesium alkoxide salt and deprotection of the silyl ether gives the desired product.

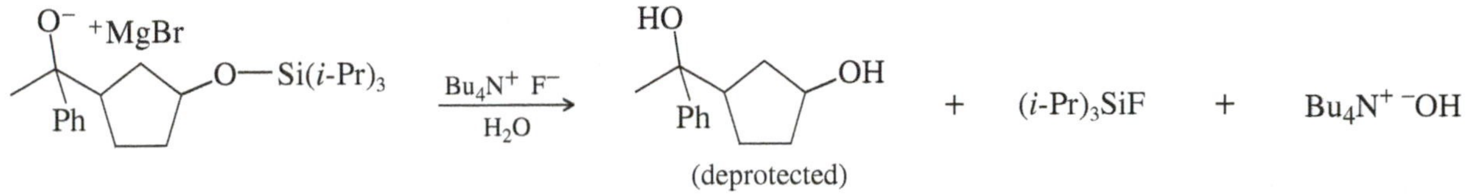

O$^-$ $^+$MgBr, Ph, O—Si(i-Pr)$_3$ $\xrightarrow[H_2O]{Bu_4N^+\ F^-}$ HO, Ph, OH (deprotected) + (i-Pr)$_3$SiF + $Bu_4N^+\ {}^-OH$

PROBLEM 14-19

Show how you would use a protecting group to convert 4-bromobutan-1-ol to hept-5-yn-1-ol.

PROBLEM-SOLVING HINT

Bond-dissociation energies show that Si—Cl bonds are weaker than Si—O bonds, so the formation of a silyl ether from a silyl chloride is generally exothermic.

Si—F bonds are much stronger than Si—O bonds, so the deprotection step with fluoride ion is also exothermic.

BDE:	kJ/mol	
$(CH_3)_3Si{-}Cl$	473	*weakest*
$(CH_3)_3Si{-}OR$	536	
$(CH_3)_3Si{-}F$	670	*strongest*

14-11 Synthesis of Epoxides

Epoxides are easily made from alkenes, and (unlike other ethers) they undergo a variety of useful synthetic reactions. For these reasons, epoxides are valuable synthetic intermediates. Here we review the **epoxidation** techniques already covered (Section 8-12) and consider in more detail the useful syntheses and reactions of epoxides.

14-11A Peroxyacid Epoxidation

Peroxyacids (sometimes called *peracids*) are used to convert alkenes to epoxides. If the reaction takes place in aqueous acid, the epoxide opens to a glycol. Therefore, to make an epoxide, we avoid strong acids. Because of its desirable solubility properties, *meta*-chloroperoxybenzoic acid (**mCPBA**) is often used for these epoxidations. mCPBA is a weakly acidic peroxyacid that is soluble in aprotic solvents such as CH_2Cl_2.

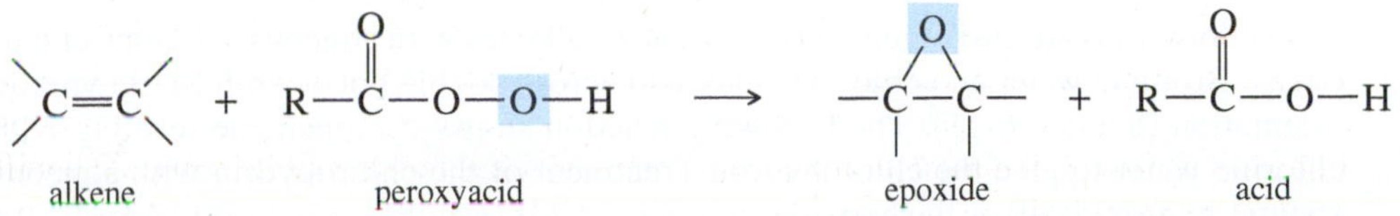

$$\underset{\text{alkene}}{>C{=}C<} + \underset{\text{peroxyacid}}{R{-}\overset{O}{\overset{\|}{C}}{-}O{-}O{-}H} \longrightarrow \underset{\text{epoxide}}{{-}\overset{O}{C{-}C}{-}} + \underset{\text{acid}}{R{-}\overset{O}{\overset{\|}{C}}{-}O{-}H}$$

Example

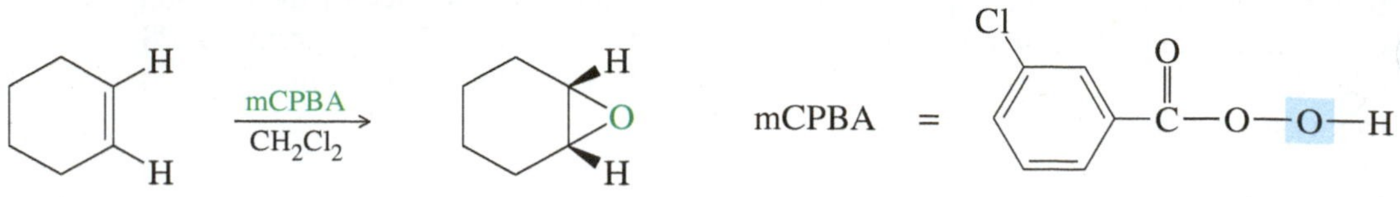

cyclohexene $\xrightarrow[CH_2Cl_2]{mCPBA}$ epoxycyclohexane (100%)

mCPBA = *meta*-chloroperoxybenzoic acid

The epoxidation takes place in a one-step, **concerted reaction** that maintains the stereochemistry of any substituents on the double bond.

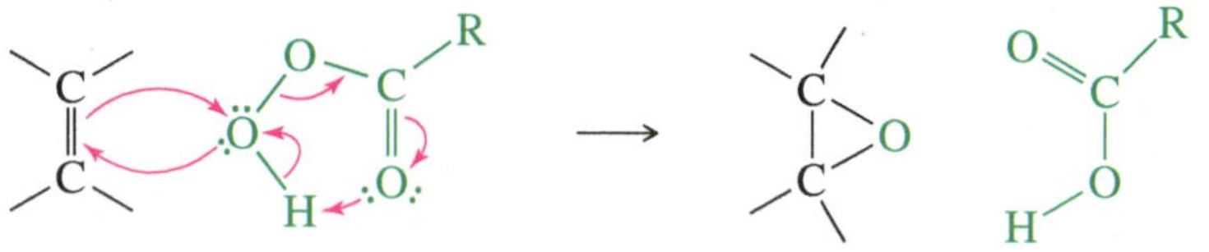

alkene + peroxyacid → epoxide + acid

Application: Antibacterial

MMPP is used in surface disinfectants for sensitive plastic and rubber equipment such as incubators. It is also being tested for use as a plaque-reducing mouthwash and toothpaste.

The peroxyacid epoxidation is quite general, with electron-rich double bonds reacting fastest. The following reactions are difficult transformations made possible by this selective, stereospecific epoxidation procedure. The second example uses magnesium monoperoxyphthalate (**MMPP**), a relatively stable water-soluble peroxyacid often used in large-scale epoxidations. These aqueous MMPP epoxidations, carried out at neutral pH to avoid opening the epoxide, avoid the large-scale use of hazardous chlorinated solvents.

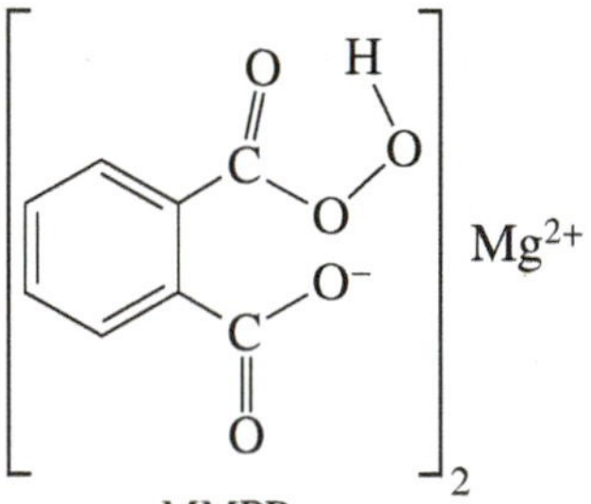

MMPP (Mg^{2+})

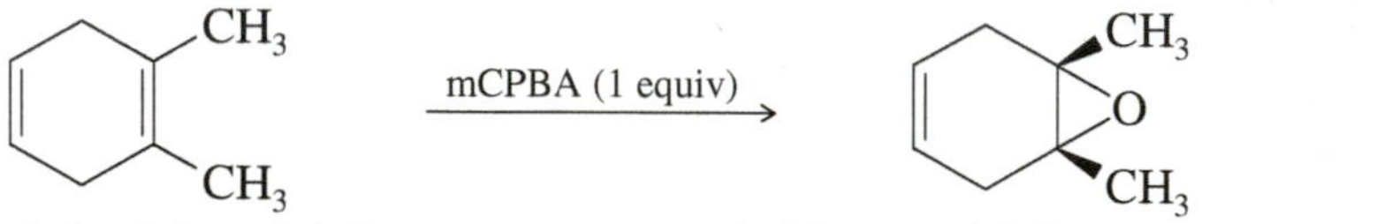

1,2-dimethylcyclohexa-1,4-diene $\xrightarrow{\text{mCPBA (1 equiv)}}$ *cis*-4,5-epoxy-4,5-dimethylcyclohexene

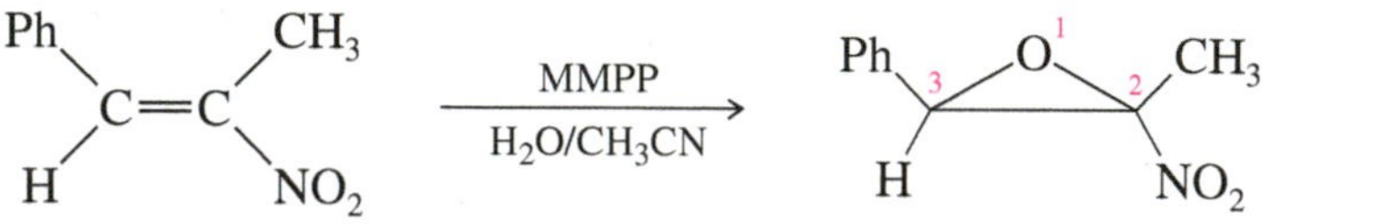

(*E*)-2-nitro-1-phenylpropene $\xrightarrow[H_2O/CH_3CN]{MMPP}$ (*E*)-2-methyl-2-nitro-3-phenyloxirane

14-11B Base-Promoted Cyclization of Halohydrins

A less common synthesis of epoxides and other cyclic ethers involves a variation of the Williamson ether synthesis. If an alkoxide ion and a halogen atom are located in the same molecule, the alkoxide may displace a halide ion and form a ring. Treatment of a **halohydrin** with base leads to an epoxide through this internal S_N2 attack.

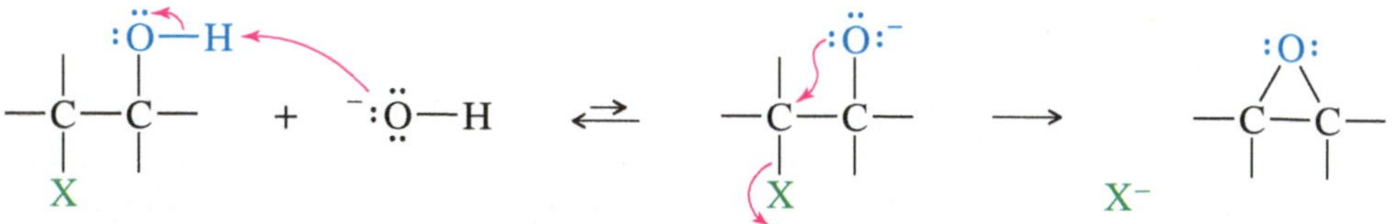

halohydrin (X = Cl, Br, I) + $^-$:Ö—H ⇌ alkoxide intermediate → epoxide + X^-

Halohydrins are easily generated by treating alkenes with aqueous solutions of halogens. Bromine water and chlorine water add across double bonds with Markovnikov orientation (Section 8-11). The following reaction shows cyclopentene reacting with chlorine water to give the chlorohydrin. Treatment of the chlorohydrin with aqueous sodium hydroxide gives the epoxide.

Formation of the chlorohydrin

cyclopentene — chlorine water ($H_2\ddot{O}$:, Cl—Cl) → chloronium ion → ($-H^+$) → *trans*-chlorohydrin (mixture of enantiomers)

Displacement of the chlorohydrin

trans-chlorohydrin ⇌ (^-OH) alkoxide → epoxide (50% overall) + Cl^-

This reaction can be used to synthesize cyclic ethers with larger rings. The difficulty lies in preventing the base (added to deprotonate the alcohol) from attacking and displacing the halide. 2,6-Lutidine, a bulky base that cannot easily attack a carbon atom, can deprotonate the hydroxy group to give a five-membered cyclic ether. Five-, six-, and seven-membered (and occasionally four-membered) cyclic ethers are formed this way.

chloro-alcohol + 2,6-lutidine (2,6-dimethylpyridine) ⇌ alkoxide → 2-methyltetrahydrofuran (85%) + Cl^-

PROBLEM 14-20

Show how you would accomplish the following transformations. Some of these examples require more than one step.

(a) 2-methylpropene ⟶ 2,2-dimethyloxirane
(b) 1-phenylethanol ⟶ 2-phenyloxirane
(c) 5-chloropent-1-ene ⟶ tetrahydropyran
(d) 5-chloropent-1-ene ⟶ 2-methyltetrahydrofuran
(e) 2-chlorohexan-1-ol ⟶ 1,2-epoxyhexane

PROBLEM 14-21

The 2001 Nobel Prize in Chemistry was awarded to three organic chemists who have developed methods for catalytic asymmetric syntheses. An asymmetric (or enantioselective) synthesis is one that converts an achiral starting material into mostly one enantiomer of a chiral product. K. Barry Sharpless (The Scripps Research Institute) developed an asymmetric epoxidation of allylic alcohols that gives excellent chemical yields and greater than 90% enantiomeric excess.

The Sharpless epoxidation uses *tert*-butyl hydroperoxide, titanium(IV) isopropoxide, and a dialkyl tartrate ester as the reagents. The following epoxidation of geraniol is typical.

(continued)

geraniol → 80% yield, > 90% e.e.

Reagents: $H_3C-C(CH_3)_2-OOH$ — *tert*-butyl hydroperoxide; titanium(IV) isopropoxide; diethyl L-tartrate ($COOCH_2CH_3$, H—C—OH, HO—C—H, $COOCH_2CH_3$)

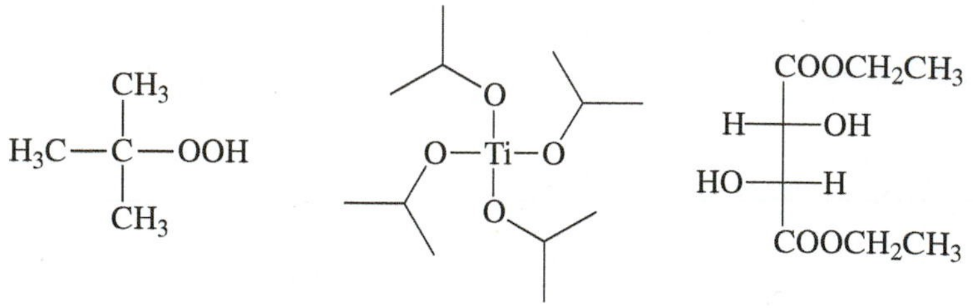

(a) Which of these reagents is most likely to be the actual oxidizing agent? That is, which reagent is reduced in the reaction? What is the likely function of the other reagents?
(b) When achiral reagents react to give a chiral product, that product is normally formed as a racemic mixture of enantiomers. How can the Sharpless epoxidation give just one nearly pure enantiomer of the product?
(c) Draw the other enantiomer of the product. What reagents would you use if you wanted to epoxidize geraniol to give this other enantiomer?

SUMMARY Epoxide Syntheses

1. *Peroxyacid epoxidation* (Section 14-11A)

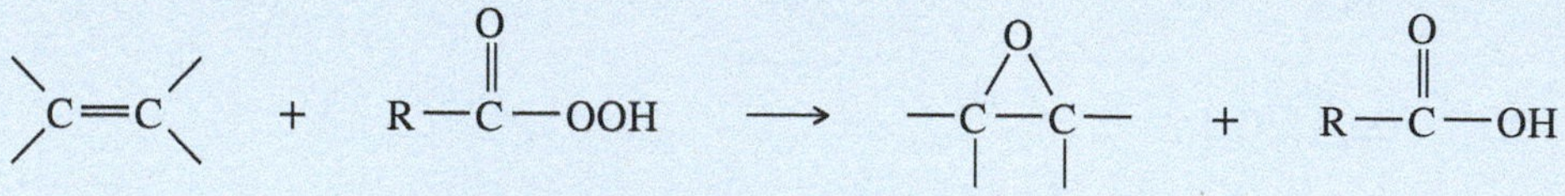

$$\text{C=C (alkene)} + \text{R-C(=O)-OOH (peroxyacid)} \longrightarrow \text{epoxide} + \text{R-C(=O)-OH (acid)}$$

2. *Base-promoted cyclization of halohydrins* (Section 14-11B)

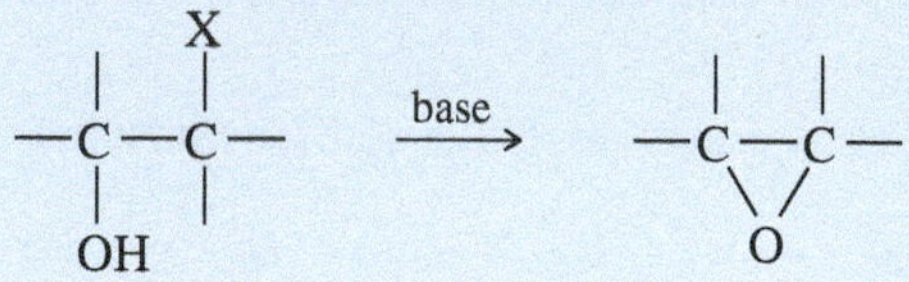

$$\text{-C(OH)-C(X)- (halohydrin)} \xrightarrow{\text{base}} \text{epoxide}$$

X = Cl, Br, I, OTs, etc.

14-12 Acid-Catalyzed Ring Opening of Epoxides

Epoxides are much more reactive than common dialkyl ethers because of the large strain energy (about 105 kJ/mol or 25 kcal/mol) associated with the three-membered ring. Unlike other ethers, epoxides react under both acidic and basic conditions. The products of acid-catalyzed opening depend primarily on the solvent used.

In Water In Section 8-13, we saw that acid-catalyzed hydrolysis of epoxides gives glycols with anti stereochemistry. The mechanism of this hydrolysis involves protonation of oxygen (forming a good leaving group), followed by S_N2 attack by water. Anti stereochemistry results from the back-side attack of water on the protonated epoxide.

MECHANISM 14-2 Acid-Catalyzed Opening of Epoxides in Water

Epoxides open in acidic solutions to form glycols.

Step 1: Protonation of the epoxide to form a strong electrophile.

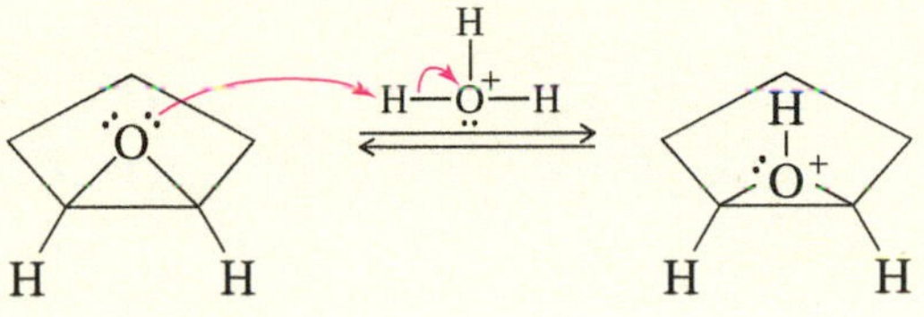

1,2-epoxycyclopentane

Step 2: Water attacks and opens the ring.

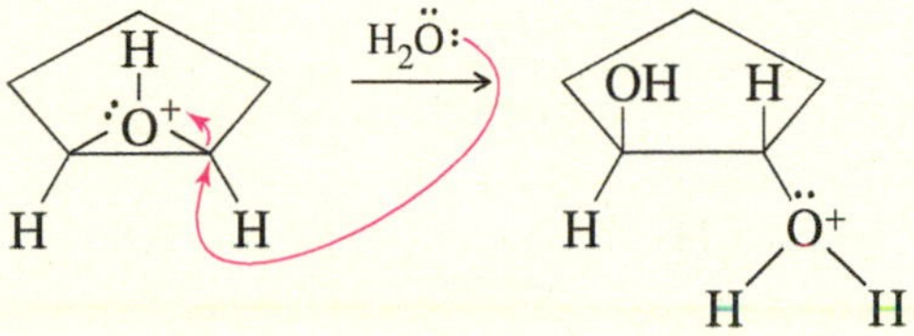

Step 3: Deprotonation to give the diol.

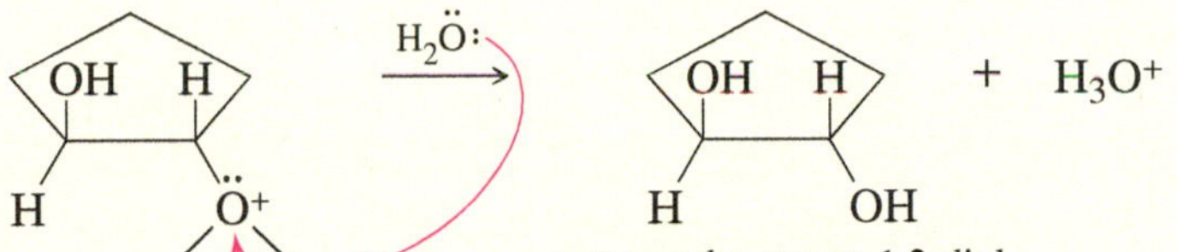

trans-cyclopentane-1,2-diol
(mixture of enantiomers)

Direct anti hydroxylation of an alkene (without isolation of the epoxide intermediate, Section 8-13) is possible by using an acidic aqueous solution of a peroxyacid. As soon as the epoxide is formed, it hydrolyzes to the glycol. Peroxyacetic acid (CH_3CO_3H) and peroxyformic acid (HCO_3H) are often used for the anti hydroxylation of alkenes.

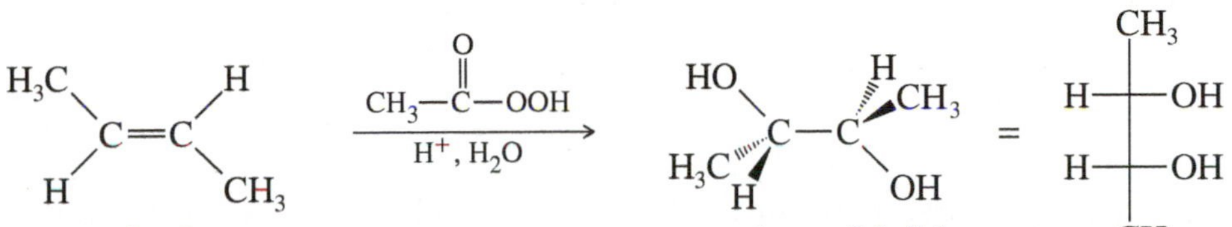

trans-but-2-ene $\xrightarrow[H^+,\ H_2O]{CH_3-C(=O)-OOH}$ *meso*-butane-2,3-diol

PROBLEM 14-22

Propose mechanisms for the epoxidation and ring-opening steps of the epoxidation and hydrolysis of *trans*-but-2-ene shown above. Predict the product of the same reaction with *cis*-but-2-ene.

In Alcohols When the acid-catalyzed opening of an epoxide takes place with an alcohol as the solvent, a molecule of alcohol acts as the nucleophile. This reaction produces an alkoxy alcohol with anti stereochemistry. This is an excellent method for making compounds with ether and alcohol functional groups on adjacent carbon atoms. For example, the acid-catalyzed opening of 1,2-epoxycyclopentane in a methanol solution gives *trans*-2-methoxycyclopentanol.

MECHANISM 14-3 Acid-Catalyzed Opening of an Epoxide in an Alcohol Solution

Epoxides open in acidic alcohol solutions to form 2-alkoxy alcohols.

Step 1: Protonation of the epoxide to form a strong electrophile.

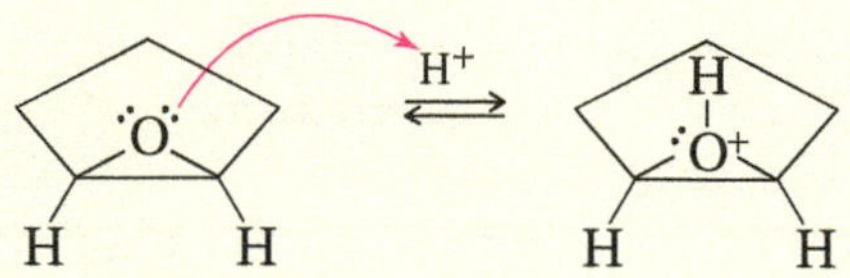

1,2-epoxycyclopentane

Step 2: The alcohol (solvent) attacks and opens the ring.

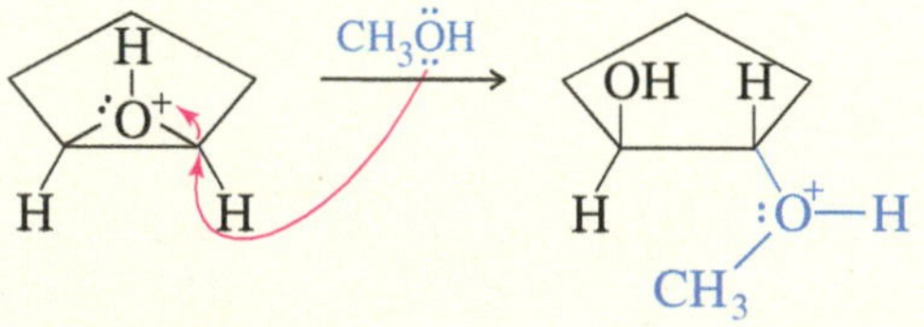

Step 3: Deprotonation to give the product, a 2-alkoxy alcohol.

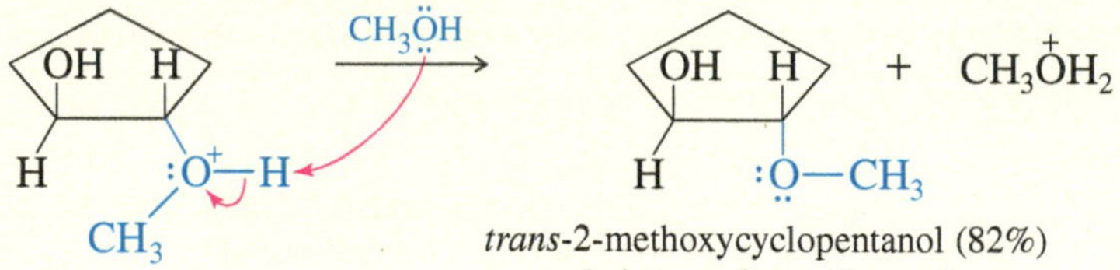

trans-2-methoxycyclopentanol (82%)
(mixture of enantiomers)

PROBLEM 14-23

Cellosolve® is the trade name for 2-ethoxyethanol, a common industrial solvent. This compound is produced in chemical plants that use ethylene as their only organic feedstock. Show how you would accomplish this industrial process.

Using Hydrohalic Acids When an epoxide reacts with a hydrohalic acid (HCl, HBr, or HI), a halide ion attacks the protonated epoxide. This reaction is analogous to the cleavage of ethers by HBr or HI. The halohydrin initially formed reacts further with HX to give a 1,2-dihalide. This is rarely a useful synthetic reaction, because the 1,2-dihalide can be made directly from the alkene by electrophilic addition of X_2.

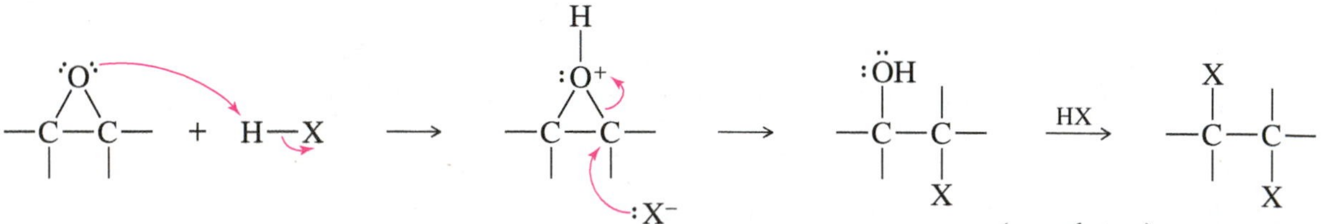

(several steps)

The Opening of Squalene-2,3-Epoxide *Steroids* are tetracyclic compounds that serve a wide variety of biological functions, including hormones (sex hormones), emulsifiers (bile acids), and membrane components (cholesterol). The biosynthesis of steroids is believed to involve an acid-catalyzed opening of squalene-2,3-epoxide (Figure 14-6). Squalene is a member of the class of natural products called *terpenes* (see Section 25-8). The enzyme *squalene epoxidase* oxidizes squalene to the epoxide, which opens and forms a carbocation that cyclizes under the control of another enzyme. The cyclized intermediate rearranges to lanosterol, which is converted to cholesterol and other steroids.

Although cyclization of squalene-2,3-epoxide is controlled by an enzyme, its mechanism is similar to the acid-catalyzed opening of other epoxides. The epoxide oxygen becomes protonated and is attacked by a nucleophile. In this case, the nucleophile is a pi bond. The initial result is a tertiary carbocation (Figure 14-7).

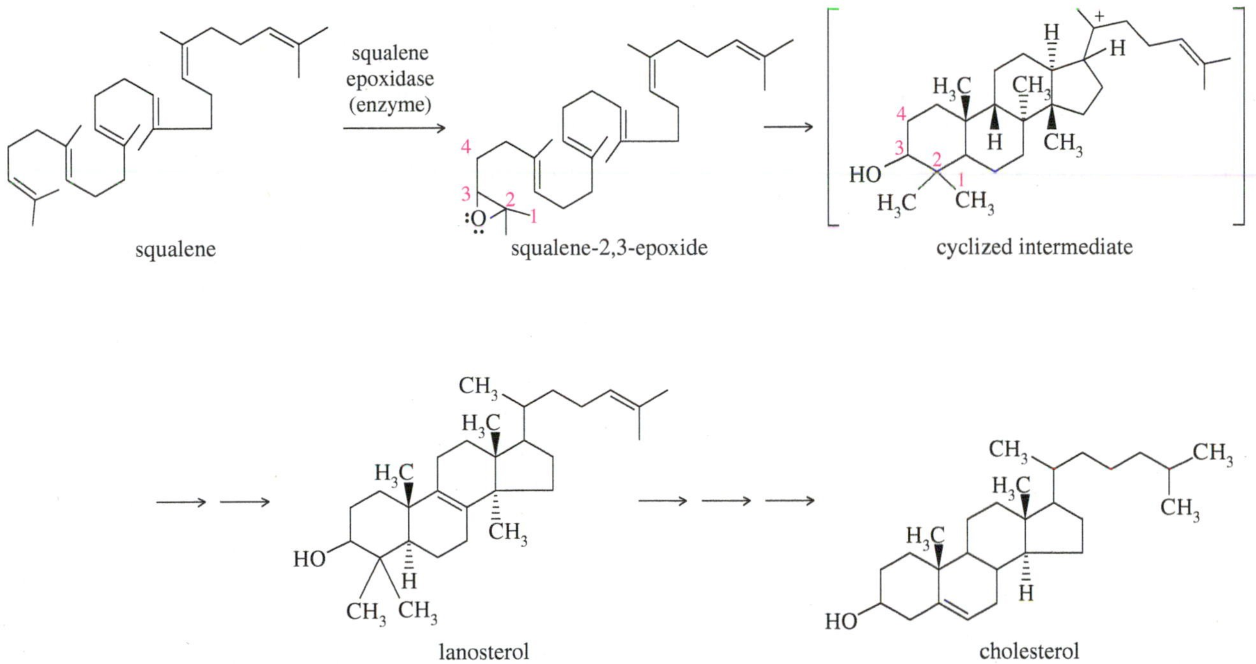

FIGURE 14-6
Role of squalene in the biosynthesis of steroids. The biosynthesis of steroids starts with epoxidation of squalene to squalene-2,3-epoxide. The opening of this epoxide promotes cyclization of the carbon skeleton under the control of an enzyme. The cyclized intermediate is converted to lanosterol and then to other steroids.

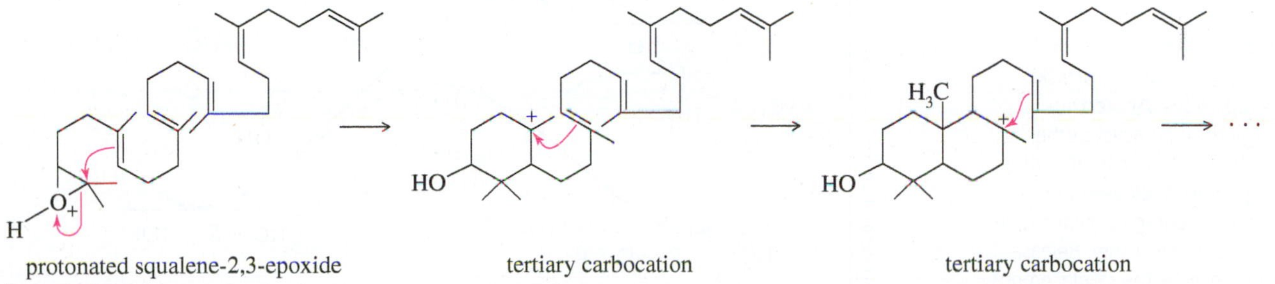

FIGURE 14-7
Cyclization of squalene epoxide begins with the acid-catalyzed opening of the epoxide. Each additional cyclization step forms another carbocation.

Application: Antifungal

Inhibitors of squalene epoxidase are used in antifungal drugs to treat athlete's foot, jock itch, ringworm, and nail infections. The drug Tinactin™ (tolnaftate) inhibits squalene epoxidase, which blocks the synthesis of the steroids the fungus needs to make its cell membrane. The defective cell membrane kills the fungus.

This initial carbocation is attacked by another double bond, leading to the formation of another ring and another tertiary carbocation. A repetition of this process leads to the cyclized intermediate shown in Figure 14-6. Note that this sequence of steps converts an achiral, acyclic starting material (squalene) into a compound with four rings and seven asymmetric carbon atoms. The enzyme-catalyzed sequence takes place with high yields and complete stereospecificity, providing a striking example of asymmetric induction in a biological system.

PROBLEM 14-24

Show the rest of the mechanism for formation of the cyclized intermediate in Figure 14-6.

14-13 Base-Catalyzed Ring Opening of Epoxides

Most ethers do not undergo nucleophilic substitutions or eliminations under basic conditions, because an alkoxide ion is a poor leaving group. Epoxides have about 105 kJ/mol (25 kcal/mol) of ring strain that is released upon ring opening, however, and this strain is enough to compensate for the poor alkoxide leaving group. Figure 14-8 compares the energy profiles for nucleophilic attack on an ether and on an epoxide. The starting epoxide is about 105 kJ/mol (25 kcal/mol) higher in energy than the ether, and its displacement has a lower activation energy.

The reaction of an epoxide with hydroxide ion leads to the same product as the acid-catalyzed opening of the epoxide: a 1,2-diol (glycol), with anti stereochemistry. In fact, either the acid-catalyzed or base-catalyzed reaction may be used to open an epoxide, but the acid-catalyzed reaction takes place under milder conditions. Unless there is an acid-sensitive functional group present, the acid-catalyzed hydrolysis is preferred.

Like hydroxide, alkoxide ions react with epoxides to form ring-opened products. For example, cyclopentene oxide reacts with sodium methoxide in methanol to give the same trans-2-methoxycyclopentanol produced in the acid-catalyzed opening in methanol.

cyclopentene oxide $\xrightarrow[CH_3OH]{CH_3-\ddot{O}:^-\ Na^+}$ *trans*-2-methoxycyclopentanol (mixture of enantiomers)

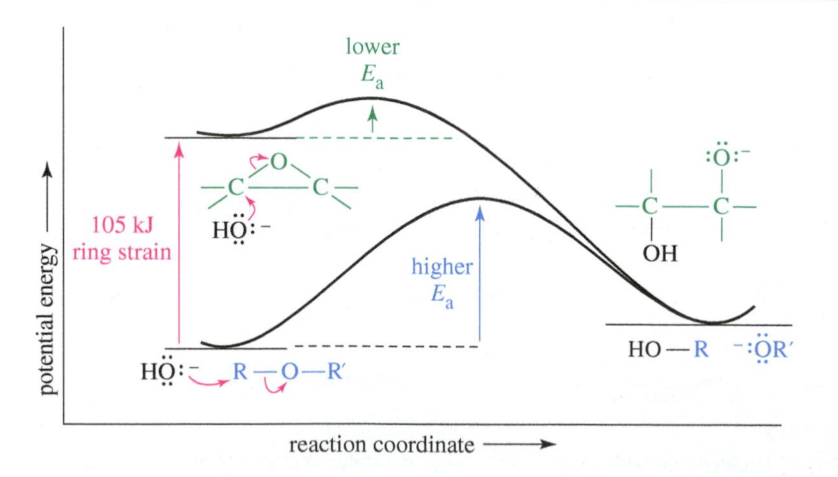

FIGURE 14-8
Energy profiles of nucleophilic attacks on ethers and epoxides. An epoxide is higher in energy than an acyclic ether by about 105 kJ/mol (25 kcal/mol) ring strain. The ring strain is released in the product, giving it an energy similar to the products from the acyclic ether. Release of the ring strain makes the displacement of an epoxide thermodynamically favorable.

MECHANISM 14-4 Base-Catalyzed Opening of Epoxides

Strong bases and nucleophiles do not attack and cleave most ethers. Epoxides are more reactive, however, because opening the epoxide relieves the strain of the three-membered ring. Strong bases can attack and open epoxides, even though the leaving group is an alkoxide.

Step 1: A strong base attacks and opens the ring to an alkoxide.

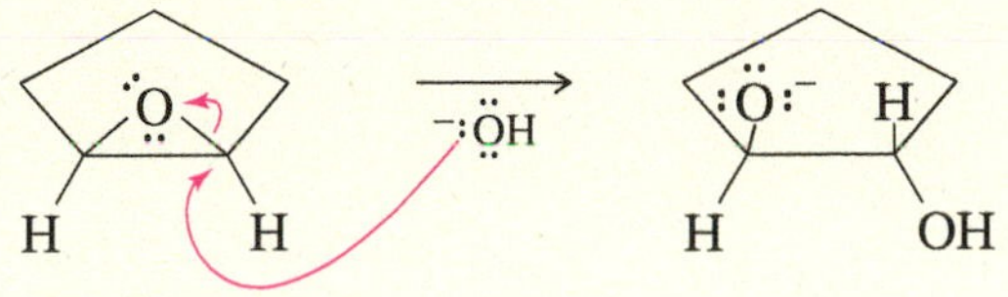

$^-$:ÖH

1,2-epoxycyclopentane

Step 2: Protonation of the alkoxide gives the diol.

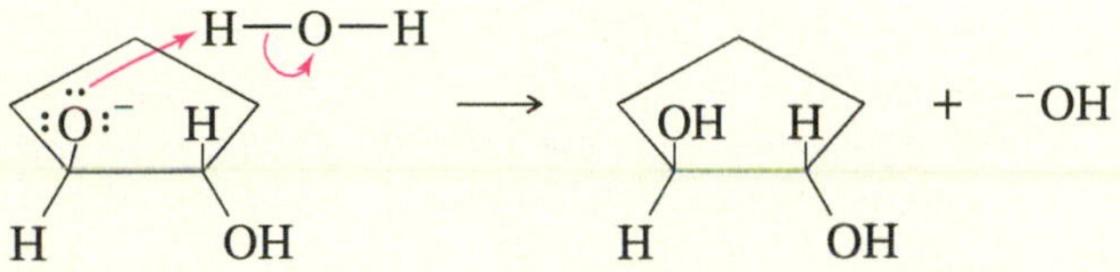

H—O—H

+ $^-$OH

trans-cyclopentane-1,2-diol
(mixture of enantiomers)

Amines can also open epoxides. Ethylene oxide reacts with aqueous ammonia to give ethanolamine, an important industrial reagent. The nitrogen atom in ethanolamine is still nucleophilic, and ethanolamine can react further to give diethanolamine and triethanolamine. Good yields of ethanolamine are achieved by using excess ammonia.

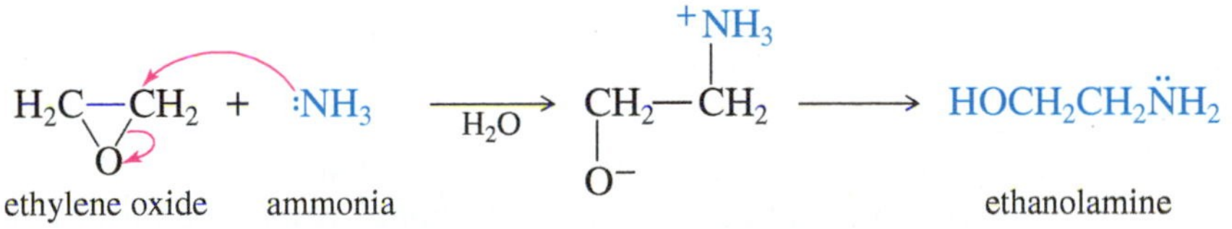

$$H_2C\text{—}CH_2 \text{ (epoxide)} + :NH_3 \xrightarrow{H_2O} {}^{-}O\text{—}CH_2\text{—}CH_2\text{—}{}^{+}NH_3 \longrightarrow HOCH_2CH_2\ddot{N}H_2$$

ethylene oxide　ammonia　ethanolamine

$$HOCH_2CH_2\ddot{N}H_2 \xrightarrow{\text{ethylene oxide}} (HOCH_2CH_2)_2\ddot{N}H \xrightarrow{\text{ethylene oxide}} (HOCH_2CH_2)_3N:$$

ethanolamine　diethanolamine　triethanolamine

PROBLEM 14-25

Propose a complete mechanism for the reaction of cyclopentene oxide with sodium methoxide in methanol.

PROBLEM 14-26

Predict the major product when each reagent reacts with ethylene oxide.

(a) $NaOCH_2CH_3$ (sodium ethoxide)
(b) $NaNH_2$ (sodium amide)
(c) NaSPh (sodium thiophenoxide)
(d) $PhNH_2$ (aniline)
(e) KCN (potassium cyanide)
(f) NaN_3 (sodium azide)

14-14 Orientation of Epoxide Ring Opening

Symmetrically substituted epoxides (such as cyclopentene oxide, above) give the same product in both the acid-catalyzed and base-catalyzed ring openings. An unsymmetrical epoxide may produce different products under acid-catalyzed and base-catalyzed conditions, however.

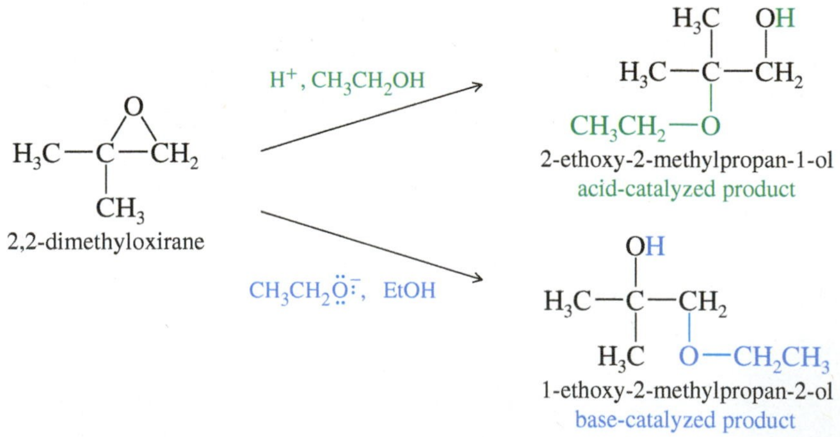

2,2-dimethyloxirane

H^+, CH_3CH_2OH → 2-ethoxy-2-methylpropan-1-ol (acid-catalyzed product)

$CH_3CH_2\ddot{O}:^-$, EtOH → 1-ethoxy-2-methylpropan-2-ol (base-catalyzed product)

Under basic conditions, the alkoxide ion simply attacks the less hindered carbon atom in an S_N2 displacement.

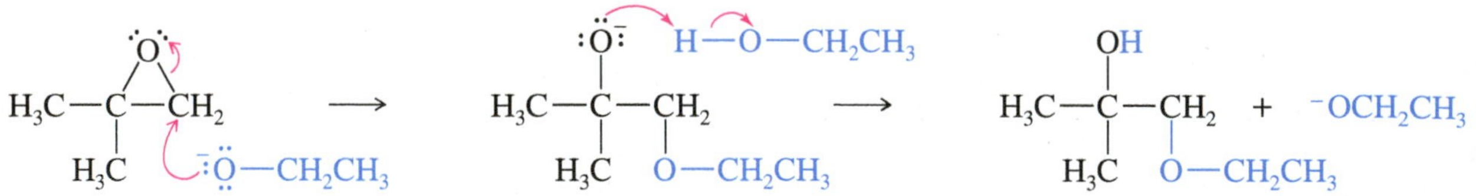

$+ \ {}^-OCH_2CH_3$

Under acidic conditions, the alcohol attacks the protonated epoxide. It might seem that the alcohol would attack at the less hindered oxirane carbon, but this is not the case. In the protonated epoxide, there is a balancing act between ring strain and the energy it costs to put some of the positive charge on the carbon atoms. We can represent this sharing of positive charge by drawing resonance forms that suggest what the cations would look like if the ring started to open. These "no-bond" resonance forms help us to visualize the charge distribution in the protonated epoxide.

PROBLEM-SOLVING HINT

In proposing mechanisms for acid-catalyzed opening of epoxides, imagine that the protonated epoxide opens to the more stable (more substituted) carbocation.

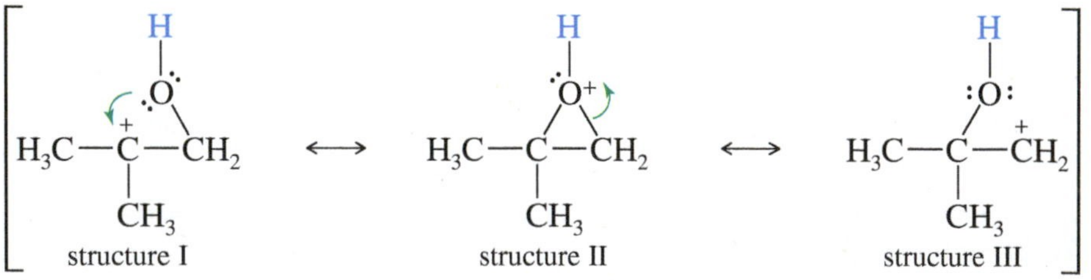

structure I ⟷ structure II ⟷ structure III

Structure II is the conventional structure for the protonated epoxide, while structures I and III show that the oxirane carbon atoms share part of the positive charge. The tertiary carbon bears a larger part of the positive charge, and it is more strongly electrophilic; that is, structure I is more important than structure III. The bond between the tertiary carbon and oxygen is weaker, implying a lower transition state energy for attack at the tertiary carbon. Attack by the weak nucleophile (ethanol in this case) is sensitive to the strength of the electrophile, and it occurs at the more electrophilic tertiary carbon.

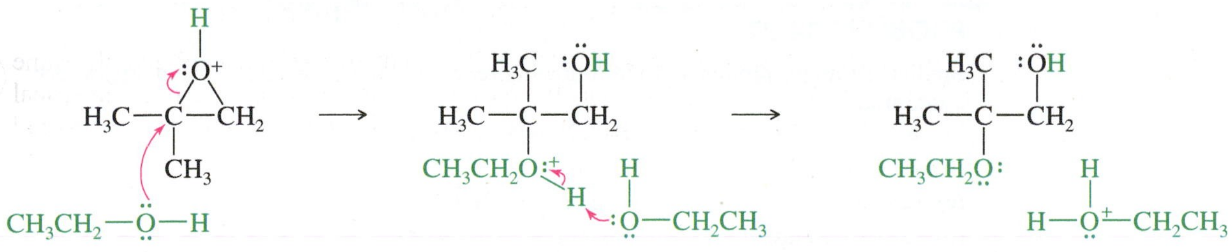

This ring opening is similar to the opening of a bromonium ion in the formation of a bromohydrin (Section 8-9) and the opening of the mercurinium ion during oxymercuration (Section 8-5). All three reactions involve the opening of an electrophilic three-membered ring by a weak nucleophile. Attack takes place at the more electrophilic carbon atom, which is usually the more substituted carbon because it can better support the positive charge. Most base-catalyzed epoxide openings, on the other hand, involve attack by a strong nucleophile at the less hindered carbon atom.

SUMMARY Orientation of Epoxide Ring Opening

Under acidic conditions where the epoxide oxygen is protonated, the nucleophile attacks the more-substituted position, as if it were an S_N1-type reaction, but still with inversion of stereochemistry.

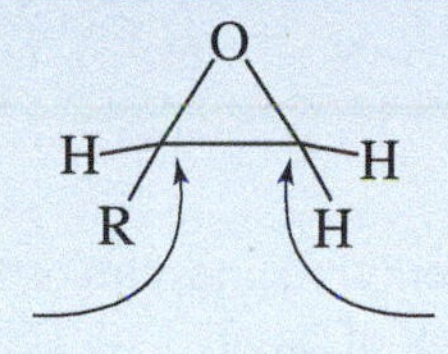

Under basic conditions, the nucleophile attacks the less-substituted position, as if it were an S_N2-type reaction.

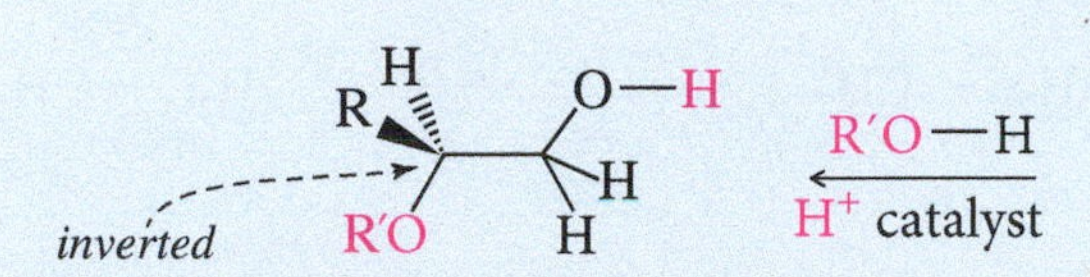

R'O—H
H^+ catalyst

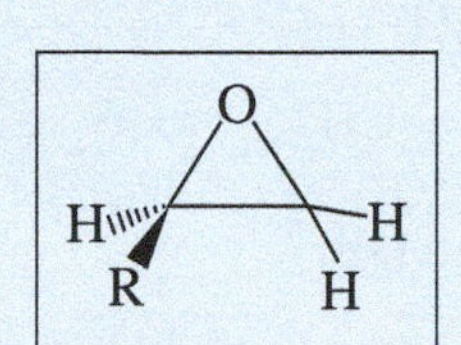

R'O⁻
R'O—H

retention of configuration

acid-catalyzed ring opening— nucleophile on the more substituted carbon; Sections 14-12, 14-14

ring opening in basic conditions— nucleophile attacks the less substituted carbon; Sections 14-13, 14-14

SOLVED PROBLEM 14-2

Predict the major products for the reaction of 1-methyl-1,2-epoxycyclopentane with
(a) sodium ethoxide in ethanol **(b)** H_2SO_4 in ethanol

SOLUTION

(a) Sodium ethoxide attacks the less hindered secondary carbon to give (*E*)-2-ethoxy-1-methylcyclopentanol.

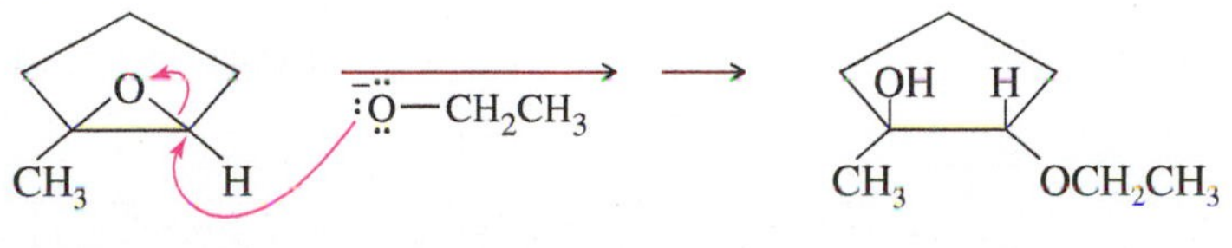

(b) Under acidic conditions, the alcohol attacks the more electrophilic tertiary carbon atom of the protonated epoxide. The product is (*E*)-2-ethoxy-2-methylcyclopentanol.

PROBLEM-SOLVING HINT

Acid-catalyzed: The nucleophile (solvent) adds to the more substituted carbon, which bears more + charge.
Base-catalyzed: The nucleophile attacks the less substituted carbon, which is less hindered.

PROBLEM 14-27

Predict the major products of the following reactions, including stereochemistry where appropriate.

(a) 2,2-dimethyloxirane + $H^+/H_2{}^{18}O$ (oxygen-labeled water)

(b) 2,2-dimethyloxirane + $H^{18}O^-/H_2{}^{18}O$

(c) (2*S*,3*R*)-2-ethyl-2,3-dimethyloxirane + CH_3O^-/CH_3OH

(d) (2*S*,3*R*)-2-ethyl-2,3-dimethyloxirane + H^+/CH_3OH

14-15 Reactions of Epoxides with Grignard and Organolithium Reagents

Like other strong nucleophiles, Grignard and organolithium reagents attack epoxides to give (after protonation) ring-opened alcohols.

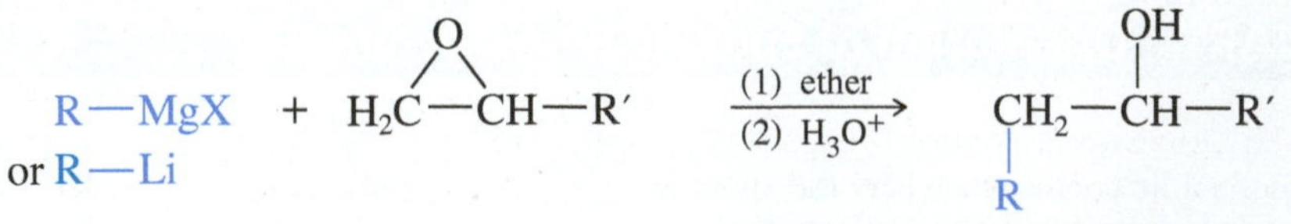

$$R{-}MgX \text{ or } R{-}Li + H_2C{-}CH{-}R' \text{ (epoxide, O bridging)} \xrightarrow[(2)\ H_3O^+]{(1)\ \text{ether}} R{-}CH_2{-}CH(OH){-}R'$$

For example, ethylmagnesium bromide reacts with oxirane (ethylene oxide) to form the magnesium salt of butan-1-ol. Protonation gives the neutral alcohol.

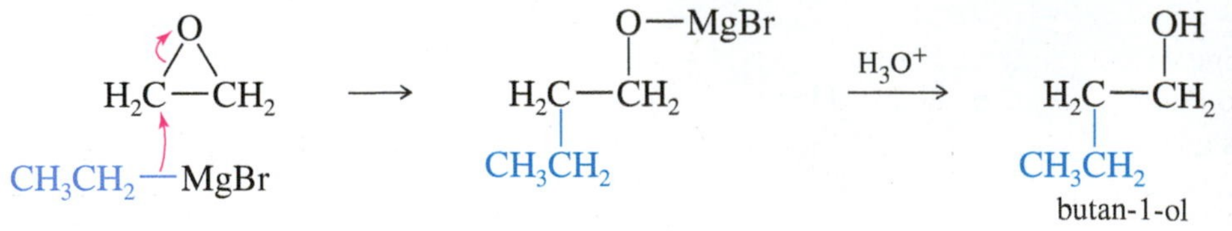

$$H_2C{-}CH_2 \text{ (oxirane)} + CH_3CH_2{-}MgBr \longrightarrow CH_3CH_2{-}H_2C{-}CH_2{-}O{-}MgBr \xrightarrow{H_3O^+} CH_3CH_2{-}H_2C{-}CH_2{-}OH$$

butan-1-ol

Substituted epoxides can be used in this reaction, with the carbanion usually attacking the less hindered epoxide carbon atom. This reaction works best if one of the oxirane carbons is unsubstituted, to allow an unhindered nucleophilic attack. Organolithium reagents (RLi) are more selective than Grignard reagents in attacking the less hindered epoxide carbon atom. Unless one carbon atom is very strongly hindered, Grignard reagents may give mixtures of products.

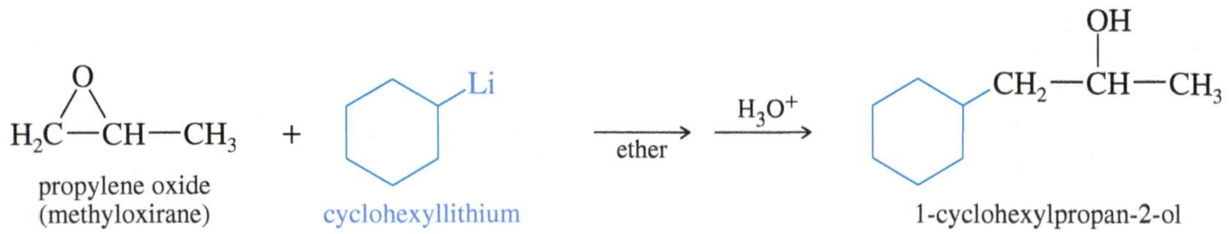

$$H_2C{-}CH{-}CH_3 \text{ (epoxide)} + \text{cyclohexyl}{-}Li \xrightarrow{\text{ether}} \xrightarrow{H_3O^+} \text{cyclohexyl}{-}CH_2{-}CH(OH){-}CH_3$$

propylene oxide (methyloxirane) + cyclohexyllithium → 1-cyclohexylpropan-2-ol

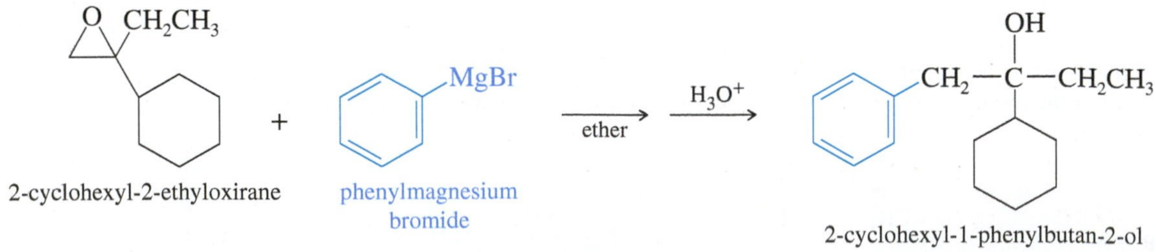

$$\text{2-cyclohexyl-2-ethyloxirane} + C_6H_5{-}MgBr \xrightarrow{\text{ether}} \xrightarrow{H_3O^+} C_6H_5{-}CH_2{-}C(OH)(C_6H_{11}){-}CH_2CH_3$$

2-cyclohexyl-2-ethyloxirane + phenylmagnesium bromide → 2-cyclohexyl-1-phenylbutan-2-ol

PROBLEM 14-28

Give the expected products of the following reactions. Include a protonation step where necessary.

(a) 2,2-dimethyloxirane + isopropylmagnesium bromide
(b) propylene oxide + *n*-butyllithium
(c) cyclopentyloxirane + ethyllithium

14-16 Epoxy Resins: The Advent of Modern Glues

The earliest glues were made of carbohydrates and proteins. Wheat paste uses the gluten in wheat, the sticky carbohydrate that holds bread together. Hide glue is a collagen-containing protein extract of animal hides, hooves, and tendons. Hide glue was used for wood and paper gluing for hundreds of years, and it is still used for fine musical instruments and other articles that must be readily taken apart without damaging the wood. Hide glue is water soluble, however, and the bond quickly fails in a damp environment. It does not fill gaps because it shrinks to a fraction of its wet volume as it dries. Glues based on casein (a milk protein) were developed to give a stronger, water-resistant bond. A casein glue (such as Elmer's®) gives a bond as strong as most woods, and it resists water for hours before it softens. However, it does not fill gaps well, and it works poorly with metals and plastics.

Workers guide the skin for the tail of a Harrier II jet into an autoclave where its layers of epoxy-coated carbon-fiber cloth will be bonded and cured under heat and pressure. The carbon-epoxy composite is used to make aircraft parts that are as strong as steel but as light as aluminum.

Imagine a glue that does not shrink at all as it hardens; it fills gaps perfectly so that pieces don't need to be fitted closely. It holds forever in water, is at least as strong as wood and plastic, and sticks to anything: wood, metal, plastic, etc. It lasts forever on the shelf without hardening, yet hardens quickly once the pieces are in place. It can be made runny so that it fills tiny voids, or thick and pasty so that it stays in place while it hardens.

This ideal glue was only a dream until the development of **epoxy** adhesives. Epoxies polymerize in place, so they match the shape of the joint perfectly and adhere to microscopic irregularities in the surfaces. There is no solvent to evaporate, so there is no shrinkage. Epoxies are bonded by ether linkages, so they are unaffected by water. Epoxies use a prepolymer that can be made as runny or as gummy as desired, and they use a hardening agent that can be modified to control the curing time. In the absence of the hardening agent, they have a long shelf life.

Manufacture and sales of epoxy resins have grown to a market of about \$20 billion worldwide, with China having become the largest manufacturer and consumer of the resins.

The most common epoxy resins use a prepolymer made from *bisphenol A* and epichlorohydrin.

$$HO-C_6H_4-C(CH_3)_2-C_6H_4-OH + H_2C\overset{O}{—}CH-CH_2Cl \longrightarrow$$

bisphenol A (BPA)   epichlorohydrin

$$H_2C\overset{O}{—}CH-CH_2-\left[O-C_6H_4-C(CH_3)_2-C_6H_4-O-CH_2-\overset{OH}{CH}-CH_2\right]_n-O-C_6H_4-C(CH_3)_2-C_6H_4-O-CH_2-CH\overset{O}{—}CH_2$$

prepolymer

Under base-catalyzed conditions, the anion of bisphenol A opens the epoxide of epichlorohydrin to give an alkoxide that snaps shut on the other end, forming another epoxide.

$$R-C_6H_4-\ddot{O}:^- + H_2C\overset{O}{—}CH-CH_2Cl \longrightarrow R-C_6H_4-O-CH_2-\overset{:\ddot{O}:^-}{CH}-CH_2Cl \longrightarrow R-C_6H_4-O-CH_2-CH\overset{O}{—}CH_2$$

This second epoxide reacts with another molecule of bisphenol A. Each molecule of bisphenol A can also react with two molecules of epichlorohydrin.

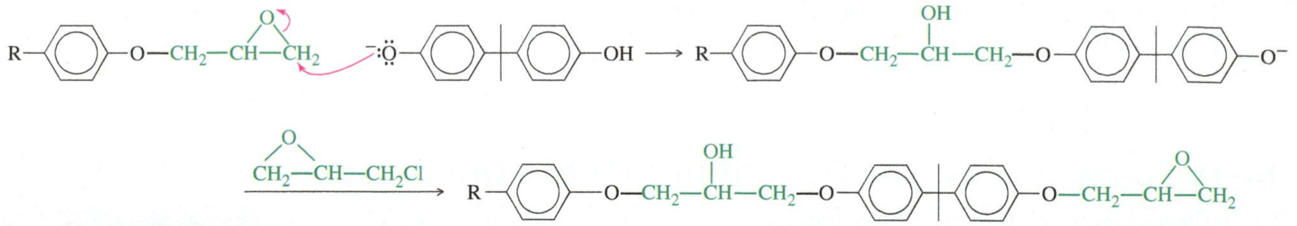

Application: Endocrine Disrupter

Bisphenol A (BPA) mimics the effects of estrogens, which can lead to health effects at high enough concentrations. BPA is used in polycarbonate bottles (see Section 21-16) and in some of the plastic linings of canned foods. Several countries have banned the sale of polycarbonate baby bottles and the use of canned food liners containing BPA because of their concerns that the polymers might hydrolyze and leach free BPA into the food or water in the container.

With exactly equal amounts of bisphenol A and epichlorohydrin, this polymerization would continue until the polymer chains were very long, and the material would be a solid polymer. In making epoxy resins, however, excess epichlorohydrin is added to form short chains with epichlorohydrins on both ends. More epichlorohydrin gives shorter chains and a runny prepolymer. Less epichlorohydrin gives longer chains (containing up to 25 epichlorohydrin/bisphenol A units) and a more viscous prepolymer.

When you buy epoxy glues, they come in two parts: the resin (prepolymer) and the hardener. The hardener can be any of a wide variety of compounds having basic or nucleophilic properties. Polyamines are the most common hardeners. The hardener can attack a terminal epoxide group, initiating a polymerization of the chain ends.

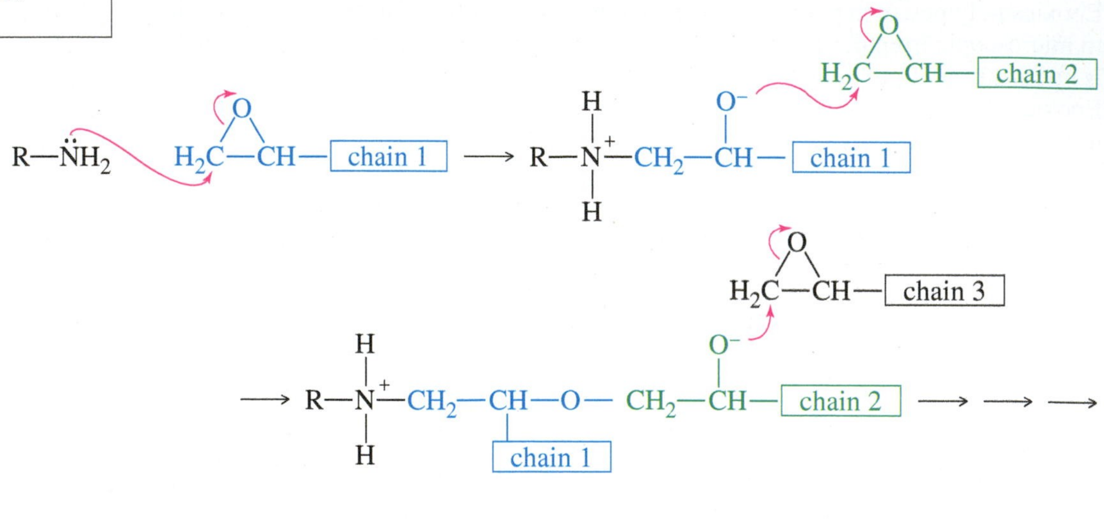

Or the hardener can deprotonate a hydroxy group from the interior of a chain, cross-linking one chain with another. The final polymer is an intricate, three-dimensional network that is strong and resistant to chemical attack.

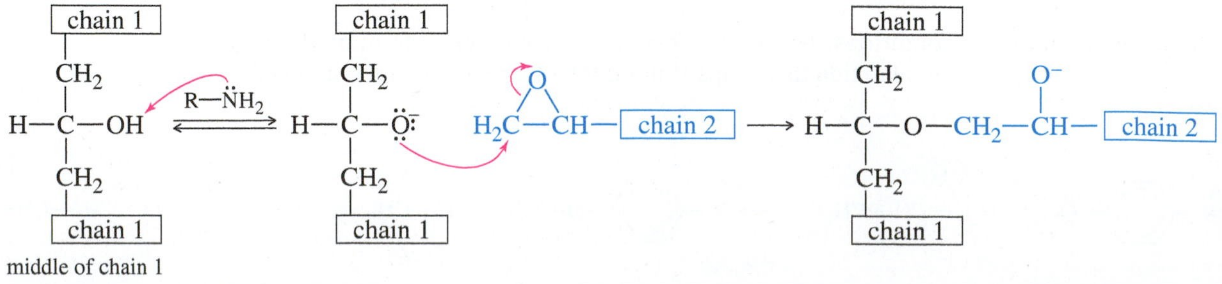

middle of chain 1

SUMMARY Reactions of Epoxides

1. *Acid-catalyzed opening* (Sections 8-13 and 14-12)
   a. *In water*

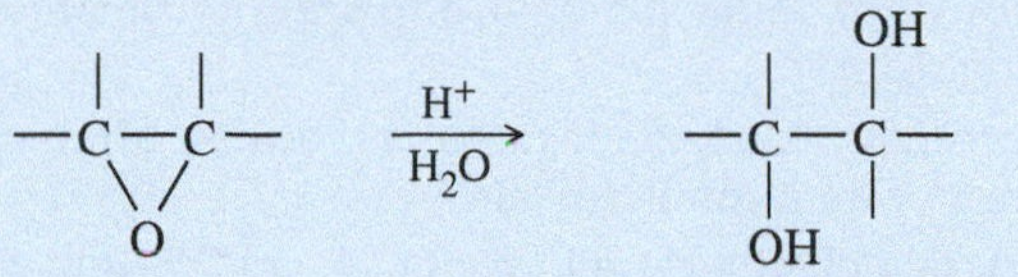

$$\text{epoxide} \xrightarrow[H_2O]{H^+} -\overset{\text{OH}}{\underset{}{C}}-\underset{\text{OH}}{C}-$$

anti stereochemistry

   b. *In alcohols*

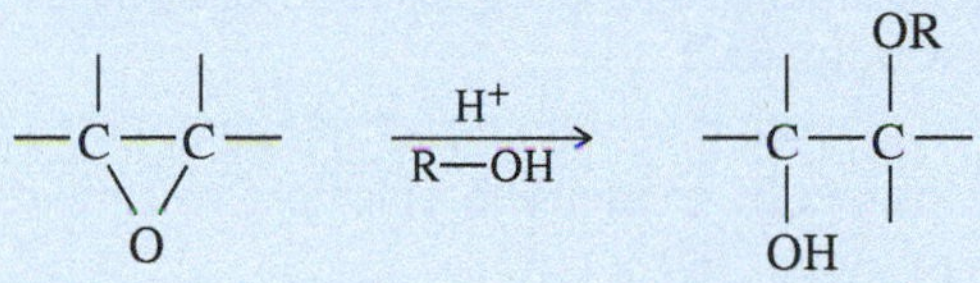

$$\text{epoxide} \xrightarrow[R-OH]{H^+} -\overset{\text{OR}}{C}-\underset{\text{OH}}{C}-$$

anti stereochemistry

The alkoxy group bonds to the more highly substituted carbon.

   c. *Using hydrohalic acids* (X = Cl, Br, I)

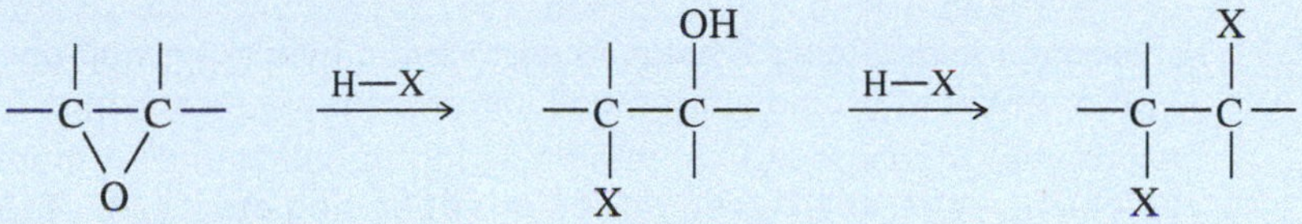

$$\text{epoxide} \xrightarrow{H-X} -\underset{X}{C}-\overset{\text{OH}}{C}- \xrightarrow{H-X} -\underset{X}{C}-\overset{X}{C}-$$

2. *Base-catalyzed opening*
   a. *With alkoxides or hydroxide* (Section 14-13)

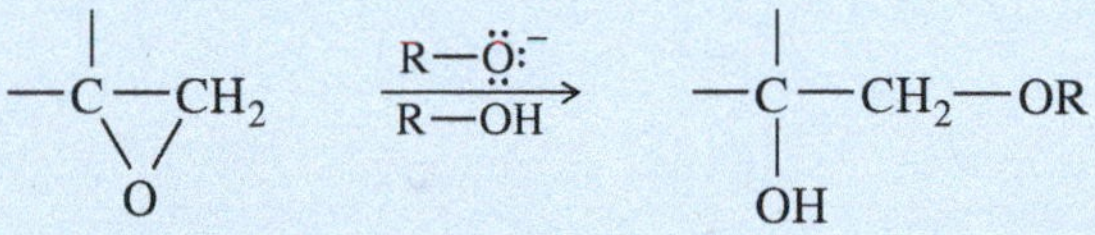

$$-C\!-\!CH_2 \ (\text{epoxide}) \xrightarrow[R-OH]{R-\ddot{O}:^-} -\underset{\text{OH}}{C}-CH_2-OR$$

The alkoxy group bonds to the less highly substituted carbon.

   b. *With organometallics* (Section 14-15)

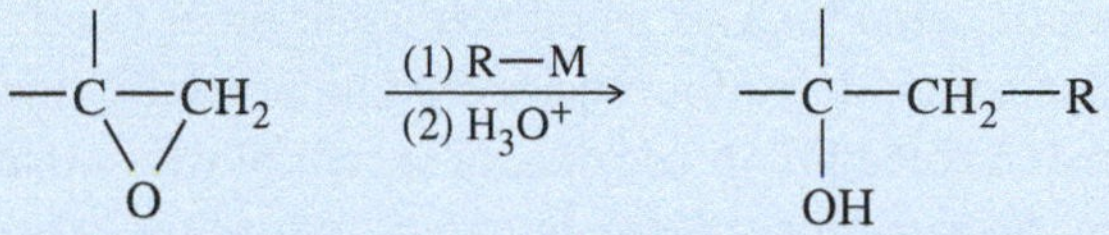

$$-C\!-\!CH_2 \ (\text{epoxide}) \xrightarrow[(2)\ H_3O^+]{(1)\ R-M} -\underset{\text{OH}}{C}-CH_2-R$$

M = Li or MgX

R bonds to the less substituted carbon

Essential Terms

alkoxy group **(alkoxyl group)** A substituent consisting of an alkyl group bonded through an oxygen atom, —O—R. (p. 677)

alkoxymercuration Addition of mercury and an alkoxy group to a double bond, usually by a solution of mercuric acetate in an alcohol. Alkoxymercuration is usually followed by sodium borohydride reduction **(demercuration)** to give an ether. (p. 684)

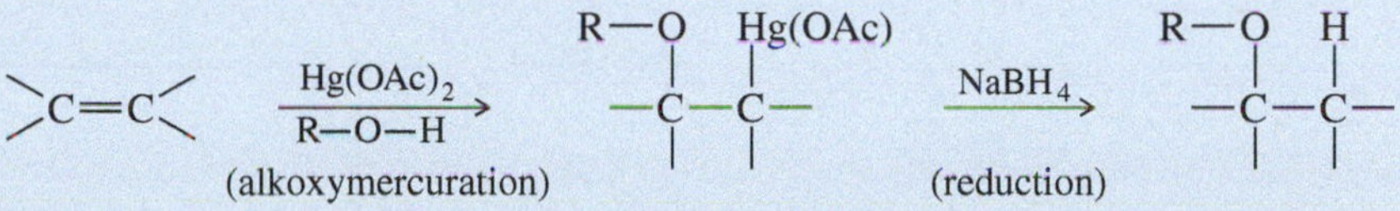

$$>C=C< \xrightarrow[R-O-H]{Hg(OAc)_2} -\overset{R-O}{C}-\overset{Hg(OAc)}{C}- \xrightarrow{NaBH_4} -\overset{R-O}{C}-\overset{H}{C}-$$

(alkoxymercuration) (reduction)

α cleavage The breaking of a bond between the first and second carbon atoms adjacent to the ether oxygen atom (or other functional group). (p. 680)

autoxidation Any oxidation that proceeds spontaneously using oxygen in the air. Autoxidation of ethers gives hydroperoxides and dialkyl peroxides. (p. 688)

concerted reaction A reaction that takes place in one step, with simultaneous bond breaking and bond forming. (p. 694)

condensation A reaction that joins two (or more) molecules, often with the loss of a small molecule such as water or an alcohol. (p. 684)

crown ether A large cyclic polyether used to complex and solvate cations in nonpolar solvents. (p. 676)

dioxane A heterocyclic ether with two oxygen atoms in a six-membered ring. (p. 679)

epoxidation Oxidation of an alkene to an epoxide. Usually accomplished by treating the alkene with a peroxyacid. (p. 693)

epoxide **(oxirane)** A compound containing a three-membered heterocyclic ether. (p. 678)

epoxy resins Polymers formed by condensing epichlorohydrin with a dihydroxy compound, most often bisphenol A. (p. 705)

ether A compound with two alkyl (or aryl) groups bonded to an oxygen atom, R—O—R′. (p. 672)

- **symmetrical ether:** An ether with two identical alkyl groups.
- **unsymmetrical ether:** An ether with two different alkyl groups.

furan The five-membered heterocyclic ether with two carbon–carbon double bonds; or a derivative of furan. (p. 679)

halohydrin A compound containing a halogen atom and a hydroxy group on adjacent carbon atoms. Chlorohydrins, bromohydrins, and iodohydrins are most common. (p. 694)

heterocyclic compound **(heterocycle)** A compound containing a ring in which one or more of the ring atoms are elements other than carbon. The noncarbon ring atoms are called **heteroatoms**. (p. 678)

heterocyclic ethers:

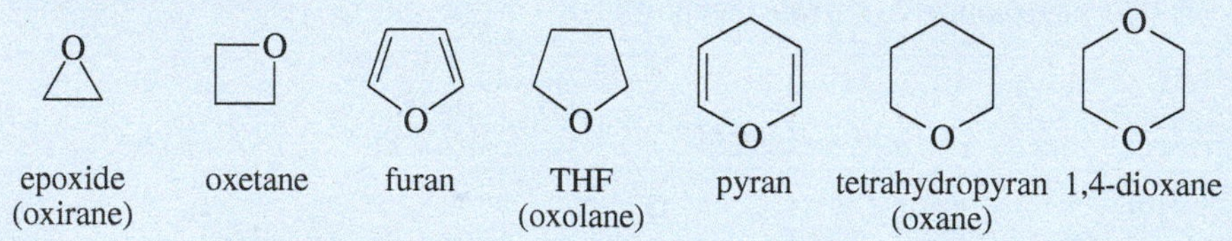

epoxide (oxirane) | oxetane | furan | THF (oxolane) | pyran | tetrahydropyran (oxane) | 1,4-dioxane

mCPBA An abbreviation for *meta*-chloroperoxybenzoic acid, a common epoxidizing agent. (p. 693)

MMPP An abbreviation for magnesium monoperoxyphthalate, a relatively stable peroxyacid often used in large-scale epoxidations. (p. 694)

oxane The systematic name for a six-membered cyclic ether (a tetrahydropyran). (p. 679)

oxetane A compound containing a four-membered heterocyclic ether. (p. 679)

oxirane The systematic name for an epoxide, or specifically for ethylene oxide. (p. 678)

oxolane The systematic name for a five-membered cyclic ether (a tetrahydrofuran). (p. 679)

peroxide Any compound containing the —O—O— linkage. The oxygen–oxygen bond is easily cleaved, and organic peroxides are prone to explosions. (p. 688)

H—O—O—H hydrogen peroxide | R—O—O—H alkyl hydroperoxide | R—O—O—R′ dialkyl peroxide

peroxyacid **(peracid)** A carboxylic acid with an extra oxygen in the hydroxy group. (p. 693)

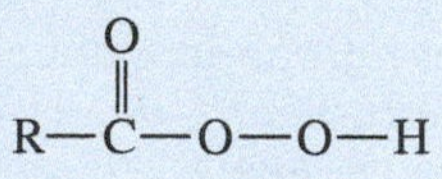

R—C(=O)—O—O—H a peroxyacid 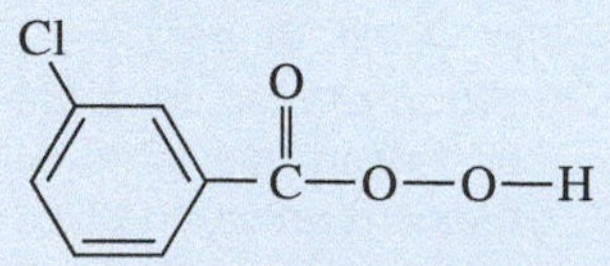 | *meta*-chloroperoxybenzoic acid (mCPBA)

protecting group A group used to prevent a sensitive functional group from reacting while another part of the molecule is being modified. The protecting group is later removed. For example, an alcohol can be converted to a silyl ether to protect it against most acids and bases and also against many oxidizing and reducing agents. Treatment with a tetraalkylammonium fluoride such as $Bu_4N^+ F^-$ deprotects the alcohol. (p. 692)

pyran The six-membered heterocyclic ether with two carbon–carbon double bonds; or a derivative of pyran. (p. 679)

silyl ether An ether of formula R'-O-SiR_3 with a substituted silicon atom replacing one of the alkyl groups of an ether. Used as protecting groups for alcohols. (p. 689)

triisopropylsilyl ether: **(TIPS ether)** A silyl ether of formula R'-O-Si(i-Pr)$_3$ commonly used to protect alcohol groups. Formed from an alcohol with TIPSCl and a tertiary amine. Deprotected using aqueous fluoride salts. (p. 692)

sulfone A compound of formula $R-SO_2-R'$ (see below). (p. 690)

sulfonium salt A salt containing a sulfur atom bonded to three alkyl groups, R_3S^+, and a counterion (see below). (p. 691)

sulfoxide A compound of formula $R-SO-R'$ (see below). (p. 690)

thioether (sulfide)	sulfonium salt	sulfoxide	sulfone
$R-S-R'$	$R-\overset{R''}{\underset{X^-}{S^+}}-R'$	$R-\overset{O}{\overset{\Vert}{S}}-R'$	$R-\overset{O}{\overset{\Vert}{\underset{O}{\underset{\Vert}{S}}}}-R'$

thioether **(sulfide)** A compound with two alkyl (or aryl) groups bonded to a sulfur atom, $R-S-R'$. (p. 689)

Williamson ether synthesis Formation of an ether by the S_N2 reaction of an alkoxide ion with an alkyl halide or tosylate. In general, the electrophile ($R'-X$) must be primary, or occasionally secondary. (p. 682)

$$R-\ddot{\underset{..}{O}}:^- \quad R'-X \longrightarrow R-O-R' + X^-$$

Essential Problem-Solving Skills in Chapter 14

Each skill is followed by problem numbers exemplifying that particular skill.

1 Draw and name ethers and heterocyclic ethers, including epoxides. Predict their relative boiling points, solubilities, and solvent properties.	Problems 14-29, 30, 31, and 32
2 Determine the structure of ethers from their spectra, and predict their characteristic absorptions and fragmentations.	Problems 14-38, 44, 51, and 52
3 Devise efficient laboratory syntheses of ethers and epoxides, including the following: **(a)** The Williamson ether synthesis **(b)** Alkoxymercuration-demercuration **(c)** Peroxyacid epoxidation **(d)** Base-promoted cyclization of halohydrins **(e)** Formation of silyl ethers	Problems 14-34, 40, 41, 42, 43, 45, 46, and 53
4 Predict the products of reactions of ethers and epoxides, including the following: **(a)** Cleavage and autoxidation of ethers **(b)** Acid- and base-promoted opening of epoxides **(c)** Reactions of epoxides with organometallic reagents **(d)** Cleavage of silyl ethers	Problems 14-33, 36, 46, 50, and 51
5 Propose mechanisms for the formation and reactions of ethers and epoxides.	Problems 14-39, 47, 48, 49, 53, and 54

Study Problems

14-29 Write structural formulas for the following compounds.

(a) ethyl isopropyl ether **(b)** di-*n*-butyl ether **(c)** 2-ethoxyoctane
(d) divinyl ether **(e)** allyl methyl ether **(f)** cyclohexene oxide
(g) *cis*-2,3-epoxyhexane **(h)** (2*R*,3*S*)-2-methoxypentan-3-ol **(i)** *trans*-2,3-dimethyloxirane

14-30 Give common names for the following compounds.

(a) $(CH_3)_2CHOCH(CH_3)CH_2CH_3$ **(b)** $(CH_3)_3COCH_2CH(CH_3)_2$ **(c)** $PhOCH_2CH_3$
(d) $ClCH_2OCH_2CH_2CH_3$ **(e)** 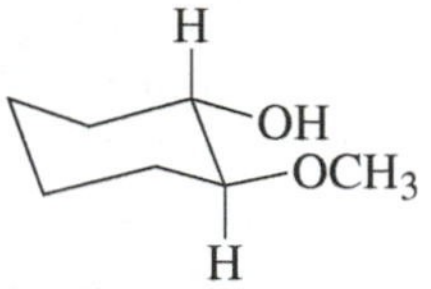 H, OH, OCH_3, H **(f)** 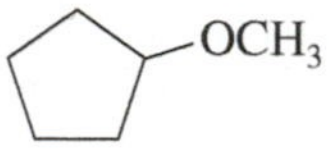 OCH_3
(g) O, CH_3 **(h)** 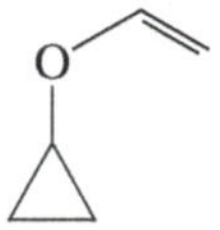 O **(i)**  O **(j)** 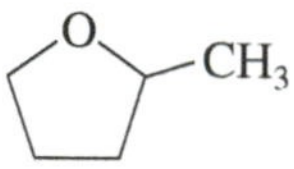 O, CH_3

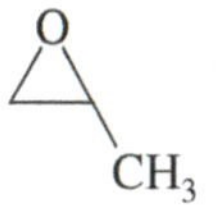

14-31 Give IUPAC names for the following compounds.

(a) $CH_3OCH(CH_3)CH_2OH$ **(b)** $PhOCH_2CH_3$ **(c)** 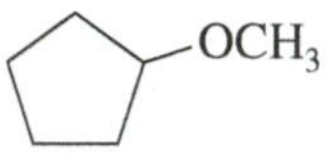 OCH_3
(d) 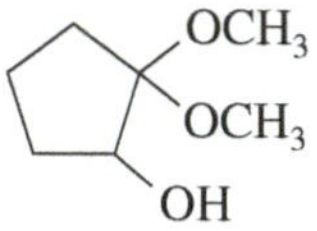 OCH_3, OCH_3, OH **(e)** 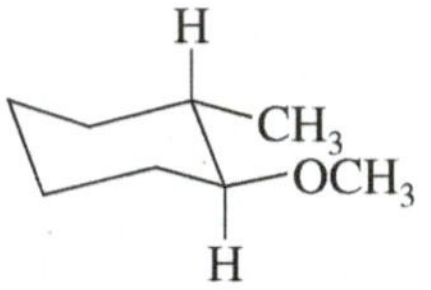 H, CH_3, OCH_3, H **(f)** 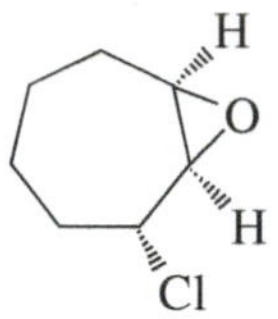 H, O, H, Cl
(g) 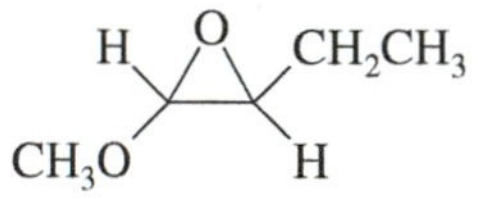 H, O, CH_2CH_3, CH_3O, H **(h)** 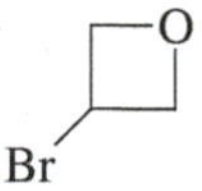 O, Br **(i)** 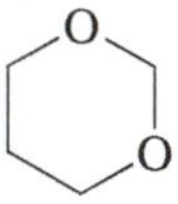 O, O

14-32 Glycerol (propane-1,2,3-triol) is a viscous syrup with molecular weight 92 g/mol, boiling point 290 °C, and density 1.24 g/mL. Transforming the three hydroxy groups into their trimethylsilyl ethers (using chlorotrimethylsilane and a tertiary amine) produces a liquid that flows easily, has molecular weight 309 g/mol, boiling point approximately 180 °C, and density 0.88 g/mL. Draw the structures of these two compounds, and explain why glycerol has a lower molecular weight but a much higher boiling point and density.

14-33 Predict the products of the following reactions.

(a) *sec*-butyl isopropyl ether + concd. HBr, heat **(b)** 2-ethoxy-2-methylpentane + concd. HBr, heat
(c) di-*n*-butyl ether + hot concd. NaOH **(d)** di-*n*-butyl ether + Na metal
(e) ethoxybenzene + concd. HI, heat **(f)** 1,2-epoxyhexane + H^+, CH_3OH
(g) *trans*-2,3-epoxyoctane + H^+, H_2O **(h)** propylene oxide + methylamine (CH_3NH_2)
(i) potassium *tert*-butoxide + *n*-butyl bromide **(j)** 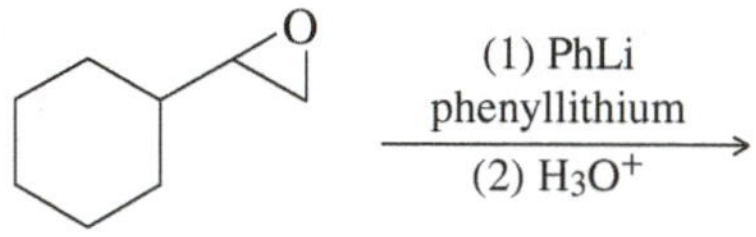 O; (1) PhLi phenyllithium; (2) H_3O^+ →
(k) 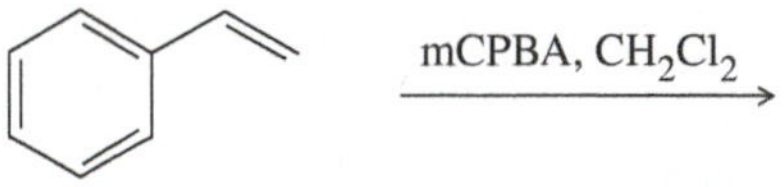 mCPBA, CH_2Cl_2 → **(l)** 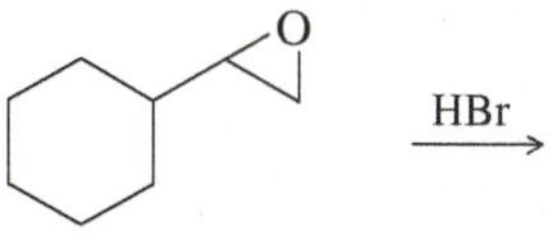 O; HBr →
(m) 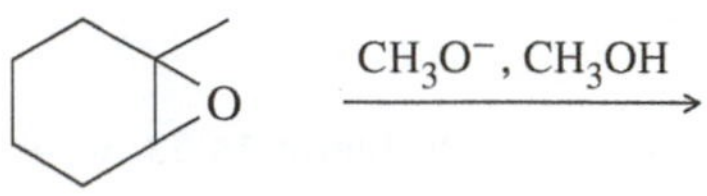 O; CH_3O^-, CH_3OH → **(n)** 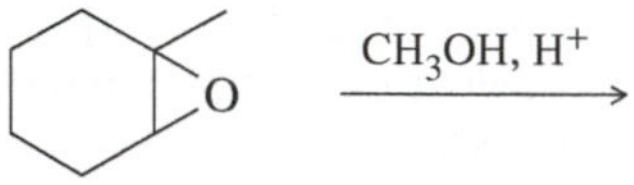 O; CH_3OH, H^+ →

14-34 Show how you would make the following ethers, using only simple alcohols and any needed reagents as your starting materials.

(a) 1-methoxybutane
(b) 2-ethoxy-2-methylpropane
(c) benzyl cyclopentyl ether
(d) *trans*-2-methoxycyclohexanol
(e) the TIPS ether of (d)
(f) cyclohexyl cyclopentyl ether

14-35 (A true story.) An inexperienced graduate student moved into a laboratory and began work. He needed some diethyl ether for a reaction, so he opened an old, rusty 1-gallon can marked "ethyl ether" and found there was half a gallon left. To purify the ether, the student set up a distillation apparatus, started a careful distillation, and went to the stockroom for the other reagents he needed. While he was at the stockroom, the student heard a muffled "boom." He quickly returned to his lab to find a worker from another laboratory putting out a fire. Most of the distillation apparatus was embedded in the ceiling.

(a) Explain what probably happened.
(b) Explain how this near-disaster might have been prevented.

14-36 Grignard reactions are often limited by steric hindrance. While Grignard reagents react in high yield with ethylene oxide and monosubstituted epoxides, yields are often lower with disubstituted epoxides. Tri- and tetrasubstituted epoxides react with difficulty, if at all.

(a) Show how to make these alcohols by a Grignard reacting with an epoxide.

(i) 2 ways

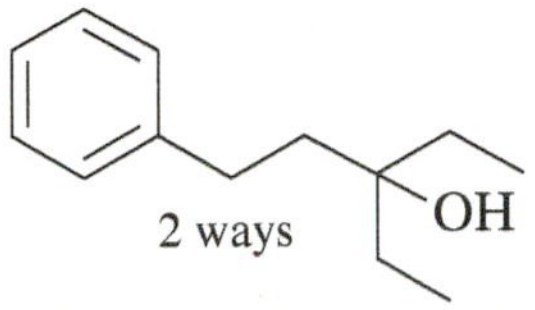

(ii)

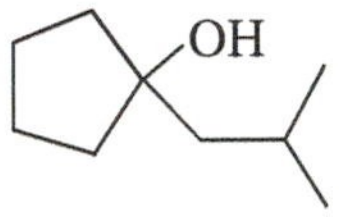

(b) These alcohols cannot be made by a Grignard plus an epoxide. Show the reagents that would be required and why that reaction would be unlikely to succeed.

(i)

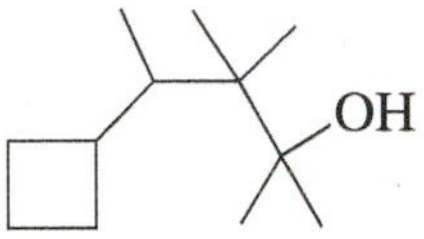

(ii)

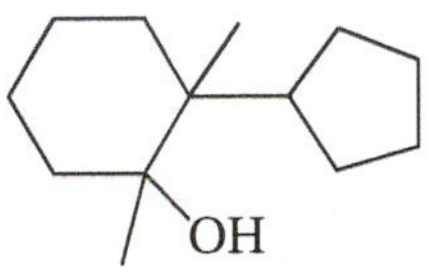

14-37 **(a)** Show how you would synthesize the pure (*R*) enantiomer of 2-butyl methyl sulfide, starting with pure (*R*)-butan-2-ol and any reagents you need.
(b) Show how you would synthesize the pure (*S*) enantiomer of the product, still starting with (*R*)-butan-2-ol and any reagents you need.

14-38 **(a)** Predict the values of m/z and the structures of the most abundant fragments you would observe in the mass spectrum of di-*n*-propyl ether.
(b) Give logical fragmentations to account for the following ions observed in the mass spectrum of 2-methoxypentane: 102, 87, 71, 59, 31.

14-39 The following reaction resembles the acid-catalyzed cyclization of squalene oxide. Propose a mechanism for this reaction.

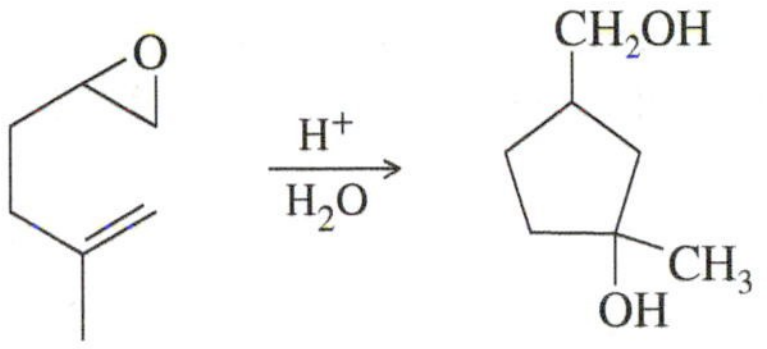

$\xrightarrow[H_2O]{H^+}$

14-40 Show how you would convert hex-1-ene to each of the following compounds. You may use any additional reagents and solvents you need.

(a) 2-methoxyhexane
(b) 1-methoxyhexane
(c) 1-methoxyhexan-2-ol
(d) 2-methoxyhexan-1-ol
(e) 1-phenylhexan-2-ol
(f) 2-methoxy-1-phenylhexane

14-41 Both $LiAlH_4$ and Grignard reagents react with carbonyl compounds to give alkoxide ion intermediates (that become protonated in an aqueous workup). Those alkoxides can react with 1° or methyl alkyl halides or tosylates to give ethers. Show how the following ethers can be formed in this two-step process. As starting materials you may use any reactants containing 7 carbons or fewer.

(a)

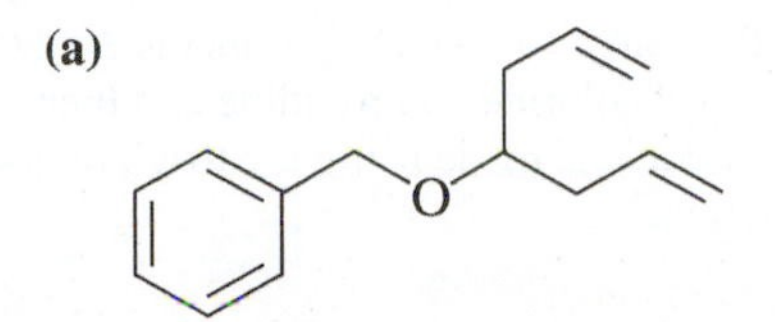

(b)

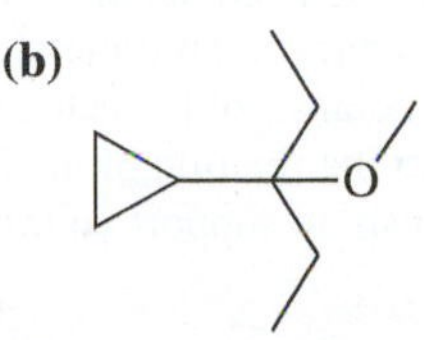

(c)

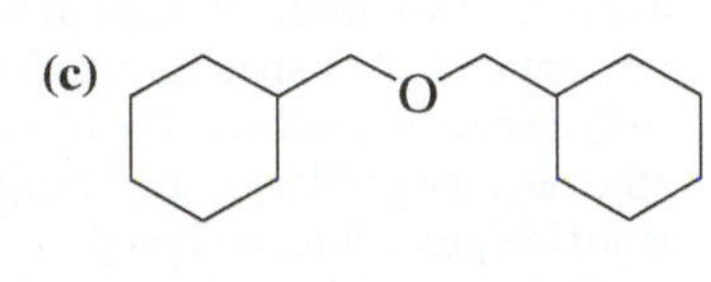

14-42 Give the structures of intermediates **A** through **H** in the following synthesis of *trans*-1-cyclohexyl-2-methoxycyclohexane.

OH; H_2SO_4, heat → **A**; **A** → (1) $Hg(OAc)_2$, CH_3OH (2) $NaBH_4$ → **B**; **B** → concd. HBr, heat → **C** (gas) + **D**; **D** → Mg, ether → **E**; **A** → mCPBA → **F**; **F** → (1) **E** (2) H_3O^+ → **G**; **G** → Na → **H**; **H** → **C** → product

14-43 Give the structures of the intermediates represented by letters in this synthesis.

OCH_3; $LiAlH_4$, ether → H_3O^+ → J; J → 1 equiv. NaOCl, TEMPO → K; J → TsCl, pyridine → L; L → KO-*t*Bu → M; M → Br_2, H_2O → N; N → TIPS–Cl, Et_3N → O; O → Mg, ether → P; P + K → ether → Q; Q → H_3O^+, $Bu_4N^+F^-$ → OH OH

14-44 (Another true story.) An organic lab student carried out the reaction of methylmagnesium iodide with acetone (CH_3COCH_3), followed by hydrolysis. During the distillation to isolate the product, she forgot to mark the vials she used to collect the fractions. She turned in a product of formula $C_4H_{10}O$ that boiled at 35 °C. The IR spectrum showed only a weak O—H stretch around 3300 cm^{-1}, and the mass spectrum showed a base peak at m/z 59. The NMR spectrum showed a quartet ($J = 7$ Hz) of area 2 at δ 3.5 and a triplet ($J = 7$ Hz) of area 3 at δ 1.3. Propose a structure for this product, explain how it corresponds to the observed spectra, and suggest how the student isolated this compound.

14-45 Show how you would synthesize the following ethers in good yield from the indicated starting materials and any additional reagents needed.

(a) cyclopentyl *n*-propyl ether from cyclopentanol and propan-1-ol
(b) *n*-butyl phenyl ether from phenol and butan-1-ol
(c) 2-ethoxyoctane from an octene
(d) 1-methoxydecane from a decene
(e) 1-ethoxy-1-methylcyclohexane from 2-methylcyclohexanol
(f) *trans*-2,3-epoxyoctane from octan-2-ol

14-46 Show how you would convert 3-bromocyclohexanol to the following diol. You may use any additional reagents you need.

HO, Br → (several steps) → HO, CH_3, OH, CH_2CH_3

14-47 There are two different ways of making 2-ethoxyoctane from octan-2-ol using the Williamson ether synthesis. When pure (−)-octan-2-ol of specific rotation −8.24° is treated with sodium metal and then ethyl iodide, the product is 2-ethoxyoctane with a specific rotation of −15.6°. When pure (−)-octan-2-ol is treated with tosyl chloride and pyridine and then with sodium ethoxide, the product is also 2-ethoxyoctane. Predict the rotation of the 2-ethoxyoctane made using the tosylation/sodium ethoxide procedure, and propose a detailed mechanism to support your prediction.

14-48 **(a)** When ethylene oxide is treated with anhydrous HBr gas, the major product is 1,2-dibromoethane. When ethylene oxide is treated with concentrated aqueous HBr, the major product is ethylene glycol. Use mechanisms to explain these results.

(b) Under base-catalyzed conditions, several molecules of propylene oxide can react to give short polymers. Propose a mechanism for the base-catalyzed formation of the following trimer.

$$3\ \underset{\text{(epoxide, O bridging C–C)}}{H_2C-CH-CH_3} \xrightarrow{^-OH} HO-CH_2-\overset{CH_3}{\overset{|}{C}H}-O-CH_2-\overset{CH_3}{\overset{|}{C}H}-O-CH_2-\overset{CH_3}{\overset{|}{C}H}-OH$$

14-49 An acid-catalyzed reaction was carried out using methyl cellosolve (2-methoxyethanol) as the solvent. When the 2-methoxyethanol was redistilled, a higher-boiling fraction (bp 162 °C) was also recovered. The mass spectrum of this fraction showed the molecular weight to be 134. The IR and NMR spectra are shown here. Determine the structure of this compound, and propose a mechanism for its formation.

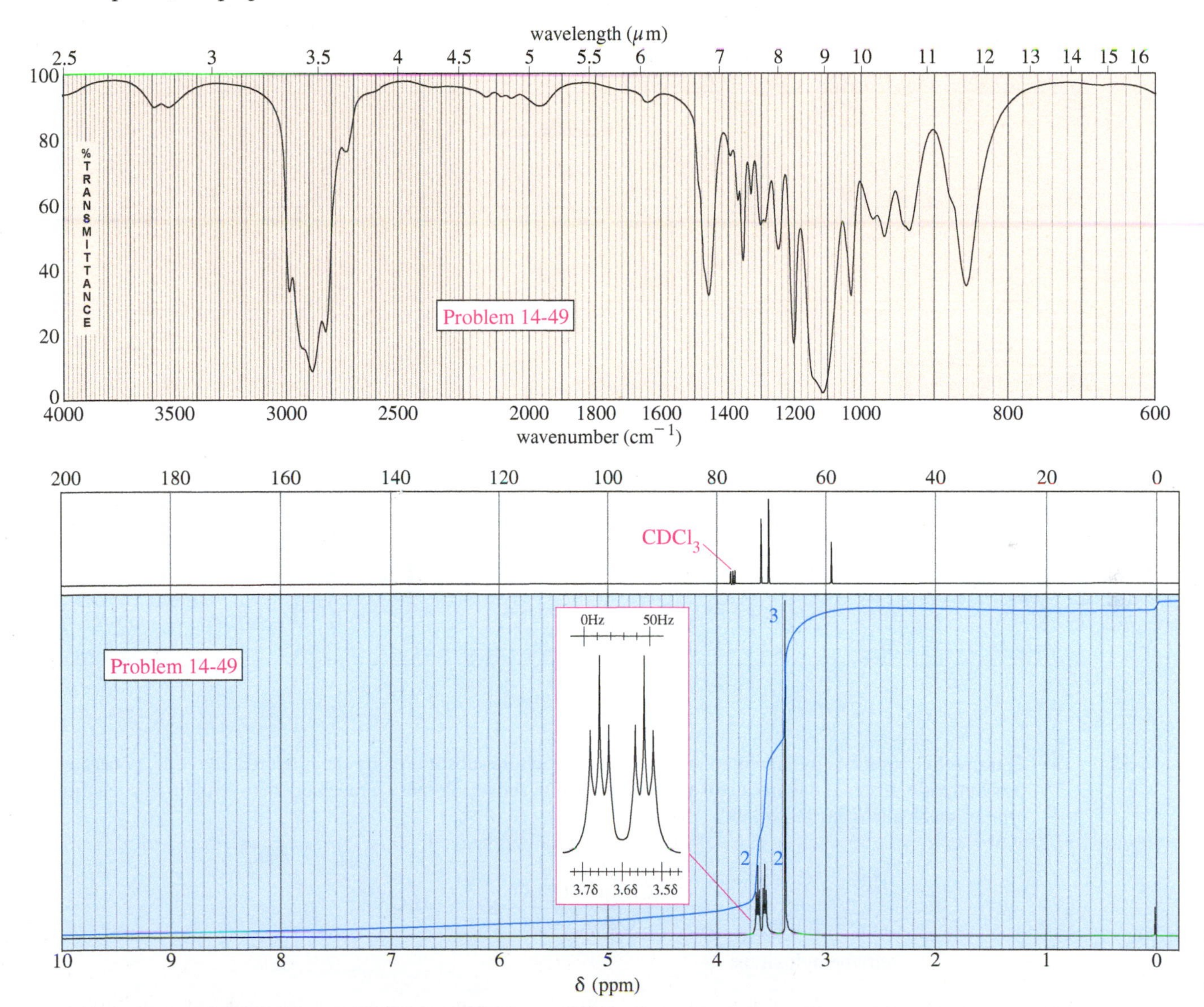

***14-50** Propylene oxide is a chiral molecule. Hydrolysis of propylene oxide gives propylene glycol, another chiral molecule.

(a) Draw the enantiomers of propylene oxide.

(b) Propose a mechanism for the acid-catalyzed hydrolysis of pure (*R*)-propylene oxide.

(c) Propose a mechanism for the base-catalyzed hydrolysis of pure (*R*)-propylene oxide.

(d) Explain why the acid-catalyzed hydrolysis of optically active propylene oxide gives a product with lower enantiomeric excess and a rotation opposite that of the product of the base-catalyzed hydrolysis.

14-51 A compound of molecular formula C_8H_8O gives the IR and NMR spectra shown here. Propose a structure, and show how it is consistent with the observed absorptions.

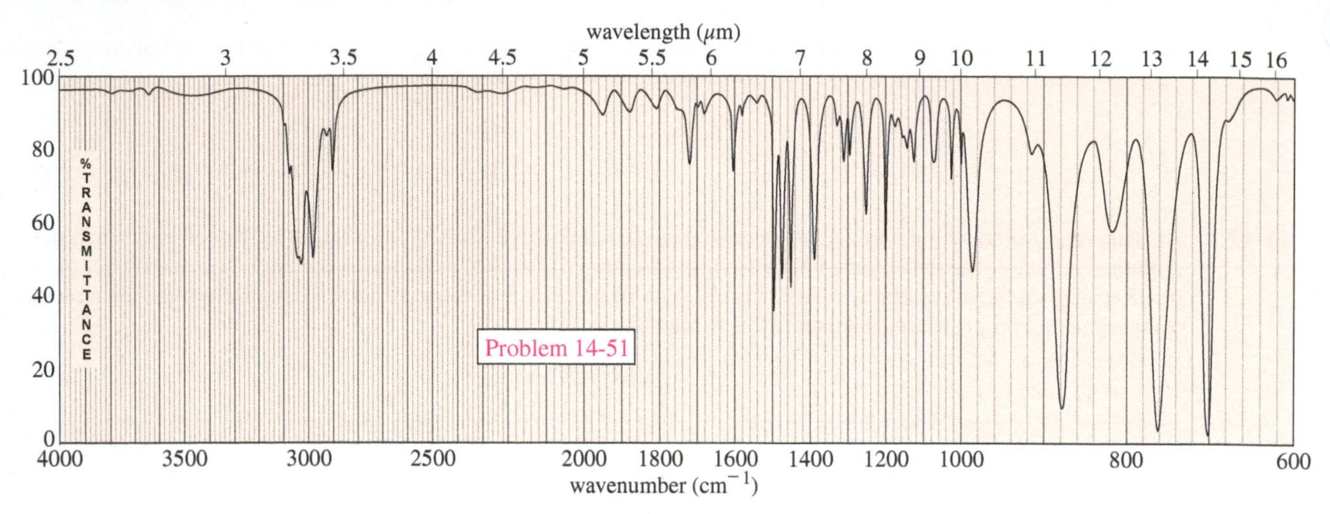

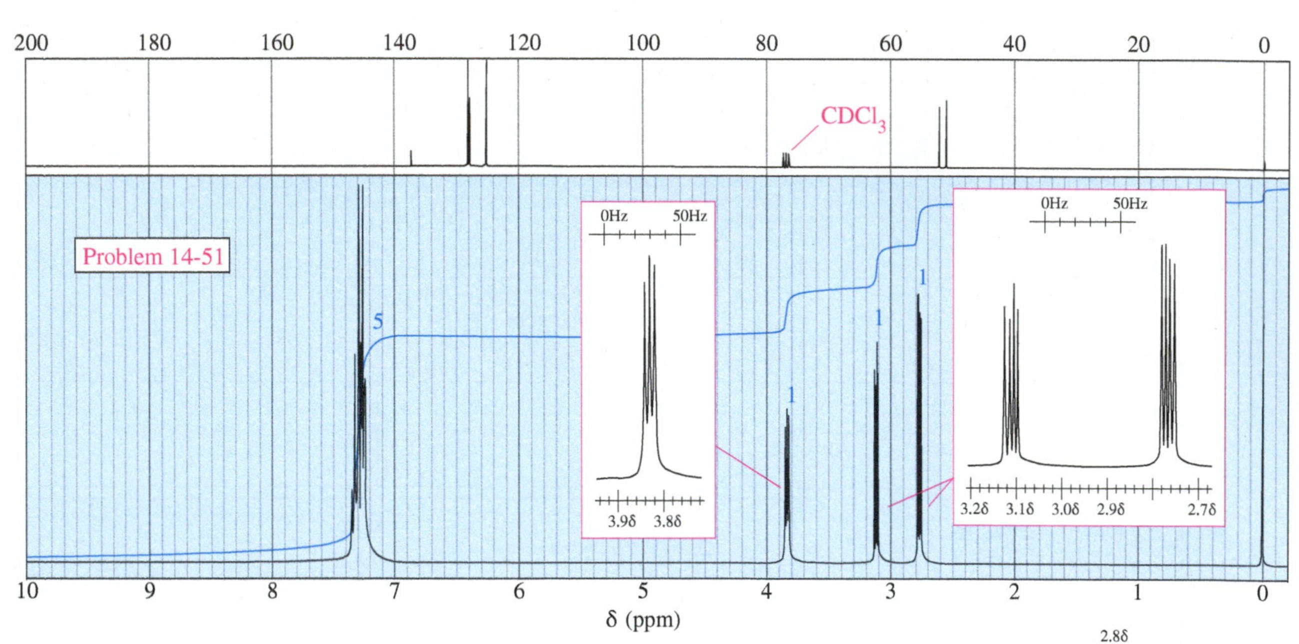

***14-52** **(a)** Tetramethyloxirane is too hindered to undergo nucleophilic substitution by the hindered alkoxide, potassium *tert*-butoxide. Instead, the product is the allylic alcohol shown. Propose a mechanism to explain this reaction. What type of mechanism does it follow?

tetramethyloxirane $\xrightarrow[(CH_3)_3C{-}OH]{(CH_3)_3C{-}O^-\ K^+}$ $H_3C{-}C(OH)(CH_3){-}C(=CH_2)CH_3$

(b) Under mild acid catalysis, 1,1-diphenyloxirane undergoes a smooth conversion to diphenylethanal (diphenylacetaldehyde). Propose a mechanism for this reaction. (Hint: Think Pinacol.)

1,1-diphenyloxirane $\xrightarrow{H^+}$ $Ph_2CH{-}CHO$

14-53 Under the right conditions, the following acid-catalyzed double cyclization proceeds in remarkably good yields. Propose a mechanism. Does this reaction resemble a biological process you have seen?

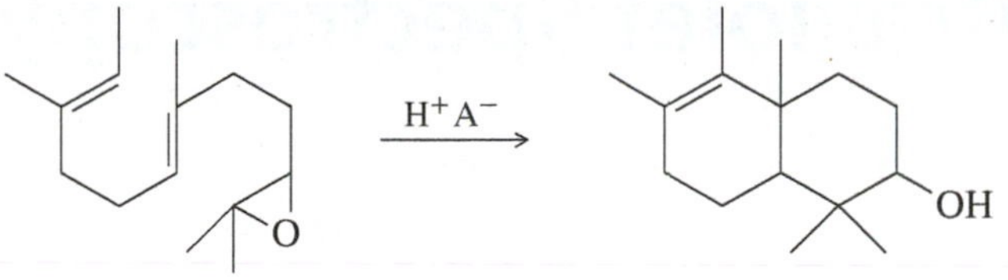

H^+A^-

***14-54** One of the crowning achievements of natural products synthesis was Bryostatin 1, published by Professor Gary Keck (University of Utah; *Journal of the American Chemical Society*, **2011**, *133*, 744–747). The Bryostatins are a family of compounds isolated from aquatic invertebrates known as Bryozoans. The compounds are of interest for a variety of biological effects, including anti-cancer activity and reversing brain damage in rodents.

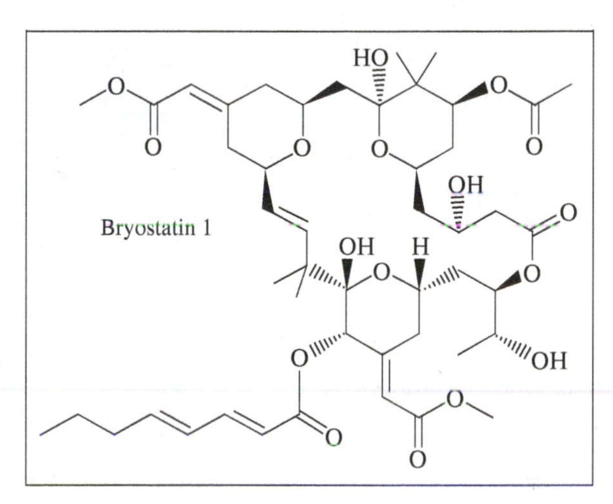
Bryostatin 1

(a) How many ether functional groups are present in Bryostatin 1?

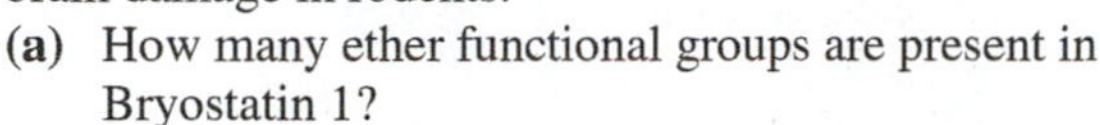

(b) Identify the other oxygen-containing functional groups.

(c) This is called a *macrolide* because it contains a large number of atoms in the large ring. How many atoms are in the large ring?

(d) How many chiral centers are in this molecule?

(e) Using the number of chiral centers you reported in part (d), calculate the number of stereoisomers possible at these chiral centers. (Ignore stereoisomers at double bonds.)

***14-55** In 2012, a group led by Professor Masayuki Satake of the University of Tokyo reported the isolation and structure determination of a toxin from a marine algal bloom that decimated the fish population off the New Zealand coast in 1998. Extensive mass spectrometry and NMR experiments ultimately led to the structure shown below, named Brevisulcenal-F. (See *Journal of the American Chemical Society*, **2012**, *134*, 4963–4968.) This structure holds the record for the largest number of fused rings, at 17.

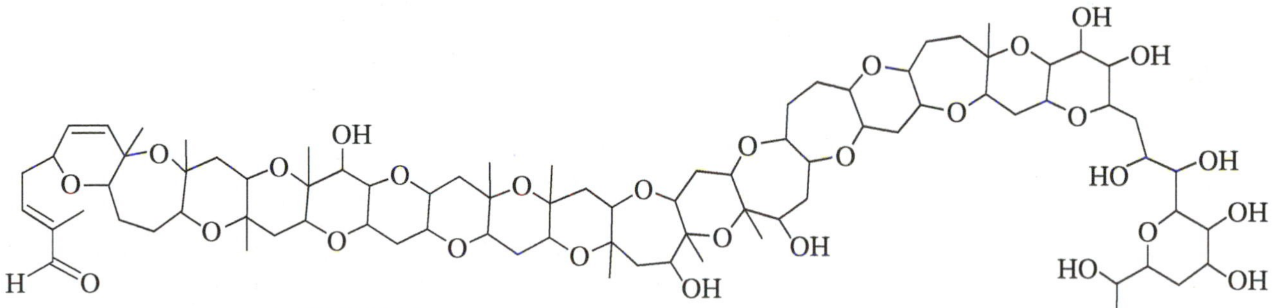

Brevisulcenal-F

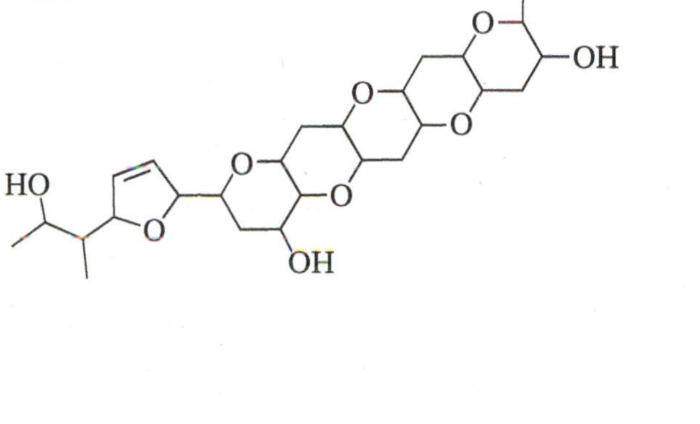

(a) How many ether groups are present?

(b) How many alcohol groups are present? Classify the alcohols as 1° or 2° or 3°.

(c) Are there any other oxygen-containing functional groups? Which, if any?

15 Conjugated Systems, Orbital Symmetry, and Ultraviolet Spectroscopy

Goals for Chapter 15

1 Explain how to construct the molecular orbitals of butadiene and other conjugated systems.

2 Draw resonance forms and propose mechanisms to explain the observed products and the enhanced rates of reactions involving conjugated reactants and resonance-stabilized intermediates.

3 Predict the products of Diels–Alder reactions, and determine which Diels–Alder reaction will give a specific synthetic product.

4 Predict which cycloadditions are thermally allowed and which are photochemically allowed by comparing the symmetry of the molecular orbitals of the reactants.

5 Use values of λ_{max} from UV–visible spectra to estimate the length of conjugated systems.

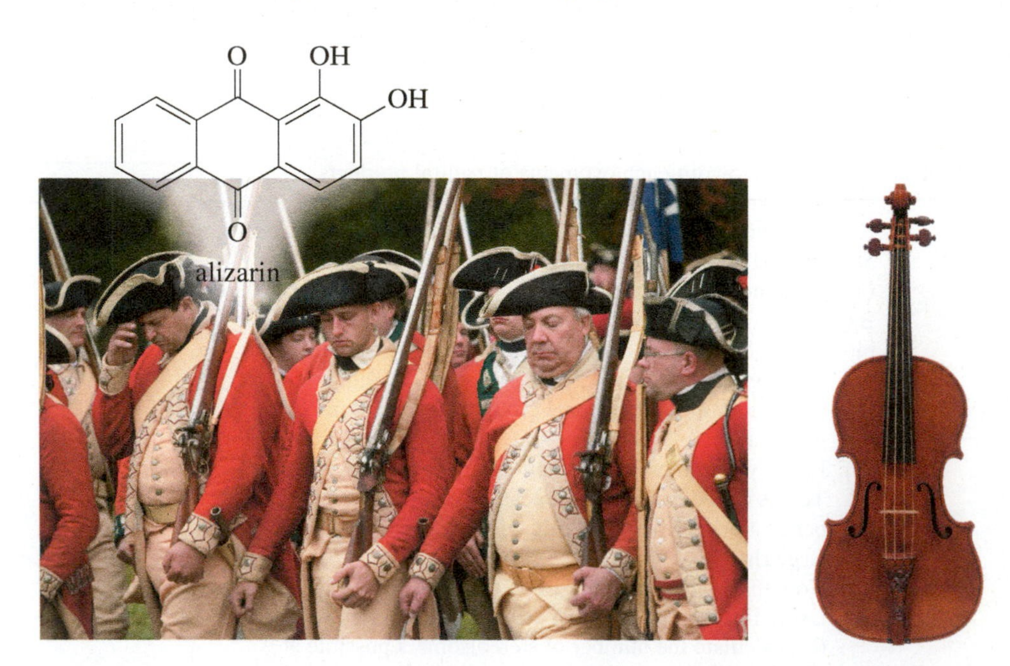

▲What did the English troops at Lexington have in common with a fine, old Italian violin? Both the fabric of the troops' coats and the varnish on the violin get their characteristic red color from an extract of the roots of the madder plant, *Rubia tinctorum*, which is Latin for "dyer's red." The principal red compound in madder is *alizarin*, or 1,2-dihydroxyanthraquinone. The color of alizarin can vary from pink to crimson to dark brown, depending on the pH and the presence of metal salts.

15-1 Introduction

Double bonds can interact with each other if they are separated by just one single bond. Such interacting double bonds are said to be **conjugated** (Section 7-8F). Double bonds with two or more single bonds separating them have little interaction and are called **isolated double bonds.** For example, penta-1,3-diene has conjugated double bonds, whereas penta-1,4-diene has isolated double bonds.

>C=C–C=C<

conjugated double bonds
(more stable than isolated double bonds)

$H_2C=CH–CH=CH–CH_3$

penta-1,3-diene

>C=C–CH_2–C=C<

isolated double bonds

$H_2C=CH–CH_2–CH=CH_2$

penta-1,4-diene

Because of the interaction between the double bonds, systems containing conjugated double bonds tend to be more stable than similar systems with isolated double bonds. In this chapter, we consider the unique properties of conjugated systems, the theoretical reasons for this extra stability, and some of the characteristic reactions of molecules containing conjugated double bonds. We also study ultraviolet spectroscopy, a tool for determining the structures of conjugated systems.

15-2 Stabilities of Dienes

In Chapter 7, we used **heats of hydrogenation** to compare the relative stabilities of alkenes. For example, the heats of hydrogenation of pent-1-ene and *trans*-pent-2-ene show that the disubstituted double bond in *trans*-pent-2-ene is 12 kJ/mol (2.9 kcal/mol) more stable than the monosubstituted double bond in pent-1-ene.

pent-1-ene $\xrightarrow[\text{Pt}]{H_2}$ pentane $\Delta H° = -126$ kJ (-30.1 kcal)

trans-pent-2-ene $\xrightarrow[\text{Pt}]{H_2}$ pentane $\Delta H° = -114$ kJ (-27.2 kcal)

When a molecule has two isolated double bonds, the heat of hydrogenation is close to the sum of the heats of hydrogenation for the individual double bonds. For example, the heat of hydrogenation of penta-1,4-diene is -252 kJ/mol (-60.2 kcal/mol), about twice that of pent-1-ene.

penta-1,4-diene $\xrightarrow[\text{Pt}]{2\,H_2}$ pentane $\Delta H° = -252$ kJ (-60.2 kcal)

For conjugated dienes, the heat of hydrogenation is less than the sum for the individual double bonds. For example, *trans*-penta-1,3-diene has a monosubstituted double bond similar to the one in pent-1-ene, and a disubstituted double bond similar to the one in pent-2-ene. The sum of the heats of hydrogenation of pent-1-ene and pent-2-ene is -240 kJ (-57.3 kcal), but the heat of hydrogenation of *trans*-penta-1,3-diene is only -225 kJ/mol (-53.7 kcal/mol), showing that the conjugated diene has about 15 kJ/mol (3.6 kcal/mol) extra stability.

pent-1-ene + *trans*-pent-2-ene $\overset{?}{=}$ predicted

Predicted: pent-1-ene -126 kJ + *trans*-pent-2-ene -114 kJ = predicted -240 kJ (-57.3 kcal)

trans-penta-1,3-diene $\xrightarrow[\text{Pt}]{2\,H_2}$ pentane

actual value -225 kJ (-53.7 kcal)

more stable by 15 kJ (3.6 kcal)

What happens if two double bonds are even closer together than in the conjugated case? Successive double bonds with no intervening single bonds are called **cumulated double bonds.** Consider penta-1,2-diene, which contains cumulated double bonds. Such 1,2-diene systems are also called **allenes,** after the simplest member of the class, propa-1,2-diene or "allene," $H_2C{=}C{=}CH_2$. The heat of hydrogenation of penta-1,2-diene is -292 kJ/mol (-69.8 kcal/mol), a larger value than any of the other pentadienes.

$H_2C{=}C{=}CH{-}CH_2CH_3$ penta-1,2-diene (ethylallene) $\xrightarrow[\text{Pt}]{2\,H_2}$ $CH_3CH_2CH_2CH_2CH_3$ pentane $\Delta H° = -292$ kJ (-69.8 kcal)

sum of pent-1-ene + pent-2-ene $\Delta H° = -240$ kJ (-57.3 kcal)

penta-1,2-diene is *less* stable by 52 kJ (12.5 kcal)

Because penta-1,2-diene has a larger heat of hydrogenation than penta-1,4-diene, we conclude that the cumulated double bonds of allenes are less stable than isolated double bonds and much less stable than conjugated double bonds. Figure 15-1 summarizes the relative stability of isolated, conjugated, and cumulated dienes and compares them with alkynes.

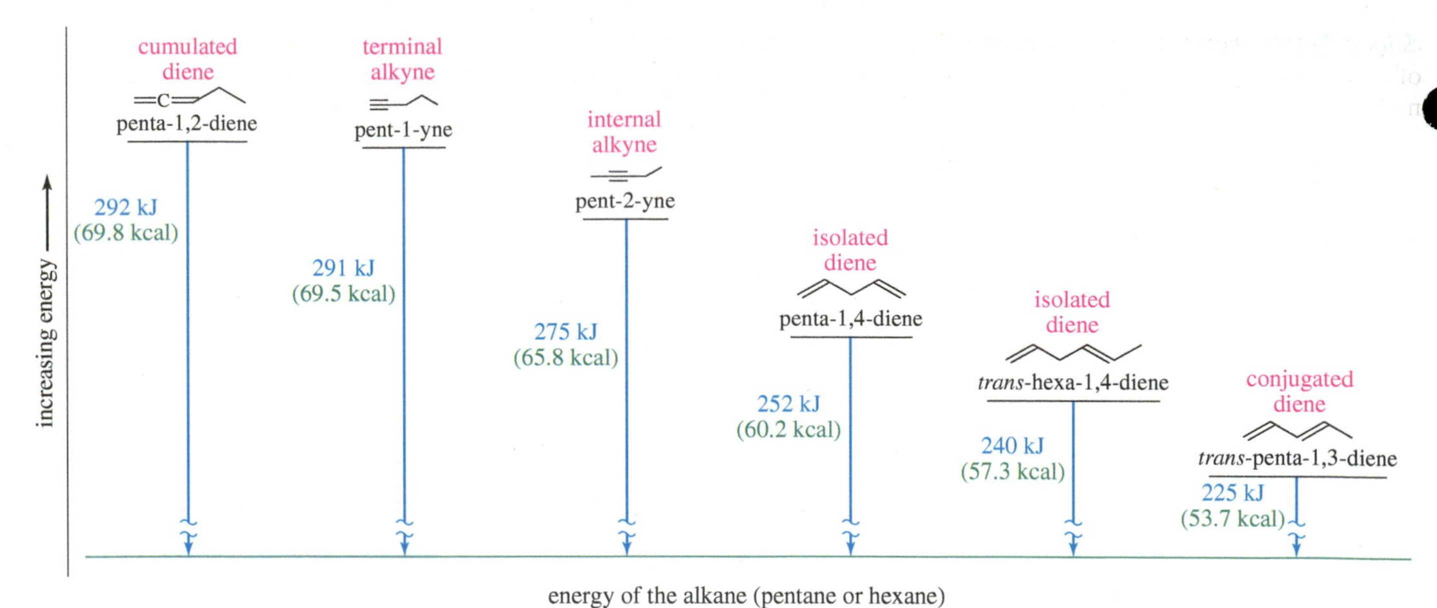

FIGURE 15-1
Relative energies of conjugated, isolated, and cumulated dienes compared with alkynes, based on molar heats of hydrogenation.

PROBLEM 15-1

Rank each group of compounds in order of increasing heat of hydrogenation.

(a) hexa-1,2-diene; hexa-1,3,5-triene; hexa-1,3-diene; hexa-1,4-diene; hexa-1,5-diene; hexa-2,4-diene

(b)

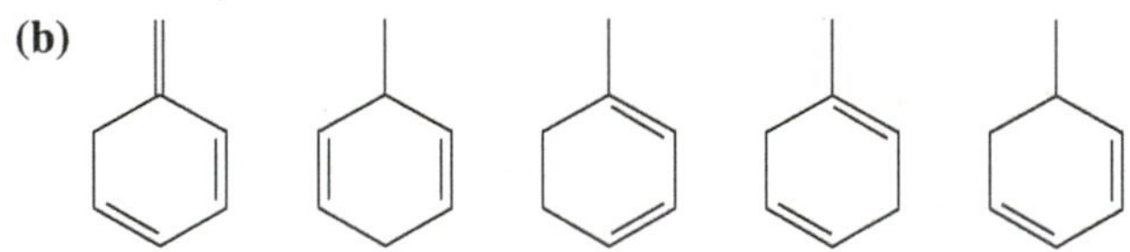

PROBLEM 15-2

In a strongly acidic solution, cyclohexa-1,4-diene tautomerizes to cyclohexa-1,3-diene. Propose a mechanism for this rearrangement, and explain why it is energetically favorable.

PROBLEM 15-3

(Review) The central carbon atom of an allene is a member of two double bonds, and it has an interesting orbital arrangement that holds the two ends of the molecule at right angles to each other.

(a) Draw an orbital diagram of allene, showing why the two ends are perpendicular.

(b) Draw the two enantiomers of penta-2,3-diene. A model may be helpful.

15-3 Molecular Orbital Picture of a Conjugated System

Figure 15-1 shows that the compound with conjugated double bonds is 15 kJ/mol (3.6 kcal/mol) more stable than a similar compound with isolated double bonds. This 15 kJ/mol of extra stability in the conjugated molecule is called the **resonance energy** of the system. (Other terms favored by some chemists are *conjugation energy*, *delocalization energy*, and *stabilization energy*.) We can best explain this extra stability of conjugated systems by examining their **molecular orbitals.** Let's begin with the molecular orbitals of the simplest conjugated diene, buta-1,3-diene.

15-3A Structure and Bonding of Buta-1,3-diene

The heat of hydrogenation of buta-1,3-diene is less than twice that of but-1-ene by about 15 kJ/mol (3.6 kcal/mol), showing that buta-1,3-diene has a resonance energy of 15 kJ/mol. Figure 15-2 shows the most stable conformation of buta-1,3-diene. Note that this conformation is planar, with the *p* orbitals on the two pi bonds aligned.

$$\underset{\text{buta-1, 3-diene}}{H_2C{=}CH{-}CH{=}CH_3} \xrightarrow{2\,H_2,\ Pt} CH_3{-}CH_2{-}CH_2{-}CH_3 \qquad \Delta H° = -237\ \text{kJ}\ (-56.6\ \text{kcal})$$

$$\underset{\text{but-1-ene}}{H_2C{=}CH{-}CH_2{-}CH_3} \xrightarrow{H_2,\ Pt} CH_3{-}CH_2{-}CH_2{-}CH_3 \qquad \Delta H° = -126\ \text{kJ}\ (-30.1\ \text{kcal})$$

$$\times 2 = -252\ \text{kJ}\ (-60.2\ \text{kcal})$$

$$\text{resonance energy of buta-1,3-diene} = 252\ \text{kJ} - 237\ \text{kJ} = 15\ \text{kJ} \quad (3.6\ \text{kcal})$$

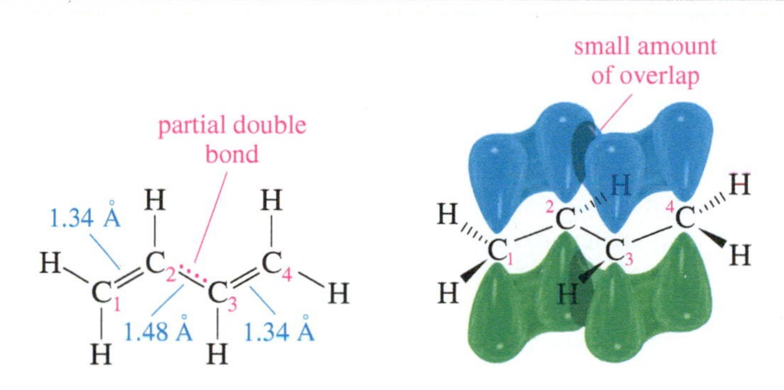

FIGURE 15-2
Structure of buta-1,3-diene in its most stable conformation. The 1.48 Å central carbon–carbon single bond is shorter than the 1.54 Å bonds typical of alkanes because of its partial double-bond character.

The C2—C3 bond in buta-1,3-diene (1.48 Å) is shorter than a carbon–carbon single bond in an alkane (1.54 Å). This bond is shortened slightly by the increased *s* character of the sp^2 hybrid orbitals, but the most important cause of this short bond is its pi bonding overlap and partial double-bond character. The planar conformation, with the *p* orbitals of the two double bonds aligned, allows overlap between the pi bonds. In effect, the electrons in the double bonds are **delocalized** over the entire molecule, creating some pi overlap and pi bonding in the C2—C3 bond. The length of this bond is intermediate between the normal length of a single bond and that of a double bond.

Lewis structures are inadequate to represent delocalized molecules such as buta-1,3-diene. To represent the bonding in conjugated systems accurately, we must consider molecular orbitals that represent the entire conjugated pi system, and not just one bond at a time.

15-3B Constructing the Molecular Orbitals of Buta-1,3-diene

All four carbon atoms of buta-1,3-diene are sp^2 hybridized, and (in the planar conformation) they all have overlapping *p* orbitals. Let's review how we constructed the pi molecular orbitals (MOs) of ethylene from the *p* atomic orbitals of the two carbon atoms (Figure 15-3). Each *p* orbital consists of two lobes, with opposite phases of the wave function in the two lobes. The plus and minus signs used in drawing these orbitals indicate the *phase of the wave function, not electrical charges*. To minimize confusion, we will color the lobes in molecular orbitals blue for the plus phase and green for the minus phase to emphasize the phase difference.

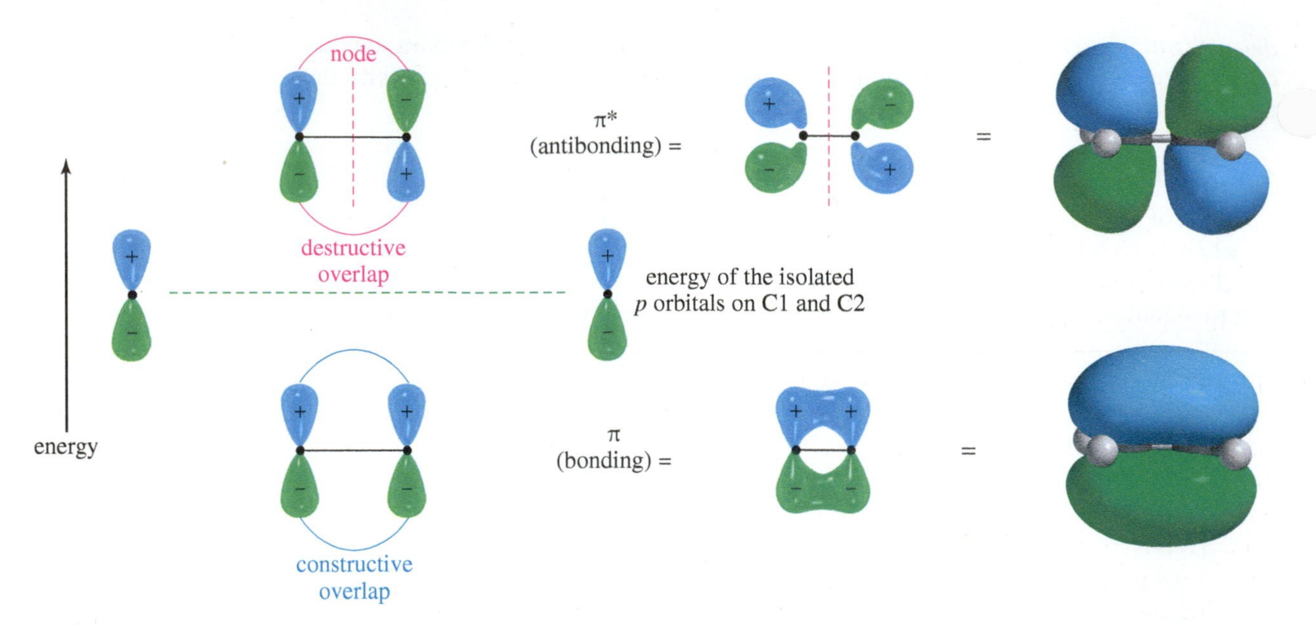

FIGURE 15-3
The pi molecular orbitals of ethylene. The pi bonding orbital is formed by constructive overlap of unhybridized *p* orbitals on the sp^2 hybrid carbon atoms. Destructive overlap of these *p* orbitals forms the antibonding pi orbital. Combination of two atomic orbitals must give exactly two molecular orbitals.

In the pi **bonding molecular orbital** of ethylene, the lobes that overlap in the bonding region between the nuclei are in phase; that is, they have the same sign (+ overlaps with +, and − overlaps with −). We call this reinforcement **constructive overlap.** Constructive overlap is an important feature of all bonding molecular orbitals.

In the pi **antibonding molecular orbital** (marked by *), on the other hand, lobes of opposite phase (with opposite signs, + with −) overlap in the bonding region. This **destructive overlap** causes cancelling of the wave function in the bonding region. Midway between the nuclei, this antibonding MO has a **node:** a region of zero electron density where the positive and negative phases exactly cancel.

Electrons have lower energy in the **bonding MO** than in the original *p* orbitals, and higher energy in the **antibonding MO.** In the ground state of ethylene, two electrons are in the bonding MO, whereas the antibonding MO is vacant. Stable molecules tend to have filled bonding MOs and empty antibonding MOs.

PROBLEM-SOLVING HINT

Stable molecules tend to have filled bonding MOs and empty antibonding MOs.

Several important principles are illustrated in Figure 15-3. Constructive overlap results in a bonding interaction; destructive overlap results in an antibonding interaction. Also, the number of molecular orbitals is always the same as the number of atomic orbitals used to form the MOs. These molecular orbitals have energies that are symmetrically distributed above and below the energy of the starting *p* orbitals. Half are bonding MOs, and half are antibonding MOs.

Now we are ready to construct the molecular orbitals of buta-1,3-diene. The *p* orbitals on C1 through C4 overlap, giving an extended system of four *p* orbitals that form four pi molecular orbitals. Two MOs are bonding, and two are antibonding. To represent the four *p* orbitals, we draw four *p* orbitals in a line. Although buta-1,3-diene is not linear, this simple straight-line representation makes it easier to draw and visualize the molecular orbitals.

represented by

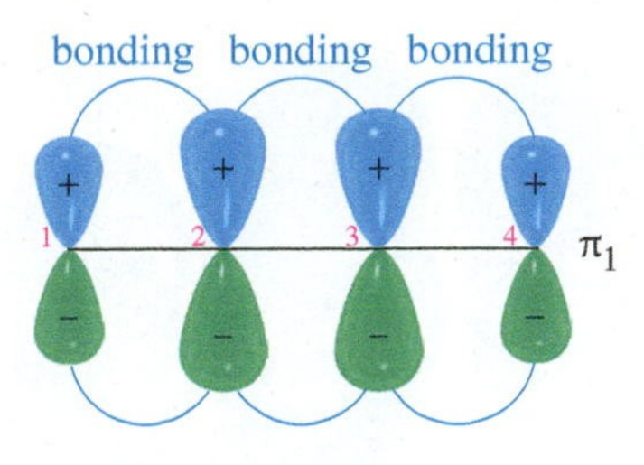

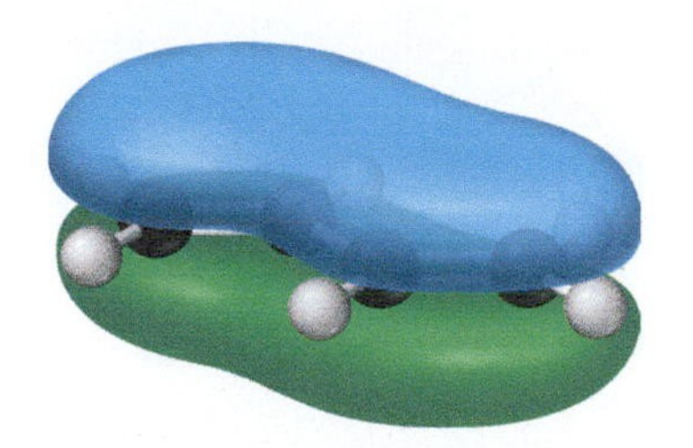

FIGURE 15-4
The π_1 bonding MO of buta-1,3-diene. This lowest-energy orbital has bonding interactions between all adjacent carbon atoms. It is labeled π_1 because it is a pi bonding orbital and it has the lowest energy.

The lowest-energy molecular orbital always consists entirely of bonding interactions. We indicate such an orbital by drawing all the positive phases of the *p* orbitals overlapping constructively on one face of the molecule, and the negative phases overlapping constructively on the other face. Figure 15-4 shows the lowest-energy MO for buta-1,3-diene. This MO places electron density on all four *p* orbitals, with slightly more on C2 and C3. (In these figures, larger and smaller *p* orbitals are used to show which atoms bear more of the electron density in a particular MO.)

This lowest-energy orbital is exceptionally stable for two reasons: There are three bonding interactions, and the electrons are delocalized over four nuclei. This orbital helps to illustrate why the conjugated system is more stable than two isolated double bonds. It also shows some pi-bond character between C2 and C3, which lowers the energy of the planar conformation and helps to explain the short C2—C3 bond length.

As with ethylene, the second molecular orbital (π_2) of butadiene (Figure 15-5) has one vertical node in the center of the molecule. This MO represents the classic picture of a diene. There are bonding interactions at the C1—C2 and C3—C4 bonds, and a (weaker) antibonding interaction between C2 and C3.

The π_2 orbital has two bonding interactions and one antibonding interaction; therefore, we expect it to be a bonding orbital (2 bonding $-$ 1 antibonding $=$ 1 bonding). It is not as strongly bonding nor as low in energy as the all-bonding π_1 orbital. Adding and subtracting bonding and antibonding interactions is not a reliable method for calculating energies of molecular orbitals. However, it is useful for predicting whether a given orbital is bonding or antibonding and for ranking orbitals in order of their energy.

The third butadiene MO (π_3^*) has two nodes (Figure 15-6). There is a bonding interaction at the C2—C3 bond, and there are two antibonding interactions, one between

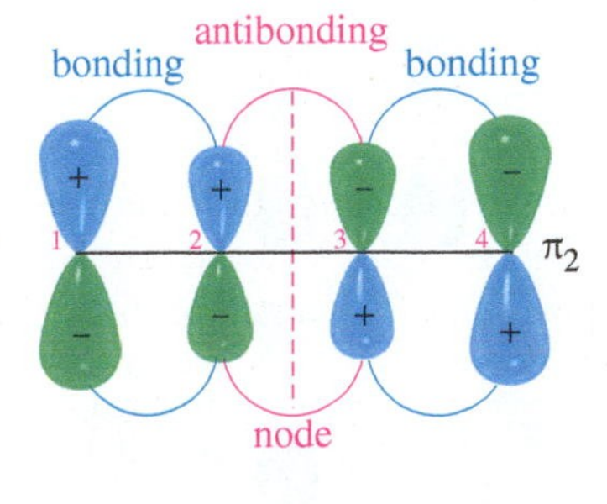

FIGURE 15-5
The π_2 bonding MO of buta-1,3-diene. The second MO has one node in the center of the molecule. There are bonding interactions at the C1—C2 and C3—C4 bonds, and there is a (weaker) antibonding interaction between C2 and C3. This π_2 orbital is bonding, but it is not bonding as strongly as π_1.

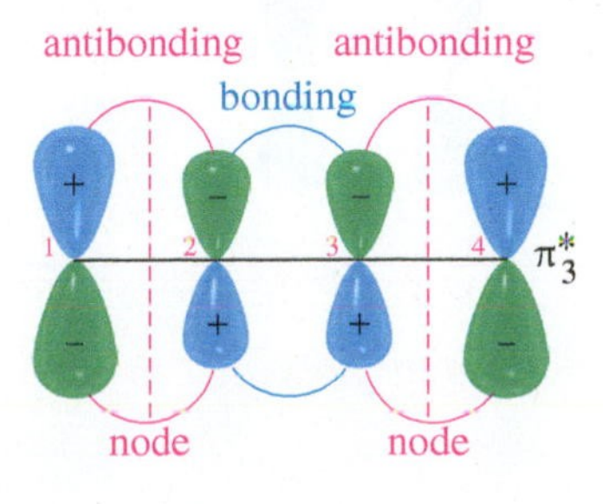

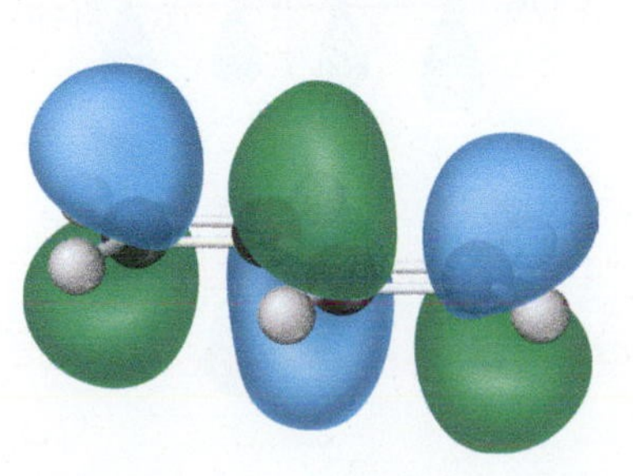

FIGURE 15-6
The π_3^* antibonding MO of buta-1,3-diene. The third MO has two nodes, giving two antibonding interactions and one bonding interaction. This is an antibonding orbital, and it is vacant in the ground state.

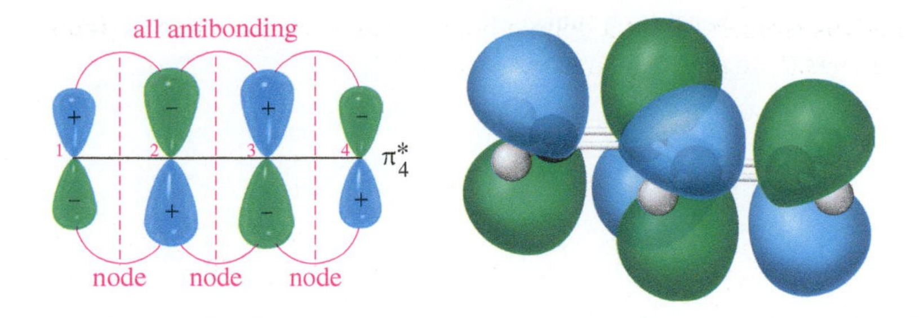

FIGURE 15-7
The π_4^* antibonding molecular orbital of buta-1,3-diene. The highest-energy MO has three nodes and three antibonding interactions. It is strongly antibonding, and it is vacant in the ground state.

C1 and C2 and the other between C3 and C4. This is an antibonding orbital (*), and it is vacant in the ground state.

The fourth, and last, molecular orbital (π_4^*) of buta-1,3-diene has three nodes and is totally antibonding (Figure 15-7). This MO has the highest energy and is unoccupied in the molecule's ground state. This highest-energy MO (π_4^*) is typical: For most systems, the highest-energy MO has antibonding interactions between all pairs of adjacent atoms.

Butadiene has four pi electrons (two electrons in each of the two double bonds in the Lewis structure) to be placed in the four MOs just described. Each MO can accommodate two electrons, and the lowest-energy MOs are filled first. Therefore, the four pi electrons go into π_1 and π_2. Figure 15-8 shows the electronic configuration of buta-1,3-diene. Both bonding MOs are filled, and both antibonding MOs are empty. Most stable molecules have this arrangement of filled bonding orbitals and vacant antibonding orbitals. Figure 15-8 also compares the relative energies of the ethylene MOs with the butadiene MOs to show that the conjugated butadiene system is slightly more stable than two ethylene double bonds.

The partial double-bond character between C2 and C3 in buta-1,3-diene explains why the molecule is most stable in a planar conformation. There are actually two planar conformations that allow overlap between C2 and C3. These conformations arise by rotation about the C2—C3 bond, and they are considered single-bond analogues

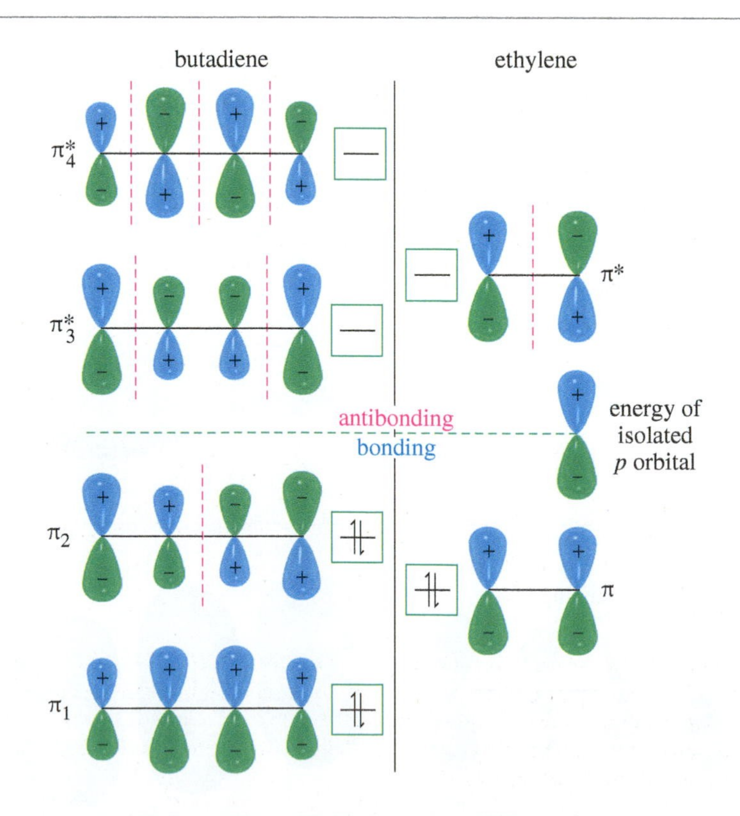

FIGURE 15-8
The electronic configurations of ethylene and buta-1,3-diene. In both buta-1,3-diene and ethylene, the bonding MOs are filled and the antibonding MOs are vacant. The average energy of the electrons is slightly lower in butadiene. This lower energy is the resonance stabilization of the conjugated diene.

of trans and cis isomers about a double bond. Thus, they are named ***s*-trans** ("single"-trans) and ***s*-cis** ("single"-cis) **conformations.**

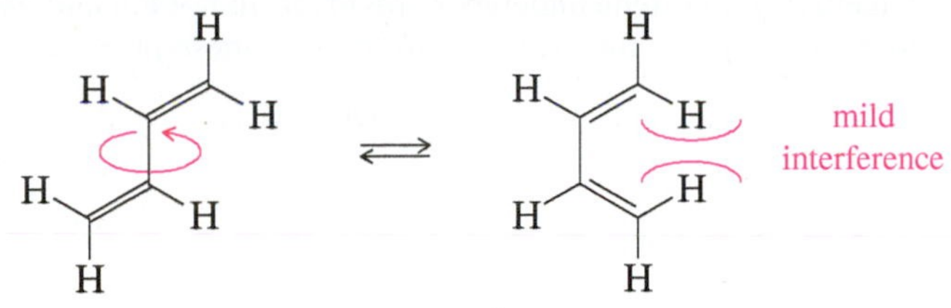

s-trans ⇌ *s*-cis (mild interference)

The *s*-trans conformation is 12 kJ/mol (2.8 kcal/mol) more stable than the *s*-cis conformation, which shows interference between the two nearby hydrogen atoms. The barrier for rotation about the C2—C3 bond from *s*-trans to *s*-cis) is only about 29 kJ/mol (about 7 kcal/mol) compared with about 250 kJ/mol (60 kcal/mol) for rotation of a double bond in an alkene. The *s*-cis and *s*-trans conformers of butadiene (and all the skew conformations in between) easily interconvert at room temperature.

15-4 Allylic Cations

Conjugated compounds undergo a variety of reactions, many of which involve intermediates that retain some of the resonance stabilization of the conjugated system. Common intermediates include allylic systems, particularly allylic cations and radicals. Allylic cations and radicals are stabilized by delocalization. First, we consider some reactions involving allylic cations and radicals, then (Section 15-8) we derive the molecular orbital picture of their bonding.

In Chapter 7, we saw that the $—CH_2—CH{=}CH_2$ group is called the **allyl group.** Many common names use this terminology.

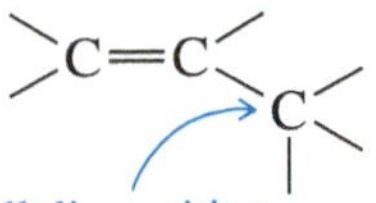

allylic position

$H_2C{=}CH—CH_2Br$ allyl bromide

$H_2C{=}CH—CH_2OH$ allyl alcohol

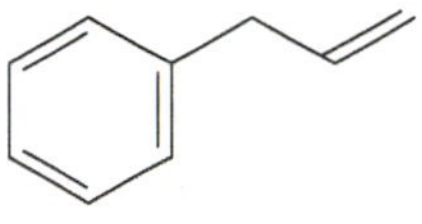

allylbenzene

When allyl bromide is heated with a good ionizing solvent, it ionizes to the **allyl cation,** an allyl group with a positive charge. More-substituted analogs are called **allylic cations.** All allylic cations are stabilized by resonance with the adjacent double bond, which delocalizes the positive charge over two carbon atoms.

$$\underset{\text{allyl bromide}}{H_2C{=}CH—CH_2—\ddot{\underset{..}{Br}}:} \rightleftarrows \underset{\text{allyl cation}}{\left[H_2C{=}CH—\overset{+}{C}H_2 \longleftrightarrow H_2\overset{+}{C}—CH{=}CH_2\right]} + :\ddot{\underset{..}{Br}}:^-$$

$H_2C{=}CH—\overset{+}{C}H—CH_3$ $(CH_3)_2C{=}CH—\overset{+}{C}H_2$ (cyclohexenyl cation)

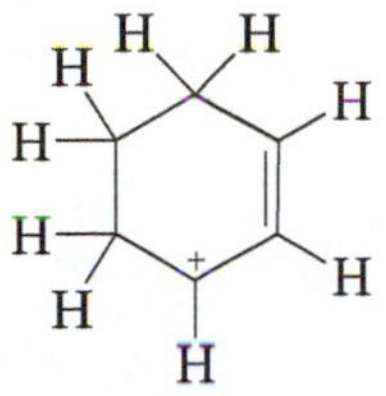

substituted allylic cations

PROBLEM 15-4

Draw another resonance form for each of the substituted allylic cations shown in the preceding figure, showing how the positive charge is shared by another carbon atom. In each case, state whether your second resonance form is a more important or less important resonance contributor than the first structure. (Which structure places the positive charge on the more-substituted carbon atom?)

PROBLEM 15-5

When 3-bromo-1-methylcyclohexene undergoes solvolysis in hot ethanol, two products are formed. Propose a mechanism that accounts for both of these products.

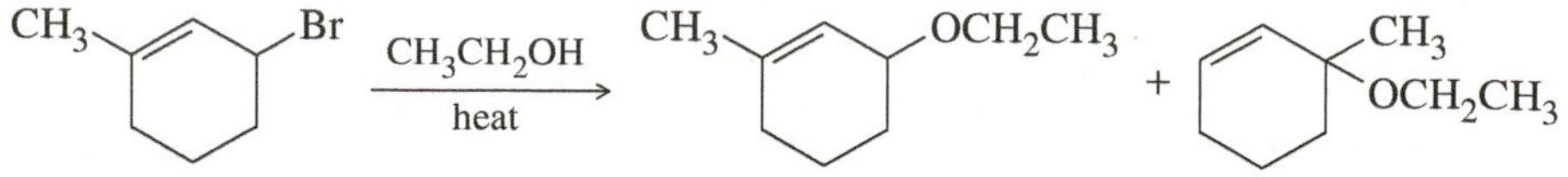

$$\text{3-bromo-1-methylcyclohexene (}CH_3\text{, Br)} \xrightarrow[\text{heat}]{CH_3CH_2OH} \text{3-ethoxy-1-methylcyclohexene (}CH_3\text{, }OCH_2CH_3\text{)} + \text{1-ethoxy-1-methylcyclohex-2-ene (}CH_3\text{, }OCH_2CH_3\text{)}$$

We can represent a delocalized ion such as the allyl cation either by resonance forms, as shown on the left in the following figure, or by a combined structure, as shown on the right. Although the combined structure is more concise, it is sometimes confusing because it attempts to convey all the information implied by two or more resonance forms.

$$\left[H_2\underset{1}{C}{=}\underset{2}{C}(H){-}\underset{3}{\overset{+}{C}}H_2 \longleftrightarrow H_2\underset{1}{\overset{+}{C}}{-}\underset{2}{C}(H){=}\underset{3}{C}H_2 \right] \quad \text{or} \quad H_2\underset{1}{\overset{\frac{1}{2}+}{C}}{\cdots}\underset{2}{C}(H){\cdots}\underset{3}{\overset{\frac{1}{2}+}{C}}H_2$$

resonance forms   combined representation

Because of its resonance stabilization, the (primary) allyl cation is about as stable as a simple secondary carbocation, such as the isopropyl cation. Most substituted allylic cations have at least one secondary carbon atom bearing part of the positive charge. They are about as stable as simple tertiary carbocations such as the *tert*-butyl cation. This resonance stabilization enhances the rates of reactions, such as the S_N1 and E1, that involve carbocation intermediates.

Stability of carbocations

$$H_3C^+ < 1° < 2°\text{, allyl} < 3°\text{, substituted allylic}$$

$$H_2\overset{\frac{1}{2}+}{C}{\cdots}CH{\cdots}\overset{\frac{1}{2}+}{C}H_2 \text{ is about as stable as } CH_3{-}\overset{+}{C}H{-}CH_3$$

$$CH_3{-}\overset{\delta+}{C}H{\cdots}CH{\cdots}\overset{\delta+}{C}H_2 \text{ is about as stable as } CH_3{-}\overset{+}{C}(CH_3)_2$$

15-5 1,2- and 1,4-Addition to Conjugated Dienes

Electrophilic additions to conjugated dienes usually involve allylic cations as intermediates. Unlike simple carbocations, an allylic cation can react with a nucleophile at either of its positive centers. Let's consider the addition of HBr to buta-1,3-diene, an electrophilic addition that produces a mixture of two constitutional isomers. One product, 3-bromobut-1-ene, results from Markovnikov addition across one of the double bonds. In the other product, 1-bromobut-2-ene, the double bond shifts to the C2—C3 position.

$$H_2C{=}CH{-}CH{=}CH_2 + HBr \longrightarrow \underset{\substack{\text{3-bromobut-1-ene}\\ \text{1,2-addition}}}{H_2C(H){-}CH(Br){-}CH{=}CH_2} + \underset{\substack{\text{1-bromobut-2-ene}\\ \text{1,4-addition}}}{H_2C(H){-}CH{=}CH{-}CH_2(Br)}$$

The first product results from electrophilic addition of HBr across a double bond. This process is called a **1,2-addition** whether or not these two carbon atoms are numbered 1 and 2 in naming the compound. In the second product, the proton and bromide ion add at the ends of the conjugated system to carbon atoms with a 1,4-relationship. Such an addition is called a **1,4-addition** whether or not these carbon atoms are numbered 1 and 4 in naming the compound.

$$\text{C=C–C=C} \xrightarrow{\text{A–B}} \underset{\text{1,2-addition}}{\text{–C(A)–C(B)–C=C}} + \underset{\text{1,4-addition}}{\text{–C(A)–C=C–C(B)–}}$$

The mechanism is similar to other electrophilic additions to alkenes. The proton is the electrophile, adding to the alkene to give the most stable carbocation. Protonation of buta-1,3-diene gives an allylic cation, which is stabilized by resonance delocalization of the positive charge over two carbon atoms. Bromide can attack this resonance-stabilized intermediate at either of the two carbon atoms sharing the positive charge. Attack at the secondary carbon gives 1,2-addition; attack at the primary carbon gives 1,4-addition.

MECHANISM 15-1 1,2- and 1,4-Addition to a Conjugated Diene

Step 1: Protonation of one of the double bonds forms a resonance-stabilized allylic cation.

$$\text{Br–H} + \text{H}_2\text{C=CH–CH=CH}_2 \longrightarrow \left[\text{H}_3\text{C–}\overset{+}{\text{C}}\text{H–CH=CH}_2 \longleftrightarrow \text{H}_3\text{C–CH=CH–}\overset{+}{\text{C}}\text{H}_2\right] \text{Br}^-$$

allylic cation

Step 2: A nucleophile attacks at either electrophilic carbon atom.

$$\left[\text{H}_3\text{C–}\overset{+}{\text{C}}\text{H–CH=CH}_2 \longleftrightarrow \text{H}_3\text{C–CH=CH–}\overset{+}{\text{C}}\text{H}_2\right] \xrightarrow{:\ddot{\text{Br}}:^-} \underset{\text{1,2-addition}}{\text{H}_3\text{C–CHBr–CH=CH}_2} \text{ and } \underset{\text{1,4-addition}}{\text{H}_3\text{C–CH=CH–CH}_2\text{–Br}}$$

(1,2 and 1,4 arrows indicate the two sites of attack)

The key to formation of these two products is the presence of a double bond in position to form a stabilized allylic cation. Molecules having such double bonds are likely to react via resonance-stabilized intermediates.

PROBLEM 15-6

Treatment of an alkyl halide with $AgNO_3$ in alcohol often promotes ionization.

$$Ag^+ + R\text{–}Cl \longrightarrow AgCl + R^+$$

When 4-chloro-2-methylhex-2-ene reacts with $AgNO_3$ in ethanol, two isomeric ethers are formed. Suggest structures, and propose a mechanism for their formation.

PROBLEM 15-7

Propose a mechanism for each reaction, showing explicitly how the observed mixtures of products are formed.

(a) 3-methylbut-2-en-1-ol + HBr → 1-bromo-3-methylbut-2-ene + 3-bromo-3-methylbut-1-ene
(b) 2-methylbut-3-en-2-ol + HBr → 1-bromo-3-methylbut-2-ene + 3-bromo-3-methylbut-1-ene
(c) cyclopenta-1,3-diene + Br_2 → 3,4-dibromocyclopent-1-ene + 3,5-dibromocyclopent-1-ene
(d) 1-chlorobut-2-ene + $AgNO_3$, H_2O → but-2-en-1-ol + but-3-en-2-ol
(e) 3-chlorobut-1-ene + $AgNO_3$, H_2O → but-2-en-1-ol + but-3-en-2-ol

15-6 Kinetic Versus Thermodynamic Control in the Addition of HBr to Buta-1,3-diene

One of the interesting peculiarities of the reaction of buta-1,3-diene with HBr is the effect of temperature on the products. If the reagents are allowed to react briefly at −80 °C, the 1,2-addition product predominates. If this reaction mixture is later allowed to warm to 40 °C, however, or if the original reaction is carried out at 40 °C, the composition favors the 1,4-addition product.

HBr + $H_2C{=}CH{-}CH{=}CH_2$

−80 °C →
- (80%) $H_2C(H){-}CH(Br){-}CH{=}CH_2$ (1,2-product)
- (20%) $H_2C(H){-}CH{=}CH{-}CH_2(Br)$ (1,4-product)

↓ 40 °C

40 °C →
- (15%) $H_2C(H){-}CH(Br){-}CH{=}CH_2$ (1,2-product)
- (85%) $H_2C(H){-}CH{=}CH{-}CH_2(Br)$ (1,4-product)

This variation in product composition reminds us that the most stable product is not always the major product. Of the two products, we expect 1-bromobut-2-ene (the 1,4-product) to be more stable, because it has the more substituted double bond. This prediction is supported by the fact that this isomer predominates when the reaction mixture is warmed to 40 °C and allowed to equilibrate.

A reaction-energy diagram for the second step of this reaction (Figure 15-9) helps to show why one product is favored at low temperatures and another at higher temperatures. The allylic cation is in the center of the diagram; it can react toward the left to give the 1,2-product or toward the right to give the 1,4-product. The initial product depends on where bromide attacks the resonance-stabilized allylic cation. Bromide can attack at either of the two carbon atoms that share the positive charge. Attack at the secondary carbon gives 1,2-addition, and attack at the primary carbon gives 1,4-addition.

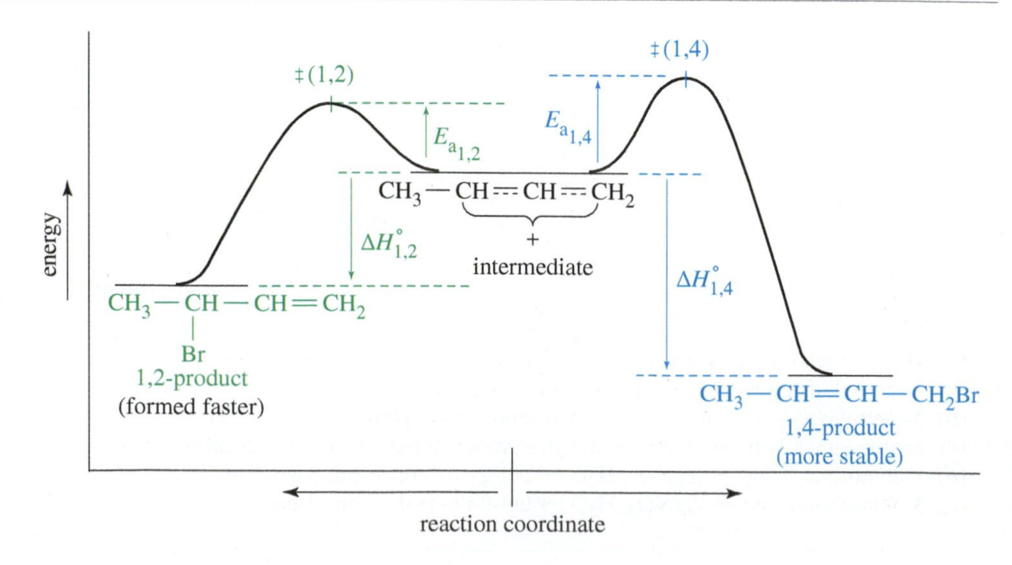

FIGURE 15-9
Reaction-energy diagram for the second step of the addition of HBr to buta-1,3-diene. The allylic carbocation (center) can react at either of its electrophilic carbon atoms. The transition state (‡) leading to 1,2-addition has a lower energy than that leading to the 1,4-product, so that the 1,2-product is formed faster (kinetic product). The 1,2-product is not as stable as the 1,4-product, however. If equilibrium is reached, the 1,4-product predominates (thermodynamic product).

$$H^+ \; H_2C{=}CH{-}CH{=}CH_2 \longrightarrow \left[H_3C{-}\overset{+}{C}H{-}CH{=}CH_2 \;\; :\ddot{Br}:^- \longleftrightarrow H_3C{-}CH{=}CH{-}\overset{+}{C}H_2 \;\; :\ddot{Br}:^- \right]$$

delocalized allylic cation

attack at secondary carbon → $H_3C{-}CH(Br){-}CH{=}CH_2$ 1,2-addition product

attack at primary carbon → $H_3C{-}CH{=}CH{-}CH_2Br$ 1,4-addition product

Kinetic Control at −80 °C The transition state for 1,2-addition has a lower energy than the transition state for 1,4-addition, giving the 1,2-addition a lower activation energy (E_a). This is not surprising, because 1,2-addition results from bromide attack at the more substituted secondary carbon, which bears more of the positive charge because it is better stabilized than the primary carbon. Because the 1,2-addition has a lower activation energy than the 1,4-addition, the 1,2-addition takes place faster (at *all* temperatures).

Attack by bromide on the allylic cation is a strongly exothermic process, so the reverse reaction has a large activation energy. At −80 °C, few collisions take place with this much energy, and the rate of the reverse reaction is practically zero. Under these conditions, the product that is formed faster predominates. Because the kinetics of the reaction determine the results, this situation is called **kinetic control** of the reaction. The 1,2-product, favored under these conditions, is called the **kinetic product.**

Thermodynamic Control at 40 °C At 40 °C, a significant fraction of molecular collisions have enough energy for reverse reactions to occur. Notice that the activation energy for the reverse of the 1,2-addition is less than that for the reverse of the 1,4-addition. Although the 1,2-product is still formed faster, it also reverts to the allylic cation faster than the 1,4-product does. At 40 °C, an equilibrium is set up, and the relative energy of each species determines its concentration. The 1,4-product is the most stable species, and it predominates. Because thermodynamics determine the results, this situation is called **thermodynamic control** (or **equilibrium control**) of the reaction. The 1,4-product, favored under these conditions, is called the **thermodynamic product.**

We will see many additional reactions whose products may be determined by kinetic control or by thermodynamic control, depending on the conditions. In general, reactions that do not reverse easily are kinetically controlled because no equilibrium is established. In kinetically controlled reactions, the product with the lowest-energy transition state predominates. Reactions that are easily reversible are thermodynamically controlled unless something happens to prevent equilibrium from being attained. In thermodynamically controlled reactions, the lowest-energy product predominates.

PROBLEM 15-8

When Br_2 is added to buta-1,3-diene at −15 °C, the product mixture contains 60% of product A and 40% of product B. When the same reaction takes place at 60 °C, the product ratio is 10% A and 90% B.

(a) Propose structures for products A and B. (*Hint:* In many cases, an allylic carbocation is more stable than a bromonium ion.)

(b) Propose a mechanism to account for formation of both A and B.

(c) Show why A predominates at −15 °C and B predominates at 60 °C.

(d) If you had a solution of pure A, and its temperature were raised to 60 °C, what would you expect to happen? Propose a mechanism to support your prediction.

15-7 Allylic Radicals

Like allylic cations, allylic radicals are stabilized by resonance delocalization. For example, Mechanism 15-2 shows the mechanism of free-radical bromination of cyclohexene. Substitution occurs entirely at the allylic position, where abstraction of a hydrogen gives a resonance-stabilized allylic radical as the intermediate.

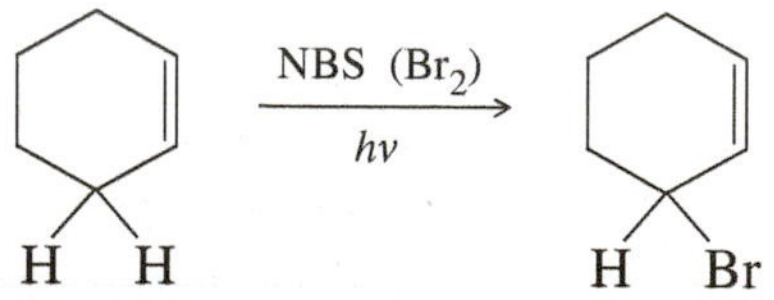

NBS (Br_2), $h\nu$

H H → H Br

MECHANISM 15-2 Free-Radical Allylic Bromination

Initiation: Formation of radicals.

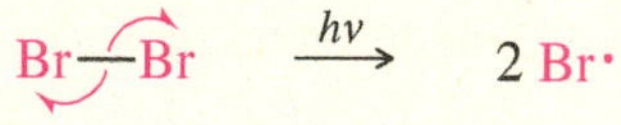

$$Br{-}Br \xrightarrow{h\nu} 2\ Br\cdot$$

Propagation: Each step consumes a radical and forms another radical leading to products.

First Propagation Step: The bromine radical abstracts an allylic hydrogen to produce an allylic radical.

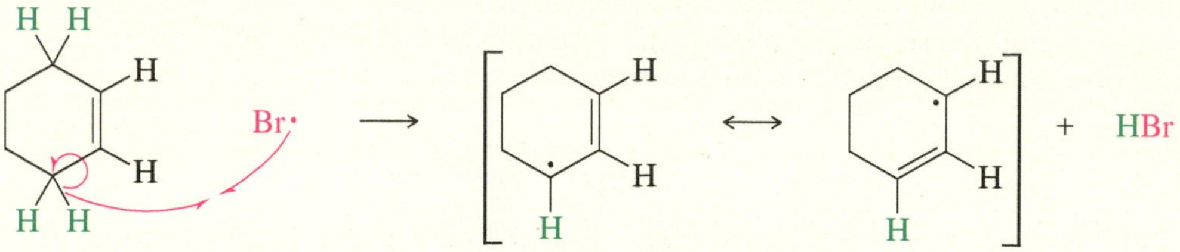

allylic hydrogens + Br· → [an allylic radical] + HBr

Second Propagation Step: The allylic radical in turn reacts with a bromine molecule to form an allyl bromide and a new bromine atom, which continues the chain.

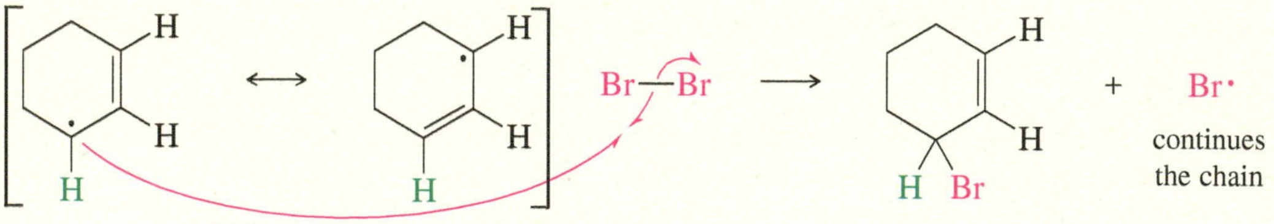

allylic radical + Br—Br → allylic bromide + Br· continues the chain

Regeneration of Br₂: NBS reacts with HBr to regenerate the molecule of bromine used in the allylic bromination step.

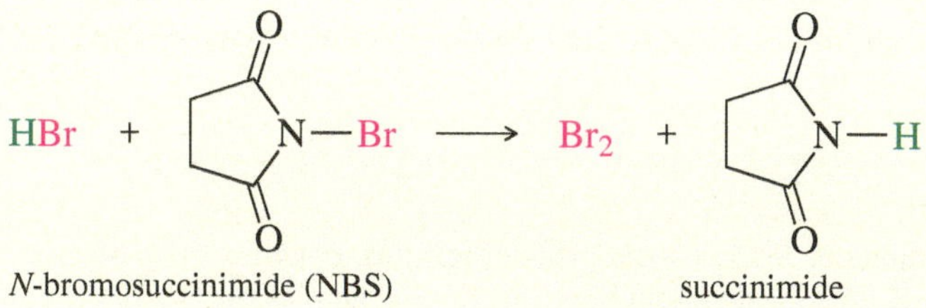

HBr + N—Br → Br_2 + N—H

N-bromosuccinimide (NBS) succinimide

Stability of Allylic Radicals Why is it that (in the first propagation step) a bromine radical abstracts only an allylic hydrogen atom, and not one from another secondary site? Abstraction of allylic hydrogens is preferred because the allylic free radical is resonance-stabilized. The bond-dissociation enthalpies required to generate several free radicals are compared below. Notice that the allyl radical (a primary free radical) is actually 31 kJ/mol (7 kcal/mol) *more* stable than the tertiary butyl radical.

Primary:	$CH_3CH_2—H$	⟶	$CH_3CH_2\cdot + H\cdot$	$\Delta H = +423$ kJ (+101 kcal)
Secondary:	$(CH_3)_2CH—H$	⟶	$(CH_3)_2CH\cdot + H\cdot$	$\Delta H = +413$ kJ (+ 99 kcal)
Tertiary:	$(CH_3)_3C—H$	⟶	$(CH_3)_3C\cdot + H\cdot$	$\Delta H = +403$ kJ (+ 96 kcal)
Allyl:	$H_2C{=}CH—CH_2—H$	⟶	$H_2C{=}CH—CH_2\cdot + H\cdot$	$\Delta H = +372$ kJ (+ 89 kcal)

The allylic cyclohex-2-enyl radical has its unpaired electron delocalized over two secondary carbon atoms, so that it is even more stable than the unsubstituted allyl radical. The second propagation step may occur at either of the radical carbons, but in this symmetrical case, either position gives 3-bromocyclohexene as the product. Less symmetrical compounds often give mixtures of products resulting from an **allylic shift:** In the product, the double bond can appear at either of the positions it occupies in the resonance forms of the allylic radical. An allylic shift in a radical reaction is similar to the 1,4-addition of an electrophilic reagent such as HBr to a diene (Section 15-5).

PROBLEM-SOLVING HINT

Stability of free radicals:
$\cdot CH_3 < 1° < 2° < 3° <$ allylic, benzylic

The following propagation steps show how a mixture of products results from the free-radical allylic bromination of but-1-ene.

$$CH_3—CH(H)—CH{=}CH_2 + Br\cdot \longrightarrow \left[CH_3—\dot{C}H—CH{=}CH_2 \longleftrightarrow CH_3—CH{=}CH—\dot{C}H_2\right] + HBr$$

resonance-stabilized allylic radical

$\downarrow Br_2$

$$CH_3—CH(Br)—CH{=}CH_2 + CH_3—CH{=}CH—CH_2Br + Br\cdot$$

(mixture)

PROBLEM 15-9

When methylenecyclohexane is treated with a low concentration of bromine under irradiation by a sunlamp, two substitution products are formed.

$$\text{methylenecyclohexane} + Br_2 \xrightarrow{h\nu} \text{two substitution products} + HBr$$

methylenecyclohexane

(a) Propose structures for these two products.
(b) Propose a mechanism to account for their formation.

Bromination Using NBS At higher concentrations, bromine *adds* across double bonds (via a bromonium ion) to give saturated dibromides (Section 8-8). In the allylic bromination just shown, bromine *substitutes* for a hydrogen atom. The key to getting substitution is to have a low concentration of bromine, together with light or free radicals to initiate the reaction. Free radicals are highly reactive, and even a small concentration of radicals can produce a fast chain reaction.

Simply adding bromine might raise the concentration too high, resulting in ionic addition of bromine across the double bond. A convenient bromine source for allylic bromination is *N*-bromosuccinimide (NBS), a brominated derivative of succinimide. Succinimide is a cyclic imide (diamide) of the four-carbon diacid succinic acid.

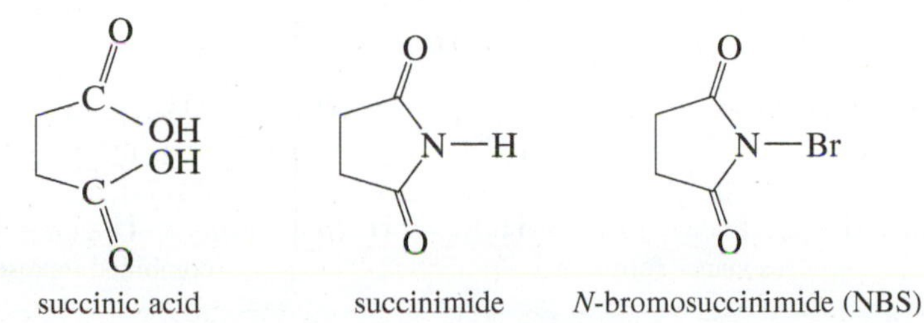

succinic acid succinimide *N*-bromosuccinimide (NBS)

NBS provides a fairly constant, low concentration of Br_2 because it reacts with HBr liberated in the substitution, converting it back into Br_2. This reaction also removes the HBr by-product, preventing it from adding across the double bond by its own free-radical chain reaction.

Step 1: Free-radical allylic substitution (Mechanism 15-2)

$$R{-}H + Br_2 \xrightarrow{h\nu} R{-}Br + HBr$$

Step 2: NBS converts the HBr by-product back into Br_2

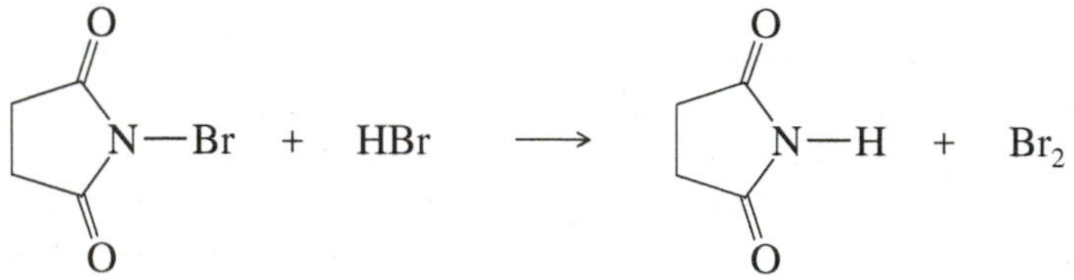

$$\text{NBS} + HBr \longrightarrow \text{succinimide (N-H)} + Br_2$$

NBS succinimide

The NBS reaction is carried out in a clever way. The allylic compound is dissolved in carbon tetrachloride, and one equivalent of NBS is added. NBS is denser than CCl_4 and not very soluble in it, so it sinks to the bottom of the CCl_4 solution. The reaction is initiated using a sunlamp for illumination or a radical initiator such as a peroxide. The NBS gradually *appears* to rise to the top of the CCl_4 layer. It is actually converted to succinimide, which is less dense than CCl_4. Once all the solid succinimide has risen to the top, the sunlamp is turned off, the solution is filtered to remove the succinimide, and the CCl_4 is evaporated to recover the product.

PROBLEM 15-10

When *N*-bromosuccinimide is added to hex-1-ene in CCl_4 and a sunlamp is shone on the mixture, three products result.

(a) Give the structures of these three products.

(b) Propose a mechanism that accounts for the formation of these three products.

PROBLEM 15-11

Predict the product(s) of light-initiated reaction with NBS in CCl_4 for the following starting materials.

(a) cyclopentene **(b)** 2,3-dimethylbut-2-ene **(c)** $C_6H_5{-}CH_3$ toluene

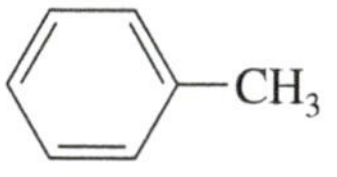

15-8 Molecular Orbitals of the Allylic System

Let's take a closer look at the electronic structure of allylic systems, using the allyl radical as our example. One resonance form shows the radical electron on C1, with a pi bond between C2 and C3. The other shows the radical electron on C3 and a pi bond between C1 and C2. These two resonance forms imply that there is half a pi bond between C1 and C2 and half a pi bond between C2 and C3, with the radical electron half on C1 and half on C3.

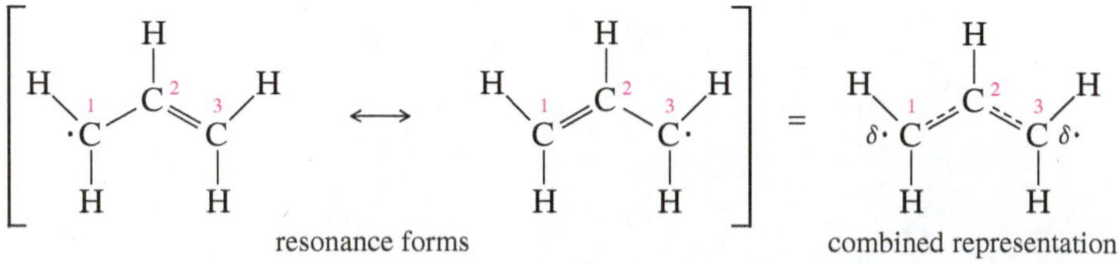

$$\left[H_2\dot{C}{-}CH{=}CH_2 \longleftrightarrow H_2C{=}CH{-}\dot{C}H_2 \right] = H_2C^{\delta\cdot}{\cdots}CH{\cdots}CH_2^{\delta\cdot}$$

resonance forms combined representation

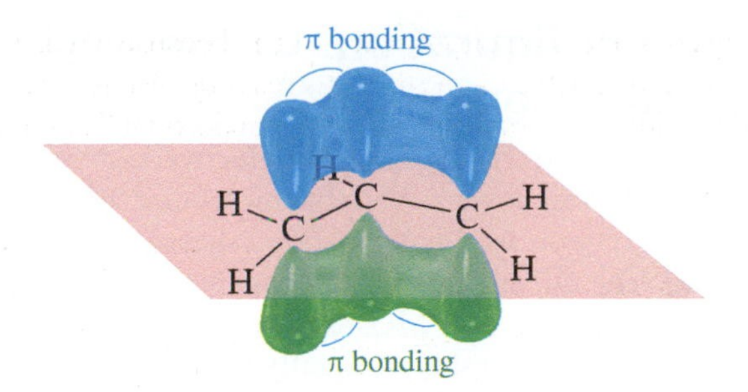

FIGURE 15-10
Geometric structure of the allyl cation, allyl radical, and allyl anion.

Remember that no resonance form has an independent existence: A compound has characteristics of all its resonance forms at the same time, but it does not "resonate" among them. The *p* orbitals of all three carbon atoms must be parallel to have simultaneous pi bonding overlap between C1 and C2 and between C2 and C3. The geometric structure of the allyl system is shown in Figure 15-10. The allyl cation, the allyl radical, and the allyl anion all have this same geometric structure, differing only in the number of pi electrons.

Just as the four *p* orbitals of buta-1,3-diene overlap to form four molecular orbitals, the three atomic *p* orbitals of the allyl system overlap to form three molecular orbitals, shown in Figure 15-11. These three MOs share several important features with the MOs of the butadiene system. The first MO is entirely bonding, the second has one node, and the third has two nodes and (because it is the highest-energy MO) is entirely antibonding.

As with butadiene, we expect that half of the MOs will be bonding, and half antibonding; however, with an odd number of MOs, they cannot be symmetrically divided. One of the MOs must appear at the middle of the energy levels, neither bonding nor antibonding: It is a **nonbonding molecular orbital.** Electrons in a nonbonding orbital have the same energy as in an isolated *p* orbital.

The structure of the nonbonding orbital (π_2) may seem strange because there is zero electron density on the center *p* orbital (C2). This is the case because π_2 must have one node, and the only symmetrical position for one node is in the center of the molecule, crossing C2. We can tell from its structure that π_2 must be nonbonding, because C1 and C3 both have zero overlap with C2. The total is zero bonding, implying a nonbonding orbital.

PROBLEM-SOLVING HINT

In drawing pi MOs, begin by assuming that some number of *p* orbitals combine to give the same number of MOs: half bonding and half antibonding. If there is an odd number of MOs, the middle one is nonbonding. The lowest-energy MO has no nodes; each higher MO has one more node.

The highest-energy MO is entirely antibonding, with a node at each overlap.

In a stable system, the bonding MOs are filled, and the antibonding MOs are empty.

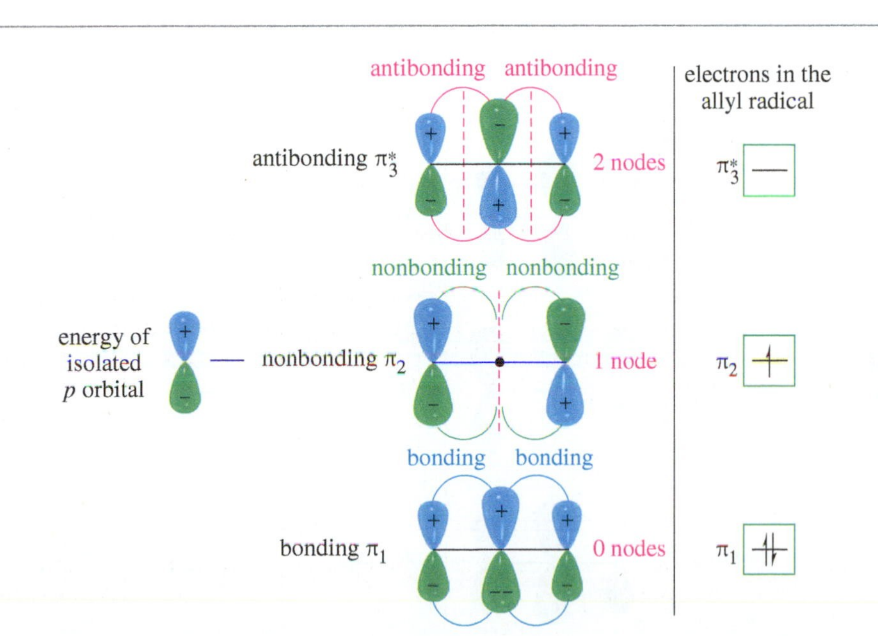

FIGURE 15-11
The three molecular orbitals of the allyl system. The lowest-energy MO (π_1) has no nodes and is entirely bonding. The intermediate orbital (π_2) is nonbonding, having one symmetrical node that coincides with the center carbon atom. The highest-energy MO (π_3^*) has two nodes and is entirely antibonding. In the allyl radical, π_1 is filled. The unpaired electron is in π_2, having its electron density entirely on C1 and C3.

15-9 Electronic Configurations of the Allyl Radical, Cation, and Anion

The right-hand column of Figure 15-11 shows the electronic structure for the allyl radical, with three pi electrons in the lowest available molecular orbitals. Two electrons are in the all-bonding MO (π_1), representing the pi bond shared between the C1—C2 bond and the C2—C3 bond. The unpaired electron goes into π_2 with zero electron density on the center carbon atom (C2). This MO representation agrees with the resonance picture showing the radical electron shared equally by C1 and C3, but not C2. Both the resonance and MO pictures successfully predict that the radical will react at either of the end carbon atoms, C1 or C3.

The electronic configuration of the allyl cation (Figure 15-12) differs from that of the allyl radical; it lacks the unpaired electron in π_2, which has half of its electron density on C1 and half on C3. In effect, we have removed half an electron from each of C1 and C3, whereas C2 remains unchanged. This MO picture is consistent with the resonance picture showing the positive charge shared by C1 and C3.

resonance forms — combined representation

Figure 15-12 also shows the electronic configuration of the allyl anion, which differs from the allyl radical in having an additional electron in π_2, the nonbonding orbital with its electron density divided between C1 and C3.

resonance forms — combined representation

This molecular orbital representation of the allyl anion is consistent with the resonance forms shown earlier, with a negative charge and a lone pair of nonbonding electrons evenly divided between C1 and C3.

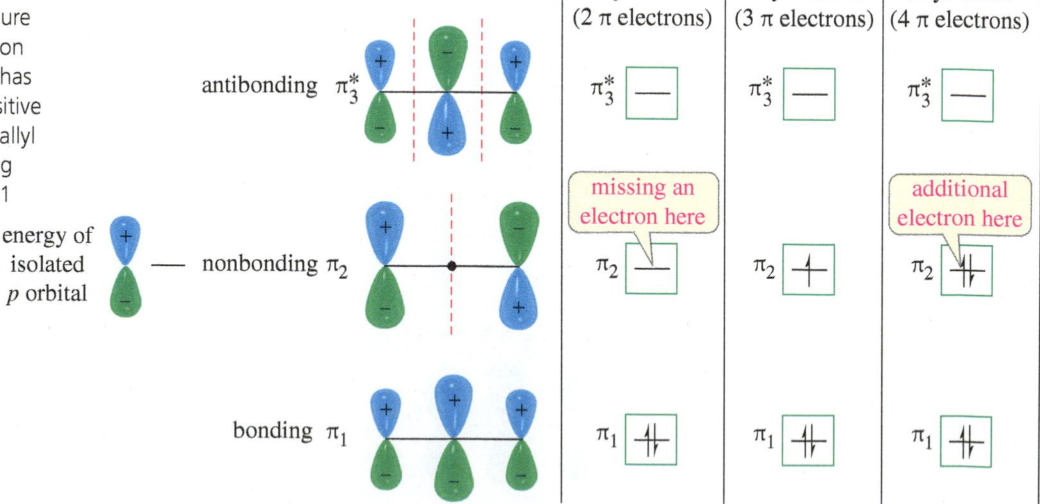

FIGURE 15-12
Comparison of the electronic structure of the allyl radical with the allyl cation and the allyl anion. The allyl cation has no electron in π_2, leaving half a positive charge on each of C1 and C3. The allyl anion has two electrons in π_2, giving half a negative charge to each of C1 and C3.

PROBLEM 15-12

Addition of 1-bromobut-2-ene to magnesium metal in dry ether results in formation of a Grignard reagent. Addition of water to this Grignard reagent gives a mixture of but-1-ene and but-2-ene (cis and trans). When the Grignard reagent is made using 3-bromobut-1-ene, addition of water produces exactly the same mixture of products in the same ratios. Explain this curious result.

15-10 S_N2 Displacement Reactions of Allylic Halides and Tosylates

Allylic halides and tosylates show enhanced reactivity toward nucleophilic displacement reactions by the S_N2 mechanism. For example, allyl bromide reacts with nucleophiles by the S_N2 mechanism about 40 times faster than *n*-propyl bromide.

Figure 15-13 shows how this rate enhancement can be explained by allylic delocalization of electrons in the transition state. The transition state for the S_N2 reaction looks like a trigonal carbon atom with a *p* orbital perpendicular to the three substituents. The electrons of the attacking nucleophile are forming a bond using one lobe of the *p* orbital while the leaving group's electrons are leaving from the other lobe. When the substrate is allylic, the transition state receives resonance stabilization through conjugation with the *p* orbitals of the pi bond. This stabilization lowers the energy of the transition state, resulting in a lower activation energy and an enhanced rate.

The enhanced reactivity of allylic halides and tosylates makes them particularly attractive as electrophiles for S_N2 reactions. Allylic halides are so reactive that they couple with Grignard and organolithium reagents, a reaction that does not work well with unactivated halides.

$$\underset{\text{allyl bromide}}{H_2C{=}CH{-}CH_2Br} + \underset{n\text{-butyllithium}}{CH_3(CH_2)_3{-}Li} \longrightarrow \underset{\text{hept-1-ene (85\%)}}{H_2C{=}CH{-}CH_2{-}(CH_2)_3CH_3} + LiBr$$

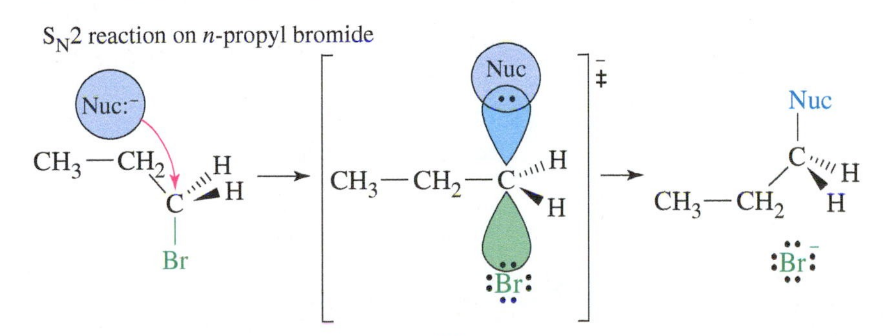

FIGURE 15-13
Allylic delocalization in the S_N2 transition state. The transition state for the S_N2 reaction of allyl bromide with a nucleophile is stabilized by conjugation of the double bond with the *p* orbital that is momentarily present on the reacting carbon atom. The resulting overlap lowers the energy of the transition state, increasing the reaction rate.

PROBLEM 15-13

Show how you might synthesize the following compounds starting with bromobenzene, and alkyl or alkenyl halides of four carbon atoms or fewer.
(a) 3-phenylprop-1-ene **(b)** 5-methylhex-2-ene ***(c)** dec-5-ene

15-11 The Diels–Alder Reaction

In 1928, German chemists Otto Diels and Kurt Alder discovered that alkenes and alkynes with electron-withdrawing groups add to conjugated dienes to form six-membered rings. The **Diels–Alder reaction** has proven to be a useful synthetic tool, providing one of the best ways to make six-membered rings with diverse functionality and controlled stereochemistry. Diels and Alder were awarded the Nobel Prize in 1950 for their work.

The Diels–Alder Reaction:

C≡N Δ (heat) C≡N

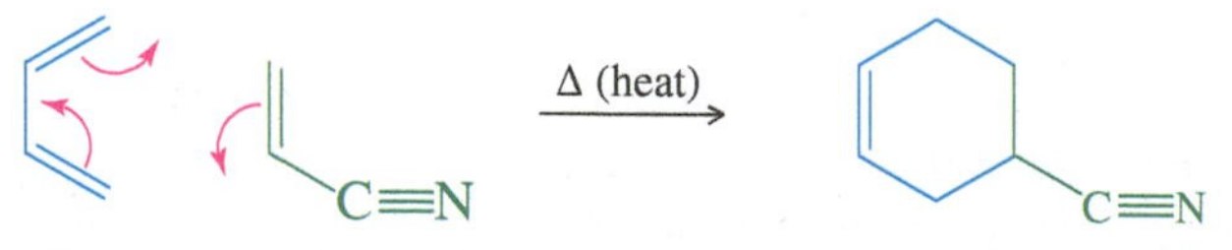

diene (4 π electrons) dienophile (2 π electrons) Diels–Alder product

The Diels–Alder reaction is called a **[4 + 2] cycloaddition** because a ring is formed by the interaction of four pi electrons in the diene with two pi electrons of the alkene or alkyne. Key Mechanism 15-3 shows the one-step mechanism of the Diels–Alder reaction. Because an electron-poor alkene or alkyne is prone to react with a diene, it is called a **dienophile** ("lover of dienes"). In effect, the Diels–Alder reaction converts two pi bonds into two sigma bonds. We can symbolize the Diels–Alder reaction by using three arrows to show the movement of three pairs of electrons. This electron movement is **concerted,** with three pairs of electrons moving simultaneously.

KEY MECHANISM 15-3 The Diels–Alder Reaction

The Diels–Alder is a one-step, concerted mechanism.

A diene reacts with an electron-poor alkene to give a new cyclohexene ring. (W is an electron-withdrawing group.)

H W C C H H Δ (heat) H C—W C—H H

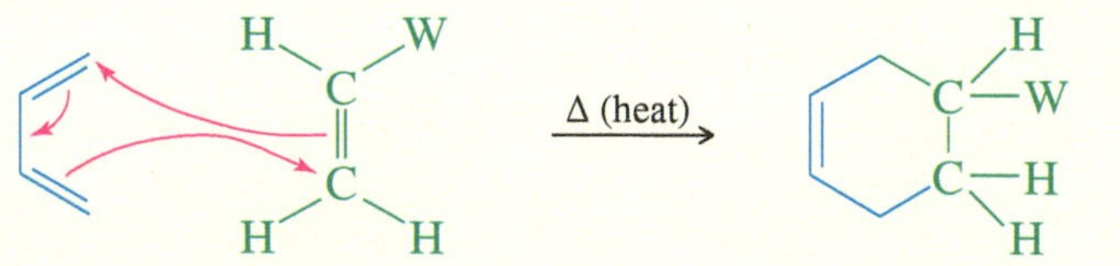

diene electron-rich dienophile electron-poor a cyclohexene ring

A diene reacts with an electron-poor alkyne to give a cyclohexadiene.

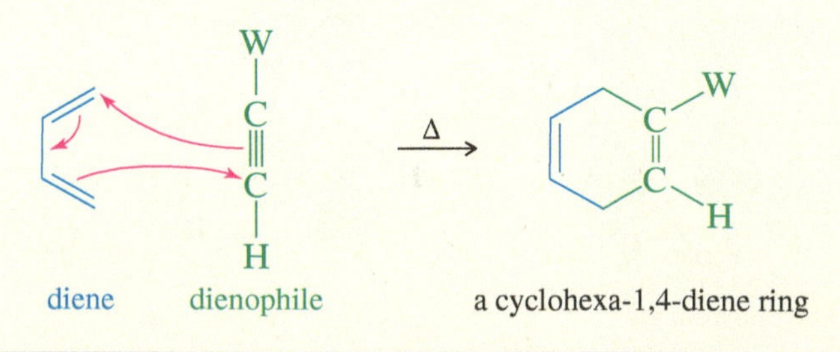

diene dienophile a cyclohexa-1,4-diene ring

EXAMPLES:

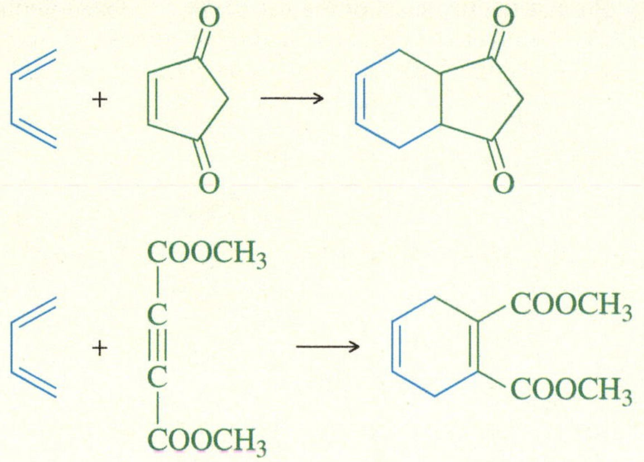

O O → O O

COOCH$_3$ C ≡ C COOCH$_3$ → COOCH$_3$ COOCH$_3$

The Diels–Alder reaction is similar to a nucleophile–electrophile reaction. The diene is electron-rich, and the dienophile is electron-poor. Simple dienes such as buta-1,3-diene are sufficiently electron-rich to be effective dienes for the Diels–Alder reaction. The presence of electron-donating (—D) groups, such as alkyl groups or alkoxy (—OR) groups, may further enhance the reactivity of the diene.

Simple alkenes and alkynes such as ethene and ethyne are poor dienophiles, however. A good dienophile generally has one or more electron-withdrawing groups (—W) pulling electron density away from the pi bond. Dienophiles commonly have carbonyl-containing (C=O) groups or cyano (—C≡N) groups to enhance their Diels–Alder reactivity. Figure 15-14 shows some representative Diels–Alder reactions involving a variety of different dienes and dienophiles.

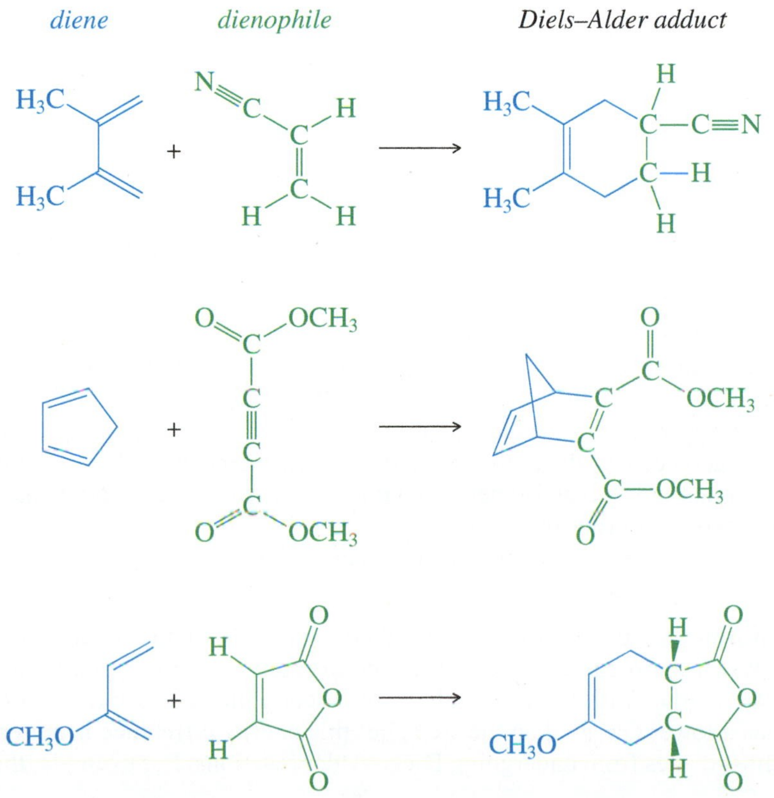

diene *dienophile* *Diels–Alder adduct*

H_3C H_3C + N≡C—C(H)=C(H)(H) → H_3C H_3C H C—C≡N C—H H

+ O=C(OCH$_3$)—C≡C—C(=O)OCH$_3$ → O C OCH$_3$ C C C—OCH$_3$ O

CH_3O + H H O O O → CH_3O H O C O C H O

FIGURE 15-14
Examples of the Diels–Alder reaction. Electron-releasing substituents activate the diene; electron-withdrawing substituents activate the dienophile.

PROBLEM-SOLVING HINT

A Diels–Alder product always contains one more ring than the reactants. The two ends of the diene form new bonds to the ends of the dienophile. The center (formerly single) bond of the diene becomes a double bond. The dienophile's double bond becomes a single bond (or its triple bond becomes a double bond).

PROBLEM 15-14

Predict the products of the following proposed Diels–Alder reactions.

(a) 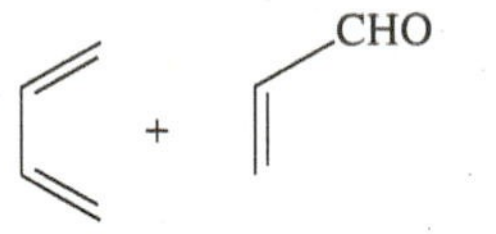 + CHO (b) 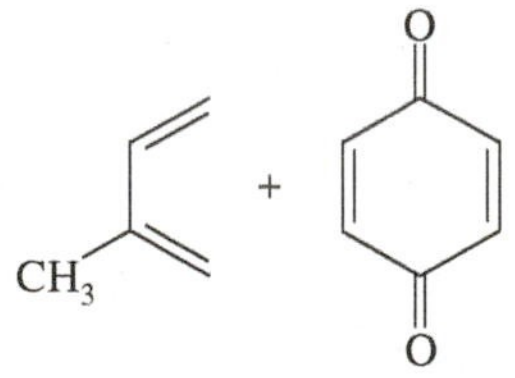 CH_3 + O O (c) 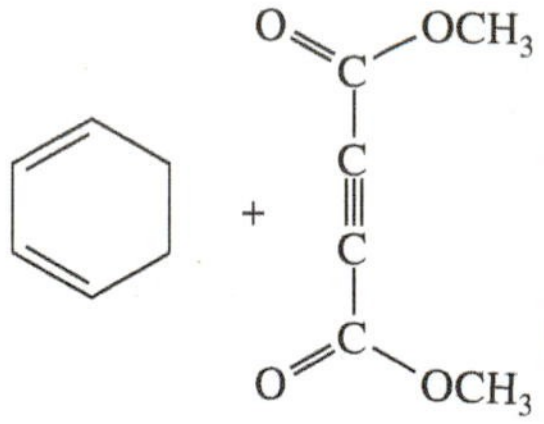 + $O{=}C(OCH_3)\text{–}C{\equiv}C\text{–}C({=}O)OCH_3$

(d) 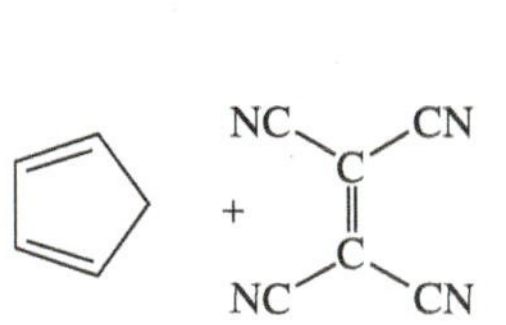 + $(NC)_2C{=}C(CN)_2$ (e) 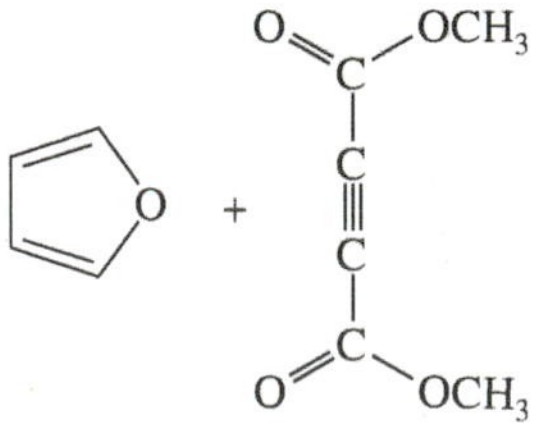 O + $O{=}C(OCH_3)\text{–}C{\equiv}C\text{–}C({=}O)OCH_3$ (f) 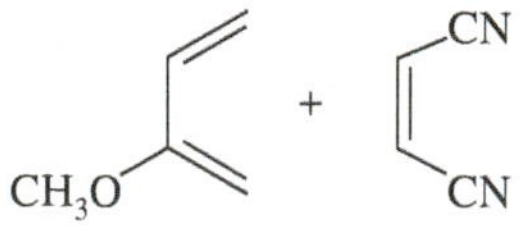 CH_3O + CN CN

PROBLEM-SOLVING HINT

To deconstruct a Diels–Alder product, look for the double bond at the center of what was the diene. Directly across the ring is the dienophile bond, usually with electron-withdrawing groups. (If a single bond, the dienophile had a double bond; if double, the dienophile had a triple bond.) Break the two bonds that join the diene and dienophile, and restore the two double bonds of the diene and the double (or triple) bond of the dienophile.

PROBLEM 15-15

What dienes and dienophiles would react to give the following Diels–Alder products?

(a) 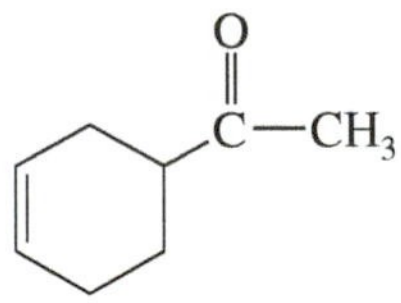 $C({=}O)\text{—}CH_3$ (b) 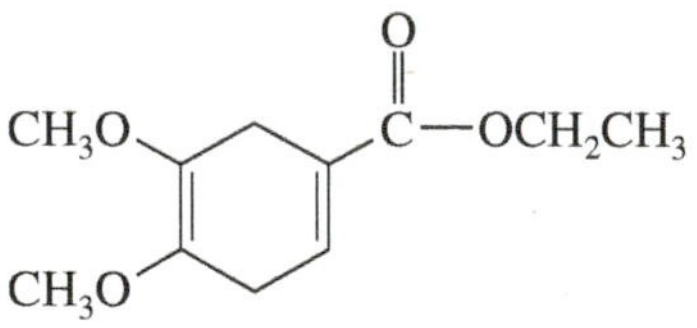 CH_3O, CH_3O, $C({=}O)\text{—}OCH_2CH_3$ (c) O, CN

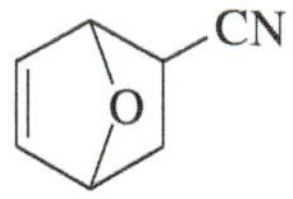

(d) 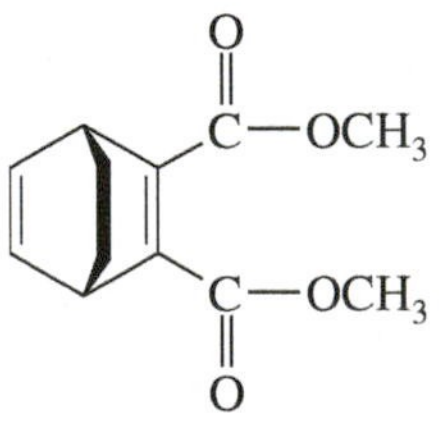 $C({=}O)\text{—}OCH_3$, $C({=}O)\text{—}OCH_3$ (e) 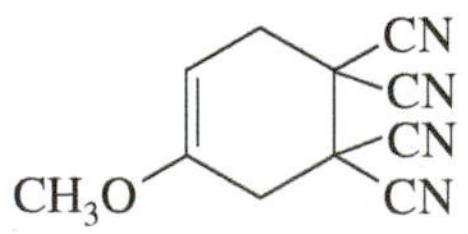 CH_3O, CN, CN, CN, CN (f) 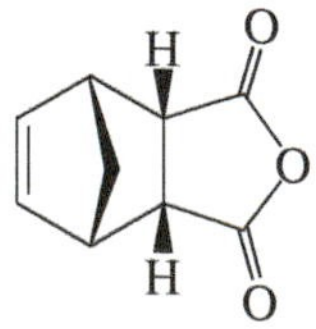 H, O, O, O, H

15-11A Stereochemical Requirements of the Diels–Alder Transition State

The mechanism of the Diels–Alder reaction is a concerted cyclic movement of six electrons: four in the diene and two in the dienophile. For the three pairs of electrons to move simultaneously, the transition state must have a geometry that allows overlap of the two end *p* orbitals of the diene with those of the dienophile. Figure 15-15 shows the required geometry of the transition state. The geometry of the Diels–Alder transition state explains why some isomers react differently from others, and it enables us to predict the stereochemistry of the products.

Three stereochemical features of the Diels–Alder reaction are controlled by the requirements of the transition state:

s-cis Conformation of the Diene The diene must be in the *s*-cis conformation to react. When the diene is in the *s*-trans conformation, the end *p* orbitals are too far apart to overlap with the *p* orbitals of the dienophile. The *s*-trans conformation usually has a lower energy than the *s*-cis, but this energy difference is not enough to prevent most dienes from undergoing Diels–Alder reactions. For example, the *s*-trans

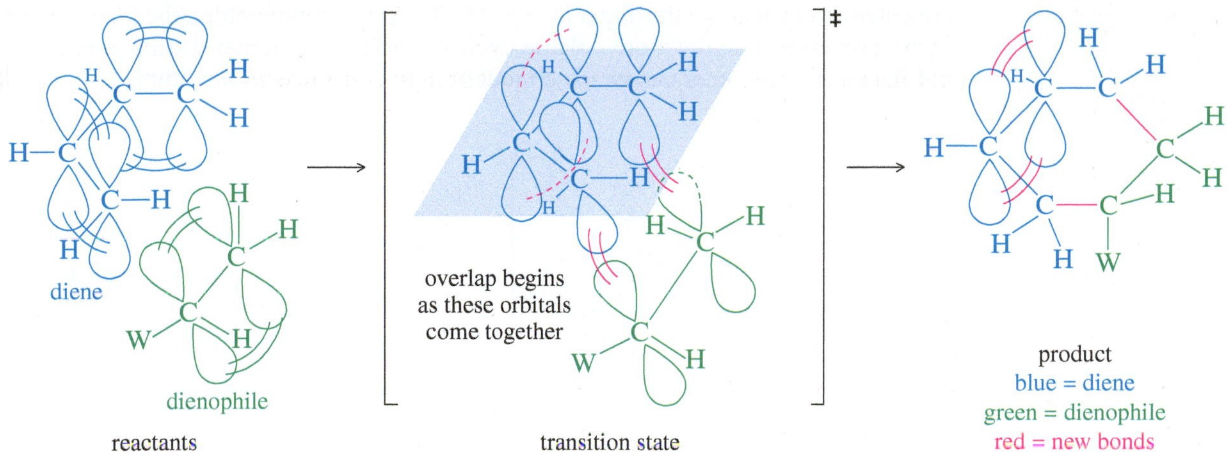

reactants

diene

dienophile

transition state

overlap begins as these orbitals come together

product

blue = diene

green = dienophile

red = new bonds

FIGURE 15-15
The geometry of the Diels–Alder transition state. The Diels–Alder reaction has a concerted mechanism, with all the bond making and bond breaking occurring in a single step. Three pairs of electrons move simultaneously, requiring a transition state with overlap between the end *p* orbitals of the diene and those of the dienophile.

conformation of butadiene is only 12 kJ/mol (2.9 kcal/mol) lower in energy than the *s*-cis conformation.

s-cis (cisoid) ⇌ *s*-trans (transoid)

12 kJ/mol more stable

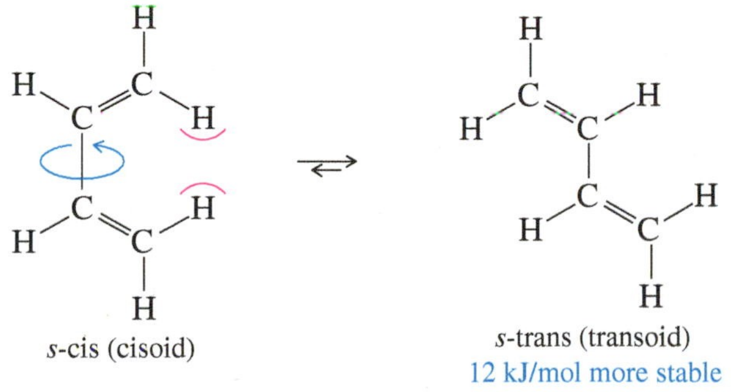

Structural features that aid or hinder the diene in achieving the *s*-cis conformation affect its ability to participate in Diels–Alder reactions. Figure 15-16 shows that dienes with functional groups that hinder the *s*-cis conformation react more slowly than butadiene. Dienes with functional groups that hinder the *s*-trans conformation react faster than butadiene.

Because cyclopentadiene is fixed in the *s*-cis conformation, it is highly reactive in the Diels–Alder reaction. It is so reactive, in fact, that at room temperature, cyclopentadiene slowly reacts with itself to form dicyclopentadiene. Cyclopentadiene

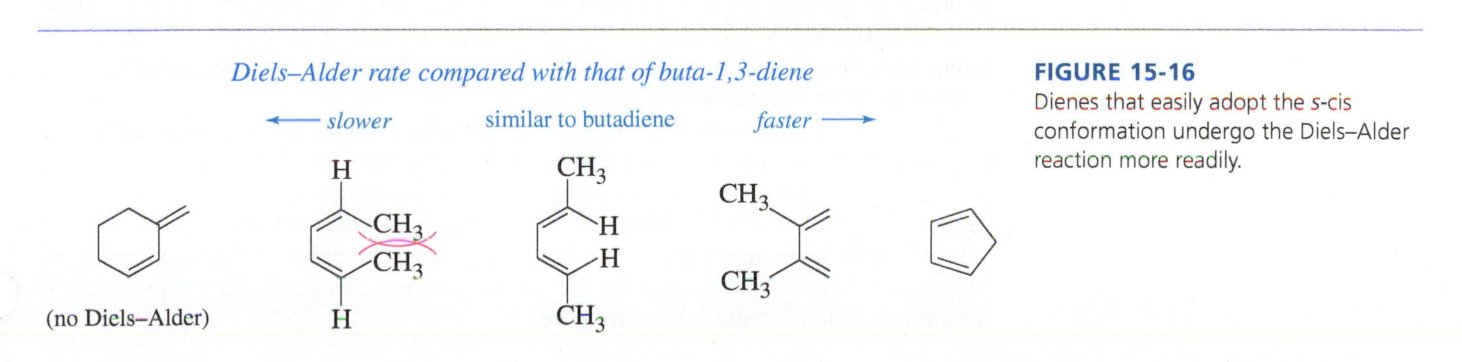

FIGURE 15-16
Dienes that easily adopt the *s*-cis conformation undergo the Diels–Alder reaction more readily.

is regenerated by heating the dimer above 200 °C. At this temperature, the Diels–Alder reaction reverses, and the more volatile cyclopentadiene monomer distills over into a cold flask. The monomer can be stored indefinitely at dry-ice temperatures.

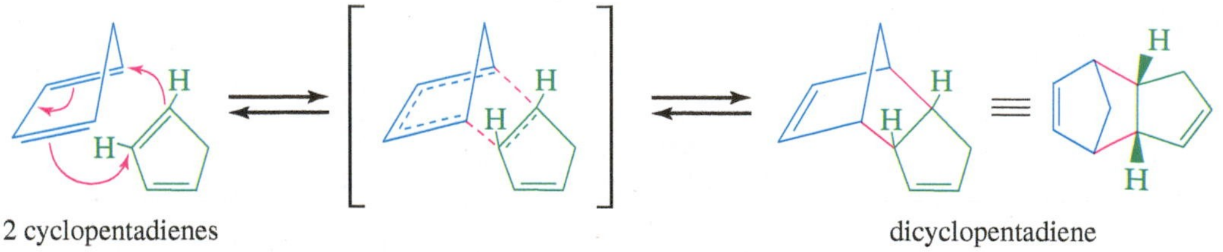

2 cyclopentadienes

dicyclopentadiene

syn Stereochemistry The Diels–Alder reaction is a syn addition with respect to both the diene and the dienophile. The dienophile adds to one face of the diene, and the diene adds to one face of the dienophile. As you can see from the transition state in Figure 15-15, there is no opportunity for any of the substituents to change their stereochemical positions during the course of the reaction. Substituents that are on the same side of the diene or dienophile will be cis on the newly formed ring. The following examples show the results of this syn addition.

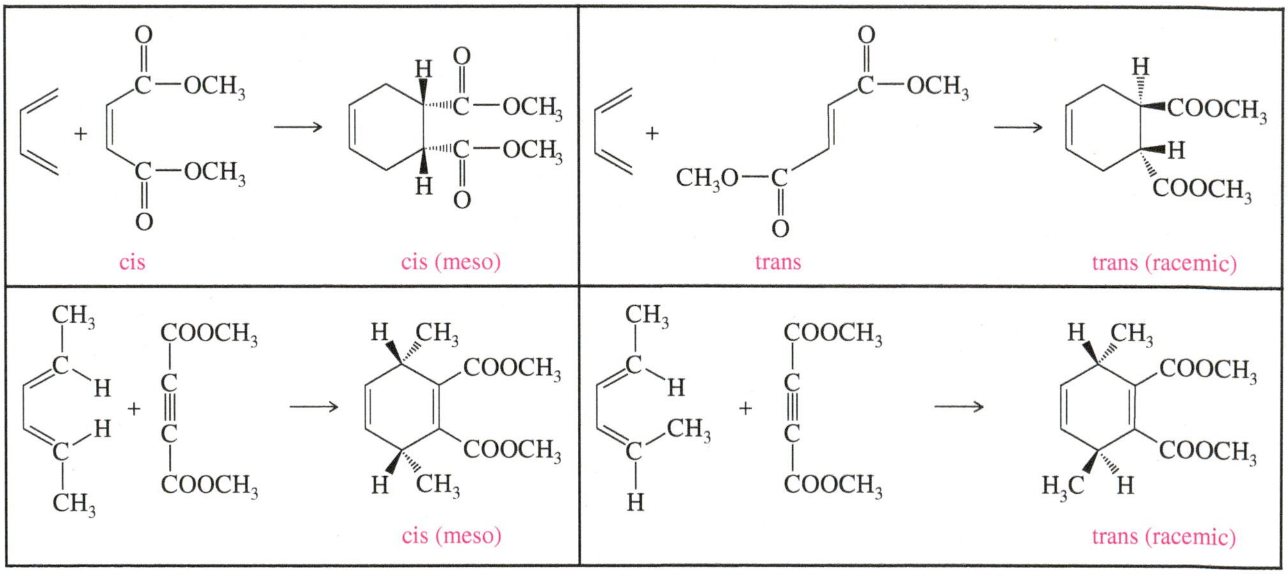

cis → cis (meso)

trans → trans (racemic)

cis (meso)

trans (racemic)

The Endo Rule When the dienophile has a pi bond in its electron-withdrawing group (as in a carbonyl group or a cyano group), the *p* orbitals in that electron-withdrawing group approach one of the central carbon atoms (C2 or C3) of the diene. This proximity results in **secondary overlap:** an overlap of the *p* orbitals of the electron-withdrawing group with the *p* orbitals of C2 and C3 of the diene (Figure 15-17). Secondary overlap helps to stabilize the transition state.

The influence of secondary overlap was first observed in reactions using cyclopentadiene to form bicyclic ring systems. In the bicyclic product (called *norbornene*), the electron-withdrawing substituent occupies the stereochemical position closest to the central atoms of the diene. This position is called the **endo position** because the substituent seems to be inside the pocket formed by the six-membered ring of norbornene. This stereochemical preference for the electron-withdrawing substituent to appear in the endo position is called the **endo rule.**

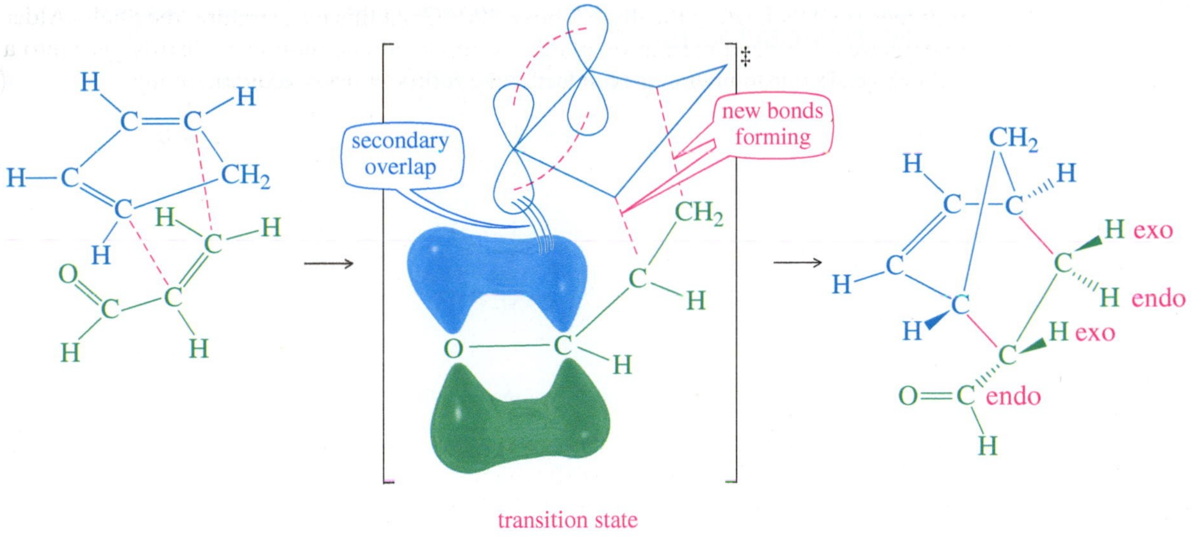

FIGURE 15-17
In most Diels–Alder reactions, there is *secondary overlap* between the *p* orbitals of the electron-withdrawing group and one of the central carbon atoms of the diene. Secondary overlap stabilizes the transition state, and it favors products having the electron-withdrawing groups in *endo* positions.

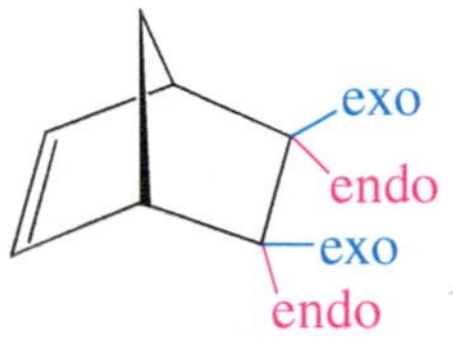

stereochemical positions of norbornene

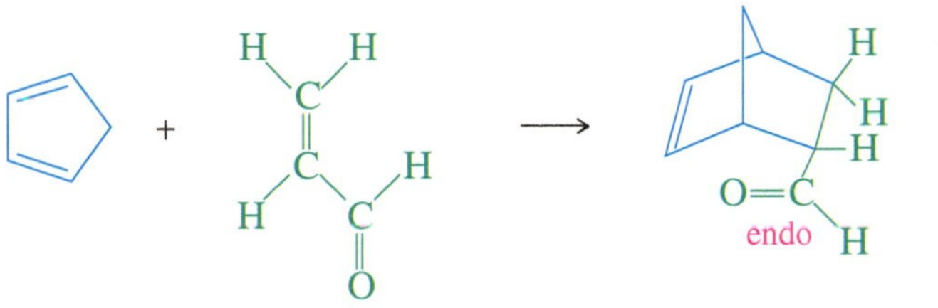

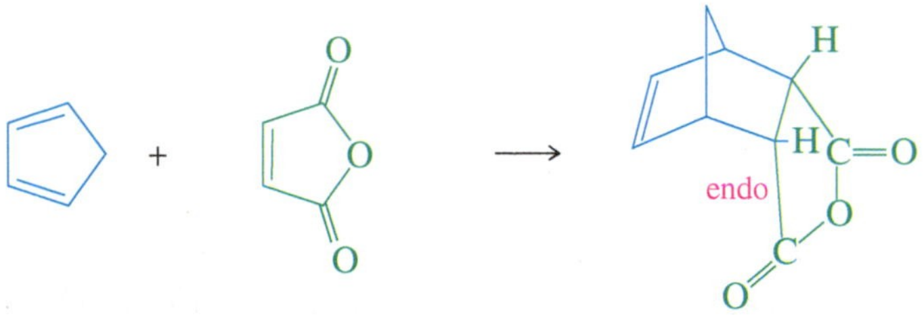

The endo rule is useful for predicting the products of many types of Diels–Alder reactions, regardless of whether they use cyclopentadiene to form norbornene systems. The following examples show the use of the endo rule with other types of Diels–Alder reactions.

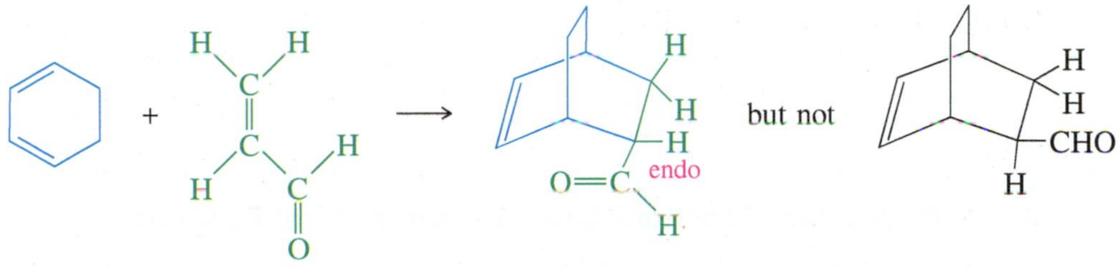

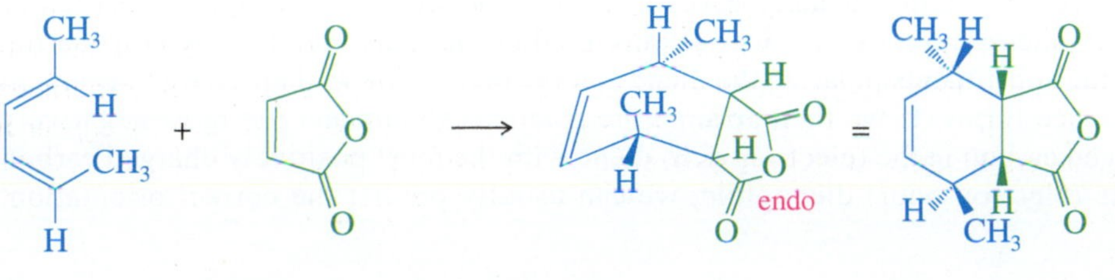

SOLVED PROBLEM 15-1

Use the endo rule to predict the product of the following cycloaddition.

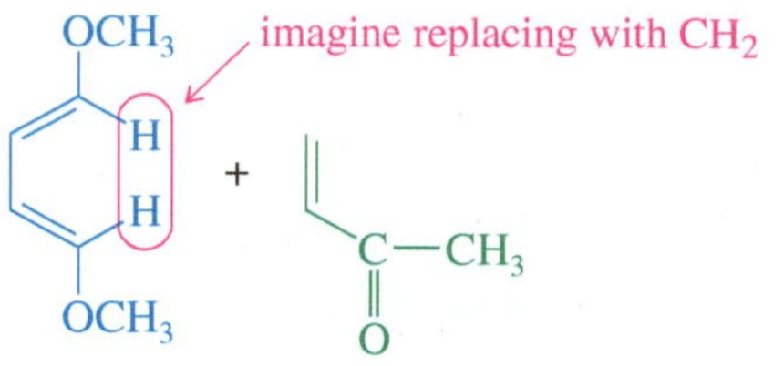

OCH_3 imagine replacing with CH_2

H H + C—CH_3 O

OCH_3

SOLUTION

Imagine this diene to be a substituted cyclopentadiene; the endo product will be formed.

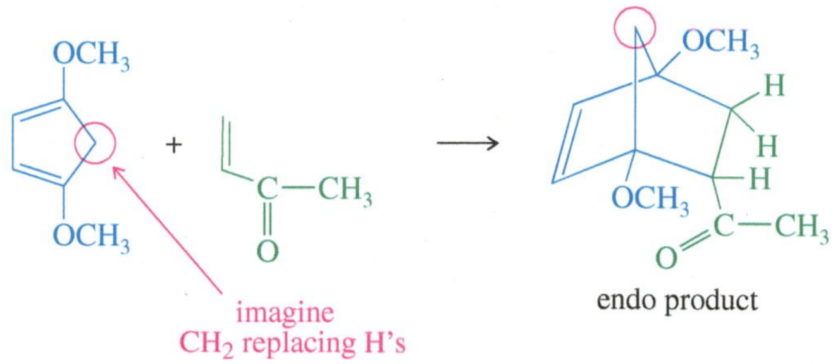

OCH_3 + C—CH_3 O → OCH_3 H H H OCH_3 C—CH_3 O

OCH_3

imagine
CH_2 replacing H's

endo product

In the imaginary reaction, we replaced the two inside hydrogens with the rest of the cyclopentadiene ring. Now we put them back and draw the actual product.

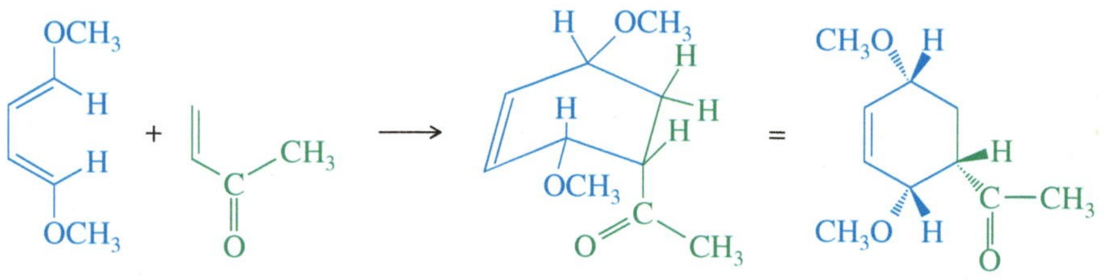

OCH_3 H H OCH_3 + C CH_3 O → H OCH_3 H H H H OCH_3 C O CH_3 = CH_3O H H C—CH_3 O CH_3O H

endo product

PROBLEM 15-16

Predict the major product for each proposed Diels–Alder reaction. Include stereochemistry where appropriate.

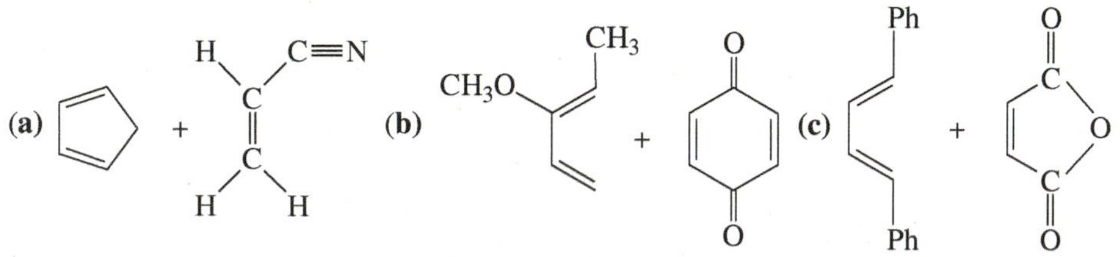

(a) + H C≡N C C H H (b) CH_3O CH_3 + O O (c) Ph Ph + O C O C O

15-11B Diels–Alder Reactions Using Unsymmetrical Reagents

Even when the diene and dienophile are both unsymmetrically substituted, the Diels–Alder reaction usually gives a single product (or a major product) rather than a random mixture. We can usually predict the major product by considering how the substituents polarize the diene and the dienophile in their charge-separated resonance forms. If we then arrange the reactants to connect the most negatively charged carbon in the (electron-rich) diene with the most positively charged carbon in the (electron-poor) dienophile, we can usually predict the correct orientation.

The following examples show that an electron-donating substituent (D) on the diene and an electron-withdrawing substituent (W) on the dienophile usually show either a 1,2- or 1,4-relationship in the product.

Formation of 1,4-product

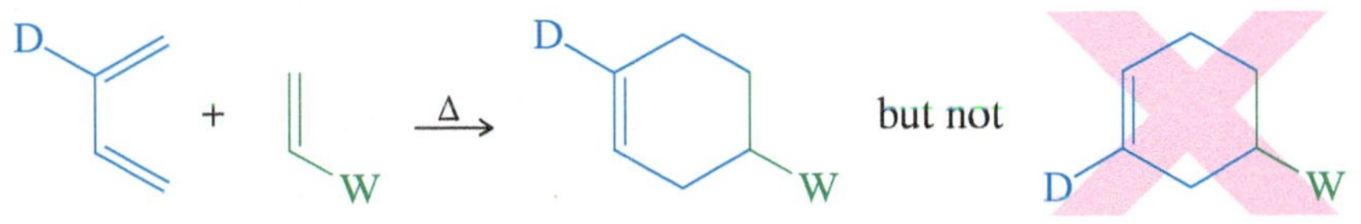

$\xrightarrow{\Delta}$ but not

1,4-product 1,3-product

Predicting this product

minus reacts with plus

diene dienophile charge-separated resonance forms 1,4-product

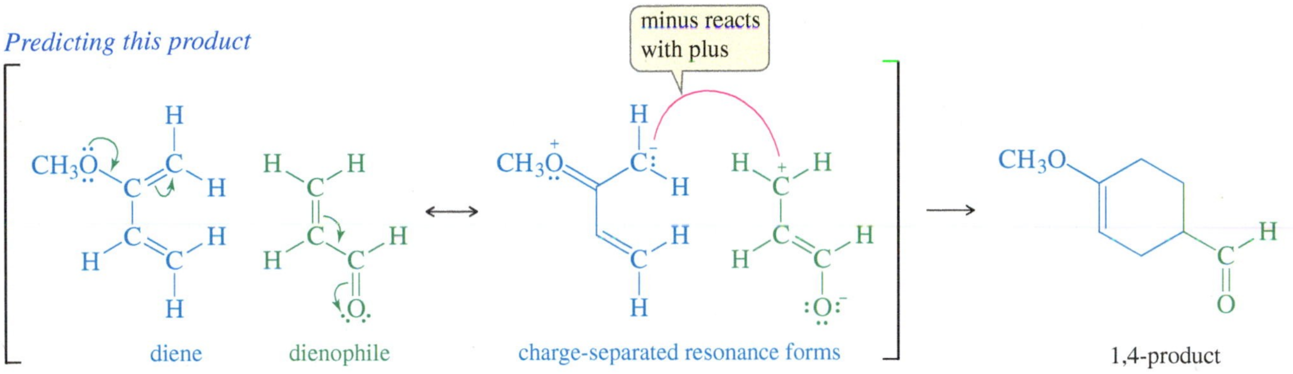

Formation of 1,2-product

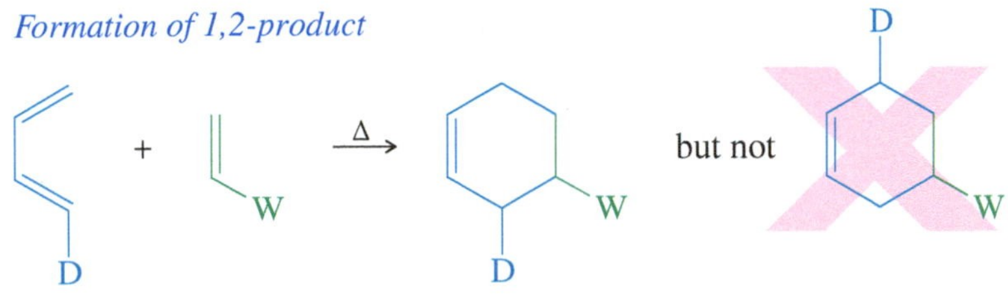

$\xrightarrow{\Delta}$ but not

1,2-product 1,3-product

Predicting this product

minus reacts with plus

diene dienophile charge-separated resonance forms 1,2-product

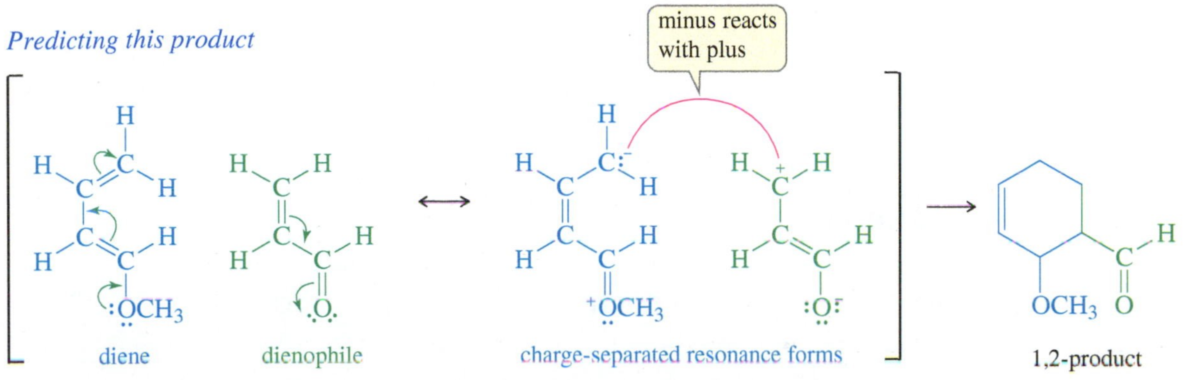

In most cases, we don't even need to draw the charge-separated resonance forms to determine which orientation of the reactants is preferred. We can predict the major products of unsymmetrical Diels–Alder reactions simply by remembering that the electron-donating groups of the diene and the electron-withdrawing groups of the dienophile usually bear either a 1,2-relationship or a 1,4-relationship in the products, but not a 1,3-relationship.

SOLVED PROBLEM 15-2

Predict the products of the following proposed Diels–Alder reactions.

(a) 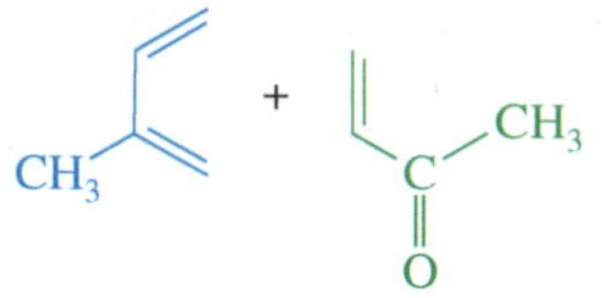 CH_3 diene + $CH_2=CH-C(=O)-CH_3$ (b) 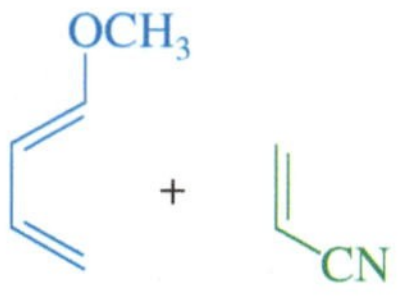 OCH_3 diene + $CH_2=CH-CN$

SOLUTION

(a) The methyl group is weakly electron-donating to the diene, and the carbonyl group is electron-withdrawing from the dienophile. The two possible orientations place these groups in a 1,4-relationship or a 1,3-relationship. We select the 1,4-relationship for our predicted product. (Experimental results show a 70:30 preference for the 1,4-product.)

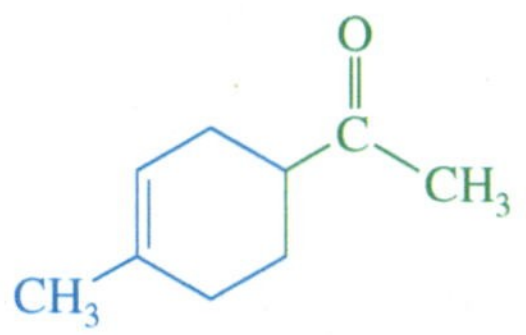

1,4-relationship (major) (70%)   1,3-relationship (minor) (30%)

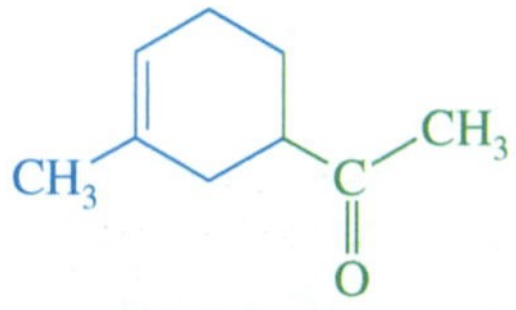

(b) The methoxy group ($-OCH_3$) is strongly electron-donating to the diene, and the cyano group ($-C{\equiv}N$) is electron-withdrawing from the dienophile. Depending on the orientation of addition, the product has either a 1,2- or a 1,3-relationship of these two groups. We select the 1,2-relationship, and the endo rule predicts cis stereochemistry of the two substituents.

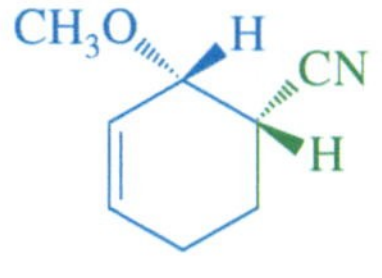

1,2-relationship (product) 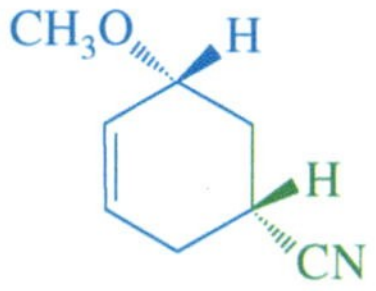   1,3-relationship (not formed)

PROBLEM 15-17

In Solved Problem 15-2, we simply predicted that the products would have a 1,2- or 1,4-relationship of the proper substituents. Draw the charge-separated resonance forms of the reactants to support these predictions.

PROBLEM 15-18

Predict the products of the following Diels–Alder reactions.

(a) 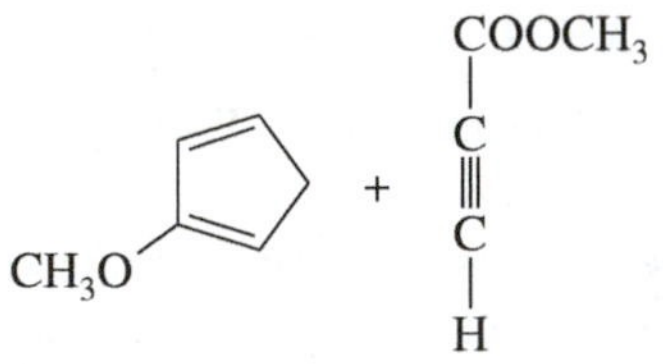 CH_3O-cyclopentadiene + $H-C{\equiv}C-COOCH_3$ (b) OCH_3-cyclohexadiene + cyclohexenone

(c) CH_3O diene + $CH_2=CH-CHO$ *(d) 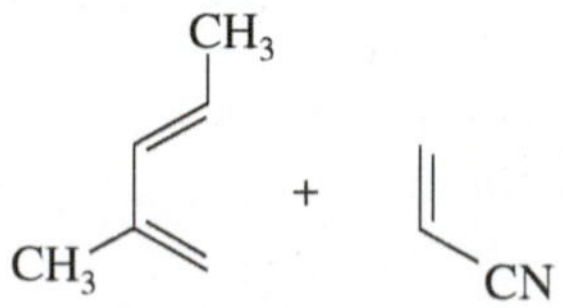 CH_3, CH_3 diene + $CH_2=CH-CN$

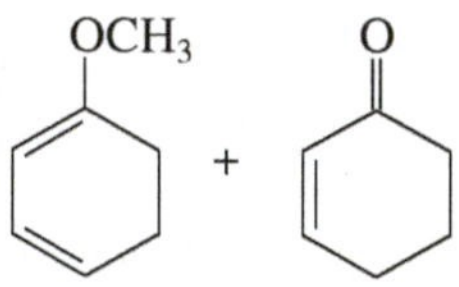

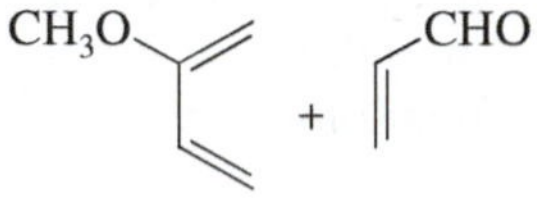

15-12 The Diels–Alder as an Example of a Pericyclic Reaction

The Diels–Alder reaction is a **cycloaddition:** Two molecules combine in a one-step, **concerted reaction** to form a new ring. Cycloadditions such as the Diels–Alder are one class of **pericyclic reactions,** which involve the concerted forming and breaking of bonds within a closed ring of interacting orbitals. Figure 15-15 (page 737) shows the closed loop of interacting orbitals in the Diels–Alder transition state. Each carbon atom of the new ring has one orbital involved in this closed loop.

A concerted pericyclic reaction has a single transition state, whose activation energy may be supplied by heat (thermal induction) or by ultraviolet light (photochemical induction). Some pericyclic reactions proceed only under thermal induction, and others proceed only under photochemical induction. Some pericyclic reactions take place under both thermal and photochemical conditions, but the two sets of conditions give different products.

For many years, pericyclic reactions were poorly understood and unpredictable. Around 1965, Robert B. Woodward and Roald Hoffmann developed a theory for predicting the results of pericyclic reactions by considering the symmetry of the molecular orbitals of the reactants and products. Their theory, called **conservation of orbital symmetry,** says that the molecular orbitals of the reactants must flow smoothly into the MOs of the products without any drastic changes in symmetry. In that case, there will be bonding interactions to help stabilize the transition state. Without these bonding interactions in the transition state, the activation energy is much higher, and the concerted cyclic reaction cannot occur. Conservation of symmetry has been used to develop "rules" to predict which pericyclic reactions are feasible and what products will result. These rules are often called the **Woodward–Hoffmann rules.**

Application: Vitamin D

The precursor to vitamin D undergoes a pericyclic ring-opening (reverse Diels–Alder) reaction in the skin after exposure to the Sun's ultraviolet rays. In cold climates, children who have little exposure to sunlight often cannot synthesize or consume enough vitamin D, and as a result they develop rickets. Fish liver oils are rich sources of vitamin D.

7-dehydrocholesterol

pericyclic reaction

provitamin D_3

15-12A Conservation of Orbital Symmetry in the Diels–Alder Reaction

We will not develop all of the Woodward–Hoffmann rules; however, we will show how the molecular orbitals can indicate whether a cycloaddition will take place. The simple Diels–Alder reaction of butadiene with ethylene serves as our first example. The molecular orbitals of butadiene and ethylene are represented in Figure 15-18.

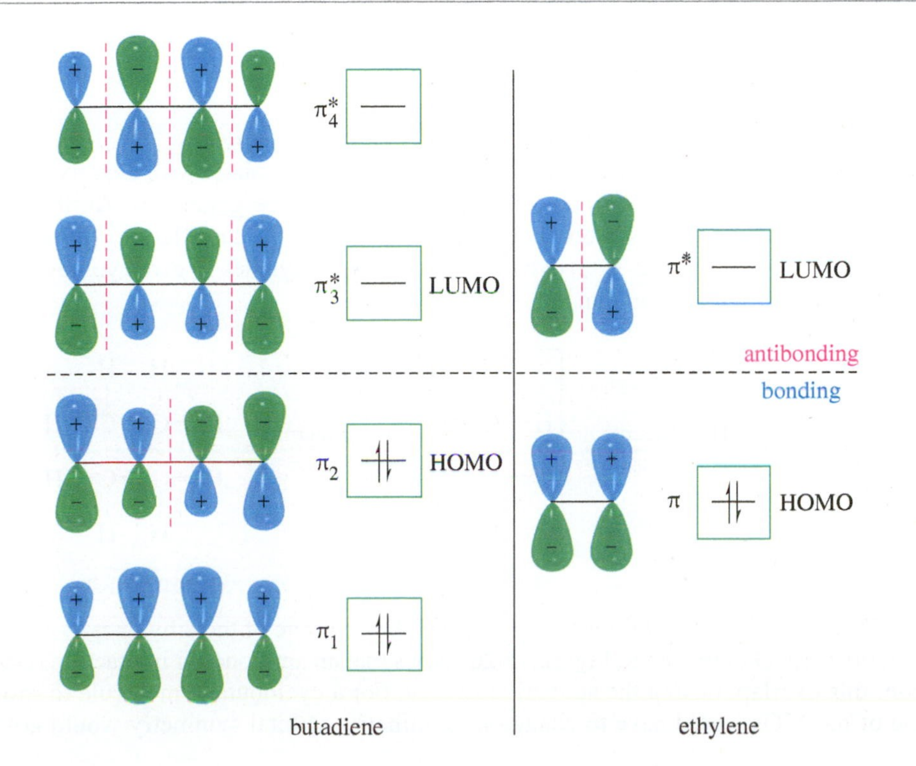

FIGURE 15-18 Molecular orbitals of butadiene and ethylene.

Butadiene, with four atomic *p* orbitals, has four molecular orbitals: two bonding MOs (filled) and two antibonding MOs (vacant). Ethylene, with two atomic *p* orbitals, has two MOs: a bonding MO (filled) and an antibonding MO (vacant).

In the Diels–Alder reaction, the diene acts as the electron-rich nucleophile, and the dienophile acts as the electron-poor electrophile. If we imagine the diene contributing a pair of electrons to the dienophile, the highest-energy electrons of the diene require the least activation energy for such a donation. The electrons in the highest-energy occupied orbital, called the **Highest Occupied Molecular Orbital (HOMO),** are the important ones because they are the most weakly held. The HOMO of butadiene is π_2, and its symmetry determines the course of the reaction.

The orbital in ethylene that receives these electrons is the lowest-energy orbital available, the **Lowest Unoccupied Molecular Orbital (LUMO).** In ethylene, the LUMO is the π^* antibonding orbital. If the electrons in the HOMO of butadiene can flow smoothly into the LUMO of ethylene, a concerted reaction can take place.

Figure 15-19 shows that the HOMO of butadiene has the correct symmetry to overlap in phase with the LUMO of ethylene. Having the correct symmetry means the orbitals that form the new bonds can overlap constructively: plus with plus and minus with minus. These bonding interactions stabilize the transition state and promote the concerted reaction. This favorable result predicts that the reaction is **symmetry-allowed.** The Diels–Alder reaction is common, and this theory correctly predicts a favorable transition state.

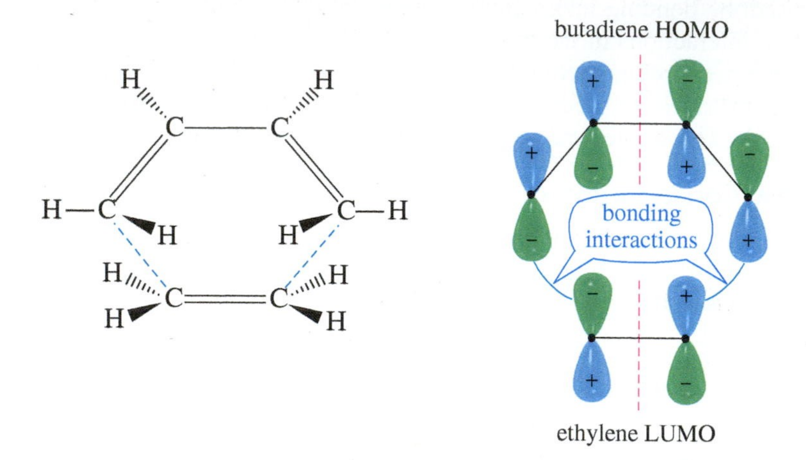

FIGURE 15-19
A symmetry-allowed reaction. The HOMO of butadiene forms a bonding overlap with the LUMO of ethylene because the orbitals have similar symmetry. This reaction is therefore symmetry-allowed.

15-12B The "Forbidden" [2 + 2] Cycloaddition

If a cycloaddition produces an overlap of positive-phase orbitals with negative-phase orbitals (destructive overlap), antibonding interactions are generated. Antibonding interactions raise the activation energy, so the reaction is classified as **symmetry-forbidden.** The thermal [2 + 2] cycloaddition of two ethylenes to give cyclobutane is a symmetry-forbidden reaction.

two ethylenes —thermal→ (transition state) —thermal→ cyclobutane

This [2 + 2] cycloaddition requires the HOMO of one of the ethylenes to overlap with the LUMO of the other. Figure 15-20 shows that an antibonding interaction results from this overlap, raising the activation energy. For a cyclobutane molecule to result, one of the MOs would have to change its symmetry. Orbital symmetry would not be

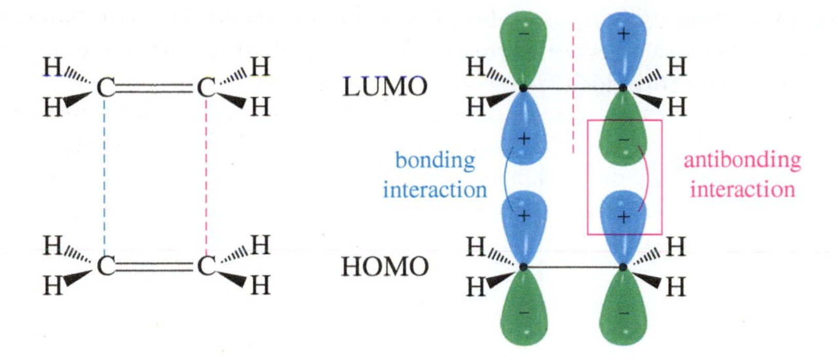

FIGURE 15-20
A symmetry-forbidden reaction. The HOMO and LUMO of two ethylene molecules have different symmetries, and they overlap to form an antibonding interaction. The concerted [2 + 2] cycloaddition is therefore symmetry-forbidden.

conserved, so the reaction is symmetry-forbidden. Such a symmetry-forbidden reaction can occasionally be made to occur; however, it cannot occur in the concerted pericyclic manner shown in the figure.

15-12C Photochemical Induction of Cycloadditions

When ultraviolet light rather than heat is used to induce pericyclic reactions, our predictions generally must be reversed. For example, the [2 + 2] cycloaddition of two ethylenes is photochemically "allowed." When a photon with the correct energy strikes ethylene, one of the pi electrons is excited to the next higher molecular orbital (Figure 15-21). This higher orbital, formerly the LUMO, is now occupied: It is the new HOMO*, the HOMO of the excited molecule.

The HOMO* of the excited ethylene molecule has the same symmetry as the LUMO of a ground-state ethylene. An excited molecule can react with a ground-state molecule to give cyclobutane (Figure 15-22). The [2 + 2] cycloaddition is

Application: Skin Cancer

Exposure of DNA to ultraviolet light induces a [2 + 2] cycloaddition reaction between the double bonds of adjacent thymines. The resulting thymine dimer, containing a cyclobutane ring, prevents DNA reproduction and can lead to the development of skin cancer.

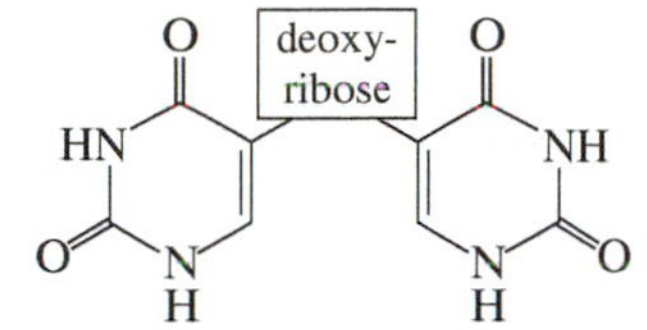

two adjacent thymine nucleotides on DNA

$h\nu$ (ultraviolet)

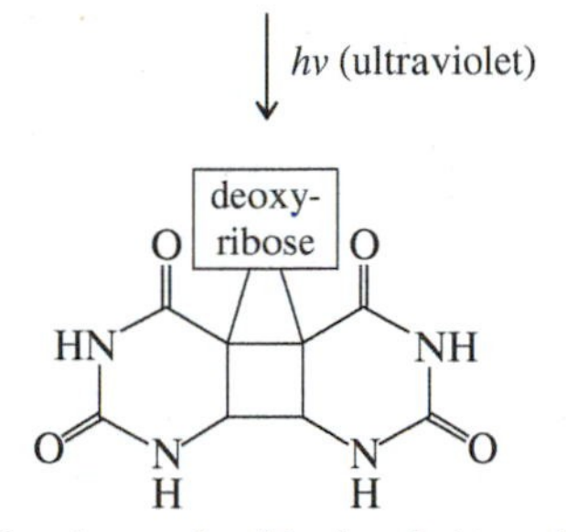

two thymine nucleotides bonded together

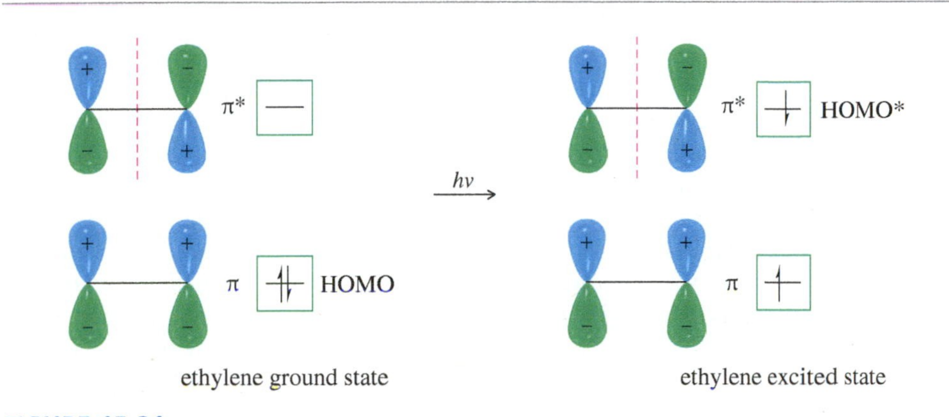

FIGURE 15-21
The effect of ultraviolet light on ethylene. Ultraviolet light excites one of the ethylene pi electrons into the antibonding orbital. The antibonding orbital is now occupied, so it is the new HOMO*.

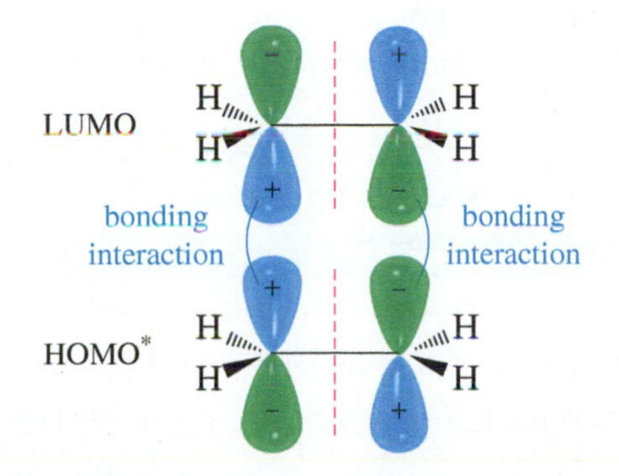

FIGURE 15-22
Photochemical [2 + 2] cycloaddition. The HOMO* of the excited ethylene overlaps favorably with the LUMO of an unexcited (ground-state) molecule. This reaction is symmetry-allowed.

therefore *photochemically allowed* but *thermally forbidden*. In most cases, photochemically allowed reactions are thermally forbidden, and thermally allowed reactions are photochemically forbidden.

two ethylenes —light (allowed)→ (transition state) —light (allowed)→ cyclobutane

PROBLEM 15-19

Show that the [4 + 2] Diels–Alder reaction is photochemically forbidden.

PROBLEM 15-20

(a) Show that the [4 + 4] cycloaddition of two butadiene molecules to give cycloocta-1,5-diene is thermally forbidden but photochemically allowed.

(b) There is a different, thermally allowed cycloaddition of two butadiene molecules. Show this reaction, and explain why it is thermally allowed. (*Hint:* Consider the dimerization of cyclopentadiene.)

15-13 Ultraviolet Absorption Spectroscopy

We have already encountered three powerful analytical techniques used by organic chemists. Infrared spectroscopy (IR, Chapter 12) observes the vibrations of molecular bonds, providing information about the nature of the bonding and the functional groups in a molecule. Nuclear magnetic resonance spectroscopy (NMR, Chapter 13) detects nuclear transitions, providing information about the electronic and molecular environment of the nuclei. From the NMR spectrum we can determine the structure of the alkyl groups present and often infer the functional groups. A mass spectrometer (MS, Chapter 12) bombards molecules with electrons, causing them to break apart in predictable ways. The masses of the molecular ion and the fragments provide a molecular weight (and perhaps a molecular formula) as well as structural information about the original compound.

We now study **ultraviolet (UV) spectroscopy,** which detects the electronic transitions of conjugated systems and provides information about the length and structure of the conjugated part of a molecule. UV spectroscopy gives more specialized information than does IR or NMR, and it is less commonly used than the other techniques.

The Cassini spacecraft took this image of Saturn's rings using the Ultraviolet Imaging Spectrograph. It shows there is more ice (turquoise) than rocks and dust (orange) in the outer parts of the rings.

15-13A Spectral Region

Ultraviolet frequencies correspond to shorter wavelengths and much higher energies than infrared (Table 15-1). The UV region is a range of frequencies just beyond the visible: *ultra*, meaning beyond, and *violet*, the highest-frequency visible light.

TABLE 15-1
Comparison of Wavelengths Used in Spectroscopy

Spectral Region	Wavelength, λ	Energy Range, kJ/mol (kcal/mol)
ultraviolet	200–400 nm ($2–4 \times 10^{-5}$ cm)	300–600 (70–140)
visible	400–800 nm ($4–8 \times 10^{-5}$ cm)	150–300 (35–70)
infrared	2.5–25 μm ($2.5–25 \times 10^{-4}$ cm)	4.6–46 (1.1–11)
NMR (radio)	0.3–5 meters	$2–40 \times 10^{-5}$ ($0.5–10 \times 10^{-5}$)

Wavelengths of the UV region are given in units of nanometers (nm; 10^{-9} m). Common UV spectrometers operate in the range of 200 to 400 nm (2×10^{-5} to 4×10^{-5} cm), corresponding to photon energies of about 300 to 600 kJ/mol (70 to 140 kcal/mol). These spectrometers often extend into the visible region (longer wavelength, lower energy) and are called **UV–visible spectrometers.** UV–visible energies correspond to electronic transitions: the energy needed to excite an electron from one molecular orbital to another.

15-13B Ultraviolet Light and Electronic Transitions

The wavelengths of UV light absorbed by a molecule are determined by the electronic energy differences between orbitals in the molecule. Sigma bonds are very stable, and the electrons in sigma bonds are usually unaffected by UV wavelengths of light above 200 nm. Pi bonds have electrons that are more easily excited into higher energy orbitals. Conjugated systems are particularly likely to have low-lying vacant orbitals, and electronic transitions into these orbitals produce characteristic ultraviolet absorptions.

Ethylene, for example, has two pi orbitals: the bonding orbital (π, the HOMO) and the antibonding orbital (π^*, the LUMO). The ground state has two electrons in the bonding orbital and none in the antibonding orbital. A photon with the right amount of energy can excite an electron from the bonding orbital (π) to the antibonding orbital (π^*). This transition from a π bonding orbital to a π^* antibonding orbital is called a *$\pi \rightarrow \pi^*$ transition* (Figure 15-23).

PROBLEM-SOLVING HINT

IR spectroscopy also detects conjugated double bonds by their lowered stretching frequencies.

1640–1680 cm^{-1}	isolated
1620–1640 cm^{-1}	conjugated
1600 cm^{-1}	aromatic

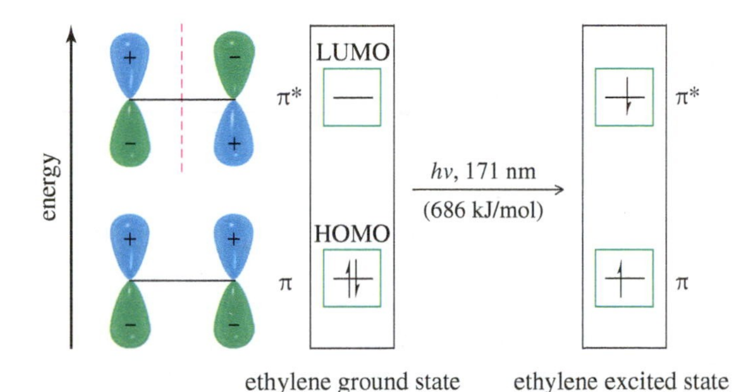

FIGURE 15-23
The absorption of a 171-nm photon excites an electron from the π bonding MO of ethylene to the π^* antibonding MO. This absorption requires light of greater energy (shorter wavelength) than the range covered by a typical UV spectrometer.

The $\pi \rightarrow \pi^*$ transition of ethylene requires absorption of light at 171 nm (686 kJ/mol, or 164 kcal/mol). Most UV spectrometers cannot detect this absorption because it is obscured by the absorption caused by oxygen in the air. In conjugated systems, however, there are electronic transitions with lower energies that correspond to wavelengths longer than 200 nm. Figure 15-24 compares the MO energies of ethylene with those of butadiene to show that the HOMO and LUMO of butadiene are closer in energy than those of ethylene. The HOMO of butadiene is higher in energy than the HOMO of ethylene, and the LUMO of butadiene is lower in energy than the LUMO of ethylene. Both differences reduce the relative energy of the $\pi_2 \rightarrow \pi_3^*$ transition. The resulting absorption is at 217 nm (540 kJ/mol, or 129 kcal/mol), which can be measured using a standard UV spectrometer.

Just as conjugated dienes absorb at longer wavelengths than simple alkenes, conjugated trienes absorb at even longer wavelengths. In general, the energy difference between HOMO and LUMO decreases as the length of conjugation increases. In hexa-1,3,5-triene, for example (Figure 15-25), the HOMO is π_3 and the LUMO is π_4^*. The HOMO in hexa-1,3,5-triene is slightly higher in energy than for buta-1,3-diene, and the LUMO is slightly lower in energy. Once again, the narrowing of the energy between the HOMO and the LUMO gives a lower-energy, longer-wavelength absorption.

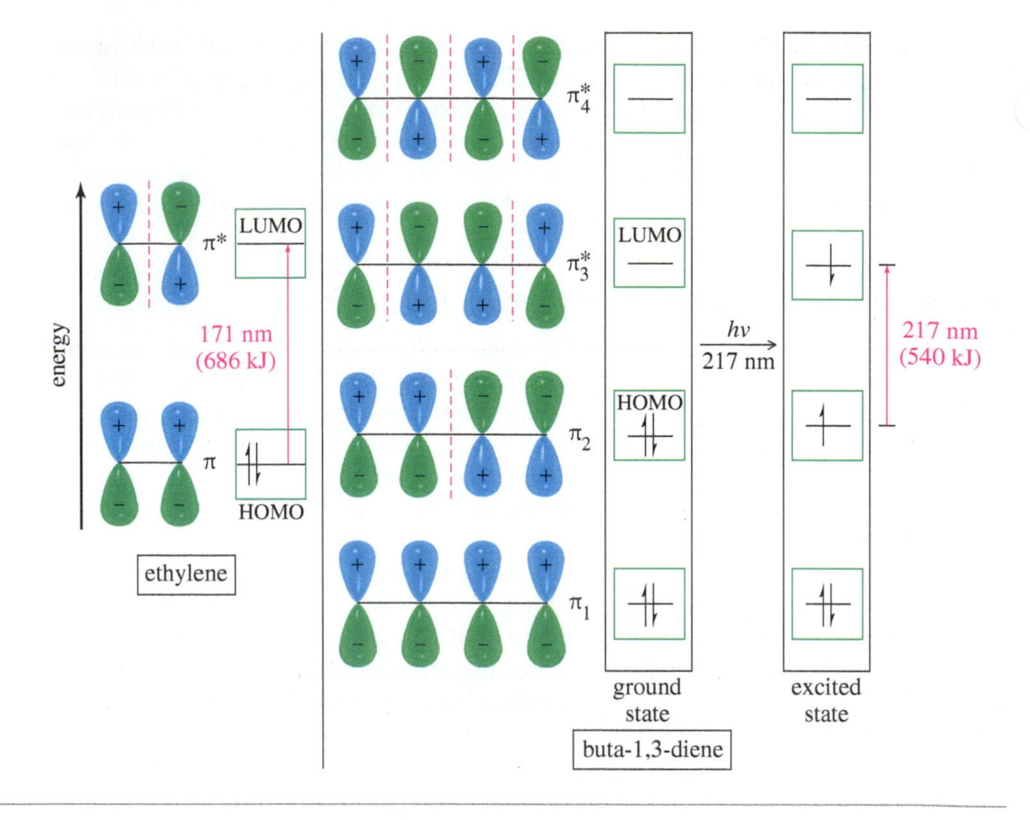

FIGURE 15-24 Comparison of HOMO-LUMO energy differences. In buta-1,3-diene, the $\pi \rightarrow \pi^*$ transition absorbs at a wavelength of 217 nm (540 kJ/mol) compared with 171 nm (686 kJ/mol) for ethylene. This longer wavelength (lower-energy) absorption results from a smaller energy difference between the HOMO and LUMO in butadiene than in ethylene.

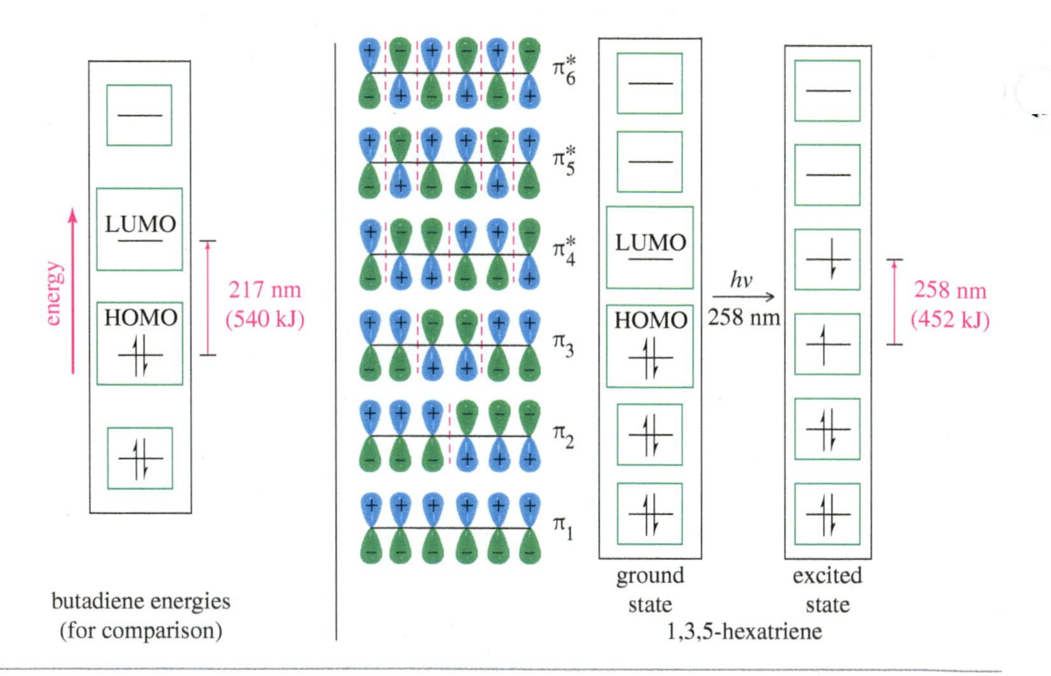

FIGURE 15-25 Hexa-1,3,5-triene has a smaller energy difference (452 kJ/mol) between its HOMO and LUMO than does buta-1,3-diene (540 kJ/mol). The $\pi \rightarrow \pi^*$ transition corresponding to this energy difference absorbs at a longer wavelength: 258 nm, compared with 217 nm for buta-1,3-diene.

The principal $\pi \rightarrow \pi^*$ transition in hexa-1,3,5-triene occurs at 258 nm (452 kJ/mol, or 108 kcal/mol).

We can summarize the effects of conjugation on the wavelength of UV absorption by stating a general rule: *A compound that contains a longer chain of conjugated double bonds absorbs light at a longer wavelength.* The trend of longer wavelengths for longer conjugated chains continues, and at seven conjugated double bonds, the absorption surpasses 400 nm and enters in the visible portion of the spectrum.

4 C═C, 290 nm

5 C═C, 334 nm

6 C═C, 362 nm

7 C═C, >400 nm

Because they have no interaction with each other, isolated double bonds do not contribute to shifting the UV absorption to longer wavelengths. Both their reactions and their UV absorptions are similar to those of simple alkenes. For example, penta-1,4-diene absorbs at 178 nm, a value that is typical of simple alkenes rather than conjugated dienes.

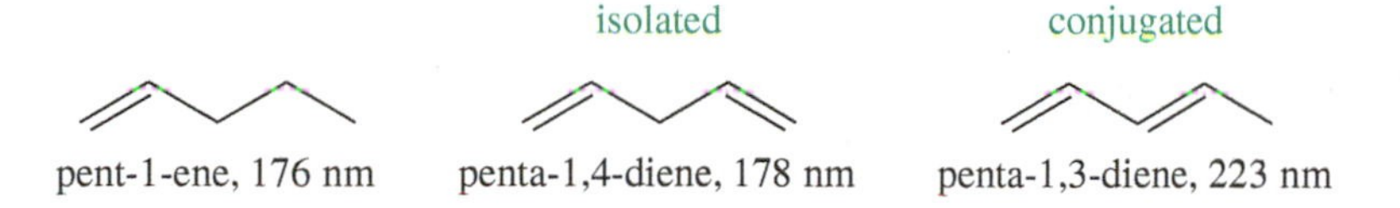

15-13C Obtaining an Ultraviolet Spectrum

To measure the ultraviolet (or UV–visible) spectrum of a compound, the sample is dissolved in a solvent (often ethanol) that does not absorb above 200 nm. The sample solution is placed in a quartz cell, and some of the solvent is placed in a **reference cell.** An ultraviolet spectrometer operates by comparing the amount of light transmitted through the sample (the **sample beam**) with the amount of light in the **reference beam.** The reference beam passes through the reference cell to compensate for any absorption of light by the cell and the solvent.

The spectrometer (Figure 15-26) has a *source* that emits all frequencies of UV light (above 200 nm). This light passes through a *monochromator*, which uses a diffraction grating or a prism to spread the light into a spectrum and select one wavelength. This single wavelength of light is split into two beams, with one beam passing through the sample cell and the other passing through the reference (solvent) cell.

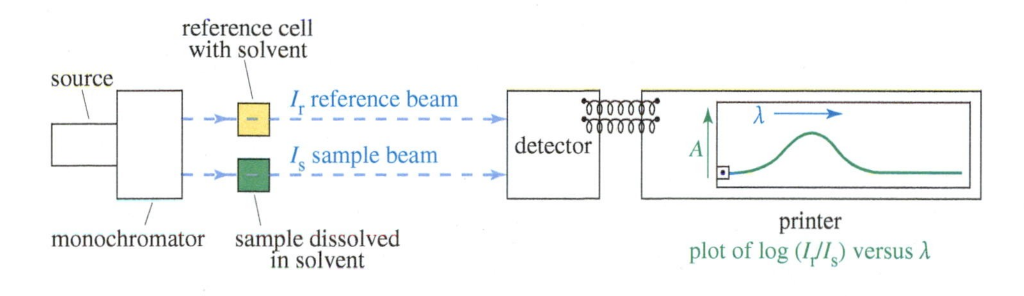

FIGURE 15-26
Schematic diagram of an ultraviolet spectrometer. In the ultraviolet spectrometer, a monochromator selects one wavelength of light, which is split into two beams. One beam passes through the sample cell, while the other passes through the reference cell. The detector measures the ratio of the two beams, and the printer plots this ratio as a function of wavelength.

The detector continuously measures the intensity ratio of the reference beam (I_r) compared with the sample beam (I_s). As the spectrometer scans the wavelengths in the UV region, a printer draws a graph (called a *spectrum*) of the absorbance of the sample as a function of the wavelength.

The *absorbance, A,* of the sample at a particular wavelength is governed by *Beer's law.*

$$\text{Beer's law: } A = \log\left(\frac{I_r}{I_s}\right) = \varepsilon c l$$

Application: Drug Analysis

The molar extinction coefficient ε, associated with a wavelength of maximum absorbance λ_{max}, is particularly useful for determining drug concentrations. For example, the concentration of tetracycline is measured at 380 nm where the molar absorptivity value is 16,200.

where

c = sample concentration in moles per liter
l = path length of light through the cell in centimeters
ε = the **molar absorptivity** (or **molar extinction coefficient**) of the sample

Molar absorptivity (ε) is a measure of how strongly the sample absorbs light at that wavelength.

If the sample absorbs light at a particular wavelength, the sample beam (I_s) is less intense than the reference beam (I_r), and the ratio I_r/I_s is greater than 1. The ratio is equal to 1 when there is no absorption. The absorbance (the logarithm of the ratio) is therefore greater than zero when the sample absorbs, and is equal to zero when it does not. A UV spectrum is a plot of A, the absorbance of the sample, as a function of the wavelength.

UV–visible spectra tend to show broad peaks and valleys. The spectral data that are most characteristic of a sample are as follows:

1. The wavelength(s) of maximum absorbance, called λ_{max}
2. The value of the molar absorptivity ε at each maximum

Because UV–visible spectra are broad and lacking in detail, they are rarely printed as actual spectra. The spectral information is given as a list of the value or values of λ_{max} together with the molar absorptivity for each value of λ_{max}.

The UV spectrum of isoprene (2-methylbuta-1,3-diene) is shown in Figure 15-27. This spectrum could be summarized as follows:

$$\lambda_{max} = 222 \text{ nm} \qquad \varepsilon = 20{,}000$$

The value of λ_{max} is read directly from the spectrum, while the molar absorptivity ε must be calculated from the concentration of the solution and the path length of the cell. For an isoprene concentration of 4×10^{-5} M and a 1-cm cell, the molar absorptivity is found by rearranging Beer's law ($A = \varepsilon cl$).

$$\varepsilon = \frac{A}{cl} = \frac{0.8}{4 \times 10^{-5}} = 20{,}000$$

Molar absorptivities in the range of 5000 to 30,000 are typical for the $\pi \rightarrow \pi^*$ transitions of conjugated polyene systems. Such large molar absorptivities are helpful, since spectra may be obtained with very small amounts of sample. On the other hand, samples and solvents for UV spectroscopy must be extremely pure. A minute impurity with a large molar absorptivity can easily obscure the spectrum of the desired compound.

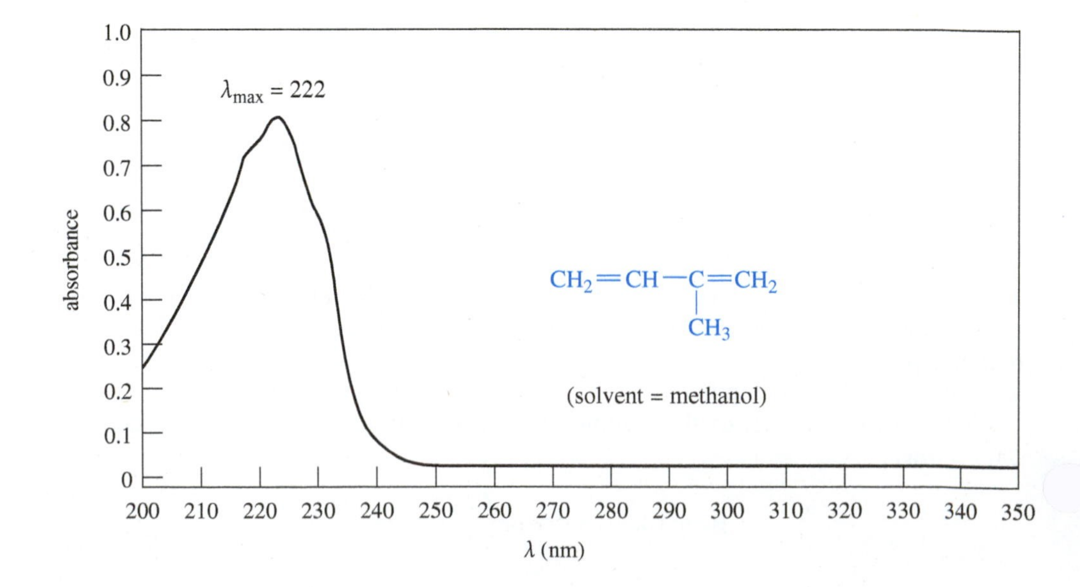

FIGURE 15-27 The UV spectrum of isoprene dissolved in methanol shows λ_{max} = 222 nm, ε = 20,000.

PROBLEM 15-21

One milligram of a compound of molecular weight 160 is dissolved in 10 mL of ethanol, and the solution is poured into a 1-cm UV cell. The UV spectrum is taken, and there is an absorption at $\lambda_{max} = 247$ nm. The maximum absorbance at 247 nm is 0.50. Calculate the value of ε for this absorption.

Application: Natural Haze

The bluish color of the Blue Ridge Mountains results from the abundant plant life there. Plants, especially conifers, release isoprene and other terpenes (alkenes composed of isoprene units, see Chapter 25) into the air around them. These alkenes react with ozone and water in the air (Chapter 8) to form particulates that scatter the blue light from the sky.

15-13D Interpreting UV–Visible Spectra

The values of λ_{max} and ε for conjugated molecules depend on the exact nature of the conjugated system and its substituents. For most purposes, we can use some simple generalizations for estimating approximate values of λ_{max} for common types of systems. Table 15-2 gives the values of λ_{max} for several types of isolated alkenes, conjugated dienes, conjugated trienes, and a conjugated tetraene.

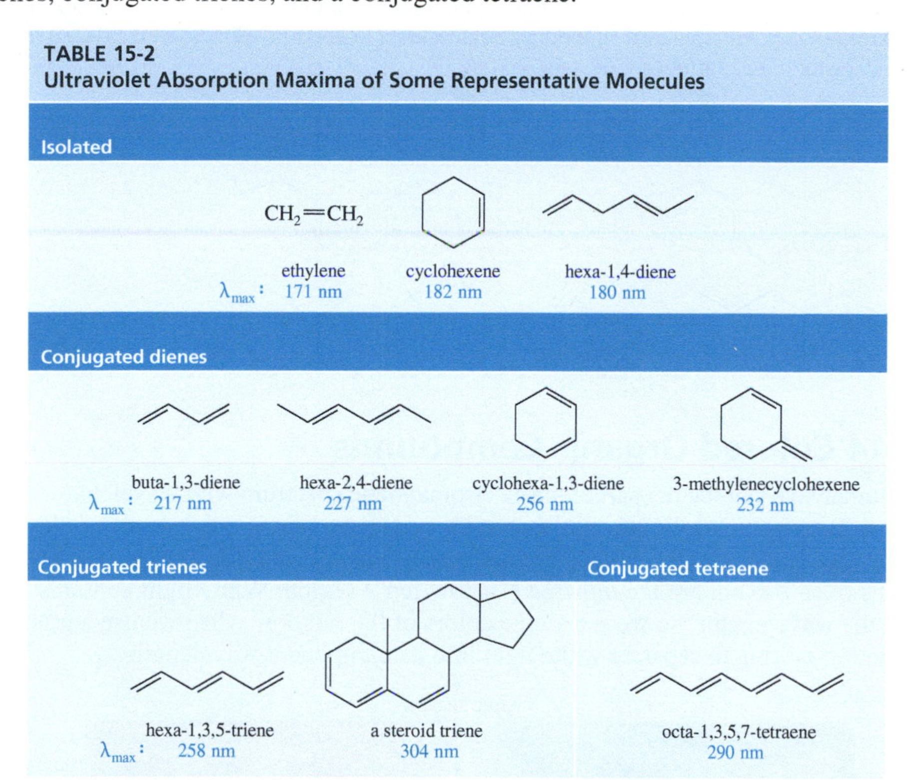

TABLE 15-2
Ultraviolet Absorption Maxima of Some Representative Molecules

Type	Compound	λ_{max}
Isolated	ethylene ($CH_2{=}CH_2$)	171 nm
Isolated	cyclohexene	182 nm
Isolated	hexa-1,4-diene	180 nm
Conjugated dienes	buta-1,3-diene	217 nm
Conjugated dienes	hexa-2,4-diene	227 nm
Conjugated dienes	cyclohexa-1,3-diene	256 nm
Conjugated dienes	3-methylenecyclohexene	232 nm
Conjugated trienes	hexa-1,3,5-triene	258 nm
Conjugated trienes	a steroid triene	304 nm
Conjugated tetraene	octa-1,3,5,7-tetraene	290 nm

The examples in Table 15-2 show that the addition of another conjugated double bond to a conjugated system has a large effect on λ_{max}. In going from ethylene (171 nm) to buta-1,3-diene (217 nm) to hexa-1,3,5-triene (258 nm) to octa-1,3,5,7-tetraene (290 nm), the values of λ_{max} increase by about 30 to 40 nm for each double bond extending the conjugated system. Alkyl groups also increase the value of λ_{max} by about 5 nm per alkyl group.

SOLVED PROBLEM 15-3

Rank the following dienes in order of increasing values of λ_{max}. (Their actual absorption maxima are 185 nm, 235 nm, 273 nm, and 300 nm.)

SOLUTION

λ_{max}: 185 nm 235 nm 273 nm 300 nm

(continued)

PROBLEM-SOLVING HINT

Some good rules of thumb:
An additional conjugated C=C increases λ_{max} about 30 to 40 nm; an additional alkyl group increases it about 5 nm. Useful base values:

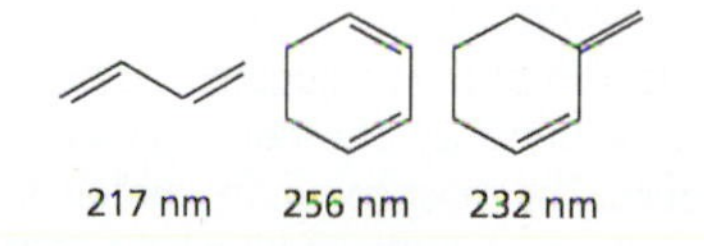

Application: UV in Synthesis

In their synthesis of vitamin B_{12}, Woodward and Eschenmoser applied the exquisite sensitivity of UV spectroscopy to follow their reactions. Using UV, they were able to detect structural changes in microgram quantities of their synthetic intermediates.

vitamin B_{12}

These compounds are an isolated diene, two conjugated dienes, and a conjugated triene. The isolated diene will have the shortest value of λ_{max} (185 nm), close to that of cyclohexene (182 nm).

The second compound looks like buta-1,3-diene (217 nm) with three additional alkyl substituents (circled). Its absorption maximum should be around (217 + 15) nm, and 235 nm must be the correct value.

The third compound looks like cyclohexa-1,3-diene (256 nm) with an additional alkyl substituent (circled) raising the value of λ_{max}, so 273 nm must be the correct value.

The fourth compound looks like cyclohexa-1,3-diene (256 nm) with an additional conjugated double bond (circled) and another alkyl group (circled). We predict a value of λ_{max} about 35 nm longer than for cyclohexa-1,3-diene, so 300 nm must be the correct value.

PROBLEM 15-22

Using the examples in Table 15-2 to guide you, match four of the following UV absorption maxima (λ_{max}) with the corresponding compounds: (1) 232 nm; (2) 256 nm; (3) 273 nm; (4) 292 nm; (5) 313 nm; (6) 353 nm.

(a) (b)

(c) (d)

15-14 Colored Organic Compounds

The human eye can see the part of the electromagnetic spectrum with wavelengths from about 400 nm (blue-violet) to a little over 700 nm (red). Wavelengths less than 400 nm are the higher-energy *ultraviolet* ("beyond violet") region, and the lower-energy wavelengths over 700 nm are the *infrared* ("below red") region. White light contains light of all the wavelengths, so we see all the colors of the rainbow when we use a prism or diffraction grating to separate white light into its component wavelengths.

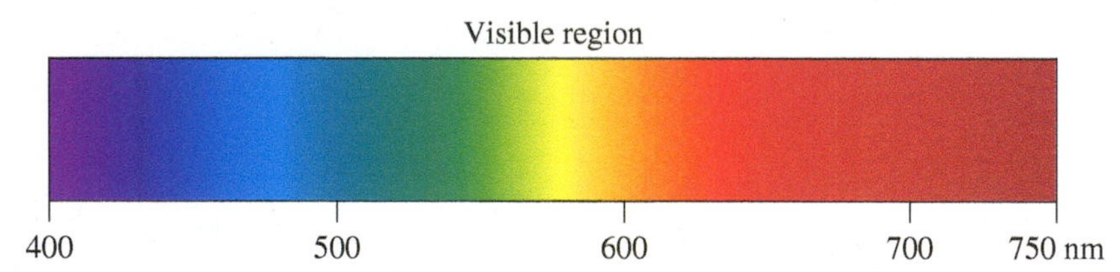

Carotene derivatives absorb different wavelengths of light, depending on the length of the conjugated system and the presence of other functional groups. The most common carotenes absorb wavelengths of blue light and appear as varieties of orange.

Colored objects do not emit their own light. They reflect part of the ambient light that falls on them. A white object reflects all of the light, but a colored object absorbs some wavelengths and reflects the rest of the spectrum. The color that we see is what is reflected, that is, the complete spectrum minus the portion that is absorbed. Therefore, we observe the opposite, or *complement*, of the light that is absorbed. For example, β-carotene, the orange color in carrots, has 11 conjugated double bonds. It absorbs strongly at 454 nm, in the blue portion of the spectrum, reflecting the orange complement.

β-carotene

β-carotene: λ_{max} = 454 nm (ε = 140,000)
absorbs blue, reflects orange

Not all color in nature arises from conjugated organic compounds. Minerals, for example, are colored by metal atoms in various oxidation states. However, virtually all of the color present in plants and animals is because of conjugated organic molecules. Figure 15-28 shows examples from some of the classes of colored compounds in living organisms. Note the extensive conjugation in each of these compounds. These brightly colored compounds have large extinction coefficients, so they require only a tiny amount to produce an intense color.

Dyes are intensely colored compounds used in fabrics, plastics, inks, and other products. Dyes were originally extracted from plants or animals and used to color cloth. For example, red carmine (page 2) was extracted from cochineal insects, and blue indigo (the dye used in blue jeans) was extracted from plant material. Both of these dyes are now synthesized in large quantities. The Romans extracted the indigo derivative Tyrian purple (imperial purple) from a sea snail and used the dye to color the robes of emperors and high-ranking senators.

Bolinus brandaris, the spiny dye-murex, is one of several species of Murex sea snails that were harvested by the millions to obtain the Tyrian purple dye they secrete.

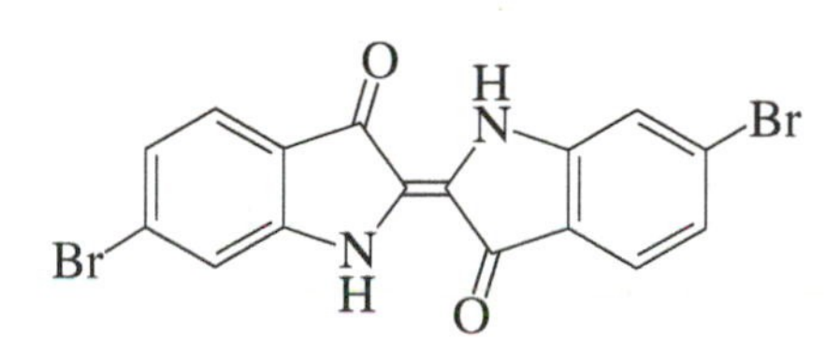

Tyrian purple
6,6'-dibromoindigo

Mauveine A

FIGURE 15-28
Colored natural products

echinochrome A 462 nm, 497 nm, 533 nm
(red–sea urchin–a naphthoquinone)

cyanidin chloride 535 nm
(crimson–cherries and flowers–an anthocyanin)

indigo 602 nm
(blue–various plants–an indigoid)

quercetin 372 nm
(red-orange–grapes and onions–a flavone)

zeaxanthin 452 nm, 483 nm
(yellow–corn–a carotenoid)

The era of synthetic dyes is credited to Sir William Henry Perkin, who accidentally synthesized the purple dye mauveine in 1856 (at age 18) while trying to make quinine. Mauveine was an inexpensive substitute for Tyrian purple. Chemists soon developed many other synthetic dyes, and by the late 1800s the dye industry had become one of the major chemical industries in Europe. Commoners could now wear all the colors that were once reserved for royalty. The dye industry also provided the commercial funding and motivation for much of the early research in organic chemistry, especially the chemistry of compounds derived from benzene.

Industrial chemists have developed thousands of commercial dyes for every imaginable use, such as fabrics, hair color, inks, toys, and foods. Many dyes are toxic, so the US government has regulated their use in foods since 1906, when the Pure Food and Drug Act created a group of dyes that were approved for coloring foods. Like all unnecessary food additives, food dyes have come under intense scrutiny to determine which ones are safest, and whether any of them cause significant side effects. Only seven dyes are currently approved for unrestricted food use in the US. Two of these food dyes are Indigo Carmine and Sunset Yellow shown below.

$Na^+ \, ^-O_3S$ … $SO_3^- \, Na^+$

FD&C blue #2
"Indigo Carmine"

$Na^+ \, ^-O_3S$—C₆H₄—N=N— (HO) … $SO_3^- \, Na^+$

FD&C yellow #6
"Sunset Yellow FCF"

This photo shows an African woman hand-dyeing cotton fabric using a natural indigo dye.

PROBLEM 15-23

Phenolphthalein is an acid–base indicator that is colorless below pH 8 and red above pH 8. Explain briefly why the first structure is colorless and the second structure is colored.

A *colorless* $\xrightarrow{HO^-}$ **B** *red*

15-15 UV-Visible Analysis in Biology and Medicine

Many of the recent advances in biology and medicine have resulted from applications of biochemistry and molecular biology based on knowledge of the physiology at the cellular and molecular levels. To understand cellular processes, we need to detect compounds at micromolar and lower concentrations, often in an intact cell, without vaporizing or destroying the sample. UV–visible spectroscopy is nondestructive and exceptionally sensitive, and it can measure small concentrations of highly conjugated metabolites such as ATP, as well as macromolecules such as proteins and nucleic acids in aqueous solutions. If we know the wavelength at maximum absorption (λ_{max}) and the molar absorptivity (ε) of the molecule of interest, we can use Beer's law to calculate minuscule concentrations of these biomolecules, often in complex mixtures.

Proteins, DNA, ATP, and many other biomolecules contain conjugated systems of pi bonds with strong characteristic absorptions in the UV region of the spectrum. Common heteroatoms with nonbonding electrons, such as oxygen and nitrogen, are frequently part of these conjugated pi systems, contributing to their unique characteristics. Table 15-3 shows some of the conjugated systems that are commonly found in biomolecules.

TABLE 15-3
UV Absorptions of Common Ring Systems Found in Biomolecules

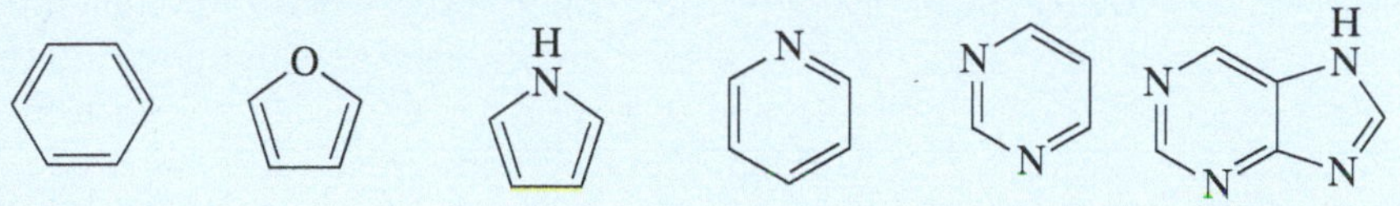

	benzene	furan	pyrrole	pyridine	pyrimidine	purine
λ_{max}	255 nm	208 nm	324 nm	256 nm	240 nm	263 nm
$\log_{10} \varepsilon$	2.4	3.9	4.47	3.1	3.4	2.9

Proteins are polymers of twenty common amino acids (Chapter 24), plus the occasional rare amino acids. Four of the twenty common amino acids contain conjugated ring systems that strongly absorb UV light, making almost all proteins detectable and quantifiable by UV analysis. The four common UV-absorbing amino acids (shown below) are phenylalanine, tyrosine, histidine, and tryptophan. The standard wavelength for measuring protein absorbance is 280 nm, where most of the absorption is due to tryptophan and tyrosine, with a small contribution from phenylalanine.

Clinical chemists analyze blood and urine to determine the concentrations of hormones, metabolites, and other substances to diagnose illnesses rapidly and accurately. The central instrument in a modern blood analyzer is a UV–visible spectrometer.

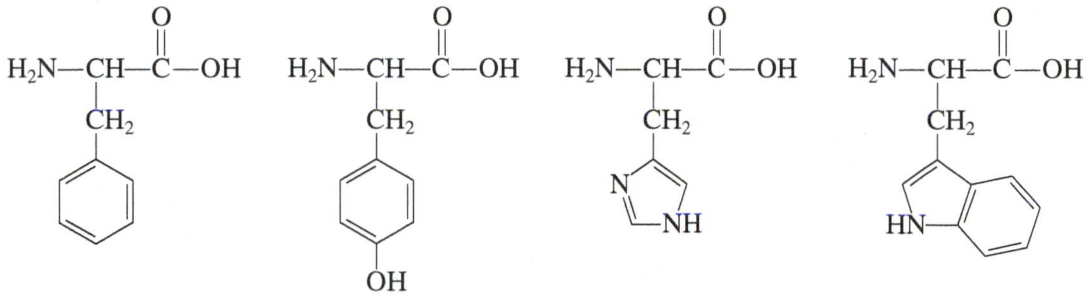

	phenylalanine	tyrosine	histidine	tryptophan
λ_{max}	257 nm	274 nm	211 nm	280 nm
$\log_{10} \varepsilon$	2.25	3.15	3.8	3.8

Depending on the UV absorption of the biomolecule to be measured, the spectrometer may detect the substance directly, or it may detect a specific "color-developing reagent" that changes its UV absorption when it reacts with the molecule of interest.

An example is the enzyme *alkaline phosphatase*, which removes the phosphate group from many types of biomolecules. Abnormal levels of alkaline phosphatase can indicate liver and bone disorders, among other conditions. Alkaline phosphatase gives a UV spectrum similar to other proteins, so other proteins would interfere with a direct analysis; however, we can measure its distinctive ability to remove phosphate groups. We add the colorless compound indoxyl phosphate, which loses its phosphate group when it reacts with the alkaline phosphatase enzyme. The product is indoxyl, which quickly dimerizes to blue indigo. By setting up the UV–visible spectrometer to measure the amount of indigo produced in a certain time period, we can calculate the amount of alkaline phosphatase present in the blood.

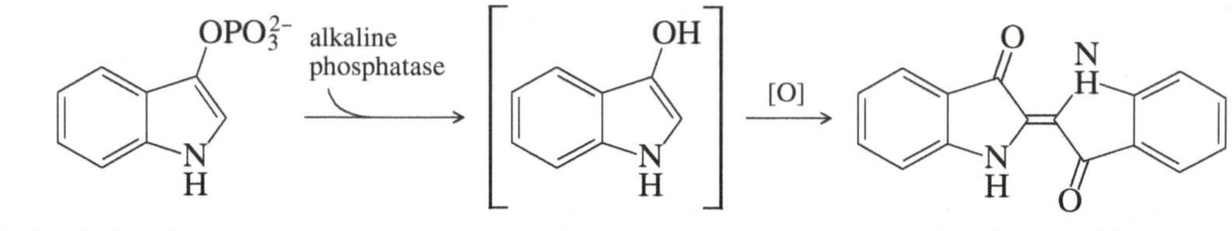

OPO_3^{2-} alkaline phosphatase → [indoxyl, OH, N H] →[O]→ indigo

indoxyl phosphate "color-developing reagent"   indoxyl   indigo, λ_{max} = 602 nm

Essential Terms

1,2-addition An addition in which two atoms or groups add to adjacent atoms. (p. 724)

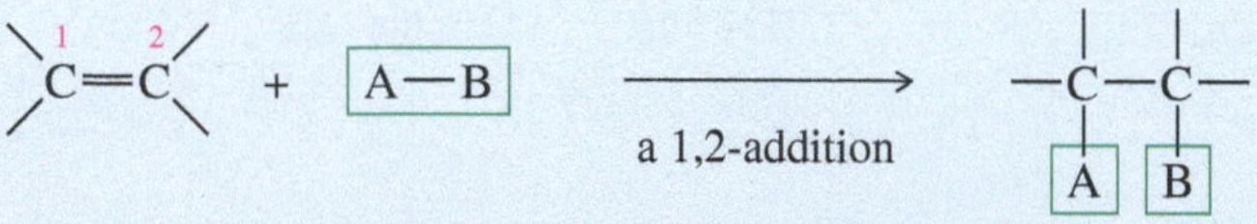

C=C (1, 2) + A—B → a 1,2-addition → —C—C— with A and B

1,4-addition An addition in which two atoms or groups add to atoms that bear a 1,4-relationship. (p. 724)

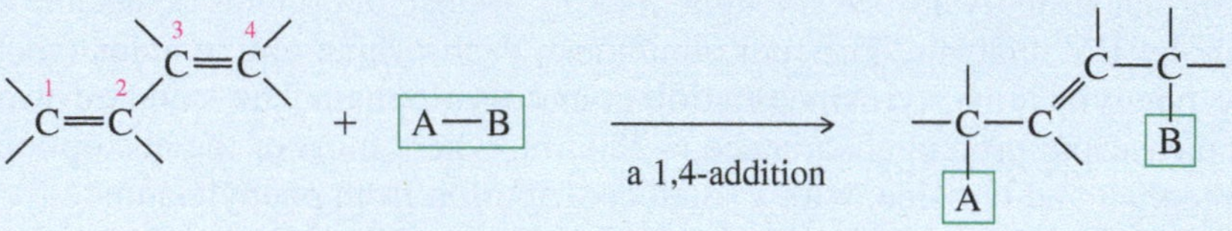

C=C—C=C (1, 2, 3, 4) + A—B → a 1,4-addition → A—C—C=C—C—B

allyl group The common name for the 2-propenyl group, $—CH_2—CH{=}CH_2$ (p. 723)

allylic position The carbon atom next to a carbon–carbon double bond. The term is used in naming compounds, such as an **allylic halide,** or in referring to reactive intermediates, such as an **allylic cation,** an **allylic radical,** or an **allylic anion.** (p. 723)

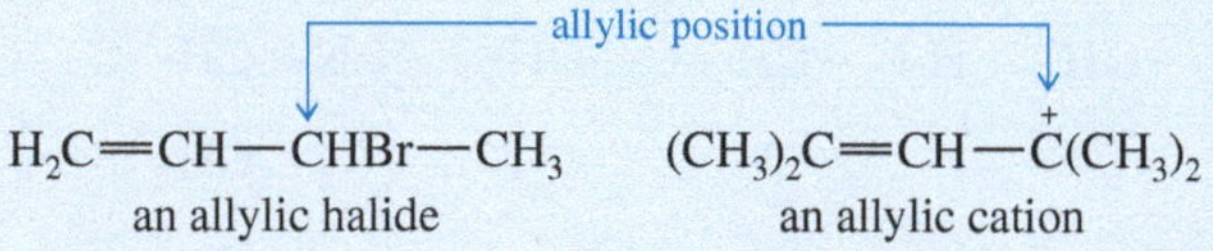

allylic position

$H_2C{=}CH—CHBr—CH_3$ an allylic halide $(CH_3)_2C{=}CH—\overset{+}{C}(CH_3)_2$ an allylic cation

allylic shift The isomerization of a double bond that occurs through the delocalization of an allylic intermediate. (p. 729)

$$H_2C{=}CH—CH_2—CH_3 \xrightarrow[hv]{NBS} H_2C{=}CH—CHBr—CH_3 + BrCH_2—CH{=}CH—CH_2$$

allylic shift product

concerted reaction A reaction in which all bond making and bond breaking occurs in the same step. The E2, S_N2, and Diels–Alder reactions are examples of concerted reactions. (p. 743)

conjugated double bonds Double bonds separated by one single bond, with interaction by overlap of the *p* orbitals in the pi bonds. (p. 716)

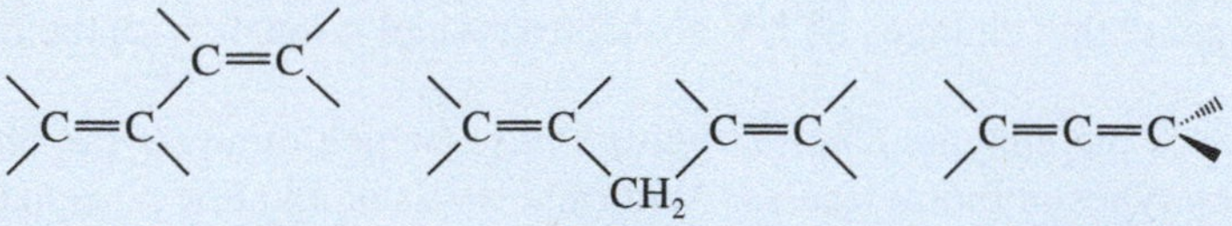

C=C—C=C conjugated   C=C—CH_2—C=C isolated   C=C=C cumulated

isolated double bonds: Double bonds separated by two or more single bonds. Isolated double bonds react inde pendently, as they do in a simple alkene. (p. 716)

cumulated double bonds: Successive double bonds with no intervening single bonds. (p. 717)

allene: **(cumulene)** A compound containing cumulated carbon–carbon double bonds. (p. 717)

conservation of orbital symmetry A theory of pericyclic reactions stating that the MOs of the reactants must flow smoothly into the MOs of the products without any drastic changes in symmetry. That is, there must be bonding interactions to help stabilize the transition state. (p. 743)

constructive overlap An overlap of orbitals that contributes to bonding. Overlap of lobes with similar phases (+ phase with + phase, or − phase with − phase) is generally constructive overlap. (p. 720)

cycloaddition A reaction of two alkenes or polyenes to form a cyclic product. Cycloadditions often take place through concerted interaction of the pi electrons in two unsaturated molecules. (p. 743)

delocalized orbital A molecular orbital that results from the combination of three or more atomic orbitals. When filled, these orbitals spread electron density over all the atoms involved. (p. 719)

destructive overlap An overlap of orbitals that contributes to antibonding. Overlap of lobes with opposite phases (+ phase with − phase) is generally destructive overlap. (p. 720)

Diels–Alder reaction A synthesis of six-membered rings by a **[4 + 2] cycloaddition.** This notation means that four pi electrons in one molecule interact with two pi electrons in the other molecule to form a new ring. (p. 734)

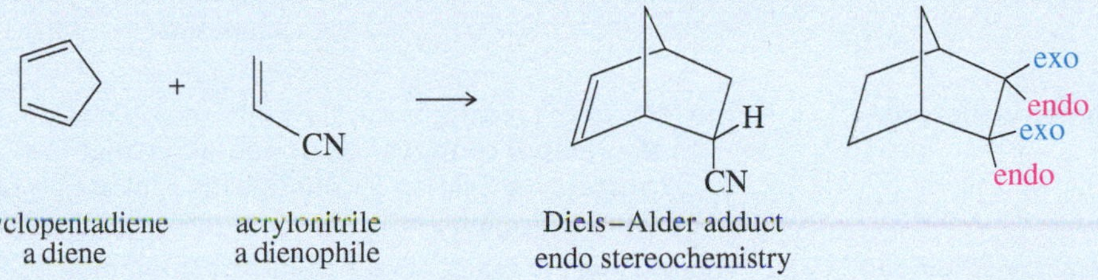

cyclopentadiene + acrylonitrile ⟶ Diels–Alder adduct

cyclopentadiene
a diene

acrylonitrile
a dienophile

Diels–Alder adduct
endo stereochemistry

- **dienophile:** The component with two pi electrons that reacts with a diene in the Diels–Alder reaction.
- **endo rule:** The stereochemical preference for electron-poor substituents on the dienophile to assume endo positions in a bicyclic Diels–Alder product. (p. 738)
- **secondary overlap:** Overlap of the *p* orbitals of the electron-withdrawing group of the dienophile with those of one of the central atoms (C2 or C3) of the diene. This overlap helps stabilize the transition state. With cyclic dienes, it favors endo products. (p. 738)

heat of hydrogenation The enthalpy of reaction that accompanies the addition of hydrogen to a mole of an unsaturated compound. (p. 717)

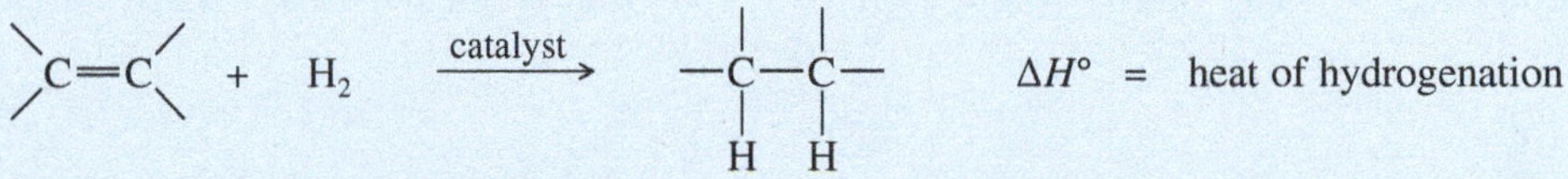

$$\mathrm{>C{=}C<} + \mathrm{H_2} \xrightarrow{\text{catalyst}} \mathrm{-\underset{H}{\overset{|}{C}}-\underset{H}{\overset{|}{C}}-} \qquad \Delta H^\circ = \text{heat of hydrogenation}$$

HOMO An acronym for **highest occupied molecular orbital.** In a photochemically excited state, this orbital is represented as HOMO*. (p. 744)

kinetic control Product distribution is governed by the rates at which the various products are formed. (p. 727)

- **kinetic product:** The product that is formed fastest; the major product under kinetic control.

LUMO An acronym for **lowest unoccupied molecular orbital.** (p. 744)

molar absorptivity, ε **(molar extinction coefficient)** A measure of how strongly a compound absorbs light at a particular wavelength. It is defined by *Beer's law*,

$$A = \log\left(\frac{I_r}{I_s}\right) = \varepsilon c l$$

where A is the absorbance, I_r and I_s are the amounts of light passing through the reference and sample beams, c is the sample concentration in moles per liter, and l is the path length of light through the cell. (p. 750)

molecular orbitals (MOs) Orbitals that include more than one atom in a molecule. Molecular orbitals can be bonding, antibonding, or nonbonding. (p. 719)

- **bonding molecular orbitals:** MOs that are lower in energy than the isolated atomic orbitals from which they are made. Electrons in these orbitals serve to hold the atoms together.
- **antibonding molecular orbitals:** MOs that are higher in energy than the isolated atomic orbitals from which they are made. Electrons in these orbitals tend to push the atoms apart.
- **nonbonding molecular orbitals:** MOs with the same energy as the isolated atomic orbitals from which they are made. Electrons in these orbitals have no effect on the bonding of the atoms. (p. 731)

node A region of a molecular orbital with zero electron density. (p. 720)

pericyclic reaction A reaction involving concerted reorganization of electrons within a closed loop of interacting orbitals. Cycloadditions are one class of pericyclic reactions. (p. 743)

reference beam A second beam in the spectrometer that passes through a **reference cell** containing only the solvent. The **sample beam** is compared with this beam to compensate for any absorption by the cell or the solvent. (p. 749)

resonance energy The extra stabilization provided by delocalization, compared with a localized structure. For dienes and polyenes, the resonance energy is the extra stability of the conjugated system compared with the energy of a compound with an equivalent number of isolated double bonds. (p. 718)

***s*-cis conformation** A cis-like conformation of a single bond in a conjugated diene or polyene. (p. 723)

***s*-trans conformation** A trans-like conformation of a single bond in a conjugated diene or polyene. (p. 723)

s-cis conformation (cisoid conformation) *s*-trans conformation (transoid conformation)

symmetry-allowed The MOs of the reactants can flow into the MOs of the products in one concerted step according to the rules of conservation of orbital symmetry. In a symmetry-allowed cycloaddition, there is constructive overlap (+ phase with + phase, − phase with − phase) between the HOMO of one molecule and the LUMO of the other. (p. 744)

symmetry-forbidden The MOs of the reactants are of incorrect symmetries to flow into those of the products in one concerted step. (p. 744)

thermodynamic control **(equilibrium control)** Product distribution is governed by the stabilities of the products. Thermodynamic control operates when the reaction mixture is allowed to come to equilibrium. (p. 727)

thermodynamic product: The most stable product; the major product under thermodynamic control.

UV–visible spectroscopy The measurement of the absorption of ultraviolet and visible light as a function of wavelength. Ultraviolet light consists of wavelengths from about 100 to 400 nm. Visible light is from about 400 nm (violet) to 750 nm (red). (p. 746)

Woodward–Hoffman rules A set of symmetry rules that predict whether a particular pericyclic reaction is symmetry-allowed or symmetry-forbidden. (p. 743)

Essential Problem-Solving Skills in Chapter 15

Each skill is followed by problem numbers exemplifying that particular skill.

1 Show how to construct the molecular orbitals of ethylene, butadiene, and the allylic system. Show the electronic configurations of ethylene, butadiene, and the allyl cation, radical, and anion.	Problems 15-36 and 37
2 Explain which reactions are enhanced by resonance stabilization of the intermediates, such as free-radical reactions and cationic reactions. Propose mechanisms to explain the enhanced rates and the observed products, and draw resonance forms of the stabilized intermediates.	Problems 15-25, 26, and 32
3 Predict the products of Diels–Alder reactions, including the orientation of cycloaddition with unsymmetrical reagents and the stereochemistry of the products.	Problems 15-29, 30, 31, 33, 36, 39, and 41
4 Predict which cycloadditions are thermally allowed and which are photochemically allowed by comparing the symmetry of the molecular orbitals of the reactants.	Problems 15-35 and 36
5 Use values of λ_{max} from UV–visible spectra to estimate the length of conjugated systems, and to distinguish between compounds that differ in their conjugated systems.	Problems 15-28, 32, 38, and 40

Study Problems

15-24 Classify the following dienes and polyenes as isolated, conjugated, cumulated, or some combination of these classifications.

(a) cycloocta-1,4-diene **(b)** cycloocta-1,3-diene **(c)** cyclodeca-1,2-diene
(d) cycloocta-1,3,5,7-tetraene **(e)** cyclohexa-1,3,5-triene (benzene) **(f)** penta-1,2,4-triene

15-25 Predict the products of the following reactions.

(a) allyl bromide + cyclohexyl magnesium bromide
(b) cyclopentadiene + anhydrous HCl
(c) 2-methylpropene + NBS, light
(d) furan + *trans*-1,2-dicyanoethylene
(e) buta-1,3-diene + bromine water
(f) hexa-1,3,5-triene + bromine in CCl_4
(g) 1-(bromomethyl)-2-methylcyclopentene, heated in methanol
(h) cyclopentadiene + methyl acrylate, $CH_2{=}CH{-}COOCH_3$
(i) cyclohexa-1,3-diene + $CH_3{-}\overset{O}{\overset{\|}{C}}{-}C{\equiv}C{-}\overset{O}{\overset{\|}{C}}{-}CH_3$

15-26 Show how the reaction of an allylic halide with a Grignard reagent might be used to synthesize the following hydrocarbons.

(a) 5-methylhex-1-ene **(b)** 2,5,5-trimethylhept-2-ene **(c)** 1-cyclopentylpent-2-ene

15-27 Draw the important resonance contributors for the following cations, anions, and radicals.

(a) $^+CH_2$ **(b)** $\dot{C}H_2$ **(c)** $COCH_3$, ^-O

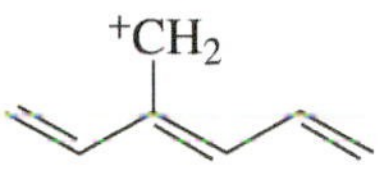

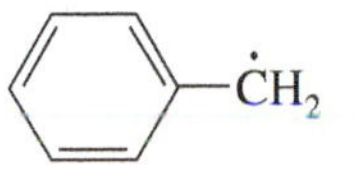

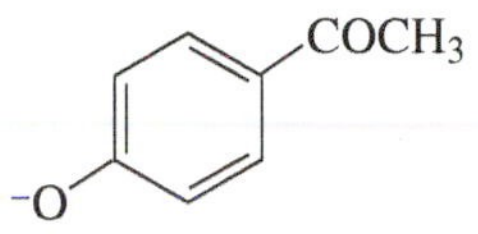

(d) H, +, H, H **(e)** $^-CH_2$, O, C, H **(f)** O, O^-

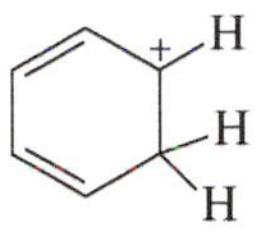

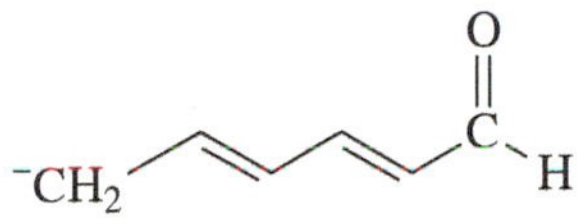

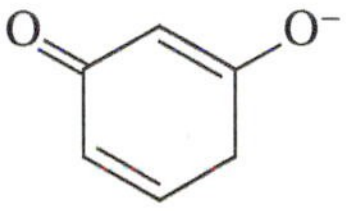

(g) O, $^+CH_2$ **(h)** OCH_3, $^+CH_2$, C, C, H, H

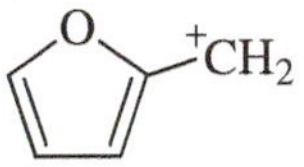

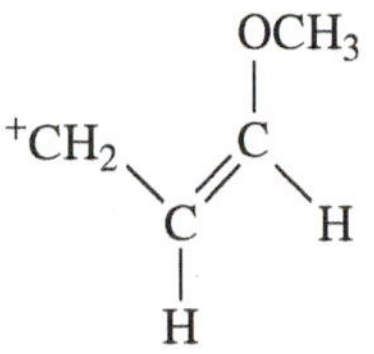

15-28 A solution was prepared using 0.0010 g of an unknown steroid (of molecular weight around 255) in 100 mL of ethanol. Some of this solution was placed in a 1-cm cell, and the UV spectrum was measured. This solution was found to have $\lambda_{max} = 235$ nm, with $A = 0.74$.

(a) Compute the value of the molar absorptivity at 235 nm.
(b) Which of the following compounds might give this spectrum?

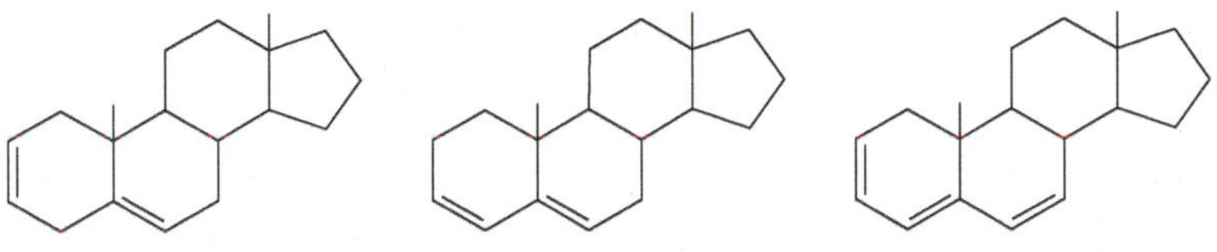

15-29 The diene lactone shown in part (a) has one electron-donating group ($-$OR) and one electron-withdrawing group (C$=$O). This diene lactone is sufficiently electron-rich to serve as the diene in a Diels–Alder reaction.

(a) What product would you expect to form when this diene reacts with methyl acetylenecarboxylate, a strong dienophile?

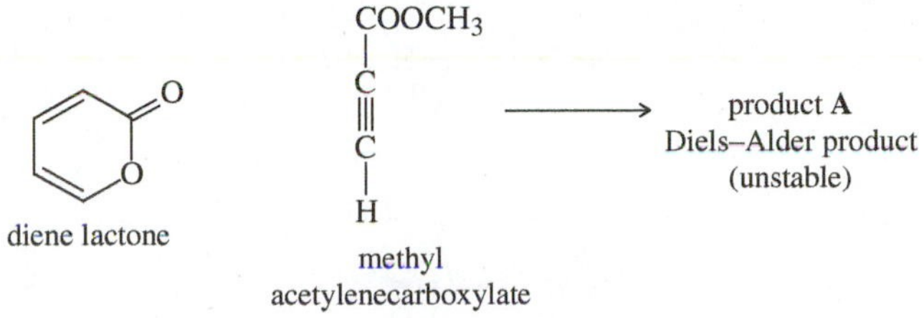

diene lactone + methyl acetylenecarboxylate ($COOCH_3{-}C{\equiv}C{-}H$) ⟶ product **A** Diels–Alder product (unstable)

(b) The Diels–Alder product **A** is not very stable. Upon mild heating, it reacts to produce CO_2 gas and methyl benzoate ($PhCOOCH_3$), a very stable product. Explain how this strongly exothermic decarboxylation takes place. (*Hint*: Under the right conditions, the Diels–Alder reaction can be reversible.)

15-30 Predict the products of the following Diels–Alder reactions. Include stereochemistry where appropriate.

(a) + C(=O)OH

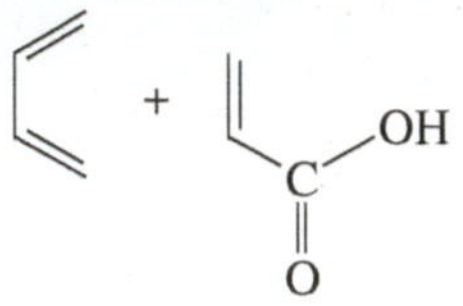

(b) + H—C≡C—$COOCH_3$

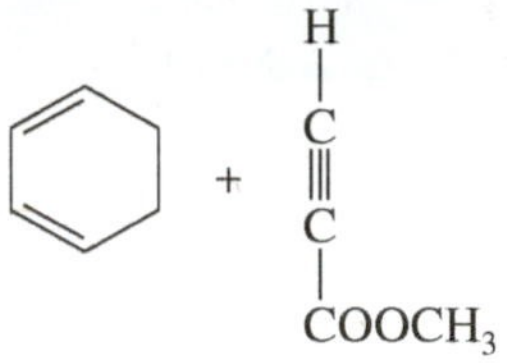

(c) + COOH

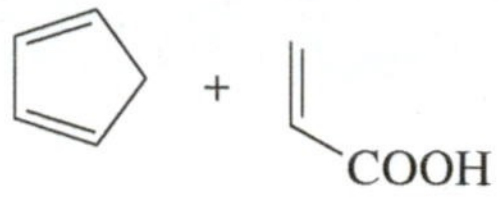

(d) OCH_3 + CHO

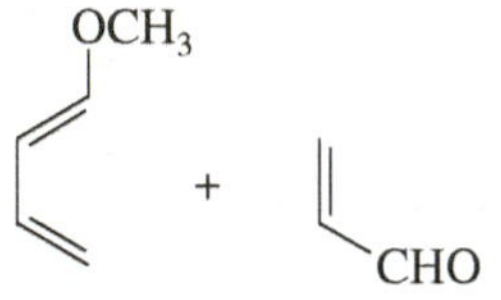

(e) + H(CN)C=C(CN)H

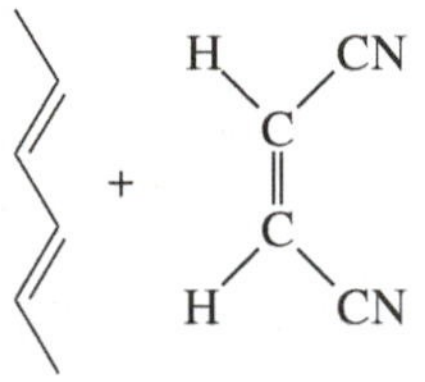

(f) + H(CN)C=C(CN)H

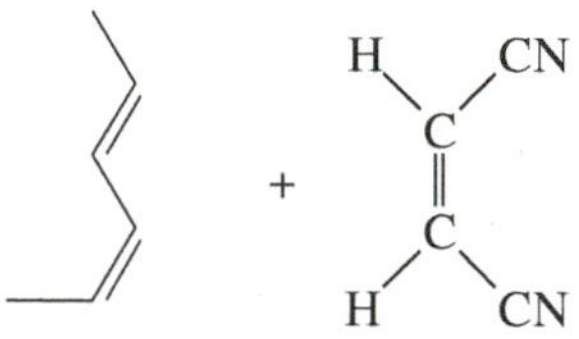

15-31 Predict the products of the following reactions, including stereochemistry where applicable.

(a) + $COOCH_3$—C≡C—$COOCH_3$ ⟶

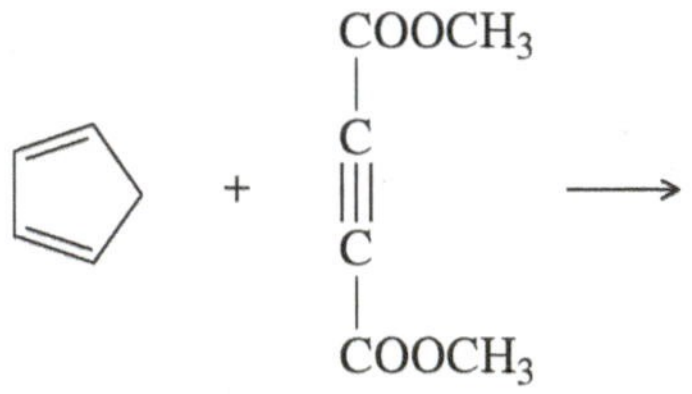

(b) O + O=⟨⟩=O ⟶

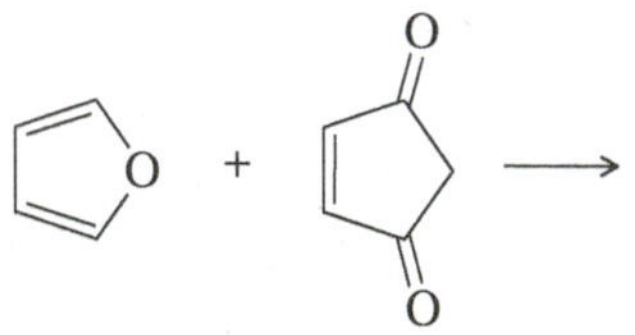

(c) + H, CH_3, H, CHO ⟶

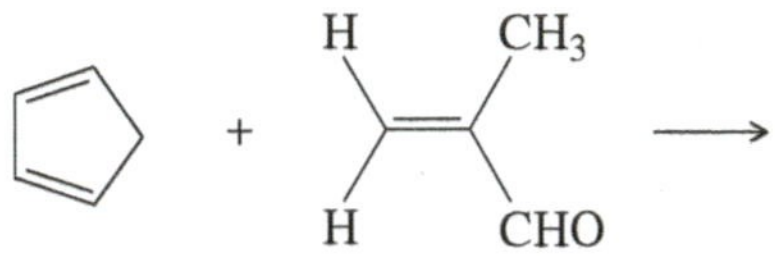

(d) H_3CO + O, O ⟶

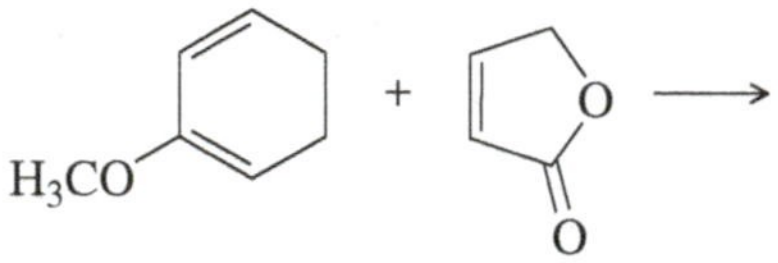

(e) $N(CH_3)_2$ + COOEt ⟶

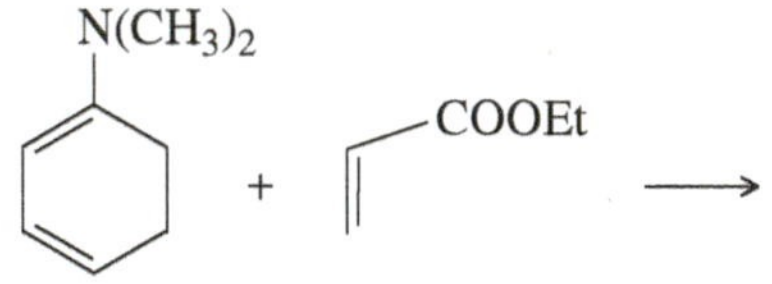

(f) OCH_3 + O ⟶

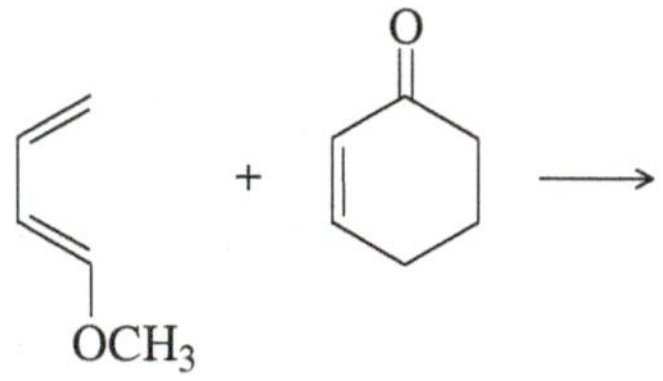

15-32 A graduate student was following a procedure to make 3-propylcyclohexa-1,4-diene. During the workup procedure, his research adviser called him into her office. By the time the student returned to his bench, the product had warmed to a higher temperature than recommended. He isolated the product, which gave the appropriate =C—H stretch in the IR, but the C=C stretch appeared around 1630 cm^{-1} as opposed to the literature value of 1650 cm^{-1} for the desired product. The mass spectrum showed the correct molecular weight, but the base peak was at M−29 rather than at M−43 as expected. Because of the anomalous IR spectrum, he took a UV spectrum that showed λ_{max} at 261 nm.

(a) Should he have his IR recalibrated or should he repeat the experiment, watching the temperature more carefully? What does the 1630 cm^{-1} absorption suggest?

(b) Draw the structure of the desired product, and propose a structure for the actual product.

(c) Show why he expected the MS base peak to be at M−43, and show how your proposed structure would give an intense peak at M−29.

15-33 Show how Diels–Alder reactions might be used to synthesize the following compounds.

(a)

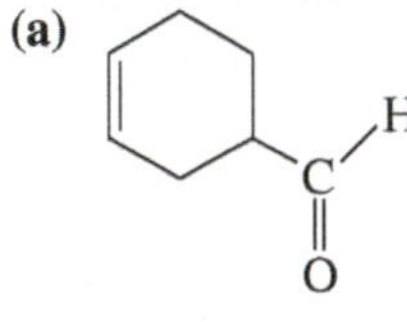

(b)

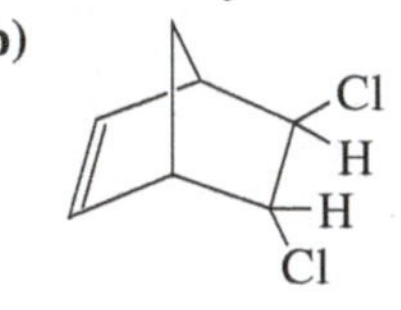

(c) CH_3, $COOCH_3$, CH_3

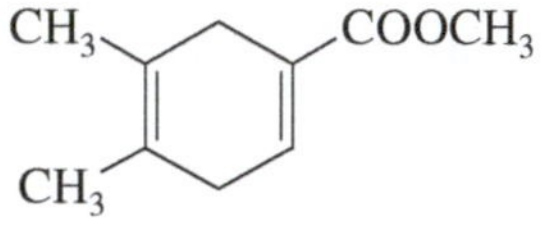

(d) H, O, H, O

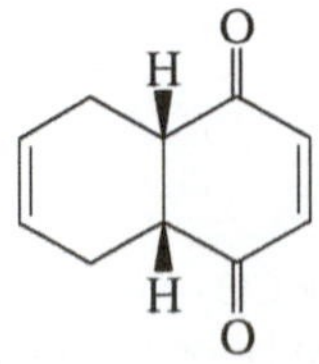

(e) H, H, O, O, O

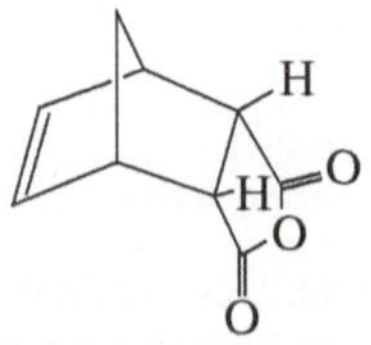

(f) H, CN, O, CN, H

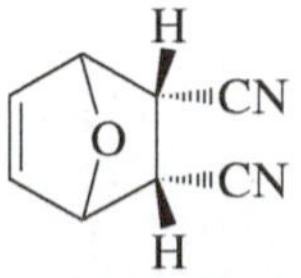

(continued)

(g) chlordane

(h) aldrin

(i)

15-34 Give the structures of the products represented by letters in this synthesis.

Part 1

cyclohexene $\xrightarrow{\text{mCPBA}}$ **A** $\xrightarrow[\text{H}_2\text{O}]{\text{NaOH}}$ **B** $\xrightarrow[\text{HOAc}]{\text{excess NaOCl}}$ **C** $\xrightarrow[\text{ether}]{\text{excess CH}_3\text{MgBr}}$ $\xrightarrow{\text{H}_3\text{O}^+}$ **D** (mixture of diastereomers) $\xrightarrow[\Delta]{\text{H}_2\text{SO}_4}$ **E** (C_8H_{12})

Part 2

$\xrightarrow[-78\ ^\circ\text{C}]{\text{O}_3}$ $\xrightarrow[\text{H}_2\text{O}]{(\text{CH}_3)_2\text{S}}$ **F** $\xrightarrow[\text{CH}_3\text{OH}]{\text{NaBH}_4}$ $\xrightarrow{\text{H}_3\text{O}^+}$ **G** $\xrightarrow[\Delta]{\text{H}_2\text{SO}_4}$ **H** ($C_3H_4O_2$)

Part 3

E (C_8H_{12}) + **H** ($C_3H_4O_2$) $\xrightarrow{\Delta}$ **I** ($C_{11}H_{16}O_2$)

15-35 Furan and maleimide undergo a Diels–Alder reaction at 25 °C to give the endo isomer of the product. When the reaction takes place at 90 °C, however, the major product is the exo isomer. Further study shows that the endo isomer of the product isomerizes to the exo isomer at 90 °C.

furan: maleimide:

(a) Draw and label the endo and exo isomers of the Diels–Alder adduct of furan and maleimide.
(b) Which isomer of the product would you usually expect from this reaction? Explain why this isomer is usually favored.
(c) Examine your answer to (b) and determine whether this answer applies to a reaction that is kinetically controlled or one that is thermodynamically controlled, or both.
(d) Explain why the endo isomer predominates when the reaction takes place at 25 °C and why the exo isomer predominates at 90 °C.

15-36 **(a)** Sketch the pi molecular orbitals of hexa-1,3,5-triene (Figure 15-25).
(b) Show the electronic configuration of the ground state of hexa-1,3,5-triene.
(c) Show what product would result from the [6 + 2] cycloaddition of hexa-1,3,5-triene with maleic anhydride.

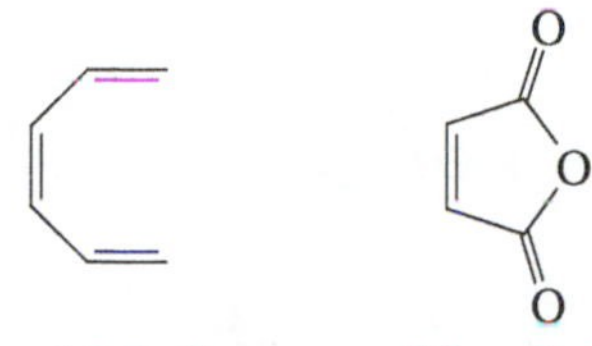

hexa-1,3,5-triene maleic anhydride

(d) Show that the [6 + 2] cyclization of hexa-1,3,5-triene with maleic anhydride is thermally forbidden but photochemically allowed.
(e) Show the Diels–Alder product that would actually result from heating hexa-1,3,5-triene with maleic anhydride.

15-37 The pentadienyl radical, $H_2C{=}CH{-}CH{=}CH{-}CH_2\cdot$, has its unpaired electron delocalized over three carbon atoms.
(a) Use resonance forms to show which three carbon atoms bear the unpaired electron.
(b) How many MOs are there in the molecular orbital picture of the pentadienyl radical?
(c) How many nodes are there in the lowest-energy MO of the pentadienyl system? How many in the highest-energy MO?
(d) Draw the MOs of the pentadienyl system in order of increasing energy.

(continued)

(e) Show how many electrons are in each MO for the pentadienyl radical (ground state).
(f) Show how your molecular orbital picture agrees with the resonance picture showing delocalization of the unpaired electron onto three carbon atoms.
(g) Remove the highest-energy electron from the pentadienyl radical to give the pentadienyl cation. Which carbon atoms share the positive charge? Does this picture agree with the resonance picture?
(h) Add an electron to the pentadienyl radical to give the pentadienyl anion. Which carbon atoms share the negative charge? Does this picture agree with the resonance picture?

***15-38** A student was studying terpene synthesis, and she wanted to make the compound shown here. First she converted 3-bromo-6-methylcyclohexene to alcohol **A.** She heated alcohol **A** with sulfuric acid and purified one of the components (compound **B**) from the resulting mixture. Compound **B** has the correct molecular formula for the desired product.
(a) Suggest how 3-bromo-6-methylcyclohexene might be converted to alcohol **A.**
(b) The UV spectrum of compound **B** shows λ_{max} at 269 nm. Is compound **B** the correct product? If not, suggest a structure for compound **B** consistent with these UV data.
(c) Propose a mechanism for the dehydration of alcohol **A** to compound **B.**

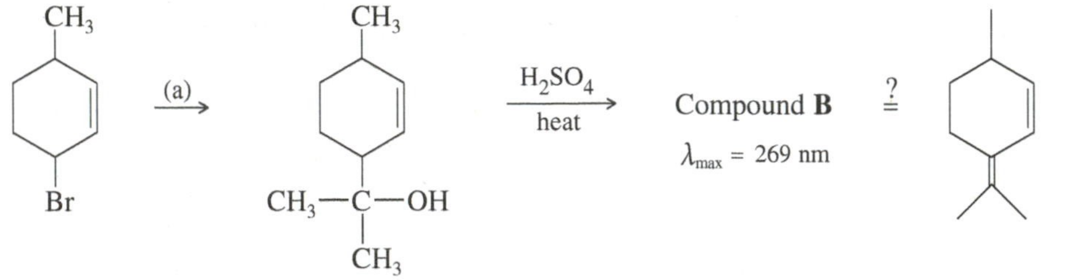

CH_3 / Br —(a)→ CH_3 / CH_3—C—OH / CH_3 (alcohol **A**) —H_2SO_4, heat→ Compound **B** (λ_{max} = 269 nm) ?= desired product

***15-39** Part of a synthesis by E. J. Corey and David Watt (Harvard University) involves the Diels–Alder cycloaddition of the following pyrone and cyclohexenone. The initial reaction gives the endo product, which loses carbon dioxide in a retro-Diels–Alder to generate a diene with predictable stereochemistry and functionality. IR and UV spectroscopy of the final product show that it contains a diene conjugated with an ester, and an unconjugated ketone. Determine the structures of the intermediate and the final product, with particular attention to their stereochemistry.

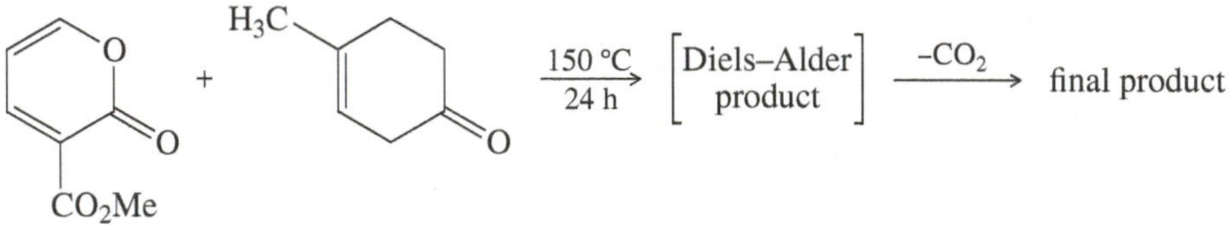

O, O, CO_2Me + H_3C, O —150 °C, 24 h→ [Diels–Alder product] —$-CO_2$→ final product

15-40 Determine whether each structure is likely to be colored or not. For those that you predict to be colored, indicate the extended conjugation by marking the series of continuous sp^2 hybridized atoms.

(a) N, O, Cl^-, $\overset{+}{N}$, CH_3, CH_3
(b) O, COOH, HO, OH
(c) OH, O, CH_3, COOH, HO, OH, OH, O
(d) OH, HO, N=N, $Na^+\ ^-O_3S$
(e) H_3CO, $N(CH_3)_2$, H_3C, O
(f) HO, C, O
(g) $Na^+\ ^-O$, I, O, I, O, I, I, $COO^-\ Na^+$
(h) NO_2, HO

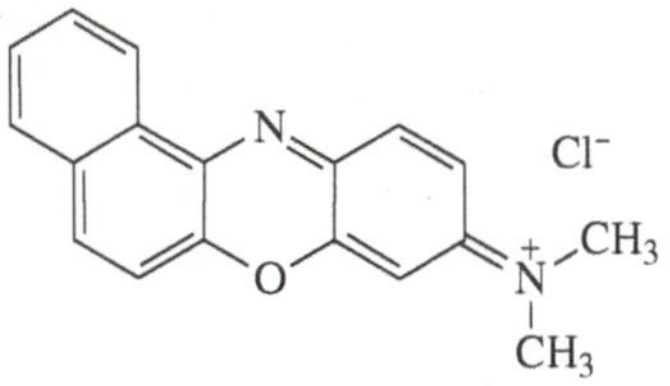

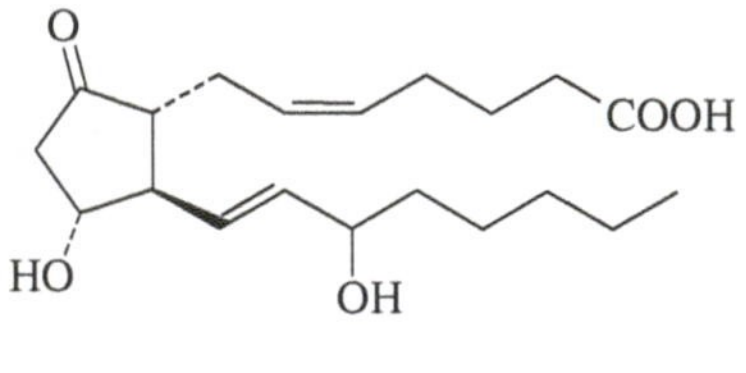

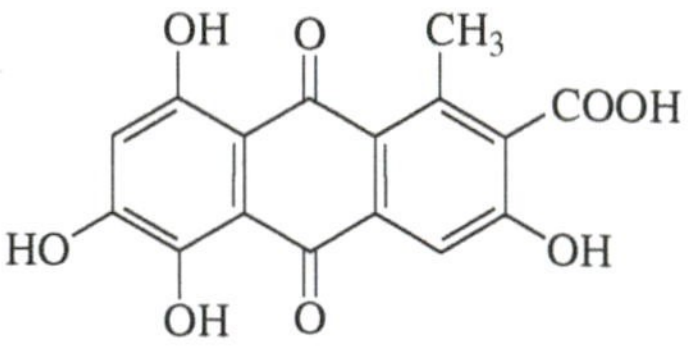

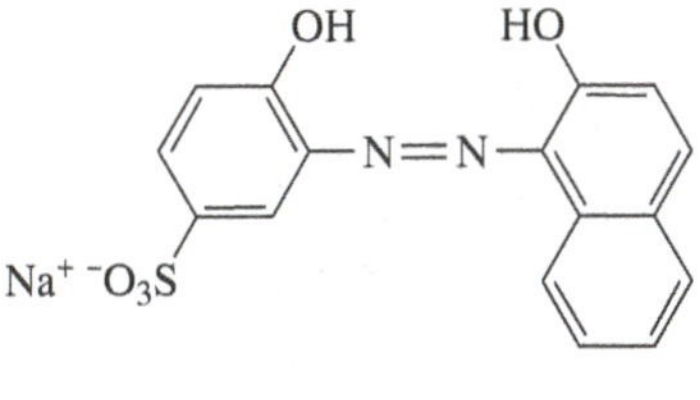

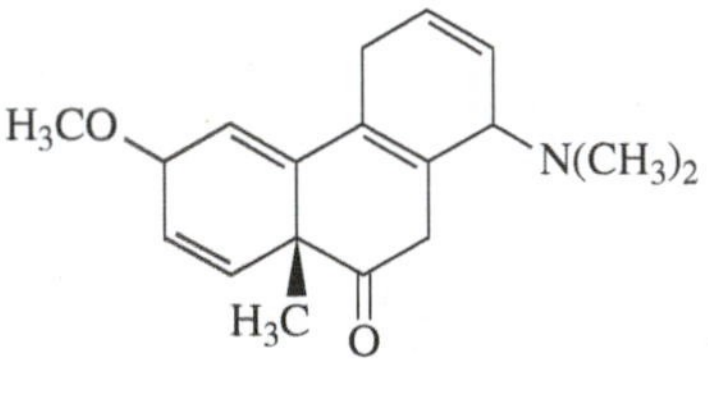

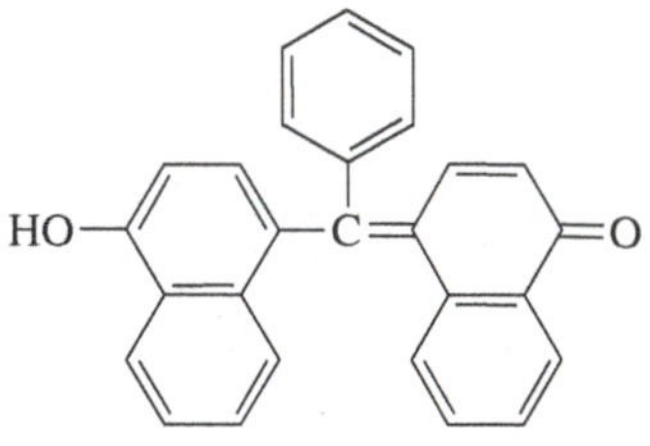

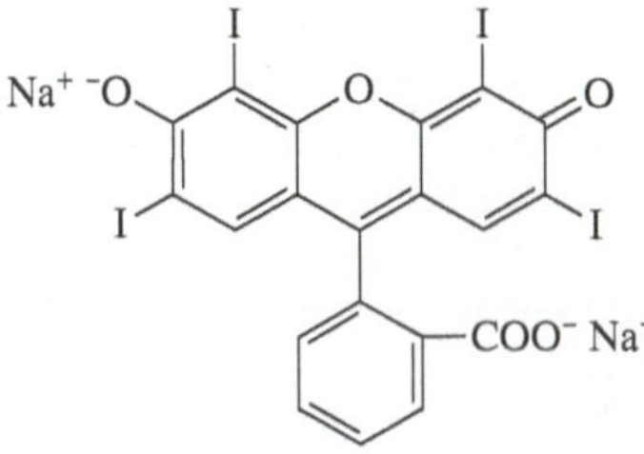

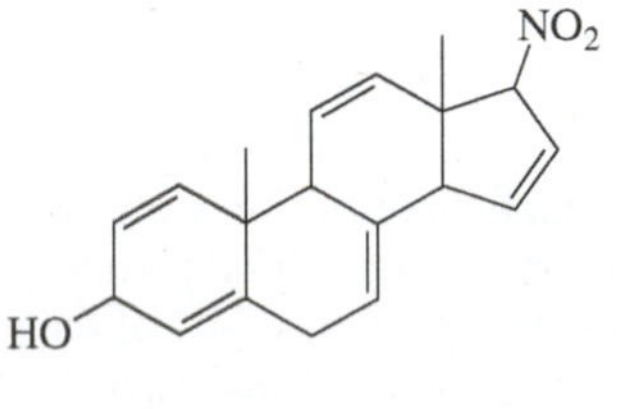

(continued)

(i)

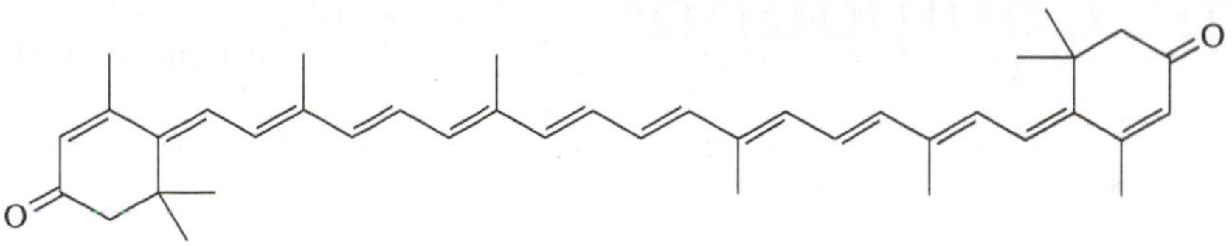

(j) $Na^+ \ ^-OOC$ ... $SO_3^- \ Na^+$ ... $Na^+ \ ^-O_3S$ ... N=N ... OH

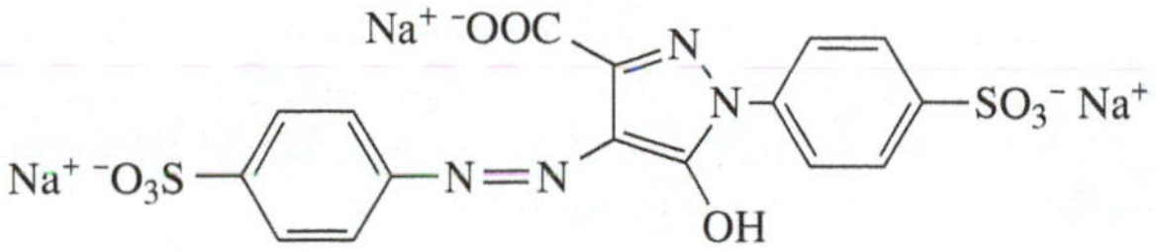

(k) Cl, Cl, Cl, Cl, OH, NH_2, O

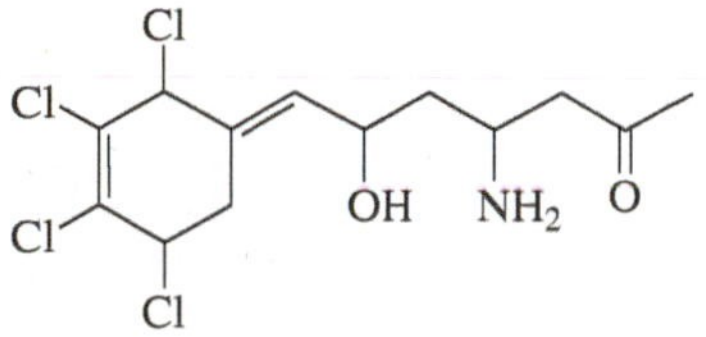

15-41 An important variation of the Diels–Alder reaction is *intramolecular*, in which the diene and the dienophile are connected. This type of Diels–Alder reaction makes two new rings. Draw the compound produced in each of these examples; try to predict stereochemistry (using models will help). In some cases, Lewis acid catalysts are used; that can be ignored for this problem.

(a) C≡N

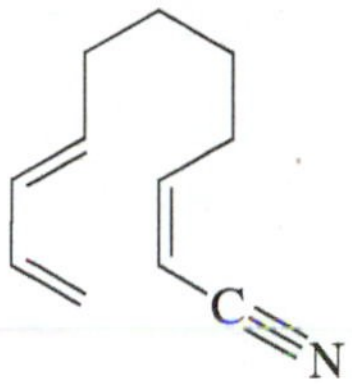

(b) O, $OSiR_3$, OR^2

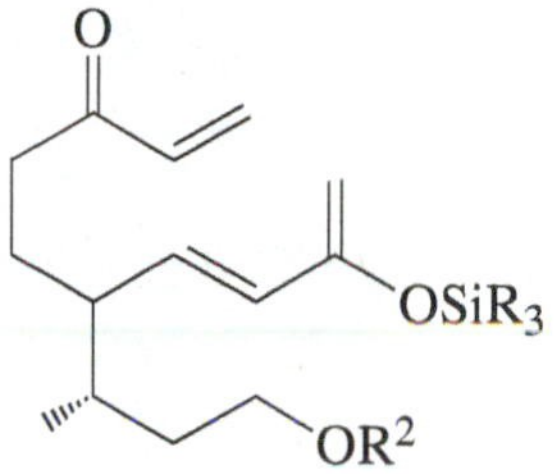

Prof. H. Miyaoka, Tokyo U.
Synthesis Letters, **2011**, 547.
(R and R^2 are different alkyl groups.)

(c) $SiMe_3$, O, O, O, O

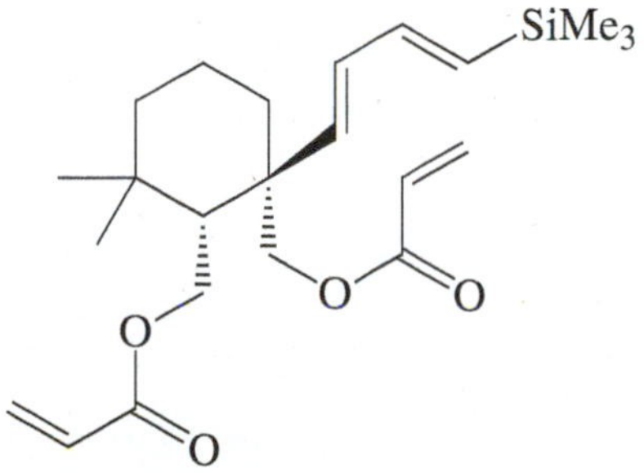

Prof. H. Zhai, Lanzhou U.
Organic Letters, **2014**, *16*, 216.

(d) O, Ar^2, C≡CH, O, Ar^1

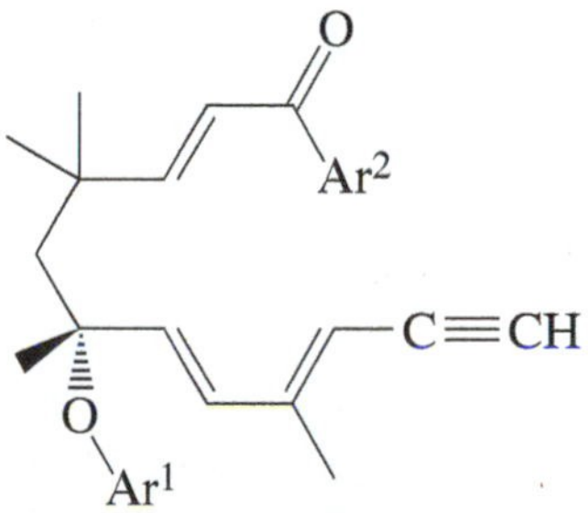

Prof. A. Kirschning, Leibnitz U. Hannover
Organic Letters, **2014**, *16*, 568.
(Ar^1 and Ar^2 are different aryl groups.)

16 Aromatic Compounds

Goals for Chapter 16

1 Determine whether Hückel's rule applies to a given structure, and predict whether the compound will be aromatic, antiaromatic, or nonaromatic.

2 Show how to construct the molecular orbitals of a conjugated cyclic system similar to benzene and cyclobutadiene.

3 Predict whether a given heterocyclic structure will be aromatic. For heterocycles containing nitrogen, determine whether nitrogen's lone pairs are used in the aromatic system, and predict whether the nitrogen atom is strongly or weakly basic.

4 Use IR, NMR, UV, and mass spectra to determine the structures of aromatic compounds. Given an aromatic compound, predict the distinguishing features of its spectra.

▶ The gold color of the stringing on this tennis racquet is characteristic of Zylon® fiber. Zylon is one of the strongest polymers known, with a breaking strength about 1.6 times that of Kevlar®. It is also heat resistant, stable to 780 °C, and highly resistant to stretching. Zylon's remarkable strength and stability result from its structure, composed entirely of aromatic rings with single bonds between them. In addition to tennis racquets, Zylon is used in protective firefighting equipment, racing suits, sailcloth, and high-strength tethers for racing and aerospace applications.

Zylon®

16-1 Introduction: The Discovery of Benzene

In 1825, Michael Faraday isolated a pure compound of boiling point 80 °C from the oily mixture that condensed from illuminating gas, the fuel burned in gaslights. Elemental analysis showed an unusually small hydrogen-to-carbon ratio of 1:1, corresponding to an empirical formula of CH. Faraday named the new compound "bicarburet of hydrogen." Eilhard Mitscherlich synthesized the same compound in 1834 by heating benzoic acid, isolated from gum benzoin, in the presence of lime. As Faraday did, Mitscherlich found that the empirical formula was CH. He also used a vapor-density measurement to determine a molecular weight of about 78, for a molecular formula of C_6H_6. Because the new compound was derived from gum benzoin, he named it benzin, now called *benzene*.

Many other compounds discovered in the 19th century seemed to be related to benzene. These compounds also had low hydrogen-to-carbon ratios as well as pleasant aromas, and they could be converted to benzene or related compounds. This group of compounds was called **aromatic** because of their pleasant odors. Other organic compounds without these properties were called **aliphatic,** meaning "fatlike." As the unusual stability of aromatic compounds was investigated, the term *aromatic* came to be applied to compounds with this stability, regardless of their odors.

16-2 The Structure and Properties of Benzene

The Kekulé Structure In 1866, Friedrich Kekulé proposed a cyclic structure for benzene with three double bonds. Considering that multiple bonds had been proposed only recently (1859), the cyclic structure with alternating single and double bonds was considered somewhat bizarre.

The **Kekulé structure** has its shortcomings, however. For example, it predicts two different 1,2-dichlorobenzenes, but only one is known to exist. Kekulé suggested (incorrectly) that a fast equilibrium interconverts the two isomers of 1,2-dichlorobenzene.

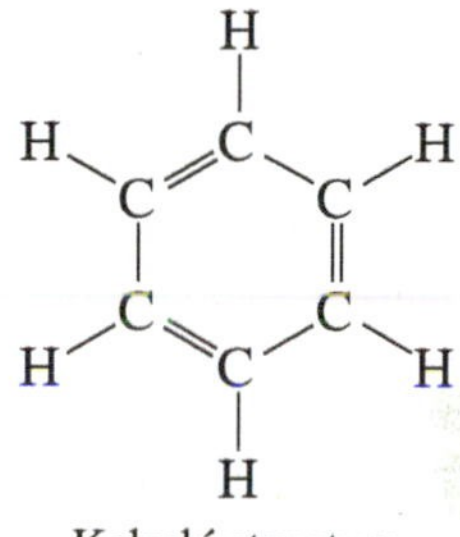

Kekulé structure of benzene

Cl Cl ?⇄ Cl Cl

1,2-dichlorobenzene

The Resonance Representation The resonance picture of benzene is a natural extension of Kekulé's hypothesis. In a Kekulé structure, the C—C single bonds would be longer than the double bonds. Spectroscopic methods have shown that the benzene ring is planar and all the bonds are the same length (1.397 Å). Because the ring is planar and the carbon nuclei are positioned at equal distances, the two Kekulé structures must differ only in the positioning of the pi electrons.

Benzene is actually a resonance hybrid of the two Kekulé structures. This representation implies that the pi electrons are delocalized, with a bond order of $1\frac{1}{2}$ between adjacent carbon atoms. The carbon–carbon bond lengths in benzene are shorter than typical single-bond lengths, yet longer than typical double-bond lengths.

all C—C bond lengths 1.397 Å

double bond 1.34 Å

single bond 1.48 Å

resonance representation

bond order = $1\frac{1}{2}$

combined representation

butadiene

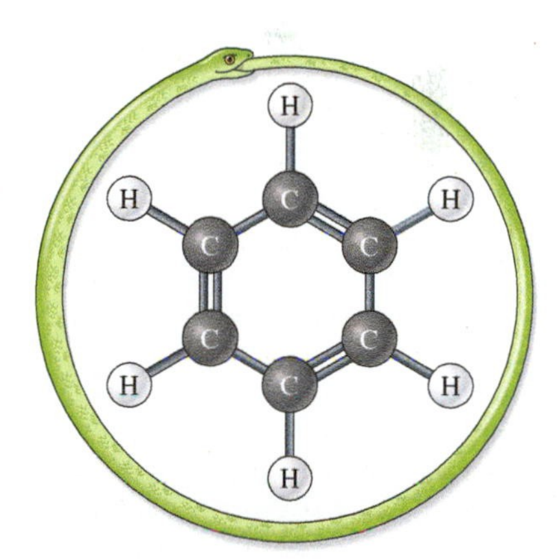

During the 1860s, benzene's multiple unsaturations were difficult to explain. An apocryphal story recounts how Kekulé fell asleep while contemplating this problem and had a dream in which the atoms danced before his eyes. They twisted in a snake-like motion, and then one of the snakes seized its own tail. When he awoke, Kekulé realized that a cyclic structure, together with three double bonds, explained benzene's formula and properties.

The resonance-delocalized picture explains most of the structural properties of benzene and its derivatives—the *benzenoid* aromatic compounds. Because the pi bonds are delocalized over the ring, we often inscribe a circle in the hexagon rather than draw three localized double bonds. This representation helps us remember there are no localized single or double bonds, and it prevents us from trying to draw supposedly different isomers that differ only in the placement of double bonds in the ring. We often use Kekulé structures in drawing reaction mechanisms, however, to show the movement of individual pairs of electrons.

PROBLEM 16-1

Write Lewis structures for the Kekulé representations of benzene. Show all the valence electrons.

Using this resonance picture, we can draw a more realistic representation of benzene (Figure 16-1). Benzene is a ring of six sp^2 hybrid carbon atoms, each bonded to one hydrogen atom. All the carbon–carbon bonds are the same length, and all the bond angles are exactly 120°. Each sp^2 carbon atom has an unhybridized p orbital perpendicular to the plane of the ring, and six electrons occupy this circle of p orbitals.

At this point, we can define an **aromatic compound** to be a cyclic compound containing some number of conjugated double bonds and having an unusually large resonance energy. Using benzene as the example, we will consider how aromatic compounds differ from aliphatic compounds. Then we will discuss why an aromatic structure confers extra stability and how we can predict aromaticity in some interesting and unusual compounds.

Application: Carcinogenesis

Long-term exposure to benzene can cause leukemia, characterized by a decrease in the number of red blood cells and an increase in the number of malfunctioning white blood cells. Metabolites of benzene may react with proteins and DNA in the bone marrow, altering the production of red and white blood cells.

The Unusual Reactions of Benzene Benzene is actually much more stable than we would expect from the simple resonance-delocalized picture. Both the Kekulé structure and the resonance-delocalized picture show that benzene is a cyclic conjugated triene. We might expect benzene to undergo the typical reactions of polyenes. In fact, its reactions are quite unusual. For example, an alkene decolorizes potassium permanganate by reacting to form a glycol (Section 8-14). The purple permanganate color disappears, and a precipitate of manganese dioxide forms. When permanganate is added to benzene, however, no reaction occurs.

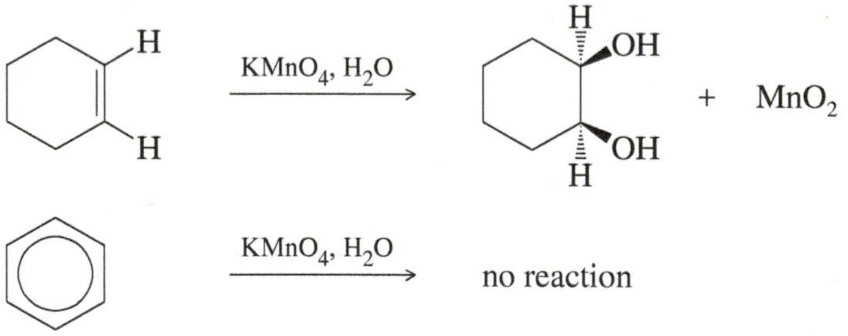

H H $\xrightarrow{KMnO_4,\ H_2O}$ H OH OH H + MnO_2

$\xrightarrow{KMnO_4,\ H_2O}$ no reaction

Most alkenes decolorize solutions of bromine in carbon tetrachloride (Section 8-10). The red bromine color disappears as bromine adds across the double bond. When bromine is added to benzene, no reaction occurs, and the red bromine color remains.

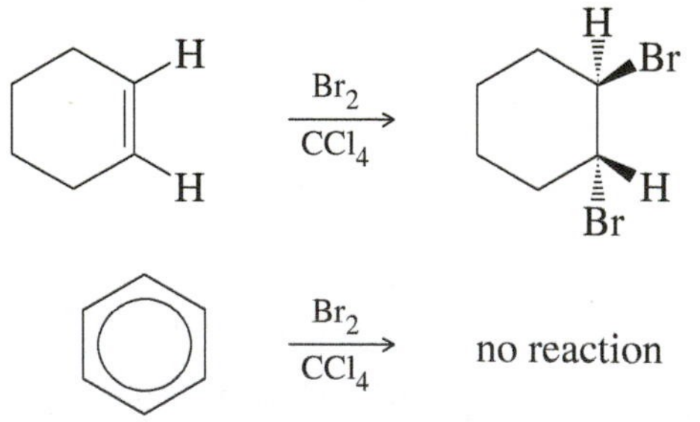

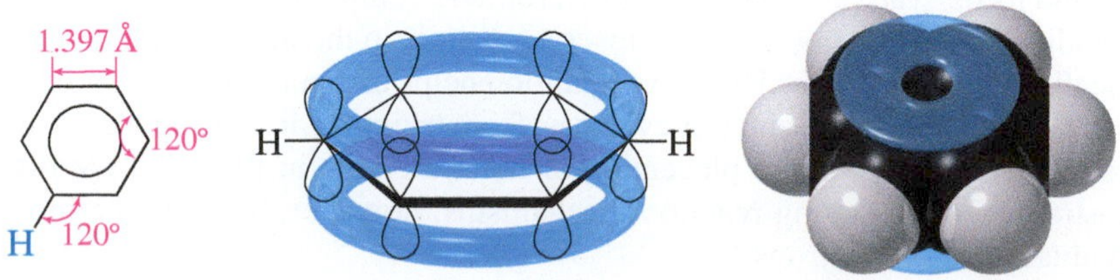

H H $\xrightarrow[CCl_4]{Br_2}$ H Br H Br

$\xrightarrow[CCl_4]{Br_2}$ no reaction

1.397 Å 120° H 120° H H

FIGURE 16-1
Benzene is a flat ring of sp^2 hybrid carbon atoms with their unhybridized p orbitals all aligned and overlapping. The ring of p orbitals contains six electrons. The carbon–carbon bond lengths are all 1.397 Å, and all the bond angles are exactly 120°.

Addition of a catalyst such as ferric bromide to the mixture of bromine and benzene causes the bromine color to disappear slowly. HBr gas is evolved as a by-product, but the expected *addition* of Br_2 does not take place. Instead, the organic product results from *substitution* of a bromine atom for a hydrogen, and all three double bonds are retained.

$$C_6H_6 \xrightarrow[CCl_4]{Br_2,\ FeBr_3} C_6H_5Br + HBr\uparrow$$

(The addition product, 5,6-dibromocyclohexa-1,3-diene, is not formed.)

The Unusual Stability of Benzene Benzene's reluctance to undergo typical alkene reactions suggests that it must be unusually stable. By comparing molar heats of hydrogenation, we can get a quantitative idea of its stability. Benzene, cyclohexene, and the cyclohexadienes all hydrogenate to form cyclohexane. Figure 16-2 shows how the experimentally determined heats of hydrogenation are used to compute the **resonance energies** of cyclohexa-1,3-diene and benzene, based on the following reasoning:

1. Hydrogenation of cyclohexene is exothermic by 120 kJ/mol (28.6 kcal/mol).
2. Hydrogenation of cyclohexa-1,4-diene is exothermic by 240 kJ/mol (57.4 kcal/mol), about twice the heat of hydrogenation of cyclohexene. The resonance energy of the isolated double bonds in cyclohexa-1,4-diene is about zero.
3. Hydrogenation of cyclohexa-1,3-diene is exothermic by 232 kJ/mol (55.4 kcal/mol), about 8 kJ (1.8 kcal) less than twice the value for cyclohexene. A resonance energy of 8 kJ (1.8 kcal) is typical for a conjugated diene.
4. Hydrogenation of benzene requires higher pressures of hydrogen and a more active catalyst. This hydrogenation is exothermic by 208 kJ/mol (49.8 kcal/mol), about 151 kJ (36.0 kcal) less than three times the value for cyclohexene.

$$\text{benzene} + 3\,H_2 \xrightarrow[\text{high pressure}]{\text{catalyst}} \text{cyclohexane}$$

$\Delta H°$	=	208 kJ/mol
3 × cyclohexene	=	359 kJ/mol
resonance energy	=	151 kJ/mol

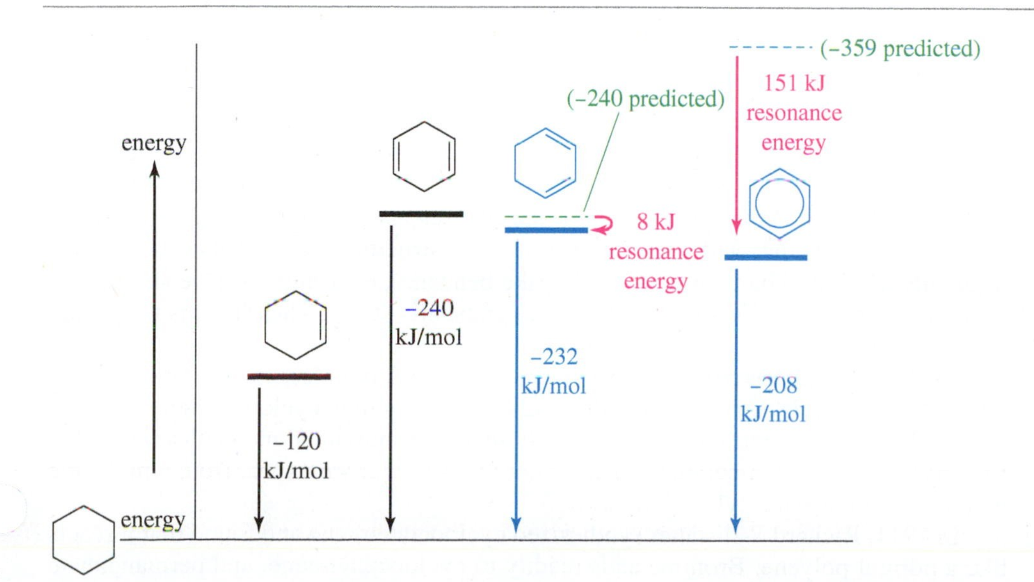

FIGURE 16-2 The molar heats of hydrogenation and the relative energies of cyclohexene, cyclohexa-1,4-diene, cyclohexa-1,3-diene, and benzene. The dashed lines represent the energies that would be predicted if every double bond had the same energy as the double bond in cyclohexene.

The huge 151 kJ/mol (36 kcal/mol) resonance energy of benzene cannot be explained by conjugation effects alone. The heat of hydrogenation for benzene is actually smaller than that for cyclohexa-1,3-diene. The hydrogenation of the *first* double bond of benzene is endothermic, the first endothermic hydrogenation we have encountered. In practice, this reaction is difficult to stop after the addition of 1 mole of H_2 because the product (cyclohexa-1,3-diene) hydrogenates more easily than benzene itself. Clearly, the benzene ring is exceptionally unreactive.

benzene + H_2 $\xrightarrow{\text{catalyst}}$ cyclohexa-1,3-diene

	$\Delta H^\circ_{\text{hydrogenation}}$	
benzene:	−208 kJ	(−49.8 kcal)
cyclohexa-1,3-diene:	−232 kJ	(−55.4 kcal)
ΔH° =	+24 kJ	(+5.6 kcal)

PROBLEM 16-2

Using the information in Figure 16-2, calculate the values of ΔH° for the following reactions:

(a) benzene + H_2 $\xrightarrow{\text{catalyst}}$ cyclohexene

(b) benzene + 2 H_2 $\xrightarrow{\text{catalyst}}$ cyclohexene

(c) cyclohexa-1,3-diene + H_2 $\xrightarrow{\text{catalyst}}$ cyclohexene

Failures of the Resonance Picture For many years, chemists assumed that benzene's large resonance energy resulted from having two identical, stable resonance structures. They thought that other hydrocarbons with analogous conjugated systems of alternating single and double bonds would show similar stability. These cyclic hydrocarbons with alternating single and double bonds are called **annulenes.** For example, benzene is the six-membered annulene, so it can be named [6]annulene. Cyclobutadiene is [4]annulene, cyclooctatetraene is [8]annulene, and larger annulenes are named similarly.

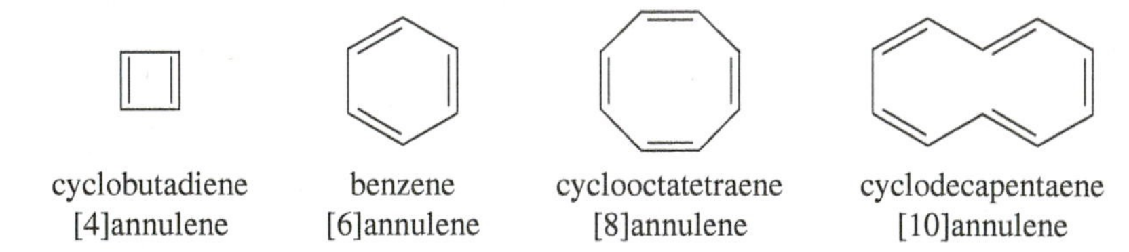

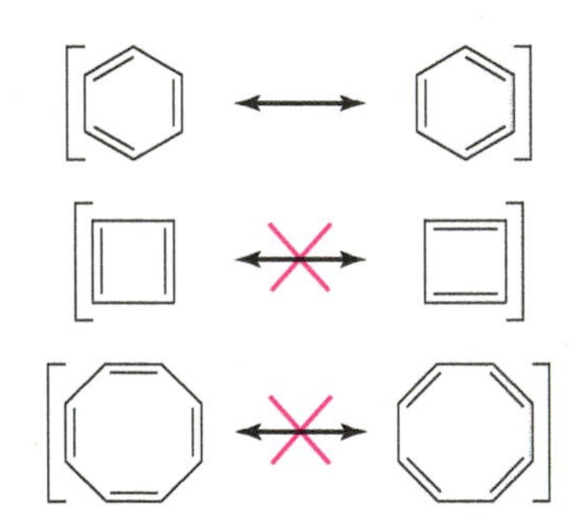

FIGURE 16-3
Cyclobutadiene and cyclooctatetraene have alternating single and double bonds similar to those of benzene. These compounds were mistakenly expected to be aromatic.

For the double bonds to be completely conjugated, the annulene must be planar so the *p* orbitals of the pi bonds can overlap. As long as an annulene is assumed to be planar, we can draw two Kekulé-like structures that seem to show a benzene-like resonance. Figure 16-3 shows proposed benzene-like resonance forms for cyclobutadiene and cyclooctatetraene. Although these resonance structures suggest that the [4] and [8]annulenes should be unusually stable (like benzene), experiments have shown that cyclobutadiene and cyclooctatetraene are not unusually stable. These results imply that the simple resonance picture is incorrect.

Cyclobutadiene has never been isolated and purified. It undergoes an extremely fast Diels–Alder dimerization. To avoid the Diels–Alder reaction, cyclobutadiene has been prepared at low concentrations in the gas phase and as individual molecules trapped in frozen argon at low temperatures. This is not the behavior we expect from a molecule with exceptional stability!

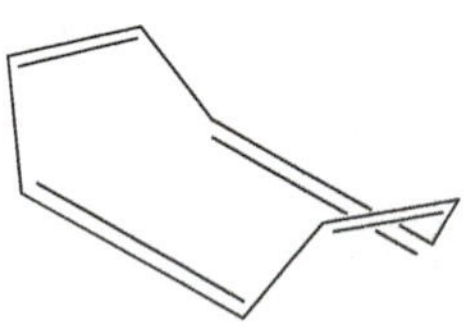

"tub" conformation of cyclooctatetraene

In 1911, Richard Willstätter synthesized cyclooctatetraene and found that it reacts like a normal polyene. Bromine adds readily to cyclooctatetraene, and permanganate oxidizes its double bonds. This evidence shows that cyclooctatetraene is much less stable than benzene. In fact, structural studies have shown that cyclooctatetraene is not planar. It is most stable in a "tub" conformation, with poor overlap between adjacent pi bonds.

PROBLEM 16-3

(a) Draw the resonance forms of benzene, cyclobutadiene, and cyclooctatetraene, showing all the carbon and hydrogen atoms.

(b) Assuming that these molecules are all planar, show how the *p* orbitals on the sp^2 hybrid carbon atoms form continuous rings of overlapping orbitals above and below the plane of the carbon atoms.

PROBLEM 16-4

Show the product of the Diels–Alder dimerization of cyclobutadiene. (This reaction is similar to the dimerization of cyclopentadiene, discussed in Section 15-11.)

16-3 The Molecular Orbitals of Benzene

Visualizing benzene as a resonance hybrid of two Kekulé structures cannot fully explain the unusual stability of the aromatic ring. As we have seen with other conjugated systems, molecular orbital theory provides the key to understanding aromaticity and predicting which compounds will have the stability of an aromatic system.

Benzene has a planar ring of six sp^2 hybrid carbon atoms, each with an unhybridized *p* orbital that overlaps with the *p* orbitals of its neighbors to form a continuous ring of orbitals above and below the plane of the carbon atoms. Six pi electrons are contained in this ring of overlapping *p* orbitals.

The six overlapping *p* orbitals create a cyclic system of molecular orbitals. Cyclic systems of molecular orbitals differ from linear systems such as buta-1,3-diene and the allyl system. A two-dimensional cyclic system requires two-dimensional MOs, with the possibility of two distinct MOs having the same energy. We can still follow the same principles in developing a molecular orbital representation for benzene, however.

1. There are six atomic *p* orbitals that overlap to form the benzene pi system. Therefore, there must be six molecular orbitals.
2. The lowest-energy molecular orbital is entirely bonding, with constructive overlap between all pairs of adjacent *p* orbitals. There are no vertical nodes in this lowest-lying MO.
3. The number of nodes increases as the MOs increase in energy.
4. The MOs should be evenly divided between bonding and antibonding MOs, with the possibility of nonbonding MOs in some cases.
5. We expect that a stable system will have filled bonding MOs and empty antibonding MOs.

PROBLEM-SOLVING HINT

These principles, used in drawing the MOs of benzene, are applicable to many MO problems.

Figure 16-4 shows the six π molecular orbitals of benzene as viewed from above, showing the sign of the top lobe of each *p* orbital. The first MO (π_1) is entirely bonding, with no nodes. It is very low in energy because it has six bonding interactions and the electrons are delocalized over all six carbon atoms. The top lobes of the *p* orbitals all have the same sign, as do the bottom lobes. The six *p* orbitals overlap to form a continuously bonding ring of electron density.

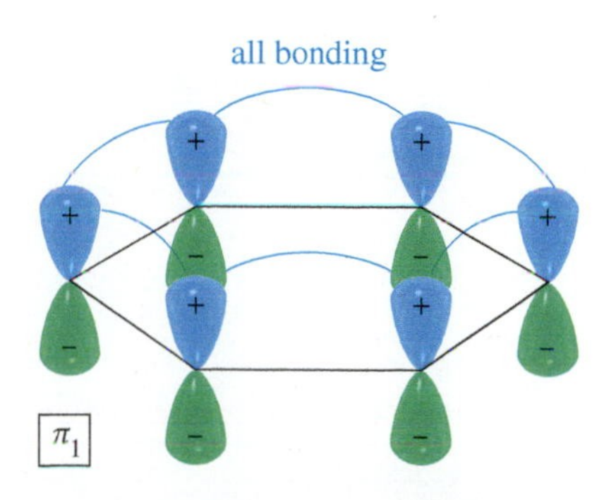

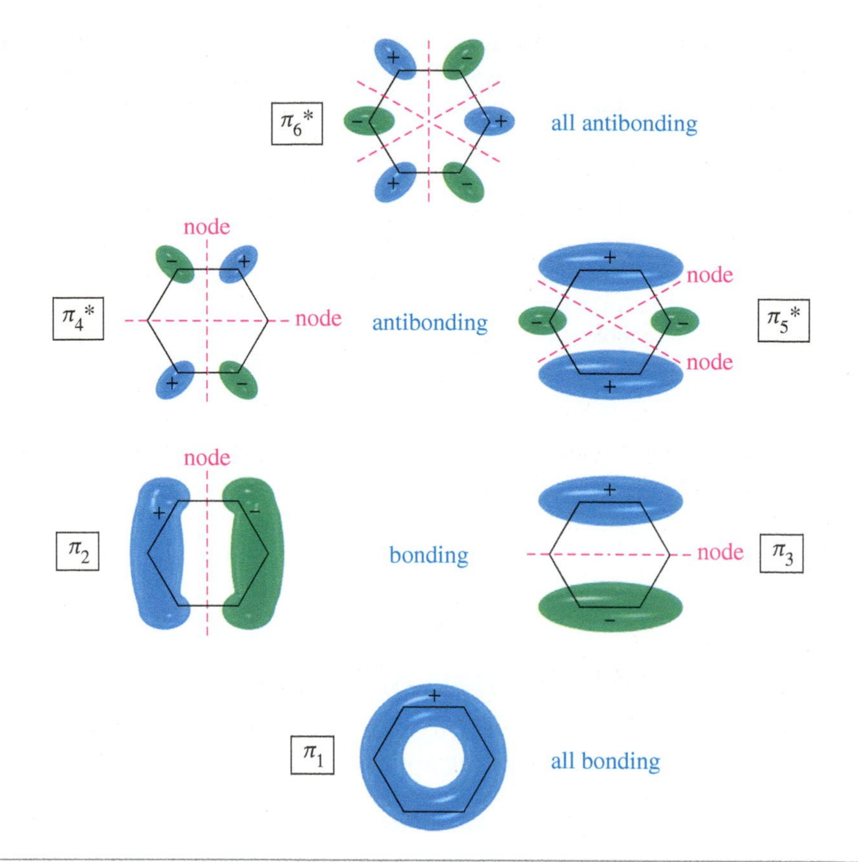

FIGURE 16-4
The six π molecular orbitals of benzene, viewed from above. The number of nodal planes increases with energy, and there are two degenerate MOs at each intermediate energy level.

In a cyclic system of overlapping *p* orbitals, the intermediate energy levels are **degenerate** (equal in energy), with two orbitals at each energy level. Both π_2 and π_3 have one nodal plane, as we expect at the second energy level. Notice that π_2 has four bonding interactions and two antibonding interactions, for a total of two net bonding interactions. Similarly, π_3 has two bonding interactions and four nonbonding interactions, also totaling two net bonding interactions. There are no antibonding interactions in π_3, but in this MO there is no electron density on the two carbon atoms that lie on the node. Although we cannot use the number of bonding and antibonding interactions as a quantitative measure of an orbital's energy, it is clear that π_2 and π_3 are bonding MOs, but not as strongly bonding as π_1.

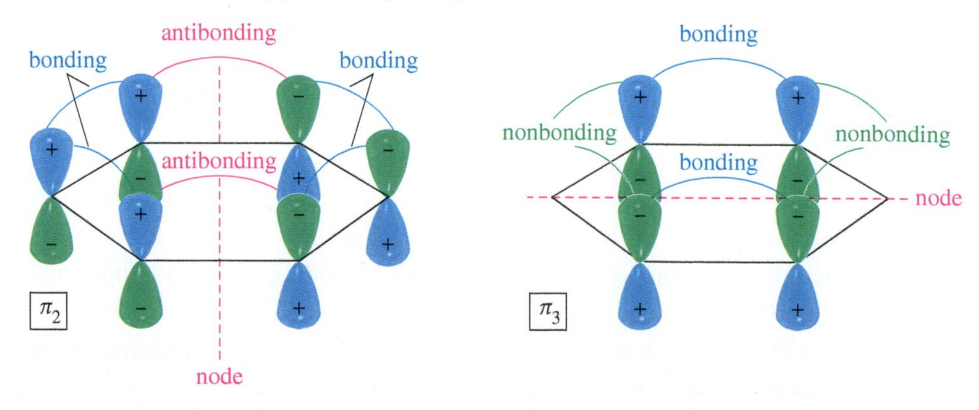

The next orbitals, π_4^* and π_5^*, are also degenerate, with two nodal planes in each. The π_4^* orbital has two antibonding interactions and four nonbonding interactions; it is an antibonding (*) orbital. Its degenerate partner, π_5^*, has four antibonding interactions and two bonding interactions, for a total of two antibonding interactions. This degenerate pair of MOs, π_4^* and π_5^*, are about as strongly antibonding as π_2 and π_3 are bonding.

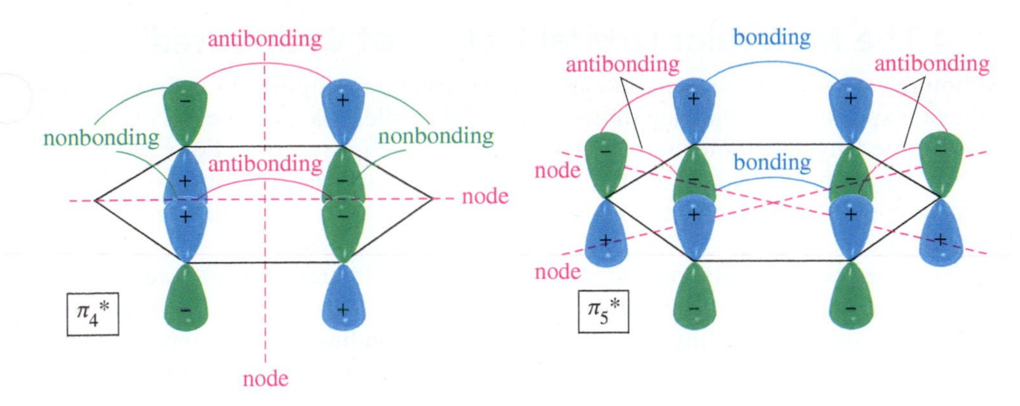

The all-antibonding π_6^* has three nodal planes. Each pair of adjacent p orbitals is out of phase and interacts destructively.

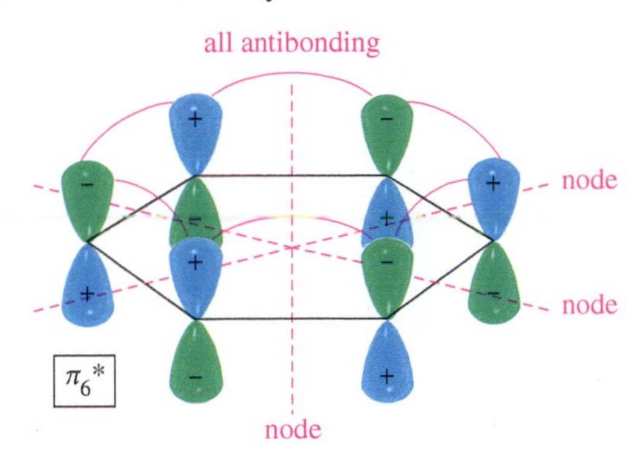

The Energy Diagram of Benzene The energy diagram of the benzene MOs (Figure 16-5) shows them to be symmetrically distributed above and below the nonbonding line (the energy of an isolated p orbital). The all-bonding and all-antibonding orbitals (π_1 and π_6^*) are lowest and highest in energy, respectively. The degenerate bonding orbitals (π_2 and π_3) are higher in energy than π_1, but still bonding. The degenerate pair π_4^* and π_5^* are antibonding, yet not as high in energy as the all-antibonding π_6^* orbital.

The Kekulé structure for benzene shows three pi bonds, representing six electrons (three pairs) involved in pi bonding. Six electrons fill the three bonding MOs of the benzene system. This electronic configuration explains the unusual stability of benzene. The first MO is all-bonding and is particularly stable. The second and third (degenerate) MOs are still strongly bonding, and all three of these bonding MOs delocalize the electrons over several nuclei. This configuration, with all the bonding MOs filled (a "closed bonding shell"), is energetically very favorable.

Application: Herbicide

People and animals cannot make the benzene nucleus, a key component of the essential amino acid phenylalanine. They must obtain phenylalanine in their diet. Plants and bacteria can synthesize the aromatic ring, however, and they synthesize their own phenylalanine. Glyphosate (the common herbicide known as Roundup®) blocks the synthesis of phenylalanine in plants. Compounds that block these pathways in bacteria are being developed as potential antibiotics.

$H_2N—CH—COOH$ with CH_2—(benzene ring) on CH

phenylalanine

HO—P(=O)(OH)—CH_2NHCH_2COOH

glyphosate

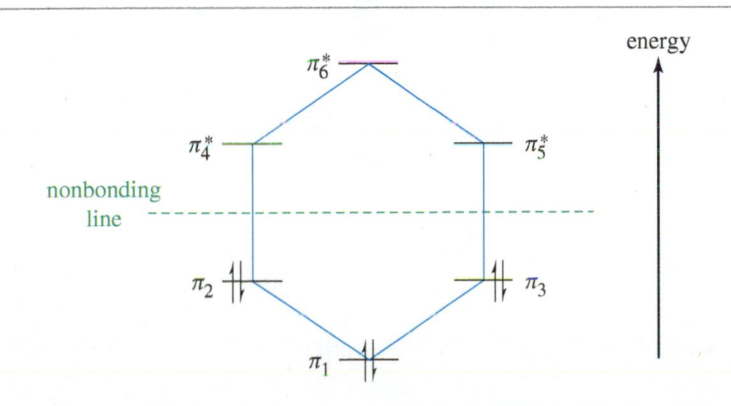

FIGURE 16-5
Energy diagram of the molecular orbitals of benzene. Benzene's six π electrons fill the three bonding orbitals, leaving the antibonding orbitals vacant.

16-4 The Molecular Orbital Picture of Cyclobutadiene

Although we can draw benzene-like resonance structures (Figure 16-3) for cyclobutadiene, experimental evidence shows that cyclobutadiene is unstable. Its instability is explained by the molecular orbitals, shown in Figure 16-6. Four sp^2 hybrid carbon atoms form the cyclobutadiene ring, and their four p orbitals overlap to form four molecular orbitals. The lowest-energy MO is π_1, the all-bonding MO with no nodes.

The next two orbitals, π_2 and π_3, are degenerate (equal energy), each having one symmetrically situated nodal plane. Each of these MOs has two bonding interactions and two antibonding interactions. The net bonding order is zero, so these two MOs are nonbonding. The final MO, π_4^*, has two nodal planes and is entirely antibonding.

Figure 16-7 is an energy diagram of the four cyclobutadiene MOs. The lowest-lying MO (π_1) is strongly bonding, and the highest-lying MO (π_4^*) is equally antibonding. The two degenerate nonbonding orbitals are intermediate in energy, falling on the nonbonding line (the energy of an isolated p orbital).

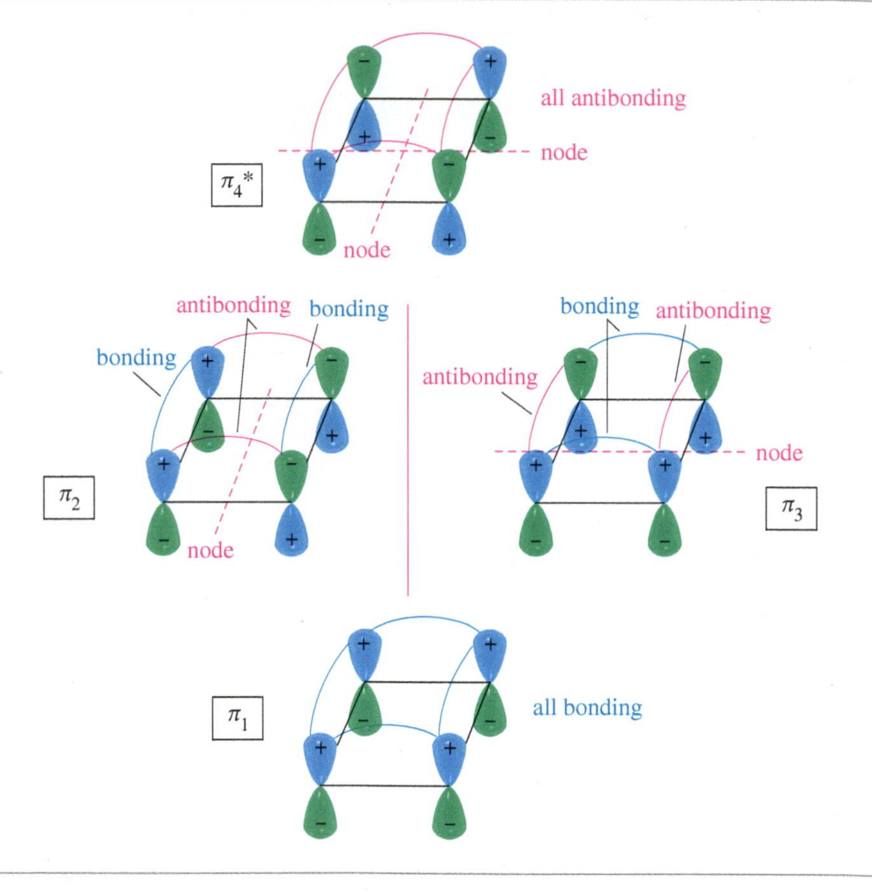

FIGURE 16-6
The pi molecular orbitals of cyclobutadiene. There are four MOs: the lowest-energy bonding orbital, the highest-energy antibonding orbital, and two degenerate nonbonding orbitals.

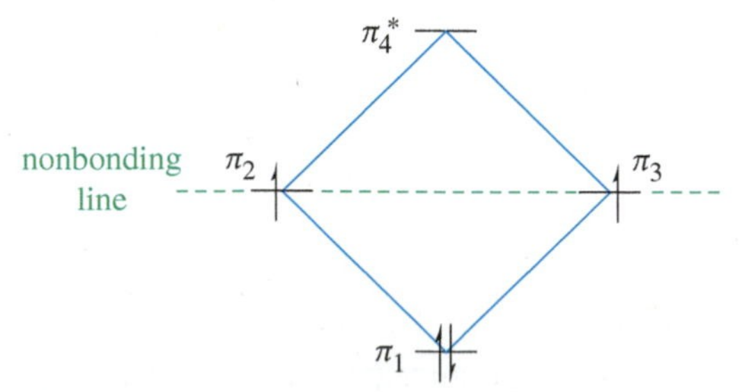

FIGURE 16-7
An electronic energy diagram of cyclobutadiene shows that two electrons are unpaired in separate nonbonding molecular orbitals.

The localized structure of cyclobutadiene shows two double bonds, implying four pi electrons. Two electrons fill π_1, the lowest-lying orbital. Once π_1 is filled, two orbitals of equal energy are available for the remaining two electrons. If the two electrons go into the same orbital, they must have paired spins and they must share the same region of space. Because electrons repel each other, less energy is required for the electrons to occupy different degenerate orbitals, with unpaired spins. This principle is another application of Hund's rule (Section 1-2).

The electronic configuration in Figure 16-7 indicates that cyclobutadiene should be unstable. Its highest-lying electrons are in nonbonding orbitals (π_2 and π_3) and are therefore very reactive. According to Hund's rule, the compound exists as a diradical (two unpaired electrons) in its ground state. Such a diradical is expected to be extremely reactive. Thus, molecular orbital theory successfully predicts the dramatic stability difference between benzene and cyclobutadiene.

The localized structure of cyclobutadiene shows two double bonds.

The MO diagram for a square cyclobutadiene suggests that the true picture is a diradical.

The Polygon Rule The patterns of molecular orbitals in benzene (Figure 16-5) and in cyclobutadiene (Figure 16-7) are similar to the patterns in other annulenes: The lowest-lying MO is the unique one with no nodes; thereafter, the molecular orbitals occur in degenerate (equal-energy) pairs until only one highest-lying MO remains. In benzene, the energy diagram looks like the hexagon of a benzene ring. In cyclobutadiene, the pattern looks like the diamond of the cyclobutadiene ring.

The **polygon rule** states that the molecular orbital energy diagram of a regular, completely conjugated cyclic system has the same regular polygonal shape as the compound, with one vertex (the all-bonding MO) at the bottom. The nonbonding line cuts horizontally through the center of the polygon. Figure 16-8 shows how the polygon rule predicts the MO energy diagrams for benzene, cyclobutadiene, and cyclooctatetraene. The pi electrons are filled into the orbitals in accordance with the aufbau principle (lowest-energy orbitals are filled first) and Hund's rule.

PROBLEM-SOLVING HINT

The polygon rule gives you a fast way to draw an electronic configuration. It also provides a quick check on molecular orbitals you might draw, to see which are bonding, antibonding, and nonbonding.

When inscribed in a circle, the polygon is sometimes called a *Frost circle* or a *Frost-Musulin diagram.*

PROBLEM 16-5

Does the MO energy diagram of cyclooctatetraene (Figure 16-8) appear to be a particularly stable or unstable configuration? Explain.

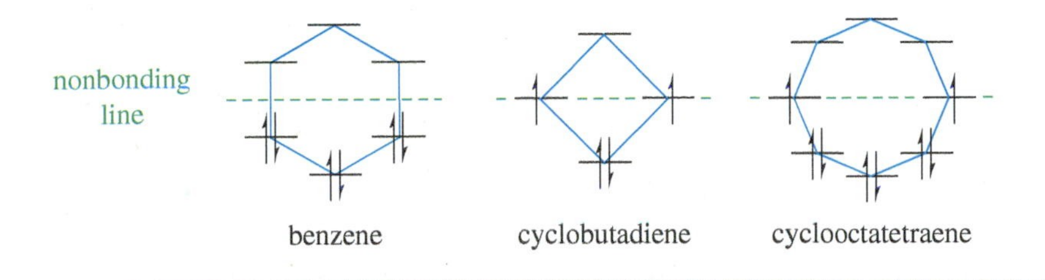

FIGURE 16-8
The polygon rule predicts that the MO energy diagrams for these annulenes will resemble the polygonal shapes of the annulenes.

16-5 Aromatic, Antiaromatic, and Nonaromatic Compounds

Our working definition of aromatic compounds has included cyclic compounds containing conjugated double bonds with unusually large resonance energies. At this point we can be more specific about the properties that are required for a compound (or an ion) to be aromatic.

Aromatic compounds are those that meet the following criteria:

1. The structure must be cyclic, containing some number of conjugated pi bonds.
2. Each atom in the ring must have an unhybridized *p* orbital. (The ring atoms are usually sp^2 hybridized or occasionally *sp* hybridized.)

3. The unhybridized *p* orbitals must overlap to form a continuous ring of parallel orbitals. In most cases, the structure must be planar (or nearly planar) for effective overlap to occur.
4. Delocalization of the pi electrons over the ring must *lower* the electronic energy.

An **antiaromatic compound** meets the first three criteria, but delocalization of the pi electrons over the ring *increases* the electronic energy.

Aromatic structures are more stable than their open-chain counterparts. For example, benzene is more stable than hexa-1,3,5-triene.

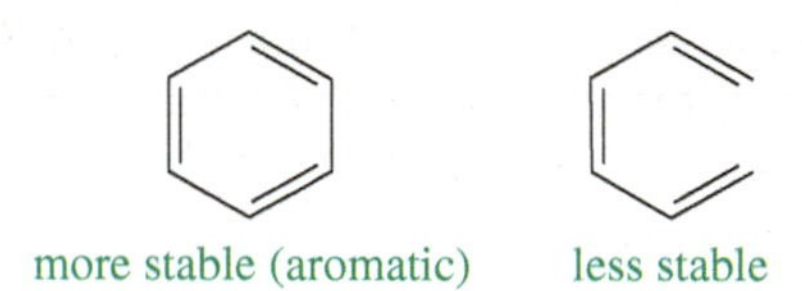

Cyclobutadiene meets the first three criteria for a continuous ring of overlapping *p* orbitals, but delocalization of the pi electrons *increases* the electronic energy. Cyclobutadiene is less stable than its open-chain counterpart (buta-1,3-diene), and it is antiaromatic.

A cyclic compound that does not have a continuous, overlapping ring of *p* orbitals cannot be aromatic or antiaromatic. It is said to be **nonaromatic,** or aliphatic. Its electronic energy is similar to that of its open-chain counterpart. For example, cyclohexa-1,3-diene is about as stable as *cis,cis*-hexa-2,4-diene.

(nonaromatic) ⟵ similar stabilities ⟶ CH_3 CH_3

16-6 Hückel's Rule

About 1931, Erich Hückel developed a shortcut for predicting which of the annulenes and related compounds are aromatic and which are antiaromatic. In using Hückel's rule, we must be certain that the compound under consideration meets the criteria for an aromatic or antiaromatic system.

> To qualify as aromatic or antiaromatic, a cyclic compound must have a continuous ring of overlapping *p* orbitals, usually in a planar conformation.

Once these criteria are met, **Hückel's rule** applies:

> Hückel's Rule: If the number of pi electrons in the cyclic system is:
> $(4N + 2)$, the system is aromatic.
> $(4N)$, the system is antiaromatic.
> N is an integer, commonly 0, 1, 2, or 3.

Common aromatic systems have 2, 6, or 10 pi electrons, for $N = 0$, 1, or 2. Antiaromatic systems might have 4, 8, or 12 pi electrons, for $N = 1$, 2, or 3.

Benzene is [6]annulene, cyclic, with a continuous ring of overlapping *p* orbitals. There are six pi electrons in benzene (three double bonds in the classical structure), so it is a $(4N+2)$ system, with $N = 1$. Hückel's rule predicts benzene to be aromatic.

As with benzene, cyclobutadiene ([4]annulene) has a continuous ring of overlapping *p* orbitals. But it has four pi electrons (two double bonds in the classical structure), which is a $(4N)$ system with $N = 1$. Hückel's rule predicts cyclobutadiene to be antiaromatic.

Cyclooctatetraene is [8]annulene, with eight pi electrons (four double bonds) in the classical structure. It is a $(4N)$ system, with $N = 2$. If Hückel's rule were applied to cyclooctatetraene, it would predict antiaromaticity. However, cyclooctatetraene is a stable hydrocarbon with a boiling point of 153 °C. It does not show the high reactivity associated with antiaromaticity, yet it is not aromatic either. Its reactions are typical of alkenes.

Cyclooctatetraene would be antiaromatic if Hückel's rule applied, so the conjugation of its double bonds is energetically unfavorable. Remember that Hückel's rule applies to a compound *only* if there is a continuous ring of overlapping *p* orbitals, usually in a planar system. Cyclooctatetraene is more flexible than cyclobutadiene, and it assumes a nonplanar "tub" conformation that avoids most of the overlap between adjacent pi bonds. Hückel's rule simply does not apply.

PROBLEM-SOLVING HINT

Hückel's rule is commonly used to determine aromaticity and antiaromaticity. A continuous, planar ring of overlapping *p* orbitals is required for the rule to apply. Otherwise, the system is nonaromatic.

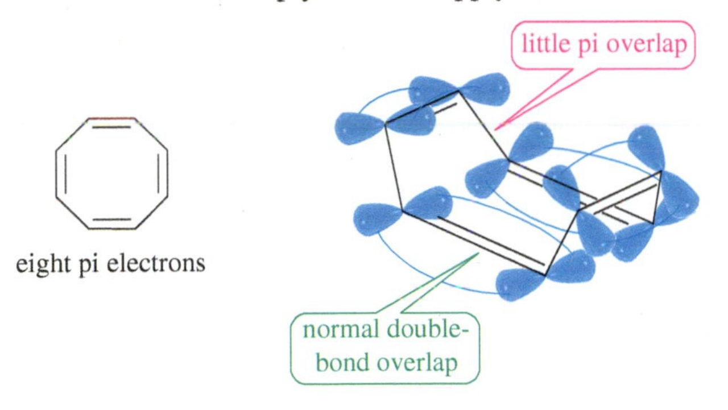

PROBLEM 16-6

Make a model of cyclooctatetraene in the tub conformation. Draw this conformation, and estimate the angle between the *p* orbitals of adjacent pi bonds.

Large-Ring Annulenes As with cyclooctatetraene, larger annulenes with $(4N)$ systems do not show antiaromaticity because they have the flexibility to adopt nonplanar conformations. Even though [12]annulene, [16]annulene, and [20]annulene are $(4N)$ systems (with $N = 3$, 4, and 5, respectively), they all react as partially conjugated polyenes.

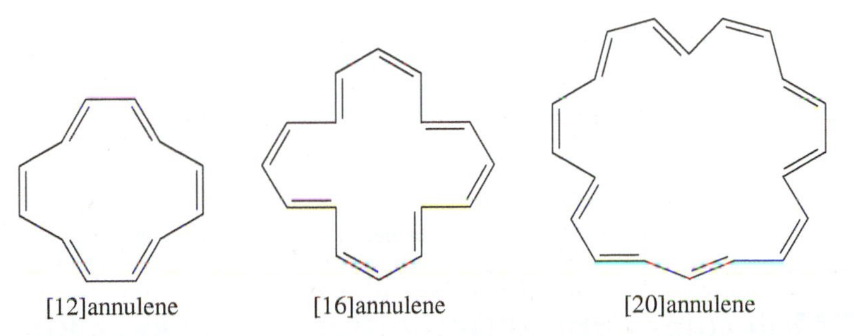

Aromaticity in the larger $(4N+2)$ annulenes depends on whether the molecule can adopt the necessary planar conformation. In the all-cis [10]annulene, the planar conformation requires an excessive amount of angle strain. The [10]annulene isomer with two trans double bonds cannot adopt a planar conformation either, because two hydrogen atoms interfere with each other. Neither of these [10]annulene isomers is aromatic, even though each has $(4N+2)$ pi electrons, with $N = 2$. If the interfering hydrogen

PROBLEM-SOLVING HINT

Predicting planarity in large, floppy annulenes is often difficult. If the annulene is *aromatic if planar*, then the resonance energy (ΔH) favors planarity, but the entropy factor ($-T\Delta S$) usually favors a more disordered conformation. Which factor prevails is hard to predict, and may depend on the temperature and solvent.

If the annulene is *antiaromatic if planar*, then both the resonance energy and the entropy favor a more disordered, nonplanar conformation unless the molecule is somehow locked in a planar conformation.

atoms in the partially trans isomer are removed, the molecule can be planar. When these hydrogen atoms are replaced with a bond, the aromatic compound naphthalene results.

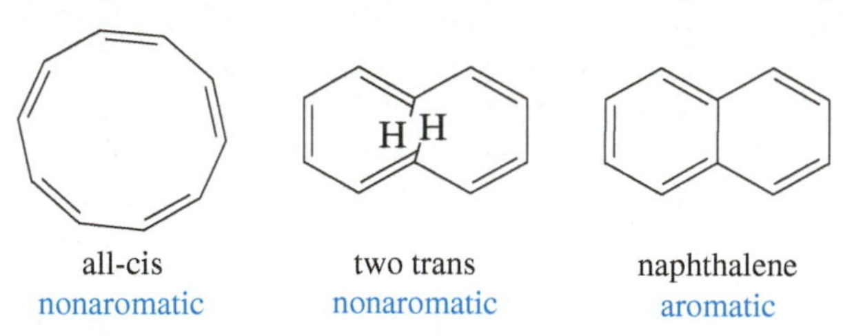

Some of the larger annulenes with ($4N+2$) pi electrons can achieve planar conformations. For example, the following [14]annulene and [18]annulene have aromatic properties.

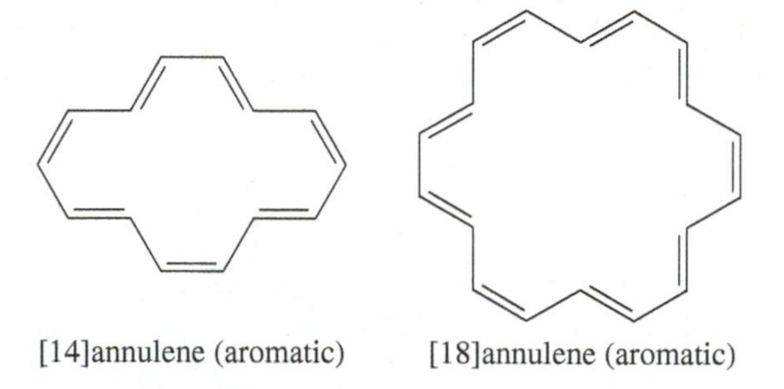

PROBLEM 16-7

Classify the following compounds as aromatic, antiaromatic, or nonaromatic.

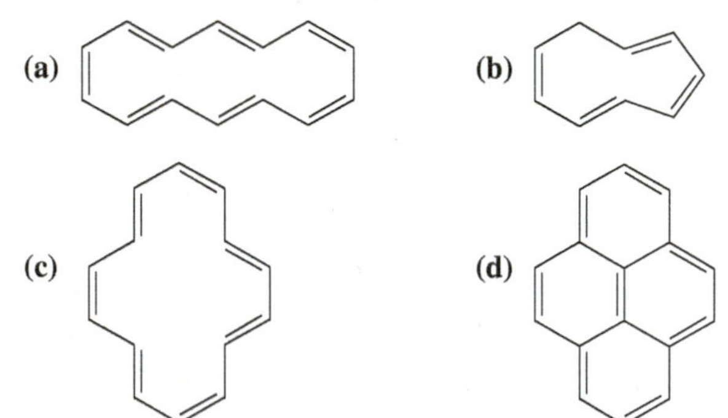

PROBLEM 16-8

One of the following compounds is much more stable than the other two. Classify each as aromatic, antiaromatic, or nonaromatic.

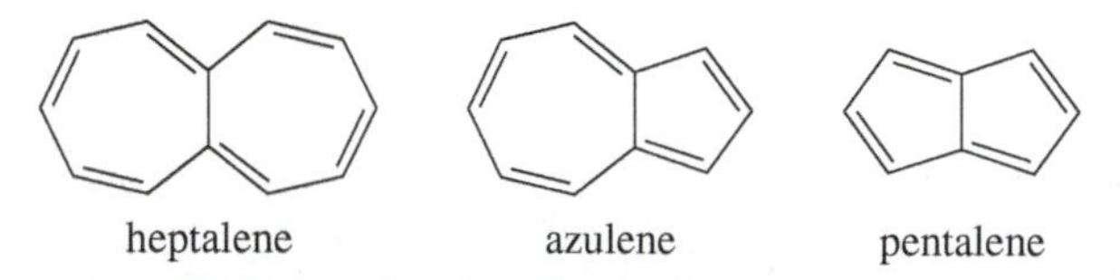

16-7 Molecular Orbital Derivation of Hückel's Rule

Benzene is aromatic because it has a filled shell of equal-energy orbitals. The degenerate orbitals π_2 and π_3 are filled, and all the electrons are paired. Cyclobutadiene, by contrast, has an open shell of electrons. There are two half-filled orbitals easily capable of donating or accepting electrons. To derive Hückel's rule, we must show under what general conditions there is a filled shell of orbitals.

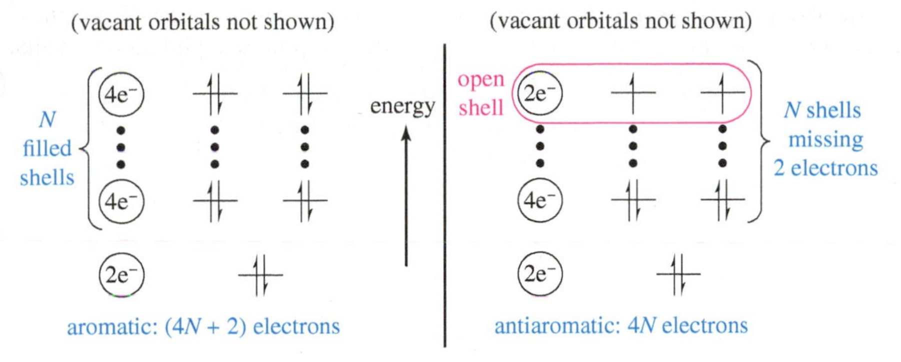

FIGURE 16-9
Pattern of molecular orbitals in a cyclic conjugated system. In a cyclic conjugated system, the lowest-lying MO is filled with two electrons. Each of the additional shells consists of two degenerate MOs, with space for four electrons. If a molecule has $(4N+2)$ pi electrons, it will have a filled shell. If it has $(4N)$ electrons, there will be two unpaired electrons in two degenerate orbitals.

Recall the pattern of MOs in a cyclic conjugated system. There is one all-bonding, lowest-lying MO, followed by degenerate pairs of bonding MOs. (There is no need to worry about the antibonding MOs because they are vacant in the ground state.) The lowest-lying MO is always filled (two electrons). Each additional shell consists of two degenerate MOs, requiring four electrons to fill a shell. Figure 16-9 shows this pattern of two electrons for the lowest orbital and then four electrons for each additional shell.

A compound has a filled shell of orbitals if it has two electrons for the lowest-lying orbital, plus $(4N)$ electrons, where N is the number of filled pairs of degenerate orbitals. The total number of pi electrons in this case is $(4N+2)$. If the system has a total of only $(4N)$ electrons, it is two electrons short of filling N pairs of degenerate orbitals. There are only two electrons in the Nth pair of degenerate orbitals. This is a half-filled shell, and Hund's rule predicts these electrons will be unpaired (a diradical).

PROBLEM-SOLVING HINT

Using the polygon rule and the following steps, draw energy diagrams for the MOs of cyclic, completely conjugated systems.

1. Draw a regular polygon with the same number of sides as the conjugated ring. Place one vertex (corresponding to the all-bonding MO) at the bottom.
2. The height of each vertex corresponds to the energy of one of the MOs. Each MO can accommodate two electrons.
3. Draw a nonbonding line horizontally through the center of the polygon. MOs below this line are bonding, and are usually filled in stable compounds. MOs above this line are antibonding, and are usually empty in stable compounds. MOs that fall on the line are nonbonding, and may or may not be filled.
4. Add the number of pi electrons in the system. Add two for each pi bond in the resonance forms, plus one for a radical or two for a nonbonding electron pair that occupies a p orbital, such as those on a carbanion, a nitrogen atom, or an oxygen atom.

PROBLEM 16-9

(a) Use the polygon rule to draw an energy diagram (as in Figures 16-5 and 16-7) for the MOs of a planar cyclooctatetraenyl system.

(b) Fill in the eight pi electrons for cyclooctatetraene. Is this electronic configuration aromatic or antiaromatic? Could the cyclooctatetraene system be aromatic if it gained or lost electrons?

***(c)** Draw pictorial representations (as in Figures 16-4 and 16-6) for the three bonding MOs and the two nonbonding MOs of cyclooctatetraene. The antibonding MOs are difficult to draw, except for the all-antibonding MO.

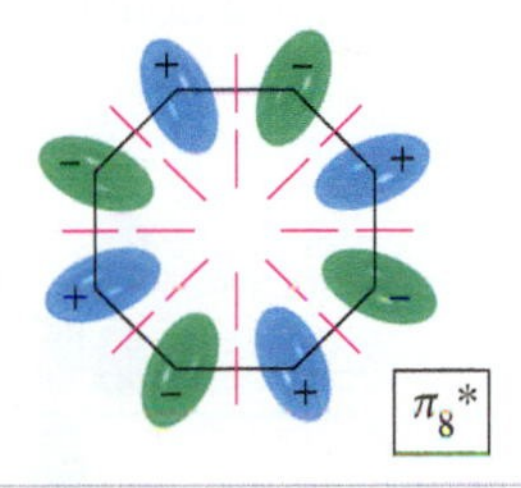

16-8 Aromatic Ions

Up to this point, we have discussed aromaticity using the annulenes as examples. Annulenes are uncharged molecules having even numbers of carbon atoms with alternating single and double bonds. Hückel's rule also applies to systems having odd numbers of carbon atoms and bearing positive or negative charges. We now consider some common aromatic ions and their antiaromatic counterparts.

16-8A The Cyclopentadienyl Ions

We can draw a five-membered ring of sp^2 hybrid carbon atoms with all the unhybridized *p* orbitals lined up to form a continuous ring. With five pi electrons, this system would be neutral, but it would be a radical because an odd number of electrons cannot all be paired. With four pi electrons (a cation), Hückel's rule predicts this system to be antiaromatic. With six pi electrons (an anion), Hückel's rule predicts aromaticity.

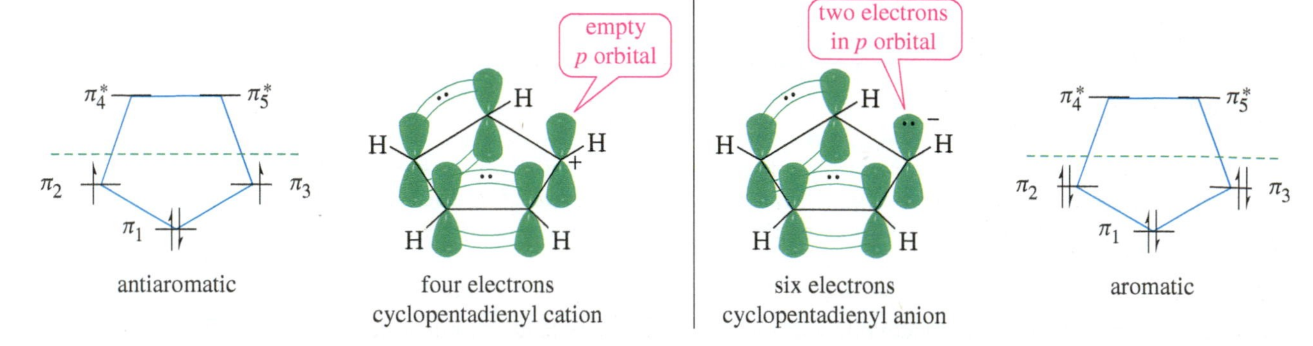

Because the cyclopentadienyl anion (six pi electrons) is aromatic, it is unusually stable compared with other carbanions. It can be formed by abstracting a proton from cyclopentadiene, which is unusually acidic for an alkene. Cyclopentadiene has a pK_a of 16, compared with a pK_a of 46 for the allylic protons in cyclohexene. In fact, cyclopentadiene is nearly as acidic as water and more acidic than many alcohols. It is entirely ionized by potassium *tert*-butoxide:

cyclopentadiene (pK_a = 16) + $^{-}OC(CH_3)_3$ ⇌ the cyclopentadienyl anion (six pi electrons) + $HOC(CH_3)_3$ (pK_a = 18)

Cyclopentadiene is unusually acidic because loss of a proton converts the nonaromatic diene to the aromatic cyclopentadienyl anion. Cyclopentadiene contains an sp^3 hybrid ($-CH_2-$) carbon atom without an unhybridized *p* orbital, so there can be no continuous ring of *p* orbitals. Deprotonation of the $-CH_2-$ group leaves an orbital occupied by a pair of electrons. This orbital can rehybridize to a *p* orbital, completing a ring of *p* orbitals containing six pi electrons: the two electrons on the deprotonated carbon, plus the four electrons in the original double bonds.

cyclopentadiene, nonaromatic (sp^3) + $^{-}OC(CH_3)_3$ ⇌ cyclopentadienyl anion, aromatic (*p* orbital) + $HOC(CH_3)_3$

more stable (aromatic) less stable

When we say the cyclopentadienyl anion is aromatic, this does not necessarily imply that it is as stable as benzene. As a carbanion, the cyclopentadienyl anion reacts readily with electrophiles. Because this ion is aromatic, however, it is more stable than the corresponding open-chain ion.

Hückel's rule predicts that the cyclopentadienyl cation, with four pi electrons, is antiaromatic. In agreement with this prediction, the cyclopentadienyl cation is not easily formed. Protonated cyclopenta-2,4-dien-1-ol does not lose water (to give the cyclopentadienyl cation), even in concentrated sulfuric acid. The antiaromatic cation is simply too unstable.

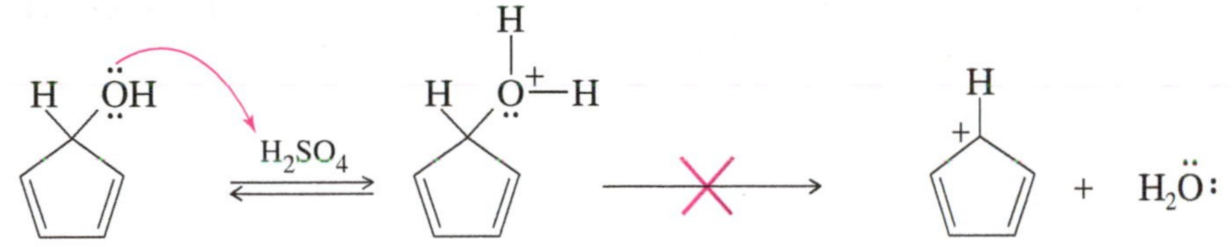

cyclopenta-2,4-dien-1-ol $\xrightleftharpoons{H_2SO_4}$ (does not occur) not formed (four pi electrons) + $H_2\ddot{O}$:

Using a simple resonance approach, we might incorrectly expect both of the cyclopentadienyl ions to be unusually stable. Shown next are resonance structures that spread the negative charge of the anion and the positive charge of the cation over all five carbon atoms of the ring. With conjugated cyclic systems such as these, the resonance approach is a poor predictor of stability. Hückel's rule, based on molecular orbital theory, is a much better predictor of stability for these aromatic and antiaromatic systems.

less stable (antiaromatic)

more stable

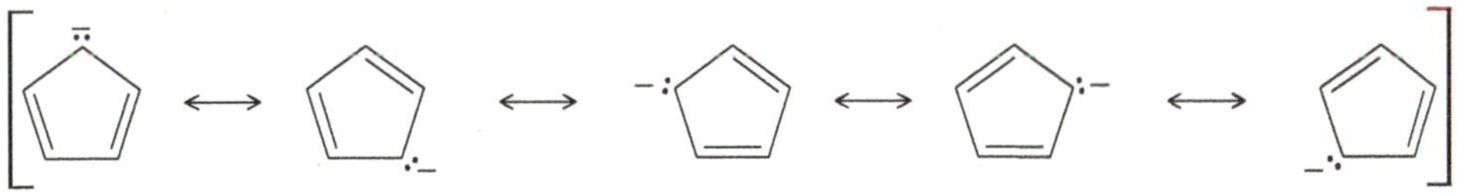

cyclopentadienyl anion: six pi electrons, aromatic

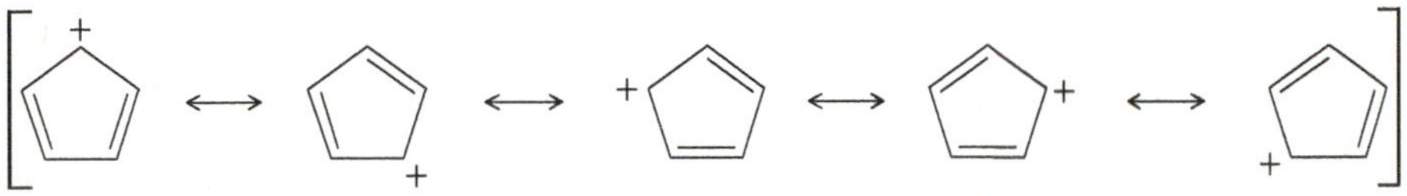

cyclopentadienyl cation: four pi electrons, antiaromatic
The resonance picture gives a misleading suggestion of stability.

PROBLEM 16-10

(a) Draw the molecular orbitals for the cyclopropenyl case.

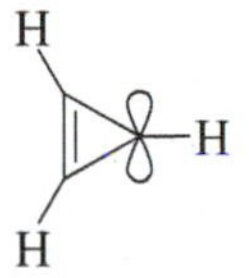

(Because there are three p orbitals, there must be three MOs: one all-bonding MO and one degenerate pair of MOs.)

(b) Draw an energy diagram for the cyclopropenyl MOs. (The polygon rule is helpful.) Label each MO as bonding, nonbonding, or antibonding, and add the nonbonding line. Notice that it goes through the approximate average of the MOs.

(c) Add electrons to your energy diagram to show the configuration of the cyclopropenyl cation and the cyclopropenyl anion. Which is aromatic and which is antiaromatic?

PROBLEM 16-11

Repeat Problem 16-10 for the cyclopentadienyl ions. Draw one all-bonding MO, then a pair of degenerate MOs, and then a final pair of degenerate MOs. Draw the energy diagram, fill in the electrons, and confirm the electronic configurations of the cyclopentadienyl cation and anion.

16-8B The Cycloheptatrienyl Ions

As with the five-membered ring, we can imagine a flat seven-membered ring with seven *p* orbitals aligned. The cation has six pi electrons, and the anion has eight pi electrons. Once again, we can draw resonance forms that seem to show either the positive charge of the cation or the negative charge of the anion delocalized over all seven atoms of the ring. By now, however, we know that the six-electron system is aromatic and the eight-electron system is antiaromatic (if it remains planar).

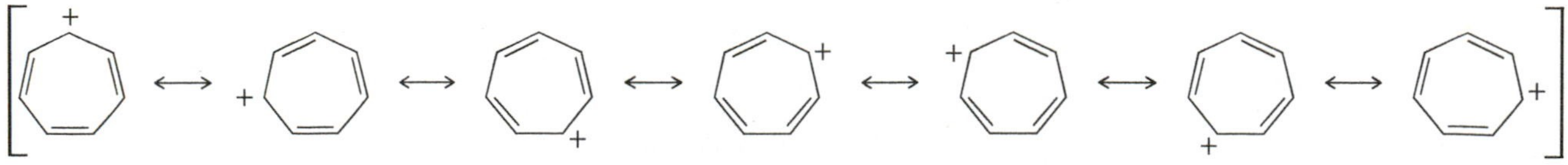

cycloheptatrienyl cation (tropylium ion): six pi electrons, aromatic

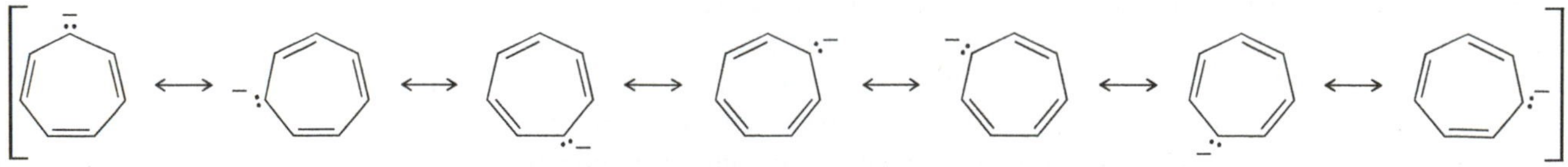

cycloheptatrienyl anion: eight pi electrons, antiaromatic (if planar)
The resonance picture gives a misleading suggestion of stability.

The cycloheptatrienyl cation is easily formed by treating the corresponding alcohol with dilute (0.01 molar) aqueous sulfuric acid. This is our first example of a hydrocarbon cation that is stable in aqueous solution.

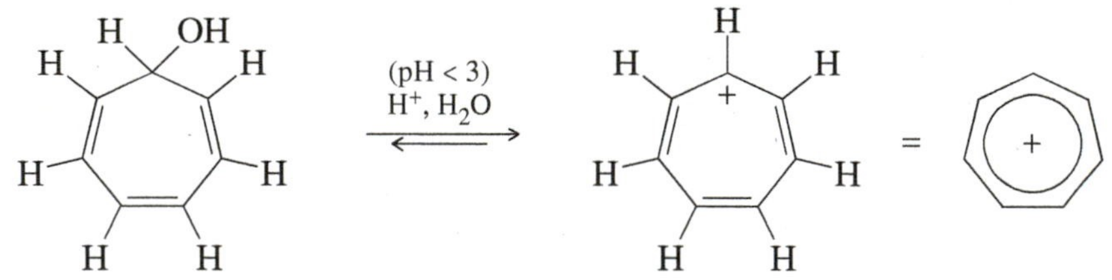

(pH < 3) H^+, H_2O

tropylium ion, six pi electrons

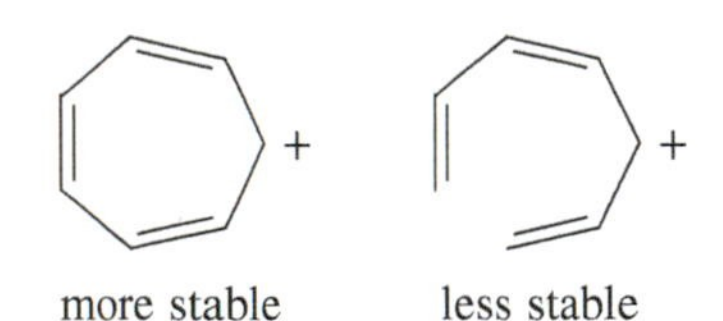

more stable (aromatic) less stable

The cycloheptatrienyl cation is called the **tropylium ion.** This aromatic ion is much less reactive than most carbocations. Some tropylium salts can be isolated and stored for months without decomposing. Nevertheless, the tropylium ion is not necessarily as stable as benzene. Its aromaticity simply implies that the cyclic ion is more stable than the corresponding open-chain ion.

Although the tropylium ion forms easily, the corresponding anion is difficult to form because it is antiaromatic. Cycloheptatriene ($pK_a = 39$) is barely more acidic than propene ($pK_a = 43$), and the anion is very reactive. This result agrees with the prediction of Hückel's rule that the cycloheptatrienyl anion is antiaromatic if it is planar.

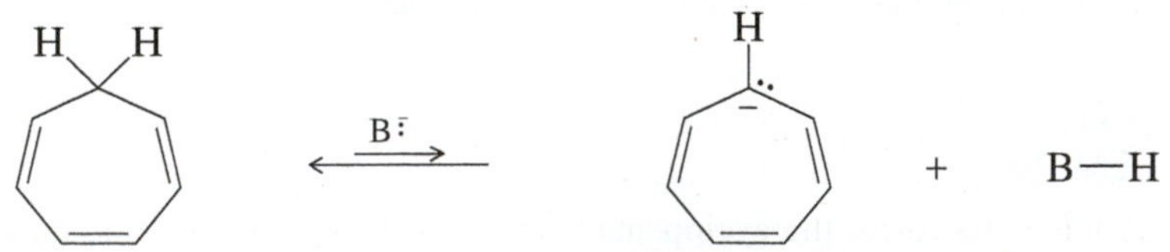

B:⁻ + B—H

cycloheptatriene
$pK_a = 39$

cycloheptatrienyl anion
eight pi electrons

16-8C The Cyclooctatetraene Dianion

We have seen that aromatic stabilization leads to unusually stable hydrocarbon anions such as the cyclopentadienyl anion. Dianions of hydrocarbons are rare and are usually much more difficult to form. Cyclooctatetraene reacts with potassium metal, however, to form an aromatic dianion.

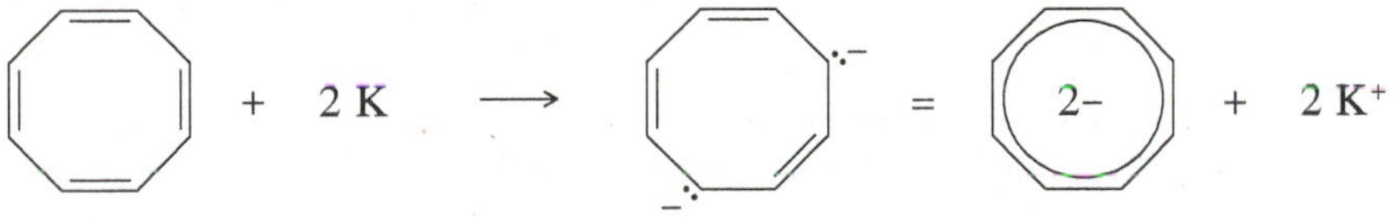

+ 2 K ⟶ = 2− + 2 K⁺

ten pi electrons

The cyclooctatetraene dianion has a planar, regular octagonal structure with C—C bond lengths of 1.40 Å close to the 1.397 Å bond lengths in benzene. Cyclooctatetraene itself has eight pi electrons, so the dianion has ten: $(4N+2)$, with $N=2$. The cyclooctatetraene dianion is easily prepared because it is aromatic.

PROBLEM 16-12

Explain why each compound or ion should be aromatic, antiaromatic, or nonaromatic.

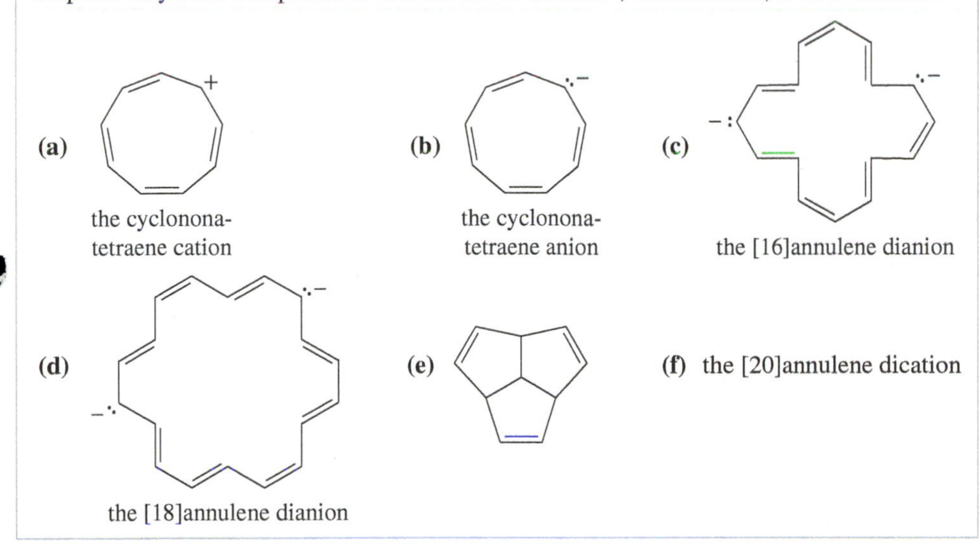

PROBLEM-SOLVING HINT

Use Hückel's rule (and the criteria for its application), rather than resonance, to determine which annulenes and ions are aromatic, antiaromatic, and nonaromatic.

PROBLEM 16-13

The following hydrocarbon has an unusually large dipole moment. Explain how a large dipole moment might arise.

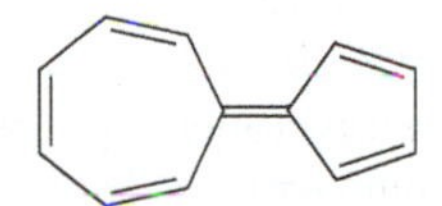

PROBLEM 16-14

When 3-chlorocyclopropene is treated with $AgBF_4$, AgCl precipitates. The organic product can be obtained as a crystalline material, soluble in polar solvents such as nitromethane but insoluble in hexane. When the crystalline material is dissolved in nitromethane containing KCl, the original 3-chlorocyclopropene is regenerated. Determine the structure of the crystalline material, and write equations for its formation and its reaction with chloride ion.

PROBLEM 16-15

The polarization of a carbonyl group can be represented by a pair of resonance structures:

$$\left[\;\rangle C{=}\ddot{O}: \;\longleftrightarrow\; \rangle\overset{+}{C}{-}\ddot{\underset{..}{O}}:^{-}\right]$$

Cyclopropenone and cycloheptatrienone are more stable than anticipated. Cyclopentadienone, however, is relatively unstable and rapidly undergoes a Diels–Alder dimerization. Explain.

cyclopropenone

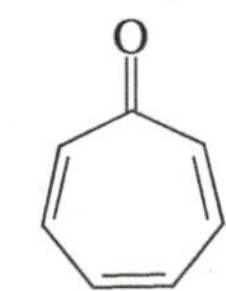

cycloheptatrienone

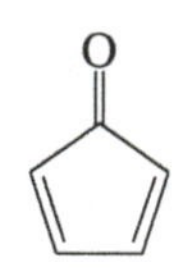

cyclopentadienone

16-8D Summary of Annulenes and Their Ions

This list summarizes applications of Hückel's rule to a variety of cyclic pi systems. These systems are classified according to the number of pi electrons: The 2, 6, and 10 π electron systems are aromatic, whereas the 4 and 8 π electron systems are antiaromatic if they are planar.

2 π electron systems (aromatic)

cyclopropenyl cation (cyclopropenium ion)

4 π electron systems (antiaromatic)

cyclobutadiene

cyclopropenyl anion

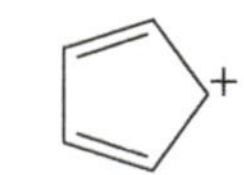

cyclopentadienyl cation

6 π electron systems (aromatic)

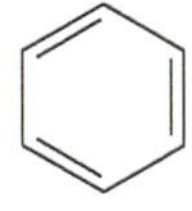

benzene

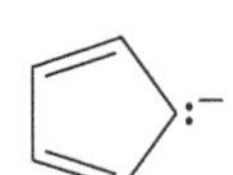

cyclopentadienyl anion (cyclopentadienide ion)

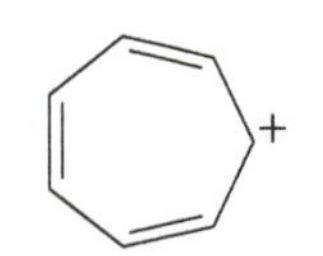

cycloheptatrienyl cation (tropylium ion)

Heterocyclic 6 π systems (aromatic)

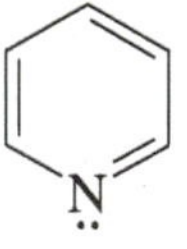

pyridine

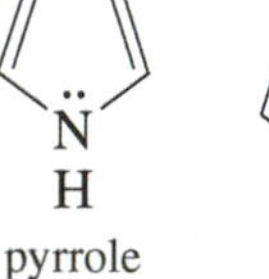

pyrrole

furan

8 π electron systems (antiaromatic if planar)

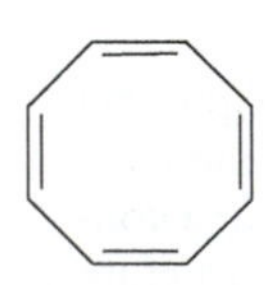

cyclooctatetraene (not planar)

cycloheptatrienyl anion

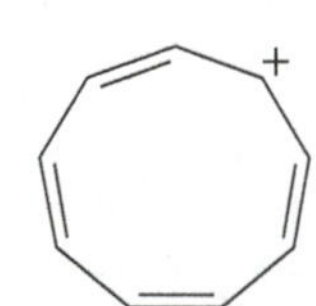

cyclononatetraenyl cation

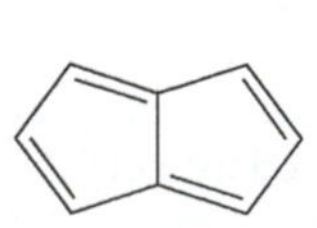

pentalene

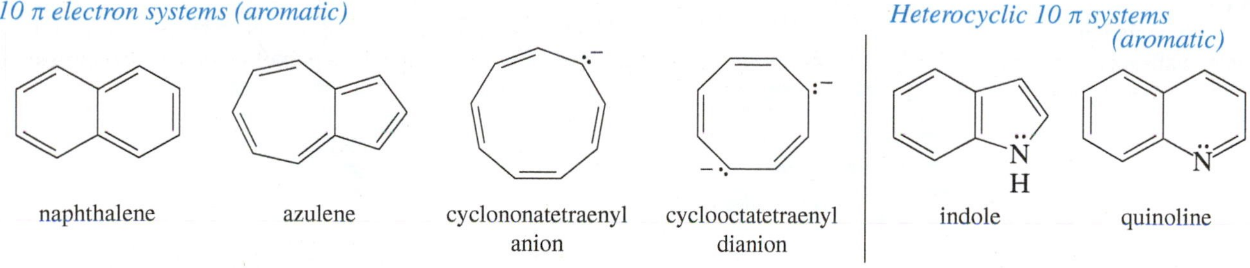

(Naphthalene can also be considered as two fused benzenes.)

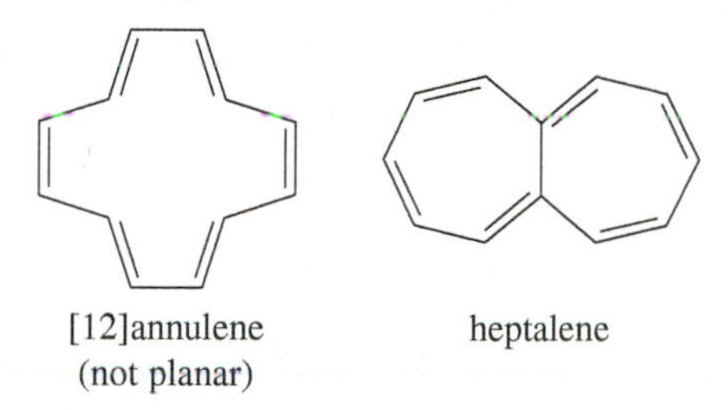

16-9 Heterocyclic Aromatic Compounds

The criteria for Hückel's rule require a ring of atoms, all with unhybridized *p* orbitals overlapping in a continuous ring. In discussing aromaticity, we have considered only compounds composed of rings of sp^2 hybrid carbon atoms. **Heterocyclic compounds,** with rings containing sp^2-hybridized atoms of other elements, can also be aromatic. Nitrogen, oxygen, and sulfur are the most common heteroatoms in heterocyclic aromatic compounds.

16-9A Pyridine

Pyridine is an aromatic nitrogen analog of benzene. It has a six-membered heterocyclic ring with six pi electrons. Pyridine has a nitrogen atom in place of one of the six C—H units of benzene, and the nonbonding pair of electrons on nitrogen replaces benzene's bond to a hydrogen atom. These nonbonding electrons are in an sp^2 hybrid orbital in the plane of the ring (Figure 16-10). They are perpendicular to the pi system and do not overlap with it.

Pyridine shows all the characteristics of aromatic compounds. It has a resonance energy of 113 kJ/mol (27 kcal/mol) and it usually undergoes substitution rather than addition. Because it has an available pair of nonbonding electrons, pyridine is basic (Figure 16-11). In an acidic solution, pyridine protonates to give the

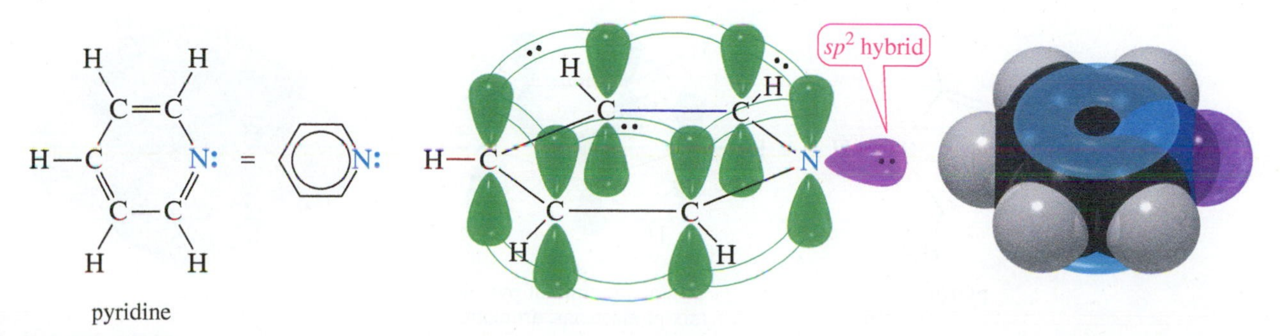

FIGURE 16-10
The pi bonding structure of pyridine. Pyridine has six delocalized electrons in its cyclic pi system. The two nonbonding electrons on nitrogen are in an sp^2 orbital, and they do not interact with the pi electrons of the ring.

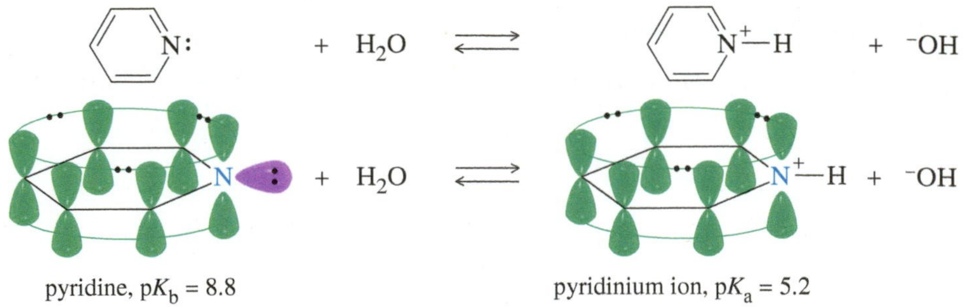

FIGURE 16-11
Pyridine is basic, with nonbonding electrons available to abstract a proton. The protonated pyridine (a pyridinium ion) is still aromatic.

pyridine, $pK_b = 8.8$ + H_2O ⇌ pyridinium ion, $pK_a = 5.2$ + ^-OH

pyridinium ion. The pyridinium ion is still aromatic because the additional proton has no effect on the electrons of the aromatic sextet: It simply bonds to pyridine's nonbonding pair of electrons.

Application: Biochemistry

Porphobilinogen, a substituted pyrrole, is the building block of the heme ring, which has many physiological functions, such as the transport and storage of oxygen. The macrocyclic (large ring) structure of heme is also aromatic, with 18 π electrons in the conjugated cyclic system.

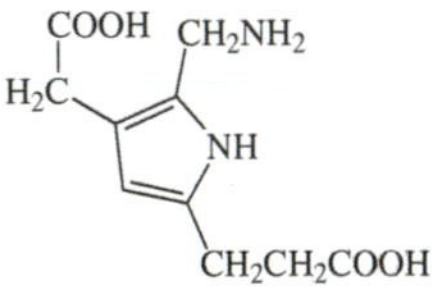

porphobilinogen

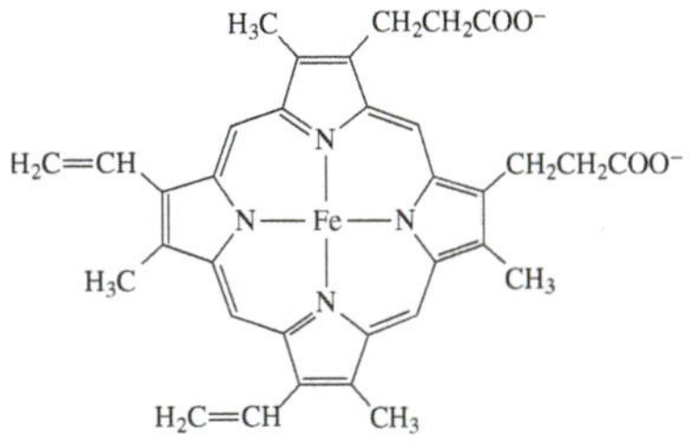

heme, found in hemoglobin and myoglobin

16-9B Pyrrole

Pyrrole is an aromatic five-membered heterocycle, with one nitrogen atom and two double bonds (Figure 16-12). Although it may seem that pyrrole has only four pi electrons, the nitrogen atom has a lone pair of electrons. The pyrrole nitrogen atom is sp^2 hybridized, and its unhybridized p orbital overlaps with the p orbitals of the carbon atoms to form a continuous ring. The lone pair on nitrogen occupies the p orbital, and (unlike the lone pair of pyridine) these electrons take part in the pi bonding system. These two electrons, added to the four pi electrons of the two double bonds, complete an aromatic sextet. Pyrrole has a resonance energy of 92 kJ/mol (22 kcal/mol).

PROBLEM 16-16

(a) Explain how pyrrole is isoelectronic with the cyclopentadienyl anion.
(b) Specifically, what is the difference between the cyclopentadienyl anion and pyrrole?
(c) Draw resonance forms to show the charge distribution on the pyrrole structure.

Pyrrole ($pK_b = 13.6$) is a much weaker base than pyridine ($pK_b = 8.8$). This difference is due to the structure of the protonated pyrrole (Figure 16-13). To form a bond to a proton requires the use of one of the electron pairs in the aromatic sextet. In the protonated pyrrole, the nitrogen atom is bonded to four different atoms (two carbon atoms and two hydrogen atoms), requiring sp^3 hybridization and leaving no unhybridized p orbital. The protonated pyrrole is nonaromatic. In fact, a sufficiently strong acid actually protonates pyrrole at the 2-position, on one of the carbon atoms of the ring, rather than on nitrogen.

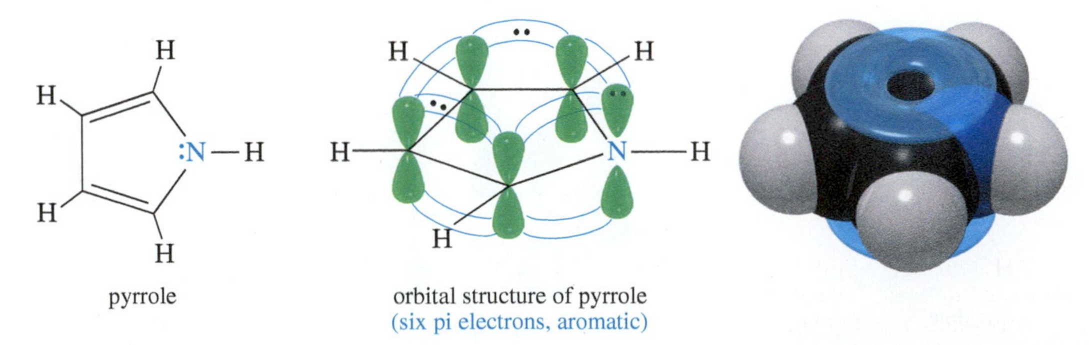

FIGURE 16-12
The pi bonding structure of pyrrole. The pyrrole nitrogen atom is sp^2 hybridized, with a lone pair of electrons in the p orbital. This p orbital overlaps with the p orbitals of the carbon atoms to form a continuous ring. Counting the four electrons of the double bonds and the two electrons in the nitrogen p orbital, there are six pi electrons.

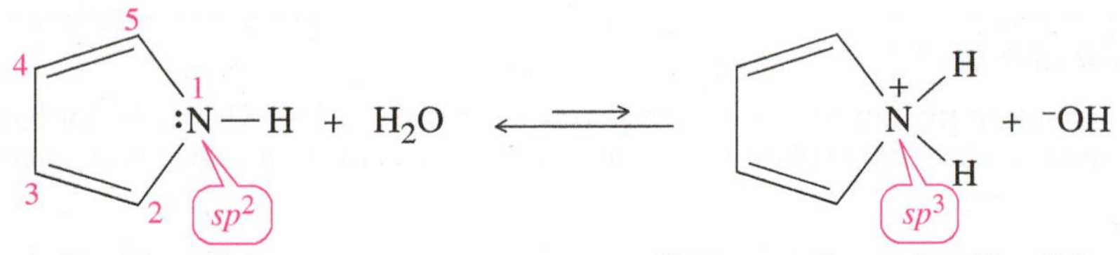

pyrrole, $pK_b = 13.6$
(weak base)

N-protonated pyrrole, $pK_a = 0.4$
(strong acid)

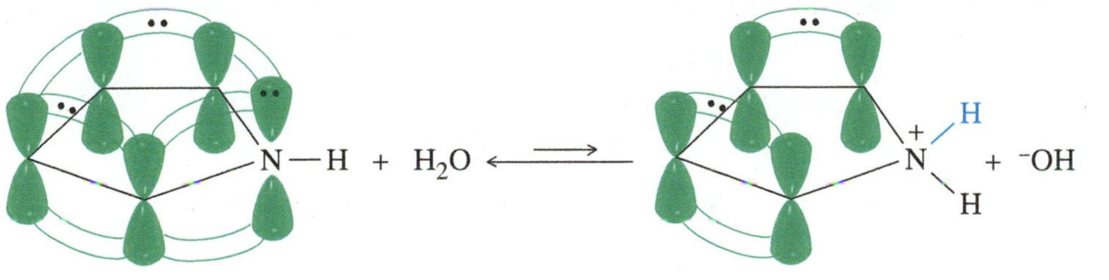

pyrrole
(aromatic)

N-protonated pyrrole
(nonaromatic)

FIGURE 16-13
Pyrrole is a very weak base. The pyrrole nitrogen atom must become sp^3 hybridized to abstract a proton. This eliminates the unhybridized *p* orbital needed for aromaticity.

16-9C Pyrimidine and Imidazole

Pyrimidine is a six-membered heterocycle with two nitrogen atoms situated in a 1,3- arrangement. Both nitrogen atoms are similar to the nitrogen in pyridine. Each has its lone pair of electrons in the sp^2 hybrid orbital in the plane of the aromatic ring. These lone pairs are not needed for the aromatic sextet, and they are basic, like the lone pair of pyridine.

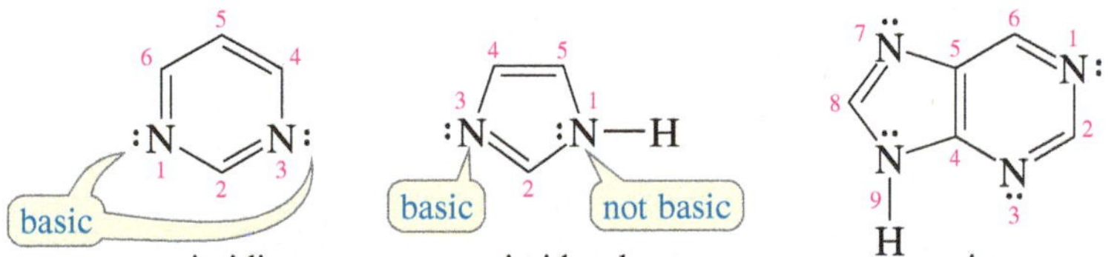

basic — pyrimidine

basic — not basic — imidazole

purine

PROBLEM-SOLVING HINT

Practice spotting basic and nonbasic nitrogen atoms. Most nonbasic (pyrrole-like) nitrogens have three single bonds, and a lone pair in a *p* orbital. Most basic (pyridine-like) nitrogens have a double bond in the ring, and their lone pair in an sp^2 hybrid orbital.

Imidazole is an aromatic five-membered heterocycle with two nitrogen atoms. The lone pair of one of the nitrogen atoms (the one not bonded to a hydrogen) is in an sp^2 orbital that is not involved in the aromatic system; this lone pair is basic. The other nitrogen uses its third sp^2 orbital to bond to hydrogen, and its lone pair is part of the aromatic sextet. As in the pyrrole nitrogen atom, this imidazole N—H nitrogen is not very basic. Once imidazole is protonated, the two nitrogens become chemically equivalent. Either nitrogen can lose a proton and return to an imidazole molecule.

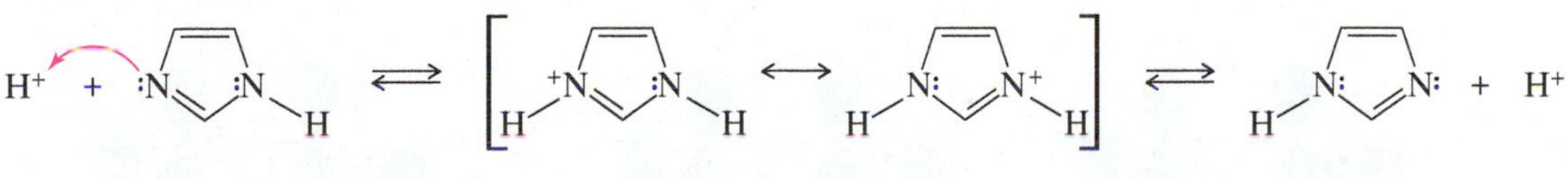

imidazole — protonated imidazole — imidazole

Purine has an imidazole ring fused to a pyrimidine ring. Purine has three basic nitrogen atoms and one pyrrole-like nitrogen.

Pyrimidine and purine derivatives serve in DNA and RNA to specify the genetic code. Imidazole derivatives enhance the catalytic activity of enzymes. We will consider these important heterocyclic derivatives in more detail in Chapters 23 and 24.

Application: Chemotherapy

Blocking DNA synthesis usually kills more cancer cells than healthy cells because cancer cells divide rapidly, which requires rapid DNA synthesis. A number of purine and pyrimidine analogs are useful as anti-cancer drugs. For example, 5-fluorouracil blocks the enzyme that produces thymidine, a key base in DNA, and kills many cancer cells as well as some healthy cells.

5-fluorouracil

Application: Sulfur in Coal

Coal contains thiophene and other organic and inorganic sulfur-containing compounds. Burning the coal releases SO_2 into the air, which contributes to acid rain. Microorganisms have evolved to use thiophene and other sulfur compounds as food sources. Such microorganisms show promise for desulfurizing coal to produce a cleaner-burning product.

PROBLEM 16-17

Show which of the nitrogen atoms in purine are basic, and which one is not basic. For the nonbasic nitrogen, explain why its nonbonding electrons are not easily available to become protonated.

PROBLEM 16-18

The proton NMR spectrum of 2-pyridone gives the chemical shifts shown.

2-pyridone (H 7.26, H 6.57, 6.15 H, 7.31 H) thymine

(a) Is 2-pyridone aromatic?

(b) Use resonance forms to explain your answer to (a). Also explain why the protons at δ 7.31 and δ 7.26 are more deshielded than the other two (δ 6.15 and δ 6.57).

(c) Thymine is one of the heterocyclic bases found in DNA. Do you expect thymine to be aromatic? Explain.

(d) The structure of 5-fluorouracil is shown in the box at the side of the page. Is 5-fluorouracil aromatic? Explain.

16-9D Furan and Thiophene

Furan is an aromatic five-membered heterocycle similar to pyrrole, but in furan the heteroatom is oxygen instead of nitrogen. The classical structure for furan (Figure 16-14) shows that the oxygen atom has two lone pairs of electrons. The oxygen atom is sp^2 hybridized, and one of the lone pairs occupies an sp^2 hybrid orbital. The other lone pair occupies the unhybridized p orbital, combining with the four electrons in the double bonds to give an aromatic sextet. Furan has a resonance energy of 67 kJ/mol (16 kcal/mol).

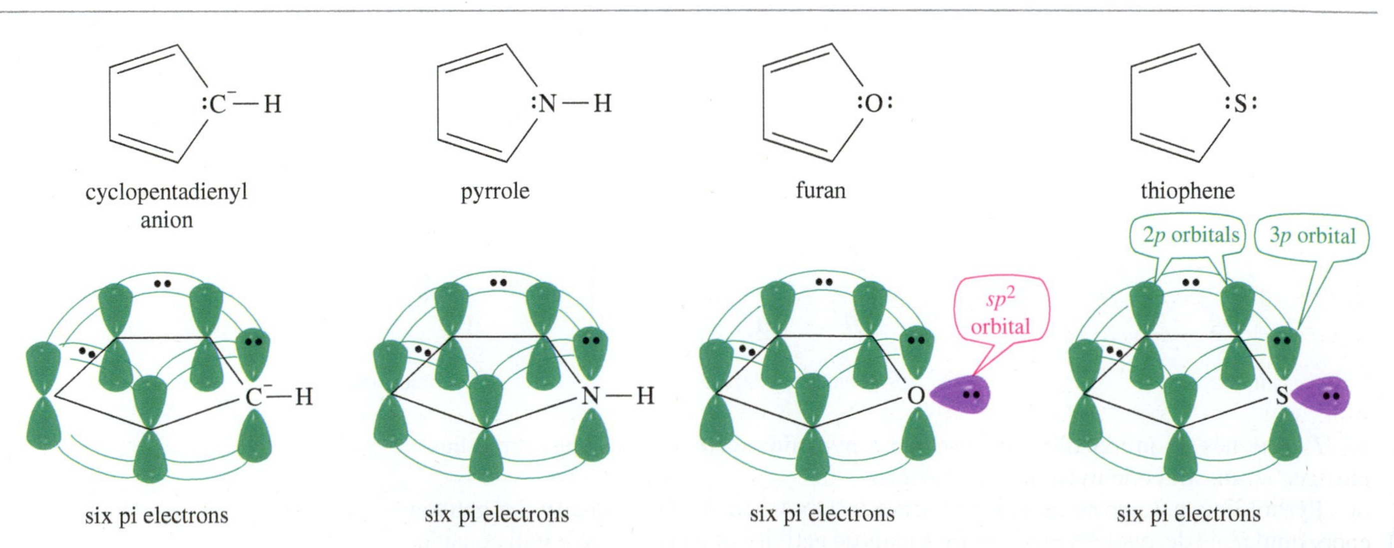

FIGURE 16-14
Pyrrole, furan, and thiophene are isoelectronic with the cyclopentadienyl anion. In furan and thiophene, the pyrrole N—H bond is replaced by a nonbonding pair of electrons in the sp^2 hybrid orbital.

Thiophene is similar to furan, with a sulfur atom in place of the furan oxygen. The bonding in thiophene is similar to that in furan, except that the sulfur atom uses an unhybridized 3*p* orbital to overlap with the 2*p* orbitals on the carbon atoms. The resonance energy of thiophene is 121 kJ/mol (29 kcal/mol).

PROBLEM 16-19

Explain why each compound is aromatic, antiaromatic, or nonaromatic.

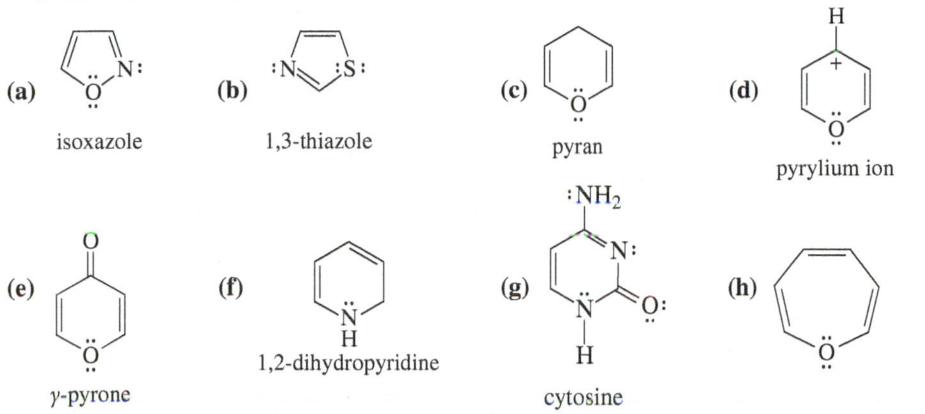

(a) isoxazole
(b) 1,3-thiazole
(c) pyran
(d) pyrylium ion
(e) γ-pyrone
(f) 1,2-dihydropyridine
(g) cytosine
(h)

PROBLEM-SOLVING HINT

A heteroatom with lone pairs (such as N, O, P, or S) can contribute at most one pair of electrons toward an aromatic system. If using one lone pair from the heteroatom gives an aromatic system, then it will usually share the electron pair. Pyrrole and furan are examples. But if using a lone pair from the heteroatom would give an antiaromatic system, then the lone pairs will not overlap with the pi system. The nitrogen in pyridine and the oxygen in Problem 16-19(h) are examples.

PROBLEM 16-20

Borazole, $B_3N_3H_6$, is an unusually stable cyclic compound. Propose a structure for borazole, and explain why it is aromatic.

16-10 Polynuclear Aromatic Hydrocarbons

The **polynuclear aromatic hydrocarbons** (abbreviated PAHs or PNAs) are composed of two or more fused benzene rings. **Fused rings** share two carbon atoms and the bond between them.

Naphthalene Naphthalene ($C_{10}H_8$) is the simplest fused aromatic compound, consisting of two fused benzene rings. We represent naphthalene by using one of the three Kekulé resonance structures or using the circle notation for the aromatic rings.

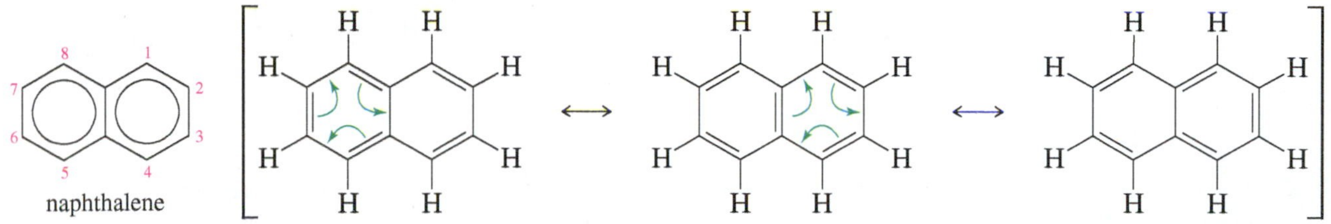

naphthalene

The two aromatic rings in naphthalene contain a total of 10 pi electrons. Two isolated aromatic rings would contain 6 pi electrons in each aromatic system, for a total of 12. The smaller amount of electron density gives naphthalene less than twice the resonance energy of benzene: 252 kJ/mol (60 kcal/mol), or 126 kJ (30 kcal), per aromatic ring, compared with benzene's resonance energy of 151 kJ/mol (36 kcal/mol).

Anthracene and Phenanthrene As the number of fused aromatic rings increases, the resonance energy per ring continues to decrease and the compounds become more reactive. Tricyclic anthracene has a resonance energy of 351 kJ/mol (84 kcal/mol), or 117 kJ (28 kcal), per aromatic ring. Phenanthrene has a slightly higher resonance energy of 381 kJ/mol (91 kcal/mol), or about 127 kJ (30 kcal), per aromatic ring. Each of these compounds has only 14 pi electrons in its three aromatic rings, compared with 18 electrons for three separate benzene rings.

anthracene phenanthrene

(Only one Kekulé structure is shown for each compound.)

Because they are not as strongly stabilized as benzene, anthracene and phenanthrene can undergo addition reactions that are more characteristic of their nonaromatic polyene relatives. Anthracene undergoes 1,4-addition at the 9- and 10-positions to give a product with two isolated, fully aromatic benzene rings. Similarly, phenanthrene undergoes 1,2-addition at the 9- and 10-positions to give a product with two fully aromatic rings. (Because they are less likely to be substituted, the bridgehead carbon atoms of fused aromatics are often left unnumbered.)

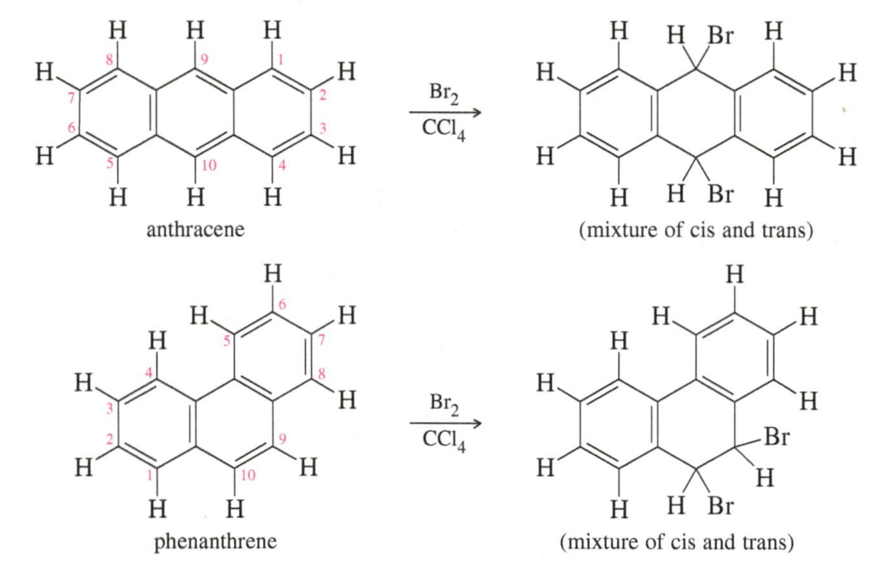

PROBLEM 16-21

(a) Draw all the Kekulé structures of anthracene and phenanthrene.
(b) Propose mechanisms for the two additions shown.
(c) In Chapter 8, most of the additions of bromine to double bonds gave entirely *anti* stereochemistry. Explain why the addition to phenanthrene gives a mixture of syn and anti stereochemistry.
(d) When the product from (c) is heated, HBr is evolved and 9-bromophenanthrene results. Propose a mechanism for this dehydrohalogenation.

The black material in diesel exhaust consists of small particles that are rich in polynuclear aromatic hydrocarbons.

Larger Polynuclear Aromatic Hydrocarbons Larger polynuclear aromatic hydrocarbons have more fused rings than anthracene and phenanthrene, and they have less resonance energy per ring and are more reactive. In drawing most of these large PAHs, we must select which Kekulé structures to use to make their rings appear aromatic. There is a high level of interest in the larger PAHs because they are formed in most combustion processes and many of them are carcinogenic (capable of causing cancer). The following three compounds, for example, are present in tobacco smoke. These compounds are so hazardous that laboratories must install special containment facilities to work with them, yet smokers expose their lung tissues to them every time they smoke a cigarette.

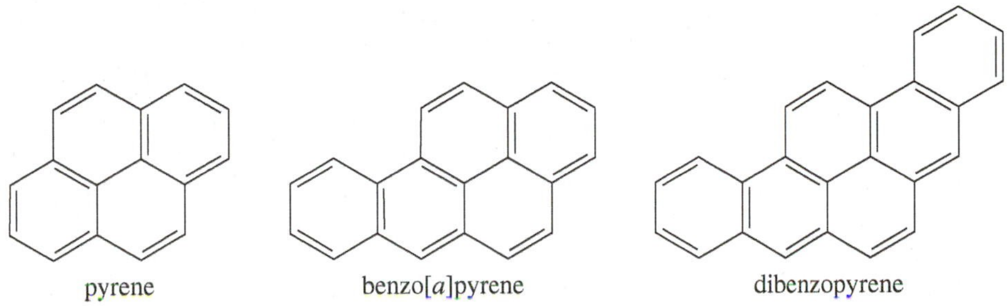

pyrene benzo[*a*]pyrene dibenzopyrene

Benzo[*a*]pyrene, one of the most thoroughly studied carcinogens, is formed whenever organic compounds undergo incomplete combustion. For example, benzo[*a*]pyrene is found in chimney soot, in broiled steaks, and in cigarette smoke. Long before our ancestors learned to use fire, they were exposed to benzo[*a*]pyrene in the smoke and ash from forest fires. Its carcinogenic effects appear to result from its epoxidation to arene oxides, which can be attacked by nucleophilic sites of DNA. The resulting DNA derivatives cannot be properly transcribed. On replication, they cause errors that produce mutations in the genes.

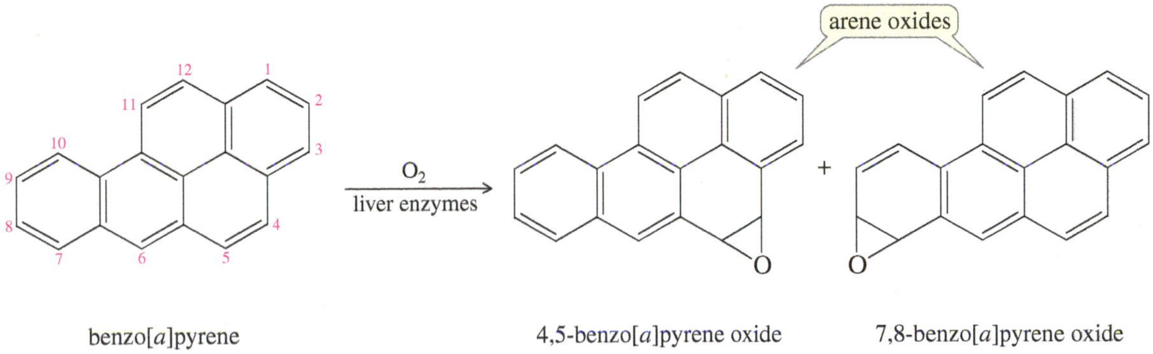

arene oxides

O_2, liver enzymes

benzo[*a*]pyrene 4,5-benzo[*a*]pyrene oxide 7,8-benzo[*a*]pyrene oxide

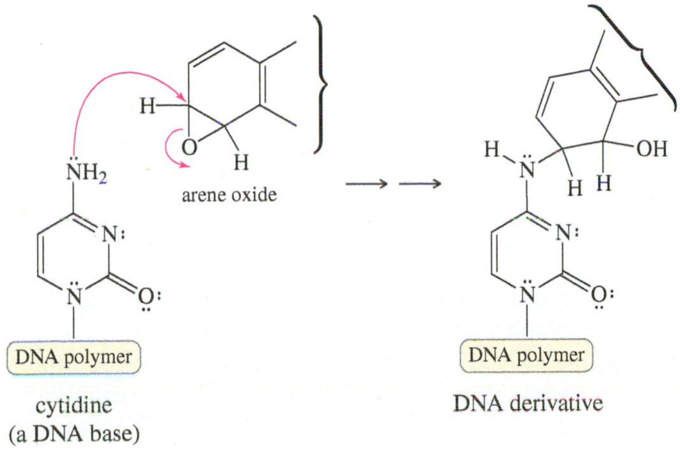

arene oxide

DNA polymer

cytidine (a DNA base)

DNA derivative

Application: Carcinogenesis

Benzo[a]pyrene in soot was the culprit for a large number of skin cancers in young boys who cleaned chimneys in the 1700s. The body transforms this compound to 4,5-benzo[a]pyrene oxide, a reactive epoxide that forms a covalent bond with DNA.

16-11 Aromatic Allotropes of Carbon

What do you get when you make an extremely large polynuclear aromatic hydrocarbon, with millions or billions of benzene rings joined together? You get graphite, one of the oldest-known forms of pure elemental carbon. Let's consider how aromaticity plays a role in the stability of both the old and the new forms of carbon.

16-11A Allotropes of Carbon: Diamond

We don't normally think of elemental carbon as an organic compound. Historically, carbon was known to exist as three **allotropes** (elemental forms with different properties): amorphous carbon, diamond, and graphite.

"Amorphous carbon" refers to charcoal, soot, coal, and carbon black. These materials are mostly microcrystalline forms of graphite. They are characterized by small particle sizes and large surface areas with partially saturated valences. These small particles readily absorb gases and solutes from solution, and they form strong, stable dispersions in polymers, such as the dispersion of carbon black in tires.

Diamond is the hardest naturally occurring substance known. Diamond has a crystalline structure containing tetrahedral carbon atoms linked together in a three-dimensional lattice (Figure 16-15). This lattice extends throughout each crystal, so that a diamond is actually one giant molecule. Diamond is an electrical insulator because the electrons are all tightly bound in sigma bonds (length 1.54 Å, typical of C—C single bonds), and they are unavailable to carry a current.

FIGURE 16-15
Structures of diamond, graphite, and graphene. Diamond is a lattice of tetrahedral carbon atoms linked in a rigid three-dimensional array. Graphite consists of planar layers of fused aromatic rings. Graphene is a single layer of graphite one atom thick.

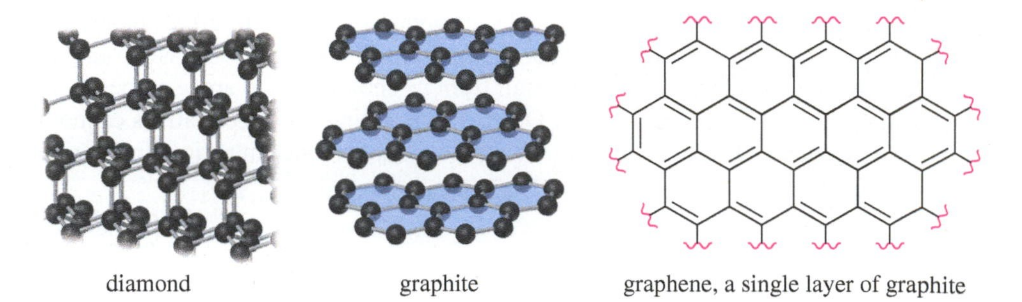

16-11B Graphite

Graphite has the layered planar structure shown in Figure 16-15. Within a layer, the C—C bond lengths are all 1.415 Å which is fairly close to the C—C bond length in benzene (1.397 Å). Between layers, the distance is 3.35 Å which is about twice the van der Waals radius for carbon, suggesting there is little or no bonding between layers. The layers can easily cleave and slide across each other, making graphite a good lubricant. This layered structure also helps to explain graphite's unusual electrical properties: It is a good electrical conductor parallel to the layers, but it resists electrical currents perpendicular to the layers.

We picture each layer of graphite as a nearly infinite lattice of fused aromatic rings. All the valences are satisfied (except at the edges), and no bonds are needed between layers. Only van der Waals forces hold the layers together, consistent with their ability to slide easily over one another. The pi electrons within a layer can conduct electrical currents parallel to the layer, but electrons cannot easily jump between layers, so graphite is resistive perpendicular to the layers.

Because of its aromaticity, graphite is slightly more stable than diamond, and the transition from diamond to graphite is slightly exothermic ($\Delta H° = -2.9$ kJ/mol, or -0.7 kcal/mol). Fortunately for those who have invested in diamonds, the favorable conversion of diamond to graphite is exceedingly slow. Diamond (3.51 g/cm^3) has a higher density than graphite (2.25 g/cm^3), implying that graphite might be converted to diamond under very high pressures. Indeed, small industrial diamonds can be synthesized by subjecting graphite to pressures over 125,000 atm and temperatures around 3000 °C, using catalysts such as Cr and Fe.

Andre Geim and Konstantin Novoselov (University of Manchester) received the 2010 Nobel Prize in Physics for producing and characterizing graphene, which is a single layer of graphite one atom thick. They used adhesive tape to pull one layer away from the surface of a piece of graphite. Single-layer graphene is transparent, strong, and an excellent electrical conductor. It has been used to make transistors, and it holds great promise for touch-screen monitors if it can ever be mass-produced in large sheets.

16-11C Fullerenes

Around 1985, Kroto, Smalley, and Curl (Rice University) isolated a molecule of formula C_{60} from the soot produced by using a laser (or an electric arc) to vaporize graphite. Molecular spectra showed that C_{60} is unusually symmetrical: It has only one type of carbon atom by ^{13}C NMR (δ 143 ppm), and there are only two types of bonds (1.39 Å and 1.45 Å). Figure 16-16 shows the structure of C_{60}, which was named **buckminsterfullerene** in honor of the American architect R. Buckminster Fuller, whose geodesic domes used similar five- and six-membered rings to form a curved roof. The C_{60} molecules are sometimes called "buckyballs," and these types of compounds (C_{60} and similar carbon clusters) are called **fullerenes.**

A soccer ball has the same structure as C_{60}, with each vertex representing a carbon atom. All the carbon atoms are chemically the same. Each carbon serves as a bridgehead for two six-membered rings and one five-membered ring. There are only two types of bonds: the bonds that are shared by a five-membered ring and a six-membered ring (1.45 Å) and the bonds shared between two six-membered rings (1.39 Å). Compare these bond lengths with a typical double bond (1.33 Å), a typical aromatic bond (1.40 Å), and a typical single bond (1.48 Å between sp^2 carbons). It appears that the six-membered rings are aromatic, but the double bonds are partially localized between the six-membered rings, (Figure 16-16). These double bonds are less reactive than typical alkene double bonds, yet they do undergo some of the addition reactions of alkenes.

Nanotubes (Figure 16-16) were discovered around 1991. These structures begin with half of the C_{60} sphere, fused to a cylinder composed entirely of fused six-membered rings (as in a layer of graphite). Nanotubes have aroused interest because they are electrically conductive only along the length of the tube and they have an enormous strength-to-weight ratio. Thousands of tons of nanotubes are produced commercially each year. They are often added to polymer resins to make the cured polymers stronger, more heat resistant, or electrically conductive. They are also used as scaffolding to promote bone growth in tissue cultures, and as tips for atomic force microscope probes.

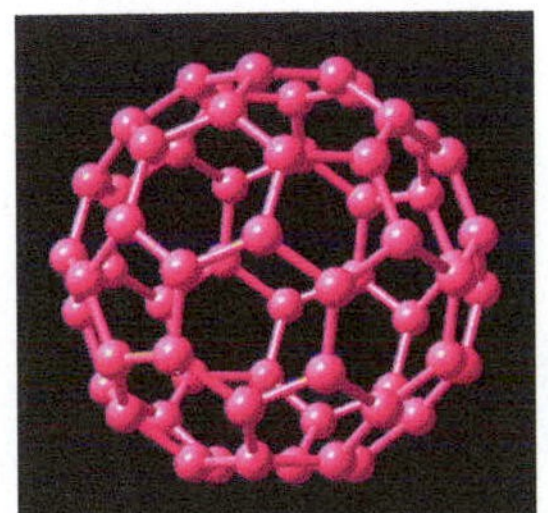

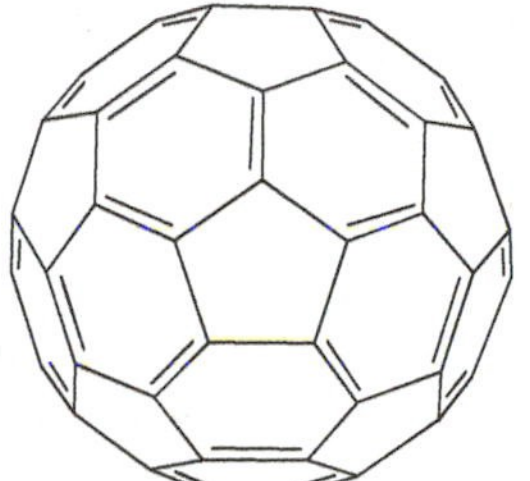

buckyball (C_{60})

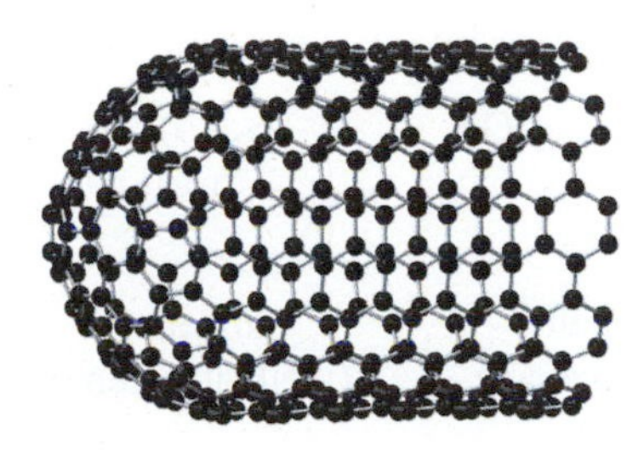

carbon nanotube

FIGURE 16-16
Structure of C_{60} and a carbon nanotube. Each carbon in C_{60} is a bridgehead carbon for a five-membered ring and two six-membered rings. A nanotube is a cylinder composed of aromatic six-membered rings similar to graphite. The end of the tube is half of a C_{60} sphere. Notice how the five-membered rings cause the structure to curve at the end of the tube.

16-12 Fused Heterocyclic Compounds

Purine is one of many fused heterocyclic compounds whose rings share two atoms and the bond between them. For example, the following compounds all contain fused heterocyclic aromatic rings:

purine

indole

benzimidazole

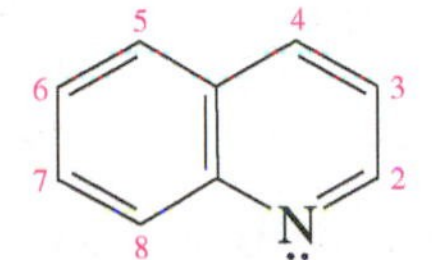

quinoline

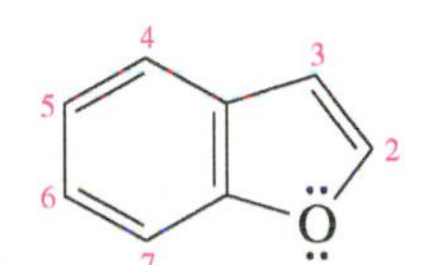

benzofuran

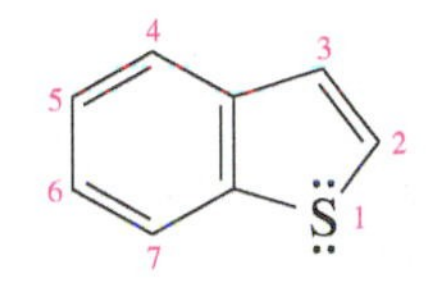

benzothiophene

L-tryptophan, an amino acid

benziodarone, a vasodilator

LSD, a hallucinogen

quinine, an antimalarial drug

FIGURE 16-17
Examples of biologically active fused heterocycles.

The properties of fused-ring heterocycles are generally similar to those of the simple heterocycles. Fused heterocyclic compounds are common in nature, and they are also used as drugs to treat a wide variety of illnesses. Figure 16-17 shows some fused heterocycles that occur naturally or are synthesized for use as drugs.

ciprofloxacin

PROBLEM 16-22

Ciprofloxacin is a member of the fluoroquinolone class of antibiotics.
(a) Which of its rings are aromatic?
(b) Which nitrogen atoms are basic?
(c) Which protons would you expect to appear between $\delta 6$ and $\delta 8$ in the proton NMR spectrum?

16-13 Nomenclature of Benzene Derivatives

Benzene derivatives have been isolated and used as industrial reagents for well over 100 years. Many of their names are rooted in the historical traditions of chemistry. The following compounds are usually called by their historical common names, and almost never by the systematic IUPAC names:

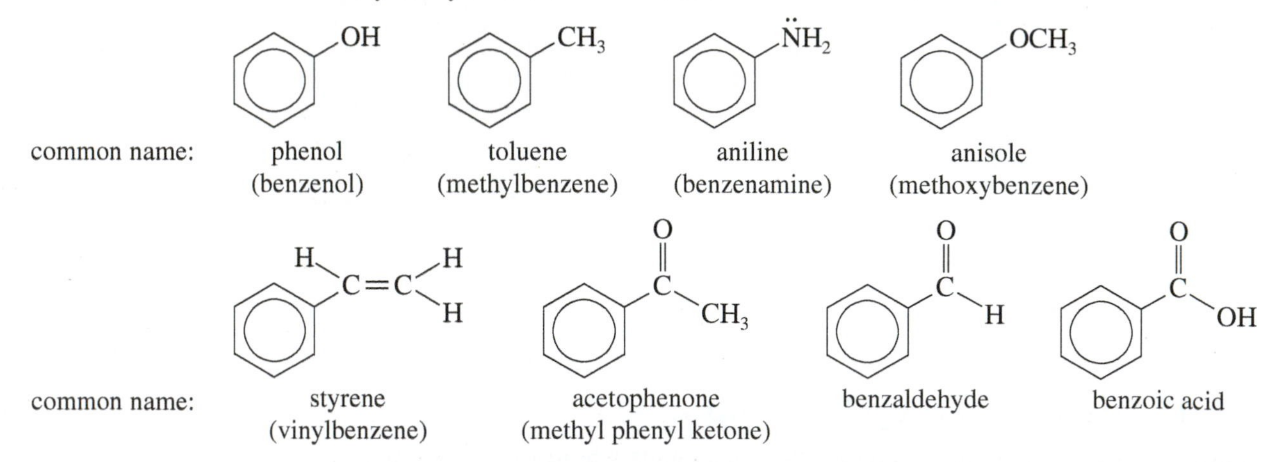

Many compounds are named as derivatives of benzene, with their substituents named just as though they were attached to an alkane.

tert-butylbenzene

nitrobenzene

ethynylbenzene
(phenylacetylene)

benzenesulfonic acid

Disubstituted benzenes are named using the prefixes **ortho-**, **meta-**, and **para-** to specify the substitution patterns. These terms are abbreviated *o*-, *m*-, and *p*-. Numbers can also be used to specify the substitution in disubstituted benzenes.

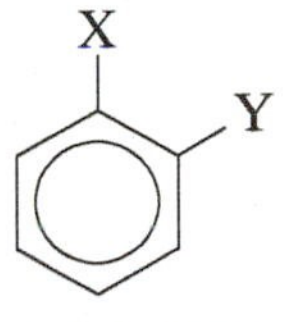

1,2 or ortho 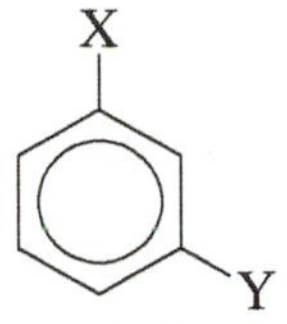   1,3 or meta 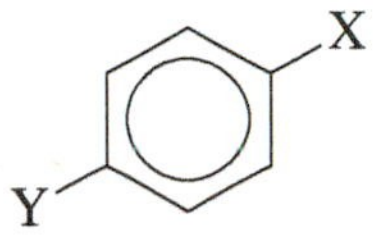   1,4 or para

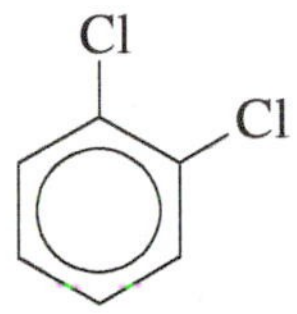

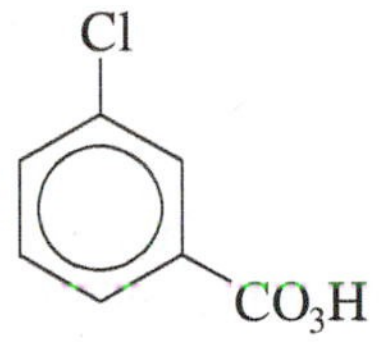

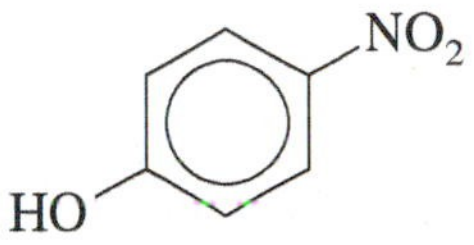

common name:	*o*-dichlorobenzene	*m*-chloroperoxybenzoic acid	*p*-nitrophenol
IUPAC name:	1,2-dichlorobenzene	3-chloroperoxybenzoic acid	4-nitrophenol

With three or more substituents on the benzene ring, numbers are used to indicate their positions. Assign the numbers as you would with a substituted cyclohexane, to give the lowest possible numbers to the substituents. The carbon atom bearing the functional group that defines the base name (as in phenol or benzoic acid) is assumed to be C1.

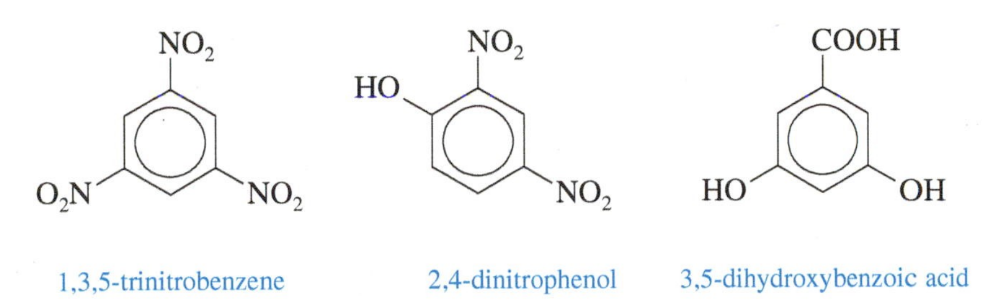

1,3,5-trinitrobenzene   2,4-dinitrophenol   3,5-dihydroxybenzoic acid

PROBLEM-SOLVING HINT

If the substitution pattern is unknown or unimportant, a structure might be drawn with ambiguous positioning. For example, the following structure might imply *ortho-*, *meta-*, or *para*-nitrophenol, or possibly a mixture of isomers.

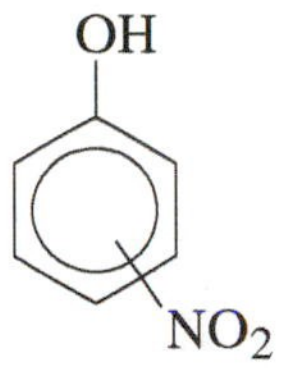

OH

NO$_2$

Many disubstituted benzenes (and polysubstituted benzenes) have historical names. Some of these are obscure, with no obvious connection to the structure of the molecule.

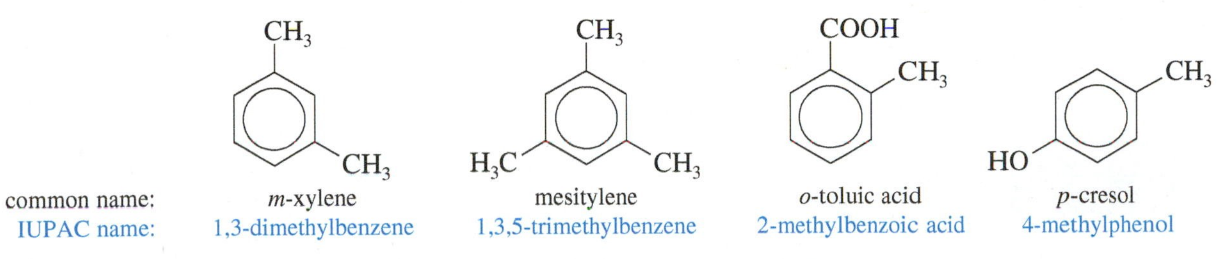

common name:	*m*-xylene	mesitylene	*o*-toluic acid	*p*-cresol
IUPAC name:	1,3-dimethylbenzene	1,3,5-trimethylbenzene	2-methylbenzoic acid	4-methylphenol

When the benzene ring is named as a substituent on another molecule, it is called a **phenyl group.** The phenyl group is used in the name just like the name of an alkyl group, and it is often abbreviated **Ph** (or ϕ) in drawing a complex structure.

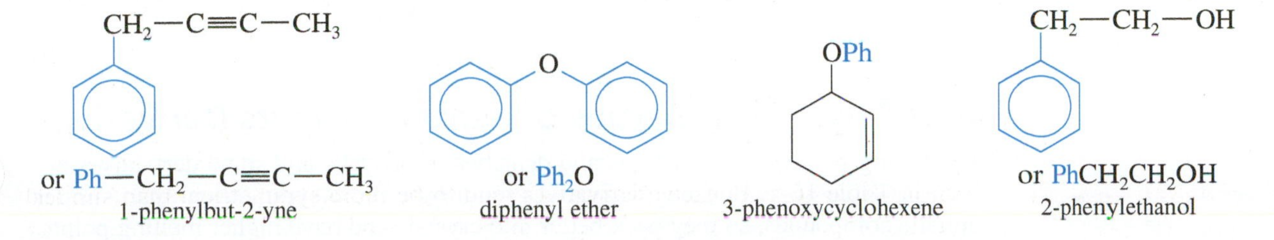

or Ph—CH$_2$—C≡C—CH$_3$
1-phenylbut-2-yne

or Ph$_2$O
diphenyl ether

3-phenoxycyclohexene

or PhCH$_2$CH$_2$OH
2-phenylethanol

The seven-carbon unit consisting of a benzene ring and a methylene ($—CH_2—$) group is often named as a **benzyl group.** Be careful not to confuse the *benzyl group* (seven carbons) with the *phenyl group* (six carbons).

PROBLEM-SOLVING HINT

A benzene ring substituent is a *phenyl group* (6 carbons).
A *benzyl group* contains an additional CH_2 group (7 carbons total).

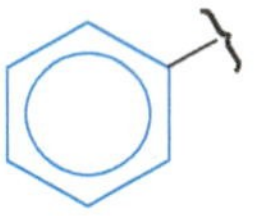

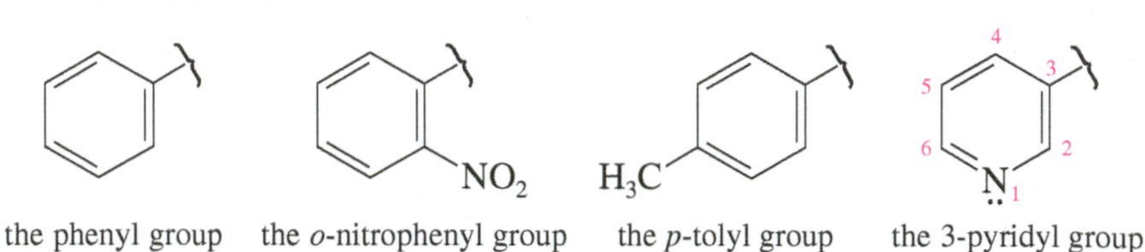

Aromatic hydrocarbons are sometimes called **arenes.** An **aryl group,** abbreviated **Ar,** is the aromatic group that remains after the removal of a hydrogen atom from an aromatic ring. The phenyl group, **Ph,** is the simplest aryl group. The generic aryl group **(Ar)** is the aromatic relative of the generic alkyl group, which we symbolize by **R.**

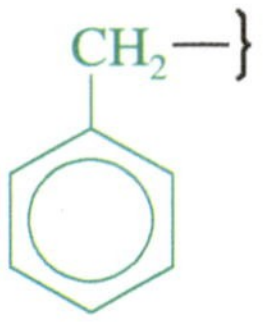

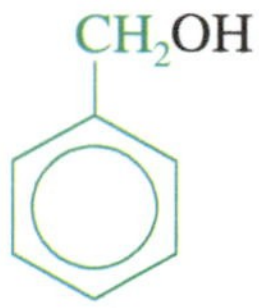

Examples of aryl groups

the phenyl group — the *o*-nitrophenyl group (NO_2) — the *p*-tolyl group (H_3C) — the 3-pyridyl group (N 1, 2, 3, 4, 5, 6)

Examples of the use of a generic aryl group

Ar—MgBr	Ar_2O or Ar—O—Ar′	Ar—NH_2	Ar—SO_3H
an arylmagnesium bromide	a diaryl ether	an arylamine	an arylsulfonic acid

Application: Capsaicin

Capsaicin is the main flavor component of chili peppers (genus *Capsicum*). It is a strong irritant to mammals, but not to birds, who eat, digest, and disperse the Capsicum seeds unharmed. Capsaicin produces a piquant burning sensation in the mouth ("heat") that many people find pleasantly spicy. Capsaicin is also used medicinally in ointments and dermal patches to relieve pain in muscles and joints.

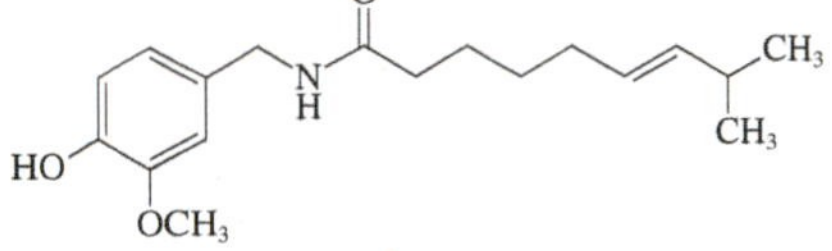

HO, OCH_3, N H, O, CH_3, CH_3

capsaicin

PROBLEM 16-23

Draw and name all the chlorinated benzenes having from one to six chlorine atoms.

PROBLEM 16-24

Name the following compounds:

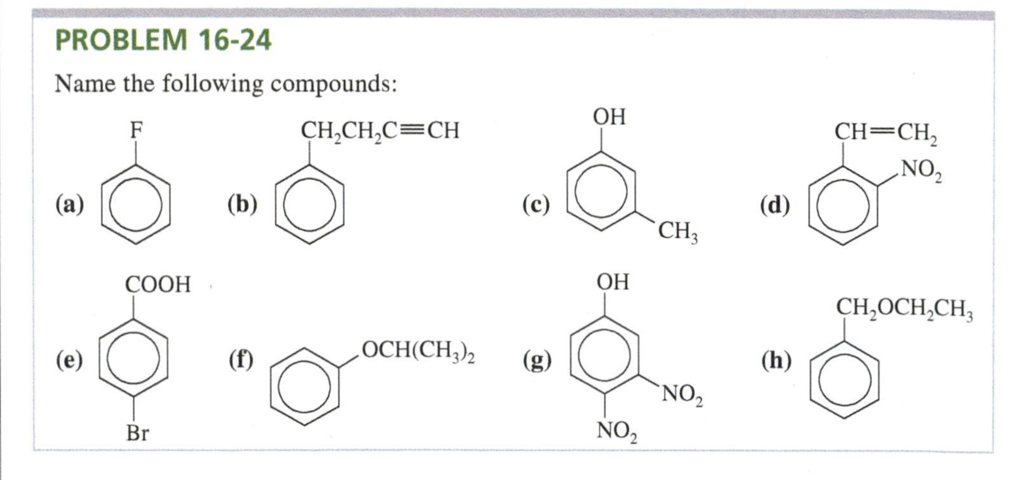

16-14 Physical Properties of Benzene and Its Derivatives

The melting points, boiling points, and densities of benzene and some derivatives are given in Table 16-1. Benzene derivatives tend to be more symmetrical than similar aliphatic compounds, so they pack better into crystals and have higher melting points.

TABLE 16-1
Physical Properties of Benzene Derivatives

Compound	mp (°C)	bp (°C)	Density (g/mL)
benzene	6	80	0.88
toluene	−95	111	0.87
ethylbenzene	−95	136	0.87
styrene	−31	146	0.91
ethynylbenzene	−45	142	0.93
fluorobenzene	−41	85	1.02
chlorobenzene	−46	132	1.11
bromobenzene	−31	156	1.49
iodobenzene	−31	188	1.83
nitrobenzene	6	211	1.20
phenol	43	182	1.07
anisole	−37	156	0.98
benzoic acid	122	249	1.31
benzyl alcohol	−15	205	1.04
aniline	−6	186	1.02
o-xylene	−26	144	0.88
m-xylene	−48	139	0.86
p-xylene	13	138	0.86
o-dichlorobenzene	−17	181	1.31
m-dichlorobenzene	−25	173	1.29
p-dichlorobenzene	54	170	1.46

Application: Anesthetic

OH

propofol

Propofol (2,6-diisopropylphenol) is a safe and effective intravenous anesthetic when used in an operating room under careful supervision. Pop star Michael Jackson died from the administration of propofol in combination with other drugs in a home setting to treat insomnia.

For example, benzene melts at 6 °C, whereas hexane melts at −95 °C. Similarly, para-disubstituted benzenes are more symmetrical than the ortho and meta isomers, and they pack better into crystals and have higher melting points.

The relative boiling points of many benzene derivatives are related to their dipole moments. For example, the dichlorobenzenes have boiling points that follow their dipole moments. Symmetrical *p*-dichlorobenzene has zero dipole moment and the lowest boiling point. *m*-Dichlorobenzene has a small dipole moment and a slightly higher boiling point. *o*-Dichlorobenzene has the largest dipole moment and the highest boiling point. Even though *p*-dichlorobenzene has the lowest boiling point, it has the highest melting point of the dichlorobenzenes because it packs best into a crystal.

Application: Pesticide

Mothballs are composed of either *p*-dichlorobenzene or naphthalene. These two compounds are used because they are crystalline solids with high vapor pressures. Vapors permeate the clothes in a sealed closet or bag, but the odor dissipates when the clothes are aired out.

o-dichlorobenzene	*m*-dichlorobenzene	*p*-dichlorobenzene
bp 181 °C	bp 173 °C	bp 170 °C
mp −17 °C	mp −25 °C	mp 54 °C

Benzene and other aromatic hydrocarbons are slightly denser than the nonaromatic analogs, but they are still less dense than water. The halogenated benzenes are denser than water. Aromatic hydrocarbons and halogenated aromatics are generally insoluble in water, although some derivatives with strongly polar functional groups (phenol, benzoic acid, etc.) are moderately soluble in water.

16-15 Spectroscopy of Aromatic Compounds

Infrared Spectroscopy (Review) Aromatic compounds are readily identified by their infrared spectra because they show a characteristic C=C stretch around 1600 cm^{-1}. This is a lower C=C stretching frequency than for isolated alkenes (1640 to 1680 cm^{-1}) or conjugated dienes (1620 to 1640 cm^{-1}) because the aromatic bond order is only about $1\frac{1}{2}$. The aromatic bond is therefore less stiff than a normal double bond, and it vibrates at a lower frequency.

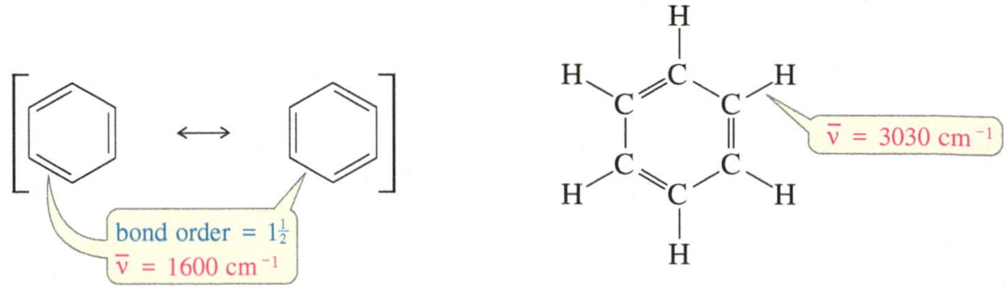

bond order = $1\frac{1}{2}$
$\bar{\nu}$ = 1600 cm^{-1}

$\bar{\nu}$ = 3030 cm^{-1}

Similarly to alkenes, aromatic compounds show unsaturated =C—H stretching just above 3000 cm^{-1} (usually around 3030 cm^{-1}). The combination of the aromatic C=C stretch around 1600 cm^{-1} and the =C—H stretch just above 3000 cm^{-1} leaves little doubt of the presence of an aromatic ring. The sample spectra labeled Compounds 4, 5, and 7 in Chapter 12 (pages 581–583) show compounds containing aromatic rings.

NMR Spectroscopy (Review) Aromatic compounds give readily identifiable ^{1}H NMR signals around δ 7 to δ 8, strongly deshielded by the aromatic ring current (Section 13-5B). In benzene, the aromatic protons absorb around δ 7.2. The signals may be moved farther downfield by electron-withdrawing groups such as carbonyl, nitro, or cyano groups, or upfield by electron-donating groups such as hydroxy, alkoxy, or amino groups.

Nonequivalent aromatic protons that are ortho or meta usually split each other. The spin-spin splitting constants are about 8 Hz for ortho protons and 2 Hz for meta protons. Figures 13-11, 13-18, 13-24, 13-29, and 13-31 show proton NMR spectra of aromatic compounds.

Aromatic carbon atoms absorb around δ 120 to δ 150 in the ^{13}C NMR spectrum. Alkene carbon atoms can also absorb in this spectral region, but the combination of ^{13}C NMR with ^{1}H NMR or IR spectroscopy usually leaves no doubt about the presence of an aromatic ring.

Mass Spectrometry The most common mass spectral fragmentation of alkylbenzene derivatives is the cleavage of a benzylic bond to give a resonance-stabilized benzylic cation. For example, in the mass spectrum of *n*-butylbenzene (Figure 16-18), the base peak is at m/z 91, from the benzyl cation. The benzyl cation may rearrange to give the aromatic tropylium ion. Alkylbenzenes frequently give ions corresponding to the tropylium ion at m/z 91.

Ultraviolet Spectroscopy The ultraviolet spectra of aromatic compounds are quite different from those of nonaromatic polyenes. For example, benzene has three absorptions in the ultraviolet region: an intense band at λ_{max} = 184 nm (ϵ = 68,000), a moderate band at λ_{max} = 204 nm (ϵ = 8800), and a characteristic low-intensity band of multiple absorptions centered around 254 nm (ϵ = 200 to 300). In the UV spectrum of benzene in Figure 16-19, the absorption at 184 nm does not appear because wavelengths shorter than 200 nm are not accessible by standard UV–visible spectrometers.

All three major bands in the benzene spectrum correspond to $\pi \rightarrow \pi^*$ transitions. The absorption at 184 nm corresponds to the energy of the transition from one of the two HOMOs to one of the two LUMOs. The weaker band at 204 nm corresponds to a "forbidden" transition that would be impossible to observe if benzene were always an unperturbed, perfectly hexagonal structure.

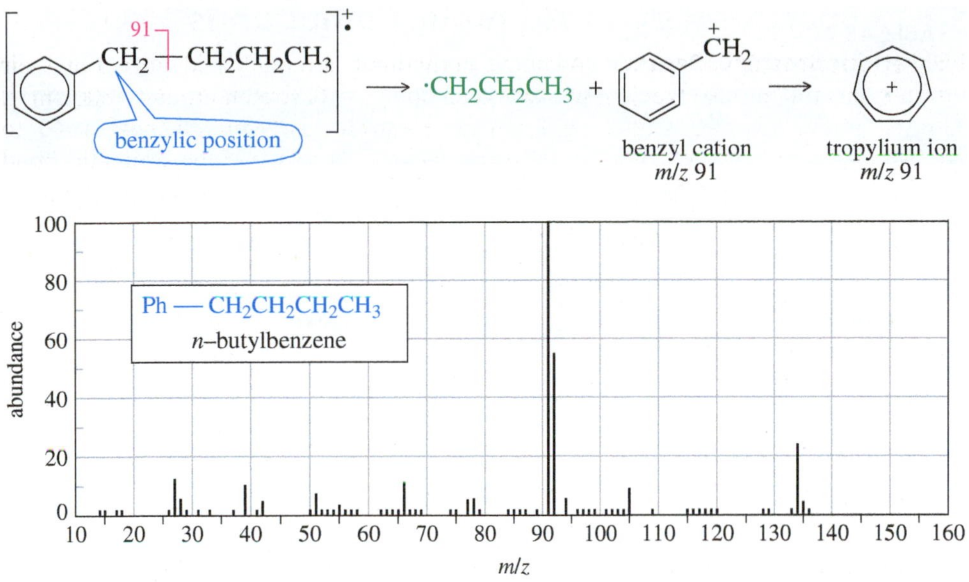

FIGURE 16-18
The mass spectrum of *n*-butylbenzene has its base peak at m/z 91, corresponding to cleavage of a benzylic bond. The fragments are a benzyl cation and a propyl radical. The benzyl cation rearranges to the tropylium ion, detected at m/z 91.

The most characteristic part of the spectrum is the band centered at 254 nm, called the **benzenoid band.** About three to six small, sharp peaks (called *fine structure*) usually appear in this band. Their molar absorptivities are weak, usually 200 to 300. These benzenoid absorptions correspond to additional forbidden transitions.

Simple benzene derivatives show most of the characteristics of benzene, including the moderate band in the 210-nm region and the benzenoid band in the 260-nm region. Alkyl and halogen substituents increase the values of λ_{max} by about 5 nm, as shown by the examples in Table 16-2. An additional conjugated double bond can increase the value of λ_{max} by about 30 nm, as shown by the UV spectrum of styrene in Figure 16-19.

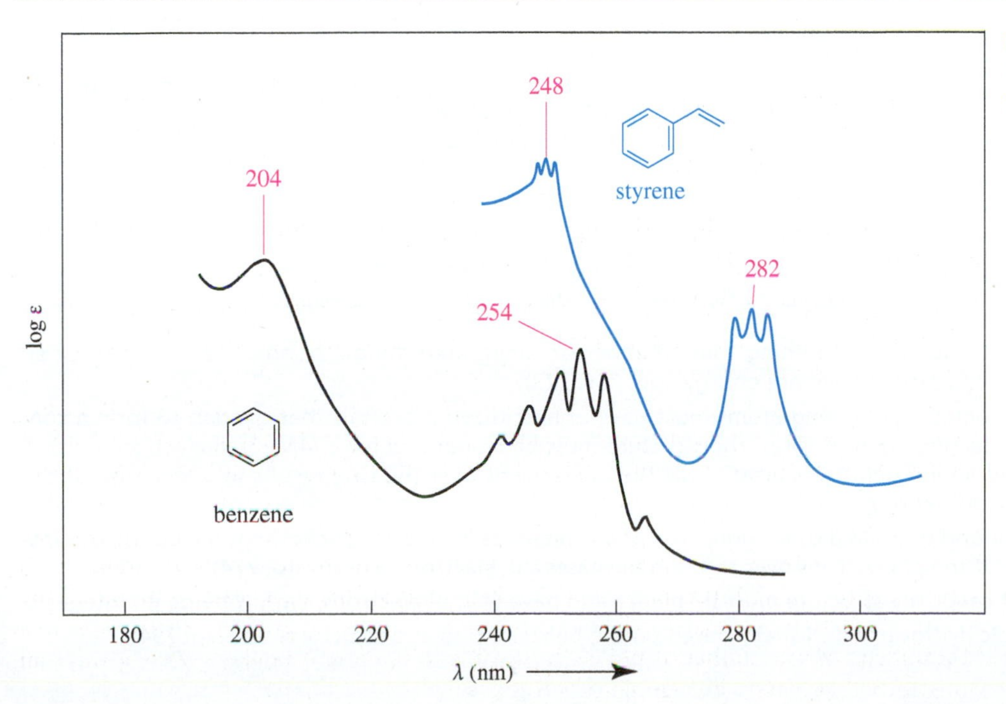

FIGURE 16-19
Ultraviolet spectra of benzene and styrene.

TABLE 16-2
Ultraviolet Spectra of Benzene and Some Derivatives

Compound	Structure	Moderate Band λ_{max}(nm)	ε	Benzenoid Band λ_{max}(nm)	ε
benzene		204	8,800	254	250
ethylbenzene	CH_2CH_3	208	7,800	260	220
m-xylene	CH_3, CH_3	212	7,300	264	300
bromobenzene	Br	210	7,500	258	170
styrene		248	15,000	282	740

PROBLEM 16-25

The UV spectrum of 1-phenylprop-2-en-1-ol shows an intense absorption at 220 nm and a weaker absorption at 258 nm. When this compound is treated with dilute sulfuric acid, it rearranges to an isomer with an intense absorption at 250 nm and a weaker absorption at 290 nm. Suggest a structure for the isomeric product and propose a mechanism for its formation.

Essential Terms

aliphatic compound An organic compound that is not aromatic. (p. 765)

allotropes Different forms of an element with different properties. For example, diamond, graphite, and fullerenes are different allotropic forms of elemental carbon. (p. 790)

annulenes Cyclic hydrocarbons with alternating single and double bonds. (p. 768)

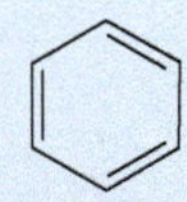
[6]annulene (benzene)

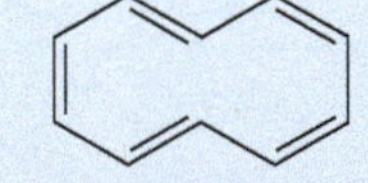
[10]annulene (cyclodecapentaene)

aromatic compound A cyclic compound containing some number of conjugated double bonds, characterized by an unusually large resonance energy. (pp. 766, 773)

To be aromatic, all its ring atoms must have unhybridized *p* orbitals that overlap to form a continuous ring. In most cases, the structure must be planar and have $(4N+2)$ pi electrons, with N being an integer. Delocalization of the pi electrons over the ring results in a lowering of the electronic energy.

antiaromatic compound A compound that has a continuous ring of *p* orbitals, as in an aromatic compound, but delocalization of the pi electrons over the ring increases the electronic energy. (p. 774)

In most cases, the structure must be planar and have $(4N)$ pi electrons, with N being an integer.

arenes Aromatic hydrocarbons, usually based on the benzene ring as a structural unit. (p. 794)

aryl group (abbreviated Ar) The aromatic group that remains after taking a hydrogen atom off an aromatic ring; the aromatic equivalent of the generic alkyl group (R). (p. 794)

benzenoid band The weak band around 250 to 270 nm in the UV spectra of benzenoid aromatics. This band is characterized by multiple sharp absorptions (fine structure). (p. 797)

benzyl group ($PhCH_2$—) The seven-carbon unit consisting of a benzene ring and a methylene group. (p. 794)

buckminsterfullerene ("buckyballs") A common name for the C_{60} molecule with the same symmetry as a soccer ball. The arrangement of five-membered and six-membered rings is similar to that in a geodesic dome. (p. 791)

degenerate orbitals Orbitals having the same energy. (p. 770)

diamond The hardest, densest, and most transparent allotrope of carbon. "A girl's best friend," according to Marilyn Monroe. (p. 790)

fullerenes A common generic term for carbon clusters similar to C_{60} (buckminsterfullerene) and compounds related to them. (p. 791)

fused rings Rings that share a common carbon–carbon bond and its two carbon atoms. (p. 787)

heterocyclic compound **(heterocycle)** A cyclic compound in which one or more of the ring atoms is not carbon. (p. 783)

  aromatic heterocycle: A heterocyclic compound that fulfills the criteria for aromaticity and has a substantial resonance energy.

Hückel's rule A cyclic molecule or ion that has a continuous ring of overlapping *p* orbitals will be

1. aromatic if the number of pi electrons is (4*N*+2), with *N* being an integer.
2. antiaromatic if the number of pi electrons is (4*N*), with *N* being an integer. (p. 774)

Kekulé structure A classical structural formula for an aromatic compound, showing localized double bonds. (p. 765)

nanotubes A common term for carbon tubes consisting of a cylinder of fused graphite-like six-membered rings and ending with half of a C_{60} sphere. (p. 791)

nonaromatic compound Neither aromatic nor antiaromatic; lacking the continuous ring of overlapping *p* orbitals required for aromaticity or antiaromaticity. (p. 774)

ortho Having a 1,2-relationship on a benzene ring. (p. 793)

meta Having a 1,3-relationship on a benzene ring. (p. 793)

para Having a 1,4-relationship on a benzene ring. (p. 793)

ortho (1,2) meta (1,3) para (1,4)

phenyl group (Ph or ϕ) The benzene ring, minus one hydrogen atom, when named as a substituent on another molecule. (p. 793)

polygon rule The energy diagram of the MOs of a regular, completely conjugated cyclic system has the same polygonal shape as the compound, with one vertex (the all-bonding MO) at the bottom. The nonbonding line cuts horizontally through the center of the polygon. (p. 773)

Energy diagrams

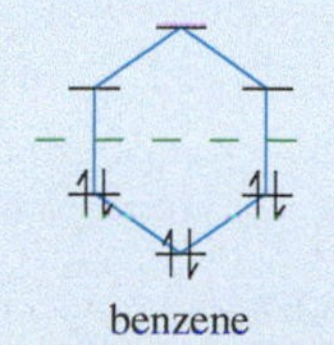

benzene

cyclobutadiene

cyclopentadienyl cation

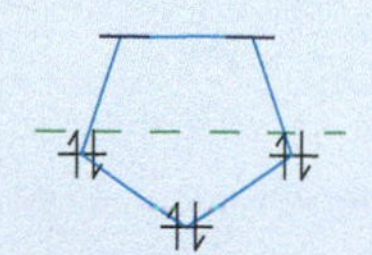

cyclopentadienyl anion

tropylium ion

polynuclear aromatic compounds Aromatic compounds with two or more fused aromatic rings. Naphthalene is a **polynuclear aromatic hydrocarbon** (PAH or PNA). Indole is a polynuclear aromatic heterocycle. (p. 787)

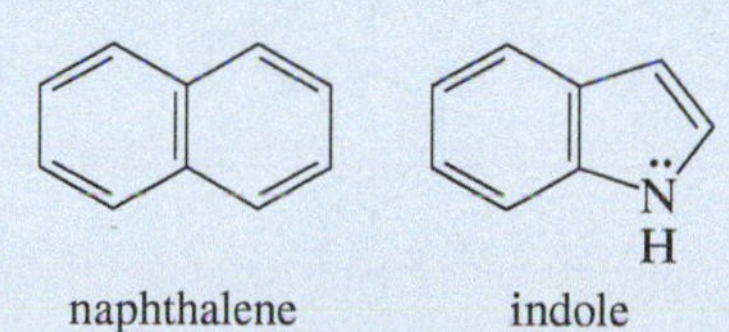

naphthalene indole

resonance energy The extra stabilization provided by delocalization, compared with a localized structure. For aromatic compounds, the resonance energy is the extra stabilization provided by the delocalization of the electrons in the aromatic ring. (p. 767)

tropylium ion The cycloheptatrienyl cation. This cation is aromatic (see the energy diagram above), and is frequently found at m/z 91 in the mass spectra of alkylbenzenes. (p. 780)

Essential Problem-Solving Skills in Chapter 16

Each skill is followed by problem numbers exemplifying that particular skill.

1	Explain how to construct the molecular orbitals of a conjugated cyclic system similar to benzene and cyclobutadiene. Use the polygon rule to draw the energy diagram, and fill in the electrons to show whether a given compound or ion is aromatic or antiaromatic.	Problems 16-11 and 16-48
2	Use Hückel's rule to predict whether a given annulene, heterocycle, or ion will be aromatic, antiaromatic, or nonaromatic.	Problems 16-29, 32, 33, 35, 40, 43, 44, 50, and 51
3	For heterocycles containing nitrogen atoms, determine whether nitrogen's lone pairs are used in the aromatic system, and predict whether the nitrogen atom is strongly or weakly basic.	Problems 16-34, 35, 43, and 44
4	Recognize fused aromatic systems such as polynuclear aromatic hydrocarbons and fused heterocyclic compounds, and use the theory of aromatic compounds to explain their properties.	Problems 16-33 and 43
5	Name aromatic compounds, and draw their structures from the names.	Problems 16-26, 27, and 28
6	Predict the properties of aromatic compounds and the effects that aromatic rings have on neighboring parts of the molecule.	Problems 16-29, 33, 36, 39, 41, and 47
7	Use IR, NMR, UV, and mass spectra to determine the structures of aromatic compounds. Given an aromatic compound, predict the distinguishing features of its spectra.	Problems 16-38, 44, 45, 46, 48, and 51

Study Problems

16-26 Draw the structure of each compound.

(a) *o*-nitroanisole **(b)** 2,4-dimethoxyphenol **(c)** *p*-aminobenzoic acid
(d) 4-nitroaniline **(e)** *m*-chlorotoluene **(f)** *p*-divinylbenzene
(g) *p*-bromostyrene **(h)** 3,5-dimethoxybenzaldehyde **(i)** tropylium chloride
(j) sodium cyclopentadienide **(k)** 2-phenylpropan-1-ol **(l)** benzyl methyl ether
(m) *p*-toluenesulfonic acid **(n)** *o*-xylene **(o)** 3-benzylpyridine

16-27 Name the following compounds:

(a)

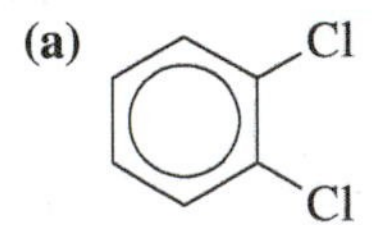

(b) NO_2 / OCH_3

(c) Br / Br / COOH

(d) CH_3O / OCH_3

(e) COOH / Cl

(f) OH / Cl / Cl / Cl

(g) CH_3 / $CHCH_2CH_3$ / CHO

(h) H / + / H / H / BF_4^-

16-28 Draw and name all the methyl, dimethyl, and trimethylbenzenes.

16-29 Four pairs of compounds are shown. In each pair, one of the compounds reacts more quickly, or with a more favorable equilibrium constant, than the less conjugated system. In each case, explain the enhanced reactivity.

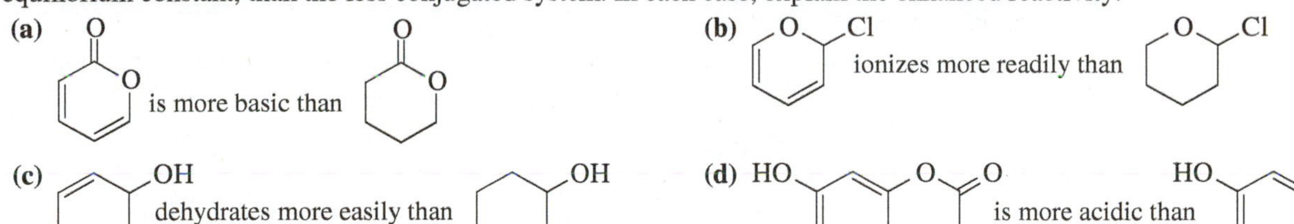

16-30 One of the following hydrocarbons is much more acidic than the others. Indicate which one, and explain why it is unusually acidic.

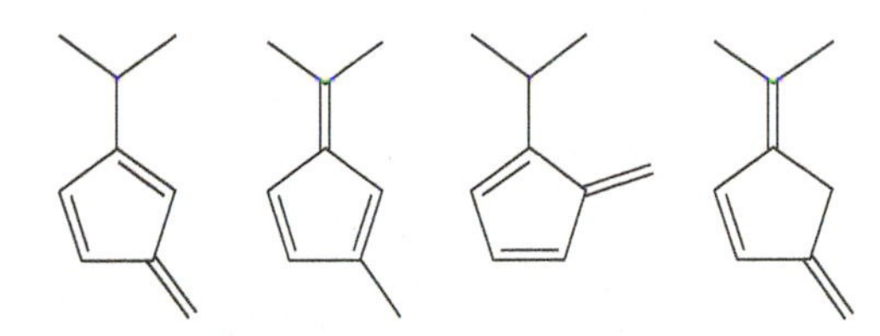

16-31 In Kekulé's time, cyclohexane was unknown, and there was no proof that benzene must be a six-membered ring. Determination of the structure relied largely on the known numbers of monosubstituted and disubstituted benzenes, together with the knowledge that benzene did not react similarly to a normal alkene. The following C_6H_6 structures were the likely candidates:

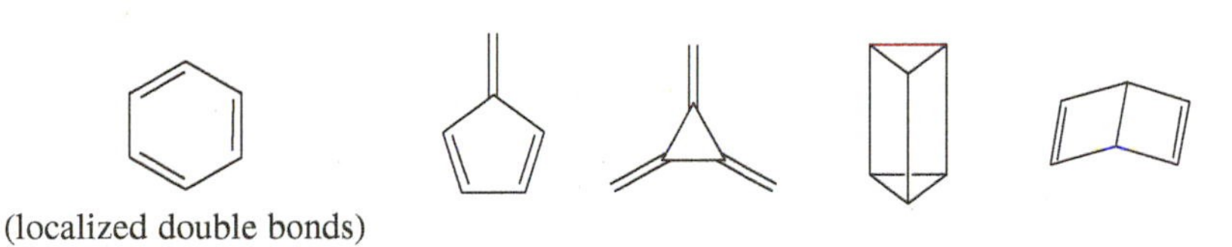

(a) Show where the six hydrogen atoms are in each structure.

(b) For each structure, draw all the possible monobrominated derivatives (C_6H_5Br) that would result from randomly substituting one hydrogen with a bromine. Benzene was known to have only one monobromo derivative.

(c) For each of the structures that had only one monobromo derivative in part (b), draw all the possible dibromo derivatives. Benzene was known to have three dibromo derivatives, but resonance theory was unknown at the time.

(d) Determine which structure was most consistent with what was known about benzene at that time: Benzene gives one monobrominated derivative and three dibrominated derivatives, and it gives negative chemical tests for an alkene.

(e) The structure that was considered the most likely structure for benzene is called *Ladenburg benzene*, after the chemist who proposed it. What factors would make Ladenburg benzene relatively unstable, in contrast with the stability observed with real benzene?

16-32 The following molecules and ions are grouped by similar structures. Classify each as aromatic, antiaromatic, or nonaromatic. For the aromatic and antiaromatic species, give the number of pi electrons in the ring.

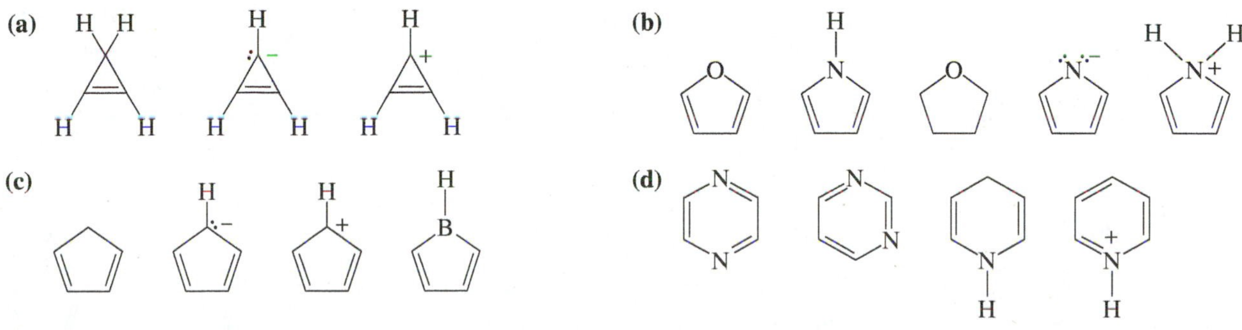

(continued)

(e) 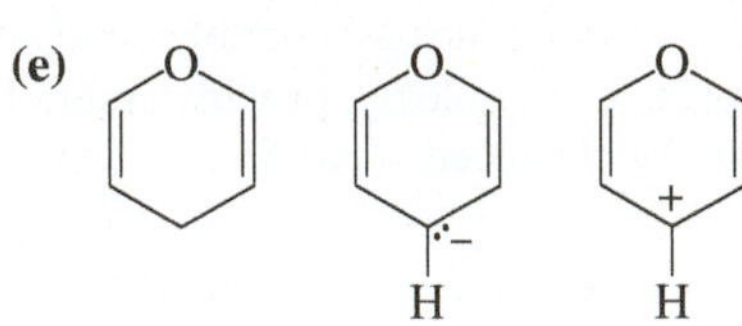

(f) 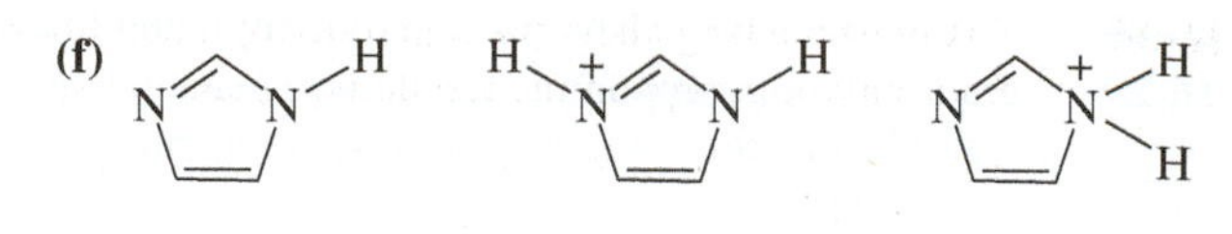

(g) 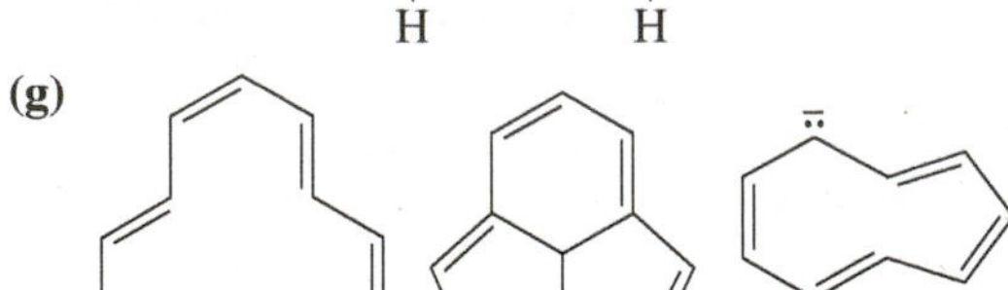

(h) 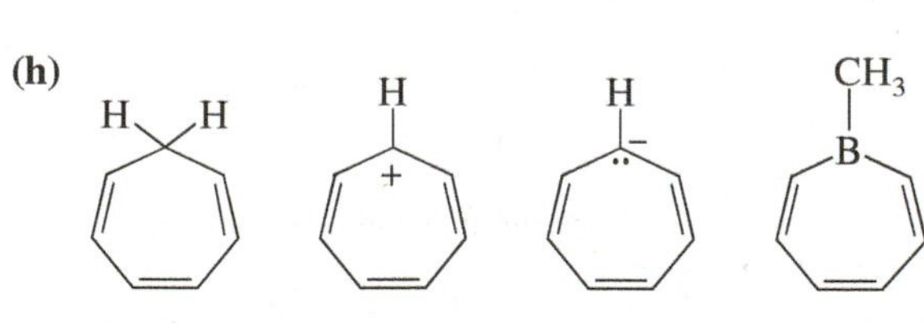

16-33 Azulene is a deep-blue hydrocarbon with resonance energy of 205 kJ/mol (49 kcal/mol). Azulene has ten pi electrons, so it might be considered as one large aromatic ring. Its electrostatic potential map shows one ring to be highly electron-rich (red) and the other to be electron-poor (blue). The dipole moment is unusually large (1.0 D) for a hydrocarbon. Show how this charge separation might arise.

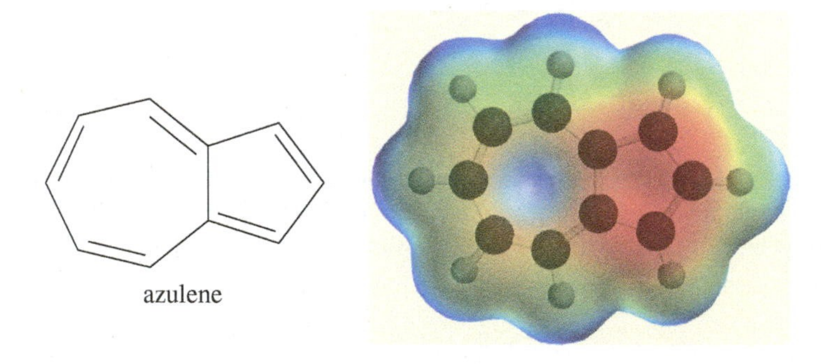

azulene

16-34 Each of the following heterocycles includes one or more nitrogen atoms. Classify each nitrogen atom as strongly basic or weakly basic, according to the availability of its lone pair of electrons.

(a) 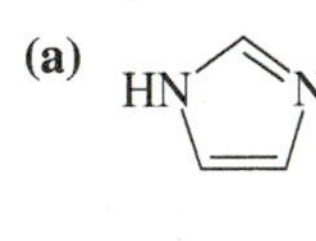
(b) 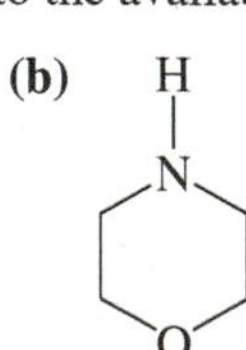
(c) 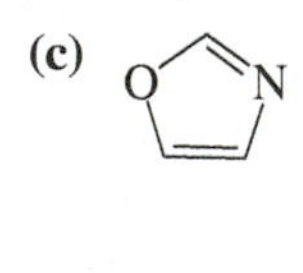
(d) 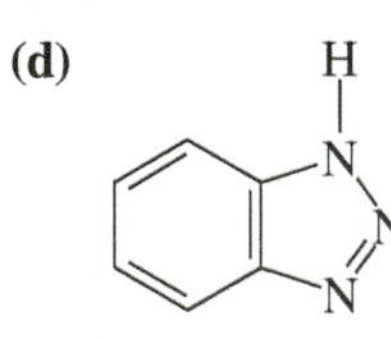
(e) 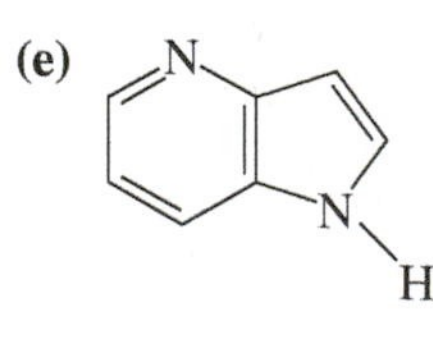

16-35 Some of the following compounds show aromatic properties, and others do not.

1. Predict which ones are likely to be aromatic, and explain why they are aromatic.
2. Predict which nitrogen atoms are more basic than water and which are less basic.

(a) 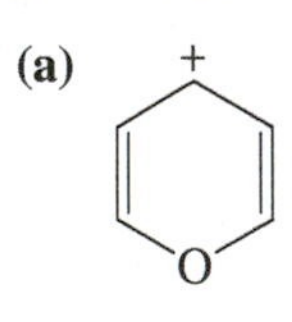
(b) 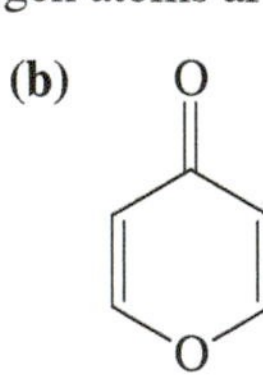
(c) 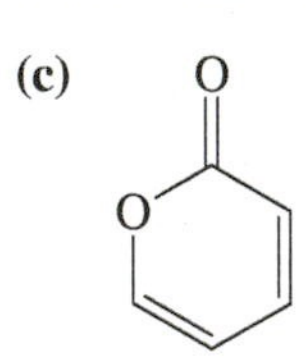
(d) 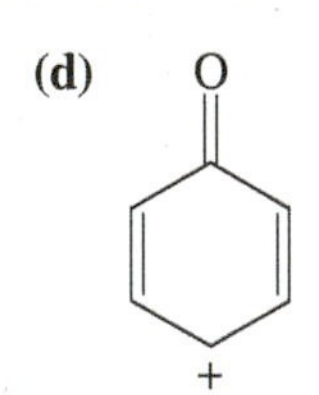
(e) 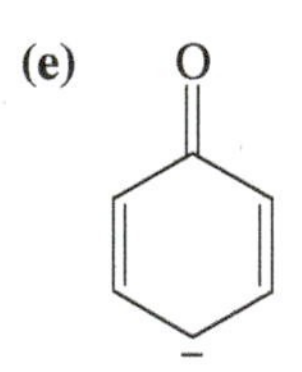

(f) 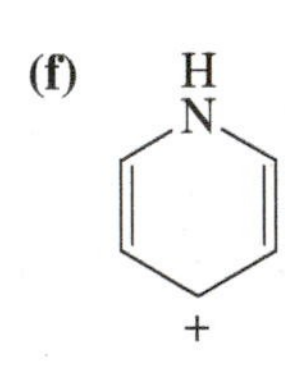
(g) 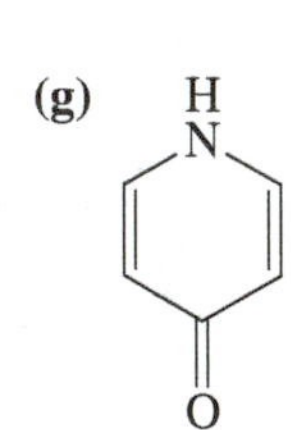
(h) 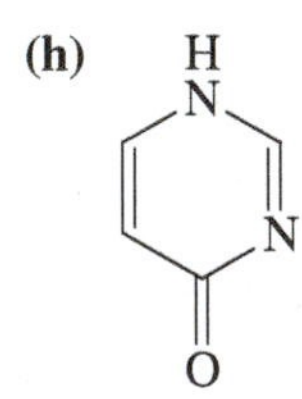
(i) 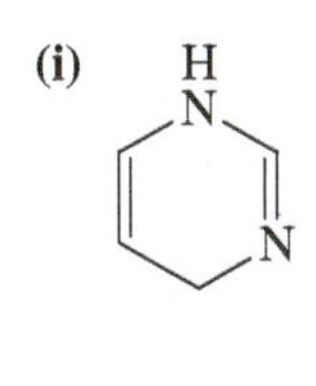
(j) 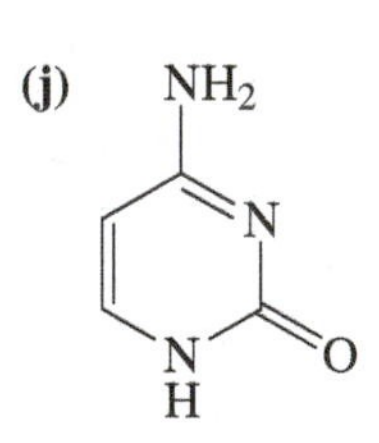

(k) 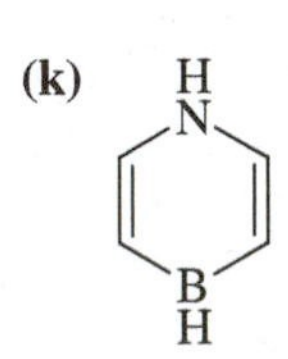
(l) 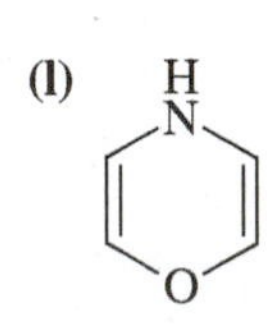
(m) 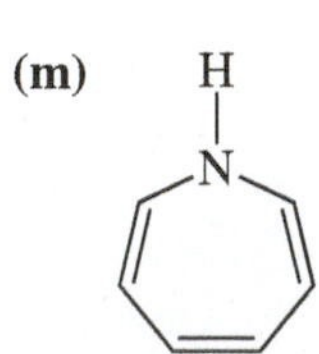
(n) 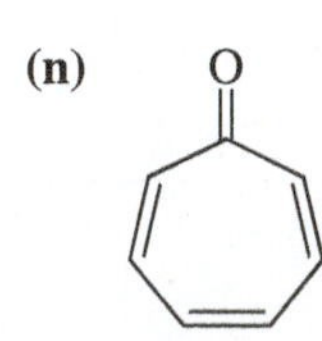
(o)

16-36 The benzene ring alters the reactivity of a neighboring group in the *benzylic position* much as a double bond alters the reactivity of groups in the allylic position.

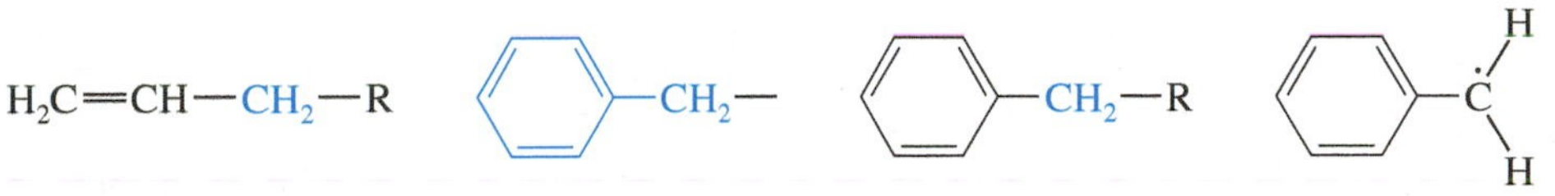

$H_2C{=}CH{-}CH_2{-}R$ — allylic position

$C_6H_5{-}CH_2{-}$ — benzyl group

$C_6H_5{-}CH_2{-}R$ — benzylic position

$C_6H_5{-}\dot{C}H_2$ — benzyl radical

Benzylic cations, anions, and radicals are all more stable than simple alkyl intermediates.

(a) Use resonance forms to show the delocalization (over four carbon atoms) of the positive charge, unpaired electron, and negative charge of the benzyl cation, radical, and anion.

(b) Toluene reacts with bromine in the presence of light to give benzyl bromide. Propose a mechanism for this reaction.

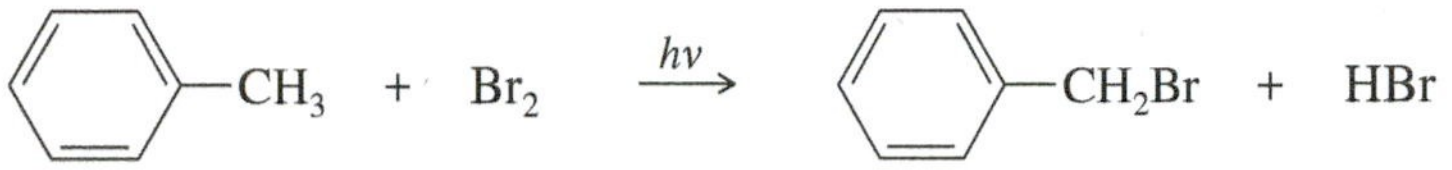

$$C_6H_5{-}CH_3 + Br_2 \xrightarrow{h\nu} C_6H_5{-}CH_2Br + HBr$$

toluene → benzyl bromide

(c) Which of the following reactions will have the faster rate and give the better yield? Use a drawing of the transition state to explain your answer.

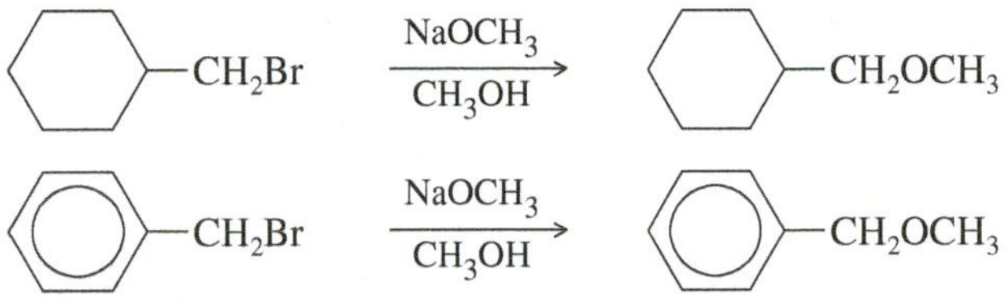

$$C_6H_{11}{-}CH_2Br \xrightarrow[CH_3OH]{NaOCH_3} C_6H_{11}{-}CH_2OCH_3$$

$$C_6H_5{-}CH_2Br \xrightarrow[CH_3OH]{NaOCH_3} C_6H_5{-}CH_2OCH_3$$

16-37 Before spectroscopy was invented, *Körner's absolute method* was used to determine whether a disubstituted benzene derivative was the ortho, meta, or para isomer. Körner's method involves adding a third group (often a nitro group) and determining how many isomers are formed. For example, when *o*-xylene is nitrated (by a method shown in Chapter 17), two isomers are formed.

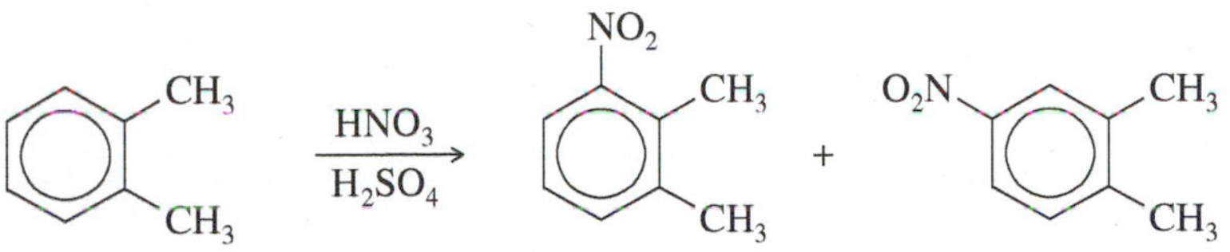

$$o\text{-}C_6H_4(CH_3)_2 \xrightarrow[H_2SO_4]{HNO_3} \text{3-nitro-}o\text{-xylene } (NO_2) + \text{4-nitro-}o\text{-xylene } (O_2N)$$

(a) How many isomers are formed by nitration of *m*-xylene?

(b) How many isomers are formed by nitration of *p*-xylene?

(c) A turn-of-the-century chemist isolated an aromatic compound of molecular formula $C_6H_4Br_2$. He carefully nitrated this compound and purified three isomers of formula $C_6H_3Br_2NO_2$. Propose structures for the original compound and the three nitrated derivatives.

16-38 For each NMR spectrum, propose a structure consistent with the spectrum and the additional information provided.

(a) Elemental analysis shows the molecular formula to be C_8H_7OCl. The IR spectrum shows a moderate absorption at 1602 cm^{-1} and a strong absorption at 1690 cm^{-1}.

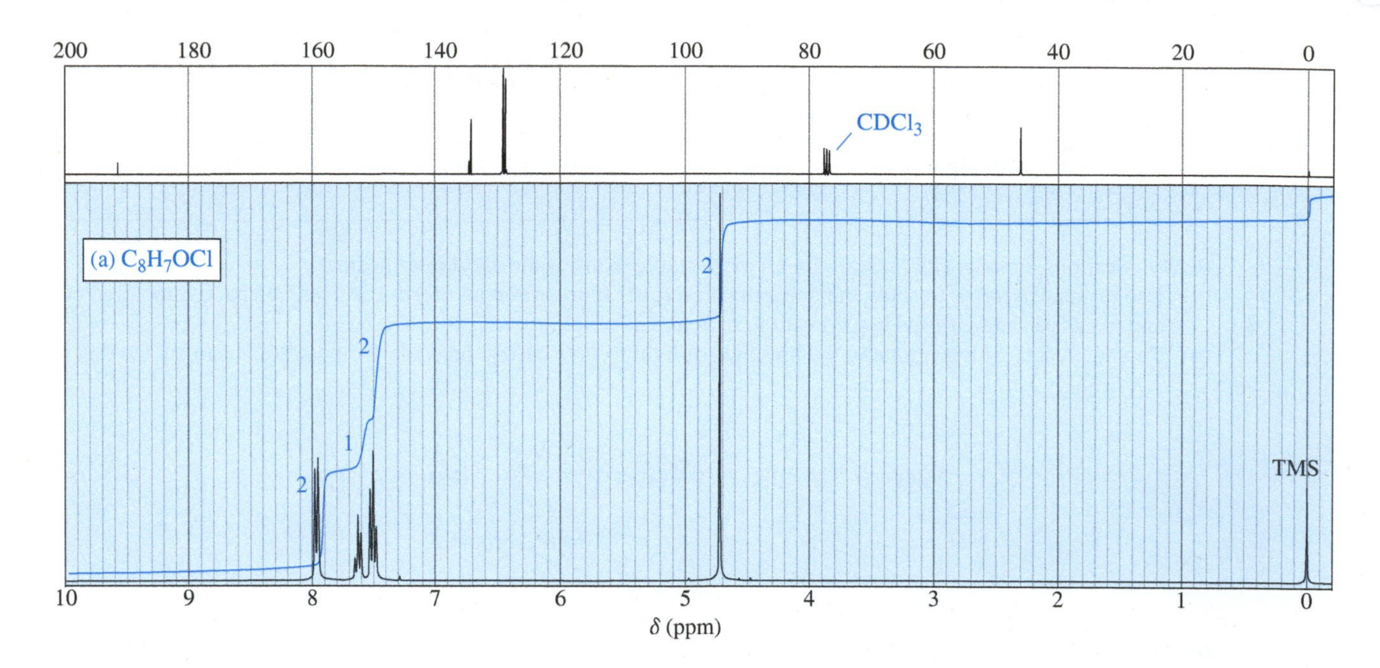

(b) The mass spectrum shows a double molecular ion of ratio 1:1 at m/z 184 and 186.

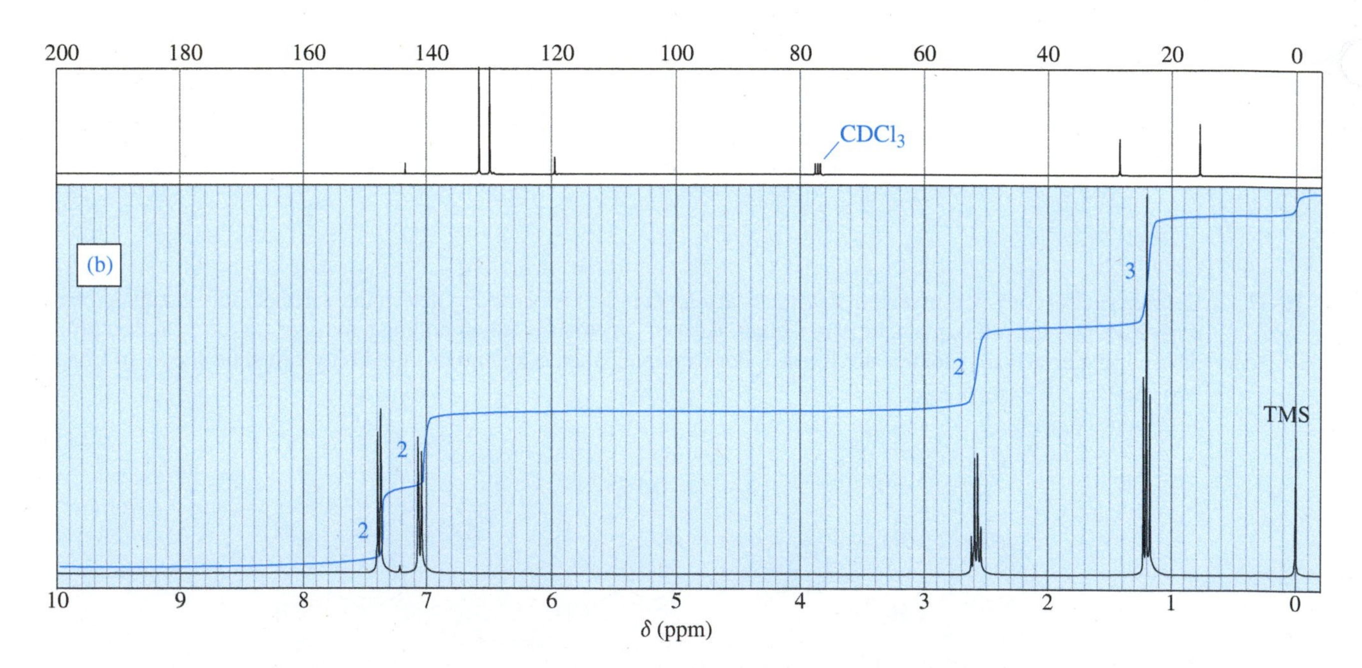

16-39 Recall (Section 16-10) that two positions of anthracene sometimes react more like polyenes than like aromatic compounds.

(a) Draw a Kekulé structure that shows how the reactive positions of anthracene are the ends of a diene, appropriate for a Diels–Alder reaction.

(b) The Diels–Alder reaction of anthracene with maleic anhydride is a common organic lab experiment. Predict the product of this Diels–Alder reaction.

O O O

maleic anhydride

16-40 Biphenyl has the following structure.

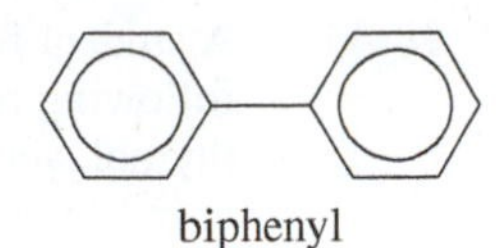

biphenyl

(a) Is biphenyl a (fused) polynuclear aromatic hydrocarbon?
(b) How many pi electrons are there in the two aromatic rings of biphenyl? How does this number compare with that for naphthalene?
(c) The heat of hydrogenation for biphenyl is about 418 kJ/mol (100 kcal/mol). Calculate the resonance energy of biphenyl.
(d) Compare the resonance energy of biphenyl with that of naphthalene and with that of two benzene rings. Explain the difference in the resonance energies of naphthalene and biphenyl.

16-41 Anions of hydrocarbons are rare, and dianions of hydrocarbons are extremely rare. The following hydrocarbon reacts with two equivalents of butyllithium to form a dianion of formula $[C_8H_6]^{2-}$. Propose a structure for this dianion, and suggest why it forms so readily.

$$\text{(hydrocarbon)} + 2\,C_4H_9Li \longrightarrow [C_8H_6]^{2-}\,(Li^+)_2 + 2\,C_4H_{10}\uparrow$$

16-42 How would you convert the following compounds to aromatic compounds?

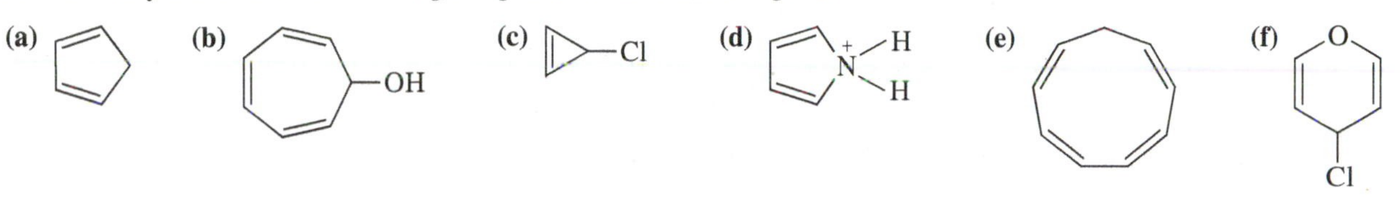

***16-43** The ribonucleosides that make up ribonucleic acid (RNA) are composed of D-ribose (a sugar) and four heterocyclic "bases." The general structure of a ribonucleoside is shown here.

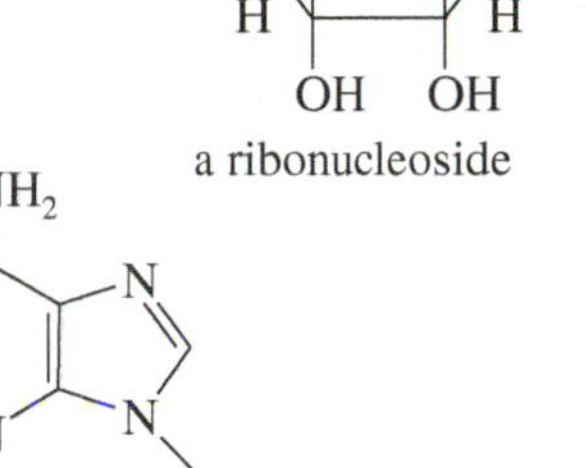

a ribonucleoside

The four heterocyclic bases are cytosine, uracil, guanine, and adenine. Cytosine and uracil are called *pyrimidine bases* because their structures resemble pyrimidine. Guanine and adenine are called *purine bases* because their structures resemble purine.

pyrimidine cytosine uracil

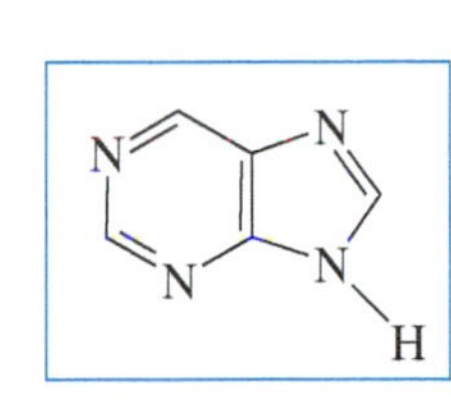

purine

guanine

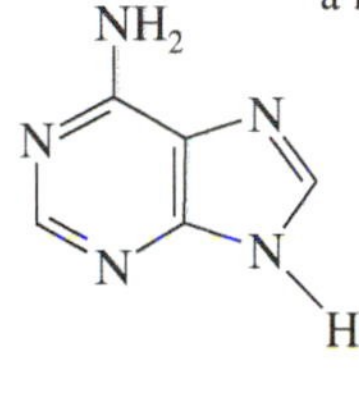

adenine

(a) Determine which rings of these bases are aromatic.
(b) Predict which nitrogen atoms are basic.
(c) Do any of these bases have easily formed tautomers that are aromatic? (Consider moving a proton from nitrogen to a carbonyl group to form a phenolic derivative.)

***16-44** Consider the following compound, which has been synthesized and characterized:

$(CH_3)_3C$, $C(CH_3)_3$, $(CH_3)_3C$ (substituted azete)

(a) Assuming this molecule is entirely conjugated, do you expect it to be aromatic, antiaromatic, or nonaromatic?
(b) Why was this molecule synthesized with three *tert*-butyl substituents? Why not make the unsubstituted compound and study it instead?
(c) Do you expect the nitrogen atom to be basic? Explain. Why doesn't nitrogen's lone pair overlap with the double bonds to give a total of six electrons in the pi system?
(d) At room temperature, the proton NMR spectrum shows only two singlets of ratio 1:2. The smaller signal remains unchanged at all temperatures. As the temperature is lowered to −110 °C, the larger signal broadens and separates into two new singlets, one on either side of the original chemical shift. At −110 °C, the spectrum consists of three separate singlets of areas 1:1:1. Explain what these NMR data indicate about the bonding in this molecule. How does your conclusion based on the NMR data agree with your prediction in part (a)?

16-45 A student found an old bottle labeled "thymol" on the stockroom shelf. After noticing a pleasant odor, she obtained the following mass, IR, and NMR spectra. The NMR peak at δ 4.8 disappears on shaking with D_2O. Propose a structure for thymol, and show how your structure is consistent with the spectra. Propose a fragmentation to explain the MS peak at m/z 135, and show why the resulting ion is relatively stable.

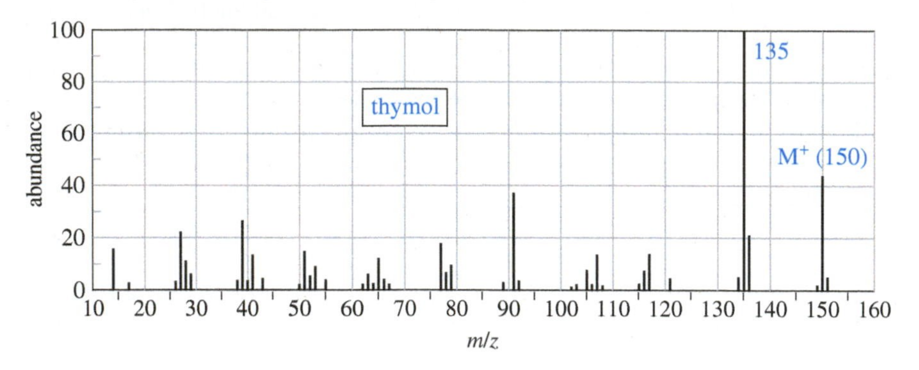

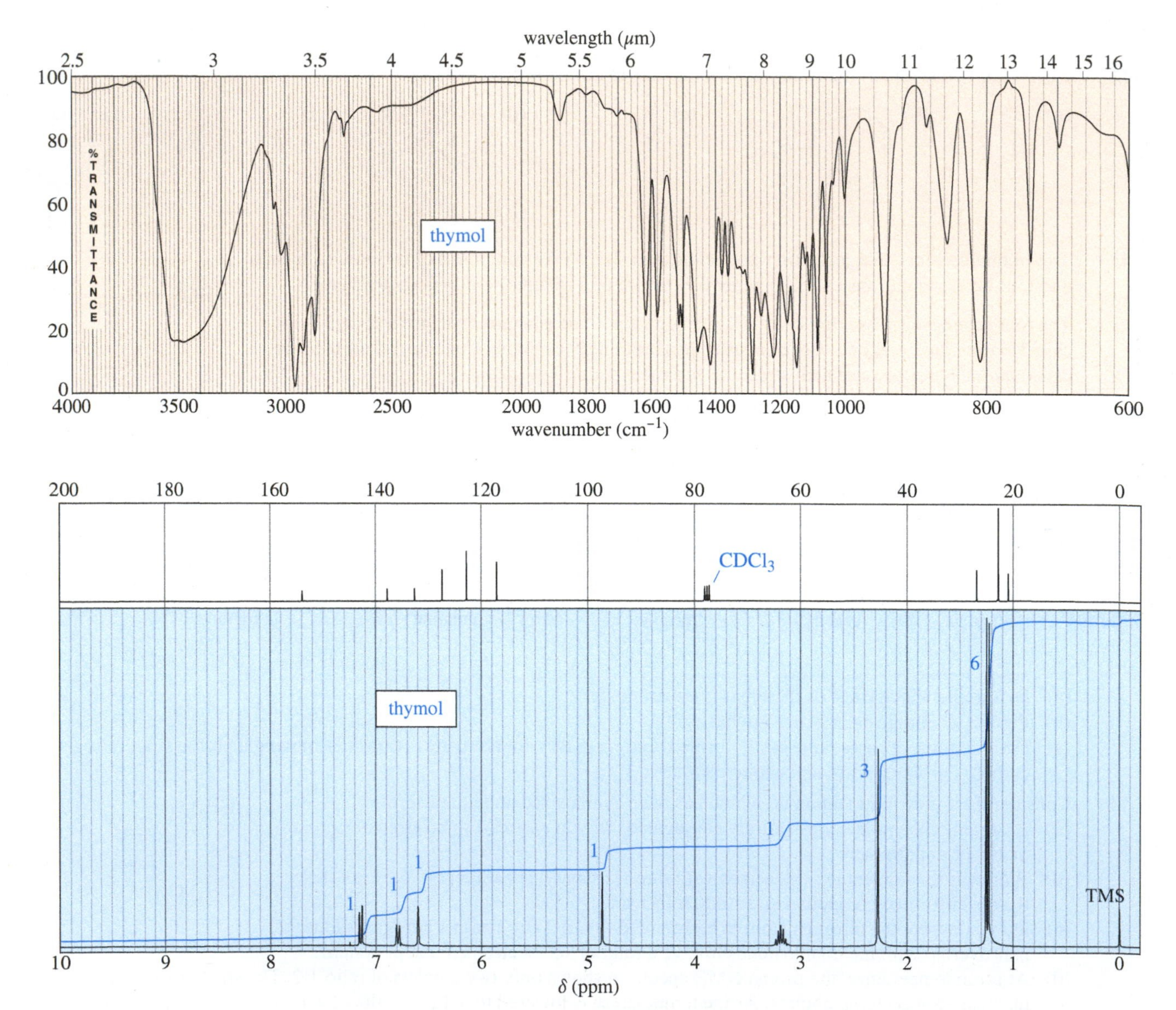

***16-46** An unknown compound gives the following mass, IR, and NMR spectra. Propose a structure, and show how it is consistent with the spectra. Show the fragmentations that give the prominent peaks at m/z 127 and 155 in the mass spectrum.

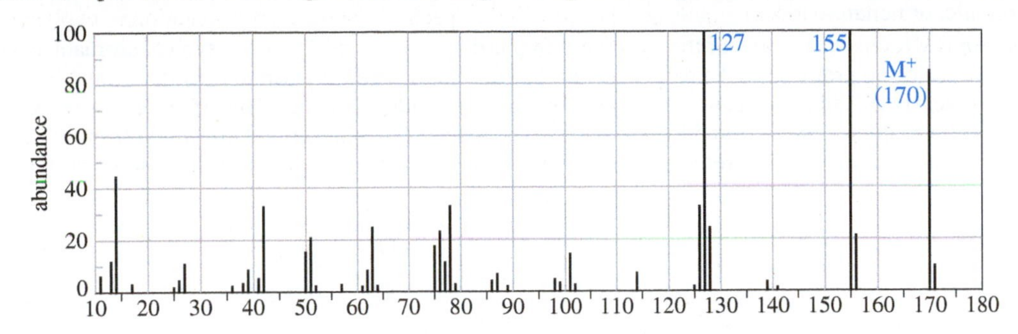

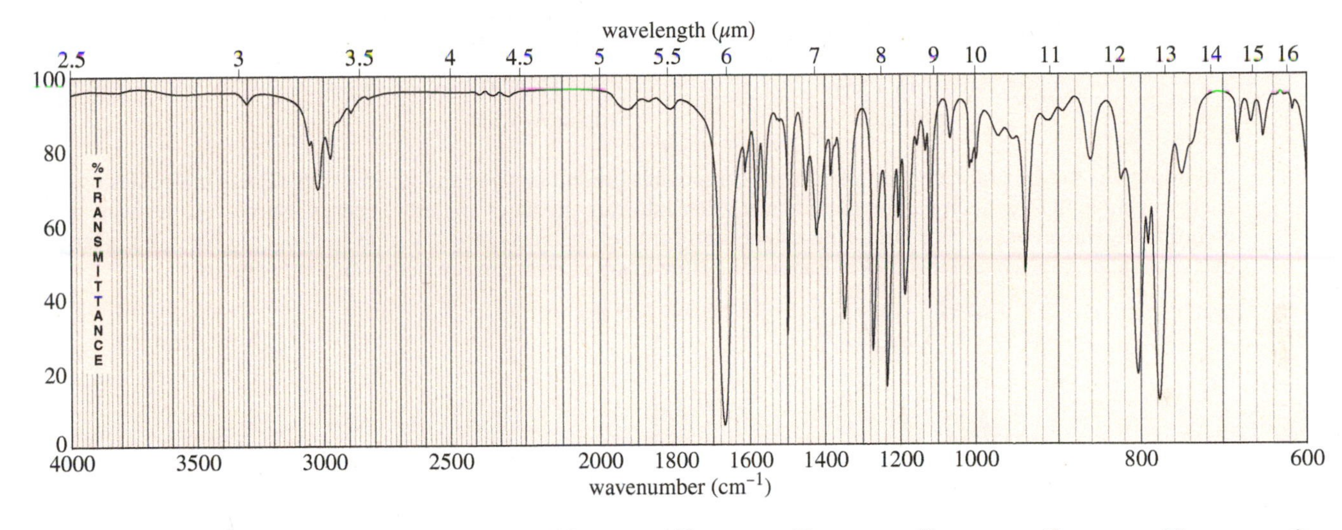

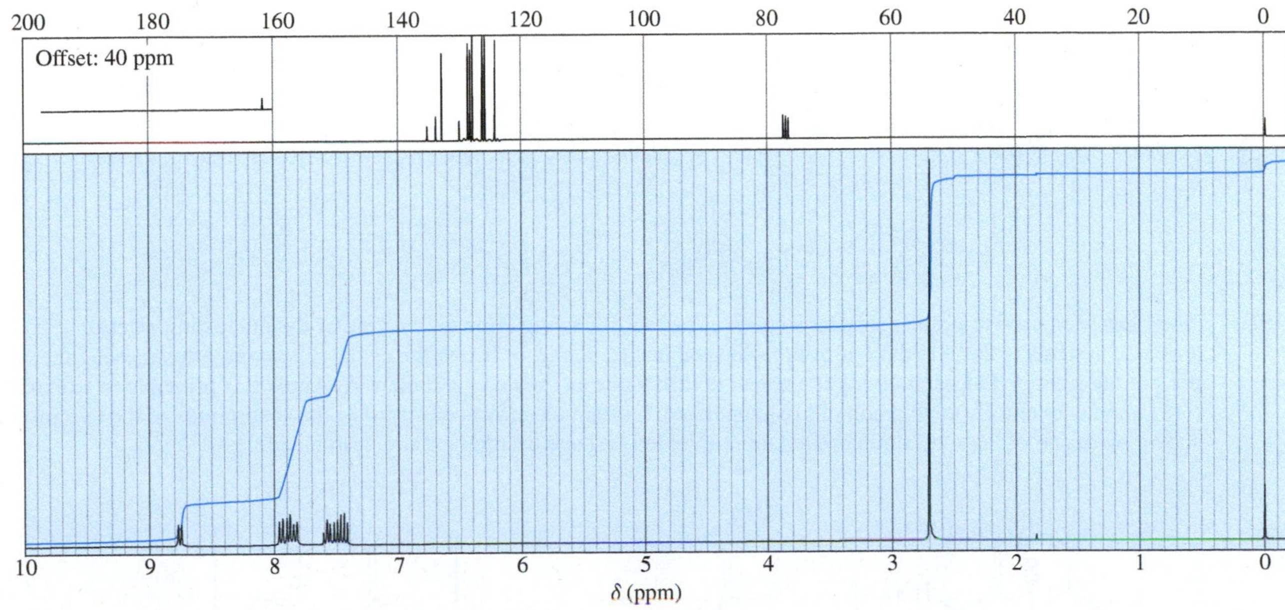

16-47 Hexahelicene seems a poor candidate for optical activity because all its carbon atoms are sp^2 hybrids and presumably flat. Nevertheless, hexahelicene has been synthesized and separated into enantiomers. Its optical rotation is enormous: $[\alpha]_D = 3700°$. Explain why hexahelicene is optically active, and speculate as to why the rotation is so large.

hexahelicene

16-48 Draw just the **bonding** π-MO's for the cycloheptatrienyl cation. Draw the energy diagram to show the relative energies of all the MO's, and show which orbitals the electrons would occupy in the ground state. Predict whether this ion is aromatic, antiaromatic, or nonaromatic.

16-49 The proton NMR chemical shifts of the hydrogens in pyridine are shown. These are typical aromatic chemical shifts, except that the ortho protons (on the carbons bonded to nitrogen) are deshielded to δ 8.60. A suitable oxidizing agent (such as a peroxyacid) can add an oxygen atom to pyridine to give pyridine *N*-oxide. The effect of this added oxygen atom is to shift the ortho protons *upfield* from δ 8.60 to δ 8.19. The meta protons are shifted *downfield* from δ 7.25 to δ 7.40. The para protons are shifted *upfield*, from δ 7.64 to δ 7.32. Explain this curious effect, shifting some protons upfield and others downfield.

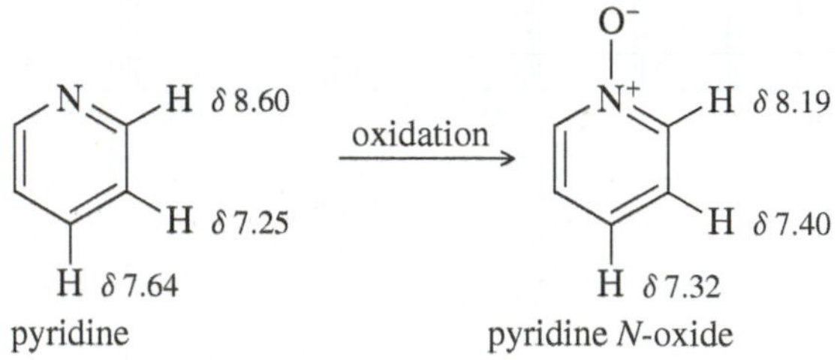

16-50 Chlorophyll is the general name for a family of compounds present in algae and green plants. These molecules use the energy in sunlight to convert carbon dioxide and water into carbohydrates and other energy sources. At the heart of chlorophyll (shown below) is a large-ring magnesium complex called a chlorin. Circle each double bond in the large cyclic conjugated pi system that makes it aromatic. How many pi electrons are in this aromatic system?

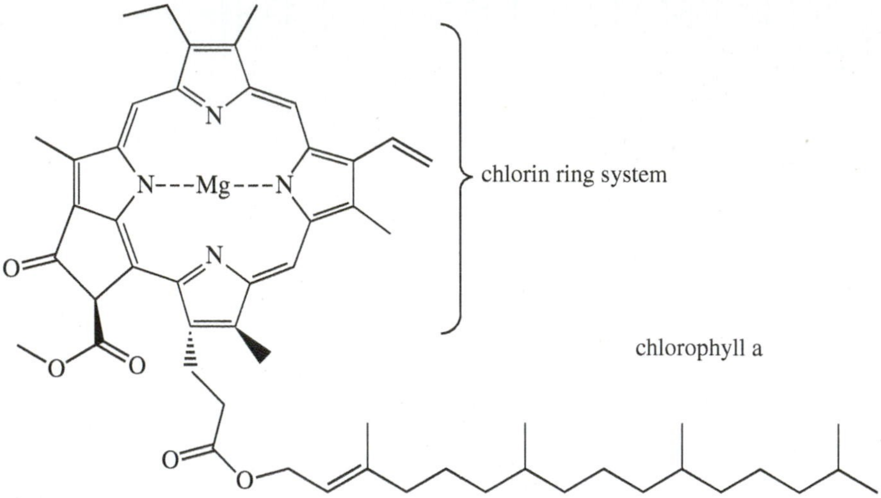

16-51 NMR has been used to probe many molecular properties, including aromaticity. One of the interesting electronic effects of aromatic systems is the shielding of protons located inside, above, or below the cyclic pi system, often moving the NMR peak to the right of TMS and giving it a negative shift value as shown by structure **A**. Apply this principle to structures **B** ([16]-annulene) and **C** (the dianion of [16]-annulene). Explain the effects observed in the absorptions of the two protons shown, with one of them becoming much more shielded and one becoming more deshielded.

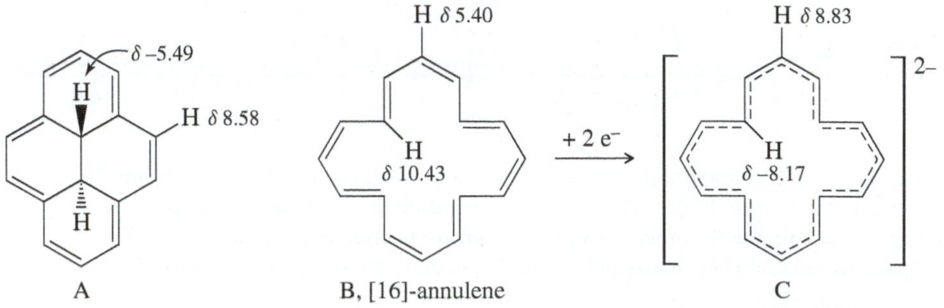

17 Reactions of Aromatic Compounds

Goals for Chapter 17

1 Understand the mechanisms of electrophilic and nucleophilic aromatic substitutions. Predict the products of these reactions and use them in syntheses.

2 Explain how substituents on the aromatic ring promote substitution at some positions but not at others.

3 Predict the coupling products of organometallic substitutions, and use them in syntheses.

4 Predict the products of oxidation and reduction of the aromatic ring, including hydrogenation, chlorination, and Birch reduction. Predict the products of the oxidation of phenols.

◀ In 1856, German chemists treated aniline and alkyl-substituted anilines with sulfuric acid, obtaining deeply colored pigments from repeated electrophilic aromatic substitutions of anilines with each other. This discovery of aniline dyes allowed chemists to make synthetic dyes much more cheaply and efficiently than dyes that were harvested from natural resources. For example, the aniline dye mauvine quickly replaced royal purple, a very expensive dye that was laboriously harvested from sea snails.

alkylbenzene $\xrightarrow[H_2SO_4]{HNO_3}$ alkylnitrobenzene (NO_2 para or ortho to R) $\xrightarrow[HCl(aq)]{Fe}$ alkylated anilines (NH_2 para or ortho to R) $\xrightarrow{H_2SO_4}$ aniline dyes

Aromatic compounds undergo many reactions, but relatively few reactions that affect the bonds in the aromatic ring itself. Most of these reactions are unique to aromatic compounds. A large part of this chapter is devoted to *electrophilic aromatic substitution*, the most important mechanism involved in the reactions of aromatic compounds. Many reactions of benzene and its derivatives are explained by minor variations of electrophilic aromatic substitution. We will study several of these reactions and then consider how substituents on the ring influence its reactivity toward electrophilic aromatic substitution and the regiochemistry seen in the products. We will also study other reactions of aromatic compounds, including nucleophilic aromatic substitution, addition reactions, reactions of side chains, and special reactions of phenols.

17-1 Electrophilic Aromatic Substitution

Like an alkene, benzene has clouds of pi electrons above and below its sigma bond framework. Although benzene's pi electrons are in a stable aromatic system, they are available to attack a strong electrophile to give a carbocation. This resonance-stabilized carbocation is called a **sigma complex** because the electrophile is joined to the benzene ring by a new sigma bond.

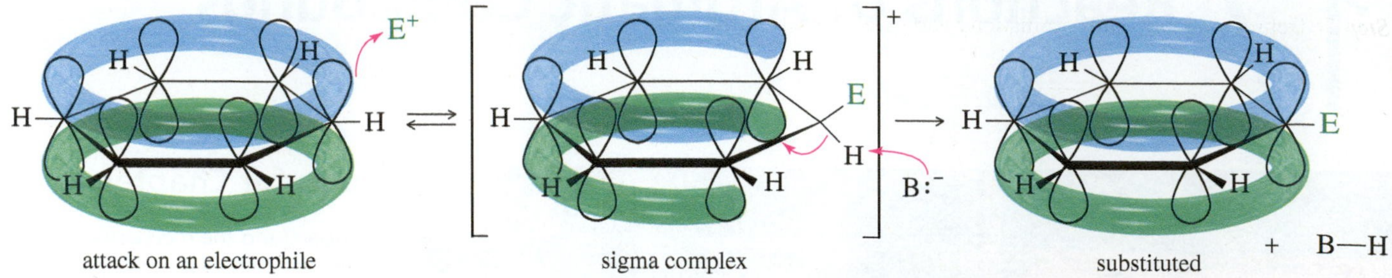

attack on an electrophile

sigma complex

substituted + B—H

The sigma complex (also called an *arenium ion*) is not aromatic because the sp^3 hybrid carbon atom interrupts the ring of *p* orbitals. Loss of aromaticity contributes to the highly endothermic nature of this first step. The sigma complex regains aromaticity either by a reversal of the first step (returning to the reactants) or by loss of the proton on the tetrahedral carbon atom, leading to the aromatic substitution product.

The overall reaction is the *substitution* of an electrophile (E^+) for a proton (H^+) on the aromatic ring: **electrophilic aromatic substitution.** This class of reactions includes substitutions by a wide variety of electrophilic reagents. Because it enables us to introduce functional groups directly onto the aromatic ring, electrophilic aromatic substitution is the most important method for synthesis of substituted aromatic compounds.

KEY MECHANISM 17-1 Electrophilic Aromatic Substitution

Step 1: Attack on the electrophile forms the sigma complex.

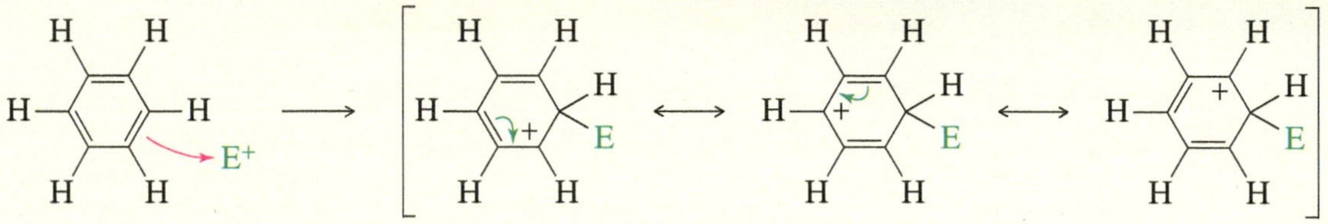

sigma complex (arenium ion)

Step 2: Loss of a proton regains aromaticity and gives the substitution product.

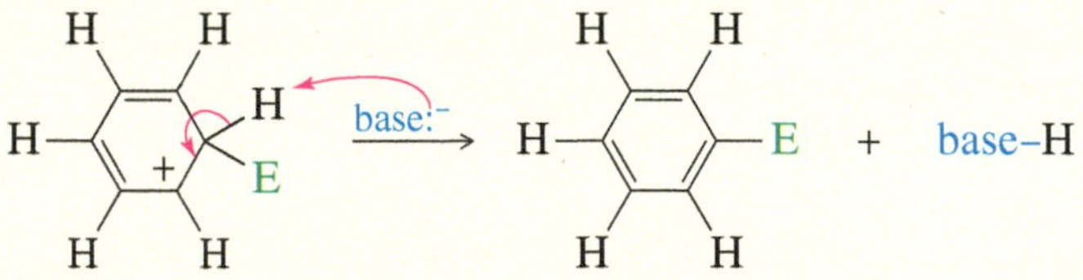

base:⁻ → + base–H

EXAMPLE: Iodination of toluene

Preliminary step: Formation of the electrophile, I^+ (the iodine cation).

$$\tfrac{1}{2}I_2 + H^+ + HNO_3 \longrightarrow I^+ + NO_2 + H_2O$$

Step 1: Attack on the electrophile forms the sigma complex.

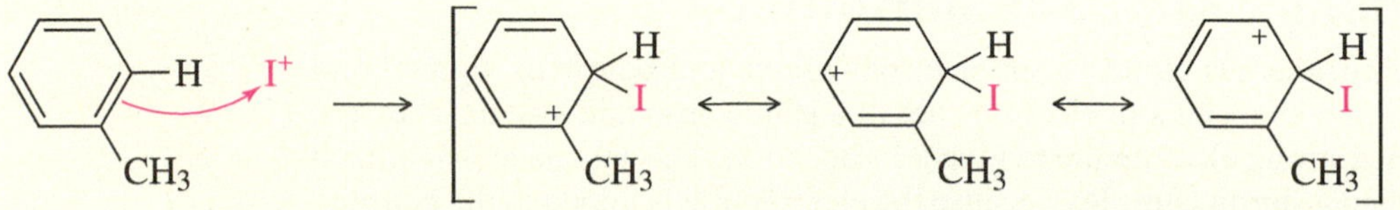

Step 2: Deprotonation regains aromaticity and gives the substitution product.

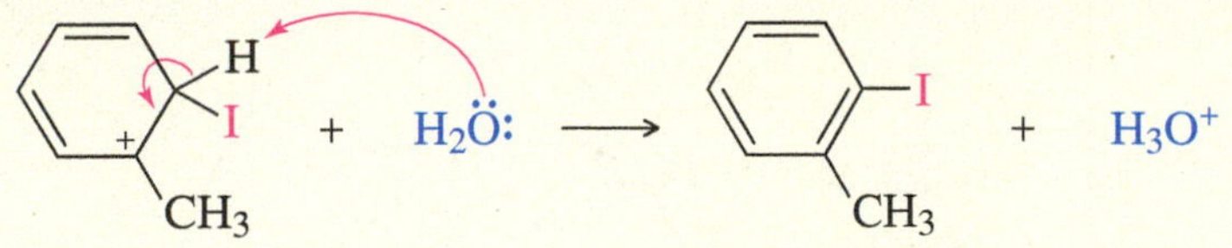

$$\text{sigma complex (H, I, } CH_3\text{)} + H_2\ddot{O}: \longrightarrow \text{iodotoluene (I, } CH_3\text{)} + H_3O^+$$

(plus other isomers)

PROBLEM 17-1

Step 2 of the iodination of benzene shows water acting as a base and removing a proton from the sigma complex. We did not consider the possibility of water acting as a nucleophile and attacking the carbocation, as in an electrophilic addition to an alkene. Draw the reaction that would occur if water reacted as a nucleophile and added to the carbocation. Explain why this type of addition is rarely observed.

17-2 Halogenation of Benzene

Bromination of Benzene Bromination follows the general mechanism for electrophilic aromatic substitution. Bromine itself is not sufficiently electrophilic to react with benzene, and the formation of Br^+ is difficult. A strong Lewis acid such as $FeBr_3$ catalyzes the reaction, however, by forming a complex with Br_2 that reacts like Br^+. Bromine donates a pair of electrons to $FeBr_3$, forming a stronger electrophile with a weakened Br—Br bond and a partial positive charge on one of the bromine atoms. Attack by benzene forms the sigma complex. Bromide ion from $FeBr_4^-$ acts as a weak base to remove a proton from the sigma complex, giving the aromatic product and HBr, and regenerating the catalyst.

MECHANISM 17-2 Bromination of Benzene

Step 1: Formation of a stronger electrophile.

$$:\ddot{Br}-\ddot{Br}: + FeBr_3 \rightleftharpoons \left[:\ddot{Br}-\overset{+}{\ddot{Br}}-\overset{-}{Fe}Br_3\right]$$

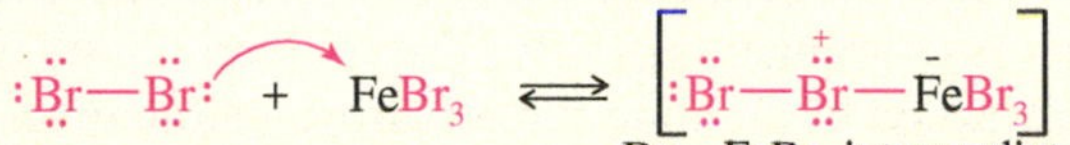

$Br_2 \cdot FeBr_3$ intermediate
(a stronger electrophile than Br_2)

Step 2: Electrophilic attack and formation of the sigma complex.

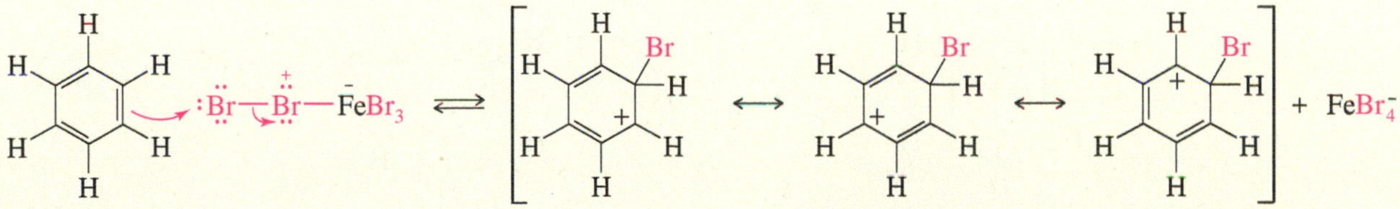

$$C_6H_6 + :\ddot{Br}-\overset{+}{\ddot{Br}}-\overset{-}{Fe}Br_3 \rightleftharpoons [\text{sigma complex resonance forms}] + FeBr_4^-$$

sigma complex

Step 3: Loss of a proton gives the products.

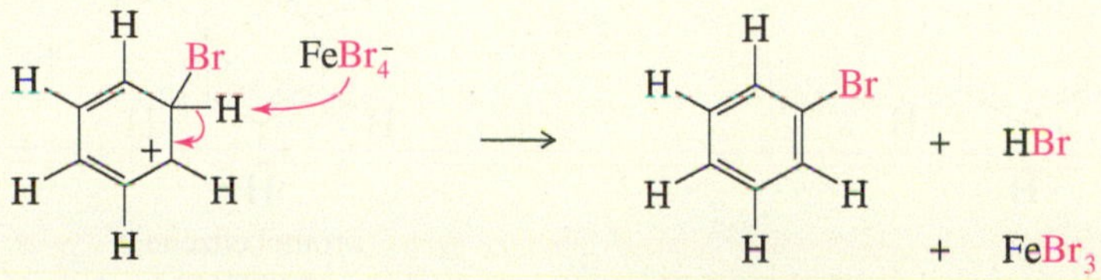

$$\text{sigma complex} + FeBr_4^- \longrightarrow C_6H_5Br + HBr + FeBr_3$$

bromobenzene

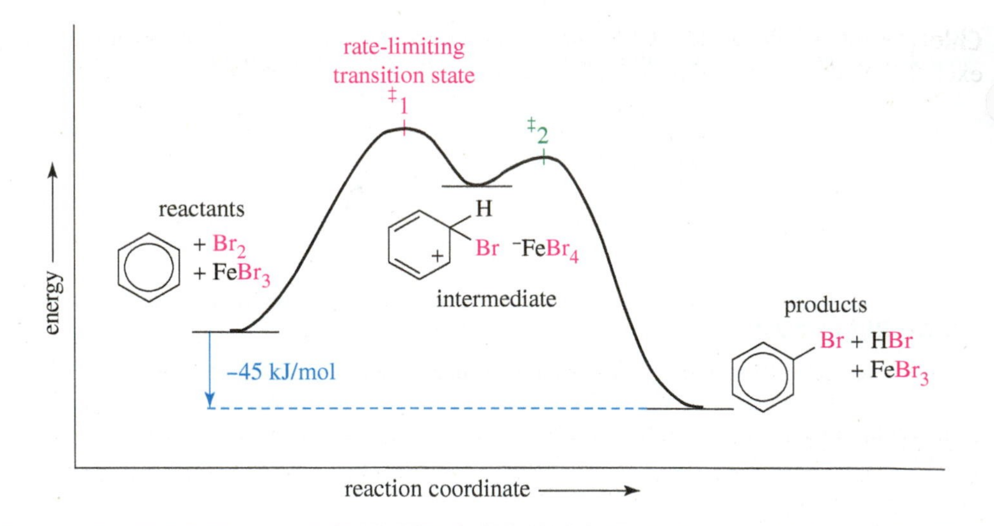

FIGURE 17-1 The energy diagram for the bromination of benzene shows that the first step is endothermic and rate-limiting and the second step is strongly exothermic.

Formation of the sigma complex is rate-limiting, and the transition state leading to it occupies the highest-energy point on the energy diagram (Figure 17-1). This step is strongly endothermic because it forms a nonaromatic carbocation. The second step is exothermic because aromaticity is regained and a molecule of HBr is evolved. The overall reaction is exothermic by 45 kJ/mol (10.8 kcal/mol).

Comparison with Alkenes Benzene is not as reactive as alkenes, which react rapidly with bromine at room temperature to give addition products (Section 8-8). For example, cyclohexene reacts to give *trans*-1,2-dibromocyclohexane. This reaction is exothermic by about 121 kJ/mol (29 kcal/mol).

cyclohexene + Br_2 ⟶ *trans*-1,2-dibromocyclohexane $\Delta H° = -121$ kJ (−29 kcal)

The analogous addition of bromine to benzene is *endothermic* because it requires the loss of aromatic stability. The addition is not seen under normal circumstances. The *substitution* of bromine for a hydrogen atom gives an aromatic product. The substitution is exothermic, but it requires a Lewis acid catalyst to convert bromine to a stronger electrophile.

benzene + Br_2 ✗⟶ 5,6-dibromocyclohexa-1,3-diene $\Delta H° = +8$ kJ (+2 kcal)

benzene + Br_2 $\xrightarrow{FeBr_3}$ bromobenzene (80%) + HBr $\Delta H° = -45$ kJ (−10.8 kcal)

Chlorination of Benzene Chlorination of benzene works much like bromination, except that aluminum chloride ($AlCl_3$) is most often used as the Lewis acid catalyst.

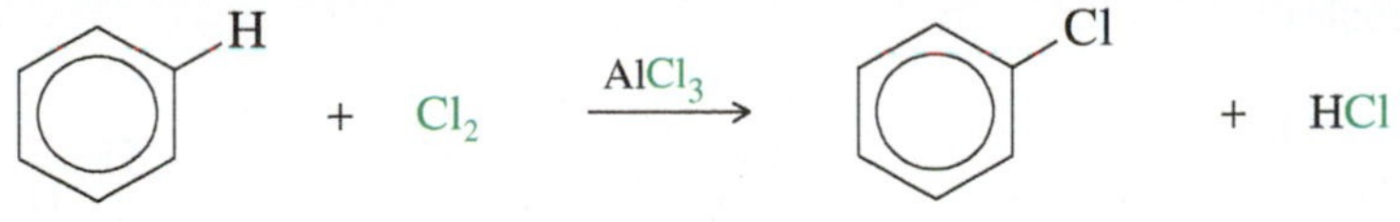

$$\text{benzene} + Cl_2 \xrightarrow{AlCl_3} \text{chlorobenzene (85\%)} + HCl$$

PROBLEM 17-2

Propose a mechanism for the aluminum chloride–catalyzed reaction of benzene with chlorine.

PROBLEM-SOLVING HINT

Note that the three resonance forms of the sigma complex show the positive charge on the three carbon atoms ortho and para to the site of substitution.

Iodination of Benzene Iodination of benzene requires an acidic oxidizing agent, such as nitric acid. Nitric acid is consumed in the reaction, so it is a reagent (an oxidant) rather than a catalyst.

$$\text{benzene} + \tfrac{1}{2}I_2 + HNO_3 \longrightarrow \text{iodobenzene (85\%)} + NO_2 + H_2O$$

Iodination probably involves an electrophilic aromatic substitution with the iodine cation (I^+) acting as the electrophile. The iodine cation results from oxidation of iodine by nitric acid.

$$H^+ + HNO_3 + \tfrac{1}{2}I_2 \longrightarrow \underset{\text{iodine cation}}{I^+} + NO_2 + H_2O$$

17-3 Nitration of Benzene

Benzene reacts with hot, concentrated nitric acid to give nitrobenzene. This sluggish reaction is hazardous because a hot mixture of concentrated nitric acid with any oxidizable material might explode. A safer and more convenient procedure uses a mixture of nitric acid and sulfuric acid. Sulfuric acid is a catalyst, allowing nitration to take place more rapidly and at lower temperatures.

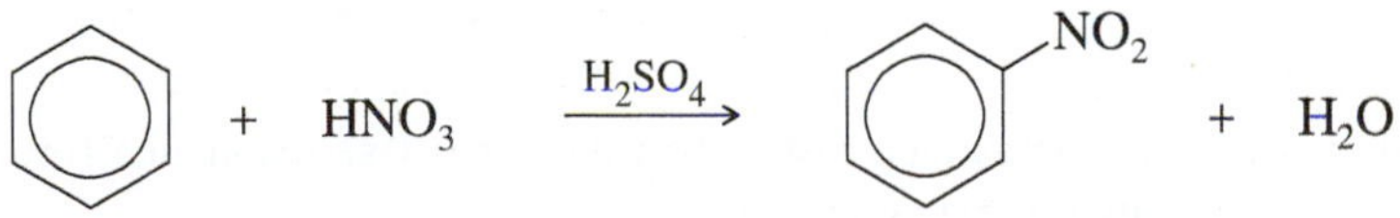

$$\text{benzene} + HNO_3 \xrightarrow{H_2SO_4} \text{nitrobenzene (85\%)} + H_2O$$

The mechanism is shown next. Sulfuric acid reacts with nitric acid to form the **nitronium ion** ($^+NO_2$), a powerful electrophile. The mechanism is similar to other sulfuric acid–catalyzed dehydrations. Sulfuric acid protonates the hydroxy group of nitric acid, allowing it to leave as water and form a nitronium ion. The nitronium ion reacts with benzene to form a sigma complex. Loss of a proton from the sigma complex gives nitrobenzene.

MECHANISM 17-3 Nitration of Benzene

Preliminary steps: Formation of the nitronium ion, NO_2^+.

Nitric acid has a hydroxy group that can become protonated and leave as water, similar to the dehydration of an alcohol.

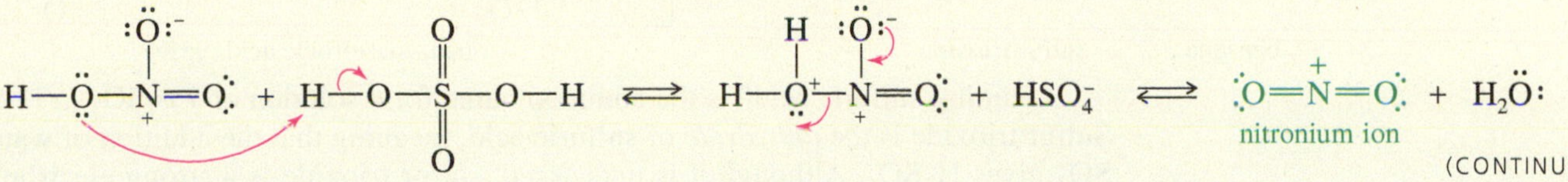

$$H{-}\ddot{O}{-}N^+(O^-){=}O + H{-}O{-}SO_2{-}O{-}H \rightleftarrows H{-}O^+(H){-}N^+(O^-){=}O + HSO_4^- \rightleftarrows \underset{\text{nitronium ion}}{O{=}N^+{=}O} + H_2\ddot{O}:$$

(CONTINUED)

Electrophilic aromatic substitution by the nitronium ion gives nitrobenzene.

Step 1: Attack on the electrophile forms the sigma complex.

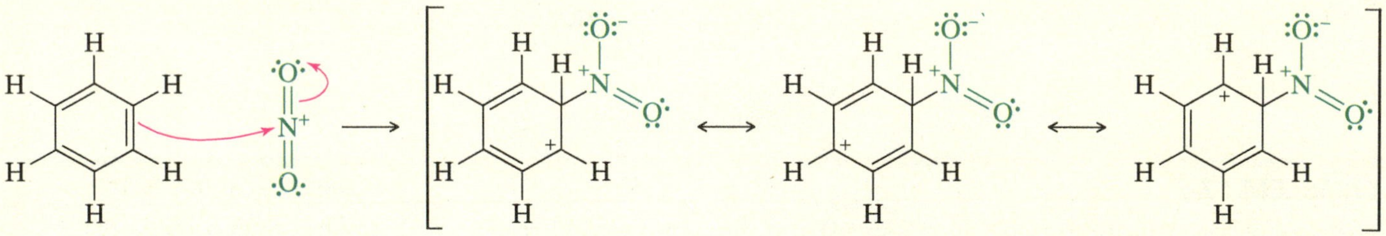

benzene nitronium ion sigma complex

Step 2: Loss of a proton gives nitrobenzene.

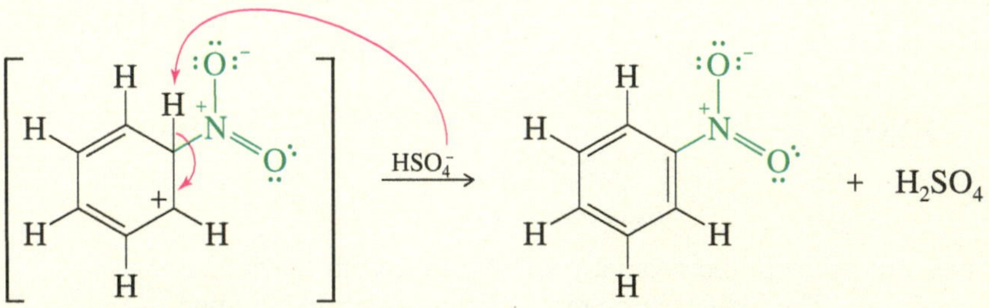

HSO_4^- + H_2SO_4

sigma complex (resonance-delocalized) nitrobenzene

Application: Drug Synthesis

p-Nitrotoluene is the starting point for the syntheses of benzocaine and procaine, two compounds used as local anesthetics. (See Section 19-20C.)

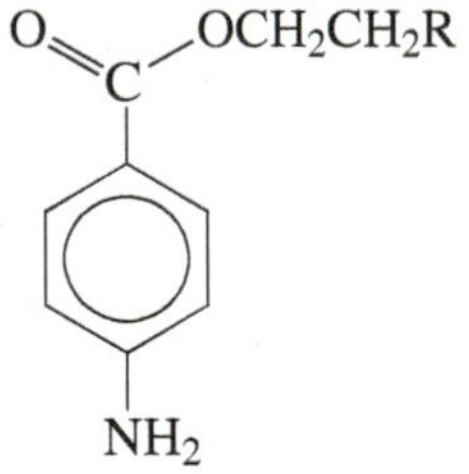

OCH_2CH_2R, NH_2

benzocaine (R = H)
procaine (R = NEt_2)

Aromatic nitro groups are easily reduced to amino ($-NH_2$) groups by treatment with an active metal such as tin, zinc, or iron in dilute acid. Nitration followed by reduction is often the best method for adding an amino group to an aromatic ring.

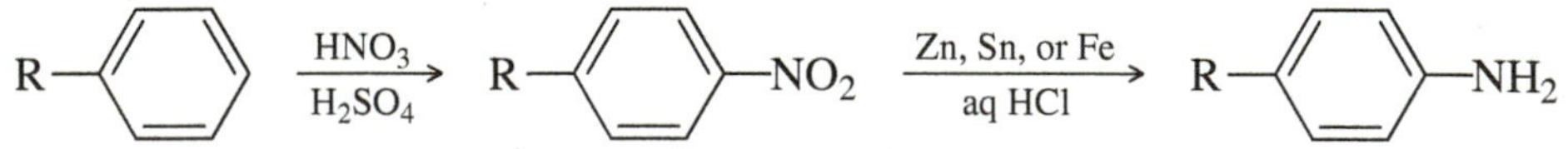

$$\text{R–C}_6\text{H}_5 \xrightarrow[H_2SO_4]{HNO_3} \text{R–C}_6\text{H}_4\text{–NO}_2 \xrightarrow[\text{aq HCl}]{\text{Zn, Sn, or Fe}} \text{R–C}_6\text{H}_4\text{–NH}_2$$

an alkylbenzene a nitrated alkylbenzene a substituted aniline

PROBLEM 17-3

p-Xylene undergoes nitration much faster than benzene. Use resonance forms of the sigma complex to explain this accelerated rate.

17-4 Sulfonation of Benzene

We have already used esters of *p*-toluenesulfonic acid as activated derivatives of alcohols with a good leaving group, the tosylate group (Section 11-5). *p*-Toluenesulfonic acid is an example of an *arylsulfonic acid* (general formula $Ar-SO_3H$), which are often used as strong acid catalysts that are soluble in nonpolar organic solvents. Arylsulfonic acids are easily synthesized by sulfonation of benzene derivatives, an electrophilic aromatic substitution using sulfur trioxide (SO_3) as the electrophile.

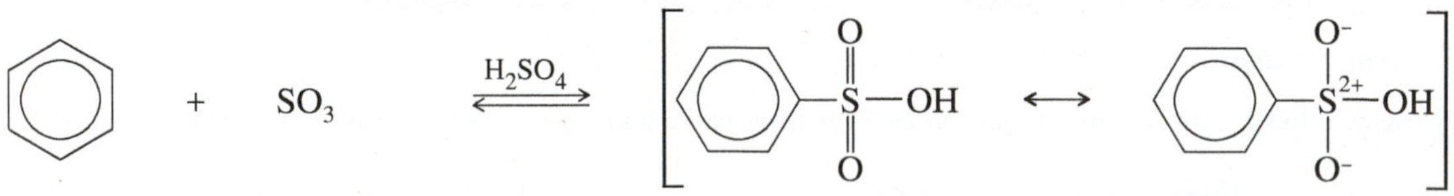

$$\text{C}_6\text{H}_6 + SO_3 \underset{}{\overset{H_2SO_4}{\rightleftharpoons}} [\text{C}_6\text{H}_5\text{–S(=O)}_2\text{–OH} \longleftrightarrow \text{C}_6\text{H}_5\text{–S}^{2+}(\text{O}^-)_2\text{–OH}]$$

benzene sulfur trioxide benzenesulfonic acid (95%)

"Fuming sulfuric acid" is the common name for a solution of 7% SO_3 in H_2SO_4. Sulfur trioxide is the *anhydride* of sulfuric acid, meaning that the addition of water to SO_3 gives H_2SO_4. Although it is uncharged, sulfur trioxide is a strong electrophile, with three sulfonyl (S═O) bonds drawing electron density away from the sulfur atom. Benzene attacks sulfur trioxide, forming a sigma complex. Loss of a proton on the tetrahedral carbon and reprotonation on oxygen gives benzenesulfonic acid.

sulfur trioxide, a powerful electrophile

MECHANISM 17-4 Sulfonation of Benzene

Sulfur trioxide is a powerful electrophile.

Step 1: Attack on the electrophile forms the sigma complex.

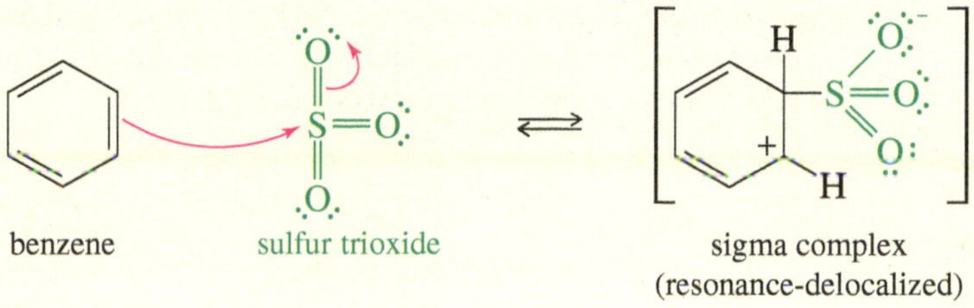

benzene sulfur trioxide sigma complex (resonance-delocalized)

Step 2: Loss of a proton regenerates an aromatic ring.

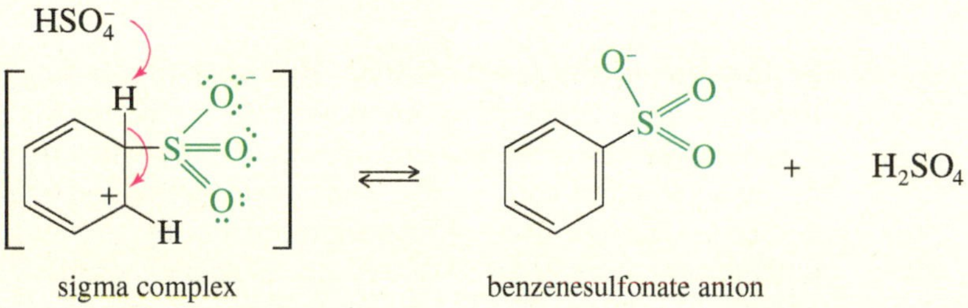

HSO_4^-

sigma complex benzenesulfonate anion + H_2SO_4

Step 3: The sulfonate group may become protonated in strong acid.

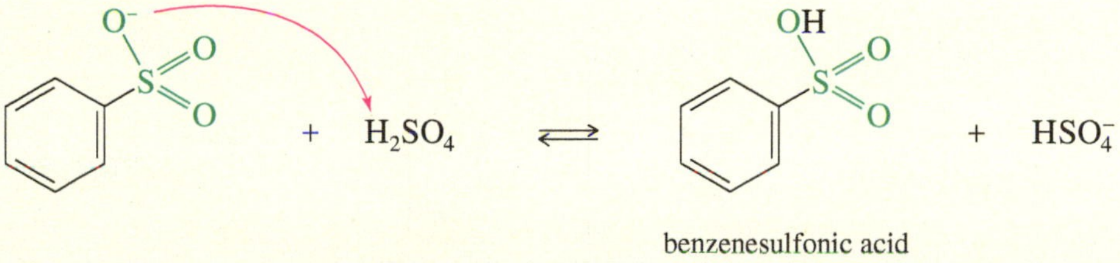

+ H_2SO_4 ⇌ benzenesulfonic acid + HSO_4^-

Sulfonation is economically important because alkylbenzene sulfonates are widely used as detergents. Sulfonation of an alkylbenzene (R = unbranched C_{10}–C_{14}) gives an alkylbenzenesulfonic acid, which is neutralized with base to give an alkylbenzene sulfonate detergent. Detergents are covered in more detail in Section 25-4.

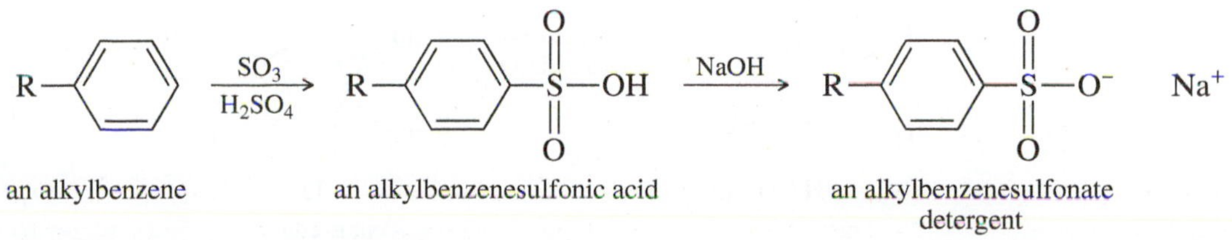

an alkylbenzene $\xrightarrow[H_2SO_4]{SO_3}$ an alkylbenzenesulfonic acid $\xrightarrow{NaOH}$ an alkylbenzenesulfonate detergent Na^+

PROBLEM 17-4

Use resonance forms to show that the dipolar sigma complex shown in the sulfonation of benzene has its positive charge delocalized over three carbon atoms and its negative charge delocalized over three oxygen atoms.

Application: Detergents

Sulfonated aromatic compounds are released into the environment from industrial and domestic uses of detergents. Environmental microbes easily metabolize alkylbenzenesulfonates with unbranched alkyl groups, so these compounds are considered to be ***biodegradable.***

The first synthetic detergents had branched alkyl groups. These branched alkylbenzenesulfonates are not easily biodegradable, and they accumulated in the environment. Lakes and rivers began to foam, and wildlife suffered from the surfactant properties of the detergents, which allowed water to wet their normally waterproof fur and feathers.

Desulfonation Sulfonation is reversible, and a sulfonic acid group may be removed from an aromatic ring by heating in dilute sulfuric acid. In practice, steam is often used as a source of both water and heat for **desulfonation.**

$$\text{benzenesulfonic acid } (C_6H_5SO_3H) + H_2O \xrightleftharpoons{H^+,\ \Delta \text{ (heat)}} \text{benzene (95\%)} + H_2SO_4$$

Desulfonation follows the same mechanistic path as sulfonation, except in the opposite order. A proton adds to a ring carbon to form a sigma complex, and then loss of sulfur trioxide gives the unsubstituted aromatic ring. Excess water removes SO_3 from the equilibrium by hydrating it to sulfuric acid.

$$C_6H_5SO_3^- + H^+ \rightleftharpoons [\text{sigma complex}] \rightleftharpoons C_6H_6 + SO_3$$

(resonance-delocalized)

$$(SO_3 + H_2O \rightleftharpoons H_2SO_4)$$

Protonation of the Aromatic Ring; Hydrogen–Deuterium Exchange Desulfonation involves protonation of an aromatic ring to form a sigma complex. Similarly, if a proton attacks benzene, the sigma complex can lose either of the two protons at the tetrahedral carbon. We can prove that a reaction has occurred by using a deuterium ion (D^+) rather than a proton and by showing that the product contains a deuterium atom in place of hydrogen. This experiment is easily accomplished by adding SO_3 to some D_2O (heavy water) to generate D_2SO_4. Benzene reacts to give a deuterated product.

$$C_6H_6 + D_3O^+ \rightleftharpoons [\text{sigma complex}] + D_2O \rightleftharpoons C_6H_5D + HD_2O^+$$

(D_3O^+)

(resonance-delocalized)

The reaction is reversible, and at equilibrium the final products reflect the D/H ratio of the solution. A large excess of deuterium gives a product with all six of the benzene hydrogens replaced by deuterium. This reaction serves as a synthesis of benzene-d_6 (C_6D_6), a common NMR solvent.

$$C_6H_6 \xrightleftharpoons{\text{large excess } D_2SO_4/D_2O} C_6D_6$$

benzene — benzene-d_6

17-5 Nitration of Toluene: The Effect of Alkyl Substitution

Up to now, we have considered only benzene as the substrate for electrophilic aromatic substitution. To synthesize more complicated aromatic compounds, we need to consider the effects other substituents might have on further substitutions. For example, toluene (methylbenzene) reacts with a mixture of nitric and sulfuric acids much like benzene does, but with some interesting differences:

1. Toluene reacts about 25 times faster than benzene under the same conditions. We say that toluene is **activated** toward electrophilic aromatic substitution and that the methyl group is an **activating group.**
2. Nitration of toluene gives a mixture of products, primarily those resulting from substitution at the ortho and para positions. Because of this preference, we say that the methyl group of toluene is an **ortho, para-director.**

toluene $\xrightarrow[H_2SO_4]{HNO_3}$ *o*-nitrotoluene (60%) + *m*-nitrotoluene (4%) + *p*-nitrotoluene (36%)

These product ratios show that the orientation of substitution is not random. If each C—H position were equally reactive, there would be equal amounts of ortho and meta substitution and half as much para substitution: 40% ortho, 40% meta, and 20% para. This is the statistical prediction based on the two ortho positions, two meta positions, and just one para position available for substitution.

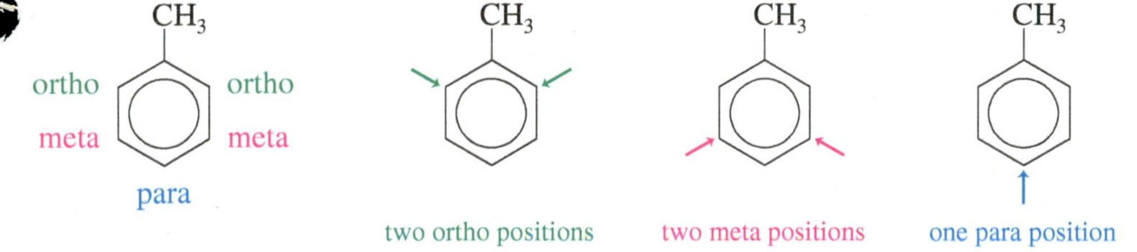

The rate-limiting step (the highest-energy transition state) for electrophilic aromatic substitution is the first step, formation of the sigma complex. This step is where the electrophile bonds to the ring, determining the substitution pattern. We can explain both the enhanced reaction rate and the preference for ortho and para substitution by considering the structures of the intermediate sigma complexes. In this endothermic reaction, the structure of the transition state leading to the sigma complex resembles the product, the sigma complex (Hammond postulate, Section 4-14). We are justified in using the stabilities of the sigma complexes to indicate the relative energies of the transition states leading to them.

When benzene reacts with the nitronium ion, the resulting sigma complex has the positive charge distributed over three secondary (2°) carbon atoms.

Benzene

benzene + $\overset{+}{N}O_2$ → [sigma complex resonance forms, each with + on a 2° carbon: 2° ↔ 2° ↔ 2°]

In ortho or para substitution of toluene, the positive charge is spread over two secondary carbons and one tertiary (3°) carbon (bearing the CH_3 group).

Application: Nitro Compounds

Aromatic nitro compounds are components of many drugs and other consumer products. For example, nitromide (3,5-dinitrobenzamide) is a potent antibacterial agent, and Ultrasüss (5-nitro-2-propoxyaniline) is 4100 times as sweet as cane sugar.

nitromide
(3,5-dinitrobenzamide)

Ultrasüss
(5-nitro-2-propoxyaniline)

Ortho attack

3° (favorable) 2° 2°

Para attack

2° 3° (favorable) 2°

Because the sigma complexes for ortho and para attack have resonance forms with tertiary carbocations, they are more stable than the sigma complex for nitration of benzene. Therefore, the ortho and para positions of toluene react faster than benzene.

The sigma complex for meta substitution has its positive charge spread over three 2° carbons; this intermediate is similar in energy to the intermediate for substitution of benzene. Therefore, meta substitution of toluene does not show the large rate enhancement seen with ortho and para substitution.

Meta attack

2° 2° 2°

The methyl group in toluene is electron-donating; it stabilizes the intermediate sigma complex and the rate-limiting transition state leading to its formation. This stabilizing effect is large when it is situated ortho or para to the site of substitution and the positive charge is delocalized onto the tertiary carbon atom. When substitution occurs at the meta position, the positive charge is not delocalized onto the tertiary carbon, and the methyl group has a smaller effect on the stability of the sigma complex. Figure 17-2 compares the reaction-energy diagrams for nitration of benzene and toluene at the ortho, meta, and para positions.

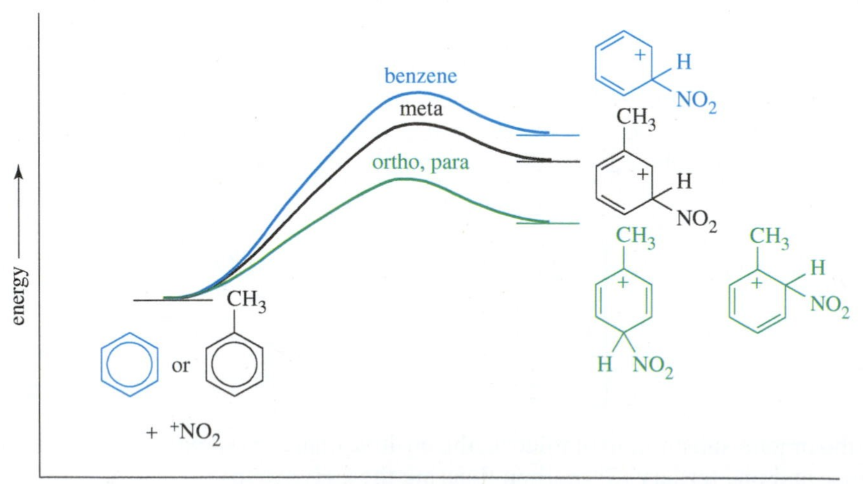

FIGURE 17-2
Energy profiles with an activating group. The methyl group of toluene stabilizes the sigma complexes and the transition states leading to them. This stabilization is most effective when the methyl group is ortho or para to the site of substitution.

17-6 Activating, Ortho, Para-Directing Substituents

The methyl group is not the only substituent that activates the benzene ring toward further substitution at the ortho and para positions. In this section, we consider other groups that share this effect, and in the next section we look at groups that have the opposite effect.

17-6A Alkyl Groups

The results observed with toluene are general for any alkylbenzene undergoing electrophilic aromatic substitution. Substitution ortho or para to the alkyl group gives a transition state and an intermediate with the positive charge shared by the tertiary carbon atom. As a result, alkylbenzenes undergo electrophilic aromatic substitution faster than benzene, and the products are predominantly ortho- and para-substituted. An alkyl group is therefore an activating substituent, and it is **ortho, para-directing.** This effect is called **inductive stabilization** because the alkyl group donates electron density through the sigma bond joining it with the benzene ring.

The reaction of ethylbenzene with bromine, catalyzed by ferric bromide, is another example of an electrophilic aromatic substitution that is enhanced by inductive stabilization. As with toluene, the rates of formation of the ortho- and para-substituted isomers are greatly enhanced with respect to the meta isomer.

ethylbenzene (CH_2CH_3 on benzene) $\xrightarrow[FeBr_3]{Br_2}$ *o*-bromo (38%) + *m*-bromo (<1%) + *p*-bromo (62%)

> **PROBLEM 17-5**
>
> **(a)** Draw a detailed mechanism for the $FeBr_3$-catalyzed reaction of ethylbenzene with bromine, and show why the sigma complex (and the transition state leading to it) is lower in energy for substitution at the ortho and para positions than it is for substitution at the meta position.
>
> **(b)** Explain why *m*-xylene undergoes nitration 100 times faster than *p*-xylene.

> **PROBLEM 17-6**
>
> Styrene (vinylbenzene) undergoes electrophilic aromatic substitution much faster than benzene, and the products are found to be primarily ortho- and para-substituted styrenes. Use resonance forms of the intermediates to explain these results.

17-6B Substituents with Nonbonding Electrons

Alkoxy Groups Anisole (methoxybenzene) undergoes nitration about 10,000 times faster than benzene and about 400 times faster than toluene. This result seems curious because oxygen is a strongly electronegative group, yet it donates electron density to stabilize the transition state and the sigma complex. Recall that the nonbonding electrons of an oxygen atom adjacent to a carbocation stabilize the positive charge through resonance.

$$\left[\;>\overset{+}{C}-\ddot{O}-R \;\longleftrightarrow\; >C=\overset{+}{O}-R \;\right]$$

only six valence electrons (on $\overset{+}{C}$) — each atom has eight valence electrons

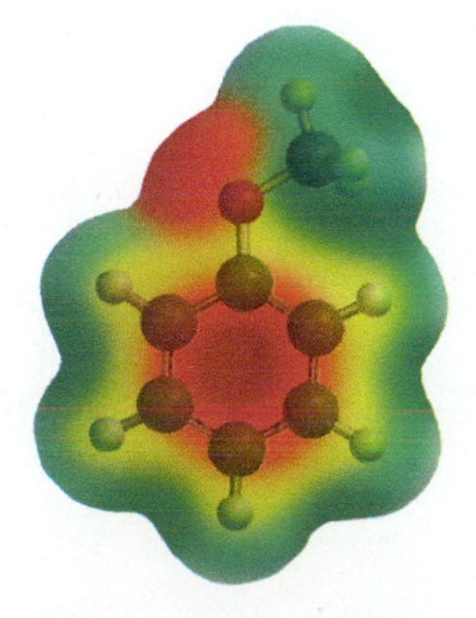

The EPM of anisole shows the aromatic ring to be electron-rich (red), consistent with the observation that anisole is strongly activated toward reactions with electrophiles.

The second resonance form puts the positive charge on the electronegative oxygen atom, but it has more covalent bonds, and it provides each atom with an octet in its valence shell. This type of stabilization is called **resonance stabilization,** and the oxygen atom is called **resonance-donating** or **pi-donating** because it donates electron density through a pi bond in one of the resonance structures. Like alkyl groups, the methoxy group of anisole preferentially activates the ortho and para positions.

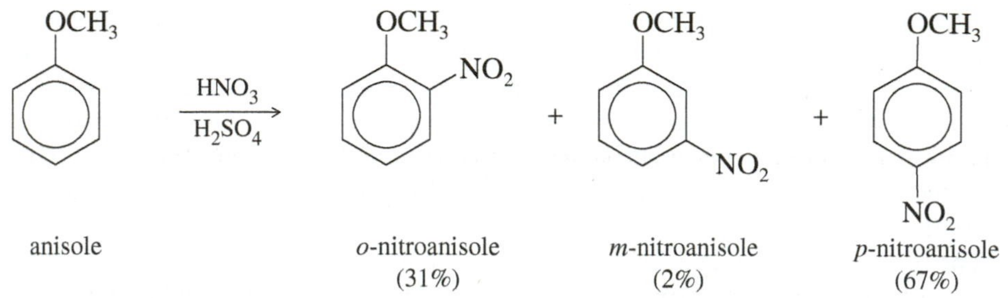

anisole $\xrightarrow[H_2SO_4]{HNO_3}$ *o*-nitroanisole (31%) + *m*-nitroanisole (2%) + *p*-nitroanisole (67%)

Resonance forms show that the methoxy group effectively stabilizes the sigma complex if it is ortho or para to the site of substitution, but not if it is meta. Resonance stabilization is provided by a pi bond between the $-OCH_3$ substituent and the ring.

Ortho attack

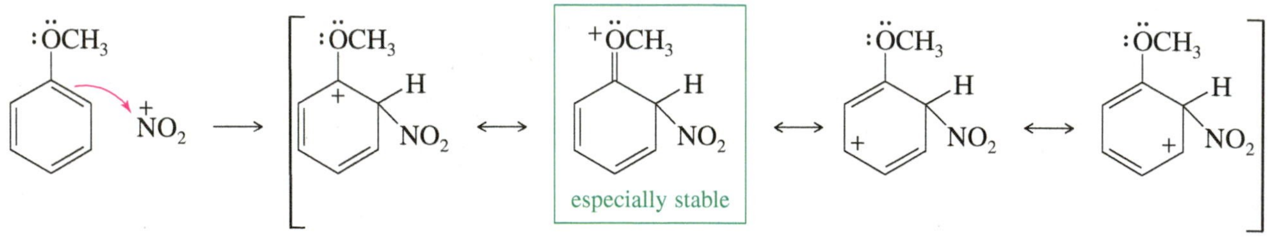

especially stable

Meta attack

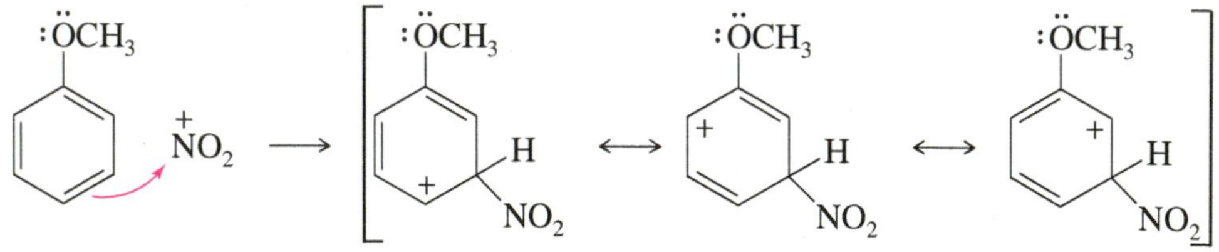

Para attack

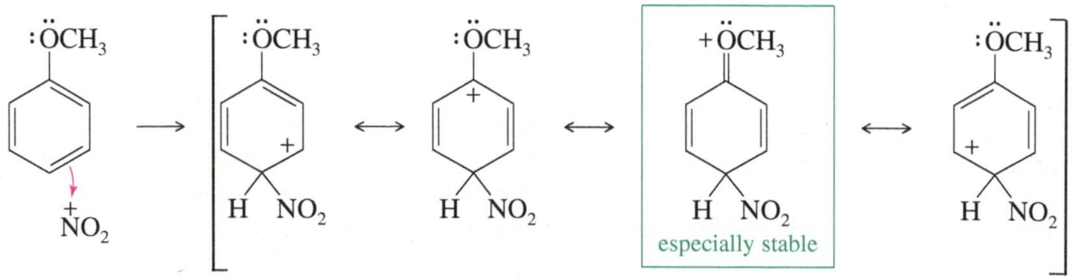

especially stable

A methoxy group is so strongly activating that anisole quickly brominates in water without a catalyst. In the presence of excess bromine, this reaction proceeds to the tribromide.

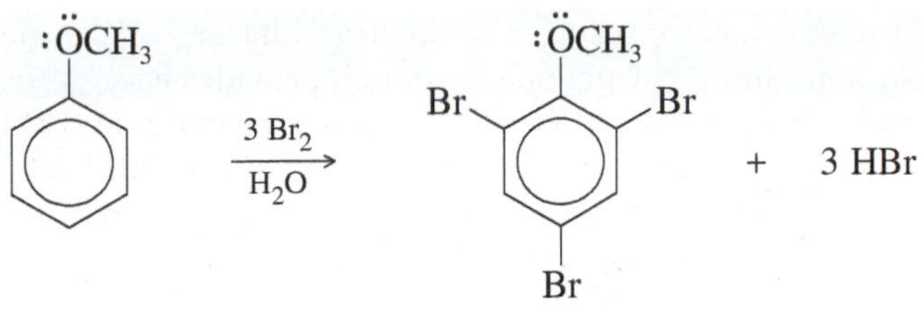

$$\text{anisole} \xrightarrow[H_2O]{3\ Br_2} \text{2,4,6-tribromoanisole (100\%)} + 3\ HBr$$

PROBLEM 17-7

Propose a mechanism for the bromination of ethoxybenzene to give *o*- and *p*-bromoethoxybenzene.

Amine Groups Like an alkoxy group, a nitrogen atom with a nonbonding pair of electrons serves as a powerful activating group. For example, aniline undergoes a fast bromination (without a catalyst) in bromine water to give the tribromide. Sodium bicarbonate is added to neutralize the HBr formed and to prevent protonation of the basic amino (—NH_2) group (see Problem 17-10).

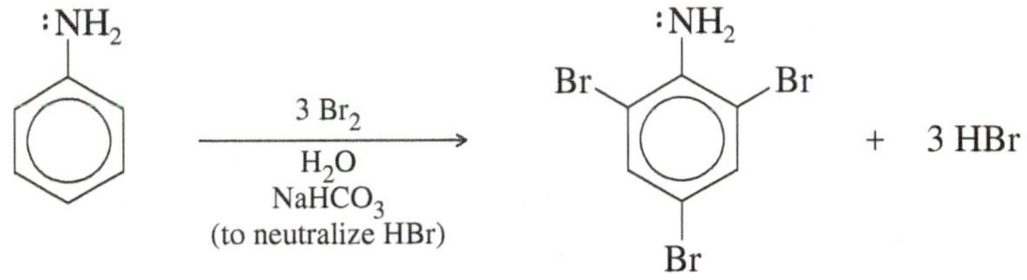

$$\text{aniline} \xrightarrow[\substack{H_2O \\ NaHCO_3 \\ \text{(to neutralize HBr)}}]{3\ Br_2} \text{2,4,6-tribromoaniline (100\%)} + 3\ HBr$$

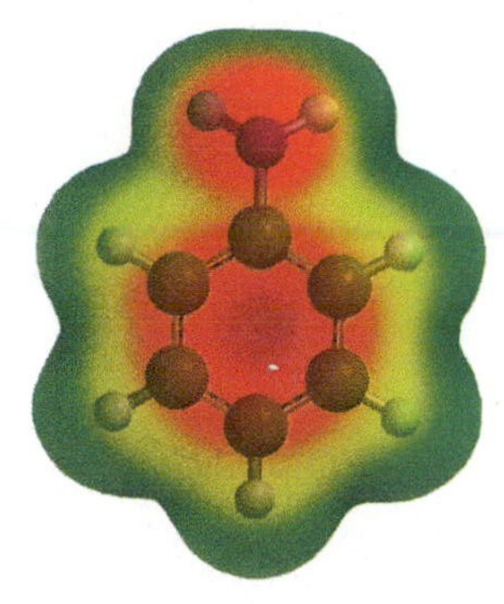

The EPM of aniline shows the aromatic ring to be even more electron-rich (red) than that of anisole.

Nitrogen's nonbonding electrons provide resonance stabilization to the sigma complex if attack takes place ortho or para to the position of the nitrogen atom.

Ortho attack

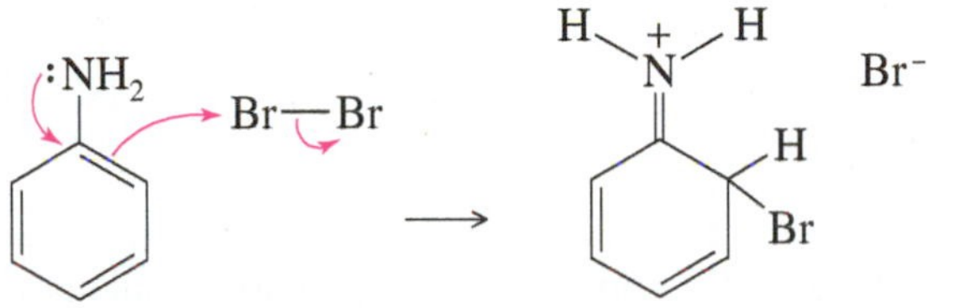

(plus other resonance forms)

Para attack

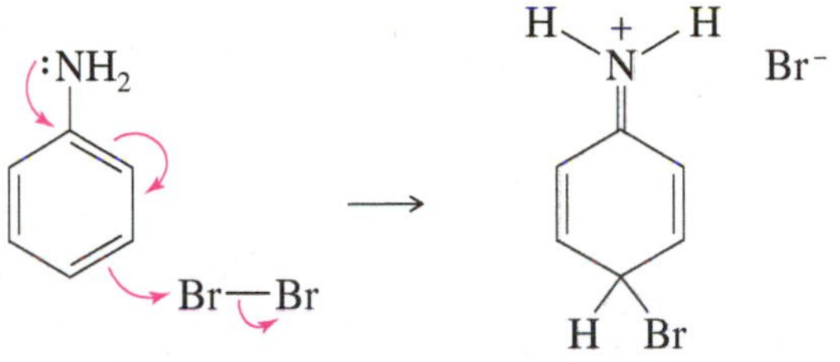

(plus other resonance forms)

PROBLEM 17-8

Draw all the resonance forms for the sigma complexes corresponding to bromination of aniline at the ortho, meta, and para positions.

Thus, any substituent with a lone pair of electrons on the atom bonded to the ring can provide resonance stabilization to a sigma complex. Several examples are illustrated next in decreasing order of their activation of an aromatic ring. All these substituents are strongly activating, and they are all ortho, para-directing.

SUMMARY Activating, Ortho, Para-Directors

Groups

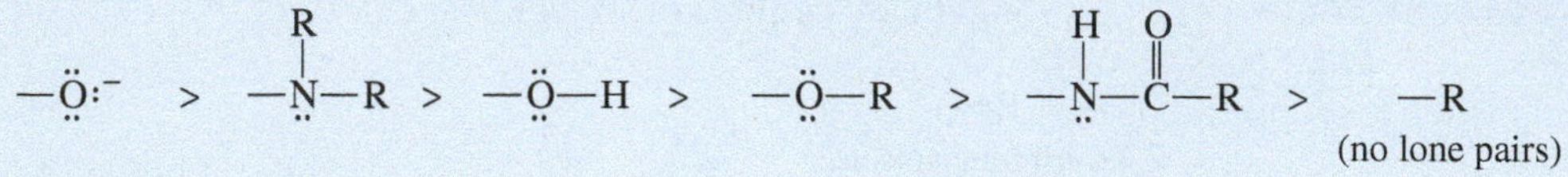

$-\ddot{O}:^- > -\ddot{N}R-R > -\ddot{O}-H > -\ddot{O}-R > -\ddot{N}H-C(=O)-R > -R$

(no lone pairs)

Compounds

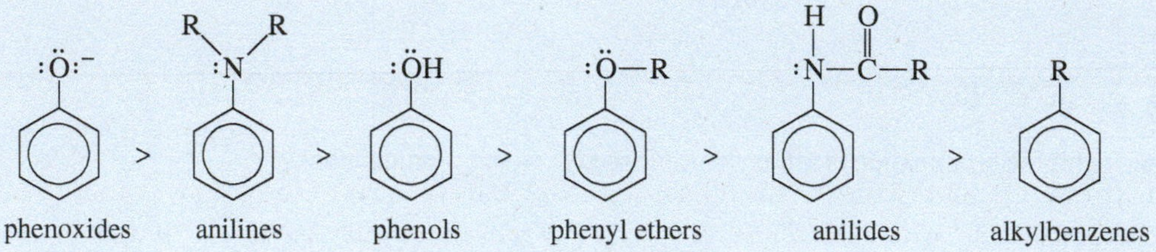

phenoxides > anilines > phenols > phenyl ethers > anilides > alkylbenzenes

> **PROBLEM 17-9**
>
> When bromine is added to two beakers, one containing phenyl isopropyl ether and the other containing cyclohexene, the bromine color in both beakers disappears. What observation could you make while performing this test that would allow you to distinguish the alkene from the aryl ether?

17-7 Deactivating, Meta-Directing Substituents

Nitrobenzene is about 100,000 times *less* reactive than benzene toward electrophilic aromatic substitution. For example, nitration of nitrobenzene requires concentrated nitric and sulfuric acids at temperatures above 100 °C. Nitration proceeds slowly, giving the meta isomer as the major product.

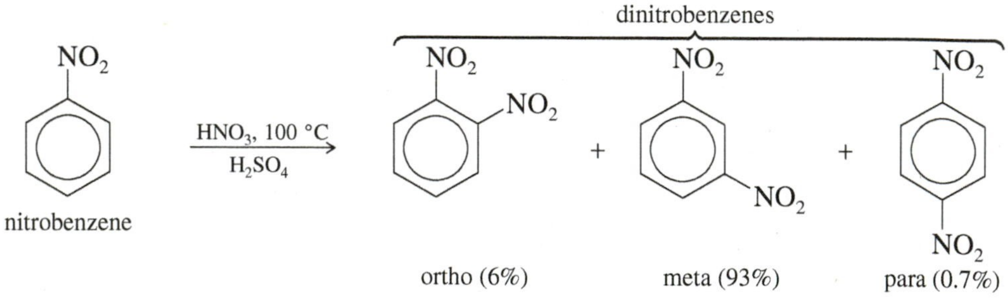

nitrobenzene $\xrightarrow[H_2SO_4]{HNO_3,\ 100\ °C}$ dinitrobenzenes: ortho (6%) + meta (93%) + para (0.7%)

These results should not be surprising. We have already seen that a substituent on a benzene ring has its greatest effect on the carbon atoms ortho and para to the substituent. An electron-donating substituent activates primarily the ortho and para positions, and an electron-withdrawing substituent (such as a nitro group) deactivates primarily the ortho and para positions.

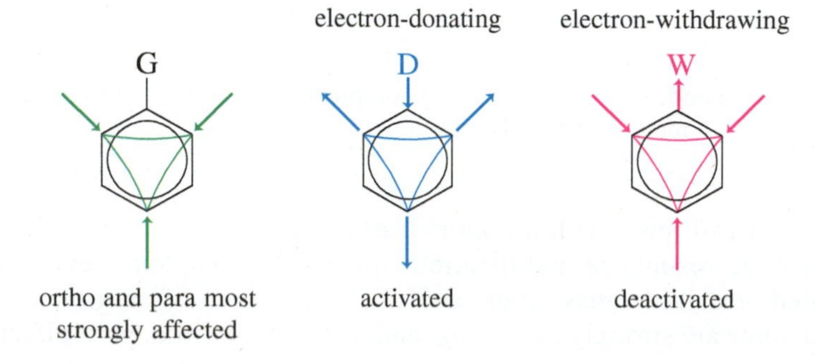

This selective deactivation leaves the meta positions the most reactive, and meta substitution is seen in the products. **Meta-directors,** often called **meta-allowing** substituents, deactivate the meta position less than the ortho and para positions, allowing meta substitution.

We can show why the nitro group is a strong **deactivating group** by considering its resonance forms. No matter how we position the electrons in a Lewis dot diagram, the nitrogen atom always has a formal positive charge.

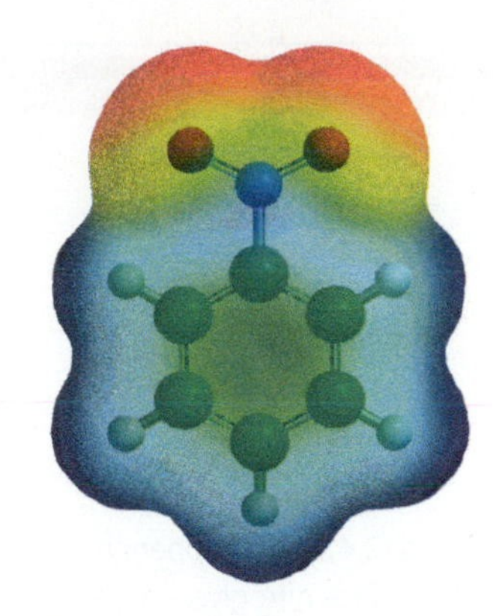

The EPM of nitrobenzene shows the aromatic ring to be electron-poor (blue tinge), consistent with the observation that nitrobenzene is deactivated toward reactions with electrophiles.

The positively charged nitrogen inductively withdraws electron density from the aromatic ring. This aromatic ring is less electron-rich than benzene, so it is deactivated toward reactions with electrophiles.

The following reactions show why this deactivating effect is strongest at the ortho and para positions. Each sigma complex has its positive charge spread over three carbon atoms. In ortho and para substitution, one of the carbon atoms bearing this positive charge is the carbon attached to the positively charged nitrogen atom of the nitro group. Because like charges repel, this close proximity of two positive charges is especially unstable.

Ortho attack

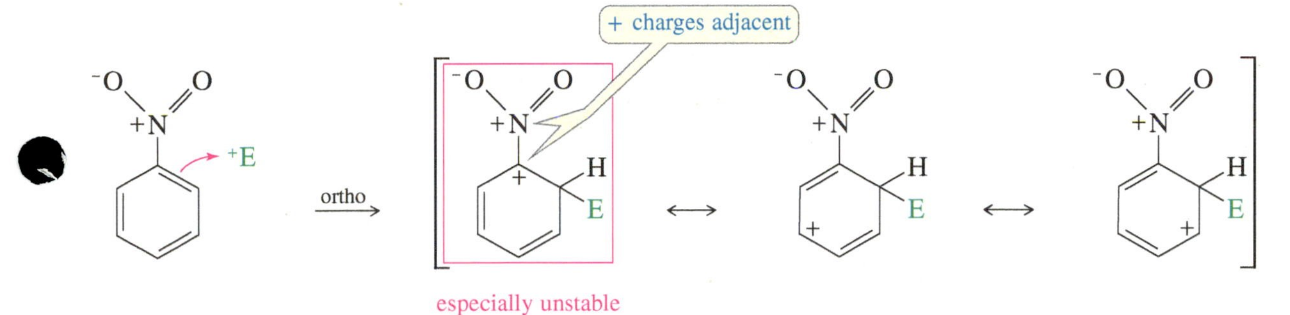

Meta attack

meta

Para attack

para

+ charges adjacent

especially unstable

Application: Safer Explosives

picric acid

Picric acid (2,4,6-trinitrophenol) was used to replace nitroglycerin (see Section 11-13B) during World War I as a safer high explosive for artillery shells. Nitroglycerin is sensitive to shock and occasionally detonated within the artillery barrel. Picric acid is made from phenol, which is activated toward EAS to introduce the three nitro groups.

TNT

TNT (2,4,6-trinitrotoluene) replaced picric acid because TNT is more difficult to detonate, making it safer to handle. The standard procedure for casting shaped charges is to melt the TNT, pour it into a mold, and allow it to re-solidify in the desired shape.

In the sigma complex for meta substitution, the carbon bonded to the nitro group does not share the positive charge of the ring. This is a more stable situation because the positive charges are farther apart. As a result, nitrobenzene reacts primarily at the meta position. We can summarize by saying that the nitro group is a deactivating group and that it is a meta-director (or meta-allower).

The energy diagram in Figure 17-3 compares the energies of the transition states and intermediates leading to ortho, meta, and para substitution of nitrobenzene with those for benzene. Notice that a higher activation energy is involved for substitution of nitrobenzene at any position, resulting in slower reaction rates than for benzene.

Just as activating substituents are all ortho, para-directors, most deactivating substituents are meta-directors. In general, deactivating substituents are groups with a positive charge (or a partial positive charge) on the atom bonded to the aromatic ring. As we saw with the nitro group, this positively charged atom repels any positive charge on the adjacent carbon atom of the ring. Of the possible sigma complexes, only the one corresponding to meta substitution avoids putting a positive charge on this ring carbon. For example, the partial positive charge on a carbonyl carbon allows substitution primarily at the meta position:

Ortho attack

acetophenone

+ charges adjacent (unfavorable)

+ charge here in other resonance forms

Meta attack

+ charge here in other resonance forms

This sigma complex does not place the positive charge on the ring carbon bearing the carbonyl group.

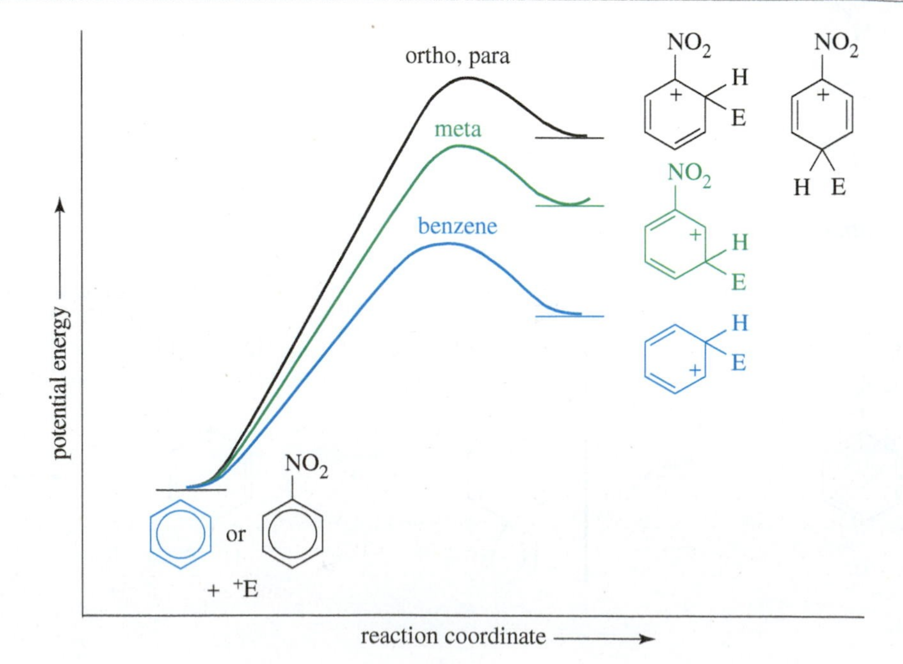

FIGURE 17-3
Energy profiles with a deactivating group. Nitrobenzene is deactivated toward electrophilic aromatic substitution at *any* position, but deactivation is strongest at the ortho and para positions. Reaction occurs at the meta position, but it is slower than the reaction with benzene.

The following summary table lists some common substituents that are deactivating and meta-directing. Resonance forms are also given to show how a positive charge arises on the atom bonded to the aromatic ring.

SUMMARY Deactivating, Meta-Directors

Group	*Resonance Forms*	*Example*
$-NO_2$ nitro	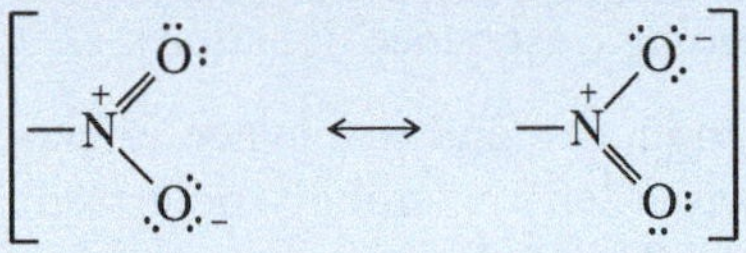 $[-\overset{+}{N}(=O)-O^- \longleftrightarrow -\overset{+}{N}(-O^-)=O]$	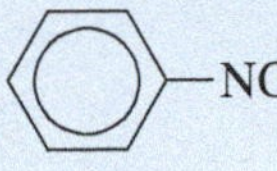 $C_6H_5-NO_2$ nitrobenzene
$-SO_3H$ sulfonic acid	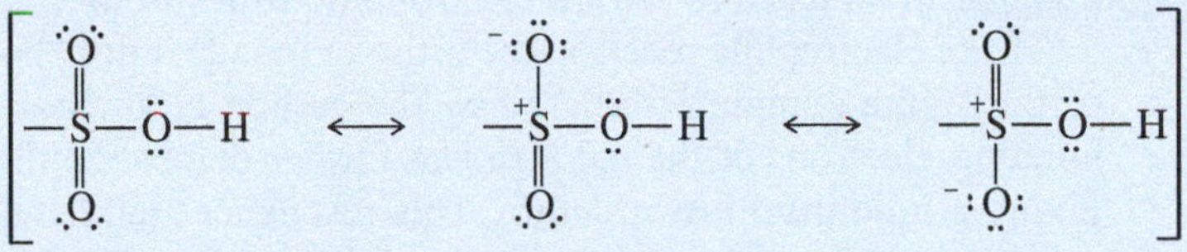 $[-S(=O)_2-O-H \longleftrightarrow -\overset{+}{S}(-O^-)(=O)-O-H \longleftrightarrow -\overset{+}{S}(=O)(-O^-)-O-H]$	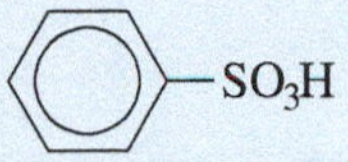 $C_6H_5-SO_3H$ benzenesulfonic acid
$-C\equiv N:$ cyano	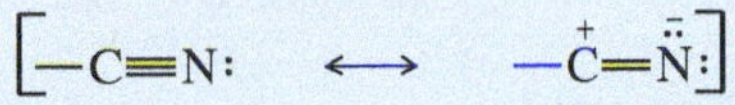 $[-C\equiv N: \longleftrightarrow -\overset{+}{C}=\overset{-}{N}:]$	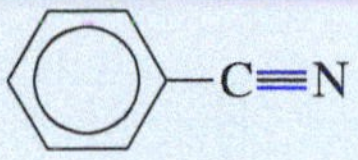 $C_6H_5-C\equiv N$ benzonitrile
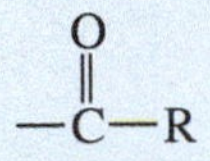 $-C(=O)-R$ ketone or aldehyde	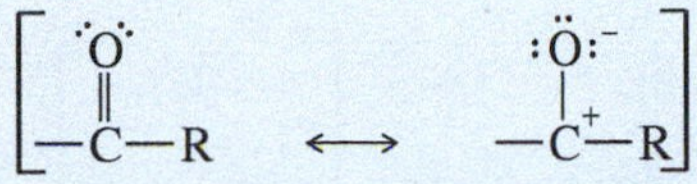 $[-C(=O)-R \longleftrightarrow -\overset{+}{C}(-O^-)-R]$	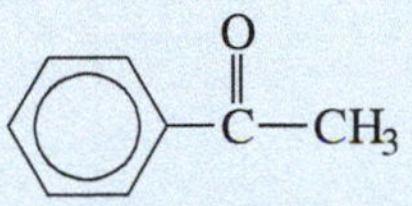 $C_6H_5-C(=O)-CH_3$ acetophenone
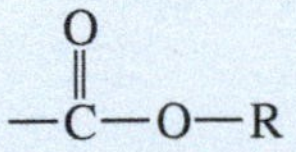 $-C(=O)-O-R$ ester	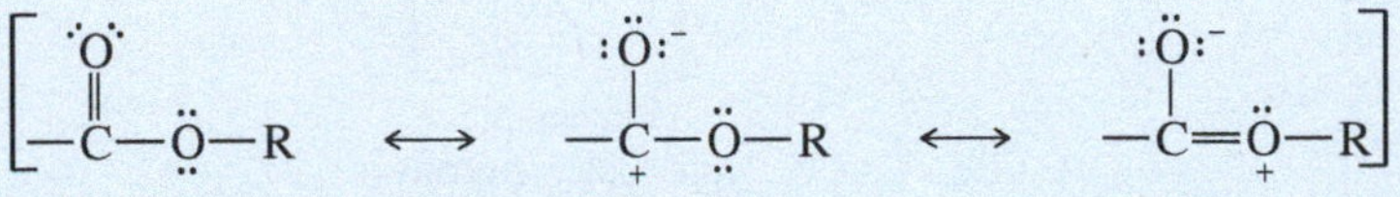 $[-C(=O)-O-R \longleftrightarrow -\overset{+}{C}(-O^-)-O-R \longleftrightarrow -C(-O^-)=\overset{+}{O}-R]$	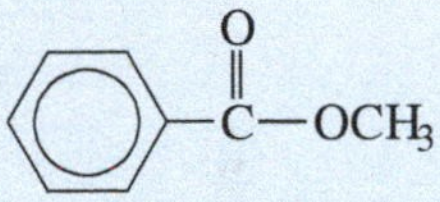 $C_6H_5-C(=O)-OCH_3$ methyl benzoate
$-\overset{+}{N}R_3$ quaternary ammonium	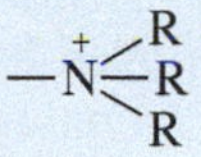 $-\overset{+}{N}R_3$	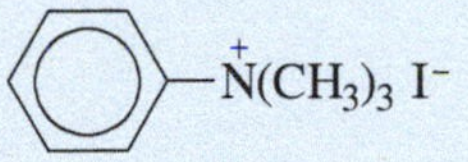 $C_6H_5-\overset{+}{N}(CH_3)_3\ I^-$ trimethylanilinium iodide

PROBLEM 17-10

In an aqueous solution containing sodium bicarbonate, aniline reacts quickly with bromine to give 2,4,6-tribromoaniline. Nitration of aniline requires very strong conditions, however, and the yields (mostly *m*-nitroaniline) are poor.

(a) What conditions are used for nitration, and what form of aniline is present under these conditions?

(b) Explain why nitration of aniline is so sluggish and why it gives mostly meta substitution.

***(c)** Although nitration of aniline is slow and gives mostly meta substitution, nitration of acetanilide ($PhNHCOCH_3$) goes quickly and gives mostly para substitution. Use resonance forms to explain this difference in reactivity.

17-8 Halogen Substituents: Deactivating, but Ortho, Para-Directing

The halobenzenes are exceptions to the general rules. Halogens are deactivating groups, yet they are ortho, para-directors. We can explain this unusual combination of properties by considering that

1. the halogens are strongly electronegative, withdrawing electron density from a carbon atom through the sigma bond (inductive withdrawal), and
2. the halogens have nonbonding electrons that can donate electron density through pi bonding (resonance donation).

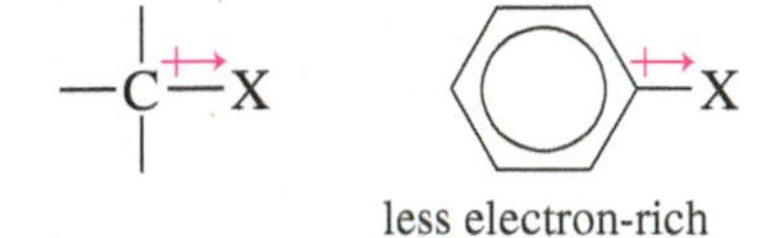

These inductive and resonance effects oppose each other. The carbon–halogen bond (shown at left) is strongly polarized, with the carbon atom at the positive end of the dipole. This polarization draws electron density away from the benzene ring, making it less reactive toward electrophilic substitution.

If an electrophile reacts at the ortho or para position, however, the positive charge of the sigma complex is shared by the carbon atom bearing the halogen. The nonbonding electrons of the halogen can further delocalize the charge onto the halogen, giving a **halonium ion** structure. This resonance stabilization allows a halogen to be pi-donating, even though it is sigma-withdrawing.

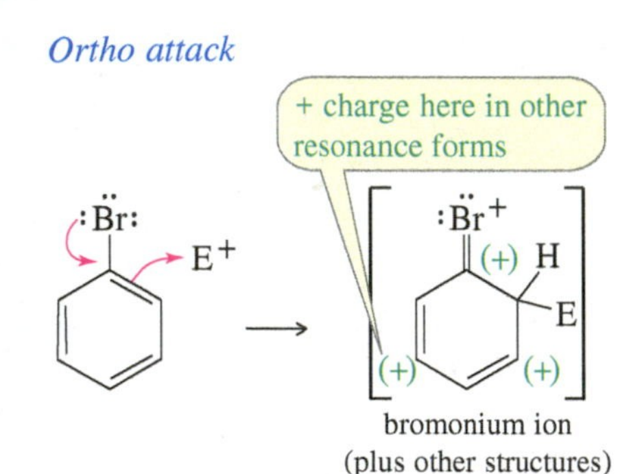

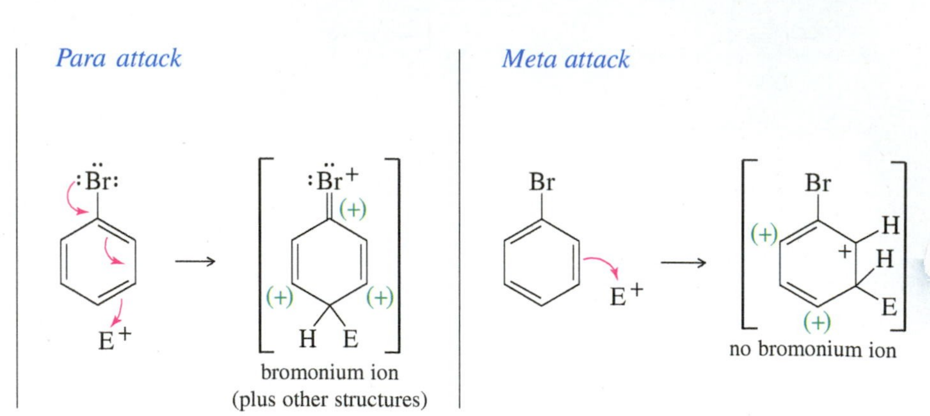

Reaction at the meta position gives a sigma complex whose positive charge is not delocalized onto the halogen-bearing carbon atom. Therefore, the meta intermediate is not stabilized by the halonium ion structure. The following reaction illustrates the preference for ortho and para substitution in the nitration of chlorobenzene.

chlorobenzene $\xrightarrow[H_2SO_4]{HNO_3}$ ortho (35%) + meta (1%) + para (64%)

(Cl-substituted rings; products bear NO_2 at the ortho, meta, and para positions, respectively.)

Figure 17-4 shows the effect of the halogen atom graphically, with an energy diagram comparing energies of the transition states and intermediates for electrophilic attack on chlorobenzene and benzene. Higher energies are required for the reactions of chlorobenzene, especially for attack at the meta position.

PROBLEM-SOLVING HINT

Remember which substituents are activating and which are deactivating. Activators are ortho, para-directing, and deactivators are meta-directing, except for the halogens.

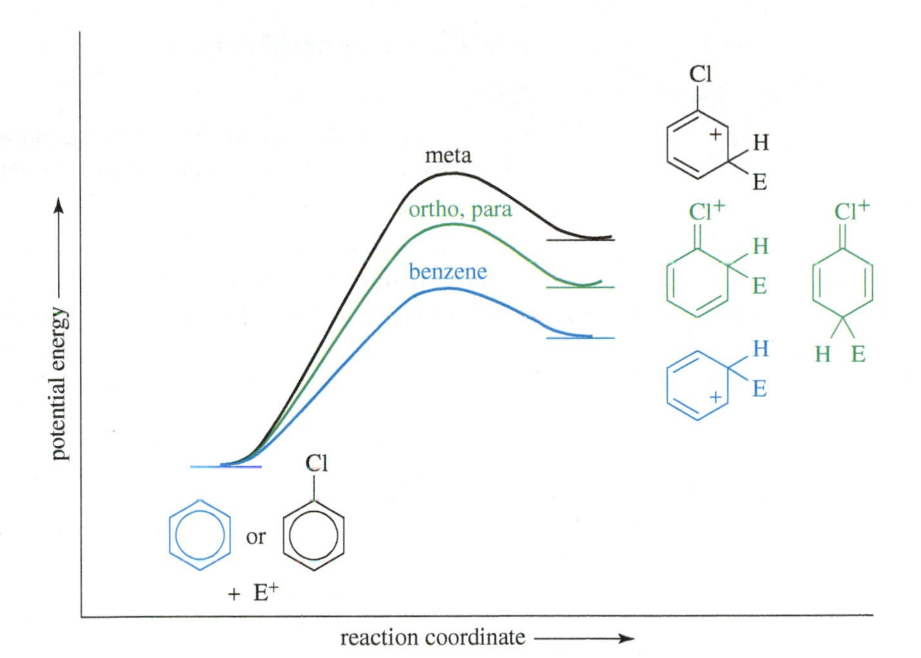

FIGURE 17-4
Energy profiles with halogen substituents. The energies of the intermediates and transition states are higher for chlorobenzene than for benzene. The highest energy results from substitution at the meta position; the energies for ortho and para substitution are slightly lower because of stabilization by the halonium ion structure.

PROBLEM 17-11

Draw all the resonance forms of the sigma complex for nitration of bromobenzene at the ortho, meta, and para positions. Point out why the intermediate for meta substitution is less stable than the other two.

SUMMARY Directing Effects of Substituents

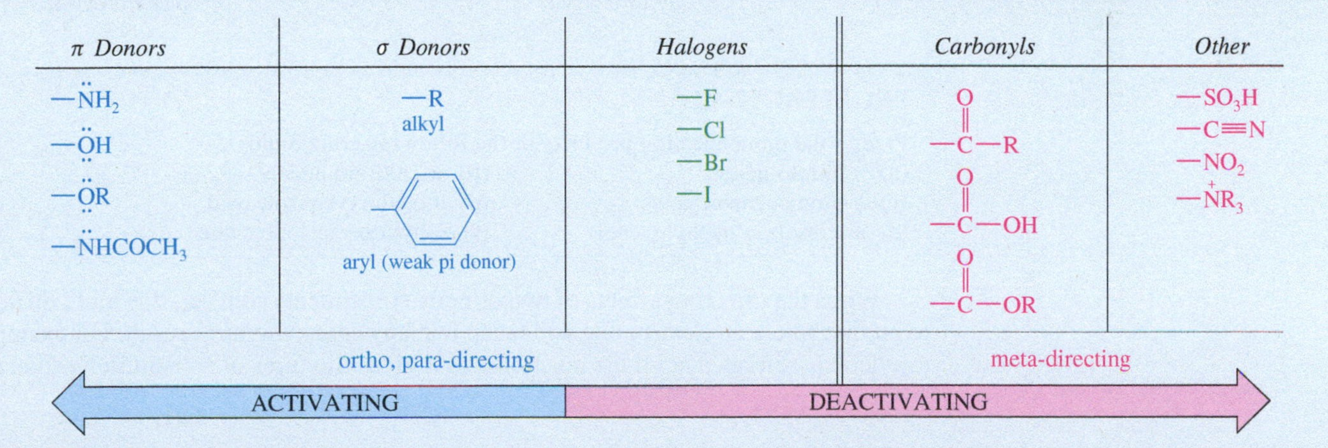

π Donors
—NH2
—OH
—OR
—NHCOCH3
σ Donors
—R
alkyl
aryl (weak pi donor)
Halogens
—F
—Cl
—Br
—I
Carbonyls
—C—R
—C—OH
—C—OR
Other
—SO3H
—C≡N
—NO2
—NR3
ortho, para-directing
meta-directing
ACTIVATING
DEACTIVATING

17-9 Effects of Multiple Substituents on Electrophilic Aromatic Substitution

Two or more substituents exert a combined effect on the reactivity of an aromatic ring. If the groups reinforce each other, the result is easy to predict. For example, we can predict that all the xylenes (dimethylbenzenes) are activated toward electrophilic substitution because the two methyl groups are both activating. In the case of a nitrobenzoic acid, both substituents are deactivating, so we predict that a nitrobenzoic acid is deactivated toward attack by an electrophile.

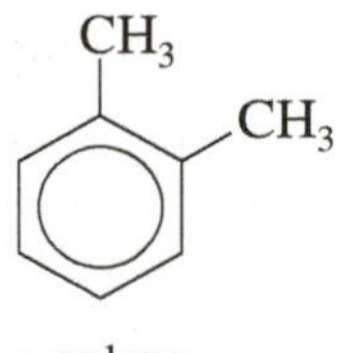

CH_3 / CH_3 — *o*-xylene, activated 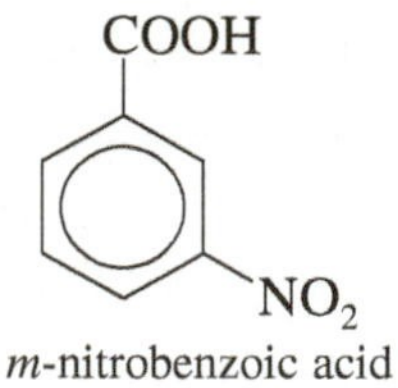
COOH / NO_2 — *m*-nitrobenzoic acid, deactivated 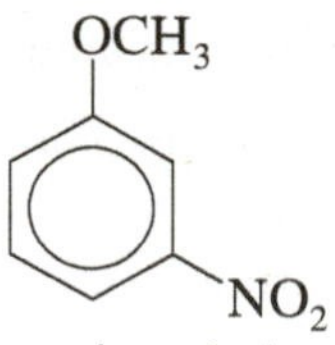
OCH_3 / NO_2 — *m*-nitroanisole, not obvious

The orientation of addition is also easy to predict in many cases. For example, in *m*-xylene there are two positions ortho to one of the methyl groups and para to the other. Electrophilic substitution occurs primarily at these two equivalent positions. There may be some substitution at the position between the two methyl groups (ortho to both), but this position is sterically hindered, and it is less reactive than the other two activated positions. In *p*-nitrotoluene, the methyl group directs an electrophile toward its ortho positions. The nitro group directs toward the same locations because they are its meta positions.

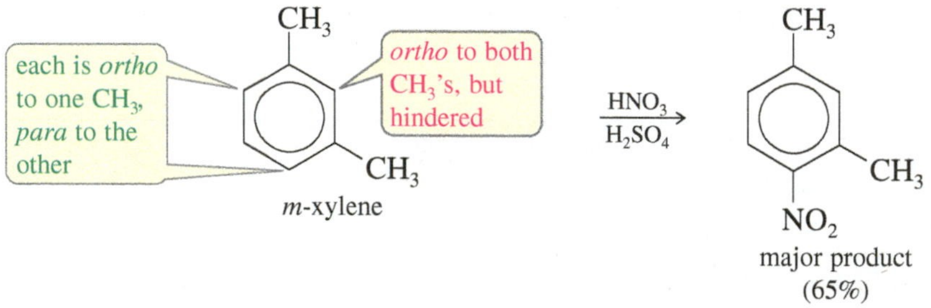

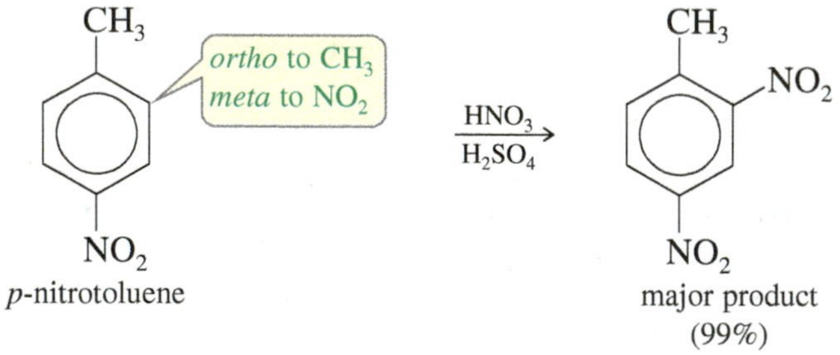

each is *ortho* to one CH_3, *para* to the other

ortho to both CH_3's, but hindered

m-xylene $\xrightarrow[H_2SO_4]{HNO_3}$ major product (65%)

ortho to CH_3 *meta* to NO_2

p-nitrotoluene $\xrightarrow[H_2SO_4]{HNO_3}$ major product (99%)

PROBLEM 17-12

Predict the mononitration products of the following compounds.

(a) *o*-nitrotoluene
(b) *m*-chlorotoluene
(c) *o*-bromobenzoic acid
(d) *p*-methoxybenzoic acid
(e) *m*-cresol (*m*-methylphenol)
(f) *o*-hydroxyacetophenone

When the directing effects of two or more substituents conflict, it is more difficult to predict where an electrophile will react. In many cases, mixtures result. For example, *o*-xylene is activated at all the positions, so it gives mixtures of substitution products.

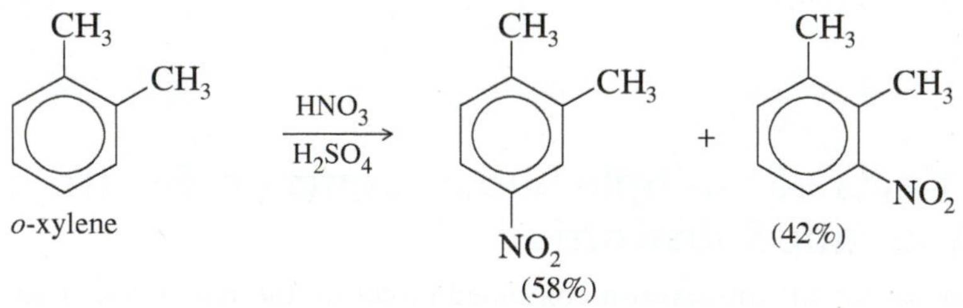

o-xylene $\xrightarrow[H_2SO_4]{HNO_3}$ (58%) + (42%)

When there is a conflict between an activating group and a deactivating group, the activating group usually directs the substitution. We can make an important generalization:

Activating groups are usually stronger directors than deactivating groups.

In fact, it is helpful to separate substituents into three classes, from strongest to weakest.

1. Powerful ortho, para-directors that stabilize the sigma complexes through resonance. Examples are —OH, —OR, and —NR_2 groups.
2. Moderate ortho, para-directors, such as alkyl groups and halogens.
3. All meta-directors.

$$-OH, -OR, -NR_2 \quad > \quad -R, -X \quad > \quad -C(=O)-R, -SO_3H, -NO_2$$

If two substituents direct an incoming electrophile toward different reaction sites, the substituent in the stronger class predominates. If both are in the same class, mixtures are likely. In the case of *m*-nitroanisole, the stronger group predominates and directs the incoming substituent. The methoxy group is a stronger director than the nitro group, and substitution occurs ortho and para to the methoxy group. Steric effects prevent much substitution at the crowded position ortho to both the methoxy group and the nitro group.

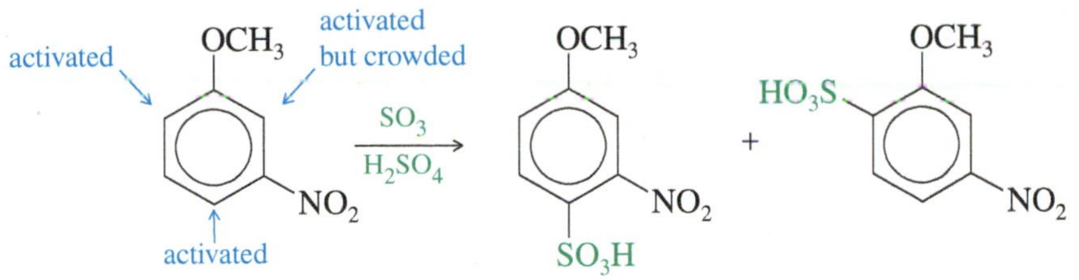

activated; activated but crowded; activated — OCH_3, NO_2 — *m*-nitroanisole $\xrightarrow[H_2SO_4]{SO_3}$ (OCH_3, NO_2, SO_3H) + (OCH_3, HO_3S, NO_2) — major products

SOLVED PROBLEM 17-1

Predict the major product(s) of bromination of *p*-chloroacetanilide.

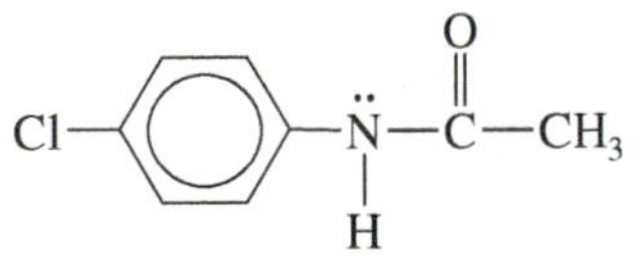

$$Cl-C_6H_4-\ddot{N}(H)-C(=O)-CH_3$$

SOLUTION

The amide group (—$NHCOCH_3$) is a strong activating and directing group because the nitrogen atom with its nonbonding pair of electrons is bonded to the aromatic ring. The amide group is a stronger director than the chlorine atom, and substitution occurs mostly at the positions ortho to the amide. Like an alkoxy group, the amide is a particularly strong activating group, and the reaction gives some of the dibrominated product.

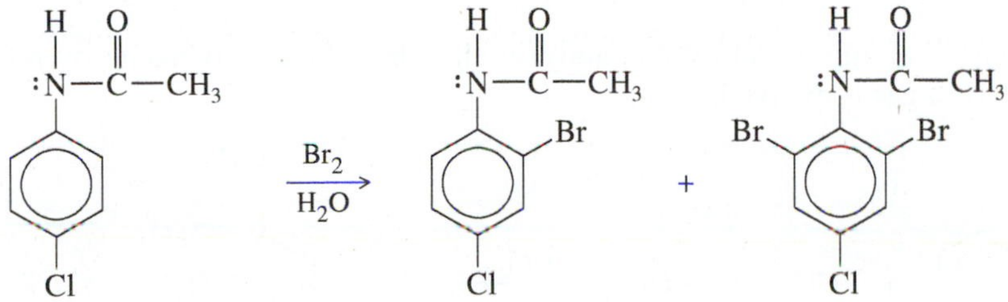

p-chloroacetanilide $\xrightarrow[H_2O]{Br_2}$ (2-bromo product: :N(H)—C(=O)—CH_3, Br, Cl) + (2,6-dibromo product: :N(H)—C(=O)—CH_3, Br, Br, Cl)

PROBLEM-SOLVING HINT

To predict products of compounds with multiple substituents, look for the most strongly activating substituent(s).

PROBLEM 17-13

Predict the mononitration products of the following aromatic compounds.

(a) *p*-methylanisole (b) *m*-nitrochlorobenzene
(c) *p*-chlorophenol (d) *m*-nitroanisole

(continued)

(e) o-methylacetanilide (structure: ring with CH_3 and $-NH-C(=O)-CH_3$)

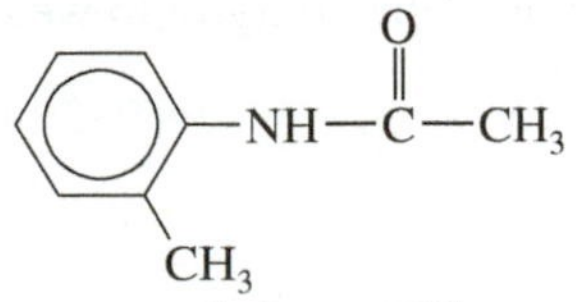

(f) $CH_3-C(=O)-NH-C_6H_4-C(=O)-NH_2$

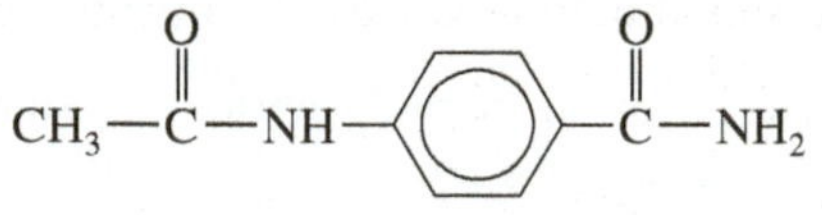

(Consider the structures of these groups. One is activating, and the other is deactivating.)

PROBLEM 17-14

Biphenyl is two benzene rings joined by a single bond. The site of substitution for a biphenyl is determined by (1) which phenyl ring is more activated (or less deactivated), and (2) which position on that ring is most reactive, using the fact that a phenyl substituent is activating and ortho, para-directing.

(a) Use resonance forms of a sigma complex to show why a phenyl substituent should be ortho, para-directing.

(b) Predict the mononitration products of the following compounds.

(i) biphenyl (ii) OH (iii) $-C(=O)-CH_3$

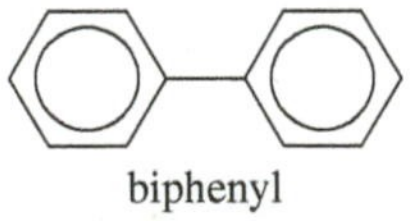

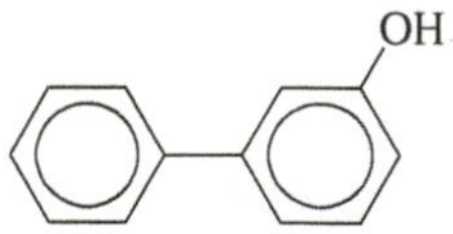

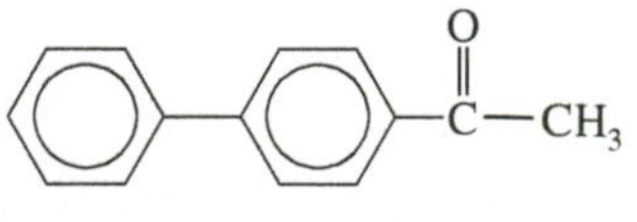

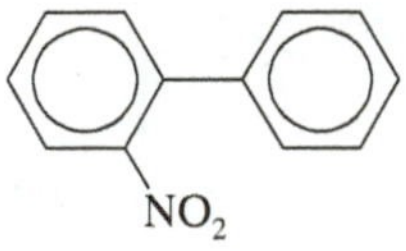

(iv) NO_2 (v) O, O (vi)

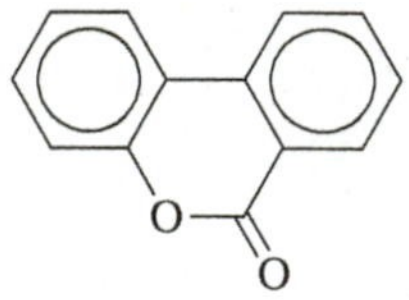

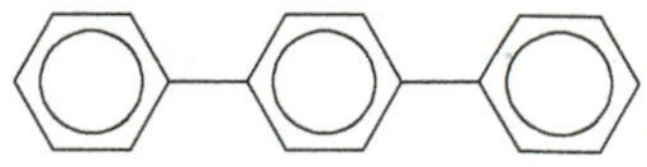

PROBLEM-SOLVING HINT

When predicting substitution products for compounds with more than one ring, first decide which ring is more activated (or less deactivated). Then consider only that ring, and decide which position is most reactive.

17-10 The Friedel–Crafts Alkylation

Carbocations are perhaps the most important electrophiles capable of substituting onto aromatic rings, because this substitution forms a new carbon–carbon bond. Reactions of carbocations with aromatic compounds were first studied in 1877 by the French alkaloid chemist Charles Friedel and his American partner, James Crafts. In the presence of Lewis acid catalysts such as aluminum chloride ($AlCl_3$) or ferric chloride ($FeCl_3$), alkyl halides were found to alkylate benzene to give alkylbenzenes. This useful reaction is called the **Friedel–Crafts alkylation.**

Friedel–Crafts alkylation

$$C_6H_5-H + R-X \xrightarrow[(AlCl_3,\ FeBr_3,\ \text{etc.})]{\text{Lewis acid}} C_6H_5-R + H-X$$

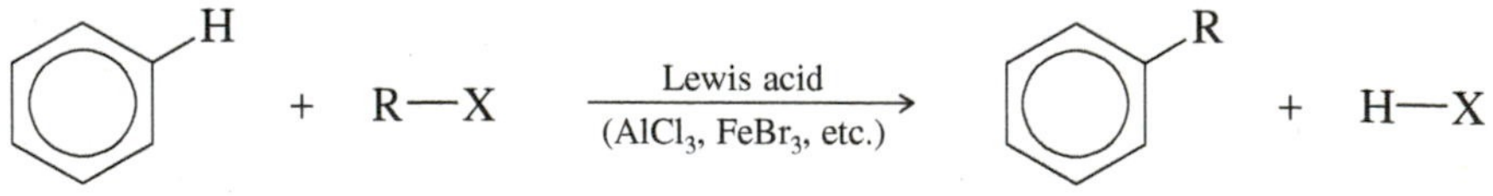

(X = Cl, Br, I)

For example, aluminum chloride catalyzes the alkylation of benzene by *tert*-butyl chloride. HCl gas is evolved.

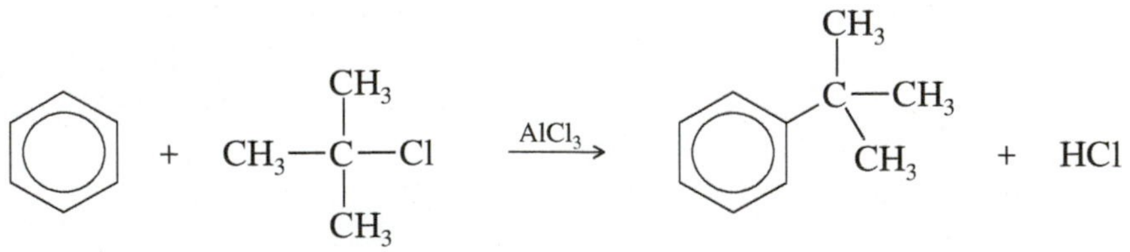

$$C_6H_6 + (CH_3)_3C-Cl \xrightarrow{AlCl_3} C_6H_5-C(CH_3)_3 + HCl$$

benzene, *tert*-butyl chloride, *tert*-butylbenzene (90%)

This alkylation is a typical electrophilic aromatic substitution, with the *tert*-butyl cation acting as the electrophile. The *tert*-butyl cation is formed by reaction of *tert*-butyl chloride with the catalyst, aluminum chloride. The *tert*-butyl cation reacts with benzene to form a sigma complex. Loss of a proton gives the product, *tert*-butylbenzene. The aluminum chloride catalyst is regenerated in the final step.

Friedel–Crafts alkylations are used with a wide variety of primary, secondary, and tertiary alkyl halides. With secondary and tertiary halides, the reacting electrophile is probably the carbocation.

$$R{-}X + AlCl_3 \rightleftharpoons R^+ + X{-}\bar{A}lCl_3$$

(R is secondary or tertiary) — reacting electrophile

MECHANISM 17-5 Friedel–Crafts Alkylation

Friedel–Crafts alkylation is an electrophilic aromatic substitution in which an alkyl cation acts as the electrophile.

EXAMPLE: Alkylation of benzene by the *tert*-butyl cation.

Step 1: Formation of a carbocation.

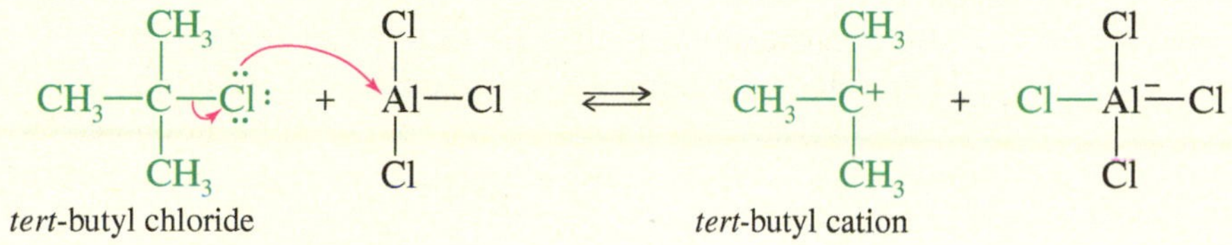

$$(CH_3)_3C{-}Cl + AlCl_3 \rightleftharpoons (CH_3)_3C^+ + Cl{-}\bar{A}lCl_3$$

tert-butyl chloride — *tert*-butyl cation

Step 2: Electrophilic attack forms a sigma complex.

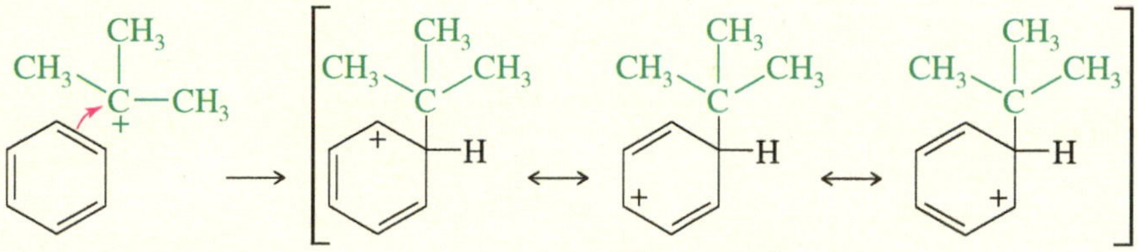

sigma complex

Step 3: Loss of a proton regenerates the aromatic ring and gives the alkylated product.

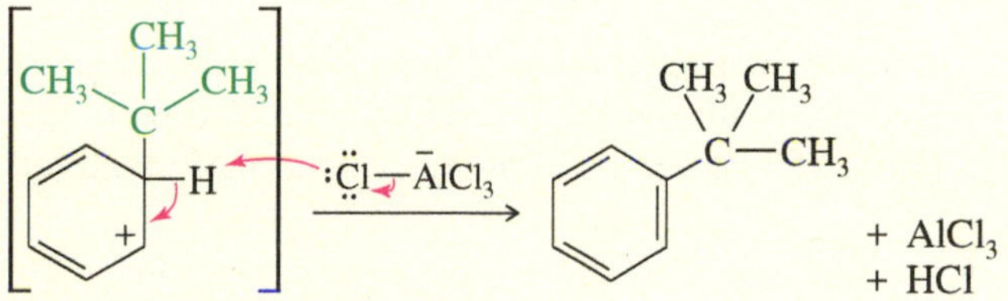

$$\text{sigma complex} \xrightarrow{:\ddot{C}l{-}\bar{A}lCl_3} C_6H_5{-}C(CH_3)_3 + AlCl_3 + HCl$$

sigma complex

With primary alkyl halides, the free primary carbocation is too unstable. The actual electrophile is likely a complex of aluminum chloride with the alkyl halide. In this complex, the carbon–halogen bond is weakened (as indicated by dashed lines), and there is considerable positive charge on the carbon atom. The mechanism for the aluminum chloride-catalyzed reaction of ethyl chloride with benzene is as follows:

$$CH_3{-}CH_2{-}Cl + AlCl_3 \rightleftharpoons CH_3{-}\overset{\delta+}{C}H_2{-}{-}{-}Cl{-}{-}{-}\overset{\delta-}{A}lCl_3$$

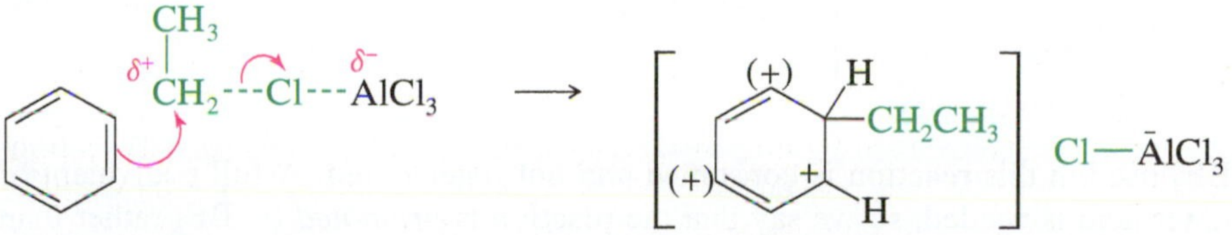

$$C_6H_6 + CH_3\overset{\delta+}{C}H_2{-}{-}{-}Cl{-}{-}{-}\overset{\delta-}{A}lCl_3 \longrightarrow [\text{sigma complex, } C_6H_6^{(+)}{-}CH_2CH_3] \; Cl{-}\bar{A}lCl_3$$

sigma complex

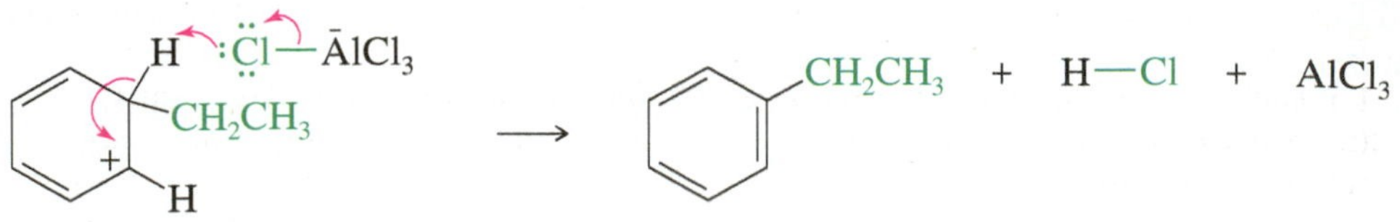

H :Cl—$\bar{A}lCl_3$ CH$_2$CH$_3$ + H ⟶ CH$_2$CH$_3$ + H—Cl + $AlCl_3$

PROBLEM 17-15

Propose products (if any) and mechanisms for the following $AlCl_3$-catalyzed reactions:
(a) chlorocyclohexane with benzene
(b) methyl chloride with anisole
*(c) 3-chloro-2,2-dimethylbutane with isopropylbenzene

Friedel–Crafts Alkylation Using Other Carbocation Sources We have seen several ways of generating carbocations, and most of these can be used for Friedel–Crafts alkylations. Two common methods are protonation of alkenes and treatment of alcohols with BF_3.

Alkenes are protonated by HF to give carbocations. Fluoride ion is a weak nucleophile and does not immediately attack the carbocation. If benzene (or an activated benzene derivative) is present, electrophilic substitution occurs. The protonation step follows Markovnikov's rule, forming the more stable carbocation, which alkylates the aromatic ring.

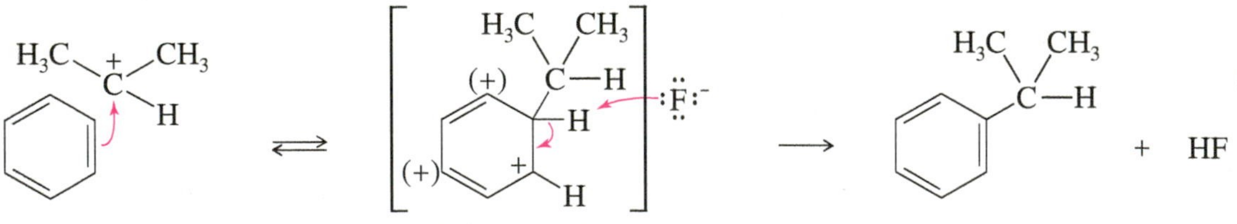

$$H_2C{=}C(CH_3)H + HF \rightleftharpoons H_3C{-}\overset{+}{C}(CH_3)H + F^-$$

sigma complex

+ HF

Alcohols are another source of carbocations for Friedel–Crafts alkylations. Alcohols commonly form carbocations when treated with Lewis acids such as boron trifluoride (BF_3). If benzene (or an activated benzene derivative) is present, substitution may occur.

Formation of the cation

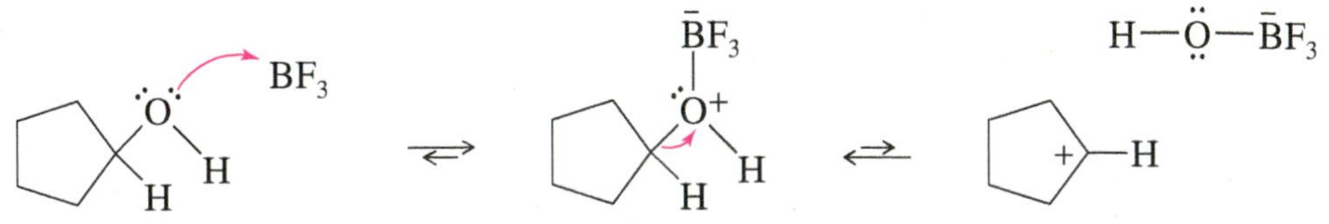

BF_3 ⇌ $\bar{B}F_3$ O+ H ⇌ H—O—$\bar{B}F_3$ + —H

Electrophilic substitution on benzene

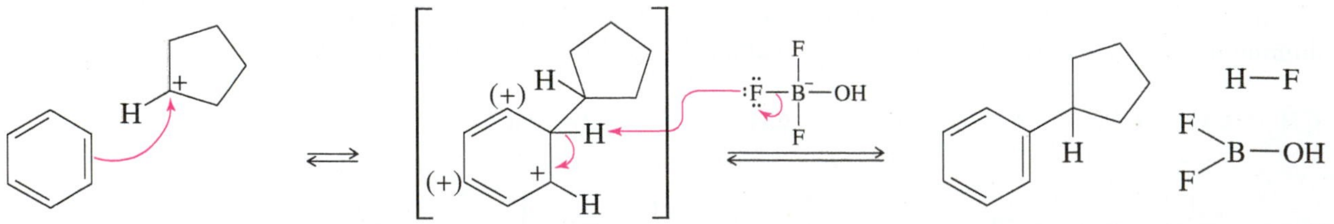

sigma complex

H—F

F$_2$B—OH

The BF_3 used in this reaction is consumed and not regenerated. A full equivalent of the Lewis acid is needed, so we say that the reaction is *promoted* by BF_3 rather than *catalyzed* by BF_3.

PROBLEM 17-16

For each reaction, show the generation of the electrophile and predict the products.

(a) benzene + cyclohexene + HF
(b) *tert*-butyl alcohol + benzene + BF_3
(c) *tert*-butylbenzene + 2-methylpropene + HF
(d) propan-2-ol + toluene + BF_3

Limitations of the Friedel–Crafts Alkylation Although the Friedel–Crafts alkylation looks good in principle, it has three major limitations that severely restrict its use.

Limitation 1 Friedel–Crafts reactions work only with benzene, activated benzene derivatives, and halobenzenes. They fail with strongly deactivated systems such as nitrobenzene, benzenesulfonic acid, and phenyl ketones. In some cases, we can get around this limitation by adding the deactivating group or changing an activating group into a deactivating group *after* the Friedel–Crafts step.

SOLVED PROBLEM 17-2

Devise a synthesis of *p*-nitro-*tert*-butylbenzene from benzene.

SOLUTION

To make *p*-nitro-*tert*-butylbenzene, we would first use a Friedel–Crafts reaction to make *tert*-butylbenzene. Nitration gives the correct product. If we were to make nitrobenzene first, the Friedel–Crafts reaction to add the *tert*-butyl group would fail.

Good

benzene $\xrightarrow[AlCl_3]{(CH_3)_3C-Cl}$ $C(CH_3)_3$-benzene $\xrightarrow[H_2SO_4]{HNO_3}$ $C(CH_3)_3$ / NO_2 (para) (plus ortho)

Bad

benzene $\xrightarrow[H_2SO_4]{HNO_3}$ NO_2-benzene (deactivated) $\xrightarrow[AlCl_3]{(CH_3)_3C-Cl}$ (reaction fails)

PROBLEM-SOLVING HINT

Friedel–Crafts reactions fail with strongly deactivated systems.

Limitation 2 Like other carbocation reactions, the Friedel–Crafts alkylation is susceptible to carbocation rearrangements. As a result, only certain alkylbenzenes can be made using the Friedel–Crafts alkylation. *tert*-Butylbenzene, isopropylbenzene, and ethylbenzene can be synthesized using the Friedel–Crafts alkylation because the corresponding cations are not prone to rearrangement. Consider what happens, however, when we try to make *n*-propylbenzene by the Friedel–Crafts alkylation.

PROBLEM-SOLVING HINT

Alkyl carbocations for Friedel–Crafts alkylations are prone to rearrangements.

Ionization with rearrangement gives isopropyl cation

$$CH_3-CH_2-CH_2-Cl + AlCl_3 \rightleftharpoons CH_3-CH(H)-\overset{\delta+}{CH_2}\cdots\overset{\delta-}{Cl}\cdots AlCl_3 \longrightarrow CH_3-\overset{+}{C}H-CH_3 + {}^-AlCl_4$$

Reaction with benzene gives isopropylbenzene

$$CH_3-\overset{+}{C}H-CH_3 \;\; {}^-AlCl_4 + \text{benzene} \longrightarrow \text{C}_6\text{H}_5-CH(CH_3)-CH_3 + HCl + AlCl_3$$

PROBLEM-SOLVING HINT

Friedel–Crafts alkylations are prone to multiple alkylations.

Limitation 3 Because alkyl groups are activating substituents, the product of the Friedel–Crafts alkylation is *more reactive* than the starting material. Multiple alkylations are hard to avoid. This limitation can be severe. If we need to make ethylbenzene, we might try adding some $AlCl_3$ to a mixture of 1 mole of ethyl chloride and 1 mole of benzene. As some ethylbenzene is formed, however, it is activated, reacting even faster than benzene itself. The product is a mixture of some (ortho and para) diethylbenzenes, some triethylbenzenes, a small amount of ethylbenzene, and some leftover benzene.

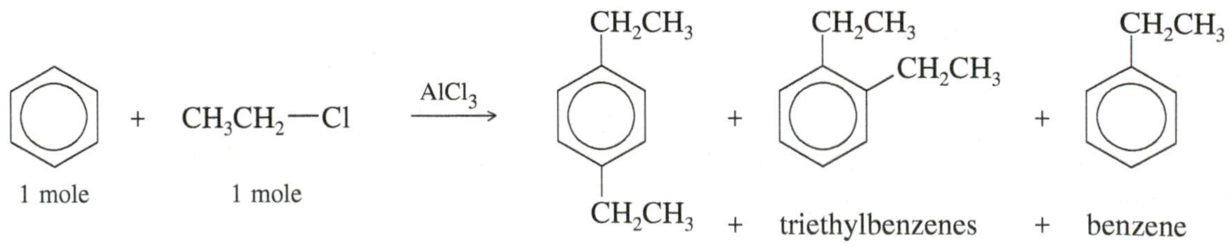

$$\text{benzene (1 mole)} + CH_3CH_2{-}Cl \text{ (1 mole)} \xrightarrow{AlCl_3} p\text{-}C_6H_4(CH_2CH_3)_2 + o\text{-}C_6H_4(CH_2CH_3)_2 + C_6H_5CH_2CH_3 + \text{triethylbenzenes} + \text{benzene}$$

The problem of overalkylation can be minimized by using a large excess of benzene. For example, if 1 mole of ethyl chloride is used with 50 moles of benzene, the concentration of ethylbenzene is always low, and the electrophile is more likely to react with benzene than with ethylbenzene. Distillation separates the product from excess benzene. This is a common industrial approach, because a continuous distillation can recycle the unreacted benzene.

In the laboratory, we must often alkylate aromatic compounds that are more expensive than benzene. Because we cannot afford to use a large excess of the starting material, a more selective method is needed. Fortunately, the Friedel–Crafts acylation, discussed in Section 17-11, introduces just one group without danger of polyalkylation or rearrangement.

PROBLEM 17-17

Predict the products (if any) of the following reactions.

(a) (excess) benzene + isobutyl chloride + $AlCl_3$
(b) (excess) toluene + butan-1-ol + BF_3
(c) (excess) nitrobenzene + 2-chloropropane + $AlCl_3$
(d) (excess) benzene + 3,3-dimethylbut-1-ene + HF

PROBLEM 17-18

Which reactions will produce the desired product in good yield? You may assume that aluminum chloride is added as a catalyst in each case. For the reactions that will not give a good yield of the desired product, predict the major products.

Reagents	Desired Product
(a) benzene + *n*-butyl bromide	*n*-butylbenzene
(b) ethylbenzene + *tert*-butyl chloride	*p*-ethyl-*tert*-butylbenzene
(c) bromobenzene + ethyl chloride	*p*-bromoethylbenzene
(d) benzamide ($PhCONH_2$) + CH_3CH_2Cl	*p*-ethylbenzamide
(e) toluene + HNO_3, H_2SO_4, heat	2,4,6-trinitrotoluene (TNT)

PROBLEM 17-19

Show how you would synthesize the following aromatic derivatives from benzene.

(a) *p*-*tert*-butylnitrobenzene **(b)** *p*-toluenesulfonic acid **(c)** *p*-chlorotoluene

17-11 The Friedel–Crafts Acylation

An **acyl group** is a carbonyl group with an alkyl group attached. Acyl groups are named systematically by dropping the final *-e* from the alkane name and adding the *-oyl* suffix. Historical names are often used for the *formyl group*, the *acetyl group*, and the *propionyl group*, however.

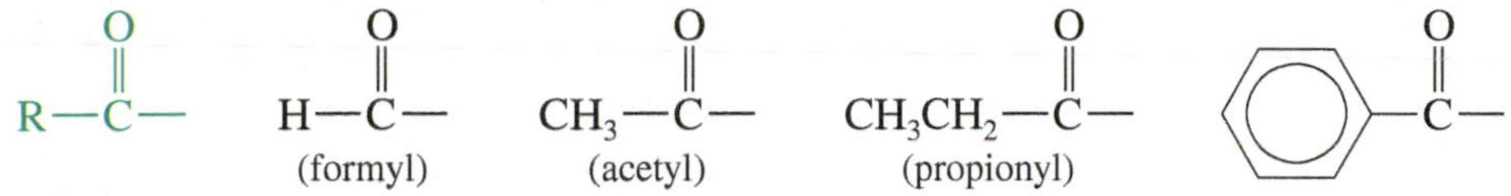

$$R-\overset{O}{\overset{\|}{C}}- \quad H-\overset{O}{\overset{\|}{C}}- \quad CH_3-\overset{O}{\overset{\|}{C}}- \quad CH_3CH_2-\overset{O}{\overset{\|}{C}}- \quad C_6H_5-\overset{O}{\overset{\|}{C}}-$$

acyl group; (formyl) methanoyl; (acetyl) ethanoyl; (propionyl) propanoyl; benzoyl

An **acyl chloride** is an acyl group bonded to a chlorine atom. Acyl chlorides are made by reaction of the corresponding carboxylic acids with thionyl chloride. Therefore, acyl chlorides are also called **acid chlorides.** We consider acyl chlorides in more detail when we study acid derivatives in Chapter 21.

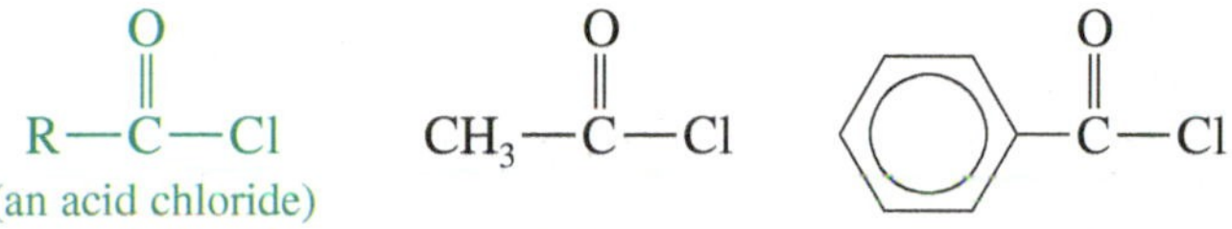

$$R-\overset{O}{\overset{\|}{C}}-Cl \qquad CH_3-\overset{O}{\overset{\|}{C}}-Cl \qquad C_6H_5-\overset{O}{\overset{\|}{C}}-Cl$$

(an acid chloride) an acyl chloride; acetyl chloride; benzoyl chloride

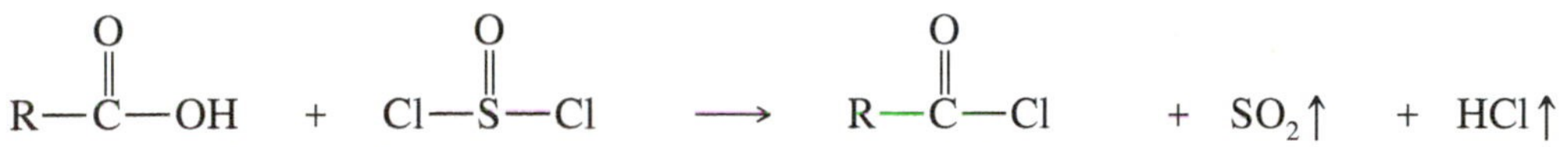

$$R-\overset{O}{\overset{\|}{C}}-OH + Cl-\overset{O}{\overset{\|}{S}}-Cl \longrightarrow R-\overset{O}{\overset{\|}{C}}-Cl + SO_2\uparrow + HCl\uparrow$$

a carboxylic acid; thionyl chloride; an acyl chloride

In the presence of aluminum chloride, an acyl chloride reacts with benzene (or an activated benzene derivative) to give a phenyl ketone: an *acylbenzene*. The **Friedel–Crafts acylation** is analogous to the Friedel–Crafts alkylation, except that the reagent is an acyl chloride instead of an alkyl halide and the product is an acylbenzene (a "phenone") instead of an alkylbenzene.

Friedel–Crafts acylation

$$C_6H_6 + R-\overset{O}{\overset{\|}{C}}-Cl \xrightarrow{AlCl_3} C_6H_5-\overset{O}{\overset{\|}{C}}-R + HCl$$

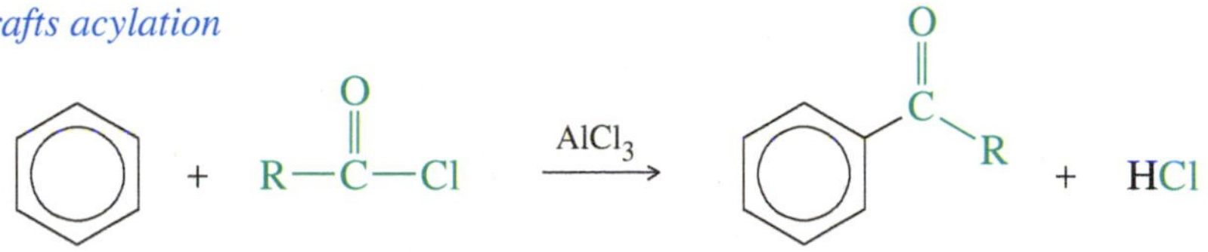

benzene; acyl halide; an acylbenzene (a phenyl ketone)

Example

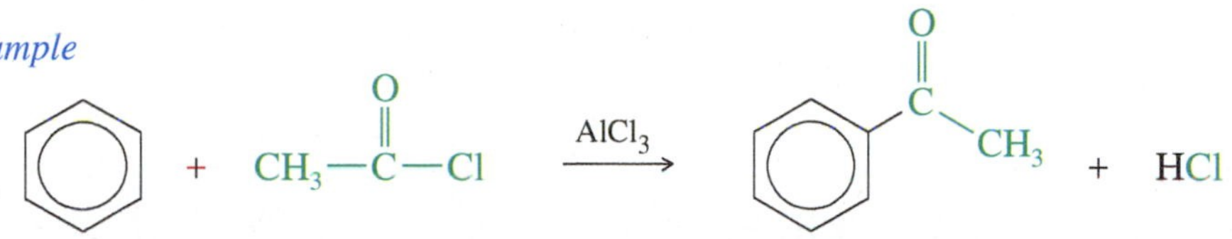

$$C_6H_6 + CH_3-\overset{O}{\overset{\|}{C}}-Cl \xrightarrow{AlCl_3} C_6H_5-\overset{O}{\overset{\|}{C}}-CH_3 + HCl$$

benzene; acetyl chloride; acetylbenzene (95%) (acetophenone)

17-11A Mechanism of Acylation

The mechanism of Friedel–Crafts acylation (shown next) resembles that for alkylation, except that the electrophile is a resonance-stabilized **acylium ion.** The acylium ion reacts with benzene or an activated benzene derivative via an electrophilic aromatic substitution to form an acylbenzene.

MECHANISM 17-6 Friedel–Crafts Acylation

Friedel–Crafts acylation is an electrophilic aromatic substitution with an acylium ion acting as the electrophile.

Step 1: Formation of an acylium ion.

$$R-C(=O)-Cl + AlCl_3 \rightleftharpoons R-C(=O)-\overset{+}{Cl}-\overset{-}{Al}Cl_3 \rightleftharpoons {}^{-}AlCl_4 + [R-\overset{+}{C}=O \longleftrightarrow R-C\equiv\overset{+}{O}:]$$

acyl chloride — complex — acylium ion

Steps 2 and 3: Electrophilic attack forms a sigma complex, and loss of a proton regenerates the aromatic system.

sigma complex — acylbenzene

$$\xrightarrow{:Cl-\overset{-}{Al}Cl_3} \text{acylbenzene} + AlCl_3 + HCl$$

Step 4: Complexation of the product. The product complex must be hydrolyzed (by water) to release the free acylbenzene.

$$\text{acylbenzene} + AlCl_3 \longrightarrow \text{product complex} \xrightarrow{\text{excess } H_2O} \text{free acylbenzene} + \text{aluminum salts}$$

The product of acylation (the acylbenzene) is a ketone. The ketone's carbonyl group has nonbonding electrons that complex with the Lewis acid ($AlCl_3$), requiring a full equivalent of $AlCl_3$ in the acylation. The initial product is the aluminum chloride complex of the acylbenzene. Addition of water hydrolyzes this complex, giving the free acylbenzene.

The electrophile in the Friedel–Crafts acylation appears to be a large, bulky complex, such as $R-\overset{+}{C}=O\ {}^{-}AlCl_4$. Para substitution usually prevails when the aromatic substrate has an ortho, para-directing group, possibly because the electrophile is too bulky for effective attack at the ortho position. For example, when ethylbenzene reacts with acetyl chloride, the major product is *p*-ethylacetophenone.

$$\text{ethylbenzene} + CH_3-C(=O)-Cl \xrightarrow[(2)\ H_2O]{(1)\ AlCl_3} \text{p-ethylacetophenone (70–80\%)}$$

ethylbenzene — acetyl chloride — *p*-ethylacetophenone (70–80%)

One of the most attractive features of the Friedel–Crafts acylation is the deactivation of the product toward further substitution. The acylbenzene has a carbonyl group (a deactivating group) bonded to the aromatic ring. Because Friedel–Crafts reactions do not occur on strongly deactivated rings, the acylation stops after one substitution.

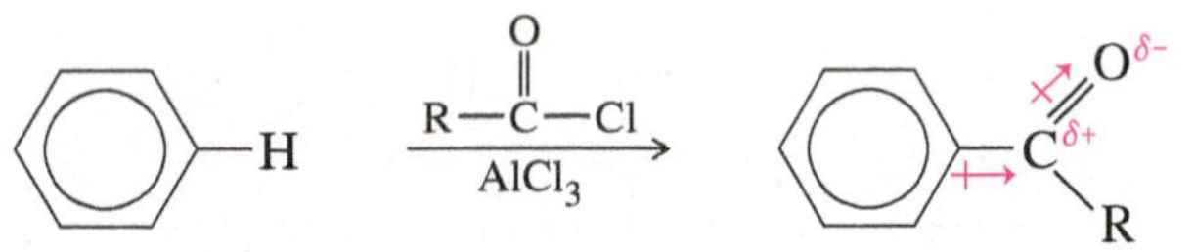

Thus, Friedel–Crafts acylation overcomes two of the three limitations of the alkylation: The acylium ion is resonance-stabilized, so that no rearrangements occur; and the acylbenzene product is deactivated, so that no further reaction occurs. Like the alkylation, however, the acylation fails with strongly deactivated aromatic rings.

SUMMARY Comparison of Friedel–Crafts Alkylation and Acylation

Alkylation	*Acylation*
The alkylation cannot be used with strongly deactivated derivatives.	Also true: Only benzene, halobenzenes, and activated derivatives are suitable.
The carbocations involved in the alkylation may rearrange.	Resonance-stabilized acylium ions are not prone to rearrangement.
Polyalkylation is commonly a problem.	The acylation forms a deactivated acylbenzene, which does not react further.

17-11B The Clemmensen Reduction: Synthesis of Alkylbenzenes

How do we synthesize alkylbenzenes that cannot be made by Friedel–Crafts alkylation? We use the Friedel–Crafts acylation to make the acylbenzene, then we reduce the acylbenzene to the alkylbenzene using the **Clemmensen reduction**: treatment with aqueous HCl and amalgamated zinc (zinc treated with mercury salts).

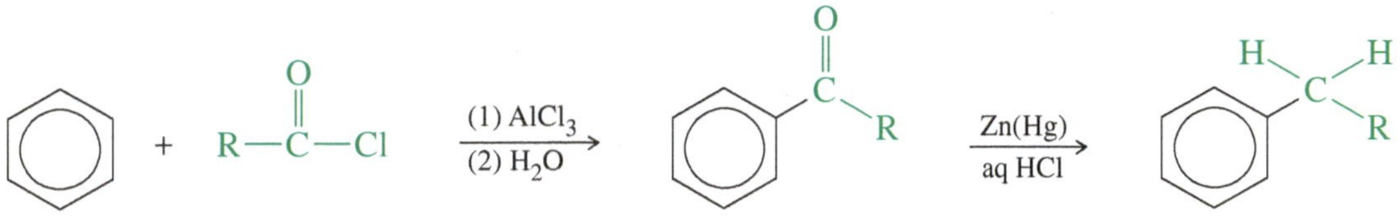

This two-step sequence can synthesize many alkylbenzenes that are impossible to make by direct alkylation. For example, we saw earlier that *n*-propylbenzene cannot be made by Friedel–Crafts alkylation. Benzene reacts with *n*-propyl chloride and $AlCl_3$ to give isopropylbenzene, together with some diisopropylbenzene. In the acylation, however, benzene reacts with propanoyl chloride and $AlCl_3$ to give ethyl phenyl ketone (propiophenone), which is easily reduced to *n*-propylbenzene.

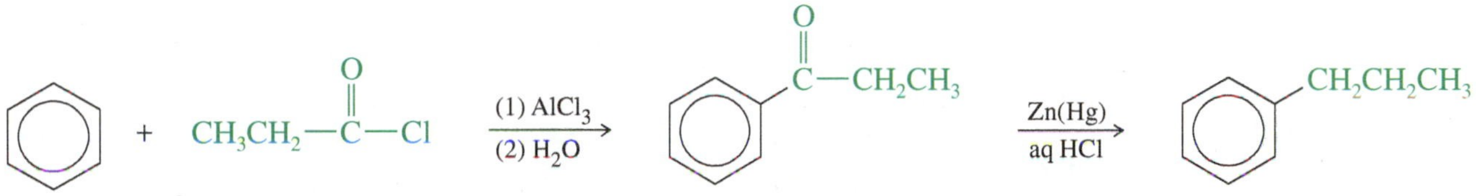

propanoyl chloride propiophenone *n*-propylbenzene

The reagents and conditions for the Clemmensen reduction are similar to those used to reduce a nitro group to an amine. Aromatic substitution followed by reduction is a valuable process for making compounds with specific substitution patterns, such as in the following synthesis.

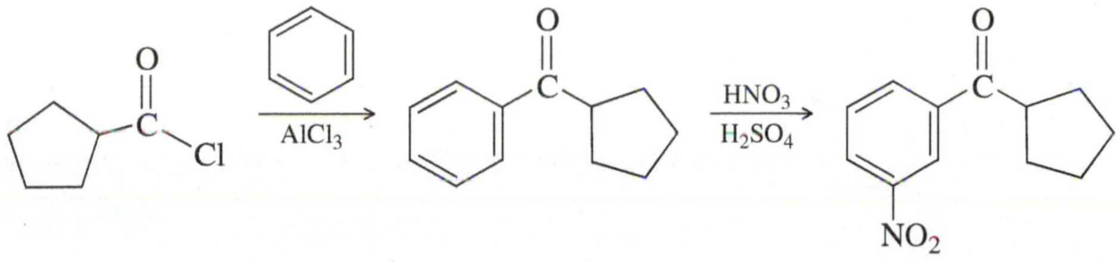

$$\text{(3-nitrophenyl) cyclopentyl ketone} \xrightarrow[\text{aq HCl}]{\text{Zn or Sn}} \text{3-(cyclopentylmethyl)anilinium } (^{+}NH_3\ Cl^-) \xrightarrow{^-OH} \text{3-(cyclopentylmethyl)aniline } (NH_2)$$

Carboxylic acids and acid anhydrides also serve as acylating agents in Friedel–Crafts reactions. We consider these acylating agents in Chapters 20 and 21 when we study the reactions of carboxylic acids and their derivatives.

17-11C The Gatterman–Koch Formylation: Synthesis of Benzaldehydes

We cannot add a formyl group to benzene by Friedel–Crafts acylation in the usual manner. The problem lies with the necessary reagent, formyl chloride, which is unstable and cannot be bought or stored.

$$\underset{\text{formyl group}}{H{-}\overset{O}{\overset{\|}{C}}{-}} \qquad \underset{\text{formyl chloride}}{\left[H{-}\overset{O}{\overset{\|}{C}}{-}Cl\right]}$$

Formylation can be accomplished by using a high-pressure mixture of carbon monoxide and HCl together with a catalyst consisting of a mixture of cuprous chloride (CuCl) and aluminum chloride. This mixture generates the formyl cation, possibly through a small concentration of formyl chloride. The reaction with benzene gives formyl benzene, better known as benzaldehyde. This reaction, called the **Gatterman–Koch synthesis,** is widely used in industry to synthesize aryl aldehydes.

$$CO + HCl \rightleftarrows \underset{\substack{\text{formyl chloride}\\ \text{(unstable)}}}{\left[H{-}\overset{O}{\overset{\|}{C}}{-}Cl\right]} \xrightarrow{AlCl_3/CuCl} \underset{\text{formyl cation}}{\left[H{-}\overset{+}{C}{=}\ddot{O}{:}\right]}\ {}^{-}AlCl_4$$

$$\text{benzene} + H{-}\overset{+}{C}{=}O \longrightarrow \underset{\text{benzaldehyde}}{Ph{-}\overset{O}{\overset{\|}{C}}{-}H} + HCl$$

PROBLEM-SOLVING HINT

Friedel–Crafts acylations are generally free from rearrangements and multiple substitution. They do not go on strongly deactivated rings, however.

PROBLEM 17-20

Show how you would use the Friedel–Crafts acylation, Clemmensen reduction, and/or Gatterman–Koch synthesis to prepare the following compounds:

(a) $Ph{-}\overset{O}{\overset{\|}{C}}{-}CH_2CH(CH_3)_2$ isobutyl phenyl ketone

(b) $Ph{-}\overset{O}{\overset{\|}{C}}{-}C(CH_3)_3$ *tert*-butyl phenyl ketone

(c) $Ph{-}\overset{O}{\overset{\|}{C}}{-}Ph$ diphenyl ketone

(d) *p*-methoxybenzaldehyde

(e) 3-methyl-1-phenylbutane

(f) 1-phenyl-2,2-dimethylpropane

(g) *n*-butylbenzene

(h) $H_3C{-}\overset{O}{\overset{\|}{C}}{-}C_6H_4{-}\underset{H}{N}{-}\overset{O}{\overset{\|}{C}}{-}CH_3$ (from benzene)

17-12 Nucleophilic Aromatic Substitution

Nucleophiles can displace halide ions from aryl halides, particularly if there are strong electron-withdrawing groups ortho or para to the halide. Because a nucleophile substitutes for a leaving group on an aromatic ring, this class of reactions is called **nucleophilic aromatic substitution.** The following examples show that both ammonia and hydroxide ion can displace chloride from 2,4-dinitrochlorobenzene:

2,4-dinitrochlorobenzene + $2\ NH_3$ $\xrightarrow{\text{(heat, pressure)}}$ 2,4-dinitroaniline (90%) + $NH_4^+\ Cl^-$

2,4-dinitrochlorobenzene $\xrightarrow[100\ ^\circ C]{2\ NaOH}$ 2,4-dinitrophenoxide ($O^-\ Na^+$) + NaCl + H_2O $\xrightarrow{H^+}$ 2,4-dinitrophenol (95%)

Nucleophilic aromatic substitution is much more restrictive in its applications than *electrophilic* aromatic substitution. In nucleophilic aromatic substitution, a strong nucleophile replaces a leaving group such as a halide. The mechanism cannot be the S_N2 mechanism because aryl halides cannot achieve the correct geometry for back-side displacement. The aromatic ring blocks approach of the nucleophile to the back of the carbon bearing the halogen.

The S_N1 mechanism cannot be involved either. Strong nucleophiles are required for nucleophilic aromatic substitution, and the reaction rate is proportional to the concentration of the nucleophile. Thus, the nucleophile must be involved in the rate-limiting step.

Electron-withdrawing substituents (such as nitro groups) *activate* the ring toward nucleophilic aromatic substitution, suggesting that the transition state is developing a negative charge on the ring. In fact, nucleophilic aromatic substitutions are difficult without at least one powerful electron-withdrawing group. (This effect is the opposite of that for *electrophilic* aromatic substitution, where electron-withdrawing substituents slow or stop the reaction.)

Nucleophilic aromatic substitutions have been studied in detail. Either of two mechanisms may be involved, depending on the reactants. One mechanism is similar to the electrophilic aromatic substitution mechanism, except that nucleophiles and carbanions are involved rather than electrophiles and carbocations. The other mechanism involves "benzyne," an interesting and unusual reactive intermediate.

17-12A The Addition–Elimination Mechanism

Consider the reaction of 2,4-dinitrochlorobenzene with sodium hydroxide, shown in Mechanism 17-7. When hydroxide (the nucleophile) attacks the carbon bearing the chlorine, a negatively charged sigma complex results. The negative charge is delocalized over the ortho and para carbons of the ring and further delocalized into the electron-withdrawing nitro groups. Loss of chloride from the sigma complex gives 2,4-dinitrophenol, which is deprotonated in this basic solution.

MECHANISM 17-7 Nucleophilic Aromatic Substitution (Addition–Elimination)

The addition–elimination mechanism requires strong electron-withdrawing groups to stabilize a negatively charged sigma complex.

Step 1: Attack by the nucleophile gives a resonance-stabilized sigma complex.

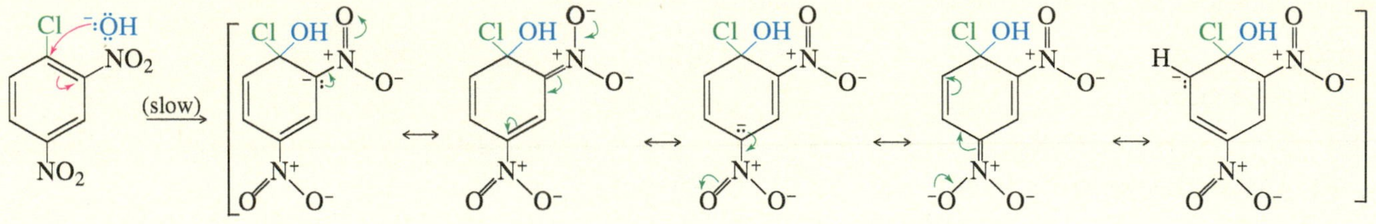

(slow)

Step 2: Loss of the leaving group gives the product.

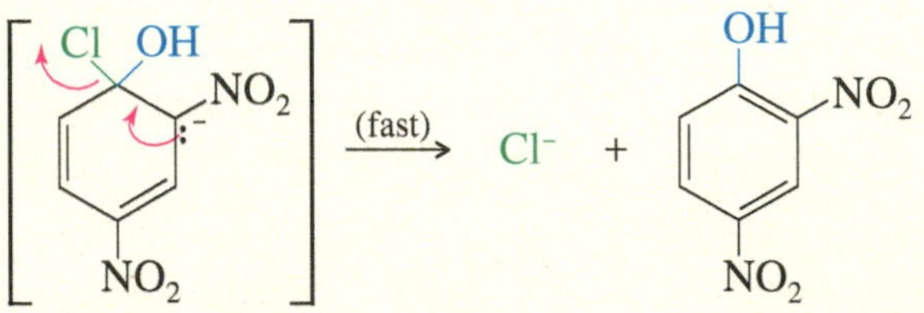

sigma complex (fast) Cl^- + a phenol

Step 3: This product (a phenol) is acidic, and is deprotonated by the base.

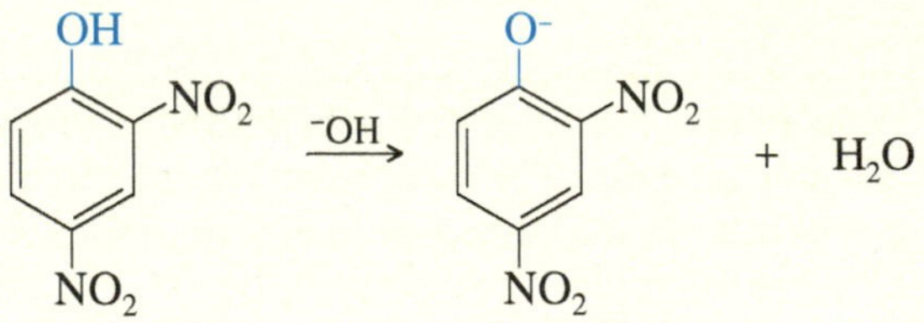

a phenol $\xrightarrow{^-OH}$ a phenoxide ion + H_2O

After the reaction is complete, acid would be added to reprotonate the phenoxide ion to give the phenol.

The resonance forms shown in the mechanism box illustrate how nitro groups ortho and para to the halogen help to stabilize the intermediate (and the transition state leading to it). Without strong resonance-withdrawing groups in these positions, formation of the negatively charged sigma complex is unlikely.

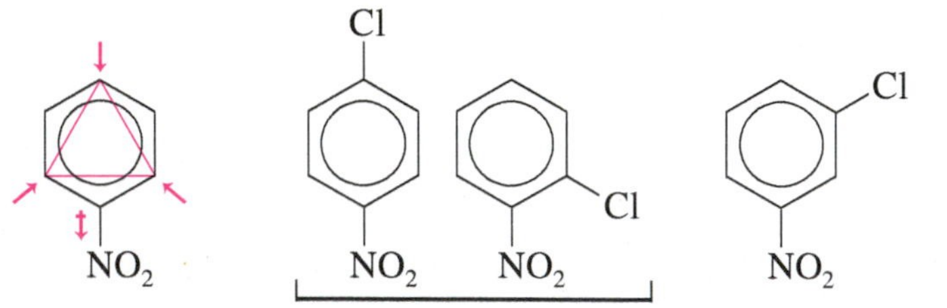

activates positions ortho and para — activated — not activated

PROBLEM 17-21

Fluoride ion is usually a poor leaving group because it is not very polarizable. Fluoride serves as the leaving group in the Sanger reagent (2,4-dinitrofluorobenzene), used in the determination of peptide structures (Chapter 24). Explain why fluoride works as a leaving group in this nucleophilic aromatic substitution, even though it is a poor leaving group in the S_N1 and S_N2 mechanisms.

2,4-dinitrofluorobenzene (Sanger reagent) + amine $\longrightarrow$ 2,4-dinitrophenyl derivative + HF

17-12B The Benzyne Mechanism: Elimination–Addition

The addition–elimination mechanism for nucleophilic aromatic substitution requires strong electron-withdrawing substituents on the aromatic ring. Under extreme conditions, however, unactivated halobenzenes react with strong bases. For example, a commercial synthesis of phenol (the "Dow process") involves treatment of chlorobenzene with sodium hydroxide and a small amount of water in a pressurized reactor at 350 °C:

chlorobenzene $\xrightarrow[H_2O]{2\ NaOH,\ 350\ °C}$ sodium phenoxide ($O^- Na^+$) + NaCl $\xrightarrow{H^+}$ phenol

Similarly, chlorobenzene reacts with sodium amide ($NaNH_2$, an extremely strong base) to give aniline, Ph—NH_2. This reaction does not require high temperatures, taking place in liquid ammonia at −33 °C.

Nucleophilic substitution of unactivated benzene derivatives occurs by a mechanism different from the addition–elimination we saw with the nitro-substituted halobenzenes. A clue to the mechanism is provided by the reaction of *p*-bromotoluene with sodium amide. The products are a 50:50 mixture of *m*- and *p*-toluidine.

p-bromotoluene $\xrightarrow[NH_3,\ -33\ °C]{Na^+\ ^-NH_2}$ *p*-toluidine (50%) + *m*-toluidine (50%)

These two products can be explained by an elimination–addition mechanism, called the **benzyne mechanism** because of the unusual intermediate. Sodium amide (or sodium hydroxide in the Dow process) reacts as a *base*, abstracting a proton. The product is a carbanion with a negative charge and a nonbonding pair of electrons localized in the sp^2 orbital that once formed the C—H bond.

$:\ddot{N}H_2$ ⇄ carbanion (sp^2 hybrid) + $:NH_3$ $\xrightarrow{\text{lose } Br^-}$ a "benzyne" (sp^2 hybrid orbitals (distorted))

The carbanion can expel bromide ion to become a neutral species. As bromide leaves with its bonding electrons, an empty sp^2 orbital remains. This orbital overlaps with the filled orbital adjacent to it, giving additional bonding between these two carbon atoms. The two sp^2 orbitals are directed 60° away from each other, so their overlap is not very effective. This reactive intermediate is called a **benzyne** because it can be symbolized by drawing a triple bond between these two carbon atoms. Triple bonds are usually linear, however, so this is a very reactive, highly strained triple bond.

Amide ion is a strong nucleophile, attacking at either end of the weak, reactive benzyne triple bond. Subsequent protonation gives toluidine. About half of the product results from attack by the amide ion at the meta carbon, and about half from attack at the para carbon.

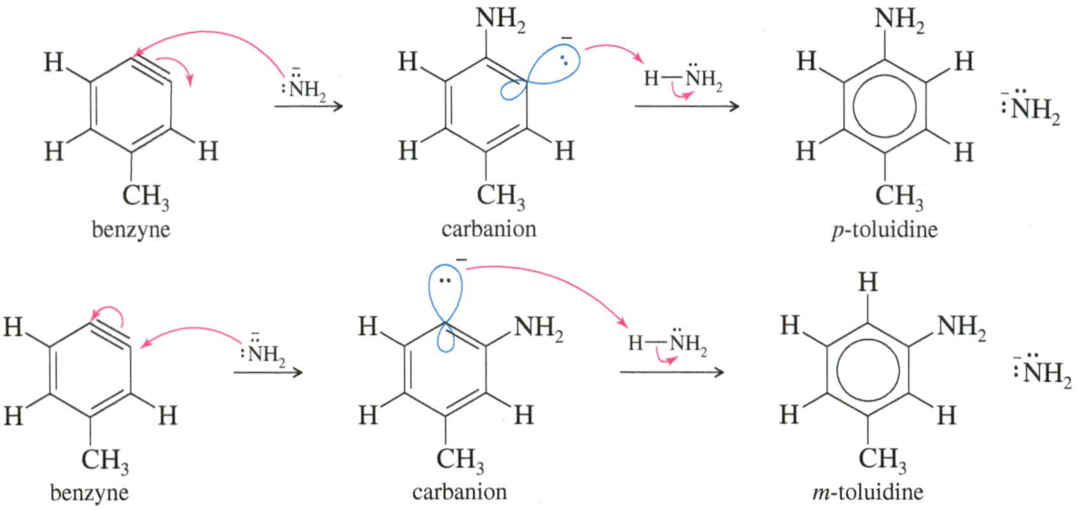

benzyne $\xrightarrow{:\ddot{N}H_2^-}$ carbanion $\xrightarrow{H-\ddot{N}H_2}$ *p*-toluidine + $:\ddot{N}H_2^-$

benzyne $\xrightarrow{:\ddot{N}H_2^-}$ carbanion $\xrightarrow{H-\ddot{N}H_2}$ *m*-toluidine + $:\ddot{N}H_2^-$

In summary, the benzyne mechanism operates when the halobenzene is unactivated toward nucleophilic aromatic substitution, and forcing conditions are used with a strong base. A two-step elimination forms a reactive benzyne intermediate. Nucleophilic attack, followed by protonation, gives the substituted product. This benzyne mechanism explains the 50:50 mixture of products obtained from the reaction of *p*-bromotoluene. The benzyne mechanism should be considered whenever a nucleophilic aromatic substitution takes place with a powerful base and without strong electron-withdrawing groups on the ring.

MECHANISM 17-8 Nucleophilic Aromatic Substitution (Benzyne Mechanism)

The benzyne mechanism (elimination–addition) is likely when the ring has no strong electron-withdrawing groups. It usually requires a powerful base or high temperatures.

Step 1: Deprotonation adjacent to the leaving group gives a carbanion.

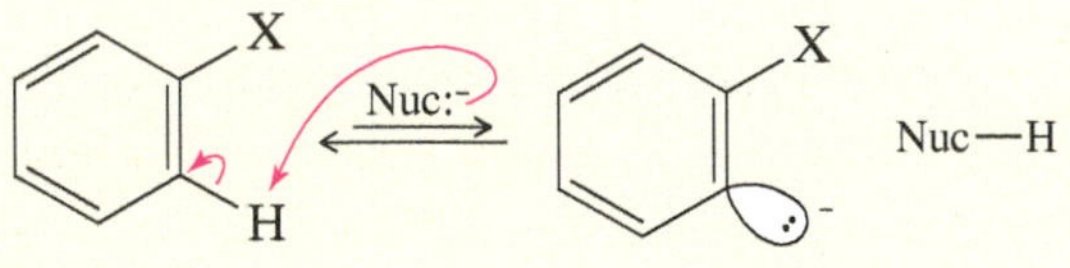

$\xrightarrow{Nuc:^-}$ Nuc—H

Step 2: The carbanion expels the leaving group to give a "benzyne" intermediate.

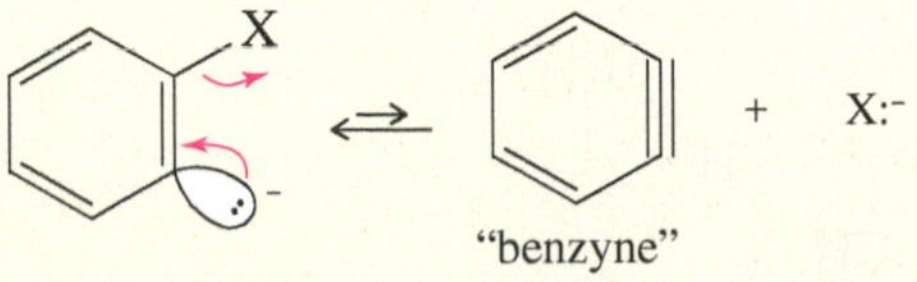

"benzyne" + $X:^-$

Step 3: The nucleophile attacks at either end of the reactive benzyne triple bond.

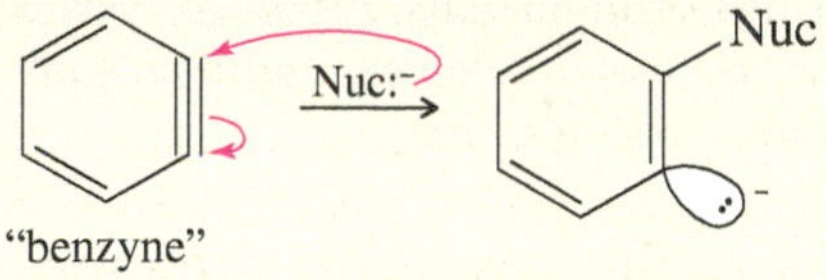

"benzyne"

Nuc:⁻

Nuc

Step 4: Reprotonation gives the product.

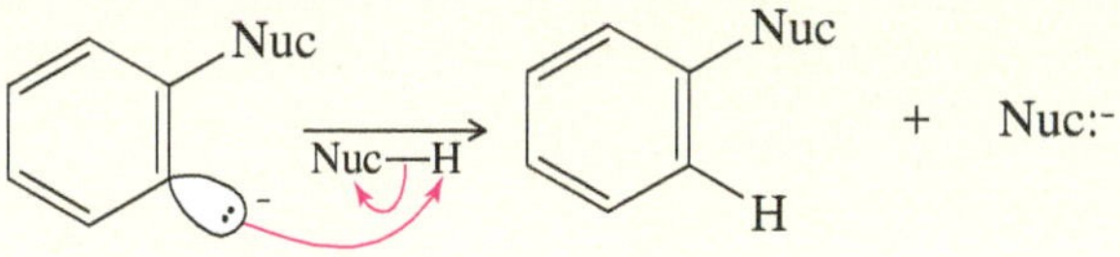

Nuc

Nuc—H

Nuc

H

+ Nuc:⁻

PROBLEM 17-22

Propose a mechanism that shows why *p*-chlorotoluene reacts with sodium hydroxide at 350 °C to give a mixture of *p*-cresol and *m*-cresol.

PROBLEM 17-23

Propose mechanisms and show the expected products of the following reactions.

(a) 2,4-dinitrochlorobenzene + sodium methoxide ($NaOCH_3$)
(b) 2,4-dimethylchlorobenzene + sodium hydroxide, 350 °C
(c) *p*-nitrobromobenzene + methylamine (CH_3-NH_2)
(d) 2,4-dinitrochlorobenzene + excess hydrazine (H_2N-NH_2)

PROBLEM-SOLVING HINT

With strong electron-withdrawing groups ortho or para, the addition–elimination mechanism is more likely. Without these activating groups, stronger conditions are required, and the benzyne mechanism is likely.

PROBLEM 17-24

The highly reactive triple bond of benzyne is a powerful dienophile. Predict the product of the Diels–Alder reaction of benzyne (from chlorobenzene and NaOH, heated) with cyclopentadiene.

17-13 Aromatic Substitutions Using Organometallic Reagents

Many useful drugs, fabrics, and plastics require the synthesis of aromatic rings with alkyl, aryl, or vinyl groups attached in the presence of multiple types of functional groups. Section 17-10 discussed the Friedel–Crafts alkylation and its limitations: (1) rearrangements and multiple alkylations are common; (2) the reaction fails on deactivated rings; and (3) it requires strong electrophiles such as $AlCl_3$ and carbocations, which are incompatible with many functional groups. To avoid these limitations, organic chemists have developed new coupling reactions (reactions that form carbon–carbon bonds) using a wide variety of methods that tolerate many other functional groups.

Some of the most successful coupling reactions use transition metals that change valences easily, adding and eliminating substituents as they pass from one oxidation state to another. Section 8-17 showed how transition-metal compounds catalyze the exchange of partners that occurs in olefin metathesis. Problem 10-21 (Section 10-9) described how organocuprate reagents (Gilman reagents) can couple with alkyl halides. Organocuprate reagents also couple with aryl and vinyl halides to make substituted benzenes and elongated alkenes. Recent research has developed many additional coupling methods using other transition metals in the reagents and catalysts.

The 2010 Nobel Prize in Chemistry was awarded to Richard F. Heck (University of Delaware), Ei-Ichi Negishi (Purdue University), and Akira Suzuki (Hokkaido University) for their development of palladium-catalyzed coupling reactions that form new carbon–carbon bonds at sp^2 hybridized carbons like those in aromatic rings and in alkenes. Most of these reactions substitute organic groups for halogen atoms on aromatic rings or in alkenes. First, we consider the use of organocuprates to couple with aromatic rings and alkenes, and then look at palladium-catalyzed reactions that form substituted aromatic rings.

17-13A Couplings Using Organocuprate Reagents

Lithium dialkylcuprate reagents (Gilman reagents) are formed by the reaction of two equivalents of an organolithium reagent with cuprous iodide. Reaction of the dialkylcuprate with an alkyl, aryl, or vinyl halide forms a new carbon–carbon bond.

PROBLEM-SOLVING HINT

Section 10-9F covered the use of lithium dialkylcuprates to couple alkyl halides, making larger alkanes.

Making the organolithium reagent:

$$\mathrm{R{-}X} + 2\,\mathrm{Li} \longrightarrow \mathrm{R{-}Li} + \mathrm{LiX}$$

(R = alkyl, vinyl, aryl)

Making the lithium organocuprate:

$$2\,\mathrm{R{-}Li} + \mathrm{CuI} \longrightarrow \mathrm{R_2CuLi} + \mathrm{LiI}$$

Coupling of the organocuprate with an alkyl, vinyl, or aryl halide:

$$\mathrm{R'{-}X} + \mathrm{R_2CuLi} \longrightarrow \mathrm{R'{-}R} + \mathrm{R{-}Cu} + \mathrm{LiX}$$

The mechanisms of these organocuprate reactions vary with the type of alkyl halide and organocuprate used, and they are not well understood. As much as we might like to think of organocuprate couplings as simple S_N2 reactions, they cannot be S_N2 because these work well with sp^2 hybrid substrates such as vinyl and aryl halides, which cannot undergo S_N2 displacement. The following examples show the wide variety of compounds that can be made by organocuprate reactions. Note that an aromatic ring can be present either in the aryl halide or in the dialkylcuprate reagent. Iodides, bromides, and chlorides can be used as the halides.

Examples:

1. An aryl halide with an alkyl or vinyl cuprate.

I + $(\text{CH}_2{=}\text{CH})_2$CuLi ⟶

iodobenzene lithium divinylcuprate styrene

2. A vinyl halide with an aryl cuprate, preserving the stereochemistry of the vinyl halide.

Br + $(\text{C}_6\text{H}_5)_2$CuLi ⟶

3. An alkyl halide with an aryl cuprate.

Cl + $(\ )_2$CuLi ⟶

OCH_3 OCH_3

4. An acyl halide with an organocuprate, giving a ketone.

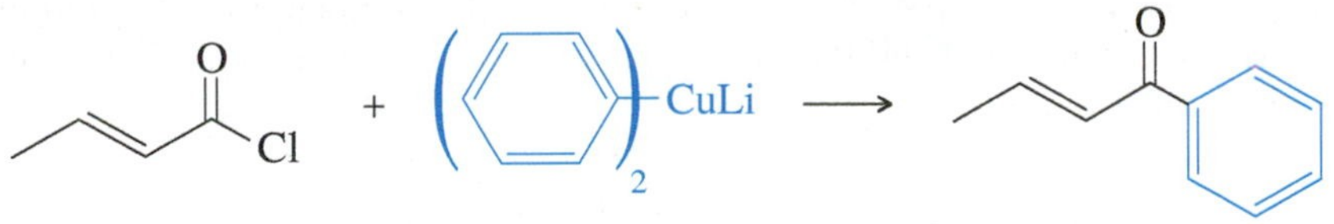

O Cl + ( )$_2$ CuLi ⟶ O

PROBLEM 17-25

What products would you expect from the following reactions?

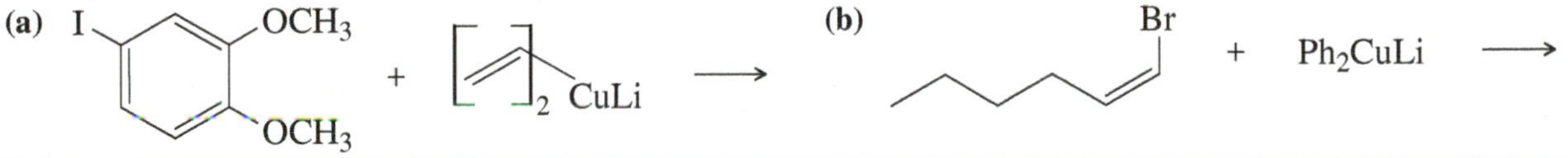

(a) I, OCH_3, OCH_3 + $[\ ]_2$ CuLi ⟶ **(b)** Br + Ph_2CuLi ⟶

PROBLEM 17-26

What organocuprate reagent would you use for the following substitutions?

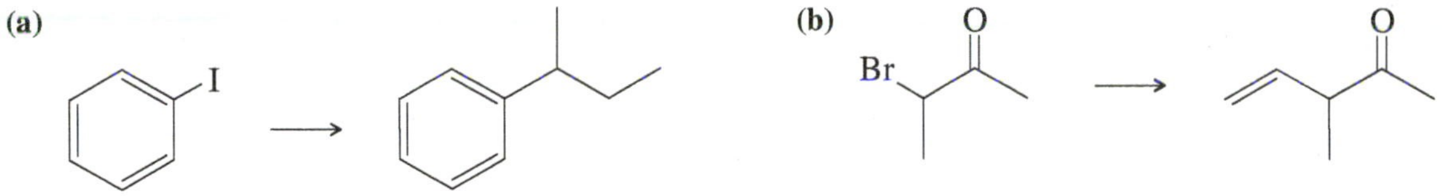

(a) I ⟶ **(b)** Br O ⟶ O

17-13B The Heck Reaction

The Heck reaction is the palladium-catalyzed coupling of an aryl or vinyl halide with an alkene to give a new C—C bond at the less substituted end of the alkene, usually with trans stereochemistry.

The Heck Reaction

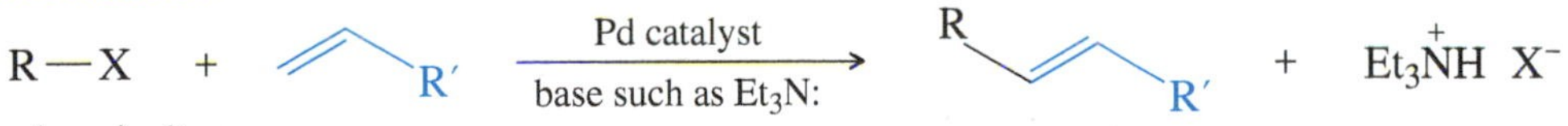

R—X + R′ $\xrightarrow[\text{base such as } Et_3N:]{\text{Pd catalyst}}$ R R′ + $Et_3\overset{+}{N}H\ X^-$

(R = aryl or vinyl)
(X = Br or I)

In most cases, the halide is a bromide or an iodide, and the alkene is typically monosubstituted. The palladium catalyst might be $Pd(OAc)_2$ or $PdCl_2$ or a variety of other palladium compounds. Only a small amount of the expensive palladium catalyst is needed. A base such as triethylamine or sodium acetate is added to neutralize the HX released in the reaction. Many reactions use triphenylphosphine (PPh_3) to complex with the palladium, which helps stabilize it and enhances its reactivity.

The Heck reaction and its variants are used routinely in drug synthesis, where the palladium catalysts can be recovered and recycled. Some Heck reactions can use water as the solvent, which eliminates the purchase and disposal of hazardous organic solvents. The following examples suggest the wide utility of the Heck reaction.

Examples of the Heck reaction:

1. An aryl halide with an aryl olefin.

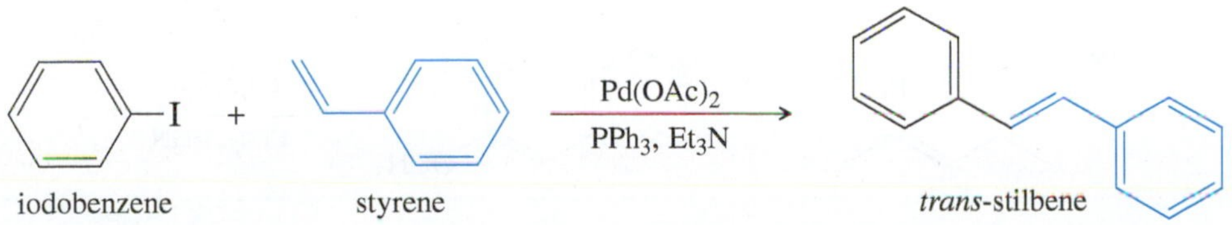

I + $\xrightarrow[PPh_3,\ Et_3N]{Pd(OAc)_2}$

iodobenzene styrene *trans*-stilbene

2. An aryl halide with a conjugated acid or ester.

bromobenzene + methyl acrylate $\xrightarrow[\text{PPh}_3,\ \text{Et}_3\text{N:}]{\text{Pd(OAc)}_2}$ methyl cinnamate

3. A vinyl halide with a conjugated nitrile.

(E)-1-iodobut-1-ene + acrylonitrile $\xrightarrow[\text{PPh}_3,\ \text{Et}_3\text{N:}]{\text{Pd(OAc)}_2}$ (2E, 4E)-hepta-2,4-dienenitrile

4. The industrial synthesis of naproxen, an over-the-counter analgesic and anti-inflammatory drug.

CH_3O–(naphthyl)–Br $\xrightarrow[\text{Pd catalyst (Heck reaction)}]{H_2C{=}CH_2}$ CH_3O–(naphthyl)–CH=CH_2 $\xrightarrow{\text{(several steps)}}$ CH_3O–(naphthyl)–CH(CH_3)COOH

5. The industrial synthesis of octyl methoxycinnamate, a common ingredient in sunscreens.

CH_3O–(phenyl)–Br + $H_2C{=}CH{-}COOH$ $\xrightarrow[\text{(Heck reaction)}]{\text{Pd catalyst}}$ CH_3O–(phenyl)–CH=CH–COOH

CH_3O–(phenyl)–CH=CH–COOH $\xrightarrow{\text{2-ethylhexan-1-ol (OH)}}$ CH_3O–(phenyl)–CH=CH–C(=O)–O–(2-ethylhexyl)

"octyl methoxycinnamate"
(E)-2-ethylhexyl-4-methoxycinnamate

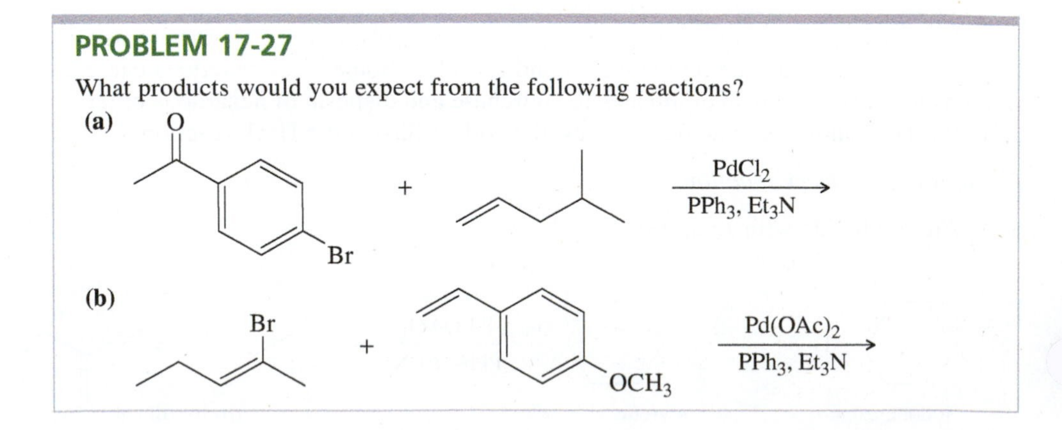

PROBLEM 17-27
What products would you expect from the following reactions?
(a)
O
+
PdCl2
PPh3, Et3N
Br
(b)
Br
+
Pd(OAc)2
PPh3, Et3N
OCH3

PROBLEM 17-28

What substituted alkene would you use in the Heck reaction to make the following products?

(a) 2-iodotoluene + ? $\xrightarrow[\text{PPh}_3,\ \text{Et}_3\text{N}]{\text{Pd(OAc)}_2}$ product with CH_3 and $C(=O)OCH_3$

(b) bromotetralin + ? $\xrightarrow[\text{PPh}_3,\ \text{Et}_3\text{N}]{\text{Pd(OAc)}_2}$ product with CN

17-13C The Suzuki Reaction (Suzuki Coupling)

The Suzuki reaction is a palladium-catalyzed substitution that couples an aryl or vinyl halide with an alkyl, alkenyl, or aryl boronic acid or boronate ester. Many variations on this fundamental reaction are possible, containing a wide variety of functional groups.

The Suzuki Reaction (Suzuki Coupling)

$$\text{R—X} + \text{R}'\text{—B(OH)}_2 \xrightarrow[\text{base such as NaOH}]{\text{Pd catalyst}} \text{R—R}' + \text{B(OH)}_3 + \text{NaX}$$

(R = aryl or vinyl)
(X = Br, I, or Cl)

a boronic acid
or a boronate ester, $R'B(OR)_2$

boric acid

Application: Greener Reagents

These types of couplings once required heavy metals and other toxic compounds. It was harder to dispose of the waste, and (in pharmaceuticals) difficult to prove that the products weren't contaminated with toxic spent reagents.

Suzuki coupling gives by-products that are much less toxic, like boric acid and borate esters. These by-products are less hazardous and easier to dispose of.

Like the Heck reaction, the Suzuki coupling can use water as the solvent. Water-based Suzuki reactions are attractive both for industrial processes and for labs that want to minimize the purchase and disposal of toxic solvents. The following examples show the variety of combinations that can be coupled using Suzuki reactions.

1. A vinyl halide with an alkenylboronate ester, preserving the stereochemistry of the reagents.

$$\text{(Z)-vinyl halide (Br)} + \text{(E)-alkenylboronate ester, B(OR)}_2 \xrightarrow[\text{NaOH}]{\text{Pd(PPh}_3)_4} \text{diene with (E, Z) stereochemistry}$$

(*Z*)-vinyl halide (*E*)-alkenylboronate ester diene with (*E*, *Z*) stereochemistry

2. An aryl halide with an arylboronic acid, using palladium on carbon and water as the solvent.

$$\text{HO—C}_6\text{H}_4\text{—Br} + \text{C}_6\text{H}_5\text{—B(OH)}_2 \xrightarrow[\text{K}_2\text{CO}_3,\ \text{H}_2\text{O}]{\text{10\% Pd/C}} \text{HO—C}_6\text{H}_4\text{—C}_6\text{H}_5$$

p-bromophenol phenylboronic acid *p*-phenylphenol

3. Synthesis of the anti-inflammatory drug flurbiprofen by an environmentally efficient Suzuki coupling that uses water as the solvent and palladium on carbon as a reusable catalyst.

sodium tetraphenylborate + (3-bromo-4-fluorophenyl) propanoic acid → flurbiprofen

Pd/C, H_2O (Suzuki coupling)

The boronate esters used for Suzuki reactions can be synthesized from commercially available alkyl-, vinyl-, and arylboronic acids. Alkyl and vinyl boronate esters are also synthesized by the hydroboration of double and triple bonds, similar to the hydroboration of alkenes and alkynes in Chapters 8 and 9. Note that the boron atom generally adds to the less substituted end of a double or triple bond. Also, the B and H add to the same side of a triple bond (syn addition) to give a trans alkenylboronate ester.

$$R'HC{=}CH_2 + H{-}B(OR)_2 \longrightarrow R'CH_2{-}CH_2{-}B(OR)_2$$

alkene + a dialkoxyborane → an alkylboronate ester

$$R{-}C{\equiv}C{-}H + H{-}B(OR)_2 \longrightarrow R'HC{=}CH{-}B(OR)_2$$

alkyne + a dialkoxyborane → a trans alkenylboronate ester

Arylboronate esters are made by a two-step process: First, we convert the aryl halide to the aryllithium compound (Chapter 10). Addition of a trialkyl borate (often trimethyl borate) allows the organolithium compound to form a carbon–boron bond and expel an alkoxide group.

$$Ph{-}Br \xrightarrow{2\ Li} Ph{-}Li \xrightarrow[\text{(trimethyl borate)}]{B(OMe)_3} Ph{-}B(OCH_3)_2 + LiOCH_3$$

bromobenzene → phenyllithium → an arylboronate ester (dimethyl phenylboronate)

The Suzuki reaction occurs by way of an interesting mechanism that is outlined in Mechanism 17-9. The first step, insertion of a palladium atom into the R—X bond to give R—Pd—X, is similar to the formation of a Grignard reagent by the insertion of Mg into the R—X bond to give R—Mg—X.

In the second step, the alkyl group from the negatively charged alkylboronate replaces the halogen on Pd, expelling a halide ion. In the final step, the two alkyl groups on Pd couple together to release the Pd atom in its original (0) oxidation state. The Pd atom can continue to catalyze more reactions, making it a true catalyst.

MECHANISM 17-9 The Suzuki Reaction

Step 1: Oxidative addition to the Pd gives Pd a higher oxidation number.

$$\text{R—X} + \text{Pd}^0 \longrightarrow \text{R—Pd}^{\text{II}}\text{—X}$$

Step 2: Exchange of ligands on Pd occurs.

$$\text{R—Pd}^{\text{II}}\text{—X} + \text{R}'\text{—}\overset{\text{OH}}{\underset{\text{OH}}{\text{B}^-}}\text{—OH} \longrightarrow \text{R—Pd}^{\text{II}}\text{—R}' + \text{HO—B(OH)}_2 + \text{X}^-$$

formed by addition of ^-OH to $R'B(OH)_2$

Step 3: Reductive elimination from the Pd gives Pd a lower oxidation number.

$$\text{R—Pd}^{\text{II}}\text{—R}' \longrightarrow \text{R—R}' + \text{Pd}^0$$

PROBLEM 17-29

What products would you expect from the following Suzuki coupling reactions?

(a) 4-tert-butylbromobenzene + 2-methylphenylboronic acid ($B(OH)_2$) $\xrightarrow[K_3PO_4]{Pd(OAc)_2}$

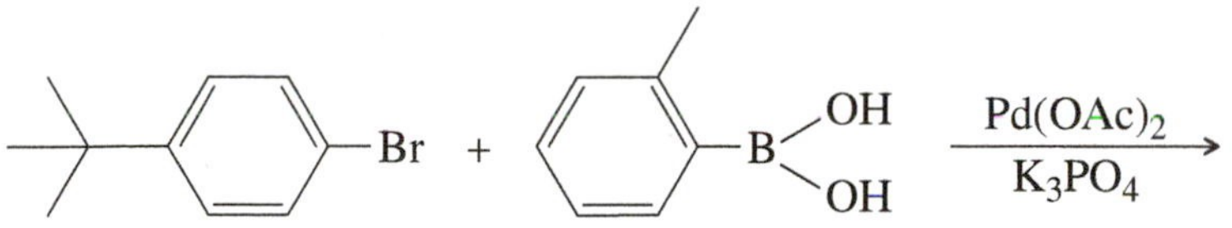

(b) bromobenzene + 3-pyridyl–$B(OR)_2$ $\xrightarrow[NaOH]{Pd(PPh_3)_4}$

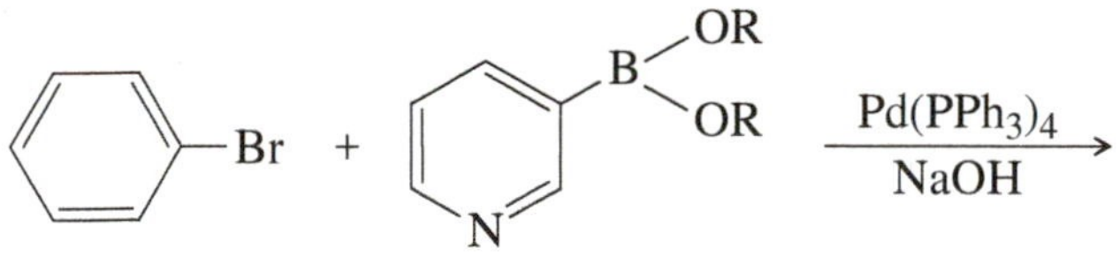

PROBLEM 17-30

Show how you would use Suzuki reactions to synthesize these products from the indicated starting materials. You may use any additional reagents you need.

(a) Br (vinyl bromide) $\longrightarrow$ (phenyl-substituted alkene)

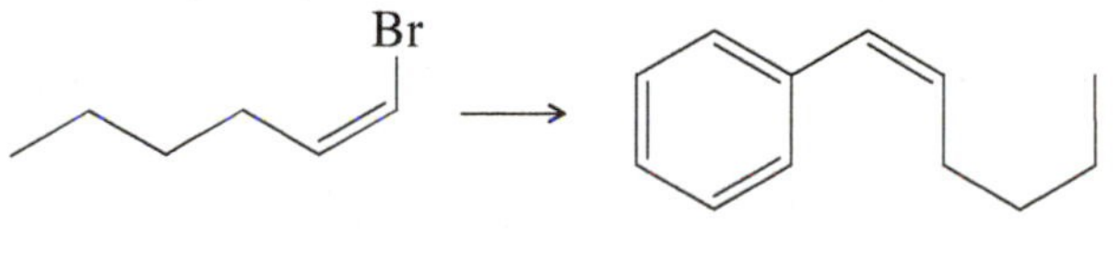

(b) Br $\longrightarrow$ (phenyl-substituted diene)

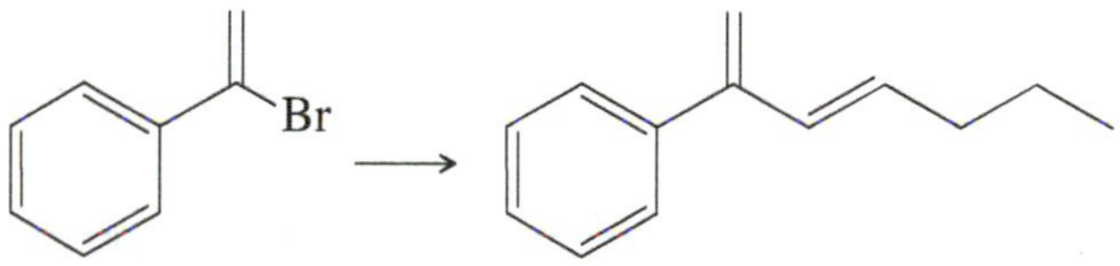

17-14 Addition Reactions of Benzene Derivatives

Although substitution is more common, aromatic compounds may undergo addition if forcing conditions are used. These additions are some of the most important industrial reactions of aromatic compounds.

17-14A Chlorination

When benzene is treated with an excess of chlorine under heat and pressure (or with irradiation by light), six chlorine atoms add to form 1,2,3,4,5,6-hexachlorocyclohexane. This product is often called *benzene hexachloride* (BHC) because it is synthesized by direct chlorination of benzene.

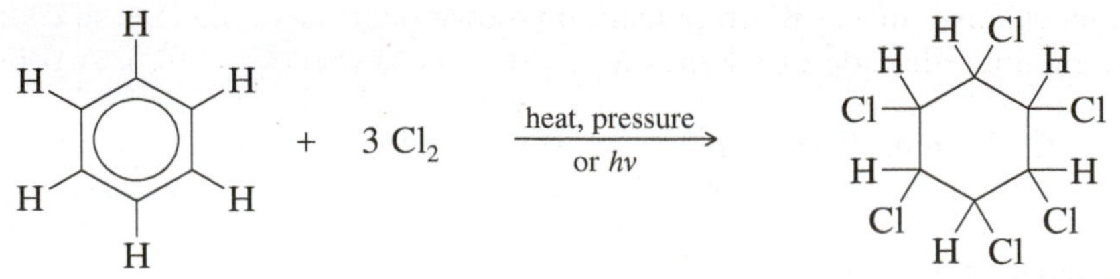

$$C_6H_6 + 3\,Cl_2 \xrightarrow[\text{or } h\nu]{\text{heat, pressure}} C_6H_6Cl_6$$

benzene → benzene hexachloride, BHC (eight isomers)

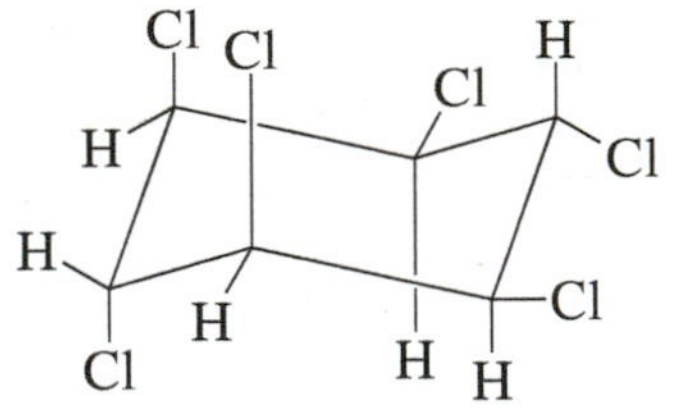

lindane

The addition of chlorine to benzene, believed to involve a free-radical mechanism, is normally impossible to stop at an intermediate stage. The first addition destroys the ring's aromaticity, and the next 2 moles of Cl_2 add very rapidly. All eight possible stereoisomers are produced in various amounts. The most important isomer for commercial purposes is the insecticide *lindane*, which is used in a shampoo to kill head lice.

17-14B Catalytic Hydrogenation of Aromatic Rings

Catalytic hydrogenation of benzene to cyclohexane takes place at elevated temperatures and pressures, often catalyzed by ruthenium or rhodium-based catalysts. Substituted benzenes react to give substituted cyclohexanes; disubstituted benzenes usually give mixtures of cis and trans isomers.

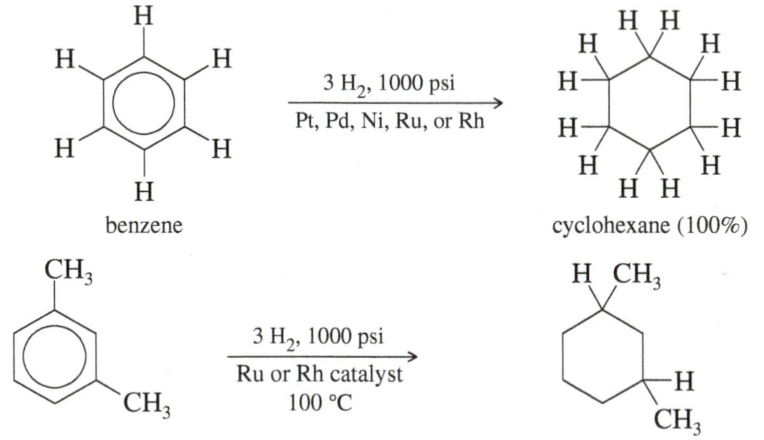

$$\text{benzene} \xrightarrow[\text{Pt, Pd, Ni, Ru, or Rh}]{3\,H_2,\ 1000\text{ psi}} \text{cyclohexane (100\%)}$$

$$m\text{-xylene} \xrightarrow[\text{Ru or Rh catalyst, } 100\ ^{\circ}\text{C}]{3\,H_2,\ 1000\text{ psi}} \text{1,3-dimethylcyclohexane (100\%)}$$

(mixture of cis and trans)

Catalytic hydrogenation of benzene is the commercial method for producing cyclohexane and substituted cyclohexane derivatives. The reduction cannot be stopped at an intermediate stage (cyclohexene or cyclohexadiene) because these alkenes are reduced faster than benzene.

17-14C Birch Reduction

In 1944, the Australian chemist A. J. Birch found that benzene derivatives are reduced to nonconjugated cyclohexa-1,4-dienes by treatment with sodium or lithium in a mixture of liquid ammonia and an alcohol. The **Birch reduction** provides a convenient method for making a wide variety of interesting and useful cyclic dienes.

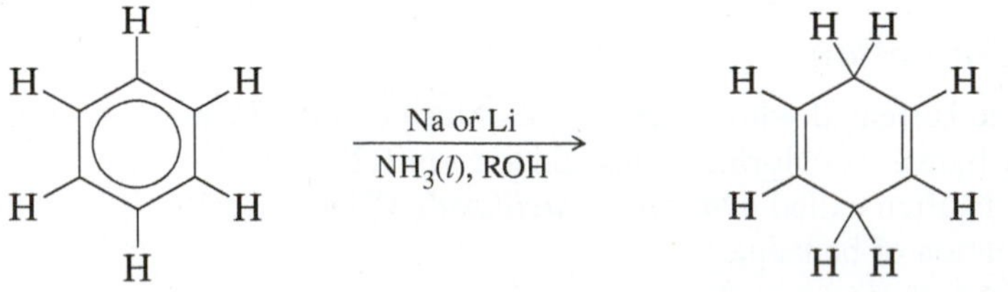

$$\text{benzene} \xrightarrow[NH_3(l),\ ROH]{\text{Na or Li}} \text{cyclohexa-1,4-diene (90\%)}$$

The mechanism of the Birch reduction (shown next) is similar to the sodium/liquid ammonia reduction of alkynes to *trans*-alkenes (Section 9-9C). A solution of sodium in liquid ammonia contains solvated electrons that can add to benzene, forming a radical anion. The strongly basic radical anion abstracts a proton from the alcohol in the solvent, giving a cyclohexadienyl radical. The radical quickly adds another solvated electron to form a cyclohexadienyl anion. Protonation of this anion gives the reduced product.

MECHANISM 17-10 The Birch Reduction

The Birch reduction involves twice adding a solvated electron, followed by a proton, to the aromatic ring.

Preceding step: Formation of solvated electrons in the ammonia solution.

$$NH_3 + Na \rightleftharpoons NH_3 \cdot e^- \text{ (deep blue solution)} + Na^+$$

solvated electron

Steps 1 and 2: Addition of an electron, followed by a proton, forms a radical.

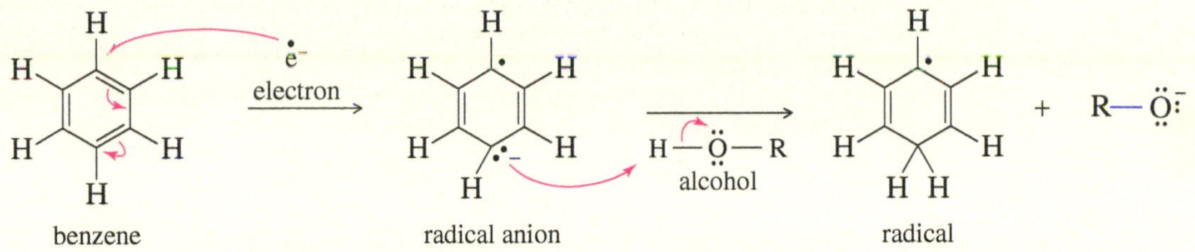

benzene + e^- (electron) → radical anion; radical anion + $H-\ddot{O}-R$ (alcohol) → radical + $R-\ddot{O}:^-$

Steps 3 and 4: Addition of a second electron, followed by a proton, gives the product.

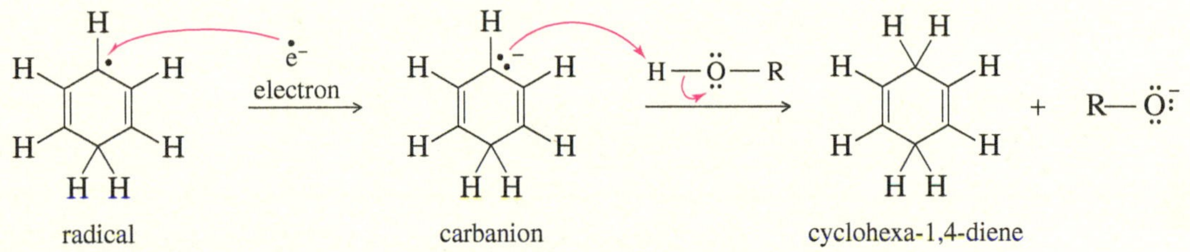

radical + e^- (electron) → carbanion; carbanion + $H-\ddot{O}-R$ → cyclohexa-1,4-diene + $R-\ddot{O}:^-$

The two carbon atoms that are reduced go through anionic intermediates. Electron-withdrawing substituents stabilize the carbanions, whereas electron-donating substituents destabilize them. Therefore, reduction takes place on carbon atoms bearing electron-withdrawing substituents (such as those containing carbonyl groups) and not on carbon atoms bearing electron-releasing substituents (such as alkyl and alkoxy groups).

A carbon bearing an electron-withdrawing carbonyl group is reduced

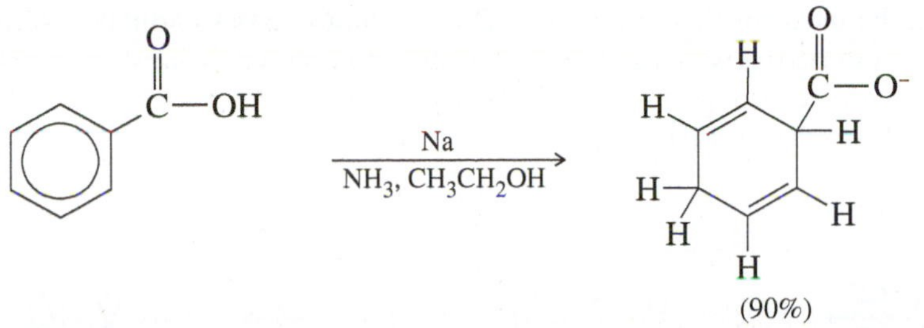

$C_6H_5COOH \xrightarrow[NH_3,\ CH_3CH_2OH]{Na}$ cyclohexa-2,5-diene-1-carboxylate ($-COO^-$) (90%)

A carbon bearing an electron-releasing alkoxy group is not reduced

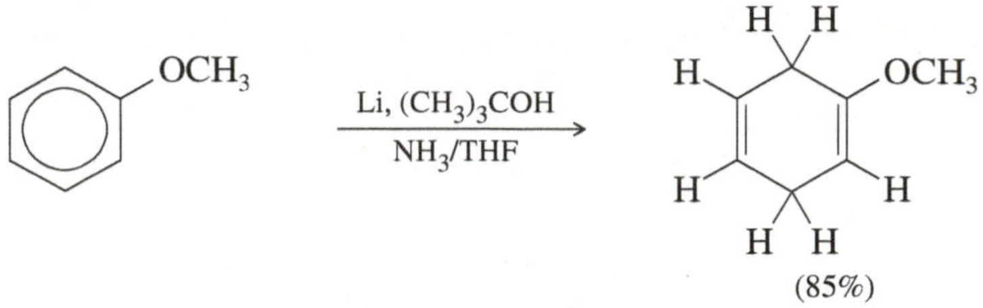

Anisole (OCH_3) $\xrightarrow[NH_3/THF]{Li,\ (CH_3)_3COH}$ 1-methoxy-1,4-cyclohexadiene (85%)

Substituents that are strongly electron-releasing ($—OCH_3$, for example) deactivate the aromatic ring toward Birch reduction. Lithium is often used with these deactivated systems, together with a cosolvent (often THF) and a weaker proton source (*tert*-butyl alcohol). The stronger reducing agent, combined with a weaker proton source, enhances the reduction.

PROBLEM 17-31

Propose mechanisms for the Birch reductions of benzoic acid and anisole just shown. Show why the observed orientation of reduction is favored in each case.

PROBLEM 17-32

Predict the major products of the following reactions.

(a) toluene + excess Cl_2 (heat, pressure)
(b) benzamide ($PhCONH_2$) + Na (liquid NH_3, CH_3CH_2OH)
(c) *o*-xylene + H_2 (1000 psi, 100 °C, Rh catalyst)
(d) *p*-xylene + Na (liquid NH_3, CH_3CH_2OH)
(e) CH_3O–naphthalene–OCH_3 $\xrightarrow[(CH_3)_3COH]{\text{excess Li},\ NH_3(l)/THF}$

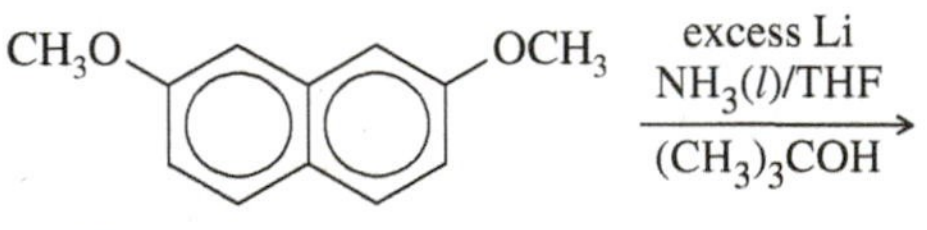

2,7-dimethoxynaphthalene

17-15 Side-Chain Reactions of Benzene Derivatives

Many reactions are not affected by the presence of a nearby benzene ring; yet others depend on the aromatic ring to promote the reaction. For example, the Clemmensen reduction is occasionally used to reduce aliphatic ketones to alkanes, but it works best reducing aryl ketones to alkylbenzenes. Several additional side-chain reactions show the effects of a nearby aromatic ring.

17-15A Permanganate Oxidation

An aromatic ring imparts extra stability to the nearest carbon atom of its side chains. The aromatic ring and *one* carbon atom of a side chain can survive a vigorous permanganate oxidation. The product is a carboxylate salt of benzoic acid. (Hot chromic acid can also be used for this oxidation.) This oxidation is occasionally useful for making benzoic acid derivatives, as long as any other functional groups are resistant to oxidation. Functional groups such as $—NO_2$, halogens, $—COOH$, and $—SO_3H$ usually survive this brutal oxidation.

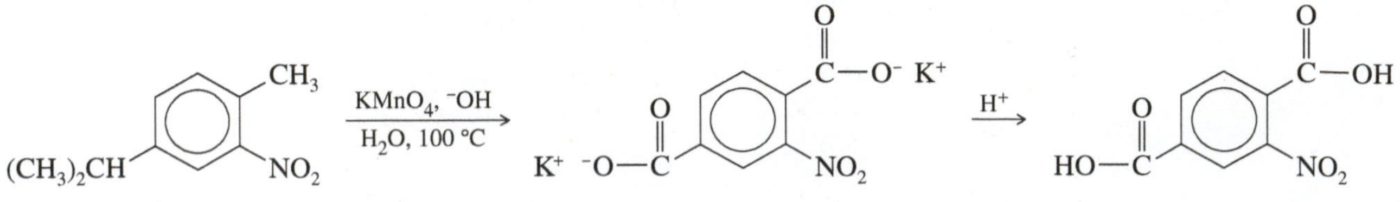

$(CH_3)_2CH$–C_6H_3(CH_3)(NO_2) $\xrightarrow[H_2O,\ 100\ °C]{KMnO_4,\ ^-OH}$ $K^+\ ^-O—C(=O)$–C_6H_3(NO_2)–$C(=O)—O^-\ K^+$ $\xrightarrow{H^+}$ $HO—C(=O)$–C_6H_3(NO_2)–$C(=O)—OH$

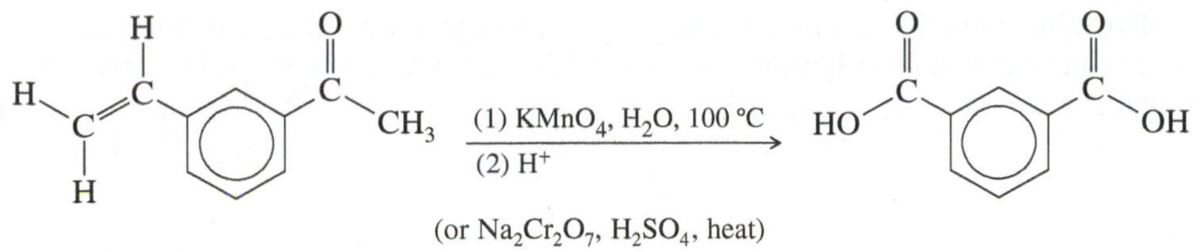

$$\xrightarrow[\text{(2) H}^+]{\text{(1) KMnO}_4\text{, H}_2\text{O, 100 °C}}$$

(or $Na_2Cr_2O_7$, H_2SO_4, heat)

PROBLEM 17-33

Predict the major products of treating the following compounds with hot, concentrated potassium permanganate, followed by acidification with dilute HCl.

(a) isopropylbenzene **(b)** *p*-xylene **(c)** (tetralin)

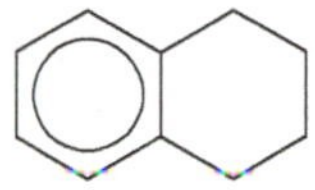

17-15B Side-Chain Halogenation

Alkylbenzenes undergo free-radical halogenation much more easily than alkanes because abstraction of a hydrogen atom at a **benzylic position** gives a resonance-stabilized benzylic radical. For example, ethylbenzene reacts with chlorine in the presence of light to give α-chloroethylbenzene. Further chlorination can occur to give a dichlorinated product.

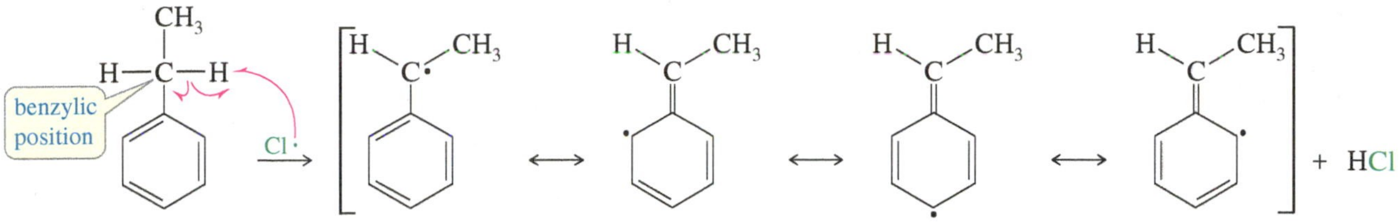

benzylic position

resonance-stabilized benzylic radical

+ HCl

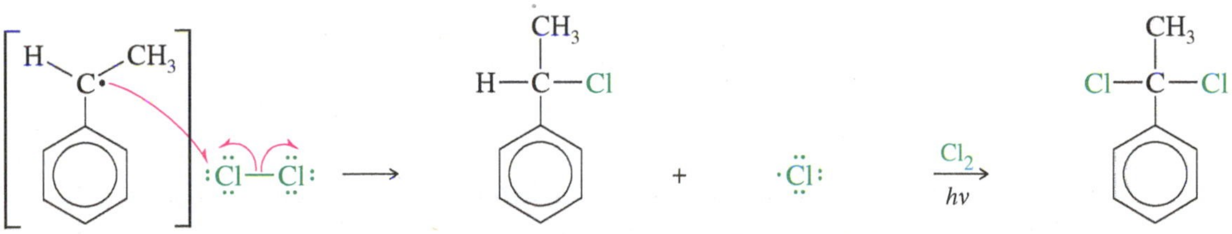

benzylic radical

α-chloroethylbenzene

chlorine radical continues the chain

dichlorinated

Although chlorination shows a preference for α substitution (the α position is the benzylic carbon bonded to the benzene ring), the chlorine radical is too reactive to give entirely benzylic substitution. Mixtures of isomers are often produced. In the chlorination of ethylbenzene, for example, there is a significant amount of substitution at the β carbon.

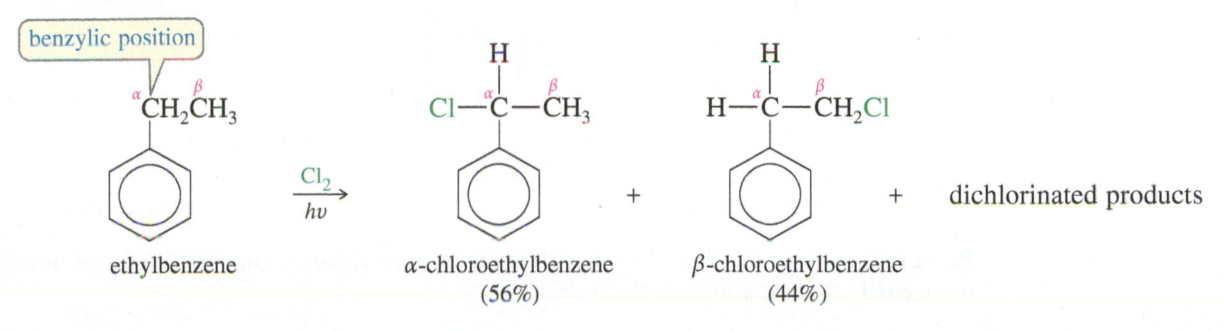

Bromine radicals are not as reactive as chlorine radicals, so bromination is more selective than chlorination (Section 4-13C). Bromine reacts exclusively at the benzylic position.

ethylbenzene ($C_6H_5CH_2CH_3$) $\xrightarrow[h\nu]{Br_2 \text{ or NBS}}$ α-bromoethylbenzene (C_6H_5–CHBr–CH_3) + (trace) α,α-dibromoethylbenzene (C_6H_5–CBr_2–CH_3)

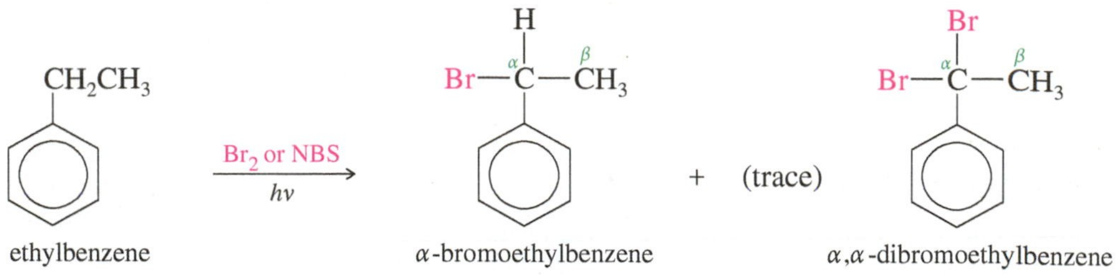

Either elemental bromine (much cheaper) or *N*-bromosuccinimide may be used as the reagent for benzylic bromination. *N*-Bromosuccinimide is preferred for allylic bromination (Section 15-7) because Br_2 can add to the double bond. This is not a problem with the relatively unreactive benzene ring unless it has powerful activating substituents.

PROBLEM-SOLVING HINT

In predicting reactions on side chains of aromatic rings, consider resonance forms that delocalize a charge or a radical electron onto the ring.

PROBLEM 17-34

Propose a mechanism for the bromination of ethylbenzene shown above.

PROBLEM 17-35

What would be the ratio of products in the reaction of chlorine with ethylbenzene if chlorine *randomly* abstracted a methyl or methylene hydrogen? What is the reactivity ratio for the benzylic hydrogens compared with the methyl hydrogens?

PROBLEM 17-36

Predict the major products when the following compounds are irradiated by light and treated with (1) 1 equivalent of Br_2 and (2) excess Br_2.

(a) isopropylbenzene **(b)** 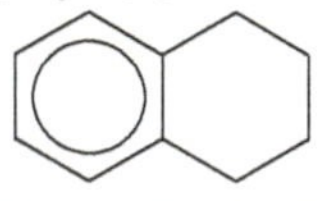 (tetralin)

17-15C Nucleophilic Substitution at the Benzylic Position

In Chapter 15, we saw that allylic halides are more reactive than most alkyl halides in both S_N1 and S_N2 reactions. Benzylic halides are also more reactive in these substitutions, for reasons similar to those for allylic halides.

First-Order Reactions First-order nucleophilic substitution requires ionization of the halide to give a carbocation. In the case of a benzylic halide, the carbocation is resonance-stabilized. For example, the 1-phenylethyl cation (2°) is about as stable as a 3° alkyl cation.

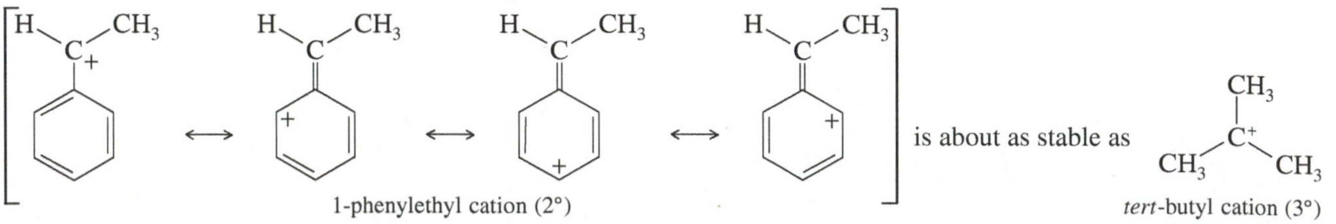

1-phenylethyl cation (2°) [four resonance forms] is about as stable as *tert*-butyl cation (3°)

Because they form relatively stable carbocations, benzyl halides undergo S_N1 reactions more easily than do most alkyl halides.

$$\text{PhCH}_2\text{—Br} \xrightarrow{CH_3CH_2OH,\ \Delta\ (heat)} \text{PhCH}_2\text{—OCH}_2CH_3$$

benzyl bromide — benzyl ethyl ether

If a benzylic cation is bonded to more than one phenyl group, the stabilizing effects are additive. An extreme example is the triphenylmethyl cation. This cation is exceptionally stable, with three phenyl groups to stabilize the positive charge. In fact, triphenylmethyl fluoroborate can be stored for years as a stable ionic solid.

$Ph_3C^+\ {}^-BF_4$

triphenylmethyl fluoroborate

PROBLEM 17-37

Propose a mechanism for the reaction of benzyl bromide with ethanol to give benzyl ethyl ether (shown above).

Second-Order Reactions Like allylic halides, benzylic halides are about 100 times more reactive than primary alkyl halides in S_N2 displacement reactions. The explanation for this enhanced reactivity is similar to that for the reactivity of allylic halides.

During S_N2 displacement on a benzylic halide, the *p* orbital partially bonds with the nucleophile and the leaving group also overlaps with the pi electrons of the ring (Figure 17-5). This stabilizing conjugation lowers the energy of the transition state, increasing the reaction rate.

S_N2 reactions of benzyl halides efficiently convert aromatic methyl groups to functional groups. Halogenation, followed by substitution, gives the functionalized product.

$$p\text{-}O_2N\text{—}C_6H_4\text{—}CH_3 \xrightarrow[h\nu]{Br_2} p\text{-}O_2N\text{—}C_6H_4\text{—}CH_2Br \xrightarrow[CH_3OH]{NaOCH_3} p\text{-}O_2N\text{—}C_6H_4\text{—}CH_2OCH_3$$

$$o\text{-}Br\text{—}C_6H_4\text{—}CH_3 \xrightarrow[h\nu]{Br_2} o\text{-}Br\text{—}C_6H_4\text{—}CH_2Br \xrightarrow[acetone]{NaCN} o\text{-}Br\text{—}C_6H_4\text{—}CH_2CN$$

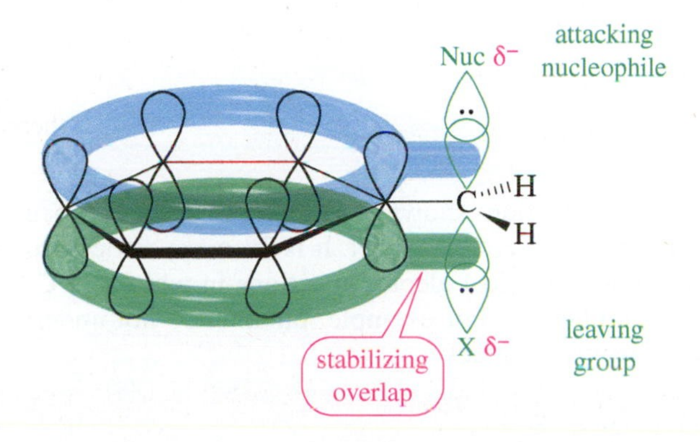

FIGURE 17-5
The transition state for S_N2 displacement of a benzylic halide is stabilized by conjugation with the pi electrons in the ring.

PROBLEM 17-38

(a) Based on what you know about the relative stabilities of alkyl cations and benzylic cations, predict the product of addition of HBr to 1-phenylpropene.
(b) Propose a mechanism for this reaction.
(c) Based on what you know about the relative stabilities of alkyl radicals and benzylic radicals, predict the product of addition of HBr to 1-phenylpropene in the presence of a free-radical initiator.
(d) Propose a mechanism for this reaction.

PROBLEM 17-39

Show how you would synthesize the following compounds, using the indicated starting materials.
(a) 3-phenylbutan-1-ol from styrene
(b) $CH_3—CH—OCH_3$ (CH bearing a p-methoxyphenyl group, OCH_3) from anisole
(c) O_2N–C₆H₄–CH_2CN (para) from toluene

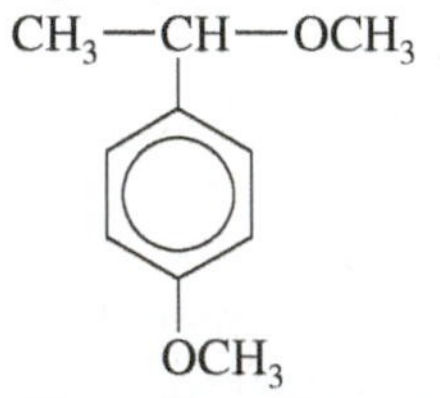

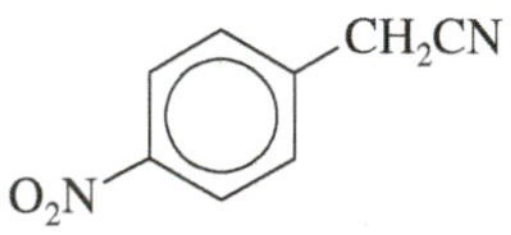

Application: Aspirin

Aspirin is an anti-inflammatory drug that blocks the synthesis of prostaglandins, which are powerful hormones that regulate smooth muscle and stimulate inflammation (Section 25-7). Aspirin also blocks synthesis of the related thromboxanes, which constrict blood vessels and promote the aggregation of platelets, the first step in formation of a blood clot. Many doctors recommend that high-risk patients take a small aspirin each day to reduce the danger of clotting in blood vessels, leading to a heart attack, stroke, or embolism.

17-16 Reactions of Phenols

Much of the chemistry of phenols is like that of aliphatic alcohols. For example, phenols can be acylated to give esters, and phenoxide ions can serve as nucleophiles in the Williamson ether synthesis (Section 14-5). Formation of phenoxide ions is particularly easy because phenols are more acidic than water; aqueous sodium hydroxide deprotonates phenols to give phenoxide ions.

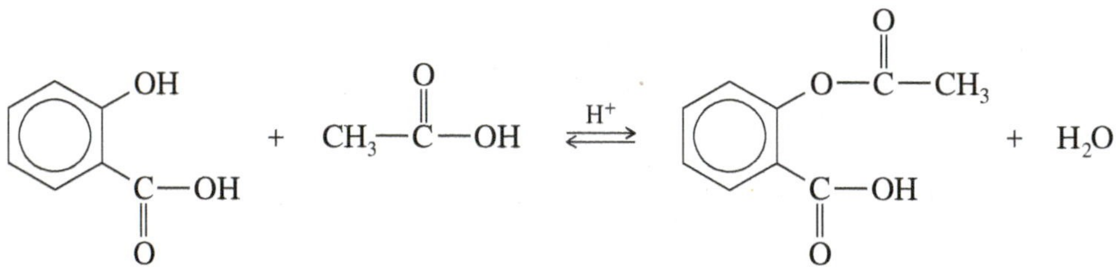

salicylic acid + acetic acid ($CH_3—C(=O)—OH$) $\xrightleftharpoons{H^+}$ acetylsalicylic acid (aspirin) ($O—C(=O)—CH_3$, $C(=O)—OH$) + H_2O

salicylic acid   acetic acid   acetylsalicylic acid (aspirin)

(p-propenylphenol, OH) $\xrightarrow[H_2O]{NaOH}$ ($O^-\ Na^+$) $\xrightarrow[\text{(or } CH_3I)]{CH_3OSO_2OCH_3 \text{ dimethyl sulfate}}$ (OCH_3)

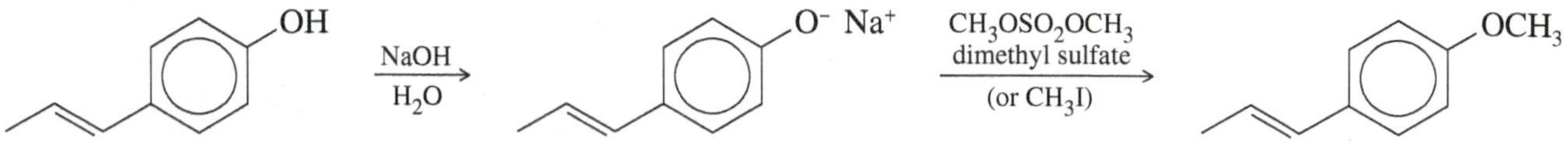

anethole (licorice flavoring)

All the alcohol-like reactions shown involve breaking of the phenolic O—H bond. This is a common way for phenols to react. It is far more difficult to break the C—O bond of a phenol, however. Most alcohol reactions in which the C—O bond breaks are not possible with phenols. For example, phenols do not undergo acid-catalyzed elimination or S_N2 back-side attack.

Phenols also undergo reactions that are not possible with aliphatic alcohols. Let's consider some reactions that are peculiar to phenols.

17-16A Oxidation of Phenols to Quinones

Phenols undergo oxidation, but they give different types of products from those seen with aliphatic alcohols. Chromic acid oxidation of a phenol gives a conjugated 1,4-diketone called a **quinone.** In the presence of air, many phenols slowly autoxidize to dark mixtures containing quinones.

m-cresol $\xrightarrow[H_2SO_4]{Na_2Cr_2O_7}$ 2-methyl-1,4-benzoquinone

Hydroquinone (benzene-1,4-diol) is easily oxidized because it already has two oxygen atoms bonded to the ring. Even very weak oxidants like silver bromide (AgBr) can oxidize hydroquinone. Silver bromide is reduced to black metallic silver in a light-sensitive reaction: Any grains of silver bromide that have been exposed to light (AgBr*) react faster than unexposed grains.

hydroquinone (benzene-1,4-diol) + 2 AgBr* ⟶ quinone (1,4-benzoquinone) + 2 Ag↓ + 2 HBr

Black-and-white photographic film is based on this reaction. A film containing small grains of silver bromide is exposed by a focused image. Where light strikes the film, the grains are activated. The film is then treated with a hydroquinone solution (the *developer*) to reduce the activated silver bromide grains, leaving black silver deposits where the film was exposed to light. The result is a negative image, with dark areas where light struck the film.

PROBLEM 17-40

The bombardier beetle defends itself by spraying a hot quinone solution from its abdomen (see photo). This solution is formed by the enzyme-catalyzed oxidation of hydroquinone by hydrogen peroxide. Write a balanced equation for this oxidation.

Quinones occur widely in nature, where they serve as biological oxidation–reduction reagents. The quinone *coenzyme Q* (CoQ) is also called *ubiquinone* because it seems ubiquitous (found everywhere) in oxygen-consuming organisms. Coenzyme Q serves as an oxidizing agent within the mitochondria of cells. The following reaction shows the reduction of coenzyme Q by NADH (the reduced form of nicotinamide adenine dinucleotide), which becomes oxidized to NAD^+.

Application: Fruit Browning

The browning of fruit is a common example of the oxidation of phenols to quinones. Apples, pears, potatoes, and so forth contain *polyphenol oxidase* (PPO), an enzyme that catalyzes the oxidation of naturally occurring derivatives of catechol (benzene-1,2-diol) by atmospheric oxygen. The products are ortho-quinones, which are unstable and quickly condense to give brown polymers.

Browning can be controlled by adding reducing agents or acidic solutions that inhibit the activity of the PPO enzyme. Solutions of sodium bisulfite, ascorbic acid (vitamin C), and lemon juice are commonly added to freshly cut fruit to retard browning.

catechol derivatives $\xrightarrow{O_2,\ PPO}$ *ortho*-quinones (unstable) ⟶ brown polymer

When threatened, the bombardier beetle mixes hydroquinone and H_2O_2 with enzymes. Peroxide oxidizes hydroquinone to quinone, and the strongly exothermic reaction heats the solution to boiling. The hot, irritating liquid sprays from the tip of the insect's abdomen.

coenzyme Q oxidized form + NADH reduced form + H^+ ⟶ coenzyme Q reduced form + NAD^+ oxidized form

$R = —(CH_2—CH{=}C(CH_3)—CH_2)_{10}—H$

17-16B Electrophilic Aromatic Substitution of Phenols

Phenols are highly reactive substrates for electrophilic aromatic substitution because the nonbonding electrons of the hydroxy group stabilize the sigma complex formed by attack at the ortho or para position (Section 17-6B). Therefore, the hydroxy group is strongly activating and ortho, para-directing. Phenols are excellent substrates for halogenation, nitration, sulfonation, and some Friedel–Crafts reactions. Because they are highly reactive, phenols are usually alkylated or acylated using relatively weak Friedel–Crafts catalysts (such as HF) to avoid overalkylation or overacylation.

$$\text{phenol} + CH_3—CH(OH)—CH_3 \xrightarrow{HF} p\text{-}CH(CH_3)_2\text{ phenol} + o\text{-}CH(CH_3)_2\text{ phenol}$$

Phenoxide ions, easily generated by treating a phenol with sodium hydroxide, are even more reactive than phenols toward electrophilic aromatic substitution. Because they are negatively charged, phenoxide ions react with positively charged electrophiles to give neutral sigma complexes whose structures resemble quinones.

$$\text{phenol} \xrightarrow[H_2O]{NaOH} \text{phenoxide ion} \xrightarrow{\text{excess } Br_2} \text{sigma complex} \xrightarrow{^-OH} p\text{-bromophenoxide} \xrightarrow{Br_2} \xrightarrow{Br_2} \text{2,4,6-tribromophenoxide}$$

phenoxide ion   sigma complex

Phenoxide ions are so strongly activated that they undergo electrophilic aromatic substitution with carbon dioxide, a weak electrophile. The carboxylation of phenoxide ion is an industrial synthesis of salicylic acid, which is then converted to aspirin, as shown on p. 856.

$$\text{phenoxide} + O{=}C{=}O \longrightarrow \text{sigma complex} \xrightarrow{^-OH} \text{salicylate} + H_2O \xrightarrow{H^+} \text{salicylic acid}$$

salicylic acid

PROBLEM 17-41

Predict the products formed when *m*-cresol (*m*-methylphenol) reacts with

(a) NaOH and then ethyl bromide
(b) acetyl chloride, $CH_3-C(=O)-Cl$
(c) bromine in CCl_4 in the dark
(d) excess bromine in CCl_4 in the light
(e) sodium dichromate in H_2SO_4
(f) two equivalents of *tert*-butyl chloride and $AlCl_3$

PROBLEM 17-42

1,4-Benzoquinone is a good Diels–Alder dienophile. Predict the products of its reaction with
(a) buta-1,3-diene **(b)** cyclopenta-1,3-diene

PROBLEM-SOLVING STRATEGY Synthesis Using Electrophilic Aromatic Substitution

Aromatic substitutions have been important synthetic tools for over a century. Most important syntheses require more than one substitution, and each new substituent affects where the next substitution will react. Therefore, any multistep sequence must be planned so that the earlier substituents direct later reactions toward the correct reaction sites.

Several factors govern these reactions:

1. The order of substitution determines the product.

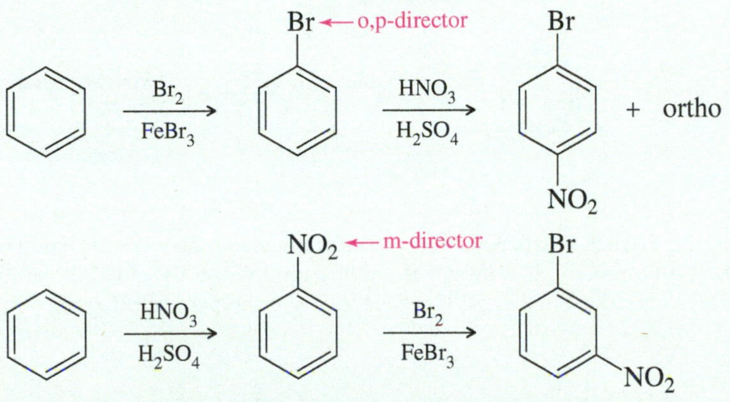

Benzene $\xrightarrow[FeBr_3]{Br_2}$ bromobenzene (Br ← o,p-director) $\xrightarrow[H_2SO_4]{HNO_3}$ *p*-bromonitrobenzene (Br, NO_2) + ortho

In ortho, para mixtures, the para isomer is usually the major product, and it can be separated from other isomers.

Benzene $\xrightarrow[H_2SO_4]{HNO_3}$ nitrobenzene (NO_2 ← m-director) $\xrightarrow[FeBr_3]{Br_2}$ *m*-bromonitrobenzene (Br, NO_2)

Conclusion: To produce the ortho or para product, attach the o,p-director first. To produce the meta product, attach the m-director first. Remember that some reactions (such as Friedel–Crafts) do not work well on strongly deactivated rings.

PROBLEM 17-43

Predict the major product at the end of each sequence.

(a) Benzene $\xrightarrow[H_2SO_4]{SO_3}$ $\xrightarrow[AlCl_3]{Cl_2}$

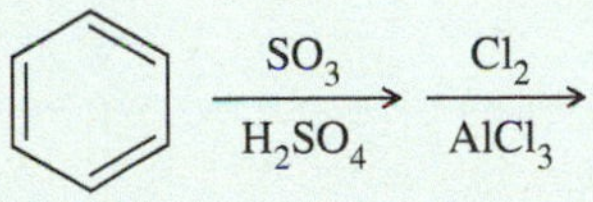

(b) Benzene $\xrightarrow[AlCl_3]{Cl_2}$ $\xrightarrow[H_2SO_4]{SO_3}$

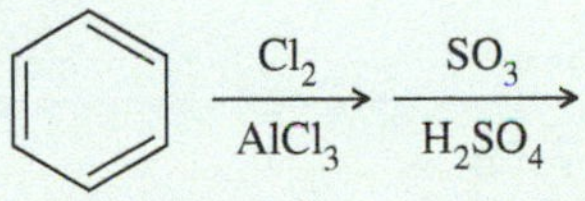

2. When the directive effects of two substituents conflict, a strongly activating group wins.

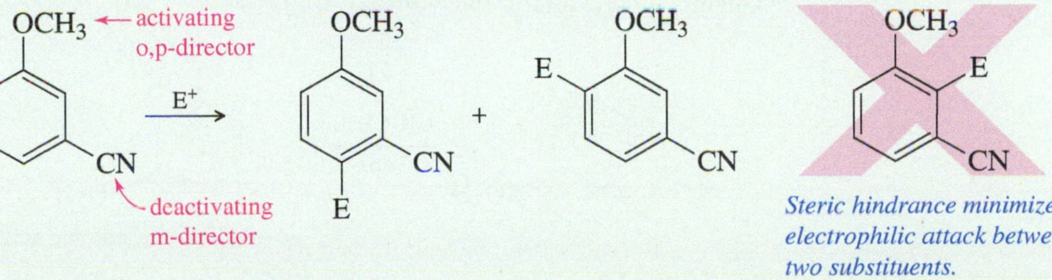

m-Methoxybenzonitrile (OCH_3 ← activating o,p-director; CN ← deactivating m-director) $\xrightarrow{E^+}$ product with E para to OCH_3 + product with E ortho to OCH_3 (away from CN); the product with E between OCH_3 and CN is crossed out.

Steric hindrance minimizes electrophilic attack between two substituents.

(continued)

PROBLEM 17-44

Predict the site(s) of electophilic attack on these compounds.

(a) OH, NO_2 (b) OCH_3, H_3C, CH_3

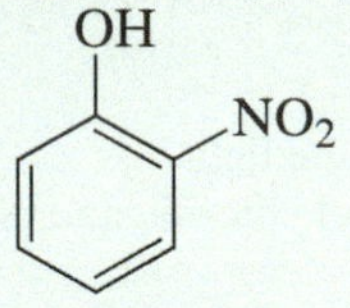

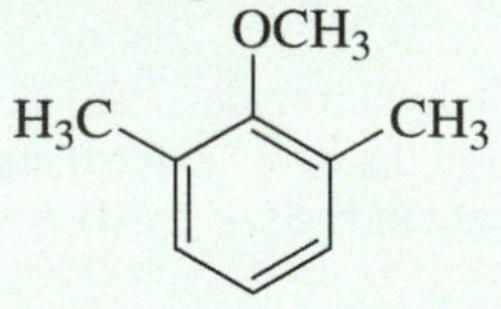

3. Friedel–Crafts reactions add acyl and alkyl groups to aromatic compounds, but they have limitations.

A. In an alkylation, a carbocation intermediate can rearrange: straight-chain alkylbenzenes cannot be produced by simple Friedel–Crafts alkylation.

+ Cl $\xrightarrow{AlCl_3}$

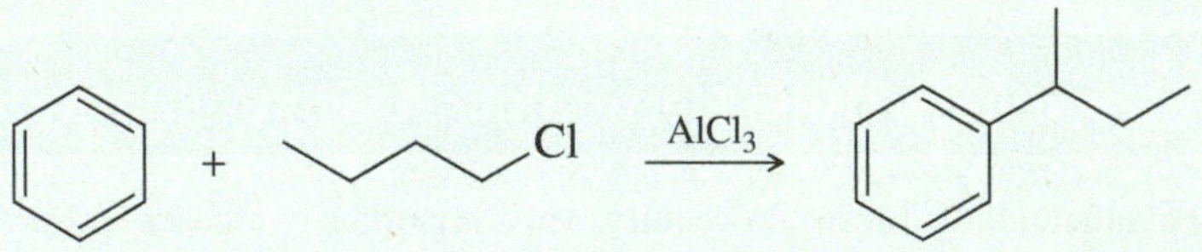

The major product results from rearrangement of the carbocation.

PROBLEM 17-45

Predict the products.

(a) + Cl $\xrightarrow{AlCl_3}$

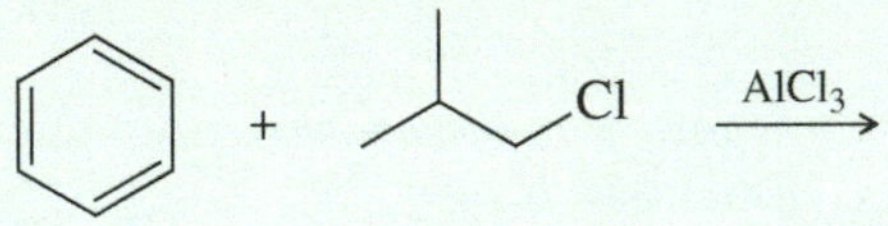

(b) + Cl $\xrightarrow{AlCl_3}$ two rearranged products (Hint: hydride shift and ring expansion.)

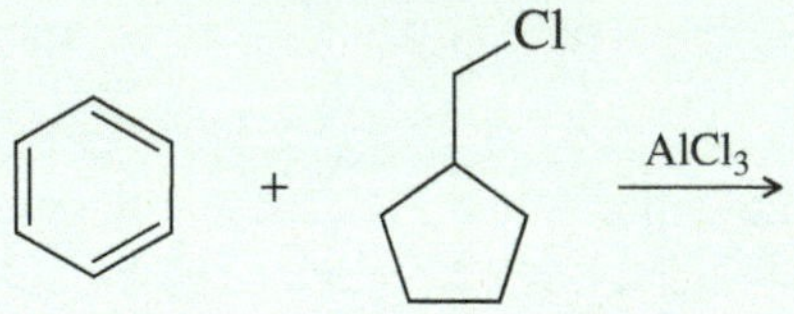

This limitation can be avoided by using the Friedel–Crafts acylation (no rearrangement, no overalkylation), followed by a Clemmensen reduction to convert the acyl group to an alkyl group. If another group is to be added, it can be added to the acylbenzene to give meta orientation, or it can be added to the alkylbenzene to give ortho,para orientation.

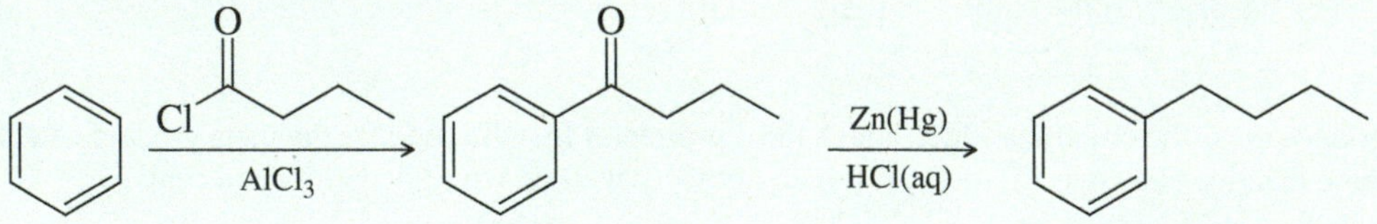

O, Cl $\xrightarrow{AlCl_3}$ O $\xrightarrow[HCl(aq)]{Zn(Hg)}$

an acylbenzene meta-director (no rearrangement)

an alkylbenzene o,p-director

PROBLEM 17-46

To synthesize *m*-ethylbenzenesulfonic acid, a student attempted the Friedel–Crafts alkylation of benzenesulfonic acid with bromoethane. Do you predict that this reaction was successful? If not, propose an alternative synthesis.

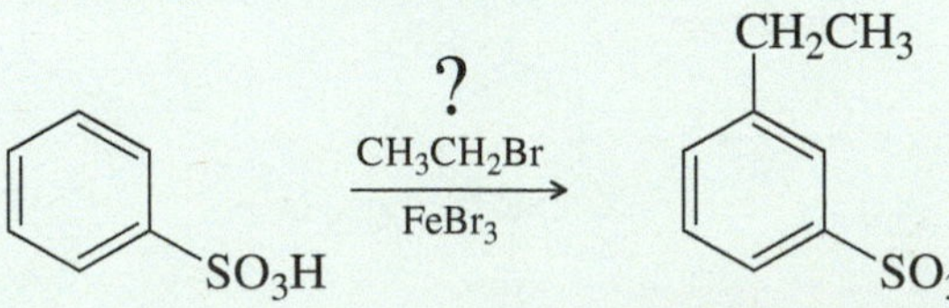

SO_3H $\xrightarrow[FeBr_3]{CH_3CH_2Br}$? CH_2CH_3, SO_3H

m-ethylbenzenesulfonic acid

B. In alkylation, substitution can occur more than once. (Each successive group increases activation.)

activating → CH_3 CH_3 CH_3 CH_3 CH_3 CH_3 CH_3 CH_3

CH_3Br / $AlCl_3$ + + + plus other isomers

Overalkylation can be minimized by using a large excess of the starting aromatic compound (benzene in this case), which is usually recycled through the process.

C. Friedel–Crafts reactions do not work on strongly deactivated benzenes, that is, any compound less reactive than halobenzenes.

Strong deactivating groups are NO_2, SO_3H, CN, CF_3, and —C(=O)—

D. Friedel–Crafts reactions do not work on benzenes with unprotected $—NH_2$ and $—NR_2$ substituents. The N: complexes with the $AlCl_3$ catalyst and becomes positively charged, thereby becoming a deactivating group.

R R N $\overset{-}{AlCl_3}$ R—N—R + strongly deactivating; F-C stops

RCl / $AlCl_3$

Amines also react with strongly acidic reagents such as HCl and H_2SO_4, protonating the nitrogen, which makes it meta-directing and strongly deactivating. Converting the basic amine to an amide (below) also protects it against this protonation.

Workaround: Convert the amine to an amide to protect it from strongly acidic (including Lewis acidic) reagents. The amide is still an activating, o,p-directing group, and it is compatible with Friedel–Crafts and many other reactions. If necessary, the amide can be removed at the end of the synthesis.

strong base weak base

NH_2 $HN—\overset{O}{\overset{\|}{C}}—CH_3$

Ac_2O / NaOAc

amine amide

O O O is called acetic anhydride, abbreviated Ac_2O.

Works in Friedel–Crafts reactions. The N in the amide does not complex with $AlCl_3$.

NH_2 $HN—\overset{O}{\overset{\|}{C}}—CH_3$ $HN—\overset{O}{\overset{\|}{C}}—CH_3$ NH_2

Ac_2O / NaOAc $(CH_3)_2CHBr$ / $AlCl_3$ H^+ or HO^- / H_2O, heat

amine amide amine

4. Other reactions may be useful.

A. A carboxylic acid (COOH) group can be attached to an aromatic ring by adding an alkyl group and then oxidizing it. Additional groups can be added ortho and para at the alkylbenzene stage, or they can be added meta after the oxidation. The oxidation is brutal, and OH and NH_2 substituents do not survive. Halogens, NO_2, and SO_3H survive the $KMnO_4$ oxidation.

R ← o,p-director COOH ← m-director

F-C acylation or alkylation $KMnO_4$ / heat

R cannot have a quaternary carbon atom bonded to the ring.

(continued)

PROBLEM 17-47

Propose a synthetic sequence of this trisubstituted benzene starting from toluene.

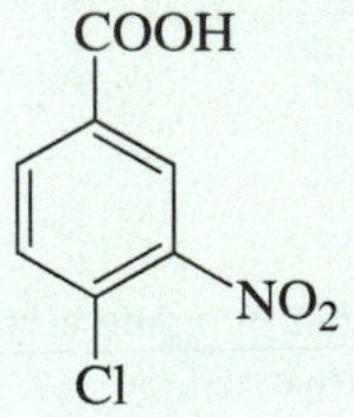

COOH

NO_2

Cl

B. Reduction of NO_2 to NH_2.

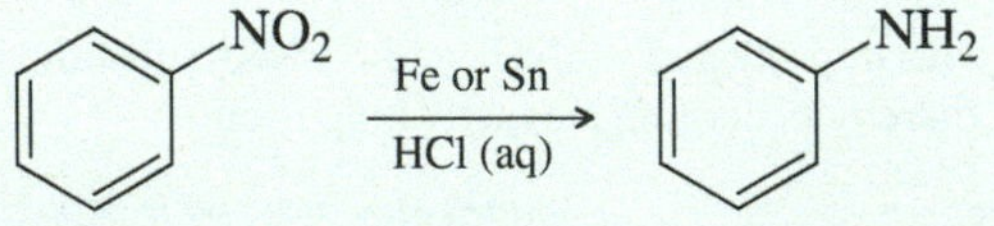

NO_2 —(Fe or Sn / HCl (aq))→ NH_2

Changes a m-director to an o,p-director.

Example:

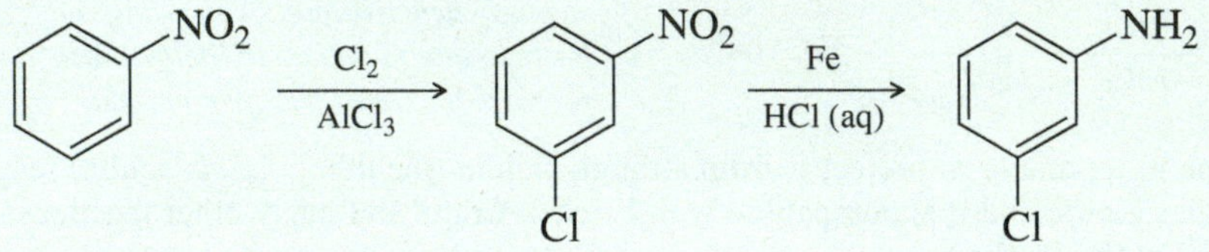

NO_2 —(Cl_2 / $AlCl_3$)→ NO_2, Cl —(Fe / HCl (aq))→ NH_2, Cl

Places two o,p-directors meta to each other.

The conditions of the Clemmensen reduction, [Zn(Hg) + HCl], or [Fe + HCl], also reduce NO_2 groups.

C. The SO_3H group as a blocking group.

The SO_3H group can be used to block a position on the ring. This is a common procedure to make ortho isomers when the para position is more reactive. Use the SO_3H group to block the para position, substitute the ortho position, and then remove the blocking group.

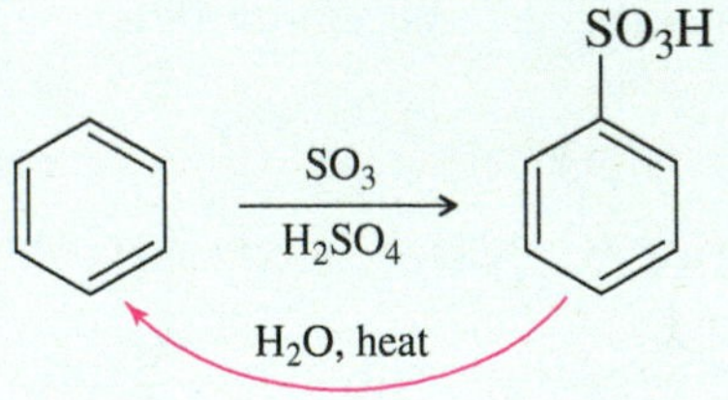

—(SO_3 / H_2SO_4)→ SO_3H

H_2O, heat

Sulfonic acid group can be removed by steam distillation.

Sulfonation is reversible. The sulfonic acid group can be removed by steam distillation.

CH_3 —(SO_3 / H_2SO_4)→ CH_3, SO_3H —(HNO_3 / H_2SO_4)→ CH_3, NO_2, SO_3H —(H_2O / heat)→ CH_3, NO_2

para is major

ortho only; not mixed with para

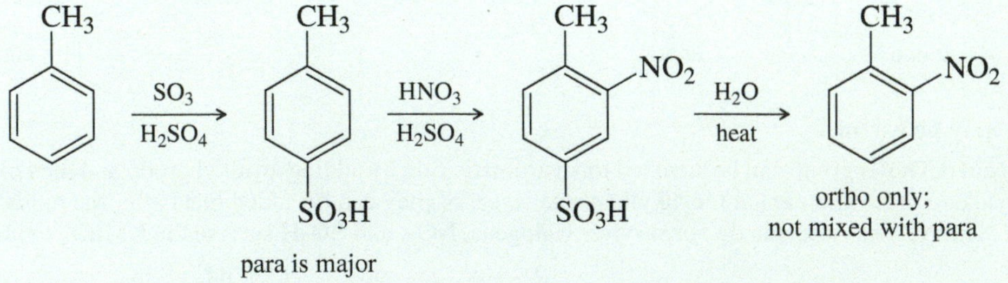

PROBLEM 17-48

Starting from toluene, propose syntheses for ortho-, meta-, and para-chlorobenzoic acid.

PROBLEM 17-49

Starting from toluene, propose a synthesis of this trisubstituted benzene.

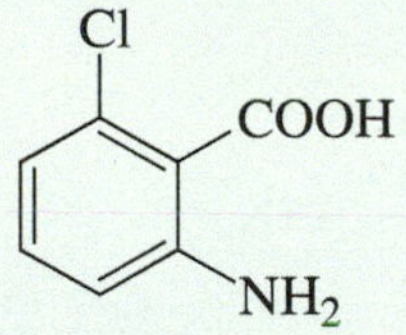

(2-amino-6-chlorobenzoic acid: benzene ring bearing Cl, COOH, NH_2)

SUMMARY Reactions of Aromatic Compounds

1. *Electrophilic aromatic substitution*
   a. *Halogenation* (Section 17-2)

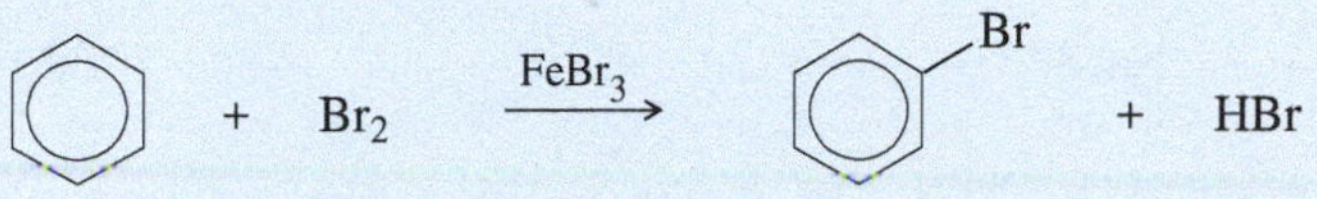

$$C_6H_6 + Br_2 \xrightarrow{FeBr_3} C_6H_5Br + HBr$$

bromobenzene

   b. *Nitration* (Section 17-3)

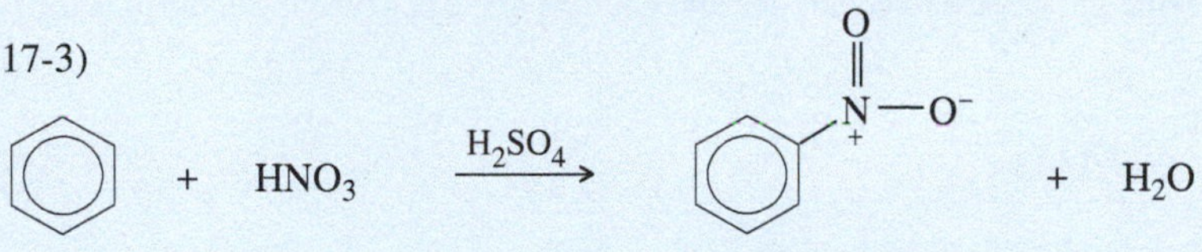

$$C_6H_6 + HNO_3 \xrightarrow{H_2SO_4} C_6H_5\overset{+}{N}(=O)-O^- + H_2O$$

nitrobenzene

Nitration followed by reduction gives anilines.

   c. *Sulfonation* (Section 17-4)

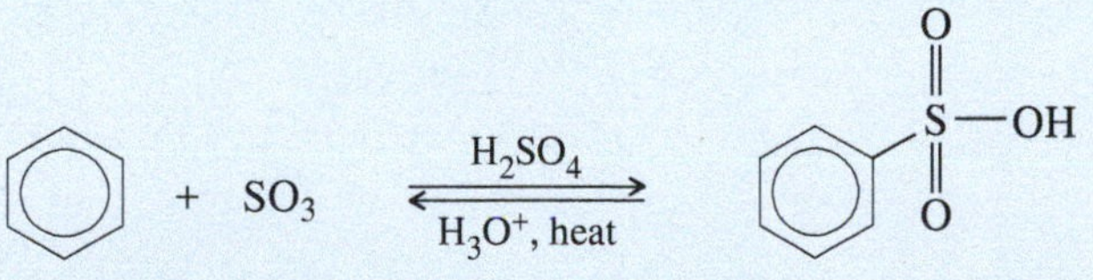

$$C_6H_6 + SO_3 \underset{H_3O^+,\ heat}{\overset{H_2SO_4}{\rightleftharpoons}} C_6H_5-S(=O)_2-OH$$

benzenesulfonic acid

   d. *Friedel–Crafts alkylation* (Section 17-10)

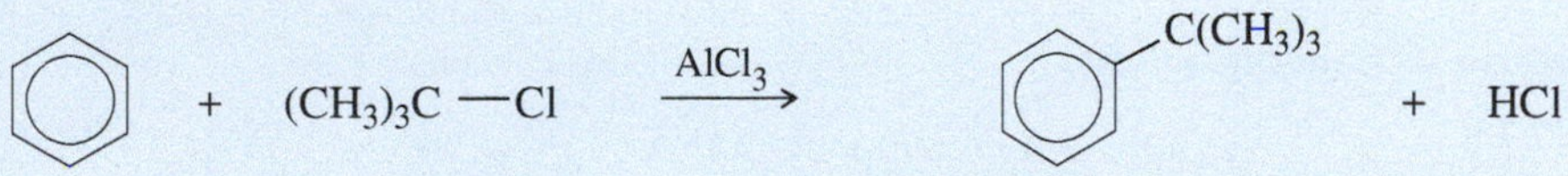

$$C_6H_6 + (CH_3)_3C-Cl \xrightarrow{AlCl_3} C_6H_5C(CH_3)_3 + HCl$$

tert-butylbenzene

   e. *Friedel–Crafts acylation* (Section 17-11)

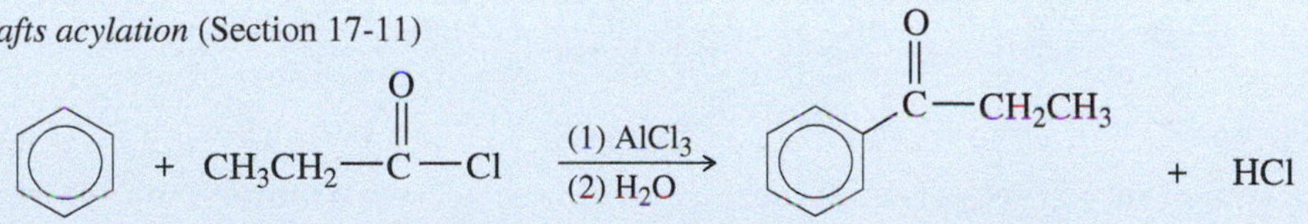

$$C_6H_6 + CH_3CH_2-C(=O)-Cl \xrightarrow[(2)\ H_2O]{(1)\ AlCl_3} C_6H_5-C(=O)-CH_2CH_3 + HCl$$

propiophenone

   f. *Gatterman–Koch synthesis* (Section 17-11C)

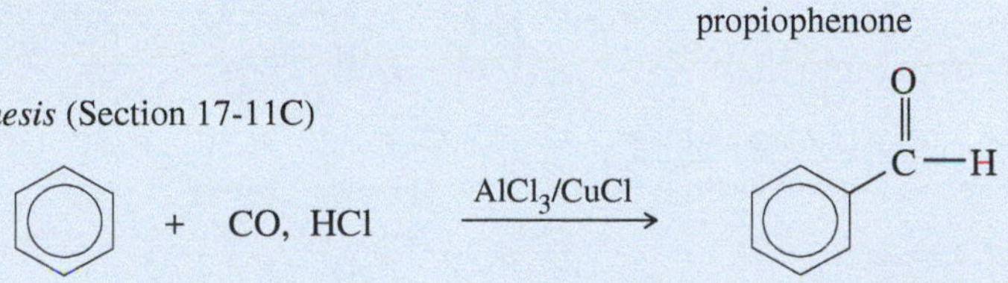

$$C_6H_6 + CO,\ HCl \xrightarrow{AlCl_3/CuCl} C_6H_5-C(=O)-H$$

benzaldehyde

(continued)

g. *Substituent effects* (Sections 17-5 through 17-9)

Activating, ortho, para-directing: —R, —ÖR, —ÖH, —Ö:⁻, —N̈R$_2$ (amines, amides)

Deactivating, ortho, para-directing: —Cl, —Br, —I

Deactivating, meta-allowing: —NO_2, —SO_3H, —$\overset{+}{N}R_3$, —C=O, —C≡N

2. *Nucleophilic aromatic substitution* (Section 17-12)

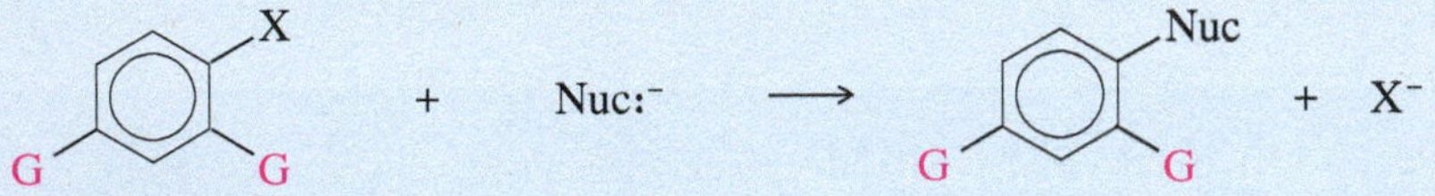

a halobenzene (G = NO_2 or other strong withdrawing group) + Nuc:⁻ (strong nucleophile) ⟶ substituted product (Nuc) + X⁻

Example

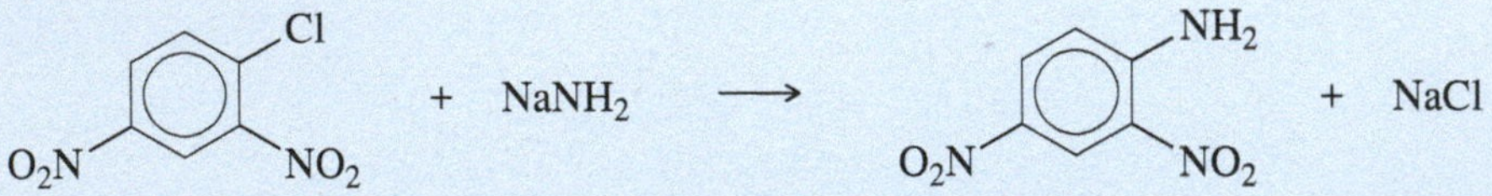

2,4-dinitrochlorobenzene + $NaNH_2$ ⟶ 2,4-dinitroaniline + NaCl

If G is not a strong electron-withdrawing group, severe conditions are required, and a benzyne mechanism is involved (Section 17–12B).

3. *Organometallic Couplings*

a. *Organocuprate couplings* (Section 17-13A)

iodobenzene + lithium divinylcuprate ($(CH_2{=}CH)_2CuLi$) ⟶ styrene

b. *Heck coupling* (Section 17-13B)

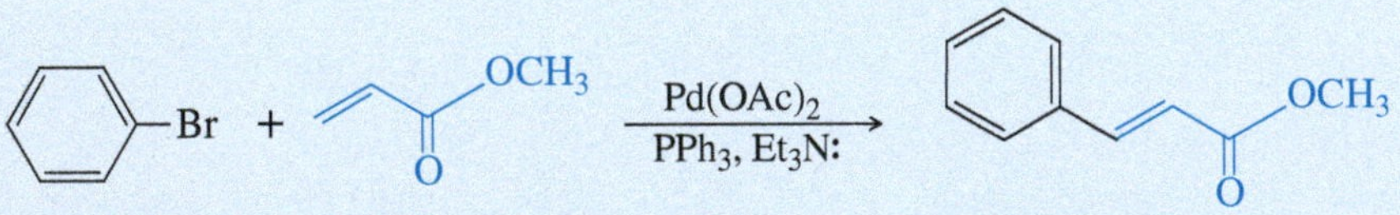

bromobenzene + methyl acrylate $\xrightarrow[\text{PPh}_3,\ \text{Et}_3\text{N:}]{\text{Pd(OAc)}_2}$ methyl cinnamate

c. *Suzuki coupling* (Section 17-13C)

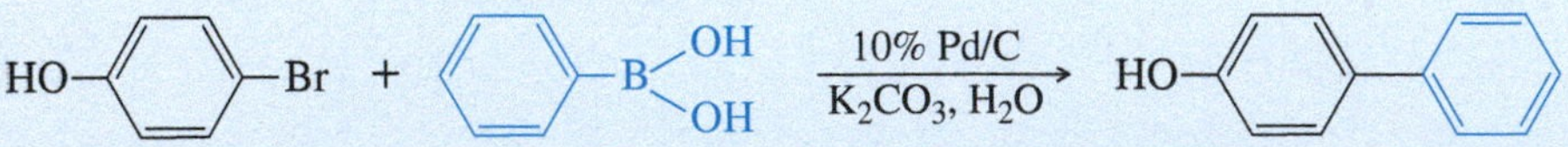

p-bromophenol + phenylboronic acid $\xrightarrow[\text{K}_2\text{CO}_3,\ \text{H}_2\text{O}]{\text{10\% Pd/C}}$ *p*-phenylphenol

4. *Addition reactions*

a. *Chlorination* (Section 17-14A)

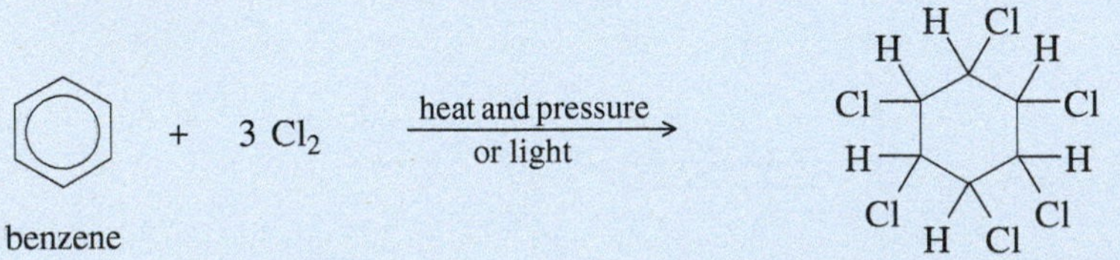

benzene + 3 Cl_2 $\xrightarrow[\text{or light}]{\text{heat and pressure}}$ benzene hexachloride (BHC)

b. *Catalytic hydrogenation* (Section 17-14B)

o-diethylbenzene + 3 H_2 $\xrightarrow[\text{100 °C, 1000 psi}]{\text{Ru or Rh catalyst}}$ 1,2-diethylcyclohexane (mixture of cis and trans)

c. *Birch reduction* (Section 17-14C)

ethylbenzene $\xrightarrow[\text{NH}_3(l),\ \text{R—OH}]{\text{Na or Li}}$ 1-ethylcyclohexa-1,4-diene

5. *Side-chain reactions*

a. *The Clemmensen reduction* (converts acylbenzenes to alkylbenzenes, Section 17-11B)

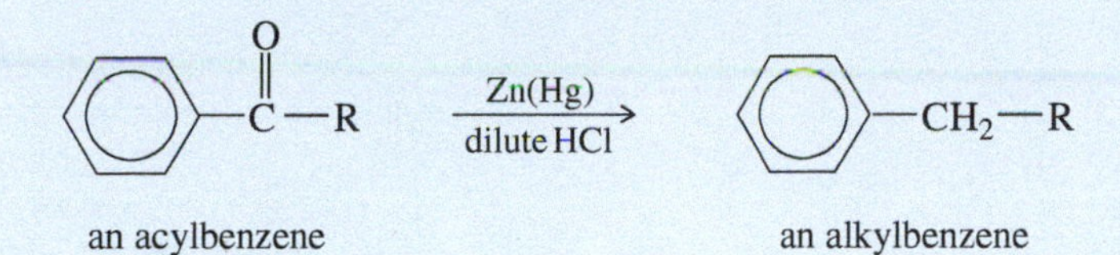

an acylbenzene → an alkylbenzene

b. *Permanganate oxidation* (Section 17-15A)

an alkylbenzene (Ph—CH_2—R) $\xrightarrow[\text{H}_2\text{O}]{\text{hot, concd. KMnO}_4}$ a benzoic acid salt (Ph—COO^- K^+)

Halogen, $-NO_2$, and $-SO_3H$ groups survive this brutal oxidation.

c. *Side-chain halogenation* (Section 17-15B)

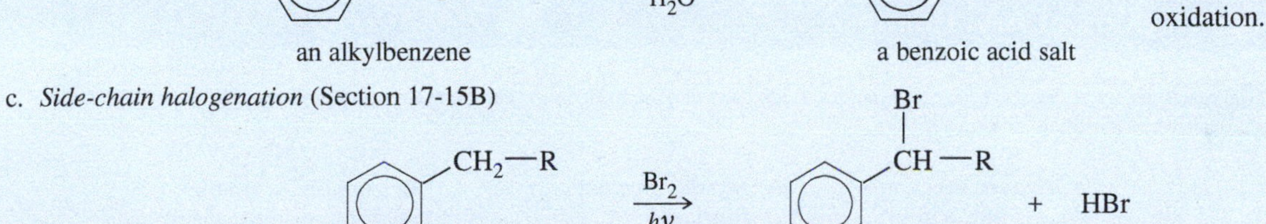

an alkylbenzene → an α-bromo alkylbenzene

d. *Nucleophilic substitution at the benzylic position* (Section 17-15C)
The benzylic position is activated toward both S_N1 and S_N2 displacements.

an α-halo alkylbenzene (Ph—CH(X)—R) + Nuc:$^-$ ⟶ Ph—CH(Nuc)—R + X^-

6. *Oxidation of phenols to quinones* (Section 17-16A)

o-chlorophenol $\xrightarrow[\text{H}_2\text{SO}_4]{\text{Na}_2\text{Cr}_2\text{O}_7}$ 2-chloro-1,4-benzoquinone

SUMMARY Electrophilic Aromatic Substitution of Benzene

A catalyst, either a protic acid like H_2SO_4, or a Lewis acid like $FeBr_3$ or $AlCl_3$, is needed for each reaction.

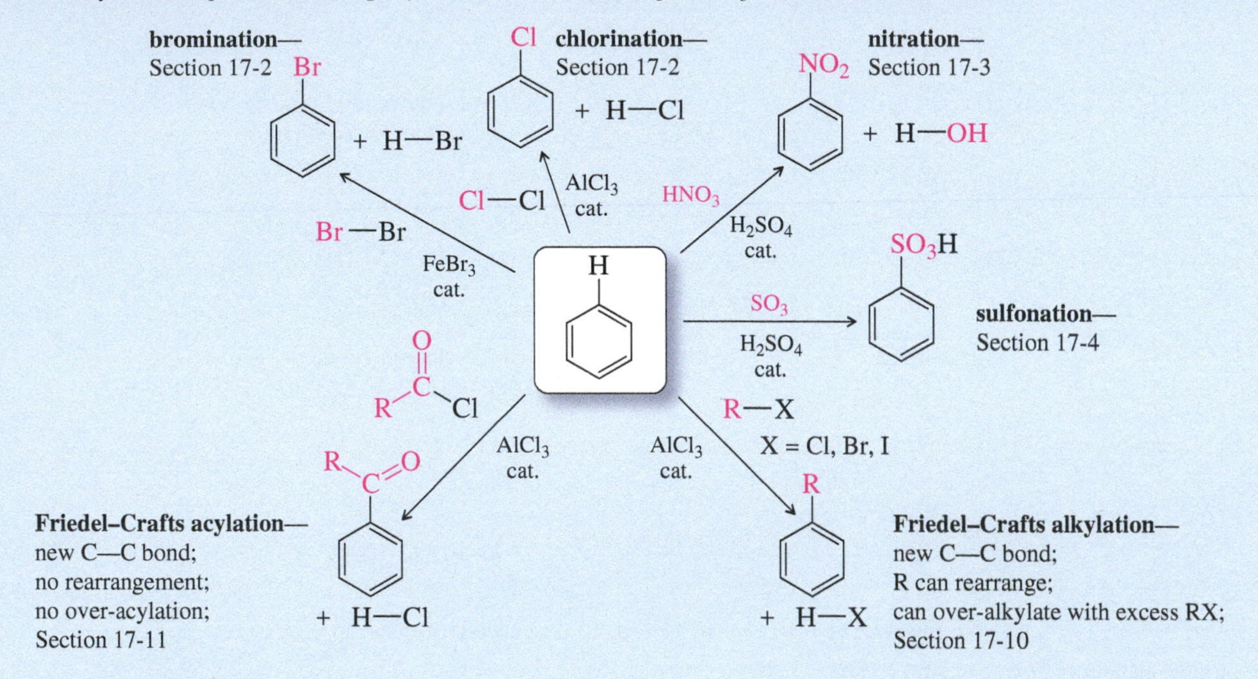

SUMMARY Substitutions of Aryl Halides

S_N1 and S_N2 reactions are not observed for aryl halides. Replacement of X from an sp^2 carbon requires special conditions, usually involving a reactive intermediate or a metal.

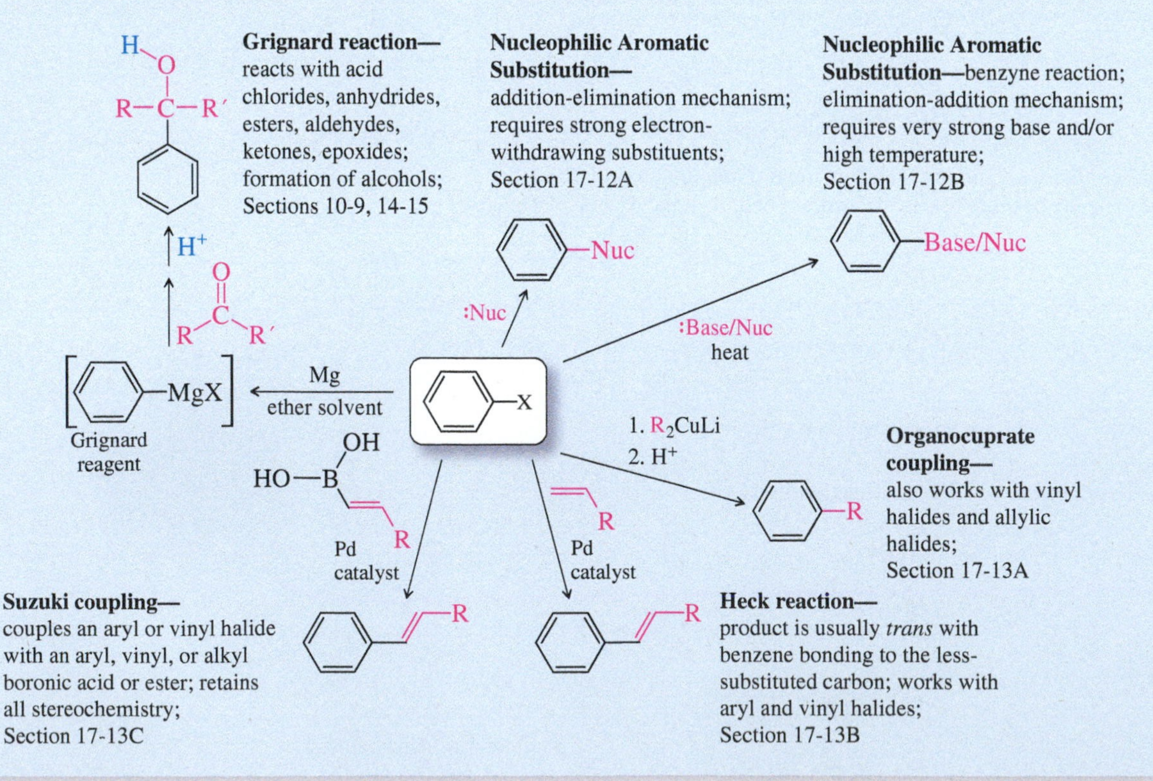

H O
R—C—R′
Grignard reaction—
reacts with acid chlorides, anhydrides, esters, aldehydes, ketones, epoxides;
formation of alcohols;
Sections 10-9, 14-15
H+
O
R C R′
MgX
Grignard reagent
Mg
ether solvent
X
Nucleophilic Aromatic Substitution—
addition-elimination mechanism;
requires strong electron-withdrawing substituents;
Section 17-12A
Nuc
:Nuc
Nucleophilic Aromatic Substitution—benzyne reaction;
elimination-addition mechanism;
requires very strong base and/or high temperature;
Section 17-12B
Base/Nuc
:Base/Nuc
heat
1. R2CuLi
2. H+
R
Organocuprate coupling—
also works with vinyl halides and allylic halides;
Section 17-13A
OH
HO—B
R
Pd catalyst
Suzuki coupling—
couples an aryl or vinyl halide with an aryl, vinyl, or alkyl boronic acid or ester; retains all stereochemistry;
Section 17-13C
R
Pd catalyst
Heck reaction—
product is usually trans with benzene bonding to the less-substituted carbon; works with aryl and vinyl halides;
Section 17-13B

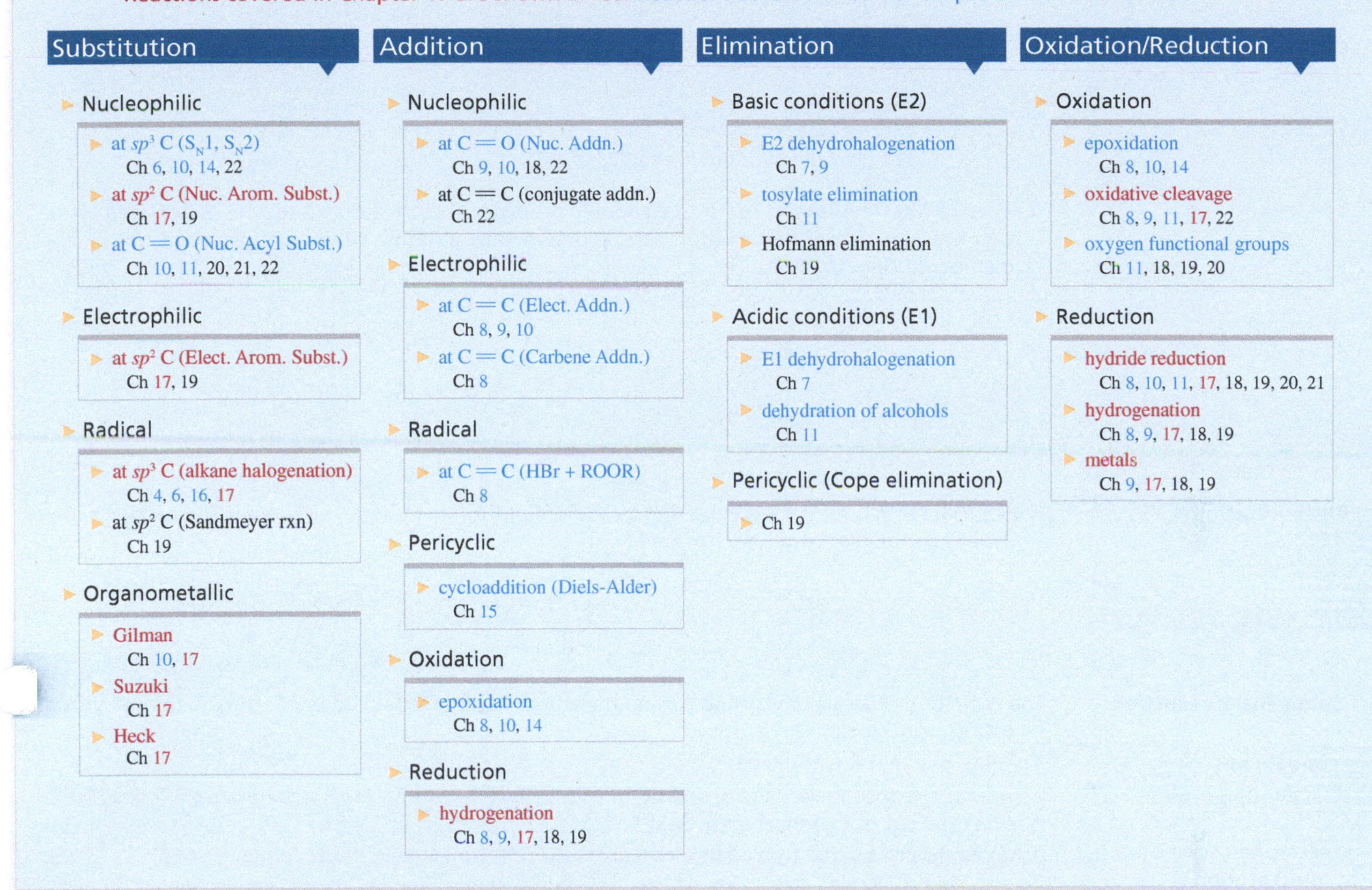

Essential Terms

activating group A substituent that makes the aromatic ring more reactive (usually toward electrophilic aromatic substitution) than benzene. (p. 817)

acyl group $(R-\overset{\overset{O}{\|}}{C}-)$ A carbonyl group with an alkyl group attached. (p. 835)

acyl chloride: **(acid chloride)** An acyl group bonded to a chlorine atom, RCOCl.

acylium ion $(R-C\equiv O^{+})$ An acyl group fragment with a positive charge. (p. 835)

alkoxy group **(alkoxyl group)** A substituent consisting of an alkyl group bonded through an oxygen atom, —O—R. (p. 819)

benzylic position The carbon atom of an alkyl group that is directly bonded to a benzene ring; the position α to a benzene ring. (p. 853)

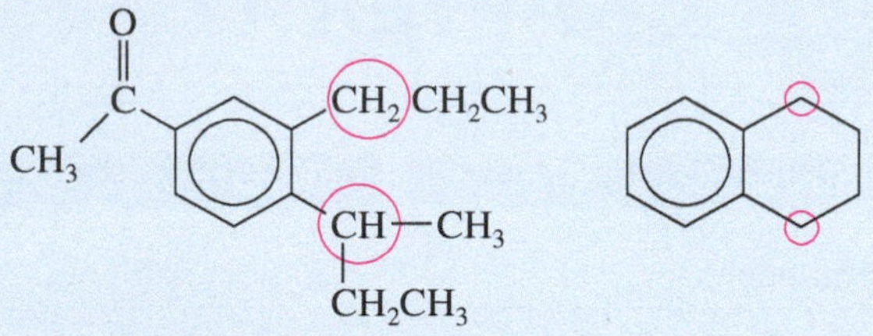

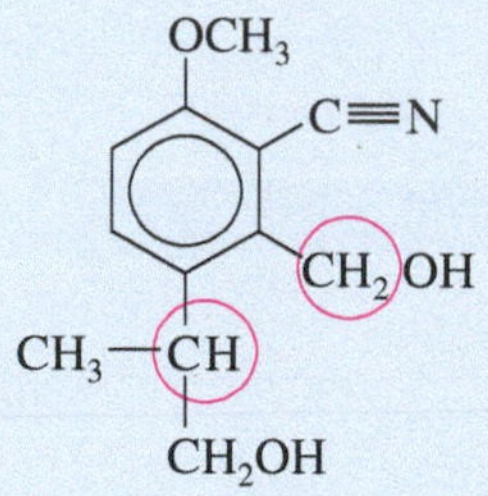

The benzylic positions are circled in red.

benzyne A reactive intermediate in some nucleophilic aromatic substitutions, benzyne is benzene with two hydrogen atoms removed. It can be drawn with a highly strained triple bond in the six-membered ring. (p. 842)

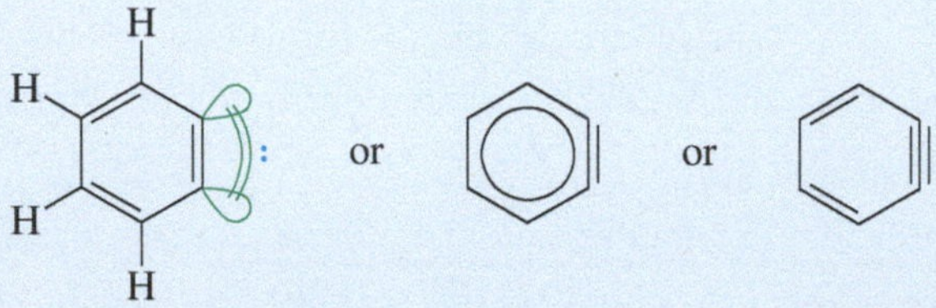

Birch reduction The partial reduction of a benzene ring by sodium or lithium in liquid ammonia. The products are usually cyclohexa-1,4-dienes. (p. 850)

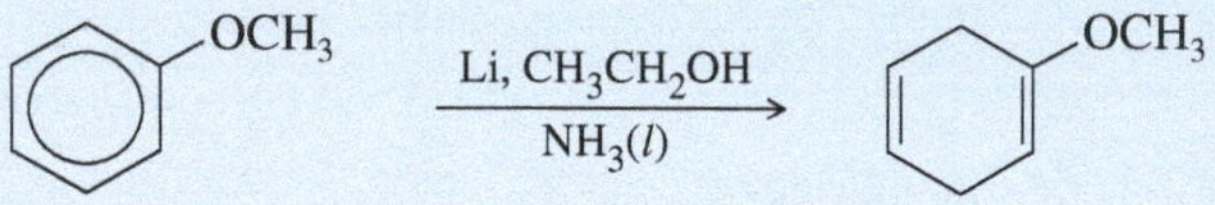

$$\text{C}_6\text{H}_5\text{OCH}_3 \xrightarrow[\text{NH}_3(l)]{\text{Li, CH}_3\text{CH}_2\text{OH}} \text{1-methoxycyclohexa-1,4-diene}$$

Clemmensen reduction The reduction of a carbonyl group to a methylene group by zinc amalgam, Zn(Hg), in dilute hydrochloric acid. (p. 837)

amalgam: An alloy of a metal with mercury.

deactivating group A substituent that makes the aromatic ring less reactive (usually toward electrophilic aromatic substitution) than benzene. (p. 823)

electrophilic aromatic substitution **(EAS)** Replacement of a hydrogen on an aromatic ring by a strong electrophile. (p. 810)

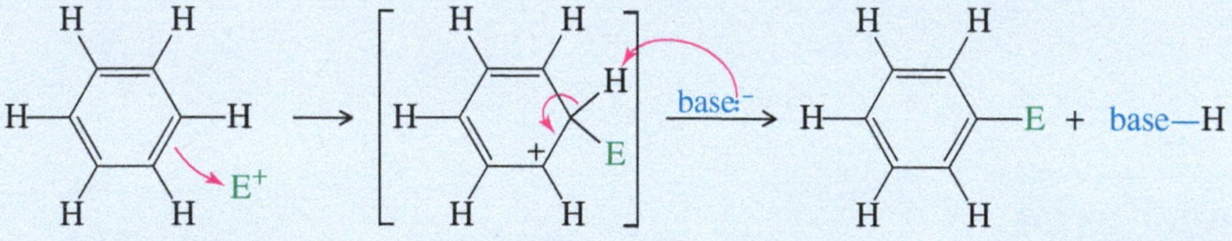

electrophilic attack — sigma complex — substituted product

$$\text{C}_6\text{H}_6 + \text{E}^+ \longrightarrow [\text{sigma complex}] \xrightarrow{\text{base:}^-} \text{C}_6\text{H}_5\text{E} + \text{base—H}$$

Friedel–Crafts acylation Formation of an acylbenzene by substitution of an acylium ion on an aromatic ring. (p. 835)

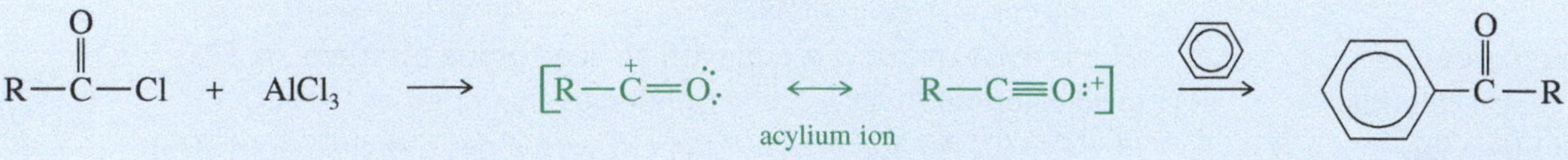

$$\text{R—C(=O)—Cl} + \text{AlCl}_3 \longrightarrow [\text{R—}\overset{+}{\text{C}}\text{=O:} \longleftrightarrow \text{R—C}\equiv\text{O:}^+] \xrightarrow{\text{C}_6\text{H}_6} \text{C}_6\text{H}_5\text{—C(=O)—R}$$

acylium ion

an acylbenzene

Friedel–Crafts alkylation Formation of an alkyl-substituted benzene derivative by substitution of an alkyl carbocation or carbocation-like species on an aromatic ring. (p. 830)

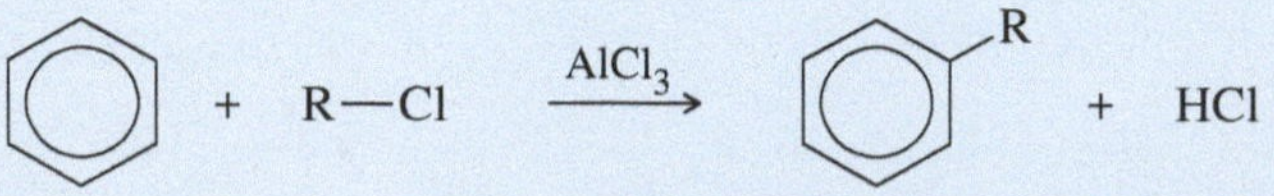

$$\text{C}_6\text{H}_6 + \text{R—Cl} \xrightarrow{\text{AlCl}_3} \text{C}_6\text{H}_5\text{R} + \text{HCl}$$

Gatterman–Koch synthesis	The synthesis of benzaldehydes by treating a benzene derivative with CO and HCl using an $AlCl_3/CuCl$ catalyst. (p. 838)
halonium ion	A positively charged ion that has a positive charge (or partial positive charge) on a halogen atom. Typically, in a halonium ion the halogen atom has two bonds and bears a formal plus charge. (Specific: chloronium ion, bromonium ion, and so forth.) (p. 826)
Heck reaction	A palladium-catalyzed coupling of an aryl or vinyl halide with an alkene to form a new C—C bond at the less substituted end of the alkene. (p. 845)
inductive stabilization	Stabilization of a reactive intermediate by donation or withdrawal of electron density through sigma bonds. (p. 819)
meta-director	**(meta-allower)** A substituent that deactivates primarily the ortho and para positions, leaving the meta position the least deactivated and most reactive. (p. 823)
nitration	Replacement of a hydrogen atom by a nitro group, $—NO_2$. (p. 813)
nitronium ion	The NO_2^+ ion, $O{=}\overset{+}{N}{=}O$. (p. 813)
nucleophilic aromatic substitution	**(NAS)** Replacement of a leaving group on an aromatic ring by a strong nucleophile. Usually takes place by an addition–elimination mechanism or by a benzyne mechanism. (p. 839)
organocuprate	A compound containing carbon–copper bonds, commonly of formula R_2CuLi (Gilman reagent). The lithium dialkylcuprate (or diarylcuprate or divinylcuprate) displaces an organohalogen bond to form a new C—C. (pp. 485, 843)
ortho, para-director	A substituent that activates primarily the ortho and para positions toward attack. (p. 817)
quinone	A derivative of a cyclohexadiene-dione. Common quinones are the 1,4-quinones (*para*-quinones); the less stable 1,2-quinones (*ortho*-quinones) are relatively uncommon. (p. 857)

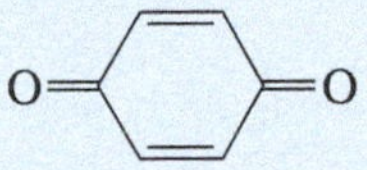

p-quinone 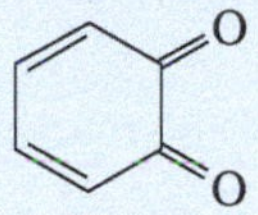 *o*-quinone

resonance stabilization	Stabilization of a reactive intermediate by donation or withdrawal of electron density through pi bonds. (p. 820)
resonance-donating:	**(pi-donating)** Capable of donating electrons through resonance involving pi bonds. (pp. 820, 826)
resonance-withdrawing:	**(pi-withdrawing)** Capable of withdrawing electron density through resonance involving pi bonds. (p. 822)

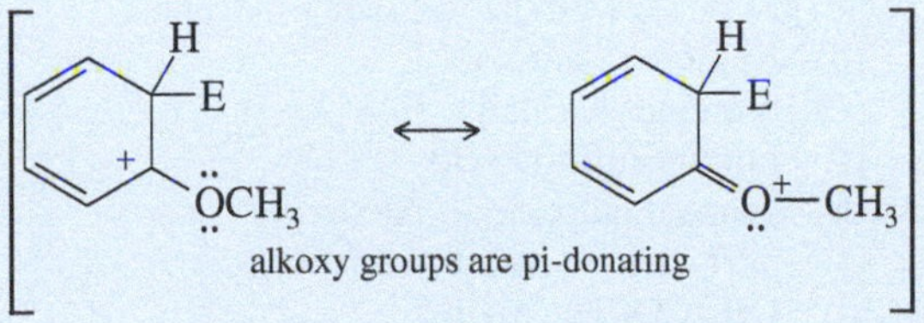

alkoxy groups are pi-donating

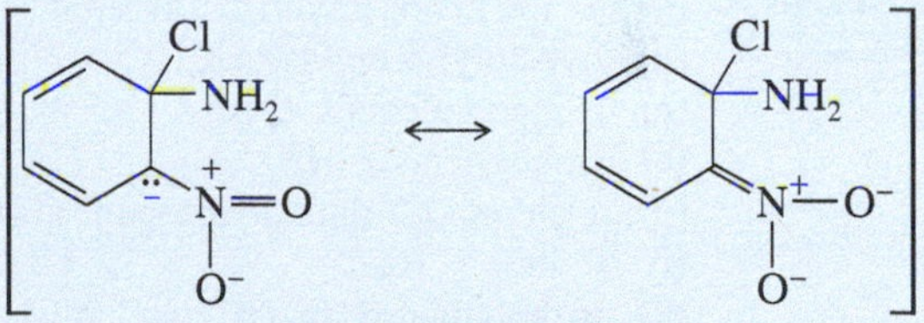

nitro groups are pi-withdrawing

sigma complex	An intermediate in electrophilic aromatic substitution or nucleophilic aromatic substitution with a sigma bond between the electrophile or nucleophile and the former aromatic ring. The sigma complex bears a delocalized positive charge in electrophilic aromatic substitution and a delocalized negative charge in nucleophilic aromatic substitution. (p. 809)
sulfonation	Replacement of a hydrogen atom by a sulfonic acid group, $—SO_3H$. (p. 814)
desulfonation:	Replacement of the $—SO_3H$ group by a hydrogen. With benzene derivatives, this is done by heating with water or steam and acid. (p. 816)
Suzuki reaction	**(Suzuki coupling)** A palladium-catalyzed substitution that couples an aryl or vinyl halide with a alkyl, aryl, or vinyl boronic acid or boronate ester. (p. 847)

Essential Problem-Solving Skills in Chapter 17

Each skill is followed by problem numbers exemplifying that particular skill.

1 Predict products and propose mechanisms for the common electrophilic aromatic substitutions: halogenation, nitration, sulfonation, and Friedel–Crafts alkylation and acylation.	Problems 17-50, 53, 54, 57, 66, 71, and 77
2 Draw resonance forms for the sigma complexes resulting from electrophilic attack on substituted aromatic rings. Explain which substituents are activating and which are deactivating, and show why they are ortho,para-directing, or meta-directing.	Problems 17-53, 54, 60, 64, 67, and 71
3 Predict the position(s) of electrophilic aromatic substitution on molecules containing substituents on one or more aromatic rings, and design syntheses that use the influence of substituents to generate the correct isomers.	Problems 17-52, 53, 54, 57, 60, 63, 67, and 76
4 Determine which nucleophilic aromatic substitutions are likely, and propose mechanisms for both the addition–elimination type and the benzyne type.	Problems 17-58, 68, 69, 70, and 73
5 Predict the coupling products of the organocuprate, Heck, and Suzuki reactions, and use these reactions in syntheses.	Problems 17-55, 61, 62, 72, and 75
6 Predict the products of the Birch reduction, hydrogenation, and chlorination of aromatic compounds, and use these reactions in syntheses.	Problems 17-51, 57, 59, and 74
7 Explain how the reactions of side chains are affected by the presence of the aromatic ring, predict the products of side-chain reactions, and use these reactions in syntheses.	Problems 17-51, 53, and 74
8 Predict the products of the reactions of phenols, and use these reactions in syntheses.	Problems 17-52, 63, 66, and 76
Problem-Solving Strategy: Synthesis Using Electrophilic Aromatic Substitution	Problems 17-50, 52, 60, 63, 72, 75, and 76

Study Problems

17-50 Predict the major products formed when benzene reacts (just once) with the following reagents.

(a) *tert*-butyl bromide, $AlCl_3$
(b) 1-chlorobutane, $AlCl_3$
(c) isobutyl alcohol + BF_3
(d) bromine + a nail
(e) isobutylene + HF
(f) fuming sulfuric acid
(g) 1-chloro-2,2-dimethylpropane + $AlCl_3$
(h) benzoyl chloride + $AlCl_3$
(i) iodine + HNO_3
(j) nitric acid + sulfuric acid
(k) carbon monoxide, HCl, and $AlCl_3$ /CuCl
(l) $CH_2(COCl)_2$, $AlCl_3$

17-51 Indane can undergo free-radical chlorination at any of the alkyl positions on the aliphatic ring.

indane

(a) Draw the possible monochlorinated products from this reaction.
(b) Draw the possible dichlorinated products from this reaction.
(c) What instrumental technique would be most helpful for determining how many products are formed, and how many of those products are monochlorinated and how many are dichlorinated?
(d) Once the products have been separated, what instrumental technique would be most helpful for determining the structures of all the dichlorinated products?

17-52 Show how you would synthesize the following compounds, starting with benzene or toluene and any necessary acyclic reagents. Assume para is the major product (and separable from ortho) in ortho, para mixtures.

(a) 1-phenyl-1-bromobutane
(b) 1-phenyl-1-methoxybutane
(c) 3-phenylpropan-1-ol
(d) ethoxybenzene
(e) 1,2-dichloro-4-nitrobenzene
(f) 1-phenylpropan-2-ol
(g) *p*-aminobenzoic acid
(h) 2-methyl-1-phenylbutan-2-ol
(i) 5-chloro-2-methylaniline
(j) 3-nitro-4-bromobenzoic acid
(k) 3-nitro-5-bromobenzoic acid
(l) 4-butylphenol
(m) 2-(4-methylphenyl)butan-2-ol

17-53 Predict the major products of the following reactions.

(a) 2,4-dinitrochlorobenzene + $NaOCH_3$
(b) phenol + *tert*-butyl chloride + $AlCl_3$
(c) nitrobenzene + fuming sulfuric acid
(d) nitrobenzene + acetyl chloride + $AlCl_3$
(e) *p*-methylanisole + acetyl chloride + $AlCl_3$
(f) *p*-methylanisole + Br_2, light
(g) 1,2-dichloro-4-nitrobenzene + $NaNH_2$
(h) *p*-nitrotoluene + Zn + dilute HCl
(i) *p*-ethylbenzenesulfonic acid + steam/H^+
(j) *p*-ethylbenzenesulfonic acid + HNO_3, H_2SO_4
(k) Ph—C(=O)—NHPh + CH_3CH_2—C(=O)—Cl, $AlCl_3$
(l) indane + hot, concd. $KMnO_4$

17-54 Predict the major products of bromination of the following compounds, using Br_2 and $FeBr_3$ in the dark.

(a) NO_2 OCH_3

(b)
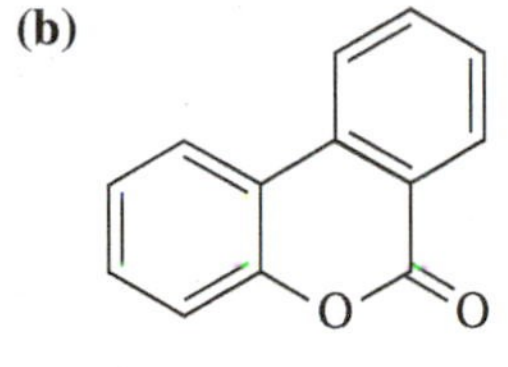

(c)
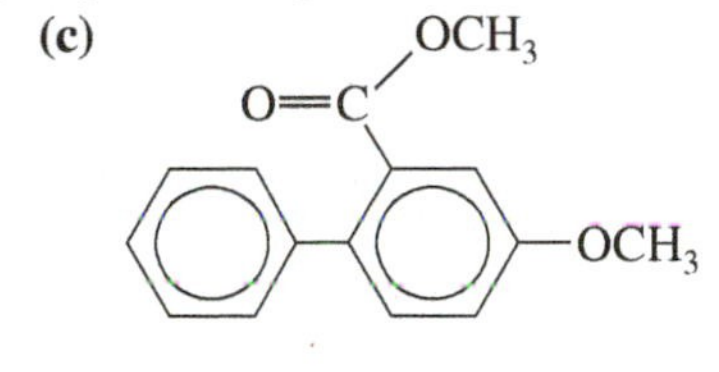

17-55 What products would you expect from the following coupling reactions?

(a) Br + [ ]$_2$ CuLi ⟶

(b) I + O OH; $PdCl_2$ / Na_2CO_3, H_2O ⟶

(c)
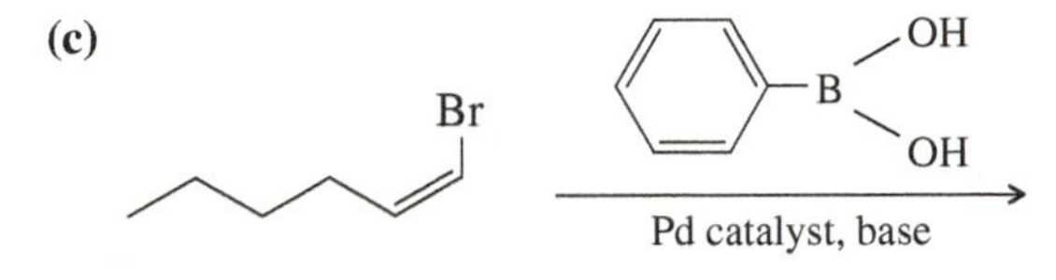

(d)
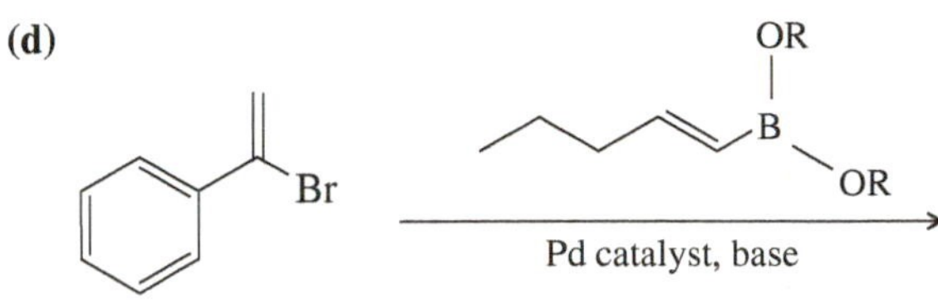

(e)
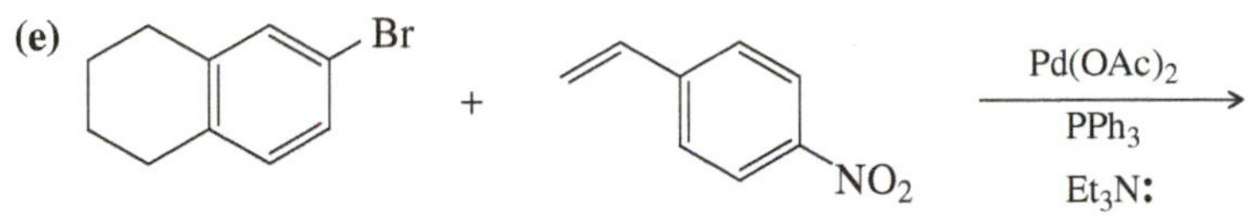

17-56 A student added 3-phenylpropanoic acid ($PhCH_2CH_2COOH$) to a molten salt consisting of a 1:1 mixture of NaCl and $AlCl_3$ maintained at 170 °C. After 5 minutes, he poured the molten mixture into water and extracted it into dichloromethane. Evaporation of the dichloromethane gave a 96% yield of the product whose spectra follow. The mass spectrum of the product shows a molecular ion at m/z 132. What is the product?

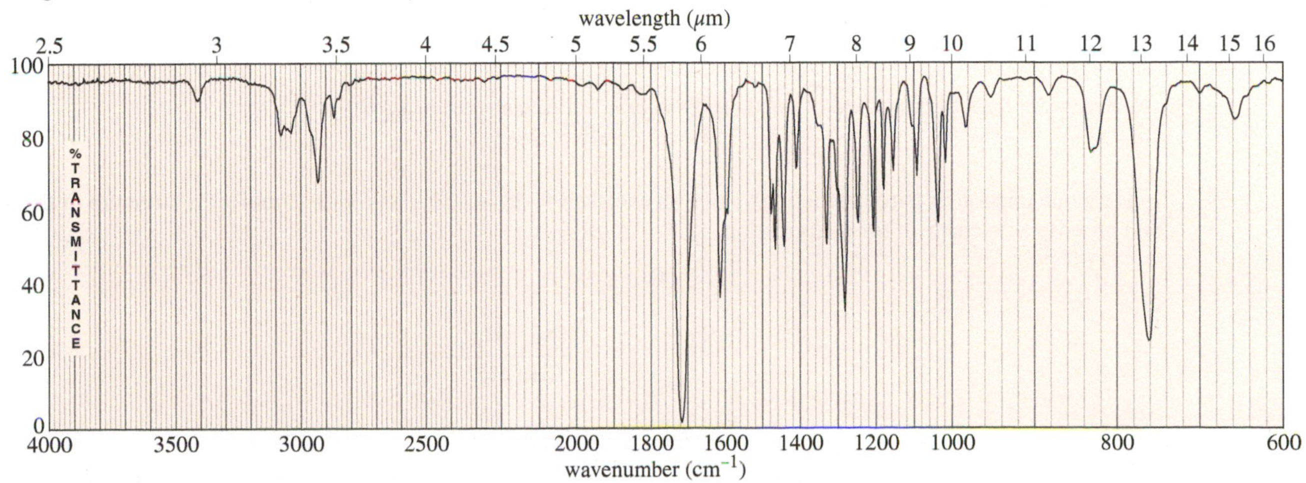

(continued)

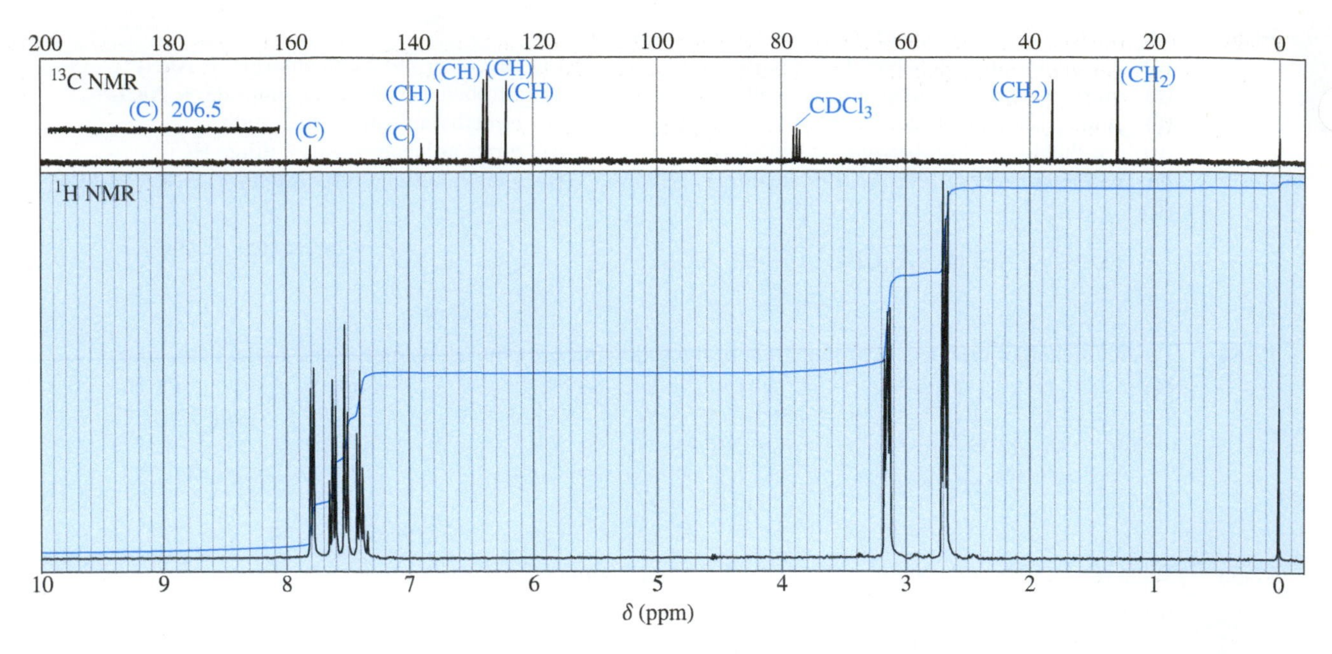

17-57 Give the structures of compounds **A** through **H** in the following series of reactions.

Benzene + (CH$_3$)$_2$CHCH$_2$COCl, $AlCl_3$ → **A** ; **A** + HNO_3, H_2SO_4 → **B**

A + Zn(Hg), HCl → **C** ; **C** + $KMnO_4$ (hot, concd.) → **D**

C + Br_2, *hν* → **E** ; **E** + NH_3 (excess) → PhCH(NH$_2$)CH$_2$CH(CH$_3$)$_2$

E + $(CH_3)_3CO^- K^+$ → **G** ; **G** + HBr → **H**

E + $NaOCH_3$ → **F**

17-58 The following compound reacts with a hot, concentrated solution of NaOH (in a sealed tube) to give a mixture of two products. Propose structures for these products, and give a mechanism to account for their formation.

5-Chloro-1,2,3,4-tetrahydronaphthalene (Cl) $\xrightarrow[350\ °C]{NaOH,\ H_2O}$ 2 products

17-59 α-Tetralone undergoes Birch reduction to give an excellent yield of a single product. Predict the structure of the product, and propose a mechanism for its formation.

α-tetralone (O) $\xrightarrow[CH_3CH_2OH]{Na,\ NH_3(l)}$

17-60 Electrophilic aromatic substitution usually occurs at the 1-position of naphthalene, also called the α position. Predict the major products of the reactions of naphthalene with the following reagents.

(a) HNO_3, H_2SO_4 **(b)** Br_2, $FeBr_3$ **(c)** CH_3CH_2COCl, $AlCl_3$
(d) isobutylene and HF **(e)** cyclohexanol and BF_3 **(f)** fuming sulfuric acid

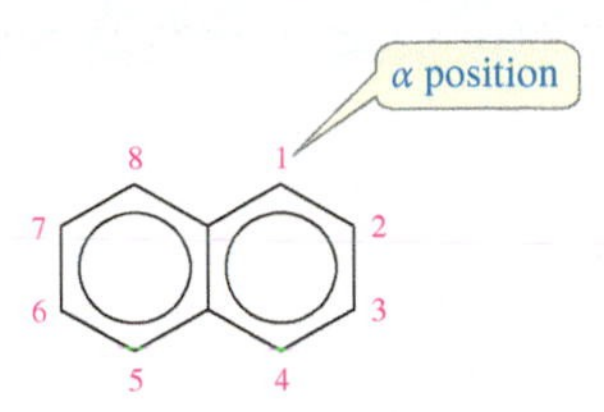

17-61 Nucleophilic aromatic substitution provides one of the common methods for making phenols. (Another method is discussed in Section 19-17.) Show how you would synthesize the following phenols, using benzene or toluene as your aromatic starting material, and explain why mixtures of products would be obtained in some cases.

(a) *p*-nitrophenol **(b)** 2,4,6-tribromophenol **(c)** *p*-chlorophenol
(d) *m*-cresol **(e)** *p*-*n*-butylphenol

17-62 Show how you would use a Suzuki reaction to synthesize Bombykol, the sex hormone of the silk moth, from *cis*-1-bromopent-1-ene and the acetylenic alcohol shown below.

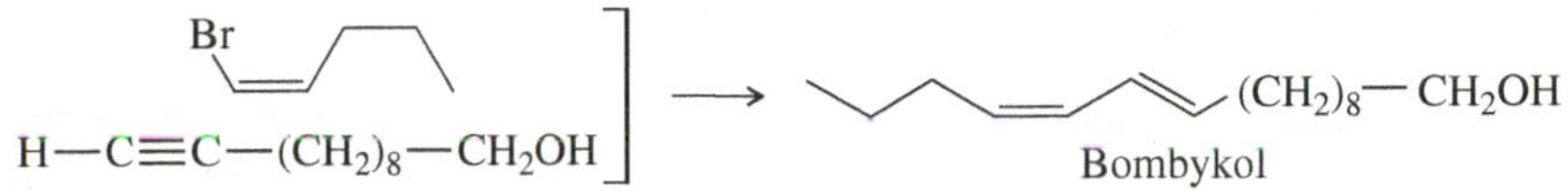

Br

$H—C{\equiv}C—(CH_2)_8—CH_2OH$ ⟶ $(CH_2)_8—CH_2OH$

Bombykol

17-63 The most common selective herbicide for killing broadleaf weeds is 2,4-dichlorophenoxyacetic acid (2,4-D). Show how you would synthesize 2,4-D from benzene, chloroacetic acid ($ClCH_2COOH$), and any necessary reagents and solvents.

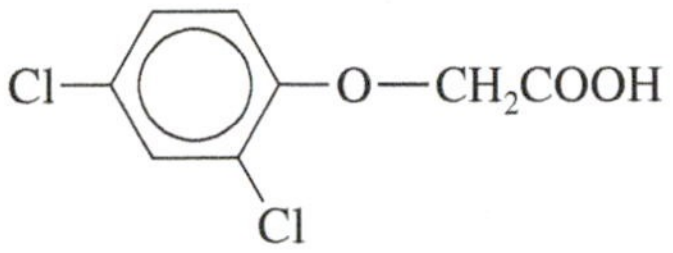

Cl— —O—CH_2COOH

Cl

2,4-dichlorophenoxyacetic acid (2,4-D)

17-64 Furan undergoes electrophilic aromatic substitution more readily than benzene; mild reagents and conditions are sufficient. For example, furan reacts with bromine to give 2-bromofuran.

(a) Propose mechanisms for the bromination of furan at the 2-position and at the 3-position. Draw the resonance forms of each sigma complex, and compare their stabilities.

(b) Explain why furan undergoes bromination (and other electrophilic aromatic substitutions) primarily at the 2-position.

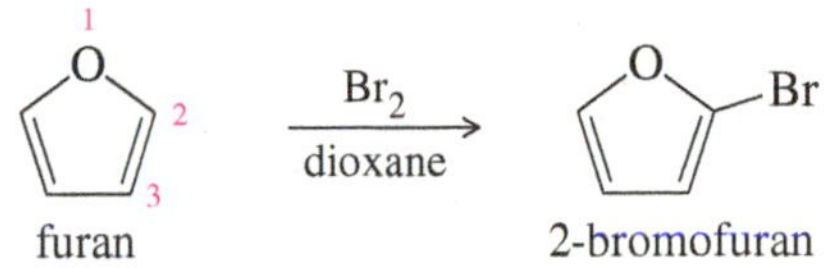

furan —Br_2, dioxane→ 2-bromofuran

17-65 **(a)** Draw the three isomers of benzenedicarboxylic acid.

(b) The isomers have melting points of 210 °C, 343 °C, and 427 °C. Nitration of the isomers at all possible positions was once used to determine their structures. The isomer that melts at 210 °C gives two mononitro isomers. The isomer that melts at 343 °C gives three mononitro isomers. The isomer that melts at 427 °C gives only one mononitro isomer. Show which isomer has which melting point.

***17-66** *Bisphenol A* is an important component of many polymers, including polycarbonates, polyurethanes, and epoxy resins. It is synthesized from phenol and acetone with HCl as a catalyst. Propose a mechanism for this reaction.

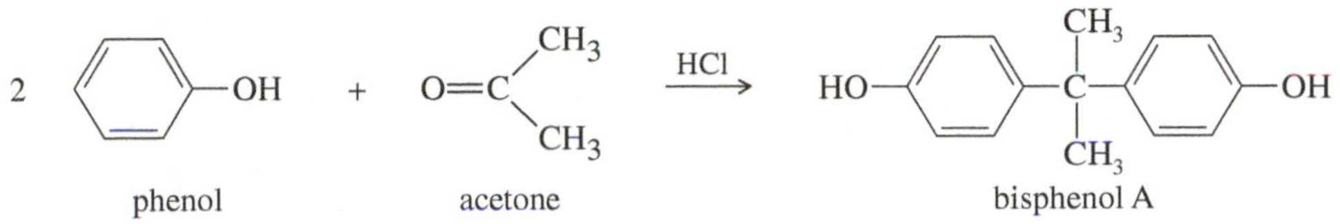

2 phenol (—OH) + acetone ($O{=}C(CH_3)_2$) —HCl→ bisphenol A (HO—C₆H₄—C(CH_3)₂—C₆H₄—OH)

17-67 Unlike most other electrophilic aromatic substitutions, sulfonation is often reversible (see Section 17-4). When one sample of toluene is sulfonated at 0 °C and another sample is sulfonated at 100 °C, the following ratios of substitution products result:

	Reaction Temperature	
Isomer of the Product	0 °C	100 °C
o-toluenesulfonic acid	43%	13%
m-toluenesulfonic acid	4%	8%
p-toluenesulfonic acid	53%	79%

(a) Explain the change in the product ratios when the temperature is increased.
(b) Predict what will happen when the product mixture from the reaction at 0 °C is heated to 100 °C.
(c) Because the SO_3H group can be added to a benzene ring and removed later, it is sometimes called a blocking group. Show how 2,6-dibromotoluene can be made from toluene using sulfonation and desulfonation as intermediate steps in the synthesis.

17-68 When 1,2-dibromo-3,5-dinitrobenzene is treated with excess NaOH at 50 °C, only one of the bromine atoms is replaced. Draw an equation for this reaction, showing the product you expect. Give a mechanism to account for the formation of your proposed product.

***17-69** When anthracene is added to the reaction of chlorobenzene with concentrated NaOH at 350 °C, an interesting Diels–Alder adduct of formula $C_{20}H_{14}$ results. The proton NMR spectrum of the product shows a singlet of area 2 around δ 3 and a broad singlet of area 12 around δ 7. Propose a structure for the product, and explain why one of the aromatic rings of anthracene reacted as a diene.

17-70 In Chapter 14, we saw that Agent Orange contains (2,4,5-trichlorophenoxy) acetic acid, called 2,4,5-T. This compound is synthesized by the partial reaction of 1,2,4,5-tetrachlorobenzene with sodium hydroxide, followed by reaction with sodium chloroacetate, $ClCH_2CO_2Na$.
(a) Draw the structures of these compounds, and write equations for these reactions.
(b) One of the impurities in the Agent Orange used in Vietnam was 2,3,7,8-tetrachlorodibenzodioxin (2,3,7,8-TCDD), often incorrectly called "dioxin." Propose a mechanism to show how 2,3,7,8-TCDD is formed in the synthesis of 2,4,5-T.
(c) Show how the TCDD contamination might be eliminated, both after the first step and on completion of the synthesis.

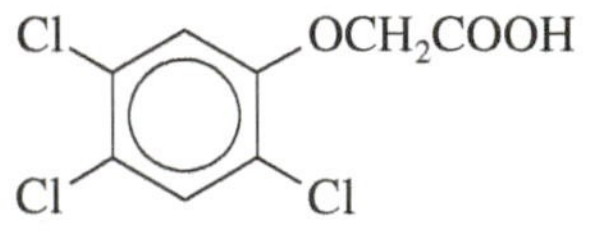

Cl, OCH_2COOH, Cl, Cl

2,4,5-T

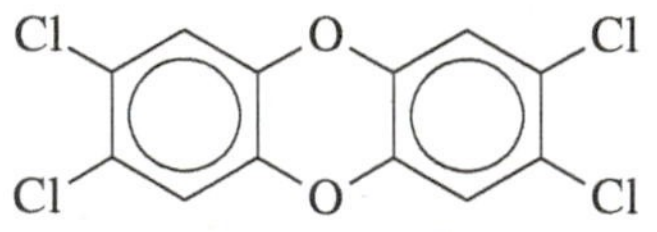

Cl, O, Cl, Cl, O, Cl

2,3,7,8-tetrachlorodibenzodioxin (TCDD)

17-71 Phenol reacts with three equivalents of bromine in CCl_4 (in the dark) to give a product of formula $C_6H_3OBr_3$. When this product is added to bromine water, a yellow solid of molecular formula $C_6H_2OBr_4$ precipitates out of the solution. The IR spectrum of the yellow precipitate shows a strong absorption (much like that of a quinone) around 1680 cm^{-1}. Propose structures for the two products.

17-72 Starting with benzene and any other reagents you need, show how you would synthesize the compound shown here. (*Hint:* Consider a Pd-catalyzed coupling for the final step.)

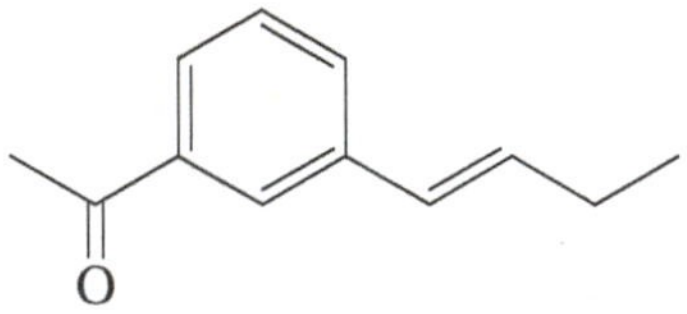

O

***17-73** A graduate student tried to make *o*-fluorophenylmagnesium bromide by adding magnesium to an ether solution of *o*-fluorobromobenzene. After obtaining puzzling results with this reaction, she repeated the reaction by using as solvent some tetrahydrofuran that contained a small amount of furan. From this reaction, she isolated a fair yield of the compound that follows. Propose a mechanism for its formation.

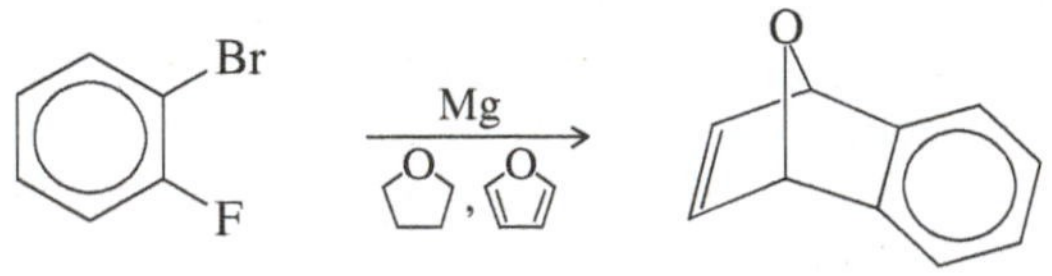

Br, F, Mg, O, O, O

*17-74 A common illicit synthesis of methamphetamine involves an interesting variation of the Birch reduction. A solution of ephedrine in alcohol is added to liquid ammonia, followed by several pieces of lithium metal. The Birch reduction usually reduces the aromatic ring (Section 17-14C), but in this case it eliminates the hydroxy group of ephedrine to give methamphetamine. Propose a mechanism, similar to that for the Birch reduction, to explain this unusual course of the reaction.

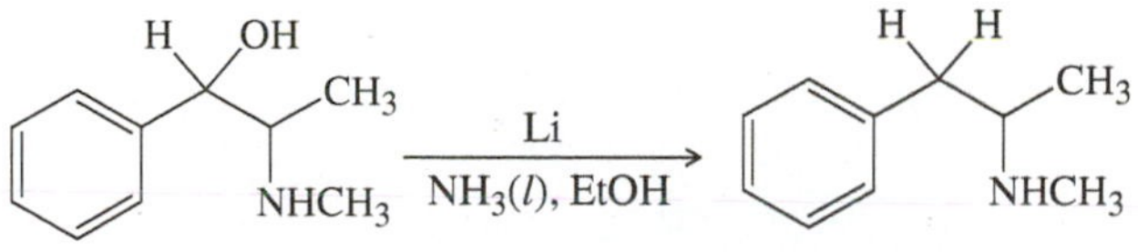

H OH CH$_3$ NHCH$_3$ — Li, NH$_3$(l), EtOH → H H CH$_3$ NHCH$_3$

ephedrine methamphetamine

17-75 Show how you would use a Suzuki reaction to synthesize the following biaryl compound. As starting materials, you may use the two indicated compounds, plus any additional reagents you need.

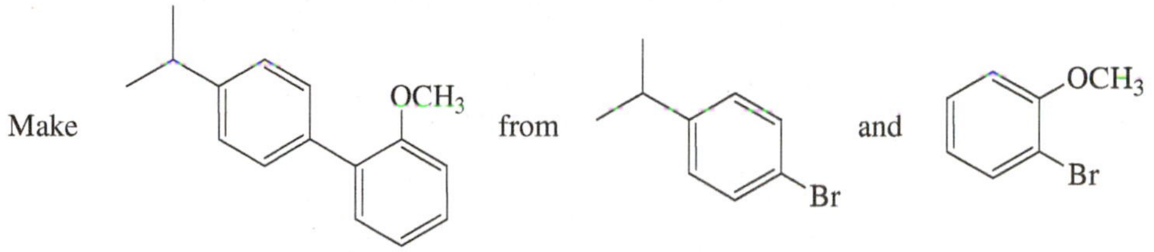

Make OCH$_3$ from Br and OCH$_3$ Br

17-76 The antioxidants BHA and BHT are commonly used as food preservatives. Show how BHA and BHT can be made from phenol and hydroquinone.

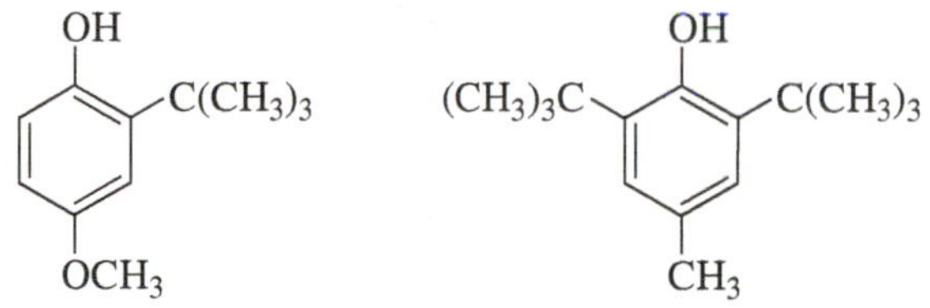

OH C(CH$_3$)$_3$ OCH$_3$ BHA

OH (CH$_3$)$_3$C C(CH$_3$)$_3$ CH$_3$ BHT

17-77 Triphenylmethanol is insoluble in water, but when it is treated with concentrated sulfuric acid, a bright yellow solution results. As this yellow solution is diluted with water, its color disappears and a precipitate of triphenylmethanol reappears. Suggest a structure for the bright yellow species, and explain this unusual behavior.

17-78 Phenolphthalein, a common nonprescription laxative, is also an acid–base indicator that is colorless in acid and red in base. Phenolphthalein is synthesized by the acid-catalyzed reaction of phthalic anhydride with 2 equivalents of phenol.

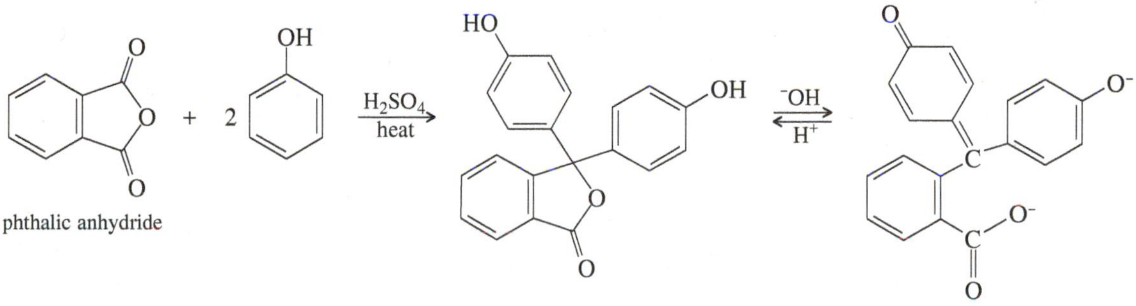

phthalic anhydride + 2 phenol (OH) — H_2SO_4, heat → phenolphthalein (HO, OH, O, O) — ^-OH / H^+ ⇌ red dianion (O, O^-, C, O^-, C, O)

(a) Propose a mechanism for the synthesis of phenolphthalein.

(b) Propose a mechanism for the conversion of phenolphthalein to its red dianion in base.

(c) Use resonance structures to show that the two phenolic oxygen atoms are equivalent (each with half a negative charge) in the red phenolphthalein dianion.

18 Ketones and Aldehydes

Goals for Chapter 18

1 Draw and name ketones and aldehydes, and use spectral information to determine their structures.

2 Propose single-step and multistep syntheses of ketones and aldehydes from compounds containing other functional groups.

3 Predict the products and propose mechanisms for the reactions of ketones and aldehydes with oxidizing and reducing agents, amines, alcohols, and phosphorus ylides.

4 Propose multistep syntheses using ketones and aldehydes as starting materials and intermediates. Protect the carbonyl group if necessary.

▲ Canthaxanthin (β,β-carotene-4,4′-dione) is one of the carotenoid pigments that give flamingos their rosy pink color. Flamingo feathers are actually white or light gray. The pink carotenoid pigments come from the brine shrimp and algae they eat, which all contain carotenes. The birds' digestive systems metabolize the carotenes, which dissolve in fats that are deposited in new feathers as they grow. Canthaxanthin is added to the food for captive flamingos in zoos to give them their characteristic pink coloration.

18-1 Carbonyl Compounds

We will study compounds containing the **carbonyl group** (C═O) in detail because they are of central importance to organic chemistry, biochemistry, and biology. Some of the common types of carbonyl compounds are listed in Table 18-1.

Carbonyl compounds are everywhere. In addition to their uses as reagents and solvents, they are constituents of fabrics, flavorings, plastics, and drugs. Naturally occurring carbonyl compounds include proteins, carbohydrates, and nucleic acids that make up all plants and animals. In the next few chapters, we will discuss the properties and reactions of simple carbonyl compounds. Then, in Chapters 23 and 24, we will apply this carbonyl chemistry to carbohydrates, nucleic acids, and proteins.

The simplest carbonyl compounds are ketones and aldehydes. A **ketone** has two alkyl (or aryl) groups bonded to the carbonyl carbon atom. An **aldehyde** has one alkyl (or

TABLE 18-1
Common Classes of Carbonyl Compounds

Class	General Formula	Class	General Formula
ketones	R—C(=O)—R′	aldehydes	R—C(=O)—H
carboxylic acids	R—C(=O)—OH	acid chlorides	R—C(=O)—Cl
esters	R—C(=O)—O—R′	amides	R—C(=O)—NH_2

aryl) group and one hydrogen atom bonded to the carbonyl carbon atom. Formaldehyde is the simplest aldehyde, with two hydrogen atoms bonded to the carbonyl group.

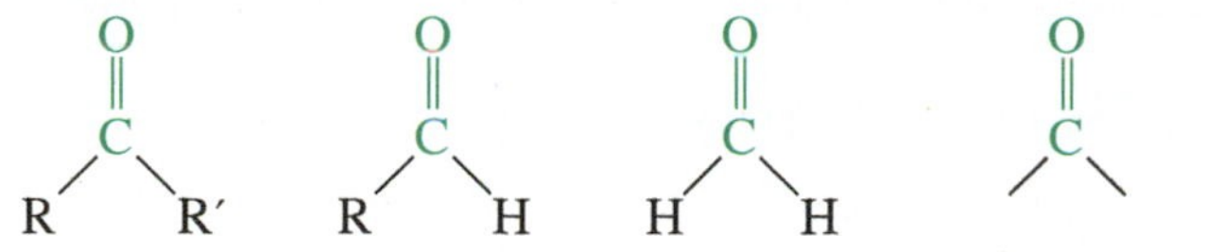

	ketone	aldehyde	formaldehyde	carbonyl group
condensed structures:	RCOR′	RCHO	CH_2O	

> *Ketone*: Two alkyl groups bonded to a carbonyl group.
> *Aldehyde*: One alkyl group and one hydrogen bonded to a carbonyl group.

Ketones and aldehydes are similar in structure, and they have similar properties. There are some differences, however, particularly in their reactions with oxidizing agents and with nucleophiles. In most cases, aldehydes are more reactive than ketones, for reasons we discuss shortly.

18-2 Structure of the Carbonyl Group

The carbonyl carbon atom is sp^2 hybridized and bonded to three other atoms through coplanar sigma bonds oriented about 120° apart. The unhybridized *p* orbital overlaps with a *p* orbital of oxygen to form a pi bond. The double bond between carbon and oxygen is similar to an alkene C=C double bond, except that the carbonyl double bond is shorter, stronger, and polarized.

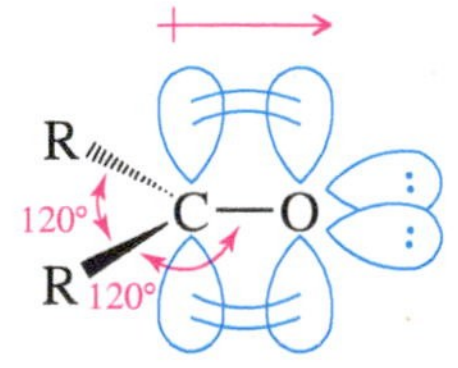

	length	*energy*
ketone C=O bond	1.23 Å	745 kJ/mol (178 kcal/mol)
alkene C=C bond	1.34 Å	611 kJ/mol (146 kcal/mol)

The double bond of the carbonyl group has a large dipole moment because oxygen is more electronegative than carbon, and the bonding electrons are not shared equally. In particular, the less tightly held pi electrons are pulled more strongly toward the oxygen atom, giving ketones and aldehydes larger dipole moments than most alkyl halides and ethers. We can use resonance forms to symbolize this unequal sharing of the pi electrons.

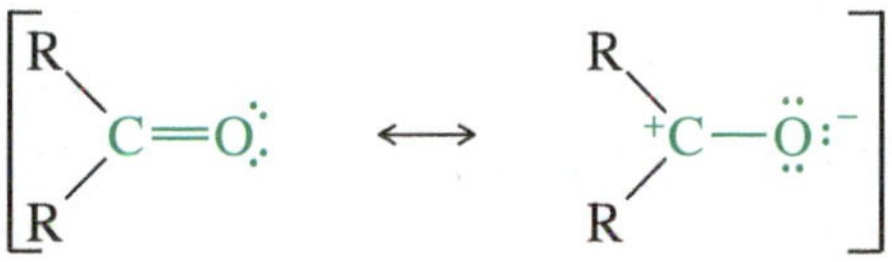

major minor

The first resonance form is more important because it involves more bonds, filled octets, and less charge separation. The contribution of the second structure is evidenced by the large dipole moments of the ketones and aldehydes shown here.

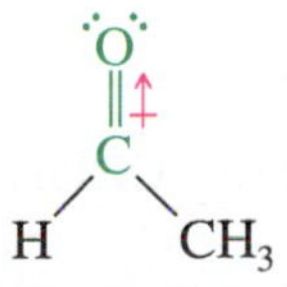

Compare with:

μ = 2.7 D	μ = 2.9 D	μ = 1.9 D	μ = 1.30 D
acetaldehyde	acetone	chloromethane	dimethyl ether

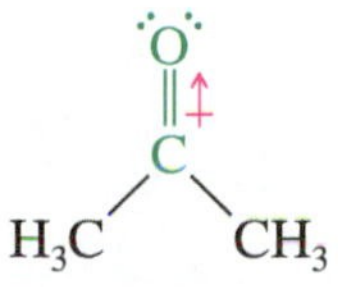

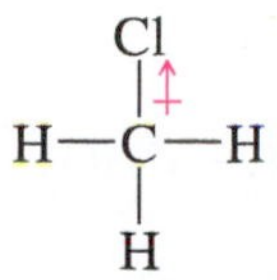

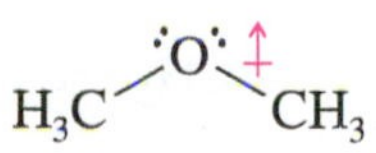

This polarization of the carbonyl group contributes to the reactivity of ketones and aldehydes: The positively polarized carbon atom acts as an electrophile (Lewis acid), and the negatively polarized oxygen acts as a nucleophile (Lewis base).

18-3 Nomenclature of Ketones and Aldehydes

IUPAC Names Systematic names of ketones are derived by replacing the final *-e* in the alkane name with *-one*. The "alkane" name becomes "alkanone." In open-chain ketones, we number the longest chain that includes the carbonyl carbon from the end closest to the carbonyl group, and we indicate the position of the carbonyl group by a number. In cyclic ketones, the carbonyl carbon atom is assigned the number 1.

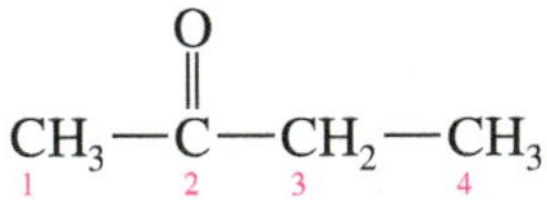

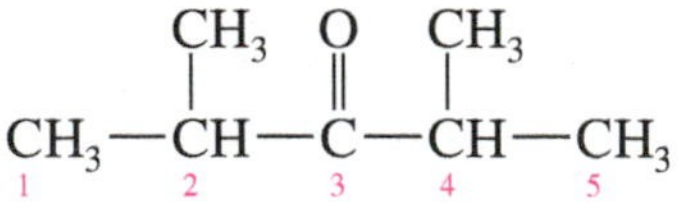

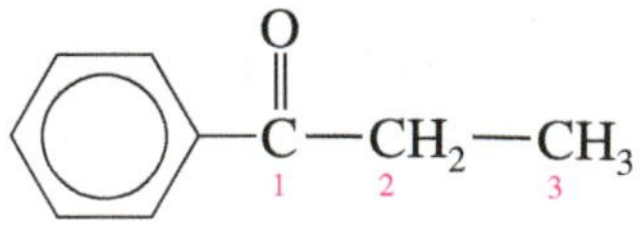

$CH_3-C(=O)-CH_2-CH_3$ $CH_3-CH(CH_3)-C(=O)-CH(CH_3)-CH_3$ $C_6H_5-C(=O)-CH_2-CH_3$

old IUPAC names in blue: 2-butanone 2,4-dimethyl-3-pentanone 1-phenyl-1-propanone
new IUPAC names in green: butan-2-one 2,4-dimethylpentan-3-one 1-phenylpropan-1-one

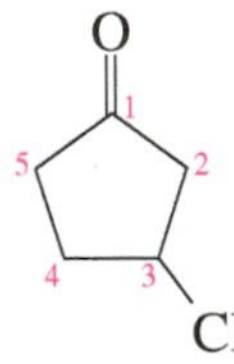

3-methylcyclopentanone

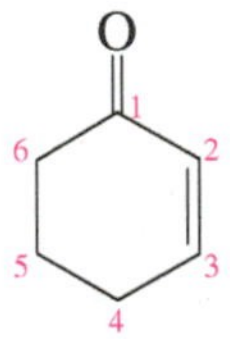

2-cyclohexenone
cyclohex-2-en-1-one

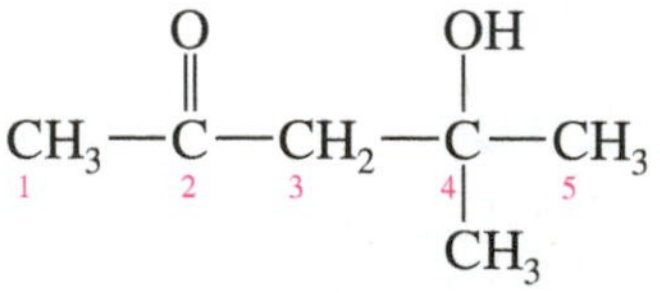

$CH_3-C(=O)-CH_2-C(OH)(CH_3)-CH_3$

4-hydroxy-4-methyl-2-pentanone
4-hydroxy-4-methylpentan-2-one

Systematic names for aldehydes are derived by replacing the final *-e* of the alkane name with *-al*. An aldehyde carbon is at the end of a chain, so it is number 1. If the aldehyde group is a substituent of a large unit (usually a ring), the suffix *carbaldehyde* is used.

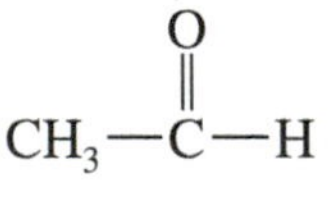

$CH_3-C(=O)-H$ ethanal

$CH_3CH_2CH_2-CH(Br)-CH(CH_3)-CH_2-C(=O)-H$ 4-bromo-3-methylheptanal

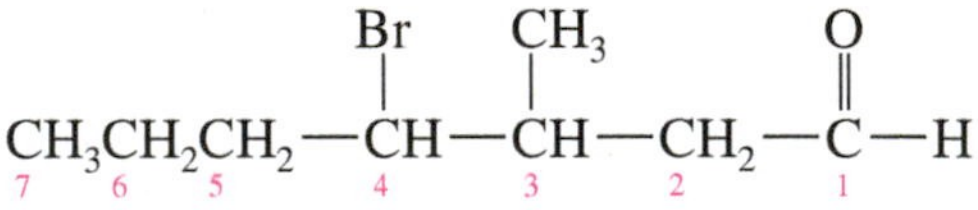

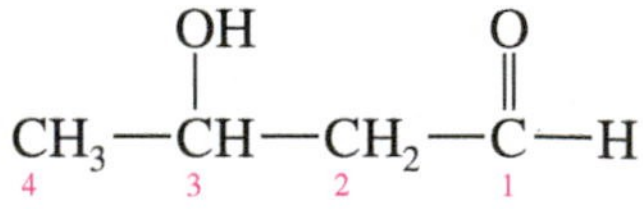

$CH_3-CH(OH)-CH_2-C(=O)-H$ 3-hydroxybutanal

$CH_3-CH_2-CH=CH-CHO$

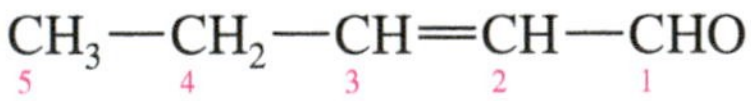

2-pentenal
pent-2-enal

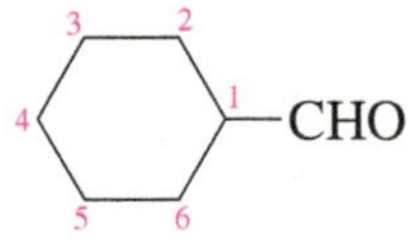

cyclohexanecarbaldehyde

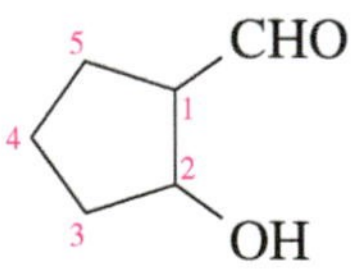

2-hydroxycyclopentane-1-carbaldehyde

Priority of Functional Groups in Naming Organic Compounds
(highest) acids
esters
aldehydes
ketones
alcohols
amines
alkenes, alkynes
alkanes
ethers
(lowest) halides

A ketone or aldehyde group can also be named as a substituent on a molecule with a higher priority functional group as its root. A ketone or aldehyde carbonyl is named by the prefix *oxo-* if it is included as part of the longest chain in the root name. When an aldehyde —CHO group is a substituent and not part of the longest chain, it is named by the prefix *formyl*. Carboxylic acids frequently contain ketone or aldehyde groups named as substituents.

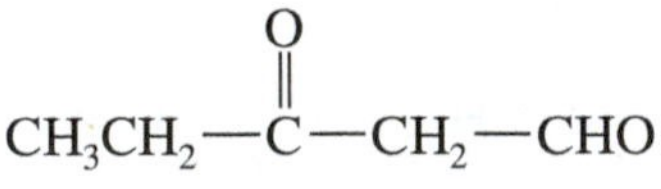

$CH_3CH_2-C(=O)-CH_2-CHO$ 3-oxopentanal

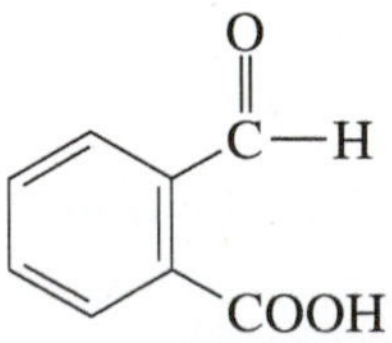

2-formylbenzoic acid

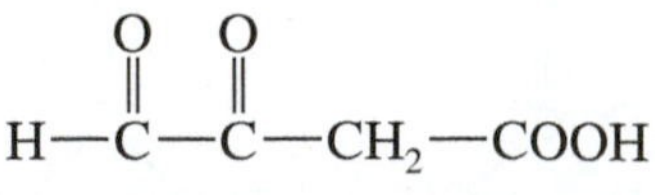

$H-C(=O)-C(=O)-CH_2-COOH$ 3,4-dioxobutanoic acid

Common Names As with other classes of compounds, ketones and aldehydes are often called by common names instead of their systematic IUPAC names. Ketone common names are formed by naming the two alkyl groups bonded to the carbonyl group. Substituent locations are given using Greek letters, beginning with the carbon *next to* the carbonyl group.

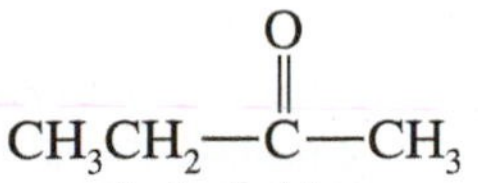

$CH_3CH_2{-}C({=}O){-}CH_3$
methyl ethyl ketone

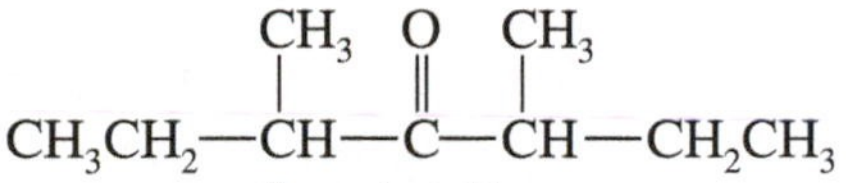

$CH_3CH_2{-}CH(CH_3){-}C({=}O){-}CH(CH_3){-}CH_2CH_3$
di-*sec*-butyl ketone

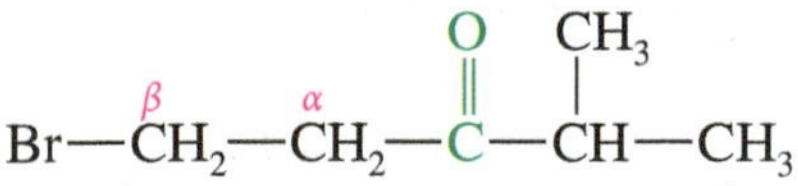

$Br{-}\overset{\beta}{C}H_2{-}\overset{\alpha}{C}H_2{-}C({=}O){-}CH(CH_3){-}CH_3$
β-bromoethyl isopropyl ketone

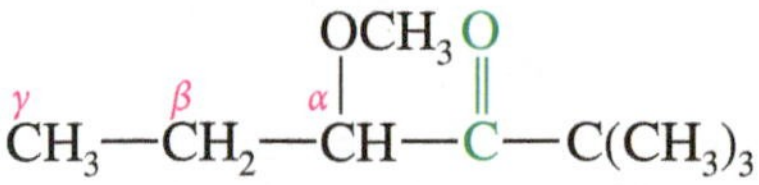

$\overset{\gamma}{C}H_3{-}\overset{\beta}{C}H_2{-}\overset{\alpha}{C}H(OCH_3){-}C({=}O){-}C(CH_3)_3$
tert-butyl α-methoxypropyl ketone

Some ketones have historical common names. Dimethyl ketone is always called *acetone*, and alkyl phenyl ketones are usually named as the acyl group followed by the suffix *-phenone*.

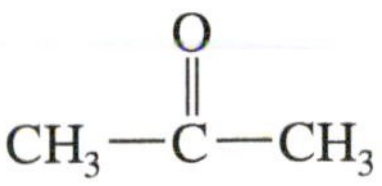

$CH_3{-}C({=}O){-}CH_3$
acetone

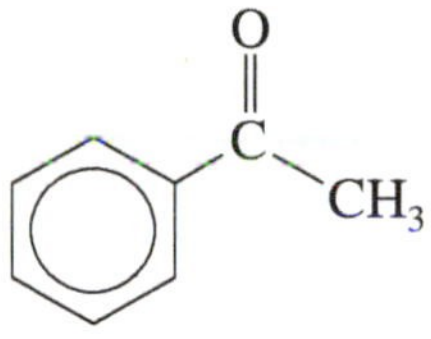

acetophenone

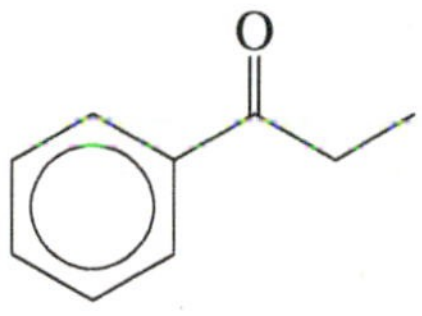

propiophenone

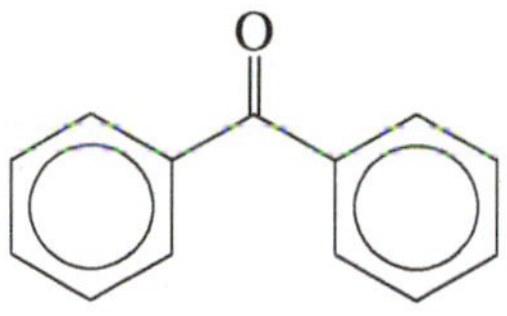

benzophenone

Common names of aldehydes are derived from the common names of the corresponding carboxylic acids (Table 18-2). These names often reflect the Latin or Greek term for the original source of the acid or the aldehyde. Greek letters are used with

TABLE 18-2
Common Names of Acids and Aldehydes

Carboxylic Acid	Derivation	Aldehyde
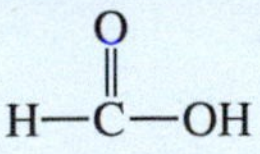 $H{-}C({=}O){-}OH$ formic acid (methanoic acid)	*formica*, "ants"	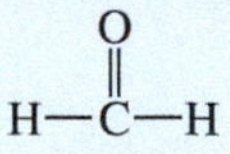 $H{-}C({=}O){-}H$ formaldehyde (methanal)
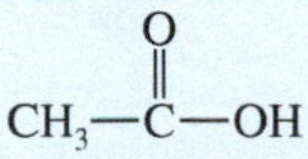 $CH_3{-}C({=}O){-}OH$ acetic acid (ethanoic acid)	*acetum*, "sour"	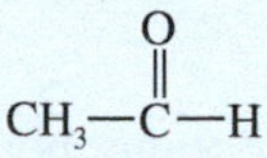 $CH_3{-}C({=}O){-}H$ acetaldehyde (ethanal)
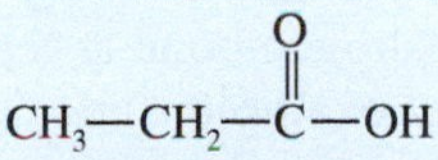 $CH_3{-}CH_2{-}C({=}O){-}OH$ propionic acid (propanoic acid)	*protos pion*, "first fat"	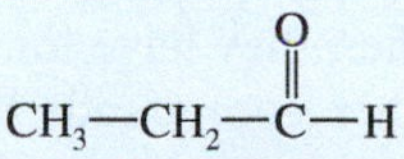 $CH_3{-}CH_2{-}C({=}O){-}H$ propionaldehyde (propanal)
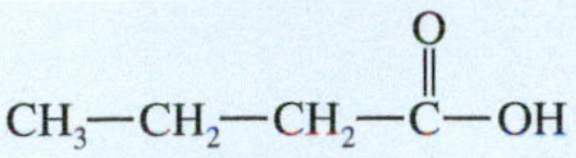 $CH_3{-}CH_2{-}CH_2{-}C({=}O){-}OH$ butyric acid (butanoic acid)	*butyrum*, "butter"	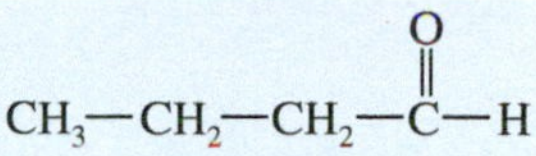 $CH_3{-}CH_2{-}CH_2{-}C({=}O){-}H$ butyraldehyde (butanal)
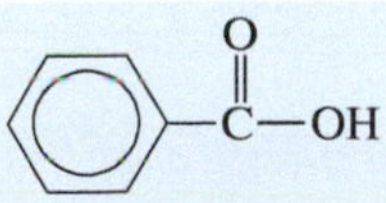 $C_6H_5{-}C({=}O){-}OH$ benzoic acid	*gum benzoin*, "blending"	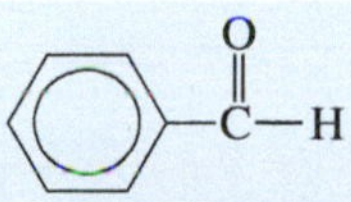 $C_6H_5{-}C({=}O){-}H$ benzaldehyde

common names of aldehydes to give the locations of substituents. The first letter (α) is given to the carbon atom *next to* the carbonyl group, which is C2 in the IUPAC name.

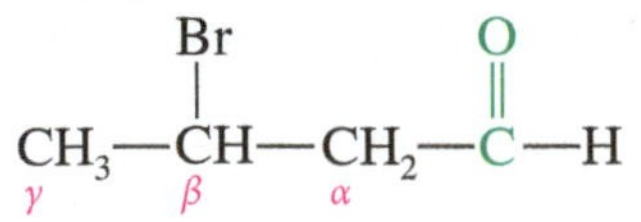

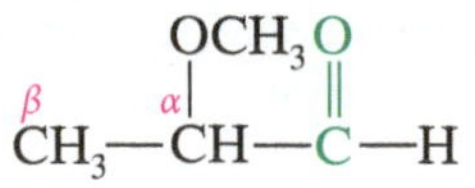

$$\underset{\gamma}{CH_3}-\underset{\beta}{CH}(Br)-\underset{\alpha}{CH_2}-C(=O)-H \qquad \underset{\beta}{CH_3}-\underset{\alpha}{CH}(OCH_3)-C(=O)-H$$

Common name:	β-bromobutyraldehyde	α-methoxypropionaldehyde
IUPAC name:	3-bromobutanal	2-methoxypropanal

PROBLEM-SOLVING HINT

The nomenclature of ketones and aldehydes is reviewed in Appendix 5, the Summary of Organic Nomenclature.

PROBLEM 18-1

Give the IUPAC name and (if possible) a common name for each compound.

(a) 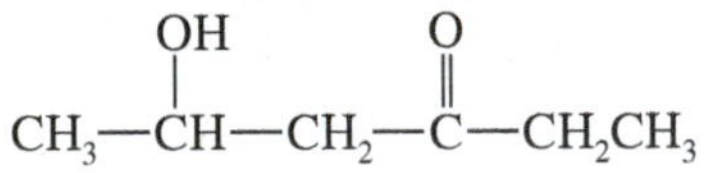 $CH_3-CH(OH)-CH_2-C(=O)-CH_2CH_3$

(b) $CH_3-CH(Ph)-CH_2-CHO$

(c) cyclohexane ring bearing OCH_3, H, H, and CHO substituents

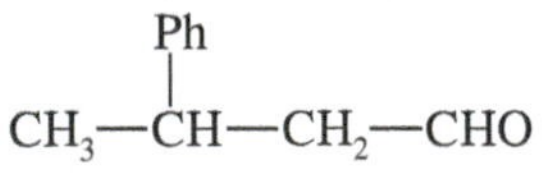

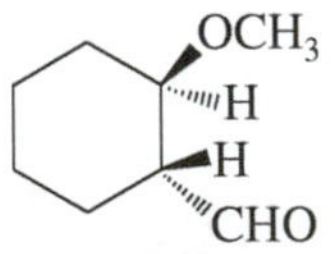

(d) 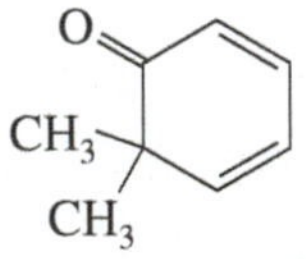 cyclohexadienone ring with O and two CH_3 groups

18-4 Physical Properties of Ketones and Aldehydes

Polarization of the carbonyl group creates dipole–dipole attractions between the molecules of ketones and aldehydes, resulting in higher boiling points than for hydrocarbons and ethers of similar molecular weights. Ketones and aldehydes have no O—H or N—H bonds, however, so their molecules cannot form hydrogen bonds with each other. Their boiling points are therefore lower than those of alcohols of similar molecular weight. The following compounds of molecular weight 58 or 60 are ranked in order of increasing boiling points. The ketone and the aldehyde are more polar and higher-boiling than the ether and the alkane, but lower-boiling than the hydrogen-bonded alcohol.

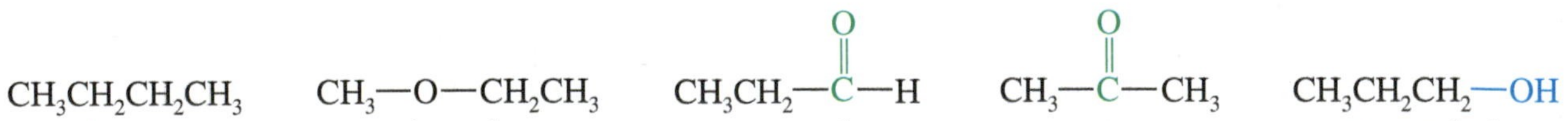

$CH_3CH_2CH_2CH_3$	$CH_3-O-CH_2CH_3$	$CH_3CH_2-C(=O)-H$	$CH_3-C(=O)-CH_3$	$CH_3CH_2CH_2-OH$
butane	methoxyethane	propanal	acetone	propan-1-ol
bp 0 °C	bp 8 °C	bp 49 °C	bp 56 °C	bp 97 °C

The melting points, boiling points, and water solubilities of some representative ketones and aldehydes are given in Table 18-3.

Although pure ketones and aldehydes cannot engage in hydrogen bonding with each other, they have lone pairs of electrons and can act as hydrogen bond acceptors with other compounds having O—H or N—H bonds. For example, the —OH hydrogen of water or an alcohol can form a hydrogen bond with the unshared electrons on a carbonyl oxygen atom.

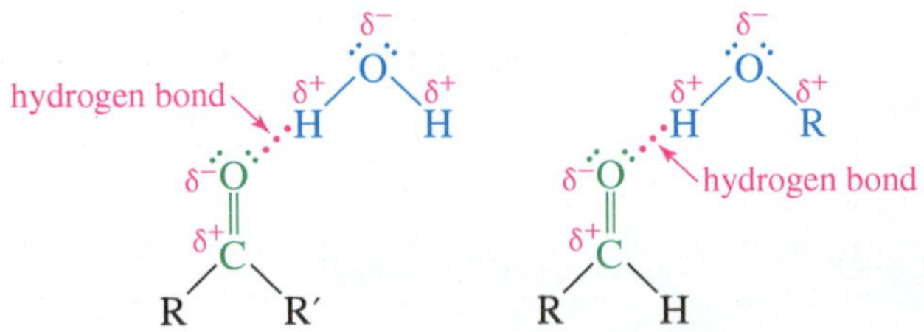

hydrogen bond

δ^- δ^+ δ^+ R R′

hydrogen bond

δ^- δ^+ R H

TABLE 18-3
Physical Properties of Ketones and Aldehydes

IUPAC Name	Common Name	Structure	mp (°C)	bp (°C)	Density (g/cm³)	H_2O Solubility (%)
		Ketones				
propan-2-one	acetone	CH_3COCH_3	−95	56	0.79	∞
butan-2-one	methyl ethyl ketone (MEK)	$CH_3COCH_2CH_3$	−86	80	0.81	25.6
pentan-2-one	methy *n*-propyl ketone	$CH_3COCH_2CH_2CH_3$	−78	102	0.81	5.5
pentan-3-one	diethyl ketone	$CH_3CH_2COCH_2CH_3$	−41	101	0.81	4.8
hexan-2-one		$CH_3CO(CH_2)_3CH_3$	−57	127	0.83	1.6
hexan-3-one		$CH_3CH_2COCH_2CH_2CH_3$	−55	124	0.82	
heptan-2-one		$CH_3CO(CH_2)_4CH_3$	−36	151	0.81	1.4
heptan-3-one		$CH_3CH_2CO(CH_2)_3CH_3$	−39	147	0.82	0.4
heptan-4-one	di-*n*-propyl ketone	$(CH_3CH_2CH_2)_2CO$	−34	144	0.82	
4-methylpent-3-en-2-one	mesityl oxide	$(CH_3)_2C{=}CHCOCH_3$	−59	131	0.86	
but-3-en-2-one	methyl vinyl ketone (MVK)	$CH_2{=}CHCOCH_3$	−6	80	0.86	
cyclohexanone			−47	157	0.94	15
acetophenone	methyl phenylketone	$C_6H_5COCH_3$	21	202	1.02	0.5
propiophenone	ethyl phenyl ketone	$C_6H_5COCH_2CH_3$	21	218	1.009	
benzophenone	diphenyl ketone	$C_6H_5COC_6H_5$	48	305	1.08	
		Aldehydes				
methanal	formaldehyde	HCHO or CH_2O	−92	−21	0.82	55
ethanal	acetaldehyde	CH_3CHO	−123	21	0.78	∞
propanal	propionaldehyde	CH_3CH_2CHO	−81	49	0.81	20
butanal	*n*-butyraldehyde	$CH_3(CH_2)_2CHO$	−97	75	0.82	7.1
2-methylpropanal	isobutyraldehyde	$(CH_3)_2CHCHO$	−66	61	0.79	11
pentanal	*n*-valeraldehyde	$CH_3(CH_2)_3CHO$	−91	103	0.82	
3-methylbutanal	isovaleraldehyde	$(CH_3)_2CHCH_2CHO$	−51	93	0.80	
hexanal	caproaldehyde	$CH_3(CH_2)_4CHO$	−56	129	0.83	0.1
heptanal	*n*-heptaldehyde	$CH_3(CH_2)_5CHO$	−45	155	0.85	0.02
propenal	acrolein	$CH_2{=}CH{-}CHO$	−88	53	0.84	30
but-2-enal	crotonaldehyde	$CH_3{-}CH{=}CH{-}CHO$	−77	104	0.86	18
benzaldehyde		C_6H_5CHO	−56	179	1.05	0.3

Because of this hydrogen bonding, ketones and aldehydes are good solvents for polar hydroxylic substances such as alcohols. They are also relatively soluble in water. Table 18-3 shows that acetaldehyde and acetone are miscible (soluble in all proportions) with water. Other ketones and aldehydes with up to four carbon atoms are fairly soluble in water. These solubility properties are similar to those of ethers and alcohols, which also engage in hydrogen bonding with water.

Formaldehyde and acetaldehyde are the most common aldehydes. Formaldehyde is a gas at room temperature, so it is often stored and used as a 40% aqueous solution called *formalin*. When dry formaldehyde is needed, it can be generated by heating one of its solid derivatives, usually *trioxane* or *paraformaldehyde*. Trioxane is a cyclic *trimer*, containing three formaldehyde units. Paraformaldehyde is a linear *polymer*, containing many formaldehyde units. These solid derivatives form spontaneously when a small amount of acid catalyst is added to pure formaldehyde.

Application: Diabetes

One symptom of untreated diabetes is the characteristic fruity smell of acetone in the patient's breath. Because diabetics cannot use carbohydrates properly, the body goes into a state called ketosis, in which it produces acetone and other ketones.

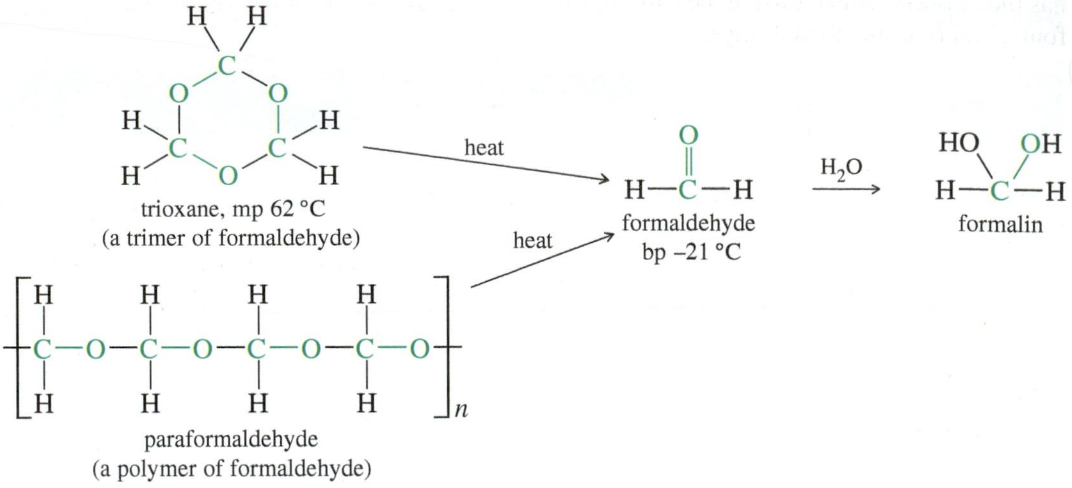

trioxane, mp 62 °C
(a trimer of formaldehyde)

heat

formaldehyde
bp −21 °C

$\xrightarrow{H_2O}$

formalin

heat

paraformaldehyde
(a polymer of formaldehyde)

Acetaldehyde boils near room temperature, and it can be handled as a liquid. Acetaldehyde is also used as a trimer (*paraldehyde*) and a tetramer (*metaldehyde*), formed from acetaldehyde under acid catalysis. Heating either of these compounds provides dry acetaldehyde. Paraldehyde is used in medicines as a sedative, and metaldehyde is used as a bait and poison for snails and slugs.

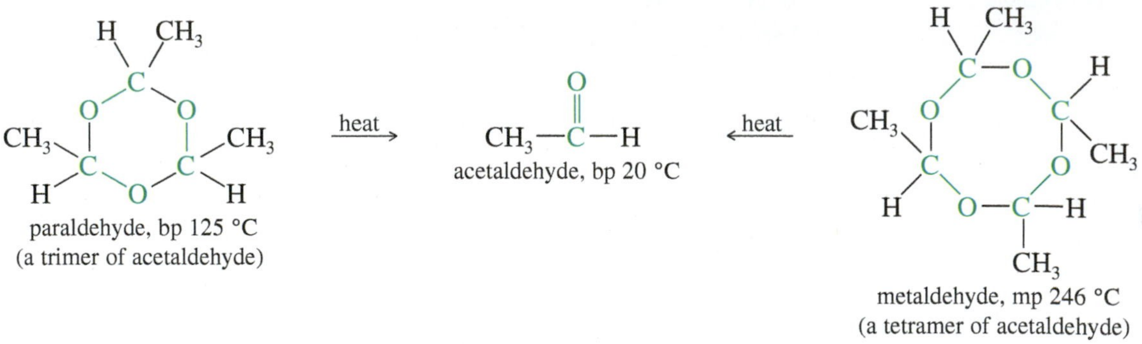

paraldehyde, bp 125 °C
(a trimer of acetaldehyde)

heat

$CH_3—C(=O)—H$
acetaldehyde, bp 20 °C

heat

metaldehyde, mp 246 °C
(a tetramer of acetaldehyde)

18-5 Spectroscopy of Ketones and Aldehydes

Chapters 12 and 13 covered some of the spectroscopic properties of ketones and aldehydes. Here we review that material and see what additional information we can obtain from spectroscopy.

18-5A Infrared Spectra of Ketones and Aldehydes

The carbonyl (C=O) stretching vibrations of simple ketones occur around 1710 cm^{-1}, and simple aldehydes around 1725 cm^{-1}. Because the carbonyl group has a large dipole moment, these absorptions are very strong. In addition to the carbonyl absorption, an aldehyde shows a set of two low-frequency C—H stretching absorptions around 2710 and 2810 cm^{-1}.

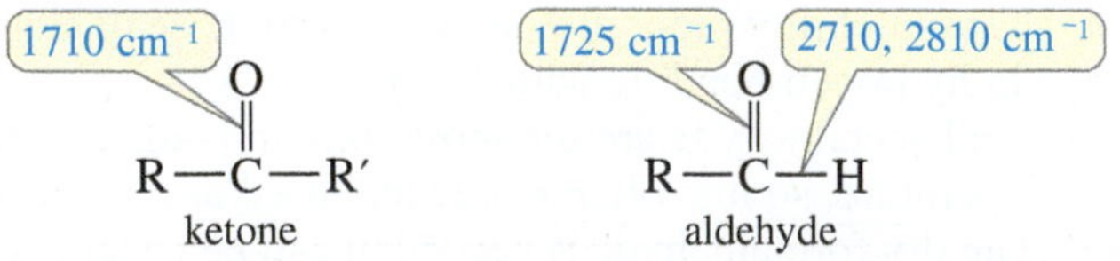

1710 cm^{-1}
R—C(=O)—R′
ketone

1725 cm^{-1} 2710, 2810 cm^{-1}
R—C(=O)—H
aldehyde

Figure 12-11 (page 572) compares the IR spectra of a simple ketone and aldehyde.

Conjugation lowers the carbonyl stretching frequencies of ketones and aldehydes because the partial pi bonding character of the single bond between the conjugated double bonds reduces the electron density of the carbonyl pi bond. The stretching frequency of this weakened carbonyl bond is lowered to about 1685 cm^{-1}. Ring strain has the opposite effect, raising the carbonyl stretching frequency in ketones with three-, four-, and five-membered rings.

about 1685 cm^{-1}

1685 cm^{-1} — acetophenone

1690 cm^{-1} — but-2-enal

1745 cm^{-1} — cyclopentanone

1815 cm^{-1} — cyclopropanone

18-5B Proton NMR Spectra of Ketones and Aldehydes

When considering the proton NMR spectra of ketones and aldehydes, we are interested primarily in the protons bonded to the carbonyl group (aldehyde protons) and the protons bonded to the adjacent carbon atom (the α carbon atom). Aldehyde protons normally appear at chemical shifts between δ 9 and δ 10. The aldehyde proton's absorption may be split ($J = 1$ to 5 Hz) if there are protons on the α carbon atom. Protons on the α carbon atom of a ketone or aldehyde usually appear at a chemical shift between δ 2.1 and δ 2.4 if there are no other electron-withdrawing groups nearby. Methyl ketones are characterized by a singlet at about δ 2.1.

α carbon: $R-CH_2-C(=O)-H$ — δ 2.4, δ 9–10 — an aldehyde

α carbon: $R-C(=O)-CH_3$ — δ 2.1 — a methyl ketone

α carbon: $R-C(=O)-CH_2R'$ — δ 2.4 — other ketones

Figure 18-1 shows the proton NMR spectrum of butanal. The aldehyde proton appears at δ 9.75, split by the protons on the α carbon atom with a small ($J = 1$ Hz) coupling constant. The α protons appear at δ 2.4, and the β and γ protons appear at increasing magnetic fields, as they are located farther from the deshielding effects of the carbonyl group.

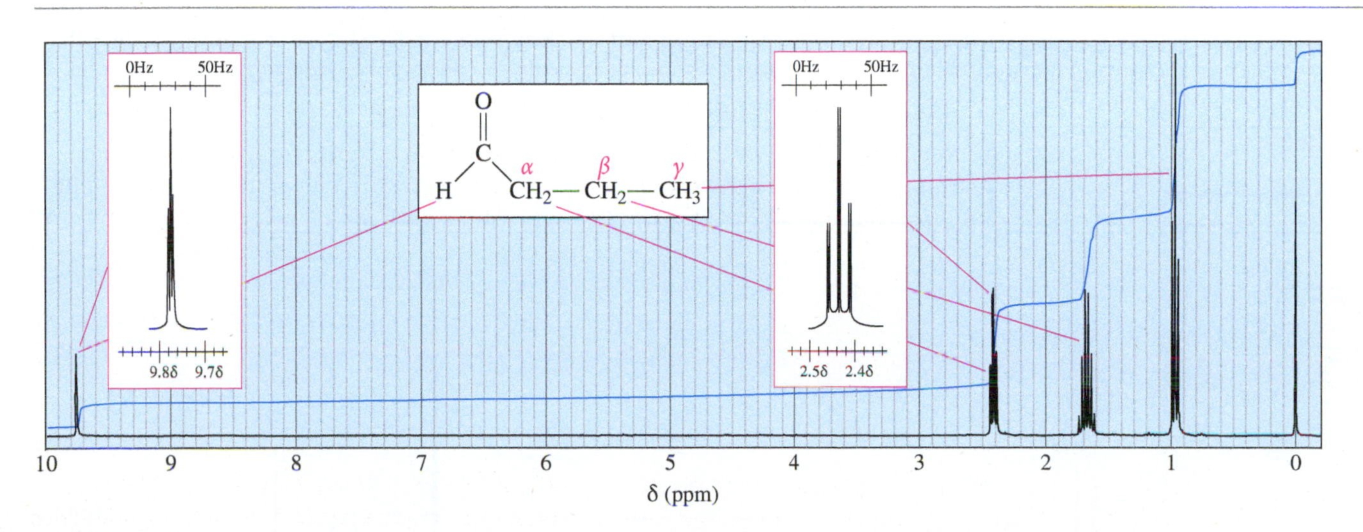

FIGURE 18-1
The proton NMR spectrum of butanal (butyraldehyde). Note the aldehyde proton at δ 9.8, split into a triplet ($J = 1$ Hz) by the two α protons. The α, β, and γ protons appear at values of δ that decrease with increasing distance from the carbonyl group.

18-5C Carbon NMR Spectra of Ketones and Aldehydes

The carbonyl carbon atoms of aldehydes and ketones have chemical shifts around 200 ppm in the carbon NMR spectrum. Because they have no hydrogens attached, ketone carbonyl carbon atoms usually give weaker absorptions than aldehydes. The α carbon atoms usually absorb at chemical shifts of about 30 to 40 ppm. Figure 18-2 shows the spin-decoupled carbon NMR spectrum of heptan-2-one, in which the carbonyl carbon absorbs at 208 ppm, and the α carbon atoms absorb at 30 ppm (methyl) and 44 ppm (methylene).

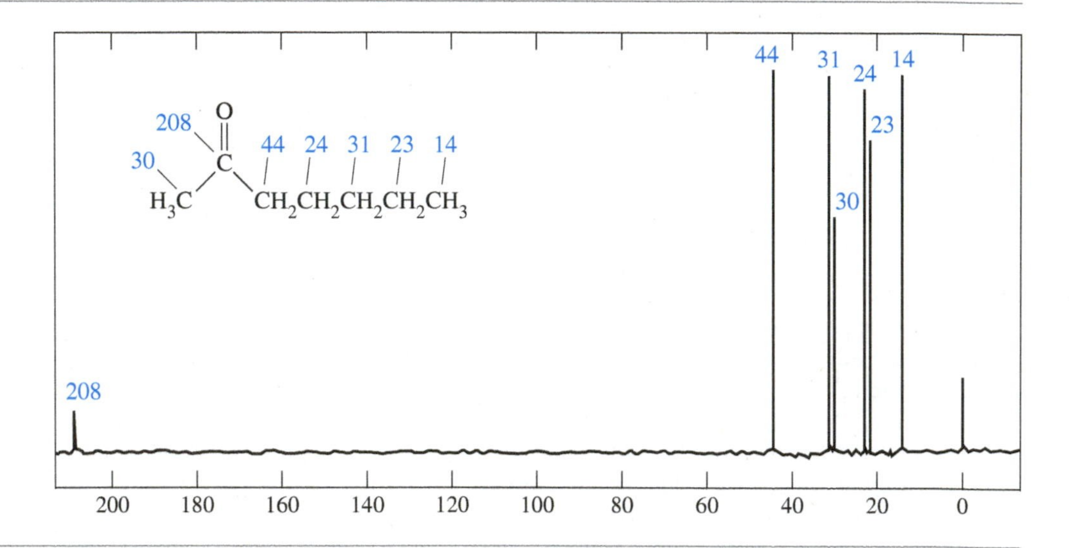

FIGURE 18-2
The spin-decoupled carbon NMR spectrum of heptan-2-one. Note the carbonyl carbon at 208 ppm and the α carbons at 30 ppm (methyl) and 44 ppm (methylene).

18-5D Mass Spectra of Ketones and Aldehydes

In the mass spectrometer, a ketone or an aldehyde may lose an alkyl group to give a resonance-stabilized acylium ion, such as the acylium ion that serves as the electrophile in the Friedel–Crafts acylation (Section 17-11).

$$\left[\mathrm{R{-}\overset{\overset{\displaystyle \ddot{O}}{\|}}{C}{-}R'}\right]^{\dot{+}} \longrightarrow [\mathrm{R{-}\overset{+}{C}{=}\ddot{O}} \longleftrightarrow \mathrm{R{-}C{\equiv}O^{+}}] + \cdot\mathrm{R'}$$

acylium ion

PROBLEM 18-2

NMR spectra for two compounds are given here, together with the molecular formulas. Each compound is a ketone or an aldehyde. In each case, show what characteristics of the spectrum imply the presence of a ketone or an aldehyde, and propose a structure for the compound.

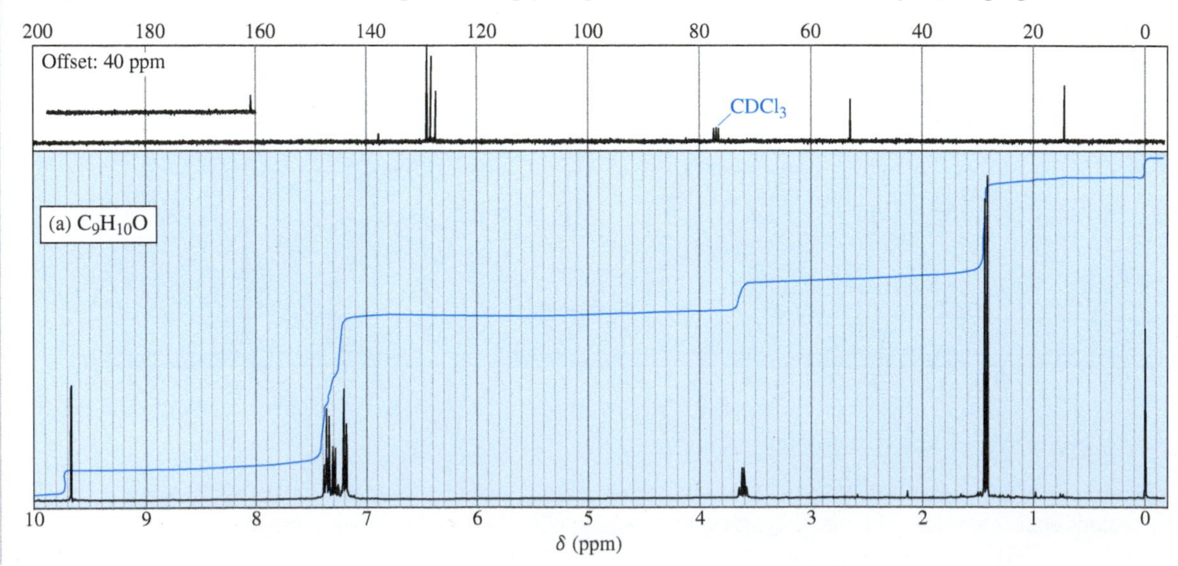

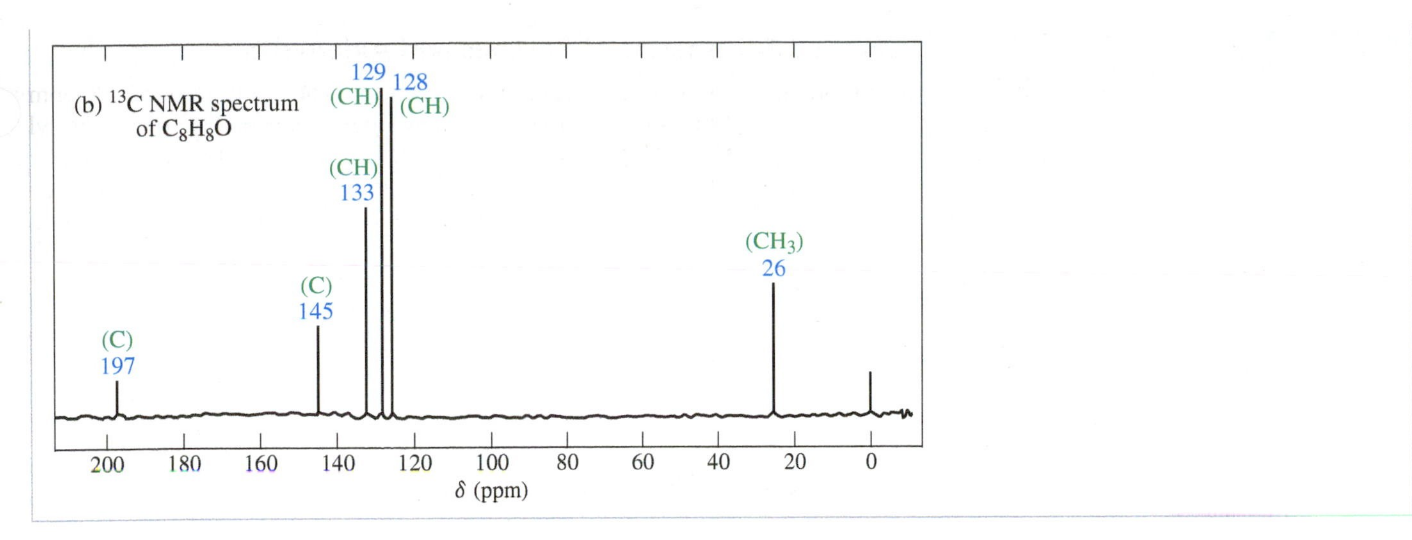

Figure 18-3 shows the mass spectrum of methyl ethyl ketone (butan-2-one). The molecular ion is prominent at m/z 72. The base peak at m/z 43 corresponds to loss of the ethyl group. Because a methyl radical is less stable than an ethyl radical, the peak corresponding to loss of the methyl group (m/z 57) is much weaker than the base peak from loss of the ethyl group.

McLafferty Rearrangement of Ketones and Aldehydes The mass spectrum of butyraldehyde (Figure 18-4) shows the peaks we expect at m/z 72 (molecular ion), m/z 57 (loss of a methyl group), and m/z 29 (loss of a propyl group). The peak at m/z 57 is from cleavage between the β and γ carbons to give a resonance-stabilized carbocation. This is also a common fragmentation with carbonyl compounds; like the other odd-numbered peaks, it results from loss of a radical. These fragmentations cannot explain the base (strongest) peak at m/z 44, however.

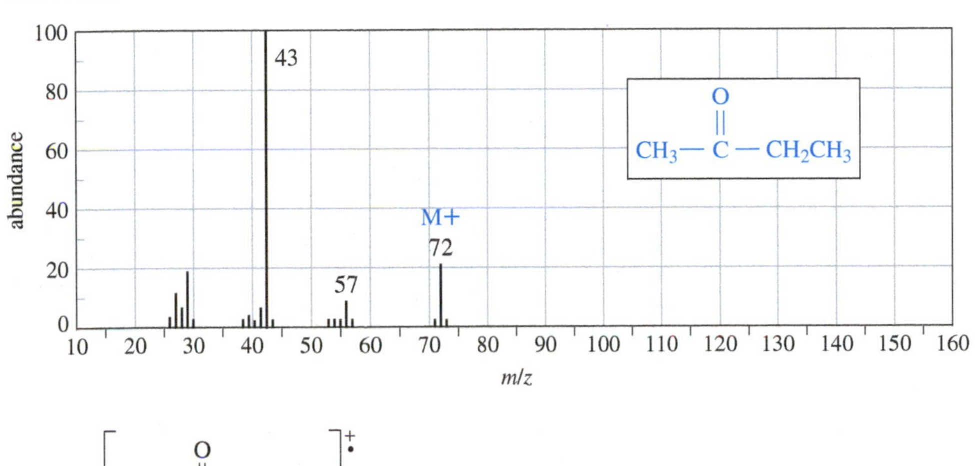

FIGURE 18-3
The mass spectrum of butan-2-one. Note the prominent molecular ion, together with a base peak from loss of an ethyl radical to give an acylium ion.

$$\left[CH_3-\overset{\overset{\displaystyle O}{\|}}{C}\,\big|\,CH_2CH_3\right]^{+\cdot} \longrightarrow CH_3-C\equiv O^+ \;+\; \cdot CH_2CH_3$$

(cleavage marked 43)

radical cation m/z 72 → acylium ion m/z 43 (base peak) + ethyl radical loss of 29

$$\left[CH_3\,\big|\,\overset{\overset{\displaystyle O}{\|}}{C}-CH_2CH_3\right]^{+\cdot} \longrightarrow CH_3CH_2-C\equiv O^+ \;+\; \cdot CH_3$$

(cleavage marked 57)

radical cation m/z 72 → acylium ion m/z 57 + methyl radical loss of 15

FIGURE 18-4
The mass spectrum of butyraldehyde shows the expected ions of masses 72, 57, and 29. The base peak at m/z 44 results from the loss of ethylene via the McLafferty rearrangement.

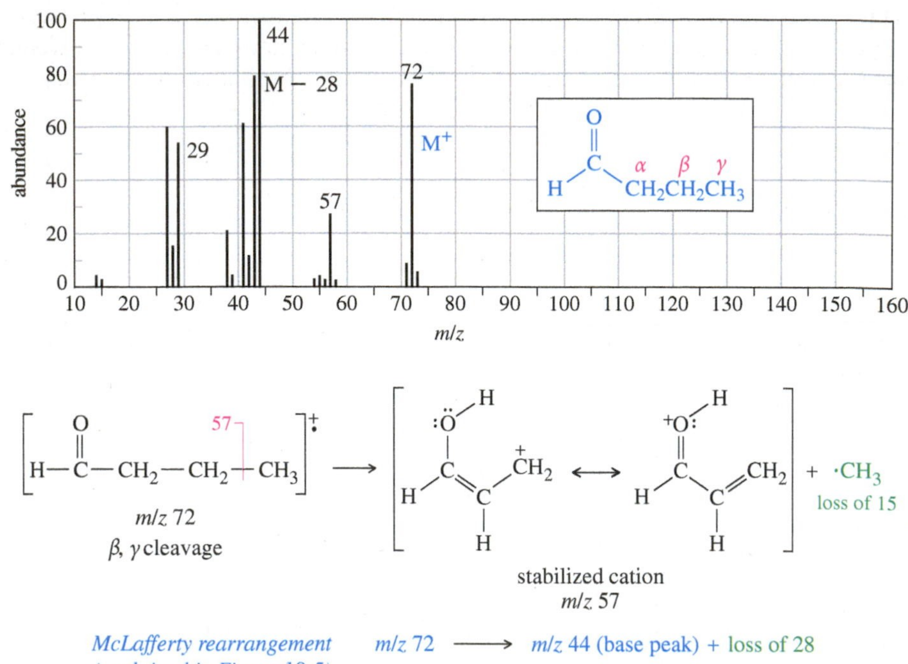

$$[\mathrm{H{-}C(=O){-}CH_2CH_2CH_3}]^{+\bullet} \longrightarrow \mathrm{H{-}C{\equiv}O^+} + \cdot\mathrm{CH_2CH_2CH_3}$$

29; m/z 72 — m/z 29 — loss of 43

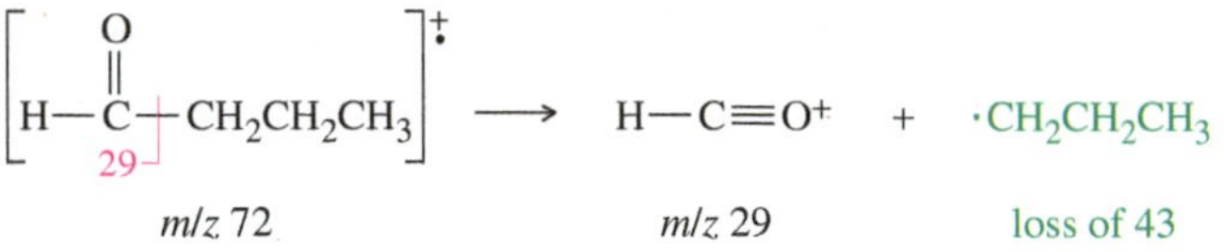

PROBLEM-SOLVING HINT

The McLafferty rearrangement is equivalent to a cleavage between the carbon atoms that are α and β to the carbonyl group, plus one mass unit for the H that is transferred. The fragment from the McLafferty rearrangement has an even-numbered mass.

The base peak is at m/z 44, from loss of a fragment of mass 28. Loss of a fragment with an even mass number corresponds to loss of a stable, neutral molecule (as when water, mass 18, is lost from an alcohol). A fragment of mass 28 corresponds to a molecule of ethylene (C_2H_4). This fragment is lost through a process called the **McLafferty rearrangement,** involving a cyclic intramolecular transfer of a hydrogen atom from the γ (gamma) carbon to the carbonyl oxygen (shown in Figure 18-5).

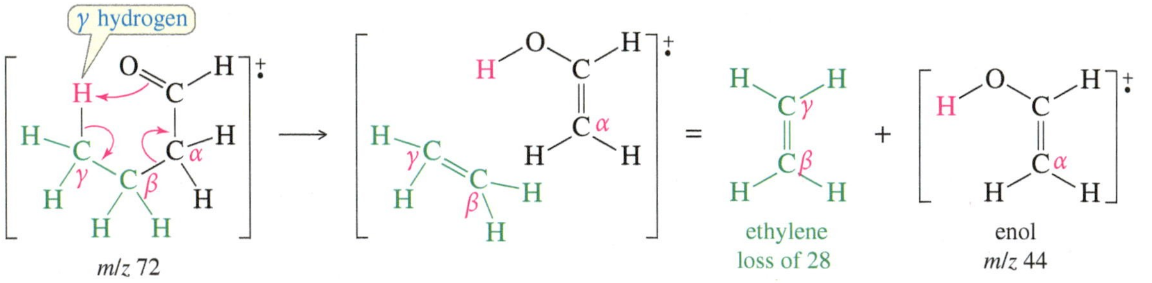

γ hydrogen

m/z 72 ⟶ [intermediate] = ethylene (loss of 28) + enol (m/z 44)

McLafferty rearrangement of butyraldehyde

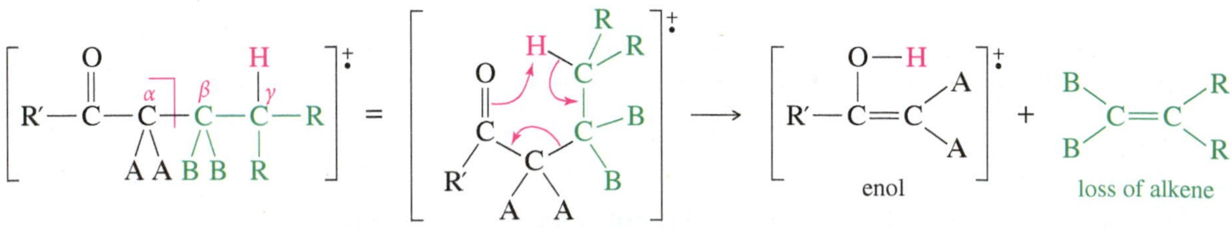

$$[\mathrm{R'{-}C(=O){-}CA_2{-}CB_2{-}CR_2H}]^{+\bullet} \longrightarrow [\mathrm{R'{-}C(OH){=}CA_2}]^{+\bullet} + \mathrm{B_2C{=}CR_2}$$

enol — loss of alkene

McLafferty rearrangement of a general ketone or aldehyde

FIGURE 18-5
Mechanism of the McLafferty rearrangement. This rearrangement may be concerted, as shown here, or the γ hydrogen may be transferred first, followed by fragmentation.

The McLafferty rearrangement is a characteristic fragmentation of ketones and aldehydes as long as they have γ hydrogens. It is equivalent to a cleavage between the α and β carbon atoms, plus one mass unit for the hydrogen that is transferred.

PROBLEM 18-3

Why were no products from the McLafferty rearrangement observed in the spectrum of butan-2-one (Figure 18-3)?

PROBLEM 18-4

Use equations to show the fragmentation leading to each numbered peak in the mass spectrum of octan-2-one.

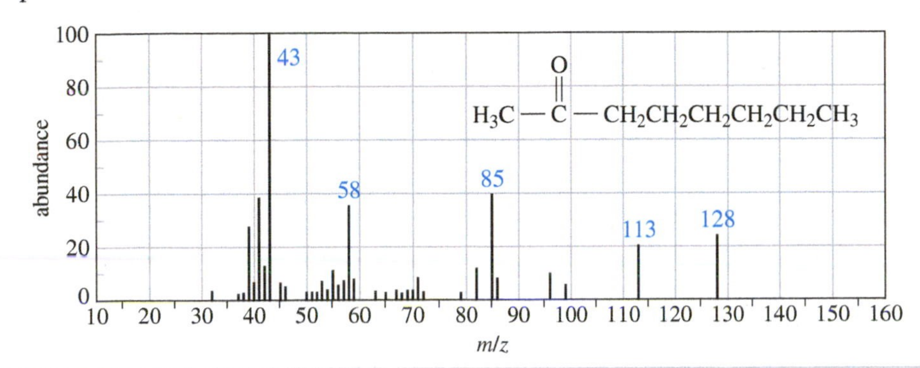

18-5E Ultraviolet Spectra of Ketones and Aldehydes

The $\pi \rightarrow \pi^*$ Transition The strongest absorptions in the ultraviolet spectra of aldehydes and ketones are the ones resulting from $\pi \rightarrow \pi^*$ electronic transitions. As with alkenes, these absorptions are observable ($\lambda_{max} > 200$ nm) only if the carbonyl double bond is conjugated with another double bond. The simplest conjugated carbonyl system is propenal, shown next. The $\pi \rightarrow \pi^*$ transition of propenal occurs at λ_{max} of 210 nm ($\varepsilon = 11{,}000$). Alkyl substitution increases the value of λ_{max} by about 10 nm per alkyl group. An additional conjugated double bond increases the value of λ_{max} by about 30 nm. Notice the large values of the molar absorptivities ($\varepsilon > 5000$), similar to those observed for the $\pi \rightarrow \pi^*$ transitions of conjugated dienes.

PROBLEM-SOLVING HINT

Conjugated carbonyl compounds have characteristic $\pi \rightarrow \pi^*$ absorptions in the UV spectrum.

Base value: 210 nm

An additional conjugated C=C increases λ_{max} about 30 nm; an additional alkyl group increases it about 10 nm.

propenal
$\lambda_{max} = 210$ nm, $\varepsilon = 11{,}000$

three alkyl groups
$\lambda_{max} = 237$ nm, $\varepsilon = 12{,}000$

three alkyl groups
$\lambda_{max} = 244$ nm, $\varepsilon = 12{,}500$

The $n \rightarrow \pi^*$ Transition A very weak band of absorptions in the ultraviolet spectra of ketones and aldehydes results from promoting one of the nonbonding electrons on oxygen to a π^* antibonding orbital. This transition involves a smaller amount of energy than the $\pi \rightarrow \pi^*$ transition, so it gives a lower frequency (longer wavelength) absorption (Figure 18-6). The $n \rightarrow \pi^*$ transitions of simple, unconjugated ketones and aldehydes give absorptions with values of λ_{max} between 280 and 300 nm. Each double bond added in conjugation with the carbonyl group increases the value of λ_{max} by about 30 nm. For example, the $n \rightarrow \pi^*$ transition of acetone occurs at λ_{max} of 280 nm ($\varepsilon = 15$).

The $n \rightarrow \pi^*$ transitions of carbonyl groups have small molar absorptivities, generally about 10 to 200. These absorptions are around 1000 times weaker than $\pi \rightarrow \pi^*$ transitions because the $n \rightarrow \pi^*$ transition corresponds to a "forbidden" electronic transition with a low probability of occurrence. The nonbonding orbitals on oxygen are

FIGURE 18-6
Comparison of the $\pi \rightarrow \pi^*$ and the $n \rightarrow \pi^*$ transitions. The $n \rightarrow \pi^*$ transition requires less energy because the nonbonding (n) electrons are higher in energy than the bonding π electrons.

"allowed" transition
$\varepsilon \cong 5000$–200,000

"forbidden" transition
$\varepsilon \cong 10$–200

perpendicular to the π^* antibonding orbitals, and there is zero overlap between these orbitals (see Figure 18-6). This forbidden transition occurs occasionally, but much less frequently than the "allowed" $\pi \rightarrow \pi^*$ transition.

PROBLEM-SOLVING HINT

Carbonyl $n \rightarrow \pi^*$ absorptions are very weak, and they are often obscured by stronger absorptions. Therefore, they are rarely as useful as the $\pi \rightarrow \pi^*$ absorptions.

PROBLEM 18-5

Oxidation of cholesterol converts the alcohol to a ketone. Under acidic or basic oxidation conditions, the C═C double bond migrates to the more stable, conjugated position. Before IR and NMR spectroscopy, chemists watched the UV spectrum of the reaction mixture to follow the oxidation. Describe how the UV spectrum of the conjugated product, cholest-4-en-3-one, differs from that of cholesterol.

cholesterol $\xrightarrow{[O]}$ cholest-4-en-3-one

Application: Formaldehyde

Many construction materials are made of, or bonded by, phenol-formaldehyde resins. Unreacted formaldehyde is toxic, causing irritation of the skin, eyes, nose, and throat. Its concentration in finished products is carefully regulated to levels below 0.1 ppm. Some construction materials, such as flooring and carpeting, have outgassed more formaldehyde than what is considered safe, and have required recalls.

18-6 Industrial Importance of Ketones and Aldehydes

In the chemical industry, ketones and aldehydes are used as solvents, starting materials, and reagents for the synthesis of other products. Although formaldehyde is well known as the formalin solution used to preserve biological specimens, most of the 4 billion kilograms of formaldehyde produced each year is used to make Bakelite®, phenol–formaldehyde resins, urea–formaldehyde glues, and other polymeric products. Acetaldehyde is used primarily as a starting material in the manufacture of acetic acid, polymers, and drugs.

Acetone is the most important commercial ketone, with over 3 billion kilograms used each year. Both acetone and methyl ethyl ketone (butan-2-one) are common industrial solvents. These ketones dissolve a wide range of organic materials, have convenient boiling points for easy distillation, and have low toxicities.

Many other ketones and aldehydes are used as flavorings and additives to foods, drugs, and other products. For example, benzaldehyde is the primary component of almond extract, and (−)-carvone gives spearmint chewing gum its minty flavor. Table 18-4 lists some simple ketones and aldehydes with well-known odors and flavors. *Pyrethrin*, isolated from pyrethrum flowers, is commercially extracted for use as a "natural" insecticide. "Natural" or not, pyrethrin can cause severe allergic reactions, nausea, vomiting, and other toxic effects in animals.

TABLE 18-4
Ketones and Aldehydes Used in Household Products

	butyraldehyde ($CH_3-CH_2-CH_2-C(=O)-H$)	vanillin	acetophenone	*trans*-cinnamaldehyde
Odor:	buttery	vanilla	pistachio	cinnamon
Uses:	margarine, foods	foods, perfumes	ice cream	candy, foods, drugs

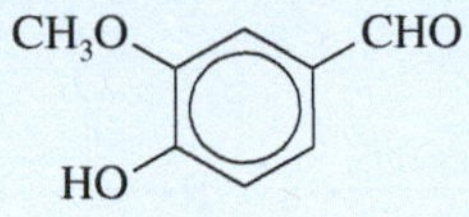

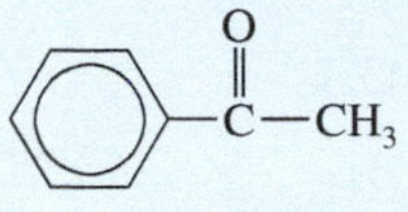

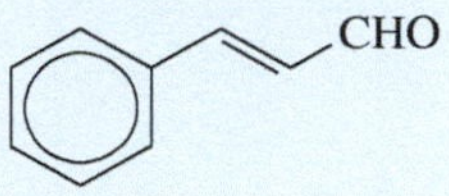

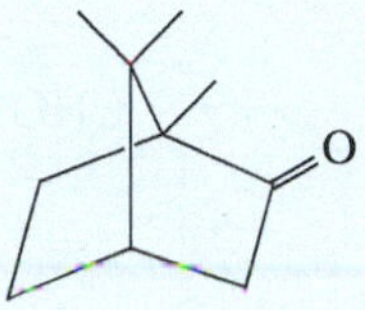

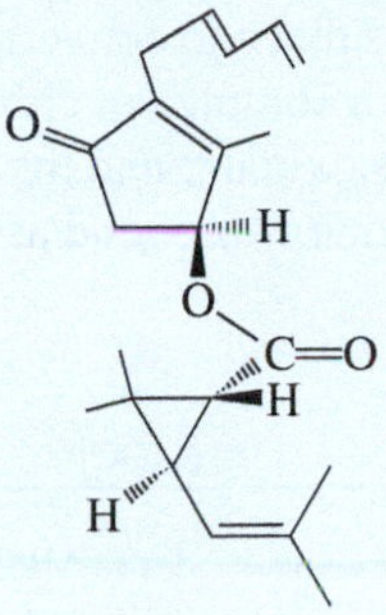

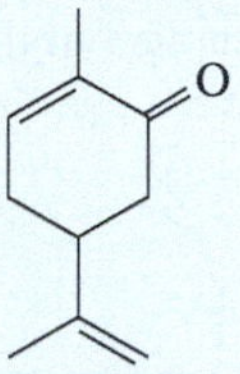

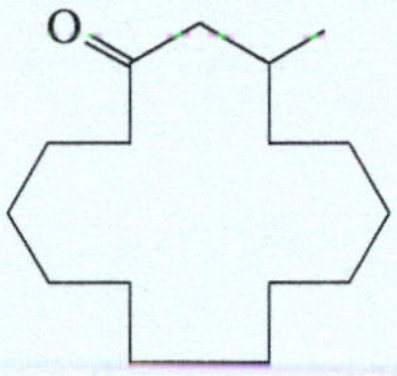

	camphor	pyrethrin	carvone	muscone
Odor:	"camphoraceous"	floral	(−) enantiomer: spearmint (+) enantiomer: caraway seed	musky aroma
Uses:	liniments, inhalants	plant insecticide	candy, toothpaste, etc.	perfumes

18-7 Review of Syntheses of Ketones and Aldehydes

In studying reactions of other functional groups, we have already encountered some of the best methods for making ketones and aldehydes. Let's review and summarize these reactions, and then consider some additional synthetic methods. A summary table of syntheses of ketones and aldehydes begins on page 925.

18-7A Ketones and Aldehydes from Oxidation of Alcohols (Section 11-2)

Ketones and aldehydes are often made by oxidizing alcohols. When we need to make a carbonyl compound, we can often use a Grignard reagent to synthesize an alcohol with the correct structure and oxidize it to the final product. In effect, you convert an aldehyde into a ketone by adding an alkyl group to give an alcohol intermediate, which is oxidized to the ketone.

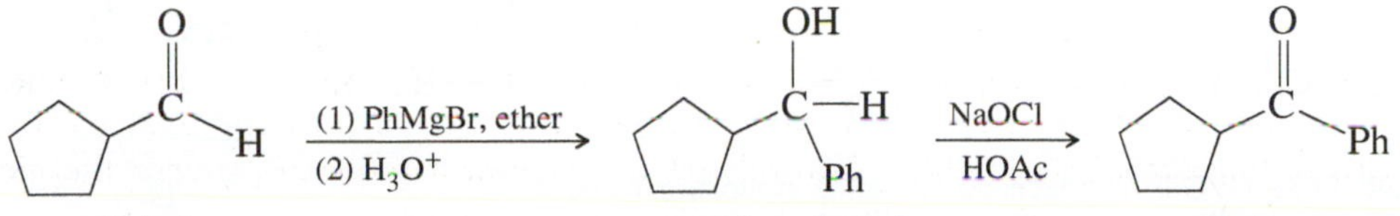

Aldehyde → secondary alcohol → ketone

$$\underset{\text{Grignard}}{R-MgX} + \underset{\text{aldehyde}}{R'-C(=O)-H} \xrightarrow{\text{ether}} \xrightarrow{H_3O^+} \underset{\text{secondary alcohol}}{R-CH(OH)-R'} \xrightarrow[\text{HOAc}]{\text{NaOCl}} \underset{\text{ketone}}{R-C(=O)-R'}$$

Sodium dichromate in sulfuric acid ("chromic acid," H_2CrO_4) is the traditional laboratory reagent for oxidizing secondary alcohols to ketones. Bleach (NaOCl) is an inexpensive, chromium-free alternative that also oxidizes secondary alcohols to ketones.

$$\underset{\text{aldehyde}}{C_5H_9-C(=O)-H} \xrightarrow[(2)\ H_3O^+]{(1)\ \text{PhMgBr, ether}} \underset{\text{secondary alcohol}}{C_5H_9-C(OH)(H)-Ph} \xrightarrow[\text{HOAc}]{\text{NaOCl}} \underset{\text{ketone}}{C_5H_9-C(=O)-Ph}$$

Primary alcohols are easily over-oxidized to carboxylic acids unless the conditions are carefully controlled.

Primary alcohols → aldehydes

$$\underset{\text{primary alcohol}}{R{-}CH_2{-}OH} \xrightarrow[-2\,H]{[\text{oxidizing agent}]} \underset{\text{aldehyde}}{R{-}C({=}O){-}H} \xrightarrow[{[O]}]{[\text{over-oxidation}]} \underset{\text{carboxylic acid}}{R{-}C({=}O){-}OH}$$

Oxidation of a primary alcohol to an aldehyde requires careful selection of an oxidizing agent to avoid over-oxidation to the carboxylic acid. One equivalent of bleach (NaOCl) plus the TEMPO catalyst, provides good yields of aldehydes without over-oxidation. Other reagents that can convert primary alcohols to aldehydes are: PCC (pyridinium chlorochromate), a complex of chromium trioxide with pyridine; the Swern oxidation, using DMSO as the oxidant; and the Dess-Martin periodinane (DMP) oxidation, using a high-valence iodine compound as the oxidant. These specialized reagents are covered in Section 11-3.

cyclohexylmethanol (CH_2OH) —NaOCl, (TEMPO)→ cyclohexanecarbaldehyde (CHO) (90%)

18-7B Ketones and Aldehydes from Ozonolysis of Alkenes (Section 8-15B)

Ozonolysis, followed by a mild reduction, cleaves alkenes to give ketones and aldehydes.

$$RHC{=}CR'R'' \xrightarrow[(2)\ (CH_3)_2S]{(1)\ O_3,\ -78\ ^\circ C} RHC{=}O + O{=}CR'R'' + (CH_3)_2S{=}O$$

Ozonolysis is useful as a synthetic method or as an analytical technique. Yields are generally good.

1-methylcyclohexene —(1) O_3, −78 °C; (2) $(CH_3)_2S$→ 6-oxoheptanal (65%)

18-7C Phenyl Ketones and Aldehydes: Friedel–Crafts Acylation (Section 17-11)

Friedel–Crafts acylation is an excellent method for making alkyl aryl ketones or diaryl ketones. It cannot be used, however, on strongly deactivated aromatic systems.

$$R{-}C({=}O){-}Cl + C_6H_5G \xrightarrow[(2)\ H_2O]{(1)\ AlCl_3} G{-}C_6H_4{-}C({=}O){-}R\ (para) + G{-}C_6H_4{-}C({=}O){-}R\ (ortho)$$

R is alkyl or aryl; G is hydrogen, an activating group, or a halogen.

$$p\text{-nitrobenzoyl chloride} + \text{benzene} \xrightarrow[(2)\ H_2O]{(1)\ AlCl_3} p\text{-nitrobenzophenone (90\%)}$$

The Gatterman–Koch synthesis is a variant of the Friedel–Crafts acylation in which carbon monoxide and HCl generate an intermediate that reacts like formyl chloride. Like Friedel–Crafts reactions, the Gatterman–Koch formylation succeeds only with benzene and activated benzene derivatives.

$$\text{toluene} \xrightarrow[AlCl_3/CuCl]{CO,\ HCl} p\text{-methylbenzaldehyde (major) (50\%)}$$

18-7D Ketones and Aldehydes from Hydration of Alkynes (Section 9-9F)

Catalyzed by Acid and Mercuric Salts Hydration of a terminal alkyne is a convenient way of making methyl ketones. This reaction is catalyzed by a combination of sulfuric acid and mercuric ion. The initial product of Markovnikov hydration is an **enol** (a vinyl alcohol), which quickly tautomerizes to its keto form. Internal alkynes can be hydrated, but mixtures of ketones often result.

$$\underset{\text{alkyne}}{R-C\equiv C-H} \xrightarrow[H_2O]{Hg^{2+},\ H_2SO_4} \underset{\text{enol (not isolated)}}{\left[(R)(HO)C=C(H)(H)\right]} \xrightarrow{H^+} \underset{\text{methyl ketone}}{R-C(=O)-CH_2-H}$$

Example

$$\underset{\text{ethynylcyclohexane}}{C_6H_{11}-C\equiv C-H} \xrightarrow[H_2O]{H_2SO_4,\ Hg^{2+}} \underset{\text{enol}}{\left[C_6H_{11}(HO)C=CH_2\right]} \xrightarrow{H^+} \underset{\text{cyclohexyl methyl ketone (90\%)}}{C_6H_{11}-C(=O)-CH_3}$$

Hydroboration–Oxidation of Alkynes Hydroboration–oxidation of an alkyne gives anti-Markovnikov addition of water across the triple bond. Di(secondary isoamyl) borane, called *disiamylborane*, is used, because this bulky borane cannot add twice across the triple bond. On oxidation of the borane, the unstable enol quickly tautomerizes to an aldehyde. (See Section 9-9F.)

$$\underset{\text{alkyne}}{R-C\equiv C-H} \xrightarrow[(2)\ H_2O_2,\ NaOH]{(1)\ Sia_2BH} \underset{\text{enol (not isolated)}}{\left[(R)(H)C=C(H)(OH)\right]} \xrightarrow{^-OH} \underset{\text{aldehyde}}{R-CH_2-C(=O)-H}$$

Example

ethynylcyclohexane $\xrightarrow[(2)\ H_2O_2,\ NaOH]{(1)\ Sia_2BH}$ cyclohexylethanal (65%)

In the following sections, we consider additional syntheses of ketones and aldehydes that we have not covered before. These syntheses form ketones and aldehydes from carboxylic acids, nitriles, acid chlorides, and alkyl halides.

SOLVED PROBLEM 18-1

Show how you would synthesize each compound from starting materials containing no more than six carbon atoms.

(a) (phenyl cyclopentyl ketone) **(b)** cyclopentyl–CH_2–CH_2–C(=O)–H

SOLUTION

(a) This compound is a ketone with 12 carbon atoms. The carbon skeleton might be assembled from two six-carbon fragments using a Grignard reaction, which gives an alcohol that is easily oxidized to the target compound.

PhMgBr + cyclopentyl–C(=O)–H $\xrightarrow[(2)\ H_3O^+]{(1)\ \text{ether solvent}}$ Ph–CH(OH)–cyclopentyl $\xrightarrow[HOAc]{NaOCl}$ target (a)

An alternative route to the target compound involves Friedel–Crafts acylation.

benzene + cyclopentyl–C(=O)–Cl $\xrightarrow{AlCl_3}$ Ph–C(=O)–cyclopentyl

(b) This compound is an aldehyde with eight carbon atoms. An aldehyde might come from oxidation of an alcohol (possibly a Grignard product) or hydroboration of an alkyne. If we use a Grignard, the restriction to six-carbon starting materials means we need to add two carbons to a methylcyclopentyl fragment, ending in a primary alcohol. Grignard addition to an epoxide does this.

cyclopentyl–CH_2MgBr $\xrightarrow[(2)\ H_3O^+]{(1)\ \text{ethylene oxide, ether}}$ cyclopentyl–CH_2–CH_2–CH_2OH $\xrightarrow[TEMPO]{NaOCl}$ target (b)

Alternatively, we could construct the carbon skeleton using acetylene as the two-carbon fragment. The resulting terminal alkyne undergoes hydroboration to the correct aldehyde.

cyclopentyl–CH_2Br $\xrightarrow{H{-}C{\equiv}C{:}^-\ Na^+}$ cyclopentyl–CH_2–C≡C–H $\xrightarrow[(2)\ H_2O_2,\ ^-OH]{(1)\ Sia_2BH}$ target (b)

PROBLEM 18-6

Show how you would synthesize each compound from starting materials containing no more than six carbon atoms.

(a) cyclohexyl–C(=O)–CH_2CH_3 (b) phenyl–C(=O)–CH_2CH_3 (c) cyclopentyl–CH_2–C(=O)–CH_3

18-8 Synthesis of Ketones from Carboxylic Acids

Organolithium reagents can be used to synthesize ketones from carboxylic acids. Organolithiums are so reactive toward carbonyls that they attack the lithium salts of carboxylate anions to give dianions. Protonation of the dianion forms the hydrate of a ketone, which quickly loses water to give the ketone (see Section 18-12).

$$\underset{\text{carboxylic acid}}{R-\overset{O}{\overset{\|}{C}}-OH} \xrightarrow{LiOH} \underset{\text{lithium carboxylate}}{R-\overset{O}{\overset{\|}{C}}-O^-\,Li^+} \xrightarrow{R'-Li} \underset{\text{dianion}}{R-\overset{O^-\,Li^+}{\overset{|}{\underset{R'}{\underset{|}{C}}}}-O^-\,Li^+} \xrightarrow{H_3O^+} \underset{\text{hydrate}}{\left[R-\overset{OH}{\overset{|}{\underset{R'}{\underset{|}{C}}}}-OH\right]} \xrightarrow{-H_2O} \underset{\text{ketone}}{R-\overset{O}{\overset{\|}{C}}-R'}$$

If the organolithium reagent is inexpensive, we can simply add two equivalents to the carboxylic acid. The first equivalent generates the carboxylate salt, and the second attacks the carbonyl group. Subsequent protonation gives the ketone.

cyclohexanecarboxylic acid $\xrightarrow{2\ \text{PhLi (phenyllithium)}}$ dianion (cyclohexyl–C(OLi)(Ph)–OLi) $\xrightarrow{H_3O^+}$ [hydrate (cyclohexyl–C(OH)(Ph)–OH)] $\xrightarrow{-H_2O}$ cyclohexyl phenyl ketone

PROBLEM 18-7

Show how you would accomplish the following synthetic conversions by adding an organolithium reagent to an acid.

(a) benzoic acid (Ph–COOH) ⟶ acetophenone (Ph–C(=O)–CH_3)

(b) bromocyclopentane ⟶ cyclopentyl methyl ketone

(c) pentanoic acid ⟶ heptan-3-one

(d) phenylacetic acid ⟶ 3,3-dimethyl-1-phenylbutan-2-one

18-9 Synthesis of Ketones and Aldehydes from Nitriles

Nitriles can also be used as starting materials for the synthesis of ketones. Discussed in Chapter 21, nitriles are compounds containing the cyano (—C≡N) functional group. Because nitrogen is more electronegative than carbon, the —C≡N triple bond is polarized like the C=O bond of the carbonyl group. Nucleophiles can add to the —C≡N triple bond by attacking the electrophilic carbon atom.

A Grignard or organolithium reagent attacks a nitrile to give the magnesium salt of an imine. Acidic hydrolysis of the imine leads to the ketone. The mechanism of this acid hydrolysis is the reverse of acid-catalyzed imine formation, covered in Section 18-14. Note that the ketone is formed during the hydrolysis *after* any excess Grignard reagent has been destroyed. Thus, the ketone is not attacked.

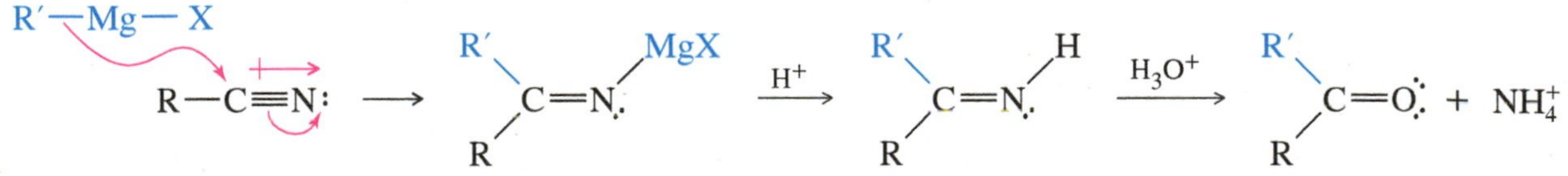

$$R'{-}Mg{-}X \quad R{-}C{\equiv}N{:} \longrightarrow R'R{>}C{=}N{-}MgX \xrightarrow{H^+} R'R{>}C{=}N{-}H \xrightarrow{H_3O^+} R'R{>}C{=}O{:} + NH_4^+$$

nucleophilic attack — Mg salt of imine — imine — ketone

Example

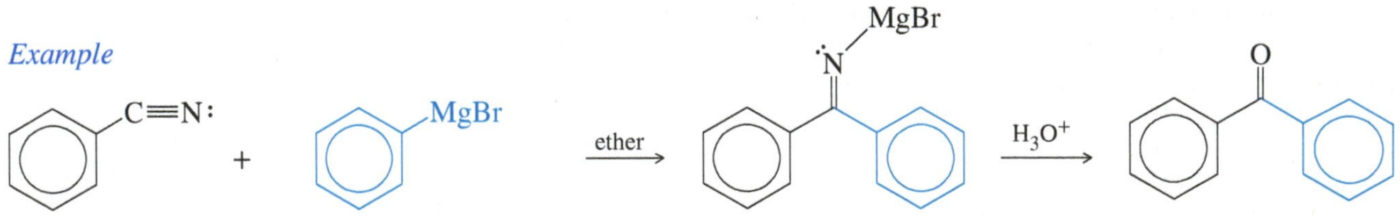

benzonitrile + phenylmagnesium bromide $\xrightarrow{\text{ether}}$ benzophenone imine (magnesium salt) $\xrightarrow{H_3O^+}$ benzophenone (80%)

Aluminum hydrides can reduce nitriles to the corresponding aldehydes. Diisobutylaluminum hydride, abbreviated $(i\text{-Bu})_2AlH$ or **DIBAL-H,** is commonly used for the reduction of nitriles. The initial reaction forms an aluminum complex that hydrolyzes in the aqueous workup.

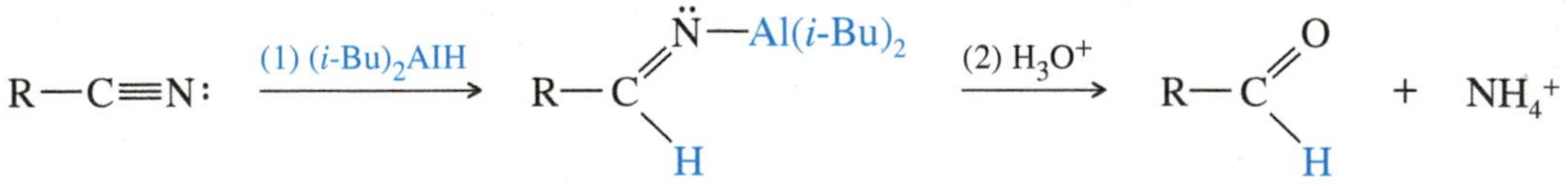

$$R{-}C{\equiv}N{:} \xrightarrow{(1)\ (i\text{-Bu})_2AlH} R{-}C(H){=}\ddot{N}{-}Al(i\text{-Bu})_2 \xrightarrow{(2)\ H_3O^+} R{-}C(H){=}O + NH_4^+$$

nitrile — aluminum complex — aldehyde

Example

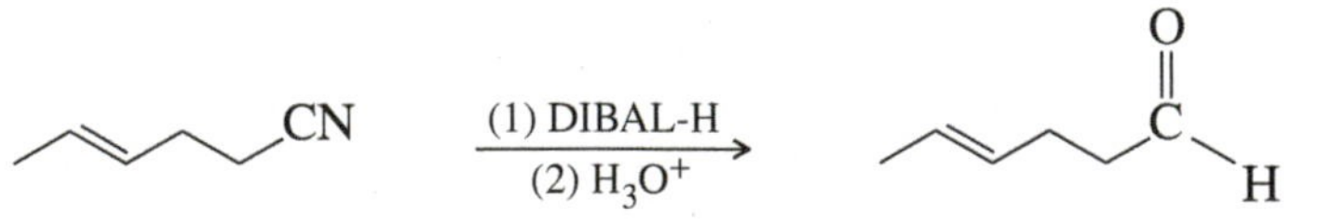

hex-4-enonitrile (CN) $\xrightarrow[(2)\ H_3O^+]{(1)\ \text{DIBAL-H}}$ hex-4-enal (C(=O)H)

$$H{-}Al(CH_2CH(CH_3)_2)_2$$

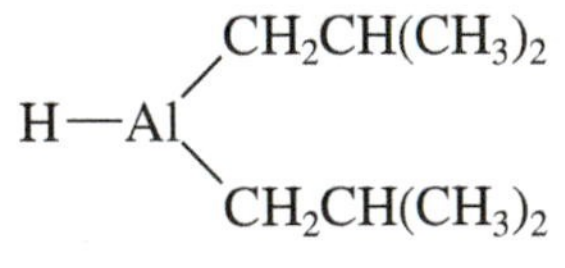

diisobutylaluminum hydride
$(i\text{-Bu})_2AlH$ or DIBAL-H

PROBLEM 18-8

Predict the products of the following reactions.

(a) $CH_3CH_2CH_2CH_2C{\equiv}N + CH_3CH_2MgBr$, then H_3O^+
(b) $CH_3CH_2CH_2CH_2C{\equiv}N$ + DIBAL-H, then H_3O^+
(c) benzyl bromide + sodium cyanide
(d) product of (c) + cyclopentylmagnesium bromide, then acidic hydrolysis
(e) product of (c) + DIBAL-H, then hydrolysis

PROBLEM 18-9

Show how the following transformations may be accomplished in good yield. You may use any additional reagents that are needed.

(a) bromobenzene → propiophenone
(b) CH_3CH_2CN → heptan-3-one
(c) benzoic acid → phenyl cyclopentyl ketone
(d) 1-bromohept-2-ene → oct-3-enal

18-10 Synthesis of Aldehydes and Ketones from Acid Chlorides and Esters

Because aldehydes are easily oxidized to acids, one might wonder whether acids are easily reduced back to aldehydes. Aldehydes tend to be more reactive than acids, however, and reducing agents that are strong enough to reduce acids also reduce aldehydes even faster.

$$R{-}C({=}O){-}OH \xrightarrow[\text{(slow)}]{LiAlH_4} [R{-}C({=}O){-}H] \xrightarrow[\text{(fast)}]{LiAlH_4} R{-}CH_2{-}O^-$$

acid; aldehyde (not isolable); alkoxide

Acids can be reduced to aldehydes by first converting them to a functional group that is easier to reduce than an aldehyde: the acid chloride. Acid chlorides (acyl chlorides) are reactive derivatives of carboxylic acids in which the acidic hydroxy group is replaced by a chlorine atom. Acid chlorides are often synthesized by treatment of carboxylic acids with thionyl chloride, $SOCl_2$.

$$R{-}C({=}O){-}OH + Cl{-}S({=}O){-}Cl \longrightarrow R{-}C({=}O){-}Cl + HCl\uparrow + SO_2\uparrow$$

acid; thionyl chloride; acid chloride

Strong reducing agents such as $LiAlH_4$ reduce acid chlorides all the way to primary alcohols. Lithium tri-*tert*-butoxyaluminum hydride is a milder reducing agent that reacts faster with acid chlorides than with aldehydes. Reduction of acid chlorides with lithium tri-*tert*-butoxyaluminum hydride gives good yields of aldehydes.

$$R{-}C({=}O){-}Cl \xrightarrow[\text{lithium tri-}tert\text{-butoxyaluminum hydride}]{Li^+\ {}^-AlH(O\text{-}t\text{-}Bu)_3} R{-}C({=}O){-}H$$

acid chloride; aldehyde

Example

$$CH_3CH(CH_3)CH_2{-}C({=}O){-}OH \xrightarrow{SOCl_2} CH_3CH(CH_3)CH_2{-}C({=}O){-}Cl \xrightarrow{Li^+\ {}^-AlH(O\text{-}t\text{-}Bu)_3} CH_3CH(CH_3)CH_2{-}C({=}O){-}H$$

isovaleric acid; isovaleroyl chloride; isovaleraldehyde (65%)

Diisobutylaluminum hydride (DIBAL-H) reduces esters directly to aldehydes at dry ice temperature, about −78 °C. Unlike $LiAlH_4$ (which reduces esters to primary alcohols), cold DIBAL-H usually does not reduce the aldehyde further. The initial reaction forms an aluminum complex that is stable toward further reduction, but hydrolyzes to the aldehyde in the aqueous workup.

$$R{-}C({=}O){-}OR' \xrightarrow[(-78\,°C)]{(1)\ (i\text{-}Bu)_2AlH} R{-}CH(O{-}Al(i\text{-}Bu)_2)(OR') \xrightarrow{(2)\ H_2O} R{-}C({=}O){-}H$$

ester; aluminum complex; aldehyde

Example

$$C_5H_9{-}C({=}O){-}OEt \xrightarrow[(2)\ H_2O]{(1)\ \text{DIBAL-H}\ (-78\,°C)} C_5H_9{-}C({=}O){-}H$$

ethyl cyclopentanecarboxylate; cyclopentanecarbaldehyde

Synthesis of Ketones Grignard and organolithium reagents react with acid chlorides much like hydride reagents: They add R^- where a hydride reagent would add H^-. As we saw in Section 10-9, Grignard and organolithium reagents add to acid chlorides to give ketones, but they add again to the ketones to give tertiary alcohols.

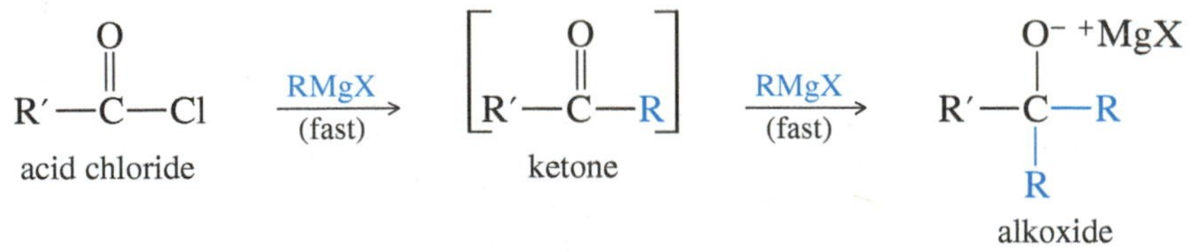

$$R'-\overset{O}{\overset{\|}{C}}-Cl \xrightarrow[\text{(fast)}]{RMgX} \left[R'-\overset{O}{\overset{\|}{C}}-R\right] \xrightarrow[\text{(fast)}]{RMgX} R'-\underset{R}{\overset{O^-\ {}^+MgX}{C}}-R$$

acid chloride — ketone — alkoxide

To stop at the ketone stage, a weaker organometallic reagent is needed: one that reacts faster with acid chlorides than with ketones. **Lithium dialkylcuprates** and other organocuprates (Sections 10-9F and 17-13a) react with acid chlorides to give good yields of a wide variety of ketones.

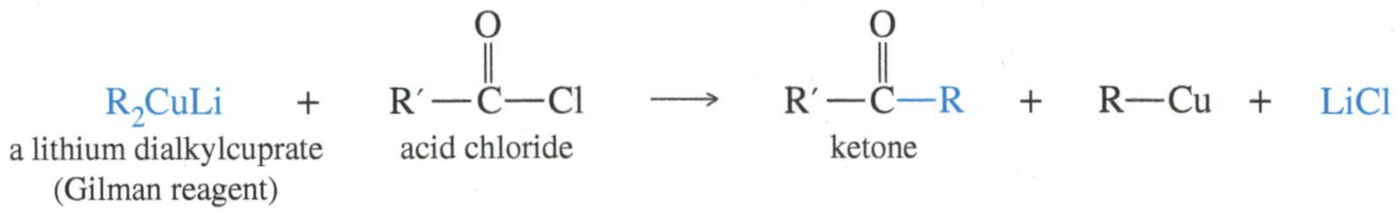

$$R_2CuLi + R'-\overset{O}{\overset{\|}{C}}-Cl \longrightarrow R'-\overset{O}{\overset{\|}{C}}-R + R-Cu + LiCl$$

a lithium dialkylcuprate (Gilman reagent) — acid chloride — ketone

The lithium dialkylcuprate is formed by the reaction of two equivalents of the corresponding organolithium reagent with cuprous iodide.

$$2\ R-Li + CuI \longrightarrow R_2CuLi + LiI$$

Example

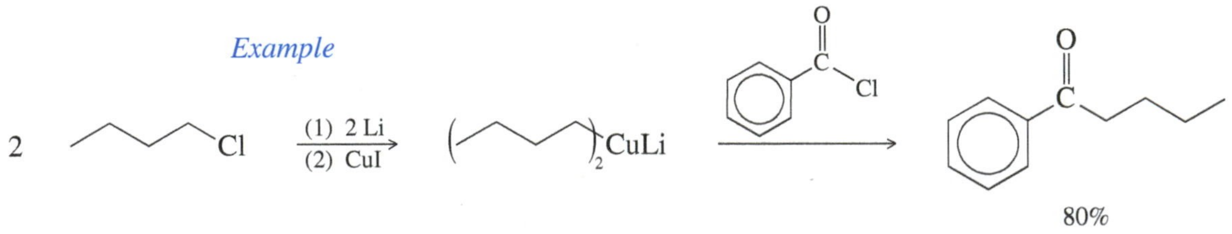

2 butyl chloride $\xrightarrow[(2)\ CuI]{(1)\ 2\ Li}$ (butyl)$_2$CuLi $\xrightarrow{\text{benzoyl chloride}}$ phenyl butyl ketone (80%)

PROBLEM 18-10

Predict the products of the following reactions:

(a) benzoyl chloride $\xrightarrow[(2)\ H_3O^+]{(1)\ LiAlH_4}$

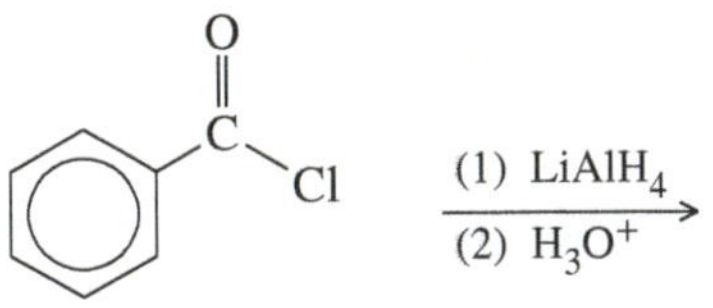

(b) benzoyl chloride $\xrightarrow{LiAlH(O\text{-}t\text{-}Bu)_3}$

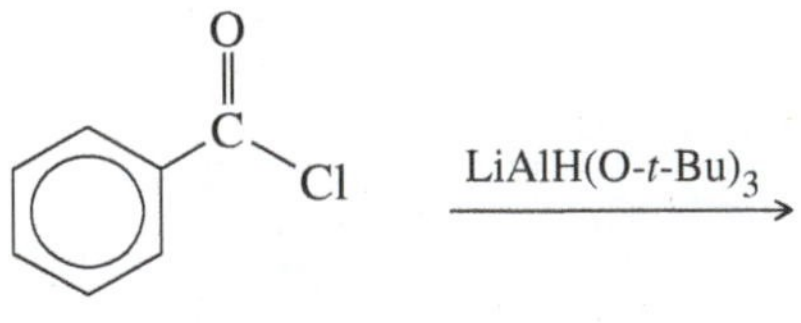

(c) pentanoyl chloride $\xrightarrow{(\text{vinyl})_2CuLi}$

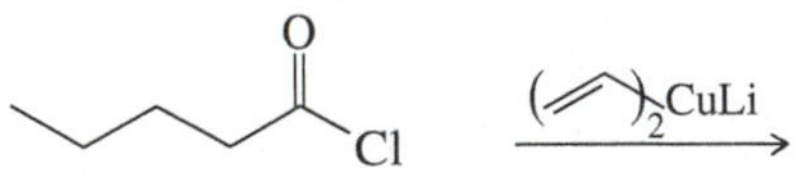

(d) butanoyl chloride $\xrightarrow[(2)\ H_3O^+]{(1)\ \text{excess } CH_2{=}CH-CH_2MgI}$

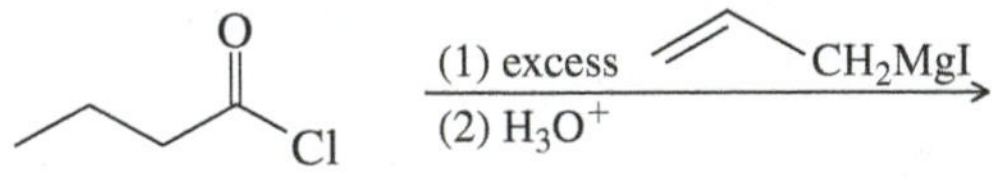

(e) methyl hex-4-enoate (OCH$_3$ ester) $\xrightarrow[(2)\ H_2O]{(1)\ \text{DIBAL-H}\ (-78\ °C)}$

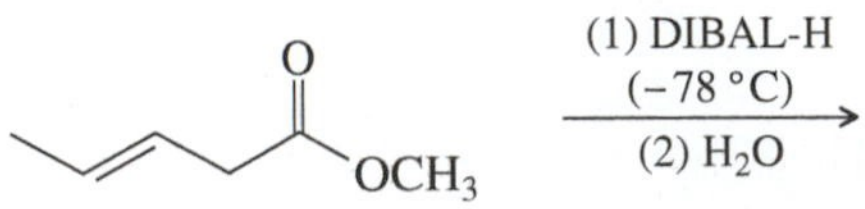

*(f) δ-valerolactone $\xrightarrow[(2)\ H_2O]{(1)\ \text{DIBAL-H}\ (-78\ °C)}$

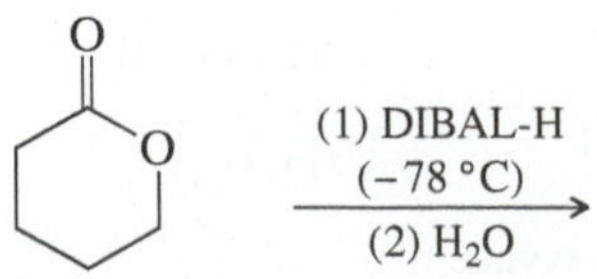

SUMMARY Syntheses of Ketones and Aldehydes

1. *Oxidation of alcohols* (Sections 11-2 and 11-3)

a. *Secondary alcohols* ⟶ *ketones*

$$\underset{\text{secondary alcohol}}{R-\overset{OH}{\overset{|}{C}H}-R'} \xrightarrow[\text{(or } H_2CrO_4)]{\text{NaOCl/HOAc}} \underset{\text{ketone}}{R-\overset{O}{\overset{\|}{C}}-R'}$$

b. *Primary alcohols* ⟶ *aldehydes*

$$\underset{\text{primary alcohol}}{R-CH_2OH} \xrightarrow[\text{(or PCC or Swern or DMP)}]{\text{NaOCl/TEMPO}} \underset{\text{aldehyde}}{R-\overset{O}{\overset{\|}{C}}-H}$$

2. *Ozonolysis of alkenes* (Section 8-15B)

$$\underset{\text{alkene}}{(R)(H)C=C(R')(R'')} \xrightarrow[\text{(2) } (CH_3)_2S]{\text{(1) } O_3, -78\ °C} \underset{\text{aldehyde}}{(R)(H)C=O} + \underset{\text{ketone}}{O=C(R')(R'')}$$

(gives aldehydes or ketones, depending on the starting alkene)

3. *Friedel–Crafts acylation* (Section 17-11)

$$R-\overset{O}{\overset{\|}{C}}-Cl + G-C_6H_5 \xrightarrow{AlCl_3} \underset{\text{aryl ketone}}{G-C_6H_4-\overset{O}{\overset{\|}{C}}-R} \quad (+\ \text{ortho})$$

R = alkyl or aryl;
G = hydrogen, an activating group, or halogen.

The Gatterman–Koch formylation (Section 17-11C)

$$HCl + CO + G-C_6H_5 \xrightarrow{AlCl_3,\ CuCl} \underset{\text{benzaldehyde derivative}}{G-C_6H_4-\overset{O}{\overset{\|}{C}}-H}$$

G = hydrogen, an activating group, or halogen.

4. *Hydration of alkynes* (Section 9-9F)

a. *Catalyzed by acid and mercuric salts (Markovnikov orientation)*

$$\underset{\text{alkyne}}{R-C\equiv C-H} \xrightarrow[H_2O]{Hg^{2+},\ H_2SO_4} \underset{\text{enol (not isolated)}}{\left[(R)(HO)C=C(H)(H)\right]} \longrightarrow \underset{\text{methyl ketone}}{R-\overset{O}{\overset{\|}{C}}-CH_3}$$

b. *Hydroboration–oxidation (anti-Markovnikov orientation)*

$$\underset{\text{alkyne}}{R-C\equiv C-H} \xrightarrow[\text{(2) } H_2O_2,\ NaOH]{\text{(1) } Sia_2BH} \underset{\text{enol (not isolated)}}{\left[(R)(H)C=C(H)(OH)\right]} \longrightarrow \underset{\text{aldehyde}}{R-CH_2-\overset{O}{\overset{\|}{C}}-H}$$

(continued)

5. *Synthesis of ketones using organolithium reagents with carboxylic acids* (Section 18-8)

$$\underset{\text{carboxylic acid}}{R-\overset{\overset{\displaystyle O}{\|}}{C}-OH} \xrightarrow{2\,R'-Li} \underset{\text{dianion}}{R-\overset{\overset{\displaystyle OLi}{|}}{\underset{\underset{\displaystyle R'}{|}}{C}}-OLi} \xrightarrow{H_3O^+} \underset{\text{ketone}}{R-\overset{\overset{\displaystyle O}{\|}}{C}-R'}$$

6. *Synthesis of ketones using organocuprates with acid chlorides* (Section 18-10)

$$\underset{\text{acid chloride}}{R'-\overset{\overset{\displaystyle O}{\|}}{C}-Cl} + \underset{\text{organocuprate}}{R_2CuLi} \longrightarrow \underset{\text{ketone}}{R'-\overset{\overset{\displaystyle O}{\|}}{C}-R}$$

7. *Synthesis of ketones from nitriles* (Section 18-9)

$$\underset{\text{nitrile}}{R-C\equiv N} + \underset{(\text{or } R'-Li)}{R'-Mg-X} \longrightarrow \underset{\text{Mg salt of imine}}{R-\overset{\overset{\displaystyle N-MgX}{\|}}{C}-R'} \xrightarrow{H_3O^+} \underset{\text{ketone}}{R-\overset{\overset{\displaystyle O}{\|}}{C}-R'}$$

8. *Aldehyde synthesis by reduction of nitriles* (Section 18-9)

$$\underset{\text{nitrile}}{R-C\equiv N:} \xrightarrow{(1)\ (i\text{-Bu})_2AlH} \underset{\text{aluminum complex}}{R-C(=\ddot{N}-Al(i\text{-Bu})_2)-H} \xrightarrow{(2)\ H_3O^+} \underset{\text{aldehyde}}{R-C(=O)-H}$$

9. *Aldehyde synthesis by reduction of acid chlorides* (Section 18-10)

$$\underset{\text{acid chloride}}{R-\overset{\overset{\displaystyle O}{\|}}{C}-Cl} \xrightarrow[(\text{or } H_2,\ Pd,\ BaSO_4,\ S)]{Li^+\ {}^-AlH(O\text{-}t\text{-Bu})_3} \underset{\text{aldehyde}}{R-\overset{\overset{\displaystyle O}{\|}}{C}-H}$$

10. *Aldehyde synthesis by reduction of esters* (Section 18-10)

$$\underset{\text{ester}}{R-C(=O)-OR'} \xrightarrow[(-78\ ^\circ C)]{(1)\ (i\text{-Bu})_2AlH} \underset{\text{aluminum complex}}{R-\overset{\overset{\displaystyle O-Al(i\text{-Bu})_2}{|}}{\underset{\underset{\displaystyle OR'}{|}}{C}}-H} \xrightarrow{(2)\ H_2O} \underset{\text{aldehyde}}{R-C(=O)-H}$$

18-11 Reactions of Ketones and Aldehydes: Introduction to Nucleophilic Addition

Ketones and aldehydes undergo many reactions to give a wide variety of useful derivatives. Their most common reaction is **nucleophilic addition,** addition of a nucleophile and a proton across the C=O double bond. The reactivity of the carbonyl group arises from the electronegativity of the oxygen atom and the resulting polarization of the carbon–oxygen double bond. The electrophilic carbonyl carbon atom is sp^2 hybridized and flat, leaving it relatively unhindered and open to attack from either face of the double bond.

As a nucleophile attacks the carbonyl group, the carbon atom changes hybridization from sp^2 to sp^3. The electrons of the pi bond are forced out to the oxygen atom to form an alkoxide anion, which protonates to give the product of nucleophilic addition.

Nuc :$^-$; nucleophilic attack ⇌ alkoxide —(H—Nuc)→ product + Nuc:$^-$

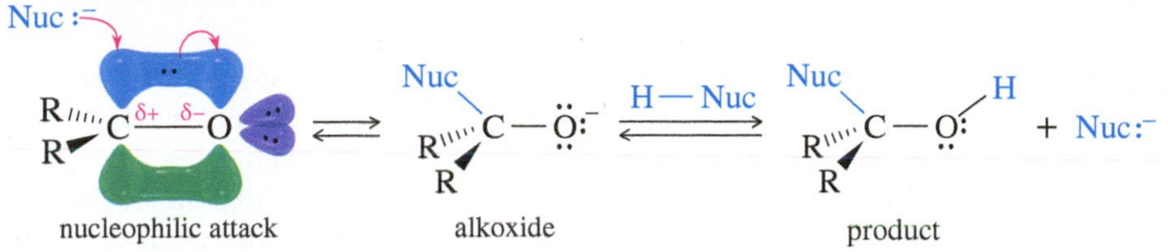

We have seen at least two examples of nucleophilic addition to ketones and aldehydes. A Grignard reagent (a strong nucleophile resembling a carbanion, R:$^-$) attacks the electrophilic carbonyl carbon atom to give an alkoxide intermediate. Subsequent protonation gives an alcohol.

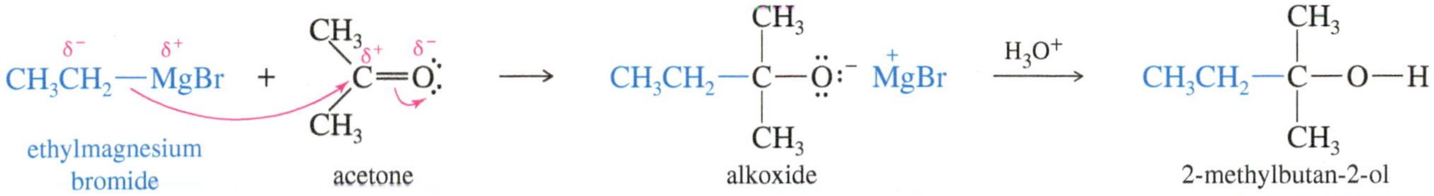

CH_3CH_2—MgBr (ethylmagnesium bromide) + $(CH_3)_2C{=}O$ (acetone) → $CH_3CH_2C(CH_3)_2$—O:$^-$ $^+$MgBr (alkoxide) —(H_3O^+)→ $CH_3CH_2C(CH_3)_2$—O—H (2-methylbutan-2-ol)

Hydride reduction of a ketone or aldehyde is another example of nucleophilic addition, with hydride ion (H:$^-$) serving as the nucleophile. Attack by hydride gives an alkoxide that protonates to form an alcohol.

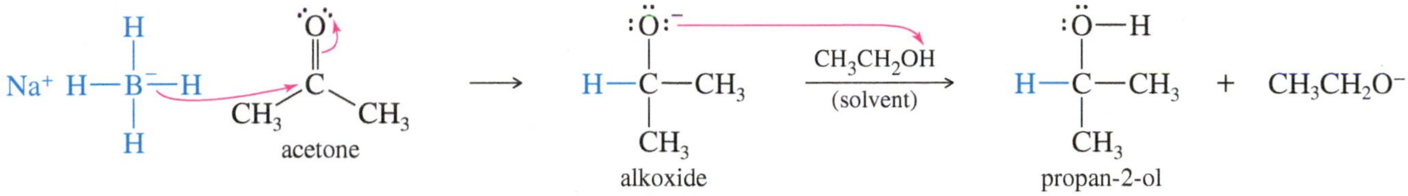

Na^+ H—$\bar{B}H_3$ + CH_3COCH_3 (acetone) → H—$C(CH_3)_2$—O:$^-$ (alkoxide) —(CH_3CH_2OH (solvent))→ H—$C(CH_3)_2$—O—H (propan-2-ol) + $CH_3CH_2O^-$

Weak nucleophiles, such as water and alcohols, can add to activated carbonyl groups under acidic conditions. A carbonyl group is a weak base, and it can become protonated in an acidic solution. A carbonyl group that is protonated (or bonded to some other electrophile) is strongly electrophilic, inviting attack by a weak nucleophile.

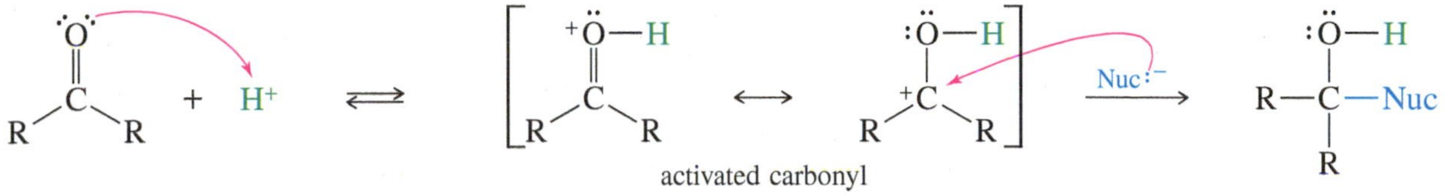

$R_2C{=}O$ + H^+ ⇌ [$R_2C{=}O^+$—H ⟷ R_2C^+—O—H] (activated carbonyl) —(Nuc:$^-$)→ R—C(R)(Nuc)—O—H

The following reaction is the acid-catalyzed nucleophilic addition of water across the carbonyl group of acetone. This hydration of a ketone or aldehyde is discussed in Section 18-12.

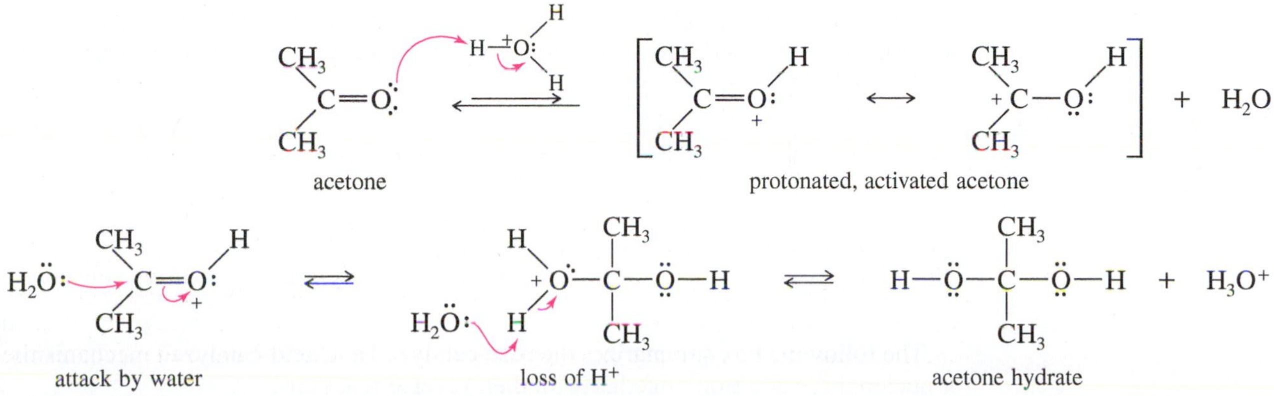

$(CH_3)_2C{=}O$ (acetone) + H_3O^+ ⇌ [$(CH_3)_2C{=}O^+$—H ⟷ $(CH_3)_2C^+$—O—H] (protonated, activated acetone) + H_2O

H_2O: + $(CH_3)_2C{=}O^+$—H (attack by water) ⇌ H_2O^+—$C(CH_3)_2$—O—H + H_2O: (loss of H^+) ⇌ H—O—$C(CH_3)_2$—O—H (acetone hydrate) + H_3O^+

In effect, the base-catalyzed addition to a carbonyl group results from nucleophilic attack of a strong nucleophile followed by protonation. Acid-catalyzed addition begins with protonation, followed by the attack of a weaker nucleophile. Many additions are reversible, with the position of the equilibrium depending on the relative stabilities of the reactants and products.

PROBLEM-SOLVING HINT

Please become familiar with these simple mechanisms. You will see many examples in the next few pages. Also, most of the important multistep mechanisms in this chapter are combinations of these simple steps.

In most cases, aldehydes are more reactive than ketones toward nucleophilic additions. They usually react more quickly than ketones, and the position of the equilibrium usually lies more toward the products than with ketones. The enhanced reactivity of aldehydes is due to an electronic effect and a steric effect. Notice that an aldehyde has only one electron-donating alkyl group, making the aldehyde carbonyl group slightly more electron-poor and electrophilic (the *electronic effect*). Also, an aldehyde has only one bulky alkyl group (compared with two in a ketone), leaving the carbonyl group more exposed toward nucleophilic attack. Especially with a bulky nucleophile, the product of attack on the aldehyde is less hindered than the product from the ketone (the *steric effect*).

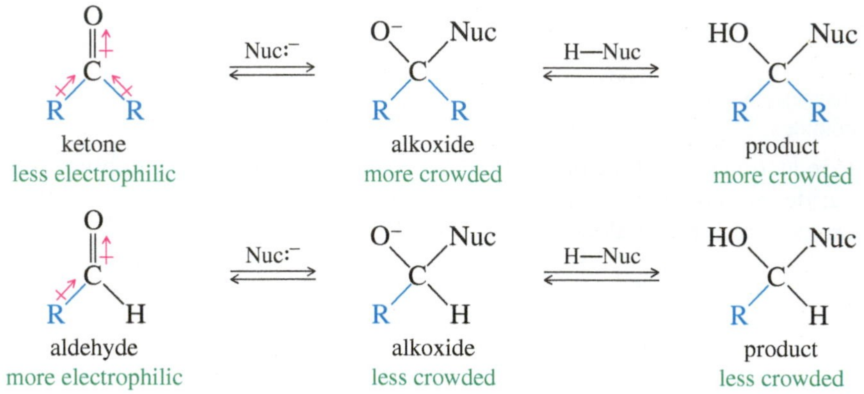

$$R_2C{=}O \xrightleftharpoons{Nuc:^-} R_2C(O^-)(Nuc) \xrightleftharpoons{H{-}Nuc} R_2C(OH)(Nuc)$$

ketone, less electrophilic → alkoxide, more crowded → product, more crowded

$$RHC{=}O \xrightleftharpoons{Nuc:^-} RHC(O^-)(Nuc) \xrightleftharpoons{H{-}Nuc} RHC(OH)(Nuc)$$

aldehyde, more electrophilic → alkoxide, less crowded → product, less crowded

PROBLEM 18-11

Show how you would accomplish the following synthetic conversions. You may use any additional reagents and solvents you need.

(a) $Ph{-}CHO \longrightarrow Ph{-}C(=O){-}Ph$

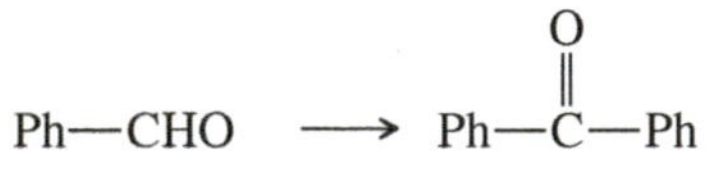

(b) $Ph{-}C(=O){-}Ph \longrightarrow Ph_3C{-}OH$

(c) $Ph{-}C(=O){-}Ph \longrightarrow Ph{-}CH(OH){-}Ph$

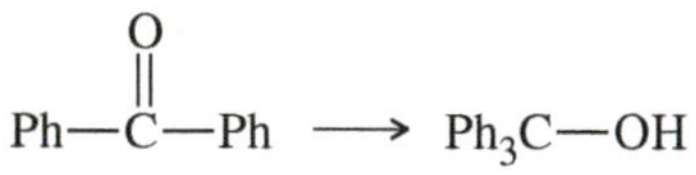

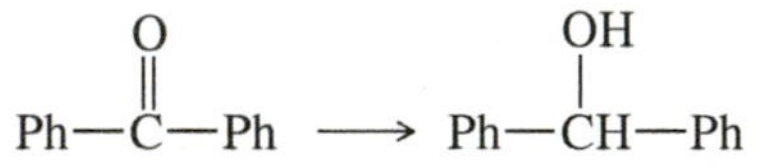

(d) $PhCHO \longrightarrow PhCH(OH)C{\equiv}CCH_2CH_3$

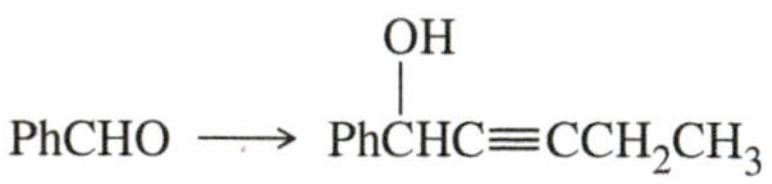

PROBLEM 18-12

Sodium triacetoxyborohydride, $NaBH(OAc)_3$, is a mild reducing agent that reduces aldehydes much more quickly than ketones. It can be used to reduce aldehydes in the presence of ketones, such as in the following reaction:

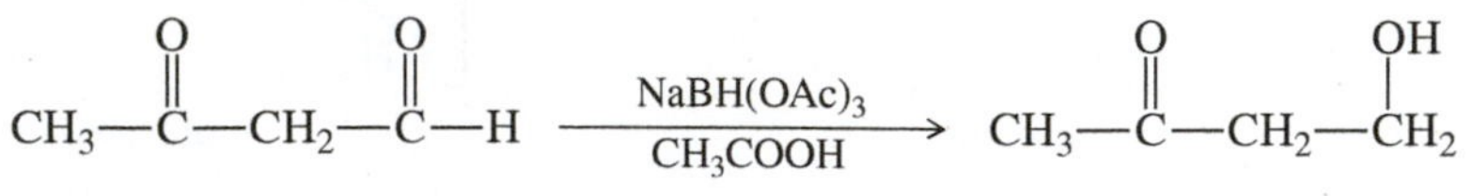

$$CH_3{-}C(=O){-}CH_2{-}C(=O){-}H \xrightarrow[CH_3COOH]{NaBH(OAc)_3} CH_3{-}C(=O){-}CH_2{-}CH_2{-}OH$$

(a) Draw a complete Lewis structure for sodium triacetoxyborohydride.
(b) Propose a mechanism for the reduction of an aldehyde by sodium triacetoxyborohydride.

The following box summarizes the base-catalyzed and acid-catalyzed mechanisms for nucleophilic addition, together with their reverse reactions.

KEY MECHANISM 18-1 Nucleophilic Additions to Carbonyl Groups

Basic Conditions (strong nucleophile)

Step 1: A strong nucleophile adds to the carbonyl group to form an alkoxide.

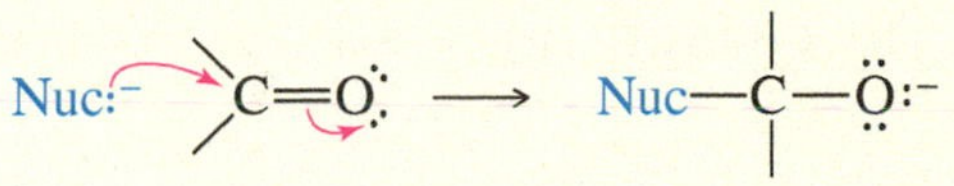

Nuc:⁻ + >C=O: ⟶ Nuc—C—O:⁻

Step 2: A weak acid protonates the alkoxide to give the addition product.

Nuc—C—O:⁻ + H—Nuc ⟶ Nuc—C—O—H   Nuc:⁻

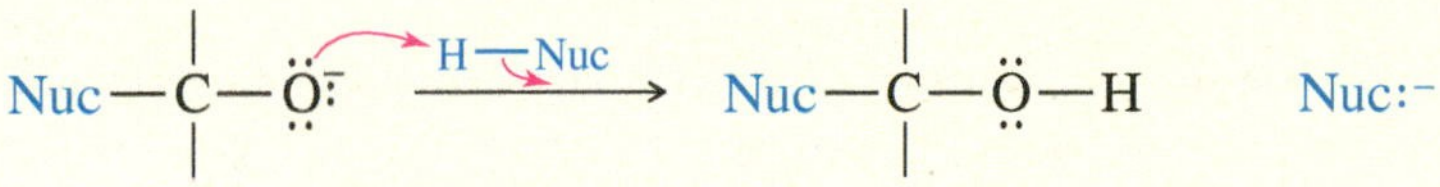

EXAMPLE: Formation of a cyanohydrin (covered in Section 18-13).

Step 1: A strong nucleophile adds to the carbonyl group to form an alkoxide.

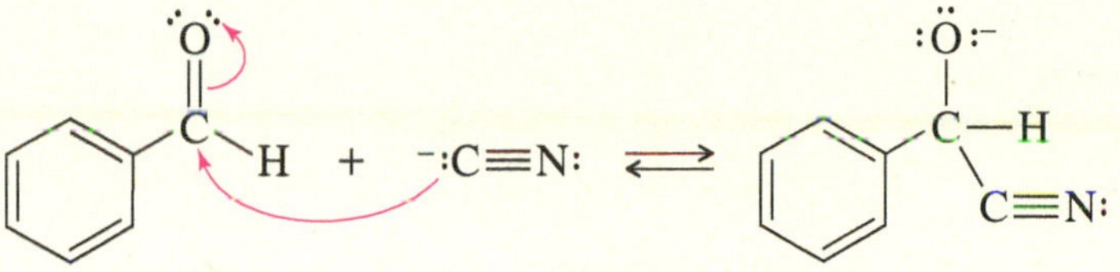

benzaldehyde + ⁻:C≡N: ⇌ Ph—CH(O:⁻)—C≡N:

Step 2: A weak acid protonates the alkoxide to give the addition product.

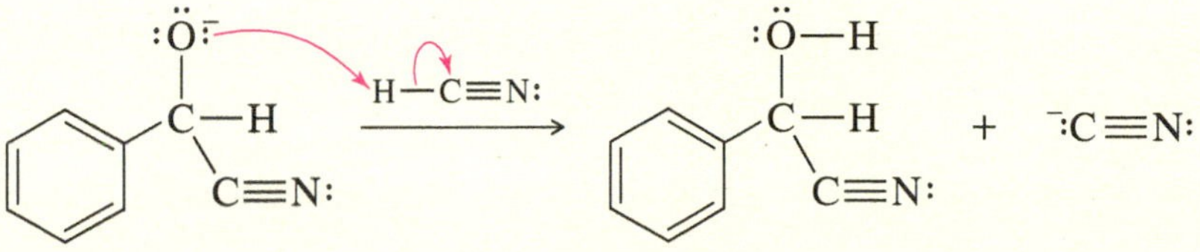

Ph—CH(O:⁻)—C≡N: + H—C≡N: ⟶ Ph—CH(OH)—C≡N: + ⁻:C≡N:

benzaldehyde cyanohydrin

Reverse reaction: Deprotonation, followed by loss of the nucleophile.

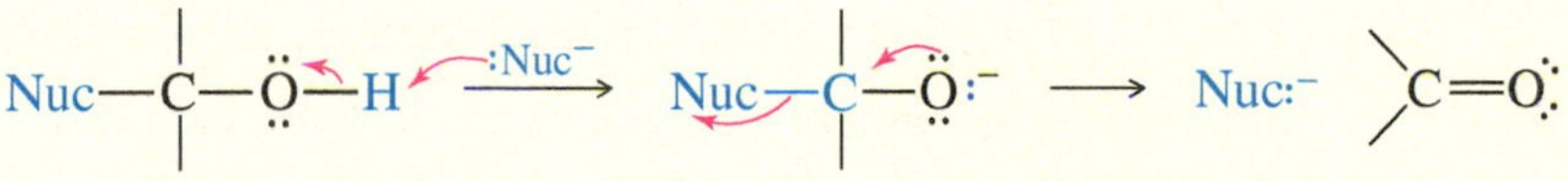

Nuc—C—O—H + :Nuc⁻ ⟶ Nuc—C—O:⁻ ⟶ Nuc:⁻ + >C=O:

PROBLEM: The formation of benzaldehyde cyanohydrin shown in the example above is reversible. Draw a mechanism for the reverse reaction.

Acidic Conditions (weak nucleophile, activated carbonyl)

Step 1: Protonation activates the carbonyl group toward nucleophilic attack.

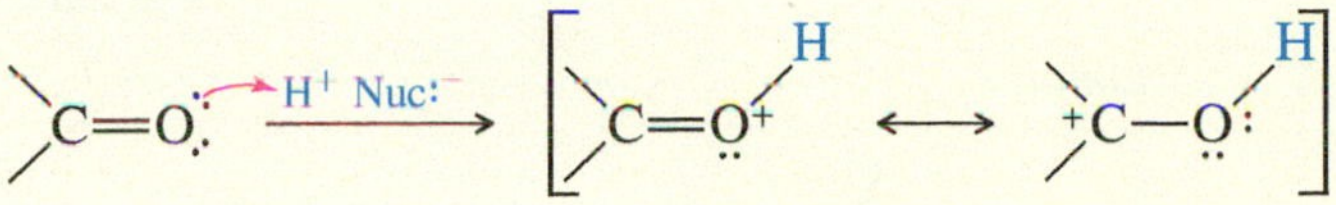

>C=O: + H⁺ Nuc:⁻ ⟶ [>C=O⁺—H ⟷ ⁺C—O:—H]

Step 2: A weak nucleophile adds to the activated (protonated) carbonyl group.

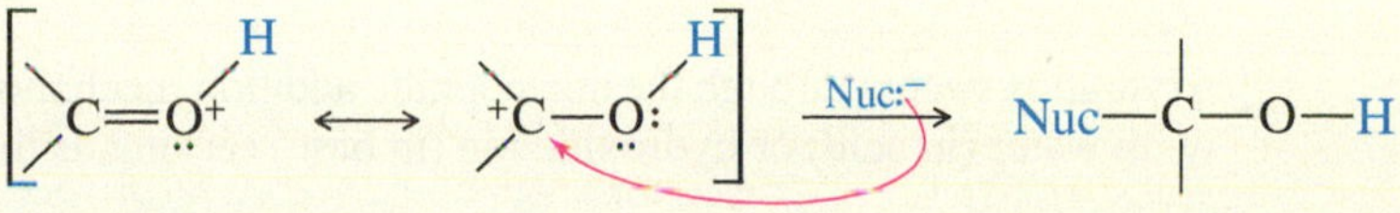

[>C=O⁺—H ⟷ ⁺C—O:—H] + Nuc:⁻ ⟶ Nuc—C—O—H

(CONTINUED)

EXAMPLE: Formation of a hemiacetal (covered in Section 18-16).

Step 1: Protonation activates the carbonyl group toward nucleophilic attack.

benzaldehyde $\xrightarrow{H^+,\ CH_3OH}$ [resonance forms of protonated benzaldehyde]

Step 2: A weak nucleophile adds to the activated (protonated) carbonyl group. Deprotonation of the product gives the hemiacetal.

a hemiacetal + $CH_3\overset{+}{O}H_2$

Reverse reaction: Loss of the weak nucleophile, followed by deprotonation.

Nuc—C—O—H ⇌ Nuc:⁻ [C=O⁺—H ⟷ ⁺C—O—H] ⇌ C=O + Nuc—H

PROBLEM: The hemiacetal formation used in the example is reversible. Draw a mechanism for the reverse reaction.

18-12 Hydration of Ketones and Aldehydes

In an aqueous solution, a ketone or an aldehyde is in equilibrium with its **hydrate,** a geminal diol. With most ketones, the equilibrium favors the unhydrated keto form of the carbonyl.

$$R'(R)C{=}O + H_2O \rightleftharpoons R'(R)C(OH)_2 \qquad K_{eq} = \frac{[\text{hydrate}]}{[\text{ketone}][H_2O]}$$

keto form — hydrate (a geminal diol)

Example

$$CH_3{-}\overset{O}{\overset{\|}{C}}{-}CH_3 + H_2O \rightleftharpoons CH_3{-}C(OH)_2{-}CH_3 \qquad K_{eq} = 0.002$$

acetone — acetone hydrate

Hydration occurs through the nucleophilic addition mechanism shown in Mechanism 18-2, with water (in acid) or hydroxide ion (in base) serving as the nucleophile.

MECHANISM 18-2 Hydration of Ketones and Aldehydes

In acid

The acid-catalyzed hydration is a typical acid-catalyzed addition to the carbonyl group. Protonation, followed by addition of water, gives a protonated product. Deprotonation gives the hydrate.

Step 1: Protonation. ***Step 2:*** Water adds. ***Step 3:*** Deprotonation.

$$R_2C{=}O \xrightleftharpoons{H_3O^+} [R_2C{=}OH^+] \xrightleftharpoons{H_2O} R_2C(OH)(OH_2^+) \xrightleftharpoons{H_2O} R_2C(OH)_2 + H_3O^+$$

In base

The base-catalyzed hydration is a perfect example of base-catalyzed addition to a carbonyl group. The strong nucleophile adds, and then protonation gives the hydrate.

Step 1: Hydroxide adds. ***Step 2:*** Protonation.

$$HO^- + R_2C{=}O \rightleftharpoons HO{-}CR_2{-}O^- \xrightleftharpoons{H{-}O{-}H} HO{-}CR_2{-}OH + {}^-OH$$

Aldehydes are more likely than ketones to form stable hydrates. The electrophilic carbonyl group of a ketone is stabilized by its two electron-donating alkyl groups, but an aldehyde carbonyl has only one stabilizing alkyl group. The partial positive charge of the aldehyde is not as well stabilized. Aldehydes are thus more electrophilic and less stable than ketones. Formaldehyde, with no electron-donating groups, is even less stable than other aldehydes.

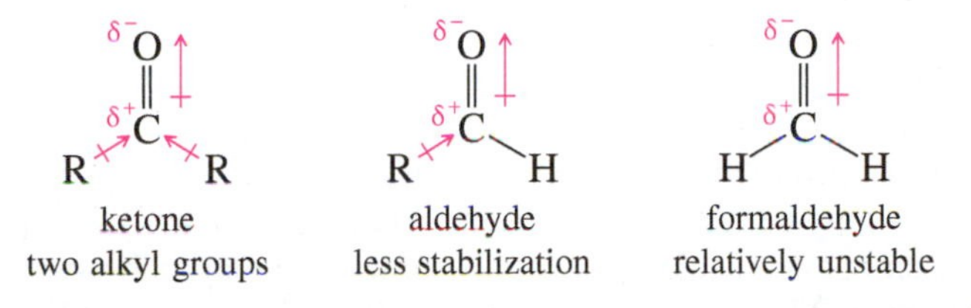

These stability effects are apparent in the equilibrium constants for hydration of ketones and aldehydes. Ketones have values of K_{eq} of about 10^{-4} to 10^{-2}. For most aldehydes, the equilibrium constant for hydration is close to 1. Formaldehyde, with no alkyl groups bonded to the carbonyl carbon, has a hydration equilibrium constant of about 40. Strongly electron-withdrawing substituents on the alkyl group of a ketone or aldehyde also destabilize the carbonyl group and favor the hydrate. Chloral (trichloroacetaldehyde) has an electron-withdrawing trichloromethyl group that favors the hydrate. Chloral forms a stable, crystalline hydrate that became famous in the movies as knockout drops or a Mickey Finn.

PROBLEM-SOLVING HINT

In basic conditions, a strong nucleophile usually adds directly to the carbonyl group. In acidic conditions, strong nucleophiles are rarely present. An acid (or Lewis acid) usually protonates the carbonyl to activate it toward attack by a weak nucleophile.

Application: Knockout Drops

The body rapidly reduces chloral (trichloroacetaldehye) to trichloroethanol, which is responsible for the drug's sleep-inducing effect.

$$CH_3{-}CH_2{-}C({=}O){-}H + H_2O \rightleftharpoons CH_3{-}CH_2{-}C(OH)_2{-}H \quad K_{eq} = 0.7$$

propanal → propanal hydrate

$$H_2C{=}O + H_2O \rightleftharpoons H_2C(OH)_2 \quad K_{eq} = 40$$

formaldehyde → formalin

$$Cl_3C{-}C({=}O){-}H + H_2O \rightleftharpoons Cl_3C{-}C(OH)_2{-}H \quad K_{eq} = 500$$

chloral → chloral hydrate

PROBLEM-SOLVING HINT

Don't be surprised to see some O—H stretch, from the hydrate, in the IR spectra of many aldehydes.

PROBLEM 18-13

Propose mechanisms for
(a) the acid-catalyzed hydration of chloral to form chloral hydrate.
(b) the base-catalyzed hydration of acetone to form acetone hydrate.

PROBLEM 18-14

Rank the following compounds in order of increasing amount of hydrate present at equilibrium.

[Structures: cyclopentyl-CH₂-CHO; cyclohexanone; cyclopentyl-CH(Br)-CHO; 2-bromocyclohexanone]

Application: Cyanohydrin defense

The almond-scented millipede (*Apheloria montana*) secretes a mixture of HCN and benzaldehyde to prevent other animals from eating it. The millipede stores mandelonitrile (benzaldehyde cyanohydrin) in a reservoir. When attacked, it discharges mandelonitrile through a reaction chamber containing enzymes that catalyze the conversion of the cyanohydrin to benzaldehyde and HCN.

18-13 Formation of Cyanohydrins

Hydrogen cyanide (H—C≡N) is a toxic, water-soluble liquid that boils at 26 °C. Because it is mildly acidic, HCN is sometimes called hydrocyanic acid.

$$H{-}C{\equiv}N: + H_2O \rightleftharpoons H_3O^+ + {}^-:C{\equiv}N: \quad pK_a = 9.2$$

The conjugate base of hydrogen cyanide is the cyanide ion ($^-$:C≡N:). Cyanide ion is a strong base and a strong nucleophile. It attacks ketones and aldehydes to give addition products called **cyanohydrins.** The mechanism is a base-catalyzed nucleophilic addition, as shown in Mechanism 18-3. Cyanide ion attacks the carbonyl group, forming an alkoxide ion that protonates to give the cyanohydrin.

Cyanohydrins may be formed using liquid HCN with a catalytic amount of sodium cyanide or potassium cyanide. HCN is highly toxic and volatile, however, and therefore dangerous to handle. Many procedures use a full equivalent of sodium or potassium cyanide (rather than HCN), dissolved in some other proton-donating solvent.

Cyanohydrin formation is reversible, and the equilibrium constant may or may not favor the cyanohydrin. These equilibrium constants follow the general reactivity trend of ketones and aldehydes:

$$\text{formaldehyde} > \text{other aldehydes} > \text{ketones}$$

Formaldehyde reacts quickly and quantitatively with HCN. Most other aldehydes have equilibrium constants that favor cyanohydrin formation. Reactions of HCN with ketones have equilibrium constants that may favor either the ketones or

MECHANISM 18-3 Formation of Cyanohydrins

Cyanohydrin formation is a perfect example of base-catalyzed addition to a carbonyl group. The strong nucleophile adds in the first step to give an alkoxide. Protonation gives the cyanohydrin.

Step 1: Cyanide adds to the carbonyl.

Step 2: Protonation gives the cyanohydrin.

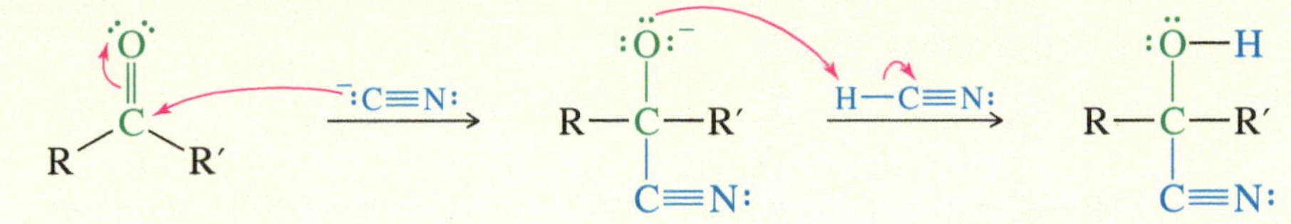

ketone or aldehyde

intermediate

cyanohydrin

EXAMPLE: Formation of benzaldehyde cyanohydrin.

Step 1: Cyanide adds to the carbonyl.

Step 2: Protonation gives the cyanohydrin.

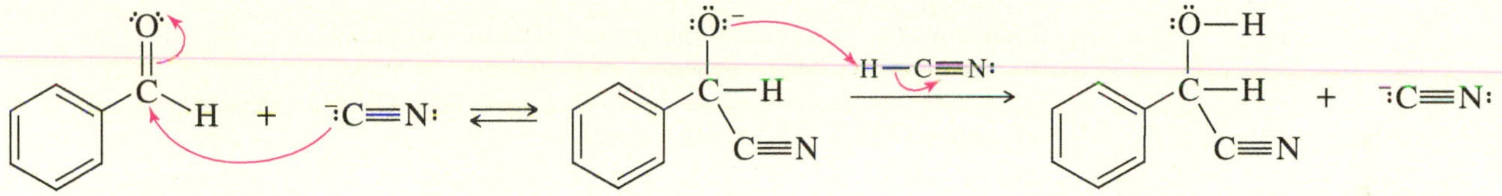

benzaldehyde

benzaldehyde cyanohydrin (mandelonitrile)

the cyanohydrins, depending on the structure. Ketones that are hindered by large alkyl groups react slowly with HCN and give poor yields of cyanohydrins.

CH_3CH_2CHO (propanal) + HCN $\xrightarrow{KCN}$ $CH_3CH_2CH(OH)CN$ (propanal cyanohydrin, 100%)

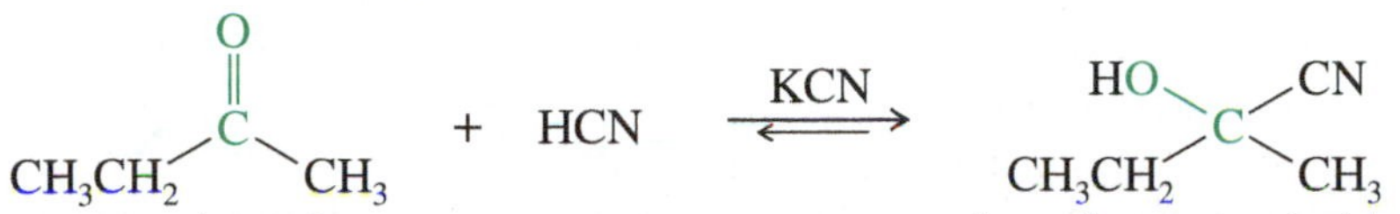

$CH_3CH_2COCH_3$ (butan-2-one) + HCN $\xrightleftharpoons{KCN}$ $CH_3CH_2C(OH)(CN)CH_3$ (butan-2-one cyanohydrin, 95%)

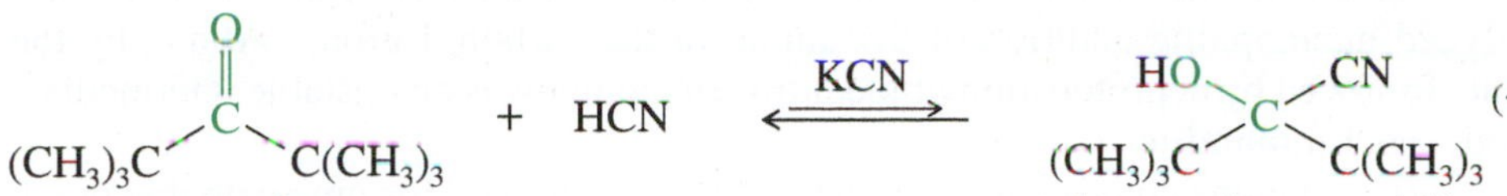

$(CH_3)_3C{-}CO{-}C(CH_3)_3$ (di-*tert*-butylketone) + HCN $\xrightleftharpoons{KCN}$ $(CH_3)_3C{-}C(OH)(CN){-}C(CH_3)_3$ (< 5%) slow reaction, poor yields

The failure with bulky ketones is largely due to steric effects. Cyanohydrin formation involves rehybridizing the sp^2 carbonyl carbon to sp^3, narrowing the angle between the alkyl groups from about 120° to about 109.5°, increasing their steric interference.

PROBLEM 18-15

Propose a mechanism for each cyanohydrin synthesis just shown.

Organic compounds containing the cyano group ($-C\equiv N$) are called **nitriles.** A cyanohydrin is therefore an α-hydroxynitrile. Nitriles hydrolyze to carboxylic acids under acidic conditions (discussed in Section 21-7D), so cyanohydrins hydrolyze to α-hydroxy acids. This is the most convenient method for making many α-hydroxy acids.

$$\underset{\text{aldehyde}}{R-\overset{\overset{\displaystyle O}{\|}}{C}-H} + HCN \xrightleftharpoons{^{-}:CN} \underset{\text{cyanohydrin}}{R-\overset{\overset{\displaystyle OH}{|}}{\underset{\underset{\displaystyle H}{|}}{C}}-CN} \xrightarrow{H_3O^+} \underset{\alpha\text{-hydroxy acid}}{R-\overset{\overset{\displaystyle OH}{|}}{\underset{\underset{\displaystyle H}{|}}{C}}-\overset{\overset{\displaystyle O}{\|}}{C}-OH}$$

PROBLEM 18-16

Show how you would accomplish the following syntheses.

(a) acetophenone → acetophenone cyanohydrin
(b) cyclopentanecarbaldehyde → 2-cyclopentyl-2-hydroxyacetic acid
(c) hexan-1-ol → 2-hydroxyheptanoic acid

18-14 Formation of Imines

Under the proper conditions, either ammonia or a primary amine reacts with a ketone or an aldehyde to form an **imine.** Imines are nitrogen analogs of ketones and aldehydes, with a carbon–nitrogen double bond in place of the carbonyl group. Imines are commonly involved as synthetic intermediates, both in biosynthesis and in industrial synthesis. One of the best methods for making amines (both in organisms and in the lab) involves making an imine, then reducing it to the amine (Section 19-18).

Like amines, imines are basic; a substituted imine is also called a **Schiff base.** Imine formation is an example of a large class of reactions called **condensations,** reactions that join two or more molecules, often with the loss of a small molecule such as water or an alcohol.

$$\underset{\text{ketone or aldehyde}}{\overset{\overset{\displaystyle O}{\|}}{C}} + \underset{\text{primary amine}}{R-\ddot{N}H_2} \xrightleftharpoons{H^+} \underset{\text{carbinolamine (hemiaminal)}}{\left[-\overset{\overset{\displaystyle OH}{|}}{\underset{\underset{\displaystyle R-\ddot{N}-H}{|}}{C}}-\right]} \rightleftharpoons \underset{\text{imine (Schiff base)}}{\underset{\underset{\displaystyle R-\ddot{N}}{\|}}{C}} + H_2O$$

The mechanism of imine formation (Key Mechanism 18-4) begins with an acid-catalyzed nucleophilic addition of the amine to the carbonyl group. Attack by the amine, followed by deprotonation of the nitrogen atom, gives an unstable intermediate called a **carbinolamine.**

Other names for a carbinolamine are hemiaminal (the IUPAC term) and α-aminoalcohol.

A carbinolamine converts to an imine by losing water and forming a double bond: a dehydration. This dehydration follows the same mechanism as the acid-catalyzed dehydration of an alcohol (Section 11-10). Protonation of the hydroxy group converts it to a good leaving group, and it leaves as water. The resulting cation is stabilized by resonance forms, including one with all octets filled and the positive charge on nitrogen. Loss of a proton gives the imine.

KEY MECHANISM 18-4 Formation of Imines

This mechanism is more easily remembered by dividing it into two parts:[1]

1. acid-catalyzed addition of the amine to the carbonyl group.
2. acid-catalyzed dehydration.

First part: Acid-catalyzed addition of the amine to the carbonyl group.

Step 1: Protonation of the carbonyl. ***Step 2:*** Addition of the amine. ***Step 3:*** Deprotonation.

carbinolamine (hemiaminal)

Second part: Acid-catalyzed dehydration.

Step 4: Protonation of the —OH group. ***Step 5:*** Loss of H_2O. ***Step 6:*** Deprotonation.

carbinolamine protonated minor major intermediate (all octets filled) imine (Schiff base)

EXAMPLE: Formation of benzaldehyde methyl imine.

First part: Acid-catalyzed addition of the amine to the carbonyl group.

Step 1: Protonation of the carbonyl. ***Step 2:*** Addition of the amine. ***Step 3:*** Deprotonation to the carbinolamine.

benzaldehyde methylamine a carbinolamine (a hemiaminal)

[1]This mechanism takes place at slightly acidic pH. The amine can act as a strong nucleophile, so the first half of this mechanism (addition to the carbonyl) may be drawn as either acid-catalyzed or as base-catalyzed. The second half (dehydration) is acid-catalyzed, so the entire mechanism is shown here as acid-catalyzed to be consistent.

(CONTINUED)

Second part: Acid-catalyzed dehydration.

Step 4: Protonation of the —OH group. ***Step 5:*** Loss of H_2O. ***Step 6:*** Deprotonation.

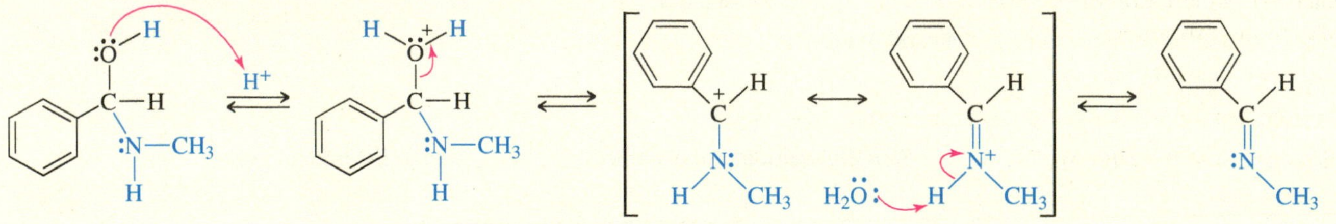

a carbinolamine ... imine

PROBLEM:

(a) What would happen if the reaction were made too acidic by the addition of too much acid?
(b) What would happen if it were too basic?

The proper pH is crucial to imine formation. The second half of the mechanism is acid-catalyzed, so the solution must be somewhat acidic. If the solution is too acidic, however, the amine becomes protonated and non-nucleophilic, inhibiting the first step. Figure 18-7 shows that the rate of imine formation is fastest around pH 4.5.

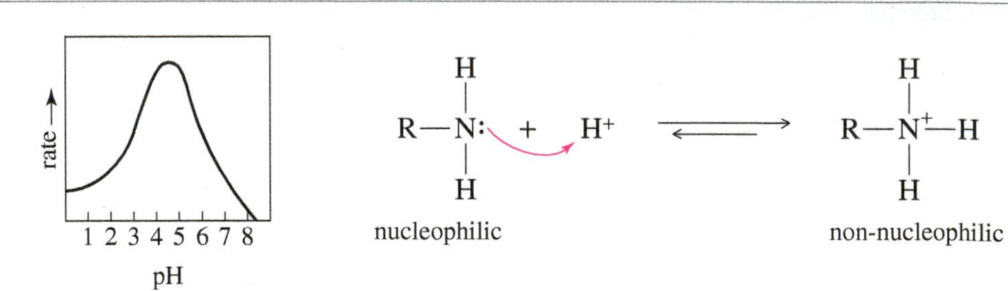

FIGURE 18-7
Although dehydration of the carbinolamine is acid-catalyzed, excess acid stops the first step by protonating the amine. Formation of the imine is fastest around pH 4.5.

The following equations show some typical imine-forming reactions. In each case, notice that the C=O group of the ketone or aldehyde is replaced by the C=N—R group of the imine.

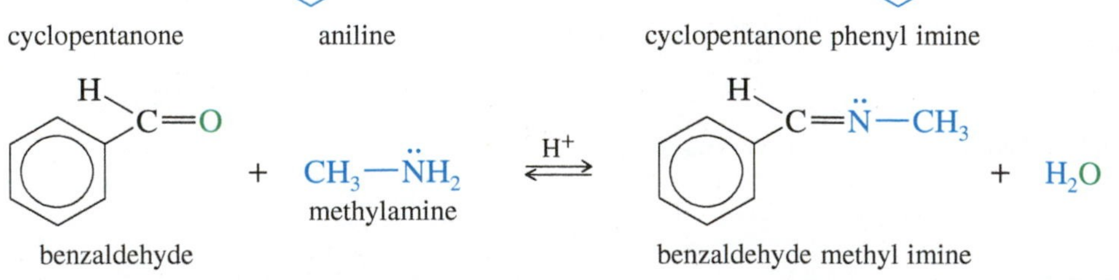

cyclohexanone + NH_3 (ammonia) $\xrightleftharpoons{H^+}$ cyclohexanone imine + H_2O

cyclopentanone + aniline $\xrightleftharpoons{H^+}$ cyclopentanone phenyl imine + H_2O

benzaldehyde + $CH_3—NH_2$ (methylamine) $\xrightleftharpoons{H^+}$ benzaldehyde methyl imine + H_2O

PROBLEM-SOLVING HINT

Imine formation is one of the important mechanisms in this chapter. It is more easily remembered as consisting of two simple mechanisms:

1. acid-catalyzed nucleophilic addition to the carbonyl, and
2. acid-catalyzed dehydration (as with an alcohol).

PROBLEM 18-17

Propose mechanisms for the three imine-forming reactions just shown.

PROBLEM 18-18

Depending on the reaction conditions, two different imines of formula C_8H_9N might be formed by the reaction of benzaldehyde with methylamine. Explain, and give the structures of the two imines.

PROBLEM 18-19

Give the structures of the carbonyl compound and the amine used to form the following imines.

(a) 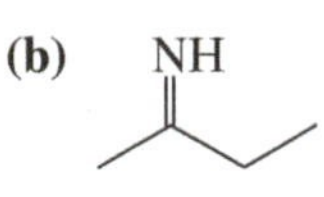 (b) (c) N=CHCH$_3$

(d) (e) (f)

Application: Toxicology

Large quantities of acetaminophen can be toxic because the body transforms it to the imine of benzoquinone. This highly reactive metabolite causes extensive liver damage and can be deadly.

acetaminophen $\xrightarrow{[O]}$ benzoquinoneimine

Imine formation is reversible, and most imines can be hydrolyzed back to the amine and the ketone or aldehyde. The Principle of Microscopic Reversibility (Section 8-4A) states that the reverse reaction taking place under the same conditions should follow the same pathway but in reverse order. Therefore, the mechanism for hydrolysis of an imine is simply the reverse of the mechanism for its formation.

benzaldehyde methyl imine $\xrightleftharpoons{H^+,\ \text{excess } H_2O}$ benzaldehyde + $CH_3-\overset{+}{N}H_3$

PROBLEM 18-20

Propose a mechanism for the hydrolysis of benzaldehyde methyl imine just shown.

18-15 Condensations with Hydroxylamine and Hydrazines

Ketones and aldehydes also condense with other ammonia derivatives, such as hydroxylamine and substituted hydrazines, to give imine derivatives. The equilibrium constants for these reactions are usually more favorable than for reactions with simple amines. Hydroxylamine reacts with ketones and aldehydes to form **oximes;** hydrazine and its derivatives react to form **hydrazones**; and semicarbazide reacts to form **semicarbazones.** The mechanisms of these reactions are similar to the mechanism of imine formation.

phenyl-2-propanone + hydroxylamine $\xrightarrow{H^+}$ phenyl-2-propanone oxime + H_2O

benzaldehyde + hydrazine $\xrightarrow{H^+}$ benzaldehyde hydrazone + H_2O

Application: Rocket Fuel

Anhydrous hydrazine is a common rocket fuel. In most cases, liquid oxygen serves as the oxidizer. The fuel and oxidizer are sprayed into the combustion chamber, where they react to generate heat and pressure, forcing the reaction products out through the rocket nozzle.

$$\text{cyclohexanone} + H_2\ddot{N}\text{—}\boxed{NH\text{—}Ph}\ (\text{phenylhydrazine}) \xrightarrow{H^+} \text{cyclohexanone phenylhydrazone } (C{=}\ddot{N}\text{—}\boxed{NH\text{—}Ph}) + H_2O$$

$$CH_3C(=O)CH_2CH_3\ (\text{butan-2-one}) + H_2\ddot{N}\text{—}\boxed{NH\text{—}C(=O)\text{—}NH_2}\ (\text{semicarbazide}) \xrightarrow{H^+} CH_3C(=\ddot{N}\text{—}\boxed{NH\text{—}C(=O)\text{—}NH_2})CH_2CH_3\ (\text{butan-2-one semicarbazone}) + H_2O$$

PROBLEM-SOLVING HINT

Please learn these common derivatives. You will see many examples, especially in the laboratory.

These derivatives are useful both as starting materials for further reactions (see Section 19-18) and for characterization and identification of the original carbonyl compounds. Oximes, semicarbazones, and phenylhydrazones are often solid compounds with characteristic melting points. Standard tables give the melting points of these derivatives for thousands of different ketones and aldehydes.

If an unknown compound forms one of these derivatives, the melting point can be compared with that in the table. If the compound's physical properties match those of a known compound and the melting point of its oxime, semicarbazone, or phenylhydrazone matches as well, we can be fairly certain of a correct identification.

SUMMARY Condensations of Amines with Ketones and Aldehydes

$$>C{=}O + H_2\ddot{N}\text{—}\boxed{Z} \underset{}{\overset{H^+}{\rightleftarrows}} >C{=}\ddot{N}\text{—}\boxed{Z} + H_2O$$

Z in Z—NH$_2$	*Reagent*	*Product*
—H	$H_2\ddot{N}$—H ammonia	>C=N̈—H an imine
—R	$H_2\ddot{N}$—R primary amine	>C=N̈—R an imine (Schiff base)
—OH	$H_2\ddot{N}$—OH hydroxylamine	>C=N̈—OH an oxime
—NH_2	$H_2\ddot{N}$—NH_2 hydrazine	>C=N̈—NH_2 a hydrazone
—NHPh	$H_2\ddot{N}$—NHPh phenylhydrazine	>C=N̈—NHPh a phenylhydrazone
—NHC(=O)NH_2	$H_2\ddot{N}$—NH—C(=O)—NH_2 semicarbazide	>C=N̈—NH—C(=O)—NH_2 a semicarbazone

PROBLEM 18-21

2,4-Dinitrophenylhydrazine is frequently used for making derivatives of ketones and aldehydes because the products (2,4-dinitrophenylhydrazones, called **2,4-DNP derivatives**) are even more likely than the phenylhydrazones to be solids with sharp melting points. Propose a mechanism for the reaction of acetone with 2,4-dinitrophenylhydrazine in a mildly acidic solution.

PROBLEM 18-22

Predict the products of the following reactions.

(a) cyclopentanone + $HONH_2$ $\xrightarrow{H^+}$

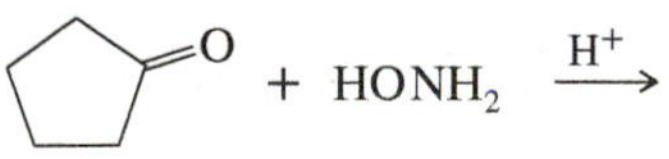

(b) α-tetralone + H_2NNH_2 $\xrightarrow{H^+}$

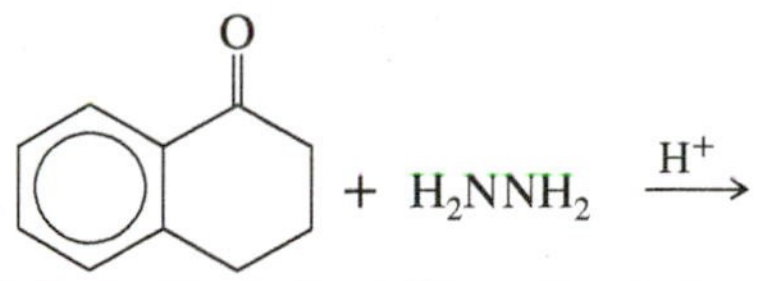

(c) Ph—C(=O)—Ph + $PhNHNH_2$ $\xrightarrow{H^+}$

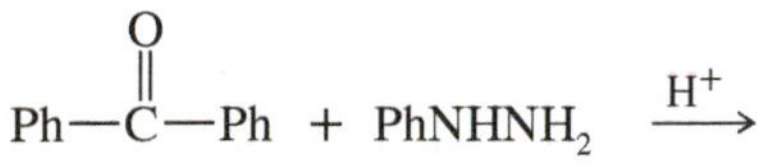

(d) PhCH=CHCHO + $H_2NC(=O)NHNH_2$ $\xrightarrow{H^+}$

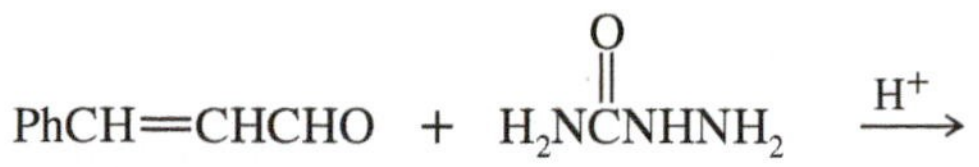

PROBLEM 18-23

Show what amines and carbonyl compounds combine to give the following derivatives.

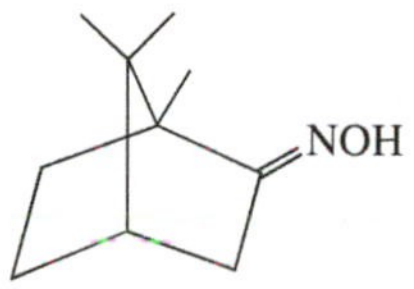

(a) Ph—CH=N—NH—C(=O)—NH$_2$

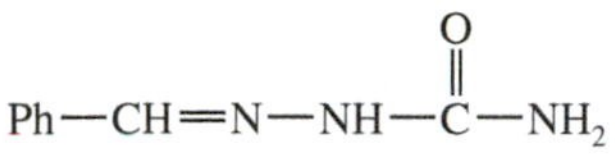

(b) NOH

(c) N—NHPh

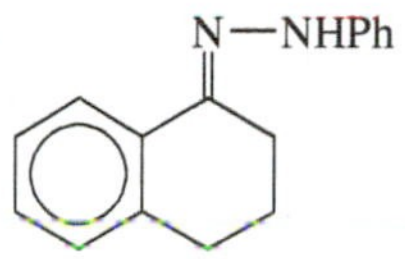

(d) NNH, O_2N, NO_2

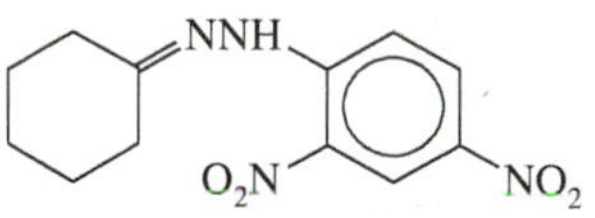

(e) N, CH_3

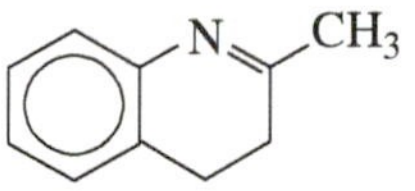

(f) N, N

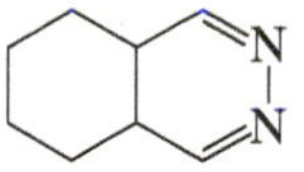

18-16 Formation of Acetals

Under acid catalysis, ketones and aldehydes react reversibly with alcohols to form acetals. An **acetal** is a derivative of a ketone or an aldehyde with two alkoxy groups on the same carbon atom, in place of the carbonyl group. An acetal derived from a ketone is often called a **ketal**.[1] Acetals are some of the most common organic compounds in the world. Table sugar, cotton fabric, and a wooden ship are all composed of acetals. We cover these common carbohydrate acetals and their polymers in Chapter 23.

In the formation of an acetal, two molecules of alcohol add to the carbonyl group, and one molecule of water is eliminated.

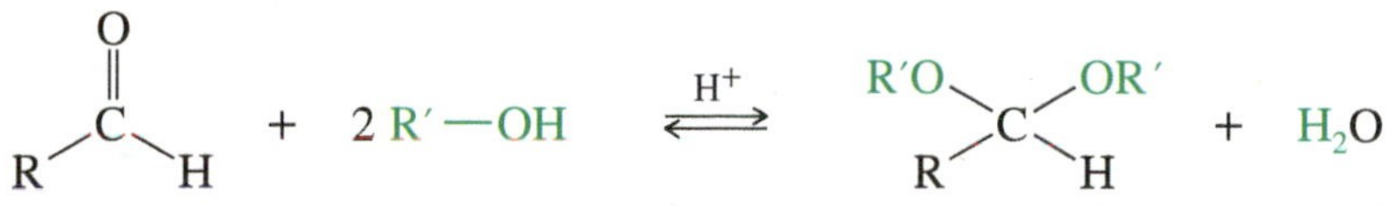

$$\underset{\text{aldehyde}}{R-C(=O)-H} + 2\,R'-OH \overset{H^+}{\rightleftharpoons} \underset{\text{acetal}}{R-C(OR')(OR')-H} + H_2O$$

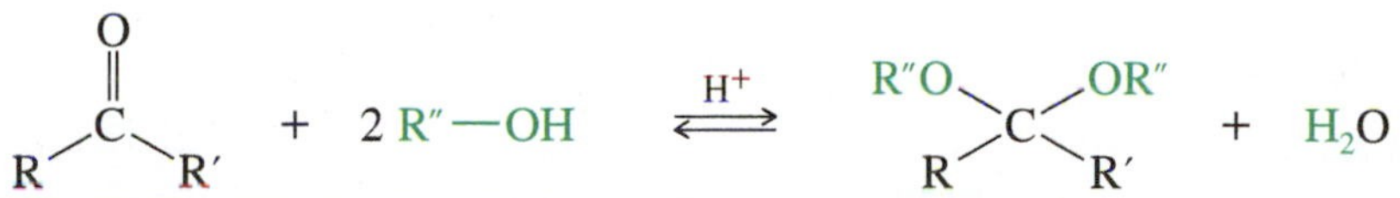

$$\underset{\text{ketone}}{R-C(=O)-R'} + 2\,R''-OH \overset{H^+}{\rightleftharpoons} \underset{\text{acetal (general), ketal (specific)}}{R-C(OR'')(OR'')-R'} + H_2O$$

Although hydration is catalyzed by either acid or base, acetal formation must be acid-catalyzed. For example, consider the reaction of cyclohexanone with methanol, catalyzed by *p*-toluenesulfonic acid.

[1]The IUPAC once dropped the term *ketal*, but has now reinstated it as a subclass of *acetals*. Therefore, an acetal of a ketone can be called either a *ketal* (specific) or an *acetal* (more general).

Overall reaction

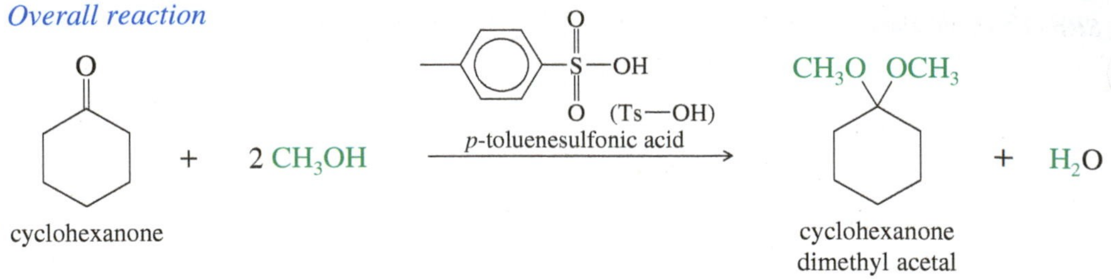

cyclohexanone + 2 CH_3OH $\xrightarrow{p\text{-toluenesulfonic acid (Ts—OH)}}$ cyclohexanone dimethyl acetal + H_2O

The mechanism for this reaction is shown in Key Mechanism 18-5. The first step is a typical acid-catalyzed addition to the carbonyl group. The acid catalyst protonates the carbonyl group, and the alcohol (a weak nucleophile) attacks the protonated, activated carbonyl. Loss of a proton from the positively charged intermediate gives a **hemiacetal.** The hemiacetal gets its name from the Greek prefix *hemi-*, meaning "half." Having added one molecule of the alcohol, the hemiacetal is halfway to becoming a "full" acetal. Like the hydrates of ketones and aldehydes, most hemiacetals are too unstable to be isolated and purified.

The second half of the mechanism converts the hemiacetal to the more stable acetal. Protonation of the hydroxy group, followed by loss of water, gives a resonance-stabilized carbocation. Attack on the carbocation by methanol, followed by loss of a proton, gives the acetal.

KEY MECHANISM 18-5 Formation of Acetals

Like imine formation, acetal formation is easily remembered by dividing it into two simple processes:

1. The first half is an acid-catalyzed addition of the alcohol to the carbonyl group.
2. The second half is an S_N1 substitution of the protonated hemiacetal.

First half: Acid-catalyzed addition of the alcohol to the carbonyl group.

Step 1: Protonation. ***Step 2:*** Alcohol adds. ***Step 3:*** Deprotonation.

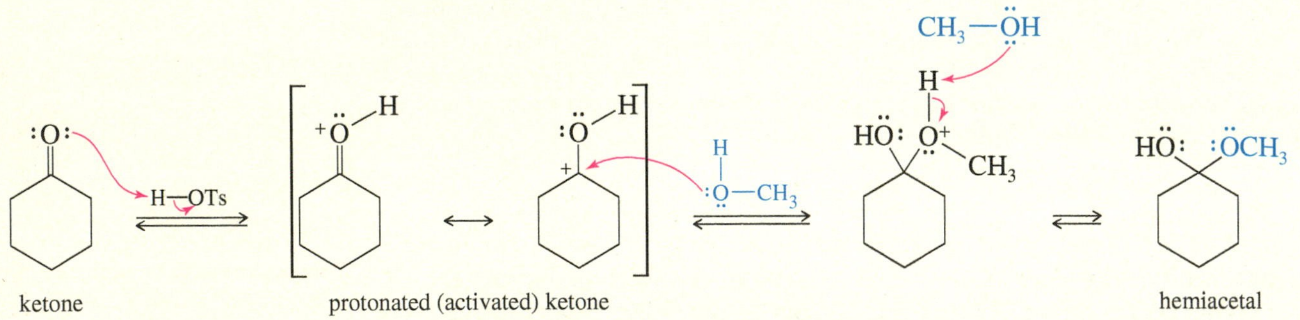

H—OTs; CH_3—OH; H—O—CH_3

ketone; protonated (activated) ketone; hemiacetal

Second half: S_N1 substitution of the protonated hemiacetal.

Step 4: Protonation. ***Step 5:*** Loss of water.

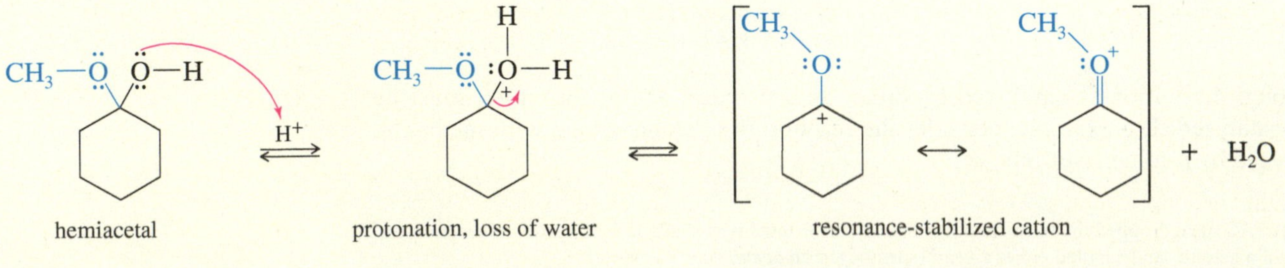

H^+; + H_2O

hemiacetal; protonation, loss of water; resonance-stabilized cation

Step 6: Second alcohol adds. ***Step 7:*** Deprotonation.

attack by methanol

acetal

PROBLEM 18-24

Propose a mechanism for the acid-catalyzed reaction of benzaldehyde with methanol to give benzaldehyde dimethyl acetal.

Because hydration is catalyzed by either acid or base, you might wonder why acetal formation is catalyzed only by acid. In fact, the first step (formation of the hemiacetal) can be base-catalyzed, involving attack by alkoxide ion followed by protonation to form the hemiacetal. The second step requires replacement of the hemiacetal —OH group by the alcohol —OR″ group. Hydroxide ion is a poor leaving group for the S_N2 reaction, so alkoxide cannot displace the —OH group. This replacement occurs under acidic conditions, however, because protonation of the —OH group and loss of water gives a resonance-stabilized cation.

Attempted base-catalyzed acetal formation

poor leaving group

attack on ketone (or aldehyde)

hemiacetal

(no S_N2 displacement)

Equilibrium of Acetal Formation Acetal formation is reversible, so the equilibrium constant controls the proportions of reactants and products that will result. For simple aldehydes, the equilibrium constants generally favor the acetal products. For example, the acid-catalyzed reaction of acetaldehyde with ethanol gives a good yield of the acetal.

With hindered aldehydes and with most ketones, the equilibrium constants favor the carbonyl compounds rather than the acetals. To enhance these reactions, the alcohol is often used as the solvent to ensure a large excess. The water formed as a by-product is removed by distillation to force the equilibrium toward the right.

Conversely, most acetals are hydrolyzed simply by shaking them with dilute acid in water. The large excess of water drives the equilibrium toward the ketone or aldehyde. The mechanism is simply the reverse of acetal formation. For example, cyclohexanone dimethyl acetal is quantitatively hydrolyzed to cyclohexanone by brief treatment with dilute aqueous acid.

PROBLEM-SOLVING HINT

Acetal formation is one of the important mechanisms in this chapter. Remember it as a two-part process consisting of these two simple mechanisms:

1. acid-catalyzed nucleophilic addition to the carbonyl group, and
2. S_N1 by protonation and loss of water, then attack by the alcohol.

Application: Topical Steroid

Fluocinolone acetonide is a steroid acetal used to treat skin conditions such as eczema and psoriasis. The acetal group decreases the water solubility of the parent steroid, enhancing its potency and allowing for a longer duration of action.

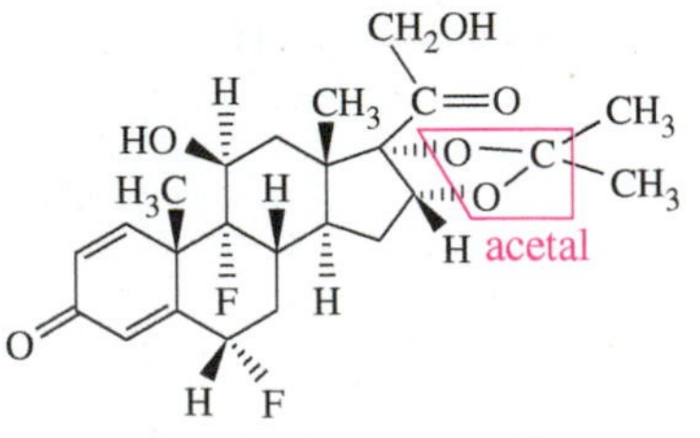

fluocinolone acetonide

$$CH_3{-}O\ O{-}CH_3 \text{ (cyclohexanone dimethyl acetal)} + H_2O \xrightleftharpoons{H^+,\ \text{excess } H_2O} \text{cyclohexanone} + 2\ CH_3OH$$

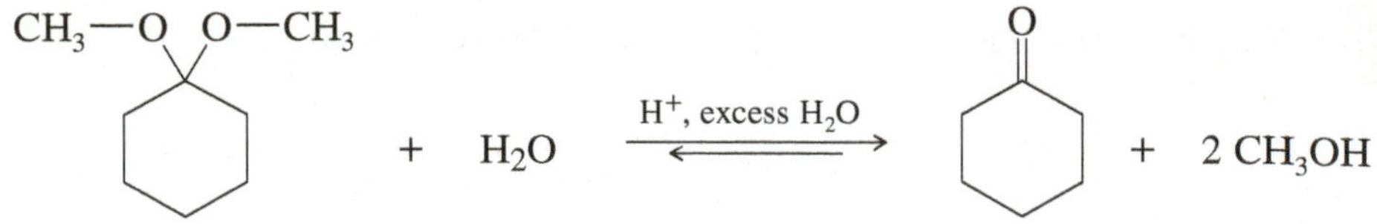

PROBLEM 18-25

Propose a mechanism for the acid-catalyzed hydrolysis of cyclohexanone dimethyl acetal.

Cyclic Acetals Formation of an acetal using a diol as the alcohol gives a cyclic acetal. Cyclic acetals often have more favorable equilibrium constants, because there is a smaller entropy loss when two molecules (a ketone and a diol) condense than when three molecules (a ketone and two molecules of an alcohol) condense. Ethylene glycol is often used to make cyclic acetals; its acetals are called **ethylene acetals** (or **ethylene ketals**).

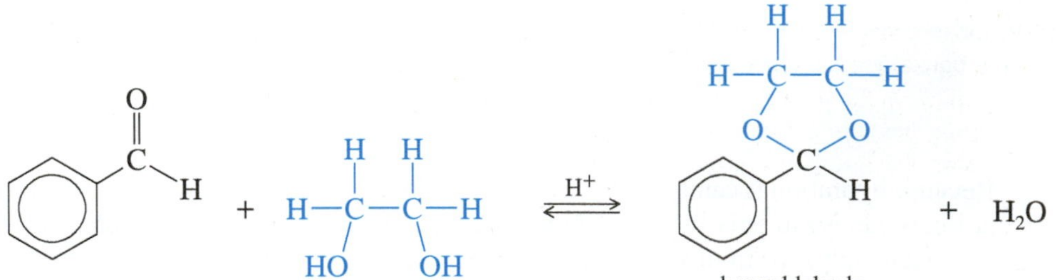

benzaldehyde + ethylene glycol $\xrightleftharpoons{H^+}$ benzaldehyde ethylene acetal + H_2O

Carbohydrates Sugars and other carbohydrates most commonly exist as cyclic acetals and hemiacetals. For example, glucose is a six-carbon sugar that is most stable as a hemiacetal. Lactose is a disaccharide (composed of two sugar units) that has one acetal and one hemiacetal. We discuss the structures of carbohydrates in detail in Chapter 23.

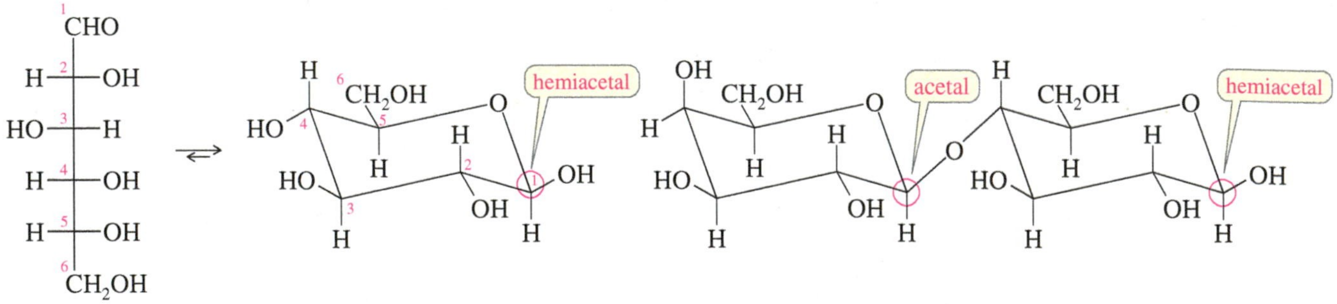

glucose (open chain) ⇌ glucose (cyclic hemiacetal)

lactose

PROBLEM-SOLVING HINT

Formation of an acetal (or hemiacetal) does not alter the oxidation state of the carbonyl carbon atom. In an acetal or hemiacetal, the carbonyl carbon atom is the one with *two* bonds to oxygen.

PROBLEM 18-26

Show what alcohols and carbonyl compounds give the following derivatives.

(a) 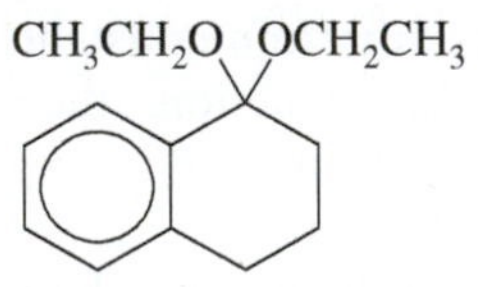 $CH_3CH_2O\ OCH_2CH_3$ (b) 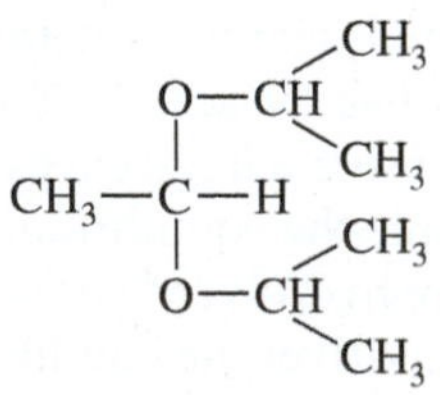 (c)

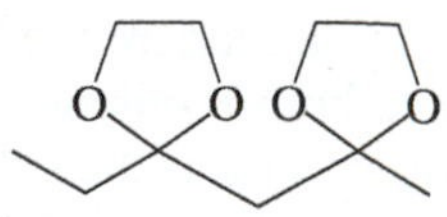

(d) 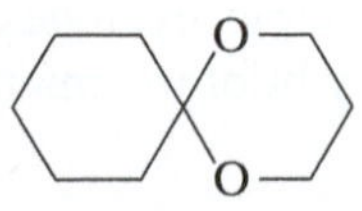 (e) 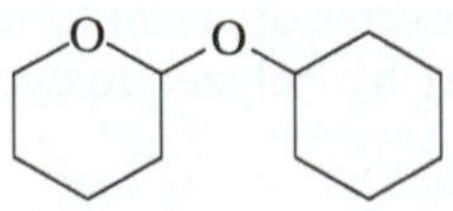 (f)

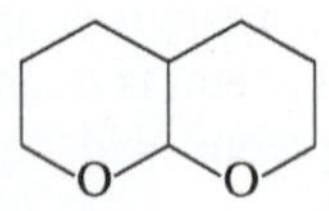

PROBLEM-SOLVING STRATEGY Proposing Reaction Mechanisms

Here we apply the general principles for proposing reaction mechanisms to the hydrolysis of an acetal. These principles were introduced in Chapters 7 and 11 and are summarized in Appendix 3A. Remember that you should draw all the bonds and substituents of each carbon atom involved in a mechanism. Show each step separately, using curved arrows to show the movement of electron pairs (from the nucleophile to the electrophile).

Our problem is to propose a mechanism for the acid-catalyzed hydrolysis of the following acetal:

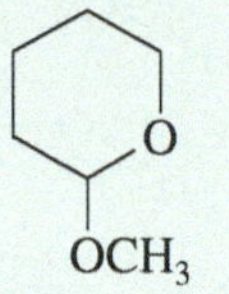

O

OCH_3

The type of mechanism is stated to be acid-catalyzed. Therefore, we assume it involves strong electrophiles and cationic intermediates (possibly carbocations), but no strong nucleophiles or strong bases and certainly no carbanions or free radicals.

1. **Consider the carbon skeletons of the reactants and products, and decide which carbon atoms in the products are likely derived from which carbon atoms in the reactants.**
First, you must decide what products are formed by hydrolysis of the acetal. In dealing with acetals and hemiacetals, any carbon atom with *two* bonds to oxygen is derived from a carbonyl group.

   Draw an equation showing all the affected atoms. The equation shows that water must somehow add (probably by a nucleophilic attack), and the ring must be cleaved.

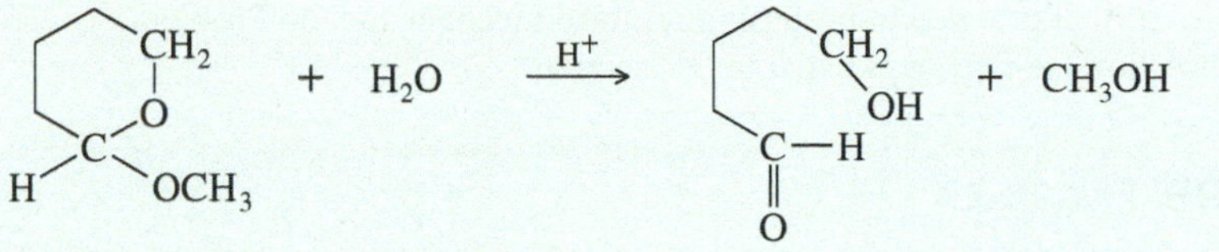

CH_2 O C H OCH_3 + H_2O $\xrightarrow{H^+}$ CH_2 OH C—H O + CH_3OH

2. **Consider whether any of the reactants is a strong enough electrophile to react without being activated. If not, consider how one of the reactants might be converted to a strong electrophile by protonation of a Lewis basic site (or complexation with a Lewis acid).**
The reactant probably will not react with water until it is activated, most likely by protonation. It can become protonated at either oxygen atom. We will arbitrarily choose the ring oxygen for protonation. The protonated compound is well suited for ring cleavage to form a stabilized (and strongly electrophilic) cation.

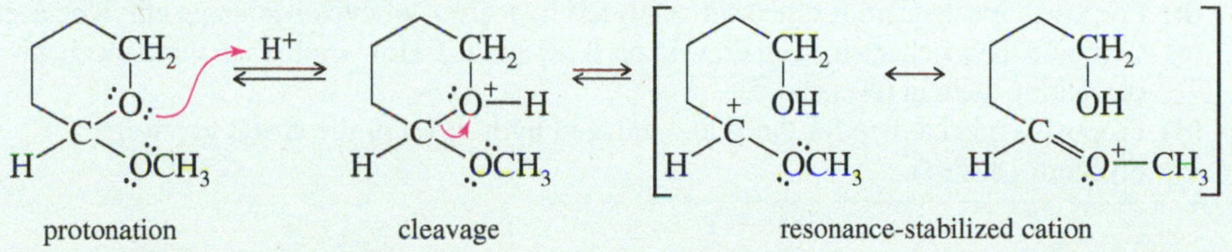

CH_2 O C H OCH_3 $\overset{H^+}{\rightleftharpoons}$ CH_2 O^+—H C H OCH_3 $\rightleftharpoons$ [CH_2 OH C^+ H OCH_3 $\longleftrightarrow$ CH_2 OH C H O^+—CH_3]

protonation cleavage resonance-stabilized cation

3. **Consider how a nucleophilic site on another reactant can attack the strong electrophile to form a bond needed in the product. Draw the product of this bond formation.**
Attack by water on the cation gives a protonated hemiacetal.

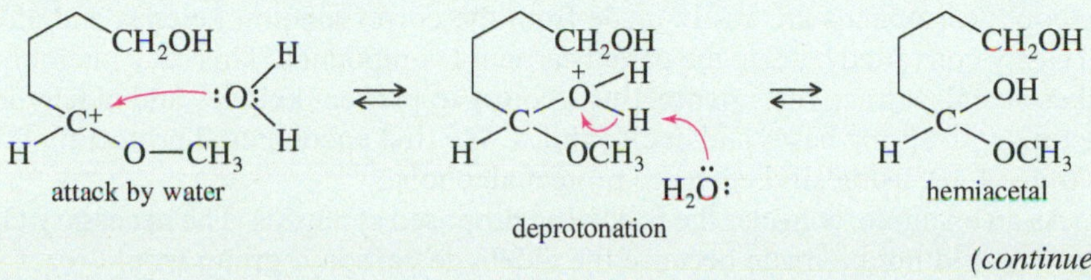

CH_2OH C^+ H O—CH_3 :O(H)H $\rightleftharpoons$ CH_2OH O^+(H)H C H OCH_3 $H_2\ddot{O}$: $\rightleftharpoons$ CH_2OH OH C H OCH_3

attack by water deprotonation hemiacetal

(continued)

PROBLEM-SOLVING HINT

To lose an —OR or —OH group under acidic conditions, consider protonating the group and losing a neutral molecule to give a carbocation.

4. Consider how the product of nucleophilic attack might be converted to the final product (if it has the right carbon skeleton) or reactivated to form another bond needed in the product.

Just as an —OH group can be lost by protonation and loss of water, the $-OCH_3$ group can be lost by protonating it and losing methanol. A protonated version of the product results.

CH_2OH H+ protonation CH_2OH OH C H O CH_3 H C OH +O—CH_3 H

CH_2OH CH_2OH deprotonation CH_2OH C+ O—H H C + O—H H C=O H H_3O^+

resonance-stabilized intermediate HOH products

5. Draw all the steps of the mechanism, using curved arrows to show the movement of electrons.

The complete mechanism is given by combining the preceding equations. You should write out the mechanism to review the steps involved.

As further practice in proposing reaction mechanisms, do Problems 18-27 and 18-28 by completing the five steps listed in this section.

PROBLEM-SOLVING HINT

The mechanism of a reverse reaction is normally the reverse of the mechanism of the forward reaction, as long as they take place under similar conditions. If you know the mechanism for formation of an acetal, you can always write the mechanism for its hydrolysis, using the same intermediates in reverse order.

PROBLEM 18-27

In the mechanism for acetal hydrolysis shown, the ring oxygen atom was protonated first, the ring was cleaved, and then the methoxy group was lost. The mechanism could also be written to show the methoxy oxygen protonating and cleaving first, followed by ring cleavage. Draw this alternative mechanism.

PROBLEM 18-28

(a) Propose a mechanism for the acid-catalyzed reaction of cyclohexanone with ethylene glycol to give cyclohexanone ethylene acetal.
(b) Propose a mechanism for the acid-catalyzed hydrolysis of cyclohexanone ethylene acetal.
(c) Compare the mechanisms you drew in parts (a) and (b). How similar are these mechanisms, comparing them in reverse order?
(d) Propose a mechanism for the acid-catalyzed hydrolysis of the acetal given in Problem 18-26(f).

18-17 Use of Acetals as Protecting Groups

Acetals hydrolyze under acidic conditions, but they are stable to strong bases and nucleophiles. Acetals are easily made from the corresponding ketones and aldehydes and easily converted back to the parent carbonyl compounds. This easy interconversion makes acetals attractive as **protecting groups** to prevent ketones and aldehydes from reacting with strong bases and nucleophiles. We first encountered protecting groups in Section 14-10, using silyl ethers to protect alcohols.

As an example, consider the following proposed synthesis. The necessary Grignard reagent could not be made because the aldehyde carbonyl group would react with its own nucleophilic organometallic group.

Proposed synthesis

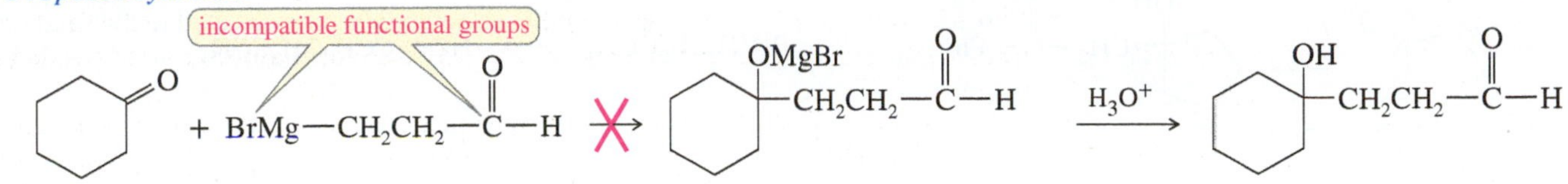

incompatible functional groups

cyclohexanone + $BrMg{-}CH_2CH_2{-}C(=O){-}H$ (impossible reagent) ✗→ [1-(OMgBr)cyclohexyl]$-CH_2CH_2-C(=O)-H$ $\xrightarrow{H_3O^+}$ [1-(OH)cyclohexyl]$-CH_2CH_2-C(=O)-H$ (target compound)

If the aldehyde is protected as an acetal, however, it is unreactive toward a Grignard reagent. The "masked" aldehyde is converted to the Grignard reagent, which is allowed to react with cyclohexanone. Dilute aqueous acid both protonates the alkoxide to give the alcohol and hydrolyzes the acetal to give the deprotected aldehyde.

Actual synthesis

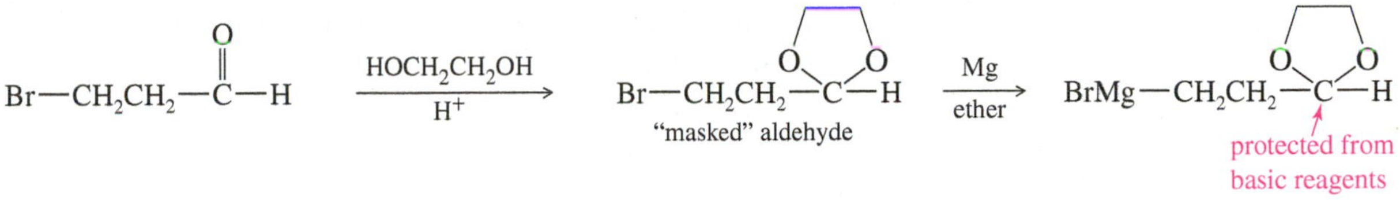

$Br{-}CH_2CH_2{-}C(=O){-}H$ $\xrightarrow[H^+]{HOCH_2CH_2OH}$ $Br{-}CH_2CH_2{-}C{-}H$ (cyclic acetal, "masked" aldehyde) $\xrightarrow[ether]{Mg}$ $BrMg{-}CH_2CH_2{-}C{-}H$ (cyclic acetal; protected from basic reagents)

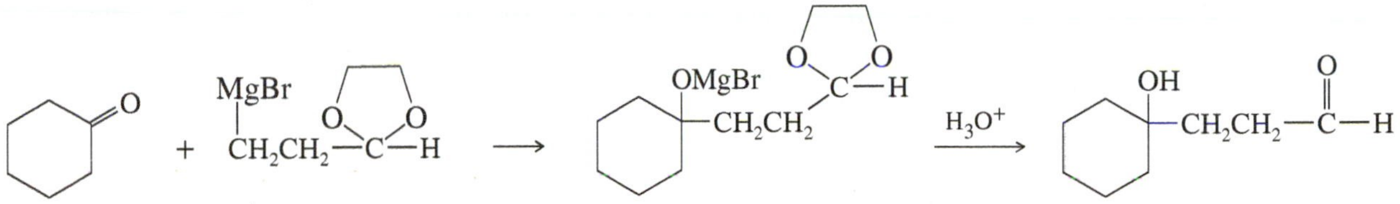

cyclohexanone + $MgBr{-}CH_2CH_2{-}C{-}H$ (cyclic acetal) → [1-(OMgBr)cyclohexyl]$-CH_2CH_2-C-H$ (cyclic acetal) $\xrightarrow{H_3O^+}$ [1-(OH)cyclohexyl]$-CH_2CH_2-C(=O)-H$ (target compound)

Selective Acetal Formation Because aldehydes form acetals more readily than ketones, we can protect an aldehyde selectively in the presence of a ketone. This selective protection leaves the ketone available for modification under neutral or basic conditions without disturbing the more reactive aldehyde group. The following example shows the reduction of a ketone in the presence of a more reactive aldehyde:

3-oxocyclohexanecarbaldehyde $\underset{H^+}{\overset{1\ equiv\ HOCH_2CH_2OH}{\rightleftharpoons}}$ acetal-protected keto compound $\xrightarrow{NaBH_4}$ alcohol (H, OH) with acetal $\xrightarrow{H_3O^+}$ 3-hydroxycyclohexanecarbaldehyde (H, OH; C(=O)H)

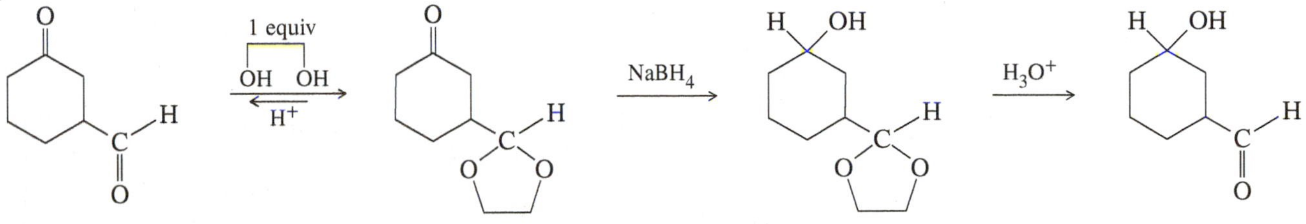

PROBLEM 18-29

Show how you would accomplish the following syntheses. You may use whatever additional reagents you need.

(a) 3-oxocyclohexane-CHO → 3-oxocyclohexane-CH_2OH

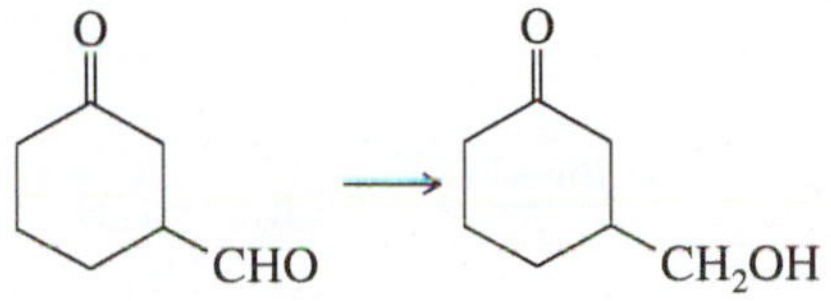

(b) 3-oxocyclohexane-CHO → 3-hydroxy-3-methyl (HO, CH_3) cyclohexane-CHO

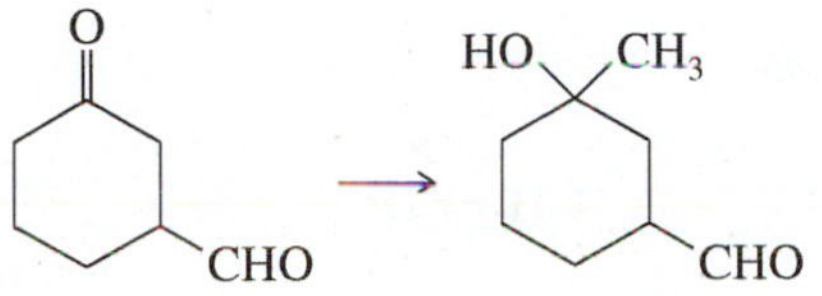

(c) 2-(CH_2Br)cyclohexanone → 2-(CH₂CH=CH-Ph)cyclohexanone

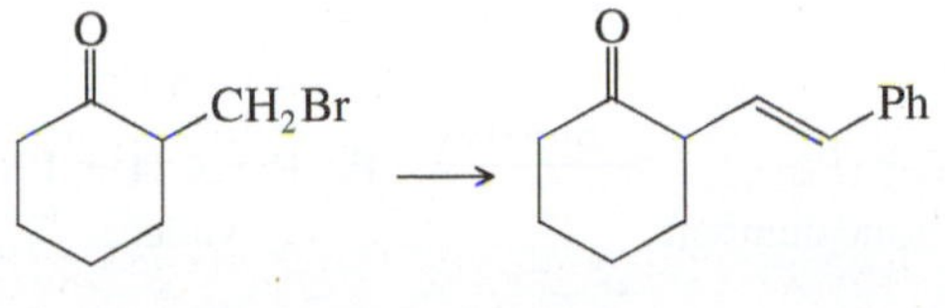

(d) 3,5-dimethylcyclohexane-1,4-dione (CH_3, CH_3) → 4-hydroxy (H, OH)-3,5-dimethylcyclohexanone (CH_3, CH_3)

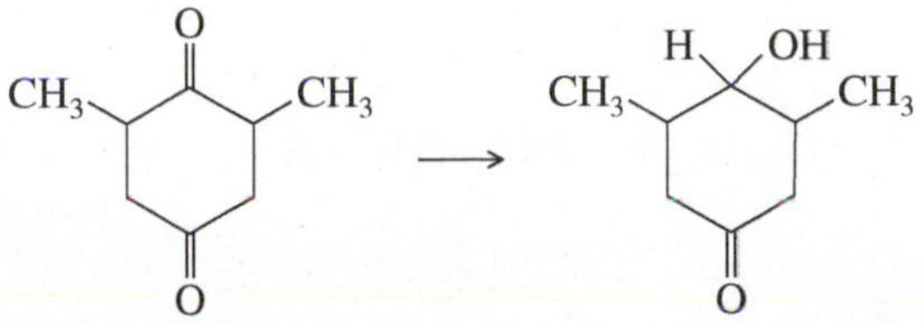

(continued)

(e) cyclopentanone → 1-(2-oxo-2-phenylethyl)cyclopentanol: $\longrightarrow$ OH, $-CH_2-\overset{O}{\overset{\|}{C}}-Ph$

(f) $BrCH_2CH_2\overset{O}{\overset{\|}{C}}CH_3 \longrightarrow HC{\equiv}CCH_2CH_2\overset{O}{\overset{\|}{C}}CH_3$

18-18 The Wittig Reaction

We have seen carbonyl groups undergo addition by a variety of carbanion-like reagents, including Grignard reagents, organolithium reagents, and acetylide ions. In 1954, Georg Wittig discovered a way of adding a phosphorus-stabilized carbanion to a ketone or aldehyde. The product is not an alcohol, however, because the intermediate undergoes elimination to an alkene. In effect, the **Wittig reaction** converts the carbonyl group of a ketone or an aldehyde into a new C=C double bond where no bond existed before. This reaction proved so useful that Wittig received the Nobel Prize in Chemistry in 1979 for this discovery.

The Wittig reaction

$$R'_2C{=}O \;+\; R(H)\ddot{C}^{-}-\overset{+}{P}Ph_3 \longrightarrow\longrightarrow R'_2C{=}C(R)H \;+\; Ph_3P{=}O$$

ketone or aldehyde; phosphorus ylide; alkene

The phosphorus-stabilized carbanion is an **ylide** (pronounced "ill′-id")—a molecule that bears no overall charge but has a negatively charged carbon atom bonded to a positively charged heteroatom. Phosphorus ylides are prepared from tri-phenylphosphine and alkyl halides in a two-step process. The first step is nucleophilic attack by triphenylphosphine on an unhindered (usually primary) alkyl halide. The product is an alkyltriphenylphosphonium salt. The phosphonium salt is treated with a strong base (usually butyllithium) to abstract a proton from the carbon atom bonded to phosphorus.

$$Ph_3P{:} + H-\underset{R}{\overset{H}{C}}-X \longrightarrow Ph_3\overset{+}{P}-\underset{R}{\overset{H}{C}}-H\;\;X^- \xrightarrow[\text{butyllithium}]{\overset{\delta^-}{C}H_2CH_2CH_2CH_3-\overset{\delta^+}{Li}} \left[Ph_3\overset{+}{P}-\ddot{C}^{-}(H)R \longleftrightarrow Ph_3P{=}C(H)R\right] + \underset{\text{butane}}{C_4H_{10}} + LiX$$

triphenylphosphine; alkyl halide; phosphonium salt; phosphorus ylide

Examples

$$Ph_3P{:} + CH_3-Br \longrightarrow \underset{\text{methyltriphenylphosphonium salt}}{Ph_3\overset{+}{P}-CH_3\;Br^-} \xrightarrow{Bu-Li} \underset{\text{ylide}}{Ph_3\overset{+}{P}-\ddot{C}H_2^-}$$

$$Ph_3P{:} + Ph-CH_2-Br \longrightarrow \underset{\text{benzyltriphenylphosphonium salt}}{Ph_3\overset{+}{P}-CH_2-Ph\;Br^-} \xrightarrow{Bu-Li} \underset{\text{ylide}}{Ph_3\overset{+}{P}-\ddot{C}H^{-}-Ph}$$

The phosphorus ylide has two resonance forms: one with a double bond between carbon and phosphorus, and another with charges on carbon and phosphorus. The double-bonded resonance form requires ten electrons in the valence shell of phosphorus, using a *d* orbital. The pi bond between carbon and phosphorus is weak, and the charged structure is the major contributor. The carbon atom actually bears a partial negative charge, balanced by a corresponding positive charge on phosphorus.

PROBLEM 18-30

Trimethylphosphine is a stronger nucleophile than triphenylphosphine, but it is rarely used to make ylides. Why is trimethylphosphine unsuitable for making most phosphorus ylides?

Because of its carbanion character, the ylide carbon atom is strongly nucleophilic. It attacks a carbonyl group to give a charge-separated intermediate called a *betaine* (pronounced "bay′-tuh-ene"). A betaine is an unusual compound because it contains a negatively charged oxygen and a positively charged phosphorus on adjacent carbon atoms. Phosphorus and oxygen form strong bonds, and the attraction of opposite charges promotes the fast formation of a four-membered *oxaphosphetane* ring. (In some cases, the oxaphosphetane may be formed directly by a cycloaddition, rather than via a betaine.)

The four-membered ring quickly collapses to give the alkene and triphenylphosphine oxide. Triphenylphosphine oxide is exceptionally stable, and the conversion of triphenylphosphine to triphenylphosphine oxide provides the driving force for the Wittig reaction.

MECHANISM 18-6 The Wittig Reaction

Step 1: The ylide attacks the carbonyl to form a betaine.

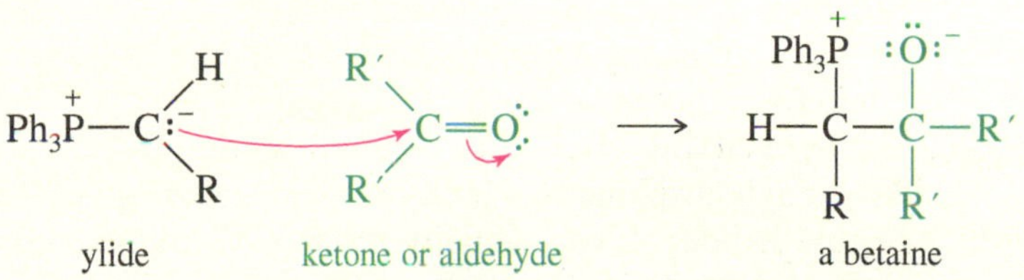

$Ph_3\overset{+}{P}-\ddot{C}^{-}(H)(R)$ + $R'_2C=O$ ⟶ $Ph_3\overset{+}{P}-C(H)(R)-C(R')_2-O^{-}$

ylide ketone or aldehyde a betaine

Step 2: The betaine closes to a four-membered ring oxaphosphetane (first P—O bond formed).

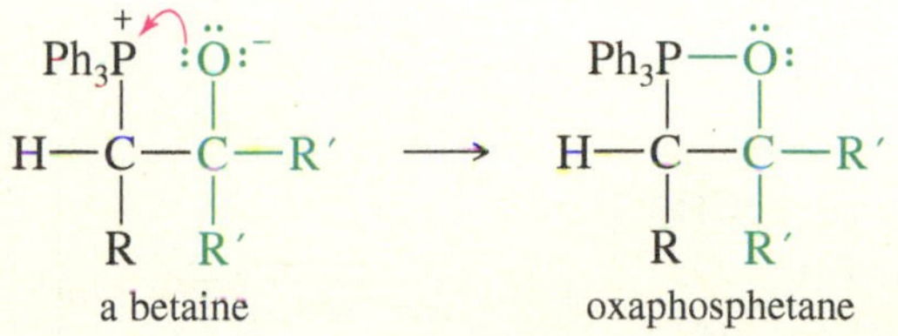

a betaine ⟶ oxaphosphetane

Step 3: The ring collapses to the products (second P—O bond formed).

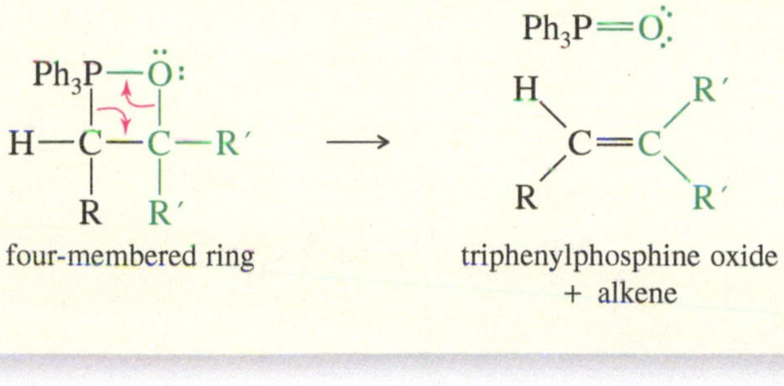

four-membered ring ⟶ $Ph_3P=O$ + $(H)(R)C=C(R')_2$

triphenylphosphine oxide + alkene

The following examples show the formation of carbon–carbon double bonds using the Wittig reaction. Mixtures of cis and trans isomers often result when geometric isomerism is possible.

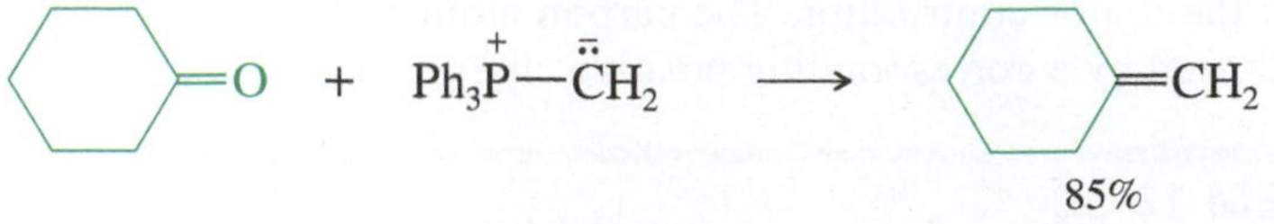

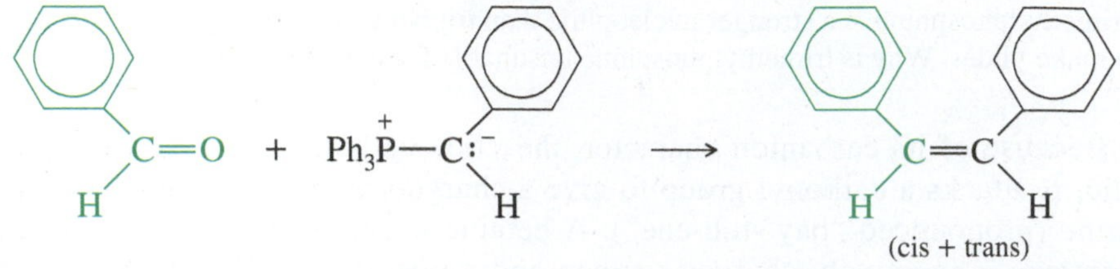

$$\text{cyclohexanone} + Ph_3\overset{+}{P}-\overset{-}{\ddot{C}}H_2 \longrightarrow \text{methylenecyclohexane} \; (85\%)$$

$$PhCH{=}O + Ph_3\overset{+}{P}-\overset{-}{C}HPh \longrightarrow PhCH{=}CHPh \; \text{(cis + trans)}$$

PROBLEM 18-31

Like other strong nucleophiles, triphenylphosphine attacks and opens epoxides. The initial product (a betaine) quickly cyclizes to an oxaphosphetane that collapses to an alkene and triphenylphosphine oxide.

(a) Show each step in the reaction of *trans*-2,3-epoxybutane with triphenylphosphine to give but-2-ene. What is the stereochemistry of the double bond in the product?

(b) Show how this sequence might be used to convert *cis*-cyclooctene to *trans*-cyclooctene.

Planning a Wittig Synthesis The Wittig reaction is a valuable synthetic tool that converts a carbonyl group to a carbon–carbon double bond. A wide variety of alkenes may be synthesized by the Wittig reaction. To determine the necessary reagents, mentally divide the target molecule at the double bond, and decide which of the two components should come from the carbonyl compound and which should come from the ylide.

In general, the ylide should come from an unhindered alkyl halide. Triphenylphosphine is a bulky reagent, reacting best with unhindered primary and methyl halides. It occasionally reacts with unhindered secondary halides, but these reactions are sluggish and often give poor yields. The following example and Solved Problem show the planning of some Wittig syntheses.

Analysis

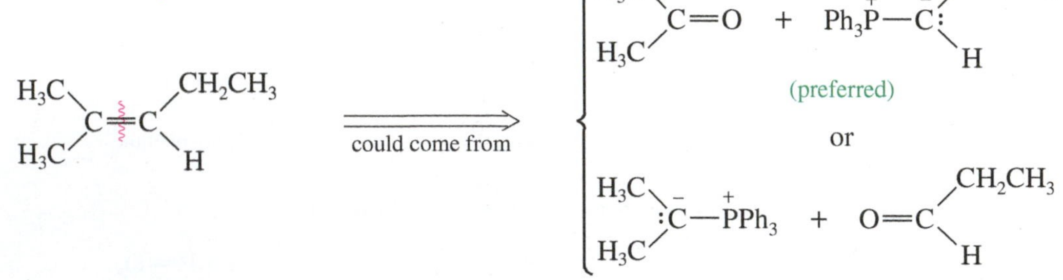

$$(H_3C)_2C{=}CH(CH_2CH_3) \overset{\text{could come from}}{\Longrightarrow} \begin{cases} (H_3C)_2C{=}O + Ph_3\overset{+}{P}-\overset{-}{C}H(CH_2CH_3) & \text{(preferred)} \\ \text{or} \\ (H_3C)_2\overset{-}{C}-\overset{+}{P}Ph_3 + O{=}CH(CH_2CH_3) \end{cases}$$

Synthesis

$$BrCH_2CH_2CH_3 \xrightarrow[\text{(2) BuLi}]{\text{(1) }Ph_3P} Ph_3\overset{+}{P}-\overset{-}{C}H(CH_2CH_3) \xrightarrow{(H_3C)_2C{=}O} (H_3C)_2C{=}CH(CH_2CH_3)$$

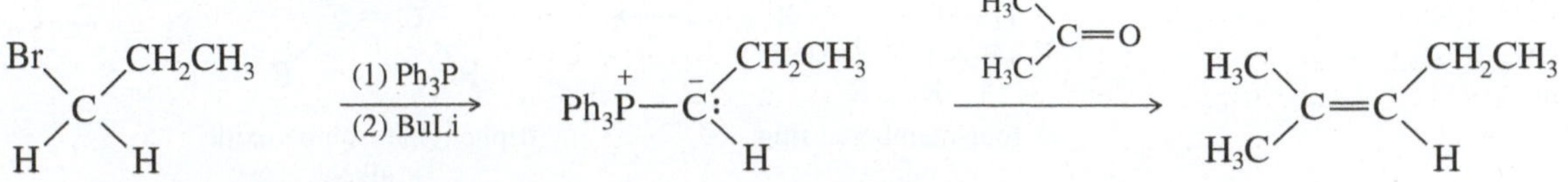

SOLVED PROBLEM 18-2

Show how you would use a Wittig reaction to synthesize 1-phenylbuta-1,3-diene.

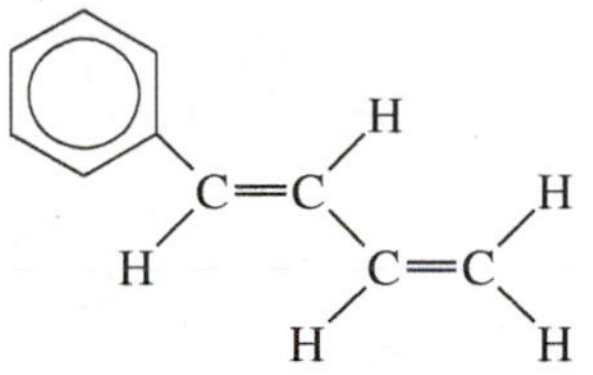

1-phenylbuta-1,3-diene

SOLUTION

This molecule has two double bonds that might be formed by Wittig reactions. The central double bond could be formed in either of two ways. Both of these syntheses will probably work, and both will produce a mixture of cis and trans isomers.

Analysis

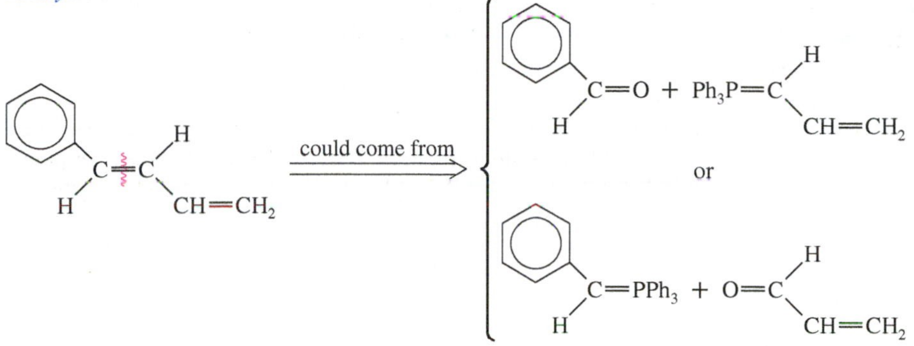

$Ph-CH=CH-CH=CH_2$ could come from $Ph-CH=O + Ph_3P=CH-CH=CH_2$ or $Ph-CH=PPh_3 + O=CH-CH=CH_2$

You should complete this solution by drawing out the syntheses indicated by this analysis (Problem 18-32).

PROBLEM 18-32

(a) Outline the syntheses indicated in Solved Problem 18-2, beginning with aldehydes and alkyl halides.
(b) Both of these syntheses of 1-phenylbuta-1,3-diene form the central double bond. Show how you would synthesize this target molecule by forming the terminal double bond.

PROBLEM-SOLVING HINT

Plan a Wittig synthesis so that the less hindered end of the double bond comes from the ylide. Remember that the ylide is made by S_N2 attack of triphenylphosphine on an unhindered alkyl halide, followed by deprotonation.

PROBLEM 18-33

Show how Wittig reactions might be used to synthesize the following compounds. In each case, start with an alkyl halide and a ketone or an aldehyde.

(a) $Ph-CH=C(CH_3)_2$
(b) $Ph-C(CH_3)=CH_2$
(c) $Ph-CH=CH-CH=CH-Ph$
(d) cyclohexylidene $=C(H)-CH_3$

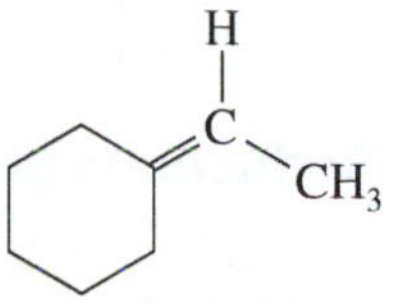

18-19 Oxidation of Aldehydes

Unlike ketones, aldehydes are easily oxidized to carboxylic acids by common oxidants such as bleach (sodium hypochlorite), chromic acid, permanganate, and peroxy acids. Aldehydes oxidize so easily that air must be excluded from their containers to avoid slow oxidation by atmospheric oxygen. Because aldehydes oxidize so easily, mild reagents such as Ag_2O can oxidize them selectively in the presence of other oxidizable functional groups.

$$R-\overset{O}{\overset{\|}{C}}-H \xrightarrow[\text{(oxidizing agent)}]{[O]} R-\overset{O}{\overset{\|}{C}}-OH$$

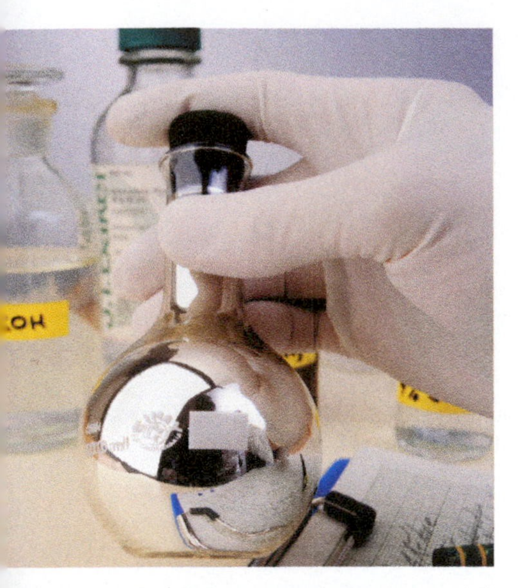
A Tollens test is usually done on a small scale, but it can also create a silver mirror on a large object.

Examples

$$CH_3-CH(CH_3)-C(=O)-H \xrightarrow[\text{dil } H_2SO_4]{Na_2Cr_2O_7} CH_3-CH(CH_3)-C(=O)-OH$$

isobutyraldehyde → isobutyric acid (90%)

$$\text{cyclohex-3-en-1-carbaldehyde} \xrightarrow[THF/H_2O]{Ag_2O} \text{cyclohex-3-en-1-carboxylic acid (97\%)}$$

Silver ion, Ag^+, oxidizes aldehydes selectively in a convenient functional-group test for aldehydes. The **Tollens test** involves adding a solution of silver–ammonia complex (the **Tollens reagent**) to the unknown compound. If an aldehyde is present, its oxidation reduces silver ion to metallic silver in the form of a black suspension or a silver mirror deposited on the inside of the container. Simple hydrocarbons, ethers, ketones, and even alcohols do not react with the Tollens reagent.

$$\underset{\text{aldehyde}}{R-C(=O)-H} + \underset{\text{Tollens reagent}}{2\,\overset{+}{Ag}(NH_3)_2} + 3\,^-OH \xrightarrow{H_2O} \underset{\text{silver}}{2\,Ag\downarrow} + \underset{\text{carboxylate}}{R-C(=O)-O^-} + 4\,NH_3 + 2\,H_2O$$

PROBLEM 18-34

Predict the major products of the following reactions.

(a) 4-hydroxycyclohexanecarbaldehyde (HO, CHO) + Ag_2O

(b) 4-hydroxycyclohexanecarbaldehyde (HO, CHO) + $K_2Cr_2O_7/H_2SO_4$

(c) 3-oxocyclohexanecarbaldehyde (CHO, O) + $\overset{+}{Ag}(NH_3)_2\ ^-OH$

(d) cyclohex-2-ene-1-carbaldehyde (CHO) + $KMnO_4$ (cold, dilute)

18-20 Reductions of Ketones and Aldehydes

We have discussed several reductions of ketones and aldehydes in earlier sections. Here, we review those reactions and then cover some additional reductions that are useful in synthesis.

18-20A Hydride Reductions (Review)

Ketones and aldehydes are most commonly reduced by sodium borohydride (see Sections 10-11 and 18-11). Sodium borohydride ($NaBH_4$) reduces ketones to secondary alcohols and aldehydes to primary alcohols. Lithium aluminum hydride ($LiAlH_4$) also accomplishes these reductions, but it is a more powerful reducing agent, and it is much more difficult to work with. Sodium borohydride is preferred for simple reductions of ketones and aldehydes. Sodium triacetoxyborohydride [$NaBH(OAc)_3$] is less reactive than $NaBH_4$, and it selectively reduces aldehydes even in the presence of ketones.

$$\underset{\text{cyclohexanecarbaldehyde}}{C_6H_{11}-C(=O)-H} \xrightarrow{NaBH_4,\ CH_3CH_2OH} \underset{\text{cyclohexylmethanol (95\%)}}{C_6H_{11}-CH_2-OH}$$

$$CH_3-\overset{O}{\overset{\|}{C}}-CH_2CH_3 \xrightarrow{NaBH_4,\ CH_3OH} CH_3-\overset{OH}{\overset{|}{C}H}-CH_2CH_3$$

butan-2-one → (±) butan-2-ol (100%)

18-20B Catalytic Hydrogenation

Like alkene double bonds, carbonyl double bonds can be reduced by catalytic hydrogenation. Catalytic hydrogenation is slower with carbonyl groups than with olefinic double bonds, however. Before sodium borohydride was available, catalytic hydrogenation was often used to reduce aldehydes and ketones, but any olefinic double bonds were reduced as well. In the laboratory, we prefer sodium borohydride over catalytic reduction because it reduces ketones and aldehydes faster than olefins, and no gas-handling equipment is required. Catalytic hydrogenation is still widely used in industry, however, because H_2 is much cheaper than $NaBH_4$, and pressure equipment is more readily available there.

The most common catalyst for catalytic hydrogenation of ketones and aldehydes is **Raney nickel.** Raney nickel is a finely divided hydrogen-bearing form of nickel made by treating a nickel–aluminum alloy with a strong sodium hydroxide solution. The aluminum in the alloy reacts to release hydrogen, leaving behind a finely divided nickel powder saturated with hydrogen. Pt and Rh catalysts are also used for hydrogenation of ketones and aldehydes.

Cyclohexene carbaldehyde $\xrightarrow[\text{(Raney nickel)}]{Ni\text{-}H_2}$ [cyclohexanecarbaldehyde] $\xrightarrow{Ni\text{-}H_2}$ cyclohexylmethanol (90%)

18-20C Deoxygenation of Ketones and Aldehydes

A **deoxygenation** replaces the carbonyl oxygen atom of a ketone or aldehyde with two hydrogen atoms, reducing the carbonyl group past the alcohol stage all the way to a methylene group. Formally, a deoxygenation is a four-electron reduction, as shown by the following equations. These equations use H_2 to symbolize the actual reducing agents, according to the general principle that one molecule of H_2 corresponds to a two-electron reduction. Formally, the deoxygenation requires two molecules of H_2, corresponding to a four-electron reduction.

2 H_2
deoxygenation
(4-*e* reduction)

$$C{=}O \xrightarrow[\text{(2-}e\text{ reduction)}]{H_2} H{-}C{-}OH \xrightarrow[\text{(2-}e\text{ reduction)}]{H_2} H{-}C{-}H + H_2O$$

In actual use, H_2 is not a good reagent for deoxygenation of ketones and aldehydes. Deoxygenation can be accomplished by either the Clemmensen reduction (under acidic conditions) or the Wolff–Kishner reduction (under basic conditions).

Clemmensen Reduction (Review) The **Clemmensen reduction** commonly converts acylbenzenes (from Friedel–Crafts acylation, Section 17-11B) to alkylbenzenes, but it also works with other ketones and aldehydes that are not sensitive to acid. The carbonyl compound is heated with an excess of amalgamated zinc (zinc treated with mercury) and hydrochloric acid. The actual reduction occurs by a complex mechanism on the surface of the zinc.

$$Ph—C(=O)—CH_2CH_3 \xrightarrow[HCl,\ H_2O]{Zn(Hg)} Ph—CH_2—CH_2CH_3$$
propiophenone → n-propylbenzene (90%)

$$CH_3—(CH_2)_5—CHO \xrightarrow[HCl,\ H_2O]{Zn(Hg)} CH_3—(CH_2)_5—CH_3$$
heptanal → n-heptane (72%)

cyclohexanone $\xrightarrow[HCl,\ H_2O]{Zn(Hg)}$ cyclohexane (75%)

Wolff–Kishner Reduction Compounds that cannot survive treatment with hot acid can be deoxygenated using the **Wolff–Kishner reduction.** The ketone or aldehyde is converted to its hydrazone, which is heated with a strong base such as KOH or potassium *tert*-butoxide. Ethylene glycol, diethylene glycol, or another high- boiling solvent is used to facilitate the high temperature (140–200 °C) needed in the second step.

$$C{=}O + H_2N—NH_2 \xrightarrow{H^+} C{=}N—NH_2 + H_2O \xrightarrow[heat]{KOH} H—C—H + H_2O + N{\equiv}N\uparrow$$
hydrazone

Examples

propiophenone $\xrightarrow{N_2H_4}$ hydrazone (NNH$_2$) $\xrightarrow[HOCH_2CH_2OCH_2CH_2OH]{KOH,\ 175\ °C}$ n-propylbenzene (82%) + N_2
(diethylene glycol)

cyclohexanone $\xrightarrow{N_2H_4}$ hydrazone (N—NH$_2$) $\xrightarrow[CH_3—S(=O)—CH_3]{t\text{-}BuO^-\ K^+}$ cyclohexane (80%) + N_2
(DMSO, a solvent)

The mechanism for formation of the hydrazone is the same as the mechanism for imine formation (Key Mechanism 18-4 in Section 18-14). The actual reduction step involves two tautomeric proton transfers from nitrogen to carbon (Mechanism 18-7). In this strongly basic solution, we expect a proton transfer from N to C to occur by loss of a proton from nitrogen, followed by reprotonation on carbon. A second deprotonation sets up the intermediate for loss of nitrogen to form a carbanion. This carbanion is quickly reprotonated to give the product.

MECHANISM 18-7 Wolff–Kishner Reduction

Formation of the Hydrazone: See Key Mechanism 18-4.

Reduction step 1: Proton transfer from N to C. (Basic conditions: Remove, then replace.)

$$R'R\,C{=}N—NH_2 \overset{^-:OH}{\rightleftharpoons} \left[R'R\,C{=}N—\ddot{N}^-—H \longleftrightarrow R'R\,{}^-\ddot{C}—N{=}N—H\right] \overset{H_2O}{\rightleftharpoons} H—C(R')(R)—N{=}N—H + {}^-OH$$

hydrazone — remove proton from N — replace proton on C

Another deprotonation enables loss of N_2

Reduction step 2: Remove second proton from N. ***Step 3:*** Lose N_2. ***Step 4:*** Protonate.

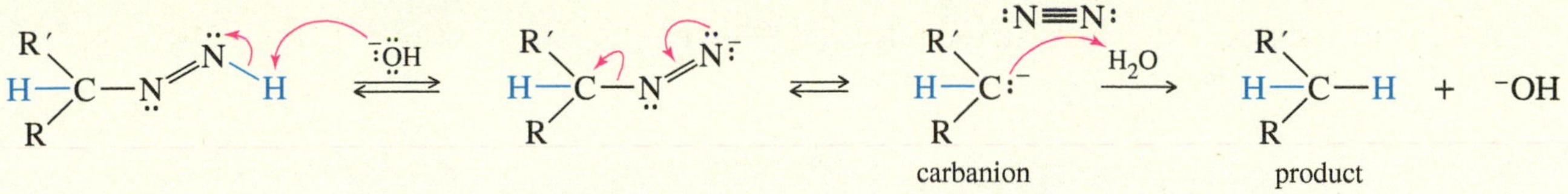

carbanion

product

PROBLEM 18-35

Propose a mechanism for both parts of the Wolff–Kishner reduction of cyclohexanone: the formation of the hydrazone, and then the base-catalyzed reduction with evolution of nitrogen gas.

PROBLEM 18-36

Predict the major products of the following reactions:

(a) $\xrightarrow[\text{HCl, } H_2O]{\text{Zn(Hg)}}$

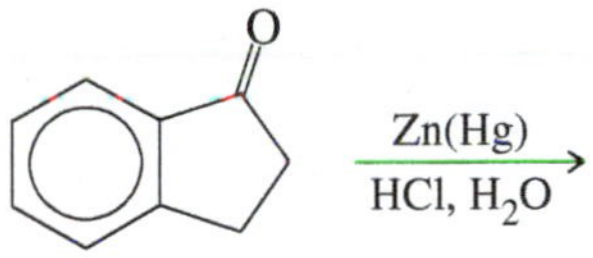

(b) $\xrightarrow[\text{(2) KOH, heat}]{\text{(1) } H_2NNH_2}$

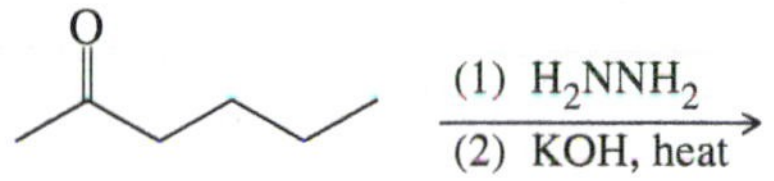

(c) $\xrightarrow[\text{(2) KOH, heat}]{\text{(1) } N_2H_4}$

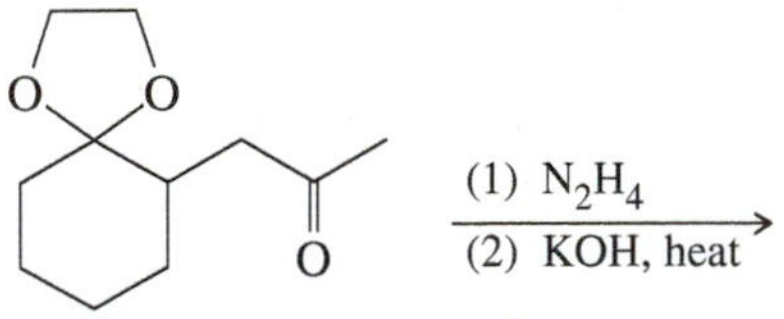

(d) $\xrightarrow[\text{HCl, } H_2O]{\text{Zn(Hg)}}$

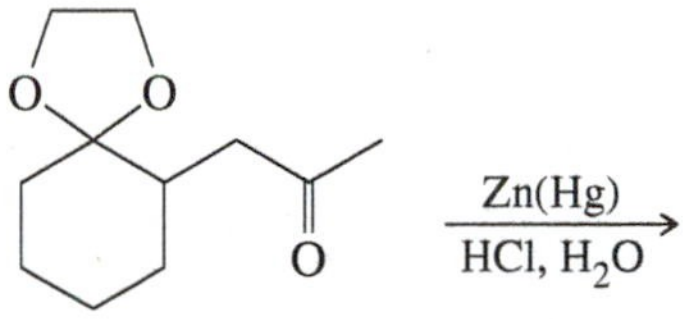

SUMMARY Reactions of Ketones and Aldehydes

1. *Addition of organometallic reagents* (Sections 9-7B and 10-9)

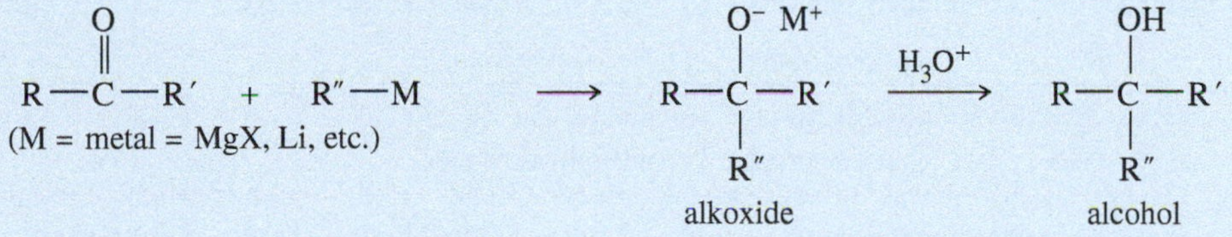

$$R{-}\overset{O}{\overset{\|}{C}}{-}R' + R''{-}M \longrightarrow R{-}\overset{O^- M^+}{\underset{R''}{C}}{-}R' \xrightarrow{H_3O^+} R{-}\overset{OH}{\underset{R''}{C}}{-}R'$$

(M = metal = MgX, Li, etc.) alkoxide alcohol

2. *Reduction* (Sections 10-12 and 18-20A)

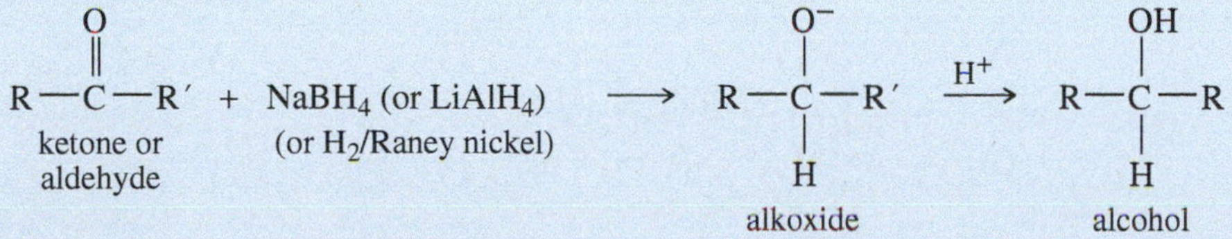

$$R{-}\overset{O}{\overset{\|}{C}}{-}R' + NaBH_4 \text{ (or } LiAlH_4) \longrightarrow R{-}\overset{O^-}{\underset{H}{C}}{-}R' \xrightarrow{H^+} R{-}\overset{OH}{\underset{H}{C}}{-}R$$

ketone or aldehyde (or H_2/Raney nickel) alkoxide alcohol

Deoxygenation reactions

a. *Clemmensen reduction* (Sections 17-11B and 18-20C)

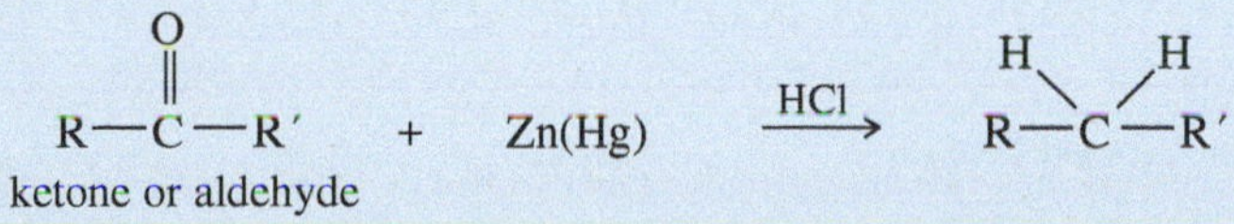

$$R{-}\overset{O}{\overset{\|}{C}}{-}R' + \text{Zn(Hg)} \xrightarrow{HCl} R{-}\overset{H\ \ H}{C}{-}R'$$

ketone or aldehyde

(continued)

b. *Wolff–Kishner reduction* (Section 18-20C)

$$\underset{\text{ketone or aldehyde}}{R{-}\overset{\overset{\displaystyle O}{\|}}{C}{-}R'} + \underset{\text{hydrazine}}{H_2N{-}NH_2} \longrightarrow \underset{\text{hydrazone}}{R{-}\overset{\overset{\displaystyle N-NH_2}{\|}}{C}{-}R'} \xrightarrow[\text{heat}]{\text{KOH}} R{-}CH_2{-}R' + H_2O + N{\equiv}N\uparrow$$

3. *Hydration* (Section 18-12)

$$\underset{\text{ketone or aldehyde}}{R{-}\overset{\overset{\displaystyle O}{\|}}{C}{-}R'} + H_2O \rightleftharpoons \underset{\text{hydrate}}{R{-}C(OH)(OH){-}R'}$$

4. *Formation of cyanohydrins* (Section 18-13)

$$\underset{\text{ketone or aldehyde}}{R{-}\overset{\overset{\displaystyle O}{\|}}{C}{-}R'} + HCN \overset{^-CN}{\rightleftharpoons} \underset{\text{cyanohydrin}}{R{-}C(OH)(CN){-}R'}$$

5. *Formation of imines* (Section 18-14)

$$\underset{\text{ketone or aldehyde}}{R{-}\overset{\overset{\displaystyle O}{\|}}{C}{-}R'} + \underset{\text{primary amine}}{R''{-}NH_2} \overset{H^+}{\rightleftharpoons} \underset{\text{imine (Schiff base)}}{R{-}\overset{\overset{\displaystyle N-R''}{\|}}{C}{-}R'} + H_2O$$

6. *Formation of oximes and hydrazones* (Section 18-15)

$$\underset{\text{ketone or aldehyde}}{R{-}\overset{\overset{\displaystyle O}{\|}}{C}{-}R'} + \underset{\text{hydroxylamine}}{H_2N{-}OH} \xrightarrow{H^+} \underset{\text{oxime}}{R{-}\overset{\overset{\displaystyle N-OH}{\|}}{C}{-}R'}$$

$$\underset{\text{ketone or aldehyde}}{R{-}\overset{\overset{\displaystyle O}{\|}}{C}{-}R'} + \underset{\text{hydrazine reagent}}{H_2N{-}NH{-}R''} \xrightarrow{H^+} \underset{\text{hydrazone derivative}}{R{-}\overset{\overset{\displaystyle N-NH-R''}{\|}}{C}{-}R'}$$

R″ =	*Reagent Name*	*Derivative Name*
—H	hydrazine	hydrazone
—Ph	phenylhydrazine	phenylhydrazone
$-\overset{\overset{\displaystyle O}{\|}}{C}-NH_2$	semicarbazide	semicarbazone

7. *Formation of acetals* (Section 18-16)

$$\underset{\text{aldehyde or ketone}}{R{-}\overset{\overset{\displaystyle O}{\|}}{C}{-}R'} + \underset{\text{alcohol}}{2\ R''{-}OH} \overset{H^+}{\rightleftharpoons} \underset{\text{acetal (or ketal)}}{R{-}C(OR'')(OR''){-}R'} + H_2O$$

8. *The Wittig reaction* (Section 18-18)

$$\underset{\text{phosphorus ylide}}{Ph_3\overset{+}{P}{-}\overset{-}{C}R_2} + \underset{\text{ketone or aldehyde}}{R'_2C{=}O} \longrightarrow \underset{\text{alkene}}{R_2C{=}CR'_2} + Ph_3P{=}O$$

9. *Oxidation of aldehydes* (Section 18-19)

$$\underset{\text{aldehyde}}{R-\overset{O}{\overset{\|}{C}}-H} \xrightarrow{\text{chromic acid, bleach, } Ag^{+}\text{, etc.}} \underset{\text{acid}}{R-\overset{O}{\overset{\|}{C}}-OH}$$

Tollens test

$$\underset{\text{aldehyde}}{R-\overset{O}{\overset{\|}{C}}-H} + \underset{\text{Tollens reagent}}{2\,Ag(NH_3)_2^{+}} + 3\,^{-}OH \xrightarrow{H_2O} \underset{\text{silver}}{2\,Ag\downarrow} + \underset{\text{carboxylate}}{R-\overset{O}{\overset{\|}{C}}-O^{-}} + 4\,NH_3 + 2\,H_2O$$

10. *Reactions of ketones and aldehydes at their α positions*
This large group of reactions is covered in Chapter 22.

Example

Aldol condensation

$$2\,CH_3-\overset{O}{\overset{\|}{C}}-H \underset{}{\overset{\text{base}}{\rightleftharpoons}} CH_3-\underset{H}{\overset{OH}{C}}-CH_2-\overset{O}{\overset{\|}{C}}-H$$

SUMMARY Nucleophilic Addition Reactions of Aldehydes and Ketones

In the central structure below, if R or R′ = H, the functional group is an aldehyde. If both R and R′ are alkyl or aryl groups, the functional group is a ketone. In general, ketones are more hindered to nucleophilic attack and react more slowly than aldehydes.

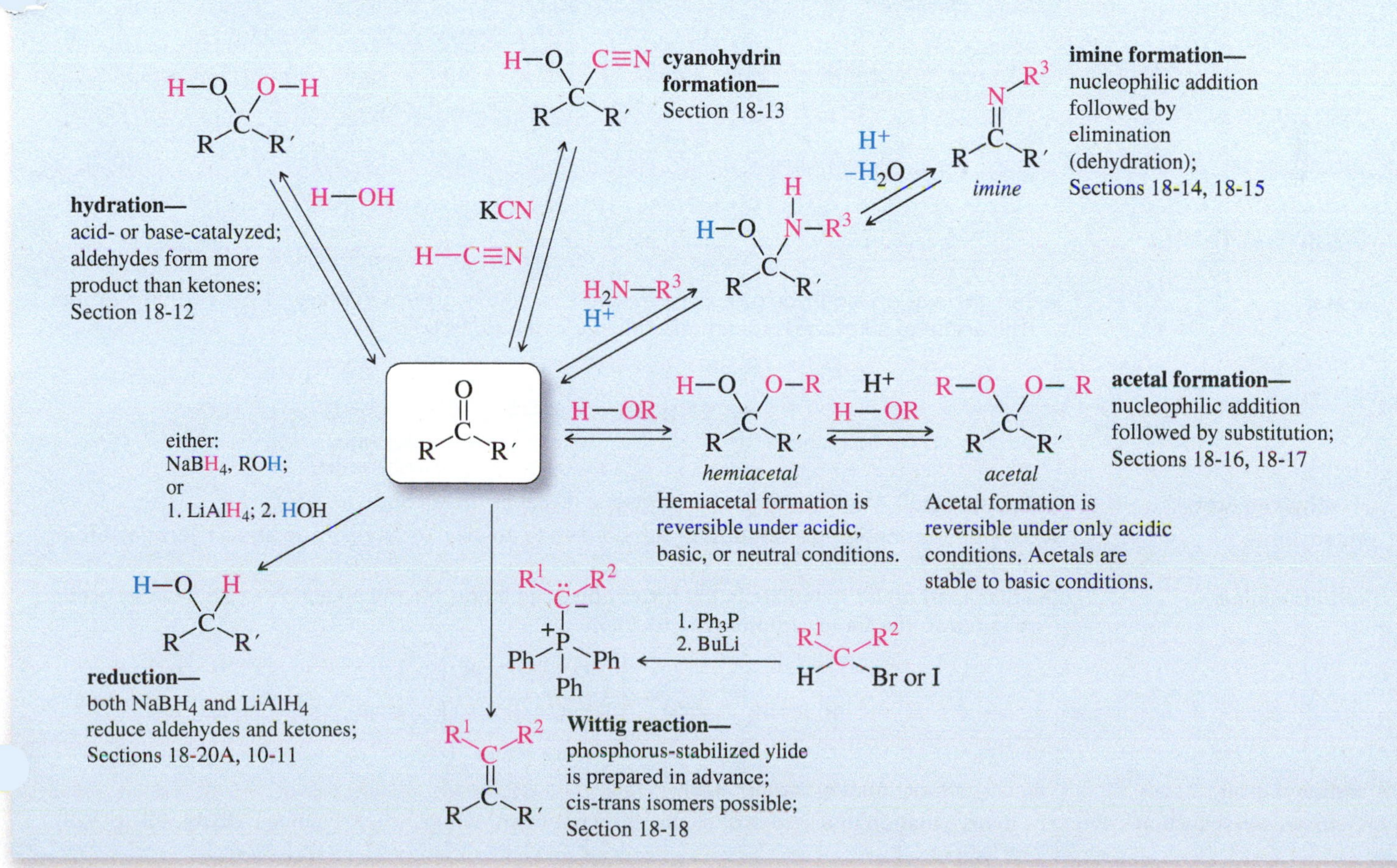

GUIDE TO ORGANIC REACTIONS IN CHAPTER 18

Reactions covered in Chapter 18 are shown in red. Reactions covered in earlier chapters are shown in blue.

Substitution

Nucleophilic
- at sp^3 C (S_N1, S_N2) Ch 6, 10, 14, 22
- at sp^2 C (Nuc. Arom. Subst.) Ch 17, 19
- at C=O (Nuc. Acyl Subst.) Ch 10, 11, 20, 21, 22

Electrophilic
- at sp^2 C (Elect. Arom. Subst.) Ch 17, 19

Radical
- at sp^3 C (alkane halogenation) Ch 4, 6, 16, 17
- at sp^2 C (Sandmeyer rxn) Ch 19

Organometallic
- Gilman Ch 10, 17
- Suzuki Ch 17
- Heck Ch 17

Addition

Nucleophilic
- at C=O (Nuc. Addn.) Ch 9, 10, 18, 22
- at C=C (conjugate addn.) Ch 22

Electrophilic
- at C=C (Elect. Addn.) Ch 8, 9, 10
- at C=C (Carbene Addn.) Ch 8

Radical
- at C=C (HBr + ROOR) Ch 8

Pericyclic
- cycloaddition (Diels-Alder) Ch 15

Oxidation
- epoxidation Ch 8, 10, 14

Reduction
- hydrogenation Ch 8, 9, 17, 18, 19

Elimination

Basic conditions (E2)
- E2 dehydrohalogenation Ch 7, 9
- tosylate elimination Ch 11
- Hofmann elimination Ch 19

Acidic conditions (E1)
- E1 dehydrohalogenation Ch 7
- dehydration of alcohols Ch 11

Pericyclic (Cope elimination)
- Ch 19

Oxidation/Reduction

Oxidation
- epoxidation Ch 8, 10, 14
- oxidative cleavage Ch 8, 9, 11, 17, 22
- oxygen functional groups Ch 11, 18, 19, 20

Reduction
- hydride reduction Ch 8, 10, 11, 17, 18, 19, 20, 21
- hydrogenation Ch 8, 9, 17, 18, 19
- metals Ch 9, 17, 18, 19

Essential Terms

acetal A derivative of an aldehyde or ketone having two alkoxy groups in place of the carbonyl group. The acetal of a ketone is sometimes called a **ketal.** (p. 911)

$$CH_3-\overset{O}{\overset{\|}{C}}-H + 2\,CH_3OH \underset{}{\overset{H^+}{\rightleftharpoons}} CH_3-\overset{CH_3O\;\;OCH_3}{C}-H + H_2O$$

acetaldehyde → acetaldehyde dimethyl acetal

ethylene acetal: **(ethylene ketal)** A cyclic acetal using ethylene glycol as the alcohol. (p. 914)

aldehyde A compound containing a carbonyl group bonded to an alkyl (or aryl) group and a hydrogen atom. (p. 876)

carbinolamine **(hemiaminal)** An intermediate in the formation of an imine, having an amine and a hydroxy group bonded to the same carbon atom. (p. 906)

$$R-\overset{O}{\overset{\|}{C}}-R + R'-NH_2 \rightleftharpoons \left[R-\overset{HO\;\;NH-R'}{C}-R\right] \rightleftharpoons R-\overset{N-R'}{\overset{\|}{C}}-R + H_2O$$

carbinolamine; imine

carbonyl group The C=O functional group. (p. 876)

Clemmensen reduction The deoxygenation of a ketone or aldehyde by treatment with zinc amalgam and dilute HCl. (p. 923)

condensation A reaction that joins two or more molecules, often with the loss of a small molecule such as water or an alcohol. (p. 906)

cyanohydrin A compound with a hydroxy group and a cyano group on the same carbon atom. Cyanohydrins are generally made by the reaction of a ketone or aldehyde with HCN. (p. 904)

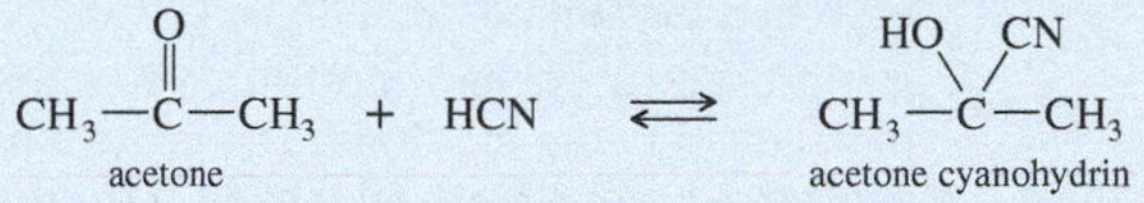

$$CH_3-\overset{O}{\overset{\|}{C}}-CH_3 + HCN \rightleftarrows CH_3-\overset{HO\ \ CN}{C}-CH_3$$

acetone — acetone cyanohydrin

deoxygenation A four-electron reduction that replaces the carbonyl oxygen atom of a ketone or aldehyde with two hydrogen atoms. The Clemmensen reduction and the Wolff–Kishner reduction are the two most common methods of deoxygenation. (p. 923)

DIBAL-H Diisobutylaluminum hydride, formula $(i\text{-Bu})_2AlH$. Used to reduce nitriles and esters selectively to aldehydes. (p. 894)

enol A vinyl alcohol. Simple enols generally tautomerize to their keto forms. (p. 891)

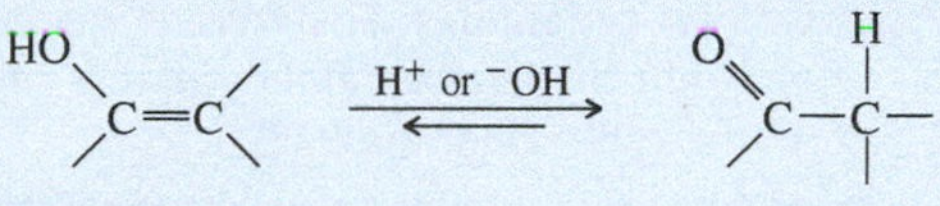

$$\text{(HO)C=C} \xrightleftharpoons{H^+ \text{ or } ^-OH} \text{O=C}-\text{C(H)}-$$

enol — keto

hemiacetal A derivative of an aldehyde or ketone similar to an acetal, but with just one alkoxy group and one hydroxy group on the former carbonyl carbon atom. (p. 912)

hemiaminal The IUPAC term for a carbinolamine, having an amine and a hydroxy group bonded to the same carbon atom. This term is analogous to the term *hemiacetal*, except with an amine instead of an alkoxy group. (p. 906)

hydrate (of an aldehyde or ketone) The geminal diol formed by addition of water across the carbonyl double bond. (p. 902)

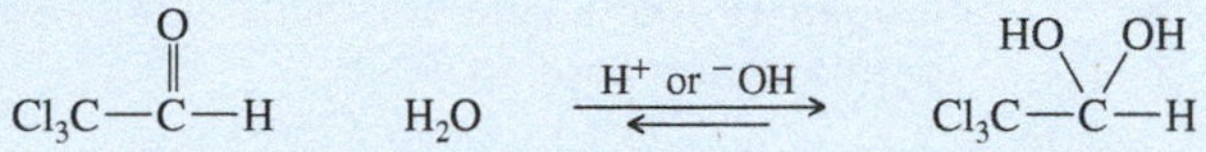

$$Cl_3C-\overset{O}{\overset{\|}{C}}-H \quad H_2O \xrightleftharpoons{H^+ \text{ or } ^-OH} Cl_3C-\overset{HO\ \ OH}{C}-H$$

chloral — chloral hydrate

hydrazone A compound containing the $C{=}N{-}NH_2$ group, formed by the reaction of a ketone or aldehyde with hydrazine. (p. 909)

2,4-DNP derivative: A hydrazone made using 2,4-dinitrophenylhydrazine. (p. 910)

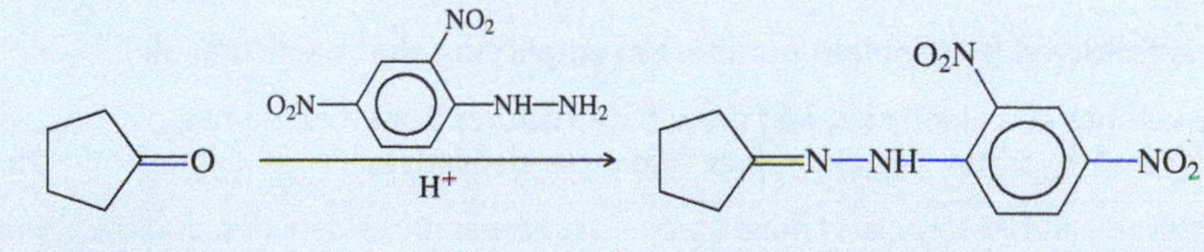

cyclopentanone $\xrightarrow[H^+]{O_2N-C_6H_3(NO_2)-NH-NH_2}$ 2,4-DNP derivative of cyclopentanone (cyclopentylidene=N—NH—$C_6H_3(NO_2)_2$)

imine A compound with a carbon–nitrogen double bond, formed by the reaction of a ketone or aldehyde with a primary amine. A substituted imine is often called a **Schiff base.** (p. 906)

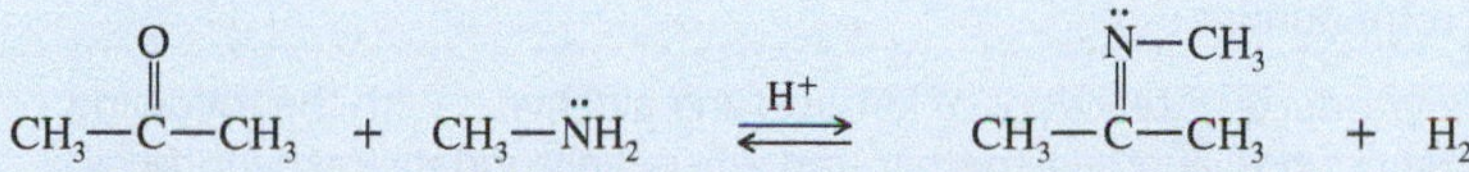

$$CH_3-\overset{O}{\overset{\|}{C}}-CH_3 + CH_3-\ddot{N}H_2 \xrightleftharpoons{H^+} CH_3-\overset{\ddot{N}-CH_3}{\overset{\|}{C}}-CH_3 + H_2O$$

acetone — methylamine — acetone methyl imine

ketal An acetal derived from a ketone. The IUPAC once abandoned this term, but later reinstated it as a subclass of acetals. (p. 911)

ketone A compound containing a carbonyl group bonded to two alkyl or aryl groups. (p. 876)

lithium dialkylcuprate **(Gilman reagent)** An organometallic reagent of formula R_2CuLi that couples with alkyl halides and acyl halides (acid chlorides). (p. 896)

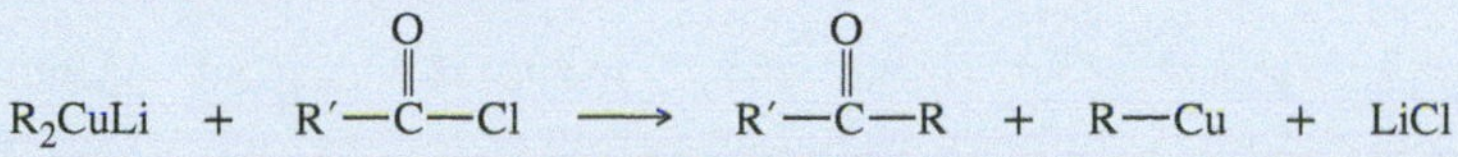

$$R_2CuLi + R'-\overset{O}{\overset{\|}{C}}-Cl \longrightarrow R'-\overset{O}{\overset{\|}{C}}-R + R-Cu + LiCl$$

McLafferty rearrangement In mass spectrometry, the loss of an alkene fragment by a cyclic rearrangement of a carbonyl compound having γ hydrogens. (p. 886)

nitrile A compound containing the cyano group, $C{\equiv}N$. (p. 906)

nucleophilic addition Addition of a reagent across a multiple bond by attack of a nucleophile at the electrophilic end of the multiple bond. As used in this chapter, nucleophilic addition is the addition of a nucleophile and a proton across the C=O bond. (p. 898)

oxime A compound containing the C=N—OH group, formed by the reaction of a ketone or aldehyde with hydroxylamine. (p. 909)

protecting group A group used to prevent a sensitive functional group from reacting while another part of the molecule is being modified. The protecting group is later removed. For example, an acetal can protect a ketone or an aldehyde from reacting under basic or neutral conditions. Dilute acid removes the acetal. (p. 916)

Raney nickel A finely divided, hydrogen-bearing form of nickel made by treating a nickel–aluminum alloy with strong sodium hydroxide. The aluminum in the alloy reacts to form hydrogen, leaving a finely divided nickel powder saturated with hydrogen. (p. 923)

semicarbazone A compound containing the C=N—NH—$CONH_2$ group, formed by the reaction of a ketone or aldehyde with semicarbazide. (p. 909)

Tollens test A test for aldehydes. The **Tollens reagent** is a silver–ammonia complex $[Ag(NH_3)_2^+ \ ^-OH]$. Tollens reagent oxidizes an aldehyde to a carboxylate salt and deposits a silver mirror on the inside of a glass container. (p. 922)

Wittig reaction Reaction of an aldehyde or ketone with a phosphorus ylide to form an alkene. One of the most versatile syntheses of alkenes. (p. 918)

$$R_2C{=}O + R'_2\ddot{C}^-{-}P^+Ph_3 \longrightarrow R_2C{=}CR'_2 + Ph_3P{=}O$$

ketone or aldehyde — phosphorus ylide — alkene

ylide: An uncharged molecule containing a carbon atom with a negative charge bonded to a heteroatom with a positive charge. A phosphorus ylide is the nucleophilic species in the Wittig reaction. (p. 918)

Wolff–Kishner reduction Deoxygenation of a ketone or aldehyde by conversion to the hydrazone, followed by treatment with a strong base. (p. 924)

Essential Problem-Solving Skills in Chapter 18

Each skill is followed by problem numbers exemplifying that particular skill.

1 Name ketones and aldehydes, and draw the structures from their names. Identify their hydrates, acetals, imines, and other derivatives.	Problems 18-37, 38, 40, 41, and 61
2 Interpret the IR, NMR, UV, and mass spectra of ketones and aldehydes, and use spectral information to determine the structures.	Problems 18-42, 43, 44, 45, 46, 71, 73, 74, 75, and 76
3 Propose single-step and multistep syntheses of ketones and aldehydes from alcohols, alkenes, alkynes, carboxylic acids, nitriles, acid chlorides, esters, and aromatic compounds.	Problems 18-55, 56, 60, and 64
4 Predict the products of reactions of ketones and aldehydes with the following types of compounds, and give mechanisms where appropriate: (a) hydride reducing agents; (b) Clemmensen and Wolff-Kishner reagents; (c) Grignard and organolithium reagents; (d) phosphorus ylides; (e) water; (f) hydrogen cyanide; (g) ammonia and primary amines; (h) hydroxylamine and hydrazine derivatives; (i) alcohols; and (j) oxidizing agents.	Problems 18-39, 48, 49, 53, 54, 65, and 66
5 Use your knowledge of the mechanisms of ketone and aldehyde reactions to propose mechanisms and products of similar reactions you have never seen before.	Problems 18-50, 58, 62, 63, 68, 69, 70, 72, 73, and 75
Problem-Solving Strategy: Proposing Reaction Mechanisms	Problems 18-50, 58, 62, 63, 68, 69, 70, 72, 73, and 75
6 Show how to convert ketones and aldehydes to other functional groups, and devise multistep syntheses using ketones and aldehydes as starting materials and intermediates.	Problems 18-47, 51, 52, 59, and 60

Study Problems

18-37 Draw structures of the following derivatives.

(a) the 2,4-dinitrophenylhydrazone of benzaldehyde
(b) the semicarbazone of cyclobutanone
(c) cyclopropanone oxime
(d) the ethylene acetal of hexan-3-one
(e) acetaldehyde dimethyl acetal
(f) the methyl hemiacetal of formaldehyde
(g) the (*E*) isomer of the ethyl imine of propiophenone
(h) the hemiacetal form of 5-hydroxypentanal

18-38 Name the following ketones and aldehydes. When possible, give both a common name and an IUPAC name.

(a) $CH_3CO(CH_2)_4CH_3$
(b) $CH_3(CH_2)_2CO(CH_2)_2CH_3$
(c) $CH_3(CH_2)_5CHO$
(d) PhCOPh
(e) $CH_3CH_2CH_2CHO$
(f) CH_3COCH_3
(g) $CH_3CH_2CHBrCH_2CH(CH_3)CHO$
(h) Ph—CH=CH—CHO
(i) $CH_3CH{=}CH{-}CH{=}CH{-}CHO$
(j) 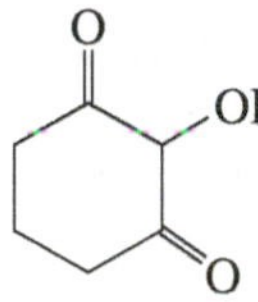 O, OH, O
(k) 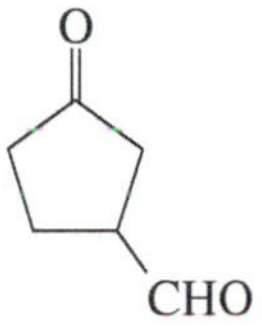 O, CHO
(l) 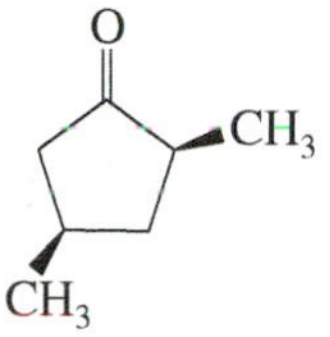 O, CH_3, CH_3

18-39 Predict the major products of the following reactions.

(a)  O; $PhNHNH_2$, H^+
(b) O, OCH_3; (1) DIBAL-H (−78 °C) (2) H_2O
(c) CN; (1) DIBAL-H (2) H_3O^+
(d) Cl, O; $(CH_2CH)_2CuLi$
(e) OCH_3, O; H_3O^+
(f) Cl, O; $LiAlH(O\text{-}t\text{-}Bu)_3$
(g) COOH; (1) 2 CH_3CH_2Li (2) H_3O^+
(h) H, C=PPh_3 + O →
(i) 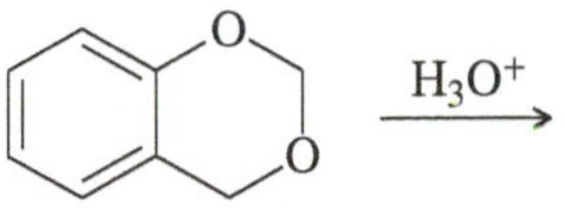 O, O; H_3O^+
(j) 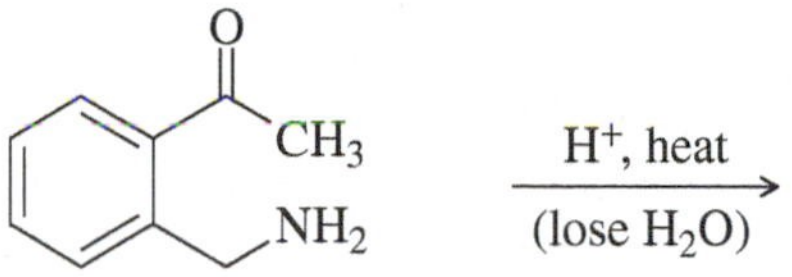 O, CH_3, NH_2; H^+, heat (lose H_2O)

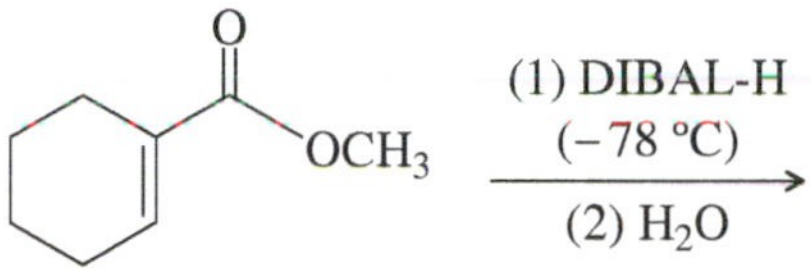

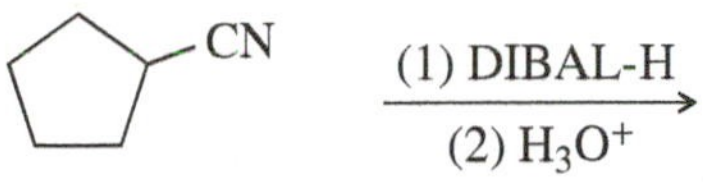

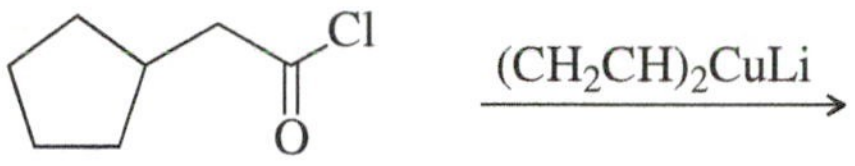

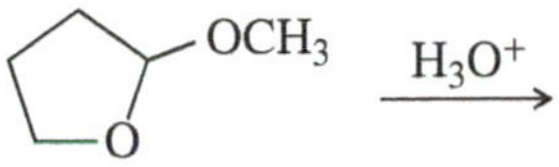

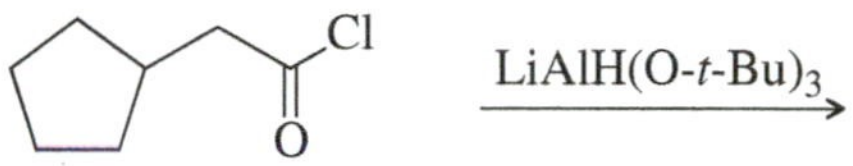

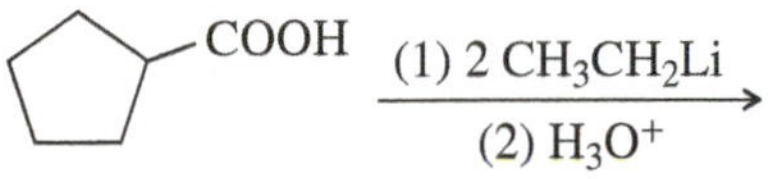

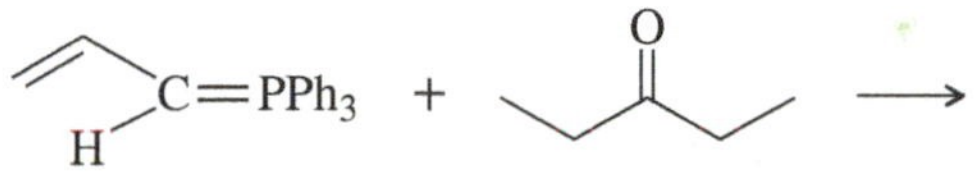

18-40 Rank the following carbonyl compounds in order of increasing equilibrium constant for hydration:

CH_3COCH_2Cl $ClCH_2CHO$ CH_2O CH_3COCH_3 CH_3CHO

18-41 Acetals can serve as protecting groups for 1,2-diols, as well as for aldehydes and ketones. When the acetal is formed from acetone plus the diol, the acetal is called an acetonide. Show the acetonides formed from these diols with acetone under acid catalysis.

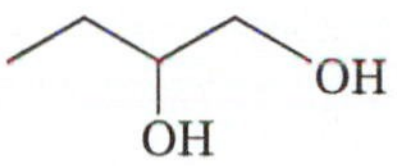

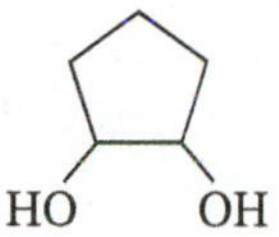

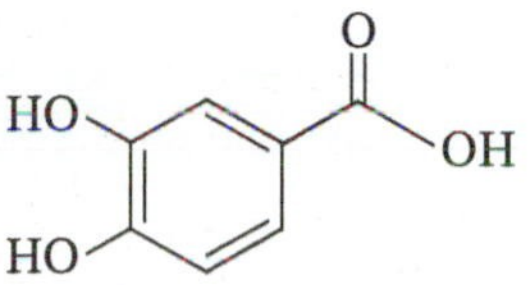

OH, OH HO, OH HO, HO, O, OH

18-42 Sketch the expected proton NMR spectrum of 3,3-dimethylbutanal.

18-43 A compound of formula $C_6H_{10}O_2$ shows only two absorptions in the proton NMR: a singlet at 2.67 ppm and a singlet at 2.15 ppm. These absorptions have areas in the ratio 2:3. The IR spectrum shows a strong absorption at 1708 cm^{-1}. Propose a structure for this compound.

18-44 The proton NMR spectrum of a compound of formula $C_{10}H_{12}O$ follows. This compound reacts with an acidic solution of 2,4-dinitrophenylhydrazine to give a crystalline derivative, but it gives a negative Tollens test. Propose a structure for this compound and give peak assignments to account for the signals in the spectrum.

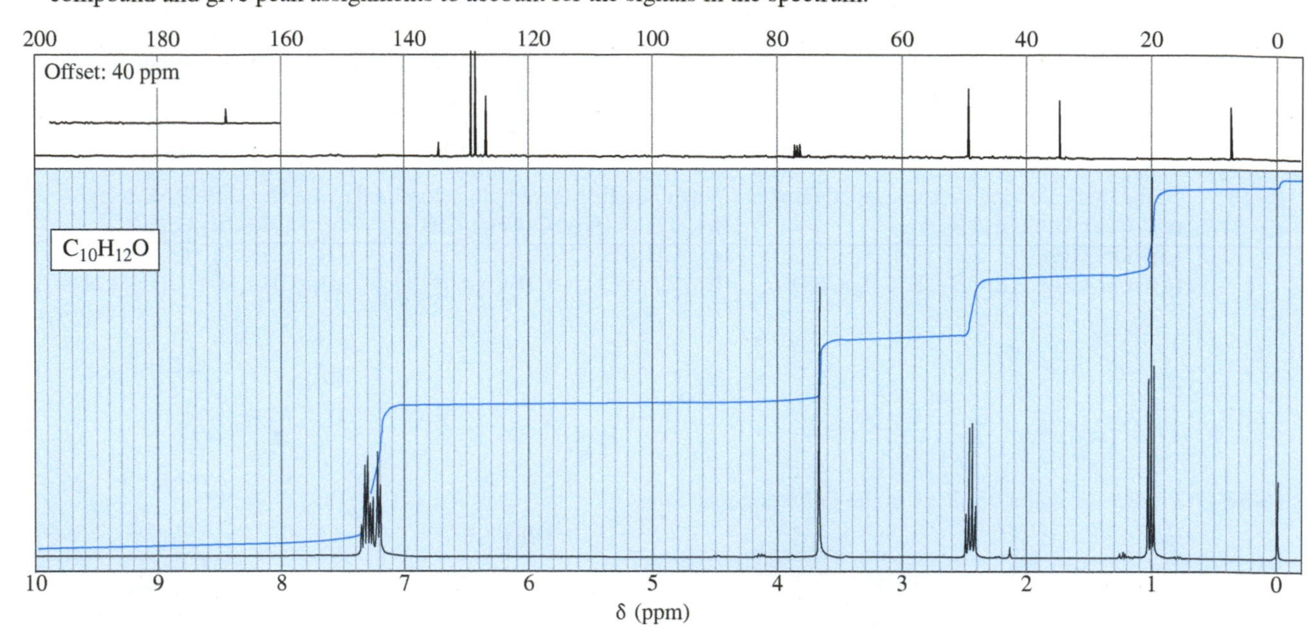

18-45 The following compounds undergo McLafferty rearrangement in the mass spectrometer. Predict the masses of the resulting charged fragments.

(a) pentanal **(b)** 3-methylhexan-2-one **(c)** 4-methylhexan-2-one

18-46 An unknown compound gives a molecular ion of m/z 70 in the mass spectrum. It reacts with semicarbazide hydrochloride to give a crystalline derivative, but it gives a negative Tollens test. The NMR and IR spectra follow. Propose a structure for this compound, and give peak assignments to account for the absorptions in the spectra. Explain why the signal at 1790 cm^{-1} in the IR spectrum appears at an unusual frequency.

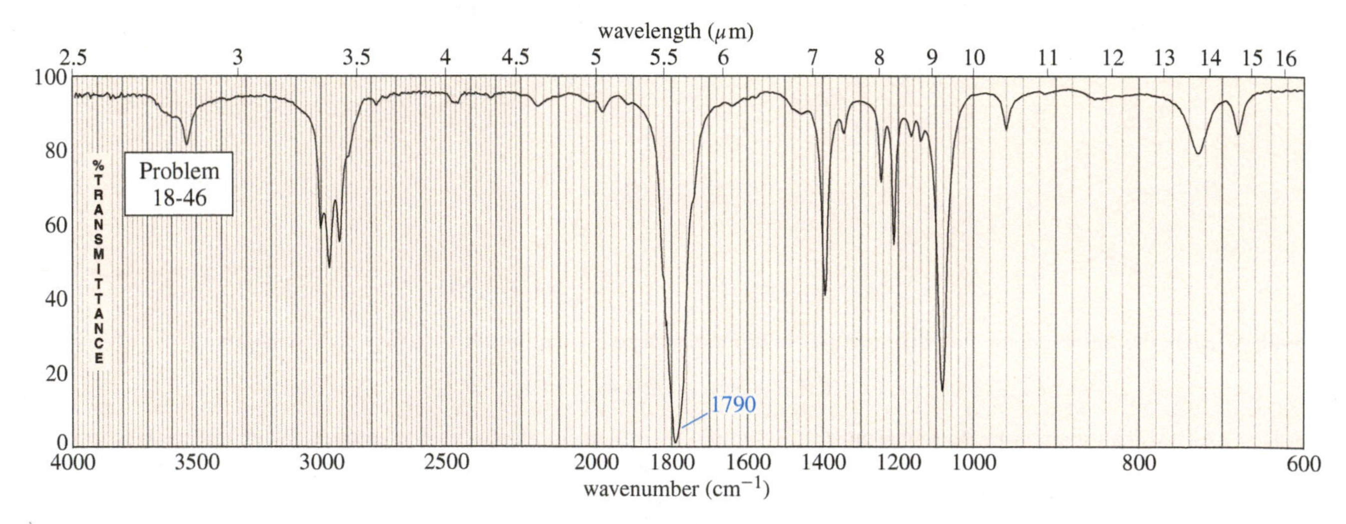

(continued)

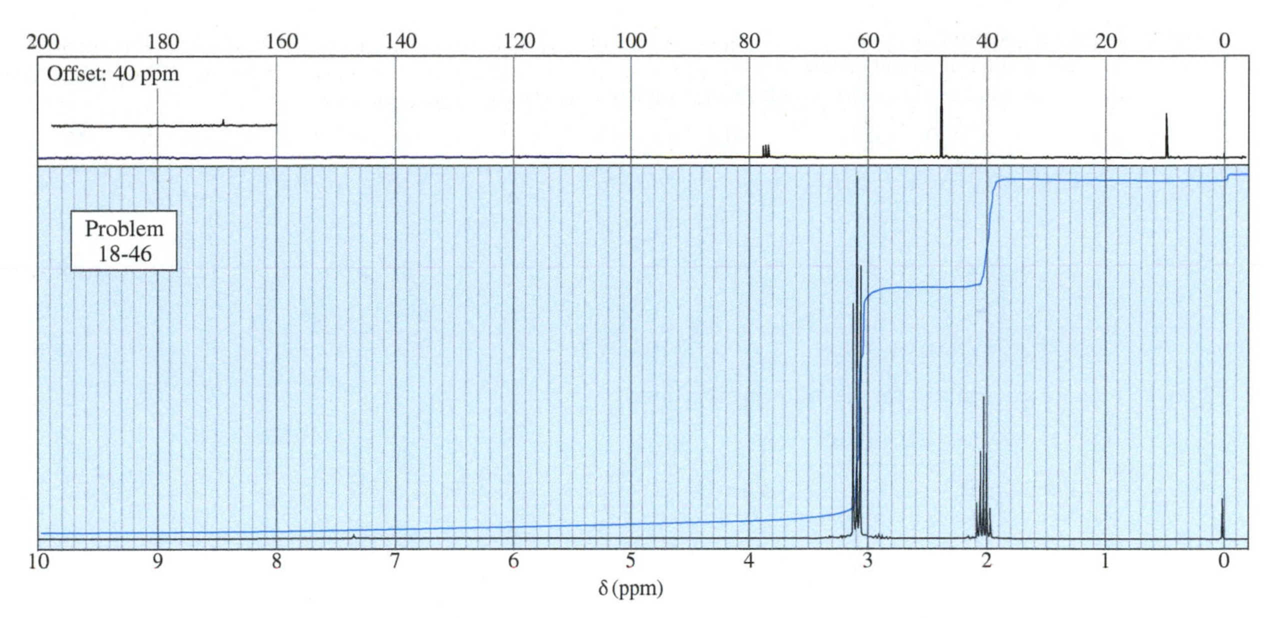

18-47 Show how you would accomplish the following synthetic conversions efficiently and in good yield. You may use any necessary additional reagents and solvents.

(a) O →

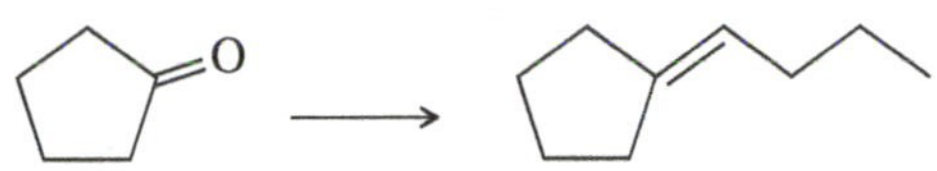

(b) OCH_3 → OCH_3 CHO

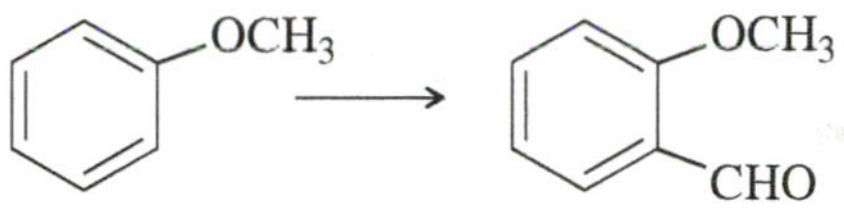

(c) CH_2Br → O H

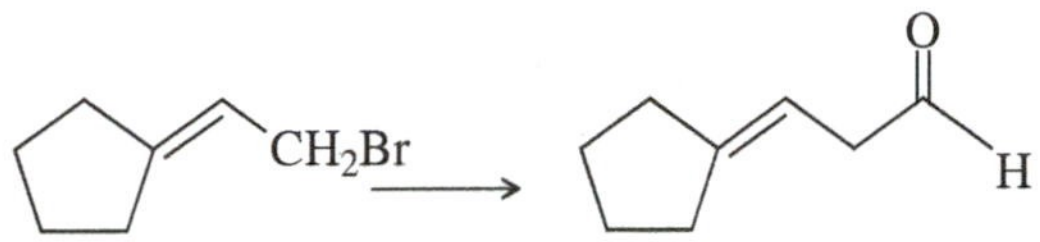

(d) CH_2Br →

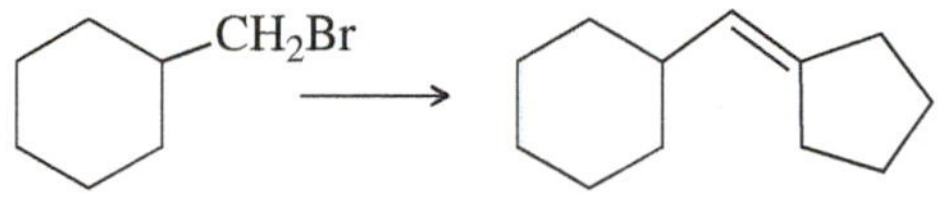

(e) COOH → O

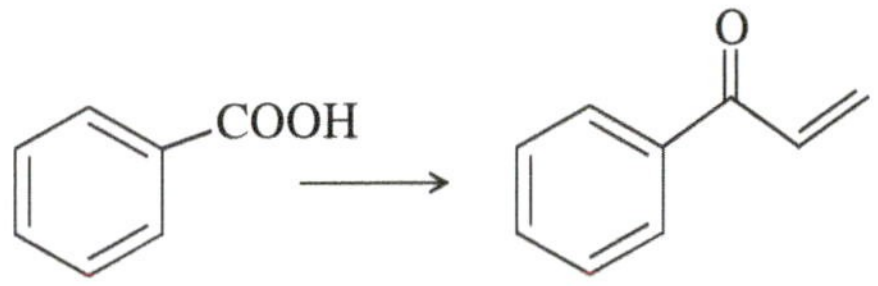

(f) O O → O H OH

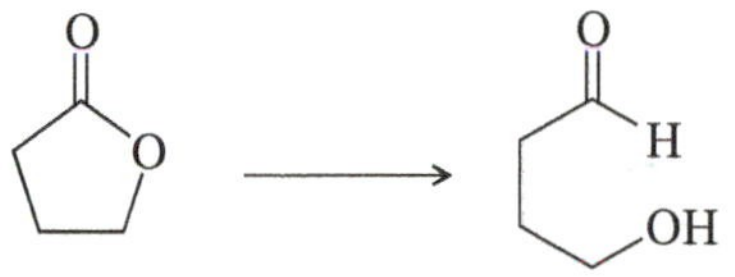

(g) COOH → O Ph

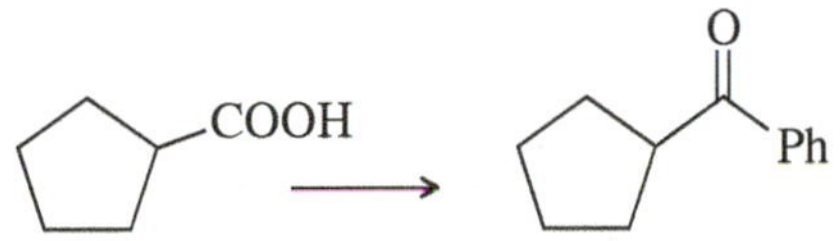

(h) CH_2OH HO CH CH_3 → CHO H_2C=C CH_3

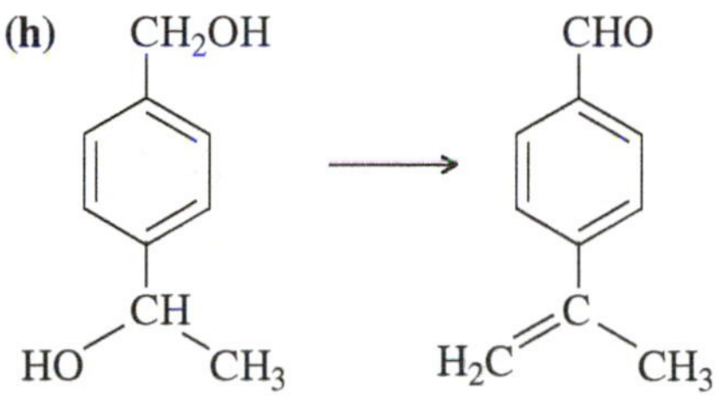

18-48 The following road-map problem centers on the structure and properties of **A,** a key intermediate in these reactions. Give structures for compounds **A** through **J.**

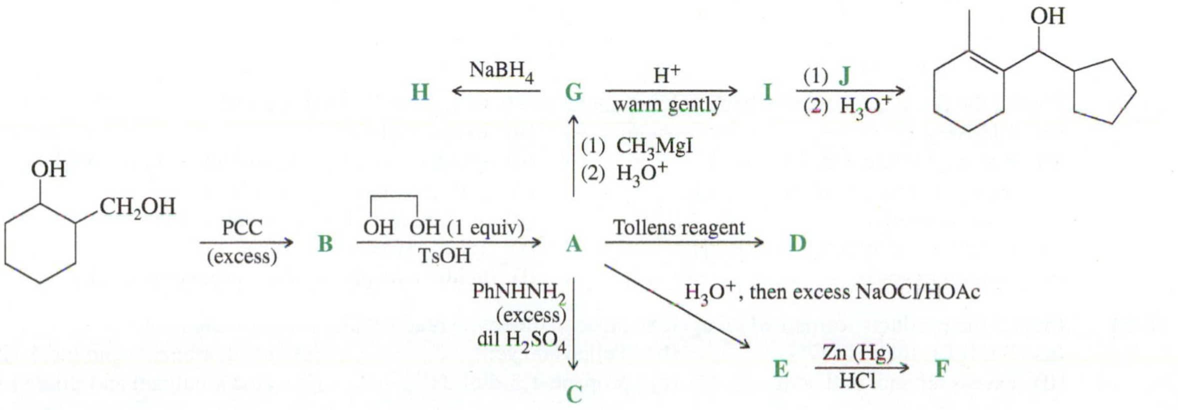

OH CH_2OH —PCC (excess)→ B —OH OH (1 equiv), TsOH→ A

A —Tollens reagent→ D

A —PhNHNH$_2$ (excess), dil H_2SO_4→ C

A —H_3O^+, then excess NaOCl/HOAc→ E —Zn (Hg), HCl→ F

A —(1) CH_3MgI (2) H_3O^+→ G

H ←$NaBH_4$— G —H^+, warm gently→ I —(1) J (2) H_3O^+→ OH

18-49 For each compound,

1. name the functional group.
2. show what compound(s) result from complete hydrolysis.

(a) 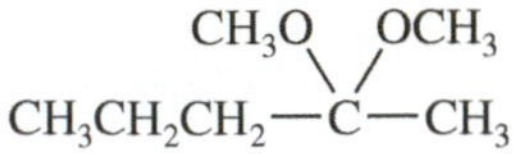 $CH_3CH_2CH_2-C(OCH_3)_2-CH_3$ (b) 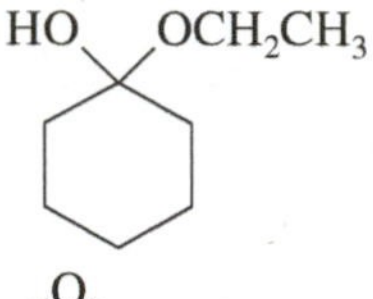 HO, OCH_2CH_3 on cyclohexane (c) 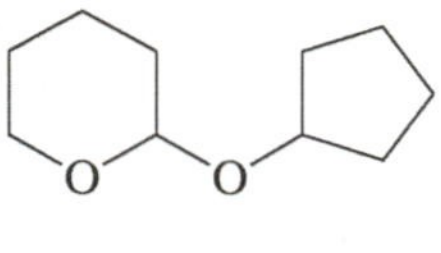 (d) 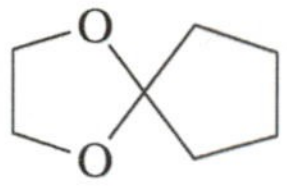 (e) 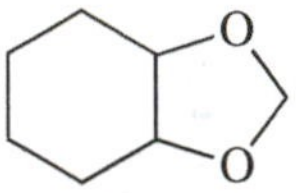 (f) 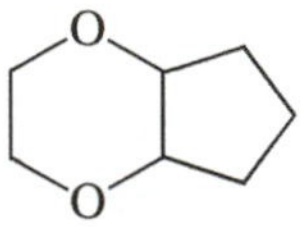 (g) 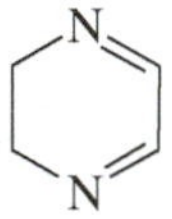 N N (h) 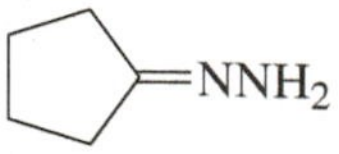 $=NNH_2$

18-50 Propose mechanisms for the following reactions.

(a) 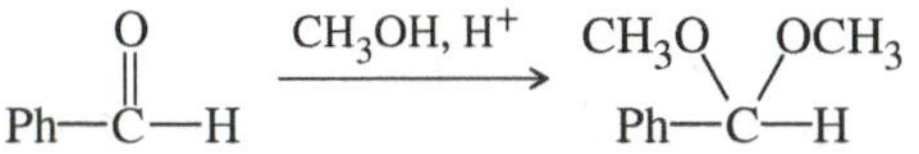 $Ph-C(=O)-H \xrightarrow{CH_3OH,\ H^+} Ph-C(OCH_3)(OCH_3)-H$

(b) 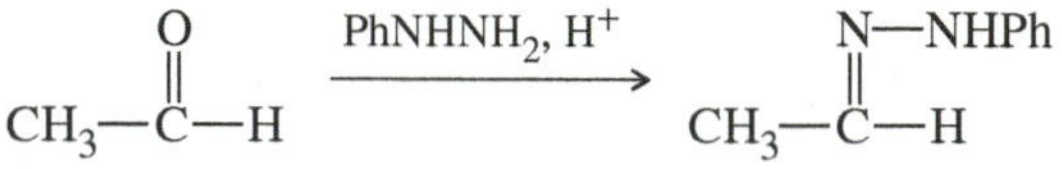 $CH_3-C(=O)-H \xrightarrow{PhNHNH_2,\ H^+} CH_3-C(=N-NHPh)-H$

(c) cyclohexanone $\xrightarrow{Ph_3P{=}CH_2}$ $=CH_2$

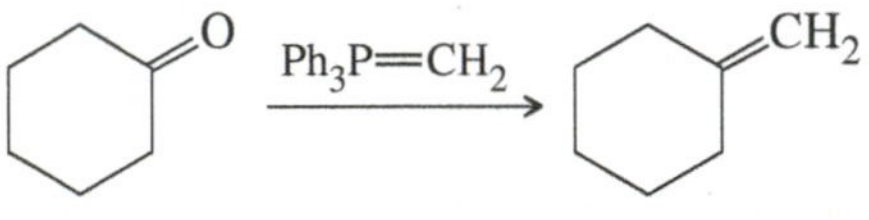

(d) 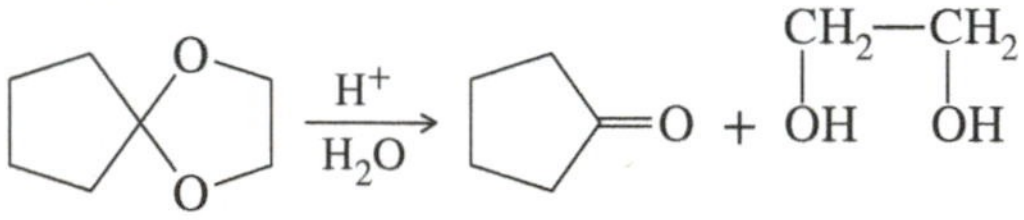 $\xrightarrow[H_2O]{H^+}$ cyclopentanone $=O$ + $CH_2(OH)-CH_2(OH)$

(e) 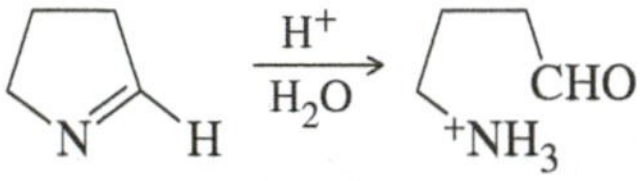 N H $\xrightarrow[H_2O]{H^+}$ CHO $^+NH_3$

(f) 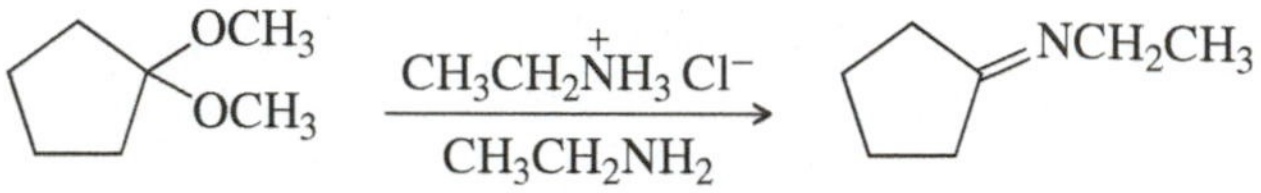 OCH_3 OCH_3 $\xrightarrow[CH_3CH_2NH_2]{CH_3CH_2\overset{+}{N}H_3\ Cl^-}$ $=NCH_2CH_3$

18-51 Show how you would accomplish the following syntheses efficiently and in good yield. You may use any necessary reagents.

(a) acetaldehyde ⟶ lactic acid, $CH_3CH(OH)COOH$

(b) 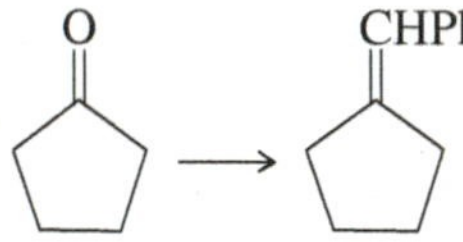 O ⟶ CHPh

(c) 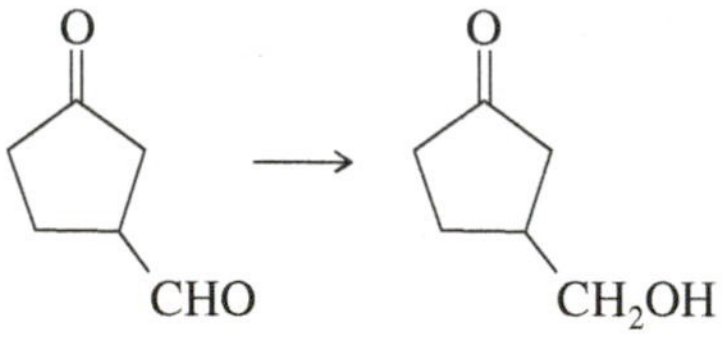 O, CHO ⟶ O, CH_2OH

(d) 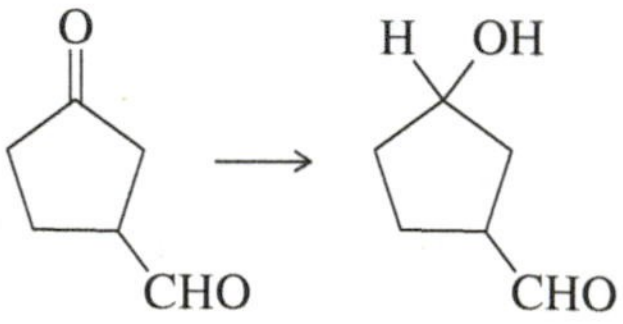 O, CHO ⟶ H OH, CHO

(e) 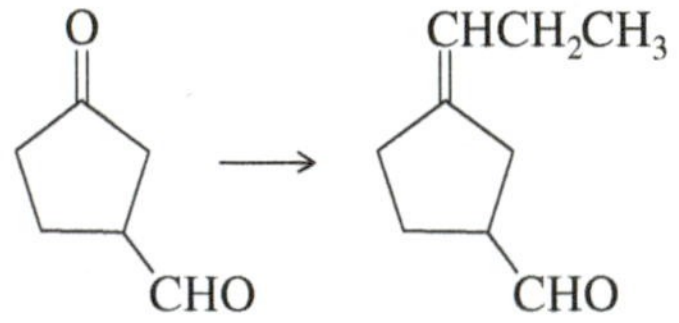 O, CHO ⟶ $CHCH_2CH_3$, CHO

(f) 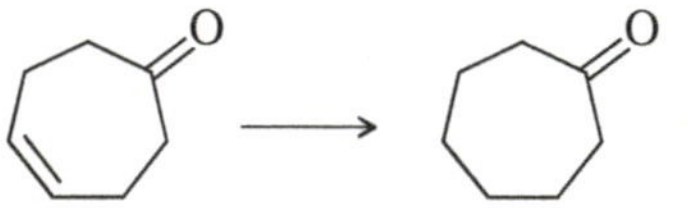 O ⟶ O

(g) 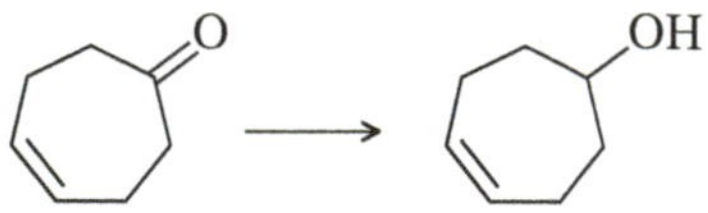 O ⟶ OH

18-52 Show how you would synthesize the following derivatives from appropriate carbonyl compounds.

(a) 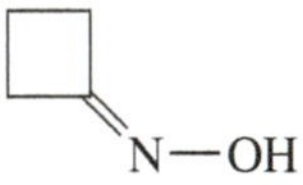 N—OH (b) 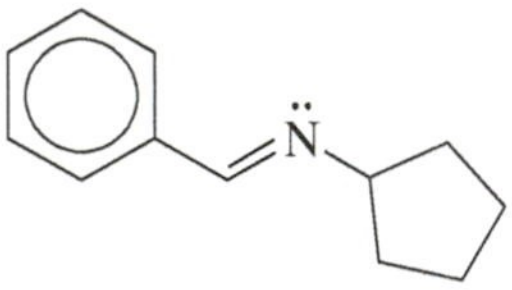 N̈ (c) 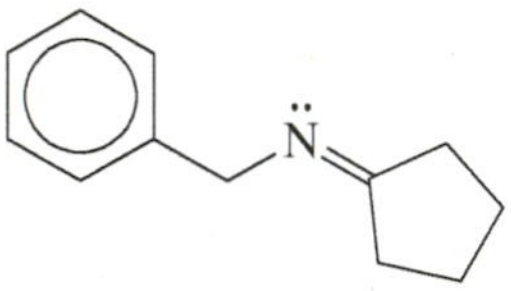 N̈ (d) 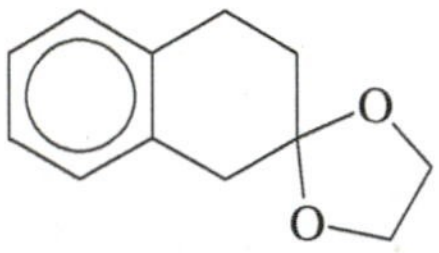 O O (e) 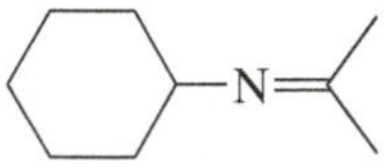 N= (f) 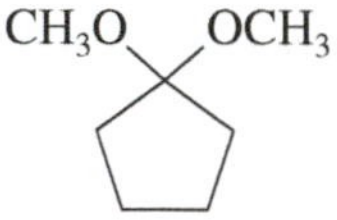 CH_3O OCH_3

18-53 Predict the products formed when cyclohexanone reacts with the following reagents.

(a) CH_3NH_2, H^+
(b) excess CH_3OH, H^+
(c) hydroxylamine and weak acid
(d) ethylene glycol and *p*-toluenesulfonic acid
(e) phenylhydrazine and weak acid
(f) PhMgBr and then mild H_3O^+
(g) Tollens reagent
(h) sodium acetylide, then mild H_3O^+
(i) hydrazine, then hot, fused KOH
(j) $Ph_3P{=}CH_2$
(k) sodium cyanide
(l) acidic hydrolysis of the product from (k)

18-54 Predict the products formed when cyclohexanecarbaldehyde reacts with the following reagents.

(a) PhMgBr, then H_3O^+
(b) Tollens reagent
(c) semicarbazide and weak acid
(d) excess ethanol and acid
(e) propane-1,3-diol, H^+
(f) zinc amalgam and dilute hydrochloric acid

18-55 Show how you would synthesize octan-2-one from each compound. You may use any necessary reagents.
(a) heptanal **(b)** oct-1-yne **(c)** 2,3-dimethylnon-2-ene
(d) octan-2-ol **(e)** heptanoic acid **(f)** $CH_3(CH_2)_5CN$

18-56 Show how you would synthesize octanal from each compound. You may use any necessary reagents.
(a) octan-1-ol **(b)** non-1-ene **(c)** oct-1-yne **(d)** 1-bromoheptane
(e) 1-bromohexane **(f)** octanoic acid **(g)** ethyl octanoate

18-57 Both $NaBH_4$ and $NaBD_4$ are commercially available, and D_2O is common and inexpensive. Show how you would synthesize the following labeled compounds, starting with butan-2-one.

(a) $CH_3-C(OH)(D)-CH_2-CH_3$ **(b)** $CH_3-C(OD)(D)-CH_2-CH_3$ **(c)** $CH_3-C(OD)(H)-CH_2-CH_3$

18-58 When $LiAlH_4$ reduces 3-methylcyclopentanone, the product mixture contains 60% *cis*-3-methylcyclopentanol and 40% *trans*-3-methylcyclopentanol. Use your models, and make three-dimensional drawings to explain this preference for the cis isomer.

18-59 The Wittig reaction is useful for placing double bonds in less stable positions. For example, the following transformation is easily accomplished using a Wittig reaction.

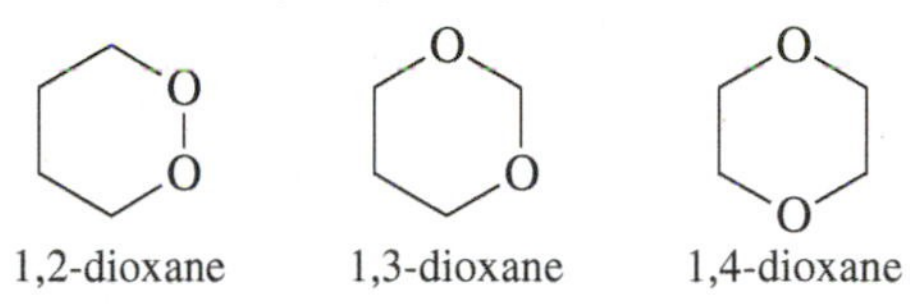

(a) Show how you would use a Wittig reaction to do this.
(b) Show how you might do this *without* using a Wittig reaction, and then explain why the Wittig reaction is a much better synthesis.

18-60 Show how you would accomplish the following syntheses.
(a) benzene ⟶ *n*-butylbenzene
(b) benzonitrile ⟶ propiophenone
(c) benzene ⟶ *p*-methoxybenzaldehyde
(d) $Ph-(CH_2)_4-OH$ ⟶ tetralone

18-61 There are three dioxane isomers: 1,2-dioxane, 1,3-dioxane, and 1,4-dioxane. One of these acts like an ether and is an excellent solvent for Grignard reactions. Another one is potentially explosive when heated. The third one quickly hydrolyzes in dilute acid. Show which isomer acts like a simple ether, and then explain why one of them is potentially explosive. Propose a mechanism for the acid hydrolysis of the third isomer.

1,2-dioxane 1,3-dioxane 1,4-dioxane

18-62 Two structures for the sugar **glucose** are shown on page 914. Interconversion of the open-chain and cyclic hemiacetal forms is catalyzed by either acid or base.
(a) Propose a mechanism for the cyclization, assuming a trace of acid is present.
(b) The cyclic hemiacetal is more stable than the open-chain form, so very little of the open-chain form is present at equilibrium. Will an aqueous solution of glucose reduce Tollens reagent and give a positive Tollens test? Explain.

18-63 Two structures of the sugar **fructose** are shown next. The cyclic structure predominates in aqueous solution.
(a) Number the carbon atoms in the cyclic structure. What is the functional group at C2 in the cyclic form?
(b) Propose a mechanism for the cyclization, assuming a trace of acid is present.

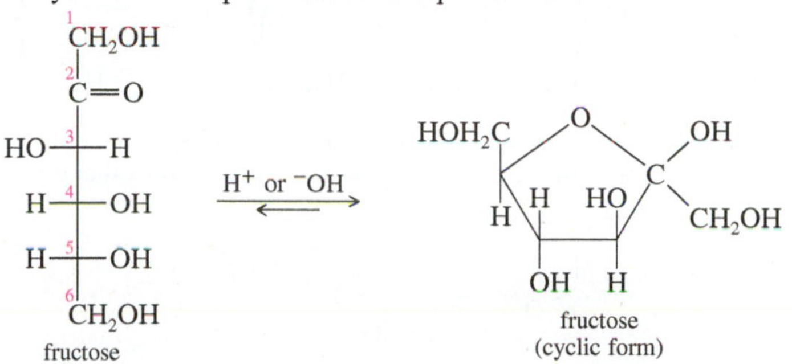

18-64 Hydration of alkynes (via oxymercuration) gives good yields of single compounds only with symmetrical or terminal alkynes. Show what the products would be from hydration of each compound.

(a) hex-3-yne **(b)** hex-2-yne **(c)** hex-1-yne
(d) cyclodecyne **(e)** 3-methylcyclodecyne

18-65 Which of the following compounds would give a positive Tollens test? (Remember that the Tollens test involves mild basic aqueous conditions.)

(a) $CH_3CH_2CH_2COCH_3$ **(b)** $CH_3CH_2CH_2CH_2CHO$ **(c)** $CH_3CH{=}CHCH{=}CHOH$
(d) $CH_3CH_2CH_2CH_2CH(OH)OCH_3$ **(e)** $CH_3CH_2CH_2CH_2CH(OCH_3)_2$ **(f)** O, OH

18-66 Solving the following road-map problem depends on determining the structure of **A,** the key intermediate. Give structures for compounds **A** through **K.**

D —$SOCl_2$→ E —$(CH_3)_2CuLi$→ A
hept-1-yne → F —(1) O_3, −78 °C; (2) $(CH_3)_2S$→ A (from G)
$CH_3(CH_2)_4MgBr$ —(1) **B**; (2) H_3O^+→ C —NaOCl, HOAc→ A
A —HCN→ H —H_3O^+→ I
A —(1) **J**; (2) H_3O^+→ OH, Ph, CH_3
A —(1) **K**; (2) H_3O^+→ OH, CH_3, O

18-67 Within each set of structures, indicate which will react fastest, and which slowest, toward nucleophilic addition in basic conditions.

(a) O, CH_3 O, H O, CH_3

(b)

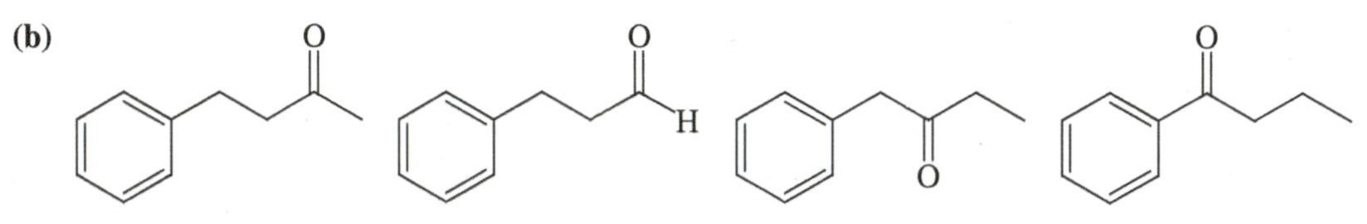

(c) F_3C–CO–CF_3 H_3C–CO–CF_3 H_3C–CO–CH_3

18-68 One of these reacts with dilute aqueous acid and the other does not. Give a mechanism for the one that reacts, and show why this mechanism does not work for the other one.

(a) O, O $\xrightarrow{H_3O^+}$? **(b)** O, O $\xrightarrow{H_3O^+}$?

18-69 Show a complete mechanism for this equilibrium established in diethyl ether with HCl gas as catalyst.

OH — N ⇌ [HCl(g), Et_2O] — O — NH

18-70 Show a complete mechanism for this reaction.

H–N ... O + CH_3OH ⇌ [HCl] N — OCH_3 + H_2O

18-71 The UV spectrum of an unknown compound shows values of λ_{max} at 225 nm (ε = 10,000) and at 318 nm (ε = 40). The mass spectrum shows a molecular ion at m/z 96 and a prominent base peak at m/z 68. The IR and NMR spectra follow. Propose a structure, and show how your structure corresponds to the observed absorptions. Propose a favorable fragmentation to account for the MS base peak at m/z 68 (loss of C_2H_4).

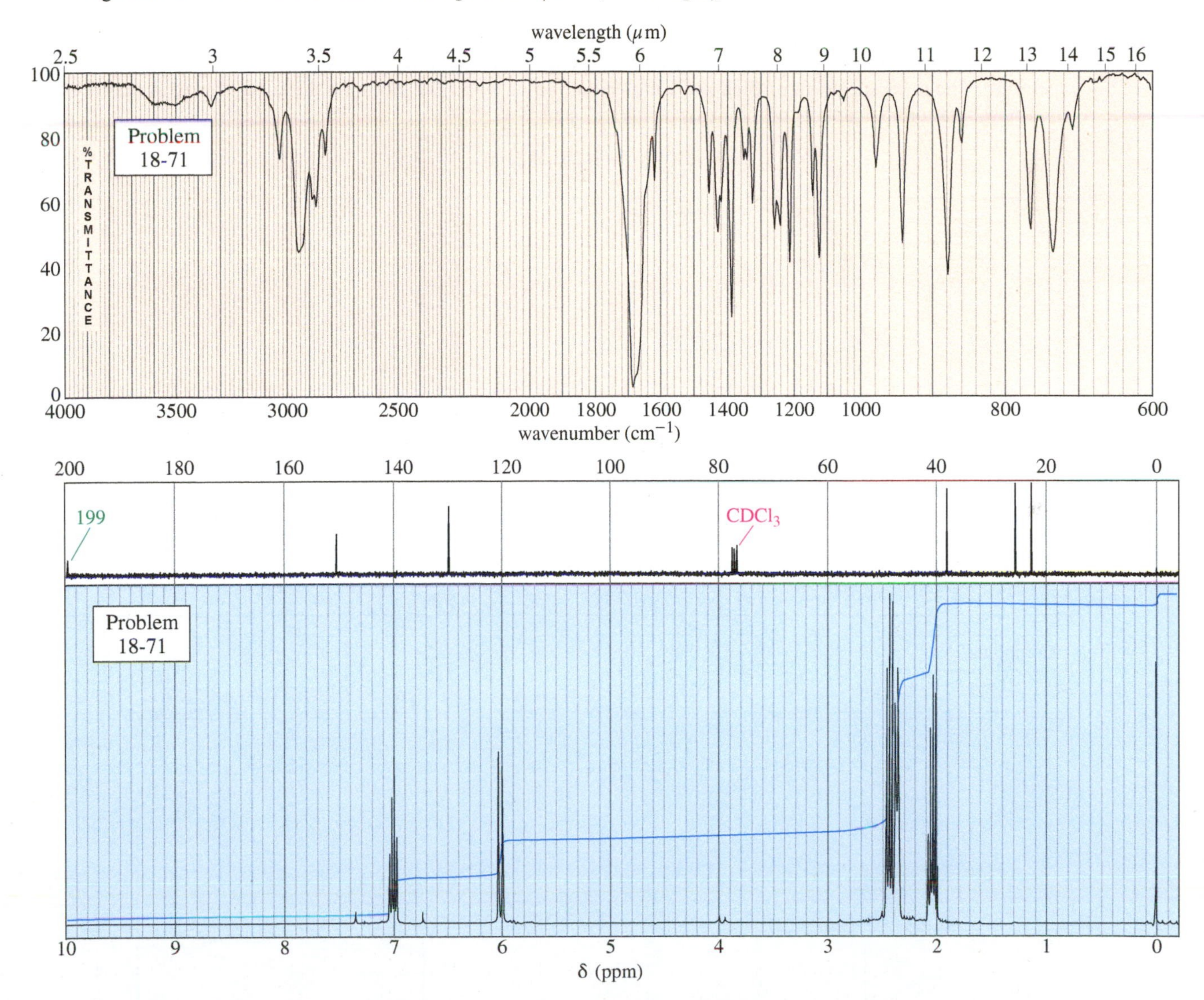

***18-72** **(a)** Simple aminoacetals hydrolyze quickly and easily in dilute acid. Propose a mechanism for hydrolysis of the following aminoacetal:

O — $N(CH_3)_2$, H → [H_3O^+] OH H — =O + $(CH_3)_2\overset{+}{N}H_2$

(continued)

(b) The nucleosides that make up DNA have heterocyclic rings linked to deoxyribose by an aminoacetal functional group. Point out the aminoacetal linkages in deoxycytidine and deoxyadenosine.

deoxycytidine   deoxyadenosine

(c) The stability of our genetic code depends on the stability of DNA. We are fortunate that the aminoacetal linkages of DNA are not easily cleaved. Show why your mechanism for part (a) does not work so well with deoxycytidine and deoxyadenosine.

18-73 The mass spectrum of unknown compound **A** shows a molecular ion at m/z 116 and prominent peaks at m/z 87 and m/z 101. Its UV spectrum shows no maximum above 200 nm. The IR and NMR spectra of **A** follow. When **A** is washed with dilute aqueous acid, extracted into dichloromethane, and the solvent evaporated, it gives a product **B. B** shows a strong carbonyl signal at 1715 cm^{-1} in the IR spectrum and a weak maximum at 274 nm (ε = 16) in the UV spectrum. The mass spectrum of **B** shows a molecular ion of m/z 72. Determine the structures of **A** and **B**, and show the fragmentation to account for the peaks at m/z 87 and 101.

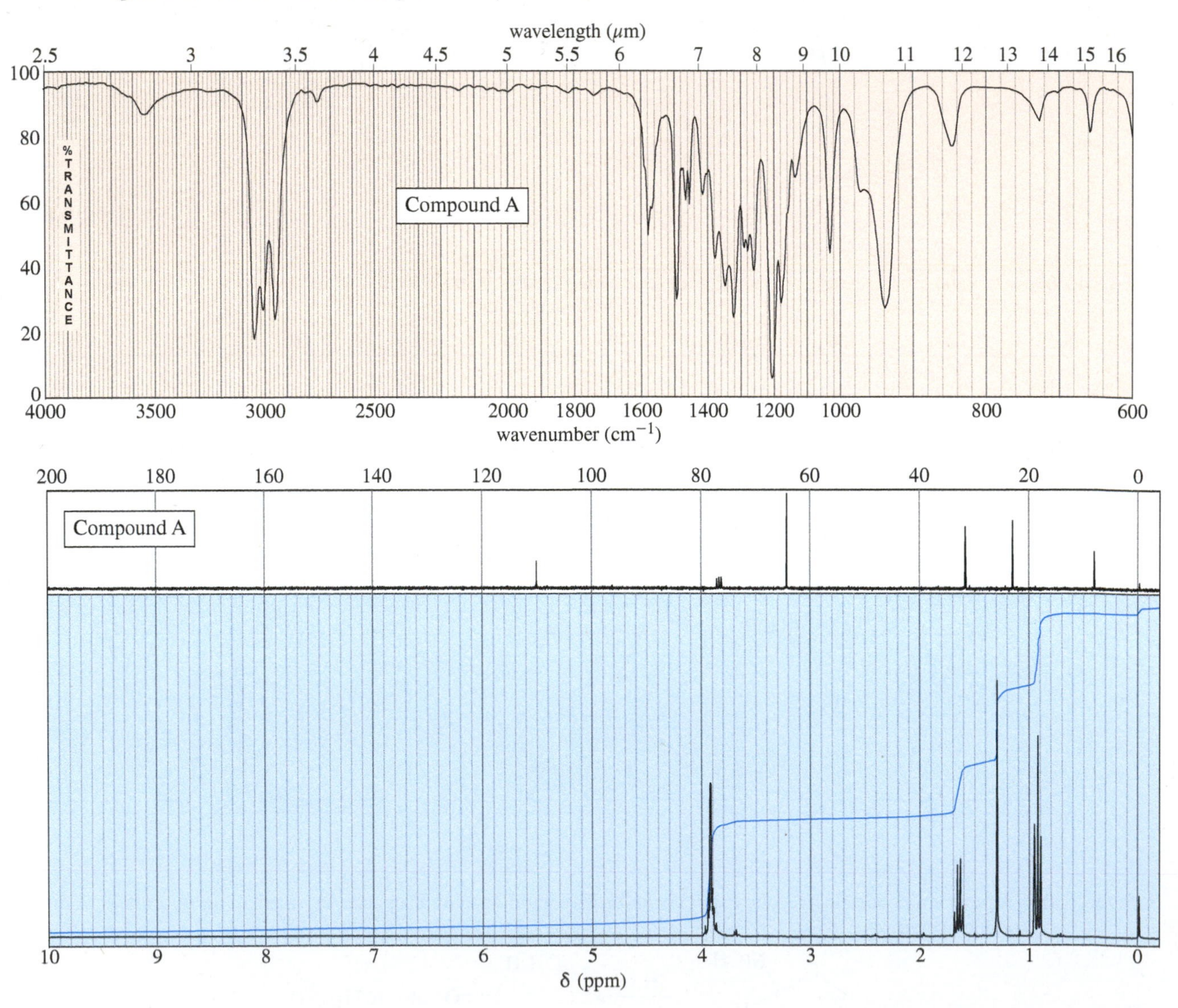

*18-74 (A true story.) The chemistry department custodian was cleaning the organic lab when an unmarked bottle fell off a shelf and smashed on the floor, leaving a puddle of volatile liquid. The custodian began to wipe up the puddle, but he was overcome with burning in his eyes and a feeling of having an electric drill thrust up his nose. He left the room and called the fire department, who used breathing equipment to go in and clean up the chemical. Three students were asked to identify the chemical quickly so that the custodian could be treated and the chemical could be handled properly. The students took IR and NMR spectra, which follow. The UV spectrum showed λ_{max} at 220 nm (ε = 16,000). The mass spectrometer was down, so no molecular weight was available. Determine the structure of this nasty compound, and show how your structure fits the spectra.

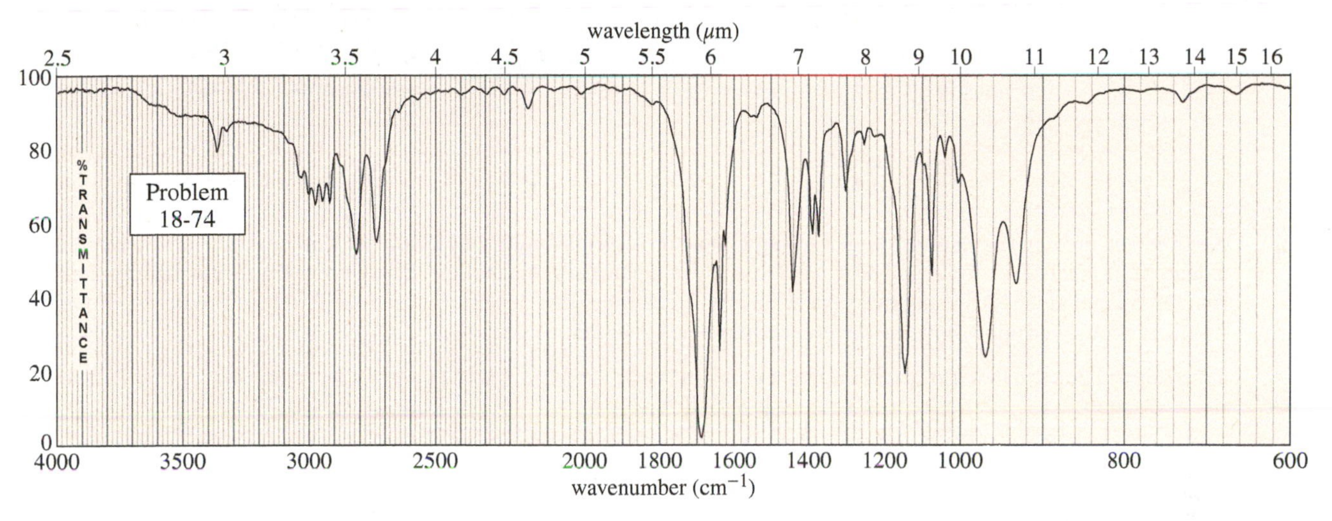

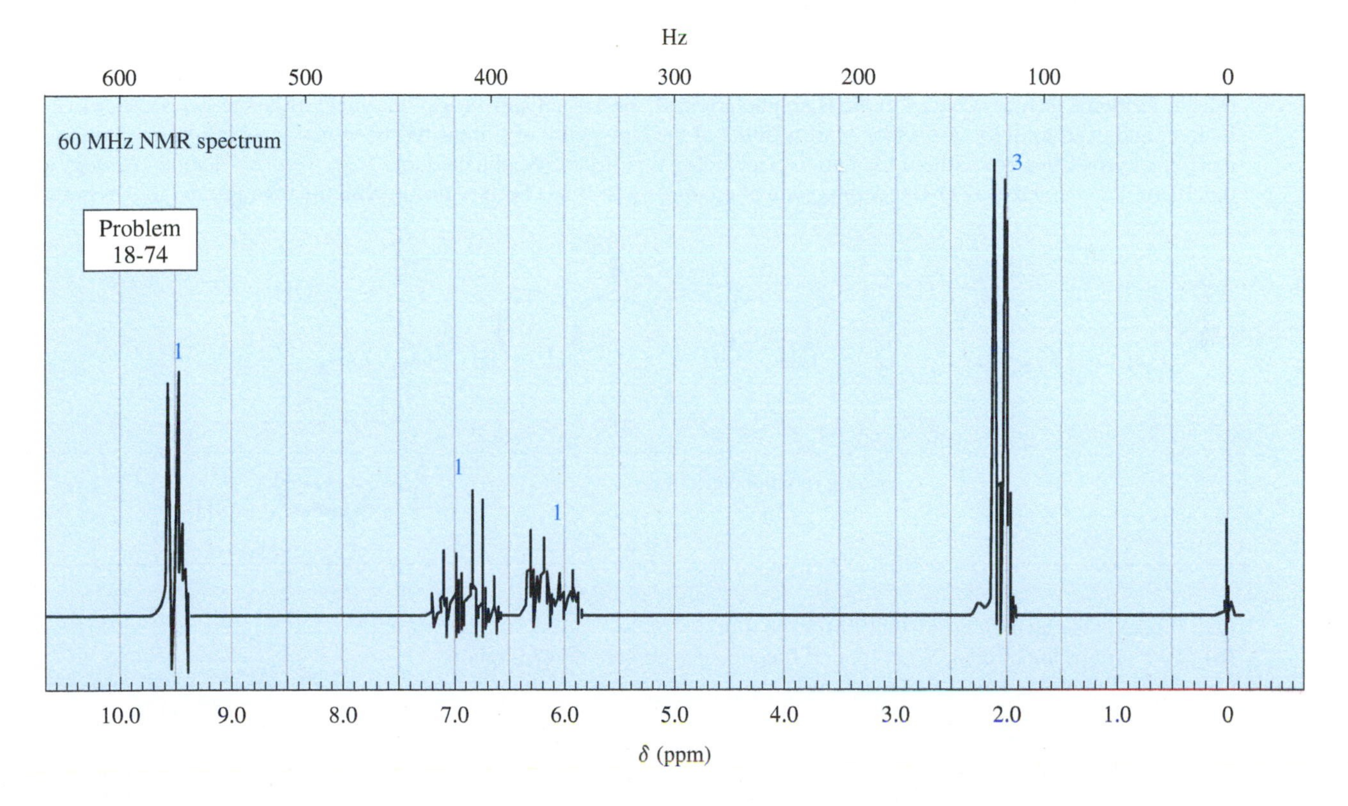

*18-75 In the absence of water, *o*-phthalaldehyde has the structure shown. Its strongest IR absorption is at 1687 cm^{-1}; the proton NMR data are shown by the structure. In the presence of water, a new compound is formed that has a strong IR absorption around 3400 cm^{-1} and no absorption in the C=O region. The proton NMR data are shown. Propose a structure of **X** consistent with this information, and suggest how **X** was formed.

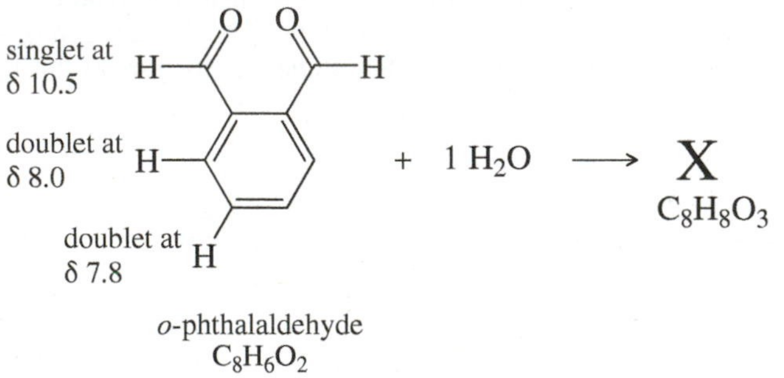

singlet at δ 10.5

doublet at δ 8.0

doublet at δ 7.8

o-phthalaldehyde $C_8H_6O_2$

+ 1 H_2O ⟶ X $C_8H_8O_3$

HNMR data
broad singlet at δ 5.0 (2H)
sharp singlet at δ 5.9 (2H)
doublet at δ 7.2 (2H)
doublet at δ 7.3 (2H)

*18-76 Assume you are a research physiologist trying to unravel a serious metabolic disorder. You have fed your lab animal Igor a deuterium-labeled substrate and now need to analyze the urinary metabolites. Show how you would differentiate these four deuterated aldehydes using mass spectrometry. Remember that deuterium has mass 2. (Hint: Predict the important fragmentations, and show how the different compounds give unique peaks.)

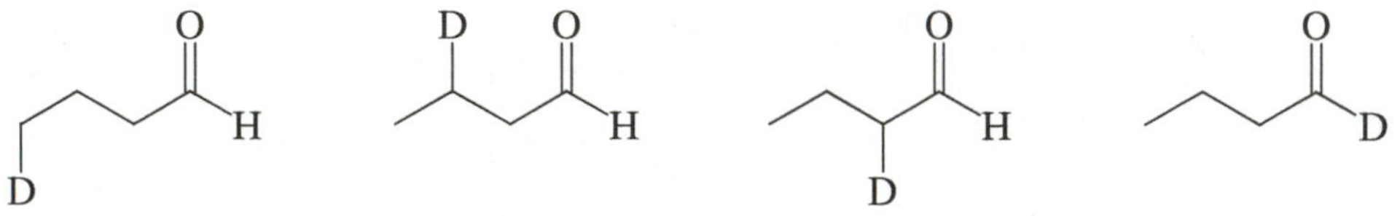

W **X** **Y** **Z**

18-77 The family of macrolide antibiotics all have large rings (macrocycle) in which an ester is what makes the ring; a cyclic ester is termed a lactone. One example is amphotericin B, used as an anti-fungal treatment of last resort because of its liver and heart toxicity. Professor Martin Burke of the University of Illinois has been making analogs to retain the antifungal properties but without the toxicity, including this structure published in 2015. (*Nature Chemical Biology* (2015) doi:10.1038/nchembio.1821) The carboxylate of amphotericin B has been replaced with the urea group (shown in red).

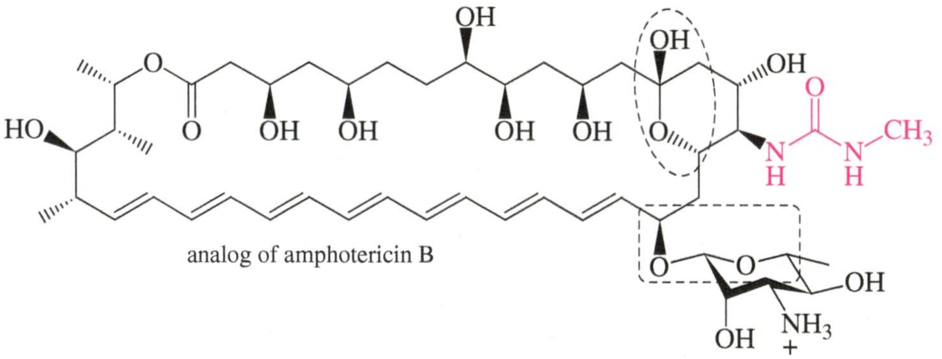

analog of amphotericin B

(a) Where is the lactone group that forms the ring?
(b) Two groups are circled. What type of functional group are they? Explain.

19 Amines

Goals for Chapter 19

1 Draw and name amines, and use spectral information to determine their structures.

2 Compare the basicity of amines with other common bases, and explain how their basicity varies with hybridization, aromaticity, resonance, and induction.

3 Describe the trends in the physical properties of amines, and contrast their physical properties with those of their salts.

4 Predict the products and propose mechanisms for the reactions of amines with ketones, aldehydes, acid chlorides, nitrous acid, alkyl halides, and oxidizing agents.

5 Propose single-step and multistep syntheses of amines from compounds containing other functional groups.

$$(CH_3)_3N^+{-}O^- \rightleftharpoons (CH_3)_3N: + \tfrac{1}{2}O_2$$

trimethylamine oxide (TMAO) — trimethylamine

▲ Amines are well-known for their fishy odors. Live fish use amine derivatives such as trimethylamine oxide (TMAO) to control the osmotic pressure in their cells. Once a fish dies, enzymes and bacteria begin to convert these compounds to trimethylamine and other degraded amines. In fish that are not fresh, these amines give the fishy odor. Some people suffer from the metabolic disease trimethylaminuria (TMAU), in which a defect in an enzyme prevents the oxidation of trimethylamine (from food digestion) to trimethylamine oxide. Trimethylamine builds up, and gives a strong fishy odor to the person's breath, sweat, and urine.

19-1 Introduction

Amines are organic derivatives of ammonia with one or more alkyl or aryl groups bonded to the nitrogen atom. As a class, amines include some of the most important biological compounds. Amines serve many functions in living organisms, such as bioregulation, neurotransmission, and defense against predators. Because of their high degree of biological activity, many amines are used as drugs and medicines. The structures and uses of some important biologically active amines are shown in Figure 19-1.

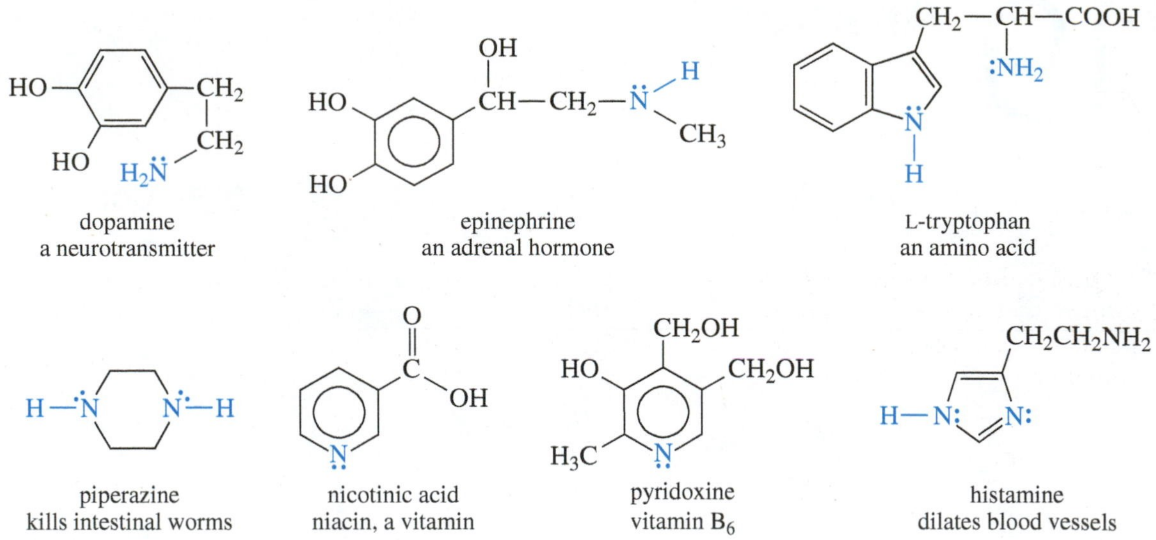

dopamine
a neurotransmitter

epinephrine
an adrenal hormone

L-tryptophan
an amino acid

piperazine
kills intestinal worms

nicotinic acid
niacin, a vitamin

pyridoxine
vitamin B_6

histamine
dilates blood vessels

FIGURE 19-1
Examples of some biologically active amines.

The *alkaloids* are an important group of biologically active amines, mostly synthesized by plants to protect them from being eaten by insects and other animals. The structures of some representative alkaloids are shown in Figure 19-2. Although some alkaloids are used medicinally (chiefly as painkillers), all alkaloids are toxic and cause death if taken in large quantities. The Greeks chose the alkaloid coniine to kill Socrates, although morphine, nicotine, or cocaine would have served equally well.

Mild cases of alkaloid poisoning can produce psychological effects that resemble peacefulness, euphoria, or hallucinations. People seeking these effects often become addicted to alkaloids. Alkaloid addiction frequently ends in death. Current estimates are over 400,000 deaths from alkaloid addiction in the United States per year, including both natural alkaloids like nicotine and cocaine, and synthetic alkaloids like methamphetamine. Most of these deaths result from addiction to nicotine in tobacco, a particularly difficult addiction to overcome.

19-2 Nomenclature of Amines

Amines are classified as **primary** (1°), **secondary** (2°), or **tertiary** (3°), corresponding to one, two, or three alkyl or aryl groups bonded to nitrogen. In a heterocyclic amine, the nitrogen atom is part of an aliphatic or aromatic ring.

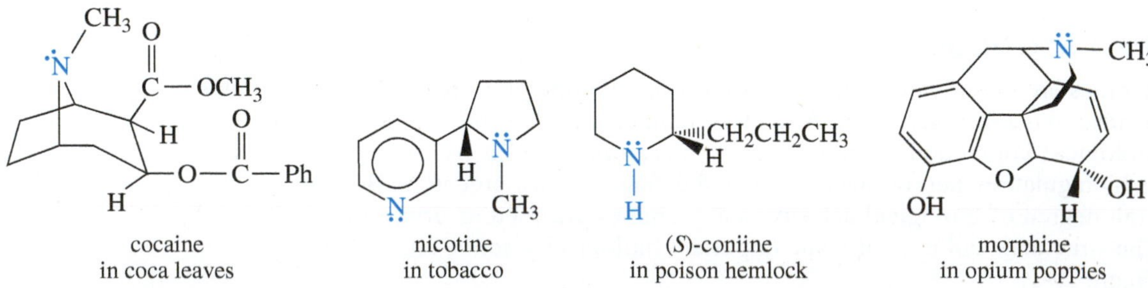

cocaine
in coca leaves

nicotine
in tobacco

(*S*)-coniine
in poison hemlock

morphine
in opium poppies

FIGURE 19-2
Some representative alkaloids.

Primary (1°) amines

cyclohexylamine (1°)

$(CH_3)_3C{-}NH_2$

tert-butylamine (1°)

Secondary (2°) amines

N-ethylaniline (2°)

piperidine (2°)

Tertiary (3°) amines

N,N-diethylaniline (3°)

quinuclidine (3°)

Quaternary ammonium salts have four alkyl or aryl bonds to a nitrogen atom. The nitrogen atom bears a positive charge, just as it does in simple ammonium salts such as ammonium chloride. The following are examples of quaternary (4°) ammonium salts:

$(CH_3CH_2)_4N^+ \; I^-$

tetraethylammonium iodide

N-butylpyridinium bromide

$CH_3{-}C(=O){-}O{-}CH_2CH_2{-}N^+(CH_3)_3$

acetylcholine, a neurotransmitter

Application: Nicotine Addiction

The structure of nicotine is shown in Figure 19-2. Like all alkaloids, nicotine is toxic, with a fatal dose of about 60–500 mg in humans. Small amounts tend to be addictive, however. A cigarette delivers about 2 mg of nicotine, which stimulates the release of dopamine in the brain and the secretion of epinephrine by the adrenal glands. Epinephrine is a stimulant, which provides the initial "kick" that smokers crave. Dopamine is one of the chemicals in the brain's reward system.

19-2A Common Names

Common names of amines are formed from the names of the alkyl groups bonded to nitrogen, followed by the suffix *-amine*. The prefixes *di-*, *tri-*, and *tetra-* are used to describe two, three, or four identical substituents.

$CH_3CH_2NH_2$
ethylamine

$(CH_3CH(CH_3)CH_2CH_2)_2NH$
diisopentylamine

$(CH_3CH_2)_2NCH_3$
diethylmethylamine

$(CH_3CH_2CH_2CH_2)_4N^+ \; Cl^-$
tetrabutylammonium chloride

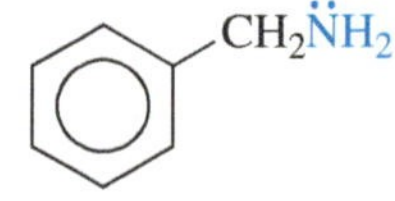

cyclohexyldimethylamine

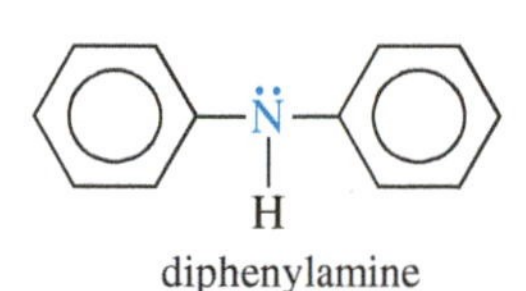

benzylamine

diphenylamine

In naming amines with more complicated structures, the $-NH_2$ group is called the **amino** group. It is treated like any other substituent, with a number or other symbol indicating its position on the ring or carbon chain.

3-aminocyclopentene
(cyclopent-2-en-1-amine)

$NH_2{-}CH_2CH_2CH_2{-}COOH$ (γ, β, α)

γ-aminobutyric acid
(4-aminobutanoic acid)

trans-3-aminocyclohexanol

p-aminobenzoic acid (PABA)

Using this system, secondary and tertiary amines are named by classifying the nitrogen atom (together with its alkyl groups) as an alkylamino group. The largest or most complicated alkyl group is taken to be the parent molecule.

$CH_3CH_2CH_2CH(N(CH_3)_2)CH_2CH_2OH$

3-(dimethylamino)hexan-1-ol

4-(ethylmethylamino)cyclohexanone

Application: Neonicotinoid Insecticides

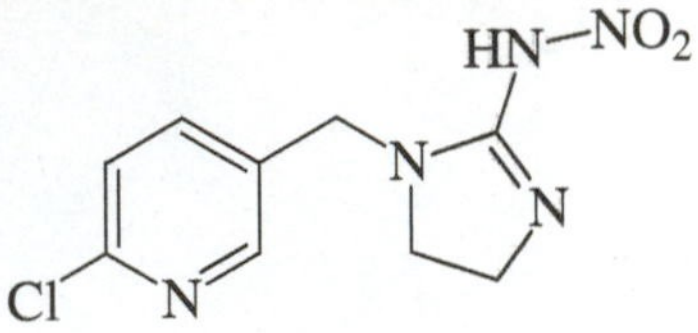

imidacloprid

Imidacloprid is one of the most widely used insecticides in the world. It is called a neonicotinoid because, like nicotine (Fig. 19-2), it blocks nerve signals and kills insects by paralysis. Imidacloprid is a systemic insecticide, meaning that it is absorbed by plants and distributed to all parts of the plant, including the pollen and nectar. Chewing and sucking insect pests are killed quickly. Pollinators such as bees and butterflies get smaller doses, but over time the sublethal doses can produce delayed mortality. The neonicotinoid insecticides are thought to be a major cause of honeybee colony collapse disorder, which kills all the bees in as many as 40% of the hives in a region.

Aromatic and heterocyclic amines are generally known by historical names. Phenylamine is called *aniline*, for example, and its derivatives are named as derivatives of aniline.

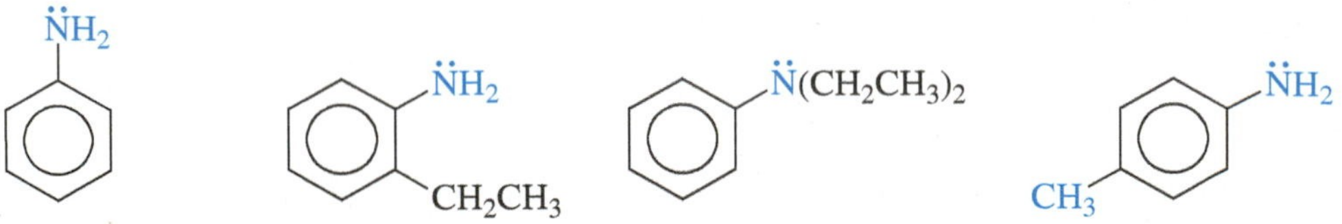

aniline | 2-ethylaniline or *o*-ethylaniline | *N*,*N*-diethylaniline | 4-methylaniline or *p*-toluidine

We first considered nitrogen heterocycles in Section 16-9. The names and structures of some common ones are shown here. The heteroatom is usually assigned position number 1.

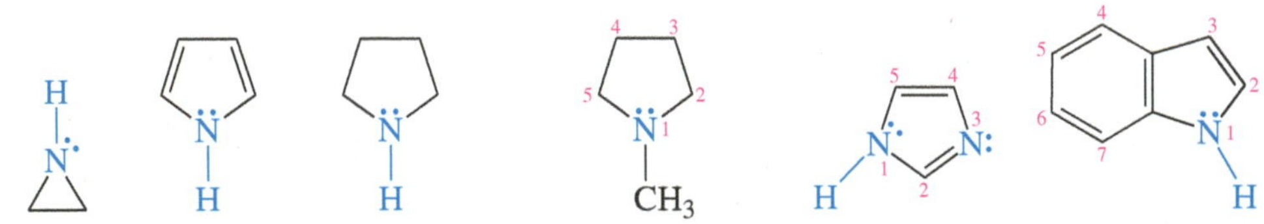

aziridine | pyrrole | pyrrolidine | 1-methylpyrrolidine (*N*-methylpyrrolidine) | imidazole | indole

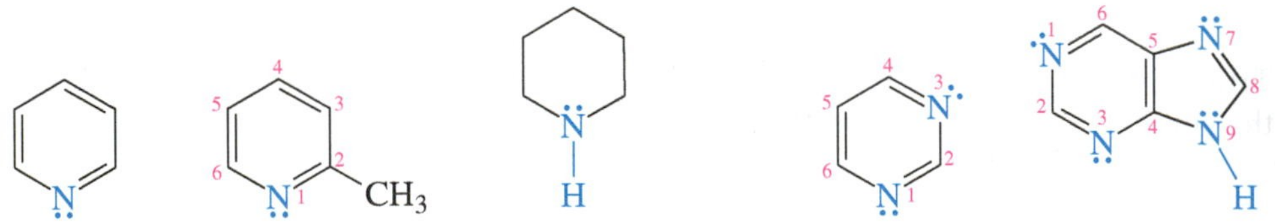

pyridine | 2-methylpyridine | piperidine | pyrimidine | purine

PROBLEM 19-1

Determine which of the heterocyclic amines just shown are aromatic. Give the reasons for your conclusions.

19-2B IUPAC Names

The IUPAC nomenclature for amines is similar to that for alcohols. The longest continuous chain of carbon atoms determines the root name. The *-e* ending in the alkane name is changed to *-amine*, and a number shows the position of the amino group along the chain. Other substituents on the carbon chain are given numbers, and the prefix *N*- is used for each substituent on nitrogen.

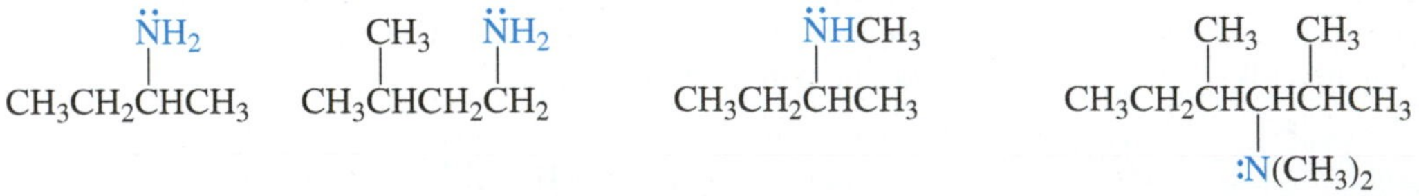

old IUPAC names in blue:	2-butanamine	3-methyl-1-butanamine	*N*-methyl-2-butanamine	*N*,*N*,2,4-tetramethyl-3-hexanamine
new IUPAC names in green:	butan-2-amine	3-methylbutan-1-amine	*N*-methylbutan-2-amine	*N*,*N*,2,4-tetramethylhexan-3-amine

PROBLEM 19-2

Draw the structures of the following compounds:

(a) *tert*-butylamine
(b) α-aminopropionaldehyde
(c) 4-(dimethylamino)pyridine
(d) 2-methylaziridine
(e) *N*-ethyl-*N*-methylhexan-3-amine
(f) *m*-chloroaniline

PROBLEM-SOLVING HINT

The nomenclature of amines is reviewed in Appendix 5, the Summary of Organic Nomenclature.

PROBLEM 19-3

Give correct names for the following amines:

(a) $CH_3—CH_2—CH_2—CH(NH_2)—CH_3$
(b) $CH_3—CH_2—CH(NHCH_3)—CH_3$

(c) benzene ring bearing NH_2 and OH
(d) pyrrole ring (N–H) bearing CH_3
(e) cyclopentane bearing two NH_2 groups and H, H
(f) cyclohexane bearing NH_2, H, and CHO, H

19-3 Structure of Amines

In Chapter 1, we saw that ammonia has a slightly distorted tetrahedral shape. A lone pair of nonbonding electrons occupies one of the tetrahedral positions. This geometry is represented by sp^3 hybridization of nitrogen, with the bulky lone pair compressing the H—N—H bond angles to 107° from the "ideal" sp^3 bond angle of 109.5°. Trimethylamine shows less angle compression because the bulky methyl groups open the angle slightly.

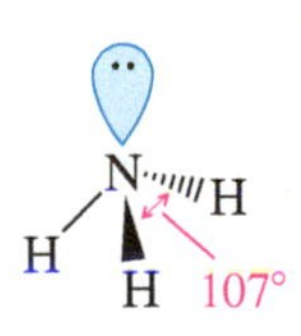

ammonia

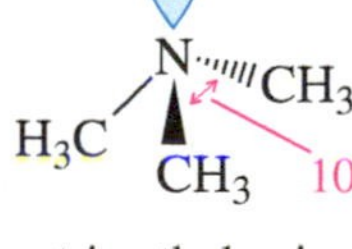

trimethylamine

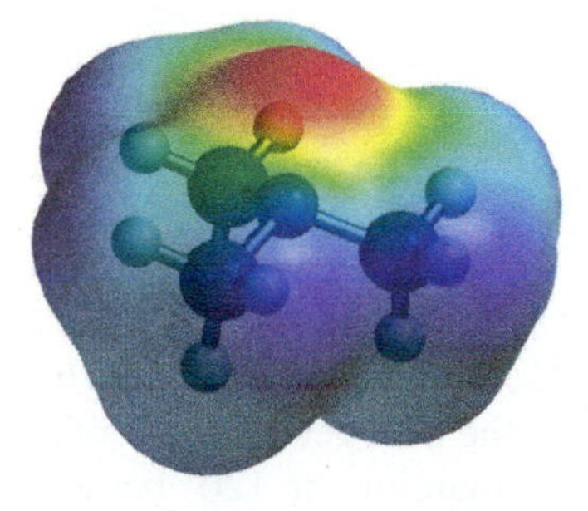

electrostatic potential map for trimethylamine

The electrostatic potential map for trimethylamine shows how the nonbonding electrons give rise to a red region (high negative potential) above the pyramidal nitrogen atom.

A tetrahedral amine with three different substituents (and a lone pair) is nonsuperimposable on its mirror image, and appears to be a chiral center. In most cases, however, we cannot resolve such an amine into two enantiomers because the enantiomers interconvert rapidly (see Figure 19-3). This interconversion takes place by **nitrogen inversion,** in which the lone pair moves from one face of the molecule to the other. The nitrogen atom is sp^2 hybridized in the transition state, and the nonbonding electrons occupy a *p* orbital. This is a fairly stable transition state, as reflected by the small activation energy of about 25 kJ/mol (6 kcal/mol). Interconversion of (*R*)- and (*S*)-ethylmethylamine occurs hundreds of times per second at room temperature. In naming the enantiomers of chiral amines, the Cahn–Ingold–Prelog convention (Section 5-3) is used, with the nonbonding electron pair having the lowest priority.

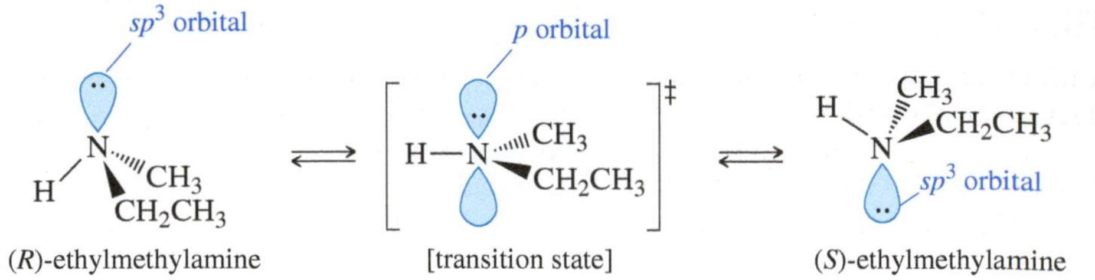

FIGURE 19-3 Nitrogen inversion interconverts the two enantiomers of a simple chiral amine. The transition state is a planar, sp^2 hybrid structure with the lone pair in a p orbital.

(*R*)-ethylmethylamine ⇌ [transition state] ⇌ (*S*)-ethylmethylamine

Although most simple amines cannot be resolved into enantiomers, several types of chiral amines can be resolved:

1. *Amines whose chirality stems from the presence of asymmetric carbon atoms.* Most chiral amines fall into this group. Nitrogen inversion is irrelevant because nitrogen is not the chirality center. For example, butan-2-amine can be resolved into enantiomers because the 2-butyl group is chiral.

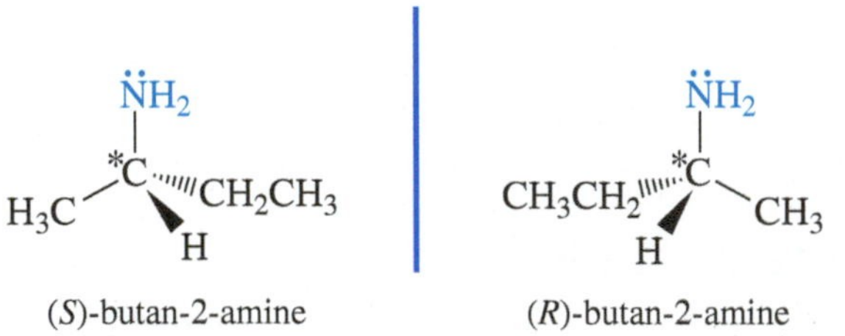

(*S*)-butan-2-amine (*R*)-butan-2-amine

2. *Quaternary ammonium salts with asymmetric nitrogen atoms.* Inversion of configuration is not possible because there is no lone pair to undergo nitrogen inversion. For example, the methyl ethyl isopropyl anilinium salts can be resolved into enantiomers.

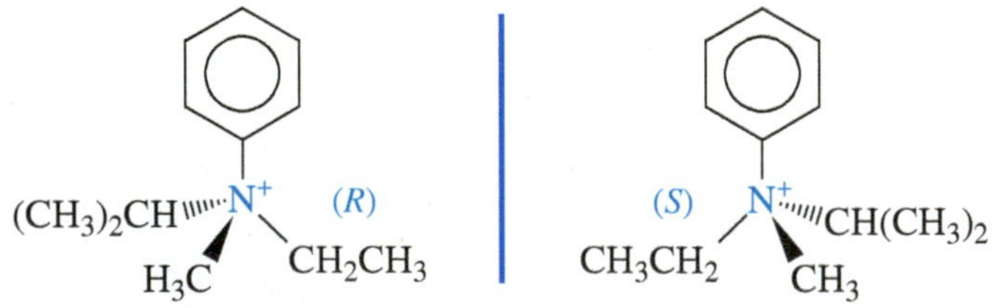

3. *Amines that cannot attain the sp^2 hybrid transition state for nitrogen inversion.* If the nitrogen atom is contained in a small ring, for example, it is prevented from attaining the 120° bond angles that facilitate inversion. Such a compound has a higher activation energy for inversion, the inversion is slow, and the enantiomers may be resolved. Chiral aziridines (three-membered rings containing a nitrogen) often may be resolved into enantiomers.

(*R*)-1,2,2-trimethylaziridine (*S*)-1,2,2-trimethylaziridine

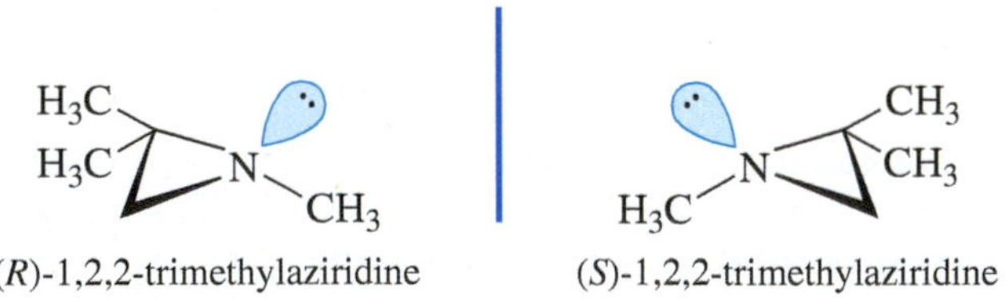

Application: Cancer Drug

Mitomycin C, an anticancer agent used to treat stomach and colon cancer, contains an aziridine ring. The aziridine functional group participates in the drug's degradation of DNA, resulting in the death of cancerous cells.

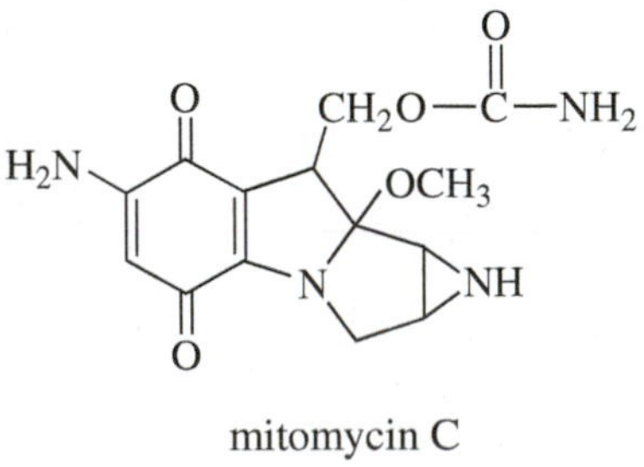

mitomycin C

PROBLEM 19-4

Which of the amines listed next is resolved into enantiomers? In each case, explain why interconversion of the enantiomers does or does not take place.

(a) *cis*-2-methylcyclohexanamine
(b) *N*-ethyl-*N*-methylcyclohexanamine
(c) *N*-methylaziridine
(d) ethylmethylanilinium iodide
(e) methylethylpropylisopropylammonium iodide

19-4 Physical Properties of Amines

Amines are strongly polar because the large dipole moment of the lone pair of electrons adds to the dipole moments of the C $\leftrightarrow$ N and H $\leftrightarrow$ N bonds. Primary and secondary amines have N—H bonds, allowing them to form hydrogen bonds. Pure tertiary amines cannot engage in hydrogen bonding because they have no N—H bonds. They can, however, accept hydrogen bonds from molecules having O—H or N—H bonds.

overall dipole moment

1° or 2° amine: hydrogen bond donor and acceptor

3° amine: hydrogen bond acceptor only

Because nitrogen is less electronegative than oxygen, the N—H bond is less polar than the O—H bond. Therefore, amines form weaker hydrogen bonds than do alcohols of similar molecular weights. Primary and secondary amines have boiling points that are lower than those of alcohols, yet higher than those of ethers of similar molecular weights. With no hydrogen bonding, tertiary amines have lower boiling points than primary and secondary amines of similar molecular weights. Table 19-1 compares the boiling points of an ether, an alcohol, and amines of similar molecular weights.

TABLE 19-1
How Hydrogen Bonding Affects Boiling Points

Compound	bp (°C)	Type	Molecular Weight
$(CH_3)_3N$:	3	tertiary amine	59
$CH_3—O—CH_2—CH_3$	8	ether	60
$CH_3—NH—CH_2—CH_3$	37	secondary amine	59
$CH_3CH_2CH_2—NH_2$	48	primary amine	59
$CH_3CH_2CH_2—OH$	97	alcohol	60

All amines, even tertiary ones, form hydrogen bonds with hydroxylic solvents such as water and alcohols. Therefore, amines tend to be soluble in alcohols, and the lower-molecular-weight amines (up to about four carbon atoms) are relatively soluble in water. Table 19-2 lists the melting points, boiling points, and water solubilities of some simple aliphatic and aromatic amines.

Perhaps the most obvious property of amines is their characteristic odor of rotting fish. Some of the diamines are particularly pungent; the following diamines have common names that describe their odors:

$H_2N—CH_2CH_2CH_2CH_2—NH_2$
putrescine
(butane-1,4-diamine)

$H_2N—CH_2CH_2CH_2CH_2CH_2—NH_2$
cadaverine
(pentane-1,5-diamine)

Application: Bacterial Decomposition

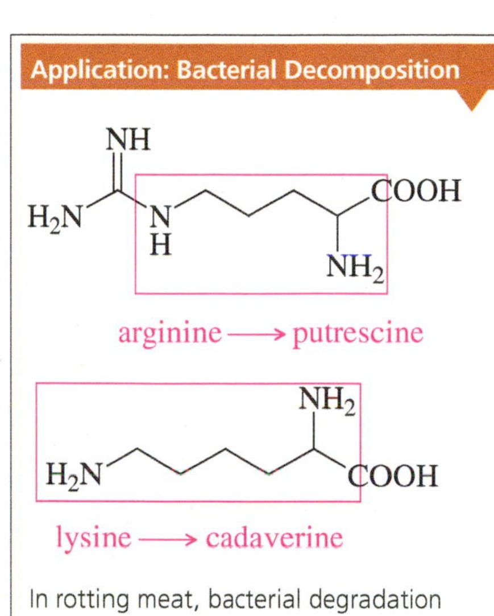

In rotting meat, bacterial degradation of the amino acids arginine and lysine produces putrescine and cadaverine, respectively.

TABLE 19-2
Physical Properties of Amines

Name	Structure	Molecular Weight	mp (°C)	bp (°C)	H_2O Solubility
		Primary amines			
methylamine	CH_3NH_2	31	−93	−7	very soluble
ethylamine	$CH_3CH_2NH_2$	45	−81	17	∞
n-propylamine	$CH_3CH_2CH_2NH_2$	59	−83	48	∞
isopropylamine	$(CH_3)_2CHNH_2$	59	−101	33	∞
n-butylamine	$CH_3CH_2CH_2CH_2NH_2$	73	−50	77	∞
cyclohexylamine	*cyclo*-$C_6H_{11}NH_2$	99	−18	134	slightly soluble
benzylamine	$C_6H_5CH_2NH_2$	107		185	∞
aniline	$C_6H_5NH_2$	93	−6	184	3.7%
		Secondary amines			
dimethylamine	$(CH_3)_2NH$	45	−96	7	very soluble
diethylamine	$(CH_3CH_2)_2NH$	73	−42	56	very soluble
di-*n*-propylamine	$(CH_3CH_2CH_2)_2NH$	101	−40	111	slightly soluble
diisopropylamine	$[(CH_3)_2CH]_2NH$	101	−61	84	slightly soluble
N-methylaniline	$C_6H_5NHCH_3$	107	−57	196	slightly soluble
diphenylamine	$(C_6H_5)_2NH$	169	54	302	insoluble
		Tertiary amines			
trimethylamine	$(CH_3)_3N$	59	−117	3.5	very soluble
triethylamine	$(CH_3CH_2)_3N$	101	−115	90	14%
tri-*n*-propylamine	$(CH_3CH_2CH_2)_3N$	143	−94	156	slightly soluble
N,N-dimethylaniline	$C_6H_5N(CH_3)_2$	121	2	194	1.4%
triphenylamine	$(C_6H_5)_3N$	251	126	225	insoluble

PROBLEM 19-5

Rank each set of compounds in order of increasing boiling points.

(a) triethylamine, di-*n*-propylamine, *n*-propyl ether **(b)** ethanol, dimethylamine, dimethyl ether
(c) diethylamine, diisopropylamine, trimethylamine

19-5 Basicity of Amines

An amine is a nucleophile (a Lewis base) because its lone pair of nonbonding electrons can form a bond with an electrophile. An amine can also act as a Brønsted–Lowry base by accepting a proton from a proton acid.

Reaction of an amine as a nucleophile

$$R{-}\ddot{N}H_2 + CH_3{-}I \longrightarrow R{-}\overset{+}{N}H_2{-}CH_3 \quad I^-$$

nucleophile electrophile new N—C bond formed

Reaction of an amine as a proton base

$$R{-}\ddot{N}H_2 + H{-}X \longrightarrow R{-}\overset{+}{N}H_2{-}H \quad X^-$$

base proton acid protonated

Because amines are fairly strong bases, their aqueous solutions are basic. An amine can abstract a proton from water, giving an ammonium ion and a hydroxide ion. The equilibrium constant for this reaction is the **base-dissociation constant** K_b for the amine (see Section 2-6B).

$$R{-}\ddot{N}H_2 + H{-}O{-}H \overset{K_b}{\rightleftharpoons} R{-}\overset{+}{N}H_3 + {}^{-}OH$$

$$K_b = \frac{[RNH_3^+][^-OH]}{[RNH_2]} \qquad pK_b = -\log_{10}K_b$$

Values of K_b for most amines are fairly small (about 10^{-3} or smaller), and the equilibrium for this dissociation lies toward the left. Nevertheless, aqueous solutions of amines are distinctly basic, and they turn litmus paper blue.

As we saw in Chapter 2, base-dissociation constants are usually listed as their negative logarithms, or pK_b values. For example, if a certain amine has $K_b = 10^{-3}$, then $pK_b = 3$. Just as we used pK_a values to indicate acid strengths (stronger acids have smaller pK_a values), we use pK_b values to compare the relative strengths of amines as proton bases.

> Stronger bases have smaller values of pK_b.

The values of pK_b for some representative amines are listed in Table 19-3.

TABLE 19-3
Basicity of Amines

Amine	K_b	pK_b	pK_a of $R_3\overset{+}{N}H$
ammonia	1.8×10^{-5}	4.74	9.26
	Primary alkyl amines		
methylamine	4.3×10^{-4}	3.36	10.64
ethylamine	4.4×10^{-4}	3.36	10.64
n-propylamine	4.7×10^{-4}	3.32	10.68
isopropylamine	4.0×10^{-4}	3.40	10.60
cyclohexylamine	4.7×10^{-4}	3.33	10.67
benzylamine	2.0×10^{-5}	4.67	9.33
	Secondary amines		
dimethylamine	5.3×10^{-4}	3.28	10.72
diethylamine	9.8×10^{-4}	3.01	10.99
di-*n*-propylamine	10.0×10^{-4}	3.00	11.00
	Tertiary amines		
trimethylamine	5.5×10^{-5}	4.26	9.74
triethylamine	5.7×10^{-4}	3.24	10.76
tri-*n*-propylamine	4.5×10^{-4}	3.35	10.65
	Aryl amines		
aniline	4.0×10^{-10}	9.40	4.60
N-methylaniline	6.1×10^{-10}	9.21	4.79
N,N-dimethylaniline	1.2×10^{-9}	8.94	5.06
p-bromoaniline	7×10^{-11}	10.2	3.8
p-methoxyaniline	2×10^{-9}	8.7	5.3
p-nitroaniline	1×10^{-13}	13.0	1.0
	Heterocyclic amines		
pyrrole	5×10^{-15}	14.3	−0.3
pyrrolidine	1.9×10^{-3}	2.73	11.27
imidazole	8.9×10^{-8}	7.05	6.95
pyridine	1.8×10^{-9}	8.75	5.25
piperidine	1.3×10^{-3}	2.88	11.12

PROBLEM-SOLVING HINT

Watch out for sources that define the pK_a of an amine differently from other compounds. Some sources cite the pK_a of the protonated amine as being the pK_a of the amine itself. For example, the actual pK_a of methylamine is around 40, corresponding to formation of CH_3NH^-, a powerful base. Some sources list the pK_a of methylamine as 10.64, which is actually the pK_a of $CH_3NH_3^+$.

Some references do not list values of K_b or pK_b for amines. Instead, they list values of K_a or pK_a for the conjugate acid, which is the ammonium ion. We can show that the product of K_a for the ammonium ion and K_b for the amine is K_W, the ion product for water, which is 10^{-14} at room temperature. This is true for any conjugate acid–base pair (see Section 2-6B).

$$\underset{\text{ammonium ion}}{R'-\overset{+}{N}H_3} + H_2O \xrightarrow{K_a} \underset{\text{amine}}{R'-\ddot{N}H_2} + H_3O^+$$

$$K_a = \frac{[RNH_2][H_3O^+]}{[RNH_3^+]} \qquad K_b = \frac{[RNH_3^+][^-OH]}{[RNH_2]}$$

$$K_a \times K_b = [H_3O^+][^-OH] = K_W = 1.0 \times 10^{-14}$$

$$pK_a + pK_b = 14 \qquad pK_b = 14 - pK_a$$

PROBLEM-SOLVING HINT

The pK_a of RNH_3^+ is the pH at which half of the molecules are protonated. At lower (more acidic) pH, the amine is mostly protonated (RNH_3^+). At higher (more basic) pH, the amine is mostly deprotonated (RNH_2).

These relationships allow us to convert values of K_a (or pK_a) for the ammonium ion and K_b (or pK_b) for the amine. They also remind us that a strongly basic amine has a weakly acidic ammonium ion and a weakly basic amine has a strongly acidic ammonium ion.

19-6 Factors that Affect Amine Basicity

Figure 19-4 shows an energy diagram for the reaction of an amine with water. On the left are the reactants: the free amine and water. On the right are the products: the ammonium ion and hydroxide ion.

Any structural feature that stabilizes the ammonium ion (relative to the free amine) shifts the reaction toward the right, making the amine a stronger base. Any feature that stabilizes the free amine (relative to the ammonium ion) shifts the reaction toward the left, making the amine a weaker base.

Substitution by Alkyl Groups As an example, consider the relative basicities of ammonia and methylamine. Alkyl groups are electron-donating toward cations, and methylamine has a methyl group to help stabilize the positive charge on nitrogen. This stabilization lowers the potential energy of the methylammonium cation, making methylamine a stronger base than ammonia. The simple alkylamines are generally stronger bases than ammonia.

$$H-\ddot{N}H_2 + H_2O \rightleftharpoons H-\overset{+}{N}H_3 + {}^-OH \qquad pK_b = 4.74 \text{ (weaker base)}$$

$$H_3C-\ddot{N}H_2 + H_2O \rightleftharpoons H_3C\rightarrow\overset{+}{N}H_3 + {}^-OH \qquad pK_b = 3.36 \text{ (stronger base)}$$

stabilized by the alkyl group

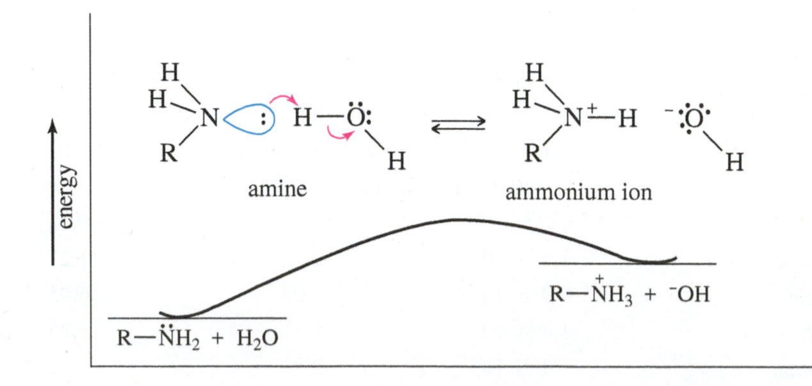

FIGURE 19-4
Potential-energy diagram of the base-dissociation reaction of an amine.

We might expect secondary amines to be stronger bases than primary amines (correct), and tertiary amines to be the strongest bases of all (incorrect). The actual situation is more complicated because of solvation effects. Because ammonium ions are charged, they are strongly solvated by water, and the energy of solvation contributes to their stability. The additional alkyl groups around the ammonium ions of secondary and tertiary amines decrease the number of water molecules that can approach closely and solvate the ions. The opposing trends of inductive stabilization and steric hindrance of solvation are balanced for secondary alkylamines, which are *slightly* stronger bases than primary or tertiary alkylamines.

Resonance Effects on Basicity Arylamines (anilines and their derivatives) are much weaker bases than simple aliphatic amines (Table 19-3). This reduced basicity is due to resonance delocalization of the nonbonding electrons in the free amine. Figure 19-5 shows how stabilization of the reactant (the free amine) makes the amine less basic. In aniline, overlap between the aromatic ring and the orbital containing nitrogen's lone pair stabilizes the lone pair and makes it less reactive. This overlap is lost in the anilinium ion, so the reactant (aniline) is stabilized compared with the product. The reaction is shifted toward the left, and aniline is less basic than most aliphatic amines.

Resonance effects also influence the basicity of pyrrole. Pyrrole is a very weak base, with a pK_b of about 15. As we saw in Chapter 15, pyrrole is aromatic because the nonbonding electrons on nitrogen are located in a p orbital, where they contribute to the aromatic sextet. When the pyrrole nitrogen is protonated, pyrrole loses its aromatic stabilization. Therefore, protonation on nitrogen is unfavorable, and pyrrole is a very weak base.

$$\text{pyrrole (aromatic)} + H_2O \underset{}{\overset{K_b}{\rightleftharpoons}} \text{protonated (not aromatic)} + {}^{-}OH \qquad K_b = 10^{-15}$$

PROBLEM-SOLVING HINT

Aromatic amines are generally less basic than aliphatic amines. This is true both when the nitrogen atom is part of the aromatic system (as in pyridine, a hybridization effect), and when the nitrogen atom is bonded to the aromatic ring (as in aniline, a resonance effect).

Hybridization Effects Our study of terminal alkynes (Section 9-6) showed that electrons are held more tightly by orbitals with more s character. This principle helps to explain why unsaturated amines tend to be weaker bases than simple aliphatic amines. In pyridine, for example, the nonbonding electrons occupy an sp^2 orbital, with greater s character and more tightly held electrons than those in the sp^3 orbital of an aliphatic amine. Pyridine's nonbonding electrons are less available for bonding to a proton. Pyridine does not lose its aromaticity on protonation, however, and it is a much stronger base than pyrrole.

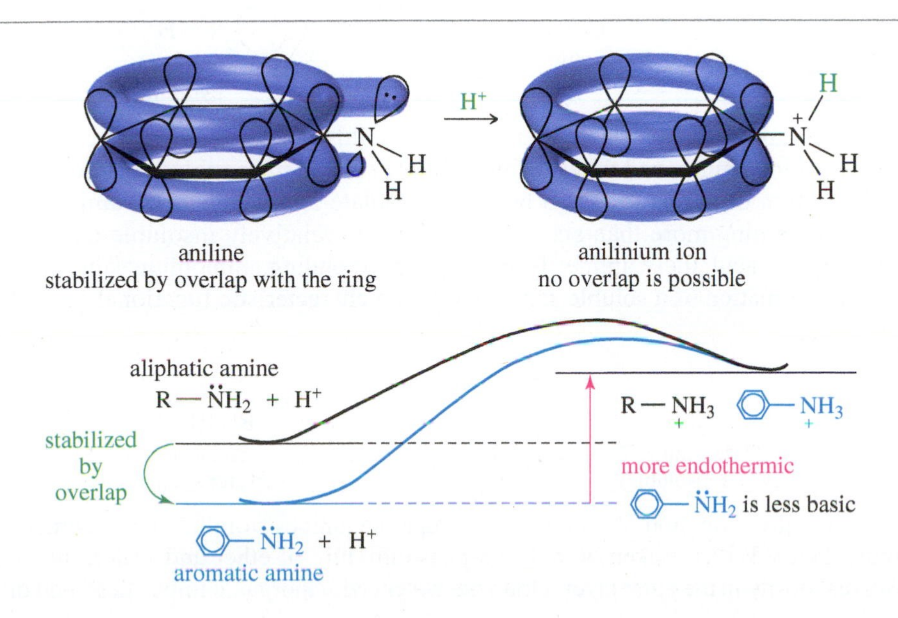

FIGURE 19-5
Aniline is stabilized by overlap of the lone pair with the aromatic ring. No such overlap is possible in the anilinium ion.

pyridine, $pK_b = 8.75$ — sp^2 hybridized (less basic)

piperidine, $pK_b = 2.88$ — sp^3 hybridized (more basic)

The effect of increased *s* character on basicity is even more pronounced in nitriles with *sp* hybridization. For example, acetonitrile has a pK_b of 24, showing that it is a very weak base. In fact, a concentrated mineral acid is required to protonate acetonitrile.

sp hybridized

$CH_3—C≡N:$ very weakly basic

$pK_b = 24$

PROBLEM 19-6

Rank each set of compounds in order of increasing basicity.
(a) NaOH, NH_3, CH_3NH_2, Ph—NH_2 **(b)** aniline, *p*-methylaniline, *p*-nitroaniline
(c) aniline, pyrrole, pyridine, piperidine **(d)** pyrrole, imidazole, 3-nitropyrrole

19-7 Salts of Amines

Protonation of an amine gives an **amine salt.** The amine salt is composed of two types of ions: the protonated amine cation (an ammonium ion) and the anion derived from the acid. Simple amine salts are named as the substituted **ammonium salts.** Salts of complex amines use the names of the amine and the acid that make up the salt.

$$CH_3CH_2CH_2—\ddot{N}H_2 + HCl \rightleftharpoons CH_3CH_2CH_2—NH_3^+\ Cl^-$$

n-propylamine + hydrochloric acid ⇌ *n*-propylammonium chloride

$$(CH_3CH_2)_3N: + H_2SO_4 \rightleftharpoons (CH_3CH_2)_3NH^+\ HSO_4^-$$

triethylamine + sulfuric acid ⇌ triethylammonium hydrogen sulfate

pyridine + acetic acid (H—O—C(=O)—CH_3) ⇌ pyridinium acetate ($\overset{+}{N}$—H $^-$O—C(=O)—CH_3)

Amine salts are ionic, high-melting, nonvolatile solids. They are more soluble in water than the parent amines, and they are only slightly soluble in nonpolar organic solvents.

Formation of amine salts can be used to isolate and characterize amines. Most amines containing more than six carbon atoms are relatively insoluble in water. In dilute aqueous acid, these amines form their corresponding ammonium salts, and they dissolve. Formation of a soluble salt is one of the characteristic functional group tests for amines.

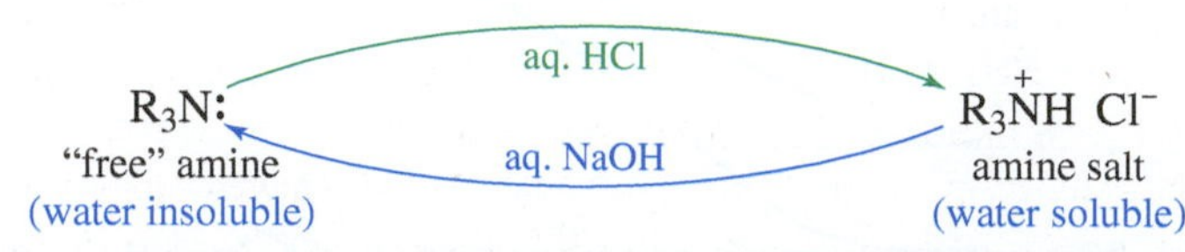

We can use the formation of amine salts to separate amines from less basic compounds (Figure 19-6). When shaken with a two-phase mixture of ether and water, the amine dissolves mostly in the ether layer. Drain the water (with inorganic impurities), add dilute

Application: Herbal Remedies

Ephedrine is the principal ingredient in Ma Huang, an herbal remedy originally used to treat asthma. The remedy is advertised as a "natural" way to enhance mood, reduce fatigue, and lose weight, but its stimulating effects on the heart can be fatal.

The term "natural" implies nothing about safety or purity. Most carcinogens, poisons, and other dangerous drugs are "natural."

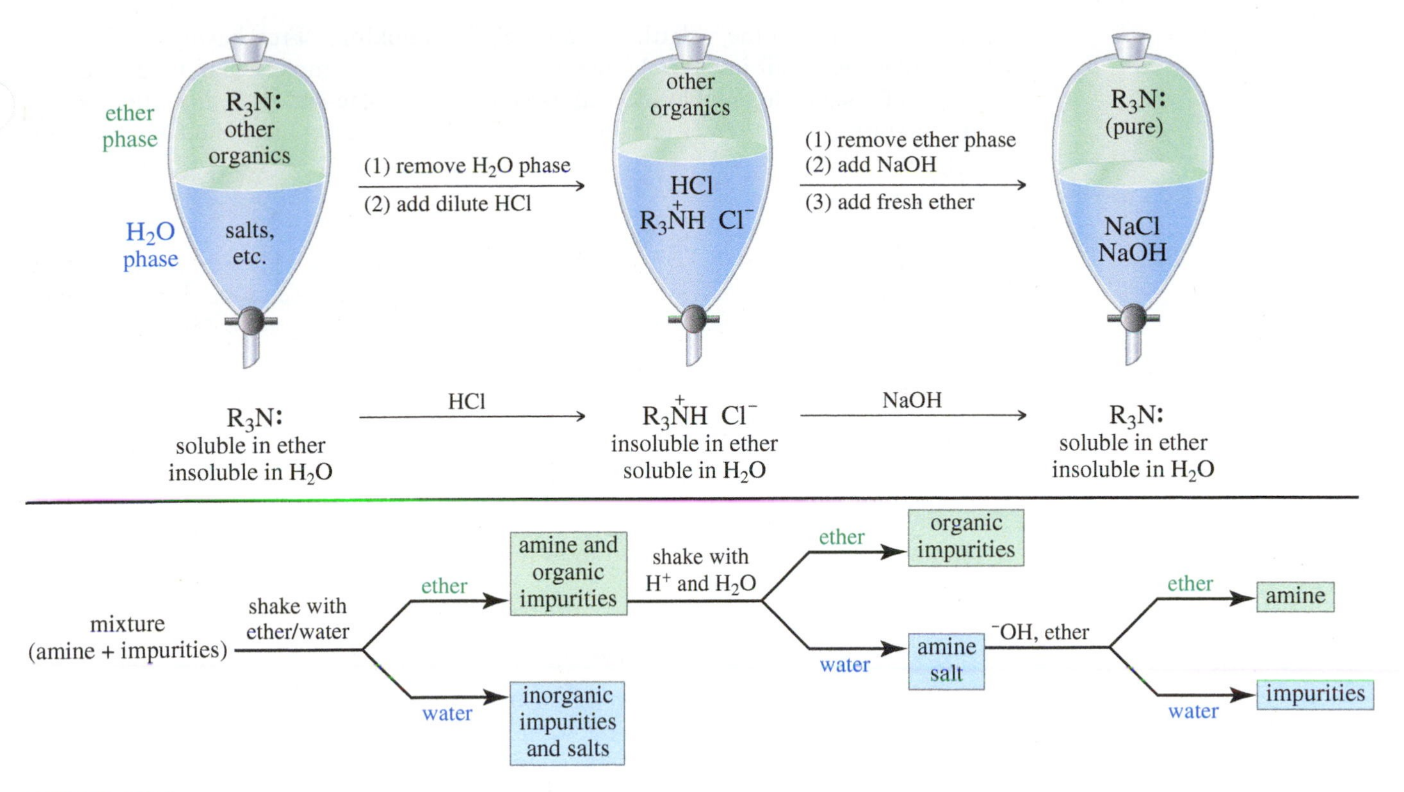

FIGURE 19-6
The basicity of an amine can be used for purification. The amine is initially more soluble in ether than in water. Addition of dilute HCl converts it to the water-soluble hydrochloride salt. Neutralization with NaOH regenerates the free amine.

acid, and the amine protonates and dissolves mostly in the aqueous phase. Drain off the ether (with the organic impurities), and add a fresh ether phase. Add dilute NaOH to make the aqueous solution alkaline, which deprotonates the amine. The purified free amine dissolves in the fresh ether phase, which is distilled to give the pure amine.

Many drugs and other biologically important amines are commonly stored and used as their salts. Amine salts are less prone to decomposition by oxidation and other reactions, and they have virtually no fishy odor. The salts are soluble in water, and they are easily converted to solutions for syrups and injectables.

As an example, the drug ephedrine is widely used in cold and allergy medications. Ephedrine melts at 79 °C, has an unpleasant fishy odor, and is oxidized by air to undesirable products. Ephedrine hydrochloride melts at 217 °C, does not oxidize easily, and has virtually no odor. Obviously, the hydrochloride salt is preferable for compounding medications.

$$\text{Ph}-\underset{\text{OH}}{\text{CH}}-\underset{\text{CH}_3}{\text{CH}}-\underset{\text{CH}_3}{\ddot{\text{N}}\text{H}} + \text{HCl} \longrightarrow \text{Ph}-\underset{\text{OH}}{\text{CH}}-\underset{\text{CH}_3}{\text{CH}}-\underset{\text{CH}_3}{\overset{+}{\text{N}}\text{H}_2}\ \text{Cl}^-$$

ephedrine
mp 79 °C, foul-smelling
easily air oxidized

ephedrine hydrochloride
mp 217 °C, no odor
stable

The chemistry of amine salts plays a large role in the illicit drug trade. Cocaine, for example, is usually smuggled and "snorted" as the hydrochloride salt, which is more stable and gives off less odor to alert the authorities. Smoking cocaine gives a more intense rush (and stronger addiction) because of fast absorption by lung tissues. But cocaine hydrochloride is not volatile; it tends to decompose before it vaporizes. Treating cocaine hydrochloride with sodium hydroxide and extracting it into ether

Cocaine hydrochloride is often divided into "lines" on a mirror and then snorted. "Crack" cocaine is sold as "rocks," commonly smoked in a crude pipe.

converts it back to the volatile "free base" for smoking. "Free-basing" cocaine is hazardous because it involves large amounts of ether. A simpler alternative is to mix a paste of cocaine hydrochloride with sodium bicarbonate and let it dry into "rocks." This mixture is called "crack cocaine" because it makes a crackling sound when heated.

NaOH or $NaHCO_3$ →
← HCl

cocaine hydrochloride — cocaine "free base"

19-8 Spectroscopy of Amines

We have discussed the general aspects of spectroscopy of amines in earlier chapters. Here, we cover more specifics and show how we combine this information to identify the presence of an amine.

19-8A Infrared Spectroscopy

The most reliable IR absorption of primary and secondary amines is the N—H stretch whose frequency appears between 3200 and 3500 cm^{-1}. Because this absorption is often broad, it is easily confused with the O—H absorption of an alcohol. In most cases, however, one or more spikes are visible in the broad N—H stretching region of an amine spectrum. Primary amines ($R—NH_2$) usually give two N—H spikes, from symmetric and antisymmetric stretching. Secondary amines ($R_2N—H$) usually give just one spike, and tertiary amines (R_3N) give no N—H absorptions.

In Figure 19-7, the characteristic N—H absorptions appear as two spikes on top of the broad N—H peak in the IR spectrum of propan-1-amine, a primary amine. Problem 19-7 contrasts the N—H stretch of a secondary amine with that of a primary amine and the O—H stretch of an alcohol.

Although an amine IR spectrum also contains absorptions resulting from vibrations of C—N bonds, these vibrations appear around 1000 to 1200 cm^{-1}, in the same region as C—C and C—O vibrations. Therefore, they are not very useful for identifying an amine.

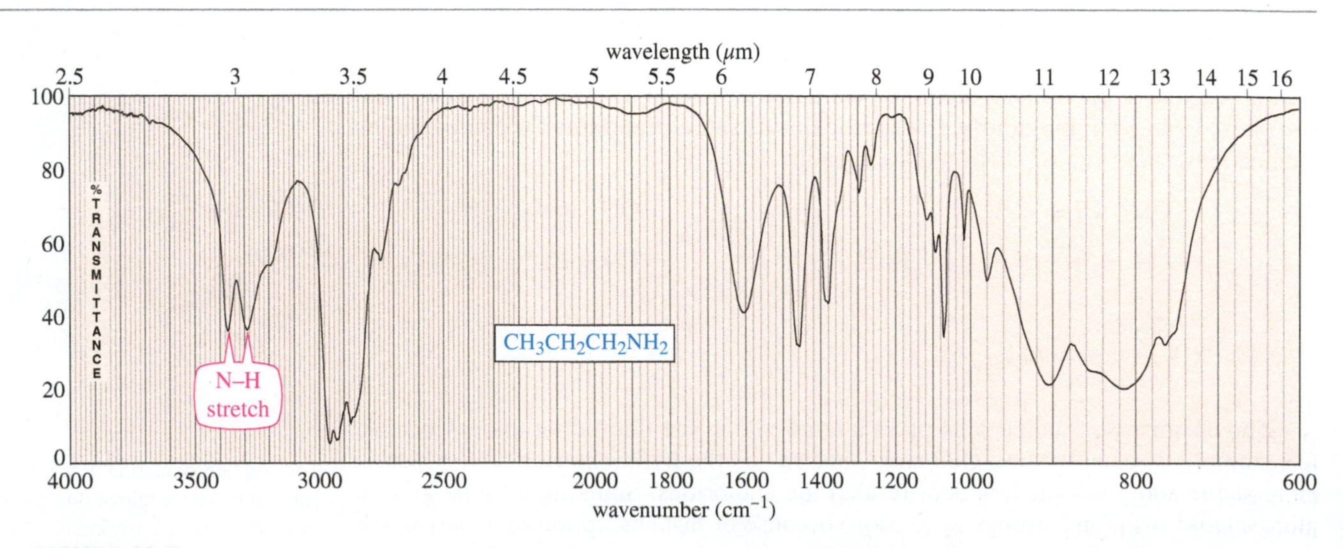

FIGURE 19-7
Infrared spectrum of propan-1-amine. Note the characteristic N—H stretching absorptions at 3300 and 3400 cm^{-1}.

PROBLEM 19-7

The following partial IR spectra correspond to a primary amine, a secondary amine, and an alcohol. Give the functional group for each spectrum.

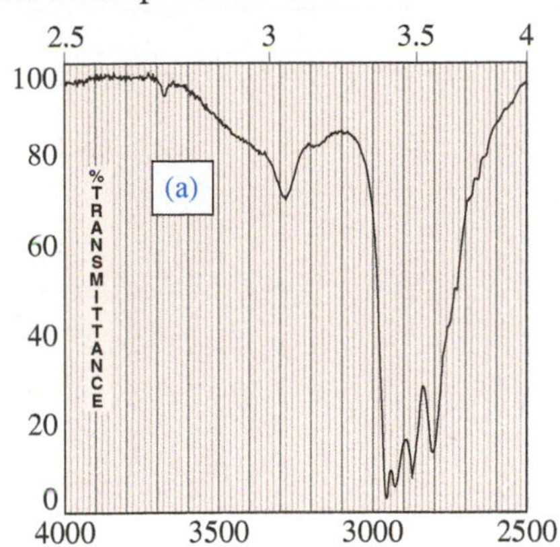

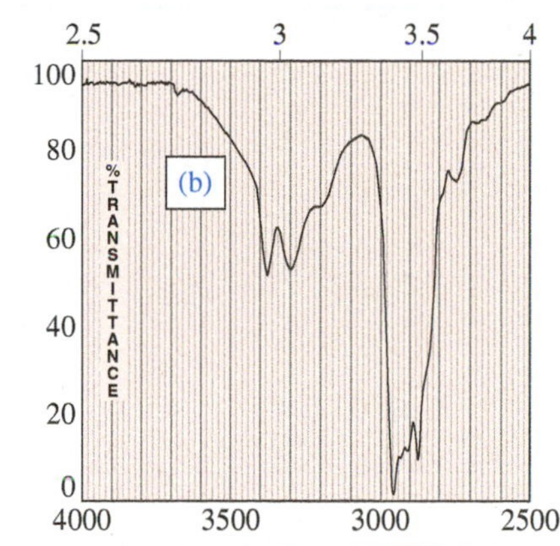

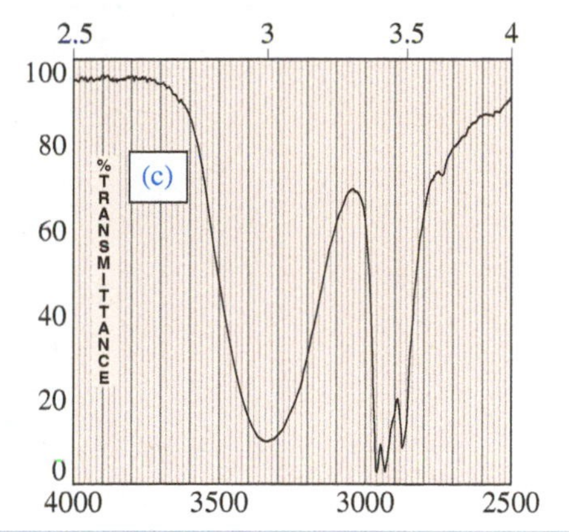

19-8B Proton NMR Spectroscopy

Like the O—H protons of alcohols, the N—H protons of amines absorb at chemical shifts that depend on the extent of hydrogen bonding. The solvent and the sample concentration influence hydrogen bonding and therefore the chemical shift. Typical N—H chemical shifts appear in the range $\delta 1$ to $\delta 4$.

Another similarity between O—H and N—H protons is their failure, in many cases, to show spin-spin splitting. In some samples, N—H protons exchange from one molecule to another at a rate that is faster than the time scale of the NMR experiment, and the N—H protons fail to show magnetic coupling. Sometimes the N—H protons of a very pure amine will show clean splitting, but these cases are rare. More commonly, the N—H protons appear as broad peaks. A broad peak should arouse suspicion of N—H protons. As with O—H protons, an absorption of N—H protons decreases or disappears after shaking the sample with D_2O.

Nitrogen is not as electronegative as oxygen and the halogens, so the protons on the α carbon atoms of amines are not as strongly deshielded. Protons on an amine's α carbon atom generally absorb between $\delta 2$ and $\delta 3$, but the exact position depends on the structure and substitution of the amine.

$CH_3—NR_2$	$R—CH_2—NR_2$	$R_2CH—NR_2$
methyl δ 2.3	methylene δ 2.7	methine δ 2.9

Protons that are beta to a nitrogen atom show a much smaller effect, usually absorbing in the range $\delta 1.1$ to $\delta 1.8$. These chemical shifts show a downfield movement of about 0.2 ppm resulting from the beta relationship. The NMR spectrum of propan-1-amine (Figure 19-8) shows these characteristic chemical shifts.

γ protons — β protons — α protons

$CH_3—CH_2—CH_2—NH_2$

δ0.9 δ1.4 δ2.6 variable (δ 1.7 in this spectrum)

19-8C Carbon NMR Spectroscopy

The α carbon atom bonded to the nitrogen of an amine usually shows a chemical shift of about 30 to 50 ppm. This range agrees with our general rule that a carbon atom shows a chemical shift about 20 times as great as the protons bonded to it. In propan-1-amine (Figure 19-8), for example, the α carbon atom absorbs at 45 ppm, whereas its protons absorb at 2.7 ppm. The β carbon is less deshielded, absorbing at 27 ppm, compared with its protons' absorption at 1.5 ppm. The γ carbon atom shows little effect from the presence of the nitrogen atom, absorbing at 11 ppm. Table 19-4 shows the carbon NMR chemical shifts of some representative amines.

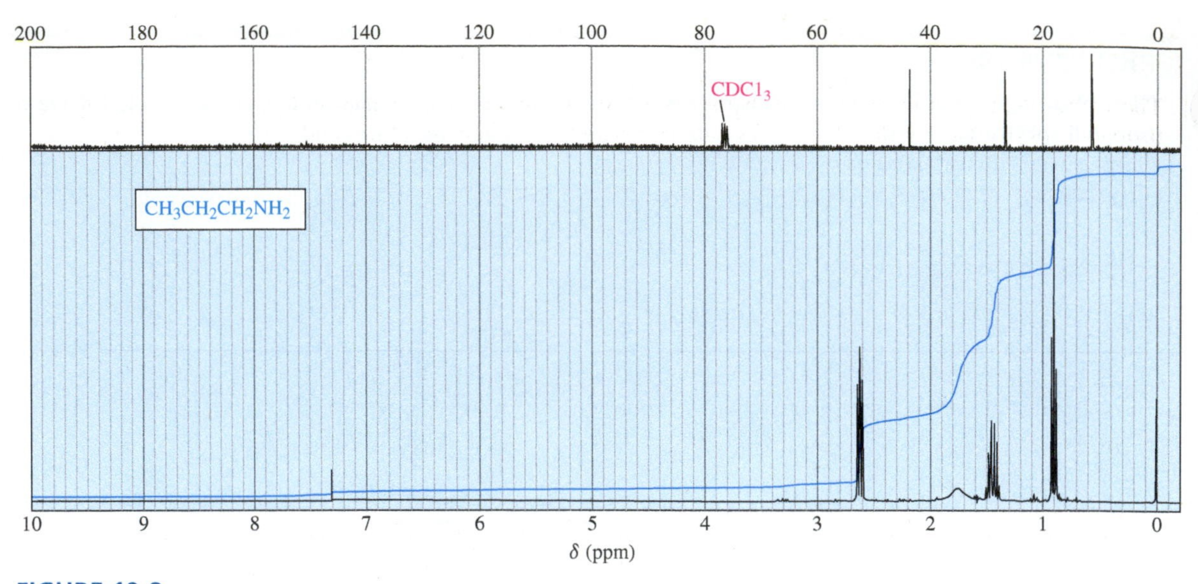

FIGURE 19-8
^{13}C and proton NMR spectra of propan-1-amine.

TABLE 19-4
Carbon NMR Chemical Shifts of Some Representative Amines

δ	γ	β	α		
			CH_3	$-NH_2$	methanamine
			26.9		
		CH_3	$-CH_2$	$-NH_2$	ethanamine
		17.7	35.9		
	CH_3	$-CH_2$	$-CH_2$	$-NH_2$	propan-1-amine
	11.1	27.3	44.9		
CH_3	$-CH_2$	$-CH_2$	$-CH_2$	$-NH_2$	butan-1-amine
14.0	20.4	36.7	42.3		

PROBLEM 19-8

The proton and ^{13}C NMR spectra of a compound of formula $C_4H_{11}N$ are shown here. Determine the structure of this amine, and give peak assignments for all of the protons in the structure.

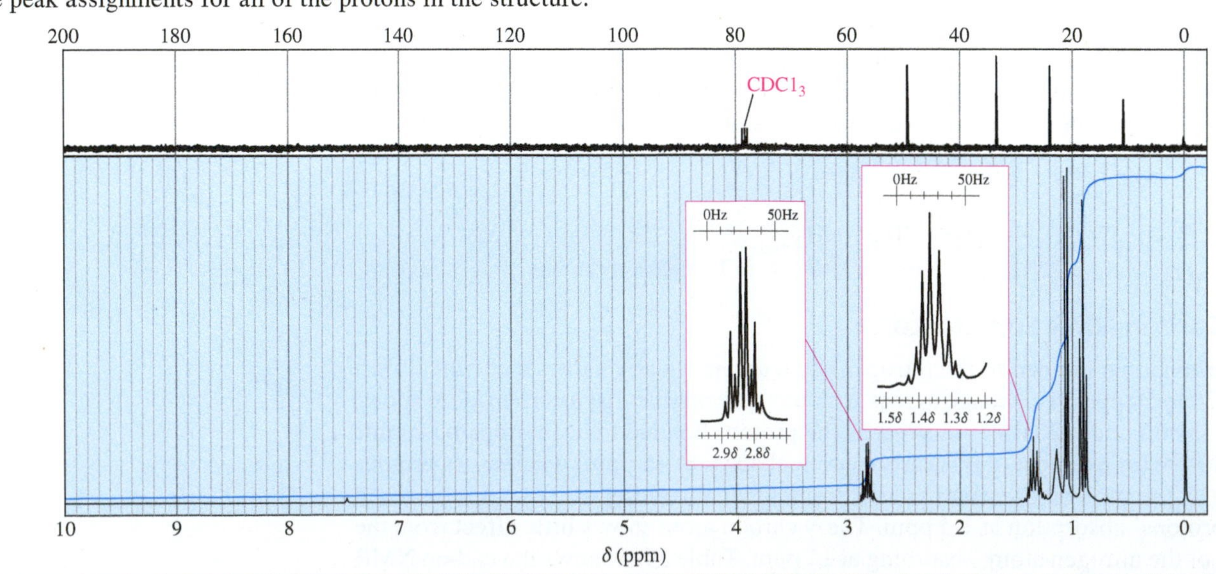

PROBLEM 19-9

The carbon NMR chemical shifts of diethylmethylamine, piperidine, propan-1-ol, and propanal follow. Determine which spectrum corresponds to each structure, and show which carbon atom(s) are responsible for each absorption.

(a) 25.9, 27.8, 47.9 **(b)** 12.4, 41.0, 51.1 **(c)** 7.9, 44.7, 201.9 **(d)** 10.0, 25.8, 63.6

19-8D Mass Spectrometry

The most obvious piece of information provided by the mass spectrum is the molecular weight. Stable compounds containing only carbon, hydrogen, oxygen, chlorine, bromine, and iodine give molecular ions with even mass numbers. Most of their fragments have odd mass numbers. This is because carbon and oxygen have even valences and even mass numbers, and hydrogen, chlorine, bromine, and iodine have odd valences and odd mass numbers.

Nitrogen has an odd valence and an even mass number. When a nitrogen atom is present in a stable molecule, the molecular weight is odd. In fact, whenever an odd number of nitrogen atoms are present in a molecule, the molecular ion has an odd mass number. Most of the fragments have even mass numbers.

The most common fragmentation of amines is α cleavage to give a resonance-stabilized cation: an *iminium* ion. This ion is simply a protonated version of an imine (Section 18-14).

$$\left[R{-}CH_2{-}\dot{N}^{+}(R)(H)\right] \longrightarrow R\cdot + \left[H_2\overset{+}{C}{-}\ddot{N}(R)(H) \longleftrightarrow H_2C{=}N^{+}(R)(H)\right]$$

α cleavage iminium ion

Figure 19-9 shows the mass spectrum of butyl propyl amine. The base peak (m/z 72) corresponds to α cleavage with loss of a propyl radical to give a resonance-stabilized iminium ion. A similar α cleavage, with loss of an ethyl radical, gives the peak at m/z 86.

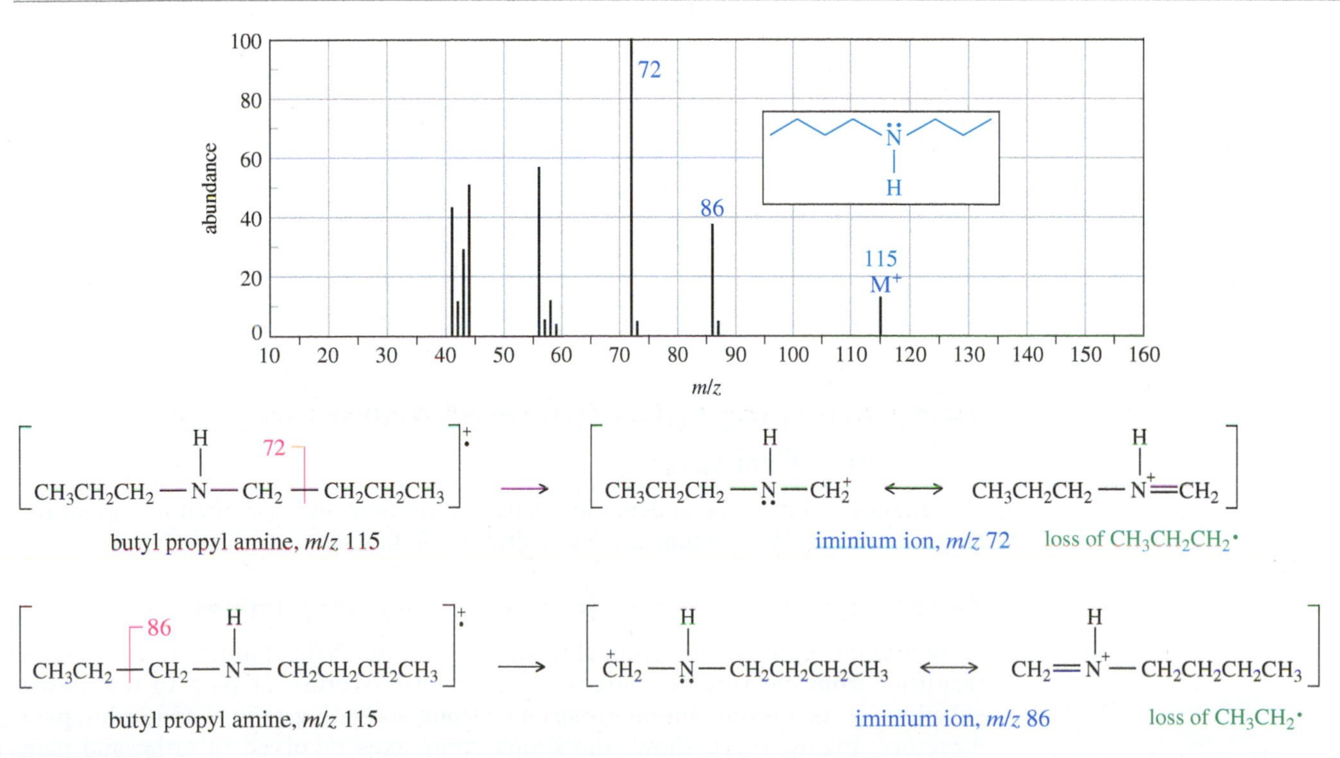

FIGURE 19-9
Mass spectrum of butyl propyl amine (*N*-propylbutan-1-amine). Note the odd mass number of the molecular ion and the even mass numbers of most fragments. The base peak corresponds to α cleavage in the butyl group, giving a propyl radical and a resonance-stabilized iminium ion.

PROBLEM 19-10

(a) Show how fragmentation occurs to give the base peak at m/z 58 in the mass spectrum of ethyl propyl amine (*N*-ethylpropan-1-amine), shown below.
(b) Show how a similar cleavage in the ethyl group gives an ion of m/z 72.
(c) Explain why the peak at m/z 72 is much weaker than the one at m/z 58.

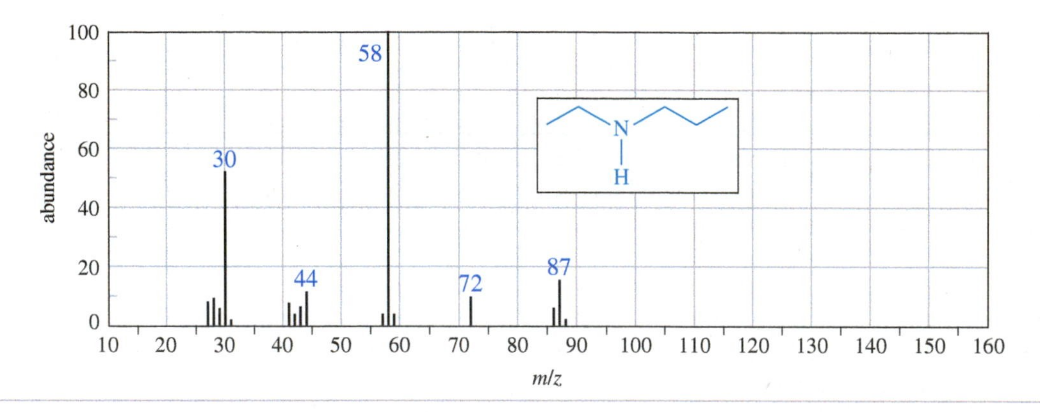

19-9 Reactions of Amines with Ketones and Aldehydes (Review)

In contrast to other functional groups, we will study the reactions of amines before we study their syntheses. This approach is better because most amine syntheses involve the reactions of amines. They begin with an amine (or ammonia) and add groups to make more-substituted amines. By studying the reactions first, we can readily understand how to use these reactions to convert simpler amines to more complicated amines.

In Section 18-14, we saw that amines attack ketones and aldehydes. When this nucleophilic attack is followed by dehydration, an imine (Schiff base) results. The analogous reaction of a hydrazine derivative gives a hydrazone, and the reaction with hydroxylamine gives an oxime. In Section 19-18, we will use these reactions to synthesize amines.

Y = H or alkyl gives an imine
Y = OH gives an oxime
Y = NHR gives a hydrazone

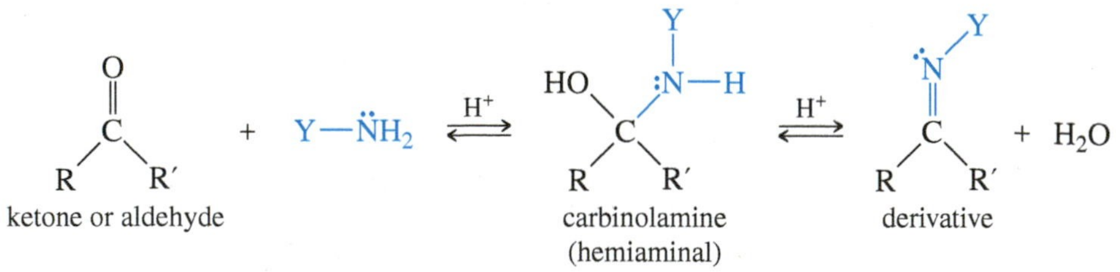

$$R{-}C({=}O){-}R' + Y{-}\ddot{N}H_2 \overset{H^+}{\rightleftharpoons} R{-}C(OH)(\ddot{N}H{-}Y){-}R' \overset{H^+}{\rightleftharpoons} R{-}C({=}\ddot{N}{-}Y){-}R' + H_2O$$

ketone or aldehyde — carbinolamine (hemiaminal) — derivative

19-10 Aromatic Substitution of Arylamines and Pyridine

Arylamines and pyridine undergo aromatic substitutions that are often important for amine synthesis. We examine these substitutions in this section.

19-10A Electrophilic Aromatic Substitution of Arylamines

In an arylamine, the nonbonding electrons on nitrogen help stabilize intermediates resulting from electrophilic attack at the positions ortho or para to the amine substituent. As a result, amino groups are strong activating groups and ortho, para directors. Figure 19-10 shows the sigma complexes involved in ortho and para substitution of aniline.

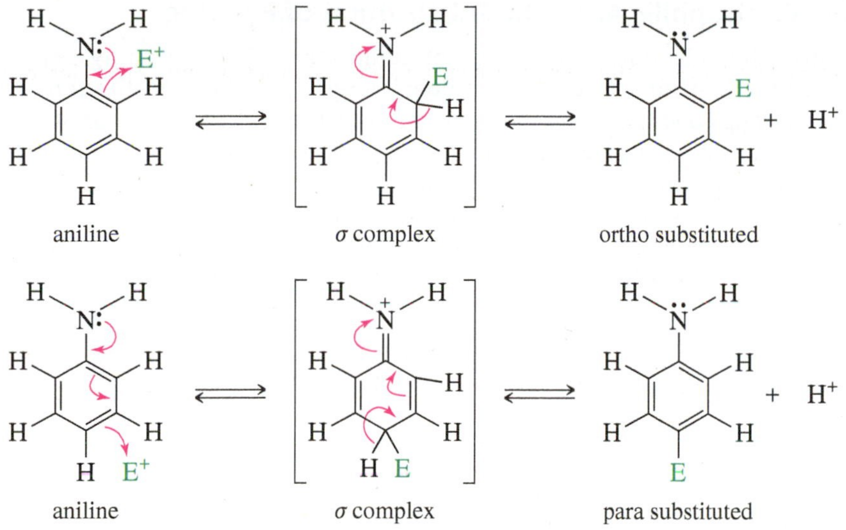

aniline ⇌ σ complex ⇌ ortho substituted + H^+

aniline ⇌ σ complex ⇌ para substituted + H^+

FIGURE 19-10
The amino group is a strong activator and ortho, para-director. The nonbonding electrons on nitrogen stabilize the σ complex when attack occurs at the ortho or para positions.

The following reactions show halogenation of aniline derivatives, which occurs readily without a catalyst. If an excess of the reagent is used, all the unsubstituted positions ortho and para to the amino group become substituted.

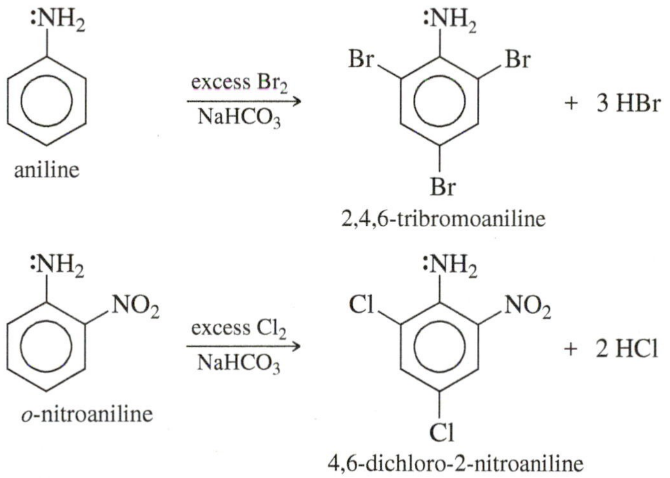

aniline $\xrightarrow[\text{NaHCO}_3]{\text{excess Br}_2}$ 2,4,6-tribromoaniline + 3 HBr

o-nitroaniline $\xrightarrow[\text{NaHCO}_3]{\text{excess Cl}_2}$ 4,6-dichloro-2-nitroaniline + 2 HCl

Care must be exercised in reactions with aniline derivatives, however. Strongly acidic reagents protonate the amino group, giving an ammonium salt that bears a full positive charge. The $-NH_3^+$ group is strongly deactivating (and meta-allowing). Therefore, strongly acidic reagents are unsuitable for substitution of anilines. Oxidizing acids (such as nitric and sulfuric acids) may oxidize the amino group, leading to decomposition and occasional violent reactions. In Section 19-12, we will see how the amino group may be acylated to decrease its basicity and permit substitution by a wide variety of electrophiles.

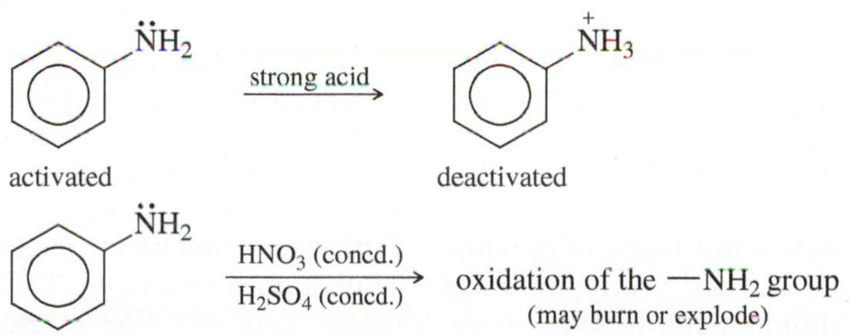

activated $\xrightarrow{\text{strong acid}}$ deactivated

$\xrightarrow[\text{H}_2\text{SO}_4\text{ (concd.)}]{\text{HNO}_3\text{ (concd.)}}$ oxidation of the $-NH_2$ group (may burn or explode)

19-10B Electrophilic Aromatic Substitution of Pyridine

In its aromatic substitution reactions, pyridine resembles a strongly deactivated benzene. Friedel–Crafts reactions fail completely, and other substitutions require unusually strong conditions. Deactivation results from the electron-withdrawing effect of the electronegative nitrogen atom. Its nonbonding electrons are perpendicular to the π system, and they cannot stabilize the positively charged intermediate. When pyridine does react, it gives substitution at the 3-position, analogous to the meta substitution shown by deactivated benzene derivatives. Mechanism 19-1 shows the substitution of pyridine at the 3-position.

MECHANISM 19-1 Electrophilic Aromatic Substitution of Pyridine

Step 1: Attack takes place at the 3-position.

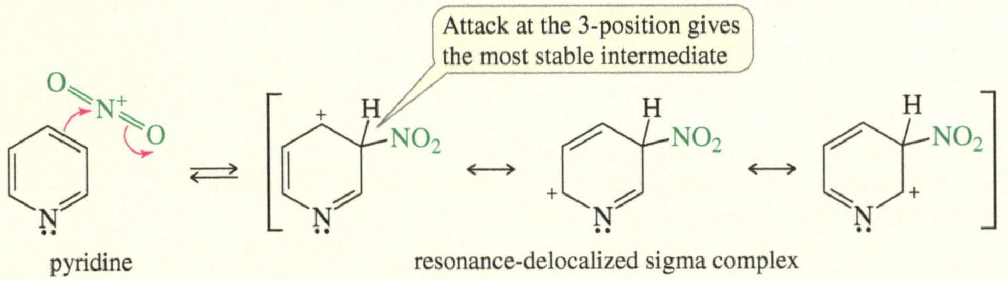

Attack at the 3-position gives the most stable intermediate

pyridine

resonance-delocalized sigma complex

Step 2: Loss of a proton gives the product.

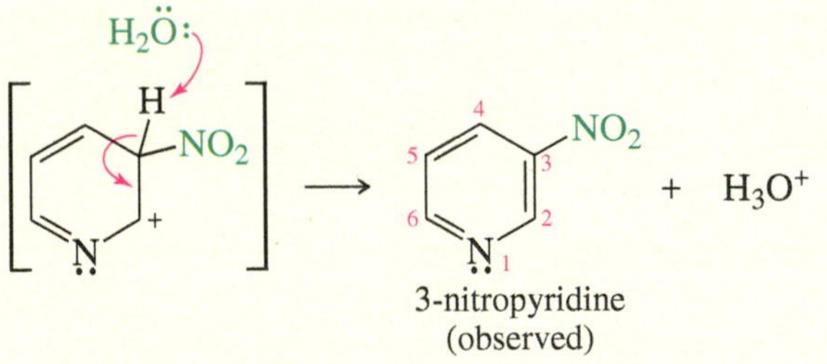

$H_2\ddot{O}:$ ... $\longrightarrow$ + H_3O^+

3-nitropyridine
(observed)

In comparison, consider the unfavorable intermediate that would be formed by attack at the 2-position:

Attack at the 2-position (or 4-position) is not observed.

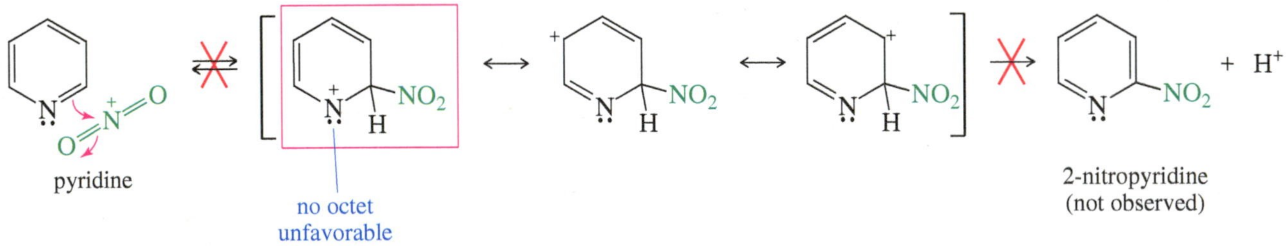

pyridine

no octet
unfavorable

+ H^+

2-nitropyridine
(not observed)

Electrophilic attack on pyridine at the 2-position gives an unstable intermediate, with one of the resonance structures showing a positive charge and only six electrons on nitrogen. In contrast, electrophilic attack at the 3-position gives a more stable intermediate with the positive charge spread over three carbon atoms and not on nitrogen.

Electrophilic substitution of pyridine is further hindered by the tendency of the nitrogen atom to attack electrophiles and take on a positive charge. The positively charged pyridinium ion is even more resistant than pyridine to electrophilic substitution.

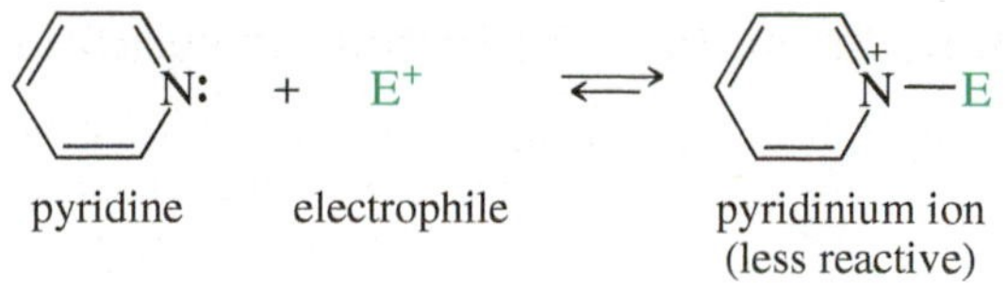

pyridine + electrophile ⇌ pyridinium ion (less reactive)

$N: + E^+ \rightleftharpoons N^+{-}E$

PROBLEM 19-11

Propose a mechanism for nitration of pyridine at the 4-position, and show why this orientation is not observed.

Two electrophilic substitutions of pyridine are shown here. Notice that these reactions require severe conditions, and the yields are poor to fair.

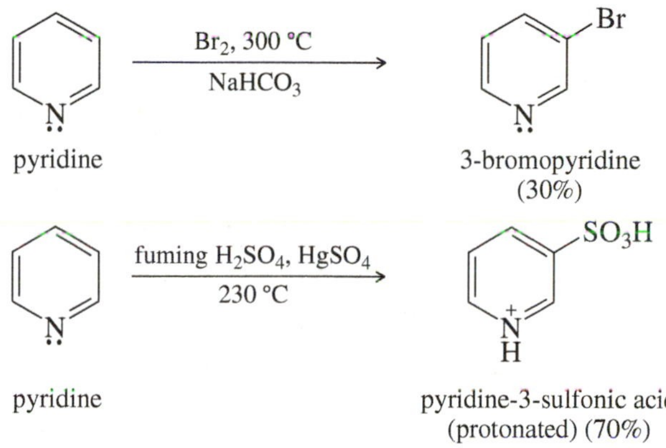

pyridine $\xrightarrow[NaHCO_3]{Br_2,\ 300\ °C}$ 3-bromopyridine (30%)

pyridine $\xrightarrow[230\ °C]{fuming\ H_2SO_4,\ HgSO_4}$ pyridine-3-sulfonic acid (protonated) (70%)

PROBLEM 19-12

Propose a mechanism for the sulfonation of pyridine, and point out why sulfonation occurs at the 3-position.

19-10C Nucleophilic Aromatic Substitution of Pyridine

Pyridine is deactivated toward electrophilic attack, but it is activated toward attack by electron-rich nucleophiles; that is, it is activated toward nucleophilic aromatic substitution. If there is a good leaving group at either the 2-position or the 4-position, a nucleophile can attack and displace the leaving group. The following reaction shows nucleophilic attack at the 2-position. The intermediate is stabilized by delocalization of the negative charge onto the electronegative nitrogen atom. This stabilization is not possible if attack occurs at the 3-position.

MECHANISM 19-2 Nucleophilic Aromatic Substitution of Pyridine

Step 1: Nucleophilic attack at the 2-position (or the 4-position) forms a stabilized intermediate.

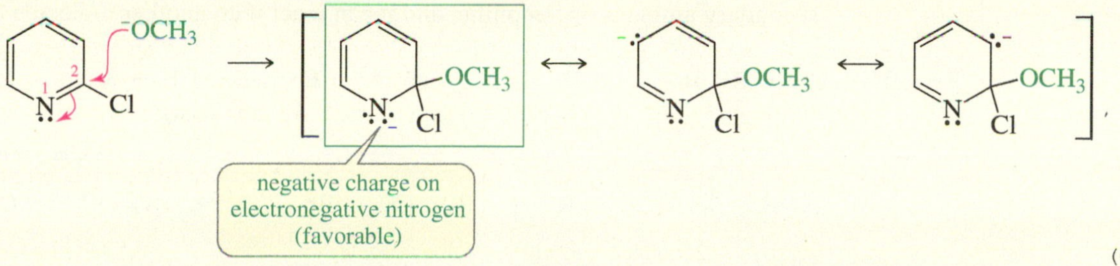

$^-OCH_3$ + 2-chloropyridine → [resonance structures with OCH_3 and Cl at C2]

negative charge on electronegative nitrogen (favorable)

(CONTINUED)

Step 2: Expulsion of the leaving group gives the product.

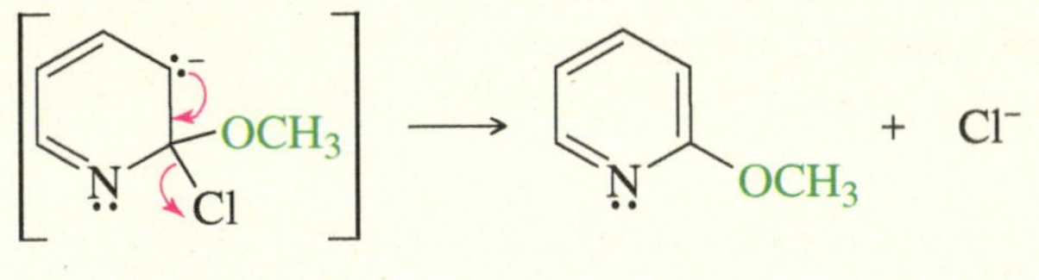

(resonance-delocalized)

Nucleophilic attack at the 3-position (not observed)

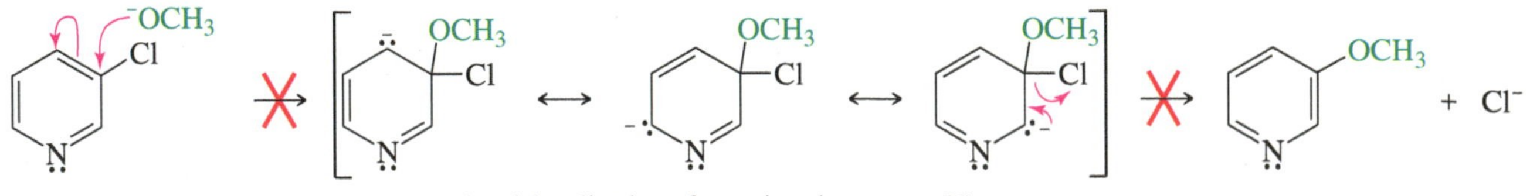

(no delocalization of negative charge onto N)

PROBLEM 19-13

We have considered nucleophilic aromatic substitution of pyridine at the 2-position and 3-position but not at the 4-position. Complete the three possible cases by showing the mechanism for the reaction of methoxide ion with 4-chloropyridine. Show how the intermediate is stabilized by delocalization of the charge onto the nitrogen atom.

PROBLEM 19-14

(a) Propose a mechanism for the reaction of 2-bromopyridine with sodium amide to give 2-aminopyridine.

(b) When 3-bromopyridine is used in this reaction, stronger reaction conditions are required and a mixture of 3-aminopyridine and 4-aminopyridine results. Propose a mechanism to explain this curious result.

19-11 Alkylation of Amines by Alkyl Halides

Amines react with primary alkyl halides to give alkylated ammonium halides. Alkylation proceeds by the S_N2 mechanism, so it is not feasible with tertiary halides because they are too hindered. Secondary halides often give poor yields, with elimination predominating over substitution.

$$\underset{\text{primary amine}}{R-\ddot{N}H_2} + \underset{\text{primary halide}}{R'-CH_2-Br} \longrightarrow \underset{\text{salt of a secondary amine}}{R-\overset{+}{N}H_2-CH_2-R'\ Br^-}$$

Unfortunately, the initially formed salt may become deprotonated. The resulting secondary amine is nucleophilic, and it can react with another molecule of the halide.

$$R-\overset{+}{N}H_2-CH_2-R'\ Br^- + R-\ddot{N}H_2 \rightleftharpoons \underset{2^\circ\text{ amine}}{R-\ddot{N}H-CH_2-R'} + R-\overset{+}{N}H_3\ Br^-$$

$$\underset{2^\circ\text{ amine}}{R-\ddot{N}H-CH_2-R'} + R'-CH_2-Br \longrightarrow \underset{\text{salt of a tertiary amine}}{R-\underset{+}{\overset{CH_2-R'}{|}}{N}H-CH_2-R'\ Br^-}$$

The difficulty of direct alkylation lies in stopping it at the desired stage. Even if just one equivalent of the halide is added, some amine molecules will react once, some will react twice, and some will react three times (to give the tetraalkylammonium salt). Others will not react at all. A complex mixture results.

Alkylation of amines can give good yields of the desired alkylated products in two types of reactions:

1. *"Exhaustive" alkylation to the tetraalkylammonium salt.* Mixtures of different alkylated products are avoided if enough alkyl halide is added to alkylate the amine as many times as possible. This **exhaustive alkylation** gives a tetraalkylammonium salt. A mild base (often $NaHCO_3$ or dilute NaOH) is added to deprotonate the intermediate alkylated amines and to neutralize the large quantities of HX formed.

$$CH_3CH_2CH_2{-}NH_2 \; + \; 3\,CH_3{-}I \xrightarrow{NaHCO_3} CH_3CH_2CH_2{-}\overset{+}{N}(CH_3)_3\; I^- \quad (90\%)$$

PROBLEM 19-15

Propose a mechanism to show the individual alkylations that form this quaternary ammonium salt.

2. *Reaction with a large excess of ammonia.* Because ammonia is inexpensive and has a low molecular weight, it is convenient to use a very large excess. Addition of a primary alkyl halide to a large excess of ammonia forms the primary amine, and the probability of dialkylation is small. Excess ammonia is simply allowed to evaporate.

$$\underset{\text{10 moles}}{\ddot{N}H_3} \; + \; \underset{\text{1 mole}}{R{-}CH_2{-}X} \longrightarrow R{-}CH_2{-}\overset{+}{N}H_3\; X^-$$

PROBLEM 19-16

Show how you would use direct alkylation to synthesize the following compounds.
(a) benzyltrimethylammonium iodide **(b)** pentan-1-amine **(c)** benzylamine

19-12 Acylation of Amines by Acid Chlorides

Primary and secondary amines react with acid halides to form amides. This reaction is a *nucleophilic acyl substitution*: the replacement of a leaving group on a carbonyl carbon by a nucleophile. We will study nucleophilic acyl substitution in detail in Chapters 20 and 21. In this case, the amine replaces chloride ion.

$$R'{-}\ddot{N}H_2 \; + \; R{-}\overset{O}{\overset{\|}{C}}{-}Cl \xrightarrow{\text{pyridine}} R{-}\overset{O}{\overset{\|}{C}}{-}\ddot{N}H{-}R' \; + \; C_5H_5\overset{+}{N}{-}H\; Cl^-$$

The amine attacks the carbonyl group of an acid chloride much like it attacks the carbonyl group of a ketone or aldehyde (Mechanism 19-3). The acid chloride is more reactive than a ketone or an aldehyde because the electronegative chlorine atom draws electron density away from the carbonyl carbon, making it more electrophilic. The chlorine atom in the tetrahedral intermediate is a good leaving group. The tetrahedral intermediate expels chloride to give the amide. A base such as pyridine or NaOH is often added to neutralize the HCl produced.

MECHANISM 19-3 Acylation of an Amine by an Acid Chloride

Step 1: A nucleophile attacks the strongly electrophilic carbonyl group of the acid chloride to form a tetrahedral intermediate.

$$\underset{\text{acid chloride}}{R-\overset{\overset{\ddot{O}}{\|}}{C}-Cl} + \underset{\text{amine}}{R'-\ddot{N}H_2} \rightleftharpoons \underset{\text{tetrahedral intermediate}}{R-\overset{:\ddot{O}:^-}{\underset{^+NH_2-R'}{C}}-Cl}$$

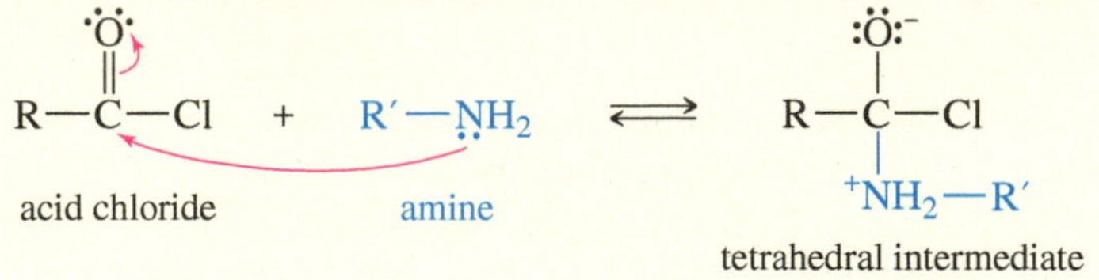

Step 2: The tetrahedral intermediate expels chloride ion.

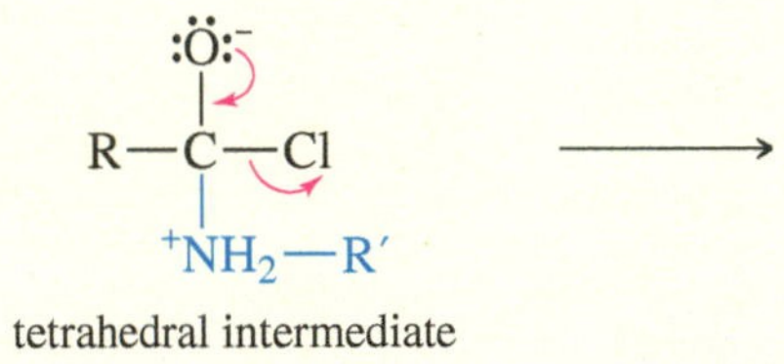

$$\underset{\text{tetrahedral intermediate}}{R-\overset{:\ddot{O}:^-}{\underset{^+NH_2-R'}{C}}-Cl} \longrightarrow R-\overset{\overset{\ddot{O}}{\|}}{C}-\overset{+}{N}H(H)-R' \quad Cl^-$$

Step 3: Loss of a proton gives the amide.

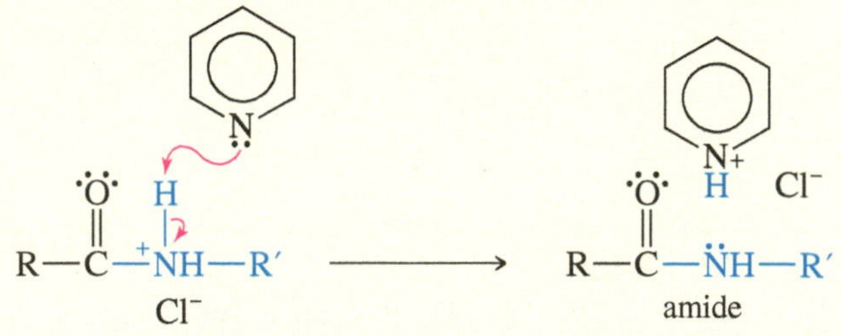

$$R-\overset{\overset{\ddot{O}}{\|}}{C}-\overset{+}{N}H(H)-R' \; Cl^- + \text{pyridine} \longrightarrow \underset{\text{amide}}{R-\overset{\overset{\ddot{O}}{\|}}{C}-\ddot{N}H-R'} + \text{pyridine-}H^+ \; Cl^-$$

Example

$$C_6H_5-\overset{\overset{O}{\|}}{C}-Cl + CH_3-NH_2 \xrightarrow{\text{pyridine}} \underset{(95\%)}{C_6H_5-\overset{\overset{O}{\|}}{C}-NHCH_3}$$

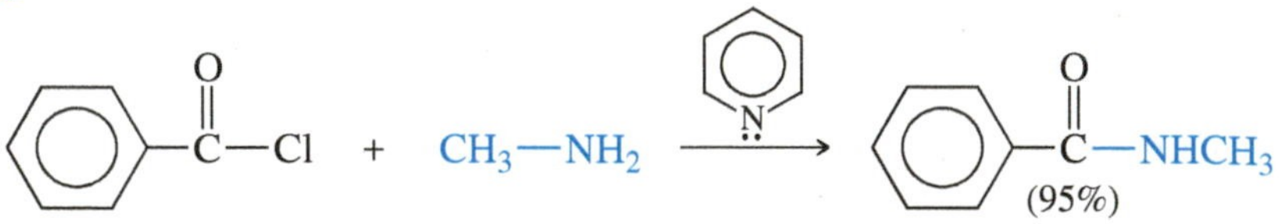

The amide produced in this reaction usually does not undergo further acylation. Amides are stabilized by a resonance structure that involves nitrogen's nonbonding electrons and places a positive charge on nitrogen. As a result, amides are much less basic and less nucleophilic than amines.

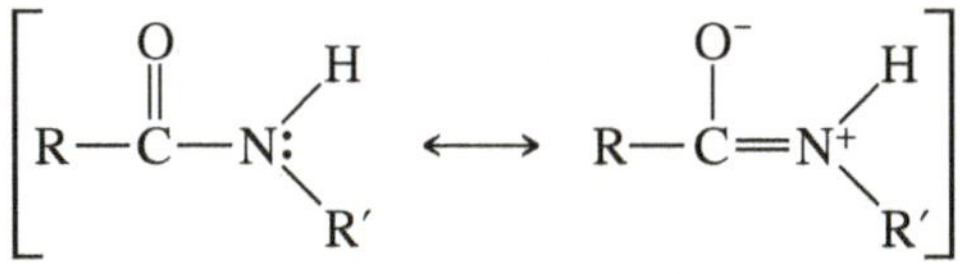

$$\left[R-\overset{\overset{O}{\|}}{C}-\ddot{N}(H)(R') \longleftrightarrow R-\overset{O^-}{\overset{|}{C}}=\overset{+}{N}(H)(R')\right]$$

resonance stabilization of an amide

The diminished basicity of amides can be used to advantage in electrophilic aromatic substitutions. For example, if the amino group of aniline is acetylated to give acetanilide, the resulting amide is still activating and ortho, para-directing. Unlike aniline, however, acetanilide may be treated with acidic (and mild oxidizing) reagents, as shown next. Aryl amino groups are frequently acylated before further substitutions are attempted on the ring, and the acyl group is removed later by acidic or basic hydrolysis (Section 21-7C).

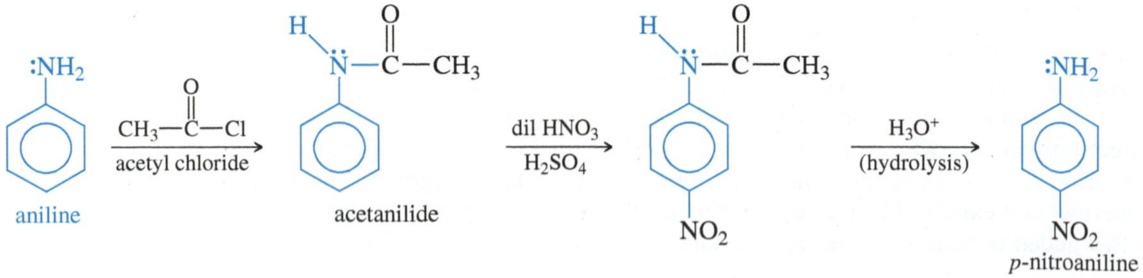

$$\underset{\text{aniline}}{C_6H_5-\ddot{N}H_2} \xrightarrow[\text{acetyl chloride}]{CH_3-\overset{\overset{O}{\|}}{C}-Cl} \underset{\text{acetanilide}}{C_6H_5-\ddot{N}H-\overset{\overset{O}{\|}}{C}-CH_3} \xrightarrow[H_2SO_4]{\text{dil } HNO_3} p\text{-}O_2N-C_6H_4-\ddot{N}H-\overset{\overset{O}{\|}}{C}-CH_3 \xrightarrow[\text{(hydrolysis)}]{H_3O^+} \underset{p\text{-nitroaniline}}{p\text{-}O_2N-C_6H_4-\ddot{N}H_2}$$

SOLVED PROBLEM 19-1

Show how you would accomplish the following synthetic conversion in good yield.

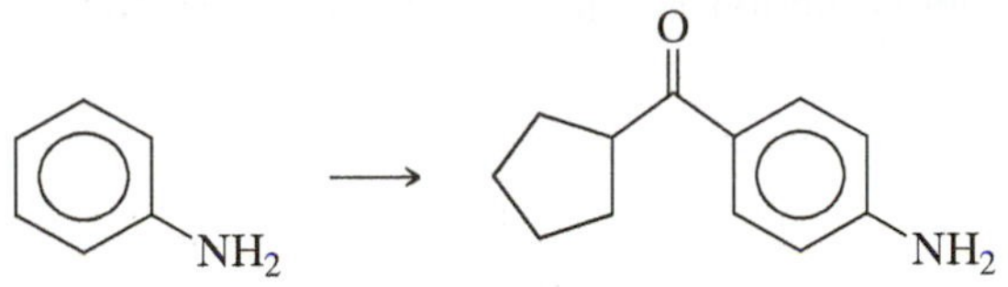

SOLUTION

An attempted Friedel–Crafts acylation on aniline would likely meet with disaster. The free amino group would attack both the acid chloride and the Lewis acid catalyst.

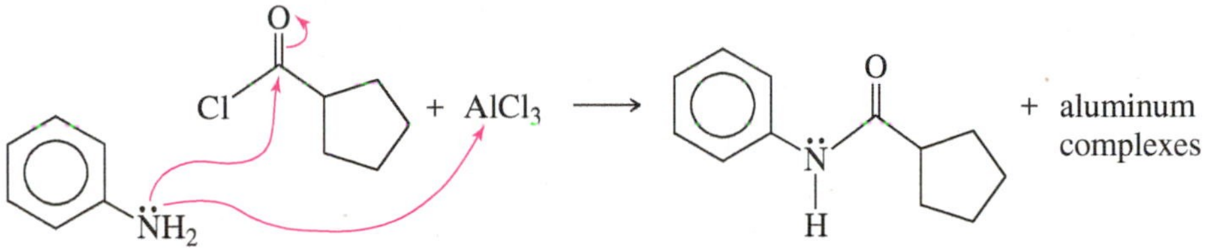

+ $AlCl_3$ ⟶ + aluminum complexes

We can control the nucleophilicity of aniline's amino group by converting it to an amide, which is still activating and ortho, para-directing for the Friedel–Crafts reaction. Acylation, followed by hydrolysis of the amide, gives the desired product.

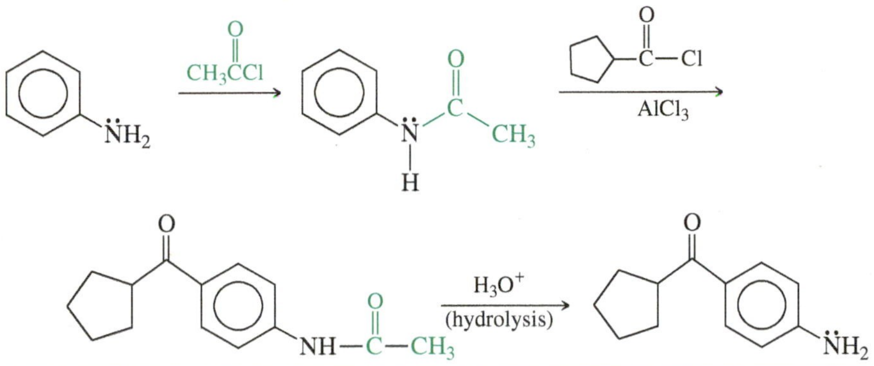

CH_3CCl ; $AlCl_3$; $NH—C—CH_3$; H_3O^+ (hydrolysis) ; NH_2

PROBLEM 19-17

Give the products expected from the following reactions.

(a) acetyl chloride + ethylamine

(b) benzoyl chloride + $(CH_3)_2NH$ dimethylamine

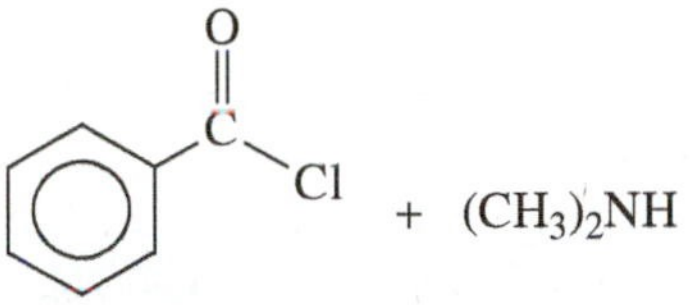

(c) $CH_3—(CH_2)_4—C(=O)—Cl$ + piperidine (NH)

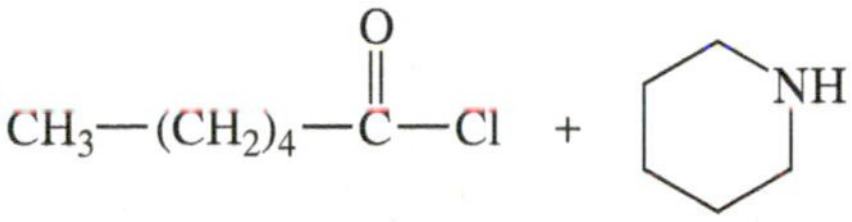

hexanoyl chloride, piperidine

19-13 Formation of Sulfonamides

Sulfonyl chlorides are the acid chlorides of sulfonic acids. Like acyl chlorides, sulfonyl chlorides are strongly electrophilic.

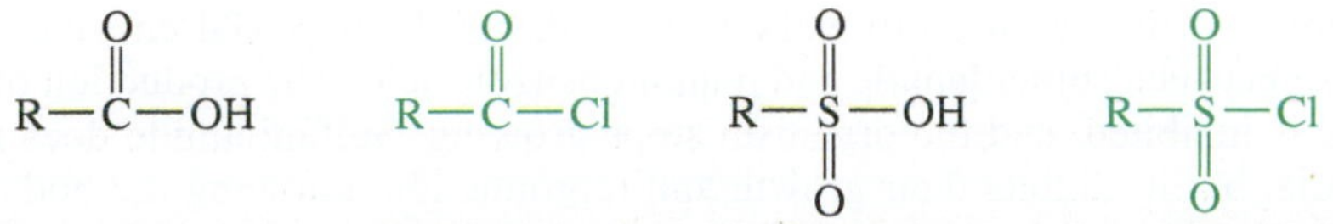

$R—C(=O)—OH$	$R—C(=O)—Cl$	$R—S(=O)_2—OH$	$R—S(=O)_2—Cl$
a carboxylic acid	an acyl chloride (acid chloride)	a sulfonic acid	a sulfonyl chloride

A primary or secondary amine attacks a sulfonyl chloride and displaces a chloride ion to give an amide. Amides of sulfonic acids are called **sulfonamides.** This reaction is similar to the formation of a sulfonate ester from a sulfonyl chloride (such as tosyl chloride) and an alcohol (Section 11-5).

$$R'{-}\ddot{N}H_2 + R{-}SO_2{-}Cl \xrightarrow{NaOH} R{-}SO_2{-}\overset{+}{N}H(H){-}R'\ Cl^- \longrightarrow R{-}SO_2{-}\ddot{N}HR' + H_2O + Cl^-$$

amine; sulfonyl chloride; sulfonamide

The *sulfa drugs* are a class of sulfonamides used as antibacterial agents. In 1936, sulfanilamide was found to be effective against streptococcal infections. Sulfanilamide is synthesized from acetanilide (having the amino group protected as an amide) by chlorosulfonation followed by treatment with ammonia. The final reaction is hydrolysis of the protecting group to give sulfanilamide.

$$\text{acetanilide} \xrightarrow{Cl{-}SO_2{-}OH} p\text{-}CH_3CONH{-}C_6H_4{-}SO_2Cl \xrightarrow[H_2O]{:NH_3} p\text{-}CH_3CONH{-}C_6H_4{-}SO_2NH_2 \xrightarrow[\text{heat}]{\text{dil HCl}} p\text{-}H_2N{-}C_6H_4{-}SO_2NH_2\ (\text{sulfanilamide})$$

Application: Early Antibiotic

During World War II, American soldiers carried first aid kits containing sulfanilamide powder and tablets. Medics sprinkled the powder on open wounds to combat infection, and the tablets were to be swallowed to prevent and treat gangrene, pneumonia, and other battlefield illnesses.

PROBLEM 19-18

What would happen in the synthesis of sulfanilamide if the amino group were not protected as an amide in the chlorosulfonation step?

The biological activity of sulfanilamide has been studied in detail. It appears that sulfanilamide is an analog of *p*-aminobenzoic acid. Streptococci use *p*-aminobenzoic acid to synthesize folic acid, an essential compound for growth and reproduction.

$$p\text{-aminobenzoic acid} \xrightarrow{\text{incorporated into}} \text{folic acid}$$

p-aminobenzoic acid — folic acid

Sulfanilamide cannot be used to make folic acid, but the bacterial enzymes cannot distinguish between sulfanilamide and *p*-aminobenzoic acid. The production of active folic acid is inhibited, and the organism stops growing. Sulfanilamide does not kill the bacteria, but it inhibits their growth and reproduction, allowing the body's own defense mechanisms to destroy the infection.

PROBLEM 19-19

Show how you would use the same sulfonyl chloride as used in the sulfanilamide synthesis to make sulfathiazole and sulfapyridine.

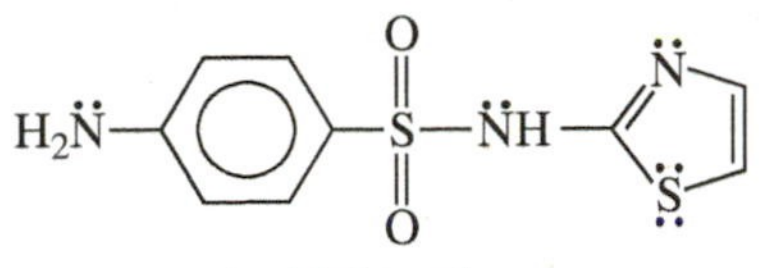

sulfathiazole

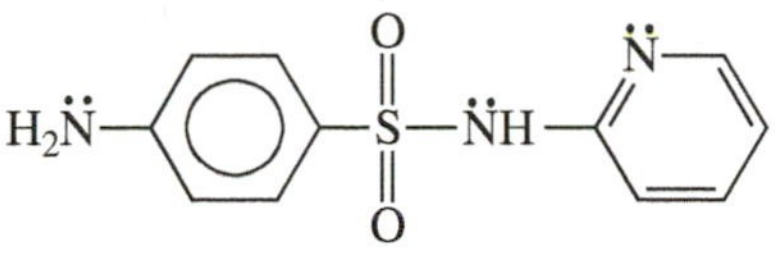

sulfapyridine

Application: Drug Resistance

The effectiveness of sulfa drugs is now limited due to bacterial resistance. One method used by resistant strains is to overproduce *p*-aminobenzoic acid, effectively diluting the drug's concentration.

19-14 Amines as Leaving Groups: The Hofmann Elimination

Amines can be converted to alkenes by elimination reactions, much like alcohols and alkyl halides undergo elimination to give alkenes (Sections 11-10 and 7-9). An amine cannot undergo elimination directly, however, because the leaving group would be an amide ion ($^-NH_2$ or ^-NHR), which is a very strong base and a poor leaving group.

An amino group can be converted to a good leaving group by exhaustive methylation, which converts it to a quaternary ammonium salt that can leave as a neutral amine. Exhaustive methylation is usually accomplished using methyl iodide.

Exhaustive methylation of an amine

$$R{-}\ddot{N}H_2 + 3\,CH_3{-}I \longrightarrow R{-}\overset{+}{N}(CH_3)_3\ I^- + 2\,HI$$

poor leaving group (R—NH$_2$); good leaving group (R—N$^+$(CH$_3$)$_3$)

Elimination of the quaternary ammonium salt generally takes place by the E2 mechanism, which requires a strong base. To provide the base, the quaternary ammonium iodide is converted to the hydroxide salt by treatment with silver oxide.

Conversion to the hydroxide salt

$$R{-}\overset{+}{N}(CH_3)_3\ I^- + \tfrac{1}{2}Ag_2O + H_2O \longrightarrow R{-}\overset{+}{N}(CH_3)_3\ {}^-OH + AgI\downarrow$$

quaternary ammonium iodide → quaternary ammonium hydroxide

Heating of the quaternary ammonium hydroxide results in E2 elimination and formation of an alkene. This elimination of a quaternary ammonium hydroxide is called the **Hofmann elimination.**

MECHANISM 19-4 Hofmann Elimination

The Hofmann elimination is a one-step, concerted E2 reaction using an amine as the leaving group.

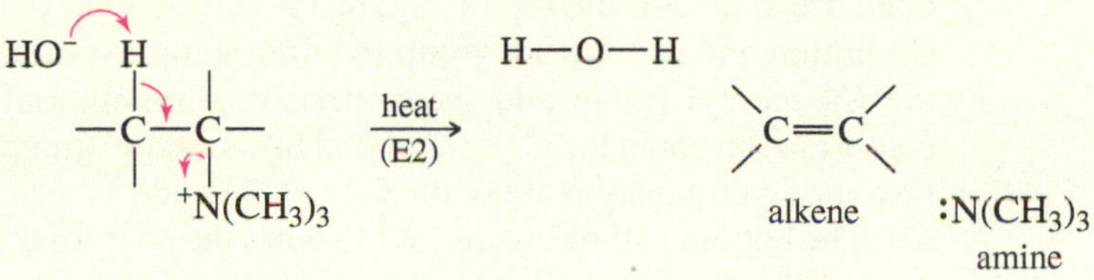

$$HO^-\ \ H{-}C{-}C{-}{}^+N(CH_3)_3 \xrightarrow[(E2)]{heat} H{-}O{-}H + \ >C{=}C< + :N(CH_3)_3$$

alkene; amine

For example, when butan-2-amine is exhaustively methylated, converted to the hydroxide salt, and heated, elimination takes place to form a mixture of but-1-ene and but-2-ene.

Exhaustive methylation and conversion to the hydroxide salt

$$\underset{\text{butan-2-amine}}{\overset{1}{CH_3}-\overset{2}{CH}(:NH_2)-\overset{3}{CH_2}-\overset{4}{CH_3}} \xrightarrow[\text{(2) } Ag_2O,\ H_2O]{\text{(1) excess } CH_3I} \underset{\text{quaternary ammonium hydroxide}}{\overset{1}{CH_3}-\overset{2}{CH}(^{+}N(CH_3)_3)-\overset{3}{CH_2}-\overset{4}{CH_3}\quad {}^{-}OH}$$

Heating and Hofmann elimination

$$\overset{1}{H_2C}(H)-\overset{2}{CH}(^{+}N(CH_3)_3)-\overset{3}{CH}(H)-\overset{4}{CH_3} \quad (^{-}\!:\!\ddot{O}H \text{ or } ^{-}\!:\!\ddot{O}H) \xrightarrow{150\ ^\circ C} \underset{\substack{\text{but-1-ene}\\ \text{Hofmann product}\\ 95\%}}{\overset{1}{H_2C}=\overset{2}{CH}-\overset{3}{CH_2}-\overset{4}{CH_3}} + \underset{\substack{\text{but-2-ene (}E\text{ and }Z\text{)}\\ \text{Zaitsev product}\\ 5\%}}{\overset{1}{CH_3}-\overset{2}{CH}=\overset{3}{CH}-\overset{4}{CH_3}} + H_2O + :N(CH_3)_3$$

In Chapter 7, we saw that eliminations of alkyl halides usually follow Zaitsev's rule; that is, the most substituted product predominates. This rule applies because the most-substituted alkene is usually the most stable. In the Hofmann elimination, however, the product is commonly the *least*-substituted alkene. We often classify an elimination as giving mostly the *Zaitsev product* (the most-substituted alkene) or the *Hofmann product* (the least-substituted alkene).

Zaitsev elimination

$$\underset{\text{2-chlorobutane}}{\overset{1}{CH_3}-\overset{2}{CH}(Cl)-\overset{3}{CH_2}-\overset{4}{CH_3}} + \underset{\text{sodium methoxide}}{Na^{+}\ {}^{-}OCH_3} \longrightarrow \underset{\substack{\text{but-1-ene}\\ \text{Hofmann product}\\ (33\%)}}{\overset{1}{H_2C}=\overset{2}{CH}-\overset{3}{CH_2}-\overset{4}{CH_3}} + \underset{\substack{\text{but-2-ene (}E\text{ and }Z\text{)}\\ \text{Zaitsev product}\\ (67\%)}}{\overset{1}{CH_3}-\overset{2}{CH}=\overset{3}{CH}-\overset{4}{CH_3}}$$

The Hofmann elimination's preference for the least-substituted alkene stems from several factors, but one of the most compelling involves the sheer bulk of the leaving group. Remember that the E2 mechanism requires an anti-coplanar arrangement of the proton and the leaving group (Section 7-9). The extremely large trialkylamine leaving group in the Hofmann elimination often interferes with this coplanar arrangement.

Figure 19-11 shows the stereochemistry of the Hofmann elimination of butan-2-amine. The methylated ammonium salt eliminates by losing trimethylamine and a proton on either C1 or C3. The possible conformations along the C2—C3 bond are shown at the top of Figure 19-11. An anti-coplanar arrangement between a C3 proton and the leaving group requires an unfavorable gauche interaction between the C4 methyl group and the bulky trimethylammonium group. The most-stable conformation about the C2—C3 bond has a methyl group in the anti-coplanar position, preventing elimination along the C2—C3 bond.

The bottom half of Figure 19-11 shows the conformations along the C1—C2 bond. *Any* of the three staggered conformations of the C1—C2 bond provides an anti relationship between one of the protons and the leaving group. The Hofmann product predominates because elimination of one of the C1 protons involves a lower-energy, more probable transition state than the crowded transition state required for Zaitsev (C2—C3) elimination.

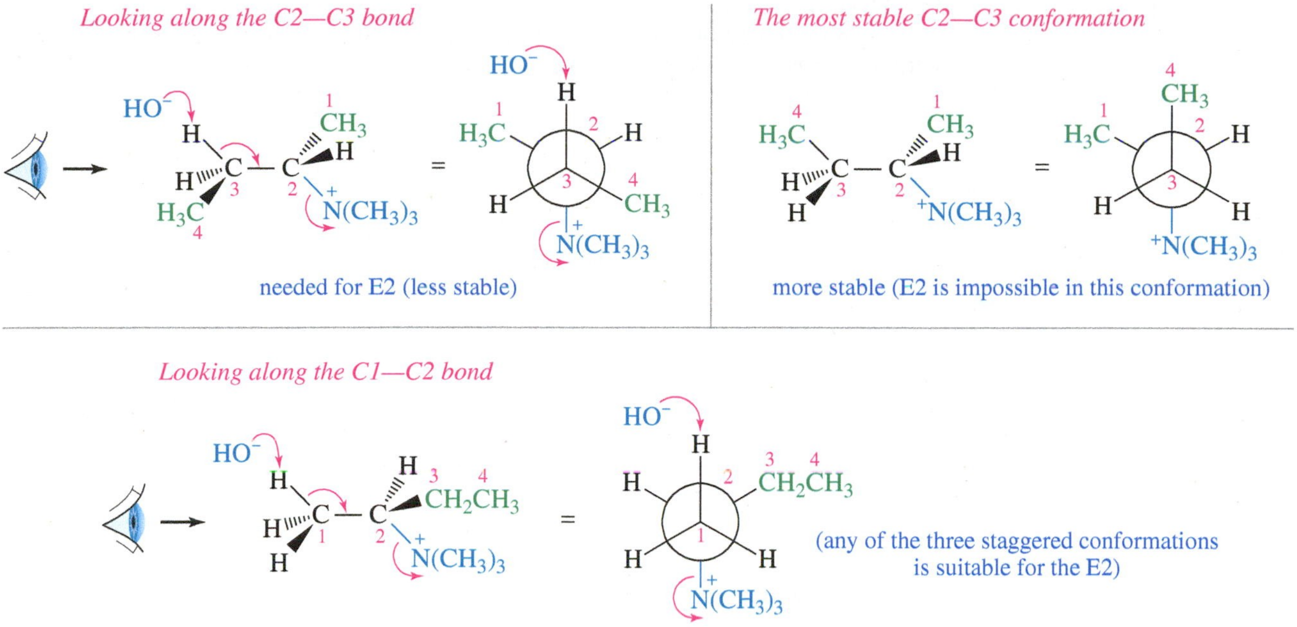

FIGURE 19-11
Hofmann elimination of exhaustively methylated butan-2-amine. The most stable conformation of the C2—C3 bond has no proton on C3 in an anti relationship to the leaving group. Along the C1—C2 bond, however, any staggered conformation has an anti relationship between a proton and the leaving group. Abstraction of a proton from C1 gives the Hofmann product.

The Hofmann elimination is frequently used to determine the structures of complex amines by converting them to simpler amines. The direction of elimination is usually predictable, giving the least-substituted alkene. Figure 19-12 shows two examples using the Hofmann elimination to simplify complex amines.

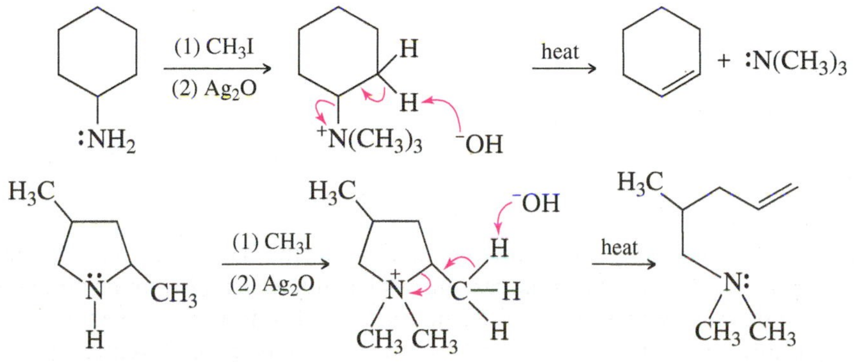

FIGURE 19-12
Examples of the Hofmann elimination. The least-substituted alkene is usually the favored product.

SOLVED PROBLEM 19-2

Predict the major product(s) formed when the following amine is treated with excess iodomethane, followed by heating with silver oxide.

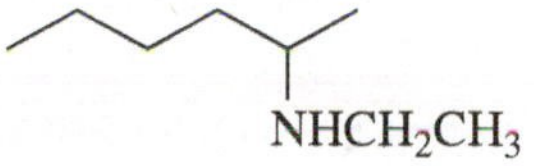

$NHCH_2CH_3$

(continued)

SOLUTION

Solving this type of problem requires finding every possible elimination of the methylated salt. In this case, the salt has the following structure:

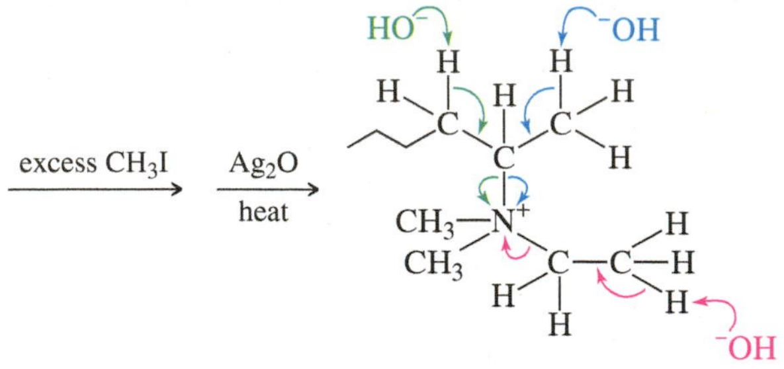

$$\xrightarrow{\text{excess } CH_3I} \xrightarrow[\text{heat}]{Ag_2O}$$ [quaternary ammonium salt with HO^- / ^-OH arrows: green, blue, and red elimination routes]

The green, blue, and red arrows show the three possible elimination routes. The corresponding products are

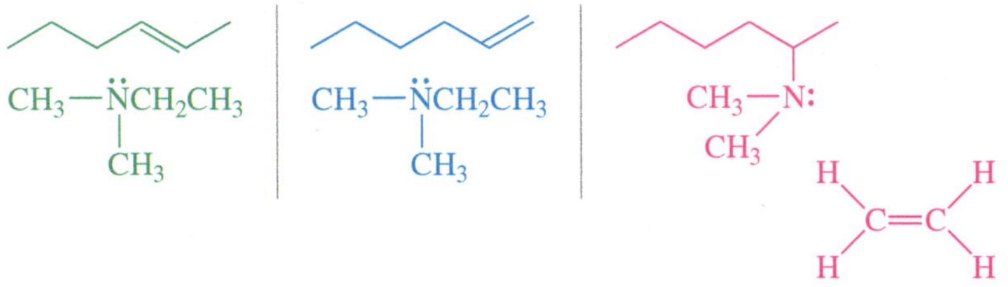

green: alkene + $CH_3-\ddot{N}CH_2CH_3$ with CH_3 | blue: alkene + $CH_3-\ddot{N}CH_2CH_3$ with CH_3 | red: amine $(CH_3)_2N:$ + $H_2C=CH_2$

The first (green) alkene has a disubstituted double bond. The second (blue) alkene is monosubstituted, and the red alkene (ethylene) has an unsubstituted double bond. We predict that the red products will be favored.

PROBLEM-SOLVING HINT

The key to solving Hofmann elimination problems is to find all possible ways the compound can eliminate. Then, the one that gives the least substituted alkene is probably favored.

PROBLEM-SOLVING HINT

Some of the stereochemical features of the Hofmann elimination are best studied using your models. Models are essential for working problems involving this elimination, such as Problem 19-20.

PROBLEM 19-20

Predict the major products formed when the following amines undergo exhaustive methylation, treatment with Ag_2O, and heating.

(a) hexan-2-amine **(b)** 2-methylpiperidine **(c)** *N*-ethylpiperidine

(d) [decahydroquinoline, N–H]

(e)

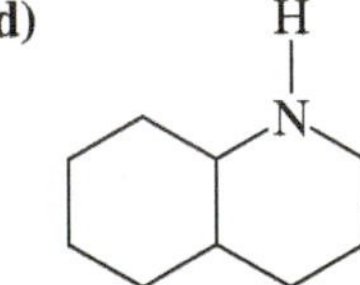

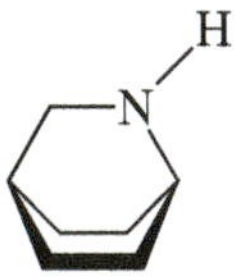

(f)

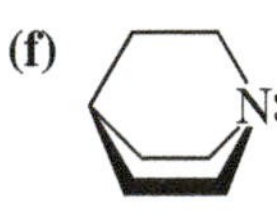

19-15 Oxidation of Amines; The Cope Elimination

Amines are notoriously easy to oxidize, and oxidation is often a side reaction in amine syntheses. Amines also oxidize during storage in contact with the air. Preventing air oxidation is one of the reasons for converting amines to their salts for storage or use as medicines.

The following partial structures show some of the bonding and oxidation states of amines:

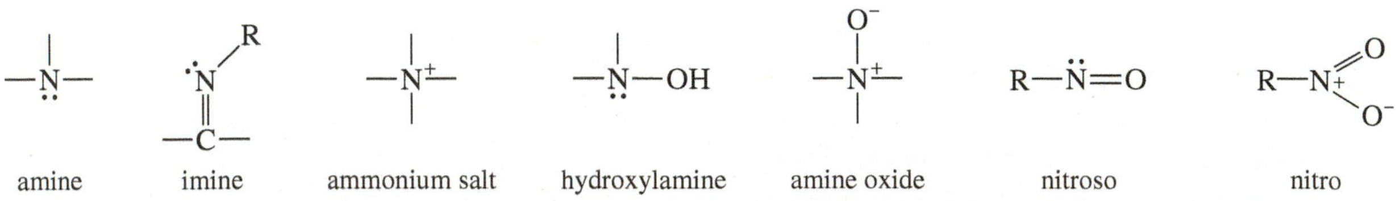

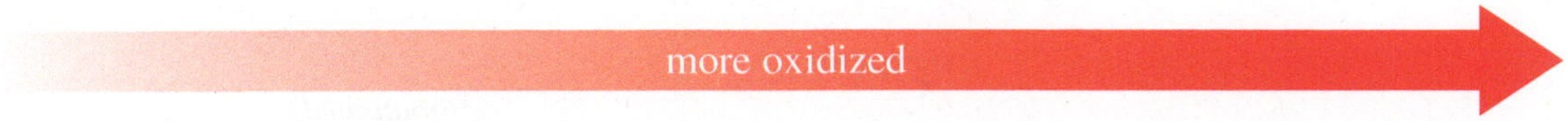

amine | imine | ammonium salt | hydroxylamine | amine oxide | nitroso ($R-\ddot{N}=O$) | nitro

more oxidized →

Depending on their specific structures, these states are generally more oxidized as you go from left to right. (Note the increasing number of bonds to oxygen.)

Most amines are oxidized by common oxidants such as H_2O_2, permanganate, and peroxyacids. Primary amines oxidize easily, but complex mixtures of products often result. The following sequence shows increasingly oxidized products of a primary amine, as it becomes more oxidized from left to right. The symbol [O] is used for a generic oxidizing agent.

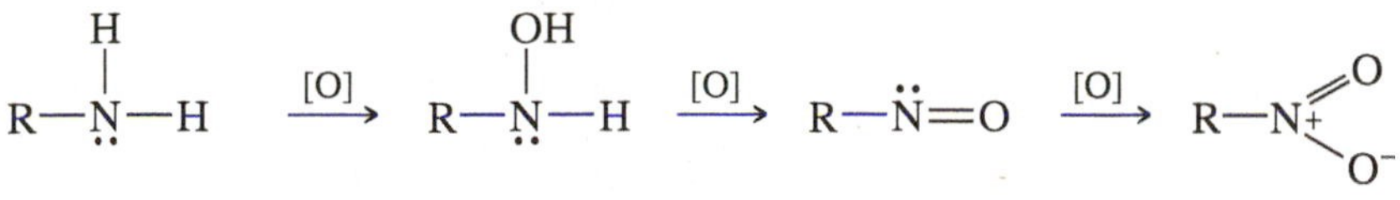

$$R{-}NH{-}H \xrightarrow{[O]} R{-}N(OH){-}H \xrightarrow{[O]} R{-}N{=}O \xrightarrow{[O]} R{-}N^{+}(=O){-}O^{-}$$

1° amine — hydroxylamine — nitroso — nitro

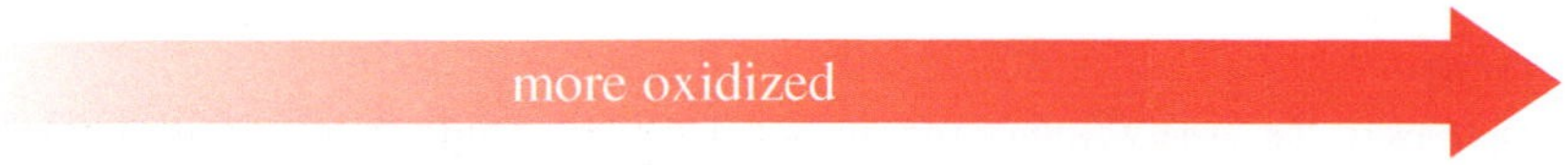

more oxidized

Secondary amines are easily oxidized to **hydroxylamines.** Side products are often formed, however, and the yields may be low. The mechanisms of amine oxidations are not well characterized, partly because many reaction paths (especially those involving free radicals) are available.

$$R_2N{-}H + H_2O_2 \longrightarrow R_2N{-}OH + H_2O$$

2° amine — a 2° hydroxylamine

Tertiary amines are oxidized to **amine oxides,** often in good yields. Either H_2O_2 or a peroxyacid may be used for this oxidation. Notice that an amine oxide must be drawn with a full positive charge on nitrogen and a negative charge on oxygen, as in nitro compounds. Because the N—O bond of the amine oxide is formed by donation of the electrons on nitrogen, this bond is often written as an arrow (N ⟶ O) in the older literature.

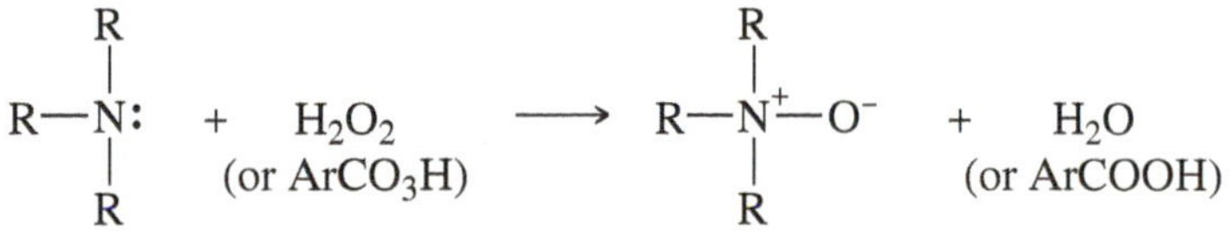

$$R_3N: + H_2O_2 \text{ (or } ArCO_3H) \longrightarrow R_3N^{+}{-}O^{-} + H_2O \text{ (or } ArCOOH)$$

3° amine — 3° amine oxide

Because of the positive charge on nitrogen, the amine oxide may undergo a **Cope elimination** (Mechanism 19-5), much like the Hofmann elimination of a quaternary ammonium salt. The amine oxide acts as its own base through a cyclic transition state, so a strong base is not needed. The Cope elimination generally gives the same orientation as Hofmann elimination, resulting in the least-substituted alkene.

Application: Biochemistry

Primary amines are oxidized in the body by monoamine oxidase (MAO). MAO converts the amine to an imine, which is hydrolyzed to yield an aldehyde and ammonia. One function of MAO is to regulate the levels of the neurotransmitters serotonin and norepinephrine. Monoamine oxidase inhibitors prevent the oxidation (and inactivation) of these neurotransmitters, thereby elevating mood. MAO inhibitors were the first antidepressants, but they are used sparingly now because of numerous side effects.

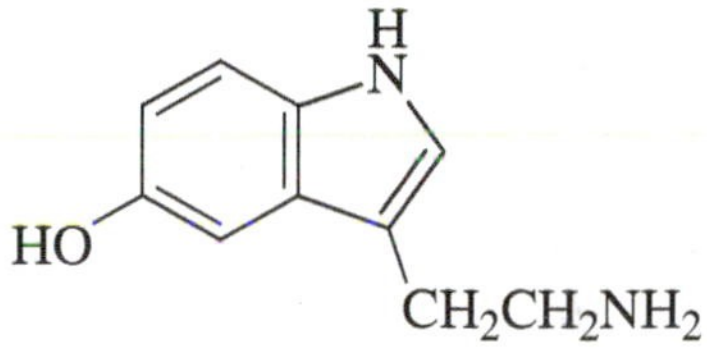

serotonin (HO-indole with $CH_2CH_2NH_2$)

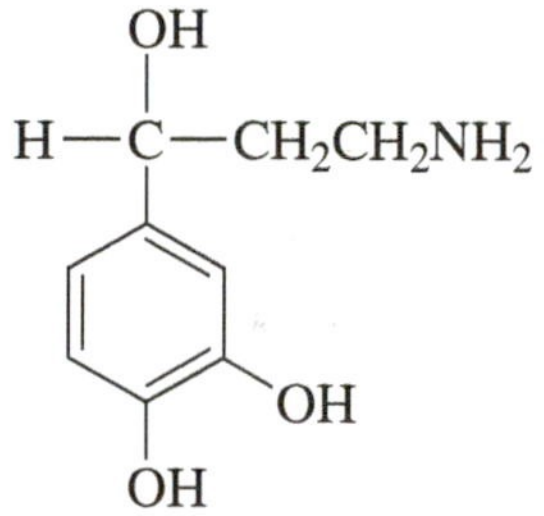

norepinephrine (H—C(OH)—$CH_2CH_2NH_2$ on a ring bearing two OH)

MECHANISM 19-5 The Cope Elimination of an Amine Oxide

The Cope elimination is a one-step, concerted internal elimination using an amine oxide as both the base and the leaving group. Syn stereochemistry is required for the Cope elimination.

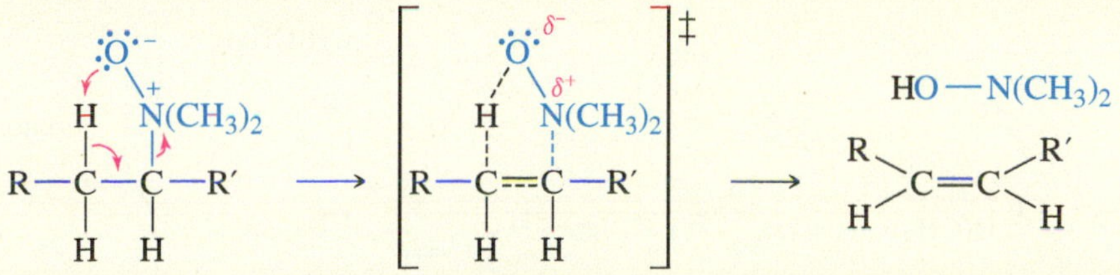

$$R{-}CH{-}CH(N^{+}(CH_3)_2O^{-}){-}R' \longrightarrow [\text{transition state}]^{\ddagger} \longrightarrow RCH{=}CHR' + HO{-}N(CH_3)_2$$

[transition state]

Cope elimination occurs under milder conditions than Hofmann elimination. It is particularly useful when a sensitive or reactive alkene must be synthesized by the elimination of an amine. Because the Cope elimination involves a cyclic transition state, it occurs with syn stereochemistry.

SOLVED PROBLEM 19-3

Predict the products expected when the following compound is treated with H_2O_2 and heated.

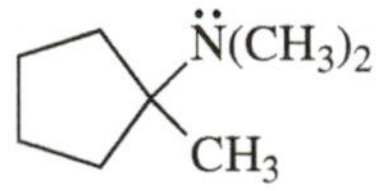

$\ddot{N}(CH_3)_2$

CH_3

SOLUTION

Oxidation converts the tertiary amine to an amine oxide. Cope elimination can give either of two alkenes. We expect the less-hindered elimination to be favored, giving the Hofmann product.

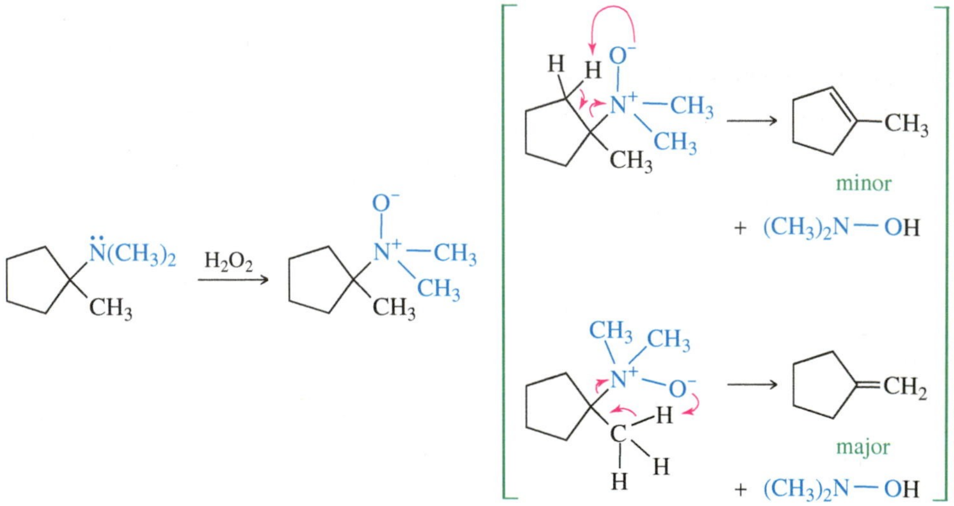

$\ddot{N}(CH_3)_2$ CH_3 $\xrightarrow{H_2O_2}$ O^- $N^+{-}CH_3$ CH_3 CH_3

H H O^- $N^+{-}CH_3$ CH_3 CH_3 $\longrightarrow$ CH_3 minor + $(CH_3)_2N{-}OH$

CH_3 CH_3 $N^+{-}O^-$ C H H H $\longrightarrow$ $=CH_2$ major + $(CH_3)_2N{-}OH$

PROBLEM 19-21

Give the products expected when the following tertiary amines are treated with a peroxyacid and heated.

(a) *N,N*-dimethylhexan-2-amine
(b) *N,N*-diethylhexan-2-amine
(c) cyclohexyldimethylamine
(d) *N*-ethylpiperidine

PROBLEM 19-22

When the (R,R) isomer of the amine shown is treated with an excess of methyl iodide, then silver oxide, then heated, the major product is the Hofmann product.

(a) Draw the structure of the major (Hofmann) product.

(b) Some Zaitsev product is also formed. It has the (E) configuration. When the same amine is treated with mCPBA and heated, the Zaitsev product has the (Z) configuration. Use stereochemical drawings of the transition states to explain these observations.

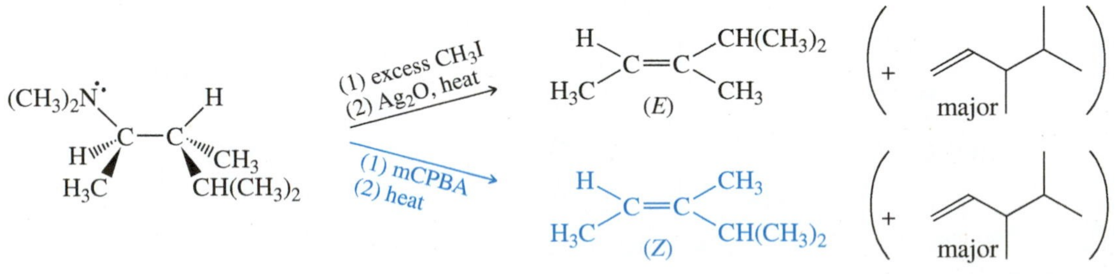

$(CH_3)_2\ddot{N}$ H H C—C CH_3 H_3C $CH(CH_3)_2$

(1) excess CH_3I (2) Ag_2O, heat $\longrightarrow$ H $CH(CH_3)_2$ C=C H_3C CH_3 (E) (+ major)

(1) mCPBA (2) heat $\longrightarrow$ H CH_3 C=C H_3C $CH(CH_3)_2$ (Z) (+ major)

19-16 Reactions of Amines with Nitrous Acid

Reactions of amines with nitrous acid (H—O—N=O) are particularly useful for synthesis. Because nitrous acid is unstable, it is generated *in situ* (in the reaction mixture) by mixing sodium nitrite ($NaNO_2$) with cold, dilute hydrochloric acid.

$$Na^+ \; {}^-{:}\ddot{O}{-}\ddot{N}{=}\ddot{O}{:} \;+\; H^+Cl^- \;\rightleftharpoons\; H{-}\ddot{O}{-}\ddot{N}{=}\ddot{O}{:} \;+\; Na^+Cl^-$$

sodium nitrite — nitrous acid

In an acidic solution, nitrous acid may protonate and lose water to give the nitrosonium ion, $^+N{=}O$. The nitrosonium ion appears to be the reactive intermediate in most reactions of amines with nitrous acid.

$$H{-}\ddot{O}{-}\ddot{N}{=}\ddot{O}{:} \;+\; H^+ \;\rightleftharpoons\; H{-}\overset{H}{\overset{|}{O^+}}{-}\ddot{N}{=}\ddot{O}{:} \;\rightleftharpoons\; H_2O \;+\; \left[{:}\overset{+}{N}{=}\ddot{O}{:} \longleftrightarrow {:}N{\equiv}\overset{+}{O}{:}\right]$$

nitrous acid — protonated nitrous acid — nitrosonium ion

Reaction with Primary Amines: Formation of Diazonium Salts Primary amines react with nitrous acid, via the nitrosonium ion, to give diazonium cations of the form $R{-}\overset{+}{N}{\equiv}N$. This procedure is called **diazotization** of an amine (Mechanism 19-6). Diazonium salts are the most useful products obtained from the reactions of amines with nitrous acid. The mechanism for diazonium salt formation begins with a nucleophilic attack on the nitrosonium ion to form an *N*-nitrosoamine.

MECHANISM 19-6 Diazotization of an Amine

Part 1: Attack on the nitrosonium ion (a strong electrophile), followed by deprotonation, gives an *N*-nitrosoamine.

$$R{-}\ddot{N}H_2 \;+\; {}^+\ddot{N}{=}\ddot{O}{:} \;\rightleftharpoons\; R{-}\overset{+}{N}H_2{-}\ddot{N}{=}\ddot{O}{:} \;\xrightarrow{H_2\ddot{O}:}\; R{-}\ddot{N}H{-}\ddot{N}{=}\ddot{O}{:} \;+\; H_3O^+$$

primary amine — nitrosonium ion — *N*-nitrosoamine

Part 2: A proton transfer (a tautomerism) from nitrogen to oxygen forms a hydroxy group and a second N—N bond.

$$R{-}\ddot{N}H{-}\ddot{N}{=}\ddot{O}{:} \;+\; H_3O^+ \;\rightleftharpoons\; \left[R{-}\ddot{N}H{-}\ddot{N}{=}\overset{+}{\ddot{O}}{-}H \longleftrightarrow R{-}\overset{+}{N}H{=}\ddot{N}{-}\ddot{O}H\right] \;+\; H_2\ddot{O}{:} \;\rightleftharpoons\; R{-}\ddot{N}{=}\ddot{N}{-}\ddot{O}H \;+\; H_3O^+$$

N-nitrosoamine — protonated *N*-nitrosoamine — second N—N bond formed

Part 3: Protonation of the hydroxy group, followed by loss of water, gives the diazonium ion.

$$R{-}\ddot{N}{=}\ddot{N}{-}\ddot{O}H \;\xrightleftharpoons{H_3O^+}\; R{-}\ddot{N}{=}\ddot{N}{-}\overset{+}{O}H_2 \;\longrightarrow\; R{-}\overset{+}{N}{\equiv}N{:} \;+\; H_2\ddot{O}{:}$$

diazonium ion

The overall diazotization reaction is

$$\underset{\text{primary amine}}{R-\ddot{N}H_2} + \underset{\text{sodium nitrite}}{NaNO_2} + 2\,HCl \longrightarrow \underset{\text{diazonium salt}}{R-\overset{+}{N}\equiv N\ \ Cl^-} + 2\,H_2O + NaCl$$

Alkanediazonium salts are unstable. They decompose to give nitrogen gas and carbocations.

$$\underset{\text{alkanediazonium cation}}{R-\overset{+}{N}\equiv N:} \longrightarrow \underset{\text{carbocation}}{R^+} + \underset{\text{nitrogen}}{:N\equiv N:}$$

The driving force for this reaction is the formation of N_2, an exceptionally stable molecule. The carbocations generated in this manner react like others we have seen: by nucleophilic attack to give substitution, by proton loss to give elimination, and by rearrangement. Because of the many competing reaction pathways, alkanediazonium salts usually decompose to give complex mixtures of products. Therefore, the diazotization of primary alkylamines is not widely used for synthesis.

Arenediazonium salts (formed from arylamines) are relatively stable, however, and they serve as intermediates in a variety of important synthetic reactions. These reactions are discussed in Section 19-17.

Reaction with Secondary Amines: Formation of *N*-Nitrosoamines Secondary amines react with the nitrosonium ion to form secondary ***N*-nitrosoamines,** sometimes called *nitrosamines*.

$$\underset{2°\text{ amine}}{R_2\ddot{N}H} + \underset{\text{nitrosonium ion}}{:\overset{+}{N}=\ddot{O}:} \rightleftharpoons R_2\overset{+}{N}H-\ddot{N}=\ddot{O}: \xrightarrow{H_2\ddot{O}:} \underset{2°\ N\text{-nitrosoamine}}{R_2\ddot{N}-\ddot{N}=\ddot{O}:} + H_3O^+$$

Secondary *N*-nitrosoamines are stable under the reaction conditions because they do not have the N—H proton needed for the tautomerism (shown in Mechanism 19-6 with a primary amine) to form a diazonium ion. The secondary *N*-nitrosoamine usually separates from the reaction mixture as an oily liquid.

Small quantities of *N*-nitrosoamines have been shown to cause cancer in laboratory animals. These findings have generated concern about the common practice of using sodium nitrite to preserve meats such as bacon, ham, and hot dogs. When the meat is eaten, sodium nitrite combines with stomach acid to form nitrous acid, which can convert amines in the food to *N*-nitrosoamines. Because nitrites are naturally present in many other foods, it is unclear just how much additional risk is involved in using sodium nitrite to preserve meats. More research is being done in this area to evaluate the risk.

The most useful reaction of amines with nitrous acid is the reaction of arylamines to form arenediazonium salts. We consider next how these diazonium salts may be used as synthetic intermediates.

PROBLEM 19-23

Predict the products from the reactions of the following amines with sodium nitrite in dilute HCl.

(a) cyclohexanamine **(b)** *N*-ethylhexan-2-amine **(c)** piperidine **(d)** aniline

19-17 Reactions of Arenediazonium Salts

In contrast to alkanediazonium salts, arenediazonium salts are relatively stable in aqueous solutions around 0–10 °C. Above these temperatures, they decompose, and they may explode if they are isolated and allowed to dry. The diazonium ($-\overset{+}{N}\equiv N$) group can be replaced by many different functional groups, including —H, —OH, —CN, and halogens.

Arenediazonium salts are formed by diazotizing a primary aromatic amine. Primary aromatic amines are commonly prepared by nitrating an aromatic ring, and then reducing the nitro group to an amino ($-NH_2$) group. In effect, by forming and diazotizing an amine, an activated aromatic position can be converted into a wide variety of functional groups. For example, toluene might be converted to a variety of substituted derivatives by using this procedure:

$$\text{toluene} \xrightarrow[H_2SO_4]{HNO_3} p\text{-}NO_2C_6H_4CH_3 \xrightarrow[(2)\ ^-OH]{(1)\ Fe,\ HCl} p\text{-}NH_2C_6H_4CH_3 \xrightarrow[HCl]{NaNO_2} p\text{-}CH_3C_6H_4\overset{+}{N}\equiv N\ Cl^- \xrightarrow{\text{various reagents}} p\text{-}ZC_6H_4CH_3$$

The following flowchart shows some of the functional groups that can be introduced via arenediazonium salts:

$Ar-\overset{+}{N}\equiv N$ →

Reagent		Products
H_3O^+, warm	Ar—OH	phenols
CuCl	Ar—Cl	aryl chlorides
CuBr	Ar—Br	aryl bromides
CuCN	Ar—C≡N	benzonitriles
HBF_4	Ar—F	aryl fluorides
KI	Ar—I	aryl iodides
H_3PO_2	Ar—H	(deamination)
H—Ar′	Ar—N=N—Ar′	azo dyes

PROBLEM-SOLVING HINT

These reactions of diazonium salts are extremely useful for solving aromatic synthesis problems.

Replacement of the Diazonium Group by Hydroxide: Hydrolysis Hydrolysis takes place when the acidic solution of an arenediazonium salt is warmed. The hydroxy group of water replaces N_2, forming a phenol. This is a useful laboratory synthesis of phenols because (unlike nucleophilic aromatic substitution) it does not require strong electron-withdrawing substituents or powerful bases and nucleophiles.

$$Ar-\overset{+}{N}\equiv N\ Cl^- \xrightarrow[H_2O]{H_2SO_4,\ heat} Ar-OH + N_2\uparrow + H^+$$

Example

$$m\text{-}H_2N{:}C_6H_4C(=O)CH_3 \xrightarrow[(2)\ H_2SO_4,\ H_2O,\ heat]{(1)\ NaNO_2,\ HCl} m\text{-}HOC_6H_4C(=O)CH_3\ (75\%)$$

Replacement of the Diazonium Group by Chloride, Bromide, and Cyanide: The Sandmeyer Reaction Copper(I) salts (cuprous salts) have a special affinity for diazonium salts. Cuprous chloride, cuprous bromide, and cuprous cyanide react with arenediazonium salts to give aryl chlorides, aryl bromides, and aryl cyanides. The use of cuprous salts to replace arenediazonium groups is called the **Sandmeyer reaction.** The Sandmeyer reaction (using cuprous cyanide) is also an excellent method for attaching another carbon substituent to an aromatic ring.

The Sandmeyer reaction

$$Ar-\overset{+}{N}\equiv N\ \ Cl^- \xrightarrow[(X = Cl, Br, C\equiv N)]{CuX} Ar-X + N_2\uparrow$$

Examples

:NH_2 (1) $NaNO_2$, HCl (2) CuCl → Cl (75%)

:NH_2 CH_3 (1) $NaNO_2$, HCl (2) CuBr → Br CH_3 (90%)

Application: Radiotracer Using An Aryl Fluoride

HO, HO, O, OH, NH_2, ^{18}F

fluorodopa or 2-fluoro-L-DOPA

Fluorodopa is L-dopa substituted with the ^{18}F isotope on the 2-position of the aromatic ring. Fluorodopa is an important radiotracer for positron emission tomography (PET) scans. For example, Parkinson's disease causes death of dopamine-producing neurons in the brain, resulting in a severe dopamine deficiency. A brain afflicted with Parkinson's disease takes up the radioactive fluorodopa more quickly than the brain of a normal patient. The synthesis of fluorodopa requires introducing a radioactive fluorine atom onto an aromatic ring. Diazotization, followed by heating of the fluoroborate salt, is one of very few synthetic methods for doing this.

Replacement of the Diazonium Group by Fluoride and Iodide When an arenediazonium salt is treated with fluoroboric acid (HBF_4), the diazonium fluoroborate precipitates out of solution. If this precipitated salt is filtered and then heated, it decomposes to give the aryl fluoride. Although this reaction requires the isolation and heating of a potentially explosive diazonium salt, it may be carried out safely if it is done carefully with the proper equipment. There are few other methods for making aryl fluorides.

$$Ar-\overset{+}{N}\equiv N\ \ Cl^- \xrightarrow{HBF_4} \underset{\text{diazonium fluoroborate}}{Ar-\overset{+}{N}\equiv N\ {}^-BF_4} \xrightarrow{heat} Ar-F + N_2\uparrow + BF_3$$

Example

:NH_2 (1) $NaNO_2$, HCl (2) HBF_4 → $N\equiv N^+$ $^-BF_4$ heat → F (50%)

Aryl iodides are formed by treating arenediazonium salts with potassium iodide. This is one of the best methods for making iodobenzene derivatives.

$$Ar-\overset{+}{N}\equiv N\ \ Cl^- \xrightarrow{KI} Ar-I + N_2\uparrow$$

Example

:NH_2 O (1) $NaNO_2$, HCl (2) KI → I O (75%)

Reduction of the Diazonium Group to Hydrogen: Deamination of Anilines Hypophosphorus acid (H_3PO_2) reacts with arenediazonium salts, replacing the diazonium group with a hydrogen. In effect, this is a reduction of the arenediazonium ion.

$$Ar-\overset{+}{N}\equiv N \quad Cl^- \xrightarrow{H_3PO_2} Ar-H + N_2\uparrow$$

Example

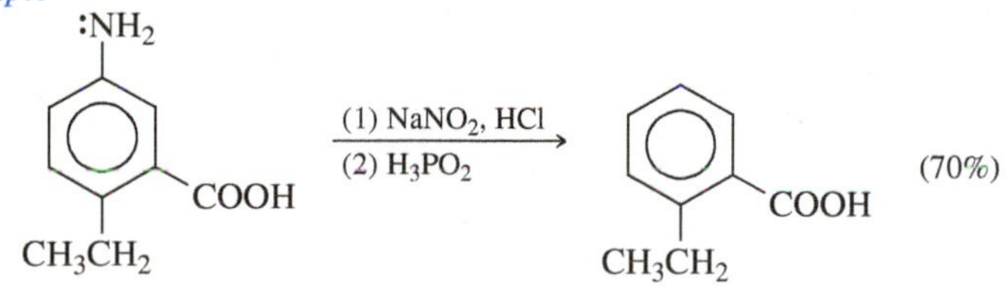

:NH_2, COOH, CH_3CH_2 —(1) $NaNO_2$, HCl; (2) H_3PO_2→ COOH, CH_3CH_2 (70%)

This reaction is sometimes used to remove an amino group that was added to activate the ring. Solved Problem 19-4 shows how one might use this technique.

SOLVED PROBLEM 19-4

Show how you would convert toluene to 3,5-dibromotoluene in good yield.

SOLUTION

Direct bromination of toluene cannot give 3,5-dibromotoluene because the methyl group activates the ortho and para positions.

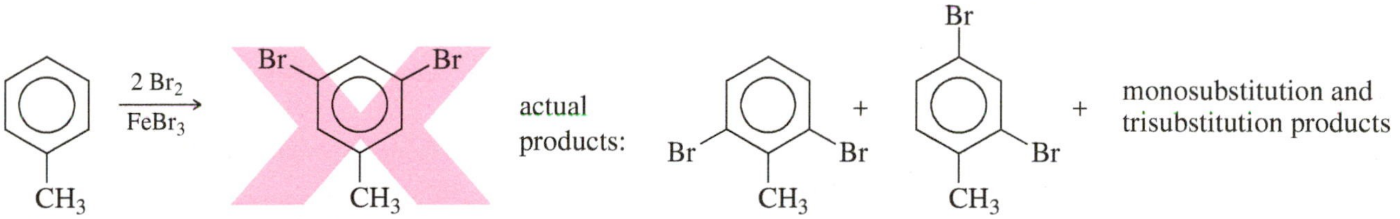

toluene (CH_3) —2 Br_2, $FeBr_3$→ no 3,5-dibromotoluene (Br, Br, CH_3)

actual products: Br, CH_3 + Br, Br, CH_3 + monosubstitution and trisubstitution products

mixture of ortho and para bromination

However, we can start by converting toluene to *p*-toluidine (*p*-methylaniline); the strongly activating amino group directs bromination to its ortho positions. Removal of the amino group (deamination) gives the desired product.

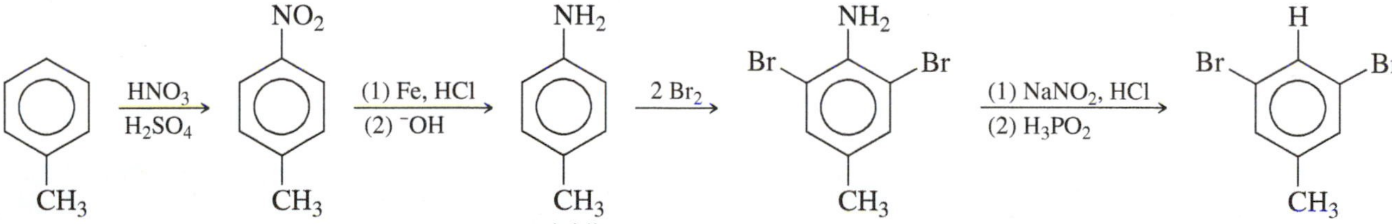

CH_3 —HNO_3, H_2SO_4→ NO_2, CH_3 —(1) Fe, HCl; (2) ^-OH→ NH_2, CH_3 (*p*-toluidine) —2 Br_2→ NH_2, Br, Br, CH_3 —(1) $NaNO_2$, HCl; (2) H_3PO_2→ H, Br, Br, CH_3 (deaminated)

Diazonium Salts as Electrophiles: Diazo Coupling Arenediazonium ions act as weak electrophiles in electrophilic aromatic substitutions. The products have the structure Ar—N=N—Ar′, containing the —N=N— **azo** linkage. For this reason, the products are called azo compounds, and the reaction is called **diazo coupling.** Because they are weak electrophiles, diazonium salts react only with strongly activated rings (such as derivatives of aniline and phenol).

$$\underset{\text{diazonium ion}}{Ar-\overset{+}{N}\equiv N} + \underset{\text{(activated)}}{H-Ar'} \longrightarrow \underset{\text{an azo compound}}{Ar-N=N-Ar'} + H^+$$

Example

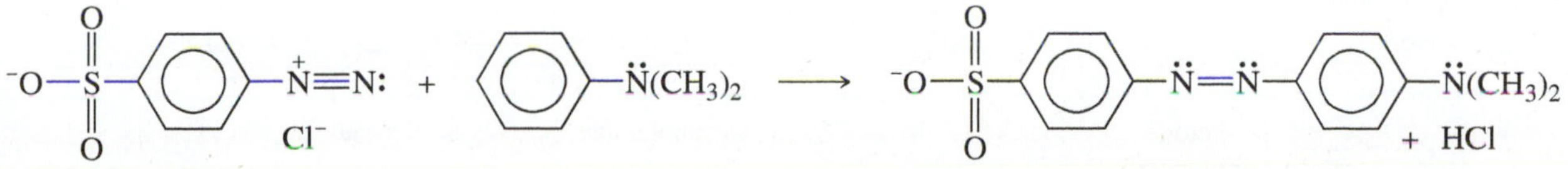

$^-O-S(=O)_2-C_6H_4-\overset{+}{N}\equiv N:$ Cl^- + $C_6H_5-\ddot{N}(CH_3)_2$ ⟶ $^-O-S(=O)_2-C_6H_4-\ddot{N}=\ddot{N}-C_6H_4-\ddot{N}(CH_3)_2$ + HCl

methyl orange (an indicator)

PROBLEM 19-24

Propose a mechanism for the synthesis of methyl orange.

Azo compounds bring two substituted aromatic rings into conjugation with an azo group, which is a strong chromophore. Therefore, most azo compounds are strongly colored, and they make excellent dyes, known as *azo dyes*. Many common azo dyes are made by diazo coupling.

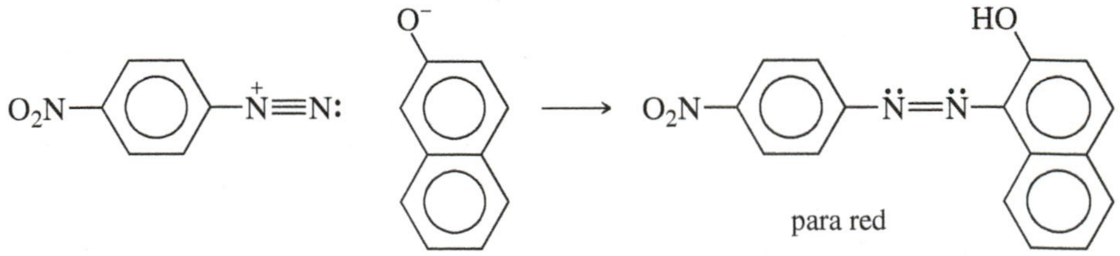

$$O_2N\text{–}C_6H_4\text{–}\overset{+}{N}{\equiv}N: \;+\; \text{2-naphthoxide } (O^-) \longrightarrow O_2N\text{–}C_6H_4\text{–}\ddot{N}{=}\ddot{N}\text{–}C_{10}H_6\text{–}OH$$

para red

Diazo coupling often takes place in basic solutions because deprotonation of the phenolic —OH groups and the sulfonic acid and carboxylic acid groups helps to activate the aromatic rings toward electrophilic aromatic substitution. Many of the common azo dyes have one or more sulfonate ($—SO_3^-$) or carboxylate ($—COO^-$) groups on the molecule to promote solubility in water and to help bind the dye to the polar surfaces of common fibers such as cotton and wool.

PROBLEM 19-25

Show how you would convert aniline to the following compounds.

(a) fluorobenzene
(b) chlorobenzene
(c) 1,3,5-trimethylbenzene
(d) bromobenzene
(e) iodobenzene
(f) benzonitrile
(g) phenol
(h) 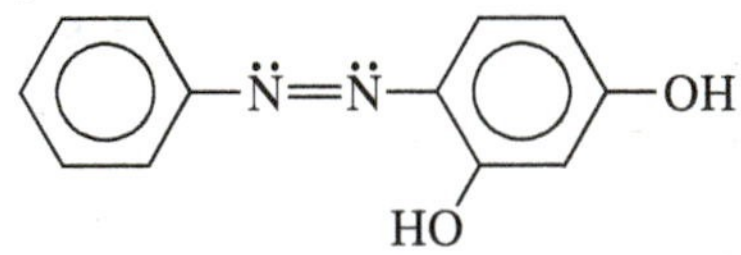 $C_6H_5\text{–}\ddot{N}{=}\ddot{N}\text{–}C_6H_3(OH)\text{–}OH$ (HO ortho to the azo group, OH para) (use aniline and resorcinol)

SUMMARY Reactions of Amines

1. *Reaction as a proton base* (Section 19-5)

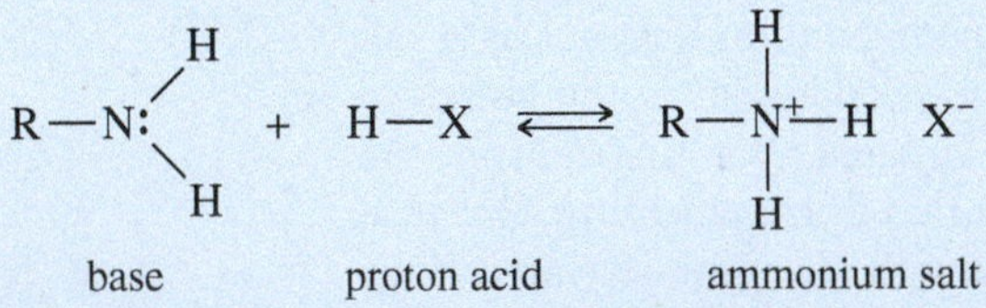

$$R\text{—}\ddot{N}H_2 + H\text{—}X \rightleftharpoons R\text{—}\overset{+}{N}H_3\; X^-$$

base   proton acid   ammonium salt

2. *Reactions with ketones and aldehydes* (Sections 18-14, 18-15, and 19-9)

Y = H or alkyl gives an imine
Y = OH gives an oxime
Y = NHR gives a hydrazone

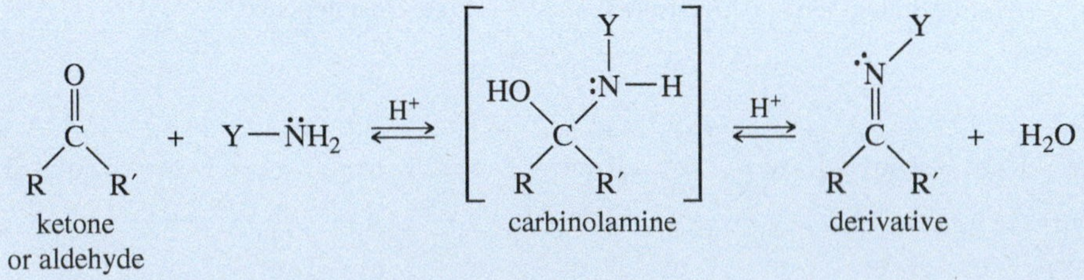

$$R\text{–}C(=O)\text{–}R' + Y\text{—}\ddot{N}H_2 \overset{H^+}{\rightleftharpoons} \left[HO\text{–}C(R)(R')\text{–}\ddot{N}(Y)\text{—}H \right] \overset{H^+}{\rightleftharpoons} R\text{–}C(=\ddot{N}\text{–}Y)\text{–}R' + H_2O$$

ketone or aldehyde   carbinolamine   derivative

3. *Alkylation* (Section 19-11)

$$\underset{\text{amine}}{R-\ddot{N}H_2} + \underset{\text{primary halide}}{R'-CH_2-Br} \longrightarrow \underset{\text{salt of alkylated amine}}{R-\overset{+}{N}H_2-CH_2-R'\ Br^-}$$

Overalkylation is common.

4. *Acylation to form amides* (Section 19-12)

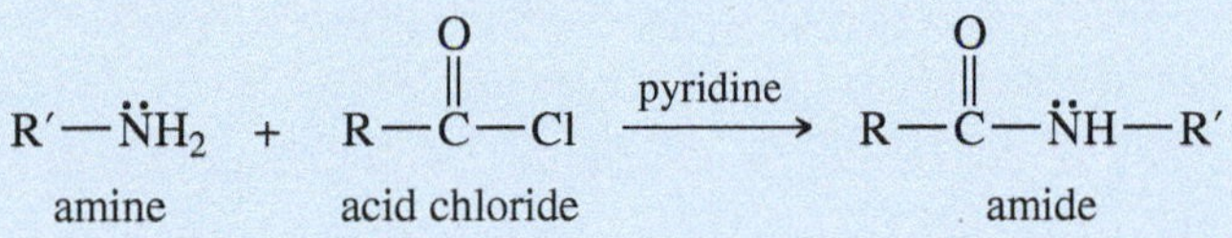

$$\underset{\text{amine}}{R'-\ddot{N}H_2} + \underset{\text{acid chloride}}{R-\overset{O}{\overset{\|}{C}}-Cl} \xrightarrow{\text{pyridine}} \underset{\text{amide}}{R-\overset{O}{\overset{\|}{C}}-\ddot{N}H-R'}$$

5. *Reaction with sulfonyl chlorides to give sulfonamides* (Section 19-13)

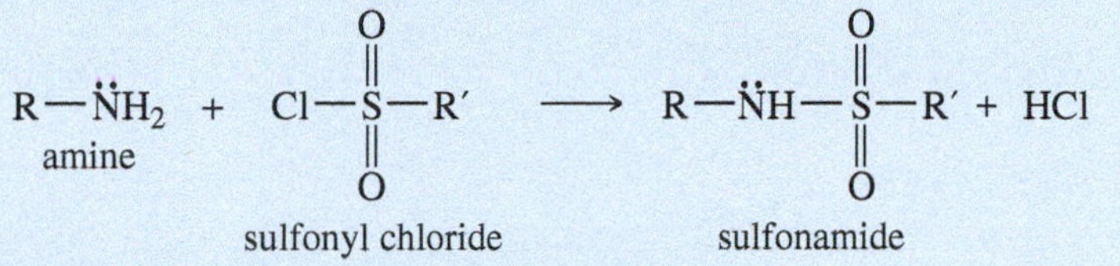

$$\underset{\text{amine}}{R-\ddot{N}H_2} + \underset{\text{sulfonyl chloride}}{Cl-\overset{O}{\overset{\|}{\underset{O}{\underset{\|}{S}}}}-R'} \longrightarrow \underset{\text{sulfonamide}}{R-\ddot{N}H-\overset{O}{\overset{\|}{\underset{O}{\underset{\|}{S}}}}-R'} + HCl$$

6. *Hofmann and Cope eliminations*

a. Hofmann elimination (Section 19-14)

Conversion to quaternary ammonium hydroxide

$$R-CH_2-CH_2-\ddot{N}H_2 \xrightarrow{3\ CH_3I} R-CH_2-CH_2-\overset{+}{N}(CH_3)_3\ I^-$$

$$\xrightarrow{Ag_2O} R-CH_2-CH_2-\overset{+}{N}(CH_3)_3\ {}^-OH$$

Elimination

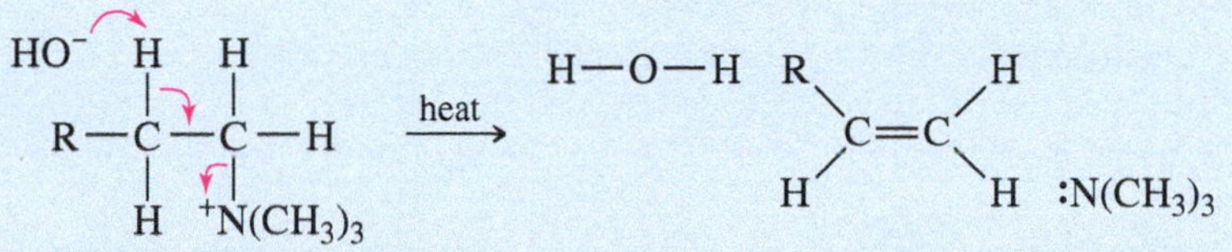

$$HO^- \quad R-CH_2(H)-CH(H)(H)-{}^+N(CH_3)_3 \xrightarrow{\text{heat}} H-O-H \quad (R)(H)C{=}C(H)(H) \quad :N(CH_3)_3$$

Hofmann elimination usually gives the least substituted alkene.

b. Cope elimination of a tertiary amine oxide (Section 19-15)

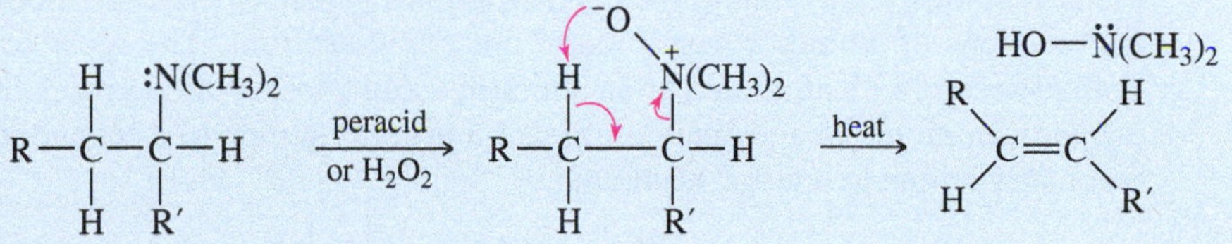

$$R-CH_2-CH(R')-:N(CH_3)_2 \xrightarrow[\text{or } H_2O_2]{\text{peracid}} R-CH_2-CH(R')-\overset{+}{N}(CH_3)_2-O^- \xrightarrow{\text{heat}} (R)(H)C{=}C(H)(R') \quad HO-\ddot{N}(CH_3)_2$$

Cope elimination also gives the least highly substituted alkene.

7. *Oxidation* (Section 19-15)

a. Secondary amines

$$\underset{2^\circ\text{ amine}}{R_2\ddot{N}-H} + H_2O_2 \longrightarrow \underset{\text{a }2^\circ\text{ hydroxylamine}}{R_2\ddot{N}-OH} + H_2O$$

b. Tertiary amines

$$\underset{3^\circ\text{ amine}}{R_3N:} + \underset{(\text{or } ArCO_3H)}{H_2O_2} \longrightarrow \underset{3^\circ\text{ amine oxide}}{R_3\overset{+}{N}-O^-} + \underset{(\text{or ArCOOH})}{H_2O}$$

(continued)

8. *Diazotization* (Section 19-16)

$$R{-}\ddot{N}H_2 \xrightarrow{NaNO_2,\ HCl} R{-}\overset{+}{N}{\equiv}N{:}\ Cl^-$$

primary alkylamine → alkanediazonium salt

$$Ar{-}\ddot{N}H_2 \xrightarrow{NaNO_2,\ HCl} Ar{-}\overset{+}{N}{\equiv}N{:}\ Cl^-$$

primary arylamine → arenediazonium salt

a. Reactions of diazonium salts (Section 19-17)

(I) Hydrolysis

$$Ar{-}\overset{+}{N}{\equiv}N{:}\ Cl^- \xrightarrow[H_2O]{H^+,\ heat} Ar{-}OH + N_2\uparrow + HCl$$

(II) The Sandmeyer reaction

$$Ar{-}\overset{+}{N}{\equiv}N{:}\ Cl^- \xrightarrow[X = Cl,\ Br,\ C{\equiv}N]{CuX} Ar{-}X + N_2\uparrow$$

(III) Replacement by fluoride or iodide

$$Ar{-}\overset{+}{N}{\equiv}N{:}\ Cl^- \xrightarrow{HBF_4} Ar{-}\overset{+}{N}{\equiv}N{:}\ BF_4^- \xrightarrow{heat} Ar{-}F + N_2\uparrow + BF_3$$

$$Ar{-}\overset{+}{N}{\equiv}N{:}\ Cl^- \xrightarrow{KI} Ar{-}I + N_2\uparrow + KCl$$

(IV) Reduction to hydrogen

$$Ar{-}\overset{+}{N}{\equiv}N{:}\ Cl^- \xrightarrow{H_3PO_2} Ar{-}H + N_2\uparrow$$

(V) Diazo coupling

$$Ar{-}\overset{+}{N}{\equiv}N{:} + H{-}Ar' \longrightarrow Ar{-}\ddot{N}{=}\ddot{N}{-}Ar' + H^+$$

diazonium ion + (activated) → an azo compound

19-18 Synthesis of Amines by Reductive Amination

Many methods are available for making amines. Most of these methods derive from the reactions of amines covered in the preceding sections. The most common amine syntheses start with ammonia or an amine and add another alkyl group. Such a process converts ammonia to a primary amine, or a primary amine to a secondary amine, or a secondary amine to a tertiary amine.

$$:NH_3 \longrightarrow\longrightarrow R{-}\ddot{N}H_2$$

ammonia → 1° amine

$$>\ddot{N}{-}H \longrightarrow\longrightarrow >\ddot{N}{-}R$$

1° or 2° amine → 2° or 3° amine

Reductive amination is the most general amine synthesis, capable of adding a primary or secondary alkyl group to an amine. Reductive amination is a two-step procedure. First, we form an imine or oxime derivative of a ketone or aldehyde, and then reduce it to the amine. In effect, reductive amination adds one alkyl group to the nitrogen atom. The product can be a primary, secondary, or tertiary amine, depending on whether the starting amine had zero, one, or two alkyl groups.

$$R{-}\ddot{N}H_2 + O{=}C< \underset{}{\overset{H^+}{\rightleftharpoons}} R{-}\ddot{N}{=}C< \xrightarrow{reduce} R{-}\underset{H}{\ddot{N}}{-}\underset{H}{C}<$$

1° amine + ketone or aldehyde ⇌ imine + H_2O → 2° amine

Primary Amines Primary amines result from condensation of hydroxylamine (zero alkyl groups) with a ketone or an aldehyde, followed by reduction of the oxime. Hydroxylamine is used in place of ammonia because most oximes are stable, easily isolated compounds. The oxime is reduced using catalytic reduction, lithium aluminum hydride, or zinc and HCl.

$$\underset{\text{ketone or aldehyde}}{R-\overset{\overset{O}{\|}}{C}-R'} \xrightarrow[H^+]{H_2\ddot{N}-OH} \underset{\text{oxime}}{R-\overset{\overset{\ddot{N}-OH}{\|}}{C}-R'} \xrightarrow{\text{reduction}} \underset{1^\circ \text{ amine}}{R-\overset{\overset{\ddot{N}H_2}{|}}{CH}-R'}$$

Examples

$$\underset{\text{pentan-2-one}}{CH_3CH_2CH_2-\overset{\overset{O}{\|}}{C}-CH_3} \xrightarrow[H^+]{H_2\ddot{N}-OH} \underset{\text{pentan-2-one oxime}}{CH_3CH_2CH_2-\overset{\overset{\ddot{N}-OH}{\|}}{C}-CH_3} \xrightarrow[Ni]{H_2} \underset{\text{pentan-2-amine}}{CH_3CH_2CH_2-\overset{\overset{\ddot{N}H_2}{|}}{CH}-CH_3}$$

$$\underset{\text{benzaldehyde}}{C_6H_5-\overset{\overset{O}{\|}}{C}-H} \xrightarrow[H^+]{H_2\ddot{N}-OH} \underset{\text{benzaldehyde oxime}}{C_6H_5-\overset{\overset{\ddot{N}-OH}{\|}}{C}-H} \xrightarrow[(2)\ H_2O]{(1)\ LiAlH_4} \underset{\text{benzylamine}}{C_6H_5-\overset{\overset{\ddot{N}H_2}{|}}{CH_2}}$$

Secondary Amines Condensation of a primary amine with a ketone or aldehyde forms an *N*-substituted imine (a Schiff base). Reduction of the imine, using either $LiAlH_4$ or $NaBH_4$, gives a secondary amine.

$$\underset{\text{ketone or aldehyde}}{R-\overset{\overset{O}{\|}}{C}-R'} \xrightarrow[H^+]{\substack{1^\circ \text{ amine} \\ R''-\ddot{N}H_2}} \underset{N\text{-substituted imine}}{R-\overset{\overset{\ddot{N}-R''}{\|}}{C}-R'} \xrightarrow{\text{reduction}} \underset{2^\circ \text{ amine}}{R-\overset{\overset{\ddot{N}HR''}{|}}{CH}-R'}$$

Example

$$\underset{\text{acetone}}{CH_3-\overset{\overset{O}{\|}}{C}-CH_3} \xrightarrow[H^+]{Ph-\ddot{N}H_2} CH_3-\overset{\overset{\ddot{N}-Ph}{\|}}{C}-CH_3 \xrightarrow[(2)\ H_2O]{(1)\ LiAlH_4} \underset{\substack{N\text{-isopropylaniline} \\ (75\%)}}{CH_3-\overset{\overset{\ddot{N}HPh}{|}}{CH}-CH_3}$$

Tertiary Amines Condensation of a secondary amine with a ketone or aldehyde gives an iminium salt. Iminium salts are frequently unstable, so they are rarely isolated. A reducing agent in the solution reduces the iminium salt to a tertiary amine. The reducing agent must reduce the iminium salt, but it must not reduce the carbonyl group of the ketone or aldehyde. Sodium triacetoxyborohydride [$NaBH(OCOCH_3)_3$ or $NaBH(OAc)_3$] is less reactive than sodium borohydride, and it reduces the imine faster than the carbonyl group. Sodium triacetoxyborohydride has largely replaced the older, more toxic reagent, sodium cyanoborohydride ($NaBH_3CN$).

$$\underset{\text{ketone or aldehyde}}{R'-\overset{\overset{O}{\|}}{C}-R''} \underset{H^+}{\overset{\substack{2^\circ \text{ amine} \\ R-\ddot{N}H-R}}{\rightleftharpoons}} \underset{\text{iminium salt}}{\left[R'-\overset{\overset{R-\overset{+}{N}-R}{\|}}{C}-R''\right]} \xrightarrow[CH_3COOH]{NaBH(OAc)_3} \underset{3^\circ \text{ amine}}{R'-\overset{\overset{R-\ddot{N}-R}{|}}{CH}-R''}$$

Example

cyclohexanone $\xrightarrow[H^+]{H\ddot{N}(CH_3)_2}$ [iminium salt] $\xrightarrow[CH_3COOH]{NaBH(OAc)_3}$ *N,N*-dimethycyclohexylamine (85%)

PROBLEM-SOLVING HINT

Reductive amination is the most useful amine synthesis: It adds a 1° or 2° alkyl group to nitrogen. Use an aldehyde to add a 1° group, and a ketone to add a 2° group.

Y—NH—H $\xrightarrow{H^+,\ R_2C{=}O}$ Y—N=CR₂ $\xrightarrow{\text{LiAlH}_4\ [\text{NaBH(OAc)}_3 \text{ to make tertiary amines}]}$ Y—NH—CHR₂

1° or 2° group added

hydroxylamine ⟶ primary amine
primary amine ⟶ secondary amine
secondary amine ⟶ tertiary amine

SOLVED PROBLEM 19-5

Show how to synthesize the following amines from the indicated starting materials.
(a) *N*-cyclopentylaniline from aniline **(b)** *N*-ethylpyrrolidine from pyrrolidine

SOLUTION

(a) This synthesis requires adding a cyclopentyl group to aniline (primary) to make a secondary amine. Cyclopentanone is the carbonyl compound.

Ph—NH—H + O=C₅H₈ $\xrightarrow{H^+}$ Ph—N=C₅H₈ $\xrightarrow[Ni]{H_2}$ Ph—NH—C₅H₉

aniline cyclopentanone

(b) This synthesis requires adding an ethyl group to a secondary amine to make a tertiary amine. The carbonyl compound is acetaldehyde. Formation of a tertiary amine by reductive amination involves an iminium intermediate, which is reduced by $NaBH(OAc)_3$ (sodium triacetoxyborohydride).

pyrrolidine N—H + CH_3—CHO (acetaldehyde) $\underset{}{\overset{H^+}{\rightleftharpoons}}$ [pyrrolidinium $\overset{+}{N}$=CH—CH_3] $\xrightarrow{NaBH(OAc)_3}$ pyrrolidine N—CH_2—CH_3

pyrrolidine acetaldehyde

PROBLEM 19-26

Show how to synthesize the following amines from the indicated starting materials by reductive amination.

(a) benzylmethylamine from benzaldehyde
(b) *N*-benzylpiperidine from piperidine
(c) *N*-cyclohexylaniline from cyclohexanone
(d) cyclohexylamine from cyclohexanone
(e) $PhCH_2CH(NH_2)CH_3$ from $PhCH_2COCH_3$
(±)-amphetamine 1-phenylpropan-2-one
(f) *N*-cyclopentylpiperidine from piperidine

19-19 Synthesis of Amines by Acylation–Reduction

The second general synthesis of amines is **acylation–reduction.** Like reductive amination, acylation–reduction adds one alkyl group to the nitrogen atom of the starting amine. Acylation of the starting amine by an acid chloride gives an amide, which is much less nucleophilic and unlikely to over-acylate (Section 19-12). Reduction of the amide by lithium aluminum hydride ($LiAlH_4$) gives the corresponding amine.

$$\underset{\text{amine}}{R-\ddot{N}H_2} + \underset{\text{acid chloride}}{Cl-\overset{O}{\overset{\|}{C}}-R'} \xrightarrow[\text{pyridine or NaOH}]{\text{acylation}} \underset{\text{amide}}{R-\ddot{N}H-\overset{O}{\overset{\|}{C}}-R'} \xrightarrow[(2)\ H_2O]{\text{reduction (1) } LiAlH_4} \underset{\text{alkylated amine}}{R-\ddot{N}H-CH_2-R'}$$

Acylation–reduction converts ammonia to a primary amine, a primary amine to a secondary amine, or a secondary amine to a tertiary amine. These reactions are quite general, with one restriction: The added alkyl group is always 1° because the carbon bonded to nitrogen is derived from the carbonyl group of the amide, reduced to a methylene ($-CH_2-$) group.

Primary amines

$$\underset{\text{acid chloride}}{R-\overset{O}{\overset{\|}{C}}-Cl} + \underset{\text{ammonia}}{\ddot{N}H_3} \longrightarrow \underset{1^\circ\text{ amide}}{R-\overset{O}{\overset{\|}{C}}-\ddot{N}H_2} \xrightarrow[(2)\ H_2O]{(1)\ LiAlH_4} \underset{1^\circ\text{ amine}}{R-CH_2-\ddot{N}H_2}$$

Example

$$\underset{\text{3-methylbutanoyl chloride}}{CH_3-\overset{CH_3}{\overset{|}{C}H}-CH_2-\overset{O}{\overset{\|}{C}}-Cl} \xrightarrow{\ddot{N}H_3} \underset{\text{3-methylbutanamide}}{CH_3-\overset{CH_3}{\overset{|}{C}H}-CH_2-\overset{O}{\overset{\|}{C}}-\ddot{N}H_2} \xrightarrow[(2)\ H_2O]{(1)\ LiAlH_4} \underset{\text{3-methylbutan-1-amine}}{CH_3-\overset{CH_3}{\overset{|}{C}H}-CH_2-CH_2-\ddot{N}H_2}$$

Secondary amines

$$\underset{\text{acid chloride}}{R-\overset{O}{\overset{\|}{C}}-Cl} + \underset{\text{primary amine}}{R'-\ddot{N}H_2} \longrightarrow \underset{N\text{-substituted amide}}{R-\overset{O}{\overset{\|}{C}}-\ddot{N}H-R'} \xrightarrow[(2)\ H_2O]{(1)\ LiAlH_4} \underset{2^\circ\text{ amine}}{R-CH_2-\ddot{N}H-R'}$$

Example

$$\underset{\text{butanoyl chloride}}{CH_3CH_2CH_2-\overset{O}{\overset{\|}{C}}-Cl} + \underset{\text{aniline}}{C_6H_5\ddot{N}H_2} \longrightarrow \underset{N\text{-phenylbutanamide}}{CH_3CH_2CH_2-\overset{O}{\overset{\|}{C}}-\ddot{N}H-C_6H_5} \xrightarrow[(2)\ H_2O]{(1)\ LiAlH_4} \underset{N\text{-butylaniline}}{CH_3CH_2CH_2-CH_2-\ddot{N}H-C_6H_5}$$

Tertiary amines

$$\underset{\text{acid chloride}}{R-\overset{O}{\overset{\|}{C}}-Cl} + \underset{\text{secondary amine}}{R'_2\ddot{N}H} \longrightarrow \underset{N,N\text{-disubstituted amide}}{R-\overset{O}{\overset{\|}{C}}-\ddot{N}R'_2} \xrightarrow[(2)\ H_2O]{(1)\ LiAlH_4} \underset{3^\circ\text{ amine}}{R-CH_2-\ddot{N}R'_2}$$

Example

$$\underset{\text{benzoyl chloride}}{C_6H_5-\overset{O}{\overset{\|}{C}}-Cl} + \underset{\text{diethylamine}}{H-\ddot{N}(CH_2CH_3)_2} \longrightarrow \underset{N,N\text{-diethylbenzamide}}{C_6H_5-\overset{O}{\overset{\|}{C}}-\ddot{N}(CH_2CH_3)_2} \xrightarrow[(2)\ H_2O]{(1)\ LiAlH_4} \underset{\text{benzyldiethylamine}}{C_6H_5-CH_2-\ddot{N}(CH_2CH_3)_2}$$

PROBLEM-SOLVING HINT

Like reductive amination, acylation–reduction adds an alkyl group to nitrogen. It is more restrictive, though, because the group added is always 1°.

$$X—\ddot{N}(Y)—H \xrightarrow{R—C(=O)—Cl} X—\ddot{N}(Y)—C(=O)—R \xrightarrow{LiAlH_4} X—\ddot{N}(Y)—CH_2—R$$

1° group added

ammonia ⟶ primary amine
primary amine ⟶ secondary amine
secondary amine ⟶ tertiary amine

SOLVED PROBLEM 19-6

Show how to synthesize *N*-ethylpyrrolidine from pyrrolidine using acylation–reduction.

SOLUTION

This synthesis requires adding an ethyl group to pyrrolidine to make a tertiary amine. The acid chloride needed will be acetyl chloride (ethanoyl chloride). Reduction of the amide gives *N*-ethylpyrrolidine.

$$\text{pyrrolidine (N—H)} + CH_3—C(=O)Cl \text{ (acetyl chloride)} \xrightarrow{\text{pyridine}} \text{N—C(=O)—}CH_3 \xrightarrow[(2)\ H_2O]{(1)\ LiAlH_4} \text{N—C(H)(H)—}CH_3$$

pyrrolidine acetyl chloride

Compare this synthesis with Solved Problem 19-5(b) to see how reductive amination and acylation–reduction can accomplish the same result.

PROBLEM 19-27

Show how to synthesize the following amines from the indicated starting materials by acylation–reduction.

(a) *N*-butylpiperidine from piperidine
(b) *N*-benzylaniline from aniline

19-20 Syntheses Limited to Primary Amines

Primary amines are the most common class of amines, and they are also used as starting materials for synthesis of secondary and tertiary amines. Many methods have been developed for making primary amines, ranging from simple alkylation of ammonia to sophisticated multistep syntheses. We will consider some of the more common syntheses.

19-20A Direct Alkylation and Gabriel Synthesis

The S_N2 reaction of amines with alkyl halides is complicated by a tendency for over-alkylation to form a mixture of monoalkylated and polyalkylated products (Section 19-11). Simple primary amines can be synthesized, however, by adding a halide or tosylate (must be a good S_N2 substrate) to a large excess of ammonia. Because there is a large excess of ammonia present, the probability that a molecule of the halide will alkylate ammonia is much larger than the probability that it will over-alkylate the amine product.

$$R—CH_2—X \; + \; \text{excess } NH_3 \longrightarrow R—CH_2—NH_2 \; + \; NH_4^+ \; X^-$$

Example

$$\underset{\text{1-bromopentane}}{CH_3CH_2CH_2CH_2CH_2—Br} \; + \; \text{excess } NH_3 \longrightarrow \underset{\text{pentan-1-amine}}{CH_3CH_2CH_2CH_2CH_2—NH_2} \; + \; NH_4^+ \; Br^-$$

PROBLEM 19-28

Addition of one equivalent of ammonia to 1-bromoheptane gives a mixture of heptan-1-amine, some dialkylamine, some trialkylamine, and even some tetraalkylammonium bromide.

(a) Give a mechanism to show how this reaction takes place, as far as the dialkylamine.
(b) How would you modify the procedure to get an acceptable yield of heptan-1-amine?

In 1887, Siegmund Gabriel (at the University of Berlin) developed the **Gabriel amine synthesis** for making primary amines without danger of over-alkylation. He used the phthalimide anion as a protected form of ammonia that cannot alkylate more than once. Phthalimide has one acidic N—H proton (pK_a 8.3) that is abstracted by potassium hydroxide to give the phthalimide anion.

phthalimide $\xrightarrow[H_2O]{KOH}$ resonance-stabilized phthalimide anion

The phthalimide anion is a strong nucleophile, displacing a halide or tosylate ion from a good S_N2 substrate. Heating the *N*-alkyl phthalimide with hydrazine displaces the primary amine, giving the very stable, aromatic hydrazide of phthalimide.

phthalimide anion $\xrightarrow[\text{alkyl halide (usually 1°)}]{R-X}$ *N*-alkyl phthalimide $\xrightarrow[\text{heat}]{H_2N-NH_2}$ phthalimide hydrazide + $R-NH_2$ primary amine

Example

$Br-CH_2CH_2CH(CH_3)CH_3$ isopentyl bromide $\xrightarrow{\text{(phthalimide anion) } K^+}$ $N-CH_2CH_2CH(CH_3)CH_3$ *N*-isopentylphthalimide $\xrightarrow[\text{heat}]{H_2N-NH_2}$ $H_2N-CH_2CH_2CH(CH_3)CH_3$ isopentylamine (95%)

PROBLEM 19-29

Show how Gabriel syntheses are used to prepare the following amines.
(a) benzylamine **(b)** hexan-1-amine **(c)** γ-aminobutyric acid

19-20B Reduction of Azides and Nitriles

Just as Gabriel used the anion of phthalimide to put the nitrogen atom into a primary amine, we can use other nucleophiles as well. We need a good nucleophile that can alkylate only once and that is easily converted to an amino group. Two good nucleophiles for introducing a nitrogen atom are the azide ion and the cyanide ion. Azide ion introduces (after reduction) an $-NH_2$ group, and cyanide ion introduces a $-CH_2-NH_2$ group.

Formation and Reduction of Azides **Azide** ion (N_3^-) is an excellent nucleophile that displaces leaving groups from unhindered primary and secondary alkyl halides and tosylates. The products are alkyl azides (RN_3), which have no tendency to react further. Azides are easily reduced to primary amines, either by $LiAlH_4$ or by catalytic hydrogenation. Alkyl azides can be explosive, so they are reduced without purification.

$$R-X + Na^+\ {}^-:N=N^+=N:^- \xrightarrow{S_N2} [R-N^-=N^+=N: \longleftrightarrow R-N^--N^+\equiv N:] \xrightarrow[\text{or } H_2/Pd]{(1)\ LiAlH_4\ (2)\ H_2O} R-NH_2$$

halide or tosylate (must be 1° or 2°); sodium azide; an alkyl azide; 1° amine

Examples

$$\text{C}_6\text{H}_5\text{—CH}_2\text{CH}_2\text{—Br} \xrightarrow[\text{S}_\text{N}2]{\text{Na}^+ \ :\bar{\text{N}}=\overset{+}{\text{N}}=\bar{\text{N}}:} \text{C}_6\text{H}_5\text{—CH}_2\text{CH}_2\text{—N}=\overset{+}{\text{N}}=\bar{\text{N}}: \xrightarrow[(2)\ \text{H}_2\text{O}]{(1)\ \text{LiAlH}_4} \text{C}_6\text{H}_5\text{—CH}_2\text{CH}_2\text{—}\ddot{\text{N}}\text{H}_2$$

1-bromo-2-phenylethane → 2-phenylethyl azide → 2-phenylethylamine (89%)

$$\text{C}_6\text{H}_{11}\text{Br} \xrightarrow{\text{NaN}_3} \text{C}_6\text{H}_{11}\text{N}_3 \xrightarrow[(2)\ \text{H}_2\text{O}]{(1)\ \text{LiAlH}_4} \text{C}_6\text{H}_{11}\text{NH}_2$$

cyclohexyl bromide → cyclohexyl azide → cyclohexylamine (54%)

Azide ion also reacts with a variety of other electrophiles. The following example shows how an azide ion opens an epoxide, and the product can be reduced to an amino alcohol:

$$\text{epoxycyclohexane} \xrightarrow{\text{Na}^+ \ :\bar{\text{N}}=\overset{+}{\text{N}}=\bar{\text{N}}:} \text{trans-2-azidocyclohexanol } (\text{N}_3, \text{H}, \text{OH}, \text{H}) \xrightarrow{\text{H}_2,\ \text{Pd}} \text{trans-2-aminocyclohexanol } (\text{NH}_2, \text{H}, \text{OH}, \text{H})$$

epoxycyclohexane

Application: Auto Airbags

Most automobile airbags contain sodium azide (NaN_3) as the propellant. When accelerometers detect a crash, the electronic controller fires the charge of sodium azide. Sodium azide decomposes explosively, generating nitrogen gas that fills the airbag.

$$2\ \text{NaN}_3 \longrightarrow 2\ \text{Na} + 3\ \text{N}_2(g)$$

Formation and Reduction of Nitriles Like the azide ion, cyanide ion ($^-$:C≡N:) is a good S_N2 nucleophile; it displaces leaving groups from unhindered primary and secondary alkyl halides and tosylates. The product is a **nitrile** (R—C≡N), which has no tendency to react further. Nitriles are reduced to primary amines by lithium aluminum hydride or by catalytic hydrogenation.

$$\text{R—X} + {}^-\text{:C}\equiv\text{N:} \longrightarrow \text{R—C}\equiv\text{N:} \xrightarrow[\text{or H}_2/\text{catalyst}]{\text{LiAlH}_4} \text{R—CH}_2\text{—}\ddot{\text{N}}\text{H}_2$$

halide or tosylate (must be 1° or 2°) — nitrile — amine (one carbon added)

Example

$$\text{CH}_3\text{CH}_2\text{CH}_2\text{Br} \xrightarrow{\text{K}^+\ {}^-\text{:C}\equiv\text{N:}} \text{CH}_3\text{CH}_2\text{CH}_2\text{—C}\equiv\text{N:} \xrightarrow[(2)\ \text{H}_2\text{O}]{(1)\ \text{LiAlH}_4} \text{CH}_3\text{CH}_2\text{CH}_2\text{—CH}_2\text{—}\ddot{\text{N}}\text{H}_2$$

1-bromopropane → butanenitrile → butan-1-amine (70%)

When the cyano (—C≡N) group is added and reduced, the resulting amine has an additional carbon atom. In effect, the cyanide substitution-reduction process is like adding $-CH_2-NH_2$. The following synthesis makes 2-phenylethylamine, which we also made by the azide synthesis:

$$\text{C}_6\text{H}_5\text{—CH}_2\text{—Br} \xrightarrow{{}^-\text{:C}\equiv\text{N:}} \text{C}_6\text{H}_5\text{—CH}_2\text{—CN:} \xrightarrow[\text{Ni}]{\text{H}_2} \text{C}_6\text{H}_5\text{—CH}_2\text{CH}_2\ddot{\text{N}}\text{H}_2$$

benzyl bromide → phenylacetonitrile → 2-phenylethylamine

Notice that the starting material in this case has one less carbon atom because the cyanide synthesis adds both a carbon and a nitrogen.

We saw (Section 18-13) that cyanide ion adds to ketones and aldehydes to form cyanohydrins. Reduction of the —C≡N group of the cyanohydrin provides a way to synthesize β-hydroxy amines.

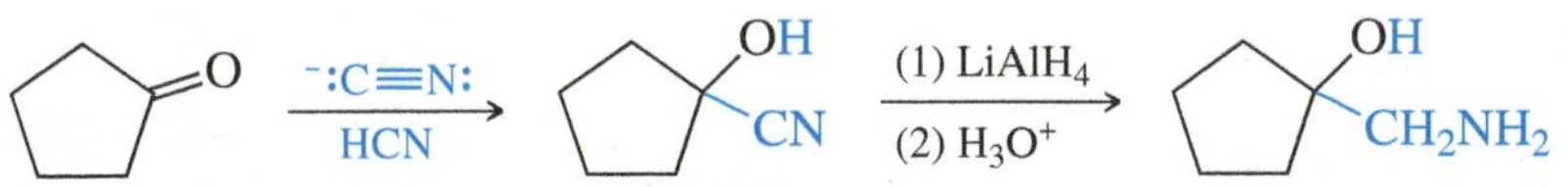

$$\text{cyclopentanone} \xrightarrow[\text{HCN}]{^-:C\equiv N:} \text{cyclopentanone cyanohydrin (1-OH, 1-CN)} \xrightarrow[(2)\ H_3O^+]{(1)\ LiAlH_4} \text{1-(aminomethyl)cyclopentanol (1-OH, 1-}CH_2NH_2\text{)}$$

cyclopentanone — cyclopentanone cyanohydrin — 1-(aminomethyl)cyclopentanol

PROBLEM 19-30

Show how you would accomplish the following synthetic conversions.

(a) benzyl bromide → benzylamine
(b) 1-bromo-2-phenylethane → 3-phenylpropan-1-amine
(c) pentanoic acid → pentan-1-amine
(d) pentanoic acid → hexan-1-amine
(e) (*R*)-2-bromobutane → (*S*)-butan-2-amine
(f) (*R*)-2-bromobutane → (*S*)-2-methylbutan-1-amine
(g) hexan-2-one → 1-amino-2-methylhexan-2-ol

PROBLEM-SOLVING HINT

To convert an alkyl halide (or alcohol, via the tosylate) to an amine, form the azide and reduce. To convert it to an amine with an additional carbon atom, form the nitrile and reduce. In either case, the alkyl group must be suitable for S_N2 displacement.

19-20C Reduction of Nitro Compounds

Both aromatic and aliphatic nitro groups are easily reduced to amino groups. The most common methods are catalytic hydrogenation and acidic reduction by an active metal. Stronger reducing agents, such as $LiAlH_4$, may also be used.

$$R{-}NO_2 \xrightarrow[\substack{or\ \text{active metal and } H^+ \\ \text{active metal = Fe, Zn, or Sn}}]{\substack{H_2/\text{catalyst} \\ \text{catalyst = Ni, Pd, or Pt}}} R{-}NH_2$$

Examples

$$o\text{-nitrotoluene} \xrightarrow{H_2,\ Ni} o\text{-toluidine (90\%)}$$

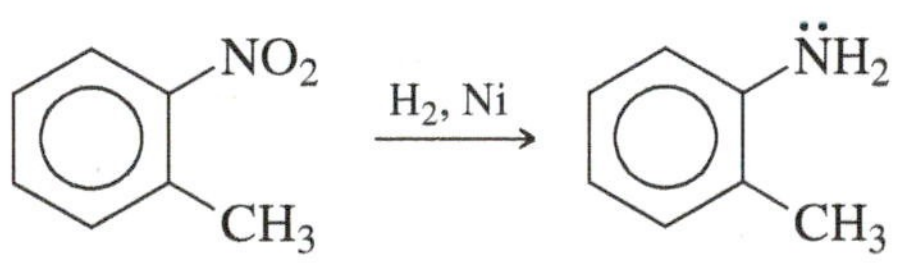

o-nitrotoluene — o-toluidine (90%)

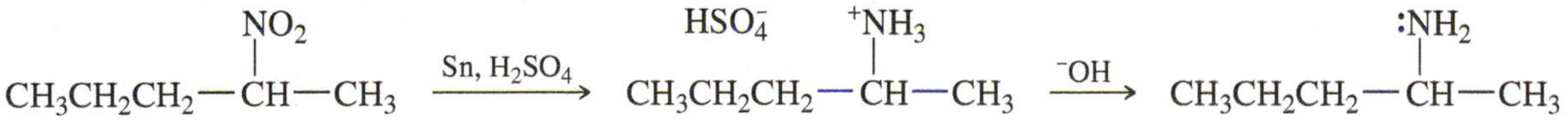

$$CH_3CH_2CH_2{-}CH(NO_2){-}CH_3 \xrightarrow{Sn,\ H_2SO_4} CH_3CH_2CH_2{-}CH(^+NH_3\ HSO_4^-){-}CH_3 \xrightarrow{^-OH} CH_3CH_2CH_2{-}CH(:NH_2){-}CH_3$$

2-nitropentane — pentan-2-amine (85%)

The most common reason for reducing aromatic nitro compounds is to make substituted anilines. Much of this chemistry was developed by the dye industry, which uses aniline derivatives for azo coupling reactions (Section 19-17) to make aniline dyes. Nitration of an aromatic ring (by electrophilic aromatic substitution) gives a nitro compound, which is reduced to the aromatic amine.

$$Ar{-}H \xrightarrow{HNO_3,\ H_2SO_4} Ar{-}NO_2 \xrightarrow{\text{reduction}} Ar{-}NH_2$$

For example, nitration followed by reduction is used in the synthesis of benzocaine (a topical anesthetic), shown below. Notice that the stable nitro group is retained through an oxidation and esterification. The final step reduces the nitro group to the relatively sensitive amine (which could not survive the oxidation step).

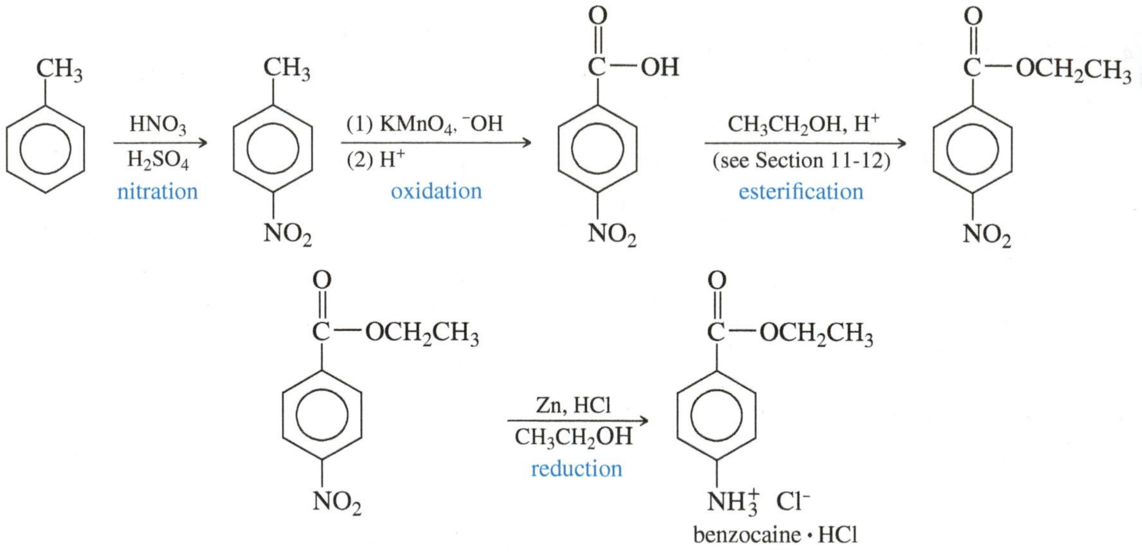

$$\text{toluene } (C_6H_5CH_3) \xrightarrow[\text{nitration}]{HNO_3,\ H_2SO_4} p\text{-}O_2N\text{-}C_6H_4\text{-}CH_3 \xrightarrow[\text{oxidation}]{(1)\ KMnO_4,\ {}^-OH;\ (2)\ H^+} p\text{-}O_2N\text{-}C_6H_4\text{-}COOH \xrightarrow[\text{esterification}]{CH_3CH_2OH,\ H^+\ (\text{see Section 11-12})} p\text{-}O_2N\text{-}C_6H_4\text{-}COOCH_2CH_3$$

$$p\text{-}O_2N\text{-}C_6H_4\text{-}COOCH_2CH_3 \xrightarrow[\text{reduction}]{Zn,\ HCl;\ CH_3CH_2OH} p\text{-}Cl^-\ H_3N^+\text{-}C_6H_4\text{-}COOCH_2CH_3$$

benzocaine • HCl

PROBLEM 19-31

Show how to prepare the following aromatic amines by aromatic nitration, followed by reduction. You may use benzene and toluene as your aromatic starting materials.

(a) aniline **(b)** *p*-bromoaniline **(c)** *m*-bromoaniline **(d)** *m*-aminobenzoic acid

SUMMARY Synthesis of Amines

1. *Reductive amination* (Section 19-18)

a. Primary amines

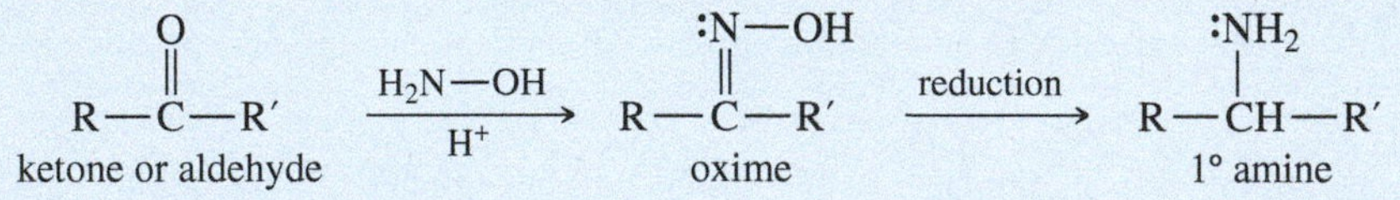

$$R-C(=O)-R' \xrightarrow[H^+]{H_2N-OH} R-C(=N-OH)-R' \xrightarrow{\text{reduction}} R-CH(NH_2)-R'$$

ketone or aldehyde → oxime → 1° amine

b. Secondary amines

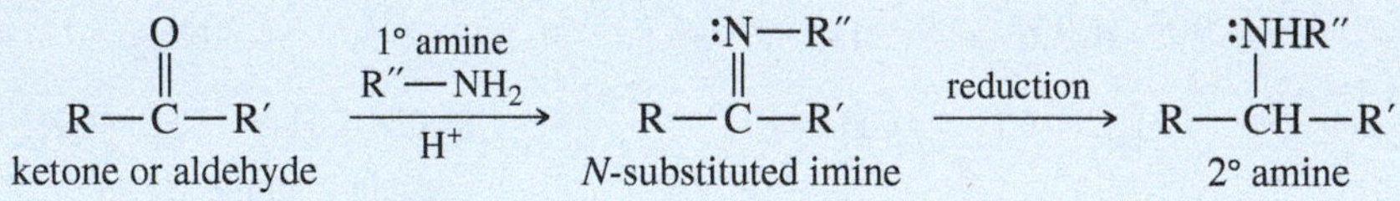

$$R-C(=O)-R' \xrightarrow[H^+]{1^\circ\text{ amine},\ R''-NH_2} R-C(=N-R'')-R' \xrightarrow{\text{reduction}} R-CH(NHR'')-R'$$

ketone or aldehyde → *N*-substituted imine → 2° amine

c. Tertiary amines

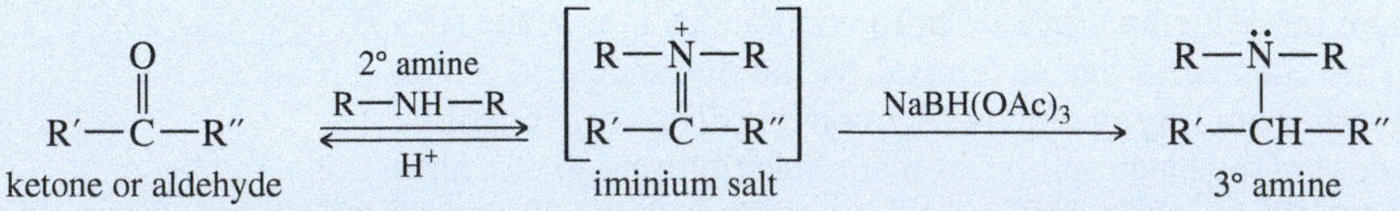

$$R'-C(=O)-R'' \underset{H^+}{\overset{2^\circ\text{ amine},\ R-NH-R}{\rightleftharpoons}} \left[R'-C(=N^+R_2)-R''\right] \xrightarrow{NaBH(OAc)_3} R'-CH(NR_2)-R''$$

ketone or aldehyde → iminium salt → 3° amine

2. *Acylation–reduction of amines* (Section 19-19)

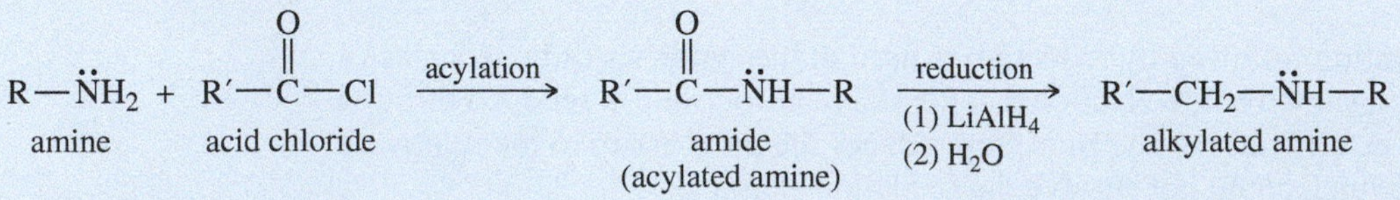

$$R-NH_2 + R'-C(=O)-Cl \xrightarrow{\text{acylation}} R'-C(=O)-NH-R \xrightarrow[(1)\ LiAlH_4;\ (2)\ H_2O]{\text{reduction}} R'-CH_2-NH-R$$

amine + acid chloride → amide (acylated amine) → alkylated amine

3. *Alkylation of ammonia* (Section 19-20A)

$$R-CH_2-X + \text{excess } NH_3 \longrightarrow R-CH_2-NH_2 + NH_4^+\ X^-$$

4. *The Gabriel synthesis of primary amines* (Section 19-20A)

$$\underset{\text{alkyl halide}}{R{-}X} \xrightarrow[\text{phthalimide anion}]{\text{phthalimide}^{-}\ K^+} \underset{N\text{-alkyl phthalimide}}{\text{phthalimide}{-}\ddot{N}{-}R} \xrightarrow[\text{heat}]{H_2NNH_2} \underset{1°\text{ amine}}{R{-}\ddot{N}H_2}$$

5. *Reduction of azides* (Section 19-20B)

$$\underset{\text{alkyl azide}}{R{-}\ddot{N}{=}\overset{+}{N}{=}\ddot{N}:^-} \xrightarrow[\text{or } H_2/Pd]{LiAlH_4} \underset{1°\text{ amine}}{R{-}\ddot{N}H_2}$$

6. *Reduction of nitriles* (Section 19-20B)

$$\underset{\text{nitrile}}{R{-}C{\equiv}N:} \xrightarrow{H_2/\text{catalyst or } LiAlH_4} \underset{1°\text{ amine}}{R{-}CH_2{-}\ddot{N}H_2}$$

7. *Reduction of nitro compounds* (Section 19-20C)

$$R{-}NO_2 \xrightarrow[\text{or active metal and } H^+]{H_2/\text{catalyst}} R{-}\ddot{N}H_2$$

catalyst = Ni, Pd, or Pt
active metal = Fe, Zn, or Sn

8. *Nucleophilic aromatic substitution* (Section 17-12)

$$R{-}\ddot{N}H_2 + Ar{-}X \longrightarrow R{-}\ddot{N}H{-}Ar + HX$$

(The aromatic ring should be activated toward nucleophilic attack.)

Essential Terms

acylation Addition of an **acyl group** ($R{-}\overset{O}{\overset{\|}{C}}{-}$), usually replacing a hydrogen atom. Acylation of an amine gives an amide. (p. 964)

$$\underset{\text{amine}}{R{-}NH_2} + \underset{\text{acid chloride}}{Cl{-}\overset{O}{\overset{\|}{C}}{-}R'} \longrightarrow \underset{\text{amide}}{R{-}NH{-}\overset{O}{\overset{\|}{C}}{-}R'} + HCl$$

acetylation: Acylation by an acetyl group ($CH_3{-}\overset{O}{\overset{\|}{C}}{-}$).

acylation–reduction A method for synthesizing amines by acylating ammonia or an amine, and then reducing the amide. (p. 982)

$$\underset{\text{amine}}{R{-}\ddot{N}H_2} + \underset{\text{acid chloride}}{R'{-}\overset{O}{\overset{\|}{C}}{-}Cl} \longrightarrow \underset{\text{amide}}{R{-}\ddot{N}H{-}\overset{O}{\overset{\|}{C}}{-}R'} \xrightarrow[(2)\ H_2O]{(1)\ LiAlH_4} \underset{\text{alkylated amine}}{R{-}\ddot{N}H{-}CH_2{-}R'}$$

amine A derivative of ammonia with one or more alkyl or aryl groups bonded to the nitrogen atom. (p. 941)

A primary amine: (1° amine) has one alkyl group bonded to nitrogen.

A secondary amine: (2° amine) has two alkyl groups bonded to nitrogen.

A tertiary amine: (3° amine) has three alkyl groups bonded to nitrogen.

$$\underset{\text{primary amine}}{R{-}\underset{..}{\overset{H}{\overset{|}{N}}}{-}H} \qquad \underset{\text{secondary amine}}{R{-}\underset{..}{\overset{H}{\overset{|}{N}}}{-}R'} \qquad \underset{\text{tertiary amine}}{R{-}\underset{..}{\overset{R''}{\overset{|}{N}}}{-}R'}$$

amino group: The —NH_2 group. If alkylated, it becomes an **alkylamino** group, —NHR or a **dialkylamino** group, —NR_2. (p. 943)

amine oxide An amine with a fourth bond to an oxygen atom. In the amine oxide, the nitrogen atom bears a positive charge, and the oxygen atom bears a negative charge. (p. 971)

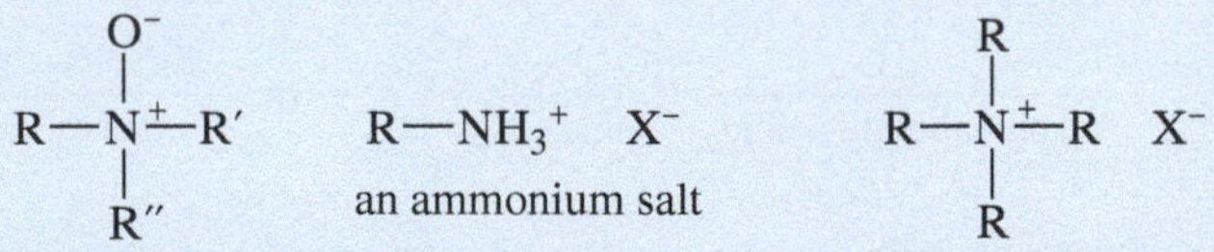

$$R-\overset{O^-}{\underset{R''}{N^+}}-R' \qquad R-NH_3^+ \quad X^- \qquad R-\overset{R}{\underset{R}{N^+}}-R \quad X^-$$

an amine oxide — an ammonium salt — a quaternary ammonium salt

ammonium salt (**amine salt**) A derivative of an amine with a positively charged nitrogen atom having four bonds. An amine is protonated by an acid to give an ammonium salt. (p. 952) A quaternary ammonium salt has a nitrogen atom bonded to four alkyl or aryl groups. (p. 943)

arenediazonium salt An aromatic compound that contains a diazonium cation, $Ar-NR_2^+$, together with its counterion, X^-. (p. 974)

azide A compound having the azido group, —N_3. (p. 985)

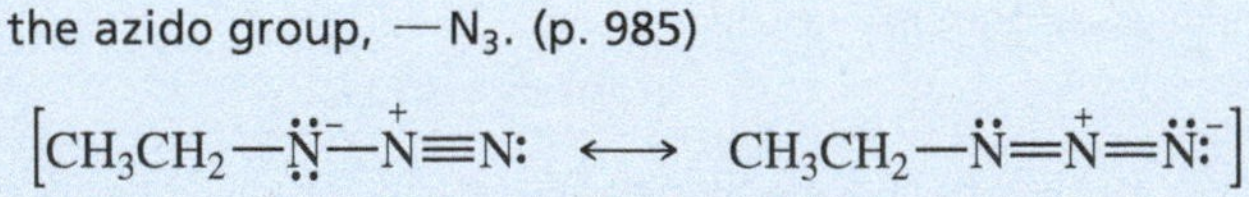

$$\left[CH_3CH_2-\ddot{\underset{..}{N}}^- -\overset{+}{N}\equiv N: \longleftrightarrow CH_3CH_2-\ddot{N}=\overset{+}{N}=\ddot{N}:^-\right]$$

ethyl azide

base-dissociation constant (K_b) A measure of the basicity of a compound such as an amine, defined as the equilibrium constant for the following reaction. The negative $\log_{10}$ of K_b is given as pK_b. (p. 948)

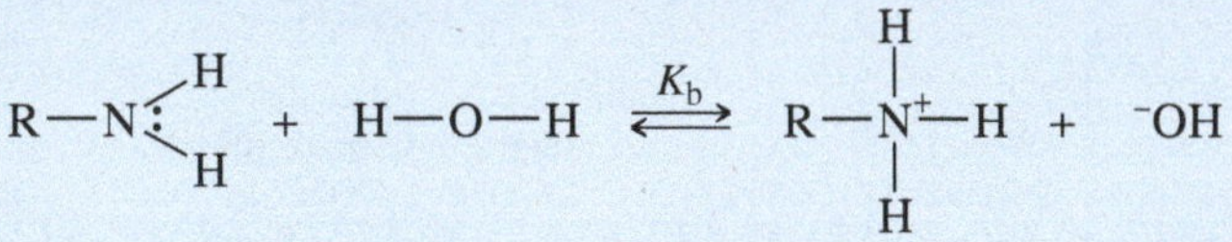

$$R-\ddot{N}H_2 + H-O-H \overset{K_b}{\rightleftharpoons} R-\overset{H}{\underset{H}{N^+}}-H + {}^-OH$$

Cope elimination A variation of the Hofmann elimination, where a tertiary amine oxide eliminates to an alkene with a hydroxylamine serving as the leaving group. (p. 971)

diazo coupling The use of a diazonium salt as an electrophile in electrophilic aromatic substitution. (p. 977)

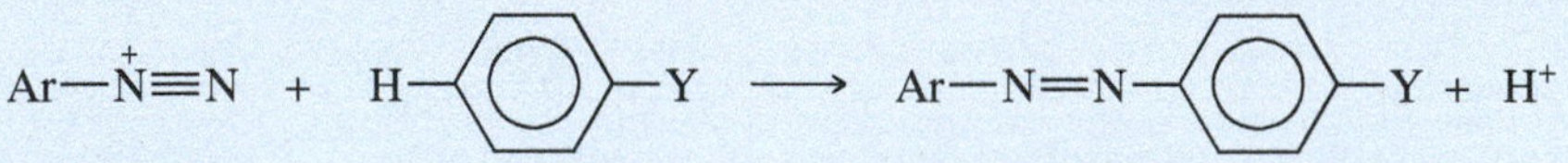

$$Ar-\overset{+}{N}\equiv N + H-C_6H_4-Y \longrightarrow Ar-N=N-C_6H_4-Y + H^+$$

diazonium ion — (activated) — an **azo** compound

diazotization of an amine The reaction of a primary amine with nitrous acid to form a diazonium salt. (p. 973)

exhaustive alkylation Treatment of an amine with an excess of an alkylating agent (often methyl iodide) to form the quaternary ammonium salt. (p. 963)

$$R-NH_2 \xrightarrow[\text{base}]{\text{excess } CH_3I} R-\overset{+}{N}(CH_3)_3 \quad I^-$$

Gabriel amine synthesis Synthesis of primary amines by alkylation of the potassium salt of phthalimide, followed by displacement of the amine by hydrazine. (p. 984)

Hofmann elimination Elimination of a quaternary ammonium hydroxide with an amine as the leaving group. The Hofmann elimination usually gives the least-substituted alkene. (p. 967)

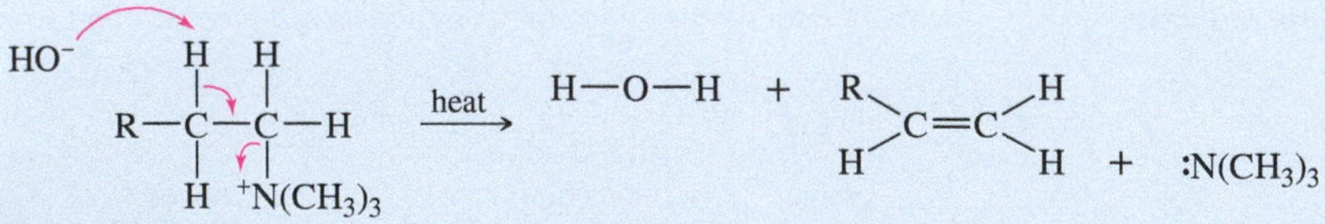

$$HO^- + R-\overset{H}{\underset{H}{C}}-\overset{H}{\underset{^+N(CH_3)_3}{C}}-H \xrightarrow{\text{heat}} H-O-H + \underset{H}{\overset{R}{}}C=C\underset{H}{\overset{H}{}} + :N(CH_3)_3$$

hydroxylamine The compound H_2NOH; or generically, an amine in which a hydroxy group is one of the three substituents bonded to nitrogen. (p. 971)

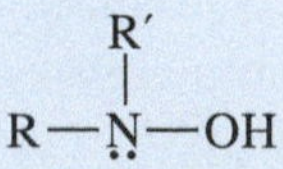

$$R-\overset{R'}{\underset{..}{N}}-OH$$

nitrile A compound of formula R—C≡N, containing the *cyano group*, —C≡N. (p. 986)

nitro compound An organic compound that contains one or more nitro (—NO_2) functional groups. (p. 987)

nitrogen inversion **(pyramidal inversion)** Inversion of configuration of a nitrogen atom in which the lone pair moves from one face of the molecule to the other. The transition state is planar, with the lone pair in a *p* orbital. (p. 945)

***N*-nitrosoamine** **(nitrosamine)** An amine with a nitroso group (—N=O) bonded to the amine nitrogen atom. The reaction of secondary amines with nitrous acid gives secondary *N*-nitrosoamines. (p. 974)

reductive amination The reduction of an imine or oxime derivative of a ketone or aldehyde. One of the most general methods for synthesis of amines. (p. 980)

$$\underset{\text{ketone or aldehyde}}{R-\overset{O}{\overset{\|}{C}}-R'} \xrightarrow[H^+]{R''-NH_2} \underset{N\text{-substituted imine}}{R-\overset{\ddot{N}-R''}{\overset{\|}{C}}-R'} \xrightarrow{\text{reduction}} \underset{2^\circ\text{ amine}}{R-\overset{\ddot{N}HR''}{\overset{|}{CH}}-R'}$$

Sandmeyer reaction Replacement of the $-\overset{+}{N}\equiv N$ group in an arenediazonium salt by a cuprous salt; usually cuprous chloride, bromide, or cyanide. (p. 976)

$$Ar-\overset{+}{N}\equiv N:\ ^-Cl \xrightarrow[(X = Cl, Br, C\equiv N)]{CuX} Ar-X + N_2\uparrow$$

sulfonamide An amide of a sulfonic acid. The nitrogen analogue of a sulfonate ester. (p. 966)

$$\underset{\text{a sulfonamide}}{R-NH-\overset{O}{\overset{\|}{\underset{O}{\underset{\|}{S}}}}-R'} \qquad \underset{\text{a }p\text{-toluenesulfonamide (a tosylamide)}}{R-NH-\overset{O}{\overset{\|}{\underset{O}{\underset{\|}{S}}}}-C_6H_4-CH_3}$$

Essential Problem-Solving Skills in Chapter 19

Each skill is followed by problem numbers exemplifying that particular skill.

1	Name amines, and draw structures from their names.	Problems 19-32, 38, 47, 48, and 52
2	Interpret the IR, NMR, and mass spectra of amines, and use the spectral information to determine the structures.	Problems 19-41, 53, 56, 58, and 59
3	Compare the basicity of amines with other common bases, and explain how their basicity varies with hybridization and aromaticity.	Problems 19-33, 34, 35, 37, 60, and 61
4	Describe the trends in the physical properties of amines, and contrast their physical properties with those of their salts.	Problems 19-36 and 37
5	Predict the products of reactions of amines with ketones and aldehydes, alkyl halides and tosylates, acid chlorides, sulfonyl chlorides, nitrous acid, and oxidizing agents. Propose mechanisms where appropriate.	Problems 19-38, 39, 46, 54, 55, 62, 63, and 64
6	Give examples using arenediazonium ion salts in diazo coupling reactions and in the synthesis of phenols and aryl chlorides, bromides, iodides, fluorides, and nitriles.	Problems 19-40, 45, and 50
7	Illustrate the uses and mechanisms of the Hofmann and Cope eliminations, and predict the major products.	Problems 19-38 and 42
8	Propose single-step and multistep syntheses of amines from other amines, ketones, aldehydes, acid chlorides, nitro compounds, alkyl halides, nitriles, and amides.	Problems 19-40, 42, 43, 44, 45, 47, 48, 49, 50, and 51
9	Show how to convert amines to other functional groups, and devise multistep syntheses using amines as starting materials and intermediates.	Problems 19-40, 42, 50, and 51

Study Problems

19-32 For each compound,

(1) classify the nitrogen-containing functional groups.

(2) provide an acceptable name.

(a) 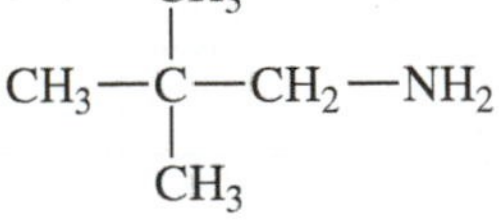 $(CH_3)_3C{-}CH_2{-}NH_2$

(b) 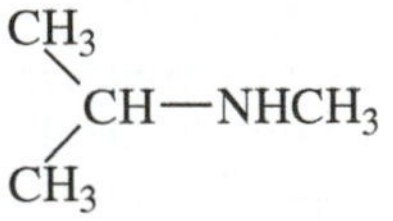 $(CH_3)_2CH{-}NHCH_3$

(c) 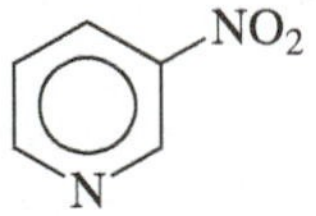 3-nitropyridine (NO_2 on pyridine ring)

(d) 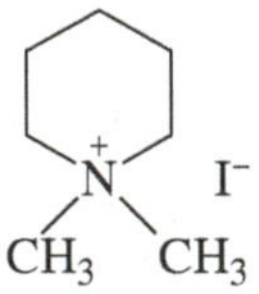 N,N-dimethylpiperidinium ($\overset{+}{N}$, CH_3, CH_3) I^-

(e) 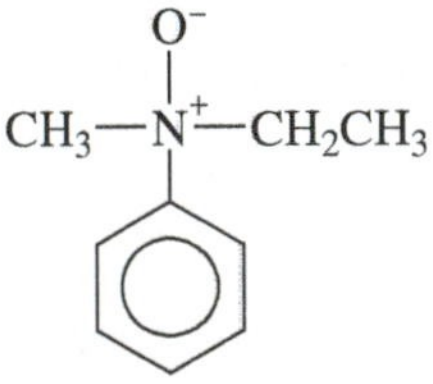 $CH_3{-}\overset{+}{N}(O^-){-}CH_2CH_3$ with phenyl on N

(f) 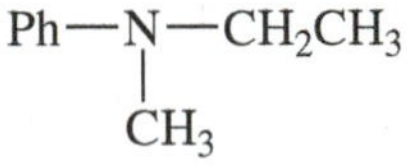 $Ph{-}N(CH_3){-}CH_2CH_3$

(g) 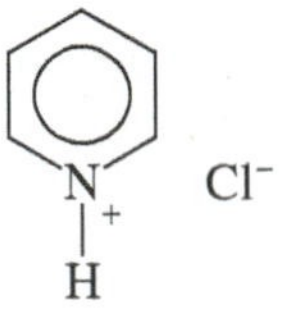 pyridinium ($\overset{+}{N}{-}H$) Cl^-

(h) $NHCH_2CH_3$, CH_2CH_3 (skeletal structure)

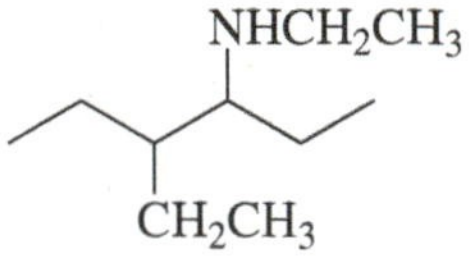

19-33 Rank the amines in each set in order of increasing basicity.

(a) 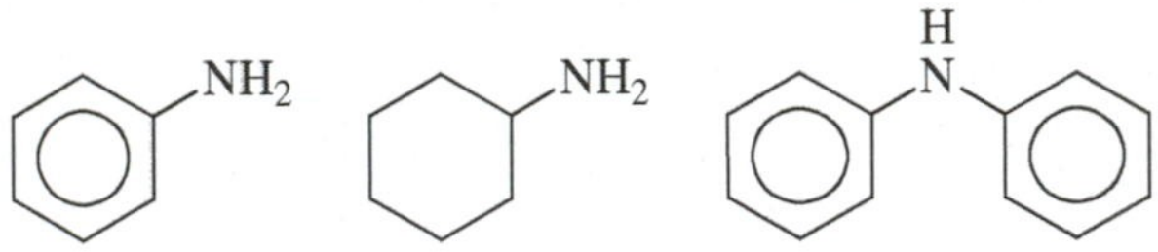 aniline (NH_2 on benzene); cyclohexylamine (NH_2); diphenylamine (H–N between two benzene rings)

(b) 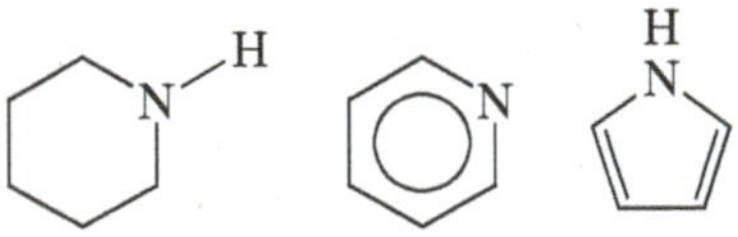 piperidine (N–H); pyridine (N); pyrrole (N–H)

(c) 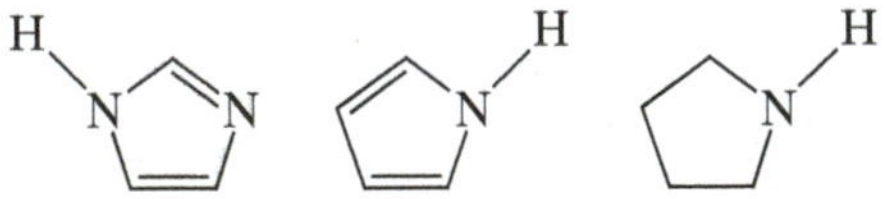 imidazole (H–N, N); pyrrole (N–H); pyrrolidine (N–H)

(d) 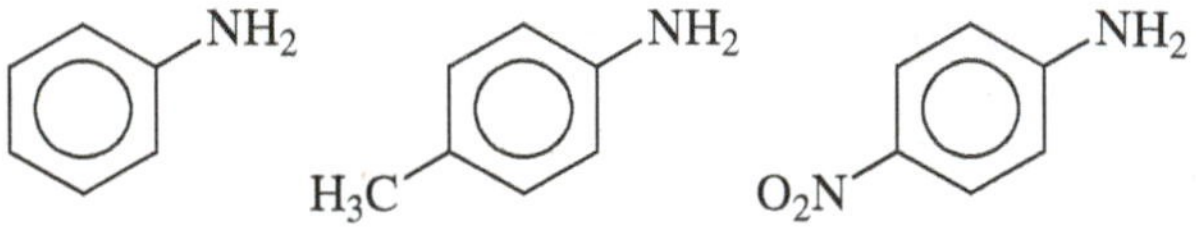 aniline (NH_2); p-toluidine (H_3C, NH_2); p-nitroaniline (O_2N, NH_2)

(e) 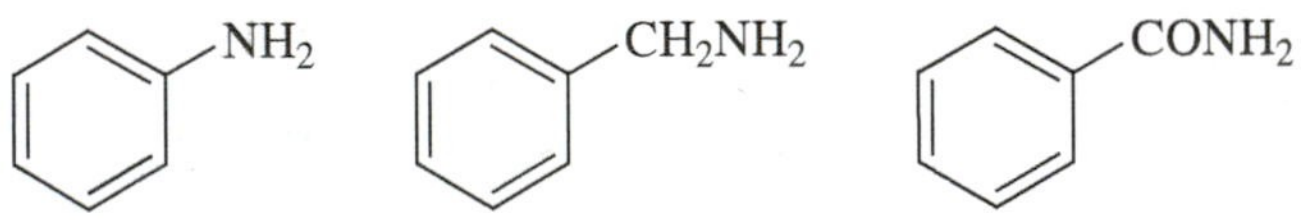 $C_6H_5NH_2$; $C_6H_5CH_2NH_2$; $C_6H_5CONH_2$

19-34 Within each structure, rank the indicated nitrogens by increasing basicity.

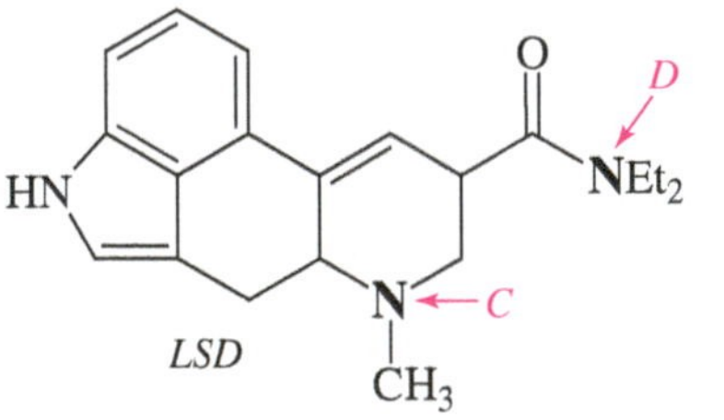

(a) 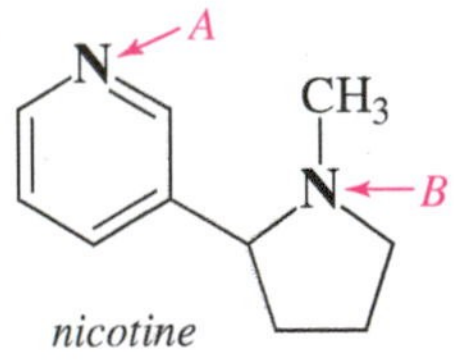 *nicotine* — A (pyridine N), B (N–CH_3 pyrrolidine N)

(b) *LSD* — HN, O, D (NEt_2), C (N–CH_3)

(c) 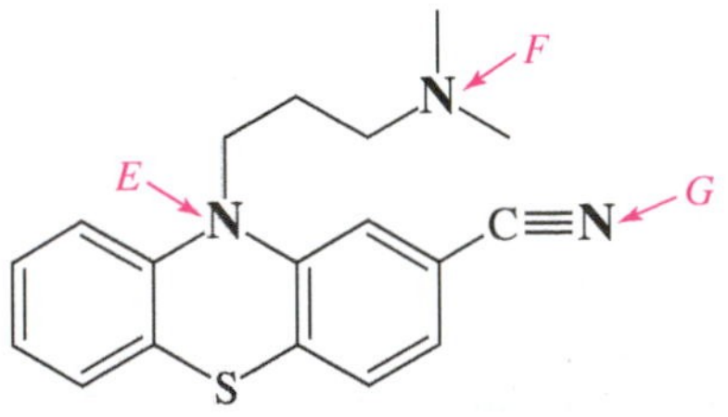 *cyamemazine, an anti-psychotic drug* — F (dimethylamino N), E (ring N), G ($C{\equiv}N$), S

(d) 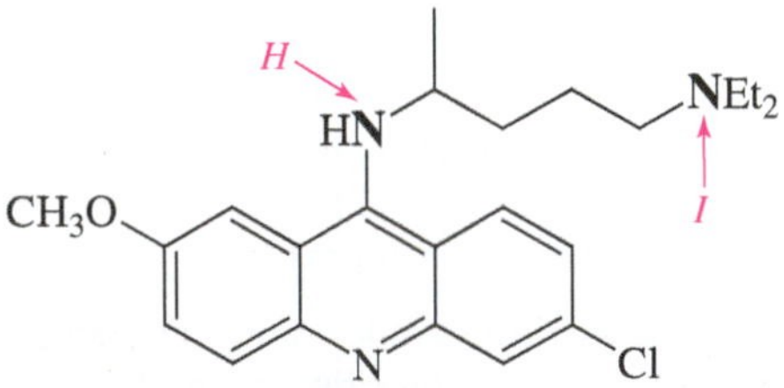 *atabrine (anti-malarial)* — H (HN), I (NEt_2), CH_3O, N, Cl

(e) 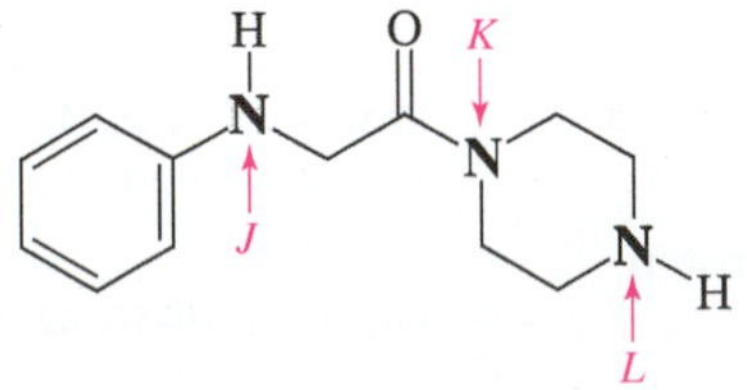 H, O, K, J, N–H, L

(f) 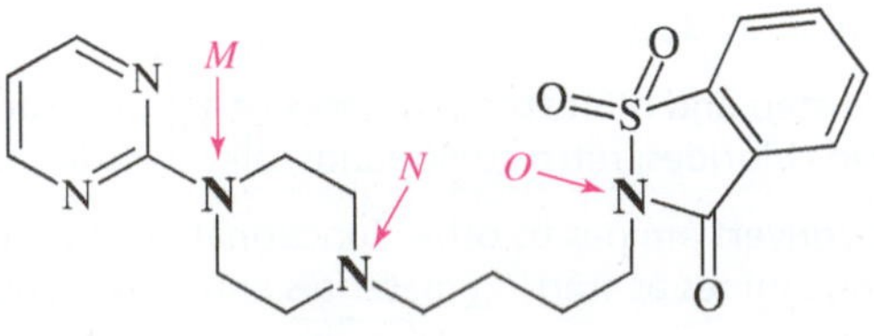 *ipsapirone, an anxiolytic drug* — N, N, M, N, O, $O{=}S$, O

19-35 In each pair of compounds, select the stronger base, and explain your choice.

(a) $HOCH_2CH_2NH_2$ or $CH_3CH_2NH_2$ **(b)** $PhNH_2$ or $PhCH_2NH_2$

(c)

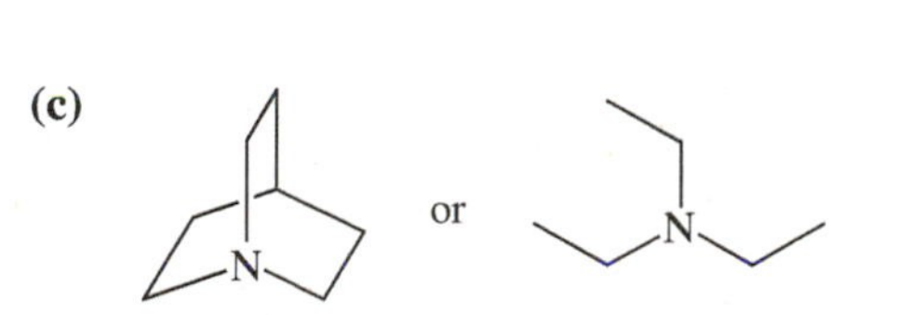

*** (d)** 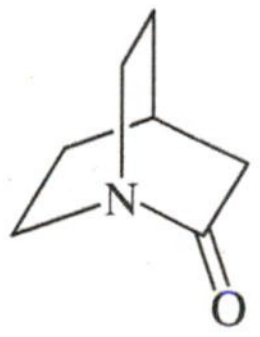 N or NH (O, O)

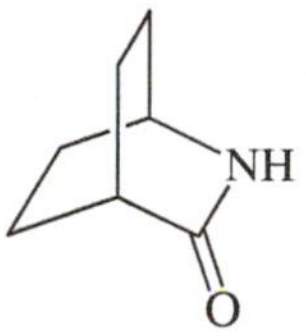

19-36 Which of the following compounds are capable of being resolved into enantiomers?

(a) *N*-ethyl-*N*-methylaniline **(b)** 2-methylpiperidine **(c)** 1-methylpiperidine **(d)** 1,2,2-trimethylaziridine

(e) 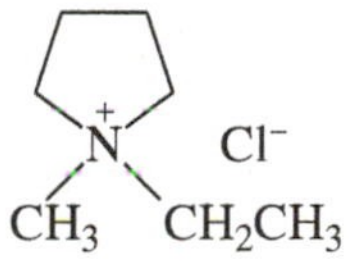 $\overset{+}{N}$ Cl^- CH_3 CH_2CH_3 **(f)** 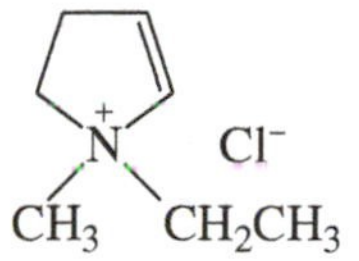 $\overset{+}{N}$ Cl^- CH_3 CH_2CH_3 **(g)** 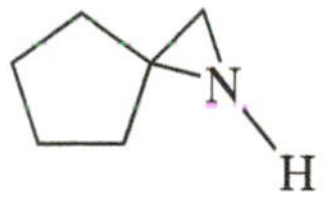 N H **(h)** N CH_3

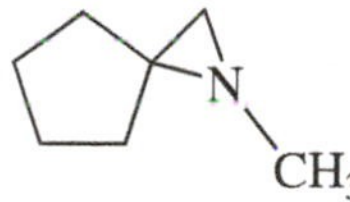

19-37 Complete the following proposed acid–base reactions, and predict whether the reactants or products are favored.

(a) 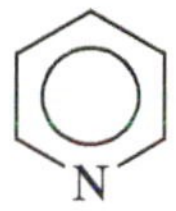 pyridine + CH_3COOH (acetic acid) $\longrightarrow$

(b) 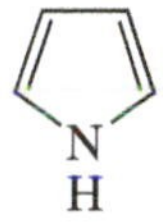 pyrrole + CH_3COOH (acetic acid) $\longrightarrow$

(c) 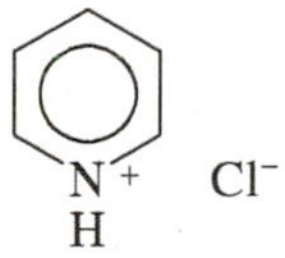 pyridinium chloride ($\overset{+}{N}H$ Cl^-) + piperidine $\longrightarrow$

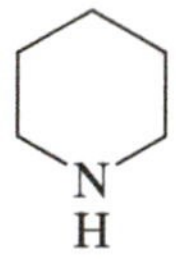

(d) 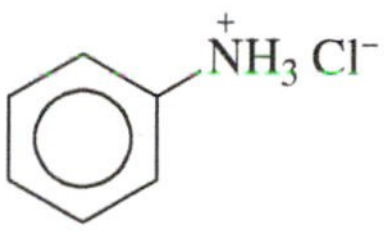 anilinium chloride ($\overset{+}{N}H_3$ Cl^-) + pyrrolidine $\longrightarrow$

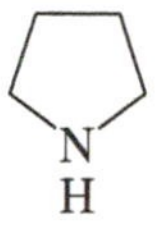

19-38 Predict the products of the following reactions:

(a) excess NH_3 + $Ph{-}CH_2\ CH_2\ CH_2\ Br$ $\longrightarrow$

(b) 1-bromopentane $\xrightarrow[(2)\ LiAlH_4\ (3)\ H_3O^+]{(1)\ NaN_3}$

(c) 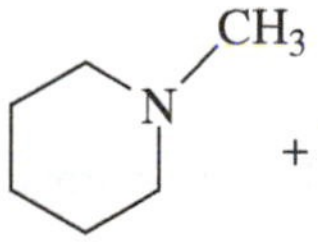 N-CH_3 (N-methylpiperidine) + H_2O_2 $\longrightarrow$

(d) product from part (c) $\xrightarrow{heat}$

(e) 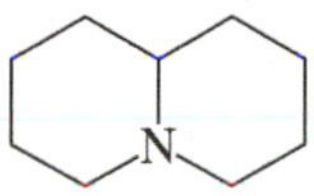 N (quinolizidine) $\xrightarrow[(3)\ heat]{(1)\ excess\ CH_3I\ (2)\ Ag_2O}$

(f) product from part (c) $\xrightarrow[(3)\ heat]{(1)\ excess\ CH_3I\ (2)\ Ag_2O}$

(g) 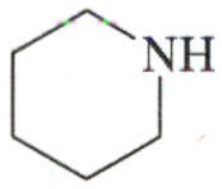 NH (piperidine) + $NaNO_2$ + HCl $\longrightarrow$

(h) 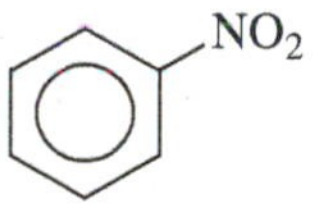 $C_6H_5NO_2$ (NO$_2$) $\xrightarrow{Zn,\ HCl}$

(i) CH_3NH_2 + (cyclohexyl)$CH_2{-}C({=}O){-}Cl$ $\xrightarrow{pyridine}$

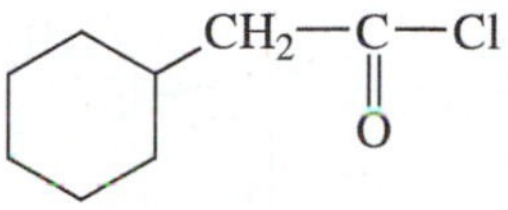

(j) product from part (i) $\xrightarrow[(2)\ H_3O^+]{(1)\ LiAlH_4}$

(k) $CH_3{-}(CH_2)_3{-}C({=}NCH_3){-}CH_2CH_3$ $\xrightarrow[(2)\ H_3O^+]{(1)\ LiAlH_4}$

(l) $Ph{-}CH_2{-}CH(CN){-}CH_3$ $\xrightarrow[(2)\ H_3O^+]{(1)\ LiAlH_4}$

19-39 Predict the products of the following reactions:

(a) butan-2-one + diethylamine $\xrightarrow{NaBH(OAc)_3}$

(b) 4-fluoropyridine $\xrightarrow{NaOCH_2CH_3}$

(c) 3-nitroaniline $\xrightarrow[(2)\ CuBr]{(1)\ HCl,\ NaNO_2}$

(d) butan-2-one $\xrightarrow[(2)\ LiAlH_4]{(1)\ KCN,\ HCN}$

(e) cyclopentanone $\xrightarrow[(2)\ LiAlH_4]{(1)\ aniline,\ H^+}$

(f) 2-bromopentane $\xrightarrow[(2)\ Ag_2O,\ heat]{(1)\ (CH_3)_3N:}$

(g) O (cyclohexanone) $\xrightarrow[-H_2O]{H^+,\ EtNH_2}$

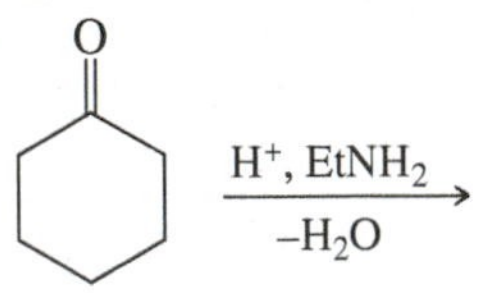

(h) O (cyclohexanone) $\xrightarrow[Na(OAc)_3BH]{Et_2NH}$

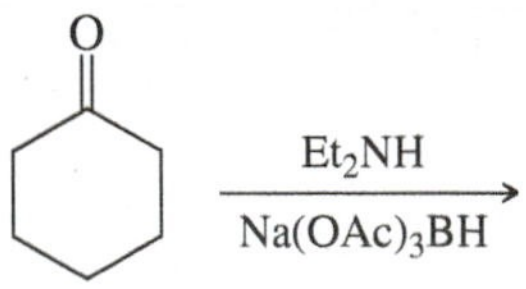

(i) D, H, NMe_2 $\xrightarrow{?}$ H

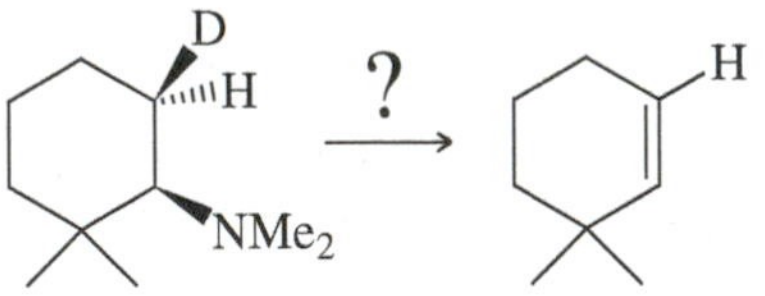

(j) D, H, NMe_2 $\xrightarrow{?}$ D

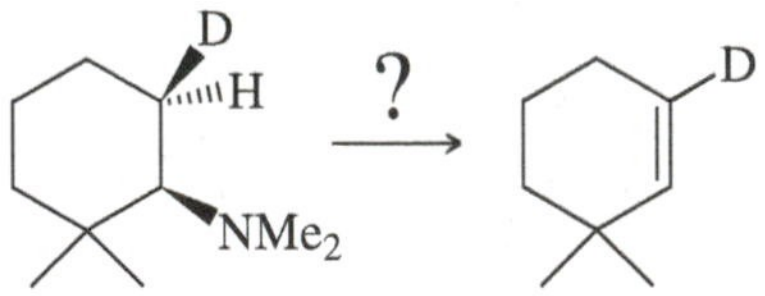

19-40 Show how *m*-toluidine can be converted to the following compounds, using any necessary reagents.

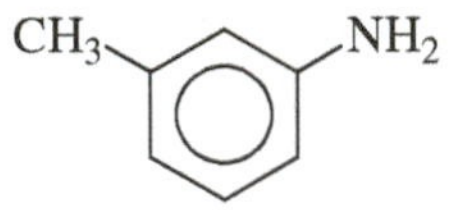

CH_3, NH_2
m-toluidine

(a) CH_3, C≡N
m-toluonitrile

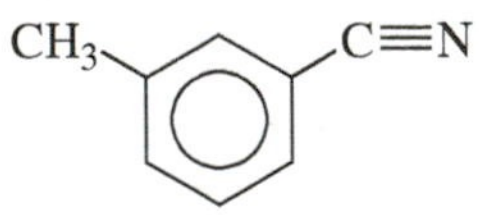

(b) CH_3, CH_2NH_2
m-methylbenzylamine

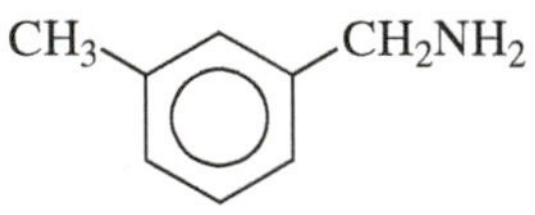

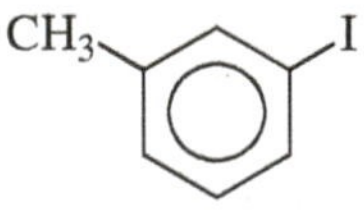

(c) CH_3, I
m-iodotoluene

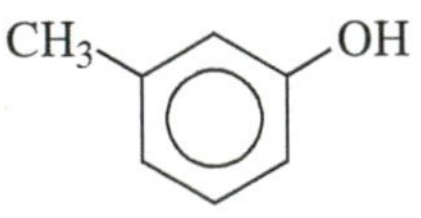

(d) CH_3, OH
m-cresol

(e) CH_3, NH_2, O_2N
3-methyl-4-nitroaniline

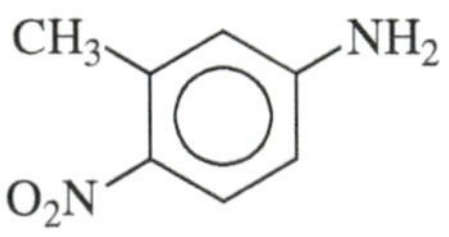

(f) H, N, CH_3
N-cyclopentyl-*m*-toluidine

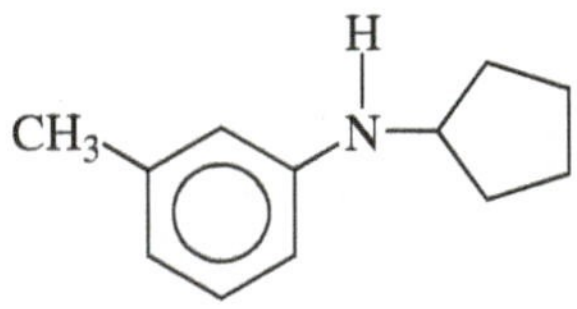

19-41 The mass spectrum of *tert*-butylamine follows shows an intense base peak at m/z 58, and very little else. Use a diagram to show the cleavage that accounts for the base peak. Suggest why no molecular ion is visible in this spectrum.

19-42 Using any necessary reagents, show how you would accomplish the following syntheses.

(a) → NH—C(=O)—(cyclopentyl)

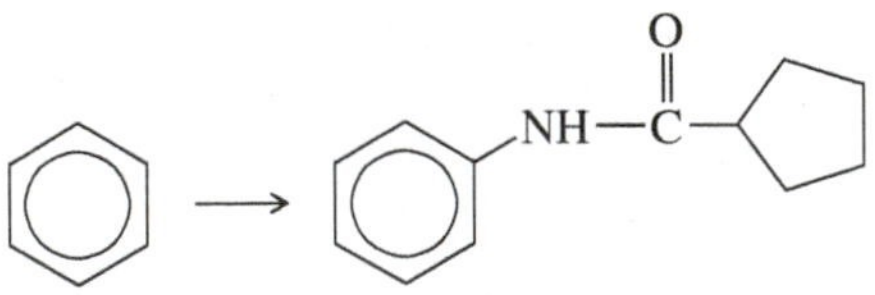

(b) NH_2 → NH—S(=O)(=O)—CH_3

(c) N—H → $N(CH_3)_2$

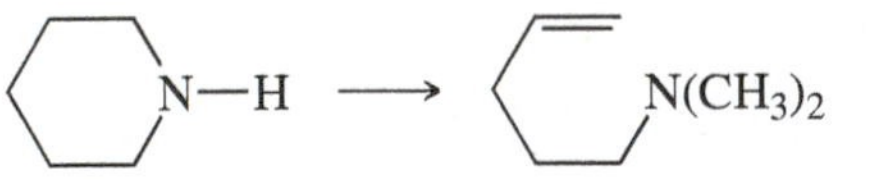

(d) O → NH_2

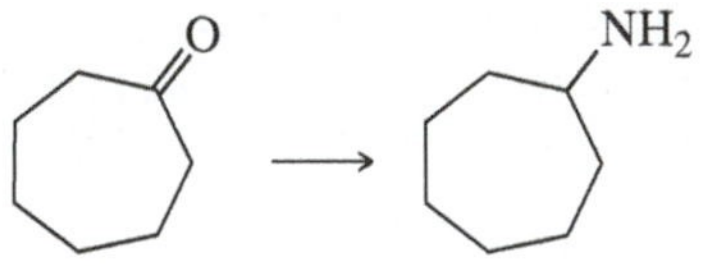

(e) N—CH_3 → O^-, N⁺, CH_3

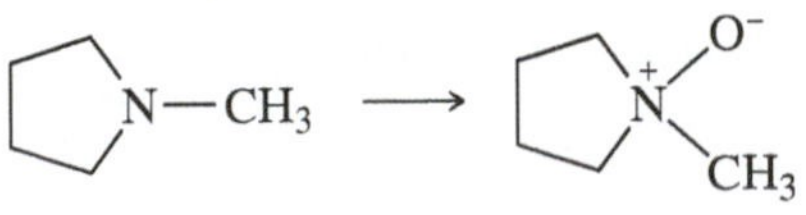

(f) N—CH_3 → OH, N, CH_3

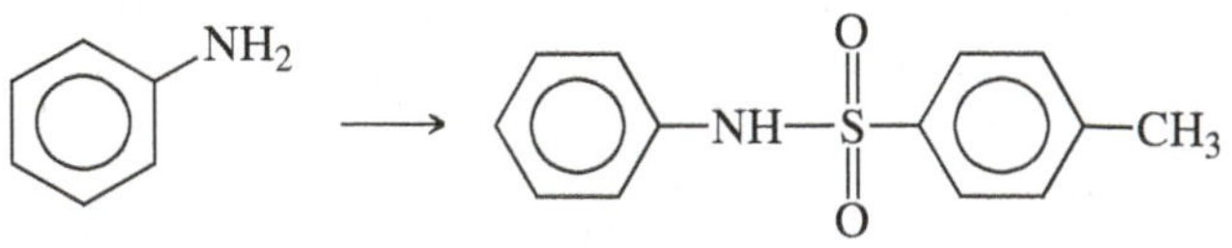

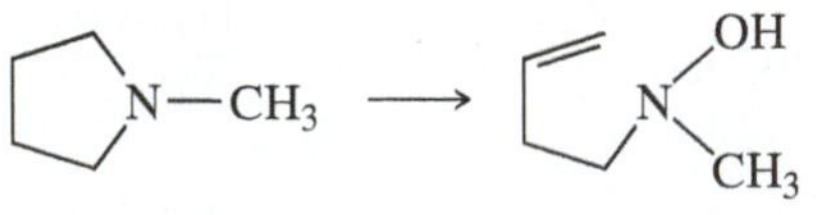

(continued)

(g) $CH_3C_6H_4COOH \longrightarrow CH_3C_6H_4C(=O)N(CH_2CH_3)_2$

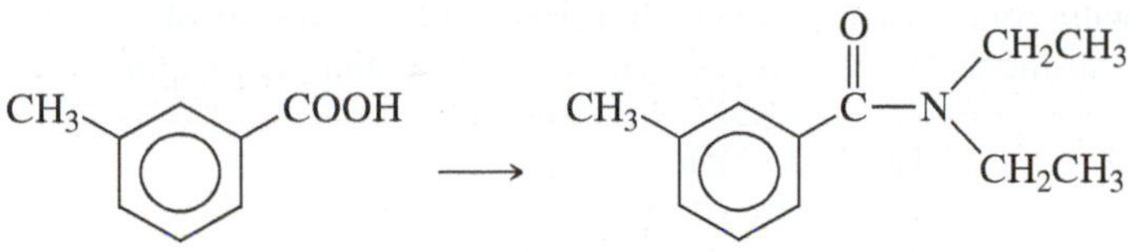

(DEET mosquito repellent)

19-43 The following drugs are synthesized using the methods in this chapter and in previous chapters. Devise a synthesis for each, starting with any compounds containing no more than six carbon atoms.
(a) Phenacetin, used with aspirin and caffeine in pain-relief medications.
(b) Methamphetamine, once considered a safe diet pill, but now known to be addictive and destructive to brain tissue.
(c) Dopamine, one of the neurotransmitters in the brain. Parkinson's disease is thought to result from a dopamine deficiency.

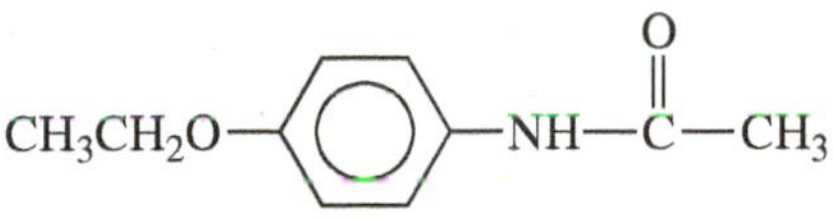

$CH_3CH_2O-C_6H_4-NH-C(=O)-CH_3$ phenacetin

$C_6H_5-CH_2-CH(NHCH_3)-CH_3$ methamphetamine

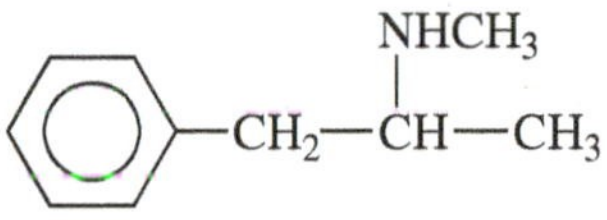

$(HO)_2C_6H_3-CH_2CH_2NH_2$ dopamine

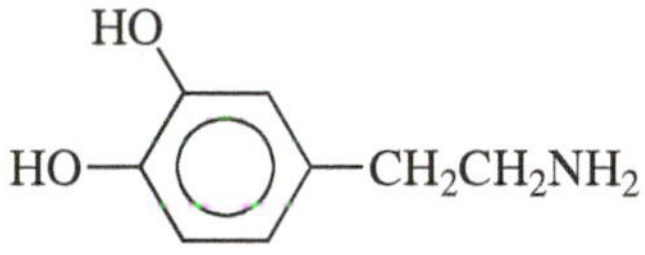

19-44 Synthesize Novocaine from benzene and any other reagents of four carbons or fewer.

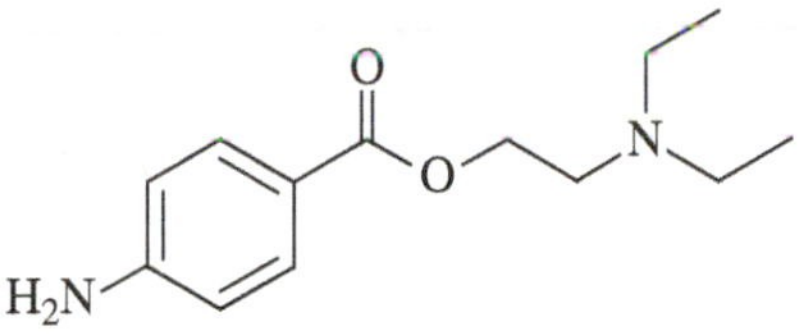

$H_2N-C_6H_4-C(=O)O-CH_2CH_2-N(CH_2CH_3)_2$

19-45 Synthesize from benzene. (*Hint:* All of these require diazonium ions.)
(a) 3-ethylbenzoic acid **(b)** 3-*n*-propylphenol
(c) 2-methyl-5-hydroxybenzoic acid **(d)** 4-methoxyaniline

19-46 Propose mechanisms for the following reactions.

(a) cyclohexanone + $(CH_3)_2NH$ $\xrightarrow[H^+]{NaBH(OAc)_3}$ cyclohexyl-$N(CH_3)_2$

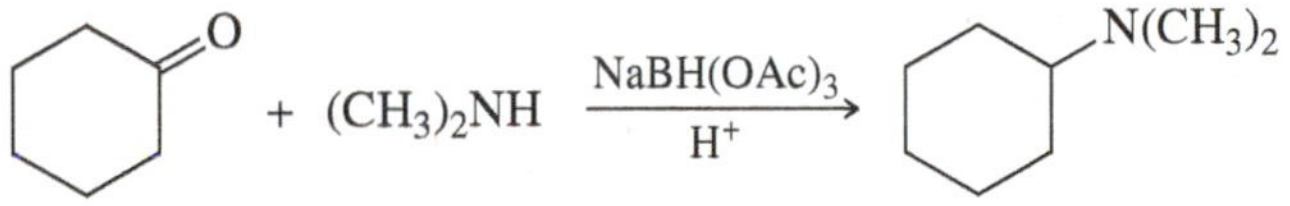

(b) 2-(2-cyanoethyl)cyclohexanone $\xrightarrow[H^+]{H_2,\ Pt}$ decahydroquinoline (N–H)

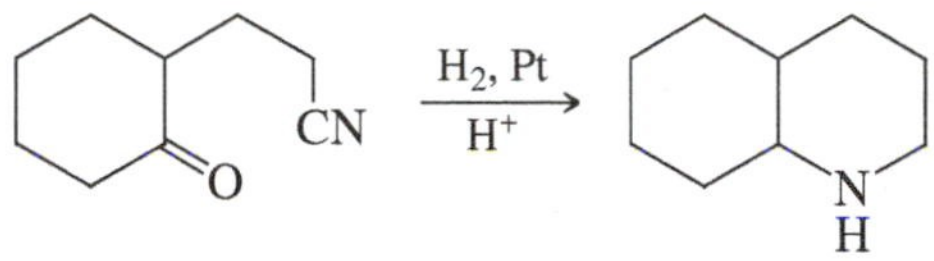

19-47 The two most general amine syntheses are the reductive amination of carbonyl compounds and the reduction of amides. Show how these techniques can be used to accomplish the following syntheses.
(a) benzoic acid → benzylamine **(b)** benzaldehyde → benzylamine
(c) pyrrolidine → *N*-ethylpyrrolidine **(d)** cyclohexanone → *N*-cyclohexylpyrrolidine
(e) $HOOC-(CH_2)_3-COOH$ → pentane-1,5-diamine (cadaverine)

19-48 Several additional amine syntheses are effectively limited to making primary amines. The reduction of azides and nitro compounds and the Gabriel synthesis leave the carbon chain unchanged. Formation and reduction of a nitrile adds one carbon atom. Show how these amine syntheses can be used for the following conversions.
(a) allyl bromide → allylamine **(b)** ethylbenzene → *p*-ethylaniline
(c) 1-bromo-3-phenylheptane → 3-phenylheptan-1-amine **(d)** 1-bromo-3-phenylheptane → 4-phenyloctan-1-amine

19-49 Show how you can synthesize the following tertiary amine three different ways, each using a different secondary amine and adding the final substituent by
(a) reductive amination (3 ways). **(b)** acylation–reduction (3 ways).

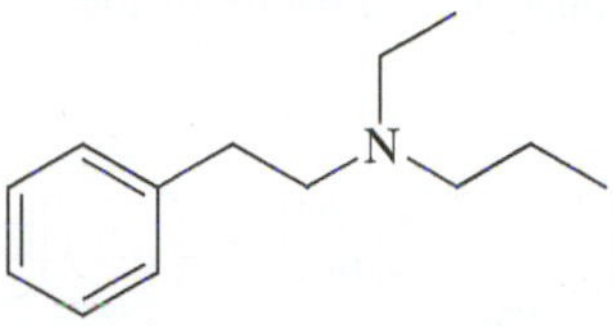

$C_6H_5CH_2CH_2-N(CH_2CH_3)(CH_2CH_2CH_3)$

19-50 Show how you can synthesize the following compounds starting with benzene, toluene, and alcohols containing no more than four carbon atoms as your organic starting materials. Assume that para is the major product (and separable from ortho) in ortho, para mixtures.

(a) pentan-1-amine
(b) *N*-methylbutan-1-amine
(c) *N*-ethyl-*N*-propylbutan-2-amine
(d) *N*-benzylpropan-1-amine
(e) C_6H_5—N=N—C_6H_4—OH (para)
(f) 3-propylaniline
(g) 4-isobutylaniline

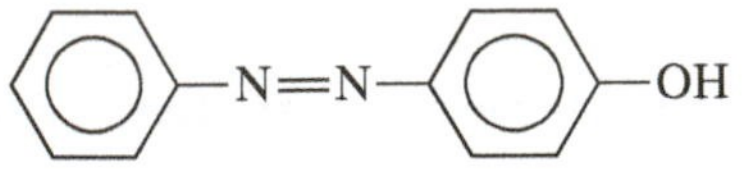

19-51 Using any necessary reagents, show how you can accomplish the following multistep syntheses.

(a) benzene ⟶ 4-butylaniline (NH_2 and $CH_2CH_2CH_2CH_3$ para on benzene ring)

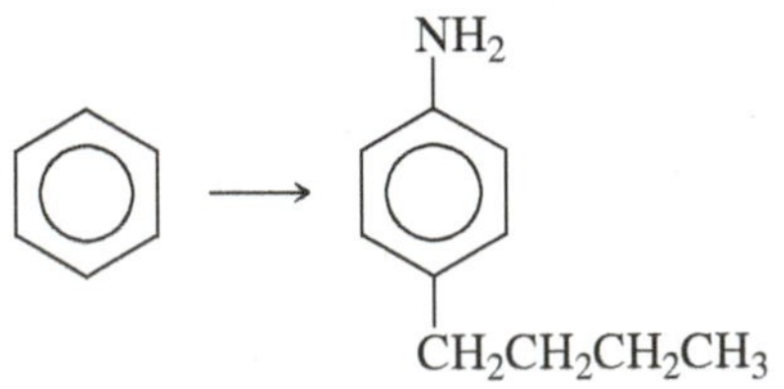

(b) quinuclidine (bicyclic amine, N) ⟶ *N*-methylpiperidine with $CH{=}CH_2$ at the 4-position (CH_3 on N)

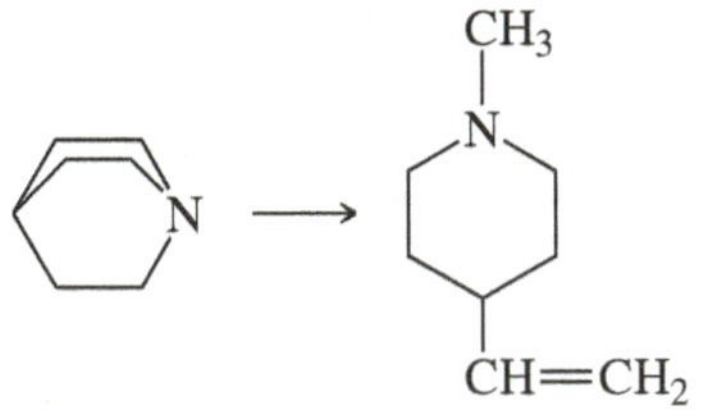

(c) biphenyl ⟶ biphenyl bearing CH_2CH_3 and CH_2NH_2 on one ring

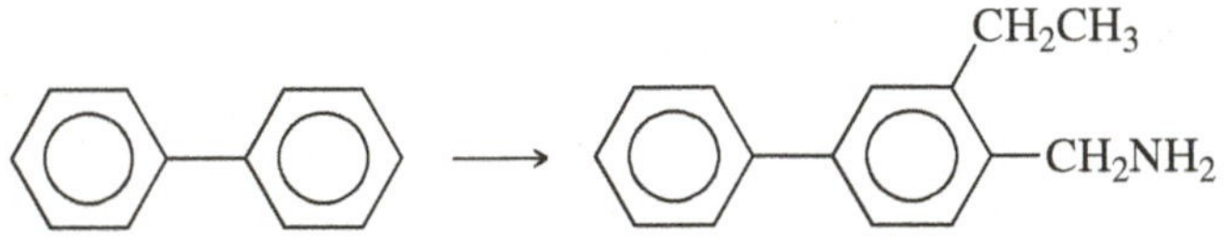

19-52 The alkaloid coniine has been isolated from hemlock and purified. Its molecular formula is $C_8H_{17}N$. Treatment of coniine with excess methyl iodide, followed by silver oxide and heating, gives the pure (*S*)-enantiomer of *N*,*N*-dimethyloct-7-ene-4-amine. Propose a complete structure for coniine, and show how this reaction gives the observed product.

19-53 A chemist is summoned to an abandoned waste-disposal site to determine the contents of a leaking, corroded barrel. The barrel reeks of an overpowering fishy odor. The chemist dons a respirator to approach the barrel and collect a sample, which she takes to her laboratory for analysis.

The mass spectrum shows a molecular ion at m/z 101, and the most abundant fragment is at m/z 86. The IR spectrum shows no absorptions above 3000 cm^{-1}, many absorptions between 2800 and 3000 cm^{-1}, no absorptions between 1500 and 2800 cm^{-1}, and a strong absorption at 1200 cm^{-1}. The proton NMR spectrum shows a triplet ($J = 7$ Hz) at δ 1.0 and a quartet ($J = 7$ Hz) at δ 2.4, with integrals of 17 spaces and 11 spaces, respectively.

(a) Show what structural information is implied by each spectrum, and propose a structure for the unknown toxic waste.
(b) Current EPA regulations restrict the disposal of liquid wastes because they tend to leak out of their containers. Propose an inexpensive method for converting this waste to a solid, relatively odorless form for reburial.
(c) Suggest how the chemist can remove the fishy smell from her clothing.

***19-54** Pyrrole undergoes electrophilic aromatic substitution more readily than benzene, and mild reagents and conditions are sufficient. These reactions normally occur at the 2-position rather than the 3-position, as shown in the following example.

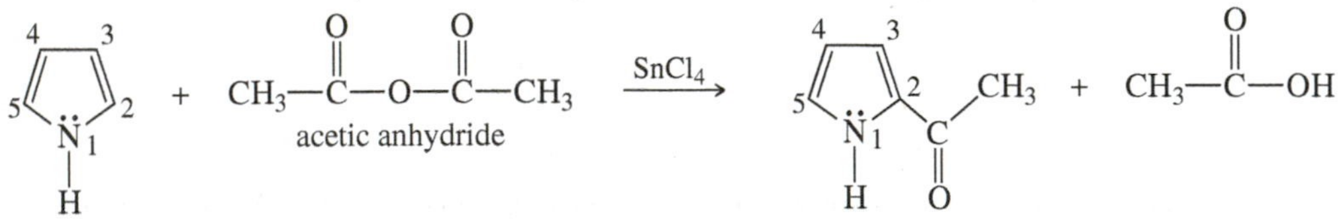

pyrrole + $CH_3-\overset{O}{\overset{\|}{C}}-O-\overset{O}{\overset{\|}{C}}-CH_3$ (acetic anhydride) $\xrightarrow{SnCl_4}$ 2-acetylpyrrole + $CH_3-\overset{O}{\overset{\|}{C}}-OH$

(a) Propose a mechanism for the acetylation of pyrrole just shown. You may begin with pyrrole and the acylium ion, $CH_3-C{\equiv}O^+$. Be careful to draw all the resonance structures of the intermediate.
(b) Explain why pyrrole reacts more readily than benzene, and also why substitution occurs primarily at the 2-position rather than the 3-position.

19-55 Section 17-12 showed how nucleophilic aromatic substitution can give aryl amines if there is a strong electron-withdrawing group ortho or para to the site of substitution. Consider the following example.

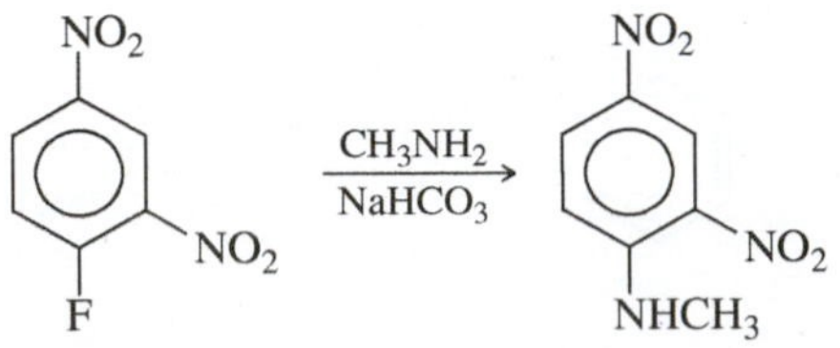

1-fluoro-2,4-dinitrobenzene $\xrightarrow[NaHCO_3]{CH_3NH_2}$ *N*-methyl-2,4-dinitroaniline ($NHCH_3$)

(continued)

(a) Propose a mechanism for this reaction.

(b) We usually think of fluoride ion as a poor leaving group. Explain why this reaction readily displaces fluoride as the leaving group.

(c) Explain why this reaction stops with the desired product, rather than reacting with another dinitrofluorobenzene.

19-56 The following spectra for **A** and **B** correspond to two structural isomers. The NMR singlet at δ 1.16 in spectrum **A** disappears when the sample is shaken with D_2O. The singlet at δ 0.6 ppm in the spectrum of **B** disappears on shaking with D_2O. Propose structures for these isomers, and show how your structures correspond to the spectra. Show what cleavage is responsible for the base peak at m/z 44 in the mass spectrum of **A** and the prominent peak at m/z 58 in the mass spectrum of **B**.

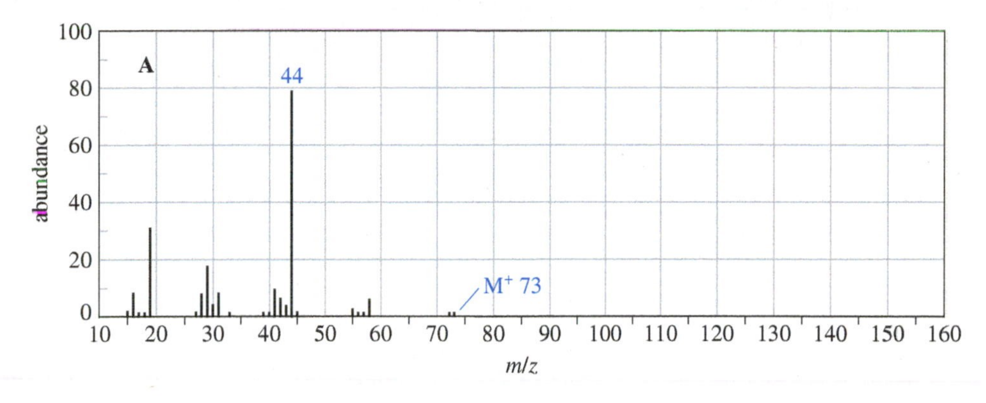

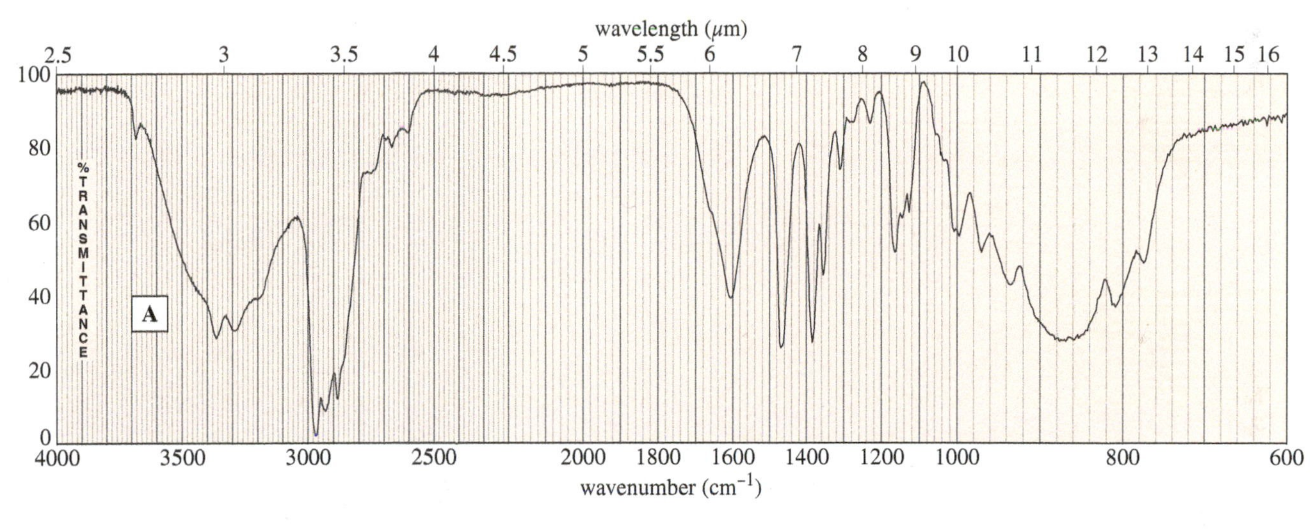

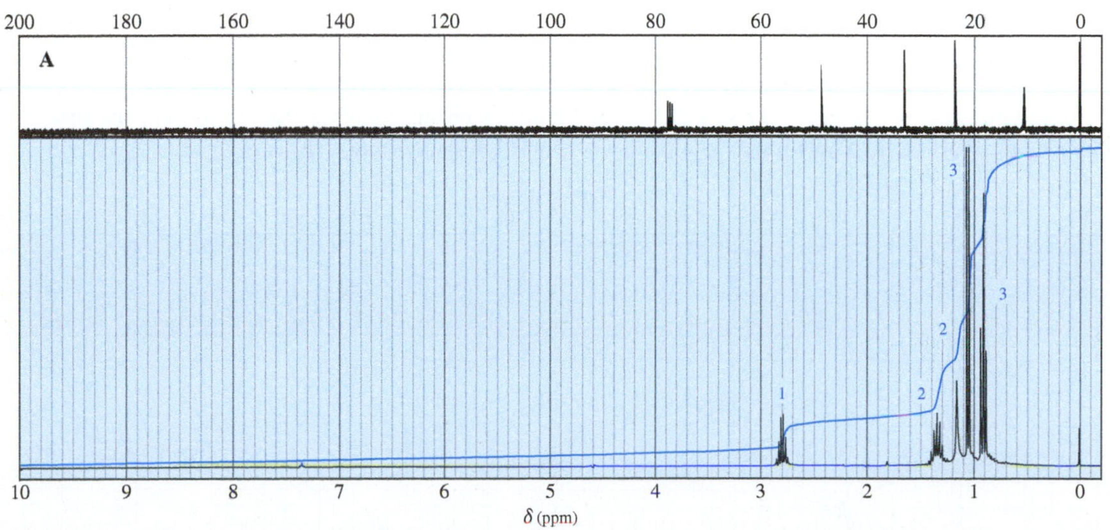

(continued)

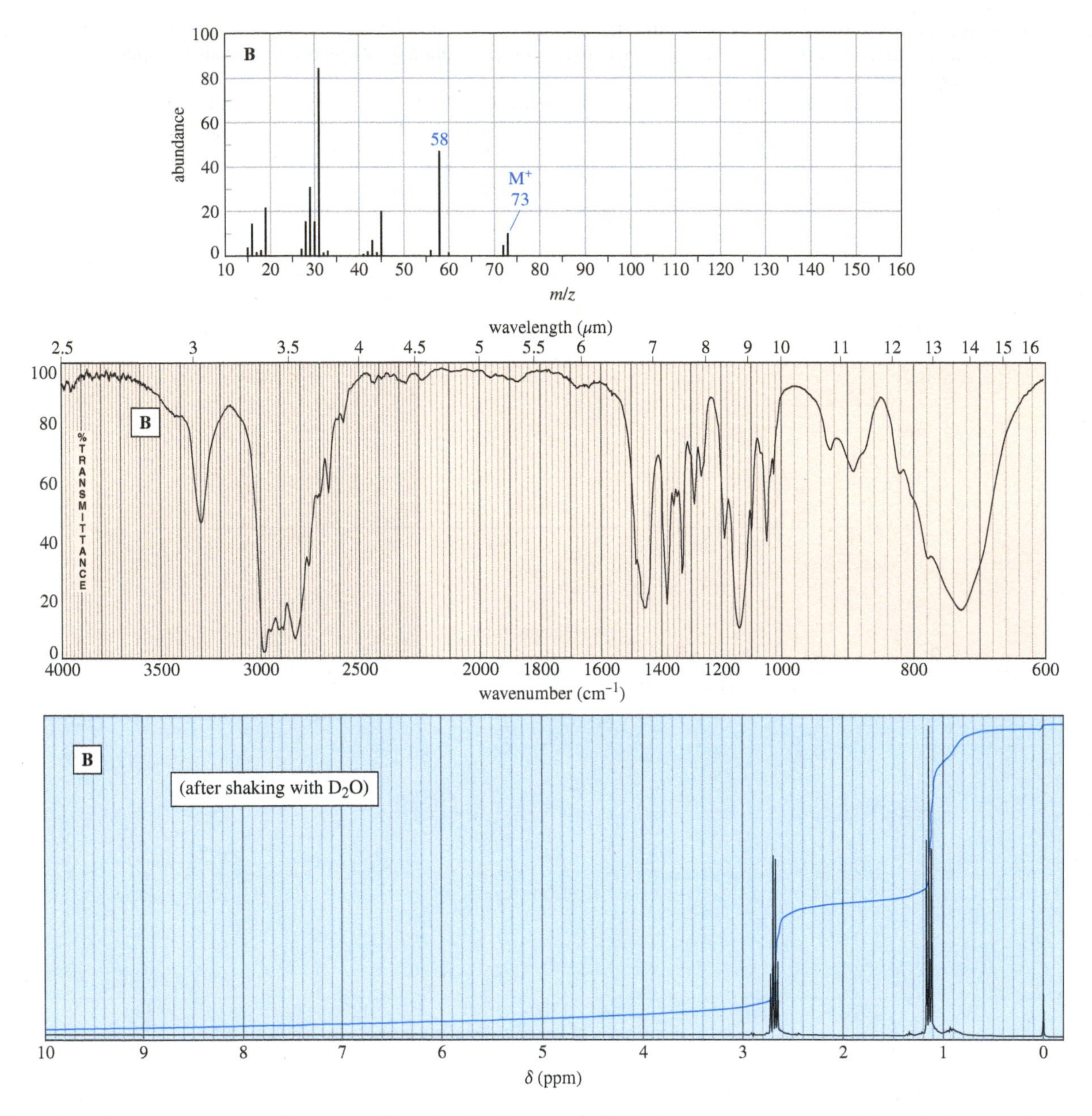

19-57 An unknown compound shows a weak molecular ion at m/z 87 in the mass spectrum, and the only large peak is at m/z 30. The IR spectrum follows. The proton NMR spectrum shows only three singlets: one of area 9 at $\delta\,0.9$, one of area 2 at $\delta\,1.0$, and one of area 2 at $\delta\,2.4$. The singlet at $\delta\,1.0$ disappears on shaking with D_2O. Determine the structure of the compound, and show the favorable fragmentation that accounts for the ion at m/z 30.

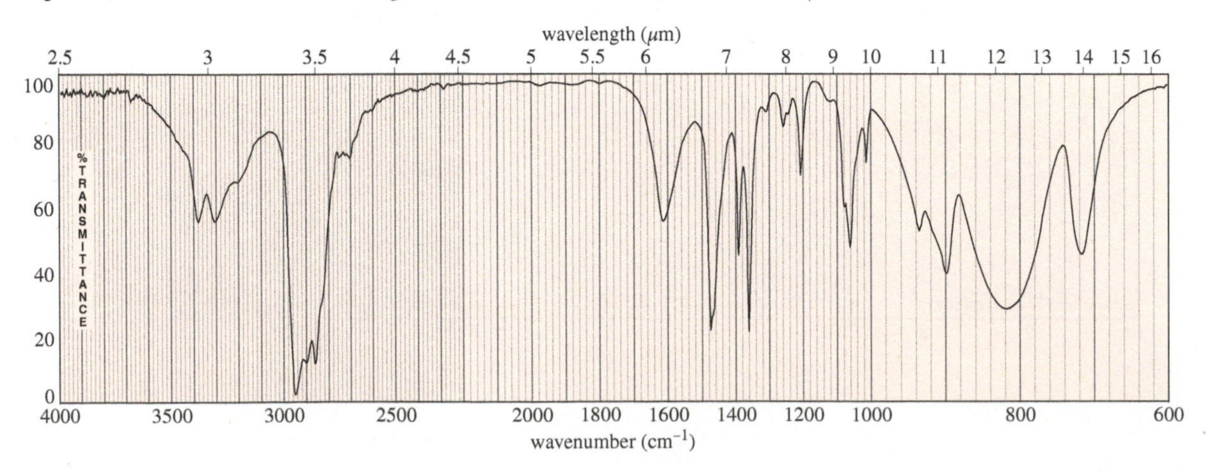

***19-58** A compound of formula $C_{11}H_{16}N_2$ gives the IR, ^1H NMR, and ^{13}C NMR spectra shown. The proton NMR peak at δ 2.0 disappears on shaking with D_2O. Propose a structure for this compound, and show how your structure accounts for the observed absorptions.

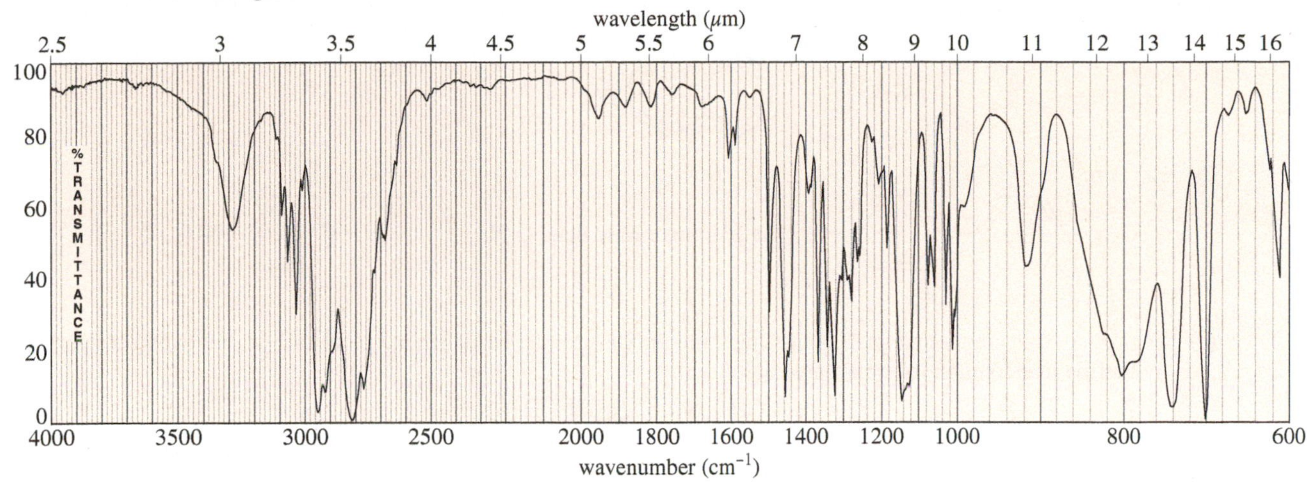

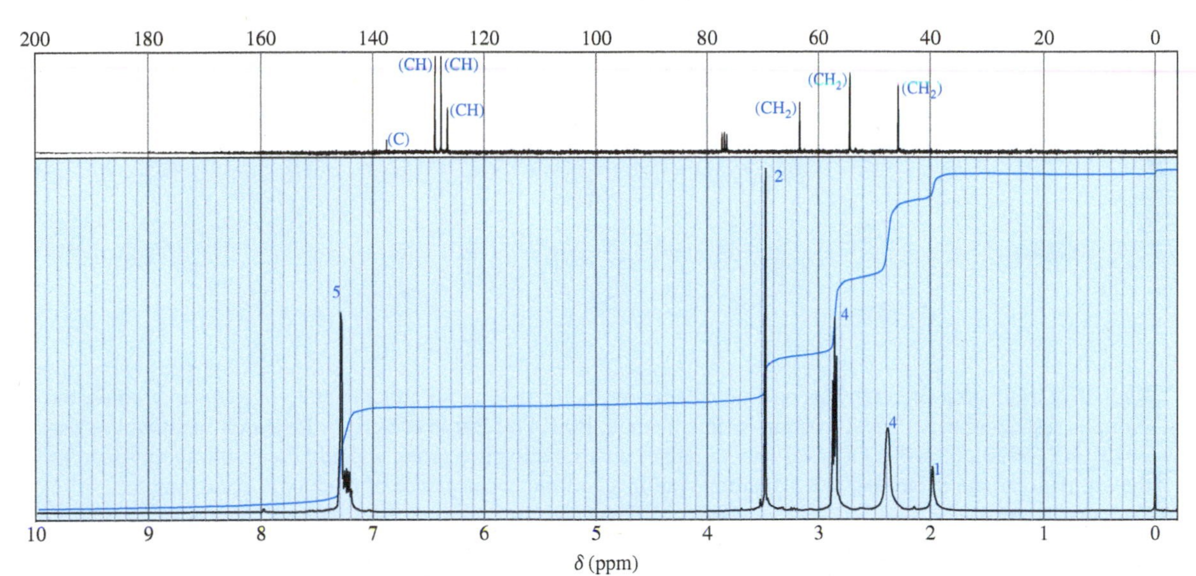

***19-59** (A true story.) A drug user responded to an ad placed by a DEA informant in a drug-culture magazine. He later flew from Colorado to Maryland, where he bought some 1-phenyl-2-propanone (P2P) from the informant. The police waited nearly a month for the suspect to synthesize something, then obtained a search warrant, and searched the residence. They found the unopened bottle of P2P; apparently, the suspect was not a good chemist and was unable to follow the instructions the informant gave him. They also found pipes and bongs with residues of marijuana and cocaine, plus a bottle of methylamine hydrochloride, some muriatic acid (dilute HCl), zinc strips, flasks, and other equipment.

(a) Assume you are consulting for the police. Show what synthesis the suspect was prepared to carry out, to provide probable cause for the charge of attempting to manufacture a controlled substance.

(b) Assume you are a member of the jury. Would you convict the defendant of attempting to manufacture a controlled substance?

19-60 **(a)** Guanidine (shown) is about as strong a base as hydroxide ion. Explain why guanidine is a much stronger base than most other amines.

(b) Show why *p*-nitroaniline is a much weaker base (3 pK_b units weaker) than aniline.

(c) Explain why *N,N*,2,6-tetramethylaniline (shown) is a much stronger base than *N,N*-dimethylaniline.

$H_2N{-}C({=}NH){-}NH_2$

guanidine

N,N,2,6-tetramethylaniline

N,N-dimethylaniline

***19-61** Basicity depends on availability of an electron pair to bond a proton. Correlate structural effects in these amines with their basicities.

(a) Explain this order:

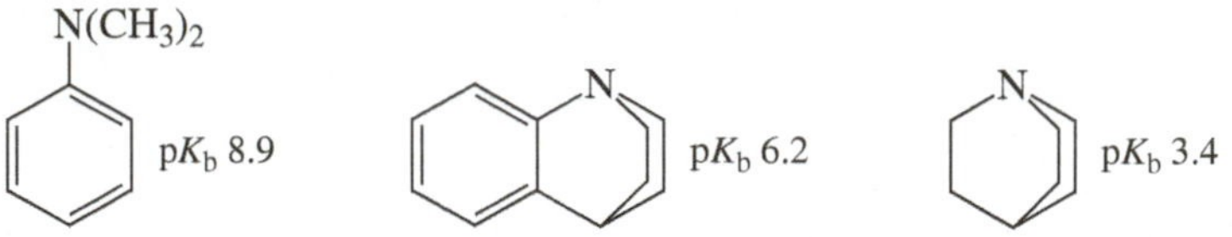

$N(CH_3)_2$ (N,N-dimethylaniline) pK_b 8.9; (benzo-fused bicyclic amine) pK_b 6.2; (quinuclidine) pK_b 3.4

(b) Explain:

O_2N–(3-methylphenyl)–(2-methylphenyl)–NH_2 is a stronger base than O_2N–(phenyl)–(phenyl)–NH_2

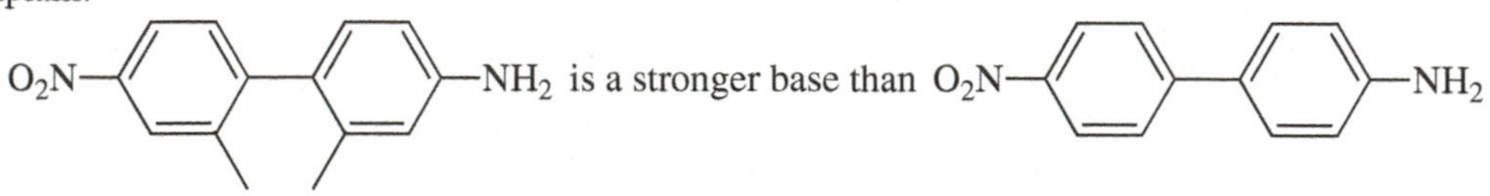

(c) The pK_b of this compound is –2.3, making it not only a stronger base than a typical aniline, but even stronger than hydroxide ion. Explain its remarkable basicity.

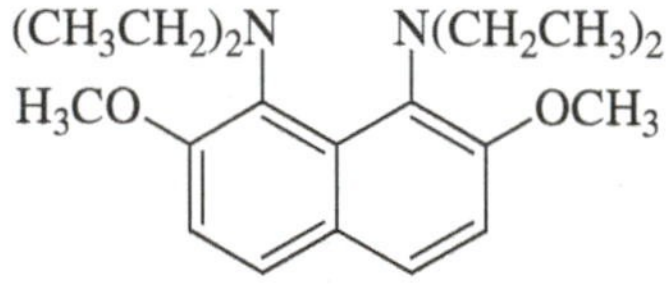

$(CH_3CH_2)_2N$ $N(CH_2CH_3)_2$
H_3CO OCH_3

***19-62** In Section 19-10B, we saw that pyridine undergoes electrophilic aromatic substitution reluctantly, requiring strong conditions and giving disappointing yields. In contrast, pyridine *N*-oxide undergoes EAS under moderate conditions, giving good yields of substitution on C2 and C4. Explain this surprising difference.

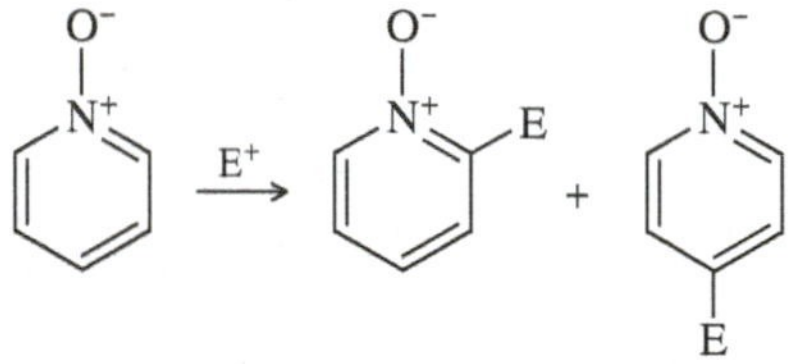

O^-–N^+ (pyridine N-oxide) $\xrightarrow{E^+}$ 2-E product + 4-E product

***19-63** Reductive amination of aldehydes and ketones is a versatile method for attaching alkyl groups to amines, but the alkyl group is restricted to a 1° or 2° carbon by this method. Prof. Phil Baran of Scripps Research Institute has reported (*Science*, **2015**, *348* (6237), 886–891) a novel way to reduce an aromatic nitro group and add the resulting amine to an alkene so that the aromatic amine is bonded to a 3° carbon—all in a continuous sequence of reactions. For example:

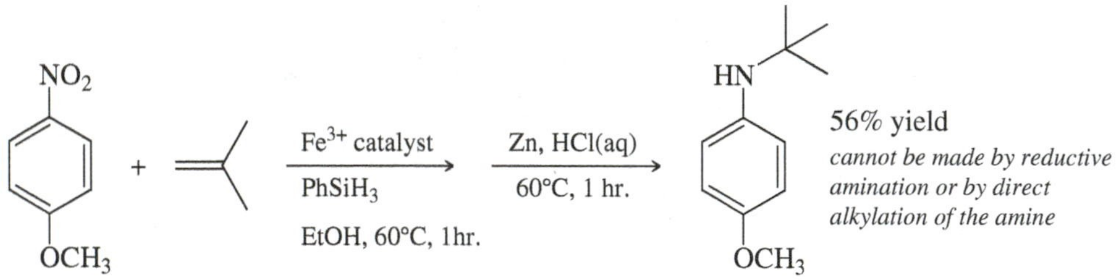

NO_2 / OCH_3 (4-nitroanisole) + (2-methylpropene) $\xrightarrow[\text{PhSiH}_3,\ \text{EtOH, 60°C, 1hr.}]{Fe^{3+}\ \text{catalyst}}$ $\xrightarrow[\text{60°C, 1 hr.}]{\text{Zn, HCl(aq)}}$ HN–(tert-butyl) / OCH_3

56% yield
cannot be made by reductive amination or by direct alkylation of the amine

Predict the products using these starting materials, all of which are reported in this paper.

(a) NO_2 (3-nitrotoluene) + (1-methylcyclohexene)

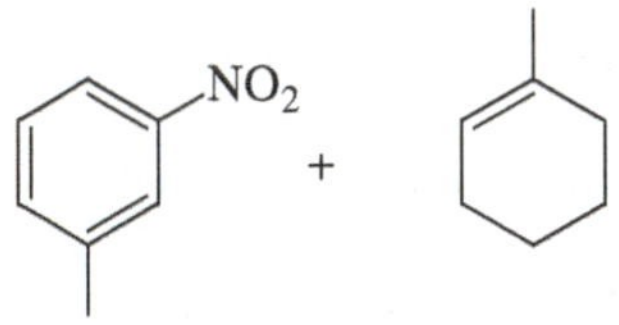

(b) NO_2 (aryl with $CH_2C(=O)CH_3$ substituent, O) + (methylenecyclopentane)

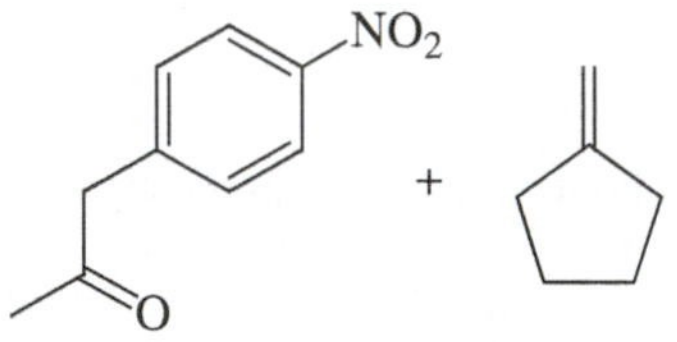

(c) Cl, NO_2 (2-chloronitrobenzene) + O (4-methyl-3-penten-2-one)

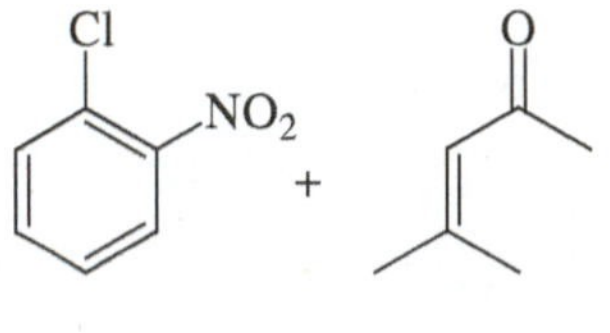

(d) NO_2, HN, HO, CN (substituted nitrobenzonitrile) + (1-heptene)

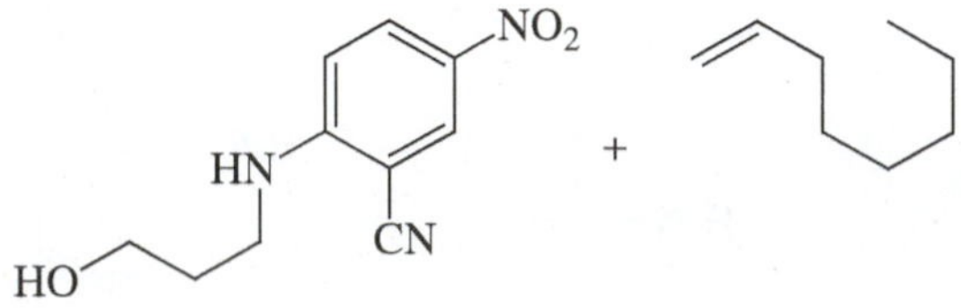

19-64 Show how the substituents containing the azo group (N=N) can facilitate both electrophilic and nucleophilic aromatic substitution.

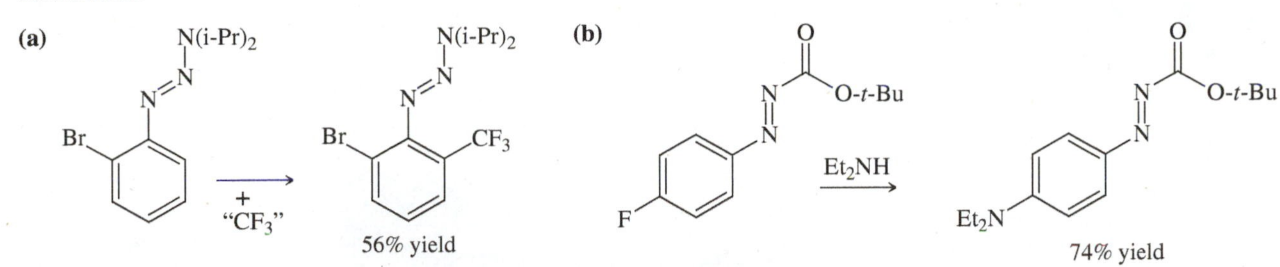

A. Hafner and S. Brase, *Angewandte Chemie Int. Ed.*, **2012**, *51*, 3713.

M. Heinrich *et al.*, *Journal of Organic Chemistry*, **2012**, *77*, 1520.

19-65 Macrolide antibiotics all have large rings (macrocycle) in which an ester makes the ring; a cyclic ester is termed a lactone. One example is erythromycin **A**, first isolated from soil bacteria in the 1950s. Over time, some pathogenic bacteria have developed resistance to erythromycin by evolving an enzymatic mechanism to cleave the macrocycle at the ketone.

To counter this resistance, chemists modified the erythromycin structure to replace the ketone with an amine that the bacteria could not detoxify. This modified antibiotic, azithromycin, trade name Zithromax®, is one of the most prescribed drugs in the world for respiratory infections.

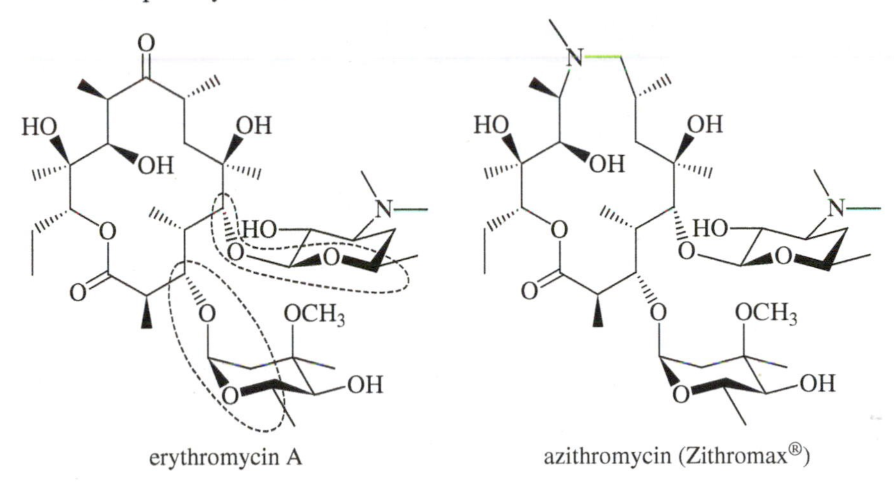

(a) Identify the lactone group in each structure that merits the classification as macrolides.
(b) Two groups are circled. What type of functional group are they? Explain.
(c) Identify the ketone in erythromycin targeted by bacteria as the site for detoxification.
(d) Identify the amine in azithromycin. What type of amine is it?
(e) From what you know about the reactivity of ketones and amines, why was an amine a good choice to be the "chemical opposite of a ketone"?

20 Carboxylic Acids

Goals for Chapter 20

1 Draw and name carboxylic acids and dicarboxylic acids, and use spectral information to determine their structures.

2 Describe the trends in the acidity and physical properties of carboxylic acids, and explain how their acidity varies with their substituents.

3 Propose single-step and multistep syntheses of carboxylic acids from compounds containing other functional groups.

4 Predict the products and propose mechanisms for the reactions of carboxylic acids with reducing agents, alcohols, amines, and organometallic reagents.

5 Propose multistep syntheses using carboxylic acids and acid chlorides as starting materials and intermediates.

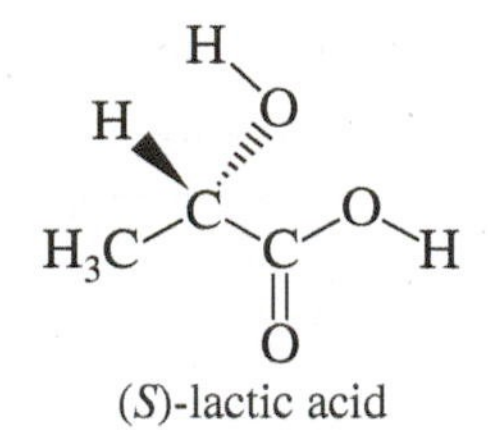

(*S*)-lactic acid

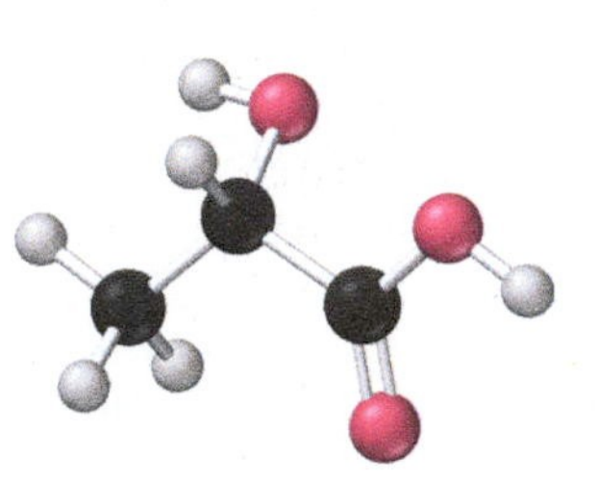

(*S*)-lactic acid

▶ When muscles must generate energy quickly, as in a sprint, they metabolize glucose to (*S*)-lactic acid. The pK_a of lactic acid is 3.86, which implies that the acid is present as its carboxylate ion (lactate ion) at a typical body pH of 7.4. Under intense exertion, the muscles produce lactate faster than it can be removed, so the lactate concentration rises from its usual value of 1–2 mmol/L to as high as 20 mmol/L. For many years, scientists believed that muscle fatigue resulted from the increased concentration of lactic acid.

20-1 Introduction

The combination of a **carb**onyl group and a hydr**oxyl** on the same carbon atom is called a **carboxyl group.** Compounds containing the carboxyl group are distinctly acidic and are called **carboxylic acids.**

$$-\overset{\overset{\displaystyle O}{\|}}{C}-O-H \qquad R-\overset{\overset{\displaystyle O}{\|}}{C}-O-H \qquad R-COOH \quad R-CO_2H$$

carboxyl group — carboxylic acid — condensed structures

Carboxylic acids are classified according to the substituent bonded to the carboxyl group. An **aliphatic acid** has an alkyl group bonded to the carboxyl group, and an **aromatic acid** has an aryl group. The simplest acid is *formic acid*, with a hydrogen atom bonded to the carboxyl group. **Fatty acids** are long-chain aliphatic acids derived from the hydrolysis of fats and oils (Section 20-6).

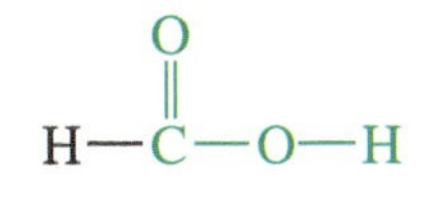

formic acid

$$CH_3-CH_2-\overset{\overset{\displaystyle O}{\|}}{C}-O-H$$

propionic acid (an aliphatic acid)

$$C_6H_5-\overset{\overset{\displaystyle O}{\|}}{C}-O-H$$

benzoic acid (an aromatic acid)

$$CH_3(CH_2)_{16}-\overset{\overset{\displaystyle O}{\|}}{C}-O-H$$

stearic acid (a fatty acid)

A carboxylic acid donates protons by heterolytic cleavage of the acidic O—H bond to give a proton and a **carboxylate ion.** We consider the ranges of acidity and the factors affecting the acidity of carboxylic acids in Section 20-4.

$$\underset{\text{carboxylic acid}}{R-\overset{\overset{\displaystyle O}{\|}}{C}-O-H} + H_2O \rightleftharpoons \underset{\text{carboxylate ion}}{R-\overset{\overset{\displaystyle O}{\|}}{C}-O^-} + H_3O^+$$

20-2 Nomenclature of Carboxylic Acids

As with most organic compounds, carboxylic acids have both common names and systematic (IUPAC) names. Because carboxylic acids have been known for hundreds of years, their common names are firmly entrenched in everyday usage. These names also appear in the names of biomolecules derived from fatty acids. Thus, learning these names is very useful for later study.

20-2A Common Names

Several aliphatic carboxylic acids have been known for centuries, and their common names reflect their historical sources. *Formic acid* was extracted from ants: *formica* in Latin. Acetic acid was isolated from vinegar, called *acetum* ("sour") in Latin. Propionic acid was considered to be the first fatty acid, and the name is derived from the Greek *protos pion* ("first fat"). Butyric acid results from the oxidation of butyraldehyde, the principal flavor of butter: *butyrum* in Latin. Caproic, caprylic, and capric acids are found in the skin secretions of goats: *caper* in Latin. The names and physical properties of some carboxylic acids are listed in Table 20-1.

In common names, the positions of substituents are named using Greek letters. Notice that the lettering begins with the carbon atom *next* to the carboxyl carbon, the α carbon. With common names, the prefix *iso-* is sometimes used for acids ending in the $-CH(CH_3)_2$ grouping.

$$-\underset{\varepsilon}{C}-\underset{\delta}{C}-\underset{\gamma}{C}-\underset{\beta}{C}-\underset{\alpha}{C}-\overset{O}{\overset{\|}{C}}-OH$$

$$\underset{\beta}{CH_3}-\underset{\alpha}{\overset{Cl}{\overset{|}{CH}}}-\overset{O}{\overset{\|}{C}}-OH$$

α-chloropropionic acid

$$\underset{\gamma}{\overset{NH_2}{\overset{|}{CH_2}}}-\underset{\beta}{CH_2}-\underset{\alpha}{CH_2}-\overset{O}{\overset{\|}{C}}-OH$$

γ-aminobutyric acid

$$\underset{\gamma}{CH_3}-\underset{\beta}{\overset{CH_3}{\overset{|}{CH}}}-\underset{\alpha}{CH_2}-\overset{O}{\overset{\|}{C}}-OH$$

isovaleric acid
(β-methylbutyric acid)

TABLE 20-1
Names and Physical Properties of Carboxylic Acids

IUPAC Name	Common Name	Formula	mp (°C)	bp (°C)	Solubility (g 100 g H_2O)
methanoic	formic	$HCOOH$	8	101	∞ (miscible)
ethanoic	acetic	CH_3COOH	17	118	∞
propanoic	propionic	CH_3CH_2COOH	−21	141	∞
prop-2-enoic	acrylic	$H_2C{=}CH{-}COOH$	14	141	∞
butanoic	butyric	$CH_3(CH_2)_2COOH$	−6	163	∞
2-methylpropanoic	isobutyric	$(CH_3)_2CHCOOH$	−46	155	23.0
trans-but-2-enoic	crotonic	$CH_3{-}CH{=}CH{-}COOH$	71	185	8.6
pentanoic	valeric	$CH_3(CH_2)_3COOH$	−34	186	3.7
2,2-dimethylpropanoic	pivalic	$(CH_3)_3C{-}COOH$	35	164	2.5
hexanoic	caproic	$CH_3(CH_2)_4COOH$	−4	206	1.0
octanoic	caprylic	$CH_3(CH_2)_6COOH$	16	240	0.7
decanoic	capric	$CH_3(CH_2)_8COOH$	31	269	0.2
dodecanoic	lauric	$CH_3(CH_2)_{10}COOH$	44		i (insoluble)
tetradecanoic	myristic	$CH_3(CH_2)_{12}COOH$	54		i
hexadecanoic	palmitic	$CH_3(CH_2)_{14}COOH$	63		i
octadecanoic	stearic	$CH_3(CH_2)_{16}COOH$	72		i
benzoic	benzoic	C_6H_5COOH	122	249	0.3

20-2B IUPAC Names

The IUPAC nomenclature for carboxylic acids uses the name of the alkane that corresponds to the longest continuous chain of carbon atoms. The final *-e* in the alkane name is replaced by the suffix *-oic acid*. The chain is numbered, *starting with the carboxyl carbon atom*, to give positions of substituents along the chain. In naming, the carboxyl group takes priority over any of the other functional groups we have discussed.

$$\underset{6}{-\text{C}}-\underset{5}{\text{C}}-\underset{4}{\text{C}}-\underset{3}{\text{C}}-\underset{2}{\text{C}}-\underset{1}{\overset{\text{O}}{\overset{\|}{\text{C}}}}-\text{OH}$$

	$\text{H}-\overset{\text{O}}{\overset{\|}{\text{C}}}-\text{OH}$	$\text{CH}_3-\overset{\text{O}}{\overset{\|}{\text{C}}}-\text{OH}$	$\underset{3}{\text{CH}_3}-\underset{2}{\text{CH}}(\text{C}_6\text{H}_{11})-\underset{1}{\overset{\text{O}}{\overset{\|}{\text{C}}}}-\text{OH}$	$\text{CH}_3\overset{\text{O}}{\overset{\|}{\text{C}}}-\overset{2}{\text{CH}}(\underset{3}{\text{CH}_2}\underset{4}{\text{CH}_2}\underset{5}{\text{CH}_3})-\underset{1}{\overset{\text{O}}{\overset{\|}{\text{C}}}}-\text{OH}$
IUPAC name:	methanoic acid	ethanoic acid	2-cyclohexylpropanoic acid	2-acetylpentanoic acid
common name:	formic acid	acetic acid	α-cyclohexylpropionic acid	α-acetylvaleric acid

	$\underset{4}{\text{CH}_2}(\text{NH}_2)-\underset{3}{\text{CH}_2}-\underset{2}{\text{CH}_2}-\underset{1}{\overset{\text{O}}{\overset{\|}{\text{C}}}}-\text{OH}$	$\underset{5}{\text{CH}_3}-\underset{4}{\text{CH}_2}-\underset{3}{\text{CH}}(\text{Ph})-\underset{2}{\text{CH}_2}-\underset{1}{\overset{\text{O}}{\overset{\|}{\text{C}}}}-\text{OH}$	$\underset{4}{\text{CH}_3}-\underset{3}{\text{CH}}(\text{CH}_3)-\underset{2}{\text{CH}_2}-\underset{1}{\overset{\text{O}}{\overset{\|}{\text{C}}}}-\text{OH}$
IUPAC name:	4-aminobutanoic acid	3-phenylpentanoic acid	3-methylbutanoic acid
common name:	γ-aminobutyric acid	β-phenylvaleric acid	isovaleric acid

Unsaturated acids are named using the name of the corresponding alkene, with the final *-e* replaced by *-oic acid*. The carbon chain is numbered starting with the carboxyl carbon, and a number gives the location of the double bond. The stereochemical terms *cis* and *trans* (and *Z* and *E*) are used as they are with other alkenes. Cycloalkanes with —COOH substituents are generally named as *cycloalkanecarboxylic acids*.

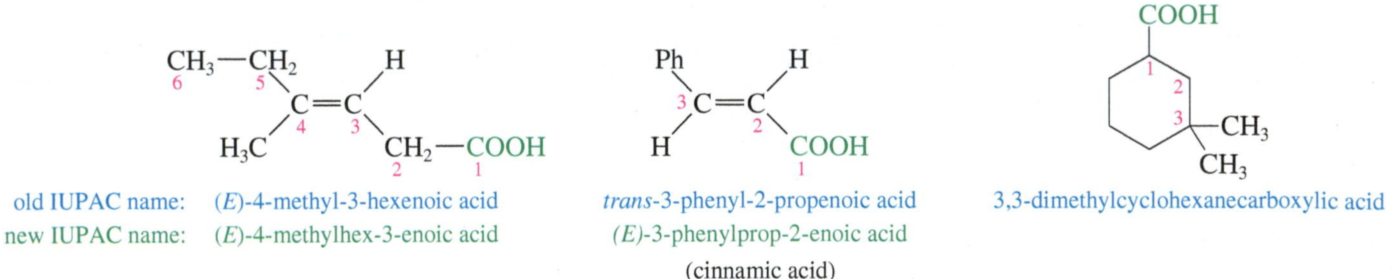

old IUPAC name:	(*E*)-4-methyl-3-hexenoic acid	*trans*-3-phenyl-2-propenoic acid	3,3-dimethylcyclohexanecarboxylic acid
new IUPAC name:	(*E*)-4-methylhex-3-enoic acid	(*E*)-3-phenylprop-2-enoic acid	
		(cinnamic acid)	

Aromatic acids of the form Ar—COOH are named as derivatives of *benzoic acid*, Ph—COOH. As with other aromatic compounds, the prefixes *ortho-*, *meta-*, and *para-* may be used to give the positions of additional substituents. Numbers are used if there are more than two substituents on the aromatic ring. Common names of substituted naphthalenes often use α and β for the 1 and 2 positions, respectively. Many aromatic acids have historical names that are unrelated to their structures.

benzoic acid	*p*-aminobenzoic acid	*o*-hydroxybenzoic acid (salicylic acid)	*p*-methylbenzoic acid (*p*-toluic acid)	α-naphthoic acid

20-2C Nomenclature of Dicarboxylic Acids

Common Names of Dicarboxylic Acids A **dicarboxylic acid** (also called a *diacid*) is a compound with two carboxyl groups. The common names of simple dicarboxylic acids are used more frequently than their systematic names. A common mnemonic for these names is "***O****h* ***m****y,* ***s****uch* ***g****ood* ***a****pple* ***p****ie,*" standing for *o*xalic, *m*alonic, *s*uccinic, *g*lutaric, *a*dipic, and *p*imelic acids. The names and physical properties of some dicarboxylic acids are given in Table 20-2.

Substituted dicarboxylic acids are given common names using Greek letters, as with the simple carboxylic acids. Greek letters are assigned beginning with the carbon atom next to the carboxyl group that is closer to the substituents.

$HO{-}C(=O){-}\underset{\alpha}{CH_2}{-}\underset{\beta}{CH}(Br){-}CH_2{-}CH_2{-}C(=O){-}OH$

β-bromoadipic acid

$HO{-}C(=O){-}\underset{\alpha}{CH}(CH_3){-}\underset{\beta}{CH}(Ph){-}CH_2{-}C(=O){-}OH$

α-methyl-β-phenylglutaric acid

Benzenoid compounds with two carboxyl groups are named **phthalic acids.** *Phthalic acid* itself is the ortho isomer. The meta isomer is called *isophthalic acid*, and the para isomer is called *terephthalic acid.*

o-phthalic acid
phthalic acid

m-phthalic acid
isophthalic acid

p-phthalic acid
terephthalic acid

IUPAC Names of Dicarboxylic Acids Aliphatic dicarboxylic acids are named simply by adding the suffix *-dioic acid* to the name of the parent alkane. For straight-chain dicarboxylic acids, the parent alkane name is determined by using the longest continuous chain that contains both carboxyl groups. The chain is numbered beginning with the carboxyl carbon atom that is closer to the substituents, and these numbers are used to give the positions of the substituents.

$HO{-}\underset{1}{C}(=O){-}\underset{2}{CH_2}{-}\underset{3}{CH}(Br){-}\underset{4}{CH_2}{-}\underset{5}{CH_2}{-}\underset{6}{C}(=O){-}OH$

3-bromohexanedioic acid

$HO{-}\underset{1}{C}(=O){-}\underset{2}{CH}(CH_3){-}\underset{3}{CH}(Ph){-}\underset{4}{CH_2}{-}\underset{5}{C}(=O){-}OH$

2-methyl-3-phenylpentanedioic acid

TABLE 20-2
Names and Physical Properties of Dicarboxylic Acids

IUPAC Name	Common Name	Formula	mp (°C)	Solubility (g 100 g H_2O)
ethanedioic	oxalic	HOOC—COOH	189 decomposes	14
propanedioic	malonic	$HOOCCH_2COOH$	136	74
butanedioic	succinic	$HOOC(CH_2)_2COOH$	185	8
pentanedioic	glutaric	$HOOC(CH_2)_3COOH$	98	64
hexanedioic	adipic	$HOOC(CH_2)_4COOH$	151	2
heptanedioic	pimelic	$HOOC(CH_2)_5COOH$	106	5
cis-but-2-enedioic	maleic	*cis*-HOOCCH=CHCOOH	130.5	79
trans-but-2-enedioic	fumaric	*trans*-HOOCCH=CHCOOH	302 sublimes	0.7
benzene-1,2-dicarboxylic	phthalic	$1,2\text{-}C_6H_4(COOH)_2$	231	0.7
benzene-1,3-dicarboxylic	isophthalic	$1,3\text{-}C_6H_4(COOH)_2$	348	0.54
benzene-1,4-dicarboxylic	terephthalic	$1,4\text{-}C_6H_4(COOH)_2$	300 sublimes	0.002

The system for naming cyclic dicarboxylic acids treats the carboxyl groups as substituents on the cyclic structure.

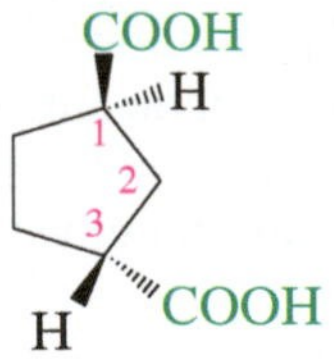

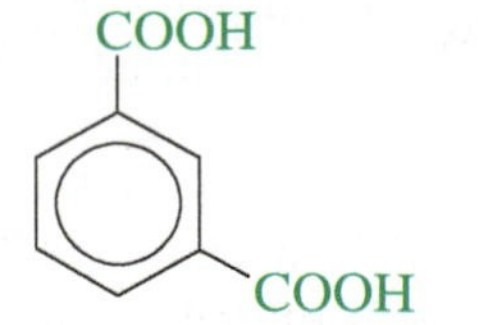

old IUPAC name: *trans*-1,3-cyclopentanedicarboxylic acid — 1,3-benzenedicarboxylic acid
new IUPAC name: (1*R*,3*R*)-cyclopentane-1,3-dicarboxylic acid — benzene-1,3-dicarboxylic acid

PROBLEM-SOLVING HINT

The nomenclature of carboxylic acids is reviewed in Appendix 5, the Summary of Organic Nomenclature.

PROBLEM 20-1

Draw the structures of the following carboxylic acids.

(a) α-methylbutyric acid
(b) 2-bromobutanoic acid
(c) 4-aminopentanoic acid
(d) *cis*-4-phenylbut-2-enoic acid
(e) *trans*-2-methylcyclohexanecarboxylic acid
(f) 2,3-dimethylfumaric acid
(g) *m*-chlorobenzoic acid
(h) 3-methylphthalic acid
(i) β-aminoadipic acid
(j) 3-chloroheptanedioic acid
(k) 4-oxoheptanoic acid
(l) phenylacetic acid

PROBLEM 20-2

Name the following carboxylic acids (when possible, give both a common name and a systematic name).

(a) I, COOH, CH_3
(b) CH_2CH_3, COOH
(c) COOH, NO_2, NO_2
(d) COOH, COOH
(e) COOH, HOOC, Cl
(f) COOH, COOH, H_3C

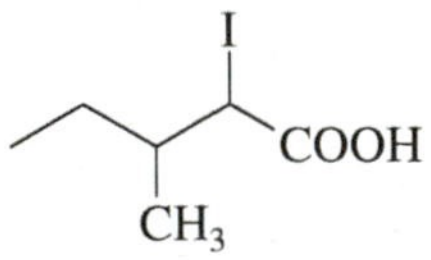

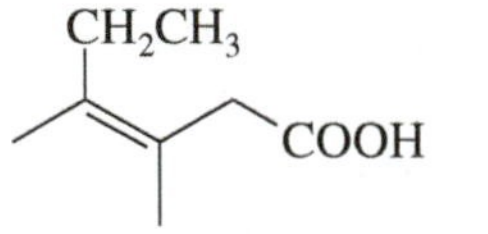

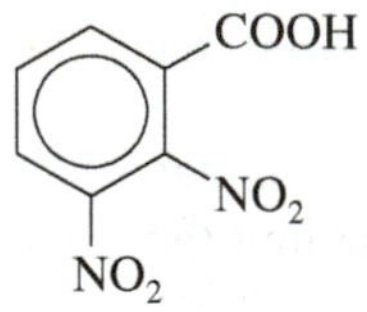

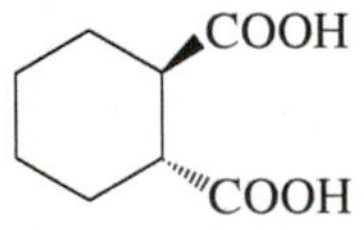

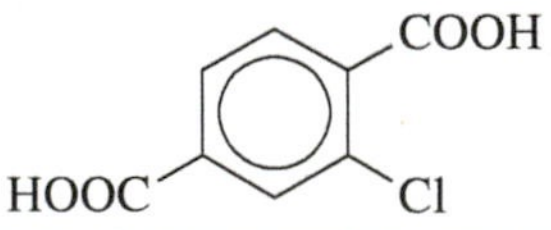

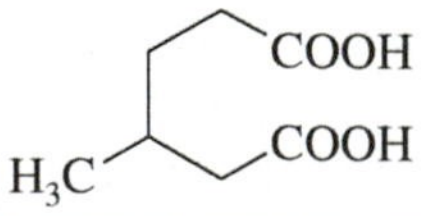

20-3 Structure and Physical Properties of Carboxylic Acids

Structure of the Carboxyl Group The structure of the most stable conformation of formic acid is shown next. The entire molecule is approximately planar. The sp^2 hybrid carbonyl carbon atom is planar, with nearly trigonal bond angles. The O—H bond also lies in this plane, eclipsed with the C=O bond.

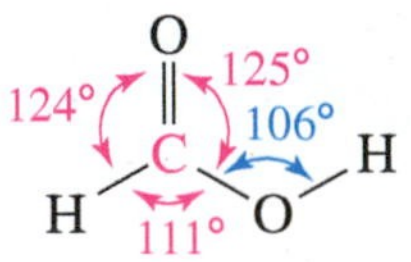

124°, 125°, 106°, 111°
bond angles

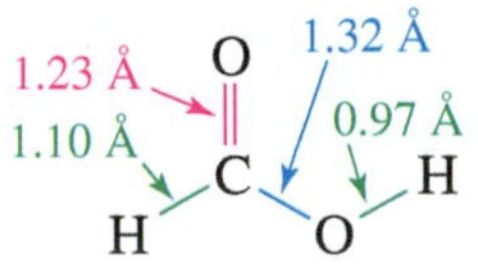

1.23 Å, 1.32 Å, 1.10 Å, 0.97 Å
bond lengths

It seems surprising that an eclipsed conformation is most stable. It appears that one of the unshared electron pairs on the hydroxy oxygen atom is delocalized into the electrophilic pi system of the carbonyl group. We can draw the following resonance forms to represent this delocalization:

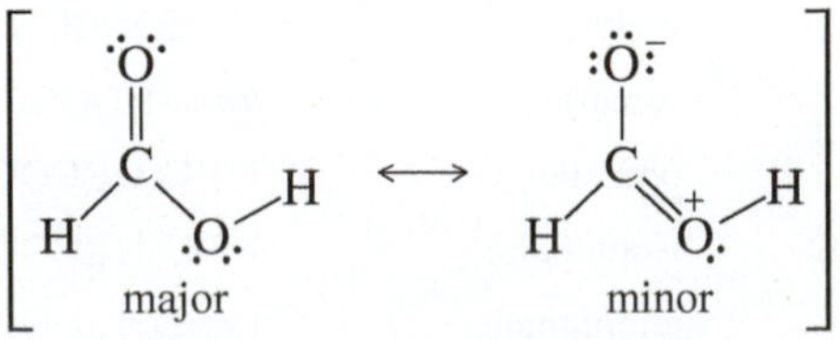

major ⟷ minor

Boiling Points Carboxylic acids boil at considerably higher temperatures than do alcohols, ketones, or aldehydes of similar molecular weights. For example, acetic acid (MW 60) boils at 118 °C, propan-1-ol (MW 60) boils at 97 °C, and propionaldehyde (MW 58) boils at 49 °C.

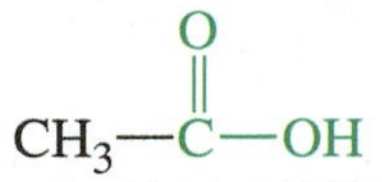

$$CH_3-\overset{\overset{\displaystyle O}{\|}}{C}-OH \qquad CH_3-CH_2-CH_2-OH \qquad CH_3-CH_2-\overset{\overset{\displaystyle O}{\|}}{C}-H$$

acetic acid, bp 118 °C 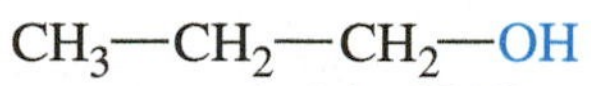 — propan-1-ol, bp 97 °C 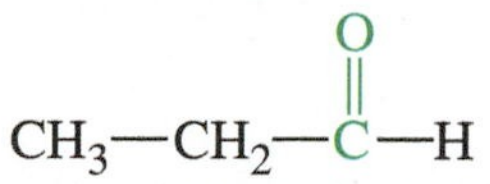 — propionaldehyde, bp 49 °C

The high boiling points of carboxylic acids result from formation of a stable, hydrogen-bonded dimer. This dimer contains an eight-membered ring joined by two hydrogen bonds, effectively doubling the molecular weight of the molecules leaving the liquid phase, and requiring more energy (higher temperature) to boil.

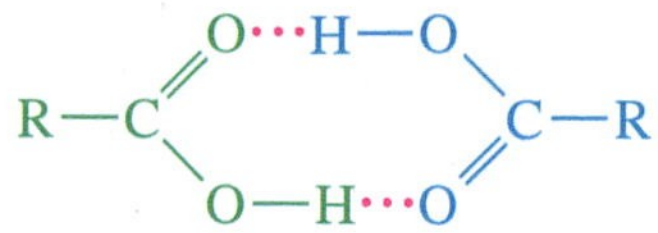

$$R-C\begin{matrix} /\!\!/ O\cdots H-O \backslash \\ \backslash O-H\cdots O /\!\!/ \end{matrix}C-R$$

hydrogen-bonded acid dimer

Melting Points The melting points of some common carboxylic acids are given in Table 20-1. Acids containing more than eight carbon atoms are generally solids, unless they contain double bonds. The presence of double bonds (especially cis double bonds) in a long chain impedes formation of a stable crystal lattice, resulting in a lower melting point. For example, both stearic acid (octadecanoic acid) and linoleic acid (*cis,cis*-octadeca-9,12-dienoic acid) have 18 carbon atoms, but stearic acid melts at 70 °C and linoleic acid melts at −5 °C.

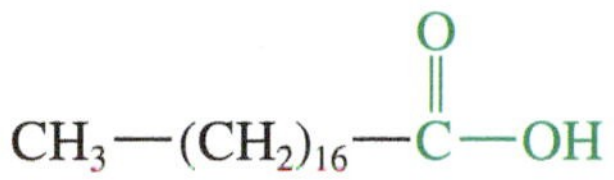

$$CH_3-(CH_2)_{16}-\overset{\overset{\displaystyle O}{\|}}{C}-OH$$

stearic acid, mp 70 °C

$$CH_3(CH_2)_4-CH=CH-CH_2-CH=CH-(CH_2)_7-\overset{\overset{\displaystyle O}{\|}}{C}-OH \quad \text{(both C=C cis)}$$

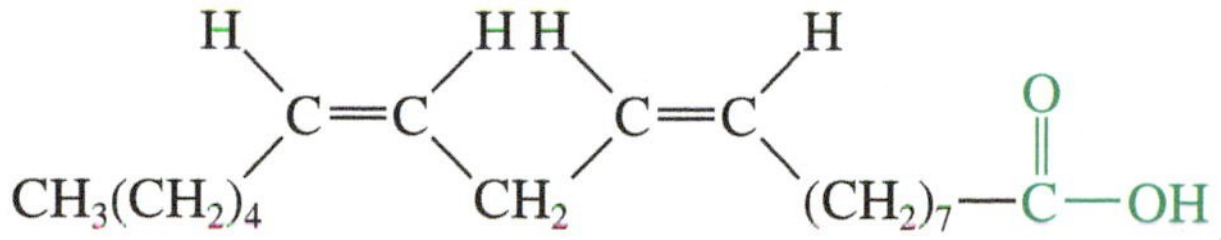

linoleic acid, mp −5 °C

The melting points of dicarboxylic acids (Table 20-2) are relatively high. With two carboxyl groups per molecule, the forces of hydrogen bonding are particularly strong in diacids; a high temperature is required to break the lattice of hydrogen bonds in the crystal and melt the diacid.

Solubilities Carboxylic acids form hydrogen bonds with water, and the lower-molecular-weight acids (up through four carbon atoms) are miscible with water. As the length of the hydrocarbon chain increases, water solubility decreases until acids with more than 10 carbon atoms are nearly insoluble in water. The water solubilities of some simple carboxylic acids and diacids are given in Tables 20-1 and 20-2.

Carboxylic acids are very soluble in alcohols because the acids form hydrogen bonds with alcohols. Also, alcohols are not as polar as water, so the longer-chain acids are more soluble in alcohols than they are in water. Most carboxylic acids are quite soluble in relatively nonpolar solvents such as chloroform because the acid continues to exist in its dimeric form in the nonpolar solvent. Thus, the hydrogen bonds of the cyclic dimer are not disrupted when the acid dissolves in a nonpolar solvent.

20-4 Acidity of Carboxylic Acids

Many of the properties and reactions of carboxylic acids relate to their acidity. Chapter 2 covered a wide variety of factors that influence the strength of acids in general. Here, we emphasize the factors that affect the relative acidity of carboxylic acids and substituted carboxylic acids.

20-4A Measurement of Acidity

A carboxylic acid may dissociate in water to give a proton and a carboxylate ion. The equilibrium constant K_a for this reaction is called the *acid-dissociation constant*. The pK_a of an acid is the negative logarithm of K_a, and we commonly use pK_a as an indication of the relative acidities of different acids (Table 20-3).

TABLE 20-3
Values of K_a and pK_a for Carboxylic Acids and Dicarboxylic Acids

Formula	Name	Values			
		Simple carboxylic acids			
		K_a (at 25 °C)	pK_a		
HCOOH	formic acid	1.77×10^{-4}	3.75		
CH_3COOH	acetic acid	1.76×10^{-5}	4.74		
CH_3CH_2COOH	propionic acid	1.34×10^{-5}	4.87		
$CH_3(CH_2)_2COOH$	butyric acid	1.54×10^{-5}	4.82		
$CH_3(CH_2)_3COOH$	pentanoic acid	1.52×10^{-5}	4.81		
$CH_3(CH_2)_4COOH$	hexanoic acid	1.31×10^{-5}	4.88		
$CH_3(CH_2)_6COOH$	octanoic acid	1.28×10^{-5}	4.89		
$CH_3(CH_2)_8COOH$	decanoic acid	1.43×10^{-5}	4.84		
C_6H_5COOH	benzoic acid	6.46×10^{-5}	4.19		
p-$CH_3C_6H_4COOH$	*p*-toluic acid	4.33×10^{-5}	4.36		
p-ClC_6H_4COOH	*p*-chlorobenzoic acid	1.04×10^{-4}	3.98		
p-$NO_2C_6H_4COOH$	*p*-nitrobenzoic acid	3.93×10^{-4}	3.41		
		Dicarboxylic acids			
		K_{a1}	pK_{a1}	K_{a2}	pK_{a2}
HOOC—COOH	oxalic	5.4×10^{-2}	1.27	5.2×10^{-5}	4.28
$HOOCCH_2COOH$	malonic	1.4×10^{-3}	2.85	2.0×10^{-6}	5.70
$HOOC(CH_2)_2COOH$	succinic	6.4×10^{-5}	4.19	2.3×10^{-6}	5.64
$HOOC(CH_2)_3COOH$	glutaric	4.5×10^{-5}	4.35	3.8×10^{-6}	5.42
$HOOC(CH_2)_4COOH$	adipic	3.7×10^{-5}	4.43	3.9×10^{-6}	5.41
cis-HOOCCH=CHCOOH	maleic	1.0×10^{-2}	2.00	5.5×10^{-7}	6.26
trans-HOOCCH=CHCOOH	fumaric	9.6×10^{-4}	3.02	4.1×10^{-5}	4.39
1,2-$C_6H_4(COOH)_2$	phthalic	1.1×10^{-3}	2.96	4.0×10^{-6}	5.40
1,3-$C_6H_4(COOH)_2$	isophthalic	2.4×10^{-4}	3.62	2.5×10^{-5}	4.60
1,4-$C_6H_4(COOH)_2$	terephthalic	2.9×10^{-4}	3.54	3.5×10^{-5}	4.46

$$R-\overset{\overset{\displaystyle O}{\|}}{C}-O-H + H_2O \rightleftharpoons R-\overset{\overset{\displaystyle O}{\|}}{C}-O^- + H_3O^+$$

$$K_a = \frac{[R-CO_2^-][H_3O^+]}{[R-CO_2H]}$$

$$pK_a = -\log_{10} K_a$$

Values of pK_a are about 5 ($K_a = 10^{-5}$) for simple carboxylic acids. For example, acetic acid has a pK_a of 4.7 ($K_a = 1.8 \times 10^{-5}$). Although carboxylic acids are not as strong as most mineral acids, they are still much more acidic than other functional groups we have studied. For example, alcohols have pK_a values in the range 16 to 18. Acetic acid (pK_a = 4.74) is about 10^{11} times as acidic as the most acidic alcohols! In fact, concentrated acetic acid causes acid burns when it comes into contact with the skin.

Dissociation of either an acid or an alcohol involves breaking an O—H bond, but dissociation of a carboxylic acid gives a carboxylate ion with the negative charge spread out equally over *two* oxygen atoms, compared with just one oxygen in an alkoxide ion (Figure 20-1). This charge delocalization makes the carboxylate ion more stable than the alkoxide ion; therefore, dissociation of a carboxylic acid to a carboxylate ion is less endothermic than dissociation of an alcohol to an alkoxide ion.

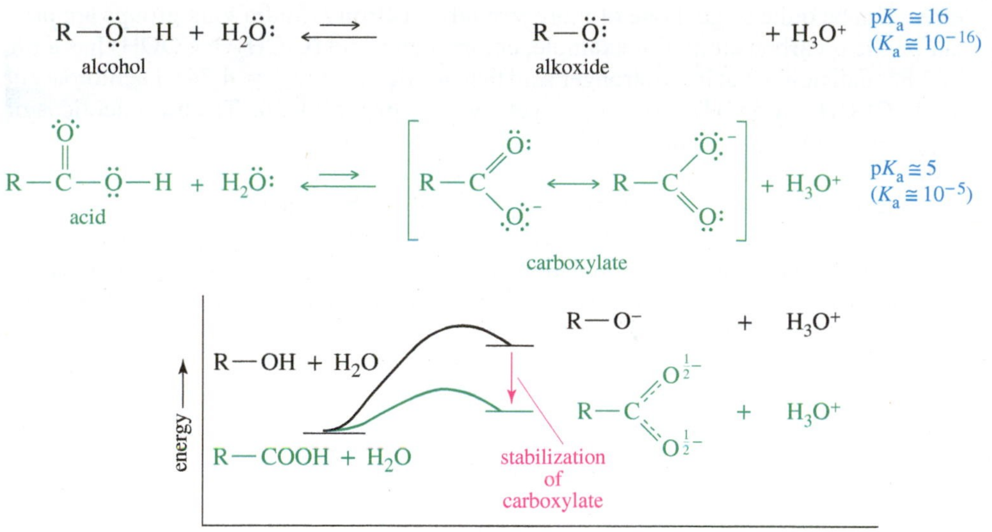

$$\text{R—O—H} + \text{H}_2\text{O} \rightleftharpoons \text{R—O}^- + \text{H}_3\text{O}^+ \qquad \text{p}K_a \cong 16 \; (K_a \cong 10^{-16})$$

alcohol — alkoxide

$$\text{R—C(=O)—O—H} + \text{H}_2\text{O} \rightleftharpoons [\text{R—C(=O)O}^- \longleftrightarrow \text{R—C(O}^-\text{)=O}] + \text{H}_3\text{O}^+ \qquad \text{p}K_a \cong 5 \; (K_a \cong 10^{-5})$$

acid — carboxylate

R—OH + H_2O → R—O^- + H_3O^+

R—COOH + H_2O → R—C($O^{\frac{1}{2}-}$)($O^{\frac{1}{2}-}$) + H_3O^+

energy

stabilization of carboxylate

FIGURE 20-1
Stability of carboxylate ions. Carboxylic acids are more acidic than alcohols because carboxylate ions are more stable than alkoxide ions. A carboxylate ion has its negative charge delocalized over two oxygen atoms, compared with only one oxygen atom bearing the negative charge in an alkoxide ion.

The carboxylate ion can be visualized either as a resonance hybrid (as in Figure 20-1) or as a conjugated system of three *p* orbitals containing four electrons. The carbon atom and the two oxygen atoms are sp^2 hybridized, and each has an unhybridized *p* orbital. Overlap of these three *p* orbitals gives a three-center π molecular orbital system. There is half of a π bond between the carbon and each oxygen atom, and there is half of a negative charge on each oxygen atom (Figure 20-2).

Table 20-3 lists pK_a values for dicarboxylic acids in addition to those for simple carboxylic acids. Diacids have two dissociation constants: K_{a1} is for the first dissociation, and K_{a2} is for the second dissociation, to give a dianion. The second carboxyl group is much less acidic than the first ($K_{a2} \ll K_{a1}$) because extra energy is required to create a second negative charge close to another, mutually repulsive, negative charge. This repulsive effect decreases as the chain gets longer.

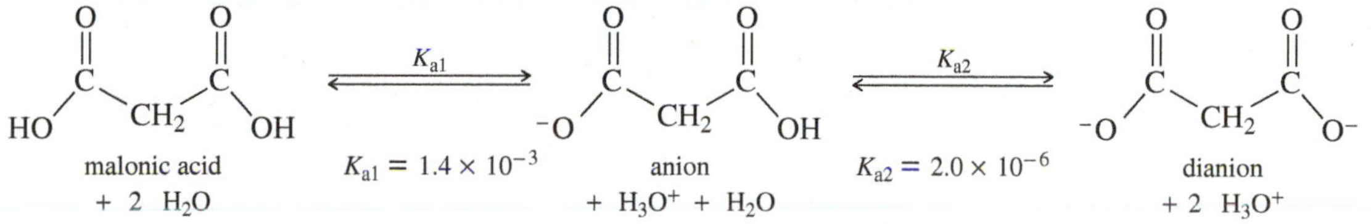

$$\text{HOOC—CH}_2\text{—COOH} \xrightleftharpoons{K_{a1}} {}^-\text{OOC—CH}_2\text{—COOH} \xrightleftharpoons{K_{a2}} {}^-\text{OOC—CH}_2\text{—COO}^-$$

malonic acid + 2 H_2O — $K_{a1} = 1.4 \times 10^{-3}$ — anion + H_3O^+ + H_2O — $K_{a2} = 2.0 \times 10^{-6}$ — dianion + 2 H_3O^+

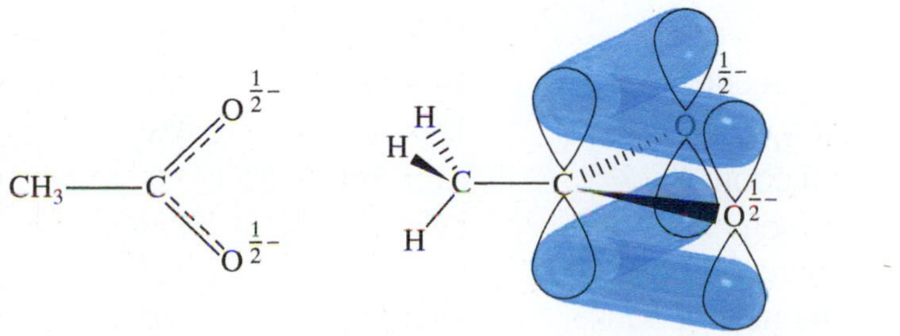

CH_3—C($O^{\frac{1}{2}-}$)($O^{\frac{1}{2}-}$)

FIGURE 20-2
Structure of the acetate ion. Each C—O bond has a bond order of $\frac{3}{2}$ from one σ bond and half a π bond. Each oxygen atom bears half of the negative charge.

20-4B Substituent Effects on Acidity

Any substituent that stabilizes the negatively charged carboxylate ion promotes dissociation and results in a stronger acid. Electronegative atoms enhance the strength of an acid by withdrawing electron density from the carboxylate ion. This inductive

PROBLEM-SOLVING HINT

These effects on acidity were covered in more detail in Sections 2-9 through 2-12. You may want to review those sections if any of this material seems unfamiliar.

effect can be quite large if one or more strongly electron-withdrawing groups are present on the α carbon atom. For example, chloroacetic acid ($ClCH_2—COOH$) has a pK_a of 2.86, indicating that it is a stronger acid than acetic acid (pK_a = 4.74). Dichloroacetic acid ($Cl_2CH—COOH$) is stronger yet, with a pK_a of 1.26. Trichloroacetic acid ($Cl_3C—COOH$) has a pK_a of 0.64, comparable in strength to some mineral acids. Table 20-4 lists values of K_a and pK_a for some substituted carboxylic acids, showing how electron-withdrawing groups enhance the strength of an acid.

The magnitude of a substituent effect depends on its distance from the carboxyl group. Substituents on the α carbon atom are most effective in increasing acid strength. More distant substituents have smaller effects on acidity, showing that inductive effects decrease rapidly with distance.

$H—CH_2—C(=O)—O—H$ acetic acid pK_a = 4.74

$Cl—CH_2—C(=O)—O—H$ chloroacetic acid pK_a = 2.86

$Cl—CHCl—C(=O)—O—H$ dichloroacetic acid pK_a = 1.26

$Cl—CCl_2—C(=O)—O—H$ trichloroacetic acid pK_a = 0.64

stronger acids →

$ClCH_2—CH_2—CH_2—C(=O)—OH$ 4-chlorobutanoic acid pK_a = 4.52

$CH_3—CHCl—CH_2—C(=O)—OH$ 3-chlorobutanoic acid pK_a = 4.05

$CH_3—CH_2—CHCl—C(=O)—OH$ 2-chlorobutanoic acid pK_a = 2.86

Substituted benzoic acids show similar trends in acidity, with electron-withdrawing groups enhancing the acid strength and electron-donating groups decreasing the acid strength. These effects are strongest for substituents in the ortho and para positions.

PROBLEM-SOLVING HINT

Electron-withdrawing groups enhance acidity because they help to stabilize the negative charge of the conjugate base (the carboxylate ion).

The amount of stabilization depends on:

1. the number of electron-withdrawing groups;
2. the strength of the electron-withdrawing groups; and
3. the distance of the electron-withdrawing groups from the COOH group.

TABLE 20-4
Values of K_a and pK_a for Substituted Carboxylic Acids

Acid	K_a	pK_a
F_3CCOOH	5.9×10^{-1}	0.23
Cl_3CCOOH	2.3×10^{-1}	0.64
$Cl_2CHCOOH$	5.5×10^{-2}	1.26
$O_2N—CH_2COOH$	2.1×10^{-2}	1.68
$NCCH_2COOH$	3.4×10^{-3}	2.46
FCH_2COOH	2.6×10^{-3}	2.59
$ClCH_2COOH$	1.4×10^{-3}	2.86
$CH_3CH_2CHClCOOH$	1.4×10^{-3}	2.86
$BrCH_2COOH$	1.3×10^{-3}	2.90
ICH_2COOH	6.7×10^{-4}	3.18
CH_3OCH_2COOH	2.9×10^{-4}	3.54
$HOCH_2COOH$	1.5×10^{-4}	3.83
$CH_3CHClCH_2COOH$	8.9×10^{-5}	4.05
PhCOOH	6.46×10^{-5}	4.19
$PhCH_2COOH$	4.9×10^{-5}	4.31
$ClCH_2CH_2CH_2COOH$	3.0×10^{-5}	4.52
CH_3COOH	1.8×10^{-5}	4.74
$CH_3CH_2CH_2COOH$	1.5×10^{-5}	4.82

stronger acids ↑

In the examples shown below, notice that a nitro substituent (electron-withdrawing) increases the strength of the acid, whereas a methoxy substituent (electron-donating) decreases the acid strength. The nitro group has a larger effect in the ortho and para positions than in the meta position.

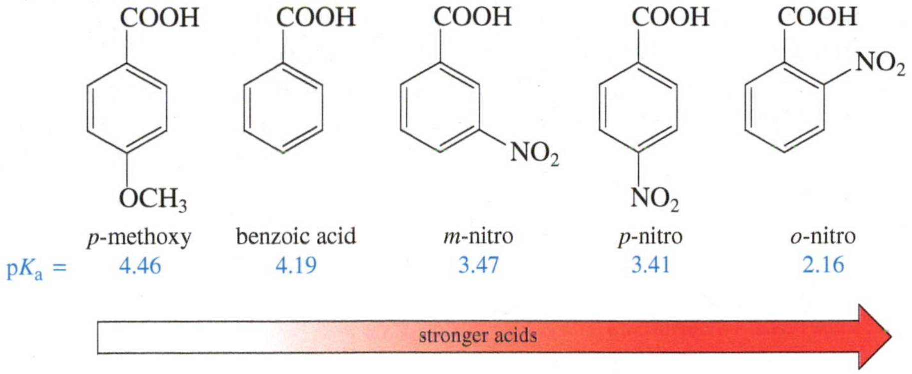

	p-methoxy	benzoic acid	m-nitro	p-nitro	o-nitro
pK_a =	4.46	4.19	3.47	3.41	2.16

stronger acids →

PROBLEM 20-3

Rank the compounds in each set in order of increasing acid strength.

(a) CH_3CH_2COOH $\quad CH_3CHBrCOOH$ $\quad CH_3CBr_2COOH$

(b) $CH_3CH_2CH_2CHBrCOOH$ $\quad CH_3CH_2CHBrCH_2COOH$ $\quad CH_3CHBrCH_2CH_2COOH$

(c) $CH_3CH(NO_2)COOH$ $\quad CH_3CH(Cl)COOH$ $\quad CH_3CH_2COOH$ $\quad CH_3CH(C{\equiv}N)COOH$

20-5 Salts of Carboxylic Acids

A strong base can completely deprotonate a carboxylic acid. The products are a carboxylate ion, the cation remaining from the base, and water. The combination of a carboxylate ion and a cation is a **salt of a carboxylic acid.**

$$\underset{\text{carboxylic acid}}{R{-}C({=}O){-}O{-}H} + \underset{\text{strong base}}{M^+\ {}^-OH} \rightleftharpoons \underset{\text{acid salt}}{R{-}C({=}O){-}O^-\ M^+} + \underset{\text{water}}{H_2O}$$

For example, sodium hydroxide deprotonates acetic acid to form sodium acetate, the sodium salt of acetic acid.

$$\underset{\text{acetic acid}}{CH_3{-}C({=}O){-}O{-}H} + \underset{\text{sodium hydroxide}}{Na^+\ {}^-OH} \rightleftharpoons \underset{\text{sodium acetate}}{CH_3{-}C({=}O){-}O^-\ Na^+} + H_2O$$

Because mineral acids are stronger than carboxylic acids, addition of a mineral acid converts a carboxylic acid salt back to the original carboxylic acid.

$$\underset{\text{acid salt}}{R{-}C({=}O){-}O^-\ M^+} + H^+ \rightleftharpoons \underset{\text{regenerated acid}}{R{-}C({=}O){-}O{-}H} + M^+$$

Example

$$\underset{\text{sodium acetate}}{CH_3{-}C({=}O){-}O^-\ Na^+} + H^+Cl^- \rightleftharpoons \underset{\text{acetic acid}}{CH_3{-}C({=}O){-}OH} + Na^+Cl^-$$

Application: Oral Drugs

The absorption of many orally administered drugs containing carboxylic acids depends on their pK_a values. Aspirin ($pK_a = 3.5$), for example, is largely absorbed from the acidic environment of the stomach (pH = 1 to 3) because it is present as the acid, which readily passes through the membranes into the blood.

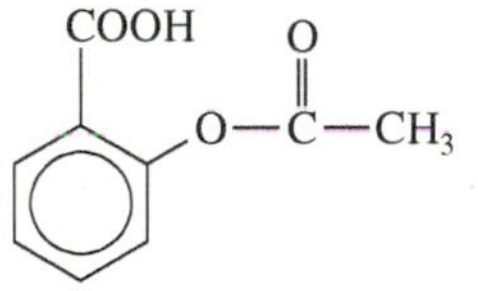

aspirin

Aspirin suppresses blood clotting and causes capillaries in the stomach lining to leak. Some manufacturers have added buffering agents (MgO and $CaCO_3$), hoping to delay the absorption of aspirin until it reaches the small intestine.

PROBLEM-SOLVING HINT

In an aqueous solution, an acid will be mostly dissociated if the pH is above (more basic than) the acid's pK_a, and mostly undissociated if the pH is below (more acidic than) the acid's pK_a.

Carboxylic acid salts have very different properties from the acids, including enhanced solubility in water and less odor. Because acids and their salts are easily interconverted, these salts serve as useful derivatives of carboxylic acids.

Nomenclature of Carboxylic Acid Salts Salts of carboxylic acids are named simply by naming the cation and then naming the carboxylate ion by replacing the *-ic acid* part of the acid name with *-ate*. The preceding example shows that sodium hydroxide reacts with acetic acid to form sodium acetate. The following examples show the formation and nomenclature of some other salts:

$$CH_3CH_2CH_2CH_2{-}\overset{\overset{\displaystyle O}{\|}}{C}{-}OH \quad + \quad LiOH \longrightarrow CH_3CH_2CH_2CH_2{-}\overset{\overset{\displaystyle O}{\|}}{C}{-}O^- \; Li^+$$

IUPAC name:	pentanoic acid	lithium hydroxide	lithium pentanoate
common name:	valeric acid		lithium valerate

$$CH_3CH_2CH_2{-}\overset{\overset{\displaystyle O}{\|}}{C}{-}OH \quad + \quad :NH_3 \longrightarrow CH_3CH_2CH_2{-}\overset{\overset{\displaystyle O}{\|}}{C}{-}O^- \; NH_4^+$$

IUPAC name:	butanoic acid	ammonia	ammonium butanoate
common name:	butyric acid		ammonium butyrate

Application: Antiseptic

Thimerosal (Merthiolate) has been used as a topical antiseptic for many years. Its antimicrobial effect is mostly due to the toxicity of the mercury atom that is bound and stabilized by the thiol group of *ortho*-mercaptobenzoic acid. The carboxylate salt of the acid is used to enhance solubility of this organomercurial compound.

Thimerosal is also used as a preservative in some contact lens solutions. Sensitivity to thimerosal, and concerns about chronic exposure to mercury, have encouraged the development of thimerosal-free antiseptics and contact-lens solutions.

COONa
$SHgCH_2CH_3$

thimerosal

Properties of Acid Salts Like the salts of amines (Section 19-7), carboxylic acid salts are solids with little odor. They generally melt at high temperatures, and they often decompose before reaching their melting points. Carboxylate salts of the alkali metals (Li^+, Na^+, K^+) and ammonium (NH_4^+) carboxylates are generally soluble in water but relatively insoluble in nonpolar organic solvents. *Soap* is a common example of carboxylate salts, consisting of the soluble sodium salts of long-chain fatty acids (Chapter 25). Carboxylate salts of most other metal ions are insoluble in water. For example, when soap is used in "hard" water containing calcium, magnesium, or iron ions, the insoluble carboxylate salts precipitate out as "hard-water scum."

$$\underset{\text{a soap}}{2\,CH_3(CH_2)_{16}{-}\overset{\overset{\displaystyle O}{\|}}{C}{-}O^-\;Na^+} + Ca^{2+} \longrightarrow \underset{\text{"hard-water scum"}}{[CH_3(CH_2)_{16}{-}\overset{\overset{\displaystyle O}{\|}}{C}{-}O]_2Ca\downarrow} + 2\,Na^+$$

Salt formation can be used to identify and purify acids. Carboxylic acids are deprotonated by the weak base sodium bicarbonate, forming the sodium salt of the acid, carbon dioxide, and water. An unknown compound that is insoluble in water, but dissolves in a sodium bicarbonate solution with a release of bubbles of carbon dioxide, is almost certainly a carboxylic acid.

$$\underset{\text{insoluble in water}}{R{-}\overset{\overset{\displaystyle O}{\|}}{C}{-}O{-}H} + NaHCO_3 \rightleftharpoons \underset{\text{water soluble}}{R{-}\overset{\overset{\displaystyle O}{\|}}{C}{-}O^-\;Na^+} + H_2O + CO_2\uparrow$$

Some purification methods take advantage of the different solubilities of acids and their salts. Nonacidic (or weakly acidic) impurities can be removed from a carboxylic acid using acid–base extractions (Figure 20-3). First, the acid is dissolved in an organic solvent such as ether and shaken with water. The acid remains in the organic phase while any water-soluble impurities are washed out. Next, the acid is washed with aqueous sodium bicarbonate, forming a salt that dissolves in the aqueous phase. Nonacidic impurities (and weakly acidic impurities such as phenols) remain in the ether phase. The phases are separated, and acidification of the aqueous phase regenerates the acid, which is insoluble in water but dissolves in a fresh portion of ether. Evaporation of the final ether layer gives the purified acid.

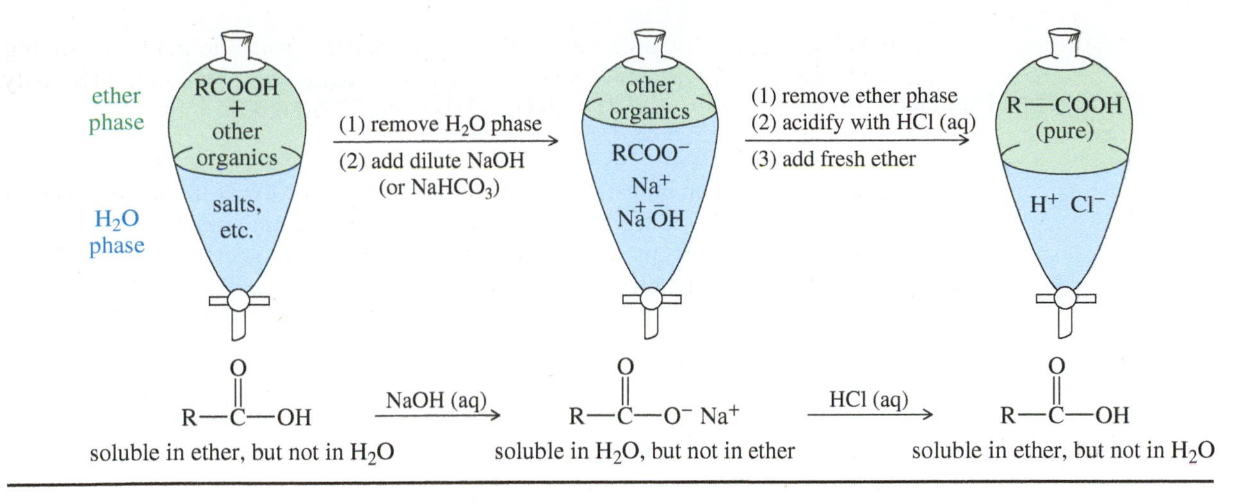

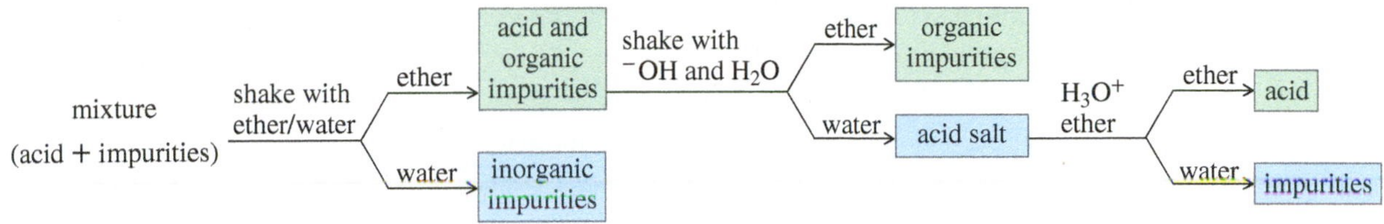

FIGURE 20-3
The solubility properties of acids and their salts may be used to remove nonacidic impurities. A carboxylic acid is more soluble in the organic phase, but its salt is more soluble in the aqueous phase. Acid–base extractions can move the acid from the ether phase into a basic aqueous phase and back into the ether phase, leaving impurities behind.

PROBLEM 20-4

Suppose you have just synthesized heptanoic acid from heptan-1-ol. The product is contaminated by sodium dichromate, sulfuric acid, heptan-1-ol, and possibly heptanal. Explain how you would use acid–base extractions to purify the heptanoic acid. Use a chart, like that in Figure 20-3, to show where the impurities are at each stage.

PROBLEM 20-5

Phenols are less acidic than carboxylic acids, with values of pK_a around 10. Phenols are deprotonated by (and therefore soluble in) solutions of sodium hydroxide but not by solutions of sodium bicarbonate. Explain how you would use extractions to isolate the three pure compounds from a mixture of *p*-cresol (*p*-methylphenol), cyclohexanone, and benzoic acid.

PROBLEM 20-6

Oxidation of a primary alcohol to an aldehyde usually gives some over-oxidation to the carboxylic acid. Assume you have used PCC to oxidize pentan-1-ol to pentanal.

(a) Show how you would use acid–base extraction to purify the pentanal.
(b) Which of the expected impurities cannot be removed from pentanal by acid–base extractions? How would you remove this impurity?

20-6 Commercial Sources of Carboxylic Acids

The most important commercial aliphatic acid is acetic acid. *Vinegar* is a 5% aqueous solution of acetic acid used in cooking and in prepared foods such as pickles, ketchup, and salad dressings. Vinegar for food is produced by fermentation of sugars and starches. An intermediate in this fermentation is ethyl alcohol. When alcoholic beverages such as wine and cider are exposed to air, the alcohol oxidizes to acetic acid. This is the source of "wine vinegar" and "cider vinegar."

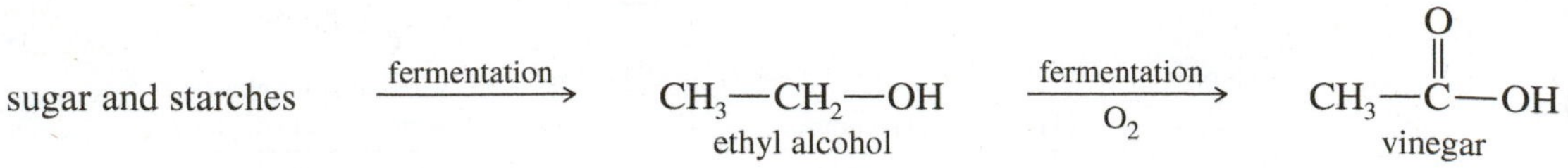

$$\text{sugar and starches} \xrightarrow{\text{fermentation}} \underset{\text{ethyl alcohol}}{CH_3—CH_2—OH} \xrightarrow[O_2]{\text{fermentation}} \underset{\text{vinegar}}{CH_3—C(=O)—OH}$$

Acetic acid is also an industrial chemical. It serves as a solvent, a starting material for synthesis, and a catalyst for a wide variety of reactions. Some industrial acetic acid is produced from ethylene, using a catalytic oxidation to form acetaldehyde, followed by another catalytic oxidation to acetic acid.

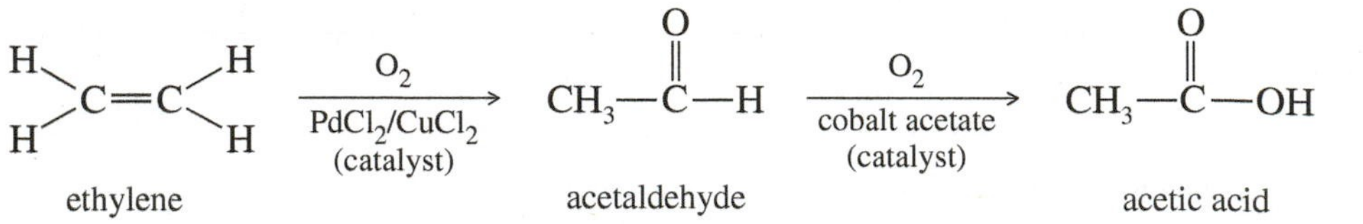

$$\underset{\text{ethylene}}{H_2C=CH_2} \xrightarrow[\substack{PdCl_2/CuCl_2\\ \text{(catalyst)}}]{O_2} \underset{\text{acetaldehyde}}{CH_3—C(=O)—H} \xrightarrow[\substack{\text{cobalt acetate}\\ \text{(catalyst)}}]{O_2} \underset{\text{acetic acid}}{CH_3—C(=O)—OH}$$

Methanol can also serve as the feedstock for an industrial synthesis of acetic acid. The rhodium-catalyzed reaction of methanol with carbon monoxide requires high pressures, so it is not suitable for a laboratory synthesis.

$$\underset{\text{methanol}}{CH_3OH} + CO \xrightarrow[\text{heat, pressure}]{\text{Rh catalyst}} \underset{\text{acetic acid}}{CH_3COOH}$$

Application: Fungicide

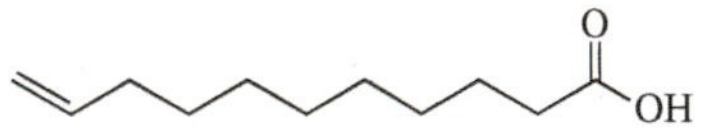

$CH_2=CH(CH_2)_8COOH$

undecylenic acid
(undec-10-enoic acid)

Undecylenic acid is a naturally occurring fungicide derived from castor oil. It is commonly used in medications for fungal skin infections such as athlete's foot and ringworm. The original medication containing undecylenic acid was named Desenex®, based on a shortened version of the chemical name.

Figure 20-4 shows how long-chain aliphatic acids are obtained from the hydrolysis of fats and oils, a reaction discussed in Chapter 25. These **fatty acids** are generally straight-chain acids with even numbers of carbon atoms ranging between about C_6 and C_{18}. The hydrolysis of animal fat gives mostly saturated fatty acids. Plant oils give large amounts of unsaturated fatty acids with one or more olefinic double bonds.

Some aromatic carboxylic acids are also commercially important. Benzoic acid is used as an ingredient in medications, a preservative in foods, and a starting material for synthesis. Benzoic acid can be produced by the oxidation of toluene with potassium permanganate, nitric acid, or other strong oxidants.

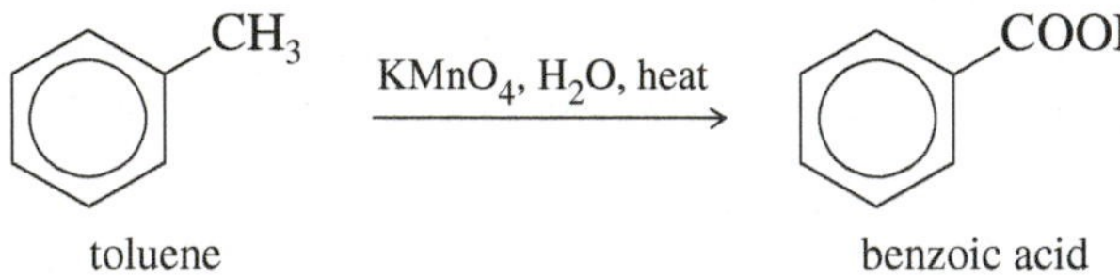

$$\underset{\text{toluene}}{C_6H_5CH_3} \xrightarrow{KMnO_4,\ H_2O,\ \text{heat}} \underset{\text{benzoic acid}}{C_6H_5COOH}$$

Two important commercial diacids are adipic acid (hexanedioic acid) and terephthalic acid (benzene-1,4-dicarboxylic acid). Adipic acid is used in making nylon 66, and terephthalic acid is used to make polyesters. The industrial synthesis of adipic acid uses benzene as the starting material. Benzene is hydrogenated to cyclohexane, whose oxidation (using a cobalt/acetic acid catalyst) gives adipic acid. Terephthalic acid is produced by the direct oxidation of *para*-xylene in acetic acid using a cobalt–molybdenum catalyst.

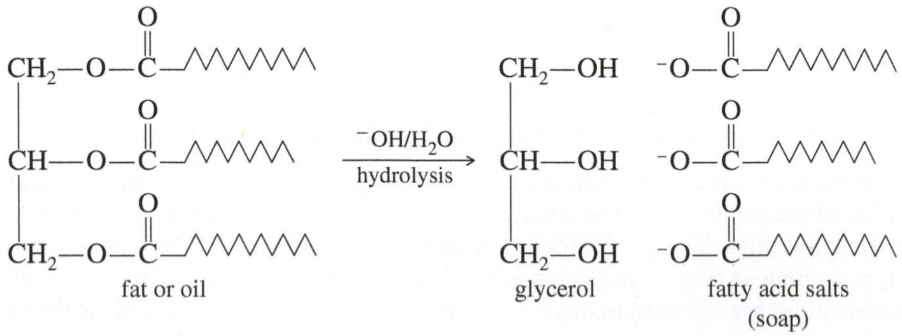

$$\underset{\text{fat or oil}}{\begin{matrix} CH_2—O—C(=O)—R \\ CH—O—C(=O)—R \\ CH_2—O—C(=O)—R \end{matrix}} \xrightarrow[\text{hydrolysis}]{^-OH/H_2O} \underset{\text{glycerol}}{\begin{matrix} CH_2—OH \\ CH—OH \\ CH_2—OH \end{matrix}} \quad \underset{\text{fatty acid salts (soap)}}{\begin{matrix} ^-O—C(=O)—R \\ ^-O—C(=O)—R \\ ^-O—C(=O)—R \end{matrix}}$$

FIGURE 20-4
Hydrolysis of a fat or an oil gives a mixture of the salts of straight-chain fatty acids. Animal fats contain primarily saturated fatty acids, whereas most vegetable oils are polyunsaturated.

benzene $\xrightarrow[\text{high pressure}]{H_2,\ Ni}$ cyclohexane $\xrightarrow[Co(OCOCH_3)_3,\ CH_3COOH]{O_2}$ adipic acid (COOH, COOH)

p-xylene (CH_3, CH_3) + $3\ O_2$ $\xrightarrow[CH_3COOH]{\text{Co/Mo catalyst}}$ terephthalic acid (COOH, COOH) + $2\ H_2O$

Application: Green Synthesis

toluene (CH_3) $\xrightarrow[O_2\ (-\ CO_2)]{\text{bacteria}}$ muconic acid (COOH, COOH) $\xrightarrow{H_2,\ \text{catalyst}}$ adipic acid (COOH, COOH)

Another synthesis of adipic acid involves the microbial degradation of toluene to muconic acid (hexa-2, 4-dienedioic acid), which undergoes catalytic hydrogenation to give adipic acid. If this process can be made economically competitive, it might produce less environmental impact than the chemical synthesis from benzene.

20-7 Spectroscopy of Carboxylic Acids

We have discussed the general aspects of spectroscopy of carboxylic acids in earlier chapters. Here we cover more specifics and show how we combine this information to identify the presence of a carboxylic acid group.

20-7A Infrared Spectroscopy

The most obvious feature in the infrared spectrum of a carboxylic acid is the intense carbonyl stretching absorption. In a saturated acid, this vibration occurs around 1710 cm^{-1}, often broadened by hydrogen bonding involving the carbonyl group. In conjugated acids, the carbonyl stretching frequency is lowered to about 1690 cm^{-1}.

R—C(=O)—O—H: C=O 1710 cm^{-1}; O—H 2500–3500 cm^{-1}

>C=C—C(=O)—O—H: C=O 1690 cm^{-1}; O—H 2500–3500 cm^{-1}

The O—H stretching vibration of a carboxylic acid absorbs in a broad band around 2500–3500 cm^{-1}. This frequency range is lower than the hydroxy stretching frequencies of water and alcohols, whose O—H groups absorb in a band centered around 3300 cm^{-1}. In the spectrum of a carboxylic acid, the broad hydroxy band appears right on top of the C—H stretching region. This overlapping of absorptions gives the 3000 cm^{-1} region a characteristic appearance of a broad peak (the O—H stretching) with sharp peaks (C—H stretching) superimposed on it. Many carboxylic acids show a shoulder or small spikes (around 2500–2700 cm^{-1}) in the broad O—H peak to the right of the C—H stretch. Figure 20-5 and Problem 20-7 show typical acid O—H stretching absorptions.

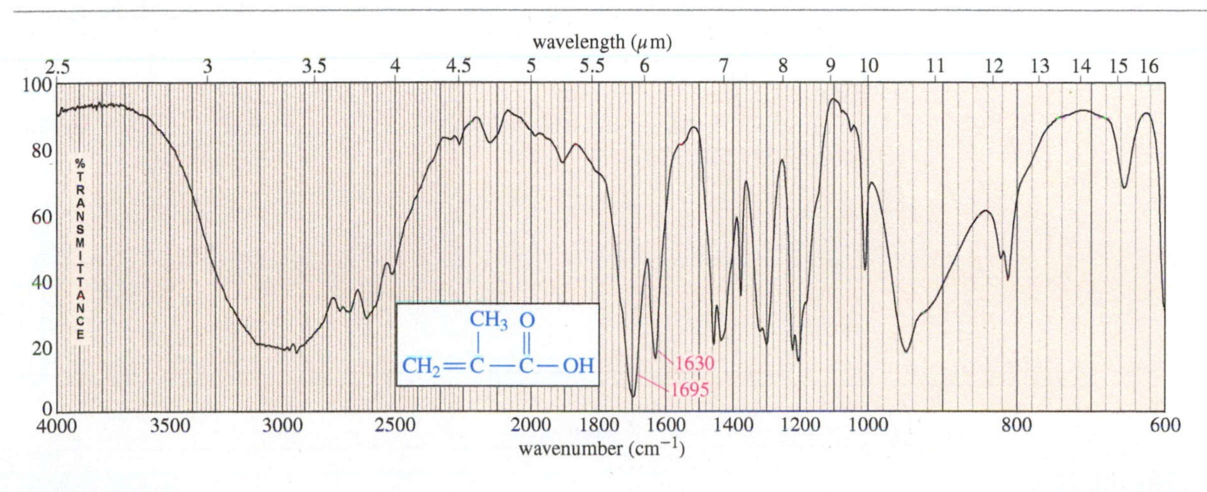

FIGURE 20-5
IR spectrum of 2-methylpropenoic acid.

The IR spectrum of 2-methylpropenoic acid (methacrylic acid) is shown in Figure 20-5. Compare this conjugated example with the spectrum of hexanoic acid (Figure 12-12, p. 573). Notice the shift in the position of the carbonyl absorptions, and notice that the conjugated, unsaturated acid has a fairly strong C=C stretching absorption around 1630 cm^{-1}, just to the right of the carbonyl absorption.

PROBLEM 20-7

The IR spectrum of *trans*-oct-2-enoic acid is shown. Point out the spectral characteristics that allow you to tell that this is a carboxylic acid, and show which features lead you to conclude that the acid is unsaturated and conjugated.

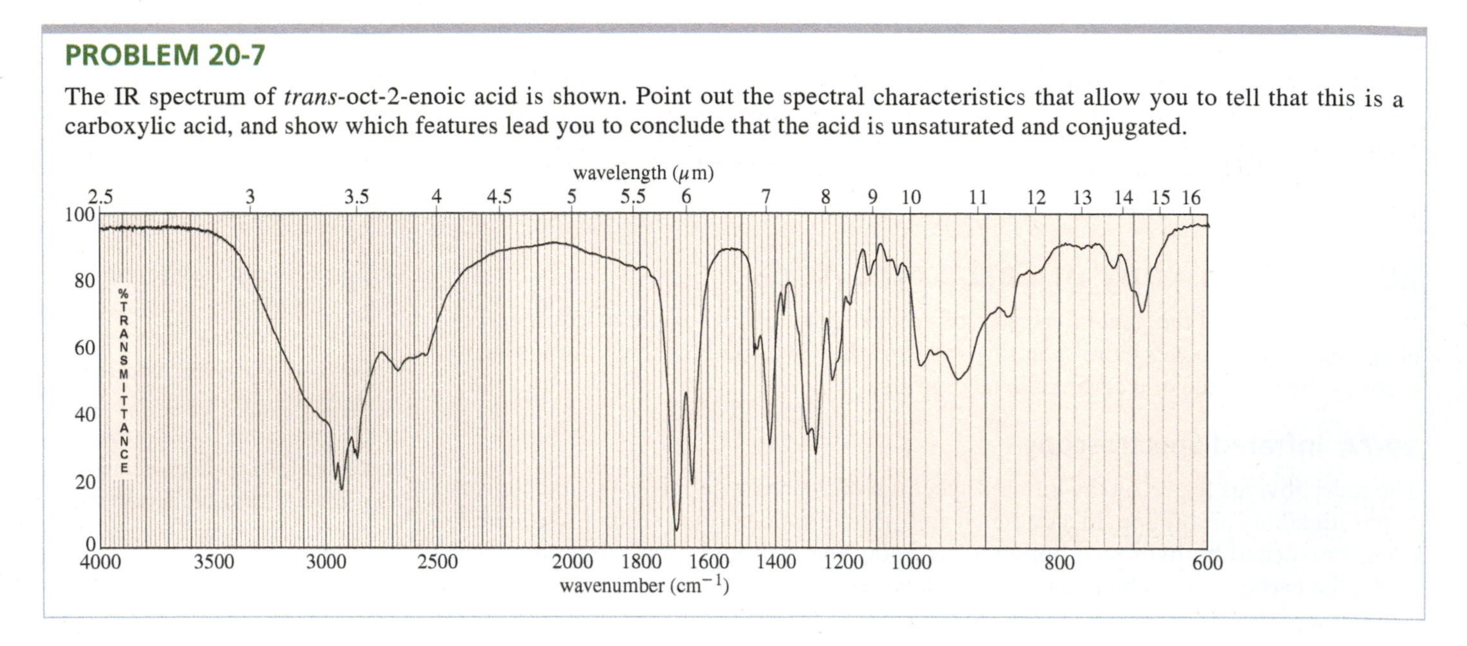

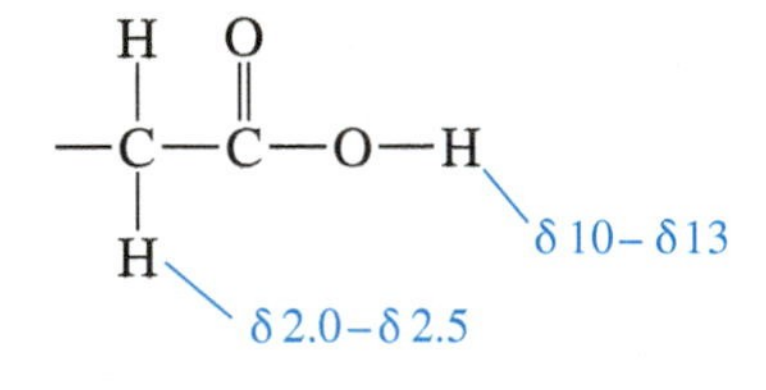

20-7B NMR Spectroscopy

Carboxylic acid protons are the most deshielded protons we have encountered, absorbing between δ10 and δ13. Depending on the solvent and the concentration, this acid proton peak may be sharp or broad, but it is always unsplit because of proton exchange.

The protons on the α carbon atom absorb between δ2.0 and δ2.5, in about the same position as the protons on a carbon atom alpha to a ketone or an aldehyde. The proton NMR spectrum of butanoic acid is shown in Figure 20-6.

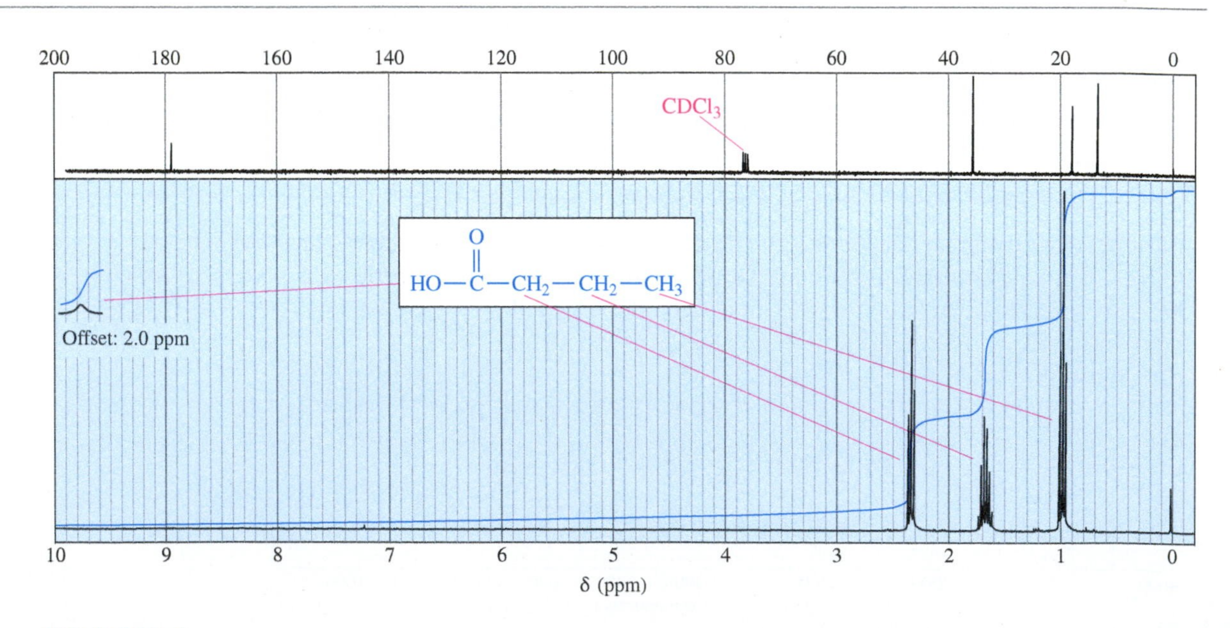

FIGURE 20-6
Proton and carbon NMR spectra of butanoic acid.

$$\underset{\substack{\delta\,11.2\\\text{singlet}}}{\text{H}}-\text{O}-\overset{\text{O}}{\overset{\|}{\text{C}}}-\underset{\substack{\delta\,2.4\\\text{triplet}}}{\text{CH}_2}-\underset{\substack{\delta\,1.6\\\text{sextet (overlapping quartet of triplets)}}}{\text{CH}_2}-\underset{\substack{\delta\,1.0\\\text{triplet}}}{\text{CH}_3}$$

The carbon NMR chemical shifts of carboxylic acids resemble those of ketones and aldehydes. The carbonyl carbon atom absorbs around 170 to 180 ppm, and the α carbon atom absorbs around 30 to 40 ppm. The chemical shifts of the carbon atoms in hexanoic acid are the following:

$$\text{HO}-\underset{181}{\overset{\text{O}}{\overset{\|}{\text{C}}}}-\underset{34}{\text{CH}_2}-\underset{25}{\text{CH}_2}-\underset{31}{\text{CH}_2}-\underset{22}{\text{CH}_2}-\underset{14\ (\text{ppm})}{\text{CH}_3}$$

PROBLEM 20-8

(a) Determine the structure of the carboxylic acid whose proton NMR spectrum appears below.
(b) Draw the NMR spectrum you would expect from the corresponding aldehyde whose oxidation gives this carboxylic acid.
(c) Point out two distinctive differences in the spectra of the aldehyde and the acid.

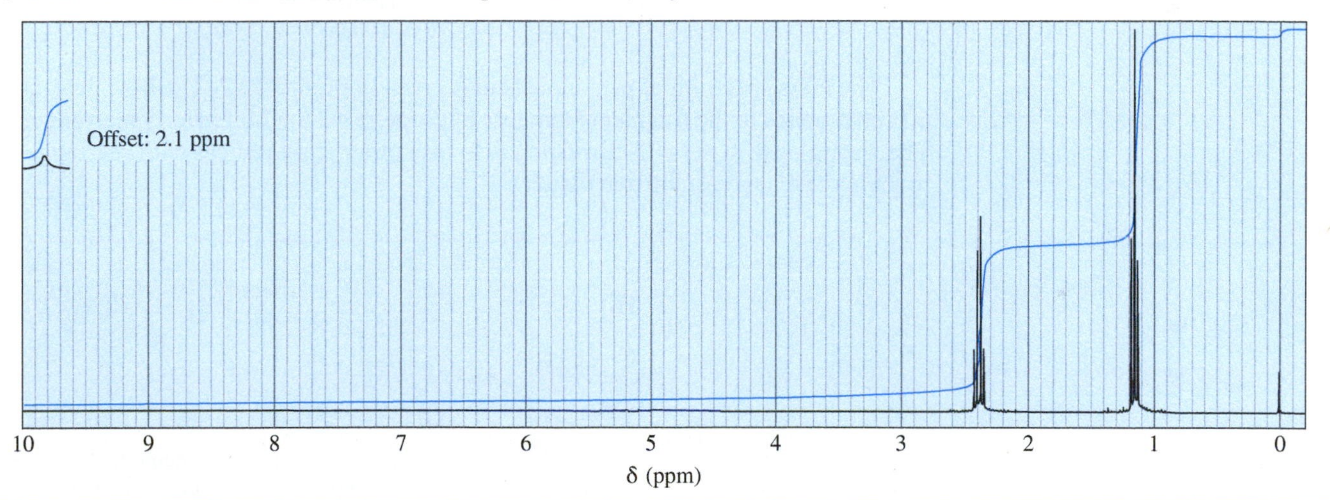

20-7C Ultraviolet Spectroscopy

Saturated carboxylic acids have a weak $n \rightarrow \pi^*$ transition that absorbs around 200 to 215 nm. This absorption corresponds to the weak transition around 270 to 300 nm in the spectra of ketones and aldehydes. The molar absorptivity is very small (about 30 to 100), and the absorption often goes unnoticed.

Conjugated acids show much stronger absorptions. One C=C double bond conjugated with the carboxyl group results in a spectrum with λ_{max} around 200 nm, but with molar absorptivity of about 10,000. A second conjugated double bond raises the value of λ_{max} to about 250 nm, as illustrated by the following examples:

$$\text{CH}_2{=}\text{CH}-\overset{\text{O}}{\overset{\|}{\text{C}}}-\text{OH} \qquad \lambda_{max} = 200\text{ nm} \qquad \varepsilon = 10{,}000$$

$$\text{CH}_3-\text{CH}{=}\text{CH}-\text{CH}{=}\text{CH}-\overset{\text{O}}{\overset{\|}{\text{C}}}-\text{OH} \qquad \lambda_{max} = 254\text{ nm} \qquad \varepsilon = 25{,}000$$

20-7D Mass Spectrometry

The molecular ion peak of a carboxylic acid is usually small because favorable modes of fragmentation are available. The most common fragmentation is loss of a molecule of an alkene (the McLafferty rearrangement, discussed in Section 18-5D). The ion that results from McLafferty rearrangement has an even-numbered mass (from loss of a molecule), as opposed to the odd-numbered ions that result from loss of fragments. Another common fragmentation is loss of an alkyl radical to give a resonance-stabilized cation with the positive charge delocalized over an allylic system and two oxygen atoms.

McLafferty rearrangement

(m/z is even)

loss of an alkyl group

resonance-stabilized cation
(m/z is odd)

The mass spectrum of pentanoic acid is given in Figure 20-7. The base peak at m/z 60 corresponds to the fragment from loss of propene via the McLafferty rearrangement. The strong peak at m/z 73 corresponds to loss of an ethyl radical with rearrangement to give a resonance-stabilized cation.

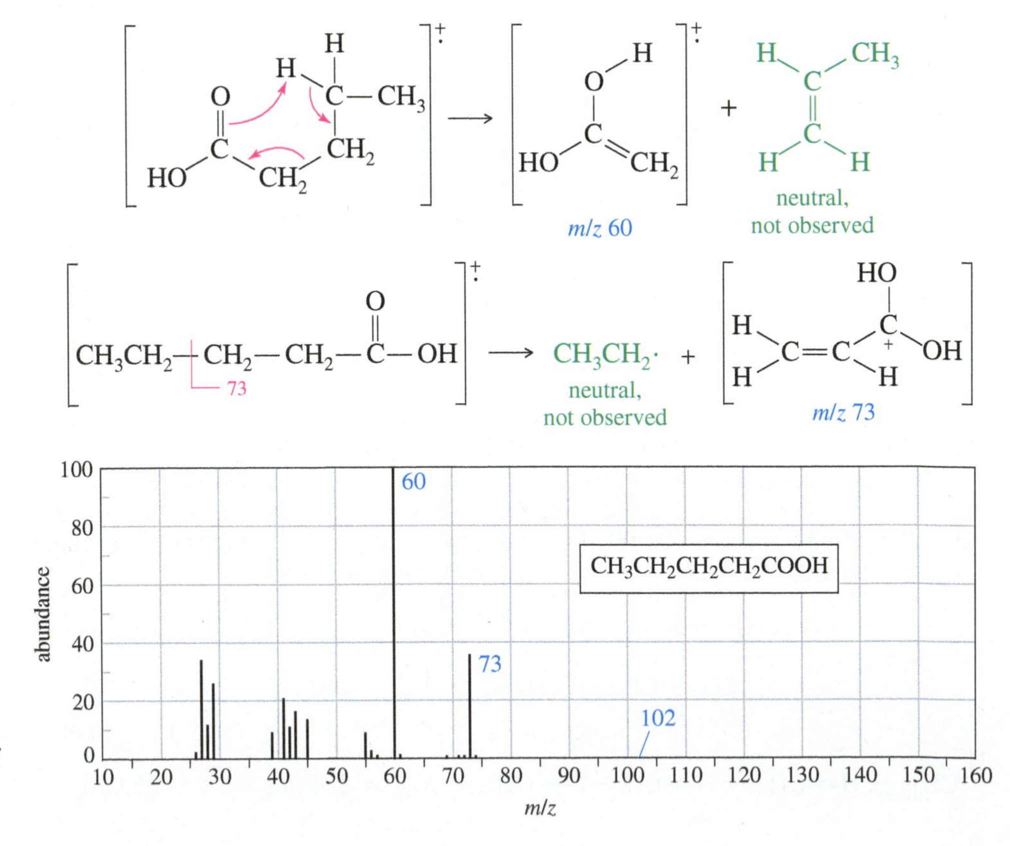

FIGURE 20-7
The mass spectrum of pentanoic acid shows a weak parent peak, an even-numbered base peak from the McLafferty rearrangement, and another strong peak from loss of an ethyl radical.

PROBLEM 20-9

Draw all four resonance forms of the fragment at m/z 73 in the mass spectrum of pentanoic acid.

PROBLEM 20-10

(a) Why do most long-chain fatty acids show a large peak in the mass spectrum at m/z 60?
(b) Use equations to explain the prominent peaks at m/z 74 and m/z 87 in the mass spectrum of 2-methylpentanoic acid.
(c) Why doesn't the mass spectrum of 2-methylpentanoic acid show a large peak at m/z 60?

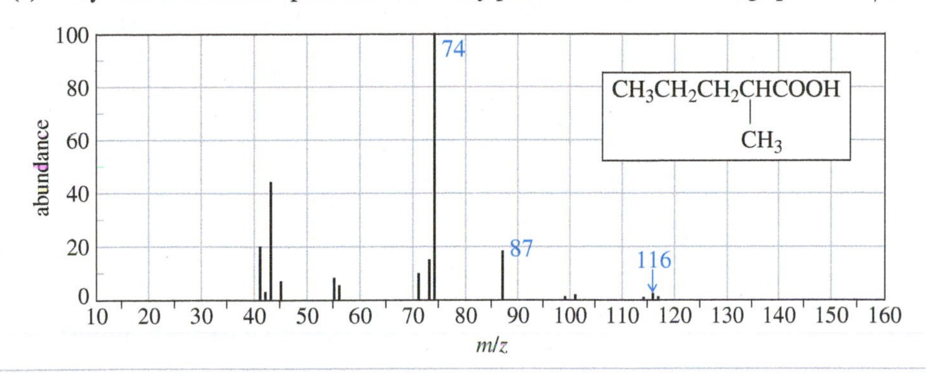

20-8 Synthesis of Carboxylic Acids

In earlier chapters, we have seen some of the most useful methods for making carboxylic acids. We review those here and discuss two other important syntheses, using Grignard reagents and nitriles as intermediates.

20-8A Review of Previous Syntheses

We have already encountered three methods for preparing carboxylic acids: (1) oxidation of alcohols and aldehydes, (2) oxidative cleavage of alkenes and alkynes, and (3) severe side-chain oxidation of alkylbenzenes.

1. Primary alcohols and aldehydes are commonly oxidized to acids by sodium hypochlorite (bleach, NaOCl), often used with TEMPO as a catalyst, or by chromic acid, H_2CrO_4 (Sections 11-2B and 18-19).

$$\underset{\text{primary alcohol}}{R-CH_2-OH} \xrightarrow[\text{(or } H_2CrO_4)]{\text{NaOCl, TEMPO}} \left[\underset{\text{aldehyde}}{R-\overset{O}{\overset{\|}{C}}-H}\right] \xrightarrow[\text{(or } H_2CrO_4)]{\text{excess NaOCl, TEMPO}} \underset{\text{carboxylic acid}}{R-\overset{O}{\overset{\|}{C}}-OH}$$

(not isolated)

Example

$$\underset{\text{3-phenylpropan-1-ol}}{Ph-CH_2-CH_2-CH_2-OH} \xrightarrow[\text{(or } H_2CrO_4)]{\text{excess NaOCl, TEMPO}} \underset{\text{3-phenylpropanoic acid}}{Ph-CH_2-CH_2-\overset{O}{\overset{\|}{C}}-OH}$$

2. Cold, dilute potassium permanganate reacts with alkenes to give glycols. Warm, concentrated permanganate solutions oxidize the glycols further, cleaving the central carbon–carbon bond. Depending on the substitution of the original double bond, ketones or acids may result (Section 8-15A).

$$\underset{\text{alkene}}{(R)(H)C{=}C(R')(R'')} \xrightarrow{\text{concd. } KMnO_4} \left[\underset{\text{glycol (not isolated)}}{H-\underset{HO}{\overset{R}{C}}-\underset{OH}{\overset{R'}{C}}-R''}\right] \longrightarrow \underset{\text{acid}}{R-COOH} + \underset{\text{ketone}}{O{=}C(R')(R'')}$$

Application: Shellfish Poisoning

Domoic acid is a toxin synthesized by several species of red algae and diatoms associated with "red tide" algal blooms. Shellfish, anchovies, and sardines feed on the poisonous organisms and accumulate the poison in their tissues. When people or marine mammals eat fish or shellfish containing domoic acid, they develop symptoms of "amnesic shellfish poisoning" (ASP), which may include tremors, seizures, brain damage, and occasionally death. Whenever an algal bloom occurs, regulatory agencies ban fishing in the affected areas to prevent the harvest of contaminated seafood.

domoic acid

Examples

$$\text{Ph(H)C{=}C(H)CH_2{-}CH_3} \xrightarrow{\text{concd. KMnO}_4} \text{Ph{-}COOH} + \text{CH_3{-}CH_2{-}COOH}$$

$$\underset{\text{cyclohexene}}{\text{C}_6\text{H}_{10}} \xrightarrow{\text{concd. KMnO}_4} \underset{\text{adipic acid}}{\text{HOOC(CH}_2)_4\text{COOH}}$$

With alkynes, either ozonolysis or a vigorous permanganate oxidation cleaves the triple bond to give carboxylic acids (Section 9-10).

$$\underset{\text{alkyne}}{\text{R{-}C{\equiv}C{-}R'}} \xrightarrow[\text{or (1) O}_3\text{ (2) H}_2\text{O}]{\text{concd. KMnO}_4} \underset{\text{(not isolated)}}{\left[\text{R{-}C(=O){-}C(=O){-}R'}\right]} \longrightarrow \underset{\text{carboxylic acids}}{\text{R{-}COOH} + \text{HOOC{-}R'}}$$

Example

$$\text{CH_3CH_2CH_2{-}C{\equiv}C{-}Ph} \xrightarrow[\text{(2) H}_2\text{O}]{\text{(1) O}_3} \text{CH_3CH_2CH_2{-}COOH} + \text{Ph{-}COOH}$$

3. Side chains of alkylbenzenes are oxidized to benzoic acid derivatives by treatment with hot potassium permanganate or hot chromic acid. Because this oxidation requires severe conditions, it is useful only for making benzoic acid derivatives with no oxidizable functional groups. Oxidation-resistant functional groups such as $-Cl$, $-NO_2$, $-SO_3H$, and $-COOH$ may be present (Section 17-15A).

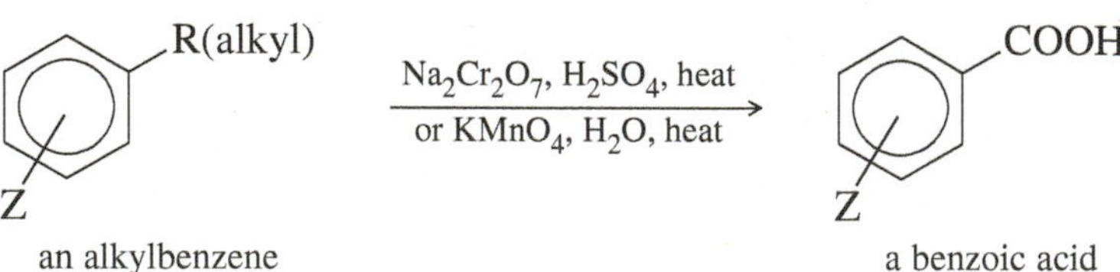

$$\underset{\substack{\text{an alkylbenzene}\\ \text{(Z must be oxidation-resistant)}}}{\text{Z{-}C}_6\text{H}_4\text{{-}R(alkyl)}} \xrightarrow[\text{or KMnO}_4\text{, H}_2\text{O, heat}]{\text{Na}_2\text{Cr}_2\text{O}_7\text{, H}_2\text{SO}_4\text{, heat}} \underset{\text{a benzoic acid}}{\text{Z{-}C}_6\text{H}_4\text{{-}COOH}}$$

Example

$$\underset{p\text{-chloroisopropylbenzene}}{\text{Cl{-}C}_6\text{H}_4\text{{-}CH(CH}_3\text{){-}CH}_3} \xrightarrow[\text{heat}]{\text{Na}_2\text{Cr}_2\text{O}_7\text{, H}_2\text{SO}_4} \underset{p\text{-chlorobenzoic acid}}{\text{Cl{-}C}_6\text{H}_4\text{{-}COOH}}$$

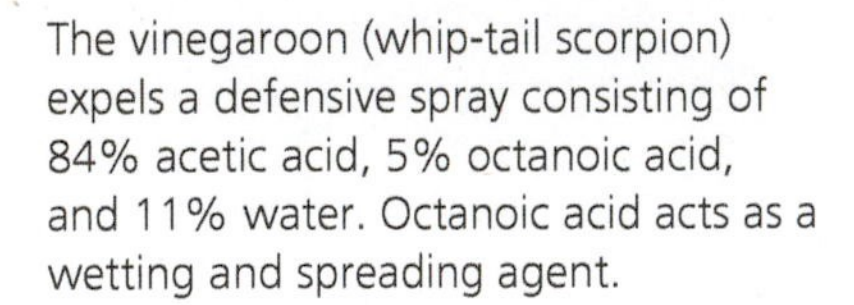

The vinegaroon (whip-tail scorpion) expels a defensive spray consisting of 84% acetic acid, 5% octanoic acid, and 11% water. Octanoic acid acts as a wetting and spreading agent.

20-8B Carboxylation of Grignard Reagents

We have seen how Grignard reagents act as strong nucleophiles, adding to the carbonyl groups of ketones and aldehydes (Section 10-9). Similarly, Grignard reagents add to carbon dioxide to form magnesium salts of carboxylic acids. Addition of dilute acid protonates these magnesium salts to give carboxylic acids. This method is useful because it converts a halide functional group to a carboxylic acid functional group with an additional carbon atom.

$$\underset{\substack{\text{(alkyl or}\\ \text{aryl halide)}}}{\text{R{-}X}} \xrightarrow[\text{ether}]{\text{Mg}} \text{R{-}MgX} \xrightarrow{\text{O{=}C{=}O}} \text{R{-}C(=O){-}O}^{-}\ {}^{+}\text{MgX} \xrightarrow{\text{H}^+} \text{R{-}C(=O){-}OH}$$

Example

$$\text{bromocyclohexane (C}_6\text{H}_{11}\text{Br)} \xrightarrow[\text{ether}]{\text{Mg}} \text{C}_6\text{H}_{11}\text{MgBr} \xrightarrow{O{=}C{=}O} \text{C}_6\text{H}_{11}\text{—C(=O)—O}^-\ {}^+\text{MgBr} \xrightarrow{H^+} \text{C}_6\text{H}_{11}\text{COOH}$$

bromocyclohexane → cyclohexanecarboxylic acid

20-8C Formation and Hydrolysis of Nitriles

Another way to convert an alkyl halide (or tosylate) to a carboxylic acid with an additional carbon atom is to displace the halide with sodium cyanide. The product is a nitrile with one additional carbon atom. Acidic or basic hydrolysis of the nitrile gives a carboxylic acid by a mechanism discussed in Chapter 21. This method is limited to halides and tosylates that are good S_N2 electrophiles: usually primary and unhindered.

$$R{-}CH_2{-}X \xrightarrow[\text{acetone}]{\text{NaCN}} R{-}CH_2{-}C{\equiv}N{:} \xrightarrow[\text{or }^-OH,\ H_2O]{H^+,\ H_2O} R{-}CH_2{-}\overset{O}{\overset{\|}{C}}{-}OH \ +\ NH_4^+$$

Example

$$C_6H_5{-}CH_2{-}Br \xrightarrow[\text{acetone}]{\text{NaCN}} C_6H_5{-}CH_2{-}C{\equiv}N \xrightarrow{H^+,\ H_2O} C_6H_5{-}CH_2{-}\overset{O}{\overset{\|}{C}}{-}OH \ +\ NH_4^+$$

benzyl bromide → phenylacetonitrile → phenylacetic acid

PROBLEM 20-11

Show how you would synthesize the following carboxylic acids, using the indicated starting materials.

(a) oct-4-yne → butanoic acid
(b) *trans*-cyclodecene → decanedioic acid
(c) bromobenzene → phenylacetic acid
(d) butan-2-ol → 2-methylbutanoic acid
(e) *p*-xylene → terephthalic acid
(f) allyl iodide → but-3-enoic acid

PROBLEM-SOLVING HINT

Oxidation of alcohols does not change the number of carbon atoms. Oxidative cleavages of alkenes and alkynes decrease the number of carbon atoms (except in cyclic cases). Carboxylation of Grignard reagents and formation and hydrolysis of nitriles increase the number of carbon atoms by one.

SUMMARY Syntheses of Carboxylic Acids

1. *Oxidation of primary alcohols and aldehydes* (Sections 11-2B and 18-19)

$$\underset{\text{primary alcohol}}{R{-}CH_2{-}OH} \xrightarrow[\text{or } H_2CrO_4]{\text{NaOCl, TEMPO}} \underset{\text{aldehyde}}{R{-}\overset{O}{\overset{\|}{C}}{-}H} \xrightarrow[\text{or } H_2CrO_4]{\text{NaOCl, TEMPO}} \underset{\text{carboxylic acid}}{R{-}\overset{O}{\overset{\|}{C}}{-}OH}$$

2. *Oxidative cleavage of alkenes and alkynes* (Sections 8-15A and 9-10)

$$\underset{\text{alkene}}{(R)(H)C{=}C(R')(R'')} \xrightarrow[\Delta,\ H_2O]{\text{concd. KMnO}_4} \underset{\text{acid}}{R{-}COOH} \ +\ \underset{\text{ketone}}{O{=}C(R')(R'')}$$

$$\underset{\text{alkyne}}{R{-}C{\equiv}C{-}R'} \xrightarrow[\text{or (1) O}_3\ \text{(2) H}_2\text{O}]{\text{concd. KMnO}_4} \underset{\text{carboxylic acids}}{R{-}COOH \ +\ HOOC{-}R'}$$

(continued)

3. *Oxidation of alkylbenzenes* (Section 17-15A)

$$\text{Z-C}_6\text{H}_4\text{-R(alkyl)} \xrightarrow[\textit{or}\ Na_2Cr_2O_7,\ H_2SO_4]{KMnO_4,\ H_2O} \text{Z-C}_6\text{H}_4\text{-COOH}$$

an alkylbenzene (Z must be oxidation-resistant) → a benzoic acid

4. *Carboxylation of Grignard reagents* (Section 20-8B)

$$R{-}X \xrightarrow[\text{ether}]{Mg} R{-}MgX \xrightarrow{O{=}C{=}O} R{-}\overset{O}{\overset{\|}{C}}{-}O^- \ \overset{+}{M}gX \xrightarrow{H_3O^+} R{-}\overset{O}{\overset{\|}{C}}{-}OH$$

alkyl or aryl halide → acid

5. *Formation and hydrolysis of nitriles* (Section 20-8C)

$$R{-}CH_2{-}X \xrightarrow[\text{acetone}]{NaCN} R{-}CH_2{-}C{\equiv}N{:} \xrightarrow[\textit{or}\ ^-OH,\ H_2O]{H^+,\ H_2O} R{-}CH_2{-}\overset{O}{\overset{\|}{C}}{-}OH$$

6. *The haloform reaction* (converts methyl ketones to acids and iodoform; Chapter 22)

$$R{-}\overset{O}{\overset{\|}{C}}{-}CH_3 \xrightarrow[^-OH]{X_2} R{-}\overset{O}{\overset{\|}{C}}{-}O^- + HCX_3$$

$X = Cl, Br, I$

7. *Malonic ester synthesis* (makes substituted acetic acids; Chapter 22)

$$CH_2(COOEt)_2 \xrightarrow[(2)\ R{-}X]{(1)\ Na^+\ ^-OCH_2CH_3} R{-}CH(COOEt)_2 \xrightarrow[(2)\ H^+,\ \text{heat}]{(1)\ ^-OH} R{-}CH_2{-}\overset{O}{\overset{\|}{C}}{-}OH + CO_2$$

20-9 Reactions of Carboxylic Acids and Derivatives; Nucleophilic Acyl Substitution

Ketones, aldehydes, and carboxylic acids all contain the carbonyl group, yet the reactions of acids are quite different from those of ketones and aldehydes. Ketones and aldehydes commonly react by nucleophilic addition to the carbonyl group; but carboxylic acids (and their derivatives) more commonly react by **nucleophilic acyl substitution,** where one nucleophile replaces another on the acyl (C═O) carbon atom.

Nucleophilic acyl substitution

$$R{-}\overset{O}{\overset{\|}{C}}{-}X + Nuc{:}^- \rightleftharpoons R{-}\overset{O}{\overset{\|}{C}}{-}Nuc + {:}X^-$$

Acid derivatives

$R{-}\overset{O}{\overset{\|}{C}}{-}OH$	$R{-}\overset{O}{\overset{\|}{C}}{-}X$	$R{-}\overset{O}{\overset{\|}{C}}{-}O{-}\overset{O}{\overset{\|}{C}}{-}R$	$R{-}\overset{O}{\overset{\|}{C}}{-}O{-}R'$	$R{-}\overset{O}{\overset{\|}{C}}{-}NH_2$
carboxylic acid	acyl halide	anhydride	ester	amide

Acid derivatives differ in the nature of the nucleophile bonded to the acyl carbon: —OH in the acid, —Cl in the acid chloride, —OR′ in the ester, and —NH_2 (or an amine) in the amide. Nucleophilic acyl substitution is the most common method for interconverting these derivatives. We will see many examples of nucleophilic acyl substitution in this chapter and in Chapter 21 ("Carboxylic Acid Derivatives"). The specific mechanisms depend on the reagents and conditions, but we can group them generally according to whether they take place under acidic or basic conditions.

Under basic conditions, a strong nucleophile can add to the carbonyl group to give a tetrahedral intermediate. This intermediate then expels the leaving group. The base-catalyzed hydrolysis of an ester to the carboxylate salt of an acid is an example of this mechanism (Mechanism 20-1). Hydroxide ion adds to the carbonyl group to give a tetrahedral intermediate. The tetrahedral intermediate stabilizes itself by expelling an alkoxide ion. The alkoxide ion quickly reacts with the acid ($pK_a = 5$) to give an alcohol ($pK_a = 16$) and a carboxylate ion.

MECHANISM 20-1 Nucleophilic Acyl Substitution in the Basic Hydrolysis of an Ester

Step 1: Hydroxide ion adds to the carbonyl group, forming a tetrahedral intermediate.

Step 2: An alkoxide ion leaves, regenerating the C=O double bond.

R—C(=O)—OR′ + ⁻:OH ⇌ R—C(—O⁻)(—OH)—OR′ ⇌ R—C(=O)—OH + ⁻:O—R′

ester + ⁻OH   tetrahedral intermediate   acid + alkoxide

Step 3: A fast, exothermic proton transfer drives the reaction to completion.

R—C(=O)—O—H + ⁻:O—R′ ⟶ R—C(=O)—O⁻ + H—O—R′

acid + alkoxide   carboxylate + alcohol

EXAMPLE: Basic hydrolysis of ethyl benzoate.

Step 1: Addition of hydroxide.

Step 2: Elimination of alkoxide.

Ph—C(=O)—OCH_2CH_3 + ⁻:OH ⇌ Ph—C(—O⁻)(—OH)—OCH_2CH_3 ⇌ Ph—C(=O)—O—H + ⁻:OCH_2CH_3

ester + ⁻OH   tetrahedral intermediate   acid + alkoxide

Step 3: Proton transfer.

Ph—C(=O)—O—H + ⁻:OCH_2CH_3 ⟶ Ph—C(=O)—O⁻ + H—O—CH_2CH_3

acid + alkoxide   carboxylate + alcohol

Nucleophilic acyl substitution also takes place in acid. Under acidic conditions, no strong nucleophile is present to attack the carbonyl group. The carbonyl group must become protonated, activating it toward nucleophilic acyl substitution. Attack by a weak nucleophile gives a tetrahedral intermediate. In most cases, the leaving group becomes protonated before it leaves, so it leaves as a neutral molecule. We now cover the Fischer esterification, a particularly useful example of an acid-catalyzed nucleophilic acyl substitution.

20-10 Condensation of Acids with Alcohols: The Fischer Esterification

The **Fischer esterification** converts carboxylic acids and alcohols directly to esters by an acid-catalyzed nucleophilic acyl substitution. The net reaction is replacement of the acid —OH group by the —OR group of the alcohol.

$$\underset{\text{acid}}{R-\overset{O}{\overset{\|}{C}}-OH} + \underset{\text{alcohol}}{R'-OH} \overset{H^+}{\rightleftharpoons} \underset{\text{ester}}{R-\overset{O}{\overset{\|}{C}}-O-R'} + H_2O$$

Examples

$$CH_3-\overset{O}{\overset{\|}{C}}-OH + CH_3CH_2-OH \underset{K_{eq}=3.38}{\overset{H_2SO_4}{\rightleftharpoons}} CH_3-\overset{O}{\overset{\|}{C}}-O-CH_2CH_3 + H_2O$$

$$\underset{\text{phthalic acid}}{C_6H_4(COOH)_2} \overset{\text{excess } CH_3OH,\ H^+}{\rightleftharpoons} \underset{\text{dimethyl phthalate}}{C_6H_4(COOCH_3)_2}$$

The Fischer esterification mechanism (Key Mechanism 20-2) is an acid-catalyzed nucleophilic acyl substitution. The carbonyl group of a carboxylic acid is not sufficiently electrophilic to be attacked by an alcohol. The acid catalyst protonates the carbonyl group and activates it toward nucleophilic attack. Attack by the alcohol, followed by loss of a proton, gives the hydrate of an ester.

Loss of water from the hydrate of the ester occurs by the same mechanism as loss of water from the hydrate of a ketone (Section 18-13). Protonation of either of the hydroxy groups allows it to leave as water, forming a resonance-stabilized cation. Loss of a proton from the second hydroxy group gives the ester. Note that this mechanism involves intermediates that are relatively stable because they are resonance-stabilized. These are indicated by brackets around one of the major resonance forms.

KEY MECHANISM 20-2 Fischer Esterification

Part 1: Acid-catalyzed addition of the alcohol to the carbonyl group.

Protonation activates the carbonyl. The alcohol adds. Deprotonation completes the reaction.

$$R-\overset{O}{\overset{\|}{C}}-OH \overset{H^+}{\rightleftharpoons} \left[R-\overset{\overset{+}{O}-H}{\overset{\|}{C}}-OH\right] \rightleftharpoons R-\overset{OH}{\overset{|}{C}}(OH)-\overset{+}{O}(H)-R' \rightleftharpoons R-\overset{OH}{\overset{|}{C}}(OH)-OR' + R'\overset{+}{O}H_2$$

(species in brackets are resonance-stabilized.) R′—O—H; ester hydrate

Part 2: Acid-catalyzed dehydration.

Protonation prepares the OH group to leave. Water leaves. Deprotonation completes the reaction.

$$R-C(OH)(OR')-OH \overset{H^+}{\rightleftharpoons} R-C(OH)(OR')-\overset{+}{O}H_2 \rightleftharpoons \left[R-C(=\overset{+}{O}H)-OR'\right] + H_2O \;\; \overset{ROH}{\rightleftharpoons} \;\; R-C(=O)-OR' + R\overset{+}{O}H_2$$

protonated ester — ester

EXAMPLE: Acid-catalyzed formation of methyl benzoate from methanol and benzoic acid.

Part 1: Acid-catalyzed addition of methanol to the carbonyl group.

Protonation activates the carbonyl. Methanol adds. Deprotonation completes the reaction.

$$Ph-C(=O)-OH \overset{H^+}{\rightleftharpoons} \left[Ph-C(=\overset{+}{O}H)-OH\right] \xrightarrow{CH_3-OH} \rightleftharpoons Ph-C(OH)(\overset{+}{O}H-CH_3)-OH \overset{CH_3-OH}{\rightleftharpoons} Ph-C(OH)(OCH_3)-OH + CH_3\overset{+}{O}H_2$$

ester hydrate

Part 2: Acid-catalyzed dehydration.

Protonation prepares the OH group to leave. Water leaves. Deprotonation completes the reaction.

$$Ph-C(OH)(OCH_3)-OH \overset{H^+}{\rightleftharpoons} Ph-C(OH)(OCH_3)-\overset{+}{O}H_2 \rightleftharpoons \left[Ph-C(=\overset{+}{O}H)-OCH_3\right] + H_2O \;\; \overset{ROH}{\rightleftharpoons} \;\; Ph-C(=O)-OCH_3 + R\overset{+}{O}H_2$$

protonated ester — methyl benzoate

QUESTION: Why can't the Fischer esterification take place under basic catalysis?

The mechanism of the Fischer esterification would seem long and complicated if you tried to memorize it, but we can understand it by breaking it down into two simpler mechanisms: (1) acid-catalyzed addition of the alcohol to the carbonyl and (2) acid-catalyzed dehydration. If you understand these mechanistic components, you can write the Fischer esterification mechanism without having to memorize it.

PROBLEM 20-12

(a) The Key Mechanism for Fischer esterification omitted some important resonance forms of the intermediates shown in brackets. Complete the mechanism by drawing all the resonance forms of these two intermediates.

(b) Propose a mechanism for the acid-catalyzed reaction of acetic acid with ethanol to give ethyl acetate.

(c) The *Principle of Microscopic Reversibility* states that a forward reaction and a reverse reaction taking place under the same conditions (as in an equilibrium) must follow the same reaction pathway in microscopic detail. The reverse of the Fischer esterification is the acid-catalyzed hydrolysis of an ester. Propose a mechanism for the acid-catalyzed hydrolysis of ethyl benzoate, $PhCOOCH_2CH_3$.

PROBLEM-SOLVING HINT

The Fischer esterification mechanism is a perfect example of acid-catalyzed nucleophilic acyl substitution. You should understand this mechanism well.

PROBLEM 20-13

Most of the Fischer esterification mechanism is identical with the mechanism of acetal formation. The difference is in the final step, where a resonance-stabilized carbocation loses a proton to give the ester. Write mechanisms for the following reactions, with the comparable steps directly above and below each other. Explain why the final step of the esterification (proton loss) cannot occur in acetal formation, and show what happens instead.

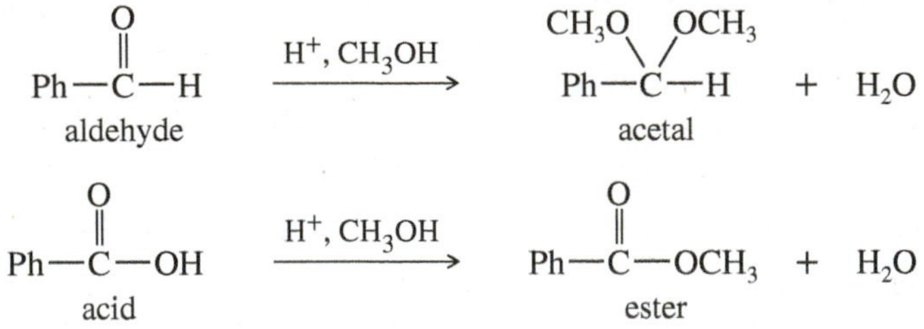

$$\underset{\text{aldehyde}}{\text{Ph}-\overset{\overset{\displaystyle O}{\|}}{C}-\text{H}} \xrightarrow{H^+,\ CH_3OH} \underset{\text{acetal}}{\text{Ph}-\overset{CH_3O\ \ OCH_3}{C}-\text{H}} + H_2O$$

$$\underset{\text{acid}}{\text{Ph}-\overset{\overset{\displaystyle O}{\|}}{C}-\text{OH}} \xrightarrow{H^+,\ CH_3OH} \underset{\text{ester}}{\text{Ph}-\overset{\overset{\displaystyle O}{\|}}{C}-OCH_3} + H_2O$$

PROBLEM 20-14

A carboxylic acid has two oxygen atoms, each with two nonbonding pairs of electrons.

(a) Draw the resonance forms of a carboxylic acid that is protonated on the hydroxy oxygen atom.

(b) Compare the resonance forms with those given previously for an acid protonated on the carbonyl oxygen atom.

(c) Explain why the carbonyl oxygen atom of a carboxylic acid is more basic than the hydroxy oxygen.

Fischer esterification is an equilibrium, and typical equilibrium constants for esterification are not very large. For example, if 1 mole of acetic acid is mixed with 1 mole of ethanol, the equilibrium mixture contains 0.65 mole each of ethyl acetate and water and 0.35 mole each of acetic acid and ethanol. Esterification using secondary and tertiary alcohols gives even smaller equilibrium constants.

Equilibrium mixture

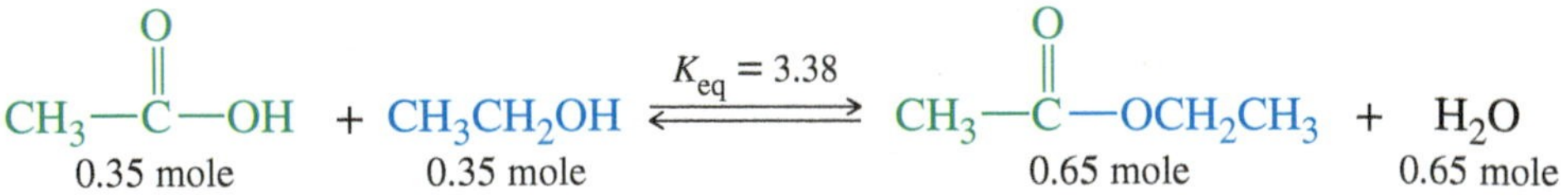

$$\underset{0.35\text{ mole}}{CH_3-\overset{\overset{\displaystyle O}{\|}}{C}-OH} + \underset{0.35\text{ mole}}{CH_3CH_2OH} \xrightleftharpoons{K_{eq} = 3.38} \underset{0.65\text{ mole}}{CH_3-\overset{\overset{\displaystyle O}{\|}}{C}-OCH_2CH_3} + \underset{0.65\text{ mole}}{H_2O}$$

Esterification may be driven to the right either by using an excess of one of the reactants or by removing one of the products. For example, in forming ethyl esters, excess ethanol is often used to drive the equilibrium as far as possible toward the ester. Alternatively, water may be removed either by distilling it out or by adding a dehydrating agent such as magnesium sulfate or **molecular sieves** (dehydrated zeolite crystals that adsorb water).

Driving the Fischer esterification toward a favorable equilibrium is rarely difficult, so this is a common method for making esters, both in the laboratory and in industry. Acid chlorides also react with alcohols to give esters (Section 20-15), but acid chlorides are more expensive, and they are more likely to promote side reactions such as dehydration of the alcohol.

PROBLEM-SOLVING HINT

In equilibrium reactions, look for ways to use an excess of a reagent or else to remove a product as it forms. Is it possible to use one of the reagents as a solvent? Can we distill off a product or drive off water?

PROBLEM 20-15

Show how Fischer esterification might be used to form the following esters. In each case, suggest a method for driving the reaction to completion.

(a) methyl salicylate **(b)** methyl formate (bp 32 °C) **(c)** ethyl phenylacetate

PROBLEM 20-16

The mechanism of the Fischer esterification was controversial until 1938, when Irving Roberts and Harold Urey of Columbia University used isotopic labeling to follow the alcohol oxygen atom through the reaction. A catalytic amount of sulfuric acid was added to a mixture of 1 mole of acetic acid and 1 mole of special methanol containing the heavy ^{18}O isotope of oxygen. After a short period, the acid was neutralized to stop the reaction, and the components of the mixture were separated.

$$CH_3-\overset{O}{\overset{\|}{C}}-O-H + CH_3-{}^{18}O-H \underset{}{\overset{H_2SO_4}{\rightleftharpoons}} CH_3-\overset{O}{\overset{\|}{C}}-O-CH_3 + H_2O$$

(a) Propose a mechanism for this reaction.
(b) Follow the labeled ^{18}O atom through your mechanism, and show where it is found in the products.
(c) The ^{18}O isotope is not radioactive. Suggest how you could experimentally determine the amounts of ^{18}O in the separated components of the mixture.

SOLVED PROBLEM 20-1

Ethyl orthoformate hydrolyzes easily in dilute acid to give formic acid and three equivalents of ethanol. Propose a mechanism for the hydrolysis of ethyl orthoformate.

$$H-C(OCH_2CH_3)_3 \xrightarrow[H_2O]{H^+} H-\overset{O}{\overset{\|}{C}}-OH + 3\ CH_3CH_2OH$$

ethyl orthoformate → formic acid + ethanol

SOLUTION

Ethyl orthoformate resembles an acetal with an extra alkoxy group, so this mechanism should resemble the hydrolysis of an acetal (Section 18-17). There are three equivalent basic sites: the three oxygen atoms. Protonation of one of these sites allows ethanol to leave, giving a resonance-stabilized cation. Attack by water gives an intermediate that resembles a hemiacetal with an extra alkoxy group. (The species in brackets are resonance-stabilized.)

$$H-C(OEt)_3 \overset{H^+}{\rightleftharpoons} H-C(OEt)_2-\overset{+}{O}(H)Et \rightleftharpoons \left[H-\overset{+}{C}(OEt)_2\right] \overset{H_2\ddot{O}:}{\rightleftharpoons} H-C(OEt)_2-\overset{+}{O}H_2 \overset{H_2\ddot{O}:}{\rightleftharpoons} H-C(OEt)_2-\ddot{O}H$$

Protonation and loss of a second ethoxy group gives an intermediate that is simply a protonated ester.

$$H-C(OEt)_2-\ddot{O}H \overset{H^+}{\rightleftharpoons} H-C(OEt)(\overset{+}{O}(H)Et)-\ddot{O}H \rightleftharpoons \left[H-C(=\overset{+}{O}-H)-\ddot{O}Et\right] \overset{H_2\ddot{O}:}{\rightleftharpoons} H-\overset{O}{\overset{\|}{C}}-\ddot{O}-Et$$

protonated ester ethyl formate

Hydrolysis of ethyl formate follows the reverse path of the Fischer esterification. This part of the mechanism is left to you as an exercise.

PROBLEM 20-17

(a) The solution given for Solved Problem 20-1 was missing some important resonance forms of the intermediates shown in brackets. Complete this mechanism by drawing all the resonance forms of these intermediates. Do your resonance forms help to explain why this reaction occurs under very mild conditions (water with a tiny trace of acid)?
(b) Finish the solution for Solved Problem 20-1 by providing a mechanism for the acid-catalyzed hydrolysis of ethyl formate.

Application: Drug Analysis

Diazomethane is often used to esterify polar or reactive compounds for mass spectrometry analysis. For example, a urine test for cocaine might involve treating the sample with diazomethane to convert benzoylecgonine, the major urinary metabolite of cocaine, to its volatile methyl ester for MS analysis.

benzoylecgonine

20-11 Esterification Using Diazomethane

Carboxylic acids are converted to their methyl esters very simply by adding an ether solution of diazomethane. The only by-product is nitrogen gas, and any excess diazomethane also evaporates. Purification of the ester usually involves only evaporation of the solvent. Yields are nearly quantitative in most cases.

$$\underset{\text{acid}}{R-\overset{O}{\overset{\|}{C}}-OH} + \underset{\text{diazomethane}}{CH_2N_2} \longrightarrow \underset{\text{methyl ester}}{R-\overset{O}{\overset{\|}{C}}-O-CH_3} + N_2\uparrow$$

Example

$$\underset{\text{cyclobutanecarboxylic acid}}{\text{cyclobutyl}-COOH} \xrightarrow{CH_2N_2} \underset{\text{methyl cyclobutanecarboxylate (100\%)}}{\text{cyclobutyl}-COOCH_3} + N_2\uparrow$$

Diazomethane is a toxic, explosive yellow gas that dissolves in ether and is fairly safe to use in ether solutions. The reaction of diazomethane with carboxylic acids probably involves transfer of the acid proton, giving a methyldiazonium salt. This diazonium salt is an excellent methylating agent, with nitrogen gas as a leaving group. Mechanism 20-3 shows this reaction.

MECHANISM 20-3 Esterification Using Diazomethane

Step 1: Proton transfer, forming a carboxylate ion and a methyldiazonium ion.

$$R-\overset{O}{\overset{\|}{C}}-O-H + \left[CH_2=\overset{+}{N}=N^{-} \longleftrightarrow {}^{-}:CH_2-\overset{+}{N}\equiv N:\right] \longrightarrow \underset{\text{carboxylate ion}}{R-\overset{O}{\overset{\|}{C}}-O:^{-}} + \underset{\text{methyldiazonium ion}}{CH_3-\overset{+}{N}\equiv N:}$$

Step 2: Nucleophilic attack on the methyl group displaces nitrogen.

$$R-\overset{O}{\overset{\|}{C}}-O:^{-} \;\; CH_3-\overset{+}{N}\equiv N: \longrightarrow R-\overset{O}{\overset{\|}{C}}-O-CH_3 + :N\equiv N:\uparrow$$

Because diazomethane is hazardous in large quantities, it is rarely used industrially or in large-scale laboratory reactions. The yields of methyl esters are excellent, however, so diazomethane is often used for small-scale esterifications of valuable and delicate carboxylic acids.

20-12 Condensation of Acids with Amines: Direct Synthesis of Amides

Amides can be synthesized directly from carboxylic acids, using heat to drive off water and force the reaction to completion. The initial acid–base reaction of a carboxylic acid with an amine gives an ammonium carboxylate salt. The carboxylate ion is a poor electrophile, and the ammonium ion is not nucleophilic, so the reaction stops at this point. Heating this salt to well above 100 °C drives off steam and forms an amide. This direct synthesis is an important industrial process, and it often works well in the laboratory.

$$\underset{\text{acid}}{R-\overset{O}{\overset{\|}{C}}-OH} + \underset{\text{amine}}{R'-\ddot{N}H_2} \rightleftarrows \underset{\text{an ammonium carboxylate salt}}{R-\overset{O}{\overset{\|}{C}}-O^- \quad H_3\overset{+}{N}-R'} \xrightarrow{\text{heat}} \underset{\text{amide}}{R-\overset{O}{\overset{\|}{C}}-\ddot{N}H-R'} + H_2O\uparrow$$

Example

$$\underset{\text{benzoic acid}}{C_6H_5COOH} + \underset{\text{ethylamine}}{CH_3CH_2\ddot{N}H_2} \longrightarrow \underset{\text{ethylammonium benzoate}}{C_6H_5COO^- \; H_3\overset{+}{N}CH_2CH_3} \xrightarrow{\text{heat}} \underset{\textit{N}\text{-ethylbenzamide}}{C_6H_5\overset{O}{\overset{\|}{C}}-\ddot{N}HCH_2CH_3} + H_2O\uparrow$$

PROBLEM 20-18

Show how to synthesize the following compounds, using appropriate carboxylic acids and amines.

(a) $CH_3-C_6H_4-\overset{O}{\overset{\|}{C}}-N(CH_2CH_3)_2$
N,N-diethyl-*meta*-toluamide (DEET insect repellent)

(b) $C_6H_5-NH-\overset{O}{\overset{\|}{C}}-CH_3$
acetanilide

(c) $H-\overset{O}{\overset{\|}{C}}-N(CH_3)_2$
N,N-dimethylformamide (DMF)

20-13 Reduction of Carboxylic Acids

Lithium aluminum hydride ($LiAlH_4$ or LAH) reduces carboxylic acids to primary alcohols. The aldehyde is an intermediate in this reduction, but it cannot be isolated because it is reduced more easily than the original acid.

$$\underset{\text{acid}}{R-\overset{O}{\overset{\|}{C}}-OH} \xrightarrow[(2)\ H_3O^+]{(1)\ LiAlH_4} \underset{\text{primary alcohol}}{R-CH_2-OH}$$

Example

$$\underset{\text{phenylacetic acid}}{C_6H_5-CH_2-\overset{O}{\overset{\|}{C}}-OH} \xrightarrow[(2)\ H_3O^+]{(1)\ LiAlH_4} \underset{\text{2-phenylethanol}}{\underset{(75\%)}{C_6H_5-CH_2-CH_2OH}}$$

Lithium aluminum hydride is a strong base, and the first step is deprotonation of the acid. Hydrogen gas is evolved, and the lithium salt results.

$$R-\overset{O}{\overset{\|}{C}}-O-H + Li^+ \; H-\overset{-}{Al}H_3 \longrightarrow H_2\uparrow + R-\overset{O}{\overset{\|}{C}}-O^- \; Li^+ + AlH_3$$

Several paths are possible for the rest of the mechanism. In one likely path, AlH_3 adds to the carbonyl group of the lithium carboxylate salt.

$$R-\overset{\ddot{O}}{\overset{\|}{C}}-\ddot{O}:^- \; Li^+ + AlH_3 \longrightarrow R-\overset{:\ddot{O}-AlH_2}{\overset{|}{C}}-H \quad (\text{with } :\ddot{O}:^- \; Li^+ \text{ on C})$$

Elimination gives an aldehyde, which is quickly reduced to a lithium alkoxide.

$$\mathrm{R{-}CH(O{-}AlH_2)(O^-\,Li^+)} \longrightarrow \underset{\text{aldehyde}}{\mathrm{R{-}CH{=}O}} + \mathrm{LiOAlH_2} \qquad \mathrm{H_3Al^-{-}H\;\;Li^+} \longrightarrow \underset{\text{lithium alkoxide}}{\mathrm{R{-}CH_2{-}O^-\,Li^+}} + \mathrm{AlH_3}$$

Water added in the second step protonates the alkoxide to the primary alcohol.

$$\mathrm{R{-}CH_2{-}O^-\;Li^+ + H_2O \longrightarrow R{-}CH_2{-}OH + LiOH}$$

Borane also reduces carboxylic acids to primary alcohols. Borane (complex with THF; see Section 8-7) reacts with the carboxyl group faster than with any other carbonyl function. It often gives excellent selectivity, as shown by the following example, where a carboxylic acid is reduced while a ketone is unaffected. ($LiAlH_4$ would also reduce the ketone.)

$$\mathrm{H_3C{-}C({=}O){-}C_6H_4{-}C({=}O){-}OH} \xrightarrow[\text{(2) } \mathrm{H_3O^+}]{\text{(1) } \mathrm{BH_3{\cdot}THF}\ (\text{or } \mathrm{B_2H_6})} \underset{(80\%)}{\mathrm{H_3C{-}C({=}O){-}C_6H_4{-}CH_2OH}}$$

Reduction to Aldehydes Reduction of carboxylic acids to aldehydes is difficult because aldehydes are more reactive than carboxylic acids toward most reducing agents. Almost any reagent that reduces acids to aldehydes also reduces aldehydes to primary alcohols. In Section 18-10, we saw that lithium tri-*tert*-butoxyaluminum hydride, $LiAlH(O\text{-}t\text{-}Bu)_3$ is a weaker reducing agent than lithium aluminum hydride. It reduces acid chlorides to aldehydes because acid chlorides are strongly activated toward nucleophilic addition of a hydride ion. Under these conditions, the aldehyde reduces more slowly and can be isolated. Therefore, reduction of an acid to an aldehyde is a two-step process: Convert the acid to the acid chloride (Section 20-15), and then reduce using lithium tri-*tert*-butoxyaluminum hydride.

$$\underset{\text{acid chloride}}{\mathrm{R{-}C({=}O){-}Cl}} + \mathrm{LiAlH(O\text{-}\mathit{t}\text{-}Bu)_3} \rightleftarrows \mathrm{R{-}CH(O^-\,Li^+){-}Cl} + \mathrm{Al(O\text{-}\mathit{t}\text{-}Bu)_3} \longrightarrow \underset{\text{aldehyde}}{\mathrm{R{-}C({=}O){-}H}} + \mathrm{LiCl}$$

Example

Step 1: Conversion to the acid chloride. *Step 2:* Reduction to the aldehyde.

$$\underset{\text{isobutyric acid}}{\mathrm{CH_3{-}CH(CH_3){-}C({=}O){-}OH}} \xrightarrow{\mathrm{SOCl_2}} \underset{\text{isobutyryl chloride}}{\mathrm{CH_3{-}CH(CH_3){-}C({=}O){-}Cl}} \xrightarrow{\mathrm{LiAlH(O\text{-}\mathit{t}\text{-}Bu)_3}} \underset{\text{isobutyraldehyde}}{\mathrm{CH_3{-}CH(CH_3){-}C({=}O){-}H}}$$

PROBLEM 20-19

Show how you would synthesize the following compounds from the appropriate carboxylic acids or acid derivatives.

(a) $C_6H_5{-}CH_2CH_2OH$ (b) $C_6H_5{-}CH_2CHO$ (c) O= (cyclopentane ring) $-CH_2OH$

20-14 Alkylation of Carboxylic Acids to Form Ketones

Carboxylic acids react with two equivalents of an organolithium reagent to give ketones. This reaction was discussed in Section 18-8.

$$R{-}C({=}O){-}O{-}H \xrightarrow[(2)\ H_2O]{(1)\ 2\ R'{-}Li} R{-}C({=}O){-}R' + R'{-}H$$

Example

$$C_6H_5{-}COOH \xrightarrow[(2)\ H_2O]{(1)\ 2\ CH_3CH_2{-}Li} C_6H_5{-}C({=}O){-}CH_2CH_3$$

benzoic acid → propiophenone

The first equivalent of the organolithium reagent simply deprotonates the acid. The second equivalent adds to the carbonyl to give a stable dianion. Hydrolysis of the dianion (by adding water) gives the hydrate of a ketone. Because the ketone is formed in a separate hydrolysis step (rather than in the presence of the organolithium reagent), overalkylation is not observed.

$$\underset{\text{carboxylic acid}}{R{-}C({=}O){-}OH} \xrightarrow{2\ R'{-}Li} R'{-}H + R{-}C({=}O){-}OLi \ (+\ R'{-}Li) \longrightarrow \underset{\text{dianion}}{R{-}C(OLi)(R'){-}OLi}$$

$$\underset{\text{dianion}}{R{-}C(OLi)(R'){-}OLi} \xrightarrow{H_3O^+} \underset{\text{hydrate of ketone}}{R{-}C(OH)(R'){-}OH} \rightleftarrows \underset{\text{ketone}}{R{-}C({=}O){-}R'} + H_2O$$

PROBLEM 20-20

Propose a mechanism for conversion of the dianion to the ketone under mildly acidic conditions.

PROBLEM 20-21

Show how the following ketones might be synthesized from the indicated acids, using any necessary reagents.

(a) propiophenone from propionic acid (two ways, using alkylation of the acid and using Friedel–Crafts acylation)

(b) methyl cyclohexyl ketone from cyclohexanecarboxylic acid

20-15 Synthesis and Use of Acid Chlorides

Halide ions are excellent leaving groups for nucleophilic acyl substitution. Therefore, acyl halides are useful intermediates for making acid derivatives. In particular, acid chlorides (acyl chlorides) are easily made and are commonly used as an activated form of a carboxylic acid. Both the carbonyl oxygen and the chlorine atom withdraw electron density from the acyl carbon atom, making it strongly electrophilic. Acid chlorides react with a wide range of nucleophiles, generally through the addition–elimination mechanism of nucleophilic acyl substitution.

an acid chloride (acyl chloride)

$$\text{R—C(=O)—Cl} + \text{Nuc:}^- \underset{}{\overset{\text{addition}}{\rightleftharpoons}} \left[\text{R—C(O}^-\text{)(Nuc)—Cl}\right] \xrightarrow{\text{elimination}} \text{R—C(=O)—Nuc} + \text{Cl}^-$$

acid chloride — tetrahedral intermediate — acid derivative

The best reagents for converting carboxylic acids to acid chlorides are thionyl chloride ($SOCl_2$) and oxalyl chloride [$(COCl)_2$] because they form gaseous by-products that do not contaminate the product. Oxalyl chloride is particularly easy to use because it boils at 62 °C and any excess is easily evaporated from the reaction mixture.

$$\text{R—C(=O)—OH} \xrightarrow[\text{or Cl—C(=O)—C(=O)—Cl}]{\text{Cl—S(=O)—Cl}} \text{R—C(=O)—Cl}$$

Examples

$$\text{CH}_3(\text{CH}_2)_7\text{CH=CH(CH}_2)_7\text{—C(=O)—OH} \xrightarrow[\text{thionyl chloride}]{\text{Cl—S(=O)—Cl}} \text{CH}_3(\text{CH}_2)_7\text{CH=CH(CH}_2)_7\text{—C(=O)—Cl} + \text{SO}_2\uparrow + \text{HCl}\uparrow$$

oleic acid — oleoyl chloride (95%)

$$\text{C}_6\text{H}_5\text{—CH}_2\text{—CH}_2\text{—C(=O)—OH} \xrightarrow[\text{oxalyl chloride}]{\text{Cl—C(=O)—C(=O)—Cl}} \text{C}_6\text{H}_5\text{—CH}_2\text{—CH}_2\text{—C(=O)—Cl} + \text{HCl}\uparrow + \text{CO}\uparrow + \text{CO}_2\uparrow$$

3-phenylpropanoic acid — 3-phenylpropanoyl chloride (95%)

The mechanisms of these reactions begin like the reaction of an alcohol with thionyl chloride. Either oxygen atom of the acid can attack sulfur, replacing chloride by a mechanism that looks like sulfur's version of nucleophilic acyl substitution. The product is an interesting, reactive chlorosulfite anhydride.

$$\text{R—C(OH)=O} + \text{SOCl}_2 \longrightarrow \left[\text{R—C(OH)=O}^+\text{—S(Cl)}_2\text{—O}^- \longleftrightarrow \text{R—C(=O}^+\text{H)—O—S(Cl)}_2\text{—O}^-\right]$$

$$\text{R—C(=O}^+\text{H)—O—S(Cl)}_2\text{—O}^- \longrightarrow \text{R—C(=O}^+\text{H)—O—S(=O)—Cl} + \text{Cl}^- \longrightarrow \text{R—C(=O)—O—S(=O)—Cl} + \text{HCl}$$

a chlorosulfite anhydride

This reactive anhydride undergoes nucleophilic acyl substitution by chloride ion to give the acid chloride.

$$\text{R—C(=O)—O—S(=O)—Cl} + \text{Cl}^- \longrightarrow \text{R—C(O}^-\text{)(Cl)—O—S(=O)—Cl} \longrightarrow \text{R—C(=O)—Cl} + \text{SO}_2 + \text{Cl}^-$$

***PROBLEM 20-22**

Propose a mechanism for the reaction of benzoic acid with oxalyl chloride. This mechanism begins like the thionyl chloride reaction, to give a reactive mixed anhydride. Nucleophilic acyl substitution by chloride ion gives a tetrahedral intermediate that eliminates a leaving group, which then fragments into carbon dioxide, carbon monoxide, and chloride ion.

Acid chlorides react with alcohols to give esters through a nucleophilic acyl substitution by the addition–elimination mechanism discussed on the previous page. Attack by the alcohol at the electrophilic carbonyl group gives a tetrahedral intermediate. Loss of chloride and deprotonation give the ester.

$$R-C(=O)-Cl + R'-\ddot{O}H \rightleftharpoons \left[R-C(O^-)(Cl)-\overset{+}{O}(H)-R' \right] \longrightarrow R-C(=O)-\overset{+}{O}(H)-R' + Cl^- \longrightarrow R-C(=O)-\ddot{O}-R' + HCl$$

ester

This reaction provides an efficient two-step method for converting a carboxylic acid to an ester. The acid is converted to the acid chloride, which reacts with an alcohol to give the ester. Pyridine or other bases are often added to neutralize the HCl generated. Otherwise, alcohols (especially tertiary alcohols) may dehydrate under strongly acidic conditions.

$$\underset{\text{acid}}{R-C(=O)-OH} \xrightarrow[\text{or } SOCl_2]{(COCl)_2} \underset{\text{acid chloride}}{R-C(=O)-Cl} \xrightarrow[\text{alcohol}]{R'-O-H} \underset{\text{ester}}{R-C(=O)-O-R'} + HCl$$

Example

$$\underset{\text{benzoic acid}}{Ph-C(=O)-OH} \xrightarrow{SOCl_2} \underset{\text{benzoyl chloride}}{Ph-C(=O)-Cl} \xrightarrow[\text{pyridine}]{CH_3CH_2-OH} \underset{\text{ethyl benzoate}}{Ph-C(=O)-O-CH_2CH_3} + C_5H_5\overset{+}{N}-H\ Cl^-$$

Ammonia and amines react with acid chlorides to give amides, also through the addition–elimination mechanism of nucleophilic acyl substitution. A carboxylic acid is efficiently converted to an amide by forming the acid chloride, which reacts with an amine to give the amide. A base such as pyridine or NaOH is often added to prevent HCl from protonating the amine.

$$\underset{\text{acid chloride}}{R-C(=O)-Cl} + \underset{\text{amine}}{R'-\ddot{N}H_2} \longrightarrow \underset{\text{amide}}{R-C(=O)-\ddot{N}H-R'} + HCl$$

Example

$$\underset{\text{acetyl chloride}}{CH_3-C(=O)-Cl} + \underset{\text{methylamine}}{CH_3-\ddot{N}H_2} \xrightarrow{NaOH} \underset{N\text{-methylacetamide}}{CH_3-C(=O)-\ddot{N}(H)-CH_3} + NaCl + H_2O$$

PROBLEM 20-23

Propose mechanisms for the nucleophilic acyl substitutions to form ethyl benzoate and *N*-methylacetamide as shown on the previous page.

PROBLEM 20-24

Show how you would use an acid chloride as an intermediate to synthesize

(a) *N*-phenylbenzamide (PhCONHPh) from benzoic acid and aniline.

(b) phenyl propionate (CH_3CH_2COOPh) from propionic acid and phenol.

SUMMARY Reactions of Carboxylic Acids

General types of reactions

$$\mathrm{R{-}C({=}O){-}OH} \longrightarrow \begin{cases} \mathrm{R{-}C({=}O){-}O^-} & \text{deprotonation} \\ \mathrm{R{-}C({=}O){-}Nuc} & \text{nucleophilic acyl substitution} \\ \mathrm{R{-}CH_2{-}OH} & \text{reduction} \\ \mathrm{R{-}H} + \mathrm{CO_2} & \text{decarboxylation (Chapter 22)} \end{cases}$$

1. *Salt formation* (Section 20-5)

$$\underset{\text{acid}}{\mathrm{R{-}C({=}O){-}OH}} + \underset{\text{strong base}}{\mathrm{M^+\ {}^-OH}} \rightleftarrows \underset{\text{salt}}{\mathrm{R{-}C({=}O){-}O^-\ M^+}} + \mathrm{H_2O}$$

2. *Conversion to esters* (Sections 20-10, 20-11, and 20-15)

Fischer esterification:

$$\underset{\text{acid}}{\mathrm{R{-}C({=}O){-}OH}} + \underset{\text{alcohol}}{\mathrm{R'{-}OH}} \underset{}{\overset{\mathrm{H^+}}{\rightleftarrows}} \underset{\text{ester}}{\mathrm{R{-}C({=}O){-}O{-}R'}} + \mathrm{H_2O}$$

$$\underset{\text{acid chloride}}{\mathrm{R{-}C({=}O){-}Cl}} + \underset{\text{alcohol}}{\mathrm{R'{-}OH}} \longrightarrow \underset{\text{ester}}{\mathrm{R{-}C({=}O){-}O{-}R'}} + \mathrm{HCl}\uparrow$$

$$\underset{\text{acid}}{\mathrm{R{-}C({=}O){-}OH}} + \underset{\text{diazomethane}}{\mathrm{CH_2N_2}} \longrightarrow \underset{\text{methyl ester}}{\mathrm{R{-}C({=}O){-}O{-}CH_3}} + \mathrm{N_2}\uparrow$$

3. *Conversion to amides* (Sections 20-12 and 20-15)

$$\underset{\text{acid}}{\mathrm{R{-}C({=}O){-}OH}} + \underset{\text{amine}}{\mathrm{R'{-}NH_2}} \rightleftarrows \underset{\text{salt}}{\mathrm{R{-}C({=}O){-}O^-\ H_3\overset{+}{N}{-}R'}} \xrightarrow{\text{heat}} \underset{\text{amide}}{\mathrm{R{-}C({=}O){-}NH{-}R'}} + \mathrm{H_2O}$$

$$\underset{\text{acid chloride}}{\mathrm{R{-}C({=}O){-}Cl}} + \underset{\text{amine}}{\mathrm{R'{-}NH_2}} \xrightarrow{\mathrm{NaOH}} \underset{\text{amide}}{\mathrm{R{-}C({=}O){-}NH{-}R'}} + \mathrm{NaCl} + \mathrm{H_2O}$$

4. *Conversion to anhydrides* (Section 21-5)

$$\underset{\text{acid chloride}}{\mathrm{R{-}C({=}O){-}Cl}} + \underset{\text{acid}}{\mathrm{HO{-}C({=}O){-}R'}} \longrightarrow \underset{\text{acid anhydride}}{\mathrm{R{-}C({=}O){-}O{-}C({=}O){-}R'}} + \mathrm{HCl}$$

5. *Reduction to primary alcohols* (Sections 10-11 and 20-13)

$$\underset{\text{acid}}{\mathrm{R{-}C({=}O){-}OH}} \xrightarrow[\substack{(2)\ \mathrm{H_3O^+} \\ \text{(or use } \mathrm{BH_3\cdot THF})}]{(1)\ \mathrm{LiAlH_4}} \underset{\text{primary alcohol}}{\mathrm{R{-}CH_2{-}OH}}$$

6. *Reduction to aldehydes* (Sections 18-10 and 20-13)

$$\underset{\text{acid}}{R-\overset{O}{\overset{\|}{C}}-OH} + \underset{\text{thionyl chloride}}{Cl-\overset{O}{\overset{\|}{S}}-Cl} \longrightarrow \underset{\text{acid chloride}}{R-\overset{O}{\overset{\|}{C}}-Cl} \xrightarrow{LiAlH(O\text{-}t\text{-}Bu)_3} \underset{\text{aldehyde}}{R-\overset{O}{\overset{\|}{C}}-H}$$

7. *Alkylation to form ketones* (Sections 18-8 and 20-14)

$$\underset{\text{lithium carboxylate}}{R-\overset{O}{\overset{\|}{C}}-O^- \; Li^+} \xrightarrow[(2)\ H_2O]{(1)\ R'-Li \text{ alkyllithium}} \underset{\text{ketone}}{R-\overset{O}{\overset{\|}{C}}-R'}$$

8. *Conversion to acid chlorides* (Section 20-15)

$$\underset{\text{acid}}{R-\overset{O}{\overset{\|}{C}}-OH} + \underset{\text{thionyl chloride}}{Cl-\overset{O}{\overset{\|}{S}}-Cl} \longrightarrow \underset{\text{acid chloride}}{R-\overset{O}{\overset{\|}{C}}-Cl} + SO_2\uparrow + HCl\uparrow$$

9. *Side-chain halogenation* (Hell–Volhard–Zelinsky reaction; Section 22-6)

$$R-CH_2-\overset{O}{\overset{\|}{C}}-OH \xrightarrow{Br_2/PBr_3} \underset{\alpha\text{-bromo acyl bromide}}{R-\overset{Br}{\overset{|}{CH}}-\overset{O}{\overset{\|}{C}}-Br} \xrightarrow{H_2O} \underset{\alpha\text{-bromoacid}}{R-\overset{Br}{\overset{|}{CH}}-\overset{O}{\overset{\|}{C}}-OH} + HBr$$

SUMMARY Reactions of Carboxylic Acids

Carboxylic acids are abundant in nature (e.g., cell biology and biochemistry) and commerce. More reactions will be discussed in Chapter 21.

Fischer esterification—
Nucleophilic Acyl Substitution (discussed in Ch 21); Section 20-10

Reduction—
both borane (BH_3) and $LiAlH_4$ reduce carboxylic acids to primary alcohols; Section 20-13

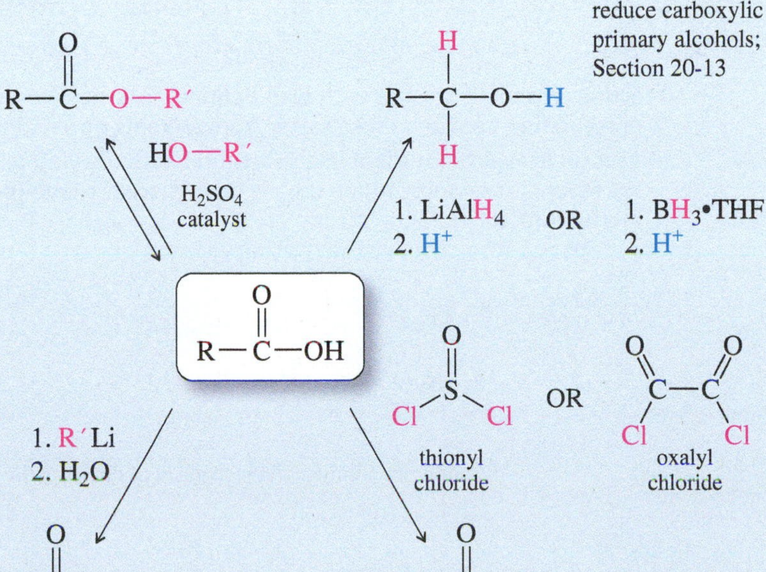

Alkylation to form ketones—
Sections 20-14, 18-8

Formation of acid chlorides—
both thionyl chloride ($SOCl_2$) and oxalyl chloride (ClCOCOCl) can be used; thionyl chloride is a federally-controlled chemical so oxalyl chloride is more common; Section 20-15

Essential Terms

acid chloride — **(acyl chloride)** An activated acid derivative in which the hydroxy group of the acid is replaced by a chlorine atom. (p. 1026)

anhydride — **(acid anhydride)** A composite of two acid molecules, with loss of water. Adding water to an anhydride regenerates the acid. A *mixed anhydride* contains two different acids. (p. 1032)

$$CH_3-C(=O)-O-C(=O)-CH_3 + H_2O \rightleftharpoons 2\ CH_3-C(=O)-OH \qquad Ph-C(=O)-O-C(=O)-CH_3$$

acetic anhydride — acetic acid — acetic benzoic anhydride

carboxyl group — The —COOH functional group of a carboxylic acid. (p. 1002)

carboxylate ion — The anion resulting from deprotonation of a carboxylic acid. (p. 1002)

carboxylation — A reaction in which a compound (usually a carboxylic acid) is formed by the addition of CO_2 to an intermediate. The addition of CO_2 to a Grignard reagent is an example of a carboxylation. (p. 1021)

carboxylic acid — Any compound containing the *carboxyl group*, —COOH. (p. 1002)

- **An aliphatic acid** has an alkyl group bonded to the carboxyl group.
- **An aromatic acid** has an aryl group bonded to the carboxyl group.
- **A dicarboxylic acid** (a diacid) has two carboxyl groups. (p. 1005)

fatty acid — A long-chain linear carboxylic acid. Some fatty acids are saturated, and others are unsaturated. (pp. 1002, 1014)

Fischer esterification — The acid-catalyzed reaction of a carboxylic acid with an alcohol to form an ester. (p. 1024)

$$R-C(=O)-O-H + R'-OH \overset{H^+}{\rightleftharpoons} R-C(=O)-O-R' + H_2O$$

acid — alcohol — ester

molecular sieves — Dehydrated zeolite crystals with well-defined pore sizes to admit molecules smaller than the pores. Often used to adsorb water from solvents or reactions. (p. 1026)

nucleophilic acyl substitution — A reaction in which a nucleophile substitutes for a leaving group on a carbonyl carbon atom. Nucleophilic acyl substitution usually takes place through the following addition–elimination mechanism. (p. 1022)

$$R-C(=\ddot{O}:)-X + Nuc:^- \rightleftharpoons R-C(-\ddot{O}:^-)(Nuc)-X \longrightarrow R-C(=\ddot{O}:)-Nuc + :X^-$$

the addition–elimination mechanism of nucleophilic acyl substitution

phthalic acids — Benzenedicarboxylic acids. *Phthalic acid* itself is the ortho isomer. The meta isomer is *isophthalic acid*, and the para isomer is *terephthalic acid*. (p. 1005)

salt of a carboxylic acid — An ionic compound containing the deprotonated anion of a carboxylic acid, called the *carboxylate ion:* $R-COO^-$. An acid salt is formed by the reaction of an acid with a base. (p. 1011)

20-37 Predict the products and propose mechanisms for the following reactions.

(a) $C_6H_5C(=O)OCH_2CH_3 \xrightarrow[\text{excess } H_2O]{H^+}$

(b) $C_6H_5C(=O)OCH_2CH_3 \xrightarrow[\text{excess } H_2O]{^-OH}$

(c) HO∕∖∕∖∕COOH $\xrightarrow[\text{remove } H_2O]{H^+}$

(d) HO∕∖∕∖∕COOH $\xrightarrow[\text{remove } H_2O]{^-OH}$

20-38 When pure (*S*)-lactic acid is esterified by racemic butan-2-ol, the product is 2-butyl lactate, with the following structure:

$$\underset{\text{lactic acid}}{CH_3-\overset{OH}{\overset{|}{C}H}-COOH} + \underset{\text{butan-2-ol}}{CH_3-\overset{OH}{\overset{|}{C}H}-CH_2CH_3} \overset{H^+}{\rightleftarrows} \underset{\text{2-butyl lactate}}{CH_3-\overset{OH}{\overset{|}{C}H}-\overset{O}{\overset{\|}{C}}-O-\overset{CH_3}{\overset{|}{C}H}-CH_2CH_3}$$

(a) Draw three-dimensional structures of the two stereoisomers formed, specifying the configuration at each asymmetric carbon atom. (Using your models may be helpful.)

(b) Determine the relationship between the two stereoisomers you have drawn.

20-39 Show how you would accomplish the following multistep syntheses. You may use any additional reagents and solvents you need.

(a) $PhCH_2CH_2OH \longrightarrow PhCH_2CH_2COOH$

(b) methylenecyclohexane (=CH_2) ⟶ 1-methylcyclohexane-1-carboxylic acid (CH_3, COOH)

(c) methylenecyclohexane (=CH_2) ⟶ cyclohexyl–CH_2COOH

(d) bromotetralone (Br) ⟶ tetralone carboxylic acid (COOH)

(e) cyclohexyl–COOH ⟶ 2,2-dicyclohexyl-1,3-dioxane (O, O)

(f) 2 cyclopentyl–C(=O)OH ⟶ cyclopentyl–CH_2–O–C(=O)–cyclopentyl

20-40 The following NMR spectra correspond to compounds of formulas (**A**) $C_9H_{10}O_2$, (**B**) $C_4H_6O_2$, and (**C**) $C_6H_{10}O_2$, respectively. Propose structures, and show how they are consistent with the observed absorptions.

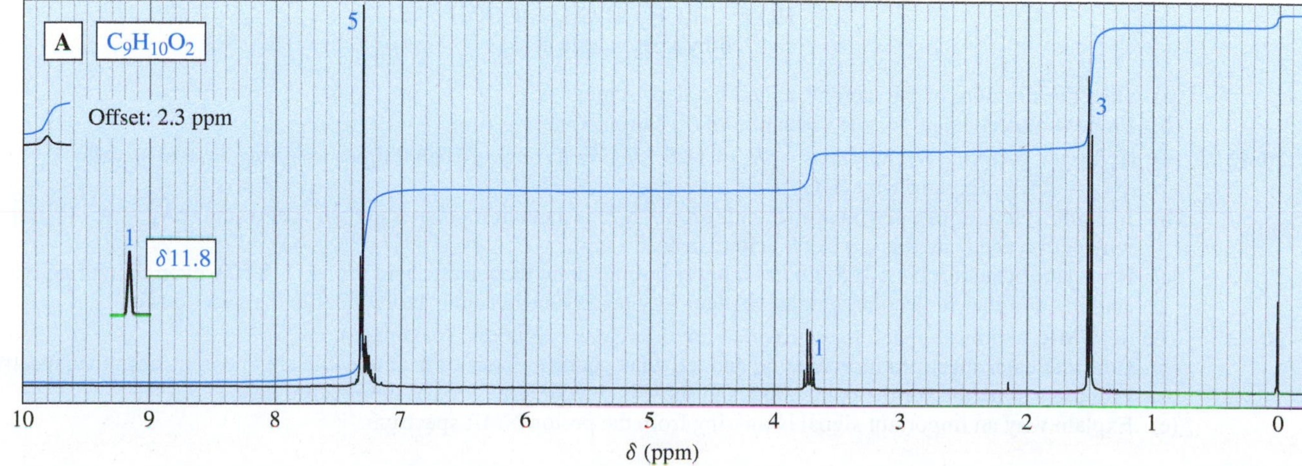

(continued)

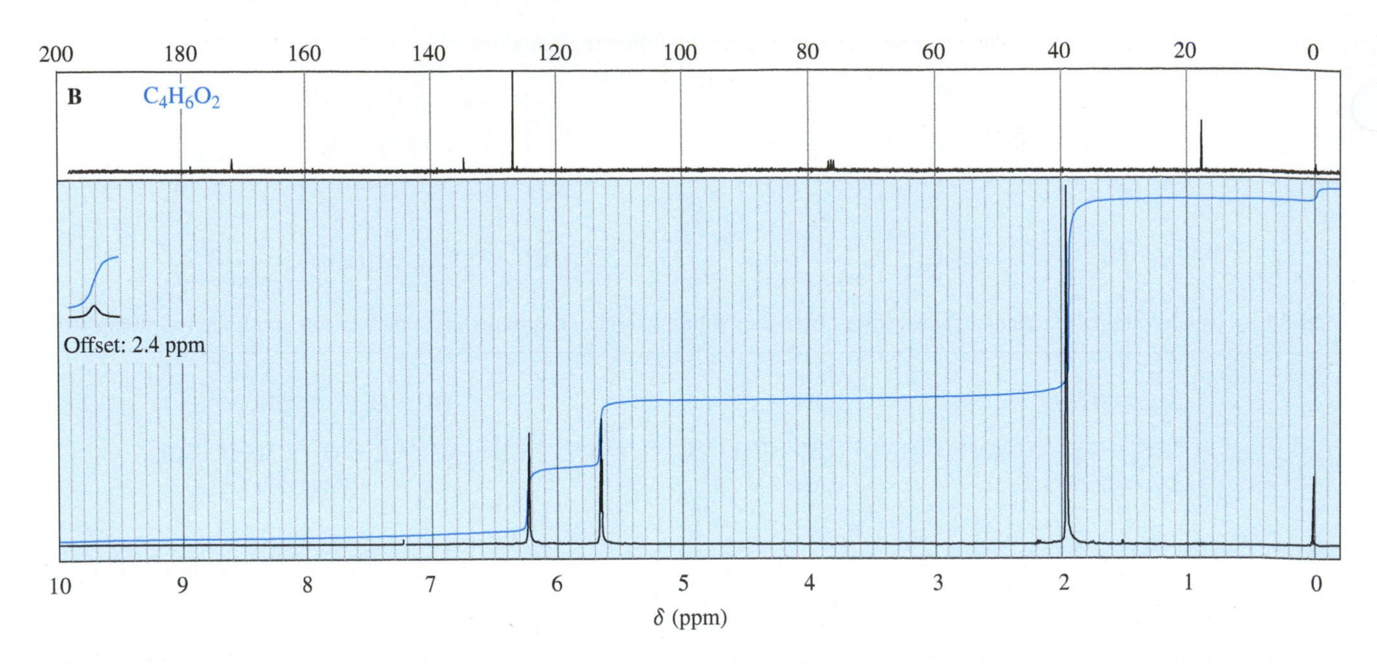

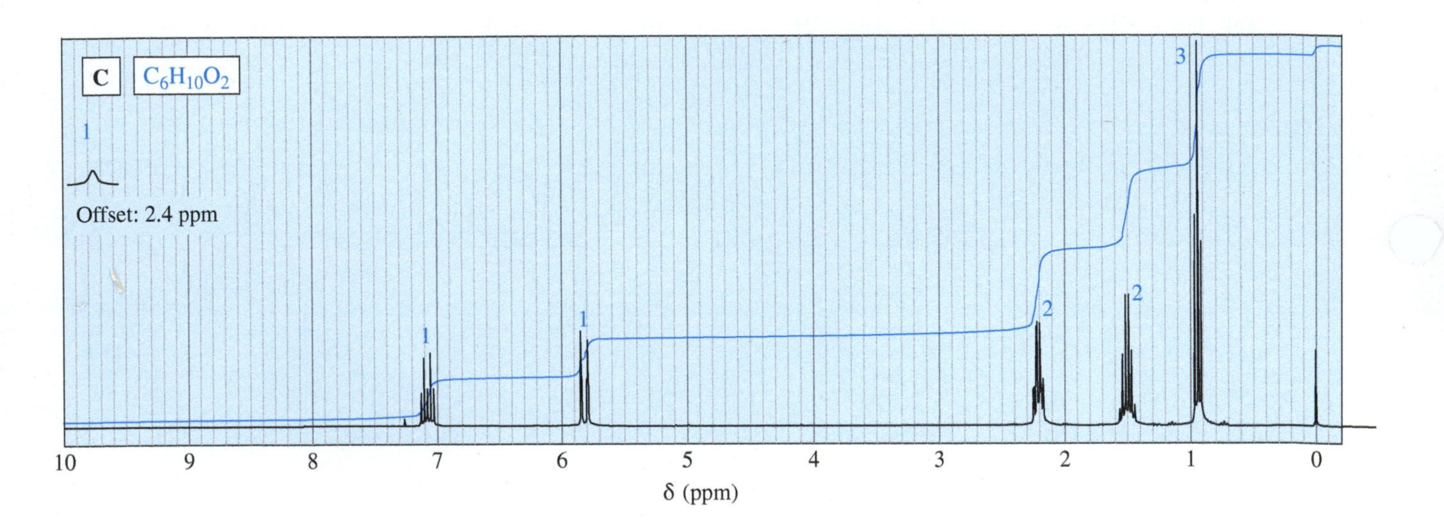

20-41 In the presence of a trace of acid, δ-hydroxyvaleric acid forms a cyclic ester (lactone).

$$HO—CH_2CH_2CH_2CH_2—COOH$$
δ-hydroxyvaleric acid

(a) Give the structure of the lactone, called δ-valerolactone.
(b) Propose a mechanism for the formation of δ-valerolactone.

20-42 **(a)** Hydrogen peroxide (HOOH) has a pK_a of 11.6, making it roughly 10,000 times as strong an acid as water ($pK_a = 15.7$). Explain why H_2O_2 is a stronger acid than H_2O.
(b) In contrast to part (a), peroxyacetic acid ($pK_a = 8.2$) is a much *weaker* acid than acetic acid ($pK_a = 4.74$). Explain why peroxyacetic acid is a weaker acid than acetic acid.
(c) Peroxyacetic acid (bp = 105 °C) has a lower boiling point than acetic acid (bp = 118 °C), even though peroxyacetic acid has a higher molecular weight. Explain why peroxyacetic acid is more volatile than acetic acid.

20-43 The IR, NMR, and mass spectra are provided for an organic compound (next page).
(a) Consider each spectrum individually, and tell what characteristics of the molecule are apparent from that spectrum.
(b) Propose a structure for the compound, and show how your structure fits the spectral data.
*__(c)__ Explain why an important signal is missing from the proton NMR spectrum.

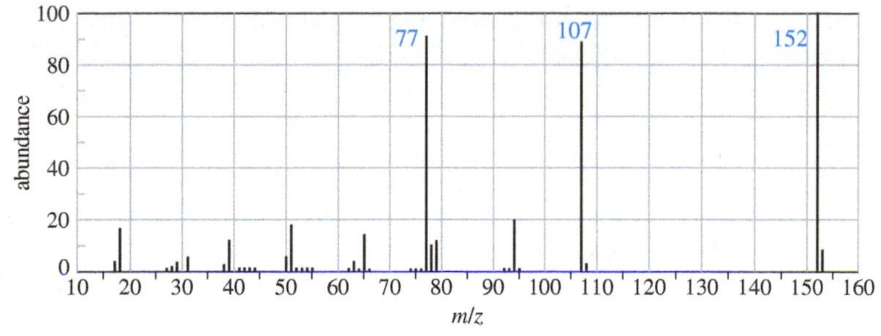

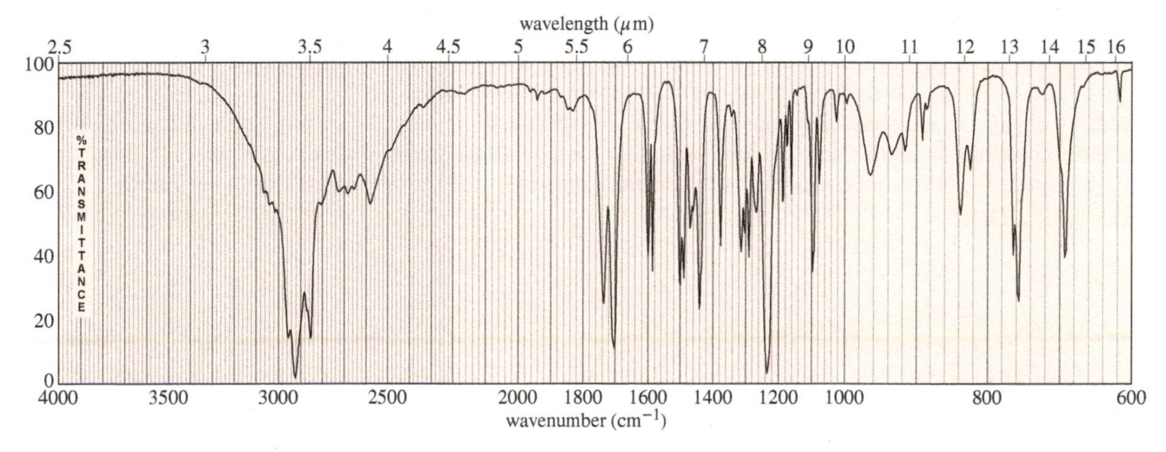

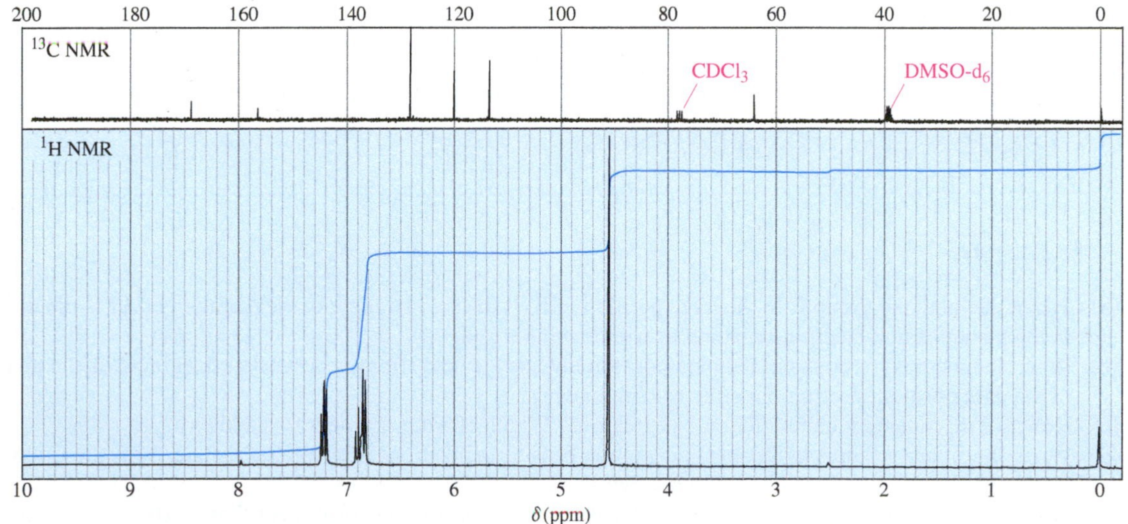

20-44 Two of the methods for converting alkyl halides to carboxylic acids are covered in Sections 20-8B and 20-8C. One is formation of a Grignard reagent followed by addition of carbon dioxide and then dilute acid. The other is substitution by cyanide ion, followed by hydrolysis of the resulting nitrile. For each of the following conversions, decide whether either or both of these methods would work, and explain why. Show the reactions you would use.

(a) (benzene ring)–CH_2Br **(b)** (benzene ring)–Br **(c)** HO–(benzene ring)–CH_2Br

(d) HO–(benzene ring)–Br **(e)** (cyclopentane)–Br **(f)** HO–(cyclopentane)–CH_2Br

20-45 (A true story) The manager of an organic chemistry stockroom prepared unknowns for a "Ketones and Aldehydes" experiment by placing two drops of the liquid unknowns in test tubes and storing the test tubes for several days until they were needed. One of the unknowns was misidentified by every student. This unknown was taken from a bottle marked "Heptaldehyde." The stockroom manager took an IR spectrum of the liquid in the bottle and found a sharp carbonyl stretch around 1720 cm^{-1} and small, sharp peaks around 2710 and 2810 cm^{-1}, as shown on the next page.

(continued)

The students complained that their spectra showed no peaks at 2710 or 2810 cm^{-1}, but a broad absorption centered over the 3000 cm^{-1} region and a carbonyl peak around 1715 cm^{-1}. They also maintained that their samples are soluble in dilute aqueous sodium hydroxide.

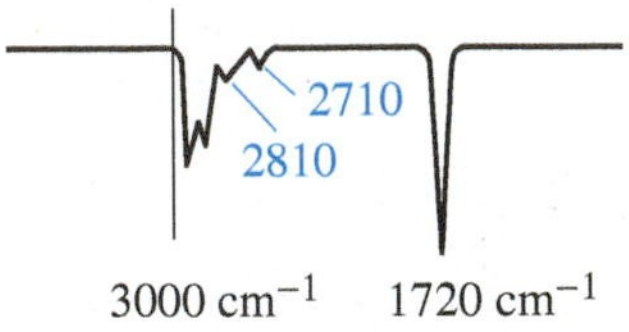

stockroom manager's spectrum

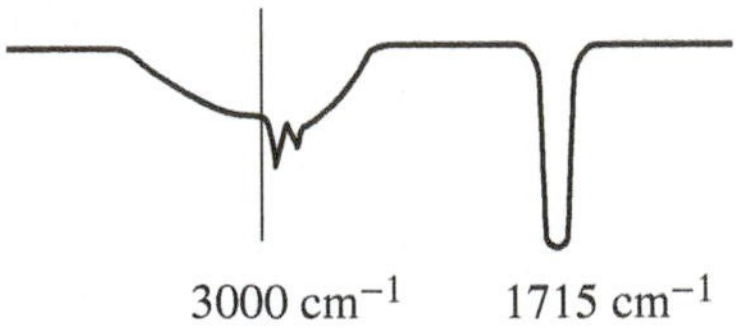

students' spectrum

(a) Identify the compound in the stockroom manager's bottle and the compound in the students' test tubes.
(b) Explain the discrepancy between the stockroom manager's spectrum and the students' results.
(c) Suggest how this misunderstanding might be prevented in the future.

20-46 Predict the major form of each compound when it is dissolved in pure water. To do this, you will need to estimate pK_a values for the compounds in (a), (b), and (c) based on similar compounds shown in Appendix 4. Values for the compound in (d) are given. Explain the differences, including why the pK_a values in part (d) are so different from the others.

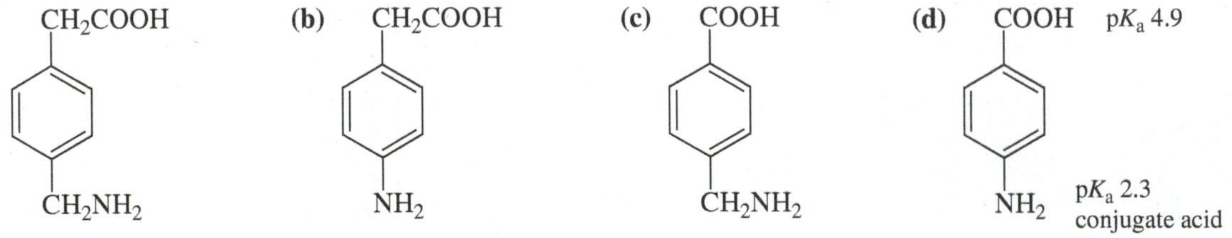

***20-47** Substituent effects: resonance and induction.

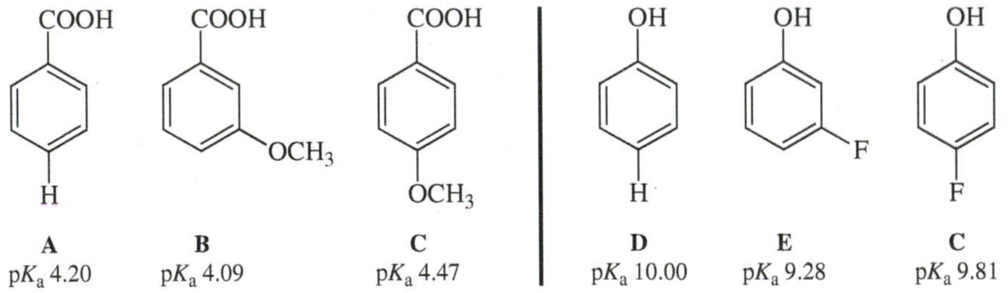

(a) When the methoxy group is in the meta position in **B**, is the acid stronger or weaker than in **A**? Is the methoxy group in the meta position electron-donating or withdrawing?
(b) When the methoxy group is in the para position in **C**, is the acid stronger or weaker than in **A**? Is the methoxy group in the para position electron-donating or withdrawing?
(c) How can this apparent contradiction be explained? Which effect is stronger for methoxy?
(d) When the fluoro group is in the meta position in **E**, is the acid stronger or weaker than in **D**? Is the fluoro group in the meta position electron-donating or withdrawing?
(e) When the fluoro group is in the para position in **F**, is the acid stronger or weaker than in **D**? Is the fluoro group in the para position electron-donating or withdrawing?
(f) Compare the results of the fluoro group with those of the above methoxy group. What must be different about the relative strength of the resonance and inductive effects for fluoro compared with methoxy?

***20-48** A student synthesized Compound **1** (below). To purify the compound, he extracted it into aqueous base and then acidified the solution to protonate the acid so that he could extract it back into ether. When he evaporated the ether, he found that his product had been converted entirely into Compound **2.**

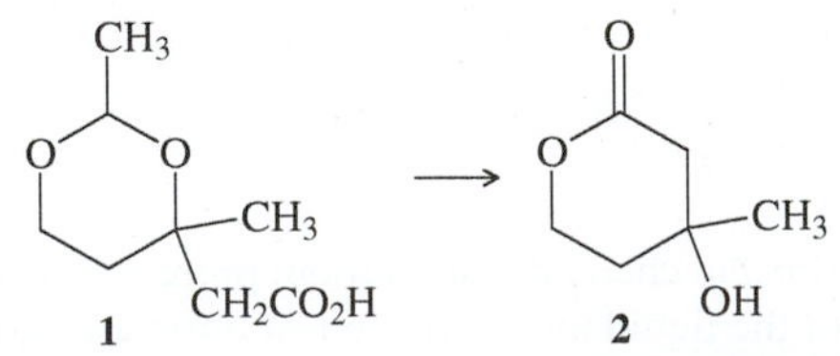

(a) What is the functional group that forms the ring in Compound **1**? In Compound **2**?
(b) How many carbon atoms are there in Compound **1**? In Compound **2**? Where did the other carbon atoms go?
(c) When did the reaction occur: When the student added the base, or when he added the acid?
(d) Propose a mechanism for the conversion of Compound **1** to Compound **2**.

21 Carboxylic Acid Derivatives

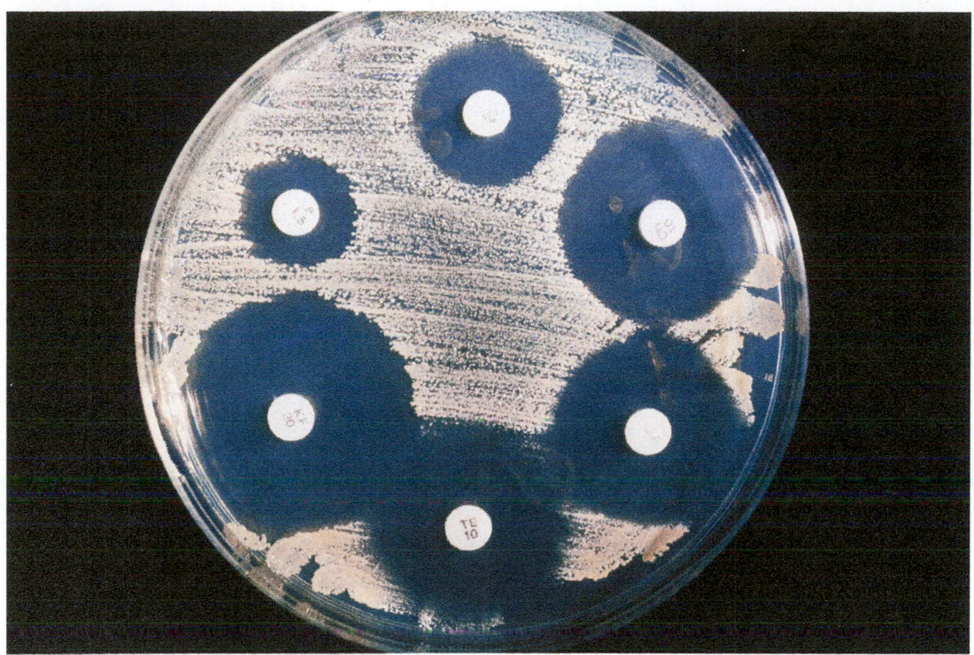

penicillin V
a penicillin

cephalexin (Keflex®)
a cephalosporin

imipenem (Primaxin®)
a carbapenem

▲ A bacterial antibiotic sensitivity test: Thin wafers containing different antibiotics are placed on an agar plate that has been inoculated with a pathogenic organism, and then incubated. The bacteria can grow around antibiotics to which they have developed a resistance. Wider rings correspond to more effective antibiotics against this organism. The active functional group in many antibiotics is a four-membered cyclic amide grouping: a β-lactam. The structures show three common types of β-lactam antibiotics. (See Section 21-13.)

Goals for Chapter 21

1 Draw and name carboxylic acid derivatives, and use spectral information to determine their structures.

2 Describe the trends in physical properties of acid derivatives, and compare the relative reactivity of esters, thioesters, amides, nitriles, anhydrides, and acid chlorides.

3 Propose single-step and multistep syntheses of acid derivatives from compounds containing other functional groups.

4 Predict the products and propose mechanisms for the reactions of carboxylic acid derivatives with reducing agents, alcohols, amines, and organometallic reagents.

5 Propose multistep syntheses using acid derivatives as starting materials and intermediates.

21-1 Introduction

Carboxylic acid derivatives are defined as *compounds with functional groups that can be converted to carboxylic acids by a simple acidic or basic hydrolysis.* The most important acid derivatives are esters, amides, and nitriles. Acid halides and anhydrides are also included in this group, although we often think of them as activated forms of the parent acids rather than completely different compounds.

	acid halide	anhydride	ester	amide	nitrile
	$R-C(=O)-X$	$R-C(=O)-O-C(=O)-R$	$R-C(=O)-O-R'$	$R-C(=O)-NH_2$	$R-C\equiv N$
Condensed structure:	RCOX	$(RCO)_2O$	RCO_2R'	$RCONH_2$	RCN

Many advances in organic chemistry involve making and using derivatives of carboxylic acids. Proteins are bonded by amide functional groups, and chemists have created synthetic amides that emulate the desirable properties of proteins. For example, the nylon in a climbing rope is a synthetic polyamide that emulates the protein in a spider's web. All the penicillin and cephalosporin antibiotics are amides that extend the antimicrobial properties of naturally occurring antibiotics.

Like amides, esters are common both in nature and in the chemical industry. Animal fats and vegetable oils are mixtures of esters, as are waxy materials such as beeswax and spermaceti. Plants often synthesize esters that give the characteristic tastes and odors to their fruits and flowers. In addition to making synthetic esters for flavors, odors, and lubricants, chemists have made synthetic polyesters such as Dacron polyester fiber used in clothing and Mylar polyester film used in magnetic recording tapes.

Some examples of naturally occurring esters and amides are shown here. Isoamyl acetate gives ripe bananas their characteristic odor, and geranyl acetate is found in the oil of roses, geraniums, and many other flowers. *N,N*-Diethyl-*meta*-toluamide (DEET®) is one of the best insect repellents known, and penicillin G is one of the antibiotics that revolutionized modern medicine.

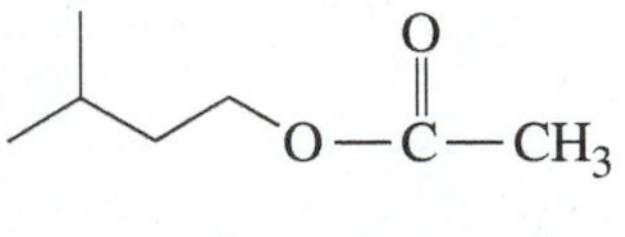

$$(CH_3)_2CHCH_2CH_2-O-\overset{O}{\overset{\|}{C}}-CH_3$$

isoamyl acetate
(banana oil)

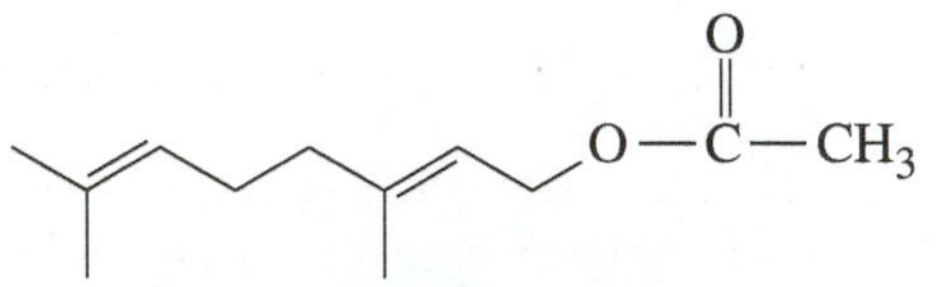

$$(CH_3)_2C{=}CHCH_2CH_2C(CH_3){=}CHCH_2-O-\overset{O}{\overset{\|}{C}}-CH_3$$

geranyl acetate
(geranium oil)

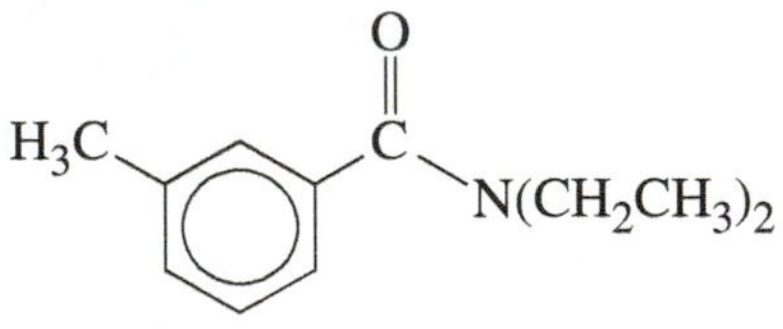

$$H_3C-C_6H_4-\overset{O}{\overset{\|}{C}}-N(CH_2CH_3)_2$$

N,N-diethyl-*meta*-toluamide

$$PhCH_2-\overset{O}{\overset{\|}{C}}-NH \quad (\text{S},\ CH_3,\ CH_3,\ N,\ O,\ COOH)$$

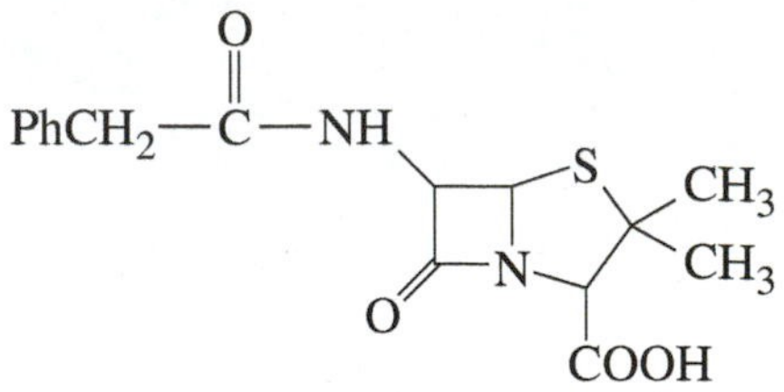

penicillin G

21-2 Structure and Nomenclature of Acid Derivatives

Most acid derivatives have structural features in common. Except for nitriles, they all have a carbonyl group bonded to an electronegative atom: Cl, O, or N. Their systematic nomenclature is logical, based on the nomenclature of their related carboxylic acids. Nitriles are included as acid derivatives because they can be hydrolyzed to give acids.

21-2A Esters of Carboxylic Acids

Esters are carboxylic acid derivatives in which the hydroxy group (—OH) is replaced by an alkoxy group (—OR). An ester is a combination of a carboxylic acid and an alcohol, with loss of a molecule of water. We have seen that esters can be formed by the Fischer esterification of an acid with an alcohol (Section 20-10).

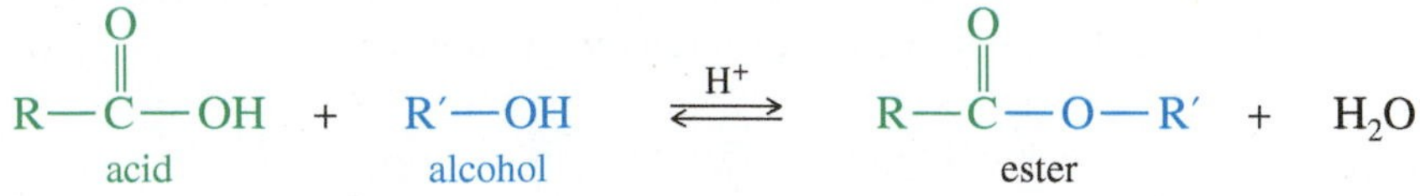

$$\underset{\text{acid}}{R-\overset{O}{\overset{\|}{C}}-OH} + \underset{\text{alcohol}}{R'-OH} \overset{H^+}{\rightleftharpoons} \underset{\text{ester}}{R-\overset{O}{\overset{\|}{C}}-O-R'} + H_2O$$

The names of esters consist of two words that reflect their composite structure. The first word is derived from the *alkyl* group of the alcohol, and the second word from the *carboxylate* group of the carboxylic acid. The IUPAC name is derived from the IUPAC names of the alkyl group and the carboxylate, and the common name is derived from the common names of each. The following examples show both the IUPAC names and the common names of some esters:

$$CH_3CH_2{-}OH + HO{-}\overset{O}{\overset{\|}{C}}{-}CH_3 \overset{H^+}{\rightleftarrows} CH_3CH_2{-}O{-}\overset{O}{\overset{\|}{C}}{-}CH_3 + H_2O$$

IUPAC name:	ethanol	ethanoic acid	ethyl ethanoate
common name:	ethyl alcohol	acetic acid	ethyl acetate

$(CH_3)_2CH{-}O{-}\overset{O}{\overset{\|}{C}}{-}H$ | Ph$-O-\overset{O}{\overset{\|}{C}}-$Ph | $CH_3{-}O{-}\overset{O}{\overset{\|}{C}}{-}CH_2{-}$Ph

IUPAC name:	1-methylethyl methanoate	phenyl benzoate	methyl 2-phenylethanoate
common name:	isopropyl formate	phenyl benzoate	methyl phenylacetate

$Ph{-}CH_2{-}O{-}\overset{O}{\overset{\|}{C}}{-}\overset{CH_3}{\overset{|}{C}H}{-}CH_3$ | $CH_3{-}O{-}\overset{O}{\overset{\|}{C}}{-}$cyclopentyl | cyclohexyl$-O-\overset{O}{\overset{\|}{C}}-H$

IUPAC name:	benzyl 2-methylpropanoate	methyl cyclopentanecarboxylate	cyclohexyl methanoate
common name:	benzyl isobutyrate	methyl cyclopentanecarboxylate	cyclohexyl formate

Lactones Cyclic esters are called **lactones.** A lactone is formed from an open-chain hydroxy acid in which the hydroxy group has reacted with the acid group to form an ester.

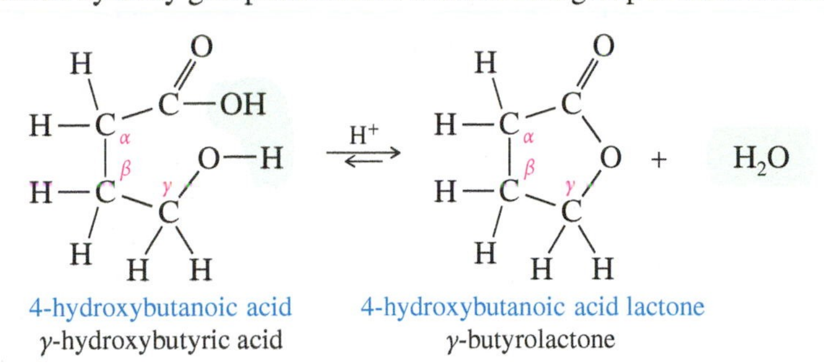

IUPAC name:	4-hydroxybutanoic acid	4-hydroxybutanoic acid lactone
common name:	γ-hydroxybutyric acid	γ-butyrolactone

The IUPAC names of lactones are derived by adding the term *lactone* at the end of the name of the parent acid. The common names of lactones, used more often than IUPAC names, are formed by changing the *-ic acid* ending of the hydroxy acid to *-olactone*. A Greek letter designates the carbon atom that bears the hydroxy group to close the ring. Substituents are named just as they are on the parent acid.

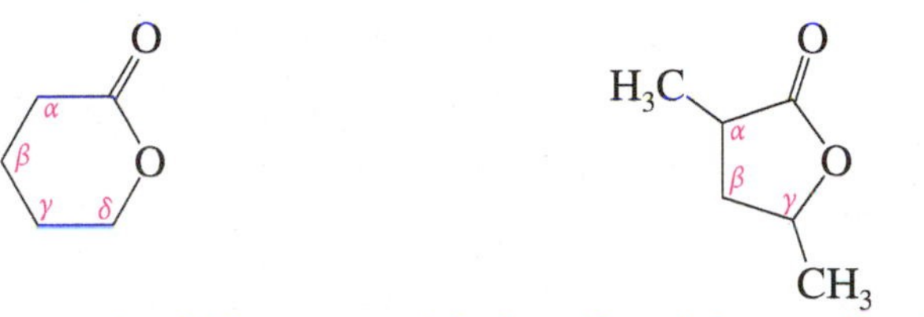

IUPAC name:	5-hydroxypentanoic acid lactone	4-hydroxy-2-methylpentanoic acid lactone
common name:	δ-valerolactone	α-methyl-γ-valerolactone

Application: Hormone antagonist

spironolactone

Spironolactone is a powerful synthetic antagonist that blocks the action of androgens (male hormones). It is used primarily to treat high blood pressure and heart failure. Spironolactone is safe enough that it can also be used to treat acne and excessive facial hair in women, caused by elevated levels of testosterone.

21-2B Amides

An **amide** is a composite of a carboxylic acid and ammonia or an amine. An acid reacts with an amine to form an ammonium carboxylate salt. When this salt is heated to well above 100 °C, water is driven off and an amide results.

$$\underset{\text{acid}}{R{-}\overset{O}{\overset{\|}{C}}{-}OH} + \underset{\text{amine}}{H_2\ddot{N}{-}R'} \longrightarrow \underset{\text{salt}}{R{-}\overset{O}{\overset{\|}{C}}{-}O^- \; H_3\overset{+}{N}{-}R'} \xrightarrow{\text{heat}} \underset{\text{amide}}{R{-}\overset{O}{\overset{\|}{C}}{-}\ddot{N}H{-}R'} + H_2O\uparrow$$

The simple amide structure shows a nonbonding pair of electrons on the nitrogen atom. Unlike amines, however, amides are only weakly basic, and we consider the amide functional group to be neutral. A concentrated strong acid is required to protonate an amide, and protonation occurs on the carbonyl oxygen atom rather than on nitrogen. This lack of basicity can be explained by picturing the amide as a resonance hybrid of the conventional structure and a structure with a double bond between carbon and nitrogen.

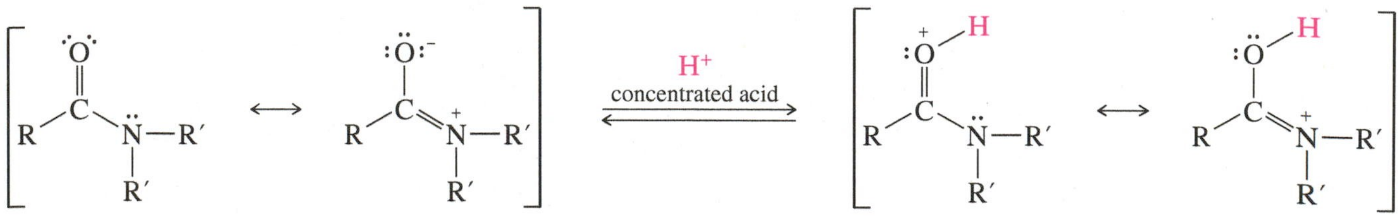

$$\left[R-C(=\ddot{O})-\ddot{N}R'_2 \longleftrightarrow R-C(-\ddot{O}:^-)=\overset{+}{N}R'_2 \right] \underset{\text{concentrated acid}}{\overset{H^+}{\rightleftharpoons}} \left[R-C(=\overset{+}{O}H)-\ddot{N}R'_2 \longleftrightarrow R-C(-\ddot{O}H)=\overset{+}{N}R'_2 \right]$$

very weakly basic — protonation on oxygen

Application: Anesthetic

Many local anesthetics are amides. Lidocaine, the prototype for this group of drugs, is the most widely used.

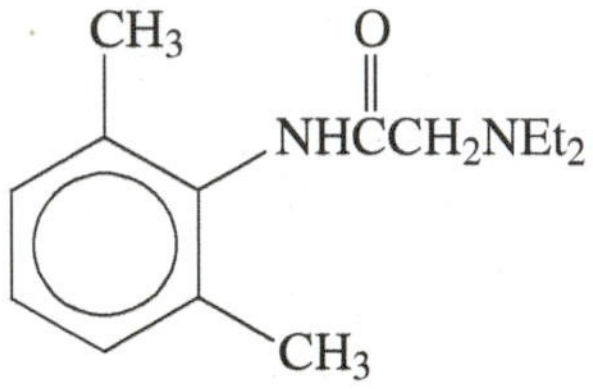

2,6-dimethylphenyl–$NHCCH_2NEt_2$ (C=O)

lidocaine

This resonance representation correctly predicts a planar amide nitrogen atom that is sp^2 hybridized to allow pi bonding with the carbonyl carbon atom. For example, formamide has a planar structure like an alkene. The C—N bond has partial double-bond character, with a rotational barrier of 75 kJ/mol (18 kcal/mol).

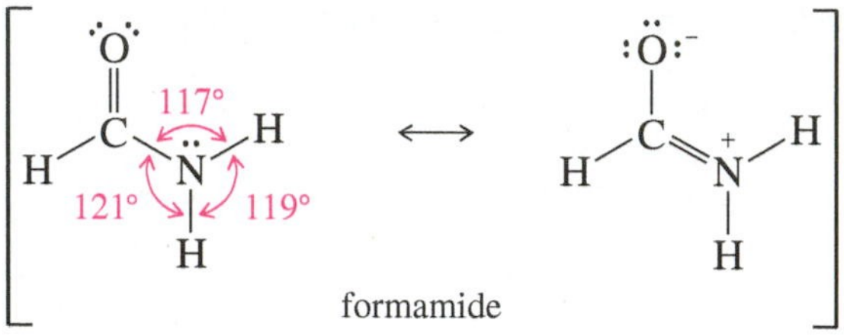

$$\left[H-C(=\ddot{O})-\ddot{N}H_2 \longleftrightarrow H-C(-\ddot{O}:^-)=\overset{+}{N}H_2 \right]$$

117°, 121°, 119°

formamide

An amide of the form R—CO—NH_2 is called a **primary amide** because there is only one carbon atom bonded to the amide nitrogen. An amide with an alkyl group on nitrogen (R—CO—NHR′) is called a **secondary amide** or an ***N*-substituted amide.** Amides with two alkyl groups on the amide nitrogen (R—CO—NR'_2) are called **tertiary amides** or ***N,N*-disubstituted amides.**

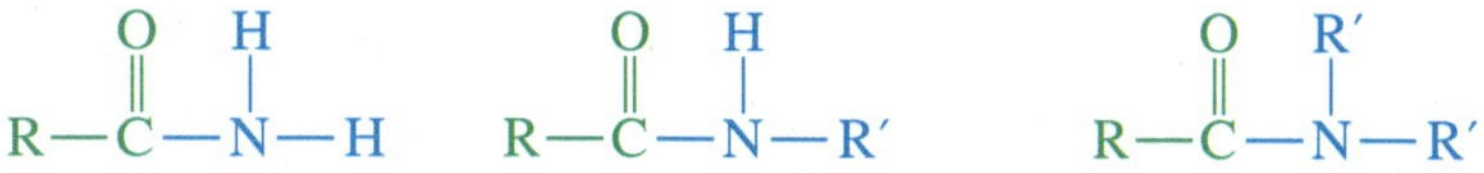

$R-C(=O)-NH-H$	$R-C(=O)-NH-R'$	$R-C(=O)-NR'-R'$
primary amide	secondary amide (*N*-substituted amide)	tertiary amide (*N,N*-disubstituted amide)

To name a primary amide, first name the corresponding acid. Drop the *-ic acid* or *-oic acid* suffix, and add the suffix *-amide*. For secondary and tertiary amides, treat the alkyl groups on nitrogen as substituents, and specify their position by the prefix *N*-.

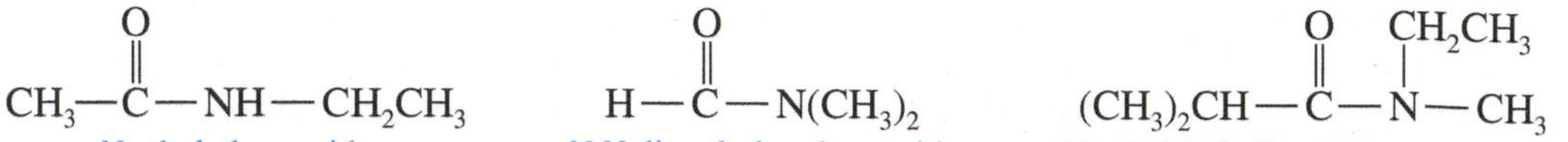

	$CH_3-C(=O)-NH-CH_2CH_3$	$H-C(=O)-N(CH_3)_2$	$(CH_3)_2CH-C(=O)-N(CH_2CH_3)-CH_3$
IUPAC name:	*N*-ethylethanamide	*N,N*-dimethylmethanamide	*N*-ethyl-*N*,2-dimethylpropanamide
common name:	*N*-ethylacetamide	*N,N*-dimethylformamide	*N*-ethyl-*N*-methylisobutyramide

For acids that are named as alkanecarboxylic acids, the amides are named by using the suffix *-carboxamide*. Some amides, such as acetanilide, have historical names that are still commonly used.

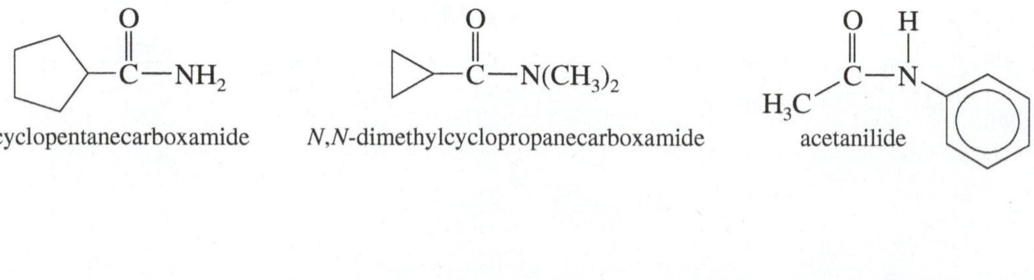

cyclopentyl–$C(=O)-NH_2$	cyclopropyl–$C(=O)-N(CH_3)_2$	$H_3C-C(=O)-NH-C_6H_5$
cyclopentanecarboxamide	*N,N*-dimethylcyclopropanecarboxamide	acetanilide

Lactams Cyclic amides are called **lactams.** Lactams are formed from amino acids, where the amino group and the carboxyl group have joined to form an amide. Lactams are named like lactones, by adding the term *lactam* at the end of the IUPAC name of the parent acid. Common names of lactams are formed by changing the *-ic acid* ending of the amino acid to *-olactam.*

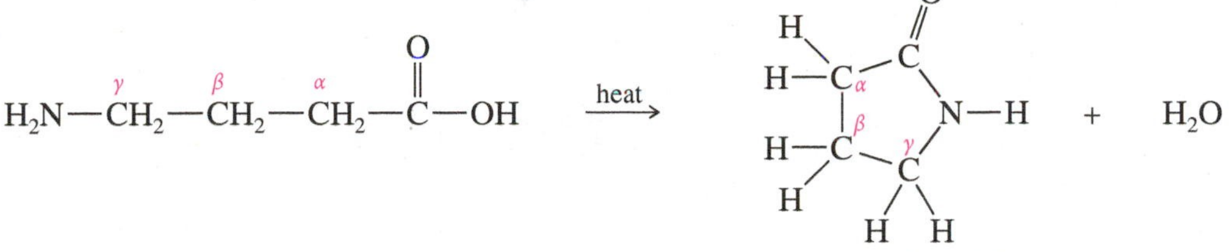

$$H_2N-\overset{\gamma}{C}H_2-\overset{\beta}{C}H_2-\overset{\alpha}{C}H_2-\overset{O}{\overset{\|}{C}}-OH \xrightarrow{\text{heat}} \text{(5-membered lactam)} + H_2O$$

IUPAC name: 4-aminobutanoic acid → 4-aminobutanoic acid lactam

common name: γ-aminobutyric acid → γ-butyrolactam

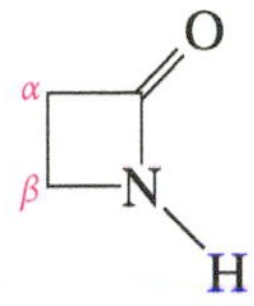

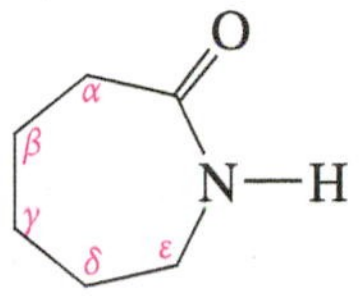

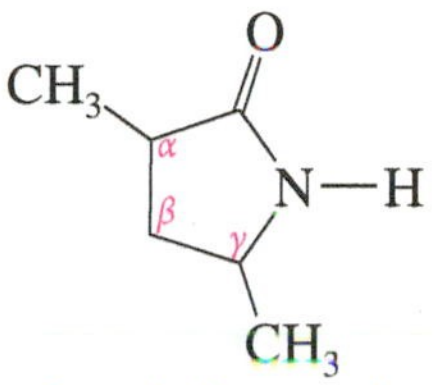

IUPAC name:	3-aminopropanoic acid lactam	6-aminohexanoic acid lactam	4-amino-2-methylpentanoic acid lactam
common name:	β-propiolactam	ε-caprolactam	α-methyl-γ-valerolactam

21-2C Nitriles

Nitriles contain the **cyano group,** $-C{\equiv}N$. Although nitriles lack the carbonyl group of carboxylic acids, they are classified as acid derivatives because they hydrolyze to give carboxylic acids and can be synthesized by dehydration of amides.

Hydrolysis to an acid

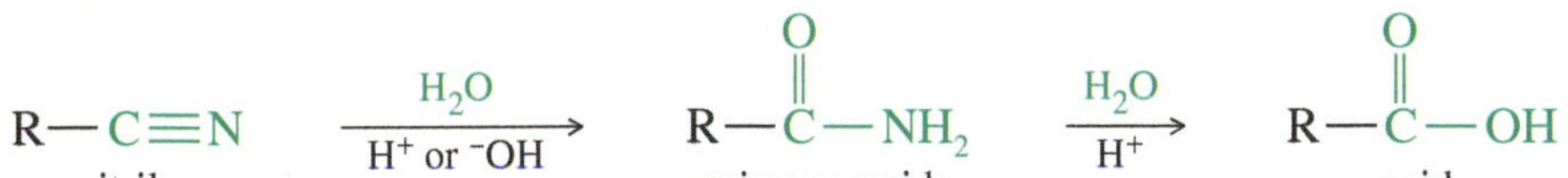

$$\underset{\text{nitrile}}{R-C{\equiv}N} \xrightarrow[H^+ \text{ or } ^-OH]{H_2O} \underset{\text{primary amide}}{R-\overset{O}{\overset{\|}{C}}-NH_2} \xrightarrow[H^+]{H_2O} \underset{\text{acid}}{R-\overset{O}{\overset{\|}{C}}-OH}$$

Synthesis from an acid

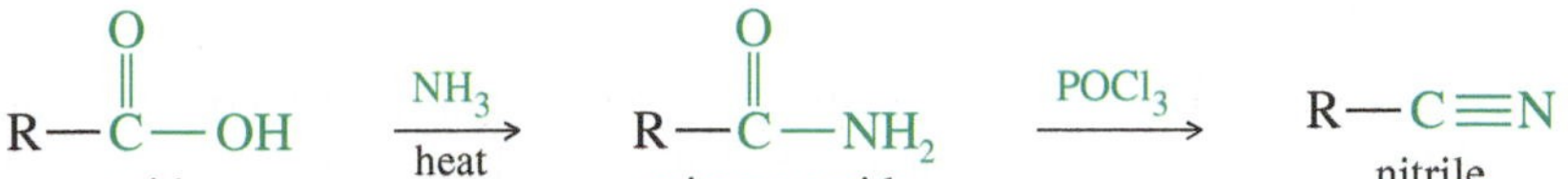

$$\underset{\text{acid}}{R-\overset{O}{\overset{\|}{C}}-OH} \xrightarrow[\text{heat}]{NH_3} \underset{\text{primary amide}}{R-\overset{O}{\overset{\|}{C}}-NH_2} \xrightarrow{POCl_3} \underset{\text{nitrile}}{R-C{\equiv}N}$$

Both the carbon atom and the nitrogen atom of the cyano group are *sp* hybridized, and the $R-C{\equiv}N$ bond angle is 180° (linear). The structure of a nitrile is similar to that of a terminal alkyne, except that the nitrogen atom of the nitrile has a lone pair of electrons in place of the acetylenic hydrogen of the alkyne. Figure 21-1 compares the structures of acetonitrile and propyne.

Although a nitrile has a lone pair of electrons on nitrogen, it is not very basic. A typical nitrile has a pK_b of about 24, requiring a concentrated solution of mineral acid to protonate the nitrile. We explain this lack of basicity by noting that the nitrile's lone pair resides in an *sp* hybrid orbital, with 50% *s* character. This orbital is close to the nucleus, and these electrons are tightly bound and relatively unreactive.

Common names of nitriles are derived from the corresponding carboxylic acids. Begin with the common name of the acid, and replace the suffix *-ic acid* with the suffix *-onitrile.* The IUPAC name is constructed from the alkane name, with the suffix *-nitrile* added.

	$CH_3—C{\equiv}N$	$CH_3—CH(Br)—CH_2—C{\equiv}N$	$CH_3—CH(OCH_3)—CH_2CH_2CH_2—C{\equiv}N$
IUPAC name:	ethanenitrile	3-bromobutanenitrile	5-methoxyhexanenitrile
common name:	acetonitrile	β-bromobutyronitrile	δ-methoxycapronitrile

FIGURE 21-1
Comparison of the electronic structures of acetonitrile and propyne (methylacetylene). In both compounds, the atoms at the ends of the triple bonds are *sp* hybridized, and the bond angles are 180°. In place of the acetylenic hydrogen atom, the nitrile has a lone pair of electrons in the *sp* orbital of nitrogen.

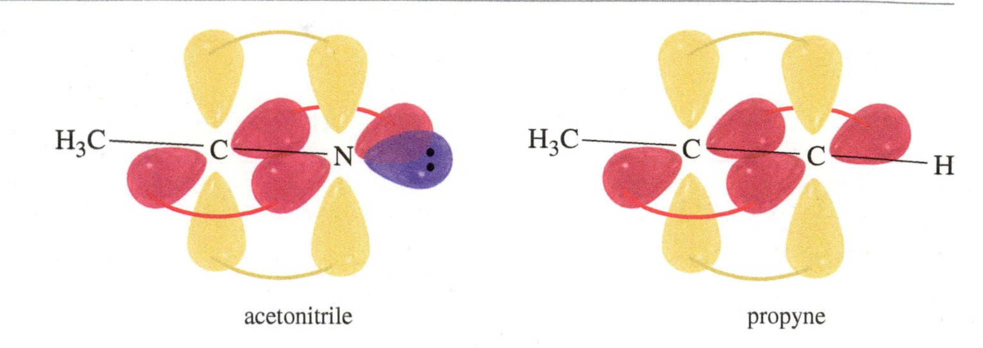

For acids that are named as alkanecarboxylic acids, the corresponding nitriles are named by using the suffix *-carbonitrile*. The —C≡N group can also be named as a substituent, the *cyano group*.

cyclopropane—CN: cyclopropanecarbonitrile

$\overset{5}{C}H_3—\overset{4}{C}H_2—\overset{3}{C}H(CN)—\overset{2}{C}H_2—\overset{1}{C}OOH$: 3-cyanopentanoic acid

21-2D Acid Halides

Acid halides, also called **acyl halides,** are activated derivatives used for the synthesis of other acyl compounds such as esters, amides, and acylbenzenes (in the Friedel–Crafts acylation). The most common acyl halides are the acid chlorides (acyl chlorides), and we will generally use acid chlorides as examples.

R—C(=O)—halogen	R—C(=O)—Cl	R—C(=O)—Br
an acid halide (acyl halide)	acid chloride (acyl chloride)	acid bromide (acyl bromide)

The halogen atom of an acyl halide inductively withdraws electron density from the carbonyl carbon, enhancing its electrophilic nature and making acyl halides particularly reactive toward nucleophilic acyl substitution. The halide ion also serves as a good leaving group.

$$R—C(=O)—Cl: + Nuc:^- \rightleftharpoons R—C(O^-)(Nuc)—Cl \longrightarrow R—C(=O)—Nuc + :Cl:^-$$

leaving group

An acid halide is named by replacing the *-ic acid* suffix of the acid name (either the common name or the IUPAC name) with *-yl* and the halide name. For acids that are named as alkanecarboxylic acids, the acid chlorides are named by using the suffix *-carbonyl chloride*.

$CH_3—C(=O)—F$	$CH_3—CH_2—C(=O)—Cl$	$CH_3—CH(Br)—CH_2—C(=O)—Br$	cyclopentyl—C(=O)—Cl
ethanoyl fluoride	propanoyl chloride	3-bromobutanoyl bromide	cyclopentanecarbonyl chloride
acetyl fluoride	propionyl chloride	β-bromobutyryl bromide	

21-2E Acid Anhydrides

The word **anhydride** means "without water." An acid anhydride contains two molecules of an acid, with loss of a molecule of water. Addition of water to an anhydride regenerates two molecules of the carboxylic acid.

$$\underset{\text{anhydride}}{R-\overset{O}{\overset{\|}{C}}-O-\overset{O}{\overset{\|}{C}}-R} + \underset{\text{water}}{H_2O} \rightleftharpoons \underset{\text{two molecules of acid}}{R-\overset{O}{\overset{\|}{C}}-OH + HO-\overset{O}{\overset{\|}{C}}-R}$$

Like acid halides, anhydrides are activated derivatives of carboxylic acids, although anhydrides are not as reactive as acid halides. In an acid chloride, the chlorine atom activates the carbonyl group and serves as a leaving group. In an anhydride, the carboxylate group serves these functions.

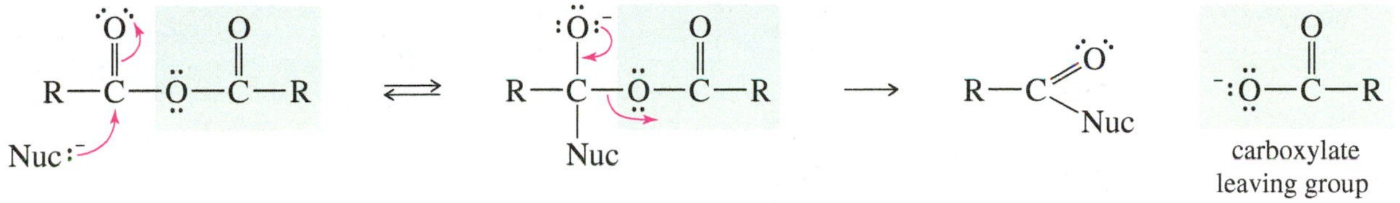

$$R-\overset{O}{\overset{\|}{C}}-O-\overset{O}{\overset{\|}{C}}-R + Nuc:^- \rightleftharpoons R-\underset{Nuc}{\overset{:O:^-}{\overset{|}{C}}}-O-\overset{O}{\overset{\|}{C}}-R \longrightarrow R-\overset{O}{\overset{\|}{C}}-Nuc \qquad \underset{\text{carboxylate leaving group}}{^-:O-\overset{O}{\overset{\|}{C}}-R}$$

Half of an anhydride's acid units are lost as leaving groups. If the acid is expensive, we would not use the anhydride as an activated form to make a derivative. The acid chloride is a more efficient alternative, using chloride as the leaving group. Anhydrides are used primarily when the necessary anhydride is cheap and readily available. Acetic anhydride, phthalic anhydride, succinic anhydride, and maleic anhydride are the ones we use most often. Diacids commonly form cyclic anhydrides, especially if a five- or six-membered ring results.

Anhydride nomenclature is very simple; the word *acid* is changed to *anhydride* in both the common name and the IUPAC name (rarely used). The following examples show the names of some common anhydrides:

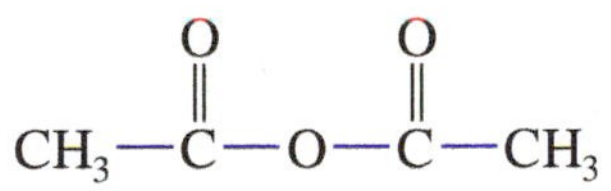

$$CH_3-\overset{O}{\overset{\|}{C}}-O-\overset{O}{\overset{\|}{C}}-CH_3 \qquad CF_3-\overset{O}{\overset{\|}{C}}-O-\overset{O}{\overset{\|}{C}}-CF_3$$

(abbreviated Ac_2O)	(abbreviated TFAA)		
ethanoic anhydride	trifluoroethanoic anhydride	benzene-1,2-dicarboxylic anhydride	but-2-enedioic anhydride
acetic anhydride	trifluoroacetic anhydride	phthalic anhydride	maleic anhydride

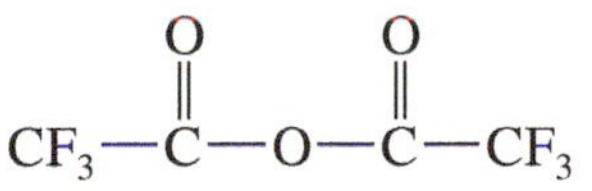

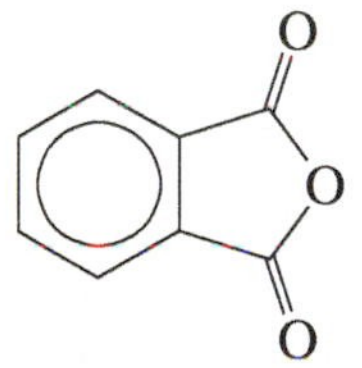

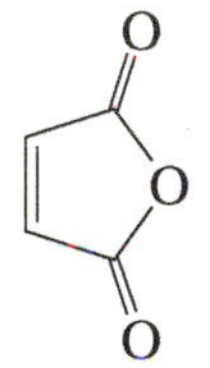

Anhydrides composed of two different acids are called **mixed anhydrides** and are named by using the names of the individual acids.

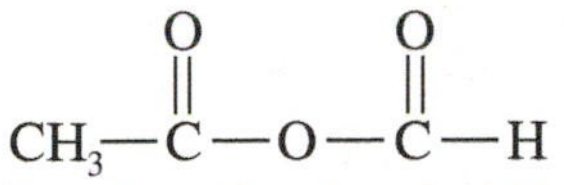

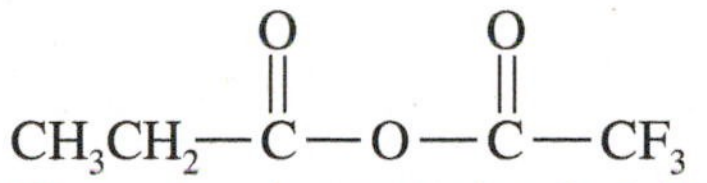

	$CH_3-\overset{O}{\overset{\|}{C}}-O-\overset{O}{\overset{\|}{C}}-H$	$CH_3CH_2-\overset{O}{\overset{\|}{C}}-O-\overset{O}{\overset{\|}{C}}-CF_3$
IUPAC name:	ethanoic methanoic anhydride	trifluoroethanoic propanoic anhydride
common name:	acetic formic anhydride	trifluoroacetic propionic anhydride

21-2F Nomenclature of Multifunctional Compounds

With all the different functional groups we have studied, it is not always obvious which functional group of a multifunctional compound is the "main" one and which groups should be named as substituents. In choosing the principal group for the root name, we use the following priorities:

acid > ester > amide > nitrile > aldehyde > ketone > alcohol > amine > alkene, alkyne

Table 21-1 summarizes these priorities, together with the suffixes used for main groups and the prefixes used for substituents. The following examples illustrate these priorities in naming multifunctional compounds:

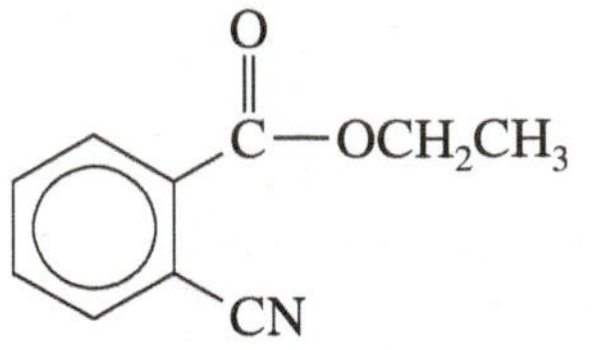

ethyl *o*-cyanobenzoate

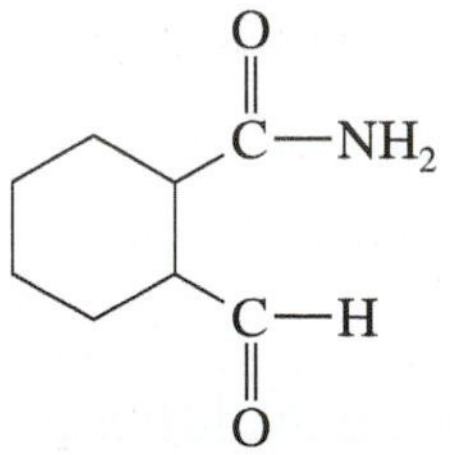

2-formylcyclohexanecarboxamide

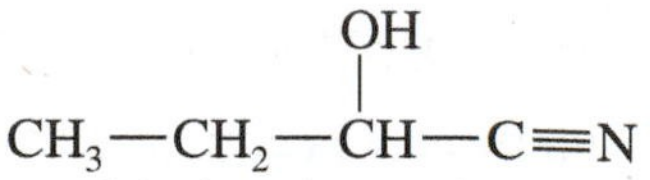

$CH_3-CH_2-CH(OH)-C\equiv N$

2-hydroxybutanenitrile

TABLE 21-1
Summary of Functional Group Nomenclature

Functional Group	Name as Main Group	Name as Substituent
	Main groups in order of decreasing priority:	
carboxylic acids	-oic acid	carboxy
esters	-oate	alkoxycarbonyl
amides	-amide	amido
nitriles	-nitrile	cyano
aldehydes	-al	formyl
ketones	-one	oxo
alcohols	-ol	hydroxy
amines	-amine	amino
alkenes	-ene	alkenyl
alkynes	-yne	alkynyl
alkanes	-ane	alkyl
ethers		alkoxy
halides		halo

PROBLEM-SOLVING HINT

The nomenclature of carboxylic acid derivatives is reviewed in Appendix 5, the Summary of Organic Nomenclature.

PROBLEM 21-1

Name the following carboxylic acid derivatives, giving both a common name and an IUPAC name where possible.

(a) $PhCOOCH_2CH(CH_3)_2$
(b) PhOCHO
(c) $PhCH(CH_3)COOCH_3$
(d) $PhNHCOCH_2CH(CH_3)_2$
(e) $CH_3CONHCH_2Ph$
(f) $CH_3CH(OH)CH_2CN$
(g) $(CH_3)_2CHCH_2COBr$
(h) $Cl_2CHCOCl$
(i) $(CH_3)_2CHCOOCHO$
(j) 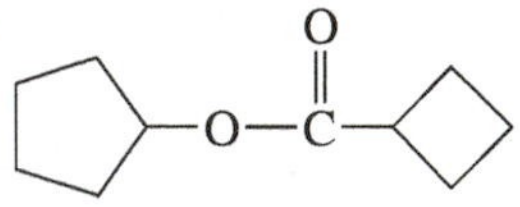
(k) 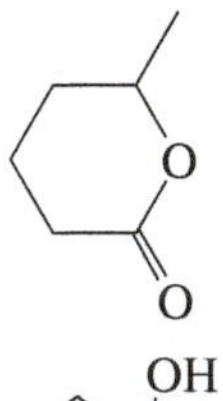
(l) PhCONH—
(m) 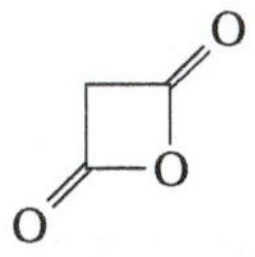
(n) OH, CN
(o) H, COOH, H, CN
(p) COCl, Br
(q) O, N—CH_3, CH_2CH_3
(r) N—C(=O)—CH_3

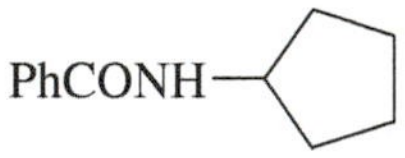

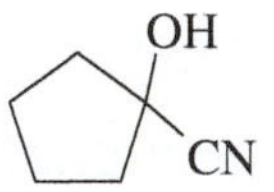

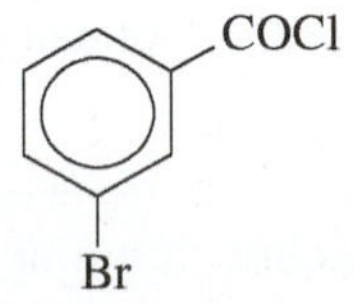

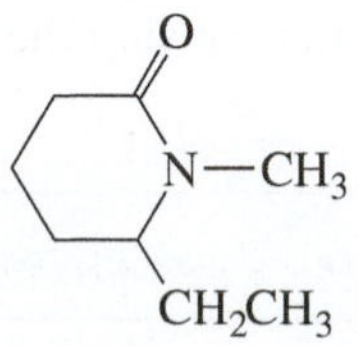

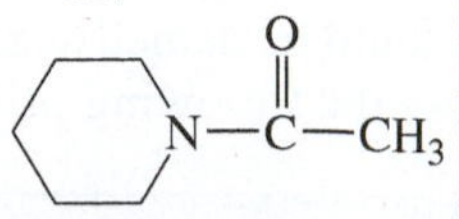

(*Hint:* Named as a piperidine derivative.)

21-3 Physical Properties of Carboxylic Acid Derivatives

The physical properties of acid derivatives depend largely on their polarity and their hydrogen-bonding properties. For example, amides show strong hydrogen bonding, which gives them unusually high boiling points and melting points.

21-3A Boiling Points and Melting Points

Figure 21-2 is a graph of the boiling points of simple acid derivatives plotted against their molecular weights. The *n*-alkanes are included for comparison. Notice that esters and acid chlorides have boiling points near those of the unbranched alkanes with similar molecular weights. These acid derivatives contain highly polar carbonyl groups, but the polarity of the carbonyl group has only a small effect on the boiling points (Section 18-4).

Carboxylic acids are strongly hydrogen bonded in the liquid phase, resulting in elevated boiling points. The stable hydrogen-bonded dimer has a higher effective molecular weight and boils at a higher temperature. Nitriles also have higher boiling points than esters and acid chlorides of similar molecular weight. This effect results from a strong dipolar association between adjacent cyano groups.

R—C(=O···H—O)(—O—H···O=)C—R

carboxylic acid dimer

$R-\overset{\delta+}{C}\equiv \overset{\delta-}{N}:$

$\overset{\delta-}{:N}\equiv \overset{\delta+}{C}-R$

dipolar association of nitriles

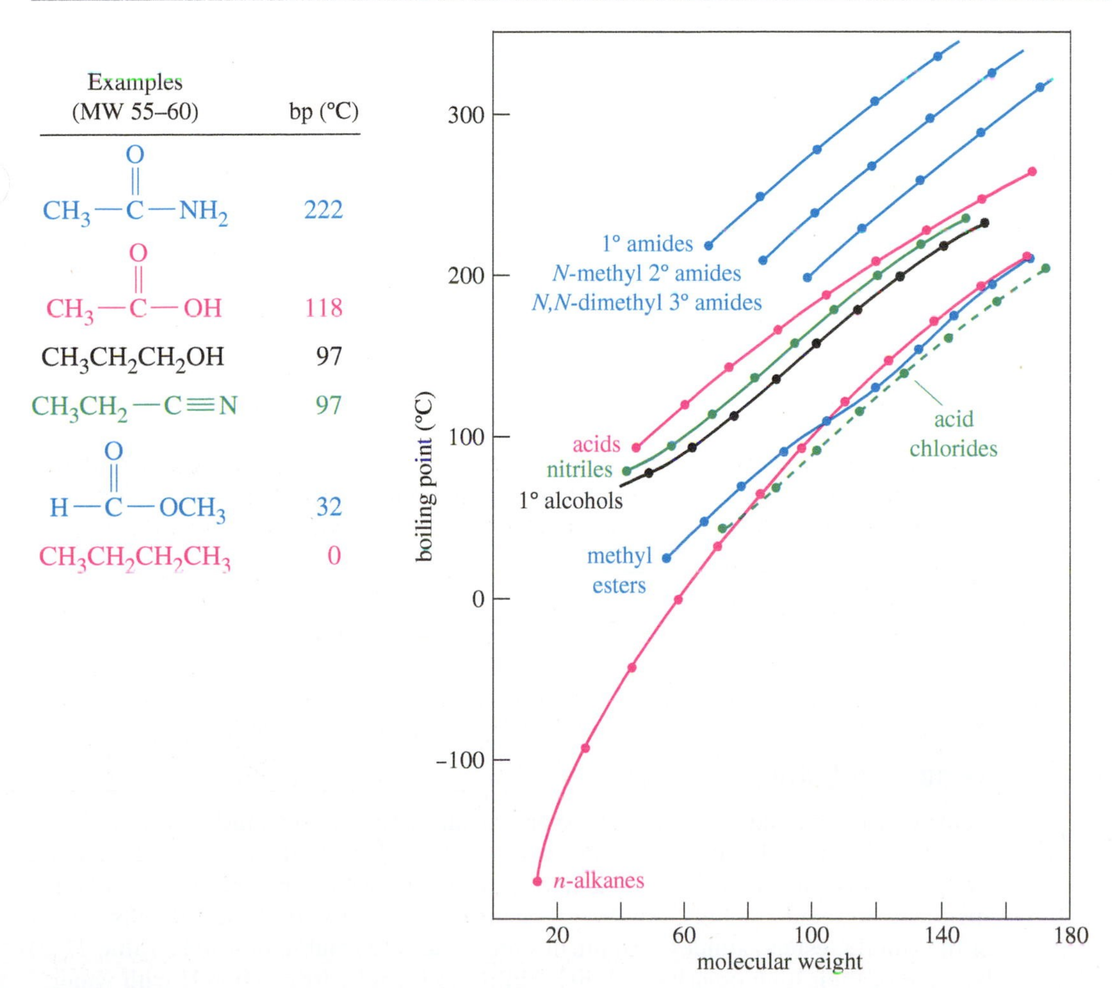

Examples (MW 55–60)	bp (°C)
$CH_3-C(=O)-NH_2$	222
$CH_3-C(=O)-OH$	118
$CH_3CH_2CH_2OH$	97
$CH_3CH_2-C\equiv N$	97
$H-C(=O)-OCH_3$	32
$CH_3CH_2CH_2CH_3$	0

FIGURE 21-2
Boiling points of acid derivatives plotted against their molecular weights. Alcohols and unbranched alkanes are included for comparison.

Amides have surprisingly high boiling points and melting points compared with other compounds of similar molecular weight. Primary and secondary amides participate in strong hydrogen bonding, shown in Figure 21-3. The resonance picture shows a partial negative charge on oxygen and a partial positive charge on nitrogen. The positively charged nitrogen polarizes the N—H bond, making the hydrogen strongly electrophilic. The negatively charged oxygen's lone pairs are particularly effective in forming hydrogen bonds to these polarized N—H hydrogens.

Pure tertiary amides lack N—H bonds, so they cannot participate in hydrogen bonding (although they are good hydrogen bond acceptors). Still, they have high boiling points, close to those of carboxylic acids of similar molecular weights. Figure 21-3 shows how a pairing of two molecules is strongly attractive, helping to stabilize the liquid phase. Vaporization disrupts this arrangement, so a higher temperature is needed for boiling.

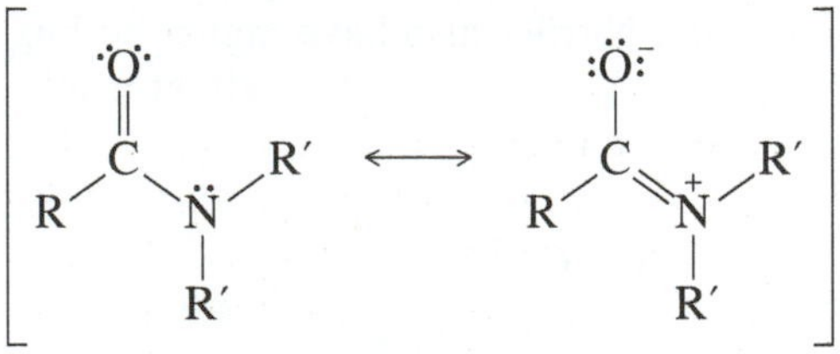

dipolar resonance in amides

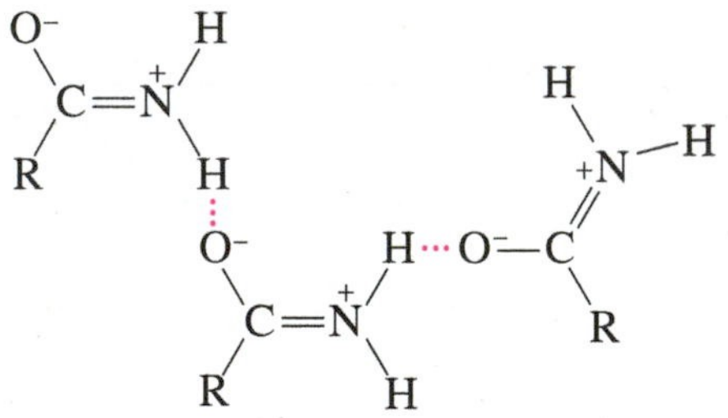

strong hydrogen bonding in amides

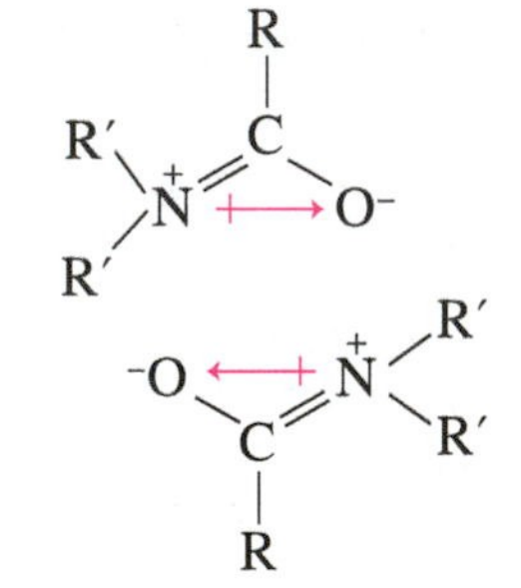

intermolecular attractions in amides

FIGURE 21-3
The resonance picture of an amide shows its strongly polar nature. Hydrogen bonds and dipolar attractions stabilize the liquid phase, resulting in higher boiling points.

Strong hydrogen bonding between molecules of primary and secondary amides also results in unusually high melting points. For example, *N*-methylacetamide (secondary, one N—H bond) has a melting point of 28 °C, which is 89° higher than the melting point (−61 °C) of its isomer dimethylformamide (tertiary, no N—H bond). With two N—H bonds to engage in hydrogen bonding, the primary amide propionamide melts at 79 °C, about 50° higher than its secondary isomer *N*-methylacetamide.

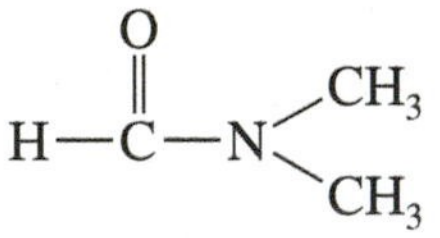

dimethylformamide	*N*-methylacetamide	propionamide
$H-C(=O)-N(CH_3)_2$	$CH_3-C(=O)-NH(CH_3)$	$CH_3CH_2-C(=O)-NH_2$
mp −61 °C	mp 28 °C	mp 79 °C

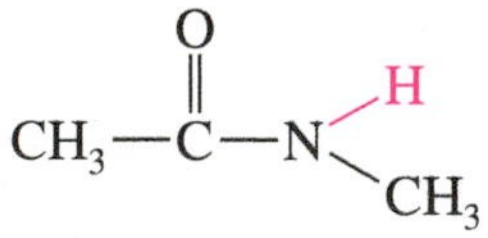

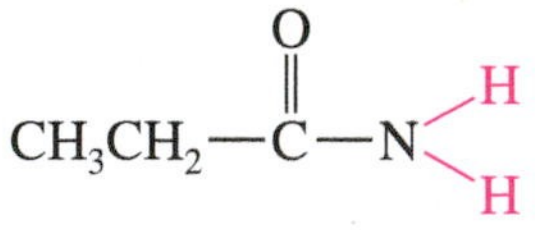

21-3B Solubility

Acid derivatives (esters, acid chlorides, anhydrides, nitriles, and amides) are soluble in common organic solvents such as alcohols, ethers, chlorinated alkanes, and aromatic hydrocarbons. Acid chlorides and anhydrides cannot be used in nucleophilic solvents such as water and alcohols, however, because they react with these solvents. Many of the smaller esters, amides, and nitriles are relatively soluble in water (Table 21-2) because of their high polarity and their ability to form hydrogen bonds with water.

Esters, tertiary amides, and nitriles are frequently used as solvents for organic reactions because they provide a polar reaction medium without O—H or N—H groups

TABLE 21-2
Esters, Amides, and Nitriles Commonly Used as Solvents for Organic Reactions

Compound	Name	mp (°C)	bp (°C)	Water Solubility
$CH_3-C(=O)-OCH_2CH_3$	ethyl acetate	-83	77	10%
$H-C(=O)-N(CH_3)_2$	dimethylformamide (DMF)	-61	153	miscible
$CH_3-C(=O)-N(CH_3)_2$	dimethylacetamide (DMA)	-20	165	miscible
$CH_3-C\equiv N$	acetonitrile	-45	82	miscible

that can donate protons or act as nucleophiles. Ethyl acetate is a moderately polar solvent with a boiling point of 77 °C, convenient for easy evaporation from a reaction mixture. Acetonitrile, dimethylformamide (DMF), and dimethylacetamide (DMA) are highly polar solvents that solvate ions almost as well as water, but without the reactivity of O—H or N—H groups. These three solvents are miscible with water and are often used in solvent mixtures with water.

21-4 Spectroscopy of Carboxylic Acid Derivatives

Several features of acid derivatives help us to distinguish them by spectroscopy. Variations in the infrared carbonyl absorptions, together with chemical shifts in the NMR, are predictable and reliable for most acid derivatives.

21-4A Infrared Spectroscopy

Different types of carbonyl groups give characteristic strong absorptions at different positions in the infrared spectrum. As a result, infrared spectroscopy is often the best method to detect and differentiate these carboxylic acid derivatives. Table 21-3 summarizes the characteristic IR absorptions of carbonyl functional groups. As

TABLE 21-3
Characteristic Carbonyl IR Stretching Absorptions

Functional Group		Frequency	Comments
ketone	$R-C(=O)-R$	C=O, 1710 cm^{-1}	lower if conjugated, higher if strained (aldehydes 1725 cm^{-1})
acid	$R-C(=O)-OH$	C=O, 1710 cm^{-1}	lower if conjugated
		O—H, 2500–3500 cm^{-1}	broad, on top of C—H stretch
ester	$R-C(=O)-O-R'$	C=O, 1735 cm^{-1}	lower if conjugated, higher if strained
amide	$R-C(=O)-N(H)-R'$	C=O, 1640–1680 cm^{-1}	
		N—H, 3200–3500 cm^{-1}	two peaks for $R-CO-NH_2$, one peak for R—CO—NHR′
acid chloride	$R-C(=O)-Cl$	C=O, 1800 cm^{-1}	very high frequency
acid anhydride	$R-C(=O)-O-C(=O)-R$	C=O, 1800 and 1750 cm^{-1}	two peaks
nitrile	$R-C\equiv N$	C≡N, 2200 cm^{-1}	just above 2200 cm^{-1}

in Chapter 12, we are using about 1710 cm^{-1} for simple ketones and acids as a standard for comparison. Appendix 2 gives a more complete table of characteristic IR frequencies.

Esters Ester carbonyl groups absorb at relatively high frequencies, about 1735 cm^{-1}. Except for strained cyclic ketones, few other functional groups absorb strongly in this region. Esters also have a C—O single-bond stretching absorption between 1000 and 1200 cm^{-1}, but many other types of bonds also absorb in this region. We do not consider this absorption to be diagnostic for an ester, but we may look for it in uncertain cases.

Conjugation lowers the carbonyl stretching frequency of an ester. Conjugated esters absorb around 1710 to 1725 cm^{-1} and might easily be confused with simple ketones (1710 cm^{-1}) and aldehydes (1725 cm^{-1}). The presence of *both* a strong carbonyl absorption in this region and a conjugated C=C absorption around 1620 to 1640 cm^{-1} suggests a conjugated ester. Compare the spectra of ethyl octanoate and methyl benzoate in Figure 21-4 to see these differences.

PROBLEM 21-2

What characteristics of the methyl benzoate spectrum rule out an aldehyde or carboxylic acid functional group giving the absorption at 1723 cm^{-1}?

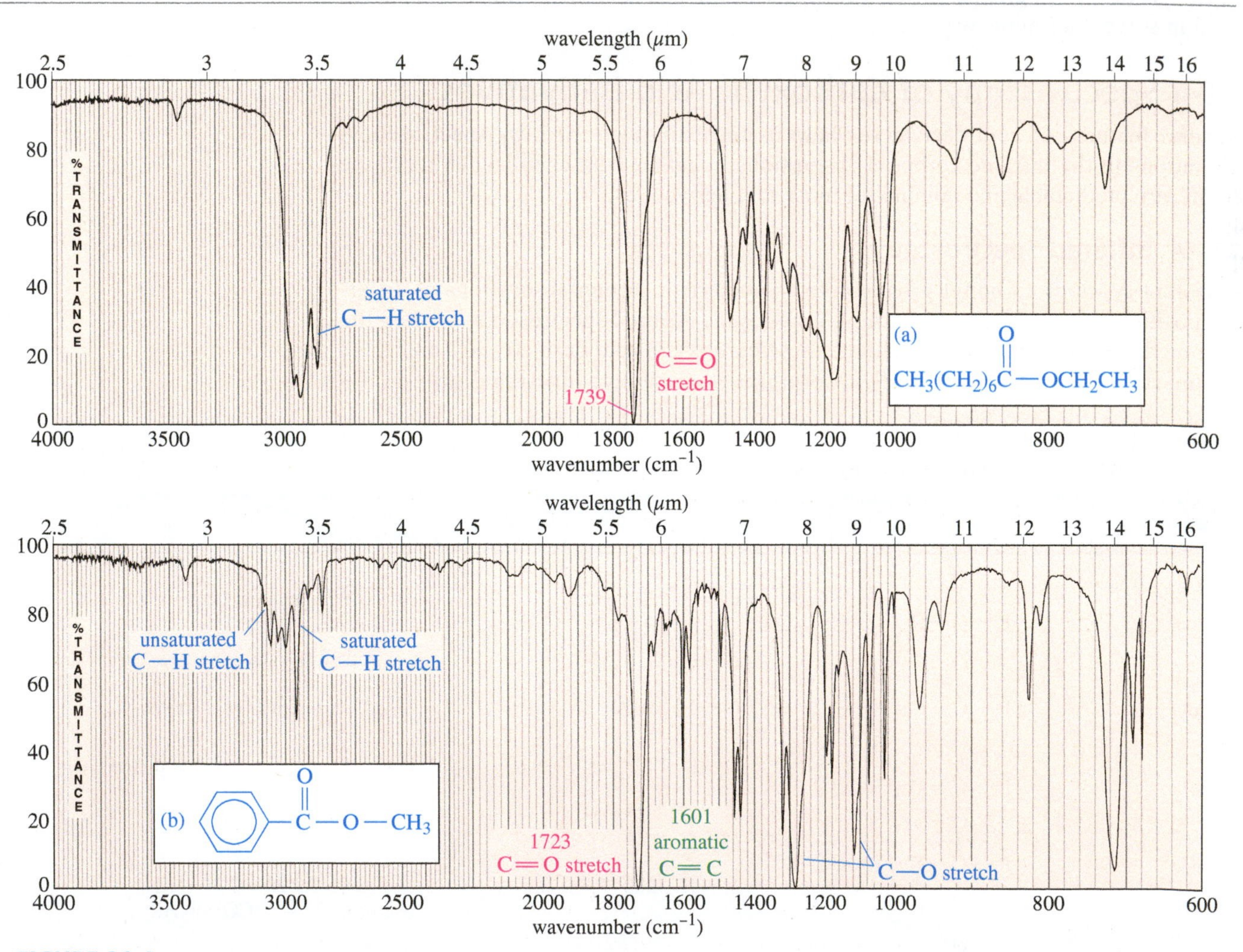

FIGURE 21-4
Infrared spectra of (a) ethyl octanoate and (b) methyl benzoate. The carbonyl stretching frequency of simple esters is around 1735 cm^{-1} and that of conjugated esters is around 1710–1725 cm^{-1}.

PROBLEM 21-3

Aldehydes, ketones, carboxylic acids, and esters all give strong carbonyl stretching absorptions in the IR spectrum. How can you use other peaks in their IR spectra to distinguish among these four common functional groups?

Amides Simple amides have much lower carbonyl stretching frequencies than the other carboxylic acid derivatives, absorbing around 1640 to 1680 cm^{-1} (often a close doublet). This low-frequency absorption agrees with the resonance picture of the amide. The C=O bond of the amide carbonyl is somewhat less than a full double bond. Because it is not as strong as the C=O bond in a simple ketone or carboxylic acid, the amide C=O has a lower stretching frequency.

less than a full double bond

Primary and secondary amides have N—H bonds that give infrared stretching absorptions in the region 3200 to 3500 cm^{-1}. These absorptions fall in the same region as the broad O—H absorption of an alcohol, but the amide N—H absorptions are usually sharper. In primary amides (R—CO—NH_2), there are two N—H bonds, and two sharp peaks occur in the region 3200 to 3500 cm^{-1}. Secondary amides (R—CO—NHR′) have only one N—H bond, and only one peak is observed in the N—H region of the spectrum. Tertiary amides (R—CO—NR'_2) have no N—H bonds, so there is no N—H absorption.

The infrared spectrum of butyramide appears in Figure 12-13a (page 576), and propanamide appears as compound 2 on page 581. Notice the strong carbonyl stretching absorption around 1630 to 1660 cm^{-1} and two N—H stretching absorptions at 3350 and 3180 cm^{-1}.

Lactones and Lactams Unstrained lactones (cyclic esters) and lactams (cyclic amides) absorb at typical frequencies for esters and amides. Ring strain raises the carbonyl absorption frequency, however. Recall that cyclic ketones with five-membered or smaller rings show a similar increase in carbonyl stretching frequency (Section 18-5A). Figure 21-5 shows the effect of ring strain on the C=O stretching frequencies of lactones and lactams.

Nitriles Nitriles show a characteristic C≡N stretching absorption around 2200 cm^{-1} in the infrared spectrum. This absorption can be distinguished from the alkyne C≡C absorption by two characteristics: Nitriles usually absorb at frequencies slightly *higher* than 2200 cm^{-1} (to the left of 2200 cm^{-1}), whereas alkynes usually absorb at frequencies slightly *lower* than 2200 cm^{-1}; and nitrile absorptions are usually stronger because the C≡N triple bond is more polar than the alkyne C≡C triple bond.

The IR spectrum of butyronitrile appears in Figure 12-14 (page 577). Notice the strong triple-bond stretching absorption at 2249 cm^{-1}. The IR spectrum of hexanenitrile (compound 3, page 581) shows C≡N stretching at 2246 cm^{-1}.

Acid Halides and Anhydrides Acid halides and anhydrides are rarely isolated as unknown compounds; but they are commonly used as reagents and intermediates, and infrared spectroscopy can confirm that an acid has been converted to a pure acid chloride or anhydride. The carbonyl stretching vibration of an acid chloride occurs at a high frequency, around 1800 cm^{-1}.

Anhydrides give *two* carbonyl stretching absorptions, one around 1800 cm^{-1} and another around 1750 cm^{-1}. Figure 21-6 shows the spectrum of propionic anhydride, with carbonyl absorptions at 1818 and 1751 cm^{-1}.

PROBLEM-SOLVING HINT

The absorptions listed in Table 21-3 are often the best spectroscopic information available for determining the functional group in an unknown acid derivative.

δ-valerolactone 1735 cm^{-1} no strain

γ-butyrolactone 1770 cm^{-1} moderate strain

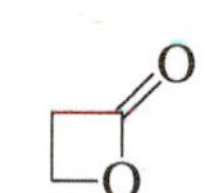

β-propiolactone 1800 cm^{-1} highly strained

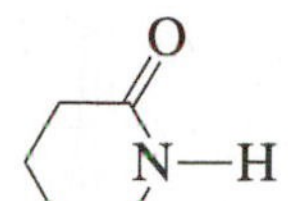

δ-valerolactam 1670 cm^{-1} no strain

γ-butyrolactam 1700 cm^{-1} moderate strain

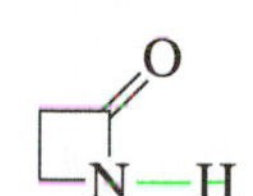

β-propiolactam 1745 cm^{-1} highly strained

FIGURE 21-5
Ring strain in a lactone or lactam increases the carbonyl stretching frequency.

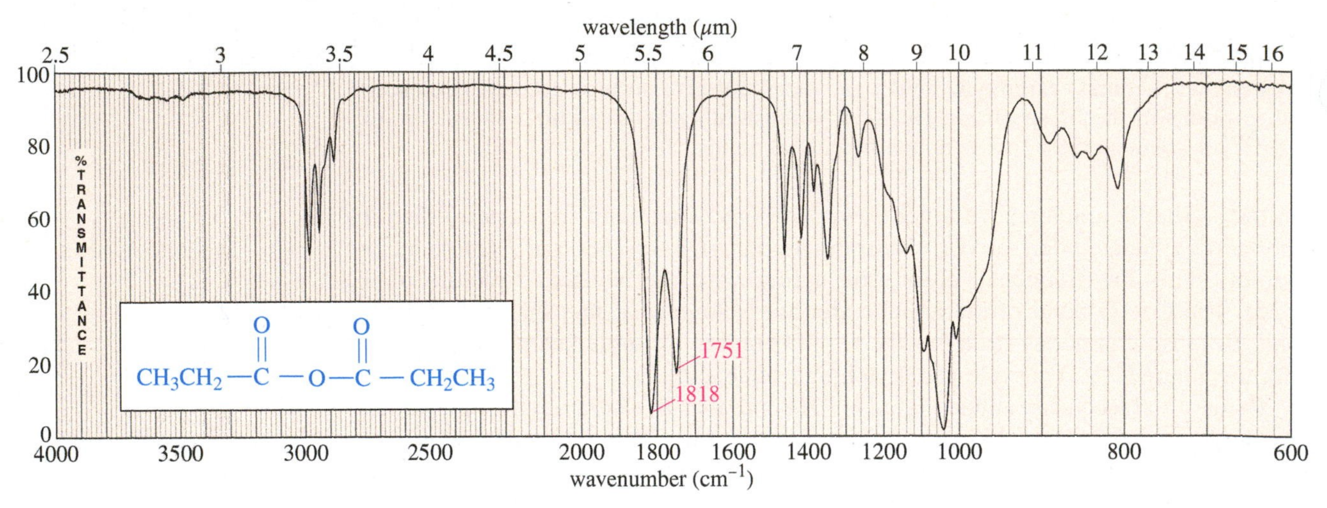

FIGURE 21-6
Infrared spectrum of propionic anhydride, showing C=O stretching absorptions at 1818 and 1751 cm^{-1}.

PROBLEM 21-4

The IR spectra shown next may include a carboxylic acid, an ester, an amide, a nitrile, an acid chloride, or an acid anhydride. Determine the functional group suggested by each spectrum, and list the specific frequencies you used to make your decision.

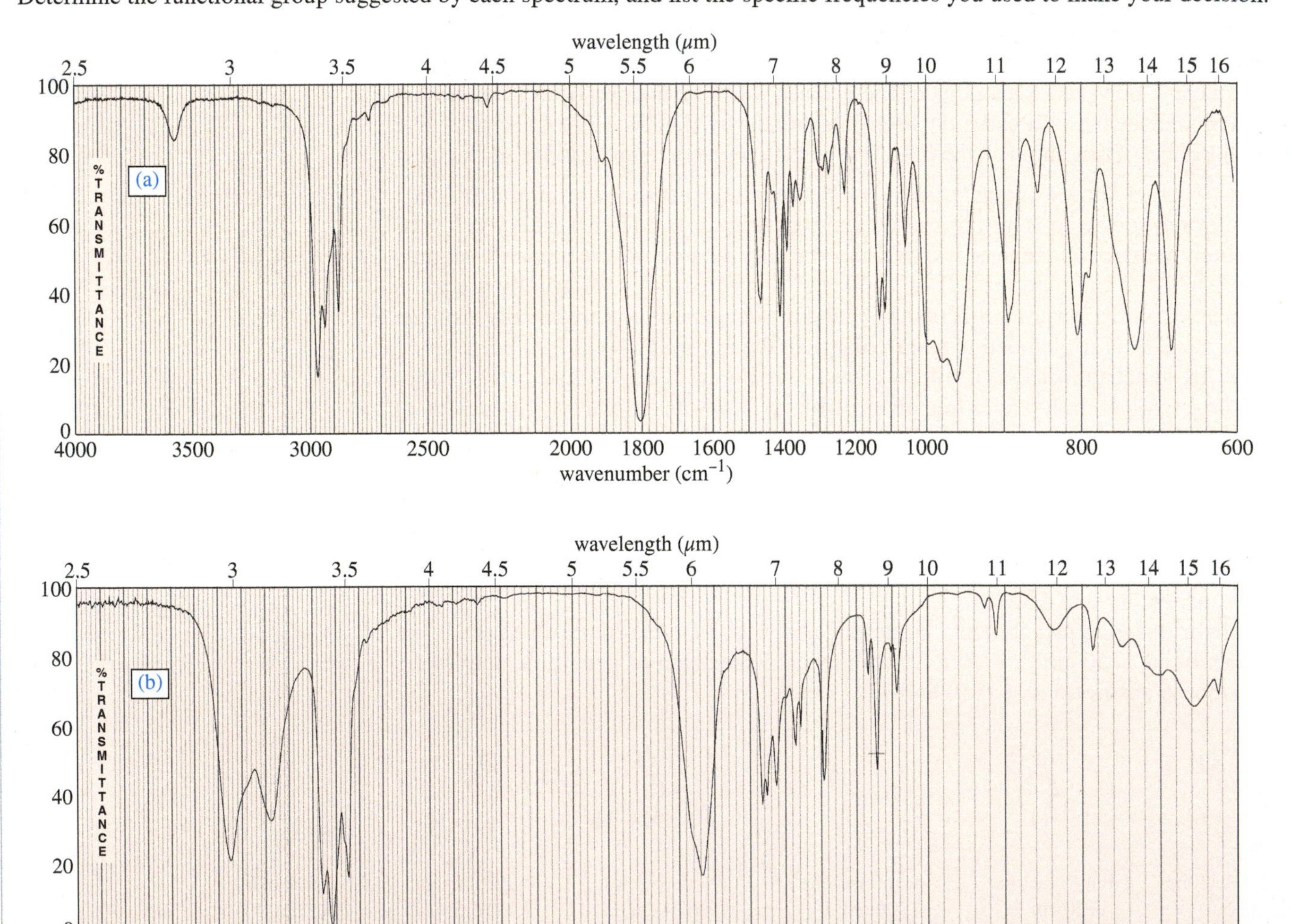

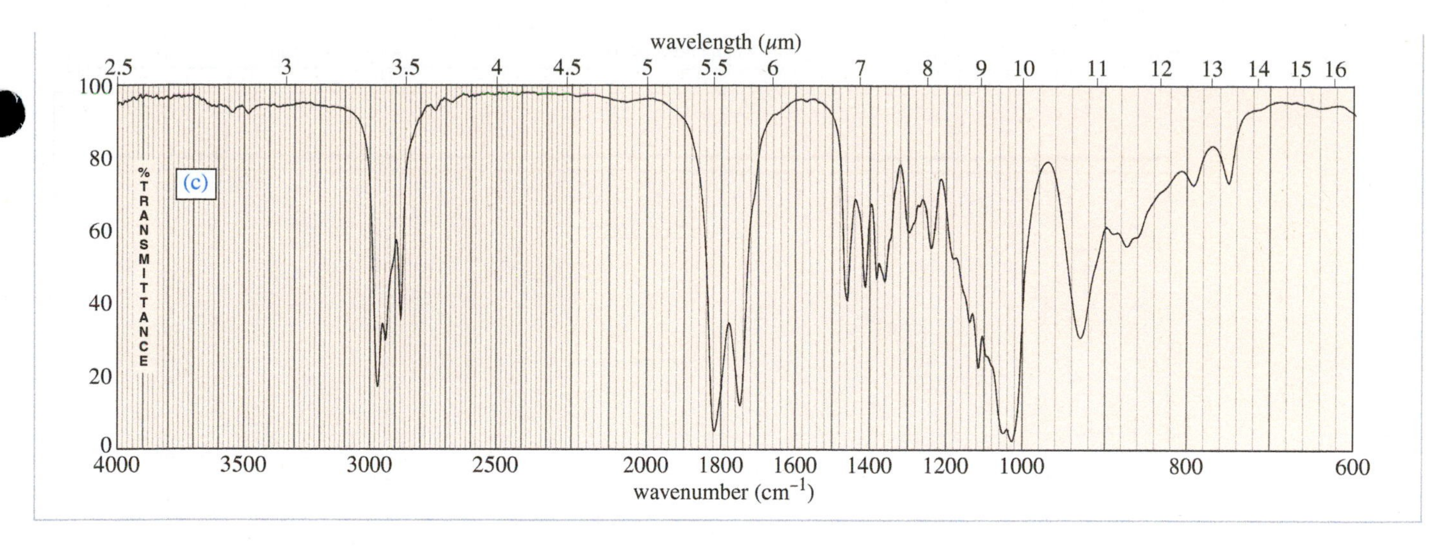

21-4B NMR Spectroscopy

NMR spectroscopy of acid derivatives is complementary to IR spectroscopy. For the most part, IR gives information about the functional groups, and NMR gives information about the alkyl groups. In many cases, the combination of IR and NMR provides enough information to determine the structure.

Proton NMR The proton chemical shifts found in acid derivatives are close to those of similar protons in ketones, aldehydes, alcohols, and amines (Figure 21-7). For example, protons alpha to a carbonyl group absorb between $\delta 2.0$ and $\delta 2.5$, whether the carbonyl group is part of a ketone, aldehyde, acid, ester, or amide. The protons of the alcohol-derived group of an ester or the amine-derived group of an amide give absorptions similar to those in the spectrum of the parent alcohol or amine.

The N—H protons of an amide, appearing between $\delta 5$ and $\delta 8$, may be broad or may show splitting, depending on concentration and solvent. Figure 13-37 (page 642) shows the NMR spectrum of ethyl carbamate, an amide with a broad N—H absorption. The *formyl* proton bonded to the carbonyl group of a formate ester or formamide resembles an aldehyde proton, but it is slightly more shielded and appears around δ 8. In a nitrile, the protons on the α carbon atom absorb around $\delta 2.5$, similar to the α protons of a carbonyl group.

The NMR spectrum of *N,N*-dimethylformamide (Figure 21-8) shows the formyl proton (H—C=O) around $\delta 8$. The two methyl groups appear as two singlets (not a spin-spin splitting doublet) near $\delta 2.9$ and $\delta 3.0$. The two singlets result from hindered rotation about the amide bond. The cisoid and transoid methyl groups interconvert slowly with respect to the NMR time scale.

O=C(H)—N(CH_3—cisoid, δ 2.9)(CH_3—transoid, δ 3.0)

Carbon NMR The carbonyl carbons of acid derivatives appear at shifts around 170 to 180 ppm, slightly more shielded than the carbonyl carbons of ketones and aldehydes. The α carbon atoms absorb around 30 to 40 ppm. The sp^3-hybridized carbons bonded to oxygen in esters absorb around 60 to 80 ppm, and those bonded to nitrogen in amides absorb around 40 to 60 ppm. The cyano carbon of a nitrile absorbs around 120 ppm.

R—CH_2—C(=O)—X: δ 2.0–δ 2.5 (alpha protons)

R—C(=O)—O—CH_2—: δ 4 (ester)

R—C(=O)—N(H)—CH_2—: δ 3 (amide); H—variable, δ 5–δ 8, broad

H—C(=O)—R: δ 9–δ 10 (aldehyde)

H—C(=O)—O—R: δ 8 (formate)

H—C(=O)—NR_2: δ 8 (formamide)

R—CH_2—C≡N: : δ 2.5 (nitrile)

FIGURE 21-7 Typical absorptions of acid derivatives in the proton NMR spectrum.

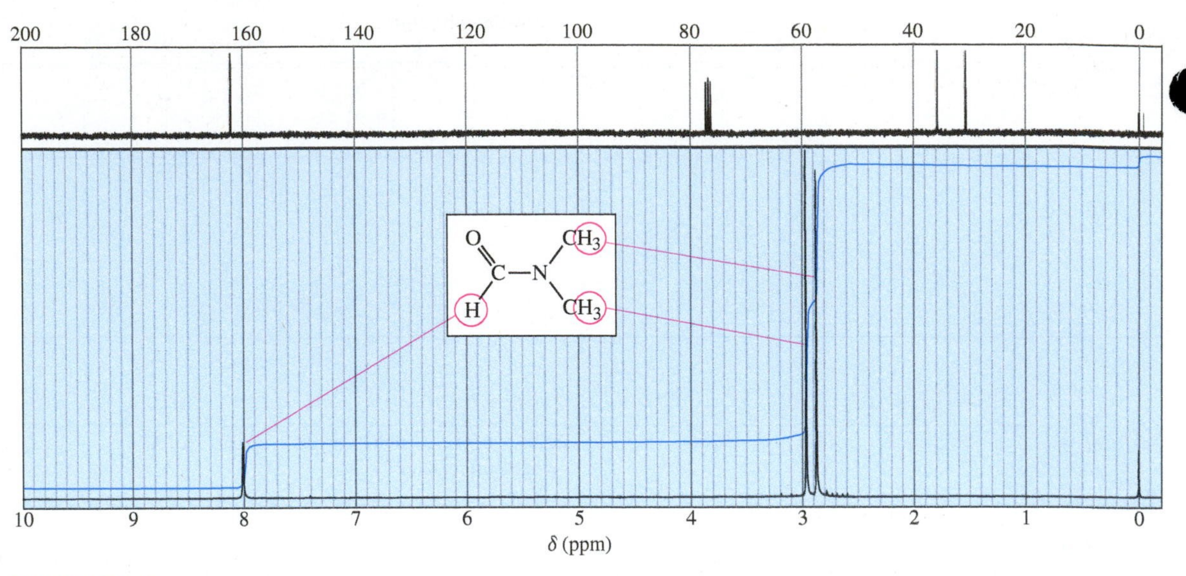

FIGURE 21-8
The proton and carbon NMR spectra of *N,N*-dimethylformamide show two methyl singlets resulting from hindered rotation about the amide bond. In both spectra, the methyl group that is transoid (in a trans-like conformation) to the carbonyl group is farther downfield than the cisoid (in a cis-like conformation) methyl group.

R—C(=O)—O—C— : C=O ~170 ppm; O—C ~70 ppm

R—C(=O)—N̈—C— : C=O ~170 ppm; N—C ~50 ppm

R—C≡N: : C≡N ~120 ppm

Figure 21-8 also shows the carbon NMR spectrum of *N,N*-dimethylformamide (DMF). Note the carbonyl carbon atom at 162 ppm and the two distinct cisoid and transoid carbons at 31 ppm and 36 ppm, respectively.

PROBLEM 21-5

For each set of IR and NMR spectra, determine the structure of the unknown compound. Explain how your proposed structure fits the spectra.

(a) C_3H_5NO **(b)** $C_5H_8O_2$

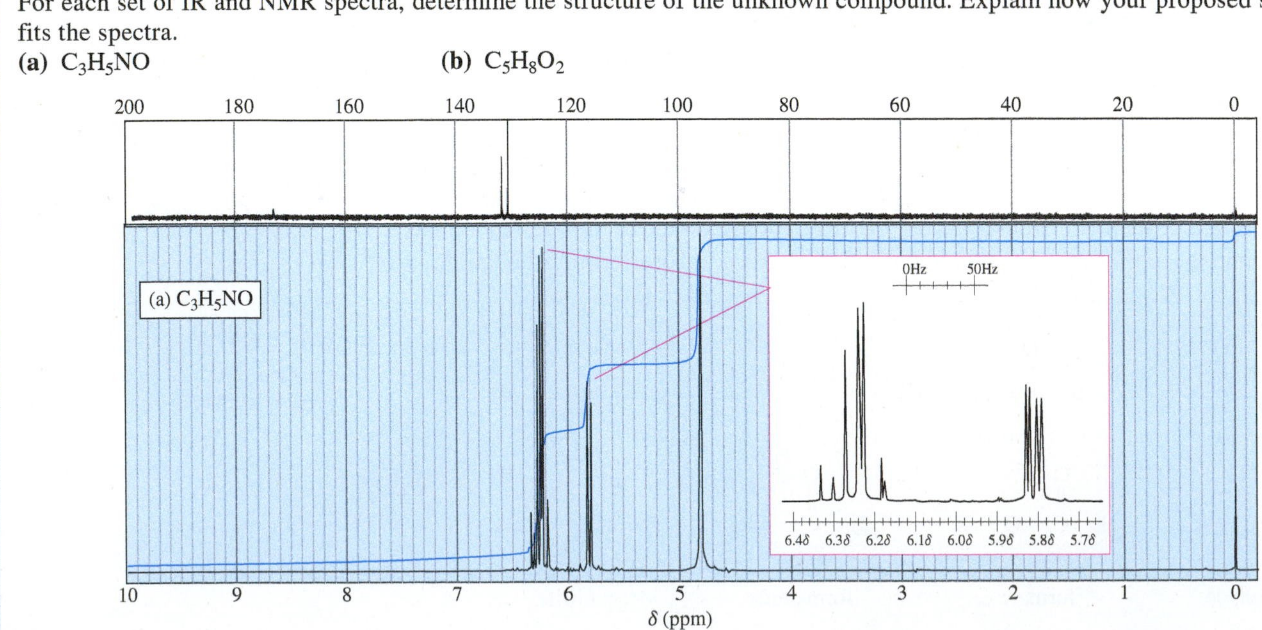

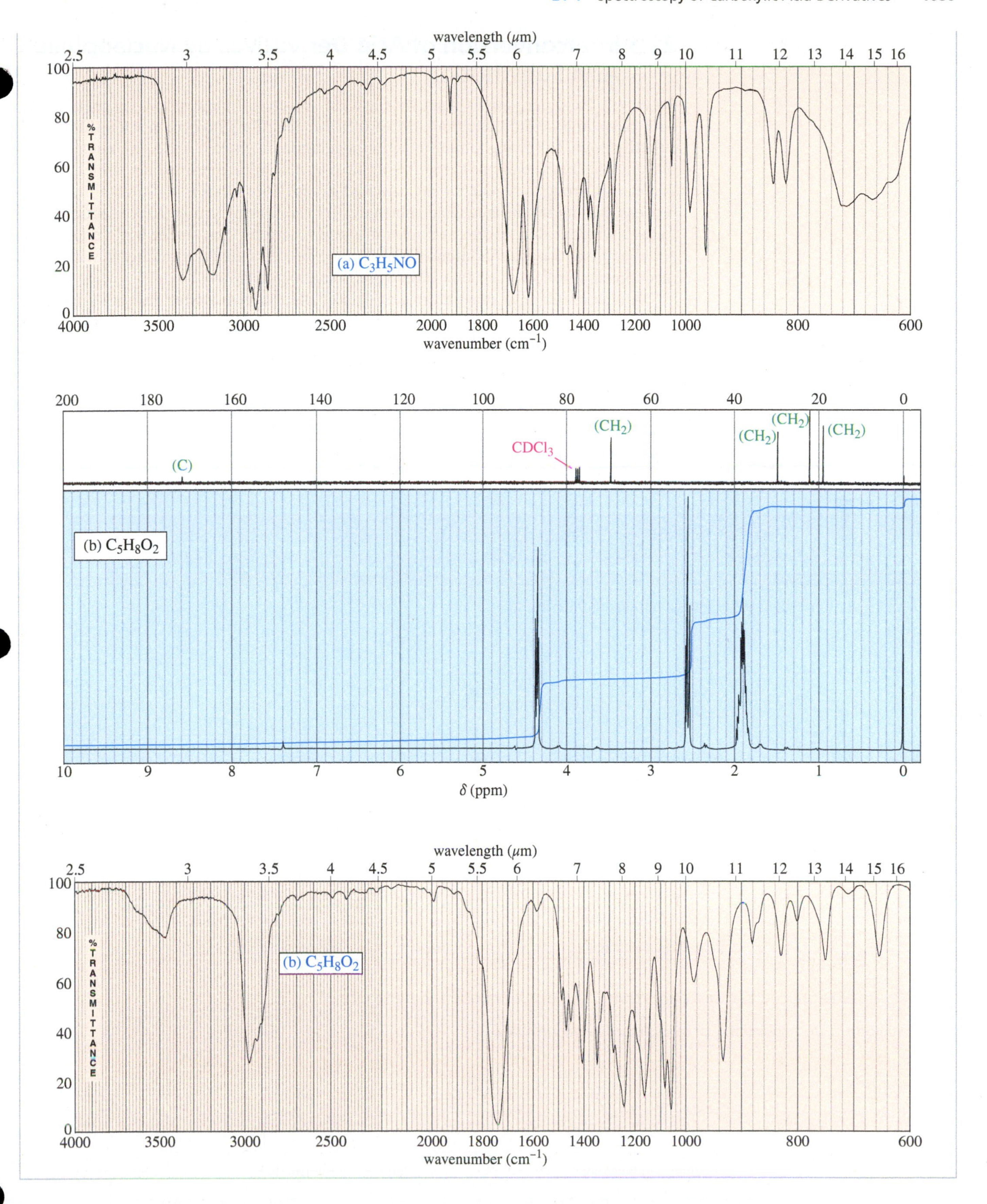

wavelength (μm)
% TRANSMITTANCE
(a) C_3H_5NO
wavenumber (cm^{-1})
(C)
CDCl₃
(CH₂)
(CH₂)
(CH₂)
(CH₂)
(b) $C_5H_8O_2$
δ (ppm)
wavelength (μm)
% TRANSMITTANCE
(b) $C_5H_8O_2$
wavenumber (cm^{-1})

21-5 Interconversion of Acid Derivatives by Nucleophilic Acyl Substitution

Preview Acid derivatives react with a wide variety of nucleophilic reagents under both basic and acidic conditions. Most of these reactions involve **nucleophilic acyl substitutions,** following similar reaction mechanisms. In each case, the nucleophilic reagent adds to the carbonyl group to produce a tetrahedral intermediate, which expels the leaving group to regenerate the carbonyl group. Through this addition–elimination process, the nucleophilic reagent substitutes for the leaving group. In the sections that follow, we consider several examples of these reactions, first under basic conditions and then under acidic conditions. In each case, we will note the similarities with other reactions that follow this same addition–elimination pathway.

Nucleophilic acyl substitutions are also called **acyl transfer** reactions because they transfer the acyl group from the leaving group to the attacking nucleophile. These reactions are common both in chemistry and in biochemistry. In many cases, we can synthesize a compound by starting with a more reactive acid derivative and treating it with an appropriate nucleophile. In biochemistry, some of the most important transformations are acyl transfer reactions used to build up or break down molecules within the cell. The following is a generalized **addition–elimination mechanism** for nucleophilic acyl substitution under basic conditions.

PROBLEM-SOLVING HINT

This mechanism applies to most of the reactions in this chapter.

KEY MECHANISM 21-1 Addition–Elimination Mechanism of Nucleophilic Acyl Substitution

Step 1: Addition of the nucleophile gives a tetrahedral intermediate.

$$\text{Nuc:}^- + \text{R}-\text{C}(=\ddot{\text{O}})-\text{Y} \rightleftarrows \left[\text{R}-\text{C}(-\text{O}^-)(\text{Nuc})-\text{Y}\right]$$

nucleophilic attack — tetrahedral intermediate

Step 2: Elimination of the leaving group regenerates the carbonyl group.

$$\left[\text{R}-\text{C}(-\text{O}^-)(\text{Nuc})-\text{Y}\right] \rightleftarrows \text{R}-\text{C}(=\ddot{\text{O}})-\text{Nuc} + \text{Y:}^-$$

tetrahedral intermediate — products — leaving group

EXAMPLE: Base-catalyzed transesterification of an ester, cyclopentyl benzoate.

Step 1: Addition of the nucleophile gives a tetrahedral intermediate.

$$\text{cyclopentyl benzoate} \xrightleftharpoons[\text{excess } CH_3OH]{^-\text{:}\ddot{\text{O}}CH_3} \text{tetrahedral intermediate}$$

cyclopentyl benzoate — tetrahedral intermediate

Step 2: Elimination of the leaving group regenerates the carbonyl group.

tetrahedral intermediate ⇌ methyl benzoate + cyclopentoxide ion

QUESTION: The reaction in the example only needs a catalytic amount of methoxide ion. Show how the catalyst is regenerated.

Depending on the nucleophile and the leaving group, we can imagine converting any acid derivative into almost any other. Not all of these reactions are practical, however. Favorable reactions generally convert a more reactive acid derivative to a less reactive one. Predicting these reactions requires a knowledge of the relative reactivity of acid derivatives.

21-5A Reactivity of Acid Derivatives

Acid derivatives differ greatly in their reactivity toward nucleophilic acyl substitution. For example, acetyl chloride reacts with water in a violently exothermic reaction, whereas acetamide is stable in boiling water. Acetamide is hydrolyzed only by boiling it in strong acid or base for several hours.

$$CH_3-C(=O)-Cl + H_2O \xrightarrow{\text{(very fast)}} CH_3-C(=O)-OH + HCl$$

$$CH_3-C(=O)-NH_2 + Na^+ \, {}^-OH \xrightarrow[\text{(slow)}]{H_2O,\ \text{boil}} CH_3-C(=O)-O^-\ Na^+ + NH_3$$

The reactivity of acid derivatives toward nucleophilic attack depends on their structure and on the nature of the attacking nucleophile. In general, reactivity follows this order:

Reactivity	*Derivative*		*Leaving group*	*Basicity*
more reactive	acid chloride	$R-C(=O)-Cl$	Cl^-	less basic
	anhydride	$R-C(=O)-O-C(=O)-R$	$^-O-C(=O)-R$	
	ester	$R-C(=O)-O-R'$	$^-O-R'$	
	amide	$R-C(=O)-NH_2$	$^-NH_2$	
less reactive	carboxylate	$R-C(=O)-O^-$	—	more basic

This order of reactivity stems partly from the basicity of the leaving groups. Strong bases are not good leaving groups, and the reactivity of the derivatives decreases as the leaving group becomes more basic.

Resonance stabilization also affects the reactivity of acid derivatives. In amides, for example, resonance stabilization is lost when a nucleophile attacks.

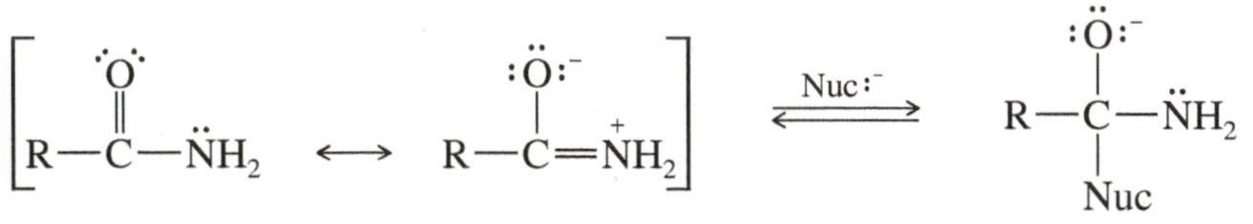

$$\left[R-\overset{\overset{\displaystyle O}{\|}}{C}-\ddot{N}H_2 \longleftrightarrow R-\overset{\overset{\displaystyle :\ddot{O}:^-}{|}}{C}=\overset{+}{N}H_2 \right] \xrightleftharpoons{\text{Nuc}:^-} R-\underset{\underset{\displaystyle \text{Nuc}}{|}}{\overset{\overset{\displaystyle :\ddot{O}:^-}{|}}{C}}-\ddot{N}H_2$$

strong resonance stabilization in amides — no resonance stabilization

A smaller amount of stabilization is present in esters.

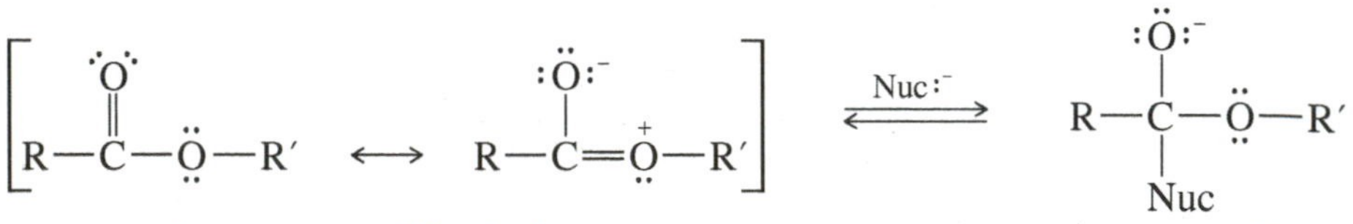

$$\left[R-\overset{\overset{\displaystyle O}{\|}}{C}-\ddot{O}-R' \longleftrightarrow R-\overset{\overset{\displaystyle :\ddot{O}:^-}{|}}{C}=\overset{+}{O}-R' \right] \xrightleftharpoons{\text{Nuc}:^-} R-\underset{\underset{\displaystyle \text{Nuc}}{|}}{\overset{\overset{\displaystyle :\ddot{O}:^-}{|}}{C}}-\ddot{O}-R'$$

weak resonance stabilization in esters — no resonance stabilization

Resonance stabilization of an anhydride is like that in an ester, but the stabilization is shared between two carbonyl groups. Each carbonyl group receives less stabilization than an ester carbonyl.

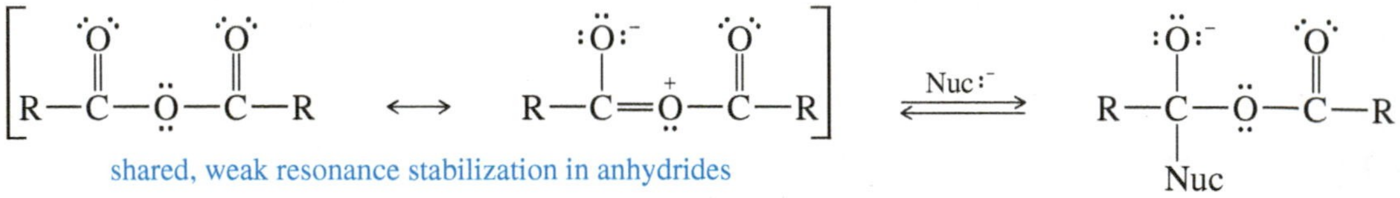

$$\left[R-\overset{\overset{\displaystyle O}{\|}}{C}-\ddot{O}-\overset{\overset{\displaystyle O}{\|}}{C}-R \longleftrightarrow R-\overset{\overset{\displaystyle :\ddot{O}:^-}{|}}{C}=\overset{+}{O}-\overset{\overset{\displaystyle O}{\|}}{C}-R \right] \xrightleftharpoons{\text{Nuc}:^-} R-\underset{\underset{\displaystyle \text{Nuc}}{|}}{\overset{\overset{\displaystyle :\ddot{O}:^-}{|}}{C}}-\ddot{O}-\overset{\overset{\displaystyle O}{\|}}{C}-R$$

shared, weak resonance stabilization in anhydrides

There is little resonance stabilization of an acid chloride, and it is quite reactive.

In general, we can easily accomplish nucleophilic acyl substitutions that convert more reactive derivatives to less reactive ones. Thus, an acid chloride is easily converted to an anhydride, ester, or amide. An anhydride is easily converted to an ester or an amide. An ester is easily converted to an amide, but an amide can be hydrolyzed only to the acid or the carboxylate ion (in basic conditions). Figure 21-9 graphically summarizes these conversions. Notice that thionyl chloride ($SOCl_2$) converts an acid to its most reactive derivative, the acid chloride (Section 20-15).

As we study these conversions of acid derivatives, it may seem that many individual mechanisms are involved. But all these mechanisms are variations on a single theme: the addition–elimination mechanism of nucleophilic acyl substitution (Key Mechanism 21-1). These reactions differ only in the nature of the nucleophile, the leaving group, and proton transfers needed before or after the actual substitution. As we study these mechanisms, watch for these differences and don't feel that you must learn each specific mechanism.

PROBLEM-SOLVING HINT

Nearly all the reactions in this chapter are nucleophilic acyl substitutions that follow the addition–elimination mechanism under acidic or basic conditions.

In the basic pathway, the nucleophile attacks the carbonyl carbon to form a tetrahedral intermediate. Then the intermediate expels the leaving group to regenerate the carbonyl group.

In the acidic pathway, the acid catalyst protonates the carbonyl group so that the weak nucleophile can add. In most cases, the leaving group also becomes protonated before it leaves, so that it can leave as a weak base rather than as a strong base.

Practice using these two pathways and recognizing when they apply. This is a much better strategy than trying to memorize the individual mechanisms.

21-5B Favorable Interconversions of Acid Derivatives

Acid chlorides are the most reactive acid derivatives, so they are easily converted to any of the other acid derivatives. Acid chlorides are often used to synthesize anhydrides, esters, and amides. Acid chlorides react with carboxylic acids (or their carboxylate salts) to form anhydrides. Either oxygen atom of the acid can attack the strongly electrophilic carbonyl group of the acid chloride to form a tetrahedral intermediate. Loss of chloride ion and a proton gives the anhydride.

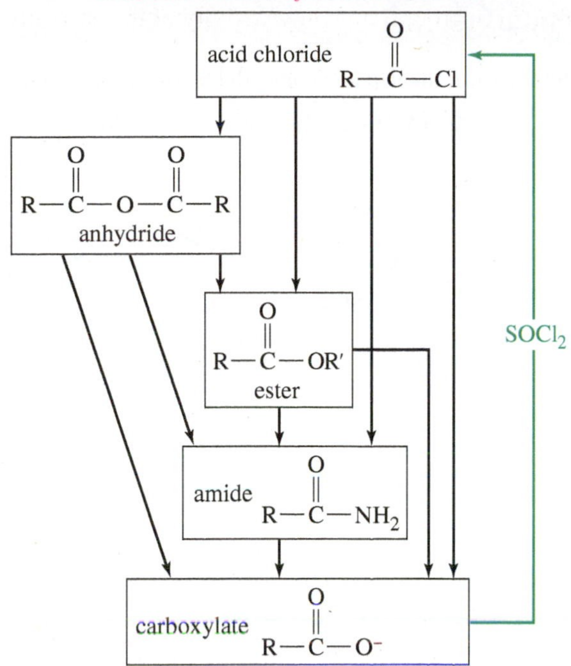

FIGURE 21-9
More reactive acid derivatives are easily converted to less reactive derivatives. A "downhill" reaction from $R-C(=O)-W$ to $R-C(=O)-Z$ generally requires Z^- or $H-Z$ as the nucleophile for nucleophilic acyl substitution.

MECHANISM 21-2 Conversion of an Acid Chloride to an Anhydride

This mechanism follows the standard pattern of an addition–elimination mechanism, ending with loss of a proton to give the final product.

Step 1: Addition of the nucleophile.

acid chloride + acid ⇌ [tetrahedral intermediate]

Step 2: Elimination of the leaving group.

tetrahedral intermediate

Step 3: Loss of a proton.

⇌ Cl^- + protonated anhydride ⇌ anhydride + HCl

Example

$$CH_3(CH_2)_5-\overset{O}{\overset{\|}{C}}-Cl + CH_3(CH_2)_5-\overset{O}{\overset{\|}{C}}-OH \longrightarrow CH_3(CH_2)_5-\overset{O}{\overset{\|}{C}}-O-\overset{O}{\overset{\|}{C}}-(CH_2)_5CH_3$$

heptanoyl chloride + heptanoic acid → heptanoic anhydride

Acid chlorides react rapidly with alcohols to give esters in a strongly exothermic reaction. This reaction requires caution to keep the temperature low to avoid dehydration of the alcohol, because acid chlorides are powerful dehydrating agents. Pyridine (or another base) is often added to the solution to neutralize the HCl by-product.

MECHANISM 21-3 Conversion of an Acid Chloride to an Ester

This is another reaction that follows the standard addition–elimination mechanism, ending with loss of a proton to give the final product.

Step 1: Addition of the nucleophile.

Step 2: Elimination of the leaving group.

Step 3: Loss of a proton.

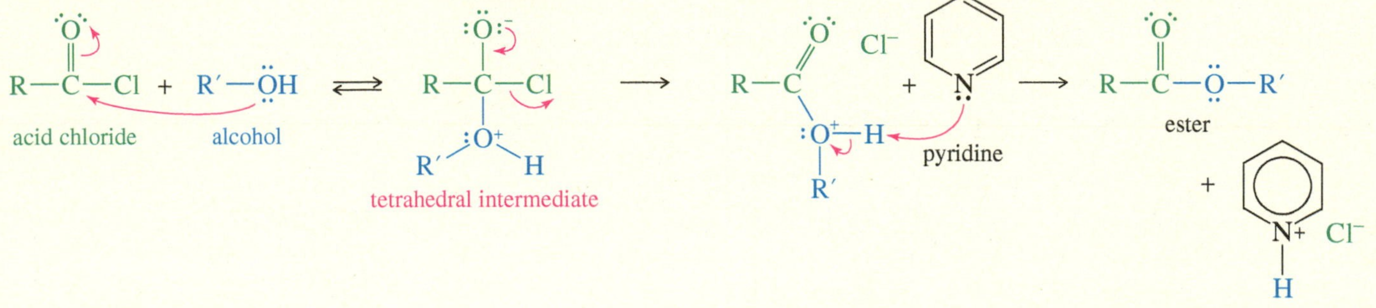

R—C(=O)—Cl (acid chloride) + R′—OH (alcohol) ⇌ R—C(O⁻)(Cl)—O⁺(H)R′ (tetrahedral intermediate) ⟶ R—C(=O)—O⁺(H)—R′ Cl⁻ + pyridine ⟶ R—C(=O)—O—R′ (ester) + pyridinium (N⁺—H) Cl⁻

Example

cyclopentanecarbonyl chloride + CH_3—CH(OH)—CH_3 (propan-2-ol) $\xrightarrow{\text{pyridine}}$ cyclopentyl—C(=O)—$OCH(CH_3)_2$ (2-propyl cyclopentanecarboxylate) + pyridinium (N⁺—H) Cl⁻

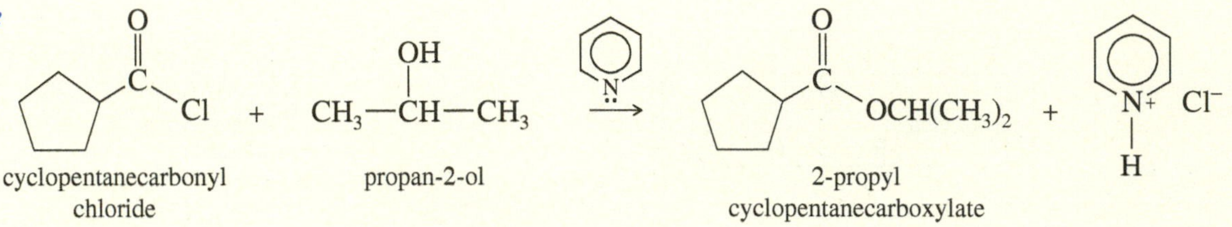

Acid chlorides react rapidly with ammonia and amines to give amides. The HCl generated by the reaction can protonate the amine starting material, so a twofold excess of the amine is required. Alternatively, a base such as pyridine or NaOH may be added with the amine to neutralize the HCl and avoid having to use a large excess of the amine.

MECHANISM 21-4 Conversion of an Acid Chloride to an Amide

This reaction also follows the steps of a standard addition–elimination mechanism, ending with loss of a proton to give the amide.

Step 1: Addition of the nucleophile.

Step 2: Elimination of the leaving group.

Step 3: Loss of a proton.

R—C(=O)—Cl (acid chloride) + 2 R'_2N—H (amine) ⇌ R—C(O⁻)(Cl)—N⁺(R'_2)—H (tetrahedral intermediate) ⟶ R—C(=O)—N⁺(R'_2)—H + R'_2NH ⟶ R—C(=O)—NR'_2 + $R'_2\overset{+}{N}H_2$ Cl⁻

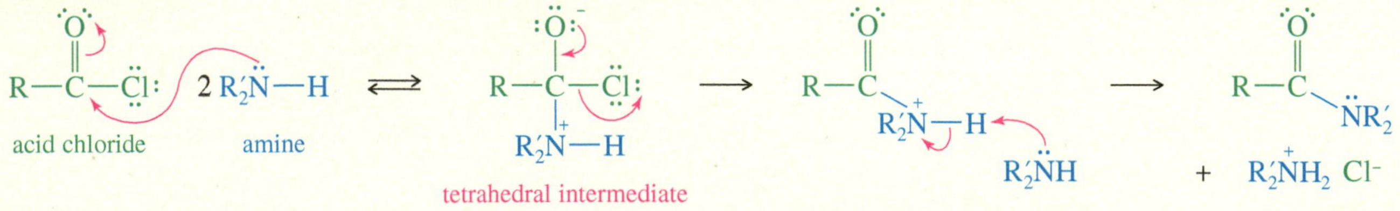

Reaction of an acid chloride with ammonia gives a primary amide. With a primary amine, this reaction gives a secondary amide; and with a secondary amine, it gives a tertiary amide.

Example

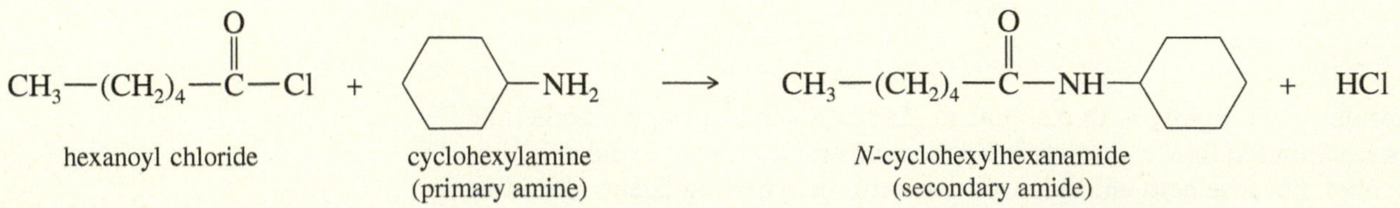

CH_3—$(CH_2)_4$—C(=O)—Cl (hexanoyl chloride) + cyclohexyl—NH_2 (cyclohexylamine (primary amine)) ⟶ CH_3—$(CH_2)_4$—C(=O)—NH—cyclohexyl (*N*-cyclohexylhexanamide (secondary amide)) + HCl

Acid anhydrides are not as reactive as acid chlorides, but they are still activated toward nucleophilic acyl substitution. An anhydride reacts with an alcohol to form an ester. Notice that one of the two acid units from the anhydride is expelled as the leaving group.

MECHANISM 21-5 Conversion of an Acid Anhydride to an Ester

This reaction follows the standard addition–elimination mechanism, ending with loss of a proton to give the ester.

Step 1: Addition of the nucleophile.

$$R-C(=O)-O-C(=O)-R + R'-\ddot{O}H \rightleftharpoons$$

anhydride alcohol

Step 2: Elimination of the leaving group.

$$R-C(-O^-)(-O^+HR')-O-C(=O)-R \longrightarrow$$

tetrahedral intermediate

Step 3: Loss of a proton.

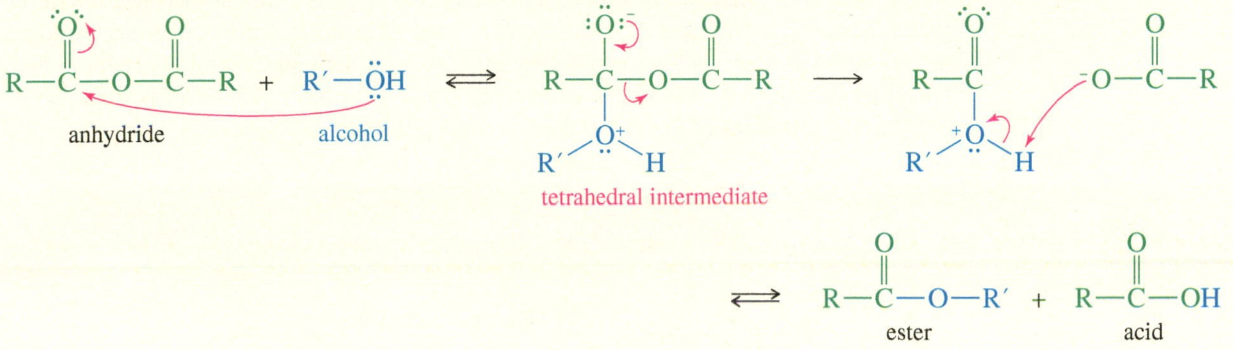

$$R-C(=O)-O^+HR' \quad {}^-O-C(=O)-R$$

$$\rightleftharpoons R-C(=O)-O-R' + R-C(=O)-OH$$

ester acid

Example

$$\text{cyclopentanol} + CH_3-C(=O)-O-C(=O)-CH_3 \longrightarrow \text{cyclopentyl}-O-C(=O)-CH_3 + AcOH$$

cyclopentanol acetic anhydride cyclopentyl acetate acetic acid

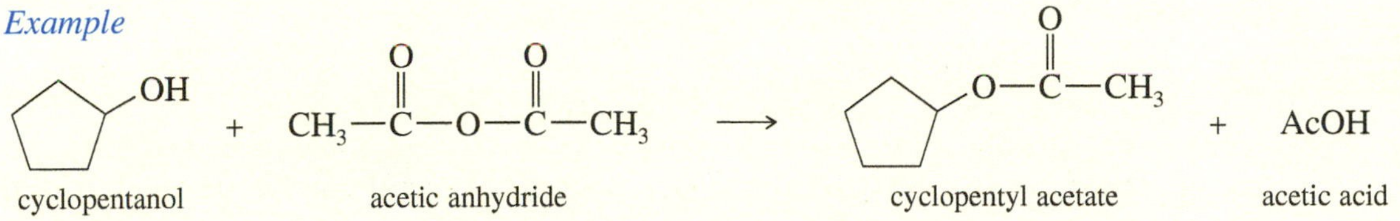

Anhydrides react quickly with ammonia and amines. Reaction of an anhydride with ammonia gives a primary amide. An anhydride reacts with a primary amine to give a secondary amide, and with a secondary amine to give a tertiary amide.

MECHANISM 21-6 Conversion of an Acid Anhydride to an Amide

This reaction follows the standard addition–elimination mechanism, ending with loss of a proton to give the amide.

Step 1: Addition of the nucleophile.

$$R-C(=O)-O-C(=O)-R + R'_2\ddot{N}H \rightleftharpoons$$

anhydride amine

Step 2: Elimination of the leaving group.

$$R-C(-O^-)(-N^+HR'_2)-O-C(=O)-R \longrightarrow$$

tetrahedral intermediate

Step 3: Loss of a proton.

$$R-C(=O)-N^+HR'_2 \quad {}^-O-C(=O)-R$$

$$\rightleftharpoons R-C(=O)-NR'_2 + R-C(=O)-OH$$

amide acid

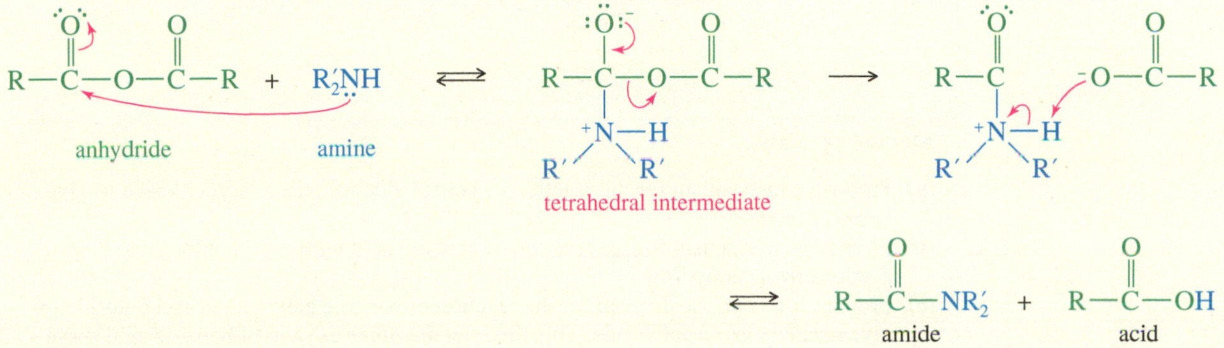

(CONTINUED)

Example

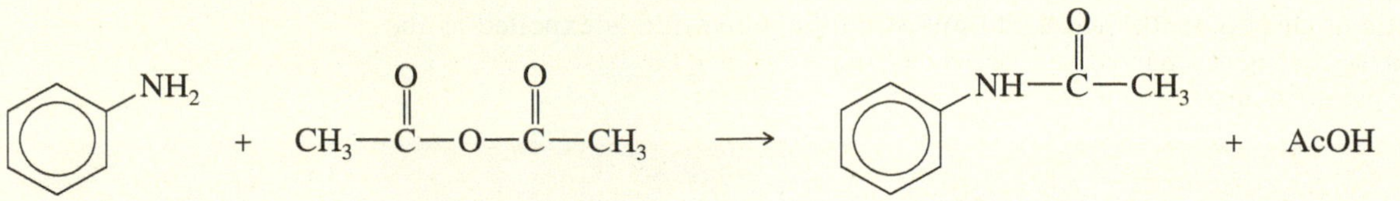

aniline + acetic anhydride ⟶ acetanilide + AcOH

$C_6H_5NH_2 + CH_3{-}C({=}O){-}O{-}C({=}O){-}CH_3 \longrightarrow C_6H_5NH{-}C({=}O){-}CH_3 + AcOH$

aniline | acetic anhydride | acetanilide | acetic acid

Esters are less reactive than anhydrides, but they can be converted to amides by heating with ammonia or an amine. This reaction is called **ammonolysis,** meaning "lysis (cleavage) by an amine." Ammonolysis using ammonia gives primary amides. Primary amines react to give secondary amides, and secondary amines react (often slowly) to give tertiary amides. In each case, the acyl group of the ester is transferred from the oxygen atom of the alcohol to the nitrogen atom of the amine.

MECHANISM 21-7 Conversion of an Ester to an Amide (Ammonolysis of an Ester)

This is yet another standard addition–elimination mechanism, ending with loss of a proton to give the amide.

Step 1: Addition of the nucleophile.

$R{-}C({=}O){-}O{-}R' + R''{-}NH_2 \rightleftarrows$

primary amine (or NH_3)

Step 2: Elimination of the leaving group.

Step 3: Loss of a proton.

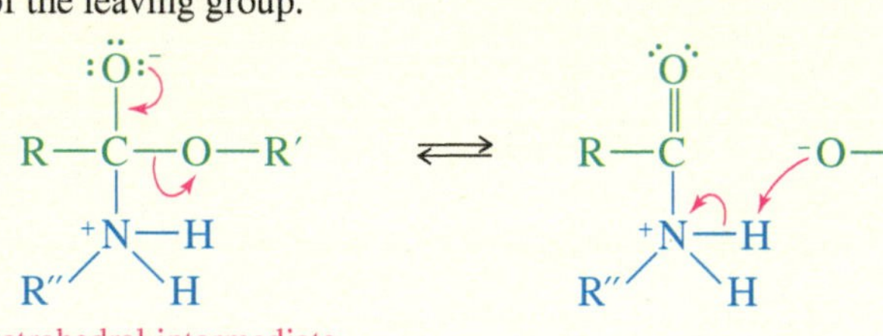

$\longrightarrow R{-}C({=}O){-}NHR'' + R'{-}OH$

amide | alcohol

Example

$H{-}C({=}O){-}O{-}CH_2CH_3 + C_6H_{11}NH_2 \longrightarrow C_6H_{11}{-}NH{-}C({=}O){-}H + CH_3CH_2{-}OH$

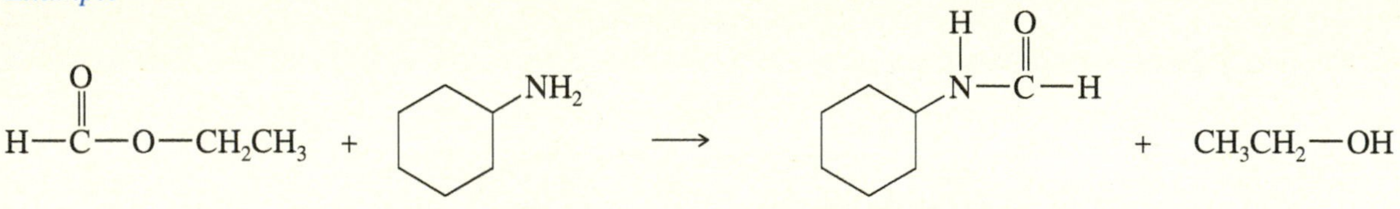

ethyl formate | cyclohexylamine | *N*-cyclohexylformamide (90%) | ethanol

PROBLEM 21-6

(a) Propose a mechanism for the reaction of benzyl alcohol with acetyl chloride to give benzyl acetate.

(b) Propose a mechanism for the reaction of benzoic acid with acetyl chloride to give acetic benzoic anhydride.

(c) Propose a *second* mechanism for the reaction of benzoic acid with acetyl chloride to give acetic benzoic anhydride. This time, let the *other* oxygen of benzoic acid serve as the nucleophile to attack the carbonyl group of acetyl chloride. Because proton transfers are fast between these oxygen atoms, it is difficult to differentiate between these two mechanisms experimentally.

(d) Propose a mechanism for the reaction of aniline with acetic anhydride to give acetanilide.

(e) Propose a mechanism for the reaction of aniline with ethyl acetate to give acetanilide. What is the leaving group in your proposed mechanism? Would this be a suitable leaving group for an S_N2 reaction?

21-5C Leaving Groups in Nucleophilic Acyl Substitutions

Loss of an alkoxide ion as a leaving group in the second step of the ammonolysis of an ester should surprise you.

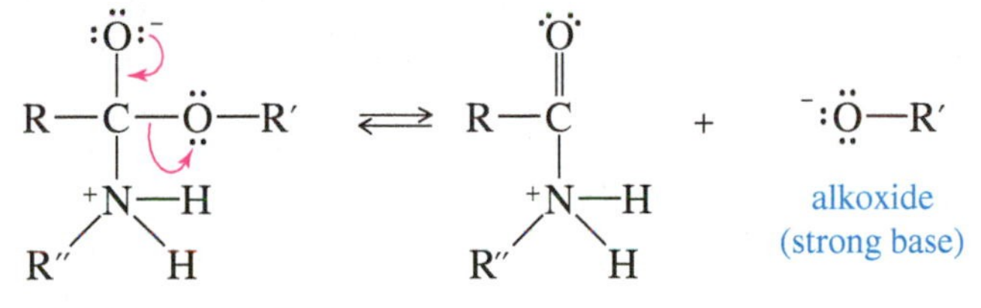

tetrahedral intermediate

alkoxide (strong base)

In our study of alkyl substitution and elimination reactions (S_N1, S_N2, E1, E2), we saw that strong bases such as hydroxide and alkoxide are poor leaving groups for these reactions. Figure 21-10 compares the acyl addition–elimination mechanism with the S_N2 mechanism. The differences in the mechanisms explain why strong bases may serve as leaving groups in acyl substitution, even though they cannot in alkyl substitution.

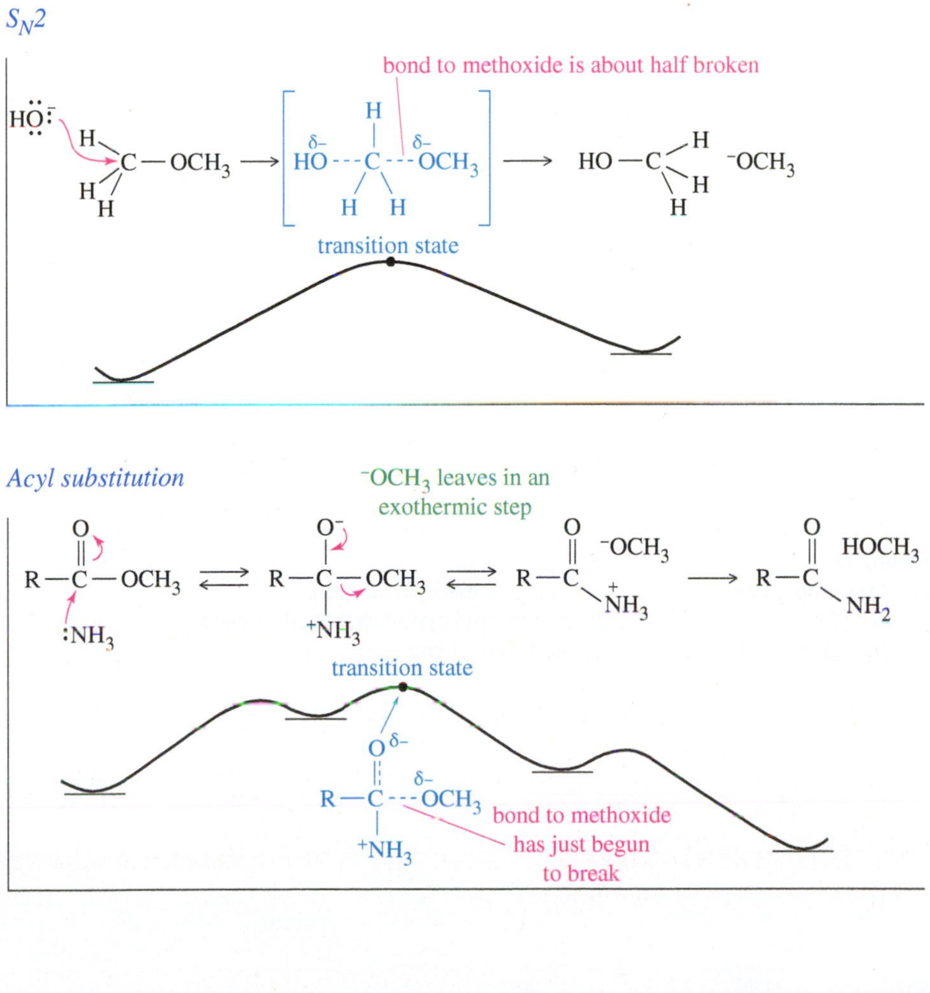

S_N2

bond to methoxide is about half broken

transition state

Acyl substitution

$^-OCH_3$ leaves in an exothermic step

transition state

bond to methoxide has just begun to break

FIGURE 21-10 Comparison of S_N2 and acyl addition–elimination reactions with methoxide as the leaving group. In the concerted S_N2, methoxide leaves in a slightly endothermic step, and the bond to methoxide is largely broken in the transition state. In the acyl substitution, methoxide leaves in an exothermic second step with a reactant-like transition state: The bond to methoxide has just begun to break in the transition state.

The S_N2 reaction's one-step mechanism is not strongly endothermic or exothermic. The bond to the leaving group is about half broken in the transition state, so the reaction rate is sensitive to the nature of the leaving group. With a poor leaving group such as alkoxide, this reaction is quite slow.

In the acyl substitution, the leaving group leaves in a separate second step. This second step is highly exothermic, and Hammond's postulate (Section 4-14) predicts that the transition state resembles the reactant: the tetrahedral intermediate. In this transition state, the bond to the leaving group has barely begun to break. The energy of the transition state (and therefore the reaction rate) is not very sensitive to the nature of the leaving group.

PROBLEM-SOLVING HINT

A strong base may serve as a leaving group if it leaves in a highly exothermic step, usually converting an unstable, negatively charged intermediate to a stable molecule.

Nucleophilic acyl substitution is our first example of a reaction with strong bases as leaving groups. We will see many additional examples of such reactions. In general, a strong base may serve as a leaving group if it leaves in a highly exothermic step, usually converting an unstable, negatively charged intermediate to a stable molecule.

PROBLEM 21-7

Which of the following proposed reactions would take place quickly under mild conditions?

(a) $$CH_3-\overset{O}{\overset{\|}{C}}-NH_2 + NaCl \longrightarrow CH_3-\overset{O}{\overset{\|}{C}}-Cl + NaNH_2$$

(b) $$Ph-\overset{O}{\overset{\|}{C}}-Cl + CH_3NH_2 \longrightarrow Ph-\overset{O}{\overset{\|}{C}}-NHCH_3 + HCl$$

(c) $$(CH_3)_2CH-\overset{O}{\overset{\|}{C}}-NH_2 + CH_3OH \longrightarrow (CH_3)_2CH-\overset{O}{\overset{\|}{C}}-OCH_3 + NH_3$$

(d) $$CH_3CH_2-\overset{O}{\overset{\|}{C}}-Cl + CH_3-\overset{O}{\overset{\|}{C}}-OH \longrightarrow CH_3CH_2-\overset{O}{\overset{\|}{C}}-O-\overset{O}{\overset{\|}{C}}-CH_3 + HCl$$

(e) $$CH_3-\overset{O}{\overset{\|}{C}}-O-\overset{O}{\overset{\|}{C}}-CH_3 + CH_3NH_2 \longrightarrow CH_3-\overset{O}{\overset{\|}{C}}-NHCH_3 + CH_3COOH$$

PROBLEM 21-8

Show how you would synthesize the following esters from appropriate acyl chlorides and alcohols.

(a) ethyl propionate
(b) phenyl 3-methylhexanoate
(c) benzyl benzoate
(d) cyclopropyl cyclohexanecarboxylate
(e) *tert*-butyl acetate
(f) diallyl succinate

PROBLEM 21-9

Show how you would use appropriate acyl chlorides and amines to synthesize the following amides.

(a) *N,N*-dimethylacetamide

(b) acetanilide ($PhNHCOCH_3$)

(c) cyclohexanecarboxamide

(d) 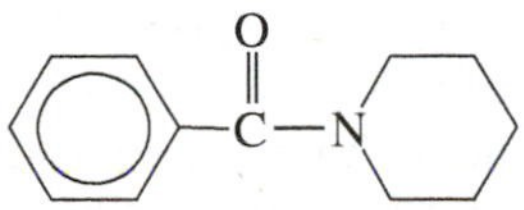 Ph—C(=O)—N(piperidine ring)

PROBLEM 21-10

(a) Show how you would use acetic anhydride and an appropriate alcohol or amine to synthesize (i) benzyl acetate and (ii) *N,N*-diethylacetamide.

(b) Propose a mechanism for each synthesis in part (a).

PROBLEM 21-11

Propose a mechanism for the reaction of benzyl acetate with methylamine. Label the attacking nucleophile and the leaving group, and draw the transition state in which the leaving group leaves.

21-6 Transesterification

Esters undergo **transesterification,** in which one alkoxy group substitutes for another, under either acidic or basic conditions. When an ester of one alcohol is treated with a different alcohol in the presence of acid or base, the two alcohol groups can interchange. An equilibrium results, and the equilibrium can be driven toward the desired ester by using a large excess of the desired alcohol or by removing the other alcohol.

Transesterification

$$R-\overset{O}{\overset{\|}{C}}-O-R' \quad + \quad \underset{\text{(large excess)}}{R''-OH} \quad \xrightleftharpoons{H^+ \text{ or } ^-OR''} \quad R-\overset{O}{\overset{\|}{C}}-O-R'' \quad + \quad R'-OH$$

Example

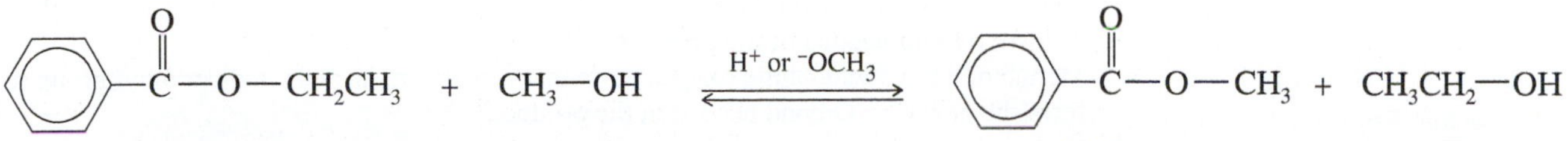

$$\underset{\text{ethyl benzoate}}{Ph-\overset{O}{\overset{\|}{C}}-O-CH_2CH_3} \quad + \quad \underset{\text{methanol}}{CH_3-OH} \quad \xrightleftharpoons{H^+ \text{ or } ^-OCH_3} \quad \underset{\text{methyl benzoate}}{Ph-\overset{O}{\overset{\|}{C}}-O-CH_3} \quad + \quad \underset{\text{ethanol}}{CH_3CH_2-OH}$$

Transesterification is possibly the simplest and best example of the acid-catalyzed and base-catalyzed nucleophilic acyl substitution mechanisms because it is an evenly balanced equilibrium with identical mechanisms for the forward and reverse reactions.

PROBLEM-SOLVING STRATEGY Proposing Reaction Mechanisms

Rather than just showing the mechanisms for acid-catalyzed and base-catalyzed transesterification, let's consider how you might work out these mechanisms as in a problem.

Base-Catalyzed Transesterification

First consider the base-catalyzed transesterification of ethyl benzoate with methanol. This is a classic example of nucleophilic acyl substitution by the addition–elimination mechanism. Methoxide ion is sufficiently nucleophilic to attack the ester carbonyl group. Ethoxide ion serves as a leaving group in a strongly exothermic second step.

$PhC(=O)OCH_2CH_3 + CH_3O^- \rightleftarrows PhC(O^-)(OCH_3)OCH_2CH_3 \rightleftarrows PhC(=O)OCH_3 + {}^-OCH_2CH_3$

nucleophilic attack — tetrahedral intermediate

Now try a base-catalyzed mechanism on your own in Problem 21-12.

PROBLEM 21-12

When ethyl 4-hydroxybutyrate is heated in the presence of a trace of a basic catalyst (sodium acetate), one of the products is a lactone. Propose a mechanism for formation of this lactone.

Acid-Catalyzed Transesterification

The acid-catalyzed reaction follows a similar mechanism, but it is more complicated because of additional proton transfers. We use the stepwise procedure to propose a mechanism for the following reaction, in which methanol replaces ethanol.

$$PhC(=O)OCH_2CH_3 + CH_3OH \overset{H^+}{\rightleftarrows} PhC(=O)OCH_3 + CH_3CH_2OH$$

1. **Consider the carbon skeletons of the reactants and products, and identify which carbon atoms in the products are likely derived from which carbon atoms in the reactants.**
   In this case, an ethoxy group is replaced by a methoxy group.
2. **Consider whether any of the reactants is a strong enough electrophile to react without being activated. If not, consider how one of the reactants might be converted to a strong electrophile by protonation of a Lewis basic site.**
   The ester carbonyl group is not a strong enough electrophile to react with methanol. Protonation converts it to a strong electrophile (shown in step 3).
3. **Consider how a nucleophilic site on another reactant can attack the strong electrophile to form a bond needed in the product.**
   Methanol has a nucleophilic oxygen atom that can attack the activated carbonyl group to form the new C—O bond needed in the product.

$$PhC(=O)OCH_2CH_3 \overset{H^+}{\rightleftarrows} [PhC(=O^+H)OCH_2CH_3] \overset{CH_3OH}{\rightleftarrows}$$

activation of C=O — (resonance-stabilized) — nucleophilic attack

Application: Biodiesel

Base-catalyzed transesterification is the process that converts waste cooking oil to biodiesel fuel.

Fats and oils are triesters of glycerol (**triglycerides**), with three long-chain fatty acids that give the molecule a high molecular weight and low volatility. A base-catalyzed transesterification (using methanol as the alcohol and NaOH as the catalyst) converts fats and oils to the methyl esters of the three individual fatty acids. With molecular weights about a third of the original triglyceride, these methyl esters are more volatile and work well in diesel engines. The mixture of fatty acid methyl esters is called **biodiesel.**

a triglyceride $\xrightarrow{3\ CH_3OH,\ NaOH\ (transesterification)}$ 3 $CH_3-O-C(=O)R$ (methyl ester) + $CH_2(OH)-CH(OH)-CH_2(OH)$ (glycerol)

Converting waste cooking oils to biodiesel is an excellent example of chemical recycling by converting a waste into a valuable product. On the other hand, converting new food-grade fats and oils to biodiesel is economically and environmentally unsound. In the absence of subsidies, food-grade oils sell for several times the price of diesel fuel. Regulations requiring biodiesel in fuels have created a large artificial demand for plant oils, especially palm oil, that has promoted the conversion of huge areas of rain forests into palm oil plantations.

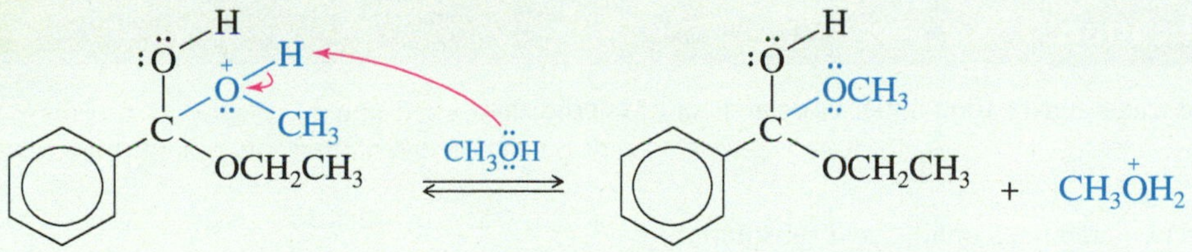

tetrahedral intermediate

4. **Consider how the product of nucleophilic attack might be converted to the final product or reactivated to form another bond needed in the product.**
The task here is to break bonds, not form them. The ethoxy group (OCH_2CH_3) must be lost. The most common mechanism for losing a group under acidic conditions is to protonate it (to make it a good leaving group), and then lose it. In fact, losing the ethoxy group is exactly the reverse of the mechanism used to gain the methoxy group.
Protonation prepares the ethoxy group to leave. When ethanol leaves, the product is simply a protonated version of the final product.

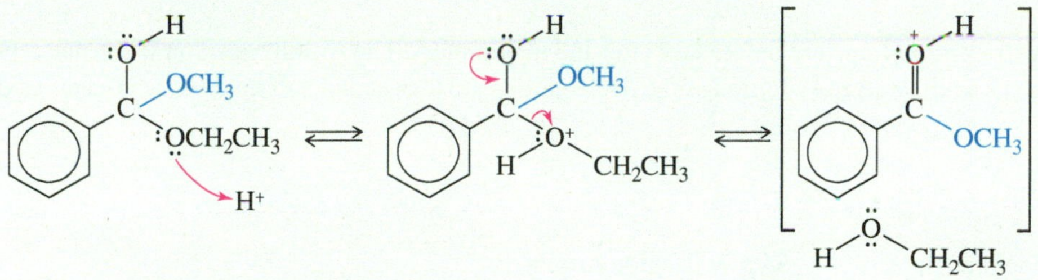

5. **Draw out all steps of the mechanism, using curved arrows to show the movement of electrons.**
Once again, this summary is left to you to help you review the mechanism.

PROBLEM-SOLVING HINT

Acid-catalyzed nucleophilic acyl substitution usually differs from the base-catalyzed reaction in two major ways:

1. The carbonyl must be protonated to activate it toward attack by a weak nucleophile.
2. In acid conditions, leaving groups are usually protonated, then lost as neutral molecules.

PROBLEM 21-13

Complete the mechanism for this acid-catalyzed transesterification by drawing out all the individual steps. Draw the important resonance contributors for each resonance-stabilized intermediate.

PROBLEM 21-14

Propose a mechanism for the following ring-opening transesterification. Use the mechanism in Problem 21-13 as a model.

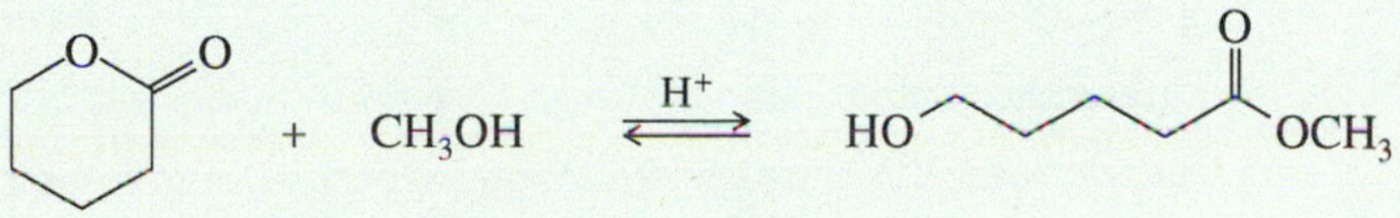

MECHANISM 21-8 Transesterification

Following is a summary of the mechanism of transesterification under basic and acidic conditions.

Base-catalyzed

Base-catalyzed transesterification is a simple two-step nucleophilic acyl substitution:

Step 1: Addition of the nucleophile. ***Step 2:*** Elimination of the leaving group.

O C OCH_3 $^-$:OR ⇌ :O:$^-$ C—OCH_3 :OR ⇌ O C OR $^-$:OCH_3

tetrahedral intermediate

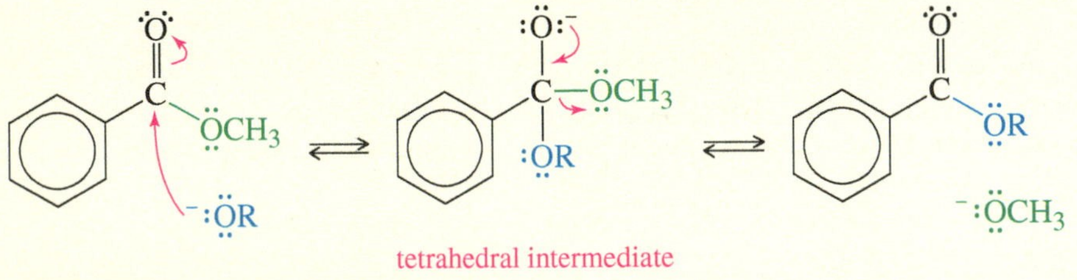

Acid-catalyzed

Acid-catalyzed transesterification requires extra proton transfers before and after the major steps. The overall reaction takes place in two stages. The first half of the reaction involves acid-catalyzed addition of the nucleophile, and the second half involves acid-catalyzed elimination of the leaving group.

First half: Acid-catalyzed addition of the nucleophile.

Step 1: Protonation of the carbonyl. ***Step 2:*** Nucleophile attack. ***Step 3:*** Deprotonation.

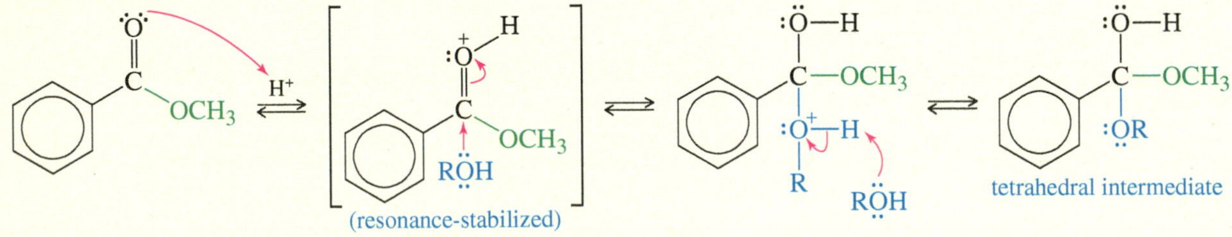

O C OCH_3 $\overset{H^+}{\rightleftharpoons}$ [:$\overset{+}{O}$—H C OCH_3 ROH] ⇌ :O—H C—OCH_3 :$\overset{+}{O}$—H R ROH ⇌ :O—H C—OCH_3 :OR

(resonance-stabilized)

tetrahedral intermediate

Second half: Acid-catalyzed elimination of the leaving group.

Step 1: Protonation of the leaving group. ***Step 2:*** Elimination of the leaving group. ***Step 3:*** Deprotonation.

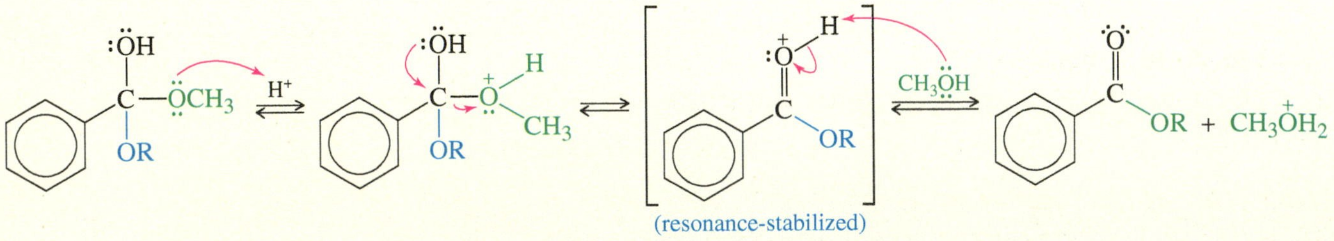

:OH C—OCH_3 OR $\overset{H^+}{\rightleftharpoons}$:OH C—$\overset{+}{O}$ H CH_3 OR ⇌ [:$\overset{+}{O}$—H C OR] $\overset{CH_3OH}{\rightleftharpoons}$ O C OR + $CH_3\overset{+}{O}H_2$

(resonance-stabilized)

Some reactions that can go as basic nucleophilic acyl substitutions actually work much better with an acid catalyst. For example, aspirin is made from salicylic acid and acetic anhydride. When these reagents are mixed, the reaction goes slowly. Addition of a drop of sulfuric acid accelerates the reaction, and it goes to completion in a minute or two.

$$\text{salicylic acid} + CH_3-\overset{O}{\overset{\|}{C}}-O-\overset{O}{\overset{\|}{C}}-CH_3 \text{ (acetic anhydride)} \xrightarrow[\text{fast}]{H_2SO_4} \text{aspirin (acetylsalicylic acid)} + CH_3COOH$$

salicylic acid

acetic anhydride

aspirin
(acetylsalicylic acid)

PROBLEM 21-15

(a) Propose a mechanism for the acid-catalyzed reaction of salicylic acid with acetic anhydride.
(b) Explain why a single drop of sulfuric acid dramatically increases the reaction rate.

21-7 Hydrolysis of Carboxylic Acid Derivatives

All acid derivatives hydrolyze to give carboxylic acids. In most cases, hydrolysis occurs under either acidic or basic conditions. The reactivity of acid derivatives toward hydrolysis varies from highly reactive acyl halides to relatively unreactive amides.

21-7A Hydrolysis of Acid Halides and Anhydrides

Acid halides and anhydrides are so reactive that they hydrolyze under neutral conditions. Hydrolysis of an acid halide or anhydride is usually an annoying side reaction that takes place on exposure to moist air. Hydrolysis can be avoided by storing acid halides and anhydrides under dry nitrogen and by using dry solvents and reagents.

$$R-\overset{O}{\overset{\|}{C}}-Cl + H_2\ddot{O}: \longrightarrow R-\underset{^+OH_2}{\overset{:O:^-}{C}}-Cl \longrightarrow R-\overset{O}{\overset{\|}{C}}-\overset{+}{O}H_2 \;\; Cl^- \;\; H_2\ddot{O}: \longrightarrow R-\overset{O}{\overset{\|}{C}}-\ddot{O}-H + H_3O^+ \; Cl^-$$

21-7B Hydrolysis of Esters

Acid-catalyzed hydrolysis of an ester is simply the reverse of the Fischer esterification equilibrium. Addition of excess water drives the equilibrium toward the acid and the alcohol.

Basic hydrolysis of esters, called **saponification,** avoids the equilibrium of the Fischer esterification. Hydroxide ion attacks the carbonyl group to give a tetrahedral intermediate. Expulsion of alkoxide ion gives the acid, and a fast proton transfer gives the carboxylate ion and the alcohol. This strongly exothermic proton transfer drives the saponification to completion. A full mole of base is consumed to deprotonate the acid.

MECHANISM 21-9 Saponification of an Ester

This is another standard addition–elimination mechanism, ending with a proton transfer to give the final products.

Step 1: Addition of the nucleophile. ***Step 2:*** Elimination of the leaving group. ***Step 3:*** Proton transfer.

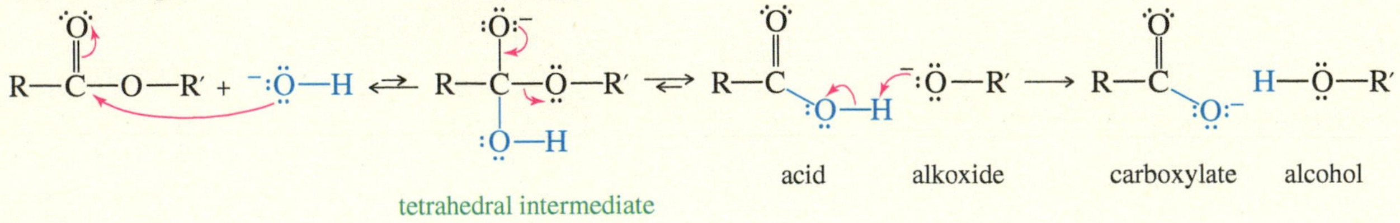

$$R-C(=O)-O-R' + {}^{-}:\ddot{O}-H \rightleftharpoons R-C(-O^{-})(-OH)-\ddot{O}-R' \rightleftharpoons R-C(=O)-OH + {}^{-}:\ddot{O}-R' \longrightarrow R-C(=O)-O^{-} + H-\ddot{O}-R'$$

tetrahedral intermediate

acid alkoxide carboxylate alcohol

Example

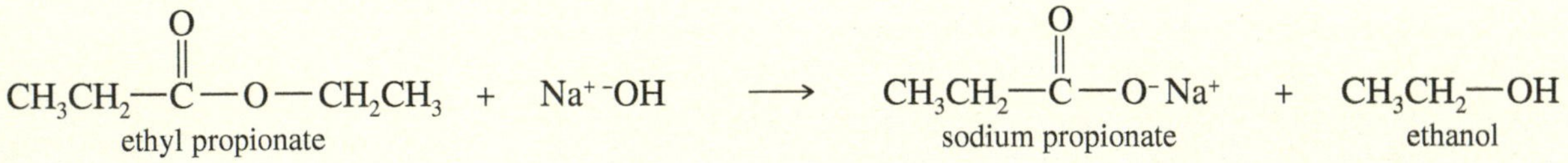

$$CH_3CH_2-C(=O)-O-CH_2CH_3 + Na^{+}\,{}^{-}OH \longrightarrow CH_3CH_2-C(=O)-O^{-}\,Na^{+} + CH_3CH_2-OH$$

ethyl propionate sodium propionate ethanol

The term *saponification* (Latin, *saponis*, "soap") literally means "the making of soap." Soap is made by the basic hydrolysis of fats, which are esters of long-chain carboxylic acids (*fatty acids*) with the triol glycerol. When sodium hydroxide hydrolyzes a fat, the resulting long-chain sodium carboxylate salts are what we know as soap. Soaps and detergents are discussed in more detail in Chapter 25.

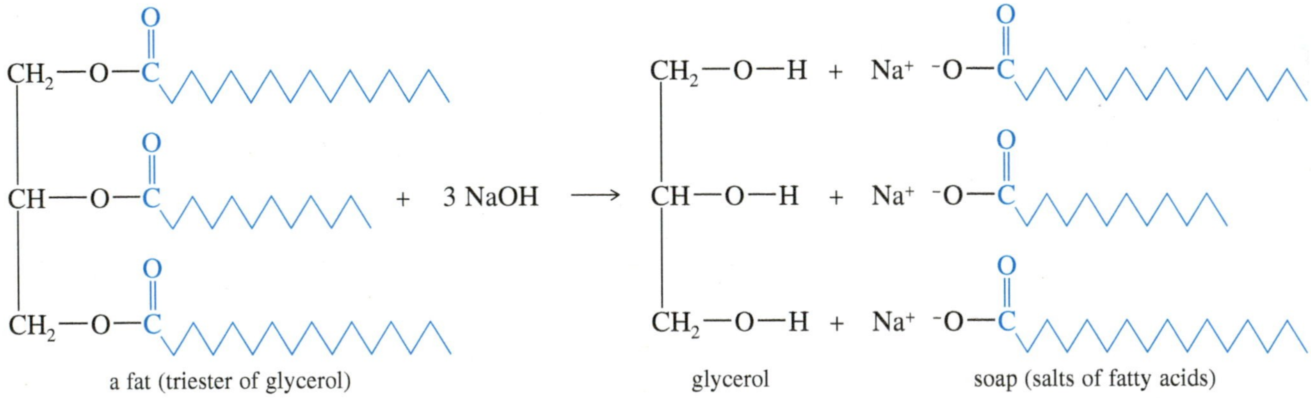

$$\begin{array}{l} CH_2-O-C(=O)R \\ CH-O-C(=O)R \\ CH_2-O-C(=O)R \end{array} + 3\,NaOH \longrightarrow \begin{array}{l} CH_2-O-H \\ CH-O-H \\ CH_2-O-H \end{array} + 3\,Na^{+}\,{}^{-}O-C(=O)R$$

a fat (triester of glycerol) glycerol soap (salts of fatty acids)

PROBLEM 21-16

Suppose we have some optically pure (*R*)-2-butyl acetate that has been "labeled" with the heavy ^{18}O isotope at one oxygen atom as shown.

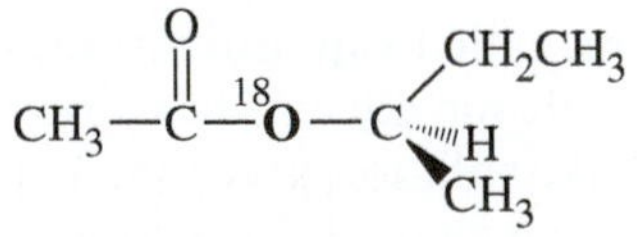

$$CH_3-C(=O)-{}^{18}O-C(H)(CH_3)(CH_2CH_3)$$

(a) Draw a mechanism for the hydrolysis of this compound under basic conditions. Predict which of the products will contain the ^{18}O label. Also predict whether the butan-2-ol product will be pure (*R*), pure (*S*), or racemized.

(b) Repeat part (a) for the acid-catalyzed hydrolysis of this compound.

(c) Explain how you would prove experimentally where the ^{18}O label appears in the products. (^{18}O is not radioactive.)

PROBLEM 21-17

(a) Explain why we speak of acidic hydrolysis of an ester as *acid-catalyzed*, but of basic hydrolysis as *base-promoted*.

(b) Soap manufacturers always use base to hydrolyze fats, and never acid. Suggest two reasons that basic hydrolysis is preferred.

PROBLEM 21-18

Propose a mechanism for the base-promoted hydrolysis of γ-butyrolactone:

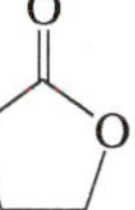

21-7C Hydrolysis of Amides

Amides hydrolyze to carboxylic acids under both acidic and basic conditions. Amides are the most stable of the acid derivatives, and stronger conditions are required for their hydrolysis than for hydrolysis of an ester. Typical hydrolysis conditions involve prolonged heating in 6 M HCl or 40% aqueous NaOH.

Basic hydrolysis

$$R-\overset{O}{\overset{\|}{C}}-NHR' + Na^{+}\,^{-}OH \xrightarrow{H_2O} R-\overset{O}{\overset{\|}{C}}-O^{-}\;Na^{+} + R'NH_2$$

Example

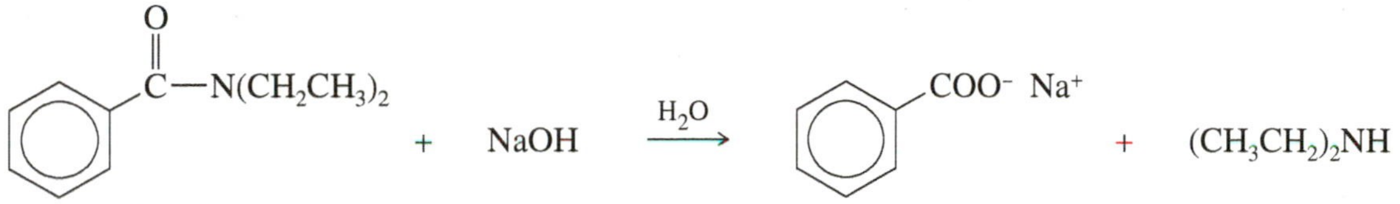

$$C_6H_5-\overset{O}{\overset{\|}{C}}-N(CH_2CH_3)_2 + NaOH \xrightarrow{H_2O} C_6H_5-COO^{-}\;Na^{+} + (CH_3CH_2)_2NH$$

N,N-diethylbenzamide — sodium benzoate — diethylamine

Acid hydrolysis

$$R-\overset{O}{\overset{\|}{C}}-NHR' + H_3O^{+} \longrightarrow R-\overset{O}{\overset{\|}{C}}-OH + R'\overset{+}{N}H_3$$

Example

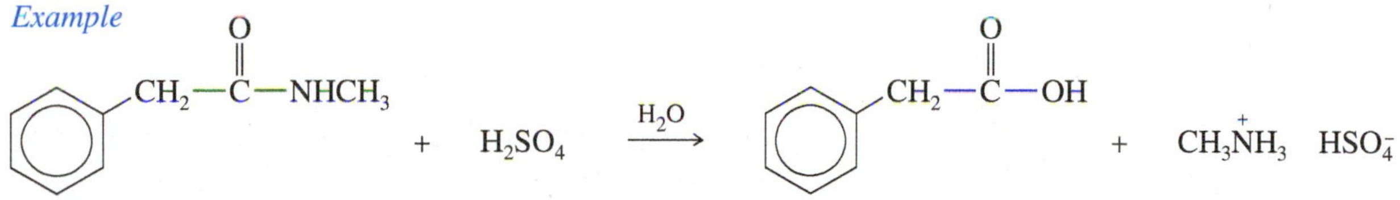

$$C_6H_5-CH_2-\overset{O}{\overset{\|}{C}}-NHCH_3 + H_2SO_4 \xrightarrow{H_2O} C_6H_5-CH_2-\overset{O}{\overset{\|}{C}}-OH + CH_3\overset{+}{N}H_3\;\;HSO_4^{-}$$

N-methyl-2-phenylacetamide — phenylacetic acid — methylammonium sulfate

The basic hydrolysis mechanism (shown next for a primary amide) is similar to that for hydrolysis of an ester. Hydroxide attacks the carbonyl to give a tetrahedral intermediate. Expulsion of an amide ion gives a carboxylic acid, which is quickly deprotonated to give the salt of the acid and ammonia.

MECHANISM 21-10 Basic Hydrolysis of an Amide

This is another standard addition–elimination mechanism, ending with a proton transfer to give the final products. The final proton transfer is extremely fast, and may occur as the very poor leaving group ($^{-}NH_2$) leaves.

Step 1: Addition of the nucleophile.

Step 2: Elimination of the leaving group.

Step 3: Proton transfer.

tetrahedral intermediate

Under acidic conditions, the mechanism of amide hydrolysis resembles the acid-catalyzed hydrolysis of an ester. Protonation of the carbonyl group activates it toward nucleophilic attack by water to give a tetrahedral intermediate. Protonation of the amino group enables it to leave as the amine. A fast exothermic proton transfer gives the acid and the protonated amine.

MECHANISM 21-11 Acidic Hydrolysis of an Amide

This mechanism takes place in two stages.

First half: Acid-catalyzed addition of the nucleophile (water).

Step 1: Protonation of the carbonyl.

Step 2: Addition of the nucleophile.

Step 3: Loss of a proton.

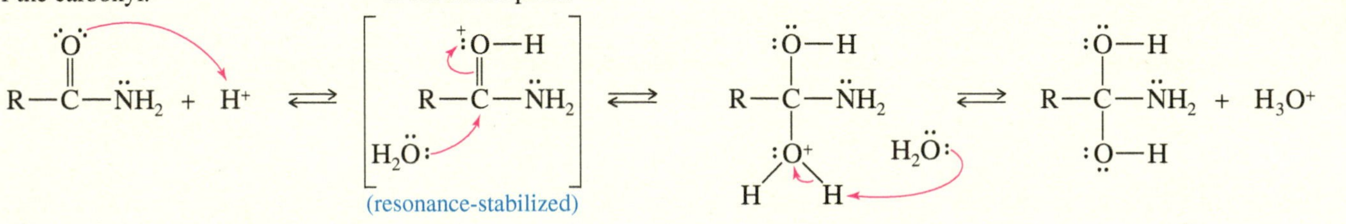

(resonance-stabilized)

Second half: Acid-catalyzed elimination of the leaving group.

Step 1: Protonation of the leaving group.

Step 2: Elimination of the leaving group.

Step 3: Deprotonation.

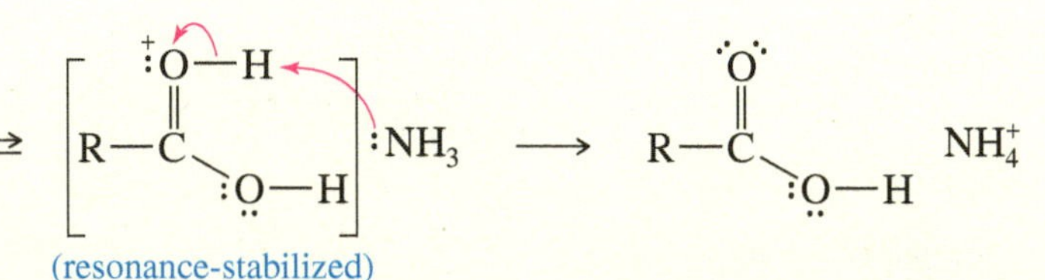

(resonance-stabilized)

PROBLEM 21-19

Draw the important resonance contributors for both resonance-stabilized cations (in brackets) in the mechanism for acid-catalyzed hydrolysis of an amide.

PROBLEM 21-20

Propose a mechanism for the hydrolysis of *N,N*-dimethylacetamide
(a) under basic conditions. **(b)** under acidic conditions.

PROBLEM 21-21

The equilibrium for hydrolysis of amides, under both acidic and basic conditions, favors the products. Use your mechanisms for the hydrolysis of *N,N*-dimethylacetamide to show which steps are sufficiently exothermic to drive the reactions to completion.

21-7D Hydrolysis of Nitriles

Nitriles are hydrolyzed to amides, and further to carboxylic acids, by heating with aqueous acid or base. Mild conditions can hydrolyze a nitrile only as far as the amide. Stronger conditions can hydrolyze it all the way to the carboxylic acid.

Basic hydrolysis of nitriles

$$\underset{\text{nitrile}}{R-C\equiv N:} + H_2O \xrightarrow[H_2O]{^-OH} \underset{1°\text{ amide}}{R-C(=O)-NH_2} \xrightarrow[H_2O]{^-OH} \underset{\text{carboxylate ion}}{R-C(=O)-O^-} + :NH_3$$

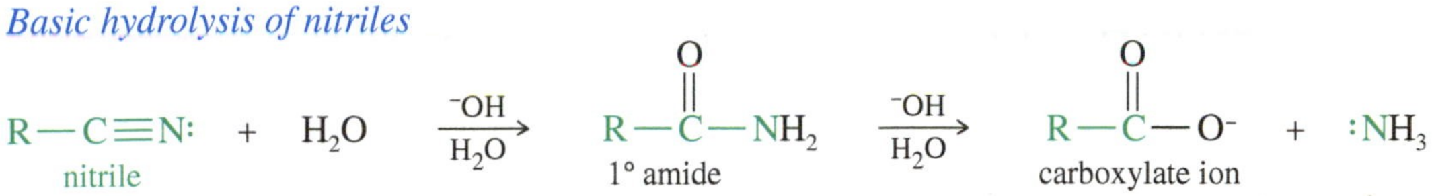

Example

$$\underset{\text{nicotinonitrile}}{\text{3-pyridyl}-C\equiv N:} \xrightarrow[H_2O/EtOH,\ 50\ °C]{NaOH} \underset{\text{nicotinamide}}{\text{3-pyridyl}-C(=O)-\ddot{N}H_2}$$

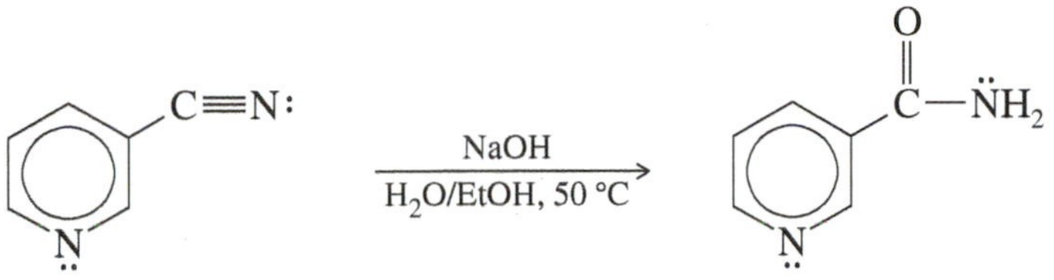

Acidic hydrolysis of nitriles

$$\underset{\text{nitrile}}{R-C\equiv N:} \xrightarrow[H_2O]{H^+} \underset{\text{primary amide}}{R-C(=O)-NH_2} \xrightarrow[H_2O]{H^+} \underset{\text{carboxylic acid}}{R-C(=O)-OH} + NH_4^+$$

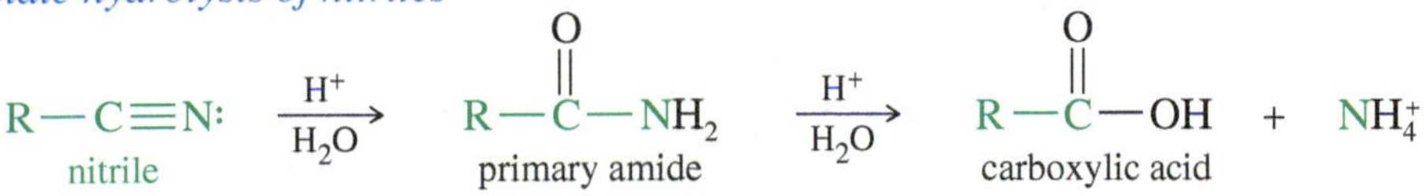

Example

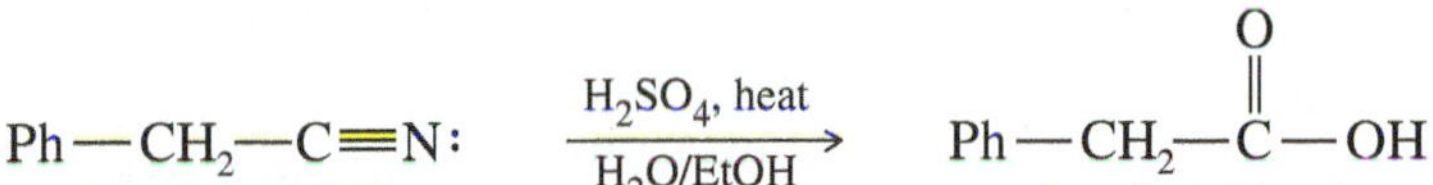

$$\underset{\text{phenylacetonitrile}}{Ph-CH_2-C\equiv N:} \xrightarrow[H_2O/EtOH]{H_2SO_4,\ \text{heat}} \underset{\text{phenylacetic acid}}{Ph-CH_2-C(=O)-OH}$$

The mechanism for basic hydrolysis begins with attack by hydroxide on the electrophilic carbon of the cyano group. Protonation gives the unstable enol tautomer of an amide. Removal of a proton from oxygen and reprotonation on nitrogen gives the amide. Further hydrolysis of the amide to the carboxylate salt involves the same base-promoted mechanism as that already discussed.

MECHANISM 21-12 Base-Catalyzed Hydrolysis of a Nitrile

Step 1: The hydroxide ion adds to the carbon of the cyano group.

Step 2: Protonation leads to the enol of an amide.

$$R-C\equiv N: \;+\; {}^{-}:\ddot{O}-H \;\rightleftarrows\; R-C(-\ddot{O}-H)=\ddot{N}:^{-} \;\xrightleftharpoons{H-\ddot{O}-H}\; R-C(-\ddot{O}-H)=\ddot{N}-H \;+\; {}^{-}:\ddot{O}-H$$

nitrile; enol tautomer of amide

Step 3: Removal and replacement of a proton (tautomerism) leads to the amide.

$$R-C(-\ddot{O}-H)=\ddot{N}-H \;\xrightleftharpoons{{}^{-}:\ddot{O}-H}\; \left[R-C(-\ddot{O}:^{-})=\ddot{N}-H \longleftrightarrow R-C(=O)-\ddot{N}^{-}-H\right] \;\xrightleftharpoons{H-\ddot{O}-H}\; R-C(=\ddot{O})-\ddot{N}H_2 \;+\; {}^{-}:\ddot{O}-H$$

enol tautomer; enolate of an amide; amide

PROBLEM 21-22

Propose a mechanism for the basic hydrolysis of benzonitrile to the benzoate ion and ammonia.

PROBLEM 21-23

The mechanism for acidic hydrolysis of a nitrile resembles the basic hydrolysis, except that the nitrile is first protonated, activating it toward attack by a weak nucleophile (water). Under acidic conditions, the proton transfer (tautomerism) involves protonation on nitrogen followed by deprotonation on oxygen. Propose a mechanism for the acid-catalyzed hydrolysis of benzonitrile to benzamide.

21-8 Reduction of Acid Derivatives

Carboxylic acids and their derivatives can be reduced to alcohols, aldehydes, and amines. Because they are relatively difficult to reduce, acid derivatives generally require a strong reducing agent such as lithium aluminum hydride ($LiAlH_4$).

21-8A Reduction to Alcohols

Lithium aluminum hydride reduces acids, acid chlorides, anhydrides, and esters to primary alcohols. (The reduction of acids was covered in Section 20-13.) Acid chlorides are more reactive than the other acid derivatives. Either lithium aluminum hydride or sodium borohydride converts acid chlorides to primary alcohols.

$$R-C(=O)-O-R' \xrightarrow{LiAlH_4} R-CH_2O^-\,Li^+ + R'-O^-\,Li^+ \xrightarrow{H_3O^+} R-CH_2OH + R'-OH$$

ester (or acid chloride); primary alkoxide; primary alcohol

Example

$$C_6H_5-CH_2-C(=O)-OCH_2CH_3 \xrightarrow[(2)\ H_3O^+]{(1)\ LiAlH_4} C_6H_5-CH_2-CH_2OH + CH_3CH_2OH$$

ethyl phenylacetate; 2-phenylethanol

Both esters and acid chlorides react through an addition–elimination mechanism to give aldehydes, which quickly reduce to alkoxides. After the reduction is complete, dilute acid is added to protonate the alkoxide.

MECHANISM 21-13 Hydride Reduction of an Ester

Nucleophilic acyl substitution gives an aldehyde, which reduces further to an alcohol.

Step 1: Addition of the nucleophile (hydride).

$$\underset{\text{ester}}{R-C(=O)-O-R'} + H-\overset{-}{Al}H_3\ Li^+ \longrightarrow \underset{\text{tetrahedral intermediate}}{R-C(H)(O^-\ Li^+)-O-R'}$$

Step 2: Elimination of alkoxide.

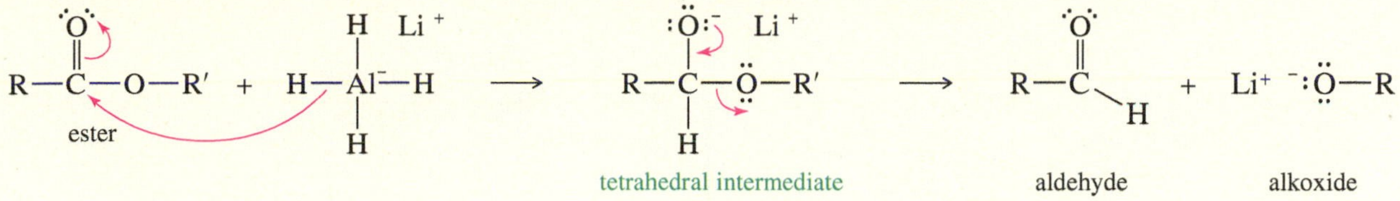

$$R-C(H)(O^-)-O-R' \longrightarrow \underset{\text{aldehyde}}{R-C(=O)-H} + \underset{\text{alkoxide}}{Li^+\ {}^-\!:\!\ddot{O}-R}$$

Step 3: Addition of a second hydride ion.

$$\underset{\text{aldehyde}}{R-C(=O)-H} + H-\overset{-}{Al}H_3\ Li^+ \longrightarrow \underset{\text{salt}}{R-CH_2-O^-\ Li^+}$$

Step 4: Add acid in the workup to protonate the alkoxide.

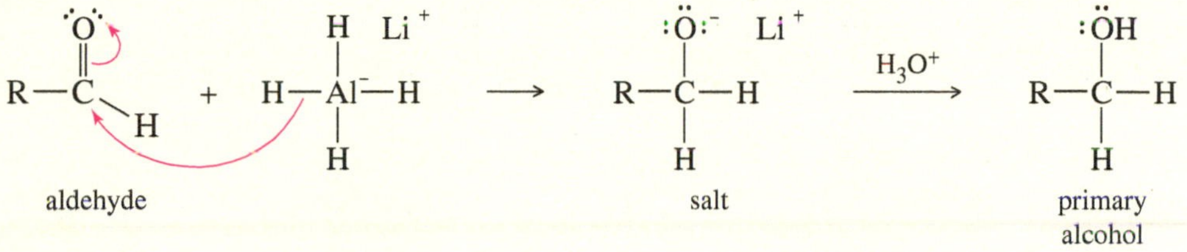

$$\underset{\text{salt}}{R-CH_2-O^-\ Li^+} \xrightarrow{H_3O^+} \underset{\text{primary alcohol}}{R-CH_2-OH}$$

PROBLEM 21-24

(a) In which step(s) of the hydride reduction of an ester does the compound undergo reduction? (*Hint*: Count the bonds to oxygen.)
(b) Propose a mechanism for the reduction of octanoyl chloride by lithium aluminum hydride.

21-8B Reduction to Aldehydes

Acid chlorides are more reactive than other acid derivatives, and they are reduced to aldehydes by mild reducing agents such as lithium tri-*tert*-butoxyaluminum hydride. Diisobutylaluminum hydride (DIBAL-H) reduces esters to aldehydes at low temperatures, and it also reduces nitriles to aldehydes. These reductions were covered in Sections 18-9 and 18-10.

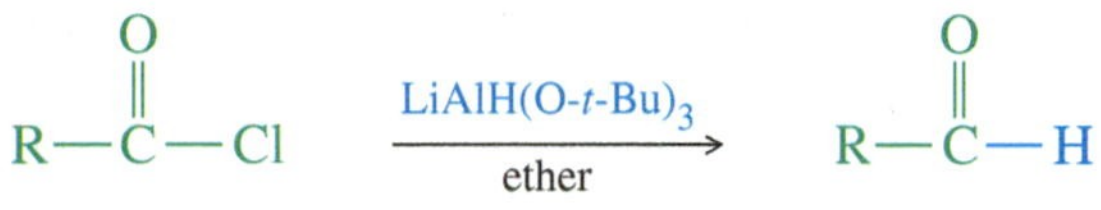

$$R-C(=O)-Cl \xrightarrow[\text{ether}]{LiAlH(O\text{-}t\text{-}Bu)_3} R-C(=O)-H$$

Example

$$\underset{\text{octanoyl chloride}}{CH_3(CH_2)_6-C(=O)-Cl} \xrightarrow{LiAlH(O\text{-}t\text{-}Bu)_3} \underset{\text{octanal}}{CH_3(CH_2)_6-C(=O)-H}$$

$$R-C(=O)-OR' \xrightarrow[-78\ ^\circ C]{(1)\ (i\text{-}Bu)_2AlH} R-C(H)(OR')-O-Al(i\text{-}Bu)_2 \xrightarrow{(2)\ H_2O} R-C(=O)-H$$

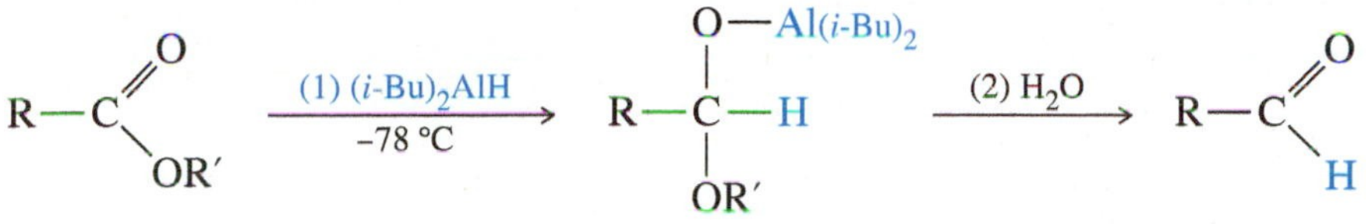

Example

$$\underset{\text{ethyl cinnamate}}{Ph-CH=CH-C(=O)-OEt} \xrightarrow[(2)\ H_2O]{(1)\ \text{DIBAL-H},\ -78\ ^\circ C} \underset{\text{cinnamaldehyde}}{Ph-CH=CH-C(=O)-H}$$

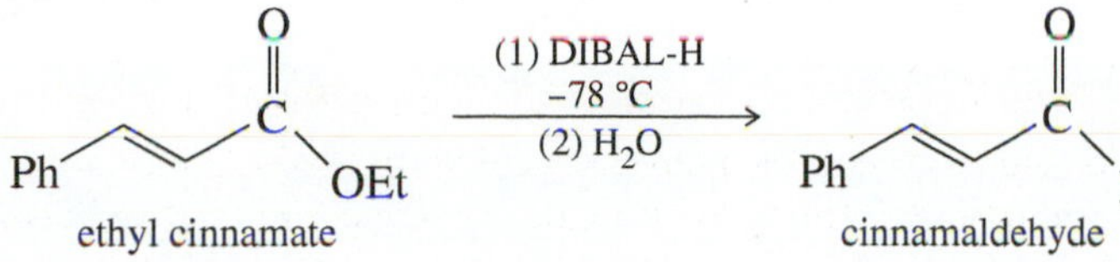

21-8C Reduction to Amines

Lithium aluminum hydride reduces amides and nitriles to amines, providing some of the best synthetic routes to amines (Sections 19-19 and 19-20B). Primary amides and nitriles are reduced to primary amines. Secondary amides are reduced to secondary amines, and tertiary amides are reduced to tertiary amines.

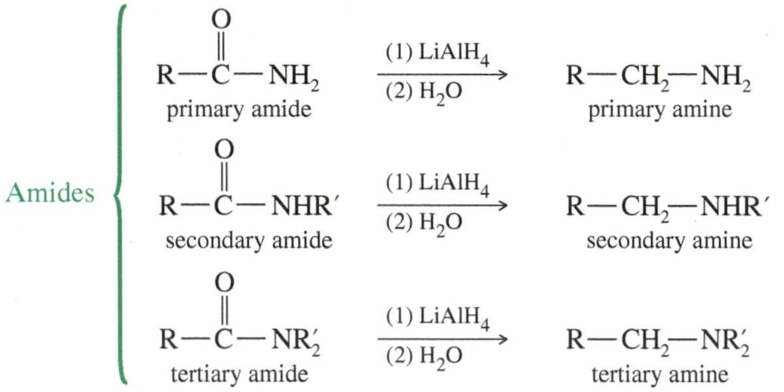

Amides

$$R-\overset{O}{\overset{\|}{C}}-NH_2 \xrightarrow[(2)\ H_2O]{(1)\ LiAlH_4} R-CH_2-NH_2$$
primary amide → primary amine

$$R-\overset{O}{\overset{\|}{C}}-NHR' \xrightarrow[(2)\ H_2O]{(1)\ LiAlH_4} R-CH_2-NHR'$$
secondary amide → secondary amine

$$R-\overset{O}{\overset{\|}{C}}-NR'_2 \xrightarrow[(2)\ H_2O]{(1)\ LiAlH_4} R-CH_2-NR'_2$$
tertiary amide → tertiary amine

Example

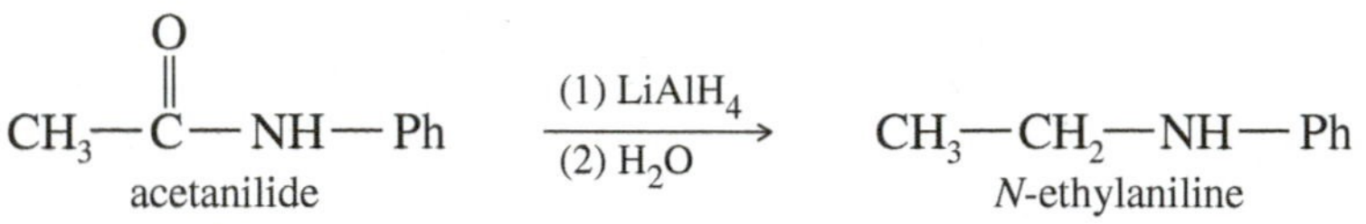

$$CH_3-\overset{O}{\overset{\|}{C}}-NH-Ph \xrightarrow[(2)\ H_2O]{(1)\ LiAlH_4} CH_3-CH_2-NH-Ph$$
acetanilide → *N*-ethylaniline

The mechanism of this reduction begins like a typical nucleophilic acyl substitution, with hydride ion adding to the carbonyl group to form a tetrahedral intermediate. The nitrogen atom is a poor leaving group, however, and the former carbonyl oxygen atom, complexed with aluminum, is a fair leaving group. The oxygen atom leaves, giving an imine or iminium salt that quickly reduces to the amine.

MECHANISM 21-14 Reduction of an Amide to an Amine

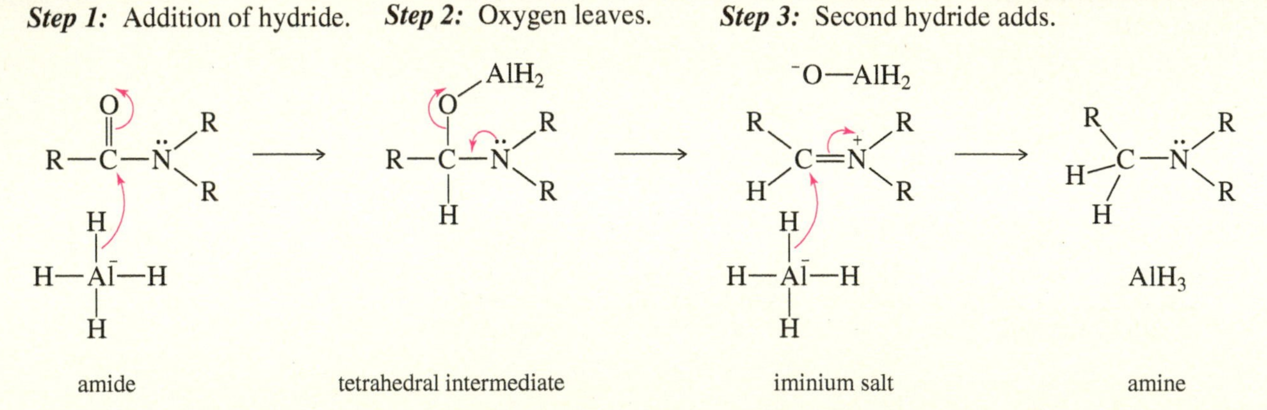

Nitriles are reduced to primary amines.

$$R-C\equiv N: \xrightarrow[or\ (1)\ LiAlH_4;\ (2)\ H_2O]{H_2/Pt} R-CH_2-NH_2$$

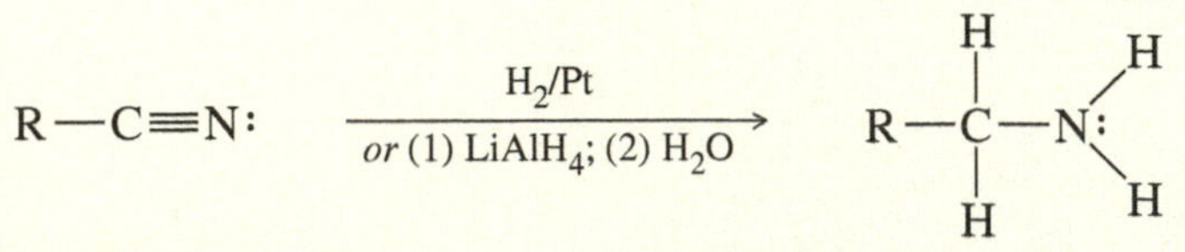

Example

$$\text{C}_6\text{H}_5-\text{CH}_2-\text{C}\equiv\text{N:} \xrightarrow[\text{(2) H}_2\text{O}]{\text{(1) LiAlH}_4} \text{C}_6\text{H}_5-\text{CH}_2-\text{CH}_2-\ddot{\text{N}}\text{H}_2$$

PROBLEM 21-25

Give the expected products of lithium aluminum hydride reduction of the following compounds (followed by hydrolysis).

(a) butyronitrile **(b)** *N*-cyclohexylacetamide **(c)** ε-caprolactam

(d) (morpholin-3-one: ring with O and N–H, C=O) **(e)** (cyclohexyl–N(CH$_3$)–C(=O)–CH$_2$CH$_3$) **(f)** (bicyclic cyclohexane–cyclopentane with CN at ring fusion)

21-9 Reactions of Acid Derivatives with Organometallic Reagents

Esters and Acid Chlorides Grignard and organolithium reagents add twice to acid chlorides and esters to give alkoxides (Section 10-9D). Protonation of the alkoxides gives alcohols.

$$\underset{\text{ester}}{\text{R}-\overset{\overset{\text{O}}{\|}}{\text{C}}-\text{OR}'} \xrightarrow[\text{(or 2 R}''\text{Li)}]{\text{2 R}''\text{MgX}} \underset{\text{alkoxide salt}}{\text{R}-\overset{\text{OMgX}}{\underset{\text{R}''}{\text{C}}}-\text{R}'' + \text{R}'\text{OMgX}} \xrightarrow{\text{H}_3\text{O}^+} \underset{\text{tertiary alcohol}}{\text{R}-\overset{\text{OH}}{\underset{\text{R}''}{\text{C}}}-\text{R}''}$$

Examples

$$\underset{\text{an ester}}{\text{Ph}-\overset{\overset{\text{O}}{\|}}{\text{C}}-\text{OEt}} + 2\ \text{PhMgBr} \longrightarrow \text{Ph}-\overset{\text{OMgBr}}{\underset{\text{Ph}}{\text{C}}}-\text{Ph} \xrightarrow{\text{H}_3\text{O}^+} \underset{\text{a 3° alcohol}}{\text{Ph}-\overset{\text{OH}}{\underset{\text{Ph}}{\text{C}}}-\text{Ph}}$$

$$\underset{\text{a formate ester}}{\text{H}-\overset{\overset{\text{O}}{\|}}{\text{C}}-\text{OEt}} + 2\ \text{C}_4\text{H}_9\text{Li} \longrightarrow \text{H}-\overset{\text{OLi}}{\underset{\text{C}_4\text{H}_9}{\text{C}}}-\text{C}_4\text{H}_9 \xrightarrow{\text{H}_3\text{O}^+} \underset{\text{a 2° alcohol}}{\text{H}-\overset{\text{OH}}{\underset{\text{C}_4\text{H}_9}{\text{C}}}-\text{C}_4\text{H}_9}$$

$$\underset{\text{an acid chloride}}{\text{CH}_3\text{CH}_2-\overset{\overset{\text{O}}{\|}}{\text{C}}-\text{Cl}} + 2\ \text{PhMgBr} \longrightarrow \text{CH}_3\text{CH}_2-\overset{\text{OMgBr}}{\underset{\text{Ph}}{\text{C}}}-\text{Ph} \xrightarrow{\text{H}_3\text{O}^+} \underset{\text{a 3° alcohol}}{\text{CH}_3\text{CH}_2-\overset{\text{OH}}{\underset{\text{Ph}}{\text{C}}}-\text{Ph}}$$

The mechanism involves nucleophilic substitution at the acyl carbon atom. Attack by the carbanion-like organometallic reagent, followed by elimination of alkoxide (from an ester) or chloride (from an acid chloride), gives a ketone. A second equivalent of the organometallic reagent adds to the ketone to give the alkoxide. Hydrolysis gives tertiary alcohols, unless the original ester is a formate (R = H), which gives a secondary alcohol. In each case, two of the groups on the product are the same, derived from the organometallic reagent.

MECHANISM 21-15 Reaction of an Ester with Two Moles of a Grignard Reagent

Step 1: Addition of the Grignard. ***Step 2:*** Elimination of alkoxide.

$$\underset{\text{ester}}{R{-}\overset{O}{\overset{\|}{C}}{-}OR'} + \underset{\text{Grignard}}{\overset{\delta-}{R''}{-}\overset{\delta+}{MgX}} \longrightarrow \underset{\text{tetrahedral intermediate}}{R{-}\overset{:\ddot{O}:^- \;{}^+MgX}{\overset{|}{C}}{-}\ddot{O}R'}\;(\text{with } R'' \text{ on C}) \longrightarrow \underset{\text{ketone}}{R{-}\overset{O}{\overset{\|}{C}}{-}R''} + R'OMgX$$

Step 3: Addition of another Grignard. ***Workup:*** Add acid to protonate the alkoxide.

$$\underset{\text{ketone}}{R{-}\overset{O}{\overset{\|}{C}}{-}R''} + R''{-}MgX \longrightarrow \underset{\text{alkoxide}}{R{-}\overset{:\ddot{O}:^- \;{}^+MgX}{\overset{|}{C}}{-}R''}\;(\text{with } R'' \text{ on C}) \xrightarrow{H_3O^+} R{-}\overset{:\ddot{O}H}{\overset{|}{C}}{-}R''\;(\text{with } R'' \text{ on C}) + MgXOH$$

Acid chlorides react just once with dialkylcuprates (Gilman reagents) to give ketones (Section 18-10).

$$R{-}\overset{O}{\overset{\|}{C}}{-}Cl + R'_2CuLi \longrightarrow R{-}\overset{O}{\overset{\|}{C}}{-}R'$$

Example

$$\text{cyclopentyl}{-}\overset{O}{\overset{\|}{C}}{-}Cl + [\text{vinyl}]_2CuLi \longrightarrow \text{cyclopentyl}{-}\overset{O}{\overset{\|}{C}}{-}CH{=}CH_2$$

Nitriles A Grignard or organolithium reagent attacks the electrophilic cyano group to form the salt of an imine. Acidic hydrolysis of the salt (in a separate step) gives the imine, which is further hydrolyzed to a ketone (Section 18-9).

Attack on the electrophilic cyano group

$$\overset{\delta-}{R'}{-}\overset{\delta+}{Mg}{-}X \quad R{-}C{\equiv}N: \longrightarrow \underset{\text{salt of imine}}{(R')(R)C{=}\ddot{N}{-}MgX} \xrightarrow[\text{Protonation}]{H_3O^+} \underset{\text{imine}}{(R')(R)C{=}\ddot{N}{-}H} \xrightarrow[\text{Acid hydrolysis}]{H_3O^+} \underset{\text{ketone}}{(R')(R)C{=}\ddot{O}:}$$

Example

$$\underset{\text{benzonitrile}}{Ph{-}C{\equiv}N:} + \underset{\text{methylmagnesium iodide}}{CH_3MgI} \longrightarrow \underset{\text{magnesium salt}}{(Ph)(CH_3)C{=}\ddot{N}{-}MgI} \xrightarrow{H_3O^+} \underset{\text{acetophenone}}{(Ph)(CH_3)C{=}\ddot{O}:}$$

PROBLEM 21-26

Draw a mechanism for the acidic hydrolysis of the magnesium salt shown above to acetophenone.

PROBLEM 21-27

Draw a mechanism for the reaction of propanoyl chloride with 2 moles of phenylmagnesium bromide.

PROBLEM 21-28

Show how you would add a Grignard reagent to an ester or a nitrile to synthesize
(a) 4-phenylheptan-4-ol. **(b)** heptan-4-ol. **(c)** pentan-2-one.

PROBLEM-SOLVING HINT

Grignards add to esters and acid chlorides to give tertiary alcohols, with one group from the ester or acid chloride and two identical groups from the Grignard. *Formate* esters give secondary alcohols, with a hydrogen from the ester and two identical groups from the Grignard.

21-10 Summary of the Chemistry of Acid Chlorides

Having discussed the reactions and mechanisms characteristic of all the common acid derivatives, we now review the syntheses and reactions of each type of compound. In addition, these sections cover any reactions that are peculiar to a specific class of acid derivatives.

Synthesis of Acid Chlorides Acid chlorides (acyl chlorides) are synthesized from the corresponding carboxylic acids using a variety of reagents. Thionyl chloride ($SOCl_2$) and oxalyl chloride $(COCl)_2$ are the most convenient reagents because they produce only gaseous side products (Section 20-15).

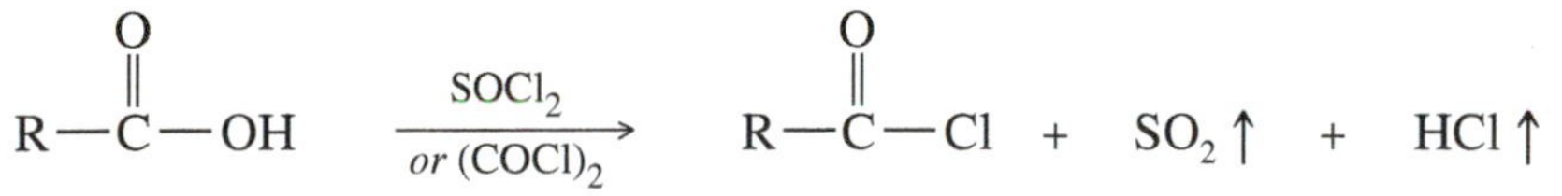

$$R-\overset{O}{\overset{\|}{C}}-OH \xrightarrow[or\ (COCl)_2]{SOCl_2} R-\overset{O}{\overset{\|}{C}}-Cl + SO_2\uparrow + HCl\uparrow$$

Reactions of Acid Chlorides Acid chlorides react quickly with water and other nucleophiles and are therefore not found in nature. Because they are the most reactive acid derivatives, acid chlorides are easily converted to other acid derivatives. Often, the best synthetic route to an ester, anhydride, or amide may involve using the acyl chloride as an intermediate.

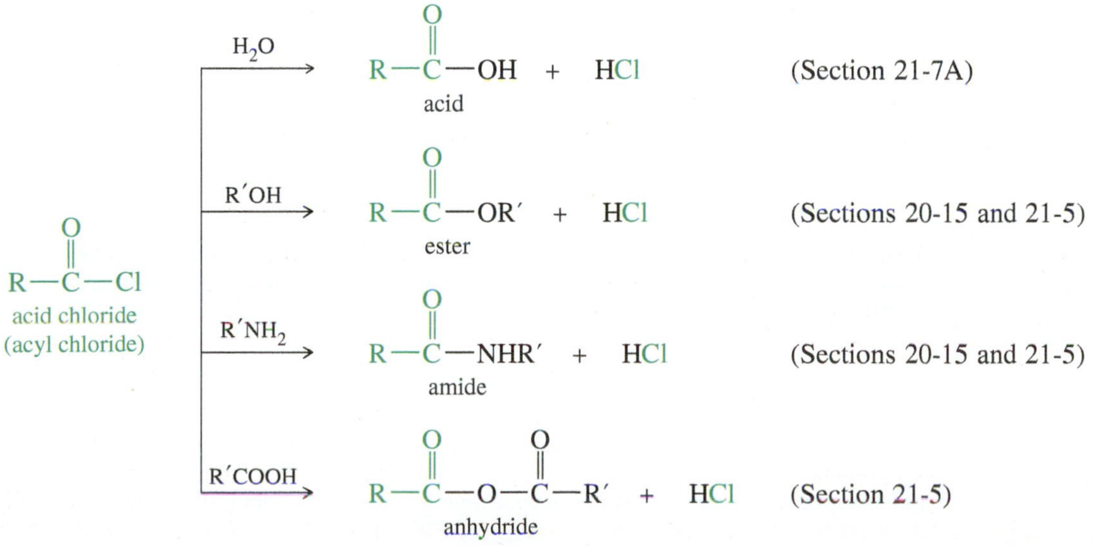

$R-\overset{O}{\overset{\|}{C}}-Cl$ acid chloride (acyl chloride)

$$\xrightarrow{H_2O} R-\overset{O}{\overset{\|}{C}}-OH + HCl \quad \text{(acid)}$$ (Section 21-7A)

$$\xrightarrow{R'OH} R-\overset{O}{\overset{\|}{C}}-OR' + HCl \quad \text{(ester)}$$ (Sections 20-15 and 21-5)

$$\xrightarrow{R'NH_2} R-\overset{O}{\overset{\|}{C}}-NHR' + HCl \quad \text{(amide)}$$ (Sections 20-15 and 21-5)

$$\xrightarrow{R'COOH} R-\overset{O}{\overset{\|}{C}}-O-\overset{O}{\overset{\|}{C}}-R' + HCl \quad \text{(anhydride)}$$ (Section 21-5)

Grignard and organolithium reagents add twice to acid chlorides to give 3° alcohols (after hydrolysis). Lithium dialkylcuprates add just once to give ketones. Sodium borohydride or lithium aluminum hydride adds hydride twice to acid chlorides, reducing them to 1° alcohols (after hydrolysis). Acid chlorides react with the weaker reducing agent tri-*tert*-butoxyaluminum hydride to give aldehydes.

$$\mathrm{R{-}C({=}O){-}Cl}\ \text{(acid chloride (acyl chloride))}$$

$$\xrightarrow[(2)\ H_2O]{(1)\ 2\ R'MgX} \mathrm{R{-}C(OH)(R'){-}R'}\ \text{(3° alcohol)} \quad \text{(Sections 10-9 and 21-9)}$$

$$\xrightarrow{R'_2CuLi} \mathrm{R{-}C({=}O){-}R'}\ \text{(ketone)} \quad \text{(Section 18-10)}$$

$$\xrightarrow[(2)\ H_2O]{(1)\ NaBH_4\ \text{or}\ LiAlH_4} \mathrm{R{-}CH_2OH}\ \text{(1° alcohol)} \quad \text{(Section 21-8A)}$$

$$\xrightarrow{LiAlH(O\text{-}t\text{-}Bu)_3} \mathrm{R{-}C({=}O){-}H}\ \text{(aldehyde)} \quad \text{(Sections 18-10 and 21-8B)}$$

Friedel–Crafts Acylation of Aromatic Rings In the presence of aluminum chloride, acyl halides acylate benzene, halobenzenes, and activated benzene derivatives. Friedel–Crafts acylation is discussed in detail in Section 17-11.

$$\mathrm{R{-}C({=}O){-}Cl} + \mathrm{C_6H_5{-}Z} \xrightarrow[(2)\ H_2O]{(1)\ AlCl_3} \mathrm{Z{-}C_6H_4{-}C({=}O){-}R}$$

(Z = H, halogen, or an activating group) — an acylbenzene

Example

$$\underset{\text{propionyl chloride}}{\mathrm{CH_3{-}CH_2{-}C({=}O){-}Cl}} + \underset{\text{anisole}}{\mathrm{CH_3O{-}C_6H_5}} \xrightarrow[(2)\ H_2O]{(1)\ AlCl_3} \underset{\text{p-methoxypropiophenone (major product)}}{\mathrm{CH_3O{-}C_6H_4{-}C({=}O){-}CH_2CH_3}}$$

PROBLEM 21-29

Draw a mechanism for the acylation of anisole by propionyl chloride. Recall that Friedel–Crafts acylation involves an acylium ion as the electrophile in electrophilic aromatic substitution.

PROBLEM 21-30

Show how Friedel–Crafts acylation might be used to synthesize the following compounds.

(a) acetophenone **(b)** benzophenone **(c)** *n*-butylbenzene

21-11 Summary of the Chemistry of Anhydrides

Like acid chlorides, anhydrides are activated acid derivatives, and they are often used for the same types of acylations. Anhydrides are not as reactive as acid chlorides, and they are occasionally found in nature. For example, cantharidin is a toxic ingredient of "Spanish fly," which is used as a vesicant ("causing burning and blistering") to destroy warts on the skin.

Because anhydrides are not as reactive as acid chlorides, they are often more selective in their reactions. Anhydrides are valuable when the appropriate acid chloride is too reactive, does not exist, or is more expensive than the corresponding anhydride.

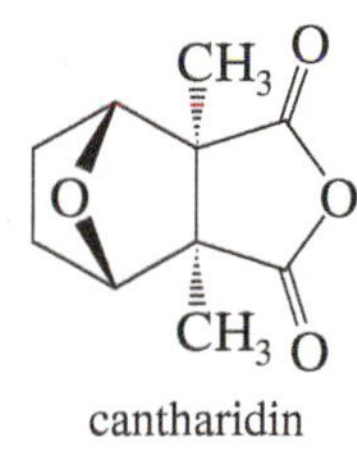

cantharidin

Acetic Anhydride Acetic anhydride is the most important carboxylic acid anhydride. It is produced at the rate of about 4 billion pounds per year, primarily for synthesis of plastics, fibers, and drugs. (See the synthesis of aspirin on p. 1073.) Acetic anhydride consists of two molecules of acetic acid, minus a molecule of water. The most common industrial synthesis begins by dehydrating acetic acid to give ketene.

$$CH_3-\overset{O}{\overset{\|}{C}}-OH \underset{}{\overset{750\ °C,\ (EtO)_3P{=}O}{\rightleftharpoons}} H_2C{=}C{=}O + H_2O$$

acetic acid → ketene

This dehydration is highly endothermic ($\Delta H = +147$ kJ/mol $= +35$ kcal/mol), but a large increase in entropy results from breaking one molecule into two. Thus, at a sufficiently high temperature (750 °C is typical), the equilibrium favors the products. Triethyl phosphate is added as a catalyst to improve the rate of the reaction.

Ketene (a gas at room temperature) is fed directly into acetic acid, where it reacts quickly and quantitatively to give acetic anhydride. This inexpensive large-scale manufacture makes acetic anhydride a convenient and inexpensive acylating reagent.

$$CH_3-\overset{O}{\overset{\|}{C}}-OH + H_2C{=}C{=}O \longrightarrow CH_3-\overset{O}{\overset{\|}{C}}-O-\overset{O}{\overset{\|}{C}}-CH_3$$

acetic acid + ketene → acetic anhydride

Blister beetles secrete cantharidin, a powerful vesicant. Crushing a blister beetle between the fingers results in severe blistering of the skin. If horses eat hay containing blister beetles, they often die from gastroenteritis and kidney failure caused by cantharidin poisoning.

General Anhydride Synthesis Other anhydrides must be made by less specialized methods. The most general method for making anhydrides is the reaction of an acid chloride with a carboxylic acid or a carboxylate salt.

$$R-\overset{O}{\overset{\|}{C}}-Cl + {}^{-}O-\overset{O}{\overset{\|}{C}}-R' \longrightarrow R-\overset{O}{\overset{\|}{C}}-O-\overset{O}{\overset{\|}{C}}-R' + Cl^-$$

acid chloride + carboxylate (or acid) → acid anhydride

Examples

$$CH_3-\overset{O}{\overset{\|}{C}}-Cl + HO-\overset{O}{\overset{\|}{C}}-Ph \xrightarrow{\text{pyridine}} CH_3-\overset{O}{\overset{\|}{C}}-O-\overset{O}{\overset{\|}{C}}-Ph + \text{pyridine}\cdot HCl$$

acetyl chloride + benzoic acid → acetic benzoic anhydride + pyridine · HCl

$$CH_3-\overset{O}{\overset{\|}{C}}-Cl + H-\overset{O}{\overset{\|}{C}}-O^-\,Na^+ \longrightarrow CH_3-\overset{O}{\overset{\|}{C}}-O-\overset{O}{\overset{\|}{C}}-H + NaCl$$

acetyl chloride + sodium formate → acetic formic anhydride

Some cyclic anhydrides are made simply by heating the corresponding diacid. A dehydrating agent, such as acetyl chloride or acetic anhydride, is occasionally added to accelerate this reaction. Because five- and six-membered cyclic anhydrides are particularly stable, the equilibrium favors the cyclic products.

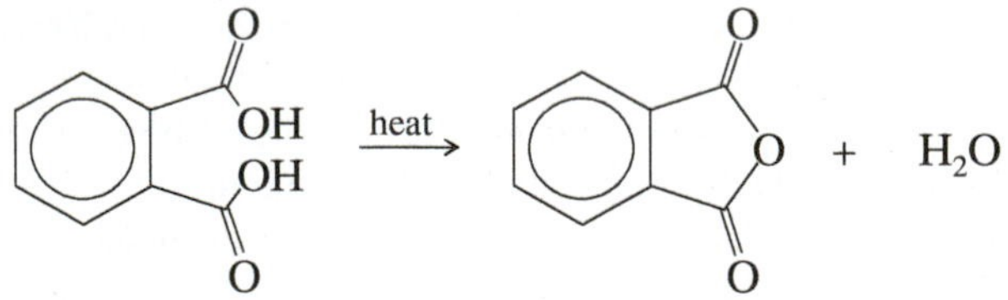

phthalic acid $\xrightarrow{\text{heat}}$ phthalic anhydride + H_2O (steam)

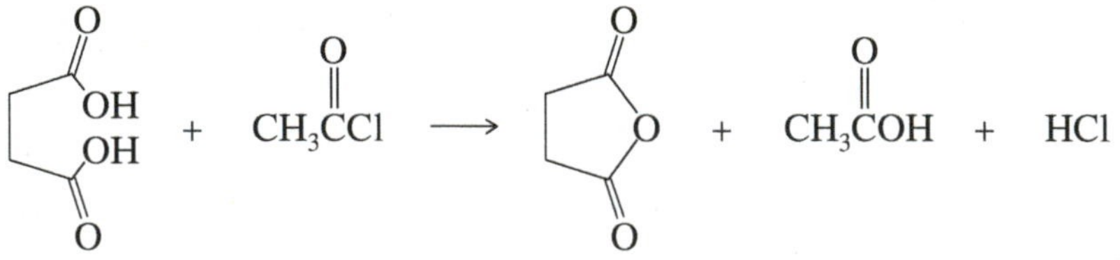

succinic acid + CH_3COCl $\longrightarrow$ succinic anhydride + CH_3COOH + HCl

Reactions of Anhydrides Anhydrides undergo many of the same reactions as acid chlorides. Like acid chlorides, anhydrides are easily converted to less reactive acid derivatives.

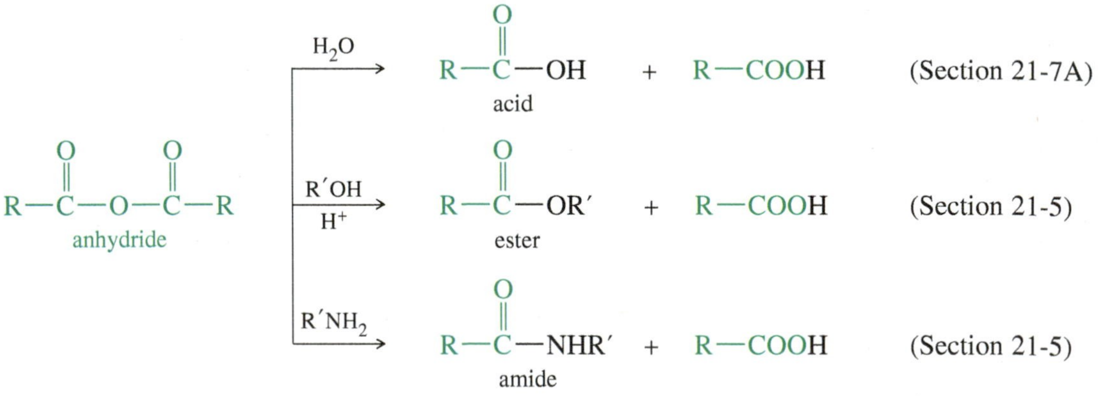

anhydride $R-CO-O-CO-R$:

- $\xrightarrow{H_2O}$ $R-CO-OH$ (acid) + $R-COOH$ (Section 21-7A)
- $\xrightarrow[H^+]{R'OH}$ $R-CO-OR'$ (ester) + $R-COOH$ (Section 21-5)
- $\xrightarrow{R'NH_2}$ $R-CO-NHR'$ (amide) + $R-COOH$ (Section 21-5)

Like acid chlorides, anhydrides participate in the Friedel–Crafts acylation. The catalyst may be aluminum chloride, polyphosphoric acid (PPA), or other acidic compounds. Cyclic anhydrides can provide additional functionality on the side chain of the aromatic product.

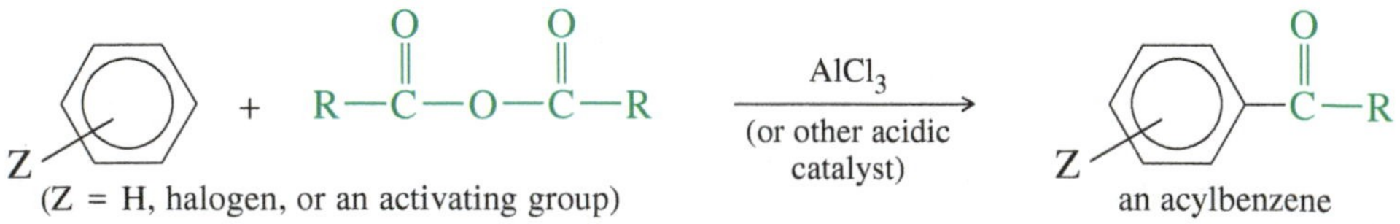

Z–benzene + $R-CO-O-CO-R$ $\xrightarrow[\text{(or other acidic catalyst)}]{AlCl_3}$ Z–benzene–CO–R

(Z = H, halogen, or an activating group) an acylbenzene

Example

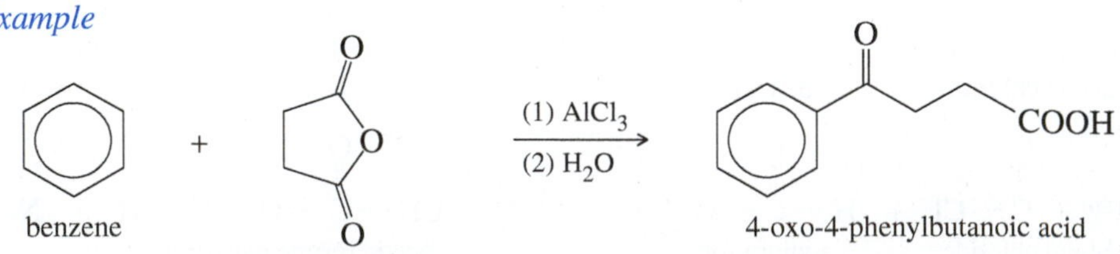

benzene + succinic anhydride $\xrightarrow[(2)\ H_2O]{(1)\ AlCl_3}$ 4-oxo-4-phenylbutanoic acid (Ph–CO–CH₂CH₂–COOH)

Most reactions of anhydrides involve loss of one of the two acid molecules as a leaving group. If a precious acid needs to be activated, converting it to the anhydride would allow only half of the acid groups to react. Converting the acid to an acid chloride would be more efficient because it would allow all the acid groups to react. There are three specific instances when anhydrides are preferred, however.

1. *Use of acetic anhydride.* Acetic anhydride is inexpensive and convenient to use, and it often gives better yields than acetyl chloride for acetylation of alcohols (to make acetate esters) and amines (to make acetamides).
2. *Use of acetic formic anhydride.* Formyl chloride (the acid chloride of formic acid) cannot be used for formylation because it quickly decomposes to CO and HCl. Acetic formic anhydride, made from sodium formate and acetyl chloride, reacts primarily at the formyl group. Lacking a bulky, electron-donating alkyl group, the formyl group is both less hindered and more electrophilic than the acetyl group. Alcohols and amines are formylated by acetic formic anhydride to give formate esters and formamides, respectively.

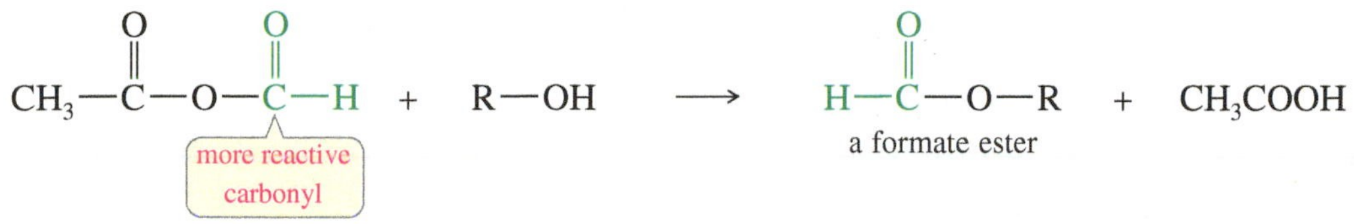

$$CH_3-\overset{O}{\overset{\|}{C}}-O-\overset{O}{\overset{\|}{C}}-H + R-OH \longrightarrow H-\overset{O}{\overset{\|}{C}}-O-R + CH_3COOH$$

more reactive carbonyl

a formate ester

$$CH_3-\overset{O}{\overset{\|}{C}}-O-\overset{O}{\overset{\|}{C}}-H + R-NH_2 \longrightarrow H-\overset{O}{\overset{\|}{C}}-NH-R + CH_3COOH$$

a formamide

3. *Use of cyclic anhydrides to make difunctional compounds.* It is often necessary to convert just one acid group of a diacid to an ester or an amide. This conversion is easily accomplished using a cyclic anhydride.

   When an alcohol or an amine reacts with a cyclic anhydride, only one of the carboxyl groups in the anhydride is converted to an ester or an amide. The other is expelled as a carboxylate ion, and a monofunctionalized derivative results.

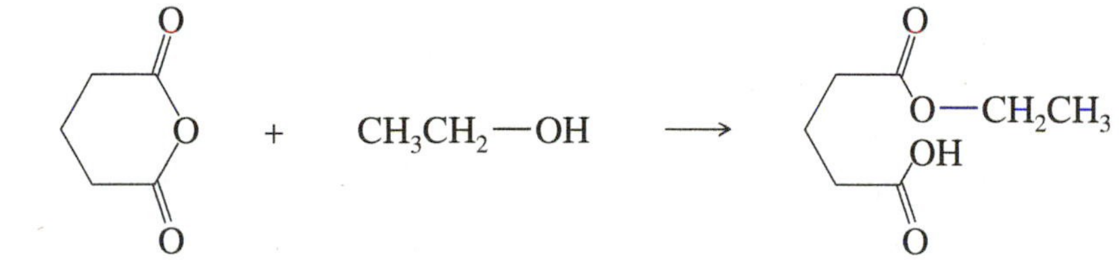

glutaric anhydride + CH_3CH_2-OH $\longrightarrow$ monoethyl ester ($O-CH_2CH_3$, OH)

glutaric anhydride

monoethyl ester

PROBLEM 21-31

(a) Give the products expected when acetic formic anhydride reacts with (i) aniline and (ii) benzyl alcohol.
(b) Propose mechanisms for these reactions.

PROBLEM 21-32

Show how you would use anhydrides to synthesize the following compounds. In each case, explain why an anhydride might be preferable to an acid chloride.

(a) *n*-octyl formate
(b) *n*-octyl acetate
(c) phthalic acid monoamide
(d) succinic acid monomethyl ester

Application: Anticoagulant

Warfarin (also named **coumadin**) is a substituted coumarin used as an anticoagulant. Warfarin slows clotting of the blood, which can prevent heart attacks and strokes in people who are prone to thrombosis (abnormal formation of blood clots). Like all medicines, warfarin becomes a poison when taken in excess. It was originally sold as a rodent poison, added to grain to kill rats and mice through internal hemorrhaging. This former rat poison is currently the most widely prescribed anticoagulant drug.

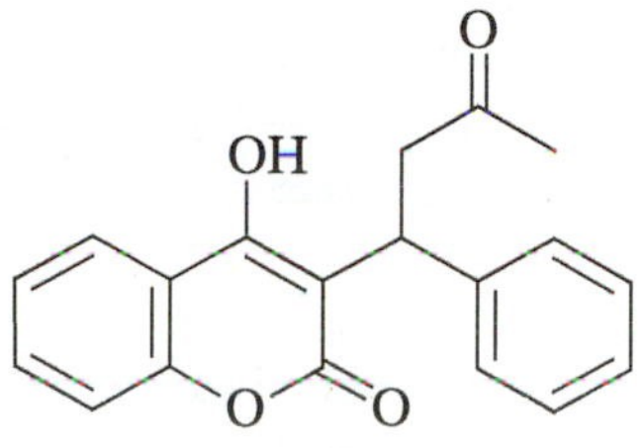

warfarin (coumadin)

21-12 Summary of the Chemistry of Esters

Esters are among the most common acid derivatives. They are found in plant oils, where they give the fruity aromas we associate with ripeness. For example, the odor of ripe bananas comes mostly from isoamyl acetate. Oil of wintergreen contains methyl salicylate, which has also been used as a medicine. Lavender oil and sweet clover contain small amounts of coumarin, which gives depth and longevity to their odors. The heads of sperm whales contain large chambers of spermaceti, a waxy ester that helps to regulate their buoyancy in the water and that may help the head to serve as a resonating chamber for communicating underwater.

isoamyl acetate (isopentyl acetate) — $(CH_3)_2CHCH_2CH_2{-}O{-}C(=O){-}CH_3$

methyl salicylate, oil of wintergreen — $C_6H_4(OH){-}C(=O){-}OCH_3$

coumarin

spermaceti (cetyl palmitate) — $CH_3(CH_2)_{14}{-}C(=O){-}O{-}(CH_2)_{15}CH_3$

Esters are widely used in industry as solvents. Ethyl acetate is a good solvent for a wide variety of compounds, and its toxicity is low compared with other solvents. Ethyl acetate is also found in household products such as cleaners, polishes, glues, and spray finishes. Ethyl butyrate and butyl butyrate were once widely used as solvents for paints and finishes, including the "butyrate dope" that was sprayed on the fabric covering of aircraft wings to make them tight and stiff. Polyesters (covered later in this section and in Chapter 26) are among the most common polymers, used in fabrics (Dacron®), films (VCR tapes), and solid plastics (soft-drink bottles).

Synthesis of Esters Esters are usually synthesized by the Fischer esterification of an acid with an alcohol or by the reaction of an acid chloride (or anhydride) with an alcohol. Methyl esters can be made by treating the acid with diazomethane. The alcohol group in an ester can be changed by transesterification, which can be catalyzed by either acid or base.

$$\underset{\text{acid}}{R{-}C(=O){-}OH} + \underset{\text{alcohol}}{R'{-}OH} \overset{H^+}{\rightleftharpoons} \underset{\text{ester}}{R{-}C(=O){-}OR'} + H_2O \quad \text{(Section 20-10)}$$

$$\underset{\text{acid chloride}}{R{-}C(=O){-}Cl} + \underset{\text{alcohol}}{R'{-}OH} \longrightarrow \underset{\text{ester}}{R{-}C(=O){-}OR'} + HCl \quad \text{(Section 20-15)}$$

$$\underset{\text{anhydride}}{R{-}C(=O){-}O{-}C(=O){-}R} + \underset{\text{alcohol}}{R'{-}OH} \xrightarrow{H^+} \underset{\text{ester}}{R{-}C(=O){-}OR'} + RCOOH \quad \text{(Section 21-5)}$$

$$\underset{\text{ester}}{R{-}C(=O){-}OR''} + \underset{\text{alcohol}}{R'{-}OH} \overset{H^+ \text{ or } ^-OR'}{\rightleftharpoons} \underset{\text{ester}}{R{-}C(=O){-}OR'} + R''OH \quad \text{(Section 21-6)}$$

$$\underset{\text{acid}}{R{-}C(=O){-}OH} + \underset{\text{diazomethane}}{CH_2N_2} \longrightarrow \underset{\text{methyl ester}}{R{-}C(=O){-}OCH_3} + N_2\uparrow \quad \text{(Section 20-11)}$$

Reactions of Esters Esters are much more stable than acid chlorides and anhydrides. For example, most esters do not react with water under neutral conditions. They hydrolyze under acidic or basic conditions, however, and an amine can displace the alkoxy group to form an amide. Lithium aluminum hydride reduces esters to primary alcohols, and Grignard and organolithium reagents add twice to give alcohols (after hydrolysis).

$$R-\overset{O}{\overset{\|}{C}}-OR' \text{ (ester)} \xrightarrow[H^+ \text{ or } ^-OH]{H_2O} R-\overset{O}{\overset{\|}{C}}-OH \text{ (acid)} + R'OH \quad \text{(Section 21-7B)}$$

$$\xrightarrow[H^+ \text{ or } ^-OR'']{R''OH} R-\overset{O}{\overset{\|}{C}}-OR'' \text{ (ester)} + R'OH \quad \text{(Section 21-6)}$$

$$\xrightarrow{R''NH_2} R-\overset{O}{\overset{\|}{C}}-NHR'' \text{ (amide)} + R'OH \quad \text{(Section 21-5)}$$

$$\xrightarrow[(2)\ H_2O]{(1)\ LiAlH_4} R-CH_2OH \text{ (1° alcohol)} + R'OH \quad \text{(Sections 10-11B and 21-8A)}$$

$$\xrightarrow[(2)\ H_2O]{(1)\ 2\ R''MgX} R-\overset{OH}{\underset{R''}{C}}-R'' \text{ (3° alcohol)} + R'OH \quad \text{(Sections 10-9D and 21-9)}$$

$$\xrightarrow[(2)\ H_2O]{(1)\ \text{DIBAL-H},\ -78\ °C} R-\overset{O}{\overset{\|}{C}}-H \text{ (aldehyde)} + R'OH \quad \text{(Sections 18-10 and 21-8)}$$

Formation of Lactones Simple lactones containing five- and six-membered rings are often more stable than the open-chain hydroxy acids. Such lactones form spontaneously under acidic conditions (via the Fischer esterification).

$$\text{(hydroxy acid: OH, COOH) 27\%} \underset{}{\overset{H^+}{\rightleftharpoons}} \text{(lactone) 73\%} + H_2O$$

Lactones that are not energetically favored may be synthesized by driving the equilibrium toward the products. For example, the ten-membered 9-hydroxynonanoic acid lactone is formed in a dilute benzene solution containing a trace of *p*-toluenesulfonic acid. The reaction is driven to completion by distilling the benzene/water azeotrope to remove water and shift the equilibrium to the right.

$$\text{9-hydroxynonanoic acid (OH, COOH)} \xrightarrow[\text{benzene}]{H^+} \text{9-hydroxynonanoic acid lactone (95\%)} + H_2O \text{ (removed)}$$

Lactones are common among natural products. For example, L-ascorbic acid (vitamin C) is necessary in the human diet to avoid the connective tissue disease known as scurvy. In acid solutions, ascorbic acid is an equilibrium mixture of the cyclic and acyclic forms, but the cyclic form predominates. Erythromycin is a member of the macrolide (large-ring lactone) group of antibiotics, which is isolated from *Streptomyces erythraeus*. It inhibits bacterial protein synthesis, thus arresting bacterial growth and development. Erythromycin is effective against a wide range of diseases, including staphylococcus, streptococcus, chlamydia, and Legionnaires' disease.

Application: Pheromone Traps

(*Z*)-11-hexadecenyl acetate

The apple fruit moth, *Argyresthia conjugella*, is a pest that bores into apples and eats them from within. One of its sex pheromones is (Z)-11-hexadecenyl acetate, which is used commercially to attract and trap the adult insects. Insect attractants are important chemicals for farming, because sticky insect traps baited with pheromones are allowed under the "organic" farming rules.

apple fruit moth

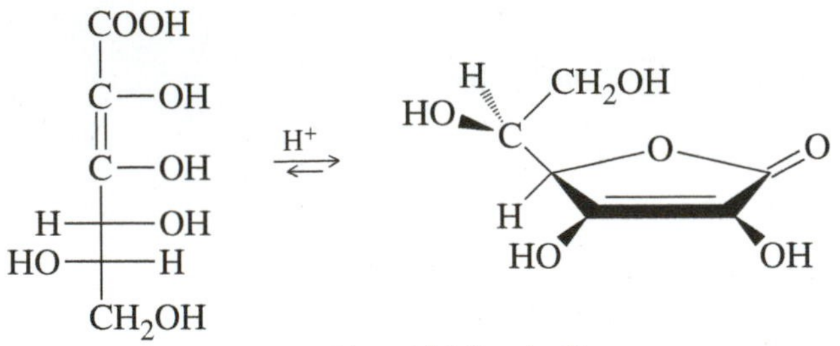

L-ascorbic acid (vitamin C)

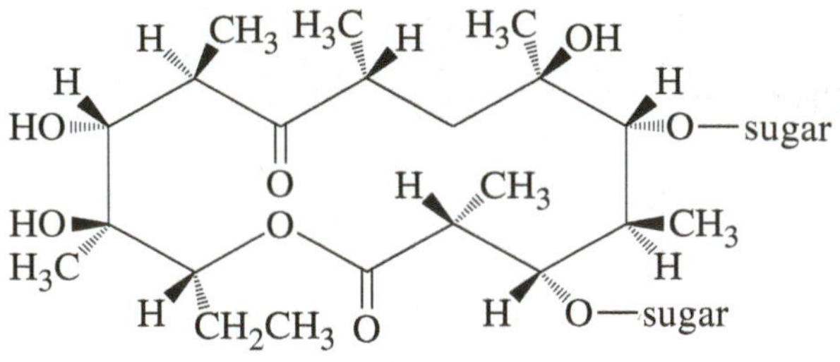

erythromycin

PROBLEM 21-33

Propose a mechanism for the formation of 9-hydroxynonanoic acid lactone, as shown in the preceding figure.

PROBLEM 21-34

Suggest the most appropriate reagent for each synthesis, and explain your choice.

(a) 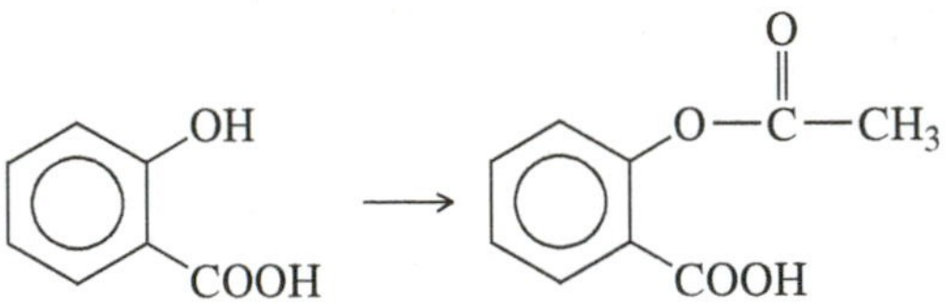 (b) 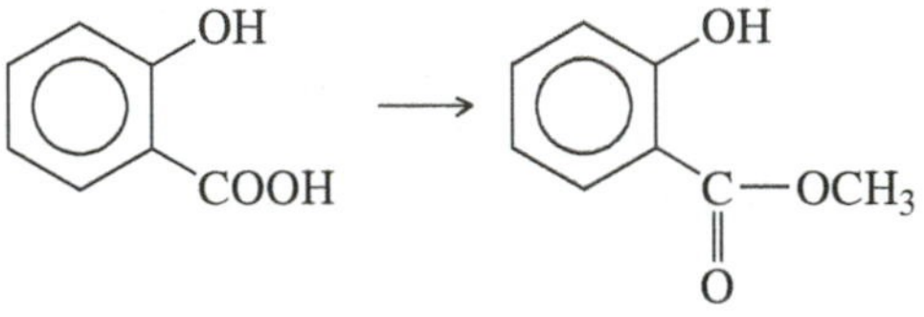 (c) 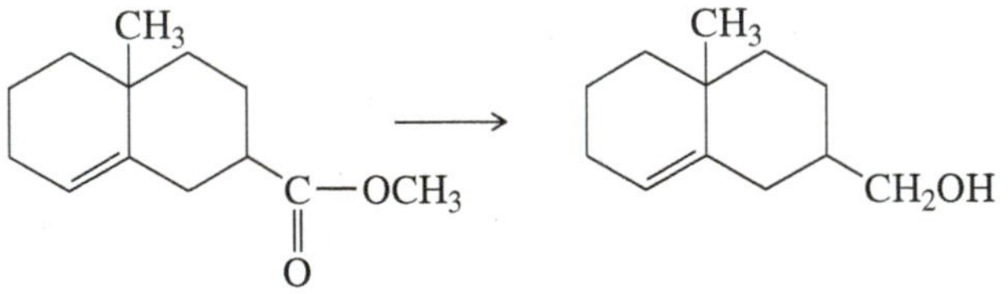 (d)

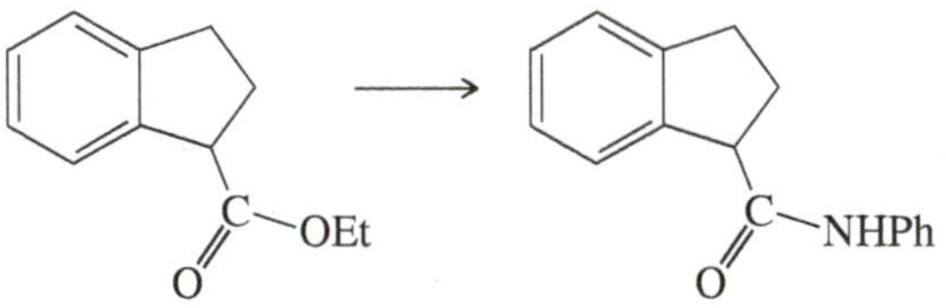

PROBLEM 21-35

Show how you would synthesize each compound, starting with an ester containing no more than eight carbon atoms. Any other necessary reagents may be used.

(a) Ph_3C-OH
(b) $(PhCH_2)_2CHOH$
(c) $PhCONHCH_2CH_3$
(d) Ph_2CHOH
(e) $PhCH_2OH$
(f) $PhCOOH$
(g) $PhCH_2COOCH(CH_3)_2$
(h) $PhCH_2-C(CH_2CH_3)_2$ with OH on the C
(i) $HO-(CH_2)_8-OH$

The Echo satellite was a 41-meter balloon made of metallized Mylar® film. It was launched in 1964 to reflect radio waves for intercontinental telephone, radio, and television. Echo was easily visible to the unaided eye as it orbited the Earth until 1969, when it reentered the atmosphere and burned up.

Polyesters Right now, you are probably using at least five things that are made from polyesters. Your clothes probably have some Dacron® polyester fiber in them, and they are almost certainly sewn with Dacron® thread. Early computers used floppy disks made of Mylar®, and the optical film in your DVD is made of Mylar®. Some of the electronics in your cell phone are probably "potted" (covered and insulated from shock) in Glyptal® polyester resin. The soft drink in your hand probably came in a plastic bottle that was blow-molded from poly(ethylene terephthalate) resin, better known as PET.

All these plastics are essentially the same compound, composed of terephthalic acid (*para*-phthalic acid) esterified with ethylene glycol. This polyester is made by a base-catalyzed transesterification of dimethyl terephthalate with ethylene glycol at a temperature around 150 °C. At this temperature, methanol escapes as a gas, driving the reaction to completion. We will study polyesters and other polymers in more detail in Chapter 26.

$$CH_3O-\overset{O}{\overset{\|}{C}}-C_6H_4-\overset{O}{\overset{\|}{C}}-OCH_3 + HO-CH_2CH_2-OH \xrightarrow[NaOCH_3]{\text{heat, loss of } CH_3OH}$$

dimethyl terephthalate; ethylene glycol

$$---\overset{O}{\overset{\|}{C}}-C_6H_4-\overset{O}{\overset{\|}{C}}\left[-O-CH_2CH_2-O-\overset{O}{\overset{\|}{C}}-C_6H_4-\overset{O}{\overset{\|}{C}}\right]_n-O-CH_2CH_2-O----$$

poly(ethylene terephthalate) or PET, also called Dacron® polyester or Mylar® film

21-13 Summary of the Chemistry of Amides

Synthesis of Amides Amides are the least reactive acid derivatives, and they can be made from any of the others. In the laboratory, amides are commonly synthesized by the reaction of an acid chloride (or anhydride) with an amine. The most common industrial synthesis involves heating an acid with an amine (at high temperatures, in the absence of oxygen) to drive off water and promote condensation. This simple industrial technique rarely works well in the laboratory, but it may succeed with the use of a coupling reagent (Section 24-10). Esters react with amines and ammonia to give amides, and the partial hydrolysis of nitriles also gives amides.

$$\underset{\text{acid}}{R-\overset{O}{\overset{\|}{C}}-OH} + \underset{\text{amine}}{R'-NH_2} \xrightarrow[(300\ °C)]{\text{heat}} \underset{\text{amide}}{R-\overset{O}{\overset{\|}{C}}-NHR'} + H_2O\uparrow \qquad \text{(Section 20-12)}$$

$$\underset{\text{acid chloride}}{R-\overset{O}{\overset{\|}{C}}-Cl} + \underset{\text{amine}}{2\ R'_2NH} \longrightarrow \underset{\text{amide}}{R-\overset{O}{\overset{\|}{C}}-NR'_2} + R'_2\overset{+}{N}H_2\ Cl^- \qquad \text{(Sections 20-15 and 21-5)}$$

$$\underset{\text{anhydride}}{R-\overset{O}{\overset{\|}{C}}-O-\overset{O}{\overset{\|}{C}}-R} + \underset{\text{amine}}{R'_2NH} \longrightarrow \underset{\text{amide}}{R-\overset{O}{\overset{\|}{C}}-NR'_2} + RCOOH \qquad \text{(Section 21-5)}$$

$$\underset{\text{ester}}{R-\overset{O}{\overset{\|}{C}}-OR''} + \underset{\text{amine}}{R'-NH_2} \longrightarrow \underset{\text{amide}}{R-\overset{O}{\overset{\|}{C}}-NHR'} + R''OH \qquad \text{(Section 21-5)}$$

$$\underset{\text{nitrile}}{R-C\equiv N} + H_2O \xrightarrow{H^+ \text{ or } ^-OH} R-\overset{O}{\overset{\|}{C}}-NH_2 \quad 1°\text{ amide} \qquad \text{(Section 21-7D)}$$

Reactions of Amides Unlike amines, amides are not appreciably basic. The $-NH_2$ group of an amide formally has a nonbonding pair of electrons. But that lone pair is involved in strong resonance with the carbonyl group, which prevents it from being basic or nucleophilic.

$$\left[R-C(=\ddot{O}:)-\ddot{N}H_2 \longleftrightarrow R-C(-\ddot{O}:^-)=\overset{+}{N}H_2 \right]$$

An amide can be protonated only in the presence of a strong acid. Even then, protonation usually occurs on the carbonyl group because of resonance stabilization of the resulting conjugate acid.

$$CH_3-C(=\ddot{O}:)-\ddot{N}H_2 + H_2SO_4 \rightleftharpoons \left[CH_3-C(=\ddot{O}^+-H)-\ddot{N}H_2 \longleftrightarrow CH_3-C(-\ddot{O}-H)=N^+H_2 \right] + HSO_4^-$$

$pK_a = -5$ $pK_a = 0$

Because amides are the most stable acid derivatives, they are not easily converted to other derivatives by nucleophilic acyl substitution. From a synthetic standpoint, their most important reaction is the reduction to amines, which is one of the best methods for synthesizing amines. Amides are hydrolyzed by a strong acid or strong base. Just as nitriles can be hydrolyzed to amides, amides can be dehydrated to nitriles.

$$R-C(=O)-NHR' \text{ (amide)} \xrightarrow[H^+ \text{ or } ^-OH]{H_2O} R-C(=O)-OH \text{ (acid)} + R'NH_2 \quad \text{(Section 21-7C)}$$

$$R-C(=O)-NHR' \text{ (amide)} \xrightarrow[(2)\ H_2O]{(1)\ LiAlH_4} R-CH_2NHR' \quad \text{(Sections 19-19 and 21-8C)}$$

$$R-C(=O)-NH_2 \text{ (1° amide)} \xrightarrow[(\text{or } P_2O_5)]{POCl_3} R-C\equiv N \text{ (nitrile)} \quad \text{(Section 21-13)}$$

Dehydration of Amides to Nitriles Strong dehydrating agents can remove the elements of water from a primary amide to give a nitrile. Dehydration of amides is one of the most common methods for synthesis of nitriles. Phosphorus pentoxide (P_2O_5) is the traditional reagent for this dehydration, but phosphorus oxychloride ($POCl_3$) sometimes gives better yields.

$$R-C(=O)-\ddot{N}H_2 \text{ (primary amide)} \xrightarrow[(\text{or } P_2O_5)]{POCl_3} R-C\equiv N: \text{ (nitrile)}$$

Example

$$CH_3CH_2CH_2CH_2-CH(CH_2CH_3)-C(=O)-\ddot{N}H_2 \text{ (2-ethylhexanamide)} \xrightarrow{P_2O_5} CH_3CH_2CH_2CH_2-CH(CH_2CH_3)-C\equiv N: \text{ (2-ethylhexanenitrile (90\%))}$$

Formation of Lactams Five-membered lactams (γ-lactams) and six-membered lactams (δ-lactams) often form upon heating or adding a dehydrating agent to the appropriate γ-amino acids and δ-amino acids. Lactams containing smaller or larger rings do not form readily under these conditions.

γ-aminobutyric acid (NH_2, COOH) $\xrightarrow{\text{heat}}$ γ-butyrolactam (NH, O) + H_2O

δ-aminovaleric acid (NH_2, COOH) $\xrightarrow{\text{heat}}$ δ-valerolactam (NH, O) + H_2O

Biological Reactivity of β-Lactams β-Lactams are unusually reactive amides and are capable of acylating a variety of nucleophiles. The considerable strain in the four-membered ring appears to be the driving force behind the unusual reactivity of β-lactams. When a β-lactam acylates a nucleophile, the ring opens and the ring strain is relieved.

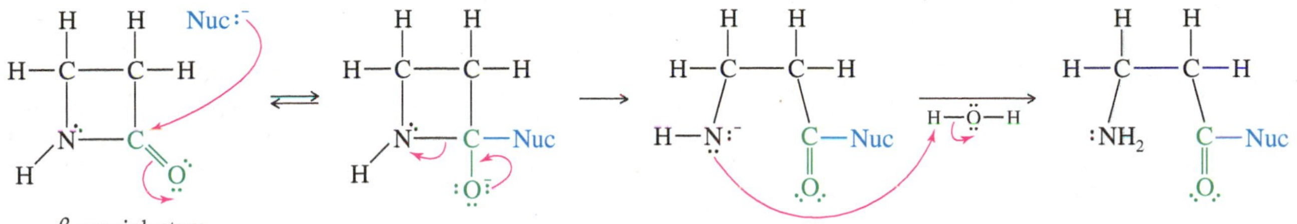

β-propiolactam

The β-lactam ring is found in three important classes of antibiotics, all isolated from fungi. *Penicillins* have a β-lactam ring fused to a five-membered ring containing a sulfur atom. *Cephalosporins* have a β-lactam ring fused to an unsaturated six-membered ring containing a sulfur atom. *Carbapenems* have a β-lactam ring fused to an unsaturated five-membered ring with a sulfur atom bonded to the ring. The structures of penicillin V, cephalexin, and imipenem illustrate these three classes of antibiotics.

penicillin V
a penicillin

cephalexin (Keflex®)
a cephalosporin

imipenem (Primaxin®)
a carbapenem

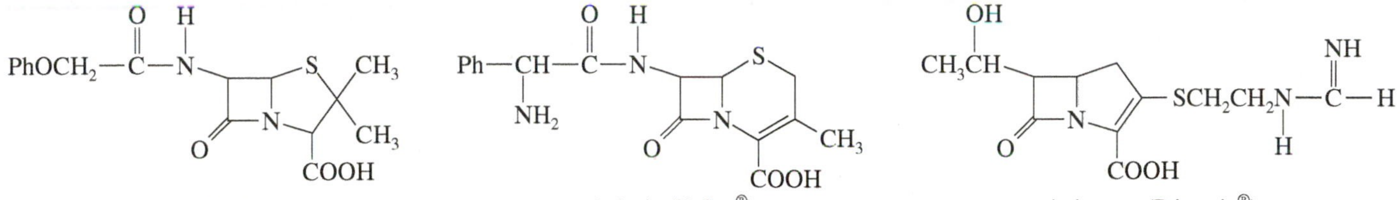

These β-lactam antibiotics apparently work by interfering with the synthesis of bacterial cell walls. Figure 21-11 shows how the carbonyl group of the β-lactam acylates a hydroxy group (from a serine residue) on one of the enzymes involved in making the cell wall. The acylated enzyme is inactive for synthesis of the cell wall protein. This acylation step is unusual because it converts an amide to an ester, an "uphill" reaction that we would assume to be endothermic. With this β-lactam, however, the strain of the four-membered ring activates the amide enough for it to acylate an alcohol to form an ester in an exothermic step.

Application: Drug Resistance

Drug-resistant bacteria inactivate β-lactam antibiotics by hydrolyzing the amide linkage of the lactam ring. Augmentin® consists of a β-lactam antibiotic (amoxicillin) and potassium clavulanate, a compound that blocks the enzyme responsible for the hydrolysis. This combination enables the amoxicillin to avoid being deactivated by the enzyme.

PROBLEM 21-36

Show how you would accomplish the following synthetic transformations. You may use any necessary reagents.

(a) *N*-ethylbenzamide → benzylethylamine
(b) ethyl benzoate → *N*-ethylbenzamide
(c) pyrrolidine → *N*-acetylpyrrolidine
(d) γ-aminobutyric acid → pyrrolidine

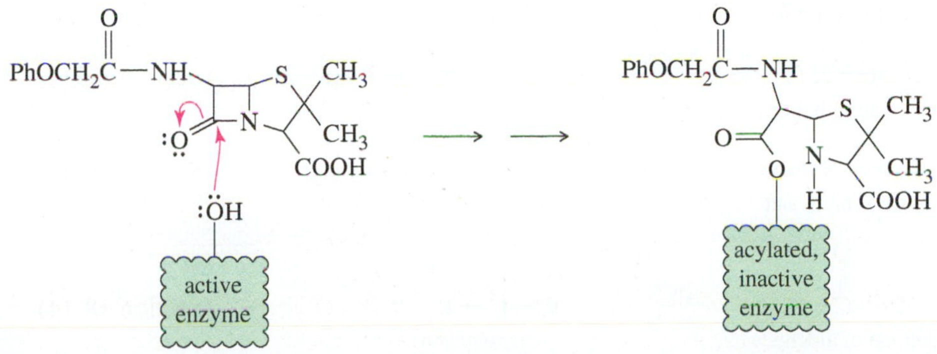

active enzyme

acylated, inactive enzyme

FIGURE 21-11
Action of β-lactam antibiotics. β-Lactam antibiotics function by acylating and inactivating one of the enzymes needed to make the bacterial cell wall.

Production of continuous-filament nylon tire cord.

PROBLEM 21-37

Show how you would accomplish the following syntheses using amides as intermediates. You may use any necessary reagents.

(a) benzoic acid → benzyldimethylamine **(b)** pyrrolidine → *N*-ethylpyrrolidine

(c) cyclopentanecarboxylic acid → cyclopentanecarbonitrile

Polyamides: Nylon The discovery of nylon in 1938 made possible a wide range of high-strength fibers, fabrics, and plastics that we take for granted today. The most common form of nylon is called nylon 6,6 because it consists of a six-carbon diacid and a six-carbon diamine in repeating blocks. Nylon 6,6 is made by mixing adipic acid and hexane-1,6-diamine (common name: *hexamethylene diamine*) to form a *nylon salt*, and then heating the salt to drive off water and form amide bonds. The molten product is extruded in continuous filaments and stretched to align the polymer chains. The combination of polymer chains aligned with the fiber, plus the strong amide hydrogen bonding between the chains, gives nylon fibers great strength. We consider nylon chemistry in more detail in Chapter 26.

$$\underset{\text{adipic acid}}{HO-\overset{O}{\overset{\|}{C}}-(CH_2)_4-\overset{O}{\overset{\|}{C}}-OH} + \underset{\text{hexamethylenediamine}}{H_2N-(CH_2)_6-NH_2} \longrightarrow \underset{\text{nylon salt}}{\begin{matrix} {}^-O-\overset{O}{\overset{\|}{C}}-(CH_2)_4-\overset{O}{\overset{\|}{C}}-O^- \\ H_3\overset{+}{N}-(CH_2)_6-\overset{+}{N}H_3 \end{matrix}}$$

heat, $-H_2O$

$$---\overset{O}{\overset{\|}{C}}-(CH_2)_4-\overset{O}{\overset{\|}{C}}\left[-NH-(CH_2)_6-NH-\overset{O}{\overset{\|}{C}}-(CH_2)_4-\overset{O}{\overset{\|}{C}}-\right]_n NH-(CH_2)_6-NH---$$

poly(hexamethylene adipamide), called nylon 6,6

21-14 Summary of the Chemistry of Nitriles

Although nitriles lack an acyl group, they are considered acid derivatives because they hydrolyze to carboxylic acids. Nitriles are frequently made from carboxylic acids (with the same number of carbons) by conversion to primary amides followed by dehydration. They are also made from primary alkyl halides and tosylates (adding one carbon) by nucleophilic substitution with cyanide ion. Aryl cyanides can be made by the Sandmeyer reaction of an aryldiazonium salt with cuprous cyanide. α-Hydroxynitriles (cyanohydrins) are made by the reaction of ketones and aldehydes with HCN.

$$\underset{\text{primary amide}}{R-\overset{O}{\overset{\|}{C}}-NH_2} \xrightarrow{POCl_3} \underset{\text{nitrile}}{R-C\equiv N} \qquad \text{(Section 21-13)}$$

$$\underset{\text{alkyl halide}}{R-X\ (1°)} \xrightarrow{NaCN} \underset{\text{nitrile}}{R-C\equiv N} + Na^+\ X^- \qquad \text{(Section 6-9)}$$

$$\underset{\text{diazonium salt}}{Ar-\overset{+}{N}\equiv N} \xrightarrow{CuCN} \underset{\text{aryl nitrile}}{Ar-C\equiv N} + N_2\uparrow \qquad \text{(Section 19-17)}$$

$$\underset{\text{ketone or aldehyde}}{R-\overset{O}{\overset{\|}{C}}-R'} \xrightarrow[KCN]{HCN} \underset{\text{cyanohydrin}}{R-\overset{HO\ \ C\equiv N}{C}-R'} \qquad \text{(Section 18-14)}$$

Reactions of Nitriles Nitriles undergo acidic or basic hydrolysis to amides, which may be further hydrolyzed to carboxylic acids. Reduction of a nitrile by lithium aluminum hydride gives a primary amine, and the reaction with a Grignard reagent gives an imine that hydrolyzes to a ketone.

$$\text{R—C}\equiv\text{N (nitrile)} \xrightarrow[\text{H}^+ \text{ or } ^-\text{OH}]{\text{H}_2\text{O}} \text{R—C(=O)—NH}_2 \text{ (amide)} \xrightarrow[\text{H}^+ \text{ or } ^-\text{OH}]{\text{H}_2\text{O}} \text{R—C(=O)—OH (acid)}$$

(Section 21-7D)

$$\text{R—C}\equiv\text{N} \xrightarrow[\text{(2) H}_2\text{O}]{\text{(1) LiAlH}_4} \text{R—CH}_2\text{NH}_2 \text{ (amine)}$$

(Sections 19-20B and 21-8C)

$$\text{R—C}\equiv\text{N} \xrightarrow{\text{R}'\text{MgX}} \text{R—C(=N—MgX)—R}' \text{ (imine salt)} \xrightarrow{\text{H}_3\text{O}^+} \text{R—C(=O)—R}' \text{ (ketone)}$$

(Sections 18-9 and 21-9)

$$\text{R—C}\equiv\text{N} \xrightarrow[\text{(2) H}_3\text{O}^+]{\text{(1) DIBAL-H, –78 °C}} \text{R—C(=O)—H (aldehyde)}$$

(Sections 18-9 and 21-8)

Application: Planetary Atmospheres

The presence of nitriles in the atmospheres of other planets is significant because they may be precursors to critical biological molecules. For example, nitriles can give rise to amino acids (Section 24-5C), the building blocks of proteins.

PROBLEM 21-38

Show how you would convert the following starting materials to the indicated nitriles:
(a) phenylacetic acid → phenylacetonitrile
(b) phenylacetic acid → 3-phenylpropionitrile
(c) *p*-chloronitrobenzene → *p*-chlorobenzonitrile

PROBLEM 21-39

Show how each transformation may be accomplished by using a nitrile as an intermediate. You may use any necessary reagents.
(a) hexan-1-ol → heptan-1-amine
(b) cyclohexanecarboxamide → cyclohexyl ethyl ketone
(c) octan-1-ol → decan-2-one

21-15 Thioesters

Most carboxylic esters are composites of carboxylic acids and alcohols. A **thioester** is formed from a carboxylic acid and a thiol. Thioesters are also called *thiol esters* to emphasize that they are derivatives of thiols.

$$\underset{\text{acid}}{\text{R—C(=O)—OH}} + \underset{\text{alcohol}}{\text{R}'\text{—OH}} \rightleftharpoons \underset{\text{ester}}{\text{R—C(=O)—O—R}'} + \text{H}_2\text{O}$$

$$\underset{\text{acid}}{\text{R—C(=O)—OH}} + \underset{\text{thiol}}{\text{R}'\text{—SH}} \rightleftharpoons \underset{\text{thioester}}{\text{R—C(=O)—S—R}'} + \text{H}_2\text{O}$$

Thioesters are more reactive toward nucleophilic acyl substitution than normal esters, but less reactive than acid chlorides and anhydrides. If we add thioesters to the order of reactivity, we have the following sequence:

Relative reactivity

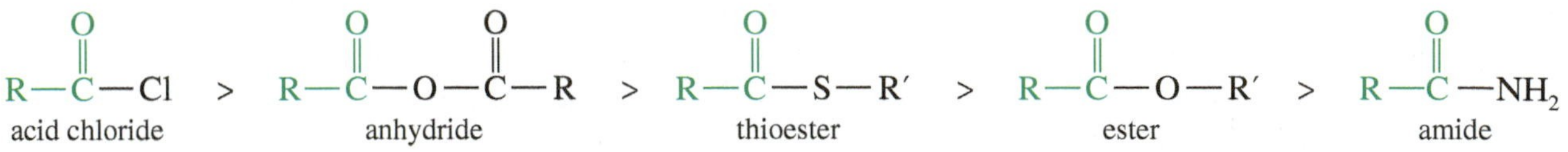

$$\underset{\text{acid chloride}}{R-\overset{O}{\overset{\|}{C}}-Cl} > \underset{\text{anhydride}}{R-\overset{O}{\overset{\|}{C}}-O-\overset{O}{\overset{\|}{C}}-R} > \underset{\text{thioester}}{R-\overset{O}{\overset{\|}{C}}-S-R'} > \underset{\text{ester}}{R-\overset{O}{\overset{\|}{C}}-O-R'} > \underset{\text{amide}}{R-\overset{O}{\overset{\|}{C}}-NH_2}$$

The enhanced reactivity of thioesters results from two major differences. First, the resonance stabilization of a thioester is less than that of an ester. In the thioester, the second resonance form involves overlap between a $2p$ orbital on carbon and a $3p$ orbital on sulfur (Figure 21-12). These orbitals are different sizes and are located at different distances from the nuclei. The overlap is weak and relatively ineffective, leaving the C—S bond of a thioester weaker than the C—O bond of an ester.

The second difference is in the leaving groups: An alkyl sulfide anion ($^{-}:\ddot{\underset{..}{S}}-R$) is a better leaving group than an alkoxide anion ($^{-}:\ddot{\underset{..}{O}}-R$) because the sulfide is less basic than an alkoxide, and the larger sulfur atom carries the negative charge spread over a larger volume of space. Sulfur is also more polarizable than oxygen, allowing more bonding as the alkyl sulfide anion is leaving (Section 6-11A).

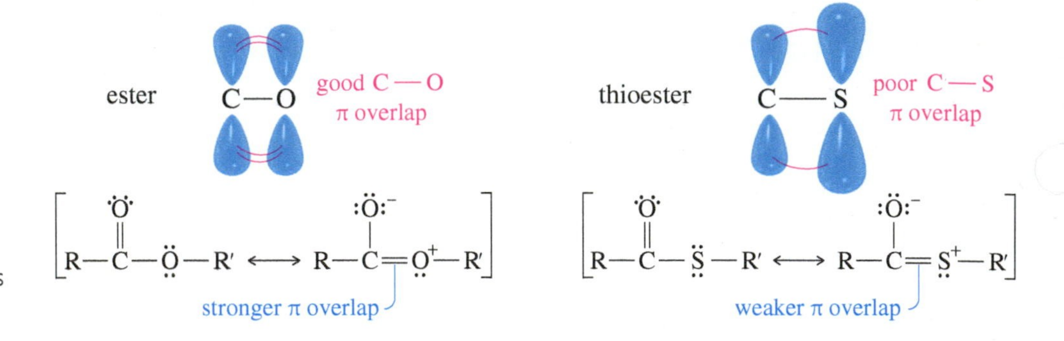

FIGURE 21-12
The resonance overlap in a thioester is not as effective as that in an ester.

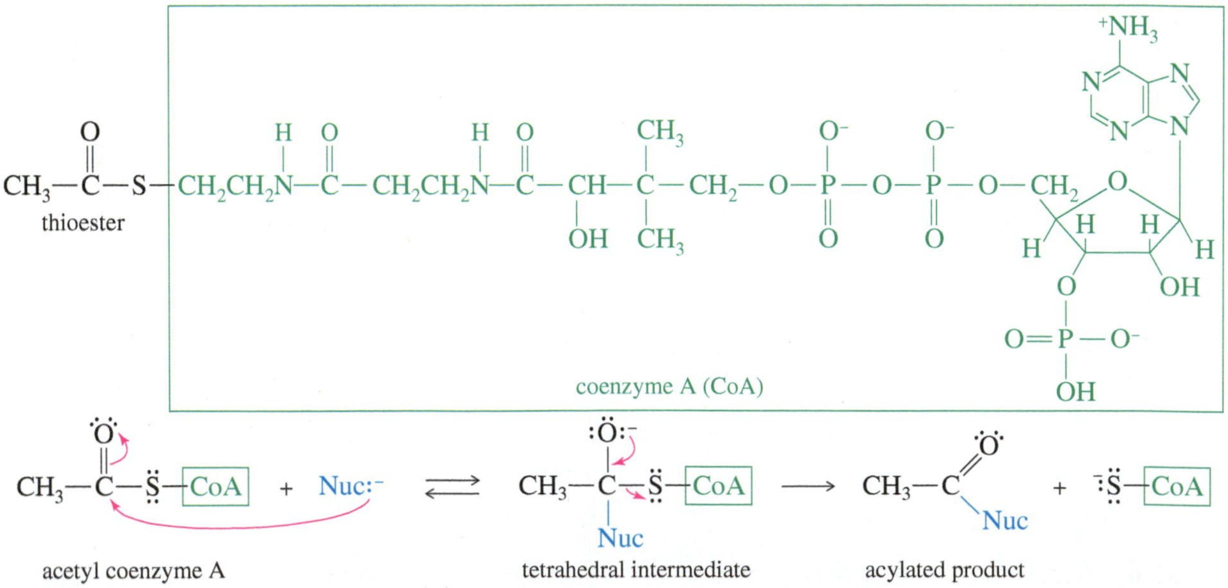

$$\underset{\text{thioester}}{CH_3-\overset{O}{\overset{\|}{C}}-S}-CH_2CH_2\overset{H}{N}-\overset{O}{\overset{\|}{C}}-CH_2CH_2\overset{H}{N}-\overset{O}{\overset{\|}{C}}-\underset{OH}{CH}-\overset{CH_3}{\underset{CH_3}{C}}-CH_2-O-\overset{O^-}{\underset{O}{P}}-O-\overset{O^-}{\underset{O}{P}}-O-CH_2-\text{(ribose-adenine)}$$

coenzyme A (CoA)

$$\underset{\text{acetyl coenzyme A}}{CH_3-\overset{O}{\overset{\|}{C}}-S-\boxed{CoA}} + Nuc:^- \rightleftharpoons \underset{\text{tetrahedral intermediate}}{CH_3-\overset{:O:^-}{\underset{Nuc}{C}}-S-\boxed{CoA}} \longrightarrow \underset{\text{acylated product}}{CH_3-\overset{O}{\overset{\|}{C}}-Nuc} + {}^-:S-\boxed{CoA}$$

FIGURE 21-13
Coenzyme A (CoA) is a thiol whose thioesters serve as biochemical acyl transfer reagents. Acetyl CoA transfers an acetyl group to a nucleophile, with coenzyme A serving as the leaving group.

Living systems need acylating reagents, but acid halides and anhydrides are too reactive for selective acylation. Also, they would hydrolyze under the aqueous conditions found in living organisms. Thioesters are not so prone to hydrolysis, yet they are excellent selective acylating reagents. For these reasons, thioesters are common acylating agents in living systems. Many biochemical acylations involve transfer of acyl groups from thioesters of coenzyme A (CoA). Figure 21-13 shows the structure of acetyl coenzyme A, together with the mechanism for transfer of the acetyl group to a nucleophile. In effect, acetyl CoA serves as a water-stable equivalent of acetyl chloride (or acetic anhydride) in living systems.

21-16 Esters and Amides of Carbonic Acid

Carbonic acid (H_2CO_3) is formed reversibly whenever carbon dioxide dissolves in water. All carbonated beverages contain carbonic acid in equilibrium with CO_2 and water.

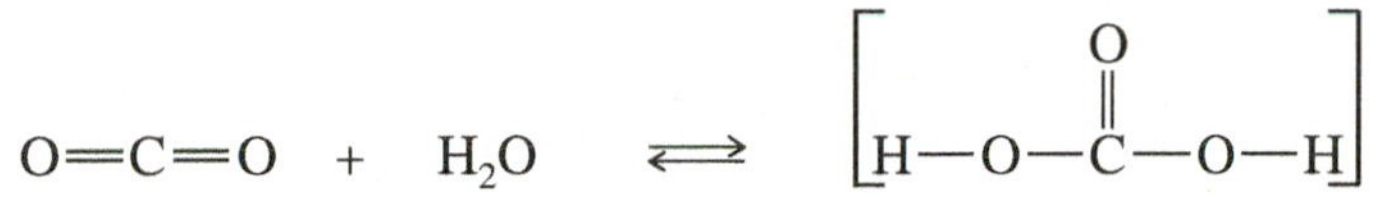

$$O{=}C{=}O \; + \; H_2O \;\rightleftharpoons\; \left[H{-}O{-}\overset{\overset{\large O}{\|}}{C}{-}O{-}H \right]$$

carbonic acid (unstable)

Although carbonic acid itself is always in equilibrium with carbon dioxide and water, it has several important stable derivatives. **Carbonate esters** are diesters of carbonic acid, with two alkoxy groups replacing the hydroxy groups of carbonic acid.

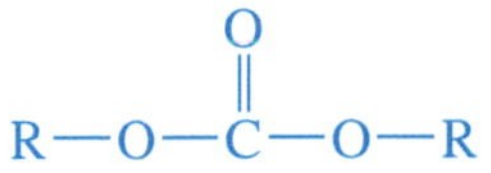

$$R{-}O{-}\overset{\overset{\large O}{\|}}{C}{-}O{-}R \qquad CH_3CH_2{-}O{-}\overset{\overset{\large O}{\|}}{C}{-}O{-}CH_2CH_3 \qquad C_6H_{11}{-}O{-}\overset{\overset{\large O}{\|}}{C}{-}O{-}CH_2CH_3$$

a carbonate ester; 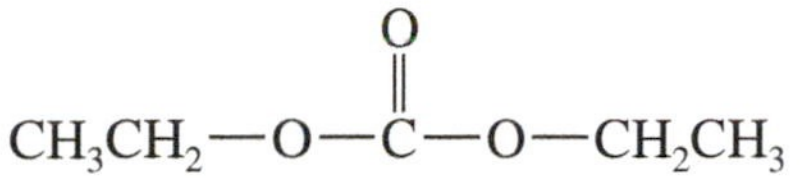 diethyl carbonate; 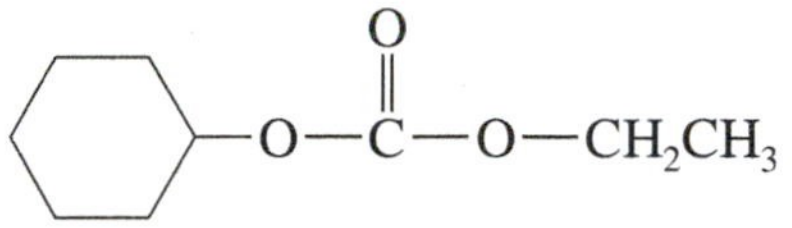 cyclohexyl ethyl carbonate

Ureas are diamides of carbonic acid, with two nitrogen atoms bonded to the carbonyl group. The unsubstituted urea, simply called *urea*, is the waste product excreted by mammals from the metabolism of excess protein. Tetramethylurea (TMU) is often used as a polar, aprotic solvent with a high boiling point (177 °C). It is highly soluble in water, so it can be washed out of a solution.

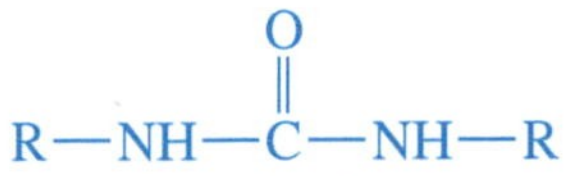

$$R{-}NH{-}\overset{\overset{\large O}{\|}}{C}{-}NH{-}R \qquad H_2N{-}\overset{\overset{\large O}{\|}}{C}{-}NH_2 \qquad (CH_3)_2N{-}\overset{\overset{\large O}{\|}}{C}{-}N(CH_3)_2$$

a substituted urea; 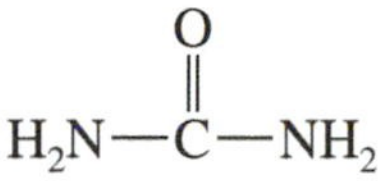 urea; 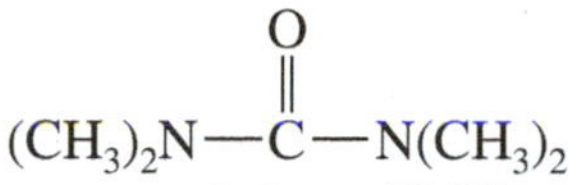 tetramethylurea (TMU)

Carbamate esters (urethanes) are the stable esters of the unstable **carbamic acid,** the monoamide of carbonic acid.

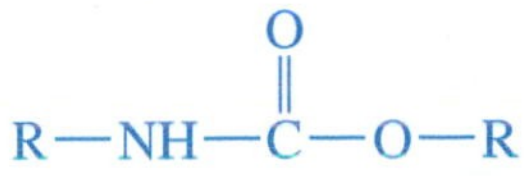

$$R{-}NH{-}\overset{\overset{\large O}{\|}}{C}{-}O{-}R \qquad \left[H_2N{-}\overset{\overset{\large O}{\|}}{C}{-}OH \right] \qquad H_2N{-}\overset{\overset{\large O}{\|}}{C}{-}OEt \qquad CH_3{-}\underset{\underset{\large H}{|}}{N}{-}\overset{\overset{\large O}{\|}}{C}{-}O{-}C_{10}H_7$$

a carbamate or urethane; 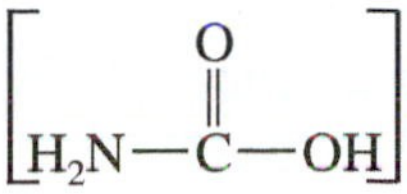 carbamic acid (unstable); ethyl carbamate; 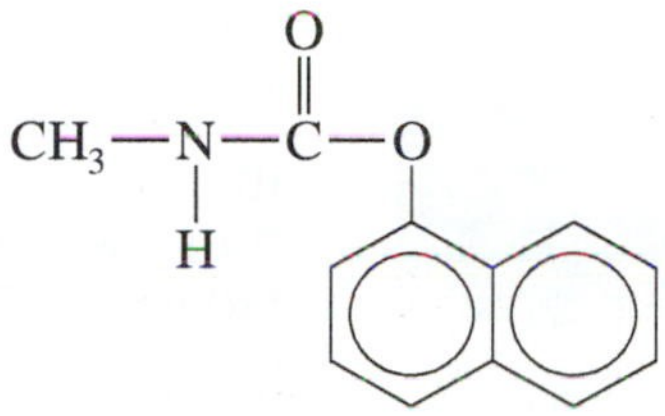 1-naphthyl-*N*-methylcarbamate (Sevin® insecticide)

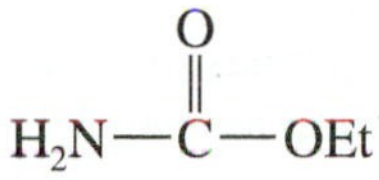

Many of these derivatives can be synthesized by nucleophilic acyl substitution from phosgene, the acid chloride of carbonic acid.

$$Cl{-}\overset{\overset{O}{\|}}{C}{-}Cl + 2\ CH_3CH_2{-}OH \longrightarrow CH_3CH_2{-}O{-}\overset{\overset{O}{\|}}{C}{-}O{-}CH_2CH_3 + 2\ HCl$$

phosgene; diethyl carbonate

$$Cl{-}\overset{\overset{O}{\|}}{C}{-}Cl \xrightarrow{CH_3CH_2OH} Cl{-}\overset{\overset{O}{\|}}{C}{-}OCH_2CH_3 \xrightarrow{C_6H_{11}{-}NH_2} C_6H_{11}{-}\underset{\underset{H}{|}}{N}{-}\overset{\overset{O}{\|}}{C}{-}OCH_2CH_3$$

ethyl *N*-cyclohexyl carbamate

$$Cl{-}\overset{\overset{O}{\|}}{C}{-}Cl + 2\ (CH_3)_2NH \longrightarrow (CH_3)_2N{-}\overset{\overset{O}{\|}}{C}{-}N(CH_3)_2 + 2\ HCl$$

tetramethylurea

Another way of making urethanes is to treat an alcohol or a phenol with an **isocyanate,** which is an anhydride of a carbamic acid. Although the carbamic acid is unstable, the urethane is stable. This is the way Sevin® insecticide is made.

$$R{-}N{=}C{=}O + HO{-}R' \longrightarrow R{-}NH{-}\overset{\overset{O}{\|}}{C}{-}O{-}R'$$

an isocyanate; alcohol; a carbamate ester (urethane)

Example

$$CH_3{-}N{=}C{=}O + \text{1-naphthol} \longrightarrow CH_3{-}\underset{\underset{H}{|}}{N}{-}\overset{\overset{O}{\|}}{C}{-}O{-}(\text{1-naphthyl})$$

methyl isocyanate; 1-naphthol; Sevin® insecticide

$$R{-}N{=}C{=}O + H_2O \longrightarrow \left[R{-}NH{-}\overset{\overset{O}{\|}}{C}{-}OH\right] \longrightarrow R{-}NH_2 + CO_2$$

an isocyanate; a carbamic acid (unstable); an amine

Before the development of tough, resilient polyurethane wheels, street roller skates used steel wheels that stopped dead when they hit the smallest pebble or crack. Rollerblades would not exist without polymer technology, both in the wheels and in the strong ABS plastic used for the uppers.

PROBLEM 21-40

Propose a mechanism for the reaction of methyl isocyanate with 1-naphthol to give Sevin® insecticide.

PROBLEM 21-41

For each heterocyclic compound,

(i) explain what type of acid derivative is present.
(ii) show what compounds would result from complete hydrolysis.
(iii) are any of the rings aromatic? Explain.

(a) [cyclohexane-fused cyclic carbonate: ring O–C(=O)–O] **(b)** [five-membered ring with S, C=O, and C=C] **(c)** [six-membered ring S–C(=O)–S]

(d) [HN–C(=O)–NH five-membered ring with C=C] **(e)** [pyrrole ring with N–H and OH on adjacent carbon] **(f)** [O–C(=O)–NH five-membered ring with C=C]

Polycarbonates and Polyurethanes The chemistry of carbonic acid derivatives is particularly important because two large classes of polymers are bonded by linkages containing these functional groups: the **polycarbonates** and the **polyurethanes.** Polycarbonates are polymers bonded by the carbonate ester linkage, and polyurethanes are polymers bonded by the carbamate ester linkage. Lexan® polycarbonate is a strong, clear polymer used in bulletproof windows and crash helmets. The diol used to make Lexan® is a phenol called *bisphenol A*, a common intermediate in polyester and polyurethane synthesis.

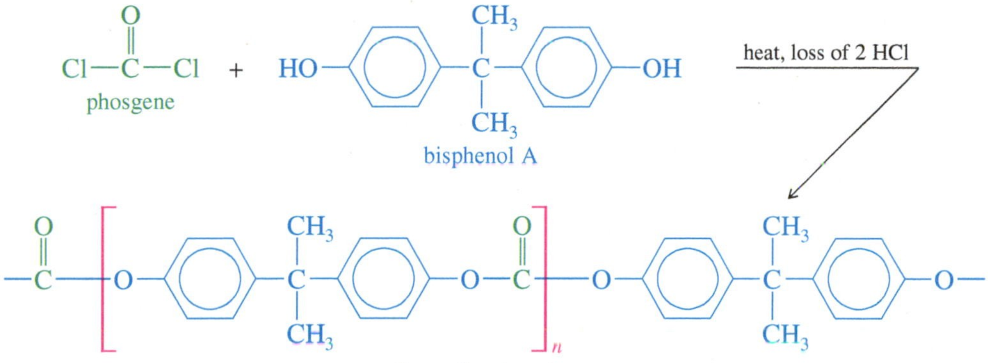

phosgene + bisphenol A —heat, loss of 2 HCl→ Lexan® polycarbonate

A polyurethane results when a diol reacts with a diisocyanate, a compound with two isocyanate groups. A common form of polyurethane is made by the reaction of ethylene glycol with *toluene diisocyanate.*

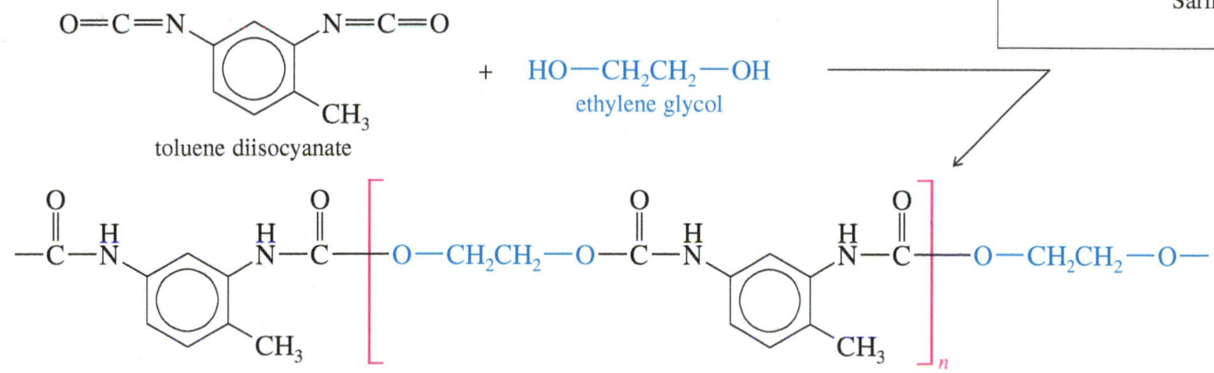

toluene diisocyanate + $HO—CH_2CH_2—OH$ (ethylene glycol) → a polyurethane

Application: Insecticide

The development of Sevin® and related insecticides resulted from studies on the alkaloid physostigmine, which has a methyl carbamate portion. These studies also led to the development of potent nerve gases such as Sarin.

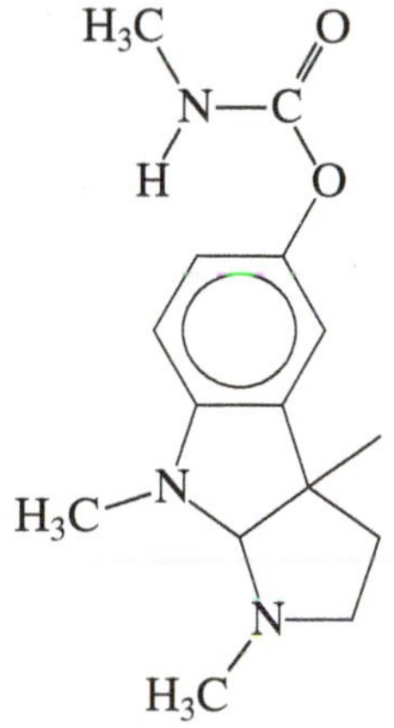

physostigmine

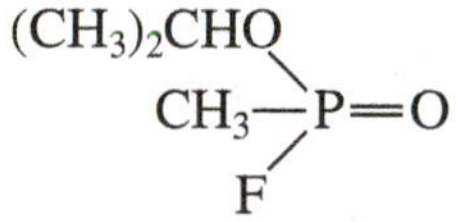

Sarin

Essential Terms

acid derivatives Compounds containing functional groups that can be converted to carboxylic acids by acidic or basic hydrolysis. (p. 1073)

Relative reactivity

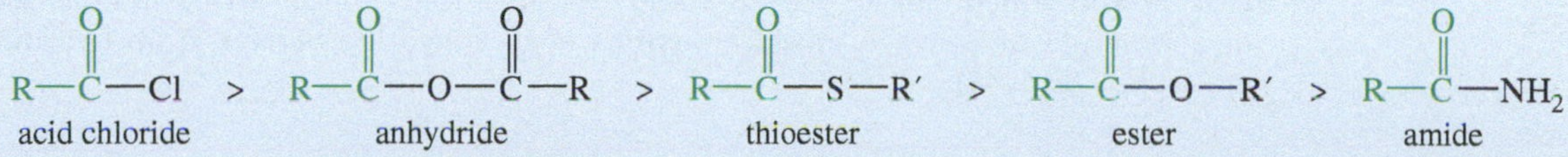

acid chloride > anhydride > thioester > ester > amide

acid halide **(acyl halide)** An activated acid derivative in which the hydroxy group of the acid is replaced by a halogen, usually chlorine. (p. 1048)

(continued)

SUMMARY Reactions of Acid Chlorides

REACTIONS OF ACID CHLORIDES

Acid chlorides are the most reactive of the carboxylic acid derivatives as shown in Figure 21-9.

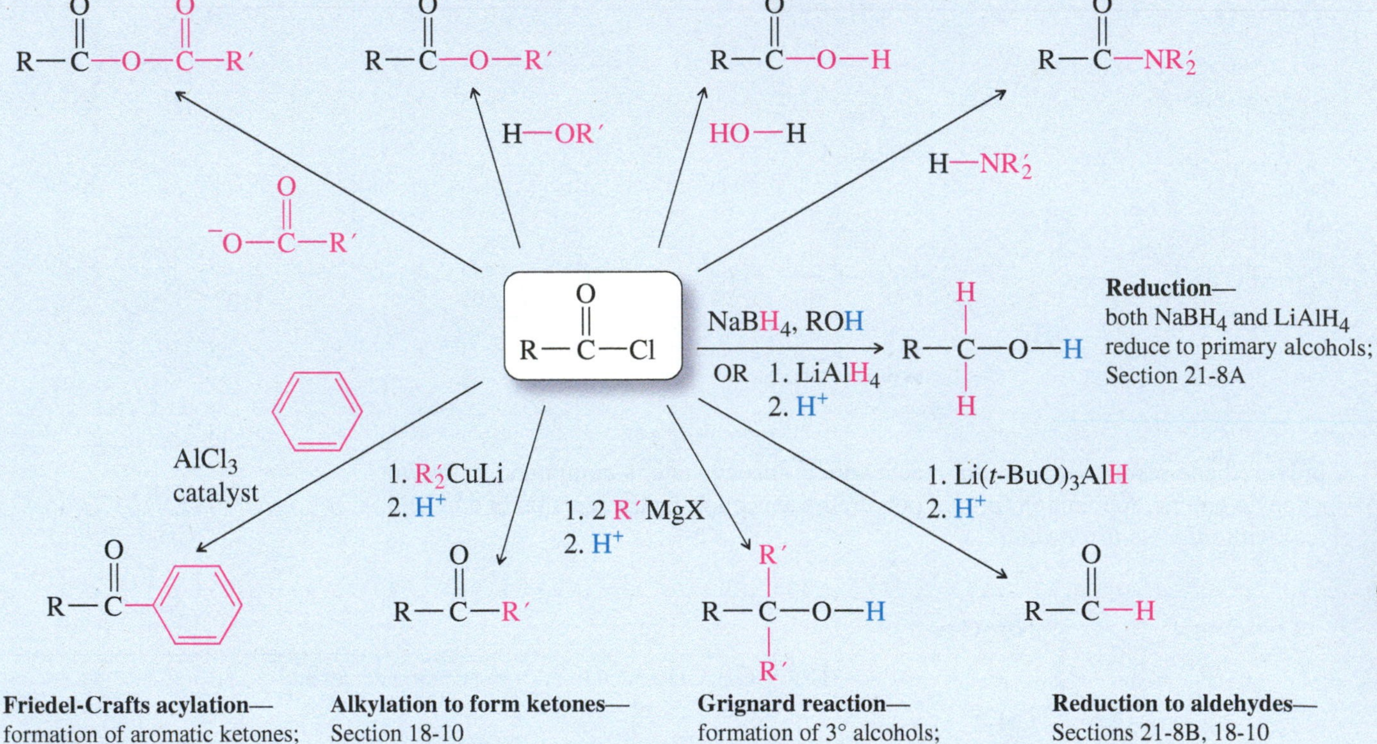

acyl transfer Another term for a *nucleophilic acyl substitution*. The term *acyl transfer* emphasizes the "transfer" of the acyl group from the leaving group to the attacking nucleophile. (p. 1060)

amide An acid derivative in which the hydroxy group of the acid is replaced by a nitrogen atom and its attached hydrogens or alkyl groups. An amide is a composite of a carboxylic acid and an amine. (p. 1045)

$$\underset{\text{primary amide}}{R-\overset{O}{\overset{\|}{C}}-NH_2} \qquad \underset{\substack{\text{secondary amide}\\(N\text{-substituted amide})}}{R-\overset{O}{\overset{\|}{C}}-\overset{H}{\overset{|}{N}}-R'} \qquad \underset{\substack{\text{tertiary amide}\\(N,N\text{-disubstituted amide})}}{R-\overset{O}{\overset{\|}{C}}-\overset{R'}{\overset{|}{N}}-R'}$$

ammonolysis of an ester Cleavage of an ester by ammonia (or an amine) to give an amide and an alcohol. (p. 1066)

anhydride **(carboxylic acid anhydride)** An activated acid derivative formed from two acid molecules with loss of a molecule of water. A **mixed anhydride** is an anhydride derived from two different acid molecules. (p. 1049)

$$2\ \underset{\text{acid}}{R-\overset{O}{\overset{\|}{C}}-OH} \rightleftarrows \underset{\text{anhydride}}{R-\overset{O}{\overset{\|}{C}}-O-\overset{O}{\overset{\|}{C}}-R} + H_2O$$

(continued)

GUIDE TO ORGANIC REACTIONS IN CHAPTER 21

Reactions covered in Chapter 21 are shown in red. Reactions covered in earlier chapters are shown in blue.

Substitution

Nucleophilic
- at sp^3 C (S_N1, S_N2) Ch 6, 10, 14, 22
- at sp^2 C (Nuc. Arom. Subst.) Ch 17, 19
- at C=O (Nuc. Acyl Subst.) Ch 10, 11, 20, 21, 22

Electrophilic
- at sp^2 C (Elect. Arom. Subst.) Ch 17, 19

Radical
- at sp^3 C (alkane halogenation) Ch 4, 6, 16, 17
- at sp^2 C (Sandmeyer rxn) Ch 19

Organometallic
- Gilman Ch 10, 17
- Suzuki Ch 17
- Heck Ch 17

Addition

Nucleophilic
- at C=O (Nuc. Addn.) Ch 9, 10, 18, 22
- at C=C (conjugate addn.) Ch 22

Electrophilic
- at C=C (Elect. Addn.) Ch 8, 9, 10
- at C=C (Carbene Addn.) Ch 8

Radical
- at C=C (HBr + ROOR) Ch 8

Pericyclic
- cycloaddition (Diels-Alder) Ch 15

Oxidation
- epoxidation Ch 8, 10, 14

Reduction
- hydrogenation Ch 8, 9, 17, 18, 19

Elimination

Basic conditions (E2)
- E2 dehydrohalogenation Ch 7, 9
- tosylate elimination Ch 11
- Hofmann elimination Ch 19

Acidic conditions (E1)
- E1 dehydrohalogenation Ch 7
- dehydration of alcohols Ch 11

Pericyclic (Cope elimination)
- Ch 19

Oxidation/Reduction

Oxidation
- epoxidation Ch 8, 10, 14
- oxidative cleavage Ch 8, 9, 11, 17, 22
- oxygen functional groups Ch 11, 18, 19, 20

Reduction
- hydride reduction Ch 8, 10, 11, 17, 18, 19, 20, 21
- hydrogenation Ch 8, 9, 17, 18, 19
- metals Ch 9, 17, 18, 19

carbamate ester See **urethane.** (p. 1097)

carbonate ester A diester of carbonic acid. (p. 1097)

carbonic acid The one-carbon dicarboxylic acid, HOCOOH. Carbonic acid is constantly in equilibrium with carbon dioxide and water. Its esters and amides are stable, however. (p. 1097)

$$[\text{H—O—C(=O)—O—H}] \qquad \text{R—O—C(=O)—O—R} \qquad \text{R—NH—C(=O)—NH—R} \qquad \text{R—NH—C(=O)—O—R}$$

carbonic acid (unstable) — a carbonate ester — a substituted urea — a carbamate or urethane

ester An acid derivative in which the hydroxy group of the acid is replaced by an alkoxy group. An ester is a composite of a carboxylic acid and an alcohol. (p. 1044)

Fischer esterification (p. 1073)

$$\underset{\text{acid}}{\text{R—C(=O)—OH}} + \underset{\text{alcohol}}{\text{R}'\text{—OH}} \overset{H^+}{\rightleftharpoons} \underset{\text{ester}}{\text{R—C(=O)—O—R}'} + H_2O$$

isocyanate A compound of formula R—N=C=O. (p. 1098)

lactam A cyclic amide. (p. 1047)

lactone A cyclic ester. (p. 1045)

(continued)

nitrile — An organic compound containing the **cyano group,** C≡N (p. 1047)

nucleophilic acyl substitution — A nucleophile substitutes for a leaving group on a carbonyl carbon atom. Nucleophilic acyl substitution usually takes place through the following **addition–elimination mechanism.** (p. 1060)

$$\mathrm{R{-}C({=}O){-}X} + \mathrm{Nuc}^{:-} \rightleftharpoons \left[\mathrm{R{-}C(O^-)(Nuc){-}X}\right] \rightleftharpoons \mathrm{R{-}C({=}O){-}Nuc} + :\mathrm{X}^-$$

addition–elimination mechanism of nucleophilic acyl substitution

polymer — A large molecule composed of many smaller units (monomers) bonded together. (p. 1088)

- **polyamide:** **(nylon)** A polymer in which the monomer units are bonded by amide linkages. (p. 1094)
- **polycarbonate:** A polymer in which the monomer units are bonded together by carbonate ester linkages. (p. 1099)
- **polyester:** A polymer in which the monomer units are bonded by ester linkages. (p. 1088)

polyurethane: A polymer in which the monomer units are bonded together by carbamate ester (urethane) linkages. (p. 1099)

saponification — Basic hydrolysis of an ester to an alcohol and a carboxylate salt. (p. 1073)

thioester — An acid derivative in which the hydroxy group of the acid is replaced by a sulfur atom and its attached alkyl or aryl group. A thioester is a composite of a carboxylic acid and a thiol. (p. 1095)

transesterification — Substitution of one alkoxy group for another in an ester. Transesterification can take place under either acidic or basic conditions. (p. 1069)

triglyceride — **(triacylglycerol)** A triester of the triol glycerol, esterified with three fatty acids. (p. 1070)

urea — A diamide of carbonic acid. (p. 1097)

urethane — **(carbamate ester)** An ester of a **carbamic acid,** RNH—COOH; a monoester, monoamide of carbonic acid. (p. 1097)

Essential Problem-Solving Skills in Chapter 21

Each skill is followed by problem numbers exemplifying that particular skill.

1 Name carboxylic acid derivatives, draw the structures from their names.	Problems 21-42 and 43
2 Compare the physical properties of acid derivatives, and explain the unusually high boiling points and melting points of amides. Compare the relative reactivity of esters, thioesters, amides, nitriles, anhydrides, and acid chlorides.	Problems 21-51, 52, 59, 60, and 62
3 Interpret the spectra of acid derivatives, and use spectral information to determine the structures. Show how the carbonyl stretching frequency in the IR depends on the structure of the acid derivative.	Problems 21-52, 61, 64, 65, 66, 67, 68, 69, and 70
4 Propose single-step and multistep syntheses of acid derivatives from compounds containing other functional groups. Propose multistep syntheses using acid derivatives as starting materials and intermediates.	Problems 21-48, 50, 53, 54, 55, and 57
5 Show how acid derivatives hydrolyze to carboxylic acids under either acidic or basic conditions. Explain why some acid derivatives (amides, for example) require much stronger conditions for hydrolysis than other derivatives.	Problems 21-47, 49, 58, 59, and 62
6 Show how acid derivatives are easily interconverted by nucleophilic acyl substitution from more reactive derivatives to less reactive derivatives under acidic or basic conditions. Show how acid chlorides serve as activated intermediates to convert acids to other acid derivatives.	Problems 21-46, 49, 51, 58, and 63
7 Predict the products, propose mechanisms for the reactions of carboxylic acid derivatives with reducing agents, alcohols, amines, and organometallic reagents such as Grignard, organolithium, and organocuprate reagents.	Problems 21-48, 50, 52, 53, 54, 55, and 60
Problem-Solving Strategy: Proposing Reaction Mechanisms	Problems 21-46, 49, 56, and 70

Study Problems

21-42 Draw structures to correspond with the following common and systematic names:

(a) phenyl formate **(b)** cyclohexyl benzoate **(c)** cyclopentyl phenylacetate
(d) *N*-butylacetamide **(e)** *N,N*-dimethylformamide **(f)** benzoic propionic anhydride
(g) benzamide **(h)** γ-hydroxyvaleronitrile **(i)** α-bromobutyryl chloride
(j) β-butyrolactone **(k)** phenyl isocyanate **(l)** cyclobutyl ethyl carbonate
(m) δ-caprolactam **(n)** trichloroacetic anhydride **(o)** ethyl *N*-methyl carbamate

21-43 Give appropriate names for the following compounds:

(a) $CH_3CH_2CH(CH_3)CH_2-C(=O)-Cl$ **(b)** $Ph-C(=O)-O-C(=O)-H$ **(c)** $CH_3-C(=O)-NH-Ph$

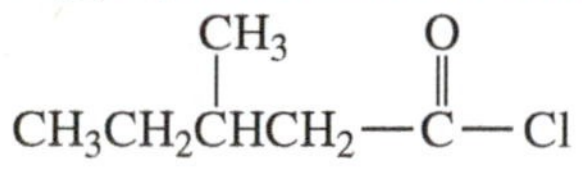

(d) $CH_3-NH-C(=O)-Ph$ **(e)** $Ph-O-C(=O)-CH_3$ **(f)** $Ph-C(=O)-O-CH_3$

(g) $C_6H_5-C{\equiv}N$ **(h)** $C_6H_5CH_2CH_2CH_2-CN$ **(i)** $CH_3O-C(=O)-C_6H_4-C(=O)-OCH_3$ (meta)

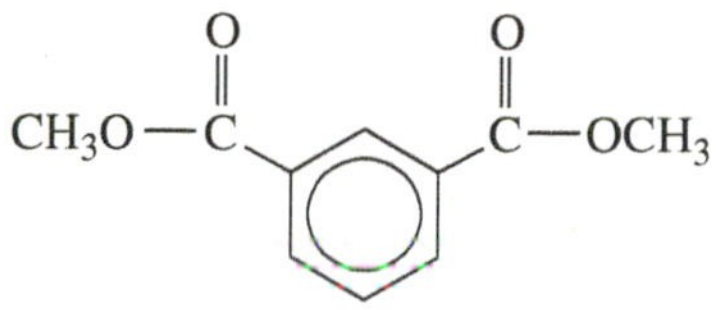

(j) CH_3 (meta) on benzene ring bearing $-C(=O)-N(CH_2CH_3)_2$ **(k)** H_3C-substituted five-membered lactone (ring O, C=O) **(l)** CH_3CH_2-substituted four-membered lactam (N–H, C=O)

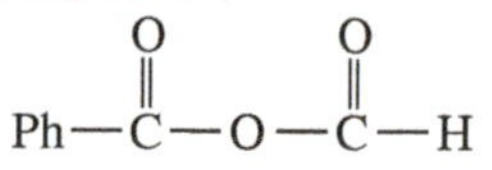

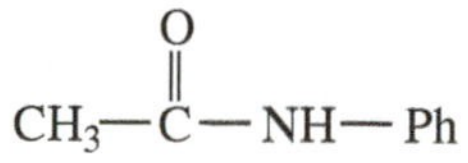

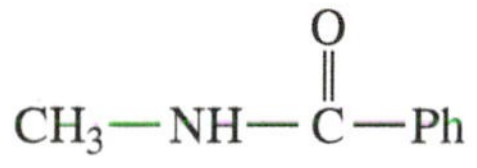

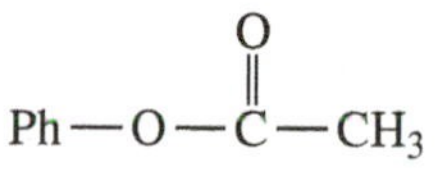

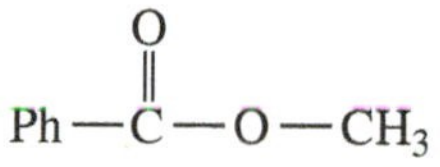

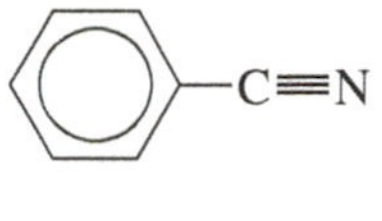

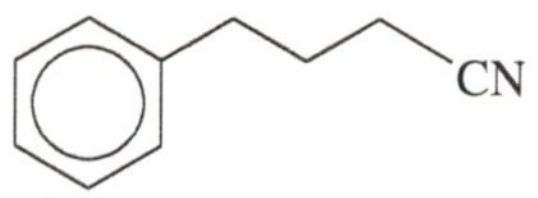

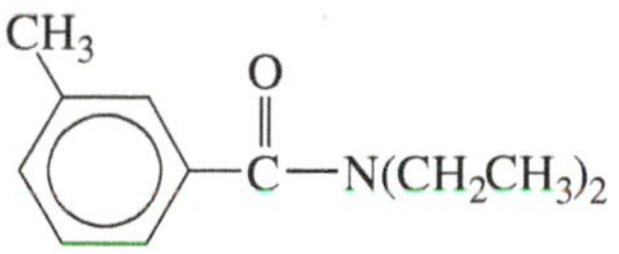

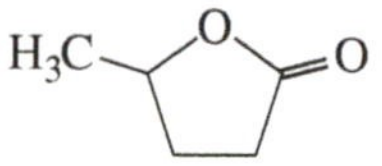

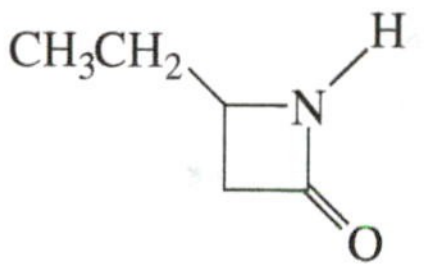

21-44 Predict the major products formed when benzoyl chloride (PhCOCl) reacts with the following reagents.

(a) ethanol **(b)** sodium acetate **(c)** aniline
(d) anisole and aluminum chloride **(e)** excess phenylmagnesium bromide, then dilute acid **(f)** $LiAlH(O\text{-}t\text{-}Bu)_3$

21-45 Predict the products of the following reactions.

(a) phenol + acetic anhydride **(b)** phenol + acetic formic anhydride
(c) aniline + phthalic anhydride **(d)** anisole + succinic anhydride and aluminum chloride
(e) $Ph-CH(OH)-CH_2-NH_2$ + 1 equivalent of acetic anhydride **(f)** $Ph-CH(OH)-CH_2-NH_2$ + excess acetic anhydride

21-46 Acid-catalyzed transesterification and Fischer esterification take place by nearly identical mechanisms. Transesterification can also take place by a base-catalyzed mechanism, but all attempts at base-catalyzed Fischer esterification (using $^-OR''$, for example) seem doomed to failure. Explain why Fischer esterification cannot be catalyzed by base.

21-47 Predict the products of saponification of the following esters.

(a) $H-C(=O)-O-Ph$ **(b)** $CH_3CH_2-C(=O)-OCH_2CH_3$ **(c)** benzo-fused six-membered lactone (dihydrocoumarin) **(d)** six-membered cyclic diester (ring with two O and two C=O)

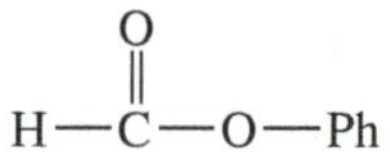

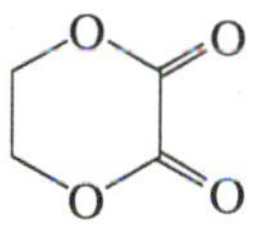

21-48 Show how you would accomplish the following syntheses in good yields.

(a) $C_6H_5-NH_2 \longrightarrow C_6H_5-NH-C(=O)-H$

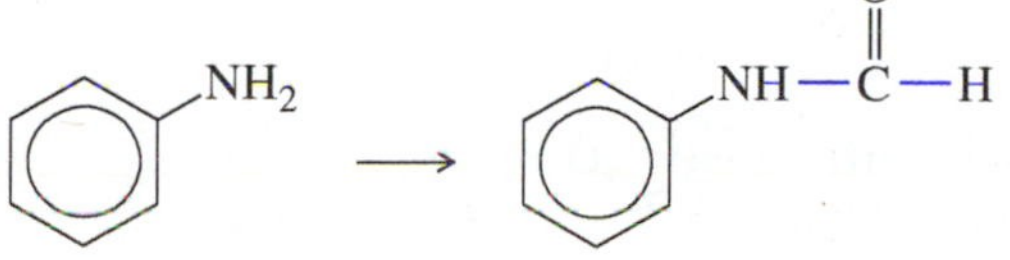

(b) $C_6H_5-COOH \longrightarrow C_6H_5-C(=O)-O-C(=O)-CH_3$

(c) *cis*-cyclohexane-1,2-diol $\longrightarrow$ fused cyclic oxalate diester (six-membered ring with two O and two C=O)

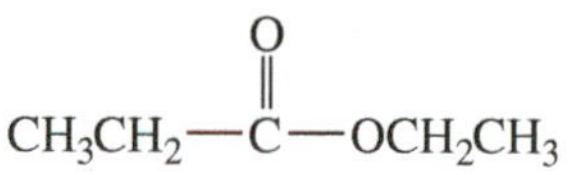

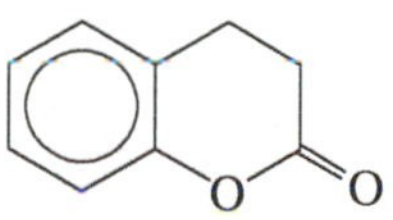

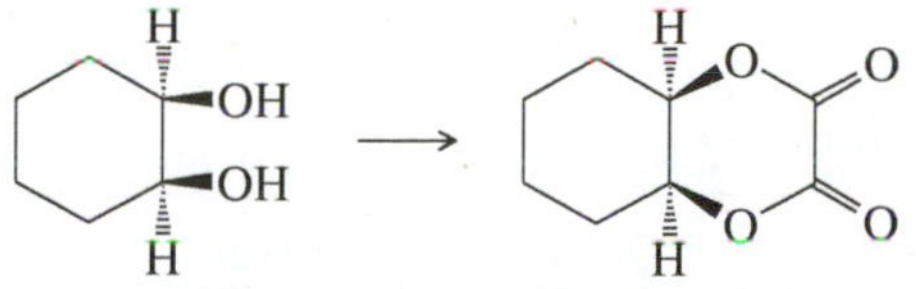

(d) phthalic acid (o-$C_6H_4(COOH)_2$) $\longrightarrow$ o-$C_6H_4(COOCH(CH_3)_2)(COOH)$

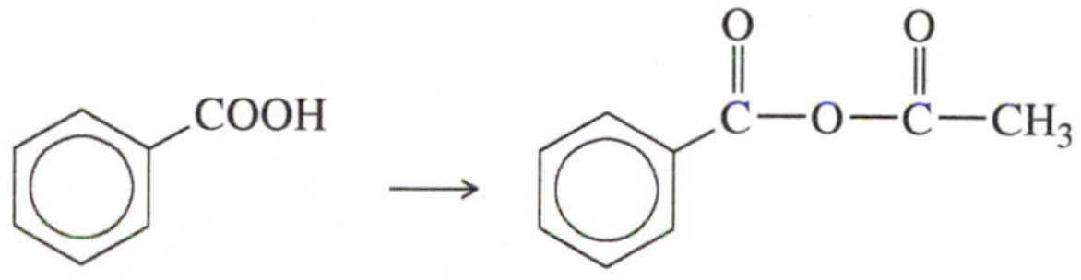

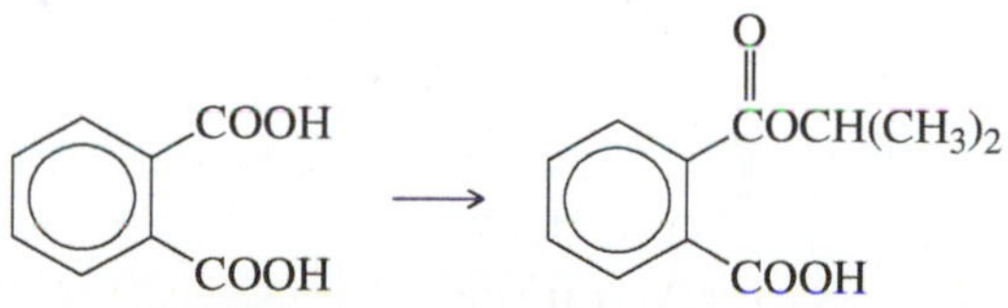

(continued)

(e) CHO, CH_2OH ⟶ (lactone)

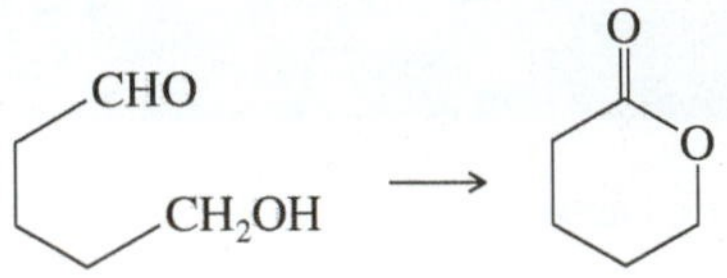

(f) CHO, COOH ⟶ (lactone)

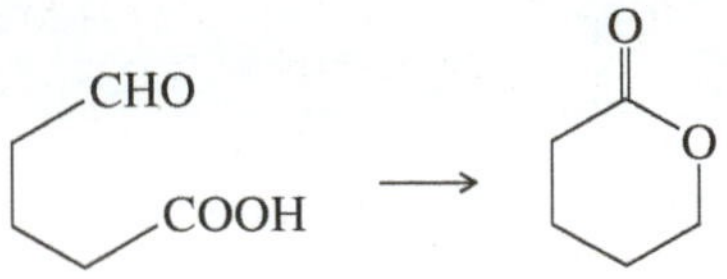

*(g) O= (cyclohexane) $COOCH_3$, $COOCH_3$ ⟶ O= (cyclohexane) CH_2OCCH_3 (O), CH_2OCCH_3 (O)

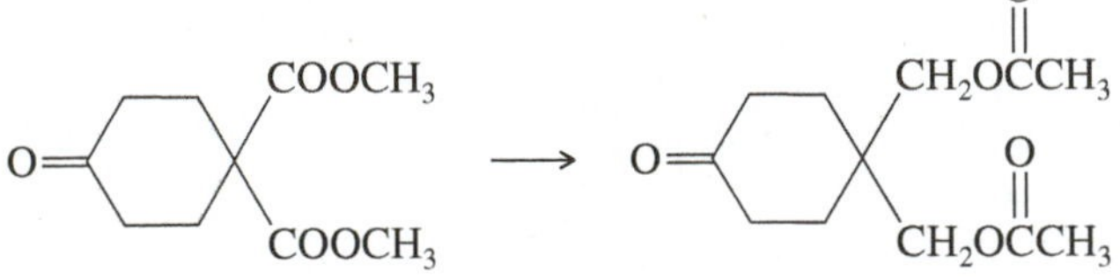

*(h) (cyclopentene) ⟶ (2-oxocyclopentane) COOH

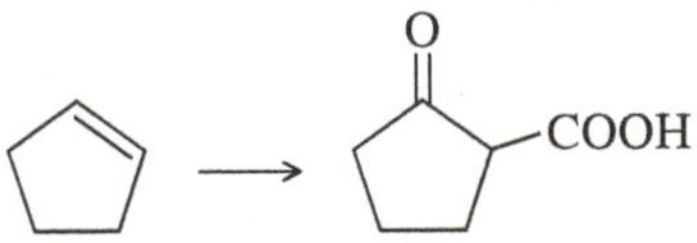

21-49 Propose mechanisms for the following reactions.

(a) $Ph-C(=O)-Cl$ + $(CH_3)_2CHOH$ ⟶ $Ph-C(=O)-OCH(CH_3)_2$

(b) $Ph-C(=O)-OCH_3$ $\xrightarrow[H_2O]{NaOH}$ $Ph-C(=O)-O^-$ + CH_3OH

(c) $Ph-C(=O)-OCH_2CH_3$ $\xrightarrow[H_2O]{H^+}$ $Ph-C(=O)-OH$ + CH_3CH_2OH

(d) (bicyclic lactone) $\xrightarrow[EtOH]{EtO^-}$ HO (cyclopentane) COOEt

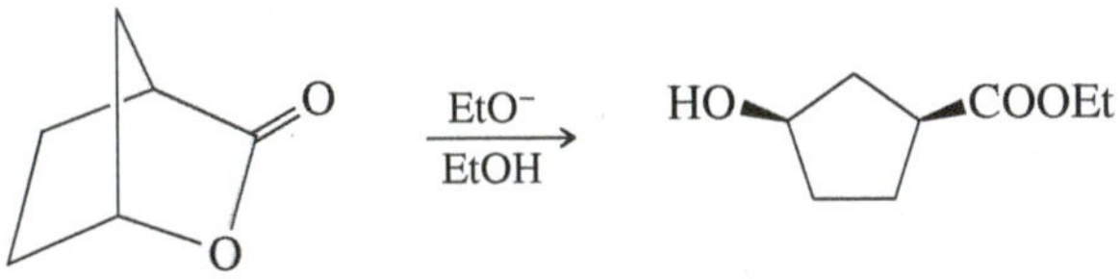

(e) H_2N ... H, O $\xrightarrow{H^+}$ (cyclic imine, N)

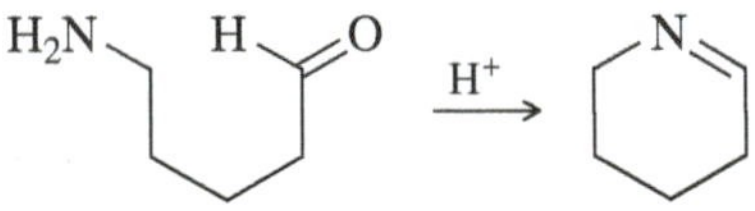

(f) H_2N ... OCH_3, O ⟶ (lactam, N–H, O)

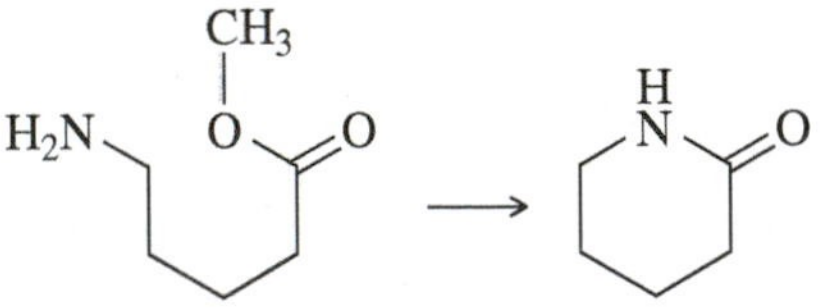

(g) $CH_3-\overset{*}{C}H(OH)-CH_2CH_3$ (*R*)-butan-2-ol $\xrightarrow[\text{(acetic anhydride)}]{Ac_2O}$ $CH_3-\overset{*}{C}H(OAc)-CH_2CH_3$ 2-butyl acetate

Does this reaction proceed with retention, inversion, or racemization of the asymmetric carbon atom?

21-50 Predict the products of the following reactions.

(a) (benzene)–C(=O)–Cl + (cyclohexane)–OH ⟶

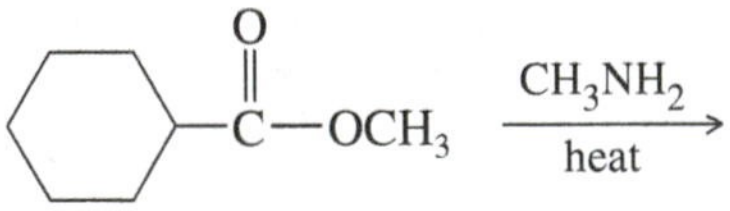

(b) (cyclohexane)$-C(=O)-OCH_3$ $\xrightarrow[heat]{CH_3NH_2}$

(c) $Ph-C(=O)-Cl$ + (pyrrolidine) N–H ⟶

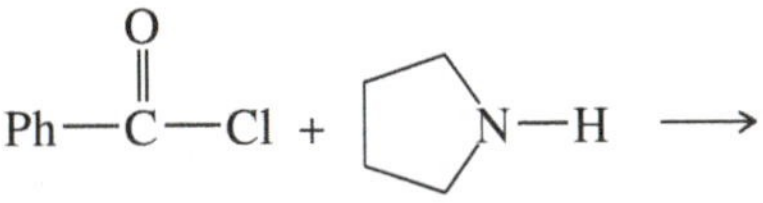

(d) (cyclic anhydride, O, O, O) + (cyclohexane)–NH_2 ⟶

(e) $Ph-C(=O)-OCH_2CH_3$ $\xrightarrow[(2)\ H_2O]{(1)\ LiAlH_4}$

(f) (lactam, O, N–H) $\xrightarrow[(2)\ H_2O]{(1)\ LiAlH_4}$

(g) (seven-membered lactone, O, O) $\xrightarrow[CH_3OH]{^-OCH_3}$

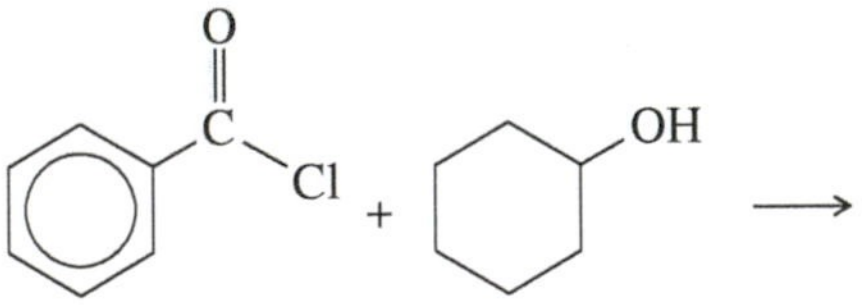

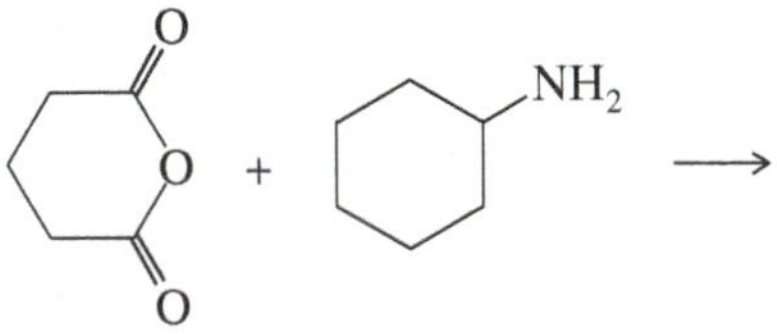

(h) (six-membered lactone, O, O) $\xrightarrow[(2)\ H_3O^+]{(1)\ \text{excess } PhMgBr}$

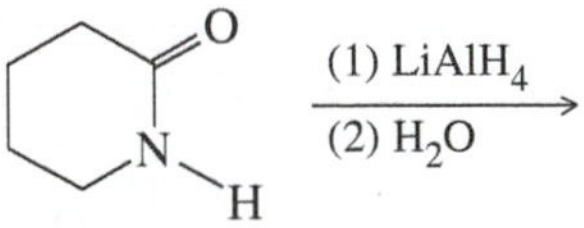

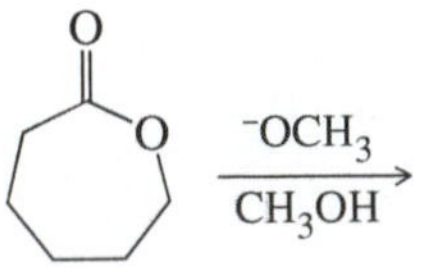

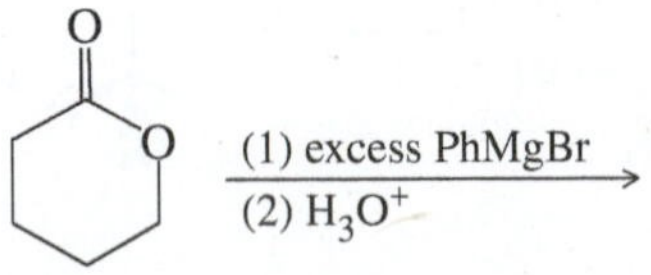

(i) cyclopentyl–C≡N $\xrightarrow[(2)\ H_3O^+]{(1)\ CH_3MgI}$

(j) caprolactam (cyclic amide, N–H) $\xrightarrow[H_2O]{NaOH}$

(k) $PhCH_2-CH(CH_3)-CH_2-C(=O)-NH_2$ $\xrightarrow[2.\ H_2O]{1.\ excess\ LiAlD_4}$

(l) cyclic ester (lactone) + $HOCH_2CH_2OH$ $\xrightarrow{H^+}$

(m) cyclic ester (lactone) $\xrightarrow[2.\ H_2O]{1.\ excess\ CH_3MgBr}$

(n) cyclic ester (lactone) $\xrightarrow[2.\ H_2O]{1.\ excess\ LiAlD_4}$

21-51 Phosgene is the acid chloride of carbonic acid. Although phosgene was used as a war gas in World War I, it is now used as a reagent for the synthesis of many useful products. Phosgene reacts like other acid chlorides, but it can react twice.

$$\left[HO-C(=O)-OH\right] \qquad Cl-C(=O)-Cl \xrightarrow{2\ Nuc:^-} Nuc-C(=O)-Nuc + 2\ Cl^-$$

carbonic acid — phosgene

(a) Predict the products formed when phosgene reacts with excess propan-2-ol.
(b) Predict the products formed when phosgene reacts with 1 equivalent of methanol, followed by 1 equivalent of aniline.
(c) *tert*-Butyloxycarbonyl chloride is an important reagent for the synthesis of peptides and proteins (Chapter 24). Show how you would use phosgene to synthesize *tert*-butyloxycarbonyl chloride.

$$CH_3-C(CH_3)_2-O-C(=O)-Cl$$

tert-butyloxycarbonyl chloride

21-52 An ether extraction of nutmeg gives large quantities of *trimyristin*, a waxy crystalline solid of melting point 57 °C. The IR spectrum of trimyristin shows a very strong absorption at 1733 cm^{-1}. Basic hydrolysis of trimyristin gives 1 equivalent of glycerol and 3 equivalents of myristic acid (tetradecanoic acid).
(a) Draw the structure of trimyristin.
(b) Predict the products formed when trimyristin is treated with lithium aluminum hydride, followed by aqueous hydrolysis of the aluminum salts.

21-53 Two widely used pain relievers are aspirin and acetaminophen. Show how you would synthesize these drugs from phenol.

aspirin: benzene ring with $O-C(=O)-CH_3$ and ortho COOH

acetaminophen: HO–benzene ring (para) with $NH-C(=O)-CH_3$

21-54 Show how you would accomplish the following syntheses. Some of these conversions may require more than one step.
(a) isopentyl alcohol → isopentyl acetate (banana oil)
(b) 3-ethylpentanoic acid → 3-ethylpentanenitrile
(c) isobutylamine → *N*-isobutylformamide
(d) ethyl acetate → 3-methylpentan-3-ol
(e) cyclohexylamine → *N*-cyclohexylacetamide
(f) bromocyclohexane → dicyclohexylmethanol
(g) dimethyl oxalate → piperazine-2,3-dione (six-membered ring with two N–H and two C=O)
(h) cyclopentyl–CH_2OH → cyclopentyl–CN

21-55 Grignard reagents add to carbonate esters as they add to other esters.

(a) Predict the major product of the following reaction.

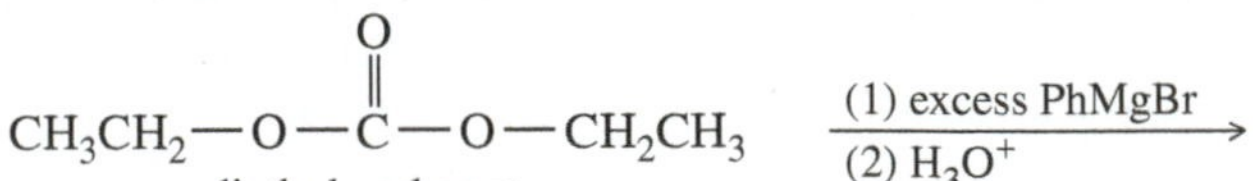

$$CH_3CH_2{-}O{-}\overset{\overset{\Large O}{\|}}{C}{-}O{-}CH_2CH_3 \xrightarrow[(2)\ H_3O^+]{(1)\ \text{excess PhMgBr}}$$

diethyl carbonate

(b) Show how you would synthesize 3-ethylpentan-3-ol using diethyl carbonate and ethyl bromide as your only organic reagents.

*(c) Diethyl carbonate is a liquid reagent that is easy to handle. In contrast, phosgene is a highly toxic and corrosive gas. Show how you might use diethyl carbonate instead of phosgene to make Lexan®. Also show how you might use diethyl carbonate instead of methyl isocyanate to make Sevin® insecticide.

***21-56** One mole of acetyl chloride is added to a liter of triethylamine, resulting in a vigorous exothermic reaction. Once the reaction mixture has cooled, 1 mole of ethanol is added. Another vigorous exothermic reaction results. The mixture is analyzed and found to contain triethylamine, ethyl acetate, and triethylammonium chloride. Propose mechanisms for the two exothermic reactions.

21-57 Show how you would accomplish the following multistep syntheses, using the indicated starting material and any necessary reagents.

(a) hept-6-en-1-ol ⟶ ε-caprolactone

(b) methoxybenzene ⟶ *p*-methoxybenzamide

(c) $C_6H_5CH_2Br \longrightarrow C_6H_5CH_2CH_2NH_2$

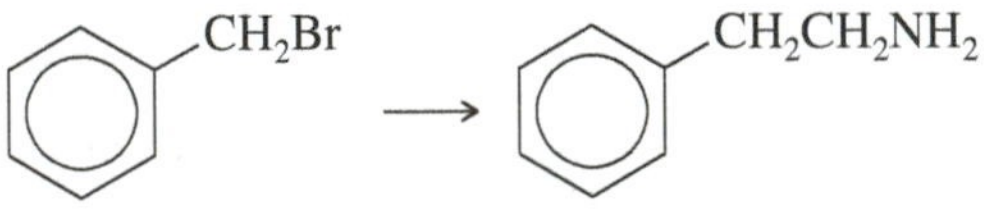

(d) gallic acid (3,4,5-trihydroxybenzoic acid: COOH; HO, OH, OH) ⟶ mescaline ($CH_2CH_2NH_2$; CH_3O, OCH_3, OCH_3)

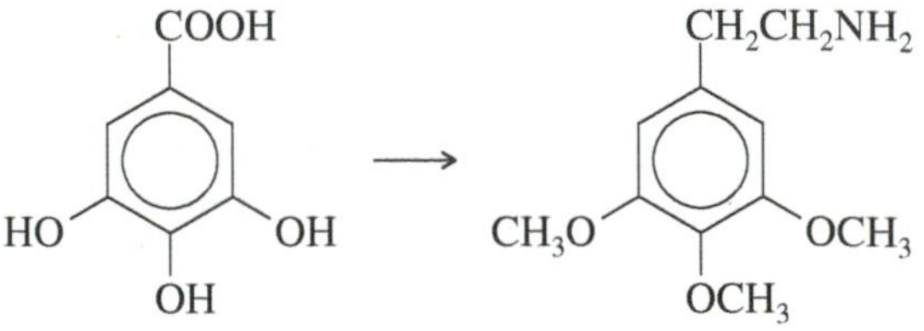

gallic acid   mescaline

***21-58** Methyl *p*-nitrobenzoate has been found to undergo saponification faster than methyl benzoate.

(a) Consider the mechanism of saponification, and explain the reasons for this rate enhancement.

(b) Would you expect methyl *p*-methoxybenzoate to undergo saponification faster or slower than methyl benzoate?

21-59 In each part, rank the compounds in order of increasing rate of nucleophilic attack at C=O by a strong nucleophile like methoxide.

(a)

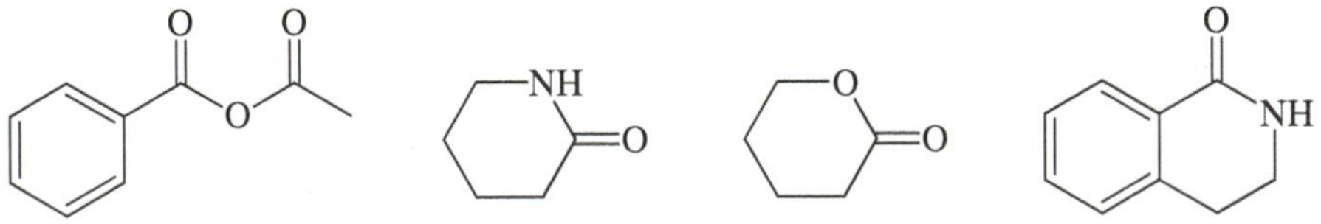

A (benzoic acetic anhydride)   B (piperidin-2-one, NH, =O)   C (tetrahydropyran-2-one, O, =O)   D (3,4-dihydroisoquinolin-1(2H)-one, O, NH)

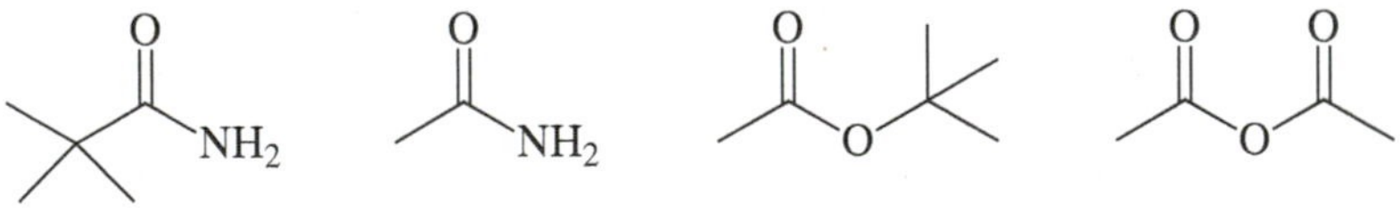

(b)

E ($(CH_3)_3CCONH_2$, NH_2)   F (CH_3CONH_2, NH_2)   G ($CH_3COOC(CH_3)_3$)   H ($CH_3COOCOCH_3$)

21-60 Explain this curious result. What does this reaction tell you about the relative reactivity of esters and ketones?

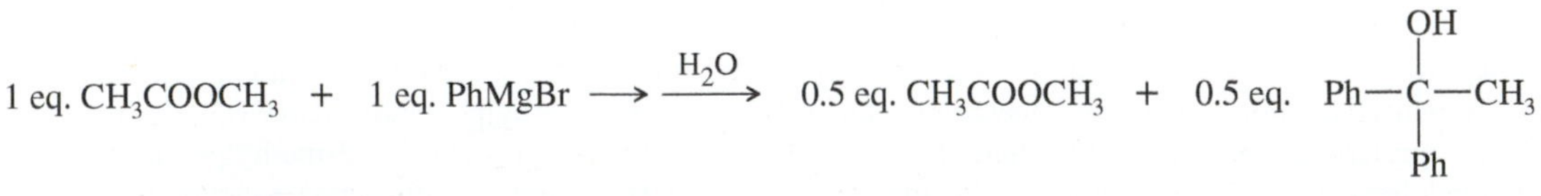

$$1\text{ eq. } CH_3COOCH_3 + 1\text{ eq. PhMgBr} \longrightarrow \xrightarrow{H_2O} 0.5\text{ eq. } CH_3COOCH_3 + 0.5\text{ eq. } Ph{-}\underset{Ph}{\overset{OH}{C}}{-}CH_3$$

21-61 A student has just added ammonia to hexanoic acid and has begun to heat the mixture when he is called away to the telephone. After a long telephone conversation, he returns to find that the mixture has overheated and turned black. He distills the volatile components and recrystallizes the solid residue. Among the components he isolates are compounds A (a liquid; molecular formula $C_6H_{11}N$) and B (a solid; molecular formula $C_6H_{13}NO$). The infrared spectrum of A shows a strong, sharp absorption at 2247 cm^{-1}. The infrared spectrum of B shows absorptions at 3390, 3200, and 1665 cm^{-1}. Determine the structures of compounds A and B.

21-62 In Section 21-16, we saw that Sevin insecticide is made by the reaction of 1-naphthol with methyl isocyanate. A Union Carbide plant in Bhopal, India, once used this process to make Sevin for use as an agricultural insecticide. On December 3, 1984, either by accident or by sabotage, a valve was opened that admitted water to a large tank of methyl isocyanate. The pressure and temperature within the tank rose dramatically, and pressure-relief valves opened to keep the tank from bursting. A large quantity of methyl isocyanate rushed out through the pressure-relief valves, and the vapors flowed with the breeze into populated areas, killing about 2500 people and injuring many more.

(a) Write an equation for the reaction that took place in the tank. Explain why the pressure and temperature rose dramatically.

(b) Propose a mechanism for the reaction you wrote in part (a).

(c) Propose an alternative synthesis of Sevin. Unfortunately, the most common alternative synthesis uses phosgene, a gas that is even more toxic than methyl isocyanate.

21-63 The structures of four useful polymers are shown, together with some of their best-known products. In each case,

(i) determine the kind of polymer (polyamide, polyester, etc.).

(ii) draw the structures of the monomers that would be released by complete hydrolysis.

(iii) suggest what monomers or stable derivatives of the monomers might be used to make these polymers.

(a) repeating unit

soft, sheer fabrics; synthetic silk

(b)

climbing ropes, violin strings

(c)

crash helmets, bulletproof "glass"

(d)

high-strength fabrics; bulletproof vests

21-64 A chemist was called to an abandoned aspirin factory to determine the contents of a badly corroded vat. Knowing that two salvage workers had become ill from breathing the fumes, she put on her breathing apparatus as soon as she noticed an overpowering odor like that of vinegar but much more pungent. She entered the building and took a sample of the contents of the vat. The mass spectrum showed a molecular weight of 102, and the NMR spectrum showed only a singlet at δ 2.15. The IR spectrum, which appears here, left no doubt about the identity of the compound. Identify the compound, and suggest a method for its safe disposal.

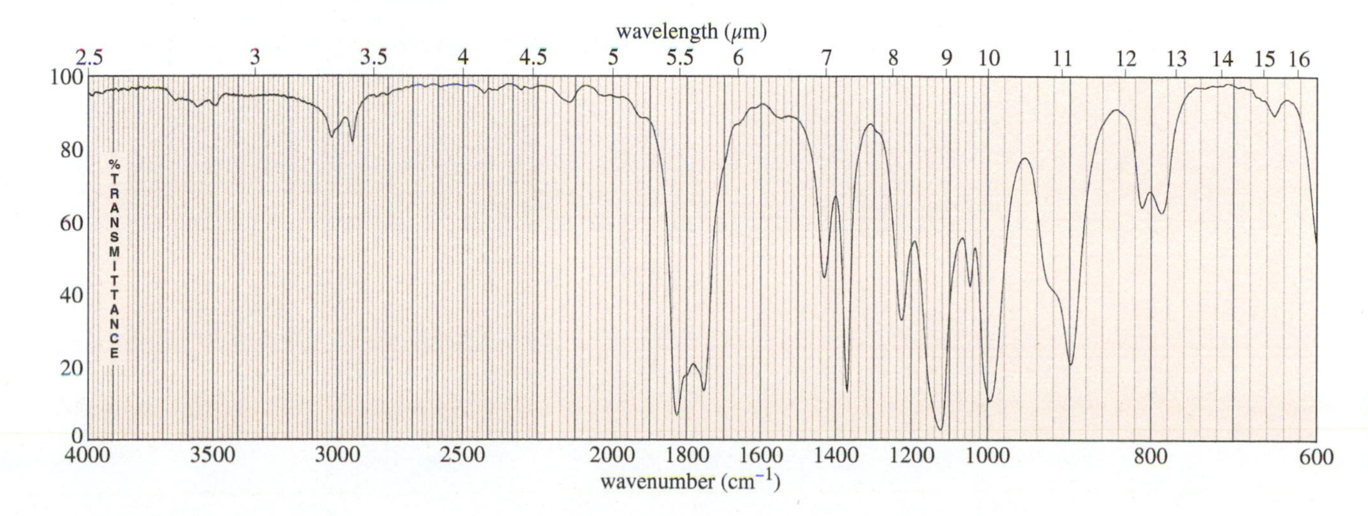

***21-65** The mass spectrum of ethyl butyrate shows two even-numbered fragments: at m/z 60 and m/z 88. These ions can be explained by a “double McLafferty” rearrangement: an initial McLafferty followed by a second rearrangement on the ion from the first. Show the cyclic transition states implicated in the two rearrangements, as well as the structures of the expected ions of masses 88 and 60. (Hint: begin with the ester side.)

ethyl butyrate, mw 116

21-66 An unknown compound gives a mass spectrum with a weak molecular ion at m/z 113 and a prominent ion at m/z 68. Its NMR and IR spectra are shown here. Determine the structure, and show how it is consistent with the observed absorptions. Propose a favorable fragmentation to explain the prominent MS peak at m/z 68.

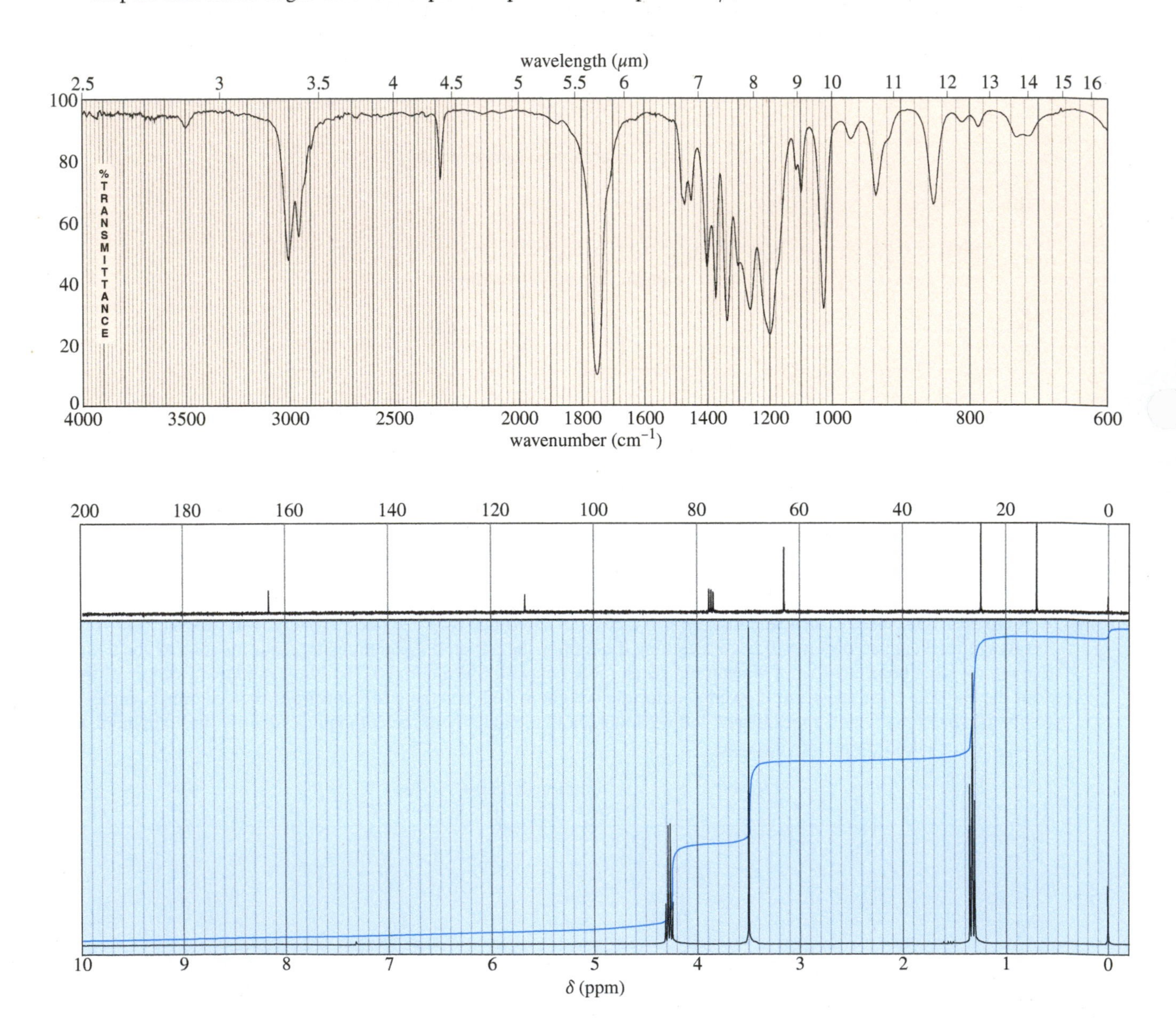

21-67 An unknown compound gives the NMR, IR, and mass spectra shown next. Propose a structure, and show how it is consistent with the observed absorptions. Show fragmentations that account for the prominent ion at m/z 69 and the smaller peak at m/z 99.

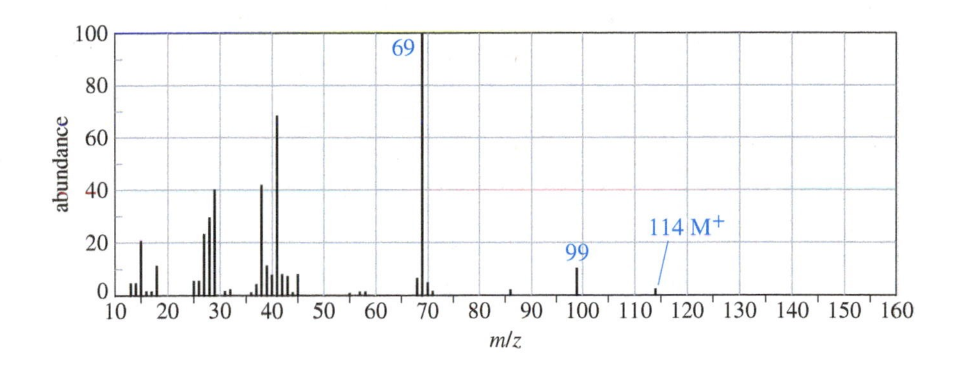

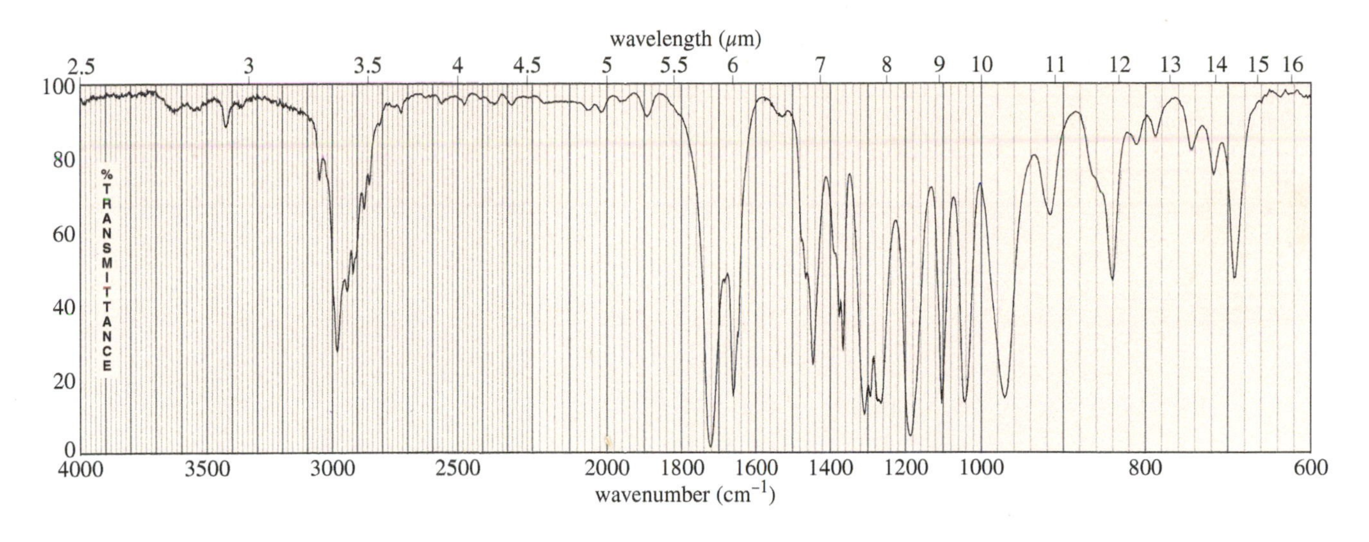

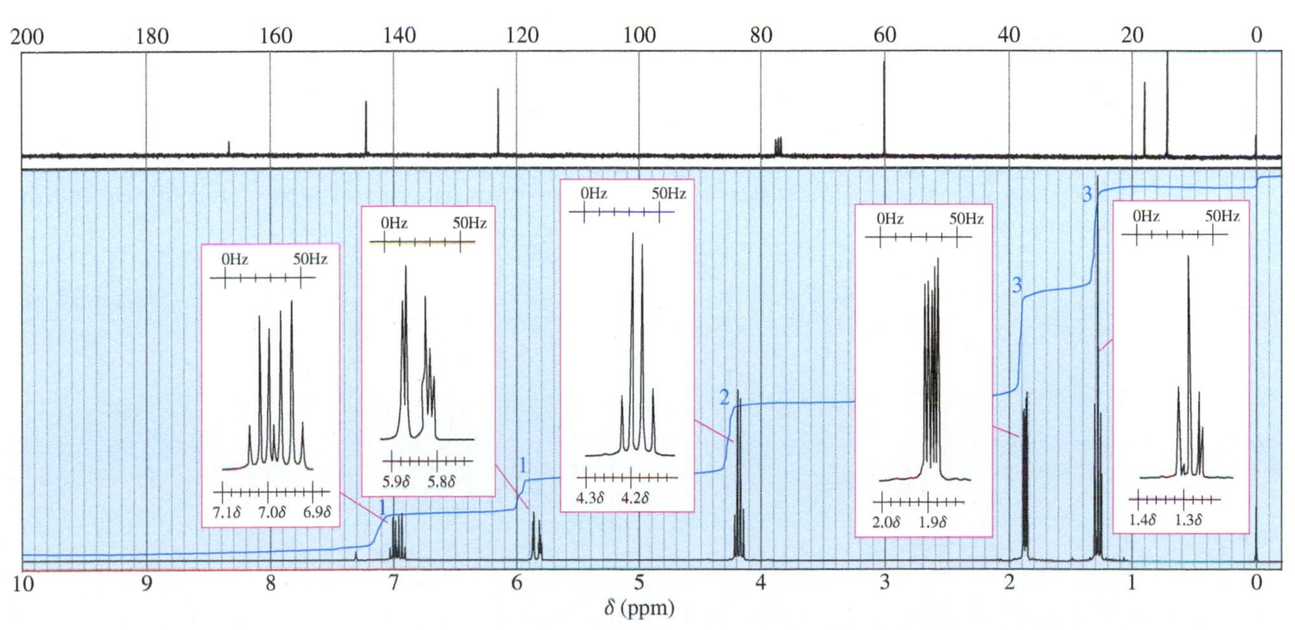

***21-68** The IR spectrum, ^{13}C NMR spectrum, and ^{1}H NMR spectrum of an unknown compound ($C_6H_8O_3$) appear next. Determine the structure, and show how it is consistent with the spectra.

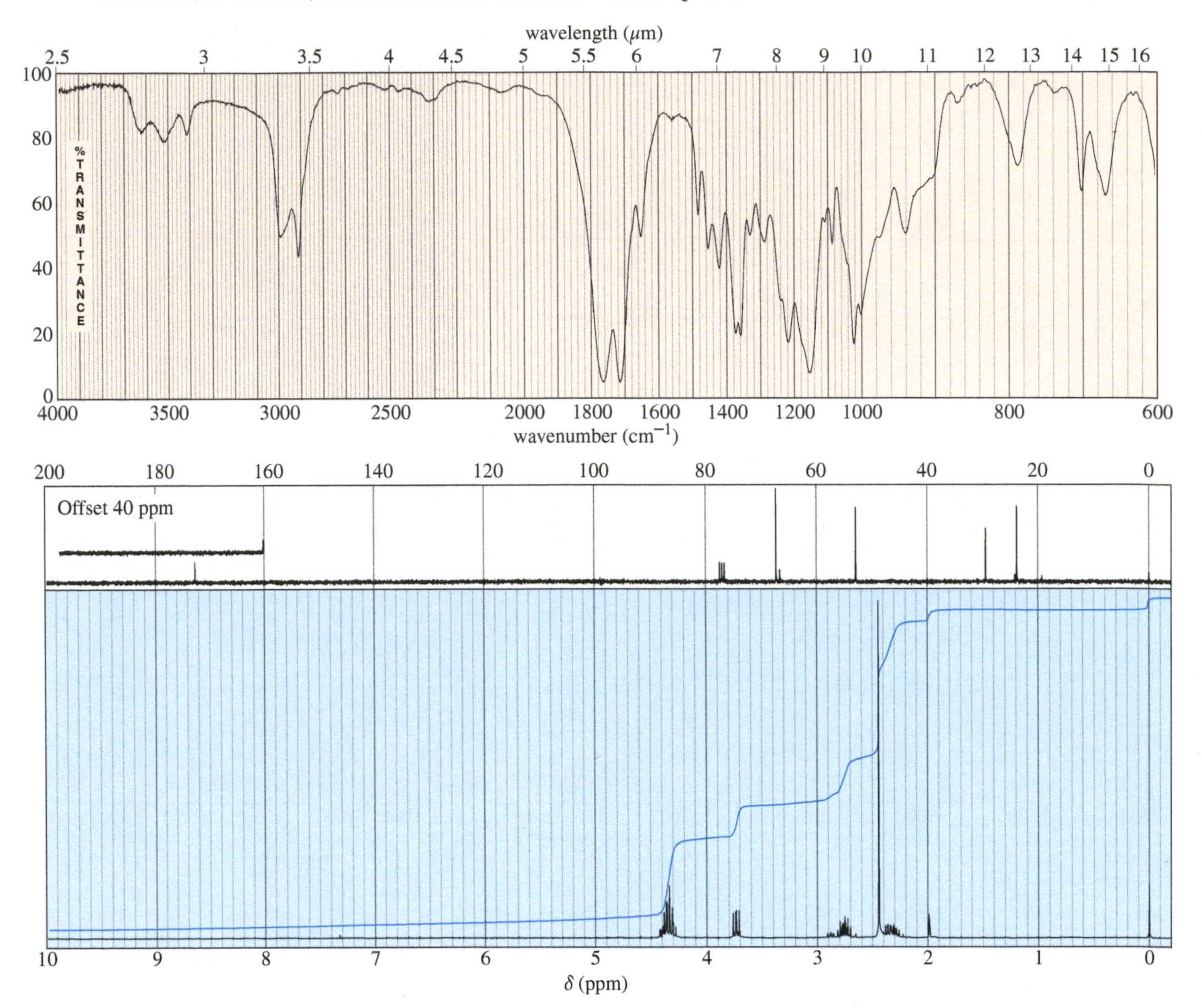

***21-69** An unknown compound of molecular formula C_5H_9NO gives the IR and NMR spectra shown here. The broad NMR peak at δ 7.55 disappears when the sample is shaken with D_2O. Propose a structure, and show how it is consistent with the absorptions in the spectra.

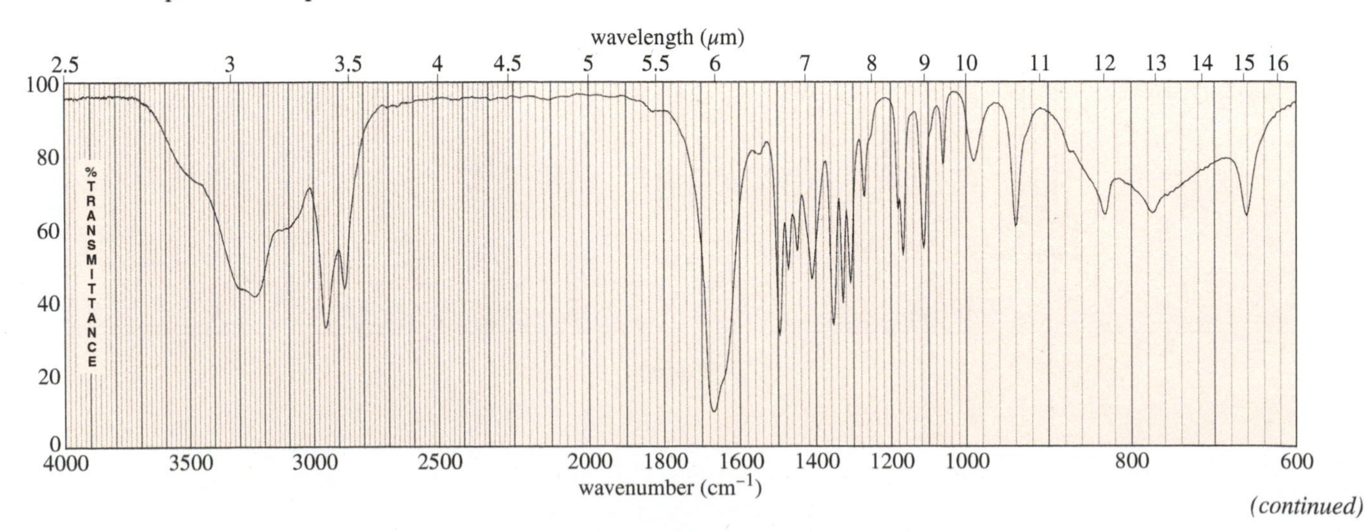

(continued)

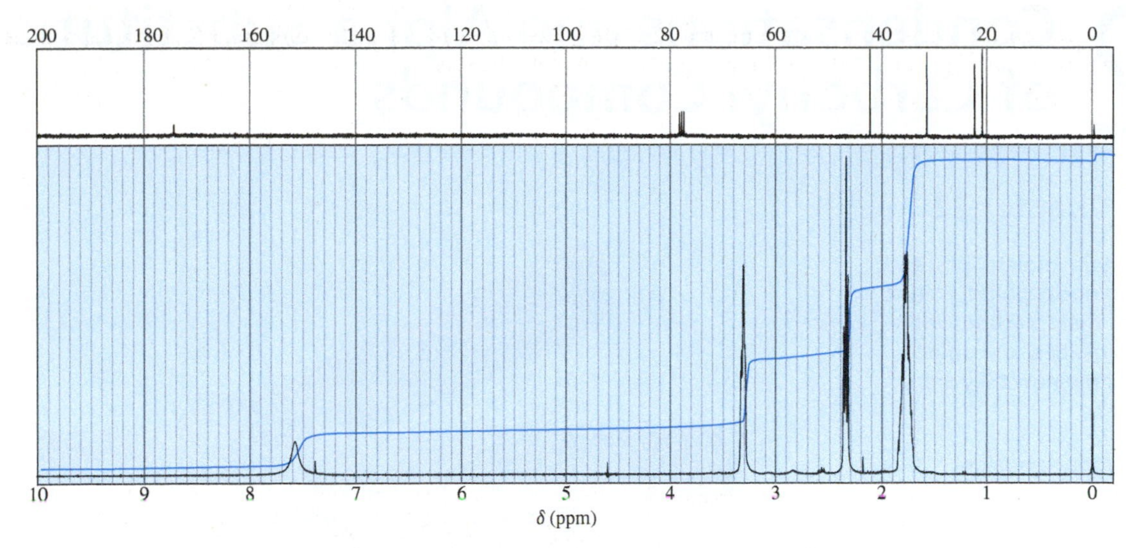

21-70 (A true story.) A chemist wanted to make the ethyl ester of 2-formylbenzoic acid, but the spectra of the product did not meet her expectations. The IR of product **Z** had an ester carbonyl, but no aldehyde absorption nor any absorption in the OH region. The HNMR data are given. What is the structure of **Z**? Propose a mechanism for its formation.

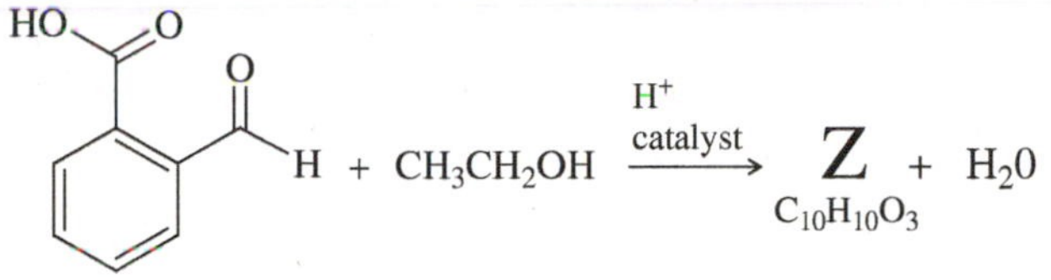

2-formylbenzoic acid + CH_3CH_2OH $\xrightarrow{H^+ \text{ catalyst}}$ **Z** + H_2O

2-formylbenzoic acid
$C_8H_6O_3$

HNMR chemical shifts:
δ 8.0 to 8.4 (4H, complex)
δ 10.4 (1H, sharp singlet)
δ 13.3 (1H, broad singlet)

Z
$C_{10}H_{10}O_3$

HNMR chemical shifts:
δ 7.5 to 7.8 (4H, complex)
δ 6.6 (1H, sharp singlet)
δ 3.9 (2H, quartet)
δ 1.1 (3H, triplet)

***21-71** Macrolide antibiotics, including erythromycin and azithromycin (Zithromax®), contain a large ring lactone. One of the largest ever reported is Gargantulide A, the structure of which was determined by the research group of Prof. William Gerwick of the Scripps Institution of Oceanography (*Organic Letters*, **2015,** *17*, 1377–1380). Isolated from a *Streptomyces* bacterium, it kills pathogenic bacteria like MRSA and *Clostridium difficile*, but it proved too toxic to the test animals to continue further testing.

(a) Identify the lactone that makes this a macrolide structure.

(b) How many rings does this structure contain? How many atoms are in the largest ring?

(c) Many complex NMR experiments were used to determine the stereochemistry of most, but not all, of the chiral centers. How many chiral centers does the structure contain?

Gargantulide A

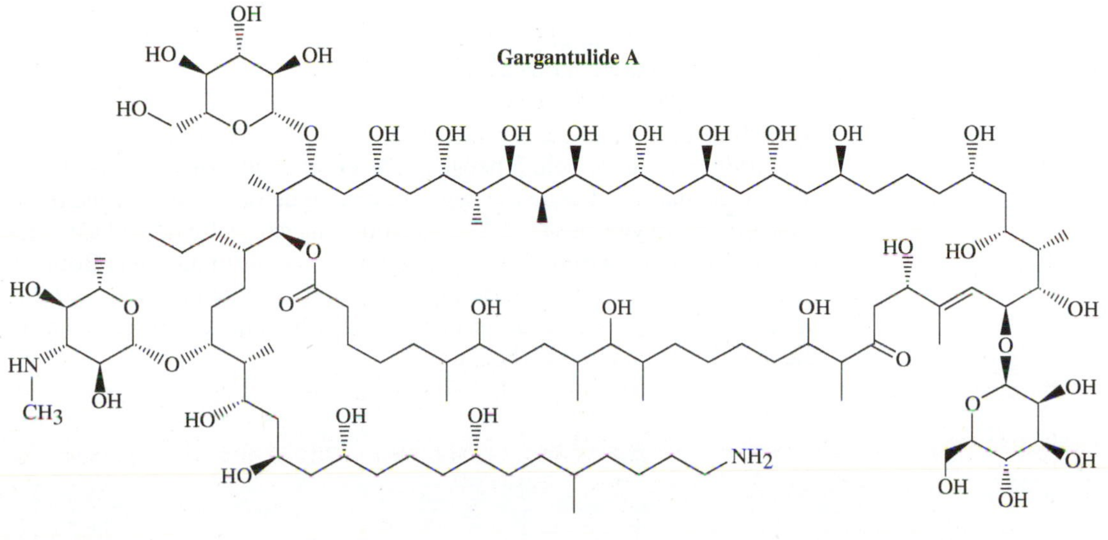

22 Condensations and Alpha Substitutions of Carbonyl Compounds

Goals for Chapter 22

1 Show how enols, enolate ions, and enamines act as nucleophiles. Predict the products of their reactions with halogens, alkyl halides, and other electrophiles. Show how they are useful in synthesis.

2 Predict the products of aldol, Claisen, and Michael condensations, and draw mechanisms to show how they are formed. Show how these condensations are useful in synthesis.

3 Recognize the products of variations of the standard condensations and propose mechanisms for them. Show how crossed Claisen condensations, malonic ester syntheses, acetoacetic ester syntheses, and Robinson annulations can synthesize a wide variety of ketone, ester, and acid derivatives.

▶ Cinnamon, obtained from the dried inner bark of *Cinnamomum verum* trees, is one of the oldest spices known. In ancient times, cinnamon was nearly as valuable as gold, and was a common gift for royalty. The principal flavor ingredient in cinnamon is cinnamaldehyde, which is easily synthesized using the aldol condensation of benzaldehyde with acetaldehyde (Section 22-9). The availability of cheap synthetic cinnamon flavoring promoted the production of cinnamon candies and other flavored products. Most chefs still prefer to use cinnamon sticks or powder made from the tree bark.

22-1 Introduction

Up to now, we have studied two of the main types of carbonyl reactions: nucleophilic addition and nucleophilic acyl substitution. In these reactions, the carbonyl group serves as an *electrophile* by accepting electrons from an attacking nucleophile. In this chapter, we consider two more types of reactions: substitution at the carbon atom next to the carbonyl group (called alpha substitution) and carbonyl condensations. Carbonyl condensations are among the most common biological methods for building up and breaking down large molecules.

Alpha (α) substitutions involve the replacement of a hydrogen atom at the **α carbon atom** (the carbon next to the carbonyl group) by some other group. The α hydrogen is more acidic because the **enolate ion** that results from its removal is resonance-stabilized, with the negative charge delocalized over the α carbon atom and the carbonyl oxygen atom. Alpha substitutions generally take place when the carbonyl compound is converted to its enolate ion or its enol tautomer. Both of these species have lost a hydrogen atom at the alpha position, and both are *nucleophilic*. Nucleophilic attack on an electrophile forms a product in which the electrophile has replaced one of the hydrogens on the α carbon atom.

MECHANISM 22-1 Alpha Substitution

Step 1: Deprotonation of α carbon to form an enolate. *Step 2:* Nucleophilic attack on an electrophile.

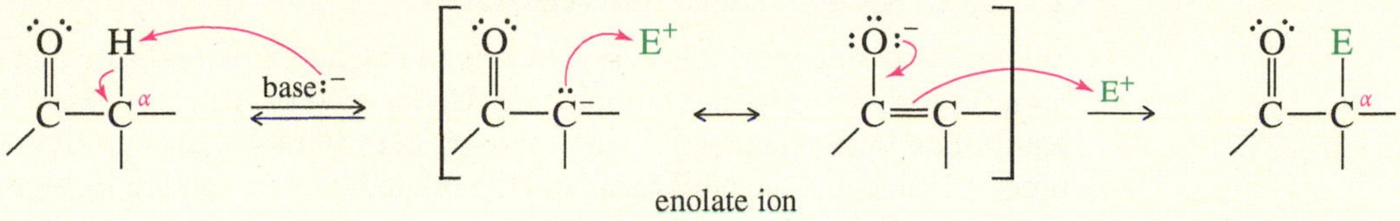

base:$^-$ E$^+$ enolate ion E$^+$ E α

Carbonyl **condensations** are alpha substitutions where the electrophile is another carbonyl compound. If the electrophile is a ketone or an aldehyde, then the enolate ion adds to that carbonyl group in a nucleophilic addition. First, the enolate ion attacks the carbonyl group to form an alkoxide. Protonation of the alkoxide gives the addition product.

PROBLEM-SOLVING HINT

In drawing mechanisms, you can show either resonance form of an enolate attacking the electrophile. Mechanism 22-1 shows both options.

MECHANISM 22-2 Addition of an Enolate to Ketones and Aldehydes (a Condensation)

Step 1: The enolate adds to the carbonyl group. *Step 2:* Protonation of the alkoxide.

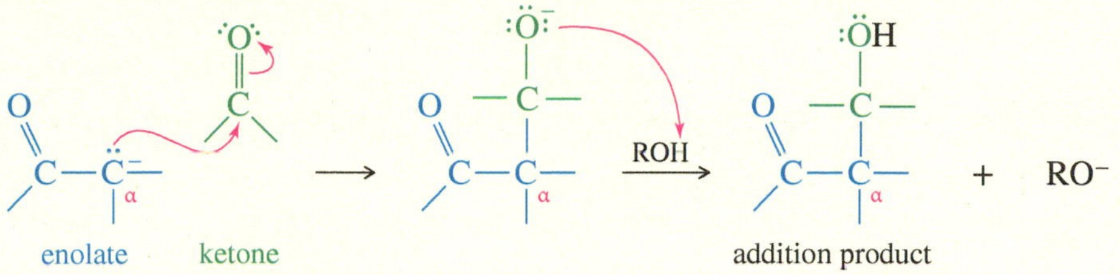

enolate ketone ROH addition product + RO$^-$

If the electrophile is an ester, then the ester undergoes a nucleophilic acyl substitution with the enolate ion serving as the nucleophile. First, the enolate adds to the ester to form a tetrahedral intermediate. Elimination of the leaving group (alkoxide) gives the substitution product.

MECHANISM 22-3 Substitution of an Enolate on an Ester (a Condensation)

Step 1: Addition of the enolate. *Step 2:* Elimination of the alkoxide.

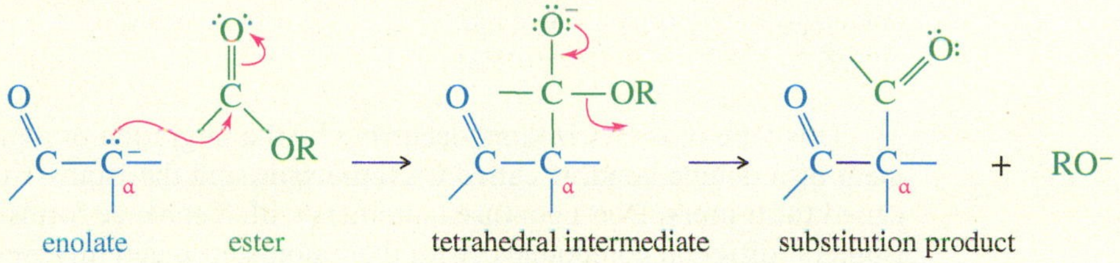

enolate ester tetrahedral intermediate substitution product + RO$^-$

Alpha substitutions and condensations of carbonyl compounds are some of the most common methods for forming carbon–carbon bonds. These types of reactions are common in biochemical pathways, particularly in the biosynthesis and metabolism of carbohydrates and fats. A wide variety of compounds can participate as nucleophiles or electrophiles (or both) in these reactions, and many useful products can be made. We begin our study of these reactions by considering the structure and formation of enols and enolate ions.

22-2 Enols and Enolate Ions

We first encountered enols (vinyl alcohols) in Chapter 9, and saw that enols are a tautomeric form of carbonyl compounds. Under acid or base catalysis, these two forms equilibrate via the keto–enol tautomerism. In most cases, the carbonyl (keto) tautomer predominates, yet the enol form and the related enolate ion are important in reactions.

22-2A Keto–Enol Tautomerism

In the presence of strong bases, ketones and aldehydes act as weak proton acids. A proton on the α carbon atom is abstracted to form a resonance-stabilized **enolate ion** with the negative charge spread over a carbon atom and an oxygen atom. Reprotonation can occur either on the α carbon (returning to the **keto** form) or on the oxygen atom, giving a vinyl alcohol, the **enol** form.

MECHANISM 22-4 Base-Catalyzed Keto–Enol Tautomerism

Step 1: Deprotonation of α C.

Step 2: Reprotonation on O.

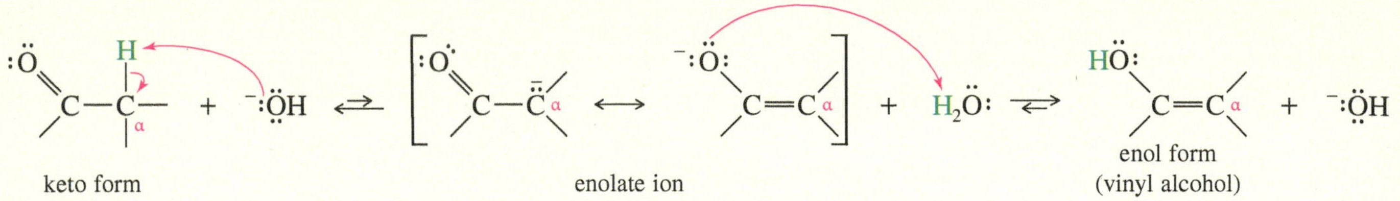

keto form + $^-$:ÖH ⇌ [enolate ion] + H_2Ö: ⇌ enol form (vinyl alcohol) + $^-$:ÖH

keto form

enolate ion

enol form
(vinyl alcohol)

In this way, base catalyzes an equilibrium between the isomeric keto and enol forms of a carbonyl compound. For simple ketones and aldehydes, the keto form predominates. Therefore, a vinyl alcohol (an enol) is best described as an alternative isomeric form of a ketone or aldehyde. In Section 9-9F, we saw that an enol intermediate, formed by hydrolysis of an alkyne, quickly isomerizes to its keto form.

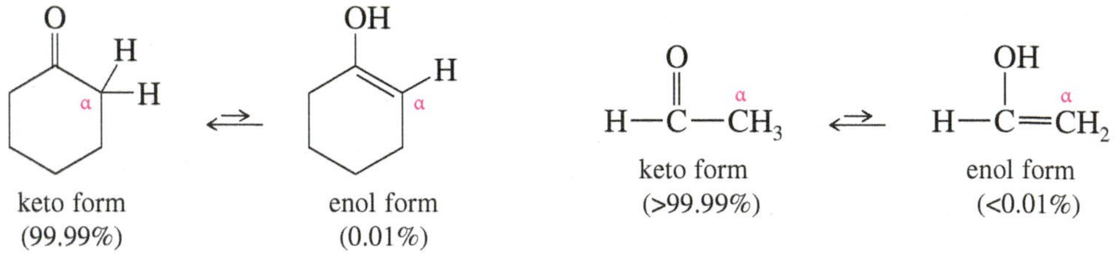

keto form (99.99%) ⇌ enol form (0.01%)

$H{-}C(=O){-}CH_3$ ⇌ $H{-}C(OH){=}CH_2$

keto form (>99.99%) ⇌ enol form (<0.01%)

This type of isomerization, occurring by the migration of a proton and the movement of a double bond, is called **tautomerism,** and the isomers that interconvert are called **tautomers.** Don't confuse tautomers with resonance forms. Tautomers are true isomers (different compounds) with their atoms arranged differently. Under the right circumstances, with no catalyst present, either individual tautomeric form may be isolated. Resonance forms are different representations of the *same* structure, with all the atoms in the same places, showing how the electrons are delocalized.

Keto–enol tautomerism is also catalyzed by acid. In acid, a proton is moved from the α carbon to oxygen by first protonating oxygen and then removing a proton from carbon.

MECHANISM 22-5 Acid-Catalyzed Keto–Enol Tautomerism

Step 1: An acid protonates the carbonyl oxygen. ***Step 2:*** Deprotonation on the α carbon gives the enol form.

keto form + H_3O^+ ⇌ [protonated carbonyl] ⇌ (with $H_2\ddot{O}$:) enol form + H_3O^+

Compare the base-catalyzed and acid-catalyzed mechanisms shown for keto–enol tautomerism. In base, the proton is removed from the α carbon, then replaced on oxygen. In acid, oxygen is protonated first, then the α carbon is deprotonated. Most proton-transfer mechanisms work this way. In base, the proton is removed from the old location, then replaced at the new location. In acid, protonation occurs at the new location, followed by deprotonation at the old location.

In addition to its mechanistic importance, keto–enol tautomerism affects the stereochemistry of ketones and aldehydes. A hydrogen atom on an α carbon may be lost and regained through keto–enol tautomerism; such a hydrogen is said to be **enolizable.** If an asymmetric carbon atom has an enolizable hydrogen atom, a trace of acid or base allows that carbon to invert its configuration, with the enol serving as the intermediate. A racemic mixture (or an equilibrium mixture of diastereomers) is the result.

enolizable hydrogens; α carbons

(*R*) configuration (originally 100%) ⇌ (H^+ or ^-OH) enol (achiral) ⇌ (*S*) configuration (50%) + (*R*) configuration (50%)

PROBLEM 22-1

Phenylacetone can form two different enols.
(a) Show the structures of these enols.
(b) Predict which enol will be present in the larger concentration at equilibrium.
(c) Propose mechanisms for the formation of the two enols in acid and in base.

PROBLEM 22-2

(a) Show each step in the mechanism of the acid-catalyzed interconversion of (*R*)- and (*S*)-3-methylpentan-2-one.
(b) When *cis*-2,4-dimethylcyclohexanone is dissolved in aqueous ethanol containing a trace of NaOH, a mixture of cis and trans isomers results. Propose a mechanism for this isomerization.

PROBLEM-SOLVING HINT

In acid, proton transfers usually occur by adding a proton in the new position, then deprotonating the old position.

In base, proton transfers usually occur by deprotonating the old position, then reprotonating at the new position.

22-2B Formation and Stability of Enolate Ions

A carbonyl group dramatically increases the acidity of the protons on the α carbon atom because deprotonation gives a resonance-stabilized enolate ion. Most of the enolate ion's negative charge resides on the electronegative oxygen atom. The pK_a for removal of an α proton from a typical ketone or aldehyde is about 20, showing that a typical ketone or aldehyde is much more acidic than an alkane or an alkene ($pK_a > 40$), or even an alkyne ($pK_a = 25$). Still, a ketone or aldehyde is less acidic than water ($pK_a = 15.7$) or an alcohol ($pK_a = 16$ to 18). When a simple ketone or aldehyde is treated with a hydroxide ion or an alkoxide ion, the equilibrium mixture contains only a small fraction of the deprotonated, enolate form.

R—C(=O)—CH(α)H—R′ + ⁻OR ⇌ [R—C(=O)—C:⁻(α)(H)—R′ (minor) ⟷ R—C(—O:⁻)=C(α)(H)—R′ (major)] + ROH

ketone or aldehyde $pK_a \approx 20$ — enolate ion — $pK_a = 16–18$

Example

cyclohexanone + ⁻:O—CH_2CH_3 ⇌ [cyclohexanone enolate resonance forms] + CH_3CH_2OH

cyclohexanone $pK_a = 19$ — ethoxide ion — cyclohexanone enolate — $pK_a = 15.9$

(equilibrium lies to the left)

Even though the equilibrium concentration of the enolate ion may be small, it serves as a useful, reactive nucleophile. When an enolate reacts with an electrophile (other than a proton), the enolate concentration decreases, and the equilibrium shifts to the right (Figure 22-1). Eventually, all the carbonyl compound reacts via a low concentration of the enolate ion.

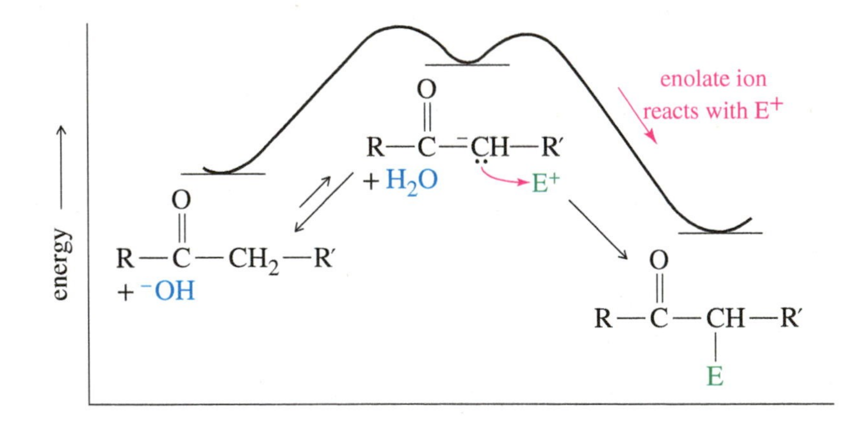

FIGURE 22-1
Reaction of the enolate ion with an electrophile removes it from the equilibrium, shifting the equilibrium to the right.

PROBLEM 22-3

Give the important resonance forms for the possible enolate ions of the following:

(a) acetone **(b)** cyclopentanone **(c)** pentane-2,4-dione

(d) [2-ethylcyclopentanone structure] **(e)** [2-acetyl-5-methylfuran structure] **(f)** [3-methylcyclohex-2-enone structure]

Sometimes this equilibrium mixture of enolate and base won't work, usually because the base (hydroxide or alkoxide) reacts with the electrophile faster than the enolate does. In these cases, we need a base that reacts completely to convert the carbonyl compound to its enolate before adding the electrophile. Although sodium hydroxide and alkoxides are not sufficiently basic, powerful bases are available to convert a carbonyl compound completely to its enolate. The most effective and useful base for this purpose is lithium diisopropylamide (LDA), the lithium salt of diisopropylamine. LDA is made by using an alkyllithium reagent to deprotonate diisopropylamine.

$$[(CH_3)_2CH]_2\ddot{N}{-}H + C_4H_9Li \longrightarrow C_4H_{10} + [(CH_3)_2CH]_2\ddot{N}{:}^- \; Li^+$$

diisopropylamine ($pK_a = 40$) + n-butyllithium ⟶ butane ($pK_a \approx 50$) + lithium diisopropylamide (LDA)

Diisopropylamine has a pK_a of about 40, showing that it is much *less* acidic than a typical ketone or aldehyde. LDA is about as basic as sodium amide ($NaNH_2$), but much less nucleophilic because it is hindered by the two bulky isopropyl groups. LDA does not easily attack a carbon atom or add to a carbonyl group. Thus, it is a powerful base, but not a strong nucleophile. When LDA reacts with a ketone, it abstracts the α proton to form the lithium salt of the enolate. We will see that these lithium enolate salts can be useful in synthesis.

$$R{-}C({=}O){-}C_\alpha H{<} + (i\text{-}C_3H_7)_2N^- \; Li^+ \rightleftharpoons R{-}C(O^- Li^+){=}C_\alpha{<} + (i\text{-}C_3H_7)_2N{-}H$$

ketone ($pK_a \approx 20$) + LDA ⇌ lithium salt of enolate + diisopropylamine ($pK_a = 40$)

(equilibrium lies far to the right)

Example

$$\text{cyclohexanone} + (i\text{-}C_3H_7)_2N^- \; Li^+ \rightleftharpoons \text{lithium enolate} + (i\text{-}C_3H_7)_2N{-}H$$

cyclohexanone ($pK_a = 19$) + LDA ⇌ lithium enolate of cyclohexanone (100%) + ($pK_a = 40$)

EPM of lithium enolate of cyclohexanone

When using LDA to make an enolate, it is common to add the carbonyl compound slowly to a cold solution of LDA, so that the enolate is formed quickly and completely, minimizing the opportunity for side reactions.

22-3 Alkylation of Enolate Ions

We have seen many reactions where nucleophiles attack unhindered alkyl halides and tosylates by the S_N2 mechanism. An enolate ion can serve as the nucleophile, becoming alkylated in the process. Because the enolate has two nucleophilic sites (the oxygen and the α carbon), it can react at either of these sites. The reaction usually takes place primarily at the α carbon, forming a new C—C bond. In effect, this is a type of α substitution, with an alkyl group substituting for an α hydrogen.

$$\left[\begin{array}{c} -\overset{\overset{\displaystyle O}{\|}}{C}-\overset{\alpha}{C}{:}^- \\ \updownarrow \\ -\overset{\overset{\displaystyle :O:^-}{|}}{C}=\overset{\alpha}{C} \end{array}\right] + R{-}CH_2{-}X \longrightarrow -\overset{\overset{\displaystyle O}{\|}}{C}-\overset{\alpha}{C}-CH_2-R + X^-$$

C-alkylation product
(more common)

$$\longrightarrow -\overset{\overset{\displaystyle O-CH_2-R}{|}}{C}=\overset{\alpha}{C} + X^-$$

O-alkylation product
(less common)

Typical bases such as sodium hydroxide or an alkoxide ion cannot be used to form enolates for alkylation because at equilibrium a large quantity of the hydroxide or alkoxide base is still present. These strongly nucleophilic bases give side reactions with the alkyl halide or tosylate. Problem 22-4 shows an example of these side reactions. Lithium diisopropylamide (LDA) avoids these side reactions. Because it is a much stronger base, LDA converts the ketone entirely to its enolate. All the LDA is consumed in forming the enolate, leaving the enolate to react without interference from the LDA. Also, LDA is a very bulky base and thus a poor nucleophile, so it generally does not react with the alkyl halide or tosylate.

$$\underset{\text{enolizable ketone}}{R-\overset{\overset{\displaystyle O}{\|}}{C}-\overset{\alpha}{\overset{\overset{\displaystyle R'}{|}}{C}H}-R'} + \underset{\text{LDA}}{(i\text{-Pr})_2N^-\ Li^+} \longrightarrow \underset{\text{enolate}}{\left[Li^+\ R-\overset{\overset{\displaystyle :O:^-}{|}}{C}=\overset{\overset{\displaystyle R'}{|}}{C}-R' \leftrightarrow Li^+\ R-\overset{\overset{\displaystyle O}{\|}}{C}-\overset{\alpha}{\overset{\overset{\displaystyle R'}{|}}{\underset{\cdot\cdot}{C}}^-}-R'\right]} + \underset{\text{diisopropylamine}}{(i\text{-Pr})_2N-H}$$

$$\underset{\text{enolate}}{R-\overset{\overset{\displaystyle O}{\|}}{C}-\overset{\alpha}{\overset{\overset{\displaystyle R'}{|}}{\underset{\cdot\cdot}{C}}^-}-R'\ \ Li^+} + \underset{\text{unhindered halide}}{R''-CH_2-X} \longrightarrow \underset{\text{alkylated}}{R-\overset{\overset{\displaystyle O}{\|}}{C}-\overset{\alpha}{\overset{\overset{\displaystyle R'}{|}}{\underset{\underset{\displaystyle CH_2-R''}{|}}{C}}}-R'} + LiX$$

Example

$$Ph-\overset{\overset{\displaystyle O}{\|}}{C}-\overset{\alpha}{\overset{\overset{\displaystyle CH_3}{|}}{C}H}-CH_3 \xrightarrow[\text{(2) }Ph-CH_2-Br]{\text{(1) LDA}} Ph-\overset{\overset{\displaystyle O}{\|}}{C}-\overset{\alpha}{\overset{\overset{\displaystyle CH_3}{|}}{\underset{\underset{\displaystyle CH_2-Ph}{|}}{C}}}-CH_3$$

Direct alkylation of enolates (using LDA) gives the best yields when only one kind of α hydrogen can be replaced by an alkyl group. If there are two different kinds of α protons that may be abstracted to give enolates, mixtures of products alkylated at the different α carbons may result. Aldehydes are not suitable for direct alkylation because they undergo side reactions when treated with LDA.

PROBLEM 22-4

A student intends to carry out the following synthesis:

cyclohexanone $\xrightarrow[\text{(2) }PhCH_2Br]{\text{(1) base}}$ 2-benzylcyclohexanone (CH_2Ph)

He adds sodium ethoxide to cyclohexanone (in ethanol solution) to make the enolate ion; then he adds benzyl bromide to alkylate the enolate ion, and heats the solution for half an hour to drive the reaction to completion.

(a) Predict the products of this reaction sequence.
(b) Suggest how this student might synthesize the correct product.

PROBLEM 22-5

Predict the major products of the following reactions.

(a) acetone $\xrightarrow[\text{(2) } CH_2{=}CHCH_2Br]{\text{(1) LDA}}$

(b) 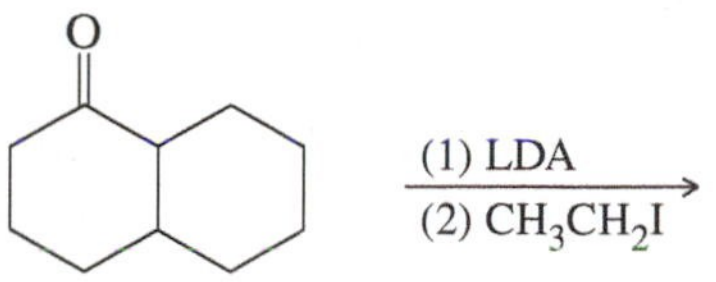 [1-decalone] $\xrightarrow[\text{(2) } CH_3CH_2I]{\text{(1) LDA}}$

(c) 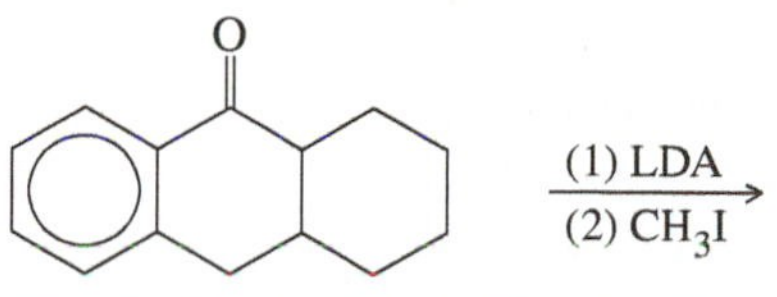 [tricyclic ketone] $\xrightarrow[\text{(2) } CH_3I]{\text{(1) LDA}}$

PROBLEM-SOLVING HINT

In drawing mechanisms, you can show either resonance form of an enolate attacking the electrophile. It is often more intuitive to show the carbanion form attacking.

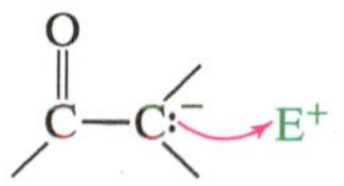

22-4 Formation and Alkylation of Enamines

A milder alternative to direct alkylation of enolate ions is the formation and alkylation of an enamine derivative. An **enamine** (a **vinyl amine**) is the nitrogen analog of an enol. The resonance picture of an enamine shows that it has some carbanion character.

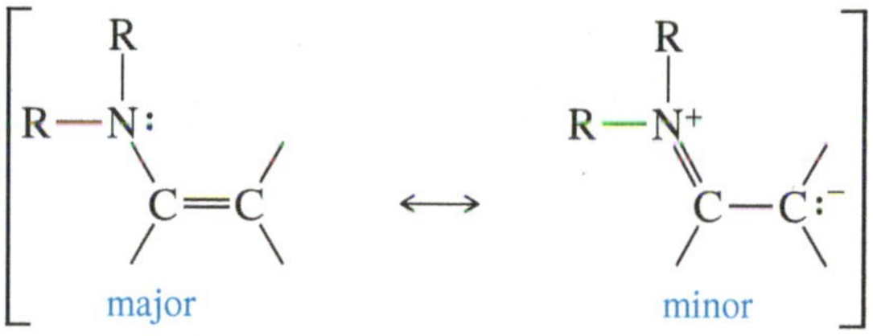

$$\left[R_2\ddot{N}{-}C{=}C \quad \text{(major)} \longleftrightarrow R_2N^+{=}C{-}\ddot{C}^- \quad \text{(minor)} \right]$$

The electrostatic potential map (EPM) of a simple enamine shows a high negative electrostatic potential (red) near the α carbon atom of the double bond. This is the nucleophilic carbon atom of the enamine.

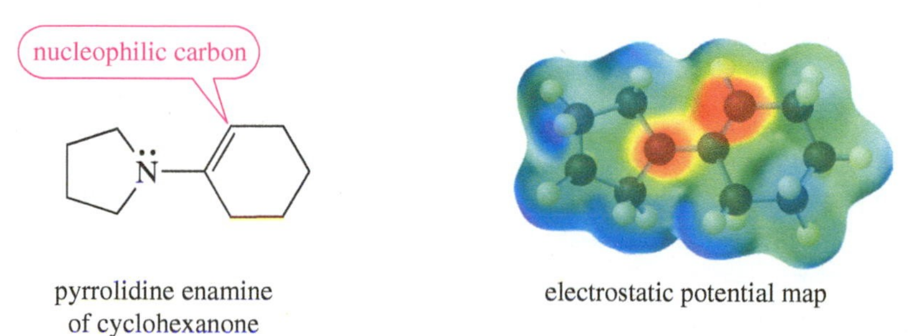

pyrrolidine enamine of cyclohexanone

electrostatic potential map

The nucleophilic carbon atom attacks an electrophile to give a resonance-stabilized cationic intermediate (an iminium ion).

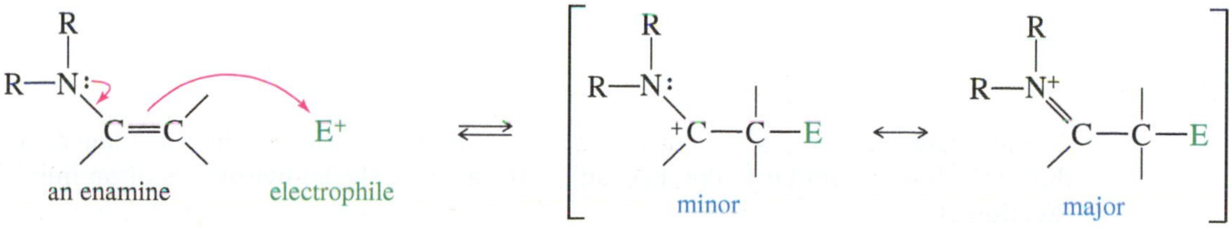

$$R_2\ddot{N}{-}C{=}C \ (\text{an enamine}) + E^+ \ (\text{electrophile}) \rightleftarrows \left[R_2\ddot{N}{-}\overset{+}{C}{-}C{-}E \ (\text{minor}) \longleftrightarrow R_2N^+{=}C{-}C{-}E \ (\text{major}) \right]$$

An enamine results from the reaction of a ketone or aldehyde with a *secondary* amine. Recall that a ketone or aldehyde reacts with a *primary* amine (Section 18-14) to form a carbinolamine, which dehydrates to give the C=N double bond of an imine. But a carbinolamine from a *secondary* amine does not form a C=N double bond because there is no proton on nitrogen to eliminate. A proton is lost from the α carbon, forming the C=C double bond of an enamine.

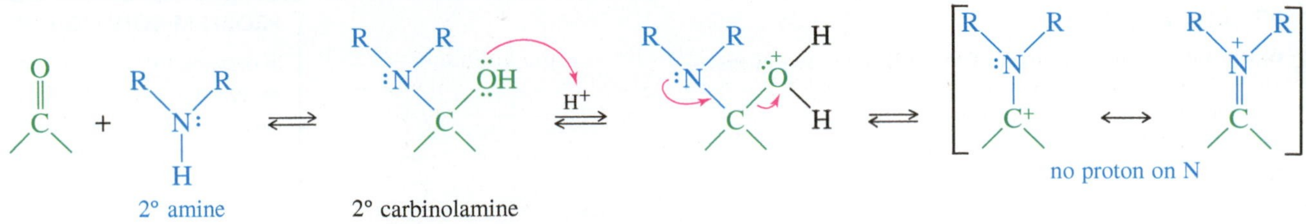

2° amine

2° carbinolamine

no proton on N

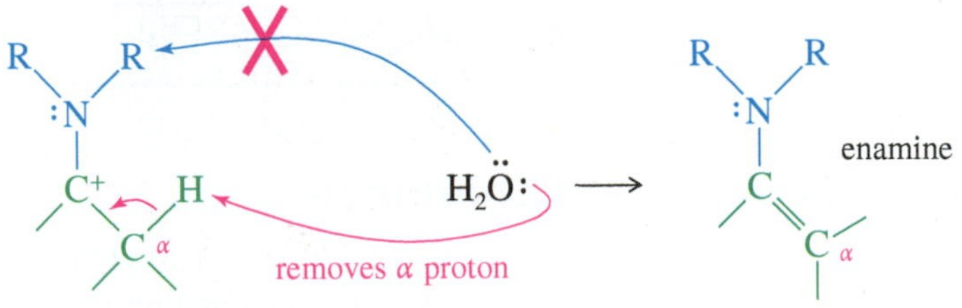

removes α proton

enamine

Example

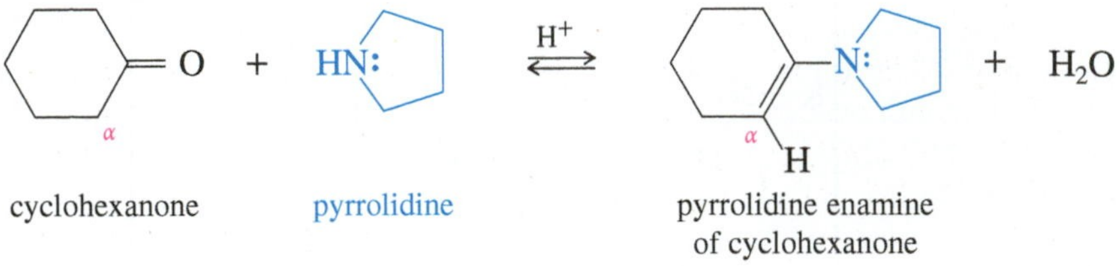

cyclohexanone + pyrrolidine $\overset{H^+}{\rightleftharpoons}$ pyrrolidine enamine of cyclohexanone + H_2O

PROBLEM 22-6

Propose a mechanism for the acid-catalyzed reaction of cyclohexanone with pyrrolidine.

Enamines are intermediate in reactivity: more reactive than an enol, but less reactive than an enolate ion. Enamine reactions occur under milder conditions than enolate reactions, so they avoid many side reactions. Enamines displace halides from reactive alkyl halides, giving alkylated iminium salts. The iminium ions are unreactive toward further alkylation or acylation. The following example shows benzyl bromide reacting with the pyrrolidine enamine of cyclohexanone.

enamine + benzyl bromide ⟶ alkylated iminium salt $\overset{H_3O^+}{\rightleftharpoons}$ alkylated ketone + pyrrolidinium

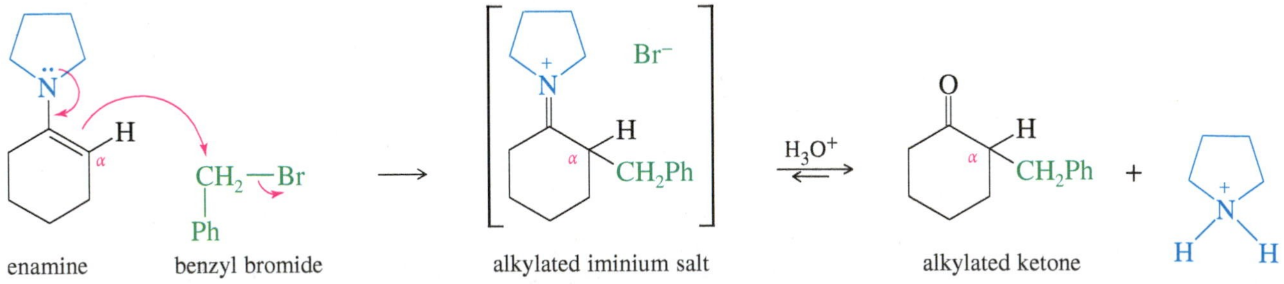

The alkylated iminium salt hydrolyzes to the alkylated ketone. The mechanism of this hydrolysis is similar to the mechanism for acid-catalyzed hydrolysis of an imine (Section 18-14).

Overall reaction

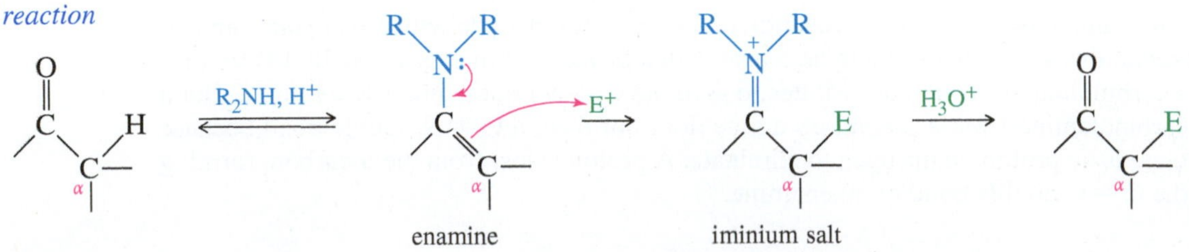

ketone $\overset{R_2NH,\ H^+}{\rightleftharpoons}$ enamine $\overset{E^+}{\longrightarrow}$ iminium salt $\overset{H_3O^+}{\longrightarrow}$ α-substituted ketone

PROBLEM 22-7

Without looking back, propose a mechanism for the hydrolysis of this iminium salt to the alkylated ketone. The first step is attack by water, followed by loss of a proton to give a carbinolamine. Protonation on nitrogen allows pyrrolidine to leave, giving the protonated ketone.

The enamine alkylation procedure is sometimes called the **Stork reaction,** after its inventor, Gilbert Stork of Columbia University. The Stork reaction can alkylate or acylate the α position of a ketone, using a variety of reactive alkyl and acyl halides. Some halides that react well with enamines to give alkylated and acylated ketone derivatives are the following:

$$\mathrm{Ph{-}CH_2{-}X} \qquad \mathrm{{>}C{=}C{-}CH_2{-}X} \qquad \mathrm{CH_3{-}X} \qquad \mathrm{R{-}C({=}O){-}Cl}$$

benzyl halides | allylic halides | methyl halides | acyl halides

The following sequence shows the acylation of an enamine to synthesize a β-diketone. The initial acylation gives an acyl iminium salt, which hydrolyzes to the β-diketone product. As we will see in Section 22-15, β-dicarbonyl compounds are easily alkylated, and they serve as useful intermediates in the synthesis of more complicated molecules.

enamine + acyl chloride ($\mathrm{Cl{-}C({=}O){-}CH_3}$) ⇌ intermediate → acyl iminium salt ($\mathrm{Cl^-}$) $\xrightarrow{H_3O^+}$ β-diketone

PROBLEM 22-8

Give the expected products of the following acid-catalyzed reactions.

(a) acetophenone + methylamine
(b) acetophenone + dimethylamine
(c) cyclohexanone + aniline
(d) cyclohexanone + piperidine

PROBLEM 22-9

Show how you would accomplish each conversion using an enamine synthesis with pyrrolidine as the secondary amine.

(a) cyclopentanone ⟶ 2-allylcyclopentanone
(b) pentan-3-one ⟶ 2-methyl-1-phenylpentan-3-one
(c)

acetophenone ⟶ $\mathrm{Ph{-}C({=}O){-}CH_2{-}C({=}O){-}Ph}$

PROBLEM-SOLVING HINT

We can summarize the overall enamine alkylation process:

1. Convert the ketone to an enamine.
2. Alkylate with a reactive alkyl (or acyl) halide.
3. Hydrolyze the iminium salt.

22-5 Alpha Halogenation of Ketones

The halogens are strong electrophiles that react rapidly with enols and enolate ions. These reactions provide efficient methods for placing halogen atoms at the alpha positions of carbonyl compounds.

22-5A Base-Promoted α Halogenation

When a ketone is treated with a halogen and base, an α-halogenation reaction occurs.

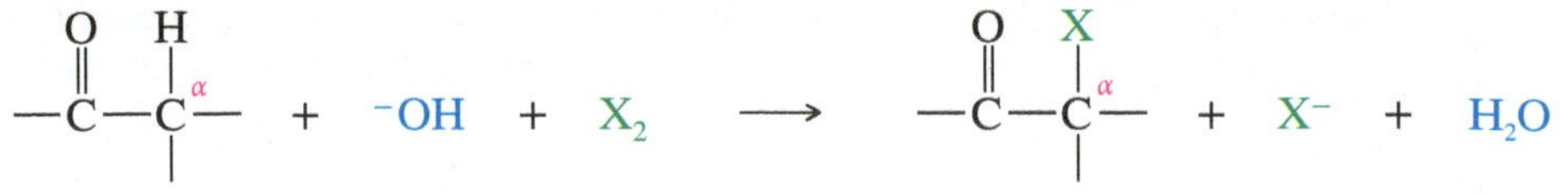

ketone + ^-OH + X_2 ⟶ α-haloketone + X^- + H_2O

($X_2 = Cl_2$, Br_2, or I_2)

Example

cyclohexanone $\xrightarrow[^-OH,\ H_2O]{Cl_2}$ 2-chlorocyclohexanone + Cl^- + H_2O

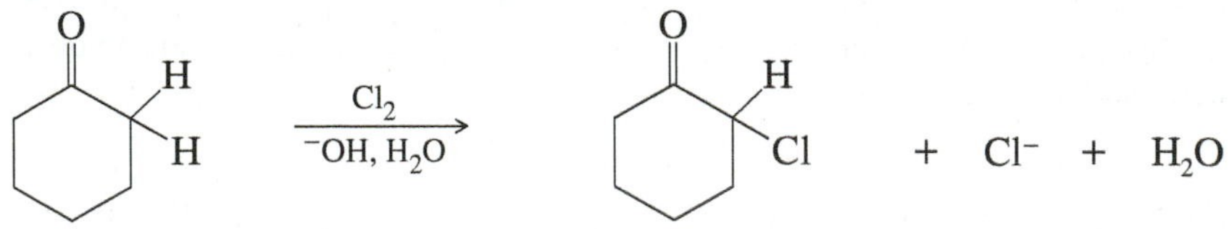

The base-promoted halogenation takes place by a nucleophilic attack of an enolate ion on the electrophilic halogen molecule. The products are the halogenated ketone and a halide ion.

MECHANISM 22-6 Base-Promoted Halogenation

Step 1: Deprotonation of the α carbon forms the enolate ion.

Step 2: The enolate ion attacks the electrophilic halogen.

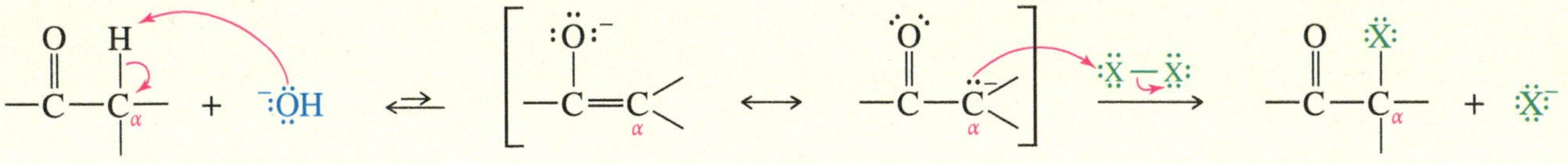

enolate ion + $H_2\ddot{O}$:

EXAMPLE: Base-promoted bromination of cyclohexanone.

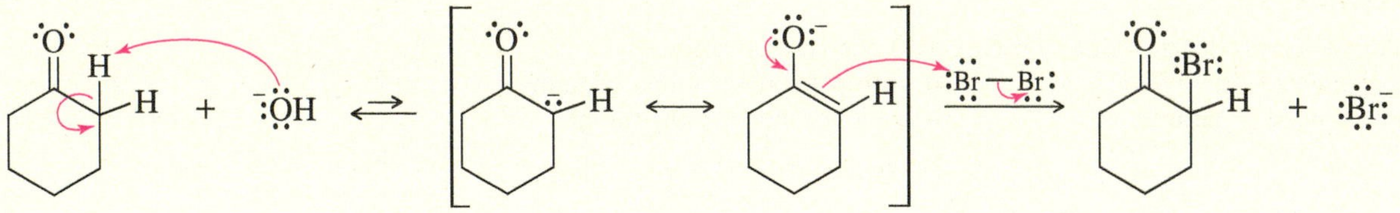

enolate ion

This reaction is called *base-promoted*, rather than base-catalyzed, because a full equivalent of the base is consumed in the reaction.

SOLVED PROBLEM 22-1

Propose a mechanism for the reaction of pentan-3-one with sodium hydroxide and bromine to give 2-bromopentan-3-one.

SOLUTION

In the presence of sodium hydroxide, a small amount of pentan-3-one is present as its enolate.

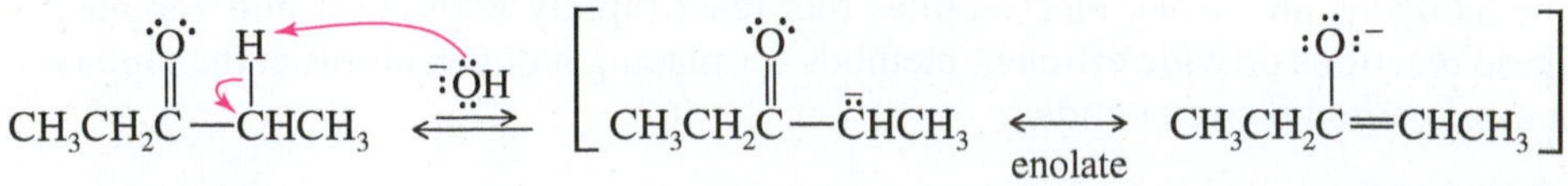

$CH_3CH_2C(=O)CHCH_3$ (H) $\rightleftarrows$ [$CH_3CH_2C(=O)\ddot{C}HCH_3$ ⟷ $CH_3CH_2C(O^-)=CHCH_3$]

enolate

(continued)

The enolate reacts with bromine to give the observed product.

$$CH_3CH_2C(=O)\text{–}\ddot{C}^-(CH_3)H + :\ddot{Br}\text{–}\ddot{Br}: \longrightarrow CH_3CH_2C(=O)\text{–}C(CH_3)(H)\text{–}\ddot{Br}: + :\ddot{Br}:^-$$

enolate α-haloketone

PROBLEM 22-10

An enolate is a very strong nucleophile. Bromine is a strong electrophile, so it can react with much weaker nucleophiles. Give mechanisms for the reactions of bromine with cyclopentene and with phenol, which are both much weaker nucleophiles than an enolate.

Multiple Halogenation In many cases, base-promoted halogenation does not stop with replacement of just one hydrogen. The product (the α-haloketone) is more reactive toward further halogenation than is the starting material, because the electron-withdrawing halogen stabilizes the enolate ion, enhancing its formation.

$$\text{–}C(=O)\text{–}C(X)(H)\text{–} + {}^-OH \rightleftarrows H_2O + \left[\text{–}C(=O)\text{–}\ddot{C}^-(X) \longleftrightarrow \text{–}C(\text{–}\ddot{O}:^-)=C(X) \right]$$

(enolate stabilized by X)

For example, bromination of pentan-3-one gives mostly 2,2-dibromopentan-3-one. After one hydrogen is replaced by bromine, the enolate ion is stabilized by both the carbonyl group and the bromine atom. A second bromination takes place faster than the first. Notice that the second substitution takes place at the same carbon atom as the first, because that carbon atom bears the enolate-stabilizing halogen.

$$CH_3CH_2\text{–}C(=O)\text{–}C(H)(Br)\text{–}CH_3 \xrightleftharpoons{{}^-OH} CH_3CH_2\text{–}C(=O)\text{–}\ddot{C}^-(Br)\text{–}CH_3 \xrightarrow{Br\text{–}Br} CH_3CH_2\text{–}C(=O)\text{–}C(Br)_2\text{–}CH_3$$

monobrominated ketone — stabilized by Br — second bromination

Because of this tendency for multiple halogenation, base-promoted halogenation is rarely used for the preparation of monohalo ketones. The acid-catalyzed procedure (discussed in Section 22-5C) is preferred.

PROBLEM 22-11

Propose a mechanism to show how acetophenone undergoes base-promoted chlorination to give trichloroacetophenone.

22-5B The Haloform Reaction

With most ketones, base-promoted halogenation continues until the α carbon atom is completely halogenated. Methyl ketones have three α protons on the methyl carbon, and they undergo halogenation three times to give trihalomethyl ketones.

$$R\text{–}C(=O)\text{–}CH_3 + 3\,X_2 + 3\,{}^-OH \longrightarrow \longrightarrow \longrightarrow R\text{–}C(=O)\text{–}CX_3 + 3\,X^- + 3\,H_2O$$

methyl ketone — trihalomethyl ketone

Application: Disinfection

The oxidizing effects of halogens are routinely used to kill bacteria, viruses, and parasites. Iodine is used as a mild disinfectant on skin for cuts and abrasions. Bromine and chlorine are used in swimming pools and water purification systems to destroy pathogens.

The photo shows the chlorine disinfection tanks used in Irvine, California to treat their wastewater before using it for irrigation.

With three electron-withdrawing halogen atoms, the trihalomethyl group can serve as a reluctant leaving group for nucleophilic acyl substitution. The trihalomethyl ketone reacts with hydroxide ion to give a tetrahedral intermediate that expels the trihalomethyl anion ($^{-}CX_3$), leaving a carboxylic acid. A fast proton exchange gives a carboxylate ion and a haloform (chloroform, $CHCl_3$; bromoform, $CHBr_3$; or iodoform, CHI_3). The overall reaction is called the **haloform reaction.**

MECHANISM 22-7 Final Steps of the Haloform Reaction

The conclusion of the haloform reaction is a nucleophilic acyl substitution, with hydroxide as the nucleophile and $^{-}CX_3$ as the leaving group.

Step 1: Hydroxide adds to the carbonyl group.

Step 2: $^{-}CX_3$ leaves.

Step 3: Fast proton transfer from the acid.

$$R-C(=O)-CX_3 + {}^{-}:\ddot{O}H \rightleftarrows R-C(O^-)(OH)-CX_3 \rightleftarrows R-C(=O)-O-H + {}^{-}:CX_3 \longrightarrow R-C(=O)-O^- + HCX_3$$

nucleophilic acyl substitution — a carboxylate ion — a haloform

The overall haloform reaction is summarized next. A methyl ketone reacts with a halogen under strongly basic conditions to give a carboxylate ion and a haloform.

$$\underset{\text{a methyl ketone}}{R-C(=O)-CH_3} \xrightarrow[(X_2 = Cl_2,\ Br_2,\ \text{or } I_2)]{\text{excess } X_2,\ {}^{-}OH} \underset{\substack{\text{a trihalomethyl ketone}\\ \text{(not isolated)}}}{\left[R-C(=O)-CX_3\right]} \xrightarrow{{}^{-}OH} \underset{\text{a carboxylate}}{R-C(=O)-O^-} + \underset{\text{a haloform}}{HCX_3}$$

Example

$$\underset{\text{butan-2-one}}{CH_3CH_2-C(=O)-CH_3} \xrightarrow[{}^{-}OH]{\text{excess } Br_2} \left[CH_3CH_2-C(=O)-CBr_3\right] \xrightarrow{{}^{-}OH} \underset{\text{propionate}}{CH_3CH_2-C(=O)-O^-} + \underset{\text{bromoform}}{HCBr_3}$$

When the halogen is iodine, the haloform product (iodoform) is a solid that separates out as a yellow precipitate. This **iodoform test** identifies methyl ketones, which halogenate three times, then lose $^{-}CI_3$, which protonates to give iodoform.

$$\underset{\text{acetophenone}}{Ph-C(=O)-CH_3} \xrightarrow[{}^{-}OH]{\text{excess } I_2} \underset{\alpha,\alpha,\alpha\text{-triiodoacetophenone}}{\left[Ph-C(=O)-CI_3\right]} \xrightarrow{{}^{-}OH} \underset{\text{benzoate}}{Ph-C(=O)-O^-} + \underset{\text{iodoform}}{HCI_3\downarrow}$$

Iodine is an oxidizing agent, and an alcohol can give a positive iodoform test if it oxidizes to a methyl ketone. The iodoform reaction can convert such an alcohol to a carboxylic acid with one less carbon atom.

$$R-CH(OH)-CH_3 + I_2 \longrightarrow R-C(=O)-CH_3 + 2\,HI \xrightarrow[{}^{-}OH]{\text{excess } I_2} \underset{\text{(one less carbon)}}{R-C(=O)-O^-} + HCI_3\downarrow$$

Example

$$CH_3(CH_2)_3-\overset{OH}{\overset{|}{C}H}-CH_3 \xrightarrow[^-OH]{I_2} CH_3(CH_2)_3-\overset{O}{\overset{\|}{C}}-CH_3 \xrightarrow[^-OH]{I_2} CH_3(CH_2)_3-\overset{O}{\overset{\|}{C}}-O^- + HCI_3\downarrow$$

hexan-2-ol → hexan-2-one → pentanoate

PROBLEM 22-12

Propose a mechanism for the reaction of cyclohexyl methyl ketone with excess bromine in the presence of sodium hydroxide.

PROBLEM 22-13

Predict the products of the following reactions.
(a) cyclopentyl methyl ketone + excess Cl_2 + excess NaOH
(b) 1-cyclopentylethanol + excess I_2 + excess NaOH
(c) propiophenone + excess Br_2 + excess NaOH

PROBLEM 22-14

Which compounds will give positive iodoform tests?
(a) 1-phenylethanol **(b)** pentan-2-one **(c)** pentan-2-ol
(d) pentan-3-one **(e)** acetone **(f)** isopropyl alcohol

22-5C Acid-Catalyzed Alpha Halogenation

The α halogenation of ketones can also be catalyzed by acid. One of the most effective procedures is to dissolve the ketone in acetic acid, which serves as both the solvent and the acid catalyst. In contrast with basic halogenation, acidic halogenation can selectively replace just one hydrogen or more than one, depending on the amount of the halogen added.

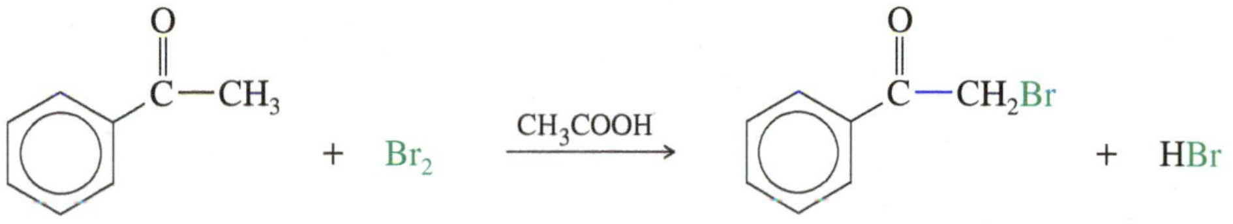

$$C_6H_5-\overset{O}{\overset{\|}{C}}-CH_3 + Br_2 \xrightarrow{CH_3COOH} C_6H_5-\overset{O}{\overset{\|}{C}}-CH_2Br + HBr$$

acetophenone → α-bromoacetophenone (70%)

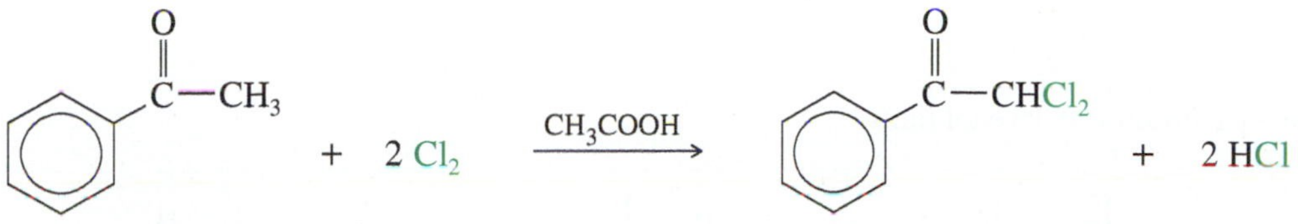

$$C_6H_5-\overset{O}{\overset{\|}{C}}-CH_3 + 2\ Cl_2 \xrightarrow{CH_3COOH} C_6H_5-\overset{O}{\overset{\|}{C}}-CHCl_2 + 2\ HCl$$

acetophenone → α,α-dichloroacetophenone

The mechanism of acid-catalyzed halogenation involves attack of the enol form on the electrophilic halogen molecule. Loss of a proton gives the α-haloketone and the hydrogen halide.

MECHANISM 22-8 Acid-Catalyzed Alpha Halogenation

Acid-catalyzed halogenation results when the enol form of the carbonyl compound serves as a nucleophile to attack the halogen (a strong electrophile). Deprotonation gives the α-haloketone.

Step 1: The enol attacks the halogen. ***Step 2:*** Deprotonation.

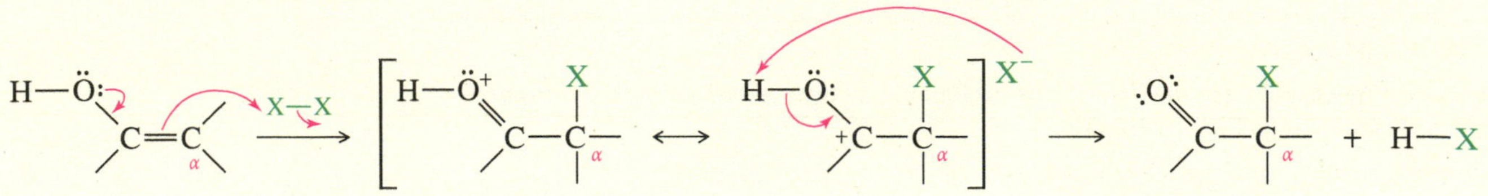

enol halogen carbocation intermediate α-haloketone

This reaction is similar to the attack of an alkene on a halogen, resulting in addition of the halogen across the double bond. The pi bond of an enol is more reactive toward halogens, however, because the carbocation that results is stabilized by resonance with the enol —OH group. Loss of the enol proton converts the intermediate to the product, an α-haloketone. We can stop the acid-catalyzed reaction at the monohalo (or dihalo) product because the halogen-substituted enol intermediate is less stable than the unsubstituted enol. Therefore, under acid-catalyzed conditions, each successive halogenation becomes slower.

Unlike ketones, aldehydes are easily oxidized, and halogens are strong oxidizing agents. Attempted halogenation of aldehydes usually results in oxidation to carboxylic acids.

PROBLEM-SOLVING HINT

Under acidic conditions, reactions at positions α to carbonyl groups often involve the enol tautomer acting as a nucleophile.

$$\underset{\text{aldehyde}}{R-\overset{O}{\overset{\|}{C}}-H} + X_2 + H_2O \longrightarrow \underset{\text{acid}}{R-\overset{O}{\overset{\|}{C}}-OH} + 2\,H-X$$

SOLVED PROBLEM 22-2

Propose a mechanism for the acid-catalyzed conversion of cyclohexanone to 2-chlorocyclohexanone.

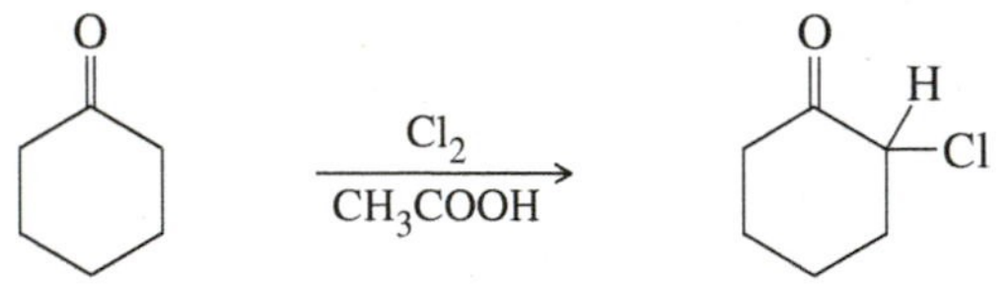

$$\text{cyclohexanone} \xrightarrow[CH_3COOH]{Cl_2} \text{2-chlorocyclohexanone (65\%)}$$

SOLUTION

Under acid catalysis, the ketone is in equilibrium with its enol form.

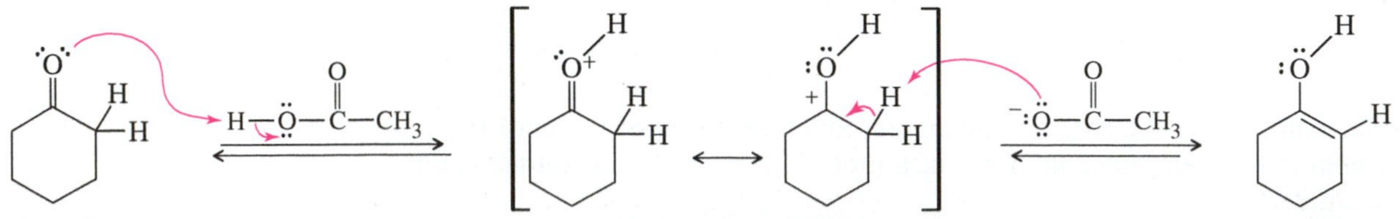

keto form stabilized intermediate enol form

(continued)

The enol acts as a weak nucleophile, attacking chlorine to give a resonance-stabilized intermediate. Loss of a proton gives the product.

$$\text{enol} \xrightarrow{:\ddot{Cl}-\ddot{Cl}:} \left[\text{oxocarbenium} \longleftrightarrow \text{resonance form}\right] \xrightarrow{:\ddot{Cl}:^-} \text{2-chlorocyclohexanone} + H-\ddot{Cl}:$$

PROBLEM 22-15

Propose a mechanism for the acid-catalyzed bromination of pentan-3-one.

PROBLEM 22-16

Acid-catalyzed halogenation is synthetically useful for converting ketones to α,β-unsaturated ketones, which are useful in Michael reactions (Section 22-18). Propose a method for converting cyclohexanone to cyclohex-2-en-1-one, an important synthetic starting material.

ketone → α,β-unsaturated; cyclohexanone → cyclohex-2-en-1-one

22-6 Alpha Bromination of Acids: The HVZ Reaction

The **Hell–Volhard–Zelinsky (HVZ) reaction** replaces a hydrogen atom with a bromine atom on the α carbon of a carboxylic acid. The resulting α-bromoacids are important for use as synthetic intermediates. The carboxylic acid is treated with bromine and phosphorus tribromide, followed by water to hydrolyze the intermediate α-bromo acyl bromide.

The HVZ reaction

$$R-CH_2-\overset{O}{\overset{\|}{C}}-OH \xrightarrow{Br_2/PBr_3} \underset{\alpha\text{-bromo acyl bromide}}{R-\overset{Br}{\overset{|}{C}H}-\overset{O}{\overset{\|}{C}}-Br} \xrightarrow{H_2O} \underset{\alpha\text{-bromoacid}}{R-\overset{Br}{\overset{|}{C}H}-\overset{O}{\overset{\|}{C}}-OH} + HBr$$

Example

$$\underset{\text{butanoic acid}}{CH_3CH_2CH_2-\overset{O}{\overset{\|}{C}}-OH} \xrightarrow{Br_2/PBr_3} \underset{\text{2-bromobutanoyl bromide}}{CH_3CH_2\overset{Br}{\overset{|}{C}H}-\overset{O}{\overset{\|}{C}}-Br} \xrightarrow{H_2O} \underset{\text{2-bromobutanoic acid}}{CH_3CH_2\overset{Br}{\overset{|}{C}H}-\overset{O}{\overset{\|}{C}}-OH} + HBr$$

The mechanism is similar to other acid-catalyzed α halogenations; the enol form of the acyl bromide serves as a nucleophilic intermediate. The first step is formation of acyl bromide, which enolizes more readily than does the acid.

$$\underset{\text{acid}}{R-CH_2-C(=O)OH} \xrightarrow{Br_2/PBr_3} \underset{\text{acyl bromide (keto form)}}{R-CH_2-C(=O)Br} \rightleftharpoons \underset{\text{enol form}}{RHC=C(OH)Br}$$

The enol is nucleophilic, attacking bromine to give the α-brominated acyl bromide.

$$:\ddot{Br}-\ddot{Br}: \;+\; \underset{\text{enol}}{RHC{=}C(OH)Br} \rightleftharpoons :\ddot{Br}:^- \;\; Br-\overset{R}{\underset{H}{C}}-C(=\overset{+}{O}H)Br \longrightarrow \underset{\alpha\text{-bromo acyl bromide}}{Br-\overset{R}{\underset{H}{C}}-C(=O)Br} + HBr$$

If a derivative of the α-bromoacid is desired, the α-bromo acyl bromide serves as an activated intermediate (similar to an acid chloride) for the synthesis of an ester, amide, or other derivative. If the α-bromoacid itself is needed, a water hydrolysis completes the synthesis.

One of the common uses for α-bromoacids is to convert them to α-amino acids. Often, simply treating them with a large excess of ammonia gives the α-amino acid directly.

$$\underset{\alpha\text{-bromoacid}}{R-\overset{Br}{CH}-\overset{O}{\overset{\|}{C}}-OH} \quad + \quad \text{excess } NH_3 \longrightarrow \underset{\alpha\text{-amino acid}}{R-\overset{NH_2}{CH}-\overset{O}{\overset{\|}{C}}-OH}$$

PROBLEM 22-17

Show the products of the reactions of these carboxylic acids with PBr_3/Br_2 before and after hydrolysis.

(a) pentanoic acid **(b)** phenylacetic acid **(c)** succinic acid **(d)** oxalic acid

22-7 The Aldol Condensation of Ketones and Aldehydes

Condensations are some of the most important enolate reactions of carbonyl compounds. Condensations combine two or more molecules, often with the loss of a small molecule such as water or an alcohol. Under basic conditions, the **aldol condensation** involves the nucleophilic addition of an enolate ion to another carbonyl group. The product, a β-hydroxy ketone or aldehyde, is called an **aldol** because it contains both an *ald*ehyde group and the hydroxy group of an alcoh*ol*. The aldol product may dehydrate to an α,β-unsaturated carbonyl compound.

The aldol condensation

$$\underset{\text{ketone or aldehyde}}{R-\overset{O}{\overset{\|}{C}}-CH_2-R' \;\; + \;\; R-\underset{O}{\underset{\|}{C}}-CH_2-R'} \xrightleftharpoons{H^+ \text{ or } ^-OH} \underset{\text{aldol product}}{R-\overset{OH}{\overset{|}{C}}(CH_2-R')-\overset{|}{CH}(R')-\underset{O}{\underset{\|}{C}}-R} \xrightarrow[H^+ \text{ or } ^-OH]{\text{heat}} \underset{\alpha,\beta\text{-unsaturated ketone or aldehyde}}{R-C(CH_2-R')=C(R')-\underset{O}{\underset{\|}{C}}-R} + H_2O$$

22-7A Base-Catalyzed Aldol Condensations

Under basic conditions, the aldol condensation occurs by a nucleophilic addition of the enolate ion (a strong nucleophile) to a carbonyl group. Protonation gives the aldol product.

KEY MECHANISM 22-9 Base-Catalyzed Aldol Condensation

The base-catalyzed aldol involves the nucleophilic addition of an enolate ion to a carbonyl group.

Step 1: A base removes an α proton to form an enolate ion.

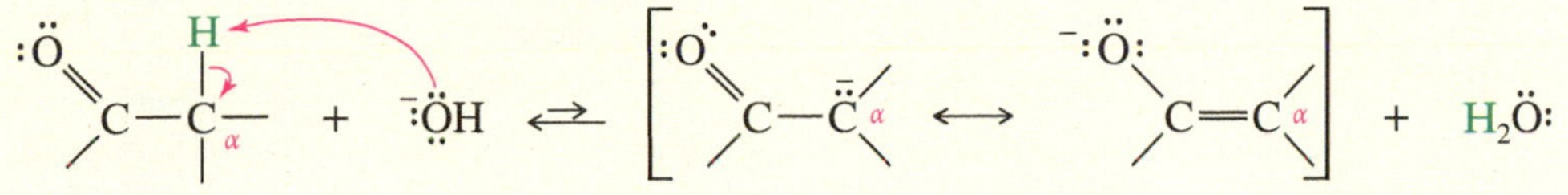

enolate ion

Step 2: The enolate ion adds to the carbonyl group.

Step 3: Protonation of the alkoxide gives the aldol product.

enolate carbonyl alkoxide aldol product

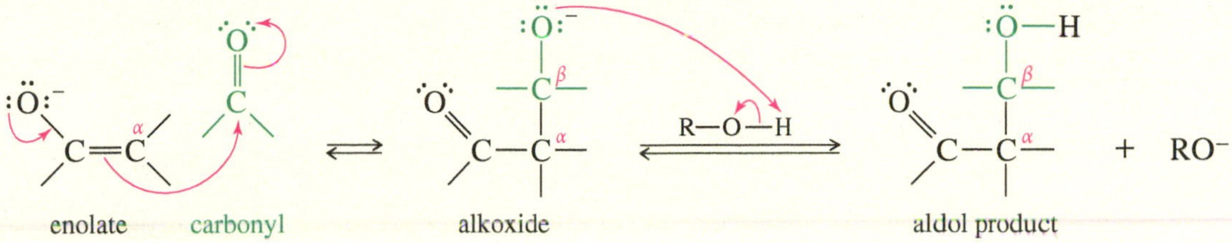

EXAMPLE: Aldol condensation of acetaldehyde.

The enolate ion of acetaldehyde attacks the carbonyl group of another acetaldehyde molecule. Protonation gives the aldol product.

Step 1: A base removes an α proton to form an enolate ion.

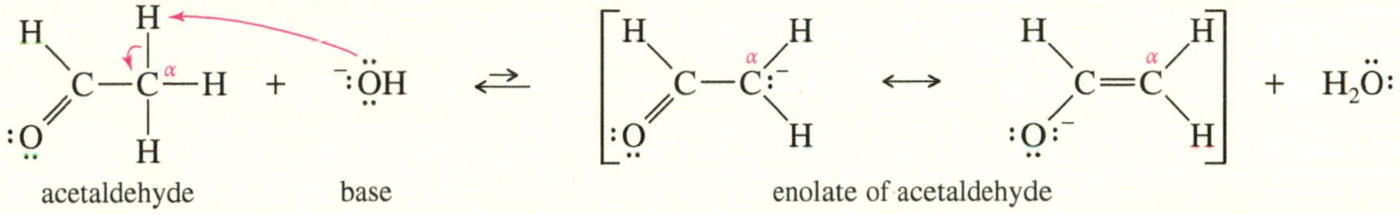

acetaldehyde base enolate of acetaldehyde

Step 2: The enolate ion adds to the carbonyl group.

Step 3: Protonation of the alkoxide gives the aldol product.

enolate acetaldehyde aldol product (50%)

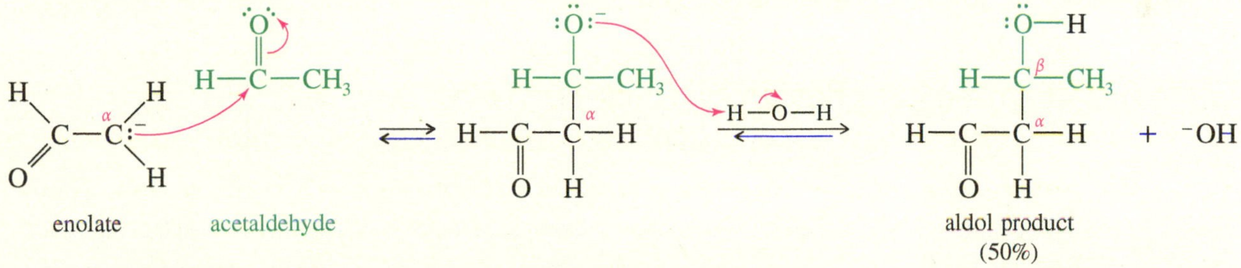

The aldol condensation is reversible, establishing an equilibrium between reactants and products. For acetaldehyde, conversion to the aldol product is about 50%. Ketones also undergo aldol condensation, but equilibrium concentrations of the products are generally small. Aldol condensations are sometimes accomplished by clever experimental methods. For example, Figure 22-2 shows how a good yield of the acetone aldol product ("diacetone alcohol") is obtained, even though the equilibrium concentration of the product is only about 1%. Acetone is boiled so it condenses into a chamber containing an insoluble basic catalyst. The reaction can take place only in the catalyst chamber. When the solution returns to the boiling flask, it contains about 1% diacetone alcohol. Diacetone alcohol is less volatile than acetone, remaining in the boiling flask while acetone boils and condenses (refluxes) in contact with the catalyst. After several hours, nearly all the acetone is converted to diacetone alcohol.

Application: Biochemistry

Aldolases are enzymes that form aldol products, most commonly in the metabolism of carbohydrates or sugars. In contrast to the chemical reaction, aldolases generate just one product stereospecifically. Hence, they are sometimes used in organic synthesis for key transformations.

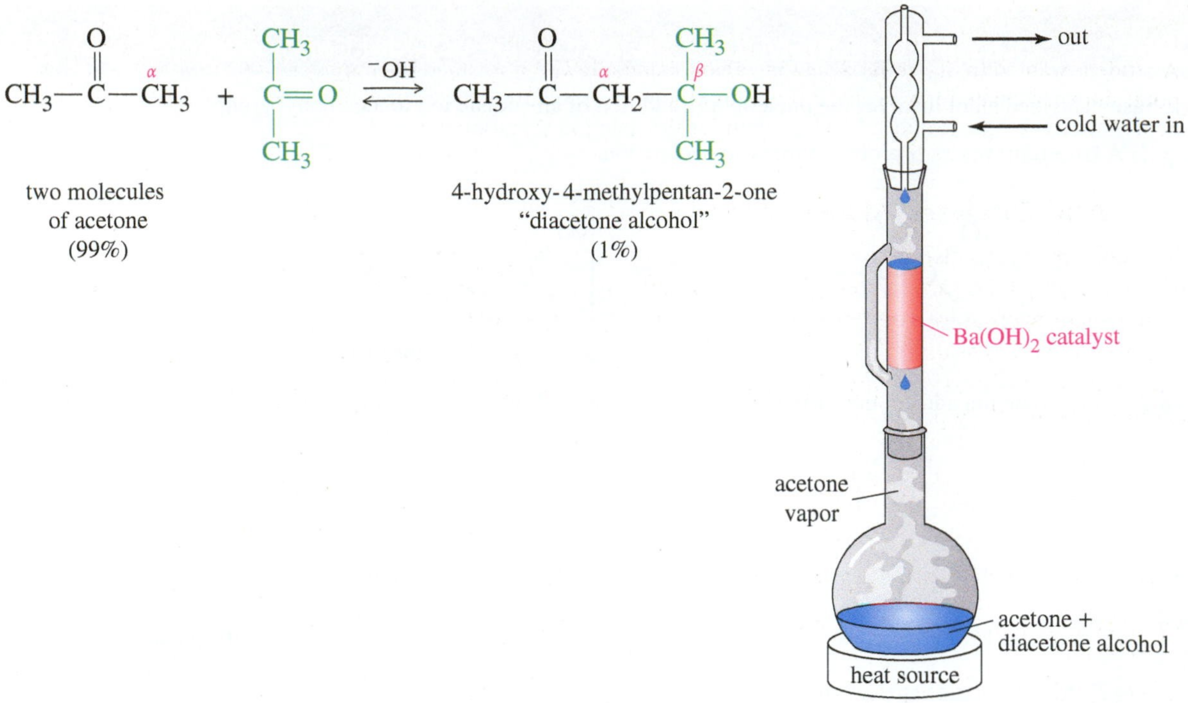

$$CH_3-\overset{O}{\overset{\|}{C}}-CH_3 + (CH_3)_2C{=}O \underset{}{\overset{^-OH}{\rightleftharpoons}} CH_3-\overset{O}{\overset{\|}{C}}-CH_2-C(CH_3)_2-OH$$

two molecules of acetone (99%)

4-hydroxy-4-methylpentan-2-one "diacetone alcohol" (1%)

out

cold water in

$Ba(OH)_2$ catalyst

acetone vapor

acetone + diacetone alcohol

heat source

FIGURE 22-2
Driving an aldol condensation to completion. The aldol condensation of acetone gives only 1% product at equilibrium, yet a clever technique gives a good yield. Acetone refluxes onto a basic catalyst such as $Ba(OH)_2$. The nonvolatile diacetone alcohol does not reflux, so its equilibrium concentration gradually increases until nearly all the acetone is converted to diacetone alcohol.

SOLVED PROBLEM 22-3

Propose a mechanism for the base-catalyzed aldol condensation of acetone (Figure 22-2).

SOLUTION

The first step is formation of the enolate to serve as a nucleophile.

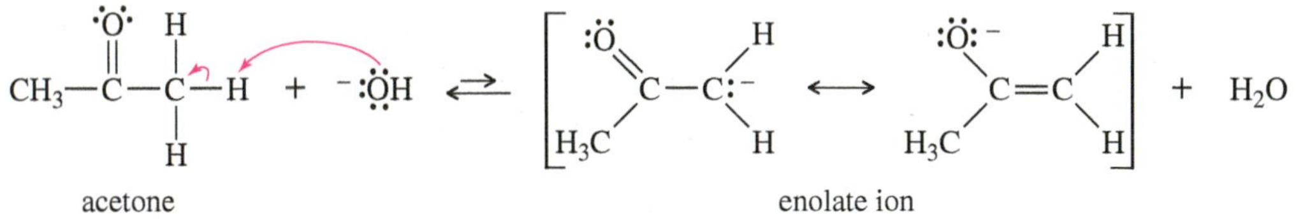

$$CH_3-\overset{O}{\overset{\|}{C}}-CH_3 + {}^-:\ddot{O}H \rightleftharpoons \left[H_3C-C(=O)-\ddot{C}H_2^- \longleftrightarrow H_3C-C(-O^-)=CH_2 \right] + H_2O$$

acetone — enolate ion

The second step is a nucleophilic attack by the enolate on another molecule of acetone. Protonation gives the aldol product.

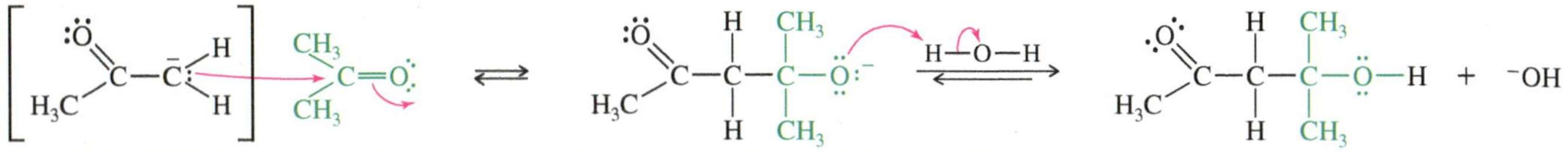

$$\left[H_3C-C(=O)-\ddot{C}H_2^- \right] + (CH_3)_2C{=}O \rightleftharpoons H_3C-C(=O)-CH_2-C(CH_3)_2-O^- \overset{H-O-H}{\rightleftharpoons} H_3C-C(=O)-CH_2-C(CH_3)_2-OH + {}^-OH$$

PROBLEM 22-18

Propose a mechanism for the aldol condensation of cyclohexanone. Do you expect the equilibrium to favor the reactant or the product?

PROBLEM 22-19

Give the expected products for the aldol condensations of
(a) propanal. **(b)** phenylacetaldehyde. **(c)** pentan-3-one.

PROBLEM 22-20

A student wanted to dry some diacetone alcohol and allowed it to stand over anhydrous potassium carbonate for a week. At the end of the week, the sample was found to contain nearly pure acetone. Propose a mechanism for the reaction that took place.

22-7B Acid-Catalyzed Aldol Condensations

Aldol condensations also take place under acidic conditions. The enol serves as a weak nucleophile to attack an activated (protonated) carbonyl group. As an example, consider the acid-catalyzed aldol condensation of acetaldehyde. The first step is formation of the enol by the acid-catalyzed keto–enol tautomerism, as discussed earlier. The enol attacks the protonated carbonyl of another acetaldehyde molecule. Loss of the enol proton gives the aldol product.

MECHANISM 22-10 Acid-Catalyzed Aldol Condensation

The acid-catalyzed aldol involves nucleophilic addition of an enol to a protonated carbonyl group.

Step 1: Formation of the enol, by protonation on O followed by deprotonation on C.

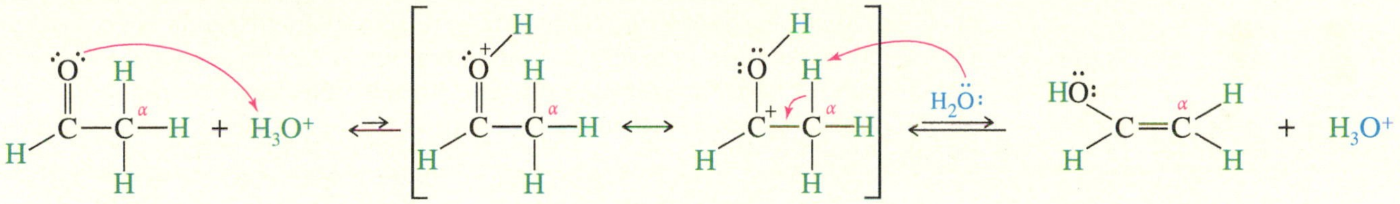

H_3O^+ $H_2\ddot{O}$: H_3O^+

keto form

protonated carbonyl

enol form

Step 2: Addition of the enol to the protonated carbonyl.

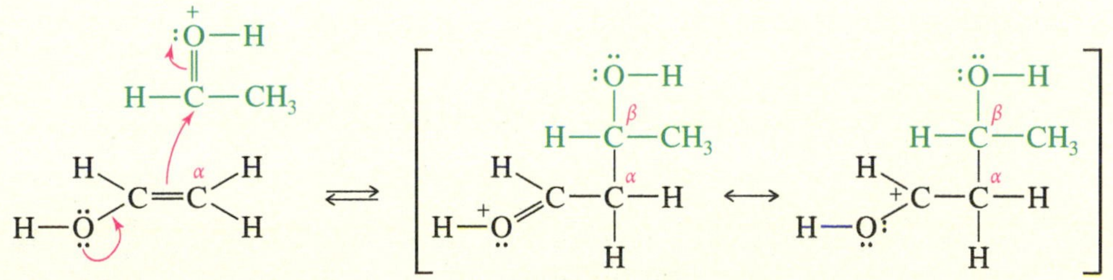

attack by enol

resonance-stabilized intermediate

Step 3: Deprotonation to give the aldol product.

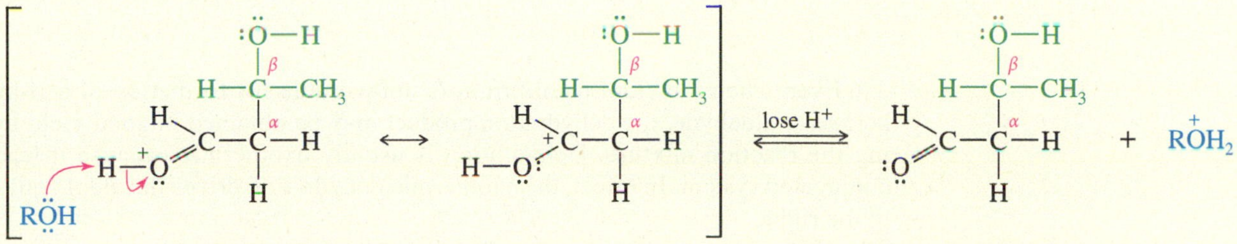

lose H^+

$R\ddot{O}H$ $R\overset{+}{O}H_2$

resonance-stabilized intermediate

aldol product

PROBLEM 22-21

Propose a complete mechanism for the acid-catalyzed aldol condensation of acetone.

22-8 Dehydration of Aldol Products

Heating a basic or acidic mixture of an aldol product leads to dehydration of the alcohol functional group. The product is a conjugated α,β-unsaturated aldehyde or ketone. Thus, an aldol condensation, followed by dehydration, forms a *new carbon–carbon double bond.* Before the Wittig reaction was discovered (Section 18-18), the aldol with dehydration was probably the best method for joining two molecules with a double bond. It is still often the cheapest and easiest method.

$$\text{diacetone alcohol} \xrightarrow[\text{heat}]{H^+ \text{ or } ^-OH} \text{4-methylpent-3-en-2-one (mesityl oxide)} + H{-}OH$$

new double bond

diacetone alcohol

4-methylpent-3-en-2-one
(mesityl oxide)

Under acidic conditions, dehydration follows a mechanism similar to those of other acid-catalyzed alcohol dehydrations (Section 11-10). We have not previously seen a base-catalyzed dehydration, however. Base-catalyzed dehydration depends on the acidity of the α proton of the aldol product. Abstraction of an α proton gives an enolate that can expel hydroxide ion to give a more stable product. Hydroxide is not a good leaving group in an E2 elimination, but it can serve as a leaving group in a strongly exothermic step like this one, which stabilizes a negatively charged intermediate. The following mechanism shows the base-catalyzed dehydration of 3-hydroxybutanal.

KEY MECHANISM 22-11 Base-Catalyzed Dehydration of an Aldol

Unlike most alcohols, aldols undergo dehydration in base. Abstraction of an α proton gives an enolate that can expel hydroxide ion to give a conjugated product.

Step 1: Formation of the enolate ion.

Step 2: Elimination of hydroxide.

removal of α proton

resonance-stabilized enolate $+ H_2O \rightleftharpoons$

conjugated system $+ {}^-OH$

Even when the aldol equilibrium is unfavorable for formation of a β-hydroxy ketone or aldehyde, the dehydration product may be obtained in good yield by heating the reaction mixture. Dehydration is usually exothermic because it leads to a conjugated system. In effect, the exothermic dehydration drives the aldol equilibrium to the right.

PROBLEM 22-22

Propose a mechanism for the dehydration of diacetone alcohol to mesityl oxide
(a) in acid. **(b)** in base.

PROBLEM 22-23

When propionaldehyde is warmed with sodium hydroxide, one of the products is 2-methylpent-2-enal. Propose a mechanism for this reaction.

PROBLEM 22-24

Predict the products of aldol condensation, followed by dehydration, of the following ketones and aldehydes.
(a) butyraldehyde **(b)** acetophenone **(c)** cyclohexanone

22-9 Crossed Aldol Condensations

When the enolate of one aldehyde (or ketone) adds to the carbonyl group of a different aldehyde or ketone, the result is called a **crossed aldol condensation.** The compounds used in the reaction must be selected carefully, or a mixture of several products will be formed.

Consider the aldol condensation between ethanal (acetaldehyde) and propanal shown below. Either of these reagents can form an enolate ion. Attack by the enolate of ethanal on propanal gives a product different from the one formed by attack of the enolate of propanal on ethanal. Also, self-condensations of ethanal and propanal continue to take place. Depending on the reaction conditions, various proportions of the four possible products result.

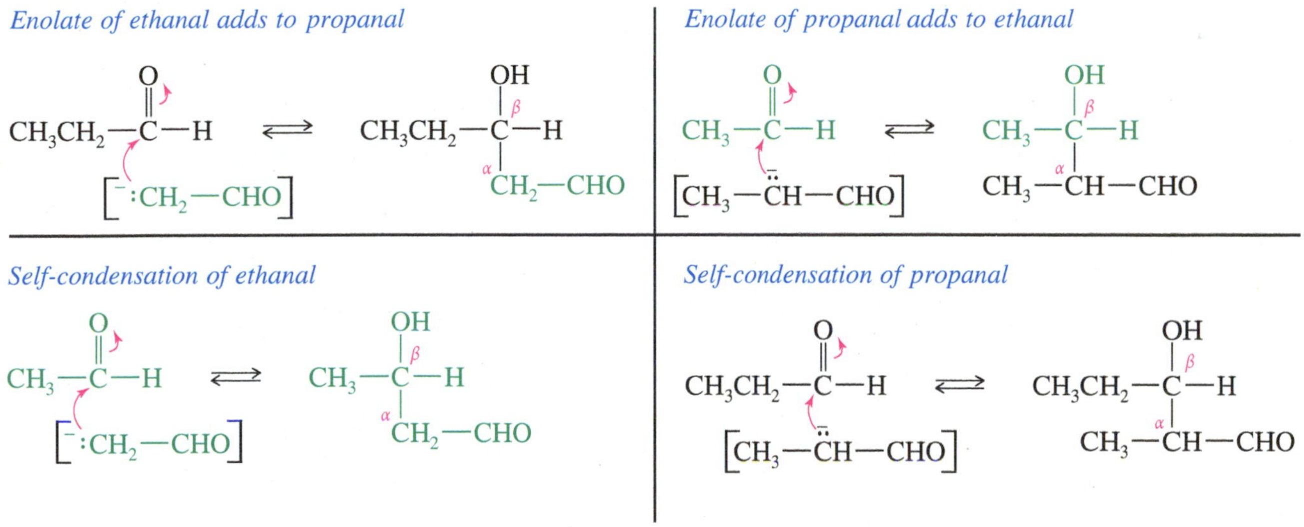

Enolate of ethanal adds to propanal

$CH_3CH_2-C(=O)-H + [^-:CH_2-CHO] \rightleftharpoons CH_3CH_2-C^{\beta}H(OH)-C^{\alpha}H_2-CHO$

Enolate of propanal adds to ethanal

$CH_3-C(=O)-H + [CH_3-\ddot{C}H^--CHO] \rightleftharpoons CH_3-C^{\beta}H(OH)-C^{\alpha}H(CH_3)-CHO$

Self-condensation of ethanal

$CH_3-C(=O)-H + [^-:CH_2-CHO] \rightleftharpoons CH_3-C^{\beta}H(OH)-C^{\alpha}H_2-CHO$

Self-condensation of propanal

$CH_3CH_2-C(=O)-H + [CH_3-\ddot{C}H^--CHO] \rightleftharpoons CH_3CH_2-C^{\beta}H(OH)-C^{\alpha}H(CH_3)-CHO$

A crossed aldol condensation can be effective if it is planned so that only one of the reactants can form an enolate ion and so that the other compound is more likely to react with the enolate. If only one of the reactants has an α hydrogen, only one enolate will be present in the solution. If the other reactant is present in excess or contains a particularly electrophilic carbonyl group, it is more likely to be attacked by the enolate ion.

The following two reactions are successful crossed aldol condensations. The aldol products may or may not undergo dehydration, depending on the reaction conditions and the structure of the products.

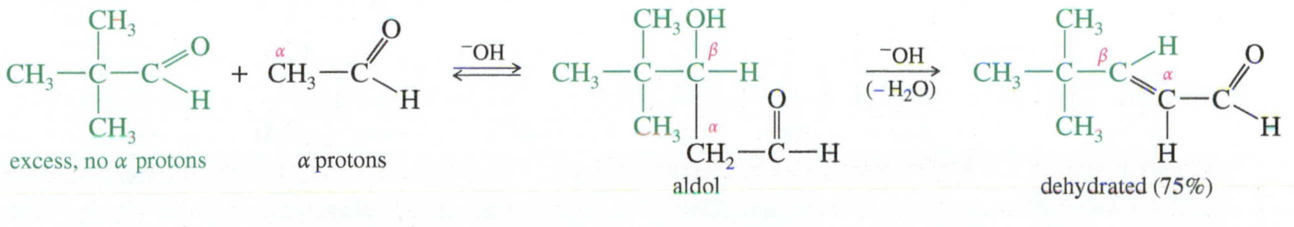

$$\underset{\text{excess, no }\alpha\text{ protons}}{(CH_3)_3C-CHO} + \underset{\alpha\text{ protons}}{CH_3^{\alpha}-CHO} \xrightleftharpoons{^-OH} \underset{\text{aldol}}{(CH_3)_3C-C^{\beta}H(OH)-C^{\alpha}H_2-C(=O)-H} \xrightarrow[(-H_2O)]{^-OH} \underset{\text{dehydrated (75\%)}}{(CH_3)_3C-C^{\beta}H=C^{\alpha}H-C(=O)H}$$

excess, no α protons + α protons $\xrightleftharpoons{^{-}OH}$ aldol $\xrightarrow[(-H_2O)]{^{-}OH}$ dehydrated (80%)

To carry out these reactions, slowly add the compound with α protons to a basic solution of the compound with no α protons. This way, the enolate ion is formed in the presence of a large excess of the other component, and the desired reaction is favored.

Lithium enolates formed using LDA (Section 22-2B) can react with other ketones and aldehydes to give crossed aldol products that could not be formed by standard base-catalyzed aldols. We can use LDA to make just the desired enolate ion, then add the compound we want to react as the electrophile. In this way, we control which enolate adds to which carbonyl group.

LDA → enolate → (CH_3CH_2—C(=O)—H) aldol product (not isolated) → ($-H_2O$) product (E + Z)

PROBLEM-SOLVING STRATEGY Proposing Reaction Mechanisms

The general principles for proposing reaction mechanisms, first introduced in Chapter 4 and summarized in Appendix 3A, are applied here to a crossed aldol condensation. This example emphasizes a base-catalyzed reaction involving strong nucleophiles. In drawing mechanisms, be careful to draw all the bonds and substituents of each carbon atom involved. Show each step separately, and draw curved arrows to show the movement of electrons from the nucleophile to the electrophile.

Our problem is to propose a mechanism for the base-catalyzed reaction of methylcyclohexanone with benzaldehyde:

$\xrightarrow{NaOCH_2CH_3}$

First, we must determine the type of mechanism. Sodium ethoxide, a strong base and a strong nucleophile, implies the reaction involves strong nucleophiles as intermediates. We expect to see strong nucleophiles and anionic intermediates (possibly stabilized carbanions), but no strong electrophiles or strong acids, and certainly no carbocations or free radicals.

1. **Consider the carbon skeletons of the reactants and products, and decide which carbon atoms in the products are likely derived from which carbon atoms in the reactants.**
   Because one of the rings is aromatic, it is clear which ring in the product is derived from which ring in the reactants. The carbon atom that bridges the two rings in the products must be derived from the carbonyl group of benzaldehyde. The two α protons from methylcyclohexanone and the carbonyl oxygen are lost as water.

lost → + H_2O

2. Consider whether any of the reactants is a strong enough nucleophile to react without being activated. If not, consider how one of the reactants might be converted to a strong nucleophile by deprotonation of an acidic site or by attack on an electrophilic site.

Neither of these reactants is a strong enough nucleophile to attack the other. If ethoxide removes an α proton from methylcyclohexanone, however, a strongly nucleophilic enolate ion results.

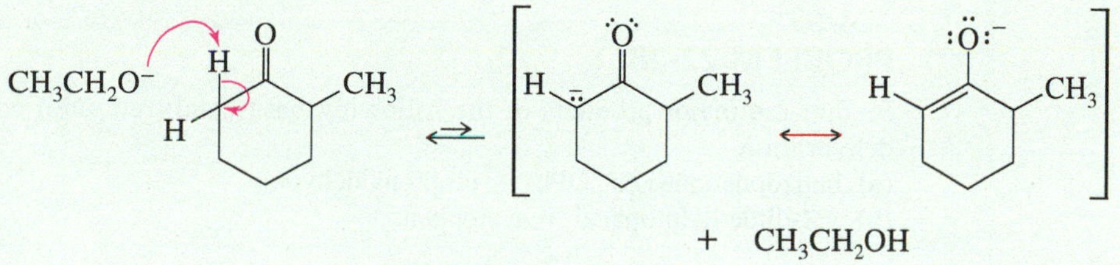

$CH_3CH_2O^-$ + CH_3CH_2OH

3. Consider how an electrophilic site on another reactant (or, in a cyclization, another part of the same molecule) can undergo attack by the strong nucleophile to form a bond needed in the product. Draw the product of this bond formation.

Attack at the electrophilic carbonyl group of benzaldehyde, followed by protonation, gives a β-hydroxy ketone (an aldol).

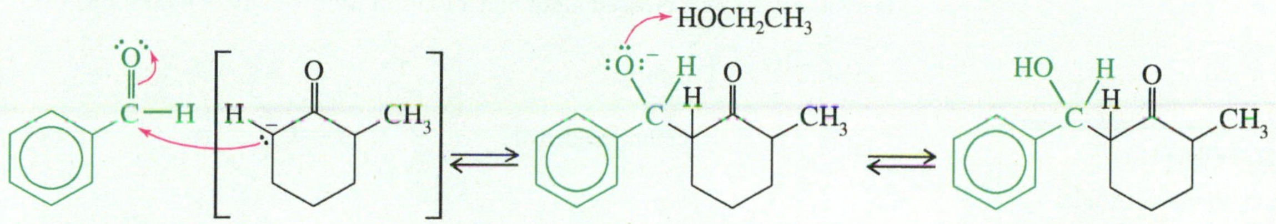

$HOCH_2CH_3$

aldol

4. Consider how the product of nucleophilic attack might be converted to the final product (if it has the right carbon skeleton) or reactivated to form another bond needed in the product.

The β-hydroxy ketone must be dehydrated to give the final product. Under these basic conditions, the usual alcohol dehydration mechanism (protonation of hydroxy, followed by loss of water) cannot occur. Removal of another proton gives an enolate ion that can lose hydroxide in a strongly exothermic step to give the final product.

$^-OCH_2CH_3$

aldol — enolate — dehydrated + OH^-

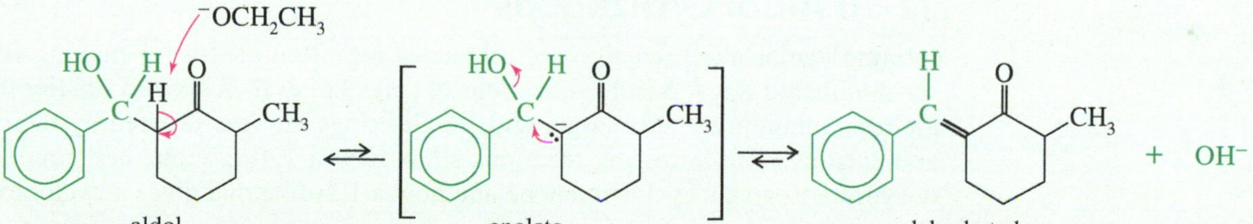

5. Draw out all the steps using curved arrows to show the movement of electrons. Be careful to show only one step at a time.

The complete mechanism is given by combining the equations shown above. We suggest you write out the mechanism as a review of the steps involved.

As further practice in proposing mechanisms for base-catalyzed reactions, do Problem 22-25 using the steps just shown.

PROBLEM 22-25

Propose mechanisms for the following base-catalyzed condensations, with dehydration.

(a) 2,2-dimethylpropanal with acetaldehyde

(b) benzaldehyde with propionaldehyde

PROBLEM 22-26

When acetone is treated with excess benzaldehyde in the presence of base, the crossed condensation adds two equivalents of benzaldehyde and expels two equivalents of water. Propose a structure for the condensation product of acetone with two molecules of benzaldehyde.

PROBLEM-SOLVING HINT

The correct mechanism for the base-catalyzed dehydration of an aldol product requires two steps:

1. deprotonation to form an enolate ion.
2. expulsion of hydroxide ion.

Do not draw a concerted E2 reaction for the dehydration of an aldol product.

PROBLEM 22-27

In the problem-solving feature above, methylcyclohexanone was seen to react at its *unsubstituted* α carbon. Try to write a mechanism for the same reaction at the methyl-substituted carbon atom, and explain why this regiochemistry is not observed.

PROBLEM 22-28

Predict the major products of the following base-catalyzed aldol condensations with dehydration.

(a) benzophenone (PhCOPh) + propionaldehyde

(b) 2,2-dimethylpropanal + acetophenone

PROBLEM 22-29

(a) Cinnamaldehyde is used in artificial cinnamon flavoring. Show how cinnamaldehyde is synthesized by a crossed aldol condensation followed by dehydration.

cinnamaldehyde

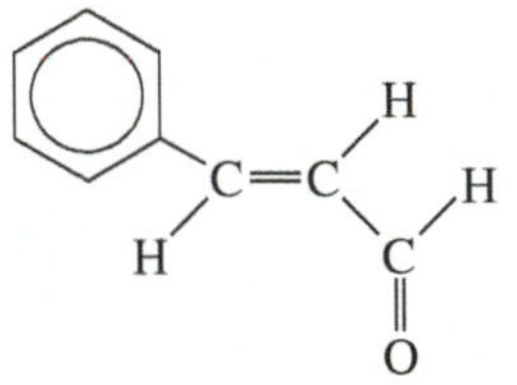

(b) Adding acetophenone slowly to a cold solution of LDA produces the enolate of acetophenone; but adding LDA slowly to a cold solution of acetophenone produces a condensation product. Show the reactions happening in each case, and explain why we observe such different results.

PROBLEM-SOLVING HINT

Practice predicting the structures of aldol products (before and after dehydration) and drawing the mechanisms. These reactions are among the most important in this chapter.

22-10 Aldol Cyclizations

Intramolecular aldol reactions of diketones are often useful for making five- and six-membered rings. Aldol cyclizations of rings larger than six and smaller than five are less common because larger and smaller rings are less favored by their energy and entropy. The following reactions show how a 1,4-diketone can condense and dehydrate to give a cyclopentenone and how a 1,5-diketone gives a cyclohexenone.

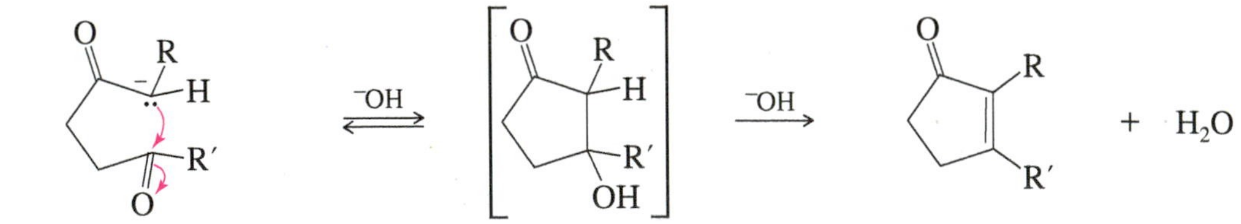

enolate of a 1,4-diketone $\xrightleftharpoons{^-OH}$ [aldol product] $\xrightarrow{^-OH}$ a cyclopentenone + H_2O

Example

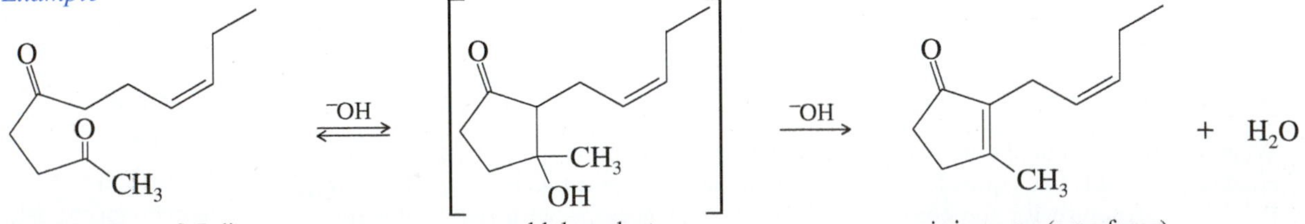

cis-undec-8-ene-2,5-dione $\xrightleftharpoons{^-OH}$ [aldol product] $\xrightarrow{^-OH}$ *cis*-jasmone (a perfume) (90%) + H_2O

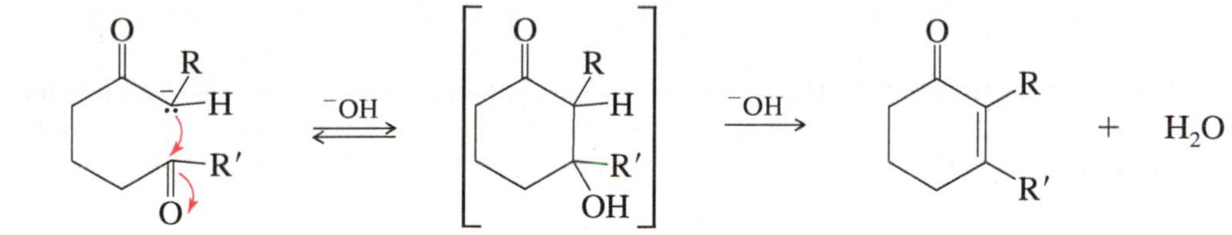

enolate of a 1,5-diketone ⇌ (^{-}OH) [aldol product] → (^{-}OH) a cyclohexenone + H_2O

Example

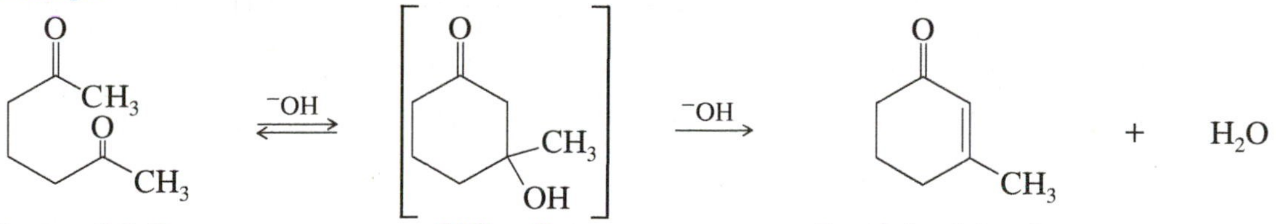

heptane-2,6-dione (a 1,5-diketone) ⇌ (^{-}OH) [aldol product] → (^{-}OH) 3-methylcyclohex-2-enone + H_2O

The following example shows how the carbonyl group of the product may be outside the ring in some cases.

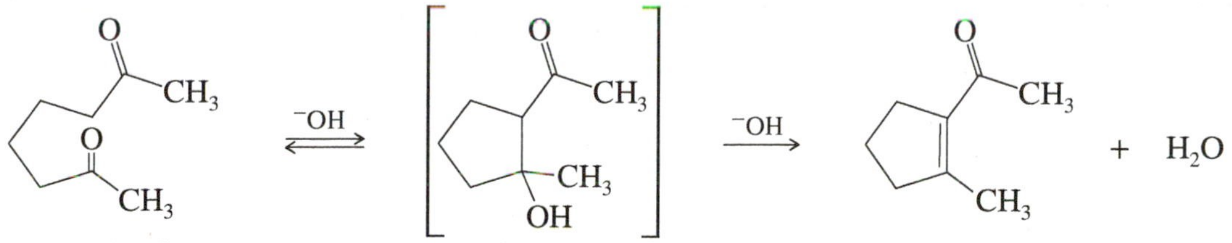

octane-2,7-dione ⇌ (^{-}OH) [aldol product] → (^{-}OH) 1-acetyl-2-methylcyclopentene + H_2O

PROBLEM 22-30

Show how octane-2,7-dione might cyclize to a cycloheptenone. Explain why ring closure to the cycloheptenone is not favored.

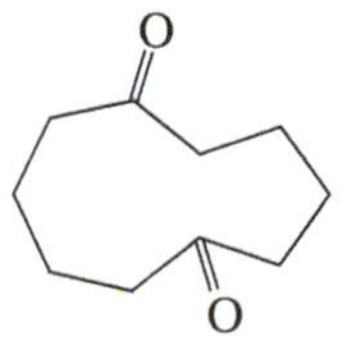

PROBLEM 22-31

When cyclodecane-1,6-dione is treated with sodium carbonate, the product gives a UV spectrum similar to that of 1-acetyl-2-methylcyclopentene. Propose a structure for the product, and give a mechanism for its formation.

cyclodecane-1,6-dione

22-11 Planning Syntheses Using Aldol Condensations

As long as we remember their limitations, aldol condensations can serve as useful synthetic reactions for making a variety of organic compounds. In particular, aldol condensations (with dehydration) form new carbon–carbon double bonds. We can use some general principles to decide whether a compound might be an aldol product and which reagents to use as starting materials.

Aldol condensations produce β-hydroxy aldehydes and ketones (aldols) and α,β-unsaturated aldehydes and ketones. If a target molecule has one of these functionalities, an aldol should be considered. To determine the starting materials, divide the structure at the α,β bond. In the case of the dehydrated product, the α,β bond is the double bond. The following analyses show the division of some aldol products into their starting materials.

$CH_3-CH_2-CH(OH)\beta \,|\, \alpha CH(CH_3)-CHO$ came from $\Longrightarrow$ CH_3-CH_2-CHO + CH_3-CH_2-CHO

break at the α,β bond — propanal — propanal

$C_6H_5-CH(OH)\beta \,|\, \alpha CH(CH_3)-CO-C_6H_5$ came from $\Longrightarrow$ C_6H_5-CHO + $CH_3CH_2-CO-C_6H_5$

break at the α,β bond — benzaldehyde — propiophenone

$C_6H_5-C(CH_3)\beta = \alpha CH-CO-C_6H_5$ came from $\Longrightarrow$ $C_6H_5-CO-CH_3$ + $CH_3-CO-C_6H_5$

break at the double bond — acetophenone — acetophenone

$C_6H_5-CH\beta = \alpha C(CH_2CH_3)-CHO$ came from $\Longrightarrow$ C_6H_5-CHO + $CH_3-CH_2-CH_2-CHO$

break at the double bond — benzaldehyde — butanal

PROBLEM 22-32

Show how each compound can be dissected into reagents joined by an aldol condensation, then decide whether the necessary aldol condensation is feasible.

(a) $CH_3CH_2CH_2-CH(OH)-CH(CH_2CH_2CH_3)-CHO$

(b) $Ph-C(OH)(CH_2CH_3)-CH(CH_3)-CO-Ph$

(c) $Ph(H)C=C(H)-CO-CH_3$

(d) 2-hydroxycyclopentyl methyl ketone ($C_5H_8(OH)-CO-CH_3$)

(e) 1-acetyl-2-methylcyclohexene (CH_3 and $-CO-CH_3$ on the ring double bond)

PROBLEM 22-33

The following compound results from base-catalyzed aldol cyclization of a 2-substituted cyclohexanone.

[bicyclic enone: fused cyclohexane–cyclohexenone, C=O]

(a) Show the diketone that would cyclize to give this product.
(b) Propose a mechanism for the cyclization.

22-12 The Claisen Ester Condensation

The α hydrogens of esters are weakly acidic, and they can be deprotonated to give enolate ions. Esters are less acidic than ketones and aldehydes because the ester carbonyl group is stabilized by resonance with the other oxygen atom. This resonance makes the carbonyl group less capable of stabilizing the negative charge of an enolate ion.

$$\left[R-\overset{\overset{\displaystyle O}{\|}}{C}-\ddot{O}-R' \longleftrightarrow R-\overset{\overset{\displaystyle :\ddot{O}:^-}{|}}{C}=\overset{+}{O}-R' \right]$$

A typical pK_a for an α proton of an ester is about 24, compared with a pK_a of about 20 for a ketone or aldehyde. Even so, strong bases do deprotonate esters.

$$\underset{\substack{\text{acetone} \\ (pK_a = 20)}}{CH_3-\overset{\overset{\displaystyle O}{\|}}{C}-CH_3} + CH_3\ddot{O}:^- \rightleftharpoons \underset{\text{enolate of acetone}}{\left[CH_3-\overset{\overset{\displaystyle \ddot{O}}{\|}}{C}-\ddot{C}H_2^- \longleftrightarrow CH_3-\overset{\overset{\displaystyle :\ddot{O}:^-}{|}}{C}=CH_2 \right]} + \underset{(pK_a = 16)}{CH_3OH}$$

$$\underset{\substack{\text{methyl acetate} \\ (pK_a = 24)}}{CH_3-O-\overset{\overset{\displaystyle O}{\|}}{C}-CH_3} + CH_3\ddot{O}:^- \rightleftharpoons \underset{\text{enolate of methyl acetate}}{\left[CH_3-O-\overset{\overset{\displaystyle \ddot{O}}{\|}}{C}-\ddot{C}H_2^- \longleftrightarrow CH_3-O-\overset{\overset{\displaystyle :\ddot{O}:^-}{|}}{C}=CH_2 \right]} + \underset{(pK_a = 16)}{CH_3OH}$$

Ester enolates are strong nucleophiles, and they undergo a wide range of interesting and useful reactions. Most of these reactions are related to the Claisen condensation, the most important of all ester condensations.

The **Claisen condensation** results when an ester molecule undergoes nucleophilic acyl substitution with an enolate ion serving as the nucleophile. First, the enolate attacks the carbonyl group, forming a tetrahedral intermediate. The intermediate has an alkoxy (—OR) group that acts as a leaving group, leaving a β-keto ester. The overall reaction combines two ester molecules to give a β-keto ester.

KEY MECHANISM 22-12 The Claisen Ester Condensation

The Claisen condensation is a nucleophilic acyl substitution on an ester, in which the attacking nucleophile is an enolate ion.

Step 1: Formation of the enolate ion.

$$R'O-\underset{\underset{\displaystyle \ddot{O}}{\|}}{C}-\overset{\overset{\displaystyle H}{|}}{\underset{\underset{\displaystyle H}{|}}{C_\alpha}}-R \overset{^-OR'}{\rightleftharpoons} \underset{\text{ester enolate ion}}{\left[R'O-\underset{\underset{\displaystyle \ddot{O}}{\|}}{C}-\underset{\underset{\displaystyle H}{|}}{\ddot{C}_\alpha^-}-R \longleftrightarrow R'O-\underset{\underset{\displaystyle :\ddot{O}:^-}{|}}{C}=\underset{\underset{\displaystyle H}{|}}{C_\alpha}-R \right]} + R'OH$$

(CONTINUED)

Step 2: Addition of the enolate to give a tetrahedral intermediate.

Step 3: Elimination of the alkoxide leaving group.

ester enolate → tetrahedral intermediate → a β-keto ester + $^{-}$:ÖR′

Notice that one molecule of the ester (deprotonated, reacting as the enolate) serves as the nucleophile to attack another molecule of the ester, which serves as the acylating reagent in this nucleophilic acyl substitution.

The β-keto ester products of Claisen condensations are more acidic than simple ketones, aldehydes, and esters because deprotonation gives an enolate whose negative charge is delocalized over both carbonyl groups. β-Keto esters have pK_a values around 11, showing they are stronger acids than water. In strong base such as ethoxide ion or hydroxide ion, the β-keto ester is rapidly and completely deprotonated.

a β-keto ester (pK_a = 11) ⇌ R′—OH (pK_a = 16–18) + [resonance-stabilized enolate ion]

Deprotonation of the β-keto ester provides a driving force for the Claisen condensation. The deprotonation is strongly exothermic, making the overall reaction exothermic and driving the reaction to completion. Because the base is consumed in the deprotonation step, a full equivalent of base must be used, and the Claisen condensation is said to be *base-promoted* rather than *base-catalyzed.* After the reaction is complete, addition of dilute acid converts the enolate back to the β-keto ester.

The following example shows the self-condensation of ethyl acetate to give ethyl acetoacetate (ethyl 3-oxobutanoate). Ethoxide is used as the base to avoid transesterification or hydrolysis of the ethyl ester (see Problem 22-34). The initial product is the enolate of ethyl acetoacetate, which is reprotonated in the final step.

2 CH_3—C(=O)—OCH_2CH_3 (ethyl acetate) $\xrightarrow[\text{sodium ethoxide}]{Na^+\ ^-OCH_2CH_3}$ [CH_3—C(=O)—$\bar{C}$H(Na^+)—C(=O)—OCH_2CH_3] (keto ester enolate) $\xrightarrow{H_3O^+}$ CH_3—C(=O)—CH_2—C(=O)—OCH_2CH_3 (ethyl acetoacetate (75%))

SOLVED PROBLEM 22-4

Propose a mechanism for the self-condensation of ethyl acetate to give ethyl acetoacetate.

SOLUTION

The first step is formation of the ester enolate. The equilibrium for this step lies far to the left; ethoxide deprotonates only a small fraction of the ester.

H—CH_2—C(=O)—OCH_2CH_3 (pK_a = 24) + $^-$:ÖCH_2CH_3 ⇌ [$^-$:CH_2—C(=O)—OCH_2CH_3] (enolate) + H—ÖCH_2CH_3 (pK_a = 16)

(continued)

The enolate ion attacks another molecule of the ester; expulsion of ethoxide ion gives ethyl acetoacetate.

$$CH_3-\overset{O}{\overset{\|}{C}}-OCH_2CH_3 + {}^{-}:CH_2-\overset{O}{\overset{\|}{C}}-OCH_2CH_3 \rightleftarrows CH_3-\overset{O^-}{\overset{|}{C}}(CH_2-\overset{O}{\overset{\|}{C}}-OCH_2CH_3)-\ddot{O}CH_2CH_3 \rightleftarrows CH_3-\overset{O}{\overset{\|}{C}}-CH_2-\overset{O}{\overset{\|}{C}}-OCH_2CH_3 + {}^{-}:\ddot{O}CH_2CH_3$$

nucleophilic attack — expulsion of ethoxide — ethyl acetoacetate

In the presence of ethoxide ion, ethyl acetoacetate is deprotonated to give its enolate. This exothermic deprotonation helps to drive the reaction to completion.

$$\underset{(pK_a = 11)}{CH_3-\overset{O}{\overset{\|}{C}}-\overset{H}{\overset{|}{C}H}-\overset{O}{\overset{\|}{C}}-OCH_2CH_3} + {}^{-}:\ddot{O}CH_2CH_3 \rightleftarrows \underset{\text{enolate}}{\left[CH_3-\overset{O}{\overset{\|}{C}}-\ddot{C}H-\overset{O}{\overset{\|}{C}}-OCH_2CH_3\right]} + \underset{(pK_a = 16)}{H-\ddot{O}CH_2CH_3}$$

When the reaction is complete, the enolate ion is reprotonated to give ethyl acetoacetate.

$$\underset{\text{enolate}}{\left[CH_3-\overset{O}{\overset{\|}{C}}-\ddot{C}H-\overset{O}{\overset{\|}{C}}-OCH_2CH_3\right]} \xrightarrow{H_3O^+} \underset{\text{ethyl acetoacetate}}{CH_3-\overset{O}{\overset{\|}{C}}-\overset{H}{\overset{|}{C}H}-\overset{O}{\overset{\|}{C}}-OCH_2CH_3}$$

PROBLEM 22-34

Ethoxide is used as the base in the condensation of ethyl acetate to avoid some unwanted side reactions. Show what side reactions would occur if the following bases were used.
(a) sodium methoxide **(b)** sodium hydroxide

Application: Biochemistry

Enzymes called polyketide synthases catalyze a series of Claisen-type reactions to generate many useful natural products, such as the antibiotic erythromycin (page 1089). These enzymes use thioesters instead of the oxygen esters.

PROBLEM 22-35

Esters with only one α hydrogen generally give poor yields in the Claisen condensation. Propose a mechanism for the Claisen condensation of ethyl isobutyrate, and explain why a poor yield is obtained.

PROBLEM 22-36

Predict the products of self-condensation of the following esters.
(a) methyl propanoate + $NaOCH_3$ **(b)** ethyl phenylacetate + $NaOCH_2CH_3$
(c) cyclopentyl$-CH_2-\overset{O}{\overset{\|}{C}}-OCH_3$ + $NaOCH_3$
(d) 2-furyl$-CH_2CH_2-$COOEt + NaOEt

SOLVED PROBLEM 22-5

Show what ester would undergo Claisen condensation to give the following β-keto ester.

$$Ph-CH_2-CH_2-\overset{O}{\overset{\|}{C}}-\underset{CH_2-Ph}{\underset{|}{CH}}-\overset{O}{\overset{\|}{C}}-OCH_3$$

(continued)

SOLUTION

First, break the structure apart at the α,β bond (α,β to the ester carbonyl). This is the bond formed in the Claisen condensation.

$$Ph{-}CH_2{-}CH_2{-}\overset{\overset{O}{\|}}{\underset{\beta}{C}}{-} \quad {-}\overset{\alpha}{C}H(CH_2{-}Ph){-}\overset{\overset{O}{\|}}{C}{-}OCH_3$$

Next, replace the α proton that was lost, and replace the alkoxy group that was lost from the carbonyl. Two molecules of methyl 3-phenylpropionate result.

$$Ph{-}CH_2{-}CH_2{-}\overset{\overset{O}{\|}}{C}{-}OCH_3 \qquad H{-}CH(CH_2{-}Ph){-}\overset{\overset{O}{\|}}{C}{-}OCH_3$$

Now draw out the reaction. Sodium methoxide is used as the base because the reactants are methyl esters.

$$2\ Ph{-}CH_2{-}CH_2{-}\overset{\overset{O}{\|}}{C}{-}OCH_3 \xrightarrow[(2)\ H_3O^+]{(1)\ Na^+\ {}^-OCH_3} Ph{-}CH_2{-}CH_2{-}\overset{\overset{O}{\|}}{C}{-}CH(CH_2{-}Ph){-}\overset{\overset{O}{\|}}{C}{-}OCH_3$$

PROBLEM-SOLVING HINT

The Claisen condensation occurs by a nucleophilic acyl substitution, with different forms of the ester acting as both the nucleophile (the enolate) and the electrophile (the ester carbonyl).

PROBLEM 22-37

Propose a mechanism for the self-condensation of methyl 3-phenylpropionate promoted by sodium methoxide.

PROBLEM 22-38

Show what esters would undergo Claisen condensation to give the following β-keto esters.

(a) $CH_3CH_2CH_2{-}C({=}O){-}CH(CH_2CH_3){-}\overset{\overset{O}{\|}}{C}{-}OCH_2CH_3$

(b) $Ph{-}CH_2{-}C({=}O){-}CH(Ph){-}\overset{\overset{O}{\|}}{C}{-}OCH_3$

(c) $(CH_3)_2CHCH_2{-}\overset{\overset{O}{\|}}{C}{-}CH(CH(CH_3)_2){-}\overset{\overset{O}{\|}}{C}{-}OEt$

22-13 The Dieckmann Condensation: A Claisen Cyclization

An internal Claisen condensation of a diester forms a ring. Such an internal Claisen cyclization is called a **Dieckmann condensation** or a **Dieckmann cyclization.** Five- and six-membered rings are easily formed by Dieckmann condensations. Rings smaller than five carbons or larger than six carbons are rarely formed by this method.

The following examples of the Dieckmann condensation show that a 1,6-diester gives a five-membered ring, and a 1,7-diester gives a six-membered ring.

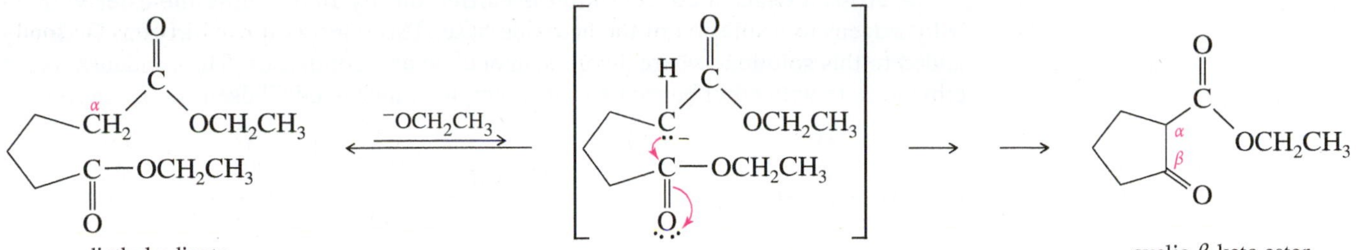

diethyl adipate
(a 1,6-diester)

$^{-}OCH_2CH_3$

cyclic β-keto ester
(80%)

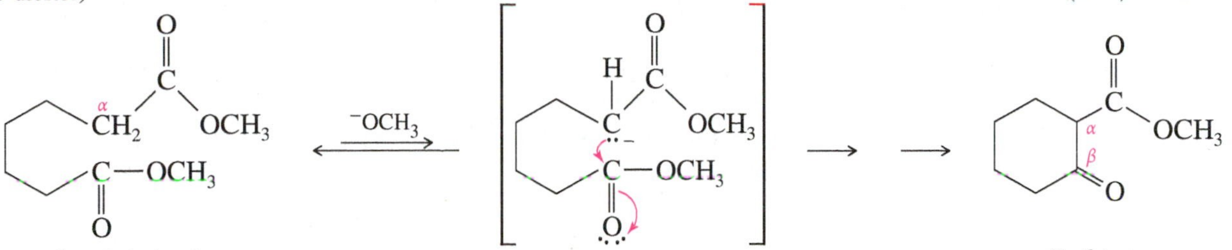

dimethyl pimelate
(a 1,7-diester)

$^{-}OCH_3$

cyclic β-keto ester

PROBLEM 22-39

Propose mechanisms for the two Dieckmann condensations just shown.

PROBLEM 22-40

Some (but not all) of the following keto esters can be formed by Dieckmann condensations. Determine which ones are possible, and draw the starting diesters.

(a) **(b)**

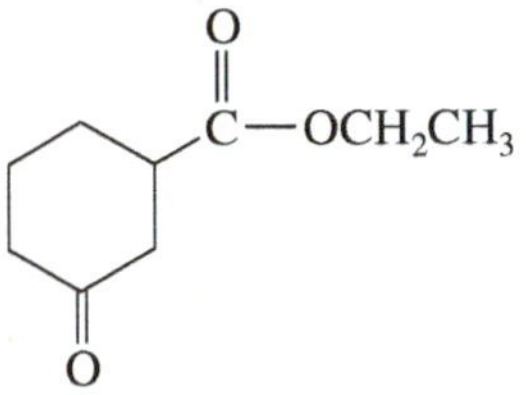

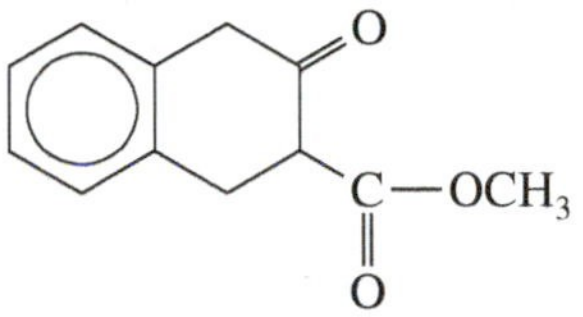

(c) 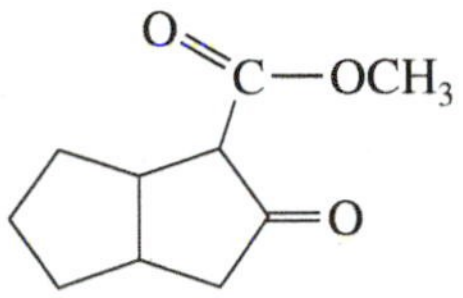 **(d)**

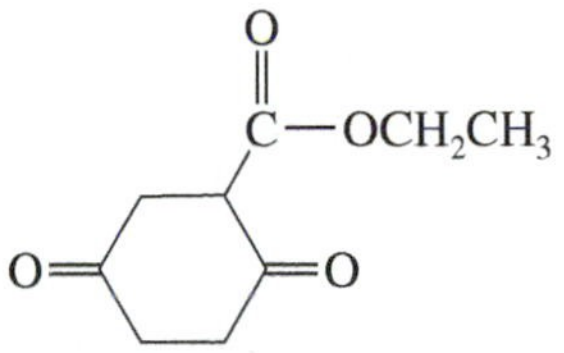

(*Hint:* Consider using a protecting group.)

22-14 Crossed Claisen Condensations

Claisen condensations can take place between different esters, particularly when only one of the esters has the α hydrogens needed to form an enolate. In a **crossed Claisen condensation,** an ester without α hydrogens serves as the electrophilic component. Some useful esters without α hydrogens are benzoate, formate, carbonate, and oxalate esters.

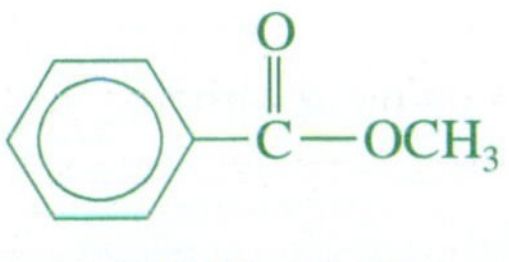

methyl benzoate

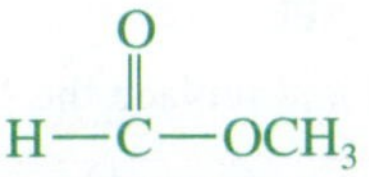

methyl formate ($H-CO-OCH_3$)

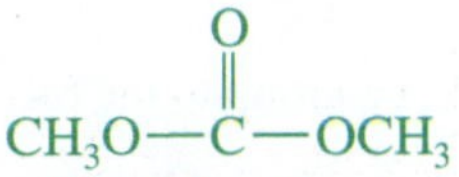

dimethyl carbonate ($CH_3O-CO-OCH_3$)

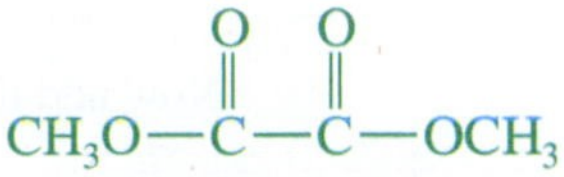

dimethyl oxalate ($CH_3O-CO-CO-OCH_3$)

A crossed Claisen condensation is carried out by first adding the ester without α hydrogens to a solution of the alkoxide base. The ester with α hydrogens is slowly added to this solution, where it forms an enolate and condenses. The condensation of ethyl acetate with ethyl benzoate is an example of a crossed Claisen condensation.

$$\text{Ph}-\overset{\text{O}}{\overset{\|}{\text{C}}}-\text{OC}_2\text{H}_5 + \overset{\alpha}{\text{CH}_3}-\overset{\text{O}}{\overset{\|}{\text{C}}}-\text{OC}_2\text{H}_5 \xrightarrow[(2)\ \text{H}_3\text{O}^+]{(1)\ \text{Na}^+\ ^-\text{OCH}_2\text{CH}_3} \text{Ph}-\underset{\beta}{\overset{\text{O}}{\overset{\|}{\text{C}}}}-\overset{\alpha}{\text{CH}_2}-\overset{\text{O}}{\overset{\|}{\text{C}}}-\text{OC}_2\text{H}_5$$

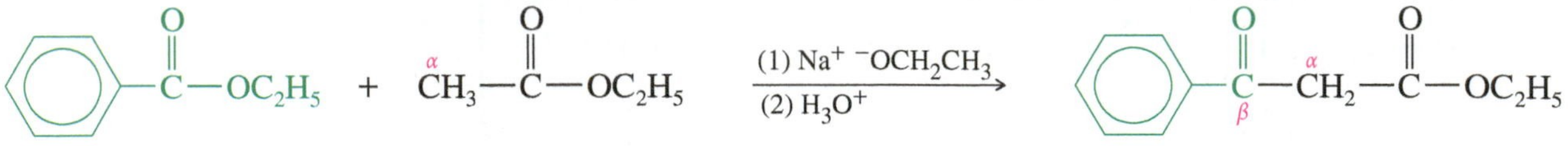

ethyl benzoate (no α hydrogens) — ethyl acetate (forms enolate) — ethyl benzoylacetate

PROBLEM 22-41

Propose a mechanism for the crossed Claisen condensation between ethyl acetate and ethyl benzoate.

Application: Biochemistry

Fatty acids are made in the body by a series of Claisen-type reactions catalyzed by a class of enzymes called fatty acid synthases. The enzymes use the thioesters of malonate and acetate as building blocks (see Figure 22-3 on page 1150).

PROBLEM 22-42

Predict the products from crossed Claisen condensation of the following pairs of esters. Indicate which combinations are poor choices for crossed Claisen condensations.

(a) $\text{Ph}-\text{CH}_2-\overset{\text{O}}{\overset{\|}{\text{C}}}-\text{OCH}_3 + \text{Ph}-\overset{\text{O}}{\overset{\|}{\text{C}}}-\text{OCH}_3 \longrightarrow$

(b) $\text{Ph}-\text{CH}_2-\overset{\text{O}}{\overset{\|}{\text{C}}}-\text{OCH}_3 + \text{CH}_3-\overset{\text{O}}{\overset{\|}{\text{C}}}-\text{OCH}_3 \longrightarrow$

(c) $\text{CH}_3-\overset{\text{O}}{\overset{\|}{\text{C}}}-\text{OC}_2\text{H}_5 + \text{C}_2\text{H}_5\text{O}-\overset{\text{O}}{\overset{\|}{\text{C}}}-\overset{\text{O}}{\overset{\|}{\text{C}}}-\text{OC}_2\text{H}_5 \longrightarrow$

(d) $\text{CH}_3-\text{CH}_2-\overset{\text{O}}{\overset{\|}{\text{C}}}-\text{OC}_2\text{H}_5 + \text{C}_2\text{H}_5\text{O}-\overset{\text{O}}{\overset{\|}{\text{C}}}-\text{OC}_2\text{H}_5 \longrightarrow$

SOLVED PROBLEM 22-6

Show how a crossed Claisen condensation might be used to prepare

$$\text{H}-\overset{\text{O}}{\overset{\|}{\text{C}}}-\underset{\text{Ph}}{\underset{|}{\text{CH}}}-\overset{\text{O}}{\overset{\|}{\text{C}}}-\text{OCH}_3$$

SOLUTION

Break the α,β bond of this β-keto ester, since that is the bond formed in the Claisen condensation.

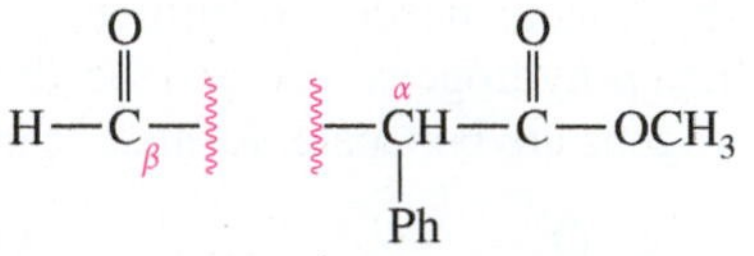

$$\text{H}-\underset{\beta}{\overset{\text{O}}{\overset{\|}{\text{C}}}}-\ \wr\wr\ -\underset{\text{Ph}}{\underset{|}{\overset{\alpha}{\text{CH}}}}-\overset{\text{O}}{\overset{\|}{\text{C}}}-\text{OCH}_3$$

Now add the alkoxy group to the carbonyl and replace the proton on the α carbon.

$$\text{H}-\overset{\text{O}}{\overset{\|}{\text{C}}}-\text{OCH}_3 \qquad \text{H}-\underset{\text{Ph}}{\underset{|}{\overset{\alpha}{\text{CH}}}}-\overset{\text{O}}{\overset{\|}{\text{C}}}-\text{OCH}_3$$

(continued)

Write out the reaction, making sure that one of the components has α hydrogens and the other does not.

$$\underset{\text{no }\alpha\text{ hydrogens}}{H{-}\overset{O}{\overset{\|}{C}}{-}OCH_3} \qquad \underset{\text{forms enolate}}{H{-}\underset{Ph}{\underset{|}{\overset{\alpha}{C}H}}{-}\overset{O}{\overset{\|}{C}}{-}OCH_3} \xrightarrow[(2)\ H_3O^+]{(1)\ Na^+\ {}^-OCH_3} H{-}\overset{O}{\overset{\|}{C}}{-}\underset{Ph}{\underset{|}{\overset{\alpha}{C}H}}{-}\overset{O}{\overset{\|}{C}}{-}OCH_3$$

PROBLEM 22-43

Show how crossed Claisen condensations could be used to prepare the following esters.

(a) $Ph{-}\overset{O}{\overset{\|}{C}}{-}\underset{CH_3}{\underset{|}{CH}}{-}\overset{O}{\overset{\|}{C}}{-}OCH_2CH_3$

(b) $Ph{-}\underset{\underset{O\quad\ O}{C(=O){-}\overset{}{C}(=O){-}OCH_3}}{\underset{|}{CH}}{-}\overset{O}{\overset{\|}{C}}{-}OCH_3$

(c) $EtO{-}\overset{O}{\overset{\|}{C}}{-}\underset{Ph}{\underset{|}{CH}}{-}\overset{O}{\overset{\|}{C}}{-}OCH_2CH_3$

(d) $(CH_3)_3C{-}\overset{O}{\overset{\|}{C}}{-}\underset{CH_2CH_2CH_3}{\underset{|}{CH}}{-}\overset{O}{\overset{\|}{C}}{-}OCH_3$

Crossed Claisen condensations between ketones and esters are also possible. Ketones are more acidic than esters, and the ketone component is more likely to deprotonate and serve as the enolate component in the condensation. The ketone enolate attacks the ester, which undergoes nucleophilic acyl substitution and thereby acylates the ketone.

$$\underset{\substack{\text{ketone, p}K_a = 20\\ \text{more acidic}}}{R{-}\overset{\alpha}{C}H_2{-}\overset{O}{\overset{\|}{C}}{-}R'} \qquad \underset{\substack{\text{ester, p}K_a = 24\\ \text{less acidic}}}{R{-}\overset{\alpha}{C}H_2{-}\overset{O}{\overset{\|}{C}}{-}OR'}$$

$$\underset{\text{ketone enolate}}{{-}\overset{O}{\overset{\|}{C}}{-}\overset{\alpha}{\underset{|}{C}}{:}^-} \quad \underset{\text{ester}}{R{-}\overset{O}{\overset{\|}{C}}{-}OR'} \rightleftarrows \underset{\text{tetrahedral intermediate}}{{-}\overset{O}{\overset{\|}{C}}{-}\overset{\alpha}{\underset{|}{C}}{-}\overset{\beta}{\underset{OR'}{\underset{|}{\overset{R}{\overset{|}{C}}}}}{-}\ddot{O}{:}^-} \longrightarrow \underset{\text{acylated ketone}}{{-}\overset{O}{\overset{\|}{C}}{-}\overset{\alpha}{\underset{|}{C}}{-}\overset{\beta}{C}(=O)R} \;+\; {}^-OR'$$

This condensation works best if the ester has no α hydrogens, so that it cannot form an enolate. Because of the difference in acidities, however, the reaction is sometimes successful between ketones and esters even when both have α hydrogens. The following examples show some crossed Claisen condensations between ketones and esters. Notice the variety of difunctional and trifunctional compounds that can be produced by appropriate choices of esters.

$$\underset{\text{acetone}}{\overset{\alpha}{C}H_3{-}\overset{O}{\overset{\|}{C}}{-}CH_3} + \underset{\text{methyl benzoate}}{C_6H_5{-}\overset{O}{\overset{\|}{C}}{-}OCH_3} \xrightarrow[(2)\ H_3O^+]{(1)\ Na^+\ {}^-OCH_3} \underset{\text{a }\beta\text{-diketone}}{C_6H_5{-}\underset{\beta}{\overset{O}{\overset{\|}{C}}}{-}\overset{\alpha}{C}H_2{-}\overset{O}{\overset{\|}{C}}{-}CH_3}$$

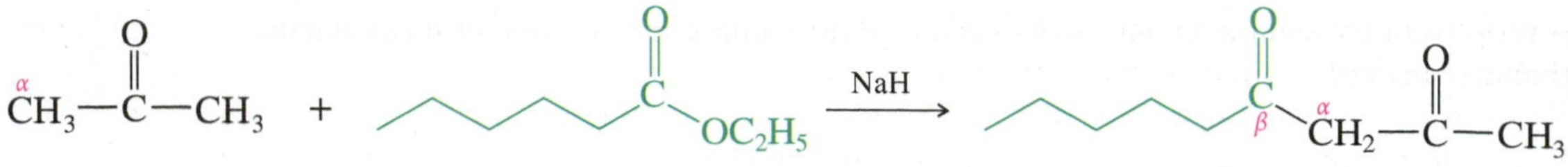

$$CH_3-\overset{O}{\overset{\|}{C}}-CH_3 \;+\; \text{ethyl hexanoate} \xrightarrow{\text{NaH}} \text{a }\beta\text{-diketone}$$

acetone — ethyl hexanoate — a β-diketone

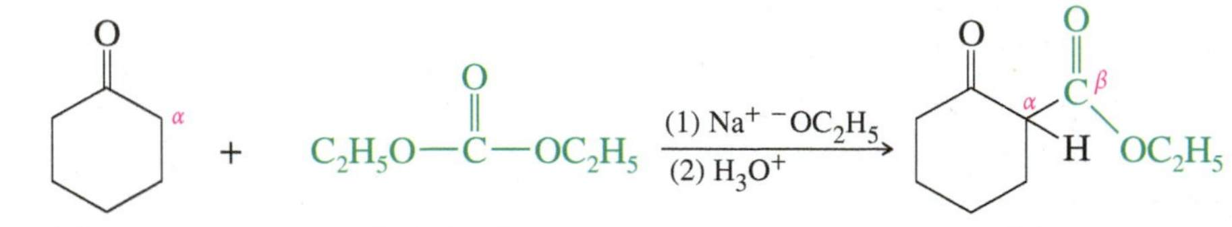

$$\text{cyclohexanone} + C_2H_5O-\overset{O}{\overset{\|}{C}}-OC_2H_5 \xrightarrow[(2)\ H_3O^+]{(1)\ Na^+\ ^-OC_2H_5} \text{a }\beta\text{-keto ester}$$

cyclohexanone — diethyl carbonate — a β-keto ester

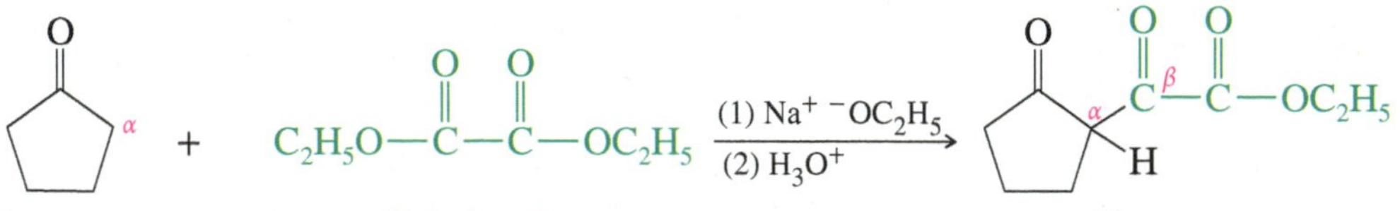

$$\text{cyclopentanone} + C_2H_5O-\overset{O}{\overset{\|}{C}}-\overset{O}{\overset{\|}{C}}-OC_2H_5 \xrightarrow[(2)\ H_3O^+]{(1)\ Na^+\ ^-OC_2H_5} \text{a diketo ester}$$

cyclopentanone — diethyl oxalate — a diketo ester

PROBLEM-SOLVING HINT

Claisen and crossed Claisen condensations are important synthetic tools and interesting mechanistic examples. Practice predicting product structures and drawing mechanisms until you gain confidence.

PROBLEM 22-44

Predict the major products of the following crossed Claisen condensations.

(a) $\text{cyclohexanone} + Ph-\overset{O}{\overset{\|}{C}}-OCH_3 \xrightarrow{NaOCH_3}$

(b) $CH_3CH_2-\overset{O}{\overset{\|}{C}}-CH_3 + CH_3CH_2O-\overset{O}{\overset{\|}{C}}-OCH_2CH_3 \xrightarrow{NaOCH_2CH_3}$

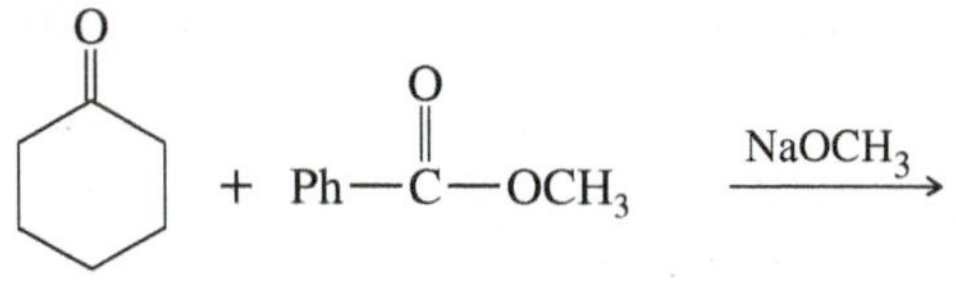

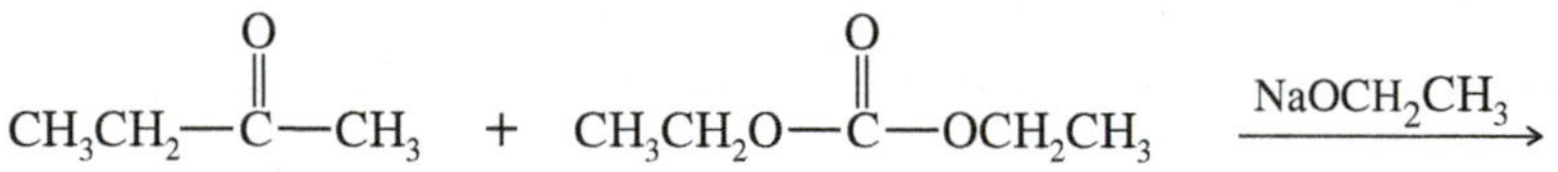

(c) $CH_3-\overset{O}{\overset{\|}{C}}-CH_2CH_2-\overset{O}{\overset{\|}{C}}-OCH_2CH_3 \xrightarrow{NaOCH_2CH_3}$

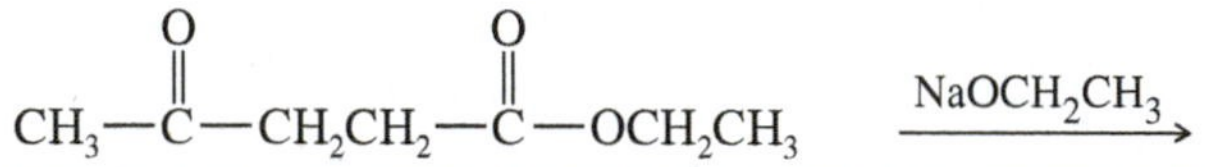

PROBLEM 22-45

Show how Claisen condensations could be used to make the following compounds.

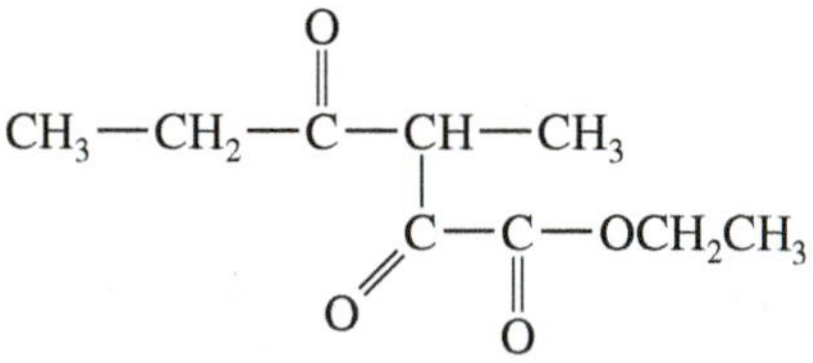

(a) 2-benzoylcyclopentanone ($\overset{O}{\overset{\|}{C}}-Ph$ on the ring carbon adjacent to C=O)

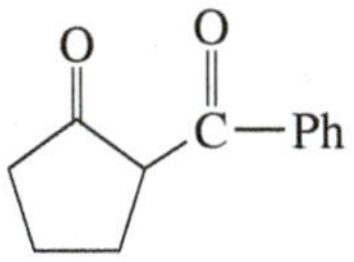

(b) $CH_3-CH_2-\overset{O}{\overset{\|}{C}}-CH(CH_3)-\overset{O}{\overset{\|}{C}}-\overset{O}{\overset{\|}{C}}-OCH_2CH_3$

(c) cyclohexane-1,3-dione

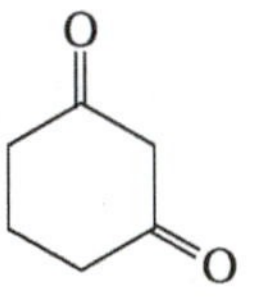

(d) cyclohexane-1,3-dione bearing $\overset{O}{\overset{\|}{C}}-OCH_2CH_3$ on C-2

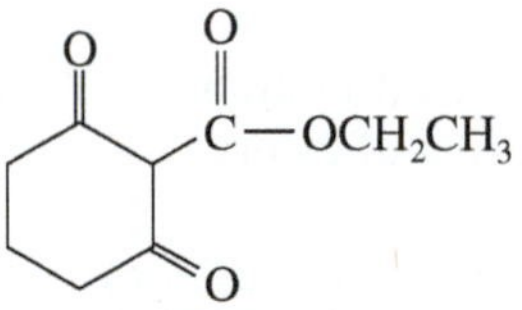

22-15 Syntheses Using β-Dicarbonyl Compounds

Many alkylation and acylation reactions are most effective using anions of β-dicarbonyl compounds that can be completely deprotonated and converted to their enolate ions by common bases such as alkoxide ions. The *malonic ester synthesis* and the *acetoacetic ester synthesis* use the enhanced acidity of the

α protons in malonic ester and acetoacetic ester to accomplish alkylations and acylations that are difficult or impossible with simple esters.

We have seen that most ester condensations use alkoxides to form enolate ions. With simple esters, only a small amount of enolate is formed. The equilibrium favors the alkoxide and the ester. The alkoxide often interferes with the desired reaction. For example, if we want an alkyl halide to alkylate an enolate, alkoxide ion in the solution will attack the alkyl halide and form an ether.

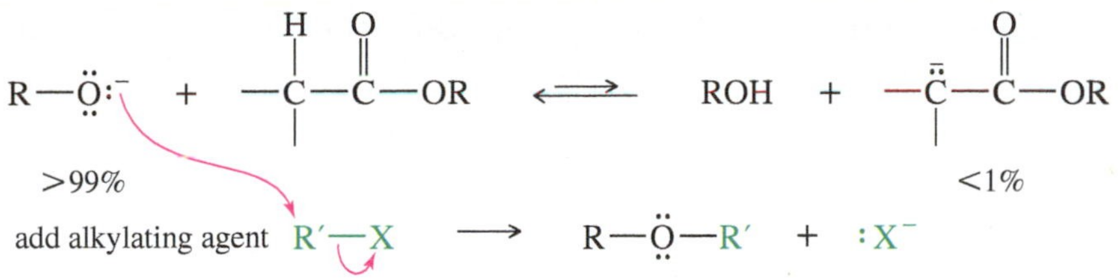

$$R-\ddot{O}:^- \;+\; -\overset{H}{\underset{|}{C}}-\overset{O}{\overset{\|}{C}}-OR \;\rightleftharpoons\; ROH \;+\; -\overset{..}{\underset{|}{C}}{}^{-}-\overset{O}{\overset{\|}{C}}-OR$$

>99%   <1%

add alkylating agent $R'-X \longrightarrow R-\ddot{O}-R' + :X^-$

In contrast, β-dicarbonyl compounds such as malonic ester and acetoacetic ester are more acidic than alcohols. They are completely deprotonated by alkoxides, and the resulting enolates are easily alkylated and acylated. At the end of the synthesis, one of the carbonyl groups can be removed by decarboxylation, leaving a compound that is difficult or impossible to make by direct alkylation or acylation of a simple ester.

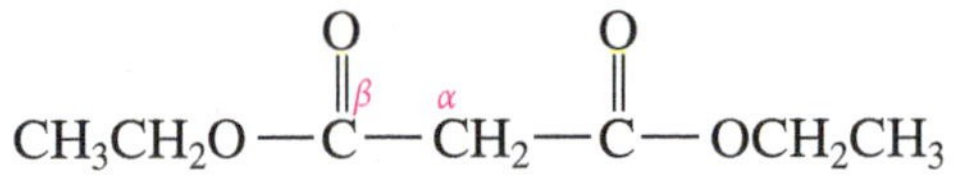

$$CH_3CH_2O-\overset{O}{\overset{\|}{C}}{}_{\beta}-\overset{\alpha}{CH_2}-\overset{O}{\overset{\|}{C}}-OCH_2CH_3$$

diethyl malonate (malonic ester)

$$CH_3-\overset{O}{\overset{\|}{C}}{}_{\beta}-\overset{\alpha}{CH_2}-\overset{O}{\overset{\|}{C}}-OCH_2CH_3$$

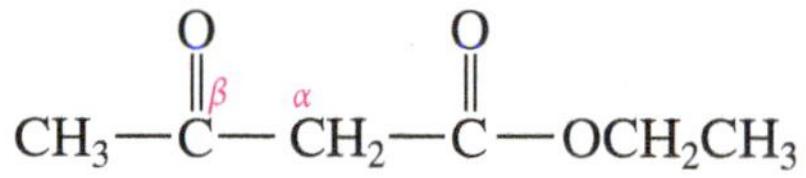

ethyl acetoacetate (acetoacetic ester)

First we compare the acidity advantages of β-dicarbonyl compounds, and then we consider how these compounds can be used in synthesis.

Acidities of β-Dicarbonyl Compounds Table 22-1 compares the acidities of some carbonyl compounds with the acidities of alcohols and water. Notice the large increase in acidity for compounds with two carbonyl groups beta to each other. The α protons of the β-dicarbonyl compounds are more acidic than the hydroxy protons of water and alcohols. This enhanced acidity results from increased stability of the enolate ion. The negative charge is delocalized over two carbonyl groups rather than just one, as shown by the resonance forms for the enolate ion of diethyl malonate (also called *malonic ester*).

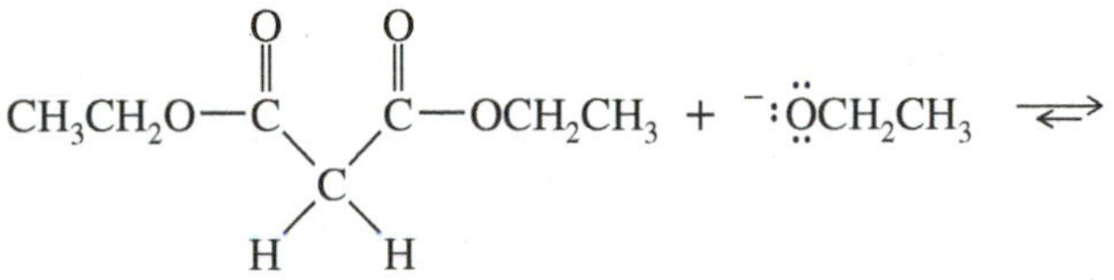

$$CH_3CH_2O-\overset{O}{\overset{\|}{C}}-CH_2-\overset{O}{\overset{\|}{C}}-OCH_2CH_3 \;+\; {}^-:\ddot{O}CH_2CH_3 \;\rightleftharpoons$$

diethyl malonate (malonic ester)
($pK_a = 13$)

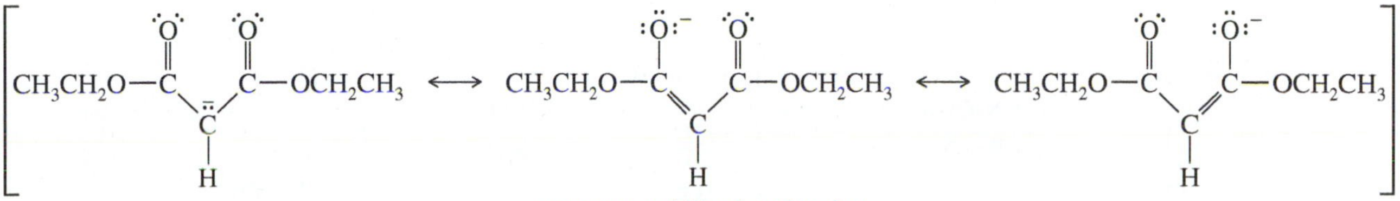

$$\left[CH_3CH_2O-\overset{O}{\overset{\|}{C}}-\overset{..}{\underset{H}{C}}{}^{-}-\overset{O}{\overset{\|}{C}}-OCH_2CH_3 \longleftrightarrow CH_3CH_2O-\overset{:\ddot{O}:^-}{\overset{|}{C}}=\underset{H}{C}-\overset{O}{\overset{\|}{C}}-OCH_2CH_3 \longleftrightarrow CH_3CH_2O-\overset{O}{\overset{\|}{C}}-\underset{H}{C}=\overset{:\ddot{O}:^-}{\overset{|}{C}}-OCH_2CH_3 \right]$$

resonance-stabilized enolate ion

PROBLEM 22-46

Show the resonance forms for the enolate ions that result when the following compounds are treated with a strong base.

(a) ethyl acetoacetate   **(b)** pentane-2,4-dione
(c) ethyl α-cyanoacetate   **(d)** nitroacetone

PROBLEM-SOLVING HINT

Table 22-1 shows that ketone carbonyl groups are more effective than ester carbonyl groups for stabilizing the negative charge of an enolate ion. This difference reflects the electron-donating nature of the alkoxy group in the ester.

TABLE 22-1
Typical Acidities of Carbonyl Compounds

Conjugate Acid	Conjugate Base	pK_a
Simple ketones and esters		
$CH_3-C(=O)-CH_3$ (α at CH_3) acetone	$[{}^-\!:CH_2-C(=O)-CH_3]$	20
$CH_3-C(=O)-OCH_2CH_3$ (α at CH_3) ethyl acetate	$[{}^-\!:CH_2-C(=O)-OCH_2CH_3]$	24
β-dicarbonyl compounds		
$CH_3-C(=O)-CH_2-C(=O)-CH_3$ (β, α) pentane-2,4-dione (acetylacetone)	$[CH_3-C(=O)-\ddot{C}H-C(=O)-CH_3]$	9
$CH_3-C(=O)-CH_2-C(=O)-OCH_2CH_3$ (β, α) ethyl acetoacetate (acetoacetic ester)	$[CH_3-C(=O)-\ddot{C}H-C(=O)-OCH_2CH_3]$	11
$CH_3CH_2O-C(=O)-CH_2-C(=O)-OCH_2CH_3$ (β, α) diethyl malonate (malonic ester)	$[CH_3CH_2O-C(=O)-\ddot{C}H-C(=O)-OCH_2CH_3]$	13
Commonly used bases (for comparison)		
$H-O-H$ water	^-OH	15.7
CH_3O-H methanol	CH_3O^-	15.5
CH_3CH_2O-H ethanol	$CH_3CH_2O^-$	15.9

22-16 The Malonic Ester Synthesis

The **malonic ester synthesis** makes substituted derivatives of acetic acid. Malonic ester (diethyl malonate) is alkylated or acylated on the more acidic carbon that is α to both carbonyl groups, and the resulting derivative is hydrolyzed and allowed to decarboxylate.

Malonic ester synthesis

$$H_2C(COOC_2H_5)_2 \xrightarrow[(2)\ R-X]{(1)\ NaOC_2H_5} R-CH(COOC_2H_5)_2 \xrightarrow[\text{heat}]{H_3O^+} R-CH_2-COOH + CO_2(g)\uparrow + 2\ C_2H_5OH$$

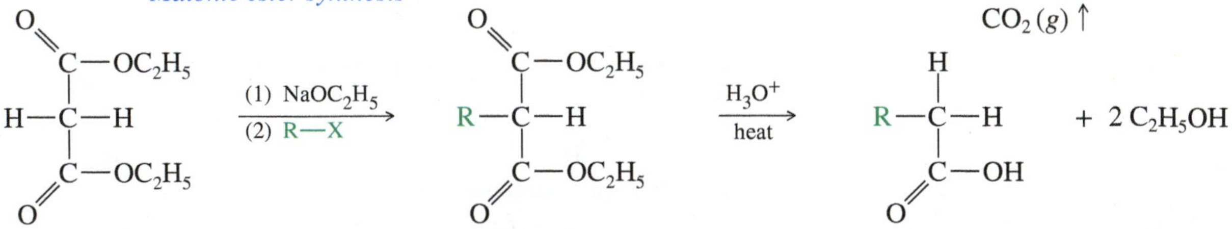

malonic ester → alkylated malonic ester → substituted acetic acid

Malonic ester is completely deprotonated by sodium ethoxide. The resulting enolate ion is alkylated by an unhindered alkyl halide, tosylate, or other electrophilic reagent. This step is an S_N2 displacement, requiring a good S_N2 substrate.

$$CH_3CH_2OC(=O)-CH_2-C(=O)OCH_2CH_3 \xrightleftharpoons{Na^+\ ^-OCH_2CH_3} \left[CH_3CH_2OC(=O)-\ddot{C}H^--C(=O)OCH_2CH_3 \quad R-X\right] \longrightarrow CH_3CH_2OC(=O)-CH(R)-C(=O)OCH_2CH_3$$

malonic ester — alkylated diethyl malonate

Hydrolysis of the alkylated diethyl malonate (a diethyl alkylmalonic ester) gives a malonic acid derivative.

$$CH_3CH_2O-C(=O)-CH(R)-C(=O)-OCH_2CH_3 \xrightarrow[H_2O]{H^+,\ heat} \left[HO-C(=O)-CH(R)-C(=O)-OH\right]$$

a diethyl alkylmalonate — an alkylmalonic acid

Any carboxylic acid with a carbonyl group in the β position is prone to decarboxylate. At the temperature of the hydrolysis, the alkylmalonic acid loses CO_2 to give a substituted derivative of acetic acid. Decarboxylation takes place through a cyclic transition state, initially giving an enol that quickly tautomerizes to the product, a substituted acetic acid.

$$\text{alkylmalonic acid} \longrightarrow CO_2 + \text{enol } (R(H)C=C(OH)_2) \xrightarrow{\text{tautomerism}} R-CH_2-C(=O)OH + CO_2\uparrow$$

alkylmalonic acid — CO_2 + enol — substituted acetic acid

The product of a malonic ester synthesis is a substituted acetic acid, with the substituent being the group used to alkylate malonic ester. In effect, the second carboxyl group is temporary, allowing the ester to be easily deprotonated and alkylated. Hydrolysis and decarboxylation remove the temporary carboxyl group, leaving the substituted acetic acid.

$$(COOC_2H_5)CH_2-C(=O)-OC_2H_5 \xrightarrow[(2)\ R-X]{(1)\ ^-OC_2H_5} R-CH(COOC_2H_5)-C(=O)-OC_2H_5 \xrightarrow[H_2O]{H^+,\ heat} R-CH_2-C(=O)-OH + CO_2\uparrow + 2\ CH_3CH_2OH$$

temporary ester group

malonic ester — alkylmalonic ester — substituted acetic acid

The alkylmalonic ester has a second acidic proton that can be removed by a base. Removing this proton and alkylating the enolate with another alkyl halide gives a dialkylated malonic ester. Hydrolysis and decarboxylation lead to a disubstituted derivative of acetic acid.

$$R-CH(COOC_2H_5)-C(=O)-OC_2H_5 \xrightarrow[(2)\ R'-X]{(1)\ NaOCH_2CH_3} R-C(R')(COOC_2H_5)-C(=O)-OC_2H_5 \xrightarrow[H_2O]{H^+,\ heat} R-CH(R')-C(=O)-OH + CO_2\uparrow + 2\ CH_3CH_2OH$$

alkylmalonic ester — dialkylmalonic ester — disubstituted acetic acid

The malonic ester synthesis is useful for making cycloalkanecarboxylic acids, some of which are not easily made by any other method. The ring is formed from a dihalide, using a double alkylation of malonic ester. The following synthesis of cyclobutanecarboxylic acid shows that a strained four-membered ring system can be generated by this ester alkylation, even though most other condensations cannot form four-membered rings.

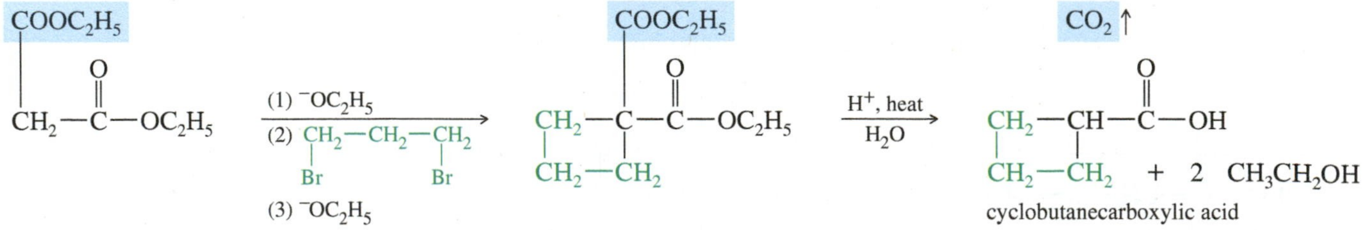

$$\mathrm{CH_2(COOC_2H_5)-C(=O)-OC_2H_5} \xrightarrow[\text{(2) } \mathrm{BrCH_2-CH_2-CH_2Br} \quad \text{(3) } ^{-}\mathrm{OC_2H_5}]{\text{(1) } ^{-}\mathrm{OC_2H_5}} \text{cyclobutane-}\mathrm{C(COOC_2H_5)-C(=O)-OC_2H_5} \xrightarrow[\mathrm{H_2O}]{\mathrm{H^+,\ heat}} \text{cyclobutane-}\mathrm{CH-C(=O)-OH} + \mathrm{CO_2}\uparrow + 2\ \mathrm{CH_3CH_2OH}$$

cyclobutanecarboxylic acid

The malonic ester synthesis might seem like an arcane technique that only an organic chemist would use. Still, it is much like the method that cells use to synthesize the long-chain fatty acids found in fats, oils, waxes, and cell membranes. Figure 22-3 outlines the steps that take place in the lengthening of a fatty acid chain by two carbon atoms at a time. The growing acid derivative (acyl-CoA) is activated as its thioester with coenzyme A (structure on page 1096). A malonic ester acylation adds two of the three carbons of malonic acid (as malonyl-CoA), with the third carbon lost in the decarboxylation. A β-keto ester results. Reduction of the ketone, followed by dehydration and reduction of the double bond, gives an acyl group that has been lengthened by two carbon atoms. The cycle is repeated until the acid has reached the necessary length, always with an even number of carbon atoms.

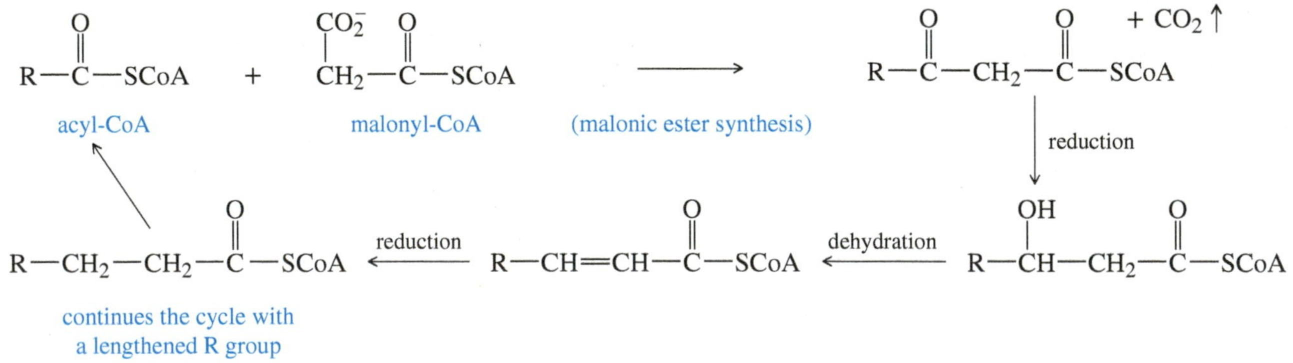

$$\underset{\text{acyl-CoA}}{\mathrm{R-C(=O)-SCoA}} + \underset{\text{malonyl-CoA}}{\mathrm{CH_2(CO_2^-)-C(=O)-SCoA}} \xrightarrow{\text{(malonic ester synthesis)}} \mathrm{R-C(=O)-CH_2-C(=O)-SCoA} + \mathrm{CO_2}\uparrow$$

$$\xrightarrow{\text{reduction}} \mathrm{R-CH(OH)-CH_2-C(=O)-SCoA} \xrightarrow{\text{dehydration}} \mathrm{R-CH{=}CH-C(=O)-SCoA} \xrightarrow{\text{reduction}} \mathrm{R-CH_2-CH_2-C(=O)-SCoA}$$

continues the cycle with a lengthened R group (→ acyl-CoA)

FIGURE 22-3
Fatty acid biosynthesis. Activated as its coenzyme A thioester, the growing fatty acid (acyl-CoA) acylates malonyl-CoA in a malonic ester synthesis. Two carbon atoms are added, with the third lost as CO_2. Enzymatic reduction, dehydration, and further reduction gives a fatty acid that has been lengthened by two carbon atoms.

PROBLEM-SOLVING HINT

A malonic ester synthesis goes through alkylation of the enolate, hydrolysis, and decarboxylation. To design a synthesis, look at the product and see what groups are added to acetic acid. Use those groups to alkylate malonic ester, then hydrolyze and decarboxylate.

SOLVED PROBLEM 22-7

Show how the malonic ester synthesis is used to prepare 2-benzylbutanoic acid.

SOLUTION

2-Benzylbutanoic acid is a substituted acetic acid having the substituents $\mathrm{Ph-CH_2-}$ and $\mathrm{CH_3CH_2-}$.

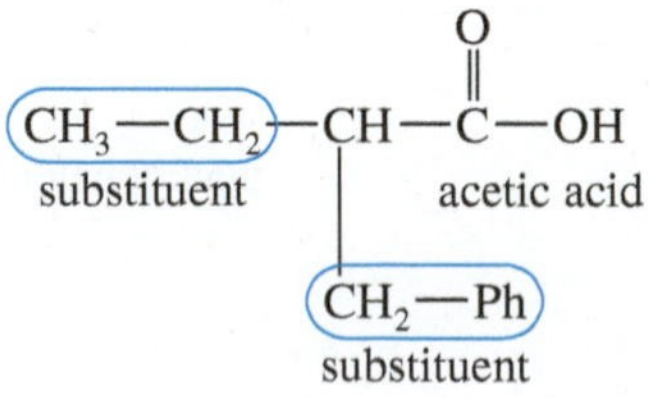

$$\underset{\text{substituent}}{\boxed{\mathrm{CH_3-CH_2}}}-\underset{\substack{|\\ \boxed{\mathrm{CH_2-Ph}}\\ \text{substituent}}}{\mathrm{CH}}-\underset{\text{acetic acid}}{\mathrm{C(=O)-OH}}$$

(continued)

Adding these substituents to the enolate of malonic ester eventually gives the correct product.

$$\underset{\text{malonic ester}}{CH_2(COOC_2H_5)-\overset{O}{\overset{\|}{C}}-OC_2H_5} \xrightarrow[\text{(2) }PhCH_2Br]{\text{(1) }NaOCH_2CH_3} CH(COOC_2H_5)(CH_2Ph)-\overset{O}{\overset{\|}{C}}-OC_2H_5 \xrightarrow[\text{(2) }CH_3CH_2Br]{\text{(1) }NaOCH_2CH_3}$$

$$\underset{\text{dialkylmalonic ester}}{CH_3CH_2-C(COOC_2H_5)(CH_2Ph)-\overset{O}{\overset{\|}{C}}-OC_2H_5} \xrightarrow[H_2O]{H^+,\text{ heat}} \underset{\text{disubstituted acetic acid}}{CH_3CH_2-CH(CH_2Ph)-\overset{O}{\overset{\|}{C}}-OH} \quad (CO_2\uparrow)$$

PROBLEM 22-47

Show how the following compounds can be made using the malonic ester synthesis.

(a) 3-phenylpropanoic acid **(b)** 2-methylpropanoic acid
(c) 4-phenylbutanoic acid **(d)** cyclopentanecarboxylic acid

PROBLEM 22-48

(a) Explain why the following substituted acetic acid cannot be formed by the malonic ester synthesis.

$$CH_2{=}CH-H_2C-C(CH_3)_2-C(=O)OH$$

(b) Sections 22-2B and 22-3 showed the use of lithium diisopropylamide (LDA) to deprotonate a ketone quantitatively. Draw the acid–base reaction between LDA and the following ester, and use estimated pK_a values to decide whether the reaction favors the reactants or products at equilibrium.

$$CH_3-\overset{CH_3}{\overset{|}{CH}}-\overset{O}{\overset{\|}{C}}-OCH_3$$

(c) Show how you might use a modern alternative to the malonic ester synthesis to make the acid shown in part (a). You may use the ester shown in part (b) as your starting material.

22-17 The Acetoacetic Ester Synthesis

The **acetoacetic ester synthesis** is similar to the malonic ester synthesis, but the final products are ketones: specifically, substituted derivatives of acetone. In the acetoacetic ester synthesis, substituents are added to the enolate ion of ethyl acetoacetate (acetoacetic ester), followed by hydrolysis and decarboxylation to produce an alkylated derivative of acetone.

$$\underset{\substack{\text{ethyl acetoacetate}\\\text{(acetoacetic ester)}}}{CH_3-\overset{O}{\overset{\|}{C}}-CH_2-\overset{O}{\overset{\|}{C}}-OC_2H_5} \xrightarrow[\text{(2) }R-X]{\text{(1) }^-OC_2H_5} \underset{\text{alkylated ester}}{CH_3-\overset{O}{\overset{\|}{C}}-\overset{R}{\overset{|}{CH}}-\overset{O}{\overset{\|}{C}}-OC_2H_5} \xrightarrow[\text{heat}]{H_3O^+} \underset{\text{substituted acetone}}{CH_3-\overset{O}{\overset{\|}{C}}-\overset{R}{\overset{|}{CH_2}}} + C_2H_5OH + CO_2\uparrow$$

Acetoacetic ester is like a molecule of acetone with a temporary ester group attached to enhance its acidity. Ethoxide ion completely deprotonates acetoacetic ester. The resulting enolate is alkylated by an unhindered alkyl halide or tosylate to give an alkylacetoacetic ester. Once again, the alkylating agent must be a good S_N2 substrate.

temporary ester group

$$\underset{\substack{\text{ethyl acetoacetate}\\(pK_a = 11)}}{H-CH(COOC_2H_5)-C(=O)-CH_3} \underset{}{\overset{^-OC_2H_5}{\rightleftharpoons}} \underset{\text{enolate ion}}{\left[^-:CH(COOC_2H_5)-C(=O)-CH_3 \right]} + C_2H_5OH \xrightarrow{R-X} \underset{\text{an alkylacetoacetic ester}}{R-CH(COOC_2H_5)-C(=O)-CH_3} + X^-$$

Acidic hydrolysis of the alkylacetoacetic ester initially gives an alkylacetoacetic acid, which is a β-keto acid. The keto group in the β position promotes decarboxylation to form a substituted version of acetone.

$$\underset{\substack{\text{alkylacetoacetic ester}\\(\text{a } \beta\text{-keto ester})}}{R-CH(COOC_2H_5)-C(=O)-CH_3} \xrightarrow[H_2O]{H^+,\ \text{heat}} \underset{\substack{\text{alkylacetoacetic acid}\\(\text{a } \beta\text{-keto acid})}}{R-CH(COOH)-C(=O)-CH_3} \xrightarrow{\text{decarboxylates}} \underset{\text{a substituted acetone}}{R-CH_2-C(=O)-CH_3} + CO_2\uparrow$$

The β-keto acid decarboxylates by the same mechanism as the alkylmalonic acid in the malonic ester synthesis. A six-membered cyclic transition state splits out carbon dioxide to give the enol form of the substituted acetone. This decarboxylation usually takes place spontaneously at the temperature of the hydrolysis.

$$\underset{\text{the } \beta\text{-keto acid}}{O{=}C(OH)-CHR-C(=O)-CH_3} \longrightarrow \underset{CO_2 + \text{enol}}{O{=}C{=}O + RCH{=}C(OH)-CH_3} \xrightarrow{\text{tautomerization}} \underset{\text{a substituted acetone}}{R-CH_2-C(=O)-CH_3} + CO_2\uparrow$$

Disubstituted acetones are formed by alkylating acetoacetic ester a second time before the hydrolysis and decarboxylation steps, as shown in the following general synthesis.

temporary ester group

$$R-CH(COOC_2H_5)-C(=O)-CH_3 \xrightarrow[(2)\ R'-X]{(1)\ ^-OC_2H_5} \underset{\text{dialkylacetoacetic ester}}{R-C(R')(COOC_2H_5)-C(=O)-CH_3} \xrightarrow[H_2O]{H^+,\ \text{heat}} \underset{\text{disubstituted acetone}}{R-CH(R')-C(=O)-CH_3} + CO_2\uparrow + CH_3CH_2OH$$

SOLVED PROBLEM 22-8

Show how the acetoacetic ester synthesis is used to make 3-propylhex-5-en-2-one.

SOLUTION

The target compound is acetone with an *n*-propyl group and an allyl group as substituents:

$$\underbrace{CH_3CH_2CH_2}_{n\text{-propyl group}}-\underset{\underset{\text{allyl group}}{\underbrace{CH_2-CH=CH_2}}}{\overset{}{CH}}-\overset{\overset{\text{acetone}}{O}}{\overset{\|}{C}}-CH_3$$

With an *n*-propyl halide and an allyl halide as the alkylating agents, the acetoacetic ester synthesis should produce 3-propylhex-5-en-2-one. Two alkylation steps give the required substitution:

$$\underset{CH_2-\overset{\overset{O}{\|}}{C}-CH_3}{\overset{COOC_2H_5}{|}} \xrightarrow[(2)\ CH_3CH_2CH_2Br]{(1)\ ^-OC_2H_5} CH_3CH_2CH_2-\overset{COOC_2H_5}{\overset{|}{CH}}-\overset{\overset{O}{\|}}{C}-CH_3 \xrightarrow[(2)\ CH_2=CH-CH_2Br]{(1)\ ^-OC_2H_5} CH_3CH_2CH_2-\underset{\underset{H_2C=CH-CH_2}{|}}{\overset{COOC_2H_5}{\overset{|}{C}}}-\overset{\overset{O}{\|}}{C}-CH_3$$

Hydrolysis proceeds with decarboxylation to give the disubstituted acetone product.

$$CH_3CH_2CH_2-\underset{\underset{H_2C=CH-CH_2}{|}}{\overset{COOC_2H_5}{\overset{|}{C}}}-\overset{\overset{O}{\|}}{C}-CH_3 \xrightarrow[H_2O]{H^+,\ heat} \left[CH_3CH_2CH_2-\underset{\underset{H_2C=CH-CH_2}{|}}{\overset{COOH}{\overset{|}{C}}}-\overset{\overset{O}{\|}}{C}-CH_3 \right]_{\beta\text{-keto acid}} \longrightarrow CO_2\uparrow + \underset{\text{3-propylhex-5-en-2-one}}{CH_3CH_2CH_2-\underset{\underset{H_2C=CH-CH_2}{|}}{CH}-\overset{\overset{O}{\|}}{C}-CH_3}$$

PROBLEM 22-49

Show the ketones that would result from hydrolysis and decarboxylation of the following β-keto esters.

(a) $PhCH_2-\underset{\underset{COOC_2H_5}{|}}{CH}-\overset{\overset{O}{\|}}{C}-CH_3$

(b) cyclobutyl carbon bearing $COOCH_2CH_3$ and $-\overset{\overset{O}{\|}}{C}-CH_3$

(c) 2-oxocyclopentyl $-\overset{\overset{O}{\|}}{C}-OCH_2CH_3$

PROBLEM-SOLVING HINT

An acetoacetic ester synthesis goes through alkylation of the enolate, hydrolysis, and decarboxylation. To design a synthesis, look at the product and see what groups are added to acetone. Use those groups to alkylate acetoacetic ester, then hydrolyze and decarboxylate.

PROBLEM 22-50

Show how the following ketones might be synthesized by using the acetoacetic ester synthesis.

(a) $PhCH_2CH_2-\overset{\overset{O}{\|}}{C}-CH_3$

(b) cyclopentyl $-\overset{\overset{O}{\|}}{C}-CH_3$

(c) $H_2C=CHCH_2\underset{}{\overset{Ph-CH_2}{\overset{|}{C}H}}-\overset{\overset{O}{\|}}{C}CH_3$

PROBLEM 22-51

(a) Although the following compound is a substituted acetone derivative, it cannot be made by the acetoacetic ester synthesis. Explain why (two reasons).

$PhCH_2-\overset{\overset{O}{\|}}{C}-\underset{\underset{CH_2CH=CH_2}{|}}{CH}-Ph$

(continued)

Application: Bacterial Synthesis

Acetone was produced during World War I by using engineered strains of *Clostridium* bacteria. These strains make an enzyme called acetoacetate decarboxylase that catalyzes the decarboxylation of acetoacetate.

(b) The use of LDA to make enolate ions (Sections 22-2B and 22-3) has provided alternatives to the acetoacetic ester synthesis. Show how you might make the compound shown in part (a), beginning with 1,3-diphenylacetone.

(c) Enamine reactions (Section 22-4) occur under relatively mild conditions, and they often give excellent yields of compounds like the one shown in part (a). Show how you might use an enamine reaction for this synthesis, beginning with 1,3-diphenylacetone.

22-18 Conjugate Additions: The Michael Reaction

α,β-Unsaturated carbonyl compounds have unusually electrophilic double bonds. The β carbon is electrophilic because it shares the partial positive charge of the carbonyl carbon atom through resonance.

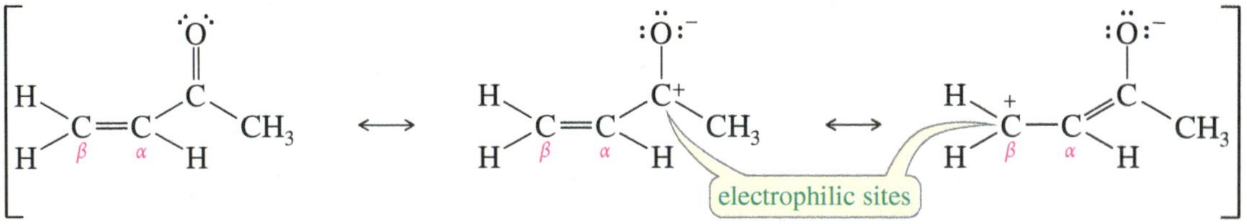

electrophilic sites

A nucleophile can attack an α,β-unsaturated carbonyl compound at either the carbonyl group or at the β position. When attack occurs at the carbonyl group, protonation of the oxygen leads to a **1,2-addition** product in which the nucleophile and the proton have added to adjacent atoms. When attack occurs at the β position, the oxygen atom is the fourth atom counting from the nucleophile, and the addition is called a **1,4-addition.** The net result of 1,4-addition is addition of the nucleophile and a hydrogen atom across a double bond that was conjugated with a carbonyl group. For this reason, 1,4-addition is often called **conjugate addition.**

Mechanism 22-13 contrasts the products and the mechanisms of 1,2-addition and 1,4-addition. Note that the intermediate in the 1,4-addition is a resonance-stabilized enolate ion.

MECHANISM 22-13 1,2-Addition and 1,4-Addition (Conjugate Addition)

1,2-addition

1,2-addition is the standard nucleophilic addition to a carbonyl group.

Step 1: Addition of the nucleophile to C=O. ***Step 2:*** Protonation of the alkoxide.

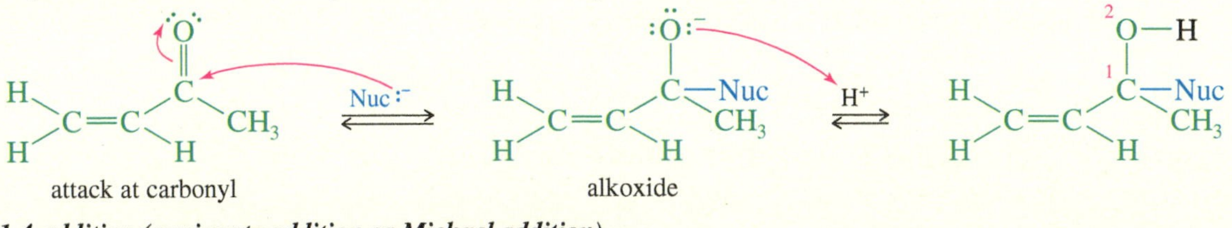

attack at carbonyl

alkoxide

1,4-addition (conjugate addition or Michael addition)

In a 1,4-addition, the nucleophile adds to the β carbon atom of an α,β-unsaturated system to give an enolate ion. Protonation may occur on oxygen to give an enol, or on carbon to give the keto form.

Step 1: Conjugate addition of the nucleophile. ***Step 2:*** Protonation of the enolate (on oxygen or on carbon).

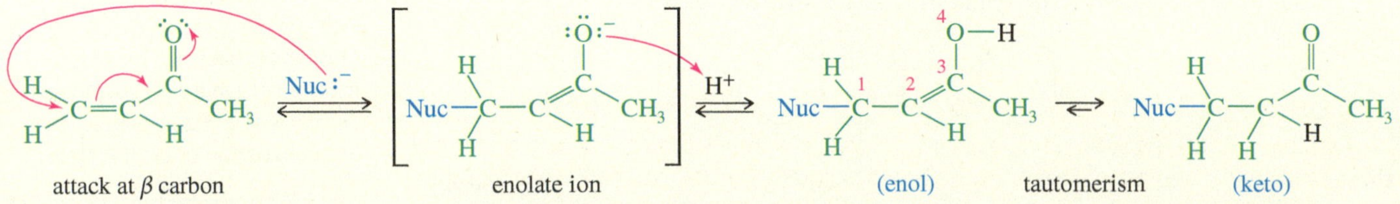

attack at β carbon

enolate ion

(enol)

tautomerism

(keto)

Conjugate addition of a carbanion to the double bond of an α,β-unsaturated carbonyl compound (or other electron-poor double bond) is called a **Michael addition.** The electrophile (the α,β-unsaturated carbonyl compound) accepts a pair of electrons, so it is called the **Michael acceptor.** The attacking nucleophile donates a pair of electrons, so it is called the **Michael donor.** A wide variety of compounds can serve as Michael donors and acceptors, as shown in Table 22-2. Common Michael donors are lithium dialkyl cuprates, enamines, and carbanions that are stabilized by two strong electron-withdrawing groups such as carbonyl groups, cyano groups, or nitro groups. Common acceptors contain a double bond conjugated with a carbonyl group, a cyano group, or a nitro group.

TABLE 22-2
Some Common Michael Donors and Michael Acceptors

Michael Donors		Michael Acceptors	
R_2CuLi	lithium dialkyl cuprate (Gilman reagents)	$H_2C{=}CH{-}C({=}O){-}H$	conjugated aldehyde
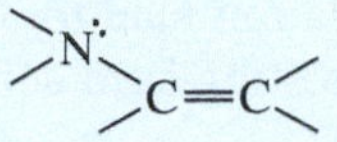 $>N{:}{-}C{=}C<$	enamine	$H_2C{=}CH{-}C({=}O){-}R$	conjugated ketone
$R{-}C({=}O){-}\ddot{C}H{-}C({=}O){-}R'$	β-diketone	$H_2C{=}CH{-}C({=}O){-}OR$	conjugated ester
$R{-}C({=}O){-}\ddot{C}H{-}C({=}O){-}OR'$	β-keto ester	$H_2C{=}CH{-}C({=}O){-}NH_2$	conjugated amide
$R{-}C({=}O){-}\ddot{C}H{-}C{\equiv}N$	β-keto nitrile	$H_2C{=}CH{-}C{\equiv}N$	conjugated nitrile
$R{-}C({=}O){-}\ddot{C}H{-}NO_2$	α-nitro ketone	$H_2C{=}CH{-}NO_2$	nitroethylene

The following example shows lithium divinylcuprate serving as a Michael donor, adding to the double bond of an α,β-unsaturated ketone. In this conjugate addition, the vinyl group adds to the β carbon atom to give an enolate ion. Protonation at the α carbon gives the product.

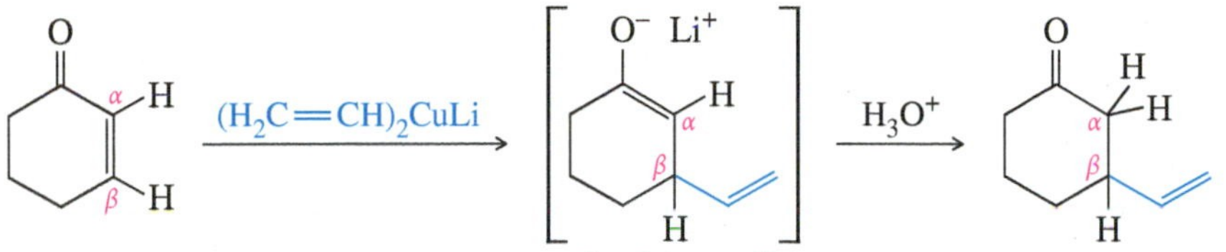

$(H_2C{=}CH)_2CuLi$ → [$O^-\ Li^+$ enolate intermediate] → H_3O^+

enolate intermediate

When using a Gilman reagent (lithium dialkylcuprate) as the Michael donor, the lithium enolate intermediate is sufficiently stable that it can undergo additional reactions. In the example shown above, further alkylation at the α position gives a product with one substituent at the α position and another at the β position.

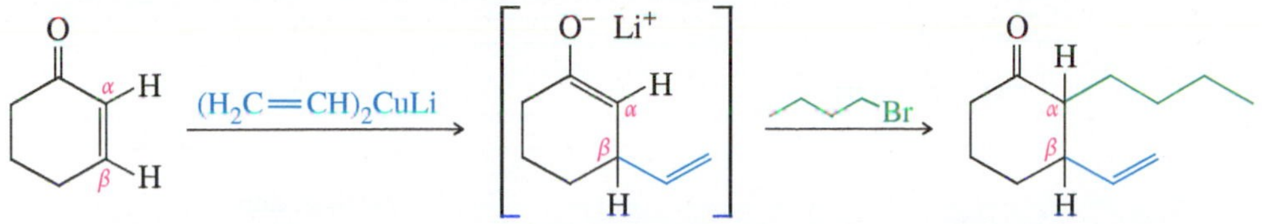

$(H_2C{=}CH)_2CuLi$ → [$O^-\ Li^+$ enolate intermediate] → Br (butyl bromide)

enolate intermediate (cis + trans)

Michael additions are useful in acetoacetic ester syntheses and malonic ester syntheses because the enolate ions of both of these esters are good Michael donors. As an example, let's consider the addition of the malonic ester enolate to methyl vinyl ketone

(MVK). The crucial step is the nucleophilic attack by the enolate at the carbon. The resulting enolate is strongly basic, and it is quickly protonated.

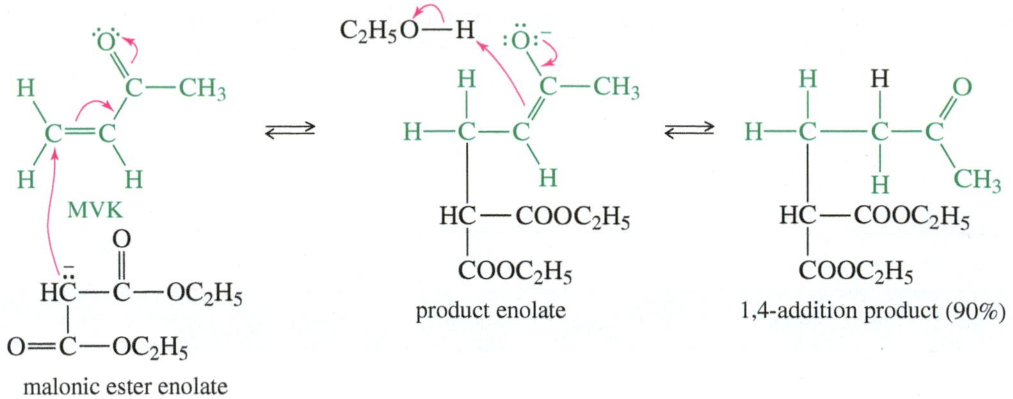

MVK

malonic ester enolate

C_2H_5O—H

product enolate

1,4-addition product (90%)

The product of this Michael addition may be treated like any other substituted malonic ester in the malonic ester synthesis. Hydrolysis and decarboxylation lead to a δ-keto acid. It is not easy to imagine other ways to synthesize this interesting keto acid.

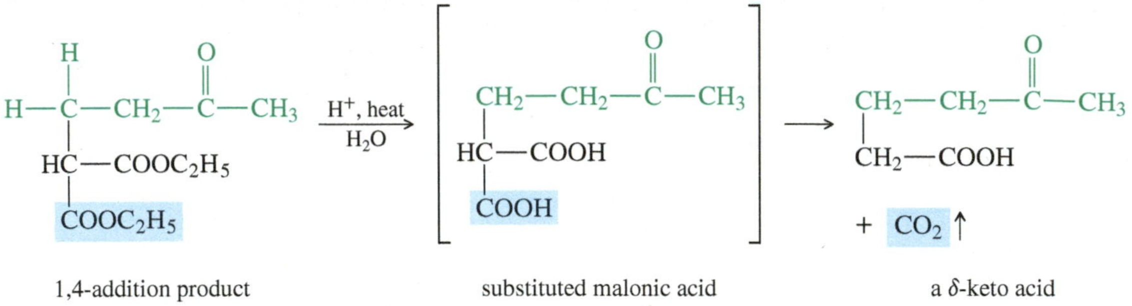

H^+, heat / H_2O

+ CO_2 ↑

1,4-addition product

substituted malonic acid

a δ-keto acid

SOLVED PROBLEM 22-9

Show how the following diketone might be synthesized using a Michael addition.

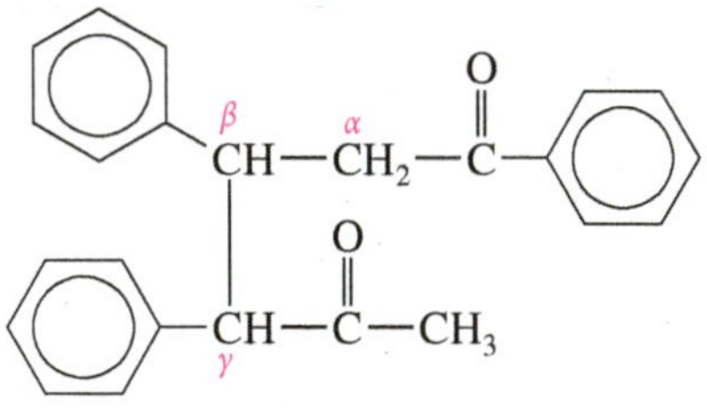

SOLUTION

A Michael addition would have formed a new bond at the β carbon of the acceptor. Therefore, we break this molecule apart at the β,γ bond.

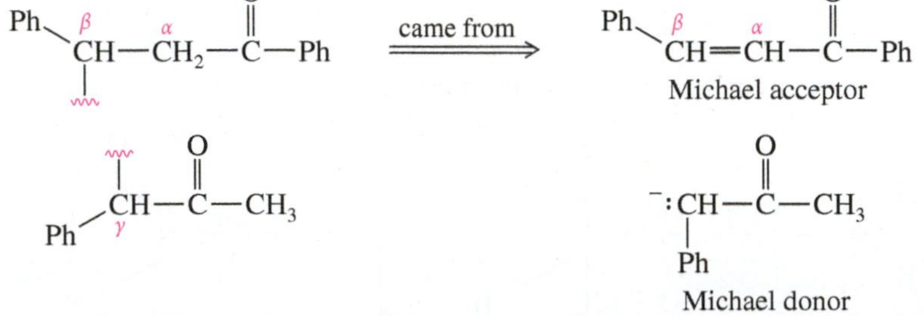

came from

Michael acceptor

Michael donor

The top fragment, where we broke the β bond, must have come from a conjugated ketone, and it must have been the Michael acceptor. The bottom fragment is a simple ketone. It is unlikely that this ketone was used without some sort of additional stabilizing group. We can add a temporary ester group to the ketone (making a substituted acetoacetic ester) and use the acetoacetic ester synthesis to give the correct product.

(continued)

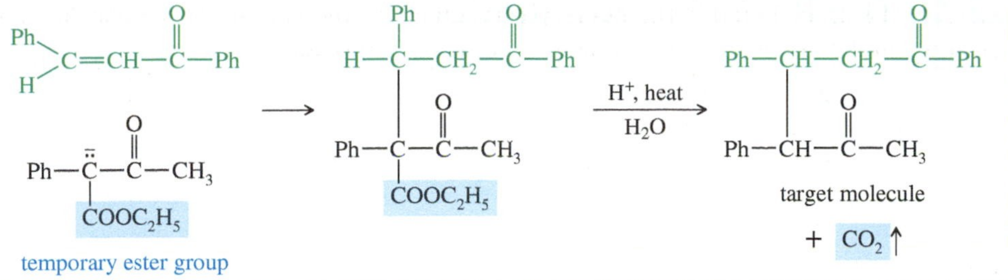

$$\text{Ph(H)C=CH-C(=O)-Ph} \;+\; \text{Ph-}\ddot{\text{C}}^{-}\text{(COOC}_2\text{H}_5\text{)-C(=O)-CH}_3 \longrightarrow \text{H-C(Ph)(-CH}_2\text{-C(=O)-Ph)-C(Ph)(COOC}_2\text{H}_5\text{)-C(=O)-CH}_3 \xrightarrow[\text{H}_2\text{O}]{\text{H}^+,\ \text{heat}} \text{Ph-CH(-CH}_2\text{-C(=O)-Ph)-CH(Ph)-C(=O)-CH}_3 + \text{CO}_2\uparrow$$

temporary ester group

target molecule

PROBLEM 22-52

In Solved Problem 22-9, the target molecule was synthesized using a Michael addition to form the bond that is β,γ to the upper carbonyl group. Another approach is to use a Michael addition to form the bond that is β,γ to the other (lower) carbonyl group. Show how you would accomplish this alternative synthesis.

PROBLEM 22-53

Show how cyclohexanone might be converted to the following δ-diketone (*Hint:* Stork).

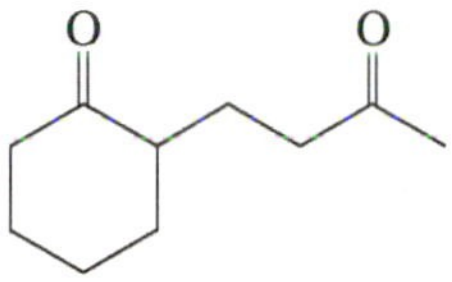

[Structure: 2-(3-oxobutyl)cyclohexanone]

PROBLEM 22-54

Show how an acetoacetic ester synthesis might be used to form a δ-diketone such as heptane-2,6-dione.

PROBLEM 22-55

Propose a mechanism for the conjugate addition of a nucleophile (Nuc:$^-$) to acrylonitrile ($H_2C{=}CHCN$) and to nitroethylene. Use resonance forms to show how the cyano and nitro groups activate the double bond toward conjugate addition.

PROBLEM 22-56

Show how the following products might be synthesized from suitable Michael donors and acceptors.

(a) $\text{Ph-CH(CH(COOCH}_2\text{CH}_3)_2\text{)-CH}_2\text{-C(=O)-OCH}_2\text{CH}_3$

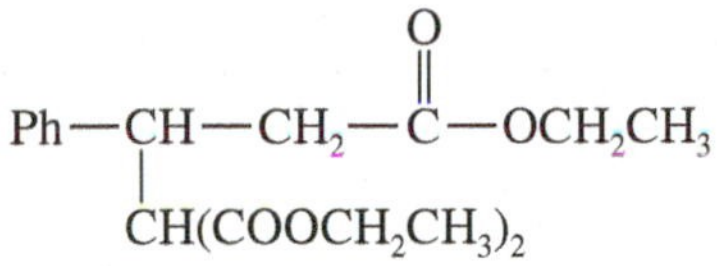

(b) $\text{CH}_2\text{-CH}_2\text{-CN}$ / $\text{CH}_2\text{-COCH}_3$ (i.e., $\text{NC-CH}_2\text{-CH}_2\text{-CH}_2\text{-COCH}_3$)

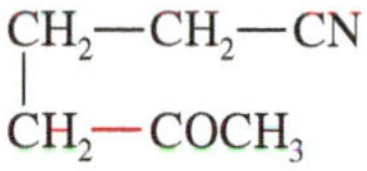

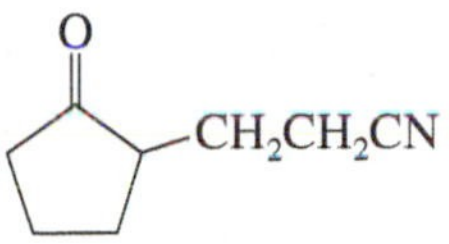

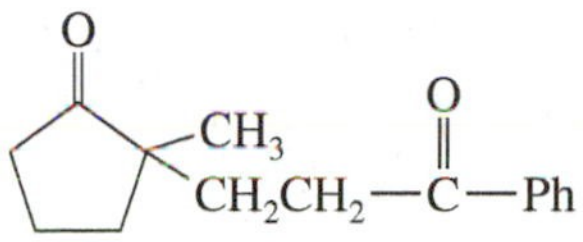

(c) 2-($\text{CH}_2\text{CH}_2\text{CN}$)cyclopentanone

(d) 2-CH_3-2-($\text{CH}_2\text{CH}_2\text{-C(=O)-Ph}$)cyclopentanone

(e) $\text{CH}_3\text{-C(=O)-CH(CH}_3\text{)-CH}_2\text{-CH}_2\text{-C(=O)-CH}_3$

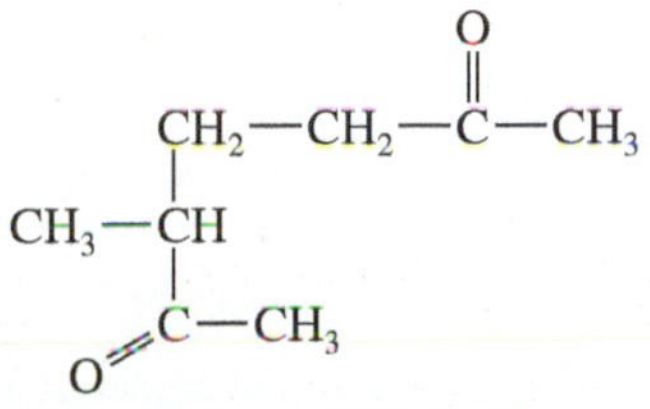

(f) 3-vinylcyclopentanone

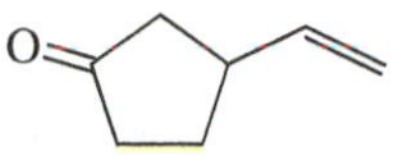

PROBLEM-SOLVING HINT

Claisen condensations usually give 1,3-dicarbonyl products, with one saturated carbon between two carbonyl groups. Michael additions commonly give 1,5-dicarbonyl products, with three saturated carbons between two carbonyl groups. When you need a compound with three carbons between two carbonyl groups, consider a Michael addition.

Application: Synthetic hormones

Several human hormones, including cortisone and estrogens used in birth control pills, have been synthesized using successive Robinson annulations (next Section) to create six-membered rings in the tetracyclic structure. The female hormone estradiol is shown below.

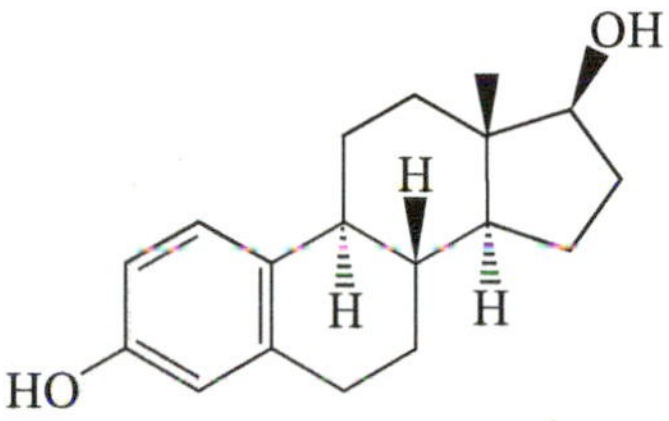

[Structure: estradiol, labels OH, H, H, H, HO]

estradiol

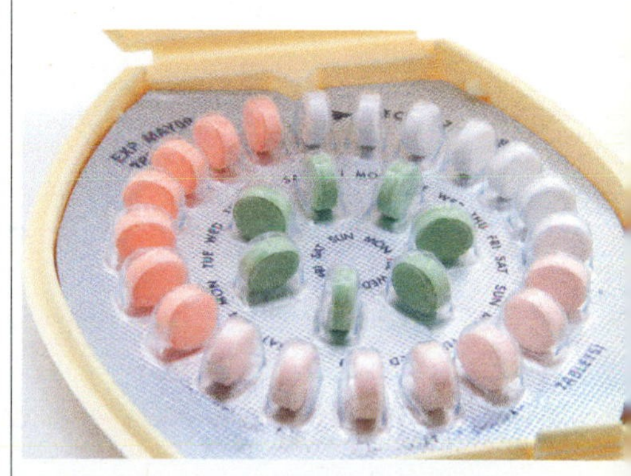

22-19 The Robinson Annulation

We have seen that Michael addition of a ketone enolate (or its enamine) to an α,β-unsaturated ketone gives a δ-diketone. If the conjugate addition takes place under strongly basic or acidic conditions, the δ-diketone undergoes a spontaneous intramolecular aldol condensation, usually with dehydration, to give a new six-membered ring: a conjugated cyclohexenone. This synthesis is called the **Robinson annulation** (ring-forming) reaction. Consider an example using a substituted cyclohexanone as the Michael donor and methyl vinyl ketone (MVK) as the Michael acceptor.

The Robinson annulation

MVK

$\xrightarrow[(^-H_2O)]{^-OH}$

new cyclohexenone (65%)

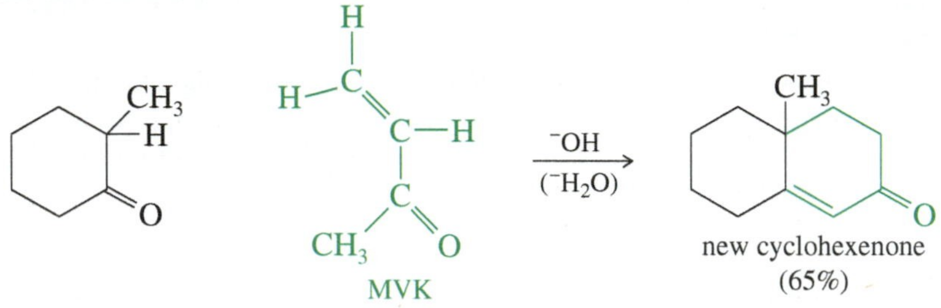

The mechanism begins with the Michael addition of the cyclohexanone enolate to MVK, forming a δ-diketone.

Step 1: Michael addition.

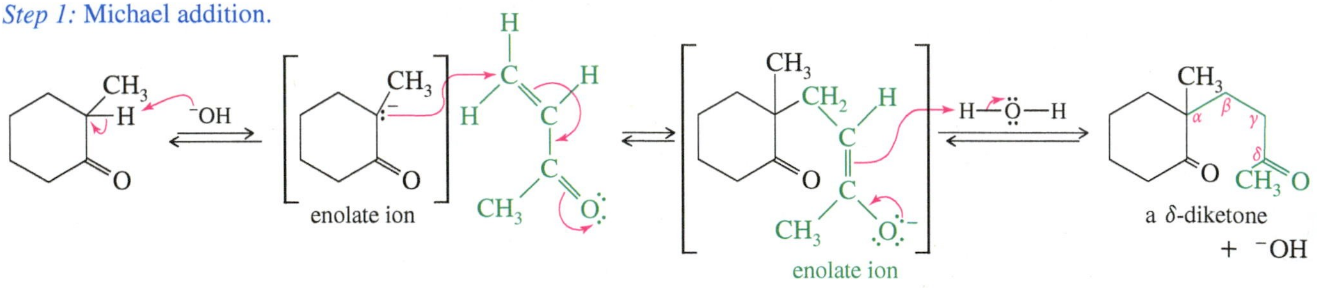

enolate ion

enolate ion

a δ-diketone + ^-OH

This δ-diketone might take part in several different aldol condensations, but it is ideally suited for a favorable one: formation of a six-membered ring. To form a six-membered ring, the enolate of the methyl ketone attacks the cyclohexanone carbonyl. The aldol product dehydrates to give a cyclohexenone.

Step 2: Cyclic aldol to form a six-membered ring.

enolate ion

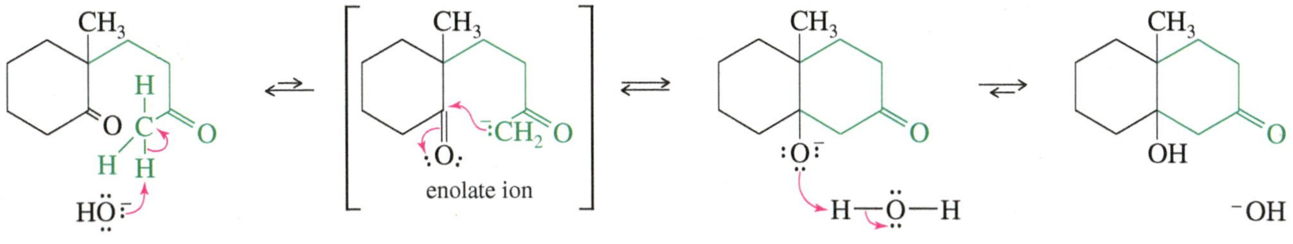

Step 3: Dehydration of the aldol product.

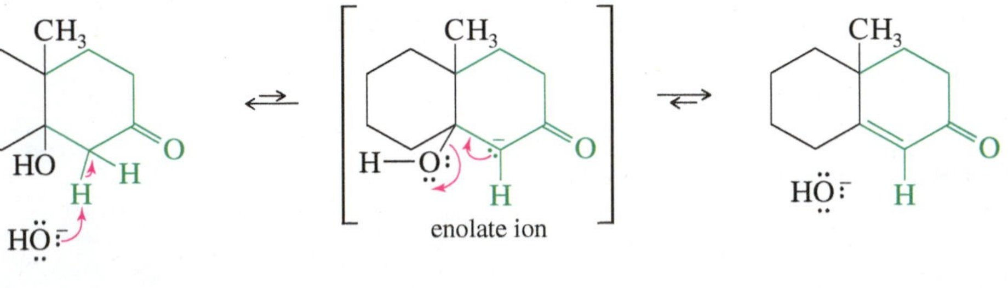

enolate ion

It is not difficult to predict the products of the Robinson annulation and to draw the mechanisms if you remember that the Michael addition is first, followed by an aldol condensation with dehydration to give a cyclohexenone.

PROBLEM-SOLVING STRATEGY Proposing Reaction Mechanisms

This problem-solving example addresses a complicated base-catalyzed reaction, using the system for proposing mechanisms summarized in Appendix 3A. The problem is to propose a mechanism for the base-catalyzed reaction of ethyl acetoacetate with methyl vinyl ketone.

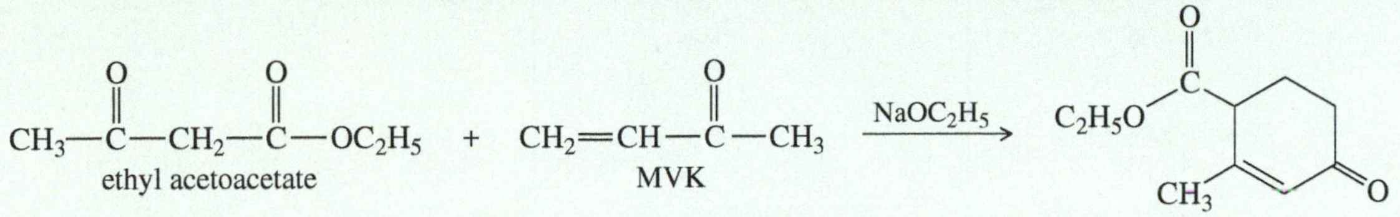

$$CH_3-\overset{O}{\overset{\|}{C}}-CH_2-\overset{O}{\overset{\|}{C}}-OC_2H_5 + CH_2{=}CH-\overset{O}{\overset{\|}{C}}-CH_3 \xrightarrow{NaOC_2H_5}$$

ethyl acetoacetate; MVK; product: C_2H_5O–C(=O)– substituted cyclohexenone bearing CH_3 and =O

First, we must determine the type of mechanism. The use of a basic catalyst suggests the reaction involves strong nucleophiles as intermediates. We expect to see anionic intermediates (possibly stabilized carbanions), but no strong electrophiles or strong acids, and no carbocations or free radicals.

1. **Consider the carbon skeletons of the reactants and products, and decide which carbon atoms in the products are likely derived from which carbon atoms in the reactants.**
   The ester group in the product must be derived from ethyl acetoacetate. The carbon from the ester (now part of the C=C double bond) should be derived from the ketone of ethyl acetoacetate. The structure of MVK can be seen in the remaining four carbons.

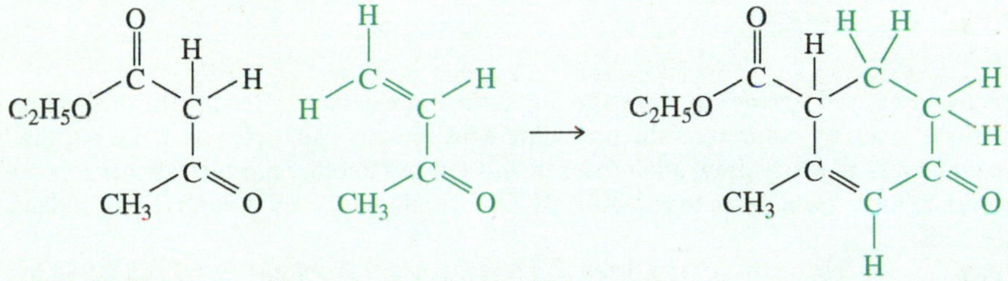

2. **Consider whether one of the reactants is a strong enough nucleophile to react without being activated. If not, consider how one of the reactants might be converted to a strong nucleophile by deprotonation of an acidic site or by attack on an electrophilic site.**
   Neither reactant is a strong enough nucleophile to attack the other. Ethyl acetoacetate is more acidic than ethanol, so ethoxide ion quickly removes a proton to give the enolate ion.

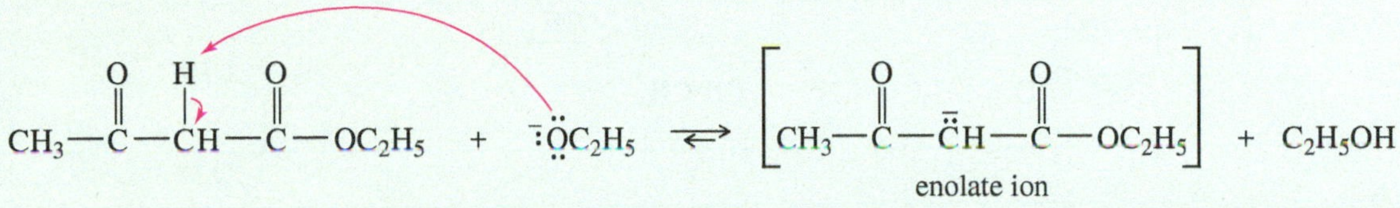

$$CH_3-\overset{O}{\overset{\|}{C}}-\overset{H}{CH}-\overset{O}{\overset{\|}{C}}-OC_2H_5 + {}^{-}:\ddot{O}C_2H_5 \rightleftharpoons \left[CH_3-\overset{O}{\overset{\|}{C}}-\bar{C}H-\overset{O}{\overset{\|}{C}}-OC_2H_5\right] + C_2H_5OH$$

enolate ion

3. **Consider how an electrophilic site on another reactant (or, in a cyclization, another part of the same molecule) can undergo attack by the strong nucleophile to form a bond needed in the product. Draw the product of this bond formation.**
   The enolate of acetoacetic ester might attack either the electrophilic double bond (Michael addition) or the carbonyl group of MVK. A Michael addition forms one of the bonds needed in the product.

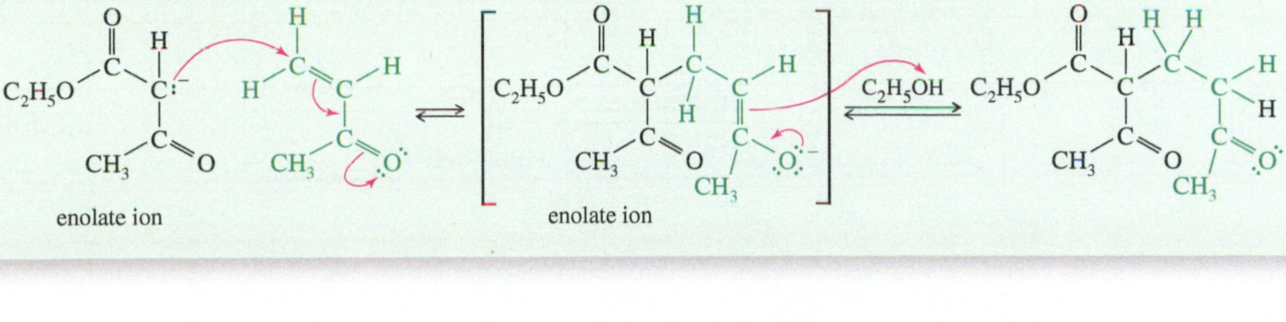

enolate ion ⇌ [enolate ion] $\xrightarrow{C_2H_5OH}$

4. Consider how the product of nucleophilic attack might be converted to the final product (if it has the right carbon skeleton) or reactivated to form another bond needed in the product.

The ketone carbonyl group of ethyl acetoacetate must be converted to a C═C double bond in the α,β position of the other ketone. This conversion corresponds to an aldol condensation with dehydration. Note that the proton we must remove is not the most acidic proton, but its removal forms the enolate that is needed to give the observed product.

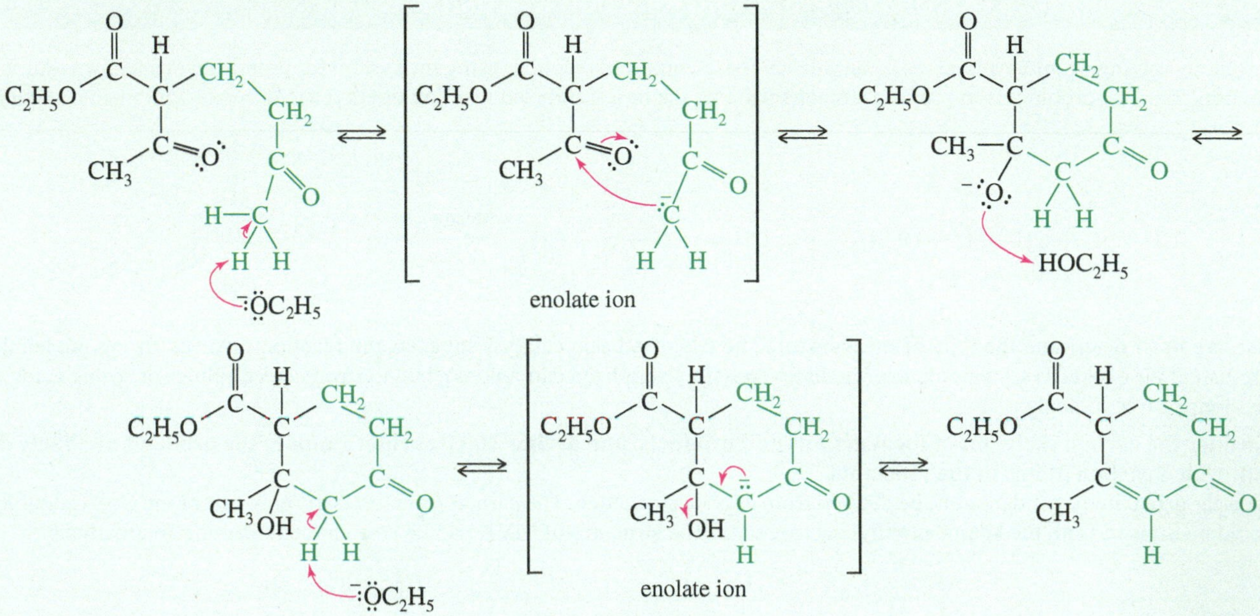

enolate ion

enolate ion

5. Draw out all the steps using curved arrows to show the movement of electrons. Be careful to show only one step at a time.

The complete mechanism is given by combining the preceding equations. We suggest you write out the mechanism as a review of the steps. Note that we could just as easily draw other mechanisms leading to other products, but that is not the point of a mechanism problem. This question asked for a mechanism to explain only this one product, even though other products are likely formed as well, and possibly in higher yields.

As further practice in proposing mechanisms for multistep condensations, try Problems 22-57 and 22-58 by using the approach shown.

PROBLEM 22-57

Propose a mechanism for the following reaction.

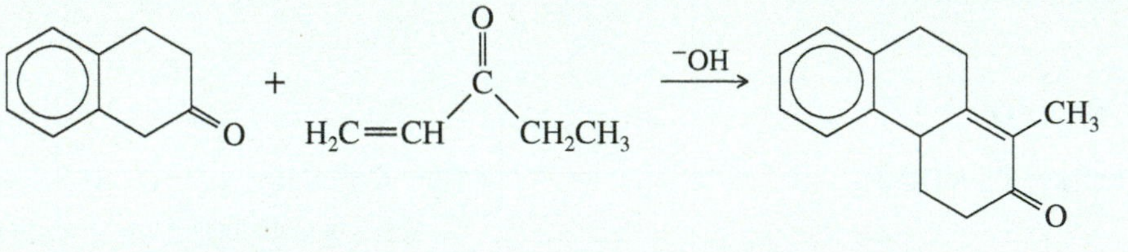

$H_2C{=}CH{-}C(=O){-}CH_2CH_3$, ^{-}OH, CH_3

PROBLEM 22-58

The base-catalyzed reaction of an aldehyde (having no α hydrogens) with an anhydride is called the *Perkin condensation*. Propose a mechanism for the following example of the Perkin condensation. (Sodium acetate serves as the base.)

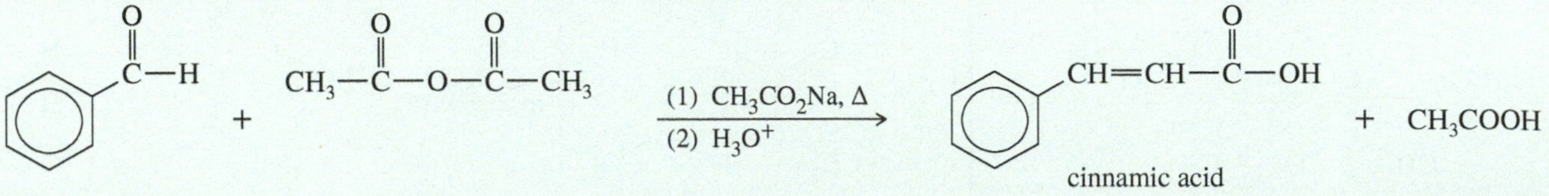

$$C_6H_5{-}CHO + CH_3{-}C(=O){-}O{-}C(=O){-}CH_3 \xrightarrow[(2)\ H_3O^+]{(1)\ CH_3CO_2Na,\ \Delta} C_6H_5{-}CH{=}CH{-}C(=O){-}OH + CH_3COOH$$

cinnamic acid

PROBLEM 22-59

Show how you would use Robinson annulations to synthesize the following compounds. Work backward, remembering that the cyclohexenone is the new ring and that the double bond of the cyclohexenone is formed by the aldol with dehydration. Take apart the double bond, then see what structures the Michael donor and acceptor must have.

(a) [bicyclic enone with CH_3 and O] **(b)** [bicyclic enone with O, CH_3 and O]

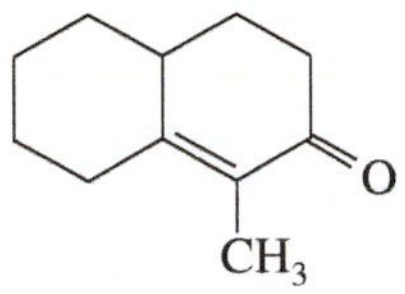

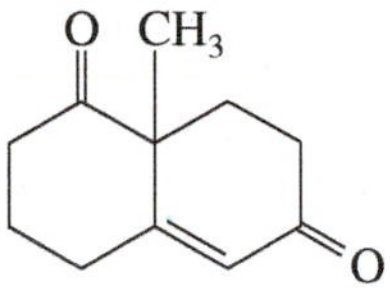

PROBLEM-SOLVING HINT

You can usually spot a product of Robinson annulation because it has a new cyclohexenone ring. The mechanism is not difficult if you remember "Michael goes first," followed by an aldol with dehydration.

SUMMARY Enolate Additions and Condensations

A complete summary of additions and condensations would be long and involved. This summary covers the major classes of condensations and related reactions.

1. *Alkylation of lithium enolates* (Section 22-3)

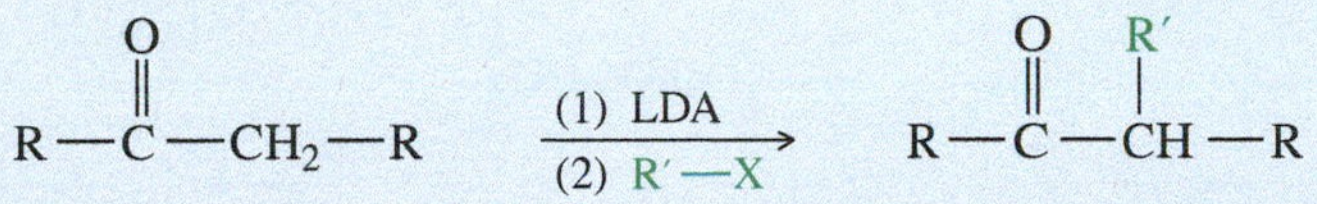

$$R-\overset{O}{\overset{\|}{C}}-CH_2-R \xrightarrow[(2)\ R'-X]{(1)\ LDA} R-\overset{O}{\overset{\|}{C}}-\overset{R'}{\overset{|}{C}H}-R$$

(LDA = lithium diisopropylamide; R′—X = unhindered 1° halide or tosylate)

2. *Alkylation of enamines (Stork reaction)* (Section 22-4)

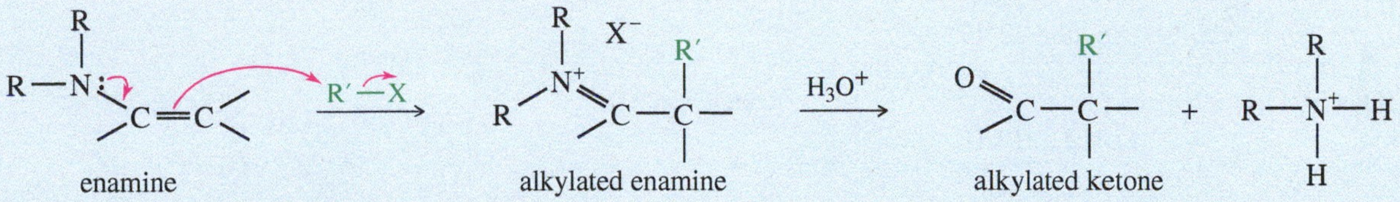

$$R_2N-C{=}C< \xrightarrow{R'-X} R_2N^+{=}C-\overset{R'}{\overset{|}{C}}-\ X^- \xrightarrow{H_3O^+} O{=}C-\overset{R'}{\overset{|}{C}}- \ + \ R_2NH_2^+$$

enamine → alkylated enamine → alkylated ketone

3. *α Halogenation* (Section 22-5)

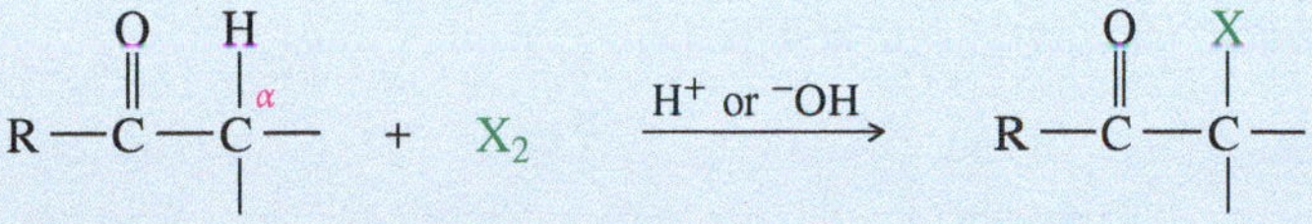

$$R-\overset{O}{\overset{\|}{C}}-\overset{H}{\overset{|}{C}}- \ + \ X_2 \xrightarrow{H^+\ or\ ^-OH} R-\overset{O}{\overset{\|}{C}}-\overset{X}{\overset{|}{C}}-$$

a. *The iodoform (or haloform) reaction* (Section 22-5B)

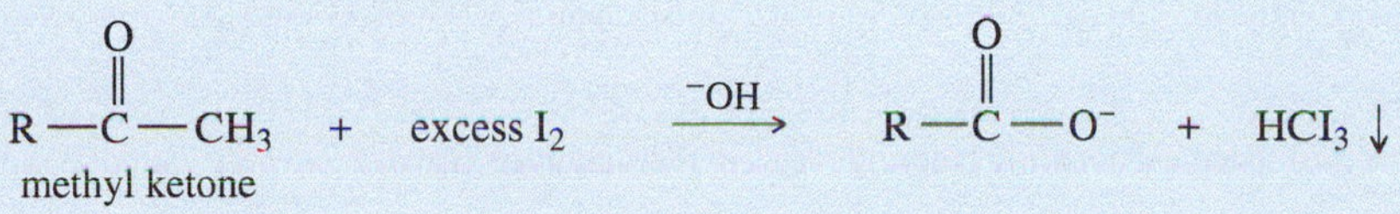

$$R-\overset{O}{\overset{\|}{C}}-CH_3 \ + \ \text{excess } I_2 \xrightarrow{^-OH} R-\overset{O}{\overset{\|}{C}}-O^- \ + \ HCI_3 \downarrow$$

methyl ketone

b. *The Hell–Volhard–Zelinsky (HVZ) reaction* (Section 22-6)

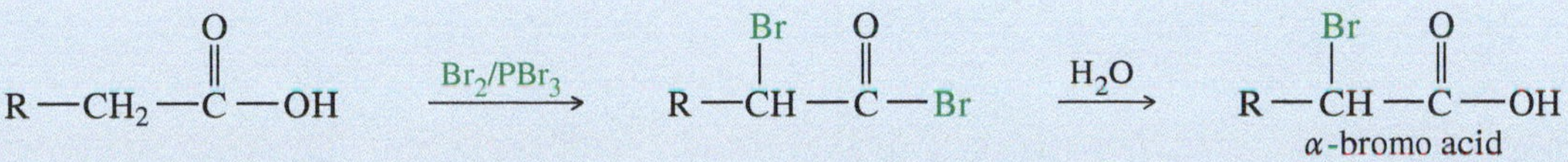

$$R-CH_2-\overset{O}{\overset{\|}{C}}-OH \xrightarrow{Br_2/PBr_3} R-\overset{Br}{\overset{|}{C}H}-\overset{O}{\overset{\|}{C}}-Br \xrightarrow{H_2O} R-\overset{Br}{\overset{|}{C}H}-\overset{O}{\overset{\|}{C}}-OH$$

α-bromo acid

(continued)

4. *The aldol condensation and subsequent dehydration* (Sections 22-7 through 22-11)

$$2\ \mathrm{R{-}CO{-}CH_2{-}R'} \underset{}{\overset{H^+ \text{ or } ^-OH}{\rightleftharpoons}} \mathrm{R{-}C(OH)(CH_2{-}R'){-}CH(R'){-}CO{-}R} \underset{H^+ \text{ or } ^-OH}{\overset{\text{heat}}{\rightleftharpoons}} \mathrm{R{-}C(CH_2{-}R'){=}C(R'){-}CO{-}R} + H_2O$$

ketone or aldehyde — aldol product — α,β-unsaturated ketone or aldehyde

5. *The Claisen ester condensation* (Sections 22-12 through 22-14)
(Cyclizations are the Dieckmann condensation.)

$$2\ \mathrm{RO{-}CO{-}CH_2{-}R'} \xrightarrow{^-OR} \mathrm{RO{-}CO{-}CH(R'){-}CO{-}CH_2{-}R'} + \mathrm{ROH}$$

The product is initially formed as its anion.

6. *The malonic ester synthesis* (Section 22-16)

$$\mathrm{H{-}CH(COOCH_2CH_3)_2} \xrightarrow[(2)\ R{-}X]{(1)\ NaOCH_2CH_3} \mathrm{R{-}CH(COOCH_2CH_3)_2} \xrightarrow[\text{heat}]{H_3O^+} \mathrm{R{-}CH_2{-}COOH} + CO_2\uparrow$$

malonic ester — substituted malonic ester — substituted acetic acid

7. *The acetoacetic ester synthesis* (Section 22-17)

$$\mathrm{H{-}CH(COOCH_2CH_3){-}C({=}O){-}CH_3} \xrightarrow[(2)\ R{-}X]{(1)\ NaOCH_2CH_3} \mathrm{R{-}CH(COOCH_2CH_3){-}C({=}O){-}CH_3} \xrightarrow[\text{heat}]{H_3O^+} \mathrm{R{-}CH_2{-}C({=}O){-}CH_3} + CO_2\uparrow$$

acetoacetic ester — substituted acetoacetic ester — substituted acetone

8. *The Michael addition (conjugate addition)* (Sections 22-18 and 22-19)

$$\mathrm{Y{-}\ddot{C}H(Z)} + \mathrm{{>}C{=}C{-}C({=}O){-}} \xrightarrow[\text{(proton source)}]{ROH} \mathrm{{-}C(CH(Y)(Z)){-}C(H){-}C({=}O){-}}$$

(Y and Z are carbonyl or other electron-withdrawing groups.)

9. *The Robinson annulation* (Section 22-19)

cyclohexanone + MVK $\xrightarrow{^-OH}$ Michael adduct $\xrightarrow{\text{aldol dehydration}}$ annulated product

SUMMARY Reactions of Stabilized Carbanions

Carbanions adjacent to strong electron-withdrawing groups like carbonyl, cyano, and nitro are stable enough to exist in solution. The general term for these stabilized carbanions is *enolates*. Enolates are nucleophiles and follow many of the nucleophilic reaction types that have been covered previously. Enolates are strong bases and usually require an acidic or aqueous workup to supply H^+.

α-Halogenation—nucleophilic substitution on a halogen molecule; Section 22-5A

$$R^1\text{-}\overset{-\cdot\cdot}{\underset{R^2}{C}}\text{-}\overset{O}{\overset{\|}{C}}\text{-}R^3 \xrightarrow{X\text{—}X,\ X = Cl, Br, I} R^1\text{-}\overset{X}{\underset{R^2}{C}}\text{-}\overset{O}{\overset{\|}{C}}\text{-}R^3$$

Alkylation—nucleophilic substitution (S_N2) on an alkyl halide; allylic, benzylic, and primary halides work best; Section 22-3

$$\text{enolate} \xrightarrow{R\text{—}X} R^1\text{-}\overset{R}{\underset{R^2}{C}}\text{-}\overset{O}{\overset{\|}{C}}\text{-}R^3$$

Aldol condensation—nucleophilic addition followed by elimination (dehydration); Sections 22-7 to 22-11

$$\text{enolate} \xrightarrow{R\text{-}CO\text{-}R'} \xrightarrow{H^+} R^1\text{-}\underset{R^2}{C}(\text{-}C(OH)RR')\text{-}\overset{O}{\overset{\|}{C}}\text{-}R^3 \xrightarrow[\text{elimination (dehydration) only if } R^1 \text{ or } R^2 = H]{H^+,\ -H_2O} R^1\text{-}C(=CRR')\text{-}\overset{O}{\overset{\|}{C}}\text{-}R^3$$

Claisen condensation—nucleophilic acyl substitution; the Dieckmann condensation is a ring-forming Claisen; Sections 22-12 to 22-14

$$\text{enolate} \xrightarrow{R\text{-}CO\text{-}LG} R^1\text{-}\underset{R^2}{C}(\text{-}C(=O)R)\text{-}\overset{O}{\overset{\|}{C}}\text{-}R^3$$

LG = leaving group, usually –OR of an ester

product is a β-dicarbonyl compound

Conjugate addition—also called the Michael reaction; Sections 22-18 and 22-19

an α,β-unsaturated carbonyl compound (or nitrile or nitro) + enolate, then H^+ → 1,5-dicarbonyl product (R^1R^2C bearing $CH_2CH_2COCH_3$ and COR^3)

Essential Terms

acetoacetic ester synthesis Alkylation or acylation of acetoacetic ester (ethyl acetoacetate), followed by hydrolysis and decarboxylation, to give substituted acetone derivatives. (p. 1151)

aldol condensation An acid- or base-catalyzed conversion of two ketone or aldehyde molecules to a β-hydroxy ketone or aldehyde (called an **aldol**). Aldol condensations often take place with subsequent dehydration to give α,β-unsaturated ketones and aldehydes. (p. 1128)

$$2\ R\text{-}\overset{O}{\overset{\|}{C}}\text{-}CH_2\text{-}R' \underset{}{\overset{H^+ \text{ or } ^-OH}{\rightleftharpoons}} R\text{-}\overset{O}{\overset{\|}{C}}\text{-}\underset{\alpha}{CH}(R')\text{-}\overset{\beta}{C}(OH)(R)\text{-}CH_2\text{-}R' \underset{H^+ \text{ or } ^-OH}{\overset{\text{heat}}{\rightleftharpoons}} R\text{-}\overset{O}{\overset{\|}{C}}\text{-}\underset{\alpha}{C}(R')\text{=}\overset{\beta}{C}(R)\text{-}CH_2\text{-}R' + H_2O$$

ketone or aldehyde — aldol product — α,β-unsaturated ketone or aldehyde

crossed aldol condensation: An aldol condensation between two different ketones or aldehydes. (p. 1133)

(continued)

GUIDE TO ORGANIC REACTIONS IN CHAPTER 22

Reactions covered in Chapter 22 are shown in red. Reactions covered in earlier chapters are shown in blue.

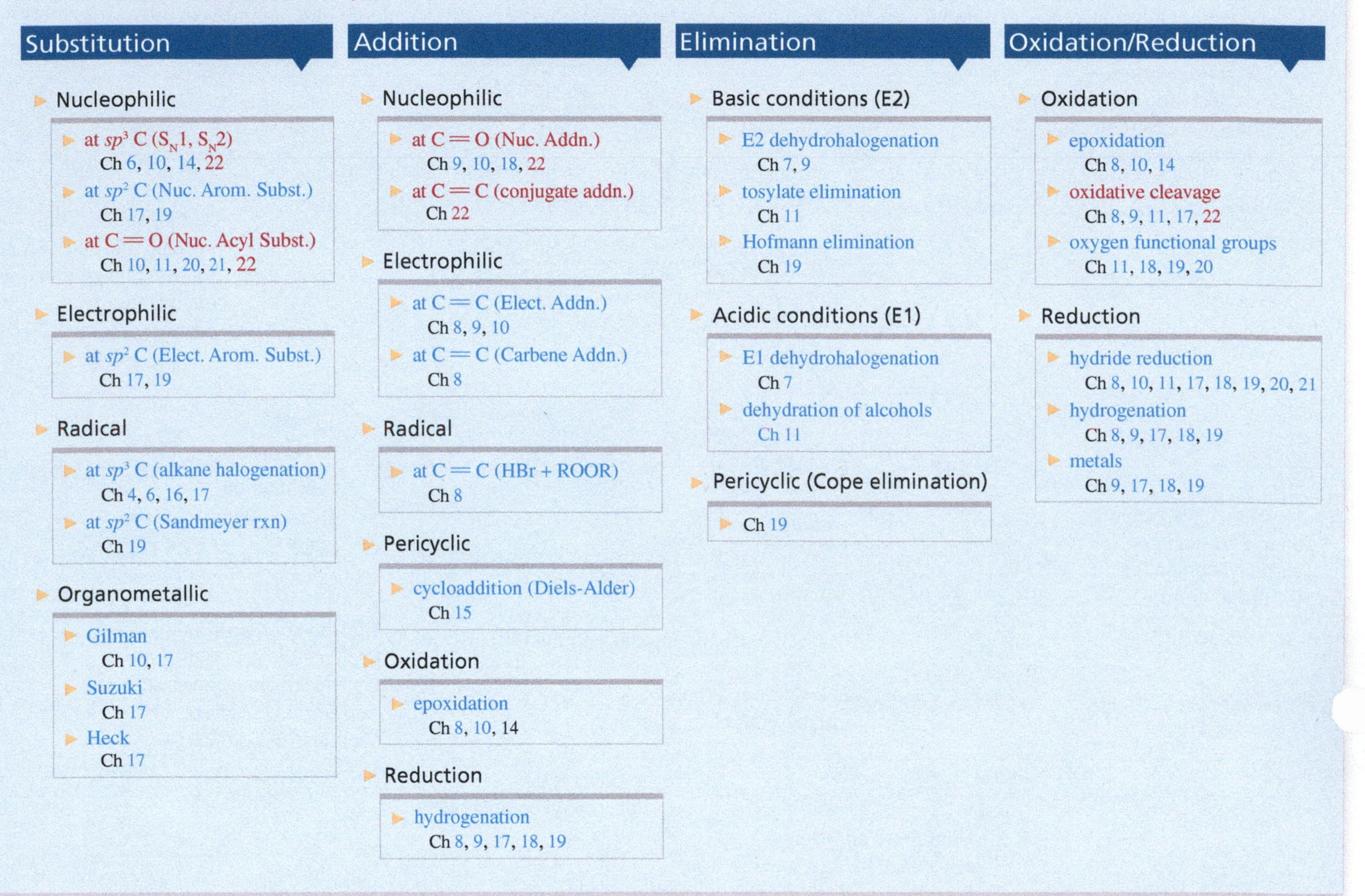

alpha (α) carbon atom The carbon atom next to a carbonyl group. The hydrogen atoms on the α carbon are called **α hydrogens** or **α protons.** (p. 1112)

alpha (α) substitution Replacement of a hydrogen atom at the α carbon atom by some other group. (p. 1112)

Claisen condensation The base-catalyzed conversion of two ester molecules to a β-keto ester. (p. 1139)

ester enolate $\longrightarrow$ tetrahedral intermediate $\longrightarrow$ a β-keto ester + $^-$OR′

crossed Claisen condensation: A Claisen condensation between two different esters or between a ketone and an ester. (p. 1143)

condensation A reaction that bonds two or more molecules, often with the loss of a small molecule such as water or an alcohol. (p. 1113)

conjugate addition **(1,4-addition)** An addition of a nucleophile to the β position of a conjugated double bond, such as that in an α,β-unsaturated ketone or ester. (p. 1154)

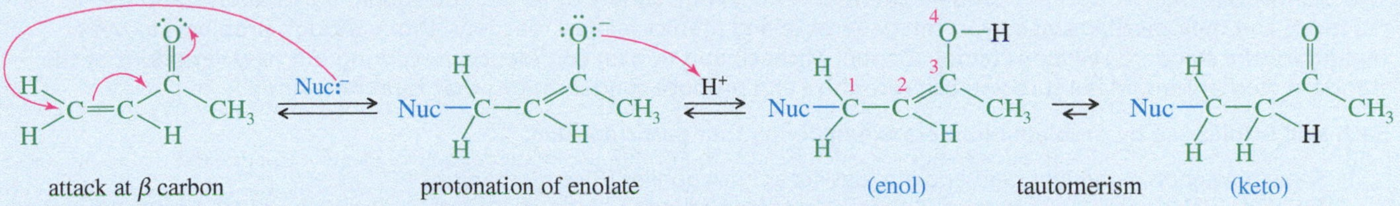

attack at β carbon — protonation of enolate — (enol) — tautomerism — (keto)

Dieckmann condensation A Claisen condensation that forms a ring. (p. 1142)

enamine A **vinyl amine,** usually generated by the acid-catalyzed reaction of a secondary amine with a ketone or an aldehyde. (p. 1119)

enol A vinyl alcohol. Simple enols usually tautomerize to their **keto** forms. (p. 1114)

enolate ion The resonance-stabilized anion formed by deprotonating the carbon atom next to a carbonyl group. (p. 1114)

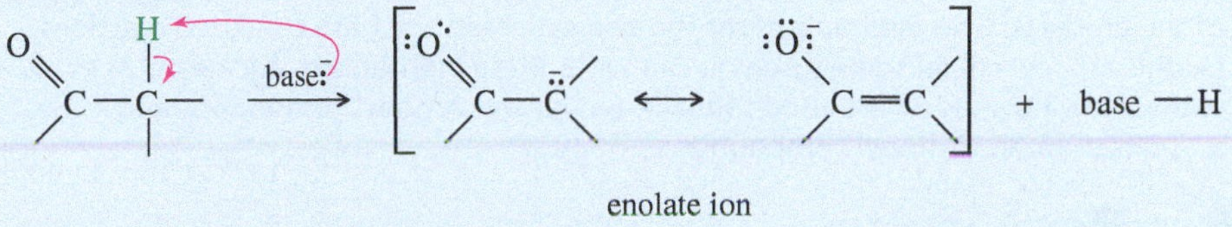

enolate ion

enolizable hydrogen (α hydrogen) A hydrogen atom on a carbon adjacent to a carbonyl group. Such a hydrogen may be lost and regained through keto–enol tautomerism, losing its stereochemistry in the process. (p. 1115)

haloform reaction The conversion of a methyl ketone to a carboxylate ion and a haloform (CHX_3) by treatment with a halogen and base. The **iodoform reaction** uses iodine to give a precipitate of solid iodoform. (p. 1124)

Hell–Volhard–Zelinsky reaction **(HVZ reaction)** Reaction of a carboxylic acid with Br_2 and PBr_3 to give an α-bromo acyl bromide, often hydrolyzed to an α-bromo acid. (p. 1127)

malonic ester synthesis Alkylation or acylation of malonic ester (diethyl malonate), followed by hydrolysis and decarboxylation, to give substituted acetic acids. (p. 1148)

Michael addition A **1,4-addition (conjugate addition)** of a resonance-stabilized carbanion (the **Michael donor**) to a conjugated double bond such as an α,β-unsaturated ketone or ester (the **Michael acceptor**). (p. 1155)

Robinson annulation Formation of a cyclohexenone ring by condensation of methyl vinyl ketone (MVK) or a substituted MVK derivative with a ketone. Robinson annulation proceeds by Michael addition to MVK, followed by an aldol condensation with dehydration. (p. 1158)

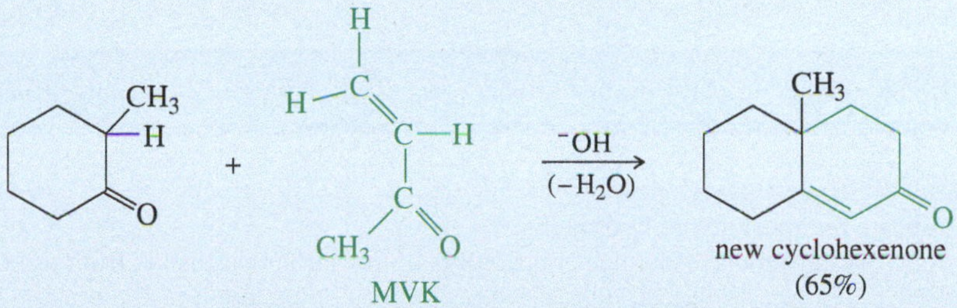

MVK — ^-OH ($-H_2O$) — new cyclohexenone (65%)

Stork reaction Alkylation or acylation of a ketone or aldehyde using its enamine derivative as the nucleophile. Acidic hydrolysis regenerates the alkylated or acylated ketone or aldehyde. (p. 1121)

tautomerism An isomerism involving the migration of a proton and the corresponding movement of a double bond. An example is the **keto–enol tautomerism** of a ketone or aldehyde with its enol form. (p. 1114)

tautomers: The isomers related by a tautomerism. (p. 1114)

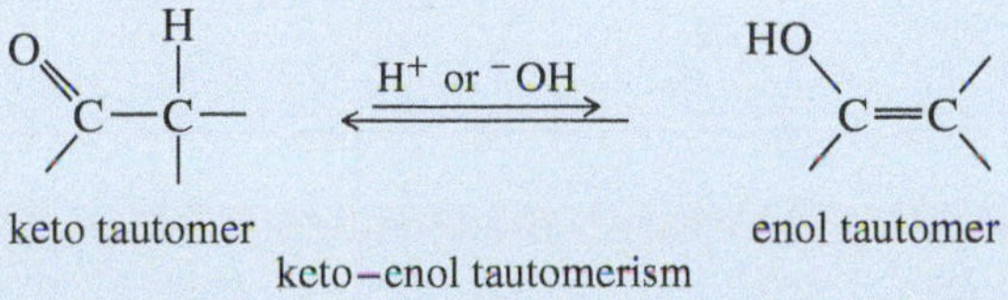

keto tautomer — H^+ or ^-OH — enol tautomer

keto–enol tautomerism

Essential Problem-Solving Skills in Chapter 22

This is a difficult chapter because condensations take on a wide variety of forms. You should try to *understand* the reactions and their mechanisms so you can generalize and predict related reactions. Work enough problems to get a feeling for the standard reactions (aldol, Claisen, Michael) and to gain confidence in working out new variations of the standard mechanisms. Make sure you can recognize and propose condensations that form new rings.

Each skill is followed by problem numbers exemplifying that particular skill.

Skill	Problems
1 Show how enols, enamines, and enolate ions act as nucleophiles. Give mechanisms for acid-catalyzed and base-catalyzed keto–enol tautomerisms.	Problems 22-60, 61, 65, 70, and 76
2 Show how alkylation and acylation of enamines and lithium enolates are used synthetically. Give mechanisms for these reactions.	Problems 22-69, 70, 76, 77, and 78
3 Give mechanisms for the acid-catalyzed and base-promoted alpha-halogenation of ketones. Explain why multiple halogenations are common with basic catalysis and give a mechanism for the haloform reaction.	Problem 22-77
4 Predict the products of aldol and crossed aldol reactions before and after dehydration of the products. Give mechanisms for the acid-catalyzed and base-catalyzed reactions. (Aldols are reversible, so be sure you can write these mechanisms *backward* as well.) Show how aldols are used to make β-hydroxy carbonyl compounds and α, β-unsaturated carbonyl compounds.	Problems 22-62, 64, 68, 69, 71, 74, 75, 77, 80, 83, 84, and 85
Problem-Solving Strategy: Proposing Reaction Mechanisms: Aldol Condensation	Problems 22-62, 64, 68, 77, 80, 83, and 84
5 Predict the products of Claisen and crossed Claisen condensations, and propose mechanisms. Show how a Claisen condensation constructs the carbon skeleton of a target molecule.	Problems 22-63, 64, 68, 69, 76, and 77
6 Show how the malonic ester synthesis makes substituted acetic acids, and how the acetoacetic ester synthesis makes substituted acetones. Give mechanisms for these reactions.	Problems 22-69, 72, 73, 74, 79, and 82
7 Predict the products of conjugate (Michael) additions, and show how to use these reactions in syntheses. Show the general mechanism of the Robinson annulation, and use it to form cyclohexenone ring systems.	Problems 22-67, 76, 77, 80, 81, and 85
Problem-Solving Strategy: Proposing Reaction Mechanisms: Robinson Annulation	Problems 22-67 and 77

Study Problems

22-60 For each molecule shown below,

1. indicate the most acidic hydrogens.

2. draw the important resonance contributors of the anion that results from removal of the most acidic hydrogen.

(a) COOH **(b)** O O **(c)** O CN **(d)** O OH

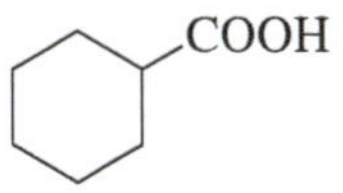

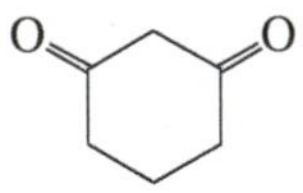

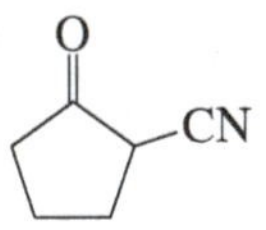

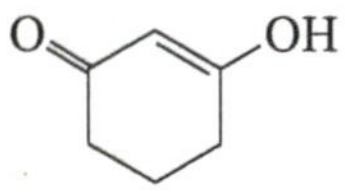

(e) O COOCH$_3$ **(f)** COOCH$_3$ C CN **(g)** $CH_3-CH=CH-\overset{O}{\overset{\|}{C}}-H$ **(h)** $CH_2=CH-CH_2-\overset{O}{\overset{\|}{C}}-H$

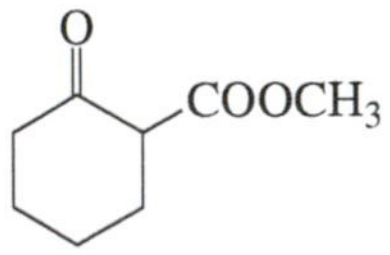

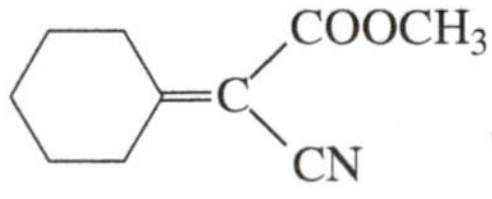

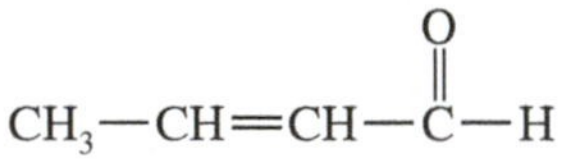

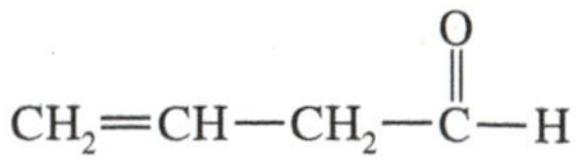

22-61 **1.** Rank the following compounds in order of increasing acidity.
2. Indicate which compounds would be more than 99% deprotonated by a solution of sodium ethoxide in ethanol.

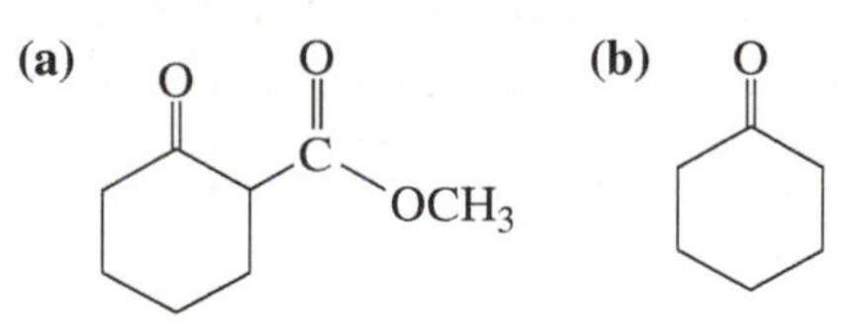

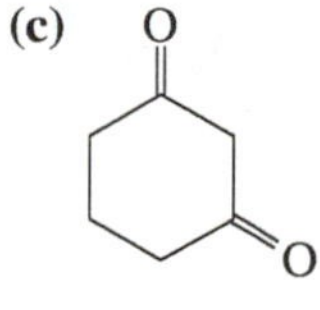

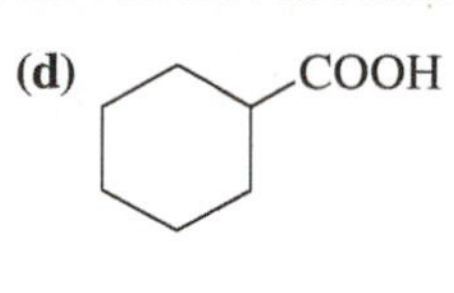

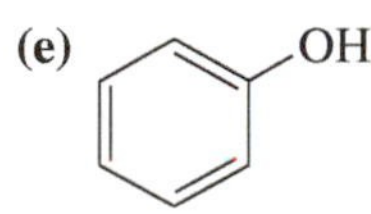

(f) OH

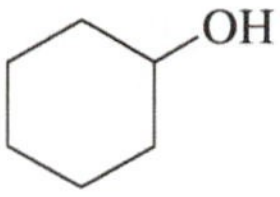

(g) O C OCH_3

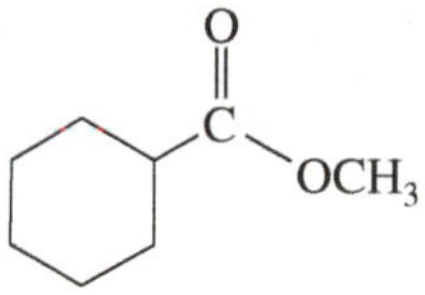

22-62 Predict the products of the following aldol condensations. Show the products both before and after dehydration.

(a) $(CH_3)_2CH{-}CH_2{-}C(=O){-}H \xrightarrow{^-OH}$

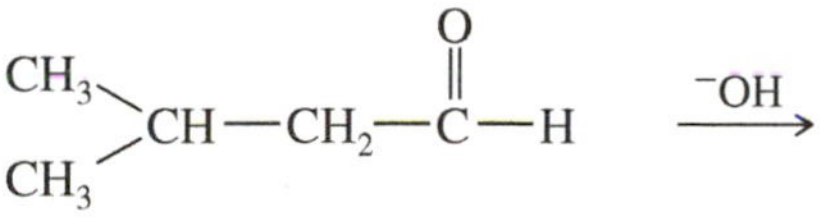

(b) O O $\xrightarrow{H^+}$

(c) $2\ Ph{-}CHO + CH_3{-}C(=O){-}CH_3 \xrightarrow{^-OH}$

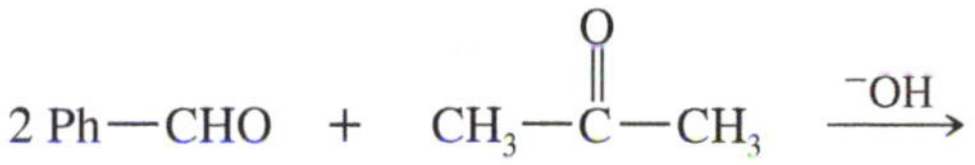

(d) $Ph{-}C(=O){-}CH_3 + (CH_3)_3C{-}C(=O){-}H \xrightarrow{^-OH}$

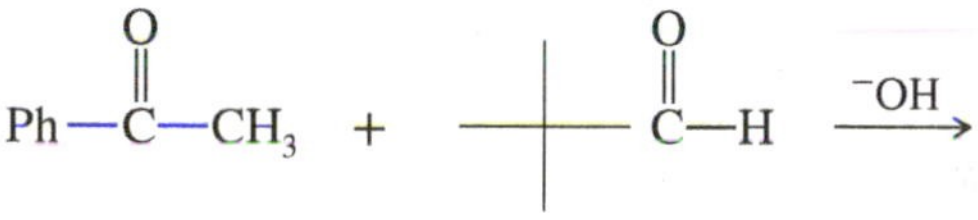

(e) O O + O CHO $\xrightarrow{^-OH}$

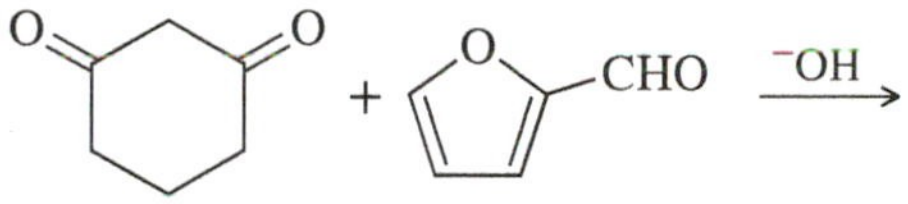

(f) O O $\xrightarrow{^-OH}$

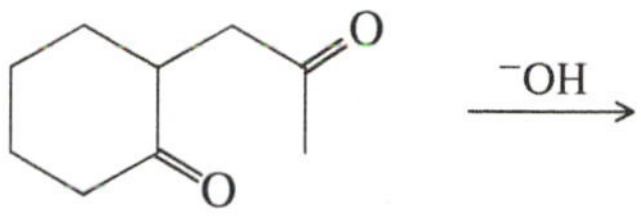

22-63 Predict the products of the following Claisen condensations.

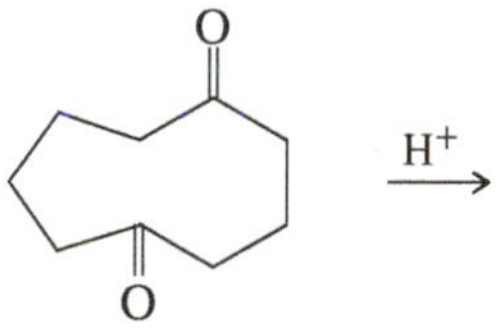

(a) $(CH_3)_2CH{-}CH_2{-}C(=O){-}OCH_3 \xrightarrow[CH_3OH]{^-OCH_3}$

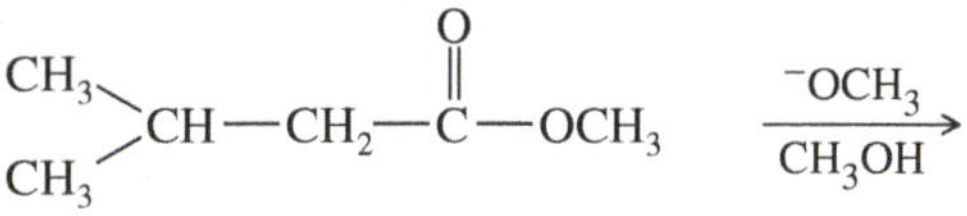

(b) $COOCH_3$ O $\xrightarrow[CH_3OH]{^-OCH_3}$

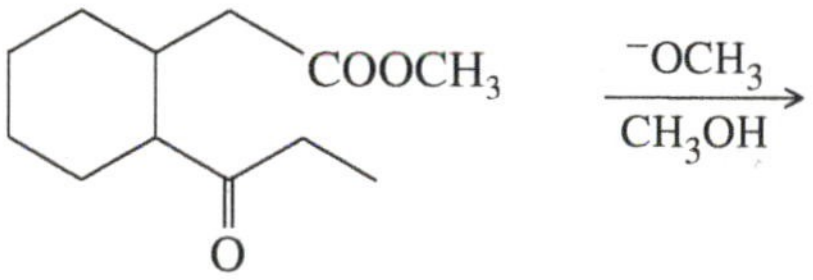

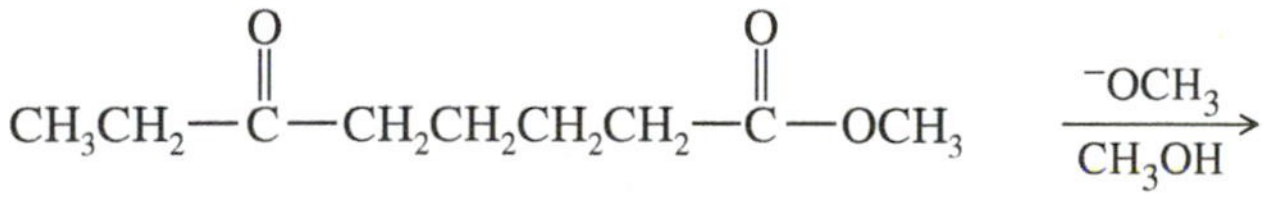

(c) $CH_3CH_2{-}C(=O){-}CH_2CH_2CH_2CH_2{-}C(=O){-}OCH_3 \xrightarrow[CH_3OH]{^-OCH_3}$ (Dieckmann)

(d) O $+\ CH_3O{-}C(=O){-}C(=O){-}OCH_3 \xrightarrow[CH_3OH]{NaOCH_3}$

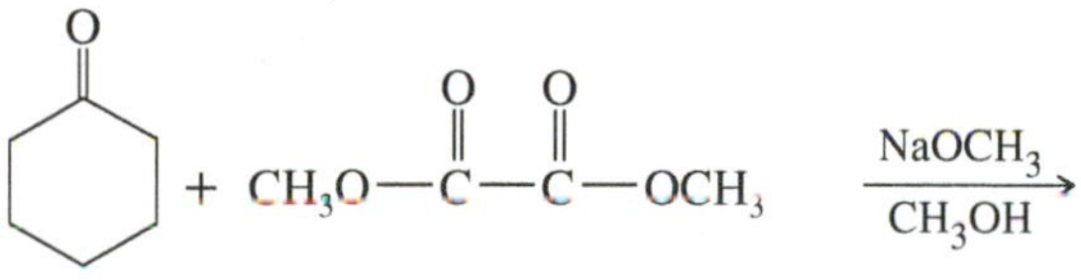

(e) $CH_2{-}C(=O){-}OCH_3$ $CH_2{-}C(=O){-}CH_3$ $\xrightarrow[CH_3OH]{NaOCH_3}$

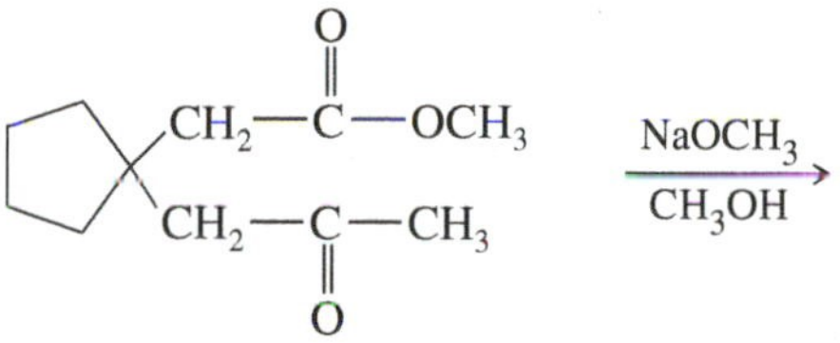

22-64 Propose mechanisms for the reactions shown in Problems 22-62 parts (a) and (b) and 22-63 parts (a) and (b).

22-65 Pentane-2,4-dione (acetylacetone) exists as a tautomeric mixture of 8% keto and 92% enol forms. Draw the stable enol tautomer, and explain its unusual stability.

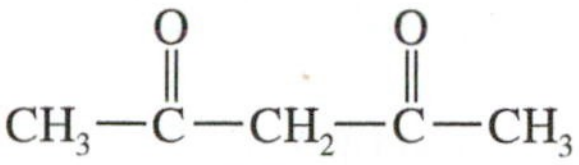

$$CH_3{-}C(=O){-}CH_2{-}C(=O){-}CH_3$$
acetylacetone

22-66 (a) Rank these compounds in order of increasing acid strength.

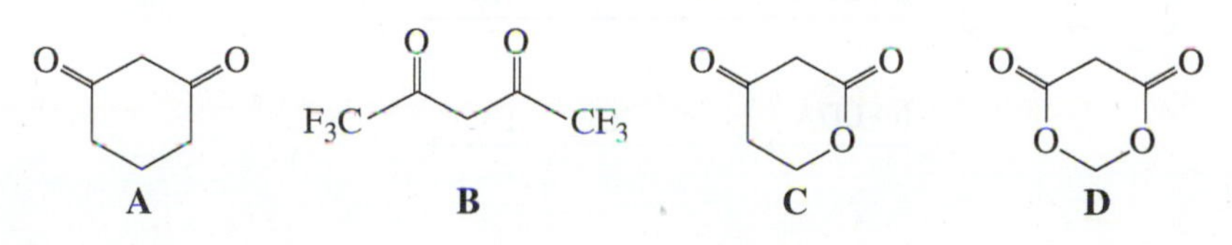

(continued)

(b) Rank these compounds in order of increasing enol content. In each case, draw the most stable enol.

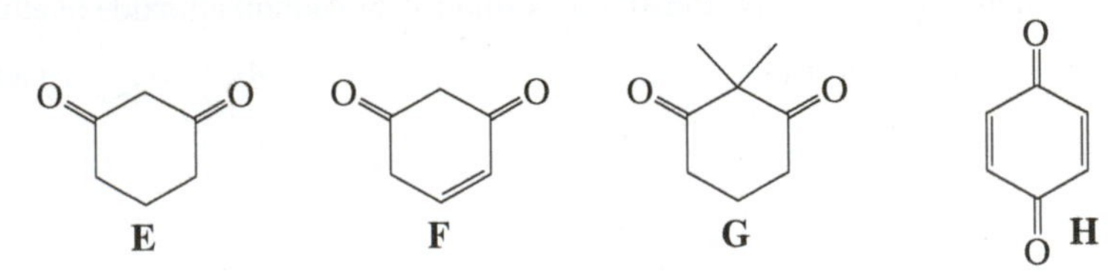

22-67 Show how you would use the Robinson annulation to synthesize the following compounds.

(a) H_3C, Ph, O

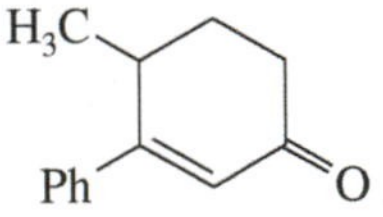

(b)

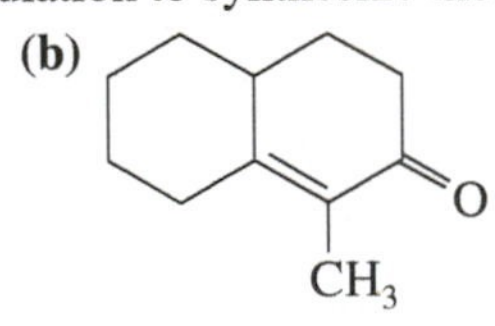

(c) O

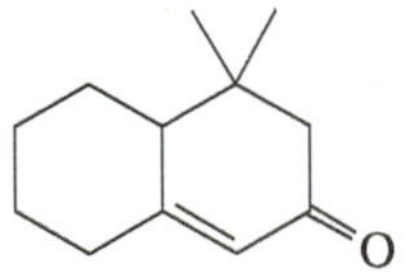

22-68 Show how you would use an aldol, Claisen, or another type of condensation to make each compound.

(a) CHO

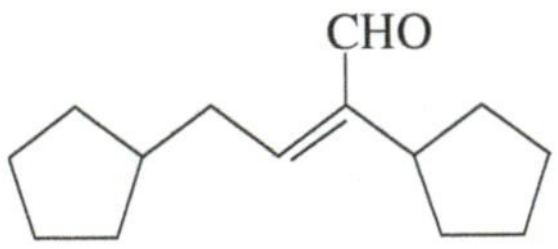

(b)

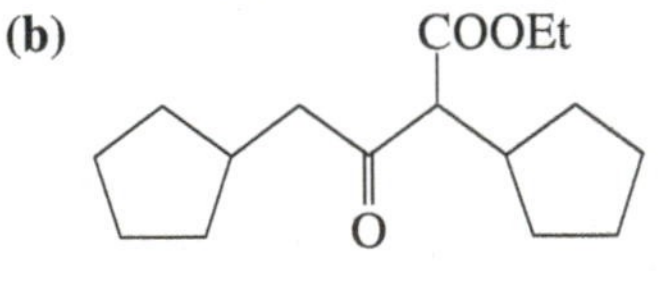

(c)

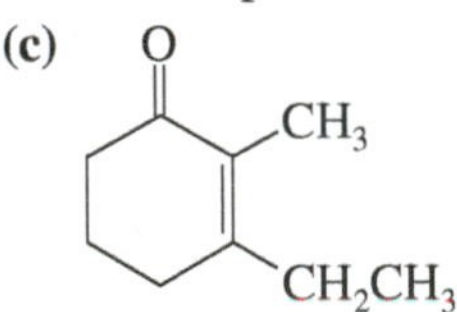

(d) O O $C(CH_3)_3$

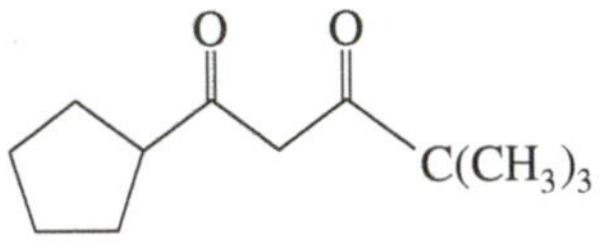

(e) O Ph

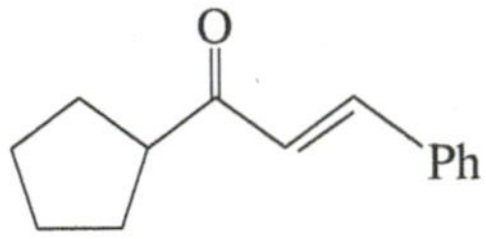

(f) O O

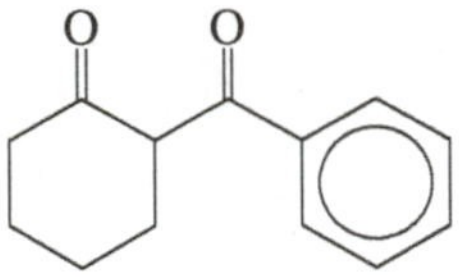

22-69 Predict the products of the following reactions.

(a) cyclopentanone + Br_2 in acetic acid

(b) 1-phenylethanol + excess I_2 in base

(c) O, H, Ph, CH_3, CH_3 $\xrightarrow[(2)\ CH_3CH_2CH_2Br]{(1)\ LDA}$

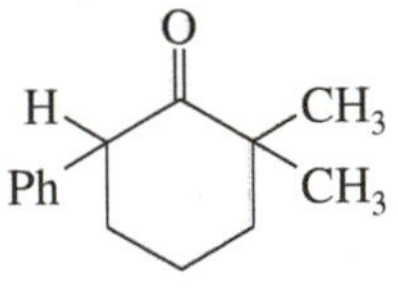

(d) H–C(=O)–Ph + cyclopentanone $\xrightarrow{^-OH}$

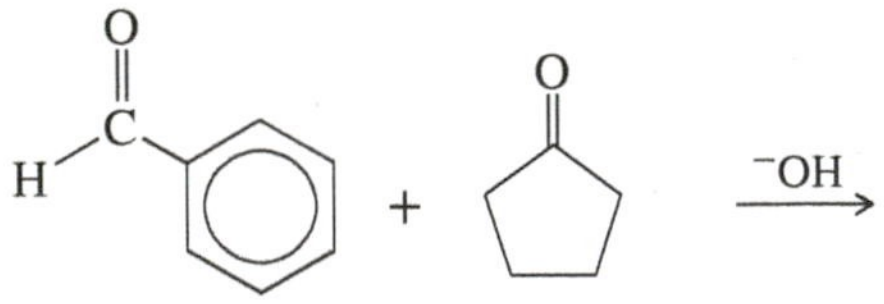

(e) N (pyrrolidine enamine), Ph $\xrightarrow[(2)\ H_3O^+]{(1)\ H_2C{=}CH{-}CH_2Br}$

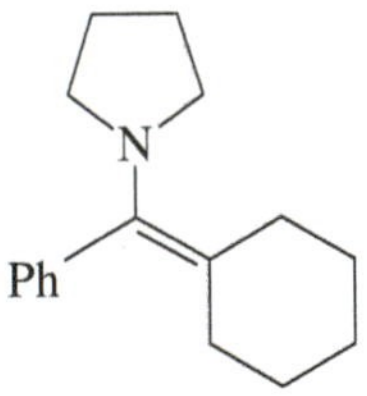

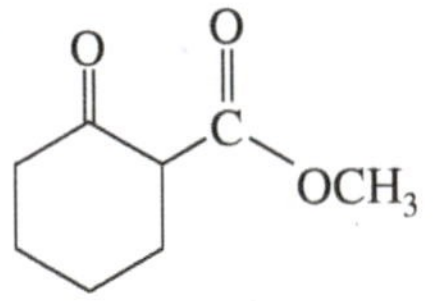

(f) O, O, C, OCH_3 $\xrightarrow[(2)\ CH_3CH_2CH_2CH_2Br]{(1)\ NaOCH_3}$

(g) product from part (f) $\xrightarrow[heat]{H_3O^+}$ (decarboxylation)

(h) $CH_3CH_2{-}\overset{O}{\overset{\|}{C}}{-}CH_2{-}\overset{O}{\overset{\|}{C}}{-}OCH_3$ $\xrightarrow[(2)\ CH_3I \quad (3)\ H_3O^+,\ heat]{(1)\ NaOCH_3}$

(i) $CH_3{-}\overset{O}{\overset{\|}{C}}{-}CH_2{-}\overset{O}{\overset{\|}{C}}{-}OCH_2CH_3$ + cyclohexenone $\xrightarrow[(2)\ H_3O^+,\ heat]{(1)\ NaOCH_2CH_3}$

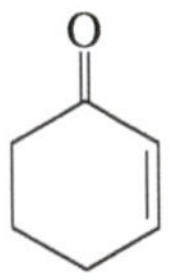

22-70 Predict the products of these reaction sequences.

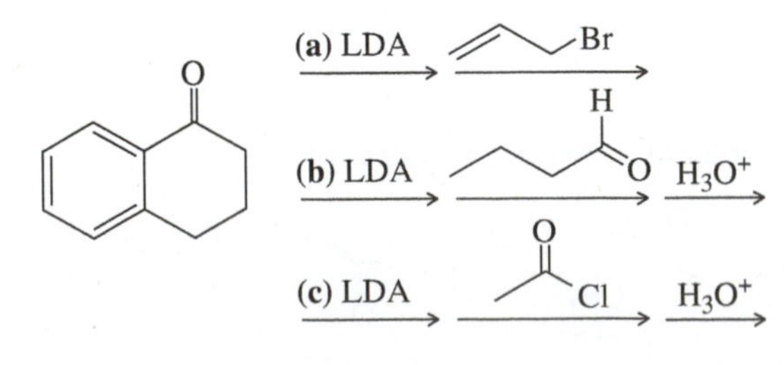

22-71 Show how you would accomplish the following conversions in good yields. You may use any necessary reagents.

(a) $CH_3-C(CH_3)_2-C(=O)-CH_3 \longrightarrow CH_3-C(CH_3)_2-C(=O)-CH_2Br$

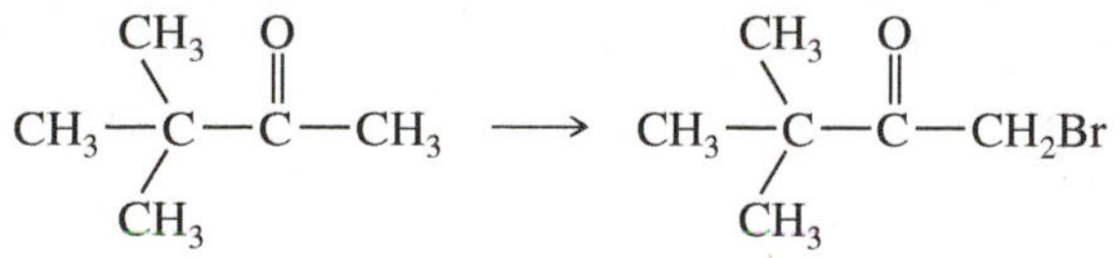

(b) cyclohexane–COOH $\longrightarrow$ cyclohexane with COOH and Br on same carbon

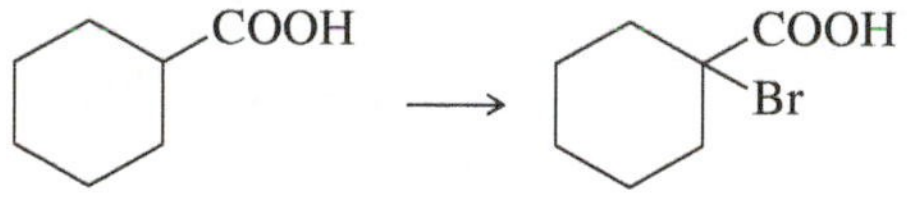

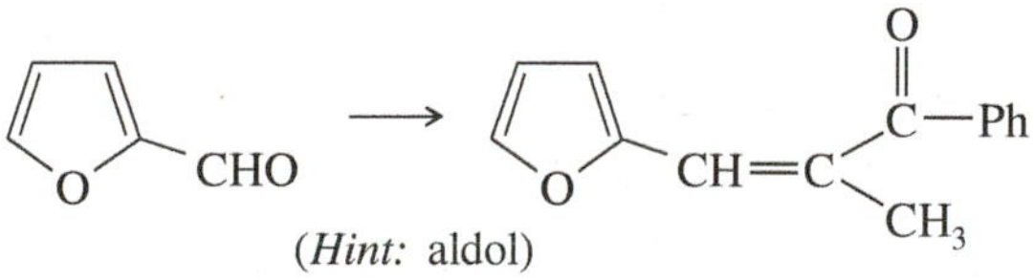

(c) $CH_3-C(CH_3)_2-C(=O)-CH_3 \longrightarrow CH_3-C(CH_3)_2-C(=O)-O^-$

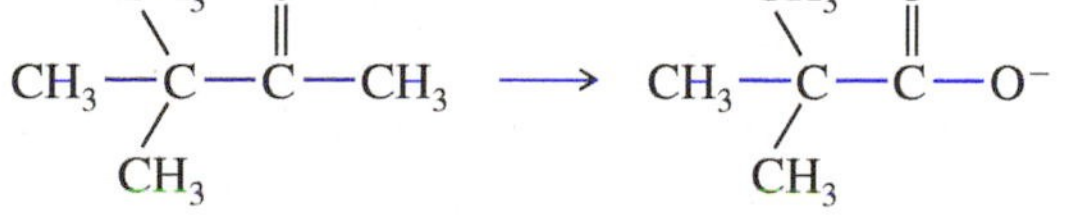

(d) $Ph-C(=O)-H \longrightarrow Ph-CH=CH-CH_3$

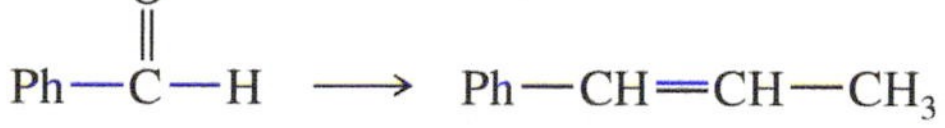

(e) furan–CHO $\longrightarrow$ furan–CH=C(CH$_3$)–C(=O)–Ph (*Hint:* aldol)

(f) cyclopentanone $\longrightarrow$ 2-acyl cyclopentanone with N-methylpyrrol-2-yl ketone

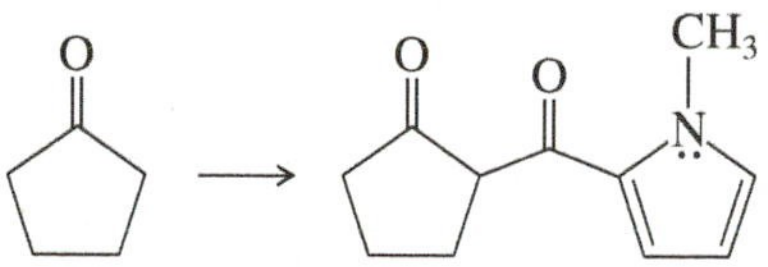

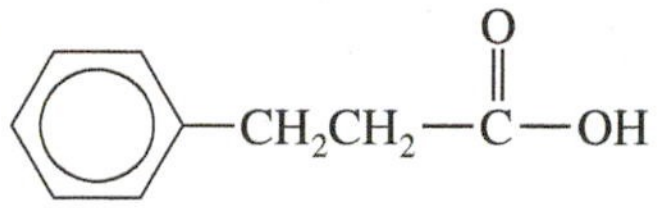

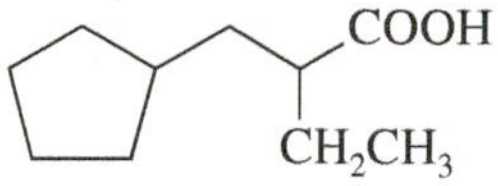

22-72 Show how you would use the malonic ester synthesis to make the following compounds.

(a) $C_6H_5-CH_2CH_2-C(=O)-OH$

(b) cyclopentyl–CH_2–CH(COOH)–CH_2CH_3

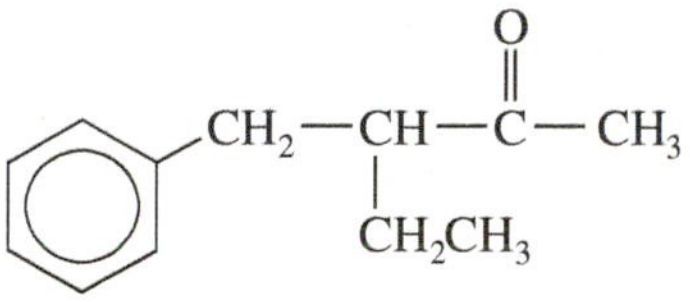

(c) cyclohexyl–C(=O)–OH

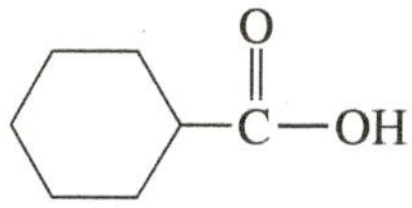

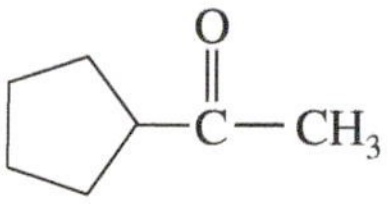

22-73 Show how you would use the acetoacetic ester synthesis to make the following compounds.

(a) $C_6H_5-CH_2-CH(CH_2CH_3)-C(=O)-CH_3$

(b) cyclopentyl–C(=O)–CH_3

* (c) 3-methylcyclohex-2-enone (CH_3)

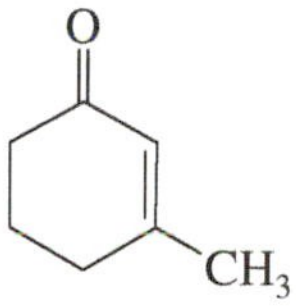

(*Hint*: Consider using heptane-2,6-dione as an intermediate.)

***22-74** The following compounds can be synthesized by aldol condensations, followed by further reactions. (In each case, work backward from the target molecule to an aldol product, and show what compounds are needed for the condensation.)

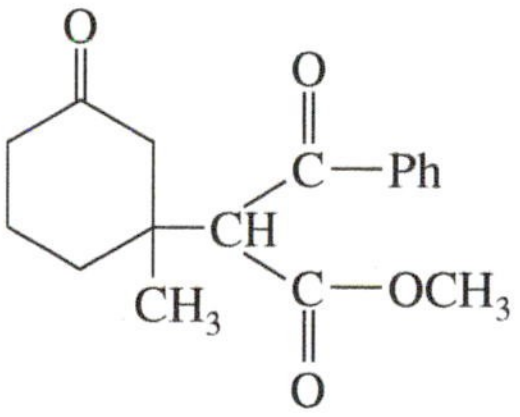

(a) $Ph-CH_2-CH_2-CH(OH)-Ph$

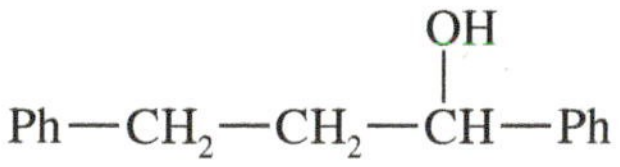

(b) 2-methyl-cyclohexanone bearing CH(C(=O)–Ph)(C(=O)–OCH_3) and CH_3 on C-2

(c) cyclopentanone with CH(Ph)–CH(NO_2)–C(=O)CH$_3$ side chain

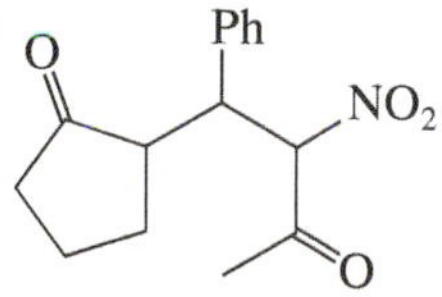

***22-75** The Knoevenagel condensation is a special case of the aldol condensation in which an active methylene compound reacts with an aldehyde or ketone, in the presence of a secondary amine as a basic catalyst, to produce a new C=C. Show the starting materials that made each of these by a Knoevenagel condensation.

(a) Ph (b) O, O, O, O (c) EtOOC, COOEt, O, H

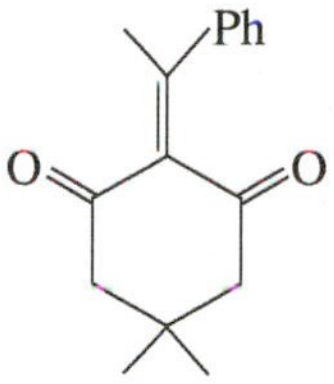

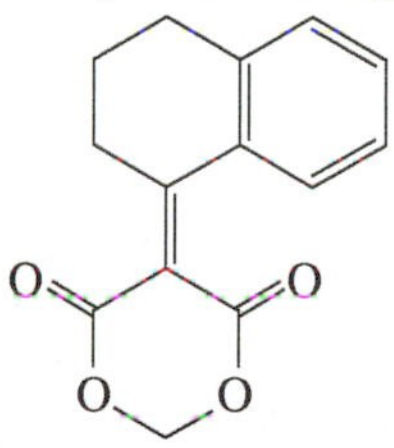

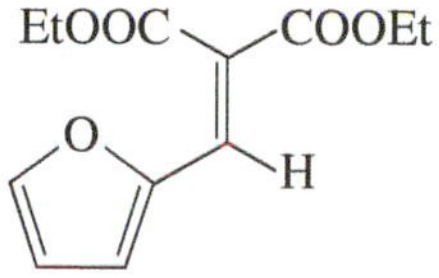

(d) EtO, O, CN (e) NO_2, O_2N (f) O, O

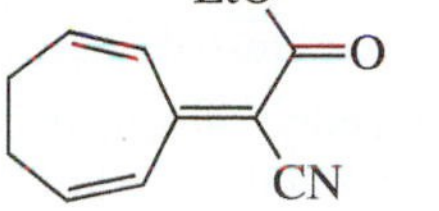

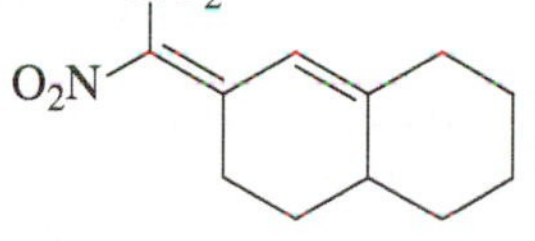

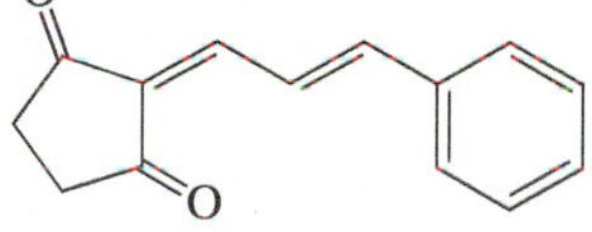

***22-76** Predict the products from this sequence of reactions.

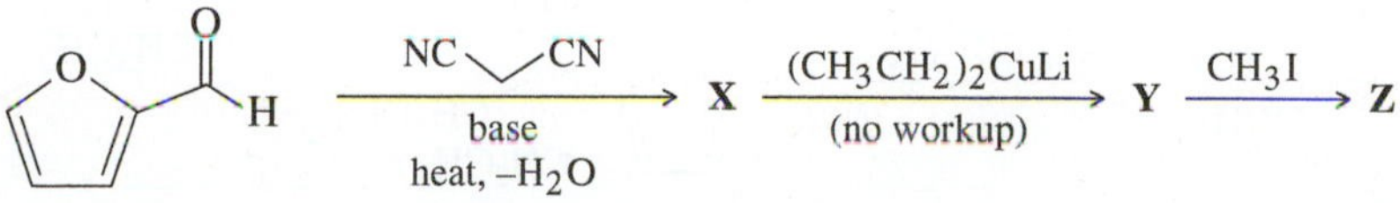

furan–CHO $\xrightarrow[\text{base, heat, } -H_2O]{NC\frown CN}$ **X** $\xrightarrow[\text{(no workup)}]{(CH_3CH_2)_2CuLi}$ **Y** $\xrightarrow{CH_3I}$ **Z**

22-77 Propose mechanisms for the following reactions.

(a) cyclohexanone + PhCHO $\xrightarrow{^-OH}$ 2-benzylidenecyclohexanone (=CHPh)

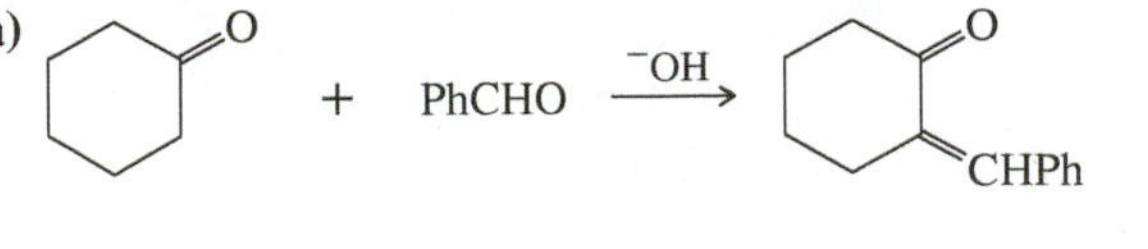

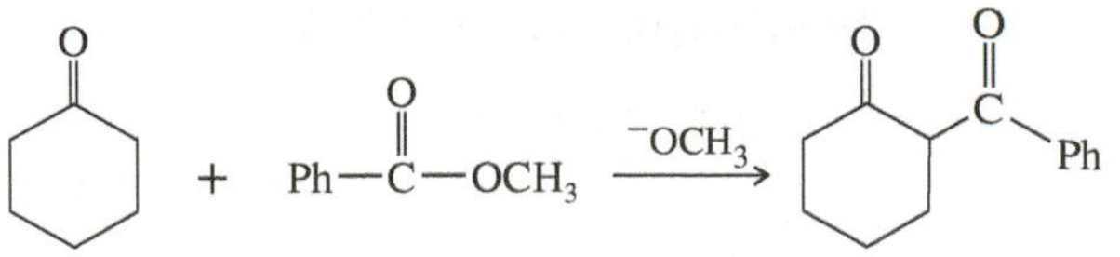

(b) cyclohexanone + $Ph-\overset{O}{\overset{\|}{C}}-OCH_3$ $\xrightarrow{^-OCH_3}$ 2-benzoylcyclohexanone ($-\overset{O}{\overset{\|}{C}}-Ph$)

(c) (OCH_3-substituted tetralone) + $H_2C{=}CH-\overset{O}{\overset{\|}{C}}-CH_2CH_3$ $\xrightarrow{^-OH}$ (tricyclic enone with H_3C and OCH_3)

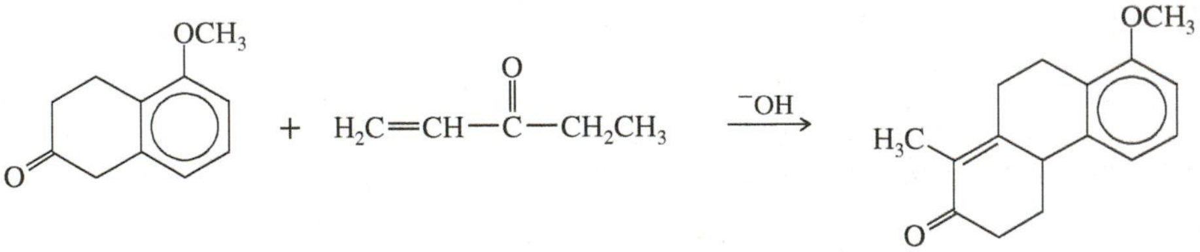

(d) (1-cyclohexenyl)pyrrolidine $\xrightarrow[(2)\ H_3O^+]{(1)\ MVK}$ (bicyclic enone)

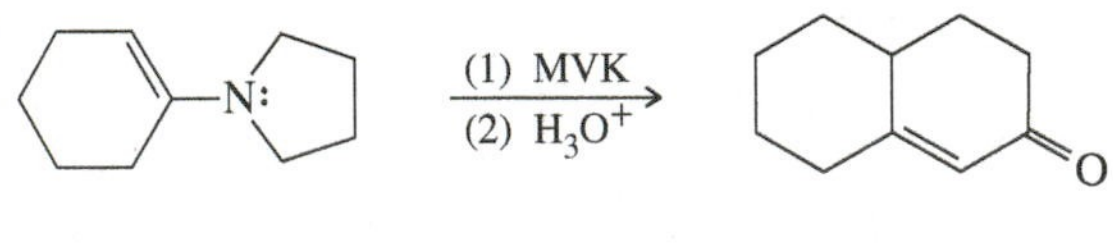

(e) acetophenone (CH_3) $\xrightarrow[AcOH]{Br_2}$ α-bromoacetophenone (CH_2Br)

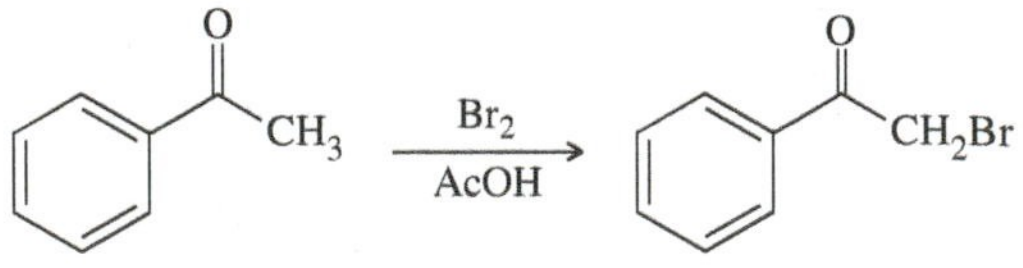

22-78 Write equations showing the expected products of the following enamine alkylation and acylation reactions. Then give the final products expected after hydrolysis of the iminium salts.

(a) pyrrolidine enamine of pentan-3-one + allyl chloride
(b) pyrrolidine enamine of acetophenone + butanoyl chloride
(c) piperidine enamine of cyclopentanone + methyl iodide
(d) piperidine enamine of cyclopentanone + methyl vinyl ketone

***22-79** Show how you would accomplish the following multistep conversions. You may use any additional reagents you need.

(a) dimethyl adipate and allyl bromide ⟶ 2-allylcyclopentanone

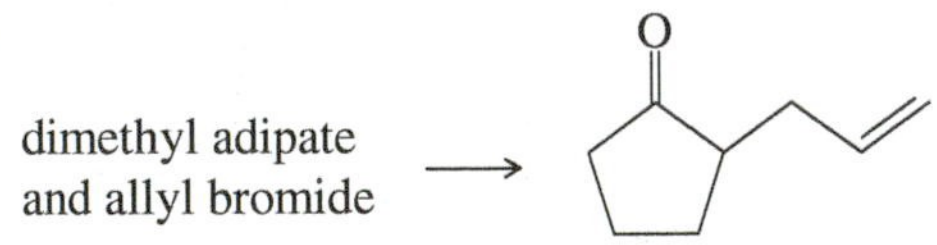

(b) cyclohexanone ⟶ 3-phenylcyclohexanone (Ph)

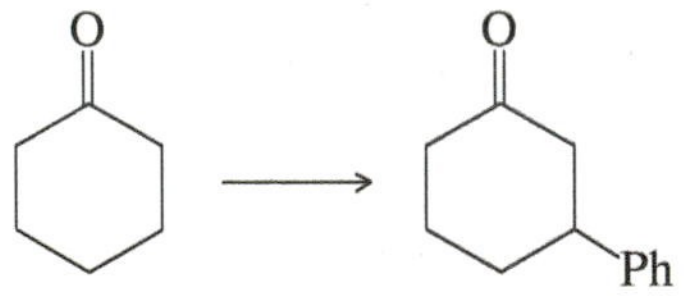

(c) indan-1-one ⟶ (tricyclic alcohol with HO and CH_3)

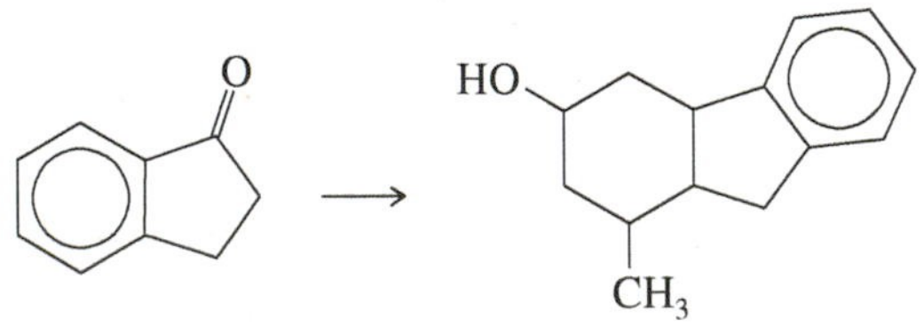

(d) cyclohexanone ⟶ (2-benzylidene-6-methyl-6-carbethoxycyclohexanone, OEt)

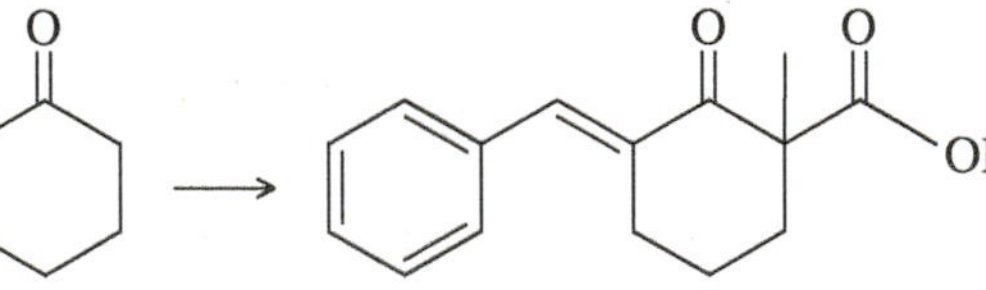

***22-80** Many of the condensations we have studied are reversible. The reverse reactions are often given the prefix *retro-*, the Latin word meaning "backward." Propose mechanisms to account for the following reactions.

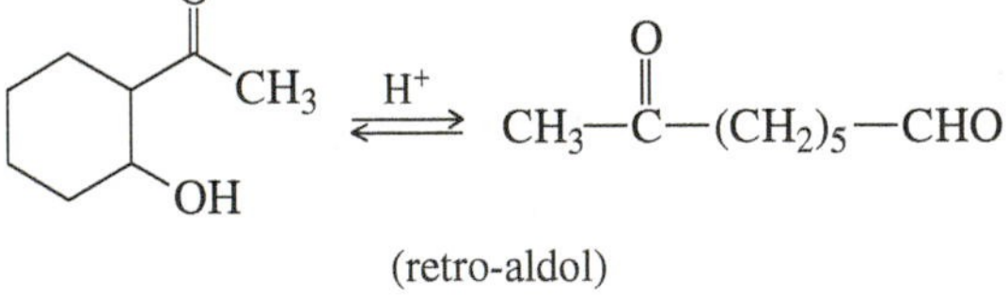

(a) (2-acetylcyclohexanol: CH_3, OH) $\overset{H^+}{\rightleftharpoons}$ $CH_3-\overset{O}{\overset{\|}{C}}-(CH_2)_5-CHO$
(retro-aldol)

(b) (cyclobutanol with acetyl, CH_3, OH) $\xrightarrow{H^+}$ 3-methylcyclohex-2-enone (CH_3)
(retro-aldol and further condensation)

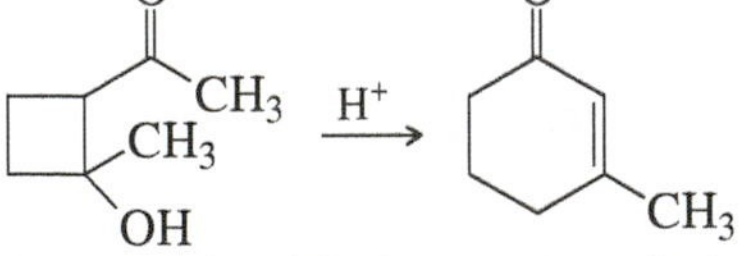

(c) (2-(2-cyanoethyl)cyclohexanone, CN) $\overset{^-OH}{\rightleftharpoons}$ cyclohexanone + $H_2C{=}CH-CN$
(retro-Michael)

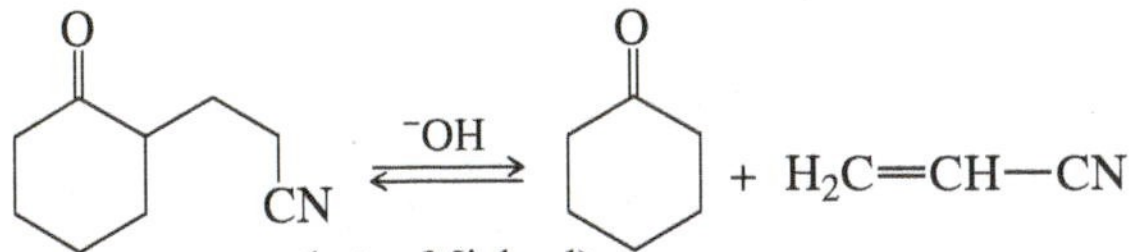

(d) (cyclohexane with CH_3, OH, $COOCH_3$) $\xrightarrow[CH_3OH]{^-OCH_3}$ 2-acetylcyclohexanone
(retro-aldol and crossed Claisen)

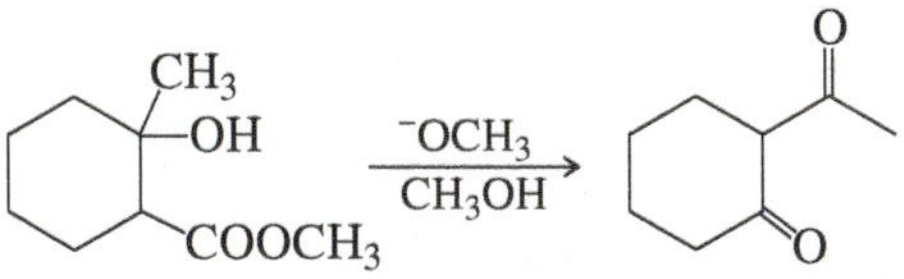

***22-81** (A true story.) Chemistry lab students added an excess of ethylmagnesium bromide to methyl furoate, expecting the Grignard reagent to add twice and form the tertiary alcohol. After water workup, they found that the product was a mixture of two compounds. One was the expected product having two ethyl groups, but the unexpected product had added three ethyl groups. Propose a mechanism to explain the formation of the unexpected product.

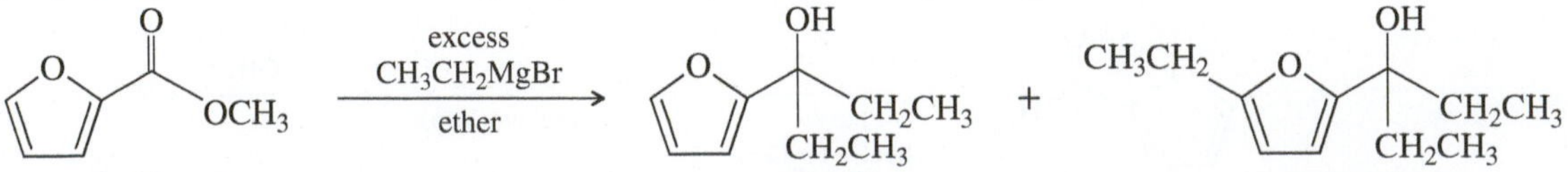

methyl furoate (OCH_3) $\xrightarrow[ether]{excess\ CH_3CH_2MgBr}$ expected (OH, CH_2CH_3, CH_2CH_3) + unexpected (CH_3CH_2-furyl, OH, CH_2CH_3, CH_2CH_3)

22-82 Propose a mechanism for the following reaction. Show the structure of the compound that results from hydrolysis and decarboxylation of the product.

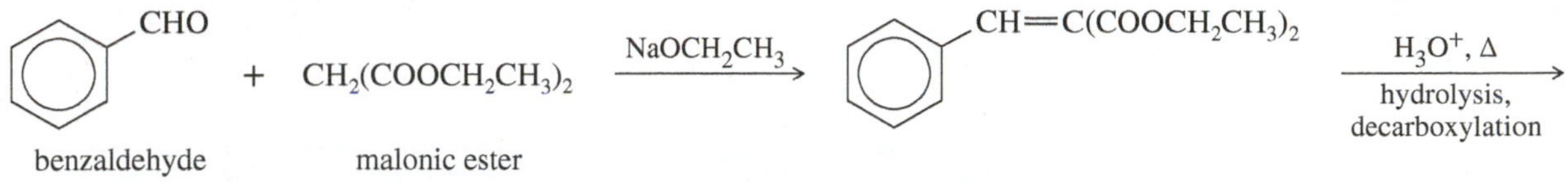

benzaldehyde (C₆H₅CHO) + $CH_2(COOCH_2CH_3)_2$ (malonic ester) $\xrightarrow{NaOCH_2CH_3}$ $C_6H_5CH{=}C(COOCH_2CH_3)_2$ $\xrightarrow[\text{hydrolysis, decarboxylation}]{H_3O^+, \Delta}$

22-83 A reaction involved in the metabolism of sugars is the splitting of fructose-1,6-diphosphate to give glyceraldehyde-3-phosphate and dihydroxyacetone phosphate. In the living system, this retro-aldol is catalyzed by an enzyme called *aldolase*; however, it can also be catalyzed by a mild base. Propose a mechanism for the base-catalyzed reaction.

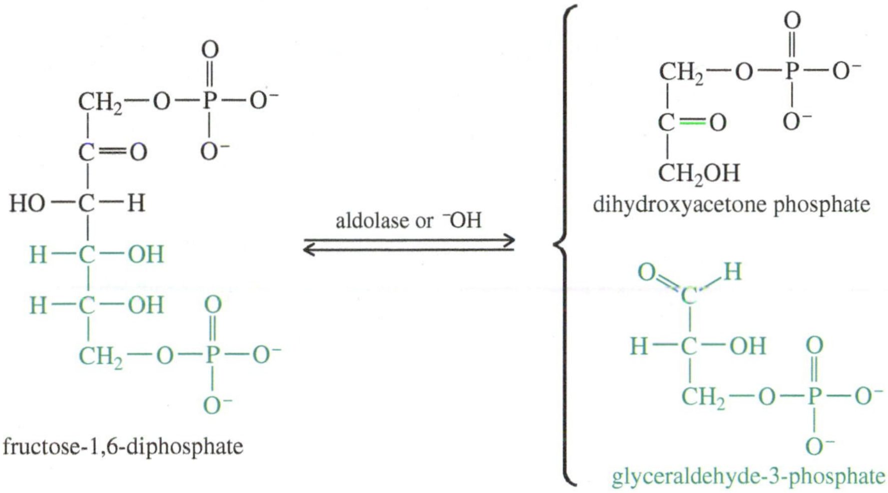

fructose-1,6-diphosphate $\underset{}{\xrightleftharpoons{\text{aldolase or }^{-}OH}}$ dihydroxyacetone phosphate + glyceraldehyde-3-phosphate

22-84 Biochemists studying the structure of collagen (a fibrous protein in connective tissue) found cross-links containing α,β-unsaturated aldehydes between protein chains. Show the structures of the side chains that react to form these cross-links, and propose a mechanism for their formation in a weakly acidic solution.

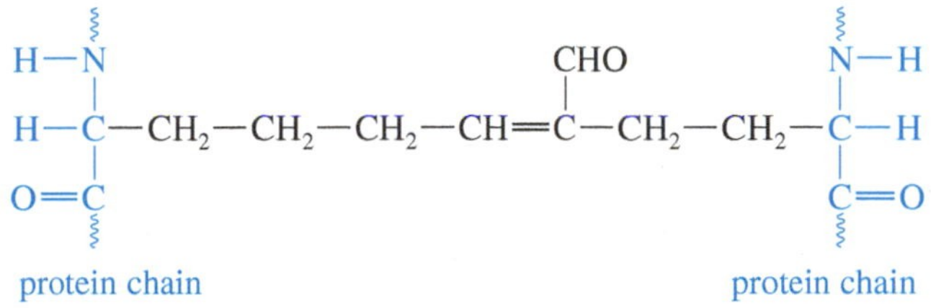

protein chain — $CH(NH)(C{=}O)-CH_2-CH_2-CH_2-CH{=}C(CHO)-CH_2-CH_2-CH(NH)(C{=}O)$ — protein chain

***22-85** Show reaction sequences (not detailed mechanisms) that explain these transformations:

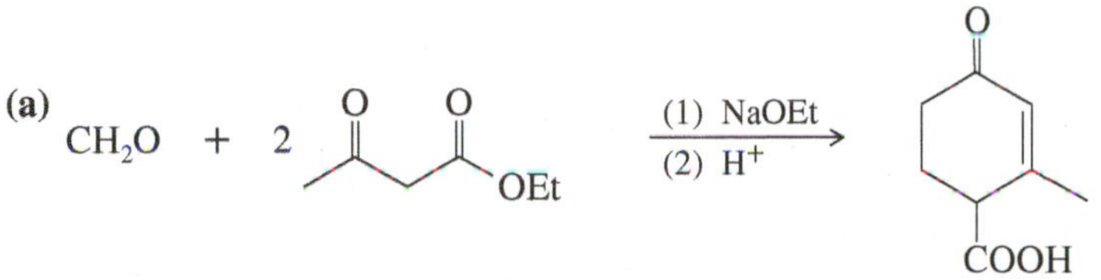

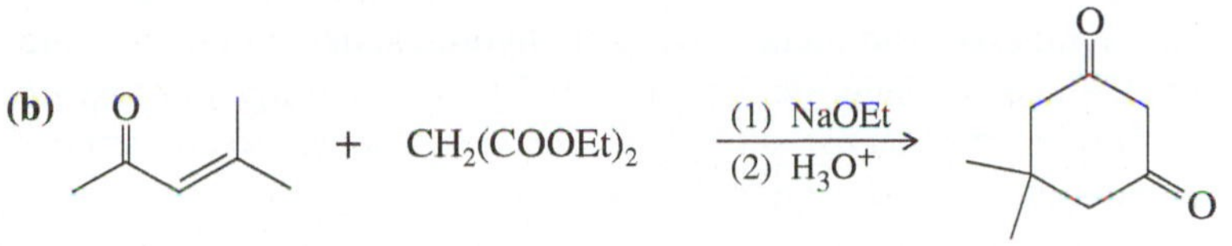

(a) CH_2O + 2 ethyl acetoacetate $\xrightarrow[(2)\ H^+]{(1)\ NaOEt}$ 3-methylcyclohex-2-enone-4-carboxylic acid (COOH)

(b) mesityl oxide + $CH_2(COOEt)_2$ $\xrightarrow[(2)\ H_3O^+]{(1)\ NaOEt}$ 5,5-dimethylcyclohexane-1,3-dione

23 Carbohydrates and Nucleic Acids

Goals for Chapter 23

1 Draw and identify the structures of glucose, its anomers, and its epimers, both as Fischer projections and as chair conformations.

2 Correctly name monosaccharides and disaccharides, and draw their structures from their names.

3 Predict the reactions of carbohydrates in acidic and basic solutions, and with oxidizing and reducing agents. Predict the reactions that convert their hydroxy groups to ethers or esters, and their carbonyl groups to acetals.

4 Draw the common types of glycosidic linkages, and identify these linkages in disaccharides and polysaccharides.

5 Recognize the structures of DNA and RNA, and draw the structures of the common ribonucleotides and deoxyribonucleotides.

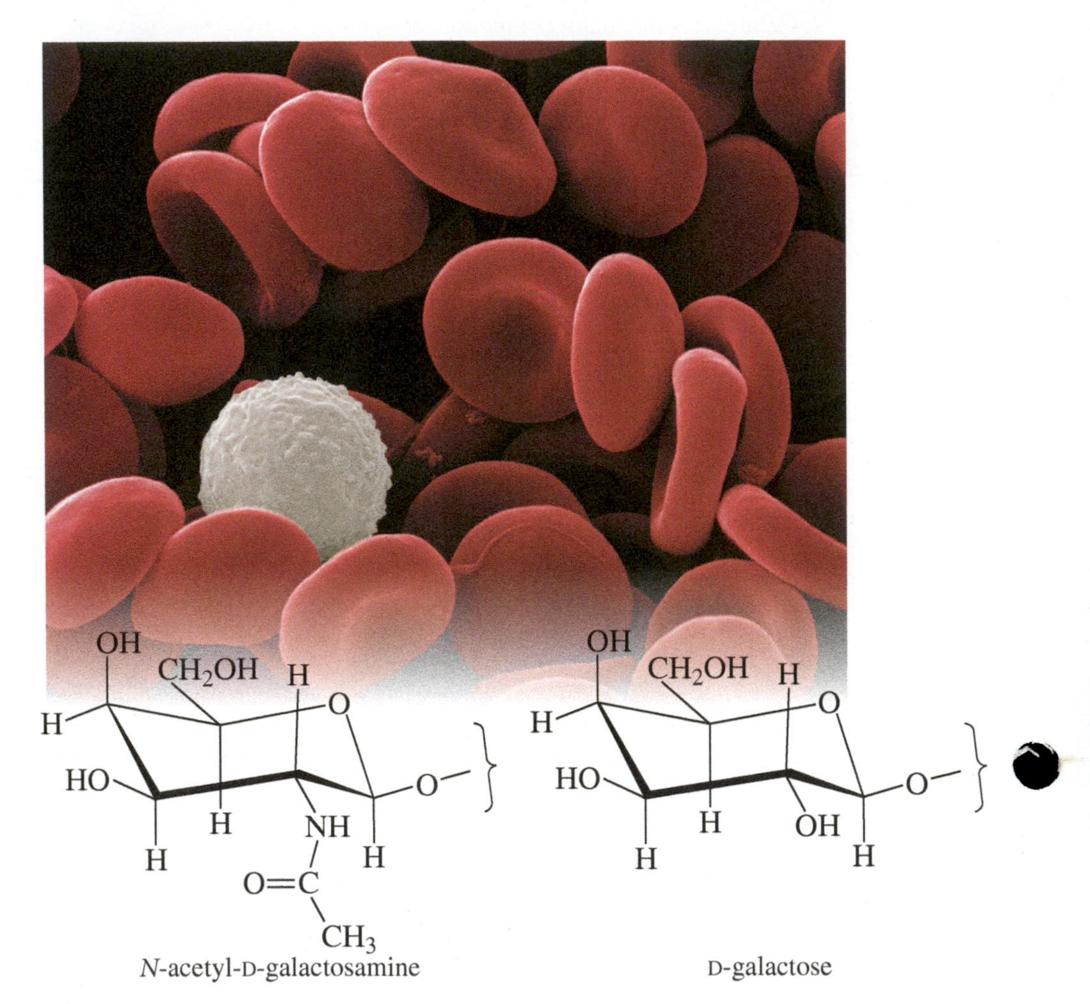

▲ Biochemists have found that cells in organisms recognize other cells by the pattern of sugar polymers (polysaccharides) on their surfaces. White blood cells identify foreign cells as pathogens if the polysaccharide sequences on the foreign cell surface do not match the usual sequences for that organism. Red blood cells also have patterns of sugar polymers that identify them as specific blood types. Type A blood cells have *N*-acetyl-D-galactosamine at the end of the sequence and Type B blood cells have D-galactose at the end. Type AB blood cells have both sequences, and Type O blood cells have neither.

23-1 Introduction

Carbohydrates are the most abundant organic compounds in nature. Nearly all plants and animals synthesize and metabolize carbohydrates, using them to store energy and deliver it to their cells. Plants synthesize carbohydrates through *photosynthesis*, a complex series of reactions that use sunlight as the energy source to convert carbon dioxide and water into glucose and oxygen. Many molecules of glucose can be linked together to form either *starch* for energy storage or *cellulose* to support the plant.

$$6\,CO_2 + 6\,H_2O \xrightarrow{\text{light}} 6\,O_2 + \underset{\text{glucose}}{C_6H_{12}O_6} \longrightarrow \text{starch, cellulose} + H_2O$$

Most living organisms oxidize glucose to carbon dioxide and water to provide the energy needed by their cells. Plants can retrieve the glucose units from starch when needed. In effect, starch is a plant's storage unit for solar energy for later use. Animals can also store glucose energy by linking many molecules together to form *glycogen*, another form of starch. *Cellulose* makes up the cell walls of plants and forms their structural framework. Cellulose is the major component of *wood*, a strong yet supple material that supports the great weight of the oak, yet allows the willow to bend with the wind.

Almost every aspect of human life involves carbohydrates in one form or another. Like other animals, we use the energy content of carbohydrates in our food to produce and store energy in our cells. In cells, they commonly form links to proteins and lipids to make glycoproteins and glycolipids, which serve important functions in the immune system, in hormones, and in cell membranes. Clothing is made from cotton and linen, two forms of cellulose. Other fabrics are made by manipulating cellulose to convert it to the semisynthetic fibers *rayon* and *cellulose acetate*. In the form of wood, we use cellulose to construct our houses and as a fuel to heat them. Even this page is made from cellulose fibers.

Carbohydrate chemistry is one of the most interesting areas of organic chemistry. Many chemists are employed by companies that use carbohydrates to make foods, building materials, and other consumer products. All biologists must understand carbohydrates, which play pivotal roles throughout the plant and animal kingdoms. At first glance, the structures and reactions of carbohydrates may seem complicated. We will learn how these structures and reactions are consistent and predictable, however, and we can study carbohydrates as easily as we study the simplest organic compounds.

23-2 Classification of Carbohydrates

Sugars have been known from antiquity for their sweet taste. We use the word ***sugar*** as a generic term for low-molecular-weight carbohydrates (monosaccharides and disaccharides), especially those that are found in foods. The term **carbohydrate** arose because most sugars have molecular formulas $C_n(H_2O)_m$, suggesting that carbon atoms are combined in some way with water. In fact, the empirical formula of most simple sugars is $C(H_2O)$. Chemists named these compounds "hydrates of carbon" or "carbohydrates" because of these molecular formulas. Our modern definition of carbohydrates includes polyhydroxyaldehydes, polyhydroxyketones, and compounds that are easily hydrolyzed to them.

Monosaccharides, or *simple sugars*, are carbohydrates that cannot be hydrolyzed to simpler compounds. Figure 23-1 shows Fischer projections of the monosaccharides *glucose* and *fructose*. Glucose is a polyhydroxyaldehyde, and fructose is a polyhydroxy ketone. Polyhydroxyaldehydes are called **aldoses** (*ald*- is for *ald*ehyde and *-ose* is the suffix for a sugar), and polyhydroxyketones are called **ketoses** (*ket*- for *ket*one, and *-ose* for sugar).

We have used Fischer projections to draw the structures of glucose and fructose because Fischer projections conveniently show the stereochemistry at all the asymmetric carbon atoms. The Fischer projection was originally developed by Emil Fischer, a carbohydrate chemist who received the Nobel Prize for his proof of the structure of glucose. Fischer developed this shorthand notation for drawing and comparing sugar structures quickly and easily. We will use Fischer projections extensively in our work with carbohydrates, so you may want to review them (Section 5-10) and make models of the structures in Figure 23-1 to review the stereochemistry implied by these structures. In aldoses, the aldehyde carbon is the most highly oxidized (and numbered 1 in the IUPAC name), so it is always at the top of the Fischer projection. In ketoses, the carbonyl group is usually the second carbon from the top.

PROBLEM-SOLVING HINT

The Fischer projection represents each asymmetric carbon atom by a cross, with the horizontal bonds projecting toward the viewer and the vertical bonds projecting away. The carbon chain is arranged along the vertical bonds, with the most oxidized end (or carbon #1 in the IUPAC name) at the top.

CHO / H—+—OH / CH_2OH = CHO / H—C*—OH / CH_2OH

For more than one asymmetric carbon atom, the Fischer projection represents a totally eclipsed conformation. This is not the most stable conformation, but it's usually the most symmetric conformation, which is most helpful for comparing stereochemistry.

PROBLEM 23-1

Draw the mirror images of glucose and fructose. Are glucose and fructose chiral? Do you expect them to be optically active?

FIGURE 23-1
Fischer projections of sugars. Glucose and fructose are monosaccharides. Glucose is an aldose (a sugar with an aldehyde group), and fructose is a ketose (a sugar with a ketone group). Carbohydrate structures are commonly drawn using Fischer projections.

glucose			fructose		
CHO	or	CHO	CH_2OH	or	CH_2OH
H–C–OH		H–OH	C=O		C=O
HO–C–H		HO–H	HO–C–H		HO–H
H–C–OH		H–OH	H–C–OH		H–OH
H–C–OH		H–OH	H–C–OH		H–OH
CH_2OH		CH_2OH	CH_2OH		CH_2OH

A **disaccharide** is a sugar that can be hydrolyzed to two monosaccharides. For example, sucrose ("table sugar") is a disaccharide that can be hydrolyzed to one molecule of glucose and one molecule of fructose.

$$1 \text{ sucrose} \xrightarrow[\text{heat}]{H_3O^+} 1 \text{ glucose} + 1 \text{ fructose}$$

Both monosaccharides and disaccharides are highly soluble in water, and most have the characteristic sweet taste we associate with sugars.

Polysaccharides are carbohydrates that can be hydrolyzed to many monosaccharide units. Polysaccharides are naturally occurring polymers (*biopolymers*) of carbohydrates. They include starch and cellulose, both biopolymers of glucose. **Starch** is a polysaccharide whose carbohydrate units are easily added to store energy or removed to provide energy to cells. The polysaccharide **cellulose** is a major structural component of plants. Hydrolysis of either starch or cellulose gives many molecules of glucose.

$$\text{starch} \xrightarrow[\text{heat}]{H_3O^+} \text{over 1000 glucose molecules}$$

$$\text{cellulose} \xrightarrow[\text{heat}]{H_3O^+} \text{over 1000 glucose molecules}$$

To understand the chemistry of these more complex carbohydrates, we must first learn the principles of carbohydrate structure and reactions, using the simplest monosaccharides as examples. Then we will apply these principles to more complex disaccharides and polysaccharides. The chemistry of carbohydrates applies the chemistry of alcohols, aldehydes, and ketones to these polyfunctional compounds. In general, the chemistry of biomolecules can be predicted by applying the chemistry of simple organic molecules with similar functional groups.

23-3 Monosaccharides

23-3A Classification of Monosaccharides

Most sugars have their own specific common names, such as glucose, fructose, galactose, and mannose. These names are not systematic, although there are simple ways to remember the common structures. We simplify the study of monosaccharides by grouping similar structures together. Three criteria guide the classification of monosaccharides:

1. The number of carbon atoms in the carbon chain
2. Whether the sugar contains a ketone or an aldehyde group
3. The stereochemical configuration of the asymmetric carbon atom farthest from the carbonyl group

As we have seen, sugars with aldehyde groups are called **aldoses,** and those with ketone groups are called **ketoses.** The number of carbon atoms in the sugar generally ranges from three to seven, designated by the terms *triose* (three carbons), *tetrose* (four carbons), *pentose* (five carbons), *hexose* (six carbons), and *heptose* (seven carbons).

Terms describing monosaccharides often reflect these first two criteria. For example, glucose has an aldehyde and contains six carbon atoms, so it is an aldohexose. Fructose also contains six carbon atoms, but it is a ketone, so it is called a ketohexose. Most ketoses have the ketone on C2, the second carbon atom of the chain. The most common naturally occurring monosaccharides are aldohexoses and aldopentoses.

an aldohexose	a ketohexose	an aldotetrose	a ketotetrose
^{1}CHO	$^{1}CH_2OH$		
$^{2}CHOH$	$^{2}C{=}O$		
$^{3}CHOH$	$^{3}CHOH$	^{1}CHO	$^{1}CH_2OH$
$^{4}CHOH$	$^{4}CHOH$	$^{2}CHOH$	$^{2}C{=}O$
$^{5}CHOH$	$^{5}CHOH$	$^{3}CHOH$	$^{3}CHOH$
$^{6}CH_2OH$	$^{6}CH_2OH$	$^{4}CH_2OH$	$^{4}CH_2OH$

PROBLEM 23-2

(a) How many asymmetric carbon atoms are there in an aldotetrose? Draw all the aldotetrose stereoisomers.
(b) How many asymmetric carbons are there in a ketotetrose? Draw all the ketotetrose stereoisomers.
(c) How many asymmetric carbons and stereoisomers are there for an aldohexose? For a ketohexose?

PROBLEM 23-3

(a) There is only one ketotriose, called *dihydroxyacetone*. Draw its structure.
(b) There is only one aldotriose, called *glyceraldehyde*. Draw the two enantiomers of glyceraldehyde.

23-3B The D and L Configurations of Sugars

Around 1880–1900, carbohydrate chemists made great strides in determining the structures of natural and synthetic sugars. They found ways to build larger sugars out of smaller ones, adding a carbon atom to convert a tetrose to a pentose and a pentose to a hexose. The opposite conversion, removing one carbon atom at a time (called a *degradation*), was also developed. A degradation could convert a hexose to a pentose, a pentose to a tetrose, and a tetrose to a triose. There is only one aldotriose, glyceraldehyde.

These chemists noticed they could start with any of the naturally occurring sugars, and degradation to glyceraldehyde always gave the dextrorotatory (+) enantiomer of glyceraldehyde. Some synthetic sugars, on the other hand, degraded to the levorotatory (−) enantiomer of glyceraldehyde. Carbohydrate chemists started using the *Fischer–Rosanoff convention*, which uses a D to designate the sugars that degrade to (+)-glyceraldehyde and an L for those that degrade to (−)-glyceraldehyde. Although these chemists did not know the absolute configurations of any of these sugars, the D and L relative configurations were useful to distinguish the naturally occurring D sugars from their unnatural L enantiomers.

We now know the absolute configurations of (+)- and (−)-glyceraldehyde. These structures serve as the configurational standards for all monosaccharides.

(+)-glyceraldehyde D series of sugars	(−)-glyceraldehyde L series of sugars
CHO	CHO
H—C—OH	HO—C—H
CH_2OH	CH_2OH

PROBLEM-SOLVING HINT

Most naturally occurring sugars are of the D series, with the OH group of the bottom asymmetric carbon on the right in the Fischer projection.

D-(+)-glucose —degrade→ D-(−)-arabinose + CO_2↑ —degrade→ D-(−)-erythrose + CO_2↑ —degrade→ D-(+)-glyceraldehyde + CO_2↑

FIGURE 23-2
Degradation to glyceraldehyde. Degradation of an aldose removes the aldehyde carbon atom to give a smaller sugar. Sugars of the D series give (+)-glyceraldehyde on degradation to the triose. Therefore, the OH group of the bottom asymmetric carbon atom of the D sugars must be on the right in the Fischer projection.

Figure 23-2 shows that degradation (covered in Section 23-10) removes the aldehyde carbon atom, and it is the *bottom asymmetric carbon* in the Fischer projection (the asymmetric carbon farthest removed from the carbonyl group) that determines which enantiomer of glyceraldehyde is formed by successive degradations.

We now know that the (+) enantiomer of glyceraldehyde has its OH group on the right in the Fischer projection, as shown in Figure 23-2. Therefore, sugars of the **D series** have the OH group of the bottom asymmetric carbon on the right in the Fischer projection. Sugars of the **L series** have the OH group of the bottom asymmetric carbon on the left. In the following examples, notice that the D or L configuration is determined by the bottom asymmetric carbon, and the enantiomer of a D sugar is always an L sugar.

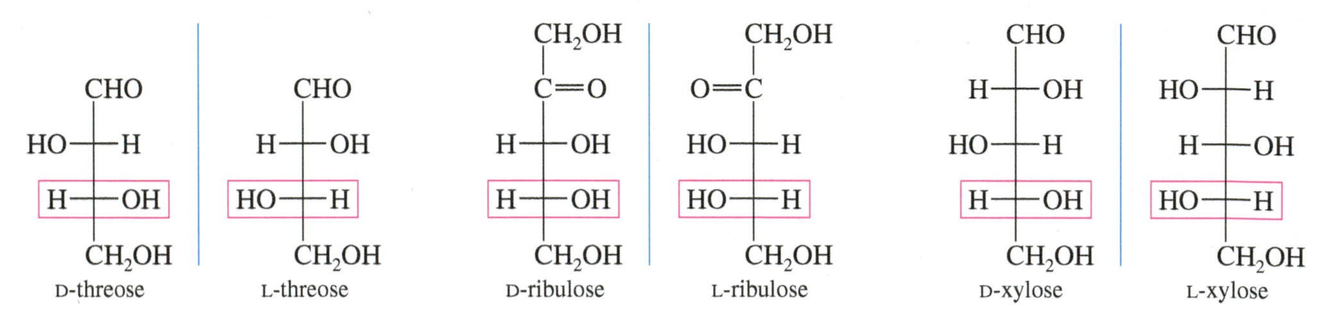

As mentioned earlier, most naturally occurring sugars have the D configuration, and most members of the D family of aldoses (up through six carbon atoms) are found in nature. Figure 23-3 shows the D family of aldoses. Notice that the D or L configuration does not tell us which way a sugar rotates the plane of polarized light. This must be determined by experiment. Some D sugars have (+) rotations, and others have (−) rotations.

On paper, the family tree of D aldoses (Figure 23-3) can be generated by starting with D-(+)-glyceraldehyde and adding another carbon at the top to generate two aldotetroses: *erythrose* with the OH group of the new asymmetric carbon on the right, and *threose* with the new OH group on the left. Adding another carbon to these aldotetroses gives four aldopentoses, and adding a sixth carbon gives eight aldohexoses.* In Section 23-11, we describe the Kiliani–Fischer synthesis, which actually adds a carbon atom and generates the pairs of elongated sugars just as we have drawn them in this family tree.

At the time the D and L system of relative configurations was introduced, chemists could not determine the absolute configurations of chiral compounds. They decided to

*Drawn in this order, the names of the four aldopentoses (ribose, arabinose, xylose, and lyxose) are remembered by the mnemonic "*Ribs are* e*x*tra *lean*." The mnemonic for the eight aldohexoses (allose, altrose, glucose, mannose, gulose, idose, galactose, and talose) is "*All altr*uists *gl*adly *ma*ke *gu*m *in gal*lon *ta*nks."

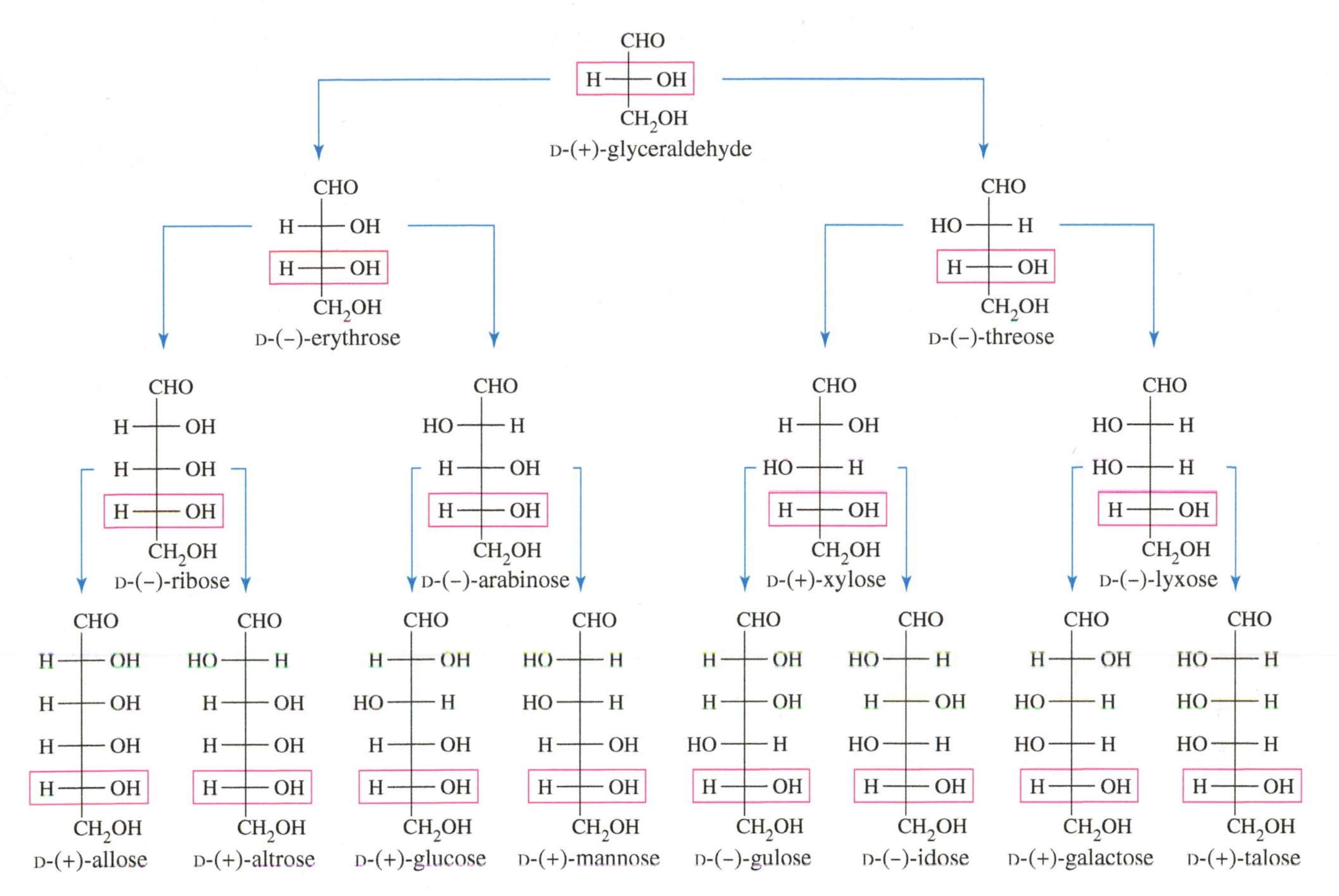

FIGURE 23-3
The D family of aldoses. All of these sugars occur naturally except for threose, lyxose, allose, and gulose.

draw the D series with the glyceraldehyde OH group on the right, and the L series with it on the left. This guess later proved to be correct, so it was not necessary to revise all the old structures.

PROBLEM 23-4

Draw and name the enantiomers of the sugars shown in Figure 23-2. Give the relative configuration (D or L) and the sign of the rotation in each case.

PROBLEM 23-5

Which configuration (R or S) does the bottom asymmetric carbon have for the D series of sugars? Which configuration for the L series?

23-3C Epimers

Many common sugars are closely related, differing only by the stereochemistry at a single carbon atom. For example, glucose and mannose differ only at C2, the first asymmetric carbon atom. Sugars that differ only by the stereochemistry at a single carbon are called **epimers,** and the carbon atom where they differ is generally stated. If the number of a carbon atom is not specified, it is assumed to be C2. Therefore, glucose and mannose are "C2 epimers" or simply "epimers." The C4 epimer of glucose is galactose, and the C2 epimer of erythrose is threose. These relationships are shown in Figure 23-4.

PROBLEM-SOLVING HINT

Use of the terms *erythro* and *threo*

The diastereomers of compounds with two adjacent chiral carbons are often called *erythro* and *threo*, based on the structures of erythrose and threose. A diastereomer is called *erythro* if its Fischer projection shows similar groups on the same side of the molecule. It is called *threo* if similar groups are on opposite sides of the Fischer projection. If the molecule is symmetric, such that one of the diastereomers is meso, then the terms *meso* and (*d,l*) are used in preference to *erythro* and *threo*.

erythro	*threo*
CH_3	CH_3
H—Cl	H—Cl
H—OH	HO—H
CH_3	CH_3

3-chlorobutan-2-ol

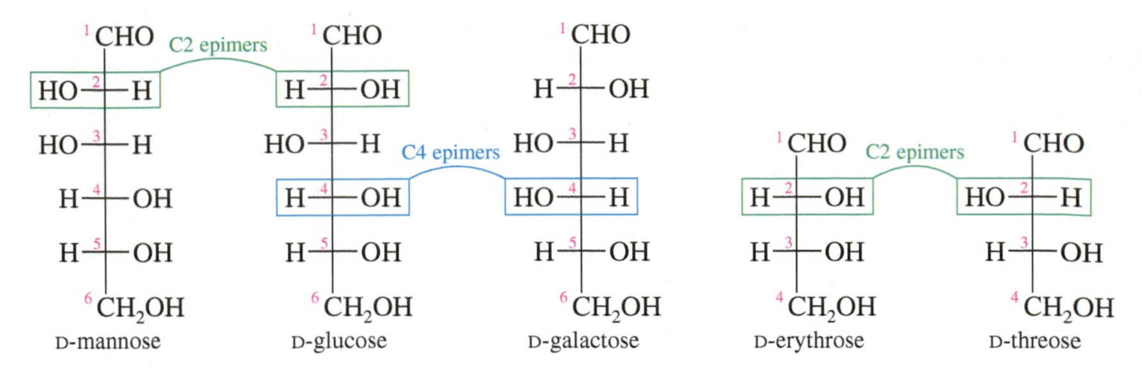

FIGURE 23-4
Epimers are sugars that differ only by the stereochemistry at a single asymmetric carbon atom. If the number of the carbon atom is not specified, it is assumed to be C2.

PROBLEM 23-6

(a) Draw D-allose, the C3 epimer of glucose.
(b) Draw D-talose, the C2 epimer of D-galactose.
(c) Draw D-idose, the C3 epimer of D-talose. Now compare your answers with Figure 23-3.
(d) Draw the C4 "epimer" of D-xylose. Notice that this "epimer" is actually an L-series sugar, and we have seen its enantiomer. Give the correct name for this L-series sugar.

23-4 Cyclic Structures of Monosaccharides

Most carbohydrates exist primarily as cyclic acetals or hemiacetals. These functional groups are made particularly stable because of the sizes and shapes of the cyclic carbohydrate structures.

Cyclic Hemiacetals In Chapter 18, we saw that an aldehyde reacts with one molecule of an alcohol to give a hemiacetal, and with a second molecule of the alcohol to give an acetal. The hemiacetal is not as stable as the acetal, and most hemiacetals decompose spontaneously to the aldehyde and the alcohol. Therefore, hemiacetals are rarely isolated.

MECHANISM 23-1 Formation of a Cyclic Hemiacetal

Step 1: Protonation of the carbonyl. ***Step 2:*** The OH group adds as a nucleophile.

δ-hydroxyaldehyde

Step 3: Deprotonation gives a cyclic hemiacetal.

derived from the OH group

$+ H_3O^+$

derived from the CHO group

cyclic hemiacetal

If the aldehyde group and the hydroxy group are part of the same molecule, a cyclic hemiacetal results. Cyclic hemiacetals are particularly stable if they result in five- or six-membered rings. In fact, five- and six-membered cyclic hemiacetals are often more stable than their open-chain forms.

The Cyclic Hemiacetal Form of Glucose Aldoses contain an aldehyde group and several hydroxy groups. The solid, crystalline form of an aldose is normally a cyclic hemiacetal. In solution, the aldose exists as an equilibrium mixture of the cyclic hemiacetal and the open-chain form. For most sugars, the equilibrium favors the cyclic hemiacetal.

Aldohexoses such as glucose can form cyclic hemiacetals containing either five-membered or six-membered rings. For most common aldohexoses, the equilibrium favors six-membered rings with a hemiacetal linkage between the aldehyde carbon and the hydroxy group on C5. Figure 23-5 shows formation of the cyclic hemiacetal of glucose. Notice that the hemiacetal has a new asymmetric carbon atom at C1. Figure 23-5 shows both possibilities at C1: The hydroxy group can be directed upward in the equatorial position, or it can be directed downward in the axial position. We discuss the stereochemistry at C1 in more detail in Section 23-5.

The cyclic structure is often drawn initially in the **Haworth projection,** which depicts the ring as being flat (of course, it is not). The Haworth projection is widely used in biology texts, but most chemists prefer to use the more realistic chair conformation. Figure 23-5 shows the cyclic form of glucose both as a Haworth projection and as a chair conformation.

Drawing Cyclic Monosaccharides Cyclic hemiacetal structures may seem complicated at first glance, but they can be drawn and recognized by following the process illustrated in Figure 23-5.

1. Mentally lay the Fischer projection on its right side. The groups that were on the right in the Fischer projection are down in the cyclic structure, and the groups that were on the left are up.
2. C5 and C6 curl back away from you. The C4—C5 bond must be rotated so that the C5 hydroxy group can form a part of the ring. For a sugar of the D series, this rotation puts the terminal $-CH_2OH$ (C6 in glucose) upward.

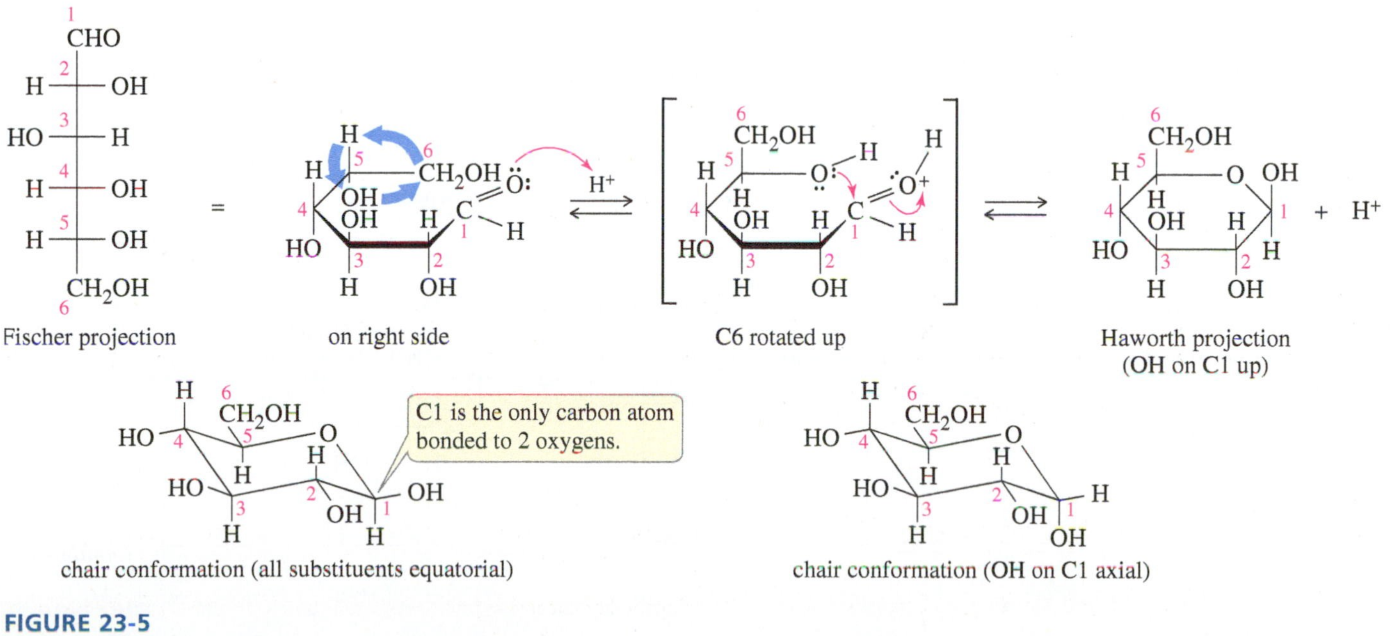

Fischer projection = on right side $\xrightarrow{H^+}$ [C6 rotated up] $\rightleftharpoons$ Haworth projection (OH on C1 up) + H^+

C1 is the only carbon atom bonded to 2 oxygens.

chair conformation (all substituents equatorial)

chair conformation (OH on C1 axial)

FIGURE 23-5
Glucose exists almost entirely as its cyclic hemiacetal form.

3. Close the ring, and draw the result. Always draw the Haworth projection or chair conformation with the oxygen at the back, right-hand corner, with C1 at the far right. C1 is easily identified because it is the hemiacetal carbon—the only carbon bonded to two oxygens. The hydroxy group on C1 can be either up or down, as discussed in Section 23-5.

Chair conformations can be drawn by recognizing the differences between the sugar in question and glucose. The following procedure is useful for drawing D-aldohexoses.

1. Draw the chair conformation puckered, as shown in Figure 23-5. The hemiacetal carbon (C1) is drawn at the far right (as the footrest), and the ring oxygen is at the back right corner.
2. Glucose has its substituents on alternating sides of the ring. In drawing the chair conformation, just put all the ring substituents in equatorial positions.
3. To draw or recognize other common sugars, notice how they differ from glucose, and make the appropriate changes.

PROBLEM-SOLVING HINT

Learn to draw glucose, both in the Fischer projection and in the chair conformation (all substituents equatorial). Draw other pyranoses by noticing the differences from glucose and changing the glucose structure as needed.

Remember the epimers of glucose (C2: mannose; C3: allose; and C4: galactose). To recognize other sugars, look for axial substituents where they differ from glucose.

SOLVED PROBLEM 23-1

Draw the cyclic hemiacetal forms of D-mannose and D-galactose both as chair conformations and as Haworth projections. Mannose is the C2 epimer of glucose, and galactose is the C4 epimer of glucose.

SOLUTION

The chair conformations are easier to draw, so we will do them first. Draw the rings exactly as we did for glucose in Figure 23-5. Number the carbon atoms, starting with the hemiacetal carbon. Mannose is the C2 epimer of glucose, so the substituent on C2 is axial, while all the others are equatorial as in glucose. Galactose is the C4 epimer of glucose, so its substituent on C4 is axial.

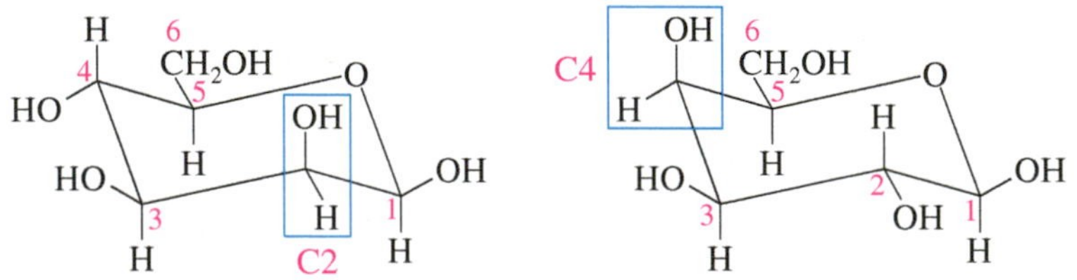

D-mannose

D-galactose

PROBLEM-SOLVING HINT

Groups on the right in the Fischer projection are down in the usual cyclic structure, and groups that were on the left in the Fischer projection are up.

The simplest way to draw Haworth structures for these two sugars is to draw the chair conformations and then draw the flat rings with the same substituents in the up and down positions. For practice, however, let's lay down the Fischer projection for galactose. You should follow along with your molecular models.

1. Lay down the Fischer projection: right ⟶ down and left ⟶ up.

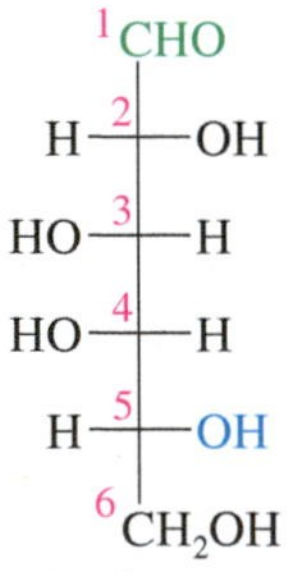

D-galactose

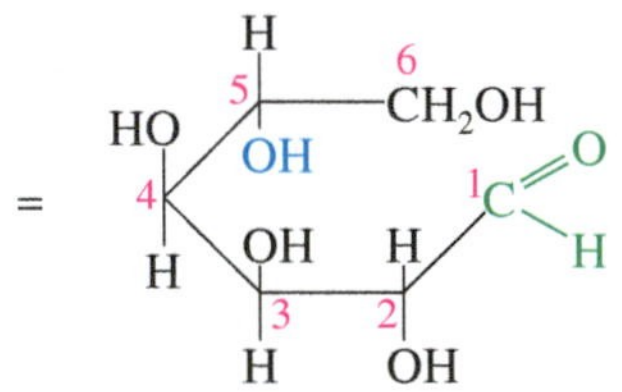

(continued)

2. Rotate the C4—C5 bond to put the C5 —OH in place. (For a D sugar, the $—CH_2OH$ goes up.)

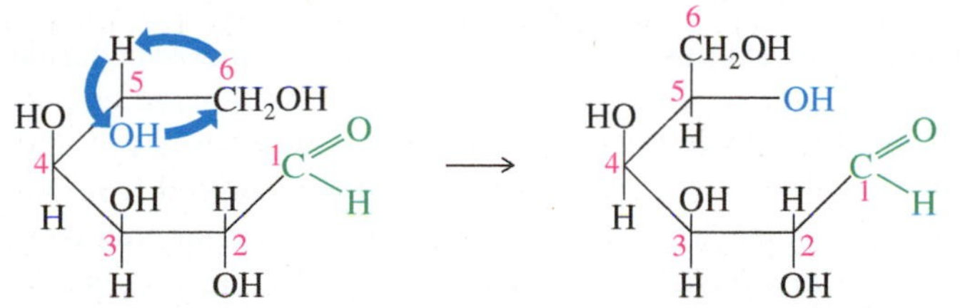

3. Close the ring, and draw the final hemiacetal. The hydroxy group on C1 can be either up or down, as discussed in Section 23-5. Sometimes this ambiguous stereochemistry is symbolized by a wavy line.

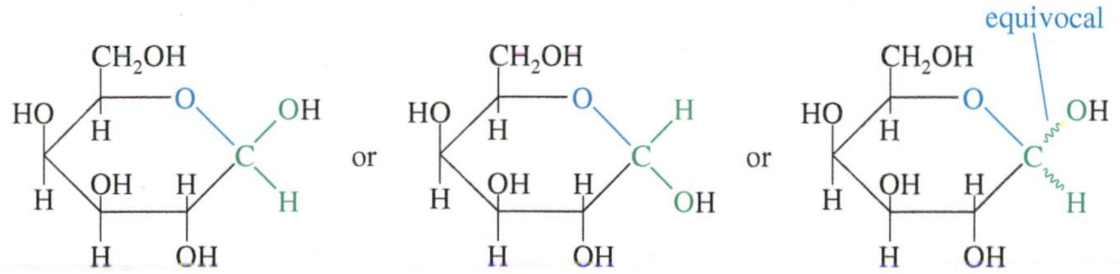

equivocal

PROBLEM 23-7

Draw the Haworth projection for the cyclic structure of D-mannose by laying down the Fischer projection.

PROBLEM 23-8

Allose is the C3 epimer of glucose. Draw the cyclic hemiacetal form of D-allose, first in the chair conformation and then in the Haworth projection.

The Five-Membered Cyclic Hemiacetal Form of Fructose Not all sugars exist as six-membered rings in their hemiacetal forms. Many aldopentoses and ketohexoses form five-membered rings. The five-membered hemiacetal ring of fructose is shown in Figure 23-6. This cyclic hemiacetal of a ketone is more specifically called a *hemiketal*.* Five-membered rings are not puckered as much as six-membered rings, so they are usually depicted as flat Haworth projections. The five-membered ring is

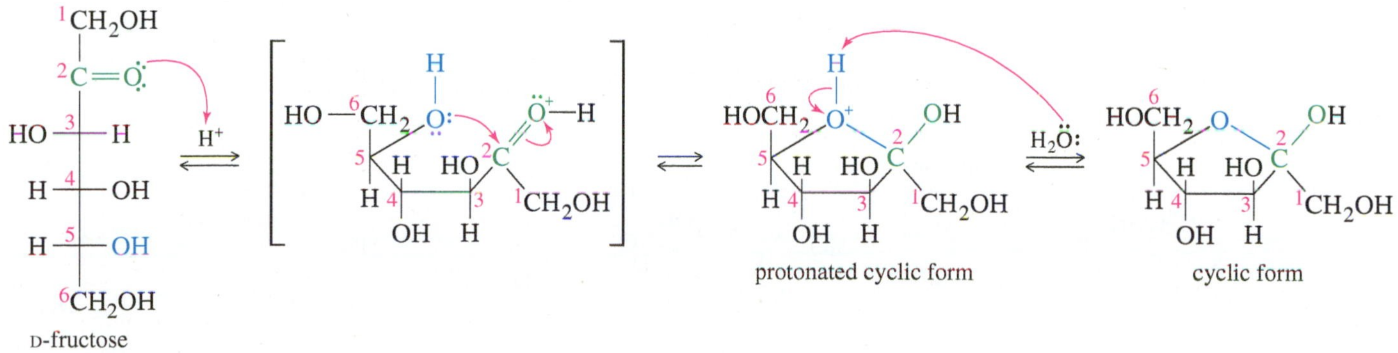

D-fructose

protonated cyclic form

cyclic form

FIGURE 23-6
Fructose forms a five-membered cyclic hemiketal.* Five-membered rings are usually represented as flat Haworth structures.

*The IUPAC has reinstated the term *ketal* as a subclass of *acetals*. Therefore, the cyclic form of fructose can be called either a *hemiketal* (specific) or a *hemiacetal* (more general).

customarily drawn with the ring oxygen in back and the hemiacetal carbon (the one bonded to two oxygens) on the right. The $-CH_2OH$ at the back left (C6) is in the up position for D-series ketohexoses.

Pyranose and Furanose Names Cyclic structures of monosaccharides are named according to their five- or six-membered rings. A six-membered cyclic hemiacetal is called a **pyranose,** derived from the name of the six-membered cyclic ether *pyran*. A five-membered cyclic hemiacetal is called a **furanose,** derived from the name of the five-membered cyclic ether *furan*. For example, the six-membered ring of glucose is called *glucopyranose*, and the five-membered ring of fructose is called *fructofuranose*. The ring is still numbered as it is in the sugar.

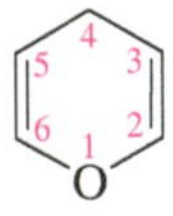

pyran

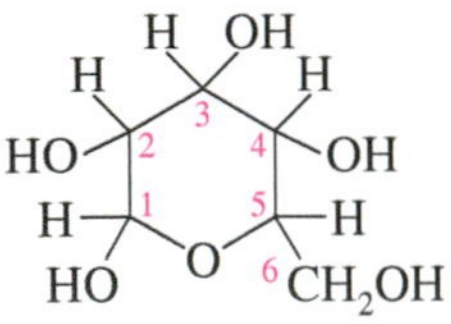

a pyranose

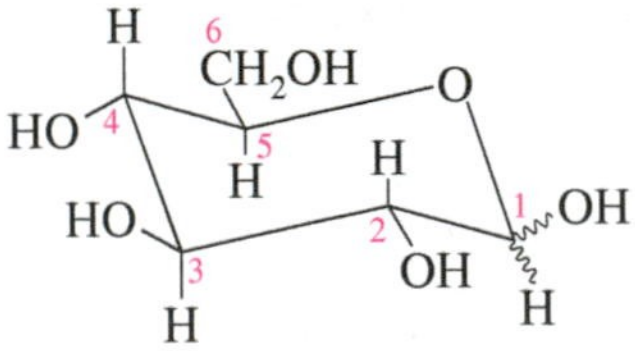

D-glucopyranose

furan

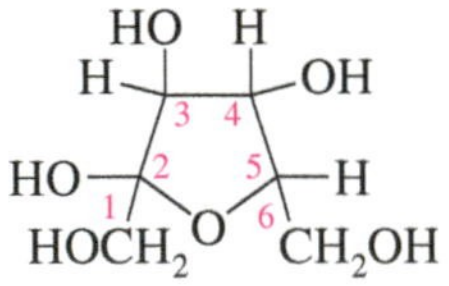

a furanose

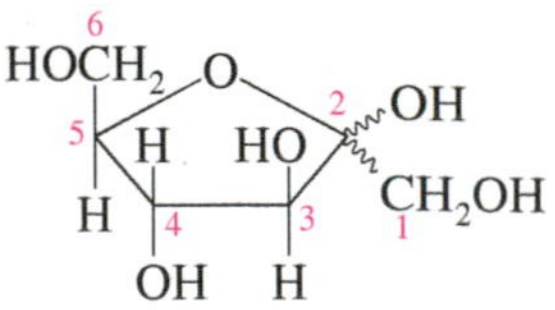

D-fructofuranose

PROBLEM 23-9

Talose is the C4 epimer of mannose. Draw the chair conformation of D-talopyranose.

PROBLEM 23-10

(a) Figure 23-2 shows that the degradation of D-glucose gives D-arabinose, an aldopentose. Arabinose is most stable in its furanose form. Draw D-arabinofuranose.
(b) Ribose, the C2 epimer of arabinose, is most stable in its furanose form. Draw D-ribofuranose.

PROBLEM 23-11

The carbonyl group in D-galactose may be isomerized from C1 to C2 by brief treatment with dilute base (by the enediol rearrangement, Section 23-7). The product is the C4 epimer of fructose. Draw the furanose structure of the product.

23-5 Anomers of Monosaccharides; Mutarotation

When a pyranose or furanose ring closes, the flat carbonyl group is converted to an asymmetric carbon in the hemiacetal. Depending on which face of the (protonated) carbonyl group is attacked, the hemiacetal $-OH$ group can be directed either up or down. These two orientations of the hemiacetal $-OH$ group give diastereomeric products called **anomers.** Figure 23-7 shows the anomers of glucose.

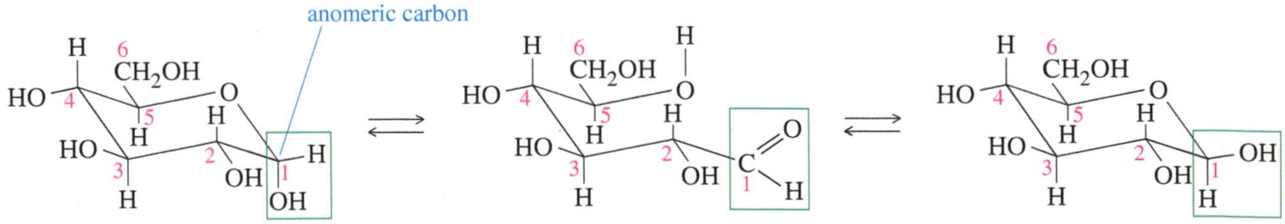

anomeric carbon

α-D-glucopyranose

open-chain form

β-D-glucopyranose

FIGURE 23-7
The anomers of glucose. The hydroxy group on the anomeric (hemiacetal) carbon is down (axial) in the α anomer and up (equatorial) in the β anomer. The β anomer of glucose has all its substituents in equatorial positions.

The hemiacetal carbon atom is called the **anomeric carbon,** easily identified as the only carbon atom bonded to two oxygens. Its —OH group is called the anomeric hydroxy group. The structure with the anomeric —OH group down (axial) is called the α (alpha) anomer, and the one with the anomeric —OH group up (equatorial) is called the β (beta) anomer. We can draw the α and β anomers of most aldohexoses by remembering that the β form of glucose (β-D-glucopyranose) has all its substituents in equatorial positions. To draw an α anomer, simply move the anomeric —OH group to the axial position.

Another way to remember the anomers is to notice that the anomeric hydroxy group is trans to the terminal $—CH_2OH$ group in the α anomer, but it is cis in the β anomer. This rule works for all sugars, from both the D and L series, as well as for furanoses. Figure 23-8 shows the two anomers of fructose, whose anomeric carbon is C2. The α anomer has the anomeric —OH group down, trans to the terminal $—CH_2OH$ group, while the β anomer has it up, cis to the terminal $—CH_2OH$.

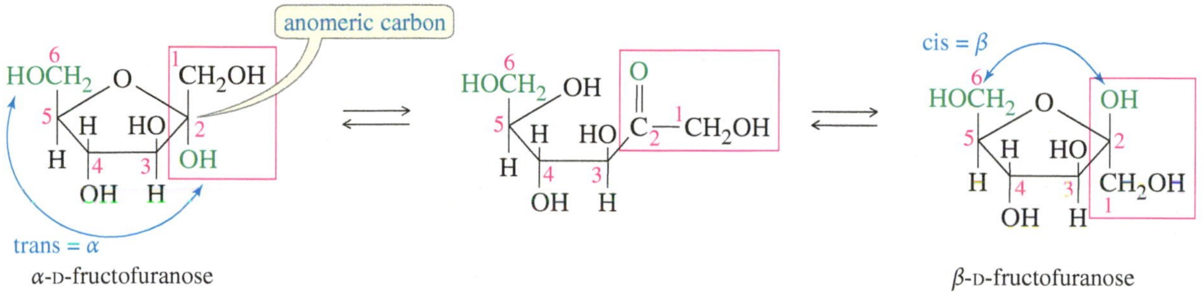

FIGURE 23-8
The α anomer of fructose has the anomeric —OH group down, trans to the terminal $—CH_2OH$ group. The β anomer has the anomeric hydroxy group up, cis to the terminal $—CH_2OH$.

PROBLEM 23-12

Draw the following monosaccharides, using chair conformations for the pyranoses and Haworth projections for the furanoses.

(a) α-D-mannopyranose (C2 epimer of glucose)
(b) β-D-galactopyranose (C4 epimer of glucose)
(c) β-D-allopyranose (C3 epimer of glucose)
(d) α-D-arabinofuranose
(e) β-D-ribofuranose (C2 epimer of arabinose)

Properties of Anomers: Mutarotation Because anomers are diastereomers, they generally have different properties. For example, α-D-glucopyranose has a melting point of 146 °C and a specific rotation of +112.2°, while β-D-glucopyranose has a melting point of 150 °C and a specific rotation of +18.7°. When glucose is crystallized from water at room temperature, pure crystalline α-D-glucopyranose results. If glucose is crystallized from water by letting the water evaporate at a temperature above 98 °C, crystals of pure β-D-glucopyranose are formed (Figure 23-9).

In each of these cases, *all* the glucose in the solution crystallizes as the favored anomer. In the solution, the two anomers are in equilibrium through a small amount of the open-chain form, and this equilibrium continues to supply more of the anomer that is crystallizing out of solution.

When one of the pure glucose anomers dissolves in water, an interesting change in the specific rotation is observed. When the α anomer dissolves, its specific rotation gradually decreases from an initial value of +112.2° to +52.6°. When the pure β anomer dissolves, its specific rotation gradually increases from +18.7° to the same value of +52.6°. This change ("mutation") in the specific rotation is called **mutarotation.** Mutarotation occurs because the two anomers interconvert in solution. When either of the pure anomers dissolves in water, its rotation gradually changes to an intermediate

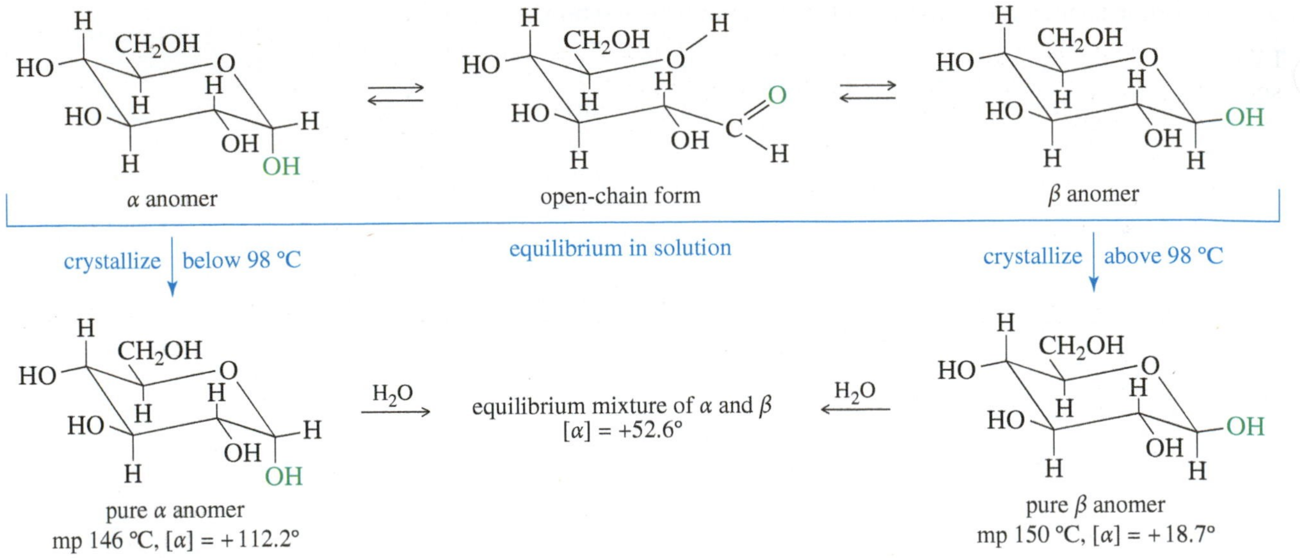

α anomer ⇌ open-chain form ⇌ β anomer

equilibrium in solution

crystallize below 98 °C → pure α anomer, mp 146 °C, $[\alpha] = +112.2°$

crystallize above 98 °C → pure β anomer, mp 150 °C, $[\alpha] = +18.7°$

pure α anomer $\xrightarrow{H_2O}$ equilibrium mixture of α and β, $[\alpha] = +52.6°$ $\xleftarrow{H_2O}$ pure β anomer

FIGURE 23-9
An aqueous solution of D-glucose contains an equilibrium mixture of α-D-glucopyranose, β-D-glucopyranose, and the intermediate open-chain form. Crystallization below 98 °C gives the α anomer, and crystallization above 98 °C gives the β anomer.

rotation that results from equilibrium concentrations of the anomers. The specific rotation of glucose is usually listed as +52.6°, the value for the equilibrium mixture of anomers. The positive sign of this rotation is the source of the name **dextrose,** an old common name for glucose.

SOLVED PROBLEM 23-2

Calculate how much of the α anomer and how much of the β anomer are present in an equilibrium mixture with a specific rotation of +52.6°.

SOLUTION

If the fraction of glucose present as the α anomer ($[\alpha] = +112.2°$) is a, the fraction present as the β anomer ($[\alpha] = +18.7°$) is b, and the rotation of the mixture is +52.6°, we have

$$a(+112.2°) + b(+18.7°) = +52.6°$$

There is very little of the open-chain form present, so the fraction present as the α anomer (a) plus the fraction present as the β anomer (b) should account for all the glucose:

$$a + b = 1 \quad \text{or} \quad b = 1 - a$$

Substituting $(1 - a)$ for b in the first equation, we have

$$a(112.2°) + (1 - a)(18.7°) = 52.6°$$

Solving this equation for a, we have $a = 0.36$, or 36%. Thus, b must be $(1 - 0.36) = 0.64$, or 64%. The amounts of the two anomers present at equilibrium are

α anomer, 36% β anomer, 64%

When we remember that the anomeric hydroxy group is axial in the α anomer and equatorial in the β anomer, it is reasonable that the more stable β anomer should predominate.

PROBLEM 23-13

Like glucose, galactose mutarotates when it dissolves in water. The specific rotation of α-D-galactopyranose is +150.7°, and that of the β anomer is +52.8°. When either of the pure anomers dissolves in water, the specific rotation gradually changes to +80.2°. Determine the percentages of the two anomers present at equilibrium.

23-6 Reactions of Monosaccharides: Reduction

Like other aldehydes and ketones, aldoses and ketoses can be reduced to the corresponding polyalcohols, called **sugar alcohols** or **alditols.** The most common reagents are sodium borohydride or catalytic hydrogenation using a nickel catalyst. Alditols are named by adding the suffix *-itol* to the root name of the sugar. The following equation shows the reduction of glucose to glucitol, sometimes called *sorbitol.*

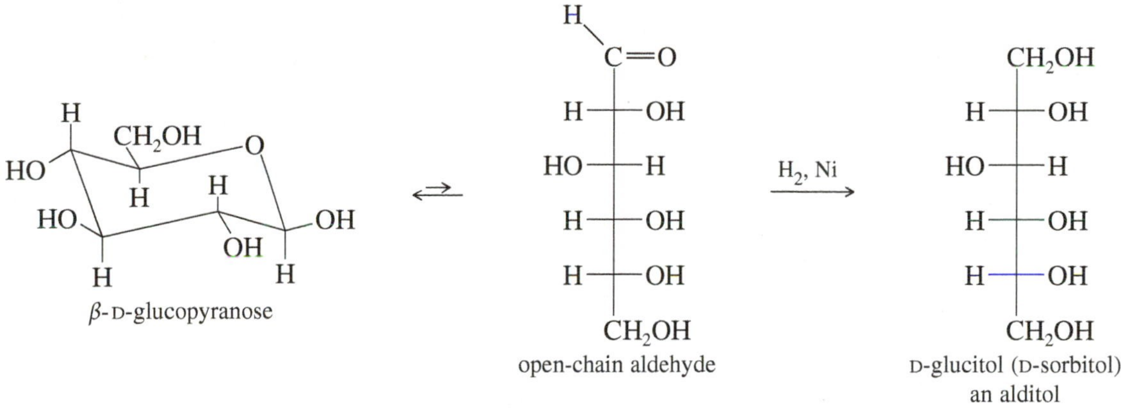

β-D-glucopyranose ⇌ open-chain aldehyde $\xrightarrow{H_2, Ni}$ D-glucitol (D-sorbitol) an alditol

Reduction of a ketose creates a new asymmetric carbon atom, formed in either of two configurations, resulting in two epimers. Figure 23-10 shows how the reduction of fructose gives a mixture of glucitol and mannitol.

Sugar alcohols are widely used in industry, primarily as food additives and sugar substitutes. Glucitol has the common name *sorbitol* because it was first isolated from the berries of the mountain ash, *Sorbus aucuparia.* Industrially, sorbitol is made by catalytic hydrogenation of glucose. Sorbitol is used as a sugar substitute, a moistening agent, and a starting material for making vitamin C. Mannitol was first isolated from plant exudates known as *mannas* (of biblical fame), the origin of the names *mannose* and *mannitol.* Mannitol is derived commercially from seaweed, or it can be made by catalytic hydrogenation of mannose. Galactitol (*dulcitol*) also can be obtained from plants, or it can be made by catalytic hydrogenation of galactose.

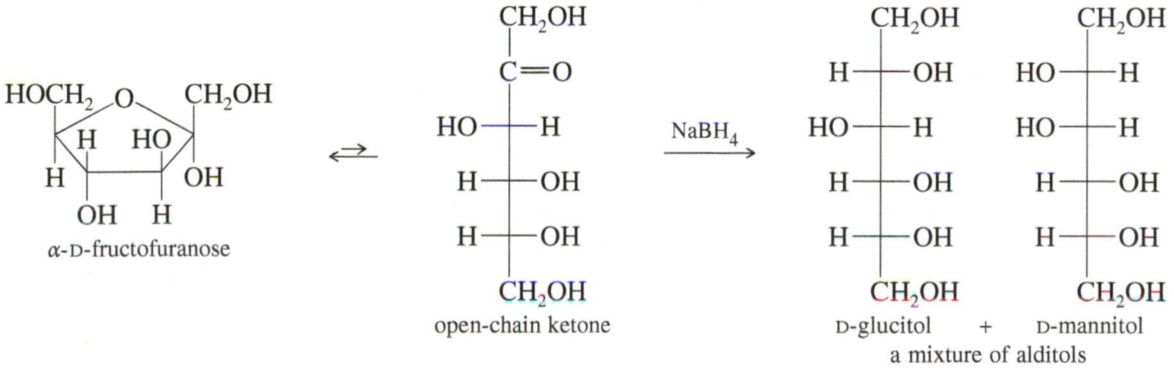

α-D-fructofuranose ⇌ open-chain ketone $\xrightarrow{NaBH_4}$ D-glucitol + D-mannitol a mixture of alditols

FIGURE 23-10
Reduction of fructose creates a new asymmetric carbon atom, which can have either configuration. The products are a mixture of glucitol and mannitol.

PROBLEM 23-14

When D-glucose is reduced with sodium borohydride, optically active glucitol results. When optically active D-galactose is reduced, however, the product is optically inactive. Explain this loss of optical activity.

PROBLEM 23-15

Emil Fischer synthesized L-gulose, an unusual aldohexose that reduces to give D-glucitol. Suggest a structure for this L sugar, and show how L-gulose gives the same alditol as D-glucose. (*Hint:* D-Glucitol has $-CH_2OH$ groups at both ends. *Either* of these primary alcohol groups might have come from reduction of an aldehyde.)

23-7 Oxidation of Monosaccharides; Reducing Sugars

Monosaccharides are oxidized by a variety of reagents. The aldehyde group of an aldose oxidizes easily. Some reagents also selectively oxidize the terminal $-CH_2OH$ group at the far end of the molecule. Oxidation is used to identify the functional groups of a sugar, to help to determine its stereochemistry, and as part of a synthesis to convert one sugar into another.

Bromine Water Bromine water oxidizes the aldehyde group of an aldose to a carboxylic acid. Bromine water is used for this oxidation because it does not oxidize the alcohol groups and it does not oxidize ketoses. Also, bromine water is acidic and does not cause epimerization or rearrangement of the carbonyl group. Because bromine water oxidizes aldoses but not ketoses, it serves as a useful test to distinguish aldoses from ketoses. The product of bromine water oxidation is an **aldonic acid** (older term: **glyconic acid**). For example, bromine water oxidizes glucose to gluconic acid.

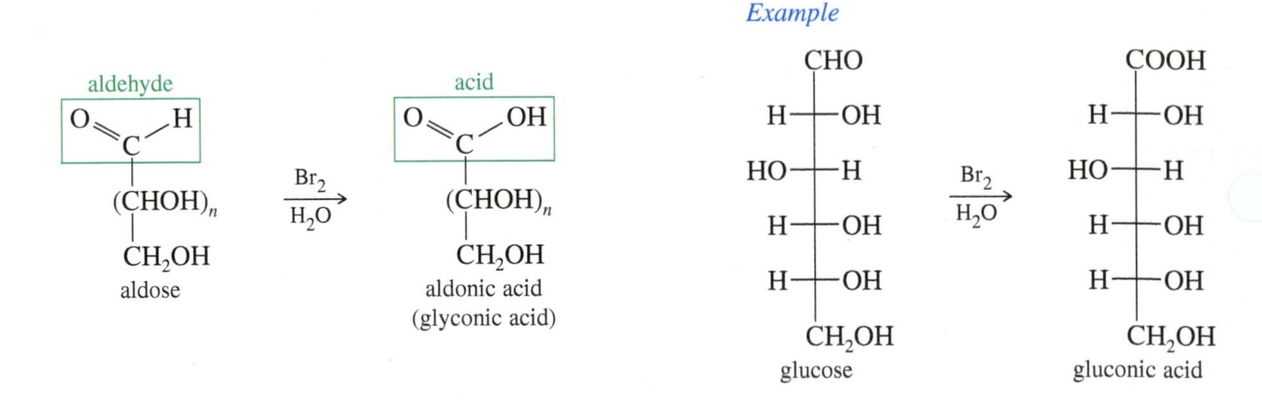

PROBLEM 23-16

Draw and name the products of bromine water oxidation of
(a) D-mannose (b) D-galactose (c) D-fructose

Nitric Acid Nitric acid is a stronger oxidizing agent than bromine water, oxidizing both the aldehyde group and the terminal $-CH_2OH$ group of an aldose to carboxylic acid groups. The resulting dicarboxylic acid is called an **aldaric acid** (older terms: **glycaric acid** or **saccharic acid**). For example, nitric acid oxidizes glucose to glucaric acid.

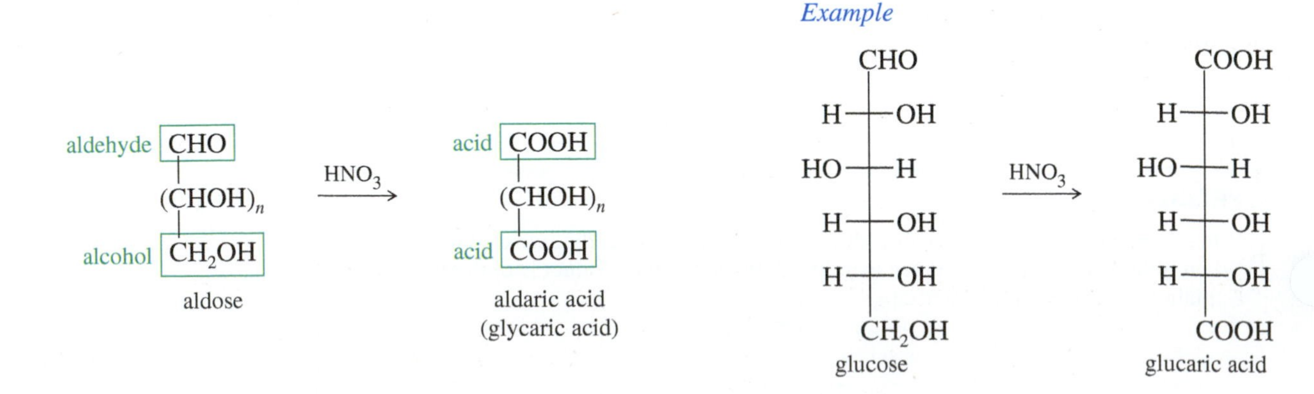

PROBLEM 23-17

Draw and name the products of nitric acid oxidation of
(a) D-mannose **(b)** D-galactose

PROBLEM 23-18

Two sugars, **A** and **B,** are known to be glucose and galactose, but it is not certain which one is which. On treatment with nitric acid, **A** gives an optically inactive aldaric acid, while **B** gives an optically active aldaric acid. Which sugar is glucose, and which is galactose?

Tollens Test **Tollens test** detects aldehydes, which react with Tollens reagent to give carboxylate ions and metallic silver, often in the form of a silver mirror on the inside of the container.

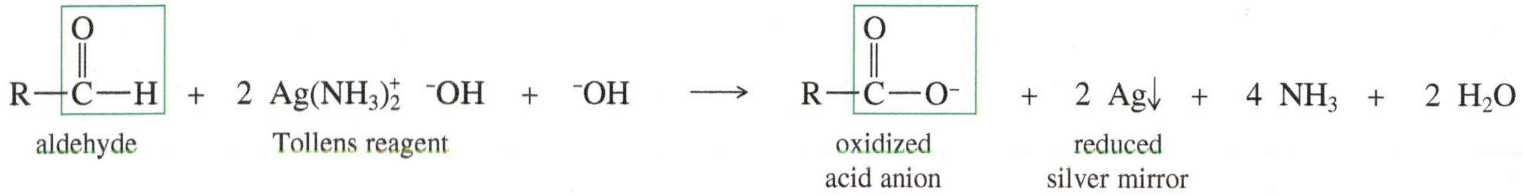

$$\underset{\text{aldehyde}}{R{-}\overset{O}{\overset{\|}{C}}{-}H} + \underset{\text{Tollens reagent}}{2\ Ag(NH_3)_2^+\ {}^-OH} + {}^-OH \longrightarrow \underset{\text{oxidized acid anion}}{R{-}\overset{O}{\overset{\|}{C}}{-}O^-} + \underset{\text{reduced silver mirror}}{2\ Ag\downarrow} + 4\ NH_3 + 2\ H_2O$$

In its open-chain form, an aldose has an aldehyde group, which reacts with Tollens reagent to give an aldonic acid and a silver mirror. This oxidation is not a good synthesis of the aldonic acid, however, because Tollens reagent is strongly basic and promotes rearrangements. Sugars that reduce Tollens reagent to give a silver mirror are called **reducing sugars.**

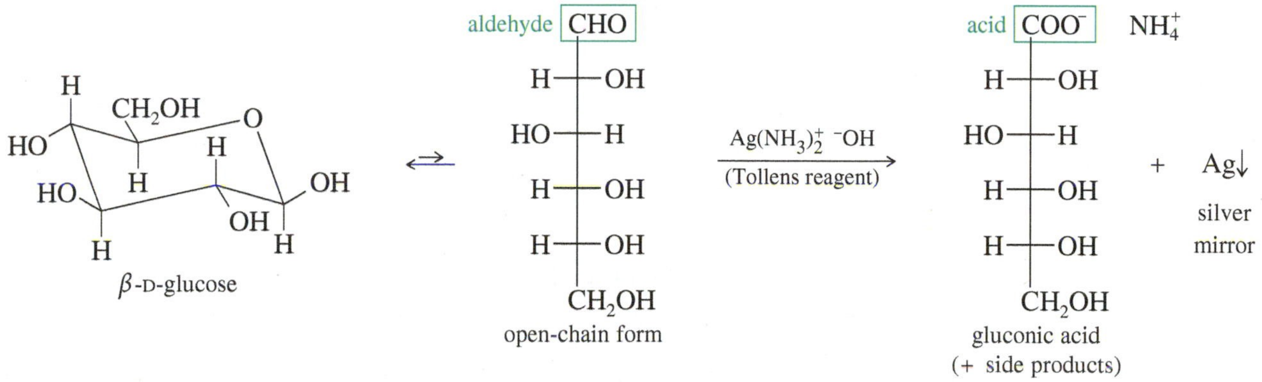

β-D-glucose ⇌ open-chain form (aldehyde CHO; H–OH; HO–H; H–OH; H–OH; CH_2OH) —$Ag(NH_3)_2^+\ {}^-OH$ (Tollens reagent)→ gluconic acid (+ side products) (acid COO^- NH_4^+; H–OH; HO–H; H–OH; H–OH; CH_2OH) + Ag↓ silver mirror

Tollens test cannot distinguish between aldoses and ketoses because the basic Tollens reagent promotes rearrangements via an enol form that is called an enediol (shown below). Under basic conditions, the open-chain form of a ketose can undergo this **enediol rearrangement** to give an aldose, which reacts to give a positive Tollens test.

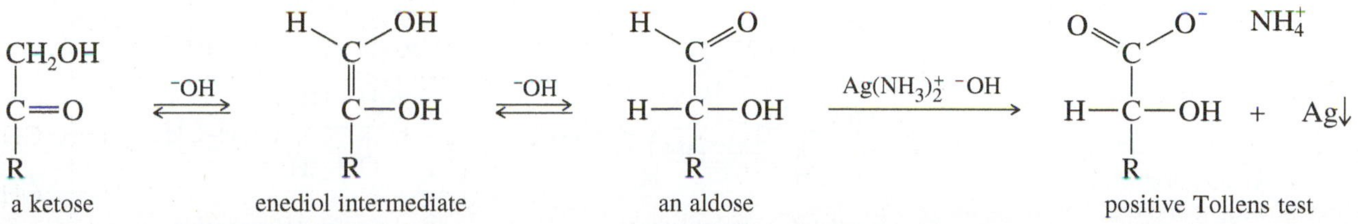

a ketose (CH_2OH–C=O–R) $\overset{^-OH}{\rightleftharpoons}$ enediol intermediate (H–C–OH = C–OH, R) $\overset{^-OH}{\rightleftharpoons}$ an aldose (H–C=O; H–C–OH; R) $\xrightarrow{Ag(NH_3)_2^+\ {}^-OH}$ positive Tollens test ($O{=}C{-}O^-$ NH_4^+; H–C–OH; R) + Ag↓

Application: Blood Glucose Meters

Diabetics must monitor their blood glucose levels several times a day. An electronic glucose meter uses test strips that contain electrodes bordering a reaction chamber. A drop of blood wicks up into the reaction chamber, which is impregnated with glucose oxidase, an enzyme that specifically catalyzes oxidation of the aldehyde group by oxygen in the air. The electrons liberated in this oxidation travel through the electrodes and produce a current that is directly proportional to the concentration of glucose in the blood sample.

PROBLEM 23-19

Except for the Tollens test, basic aqueous conditions are generally avoided with sugars because they lead to fast isomerizations.

(a) Under basic conditions, the proton alpha to the aldehyde (or ketone) carbonyl group is reversibly removed, and the resulting enolate ion is no longer asymmetric. Reprotonation can occur on either face of the enolate, giving either the original structure or its epimer. Because a mixture of epimers results, this process is called **epimerization.** Propose a mechanism for the base-catalyzed equilibration of glucose to a mixture of glucose and its C2 epimer, mannose.

*__(b)__ Propose a mechanism for the isomerization of a ketose to an aldose, via the enediol intermediate, shown immediately above. Note that the enediol has two —OH protons, and removing one or the other gives two different enolate ions.

23-8 Nonreducing Sugars: Formation of Glycosides

What good is the Tollens test if it doesn't distinguish between aldoses and ketoses? The answer lies in the fact that Tollens reagent must react with the open-chain form of the sugar, which has a free aldehyde or ketone. If the cyclic form cannot open to the free carbonyl compound, the sugar does not react with Tollens reagent. Hemiacetals are easily opened, but an acetal is stable under neutral or basic conditions (Section 18-16). If the carbonyl group is in the form of a cyclic acetal, the cyclic form cannot open to the free carbonyl compound, and the sugar gives a negative Tollens test (Figure 23-11).

Sugars in the form of acetals are called **glycosides,** and their names end in the *-oside* suffix. For example, a glycoside of glucose would be a **glucoside,** and if it were a six-membered ring, it would be a *glucopyranoside*. Similarly, a glycoside of ribose would be a *riboside*, and if it were a five-membered ring, it would be a *ribofuranoside*. In general, a sugar whose name ends with the suffix *-ose* is a reducing sugar, and one whose name ends with *-oside* is nonreducing. Because they exist as stable acetals rather than hemiacetals, glycosides cannot spontaneously open to their open-chain forms, and they do not mutarotate. They are locked in a particular anomeric form.

We can summarize by saying that Tollens test distinguishes between reducing sugars and nonreducing sugars: Reducing sugars (aldoses and ketoses) are hemiacetals, and they mutarotate. Nonreducing sugars (glycosides) are acetals, and they do not mutarotate.

CH_2OH acetal; H, O, HO, H, OR, H, H, OH, OH H

$\xrightarrow{Ag(NH_3)_2^+ \ ^-OH}$ no reaction

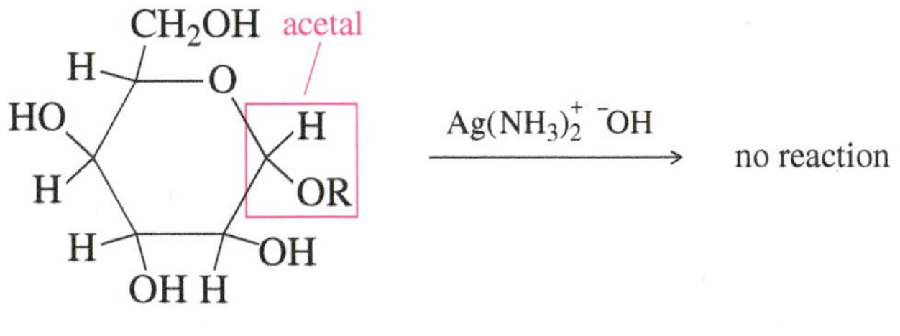

a glycoside

Examples of nonreducing sugars

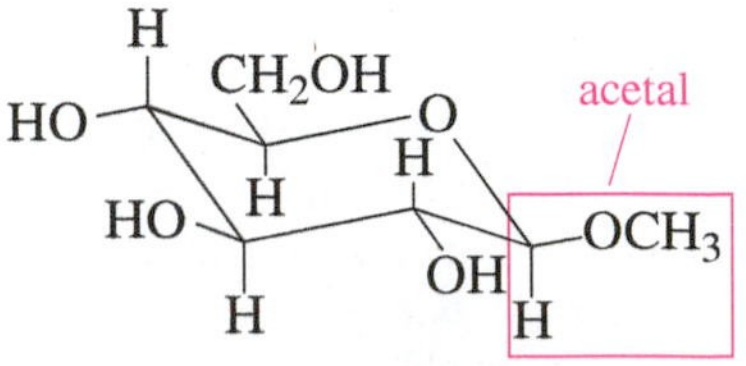

H, CH_2OH, HO, O, H, HO, H, acetal, OCH_3, OH, H, H

methyl β-D-glucopyranoside
(or methyl β-D-glucoside)

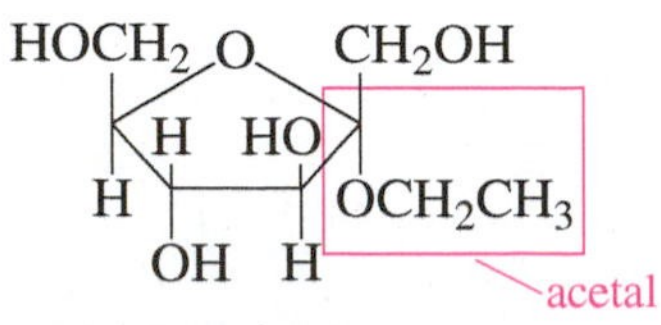

$HOCH_2$, O, CH_2OH, H, HO, H, OCH_2CH_3, OH, H, acetal

ethyl α-D-fructofuranoside
(or ethyl α-D-fructoside)

FIGURE 23-11
Glycosides. Sugars that are full acetals are stable to Tollens reagent and are nonreducing sugars. Such sugars are called glycosides.

PROBLEM 23-20

Which of the following are reducing sugars? Comment on the common name *sucrose* for table sugar.

(a) methyl α-D-galactopyranoside
(b) β-L-idopyranose (an aldohexose)
(c) α-D-allopyranose
(d) ethyl β-D-ribofuranoside

(e)

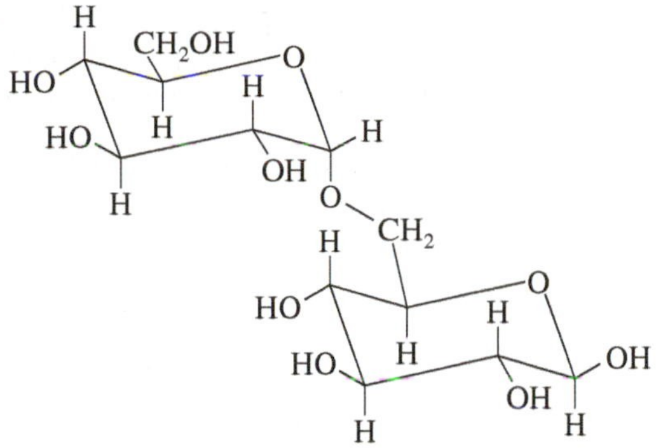

(f)

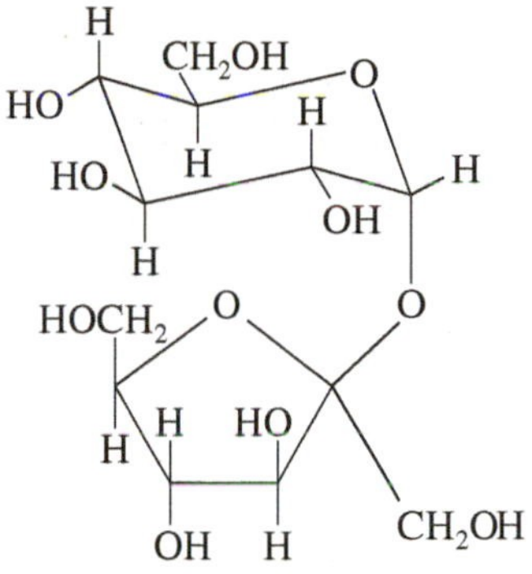

sucrose

PROBLEM 23-21

Draw the structures of the compounds named in Problem 23-20 parts (a), (c), and (d). Allose is the C3 epimer of glucose, and ribose is the C2 epimer of arabinose.

Formation of Glycosides Recall that aldehydes and ketones are converted to acetals by treatment with an alcohol and a trace of acid catalyst (Section 18-16). These conditions also convert aldoses and ketoses to the acetals we call glycosides. Regardless of the anomer used as the starting material, both anomers of the glycoside are formed (as an equilibrium mixture) under these acidic conditions. The more stable anomer predominates. For example, the acid-catalyzed reaction of glucose with methanol gives a mixture of methyl glucosides.

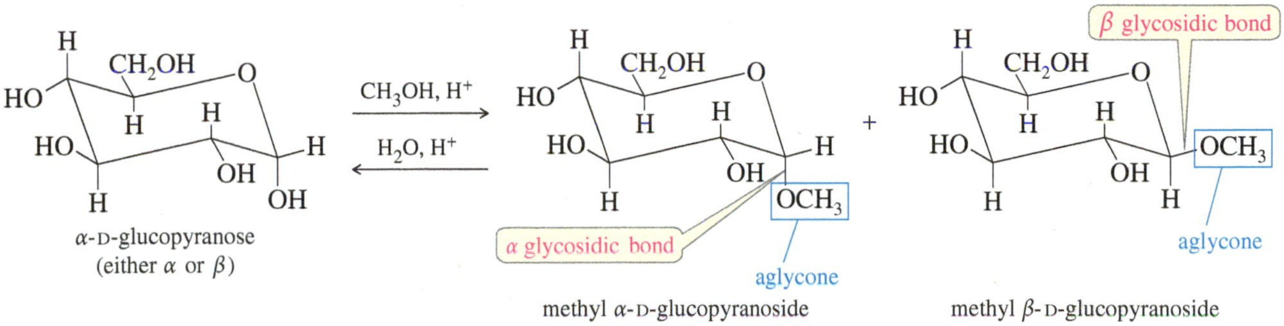

Like other acetals, glycosides are stable to basic conditions, but they hydrolyze in aqueous acid to a free sugar and an alcohol. Glycosides are stable with basic reagents and in basic solutions.

An **aglycone** is the group bonded to the anomeric carbon atom of a glycoside. For example, methanol is the aglycone in a methyl glycoside. Many aglycones are bonded through an oxygen atom, but others are bonded through a nitrogen atom or some other heteroatom. Figure 23-12 shows the structures of some glycosides with interesting aglycones.

Disaccharides and polysaccharides are glycosides in which the alcohol forming the glycoside bond is an —OH group of another monosaccharide. We will consider disaccharides and polysaccharides in Sections 23-12 and 23-13.

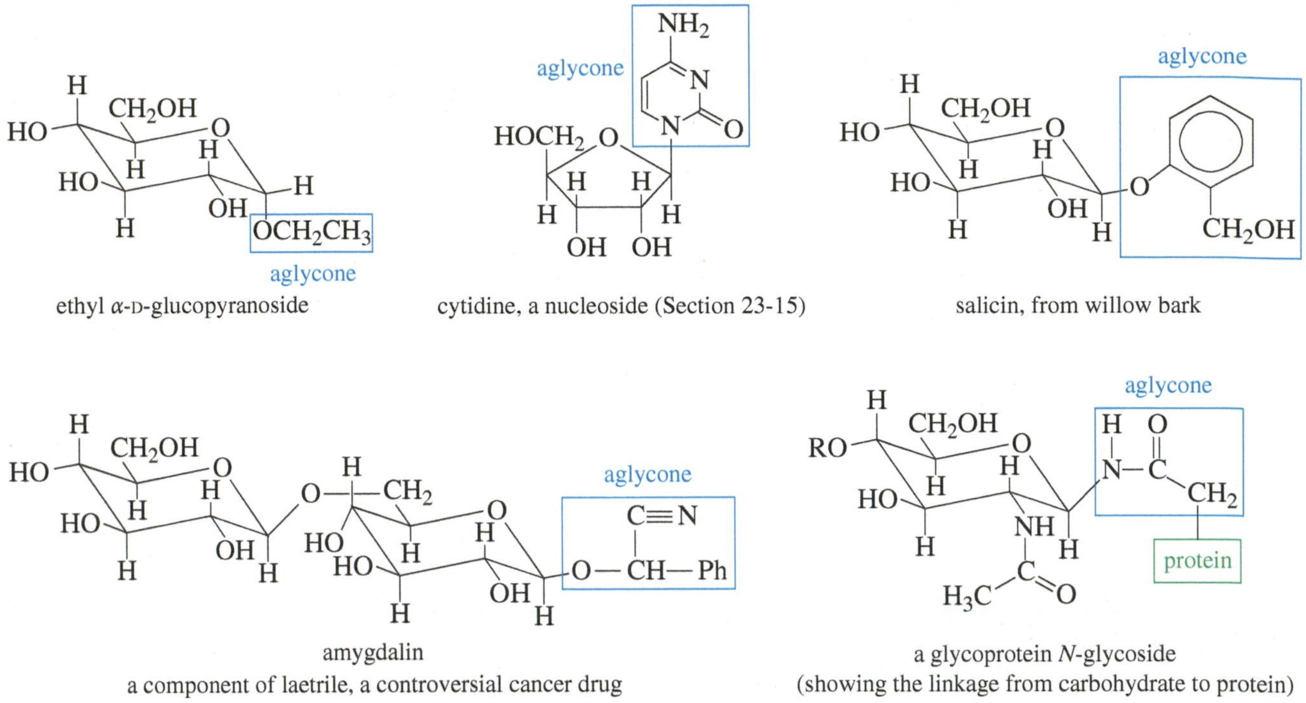

ethyl α-D-glucopyranoside

cytidine, a nucleoside (Section 23-15)

salicin, from willow bark

amygdalin
a component of laetrile, a controversial cancer drug

a glycoprotein *N*-glycoside
(showing the linkage from carbohydrate to protein)

FIGURE 23-12
Aglycones. The group bonded to the anomeric carbon of a glycoside is called an aglycone. Some aglycones are bonded through an oxygen atom (a true acetal), and others are bonded through other atoms such as nitrogen (an aminoglycoside).

PROBLEM 23-22

The mechanism of glycoside formation is the same as the second part of the mechanism for acetal formation. Propose a mechanism for the formation of methyl β-D-glucopyranoside.

Application: Diabetes

Many diabetics have long-standing elevated blood glucose levels. In the open-chain form, glucose condenses with the amino groups of proteins. This glycosylation of proteins may cause some of the chronic effects of diabetes.

PROBLEM 23-23

Show the products that result from hydrolysis of amygdalin in dilute acid. Can you suggest why amygdalin might be toxic to tumor (and possibly other) cells?

PROBLEM 23-24

Treatment of either anomer of fructose with excess ethanol in the presence of a trace of HCl gives a mixture of the α and β anomers of ethyl-D-fructofuranoside. Draw the starting materials, reagents, and products for this reaction. Circle the aglycone in each product.

23-9 Ether and Ester Formation

Because they contain several hydroxy groups, sugars are very soluble in water and rather insoluble in organic solvents. Sugars are difficult to recrystallize from water because they often form supersaturated syrups like honey and molasses. If the hydroxy groups are alkylated to form ethers, sugars behave like simpler organic compounds. The ethers are soluble in organic solvents, and they are more easily purified by recrystallization and simple chromatographic methods.

Treating a sugar with methyl iodide and silver oxide converts the hydroxy groups to methyl ethers. Silver oxide polarizes the $H_3C—I$ bond, making the methyl carbon sugar hydroxy group — polarized CH_3I — ether

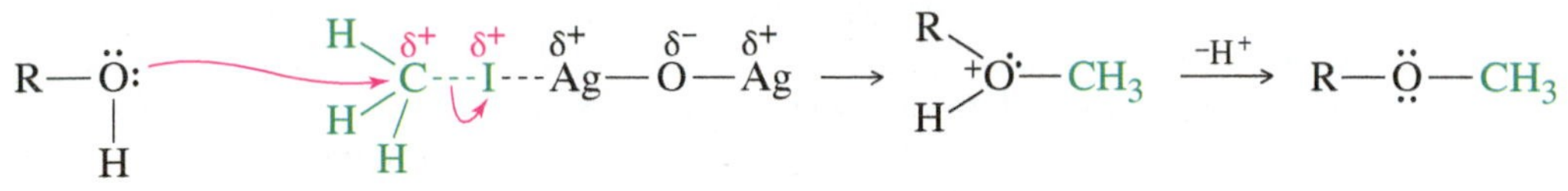

Example

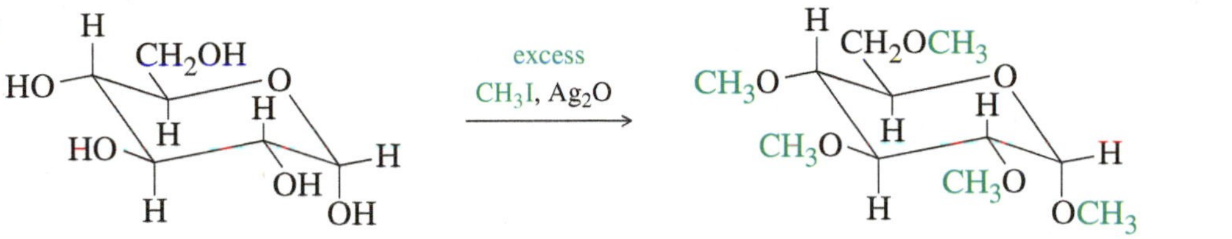

α-D-glucopyranose — excess CH_3I, Ag_2O → methyl 2,3,4,6-tetra-*O*-methyl-α-D-glucopyranoside

FIGURE 23-13
Formation of methyl ethers. Treatment of an aldose or a ketose with methyl iodide and silver oxide gives the totally methylated ether. If the conditions are carefully controlled, the stereochemistry at the anomeric carbon is usually preserved.

strongly electrophilic. Attack by the carbohydrate —OH group, followed by deprotonation, gives the ether. Figure 23-13 shows that the anomeric hydroxy group is also converted to an ether. If the conditions are carefully controlled, the hemiacetal C—O bond is not broken, and the configuration at the anomeric carbon is preserved.

The Williamson ether synthesis is the most common method for forming simple ethers, but it involves a strongly basic alkoxide ion. Under these basic conditions, a simple sugar would isomerize and decompose. A modified Williamson method may be used if the sugar is first converted to a glycoside (by treatment with an alcohol and an acid catalyst). The glycoside is an acetal, stable to base. Treatment of a glycoside with sodium hydroxide and methyl iodide or dimethyl sulfate gives the methylated carbohydrate.

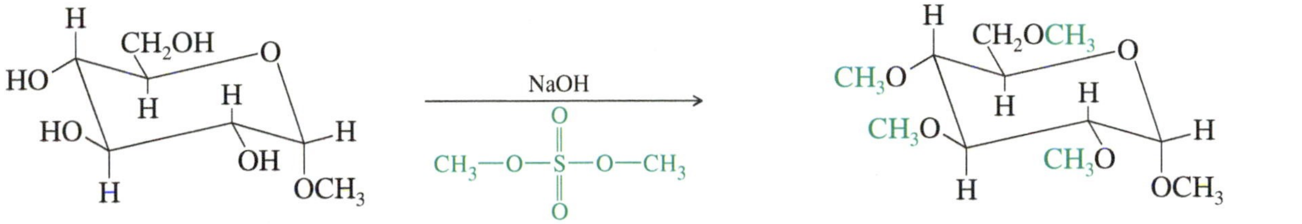

methyl α-D-glucopyranoside (stable to base) — NaOH, $CH_3—O—SO_2—O—CH_3$ (dimethyl sulfate) → methyl 2,3,4,6-tetra-*O*-methyl-α-D-glucopyranoside

We can use methylation, followed by hydrolysis, to determine the ring sizes of sugars. For example, methylation of β-D-glucose gives the pentamethyl derivative: methyl 2,3,4,6-tetra-*O*-methyl-β-D-glucopyranoside, shown below. The five methyl groups are not the same, however. Four are methyl ethers, but one is the glycosidic methyl group of an acetal.

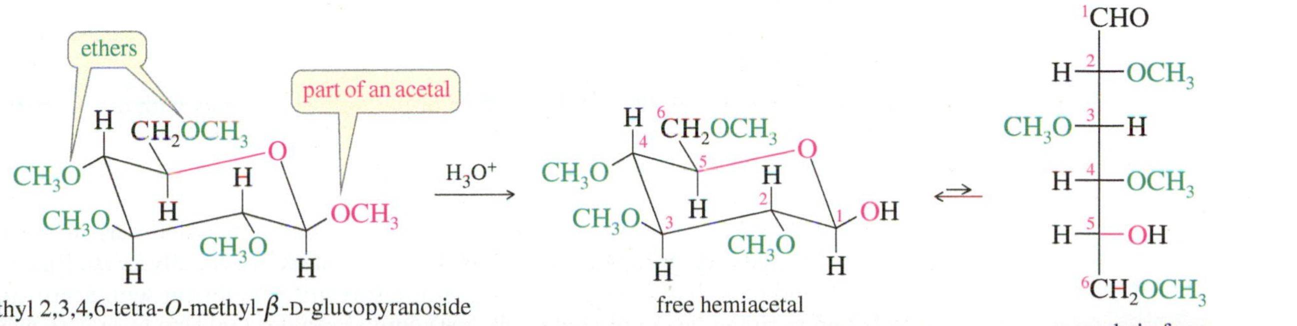

ethers — part of an acetal

methyl 2,3,4,6-tetra-*O*-methyl-β-D-glucopyranoside — H_3O^+ → free hemiacetal ⇌ open-chain form 2,3,4,6,-*O*-tetramethyl-D-glucose

Acetals are easily hydrolyzed by dilute acid, but ethers are stable under these conditions. Treatment of the pentamethyl glucose derivative with dilute acid hydrolyzes only the acetal methyl group. We can use NMR spectroscopy to determine that the free hydroxy group is on C5 of the hydrolyzed product, showing that the cyclic form of β-D-glucose has the six-membered ring of a pyranose.

PROBLEM 23-25

(a) Show the product that results when fructose is treated with an excess of methyl iodide and silver oxide.
(b) Show what happens when the product of part (a) is hydrolyzed using dilute acid.
(c) What do the results of parts (a) and (b) imply about the hemiacetal structure of fructose?

We can also easily convert hydroxy groups to silyl ethers. Section 14-10B covered the use of the triisopropylsilyl (TIPS) protecting group for alcohols. Similarly, sugars can be converted to their silyl ethers by treatment with a silyl chloride, such as chlorotrimethylsilane (TMSCl), and a tertiary amine, such as triethylamine.

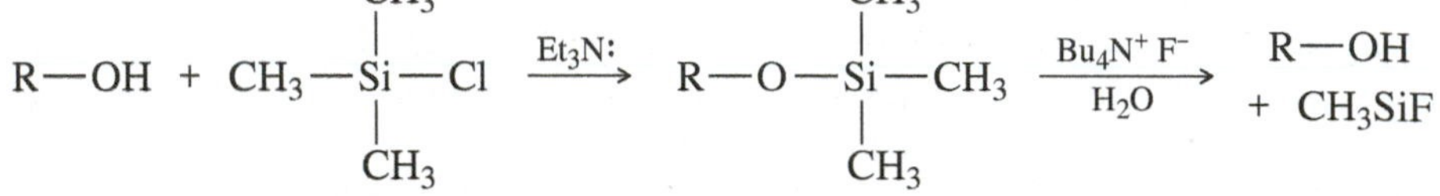

$$R{-}OH + CH_3{-}Si(CH_3)_2{-}Cl \xrightarrow{Et_3N:} R{-}O{-}Si(CH_3)_2{-}CH_3 \xrightarrow[H_2O]{Bu_4N^+\,F^-} R{-}OH + CH_3SiF$$

chlorotrimethylsilane $(CH_3)_3SiCl$ (TMSCl) — TMS ether (R—O—TMS) — (deprotected)

Sugars are most commonly converted to their silyl ethers to make them easier to handle and sufficiently volatile for gas chromatography and mass spectrometry. For example, glucose would be more likely to char and decompose inside the injector of a gas chromatograph, rather than to vaporize and flow through the column with the gas phase. The trimethylsilyl derivative of glucose is more volatile, however, and it vaporizes at a low enough temperature to survive gas chromatography and mass spectrometry.

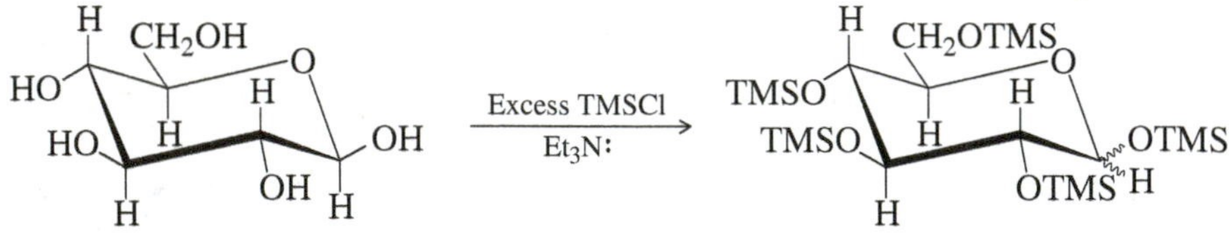

glucose (water soluble, not volatile) —Excess TMSCl, Et_3N:→ TMS derivative (organic-soluble, volatile)

PROBLEM 23-26

Propose a mechanism for methylation of any one of the hydroxy groups of methyl α-D-glucopyranoside, using NaOH and dimethyl sulfate.

PROBLEM 23-27

Draw the expected product of the reaction of the following sugars with excess methyl iodide and silver oxide.
(a) α-D-fructofuranose **(b)** β-D-galactopyranose

Application: Drug Excretion

Oxidation of glucose at C6 produces glucuronic acid. In the body, a common method for metabolizing drugs is to attach glucuronic acid. The resulting glucuronide derivative is water-soluble and easily excreted in the urine.

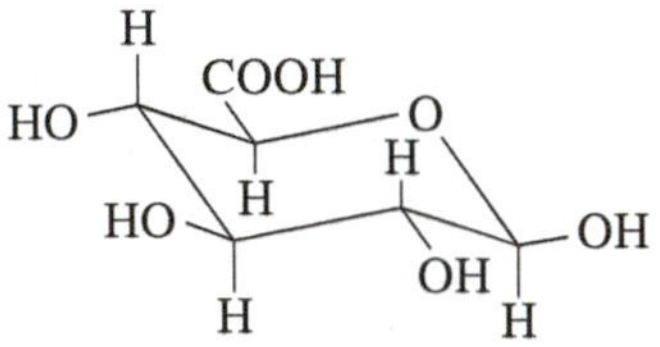

glucuronic acid

Ester Formation Another way to convert sugars to easily handled derivatives is to acylate the hydroxy groups to form esters. Sugar esters are readily crystallized and purified, and they dissolve in common organic solvents. Treatment with acetic anhydride and pyridine (as a mild basic catalyst) converts sugar hydroxy groups to acetate esters, as shown in Figure 23-14. This reaction acetylates all the hydroxy groups, including that of the hemiacetal on the anomeric carbon. The anomeric C—O bond sugar

acetate ester

Example

excess $(CH_3CO)_2O$ pyridine

β-D-fructofuranose

penta-O-acetyl-β-D-fructofuranoside

FIGURE 23-14
Formation of acetate esters. Acetic anhydride and pyridine convert all the hydroxy groups of a sugar to acetate esters. The stereochemistry at the anomeric carbon is usually preserved.

is not broken in the acylation, and the stereochemistry of the anomeric carbon atom is usually preserved. If we start with a pure α anomer or a pure β anomer, the product is the corresponding anomer of the acetate.

PROBLEM 23-28

Predict the products formed when the following sugars react with excess acetic anhydride and pyridine.

(a) α-D-glucopyranose **(b)** β-D-ribofuranose

23-10 Chain Shortening: The Ruff Degradation

In our discussion of D and L sugars, we briefly mentioned a method for shortening the chain of an aldose by removing the aldehyde carbon at the top of the Fischer projection. Such a reaction, removing one of the carbon atoms, is called a **degradation.**

The most common method used to shorten sugar chains is the **Ruff degradation,** developed by Otto Ruff, a prominent German chemist around the turn of the twentieth century. The Ruff degradation is a two-step process that begins with a bromine–water oxidation of the aldose to its aldonic acid. Treatment of the aldonic acid with hydrogen peroxide and ferric sulfate oxidizes the carboxyl group to CO_2 and gives an aldose with one fewer carbon atom. The Ruff degradation is used mainly for structure determination and synthesis of new sugars.

Ruff degradation

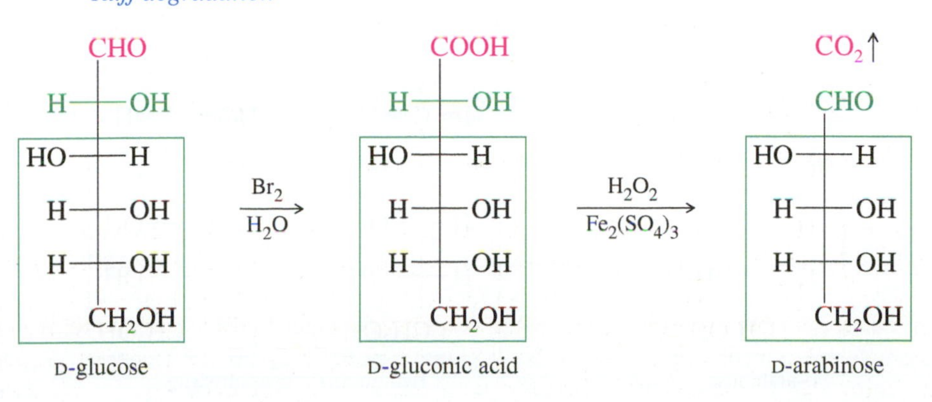

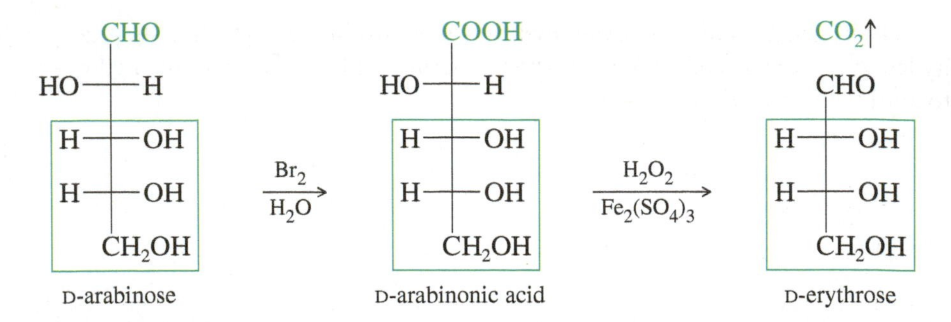

D-arabinose D-arabinonic acid D-erythrose

PROBLEM 23-29

Show that Ruff degradation of D-mannose gives the same aldopentose (D-arabinose) as does D-glucose.

PROBLEM 23-30

D-Lyxose is formed by Ruff degradation of galactose. Give the structure of D-lyxose. Ruff degradation of D-lyxose gives D-threose. Give the structure of D-threose.

PROBLEM 23-31

D-Altrose is an aldohexose. Ruff degradation of D-altrose gives the same aldopentose as does degradation of D-allose, the C3 epimer of glucose. Give the structure of D-altrose.

23-11 Chain Lengthening: The Kiliani–Fischer Synthesis

The **Kiliani–Fischer synthesis** lengthens an aldose carbon chain by adding one carbon atom to the aldehyde end of the aldose. The result of this process is a chain-lengthened sugar with a new carbon atom at C1 and the former aldehyde group (the former C1) now at C2. This synthesis is useful both for determining the structure of existing sugars and for synthesizing new sugars.

The Kiliani–Fischer synthesis

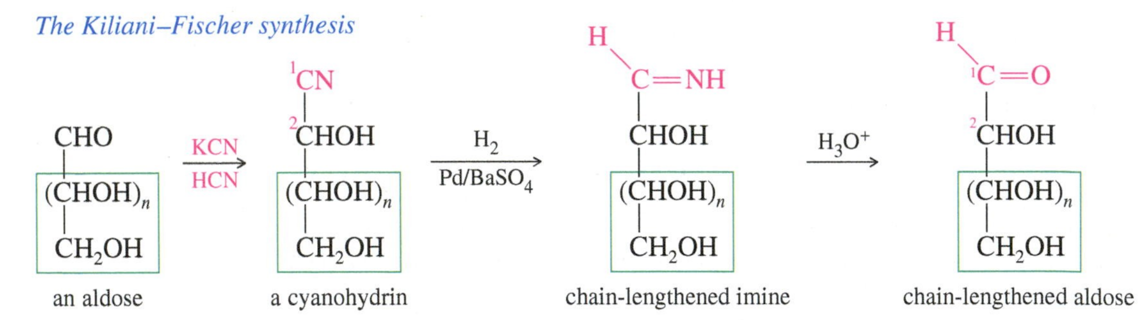

an aldose a cyanohydrin chain-lengthened imine chain-lengthened aldose

The aldehyde carbon atom is made asymmetric in the first step with the formation of the cyanohydrin. Two epimeric cyanohydrins result. For example, D-arabinose reacts with HCN to give the following cyanohydrins.

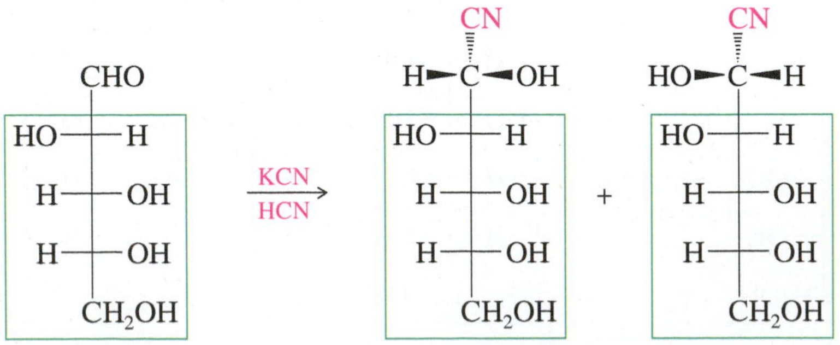

CHO
HO—H
H—OH
H—OH
CH_2OH

KCN / HCN →

CN
H—C—OH
HO—H
H—OH
H—OH
CH_2OH

+

CN
HO—C—H
HO—H
H—OH
H—OH
CH_2OH

D-arabinose two epimeric cyanohydrins

Hydrogenation of these cyanohydrins gives two imines, which hydrolyze to aldehydes. A poisoned catalyst of palladium on barium sulfate is used for the hydrogenation to avoid overreduction.

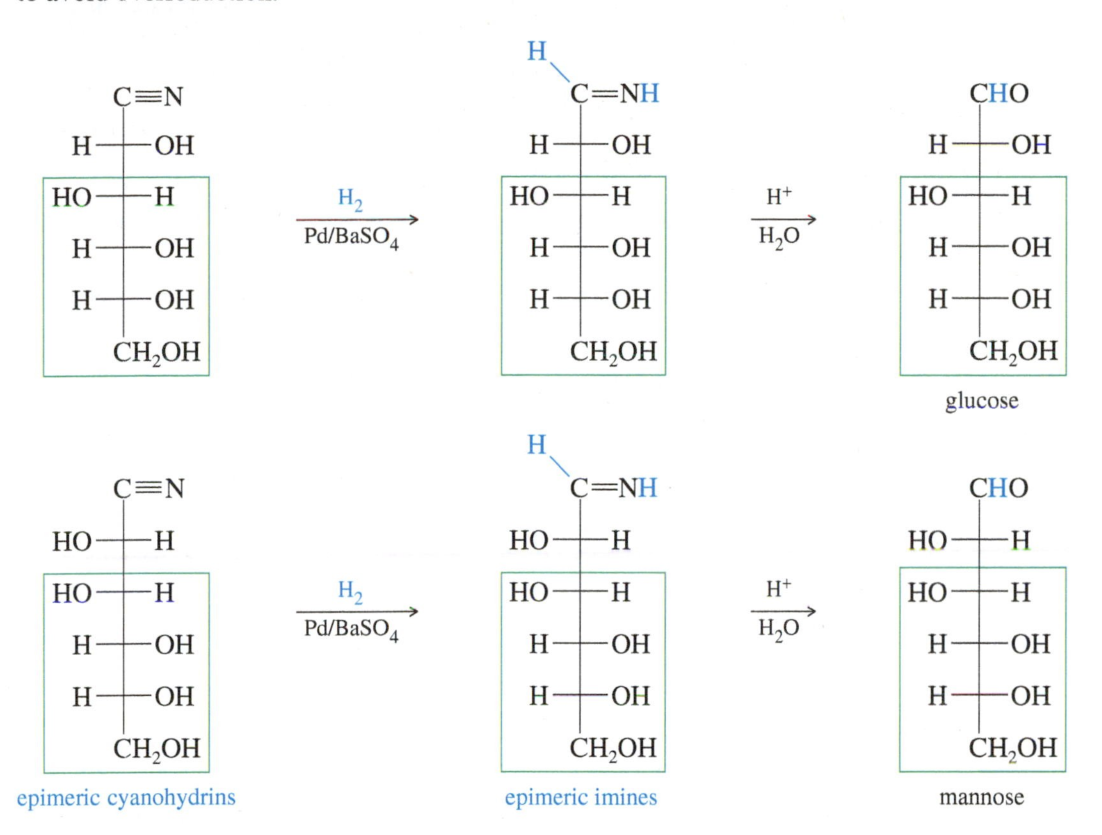

The Kiliani–Fischer synthesis accomplishes the opposite of the Ruff degradation. Ruff degradation of either of two C2 epimers gives the same shortened aldose, and the Kiliani–Fischer synthesis converts this shortened aldose back into a mixture of the same two C2 epimers. For example, glucose and mannose both undergo Ruff degradation to give arabinose. Conversely, the Kiliani–Fischer synthesis converts arabinose into a mixture of glucose and mannose.

PROBLEM 23-32

Ruff degradation of D-arabinose gives D-erythrose. The Kiliani–Fischer synthesis converts D-erythrose to a mixture of D-arabinose and D-ribose. Draw out these reactions, and give the structure of D-ribose.

PROBLEM 23-33

The *Wohl degradation*, an alternative to the Ruff degradation, is nearly the reverse of the Kiliani–Fischer synthesis. The aldose carbonyl group is converted to the oxime, which is dehydrated by acetic anhydride to the nitrile (a cyanohydrin). Cyanohydrin formation is reversible, and a basic hydrolysis allows the cyanohydrin to lose HCN. Using the following sequence of reagents, give equations for the individual reactions in the Wohl degradation of D-arabinose to D-erythrose. Mechanisms are not required.

(1) hydroxylamine hydrochloride **(2)** acetic anhydride **(3)** ^{-}OH, H_2O

PROBLEM-SOLVING HINT

In working this type of problem, it is often easier to start with the smallest structure mentioned (often glyceraldehyde) and work backward to larger structures. Write out all possible structures, and use the clues to eliminate the wrong ones.

PROBLEM 23-34

Aldohexoses **A** and **B** both undergo Ruff degradation to give aldopentose **C.** On treatment with warm nitric acid, aldopentose **C** gives an optically active aldaric acid. **B** also reacts with warm nitric acid to give an optically active aldaric acid, but **A** reacts to give an optically inactive aldaric acid. Aldopentose **C** is degraded to aldotetrose **D,** which gives optically active tartaric acid when it is treated with nitric acid. Aldotetrose **D** is degraded to (+)-glyceraldehyde. Deduce the structures of sugars **A, B, C,** and **D,** and use Figure 23-3 to determine the correct names of these sugars.

PROBLEM 23-35

In 1891, Emil Fischer determined the structures of glucose and the seven other D-aldohexoses using only simple chemical reactions and clever reasoning about stereochemistry and symmetry. He received the Nobel Prize for this work in 1902. Fischer had determined that D-glucose is an aldohexose, and he used Ruff degradations to degrade it to (+)-glyceraldehyde. Therefore, the eight D-aldohexose structures shown in Figure 23-3 are the possible structures for glucose.

Pretend that no names are shown in Figure 23-3 except for glyceraldehyde, and use the following results to prove which of these structures represent glucose, mannose, arabinose, and erythrose.

(a) Upon Ruff degradation, glucose and mannose give the same aldopentose: arabinose. Nitric acid oxidation of arabinose gives an optically active aldaric acid. What are the *two* possible structures of arabinose?

(b) Upon Ruff degradation, arabinose gives the aldotetrose erythrose. Nitric acid oxidation of erythrose gives an optically inactive aldaric acid, *meso*-tartaric acid. What is the structure of erythrose?

(c) Which of the two possible structures of arabinose is correct? What are the possible structures of glucose and mannose?

(d) Fischer's genius was needed to distinguish between glucose and mannose. He developed a series of reactions to convert the aldehyde group of an aldose to an alcohol while converting the terminal alcohol to an aldehyde. In effect, he swapped the functional groups on the ends. When he interchanged the functional groups on D-mannose, he was astonished to find that the product was still D-mannose. Show how this information completes the proof of the mannose structure, and show how it implies the correct glucose structure.

(e) When Fischer interchanged the functional groups on D-glucose, the product was an unnatural L sugar. Show which unnatural sugar he must have formed, and show how it completes the proof of the glucose structure.

SUMMARY Reactions of Sugars

1. *Reduction* (Section 23-6)

$$\underset{\text{aldose}}{\begin{array}{c} CHO \\ | \\ (CHOH)_n \\ | \\ CH_2OH \end{array}} \xrightarrow[\text{or } H_2/Ni]{NaBH_4} \underset{\text{alditol}}{\begin{array}{c} CH_2OH \\ | \\ (CHOH)_n \\ | \\ CH_2OH \end{array}}$$

2. *Oxidation* (Section 23-7)

a. *To aldonic acids (glyconic acids) by bromine water*

$$\underset{\text{aldose}}{\begin{array}{c} CHO \\ | \\ (CHOH)_n \\ | \\ CH_2OH \end{array}} \xrightarrow[H_2O]{Br_2} \underset{\text{aldonic acid}}{\begin{array}{c} COOH \\ | \\ (CHOH)_n \\ | \\ CH_2OH \end{array}}$$

b. *To aldaric acids (glycaric acids) by nitric acid*

CHO–$(CHOH)_n$–CH_2OH (aldose) $\xrightarrow{HNO_3}$ COOH–$(CHOH)_n$–COOH (aldaric acid)

c. *Tollens test for reducing sugars*

CHO–CHOH–$(CHOH)_n$–CH_2OH (aldose) or CH_2OH–C=O–$(CHOH)_n$–CH_2OH (ketose) $\xrightarrow{Ag(NH_3)_2OH}$ COO^-–CHOH–$(CHOH)_n$–CH_2OH + rearrangement + Ag (silver mirror)

3. *Glycoside formation (conversion to an acetal)* (Section 23-8)

(either anomer) $\xrightarrow[H^+]{CH_3OH}$ a methyl glycoside (OCH_3, $(\alpha + \beta)$)

(more stable anomer predominates)

4. *Alkylation to give ethers* (Section 23-9)

$\xrightarrow[Ag_2O]{\text{excess } CH_3I}$ (CH_3O, CH_3O, CH_2OH_3, CH_3O, OCH_3)

(gives the same anomer as the starting material)

Treatment with a silyl chloride (such as TMSCl) and a tertiary amine converts sugars to their silyl ethers. Fluoride salts such as aqueous Bu_4NF hydrolyze the silyl ethers.

Treatment with a silyl chloride (such as TMSCl) and a tertiary amine converts sugars to their silyl ethers. Fluoride salts such as aqueous Bu_4NF hydrolyze the silyl ethers.

5. *Acylation to give esters* (Section 23-9)

$\xrightarrow[\text{pyridine}]{\text{excess } Ac_2O}$ (CH_2OAc, AcO, AcO, AcO, OAc)

(gives the same anomer as the starting materials)

6. *Ruff degradation* (Section 23-10)

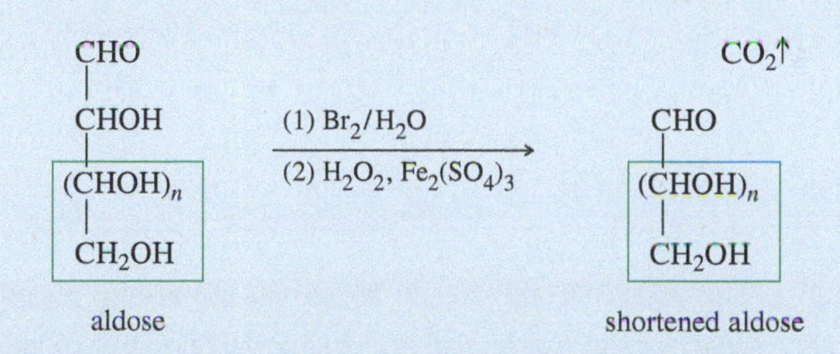

(continued)

7. *Kiliani–Fischer synthesis* (Section 23-11)

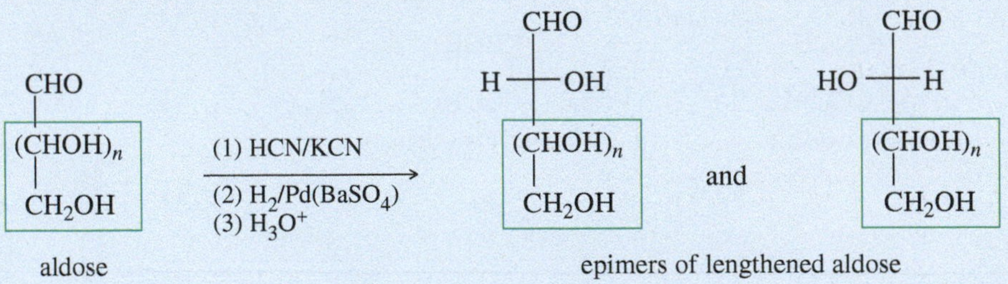

CHO–$(CHOH)_n$–CH_2OH (aldose) $\xrightarrow[(2)\ H_2/Pd(BaSO_4)\ (3)\ H_3O^+]{(1)\ HCN/KCN}$ CHO–CH(OH)–$(CHOH)_n$–CH_2OH (H–C–OH) and CHO–CH(OH)–$(CHOH)_n$–CH_2OH (HO–C–H)

aldose

epimers of lengthened aldose

23-12 Disaccharides

As we have seen, the anomeric carbon of a sugar can react with the hydroxy group of an alcohol to give an acetal called a *glycoside*. If the hydroxy group is part of another sugar molecule, then the glycoside product is a **disaccharide,** a sugar composed of two monosaccharide units (Figure 23-15).

In principle, the anomeric carbon can react with *any* of the hydroxy groups of another sugar to form a disaccharide. In naturally occurring disaccharides, however, there are three common glycosidic bonding arrangements.

1. A 1,4′ link. The anomeric carbon is bonded to the oxygen atom on C4 of the second sugar. The prime symbol (′) in 1,4′ indicates that C4 is on the second sugar.
2. A 1,6′ link. The anomeric carbon is bonded to the oxygen atom on C6 of the second sugar.
3. A 1,1′ link. The anomeric carbon of the first sugar is bonded through an oxygen atom to the anomeric carbon of the second sugar.

We will consider some naturally occurring disaccharides with these common glycosidic linkages.

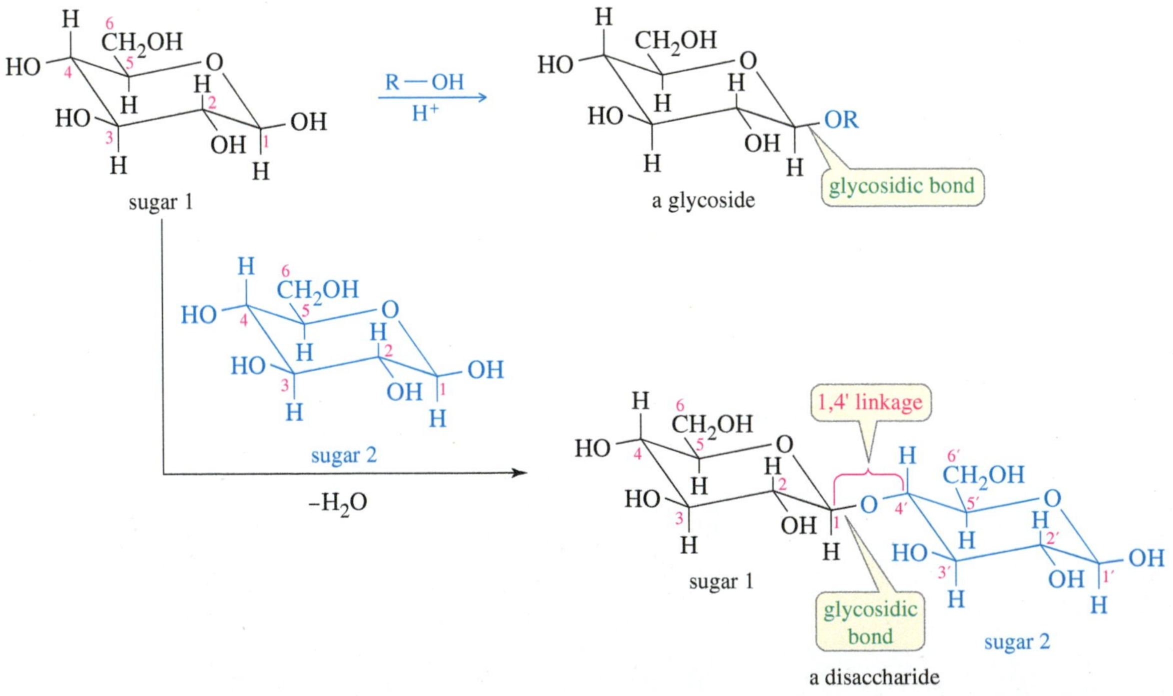

sugar 1 $\xrightarrow[H^+]{R—OH}$ a glycoside (glycosidic bond)

sugar 1 + sugar 2 $\xrightarrow{-H_2O}$ a disaccharide (1,4′ linkage; glycosidic bond)

FIGURE 23-15
Disaccharides. A sugar reacts with an alcohol to give an acetal called a glycoside. When the alcohol is part of another sugar, the product is a disaccharide.

23-12A The 1,4′ Linkage: Cellobiose, Maltose, and Lactose

The most common glycosidic linkage is the 1,4′ link. The anomeric carbon of one sugar is bonded to the oxygen atom on C4 of the second ring.

Cellobiose: A β-1,4′ Glucosidic Linkage *Cellobiose*, the disaccharide obtained by partial hydrolysis of cellulose, contains a 1,4′ linkage. In cellobiose, the anomeric carbon of one glucose unit is linked through an equatorial (β) carbon–oxygen bond to C4 of another glucose unit. This β-1,4′ linkage from a *glucose* acetal is called a β-1,4′ ***gluco*sidic linkage.**

Cellobiose, 4-O-(β-D-glucopyranosyl)-β-D-glucopyranose or 4-O-(β-D-glucopyranosyl)-D-glucopyranose

β-glucosidic linkage

β-glucosidic linkage

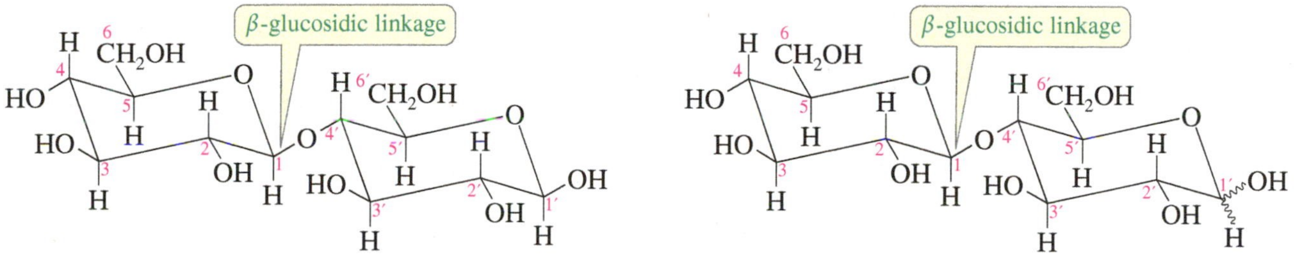

Two alternative ways of drawing and naming cellobiose

The complete name for cellobiose, 4-*O*-(β-D-glucopyranosyl)-β-D-glucopyranose, gives its structure. This name says that a β-D-glucopyranose ring (the right-hand ring) is substituted in its 4-position by an oxygen attached to a (β-D-glucopyranosyl) ring, drawn on the left. The name in parentheses says the substituent is a β-glucose, and the -*syl* ending indicates that this ring is a glycoside. The left ring with the -*syl* ending is an acetal and cannot mutarotate, while the right ring with the -*ose* ending is a hemiacetal and can mutarotate. Because cellobiose has a glucose unit in the hemiacetal form (and therefore is in equilibrium with its open-chain aldehyde form), it is a reducing sugar. Once again, the -*ose* ending indicates a mutarotating, reducing sugar.

Mutarotating sugars are often shown with a wavy line to the free anomeric hydroxy group, signifying that they can exist as an equilibrium mixture of the two anomers. Their names are often given without specifying the stereochemistry of this mutarotating hydroxy group, as in 4-*O*-(β-D-glucopyranosyl)-D-glucopyranose.

Maltose: An α-1,4′ Glucosidic Linkage *Maltose* is a disaccharide formed when starch is treated with sprouted barley, called *malt*. This malting process is the first step in brewing beer, converting polysaccharides to disaccharides and monosaccharides that ferment more easily. Like cellobiose, maltose contains a 1,4′ glycosidic linkage between two glucose units. The difference in maltose is that the stereochemistry of the glucosidic linkage is α rather than β.

Maltose, 4-O-(α-D-glucopyranosyl)-D-glucopyranose

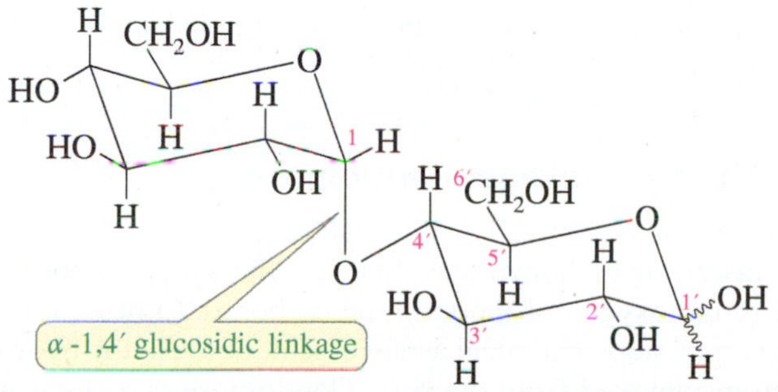

α-1,4′ glucosidic linkage

Like cellobiose, maltose has a free hemiacetal ring (on the right). This hemiacetal is in equilibrium with its open-chain form, and it mutarotates and can exist in either the α or β anomeric form. Because maltose exists in equilibrium with an open-chain aldehyde, it reduces Tollens reagent, and maltose is a reducing sugar.

PROBLEM 23-36

Draw the structures of the individual mutarotating α and β anomers of maltose.

PROBLEM 23-37

Give an equation to show the reduction of Tollens reagent by maltose.

Lactose: A β-1,4′ Galactosidic Linkage *Lactose* is similar to cellobiose, except that the glycoside (the left ring) in lactose is galactose rather than glucose. Lactose is composed of one galactose unit and one glucose unit. The two rings are linked by a β-glycosidic bond of the galactose acetal to the 4-position on the glucose ring: a β-1,4′ ***galacto*sidic linkage.**

Lactose, 4-O-(β-D-galactopyranosyl)-D-glucopyranose

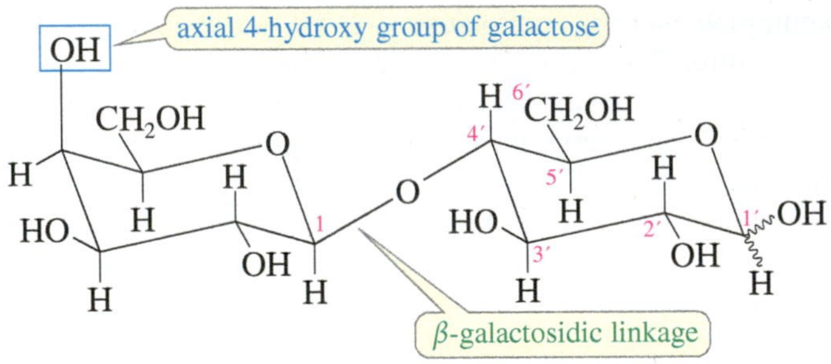

Lactose occurs naturally in the milk of mammals, including cows and humans. Hydrolysis of lactose requires a β-galactosidase enzyme (sometimes called lactase). Some humans synthesize a β-galactosidase, but others do not. This enzyme is present in the digestive fluids of normal infants to hydrolyze their mother's milk. Once the child stops drinking milk, production of the enzyme gradually stops. In most parts of the world, people do not use milk products after early childhood, and the adult population can no longer digest lactose. Consumption of milk or milk products can cause digestive discomfort in *lactose-intolerant* people who lack the β-galactosidase enzyme. Lactose-intolerant infants must drink soybean milk or another lactose-free formula.

PROBLEM 23-38

Does lactose mutarotate? Is it a reducing sugar? Explain. Draw the two anomeric forms of lactose.

23-12B The 1,6′ Linkage: Gentiobiose

In addition to the common 1,4′ glycosidic linkage, the 1,6′ linkage is also found in naturally occurring carbohydrates. In a 1,6′ linkage, the anomeric carbon of one sugar is linked to the oxygen of the terminal carbon (C6) of another. This linkage gives a different sort of stereochemical arrangement, because the hydroxy group on C6 is one carbon atom removed from the ring. Gentiobiose is a sugar with two glucose units joined by a β-1,6′ glucosidic linkage.

Gentiobiose, 6-O-(β-D-glucopyranosyl)-D-glucopyranose

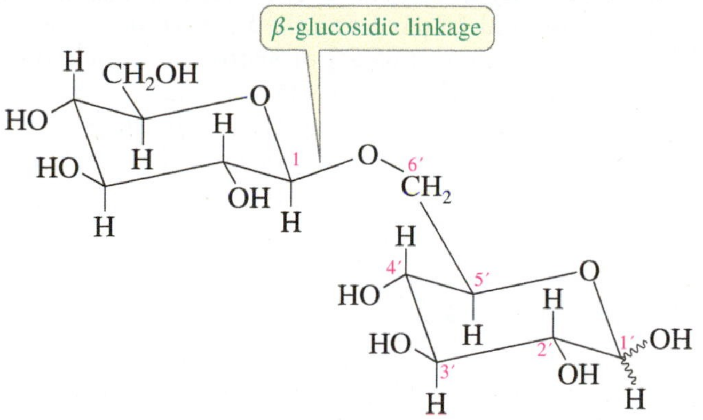

β-glucosidic linkage

Although the 1,6′ linkage is rare in disaccharides, it is commonly found as a branch point in polysaccharides. For example, branching in amylopectin (insoluble starch) occurs at 1,6′ linkages, as discussed in Section 23-13B.

PROBLEM 23-39

Is gentiobiose a reducing sugar? Does it mutarotate? Explain your reasoning.

23-12C Linkage of Two Anomeric Carbons: Sucrose

Some sugars are joined by a direct glycosidic linkage between their anomeric carbon atoms: a 1,1′ linkage. *Sucrose* (common table sugar), for example, is composed of one glucose unit and one fructose unit bonded by an oxygen atom linking their anomeric carbon atoms. (Because fructose is a ketose and its anomeric carbon is C2, this is actually a 1,2′ linkage.) Notice that the linkage is in the α position with respect to the glucose ring and in the β position with respect to the fructose ring.

Sucrose, α-D-glucopyranosyl-β-D-fructofuranoside
(or β-D-fructofuranosyl-α-D-glucopyranoside)

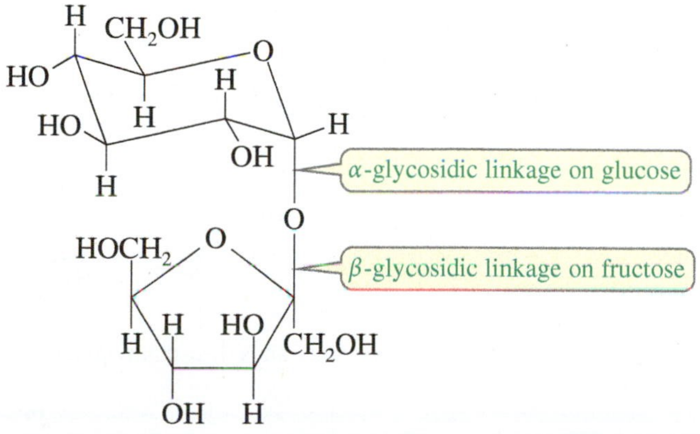

α-glycosidic linkage on glucose

β-glycosidic linkage on fructose

Both monosaccharide units in sucrose are present as acetals, or glycosides. Neither ring is in equilibrium with its open-chain aldehyde or ketone form, so sucrose does not reduce Tollens reagent and it cannot mutarotate. Because both units are glycosides, the systematic name for sucrose can list either of the two glycosides as being a substituent on the other. Both systematic names end in the *-oside* suffix, indicating a nonmutarotating, nonreducing sugar. Like many other common names, *sucrose* ends in the *-ose* ending even though it is a nonreducing sugar. Common names are not reliable indicators of the properties of sugars.

Application: Sucrose Stability

Sucrose (a nonreducing sugar) is not as easily oxidized as a reducing sugar, so it is much more useful for preserving foods such as jams and jellies. A reducing sugar like glucose would oxidize and spoil.

Sucrose is hydrolyzed by enzymes called *invertases*, found in honeybees and yeasts, that specifically hydrolyze the β-D-fructofuranoside linkage. The resulting mixture of glucose and fructose is called *invert sugar* because hydrolysis converts the positive rotation [+66.5°] of sucrose to a negative rotation that is the average of glucose [+52.7°] and fructose [−92.4°]. The most common form of invert sugar is honey, a supersaturated mixture of glucose and fructose hydrolyzed from sucrose by the invertase enzyme of honeybees. Glucose and fructose were once called *dextrose* and *levulose*, respectively, according to their opposite signs of rotation.

SOLVED PROBLEM 23-3

An unknown carbohydrate of formula $C_{12}H_{22}O_{11}$ reacts with Tollens reagent to form a silver mirror. An α-glycosidase has no effect on the carbohydrate, but a β-galactosidase hydrolyzes it to D-galactose and D-mannose. When the carbohydrate is methylated (using methyl iodide and silver oxide) and then hydrolyzed with dilute HCl, the products are 2,3,4,6-tetra-*O*-methylgalactose and 2,3,4-tri-*O*-methylmannose. Propose a structure for this unknown carbohydrate.

SOLUTION

The formula shows that this is a disaccharide composed of two hexoses. Hydrolysis gives D-galactose and D-mannose, identifying the two hexoses. Hydrolysis requires a β-galactosidase, showing that galactose and mannose are linked by a β-galactosyl linkage. Since the original carbohydrate is a reducing sugar, one of the hexoses must be in a free hemiacetal form. Galactose is present as a glycoside; thus, mannose must be present in its hemiacetal form. The unknown carbohydrate must be a (β-galactosyl)-mannose.

The methylation/hydrolysis procedure shows the point of attachment of the glycosidic bond to mannose and also confirms the size of the six-membered rings. In galactose, all the hydroxy groups are methylated except C1 and C5. C1 is the anomeric carbon, and the C5 oxygen is used to form the hemiacetal of the pyranose ring. In mannose, all the hydroxy groups are methylated except C1, C5, and C6. The C5 oxygen is used to form the pyranose ring (the C6 oxygen would form a less stable seven-membered ring); therefore, the oxygen on C6 must be involved in the glycosidic linkage. The structure and systematic name are shown here.

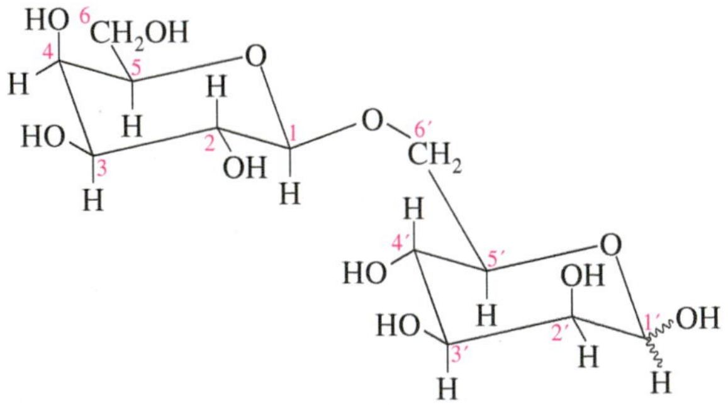

6-*O*-(β-D-galactopyranosyl)-D-mannopyranose

Application: Blood Types

The primary differences among the blood types O, A, B, and AB involve antigenic carbohydrates on the surface of the red blood cells. Type O cells have no antigenic carbohydrates on the surface, while Type A cells have an antigenic *N*-acetylgalactosamine, and Type B cells have an antigenic galactose. Type AB cells have both galactose and *N*-acetylgalactosamine antigens.

Type O is called the "universal donor" because the cells have no antigen to provoke a deadly antigen-antibody reaction. If the other blood factors (Rh factor, for example) are compatible, type O can be donated to people with the other blood types.

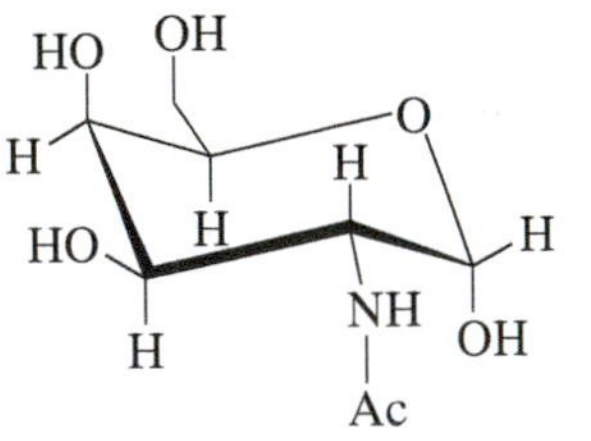

N-acetylgalactosamine

PROBLEM 23-40

Trehalose is a nonreducing disaccharide ($C_{12}H_{22}O_{11}$) isolated from the poisonous mushroom *Amanita muscaria*. Treatment with an α-glucosidase converts trehalose to two molecules of glucose, but no reaction occurs when trehalose is treated with a β-glucosidase. When trehalose is methylated by dimethyl sulfate in mild base and then hydrolyzed, the only product is 2,3,4,6-tetra-*O*-methylglucose. Propose a complete structure and systematic name for trehalose.

PROBLEM 23-41

Raffinose is a trisaccharide ($C_{18}H_{32}O_{16}$) isolated from cottonseed meal. Raffinose does not reduce Tollens reagent, and it does not mutarotate. Complete hydrolysis of raffinose gives D-glucose, D-fructose, and D-galactose. When raffinose is treated with invertase, the products are D-fructose and a reducing disaccharide called *melibiose*. Raffinose is unaffected by treatment with a β-galactosidase, but an α-galactosidase hydrolyzes it to D-galactose and sucrose. When raffinose is treated with dimethyl sulfate and base followed by hydrolysis, the products are 2,3,4-tri-*O*-methylglucose, 1,3,4,6-tetra-*O*-methylfructose, and 2,3,4,6-tetra-*O*-methylgalactose. Determine the complete structures of raffinose and melibiose, and give a systematic name for melibiose.

23-13 Polysaccharides

Polysaccharides (also called **glycans**) are carbohydrates that contain many monosaccharide units joined by glycosidic bonds. They are one class of *biopolymers*, or naturally occurring polymers. Smaller polysaccharides, containing about three to ten monosaccharide units, are sometimes called **oligosaccharides.** Most polysaccharides have hundreds or thousands of simple sugar units linked together into long polymer chains. Except for units at the ends of chains, all the anomeric carbon atoms of polysaccharides are involved in acetal glycosidic links. Therefore, polysaccharides give no noticeable reaction with Tollens reagent, and they do not mutarotate.

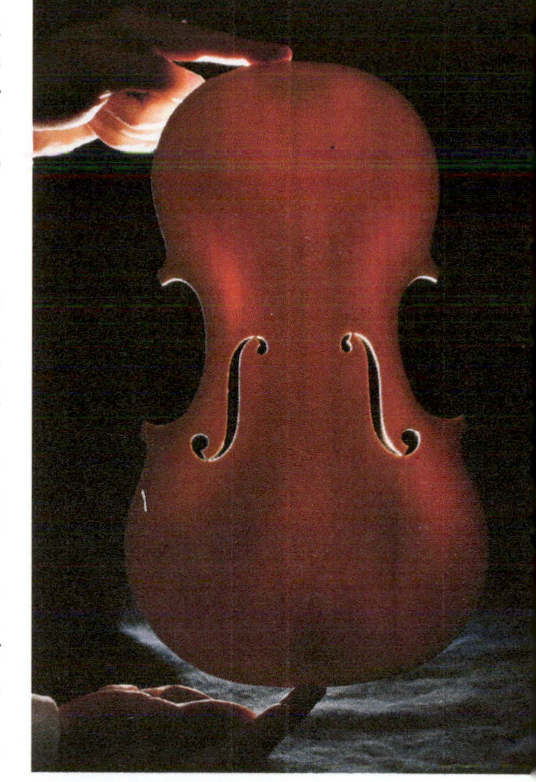

The acoustic properties of cellulose have never been surpassed by other substances. Here, a luthier holds a spruce front plate up to a light to show how the thickness is carefully graduated to enhance a pleasing sound.

23-13A Cellulose

Cellulose, a polymer of D-glucose, is the most abundant organic material. Cellulose is synthesized by plants as a structural material to support the weight of the plant. Long cellulose molecules, called *microfibrils*, are held in bundles by hydrogen bonding between the many —OH groups of the glucose rings. About 50% of dry wood and about 90% of cotton fiber is cellulose.

Cellulose is composed of D-glucose units linked by β-1,4′ glycosidic bonds. This bonding arrangement (like that in cellobiose) is rather rigid and very stable, giving cellulose desirable properties for a structural material. Figure 23-16 shows a partial structure of cellulose.

Humans and other mammals lack the β-glucosidase enzyme needed to hydrolyze cellulose, so they cannot use it directly for food. Several groups of bacteria and protozoa can hydrolyze cellulose, however. Termites and ruminants maintain colonies of these bacteria in their digestive tracts. When a cow eats hay, these bacteria convert about 20% to 30% of the cellulose to digestible carbohydrates.

Rayon is a fiber made from cellulose that has been converted to a soluble derivative and then regenerated. In the common *viscose process*, wood pulp is treated with carbon disulfide and sodium hydroxide to convert the free hydroxy groups to xanthates, which are soluble in water. The viscous solution (called *viscose*) is forced through a spinneret into an aqueous sodium bisulfate solution, where a fiber of insoluble cellulose is regenerated. Alternatively, the viscose solution can be extruded in sheets to give *cellophane* film. Rayon and cotton are both cellulose, yet rayon thread can be much stronger because it consists of long, continuously extruded fibers, rather than short cotton fibers spun together.

β-glucosidic linkage

FIGURE 23-16 Partial structure of cellulose. Cellulose is a β-1,4′ polymer of D-glucose, systematically named poly(1,4′-*O*-β-D-glucopyranoside).

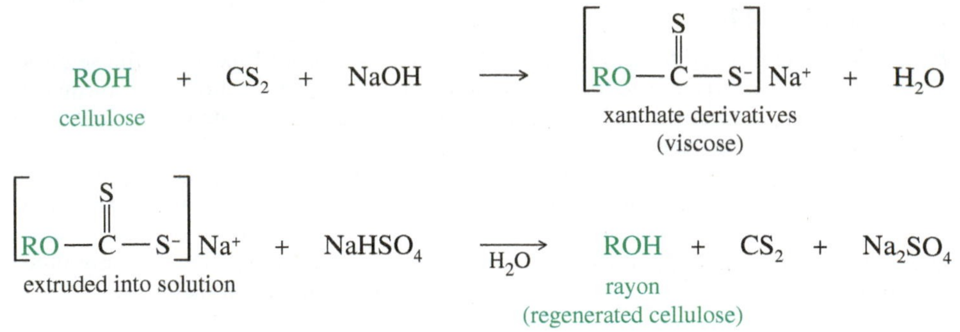

$$\underset{\text{cellulose}}{ROH} + CS_2 + NaOH \longrightarrow \underset{\substack{\text{xanthate derivatives}\\\text{(viscose)}}}{\left[RO{-}\overset{\overset{\displaystyle S}{\|}}{C}{-}S^-\right]Na^+} + H_2O$$

$$\underset{\text{extruded into solution}}{\left[RO{-}\overset{\overset{\displaystyle S}{\|}}{C}{-}S^-\right]Na^+} + NaHSO_4 \xrightarrow[H_2O]{} \underset{\substack{\text{rayon}\\\text{(regenerated cellulose)}}}{ROH} + CS_2 + Na_2SO_4$$

PROBLEM 23-42

Cellulose is converted to *cellulose acetate* by treatment with acetic anhydride and pyridine. Cellulose acetate is soluble in common organic solvents, and it is easily dissolved and spun into fibers. Show the structure of cellulose acetate.

23-13B Starches: Amylose, Amylopectin, and Glycogen

Plants use starch granules for storing energy. When the granules are dried and ground up, different types of starches can be separated by mixing them with hot water. About 20% of the starch is water-soluble *amylose*, and the remaining 80% is water-insoluble *amylopectin*. When starch is treated with dilute acid or appropriate enzymes, it is progressively hydrolyzed to maltose and then to glucose.

Application: Dental Plaque

Oral bacteria convert glucose, fructose, sucrose, and other common sugars into a polysaccharide called dextran. Dextran is an essential component of the plaque that forms around teeth and protects bacteria from the antibacterial components in saliva. The dextran chain consists of glucose molecules linked by α-1,6′ glucosidic linkages, with branches at α-1,3′ linkages. Candy makers use glucitol ("sorbitol") and mannitol to sweeten "sugarless" candies and gum because bacteria cannot easily convert these sugar alcohols to the glucose they need to make dextran.

Amylose Like cellulose, **amylose** is a linear polymer of glucose with 1,4′ glycosidic linkages. The difference is in the stereochemistry of the linkage. Amylose has α-1,4′ links, while cellulose has β-1,4′ links. A partial structure of amylose is shown in Figure 23-17.

The subtle stereochemical difference between cellulose and amylose results in some striking physical and chemical differences. The α linkage in amylose kinks the polymer chain into a helical structure. This kinking increases hydrogen bonding with water and lends additional solubility. As a result, amylose is soluble in water, and cellulose is not. Cellulose is stiff and sturdy, but amylose is not. Unlike cellulose, amylose is an excellent food source. The α-1,4′ glucosidic linkage is easily hydrolyzed by an α-glucosidase enzyme, found in all animals.

The helical structure of amylose also serves as the basis for an interesting and useful reaction. The inside of the helix is just the right size and polarity to accept an iodine (I_2) molecule. When iodine is lodged within this helix, a deep blue starch-iodine

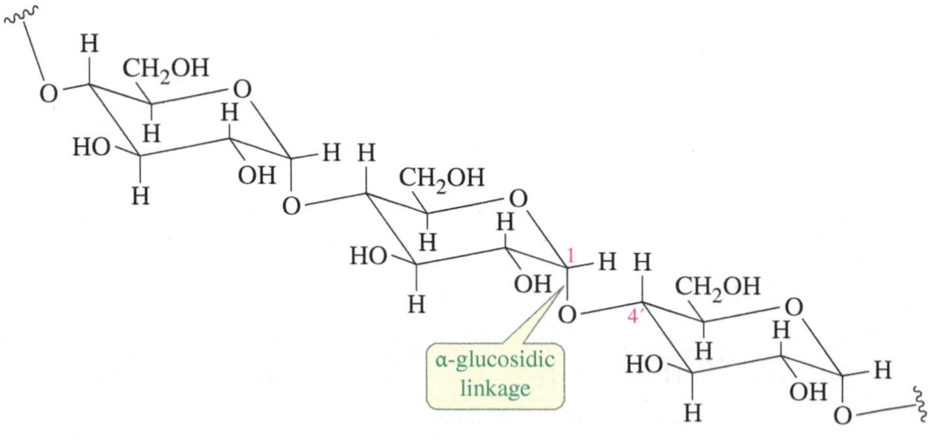

FIGURE 23-17
Partial structure of amylose. Amylose is an α-1,4′ polymer of glucose, systematically named poly(1,4′-*O*-α-D-glucopyranoside). Amylose differs from cellulose only in the stereochemistry of the glycosidic linkage.

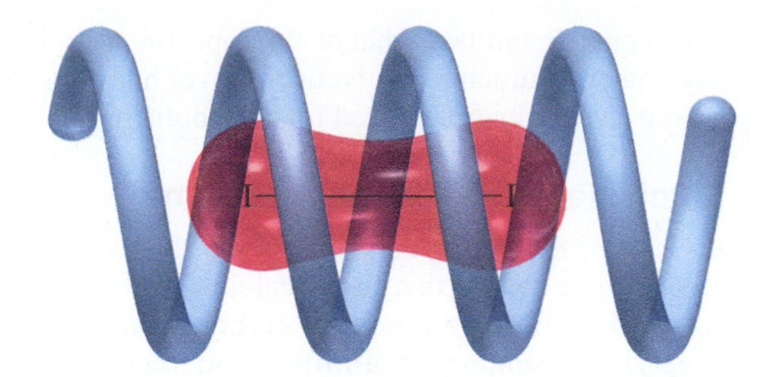

FIGURE 23-18
The starch-iodine complex of amylose. The amylose helix forms a dark blue charge-transfer complex with molecular iodine.

complex results (Figure 23-18). This is the basis of the *starch-iodide* test for oxidizers. The material to be tested is added to an aqueous solution of amylose and potassium iodide. If the material is an oxidizer, some of the iodide (I^-) is oxidized to iodine (I_2), which forms the blue complex with amylose.

Amylopectin **Amylopectin,** the insoluble fraction of starch, is also primarily an α-1,4′ polymer of glucose. The difference between amylose and amylopectin lies in the branched nature of amylopectin, with a branch point about every 20 to 30 glucose units. Another chain starts at each branch point, connected to the main chain by an α-1,6′ glycosidic linkage. A partial structure of amylopectin, including one branch point, is shown in Figure 23-19.

Glycogen **Glycogen** is the carbohydrate that animals use to store glucose for readily available energy. A large amount of glycogen is stored in the muscles themselves, ready for immediate hydrolysis and metabolism. Additional glycogen is stored in the liver, where it can be hydrolyzed to glucose for secretion into the bloodstream, providing an athlete with a "second wind."

Application: Low-Carb Diets

Low-carbohydrate diets restrict the intake of carbohydrates, sometimes resulting in rapid weight loss. The weight is lost because glycogen and fatty acids are burned to maintain blood glucose levels.

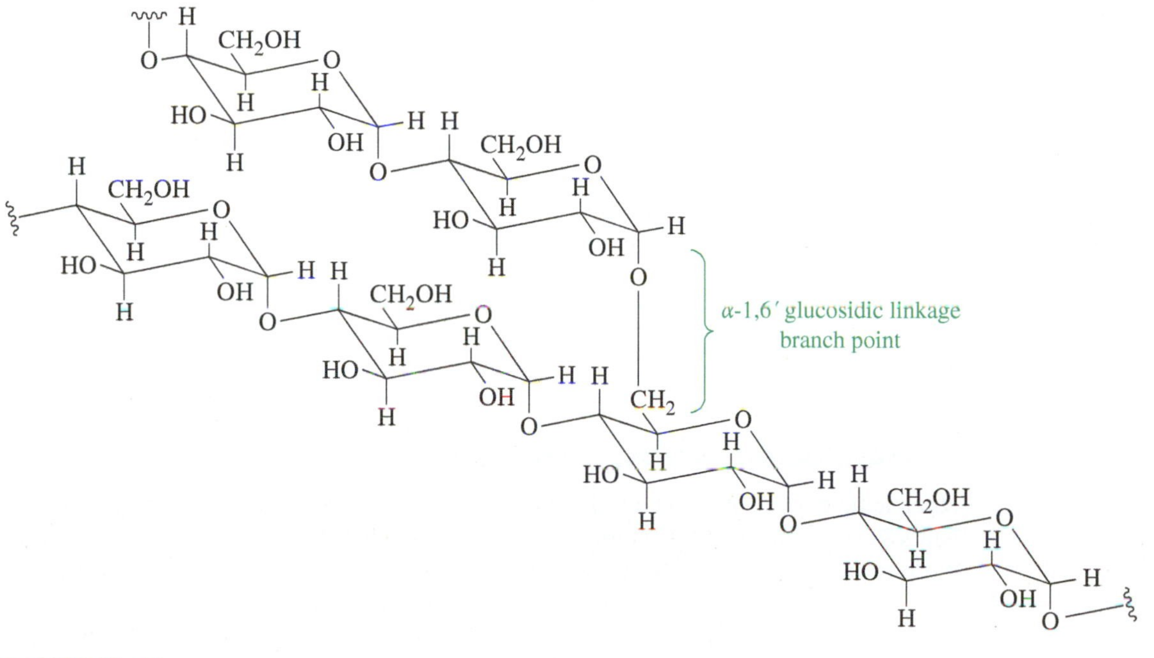

FIGURE 23-19
Partial structure of amylopectin. Amylopectin is a branched α-1,4′ polymer of glucose. At the branch points, there is a single α-1,6′ linkage that provides the attachment point for another chain. Glycogen has a similar structure, except that its branching is more extensive.

This cicada is shedding its nymphal exoskeleton. Chitin lends strength and rigidity to the exoskeletons of insects, but it cannot grow and change shape with the insect.

The structure of glycogen is similar to that of amylopectin, but with more extensive branching. The highly branched structure of glycogen leaves many end groups available for quick hydrolysis to provide glucose needed for metabolism.

23-13C Chitin: A Polymer of *N*-Acetylglucosamine

Chitin (pronounced *kī'-t'n*, rhymes with Titan) forms the exoskeletons of insects. In crustaceans, chitin forms a matrix that binds calcium carbonate crystals into the exoskeleton. Chitin is different from the other carbohydrates we have studied. It is a polymer of *N*-acetylglucosamine, an **amino sugar** (actually an amide) that is common in living organisms. In *N*-acetylglucosamine, the hydroxy group on C2 of glucose is replaced by an amino group (forming glucosamine), and that amino group is acetylated.

N-Acetylglucosamine, or 2-acetamido-2-deoxy-D-glucose

Chitin is bonded like cellulose, except using *N*-acetylglucosamine instead of glucose. Like other amides, *N*-acetylglucosamine forms exceptionally strong hydrogen bonds between the amide carbonyl groups and N—H protons. The glycosidic bonds are β-1,4′ links, giving chitin structural rigidity, strength, and stability that exceed even those of cellulose. Unfortunately, this strong, rigid polymer cannot easily expand, so it must be shed periodically by molting as the animal grows.

Chitin, or poly (1,4′-O-β-2-acetamido-2-deoxy-D-glucopyranoside), a β-1,4-linked polymer of N-acetylglucosamine

β-glycosidic linkage

Application: Insecticide

Chitin synthase inhibitors are used commercially as insecticides because they prevent the formation of a new exoskeleton and the shedding of the old one. The insect becomes trapped in an old exoskeleton that cannot grow. These inhibitors are highly toxic to insects and crustaceans, but relatively nontoxic to mammals. The most common chitin synthase inhibitors are substituted benzoyl-ureas such as diflubenzuron, which was first registered as an insecticide in 1976.

diflubenzuron or Dimilin™

23-14 Nucleic Acids: Introduction

Nucleic acids are substituted polymers of the aldopentose ribose that carry an organism's genetic information. A tiny amount of DNA in a fertilized egg cell determines the physical characteristics of the fully developed animal. The difference between a frog and a human is encoded in a relatively small part of this DNA. Each cell carries a complete set of genetic instructions that determine the type of cell, what its function will be, when it will grow and divide, and how it will synthesize all the structural proteins, enzymes, fats, carbohydrates, and other substances the cell and the organism need to survive.

The two major classes of nucleic acids are **ribonucleic acids (RNA)** and **deoxyribonucleic acids (DNA).** In a typical cell, DNA is found primarily in the nucleus, where it carries the permanent genetic code. The molecules of DNA are huge, with molecular weights up to 50 billion. When the cell divides, DNA replicates to form two copies for the daughter cells. DNA is relatively stable, providing a medium for transmission of genetic information from one generation to the next.

RNA molecules are typically much smaller than DNA molecules, and they are more easily hydrolyzed and broken down. RNA commonly serves as a working copy of the nuclear DNA being decoded. Nuclear DNA directs the synthesis of *messenger RNA*, which leaves the nucleus to serve as a template for the construction of protein molecules in the ribosomes. After it has served its purpose, the messenger RNA is then enzymatically cleaved to its component parts, which become available for assembly into new RNA molecules to direct other syntheses.

The backbone of a nucleic acid is a polymer of ribofuranoside rings (five-membered rings of the sugar ribose) linked by phosphate ester groups. Each ribose unit carries a heterocyclic *base* that provides part of the information needed to specify a particular amino acid in protein synthesis. Figure 23-20 shows the ribose-phosphate backbone of RNA.

DNA and RNA each contain four monomers, called **nucleotides,** that differ in the structure of the bases bonded to the ribose units. Yet this deceptively simple structure encodes complex information just as the 0 and 1 bits used by a computer encode complex programs. First we consider the structure of individual nucleotides, then the bonding of these monomers into single-stranded nucleic acids, and finally the base pairing that binds two strands into the double helix of nuclear DNA.

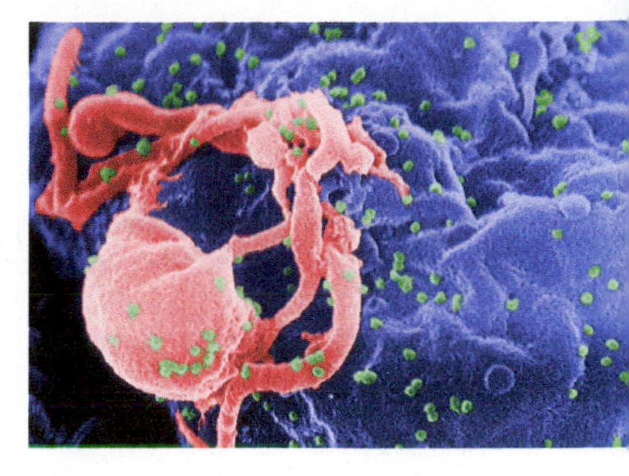

HIV (the AIDS virus, in green) is shown here attacking a T-4 lymphocyte (in red). HIV is an RNA virus whose genetic material must be translated to DNA before inserting itself into the host cell's DNA. Several of the anti-AIDS drugs are directed toward stopping this reverse transcription of RNA to DNA. (Magnification 1000X)

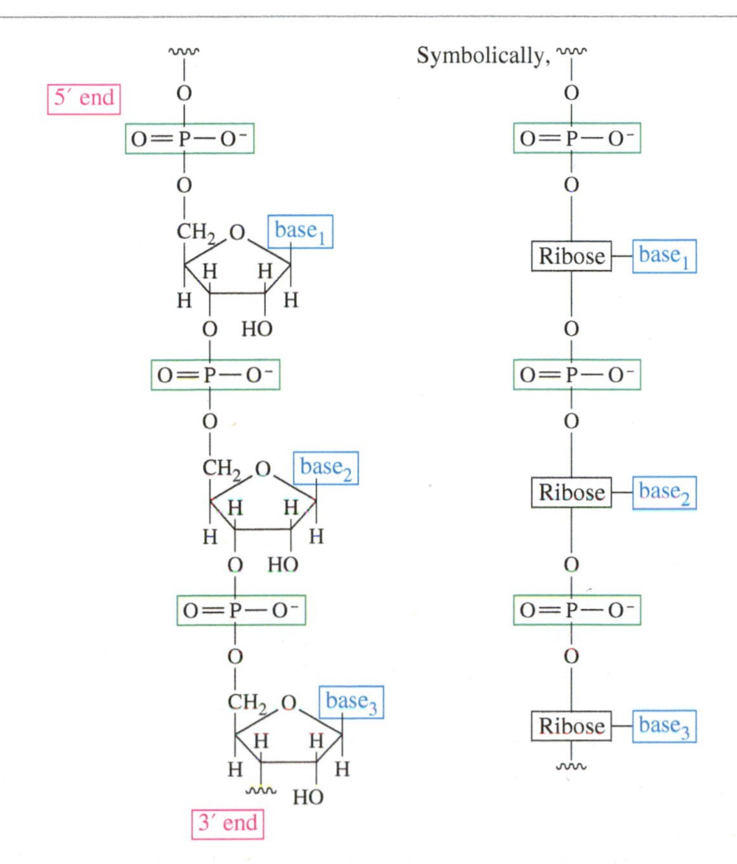

FIGURE 23-20
A short segment of the RNA polymer. Nucleic acids are assembled on a backbone made up of ribofuranoside units linked by phosphate esters.

23-15 Ribonucleosides and Ribonucleotides

Ribonucleosides are components of RNA based on glycosides of the furanose form of D-ribose. We have seen (Section 23-8) that a glycoside may have an aglycone (the substituent on the anomeric carbon) bonded by a nitrogen atom. A ribonucleoside is a β-D-ribofuranoside (a β-glucosidase of D-ribofuranose) whose aglycone is a heterocyclic nitrogen base. The following structures show the open-chain and furanose forms of ribose, and a ribonucleoside with a generic base bonded through a nitrogen atom.

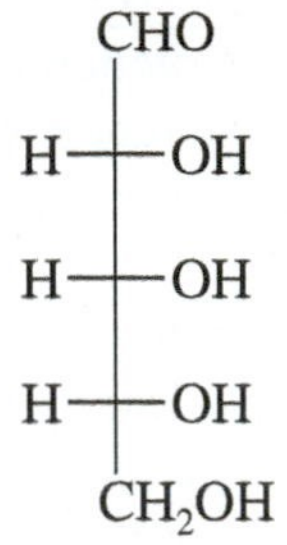

Application: Gout

Mammals metabolize purines to form several waste products, including uric acid. Gout is caused by elevated levels of uric acid in the body, causing crystals of urate salts to precipitate in the joints.

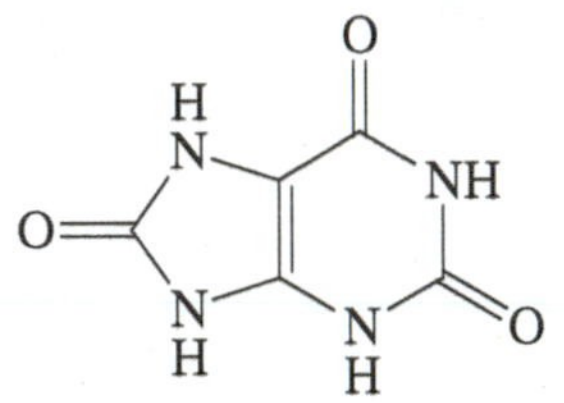

uric acid

D-ribose

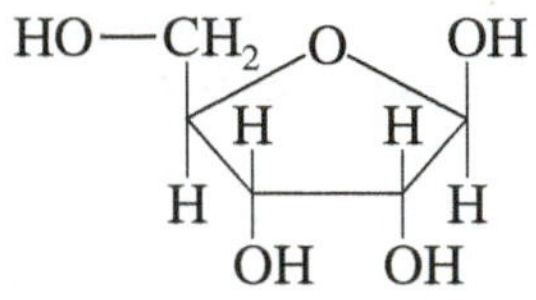

β-D-ribofuranose

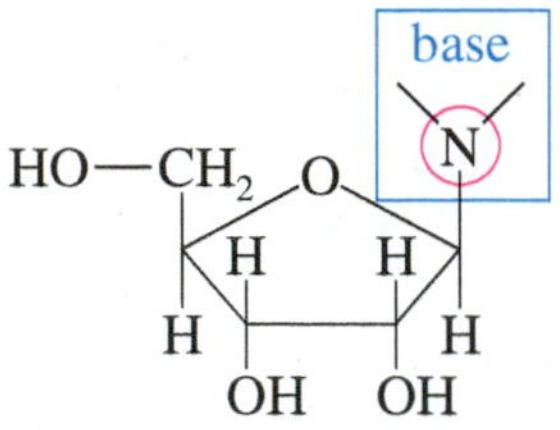

a ribonucleoside

The four bases commonly found in RNA are divided into two classes: The monocyclic compounds cytosine and uracil are called *pyrimidine bases* because they resemble substituted pyrimidines, and the bicyclic compounds adenine and guanine are called *purine bases* because they resemble the bicyclic heterocycle purine (Section 16-9C).

pyrimidine 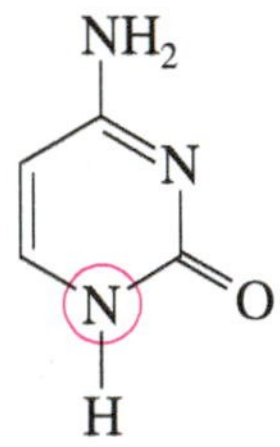　cytosine (C) 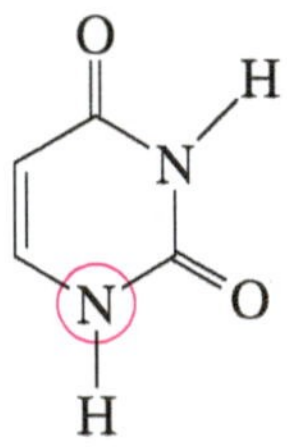　uracil (U)　adenine (A) 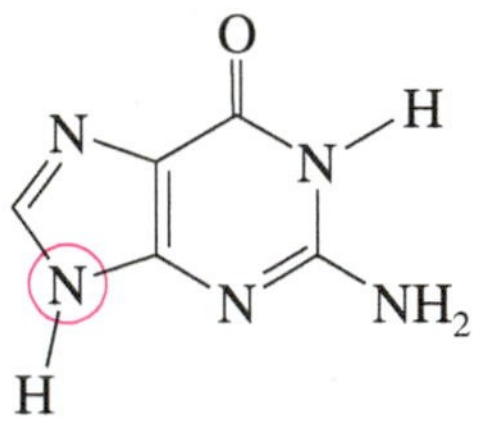　guanine (G) 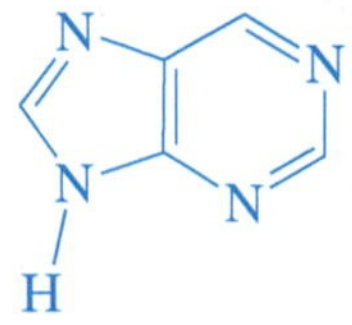　purine

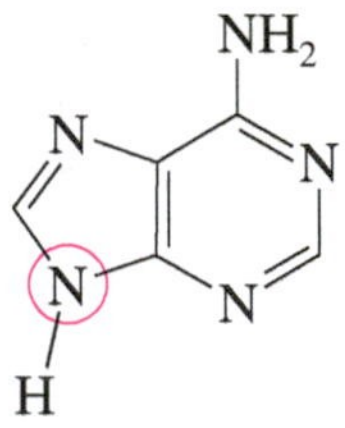

pyrimidine bases　purine bases

When bonded to ribose through the circled nitrogen atoms, the four heterocyclic bases make up the four ribonucleosides cytidine, uridine, adenosine, and guanosine (Figure 23-21). Notice that the two ring systems (the base and the sugar) are numbered separately, and the carbons of the sugar are given primed numbers. For example, the 3′ carbon of cytidine is C3 of the ribose ring.

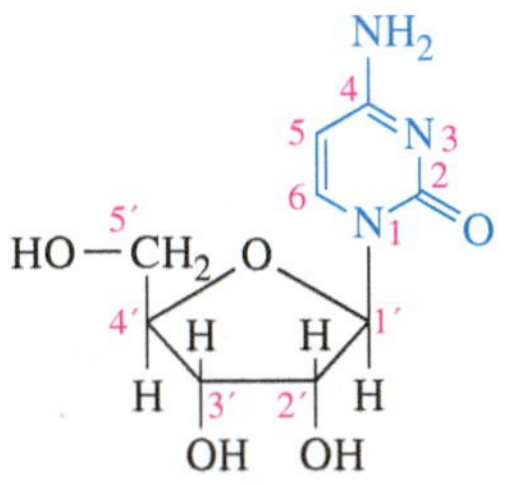

cytidine (C) 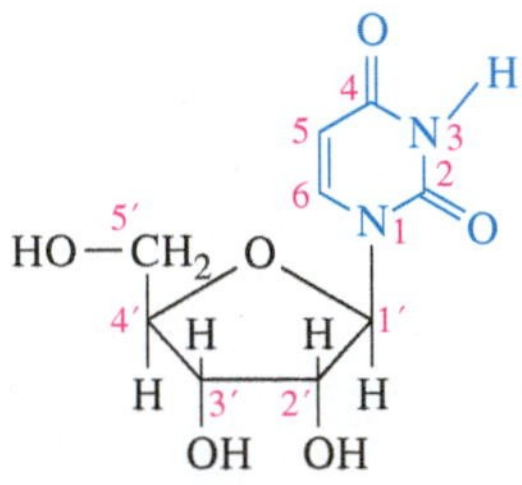　uridine (U) 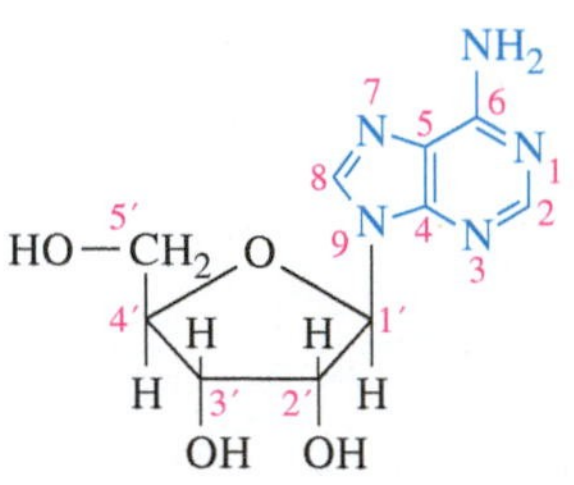　adenosine (A) 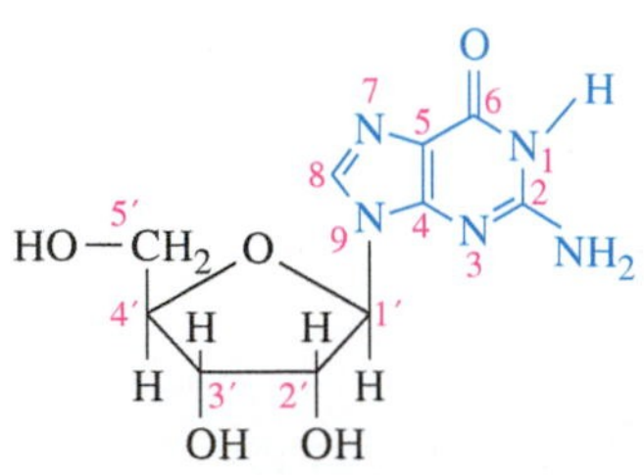　guanosine (G)

FIGURE 23-21
The four common ribonucleosides are cytidine, uridine, adenosine, and guanosine.

PROBLEM 23-43

Cytosine, uracil, and guanine have tautomeric forms with aromatic hydroxy groups. Draw these tautomeric forms.

PROBLEM 23-44

(a) An aliphatic aminoglycoside is relatively stable to base, but it is quickly hydrolyzed by dilute acid. Propose a mechanism for the acid-catalyzed hydrolysis.

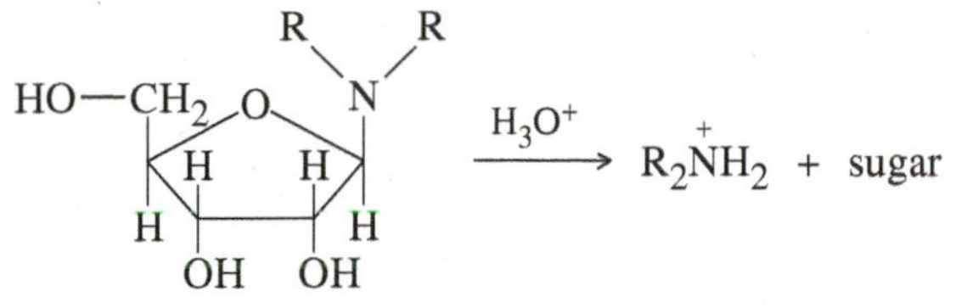

$$\text{an aliphatic riboside} \xrightarrow{H_3O^+} R_2\overset{+}{N}H_2 + \text{sugar}$$

an aliphatic riboside

(b) Ribonucleosides are not so easily hydrolyzed, requiring relatively strong acid. Using your mechanism for part (a), show why cytidine and adenosine (for example) are not so readily hydrolyzed. Explain why this stability is important for living organisms.

Ribonucleotides Ribonucleic acid consists of ribonucleosides bonded together into a polymer. This polymer cannot be bonded by glycosidic linkages like those of other polysaccharides because the glycosidic bonds are already used to attach the heterocyclic bases. Instead, the ribonucleoside units are linked by phosphate esters. The 5′-hydroxy group of each ribofuranoside is esterified by phosphoric acid. A ribonucleoside that is phosphorylated at its 5′ carbon is called a **ribonucleotide** ("tied" to phosphate). The four common ribonucleotides, shown in Figure 23-22, are simply phosphorylated versions of the four common ribonucleosides.

The phosphate groups of these ribonucleotides can exist in any of three ionization states, depending on the pH of the solution. At the nearly neutral pH of most organisms (pH = 7.4), there is one proton on the phosphate group. By convention, however, these groups are usually written completely ionized.

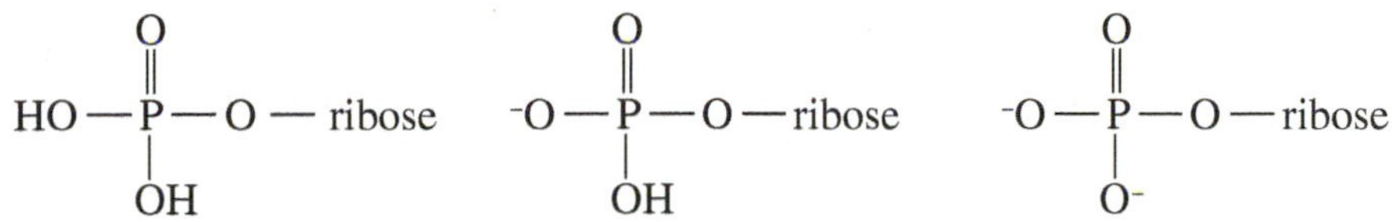

$HO{-}P(=O)(OH){-}O{-}\text{ribose}$ — in acid

$^{-}O{-}P(=O)(OH){-}O{-}\text{ribose}$ — nearly neutral

$^{-}O{-}P(=O)(O^{-}){-}O{-}\text{ribose}$ — in base (usually written)

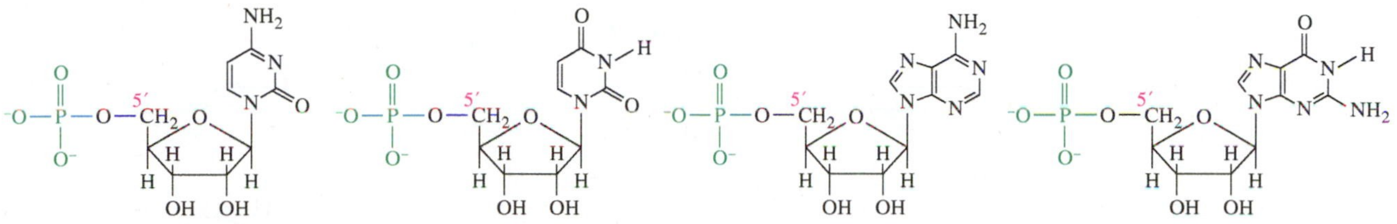

cytidine monophosphate, CMP (cytidylic acid)

uridine monophosphate, UMP (uridylic acid)

adenosine monophosphate, AMP (adenylic acid)

guanosine monophosphate, GMP (guanidylic acid)

FIGURE 23-22
Four common ribonucleotides. These are ribonucleosides esterified by phosphoric acid at their 5′ position, the $-CH_2OH$ at the end of the ribose chain.

23-16 The Structures of RNA and DNA

Ribonucleic acid (RNA) and deoxyribonucleic acid (DNA) are both biopolymers of nucleic acids, but they have minor structural differences that lead to major functional differences. All living cells use DNA as the primary genetic material that is passed from one generation to another. DNA directs and controls the synthesis of RNA, which serves as a short-lived copy of part of the much larger DNA molecule. Then, the cellular machinery translates the nucleotide sequence of the RNA molecule into a sequence of amino acids needed to make a protein.

23-16A The Structure of Ribonucleic Acid

Figure 23-23 shows how the individual ribonucleotide units are bonded into the RNA polymer. Each nucleotide has a phosphate group on its 5′ carbon (the end carbon of ribose) and a hydroxy group on the 3′ carbon. Two nucleotides are joined by a phosphate ester linkage between the 5′-phosphate group of one nucleotide and the 3′-phosphate group of another.

The RNA polymer consists of many nucleotide units bonded this way, with a phosphate ester linking the 5′ end of one nucleoside to the 3′ end of another. A molecule of RNA always has two ends (unless it is in the form of a large ring). One end has a free 3′ group, and the other end has a free 5′ group. We refer to the ends as the 3′ *end* and the 5′ *end*, and we refer to directions of replication as the 3′ → 5′ *direction* and the 5′ → 3′ *direction*. Figures 23-20 and 23-23 show short segments of RNA with the 3′ end and the 5′ end labeled.

23-16B Deoxyribose and the Structure of Deoxyribonucleic Acid

All our descriptions of ribonucleosides, ribonucleotides, and ribonucleic acid also apply to the components of DNA. The principal difference between RNA and DNA is the presence of D-2-deoxyribose as the sugar in DNA instead of the D-ribose found in RNA. The prefix *deoxy*- means that an oxygen atom is missing, and the number 2 means it is missing from C2.

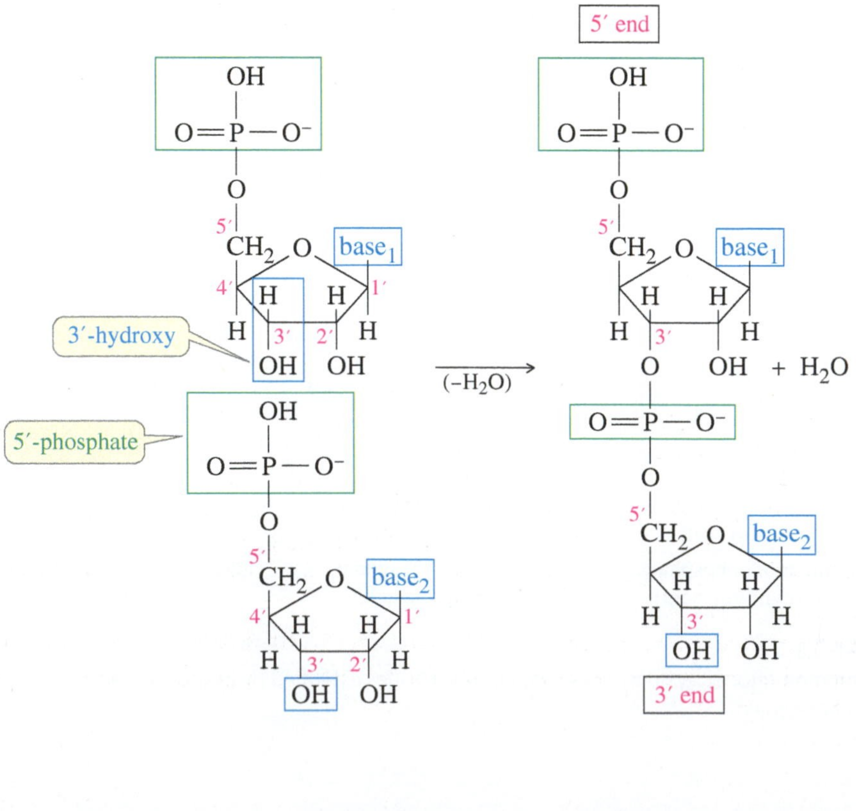

FIGURE 23-23
Phosphate linkage of nucleotides in RNA. Two nucleotides are joined by a phosphate linkage between the 5′-phosphate group of one and the 3′-hydroxy group of the other.

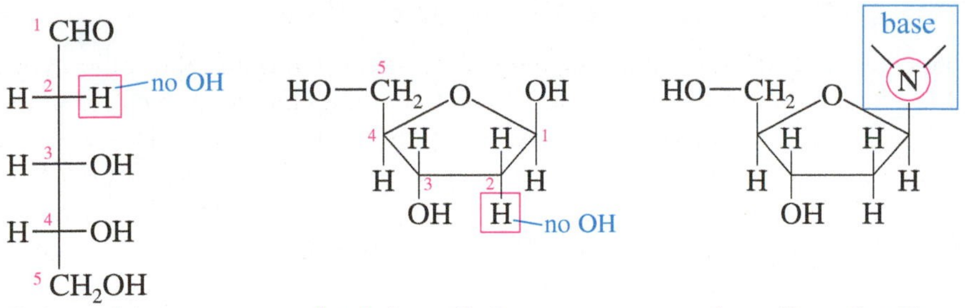

D-2-deoxyribose β-D-2-deoxyribofuranose a deoxyribonucleoside

Another key difference between RNA and DNA is the presence of thymine in DNA instead of the uracil in RNA. Thymine is simply uracil with an additional methyl group. The four common bases of DNA are cytosine, thymine, adenine, and guanine.

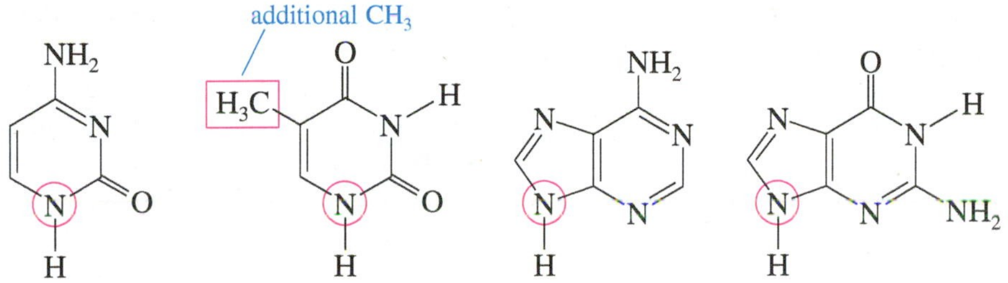

cytosine (C) thymine (T) adenine (A) guanine (G)

pyrimidine bases purine bases

These four bases are incorporated into deoxyribonucleosides and deoxyribonucleotides similar to the bases in ribonucleosides and ribonucleotides. The following structures show the common nucleosides that make up DNA. The corresponding nucleotides are simply the same structures with phosphate groups at the 5′ positions.

The structure of the DNA polymer is similar to that of RNA, except there are no hydroxy groups on the 2′ carbon atoms of the ribose rings. The alternating deoxyribose rings and phosphates act as the backbone, while the bases attached to the deoxyribose units carry the genetic information. The sequence of nucleotides is called the **primary structure** of the DNA strand.

Four common deoxyribonucleosides that make up DNA

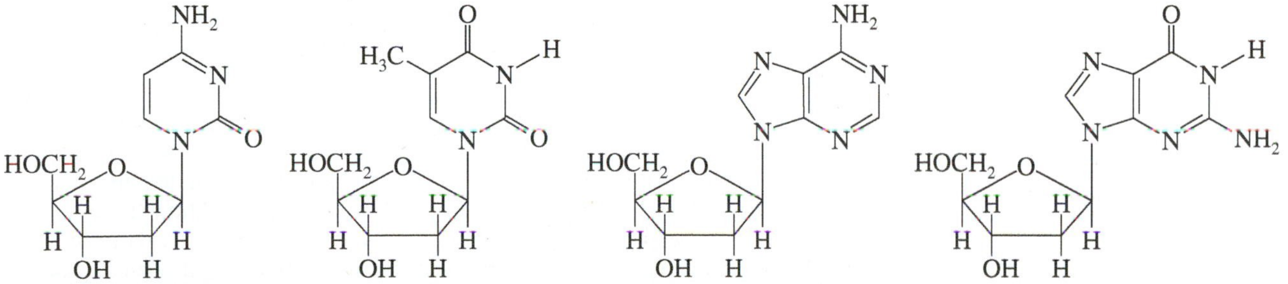

deoxycytidine deoxythymidine deoxyadenosine deoxyguanosine

23-16C Base Pairing

Having discussed the primary structure of DNA and RNA, we now consider how the nucleotide sequence is reproduced or transcribed into another molecule. This information transfer takes place by an interesting hydrogen-bonding interaction between specific pairs of bases.

Each pyrimidine base forms a stable hydrogen-bonded pair with only one of the two purine bases (Figure 23-24). Cytosine forms a base pair, joined by three hydrogen

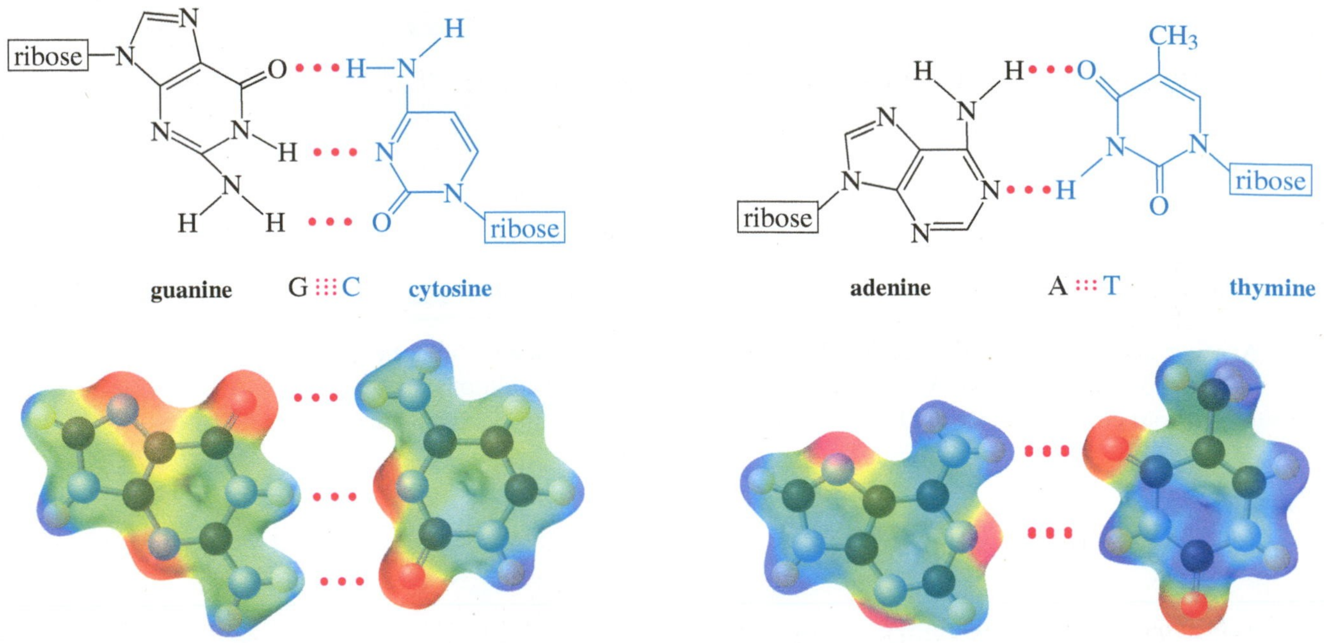

guanine G:::C cytosine

adenine A:::T thymine

FIGURE 23-24
Base pairing in DNA and RNA. Each purine base forms a stable hydrogen-bonded pair with a specific pyrimidine base. Guanine forms a base pair with three hydrogen bonds to cytosine, and adenine forms a base pair with two hydrogen bonds to thymine (or uracil in RNA). The electrostatic potential maps show that hydrogen bonding takes place between electron-poor hydrogen atoms (blue and purple regions) and electron-rich nitrogen or oxygen atoms (red regions). (In these drawings, "ribose" means β-D-2-deoxyribofuranoside in DNA and β-D-ribofuranoside in RNA.)

bonds, with guanine. Thymine (or uracil in RNA) forms a base pair with adenine, joined by two hydrogen bonds. The lengths of all these hydrogen bonds have to be the same in order to form a stable double-stranded structure. Guanine is said to be *complementary* to cytosine, and adenine is complementary to thymine. This base pairing was first suspected in 1950, when Erwin Chargaff of Columbia University noticed that various DNAs, taken from a wide variety of species, had about equal amounts of adenine and thymine and about equal amounts of guanine and cytosine.

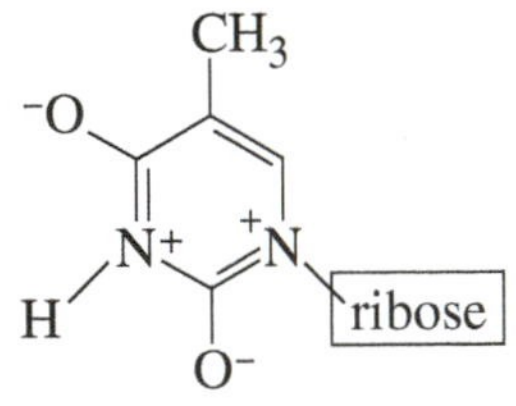

thymine

PROBLEM 23-45

All of the rings of the four heterocyclic bases are aromatic. This is more apparent when the polar resonance forms of the amide groups are drawn, as is done for thymine at left. Redraw the hydrogen-bonded guanine-cytosine and adenine-thymine pairs shown in Figure 23-24, using the polar resonance forms of the amides. Show how these forms help to explain why the hydrogen bonds involved in these pairings are particularly strong. Remember that a hydrogen bond arises between an electron-deficient hydrogen atom and an electron-rich pair of nonbonding electrons.

23-16D The Double Helix of DNA

In 1953, James D. Watson and Francis C. Crick used X-ray diffraction patterns of DNA fibers to determine the molecular structure and conformation of DNA. They found that DNA contains two complementary polynucleotide chains held together by hydrogen bonds between the paired bases. Figure 23-25 shows a portion of the double strand of DNA, with each base paired with its complement. The two strands are *antiparallel*: One strand is arranged $3' \rightarrow 5'$ from left to right, while the other runs in the opposite direction, $5' \rightarrow 3'$ from left to right.

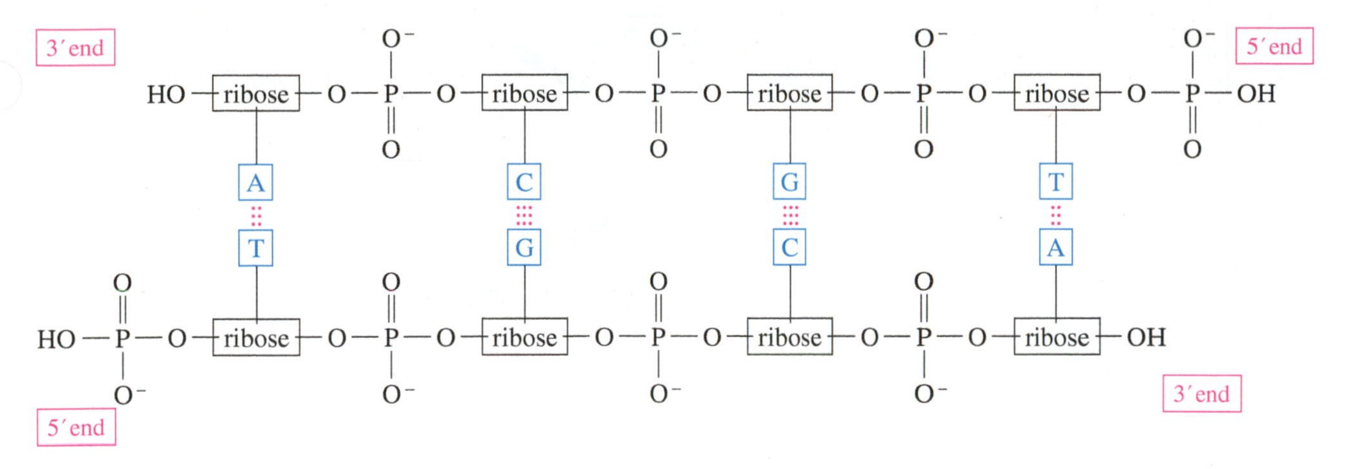

FIGURE 23-25
Antiparallel strands of DNA. DNA usually consists of two complementary strands, with all the base pairs hydrogen bonded together. The two strands are antiparallel, running in opposite directions. (In these drawings of DNA, "ribose" means β-D-2-deoxyribofuranoside.)

Watson and Crick also found that the two complementary strands of DNA are coiled into a helical conformation about 20 Å in diameter, with both chains coiled around the same axis. The helix makes a complete turn for every ten residues, or about one turn in every 34 Å of length. Figure 23-26 shows the double helix of DNA. In this drawing, the two sugar-phosphate backbones form the vertical double helix with the heterocyclic bases stacked horizontally in the center. Attractive stacking forces between the pi clouds of the aromatic pyrimidine and purine bases are substantial, further helping to stabilize the helical arrangement.

When DNA undergoes replication (in preparation for cell division), an enzyme uncoils part of the double strand. Individual nucleotides naturally hydrogen bond to their complements on the uncoiled part of the original strand, and a *DNA polymerase* enzyme couples the nucleotides to form a new strand. This process is depicted schematically in Figure 23-27. A similar process transcribes DNA into a complementary molecule of messenger RNA for use by ribosomes as a template for protein synthesis.

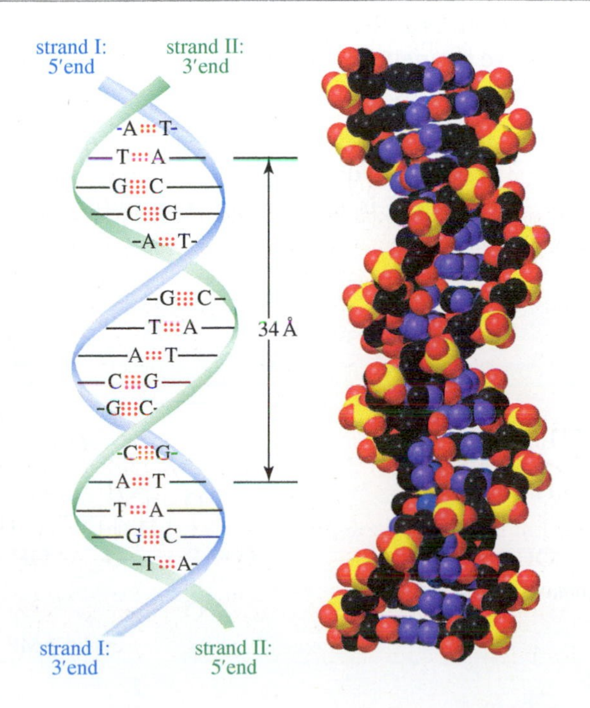

FIGURE 23-26
Double helix of DNA. Two complementary strands are joined by hydrogen bonds between the base pairs. This double strand coils into a helical arrangement.

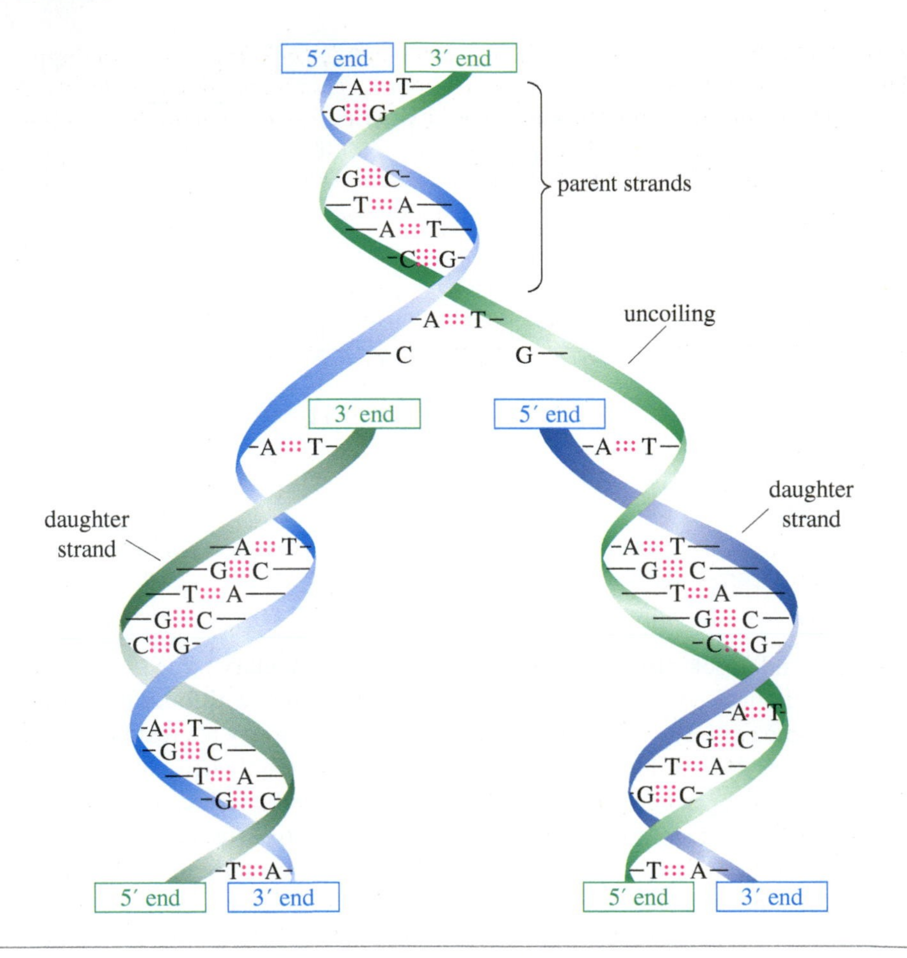

FIGURE 23-27
Replication of the double strand of DNA. A new strand is assembled on each of the original strands, with the DNA polymerase enzyme forming the phosphate ester bonds of the backbone.

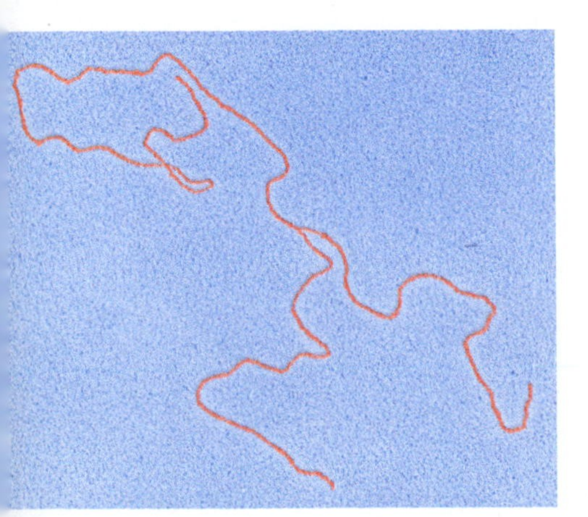

Electron micrograph of double-stranded DNA that has partially uncoiled to show the individual strands. (Magnification 13,000X)

A great deal is known about replication of DNA and translation of the DNA/RNA sequence of bases into proteins. These exciting aspects of nucleic acid chemistry are part of the field of *molecular biology*, and they are covered in detail in biochemistry courses.

23-17 Additional Functions of Nucleotides

We generally think of nucleotides as the monomers that form DNA and RNA, yet these versatile biomolecules serve a variety of additional functions. Here we briefly consider a few additional uses of nucleotides.

AMP: A Regulatory Hormone Adenosine monophosphate (AMP) also occurs in a cyclic form, where the 3′- and 5′-hydroxy groups are both esterified by the same phosphate group. This *cyclic AMP* is involved in transmitting and amplifying the chemical signals of other hormones.

Application: Genetic Disease

Adenosine deaminase replaces the C6 amino group with a hydroxy group, an important step in purine metabolism. A genetic deficiency of the enzyme causes a severe immunodeficiency, called "baby in a bubble syndrome." The child must live in a sterile environment until he or she can receive a bone marrow transplant to provide a new immune system.

adenosine monophosphate (AMP)

cyclic AMP

NAD: A Coenzyme Nicotinamide adenine dinucleotide (NAD) is one of the principal oxidation–reduction reagents in biological systems. This nucleotide has the structure of two D-ribose rings (a *di*nucleotide) linked by their 5′ phosphates. The aglycone of one ribose is nicotinamide, and the aglycone of the other is adenine. A dietary deficiency of nicotinic acid (niacin) leads to the disease called *pellagra*, caused by the inability to synthesize enough nicotinamide adenine dinucleotide.

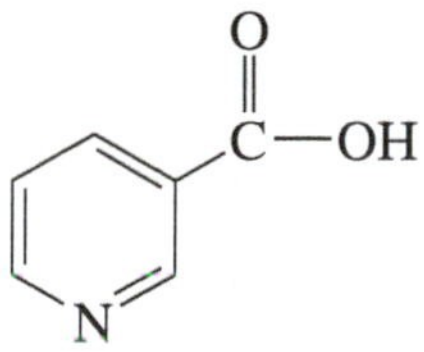

nicotinic acid (niacin)

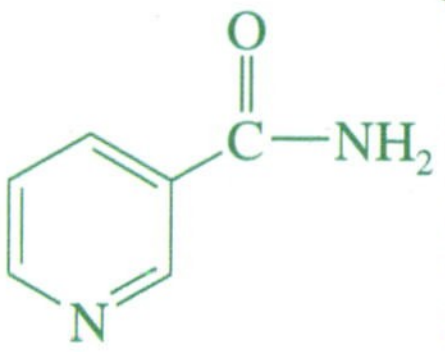

nicotinamide

adenine

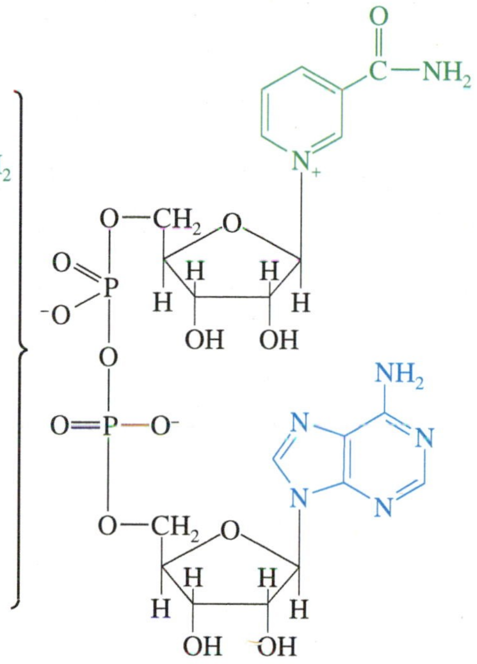

NAD^+
nicotinamide adenine dinucleotide

The following equation shows how NAD^+ serves as the oxidizing agent in the biological oxidation of an alcohol. Just the nicotinamide portion of NAD shown takes part in the reaction. The enzyme that catalyzes this reaction is called alcohol dehydrogenase (ADH).

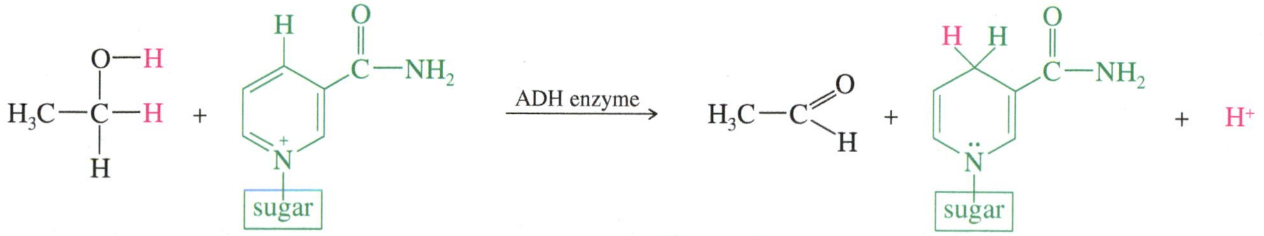

ethanol + NAD^+ —ADH enzyme→ acetaldehyde + NADH (reduced) + H^+

Application: Insomnia

Adenosine may also act as a neurotransmitter that induces sleep. Caffeine blocks the adenosine receptor, resulting in wakefulness or insomnia.

ATP: An Energy Source When glucose is oxidized in the living cell, the energy released is used to synthesize *adenosine triphosphate* (ATP), an anhydride of phosphoric acid. As with most anhydrides, hydrolysis of ATP is highly exothermic. The hydrolysis products are adenosine diphosphate (ADP) and inorganic phosphate.

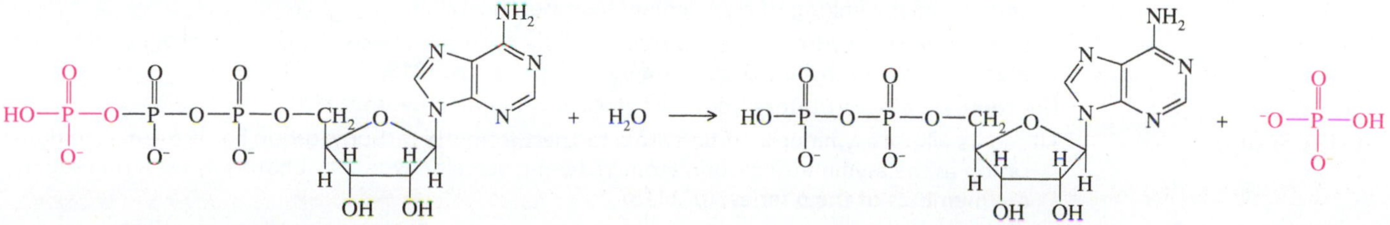

adenosine triphosphate (ATP) + H_2O ⟶ adenosine diphosphate (ADP) + phosphate

$\Delta H° = -31$ kJ/mol (−7.3 kcal/mol)

The highly exothermic nature of ATP hydrolysis is largely explained by the heats of hydration of the products. ADP is hydrated about as well as ATP, but inorganic phosphate has a large heat of hydration. Hydrolysis also reduces the electrostatic repulsion of the three negatively charged phosphate groups in ATP. Hydrolysis of adenosine triphosphate (ATP) liberates 31 kJ (7.3 kcal) of energy per mole of ATP. This is the energy that muscle cells use to contract and all cells use to drive their endothermic chemical processes.

Essential Terms

aglycone A nonsugar residue bonded to the anomeric carbon of a glycoside (the acetal form of a sugar). Aglycones are commonly bonded to the sugar through oxygen or nitrogen. (p. 1189)

aldaric acid **(glycaric acid, saccharic acid)** A dicarboxylic acid formed by oxidation of both end carbon atoms of a monosaccharide. (p. 1186)

alditol **(sugar alcohol)** A polyalcohol formed by reduction of the carbonyl group of a monosaccharide. (p. 1185)

aldonic acid **(glyconic acid)** A monocarboxylic acid formed by oxidation of the aldehyde group of an aldose. (p. 1186)

aldose A monosaccharide containing an aldehyde carbonyl group. (p. 1173)

amino sugar A sugar (such as glucosamine) in which a hydroxy group is replaced by an amino group. (p. 1206)

anomeric carbon The hemiacetal carbon in the cyclic form of a sugar (the carbonyl carbon in the open-chain form). The anomeric carbon is easily identified because it is the only carbon with two bonds to oxygen atoms. (p. 1183)

anomers Sugar stereoisomers that differ in configuration only at the anomeric carbon. Anomers are classified as α or β depending on whether the anomeric hydroxy group (or the aglycone in a glycoside) is trans (α) or cis (β) to the terminal $-CH_2OH$. (p. 1182)

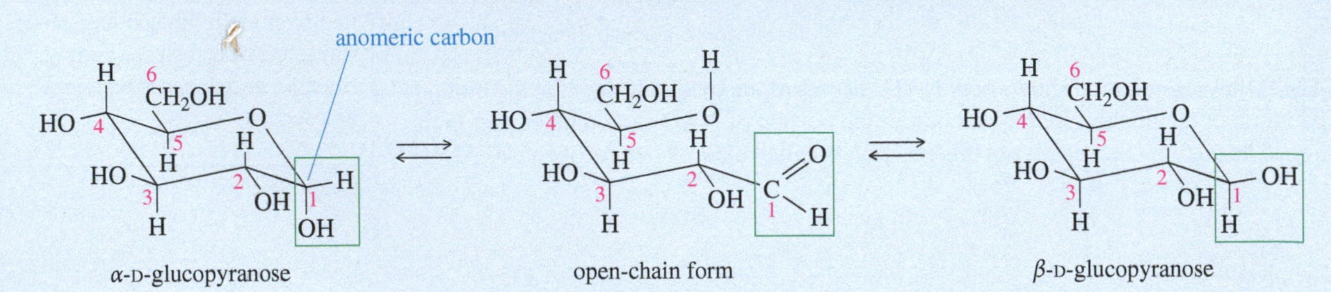

α-D-glucopyranose   open-chain form   β-D-glucopyranose

carbohydrates **(sugars)** Polyhydroxy aldehydes and ketones, including their derivatives and polymers. Many have formula $C_n(H_2O)_m$ from which they received the name "hydrates of carbon" or "carbohydrates." (p. 1173)

cellulose A linear β-1,4′ polymer of D-glucopyranose. Cellulose forms the cell walls of plants and is the major constituent of wood and cotton. (p. 1174)

chitin A β-1,4′ polymer of *N*-acetylglucosamine that lends strength and rigidity to the exoskeletons of insects and crustaceans. (p. 1206)

degradation A reaction that causes loss of a carbon atom. (p. 1193)

deoxyribonucleic acid **(DNA)** A biopolymer of deoxyribonucleotides that serves as a template for the synthesis of ribonucleic acid. DNA is also the template for its own replication, through uncoiling and the pairing and enzymatic linking of complementary bases. (p. 1207)

deoxy sugar A sugar in which a hydroxy group is replaced by a hydrogen. Deoxy sugars are recognized by the presence of a methylene group or a methyl group. (p. 1210)

dextrose The common dextrorotatory isomer of glucose, D-(+)-glucose. (p. 1184)

D series of sugars All sugars whose asymmetric carbon atom farthest from the carbonyl group has the same configuration as the asymmetric carbon atom in D-(+)-glyceraldehyde. Most naturally occurring sugars are members of the D series. (p. 1175)

disaccharide A carbohydrate whose hydrolysis gives two monosaccharide molecules. (p. 1174)

enediol rearrangement A base-catalyzed tautomerization that interconverts aldoses and ketoses with an enediol as an intermediate. This enolization also epimerizes C2 and other carbon atoms. (p. 1187)

epimers Two diastereomeric sugars differing only in the configuration at a single asymmetric carbon atom. The epimeric carbon atom is usually specified, as in "C4 epimers." If no epimeric carbon is specified, it is assumed to be C2. The interconversion of epimers is called **epimerization.** (pp. 1177, 1188)

erythro and threo Diastereomers having similar groups on the same side (erythro) or on opposite sides (threo) of the Fischer projection. This terminology was adapted from the names of the aldotetroses *erythrose* and *threose*. (p. 1177)

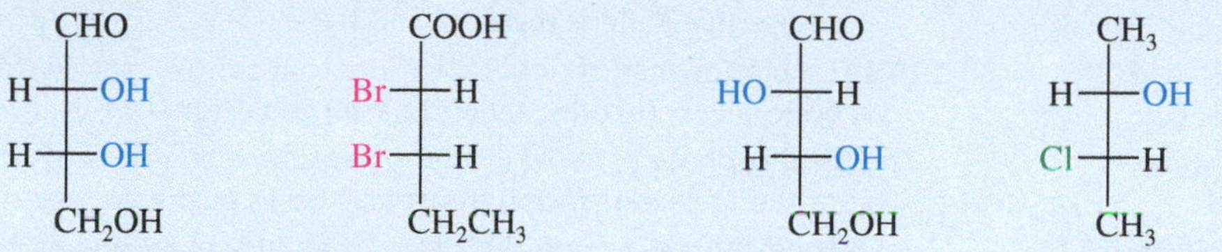

D-erythrose | *erythro*-2,3-dibromopentanoic acid | D-threose | *threo*-3-chlorobutan-2-ol

furanose A five-membered cyclic hemiacetal form of a sugar. (p. 1182)

furanoside A five-membered cyclic glycoside. (p. 1188)

glucoside A glycoside derived from glucose. (p. 1188)

glycoside A cyclic acetal form of a sugar. Glycosides are stable to base, and they are nonreducing sugars. Glycosides are generally **furanosides** (five-membered) or **pyranosides** (six-membered), and they exist in anomeric α and β forms. (p. 1188)

glycosidic linkage A general term for an acetal bond from an anomeric carbon joining two monosaccharide units. (pp. 1199, 1200)

- **glucosidic linkage:** A glycosidic linkage using an acetal bond from the anomeric carbon of glucose.
- **galactosidic linkage:** A glycosidic linkage using an acetal bond from the anomeric carbon of galactose.

Haworth projection A flat-ring representation of a cyclic sugar. The Haworth projection does not show the axial and equatorial positions of a pyranose, but it does show the cis and trans relationships. (p. 1179)

ketose A monosaccharide containing a ketone carbonyl group. (p. 1173)

Kiliani–Fischer synthesis A method for elongating an aldose at the aldehyde end. The aldose is converted into two epimeric aldoses with an additional carbon atom. For example, Kiliani–Fischer synthesis converts D-arabinose to a mixture of D-glucose and D-mannose. (p. 1194)

L series of sugars All sugars whose asymmetric carbon atom farthest from the carbonyl group has the same configuration as the asymmetric carbon atom in L-(−)-glyceraldehyde. Sugars of the L series are not common in nature. (p. 1175)

monosaccharide A carbohydrate that does not undergo hydrolysis of glycosidic bonds to give smaller sugar molecules. (p. 1173)

mutarotation A spontaneous change in optical rotation that occurs when a pure anomer of a sugar in its hemiacetal form equilibrates with the other anomer to give an equilibrium mixture with an averaged value of the optical rotation. (p. 1183)

nucleoside An *N*-glycoside of β-D-ribofuranose or β-D-deoxyribofuranose, where the aglycone is one of several derivatives of pyrimidine or purine. (p. 1208)

nucleotide A 5′-phosphate ester of a nucleoside. (p. 1207)

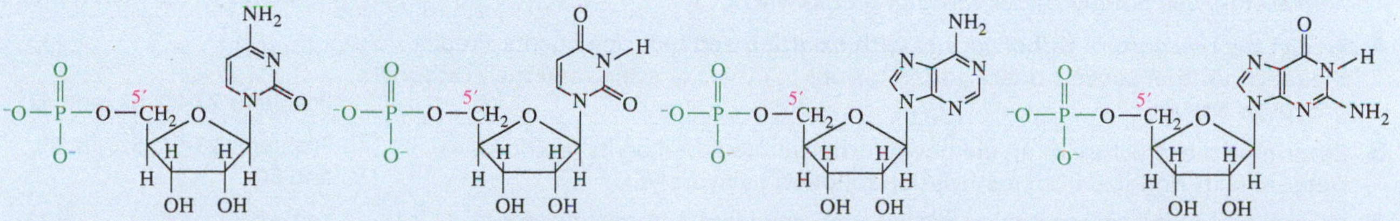

cytidine monophosphate, CMP (cytidylic acid) | uridine monophosphate, UMP (uridylic acid) | adenosine monophosphate, AMP (adenylic acid) | guanosine monophosphate, GMP (guanidylic acid)

The four common ribonucleotides of RNA

oligosaccharide A carbohydrate whose hydrolysis gives about two to ten monosaccharide units, but not as many as a polysaccharide. (p. 1203)

polysaccharide **(glycan)** A carbohydrate whose hydrolysis gives many monosaccharide molecules. (p. 1174)

primary structure The primary structure of a nucleic acid is the sequence of nucleotides forming the polymer. This sequence determines the genetic characteristics of the nucleic acid. (p. 1211)

pyranose A six-membered cyclic hemiacetal form of a sugar. (p. 1182)

pyranoside A six-membered cyclic glycoside. (p. 1188)

rayon A commercial fiber made from regenerated cellulose. (p. 1203)

reducing sugar Any sugar that gives a positive Tollens test. Both ketoses and aldoses (in their hemiacetal forms) give positive Tollens tests. (p. 1187)

ribonucleic acid **(RNA)** A biopolymer of ribonucleotides that controls the synthesis of proteins. The synthesis of RNA is generally controlled by and patterned after DNA in the cell. (p. 1207)

ribonucleotide The 5′-phosphate ester of a **ribonucleoside,** a component of RNA based on β-D-ribofuranose and containing one of four heterocyclic bases as the aglycone. (p. 1209)

Ruff degradation A method for shortening the chain of an aldose by one carbon atom by treatment with bromine water, followed by hydrogen peroxide and $Fe_2(SO_4)_3$. (p. 1193)

starches A class of α-1,4′ polymers of glucose used for carbohydrate storage in plants and animals. (p. 1204)

- **amylose:** A linear α-1,4′ polymer of D-glucopyranose used for carbohydrate storage in plants.
- **amylopectin:** A branched α-1,4′ polymer of D-glucopyranose used for carbohydrate storage in plants. Branching occurs at α-1,6′ glycosidic linkages.
- **glycogen:** An extensively branched α-1,4′ polymer of D-glucopyranose used for carbohydrate storage in animals. Branching occurs at α-1,6′ glycosidic linkages.

sugar **(saccharide)** A generic term for low-molecular-weight carbohydrates (mostly monosaccharides and disaccharides), especially those that are found in foods. A **simple sugar** is a monosaccharide. (p. 1173)

Tollens test A test for reducing sugars, employing the same silver–ammonia complex used as a test for aldehydes. A positive test gives a silver precipitate, often in the form of a silver mirror. Tollens reagent is basic, and it promotes enediol rearrangements that interconvert ketoses and aldoses. Therefore, both aldoses and ketoses give positive Tollens tests if they are in their hemiacetal forms, in equilibrium with open-chain carbonyl structures. (p. 1187)

Essential Problem-Solving Skills in Chapter 23

Each skill is followed by problem numbers exemplifying that particular skill.

1	Draw the Fischer projections and the chair conformations of the anomers and epimers of glucose from memory. Identify and name these sugars based on how they differ from the structure of glucose.	Problems 23-46 and 47
2	Correctly name monosaccharides and disaccharides, and draw their structures from their names.	Problems 23-47, 48, 49, 53, and 54
3	Predict which carbohydrates mutarotate, which reduce Tollens reagent, and which undergo epimerization and isomerization under basic conditions. (Those with free hemiacetals will, but glycosides with full acetals will not.)	Problems 23-52, 56, 57, and 60
4	Predict the reactions of carbohydrates with oxidizing and reducing agents. Predict the reactions that convert their hydroxy groups to ethers or esters, and their carbonyl groups to acetals.	Problems 23-50, 51, and 52
5	Determine the structure of an unknown carbohydrate based on its reactions. Determine its ring size from methylation followed by hydrolysis.	Problems 23-51, 57, 58, 59, and 60
6	Draw the common types of glycosidic linkages, and identify these linkages in disaccharides and polysaccharides.	Problems 23-53, 54, 55, and 58
7	Recognize the structures of DNA and RNA, and draw the structures of the common ribonucleotides and deoxyribonucleotides.	Problems 23-63, 64, and 66

Study Problems

23-46 Glucose is the most abundant monosaccharide. From memory, draw glucose in
(a) the Fischer projection of the open chain.
(b) the most stable chair conformation of the most stable pyranose anomer.
(c) the Haworth projection of the most stable pyranose anomer.

23-47 Without referring to the chapter, draw the chair conformations of
(a) β-D-mannopyranose (the C2 epimer of glucose).
(b) α-D-allopyranose (the C3 epimer of glucose).
(c) β-D-galactopyranose (the C4 epimer of glucose).
(d) *N*-acetylglucosamine, glucose with the C2 oxygen atom replaced by an acetylated amino group.

23-48 Use Figure 23-3 (the D family of aldoses) to name the following aldoses.
(a) the C2 epimer of D-arabinose **(b)** the C3 epimer of D-mannose **(c)** the C3 epimer of D-threose
(d) the enantiomer of D-galactose **(e)** the C5 epimer of D-glucose

23-49 Classify the following monosaccharides. (*Examples:* D-aldohexose, L-ketotetrose.)
(a) (+)-glucose **(b)** (−)-arabinose **(c)** L-fructose

(d)
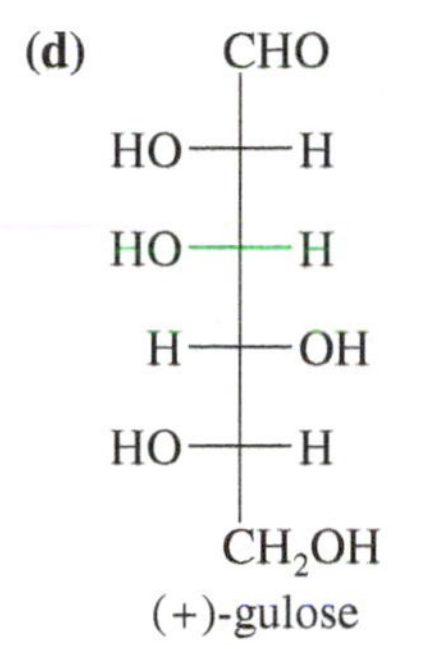

(+)-gulose

(e)
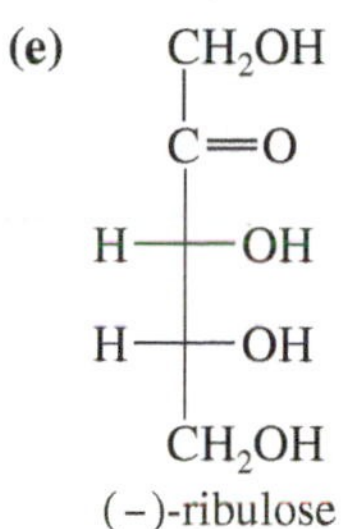

(−)-ribulose

(f)
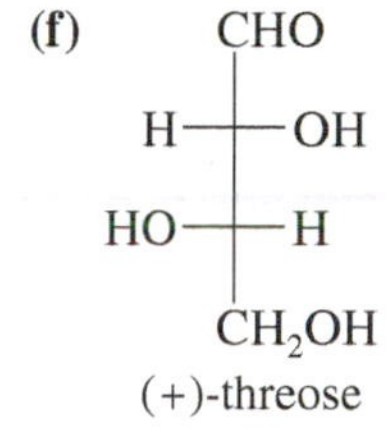

(+)-threose

(g)
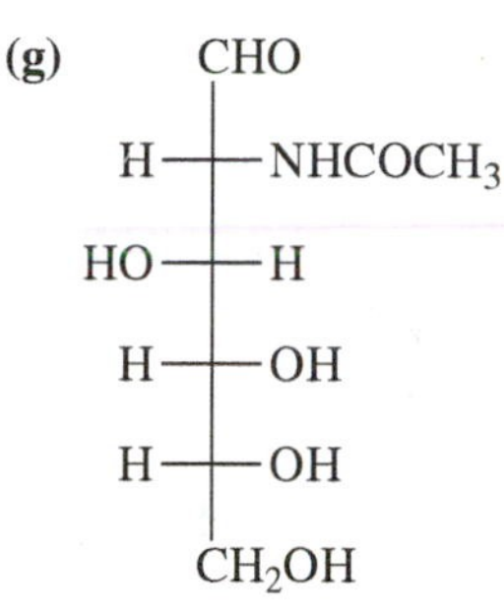

N-acetylglucosamine

23-50 **(a)** Give the products expected when (+)-glyceraldehyde reacts with HCN.
(b) What is the relationship between the products? How might they be separated?
(c) Are the products optically active? Explain.

23-51 The relative configurations of the stereoisomers of tartaric acid were established by the following syntheses:
(1) D-(+)-glyceraldehyde $\xrightarrow{\text{HCN}}$ diastereomers **A** and **B** (separated)
(2) Hydrolysis of **A** and **B** using aqueous $Ba(OH)_2$ gave **C** and **D,** respectively.
(3) HNO_3 oxidation of **C** and **D** gave (−)-tartaric acid and *meso*-tartaric acid, respectively.
(a) You know the absolute configuration of D-(+)-glyceraldehyde. Use Fischer projections to show the absolute configurations of products **A, B, C,** and **D.**
(b) Show the absolute configurations of the three stereoisomers of tartaric acid: (+)-tartaric acid, (−)-tartaric acid, and *meso*-tartaric acid.

23-52 Predict the products obtained when D-galactose reacts with each reagent.
(a) Br_2 and H_2O **(b)** NaOH, H_2O **(c)** CH_3OH, H^+ **(d)** $Ag(NH_3)_2^+ \ ^-OH$
(e) H_2, Ni **(f)** excess Ac_2O and pyridine **(g)** excess CH_3I, Ag_2O **(h)** $NaBH_4$
(i) Br_2, H_2O, then H_2O_2 and $Fe_2(SO_4)_3$ **(j)** (1) KCN/HCN; (2) H_2, Pd/$BaSO_4$; (3) H_3O^+ **(k)** excess HIO_4

23-53 Draw the following sugar derivatives.
(a) methyl β-D-glucopyranoside **(b)** 2,3,4,6-tetra-*O*-methyl-D-mannopyranose
(c) 1,3,6-tri-*O*-methyl-D-fructofuranose **(d)** methyl 2,3,4,6-tetra-*O*-methyl-β-D-galactopyranoside

23-54 Draw the structures (using chair conformations of pyranoses) of the following disaccharides.
(a) 4-*O*-(α-D-glucopyranosyl)-D-galactopyranose
(b) α-D-fructofuranosyl-β-D-mannopyranoside
(c) 6-*O*-(β-D-galactopyranosyl)-D-glucopyranose

23-55 Erwin Chargaff's discovery that DNA contains equimolar amounts of guanine and cytosine and also equimolar amounts of adenine and thymine has come to be known as *Chargaff's rule*:

$$G = C \text{ and } A = T$$

(a) Does Chargaff's rule imply that equal amounts of guanine and adenine are present in DNA? That is, does G = A?
(b) Does Chargaff's rule imply that the sum of the purine residues equals the sum of the pyrimidine residues? That is, does A + G = C + T?
(c) Does Chargaff's rule apply only to double-stranded DNA, or would it also apply to each individual strand if the double helical strand were separated into its two complementary strands?

23-56 Which of the sugars mentioned in Problems 23-53 and 23-54 are reducing sugars? Which ones would undergo mutarotation?

23-57 After a series of Kiliani–Fischer syntheses on (+)-glyceraldehyde, an unknown sugar is isolated from the reaction mixture. The following experimental information is obtained:

(1) Molecular formula $C_6H_{12}O_6$
(2) Undergoes mutarotation.
(3) Reacts with bromine water to give an aldonic acid.
(4) Reacts with HNO_3 to give an optically active aldaric acid.
(5) Ruff degradation followed by HNO_3 oxidation gives an optically inactive aldaric acid.
(6) Two Ruff degradations followed by HNO_3 oxidation give *meso*-tartaric acid.
(7) When the original sugar is treated with CH_3I and Ag_2O, a pentamethyl derivative is formed. Hydrolysis gives a tetramethyl derivative with a free hydroxy group on C5.

(a) Draw a Fischer projection for the open-chain form of this unknown sugar. Use Figure 23-3 to name the sugar.
(b) Draw the most stable conformation of the most stable cyclic hemiacetal form of this sugar, and give the structure a complete systematic name.

23-58 An unknown reducing disaccharide is found to be unaffected by invertase enzymes. Treatment with an α-galactosidase cleaves the disaccharide to give one molecule of D-fructose and one molecule of D-galactose. When the disaccharide is treated with excess iodomethane and silver oxide and then hydrolyzed in dilute acid, the products are 2,3,4,6-tetra-*O*-methylgalactose and 1,3,4-tri-*O*-methylfructose. Propose a structure for this disaccharide, and give its complete systematic name.

23-59 **(a)** Which of the D-aldopentoses will give optically active aldaric acids on oxidation with HNO_3 ?
(b) Which of the D-aldotetroses will give optically active aldaric acids on oxidation with HNO_3 ?
(c) Sugar **X** is known to be a D-aldohexose. On oxidation with HNO_3, **X** gives an optically inactive aldaric acid. When **X** is degraded to an aldopentose, oxidation of the aldopentose gives an optically active aldaric acid. Determine the structure of **X.**
(d) Even though sugar **X** gives an optically inactive aldaric acid, the pentose formed by degradation gives an optically active aldaric acid. Does this finding contradict the principle that optically inactive reagents cannot form optically active products?
(e) Show what product results if the aldopentose formed from degradation of **X** is further degraded to an aldotetrose. Does HNO_3 oxidize this aldotetrose to an optically active aldaric acid?

23-60 When the gum of the shrub *Sterculia setigera* is subjected to acidic hydrolysis, one of the water-soluble components of the hydrolysate is found to be tagatose. The following information is known about tagatose:

(1) Molecular formula $C_6H_{12}O_6$
(2) Undergoes mutarotation.
(3) Does not react with bromine water.
(4) Reduces Tollens reagent to give D-galactonic acid and D-talonic acid.
(5) Methylation of tagatose (using excess CH_3I and Ag_2O) followed by acidic hydrolysis gives 1,3,4,5-tetra-*O*-methyltagatose.

(a) Draw a Fischer projection structure for the open-chain form of tagatose.
(b) Draw the most stable conformation of the most stable cyclic hemiacetal form of tagatose.

***23-61** Some protecting groups can block two OH groups of a carbohydrate at the same time. One such group is shown here, protecting the 4-OH and 6-OH groups of β-D-glucose.

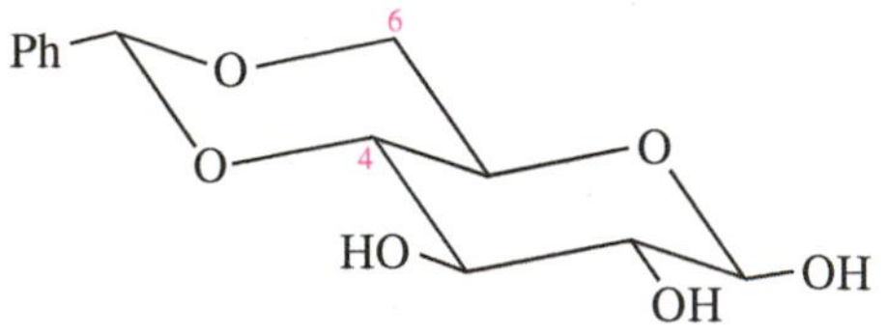

Ph O O 6 4 O HO OH OH

(a) What type of functional group is involved in this blocking group?
(b) What did glucose react with to form this protected compound?
(c) When this blocking group is added to glucose, a new chiral center is formed. Where is it?
Draw the stereoisomer that has the other configuration at this chiral center. What is the relationship between these two stereoisomers of the protected compound?
(d) Which of the two stereoisomers in part (c) do you expect to be the major product? Why?
(e) A similar protecting group, called an acetonide, can block reaction at the 2′ and 3′ oxygens of a ribonucleoside. This protected derivative is formed by the reaction of the nucleoside with acetone under acid catalysis. From this information, draw the protected product formed by the reaction.

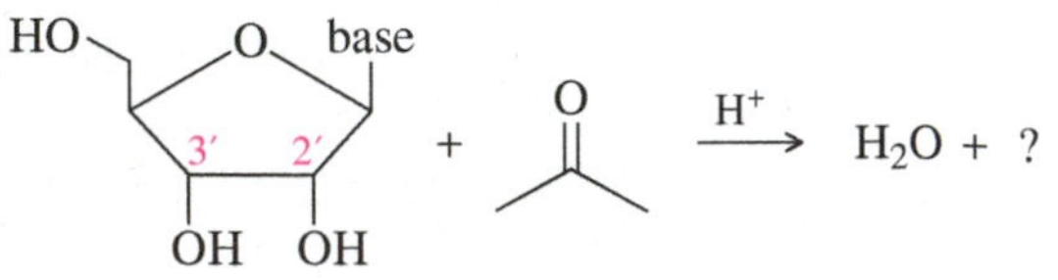

HO O base 3′ 2′ OH OH + O $\xrightarrow{H^+}$ H_2O + ?

23-62 An important protecting group developed specifically for polyhydroxy compounds like nucleosides is the tetraisopropyldisiloxanyl group, abbreviated TIPDS, that can protect two alcohol groups in a molecule.

i-Pr, *i*-Pr, *i*-Pr—Si—O—Si—*i*-Pr, Cl, Cl + HO, base, O, H, H, H, H, OH, OH $\xrightarrow{Et_3N}$

TIPDS chloride   a ribonucleoside

(a) The TIPDS group is somewhat hindered around the Si atoms by the isopropyl groups. Which OH is more likely to react first with TIPDS chloride? Show the product with the TIPDS group on one oxygen.

(b) Once the TIPDS group is attached at the first oxygen, it reaches around to the next closest oxygen. Show the final product with two oxygens protected.

(c) The unprotected hydroxy group can now undergo reactions without affecting the protected oxygens. Show the product after the protected nucleoside from (b) is treated with tosyl chloride and pyridine, followed by NaBr, ending with deprotection with Bu_4NF.

23-63 Draw the structures of the following nucleotides.

(a) guanosine triphosphate (GTP)

(b) deoxycytidine monophosphate (dCMP)

(c) cyclic guanosine monophosphate (cGMP)

23-64 Draw the structure of a four-residue segment of DNA with the following sequence.

(3′end) G-T-A-C (5′ end)

23-65 Retroviruses like HIV, the pathogen responsible for AIDS, incorporate an RNA template that is copied into DNA during infection. The *reverse transcriptase* enzyme that copies RNA into DNA is relatively nonselective and error-prone, leading to a high mutation rate. Its lack of selectivity is exploited by the anti-HIV drug AZT (3′-azido-2′,3′-dideoxythymidine), which becomes phosphorylated and is incorporated by reverse transcriptase into DNA, where it acts as a chain terminator. Mammalian DNA polymerases are more selective, having a low affinity for AZT, so its toxicity is relatively low.

(a) Draw the structures of AZT and natural deoxythymidine.

(b) Draw the structure of AZT 5′-triphosphate, the derivative that inhibits reverse transcriptase.

***23-66** Exposure to nitrous acid (see Section 19-16), sometimes found in cells, can convert cytosine to uracil.

(a) Propose a mechanism for this conversion.

(b) Explain how this conversion would be mutagenic upon replication.

(c) DNA generally includes thymine, rather than uracil (found in RNA). Based on this fact, explain why the nitrous-acid-induced mutation of cytosine to uracil is more easily repaired in DNA than it is in RNA.

***23-67** H. G. Khorana won the Nobel Prize in Medicine in 1968 for developing the synthesis of DNA and RNA and for helping to unravel the genetic code. Part of the chemistry he developed was the use of selective protecting groups for the 5′ OH group of nucleosides.

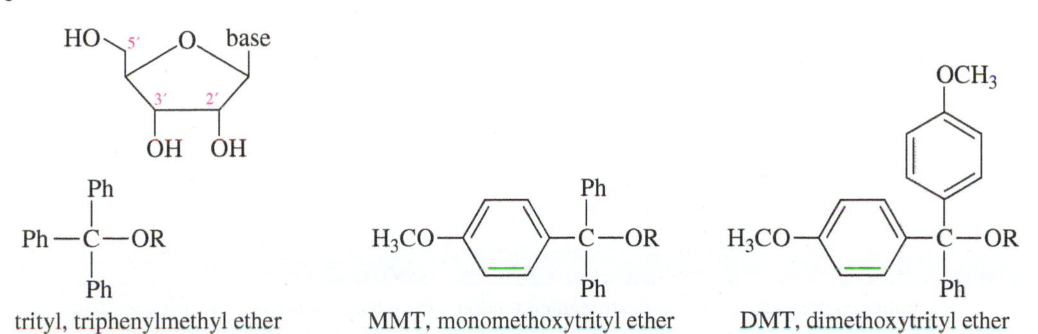

trityl, triphenylmethyl ether   MMT, monomethoxytrityl ether   DMT, dimethoxytrityl ether

The trityl ether derivative of just the 5′ OH group is obtained by reaction of the nucleoside with trityl chloride, MMT chloride, or DMT chloride and a base like Et_3N. The trityl ether derivative can be removed in dilute aqueous acid. DMT derivatives hydrolyze fastest, followed by MMT derivatives, and trityl derivatives slowest.

(a) Draw the product with the trityl derivative on the 5′ oxygen.

(b) Explain why the trityl derivative is selective for the 5′ OH group. Why doesn't it react at 2′ or 3′?

(c) Why is the DMT group easiest to remove under dilute acid conditions? Why does the solution instantly turn orange when acid is added to a DMT derivative?

24 Amino Acids, Peptides, and Proteins

Goals for Chapter 24

1 Name amino acids and peptides, and draw the structures from their names. Explain why the naturally occurring amino acids are called L-amino acids.

2 Identify which amino acids are acidic, which are basic, and which are neutral. Use the isoelectric point to predict the charge on an amino acid at a given pH.

3 Show how to synthesize amino acids from simpler compounds, and show how to combine amino acids in the proper sequence to synthesize a peptide.

4 Use information from terminal residue analysis and partial hydrolysis to determine the structure of an unknown peptide.

5 Identify the levels of protein structure, and explain how a protein's structure affects its properties.

◀ Spider web is composed mostly of fibroin, a protein with pleated-sheet secondary structure (Section 24-12). The pleated-sheet arrangement allows for multiple hydrogen bonds between molecules, conferring great strength. Spider silk has a tensile strength about the same as steel, but it weighs six times less.

24-1 Introduction

Proteins are the most abundant organic molecules in animals, playing important roles in all aspects of cell structure and function. Proteins are biopolymers of ***α*-amino acids,** so named because the amino group is bonded to the α carbon atom, next to the carbonyl group. The physical and chemical properties of a protein are determined by its constituent amino acids. The individual amino acid subunits are joined by amide linkages called **peptide bonds.** Figure 24-1 (next page) shows the general structure of an α-amino acid and a protein.

Proteins have an amazing range of structural and catalytic properties as a result of their varying amino acid composition. Because of this versatility, proteins serve an astonishing variety of functions in living organisms. Some of the functions of the major classes of proteins are outlined in Table 24-1.

TABLE 24-1
Examples of Protein Functions

Class of Protein	Example	Function of Example
structural proteins	collagen, keratin	strengthen tendons, skin, hair, nails
enzymes	DNA polymerase	replicates and repairs DNA
transport proteins	hemoglobin	transports O_2 to the cells
contractile proteins	actin, myosin	cause contraction of muscles
protective proteins	antibodies	complex with foreign proteins
hormones	insulin	regulates glucose metabolism
toxins	snake venoms	incapacitate prey

PROBLEM-SOLVING HINT

Proteins are the third type of biopolymer described in this book. The three types of biopolymers are the following:

1. Polysaccharides (glycans, starches) are polymers of simple sugars.
2. DNA and RNA are polymers of nucleic acids.
3. Peptides and proteins are polymers of amino acids.

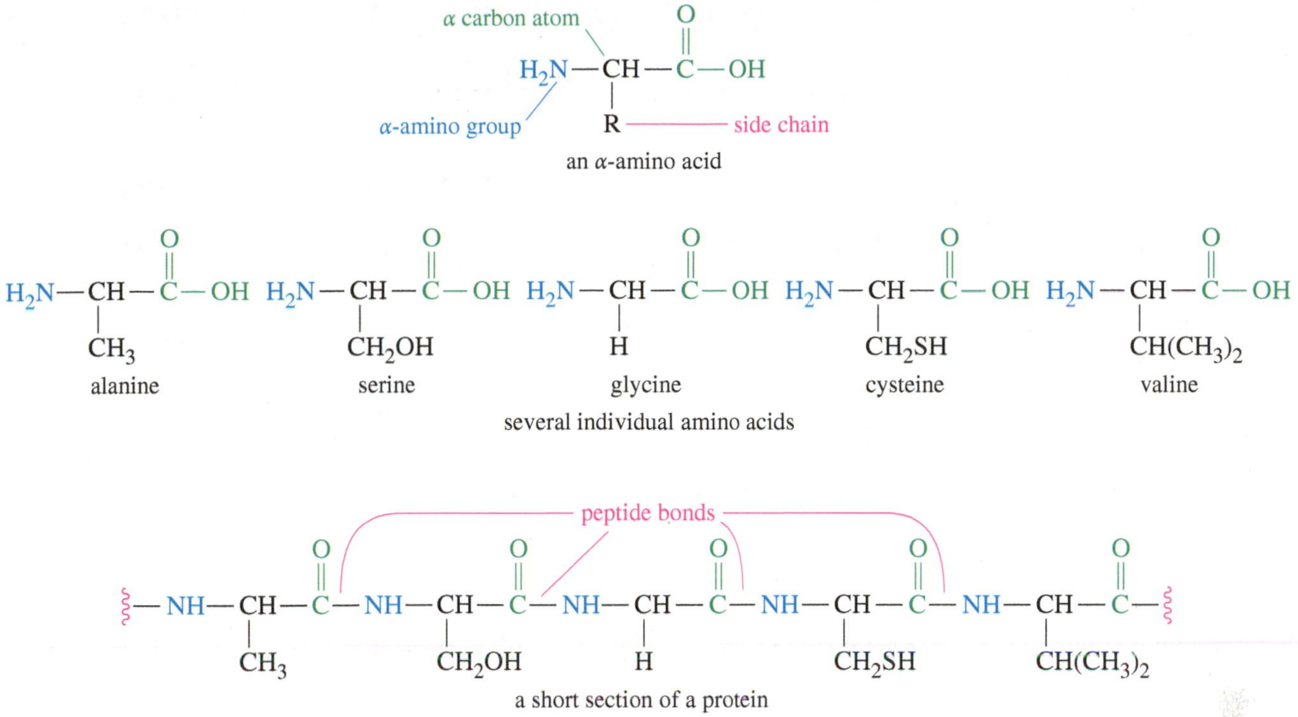

α carbon atom

$$H_2N{-}CH{-}\overset{\displaystyle O}{\overset{\|}{C}}{-}OH$$
$$\quad\ \ |$$
$$\quad\ \ R$$

α-amino group — side chain

an α-amino acid

$H_2N{-}CH(CH_3){-}C({=}O){-}OH$	$H_2N{-}CH(CH_2OH){-}C({=}O){-}OH$	$H_2N{-}CH(H){-}C({=}O){-}OH$	$H_2N{-}CH(CH_2SH){-}C({=}O){-}OH$	$H_2N{-}CH(CH(CH_3)_2){-}C({=}O){-}OH$
alanine	serine	glycine	cysteine	valine

several individual amino acids

peptide bonds

$$\sim NH{-}\underset{CH_3}{CH}{-}\overset{O}{\overset{\|}{C}}{-}NH{-}\underset{CH_2OH}{CH}{-}\overset{O}{\overset{\|}{C}}{-}NH{-}\underset{H}{CH}{-}\overset{O}{\overset{\|}{C}}{-}NH{-}\underset{CH_2SH}{CH}{-}\overset{O}{\overset{\|}{C}}{-}NH{-}\underset{CH(CH_3)_2}{CH}{-}\overset{O}{\overset{\|}{C}}{-}\sim$$

a short section of a protein

FIGURE 24-1
Structure of a general protein and its constituent amino acids. The amino acids are joined by amide linkages called peptide bonds.

The study of proteins is one of the major branches of biochemistry, and there is no clear division between the organic chemistry of proteins and their biochemistry. In this chapter, we begin the study of proteins by learning about their constituents, the amino acids. We also discuss how amino acid monomers are linked into the protein polymer, and how the properties of a protein depend on those of its constituent amino acids. These concepts are needed for the further study of protein structure and function in a biochemistry course.

24-2 Structure and Stereochemistry of the α-Amino Acids

The term **amino acid** might mean any molecule containing both an amino group and any type of acid group; however, the term is almost always used to refer to an α-amino carboxylic acid. The simplest α-amino acid is aminoacetic acid, called *glycine*. Other common amino acids have side chains (symbolized by R) substituted on the α carbon atom. For example, alanine is the amino acid with a methyl side chain.

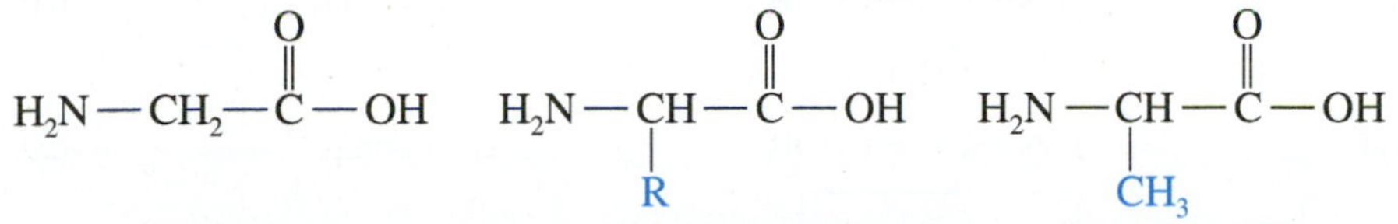

$H_2N{-}CH_2{-}C({=}O){-}OH$	$H_2N{-}CH(R){-}C({=}O){-}OH$	$H_2N{-}CH(CH_3){-}C({=}O){-}OH$
glycine	a substituted amino acid	alanine ($R = CH_3$)

Except for glycine, the α-amino acids are all chiral. In all of the chiral amino acids, the chiral center is the asymmetric α carbon atom. Nearly all the naturally occurring amino acids are found to have the (*S*) configuration at the α carbon atom. Figure 24-2 shows a Fischer projection of the (*S*) enantiomer of alanine, with the carbon chain along the vertical and the carbonyl carbon at the top. Notice that the configuration of (*S*)-alanine is similar to that of L-(−)-glyceraldehyde, with the amino group on the left in the Fischer projection. Because their stereochemistry is similar to that of L-(−)-glyceraldehyde, the naturally occurring (*S*)-amino acids are classified as **L-amino acids.**

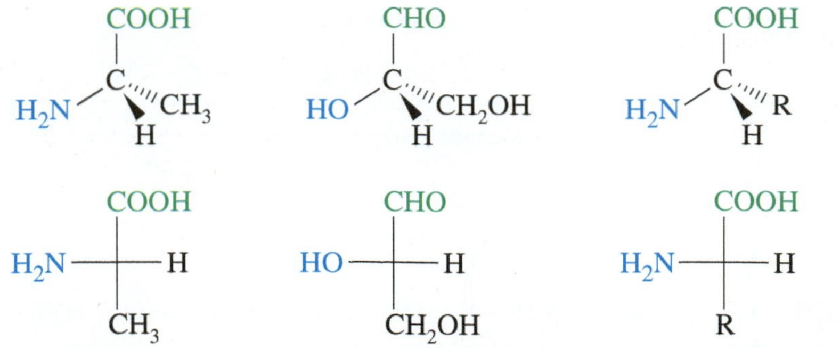

FIGURE 24-2
Almost all the naturally occurring amino acids have the (S) configuration. They are called L-amino acids because their stereochemistry resembles that of L-(−)-glyceraldehyde.

Application: Antibiotics

Bacteria require specific enzymes, called racemases, to interconvert D- and L-amino acids. Mammals do not use D-amino acids, so compounds that block racemases do not affect mammals and show promise as antibiotics.

Although D-amino acids are occasionally found in nature, we usually assume the amino acids under discussion are the common L-amino acids. Remember once again that the D and L nomenclature, like the *R* and *S* designation, gives the configuration of the asymmetric carbon atom. It does not imply the sign of the optical rotation, (+) or (−), which must be determined experimentally.

Amino acids combine many of the properties and reactions of both amines and carboxylic acids. The combination of a basic amino group and an acidic carboxyl group in the same molecule also results in some unique properties and reactions. The side chains of some amino acids have additional functional groups that lend interesting properties and undergo reactions of their own.

24-2A The Standard Amino Acids of Proteins

The **standard amino acids** are 20 common α-amino acids that are found in nearly all proteins. The standard amino acids differ from each other in the structure of the side chains bonded to their α carbon atoms. All the standard amino acids are L-amino acids. Table 24-2 shows the 20 standard amino acids, grouped according to the chemical properties of their side chains. Each amino acid is given a three-letter abbreviation and a one-letter symbol (green) for use in writing protein structures.

TABLE 24-2
The Twenty Standard Amino Acids

Name	Symbol	Abbreviation	Structure	Functional Group in Side Chain	Isoelectric Point
side chain is nonpolar, H or alkyl					
glycine	G	Gly	$H_2N—CH—COOH$ with side chain H	none	6.0
alanine	A	Ala	$H_2N—CH—COOH$ with side chain CH_3	alkyl group	6.0
*valine	V	Val	$H_2N—CH—COOH$ with side chain $CH(CH_3)_2$	alkyl group	6.0
*leucine	L	Leu	$H_2N—CH—COOH$ with side chain $CH_2—CH(CH_3)—CH_3$	alkyl group	6.0
*isoleucine	I	Ile	$H_2N—CH—COOH$ with side chain $CH_3—CH—CH_2CH_3$	alkyl group	6.0

TABLE 24-2 *(continued)*

Name	Symbol	Abbreviation	Structure	Functional Group in Side Chain	Isoelectric Point
*phenylalanine	F	Phe	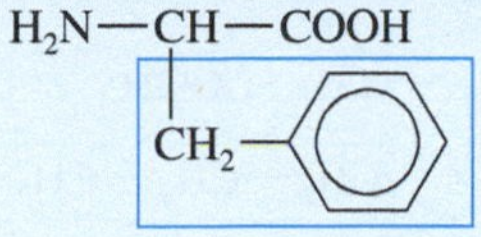 $H_2N-CH-COOH$ $CH_2-C_6H_5$ (phenyl ring)	aromatic group	5.5
proline	P	Pro	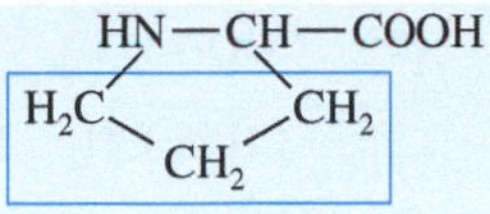 $HN-CH-COOH$ H_2C, CH_2, CH_2 (ring)	rigid cyclic structure	6.3
side chain contains an —OH					
serine	S	Ser	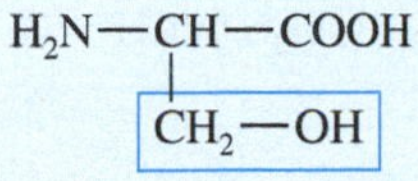 $H_2N-CH-COOH$ CH_2-OH	hydroxy group	5.7
*threonine	T	Thr	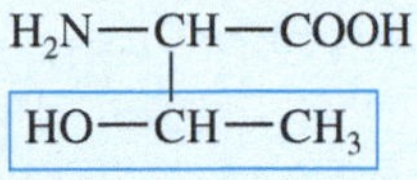 $H_2N-CH-COOH$ $HO-CH-CH_3$	hydroxy group	5.6
tyrosine	Y	Tyr	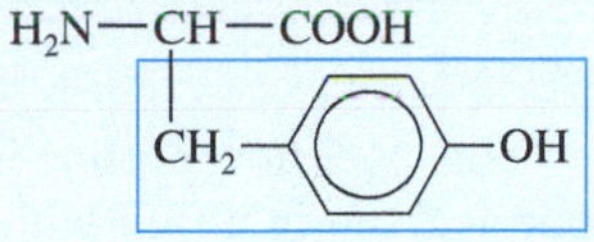 $H_2N-CH-COOH$ $CH_2-C_6H_4-OH$ (ring)	phenolic OH group	5.7
side chain contains sulfur					
cysteine	C	Cys	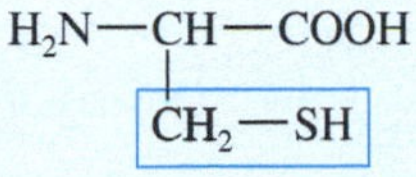 $H_2N-CH-COOH$ CH_2-SH	thiol	5.0
*methionine	M	Met	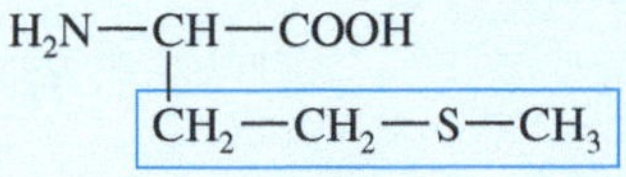 $H_2N-CH-COOH$ $CH_2-CH_2-S-CH_3$	sulfide	5.7
side chain contains nonbasic nitrogen					
asparagine	N	Asn	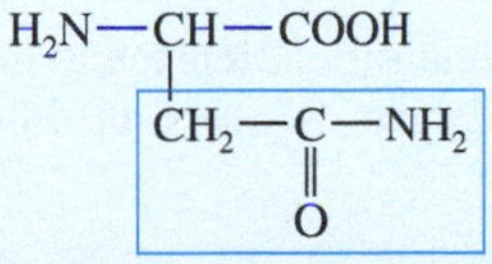 $H_2N-CH-COOH$ $CH_2-C(=O)-NH_2$	amide	5.4
glutamine	Q	Gln	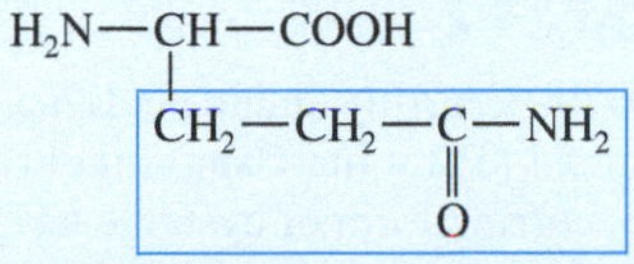 $H_2N-CH-COOH$ $CH_2-CH_2-C(=O)-NH_2$	amide	5.7
*tryptophan	W	Trp	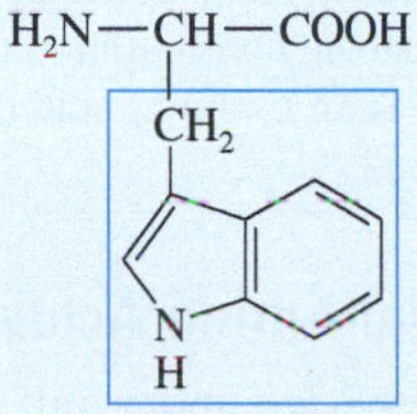 $H_2N-CH-COOH$ CH_2–indole (N–H)	indole	5.9
side chain is acidic					
aspartic acid	D	Asp	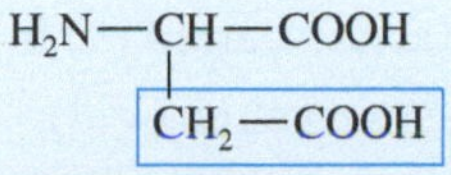 $H_2N-CH-COOH$ CH_2-COOH	carboxylic acid	2.8
glutamic acid	E	Glu	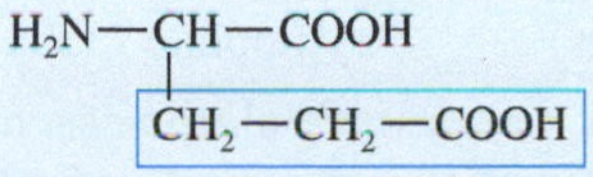 $H_2N-CH-COOH$ CH_2-CH_2-COOH	carboxylic acid	3.2

(continued)

TABLE 24-2 (*continued*)

Name	Symbol	Abbreviation	Structure	Functional Group in Side Chain	Isoelectric Point
side chain is basic					
*lysine	K	Lys	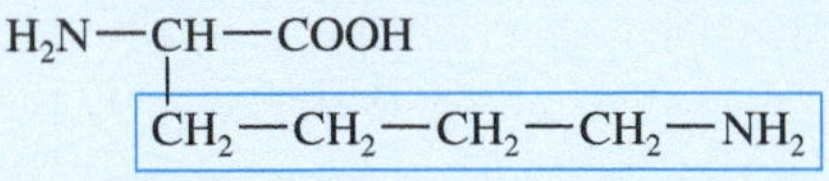 $H_2N-CH(-CH_2-CH_2-CH_2-CH_2-NH_2)-COOH$	amino group	9.7
*arginine	R	Arg	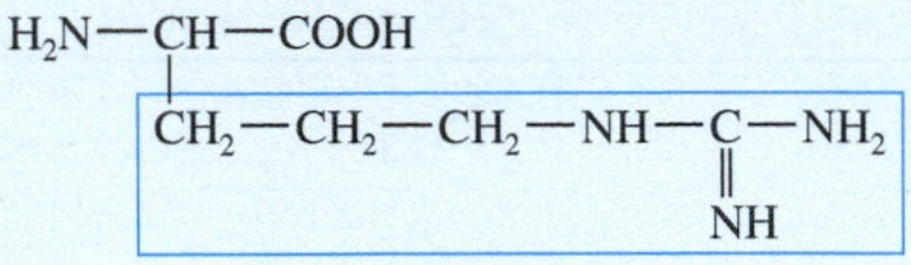 $H_2N-CH(-CH_2-CH_2-CH_2-NH-C(=NH)-NH_2)-COOH$	guanidino group	10.8
*histidine	H	His	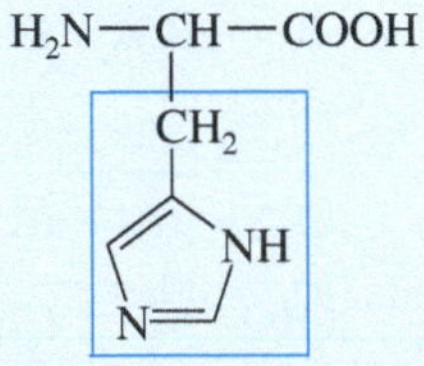 $H_2N-CH(-CH_2-\text{imidazole ring (NH, N=)})-COOH$	imidazole ring	7.6

*essential amino acid

Notice in Table 24-2 how proline is different from the other standard amino acids. Its amino group is fixed in a ring with its α carbon atom. This cyclic structure lends additional strength and rigidity to proline-containing peptides.

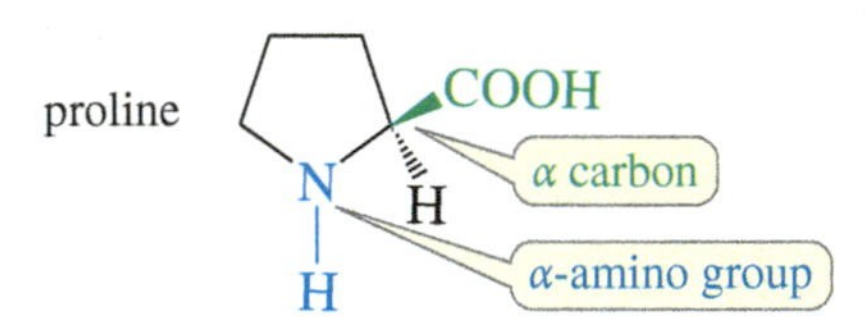

PROBLEM 24-1

Draw three-dimensional representations of the following amino acids.
(a) L-phenylalanine **(b)** L-histidine **(c)** D-serine **(d)** L-tryptophan

PROBLEM 24-2

Most naturally occurring amino acids have chiral centers (the asymmetric α carbon atoms) that are named (*S*) by the Cahn–Ingold–Prelog convention (Section 5-3). The common naturally occurring form of cysteine has a chiral center that is named (*R*), however.

(a) What is the relationship between (*R*)-cysteine and (*S*)-alanine? Do they have the opposite three-dimensional configuration (as the names might suggest) or the same configuration?

(b) (*S*)-Alanine is an L-amino acid (Figure 24-2). Is (*R*)-cysteine a D-amino acid or an L-amino acid?

24-2B Essential Amino Acids

Humans can synthesize about half of the amino acids needed to make proteins. Other amino acids, called the **essential amino acids,** must be provided in the diet. The ten essential amino acids, starred (*) in Table 24-2, are the following:

arginine (Arg)	valine (Val)	methionine (Met)	leucine (Leu)
threonine (Thr)	phenylalanine (Phe)	histidine (His)	isoleucine (Ile)
lysine (Lys)	tryptophan (Trp)		

Proteins that provide all the essential amino acids in about the right proportions for human nutrition are called **complete proteins.** Examples of complete proteins are those in meat, fish, milk, and eggs. About 50 g of complete protein per day is adequate for adult humans.

Proteins that are severely deficient in one or more of the essential amino acids are called **incomplete proteins.** If the protein in a person's diet comes mostly from one incomplete source, the amount of human protein that can be synthesized is limited by the amounts of the deficient amino acids. Plant proteins are generally incomplete. Rice, corn, and wheat are all deficient in lysine. Rice also lacks threonine, and corn also lacks tryptophan. Beans, peas, and other legumes have the most complete proteins among the common plants, but they are deficient in methionine.

Vegetarians can achieve an adequate intake of the essential amino acids if they eat many different plant foods. Plant proteins can be chosen to be complementary, with some foods supplying amino acids that others lack. For example, rice and beans are often combined because rice is low in lysine but high in methionine, and beans are low in methionine but high in lysine. An alternative is to supplement the vegetarian diet with a rich source of complete protein such as milk or eggs.

Application: Gelatin

Gelatin is made from collagen, which is a structural protein composed primarily of glycine, proline, and hydroxyproline. As a result, gelatin has low nutritional value because it lacks many of the essential amino acids.

PROBLEM 24-3

The herbicide *glyphosate* (Roundup®) kills plants by inhibiting an enzyme needed for synthesis of phenylalanine. Deprived of phenylalanine, the plant cannot make the proteins it needs, and it gradually weakens and dies. Although a small amount of glyphosate is deadly to a plant, its human toxicity is quite low. Suggest why this powerful herbicide has little effect on humans.

24-2C Rare and Unusual Amino Acids

In addition to the standard amino acids, other amino acids are found in protein in smaller quantities. For example, 4-hydroxyproline and 5-hydroxylysine are hydroxylated versions of standard amino acids. These are called *rare* amino acids, even though they are commonly found in collagen.

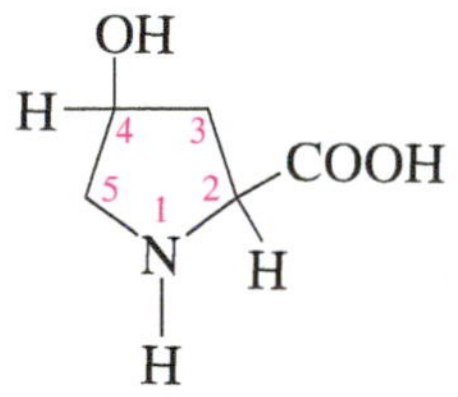

4-hydroxyproline

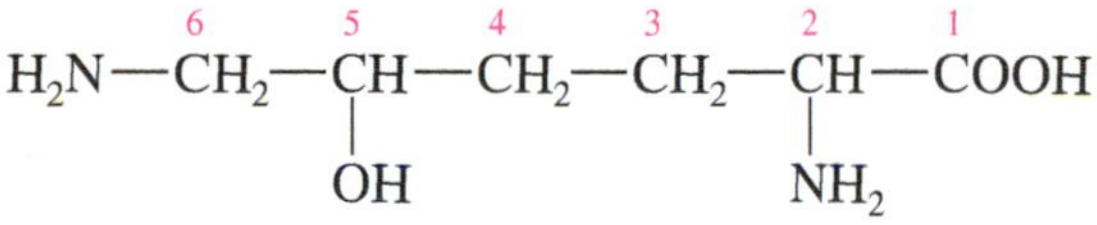

$$H_2N{-}CH_2{-}CH(OH){-}CH_2{-}CH_2{-}CH(NH_2){-}COOH$$

5-hydroxylysine

Some of the less common D enantiomers of amino acids are also found in nature. For example, D-glutamic acid is found in the cell walls of many bacteria, and D-serine is found in earthworms. Some naturally occurring amino acids are not α-amino acids: γ-Aminobutyric acid (GABA) is one of the neurotransmitters in the brain, and β-alanine is a constituent of the vitamin pantothenic acid.

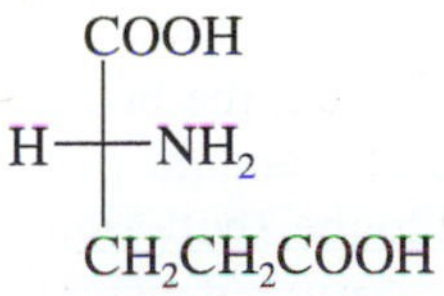

COOH / H—C—NH₂ / CH₂CH₂COOH: D-glutamic acid; COOH / H—C—NH₂ / CH₂OH: D-serine

$$\overset{\gamma}{C}H_2(NH_2){-}\overset{\beta}{C}H_2{-}\overset{\alpha}{C}H_2{-}COOH$$

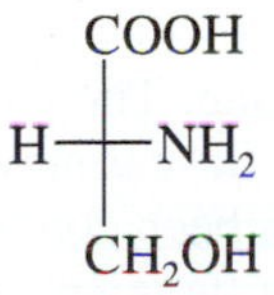

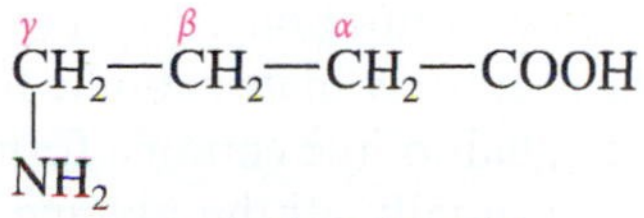

γ-aminobutyric acid

$$\overset{\beta}{C}H_2(NH_2){-}\overset{\alpha}{C}H_2{-}COOH$$

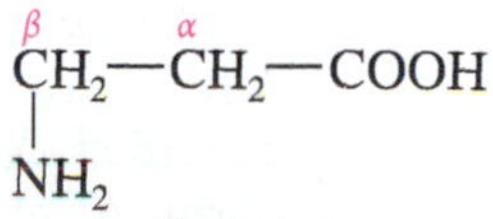

β-alanine

24-3 Acid–Base Properties of Amino Acids

Although we commonly write amino acids with an intact carboxyl (—COOH) group and amino (—NH_2) group, their actual structure is ionic and depends on the pH. The carboxyl group loses a proton, giving a carboxylate ion, and the amino group is protonated to an ammonium ion. This structure is called a **dipolar ion** or a **zwitterion** (German for "dipolar ion").

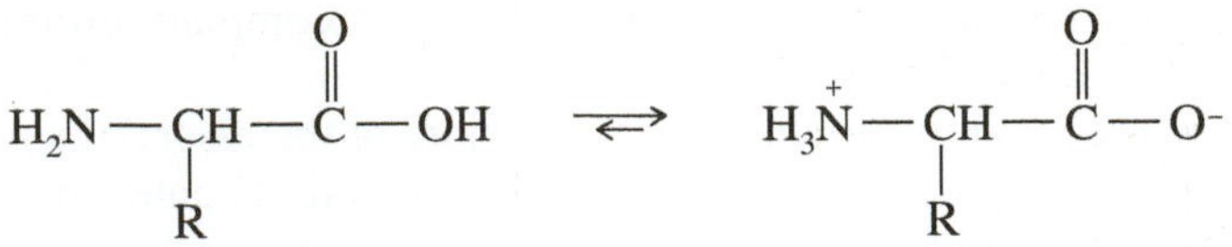

$$H_2N-CH(R)-C(=O)-OH \rightleftharpoons H_3\overset{+}{N}-CH(R)-C(=O)-O^-$$

uncharged structure (minor component) — dipolar ion, or zwitterion (major component)

The dipolar nature of amino acids gives them some unusual properties:

1. Amino acids have **high melting points,** generally over 200 °C.

$$H_3\overset{+}{N}-CH_2-COO^-$$

glycine, mp 262 °C

2. Amino acids are more **soluble in water** than they are in ether, dichloromethane, and other common organic solvents.
3. Amino acids have much **larger dipole moments** (μ) than simple amines or simple acids.

$H_3\overset{+}{N}-CH_2-COO^-$	$CH_3-CH_2-CH_2-NH_2$	CH_3-CH_2-COOH
glycine, $\mu = 14$ D	propylamine, $\mu = 1.4$ D	propionic acid, $\mu = 1.7$ D

4. Amino acids are **less acidic than most carboxylic acids** and **less basic than most amines.** In fact, the acidic part of the amino acid molecule is the $-NH_3^+$ group, not a $-COOH$ group. The basic part is the $-COO^-$ group, and not a free $-NH_2$ group.

$R-COOH$	$R-NH_2$	$H_3\overset{+}{N}-CH(R)-COO^-$
$pK_a = 5$	$pK_b = 4$	$pK_a = 10$, $pK_b = 12$

Because amino acids contain both acidic ($-NH_3^+$) and basic ($-COO^-$) groups, they are *amphoteric* (having both acidic and basic properties). The predominant form of the amino acid depends on the pH of the solution. In an acidic solution, the $-COO^-$ group is protonated to a free $-COOH$ group, and the molecule has an overall positive charge. As the pH is raised, the $-COOH$ loses its proton at about pH 2. This point is called pK_{a1}, the first acid-dissociation constant. As the pH is raised further, the $-NH_3^+$ group loses its proton at about pH 9 or 10. This point is called pK_{a2}, the second acid-dissociation constant. Above this pH, the molecule has an overall negative charge.

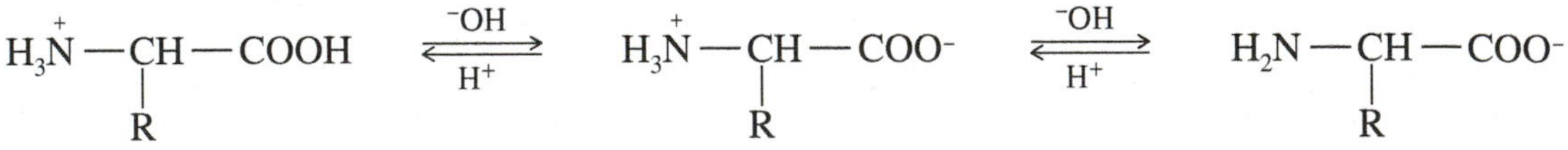

$$H_3\overset{+}{N}-CH(R)-COOH \underset{H^+}{\overset{^-OH}{\rightleftharpoons}} H_3\overset{+}{N}-CH(R)-COO^- \underset{H^+}{\overset{^-OH}{\rightleftharpoons}} H_2N-CH(R)-COO^-$$

cationic in acid — $pK_{a1} \approx 2$ — neutral — $pK_{a2} \approx 9\text{–}10$ — anionic in base

Figure 24-3 shows a titration curve for glycine. The curve starts at the bottom left, where glycine is entirely in its cationic form. Base is slowly added, and the pH is recorded. At pH 2.3, half of the cationic form has been converted to the zwitterionic form. At pH 6.0, essentially all the glycine is in the zwitterionic form. At pH 9.6, half of the zwitterionic form has been converted to the basic form. From this graph, we can see that glycine is mostly in the cationic form at pH values below 2.3, mostly in the zwitterionic form at pH values between 2.3 and 9.6, and mostly in the anionic form at pH values above 9.6. By varying the pH of the solution, we can control the charge on the molecule. This ability to control the charge of an amino acid is useful for separating and identifying amino acids by electrophoresis, as described in Section 24-4.

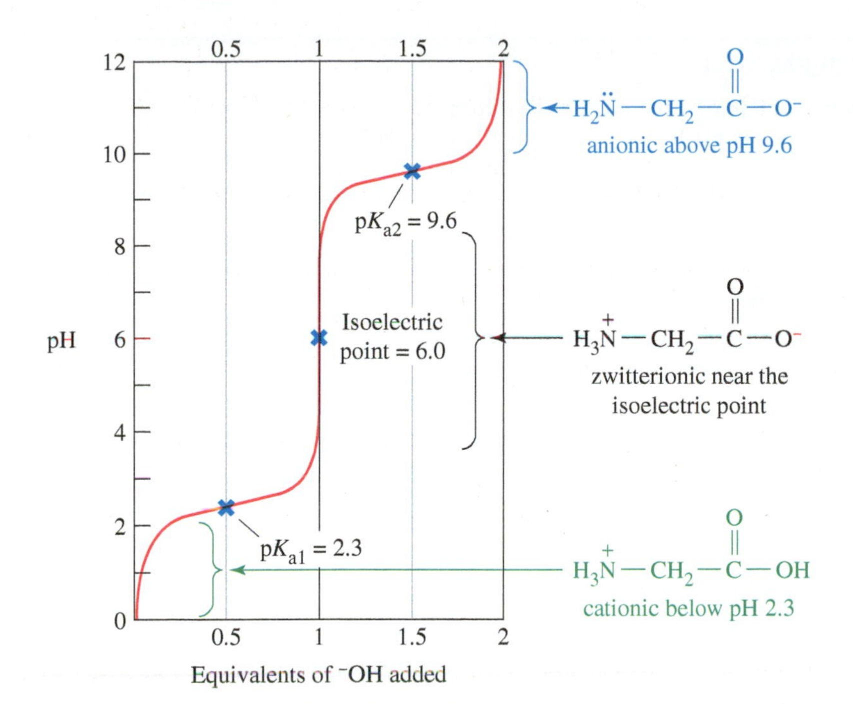

FIGURE 24-3
A titration curve for glycine. The pH controls the charge on glycine: cationic below pH 2.3; zwitterionic between pH 2.3 and 9.6; and anionic above pH 9.6. The isoelectric pH is 6.0.

24-4 Isoelectric Points and Electrophoresis

An amino acid bears a positive charge in acidic solution (low pH) and a negative charge in basic solution (high pH). There must be an intermediate pH where the amino acid is evenly balanced between the two forms, as the dipolar zwitterion with a net charge of zero. This pH is called the **isoelectric pH** or the **isoelectric point,** abbreviated **pI.**

$$\underset{\substack{\text{low pH} \\ \text{(cationic in acid)}}}{H_3\overset{+}{N}-\underset{\substack{| \\ R}}{CH}-COOH} \underset{H^+}{\overset{^-OH}{\rightleftharpoons}} \underset{\substack{\text{isoelectric pH} \\ \text{(neutral)}}}{H_3\overset{+}{N}-\underset{\substack{| \\ R}}{CH}-COO^-} \underset{H^+}{\overset{^-OH}{\rightleftharpoons}} \underset{\substack{\text{high pH} \\ \text{(anionic in base)}}}{H_2N-\underset{\substack{| \\ R}}{CH}-COO^-}$$

The isoelectric points of the standard amino acids are given in Table 24-2. Notice that the isoelectric pH depends on the amino acid structure in a predictable way.

acidic amino acids:	aspartic acid (2.8), glutamic acid (3.2)
neutral amino acids:	(5.0 to 6.3)
basic amino acids:	lysine (9.7), arginine (10.8), histidine (7.6)

The side chains of aspartic acid and glutamic acid contain acidic carboxyl groups. These amino acids have acidic isoelectric points around pH 3. An acidic solution is needed to prevent deprotonation of the second carboxylic acid group and to keep the amino acid in its neutral isoelectric state.

Basic amino acids (histidine, lysine, and arginine) have isoelectric points at pH values of 7.6, 9.7, and 10.8, respectively. These values reflect the weak basicity of the imidazole ring, the intermediate basicity of an amino group, and the strong basicity of the guanidino group. A basic solution is needed in each case to prevent protonation of the basic side chain to keep the amino acid electrically neutral.

The other amino acids are considered neutral, with no strongly acidic or basic side chains. Their isoelectric points are slightly acidic (from about 5 to 6) because the $-NH_3^+$ group is slightly more acidic than the $-COO^-$ group is basic.

PROBLEM 24-4

Draw the structure of the predominant form of
(a) isoleucine at pH 11. **(b)** proline at pH 2.
(c) arginine at pH 7. **(d)** glutamic acid at pH 7.
(e) a mixture of alanine, lysine, and aspartic acid at (i) pH 6; (ii) pH 11; (iii) pH 2.

PROBLEM-SOLVING HINT

At its isoelectric point (pI), an amino acid has a net charge of zero, with NH_3^+ and COO^- balancing each other. In a more acidic solution (lower pH), the carboxyl group becomes protonated, and the net charge is positive. In a more basic solution (higher pH), the amino group loses its proton, and the net charge is negative.

PROBLEM 24-5

Draw the resonance forms of a protonated guanidino group, and explain why arginine has such a strongly basic isoelectric point.

PROBLEM 24-6

Although tryptophan contains a heterocyclic amine, it is considered a neutral amino acid. Explain why the indole nitrogen of tryptophan is more weakly basic than one of the imidazole nitrogens of histidine.

Electrophoresis uses differences in isoelectric points to separate mixtures of amino acids (Figure 24-4). A streak of the amino acid mixture is placed in the center of a layer of acrylamide gel or a piece of filter paper wet with a buffer solution. Two electrodes are placed in contact with the edges of the gel or paper, and a potential of several thousand volts is applied across the electrodes. Positively charged (cationic) amino acids are attracted to the negative electrode (the cathode), and negatively charged (anionic) amino acids are attracted to the positive electrode (the anode). An amino acid at its isoelectric point has no net charge, so it does not move.

As an example, consider a mixture of alanine, lysine, and aspartic acid in a buffer solution at pH 6. Alanine is at its isoelectric point, in its dipolar zwitterionic form with a net charge of zero. A pH of 6 is more acidic than the isoelectric pH for lysine (9.7), so lysine is in the cationic form. Aspartic acid has an isoelectric pH of 2.8, so it is in the anionic form.

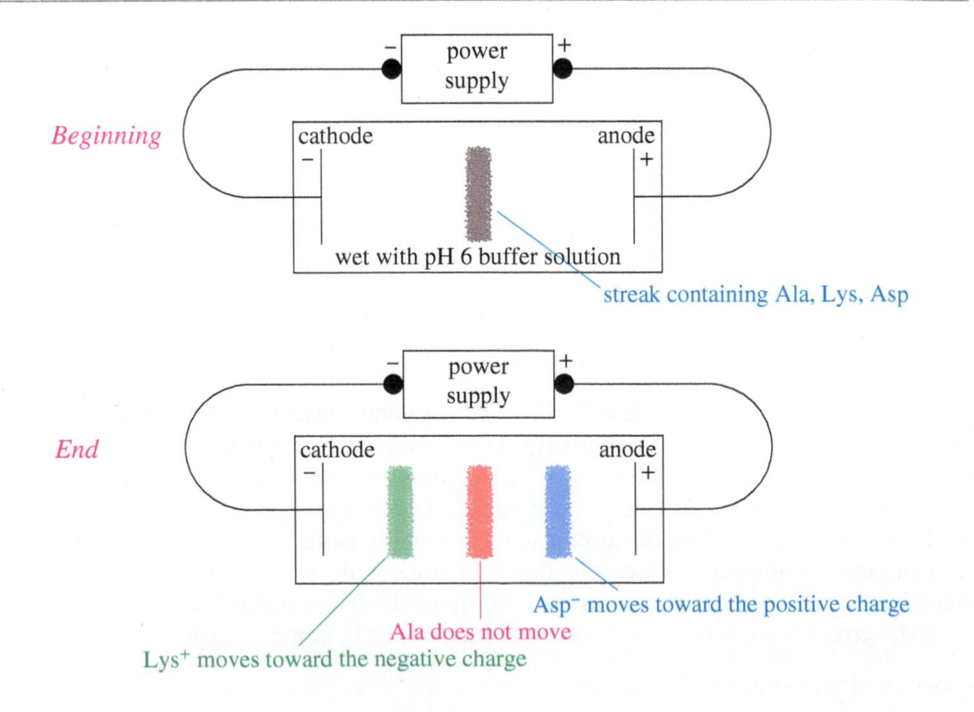

FIGURE 24-4
A simplified picture of the electrophoretic separation of alanine, lysine, and aspartic acid at pH 6. Cationic lysine is attracted to the cathode; anionic aspartic acid is attracted to the anode. Alanine is at its isoelectric point, so it does not move.

Structure at pH 6

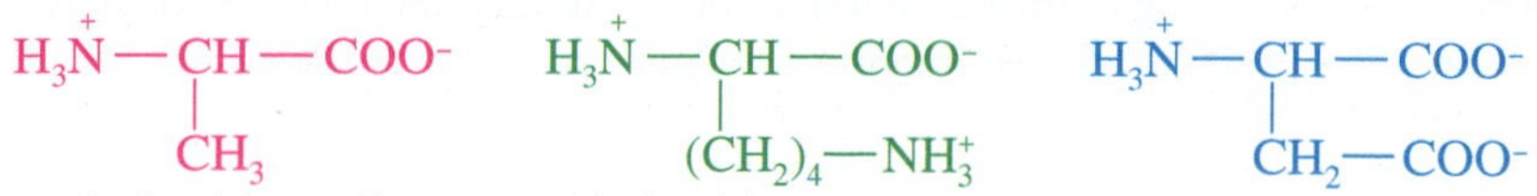

$$H_3\overset{+}{N}-\underset{CH_3}{\underset{|}{CH}}-COO^- \qquad H_3\overset{+}{N}-\underset{(CH_2)_4-\overset{+}{N}H_3}{\underset{|}{CH}}-COO^- \qquad H_3\overset{+}{N}-\underset{CH_2-COO^-}{\underset{|}{CH}}-COO^-$$

alanine (charge 0) lysine (charge +1) aspartic acid (charge −1)

When a voltage is applied to a mixture of alanine, lysine, and aspartic acid at pH 6, alanine does not move. Lysine moves toward the negatively charged cathode, and aspartic acid moves toward the positively charged anode (Figure 24-4). After a period of time, the separated amino acids are recovered by cutting the paper or scraping the bands out of the gel. If electrophoresis is being used as an analytical technique (to determine the amino acids present in the mixture), the paper or gel is treated with a reagent such as ninhydrin (Section 24-7C) to make the bands visible. Then the amino acids are identified by comparing their positions with those of standards.

PROBLEM 24-7

Draw the electrophoretic separation of Ala, Lys, and Asp at pH 9.7.

PROBLEM 24-8

Draw the electrophoretic separation of Trp, Cys, and His at pH 6.0.

24-5 Synthesis of Amino Acids

Naturally occurring amino acids can be obtained by hydrolyzing proteins and separating the amino acid mixture. Even so, it is often less expensive to synthesize the pure amino acid. In some cases, an unusual amino acid or an unnatural enantiomer is needed, and it must be synthesized. In this chapter, we consider four methods for making amino acids. All these methods are extensions of reactions we have already studied.

24-5A Reductive Amination

Reductive amination of ketones and aldehydes is one of the best methods for synthesizing amines (Section 19-18). It also forms amino acids. When an α-ketoacid is treated with ammonia, the ketone reacts to form an imine. The imine is reduced to an amine by hydrogen and a palladium catalyst. Under these conditions, the carboxylic acid is not reduced.

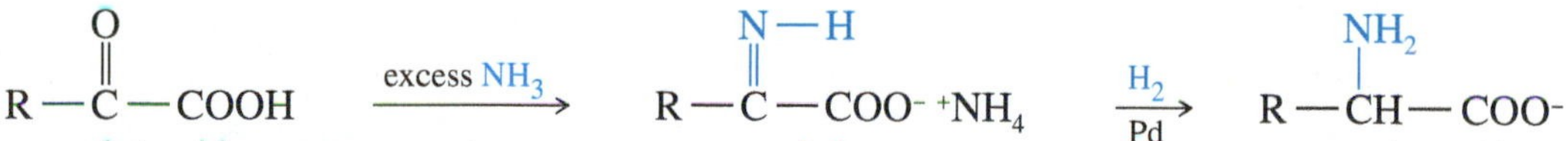

$$\underset{\alpha\text{-ketoacid}}{R-\overset{O}{\overset{\|}{C}}-COOH} \xrightarrow{\text{excess } NH_3} \underset{\text{imine}}{R-\overset{N-H}{\overset{\|}{C}}-COO^-\ {}^+NH_4} \xrightarrow[Pd]{H_2} \underset{\alpha\text{-amino acid}}{R-\overset{NH_2}{\overset{|}{CH}}-COO^-}$$

This entire synthesis is accomplished in one step by treating the α-ketoacid with ammonia and hydrogen in the presence of a palladium catalyst. The product is a racemic α-amino acid. The following reaction shows the synthesis of racemic phenylalanine from 3-phenyl-2-oxopropanoic acid.

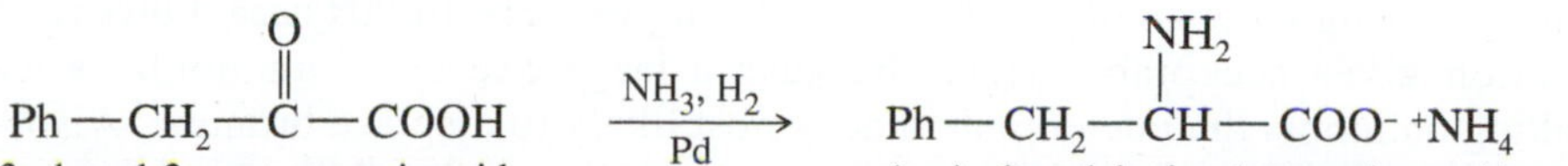

$$\underset{\text{3-phenyl-2-oxopropanoic acid}}{Ph-CH_2-\overset{O}{\overset{\|}{C}}-COOH} \xrightarrow[Pd]{NH_3,\ H_2} \underset{\text{(D,L)-phenylalanine (ammonium salt) (30\%)}}{Ph-CH_2-\overset{NH_2}{\overset{|}{CH}}-COO^-\ {}^+NH_4}$$

We call reductive amination a **biomimetic** ("mimicking the biological process") synthesis because it resembles the biological synthesis of amino acids. The biosynthesis begins with reductive amination of α-ketoglutaric acid (an intermediate in the metabolism of carbohydrates), using ammonium ion as the aminating agent and NADH as the reducing agent. The product of this enzyme-catalyzed reaction is the pure L enantiomer of glutamic acid.

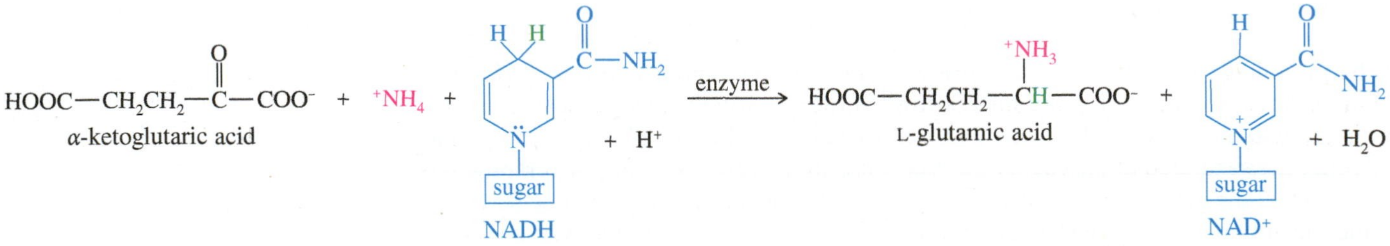

$$\underset{\alpha\text{-ketoglutaric acid}}{HOOC-CH_2CH_2-C(=O)-COO^-} + {}^+NH_4 + \underset{\text{NADH}}{\text{NADH}} + H^+ \xrightarrow{\text{enzyme}} \underset{\text{L-glutamic acid}}{HOOC-CH_2CH_2-CH({}^+NH_3)-COO^-} + \underset{\text{NAD}^+}{\text{NAD}^+} + H_2O$$

Biosynthesis of other amino acids uses L-glutamic acid as the source of the amino group. Such a reaction, moving an amino group from one molecule to another, is called a **transamination,** and the enzymes that catalyze these reactions are called *transaminases*. For example, the following reaction shows the biosynthesis of aspartic acid using glutamic acid as the nitrogen source. Once again, the enzyme-catalyzed biosynthesis gives the pure L enantiomer of the product.

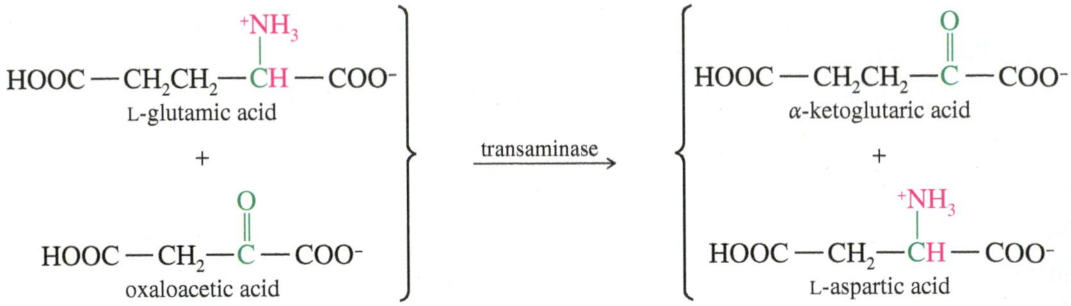

$$\left\{\begin{matrix} \underset{\text{L-glutamic acid}}{HOOC-CH_2CH_2-CH({}^+NH_3)-COO^-} \\ + \\ \underset{\text{oxaloacetic acid}}{HOOC-CH_2-C(=O)-COO^-} \end{matrix}\right\} \xrightarrow{\text{transaminase}} \left\{\begin{matrix} \underset{\alpha\text{-ketoglutaric acid}}{HOOC-CH_2CH_2-C(=O)-COO^-} \\ + \\ \underset{\text{L-aspartic acid}}{HOOC-CH_2-CH({}^+NH_3)-COO^-} \end{matrix}\right\}$$

PROBLEM 24-9

Show how the following amino acids might be formed in the laboratory by reductive amination of the appropriate α-ketoacid.

(a) alanine **(b)** leucine **(c)** serine **(d)** glutamine

24-5B Amination of an α-Halo Acid

The Hell–Volhard–Zelinsky reaction (Section 22-6) is an effective method for introducing bromine at the α position of a carboxylic acid. The racemic α-bromo acid is converted to a racemic α-amino acid by direct amination, using a large excess of ammonia.

$$\underset{\text{carboxylic acid}}{R-CH_2-C(=O)-OH} \xrightarrow[\text{(2) } H_2O]{\text{(1) } Br_2/PBr_3} \underset{\alpha\text{-bromo acid}}{R-CH(Br)-C(=O)-OH} \xrightarrow[\text{(large excess)}]{NH_3} \underset{\substack{(\text{D,L})\text{-}\alpha\text{-amino acid} \\ \text{(ammonium salt)}}}{R-CH(NH_2)-C(=O)-O^-\ {}^+NH_4}$$

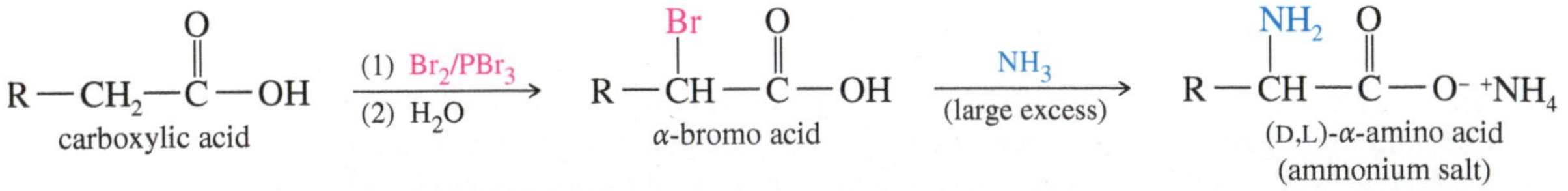

In Section 19-11, we saw that direct alkylation is often a poor synthesis of amines, giving large amounts of overalkylated products. In this case, however, the reaction gives acceptable yields because a large excess of ammonia is used, making ammonia the nucleophile that is most likely to displace bromine. Also, the adjacent carboxylate ion in the product reduces the nucleophilicity of the amino group. The following sequence shows bromination of 3-phenylpropanoic acid, followed by displacement of bromide ion, to form the ammonium salt of racemic phenylalanine.

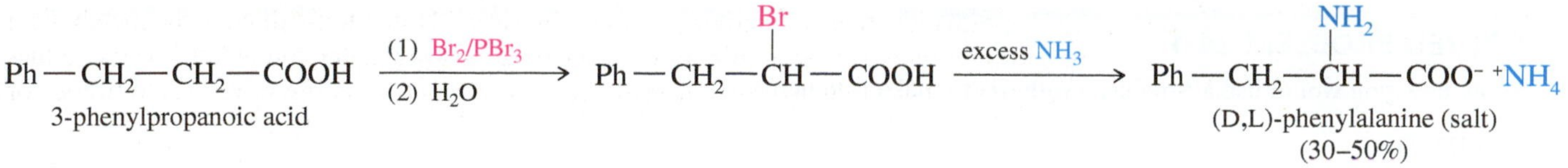

$$\underset{\text{3-phenylpropanoic acid}}{Ph-CH_2-CH_2-COOH} \xrightarrow[(2)\ H_2O]{(1)\ Br_2/PBr_3} Ph-CH_2-\overset{Br}{\overset{|}{C}H}-COOH \xrightarrow{\text{excess } NH_3} \underset{\substack{\text{(D,L)-phenylalanine (salt)} \\ (30\text{–}50\%)}}{Ph-CH_2-\overset{NH_2}{\overset{|}{C}H}-COO^- \ {}^+NH_4}$$

PROBLEM 24-10

Show how you would use bromination followed by amination to synthesize the following amino acids.
(a) glycine **(b)** leucine **(c)** glutamic acid

24-5C The Strecker Synthesis

The first known synthesis of an amino acid occurred in 1850 in the laboratory of Adolph Strecker in Tübingen, Germany. Strecker added acetaldehyde to an aqueous solution of ammonia and HCN. The product was α-amino propionitrile, which Strecker hydrolyzed to racemic alanine.

The Strecker synthesis of alanine

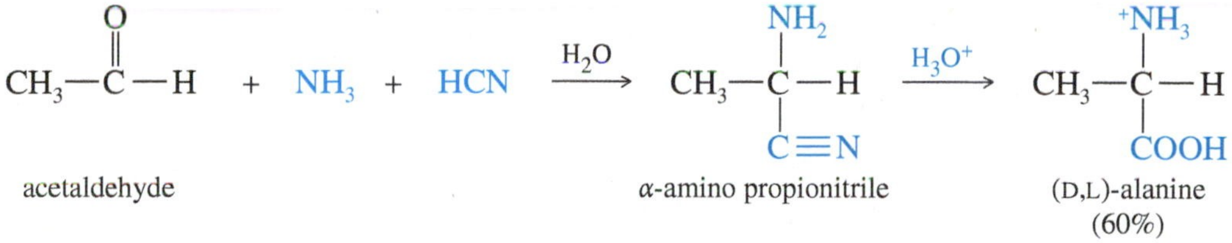

$$\underset{\text{acetaldehyde}}{CH_3-\overset{O}{\overset{\|}{C}}-H} + NH_3 + HCN \xrightarrow{H_2O} \underset{\alpha\text{-amino propionitrile}}{CH_3-\overset{NH_2}{\overset{|}{\underset{C\equiv N}{\underset{|}{C}}}}-H} \xrightarrow{H_3O^+} \underset{\substack{\text{(D,L)-alanine} \\ (60\%)}}{CH_3-\overset{^+NH_3}{\overset{|}{\underset{COOH}{\underset{|}{C}}}}-H}$$

The **Strecker synthesis** can form a large number of amino acids from appropriate aldehydes. The mechanism is shown next. First, the aldehyde reacts with ammonia to give an imine. The imine is a nitrogen analogue of a carbonyl group, and it is electrophilic when protonated. Attack of cyanide ion on the protonated imine gives the α-amino nitrile. This mechanism is similar to that for formation of a cyanohydrin (Section 18-14), except that in the Strecker synthesis cyanide ion attacks an imine rather than the aldehyde itself.

Step 1: The aldehyde reacts with ammonia to form the imine (mechanism in Section 18-14).

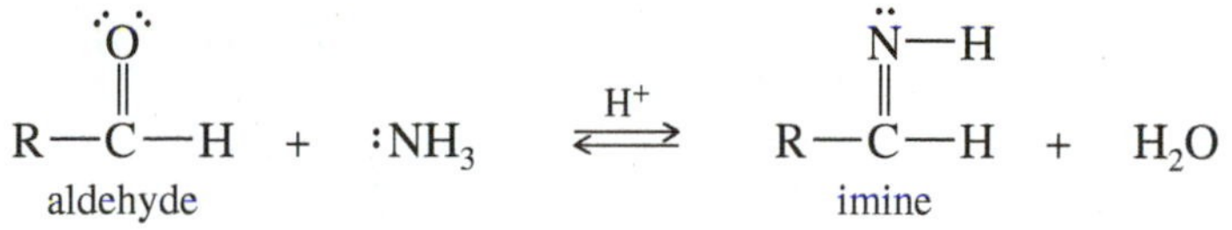

$$\underset{\text{aldehyde}}{R-\overset{\ddot{O}:}{\overset{\|}{C}}-H} + :NH_3 \overset{H^+}{\rightleftharpoons} \underset{\text{imine}}{R-\overset{\ddot{N}-H}{\overset{\|}{C}}-H} + H_2O$$

Step 2: Cyanide ion attacks the protonated imine.

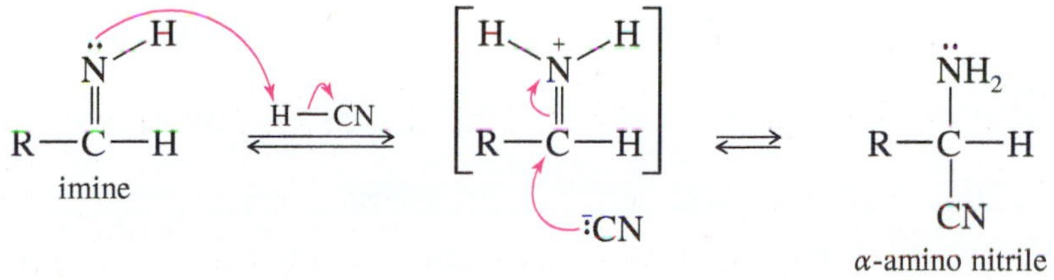

$$\underset{\text{imine}}{R-\overset{\ddot{N}-H}{\overset{\|}{C}}-H} \overset{H-CN}{\rightleftharpoons} \left[R-\overset{H-\overset{+}{N}-H}{\overset{\|}{C}}-H\right] \xrightarrow{:CN^-} \rightleftharpoons \underset{\alpha\text{-amino nitrile}}{R-\overset{\ddot{N}H_2}{\overset{|}{\underset{CN}{\underset{|}{C}}}}-H}$$

In a separate step, hydrolysis of the α-amino nitrile (Section 21-7D) gives an α-amino acid.

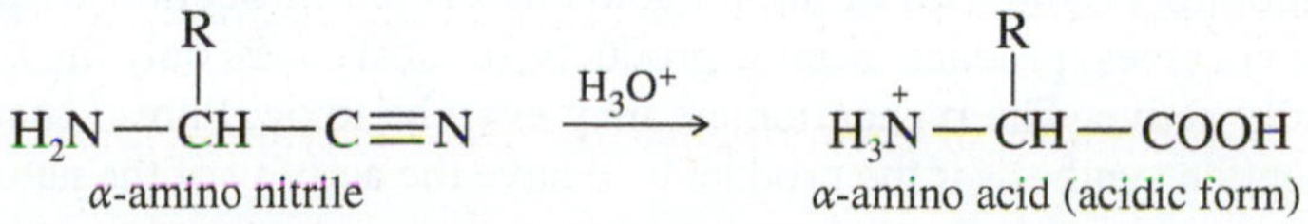

$$\underset{\alpha\text{-amino nitrile}}{H_2N-\overset{R}{\overset{|}{C}H}-C\equiv N} \xrightarrow{H_3O^+} \underset{\alpha\text{-amino acid (acidic form)}}{H_3\overset{+}{N}-\overset{R}{\overset{|}{C}H}-COOH}$$

PROBLEM-SOLVING HINT

In the Strecker synthesis, the aldehyde carbon becomes the α carbon of the amino acid. Therefore, you begin with R—CHO, where R is the side chain of the amino acid.

SOLVED PROBLEM 24-1

Show how you would use a Strecker synthesis to make isoleucine.

SOLUTION

Isoleucine has a *sec*-butyl group for its side chain. Recall that CH_3—CHO undergoes Strecker synthesis to give alanine, with CH_3 as the side chain. Therefore, *sec*-butyl—CHO should give isoleucine.

$$CH_3CH_2CH(CH_3)-C(=O)-H \xrightarrow[H_2O]{NH_3,\ HCN} CH_3CH_2CH(CH_3)-C(NH_2)(C\equiv N)-H \xrightarrow{H_3O^+} CH_3CH_2CH(CH_3)-C(^+NH_3)(COOH)-H$$

sec-butyl—CHO (2-methylbutanal) → (D,L)-isoleucine

PROBLEM 24-11

(a) Show how you would use a Strecker synthesis to make phenylalanine.
(b) Propose a mechanism for each step in the synthesis in part (a).

PROBLEM 24-12

Show how you would use a Strecker synthesis to make
(a) leucine. **(b)** valine. **(c)** aspartic acid.

SUMMARY Syntheses of Amino Acids

1. *Reductive amination* (Section 24-5A)

$$R-C(=O)-COOH \xrightarrow{\text{excess } NH_3} R-C(=N-H)-COO^-\ ^+NH_4 \xrightarrow[Pd]{H_2} R-CH(NH_2)-COO^-$$

α-ketoacid → imine → α-amino acid

2. *Amination of an α-haloacid* (Section 24-5B)

$$R-CH_2-C(=O)-OH \xrightarrow[(2)\ H_2O]{(1)\ Br_2/PBr_3} R-CH(Br)-C(=O)-OH \xrightarrow[\text{(large excess)}]{NH_3} R-CH(NH_2)-C(=O)-O^-\ ^+NH_4$$

carboxylic acid → α-bromo acid → (D,L)-α-amino salt (ammonium salt)

3. *The Strecker synthesis* (Section 24-5C)

$$R-C(=O)-H + NH_3 + HCN \xrightarrow{H_2O} R-C(NH_2)(C\equiv N)-H \xrightarrow{H_3O^+} R-C(^+NH_3)(COOH)-H$$

aldehyde → α-amino nitrile → α-amino acid

24-6 Resolution of Amino Acids

All of the laboratory syntheses of amino acids described in Section 24-5 (except for those that use enzymes) produce racemic products. In most cases, only the L enantiomers are biologically active. The D enantiomers may even be toxic. Pure L enantiomers are needed for peptide synthesis if the product is to have the activity of the natural material. Therefore, we must be able to resolve a racemic amino acid into its enantiomers.

In many cases, amino acids can be resolved by the methods we have already discussed (Section 5-16). If a racemic amino acid is converted to a salt with an optically pure chiral acid or base, two diastereomeric salts are formed. These salts can be separated by physical means such as selective crystallization or chromatography. Pure enantiomers are then regenerated from the separated diastereomeric salts. Strychnine and brucine are naturally occurring optically active bases, and tartaric acid is used as an optically active acid for resolving racemic mixtures.

Enzymatic resolution is also used to separate the enantiomers of amino acids. Enzymes are chiral molecules with specific catalytic activities. For example, when an acylated amino acid is treated with an enzyme like hog kidney acylase or carboxypeptidase, the enzyme cleaves the acyl group from just the molecules having the natural (L) configuration. The enzyme does not recognize D-amino acids, so they are unaffected. The resulting mixture of acylated D-amino acid and deacylated L-amino acid is easily separated. Figure 24-5 shows how this selective enzymatic deacylation is accomplished.

PROBLEM 24-13

Suggest how you would separate the free L-amino acid from its acylated D enantiomer in Figure 24-5.

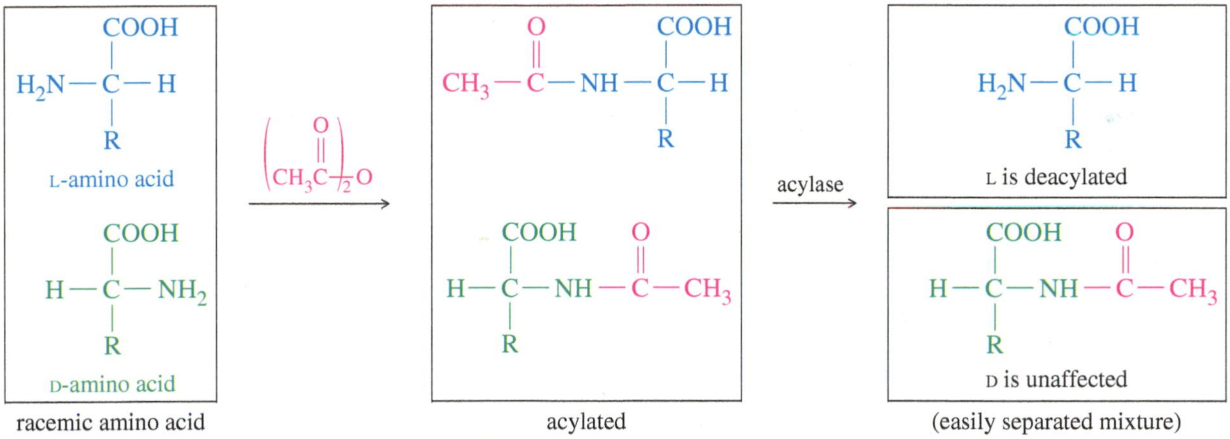

L-amino acid ($H_2N—C(COOH)(R)—H$)

D-amino acid ($H—C(COOH)(R)—NH_2$)

racemic amino acid

$\xrightarrow{(CH_3CO)_2O}$

$CH_3—C(=O)—NH—C(COOH)(R)—H$

$H—C(COOH)(R)—NH—C(=O)—CH_3$

acylated

$\xrightarrow{\text{acylase}}$

$H_2N—C(COOH)(R)—H$

L is deacylated

$H—C(COOH)(R)—NH—C(=O)—CH_3$

D is unaffected

(easily separated mixture)

FIGURE 24-5
Selective enzymatic deacylation. An acylase enzyme (such as hog kidney acylase or carboxypeptidase) deacylates only the natural L-amino acid.

24-7 Reactions of Amino Acids

Amino acids undergo many of the standard reactions of both amines and carboxylic acids. Conditions for some of these reactions must be carefully selected, however, so that the amino group does not interfere with a carboxyl group reaction, and vice versa. We will consider two of the most useful reactions: esterification of the carboxyl group and acylation of the amino group. These reactions are often used to protect either the carboxyl group or the amino group while the other group is being modified or coupled to another amino acid. Amino acids also undergo reactions that are specific to the α-amino acid structure. One of these unique amino acid reactions is the formation of a colored product on treatment with ninhydrin, discussed in Section 24-7C.

24-7A Esterification of the Carboxyl Group

Like monofunctional carboxylic acids, amino acids are esterified by treatment with a large excess of an alcohol and an acidic catalyst (often gaseous HCl). Under these acidic conditions, the amino group is present in its protonated ($—NH_3^+$) form, so it does not interfere with esterification. The following example illustrates esterification of an amino acid.

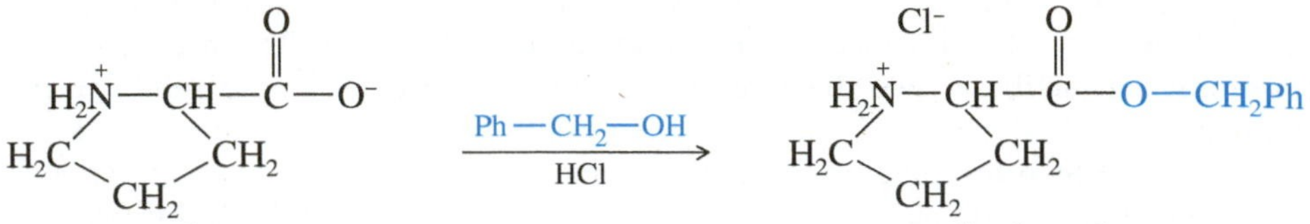

$$\text{proline} \xrightarrow[\text{HCl}]{Ph-CH_2-OH} \text{proline benzyl ester (90\%)}$$

Esters of amino acids are often used as protected derivatives to prevent the carboxyl group from reacting in some undesired manner. Methyl, ethyl, and benzyl esters are the most common protecting groups. Aqueous acid hydrolyzes the ester and regenerates the free amino acid.

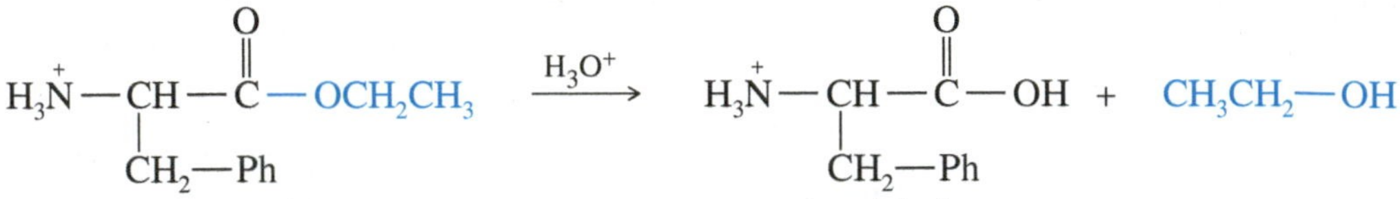

$$H_3\overset{+}{N}-CH(CH_2-Ph)-C(=O)-OCH_2CH_3 \xrightarrow{H_3O^+} H_3\overset{+}{N}-CH(CH_2-Ph)-C(=O)-OH + CH_3CH_2-OH$$

phenylalanine ethyl ester → phenylalanine

Benzyl esters are particularly useful as protecting groups because they can be removed either by acidic hydrolysis or by neutral **hydrogenolysis** ("breaking apart by addition of hydrogen"). Catalytic hydrogenation cleaves the benzyl ester, converting the benzyl group to toluene and leaving the deprotected amino acid. Although the mechanism of this hydrogenolysis is not well known, it apparently hinges on the ease of formation of benzylic intermediates.

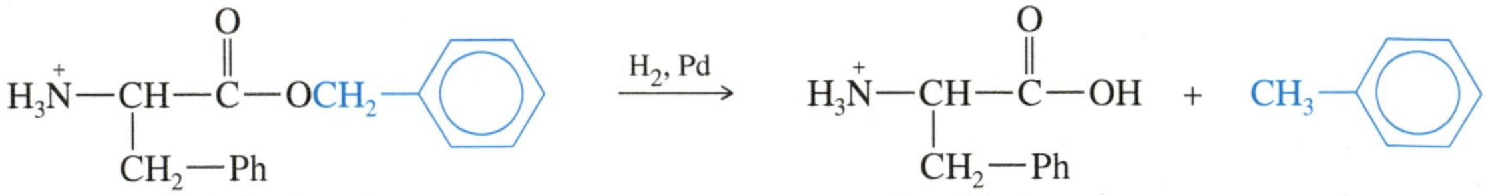

$$H_3\overset{+}{N}-CH(CH_2-Ph)-C(=O)-OCH_2-C_6H_5 \xrightarrow{H_2,\ Pd} H_3\overset{+}{N}-CH(CH_2-Ph)-C(=O)-OH + CH_3-C_6H_5$$

phenylalanine benzyl ester → phenylalanine + toluene

PROBLEM 24-14

Propose a mechanism for the acid-catalyzed hydrolysis of phenylalanine ethyl ester.

PROBLEM 24-15

Give equations for the formation and hydrogenolysis of glutamine benzyl ester.

Application: Allergy

Decarboxylation is an important reaction of amino acids in many biological processes. Histamine, which causes runny noses and itchy eyes, is synthesized in the body by decarboxylation of histidine (shown at right). The enzyme that catalyzes this reaction is called histidine decarboxylase.

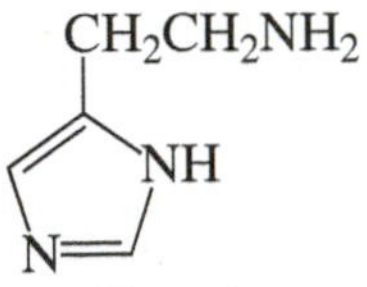

$CH_2CH_2NH_2$ (on imidazole ring: NH, N)

histamine

24-7B Acylation of the Amino Group: Formation of Amides

Just as an alcohol esterifies the carboxyl group of an amino acid, an acylating agent converts the amino group to an amide. Acylation of the amino group is often done to protect it from unwanted nucleophilic reactions. A wide variety of acid chlorides and anhydrides are used for acylation. Benzyl chloroformate acylates the amino group to give a benzyloxycarbonyl derivative, often used as a protecting group in peptide synthesis (Section 24-10).

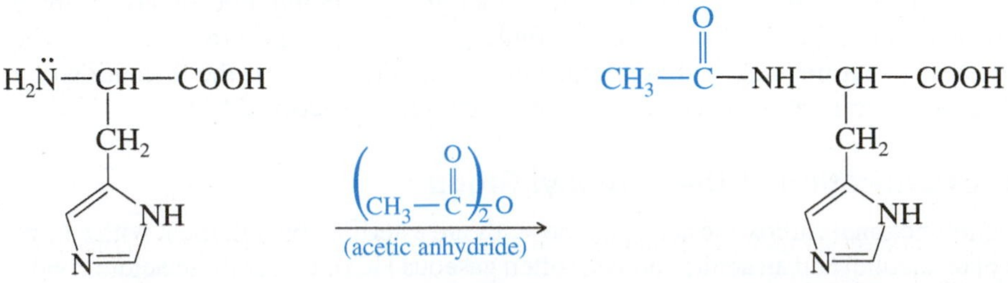

$$H_2\ddot{N}-CH(CH_2\text{-imidazolyl})-COOH \xrightarrow{(CH_3-C(=O))_2O} CH_3-C(=O)-NH-CH(CH_2\text{-imidazolyl})-COOH$$

histidine → (acetic anhydride) → *N*-acetylhistidine

$$H_2\ddot{N}-CH(CH_2CH(CH_3)_2)-COOH \xrightarrow[\text{(benzyl chloroformate)}]{PhCH_2OC(=O)-Cl} PhCH_2O-C(=O)-NH-CH(CH_2CH(CH_3)_2)-COOH$$

leucine → *N*-benzyloxycarbonyl leucine (90%)

The amino group of the *N*-benzyloxycarbonyl derivative is protected as the amide half of a carbamate ester (a urethane, Section 21-16), which is more easily hydrolyzed than most other amides. In addition, the ester half of this urethane is a benzyl ester that undergoes hydrogenolysis. Catalytic hydrogenolysis of the *N*-benzyloxycarbonyl amino acid gives an unstable carbamic acid that quickly decarboxylates to give the deprotected amino acid.

$$C_6H_5CH_2-O-C(=O)-N(H)-CH(CH_2CH(CH_3)_2)-COOH \xrightarrow{H_2,\ Pd} C_6H_5CH_3 + \left[HO-C(=O)-N(H)-CH(CH_2CH(CH_3)_2)-COOH\right] \longrightarrow CO_2\uparrow + H_2N-CH(CH_2CH(CH_3)_2)-COOH$$

N-benzyloxycarbonyl leucine; toluene; a carbamic acid; leucine

PROBLEM 24-16

Give equations for the formation and hydrogenolysis of *N*-benzyloxycarbonyl methionine.

24-7C Reaction with Ninhydrin

Ninhydrin is a common reagent for visualizing spots or bands of amino acids that have been separated by chromatography or electrophoresis. When ninhydrin reacts with an amino acid, one of the products is a deep violet, resonance-stabilized anion called *Ruhemann's purple*. Ninhydrin produces this same purple dye regardless of the structure of the original amino acid. The side chain of the amino acid is lost as an aldehyde.

Reaction of an amino acid with ninhydrin

$$H_2N-CH(R)-COOH + 2\ \text{ninhydrin} \xrightarrow{\text{pyridine}} \text{Ruhemann's purple} + CO_2\uparrow + R-CHO$$

amino acid; ninhydrin; Ruhemann's purple

The reaction of amino acids with ninhydrin can detect amino acids on a wide variety of substrates. For example, if a kidnapper touches a ransom note with his fingers, the dermal ridges on his fingers leave traces of amino acids from skin secretions. Treatment of the paper with ninhydrin and pyridine causes these secretions to turn purple, forming a visible fingerprint.

PROBLEM 24-17

Use resonance forms to show delocalization of the negative charge in the Ruhemann's purple anion.

SUMMARY Reactions of Amino Acids

1. *Esterification of the carboxyl group* (Section 24-7A)

$$\underset{\text{amino acid}}{H_3\overset{+}{N}-\underset{R}{CH}-\overset{O}{\overset{\|}{C}}-O^-} + \underset{\text{alcohol}}{R'-OH} \xrightarrow{H^+} \underset{\text{amino ester}}{H_3\overset{+}{N}-\underset{R}{CH}-\overset{O}{\overset{\|}{C}}-O-R'} + H_2O$$

2. *Acylation of the amino group: formation of amides* (Section 24-7B)

$$\underset{\text{amino acid}}{H_2N-\underset{R}{CH}-\overset{O}{\overset{\|}{C}}-OH} + \underset{\text{acylating agent}}{R'-\overset{O}{\overset{\|}{C}}-X} \longrightarrow \underset{\text{acylated amino acid}}{R'-\overset{O}{\overset{\|}{C}}-NH-\underset{R}{CH}-\overset{O}{\overset{\|}{C}}-OH} + H-X$$

3. *Reaction with ninhydrin* (Section 24-7C)

$$\underset{\text{amino acid}}{H_2N-\underset{R}{CH}-COOH} + 2\ \text{ninhydrin} \xrightarrow{\text{pyridine}} \text{Ruhemann's purple} + R-CHO + CO_2\uparrow$$

4. *Formation of peptide bonds* (Sections 24-8 and 24-10)

$$H_3\overset{+}{N}-\underset{R^1}{CH}-\overset{O}{\overset{\|}{C}}-O^- + H_3\overset{+}{N}-\underset{R^2}{CH}-\overset{O}{\overset{\|}{C}}-O^- \xrightarrow{\text{loss of } H_2O} H_3\overset{+}{N}-\underset{R^1}{CH}-\overset{O}{\overset{\|}{C}}-NH-\underset{R^2}{CH}-\overset{O}{\overset{\|}{C}}-O^-$$

peptide bond

Amino acids also undergo many other common reactions of amines and acids.

24-8 Structure and Nomenclature of Peptides and Proteins

Amino acids are the monomers that combine to form polymers: peptides and proteins. Peptide hormones and the various types of proteins (Section 24-11) are essential to life at the molecular level.

24-8A Peptide Structure

The most important reaction of amino acids is the formation of peptide bonds. Amines and acids can condense, with the loss of water, to form amides. Industrial processes often make amides simply by mixing the acid and the amine, and then heating the mixture to drive off water.

$$\underset{\text{acid}}{R-\overset{O}{\overset{\|}{C}}-OH} + \underset{\text{amine}}{H_2\ddot{N}-R'} \longrightarrow \underset{\text{salt}}{R-\overset{O}{\overset{\|}{C}}-O^-\ H_3\overset{+}{N}-R'} \xrightarrow{\text{heat}} \underset{\text{amide}}{R-\overset{O}{\overset{\|}{C}}-\ddot{N}H-R'} + H_2O$$

Recall from Section 21-13 that amides are the most stable acid derivatives. This stability is partly due to the strong resonance interaction between the nonbonding electrons on nitrogen and the carbonyl group. The amide nitrogen is no longer a strong base, and the C—N bond has restricted rotation because of its partial double-bond character. Figure 24-6 shows the resonance forms we use to explain the partial double-bond

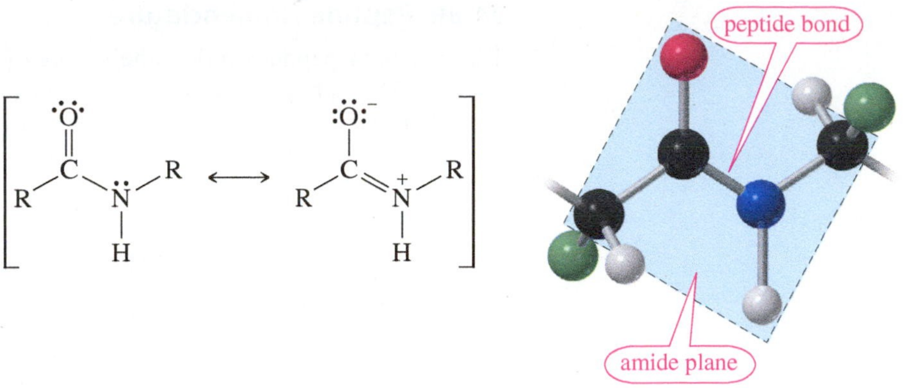

FIGURE 24-6
Resonance stabilization of an amide accounts for its enhanced stability, the weak basicity of the nitrogen atom, and the restricted rotation of the C—N bond. In a peptide, the amide bond is called a peptide bond. It holds six atoms in a plane: the C and O of the carbonyl, the N and its H, and the two associated α carbon atoms.

character and restricted rotation of an amide bond. In a peptide, this partial double-bond character results in six atoms being held rather rigidly in a plane.

Having both an amino group and a carboxyl group, an amino acid is ideally suited to form an amide linkage. Under the proper conditions, the amino group of one molecule condenses with the carboxyl group of another. The product is an amide called a *dipeptide* because it consists of two amino acids. The amide linkage between the amino acids is called a **peptide bond.** Although it has a special name, a peptide bond is just like other amide bonds we have studied.

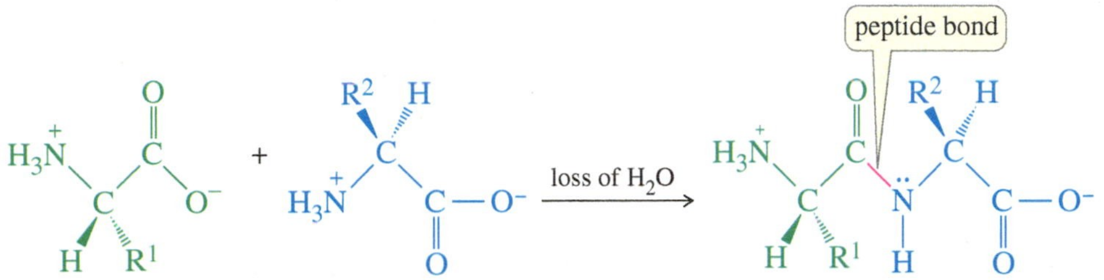

In this manner, any number of amino acids can be bonded in a continuous chain. A **peptide** is a compound containing two or more amino acids linked by amide bonds between the amino group of each amino acid and the carboxyl group of the neighboring amino acid. Each amino acid unit in the peptide is called a **residue.** A **polypeptide** is a peptide containing many amino acid residues but usually having a molecular weight of less than about 5000. **Proteins** contain more amino acid units, with molecular weights ranging from about 5000 to about 40,000,000. The term **oligopeptide** is occasionally used for peptides containing about 4 to 10 amino acid residues. Figure 24-7 shows the structure of the nonapeptide bradykinin, a human hormone that helps to control blood pressure.

The end of the peptide with the free amino group ($-NH_3^+$) is called the **N-terminal end** or the **N terminus,** and the end with the free carboxyl group ($-COO^-$) is called the **C-terminal end** or the **C terminus.** Peptide structures are generally drawn with the N terminus at the left and the C terminus at the right, as bradykinin is drawn in Figure 24-7.

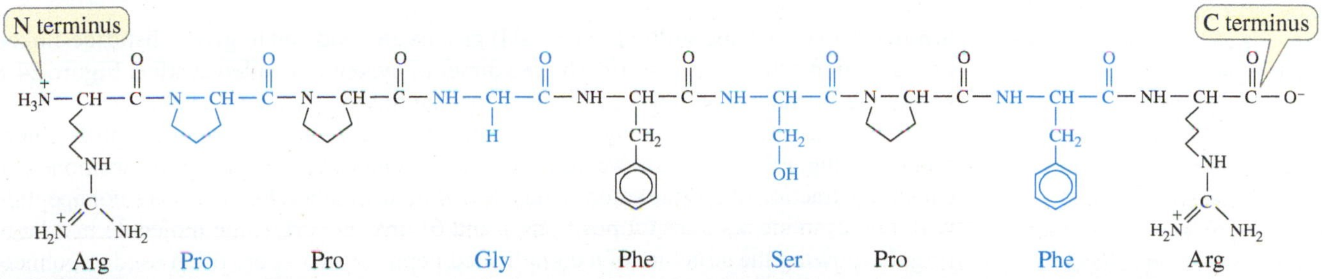

FIGURE 24-7
The human hormone bradykinin is a nonapeptide with a free $-NH_3^+$ at its N terminus and a free $-COO^-$ at its C terminus.

24-8B Peptide Nomenclature

The names of peptides reflect the names of the amino acid residues involved in the amide linkages, beginning at the N terminus. All except the last are given the *-yl* suffix of acyl groups. For example, the following dipeptide is named alanylserine. The alanine residue has the *-yl* suffix because it has acylated the nitrogen of serine.

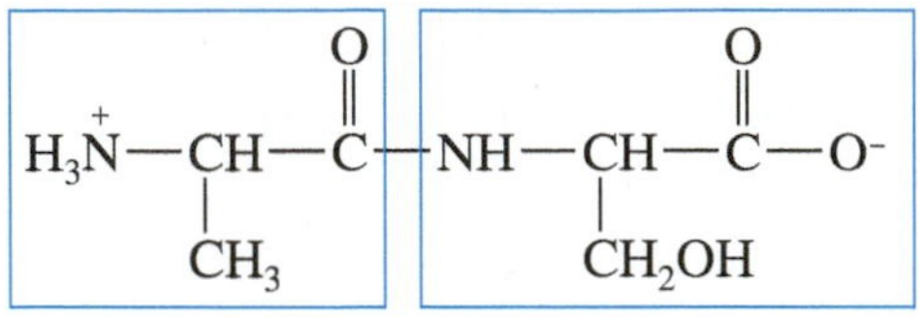

$$\underbrace{H_3\overset{+}{N}-\underset{\underset{CH_3}{|}}{CH}-\overset{\overset{O}{\|}}{C}}_{\text{alanyl}}-\underbrace{NH-\underset{\underset{CH_2OH}{|}}{CH}-\overset{\overset{O}{\|}}{C}-O^-}_{\text{serine}}$$

Ala-Ser

Bradykinin (Figure 24-7) is named as follows (without any spaces):

arginyl prolyl prolyl glycyl phenylalanyl seryl prolyl phenylalanyl arginine

This is a cumbersome and awkward name. A shorthand system is more convenient, representing each amino acid by its three-letter abbreviation. These abbreviations, given in Table 24-2, are generally the first three letters of the name. Once again, the amino acids are arranged from the N terminus at the left to the C terminus at the right. Bradykinin has the following abbreviated name:

Arg-Pro-Pro-Gly-Phe-Ser-Pro-Phe-Arg

Single-letter symbols (also given in Table 24-2) are widely used as well. Using single letters, we symbolize bradykinin by

RPPGFSPFR

PROBLEM 24-18

Draw the complete structures of the following peptides:
(a) Thr-Phe-Met **(b)** serylarginylglycylphenylalanine **(c)** IMQDK **(d)** ELVIS

24-8C Disulfide Linkages

Amide linkages (peptide bonds) form the backbone of the amino acid chains we call peptides and proteins. A second kind of covalent bond is possible between any cysteine residues present. Cysteine residues can form **disulfide bridges** (also called **disulfide linkages**) that can join two chains or link a single chain into a ring.

Mild oxidation joins two molecules of a thiol into a disulfide, forming a disulfide linkage between the two thiol molecules. This reaction is reversible, and a mild reduction cleaves the disulfide.

$$\underbrace{R-SH \; + \; HS-R}_{\text{two molecules of thiol}} \underset{[\text{reduction}]}{\overset{[\text{oxidation}]}{\rightleftharpoons}} \underbrace{R-S-S-R}_{\text{disulfide}} \; + \; H_2O$$

Similarly, two cysteine sulfhydryl (—SH) groups are oxidized to give a disulfide-linked pair of amino acids. This disulfide-linked dimer of cysteine is called cystine. Figure 24-8 shows the formation of a cystine disulfide bridge linking two peptide chains.

Two cysteine residues may form a disulfide bridge within a single peptide chain, making a ring. Figure 24-9 shows the structure of human oxytocin, a peptide hormone that causes contraction of uterine smooth muscle and induces labor. Oxytocin is a nonapeptide with two cysteine residues (at positions 1 and 6) linking part of the molecule in a large ring. In drawing the structure of a complicated peptide, arrows are often used to connect the amino acids, showing the direction from N terminus to C terminus. Notice that the C terminus of oxytocin is a primary amide ($Gly \cdot NH_2$) rather than a free carboxyl group.

Application: Hormone

Orexin A (from the Greek ***orexis***, "appetite") is a 33-amino-acid neuropeptide with two intramolecular disulfide bridges. Orexin A is a powerful stimulant for food intake and gastric juice secretion. Scientists are studying orexin A to learn more about the regulation of appetite and eating, hoping to learn more about causes and potential treatments for anorexia nervosa.

FIGURE 24-8
Cystine, a dimer of cysteine, results when two cysteine residues are oxidized to form a disulfide bridge.

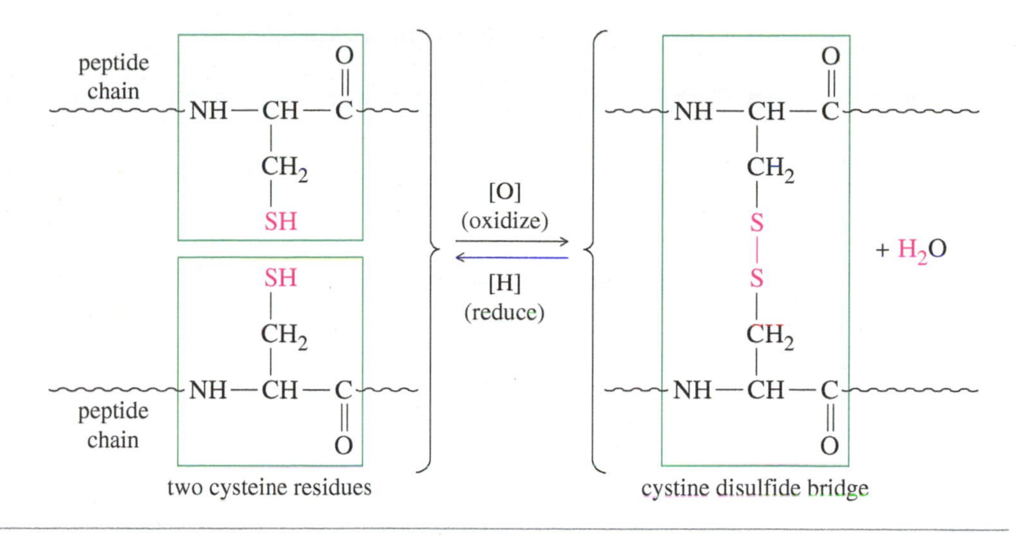

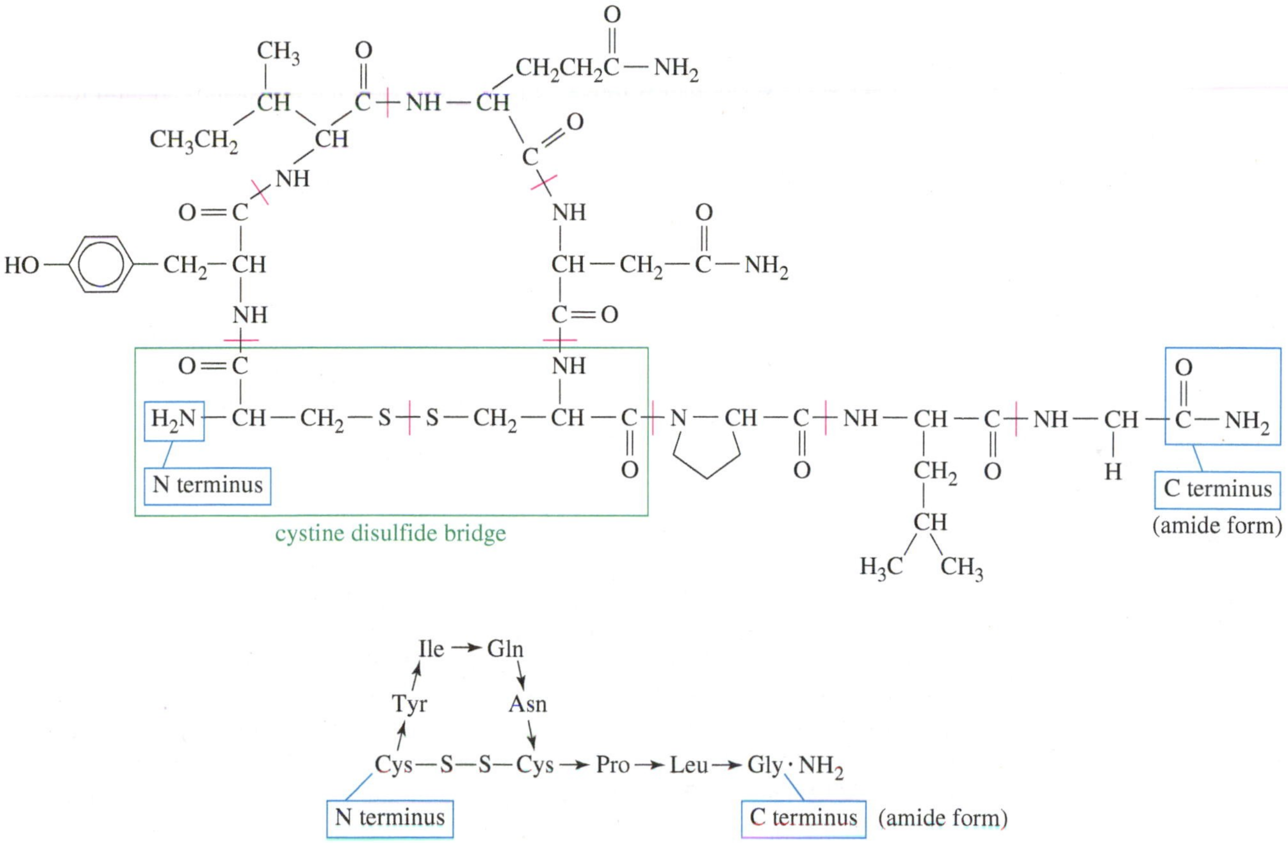

N terminus
cystine disulfide bridge
C terminus
(amide form)

Ile → Gln
Tyr Asn
Cys—S—S—Cys → Pro → Leu → Gly · NH_2
N terminus
C terminus (amide form)

FIGURE 24-9
Structure of human oxytocin. A disulfide linkage holds part of the molecule in a large ring.

Figure 24-10 shows the structure of insulin, a more complex peptide hormone that regulates glucose metabolism. Insulin is composed of two separate peptide chains: the *A chain*, containing 21 amino acid residues, and the *B chain*, containing 30. The A and B chains are joined at two positions by disulfide bridges, and the A chain has an additional disulfide bond that holds six amino acid residues in a ring. The C-terminal amino acids of both chains occur as primary amides.

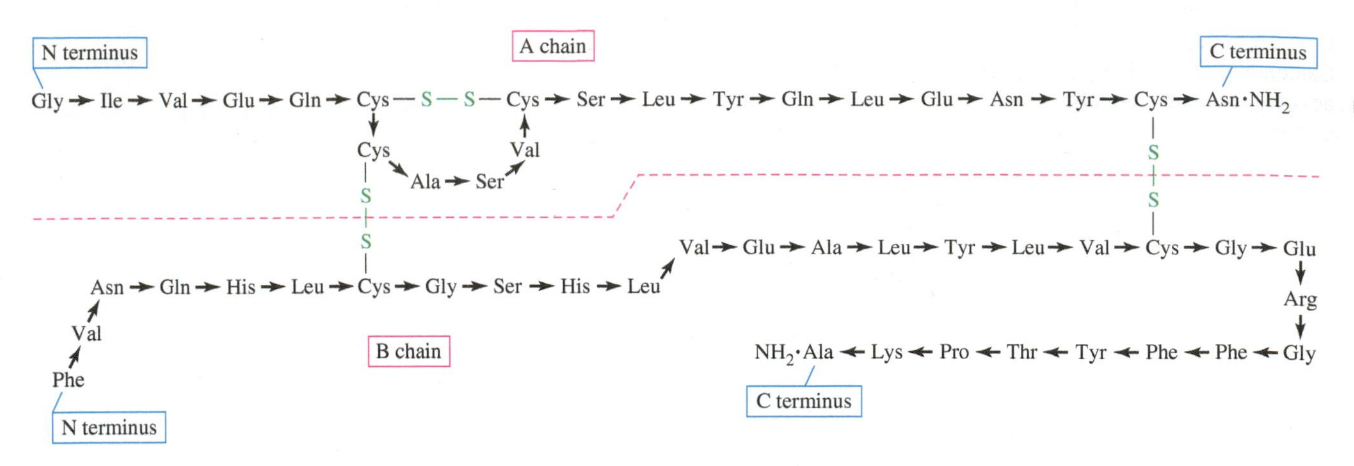

FIGURE 24-10
Structure of insulin. Two chains are joined at two positions by disulfide bridges, and a third disulfide bond holds the A chain in a ring.

Application: Permanent Wave

Reduction of the disulfide bridges in hair, followed by re-oxidation while held by curlers, re-forms the disulfide bridges in new positions. These new disulfide bridges hold the hair in the curved shape of the curlers.

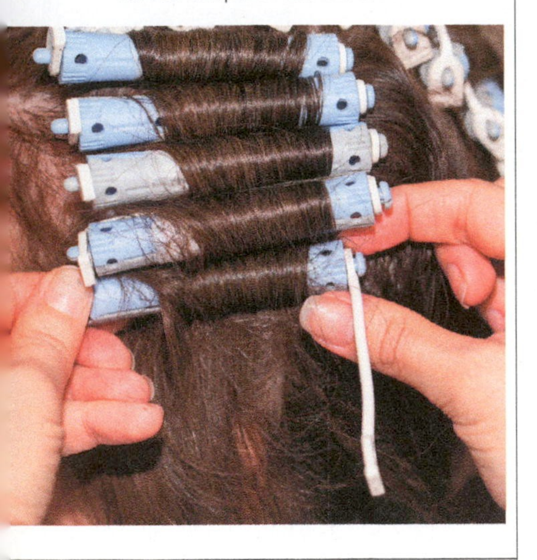

Disulfide bridges are commonly manipulated in the process of giving hair a *permanent wave*. Hair is composed of protein, which is made rigid and tough partly by disulfide bonds. When hair is treated with a solution of a thiol such as sodium thioglycolate ($HS—CH_2—COONa$), the disulfide bridges are reduced and cleaved. The hair is wrapped around curlers, and the disulfide bonds are allowed to re-form, either by air oxidation or by application of a *neutralizer* (dilute H_2O_2). The disulfide bonds re-form in new positions, holding the hair in the bent conformation enforced by the curlers.

24-9 Peptide Structure Determination

Insulin is a relatively simple protein, yet it is a complicated organic structure. How is it possible to determine the complete structure of a protein with hundreds of amino acid residues and a molecular weight of many thousands? Chemists have developed clever ways to determine the exact sequence of amino acids in a protein. We will consider some of the most common methods.

24-9A Cleavage of Disulfide Linkages

The first step in structure determination is to break all the disulfide bonds, opening any disulfide-linked rings and separating the individual peptide chains. The individual peptide chains are then purified and analyzed separately.

Cystine bridges are easily cleaved by reducing them to the thiol (cysteine) form. These reduced cysteine residues have a tendency to reoxidize and re-form disulfide bridges, however. A more permanent cleavage involves oxidizing the disulfide linkages with peroxyformic acid (Figure 24-11). This oxidation converts the disulfide bridges to sulfonic acid ($—SO_3H$) groups. The oxidized cysteine units are called *cysteic acid* residues.

24-9B Determination of the Amino Acid Composition

Once the disulfide bridges have been broken and the individual peptide chains have been separated and purified, the structure of each chain must be determined. The first step is to determine which amino acids are present and in what proportions. To analyze the amino acid composition, the peptide chain is completely hydrolyzed by boiling it for 24 hours in 6 M HCl. The resulting mixture of amino acids (the *hydrolysate*) is placed on the column of an *amino acid analyzer*, diagrammed in Figure 24-12.

In the amino acid analyzer, the components of the hydrolysate are dissolved in an aqueous buffer solution and separated by passing them down an ion-exchange column. The solution emerging from the column is mixed with ninhydrin, which reacts with amino acids to give the purple ninhydrin color. The absorption of light is recorded and printed out as a function of time.

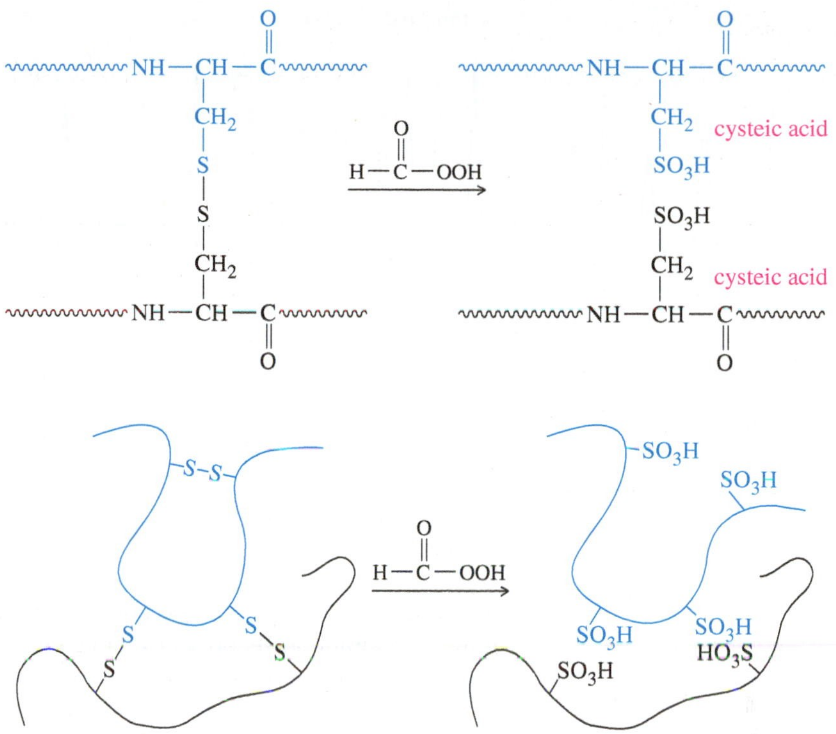

FIGURE 24-11
Oxidation of a protein by peroxyformic acid cleaves all the disulfide linkages by oxidizing cystine to cysteic acid.

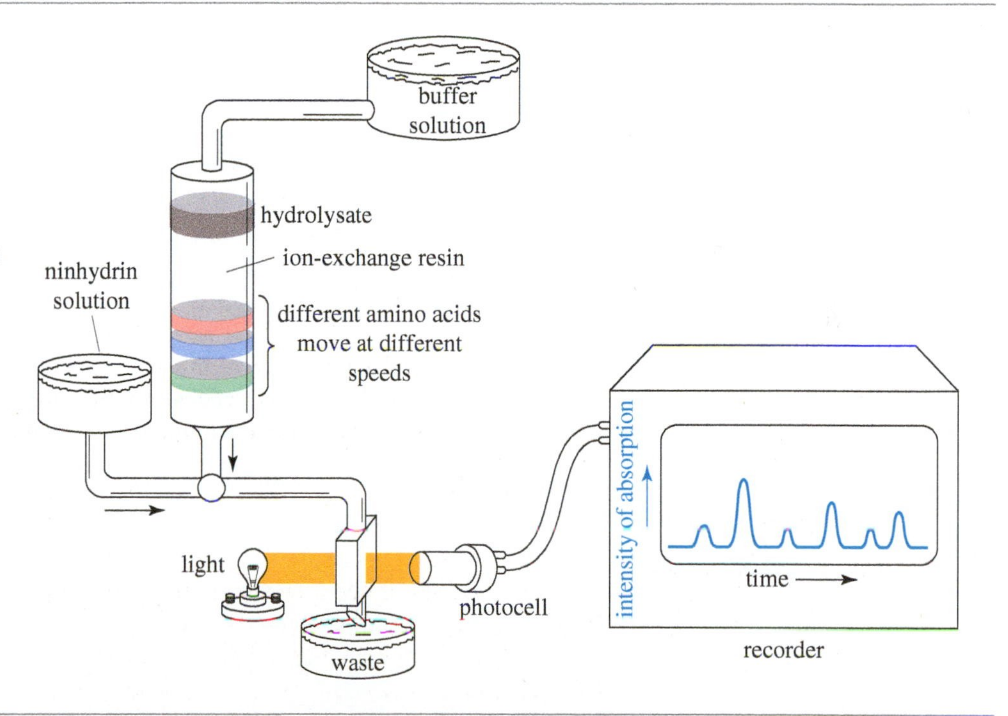

FIGURE 24-12
In an amino acid analyzer, the hydrolysate passes through an ion-exchange column. The solution emerging from the column is treated with ninhydrin, and its absorbance is recorded as a function of time. Each amino acid is identified by the retention time required to pass through the column.

The time required for each amino acid to pass through the column (its *retention time*) depends on how strongly that amino acid interacts with the ion-exchange resin. The retention time of each amino acid is known from standardization with pure amino acids. The amino acids present in the sample are identified by comparing their retention times with the known values. The area under each peak is nearly proportional to the amount of the amino acid producing that peak, so we can determine the relative amounts of amino acids present.

Figure 24-13 shows a standard trace of an equimolar mixture of amino acids, followed by the trace produced by the hydrolysate from human bradykinin (Arg-Pro-Pro-Gly-Phe-Ser-Pro-Phe-Arg).

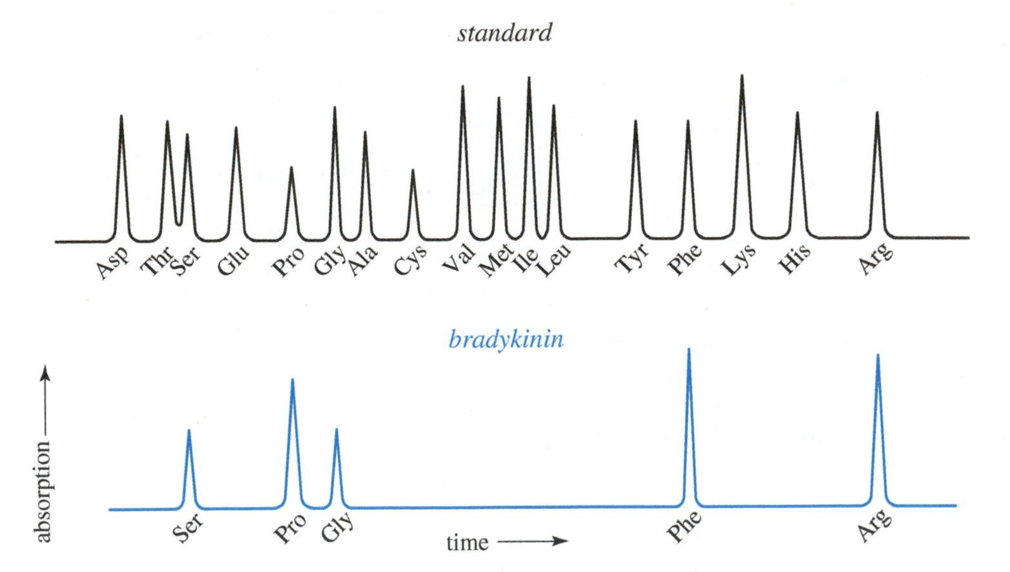

FIGURE 24-13
Use of an amino acid analyzer to determine the composition of human bradykinin. The bradykinin peaks for Pro, Arg, and Phe are larger than those in the standard equimolar mixture because bradykinin has three Pro residues, two Arg residues, and two Phe residues.

Sequencing the Peptide: Terminal Residue Analysis The amino acid analyzer determines the amino acids present in a peptide, but it does not reveal their **sequence:** the order in which they are linked together. The peptide sequence is destroyed in the hydrolysis step. To determine the amino acid sequence, we must cleave just one amino acid from the chain and leave the rest of the chain intact. The cleaved amino acid can be separated and identified, and the process can be repeated on the rest of the chain. The amino acid may be cleaved from either end of the peptide (either the N terminus or the C terminus), so we will consider one method used for each end. This general method for peptide sequencing is called **terminal residue analysis.**

24-9C Sequencing from the N Terminus: The Edman Degradation

The most efficient method for sequencing peptides is the **Edman degradation.** A peptide is treated with phenyl isothiocyanate, followed by acid hydrolysis. The products are the shortened peptide chain and a heterocyclic derivative of the N-terminal amino acid called a *phenylthiohydantoin.*

This reaction takes place in three stages. First, the free amino group of the N-terminal amino acid reacts with phenylisothiocyanate to form a phenylthiourea. Second, the phenylthiourea cyclizes to a thiazolinone and expels the shortened peptide chain. Third, the thiazolinone isomerizes to the more stable phenylthiohydantoin.

Step 1: Nucleophilic attack by the free amino group on phenyl isothiocyanate, followed by a proton transfer, gives a phenylthiourea.

Ph—N=C=S: + H_2N—CH(R^1)—C(=O)—NH—peptide ⇌ Ph—N=C(—S:⁻)—H_2N^+—CH(R^1)—C(=O)—NH—peptide ⇌ Ph—NH—C(=S)—HN—CH(R^1)—C(=O)—NH—peptide

a phenylthiourea

Step 2: Treatment with HCl induces cyclization to a thiazolinone and expulsion of the shortened peptide chain.

protonated phenylthiourea ⇌ ⇌ ⇌ a thiazolinone + H_2N—peptide + H_3O^+

Step 3: In acid, the thiazolinone isomerizes to the more stable phenylthiohydantoin.

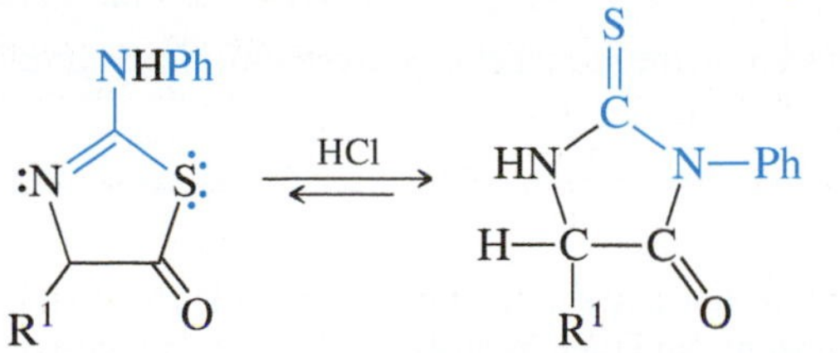

NHPh, :N, S:, R^1, O — HCl ⇌ — S, C, HN, N—Ph, H—C—C, R^1, O

thiazolinone   a phenylthiohydantoin

The phenylthiohydantoin derivative is identified by chromatography, by comparing it with phenylthiohydantoin derivatives of the standard amino acids. This gives the identity of the original N-terminal amino acid. The rest of the peptide is cleaved intact, and further Edman degradations are used to identify additional amino acids in the chain. This process is well suited to automation, and several types of automatic sequencers have been developed.

Figure 24-14 shows the first two steps in the sequencing of oxytocin. Before sequencing, the oxytocin sample is treated with peroxyformic acid to convert the disulfide bridge to cysteic acid residues.

Step 1: Cleavage and determination of the N-terminal amino acid

$H_2\ddot{N}$—CH(—CH_2—SO_3H)—C(=O)—NH—Tyr—Ile—Gln—peptide —(1) Ph—N=C=S; (2) H_3O^+→ HN, C=S, N—Ph, CH—C=O, CH_2SO_3H + $H_2\ddot{N}$—Tyr—Ile—Gln—peptide

cysteic acid   cysteic acid phenylthiohydantoin

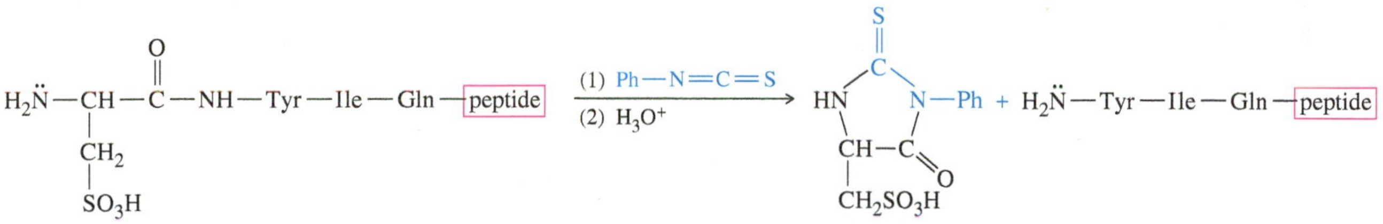

Step 2: Cleavage and determination of the second amino acid (the new N-terminal amino acid)

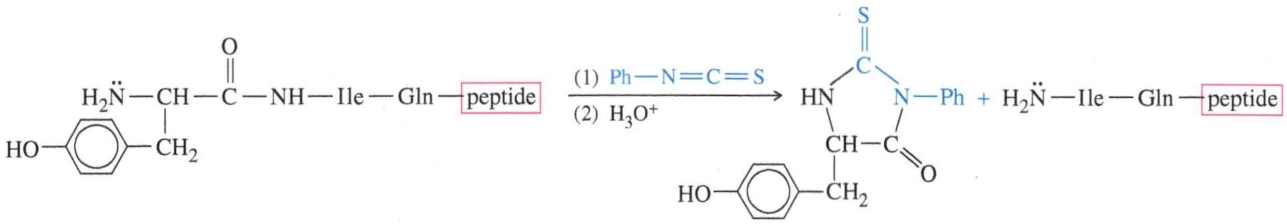

$H_2\ddot{N}$—CH(—CH_2—C₆H₄—OH)—C(=O)—NH—Ile—Gln—peptide —(1) Ph—N=C=S; (2) H_3O^+→ HN, C=S, N—Ph, CH—C=O, HO—C₆H₄—CH_2 + $H_2\ddot{N}$—Ile—Gln—peptide

tyrosine phenylthiohydantoin

FIGURE 24-14
The first two steps in sequencing oxytocin. Each Edman degradation cleaves the N-terminal amino acid and forms its phenylthiohydantoin derivative. The shortened peptide is available for the next step.

In theory, Edman degradations could sequence a peptide of any length. In practice, however, the repeated cycles of degradation cause some internal hydrolysis of the peptide, with loss of sample and accumulation of by-products. After about 50 cycles of degradation, further accurate analysis becomes impossible. A small peptide such as bradykinin can be completely determined by Edman degradation, but larger proteins must be broken into smaller fragments (Section 24-9E) before they can be completely sequenced.

PROBLEM 24-19

Draw the structure of the phenylthiohydantoin derivatives of
(a) alanine. **(b)** tryptophan. **(c)** lysine. **(d)** proline.

PROBLEM 24-20

Show the third and fourth steps in the sequencing of oxytocin. Use Figure 24-14 as a guide.

PROBLEM 24-21

The **Sanger method** for N-terminus determination is a less common alternative to the Edman degradation. In the Sanger method, the peptide is treated with the Sanger reagent, 2,4-dinitrofluorobenzene, and then hydrolyzed by reaction with 6 M aqueous HCl. The N-terminal amino acid is recovered as its 2,4-dinitrophenyl derivative and identified.

The Sanger method

O_2N–(ring)–F + $H_2\ddot{N}$—CH(R^1)—C(=O)—NH—peptide ⟶

2,4-dinitrofluorobenzene (Sanger reagent) (NO_2); peptide

O_2N–(ring, NO_2)–NH—CH(R^1)—C(=O)—NH—peptide (derivative) $\xrightarrow{\text{6 M HCl, heat}}$ O_2N–(ring, NO_2)–NH—CH(R^1)—COOH + amino acids

2,4-dinitrophenyl derivative

(a) Propose a mechanism for the reaction of the N terminus of the peptide with 2,4-dinitrofluorobenzene.
(b) Explain why the Edman degradation is usually preferred over the Sanger method.

24-9D C-Terminal Residue Analysis

There is no efficient method for sequencing several amino acids of a peptide starting from the C terminus. In many cases, however, the C-terminal amino acid can be identified using the enzyme *carboxypeptidase*, which cleaves the C-terminal peptide bond. The products are the free C-terminal amino acid and a shortened peptide. Further reaction cleaves the second amino acid that has now become the new C terminus of the shortened peptide. Eventually, the entire peptide is hydrolyzed to its individual amino acids.

peptide—NH—CH(R^{n-1})—C(=O)—NH—CH(R^n)—C(=O)—OH $\xrightarrow[H_2O]{\text{carboxypeptidase}}$ peptide—NH—CH(R^{n-1})—C(=O)—OH + H_2N—CH(R^n)—C(=O)—OH

(further cleavage) free amino acid

A peptide is incubated with the carboxypeptidase enzyme, and the appearance of free amino acids is monitored. In theory, the amino acid whose concentration increases first should be the C terminus, and the next amino acid to appear should be the second residue from the end. In practice, different amino acids are cleaved at different rates, making it difficult to determine amino acids past the C terminus and occasionally the second residue in the chain.

Application: Blood Clotting

The selective enzymatic cleavage of proteins is critical to many biological processes. For example, the clotting of blood depends on the enzyme thrombin cleaving fibrinogen at specific points to produce fibrin, the protein that forms a clot.

24-9E Breaking the Peptide into Shorter Chains: Partial Hydrolysis

Before a large protein can be sequenced, it must be broken into smaller chains, not longer than about 30 amino acids. Each of these shortened chains is sequenced, and then the entire structure of the protein is deduced by fitting the short chains together like pieces of a jigsaw puzzle.

Partial cleavage can be accomplished either by using dilute acid with a shortened reaction time or by using enzymes, such as *trypsin* and *chymotrypsin*, that break bonds between specific amino acids. The acid-catalyzed cleavage is not very selective, leading to a mixture of short fragments resulting from cleavage at various positions. Enzymes are more selective, giving cleavage at predictable points in the chain.

TRYPSIN: Cleaves the chain at the carboxyl groups of the basic amino acids lysine and arginine.

CHYMOTRYPSIN: Cleaves the chain at the carboxyl groups of the aromatic amino acids phenylalanine, tyrosine, and tryptophan.

Let's use oxytocin (Figure 24-9) as an example to illustrate the use of partial hydrolysis. Oxytocin could be sequenced directly by C-terminal analysis and a series of Edman degradations, but it provides a simple example of how a structure can be pieced together from fragments. Acid-catalyzed partial hydrolysis of oxytocin (after cleavage of the disulfide bridge) gives a mixture that includes the following peptides:

Ile-Gln-Asn-Cys   Gln-Asn-Cys-Pro   Pro-Leu-Gly · NH_2 Cys-Tyr-Ile-Gln-Asn   Cys-Pro-Leu-Gly

When we match the overlapping regions of these fragments, the complete sequence of oxytocin appears:

Cys-Tyr-Ile-Gln-Asn
Ile-Gln-Asn-Cys
Gln-Asn-Cys-Pro
Cys-Pro-Leu-Gly
Pro-Leu-Gly · NH_2

Complete structure

Cys-Tyr-Ile-Gln-Asn-Cys-Pro-Leu-Gly · NH_2

The two Cys residues in oxytocin may be involved in disulfide bridges, either linking two of these peptide units or forming a ring. By measuring the molecular weight of oxytocin, we can show that it contains just one of these peptide units; therefore, the Cys residues must link the molecule in a ring.

Application: Meat Tenderizer

Proteolytic (protein-cleaving) enzymes also have applications in consumer products. For example, papain (from papaya extract) serves as a meat tenderizer. It cleaves the fibrous proteins, making the meat less tough.

PROBLEM 24-22

Show where trypsin and chymotrypsin would cleave the following peptide.

Tyr-Ile-Gln-Arg-Leu-Gly-Phe-Lys-Asn-Trp-Phe-Gly-Ala-Lys-Gly-Gln-Gln · NH_2

PROBLEM 24-23

After treatment with peroxyformic acid, the peptide hormone vasopressin is partially hydrolyzed. The following fragments are recovered. Propose a structure for vasopressin.

Phe-Gln-Asn   Pro-Arg-Gly · NH_2 Cys-Tyr-Phe
Asn-Cys-Pro-Arg   Tyr-Phe-Gln-Asn

24-10 Laboratory Peptide Synthesis

Total synthesis of peptides is rarely an economical method for their commercial production. Important peptides are usually derived from biological sources. For example, insulin for diabetics was originally taken from pig pancreas. Now, recombinant DNA techniques have improved the quality and availability of peptide pharmaceuticals. It is possible to extract the piece of DNA that contains the code for a particular protein,

Application: GMO Proteins

Genetically-modified organisms produce a growing number of life-saving pharmaceutical products. Important GMO proteins include insulin from bacteria and antithrombin (anticoagulant) from goat's milk. Hemophiliac factor VIII (to allow clotting) and tissue plasminogen activator (tPA, to dissolve clots) are proteins produced by genetically modified cultured cells. Vaccines for diseases such as hepatitis, plague, cholera, influenza, dysentery, and Ebola are produced by genetically modified yeasts and plants such as tobacco.

insert it into a bacterium, and induce the bacterium to produce the protein. Genetically modified strains of *Escherichia coli* have been developed to produce human insulin that does not cause dangerous reactions in people who are allergic to pork products.

Laboratory peptide synthesis is still an important area of chemistry, however. If the synthetic peptide turns out to be the same as the natural peptide, it proves that the proposed structure is correct. Also, the synthesis provides a larger amount of the material for further biological testing. Synthetic peptides can be made with altered amino acid sequences to compare their biological activity with the natural peptides. These comparisons can point out the critical areas of the peptide, which may suggest causes and treatments for genetic diseases involving similar abnormal peptides.

Peptide synthesis requires the formation of amide bonds between the proper amino acids in the proper sequence. With simple acids and amines, we would form an amide bond simply by converting the acid to an activated derivative (such as an acyl halide or anhydride) and adding the amine.

$$\text{R—C(=O)—X} + \text{H}_2\ddot{\text{N}}\text{—R}' \longrightarrow \text{R—C(=O)—NH—R}' + \text{H—X}$$

(X is a good leaving group, preferably electron-withdrawing)

Amide formation is not so easy with amino acids, however. Each amino acid has both an amino group and a carboxyl group. If we activate the carboxyl group, it reacts with its own amino group. If we mix some amino acids and add a reagent to make them couple, they form every conceivable sequence. Also, some amino acids have side chains that might interfere with peptide formation. For example, glutamic acid has an extra carboxyl group, and lysine has an extra amino group. As a result, peptide synthesis always involves both activating reagents to form the desired peptide bonds and protecting groups to block the formation of undesired bonds.

Chemists have developed many specific techniques for synthesizing peptides. Before 1963, they used typical solution techniques. These *solution-phase methods* involved adding reagents to solutions of growing peptide chains and purifying the products as needed. The chemists would apply protecting groups, add activating groups, couple an amino acid to the growing peptide, and then laboriously purify the product. In 1963, Robert Bruce Merrifield of Rockefeller University developed a method for synthesizing peptides without having to purify the intermediates. He did this by attaching the growing peptide chains to solid polystyrene beads. After each amino acid is added, the excess reagents are washed away by rinsing the beads with solvent. This *solid-phase method* lends itself to automation, and Merrifield built a machine that can add several amino acid units while running unattended. Using this machine, Merrifield synthesized ribonuclease (124 amino acids) in just 6 weeks, obtaining an overall yield of 17%. Merrifield's work in solid-phase peptide synthesis won the Nobel Prize in Chemistry in 1984.

Merrifield's method synthesizes the peptide starting from the C terminus and working toward the N terminus, which is the opposite of the way we usually draw peptides. First, the protected C-terminal amino acid is attached to the polymer bead. For this reaction, its $-NH_2$ group must be protected so that it does not react. The protecting group is removed, and then the next protected amino acid is coupled to the first. Many more deprotection and coupling reactions occur until the entire peptide is formed. At that point, it is cleaved from the polymer bead. Here is a summary of the process:

1. *Attach the protected C-terminal amino acid to the bead, and then deprotect.*

$$\boxed{\text{protecting group}}\text{—NH—CH(R}^n\text{)—C(=O)—O}^- + \bullet \longrightarrow \boxed{\text{protecting group}}\text{—NH—CH(R}^n\text{)—C(=O)—O—}\bullet \xrightarrow{\text{deprotect}} \text{H}_2\text{N—CH(R}^n\text{)—C(=O)—O—}\bullet$$

2. *Couple the next protected amino acid, and then deprotect.* (Repeat many times.)

$$\left[\begin{array}{c} H_2N{-}CH(R^n){-}C(=O){-}O{-}\text{(bead)} \\ \text{protecting group}{-}NH{-}CH(R^{n-1}){-}C(=O){-}O^- \\ \text{next amino acid} \end{array}\right] \xrightarrow[\text{2. deprotect}]{\text{1. coupling agent}} \underset{\text{elongated peptide}}{H_2N{-}CH(R^{n-1}){-}C(=O){-}NH{-}CH(R^n){-}C(=O){-}O{-}\text{(bead)}} \xrightarrow{\text{(repeat)}}$$

3. *After adding all the residues, cleave the finished peptide from the bead.*

$$\underset{\text{first amino acid}}{H_2N{-}CH(R^1){-}C(=O)}{-}\boxed{\text{many amino acid residues}}{-}NH{-}CH(R^n){-}C(=O){-}O{-}\text{(bead)} \xrightarrow[\text{cleave}]{\text{HF}} \underset{\text{final peptide}}{H_2N{-}CH(R^1){-}C(=O){-}\boxed{\text{many amino acid residues}}{-}NH{-}CH(R^n){-}C(=O){-}OH}$$

Many different reagents are available for each of these steps, but we will consider only one set of reagents. The general principles are the same regardless of the specific reagents.

24-10A The Individual Reactions

Three reactions are crucial for solid-phase peptide synthesis. These reactions attach the first protected amino acid to the solid support, deprotect each amino group for its turn to react, and form the peptide bonds between the amino acids.

Attaching the Peptide to the Solid Support The solid-phase synthesis is done in the reverse direction, right-to-left as we draw the peptide. It starts with the C terminus and builds toward the N terminus. The first step is to attach the *last* amino acid (the C terminus) to the solid support.

The solid support is a special polystyrene bead in which some of the aromatic rings have chloromethyl groups. This polymer, often called the *Merrifield resin*, is made by copolymerizing styrene with a few percent of *p*-(chloromethyl)styrene.

Formation of the Merrifield resin

$$\underset{\text{styrene}}{C_6H_5{-}CH{=}CH_2} + \underset{p\text{-(chloromethyl)styrene}}{ClCH_2{-}C_6H_4{-}CH{=}CH_2} \longrightarrow \underset{\text{polymer}}{{-}CH(C_6H_5){-}CH_2{-}CH(C_6H_4CH_2Cl){-}CH_2{-}CH(C_6H_5){-}CH_2{-}} = \underset{\text{abbreviation}}{ClCH_2{-}C_6H_4{-}\text{(P)}}$$

Like other benzyl halides, the chloromethyl groups on the polymer are reactive toward S_N2 attack. The carboxyl group of an N-protected amino acid displaces chloride, giving an amino acid ester of the polymer. In effect, the polymer serves as the alcohol part of an ester protecting group for the carboxyl end of the C-terminal amino acid. The amino group must be protected, or it would attack the chloromethyl groups.

Attachment of the C-terminal amino acid

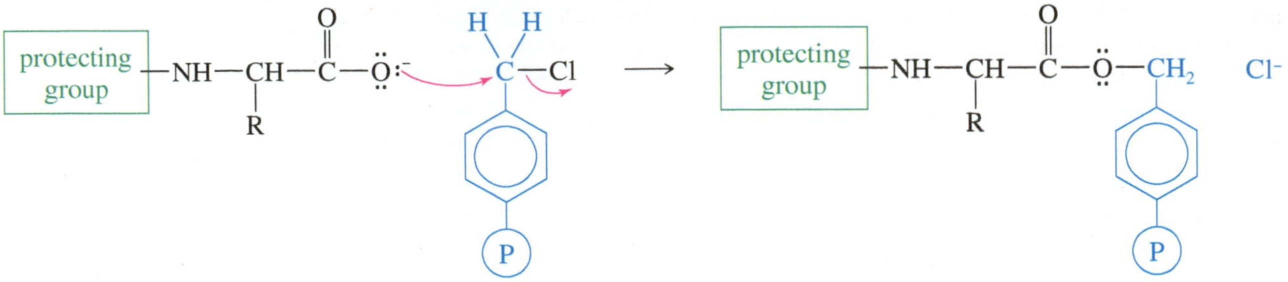

Once the C-terminal amino acid is fixed to the polymer, the chain is built on the amino group of this amino acid.

Using the 9-Fluorenylmethoxycarbonyl (Fmoc) Protecting Group The most common protecting group for the $-NH_2$ groups of amino acids and peptides is currently the Fmoc group. Its structure is shown below, with the structure of a protected amino acid. We will generally refer to it by its abbreviation, Fmoc.

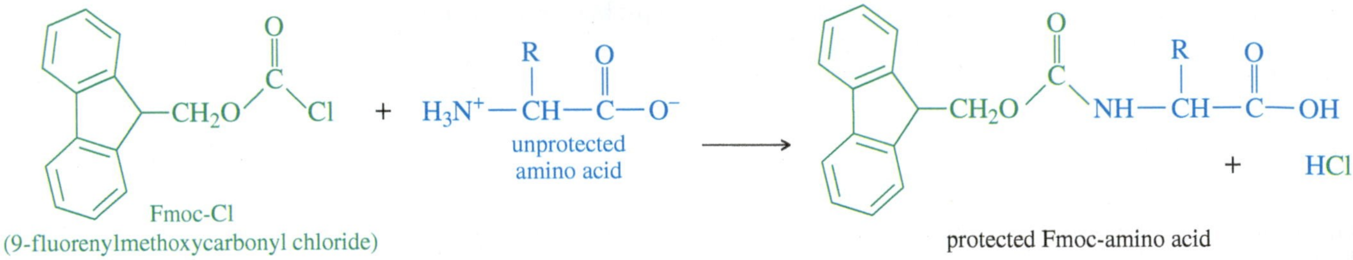

Fmoc-Cl
(9-fluorenylmethoxycarbonyl chloride)

unprotected amino acid

protected Fmoc-amino acid

The protected amino acid no longer has a nucleophilic $-NH_2$ group; that group is now an amide, part of a urethane (carbamate ester). The Fmoc protecting group is easily removed by very mildly basic conditions. The most common reagent for its removal is a solution of piperidine in DMF (*N,N*-dimethylformamide).

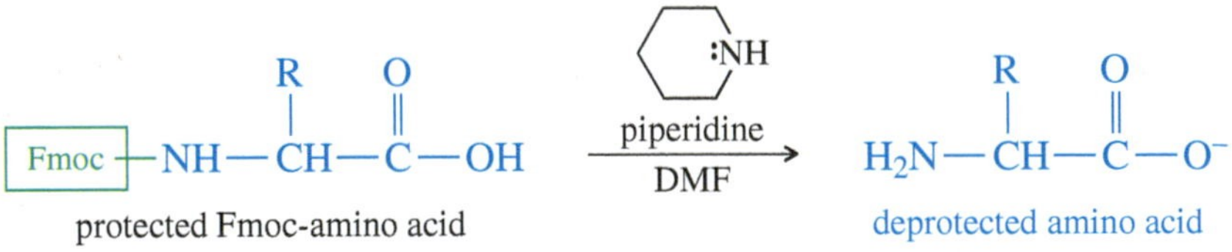

protected Fmoc-amino acid — piperidine, DMF → deprotected amino acid

Chemists who synthesize peptides generally do not make their own Fmoc-protected amino acids. Because they use all their amino acids in protected form, they buy and use commercially available Fmoc amino acids.

Use of DCC as a Peptide Coupling Agent The final reaction needed for the Merrifield procedure is the peptide bond-forming condensation. When a mixture of an amine and an acid is treated with *N,N'*-dicyclohexylcarbodiimide (abbreviated DCC), the amine and the acid couple to form an amide. The molecule of water lost in this condensation converts DCC to *N,N'*-dicyclohexylurea (DCU).

acid + amine + *N,N'*-dicyclohexylcarbodiimide (DCC) → amide + *N,N'*-dicyclohexylurea (DCU)

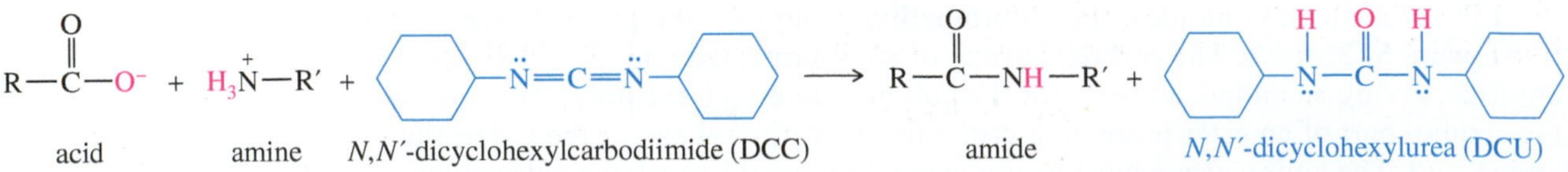

The mechanism for DCC coupling is not as complicated as it may seem. The carboxylate ion adds to the strongly electrophilic carbon of the diimide, giving an activated acyl derivative of the acid. This activated derivative reacts readily with the amine to give the amide. In the final step, DCU serves as an excellent leaving group. The cyclohexane rings are miniaturized for clarity.

Formation of an activated acyl derivative

$$\text{C}_6\text{H}_{11}\text{–N=C=N–C}_6\text{H}_{11} + \text{R–C(=O)–O}^- \longrightarrow \text{R–C(=O)–O–C(=N–C}_6\text{H}_{11})\text{–N}^-\text{–C}_6\text{H}_{11} \xrightarrow{\text{H–}\overset{+}{\text{N}}\text{H}_2\text{–R}'} \underset{\text{activated}}{\text{R–C(=O)–O–C(=N–C}_6\text{H}_{11})\text{–NH–C}_6\text{H}_{11}} + \text{H}_2\text{N–R}'$$

Coupling with the amine and loss of DCU

$$\text{R–C(=O)–O–C(=N–C}_6\text{H}_{11})\text{–NH–C}_6\text{H}_{11} + \text{R}'\text{–NH}_2 \longrightarrow \text{R–C(O}^-)(\overset{+}{\text{N}}\text{H}_2\text{–R}')\text{–O–C(=N–C}_6\text{H}_{11})\text{–NH–C}_6\text{H}_{11} \longrightarrow \text{R–C(=O)–}\overset{+}{\text{N}}\text{H}_2\text{R}' + {}^-\text{O–C(=N–C}_6\text{H}_{11})\text{–NH–C}_6\text{H}_{11} \longrightarrow \underset{\text{amide (peptide)}}{\text{R–C(=O)–NHR}'} + \text{DCU}$$

At the completion of the synthesis, the ester bond to the polymer is cleaved by anhydrous HF. Because this is an ester bond, it is more easily cleaved than the amide bonds of the peptide.

Cleavage of the finished peptide

$$\boxed{\text{peptide}}\text{–C(=O)–O–CH}_2\text{–C}_6\text{H}_4\text{–(P)} \xrightarrow{\text{HF}} \boxed{\text{peptide}}\text{–C(=O)–OH} + \text{CH}_2\text{F–C}_6\text{H}_4\text{–(P)}$$

PROBLEM-SOLVING HINT

Remember that solid-phase peptide synthesis goes C → N.

1. Attach the Fmoc-protected C terminus to the bead first.
2. Couple each amino acid by removing the Fmoc group from the N terminus, and then add the next Fmoc-protected amino acid with DCC.
3. Cleave (HF) the finished peptide from the bead.

PROBLEM 24-24

Propose a mechanism for the coupling of acetic acid and aniline using DCC as a coupling agent.

24-10B An Example of Solid-Phase Peptide Synthesis

Now we present an example to illustrate how these procedures are combined in the Merrifield solid-phase peptide synthesis. We will consider the synthesis of a simple tripeptide that has no problematic side chains requiring additional protecting groups.

Ala-Val-Phe

The solid-phase synthesis proceeds in the opposite direction from the way we write the peptide structure. The first step is attachment of the N-protected C-terminal amino acid (Fmoc-phenylalanine) to the polymer.

Fmoc-Phe + CH_2—Cl (on polymer-bound benzene ring) ⟶ Fmoc-Phe—(P)

Piperidine cleaves the Fmoc protecting group of phenylalanine so that its amino group can be coupled with the next amino acid.

Fmoc-Phe—(P) —piperidine, DMF→ Phe—(P)

The second amino acid (valine) is added in its N-protected Fmoc form so that it cannot couple with itself. Addition of DCC couples the valine carboxyl group with the free —NH_2 group of phenylalanine.

Fmoc-Val + Phe—(P) —DCC→ Fmoc-Val-Phe—(P) + DCU

To couple the final amino acid (alanine), the chain is first deprotected by treatment with piperidine. Then the N-protected Fmoc-alanine and DCC are added.

Step 1: Deprotection

Fmoc-Val-Phe—(P) —piperidine, DMF→ Val-Phe—(P)

Step 2: Coupling

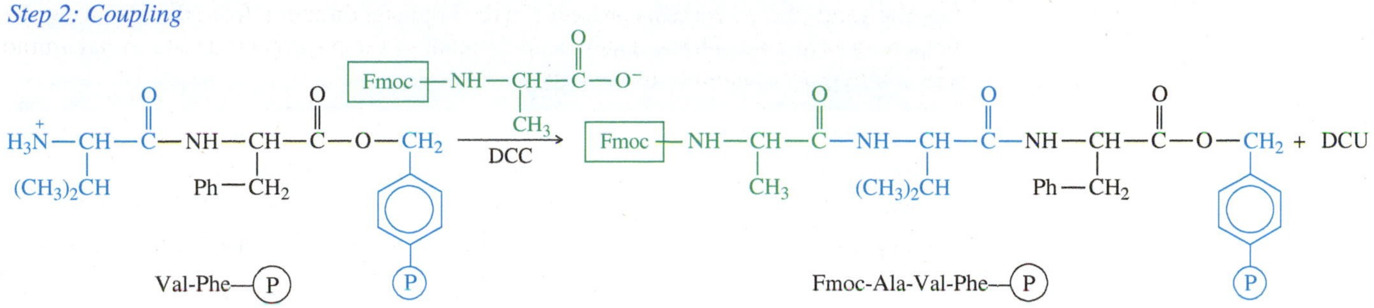

$$\mathrm{H_3\overset{+}{N}{-}CH(CH(CH_3)_2){-}C(=O){-}NH{-}CH(CH_2Ph){-}C(=O){-}O{-}CH_2{-}C_6H_4{-}(P)} \quad \text{(Val-Phe—(P))} \;+\; \boxed{\mathrm{Fmoc}}\mathrm{{-}NH{-}CH(CH_3){-}C(=O){-}O^-} \xrightarrow{\text{DCC}} \boxed{\mathrm{Fmoc}}\mathrm{{-}NH{-}CH(CH_3){-}C(=O){-}NH{-}CH(CH(CH_3)_2){-}C(=O){-}NH{-}CH(CH_2Ph){-}C(=O){-}O{-}CH_2{-}C_6H_4{-}(P)} \;+\; \text{DCU}$$

Fmoc-Ala-Val-Phe—(P)

If we were making a longer peptide, the addition of each subsequent amino acid would require the repetition of two steps:

1. Use piperidine in DMF to deprotect the amino group at the end of the growing chain.
2. Add the next Fmoc-amino acid, using DCC as a coupling agent.

Once the peptide is completed, the final Fmoc protecting group must be removed, and the peptide must be cleaved from the polymer. Anhydrous HF cleaves the ester linkage that bonds the peptide to the polymer, and it also removes the Fmoc protecting group. In our example, the following reaction occurs:

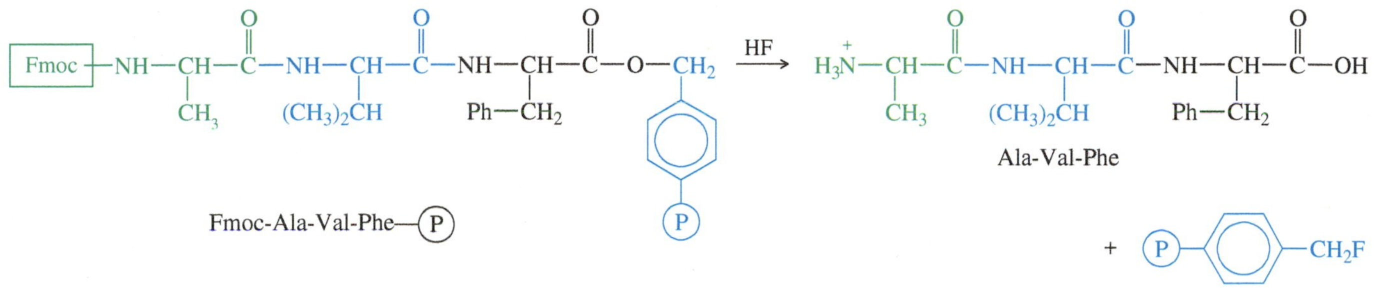

$$\boxed{\mathrm{Fmoc}}\mathrm{{-}NH{-}CH(CH_3){-}C(=O){-}NH{-}CH(CH(CH_3)_2){-}C(=O){-}NH{-}CH(CH_2Ph){-}C(=O){-}O{-}CH_2{-}C_6H_4{-}(P)} \xrightarrow{\text{HF}} \mathrm{H_3\overset{+}{N}{-}CH(CH_3){-}C(=O){-}NH{-}CH(CH(CH_3)_2){-}C(=O){-}NH{-}CH(CH_2Ph){-}C(=O){-}OH} \;+\; \mathrm{(P){-}C_6H_4{-}CH_2F}$$

Fmoc-Ala-Val-Phe—(P) → Ala-Val-Phe

PROBLEM 24-25

Show how you would synthesize Leu-Gly-Ala-Val-Phe starting with Fmoc-Ala-Val-Phe—(P).

PROBLEM 24-26

Show how solid-phase peptide synthesis would be used to make Ile-Gly-Asn.

24-11 Classification of Proteins

Proteins may be classified according to their chemical composition, their shape, or their function. Protein composition and function are treated in detail in a biochemistry course. For now, we briefly survey the types of proteins and their general classifications.

Proteins are grouped into *simple* and *conjugated* proteins according to their chemical composition. **Simple proteins** are those that hydrolyze to give only amino acids. All the protein structures we have considered so far are simple proteins. Examples are insulin, ribonuclease, oxytocin, and bradykinin. **Conjugated proteins** are bonded to a nonprotein **prosthetic group** such as a sugar, a nucleic acid, a lipid, or some other group. Table 24-3 lists some examples of conjugated proteins.

TABLE 24-3
Classes of Conjugated Proteins

Class	Prosthetic Group	Examples
glycoproteins	carbohydrates	γ-globulin, interferon
nucleoproteins	nucleic acids	ribosomes, viruses
lipoproteins	fats, cholesterol	high-density lipoprotein
metalloproteins	a complexed metal	hemoglobin, cytochromes

Proteins are classified as *fibrous* or *globular* depending on whether they form long filaments or coil up on themselves. **Fibrous proteins** are stringy, tough, and usually insoluble in water. They function primarily as structural parts of the organism. Examples of fibrous proteins are α-keratin in hooves and fingernails, and collagen in tendons. **Globular proteins** are folded into roughly spherical shapes. They usually function as enzymes, hormones, or transport proteins. **Enzymes** are protein-containing biological catalysts; an example is ribonuclease, which cleaves RNA. Hormones help to regulate processes in the body. An example is insulin, which regulates glucose levels in the blood and its uptake by cells. Transport proteins bind to specific molecules and transport them in the blood or through the cell membrane. An example is hemoglobin, which transports oxygen in the blood from the lungs to the tissues.

24-12 Levels of Protein Structure

Proteins show four distinct levels of structure:

1. the covalent-bonded structure;
2. the hydrogen-bonded local arrangement into an α-helix or a pleated sheet;
3. the complete 3-dimensional conformation; and
4. the association of two or more peptide units in the active protein.

Each level has a role in determining the overall function of the protein.

24-12A Primary Structure

Up to now, we have discussed the *primary structure* of proteins. The **primary structure** is the covalently bonded structure of the molecule. This definition includes the sequence of amino acids, together with any disulfide bridges. All the properties of the protein are determined, directly or indirectly, by the primary structure. Any folding, hydrogen bonding, or catalytic activity depends on the proper primary structure.

24-12B Secondary Structure

Although we often think of peptide chains as linear structures, they tend to form orderly hydrogen-bonded arrangements. In particular, the carbonyl oxygen atoms form hydrogen bonds with the amide (N—H) hydrogens. This tendency leads to orderly patterns of hydrogen bonding: the **α helix** and the **pleated sheet.** These hydrogen-bonded arrangements, if present, are called the **secondary structure** of the protein.

When a peptide chain winds into a helical coil, each carbonyl oxygen can hydrogen-bond with an N—H hydrogen on the next turn of the coil. Many proteins wind into an α helix (a helix that looks like the thread on a right-handed screw) with the side chains positioned on the outside of the helix. For example, the fibrous protein α keratin is arranged in the α-helical structure, and most globular proteins contain segments of α helix. Figure 24-15 shows the α-helical arrangement.

Segments of peptides can also form orderly arrangements of hydrogen bonds by lining up side-by-side. In this arrangement, each carbonyl group on one chain forms a hydrogen bond with an N—H hydrogen on an adjacent chain. This arrangement may involve many peptide molecules lined up side-by-side, resulting in a two-dimensional *sheet*.

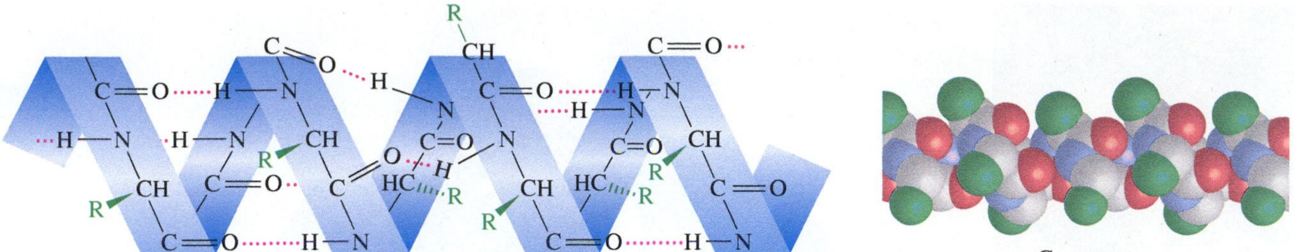

C = gray
N = blue
O = red
R = green

FIGURE 24-15
The α helical arrangement. The peptide chain curls into a helix so that each peptide carbonyl group is hydrogen-bonded to an N—H hydrogen on the next turn of the helix. Side chains are symbolized by green atoms in the space-filling structure.

The bond angles between amino acid units are such that the sheet is *pleated* (creased), with the amino acid side chains arranged on alternating sides of the sheet. Silk fibroin, the principal fibrous protein in the silks of insects and arachnids, has a pleated-sheet secondary structure. Figure 24-16 shows the pleated-sheet structure.

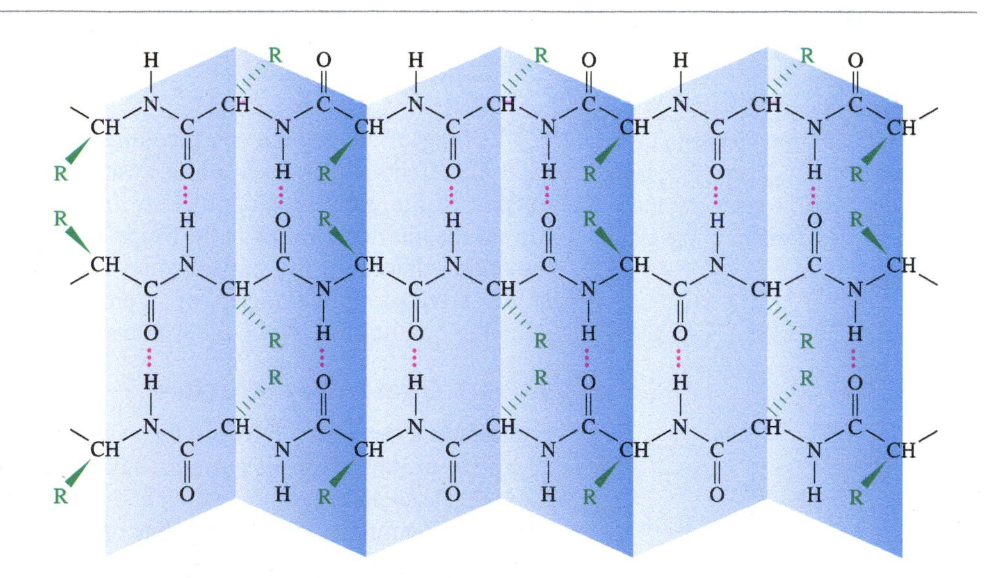

FIGURE 24-16
The pleated-sheet arrangement. Each peptide carbonyl group is hydrogen-bonded to an N—H hydrogen on an adjacent peptide chain.

A protein may or may not have the same secondary structure throughout its length. Some parts may be curled into an α helix, while other parts are lined up in a pleated sheet. Parts of the chain may have no orderly secondary structure at all. Such a structureless region is called a **random coil.** Most globular proteins, for example, contain segments of α helix or pleated sheet separated by kinks of random coil, allowing the molecule to fold into its globular shape.

24-12C Tertiary Structure

The **tertiary structure** of a protein is its complete three-dimensional conformation. Think of the secondary structure as a spatial pattern in a local region of the molecule. Parts of the protein may have the α-helical structure, while other parts may have the pleated-sheet structure, and still other parts may be random coils. The tertiary structure includes all the secondary structure and all the kinks and folds in between. The tertiary structure of a typical globular protein is represented in Figure 24-17.

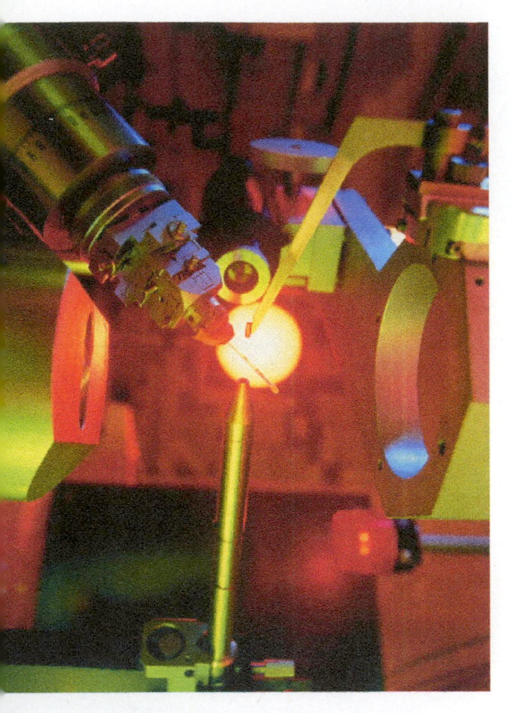

Tertiary structures of proteins are determined by X-ray crystallography. A single crystal of the protein is bombarded with X rays, whose wavelengths are appropriate to be diffracted by the regular atomic spacings in the crystal. A computer then determines the locations of the atoms in the crystal.

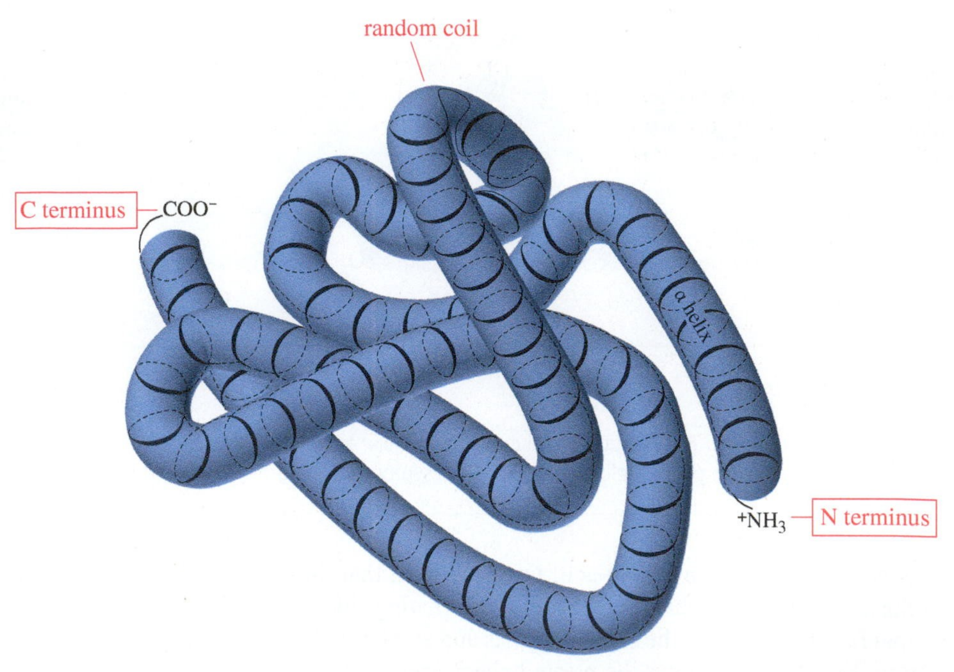

FIGURE 24-17
The tertiary structure of a typical globular protein includes segments of α helix with segments of random coil at the points where the helix is folded.

Coiling of an enzyme can give three-dimensional shapes that produce important catalytic effects. Polar, *hydrophilic* (water-loving) side chains are oriented toward the outside of the globule. Nonpolar, *hydrophobic* (water-hating) groups are arranged toward the interior. Coiling in the proper conformation creates an enzyme's **active site,** the region that binds the substrate and catalyzes the reaction. A reaction taking place at the active site in the interior of an enzyme may occur under essentially anhydrous, nonpolar conditions—while the whole system is dissolved in water!

24-12D Quaternary Structure

Quaternary structure refers to the association of two or more peptide chains in the complete protein. Not all proteins have quaternary structure. The ones that do are those that associate together in their active form. For example, hemoglobin, the oxygen carrier in mammalian blood, consists of four peptide chains fitted together to form a globular protein. Figure 24-18 summarizes the four levels of protein structure.

24-13 Protein Denaturation

For a protein to be biologically active, it must have the correct structure at all levels. The sequence of amino acids must be right, with the correct disulfide bridges linking the cysteines on the chains. The secondary and tertiary structures are important as well. The protein must be folded into its natural conformation, with the appropriate areas of α helix and pleated sheet. For an enzyme, the active site must have the right conformation, with the necessary side-chain functional groups in the correct positions. Conjugated proteins must have the right prosthetic groups, and multichain proteins must have the right combination of individual peptides.

With the exception of the covalent primary structure, all these levels of structure are maintained by weak solvation and hydrogen-bonding forces. Small changes in the environment can cause a chemical or conformational change resulting in **denaturation:** disruption of the normal structure and loss of biological activity. Many factors can cause denaturation, but the most common ones are heat and pH.

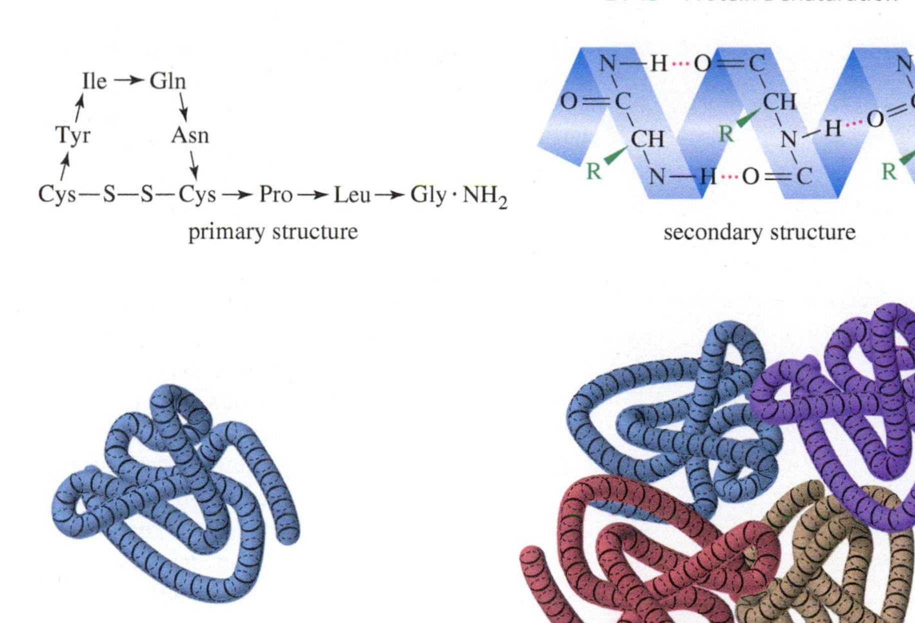

FIGURE 24-18
A schematic comparison of the levels of protein structure. Primary structure is the covalently bonded structure, including the amino acid sequence and any disulfide bridges. Secondary structure refers to the areas of α helix, pleated sheet, or random coil. Tertiary structure refers to the overall conformation of the molecule. Quaternary structure refers to the association of two or more peptide chains in the active protein.

24-13A Reversible and Irreversible Denaturation

The cooking of egg white is an example of protein denaturation by high temperature. Egg white contains soluble globular proteins called *albumins*. When egg white is heated, the albumins unfold and coagulate to produce a solid rubbery mass. Different proteins have different abilities to resist the denaturing effect of heat. Egg albumin is quite sensitive to heat, but bacteria that live in geothermal hot springs have developed proteins that retain their activity in boiling water.

When a protein is subjected to an acidic pH, some of the side-chain carboxyl groups become protonated and lose their ionic charge. Conformational changes result, leading to denaturation. In a basic solution, amino groups become deprotonated, similarly losing their ionic charge, causing conformational changes and denaturation.

Irreversible denaturation of egg albumin. The egg white does not become clear and runny again when it cools.

Milk turns sour because of the bacterial conversion of carbohydrates to lactic acid. When the pH becomes strongly acidic, soluble proteins in milk are denatured and precipitate. This process is called *curdling*. Some proteins are more resistant to acidic and basic conditions than others. For example, most digestive enzymes such as amylase and trypsin remain active under acidic conditions in the stomach, even at a pH of about 1.

In many cases, denaturation is irreversible. When cooked egg white is cooled, it does not become uncooked. Curdled milk does not uncurdle when it is neutralized. Denaturation may be reversible, however, if the protein has undergone only mild denaturing conditions. For example, a protein can be *salted out* of solution by a high salt concentration, which denatures and precipitates the protein. When the precipitated protein is redissolved in a solution with a lower salt concentration, it usually regains its activity together with its natural conformation.

24-13B Prion Diseases

Up through 1980, people thought that all infectious diseases were caused by microbes of some sort. They knew about diseases caused by viruses, bacteria, protozoa, and fungi. There were some strange diseases, however, for which no one had isolated and cultured the pathogen. *Creutzfeldt–Jakob Disease* (CJD) in humans, *scrapie* in sheep, and *transmissible encephalopathy* in mink (TME) all involved a slow, gradual loss

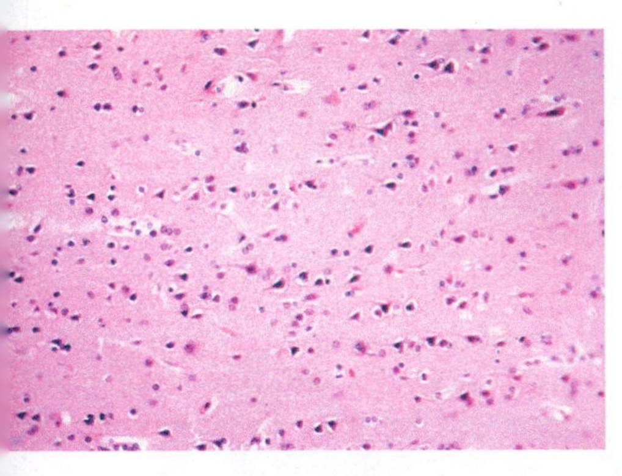

Micrograph of normal human brain tissue. The nuclei of neurons appear as dark spots.

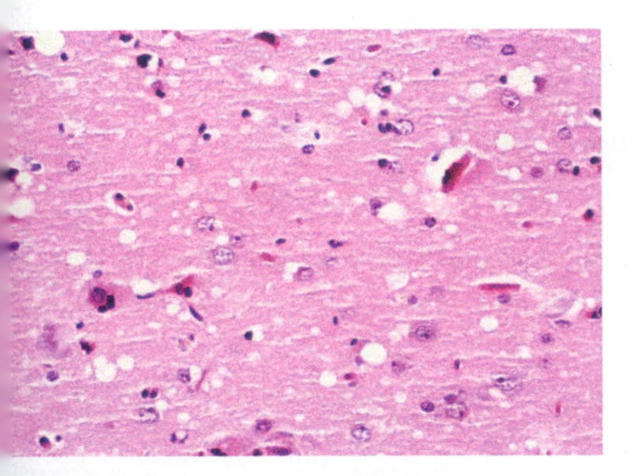

Brain tissue of a patient infected with vCJD. Note the formation of (white) vacuole spaces and (dark, irregular) plaques of prion protein. (Magnification 200X)

of mental function and eventual death. The brains of the victims all showed unusual plaques of amyloid protein surrounded by spongelike tissue.

Workers studying these diseases thought there was an infectious agent involved (as opposed to genetic or environmental causes) because they knew that scrapie and TME could be spread by feeding healthy animals the ground-up remains of sick animals. They had also studied *kuru*, a disease much like CJD among tribes in which family members showed their respect for the dead by eating their brains. These diseases were generally attributed to "slow viruses" that were yet to be isolated.

In the 1980s, neurologist Stanley B. Prusiner (of the University of California at San Francisco) made a homogenate of scrapie-infected sheep brains; systematically separated out all the cell fragments, bacteria, and viruses; and found that the remaining material was still infectious. He separated out the proteins and found a protein fraction that was still infectious. He suggested that scrapie (and presumably similar diseases) is caused by a protein infectious agent that he called **prion protein.** This conclusion contradicted the established principle that contagious diseases require a living pathogen. Many skeptical workers repeated Prusiner's work in hopes of finding viral contaminants in the infectious fractions, and most of them finally came to the same conclusion. Prusiner received the 1998 Nobel Prize in Medicine or Physiology for this work.

Since Prusiner's work, prion diseases have become more important because of their threat to humans. Beginning in 1996, some cows in the United Kingdom developed "mad cow disease" and would threaten other animals, wave their heads, fall down, and eventually die. The disease, called *bovine spongiform encephalopathy*, or BSE, was probably transmitted to cattle by feeding them the remains of scrapie-infected sheep. The most frightening aspect of the BSE outbreak was that people could contract a fatal disease, called *new-variant Creutzfeldt–Jakob Disease* (vCJD) from eating the infected meat. Since that time, a similar disease, called *chronic wasting disease*, or CWD, has been found in wild deer and elk in the Rocky Mountains. All of these (presumed) prion diseases are now classified as *transmissible spongiform encephalopathies*, or TSEs.

The most widely accepted theory of prion diseases suggests that the infectious prion protein has the same primary structure as a normal protein found in nerve cells, but it differs in its tertiary structure. In effect, it is a misfolded, denatured version of a normal protein that polymerizes to form the amyloid protein plaques seen in the brains of infected animals. When an animal ingests infected food, the polymerized protein resists digestion. Because it is simply a misfolded version of a normal protein, the infectious prion does not provoke the host's immune system to attack the pathogen.

When the abnormal prion interacts with the normal version of the protein on the membranes of nerve cells, the abnormal protein somehow induces the normal molecules to change their shape. This is the part of the process we know the least about. (We might think of it like crystallization, in which a seed crystal induces other molecules to crystallize in the same conformation and crystal form.) These newly misfolded protein molecules then induce more molecules to change shape. The polymerized abnormal protein cannot be broken down by the usual protease enzymes, so it builds up in the brain and causes the plaques and spongy tissue associated with TSEs.

We once thought that a protein with the correct primary structure, placed in the right physiological solution, would naturally fold into the correct tertiary structure and stay that way. We were wrong. We now know that protein folding is a carefully controlled process in which enzymes and *chaperone proteins* promote correct folding as the protein is synthesized. Prion diseases have shown that there are many factors that cause proteins to fold into natural or unnatural conformations, and that the folding of the protein can have major effects on its biological properties within an organism.

Essential Terms

active site The region of an enzyme that binds the substrate and catalyzes the reaction. (p. 1256)

amino acid Literally, any molecule containing both an amino group ($-NH_2$) and a carboxyl group ($-COOH$). The term usually means an **α-amino acid,** with the amino group on the carbon atom next to the carboxyl group. (p. 1223)

biomimetic synthesis A laboratory synthesis that is patterned after a biological synthesis. For example, the synthesis of amino acids by reductive amination resembles the biosynthesis of glutamic acid. (p. 1231)

complete proteins Proteins that provide all the essential amino acids in about the right proportions for human nutrition. Examples include those in meat, fish, milk, and eggs. **Incomplete proteins** are severely deficient in one or more of the essential amino acids. Most plant proteins are incomplete. (p. 1226)

conjugated protein A protein that contains a nonprotein prosthetic group such as a sugar, nucleic acid, lipid, or metal ion. (p. 1253)

C terminus **(C-terminal end)** The end of the peptide chain with a free or derivatized carboxyl group. As the peptide is written, the C terminus is usually on the right. The amino group of the C-terminal amino acid links it to the rest of the peptide. (p. 1239)

denaturation An unnatural alteration of the conformation or the ionic state of a protein. Denaturation generally results in precipitation of the protein and loss of its biological activity. Denaturation may be reversible, as in salting out a protein, or irreversible, as in cooking an egg. (p. 1256)

disulfide linkage **(disulfide bridge)** A bond between two cysteine residues formed by mild oxidation of their thiol groups to a disulfide. (p. 1240)

Edman degradation A method for removing and identifying the N-terminal amino acid from a peptide without destroying the rest of the peptide chain. The peptide is treated with phenyl isothiocyanate, followed by a mild acid hydrolysis to convert the N-terminal amino acid to its phenylthiohydantoin derivative. The Edman degradation can be used repeatedly to determine the sequence of many residues beginning at the N terminus. (p. 1244)

electrophoresis A procedure for separating charged molecules by their migration in a strong electric field. The direction and rate of migration are governed largely by the average charge on the molecules. (p. 1230)

enzymatic resolution The use of enzymes to separate enantiomers. For example, the enantiomers of an amino acid can be acylated and then treated with hog kidney acylase. The enzyme hydrolyzes the acyl group from the natural L-amino acid, but it does not react with the D-amino acid. The resulting mixture of the free L-amino acid and the acylated D-amino acid is easily separated. (p. 1235)

enzyme A protein-containing biological catalyst. Many enzymes also include *prosthetic groups*, nonprotein constituents that are essential to the enzyme's catalytic activity. (p. 1254)

essential amino acids Ten standard amino acids that are not biosynthesized by humans and must be provided in the diet. (p. 1226)

fibrous proteins A class of proteins that are stringy, tough, thread-like, and usually insoluble in water. (p. 1254)

globular proteins A class of proteins that are relatively spherical in shape. Globular proteins generally have lower molecular weights and are more soluble in water than fibrous proteins. (p. 1254)

α helix A helical peptide conformation in which the carbonyl groups on one turn of the helix are hydrogen-bonded to N—H hydrogens on the next turn. Extensive hydrogen bonding stabilizes this helical arrangement. (p. 1254)

hydrogenolysis Cleavage of a bond by the addition of hydrogen. For example, catalytic hydrogenolysis cleaves benzyl esters. (p. 1236)

$$\underset{\text{benzyl ester}}{R-\overset{\overset{\displaystyle O}{\|}}{C}-O-CH_2-C_6H_5} \xrightarrow{H_2,\ Pd} \underset{\text{acid}}{R-\overset{\overset{\displaystyle O}{\|}}{C}-O-H} + \underset{\text{toluene}}{H-CH_2-C_6H_5}$$

isoelectric point, pI **(isoelectric pH)** The pH at which an amino acid (or protein) does not move under electrophoresis. This is the pH where the average charge on its molecules is zero, with most of the molecules in their zwitterionic form. (p. 1229)

L-amino acid An amino acid having a stereochemical configuration similar to that of L-(−)-glyceraldehyde. Most naturally occurring amino acids have the L configuration. (p. 1223)

COOH	CHO	COOH
H_2N—C—H	HO—C—H	H_2N—C—H
CH_3	CH_2OH	R
L-alanine (*S*)-alanine	L-(−)-glyceraldehyde (*S*)-glyceraldehyde	an L-amino acid (*S*) configuration

N terminus **(N-terminal end)** The end of the peptide chain with a free or derivatized amino group. As the peptide is written, the N terminus is usually on the left. The carboxyl group of the N-terminal amino acid links it to the rest of the peptide. (p. 1239)

oligopeptide A small polypeptide, containing about four to ten amino acid residues. (p. 1239)

peptide Any polymer of amino acids linked by amide bonds between the amino group of each amino acid and the carboxyl group of the neighboring amino acid. The terms *dipeptide, tripeptide*, etc. may specify the number of amino acids in the peptide. (p. 1239)

peptide bonds Amide linkages between amino acids. (p. 1222)

pleated sheet A two-dimensional peptide conformation with the peptide chains lined up side-by-side. The carbonyl groups on each peptide chain are hydrogen-bonded to N—H hydrogens on the adjacent chain, and the side chains are arranged on alternating sides of the sheet. (p. 1254)

polypeptide A peptide containing many amino acid residues. Although proteins are polypeptides, the term *polypeptide* is commonly used for molecules with lower molecular weights than proteins. (p. 1239)

primary structure The covalently bonded structure of a protein; the sequence of amino acids, together with any disulfide bridges. (p. 1254)

prion protein A protein infectious agent that is thought to promote misfolding and polymerization of normal protein molecules, leading to amyloid plaques and destruction of nerve tissue. (p. 1258)

prosthetic group The nonprotein part of a conjugated protein. Examples of prosthetic groups are sugars, lipids, nucleic acids, and metal complexes. (p. 1253)

protein A biopolymer of amino acids. Proteins are polypeptides with molecular weights higher than about 5000 amu. (p. 1239)

quaternary structure The association of two or more peptide chains into a composite protein. (p. 1256)

random coil A type of protein secondary structure in which the chain is neither curled into an α helix nor lined up in a pleated sheet. In a globular protein, the kinks that fold the molecule into its globular shape are usually segments of random coil. (p. 1255)

residue An amino acid unit of a peptide. (p. 1239)

Sanger method A method for determining the N-terminal amino acid of a peptide. The peptide is treated with 2,4-dinitrofluorobenzene (Sanger's reagent), and then completely hydrolyzed. The derivatized amino acid is easily identified, but the rest of the peptide is destroyed in the hydrolysis. (p. 1246)

secondary structure The local hydrogen-bonded arrangement of a protein. The secondary structure is generally the α helix, pleated sheet, or random coil. (p. 1254)

sequence As a noun, the order in which amino acids are linked together in a peptide. As a verb, to determine the sequence of a peptide. (p. 1244)

simple proteins Proteins composed of only amino acids (having no prosthetic groups). (p. 1253)

solid-phase peptide synthesis A method in which the C-terminal amino acid is attached to a solid support (polystyrene beads) and the peptide is synthesized in the C → N direction by successive coupling of protected amino acids. When the peptide is complete, it is cleaved from the solid support. (p. 1248)

standard amino acids The 20 α-amino acids found in nearly all naturally occurring proteins. (p. 1224)

Strecker synthesis Synthesis of α-amino acids by reaction of an aldehyde with ammonia and cyanide ion, followed by hydrolysis of the intermediate α-amino nitrile. (p. 1233)

$$\underset{\text{aldehyde}}{R-\overset{O}{\overset{\|}{C}}-H} + NH_3 + HCN \xrightarrow{H_2O} \underset{\alpha\text{-amino nitrile}}{R-\underset{C\equiv N}{\overset{NH_2}{C}}-H} \xrightarrow{H_3O^+} \underset{\alpha\text{-amino acid}}{R-\underset{COOH}{\overset{^+NH_3}{C}}-H}$$

Strecker synthesis of an amino acid

terminal residue analysis Sequencing a peptide by removing and identifying the residue at the N terminus or at the C terminus. (p. 1244)

tertiary structure The complete three-dimensional conformation of a protein. (p. 1255)

transamination Transfer of an amino group from one molecule to another. Transamination is a common method for the biosynthesis of amino acids, often involving glutamic acid as the source of the amino group. (p. 1232)

zwitterion **(dipolar ion)** A structure with an overall charge of zero but having a positively charged substituent and a negatively charged substituent. Most amino acids exist in zwitterionic forms. (p. 1227)

$$H_2N{-}CH(R){-}C(=O){-}OH \rightleftharpoons H_3\overset{+}{N}{-}CH(R){-}C(=O){-}O^-$$

uncharged structure (minor component) — dipolar ion, or zwitterion (major component)

Essential Problem-Solving Skills in Chapter 24

Each skill is followed by problem numbers exemplifying that particular skill.

1	Correctly name amino acids and peptides, and draw the structures from their names.	Problems 24-28, 34, and 35
2	Use perspective drawings and Fischer projections to show the stereochemistry of D- and L-amino acids. Explain why the naturally occurring amino acids are called L-amino acids.	Problems 24-32 and 46
3	Explain which amino acids are acidic, which are basic, and which are neutral. Use the isoelectric point to predict whether a given amino acid will be positively charged, negatively charged, or neutral at a given pH.	Problems 24-27 and 34
4	Show how to make a given amino acid using one of the following syntheses: reductive amination, HVZ followed by ammonia, and the Strecker synthesis.	Problems 24-29, 30, 32, and 33
5	Predict products of the acylation and esterification of amino acids, and their reaction with ninhydrin.	Problems 24-29 and 31
6	Use information from terminal residue analysis and partial hydrolysis to determine the structure of an unknown peptide.	Problems 24-36, 37, 39, 43, and 44
7	Show how you would use solid-phase synthesis to make a given peptide. Use appropriate protecting groups to prevent unwanted couplings.	Problems 24-38 and 45
8	Discuss and identify the four levels of protein structure (primary, secondary, tertiary, and quaternary). Explain how the structure of a protein affects its properties and how denaturation changes the structure.	

Study Problems

24-27 **(a)** The isoelectric point (pI) of phenylalanine is pH 5.5. Draw the structure of the major form of phenylalanine at pH values of 1, 5.5, and 11.

(b) The isoelectric point of histidine is pH 7.6. Draw the structures of the major forms of histidine at pH values of 1, 4, 7.6, and 11. Explain why the nitrogen in the histidine ring is a weaker base than the α-amino group.

(c) The isoelectric point of glutamic acid is pH 3.2. Draw the structures of the major forms of glutamic acid at pH values of 1, 3.2, 7, and 11. Explain why the side-chain carboxylic acid is a weaker acid than the acid group next to the α-carbon atom.

24-28 Draw the complete structure of the following peptide.

Ser-Gln-Met · NH_2

24-29 Predict the products of the following reactions.

(a) Ile + (ninhydrin: indane-1,3-dione with C(OH)$_2$ at C-2) $\xrightarrow[\text{heat}]{\text{pyridine}}$

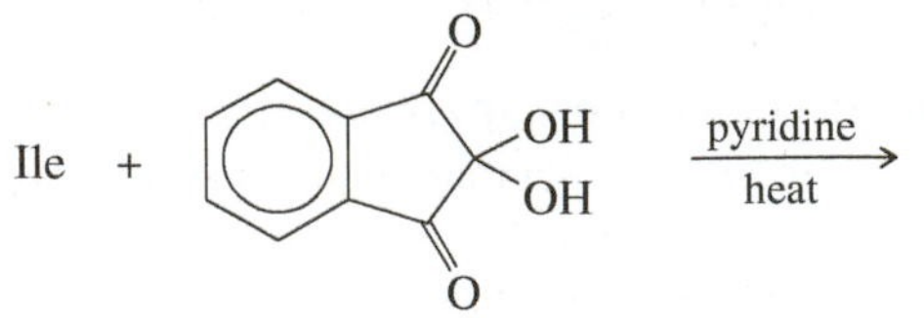

(b) $Ph-CH_2-O-\overset{O}{\overset{\|}{C}}-NH-\overset{CH_3}{\overset{|}{CH}}-COOH \xrightarrow{H_2,\ Pd}$

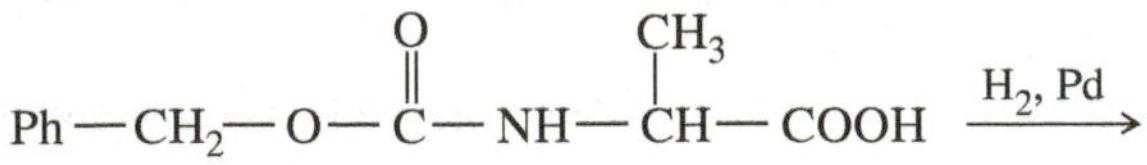

(c) Lys + excess $(CH_3CO)_2O \longrightarrow$

(d) (D,L)-proline $\xrightarrow[\text{(2) hog kidney acylase, } H_2O]{\text{(1) excess } Ac_2O}$

(e) $CH_3CH_2-\overset{CHO}{\overset{|}{CH}}-CH_3 \xrightarrow[H_2O]{NH_3,\ HCN}$

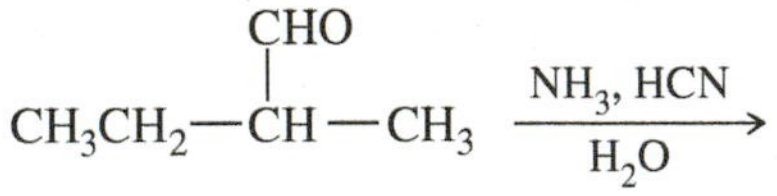

(f) product from part (e) $\xrightarrow{H_3O^+}$

(g) 4-methylpentanoic acid + $Br_2/PBr_3 \longrightarrow$

(h) product from part (g) + excess $NH_3 \longrightarrow$

24-30 Show how you would synthesize any of the standard amino acids from each starting material. You may use any necessary reagents.

(a) $(CH_3)_2CH-\overset{O}{\overset{\|}{C}}-COOH$

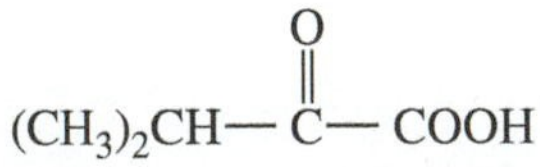

(b) $CH_3-\underset{CH_2CH_3}{\underset{|}{CH}}-CH_2-COOH$

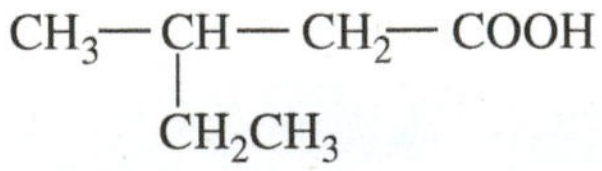

(c) $(CH_3)_2\,CH-CH_2-CHO$

24-31 Show how you would convert alanine to the following derivatives. Show the structure of the product in each case.

(a) alanine isopropyl ester
(b) *N*-benzoylalanine
(c) *N*-benzyloxycarbonyl alanine
(d) *tert*-butyloxycarbonyl alanine

24-32 Suggest a method for the synthesis of the unnatural D enantiomer of alanine from the readily available L enantiomer of lactic acid.

$$CH_3-CHOH-COOH$$
lactic acid

24-33 Show how you would use the Strecker synthesis to make tryptophan. What stereochemistry would you expect in your synthetic product?

24-34 Write the complete structures for the following peptides. Tell whether each peptide is acidic, basic, or neutral.

(a) methionylthreonine
(b) threonylmethionine
(c) arginylaspartyllysine
(d) Glu-Cys-Gln

24-35 The following structure is drawn in an unconventional manner.

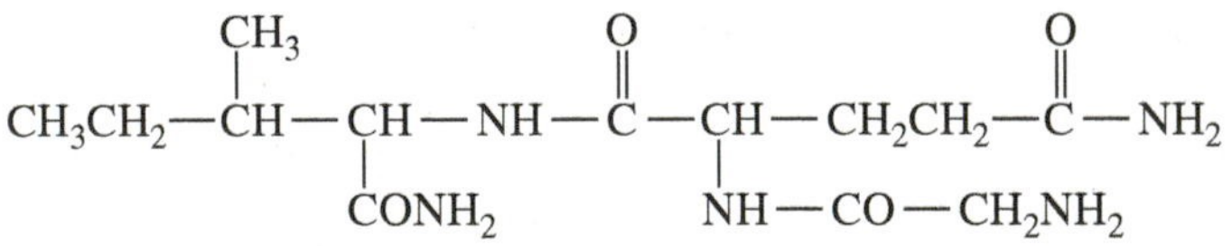

$$CH_3CH_2-\overset{CH_3}{\overset{|}{CH}}-\underset{CONH_2}{\underset{|}{CH}}-NH-\overset{O}{\overset{\|}{C}}-\underset{NH-CO-CH_2NH_2}{\underset{|}{CH}}-CH_2CH_2-\overset{O}{\overset{\|}{C}}-NH_2$$

(a) Label the N terminus and the C terminus.
(b) Label the peptide bonds.
(c) Identify and label each amino acid present.
(d) Give the full name and the abbreviated name.

24-36 *Aspartame* (Nutrasweet®) is a remarkably sweet-tasting dipeptide ester. Complete hydrolysis of aspartame gives phenyl alanine, aspartic acid, and methanol. Mild incubation with carboxypeptidase has no effect on aspartame. Treatment of aspartame with phenyl isothiocyanate, followed by mild hydrolysis, gives the phenylthiohydantoin of aspartic acid. Propose a structure for aspartame.

24-37 A molecular weight determination has shown that an unknown peptide is a pentapeptide, and an amino acid analysis shows that it contains the following residues: one Gly, two Ala, one Met, one Phe. Treatment of the original pentapeptide with carboxypeptidase gives alanine as the first free amino acid released. Sequential treatment of the pentapeptide with phenyl isothiocyanate followed by mild hydrolysis gives the following derivatives:

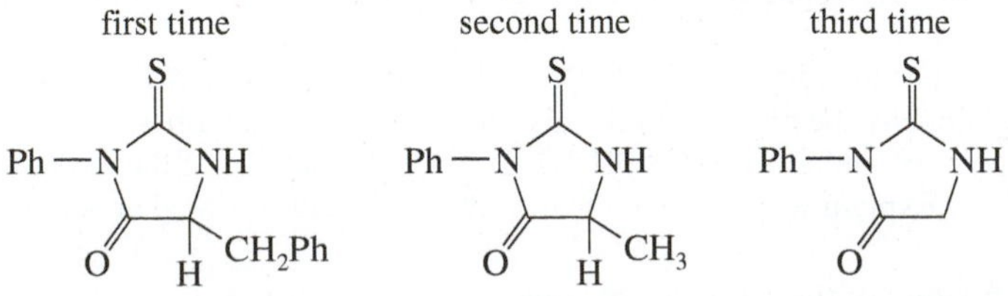

first time: phenylthiohydantoin (Ph—N, NH, C=S, C=O) with H and CH_2Ph substituents

second time: phenylthiohydantoin (Ph—N, NH, C=S, C=O) with H and CH_3 substituents

third time: phenylthiohydantoin (Ph—N, NH, C=S, C=O), unsubstituted CH_2

Propose a structure for the unknown pentapeptide.

24-38 Show the steps and intermediates in the synthesis of Leu-Ala-Phe by the solid-phase process.

24-39 Peptides often have functional groups other than free amino groups at the N terminus and other than carboxyl groups at the C terminus.

(a) A tetrapeptide is hydrolyzed by heating with 6 M HCl, and the hydrolysate is found to contain Ala, Phe, Val, and Glu. When the hydrolysate is neutralized, the odor of ammonia is detected. Explain where this ammonia might have been incorporated in the original peptide.

(b) The tripeptide *thyrotropic hormone releasing factor* (TRF) has the full name pyroglutamylhistidylprolinamide. The structure appears here. Explain the functional groups at the N terminus and at the C terminus.

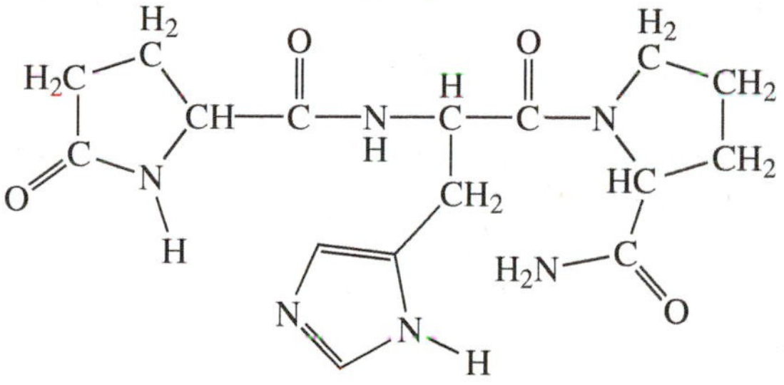

(c) On acidic hydrolysis, an unknown pentapeptide gives glycine, alanine, valine, leucine, and isoleucine. No odor of ammonia is detected when the hydrolysate is neutralized. Reaction with phenyl isothiocyanate followed by mild hydrolysis gives *no* phenylthiohydantoin derivative. Incubation with carboxypeptidase has no effect. Explain these findings.

24-40 Lipoic acid is often found near the active sites of enzymes, usually bound to the peptide by a long, flexible amide linkage with a lysine residue.

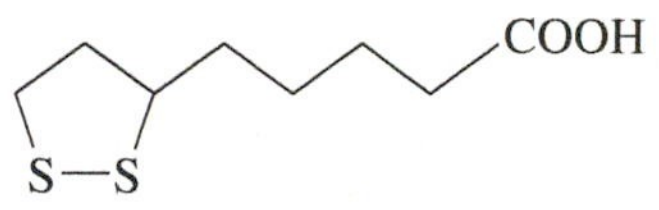

lipoic acid 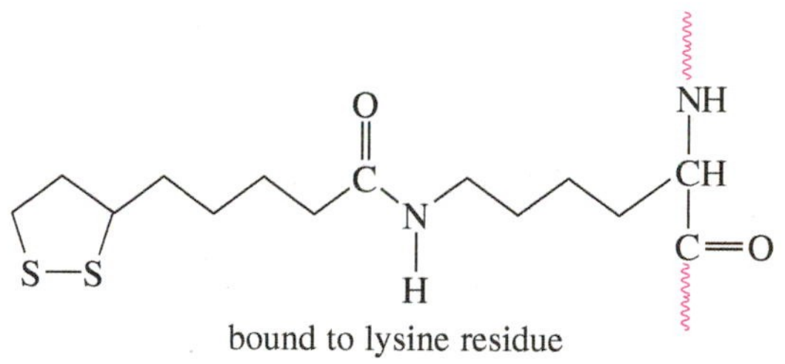   bound to lysine residue

(a) Is lipoic acid a mild oxidizing agent or a mild reducing agent? Draw it in both its oxidized and reduced forms.

(b) Show how lipoic acid might react with two Cys residues to form a disulfide bridge.

(c) Give a balanced equation for the hypothetical oxidation or reduction, as you predicted in part (a), of an aldehyde by lipoic acid.

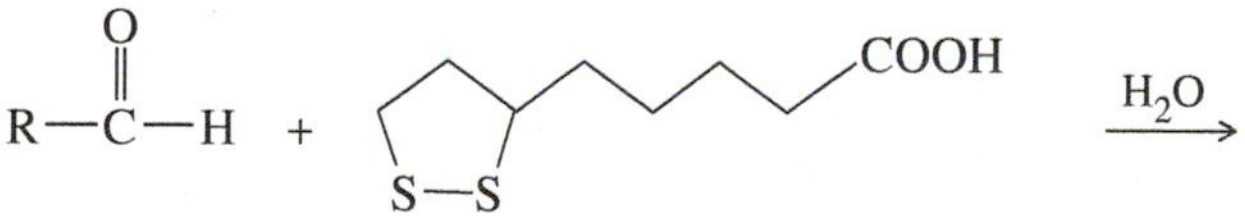

24-41 Histidine is an important catalytic residue found at the active sites of many enzymes. In many cases, histidine appears to remove protons or to transfer protons from one location to another.

(a) Show which nitrogen atom of the histidine heterocycle is basic and which is not.

(b) Use resonance forms to show why the protonated form of histidine is a particularly stable cation.

(c) Show the structure that results when histidine accepts a proton on the basic nitrogen of the heterocycle and then is deprotonated on the other heterocyclic nitrogen. Explain how histidine might function as a pipeline to transfer protons between sites within an enzyme and its substrate.

24-42 Metabolism of arginine produces urea and the rare amino acid *ornithine*. Ornithine has an isoelectric point close to 10. Propose a structure for ornithine.

24-43 Glutathione (GSH) is a tripeptide that serves as a mild reducing agent to detoxify peroxides and maintain the cysteine residues of hemoglobin and other red blood cell proteins in the reduced state. Complete hydrolysis of glutathione gives Gly, Glu, and Cys. Treatment of glutathione with carboxypeptidase gives glycine as the first free amino acid released. Treatment of glutathione with 2,4-dinitrofluorobenzene (Sanger reagent, Problem 24-21, page 1246), followed by complete hydrolysis, gives the 2,4-dinitrophenyl derivative of glutamic acid. Treatment of glutathione with phenyl isothiocyanate does not give a recognizable phenylthiohydantoin, however.

(a) Propose a structure for glutathione consistent with this information. Why would glutathione fail to give a normal product from Edman degradation, even though it gives a normal product from the Sanger reagent followed by hydrolysis?

(b) Oxidation of glutathione forms glutathione disulfide (GSSG). Propose a structure for glutathione disulfide, and write a balanced equation for the reaction of glutathione with hydrogen peroxide.

24-44 Complete hydrolysis of an unknown basic decapeptide gives Gly, Ala, Leu, Ile, Phe, Tyr, Glu, Arg, Lys, and Ser. Terminal residue analysis shows that the N terminus is Ala and the C terminus is Ile. Incubation of the decapeptide with chymotrypsin gives two tripeptides, **A** and **B,** and a tetrapeptide, **C.** Amino acid analysis shows that peptide **A** contains Gly, Glu, Tyr, and NH_3; peptide **B** contains Ala, Phe, and Lys; and peptide **C** contains Leu, Ile, Ser, and Arg. Terminal residue analysis gives the following results.

	N terminus	C terminus
A	Gln	Tyr
B	Ala	Phe
C	Arg	Ile

Incubation of the decapeptide with trypsin gives a dipeptide **D,** a pentapeptide **E,** and a tripeptide **F.** Terminal residue analysis of **F** shows that the N terminus is Ser and the C terminus is Ile. Propose a structure for the decapeptide and for fragments **A** through **F.**

24-45 There are many methods for activating a carboxylic acid in preparation for coupling with an amine. The following method converts the acid to an *N*-hydroxysuccinimide (NHS) ester.

$$R{-}COOH + F_3C{-}C(=O){-}O{-}N(\text{succinimide}) \xrightarrow{Et_3N} R{-}C(=O){-}O{-}N(\text{succinimide}) + F_3C{-}COOH$$

NHS ester

(a) Explain why an NHS ester is much more reactive than a simple alkyl ester.
(b) Propose a mechanism for the reaction shown.
(c) Propose a mechanism for the reaction of the NHS ester with an amine, $R{-}NH_2$.

24-46 Sometimes chemists need the unnatural D enantiomer of an amino acid, often as part of a drug or an insecticide. Most L-amino acids are isolated from proteins, but the D-amino acids are rarely found in natural proteins. D-amino acids can be synthesized from the corresponding L-amino acids. The following synthetic scheme is one of the possible methods.

$$\text{L configuration amino acid (COOH, } H_2N\text{, R, H)} \xrightarrow[HCl]{NaNO_2} \text{intermediate 1} \xrightarrow{NaN_3} \text{intermediate 2} \xrightarrow[Pd]{H_2} \text{D configuration amino acid (COOH, R, H, } NH_2)$$

L configuration D configuration

(a) Draw the structures of intermediates 1 and 2 in this scheme.
(b) How do we know that the product is entirely the unnatural D configuration?

24-47 A student took the proton NMR spectrum of phenylalanine in D_2O solution, and had the instrument suppress the DOH solvent peak. The spectrum is shown below. The integrated relative areas of the peaks are 5:1:1:1.
(a) Draw the structure of phenylalanine as it exists in D_2O solution. (There is a large excess of D_2O, and any exchangeable protons in phenylalanine will exchange with the solvent.)
(b) Assign the peaks in the spectrum to the protons in the structure.
(c) Why don't we see the $-NH_2$ or $-COOH$ protons in the spectrum?
(d) What is the relationship between the two protons that generate nearly mirror-image multiplets at 3.1 and 3.3?

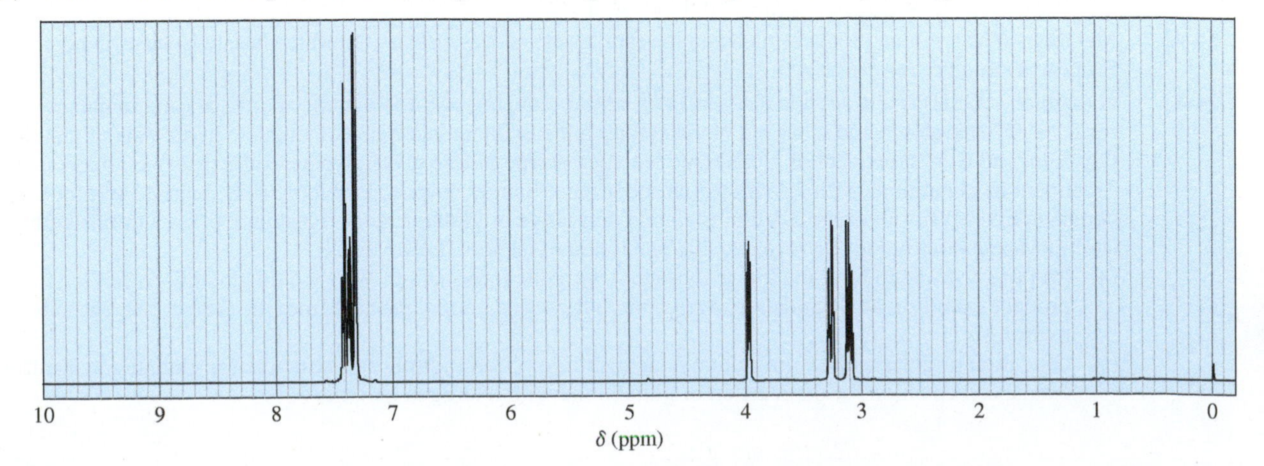

25 Lipids

◀ Whale oil was the most important product of 19th-century commercial whaling. In their quest for oil, whalers hunted many species nearly to extinction. The oil of baleen whales, such as the humpback whale (*Megaptera novaeangliae*) pictured, is composed entirely of triglycerides. Whale oil was used primarily as a lamp oil until it was replaced by kerosene, which was cheaper and easier to obtain. Whale oil was also used as a specialized lubricant for moving mechanical parts, including early automatic transmission fluid. Synthetic oils and lubricants are now available for all of these applications, and there is no need for whale oil or other whale products.

Goals for Chapter 25

1 Classify simple and complex lipids. Identify waxes, triglycerides, phospholipids, steroids, prostaglandins, and terpenes.

2 Explain how unsaturations affect the properties of fats and oils. Compare the properties of saturated fats with those of polyunsaturated oils and partially hydrogenated vegetable oils.

3 Predict the reactions of lipids under basic hydrolysis and with standard organic reagents.

4 Compare soaps and detergents, and explain how they emulsify nonpolar substances in water.

25-1 Introduction

What do the following have in common? An athlete is disqualified from the Olympics for illegal use of anabolic steroids. You spray a bread pan with canola oil to keep the bread from sticking. Your mother is rushed to surgery to remove a gallbladder packed with cholesterol. You wax your shiny new car with carnauba wax. Your father is treated with a prostaglandin to lower his blood pressure. An artist uses turpentine to clean her brushes after painting the brilliant autumn colors.

All these actions involve the use, misuse, or manipulation of lipids. Steroids, prostaglandins, fats, oil, waxes, terpenes, and even the colorful carotenes in the falling leaves are all lipids. In our study of organic chemistry, we have usually classified compounds according to their functional groups. Lipids, however, are classified by their solubility: **Lipids** are substances that can be extracted from cells and tissues by nonpolar organic solvents.

Lipids include many types of compounds containing a wide variety of functional groups. You could easily prepare a solution of lipids by grinding a T-bone steak in a blender and then extracting the puree with chloroform or diethyl ether. The resulting solution of lipids would contain a multitude of compounds, many with complex structures. To facilitate the study of lipids, chemists have divided this large family into two major classes: complex lipids and simple lipids.

Complex lipids are those that are easily hydrolyzed to simpler constituents. Most complex lipids are esters of long-chain carboxylic acids called *fatty acids*. The two major groups of fatty acid esters are *waxes* and *glycerides*. Waxes are esters of long-chain alcohols, and glycerides are esters of glycerol.

Simple lipids are those that are not easily hydrolyzed by aqueous acid or base. This term often seems inappropriate because many so-called "simple" lipids are quite complex molecules. We will consider three important groups of simple lipids: steroids, prostaglandins, and terpenes. Figure 25-1 shows some examples of complex and simple lipids.

FIGURE 25-1
Examples of lipids. Complex lipids contain ester functional groups that can be hydrolyzed to acids and alcohols. Simple lipids are not easily hydrolyzed.

Examples of complex lipids

$CH_2{-}O{-}C(=O){-}(CH_2)_{16}CH_3$
$CH{-}O{-}C(=O){-}(CH_2)_{16}CH_3$
$CH_2{-}O{-}C(=O){-}(CH_2)_{16}CH_3$

tristearin, a fat

$CH_3(CH_2)_{15}{-}O{-}C(=O){-}(CH_2)_{14}CH_3$

spermaceti (cetyl palmitate), a wax

Examples of simple lipids

H_3C, H_3C, HO

cholesterol, a steroid

α-pinene, a terpene

25-2 Waxes

Plant leaves often have a wax coating to prevent excessive loss of water.

Waxes are esters of long-chain fatty acids with long-chain alcohols. They occur widely in nature and serve a number of purposes in plants and animals. *Spermaceti* (Figure 25-1), found in the head of the sperm whale, probably helps to regulate the animal's buoyancy for deep diving. It may also serve to amplify high-frequency sounds for locating prey. *Beeswax* is a mixture of waxes, hydrocarbons, and alcohols that bees use to form their honeycomb. *Carnauba wax* is a mixture of waxes of very high molecular weights. The carnauba plant secretes this waxy material to coat its leaves to prevent excessive loss of water by evaporation. Waxes are also found in the protective coatings of insects' exoskeletons, mammals' fur, and birds' feathers. In contrast to these waxes, the "paraffin wax" used to seal preserves is not a true wax; rather, it is a mixture of high-molecular-weight alkanes.

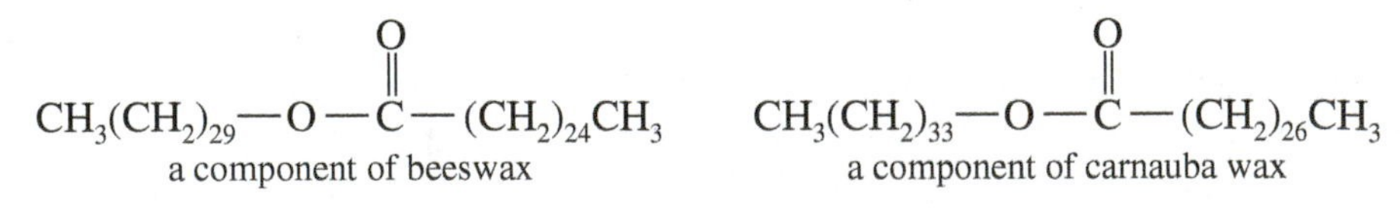

For many years, natural waxes were used in making cosmetics, adhesives, varnishes, and waterproofing materials. Synthetic materials have now replaced natural waxes for most of these uses.

25-3 Triglycerides

Glycerides are simply fatty acid esters of the triol *glycerol*. The most common glycerides are **triglycerides (triacylglycerols),** in which all three of the glycerol —OH groups have been esterified by fatty acids. For example, tristearin (Figure 25-1) is a component of beef fat in which all three —OH groups of glycerol are esterified by stearic acid, $CH_3(CH_2)_{16}COOH$. Most naturally occurring triglycerides are *mixed triglycerides*, containing two or three different fatty acids.

Triglycerides are commonly called **fats** if they are solid at room temperature and **oils** if they are liquid at room temperature. Most triglycerides derived from mammals are fats, such as beef tallow or lard. Although these fats are solid at room temperature, the warm body temperature of the living animal keeps them somewhat fluid, allowing for movement. In plants and cold-blooded animals, triglycerides are generally oils, such as corn oil, peanut oil, or fish oil. A fish requires liquid oils rather than solid fats because it would have difficulty moving if its triglycerides solidified whenever it swam in a cold stream.

Fats and oils are commonly used for long-term energy storage in plants and animals. Fat is a more efficient source of long-term energy than carbohydrates because metabolism of a gram of fat releases about 9 food Calories (kcal), but each gram of sugar, starch, or protein releases only about 4 food Calories of energy. An average 70-kg adult male stores about 4000 kJ (about 1000 kcal) of readily available energy as glycogen (0.2 kg), and about 600,000 kJ (about 140,000 kcal) of long-term energy as fat (15 kg)—enough to supply his resting metabolic needs for nearly 3 months!

The **fatty acids** of common triglycerides are long, unbranched carboxylic acids with about 12 to 20 carbon atoms. Most fatty acids contain even numbers of carbon atoms because they are derived from two-carbon acetic acid units. Some of the common fatty acids have saturated carbon chains, while others have one or more carbon–carbon double bonds. Table 25-1 shows the structures of some common fatty acids derived from fats and oils.

PROBLEM 25-1

Trimyristin, a solid fat present in nutmeg, is hydrolyzed to give one equivalent of glycerol and three equivalents of myristic acid. Give the structure of trimyristin.

Table 25-1 shows that saturated fatty acids have melting points that increase gradually with their molecular weights. The presence of a cis double bond lowers the melting point, however. Notice that the 18-carbon saturated acid (stearic acid) has a melting point of 70 °C, while the 18-carbon acid with a cis double bond (oleic acid) has a melting point of 4 °C. This lowering of the melting point results from the unsaturated acid's "kink" at the position of the double bond (Figure 25-2). Kinked molecules cannot pack as tightly together in a solid as the uniform zigzag chains of a saturated acid.

A second double bond lowers the melting point further (linoleic acid, mp −5 °C), and a third double bond lowers it still further (linolenic acid, mp −11 °C). The trans double bonds in eleostearic acid (mp 49 °C) have a smaller effect on the melting point than the cis double bonds of linolenic acid. The geometry of a trans double bond is similar to the zigzag conformation of a saturated acid, so it does not kink the chain as much as a cis double bond.

TABLE 25-1
Structures and Melting Points of Some Common Fatty Acids

Name	Carbons	Structure	Melting Point (°C)
Saturated acids			
lauric acid	12	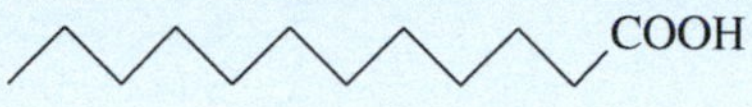 COOH	44
myristic acid	14	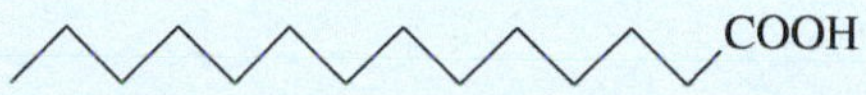 COOH	59
palmitic acid	16	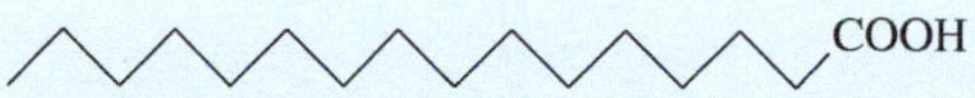 COOH	64
stearic acid	18	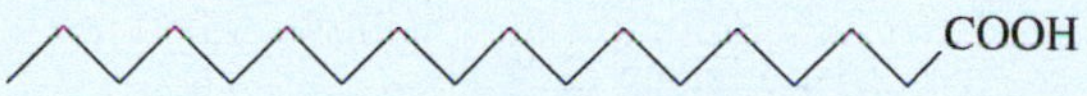 COOH	70
arachidic acid	20	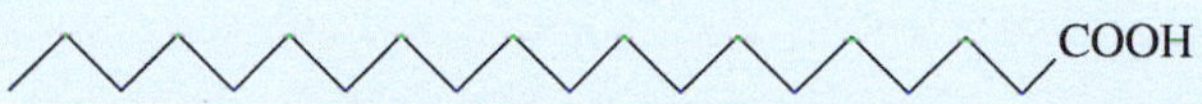 COOH	76
Unsaturated acids			
oleic acid	18	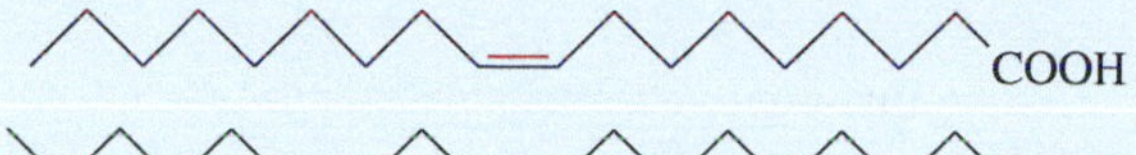 COOH	4
linoleic acid	18	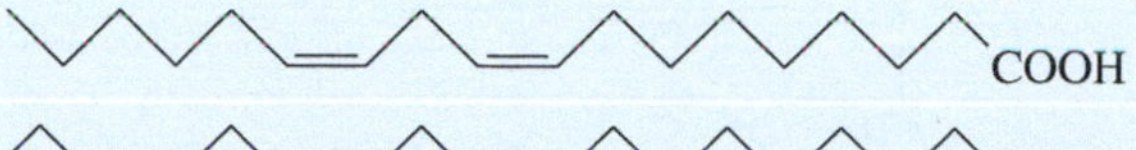 COOH	-5
linolenic acid	18	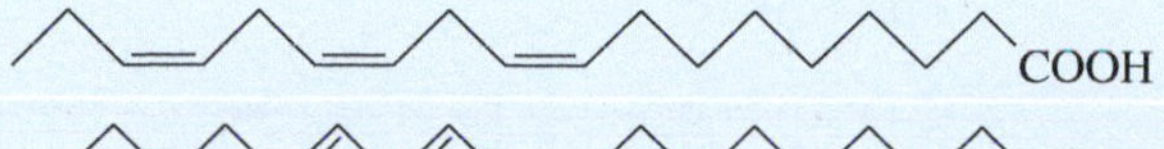 COOH	-11
eleostearic acid	18	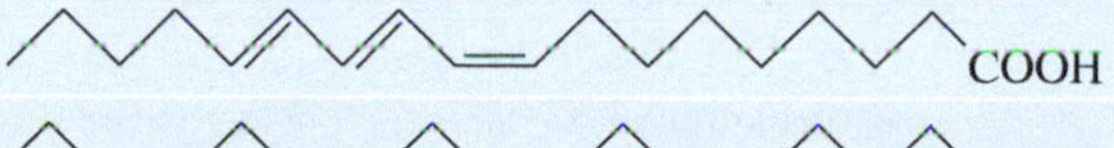 COOH	49
arachidonic acid	20	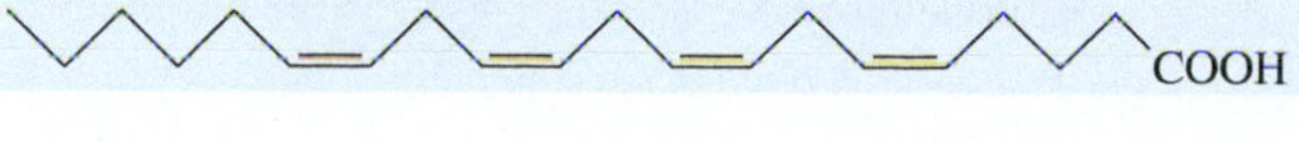 COOH	-49

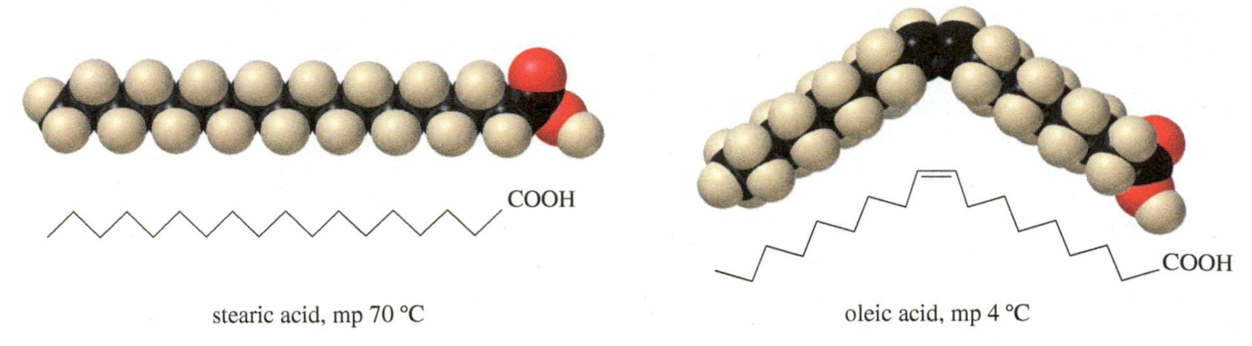

FIGURE 25-2
Comparison of stearic acid and oleic acid. The cis double bond in oleic acid lowers the melting point by 66 °C.

The melting points of fats and oils also depend on the degree of unsaturation (especially cis double bonds) in their fatty acids. A triglyceride derived from saturated fatty acids has a higher melting point because it packs more easily into a solid lattice than a triglyceride derived from kinked, unsaturated fatty acids. Figure 25-3 shows typical conformations of triglycerides containing saturated and unsaturated fatty acids. Tristearin (mp 72 °C) is a **saturated fat** that packs well in a solid lattice. Triolein (mp −4 °C) has the same number of carbon atoms as tristearin, but triolein has three cis double bonds causing kinked conformations that prevent optimum packing in the solid state.

Most saturated triglycerides are *fats* because they are solid at room temperature. Most triglycerides with several unsaturations are *oils* because they are liquid at room temperature. The term **polyunsaturated** simply means there are several double bonds in the fatty acids of the triglyceride.

Most naturally occurring fats and oils are mixtures of triglycerides containing a variety of saturated and unsaturated fatty acids. Even the individual triglycerides are often mixed, containing two or three different fatty acids. In general, oils from plants and cold-blooded animals contain more unsaturations than fats from warm-blooded animals. Table 25-2 gives the approximate composition of the fatty acids obtained from hydrolysis of some common fats and oils.

FIGURE 25-3
Unsaturated triglycerides have lower melting points because their unsaturated fatty acids do not pack as well in a solid lattice.

tristearin, mp 72 °C

triolein, mp −4 °C

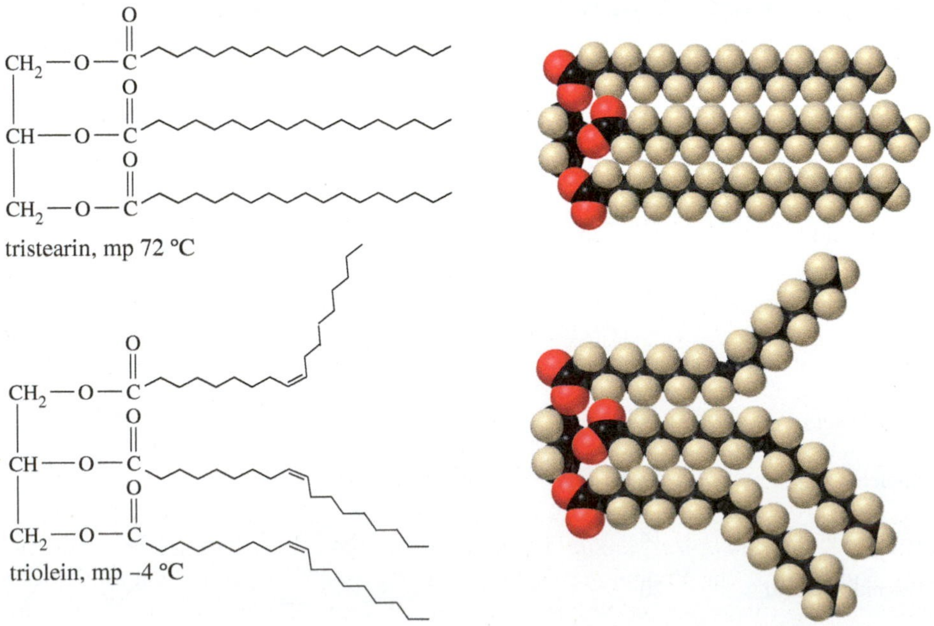

TABLE 25-2
Fatty Acid Composition of Some Fats and Oils, Percent by Weight

	Saturated Fatty Acids				Unsaturated Fatty Acids		
Source	**Lauric**	**Myristic**	**Palmitic**	**Stearic**	**Oleic**	**Linoleic**	**Linolenic**
beef fat	0	6	27	14	49	2	0
lard	0	1	24	9	47	10	0
human fat	1	3	27	8	48	10	0
herring oil	0	5	14	3	0	0	30*
corn oil	0	1	10	3	50	34	0
olive oil	0	0.1	7	2	84	5	2
soybean oil	0.2	0.1	10	2	29	51	7
canola oil	0	0	2	7	54	30	7
linseed oil	0	0.2	7	1	20	20	52

* Contains large amounts of even more highly unsaturated fatty acids.

25-3A Hydrogenation of Triglycerides; Trans Fats

For many years, *lard* (a soft, white solid obtained by rendering pig fat) was commonly used for cooking and baking. Although vegetable oil could be produced more cheaply and in greater quantities, consumers were unwilling to use vegetable oils because they were accustomed to using white, creamy lard. Then vegetable oils were treated with hydrogen gas and a nickel catalyst, reducing some of the double bonds to produce a creamy, white *vegetable shortening* that resembles lard. This "partially hydrogenated vegetable oil" largely replaced lard for cooking and baking. *Margarine* is a similar material with butyraldehyde added to give it a taste like that of butter.

More recently, consumers have learned that polyunsaturated vegetable oils may be more healthful, prompting many to switch to natural vegetable oils. Consumers are also concerned with the presence of unnatural trans fatty acids in "partially hydrogenated vegetable oil." During the hydrogenation process, the catalyst lowers the activation energy of both the forward (hydrogenation) and reverse (dehydrogenation) processes. The naturally occurring cis double bonds in vegetable oils can hydrogenate and the products can dehydrogenate. The double bonds end up in random positions, with either cis or trans stereochemistry. The white, creamy product has fewer double bonds overall, but some of the remaining double bonds may be in positions or stereochemical configurations that do not occur in nature.

Based on the latest research, the FDA has concluded that **trans fats** (fats containing trans fatty acids) are associated with an increased risk of coronary heart disease. The FDA now requires listing the amounts of trans fats on food labels, and it recommends a nationwide ban on trans fats. An increasing number of national and local governments have already banned or restricted the use of partially hydrogenated vegetable oils containing trans fats in food.

PROBLEM 25-2

Give an equation for the complete hydrogenation of trilinolein using an excess of hydrogen. Name the product, and predict approximate melting points for the starting material and the product.

25-3B Transesterification of Fats and Oils to Biodiesel

Most diesel engines can run on cooking oil once they are warm, but cooking oil is not sufficiently volatile to start a cold diesel engine. A base-catalyzed transesterification, using methanol as the alcohol and NaOH as the catalyst, converts fats and oils to the methyl esters of the three individual fatty acids. With molecular weights about a third of the original triglyceride, these methyl esters are more volatile and work well in diesel engines. The mixture of fatty acid methyl esters is called **biodiesel.**

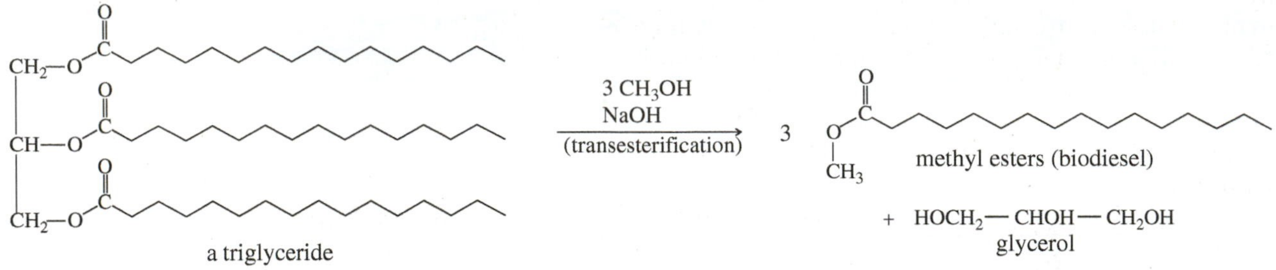

a triglyceride $\xrightarrow[\text{(transesterification)}]{3\ CH_3OH,\ NaOH}$ 3 methyl esters (biodiesel) + $HOCH_2—CHOH—CH_2OH$ glycerol

Biodiesel potentially offers environmental advantages over conventional diesel fuel. Most important, it converts waste cooking oil into a useful product, reducing the amount of waste going into landfills and replacing some of the petroleum that must be burned. Also, biodiesel comes from biomass that has been recently synthesized from atmospheric carbon dioxide, so the cycle of its production and use might not add as much carbon dioxide to the atmosphere as does the burning of petroleum-based diesel fuels.

Several countries have enacted laws mandating the use of biodiesel in blended diesel fuels, hoping to slow the increase in atmospheric carbon dioxide that is thought to contribute to global warming. But complex problems rarely have such simple solutions. The supply of waste fats and oils is not enough to produce the biodiesel required by these laws. Converting new food-grade fats and oils to biodiesel is economically unsound because food-grade oils sell for several times the price of diesel fuel. Fuel suppliers have turned to the world market for vegetable oils, encouraging the clearing of rain forests in tropical countries to produce palm oil and soybean oil for transesterification to biodiesel.

PROBLEM 25-3

Give an equation for the complete transesterification of triolein using an excess of methanol as the alcohol and sodium hydroxide as the catalyst.

25-4 Saponification of Fats and Oils: Soaps and Detergents

Saponification is the base-promoted hydrolysis of the ester linkages in fats and oils (review Section 21-7B). One of the products is soap, and the word *saponification* is derived from the Latin word *saponis*, meaning "soap." Saponification was discovered before 500 B.C., when people found that a curdy material resulted when animal fat was heated with wood ashes. Alkaline substances in the ashes promote hydrolysis of the ester linkages of the fat. Soap is currently made by boiling animal fat or vegetable oil with a solution of sodium hydroxide. The following reaction shows the formation of soap from tristearin, a component of beef fat.

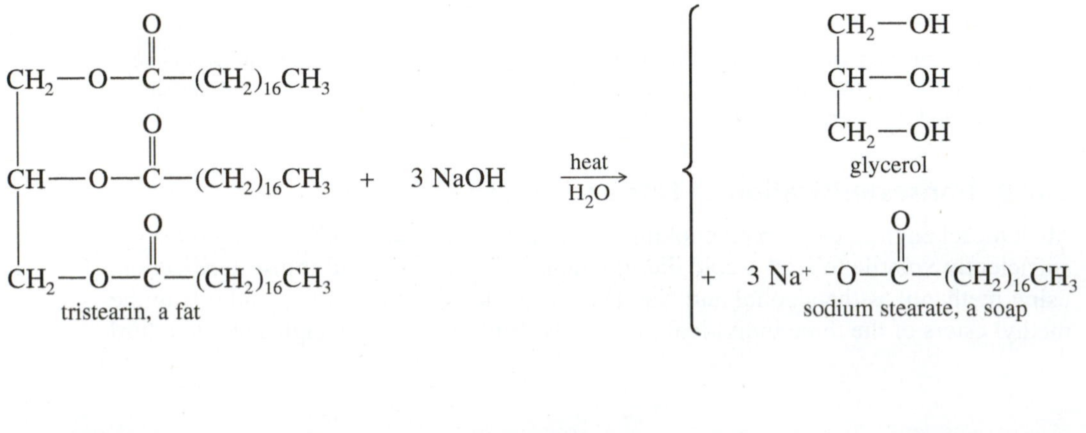

$$\begin{array}{l} CH_2—O—C(=O)—(CH_2)_{16}CH_3 \\ CH—O—C(=O)—(CH_2)_{16}CH_3 \\ CH_2—O—C(=O)—(CH_2)_{16}CH_3 \end{array} + 3\ NaOH \xrightarrow[H_2O]{\text{heat}} \begin{array}{l} CH_2—OH \\ CH—OH \\ CH_2—OH \end{array} + 3\ Na^+\ {}^-O—C(=O)—(CH_2)_{16}CH_3$$

tristearin, a fat — glycerol — sodium stearate, a soap

Chemically, a **soap** is the sodium or potassium salt of a fatty acid. The negatively charged carboxylate group is **hydrophilic** ("attracted to water"), and the long hydrocarbon chain is **hydrophobic** ("repelled by water") and **lipophilic** ("attracted to oils"). The electrostatic potential map of the stearate ion is shown in Figure 25-4. Notice the high electron density (red) around the negatively charged carboxylate end of the molecule. The carboxylate oxygen atoms share the negative charge and participate in strong hydrogen bonding with water molecules. The rest of the molecule (green) is a hydrocarbon chain that cannot participate in hydrogen bonding with water.

In water, soap forms a cloudy solution of **micelles:** clusters of about 100 to 200 soap molecules with their polar "heads" (the carboxylate groups) on the surface of the cluster and their hydrophobic "tails" (the hydrocarbon chains) enclosed within. The micelle (Figure 25-4) is an energetically stable particle because the hydrophilic groups are hydrogen-bonded to the surrounding water, while the hydrophobic groups are shielded within the interior of the micelle, interacting with other hydrophobic groups.

Soaps are useful cleaning agents because of the different affinities of a soap molecule's two ends. Greasy dirt is not easily removed by pure water because grease is hydrophobic and insoluble in water. The long hydrocarbon chain of a soap molecule dissolves in the grease, with the hydrophilic head at the surface of the grease droplet. Once the surface of the grease droplet is covered by many soap molecules, a micelle can form around it with a tiny grease droplet at its center. This grease droplet is easily

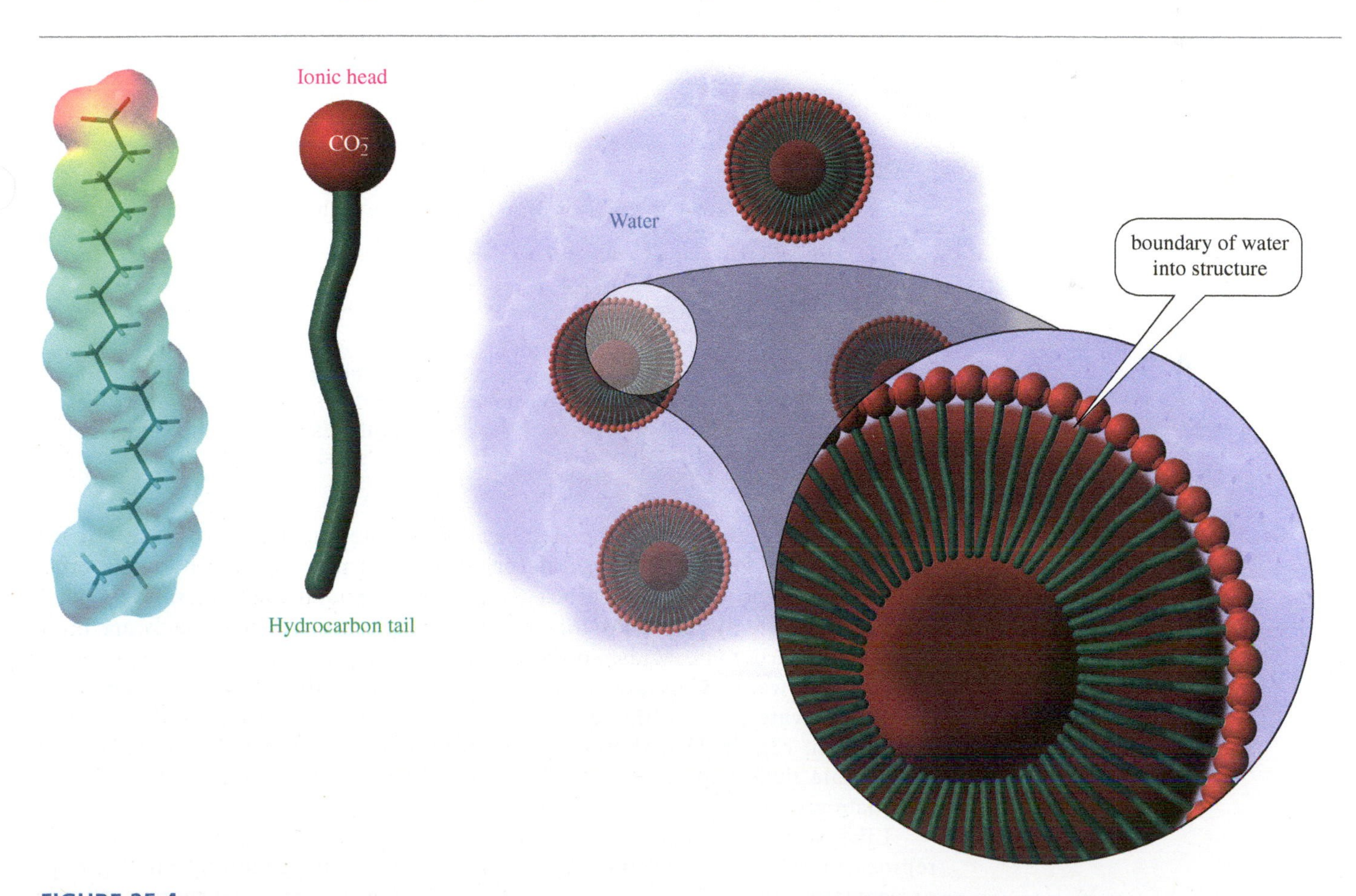

FIGURE 25-4
Aggregation of soap in micelles. The electrostatic potential map of a soap molecule shows high electron density in the negatively charged head and medium electron density (green) in the hydrocarbon tail. In water, soap forms a cloudy solution of micelles, with the hydrophilic heads in contact with water and the hydrophobic tails clustered in the interior. The Na^+ ions (not shown) are dissolved in the water surrounding the micelle.

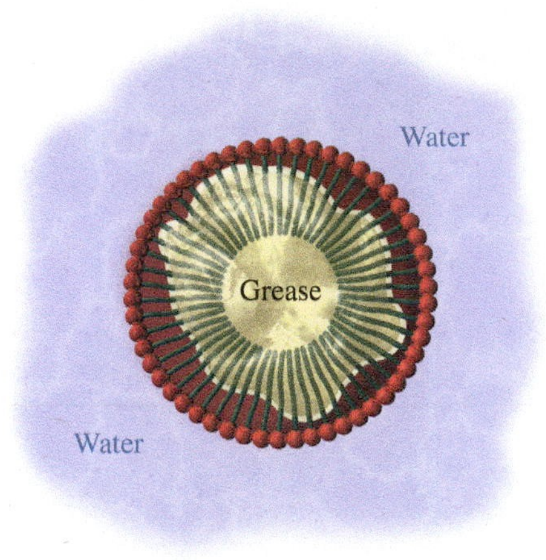

FIGURE 25-5
Emulsification of grease. In a soapy solution, grease is emulsified by forming micelles coated by the hydrophilic carboxylate groups of the soap.

suspended in water because it is covered by the hydrophilic carboxylate groups of the soap (Figure 25-5). The resulting mixture of two insoluble phases (grease and water), with one phase dispersed throughout the other in small droplets, is called an **emulsion.** We say the grease has been **emulsified** by the soapy solution. When the wash water is rinsed away, the grease goes with it.

The usefulness of soaps is limited by their tendency to precipitate out of solution in hard water. **Hard water** is water that is acidic or that contains ions of calcium, magnesium, or iron. In acidic water (such as the "acid rain" of environmental concern), soap molecules are protonated to the free fatty acids. Without the ionized carboxylate group, the fatty acid floats to the top as a greasy "acid scum" precipitate.

$$\underset{\text{a soap}}{CH_3(CH_2)_n{-}\overset{\overset{\displaystyle O}{\|}}{C}{-}O^-\,Na^+} + H^+ \longrightarrow \underset{\text{acid scum}}{CH_3(CH_2)_n{-}\overset{\overset{\displaystyle O}{\|}}{C}{-}OH\downarrow} + Na^+$$

Many areas have household water containing calcium, magnesium, and iron ions. Although these mineral-rich waters can be healthful for drinking, the ions react with soaps to form insoluble salts called *hard-water scum.* The following equation shows the reaction of a soap with calcium, common in areas where water comes in contact with limestone rocks.

$$2\,\underset{\text{a soap}}{CH_3(CH_2)_n{-}\overset{\overset{\displaystyle O}{\|}}{C}{-}O^-\,Na^+} + Ca^{2+} \longrightarrow \underset{\text{hard-water scum}}{[CH_3(CH_2)_n{-}\overset{\overset{\displaystyle O}{\|}}{C}{-}O]_2Ca\downarrow} + 2\,Na^+$$

PROBLEM 25-4

Give equations to show the reactions of sodium stearate with the following.
(a) Ca^{2+} **(b)** Mg^{2+} **(c)** Fe^{3+}

PROBLEM 25-5

Several commercial laundry soaps contain water-softening agents, usually sodium carbonate (Na_2CO_3) or sodium phosphate (Na_3PO_4 or Na_2HPO_4). Explain how these water-softening agents allow soaps to be used in water that is hard by virtue of its
(a) low pH. **(b)** dissolved Ca^{2+}, Mg^{2+}, and Fe^{3+} salts.

Soaps precipitate in hard water because of the chemical properties of the carboxylic acid group. **Synthetic detergents** avoid precipitation by using other functional groups in place of carboxylic acid salts. Sodium salts of sulfonic acids are the most widely used class of synthetic detergents. Sulfonic acids are more acidic than carboxylic acids, so their salts are not protonated, even in strongly acidic wash water. Calcium, magnesium, and iron salts of sulfonic acids are soluble in water, so sulfonate salts can be used in hard water without forming a scum. Figure 25-6 shows the structure and electrostatic potential map of a sulfonate detergent, with red (electron-rich) regions around the hydrophilic sulfonate group.

Like soaps, synthetic detergents combine hydrophilic and hydrophobic regions in the same molecule. Hydrophobic regions are generally alkyl groups or aromatic rings. Hydrophilic regions may contain anionic groups, cationic groups, or nonionic groups containing several oxygen atoms or other hydrogen-bonding atoms. Figure 25-6 shows examples of anionic, cationic, and nonionic detergents.

An alkylbenzenesulfonate detergent

$$\text{(alkyl)}-C_6H_4-SO_3^-\ Na^+ + \left\{\begin{array}{c} H^+ \\ Ca^{2+} \\ Mg^{2+} \\ Fe^{3+} \end{array}\right\} \longrightarrow \text{no precipitate}$$

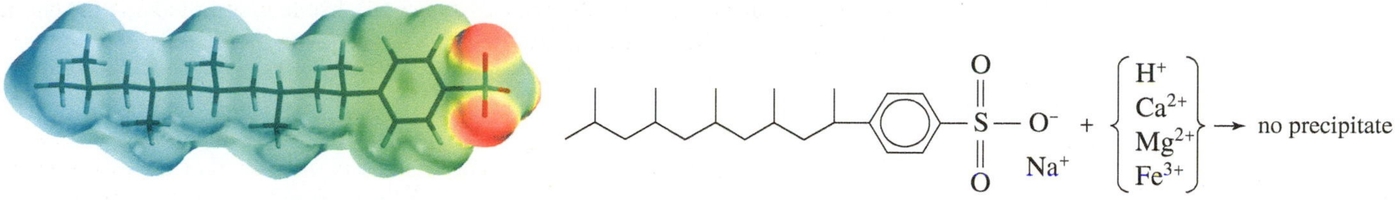

Examples of other types of detergents

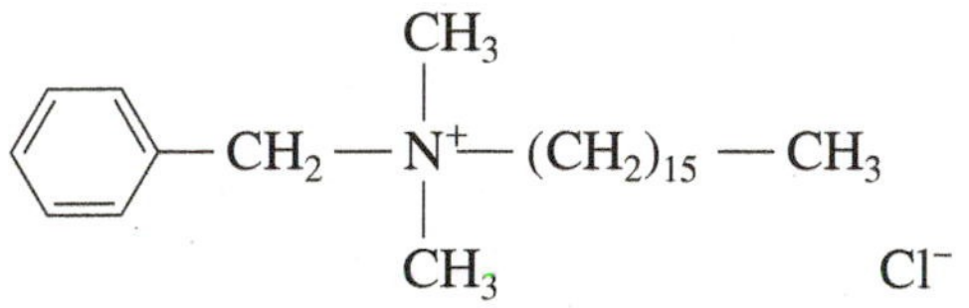

$$C_6H_5-CH_2-\overset{CH_3}{\underset{CH_3}{N^+}}-(CH_2)_{15}-CH_3 \quad Cl^-$$

benzylcetyldimethylammonium chloride
(benzalkonium chloride)

$$C_9H_{19}-C_6H_4-(OCH_2CH_2)_9-OH$$

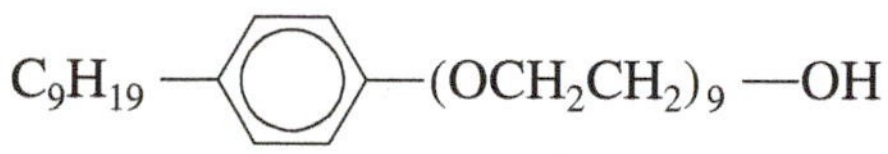

Nonoxynol®, Ortho Pharmaceuticals

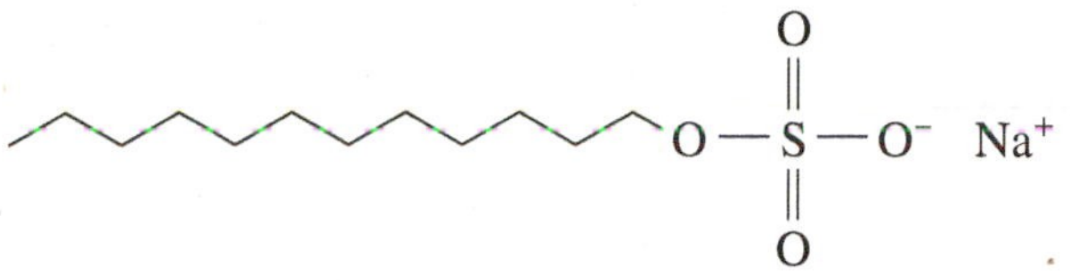

$$CH_3(CH_2)_{11}-O-SO_2-O^-\ Na^+$$

sodium dodecyl sulfate
(sodium lauryl sulfate)

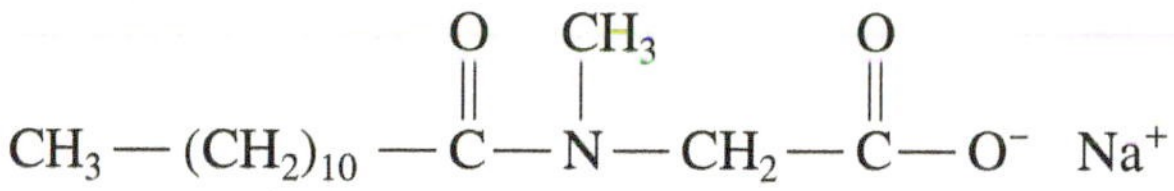

$$CH_3-(CH_2)_{10}-\overset{O}{\overset{\|}{C}}-\overset{CH_3}{N}-CH_2-\overset{O}{\overset{\|}{C}}-O^-\ Na^+$$

N-lauroyl-*N*-methylglycine, sodium salt
Gardol®, Colgate-Palmolive Co.

FIGURE 25-6
Synthetic detergents may have anionic, cationic, or nonionic hydrophilic functional groups. Of these detergents, only Gardol® is a carboxylate salt and forms a precipitate in hard water.

PROBLEM 25-6

Point out the hydrophilic and hydrophobic regions in the structures of benzalkonium chloride, Nonoxynol®, and Gardol® (Figure 25-6).

PROBLEM 25-7

The synthesis of the alkylbenzenesulfonate detergent shown in Figure 25-6 begins with the partial polymerization of propylene to give a pentamer.

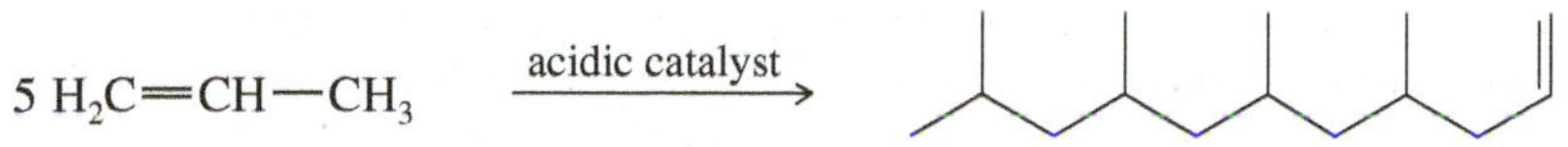

$$5\ H_2C{=}CH-CH_3 \xrightarrow{\text{acidic catalyst}} \text{a pentamer}$$

Show how aromatic substitution reactions can convert this pentamer to the final synthetic detergent.

25-5 Phospholipids

Phospholipids are lipids that contain groups derived from phosphoric acid. The most common phospholipids are **phosphoglycerides,** which are closely related to common fats and oils. A phosphoglyceride generally has a phosphoric acid group in place of one of the fatty acids of a triglyceride. The simplest class of phosphoglycerides is **phosphatidic acids;** these consist of glycerol esterified by two fatty acids and one phosphoric acid group. Although it is often drawn in its acid form, a phosphatidic acid is actually deprotonated at neutral pH.

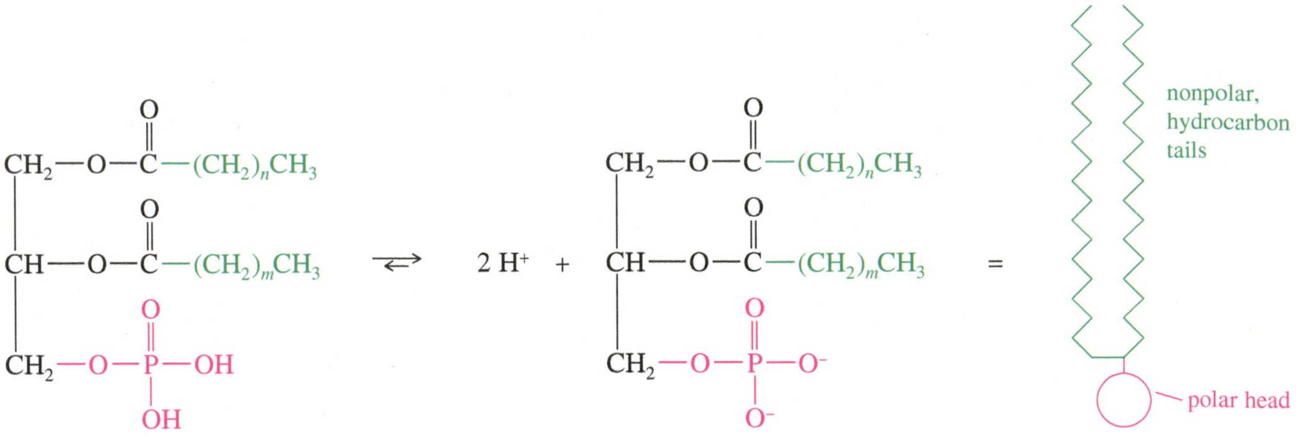

$$\begin{array}{l} CH_2{-}O{-}\overset{O}{\overset{\|}{C}}{-}(CH_2)_nCH_3 \\ | \\ CH{-}O{-}\overset{O}{\overset{\|}{C}}{-}(CH_2)_mCH_3 \\ | \\ CH_2{-}O{-}\overset{O}{\overset{\|}{P}}{-}OH \\ \qquad\qquad | \\ \qquad\quad\; OH \end{array} \rightleftharpoons 2\,H^+ + \begin{array}{l} CH_2{-}O{-}\overset{O}{\overset{\|}{C}}{-}(CH_2)_nCH_3 \\ | \\ CH{-}O{-}\overset{O}{\overset{\|}{C}}{-}(CH_2)_mCH_3 \\ | \\ CH_2{-}O{-}\overset{O}{\overset{\|}{P}}{-}O^- \\ \qquad\qquad | \\ \qquad\quad\; O^- \end{array} =$$

a phosphatidic acid

ionized form

nonpolar, hydrocarbon tails

polar head

schematic representation

PROBLEM 25-8

Show that a phosphatidic acid is chiral, even though none of its fatty acids are chiral. Where is the asymmetric carbon atom?

Application: Respiratory Distress

Dipalmitoyl phosphatidyl choline is one component of human lung surfactant, which coats the inner surfaces of the lung membranes and prevents them from clinging together and collapsing. Premature infants often produce little or no lung surfactant, which can lead to collapsed alveoli and other symptoms of infant respiratory distress syndrome (IRDS).

Many phospholipids contain an additional alcohol esterified to the phosphoric acid group. **Cephalins** are esters of ethanolamine, and **lecithins** are esters of choline. Both cephalins and lecithins are widely found in plant and animal tissues.

$HO{-}CH_2CH_2{-}NH_2$

ethanolamine

$HO{-}CH_2CH_2{-}\overset{+}{N}(CH_3)_3$

choline

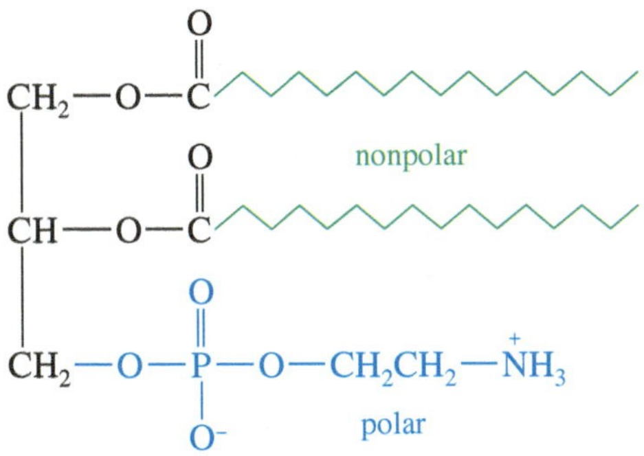

$$\begin{array}{l} CH_2{-}O{-}\overset{O}{\overset{\|}{C}}{\sim\sim\sim\sim} \\ | \qquad\qquad\quad \text{nonpolar} \\ CH{-}O{-}\overset{O}{\overset{\|}{C}}{\sim\sim\sim\sim} \\ | \\ CH_2{-}O{-}\overset{O}{\overset{\|}{P}}{-}O{-}CH_2CH_2{-}\overset{+}{N}H_3 \\ \qquad\qquad | \qquad\quad \text{polar} \\ \qquad\quad\; O^- \end{array}$$

a cephalin,
or phosphatidyl ethanolamine

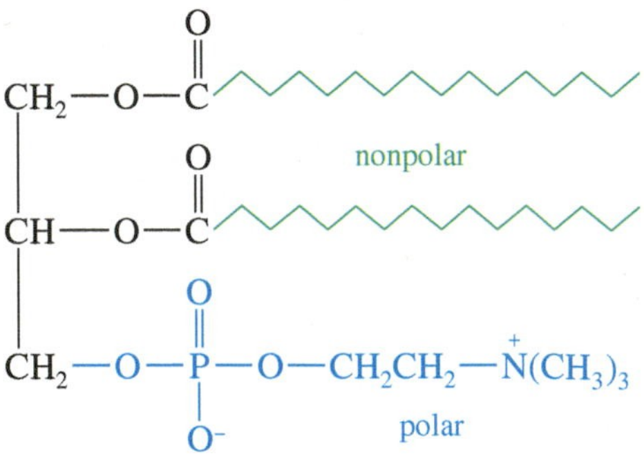

$$\begin{array}{l} CH_2{-}O{-}\overset{O}{\overset{\|}{C}}{\sim\sim\sim\sim} \\ | \qquad\qquad\quad \text{nonpolar} \\ CH{-}O{-}\overset{O}{\overset{\|}{C}}{\sim\sim\sim\sim} \\ | \\ CH_2{-}O{-}\overset{O}{\overset{\|}{P}}{-}O{-}CH_2CH_2{-}\overset{+}{N}(CH_3)_3 \\ \qquad\qquad | \qquad\quad \text{polar} \\ \qquad\quad\; O^- \end{array}$$

a lecithin,
or phosphatidyl choline

Like phosphatidic acids, lecithins and cephalins contain a polar "head" and two long, nonpolar hydrocarbon "tails." This soap-like structure gives phospholipids some interesting properties. Like soaps, they form micelles and other aggregations with their polar heads on the outside and their nonpolar tails protected on the inside.

Another stable form of aggregation is a **lipid bilayer,** which forms animal cell membranes (Figure 25-7). In a lipid bilayer, the hydrophilic heads coat the two surfaces of a membrane, and the hydrophobic tails are protected within. Cell membranes contain phosphoglycerides oriented in a lipid bilayer, forming a barrier that restricts the flow of water and dissolved substances. Studies have shown that the two tails of phospholipids occupy just the right amount of space to make the strongest lipid bilayers. The single tails of soaps leave too much space, and the triple tails of triglycerides leave too little space between molecules for optimum bilayers.

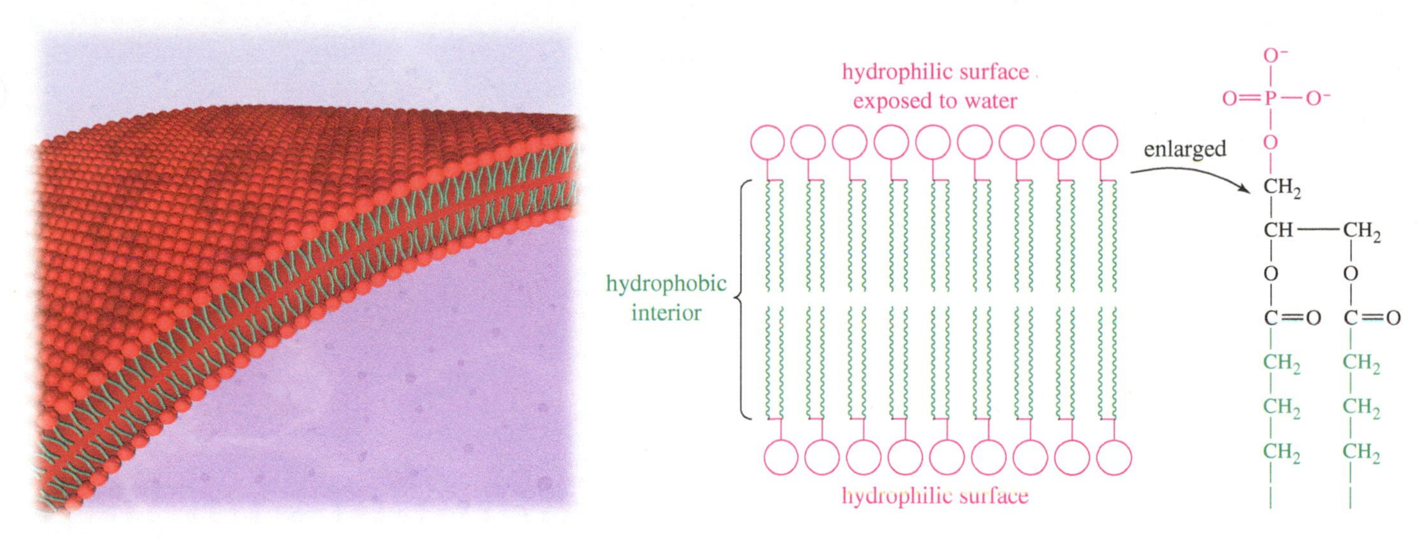

FIGURE 25-7
Structure of a lipid bilayer membrane. Phosphoglycerides can aggregate into a bilayer membrane with their polar heads exposed to the aqueous solution and the hydrocarbon tails protected within. This lipid bilayer is an important part of the cell membrane.

25-6 Steroids

Steroids are complicated polycyclic molecules found in all plants and animals. They are classified as *simple lipids* because they do not undergo hydrolysis like fats, oils, and waxes do. Steroids encompass a wide variety of compounds, including hormones, emulsifiers, and components of membranes. **Steroids** are compounds whose structures are based on the tetracyclic androstane ring system, shown here. The four rings are designated A, B, C, and D, beginning with the ring at lower left, and the carbon atoms are numbered beginning with the A ring and ending with the two "angular" (axial) methyl groups.

androstane

We have seen (Section 3-16B) that fused ring systems can have either trans or cis stereochemistry at each ring junction. In the androstane structure shown above, all three of the ring junctions are trans. A simpler example is the geometric isomerism of *trans*- and *cis*-decalin shown in Figure 25-8. If you make models of these isomers, you will find that the trans isomer is quite rigid and flat (aside from the ring puckering). In contrast, the cis isomer is relatively flexible, with the two rings situated at a sharp angle to each other.

Each of the three ring junctions is trans in androstane. Most steroids have this all-trans structure, which results in a stiff, nearly flat molecule with the two axial methyl groups perpendicular to the plane. In some steroids, the junction between rings A and B is cis, requiring the A ring to fold down below the rest of the ring system. Figure 25-9 shows the androstane ring system with both trans and cis A-B ring junctions. The B-C and C-D ring junctions are nearly always trans in natural steroids.

FIGURE 25-8
Cis-trans isomers of decalin. In *trans*-decalin, the two bonds to the second ring are trans to one another, and the hydrogens on the junction are also trans. In *cis*-decalin, the bonds to the second ring are cis, and the junction hydrogens are also cis.

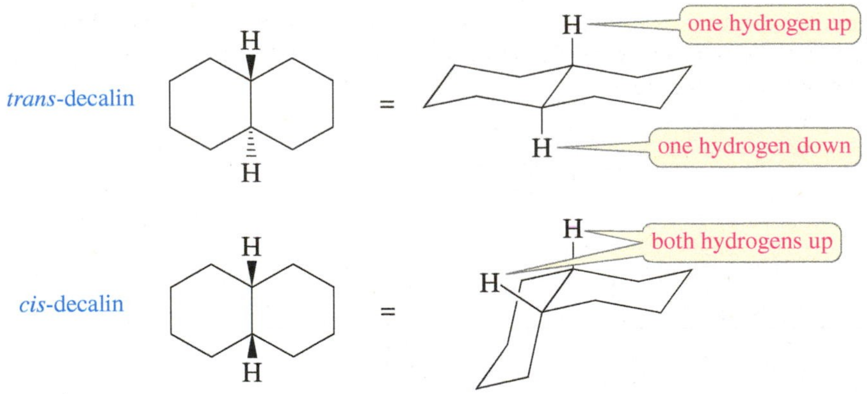

trans-decalin
one hydrogen up
one hydrogen down
cis-decalin
both hydrogens up

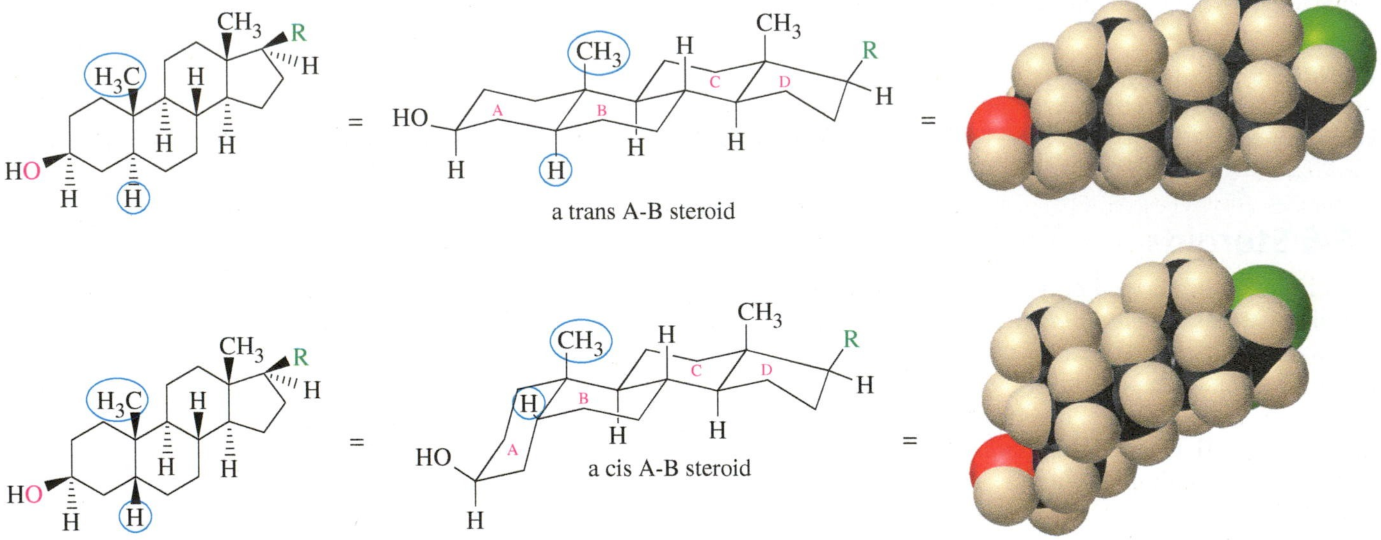

a trans A-B steroid

a cis A-B steroid

FIGURE 25-9
Common steroids may have either a cis or a trans A-B ring junction. The other ring junctions are normally trans.

Application: Statins

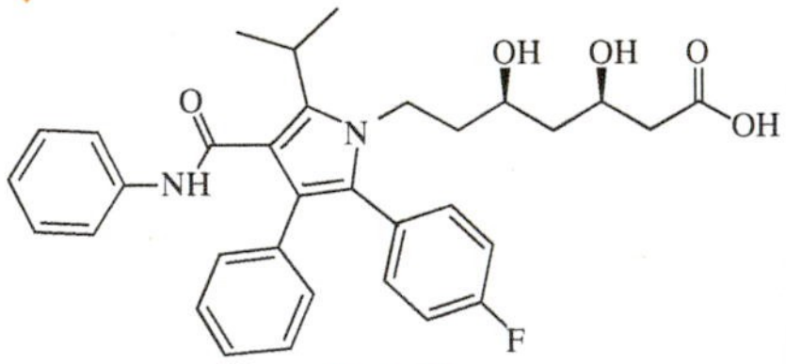

Lipitor®
(atorvastatin)

High levels of cholesterol in the blood cannot be corrected by diet alone. Most of the excess cholesterol is *endogenous* cholesterol, which is produced by the liver. Statins such as Lipitor® were developed in the 1970s and 1980s to interrupt the metabolic pathway that produces cholesterol. At one time, Lipitor® was the best-selling prescription drug in the United States.

Most steroids have an oxygen functional group (=O or —OH) at C3 and some kind of side chain or other functional group at C17. Many also have a double bond from C5 to either C4 or C6. The structures of androsterone and cholesterol serve as examples. Androsterone, a male sex hormone, is based on the simple androstane ring system. Cholesterol is a common biological intermediate and is believed to be the biosynthetic precursor to other steroids. It has a side chain at C17 and a double bond between C5 and C6.

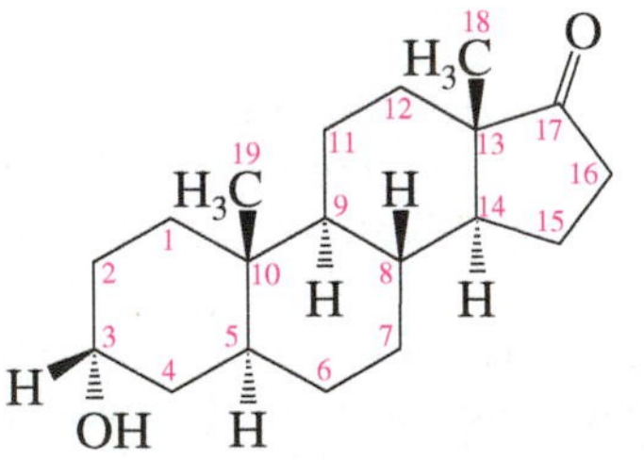

androsterone

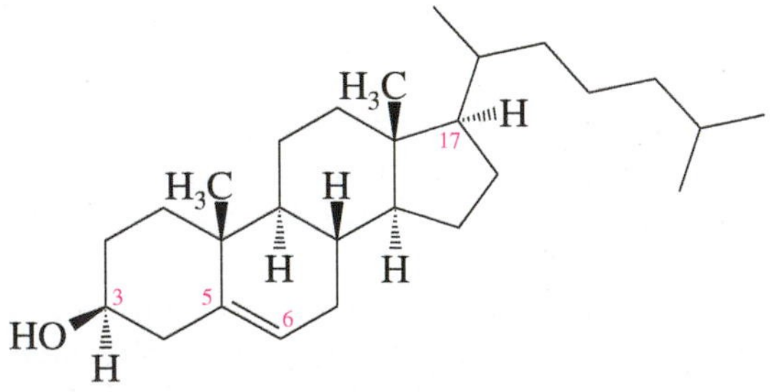

cholesterol

The principal sex hormones have been characterized and studied extensively. Testosterone is the most potent of the natural male sex hormones, and estradiol is the most potent natural female hormone. Notice that the female sex hormone differs from the male hormone by its aromatic A ring. For the A ring to be aromatic, the

C19 methyl group must be lost. In mammals, testosterone is converted to estradiol in the female's ovaries, where enzymes remove C19 and two hydrogen atoms to give the aromatic A ring.

C19 methyl

testosterone

estradiol

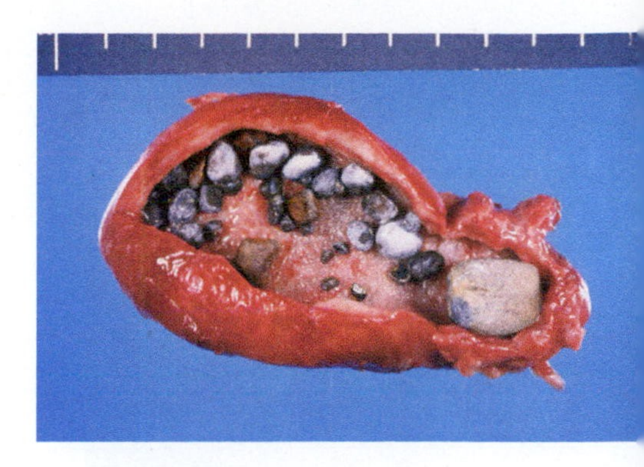

These gallstones, shown here within the gallbladder, are composed mostly of cholesterol.

PROBLEM 25-9

How would you use a simple extraction to separate a mixture of testosterone and estradiol?

When steroid hormones were first isolated, people believed that no synthetic hormone could rival the astonishing potency of natural steroids. In the past 50 years, however, many synthetic steroids have been developed. Some of these synthetic hormones are hundreds or thousands of times more potent than natural steroids. One example is ethynyl estradiol, a synthetic female hormone that is more potent than estradiol. Ethynyl estradiol is a common ingredient in oral contraceptives.

ethynyl estradiol

Application: Hormone Receptors

The biological activities of steroid hormones result from interactions with specific receptors. For example, estradiol interacts with the estrogen receptor. Some industrial chemicals, such as DDT, bisphenol A, and polychlorinated biphenyls, may possibly interact with these receptors and cause hormonal effects on people and wildlife.

Some of the most important physiological steroids are the adrenocortical hormones, synthesized by the adrenal cortex. Most of these hormones have either a carbonyl group or a hydroxy group at C11 of the steroid skeleton. The principal adrenocortical hormone is cortisol, used for the treatment of inflammatory diseases of the skin (psoriasis), the joints (rheumatoid arthritis), and the lungs (asthma). Figure 25-10 compares the structure of natural cortisol with two synthetic corticoids: fluocinolone acetonide, a fluorinated synthetic hormone that is more potent than cortisol for treating skin inflammation; and beclomethasone, a chlorinated synthetic hormone that is more potent than cortisol for treating asthma.

cortisol

fluocinolone acetonide

beclomethasone

FIGURE 25-10
Cortisol is the major natural hormone of the adrenal cortex. Fluocinolone acetonide is more potent for treating skin inflammation, and beclomethasone is more potent for treating asthma.

Application: Steroids for Oral Contraceptives

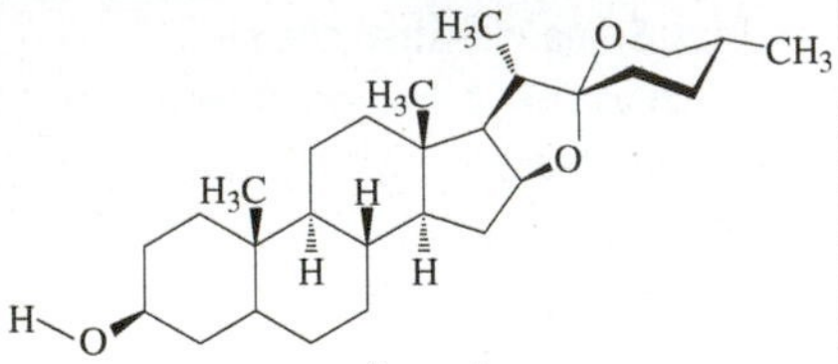

diosgenin

When oral contraceptives first became available, the most efficient routes to steroid hormones began with the steroid diosgenin, extracted from the roots of the wild Mexican barbasco yam *Dioscorea mexicana*. Several chemists from the United States founded the Syntex chemical company in Mexico, extracted diosgenin from barbasco, and converted it to progesterone, cortisone, testosterone, estrone, and estradiol. Syntex became the world's largest supplier of hormones for oral contraceptives. The market for barbasco collapsed in the 1970s, when steroids were found in soybeans, a cheaper and more reliable source of precursors.

PROBLEM 25-10

Draw each molecule in a stable chair conformation, and indicate whether each red group is axial or equatorial.

(a)

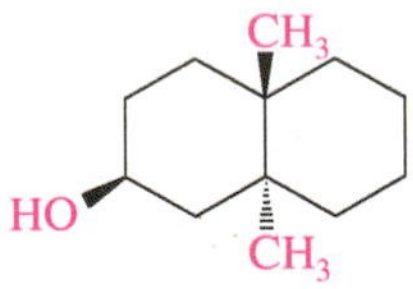

(b)

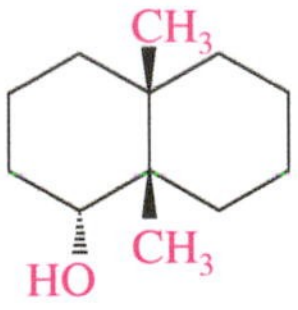

(c)

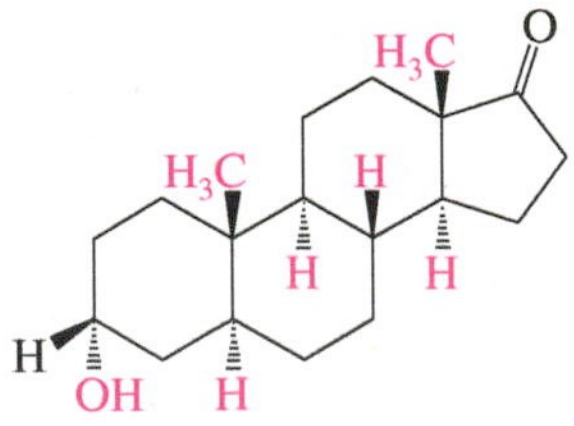

androsterone

(d)

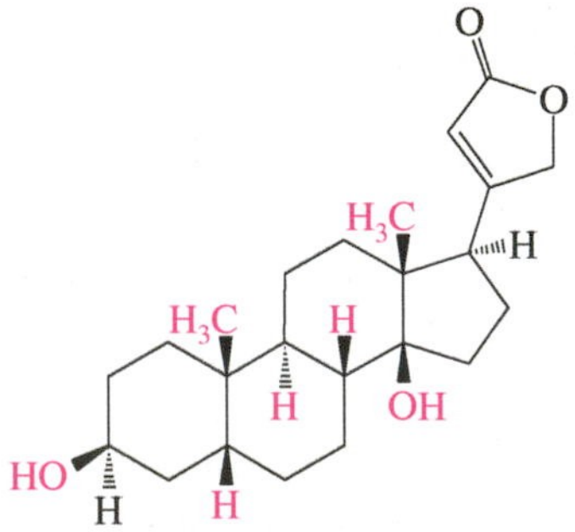

digitoxigenin, a cardiac stimulant

25-7 Prostaglandins

Prostaglandins are fatty acid derivatives that are even more powerful biochemical regulators than steroids. They are called prostaglandins because they were first isolated from secretions of the prostate gland. They were later found to be present in all body tissues and fluids, usually in minute quantities. Prostaglandins affect many body systems, including the nervous system, smooth muscle, blood, and the reproductive system. They play important roles in regulating such diverse functions as blood pressure, blood clotting, the allergic inflammatory response, activity of the digestive system, and the onset of labor.

Prostaglandins have a cyclopentane ring with two long side chains trans to each other, with one side chain ending in a carboxylic acid. Most prostaglandins have 20 carbon atoms, numbered as follows:

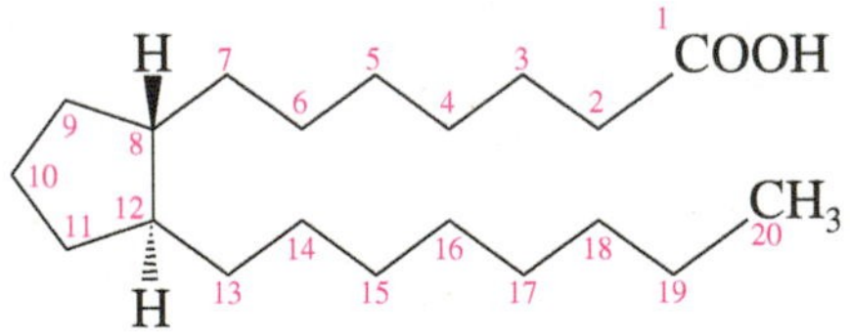

Many prostaglandins have hydroxy groups on C11 and C15 and a trans double bond between C13 and C14. They also have a carbonyl group or a hydroxy group on C9. If there is a carbonyl group at C9, the prostaglandin is a member of the *E series*. If there is a hydroxy group at C9, it is a member of the *F series*, and the symbol α means that the hydroxy group is directed down. Many prostaglandins have a cis double bond between C5 and C6. The number of double bonds is also given in the name, as shown here for two common prostaglandins.

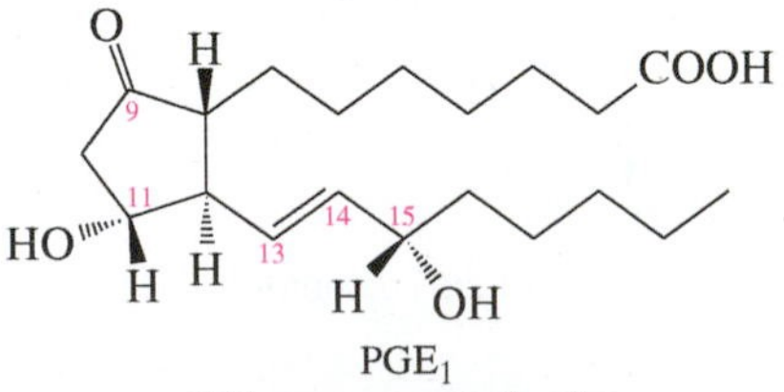

PGE_1
(PG means prostaglandin;
E means ketone at C9;
1 means one C=C double bond)

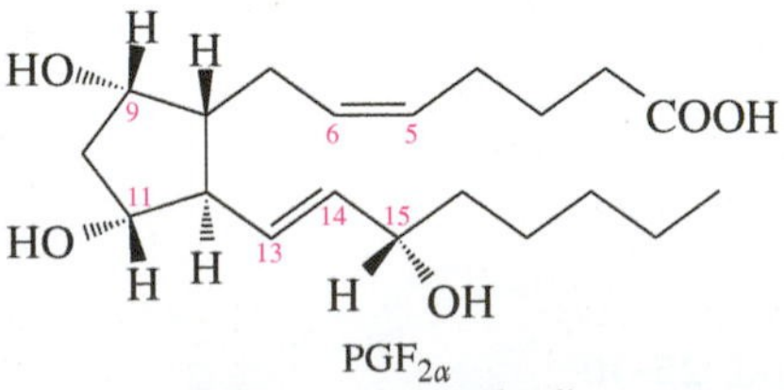

$PGF_{2\alpha}$
(PG means prostaglandin;
F means hydroxy at C9, and α means down;
2 means two C=C double bonds)

arachidonic acid $\xrightarrow[\text{enzyme}]{2\,O_2}$ $\xrightarrow{\text{reduction}}$ $PGF_{2\alpha}$

FIGURE 25-11
Biosynthesis of prostaglandins begins with an enzyme-catalyzed oxidative cyclization of arachidonic acid.

Prostaglandins are derived from arachidonic acid, a 20-carbon fatty acid with four cis double bonds. Figure 25-11 shows schematically how an enzyme oxidizes and cyclizes arachidonic acid to give the prostaglandin skeleton. One of the functions of aspirin is to inhibit this enzymatic prostaglandin synthesis and alleviate the inflammatory response.

Prostaglandins are difficult to isolate from animal tissues because they are present in extremely small concentrations and they are quickly degraded. Although they have been made by total synthesis, the process is long and difficult, and only a small amount of product is obtained. There was no way to obtain commercial quantities of prostaglandins until prostaglandin A2 was found to occur naturally in about 1% concentrations in the gorgonian coral *Plexaura homomalla*. This coral prostaglandin now serves as a starting material for short, efficient syntheses of medically useful prostaglandins.

The gorgonian coral *Plexaura homomalla* is a source of prostaglandin A2. This compound can be converted to medically useful prostaglandins that regulate mammalian reproduction, blood pressure, and digestion.

Application: Ulcer Drug

Despite their range of activities, many naturally occurring prostaglandins would make poor drugs because they are rapidly converted to inactive products. Synthetic modifications can prolong their activity. Misoprostol is a stable synthetic derivative of PGE_1 that is used for the treatment of ulcers.

misoprostol

25-8 Terpenes

Terpenes are a diverse family of compounds with carbon skeletons composed of five-carbon isopentyl (isoprene) units. Terpenes are commonly isolated from the **essential oils** of plants, the fragrant oils that are concentrated from plant material by steam distillation. The term *essential oils* literally means "oils derived from the essence" of plants. They often have pleasant tastes or aromas, and they are widely used as flavorings, deodorants, and medicines. Figure 25-12 shows the structures of four terpenes that are isolated from essential oils.

25-8A Characteristics and Nomenclature of Terpenes

Hundreds of essential oils were used as perfumes, flavorings, and medicines for centuries before chemists were capable of studying the mixtures. In 1818, it was found that oil of turpentine has a C:H ratio of 5:8, and many other essential oils have similar C:H ratios. This group of piney-smelling natural products with similar C:H ratios came to be known as **terpenes.**

In 1887, German chemist Otto Wallach determined the structures of several terpenes and discovered that all of them are formally composed of two or more five-carbon units of **isoprene** (2-methylbuta-1,3-diene). The isoprene unit maintains its isopentyl structure in a terpene, usually with modification of the isoprene double bonds.

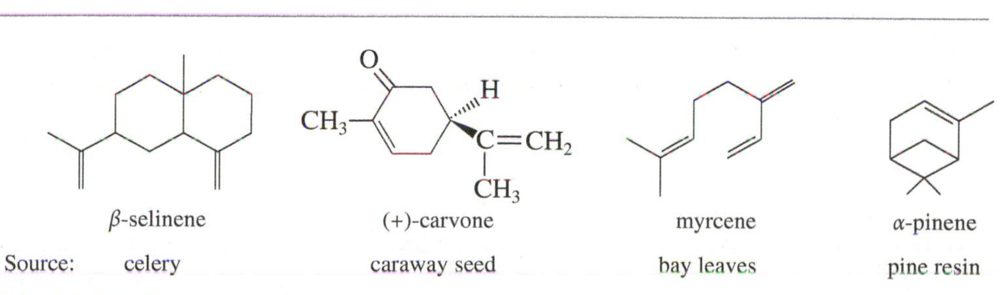

FIGURE 25-12
Many terpenes are derived from the essential oils of fragrant plants.

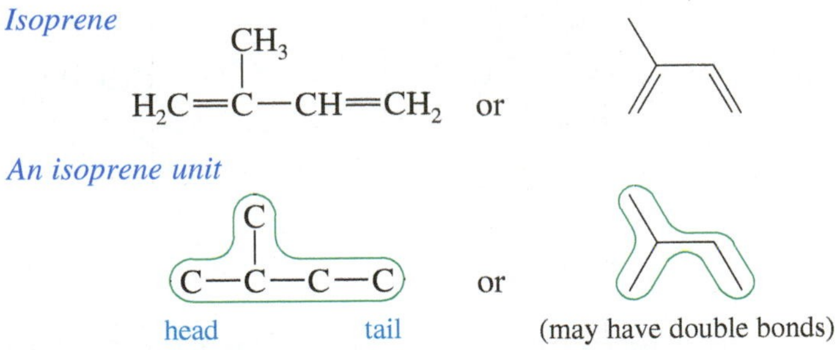

Isoprene

$$H_2C{=}\underset{}{\overset{CH_3}{\overset{|}{C}}}{-}CH{=}CH_2 \quad \text{or}$$

An isoprene unit

$$C{-}\overset{C}{\overset{|}{C}}{-}C{-}C \quad \text{or}$$

head tail (may have double bonds)

The isoprene molecule and the isoprene unit are said to have a "head" (the branched end) and a "tail" (the unbranched ethyl group). Myrcene can be divided into two isoprene units, with the head of one unit bonded to the tail of the other.

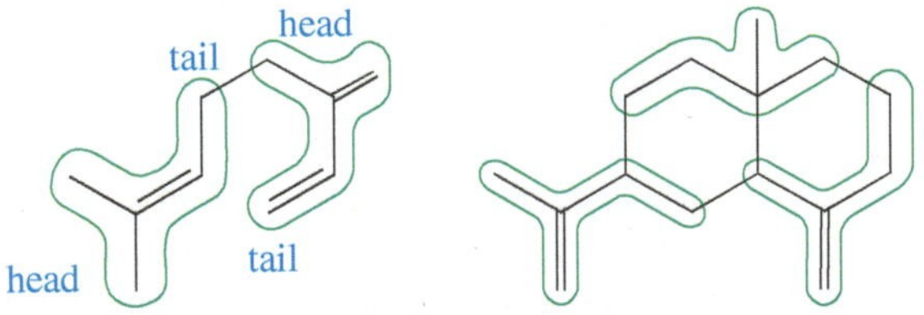

head tail head tail

myrcene β-selinene

β-Selinene has a more complicated structure, with two rings and a total of 15 carbon atoms. Nevertheless, β-selinene is composed of three isoprene units. Once again, these three units are bonded head to tail, although the additional bonds used to form the rings make the head-to-tail arrangement more difficult to see.

Many terpenes contain additional functional groups, especially carboxyl groups and hydroxy groups. A terpene aldehyde, a terpene alcohol, a terpene ketone, and a terpene acid are shown next.

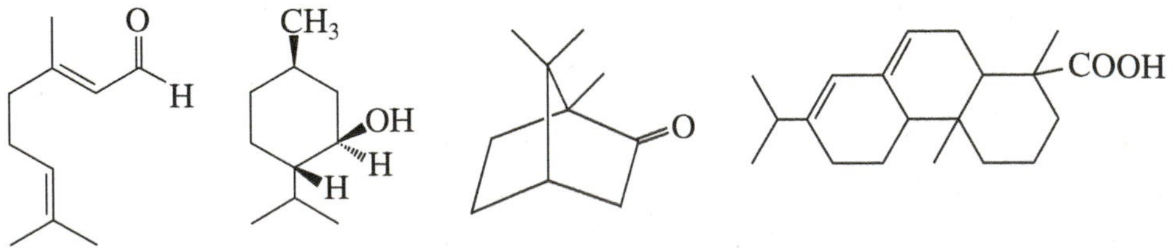

geranial menthol camphor abietic acid

PROBLEM 25-11

Circle the isoprene units in geranial, menthol, camphor, and abietic acid.

25-8B Classification of Terpenes

Class Name	Carbons
monoterpenes	10
sesquiterpenes	15
diterpenes	20
triterpenes	30
tetraterpenes	40

Terpenes are classified according to the number of carbon atoms they contain, in units of 10. A terpene with 10 carbon atoms (2 isoprene units) is called a **monoterpene,** one with 20 carbon atoms (4 isoprene units) is a **diterpene,** and so on. Terpenes with 15 carbon atoms (3 isoprene units) are called **sesquiterpenes,** meaning that they have $1\frac{1}{2}$ times 10 carbon atoms. Myrcene, geranial, menthol, and camphor are monoterpenes, β-selinene is a sesquiterpene, abietic acid is a diterpene, and squalene (Figure 25-13) is a **triterpene.**

Carotenes, with 40 carbon atoms, are **tetraterpenes.** Their extended system of conjugated double bonds moves the intense $\pi \rightarrow \pi^*$ ultraviolet absorption into the visible region, making them brightly colored. Carotenes are responsible for the pigmentation of carrots, tomatoes, and squash, and they help to give tree leaves their fiery colors in autumn. β-Carotene is the most common carotene isomer. It can be divided into two head-to-tail diterpenes, linked tail to tail.

squalene → → cyclized intermediate

→ lanosterol → → → cholesterol

FIGURE 25-13
Cholesterol is a triterpenoid that has lost three (blue) carbon atoms from the original six isoprene units of squalene. Another carbon atom (red) has migrated to form the axial methyl group between rings C and D.

β-carotene: λ_{max} = 454 nm, ε = 140,000

PROBLEM 25-12

Circle the eight isoprene units in β-carotene.

Carotenes are believed to be biological precursors of retinol, commonly known as vitamin A. If a molecule of β-carotene is split in half at the tail-to-tail linkage, each of the diterpene fragments can be converted to retinol.

carotene $\xrightarrow[\text{enzyme}]{2\ H_2O}$ 2 retinol (vitamin A)

PROBLEM 25-13

(a) Circle the isoprene units in the following terpenes.
(b) Classify each of these as a monoterpene, diterpene, etc.

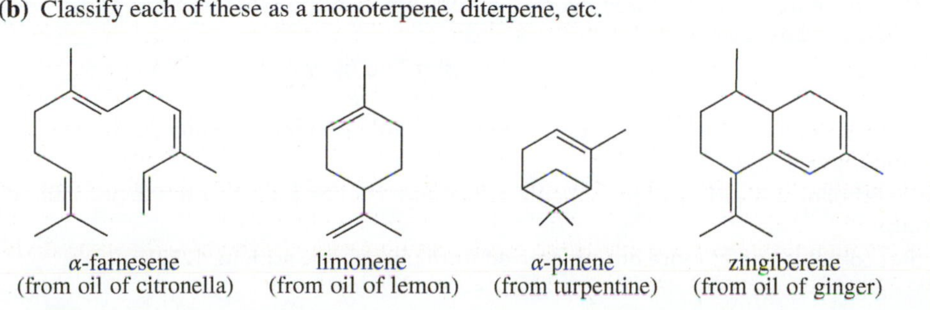

25-8C Terpenoids

Many natural products are derived from terpenes, even though they do not have carbon skeletons composed exclusively of C_5 isoprene units. These terpene-like compounds are called **terpenoids.** They may have been altered through rearrangements, loss of carbon atoms, or introduction of additional carbon atoms. Cholesterol is an example of a terpenoid that has lost some of the isoprenoid carbon atoms.

Figure 25-13 shows that cholesterol is a triterpenoid, formed from six isoprene units with loss of three carbon atoms. The six isoprene units are bonded head to tail, with the exception of one tail-to-tail linkage. The triterpene precursor of cholesterol is believed to be squalene. We can envision an acid-catalyzed cyclization of squalene to give an intermediate that is later converted to cholesterol with loss of three carbon atoms. Possible mechanisms are outlined in Figures 14-6 and 14-7 (page 699).

Essential Terms

biodiesel A mixture of methyl or ethyl fatty acid esters that is produced from fats and oils (triglycerides) by transesterification with methanol or ethanol. This mixture can be burned in most diesel engines without modifications. (p. 1270)

detergent **(synthetic detergent)** A synthesized compound that acts as an emulsifying agent. Some of the common classes of synthetic detergents are alkylbenzenesulfonate salts, alkyl sulfate salts, alkylammonium salts, and nonionic detergents containing several hydroxy groups or ether linkages. (p. 1272)

emulsify To promote formation of an emulsion. (p. 1272)

emulsion A mixture of two immiscible liquids, one dispersed throughout the other in small droplets. (p. 1272)

essential oils Fragrant oils *(essences)* that are concentrated from plant material, usually by steam distillation. (p. 1279)

fat A fatty acid triester of glycerol (a triglyceride) that is solid at room temperature. (p. 1266)

fatty acid A long-chain carboxylic acid. Most naturally occurring fatty acids contain even numbers of carbon atoms between 12 and 20. (p. 1267)

glyceride A fatty acid ester of glycerol. (p. 1266)

hard water Water that contains acids or ions (such as Ca^{2+}, Mg^{2+}, or Fe^{3+}) that react with soaps to form precipitates. (p. 1272)

hydrophilic Attracted to water; polar. (p. 1271)

hydrophobic Repelled by water; usually nonpolar and **lipophilic** (soluble in oils and in nonpolar solvents). (p. 1271)

isoprene The common name for 2-methylbuta-1,3-diene, the structural building block for terpenes. (p. 1279)

lipid bilayer An aggregation of phosphoglycerides with the hydrophilic heads forming the two surfaces of a planar structure and the hydrophobic tails protected within. A lipid bilayer forms part of the animal cell membrane. (p. 1274)

lipids Substances that can be extracted from cells and tissues by nonpolar organic solvents. (p. 1265)
- **complex lipids:** Lipids that are easily hydrolyzed to simpler constituents, usually by saponification of an ester.
- **simple lipids:** Lipids that are not easily hydrolyzed to simpler constituents.

micelle A cluster of molecules of a soap, phospholipid, or other emulsifying agent suspended in a solvent, usually water. The hydrophilic heads of the molecules are in contact with the solvent, and the hydrophobic tails are enclosed within the cluster. The micelle may or may not contain an oil droplet. (p. 1271)

oil A fatty acid triester of glycerol (a triglyceride) that is liquid at room temperature. (p. 1266)

phosphoglyceride An ester of glycerol in which the three hydroxy groups are esterified by two fatty acids and a phosphoric acid derivative. (p. 1273)
- **phosphatidic acids:** A variety of phosphoglycerides consisting of glycerol esterified by two fatty acids and one free phosphoric acid group.
- **cephalins:** **(phosphatidyl ethanolamines)** A variety of phosphoglycerides with ethanolamine esterified to the phosphoric acid group.
- **lecithins:** **(phosphatidyl cholines)** A variety of phosphoglycerides with choline esterified to the phosphoric acid group.

phospholipid Any lipid that contains one or more groups derived from phosphoric acid. (p. 1273)

polyunsaturated Containing multiple carbon–carbon double bonds. Usually applied to fish oils and vegetable oils that contain, on average, several double bonds per triglyceride molecule. (p. 1268)

prostaglandins A class of biochemical regulators consisting of a 20-carbon carboxylic acid containing a cyclopentane ring and various other functional groups. (p. 1278)

saponification Base-promoted hydrolysis of an ester. Originally used to describe the hydrolysis of fats to make soap. (p. 1270)

saturated fats Fatty acid triesters of glycerol containing few or no carbon–carbon double bonds (containing primarily saturated fatty acids). Butter, lard, and tallow contain large amounts of saturated fats. (p. 1268)

soap The sodium or potassium salts of fatty acids. (p. 1271)

steroid A compound whose structure is based on the tetracyclic androstane ring system. (p. 1275)

terpenes A diverse family of compounds with carbon skeletons composed of two or more 5-carbon isoprene units. **Monoterpenes** contain 10 carbon atoms, **sesquiterpenes** contain 15, **diterpenes** contain 20, **triterpenes** contain 30, and **tetraterpenes** contain 40 carbons. (p. 1279)

terpenoids A family of compounds including both terpenes and compounds of terpene origin whose carbon skeletons have been altered or rearranged. (p. 1282)

trans fats Fatty acid triesters of glycerol containing the unnatural trans isomers of fatty acids. Trans fats are often formed as by-products in the partial hydrogenation of vegetable oils to produce margarine and vegetable shortening. (p. 1269)

triglyceride **(triacylglycerol)** A fatty acid triester of glycerol. Triglycerides that are solid at room temperature are *fats*, and those that are liquid are *oils*. (p. 1266)

wax An ester of a long-chain fatty acid with a long-chain alcohol. (p. 1266)

Essential Problem-Solving Skills in Chapter 25

Each skill is followed by problem numbers exemplifying that particular skill.

1	Classify simple and complex lipids. Identify waxes, triglycerides, phospholipids, steroids, prostaglandins, and terpenes.	Problems 25-14, 17, 18, 22, and 23
2	Explain how unsaturations affect the properties of fats and oils. Compare the properties of saturated fats with those of polyunsaturated oils and partially hydrogenated vegetable oils.	Problems 25-20, 29, and 32
3	Identify the isoprene units in terpenes, and classify them according to the number of carbon atoms they contain.	Problems 25-27, 30, and 31
4	Predict the reactions of lipids under basic hydrolysis and with standard organic reagents. Show the reactions of the ester and olefinic groups of glycerides and the carboxyl groups of fatty acids.	Problems 25-15, 16, 18, 19, 20, 21, 23, and 24
5	Compare soaps and detergents, and explain how they emulsify nonpolar substances in water.	Problems 25-19 and 26

Study Problems

25-14 Draw the structure of an example of each of the following types of lipids:
(a) a saturated fat **(b)** a polyunsaturated oil **(c)** a wax
(d) a soap **(e)** a detergent **(f)** a phospholipid
(g) a prostaglandin **(h)** a steroid **(i)** a sesquiterpene

25-15 Predict the products obtained from the reaction of triolein with the following reagents.
(a) NaOH in water **(b)** H_2 and a nickel catalyst **(c)** Br_2 in CCl_4
(d) ozone, then dimethyl sulfide **(e)** warm $KMnO_4$ in water **(f)** $CH_2I_2/Zn(Cu)$

25-16 Show how you would convert oleic acid to the following fatty acid derivatives.
(a) octadecan-1-ol **(b)** stearic acid **(c)** octadecyl stearate
(d) nonanal **(e)** nonanedioic acid **(f)** 2,9,10-tribromostearic acid

25-17 Give the general classification of each compound.

(a) glyceryl tripalmitate

(b) $CH_3-(CH_2)_{10}-CH_2-O-SO_2-O^- Na^+$

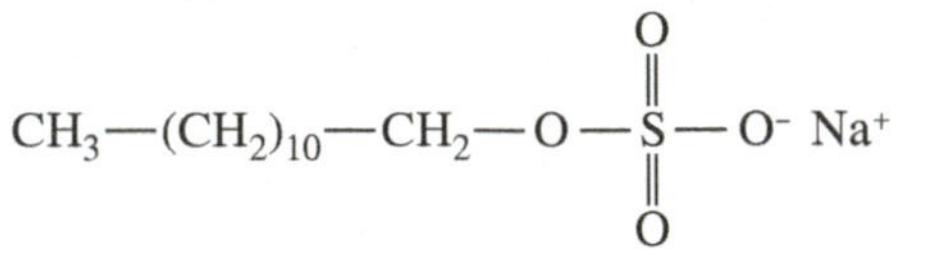

sodium lauryl sulfate (in shampoo)

(c) $CH_3-(CH_2)_{13}-O-C(=O)-(CH_2)_{16}-CH_3$

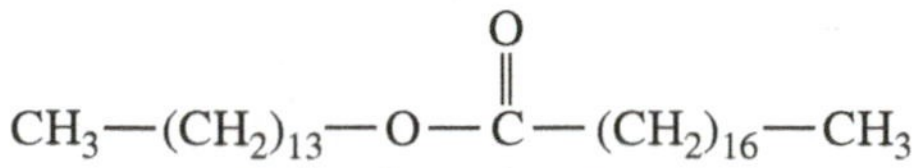

tetradecyl octadecanoate

(d) caryophyllene (from cloves)

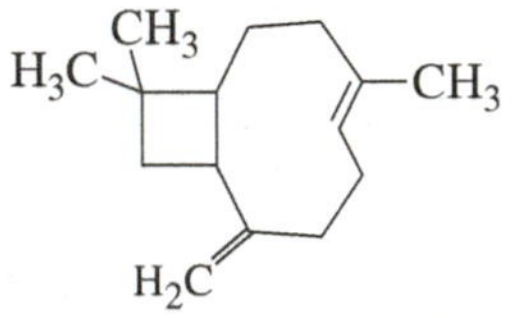

(e) PGA_2

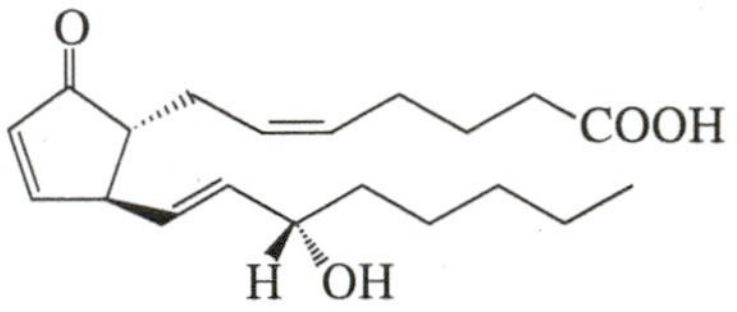

(f) norethindrone (a synthetic hormone)

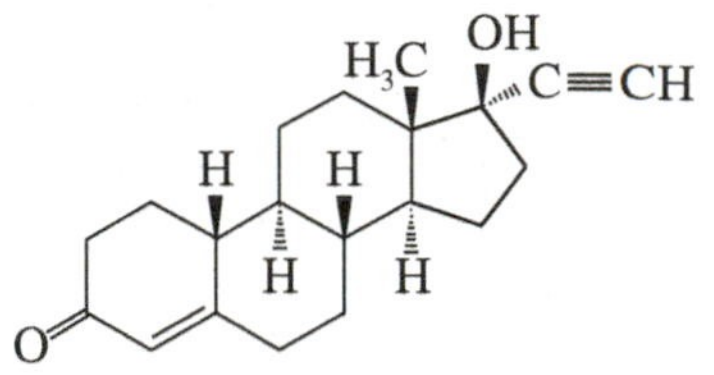

25-18 Phospholipids undergo saponification much like triglycerides. Draw the structure of a phospholipid meeting the following criteria. Then draw the products that would result from its saponification.

(a) a cephalin containing stearic acid and oleic acid
(b) a lecithin containing palmitic acid

25-19 Some of the earliest synthetic detergents were the sodium alkyl sulfates, $CH_3(CH_2)_nCH_2-OSO_3^- Na^+$. Show how you would make sodium octadecylsulfate using tristearin as your organic starting material.

25-20 Which of the following chemical reactions could be used to distinguish between a polyunsaturated vegetable oil and a petroleum oil containing a mixture of saturated and unsaturated hydrocarbons? Explain your reasoning.

(a) addition of bromine in CCl_4
(b) hydrogenation
(c) saponification
(d) ozonolysis

25-21 How would you use simple chemical tests to distinguish between the following pairs of compounds?

(a) sodium stearate and *p*-dodecylbenzenesulfonate
(b) beeswax and "paraffin wax"
(c) trimyristin and myristic acid
(d) trimyristin and triolein

25-22 A triglyceride can be optically active if it contains two or more different fatty acids.

(a) Draw the structure of an optically active triglyceride containing one equivalent of myristic acid and two equivalents of oleic acid.
(b) Draw the structure of an optically inactive triglyceride with the same fatty acid composition.

25-23 Draw the structure of an optically active triglyceride containing one equivalent of stearic acid and two equivalents of oleic acid. Draw the products expected when this triglyceride reacts with the following reagents. In each case, predict whether the products will be optically active.

(a) H_2 and a nickel catalyst
(b) Br_2 in CCl_4
(c) hot aqueous NaOH
(d) ozone followed by $(CH_3)_2S$

25-24 The structure of limonene appears in Problem 25-13. Predict the products formed when limonene reacts with the following reagents.

(a) excess HBr
(b) excess HBr, peroxides
(c) excess Br_2 in CCl_4
(d) ozone, followed by dimethyl sulfide
(e) warm, concentrated $KMnO_4$
(f) $BH_3 \cdot THF$, followed by basic H_2O_2

***25-25** *Olestra*® is a fat-based fat substitute that became available in snack foods such as potato chips in 1998. Previous fat substitutes were carbohydrate-based or protein-based mixtures that did not give as good a sensation in the mouth, and are not suitable for frying. With Olestra®, the glycerol molecule of a fat is replaced by sucrose (p. 1201). In Olestra®, the sucrose molecule has six, seven, or (most commonly) eight fatty acids esterified to its hydroxy groups. The fatty acids come from hydrolysis of vegetable oils such as soybean, corn, palm, coconut, and cottonseed oils. This unnaturally bulky, fat-like molecule does not pass through the intestinal walls, and digestive enzymes cannot get close to the sucrose center to bind it to their active sites. Olestra® passes through the digestive system unchanged, and it provides zero calories. Draw a typical Olestra® molecule, using any fatty acids that are commonly found in vegetable oils.

25-26 Cholic acid, a major constituent of bile, has the structure shown.

(a) Draw the structure of cholic acid, showing the rings in their chair conformations, and label each methyl group and hydroxy group as axial or equatorial. (Making a model may be helpful.)
(b) Cholic acid is secreted in bile as an amide linked to the amino group of glycine. This cholic acid–amino acid combination acts as an emulsifying agent to disperse lipids in the intestines for easier digestion. Draw the structure of the cholic acid–glycine combination, and explain why it is a good emulsifying agent.

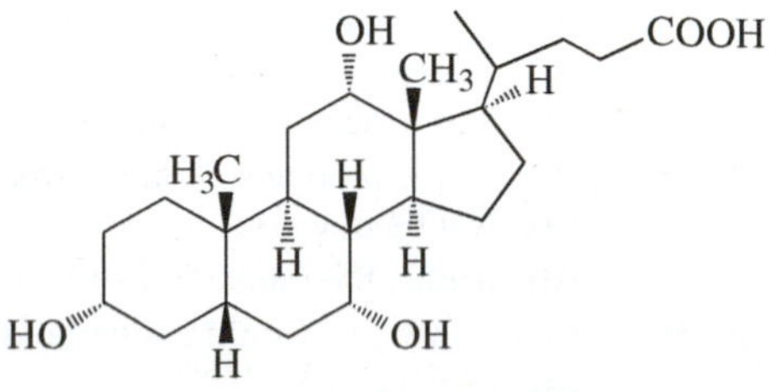

25-27 Carefully circle the isoprene units in the following terpenes, and label each compound as a monoterpene, sesquiterpene, or diterpene.

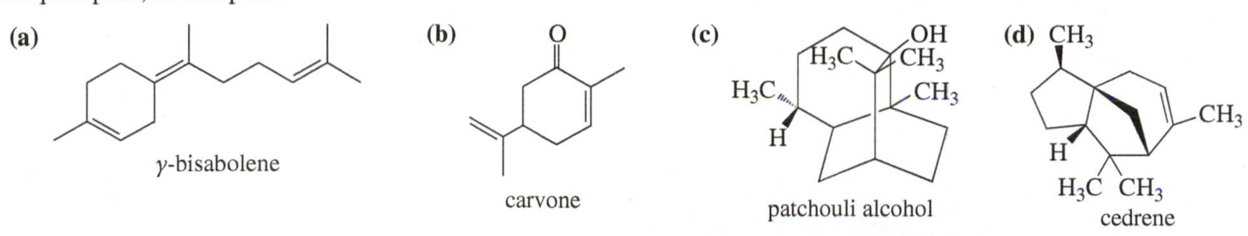

25-28 When an extract of parsley seed is saponified and acidified, one of the fatty acids isolated is *petroselenic acid*, formula $C_{18}H_{34}O_2$. Hydrogenation of petroselenic acid gives pure stearic acid. When petroselenic acid is treated with warm potassium permanganate followed by acidification, the only organic products are dodecanoic acid and adipic acid. The NMR spectrum shows absorptions of vinyl protons split by coupling constants of 7 Hz and 10 Hz. Propose a structure for petroselenic acid, and show how your structure is consistent with these observations.

25-29 The long-term health effects of eating partially hydrogenated vegetable oils concern some nutritionists because many unnatural fatty acids are produced. Consider the partial hydrogenation of linolenic acid by the addition of one or two equivalents of hydrogen. Show how this partial hydrogenation can produce at least three different fatty acids we have not seen before.

25-30 Two naturally occurring lactones are shown. For each compound, determine
(a) whether the compound is a terpene. If so, circle the isoprene units.
(b) whether the compound is aromatic, and explain your reasoning.
(c) Show the product resulting from saponification with aqueous NaOH.

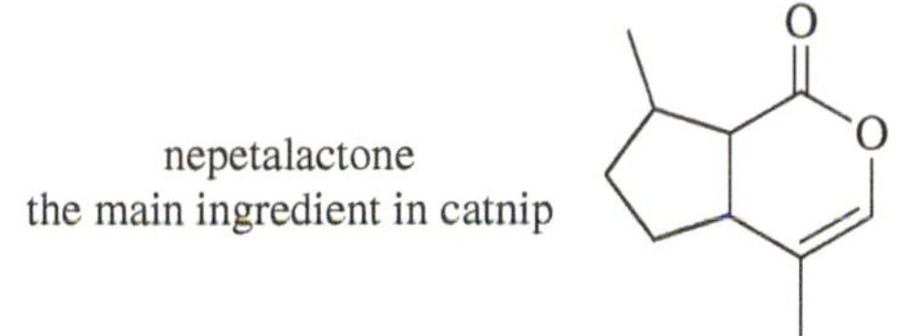

nepetalactone
the main ingredient in catnip

This compound, generated in the smoke from burning plants, promotes seed germination in plants that require fire to reproduce.

25-31 The following five compounds are found in Vicks Vapo-Rub®.
(a) Which are terpenes? Circle the isoprene units of the terpenes.
(b) Do you expect Vicks Vapo-Rub to be optically active? Explain.

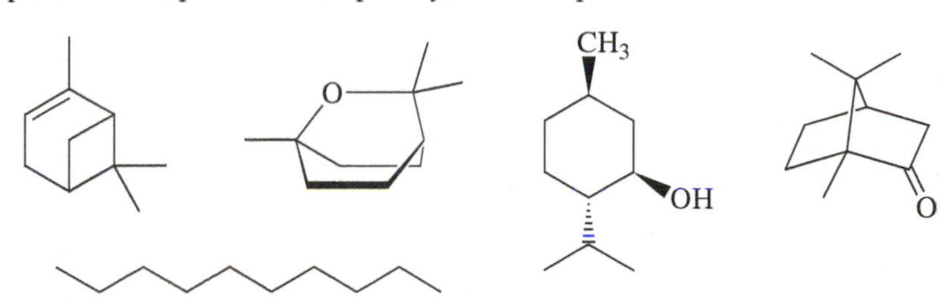

25-32 Oils containing highly unsaturated acids like linolenic acid undergo oxidation in air. This reaction, called oxidative rancidity, is accelerated by heat, explaining why saturated fats are preferred for deep fat frying.

COOH

$\cdot\ddot{O}-\ddot{O}\cdot$

O—OH

COOH

(a) Molecular oxygen is a diradical. What type of mechanism does a diradical suggest for this reaction?
(b) Why is the position shown (carbon-11) a likely site for attack?
(c) Propose a plausible mechanism for this reaction.
(d) BHA and BHT are antioxidants added to foods to interrupt the oxidation mechanism. Suggest how these molecules might work as antioxidants.

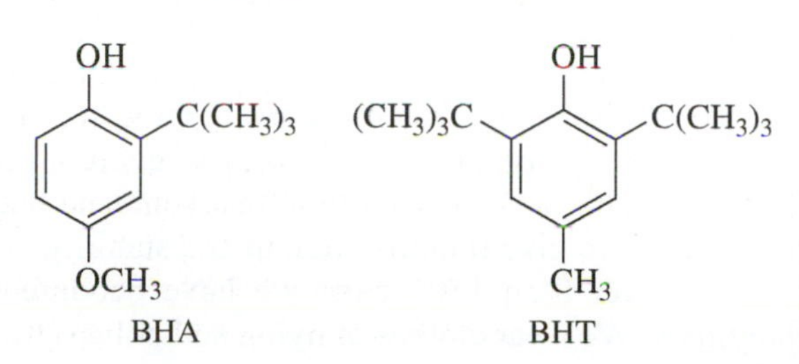

26 Synthetic Polymers

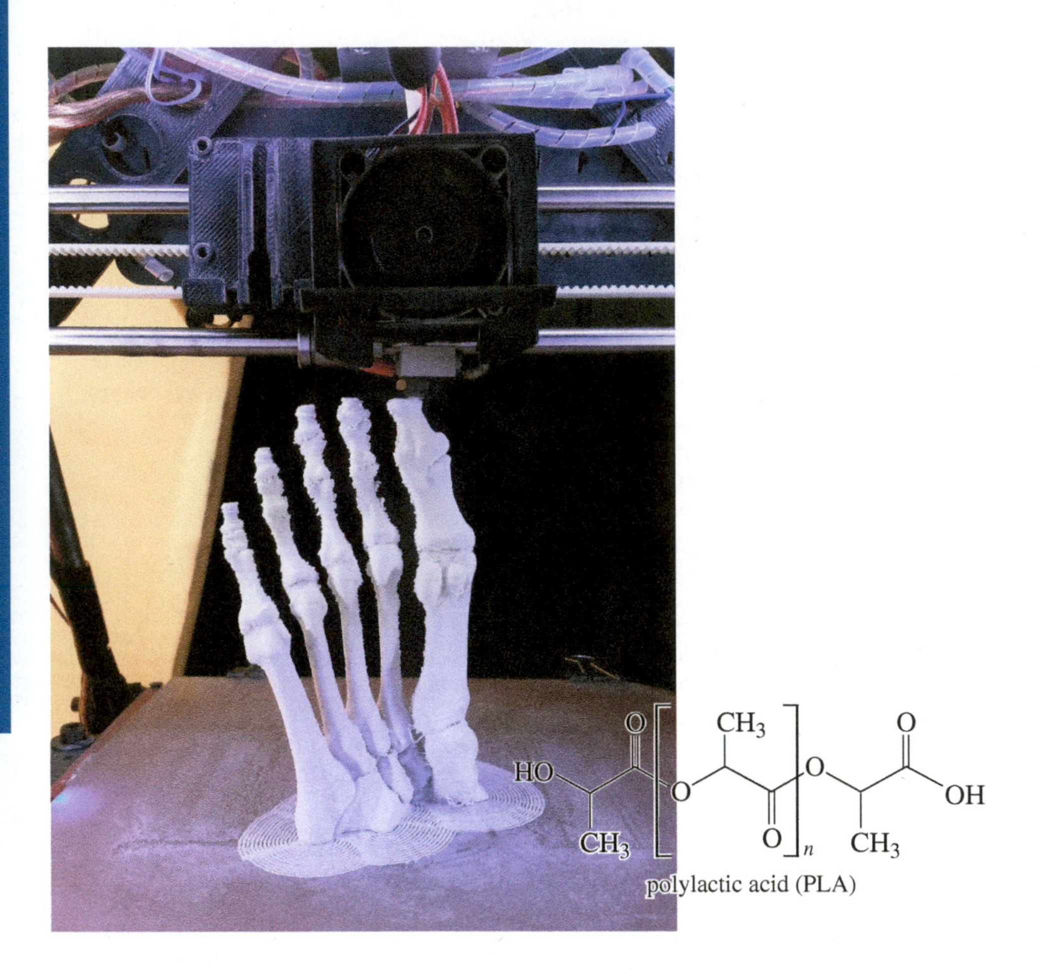

Goals for Chapter 26

1 Explain the differences between addition and condensation polymers. Show the differences between the chain-growth and step-growth mechanisms of polymerization.

2 Given a polymer, determine the structure(s) of the monomer(s). Given the monomer(s), predict the structure of the polymer.

3 Propose mechanisms for the formation of addition and condensation polymers, and use those mechanisms to explain why the polymers have the observed molecular structures.

4 Predict the general characteristics of a polymer based on its structure, cross-linking, and presence of plasticizers. Explain how its characteristics change as the polymer is heated.

◀ 3D printing is an additive process that lays down successive layers of material under computer control. This photo shows a hot printer head sculpting the bones of a human foot by extruding a molten filament of a thermoplastic polymer that hardens in the desired shape. Polylactic acid (PLA), made from plant materials such as cornstarch, is a common filament used in 3D printers. This polyester has a convenient melting point (150–160 °C) and readily forms fully fused products.

26-1 Introduction

People have always used polymers. Prehistoric tools and shelters made from wood and straw derive their strength and resilience from cellulose, a biopolymer of glucose. Clothing made from the hides and hair of animals is made strong and supple by proteins, which are biopolymers of amino acids. After people learned to use fire, they made ceramic pottery and glass, using naturally occurring inorganic polymers.

A **polymer** is a large molecule composed of many smaller repeating units (the **monomers**) bonded together. Today when we speak of polymers, we generally mean *synthetic organic polymers* rather than natural organic biopolymers such as DNA, cellulose, and protein, or inorganic polymers such as glass and concrete. The first synthetic organic polymer was made in 1838, when vinyl chloride was accidentally polymerized. Polystyrene was discovered in 1839, shortly after styrene was synthesized and purified. The discovery of polystyrene was inevitable, since styrene polymerizes spontaneously unless a stabilizer is added.

Also in 1839, Charles Goodyear (of tire and blimp fame) discovered how to convert the gummy polymeric sap of the rubber tree to a strong, stretchy material by heating it with sulfur. *Vulcanized rubber* quickly revolutionized the making of boots, tires, and rainwear. This was the first time that someone had artificially cross-linked a natural biopolymer to give it more strength and stability.

In fewer than 150 years, we have become literally surrounded by synthetic polymers. We wear clothes of nylon and polyester, we walk on polypropylene carpets, we drive cars with ABS plastic fenders and synthetic rubber tires, and we use artificial hearts and other organs made of silicone polymers. Our pens and computers, our toys and our televisions are made largely of plastics.

Articles that are not made from synthetic polymers are often held together or coated with polymers. A bookcase may be made from wood, but the wood is bonded by a phenol-formaldehyde polymer and painted with a latex polymer. Each year, about *400 billion* pounds of synthetic organic polymers are produced worldwide, mostly for use in consumer products. Large numbers of organic chemists are employed to develop and produce these polymers.

In this chapter, we discuss some of the fundamental principles of polymer chemistry. We begin with a survey of the different kinds of polymers and then consider the reactions used to induce **polymerization.** Finally, we discuss some of the structural characteristics that determine the physical properties of a polymer.

Classes of Synthetic Polymers The two major classes of synthetic polymers are chain-growth polymers and step-growth polymers. In a **chain-growth polymerization,** elongation of the chain takes place only at the end of the chain. This occurs by the rapid addition of one monomer at a time to a reactive intermediate (cation, radical, or anion) at the growing end of the chain. Chain-growth polymers are usually **addition polymers,** which result from monomers adding together without the loss of any molecules. Monomers for chain-growth polymerization are commonly alkenes, and polymerization involves successive additions across the double bonds. Poly(vinyl chloride), widely used as a synthetic leather, is a chain-growth addition polymer that is often made by free-radical polymerization.

$$\text{growing chain} \;-CH_2-\dot{C}HCl \;+\; \underset{\text{vinyl chloride}}{CH_2=CHCl} \longrightarrow \underset{\text{elongated chain}}{-CH_2-CHCl-CH_2-\dot{C}HCl} \longrightarrow \longrightarrow \underset{\text{poly(vinyl chloride)}}{-CH_2-CHCl-[CH_2-CHCl]_n-CH_2-CHCl-}$$

repeating unit

In a **step-growth polymerization,** any two monomers having the correct functionality can react with each other, or two polymer chains can combine. Most step-growth polymers are **condensation polymers,** bonded by some kind of condensation (bond formation with loss of a small molecule) between the monomers or the polymer segments. The most common condensations involve the formation of amides and esters. Dacron polyester is an example of a step-growth condensation polymer.

$$\underset{\text{dimethyl terephthalate}}{CH_3-O-C(=O)-C_6H_4-C(=O)-O-CH_3} \;+\; \underset{\text{ethylene glycol}}{HO-CH_2-CH_2-OH}$$

$$\longrightarrow \underset{\text{Dacron}^{®}\text{ polyester}}{CH_3-O-C(=O)-C_6H_4-C(=O)-[O-CH_2-CH_2-O-C(=O)-C_6H_4-C(=O)]_n-O-CH_2-CH_2-OH} \;+\; 2n\, CH_3OH$$

repeating unit

26-2 Chain-Growth Polymers

Many alkenes undergo chain-growth polymerization when treated with small amounts of suitable initiators. The products are addition polymers, resulting from repeated additions across the double bonds of the monomers. Table 26-1 shows some of the most common addition polymers, all made from substituted alkenes. The chain-growth

TABLE 26-1
Some of the Most Important Addition Polymers

Polymer	Polymer Uses	Monomer	Polymer Repeating Unit
polyethylene	bottles, bags, films	$H_2C{=}CH_2$	$-[CH_2-CH_2]_n-$
polypropylene	plastics, olefin fibers	$H_2C{=}CH(CH_3)$	$-[CH_2-CH(CH_3)]_n-$
polystyrene	plastics, foam insulation	$H_2C{=}CH(C_6H_5)$	$-[CH_2-CH(C_6H_5)]_n-$
poly(isobutylene)	specialized rubbers	$H_2C{=}C(CH_3)_2$	$-[CH_2-C(CH_3)_2]_n-$
poly(vinyl chloride)	vinyl plastics, films, water pipes	$H_2C{=}CHCl$	$-[CH_2-CH(Cl)]_n-$
poly(acrylonitrile)	Orlon®, Acrilan® fibers	$H_2C{=}CH(C{\equiv}N)$	$-[CH_2-CH(CN)]_n-$
poly(methyl α-methacrylate)	acrylic fibers, Plexiglas®, Lucite® paints	$H_2C{=}C(CH_3)-C(=O)-OCH_3$	$-[CH_2-C(CH_3)(COOCH_3)]_n-$
poly(methyl α-cyanoacrylate)	"super" glues	$H_2C{=}C(C{\equiv}N)-C(=O)-OCH_3$	$-[CH_2-C(CN)(COOCH_3)]_n-$
poly(tetrafluoroethylene)	Teflon® coatings, PTFE plastics	$F_2C{=}CF_2$	$-[CF_2-CF_2]_n-$

mechanism involves addition of the reactive end of the growing chain across the double bond of the monomer. Depending on the monomer and the initiator used, the reactive intermediates may be free radicals, carbocations, or carbanions. Although these three types of chain-growth polymerizations are similar, we consider them individually.

26-2A Free-Radical Polymerization

Free-radical polymerization results when a suitable alkene is heated with a radical initiator. For example, styrene polymerizes to polystyrene when it is heated to 100 °C in the presence of benzoyl peroxide. This chain-growth polymerization is a free-radical chain reaction. Benzoyl peroxide cleaves when heated to give two carboxyl radicals, which quickly decarboxylate to give phenyl radicals.

benzoyl peroxide $\xrightarrow{\text{heat}}$ 2 carboxyl radicals $\longrightarrow$ 2 phenyl radicals + 2 CO_2

A phenyl radical adds to styrene to give a resonance-stabilized benzylic radical. This reaction starts the growth of the polymer chain. Each propagation step adds another molecule of styrene to the growing chain. This addition takes place with the orientation that gives another resonance-stabilized benzylic radical.

Chain growth may continue with the addition of several hundred or several thousand styrene units. The length of a polymer chain depends on the number of additions of monomers that occur before a termination step stops the process. Strong polymers with high molecular weights result from conditions that favor fast chain growth and minimize termination steps. Eventually the chain reaction stops, either by the coupling of two chains or by reaction with an impurity (such as oxygen) or simply by running out of monomer.

MECHANISM 26-1 Free-Radical Polymerization

Initiation step: The initiator forms a radical that reacts with the monomer to start the chain.

benzoyl peroxide $\xrightarrow[(-CO_2)]{\text{heat}}$ phenyl radicals + styrene $\xrightarrow{\text{heat}}$ benzylic radical

Propagation step: Another molecule of monomer adds to the chain.

growing chain + styrene $\longrightarrow$ elongated chain $\xrightarrow{\text{many more styrene molecules}}$ polystyrene

n = about 100 to 10,000

PROBLEM 26-1

Show the intermediate that would result if the growing chain added to the other end of the styrene double bond. Explain why the final polymer has phenyl groups substituted on every other carbon atom rather than randomly distributed.

Application: Bacterial Polymers

In the presence of limited nutrients, bacteria can be induced to make polyhydroxybutyrates and valerates, which are processed into a copolymer known as Biopol™. Biopol has properties similar to polypropylene, but it is biodegradable and obtained from nonpetroleum sources.

Ethylene and propylene also polymerize by free-radical, chain-growth polymerization. With ethylene, the free-radical intermediates are less stable, so stronger reaction conditions are required. Ethylene is commonly polymerized by free-radical initiators at pressures around 3000 atm and temperatures of about 200 °C. The product, called *low-density polyethylene*, is the material commonly used in stretchy polyethylene grocery bags.

PROBLEM 26-2

Propose a mechanism for reaction of the first three propylene units in the polymerization of propylene in the presence of benzoyl peroxide.

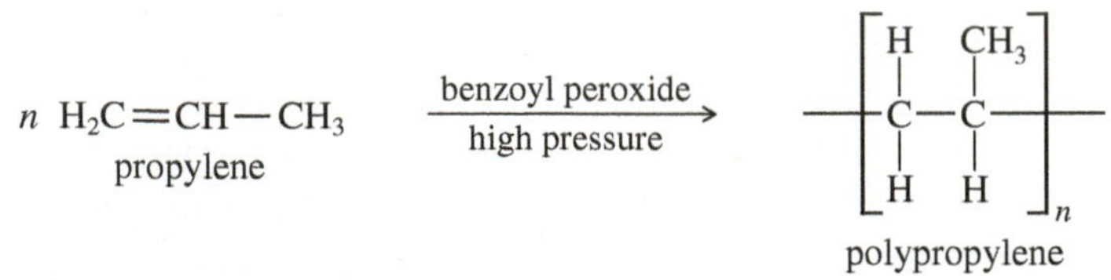

Chain Branching by Hydrogen Abstraction Low-density polyethylene is soft and flimsy because it has a highly branched, amorphous structure. (High-density polyethylene, discussed in Section 26-4, is much stronger because of the orderly structure of unbranched linear polymer chains.) Chain branching in low-density polyethylene results from abstraction of a hydrogen atom in the middle of a chain by the free radical at the end of a chain. A new chain grows from the point of the free radical in the middle of the chain. Figure 26-1 shows abstraction of a hydrogen from a polyethylene chain and the first step in the growth of a branch chain at that point.

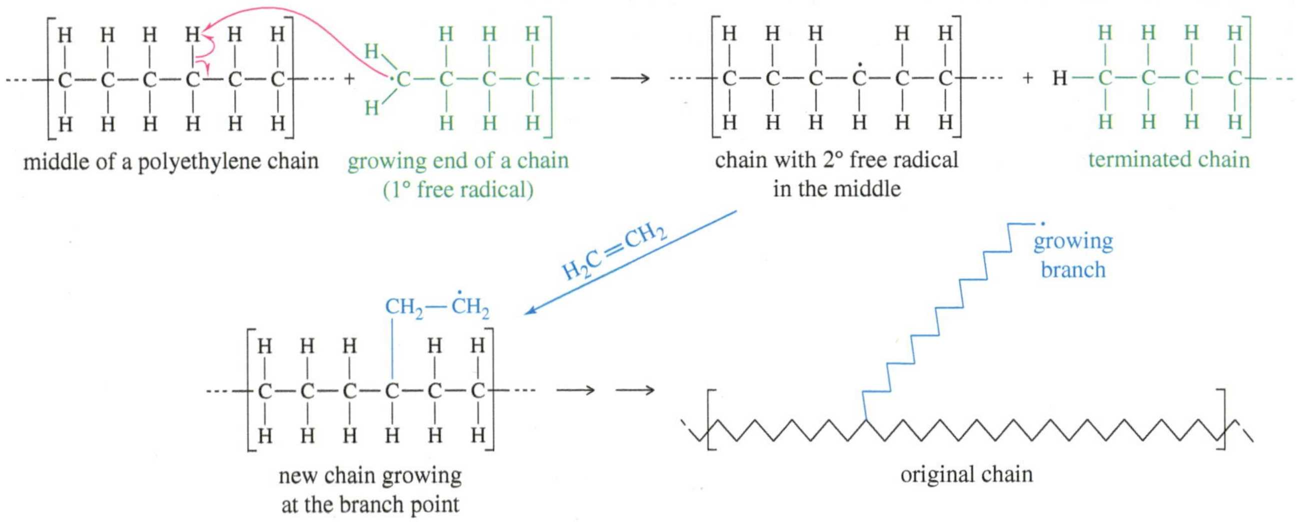

FIGURE 26-1
Chain branching in radical polymerization. Chain branching occurs when the growing end of a chain abstracts a hydrogen atom from the middle of a chain. A new branch grows off the chain at that point.

PROBLEM 26-3

Give a mechanism, using Figure 26-1 as a guide, showing chain branching during the free-radical polymerization of styrene. There are two types of aliphatic hydrogens in the polystyrene chain. Which type is more likely to be abstracted?

26-2B Cationic Polymerization

Cationic polymerization occurs by a mechanism similar to the free-radical process, except that it involves carbocation intermediates. Strongly acidic catalysts are used to initiate cationic polymerization. BF_3 is a particularly effective catalyst, requiring a trace of water or methanol as a co-catalyst. Even when the reagents are carefully dried, there is enough water present for the first initiation step of the mechanism shown in Mechanism 26-2.

MECHANISM 26-2 Cationic Polymerization

Initiation steps: The acidic catalyst protonates the monomer, starting the chain.

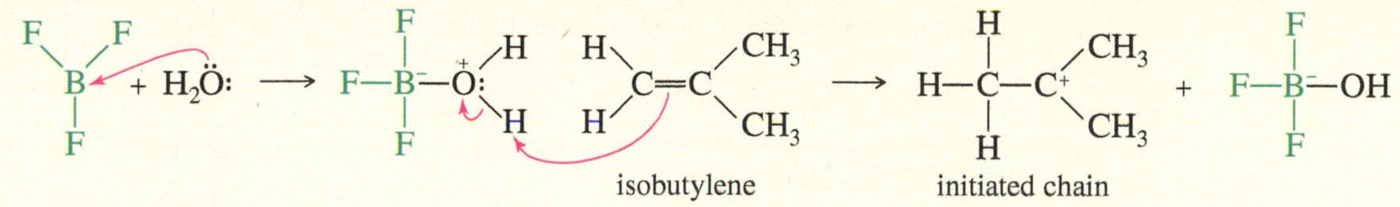

$$BF_3 + H_2\ddot{O}: \longrightarrow F_3\overset{-}{B}-\overset{+}{O}H_2 + (CH_3)_2C{=}CH_2 \longrightarrow H_3C-\overset{+}{C}(CH_3)_2 + F_3\overset{-}{B}-OH$$

isobutylene; initiated chain

Propagation step: Another molecule of monomer adds to the cationic end of the chain.

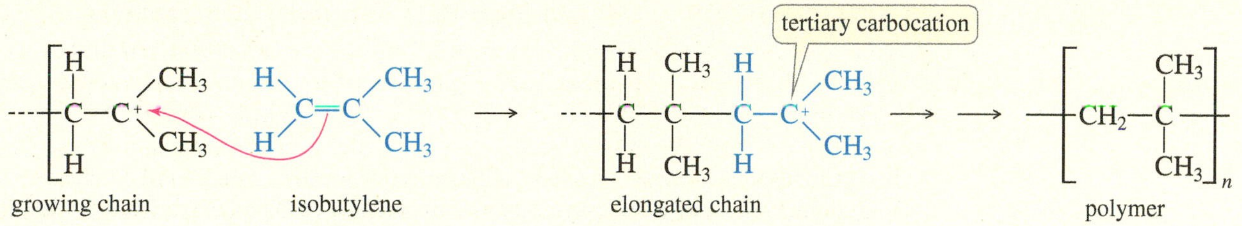

$$\cdots CH_2-\overset{+}{C}(CH_3)_2 + CH_2{=}C(CH_3)_2 \longrightarrow \cdots CH_2-C(CH_3)_2-CH_2-\overset{+}{C}(CH_3)_2 \longrightarrow \longrightarrow \left[CH_2-C(CH_3)_2\right]_n$$

growing chain; isobutylene; elongated chain (tertiary carbocation); polymer

A major difference between cationic and free-radical polymerization is that the cationic process needs a monomer that forms a relatively stable carbocation when it reacts with the cationic end of the growing chain. Some monomers form more stable intermediates than others. For example, styrene and isobutylene undergo cationic polymerization easily, while ethylene and acrylonitrile do not polymerize well under these conditions. Figure 26-2 compares the intermediates involved in these cationic polymerizations.

Good monomers for cationic polymerization

$$R^+ + CH_2{=}CH(C_6H_5) \longrightarrow R-CH_2-\overset{+}{C}H(C_6H_5) \quad \text{(good)}$$

growing chain; styrene; benzylic carbocation

$$R^+ + CH_2{=}C(CH_3)_2 \longrightarrow R-CH_2-\overset{+}{C}(CH_3)_2 \quad \text{(good)}$$

growing chain; isobutylene; tertiary carbocation

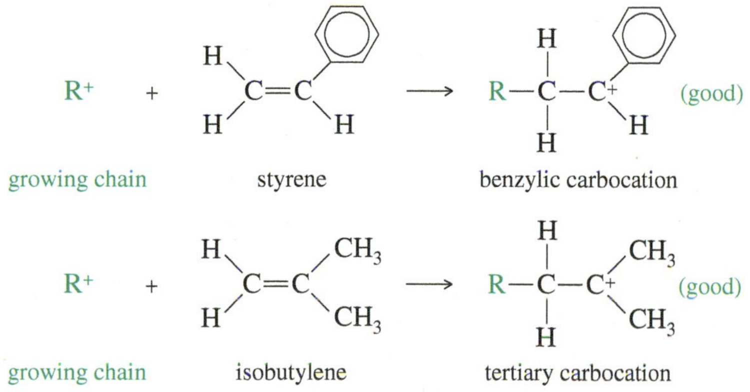

Poor monomers for cationic polymerization

$$R^+ + CH_2{=}CH_2 \longrightarrow R-CH_2-\overset{+}{C}H_2 \quad \text{(poor)}$$

growing chain; ethylene; primary carbocation

$$R^+ + CH_2{=}CH(CN) \longrightarrow R-CH_2-\overset{+}{C}H(CN) \quad \text{(poor)}$$

growing chain; acrylonitrile; destabilized carbocation

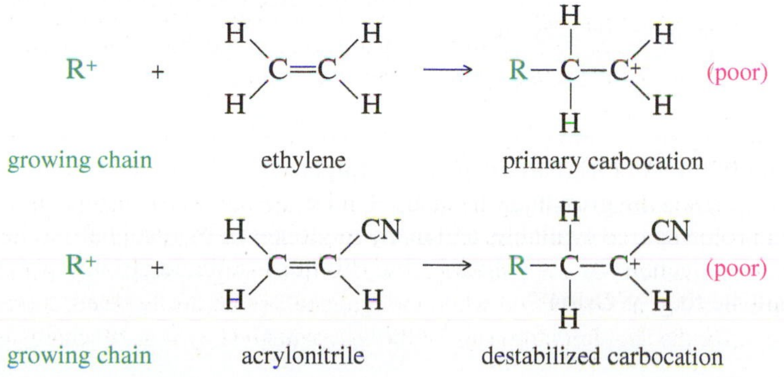

FIGURE 26-2
Cationic polymerization requires relatively stable carbocation intermediates.

PROBLEM 26-4

The mechanism given for cationic polymerization of isobutylene (Mechanism 26-2) shows that all the monomer molecules add with the same orientation, giving a polymer with methyl groups on alternate carbon atoms of the chain. Explain why no isobutylene molecules add with the opposite orientation.

PROBLEM 26-5

Suggest which of the following monomers might polymerize well on treatment with BF_3.
(a) vinyl chloride **(b)** vinyl acetate **(c)** methyl α-cyanoacrylate

PROBLEM 26-6

Chain branching occurs in cationic polymerization much as it does in free-radical polymerization. Propose a mechanism to show how branching occurs in the cationic polymerization of styrene. Suggest why isobutylene might be a better monomer for cationic polymerization than styrene.

26-2C Anionic Polymerization

Anionic polymerization occurs through carbanion intermediates. Effective anionic polymerization requires a monomer that gives a stabilized carbanion when it reacts with the anionic end of the growing chain. A good monomer for anionic polymerization should contain at least one strong electron-withdrawing group such as a carbonyl group, a cyano group, or a nitro group. The following reaction shows the chain-lengthening step in the polymerization of methyl acrylate. Notice that the chain-growth step of an anionic polymerization is simply a conjugate addition to a Michael acceptor (Section 22-18).

Chain-growth step in anionic polymerization

growing chain + methyl acrylate → stabilized anion → → polymer $\left[-CH_2-CH(COOCH_3)- \right]_n$

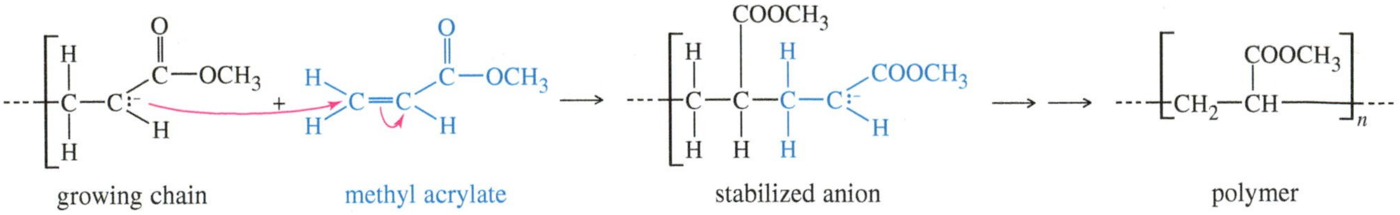

PROBLEM 26-7

Draw the important resonance forms of the stabilized anion formed in the anionic polymerization of methyl acrylate.

Anionic polymerization is usually initiated by a strong carbanion-like reagent such as an organolithium or Grignard reagent. Conjugate addition of the initiator to a monomer molecule starts the growth of the chain. Under the polymerization conditions, there is no good proton source available, and many monomer units react before the carbanion is protonated. Mechanism 26-3 shows a butyllithium-initiated anionic polymerization of acrylonitrile to give Orlon®.

MECHANISM 26-3 Anionic Polymerization

Initiation step: The initiator adds to the monomer to form an anion.

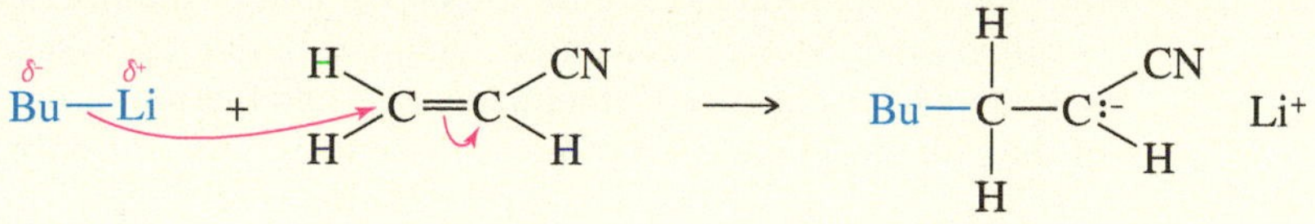

$$\overset{\delta-}{\text{Bu}}-\overset{\delta+}{\text{Li}} + H_2C{=}CH(CN) \longrightarrow \text{Bu}-CH_2-\ddot{C}^{-}H(CN) \quad Li^+$$

butyllithium　　acrylonitrile　　stabilized anion

Propagation step: Another molecule of monomer adds to the chain.

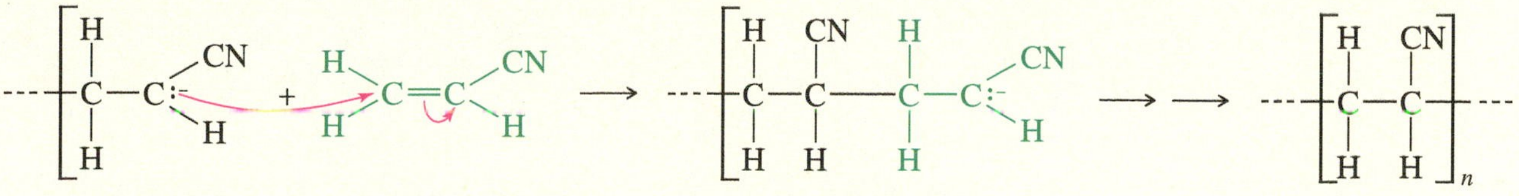

$$\cdots[CH_2]-\ddot{C}^{-}H(CN) + H_2C{=}CH(CN) \longrightarrow \cdots[CH_2-CH(CN)]-CH_2-\ddot{C}^{-}H(CN) \longrightarrow \longrightarrow \cdots[CH_2-CH(CN)]_n\cdots$$

growing chain　　acrylonitrile　　elongated chain　　polymer

PROBLEM 26-8

Methyl α-cyanoacrylate (Super Glue) is easily polymerized, even by weak bases. Draw a mechanism for its base-catalyzed polymerization, and explain why this polymerization goes so quickly and easily.

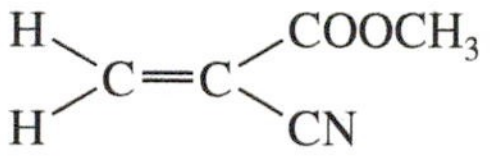

$$H_2C{=}C(COOCH_3)(CN)$$

methyl α-cyanoacrylate

PROBLEM 26-9

Chain branching is not as common with anionic polymerization as it is with free-radical polymerization and cationic polymerization.

(a) Propose a mechanism for chain branching in the polymerization of acrylonitrile.

(b) Compare the relative stabilities of the intermediates in this mechanism with those you drew for chain branching in the cationic polymerization of styrene (Problem 26-6). Explain why chain branching is less common in this anionic polymerization.

26-3 Stereochemistry of Polymers

Chain-growth polymerization of alkenes usually gives a head-to-tail bonding arrangement, with any substituent(s) appearing on alternate carbons of the polymer chain. This bonding arrangement is shown here for a generic polyalkene. Although the polymer backbone is joined by single bonds (and can undergo conformational changes), it is shown in the most stable all-anti conformation.

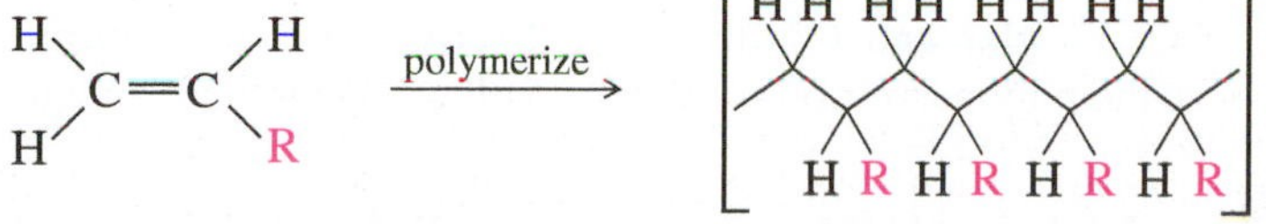

$$H_2C{=}CHR \xrightarrow{\text{polymerize}} [\,CH_2-CHR-CH_2-CHR-CH_2-CHR-CH_2-CHR\,]$$

The stereochemistry of the side groups (R) in the polymer has a major effect on the polymer's properties. The polymer has many chiral centers, raising the possibility of millions of stereoisomers. Polymers are grouped into three classes, according to their predominant stereochemistry. If the side groups are generally on the same side of the polymer backbone, the polymer is called **isotactic** (Greek, *iso*, meaning "same," and *tactic*, meaning "order"). If the side groups generally alternate from one side to the other, the polymer is called **syndiotactic** (Greek, meaning "alternating order"). If the side groups occur randomly on either side of the polymer backbone, the polymer is called **atactic** (Greek, meaning "no order"). In most cases, isotactic and syndiotactic polymers have enhanced strength, clarity, and thermal properties over the atactic form of the polymer. Figure 26-3 shows these three types of polymers.

An isotactic polymer (side groups on the same side of the backbone)

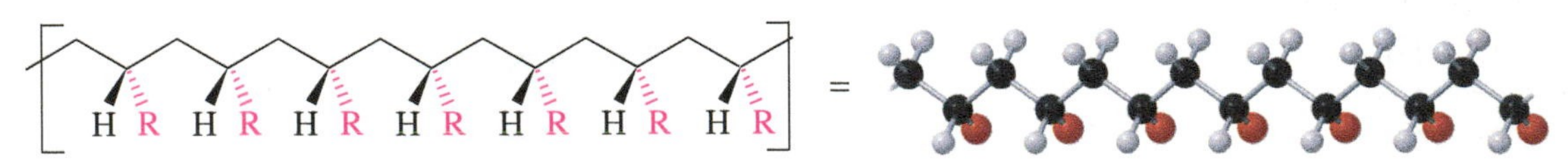

A syndiotactic polymer (side groups on alternating sides of the backbone)

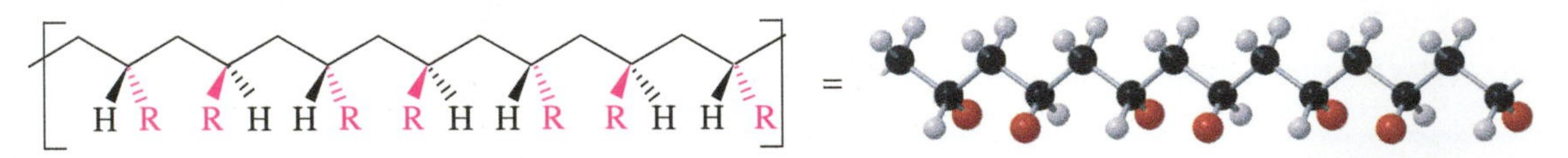

An atactic polymer (side groups on random sides of the backbone)

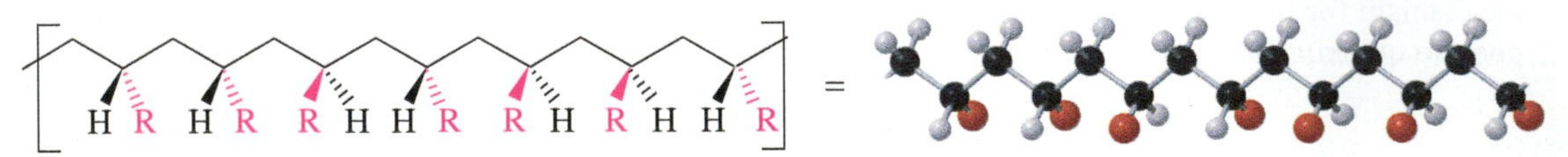

FIGURE 26-3
Three stereochemical types of addition polymers.

PROBLEM 26-10

Draw the structures of isotactic poly(acrylonitrile) and syndiotactic polystyrene.

26-4 Stereochemical Control of Polymerization: Ziegler–Natta Catalysts

For any particular polymer, the three stereochemical forms have distinct properties. In most cases, the stereoregular isotactic and syndiotactic polymers are stronger and stiffer because of their greater crystallinity (a regular packing arrangement). The conditions used for polymerization often control the stereochemistry of the polymer. Anionic polymerizations are the most stereoselective; they usually give isotactic or syndiotactic polymers, depending on the nature of the side group. Cationic polymerizations are often stereoselective, depending on the catalysts and conditions used. Free-radical polymerization is nearly random, resulting in branched, atactic polymers.

In 1953, Karl Ziegler and Giulio Natta discovered that aluminum–titanium initiators catalyze the polymerization of alkenes, with two major advantages over other catalysts:

1. The polymerization is highly stereoselective. Either the isotactic form or the syndiotactic form may be made by selecting the proper Ziegler–Natta catalyst.
2. Because the intermediates are stabilized by the catalyst, very little hydrogen abstraction occurs. The resulting polymers are linear with almost no branching.

A **Ziegler–Natta catalyst** is an organometallic complex, often containing titanium and aluminum. A typical catalyst is formed by adding a solution of $TiCl_4$ (titanium tetrachloride) to a solution of $(CH_3CH_2)_3Al$ (triethyl aluminum). This mixture is then "aged" by heating it for about an hour. The precise structure of the active catalyst is not known, but the titanium atom appears to form a complex with both the growing polymer chain and a molecule of monomer. The monomer attaches to the end of the chain (which remains complexed to the catalyst), leaving the titanium atom with a free site for complexation to the next molecule of monomer.

With a Ziegler–Natta catalyst, a *high-density polyethylene* (or *linear* polyethylene) can be produced with almost no chain branching and with much greater strength than common low-density polyethylene. Many other polymers are produced with improved properties using Ziegler–Natta catalysts. In 1963, Ziegler and Natta received the Nobel Prize for their work, which had revolutionized the polymer industry in only ten years.

Application: Plumbing Pipe

The residential plumbing industry is moving away from copper (expensive and rigid) to PEX pipe, a cross-linked high-density polyethylene. PEX is durable, flexible, and easy to connect, and it does not accumulate hard water deposits. It also avoids most of the cracking and leaking that is common with older PVC plumbing.

26-5 Natural and Synthetic Rubbers

Natural rubber is isolated from a white fluid, called **latex,** that exudes from cuts in the bark of *Hevea brasiliensis*, the South American rubber tree. Many other plants secrete this polymer as well. The name **rubber** was first used by Joseph Priestly, who used the crude material to "rub out" errors in his pencil writing. Natural rubber is soft and sticky. An enterprising Scotsman named Charles Macintosh found that rubber makes a good waterproof coating for raincoats. Natural rubber is not strong or elastic, however, so its uses were limited to waterproofing cloth and other strong materials.

Structure of Natural Rubber Like many other plant products, natural rubber is a terpene composed of isoprene units (Section 25-8). If we imagine lining up many molecules of isoprene in the *s*-cis conformation and moving pairs of electrons as shown in the following figure, we would produce a structure similar to natural rubber. This polymer results from 1,4-addition to each isoprene molecule, with all the double bonds in the cis configuration. Another name for natural rubber is *cis*-1,4-polyisoprene.

Imaginary polymerization of isoprene units

Natural rubber

White latex drips out of cuts in the bark of a rubber tree in a Malaysian rubber plantation.

The cis double bonds in natural rubber force it to assume a kinked conformation that may be stretched and still return to its shorter, kinked structure when released. Unfortunately, when we pull on a mass of natural rubber, the chains slide by each other and the material pulls apart. This is why natural rubber is not suitable for uses requiring strength or durability.

Vulcanization: Cross-Linking of Rubber In 1839, Charles Goodyear accidentally dropped a mixture of natural rubber and sulfur onto a hot stove. He was surprised to find that the rubber had become strong and elastic. This discovery led to the process that Goodyear called **vulcanization,** after the Roman god of fire and the volcano. Vulcanized rubber has much greater toughness and elasticity than natural rubber. It withstands relatively high temperatures without softening, and it remains elastic and flexible when cold.

Vulcanization also allows the casting of complicated shapes such as rubber tires. Natural rubber is putty-like, and it is easily mixed with sulfur, formed around the tire cord, and placed into a mold. The mold is closed and heated, and the gooey mass of string and rubber is vulcanized into a strong, elastic tire carcass.

On a molecular level, vulcanization causes cross-linking of the *cis*-1,4-polyisoprene chains through disulfide (—S—S—) bonds, similar to the cystine bridges that link peptides (Section 24-8C). In vulcanized rubber, the polymer chains are linked together so they can no longer slip past each other. When the material is stressed, the chains stretch, but cross-linking prevents tearing. When the stress is released, the chains return to their shortened, kinked conformations as the rubber snaps back. Figure 26-4 shows the structure of rubber before and after vulcanization.

Rubber can be prepared with a wide range of physical properties by controlling the amount of sulfur used in vulcanization. Low-sulfur rubber, made with about 1% to 2% sulfur, is soft and stretchy. It is good for rubber bands and inner tubes. Medium-sulfur rubber (about 2% to 5% sulfur) is somewhat harder, but still flexible, making good tires. High-sulfur rubber (10% to 30% sulfur) is called *hard rubber* and was once used as a hard synthetic plastic. Using more sulfur in the mixture increases the number of disulfide cross-links, as well as the frequency of bridges containing three or more sulfur atoms.

Wallace Carothers, the inventor of nylon, stretches a piece of synthetic rubber in his laboratory at the DuPont company.

PROBLEM 26-11

(a) Draw the structure of gutta-percha, a natural rubber with all its double bonds in the trans configuration.
(b) Suggest why gutta-percha is not very elastic, even after it is vulcanized.

Synthetic Rubber There are many different formulations for **synthetic rubbers,** but the simplest is a polymer of buta-1,3-diene. Specialized Ziegler–Natta catalysts can produce buta-1,3-diene polymers where 1,4-addition has occurred on each butadiene unit and the remaining double bonds are all cis. This polymer has properties similar to those of natural rubber, and it can be vulcanized in the same way.

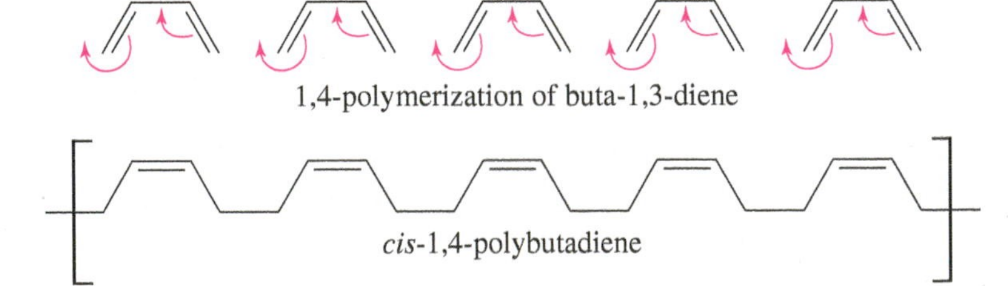

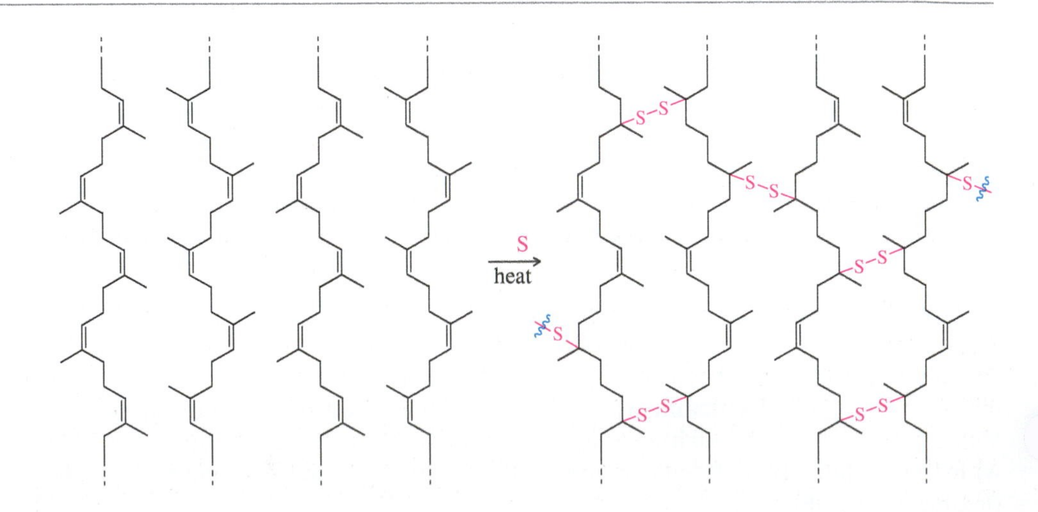

FIGURE 26-4
Vulcanized rubber has disulfide cross-links between the polyisoprene chains. Cross-linking forms a stronger, elastic material that does not pull apart when it is stretched.

26-6 Copolymers of Two or More Monomers

All the polymers we have discussed are **homopolymers,** polymers made up of identical monomer units. Many polymeric materials are **copolymers,** made by polymerizing two or more different monomers together. In many cases, monomers are chosen so that they add selectively in an alternating manner. For example, when a mixture of vinyl chloride and vinylidene chloride (1,1-dichloroethylene) is induced to polymerize, the growing chain preferentially adds the monomer that is *not* at the end of the chain. This selective reaction gives the alternating copolymer *Saran*®, used as a film for wrapping food.

Overall reaction

$$\underset{\text{vinyl chloride}}{\mathrm{H_2C{=}CHCl}} + \underset{\text{vinylidene chloride}}{\mathrm{H_2C{=}CCl_2}} \longrightarrow \underset{\text{Saran}}{\left[-\mathrm{CH_2}-\underset{\mathrm{H}}{\overset{\mathrm{Cl}}{\mathrm{C}}}-\mathrm{CH_2}-\underset{\mathrm{Cl}}{\overset{\mathrm{Cl}}{\mathrm{C}}}-\right]_n}$$

Three or more monomers may combine to give polymers with desired properties. For example, acrylonitrile, butadiene, and styrene are polymerized to give ABS plastic, a strong, tough, and resilient material used for bumpers, crash helmets, and other articles that must withstand heavy impacts.

PROBLEM 26-12

(a) Isobutylene and isoprene copolymerize to give "butyl rubber." Draw the structure of the repeating unit in butyl rubber, assuming that the two monomers alternate.

(b) Styrene and butadiene copolymerize to form styrene-butadiene rubber (SBR) for automobile tires. Draw the structure of the repeating unit in SBR, assuming that the two monomers alternate.

26-7 Step-Growth Polymers

Step-growth polymers are formed by difunctional molecules, any two of which can react to form bonds between them. Monomer molecules may react to form dimers, and dimers may react together to give tetramers. These short chains containing a few monomer units are called **oligomers.** The oligomers can elongate both by adding monomer molecules at either end, or by shorter chains reacting together to form longer chains. Each reaction is an individual step in the growth of the polymer, and there is no chain reaction.

Most step-growth polymerizations involve condensations to form acid-derivative linkages, such as ester or amide bonds, between the monomers. Such polymers are called **condensation polymers,** and they include most polyesters, polyamides, and polycarbonates. We will discuss the four most common types of step-growth polymers: polyamides, polyesters, polycarbonates, and polyurethanes.

26-7A Polyamides: Nylon

When Wallace Carothers of DuPont discovered nylon in 1935, he opened the door to a new age of fibers and textiles. At that time, thread used for clothing was made of spun animal and plant fibers. These fibers were held together by friction or sizing, but they were weak and subject to unraveling and rotting. Silk (a protein) was the strongest fiber known at the time, and Carothers reasoned that a polymer bonded by amide linkages might approach the strength of silk. Nylon proved to be a completely new type of fiber, with remarkable strength and durability. It can be melted and extruded into a strong, continuous fiber, and it cannot rot. Thread spun from continuous nylon fibers is so much stronger than natural materials that it can be made much thinner. Availability of

Scanning electron micrograph of the material in a nylon stocking. Sheer stockings require long, continuous fibers of small diameter and enormous strength. (Magnification 150X)

this strong, thin thread made possible stronger ropes, sheer fabrics, and nearly invisible women's stockings that came to be called "nylons."

Nylon is the common name for polyamides. **Polyamides** are generally made from reactions of diacids with diamines. The most common polyamide is called nylon 6,6 because it is made by reaction of a six-carbon diacid (adipic acid) with a six-carbon diamine. The six-carbon diamine, systematically named *hexane-1,6-diamine*, is commonly called *hexamethylene diamine*. When adipic acid is mixed with hexamethylene diamine, a proton-transfer reaction gives a white solid called *nylon salt*. When nylon salt is heated to 250 °C, water is driven off as a gas, and molten nylon results. Molten nylon is cast into a solid shape or extruded through a spinneret to produce a fiber.

$$\underset{\text{adipic acid}}{HO-\overset{\overset{O}{\|}}{C}-(CH_2)_4-\overset{\overset{O}{\|}}{C}-OH} + \underset{\text{hexamethylene diamine}}{H_2N-(CH_2)_6-NH_2} \longrightarrow \underset{\text{nylon salt}}{\begin{matrix} {}^-O-\overset{\overset{O}{\|}}{C}-(CH_2)_4-\overset{\overset{O}{\|}}{C}-O^- \\ H_3\overset{+}{N}-(CH_2)_6-\overset{+}{N}H_3 \end{matrix}}$$

heat, $-H_2O$

$$\text{---}\overset{\overset{O}{\|}}{C}-(CH_2)_4-\overset{\overset{O}{\|}}{C}\left[NH-(CH_2)_6-NH-\overset{\overset{O}{\|}}{C}-(CH_2)_4-\overset{\overset{O}{\|}}{C}\right]_n NH-(CH_2)_6-NH\text{---}$$

poly(hexamethylene adipamide), called nylon 6,6

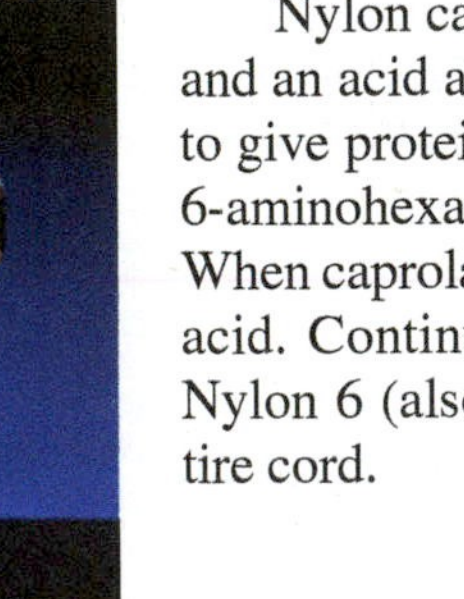

Kevlar body armor works by "catching" a bullet in a multilayer web of woven fabrics. Kevlar's extraordinary strength resists tearing and cutting, allowing the stressed fibers to absorb and disperse the impact to other fibers in the fabric.

Nylon can also be made from a single monomer having an amino group at one end and an acid at the other. This reaction is similar to the polymerization of α-amino acids to give proteins. Nylon 6 is a polymer of this type, made from a six-carbon amino acid: 6-aminohexanoic acid (ε-aminocaproic acid). This synthesis starts with ε-caprolactam. When caprolactam is heated with a trace of water, some of it hydrolyzes to the free amino acid. Continued heating gives condensation and polymerization to molten nylon 6. Nylon 6 (also called *Perlon*®) is used for making strong, flexible fibers for ropes and tire cord.

$$\underset{\varepsilon\text{-caprolactam}}{\text{(cyclic lactam, C=O, N–H)}} \xrightleftharpoons{H_2O,\ \text{heat}} \underset{\varepsilon\text{-aminocaproic acid}}{H_3\overset{+}{N}-(CH_2)_5-\overset{\overset{O}{\|}}{C}-O^-}$$

heat, $-H_2O$

$$\text{---}NH-(CH_2)_5-\overset{\overset{O}{\|}}{C}-NH-(CH_2)_5-\overset{\overset{O}{\|}}{C}\left[NH-(CH_2)_5-\overset{\overset{O}{\|}}{C}\right]_n NH-(CH_2)_5-\overset{\overset{O}{\|}}{C}\text{---}$$

poly(6-aminohexanoic acid), called nylon 6 or Perlon

PROBLEM 26-13

(a) *Nomex*®, a strong fire-resistant fabric, is a polyamide made from *meta*-phthalic acid and *meta*-diaminobenzene. Draw the structure of Nomex.

(b) *Kevlar*®, made from terephthalic acid (*para*-phthalic acid) and *para*-diaminobenzene, is used in making tire cord and bulletproof vests. Draw the structure of Kevlar.

26-7B Polyesters

The introduction of **polyester** fibers has brought about major changes in the way we care for our clothing. Nearly all modern permanent-press fabrics owe their wrinkle-free behavior to polyester, often blended with other fibers. These polyester blends have reduced or eliminated the need for starching and ironing clothes to achieve a wrinkle-free surface that holds its shape.

The most common polyester is *Dacron®*, the polymer of terephthalic acid (*para*-phthalic acid or benzene-1,4-dicarboxylic acid) with ethylene glycol. In principle, this polymer might be made by mixing the diacid with the glycol and heating the mixture to drive off water. In practice, however, a better product is obtained using a transesterification process (Section 21-6). The dimethyl ester of terephthalic acid is heated to about 150 °C with ethylene glycol. Methanol is evolved as a gas, driving the reaction to completion. The molten product is spun into Dacron fiber or cast into *Mylar®* film.

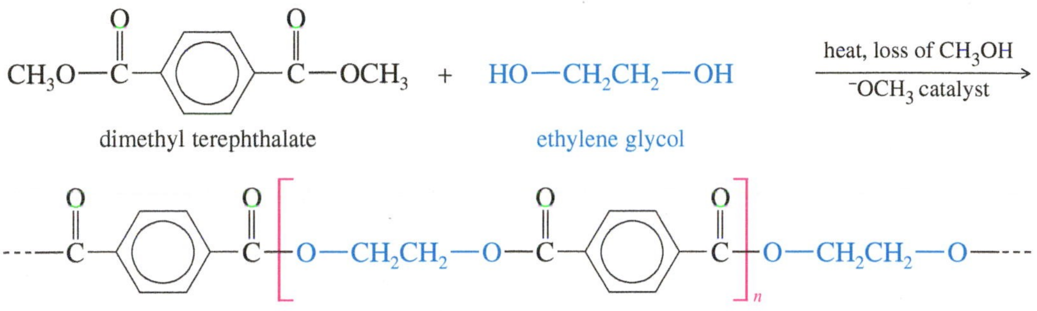

$$CH_3O-\overset{O}{\overset{\|}{C}}-C_6H_4-\overset{O}{\overset{\|}{C}}-OCH_3 \;+\; HO-CH_2CH_2-OH \xrightarrow[\;^{-}OCH_3\text{ catalyst}]{\text{heat, loss of }CH_3OH}$$

dimethyl terephthalate ethylene glycol

$$\cdots-\overset{O}{\overset{\|}{C}}-C_6H_4-\overset{O}{\overset{\|}{C}}\left[-O-CH_2CH_2-O-\overset{O}{\overset{\|}{C}}-C_6H_4-\overset{O}{\overset{\|}{C}}-\right]_n O-CH_2CH_2-O-\cdots$$

poly(ethylene terephthalate) or PET, also called Dacron polyester or Mylar film

Dacron fiber is used to make fabric and tire cord, and Mylar film is used to make magnetic recording tape. Mylar film is strong, flexible, and resistant to ultraviolet degradation. Aluminized Mylar was used to make the Echo satellites, (page 1090), huge balloons that were put into orbit around the Earth as giant reflectors in the early 1960s. Poly(ethylene terephthalate) is also blow-molded to make plastic soft-drink bottles that are sold by the billions each year.

PROBLEM 26-14

Kodel® polyester is formed by transesterification of dimethyl terephthalate with 1,4-di(hydroxy methyl)cyclohexane. Draw the structure of Kodel.

PROBLEM 26-15

Glyptal® resin makes a strong, solid polymer matrix for electronic parts. Glyptal is made from terephthalic acid and glycerol. Draw the structure of Glyptal, and explain its remarkable strength and rigidity.

Application: Bio-Absorbable Materials

A polyester of ε-caprolactone is used to make bio-absorbable materials for use in the body. For example, suture thread for surgical stitches can be made from poly (ε-caprolactone).

26-7C Polycarbonates

A *carbonate ester* is simply an ester of carbonic acid. Carbonic acid itself exists in equilibrium with carbon dioxide and water, but its esters are quite stable (Section 21-16).

$$HO-\overset{O}{\overset{\|}{C}}-OH \rightleftharpoons CO_2 + H_2O \qquad\qquad R-O-\overset{O}{\overset{\|}{C}}-O-R'$$

carbonic acid a carbonate ester

Carbonic acid is a diacid; with suitable diols, it can form polyesters. For example, when phosgene (the acid chloride of carbonic acid) reacts with a diol, the product is a **poly(carbonate ester).** The following equation shows the synthesis of *Lexan®* polycarbonate: a strong, clear, and colorless material that is used for

Application: Medical Polymers

Polycarbonate is a tough, clear material that withstands repeated sterilization. These properties account for its wide use in medical devices such as blood filters, surgical instruments, and intravenous line components.

bulletproof windows and crash helmets. The diol used to make Lexan is a phenol called *bisphenol A*, a common intermediate in polyester and polyurethane synthesis.

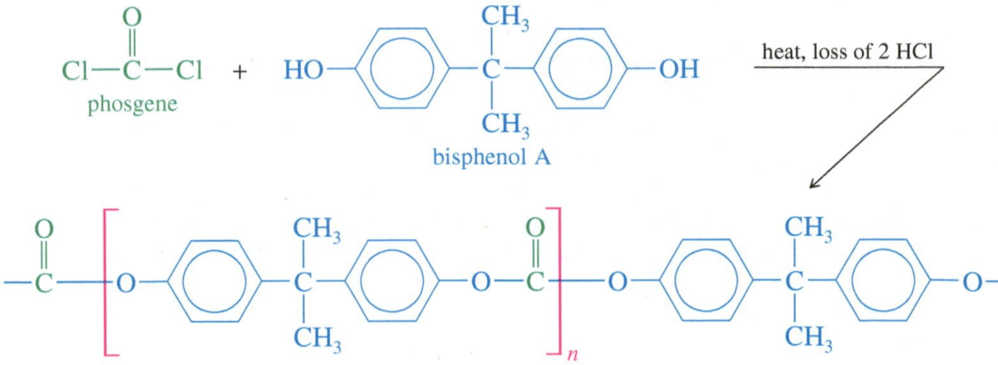

$$Cl-C(=O)-Cl \text{ (phosgene)} + HO-C_6H_4-C(CH_3)_2-C_6H_4-OH \text{ (bisphenol A)} \xrightarrow{\text{heat, loss of 2 HCl}} \text{Lexan polycarbonate}$$

$$-C(=O)-[O-C_6H_4-C(CH_3)_2-C_6H_4-O-C(=O)]_n-O-C_6H_4-C(CH_3)_2-C_6H_4-O-$$

Lexan polycarbonate

PROBLEM 26-16

(a) Propose a mechanism for the reaction of bisphenol A with phosgene.
(b) Diethyl carbonate serves as a less-toxic alternative to phosgene for making Lexan. Propose a mechanism for the transesterification of diethyl carbonate with bisphenol A, catalyzed by a trace of sodium ethoxide. What small molecule is given off in this condensation?

Application: Latex-Free Gloves

Latex surgical gloves cause severe allergies for some people, especially health-care workers who must wear gloves for most of their working day. Latex gloves also deteriorate over time or in contact with oil-based creams and ointments.

Latex-free gloves are often made of polyurethane, which causes fewer allergic reactions. These polyurethane materials can be used with oil- or water-based creams and ointments, and they deteriorate less in storage.

PROBLEM 26-17

Bisphenol A is made on a large scale by a condensation of phenol with acetone. Suggest an appropriate catalyst, and propose a mechanism for this reaction. (*Hint:* This is a condensation because three molecules are joined with loss of water. The mechanism belongs to another class of reactions, though.)

26-7D Polyurethanes

A *urethane* (Section 21-16) is an ester of a carbamic acid (R—NH—COOH), a half-amide of carbonic acid. Carbamic acids themselves are unstable, quickly decomposing to amines and CO_2. Their esters (urethanes) are quite stable, however.

$$\underset{\text{a carbamic acid}}{R-NH-C(=O)-OH} \longrightarrow \underset{\text{amine}}{R-NH_2} + CO_2\uparrow \qquad \underset{\text{a urethane or carbamate ester}}{R-NH-C(=O)-O-R'}$$

Because carbamic acids are unstable, normal esterification procedures cannot be used to form urethanes. Urethanes are most commonly made by treating an *isocyanate* with an alcohol or a phenol. The reaction is highly exothermic, and it gives a quantitative yield of a carbamate ester.

$$\underset{\text{isocyanate}}{R-N=C=O} + \underset{\text{alcohol}}{HO-R'} \longrightarrow \underset{\text{carbamate ester (urethane)}}{R-NH-C(=O)-O-R'}$$

Example

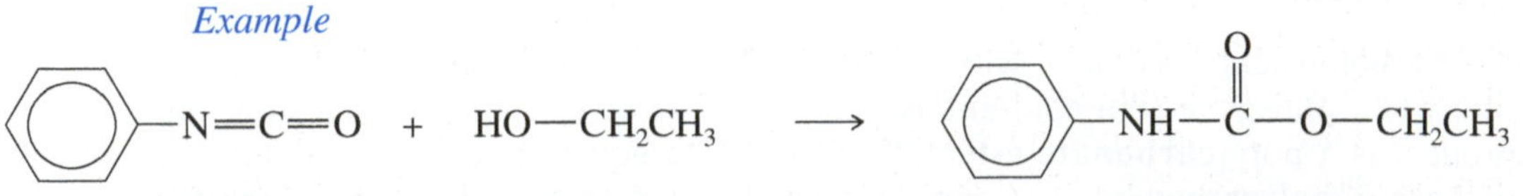

$$\underset{\text{phenyl isocyanate}}{C_6H_5-N=C=O} + \underset{\text{ethanol}}{HO-CH_2CH_3} \longrightarrow \underset{\text{ethyl-}N\text{-phenylcarbamate}}{C_6H_5-NH-C(=O)-O-CH_2CH_3}$$

PROBLEM 26-18

Propose a mechanism for the reaction of phenyl isocyanate with ethanol.

A **polyurethane** results when a diol reacts with a diisocyanate, a compound with two isocyanate groups. The compound shown next, commonly called *toluene diisocyanate*, is frequently used for making polyurethanes. When ethylene glycol or another diol is added to toluene diisocyanate, a rapid exothermic reaction gives the polyurethane. Low-boiling liquids such as butane are often added to the reaction mixture. Heat evolved by the polymerization vaporizes the volatile liquid, producing bubbles that convert the viscous polymer to a frothy mass of polyurethane foam.

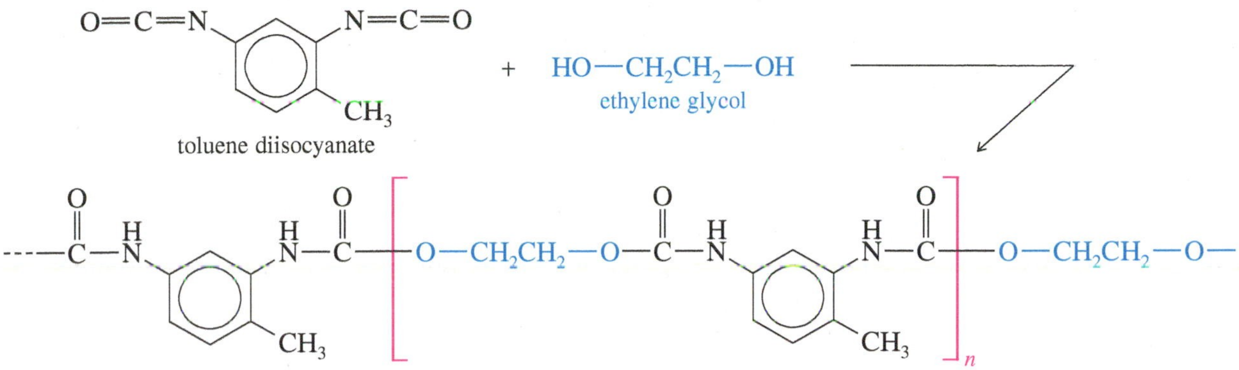

$O{=}C{=}N-C_6H_3(CH_3)-N{=}C{=}O$ + $HO-CH_2CH_2-OH$

toluene diisocyanate

ethylene glycol

$$\cdots-\overset{O}{\overset{\|}{C}}-\overset{H}{N}-C_6H_3(CH_3)-\overset{H}{N}-\overset{O}{\overset{\|}{C}}-\left[O-CH_2CH_2-O-\overset{O}{\overset{\|}{C}}-\overset{H}{N}-C_6H_3(CH_3)-\overset{H}{N}-\overset{O}{\overset{\|}{C}}\right]_n-O-CH_2CH_2-O-$$

a polyurethane

PROBLEM 26-19

Explain why the addition of a small amount of glycerol to the polymerization mixture gives a stiffer urethane foam.

PROBLEM 26-20

Give the structure of the polyurethane formed by the reaction of toluene diisocyanate with bisphenol A.

26-8 Polymer Structure and Properties

Although polymers are very large molecules, we can explain their chemical and physical properties in terms of what we already know about smaller molecules. For example, when you spill a base on your polyester slacks, the fabric is weakened because the base hydrolyzes some of the ester linkages. The physical properties of polymers can also be explained using concepts we have already encountered. Although polymers do not crystallize or melt quite like smaller molecules, we can detect crystalline regions in a polymer, and we can measure the temperature at which these *crystallites* melt. In this section, we consider briefly some of the important aspects of polymer crystallinity and thermal behavior.

26-8A Polymer Crystallinity

Polymers rarely form the large crystals characteristic of other organic compounds, but many do form microscopic crystalline regions called **crystallites.** A highly regular polymer that packs well into a crystal lattice will be highly crystalline, and it will generally be denser, stronger, and more rigid than a similar polymer with a lower degree of **crystallinity.** Figure 26-5 shows how the polymer chains are arranged in parallel lines in crystalline areas within a polymer.

Polyethylene provides an example of how crystallinity affects a polymer's physical properties. Free-radical polymerization gives a highly branched, low-density

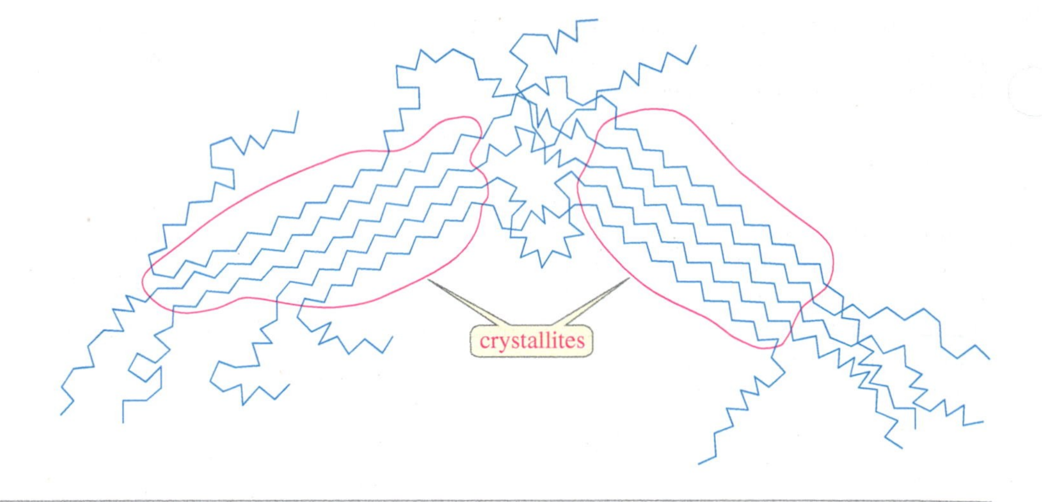

FIGURE 26-5
Crystallites are areas of crystalline structure within the large mass of a solid polymer.

polyethylene that forms very small crystallites because the random chain branching destroys the regularity of the crystallites. An unbranched, high-density polyethylene is made using a Ziegler–Natta catalyst. The *linear* structure of the high-density material packs more easily into a crystal lattice, so it forms larger and stronger crystallites. We say that high-density polyethylene has a higher degree of crystallinity, and it is therefore denser, stronger, and more rigid than low-density polyethylene.

Stereochemistry also affects the crystallinity of a polymer. Stereoregular isotactic and syndiotactic polymers are generally more crystalline than atactic polymers. By careful choice of catalysts, we can make a linear polymer with either isotactic or syndiotactic stereochemistry.

26-8B Thermal Properties

Application: Controlled-Release Drugs

Drugs are sometimes enclosed in water-soluble polymers to control the rate of drug release. The polymer breaks down in the body over time at a predictable rate and gradually releases the drug.

At low temperatures, long-chain polymers are *glasses*. They are solid and unyielding, and a strong impact causes them to fracture. As the temperature is raised, the polymer goes through a **glass transition temperature,** abbreviated T_g. Above T_g, a highly crystalline polymer becomes flexible and moldable. We say it is a **thermoplastic** because application of heat makes it plastic (moldable). As the temperature is raised further, the polymer reaches the **crystalline melting temperature,** abbreviated T_m. At this temperature, crystallites melt and the individual molecules can slide past one another.

Above T_m, the polymer is a viscous liquid and can be extruded through spinnerets to form fibers. The fibers are immediately cooled in water to form crystallites and then stretched (drawn) to orient the crystallites along the fiber, increasing its strength.

Long-chain polymers with low crystallinity (called **amorphous polymers**) become rubbery when heated above the glass transition temperature. Further heating causes them to grow gummier and less solid until they become viscous liquids without definite melting points. Figure 26-6 compares the thermal properties of crystalline and amorphous long-chain polymers.

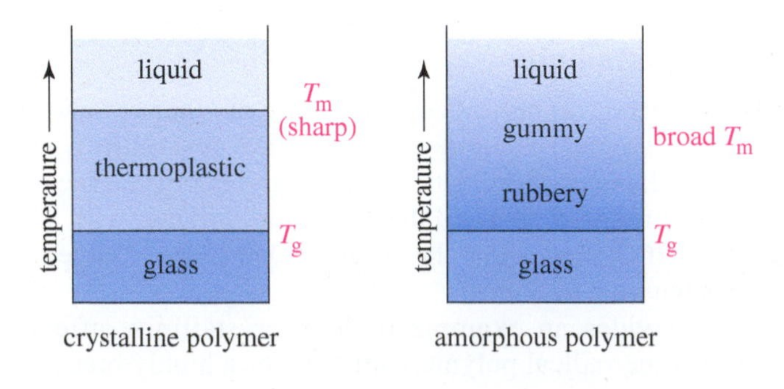

FIGURE 26-6
Crystalline and amorphous long-chain polymers show different physical properties when they are heated.

These phase transitions apply only to long-chain polymers. Cross-linked polymers are more likely to stay rubbery, and they may not melt until the temperature is so high that the polymer begins to decompose.

26-8C Plasticizers

In many cases, a polymer has desirable properties for a particular use, but it is too brittle—either because its glass transition temperature (T_g) is above room temperature or because the polymer is too highly crystalline. In such cases, addition of a **plasticizer** often makes the polymer more flexible. A plasticizer is a nonvolatile liquid that dissolves in the polymer, lowering the attractions between the polymer chains and allowing them to slide by one another. The overall effect of the plasticizer is to reduce the crystallinity of the polymer and lower its glass transition temperature (T_g).

A common example of a plasticized polymer is poly(vinyl chloride). The common atactic form has a T_g of about 80 °C, well above room temperature. Without a plasticizer, "vinyl" is stiff and brittle. Dibutyl phthalate (see the structure at right) is added to the polymer to lower its glass transition temperature to about 0 °C. This plasticized material is the flexible, somewhat stretchy film we think of as vinyl raincoats, shoes, and even inflatable boats. Dibutyl phthalate is slightly volatile, however, and it gradually evaporates. The soft, plasticized vinyl gradually loses its plasticizer and becomes hard and brittle.

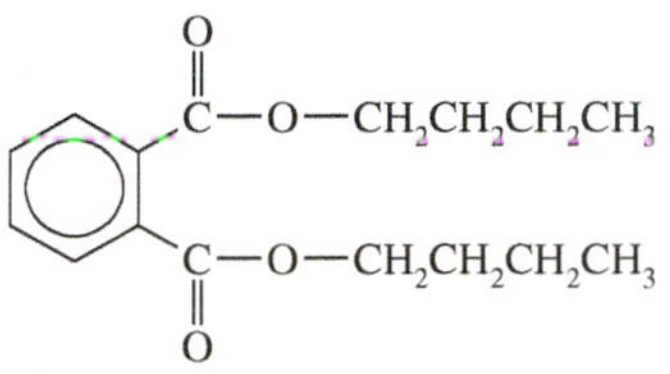

$C(=O)-O-CH_2CH_2CH_2CH_3$

$C(=O)-O-CH_2CH_2CH_2CH_3$

dibutyl phthalate

26-9 Recycling of Plastics

Synthetic polymers are useful in consumer products because they do not readily rot, oxidize, or hydrolyze. They are waterproof, flexible, and durable. These same properties are problematic when plastic consumer products are discarded in landfills. The volume is enormous: between 10% and 15% of solid municipal waste.

In many cases, however, these discarded polymers can serve as a feedstock for making new plastic products because most polymers can be re-melted and recycled. Some of the ones that cannot be re-melted can be recycled by using chemical reactions to revert them to their monomers, but this process is much more expensive than simply re-melting them.

Recycled plastics always contain small amounts of impurities such as adhesives, labels, and possibly small amounts of other polymers. For that reason, recycled plastics have less strength and uniformity than newly made plastics, and they are not used where strength is crucial. For example, recycled plastics are not often used for storing food products, and they are never used for aviation components.

In order to recycle plastics economically, the feedstock must consist of just one nearly pure polymer. The most important stage of recycling is sorting the plastics by their types. Table 26-2 shows how plastics are classified according to a number. These numbers and their recycling symbols are molded into most plastic consumer products.

Once the plastics are sorted by type, they are shredded into small chips. Then the chips are washed and floated in a solution that allows metal fragments to sink. Finally, the chips are shipped to a manufacturing facility to be re-melted and molded for a new useful life.

TABLE 26-2
Classes of Polymers for Recycling

Polymer	Symbol	New Uses	Recycled Uses
PET poly(ethylene terephthalate)	 01 PET	Polyester textiles, tire cords, soft drink and water bottles	Polyester fill fibers, carpet, paneling, strapping, furniture
HDPE high-density polyethylene	 02 HDPE	Fuel tanks, bottles, grocery bags, plumbing pipe, boats, labware	Plastic lumber, trash cans, planters, mailboxes

(continued)

TABLE 26-2 *(Continued)*

Polymer	Symbol	New Uses	Recycled Uses
PVC poly(vinyl chloride)	03 PVC	Window frames, raincoats, siding, plumbing pipe, gutters, inflatables	Plumbing drain pipes, siding, gutters, toys, plastic lumber
LDPE low-density polyethylene	04 LDPE	Filmy plastic bags, trash bags, packaging, squeezable bottles	Plastic film and bags
PP polypropylene	05 PP	"Olefin" carpet, rope, dishware, packaging, auto parts, fishing nets	Carpet, rope, tarpaulins
PS polystyrene	06 PS	Toys, electronic gadgets, optics, styrofoam insulation	Clothes hangers, insulation, plastic lumber, containers
Other: Nylons, polycarbonates, acrylics, polyurethanes, ABS, etc.	07 O	All other uses	Not usually recycled

Essential Terms

addition polymer A polymer that results from the addition reactions of alkenes, dienes, or other compounds with double and triple bonds. Most addition polymers form by a chain-growth process. (p. 1287)

amorphous polymer A long-chain polymer with low crystallinity. (p. 1302)

anionic polymerization The process of forming an addition polymer by chain-growth polymerization involving an anion at the end of the growing chain. (p. 1292)

atactic polymer A polymer with the side groups on random sides of the polymer backbone. (p. 1294)

cationic polymerization The process of forming an addition polymer by chain-growth polymerization involving a cation at the end of the growing chain. (p. 1290)

chain-growth polymerization The rapid addition of one monomer at a time to a growing polymer chain, usually with a reactive intermediate (cation, radical, or anion) at the growing end of the chain. Most chain-growth polymers are addition polymers of alkenes and dienes. (p. 1287)

condensation polymer A polymer that results from condensation (bond formation with loss of a small molecule) between the monomers or the polymer segments. Most condensation polymers form by a step-growth process that forms ester or amide linkages between any two molecules, not necessarily at the end of a growing chain. (p. 1287)

copolymer A polymer made from two or more different monomers. (p. 1297)

crystalline melting temperature (T_m) The temperature at which melting of the crystallites in a highly crystalline polymer occurs. Above T_m, the polymer is a viscous liquid. (p. 1302)

crystallinity The relative amount of the polymer that is included in crystallites, and the relative sizes of the crystallites. (p. 1301)

crystallites Microscopic crystalline regions found within a solid polymer below the crystalline melting temperature. (p. 1301)

free-radical polymerization The process of forming an addition polymer by chain-growth polymerization involving a free radical at the end of the growing chain. (p. 1288)

glass transition temperature (T_g) The temperature above which a polymer becomes rubbery or flexible. (p. 1302)

homopolymer A polymer made from identical monomer units. (p. 1297)

isotactic polymer A polymer with all the side groups on the same side of the polymer backbone. (p. 1294)

monomer One of the small molecules that bond together to form a polymer. (p. 1286)

nylon The common name for polyamides. (p. 1298)

oligomer A small polymer chain that consists of a few monomer units up to a few dozen monomer units. (p. 1297)

plasticizer A nonvolatile liquid that is added to a polymer to make it more flexible and less brittle below its glass transition temperature. In effect, a plasticizer reduces the crystallinity of a polymer and lowers T_g. (p. 1303)

polyamide **(nylon)** A polymer whose repeating monomer units are bonded by amide linkages, much like the peptide linkages in protein. (p. 1298)

polycarbonate A polymer whose repeating monomer units are bonded by carbonate ester linkages. (p. 1299)

polyester A polymer whose repeating monomer units are bonded by carboxylate ester linkages. (p. 1299)

polymer A large molecule composed of many smaller units (monomers) bonded together. (p. 1286)

polymerization The process of linking monomer molecules into a polymer. (p. 1287)

polyurethane A polymer whose repeating monomer units are bonded by urethane (carbamate ester) linkages. (p. 1301)

rubber A natural polymer isolated from **latex** that exudes from cuts in the bark of the South American rubber tree. Alternatively, synthetic polymers with rubber-like properties are called **synthetic rubber.** (p. 1295)

step-growth polymerization A polymerization in which any two molecules having the correct functionality can react with each other, or two polymer chains can combine. Most step-growth polymers are condensation polymers, resulting from the formation of ester or amide linkages between the monomers. (p. 1287)

syndiotactic polymer A polymer with the side groups on alternating sides of the polymer backbone. (p. 1294)

thermoplastic A polymer that becomes moldable at high temperature. (p. 1302)

vulcanization Heating of natural or synthetic rubber with sulfur to form disulfide cross-links. Cross-linking adds durability and elasticity to rubber. (p. 1295)

Ziegler–Natta catalyst Any one of a group of addition polymerization catalysts involving titanium–aluminum complexes. Ziegler–Natta catalysts produce stereoregular (either isotactic or syndiotactic) polymers in most cases. (p. 1295)

Essential Problem-Solving Skills in Chapter 26

Each skill is followed by problem numbers exemplifying that particular skill.

1 Given the structure of a polymer, determine whether it is a chain-growth polymer or a step-growth polymer, and draw the structure of the monomer(s).	Problems 26-22, 23, 24, 25, 26, 27, 36, and 37
2 Given the structures of the monomers, predict whether the polymer will be a chain-growth polymer or a step-growth polymer, and draw the structure of the polymer chain.	Problems 26-21, 28, 29, 31, 32, 33, and 35
3 Use mechanisms to show how monomers polymerize under acidic, basic, or free-radical conditions. For chain-growth polymerization, determine whether the reactive end is more stable as a cation (acidic conditions), anion (basic conditions), or free radical (radical initiator). For step-growth polymerization, consider the mechanism of the condensation.	Problems 26-21, 26, 28, 33, and 34
4 Predict the general characteristics (strength, elasticity, crystallinity, chemical reactivity) of a polymer based on its structure, and explain how its physical characteristics change as it is heated past T_g and T_m.	Problems 26-27, 29, 30, and 32
5 Explain how chain branching, cross-linking, stereochemistry (isotactic, syndiotactic, or atactic), and plasticizers affect the properties of polymers.	Problems 26-10, 11, and 15

Study Problems

26-21 Polyisobutylene is one of the components of butyl rubber used for making inner tubes.
(a) Give the structure of polyisobutylene.
(b) Is this an addition polymer or a condensation polymer?
(c) What conditions (cationic, anionic, free-radical) would be most appropriate for polymerization of isobutylene? Explain your answer.

26-22 Polychloroprene, commonly known as *neoprene*, is widely used in wetsuits and in rubber parts that must withstand exposure to gasoline or other solvents.
(a) Is neoprene an addition polymer or a condensation polymer?
(b) What monomer is used to make this synthetic rubber?

$$\left(CH_2-\overset{H}{C}=\overset{Cl}{C}-CH_2\right)_n$$

polychloroprene (neoprene)

26-23 Poly(trimethylene carbamate) is used in high-quality synthetic leather. It has the structure shown.
(a) What type of polymer is poly(trimethylene carbamate)?
(b) Is this a chain-growth polymer or a step-growth polymer?
(c) Draw the products that would be formed if the polymer were completely hydrolyzed under acidic or basic conditions.

$$\left(CH_2CH_2CH_2-\overset{H}{N}-\overset{O}{\overset{\|}{C}}-O\right)_n$$

poly(trimethylene carbamate)

26-24 Poly(butylene terephthalate) is a hydrophobic plastic material widely used in automotive ignition systems.
(a) What type of polymer is poly(butylene terephthalate)?
(b) Is this an addition polymer or a condensation polymer?
(c) Suggest what monomers might be used to synthesize this polymer and how the polymerization might be accomplished.

$$\left(CH_2CH_2CH_2CH_2-O-\overset{O}{\overset{\|}{C}}-C_6H_4-\overset{O}{\overset{\|}{C}}-O\right)_n$$

poly(butylene terephthalate)

26-25 *Urylon* fibers are used in premium fishing nets because the polymer is relatively stable to UV light and aqueous acid and base. Urylon has the structure shown.
(a) What functional group is contained in the Urylon structure?
(b) Is Urylon a chain-growth polymer or a step-growth polymer?
(c) Draw the products that would be formed if the polymer were completely hydrolyzed under acidic or basic conditions.

$$\left((CH_2)_9-\overset{H}{N}-\overset{O}{\overset{\|}{C}}-\overset{H}{N}\right)_n$$

Urylon

26-26 Polyethylene glycol, or *Carbowax*® [$(-O-CH_2-CH_2-)_n$], is widely used as a binder, thickening agent, and packaging additive for foods.
(a) What type of polymer is polyethylene glycol? (We have not seen this type of polymer before.)
(b) The systematic name for polyethylene glycol is poly(ethylene oxide). What monomer would you use to make polyethylene glycol?
(c) What conditions (free-radical initiator, acid catalyst, basic catalyst, etc.) would you consider using in this polymerization?
(d) Propose a polymerization mechanism as far as the tetramer.

26-27 Ring-opening metathesis polymerization (ROMP, see Section 8-17) is a promising new technique for polymerizing cyclic olefins. In its simplest form, the reaction involves a cycloalkene (preferably with some ring strain to drive the reaction) whose double bonds undergo metathesis (a trading of partners at the ends of the double bonds) to give a polymer containing both single and double bonds.

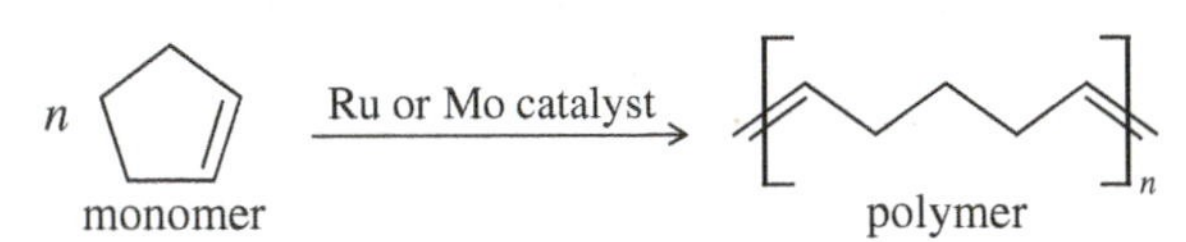

ring-opening metathesis polymerization (ROMP)

Show which monomers might have reacted to produce the following polymers:

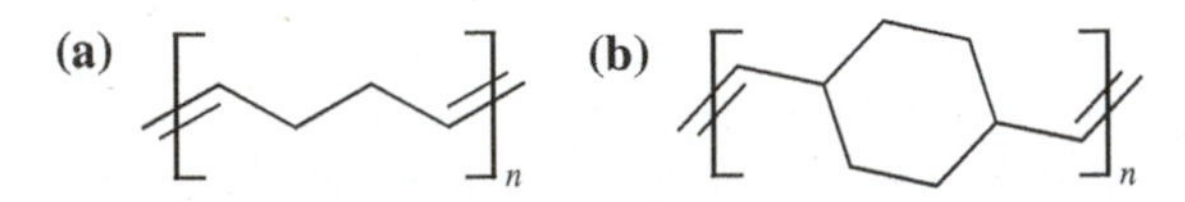

26-28 Polyoxymethylene (polyformaldehyde) is the tough, self-lubricating *Delrin*® plastic used in gear wheels.
(a) Give the structure of polyformaldehyde.
(b) Formaldehyde is polymerized using an acidic catalyst. Using H^+ as a catalyst, propose a mechanism for the polymerization as far as the trimer.
(c) Is Delrin an addition polymer or a condensation polymer?

26-29 The 2000 Nobel Prize in Chemistry was awarded for work on polyacetylenes. Acetylene can be polymerized using a Ziegler–Natta catalyst. The cis or trans stereochemistry of the products can be controlled by careful selection and preparation of the catalyst. The resulting polyacetylene is an electrical semiconductor with a metallic appearance. *cis*-Polyacetylene has a copper color, and *trans*-polyacetylene is silver.

(a) Draw the structures of *cis*- and *trans*-polyacetylene.
(b) Use your structures to show why these polymers conduct electricity.
(c) It is possible to prepare polyacetylene films whose electrical conductivity is *anisotropic.* That is, the conductivity is higher in some directions than in others. Explain how this unusual behavior is possible.

26-30 Use chemical equations to show how the following accidents cause injury to the clothing involved (not to mention the skin under the clothing!).

(a) An industrial chemist spills aqueous H_2SO_4 on her nylon stockings but fails to wash it off immediately.
(b) An organic laboratory student spills aqueous NaOH on his polyester slacks.

26-31 Poly(vinyl alcohol), a hydrophilic polymer used in aqueous adhesives, is made by polymerizing vinyl acetate and then hydrolyzing the ester linkages.

(a) Give the structures of poly(vinyl acetate) and poly(vinyl alcohol).
(b) Vinyl acetate is an ester. Is poly(vinyl acetate) therefore a polyester? Explain.
(c) We have seen that basic hydrolysis destroys the Dacron polymer. Poly(vinyl acetate) is converted to poly(vinyl alcohol) by a basic hydrolysis of the ester groups. Why doesn't the hydrolysis destroy the poly(vinyl alcohol) polymer?
(d) Why is poly(vinyl alcohol) made by this circuitous route? Why not just polymerize vinyl alcohol?

26-32 In reference to cloth or fiber, the term *acetate* usually means *cellulose acetate,* a semisynthetic polymer made by treating cellulose with acetic anhydride. Cellulose acetate is spun into yarn by dissolving it in acetone or methylene chloride and forcing the solution through spinnerets into warm air, where the solvent evaporates.

(a) Draw the structure of cellulose acetate.
(b) Explain why cellulose acetate is soluble in organic solvents, even though cellulose is not.
(c) (A true story) An organic chemistry student wore a long-sleeved acetate blouse to the laboratory. She was rinsing a warm separatory funnel with acetone when the pressure rose and blew out the stopper. Her right arm was drenched with acetone, but she was unconcerned because acetone is not very toxic. About ten minutes later, the right arm of the student's blouse disintegrated into a pile of white fluff, leaving her with a ragged short sleeve and the tatters of a cuff remaining around her wrist. Explain how a substance as innocuous as acetone ruined the student's blouse.
(d) Predict what usually happens when students wear polyvinyl chloride shoes to the organic laboratory.

***26-33** One of the earliest commercial plastics was *Bakelite*®, formed by the reaction of phenol with a little more than one equivalent of formaldehyde under acidic or basic conditions. Baeyer first discovered this reaction in 1872, and practical methods for casting and molding Bakelite were developed around 1909. Phenol-formaldehyde plastics and resins (also called *phenolics*) are highly cross-linked because each phenol ring has three sites (two ortho and one para) that can be linked by condensation with formaldehyde. Suggest a general structure for a phenol-formaldehyde resin, and propose a mechanism for its formation under acidic conditions. (*Hint:* Condensation of phenol with formaldehyde resembles the condensation of phenol with acetone, used in Problem 26-17, to make bisphenol A.)

***26-34** Plywood and particle board are often glued with cheap, waterproof urea-formaldehyde resins. Two to three moles of formaldehyde are mixed with one mole of urea and a little ammonia as a basic catalyst. The reaction is allowed to proceed until the mixture becomes syrupy, and then it is applied to the wood surface. The wood surfaces are held together under heat and pressure, while polymerization continues and cross-linking takes place. Propose a mechanism for the base-catalyzed condensation of urea with formaldehyde to give a linear polymer, and then show how further condensation leads to cross-linking. (*Hint:* The carbonyl group lends acidity to the N—H protons of urea. A first condensation with formaldehyde leads to an imine, which is weakly electrophilic and reacts with another deprotonated urea.)

26-35 The polyester named Lactomer® is an alternating copolymer of lactic acid and glycolic acid. Lactomer is used for absorbable suture material because stitches of Lactomer hydrolyze slowly over a two-week period and do not have to be removed. The hydrolysis products, lactic acid and glycolic acid, are normal metabolites and do not provoke an inflammatory response. Draw the structure of the Lactomer polymer.

HO–CH₂–C(=O)–OH
glycolic acid

HO–CH(CH₃)–C(=O)–OH
lactic acid

26-36 Compare the molecular structures of cotton and polypropylene, the two major components of thermal underwear. One of these gets wet easily and holds the water in contact with the skin. The other one does not get wet, but wicks the water away from the skin and feels relatively dry to the touch. Explain the difference in how these two fabrics respond to moisture.

Appendices

APPENDIX 1A
NMR: Spin-Spin Coupling Constants

Type	*J*, Hz	Type	*J*, Hz
$>C(H)H$ (geminal)	12–15	$>C=C(H)-C(H)<$	4–10
$>CH-CH<$ with free rotation	2–9 ~7	$H-C=C-C-H$ (allylic)	0.5–2.5
$H-C-(-C-)_n-C-H$	~0	$H-C=C-C-H$	~0
CH_3-CH_2-X	6.5–7.5	$>C=CH-CH=C<$	9–13
$(CH_3)_2CH-X$	5.5–7.0	$>CH-C\equiv C-H$	2–3
H–C–C–H on cyclohexane ring (X, Y substituents)	a,a 5–10 a,e 2–4 e,e 2–4	$>CH-C(H)=O$	1–3
$>C=C(H)H$	0.5–3	$>C=C(H)-C(H)=O$	6–8
$H(C=C)H$ (cis)	7–12	benzene ring	ortho 6–9 meta 1–3 para 0–1
$H(C=C)H$ (trans)	13–18		

a = axial, e = equatorial

APPENDIX 1B
NMR: Proton Chemical Shifts

Structural type	δ Value and range[a] (14 13 12 11 10 9 8 7 6 5 4 3 2 1 0)
TMS, 0.000	
$-CH_2-$, cyclopropane	
CH_4	
ROH, monomer, very dilute solution	
$CH_3-\overset{\vert}{\underset{\vert}{C}}-$(saturated)	
R_2NH[b], 0.1–0.9 mole fraction in an inert solvent	
$CH_3-\overset{\vert}{\underset{\vert}{C}}-\overset{\vert}{\underset{\vert}{C}}-X$ (X = Cl, Br, I, OH, OR, C=O, N)	
$-CH_2-$(saturated)	
RSH[b]	
RNH_2[b], 0.1–0.9 mole fraction in an inert solvent	
$-\overset{\vert}{\underset{\vert}{C}}-H$ (saturated)	
$CH_3-\overset{\vert}{\underset{\vert}{C}}-X$ (X = F, Cl, Br, I, OH, OR, OAr, N)	
CH_3 on C=C (vinylic methyl)	
$CH_3-\overset{\vert}{C}=O$	
CH_3Ar	
CH_3-S-	
$CH_3-N<$	
$H-C\equiv C-$, nonconjugated	
$H-C\equiv C-$, conjugated	
$H-\overset{\vert}{\underset{\vert}{C}}-X$ (X = F, Cl, Br, I, O)	
ArSH[b]	
CH_3-O-	
$ArNH_2$[b], ArNHR[b], and Ar_2NH[b]	

[a] Normally, absorptions for the functional groups indicated will be found within the range shown in black. Occasionally, a functional group will absorb outside this range. Approximate limits are indicated by extended outlines.

[b] Absorption positions of these groups are concentration-dependent and are shifted to lower δ values in more dilute solutions.

APPENDIX 1B
NMR: Proton Chemical Shifts

Structural type	δ Value and range[a] (14 13 12 11 10 9 8 7 6 5 4 3 2 1 0)
ROH[b], 0.1–0.9, mole fraction in an inert solvent	
$CH_2{=}C<$, nonconjugated	
$H{-}C{=}C<$, acyclic, nonconjugated	
$H{-}C{=}C<$, cyclic, nonconjugated	
$CH_2{=}C<$, conjugated	
ArOH[b], polymeric association	
$H{-}C{=}C<$, conjugated	
$H{-}C{=}C<$, acyclic, conjugated	
$H{-}N{-}C(=O)$	
ArH, benzenoid	
ArH, nonbenzenoid	
RNH_3^+, $R_2NH_2^+$, and R_3NH^+, (trifluoroacetic acid solution)	
$H{-}C(=O){-}N<$	
$H{-}C(=O){-}O{-}$	
$ArNH_3^+$, $ArRNH_2^+$, and ArR_2NH^+, (trifluoroacetic acid solution)	
$>C{=}N{-}OH$[b]	
RCHO, aliphatic, α, β-unsaturated	
RCHO, aliphatic	
ArCHO	
ArOH, intermolecularly bonded	
$-SO_3H$	
RCO_2H, dimer, in nonpolar solvents	

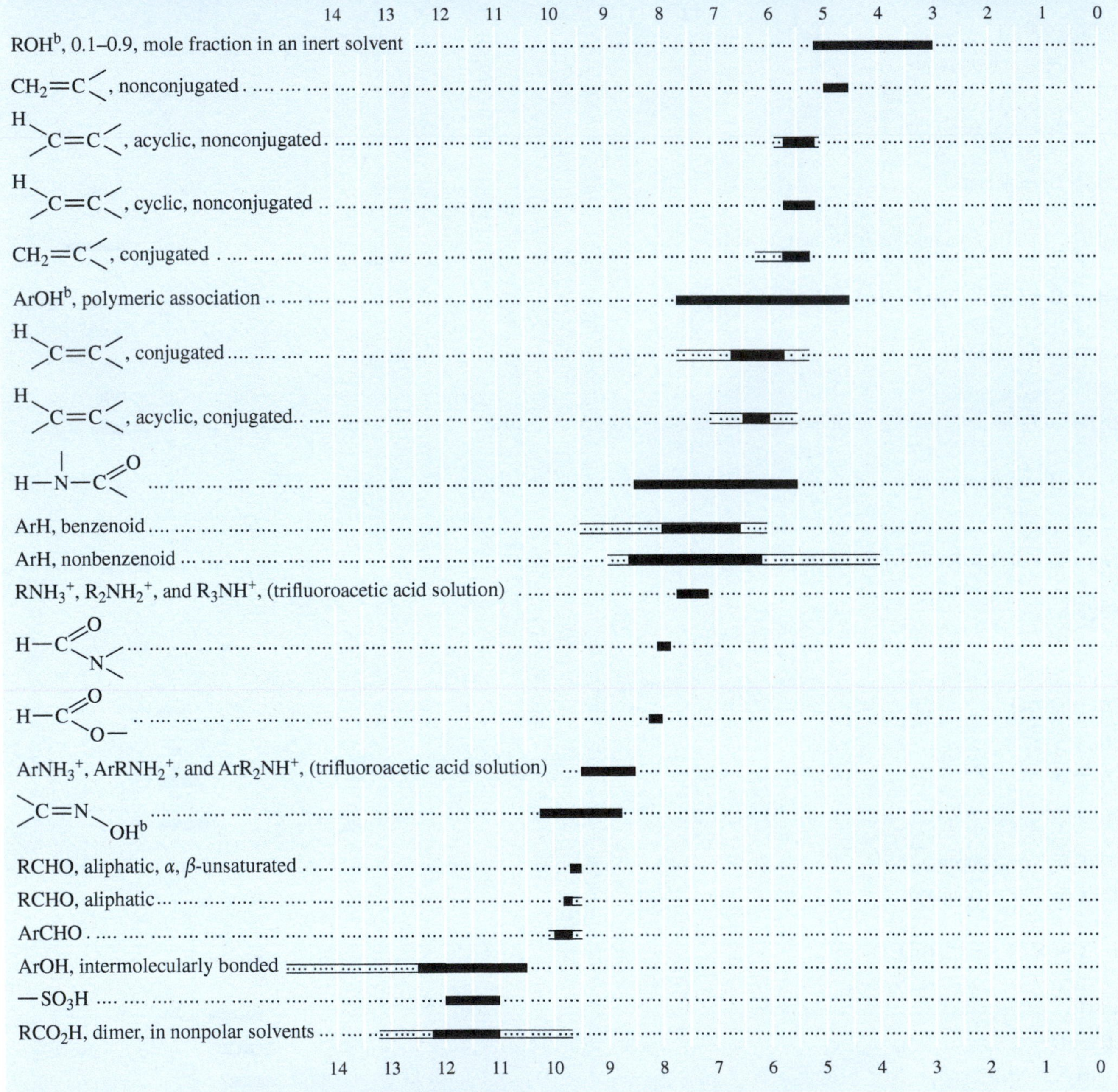

14 13 12 11 10 9 8 7 6 5 4 3 2 1 0

[a] Normally, absorptions for the functional groups indicated will be found within the range shown in black. Occasionally, a functional group will absorb outside this range. Approximate limits are indicated by extended outlines.

[b] Absorption positions of these groups are concentration-dependent and are shifted to lower δ values in more dilute solutions.

APPENDIX 1C
NMR: ^{13}C Chemical Shifts in Organic Compounds*

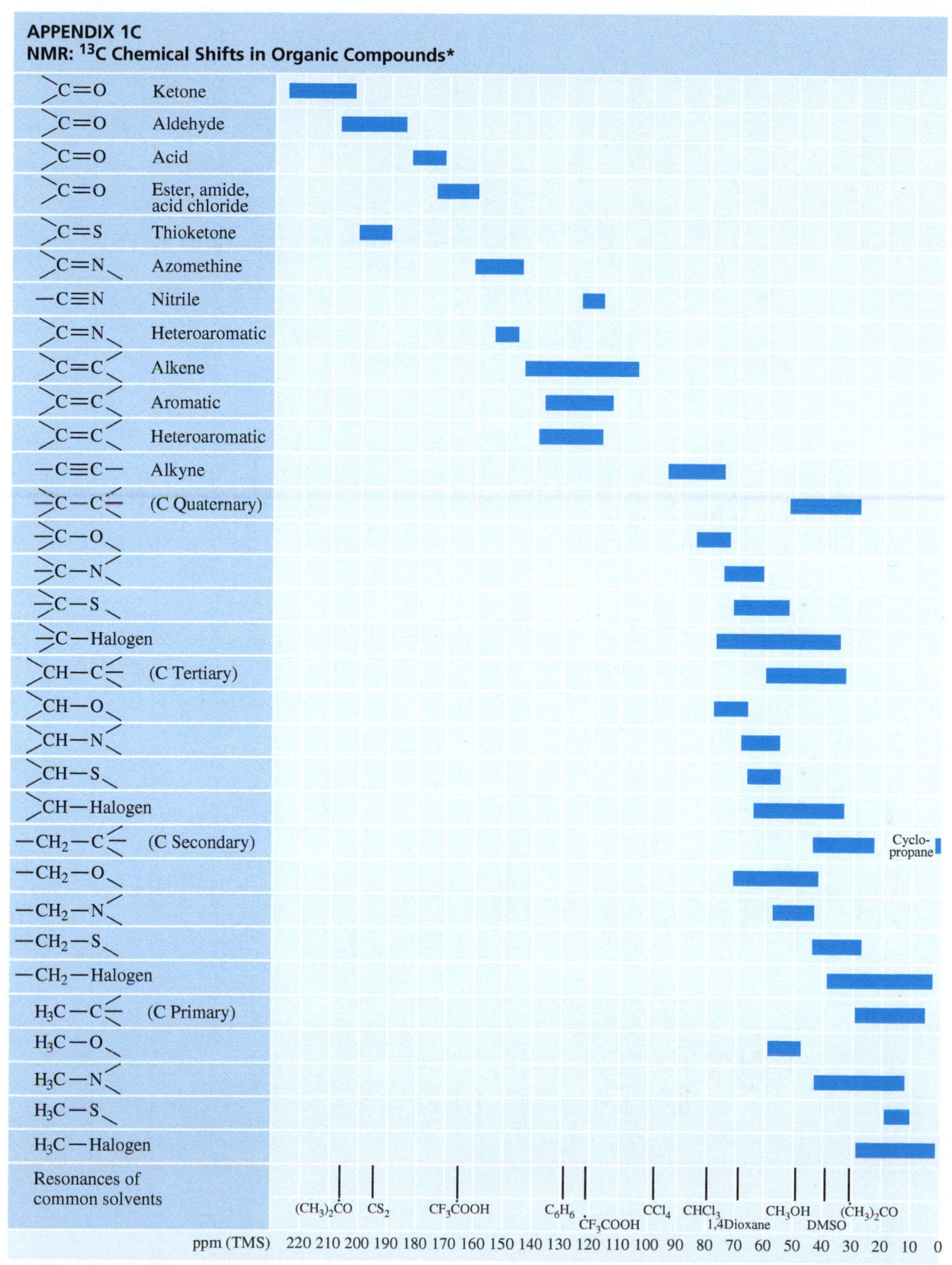

*Relative to internal tetramethylsilane.

APPENDIX 2A

IR: Characteristic Infrared Group Frequencies (s = strong, m = medium, w = weak; overtone bands are marked 2ν)

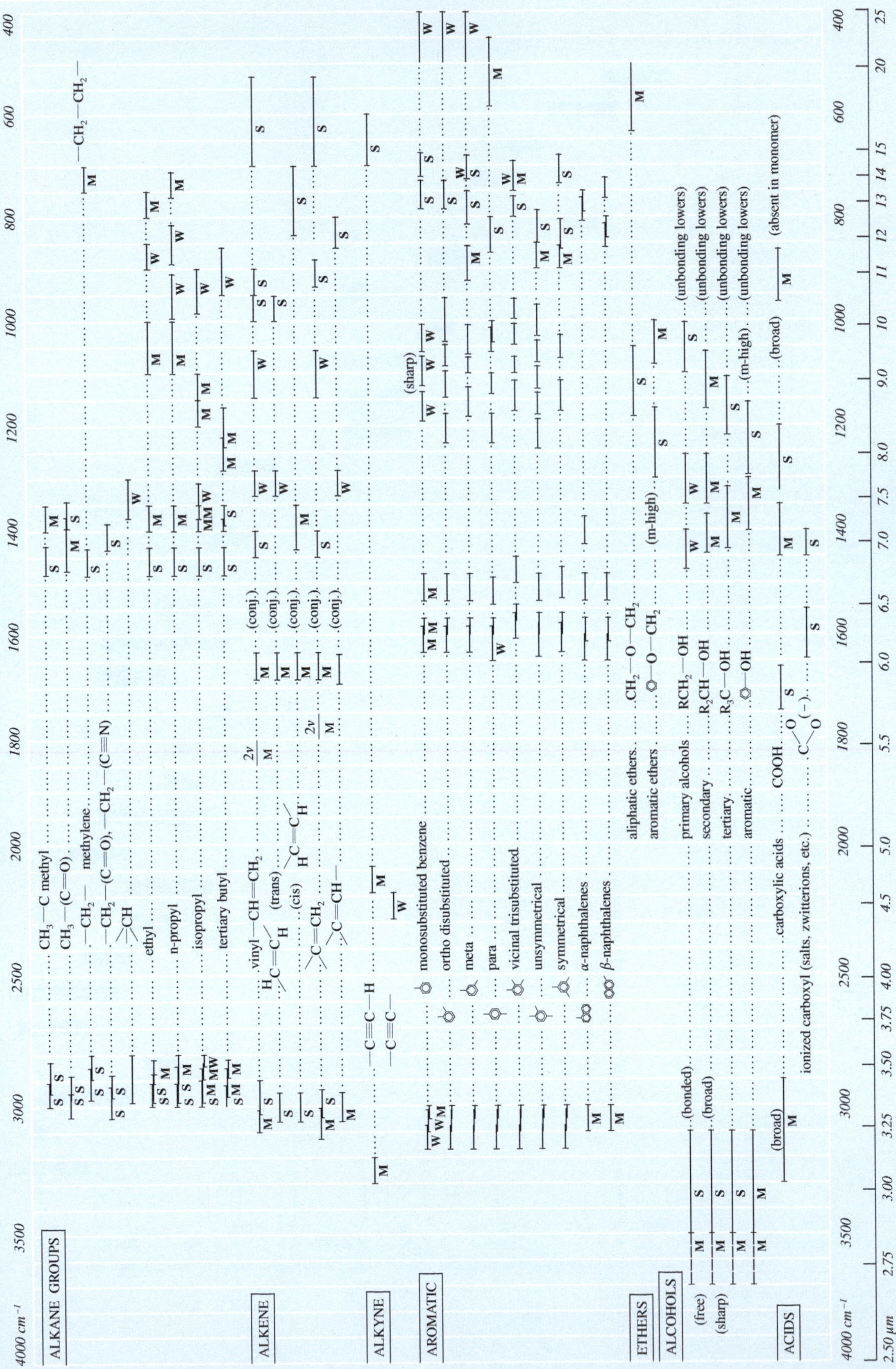

4000 cm^{-1} 3500 3000 2500 2000 1800 1600 1400 1200 1000 800 600 400

ALKANE GROUPS

- CH_3—C methyl
- CH_3—(C=O)
- —CH_2— methylene
- —CH_2—(C=O), —CH_2—(C≡N)
- >CH
- ethyl
- n-propyl
- isopropyl
- tertiary butyl
- —CH_2—CH_2—

ALKENE

- vinyl —CH=CH_2 (conj.)
- HC=CH (trans) (conj.)
- HC=CH (cis) (conj.)
- >C=CH_2 (conj.)
- >C=CH— (conj.)

ALKYNE

- —C≡C—H
- —C≡C—

AROMATIC

- monosubstituted benzene (sharp)
- ortho disubstituted
- meta
- para
- vicinal trisubstituted
- unsymmetrical
- symmetrical
- α-naphthalenes
- β-naphthalenes

ETHERS

- aliphatic ethers CH_2—O—CH_2
- aromatic ethers ⌬—O—CH_2 (m-high)

ALCOHOLS

- (free) (sharp) (bonded) (broad)
- primary alcohols RCH_2—OH (unbonding lowers)
- secondary R_2CH—OH (unbonding lowers)
- tertiary R_3C—OH (unbonding lowers)
- aromatic ⌬—OH (m-high) (unbonding lowers)

ACIDS

- carboxylic acids COOH (broad) (broad) (absent in monomer)
- ionized carboxyl (salts, zwitterions, etc.) $C\langle{}^{O}_{O}(-)$

4000 cm^{-1} 3500 3000 2500 2000 1800 1600 1400 1200 1000 800 600 400

2.50 μm 2.75 3.00 3.25 3.50 3.75 4.00 4.5 5.0 5.5 6.0 6.5 7.0 7.5 8.0 9.0 10 11 12 13 14 15 20 25

Courtesy of N. B. Colthup, Stanford Research Laboratories, American Cyanamid Company, and the editor of the *Journal of the Optical Society.*

APPENDIX 2A

IR: Characteristic Infrared Group Frequencies (s = strong, m = medium, w = weak; overtone bands are marked 2ν)

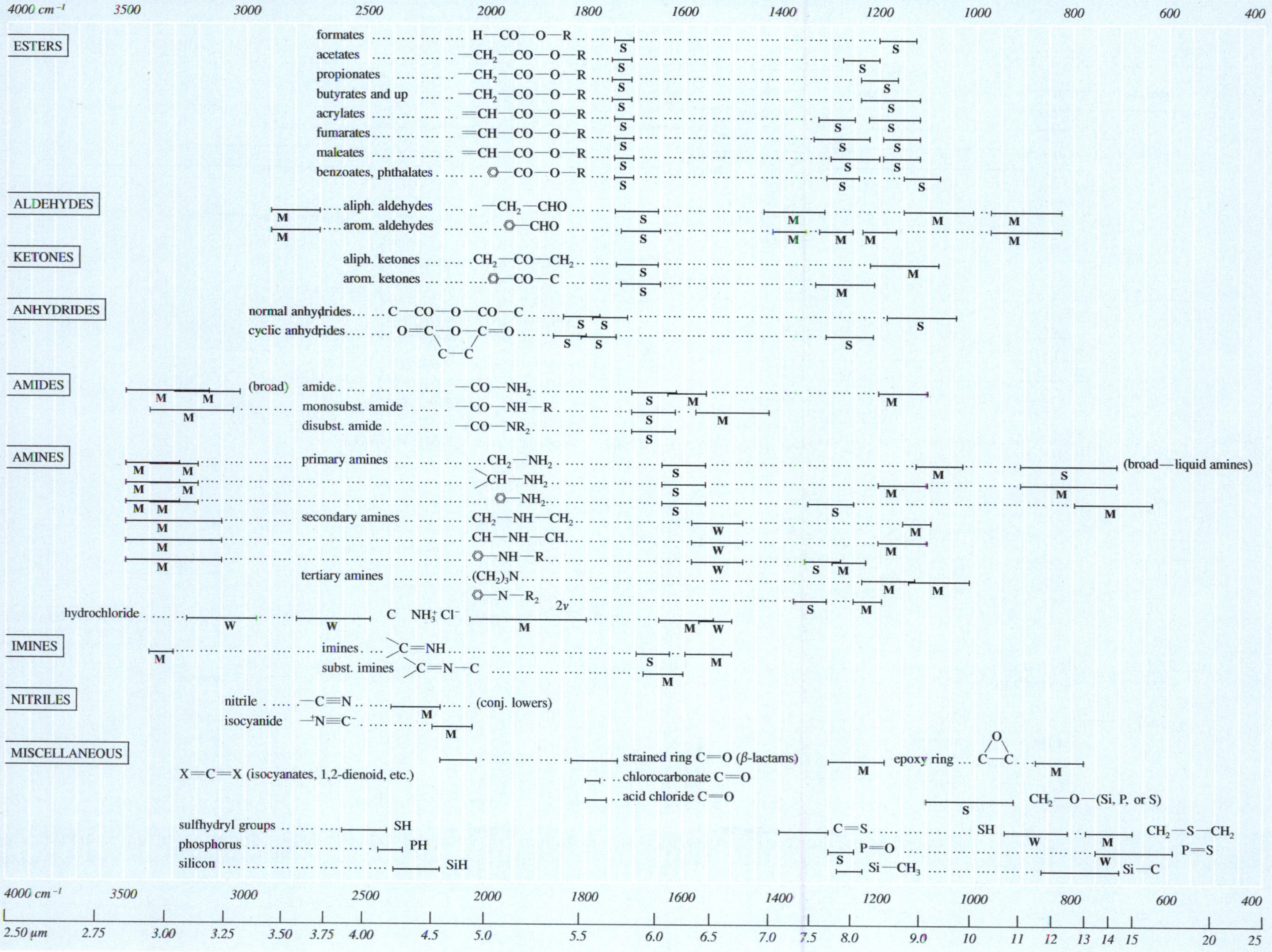

4000 cm⁻¹ 3500 3000 2500 2000 1800 1600 1400 1200 1000 800 600 400

ESTERS

- formates …. H—CO—O—R …. S …. S
- acetates …. $-CH_2-CO-O-R$ …. S …. S
- propionates …. $-CH_2-CO-O-R$ …. S …. S
- butyrates and up …. $-CH_2-CO-O-R$ …. S …. S
- acrylates …. $=CH-CO-O-R$ …. S …. S
- fumarates …. $=CH-CO-O-R$ …. S …. S S
- maleates …. $=CH-CO-O-R$ …. S …. S S
- benzoates, phthalates …. ⌬—CO—O—R …. S …. S S S

ALDEHYDES

- M …. aliph. aldehydes …. $-CH_2-CHO$ …. S …. M …. M …. M
- M …. arom. aldehydes …. ⌬—CHO …. S …. M M M …. M

KETONES

- aliph. ketones …. $CH_2-CO-CH_2$ …. S …. M
- arom. ketones …. ⌬—CO—C …. S …. M

ANHYDRIDES

- normal anhydrides …. C—CO—O—CO—C …. S S …. S
- cyclic anhydrides …. O=C—O—C=O (ring C—C) …. S S …. S

AMIDES

- M M (broad) amide …. $-CO-NH_2$ …. S M …. M
- M …. monosubst. amide …. —CO—NH—R …. S …. M
- disubst. amide …. $-CO-NR_2$ …. S

AMINES

- M M …. primary amines …. CH_2-NH_2 …. S …. M …. S (broad—liquid amines)
- M M …. $\rangle CH-NH_2$ …. S …. M …. M
- M M …. ⌬—NH_2 …. S …. S …. M
- M …. secondary amines …. $CH_2-NH-CH_2$ …. W …. M
- M …. CH—NH—CH …. W …. M
- M …. ⌬—NH—R …. W …. S M
- tertiary amines …. $(CH_2)_3N$ …. M M
- ⌬—N—R_2 …. 2ν …. S …. M
- hydrochloride …. W …. W C $NH_3^+Cl^-$ …. M …. M W

IMINES

- M …. imines …. $\rangle C=NH$ …. S …. M
- subst. imines …. $\rangle C=N-C$ …. M

NITRILES

- nitrile …. —C≡N …. M …. (conj. lowers)
- isocyanide …. $-^{+}N\equiv C^{-}$ …. M

MISCELLANEOUS

- X=C=X (isocyanates, 1,2-dienoid, etc.) …. strained ring C=O (β-lactams) …. M …. epoxy ring …. C—C (with O) …. M
- …. chlorocarbonate C=O
- …. acid chloride C=O …. S …. CH_2-O-(Si, P, or S)
- sulfhydryl groups …. SH …. C=S …. SH …. W …. M …. CH_2-S-CH_2
- phosphorus …. PH …. S …. P=O …. W …. P=S
- silicon …. SiH …. Si—CH_3 …. Si—C

4000 cm⁻¹ 3500 3000 2500 2000 1800 1600 1400 1200 1000 800 600 400

2.50 μm 2.75 3.00 3.25 3.50 3.75 4.00 4.5 5.0 5.5 6.0 6.5 7.0 7.5 8.0 9.0 10 11 12 13 14 15 20 25

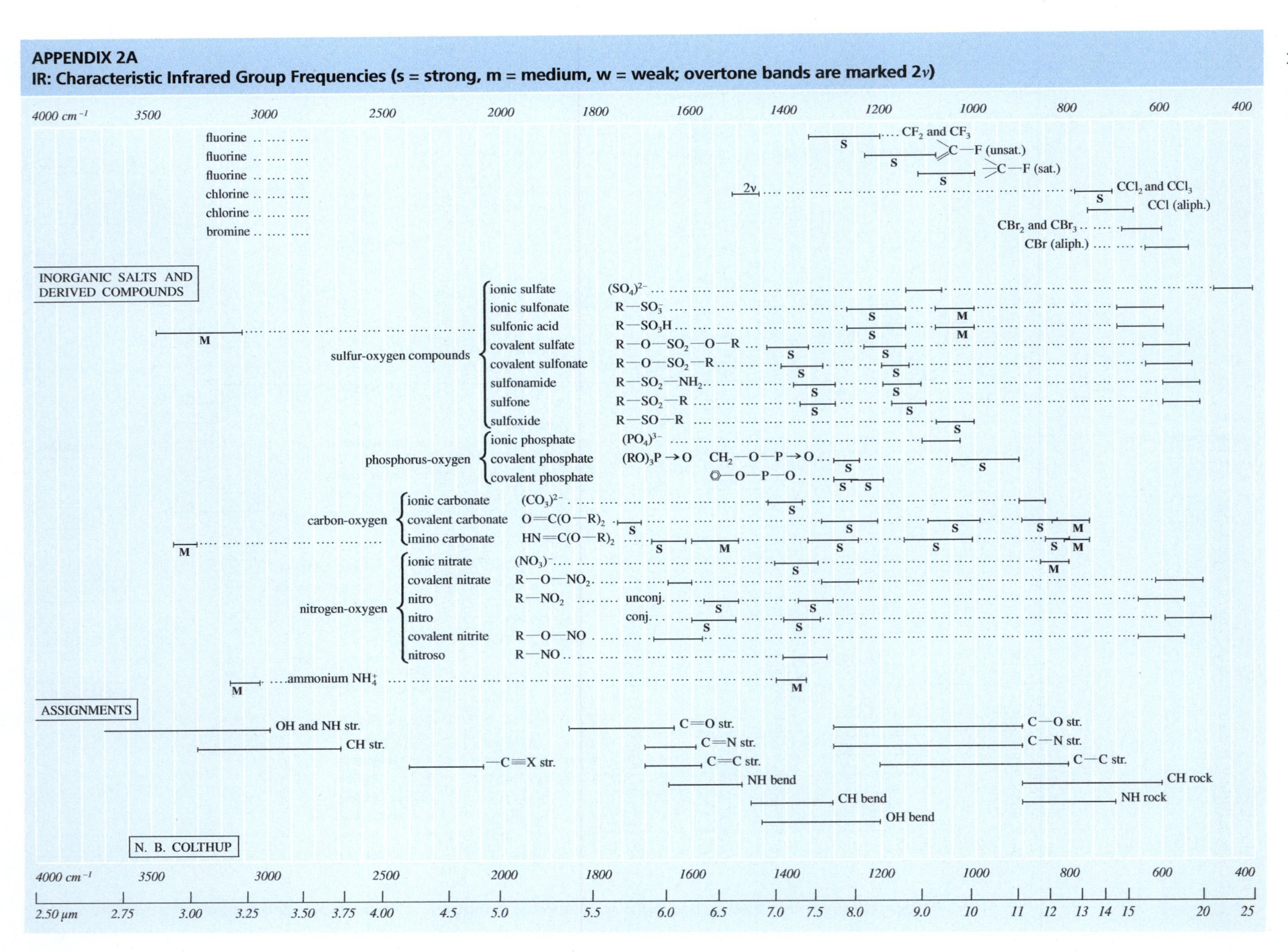
APPENDIX 2A
IR: Characteristic Infrared Group Frequencies (s = strong, m = medium, w = weak; overtone bands are marked 2ν)
4000 cm⁻¹
3500
3000
2500
2000
1800
1600
1400
1200
1000
800
600
400
fluorine
fluorine
fluorine
chlorine
chlorine
bromine
CF₂ and CF₃
C—F (unsat.)
C—F (sat.)
2ν
CCl₂ and CCl₃
CCl (aliph.)
CBr₂ and CBr₃
CBr (aliph.)
INORGANIC SALTS AND DERIVED COMPOUNDS
sulfur-oxygen compounds
ionic sulfate
ionic sulfonate
sulfonic acid
covalent sulfate
covalent sulfonate
sulfonamide
sulfone
sulfoxide
(SO₄)²⁻
R—SO₃⁻
R—SO₃H
R—O—SO₂—O—R
R—O—SO₂—R
R—SO₂—NH₂
R—SO₂—R
R—SO—R
phosphorus-oxygen
ionic phosphate
covalent phosphate
covalent phosphate
(PO₄)³⁻
(RO)₃P→O
CH₂—O—P→O
—O—P—O
carbon-oxygen
ionic carbonate
covalent carbonate
imino carbonate
(CO₃)²⁻
O=C(O—R)₂
HN=C(O—R)₂
nitrogen-oxygen
ionic nitrate
covalent nitrate
nitro
nitro
covalent nitrite
nitroso
(NO₃)⁻
R—O—NO₂
R—NO₂
unconj.
conj.
R—O—NO
R—NO
ammonium NH₄⁺
ASSIGNMENTS
OH and NH str.
CH str.
—C≡X str.
C=O str.
C=N str.
C=C str.
NH bend
CH bend
OH bend
C—O str.
C—N str.
C—C str.
CH rock
NH rock
N. B. COLTHUP
2.50 μm
2.75
3.00
3.25
3.50
3.75
4.00
4.5
5.0
5.5
6.0
6.5
7.0
7.5
8.0
9.0
10
11
12
13
14
15
20
25

APPENDIX 2B
IR: Characteristic Infrared Absorptions of Functional Groups

Group	Intensity[a]	Range (cm^{-1})
A. Hydrocarbon chromophore		
1. C—H stretching		
a. Alkane	m–s	2962–2853
b. Alkene, monosubstituted	m	3040–3010
(vinyl)	and m	3095–3075
Alkene, disubstituted, cis	m	3040–3010
Alkene, disubstituted, trans	m	3040–3010
Alkene, disubstituted, gem	m	3095–3075
Alkene, trisubstituted	m	3040–3010
c. Alkyne	s	~3300
d. Aromatic	v	~3030
2. C—H bending		
a. Alkane, C—H	w	~1340
Alkane, $—CH_2—$	m	1485–1445
Alkane, $—CH_3$	m	1470–1430
	and s	1380–1370
Alkane, *gem*-dimethyl	s	1385–1380
	and s	1370–1365
Alkane, *tert*-butyl	m	1395–1385
	and s	~1365
b. Alkene, monosubstituted (vinyl)	s	995–985
	s	915–905
	and s	1420–1410
Alkene, disubstituted, cis	s	~690
Alkene, disubstituted, trans	s	970–960
	and m	1310–1295
Alkene, disubstituted, gem	s	895-885
	and s	1420–1410
Alkene, trisubstituted	s	840–790
c. Alkyne	s	~630
d. Aromatic, substitution type:[b]		
Five adjacent	v, s	~750
hydrogen atoms	and v, s	~700
Four adjacent hydrogen atoms	v, s	~750
Three adjacent hydrogen atoms	v, m	~780
Two adjacent hydrogen atoms	v, m	~830
One hydrogen atom	v, m	~880
3. C—C multiple bond stretching		
a. Alkene, nonconjugated	v	1680–1620
Alkene, monosubstituted (vinyl)	m	~1645
Alkene, disubstituted, cis	m	~1658
Alkene, disubstituted, trans	m	~1675
Alkene, disubstituted, gem	m	~1653
Alkene, trisubstituted	m	~1669
Alkene, tetrasubstituted	w	~1669
Diene	w	~1650
	and w	~1600
b. Alkyne, monosubstituted	m	2140–2100
Alkyne, disubstituted	v, w	2260–2190
c. Allene	m	~1960
	and m	~1060
d. Aromatic	v	~1600
	v	~1580
	m	~1500
	and m	~1450

Group	Intensity[a]	Range (cm^{-1})
B. Carbonyl chromophore		
1. Ketone stretching vibrations		
a. Saturated, acyclic	s	1725–1705
b. Saturated, cyclic:		
6-Membered ring (and higher)	s	1725–1705
5-Membered ring	s	1750–1740
4-Membered ring	s	~1775
c. α, β-Unsaturated, acyclic	s	1685–1665
d. α, β-Unsaturated, cyclic:		
6-Membered ring (and higher)	s	1685–1665
5-Membered ring	s	1725–1708
e. $\alpha, \beta, \alpha', \beta'$-Unsaturated, acyclic	s	1670–1663
f. Aryl	s	1700–1680
g. Diaryl	s	1670–1660
h. β-Diketones	s	1730–1710
i. β-Diketones (enolic)	s	1640–1540
j. 1,4-Quinones	s	1690–1660
k. Ketenes	s	~2150
2. Aldehyes		
a. Carbonyl stretching vibrations:		
Saturated, aliphatic	s	1740–1720
α, β-Unsaturated, aliphatic	s	1705–1680
$\alpha, \beta, \gamma, \delta$-Unsaturated, aliphatic	s	1680–1660
Aryl	s	1715–1695
b. C—H stretching vibrations,		
two bands	w	2900–2820
	and w	2775–2700
3. Ester stretching vibrations		
a. Saturated, acyclic	s	1750–1735
b. Saturated, cyclic:		
δ-Lactones (and larger rings)	s	1750–1735
γ-Lactones	s	1780–1760
β-Lactones	s	~1820
c. Unsaturated:		
vinyl ester type	s	1800–1770
α, β-Unsaturated and aryl	s	1730–1717
α, β-Unsaturated δ-lactone	s	1730-1717
α, β-Unsaturated, γ-lactone	s	1760–1740
α, γ-Unsaturated, γ-lactone	s	~1800
d. α-Ketoesters	s	1755–1740
e. β-Ketoesters (enolic)	s	~1650
f. Carbonates	s	1780–1740
g. Thioesters	s	~1690
4. Carboxylic acids		
a. Carbonyl stretching vibrations:		
Saturated aliphatic	s	1725–1700
α, β-Unsaturated aliphatic	s	1715–1690
Aryl	s	1700–1680
b. Hydroxyl stretching (bonded),		
several bands	w	2700–2500
c. Carboxylate anion	s	1610–1550
stretching	and s	1400–1300
5. Anhydride stretching vibrations		
a. Saturated, acyclic	s	1850–1800
	and s	1790–1740

(Continued)

APPENDIX 2B
IR: Characteristic Infrared Absorptions of Functional Groups *(continued)*

Group	Intensity[a]	Range (cm^{-1})
b. α, β-Unsaturated and aryl,	s	1830–1780
acyclic anhydrides	and s	1770–1720
c. Saturated, 5-membered	s	1870–1820
ring anhydrides	and s	1800–1750
d. α, β-Unsaturated,	s	1850–1800
5-membered ring	and s	1830–1780
6. Acyl halide stretching vibrations		
a. Acyl fluorides	s	~1850
b. Acyl chlorides	s	~1795
c. Acyl bromides	s	~1810
d. α, β-Unsaturated and aryl	s	1780–1750
	and m	1750–1720
7. Amides		
a. Carbonyl stretching vibrations:		
Primary, solid and concentrated solution	s	~1650
Primary, dilute solution	s	~1690
Secondary, solid and concentrated solution	s	1680–1630
Secondary, dilute solution	s	1700–1670
Tertiary, solid and all solutions	s	1670–1630
Cyclic, δ-lactams	s	~1680
Cyclic, γ-lactams	s	~1700
Cyclic, γ-lactams, fused to another ring	s	1750–1700
Cyclic, β-lactams	s	1760–1730
Cyclic, β-lactams, fused to another ring, dilute solution	s	1780–1770
Ureas, acyclic	s	~1600
Ureas, cyclic, 6-Membered ring	s	~1640
Ureas, cyclic, 5-Membered ring	s	~1720
Urethanes	s	1740–1690
Imides, acyclic	s	~1710
	and s	~1700
Imides, cyclic,	s	~1710
6-membered ring	and s	1700
Imides, cyclic, α, β-unsaturated,	s	~1730
6-membered ring	and s	~1670
Imides, cyclic, 5-membered	s	~1770
ring	and s	~1700
Imides, cyclic, α, β-unsaturated,	s	~1790
5-membered ring	and s	~1710
b. N—H stretching vibrations:		
Primary, free; two bands	m	~3500
	and m	~3400
Primary, bonded:	m	~3350
two bands	and m	~3180
Secondary, free; one band	m	~3430
Secondary, bonded; one band	m	3320–3140
c. N—H bending vibrations:		
Primary amides, dilute solution	s	1620–1590
Secondary amides	s	1550–1510

Group	Intensity[a]	Range (cm^{-1})
C. Miscellaneous chromophoric groups		
1. Alcohols and phenols		
a. O—H stretching vibrations:		
Free O—H	v, sh	3650–3590
Intermolecuarly hydrogen bonded O—H change or dilution		
Single bridge compounds	v, sh	3550–3450
Polymeric association	s, b	3400–3200
Intramolecularly hydrogen bonded (no change on dilution)		
Single-bridge compounds	v, sh	3570–3450
Chelate compounds	w, b	3200–2500
b. O—H bending and C—O stretching vibrations:		
Primary alcohols	s	~1050
	and s	1350–1260
Secondary alcohols	s	~1100
	and s	1350–1260
Tertiary alcohols	s	~1150
	and s	1410–1310
Phenols	s	~1200
	and s	1410–1310
2. Amines		
a. N—H stretching vibrations:		
Primary, free; two bands	m	~3500
	and m	~3400
Secondary, free; one band	m	3500–3310
Imines (=N—H); one band	m	3400–3300
Amine salts	m	3130–3030
b. N—H bending vibrations:		
Primary	s-m	1650–1590
Secondary	w	1650–1550
Amine salts	s	1600–1575
	and s	~1500
c. C—N vibrations:		
Aromatic, primary	s	1340–1250
Aromatic, secondary	s	1350–1280
Aromatic, tertiary	s	1360–1310
Aliphiatic	w	1220–1020
	and w	~1410
3. Unsaturated nitrogen compounds		
a. C≡N stretching vibrations:		
Alkyl nitriles	m	2260–2240
α, β-Unsaturated alkyl nitriles	m	2235–2215
Aryl nitriles	m	2240–2220
Isocyanates	m	2275–2240
Isocyanides	m	2220–2070
b. >C=N— stretching vibrations (imines, oximes)		
stretching vibrations (imines, oximes)		
Alkyl compounds	v	1690–1640
α, β-Unsaturated compounds	v	1660–1630
c. —N=N— stretching vibrations, azo compounds	v	1630–1575
d. —N=C=N— stretching vibrations, diimide	s	2155–2130
e. —N_3 stretching vibrations,	s	2160–2120
azides	and w	1340–1180

(Continued)

APPENDIX 2B
IR: Characteristic Infrared Absorptions of Functional Groups *(continued)*

Group	Intensity[a]	Range (cm^{-1})
f. C—NO_2 nitro compounds:		
Aromatic	s	1570–1500
(Aromatic intro compounds)	and s	1370–1300
Aliphatic	s	1570–1550
	and s	1380–1370
g. O—NO_2, nitrates	s	1650–1600
	and s	1300–1250
h. C—NO, nitroso compound	s	1600–1500
i. O—NO, nitrites	s	1680–1650
	and s	1625–1610
4. Halogen compounds, C—X stretching vibrations		
a. C—F	s	1400–1000
b. C—Cl	s	800–600
c. C—Br	s	600–500
d. C—I	s	~500
5. Sulfur compounds		
a. S—H stretching vibrations	w	2600–2550
b. C═S stretching vibrations	s	1200–1050
c. S═O stretching vibrations		
Sulfoxides	s	1070–1030
Sulfones	s	1160–1140
	and s	1350–1300
Sulfites	s	1230–1150
	and s	1430–1350
Sulfonyl chlorides	s	1185–1165
	and s	1370–1340
Sufonamides	s	1180–1140
	and s	1350–1300
Sulfonic acids	s	1210–1150
	s	1060–1030
	and s	~650
Thioesters (C═O)S	s	~1690

[a] Abbreviations: s = strong, m = medium, w = weak, v = variable, b = broad, sh = sharp, ~ = approximately
[b] Substituted benzenes also show weak bands in the 2000–1670 cm^{-1} region

APPENDIX 3A
Methods and Suggestions for Proposing Mechanisms

In this appendix, we consider how an organic chemist systematically approaches a mechanism problem. Although there is no "formula" for solving all mechanism problems, this stepwise method should provide a starting point for you to begin building experience and confidence. Solved problems that apply this approach appear on pages 183, 346, 531, 915, 1070, 1134, and 1159.

Determining the Type of Mechanism

First, determine what conditions or catalysts are involved. In general, reactions may be classified as (a) involving strong electrophiles (includes acid-catalyzed reactions), (b) involving strong nucleophiles (includes base-catalyzed reactions), or (c) involving free radicals. These three types of mechanisms are quite distinct, and you should first try to determine which type is involved. If uncertain, you can develop more than one type of mechanism and see which one fits the facts better.

(a) In the presence of a strong acid or a reactant that can give a strong electrophile, the mechanism probably involves strong electrophiles as intermediates. Acid-catalyzed reactions and reactions involving carbocations (such as the S_N1, E1, and most alcohol dehydrations) generally fall in this category.

(b) In the presence of a strong base or strong nucleophile, the mechanism probably involves strong nucleophiles as intermediates. Base-catalyzed reactions and those whose rates depend on base strength (such as S_N2 and E2) generally fall in this category.

(c) Free-radical reactions usually require a free-radical initiator such as chlorine, bromine, NBS, AIBN, or a peroxide. In most free-radical reactions, there is no need for a strong acid or base.

Points to Watch in All Mechanisms

Once you have determined which type of mechanism is likely, some general principles can guide you in proposing a mechanism. Regardless of the type of mechanism, however, you should follow three general rules in proposing a mechanism:

1. **Draw all bonds and all substituents of each carbon atom affected throughout the mechanism. Do not use condensed or line–angle formulas for reaction sites.** Three-bonded carbon atoms are most likely reactive intermediates: carbocations in reactions involving strong electrophiles, carbanions in reactions involving strong nucleophiles, and free radicals in radical reactions. If you draw condensed formulas or line–angle formulas, you might misplace a hydrogen atom and show a reactive species on the wrong carbon.

2. **Show only one step at a time. Do not show two or three bonds changing position in one step unless the changes really are concerted (take place simultaneously).** For example, three pairs of electrons really do move in one step in the Diels–Alder reaction; but in the dehydration of an alcohol, protonation of the hydroxyl group and loss of water are two separate steps.
3. **Use curved arrows to show *movement of electrons*, always from the nucleophile (electron donor) to the electrophile (electron acceptor).** For example, a proton has no electrons to donate, so a curved arrow should never be drawn from H^+ to anything. When an alkene is protonated, the arrow should go from the electrons of the double bond to the proton. Don't try to use curved arrows to "point out" where the proton (or other reagent) goes. In a free-radical reaction, half-headed arrows show single electrons coming together to form bonds or separating to give other radicals.

Approaches to Specific Types of Mechanisms

Reactions Involving Strong Electrophiles General principles: When a strong acid or electrophile is present, expect intermediates that are strong acids and strong electro-philes. Cationic intermediates are common, but avoid drawing any ion with more than one + charge. Carbocations, protonated (three-bonded) oxygen atoms, protonated (four-bonded) nitrogen atoms, and other strong acids might be involved. Any bases and nucleophiles in such a reaction are generally weak. Avoid drawing carbanions, hydroxide ions, alkoxide ions, and other strong bases. They are unlikely to coexist with strong acids and strong electrophiles.

Functional groups are often converted to carbocations or other strong electrophiles by protonation or by reaction with a strong electrophile; then the carbocation or other strong electrophile reacts with a weak nucleophile such as an alkene or the solvent.

1. Consider the carbon skeletons of the reactants and products, and identify which carbon atoms in the products are most likely derived from which carbon atoms in the reactants.
2. Consider whether any of the reactants is a sufficiently strong electrophile to react without being activated. If not, consider how one of the reactants might be converted to a strong electrophile by protonation of a basic site, complexation with a Lewis acid, or ionization.
3. Consider how a nucleophilic site on another reactant (or, in a cyclization, another part of the same molecule) can attack this strong electrophile to form a bond needed in the product. Draw the product of this bond formation.
   - If the intermediate is a carbocation, consider whether it is likely to rearrange to form a bond in the product.
   - If there is no possible nucleophilic attack that leads in the direction of the product, consider other ways of converting one of the reactants to a strong electrophile.
4. Consider how the product of nucleophilic attack might be converted to the final product (if it has the right carbon skeleton) or reactivated to form another bond needed in the product.
5. Draw out all the steps, using curved arrows to show movement of electrons. Be careful to show only one step at a time.

Reactions Involving Strong Nucleophiles General principles: When a strong base or nucleophile is present, expect intermediates that are strong bases and strong nucleophiles. Anionic intermediates are common, but avoid drawing any ions with more than one negative charge. Alkoxide ions, hydroxide ions, stabilized carbanions, and other strong bases might be involved. Any acids and electrophiles in such a reaction are generally weak. Avoid drawing carbocations, free H^+, protonated carbonyl groups, protonated hydroxyl groups, and other strong acids. They are unlikely to coexist with strong bases and strong nucleophiles.

Functional groups are often converted to strong nucleophiles by deprotonation of the group itself; by deprotonation of the alpha position of a carbonyl group, nitro group, or nitrile; or by attack of another strong nucleophile. Then the resulting carbanion or other nucleophile reacts with a weak electrophile such as a carbonyl group, an alkyl halide, or the double bond of a Michael acceptor.

1. Consider the carbon skeletons of the reactants and products, and identify which carbon atoms in the products are most likely derived from which carbon atoms in the reactants.
2. Consider whether any of the reactants is a sufficiently strong nucleophile to react without being activated. If not, consider how one of the reactants might be converted to a strong nucleophile by deprotonation of an acidic site or by attack on an electrophilic site.
3. Consider how an electrophilic site on another reactant (or, in a cyclization, another part of the same molecule) can undergo attack by the strong nucleophile to form a bond needed in the product. Draw the product of this bond formation.
   - If no appropriate electrophilic site can be found, consider another way of converting one of the reactants to a strong nucleophile.

4. Consider how the product of nucleophilic attack might be converted to the final product (if it has the right carbon skeleton) or reactivated to form another bond needed in the product.
5. Draw out all the steps, using curved arrows to show movement of electrons. Be careful to show only one step at a time.

Reactions Involving Free Radicals General principles: Free-radical reactions generally proceed by chain-reaction mechanisms, using an initiator with an easily broken bond (such as chlorine, bromine, or a peroxide) to start the chain reaction. In drawing the mechanism, expect free-radical intermediates (especially highly substituted or resonance-stabilized intermediates). Cationic intermediates and anionic intermediates are not usually involved. Watch for the most stable free radicals, and avoid high-energy radicals such as hydrogen atoms.

Initiation

1. Draw a step involving homolytic (free-radical) cleavage of the weak bond in the initiator to give two radicals.
2. Draw a reaction of the initiator radical with one of the starting materials to give a free-radical version of the starting material.

   The initiator might abstract a hydrogen atom or add to a double bond, depending on what reaction leads toward the observed product. You might want to consider bond-dissociation energies to see which reaction is energetically favored.

Propagation

1. Draw a reaction of the free-radical version of the starting material with another starting material molecule to form a bond needed in the product and generate a new radical intermediate. Two or more propagation steps may be needed to give the entire chain reaction.

Termination

1. Draw termination steps showing the recombination or destruction of radicals. Termination steps are side reactions rather than part of the product-forming mechanism. Reaction of any two free radicals to give a stable molecule is a termination step, as is a collision of a free radical with the container.

APPENDIX 3B
Suggestions for Developing Multistep Syntheses

In this appendix, we consider how an organic chemist systematically approaches a multistep synthesis problem. As with mechanism problems, there is no reliable formula that can be used to solve all synthesis problems, yet students need guidance in how they should begin.

In a multistep synthesis problem, the solution is rarely immediately apparent. A synthesis is best developed systematically, working backward (in the *retrosynthetic* direction) and considering alternative ways of solving each stage of the synthesis. A strict retrosynthetic approach requires considering all possibilities for the final step, evaluating each reaction, and then evaluating every way of making each of the possible precursors.

This exhaustive approach is very time-consuming. It works well on a large computer, but most organic chemists solve problems more directly by attacking the crux of the problem: steps that build the carbon skeleton. Once the carbon skeleton is assembled (with usable functionality), converting the functional groups to those required in the target molecule is relatively easy.

The following steps suggest a systematic approach to developing a multistep synthesis. These steps should help you organize your thoughts and approach syntheses like many organic chemists do: in a generally retrosynthetic direction, but with primary emphasis on the crucial steps that form the carbon skeleton of the target molecule. Solved problems that apply this approach appear on pages 410, 452, and 543.

1. Review the functional groups and carbon skeleton of the target compound, considering what kinds of reactions might be used to create them.
2. Review the functional groups and carbon skeletons of the starting materials (if specified), and see how their skeletons might fit together into the skeleton of the target compound.
3. Compare methods for assembling the carbon skeleton of the target compound. Which ones produce a key intermediate with the correct carbon skeleton and functional groups correctly positioned for conversion to the functionality in the target molecule?

   Also notice what functional groups are required in the reactants for the skeleton-forming steps and whether they are easily accessible from the specified starting materials.

4. Write down the steps involved in assembling the key intermediate with the correct carbon skeleton.
5. Compare methods for converting the key intermediate's functional groups to those in the target compound, and select reactions that are likely to give the correct product. Reactive functional groups are often added late in a synthesis to prevent them from interfering with earlier steps.
6. Working backward through as many steps as necessary, compare methods for synthesizing the reactants needed for assembly of the key intermediate. (This process may require writing several possible reaction sequences and evaluating them, keeping in mind the specified starting materials.)
7. Summarize the complete synthesis in the forward direction, including all steps and all reagents, and check it for errors and omissions.

APPENDIX 4
pK_a Values for Representative Compounds

Compound	pK_a
$CH_3C\equiv\overset{+}{N}H$	−10.1
HI	−10
HBr	−9
$CH_3CH{=}\overset{+}{O}H$	−8
$CH_3C({=}\overset{+}{O}H)CH_3$	−7.3
HCl	−7
$CH_3\overset{+}{S}H_2$	−6.8
$CH_3C({=}\overset{+}{O}H)OCH_3$	−6.5
$CH_3C({=}\overset{+}{O}H)OH$	−6.1
H_2SO_4	−5
$CH_3CH_2\overset{+}{O}(H)CH_2CH_3$	−3.6
$CH_3CH_2\overset{+}{O}(H)H$	−2.4
$CH_3\overset{+}{O}(H)H$	−2.5
H_3O^+	−1.7
HNO_3	−1.3
CH_3SO_3H	−1.2
$C_6H_5-SO_3H$	−0.60
$CH_3C({=}\overset{+}{O}H)NH_2$	0.0
F_3CCOH (with C=O)	0.2
$C_4H_4\overset{+}{N}H(H)$ (protonated pyrrole)	0.4
Cl_3CCOH (with C=O)	0.64
$C_5H_5\overset{+}{N}-OH$	0.79
$O_2N-C_6H_4-\overset{+}{N}H_3$	1.0
$C_4H_4N_2H^+$ (ring, N and $\overset{+}{N}H$)	1.0
Cl_2CHCOH (with C=O)	1.3
HSO_4^-	2.0
H_3PO_4	2.1
$C_5H_4N_4H^+$ (ring, $H\overset{+}{N}$, N, N, NH)	2.5
FCH_2COH (with C=O)	2.7
$ClCH_2COH$ (with C=O)	2.8
$BrCH_2COH$ (with C=O)	2.9
HF	3.17
ICH_2COH (with C=O)	3.2
HNO_2	3.4
$O_2N-C_6H_4-COH$ (with C=O)	3.4
$HCOH$ (with C=O)	3.76
$Br-C_6H_4-COH$ (with C=O)	3.8
$Br-C_6H_4-\overset{+}{N}H_3$	3.9
C_6H_5-COH (with C=O)	4.2
$CH_3-C_6H_4-COH$ (with C=O)	4.3
$CH_3O-C_6H_4-COH$ (with C=O)	4.5
$C_6H_5-\overset{+}{N}H_3$	4.60
CH_3COH (with C=O)	4.74
$C_9H_7\overset{+}{N}H$ (fused ring, $\overset{+}{N}H$)	4.9
$CH_3-C_6H_4-\overset{+}{N}H_3$	5.1
$C_5H_5\overset{+}{N}H$	5.25
$CH_3O-C_6H_4-\overset{+}{N}H_3$	5.3
$CH_3C(CH_3){=}\overset{+}{N}HCH_3$	5.5
CH_3CCH_2CH (with two C=O)	5.9
$HO\overset{+}{N}H_3$	6.0
H_2CO_3	6.4
$C_3H_3(HN)(\overset{+}{N}H)$ (ring, HN and NH)	6.95
H_2S	7.0
$O_2N-C_6H_4-OH$	7.1
$H_2PO_4^-$	7.2
C_6H_5-SH	7.8

APPENDIX 4
pK_a Values for Representative Compounds *(continued)*

Compound	pK_a
$(CH_2)_2\overset{+}{N}H_2$ (aziridinium ring)	8.0
$H_2N\overset{+}{N}H_3$	8.1
$CH_3\overset{O}{\overset{\|}{C}}OOH$	8.2
$O(CH_2CH_2)_2\overset{+}{N}H_2$ (morpholinium ring)	8.36
$CH_3CH_2NO_2$	8.6
$CH_3\overset{O}{\overset{\|}{C}}CH_2\overset{O}{\overset{\|}{C}}CH_3$	8.9
$C_5H_4N_4$ (purine, N–H)	8.9
$HC{\equiv}N$	9.22
$\overset{+}{N}H_4$	9.26
$Cl-C_6H_4-OH$	9.4
$HOCH_2CH_2\overset{+}{N}H_3$	9.5
$(CH_3)_3\overset{+}{N}H$	9.74
$H_3\overset{+}{N}CH_2\overset{O}{\overset{\|}{C}}O^-$	9.8
C_6H_5-OH	10.0
$CH_3-C_6H_4-OH$	10.2
HCO_3^-	10.2
CH_3NO_2	10.2
$H_2N-C_6H_4-OH$	10.3
CH_3CH_2SH	10.5
$CH_3CH_2\overset{+}{N}H_3$	10.64
$CH_3\overset{+}{N}H_3$	10.64
$C_6H_{11}-\overset{+}{N}H_3$	10.67
$CH_3\overset{O}{\overset{\|}{C}}CH_2\overset{O}{\overset{\|}{C}}OEt$	10.7
$(CH_3)_2\overset{+}{N}H_2$	10.72
$(CH_2)_5\overset{+}{N}H_2$ (piperidinium ring)	11.12
$(CH_2)_4\overset{+}{N}H_2$ (pyrrolidinium ring)	11.27
HPO_4^{2-}	12.3
CF_3CH_2OH	12.4
$EtO\overset{O}{\overset{\|}{C}}CH_2\overset{O}{\overset{\|}{C}}OEt$	13.3
$HC{\equiv}CCH_2OH$	13.5
$H_2N\overset{O}{\overset{\|}{C}}NH_2$	13.7
$CH_3\overset{+}{N}(CH_3)_2CH_2CH_2OH$	13.9
$C_3H_4N_2$ (imidazole ring, N / NH)	14.4
CH_3OH	15.5
H_2O	15.7
CH_3CH_2OH	15.9
$CH_3\overset{O}{\overset{\|}{C}}NH_2$	16
$C_6H_5-\overset{O}{\overset{\|}{C}}CH_3$	16.0
C_4H_4NH (pyrrole ring, N–H)	~17
$CH_3\overset{O}{\overset{\|}{C}}H$	17
$(CH_3)_3COH$	18
$CH_3\overset{O}{\overset{\|}{C}}CH_3$	20
$CH_3\overset{O}{\overset{\|}{C}}OEt$	24.5
$HC{\equiv}CH$	25
$CH_3C{\equiv}N$	25
$CH_3\overset{O}{\overset{\|}{C}}N(CH_3)_2$	30
NH_3	36
$(CH_2)_4NH$ (pyrrolidine ring, N–H)	36
CH_3NH_2	40
$[(CH_3)_2CH]_2NH$	40
$C_6H_5-CH_3$	41
C_6H_6	43
$CH_2{=}CHCH_3$	43
$CH_2{=}CH_2$	44
C_3H_6 (cyclopropane ring)	46
CH_4	50
CH_3CH_3	50
$CH_3CH_2CH_2CH_3$	50

APPENDIX 5

Summary of Organic Nomenclature

An earlier version of this document was created in the 1970s by various faculty members at California Polytechnic State University. This version has been revised and expanded by Jan William Simek.

Introduction

The purpose of the IUPAC system of nomenclature is to establish an international standard of naming compounds to facilitate communication. The goal of the system is to give each structure a unique and unambiguous name, and to correlate each name with a unique and unambiguous structure.

I. Fundamental Principle

IUPAC nomenclature is based on naming a molecule's longest chain of carbons connected by single bonds, whether in a continuous chain or in a ring. All deviations—either multiple bonds or atoms other than carbon and hydrogen—are indicated by prefixes or suffixes according to a specific set of priorities.

II. Alkanes and Cycloalkanes (also called "aliphatic" compounds)

Alkanes are the family of saturated hydrocarbons—that is, molecules containing carbon and hydrogen connected by single bonds only. These molecules can be in continuous chains (called linear or acyclic) or in rings (called cyclic or alicyclic). The names of alkanes and cycloalkanes are the root names of organic compounds. Beginning with the five-carbon alkane, the number of carbons in the chain is indicated by the Greek or Latin prefix. Rings are designated by the prefix *cyclo-*. (In the geometrical symbols for rings, each apex represents a carbon with the number of hydrogens required to fill its valence.)

C_1	CH_4	methane	C_{12}	$CH_3[CH_2]_{10}CH_3$	dodecane
C_2	CH_3CH_3	ethane	C_{13}	$CH_3[CH_2]_{11}CH_3$	tridecane
C_3	$CH_3CH_2CH_3$	propane	C_{14}	$CH_3[CH_2]_{12}CH_3$	tetradecane
C_4	$CH_3[CH_2]_2CH_3$	butane	C_{20}	$CH_3[CH_2]_{18}CH_3$	icosane
C_5	$CH_3[CH_2]_3CH_3$	pentane	C_{21}	$CH_3[CH_2]_{19}CH_3$	henicosane
C_6	$CH_3[CH_2]_4CH_3$	hexane	C_{22}	$CH_3[CH_2]_{20}CH_3$	docosane
C_7	$CH_3[CH_2]_5CH_3$	heptane	C_{23}	$CH_3[CH_2]_{21}CH_3$	tricosane
C_8	$CH_3[CH_2]_6CH_3$	octane	C_{30}	$CH_3[CH_2]_{28}CH_3$	triacontane
C_9	$CH_3[CH_2]_7CH_3$	nonane	C_{31}	$CH_3[CH_2]_{29}CH_3$	hentriacontane
C_{10}	$CH_3[CH_2]_8CH_3$	decane	C_{40}	$CH_3[CH_2]_{38}CH_3$	tetracontane
C_{11}	$CH_3[CH_2]_9CH_3$	undecane	C_{50}	$CH_3[CH_2]_{48}CH_3$	pentacontane

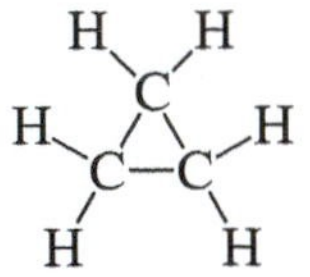

cyclopropane  cyclobutane 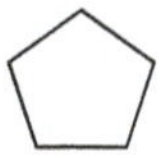 cyclopentane  cyclohexane 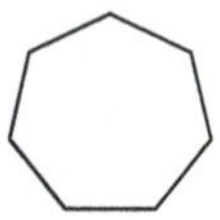 cycloheptane 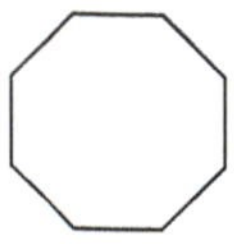 cyclooctane

The IUPAC system of nomenclature is frequently revised. The 1993 guidelines place the position number close to the functional group designation; however, you should be able to use and recognize names in either the old or the new style. Ask your instructor which system to use.

III. Nomenclature of Molecules Containing Substituents and Functional Groups

A. Priorities of Substituents and Functional Groups

LISTED HERE FROM HIGHEST TO LOWEST PRIORITY, except that the substituents within Group C have equivalent priority.

Group A—Functional Groups Named by Prefix or Suffix

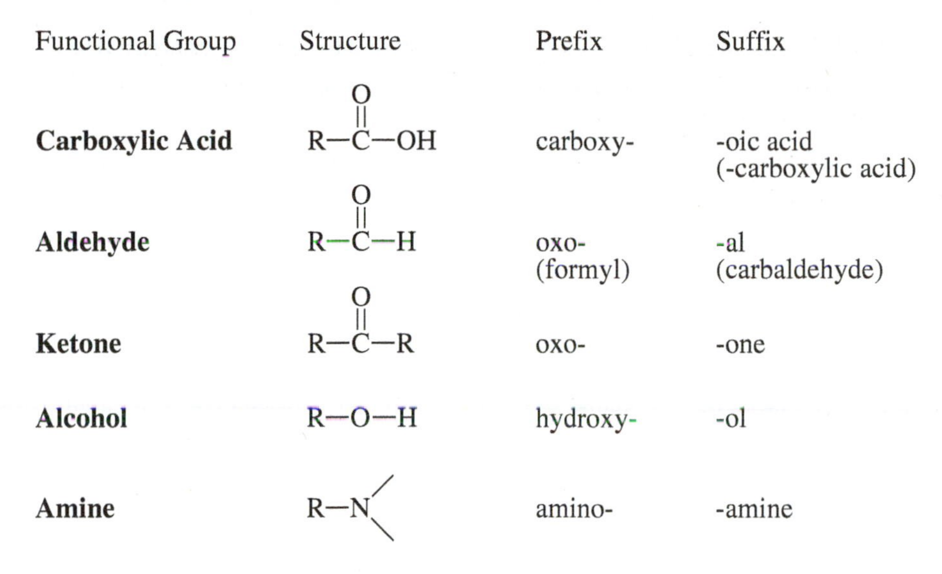

Functional Group	Structure	Prefix	Suffix
Carboxylic Acid	R—C(=O)—OH	carboxy-	-oic acid (-carboxylic acid)
Aldehyde	R—C(=O)—H	oxo- (formyl)	-al (carbaldehyde)
Ketone	R—C(=O)—R	oxo-	-one
Alcohol	R—O—H	hydroxy-	-ol
Amine	R—N<	amino-	-amine

Group B—Functional Groups Named by Suffix only

Functional Group	Structure	Prefix	Suffix
Alkene	>C=C<	---------	-ene
Alkyne	—C≡C—	---------	-yne

Group C—Substituent Groups Named by Prefix Only

Substituent	Structure	Prefix	Suffix
Alkyl (see below)	R—	alkyl-	--------
Alkoxy	R—O—	alkoxy-	--------

Alkoxy groups take the name of the alkyl group (like methyl or ethyl), drop the *-yl*, and add *-oxy*. CH_3O is methoxy; CH_3CH_2O is ethoxy.

Halogen	F—	fluoro-	--------
	Cl—	chloro-	--------
	Br—	bromo-	--------
	I—	iodo-	--------

Miscellaneous substituents and their prefixes

$—NO_2$	$—CH{=}CH_2$	$—CH_2CH{=}CH_2$	—C₆H₅ (benzene ring)
nitro	vinyl	allyl	phenyl

Common alkyl groups—replace *-ane* ending of alkane name with *-yl*. Alternate names for complex substituents are given in brackets.

$—CH_3$	$—CH_2CH_3$	$—CH_2CH_2CH_3$	$—CH_2CH_2CH_2CH_3$
methyl	ethyl	propyl (*n*-propyl)	butyl (*n*-butyl)

$—CH(CH_3)_2$	$—CH(CH_3)CH_2CH_3$	$—CH_2—CH(CH_3)_2$	$—C(CH_3)_3$
isopropyl [1-methylethyl]	*sec*-butyl [1-methylpropyl]	isobutyl [2-methylpropyl]	*tert*-butyl [1,1-dimethylethyl]

B. Naming Substituted Alkanes and Cycloalkanes—Group C Substituents Only

Organic compounds containing substituents from Group C are named following this sequence of steps, as indicated in the following examples:

- Step 1. Find the longest continuous carbon chain. Determine the root name for this parent chain. In cyclic compounds, the ring is usually considered the parent chain, unless it is attached to a longer chain of carbons; indicate a ring with the prefix *cyclo-* before the root name. (When there are two longest chains of equal length, use the chain with the greater number of substituents.)
- Step 2. Number the chain in the direction such that the position number of the first substituent is the smaller number. If the first substituents have the same number, then number so that the second substituent has the smaller number, etc.
- Step 3. Determine the name and position number of each substituent. (Each substituent on a nitrogen is designated with an "*N*" instead of a number; see Section **III**.D.1.)
- Step 4. Indicate the number of identical groups by the prefixes *di-*, *tri-*, *tetra-*, etc.
- Step 5. Place the position numbers and names of the substituent groups, in alphabetical order, before the root name. In alphabetizing, ignore prefixes like *sec-*, *tert-*, *di-*, *tri-*, etc., but include *iso-* and *cyclo-*. Always include a position number for each substituent, regardless of redundancies. In case of ties, where numbering could begin with either of two carbons, begin with the carbon closer to the one with more substituents, or else the carbon with the substituent whose name is earlier in the alphabet.

$CH_3—CH(Cl)—CH(Br)—C(CH_3)_2—CH(CH_2CH_2CH_3)—CH_2CH_3$

3-bromo-2-chloro-5-ethyl-4,4-dimethyloctane

$CH_3—CH(CH_3)—CH(CH_2CH_2CH_3)—CH(F)—CHCH_3(CH_3)$

3-fluoro-4-isopropyl-2-methylheptane

$H_3C—CHCH_2CH_3$ (on cyclohexane C1), NO_2 (on C3)

1-*sec*-butyl-3-nitrocyclohexane (Numbering is determined by the alphabetical order of substituents, "b" of "butyl" before "n" of "nitro".)

1-bromo-3-fluorocyclobutane

3-bromo-1,1-difluorocyclobutane

C. Naming Molecules Containing Functional Groups from Group B—Suffix Only

1. Alkenes—Follow the same steps as for alkanes, except for the following:

(a) Number the chain of carbons *that includes the* C=C so that the C=C has the lower position number, because it has a higher priority than any substituents.

(**b**) Change *-ane* to *-ene* and assign a position number to the first carbon of the C═C; place the position number just before the name of the functional group(s).

(**c**) Designate geometrical isomers with a *cis-trans* or *E-Z* prefix (described later).

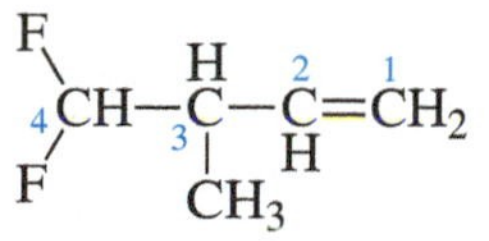

4,4-difluoro-3-methylbut-1-ene

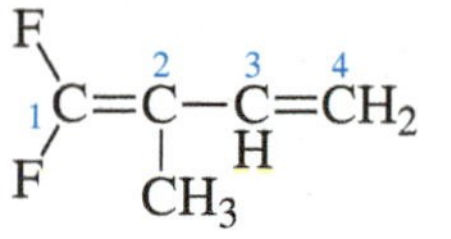

1,1-difluoro-2-methyl-buta-1,3-diene

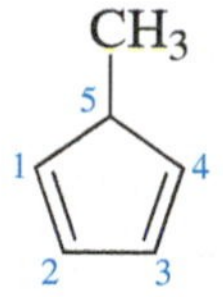

5-methylcyclopenta-1,3-diene

Special case: When the chain cannot include an alkene, a substituent name is used.

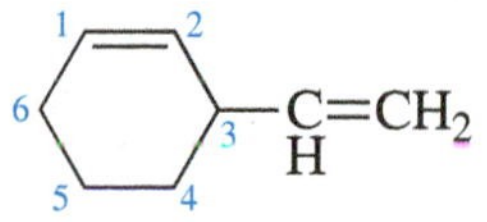

3-vinylcyclohex-1-ene

Numbering must be on EITHER a ring OR a chain, but not both.

2. Alkynes—Follow the same steps as for alkanes, except for the following:

(**a**) Number the chain of carbons *that includes the* C≡C so that the alkyne has the lower position number.

(**b**) Change *-ane* to *-yne* and assign a position number to the first carbon of the C≡C; place the position number just before the name of the functional group(s).

Note: The Group B functional groups (alkene and alkyne) are considered to have equal priority: in a molecule with both an *-ene* and an *-yne*, whichever is closer to the end of the chain determines the direction of numbering. In the case in which each would have the same position number, the alkene takes the lower number. In the name, *ene* comes before *yne* because of alphabetization.

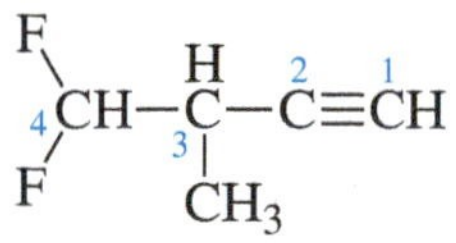

4,4-difluoro-3-methylbut-1-yne

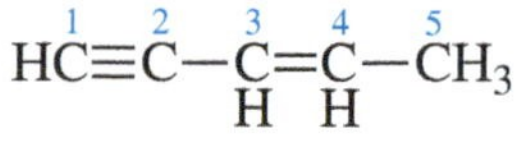

pent-3-en-1-yne
("yne" closer to end of chain)

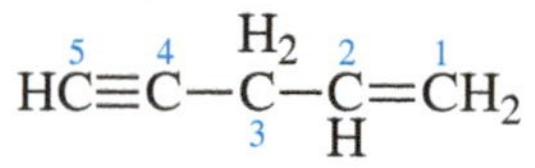

pent-1-en-4-yne
(The "ene" and "yne" have equal priority unless they have the same position number, when "ene" takes the lower number.)

(Notes: 1. An "e" is dropped if the letter following it is a vowel: "pent-3-en-1-yne," not "pent-3-ene-1-yne." 2. An "a" is added if inclusion of *di-*, *tri-*, and so on, would put two consonants together: "buta-1,3-diene," not "but- 1,3-diene.")

D. Naming Molecules Containing Functional Groups from Group A—Prefix or Suffix

In naming molecules containing one or more of the functional groups in Group A, the group of highest priority is indicated by suffix; the others are indicated by prefix, with priority equivalent to any other substituents. The table in Section **III**.A. defines the priorities; they are discussed next in order of increasing priority.

Now that the functional groups and substituents from Groups A, B, and C have been described, a modified set of steps for naming organic compounds can be applied to all simple structures:

- Step 1. Find the highest priority functional group. Determine and name the longest continuous carbon chain that includes this group.
- Step 2. Number the chain so that the highest priority functional group is assigned the lower number. (The number "1" is often omitted when there is no confusion about where the group must be. Aldehydes and carboxylic acids must be at the first carbon of a chain, so a "1" is rarely used with those functional groups.)
- Step 3. If the carbon chain includes multiple bonds (Group B), replace *-ane* with *-ene* for an alkene or *-yne* for an alkyne. Designate the position of the multiple bond with the number of the first carbon of the multiple bond.
- Step 4. If the molecule includes Group A functional groups, replace the last "e" with the suffix of the highest priority functional group, and include its position number just before the name of the highest priority functional group.
- Step 5. Indicate all Group C substituents and Group A functional groups of lower priority with a prefix. Place the prefixes, with appropriate position numbers, in alphabetical order before the root name.

1. Amines: prefix: *amino-*; suffix: *-amine*—substituents on nitrogen denoted by *N-*

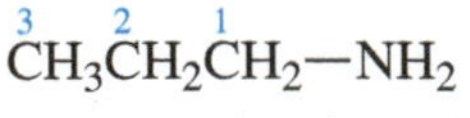

$\overset{3}{C}H_3\overset{2}{C}H_2\overset{1}{C}H_2-NH_2$

propan-1-amine

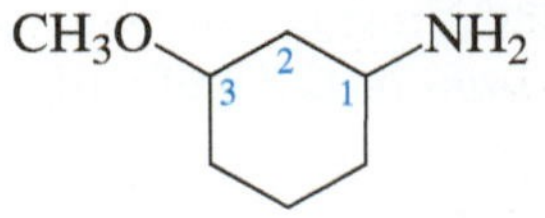

3-methoxycyclohexan-1-amine
("1" is optional in this case.)

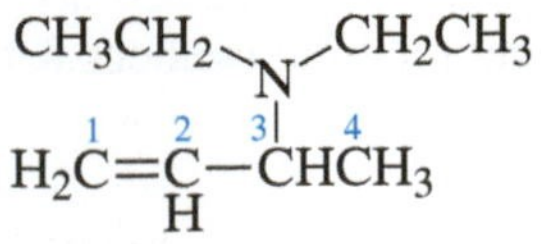

N,N-diethylbut-3-en-2-amine

2. Alcohols: prefix: *hydroxy-*; suffix: *-ol*

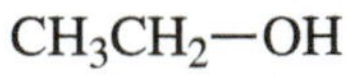

CH_3CH_2-OH

ethanol

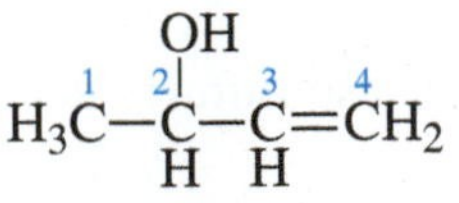

but-3-en-2-ol

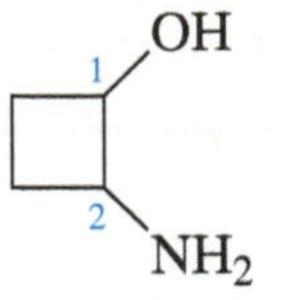

2-aminocyclobutan-1-ol
("1" is optional in this case.)

3. Ketones: prefix: *oxo-*; suffix: *-one* (pronounced "own")

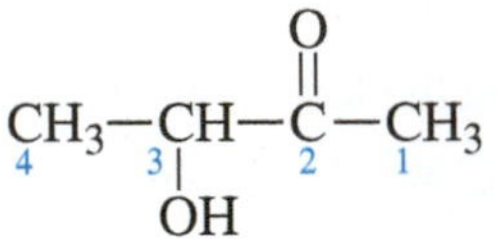

3-hydroxybutan-2-one

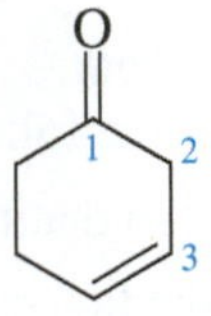

cyclohex-3-en-1-one
("1" is optional in this case.)

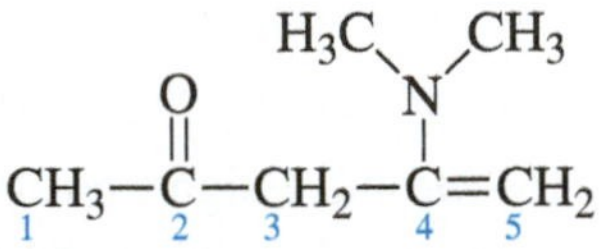

4-(*N,N*-dimethylamino)pent-4-en-2-one

4. Aldehydes: prefix: *oxo-* or *formyl-* (O=CH−); suffix: *-al* (abbreviation: —CHO)
An aldehyde can only be on carbon 1, so the "1" is generally omitted from the name.

methanal;
formaldehyde

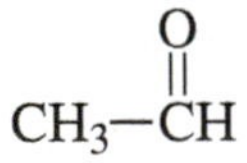

ethanal;
acetaldehyde

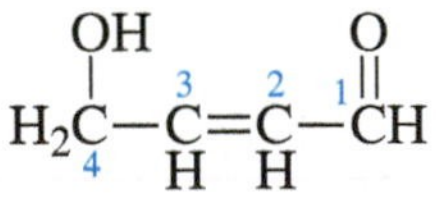

4-hydroxybut-2-enal

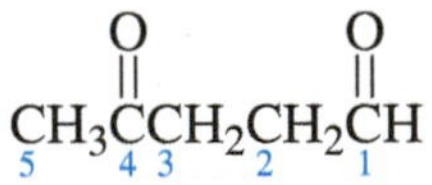

4-oxopentanal

Special case: When the chain cannot include the carbon of the aldehyde, the suffix *-carbaldehyde* is used:

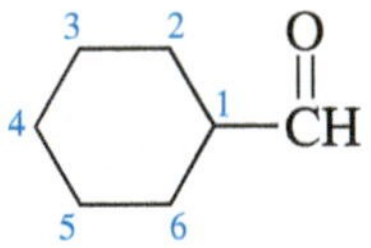

cyclohexanecarbaldehyde

5. Carboxylic Acids: prefix: *carboxy-*; suffix: *-oic acid* (abbreviation: —COOH)
A carboxylic acid can only be on carbon 1, so the "1" is generally omitted from the name.
(Note: Chemists traditionally use, and IUPAC accepts, the names "formic acid" and "acetic acid" in place of "methanoic acid" and "ethanoic acid.")

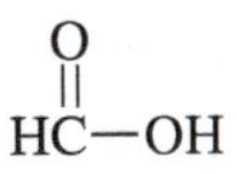

methanoic acid;
formic acid

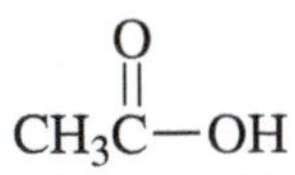

ethanoic acid;
acetic acid

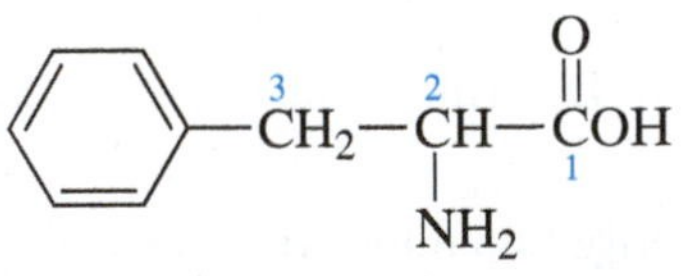

2-amino-3-phenylpropanoic acid

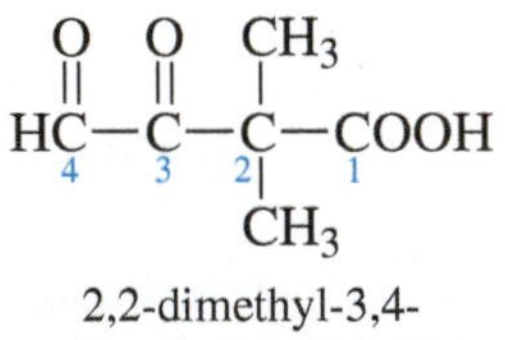

2,2-dimethyl-3,4-dioxobutanoic acid

Special case: When the chain numbering cannot include the carbon of the carboxylic acid, the suffix "carboxylic acid" is used:

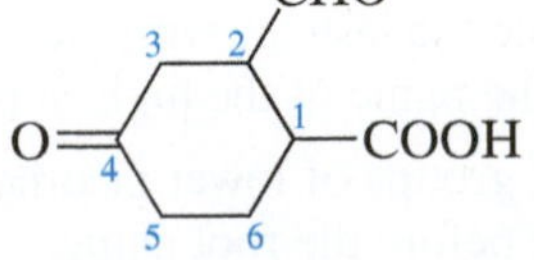

2-formyl-4-oxocyclohexanecarboxylic acid
("Formyl" is used to indicate an aldehyde as a substituent when its carbon cannot be in the chain numbering.)

E. Naming Carboxylic Acid Derivatives

The six common groups derived from carboxylic acids are, in decreasing priority after carboxylic acids, salts, anhydrides, esters, acyl halides, amides, and nitriles.

1. Salts of Carboxylic Acids

Salts are named with cation first, followed by the anion name of the carboxylic acid, where *-ic acid* is replaced by *-ate*:

acet**ic acid**	becomes	acet**ate**
butano**ic acid**	becomes	butano**ate**
cyclohexanecarboxyl**ic acid**	becomes	cyclohexanecarboxyl**ate**

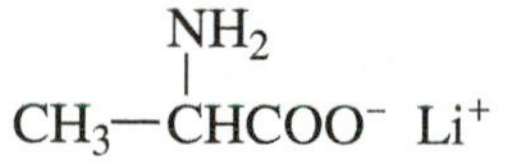

$CH_3-CH(NH_2)COO^-\ Li^+$
lithium 2-aminopropanoate

$ClCH_2-COO^-\ Na^+$
sodium chloroacetate

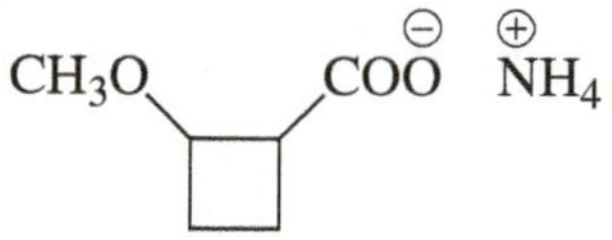

CH_3O, $COO^{\ominus}\ NH_4^{\oplus}$
ammonium 2-methoxy-cyclobutanecarboxylate

2. Anhydrides: *-oic acid* is replaced by *-oic anhydride*:

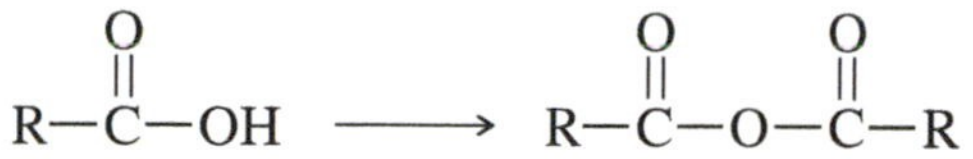

$$R-\overset{O}{\overset{\|}{C}}-OH \longrightarrow R-\overset{O}{\overset{\|}{C}}-O-\overset{O}{\overset{\|}{C}}-R$$

alkanoic acid → alkanoic anhydride

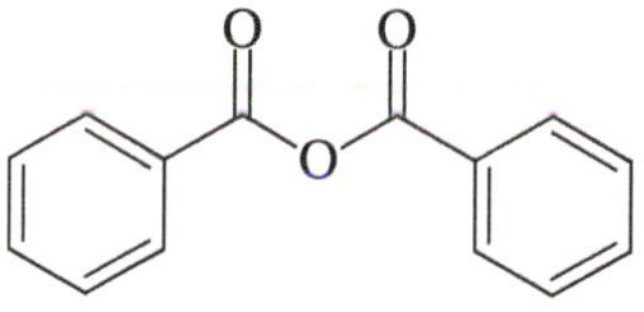

benzoic anhydride

3. Esters

Esters are named as "organic salts"; that is, the alkyl name comes first, followed by the name of the carboxylate anion. (common abbreviation: —COOR)

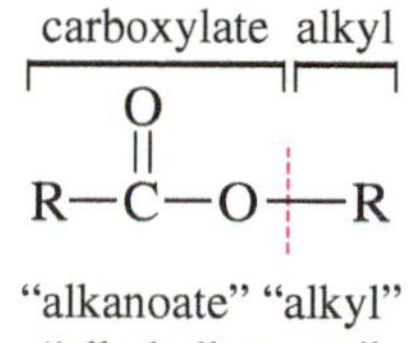

"alkanoate" "alkyl"
"alkyl alkanoate"

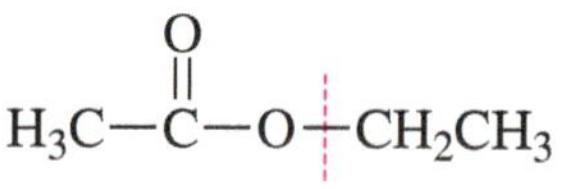

$H_3C-\overset{O}{\overset{\|}{C}}-O-CH_2CH_3$
ethyl acetate

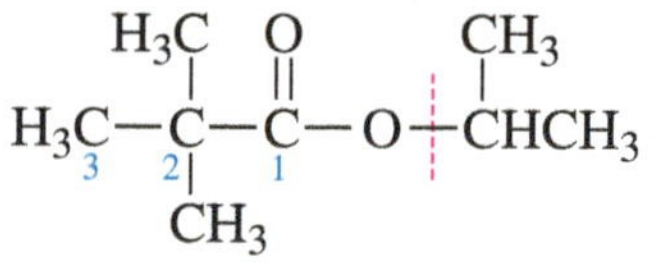

$H_3C-C(CH_3)(CH_3)-\overset{O}{\overset{\|}{C}}-O-CH(CH_3)CH_3$ (carbons numbered 3, 2, 1)
isopropyl 2,2-dimethylpropanoate

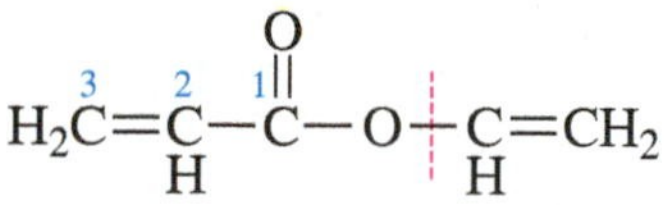

$H_2C{=}CH-\overset{O}{\overset{\|}{C}}-O-CH{=}CH_2$ (carbons numbered 3, 2, 1)
vinyl prop-2-enoate

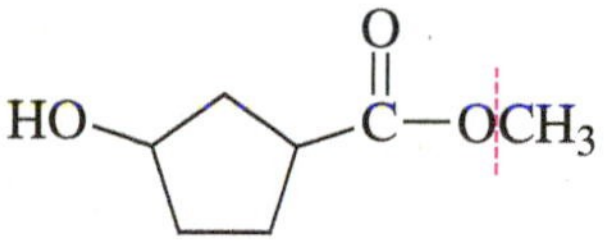

HO, $\overset{O}{\overset{\|}{C}}-OCH_3$
methyl 3-hydroxycyclo-pentanecarboxylate

$-CH_2COO-$
cyclohexyl 2-phenylacetate

4. Acyl Halides: *-oic acid* replaced by *-oyl halide*:

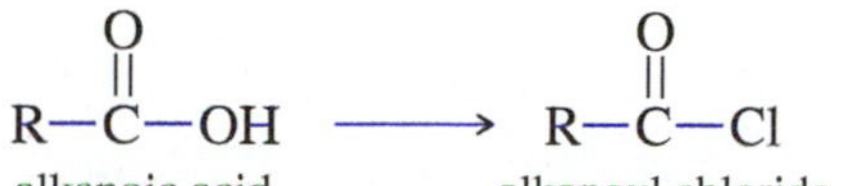

$$R-\overset{O}{\overset{\|}{C}}-OH \longrightarrow R-\overset{O}{\overset{\|}{C}}-Cl$$

alkanoic acid → alkanoyl chloride

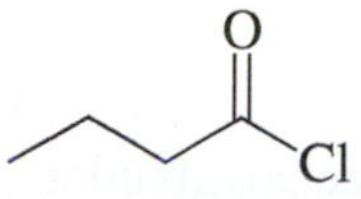

O, Cl
butanoyl chloride

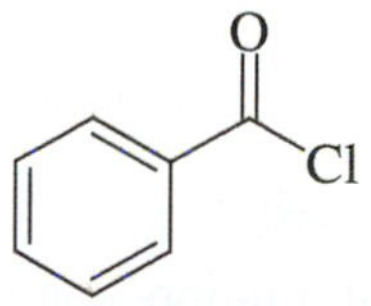

O, Cl
benzoyl chloride

5. Amides: *-oic acid* is replaced by *-amide*:

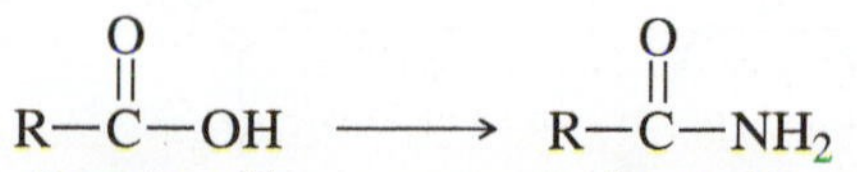

$$R-\overset{O}{\overset{\|}{C}}-OH \longrightarrow R-\overset{O}{\overset{\|}{C}}-NH_2$$

alkanoic acid → alkanamide

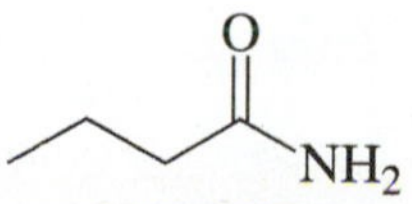

O, NH_2
butanamide

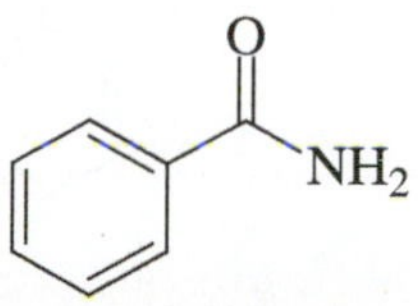

O, NH_2
benzamide

6. Nitriles: *-oic acid* is replaced by *-enitrile*:

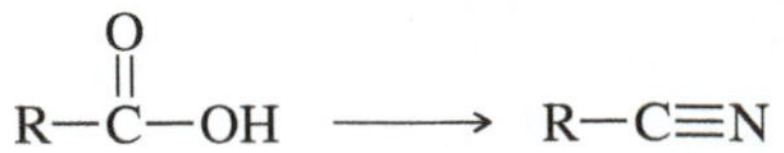

alkanoic acid alkanenitrile

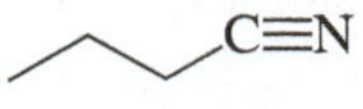

butanenitrile

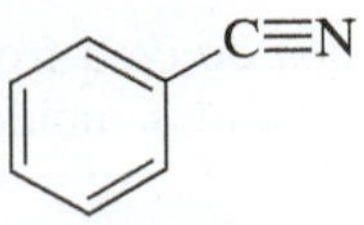

benzonitrile
(common spelling differs from IUPAC)

IV. Nomenclature of Aromatic Compounds

"Aromatic" compounds are those derived from benzene and similar ring systems. As with the aliphatic nomenclature described earlier, the process is to determine the root name of the parent ring; determine priority, name, and position number of substituents; and assemble the name in alphabetical order.

Functional group priorities are the same in aliphatic and aromatic nomenclature. See Section **III.A** for the list of priorities.

A. Common Parent Ring Systems

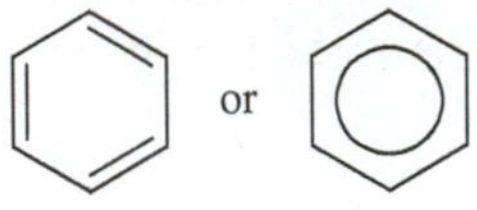

benzene

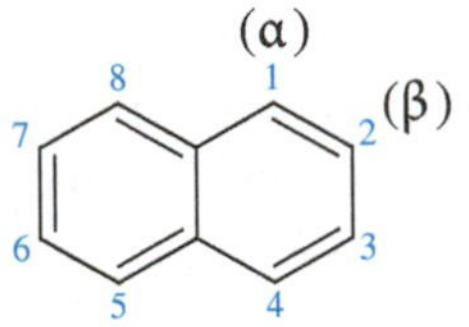

napthalene

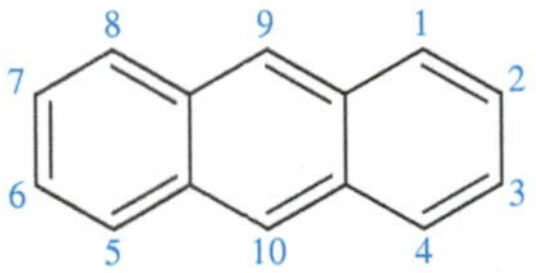

anthracene

B. Monosubstituted Benzenes

1. Most substituents keep their designation, followed by the word "benzene":

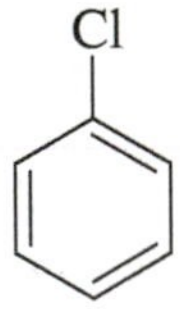

chlorobenzene

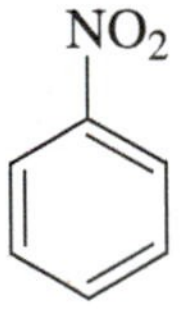

nitrobenzene

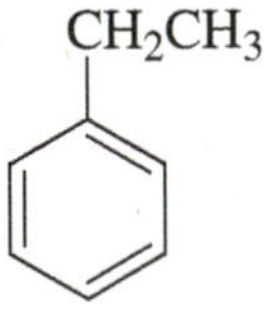

ethylbenzene

2. Some common substituents change the root name of the ring. IUPAC accepts these as root names, listed here in decreasing priority (same as Section **III.A**, Group A):

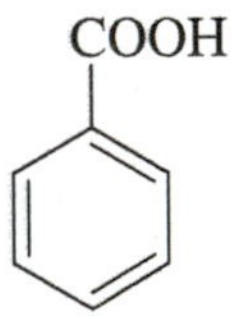

benzoic acid

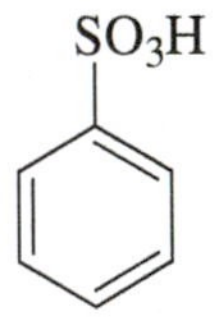

benzene-sulfonic acid

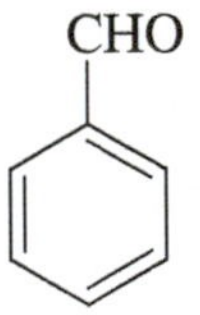

benzaldehyde

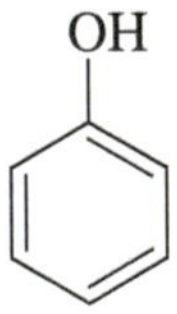

phenol

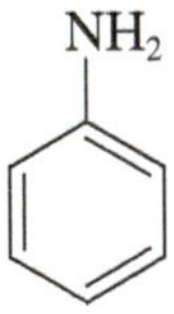

aniline

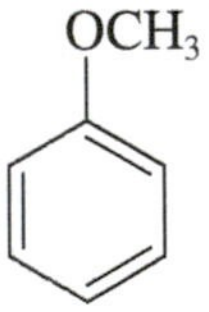

anisole

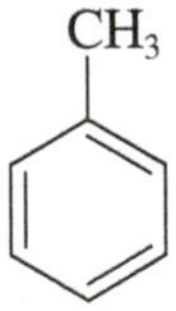

toluene

C. Disubstituted Benzenes

1. Designation of substitution—only three possibilities:

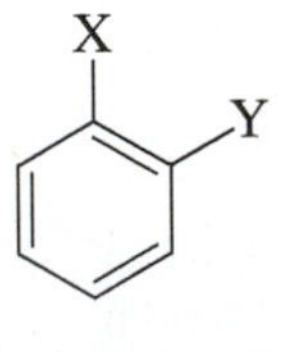

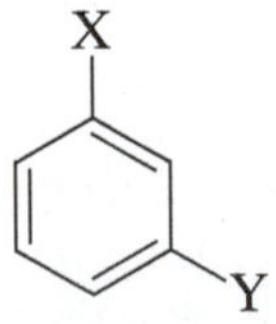

common:	*ortho-* (*o-*)	*meta-* (*m-*)	*para-* (*p-*)
IUPAC:	1,2-	1,3-	1,4-

2. Naming disubstituted benzenes—priorities from Section **III.A**, Group A, determine root name and substituents.

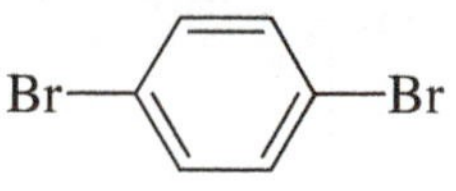

p-dibromobenzene
1,4-dibromobenzene

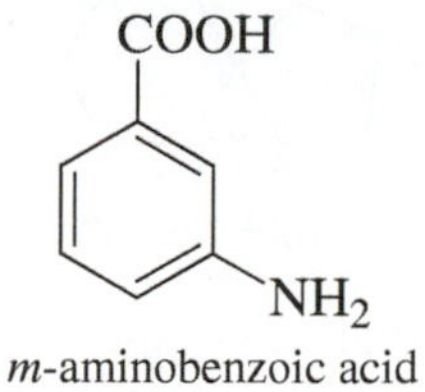

m-aminobenzoic acid
3-aminobenzoic acid

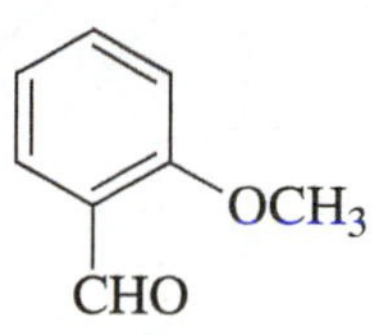

o-methoxybenzaldehyde
2-methoxybenzaldehyde

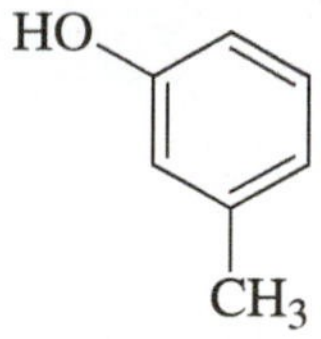

m-methylphenol
3-methylphenol

D. Polysubstituted Benzenes—must use numbers to indicate substituent position

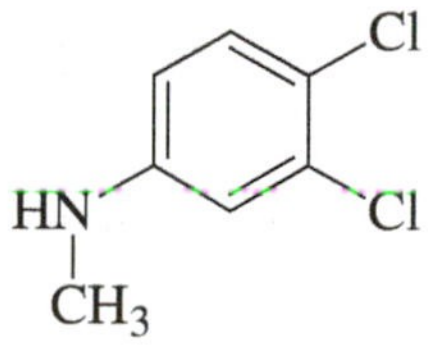

3,4-dichloro-*N*-methylaniline

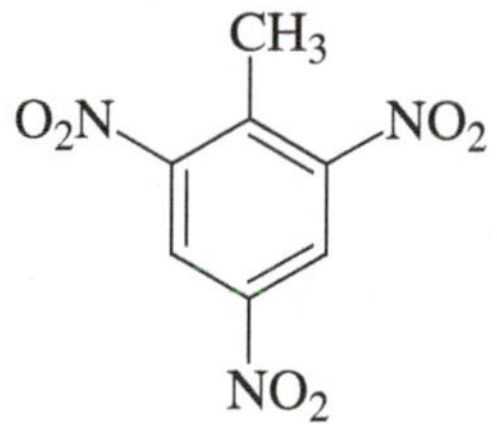

2,4,6-trinitrotoluene
(TNT)

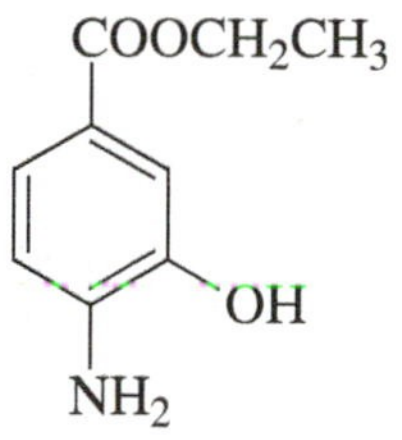

ethyl 4-amino-3-hydroxybenzoate

E. Aromatic Ketones

A special group of aromatic compounds are ketones where the carbonyl is attached to at least one benzene ring. Such compounds are named as "phenones"; the prefix depends on the size and nature of the group on the other side of the carbonyl. These are the common examples:

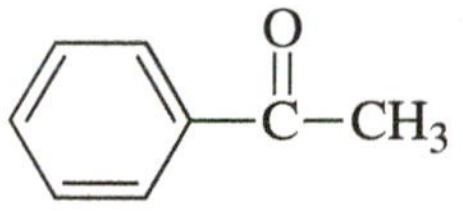

acetophenone

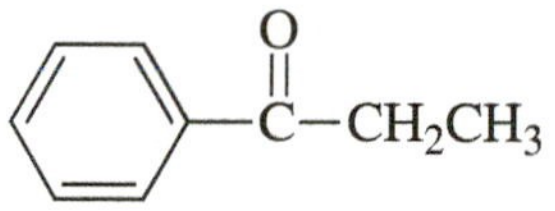

propiophenone

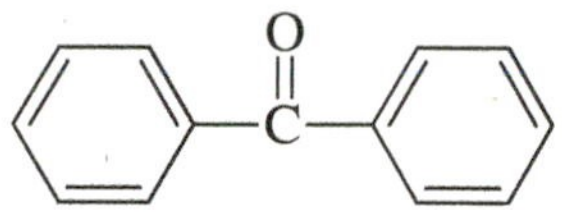

benzophenone

V. Nomenclature of Bicyclic Compounds

"Bicyclic" compounds are those that contain two rings. There are four possible arrangements of two rings that depend on how many atoms are shared by the two rings. The first arrangement in which the rings do not share any atoms does not use any special nomenclature, but the other types require a method to designate how the rings are put together. Once the ring system is named, then functional groups and substituents follow the standard rules described earlier.

Type 1. Two rings with no common atoms

These follow the standard rules of choosing one parent ring system and describing the other ring as a substituent.

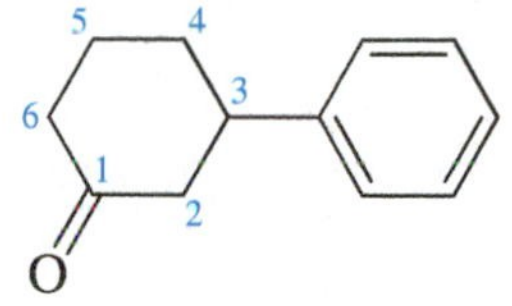

Ketone is the highest priority functional group; phenyl is the substituent.
⟶ 3-phenylcyclohexan-1-one ("1" could be omitted here.)

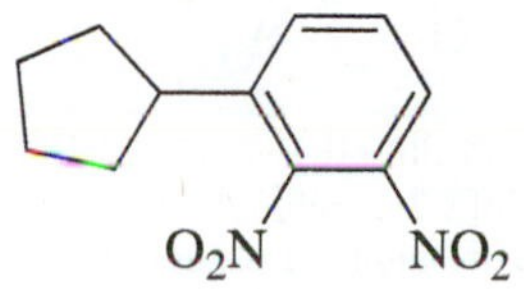

Benzene is the parent ring system, as it is larger than cyclopentane and it has three substituents.
⟶ 1-cyclopentyl-2,3-dinitrobenzene

Type 2. Two rings with one common atom—spiro ring system

The ring system in spiro compounds is indicated by the word "spiro" (instead of "cyclo"), followed by brackets indicating how many atoms are contained in each path around the rings, ending with the alkane name describing how many carbons are in the ring systems, including the spiro carbon. (If any atoms are not carbons, see section **VI.**) Numbering follows the smaller path first, passing through the spiro carbon and around the second ring.

spiro[3.4]octane spiro[4.5]decane

Substituents and functional groups are indicated in the usual ways. Spiro ring systems are always numbered smaller before larger, and they are numbered in such a way as to give the highest priority functional group the lower position number.

CH_3 CH_3

spiro[3.4]oct-5-ene 7,7-dimethylspiro[4.5]decan-2-one

Type 3. Two rings with two common atoms—fused ring system

Two rings that share two common atoms are called fused rings. This ring system and the next type called bridged rings share the same designation of ring system. Each of the two common atoms is called a bridgehead atom, and there are three paths between the two bridgehead atoms. In contrast with naming the spiro rings, the *longer* path is counted first, and then the shorter, and then the shortest. In fused rings, the shortest path is always a zero, meaning zero atoms between the two bridgehead atoms. Numbering starts at a bridgehead, continues around the largest ring, and then goes through the other bridgehead and around the shorter ring. (In these structures, bridgeheads are marked with a dark circle for clarity.)

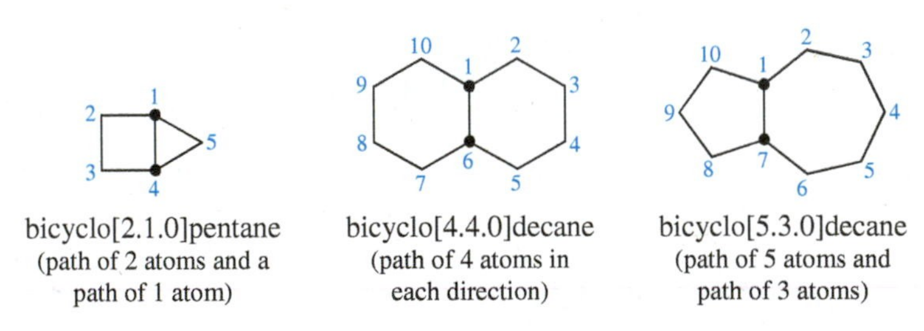

bicyclo[2.1.0]pentane (path of 2 atoms and a path of 1 atom)

bicyclo[4.4.0]decane (path of 4 atoms in each direction)

bicyclo[5.3.0]decane (path of 5 atoms and path of 3 atoms)

Substituents and functional groups are indicated in the usual ways. Fused ring systems are always numbered larger before smaller, and they are numbered in such a way as to give the highest priority functional group the lower position number.

Type 4. Two rings with more than two common atoms—bridged ring system

Two rings that share more than two common atoms are called bridged rings. Bridged rings share the same designation of ring system as Type 3, in which there are three paths between the two bridgehead atoms. The longer path is counted first, and then the medium, and then the shortest. Numbering starts at a bridgehead, continues around the largest ring, goes through the other bridgehead and around the medium path, and ends with the shortest path numbered from the original bridgehead atom. (In these structures, bridgeheads are marked with a dark circle for clarity.)

H_3C C_3H Br Br O HO

5,5-dibromo-bicyclo[2.1.1]hexane (paths of 2 atoms, 1 atom, and 1 atom)

bicyclo[2.2.2]oct-5-en-2-one (three paths of 2 atoms)

8,8-dimethyl-bicyclo[3.2.1]octan-1-ol (paths of 3 atoms, 2 atoms, and 1 atom)

VI. Replacement Nomenclature of Heteroatoms

The term "heteroatom" applies to any atom other than carbon or hydrogen. It is common for heteroatoms to appear in locations that are inconvenient to name following basic rules, so a simple system called "replacement nomenclature" has been devised. The fundamental principle is to name a compound as if it contained only carbons in the skeleton, plus any functional groups or substituents, and then indicate which carbons are "replaced" by heteroatoms. The prefixes used to indicate these substitutions are listed here *in decreasing priority and listed in this order in the name*:

Element	Prefix
O	oxa-
S	thia-
N	aza-
P	phospha-
Si	sila-
B	bora-

Example

OH

nonan-1-ol

NH BH H_2Si S OH

2-thia-8-aza-4-sila-6-boranonan-1-ol

In the example, note that the (imaginary) compound no longer has nine carbons, even though the name still includes "nonan." The heteroatoms have replaced carbons, but the compound is named as if it still had those carbons.

Where the replacement system is particularly useful is in polycyclic compounds. This example is a commercially available and synthetically useful reagent that uses this system.

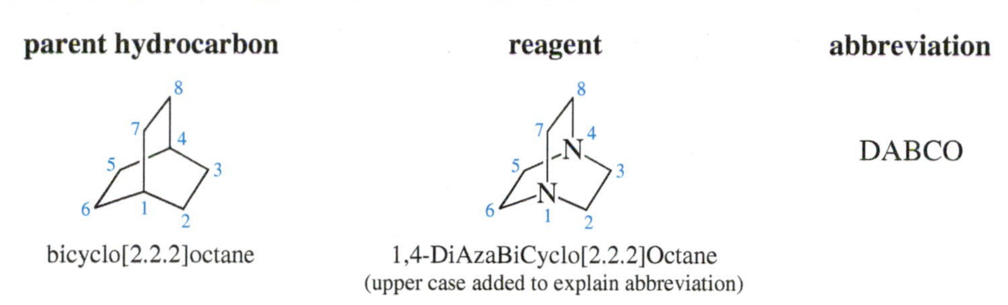

bicyclo[2.2.2]octane

1,4-DiAzaBiCyclo[2.2.2]Octane
(upper case added to explain abbreviation)

VII. Designation of Stereochemistry

Compounds that exhibit stereoisomerism, whether geometric isomers around double bonds, substituent groups on rings, or molecules with asymmetric tetrahedral atoms (usually carbons), require systems to designate relative and absolute orientation of the groups. These systems have been discussed in the text and are only exemplified here.

Stereochemistry around double bonds uses *cis-trans* or the *E-Z* designation of the Cahn-Ingold-Prelog sytem. (See Section 7-5.)

Z

O Cl CH_3 H_3C

(*Z*)-3-chloropent-3-en-2-one

Relative positions of substituents around a ring use the *cis-trans* designation. (See Section 3-11.)

Br H H CH_3

cis-3-methyl-1-bromocyclohexane

Absolute configuration of chiral centers uses the R-S designation of the Cahn-Ingold-Prelog system. (See Section 5-3.)

Br H H CH_3

(1*R*,3*S*)-3-methyl-1-bromocyclohexane

Answers to Selected Problems

These short answers are sometimes incomplete, but they should put you on the right track. Complete answers to all problems are found in the *Solutions Manual*.

CHAPTER 1

1.5. (a) C—Cl; (b) C—O; (c) C—N; (d) C—S; (e) C—B; (f) N—Cl; (g) N—O; (h) N—S; (i) N—B; (j) B—Cl. **1.6.** (a) +1 on O; (b) +1 on N, −1 on Cl; (c) +1 on N, −1 on Cl; (d) +1 on Na, −1 on O; (e) +1 on C; (f) −1 on C; (g) +1 on Na, −1 on B; (h) +1 on Na, −1 on B; (i) +1 on O, −1 on B; (j) +1 on N; (k) +1 on K, −1 on O; (l) +1 on O. **1.14.** (a) CH_2O, $C_3H_6O_3$; (b) $C_2H_5NO_2$, same; (c) C_2H_4ClNO, same; (d) C_2H_3Cl, $C_4H_6Cl_2$. **1.16.** sp^3; Two lone pairs compress the bond angle to 104.5°. **1.18.** Methyl carbon; sp^3, about 109.5°. Nitrile carbon *sp*, 180°. Nitrile nitrogen *sp*, no bond angle. **1.20.** The central carbon is *sp*, with two unhybridized *p* orbitals at right angles. Each terminal $=CH_2$ group must be aligned with one of these *p* orbitals. **1.23.** $CH_3-CH=N-CH_3$ shows cis-trans isomerism about the C=N double bond, but $(CH_3)_2C=N-CH_3$ has two identical substituents on the C=N carbon atom, and there are no cis-trans isomers. **1.25.** (a) constitutional isomers; (b) cis-trans isomers; (c) constitutional isomers; (d) same compound; (e) same compound; (f) same compound; (g) not isomers; (h) constitutional isomers; (i) same compound; (j) constitutional isomers; (k) constitutional isomers. **1.27.** (a) carbon; (b) oxygen; (c) phosphorus; (d) chlorine. **1.34.** The following are condensed structures that you should convert to Lewis structures. (a) $CH_3CH_2CH_2CH_3$ and $CH_3CH(CH_3)_2$; (c) $CH_3CH_2NH_2$ and CH_3NHCH_3; (e) $CH_2(CH_2OH)_2$ and $CH_3CHOHCH_2OH$ and $CH_3OCH_2OCH_3$ and others; (f) $CH_2=CHOH$ and CH_3CHO. **1.38.** (a) C_5H_5N; (b) C_4H_9N; (c) C_4H_4O; (d) $C_4H_9NO_2$; (e) $C_{11}H_{19}NO$; (f) $C_6H_{12}O$; (g) $C_7H_8O_3S$; (h) $C_7H_8O_3$. **1.41.** (a) different compounds; (b) resonance forms; (c) different compounds; (d) different compounds; (e) resonance forms; (f) resonance forms; (g) resonance forms; (h) different compounds; (i) different compounds; (j) resonance forms; (k) resonance forms; (l) resonance forms. **1.45.** (a) second; (b) first; (c) first; (d) second; (e) first; (f) second. **1.47.** Empirical formula C_3H_6O; molecular formula $C_6H_{12}O_2$. **1.48.** (a) $C_9H_{12}O$; (b) $C_{18}H_{24}O_2$. **1.49.** No stereoisomers. **1.50.** Cyclopropane has bond angles of 60°, compared with the 109.5° bond angle of an unstrained alkane. **1.53.** Urea must have two sp^2-hybridized nitrogen atoms because they are involved in pi-bonding in the other resonance forms. **1.58.** (a), (e), and (f). **1.59.** (a) constitutional isomers; (b) constitutional isomers; (c) cis-trans isomers; (d) constitutional isomers; (e) cis-trans isomers; (f) same compound; (g) cis-trans isomers; (h) constitutional isomers.

CHAPTER 2

2.2. The N—F dipole moments oppose the dipole moment of the lone pair. **2.4.** *trans* has zero dipole moment because the bond dipole moments cancel. **2.7.** (a) $CH_3CH_2OCH_2CH_3$; (c) $CH_3CH_2NHCH_3$; (d) CH_3CH_2OH; (e) CH_3COCH_3. **2.8.** (a) 0.209; (b) 13.88. **2.10.** (a) favors products; (b) favors reactants; (c) through (j) all favor products. **2.24.** (a) alkane; (b) alkene; (c) alkyne; (d) cycloalkyne and cycloalkene; (e) cycloalkane and alkene; (f) aromatic hydrocarbon and alkyne; (g) cycloalkene and alkene; (h) cycloalkane and alkane; (i) aromatic hydrocarbon and cycloalkene. **2.25.** (a) aldehyde and alkene; (b) alcohol; (c) ketone; (d) ether and alkene; (e) carboxylic acid; (f) ether and alkene; (g) ketone and alkene; (h) aldehyde; (i) alcohol. **2.26.** (a) amide; (b) amine; (c) ester; (d) acid chloride and alkene; (e) ether; (f) nitrile; (g) carboxylic acid; (h) cyclic ester and alkene; (i) ketone, cyclic ether; (j) cyclic amine; (k) cyclic amide; (l) amide; (m) cyclic ester; (n) aldehyde, cyclic amine; (o) ketone, cycloalkene **2.30.** CO_2 is *sp*-hybridized and linear; the bond dipole moments cancel. The sulfur atom in SO_2 is sp^2-hybridized and bent; the bond dipole moments do not cancel. **2.31.** (a), (c), (h), and (l) can form hydrogen bonds in the pure state. These four plus (b), (d), (g), (i), (j), and (k) can form hydrogen bonds with water. **2.33.** Both can form H-bonds with water, but only the alcohol can form H-bonds with itself. **2.40.** (b) The =NH nitrogen atom is the most basic. **2.49.** (a) $CH_3CH_2O^- Li^+ + CH_4$; (b) methane; CH_3Li is a very strong base. **2.56.** (a) cyclic ether; (b) cycloalkene, carboxylic acid; (c) alkene, aldehyde; (d) aromatic, ketone; (e) alkene, cyclic ester; (f) cyclic amide; (g) aromatic, nitrile, ether; (h) amine, ester; (i) amine, alcohol, carboxylic acid.

CHAPTER 3

3.1. (a) $C_{28}H_{58}$; (b) $C_{44}H_{90}$. **3.2.** (a) 3-methylpentane; (b) 2-bromo-3-methylpentane; (c) 5-ethyl-2-methyl-4-propylheptane; (d) 4-isopropyl-2-methyldecane. **3.4.** (a) 2-methylbutane; (b) 2,2-dimethylpropane; (c) 3-ethyl-2-methylhexane; (d) 2,4-dimethylhexane; (e) 3-ethyl-2,2,4,5-tetramethylhexane; (f) 4-*tert*-butyl-3-methylheptane. **3.9.** (a) $C_{12}H_{26}$; (b) $C_{15}H_{32}$. **3.10.** (a) hexane < octane < decane; (b) $(CH_3)_3C-C(CH_3)_3 < CH_3CH_2C(CH_3)_2CH_2CH_2CH_3 <$ octane. **3.15.** (a) 1,1-dimethyl-3-(1-methylpropyl)cyclopentane or 3-*sec*-butyl-1,1-dimethylcyclopentane; (b) 3-cyclopropyl-1,1-dimethylcyclohexane; (c) 4-cyclobutylnonane. **3.17.** (b), (c), and (d). **3.18.** (a) *cis*-1-methyl-3-propylcyclobutane; (b) *trans*-1-*tert*-butyl-3-ethylcyclohexane; (c) *trans*-1,2-dimethylcyclopropane. **3.19.** Trans is more stable. In the cis isomer, the methyl groups are nearly eclipsed. **3.26.** (a) *cis*-1,3-dimethylcyclohexane; (b) *cis*-1,4-dimethylcyclohexane; (c) *trans*-1,2-dimethylcyclohexane; (d) *cis*-1,3-dimethylcyclohexane; (e) *cis*-1,3-dimethylcyclohexane; (f) *trans*-1,4-dimethylcyclohexane. **3.31.** (a) bicyclo[3.1.0]hexane; (b) bicyclo[3.3.1]nonane; (c) bicyclo[2.2.2]octane; (d) bicyclo[3.1.1]heptane. **3.34.** (a) All except the third (isobutane) are *n*-butane. (b) The first and fourth structures are *cis*-but-2-ene. The second and fifth structures are but-1-ene. The third structure is *trans*-but-2-ene, and the last structure is 2-methylpropene. (c) The first and second are *cis*-1,2-dimethylcyclopentane. The third and fourth are *trans*-1,2-dimethylcyclopentane. The fifth is *cis*-1,3-dimethylcyclopentane. (f) The first, second, and fourth structures are 2,3-dimethylbutane. The third and fifth structures are 2,2-dimethylbutane. **3.39.** (a) 3-ethyl-2,2,6-trimethylheptane; (b) 3-ethyl-2,6,7-trimethyloctane; (c) 3,7-diethyl-2,2,8-trimethyldecane; (d) 1,1-diethyl-2-methylcyclobutane; (e) bicyclo[4.1.0]heptane; (f) *cis*-1-ethyl-3-propylcyclopentane; (g) (1,1-diethylpropyl)cyclohexane; (h) *cis*-1-ethyl-4-isopropylcyclodecane. **3.41.** (a) should be 3-methylhexane; (b) 3-ethyl-2-methylhexane; (c) 2-chloro-3-methylhexane; (d) 2,2-dimethylbutane; (e) *sec*-butylcyclohexane or (1-methylpropyl)cyclohexane; (f) should be *cis* or *trans*-1,2-diethylcyclopentane. **3.42.** (a) octane; (b) nonane; (c) nonane. **3.47.** The trans isomer is more stable, because both of the bonds to the second cyclohexane ring are in equatorial positions.

CHAPTER 4

4.3. (a) One photon of light would be needed for every molecule of product formed (the quantum yield would be 1); (b) Methane does not absorb the visible light that initiates the reaction, and the quantum yield would be 1. **4.4.** (a) Hexane has three different kinds of hydrogen atoms, but cyclohexane has only one type. (b) Large excess of cyclohexane. **4.5.** (a) $K_{eq} = 2.3$; (b) $[CH_3Br] = [H_2S] = 0.40$ M, $[CH_3SH] = [HBr] = 0.60$ M. **4.8.** (a) positive; (b) negative; (c) not easy to predict. **4.10.** (a) initiation +190 kJ/mole; propagation +73 kJ/mole and −112 kJ/mole; (b) overall −39 kJ/mole. **4.11.** (a) first order; (b) zeroth order; (c) first order overall. **4.13.** (a) zero, zero, zeroth order overall; (b) rate = k_r; (c) increase the surface area of the platinum catalyst. **4.14.** (b) +10 kJ/mole; (c) −7 kJ/mole. **4.15.** (c) +114 kJ/mole. **4.17.** (a) initiation +149 kJ/mole; propagation +141 kJ/mole and −92 kJ/mole; (b) overall +49 kJ/mole;

(c) low rate and very unfavorable equilibrium constant. **4.18.** 1°:2° ratio of 6:2, product ratio of 75% 1° and 25% 2°. **4.22.** (a) The combustion of isooctane involves highly branched, more stable tertiary free radicals that react less explosively. (b) *tert*-butyl alcohol forms relatively stable alkoxy radicals that react less explosively. **4.29.** Stability: (d) res 3° > (c) 3°> (b) 2°> (a) 1°. **4.30.** Stability: (d) res 3° > (c) 3°> (b) 2°> (a) 1°. **4.39.** rate = $k_r[H^+][(CH_3)_3C{-}OH]$; second order overall. **4.42.** $PhCH_2\cdot > CH_2{=}CHCH_2\cdot > (CH_3)_3C\cdot > (CH_3)_2CH\cdot > CH_3CH_2\cdot > CH_3\cdot$.

CHAPTER 5

5.1. chiral: corkscrew, desk, screw-cap bottle, rifle, knot, left-handed can opener. **5.2.** (b), (d), (e), and (f) are chiral. **5.3.** (a) chiral, one C*; (b) achiral, no C*; (c) chiral, one C*; (d) chiral, one C*; (e) achiral, no C*; (f) achiral, two C*; (g) chiral, one C*; (h) chiral, two C*; (i) chiral, two C*. **5.5.** (a) mirror, achiral; (b) mirror, achiral; (c) chiral, no mirror; (d) chiral, no mirror; (e) chiral, no mirror; (f) chiral, no mirror; (g) mirror, achiral; (h) mirror, achiral. **5.6.** (a) (*R*); (b) (*S*); (c) (*R*); (d) (*S*), (*S*); (e) (*R*), (*S*); (f) (*R*), (*S*); (g) (*R*), (*R*); (h) (*R*); (i) (*S*). **5.8.** +8.7°. **5.10.** Dilute the sample. If clockwise, will make less clockwise, and vice-versa. **5.12.** e.e. = 33.3%. Specific rotation = 33.3% of +13.5° = +4.5°. **5.15.** (a), (b), (e), and (f) are chiral. Only (e) has asymmetric carbons. **5.16.** (a) enantiomer, enantiomer, same; (b) same, enantiomer, enantiomer; (c) enantiomer, same, same. **5.18.** (a), (d), and (f) are chiral. The others have internal mirror planes. **5.19.** (from 5–18) (a) (*R*); (b) none; (c) none; (d) (2*R*), (3*R*); (e) (2*S*), (3*R*); (f) (2*R*), (3*R*); (new ones) (g) (*R*); (h) (*S*); (i) (*S*). **5.20.** (a) enantiomers; (b) diastereomers; (c) diastereomers; (d) constitutional isomers; (e) enantiomers; (f) diastereomers; (g) enantiomers; (h) enantiomers; (i) diastereomers. **5.23.** (a), (b), and (d) are pairs of diastereomers and could theoretically be separated by their physical properties. **5.30.** (a) same compound; (b) enantiomers; (c) enantiomers; (d) enantiomers; (e) diastereomers; (f) diastereomers; (g) enantiomers; (h) same compound; (i) enantiomers. **5.34.** (b) −15.90°; (c) −7.95°/−15.90° = 50% e.e. Composition is 75% (*R*) and 25% (*S*).

CHAPTER 6

6.1. (a) vinyl halide; (b) alkyl halide; (c) alkyl halide; (d) alkyl halide; (e) vinyl halide; (f) aryl halide. **6.5.** (a) ethyl chloride; (b) 1-bromopropane; (c) *cis*-2,3-dibromobut-2-ene; (d) *cis*-1,2-dichlorocyclobutane. **6.7.** Water is denser than hexane, so water forms the lower layer. Chloroform is denser than water, so chloroform forms the lower layer. Water and ethanol are miscible, so they form only one phase. **6.11.** (a) substitution; (b) elimination; (c) elimination, also a reduction. **6.13.** (a) 0.02 mol/L per second. **6.14.** (a) $(CH_3)_3COCH_2CH_3$; (b) $HC{\equiv}CCH_2CH_2CH_2CH_3$; (c) $(CH_3)_2CHCH_2NH_2$; (d) $CH_3CH_2CH_2C{\equiv}N$; (e) 1-iodopentane; (f) 1-fluoropentane. **6.16.** (a) $(CH_3CH_2)_2NH$, less hindered; (b) $(CH_3)_2S$, S more polarizable; (c) PH_3, P more polarizable; (d) CH_3S^-, neg. charged; (e) $(CH_3)_3N$, N less electronegative; (f) acetate is better: more basic, no inductive effect from F; (g) $CH_3CH_2CH_2O^-$, less hindered; (h) I^-, more polarizable. **6.18.** methyl iodide > methyl chloride > ethyl chloride > isopropyl bromide ≫ neopentyl bromide, *tert*-butyl iodide. **6.19.** (a) 2-methyl-1-iodopropane; (b) cyclohexyl bromide; (c) isopropyl bromide; (d) 2-chlorobutane; (e) 1-iodobutane. **6.23.** (a) 2-bromopropane; (b) 2-bromo-2-methylbutane; (c) allyl bromide; (d) 2-bromopropane; (e) 2-iodo-2-methylbutane; (f) 2-bromo-2-methylbutane. **6.27.** (a) $(CH_3)_2C(OCOCH_3)CH_2CH_3$, first order; (b) 1-methoxy-2-methylpropane, second order; (c) 1-ethoxy-1-methylcyclohexane, first order; (d) methoxycyclohexane, first order; (e) ethoxycyclohexane, second order. **6.32.** (a) 2-bromo-2-methylpentane; (b) 1 chloro-1-methylcyclohexane; (c) 1,1-dichloro-3-fluorocycloheptane; (d) 4-(2-bromoethyl)-3-(fluoromethyl)-2-methylheptane; (e) 4,4-dichloro-5-cyclopropyl-1-iodoheptane; (f) *cis*-1,2-dichloro-1-methylcyclohexane. **6.33.** (a) 1-chlorobutane; (b) 1-iodobutane; (c) 4-chloro-2,2-dimethylpentane; (d) 1-bromo-2,2-dimethylpentane; (e) chloromethylcyclohexane; (f) 2-methyl-1-bromopropane. **6.34.** (a) *tert*-butyl chloride; (b) 2-chlorobutane; (c) bromocyclohexane; (d) iodocyclohexane; (e) $PhCHBrCH_3$; (f) 3-bromocyclohexene. **6.37.** (a) rate doubles; (b) rate multiplied by six; (c) rate increases. **6.44.** (a) (*R*)-2-cyanobutane (inversion); (b) (2*S*,3*R*)-3-methylpentan-2-ol (inversion); (c) racemic mixture of 3-ethoxy-2,3-dimethylpentanes (racemization). **6.45.** (a) diethyl ether; (b) $PhCH_2CH_2CN$; (c) $PhSCH_2CH_3$; (d) 1-dodecyne; (e) N-methylpyridinium iodide; (f) $(CH_3)_3CCH_2CH_2NH_2$; (g) tetrahydrofuran; (h) *cis*-4-methylcyclohexanol. **6.48.** (a) o.p. = e.e. = 15.58/15.90 = 98% (99% (*S*) and 1% (*R*)); (b) The e.e. of (*S*) decreases twice as fast as radioactive iodide substitutes, thus gives the (*R*) enantiomer; implies the S_N2 mechanism. **6.51.** NBS provides low conc. Br_2 for free-radical bromination. Abstraction of one of the CH_2 hydrogens gives a resonance-stabilized free radical; product $PhCHBrCH_3$.

CHAPTER 7

7.4. (a) two; (b) one; (c) three; (d) four; (e) five. **7.5.** (a) 4-methylpent-1-ene; (b) 2-ethylhex-1-ene; (c) penta-1,4-diene; (d) penta-1,2,4-triene; (e) 2,5-dimethylcyclopenta-1,3-diene; (f) 3-methylene-4-vinylcyclohex-1-ene; (g) allylbenzene or 3-phenylpropene. **7.6.** (1) (a), (c), (d), and (f) show geometric isomerism. **7.7.** (a) *cis*-3,4-dimethylpent-2-ene; (b) 3-ethylhexa-1,4-diene; (c) 1-methylcyclopentene; (d) give positions of double bonds; (e) specify cis or trans; (f) (*E*) or (*Z*), not *cis*. **7.11.** (a) *cis*-1,2-dibromoethene; (b) *cis* (*trans* has zero dipole moment); (c) 1,2-dichlorocyclohexene. **7.12.** 2,3-dimethylbut-2-ene is more stable by 6 kJ/mole. **7.14.** (a) stable; (b) unstable; (c) stable; (d) stable; (e) unstable (maybe stable cold); (f) stable; (g) unstable; (h) stable; (i) unstable (maybe stable cold). **7.23.** 3-methylbut-1-ene by E2 (minor); 2-methylbut-2-ene by E2 (major); and 2-ethoxy-3-methylbutane (trace) by S_N2. **7.29.** There is no hydrogen *trans* to the bromide leaving group. **7.36.** (a) $\Delta G > 0$, disfavored; (b) $\Delta G < 0$, favored; **7.37.** (a) strong bases and nucleophiles; (b) strong acids and electrophiles; (c) free-radical chain reaction; (d) strong acids and electrophiles. **7.42.** (b), (c), (e) and (f) show geometric isomerism. **7.43.** (a) 2-ethylpent-1-ene; (b) 3-ethylpent-2-ene; (c) (3*E*,5*E*)-2,6-dimethylocta-1,3,5-triene; (d) (*E*)-4-ethylhept-3-ene; (e) 1-cyclohexylcyclohexa-1,3-diene; (f) (3*Z*,5*Z*)-6-chloro-3-(chloromethyl)octa-1,3,5-triene. **7.51.** (a) a 1-halohexane; (b) a *tert*-butyl halide; (c) a 3-halopentane; (d) a halomethylcyclohexane; (e) a 4-halocyclohexane (preferably *cis*). **7.52.** (a) pent-2-ene; (b) 1-methylcyclopentene; (c) 1-methylcyclohexene; (d) 2-methylbut-2-ene; (rearrangement). **7.53.** (a) cyclopentene; (b) 2-methylbut-2-ene (major) and 2-methylbut-1-ene (minor); (c) 1-methylcyclopentene (major), methylenecyclopentane (minor), possibly 3-methylcyclopentene (minor). **7.73.** E1 with rearrangement by an alkyl shift followed by a hydride shift. The Zaitsev product violates Bredt's rule.

CHAPTER 8

8.1. (a) 2-chloropentane; (b) 2-chloro-2-methylpropane; (c) 1-iodo-1-methylcyclohexane; (d) mixture of *cis*- and *trans*-1-bromo-3-methyl- and 1-bromo-4-methylcyclohexane. **8.3.** (a) 1-bromo-2-methylcyclopentane; (b) 2-bromo-1-phenylpropane. **8.6.** (a) 1-methylcyclopentanol; (b) 2-phenylpropan-2-ol; (c) 1-phenylcyclohexanol. **8.10.** (b) propan-1-ol; (d) 2-methylpentan-3-ol; (f) *trans*-2-methylcyclohexanol. **8.13.** (a) *trans*-2-methylcycloheptanol; (b) mostly 4.4-dimethylpentan-2-ol; (c) —OH *exo* on the less substituted carbon. **8.16.** (a) The carbocation can be attacked from either face. **8.22.** (a) Cl_2/H_2O; (b) KOH/heat, then Cl_2/H_2O; (c) H_2SO_4/heat, then Cl_2/H_2O. **8.28.** (a) CH_2I_2 + Zn(Cu); (b) CH_2Br_2, NaOH, H_2O; (c) dehydrate (H_2SO_4), then $CHCl_3$, NaOH/H_2O. **8.34.** (a) *cis*-cyclohexane-1,2-diol; (b) *trans*-cyclohexane-1,2-diol; (c) and (f) (*R,S*)-pentane-2,3-diol (+enantiomer); (d) and (e) (*S,S*)-pentane-2,3-diol (+enantiomer). **8.35.** (a) OsO_4/H_2O_2; (b) CH_3CO_3H/H_3O^+; (c) CH_3CO_3H/H_3O^+; (d) OsO_4/H_2O_2. **8.49.** (a) 1-methylcyclohexene, RCO_3H/H_3O^+; (b) cyclooctene, OsO_4/H_2O_2; (c) *trans*-cyclodecene, Br_2; (d) cyclohexene, Cl_2/H_2O. **8.62.** $CH_3(CH_2)_{12}CH{=}CH(CH_2)_7CH_3$, cis or trans unknown.

CHAPTER 9

9.3. decomposition to its elements, C and H_2. **9.4.** Treat the mixture with $NaNH_2$ to remove the hex-1-yne. **9.5.** (a) $Na^+\ {}^-C{\equiv}CH$ and NH_3; (b) $Li^+\ {}^-C{\equiv}CH$ and CH_4; (c) no reaction; (d) no reaction; (e) acetylene + $NaOCH_3$; (f) acetylene + NaOH; (g) no reaction;

(h) no reaction; (i) NH_3 + $NaOCH_3$. **9.7.** (a) $NaNH_2$; butyl halide; (b) $NaNH_2$; propyl halide; $NaNH_2$; methyl halide; (c) $NaNH_2$; ethyl halide; repeat; (d) S_N2 on *sec*-butyl halide is unfavorable; (e) $NaNH_2$; isobutyl halide (low yield); $NaNH_2$; methyl halide; (f) $NaNH_2$ added for second substitution on 1,8-dibromooctane might attack the halide. **9.8.** (a) sodium acetylide + formaldehyde; (b) sodium acetylide + CH_3I, then $NaNH_2$, then $CH_3CH_2CH_2CHO$; (c) sodium acetylide + $PhCOCH_3$; (d) sodium acetylide + CH_3I, then $NaNH_2$, then $CH_3CH_2COCH_3$. **9.12.** (a) H_2, Lindlar; (b) Na, NH_3; (c) Add halogen, dehydrohalogenate to the alkyne, Na, NH_3; (d) $NaNH_2$, then EtBr, then H_2 with Lindlar. **9.18.** (a) Cl_2; (b) HBr, peroxides; (c) HBr, no peroxides; (d) excess Br_2; (e) reduce to hex-1-ene, add HBr; (f) excess HBr. **9.20.** (a) The two ends of the triple bond are equivalent; (b) The two ends of the triple bond are not equivalent, yet not sufficiently different for good selectivity. **9.21.** (a) hexan-2-one; hexanal; (b) mixtures of hexan-2-one and hexan-3-one; (c) hexan-3-one for both; (d) cyclodecanone for both. **9.24.** (a) $CH_3C{\equiv}C(CH_2)_4C{\equiv}CCH_3$. **9.27.** (a) ethylmethylacetylene; (b) phenylacetylene; (c) *sec*-butyl-*n*-propylacetylene; (d) *sec*-butyl-*tert*-butylacetylene. **9.38.** cyclohexa-1,3-diene with ($HC{\equiv}C{-}CH{=}CH{-}$) at the 1 position (cis or trans).

CHAPTER 10

10.1. (a) 2-phenylbutan-2-ol; (b) (*E*)-5-bromohept-3-en-2-ol; (c) 4-methylcyclohex-3-en-1-ol; (d) *trans*-2-methylcyclohexanol; (e) (*E*)-2-chloro-3-methylpent-2-en-1-ol; (f) (2*R*,3*S*)-2-bromohexan-3-ol. **10.4.** (a) 8,8-dimethylnonane-2,7-diol; (b) octane-1,8-diol; (c) *cis*-cyclohex-2-ene-1,4-diol; (d) 3-cyclopentylheptane-2,4-diol; (e) *trans*-cyclobutane-1,3-diol. **10.5.** (a) cyclohexanol; more compact; (b) 4-methylphenol; more compact, stronger H-bonds; (c) 3-ethylhexan-3-ol; more spherical; (d) cyclooctane-1,4-diol; more OH groups per carbon; (e) enantiomers; equal solubility. **10.7.** (a) methanol; less substituted; (b) 2-chloropropan-1-ol; chlorine closer to the OH group; (c) 2,2-dichloroethanol; two chlorines to stabilize the alkoxide; (d) 2,2-difluoropropan-1-ol; F is more electronegative than Cl, stabilizing the alkoxide. **10.9.** The anions of 2-nitrophenol and 4-nitrophenol (but not 3-nitrophenol) are stabilized by resonance with the nitro group. **10.10.** (a) The phenol (left) is deprotonated by sodium hydroxide; it dissolves; (b) In a separatory funnel, the alcohol (right) will go into an ether layer and the phenolic compound will go into an aqueous sodium hydroxide layer. **10.11.** (b), (f), (g), (h). **10.15.** (a) three ways: (i) CH_3CH_2MgBr + $PhCOCH_2CH_2CH_3$; (ii) PhMgBr + $CH_3CH_2COCH_2CH_2CH_3$; (iii) $CH_3CH_2CH_2MgBr$ + $PhCOCH_2CH_3$; (b) PhMgBr + PhCOPh; (c) EtMgBr + cyclopentanone; (d) c-C_5H_9MgBr + pentan-2-one. **10.17.** (a) 2 PhMgBr + PhCOCl; (b) 2 CH_3CH_2MgBr + $(CH_3)_2CHCOCl$; (c) 2 *c*-HxMgBr + PhCOCl. **10.19.** (a) PhMgBr + ethylene oxide; (b) $(CH_3)_2CHCH_2MgBr$ + ethylene oxide; (c) 2-methylcyclohexylmagnesium bromide + ethylene oxide. **10.23.** (a) Grignard removes NH proton; (b) Grignard attacks ester; (c) Water will destroy Grignard; (d) Grignard removes OH proton. **10.26.** (a) heptanoic acid + $LiAlH_4$; or heptanal + $NaBH_4$; (b) heptan-2-one + $NaBH_4$; (c) 2-methylhexan-3-one + $NaBH_4$; (d) ketoester + $NaBH_4$. **10.33.** (a) hexan-1-ol, larger surface area; (b) hexan-2-ol, hydrogen-bonded; (c) hexane-1,5-diol, two OH groups; (d) hexan-2-ol. **10.36.** (a) cyclohexylmethanol; (b) 2-cyclopentylpentan-2-ol; (c) 2-methyl-1-phenylpropan-1-ol; (d) methane + 3-hydroxycyclohexanone; (e) cyclopentylmethanol; (f) triphenylmethanol; (g) $Ph_2C(OH)(CH_2)_4OH$; (h) 1,3-dicyclopentyl-2-phenylpropan-2-ol; (i) 2-cyclopentylethanol; (j) reduction of just the ketone, but not the ester; (k) 3-(2-hydroxyethyl) cyclohexanol from reduction of ketone and ester; (l) the tertiary alcohol from Markovnikov orientation of addition of H—OH; (m) the secondary alcohol from anti-Markovnikov orientation of addition of H—OH; (n) (2*S*,3*S*)-hexane-2,3-diol (+ enantiomer); (o) (2*S*,3*R*)-hexane-2,3-diol (+ enantiomer); (p) hepta-1,4-diene. **10.39.** (a) EtMgBr; (b) Grignard with formaldehyde; (c) *c*-HxMgBr; (d) cyclohexylmagnesium bromide with ethylene oxide; (e) PhMgBr with formaldehyde; (f) 2 CH_3MgI; (g) cyclopentylmagnesium bromide; (h) hexylmagnesium bromide with ethylene oxide.

CHAPTER 11

11.1. (a) oxidation, oxidation; (b) oxidation, oxidation, reduction, oxidation; (c) neither (C2 is oxidation, C3 reduction); (d) reduction; (e) oxidation; (f) neither; (g) neither; (h) neither; then reduction; (i) oxidation; (j) oxidation then neither; (k) oxidation; (l) reduction then oxidation, no net change. **11.6.** Cr reagents: (a) PCC; (b) chromic acid; (c) chromic acid or Jones reagent; (d) oxidize, add Grignard; (f) dehydrate, hydroborate, oxidize (chromic acid or Jones reagent). **11.7.** An alcoholic has more alcohol dehydrogenase. More ethanol is needed to tie up this larger amount of enzyme. **11.8.** CH_3COCHO (pyruvaldehyde) and $CH_3COCOOH$ (pyruvic acid). **11.10.** Treat the tosylate with (a) bromide; (b) ammonia; (c) ethoxide; (d) cyanide. **11.14.** (a) chromic acid or Lucas reagent; (b) chromic acid or Lucas reagent; (c) Lucas reagent only; (d) Lucas reagent only; allyl alcohol forms a resonance-stabilized carbocation; (e) Lucas reagent only. **11.19.** (a) thionyl chloride (retention); (b) tosylate (retention), then S_N2 using chloride ion (inversion). **11.20.** resonance-delocalized cation, positive charge spread over two carbons. **11.22.** (a) 2-methylbut-2-ene (+ 2-methylbut-1-ene); (b) pent-2-ene (+ pent-1-ene); (c) pent-2-ene (+ pent-1-ene); (d) *c*-$Hx{=}C(CH_3)_2$ (+ 1-isopropylcyclohexene); (e) 1-methylcyclohexene (+ 3-methylcyclohexene). **11.25.** Using R—OH and R′—OH will form R—O—R, R′—O—R′, and R—O—R′. **11.31.** (a) $CH_3CH_2CH_2COCl$ + propan-1-ol; (b) CH_3CH_2COCl + butan-1-ol; (c) $(CH_3)_2CHCOCl$ + *p*-methylphenol; (d) PhCOCl + cyclopropanol. **11.33.** An acidic solution (to protonate the alcohol) would protonate methoxide ion. **11.34.** (a) the alkoxide of cyclohexanol and an ethyl halide or tosylate; (b) dehydration of cyclohexanol. **11.42.** (a) Na, then ethyl bromide; (b) NaOH, then PCC to aldehyde; Grignard, then dehydrate; (c) Mg in ether, then $CH_3CH_2CH_2CHO$, then oxidize; (d) PCC, then EtMgBr. **11.46.** (a) thionyl chloride; (b) make tosylate, displace with bromide; (c) make tosylate, displace with hydroxide. **11.52.** Compound **A** is butan-2-ol. **11.59.** **X** is but-3-en-1-ol; **Y** is tetrahydrofuran (5-membered cyclic ether).

CHAPTER 12

12.3. (a) alkene; (b) alkane; (c) terminal alkyne. **12.4.** (a) amine (secondary); (b) acid; (c) alcohol. **12.5.** (a) conjugated ketone; (b) ester; (c) primary amide. **12.6.** (a) 3080 C—H; 1642 C=C *alkene*; (b) 2712, 2814—CHO; 1691 carbonyl-*aldehyde*; (c) over-inflated C—H region —COOH; 1703 carbonyl (maybe conjugated); 1650 C=C (maybe conjugated)-*conjugated acid*; (d) 1742 ester (or strained ketone)-*ester*. **12.7.** (a) bromine (C_6H_5Br); (b) iodine (C_2H_5I); (c) chlorine (C_4H_7Cl); (d) nitrogen ($C_7H_{17}N$). **12.8.** the isobutyl cation, $(CH_3)_2CHCH_2^+$. **12.11.** 126: loss of water; 111: allylic cleavage; 87: cleavage next to alcohol. **12.14.** (a) about 1660 and 1710; the carbonyl is much stronger; (b) about 1660 for both; the ether is much stronger; (c) about 1660 for both; the imine is much stronger; (d) about 1660 for both; the terminal alkene is stronger. **12.16.** (a) $CH_2{=}C(CH_3)COOH$; (b) $(CH_3)_2CHCOCH_3$; (c) $PhCH_2C{\equiv}N$; (d) $PhCH_2CH_2OH$. **12.17.** (a) 86, 71, 43; (b) 98, 69; (c) 84, 87, 45. **12.20.** (a) 1-bromobutane. **12.23.** (c) oct-1-yne.

CHAPTER 13

13.1. (a) δ2.17; (b) δ2.17; (c) 130 Hz. **13.3.** (a) three; (b) two; (c) three; (d) two; (e) three; (f) five. **13.6.** (a) 2-methylbut-3-yn-2-ol; (b) *p*-dimethoxybenzene; (c) 1,2-dibromo-2-methylpropane. **13.10.** *trans* CHCl=CHCN. **13.11.** (a) 1-chloropropane; (b) methyl *p*-methylbenzoate, $CH_3C_6H_4COOCH_3$. **13.14.** (a) H^a, δ9.7 (doublet); H^b, δ6.7 (multiplet); H^c, δ7.5 (doublet); (b) J_{ab} = 8 Hz, J_{bc} = 18 Hz (approx). **13.18.** (a) Five; the two hydrogens on C3 are diastereotopic. (b) Six; all the CH_2 groups have diastereotopic hydrogens. (c) Six; three on the Ph, and the CH_2 hydrogens are diastereotopic. (d) Three; the hydrogens cis and trans to the Cl are diastereotopic. **13.21.** (a) butane-1,3-diol; (b) $H_2NCH_2CH_2OH$. **13.24.** (a) $(CH_3)_2CHCOOH$; (b) $PhCH_2CH_2CHO$; (c) $CH_3COCOCH_2CH_3$; (d) $CH_2{=}CHCH(OH)CH_3$; (e) $CH_3CH_2C(OH)(CH_3)CH(CH_3)_2$. **13.29.** (a) allyl alcohol, $H_2C{=}CHCH_2OH$. **13.30.** (a) 4-hydroxybutanoic acid lactone (cyclic ester). **13.31.** (a) cyclohexene. **13.32.** isobutyl bromide. **13.36.** (a) isopropyl alcohol. **13.38.** (a) $PhCH_2CH_2OCOCH_3$. **13.42.** 1,1,2-trichloropropane. **13.45. A** is 2-methylbut-2-ene (Zaitsev product); **B** is 2-methylbut-1-ene. **13.47.** $PhCH_2CN$.

CHAPTER 14

14.2. $(CH_3CH_2)_2\overset{+}{O}-\overset{-}{Al}Cl_3$. **14.4.** (a) methoxyethene; methyl vinyl ether; (b) ethyl isopropyl ether; 2-ethoxypropane; (c) 2-chloroethyl methyl ether; 1-chloro-2-methoxyethane; (d) 2-ethoxy-2,3-dimethylpentane; (e) 1,1-dimethoxycyclopentane; (f) *trans*-2-methoxycyclohexanol; (g) cyclopropyl methyl ether; methoxycyclopropane. **14.6.** (a) dihydropyran; (b) 2-chloro-1,4-dioxane; (c) 3-isopropylpyran; (d) *trans*-2,3-diethyloxirane or *trans*-3,4-epoxyhexane; (e) 3-bromo-2-ethoxyfuran; (f) 3-bromo-2,2-dimethyloxetane. **14.11.** Intermolecular condensation of a mixture of methanol and ethanol would produce a mixture of diethyl ether, dimethyl ether, and ethyl methyl ether. **14.13.** Intermolecular condensation might work for (a). Use the Williamson for (b). Alkoxymercuration is best for (c). **14.15.** (a) bromocyclohexane and ethyl bromide; (b) 1,5-diiodopentane; (c) phenol and methyl bromide; (e) phenol, ethyl bromide, and 1,4-dibromo-2-methylbutane. **14.23.** Epoxidation of ethylene gives ethylene oxide, and catalytic hydration of ethylene gives ethanol. Acid-catalyzed opening of the epoxide in ethanol gives cellosolve. **14.26.** (a) $CH_3CH_2OCH_2CH_2O^-\ Na^+$; (b) $H_2NCH_2CH_2O^-\ Na^+$; (c) $Ph-SCH_2CH_2O^-\ Na^+$; (d) $PhNHCH_2CH_2OH$; (e) $N{\equiv}C-CH_2CH_2O^-\ K^+$; (f) $N_3CH_2CH_2O^-\ Na^+$. **14.27.** (a) 2-methylpropane-1,2-diol, ^{18}O at the C2 hydroxyl group; (b) 2-methylpropane-1,2-diol, ^{18}O at the C1 hydroxyl group; (c) (2*S*,3*S*)-2-methoxy-3-methylpentan-3-ol; (d) (2*R*,3*R*)-3-methoxy-3-methylpentan-2-ol. **14.35.** (a) The old ether had autoxidized to form peroxides. On distillation, the peroxides were heated and concentrated, and they detonated. (b) Discard the old ether or treat it to reduce the peroxides. **14.40.** (c) epoxide + $NaOCH_3$ in methanol; (d) epoxide + methanol, H^+. **14.47.** Sodium then ethyl iodide gives retention of configuration. Tosylation gives retention, then the Williamson gives inversion. Second product +15.6°. **14.49.** $(CH_3OCH_2CH_2)_2O$. **14.51.** phenyloxirane.

CHAPTER 15

15.1. (a) hexa-2,4-diene < hexa-1,3-diene < hexa-1,4-diene < hexa-1,5-diene < hexa-1,2-diene < hexa-1,3,5-triene; (b) third < fifth < fourth < second < first. **15.8.** (a) **A** is 3,4-dibromobut-1-ene; **B** is 1,4-dibromobut-2-ene; (c) Hint: **A** is the kinetic product, **B** is the thermodynamic product; (d) Isomerization to an equilibrium mixture. 10% **A** and 90% **B**. **15.9.** (a) 1-(bromomethyl)cyclohexene and 2-bromo-1-methylenecyclohexane. **15.11.** (a) 3-bromocyclopentene; (c) $PhCH_2Br$. **15.12.** Both generate the same allylic carbanion. **15.13.** In this reaction, alkyllithiums or Grignard reagents can be used interchangeably. (a) allyl bromide + phenyllithium; (b) isopropyllithium + 1-bromobut-2-ene; (c) 1,4-dibromobut-2-ene + two equivalents of propyllithium. **15.20.** (b) [4 + 2] cycloaddition of one butadiene with just one of the double bonds of another butadiene. **15.21.** 800. **15.22.** (a) 353 nm; (b) 313 nm; (c) 232 nm; (d) 292 nm. **15.24.** (a) isolated; (b) conjugated; (c) cumulated; (d) conjugated; (e) conjugated; (f) cumulated and conjugated. **15.25.** (a) allylcyclohexane; (b) 3-chlorocyclopentene; (c) 3-bromo-2-methylpropene; (e) 4-bromobut-2-en-1-ol and 1-bromobut-3-en-2-ol; (f) 5,6-dibromohexa-1,3-diene, 1,6-dibromohexa-2,4-diene, and 3,6-dibromohexa-1,4-diene (minor); (g) 1-(methoxymethyl)-2-methylcyclopentene and 1-methoxy-1-methyl-2-methylenecyclopentane; (h) and (i) Diels–Alder adducts. **15.26.** (a) allyl bromide + isobutyl Grignard; (b) 1-bromo-3-methylbut-2-ene + $CH_3CH_2C(CH_3)_2MgBr$; (c) cyclopentyl-MgBr + 1-bromopent-2-ene. **15.28.** (a) 19,000; (b) second structure. **15.32.** (a) The product isomerized, 1630 suggests conjugated; the UV spectrum supports conjugation; (b) 2-propylcyclohexa-1,3-diene.

CHAPTER 16

16.2. (a) +31.8 kJ/mole; (b) −88.6 kJ/mole; (c) −112.0 kJ/mole. **16.5.** Two of the eight pi electrons are unpaired in two non-bonding orbitals, an unstable configuration. **16.7.** (a) nonaromatic (internal H's prevent planarity); (b) nonaromatic (one ring atom has no *p* orbital); (c) aromatic, [14]annulene; (d) aromatic (in the outer system). **16.8.** Azulene is aromatic, but the other two are antiaromatic. **16.10.** The cation (cyclopropenium ion) is aromatic; the anion is antiaromatic. **16.12.** (a) antiaromatic if planar; (b) aromatic if planar; (c) aromatic if planar; (d) antiaromatic if planar; (e) nonaromatic; (f) aromatic if planar. **16.14.** cyclopropenium fluoroborate. **16.19.** (a) aromatic; (b) aromatic; (c) nonaromatic; (d) aromatic; (e) aromatic; (f) nonaromatic; (g) aromatic; (h) not aromatic. **16.24.** (a) fluorobenzene; (b) 4-phenylbut-1-yne; (c) 3-methylphenol or *m*-cresol; (d) *o*-nitrostyrene; (e) *p*-bromobenzoic acid; (f) isopropyl phenyl ether; (g) 3,4-dinitrophenol; (h) benzyl ethyl ether. **16.25.** 3-phenylprop-2-en-1-ol. **16.27.** (a) *o*-dichlorobenzene; (b) *p*-nitroanisole; (c) 2,3-dibromobenzoic acid; (d) 2,7-dimethoxynaphthalene; (e) *m*-chlorobenzoic acid; (f) 2,4,6-trichlorophenol; (g) 2-*sec*-butylbenzaldehyde; (h) cyclopropenium tetrafluoroborate. **16.30.** The second is deprotonated to an aromatic cyclopentadienyl anion. **16.31.** (d), (e) The fourth structure, with two three-membered rings, was considered the most likely and was called Ladenburg benzene. **16.37.** (a) three; (b) one; (c) *meta*-dibromobenzene. **16.38.** (a) α-chloroacetophenone; (b) 4-bromo-1-ethylbenzene. **16.45.** 2-isopropyl-5-methylphenol.

CHAPTER 17

17.3. The sigma complex for *p*-xylene has the + charge on two 2° carbons and one 3° carbon, compared with three 2° carbons in benzene. **17.9.** Bromine *adds* to the alkene but *substitutes* on the aryl ether, evolving gaseous HBr. **17.10.** Strong acid is used for nitration, and the amino group of aniline is protonated to a deactivating $-NH_3^+$ group. **17.12.** (a) 2,4- and 2,6-dinitrotoluene; (b) 3-chloro-4-nitrotoluene and 5-chloro-2-nitrotoluene; (c) 3- and 5-nitro-2-bromobenzoic acid; (d) 4-methoxy-3-nitrobenzoic acid; (e) 5-methyl-2-nitrophenol and 3-methyl-4-nitrophenol. **17.15.** (a) phenylcyclohexane; (b) *o*- and *p*-methylanisole, with overalkylation products; (c) 1-isopropyl-4-(1,1,2-trimethylpropyl)benzene. **17.16.** (a) phenylcyclohexane; (b) *tert*-butylbenzene; (c) *p*-di-*tert*-butylbenzene; (d) *o*- and *p*-isopropyltoluene. **17.17.** (a) *tert*-butylbenzene; (b) 2- and 4-*sec*-butyltoluene; (c) no reaction; (d) (1,1,2-trimethylpropyl)benzene. **17.18.** (a) *sec*-butylbenzene and others; (b) OK; (c) +disub, trisub; (d) No, deactivated; (e) OK. **17.20.** (a) $(CH_3)_2CHCH_2COCl$, benzene, $AlCl_3$; (b) $(CH_3)_3CCOCl$, benzene, $AlCl_3$; (c) PhCOCl, benzene, $AlCl_3$; (d) CO/HCl, $AlCl_3/CuCl$, anisole; (f) Clemmensen on (b); (g) $CH_3(CH_2)_2COCl$, benzene, $AlCl_3$ then Clemmensen. **17.21.** Fluoride leaves in a fast exothermic step; the C—F bond is only slightly weakened in the reactant-like transition state (Hammond postulate). **17.23.** (a) 2,4-dinitroanisole; (b) 2,4- and 3,5-dimethylphenol; (c) *N*-methyl-4-nitroaniline; (d) 2,4-dinitrophenylhydrazine. **17.32.** (a) 1,2,3,4,5,6-hexachloro-1-(trichloromethyl)cyclohexane; (c) *cis*- and *trans*-1,2- dimethylcyclohexane; (d) 1,4-dimethylcyclohexa-1,4-diene. **17.33.** (a) benzoic acid; (b) terephthalic acid (benzene-1,4-dicarboxylic acid); (c) phthalic acid (benzene-1,2-dicarboxylic acid). **17.35.** 60% beta, 40% alpha; reactivity ratio = 1.9 to 1. **17.38.** (a) 1-bromo-1-phenylpropane. **17.39.** (a) HBr, then Grignard with ethylene oxide; (b) CH_3COCl and $AlCl_3$, then Clemmensen, Br_2 and light, then $^-OCH_3$; (c) nitrate, then Br_2 and light, then NaCN. **17.41.** (a) 3-ethoxytoluene; (b) *m*-tolyl acetate; (c) 2,4,6-tribromo-3-methylphenol; (d) 2,4,6-tribromo-3-(tribromomethyl)phenol; (e) 2-methyl-1,4-benzoquinone; (f) 2,4-di-*tert*-butyl-5-methylphenol. **17.56.** indanone. **17.67.** kinetic control at 0 °C, thermodynamic control at 100 °C.

CHAPTER 18

18.1. (a) 5-hydroxyhexan-3-one; ethyl β-hydroxypropyl ketone; (b) 3-phenylbutanal: β-phenylbutyraldehyde; (c) *trans*-2-methoxycyclohexanecarbaldehyde; (d) 6,6-dimethylcyclohexa-2,4-dienone. **18.2.** (a) 2-phenylpropanal; (b) acetophenone. **18.3.** No γ-hydrogens. **18.8.** (a) heptan-3-one; (b) pentanal; (c) phenylacetonitrile; (d) benzyl cyclopentyl ketone; (e) phenylacetaldehyde. **18.10.** (a) benzyl alcohol; (b) benzaldehyde; (c) hept-1-en-3-one; (d) 4-propylhepta-1,6-dien-4-ol; (e) pent-3-enal; (f) 5-hydroxypentanal. **18.14.** second $<$ fourth $<$ first $<$ third. **18.18.** *Z* and *E* isomers. **18.19.** (a) cyclohexanone and methylamine; (b) butan-2-one and ammonia; (c) acetaldehyde and aniline; (d) 6-aminohexan-2-one. **18.23.** (a) benzaldehyde and semicarbazide; (b) camphor and hydroxylamine; (c) tetralone and phenylhydrazine; (d) cyclohexanone and 2,4-DNP; (e) 4-(o-aminophenyl)-butan-2-one. **18.26.** (a) tetralone and ethanol; (b) acetaldehyde and propan-2-ol; (c) hexane-2,4-dione and ethanediol; (d) cyclohexanone and propane-1,3-diol; (e) 5-hydroxypentanal and cyclohexanol; (f) $(HOCH_2CH_2CH_2)_2CHCHO$. **18.30.** $[(CH_3)_3P-R]^+$ could lose a proton from a CH_3. **18.33.** (a) Wittig of $PhCH_2Br$ + acetone; (b) Wittig of $CH_3I + PhCOCH_3$; (c) Wittig of $PhCH_2Br + PhCH{=}CHCHO$; (d) Wittig of EtBr + cyclohexanone. **18.34.** (a) 4-hydroxycyclohexanecarboxylic acid; (b) 4-oxocyclohexanecarboxylic acid; (c) 3-oxocyclohexanecarboxylic acid; (d) *cis*-3,4-dihydroxycyclohexanecarboxylic acid. **18.36.** (a) indane; (b) hexane; (c) ethylene ketal of 2-propylcyclohexanone; (d) propylcyclohexane **18.43.** (a) hexane-2,5-dione. **18.44.** 1-phenyl-butan-2-one (benzyl ethyl ketone). **18.46.** cyclobutanone. **18.52.** (all H^+ cat.) (a) cyclobutanone and hydroxylamine; (b) benzaldehyde and cyclopentylamine; (c) benzylamine and cyclopentanone; (d) β-tetralone and ethylene glycol; (e) cyclohexylamine and acetone; (f) cyclopentanone and methanol. **18.57.** (a) $NaBD_4$, then H_2O; (b) $NaBD_4$, then D_2O; (c) $NaBH_4$, then D_2O. **18.60.** (a) $CH_3CH_2CH_2COCl$ and $AlCl_3$, then Clemmensen; (b) EtMgBr, then H_3O^+; (c) $Cl_2/FeCl_3$, then Dow process to phenol; NaOH, CH_3I, then Gatterman; (d) oxidize to the acid, $SOCl_2$, then $AlCl_3$. **18.64.** (a) hexan-3-one; (b) hexan-2-one and hexan-3-one; (c) hexan-2-one; (d) cyclodecanone; (e) 2- and 3-methylcyclodecanone. **18.66.** **A** heptan-2-one. **18.73.** **A** is the ethylene acetal of butan-2-one; **B** is butan-2-one. **18.74.** *trans*-but-2-enal (crotonaldehyde).

CHAPTER 19

19.1. Pyridine, 2-methylpyridine, pyrimidine, pyrrole, imidazole, indole, and purine are aromatic. **19.3.** (a) pentan-2-amine; (b) *N*-methylbutan-2-amine; (c) *m*-aminophenol; (d) 3-methylpyrrole; (e) *trans*-cyclopentane-1,2-diamine; (f) *cis*-3-aminocyclohexanecarbaldehyde. **19.4.** (a) resolvable (chiral carbons); (b) not resolvable (N inverts); (c) symmetric; (d) not resolvable; proton on N is removable; (e) resolvable (chiral quat. salt). **19.6.** (a) aniline $<$ ammonia $<$ methylamine $<$ NaOH; (b) *p*-nitroaniline $<$ aniline $<$ *p*-methylaniline; (c) pyrrole $<$ aniline $<$ pyridine $<$ piperidine; (d) 3-nitropyrrole $<$ pyrrole $<$ imidazole. **19.7.** (a) secondary amine; (b) primary amine; (c) alcohol. **19.8.** butan-2-amine **19.9.** (a) piperidine; (b) diethylmethylamine; (c) propanal; (d) propan-1-ol. **19.16.** (a) benzylamine + excess CH_3I; (b) 1-bromopentane + excess NH_3; (c) benzyl bromide + excess NH_3. **19.17.** (a) $CH_3CONHCH_2CH_3$; (b) $PhCON(CH_3)_2$; (c) *N*-hexanoylpiperidine. **19.23.** (a) cyclohexanediazonium chloride (then cyclohexanol and cyclohexene); (b) *N*-nitroso-*N*-ethylhexan-2-amine; (c) N-nitrosopiperidine; (d) benzenediazonium chloride. **19.25.** (a) diazotize, then HBF_4, heat; (b) diazotize, then CuCl; (c) protect (CH_3COCl), then 3 $CH_3I/AlCl_3$, H_3O^+, diazotize, H_3PO_2; (d) diazotize, then CuBr; (e) diazotize, then KI; (f) diazotize, then CuCN; (g) diazotize, then H_2SO_4, H_2O, heat; (h) diazotize, then couple with resorcinol. **19.26.** (a) CH_3NH_2, $NaBH(OAc)_3$; (b) PhCHO, $NaBH(OAc)_3$; (c) aniline/H^+, then $LiAlH_4$; (d) H_2NOH/H^+, then $LiAlH_4$; (e) H_2NOH/H^+, then $LiAlH_4$; (f) piperidine + cyclopentanone + $NaBH(OAc)_3$. **19.31.** (a) nitrate, reduce; (b) brominate, then nitrate and reduce; (c) nitrate, then brominate and reduce; (d) oxidize toluene, then nitrate and reduce. **19.36.** only (b), (d), (f), and (h). **19.53.** (a) triethylamine; (b) An acid converts it to a solid ammonium salt. (c) Rinse the clothes with diluted vinegar (acetic acid). **19.56.** **A** is butan-2-amine; **B** is diethylamine. **19.57.** 2,2-dimethylpropan-1-amine.

CHAPTER 20

20.2. (a) 2-iodo-3-methylpentanoic acid; α-iodo-β-methylvaleric acid; (b) (*Z*)-3,4-dimethylhex-3-enoic acid; (c) 2,3-dinitrobenzoic acid; (d) *trans*-cyclohexane-1,2-dicarboxylic acid; (e) 2-chlorobenzene-1,4-dicarboxylic acid; 2-chloroterephthalic acid; (f) 3-methylhexanedioic acid; β-methyladipic acid. **20.3.** (a) first, second, third; (b) third, second, first; (c) third, second, fourth, first. **20.7.** Broad acid OH centered around 3000; conjugated carbonyl about 1690; C=C about 1650. **20.8.** (a) propanoic acid; (b) —CHO proton triplet between δ9 and δ10. **20.11.** (a) $KMnO_4$; (b) $KMnO_4$; (c) PhMgBr + ethylene oxide, oxidize; (d) PBr_3, Grignard, CO_2; (e) conc. $KMnO_4$, heat; (f) KCN, then H_3O^+. **20.15.** (a) methanol and salicylic acid, H^+; methanol solvent, dehydrating agent; (b) methanol and formic acid, H^+, distill product as it forms; (c) ethanol and phenylacetic acid, H^+, ethanol solvent, dehydrating agent. **20.16.** (a) see Fischer esterification; (b) $C-{}^{18}O-CH_3$; (c) mass spectrometry. **20.19.** (a) phenylacetic acid and $LiAlH_4$; (b) phenylacetic acid and $LiAlH_4$, then PCC; (c) 3-oxocyclopentanecarboxylic acid + B_2H_6, then H_3O^+. **20.21.** (a) benzene + CH_3CH_2COCl, $AlCl_3$; or propionic acid + 2 PhLi, then H_3O^+; (b) Add 2 CH_3Li, then H_3O^+. **20.36.** (a) Grignard + CO_2; or KCN, then H_3O^+; (b) conc. $KMnO_4$, heat; (c) Ag^+; (d) $SOCl_2$, then $LiAlH(O\text{-}t\text{-}Bu)_3$; or $LiAlH_4$, then PCC; (e) CH_3OH, H^+; or CH_2N_2; (f) $LiAlH_4$ or B_2H_6; (g) $SOCl_2$, then excess CH_3NH_2. **20.38.** diastereomers. **20.40.** (a) 2-phenylpropanoic acid; (b) 2-methylpropenoic acid; (c) *trans*-hex-2-enoic acid. **20.43.** phenoxyacetic acid. **20.45.** (a) stockroom; heptaldehyde; students; heptanoic acid; (b) air oxidation; (c) prepare fresh samples immediately before using.

CHAPTER 21

21.2. No aldehyde C—H at 2700 and 2800; no acid O—H centered at 3000. **21.4.** (a) acid chloride C=O at 1810; (b) primary amide $H_2C{=}CHCONH_2$ at 1650, two N—H around 3300; (c) anhydride C=O double absorption at 1750 and 1820. **21.5.** (a) acrylamide, (b) 5-hydroxypentanoic acid lactone **21.8.** (a) ethanol, propionyl chloride; (b) phenol, 3-methylhexanoyl chloride; (c) benzyl alcohol, benzoyl chloride; (d) cyclopropanol, cyclohexanecarbonyl chloride; (e) *tert*-butyl alcohol, acetyl chloride; (f) allyl alcohol, succinoyl chloride. **21.9.** (a) dimethylamine, acetyl chloride; (b) aniline, acetyl chloride; (c) ammonia, cyclohexanecarbonyl chloride; (d) piperidine, benzoyl chloride. **21.10.** (i) $PhCH_2OH$; (ii) Et_2NH. **21.25.** (a) butan-1-amine; (b) cyclohexylethylamine; (c) $(CH_2)_6NH$ (7-membered ring); (d) morpholine; (e) cyclohexylmethylpropylamine. **21.30.** (a) benzene + acetyl chloride; (b) benzene + benzoyl chloride; (c) benzene + butyryl chloride, then Clemmensen. **21.32.** (a) *n*-octyl alcohol, acetic formic anhydride (formyl chloride is unavailable); (b) *n*-octyl alcohol, acetic anhydride (cheap, easy to use); (c) phthalic anhydride, ammonia (anhydride forms monoamide); (d) succinic anhydride, methanol (anhydride forms monoester). **21.34.** (a) acetic anhydride; (b) methanol, H^+; (c) $LiAlH_4$, then protonate; (d) $PhNH_2$, warm. **21.37.** (a) $SOCl_2$, then $HN(CH_3)_2$, then $LiAlH_4$; (b) acetic anhydride, then $LiAlH_4$. **21.38.** (a) $SOCl_2$, then NH_3, then $POCl_3$; (b) $LiAlH_4$, make tosylate, NaCN; (c) Fe/HCl, diazotize, CuCN. **21.44.** (a) ethyl benzoate; (b) acetic benzoic anhydride; (c) PhCONHPh; (d) 4-methoxybenzophenone; (e) Ph_3COH; (f) benzaldehyde. **21.47.** (after H^+) (a) HCOOH + PhOH; (b) CH_3CH_2COOH + CH_3CH_2OH; (c) 3-(*o*-hydroxyphenyl)propanoic acid; (d) $(CH_2OH)_2$ + $(COOH)_2$. **21.48.** (a) acetic formic anhydride; (b) $SOCl_2$, then CH_3COONa; (c) oxalyl chloride; (d) H^+ and heat to form anhydride, then one equivalent of $(CH_3)_2CHOH$; (e) oxidize aldehyde with Ag^+, then form lactone with H^+; (f) $NaBH_4$ to reduce aldehyde, then H^+ to form lactone. **21.55.** (a) Ph_3COH; (b) 3 EtMgBr + EtOCOOEt, then H_3O^+. **21.61.** **A** is hexanenitrile; **B** is hexanamide. **21.64.** Acetic anhydride; add water to hydrolyze it to dilute acetic acid. **21.66.** $CH_3CH_2OCOCH_2CN$. **21.69.** δ-valerolactam.

CHAPTER 22

22.8. (a) $PhC(NCH_3)CH_3$; (b) $CH_2{=}C(Ph)NMe_2$; (c) cyclohexanone phenyl imine; (d) piperidine enamine of cyclohexanone.
22.9. (a) enamine + allyl bromide; (b) enamine + $PhCH_2Br$; (c) enamine + PhCOCl. **22.13.** (a), (b) cyclopentanecarboxylate and chloroform/iodoform; (c) $PhCOCBr_2CH_3$. **22.19.** (a) 3-hydroxy-2-methylpentanal; (b) 3-hydroxy-2,4-diphenylbutanal. **22.20.** retro-aldol, reverse of aldol condensation. **22.24.** (a) 2-ethylhex-2-enal; (b) 1,3-diphenylbut-2-en-1-one; (c) 2-cyclohexylidenecyclohexanone.
22.26. $PhCH{=}CHCOCH{=}CHPh$, "dibenzalacetone".
22.28. (a) 2-methyl-3,3-diphenylprop-2-enal; (b) 4,4-dimethyl-1-phenylpent-2-en-1-one. **22.29.** benzaldehyde and acetaldehyde.
22.32. (a) butanal and pentanal (no); (b) two $PhCOCH_2CH_3$ (yes); (c) acetone and PhCHO (yes); (d) 6-oxoheptanal (yes, but also attack by enolate of aldehyde); (e) nonane-2,8-dione (yes). **22.34.** (a) transesterification to a mixture of methyl and ethyl esters; (b) saponification.
22.35. no second alpha proton to form the final enolate to drive the reaction to completion. **22.36.** (a) methyl 2-methyl-3-oxopentanoate; (b) ethyl 2,4-diphenyl-3-oxobutanoate. **22.37.** methyl 2-benzyl-5-phenyl-3-oxopentanoate **22.38.** (a) ethyl butyrate; (b) methyl phenylacetate; (c) ethyl 3-methylbutanoate, or common name: ethyl isovalerate. **22.42.** (a) $PhCO{-}CH(Ph)COOCH_3$; (b) poor choice, four products; (c) $EtOCOCO{-}CH_2COOEt$; (d) $EtOCOCH(CH_3)COOEt$.
22.43. (a) PhCOOEt + CH_3CH_2COOEt; (b) $PhCH_2COOMe$ + MeOCOCOOMe; (c) $(EtO)_2C{=}O$ + $PhCH_2COOEt$; (d) $(CH_3)_3CCOOMe$ + $CH_3(CH_2)_3COOMe$. **22.47.** Alkylate malonic ester with; (a) $PhCH_2Br$; (b) CH_3I twice; (c) $PhCH_2CH_2Br$; (d) $Br(CH_2)_4Br$ (twice). **22.49.** (a) 4-phenylbutan-2-one; (b) cyclobutyl methyl ketone; (c) cyclopentanone. **22.50.** Alkylate acetoacetic ester with: (a) $PhCH_2Br$; (b) $Br(CH_2)_4Br$ (twice); (c) $PhCH_2Br$, then $CH_2{=}CHCH_2Br$. **22.53.** Alkylate the enamine of cyclohexanone with MVK.
22.56. (a) malonic ester anion + ethyl cinnamate; (b) acetoacetic ester anion + acrylonitrile, then H_3O^+; (c) enamine of cyclopentanone + acrylonitrile, then H_3O^+; (d) enamine of 2-methylcyclopentanone + $PhCOCH{=}CH_2$, then H_3O^+; (e) alkylate acetoacetic ester with CH_3I, then MVK, then H_3O^+; (f) cyclopent-2-enone + $(CH_2{=}CH)_2CuLi$. **22.61.** (1) $g < b < f < a < e < c < d$; (2) a, c, d, e. **22.67.** (a) EtCOPh + MVK; (b) cyclohexanone and ethyl vinyl ketone; (c) cyclohexanone and $(CH_3)_2C{=}CHCOCH_3$.
22.72. Alkylate with: (a) $PhCH_2Br$; (b) CH_3CH_2Br, then (bromomethyl) cyclopentane; (c) $Br(CH_2)_5Br$, alkylate on each end to make a cyclohexane ring. **22.73.** Alkylate with: (a) CH_3CH_2Br, then $PhCH_2Br$; (b) $Br(CH_2)_4Br$; (c) MVK (hydrolysis, decarboxylation, then aldol gives product). **22.79.** (a) Dieckmann of dimethyl adipate, alkylation by allyl bromide, hydrolysis and decarboxylation; (c) Robinson with $CH_3CH{=}CHCOCH_3$, then reduction; (d) form enamine or enolate, acylate with ClCOOEt, methylate with CH_3I, do aldol with benzaldehyde.

CHAPTER 23

23.2. (a) two C*, two pairs of enantiomers; (b) one C*, one pair of enantiomers; (c) four C*, eight pairs of enantiomers; three C*, four pairs of enantiomers. **23.5.** (*R*) for D series, (*S*) for L series.
23.13. 28% alpha, 72% beta. **23.15.** L-gulose has the same structure as D-glucose, but with the CHO and CH_2OH ends interchanged.
23.16. (a) D-mannonic acid; (b) D-galactonic acid; (c) Br_2 does not oxidize ketoses. **23.17.** (a) D-mannaric acid; (b) D-galactaric acid. **23.18.** **A** is galactose; **B** is glucose. **23.20.** (a) non-reducing; (b) reducing; (c) reducing; (d) non-reducing; (e) reducing; (f) "sucrose" is nonreducing; should have "-oside" ending. **23.23.** glucose, benzaldehyde, and HCN (toxic). **23.34.** **A** = D-galactose; **B** = D-talose; **C** = D-lyxose; **D** = D-threose. **23.38.** reducing and mutarotating.
23.39. reducing and mutarotating. **23.40.** Trehalose is α-D-glucopyranosyl-α-D-glucopyranoside. **23.41.** Melibiose is 6-O-(α-D-galactopyranosyl)-D-glucopyranose. **23.48.** (a) D-ribose; (b) D-altrose; (c) L-erythrose; (d) L-galactose; (e) L-idose.
23.55. (a) no; (b) yes; (c) Only applies to double stranded DNA. **23.59.** (a) D-arabinose and D-lyxose; (b) D-threose; (c) **X** = D-galactose; (d) No; the optically active hexose is degraded to an optically active pentose that is oxidized to an optically active aldaric acid; (e) D-threose gives an optically active aldaric acid. **23.60.** (a) D-tagatose is a ketohexose, the C4 epimer of D-fructose; (b) A pyranose with the anomeric carbon (C2) bonded to the oxygen atom of C6.

CHAPTER 24

24.6. As in pyrrole, the lone pair on the indole N is part of the aromatic sextet. One N in histidine is like that in pyridine, with the lone pair in an sp^2 hybrid orbital. **24.9.** Reductive amination of (a) $CH_3COCOOH$; (b) $(CH_3)_2CHCH_2COCOOH$; (c) $HOCH_2COCOOH$; (d) $H_2NCOCH_2CH_2COCOOH$. **24.10.** Start with (a) CH_3COOH; (b) $(CH_3)_2CHCH_2CH_2COOH$; (c) $HOOCCH_2CH_2CH_2COOH$.
24.13. The free amino group of the deacylated L enantiomer should become protonated (and soluble) in dilute acid. **24.21.** (a) nucleophilic aromatic substitution; (b) Edman cleaves only the N-terminal amino acid, leaving the rest of the chain intact for further degradation.
24.23. Cys-Tyr-Phe-Gln-Asn-Cys-Pro-Arg-Gly • NH_2. **24.25.** Add piperidine/DMF then Fmoc-Gly and DCC, then piperidine/DMF, then Fmoc-Leu and DCC, then HF. **24.29.** (a) Ruhemann's purple; (b) alanine; (c) $CH_3CONH(CH_2)_4CH(COOH)NHCOCH_3$; (d) L-proline and N-acetyl-D-proline; (e) $CH_3CH_2CH(CH_3)CH(NH_2)CN$; (f) isoleucine; (g) 2-bromo-4-methylpentanoic acid (after water workup); (h) 2-amino-4-methylpentanoic acid or leucine. **24.30.** (a) $NH_3/H_2/Pd$; (b) Br_2/PBr_3, H_2O, excess NH_3; (c) $NH_3/HCN/H_2O$, H_3O^+;
24.32. Convert the alcohol to a tosylate and displace with excess ammonia. **24.36.** aspartylphenylalanine methyl ester. **24.37.** Phe-Ala-Gly-Met-Ala. **24.39.** (a) C-terminal amide ($CONH_2$), or amide (Gln) of Glu; (b) The N-terminal Glu is a cyclic amide (a "pyroglutamyl" group) that effectively blocks the N-terminus. The C-terminal Pro is an amide; (c) cyclic pentapeptide. **24.42.** Ornithine is $H_2N(CH_2)_3CH(NH_2)COOH$, a homolog of lysine, with a similar **pI**. **24.44.** Ala-Lys-Phe-Gln-Gly-Tyr-Arg-Ser-Leu-Ile.

CHAPTER 25

25.2. Hydrogenation of trilinolein (m.p. below −4 °C) gives tristearin (m.p. 72 °C). **25.9.** Estradiol is a phenol, soluble in aqueous sodium hydroxide. **25.13.** (1) sesquiterpene; (2) monoterpene; (3) monoterpene; (4) sesquiterpene. **25.16.** (a) H_2/Ni, $LiAlH_4$; (b) H_2/Ni; (c) stearic acid from (d) add $SOCl_2$, then octadecan-1-ol; (d) O_3, then $(CH_3)_2S$; (e) $KMnO_4$, then H^+; (f) Br_2/PBr_3, then H_2O. **25.17** (a) a triglyceride (a fat); (b) an alkyl sulfate detergent; (c) a wax; (d) a sesquiterpene; (e) a prostaglandin; (f) a steroid. **25.19.** reduce ($LiAlH_4$), esterify with sulfuric acid. **25.21.** Sodium stearate precipitates in dilute acid or Ca^{2+}; (b) Paraffin "wax" does not saponify; (c) Myristic acid shows acidic properties when treated with base; (d) Triolein decolorizes Br_2 in CCl_4. **25.28.** Petroselenic acid is *cis*-octadec-6-enoic acid.

CHAPTER 26

26.1. The radical intermediates would not be benzylic if they added with the other orientation. **26.3.** The benzylic hydrogens are more likely to be abstracted. **26.4.** They all add to give the more highly substituted carbocation. **26.5.** (a) is possible; (b) is very good; (c) is terrible. **26.6.** The cation at the end of a chain abstracts hydride from a benzylic position in the middle of a chain. In isobutylene, a tertiary cation would have to abstract a hydride from a secondary position: unlikely. **26.15.** The third hydroxyl group of glycerol allows for profuse cross-linking of the chains (with a terephthalic acid linking two of these hydroxyl groups), giving a very rigid polyester. **26.19.** Glycerol allows profuse cross-linking, as in Problem 26-15. **26.23.** (a) a polyurethane; (b) step-growth polymer; (c) $HO(CH_2)_3NH_2$ and CO_2. **26.24.** (a) a polyester; (b) condensation polymer; (c) dimethyl terephthalate and butane-1,4-diol; transesterification. **26.25.** (a) a polyurea; (b) step-growth polymer; (c) $H_2N(CH_2)_9NH_2$ and CO_2. **26.26.** (a) polyether (addition polymer); (b) ethylene oxide; (c) base catalyst. **26.28.** (a) $-CH_2-O-[CH_2-O]_n-$; (c) addition polymer. **26.31.** (b) and (c) No to both. Poly(vinyl acetate) is an addition polymer. The ester bonds are not in the main polymer chain; (d) Vinyl alcohol (the enol form of acetaldehyde) is not stable.

Photo Credits

page Multi:	Pearson Education, Inc.
page 1 (top):	Darwin Dale/Science Source
page 2 (bottom):	Getty Images
page 2 (bottom):	Nature's Images/Science Source
page 2 (left):	SARAWUT KUNDEJ/Shutterstock
page 2 (right):	John Glover/DK Images
page 3 (top):	Peter Jordan/Newscom
page 29 (right):	George Grall/Getty Images
page 47 (top):	Richard Megna/Fundamental Photographs
page 55 (top):	Lightspring/Shutterstock
page 61 (right):	Meckes/Ottawa/Science Source
page 61 (right):	paytai/Shutterstock
page 66 (left):	Filipe B. Varela/Shutterstock
page 107 (centre):	NASA/JPL-Caltech/USGS
page 117 (right):	Sean Gardner/Reuters
page 119 (top):	Paul Franklin/DK Images
page 120 (top):	Peter Jordan/Alamy
page 121 (right):	Courtesy of the National Research Council of Canada.
page 122 (left):	AP Images
page 201 (top):	Niels Poulsen DK/Alamy
page 239 (right):	Superstock/Superstock
page 239 (right):	David Parker/Science Source
page 247 (top):	Scott Johnson/Kwajalein Underwater
page 252 (left):	Sclepro/Science Photo Library/Alamy
page 296 (right):	Sascha Janson/Alamy
page 345 (top):	Kellogg Brown & Root Inc.
page 359 (left):	JACOPIN/BSIP/Custom Medical Stock Photo
page 383 (right):	Pearson Education, Inc.
page 389 (right):	Parr Instrument Company
page 406 (left):	Courtesy of Hagley Museum and Library.
page 410 (top):	Barry Freeman, Auburn University, Bugwood.org
page 428 (right):	Thomas Marent/Dorling Kindersley, Ltd
page 432 (top):	Arhendrix/Fotolia
page 433 (top):	Shutterstock
page 462 (top):	Leroy G. Wade
page 471 (top):	Nestor A. Vasquez/Shutterstock
page 496 (left):	Holly Kuchera/Shutterstock
page 497 (right):	Postsmth/Fotolia
page 507 (top):	krmk/Fotolia
page 513 (right):	Mira/Alamy
page 541 (top):	Science Source
page 543 (top):	Richard Megna/Fundamental Photographs
page 558 (right):	Jens Mittag/Amy Werner
page 587 (right):	Kratos Analytical Inc.
page 607 (left):	Simon Fraser/Science Source
page 659 (top):	Gustoimages/Science Source
page 659 (top):	Medical Body Scans/Science Source
page 672 (top):	NASA
page 673 (right):	SSPL/The Image Works
page 705 (top):	Bettmann/Corbis
page 716 (top):	Visions of America/UIG/Getty Images
page 716 (top):	Robert Bailey/Epa /Landov
page 746 (left):	NASA
page 751 (top):	Jon Bilous/123RF GB Ltd
page 752 (left):	Smileus/Shutterstock
page 753 (top):	AGE Fotostock
page 754 (left):	Trevor Kittelty/Shutterstock
page 764 (top):	danmorgan12/Fotolia
page 788 (left):	Edmond Jan Hoorick/Getty Images
page 794 (left):	Pilipphoto/Fotolia
page 809 (top):	2/Travel Ink/Corbis
page 857 (right):	Nature Production/Nature Picture Library
page 876 (top):	Nazzu/Fotolia
page 888 (left):	Nagy-Bagoly Arpad/123RF
page 941 (top):	Javarman/Fotolia
page 943 (top):	Danmir12/Fotolia
page 944 (top):	Subrata Chakraborty/123RF
page 953 (right):	SuperStock
page 1002 (top):	Warrengoldswain/123RF
page 1019 (top):	Europics/Newscom
page 1020 (left):	WildLife GmbH/Alamy
page 1043 (top):	Biophoto Associates/Science Source/Getty Images
page 1085 (right):	Vblinov/Shutterstock
page 1089 (right):	Ian Kimber.
page 1090 (left):	NASA
page 1094 (top):	Ted Horowitz/Alamy
page 1098 (left):	Maridav/Shutterstock
page 1112 (right):	Nik Wheeler/Encyclopedia/Corbis
page 1112 (top):	Valery121283/Fotolia
page 1123 (top):	GIPhotoStock X/Alamy
page 1157 (right):	Kim Reinick/123RF
page 1172 (top):	David Spears FRPS FRMS/Latitude/Corbis
page 1203 (right):	Tim Shaffer/Reuters
page 1206 (top):	Auscape/Universal Images Group/AGE Fotostock
page 1207 (top):	Centers for Disease Control and Prevention (CDC)
page 1214 (top):	Dr. Gopal Murti/Science Source
page 1222 (top):	B Lamm/Blickwinkel/AGE Fotostock
page 1242 (right):	Schankz/Shutterstock
page 1256 (top):	Science Source
page 1257 (left):	Silver Burdett Ginn/Pearson Education, Inc.
page 1258 (top):	Michael Abbey/Science Source
page 1258 (top):	Ralph C. Eagle, Jr./Science Source
page 1265 (top):	Fred Buyle/Corbis Sports Premium/Corbis
page 1266 (left):	Pamela Long/Getty Images
page 1277 (top):	Biophoto Associates/Science Source
page 1278 (top):	Pinkannjoh/Blickwinkel/AGE Fotostock
page 1279 (top):	Charles Sheppard
page 1286 (top):	Dario Sabljak/Fotolia
page 1295 (right):	Thien Eu/Shutterstock
page 1296 (left):	E.I. DuPont de Nemours & Co., Inc.
page 1297 (right):	Science Source
page 1298 (left):	DuPont Corporation, Texas

Index

B

C

D

E

J

K

L

M

P

Q

R

S

Typical Values of Proton NMR Chemical Shifts

Type of Proton		Approximate δ
alkane (—CH_3)	methyl	0.9
alkane (—CH_2—)	methylene	1.3
alkane (—CH—)	methine	1.4
—C(=O)—CH_3	methyl ketone	2.1
—C≡C—H	acetylenic	2.5
R—CH_2—X	(X = halogen, —O—)	3–4
C=C—H	vinyl	5–6
C=C—CH_3	allylic	1.7
Ph—H	aromatic	7.2
Ph—CH_3	benzylic	2.3
R—CHO	aldehyde	9–10
R—COOH	acid	10–12
R—OH	alcohol	variable, about 2–5
Ar—OH	phenol	variable, about 4–7
R—NH_2	amine	variable, about 1.5–4

These values are approximate, because all chemical shifts are affected by neighboring substituents. The numbers given here assume that alkyl groups are the only other substituents present. A more complete table of chemical shifts appears in Appendix 1.

Summary of Functional Group Nomenclature

Functional Group	Name as Main Group	Name as Substituent
Main groups in order of decreasing priority		
carboxylic acids	-oic acid	carboxy
esters	-oate	alkoxycarbonyl
amides	-amide	amido
nitriles	-nitrile	cyano
aldehydes	-al	formyl
ketones	-one	oxo
alcohols	-ol	hydroxy
amines	-amine	amino
alkenes	-ene	alkenyl
alkynes	-yne	alkynyl
alkanes	-ane	alkyl
ethers		alkoxy
halides		halo

Typical Values of IR Stretching Frequencies

Frequency (cm^{-1})	Functional Group		Comments
3300	alcohol	O—H	always broad
	amine, amide	N—H	may be broad, sharp, or broad with spikes
	alkyne	≡C—H	always sharp, usually strong
3000	alkane	—C—H	just below 3000 cm^{-1}
	alkene	=C—H	just above 3000 cm^{-1}
	acid	O—H	very broad 2500–3500 cm^{-1}
2200	alkyne	—C≡C—	just below 2200 cm^{-1}
	nitrile	—C≡N	just above 2200 cm^{-1}
1710 (very strong)	carbonyl	C=O	ketones, acids about 1710 cm^{-1} aldehydes about 1725 cm^{-1} esters higher, about 1735 cm^{-1} amides lower, about 1650 cm^{-1} conjugation lowers frequency
1660	alkene	C=C	conjugation lowers frequency aromatic C=C about 1600 cm^{-1}
	imine	C=N	stronger than C=C
	amide	C=O	stronger than C=C (see above)

Ethers, esters, and alcohols also show C—O stretching between 1000 and 1200 cm^{-1}.

More complete tables of IR frequencies appear in Appendices 2A and 2B.

Periodic Table of the Elements

6 C 12.01 Carbon	
6	Atomic number
C	Element symbol
12.01	Atomic weight*
Carbon	Element name

PERIOD	1A	2A	3B	4B	5B	6B	7B	8B	8B	8B	1B	2 B	3A	4A	5A	6A	7A	Noble gases 8A
1	1 **H** 1.008 Hydrogen																	2 **He** 4.003 Helium
2	3 **Li** 6.941 Lithium	4 **Be** 9.012 Beryllium											5 **B** 10.81 Boron	6 **C** 12.01 Carbon	7 **N** 14.01 Nitrogen	8 **O** 16.00 Oxygen	9 **F** 19.00 Fluorine	10 **Ne** 20.18 Neon
3	11 **Na** 22.99 Sodium	12 **Mg** 24.31 Magnesium											13 **Al** 26.98 Aluminum	14 **Si** 28.09 Silicon	15 **P** 30.97 Phosphorus	16 **S** 32.07 Sulfur	17 **Cl** 35.45 Chlorine	18 **Ar** 39.95 Argon
4	19 **K** 39.10 Potassium	20 **Ca** 40.08 Calcium	21 **Sc** 44.96 Scandium	22 **Ti** 47.867 Titanium	23 **V** 50.94 Vanadium	24 **Cr** 52.00 Chromium	25 **Mn** 54.94 Manganese	26 **Fe** 55.85 Iron	27 **Co** 58.93 Cobalt	28 **Ni** 58.70 Nickel	29 **Cu** 63.546 Copper	30 **Zn** 65.39 Zinc	31 **Ga** 69.72 Gallium	32 **Ge** 72.64 Germanium	33 **As** 74.92 Arsenic	34 **Se** 78.96 Selenium	35 **Br** 79.90 Bromine	36 **Kr** 83.80 Krypton
5	37 **Rb** 85.47 Rubidium	38 **Sr** 87.62 Strontium	39 **Y** 88.91 Yttrium	40 **Zr** 91.22 Zirconium	41 **Nb** 92.90 Niobium	42 **Mo** 95.94 Molybdenum	43 **Tc** (98) Technetium	44 **Ru** 101.07 Ruthenium	45 **Rh** 102.9 Rhodium	46 **Pd** 106.4 Palladium	47 **Ag** 107.9 Silver	48 **Cd** 112.4 Cadmium	49 **In** 114.8 Indium	50 **Sn** 118.7 Tin	51 **Sb** 121.8 Antimony	52 **Te** 127.6 Tellurium	53 **I** 126.9 Iodine	54 **Xe** 131.3 Xenon
6	55 **Cs** 132.9 Cesium	56 **Ba** 137.327 Barium	71 **Lu** 175.0 Lutetium	72 **Hf** 178.5 Hafnium	73 **Ta** 180.9479 Tantalum	74 **W** 183.84 Tungsten	75 **Re** 186.2 Rhenium	76 **Os** 190.23 Osmium	77 **Ir** 192.2 Iridium	78 **Pt** 195.1 Platinum	79 **Au** 197.0 Gold	80 **Hg** 200.6 Mercury	81 **Tl** 204.4 Thallium	82 **Pb** 207.2 Lead	83 **Bi** 209.0 Bismuth	84 **Po** (209) Polonium	85 **At** (210) Astatine	86 **Rn** (222) Radon
7	87 **Fr** (223) Francium	88 **Ra** (226) Radium	103 **Lr** (262) Lawrencium	104 **Rf** (261) Rutherfordium	105 **Db** (262) Dubnium	106 **Sg** (266) Seaborgium	107 **Bh** (264) Bohrium	108 **Hs** (269) Hassium	109 **Mt** (268) Meitnerium	110 **Ds** (281) Darmstadtium	111 **Rg** (272) Roentgenium	112 **Cn** (285) Copernicium	113 (284)	114 **Fl** (289) Flerovium	115 (288)	116 **Lv** (292) Livermorium	117 (294)	118 (294)

Lanthanide series:	57 **La** 138.9 Lanthanum	58 **Ce** 140.1 Cerium	59 **Pr** 140.9 Praseodymium	60 **Nd** 144.2 Neodymium	61 **Pm** (145) Promethium	62 **Sm** 150.4 Samarium	63 **Eu** 152.0 Europium	64 **Gd** 157.3 Gadolinium	65 **Tb** 158.9 Terbium	66 **Dy** 162.5 Dysprosium	67 **Ho** 164.9 Holmium	68 **Er** 167.3 Erbium	69 **Tm** 168.9 Thulium	70 **Yb** 173.0 Ytterbium
Actinide series:	89 **Ac** (227) Actinium	90 **Th** 232.0 Thorium	91 **Pa** (231) Protactinium	92 **U** 238.0 Uranium	93 **Np** (237) Neptunium	94 **Pu** (244) Plutonium	95 **Am** (243) Americium	96 **Cm** (247) Curium	97 **Bk** (247) Berkelium	98 **Cf** (251) Californium	99 **Es** (252) Einsteinium	100 **Fm** (257) Fermium	101 **Md** (258.10) Mendelevium	102 **No** (259) Nobelium

*Numbers in parentheses are mass numbers of the most stable or best-known isotope of radioactive elements.